Molecular Biology of Membrane Transport Disorders

Edited by

Stanley G. Schultz
University of Texas Medical School at Houston
Houston, Texas

Thomas E. Andreoli
University of Arkansas College of Medicine
Little Rock, Arkansas

Arthur M. Brown
Case Western Reserve University
Cleveland, Ohio

Douglas M. Fambrough
The Johns Hopkins University
Baltimore, Maryland

Joseph F. Hoffman
Yale University School of Medicine
New Haven, Connecticut

and

Michael J. Welsh
University of Iowa College of Medicine
Iowa City, Iowa

Plenum Press • New York and London

Library of Congress Cataloging-in-Publication Data

Molecular biology of membrane transport disorders / edited by Stanley G. Schultz ... [et al.].
p. cm.
Includes bibliographical references and index.
ISBN 0-306-45164-6
1. Membrane disorders. 2. Pathology, Molecular. I. Schultz, Stanley G.
RB113.M5855 1996
616.07--dc20 96-21633
CIP

ISBN 0-306-45164-6

A Division of Plenum Publishing Corporation
233 Spring Street, New York, N. Y. 10013

10 9 8 7 6 5 4 3 2 1

Printed in the United States of America

Preface

When the six of us gathered to start planning for what was to be the *Third Edition* of *Physiology of Membrane Disorders,* it was clear that since 1986, when the *Second Edition* appeared, the field had experienced the dawning of a new era dominated by a change in focus from phenomenology to underlying mechanisms propelled by the power of molecular biology. In 1985, detailed molecular information was available for only three membrane transporters: the *lac* permease, bacterial rhodopsin, and the acetylcholine receptor. During the decade that has since elapsed, almost all of the major ion channels and transport proteins have been cloned, sequenced, mutagenized, and expressed in homologous as well as heterologous cells. Few, if any, of the transporters that were identified during the previous era have escaped the probings of the new molecular technologies and, in many instances, considerable insight has been gained into their mechanisms of function in health and disease. Indeed, in some instances novel, unexpected transporters have emerged that have yet to have their functions identified.

The decision to adopt the new title *Molecular Biology of Membrane Transport Disorders* was a natural outgrowth of these considerations. The seminal purpose of this volume, however, remains unchanged from, and is a logical extension of, that of its predecessor; namely, to provide a frame of reference for the understanding of the pathogenesis of clinical conditions that involve derangements in membrane transport processes at the molecular level—a goal that ten years ago could only have been contemplated but that is now within our grasp. Accordingly, this volume is structured similarly to *PMD* but, because of length limitations, it focuses on areas where our understanding has advanced to the molecular level; no attempt has been made to duplicate sections of the *Second Edition* of *PMD* that are as timely today as they were when they were written.

Stanley G. Schultz
Thomas E. Andreoli
Arthur M. Brown
Douglas M. Fambrough
Joseph F. Hoffman
Michael J. Welsh

Contributors

Robert J. Alpern • Department of Internal Medicine, The University of Texas, Southwestern Medical Center, Dallas, Texas 75235

Suresh V. Ambudkar • Department of Medicine, The Johns Hopkins University School of Medicine, Baltimore, Maryland 21218

Thomas E. Andreoli • Division of Nephrology and Department of Internal Medicine, University of Arkansas College of Medicine, Little Rock, Arkansas 72205

Yvonne L. O. Balba • Department of Pharmacology, University of Texas–Houston Medical School, Houston, Texas 77225

Krister Bamberg • Membrane Biology Laboratory, Department of Medicine, University of California, Los Angeles, and Wadsworth VA Medical Center, Los Angeles, California 90073

Laurence Bianchini • Department of Biochemistry, University of Nice, Nice, France

Lutz Birnbaumer • Departments of Anesthesiology and Biological Chemistry, School of Medicine, and the Molecular Biology Institute, University of California at Los Angeles, Los Angeles, California 90095

Mariel Birnbaumer • Departments of Anesthesiology and Biological Chemistry, School of Medicine, and the Molecular Biology Institute, University of California at Los Angeles, Los Angeles, California 90095

Robert J. Bloch • Department of Physiology, University of Maryland, School of Medicine, Baltimore, Maryland 21201

Steven R. Brant • Departments of Medicine and Physiology, GI Unit, The Johns Hopkins University School of Medicine, Baltimore, Maryland 21205

Thomas J. Burke • Department of Medicine, University of Colorado Medical School, Denver, Colorado 80262

William B. Busa • Department of Biology, The Johns Hopkins University, Baltimore, Maryland 21218

William A. Catterall • Department of Pharmacology, University of Washington, Seattle, Washington 98195

Marilyn M. Cornwell • Fred Hutchinson Cancer Research Center, Seattle, Washington 98195

Jonathan Covault • Department of Physiology and Neurobiology, University of Connecticut, Storrs, Connecticut 06269

Mark G. Darlison • Institute for Cell Biochemistry and Clinical Neurobiology, University Hospital Eppendorf, University of Hamburg, 20246 Hamburg, Germany

David C. Dawson • Department of Physiology, The University of Michigan Medical School, Ann Arbor, Michigan 48109

Patricia V. Donnelly • Department of Pharmacology, University of Texas–Houston Medical School, Houston, Texas 77225

Mark Donowitz • Departments of Medicine and Physiology, GI Unit, The Johns Hopkins University School of Medicine, Baltimore, Maryland 21205

William P. Dubinsky • Department of Integrative Biology, The University of Texas–Houston Medical School, Houston, Texas 77225

Douglas M. Fambrough • Department of Biology, The Johns Hopkins University, Baltimore, Maryland 21218

Ursula A. Germann • Vertex Pharmaceuticals Incorporated, Cambridge, Massachusetts

Michael M. Gottesman • Laboratory of Cell Biology, National Cancer Institute, National Institutes of Health, Bethesda, Maryland 20892

Robert C. Griggs • Department of Neurology, University of Rochester, Rochester, New York 14642

Robert J. Harvey • Institute for Cell Biology and Clinical Neurobiology, University Hospital Eppendorf, University of Hamburg, 20246 Hamburg, Germany

Sandra Hoogerwerf • Departments of Medicine and Physiology, GI Unit, The Johns Hopkins University School of Medicine, Baltimore, Maryland 21205

C. K. Ifune • Washington University School of Medicine, St. Louis, Missouri 63110

Giuseppe Inesi • Department of Biological Chemistry, The University of Maryland School of Medicine, Baltimore, Maryland 21201

Laurinda A. Jaffe • Department of Physiology, University of Connecticut Health Center, Farmington, Connecticut 06032

H. Ronald Kaback • Howard Hughes Medical Institute, Departments of Physiology and Microbiology and Molecular Genetics, Molecular Biology Institute, University of California, Los Angeles, Los Angeles, California 90024

Henry J. Kaminski • Department of Neurology, Case Western Reserve University School of Medicine, Cleveland Veterans Affairs Medical Center, University Hospitals of Cleveland, Cleveland, Ohio 44106

Ronald S. Kaplan • Department of Pharmacology, College of Medicine, University of South Alabama, Mobile, Alabama 36688

Robert S. Kass • Department of Pharmacology, College of Physicians and Surgeons, Columbia University, New York, New York 10032

Victoria P. Knutson • Department of Pharmacology, University of Texas–Houston Medical School, Houston, Texas 77225

Freddie Kokkie • Departments of Medicine and Physiology, GI Unit, The Johns Hopkins University School of Medicine, Baltimore, Maryland 21205

Roger Lester • Department of Internal Medicine, University of Arkansas for Medical Sciences, and McClellan VA Hospital, Little Rock, Arkansas 77205

Susan A. Levine • Departments of Medicine and Physiology, GI Unit, The Johns Hopkins University School of Medicine, Baltimore, Maryland 21205

Jennifer Lippincott-Schwartz • Cell Biology and Metabolism Branch, National Institute of Child Health and Development, National Institutes of Health, Bethesda, Maryland 20892

Maria M. Lopez-Reyes • Department of Pharmacology, University of Texas–Houston Medical School, Houston, Texas 77225

David H. MacLennan • Banting and Best Department of Medical Research, University of Toronto, C. H. Best Institute, Toronto, Ontario M5G 1L6, Canada

Otilia Mayorga-Wark • Department of Integrative Biology, The University of Texas–Houston Medical School, Houston, Texas 77225

Orson W. Moe • Veterans Administration Medical Center, Dallas, Texas 75216

Jami Montgomery • Departments of Medicine and Physiology, GI Unit, The Johns Hopkins University School of Medicine, Baltimore, Maryland 21205

Samir Nath • Departments of Medicine and Physiology, GI Unit, The Johns Hopkins University School of Medicine, Baltimore, Maryland 21205

Lawrence G. Palmer • Departments of Physiology and Biophysics, Cornell University Medical College, New York, New York 10021

John C. Parker • Department of Medicine, University of North Carolina, Clinical Research Center, Chapel Hill, North Carolina 27599

Ira Pastan • Laboratory of Molecular Biology, National Cancer Institute, National Institutes of Health, Bethesda, Maryland 20892

Michael S. Phillips • Banting and Best Department of Medical Research, University of Toronto, C. H. Best Institute, Toronto, Ontario M5G 1L6, Canada

Jacques Pouysségur • Department of Biochemistry, University of Nice, Nice, France

Christian Prinz • Membrane Biology Laboratory, Department of Medicine, University of California, Los Angeles, and Wadsworth VA Medical Center, Los Angeles, California 90073

Louis Ptáček • Department of Neurology, Human Molecular Biology and Genetics, The University of Utah, Salt Lake City, Utah 84132

W. Brian Reeves • Division of Nephrology and Department of Internal Medicine, University of Arkansas College of Medicine, Little Rock, Arkansas 72205

Robert L. Ruff • Departments of Neurology and Neuroscience, Case Western Reserve University School of Medicine, Cleveland Veterans Affairs Medical Center, University Hospitals of Cleveland, Cleveland, Ohio 44106

George Sachs • Membrane Biology Laboratory, Department of Medicine, University of California, Los Angeles, and Wadsworth VA Medical Center, Los Angeles, California 90073

John R. Sachs • Department of Medicine, State University of New York at Stony Brook, Stony Brook, New York 11794

Henry Sackin • Department of Physiology and Biophysics, Cornell University Medical College, New York, New York 10021

Robert W. Schrier • Department of Medicine, University of Colorado Medical School, Denver, Colorado 80262

Joseph H. Sellin • Departments of Medicine and Integrative Biology, The University of Texas–Houston Medical School, Houston, Texas 77030

Jai Moo Shin • Membrane Biology Laboratory, Department of Medicine, University of California, Los Angeles, and Wadsworth VA Medical Center, Los Angeles, California 90073

Joe Henry Steinbach • Department of Anesthesiology, Washington University School of Medicine, St. Louis, Missouri 63110

Chung-Ming Tse • Departments of Medicine and Physiology, GI Unit, The Johns Hopkins University School of Medicine, Baltimore, Maryland 21205

Michael J. Welsh • Howard Hughes Medical Institute, Departments of Internal Medicine and Physiology and Biophysics, University of Iowa College of Medicine, Iowa City, Iowa 52242

Jeannie Yip • Departments of Medicine and Physiology, GI Unit, The Johns Hopkins University School of Medicine, Baltimore, Maryland 21205

C. H. Chris Yun • Departments of Medicine and Physiology, GI Unit, The Johns Hopkins University School of Medicine, Baltimore, Maryland 21205

Yilin Zhang • Banting and Best Department of Medical Research, University of Toronto, C. H. Best Institute, Toronto, Ontario M5G 1L6, Canada

Piotr Zimniak • Departments of Internal Medicine and of Biochemistry and Molecular Biology, University of Arkansas for Medical Sciences, and McClellan VA Hospital, Little Rock, Arkansas 72205

Contents

3. The Membrane-Associated Cytoskeleton and Exoskeleton

Robert J. Bloch

4. Methods of Reconstitution of Ion Channels

William P. Dubinsky and Otilia Mayorga-Wark

5. Permeability and Conductance in Ion Channels: A Primer

David C. Dawson

24. Cell Death

Thomas J. Burke and Robert W. Schrier

25. Genetic Variants of Erythrocytes

John C. Parker

Molecular Biology of Membrane Transport Disorders

1

Membrane Traffic and Compartmentalization within the Secretory Pathway

Jennifer Lippincott-Schwartz

1.1. INTRODUCTION

A wide variety of processes necessary for cell growth and homeostasis are carried out within the secretory pathway of eukaryotic cells. These include biosynthesis and assembly of protein complexes, lipid biosynthesis and metabolism, and protein processing, degradation, and secretion. These functions occur within distinct membrane-bound compartments, or organelles, which communicate by membrane transport pathways. Movement of protein and lipid through these pathways involves the formation, targeting, and fusion of membrane-bound transport intermediates. Regulation of the organization and function of this membrane system is crucial for enabling cells to selectively transport and localize protein and lipid in response to changing cellular needs.

Over the past decade, much research has focused on understanding how organelles of the secretory system and the membrane transport pathways that connect them are organized and regulated. A wide variety of approaches have revealed that the membranes of the secretory system are extremely dynamic, with bidirectional membrane transport pathways and overlapping compartmental boundaries. Moreover, the characteristics of individual organelles have been shown to be intimately tied to the properties of membrane traffic and sorting, with imbalances in membrane flow or aberrations in membrane sorting having profound consequences for organelle structure and identity. The molecular basis for these characteristics of the secretory system have only just begun to be identified, with interacting classes of molecules identified for vesicle formation, sorting, targeting, and fusion steps. In this chapter recent progress toward understanding the mechanisms of membrane traffic and compartmentalization within the secretory pathway will be discussed.

1.2. COMPARTMENTAL ORGANIZATION

The secretory pathway of eukaryotes is comprised of several morphologically and functionally distinct compartments [including the endoplasmic reticulum (ER), intermediate compartment (IC), and Golgi complex] through which newly synthesized protein and lipid move en route to the cell surface (see Figure 1.1). Whereas the ER serves as the port of entry for all membrane and protein into the secretory pathway, the IC and Golgi serve as the sites where membranes exported from the ER converge and are processed/sorted before transport to different destinations. In order to function efficiently as a transport system, the membrane traffic which interconnects these compartments must be both highly regulated and fine-tuned to the synthetic and transport activities required for cell growth.

1.2.1. The Endoplasmic Reticulum

The ER is the starting point in the secretory pathway. Comprised of an extensive array of interconnecting membrane tubules and cisternae which extend throughout the cell,[1] the ER is specialized for lipid biosynthesis and metabolism, as well as for protein synthesis, folding, assembly, degradation, and transport.[2,3] Protein biosynthesis, processing, and folding occur within the rough ER whose outer membrane is studded with ribosomes. Here, a variety of specific resident ER components ensure the proper entry, folding, and processing of newly synthesized proteins. These include signal peptidase, which cleaves off signal sequences from nascent polypeptides; glycosyl transferase, which adds oligosaccharide chains to specific residues on the polypeptide; BiP and protein disulfide isomerase (PDI), which act as chaperones to facilitate

Jennifer Lippincott-Schwartz • Cell Biology and Metabolism Branch, National Institute of Child Health and Human Development, National Institutes of Health, Bethesda, Maryland 20892

Molecular Biology of Membrane Transport Disorders, edited by S.G. Schultz *et al.* Plenum Press, New York, 1996

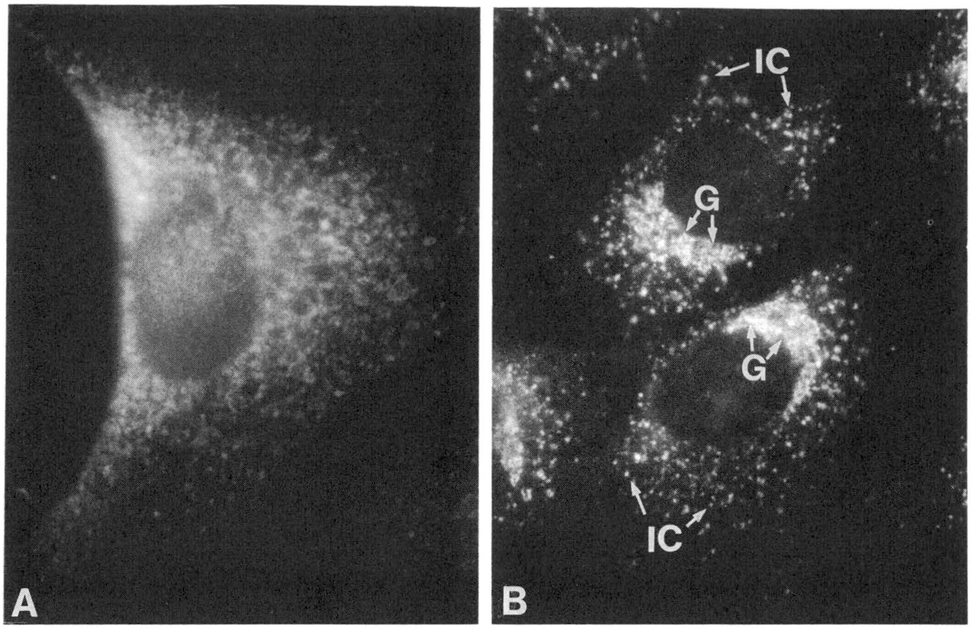

Figure 1.1. Distribution of the ER labeled with antibodies specific to ER proteins (panel A), and the IC and Golgi complex (G) labeled with antibodies to βCOP (panel B).

folding.[3] Folding and assembly of proteins is also facilitated by the specialized luminal environment of the ER, which is oxidizing (promoting disulfide bond formation) and has high free calcium levels.[4] Proper folding and assembly of proteins is required for their targeting to correct final intracellular destinations and for their proper functioning there.[5]

1.2.2. The Intermediate Compartment

Once a protein is properly folded and assembled, it can exit the ER. This is mediated by membrane transport vesicles that bud off from specialized regions of the ER known as transitional elements. These regions consist of clusters of elaborate smooth ER tubules that can be found in regions of the ER near the central Golgi complex as well as at multiple peripheral sites.[6,7] ER-derived transport vesicles are not long-lived but rapidly fuse to produce the IC, which are pleiomorphic structures (between 200 and 500 nm in diameter) that are scattered throughout the cytoplasm.[8,9] ICs consist of tubular cisternae interconnected with vacuolar elements of variable size and form, depending on conditions of temperature and microtubule status.[7–9] In many respects, ICs resemble endosomes which are also pleiomorphic, tubulovesicular structures.[10] The presence within ICs of vacuolar and tubular domains raises the possibility that they can segregate membrane area and internal volume, analogous to sorting endosomes.[9] This could play a role in recycling of membrane back to the ER.

Membranes of the IC are enriched in several proteins, including three proteins of 63, 53, and 58 kDa (whose functions are unknown), as well as the small GTP-binding proteins, rab 1b and 2, and a peripheral membrane protein of 110 kDa, known as βCOP.[8,9,11] Additional polypeptides, different from those of the ER and the Golgi complex, have also been isolated from IC fractions off sucrose gradients.[8] These proteins are likely to participate in the biochemical processes that have been proposed to occur in the IC, including fatty acylation of proteins involved in the first step in *O*-glycosylation and in the generation of the mannose-6-phosphate signal for lysosomal proteins.

1.2.3. The Golgi Complex

Protein and lipid accumulated within the IC next move into the Golgi complex, which usually appears as a set of three to ten flattened cisternae with dilated rims arranged as a stack.[12] Surrounding this stack are an array of small vesicles and a network of tubules that are thought to be involved in transport into and from this organelle. In contrast to the IC, which appears as numerous discrete structures scattered throughout the cytoplasm, the Golgi complex usually appears as one interconnected structure that is localized near the microtubule organizing center (MTOC) as a result of its interaction with microtubules.[13–15]

That all membrane traveling from the ER to the plasma membrane travels through the Golgi complex has led to the idea that its stack of flattened membranes acts as a "countercurrent" fractionation system to separate proteins destined for the plasma membrane from those to be retained in the ER.[16] Under such a model, molecules not destined for the plasma membrane or other post-Golgi sites would be recycled to the ER. The molecular mechanism(s) for sorting of molecules within the Golgi is under current investigation and is proposed to involve protein retention and sorting signals[17] as well as lateral partitioning of proteins between coexisting lipid domains.[16] The Golgi complex performs several other important functions within cells including glycosylation and processing of glycoproteins and glycolipids, and sorting of membrane and soluble components that exit the Golgi to different destinations.

The diverse roles of the Golgi complex are reflected in its overall organization.[18,19] Newly synthesized membrane and secretory components first enter the Golgi at the *cis* Golgi network (CGN), which is closest to the ER. Appearing as an array of tubules and associated vesicles, the CGN sorts molecules to be returned to the ER from those moving by bulk flow through the secretory pathway. Recycling of membrane and protein back to the ER is crucial for maintaining the surface area of the ER in the face of continuous outflow of membrane and for retrieving escaped ER resident proteins.[20] It is also likely to provide a quality control for transport of molecules out of the ER–Golgi system, enabling molecules that are incorrectly folded or assembled to return to the environment of the ER where they can be refolded and assembled properly.[21,22] Retrieval of escaped soluble ER resident proteins from the Golgi is thought to occur by a mechanism involving association of a C-terminal tetrapeptide sequence, KDEL (or HDEL), on these proteins with the receptor protein, ERD-2, which returns them to the ER.[20]

From the CGN, molecules traverse through the Golgi stacks where most protein and lipid modifications occur. These include processing of *N*- and *O*-linked glycoproteins, phosphorylation of glycoproteins, elongation of glycosylaminoglycans and glycolipids, and addition of lipid to secretory lipoproteins.[23] The cisternal stacklike appearance of Golgi stacks is likely to enhance the efficiency of the enzymes catalyzing these reactions by minimizing luminal volume and increasing membrane surface area. Moreover, the polarized distribution of these enzymes might facilitate the ordered sequence through which these proteins are processed as they move through the stack.

Exit out of the Golgi occurs at the opposite face of the CGN known as the *trans* Golgi network (TGN). Here, protein and lipid are sorted to different final destinations. Morphologically the TGN appears as a saccular tubular network that varies in size and shape in different cell types.[24] In addition to sorting to post-Golgi compartments, the TGN is also involved in late protein modification events, including galactose α-2,6-sialylation, tyrosine sulfation, proteolytic cleavages at dibasic residues on prohormones, and in the condensation of molecules to produce secretory granules.[25–27] From the TGN, membrane and protein can be transported to the plasma membrane, endosomes, secretory granules, or lysosomes.

1.3. CHARACTERISTICS OF TRANSPORT INTERMEDIATES

Elegant work from the 1960s and 1970s defined the compartmental organization of the secretory pathway, its directionality, and the role of membrane-bound transport intermediates.[28] Recent studies have focused on the characteristics of transport intermediates including what determines the specificity of their membrane and soluble cargo and their morphological and functional properties.

1.3.1. Cargo Specificity: Bulk Flow versus Concentration

The failure to identify discrete structures on proteins that signal entry into transport intermediates has given rise to the nonselective or "bulk flow" model for transport. According to this model, protein and lipid move by default into transport intermediates as they move through the secretory pathway to the plasma membranes.[29] The degree to which proteins are retarded from this "secretory bulk flow" (for example, by their adsorption to the luminal matrix of organelles) would determine the rate at which they pass through the system. Although this model is widely accepted for transport of luminal proteins, recent work by Balch and colleagues[30] suggests that it may not hold for membrane proteins. Using quantitative immunoelectron microscopy, these authors demonstrated that a cell surface-directed protein, vesicular stomatitis virus G glycoprotein (VSV G), is concentrated five- to tenfold along the surface of the ER at the site of vesicle budding. Since concentration of VSV G is unlikely to result from passive movement of VSV G into defined membrane regions specialized for export, Balch *et al.* have proposed an indirect or direct interaction of VSV G with a transport machinery. This would involve positive sorting signals to promote concentration. Although still unknown, such signals might involve a process whereby membrane cargo triggers active recruitment of cytosolic components onto membrane, leading to the formation of nascent budding sites.

Positive sorting signals are not thought to be involved in the process whereby luminal proteins are transported. Nevertheless, recent quantitative immunoelectron microscopy[31] has shown that certain secretory proteins are concentrated 50-fold during transport between the ER and the first *cis*-most Golgi cisternae, with a 6- to 10-fold concentration over the rest of the secretory pathway. This could be related to a mechanism which retains these molecules during transport, for example, by their affinity to membrane-anchored glycoconjugates at various sites. Alternatively, concentration could be achieved by the selective removal of membrane and fluid (by recycling to the ER) from forward-moving transport intermediates. This latter interpretation requires a yet unidentified mechanism for segregating recycling components from forward-moving transport intermediates.

1.3.2 Morphological and Functional Characteristics of Transport Intermediates: Roles of Vesicles, Organelles, and Tubules

The idea that secretory transport is mediated by a series of carrier vesicles that bud from one compartment and then fuse with the next in series along the secretory pathway has received widespread support from *in vitro* studies reconstituting transport in mammalian cells as well as from genetic studies in yeast.[29,32] When isolated Golgi stacks from mammalian cells are incubated with cytosol and ATP, for instance, numerous 75-nm-diameter vesicles appear.[29] These are presumed to represent transport vesicles since they contain transported proteins. Consistent with this, when Golgi stacks and their associated vesicles are incubated with acceptor Golgi membranes, vectorial transport (presumed to be mediated by vesicles shuttling between Golgi cisternae) can be monitored by the maturation of saccharide chains on glycoprotein cargo.

Vesicle-mediated traffic is also well documented in yeast. Here, large accumulations of secretory vesicles derived from the ER have been observed in various yeast mutants defective in secretion.[32] These ER-derived vesicles and their associated cargo can be isolated by differential centrifugation and density gradient fractionation. When they are subsequently washed and added to acceptor Golgi membrane in the presence of ATP and cytosol, targeting to and fusion with acceptor membranes occurs.

Although it is indisputable from the above studies that small vesicles can mediate traffic between organelles under certain circumstances, the question is whether they are the only mode of membrane transport through the secretory pathway. This question is particularly relevant when considering transport from the IC to the Golgi complex and between Golgi subcompartments. Recent *in vivo* morphological studies have suggested that cargo-enriched IC structures themselves might translocate into the Golgi region and fuse with Golgi membrane there.[7,10] This would be in contrast to the view that IC elements are stable structures from which vesicles targeted to the Golgi bud. Support for the "organelle transport" model has come from morphological experiments following the fate of cargo accumulated within IC structures at 16°C, which impedes membrane transport into the Golgi. On warming to 37°C, IC elements and their associated cargo were found to translocate as discrete entities into the central Golgi region, consistent with the view that ICs are mobile transport structures.[7]

That a similar nonvesicular mechanism may mediate transport between Golgi cisternae has been suggested by Clermont and colleagues.[33] They performed *in vivo* studies following the movement of the secretory protein, casein, as it passes through the Golgi in secretory cells of the mammary gland. Unlike most secretory proteins, casein assembles into large submicellar structures of approximately 20 nm diameter as it moves from the ER into the Golgi complex. These structures are easily observed in fixed tissue without using permeabilization and immunolabeling techniques. Using high-resolution electron microscopy and stereoimaging, Clermont *et al.* found casein submicelles throughout all cisternae of the Golgi except where cisternal lumina were too narrow to accommodate them. Strikingly, casein submicelles were excluded from the vesicles closely surrounding Golgi cisternae. These vesicles have previously been thought to be transport intermediates for forward traffic through the Golgi complex. Since it seems unlikely that casein submicelles in each cisterna repeatedly disintegrate into subcomponents for vesicle transport only to be reconstituted as submicelles in succeeding cisternae, Clermont *et al.* argue that these vesicles do not serve as carriers for forward movement of proteins like casein through the Golgi stack. A similar conclusion was reached by Melkonian *et al.*[34] in their electron microscopic studies of scale formation in the Golgi stacks of algae. There, Golgi cisternae all contained scales at various stages of formation, but the vesicles seen at their edges were devoid of the electron-dense scale material.

How might secretory components that are too large to enter transport vesicles move forward through the Golgi complex? If the IC is a transport intermediate that continuously matures by recycling of selected components as it moves toward the Golgi region, it is possible that Golgi cisternae might transport cargo by a similar maturation process. According to this view, IC elements moving into the Golgi region would continuously replace existing cisternae of the CGN as these mature into Golgi stacks.[9,10] A polarized cisternal array or stack would arise as cisternae continued to form and mature at the *cis* face of the cisternal stack. Transport of proteins forward through the stack would not require formation of transport vesicles, but could be mediated by cisternae themselves through "cisternal progression." The "resident" distribution of enzymes within the Golgi stack would reflect their steady-state distribution within cisternae as these enzymes cycled from cisternae near the *trans* face of the stack into cisternae located near the *cis* face.

The "cisternal pregression" model for movement of proteins through the Golgi complex contrasts with the more popular vesicular transport model where Golgi cisternae are considered to be relatively stable structures that communicate via small vesicles.[29] Since existing experimental approaches can only indirectly study transport through the Golgi, which of these models proves correct will be difficult to resolve. A crucial difference between these two models is the role of retrograde (Golgi-to-ER) transport. For the vesicular transport model, retrograde transport is not necessarily linked to forward traffic or required for maintenance of Golgi identity. This is because Golgi cisternae, according to this view, are stable structures which accept incoming vesicles and produce outflowing vesicles. In the organelle transport model, by contrast, retrograde traffic is crucial both for forward traffic and for maintaining cisterna identity. This is because "resident" Golgi enzymes must be continually recycled to maintain their steady state within progressing cisternae. Without such cycling, cisternae would rapidly lose their identity and ability to function in glycoprotein and glycolipid processing reactions.

Recent studies have shown that membrane tubules, in addition to vesicles and vesicular structures (i.e., organelles), can function to transport membrane components within the cell. Real-time imaging of living cells labeled with fluorescent lipids that preferentially label ER (using $DiOC_6$)[1] or Golgi membranes (using NBD-ceramide)[35] have shown the formation of extensive tubular interconnections both within the ER and between Golgi cisternae. These tubular connections can be observed in *in vitro* preparations of ER and Golgi stacks after incubation with ATP and cytosol.[36] That such tubular connections can mediate transport between Golgi cisternae and between the ER and Golgi has been demonstrated by studies with the fungal metabolite, brefeldin A (BFA).[37,38]

In the presence of BFA, vesicle-mediated protein transport out of the ER is blocked and tubular interconnections between Golgi cisternae become even more exaggerated. This results in the fusion of Golgi tubular networks with the ER and the mixing of ER and Golgi components. Despite no forward traffic out of the ER in the presence of BFA, *in vitro* assays measuring intra-Golgi transport revealed no change in the rate of transport-coupled processing of glycoprotein compared to untreated cells.[39] These observations suggest, therefore, that tubules can mediate traffic between organelles in the absence of vesicles.

Because membrane tubules in BFA-treated cells are utilized for carrying membrane back to the ER, it has been proposed that tubules normally mediate retrograde traffic.[9,38] Under such a view, the vacuolar domains of transport intermediates (arising from fused ER-derived vesicles) would carry bulk flow luminal contents forward in the secretory pathway, while tubular domains would recycle membrane and specific proteins back to the ER. An attractive feature of this model is that it suggests a simple mechanism based on segregation of structures that maximize surface-to-volume ratio (tubules) from structures that minimize surface-to-volume ratio (vacuoles) for sorting of membrane and luminal content. This would be analogous to sorting of fluid-phase molecules from recycling membrane receptors in the endosomal system.[40]

The above studies have indicated that transport intermediates of the secretory pathway are much more complex than originally thought. Not only do they appear in diverse forms, from small vesicles to elongated tubules arising from larger membrane structures, but they are able to sort molecules in both forward and reverse directions along the secretory pathway (see Figure 1.2). These studies have also emphasized the intimate relationship between membrane traffic and compartmentalization. Organelles serve not only to generate/ consume smaller transport intermediates, but since they translocate through the cytoplasm, organelles themselves can behave as transport intermediates. Further knowledge of the respective roles of vesicles, organelles, and tubules in mediating transport, therefore, will be crucial for interpreting information about the molecular components and mechanisms of transport obtained from genetic and biochemical studies.

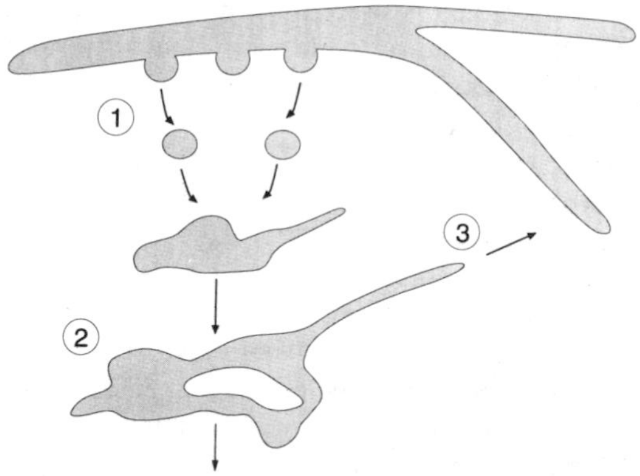

Figure 1.2. Diverse modes of membrane transport. Protein transport can be mediated by vesicles (1), which bud from donor structures, as well as by larger, pleiomorphic membrane transport intermediates (2), which arise from fusion of vesicles. These membrane transport intermediates undergo morphological transformations to produce both tubule processes (3) and vacuolar domains which transport membrane to alternative sites.

1.4. MOLECULAR MACHINERY FOR TRANSPORT

Insight into the molecular machinery that underlies the formation, targeting, and fusion of transport intermediates has come from genetic and biochemical approaches. In *Saccharomyces cereviseae,* for instance, at least 15 genes have been identified which are required for traffic between the ER and Golgi complex, including *SAR 1, SEC 12, SEC 13, SEC 16-23, Bos 1, BET 1, YPT1, and SEC 7.*[32] Many of these proteins have been isolated biochemically from mammalian cells using ER-to-Golgi assays in cell-free systems, indicating a conservation across species of components involved in secretory transport.[29]

1.4.1. Vesicle Formation

Recent studies in yeast have indicated that export from the ER is initiated by the activity of the ras-like GTP-binding protein, sar 1p, which reversibly binds to ER membranes.[41] Once bound to the ER, sar 1p is converted to its GTP-bound form by the resident ER protein sec 12p, which is a GDP/GTP exchange protein.[42] On activation by sec 12p, transport vesicles carrying sar 1p and sec 23/sec 24p bud off from the ER.[43] Within these vesicles sec 23p acts as a GTPase-activating protein (GAP) to convert sar 1p to its GDP-bound form.[44] This releases sar 1p from vesicles, allowing it to rebind to ER membrane, whereupon it can promote another cycle of vesicle budding (see Figure 1.3A). In addition to these molecules, sec 13p and sec 31p have been shown also to participate in promoting vesicle budding and release from the ER.[44] These molecules, when combined with sar 1p and sec 23/24p and then added to ER membranes, produce fusion-competent transport vesicles, suggesting they are all that is needed for vesicle formation. Whether they are all that is needed to mediate concentration of membrane cargo into vesicles is not known. In the above studies in yeast, only transport of soluble cargo (which enters vesicles by bulk flow) was monitored. *In vitro* assays measuring the transport of a *membrane* marker in mammalian cells, however, have revealed the requirement of two additional components—coatomer and ARF (ADP-riboslyation factor)—in vesicle formation and transport.

Coatomers were first identified by Rothman and colleagues in an *in vitro* assay that reconstitutes intra-Golgi transport of the membrane protein, VSV G.[29] Using GTPγS, a chemical reagent that blocks transport, Golgi-derived vesicles carrying an electron-dense protein coat on their surface accumulated. The coat structure on these vesicles was reminiscent of the clathrin coat that surrounds endocytotic vesicles, but lacked its well-defined geometry. When these nonclathrin "coated" vesicles were purified, their protein coat was found to contain a number of stoichiometric components (called

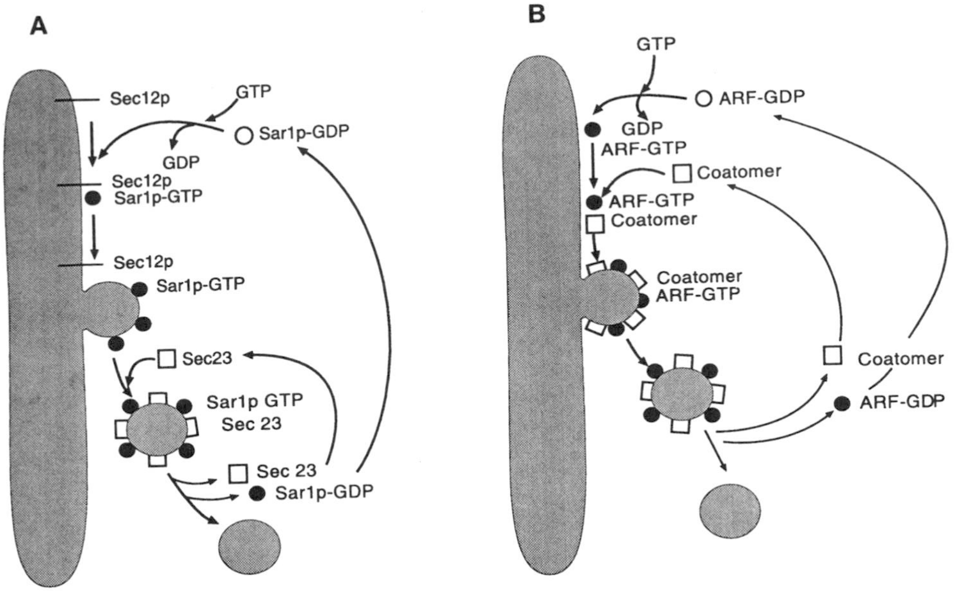

Figure 1.3. Models for vesicle budding derived from yeast genetics (A) and from mammalian *in vitro* systems (B).

COPs) which form a larger "coatomer" complex.[46] One of the COPs, a 110-kDa protein called βCOP, had a sequence that was found to be weakly homologous to β-adaptin, a component of the clathrin coat.[47] Because coatomer complexes could be readily purified from cytosol as well as from vesicles, it was proposed that coatomer cycles between a cytosol and membrane association.[46]

An additional protein identified from coatomers, ARF,[48] which previously was characterized as an ADP-ribosylation factor, was found to mediate the cycle of coatomer association/dissociation from membrane.[49] ARF is a ras-like GTP-binding protein of 21 kDa, which associates with Golgi membranes in a GTP-bound state via a myristic acid residue at its amino terminus.[50,51] Since membrane association of ARF–GTP is a prerequisite for coatomer association with Golgi membranes,[51,52] it has been hypothesized that membrane-bound ARF induces cytosolic coatomers to aggregate on Golgi membranes, triggering the formation of transport vesicles (see Figure 1.3B). The recent finding showing that exogenous ARF 1 added to cells activates phospholipase D (PLO)[53,54] indicates that ARF can also influence the phospholipid content of membrane. Whether PLD activation and coatomer assembly by ARF represent related steps in a single process or diverging pathways in ARF signaling is unknown. These studies, nevertheless, suggest that the processes whereby membrane lipid content is regulated play an essential role in membrane dynamics, organelle integrity, and intracellular traffic.

As yet, there is no direct proof that ARF and coatomer are required for vesicle budding. Several observations are consistent with this possibility, however. First, electron micrographs of isolated Golgi membranes incubated with cytosol (containing coatomer and ARF) and GTPγS showed numerous coated buds and coated vesicles forming from cisternae.[29] If cytosol was depleted of coatomer and incubated with Golgi membranes, no coated buds or vesicles were apparent. Furthermore, addition of purified coatomer reversed this phenotype, resulting in coated vesicle formation. Second, a peptide that blocks ARF activity was found to block both coated vesicle budding and transport-coupled glycosylation of VSV G protein when added to these transport assays.[55,56] Finally, the finding that BFA inhibits ARF activity (by preventing exchange of GDP bound to ARF for GTP)[57,58] has enabled researchers to explore the consequences of ARF inactivation on coat assembly and vesicle budding. BFA treatment was found to block binding of coatomers to membranes and simultaneously to block coated vesicle formation within cells.[38]

Although these observations make a strong argument for a role of coatomer and ARF in coated vesicle budding, the fact that in yeast, vesicles can be produced in the absence of these factors,[45] raises the question of whether coat binding serves other roles. A clue to this role has come from studies with BFA.[38] While preventing "COP"-coated vesicles from forming, BFA also induces retrograde tubules resulting in the net redistribution of Golgi membrane back to the ER. This suggests that an additional role of ARF and coatomer binding is to regulate retrograde transport. One way this could be accomplished is if COPs prevented proteins from entering tubules, perhaps by sequestering them to "coated" membrane domains. The same sorting mechanism could also serve to sequester newly synthesized membrane proteins to ER exit sites where coatomer and ARF have bound to nascent buds. In the absence of COP binding, membrane proteins destined for the plasma membrane would be incapable of sorting to exit sites within the ER. Likewise, membrane proteins in the Golgi would no longer be retained in nontubular domains and instead would be free to diffuse into tubule structures that recycle membrane to the ER. Coatomer binding under this model would serve both to segregate membrane proteins to sites of budding in the ER and to retain them within these "coated" regions as vesicles fuse into larger transport intermediates.

1.4.2. Membrane Targeting and Fusion

Once transport intermediates have formed (with their appropriate membrane and soluble cargo), they target and fuse to other membranes. The class of proteins in the ypt1/rab family of GTP binding proteins has been implicated in these processes.[59] Although their precise functions are unknown, it has been suggested that through a cycle of binding and hydrolysis of GTP, they serve as molecular switches for catalyzing the assembly of protein complexes involved in regulated assembly (or activation) of the machinery for fusion.[60] Proteins of the rab family are thought to fulfill this function as they undergo a cycle of cytosol and membrane localization. Different members of the rab family have distinctive membrane localizations. For example, rab 1b and rab 2 are associated with the Golgi and IC, whereas rab 5 is associated with early endosomes and the plasma membrane.[59]

Despite their diverse cellular localizations, rab family members all share similar "functional domains" including a nucleo-binding domain, an effector domain where GTPase-activating proteins (GAPs) bind, and a C-terminal region where geranyl-geranyl groups are posttranslationally added.[59] Knowledge of the domain structure of rab proteins has led to the construction of mutants and the generation of inhibitory antibodies to study the mechanism of rab action during membrane traffic. Cells expressing rab 1b and rab 2 proteins with mutations in the guanine nucleotide-binding domain, for example, were shown to be blocked in ER-to-Golgi transport and accumulate the rab proteins in their GDP-bound, or inactive, forms.[61] Moreover, in yeast, addition of antibodies to the yeast homologue of rab 1b, YPT1, resulted in a block in ER-to-Golgi transport and an accumulation of vesicles containing YPT1 on their surfaces.[59] Rab proteins, therefore, appear to be loaded onto transport vesicles and remain there until targeting and fusion is complete.

Recent work has focused on identifying accessory proteins whose role is to stimulate rates of GTP hydrolysis via GAPs or rates of dissociation via guanine nucleotide exchange factors (GEFs) that are specific to rab 1b and rab 2. These studies, in turn, have led to the identification of a GDP dissociation inhibitor (GDI)[62] and rabphilin, which is proposed to serve as a rab receptor.[63] These accessory proteins are proposed to function together to control each step of a functional cycle that couples GTP binding and hydrolysis on rab to its membrane attachment, function, and recycling. Thus, the site of membrane attachment of rab proteins might be dictated by the localization of exchange proteins, while the site of dissociation of rabs from membranes could be dictated by the location of GAPs.

Membrane fusion appears to be mediated by additional components that function downstream of the rab proteins. Several recently isolated soluble and membrane proteins have been proposed to form a putative fusion machinery. The first of these proteins to be isolated was an *N*-ethylmaleimide (NEM)-sensitive factor (NSF), which exists as a cytosolic tetramer of 76-kDa subunits.[64] The observation that NSF is required for a large variety of membrane fusion reactions and that sec 18 (the yeast homologue of NSF) also acts at multiple steps of transport have led to the hypothesis that NSF is a general fusion factor.[29] In order to associate with membranes, NSF was found to require an additional cytosolic protein termed soluble NSF attachment protein (SNAP).[65] These proteins, in turn, interact with membrane proteins called SNAREs (SNAP receptors). Identification of three such SNARE molecules, synaptobrevein, syntaxin, and SNAP 25, using an affinity column consisting of NSF–SNAP complex, has given rise to the SNARE hypothesis for membrane fusion[66] (see Figure 1.4). According to this model, v-SNAREs on carrier membranes would interact with a second set of integral membrane proteins on the target membrane, termed t-SNAREs, and through these protein–protein interactions specificity of membrane–target recognition would be established. The paradigm for this model has come from studies of synaptic fusion where synaptophysin/synaptogamin on donor vesicle membrane has been shown to interact with syntaxin on target membrane during synaptic vesicle fusion with the plasma membrane.[67] According to the SNARE hypothesis, soluble NSF/SNAPs would form a bridge to initiate the interaction between v- and t-SNAREs on membranes.[66] Since after membrane fusion v- and t-SNAREs both reside in the same membrane, mechanisms must exit for regulating SNARE activity. Moreover, there must be processes for routing v-SNAREs back to donor membranes if these molecules are to be reutilized in subsequent fusion events. Rabs or ARFs could influence these processes, since both of these molecules are recruited onto membranes from the cytosol by a cycle of GTP binding. Identifying how these and other factors regulate SNARE activity are questions for future work.

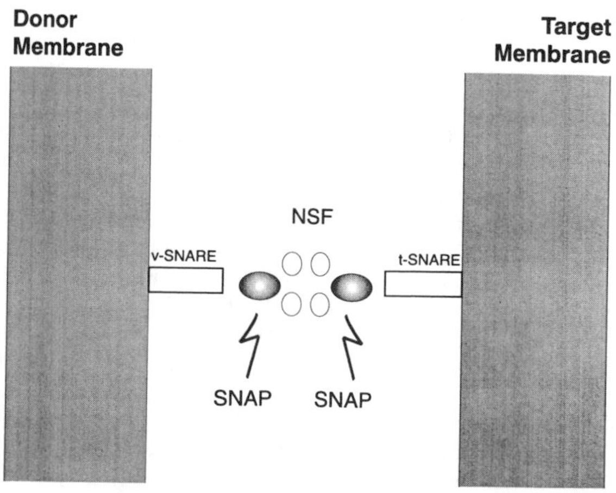

Figure 1.4. Membrane fusion machinery: the SNARE hypothesis.

1.5. CONCLUSION

Our understanding of the mechanisms of membrane traffic and compartmentalization within the secretory pathway has emerged from a variety of morphological, biochemical, and genetic-based research. These studies have revealed a fundamental relationship between membrane traffic and organelle biogenesis and maintenance, with perturbations in membrane traffic quickly leading to changes in organelle structure and identity. Dissection of the biochemical basis of transport and compartmentalization within this system has revealed crucial roles for cytosolic protein complexes which undergo regulated assembly/disassembly with membranes. Future work will need to address how the activity and localization of these cytosolic protein complexes are regulated, and by what mechanism they transform membrane shape and dynamics for the purpose of protein transport and compartmental function.

REFERENCES

1. Terasaki, M., Chen, L. B., and Fujiwara, K. (1986). Microtubules and the endoplasmic reticulum are highly interdependent structures. *J. Cell Biol.* **103:**1557–1568.
2. Koch, G. L. (1992). The endoplasmic reticulum. In *Fundamentals of Medical Cell Biology,* Vol. 4, JAI Press, Greenwich, CT, pp. 397–420.
3. Sitia, R., and Meldolesi, J. (1993). Endoplasmic reticulum: A dynamic patchwork of specialized subregions. *Mol. Biol. Cell* **3:**1067–1072.
4. Hurtley, S. M., and Helenius, A. (1989). Protein oligomerization in the endoplasmic reticulum. *Annu. Rev. Cell Biol.* **5:**277–307.
5. DeSilva, A., Braakman, I., and Helenius, A. (1993). Posttranslational folding of VSV G protein in the ER: Involvement of noncovalent and covalent complexes. *J. Cell Biol.* **120:**647–655.
6. Saraste, J., and Kuismanen, E. (1984). Pre- and post-Golgi vacuoles operate in the transport of Semlike Forest virus membrane glycoproteins to the cell surface. *Cell* **38:**535–549.
7. Saraste, J., and Svensson, K. (1991). Distribution of the intermediate elements operating in ER to Golgi transport. *J. Cell Sci.* **100:**415–430.
8. Hauri, H.-P., and Schweizer, A. (1992). Relationship of the ER–Golgi intermediate compartment to endoplasmic reticulum and Golgi apparatus. *Curr. Opin. Cell Biol.* **4:**600–608.
9. Lippincott-Schwartz, J. (1993). Bidirectional membrane traffic between the endoplasmic reticulum and Golgi apparatus. *Trends Cell Biol.* **3:**81–88.
10. Saraste, J., and Kuismanen, E. (1992). Pathways of protein sorting and membrane traffic between the rough ER and the Golgi complex. *Semin. Cell Biol.* **3:**343–355.
11. Tisdale, E. J., Bourne, J. R., Khosravi-Far, R., Der, C. J., and Balch, W. E. (1992). GTP-binding mutants of rab1 and rab2 are potent inhibitors of vesicular transport from the endoplasmic reticulum to the Golgi complex. *J. Cell Biol.* **119:**749–761.
12. Rambourg, A., and Clermont, Y. (1990). Three-dimensional electron microscopy: Structure of the Golgi apparatus. *Eur. J. Cell Biol.* **51:** 189–200.
13. Thyberg, J., and Moskalewski, S. (1985). Microtubules and the organization of the Golgi complex. *Exp. Cell Res.* **159:**1–16.
14. Rogalski, A., Bergmann, J., and Singer, S. J. (1984). Effect of microtubule assembly status on the intracellular processing and surface expression of an integral protein of the plasma membrane. *J. Cell Biol.* **99:**1101–1109.
15. Turner, J. R., and Tartakoff, A. M. (1989). The response of the Golgi complex to microtubule alterations: Roles of metabolic energy and membrane traffic in Golgi complex organization. *J. Cell Biol.* **109:** 2081–2088.
16. Bretscher, M. S., and Munro, S. (1993). Cholesterol and the Golgi apparatus. *Science* **261:**1280–1281.
17. Weisz, O. A., Swift, A. M., and Machamer, C. E. (1993). Oligomerization of a membrane protein correlates with its retention in the Golgi complex. *J. Cell Biol.* **122:**1185–1196.
18. Farquhar, M. G. (1991). Protein traffic through the Golgi complex. In *Intracellular Trafficking of Proteins* (C. J. Steer and J. A. Hanover, eds.), Cambridge University Press, London. pp. 431–471.
19. Mellman, I., and Simons, K. (1992). The Golgi complex: In vitro veritas? *Cell* **68:**829–840.
20. Pelham, H. R. B. (1991). Recycling of proteins between the endoplasmic reticulum and Golgi complex. *Curr. Opin. Cell Biol.* **3:**585–591.
21. Hsu, V. W., Yuan, L. C., Nuchtern, J. G., Lippincott-Schwartz, J., Hammerling, G. J., and Klausner, R. D. (1991). A recycling pathway between the endoplasmic reticulum and the Golgi apparatus for retention of unassembled MHC class I molecules. *Nature* **352:**441–444.
22. Jackson, M. R., Nilsson, T., and Peterson, P. A. (1993). Retrieval of transmembrane proteins to the endoplasmic reticulum. *J. Cell Biol.* **121:**317–333.
23. Tartakoff, A. M., and Turner, J. R. (1992). The Golgi complex. In *Fundamentals of Medical Cell Biology,* Vol. 4, JAI Press, Greenwich, CT, pp. 283–304.
24. Griffiths, G., and Simons, K. (1986). The trans Golgi network: Sorting at the exit site of the Golgi complex. *Science* **234:**438–443.
25. Griffiths, G., Pfeiffer, S., Simons, K., and Matlin, K. (1985). Exit of newly synthesized membrane proteins from the trans cisterna of the Golgi complex to the plasma membrane. *J. Cell Biol.* **101:**949–964.
26. Huttner, W. B., and Baeuerle, P. A. (1988). Protein sulfation on tyrosine. *Mod. Cell Biol.* **6:**97–140.
27. Sossin, W. S., Fisher, J. M., and Scheller, R. H. (1990). Sorting within the regulated secretory pathway occurs in the trans-Golgi network. *J. Cell Biol.* **110:**1–12.
28. Palade, G. E. (1975). Intracellular aspects of the process of protein secretion. *Science* **189:**347–358.
29. Rothman, J. E., and Orci, L. (1992). Molecular dissection of the secretory pathway. *Nature* **355:**409–415.
30. Balch, W. E., McCaffery, J. M., Plutner, H., and Farquhar, M. G. (1994). VSV G is sorted and concentrated during export from the ER. *Cell* **76:**841–852.
31. Slot, J. W., Oprins, A., and Geuze, H. J. (1993). Non-parallel concentration of pancreatic secretory proteins during intracellular transport. *Mol. Biol. Cell* **4:**207a.
32. Schekman, R. (1992). Genetic and biochemical analysis of vesicular traffic in yeast. *Curr. Opin. Cell Biol.* **4:**587–592.
33. Clermont, Y., Xia, L., Rambourg, A., Turner, J. D., and Hermo, L. (1939). Transport of casein submicelles and formation of secretion granules in the Golgi apparatus of epithelial cells of the lactating mammary gland of the rat. *Anat. Rec.* **235:**363–373.
34. Melkonian, M., Becker, B., and Becker, D. (1991). Scale formation in algae. *J. Electron Microsc. Tech.* **17:**165–178.
35. Cooper, M. S., Cornell-Bell, A. H., Chernjavsky, A., Dani, J. W., and Smith, S. J. (1990). Tubulovesicular processes emerge from the trans-Golgi cisternae, extend along microtubules, and interlink adjacent trans-Golgi elements into a reticulum. *Cell* **61:**135–145.
36. Weidman, P., Roth, R., and Heuser, J. (1993). Golgi membrane dynamics imaged by freeze etch electron microscopy: Views of different membrane coatings involved in tubulation versus vesiculation. *Cell* **75:**123–133.
37. Lippincott-Schwartz, J., Donaldson, J. G., Schweizer, A., Berger, E. G., Hauri, H.-P., Yuan, L. C., and Klausner, R. D. (1990). Microtubule-dependent retrograde transport of proteins into the ER in the

presence of brefeldin A suggests in ER recycling pathway. *Cell* **60:**821–836.
38. Klausner, R. D., Donaldson, J. G., and Lippincott-Schwartz, J. (1992). Brefeldin A: Insights into the control of membrane traffic and organelle structure. *J. Cell Biol.* **116:**1071–1080.
39. Orci, L., Tagaya, M., Amherdt, M., Perrelet, A., Donaldson, J. G., Lippincott-Schwartz, J., Klausner, R. D., and Rothman, J. E. (1991). Brefeldin A, a drug that blocks secretion, prevents the assembly of non-clathrin-coated buds on Golgi cisternae. *Cell* **64:**1183–1195.
40. Geuze, H. J., Slot, J. W., Strous, G. J., Lodish, H. F., and Schwartz, A. L. (1983). Intracellular site of asialoglycoprotein receptor–ligand uncoupling: Double label immuno-electron microscopy during receptor-mediated endocytosis. *Cell* **32:**277–287.
41. Nakano, A., and Muramatsu, M. (1989). A novel GTP-binding protein, Sar 1p, is involved in transport from the endoplasmic reticulum to the Golgi apparatus. *J. Cell Biol.* **109:**2677–2691.
42. Barlow, C., and Schekman, R. (1993). SEC12 encodes a guanine nucleotide exchange factor essential for transport vesicle budding from the ER. *Nature* **365:**347–349.
43. Oka, T., and Nakano, A. (1994). Inhibition of GTP hydrolysis by Sar1p causes accumulation of vesicles that are a functional intermediate of the ER-to-Golgi transport in yeast. *J. Cell Biol.* **124:**425–434.
44. Yoshihisa, T., Barlowe, C., and Schekman, R. (1993). Requirements for a GTPase-activating protein in vesicle budding from the endoplasmic reticulum. *Science* **259:**1466–1468.
45. Salama, N. R., Yeung, T., and Schekman, R. W. (1993). *EMBO J.* The sec13p complex and reconstitution of vesicle budding from the ER with purified cytosolic proteins. **12:**4073–4082.
46. Waters, M. G., Serafini, T., and Rothman, J. E. (1991). "Coatomer": A cytosolic protein complex containing subunits of non-clathrin coated Golgi transport vesicles. *Nature* **339:**355–359.
47. Duden, R., Griffiths, G., Frank, R., Argos, P., and Kreis, T. E. (1991). βCOP, a 110 kD protein associated with non-clathrin coated vesicles and the Golgi complex, shows homology to β-adaptin. *Cell* **64:** 649–665.
48. Serafini, T., Orci, L., Amherdt, M., Brunner, M., Kahn, R. A., Rothman, J. E., and Wieland, F. T. (1991). A coat subunit of Golgi-derived non-clathrin-coated vesicles with homology to the clathrin-coated vesicle coat protein β-adaptin. *Nature* **349:**215–220.
49. Donaldson, J. G., Cassel, D., Kahn, R. A., and Klausner, R. D. (1992). ADP-ribosylation factor, a small GTP-binding protein is required for binding of the coatomer protein bCOP to Golgi membranes. *Proc. Natl. Acad. Sci. USA* **89:**6408–6412.
50. Kahn, R. A. (1991). Fluoride is not an activator of the smaller (20–25 kDa) GTP-binding proteins. *J. Biol. Chem.* **266:**15595–15597.
51. Donaldson, J. G., and Klausner, R. D. (1994). ARF: A key regulatory switch in membrane traffic and organelle structure. *Curr. Opin. Cell Biol.,* **6:**527–532.
52. Palmer, D. J., Helms, J. B., Beckers, C. J. M., Orci, L., and Rothman, J. E. (1993). Binding of coatomer to Golgi membranes requires ADP-ribosylation factor. *J. Biol. Chem.* **268:**12083–12089.
53. Brown, H. A., Gutowski, S., Moomaw, C. R., Slaughter, C., and Sternweis, P. C. (1993). ADP-ribosylation factor, a small GTP-dependent regulatory protein, stimulates phospholipase D activity. *Cell* **75:**1137–1144.
54. Cockcroft, S., Thomas, G. M. H., Fensome, A., Geny, B., Cunningham, E., Gout, I., Hiles, I., Totty, N. F., Troung, O., and Hsuan, J. J. (1994). Phospholipase D: A downstream effector of ARF in granulocytes. *Science* **263:**523–526.
55. Zhang, C., Rosenwald, A. G., Willingham, M. C., Skuntz, S., Clark, J., and Kahn, R. A. (1994). Expression of a dominant allele of human ARF1 inhibits membrane traffic in vivo. *J. Cell Biol.* **124:**289–300.
56. Dascher, C., and Balch, W. E. (1994). Dominant inhibitory mutants of ARF1 inhibit ER to Golgi transport and trigger disassembly of the Golgi apparatus. *J. Biol. Chem.* **269:**1437–1448.
57. Donaldson, J. G., Finazzi, D., and Klausner, R. D. (1992). Brefeldin A inhibits Golgi membrane-catalyzed exchange of guanine nucleotide onto ARF protein. *Nature* **360:**350–352.
58. Helms, J. B., and Rothman, J. E. (1992). Inhibition by brefeldin A of a Golgi membrane enzyme that catalyses exchange of guanine nucleotide bound to ARF. *Nature* **360:**352–354.
59. Ferro-Novick, S., and Novick, P. (1993). The role of GTP-binding proteins in transport along the exocytic pathway. *Annu. Rev. Cell Biol.* **9:**575–599.
60. Novick, P., and Brennwald, P. (1993). Friends and family: The role of the rab GTPase in vesicular traffic. *Cell* **75:**597–601.
61. Nuoffer, C., Davidson, H. W., Matteson, J., Meinkoth, J., and Balch, W. E. (1994). A GDP-bound form of rab1 inhibits protein export from the endoplasmic reticulum and transport between Golgi compartments. *J. Cell Biol.* **125:**225–237.
62. Takai, Y., Kaibuchi, K., Kikuchi, A., and Kawata, M. (1992). Small GTP-binding proteins. *Int. Rev. Cytol.* **133:**187–230.
63. Shiriataki, H., Kaibuchi, K., Yamaguichi, T., Wada, K., Horiuchi, H., and Takai, Y. (1992). A possible target protein for smg-25A/rab3A small GTP-binding protein. *J. Biol. Chem.* **267:**10946–10949.
64. Malhotra, V., Serafini, T., Orci, L., Shepard, J. C., and Rothman, J. E. (1989). Purification of a novel class of coated vesicles mediating biosynthetic protein transport through the Golgi stack. *Cell* **58:**329–336.
65. Clary, D. O., Griff, I. C., and Rothman, J. E. (1990). SNAPs, a family of NSF attachment proteins involved in intracellular membrane fusion in animals and yeast. *Cell* **61:**709–721.
66. Sollner, T., Whiteheart, S. W., Brunner, M., Erdjument-Bromage, H., Geromanos, S., Tempst, P., and Rothman, J. E. (1993). SNAP receptors implicated in vesicle targeting and fusion. *Nature* **362:**318–324.
67. Sudhof, T. C., Camilli, P. D., Niemann, H., and Jahn, R. (1993). Membrane fusion machinery: Insights from synaptic proteins. *Cell* **75:**1–4.

2

Cell Adhesion

Jonathan Covault

2.1. INTRODUCTION

The central role of cell adhesion in multicellular organisms is self-evident. In order to form functional tissues, cells must physically adhere to one another and to extracellular matrix (ECM) materials. Somewhat more subtle is the role that regulated and selective cell adhesion plays in the development, function, and regeneration of vertebrate organisms. Studies of a variety of biological phenomena by zoologists early in this century led to the formulation of quite specific models of selective cell and tissue affinities[1–3] including their hypothetical molecular basis (Figure 2.1). As reflected in this early molecular diagram of the cell membrane, little was known about molecular details of cell adhesion. Nonetheless, as noted in the legend, several of the molecular principles envisioned by Weiss half a century ago have now been given molecular form. During the past 15–20 years, the application of immunologic and molecular genetic tools to these problems has begun to give more detailed form to the hypothetical black bar molecules in Figure 2.1. The list of identified cell adhesion molecules (CAMs) has grown from a handful in 1980 to nearly 100 today. Our goal in this chapter is to outline the steps that have been used to identify CAMs, to describe salient features of this group of membrane molecules, and to give an overview of what is currently known about their roles in a variety of cellular interactions. To accomplish the latter we have chosen to describe in moderate detail the involvement of CAMs in three well-described systems rather than to catalog the many examples of CAM involvement in cellular processes. Finally, our emerging understanding of CAM function or dysfunction in several specific disease processes will be considered.

2.2. EARLY STUDIES OF CELL ADHESION

In contrast to enzyme or ion channel functional activities, cell adhesion is a rather imprecise event to quantitate. Early progress in identification of CAMs relied on the development of convenient assays for selective cell–cell adhesion. The earliest model system dates to the work of Wilson[4] who showed that heterogeneous suspensions of cells mechanically dissociated from two differently colored species of marine sponge (e.g., the red *Microciona* and yellow *Stylotella* species) reassociated into species-specific cell aggregates. Subsequently, Moscona, Roth, and others developed rotary suspension cell reaggregation assays utilizing enzymatically dissociated embryonic vertebrate cells to demonstrate selective cell–cell adhesion.[5,6] In Roth's system, dissociated cells from various embryonic tissues were radiolabeled by preincubation with either ^{3}H- or ^{14}C-labeled metabolic precursors. The binding of these labeled tracer cells to larger homotypic versus heterotypic cellular aggregates could then easily be scored. Table 2.1 summarizes the results from such an experiment. Results from this early work clearly demonstrated the existence of distinct cell adhesion specificities: dissociated cells adhere preferentially to homologous cells rather than cells from other organs.

Numerous modifications of these basic reaggregation assays combined with immunologic techniques have been used to identify a variety of CAMs. Other researchers have used traditional biochemical methods to purify cell surface components which mediate adhesion. One particularly useful approach pioneered by Edelman's group involves production of polyspecific antisera to the entire cell surface of a particular cell type which could selectively block adhesion of cells in rotary suspension cultures or growing as monolayers. By fractionating cell surface components using traditional biochemical techniques, glycoprotein fractions could be purified which neutralized the antiadhesive effects of the original polyspecific antisera and thereby indirectly implicated the purified component in cell adhesion. Such an approach was used to identify many of the earliest and now best-characterized adhesion molecules including NCAM, NgCAM, L-CAM/-uvomorulin/cell-CAM 120/80, and cell–substratum adhesion molecules.[7–15] While little used today to identify new adhesion molecules, these basic *in vitro* cell aggregation assays are finding important utility in showing that newly identified putative CAMs are indeed capable of promoting cell adhesion. In this case cDNAs encoding individual candidate adhesion molecules are transfected into poorly adherent cells such as

Jonathan Covault • Department of Physiology and Neurobiology, University of Connecticut, Storrs, Connecticut 06269.
Molecular Biology of Membrane Transport Disorders, edited by S.G. Schultz *et al.* Plenum Press, New York, 1996

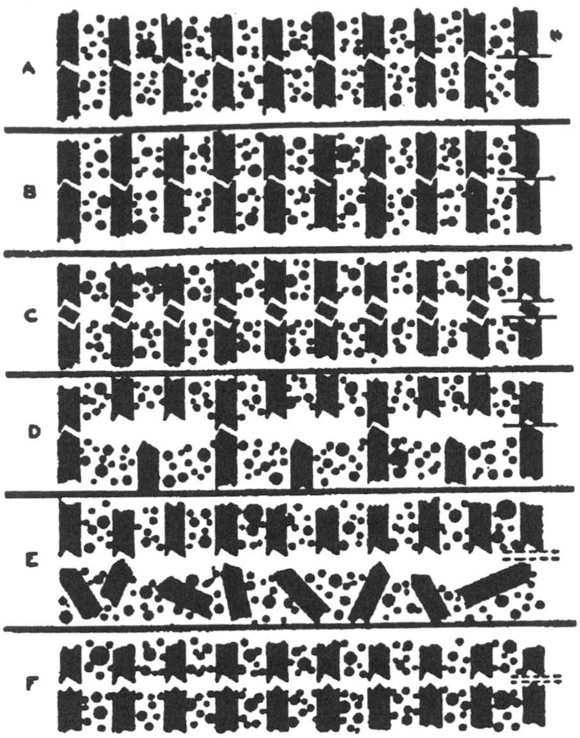

Figure 2.1. "Diagram of hypothetical molecular configurations in adjacent surfaces of two cells" (from Weiss, 1947[2]). The solid, vertically oriented bars are Weiss's representation of cell adhesion molecules with binding sites at their extracellular terminus. Filled circles represent the lipid component of the cell membrane and reflect the primitive molecular understanding of the membrane in the 1940s. The solid versus hatched lines at the right of each panel indicate adhesion versus lack of adhesion, respectively, between apposing cell surfaces, while a singe versus double line indicates close versus slightly distant cellular apposition. Panel A illustrates two different cell populations with "complementary affinity" cell surface ligands which today would be termed heterophilic (molecule) heterotypic (cell type) adhesion. Panel B illustrates "linking between two identical systems—homonomic affinity," each containing a mixture of two mutually adhesive molecules which would now be termed heterophilic homotypic cell adhesion. Panel C illustrates the involvement of "an intermediary substance of complementary configuration" promoting cell–cell adhesion which is illustrated in modern terms by the involvement of multivalent substances such as von Willebrand factor or the extracellular matrix components heparan sulfate, fibronectin, collagen, and laminin as intermediates in cell–cell adhesion. Panel D "is intended to show how affinity can vary in strength" by the modulation of adhesion molecule cell surface density, something that has now been demonstrated experimentally. Panel E illustrates that the "surface orientation of complementary molecules . . . becomes an indispensable prerequisite" for effective cell adhesion. Molecular studies have now confirmed that modulation of the molecular conformation and attendant binding affinity of several cell adhesion molecules is indeed central to the regulation of their physiologic function. Panel F diagrams "detachment (or non-attachment) due to lack of correspondence of shape between the key members of two frontier" cell populations. "Whether or not two systems, which differentiate in contiguity, remain attached or become separated, will, therefore, depend on whether their respective frontier populations undergo parallel or divergent changes" in their population of adhesion molecules. This principle has been molecularly illustrated during the separation of the neural tube ectoderm from the adjacent epidermal ectoderm. An initially uniform expression of the cell adhesion molecules NCAM and E-cadherin by dorsal ectodermal cells is replaced by the selective expression of NCAM by differentiating neural ectoderm and E-cadherin by the adjacent lateral epidermal ectoderm.[261] (Illustration and quotations from Ref. 2 with permission.)

mouse L cells and the conferral of *de novo* adhesiveness is scored by way of short-term cell aggregation assays.[16] A further refinement of Roth's mixed cell reaggregation experiment using fluorescent labeling of native versus cDNA-transfected cells has been used to distinguish homophilic[17] versus heterophilic[18] binding and to demonstrate the ability of two different CAMs to produce cell sorting.[19]

A second generation of CAMs were identified by use of monoclonal antibody (mAb) technology. Panels of mAbs generated against cell surface components have been screened for those that either block adhesion or that stain subsets of cells. It appears that most moderately abundant cell surface glycoproteins with a restricted cellular distribution are in fact CAMs. This correlation provides an important tool as it is a relatively easy, although tedious, task to screen hundreds to thousands of mAbs on tissue sections for those that stain subsets of cells. More recently, with the recognition that individual CAMs are members of larger gene families, molecular genetic approaches have been employed to screen mRNA populations for sequences related to known adhesion molecules in order to identify new CAMs.

Table 2.1. Adhesion of Radiolabeled Cells to Unlabeled Cell Aggregates[a,b]

	Radiolabeled cell type		
Unlabeled aggregate	Cartilage	Liver	Skeletal muscle
Cartilage	100	6	48
Liver	10	100	0
Skeletal muscle	38	49	100

[a]Binding of labeled cells to tissue aggregates expressed as a percentage relative to homotypic binding.
[b]Data from Ref. 6.

2.3. CELL ADHESION MOLECULE FAMILIES

The majority of molecularly characterized CAMs derive from four multigene families—cadherins, immunoglobulin (Ig)-like CAMs, integrins, and selectins—each encoding closely related glycoproteins. A number of CAMs fall outside of these four groups and may represent members of additional CAM families. Specific CAMs together with some of their characteristics are listed in Table 2.2; representatives of each family are diagrammed in Figure 2.2. For more detailed information, several recent reviews[20–25] are available in addition to the selected references noted in Table 2.2. As a group, CAMs share several features. All are glycoproteins, ranging in size from 80 to 200 kDa, with abundant carbohydrate decorations comprising 10–60% of their total apparent molecular mass as determined by SDS gel electrophoresis. Most are transmembrane proteins with a large extracellular domain, a single transmembrane domain, and typically a comparatively short carboxyl-terminal cytoplasmic domain (30–120 amino acids). Several are tethered to the membrane by way of glycosyl-phosphatidylinositol (GPI) lipid linkages, providing high lateral diffusion capability as well as potential rapid release from the cell membrane by extracellular phospholipase activities. The ability to enzymatically release CAMs from the membrane might serve to allow dynamic regulation (decrease) of cell–cell adhesion by GPI-linked CAMs, provide a supply of released CAMs for deposition in ECMs, or produce a soluble CAM ligand which could disrupt cell–cell adhesion on neighboring cells. That one or more of these mechanisms may in fact have a physiologic role is suggested by the finding of high levels of soluble axonin-1, a GPI-linked CAM, in association with specific axon tracts and CSF fluids.[26] Although lacking transmembrane and intracellular domains, GPI-linked cell surface proteins can nonetheless be involved in generation of transmembrane signaling events presumably by complexing with transmembrane proteins.[27,28]

Multiple isoforms of individual CAMs are frequently produced by way of stage- and cell type-specific variation in mRNA splicing. NCAM provides the best-studied example.[29–32] At least three major variations in core polypeptide length and cellular association are produced by use of alternate carboxyl-terminal domains. The shortest variant, incorporating exons 1–15, generates a 120-kDa form of NCAM which is GPI-linked to the membrane. A second variant, utilizing exons 1–14, 16, and 18, has an apparent size of 140 kDa, a transmembrane domain and a small cytoplasmic domain. The third, a 180-kDa form, is encoded by exons 1–14 and 16–18 and contains a large cytoplasmic domain which, unlike the 140-kDa form, binds to the cytoskeletal protein fodrin.[33] In addition to these variations in cell association, which have considerable impact on NCAM function, several smaller variations of extracellular domains are produced by splicing variants incorporating small exons located between exons 7 and 8 and between exons 12 and 13. The exon 12–13 variant, which is frequently seen in muscle and glial cells, is thought to introduce a kink in the NCAM polypeptide, decreasing its effective extension from the cell surface. Both the macro and micro NCAM splice variants show tissue and developmental specificity in their expression.

As a group CAMs are about evenly divided between those that utilize homophilic versus heterophilic binding to produce intercellular adhesion. In several cases, molecular studies have identified regions of individual CAMs essential for intermolecular binding. In all cases to date the intercellular binding domain is positioned, perhaps not unexpectedly, at the distal amino-terminal end of the molecule. Electron microscopic images have been reported for three CAMs.[34–37] The typical appearance is that of 2- to 4-nm-thick rods 19–50 nm long, often with a single, apparently flexible, 90° bend. Preparations of the homophilic adhesion molecule NCAM frequently appear as dimers or trimers with molecular contacts at their extreme, formerly extracellular ends confirming biochemical studies of the NCAM binding site. Thus, although prepared about 50 years ago, Weiss's sketches (Figure 2.1) of hypothetical CAMs depicted as rods with CAM bind sites at their distal ends turn out to be perhaps more accurate than even Weiss might have imagined.

2.3.1. Cadherins

The cadherin family of adhesion molecules mediate Ca^{2+}-dependent adhesion, hence the name cadherin. Most cadherins are named using a prefix letter denoting the tissue in which that cadherin was first characterized. Cadherins are capable of producing relatively strong cell–cell adhesion. They are generally considered to be the class of CAMs most important for the formation of long-lasting cell–cell adhesion. With the exception of desmocollins and desmogleins, which interact with intermediate filaments, cadherins are typically colocalized with intracellular actin bundles. E-cadherin, perhaps the best characterized of this group, is a key component of adherens-type cell–cell junctions. The extracellular domain, comprising three-quarters of the molecule, contains three internally repeated domains each predicted to contain a series of six antiparallel β strands and a pair of Ca^{2+}-binding sites[38] (Figure 2.2A). The two amino-terminal domains show the greatest homology between individual cadherins with approximately 70% identity between members.[21] In addition, E-, N-, B-, R-, and M-cadherin have highly conserved cytoplasmic domains essential for cell binding function.[39] This intracellular conserved domain is thought to mediate indirect cadherin interaction with actin filaments via a set of three intermediary cytoskeletal proteins termed α, β, and γ catenin.[40–42] Cadherin adhesion activity is dependent on the integrity of this catenin linkage of cadherins to the cytoskeleton: transfection of catenin cDNA into cadherin-positive but catenin-deficient cells causes cell aggregation while mutation of the catenin binding domain of cadherins abolishes adhesion.[39,41,43] T-cadherin, a truncated cadherin, is GPI-linked to the membrane and has no cytoplasmic domain while the desmoglein and desmocollin cadherins have a completely different cytosolic domain sequence which may reflect their association with intermediate as opposed to actin filaments.

The intercellular binding activity of cadherins appears to be localized in the amino-terminal quarter of the molecule. Cadherin subtype-specific antibodies which block adhesion bind here, and exchange of an amino-terminal, 113-amino-acid domain can interconvert the binding specificities of E- and P-cadherin.[44] Second, a nearly invariant tripeptide sequence, histidine–alanine–valine, is found in this region and may be involved in homophilic binding. Synthetic peptides containing this sequence have been shown to block cadherin function.[45] Finally, site-specific, single amino acid mutations of the putative Ca^{2+}-binding site within the first amino-terminal repeating domain inactivate the adhesive function of E-cadherin.[46]

The list of cadherins will undoubtedly continue to grow. In addition to the cadherins listed in Table 2.2, eight other, less well-characterized cadherins, denoted cadherin-4 through -11, have been identified by use of polymerase chain reaction amplification of brain RNA using primers encompassing the highly conserved cytoplasmic domains shared by N-, E-, and

text continues on page 20

Table 2.2. Four Families of Cell Adhesion Molecules

Adhesion molecule	Other names	Expressed by	Ligands	Key features and putative functions	Selected references
Cadherins					
B-cadherin		Subsets of CNS neurons, choroid plexus, developing organs and tissues	Unknown		262
Desmocollins	Dsc 1, 2, and 3	Epithelial cell desmosomes	Unknown	Family of three related genes. The desmocollin cytoplasmic region is sufficient to induce the formation of desmosomelike plaques and anchorage of intermediate filament bundles	102, 263–266
Desmogleins	Dsg 1, 2, and 3 (PVA)	Epithelial cell desmosomes	Unknown	Family of three related genes. Contain a repeat of a 29-aa sequence within a cytoplasmic domain. Autoantibodies to Dsg 3 (PVA) produce a life-threatening skin disease, pemphigus vulgaris.	102,103, 267–269
E-cadherin	Uvomorulin, L-CAM, cell-CAM 120/80	Epithelia cells, subset of DRG sensory neurons. Blastomeres of early embryo	E-cadherin	Segregation of tissue layers. Transmembrane component of adherens junctions. Tumor suppressor	38,99,100, 194,198, 261,270
M-cadherin		Developing myotubes	Unknown	May participate together with N-cadherin and NCAM in promoting myoblast fusion with developing myotubes	271
N-cadherin	A-CAM	Developing neural tube, urogenital system, neural crest derivatives, CNS glial cells, cardiac, embryonic, and denervated skeletal muscle	N-cadherin	Regulation of morphogenetic cell movements during somitogenesis and formation of the early nervous and urogenital systems. Stimulation of axon growth. Muscle cell adhesion/fusion	101,159, 272–274
P-cadherin		Placenta, developing endoderm and ectoderm	P-cadherin	Dynamic patterns of expression which correlate with morphogenetic events	275,276
R-cadherin		Retina neuron and glial cells	R- and N-cadherin	Organization of retinal components during late stages of retinal morphogenesis	277
T-cadherin (truncated)	Heart, skeletal muscle, kidney, subsets of neurons. Rostral somitic sclerotome	Unknown		A "truncated" cadherin lacking a cytoplasmic domain. CPI-linked to the membrane. Restricted expression to the rostral half of somatic sclerotomes suggests a	278,279

				potential role in segmentation of the peripheral nervous system	
Ig-related CAMs					
ABGPs (rat) (ankyrin-binding glycoproteins)		Postnatal and adult brain	Unknown	Six Ig C2 type domains/four FNIII repeats proline/threonine-rich *O*-glycosylation domain. Binds 1:1 with high affinity to the cytoskeletal protein ankyrin. Stabilization of neural adhesion via strong cytoskeletal binding.	57
Axonin-1 (chick)	TAG-1 (rat); TAX-1 (human)	Transiently by subsets of developing neurons/axons	TAG-1/axonin-1; G4/NgCAM	Six Ig C2 domains/four FNIII repeats. GPI-membrane anchor. Large amounts are secreted/shed by developing axons. Involved in selective axons fasciculation and growth	26, 280–282
BGP (biliary glycoprotein)	MHVR	Liver, intestinal epithelial cells	BGP	Member of CEA-related gene family. Multiple isoforms containing 2–3 Ig domains result from alternative splicing. Ca-dependent homophilic adhesion, unknown functional role. Receptor for murine hepatitis virus	175,176, 283–285
CD2	LFA-2, erythrocyte rosette receptor	T lymphocytes	LFA-3		286,287
CD4		Helper T lymphocytes, macrophages, subsets of neurons and glial cells	Class II MHC antigen	Two V and one C2 Ig domain. Participants with T-cell receptor in T-cell activation by increasing the avidity of T-cell adhesion to B cells and macrophages and by CD4 stimulation of tyrosine kinase activity. Receptor for human immunodeficiency virus	288,289
CD8		Killer T lymphocytes	Class I MHC antigen	Heterodimer each containing one V-type Ig domain. Participates with T-cell receptor in T-cell activation by increasing the avidity of T-cell adhesion to target cells and by CD8 stimulation of tyrosine kinase activity.	288,290
CEA (carcino-embryonic antigen)		Entire surface membrane in colonic carcinomas and developing intestinal epithelia, restricted to luminal microvillus membrane in adult	CEA	One V and six C2-type Ig domains. GPI-membrane anchor. Overexpressed by colon carcinomas. Weak intercellular adhesion; intercellular expression may prevent formation of close intercellular apposition/adhesion	205,291, 292

(Continued)

Table 2.2. Continued

Adhesion molecule	Other names	Expressed by	Ligands	Key features and putative functions	Selected references
F11 (chick)	F3 (mouse); contactin (chick)	Subsets of developing and adult neurons.	NgCAM, restrictin (ECM glycoprotein), J1-160/180	Six Ig domains/four FNIII repeats. GPI-membrane anchor. Appears late in development. Involved in selective axon fasciculation	75,76 293–296
ICAM-1	CD54	Endothelial and many epithelial cells. Low levels in basal state but strongly induced by inflammatory cytokines	LFA-1; Mac-1	Five Ig C2 domains. Highest relative affinity of the ICAM family for LFA-1 binding. LFA-1 binding maps to first Ig domain. Mg-dependent adhesion. Adhesion of leukocytes to activated endothelia and epithelia. Human rhinovirus receptor	170–172, 297,298
ICAM-2		Constitutively expressed on resting endothelial cells	LFA-1	Two Ig C2 domains. Mg-dependent adhesion. Recirculation of LFA-1-positive lymphocytes through tissue endothelium.	299,300
ICAM-3	ICAM-R	Resting and stimulated lymphocytes, monocytes, and neutrophils. *Not* expressed by endothelial cells	LFA-1	Five Ig C2 domains. 37-aa cytoplasmic domain. Initial lymphocyte–lymphocyte adhesion	301–304
L1 (mouse)	NILE (rat, NGF inducible large external glycoprotein)	Unmyelinated axons of postmitotic neurons. Schwann cells	L1	Six Ig domains/five FNIII repeats. Cytoplasmic domain 50% identity with rat ankyrin binding domain. Involved in axon fasciculation, stimulation of neurite growth, and cerebellar granule cell migration	305–308
LFA-3 (lymphocyte function-associated antigen-3)	CD58, Pan-CAM	Widespread distribution on both hemopoietic and non-hemopoietic cells	CD2		286,309, 310
MAG (myelin-associated glycoprotein)		Oligodendroglia, myelinating Schwann cells particularly at the periaxonal membrane	MAG; collagen; heparan sulfate	Five Ig domains. Involved in neuron–oligodendroglial adhesion, may be initiate of myelination. Axon–myelin sheath adhesion, maintenance of 12–15-nm periaxonal spacing	311–315
NCAM	CD56	Many embryonic cell types, adult neurons (cell body, dendrites, and axons), and denervated muscle	NCAM, heparan sulfate	Five Ig domains/two FNIII repeats. Transmembrane and GPI-anchor forms. Contains variable amounts of an unusual polysialic acid which can modulate cell contact. Involved in histogenesis in derivatives of all germ layers	8,30,32, 34,35, 51,148, 166,261, 316

Neurofascin (chick)		Subsets of developing axon tracts	Unknown	Six Ig C2-type domains/four FNIII repeats, proline/threonine-rich domain. Cytoplasmic region containing ankyrin-binding domain. Selective axon fasciculation	317,318
NgCAM (chick, neuron–glia CAM)	G4; 8D9	Unmyelinated axons of postmitotic neurons. Schwann cells	NgCAM; axonin-1; uncharacterized glial ligand	Six Ig domains/five FNIII repeats. May be chick homologue of L1 but large sequence divergence. Involved in axon fasciculation, stimulation of axon growth and neuronal migration	10, 319–323
NrCAM (chick, NgCAM related)	Bravo	Subset of embryonic and posthatch neuron cell bodies and processes. Muller glial cells	Unknown	Six Ig C2-type domains/five FNIII repeats. Proteolytic site in FNIII domain 3. Cytoplasmic region containing ankyrin-binding domain. Axon fasciculation, neuron–neuron and neuron–glial adhesion	324,325
P_o		Myelinating Schwann cells	Po	Single Ig V-type domain. 69-aa highly basic cytoplasmic domain. 50% of PNS myelin protein. Adhesion between external (and perhaps internal) leaflets of compact myelin Schwann cell membranes	17,326
PECAM-1 (platelet–endothelial CAM-1)	CD31; endoCAM	Endothelial intercellular junctions, platelets and leukocytes	Unknown, PECAM-1 likely	Six Ig C2 domains. Ca-dependent adhesion. Endothelial cell–cell adhesion; platelet aggregation/adhesion to endothelia	327–329
SC1/GRASP (Ig-related restricted axonal surface Protein)	BEN, JC7, F84.1 (rat)	Transient expression on subsets of developing neurons and a variety of developing epithelia. Persistent expression on subsets of glial cells late in development	SC1/GRASP	Two V and three C2-type Ig domains. Decreased expression in motoneurons following target contact. Selective axon fasciculation and growth	160, 330–332
VCAM-1 (vascular CAM-1)	INCAM-110	Endothelial cells—constitutively on PP-HEV, induced by chronic inflammation elsewhere	$\alpha4/\beta1$ and $\alpha4/\beta p$ integrins	Ca-dependent adhesion involved in tissue-selective lymphocyte recirculation	134,333
Integrins					
Leukocyte adhesins (Leu-CAM)					
β subunit / α subunit					
β_2 (CD18) / α_L (CD11a)	LFA-1, CD11a/CD18	Lymphocytes, monocytes	ICAM-1, ICAM-2, ICAM-3	Requires cytokine activation. Leukocyte adhesion to activated endothelium. Adhesion of cytotoxic T lymphocytes to target cells and activation of T lymphocytes	133, 334–338

(Continued)

Table 2.2. Continued

Adhesion molecule		Other names	Expressed by	Ligands	Key features and putative functions	Selected references
Integrins						
Leukocyte adhesins (Leu-CAM) *(cont.)*						
β subunit	α subunit					
β_2	α_M (CD11b)	Mac-1, CR3 (complement receptor 3)	Neutrophils, monocytes	ICAM-1, inactivated complement component C3b, fibrinogen, factor X	Neutrophil aggregation, neutrophil–endothelial adhesion, phagocytosis of opsonized particles	337, 339–341
β_2	α_X (CD11c)	p150,95	Macrophages, monocytes	Inactivated complement component C3b	Monocyte–endothelial adhesion, adhesion of cytotoxic T lymphocytes to target cells	342,343
VLA proteins						
β_1 (CD29)	α_1	VLA-1	Endothelial cells, neurons, fibroblasts, smooth muscle cells	Laminin (cross region), collagens		344
β_1	α_2 (CD49b)	VLA-2, ECMRII	Endothelial cells, platelets, phagocytes, fibroblasts, epidermal cells	Collagens, laminin	Echovirus 1 receptor	183,345
β_1	α_3	VLA-3, ECMRI	Many cell types	Laminin, collagens, fibronectin (RGD)		345,346
β_1	α_4 (CD49d)	VLA-4, PP-homing receptor	Lymphocytes, monocytes, eosinophils, fibroblasts	Type III connecting segment of fibronectin via EILDV sequence, PP-HEV (VCAM-1)	Leukocyte homing to areas of chronic inflammation, fibroblast migration and attachment in wound repair	117,347, 349,351
β_1	α_5 (CD49e)	VLA-5, FNR (fibronectin receptor)	Endothelial cells, platelets, phagocytes	Fibronectin (RGD)	Induction of specific gene expression	58,346, 352,353
β_1	α_6 (CD49f)	VLA-6	Neurons	Laminin	Stimulation of neurite outgrowth on laminin substrates	354,355
$\beta_P(\beta_7)$	α_4	VLA-4 (alt), PP-homing receptor	Lymphocytes	Fibronectin via EILDV sequence,		356–359

				VCAM-1, PP-HEV		
β_4	α_6	VLA-6 (alt)		Laminin	Component of hemidesmosomes	360–362
Cytoadhesins						
β_3 (CD61)	α_{IIb} (CD41)	IIb/IIIa	Platelets	Fibrinogen, fibronectin, von Willebrand factor, thrombospondin, (RGD)	Platelet aggregation and thrombus formation. Absent in Glanzmann's thrombasthenia platelets	363–365
β_3	α_V (CD51)	VNR (vitronectin receptor)	Endothelial cells	Vitronectin, fibrinogen, fibronectin, von Willebrand factor, thrombospondin, (RGD)	Endothelial cell adhesion to thrombus components required in the formation of hemostatic plug and wound healing in damaged endothelium. Adenovirus endocytosis	366–369
β_5	α_V	VNR (alt)		Vitronectin, fibronectin, (RGD)		371
Selectins						
L-selectin		Homing receptor; LECCAM-1; LECAM-1, MEL-14; Leu-8	Leukocytes (constitutive)	Mucin-like CAM GlyCAM-1; sialyl Lewis^x-like glycoprotein; fucosylated and sulfated carbohydrates; peripheral lymph node addressin	Homing receptor expressed by circulating lymphocytes required for their recognition and binding to specific glycoconjugates present on HEVs of peripheral lymph nodes. Acts as a ligand for P- and E-selectin-mediated leukocyte rolling on activated endothelium. Decreased expression on cell activation	64–66, 71,73, 128,136
E-selectin		ELAM-1 (endothelial leukocyte adhesion molecule-1); LECCAM-2; skin vascular addressin	Endothelium (transcriptionally activated)	150-kDa neutrophil glycoprotein; sialyl Lewis^X-like glycoconguates; CLA (cutaneous lymphocyte antigen)	Increased expression over a few hours in endothelial cells exposed to inflammatory stimuli. Participate in neutrophil binding to activated endothelium	67,68, 70,74
P-Selectin		GMP-140 (granule membrane protein-140 kDa); PADGEM (platelet activation-dependent granule external membrane protein); CD62; LECCAM-3	Activated platelets and endothelium	120-kDa neutrophil glycoprotein; sialyl Lewis^x-like glycoconjugates and sulfated glycans	Surface expression increases within minutes after thrombotic activation of platelets or cytokine activation of endothelium via mobilization of P-selectin storage granules Binding of leukocytes to platelet thrombi and to inflamed endothelia	63,69, 72,130, 370

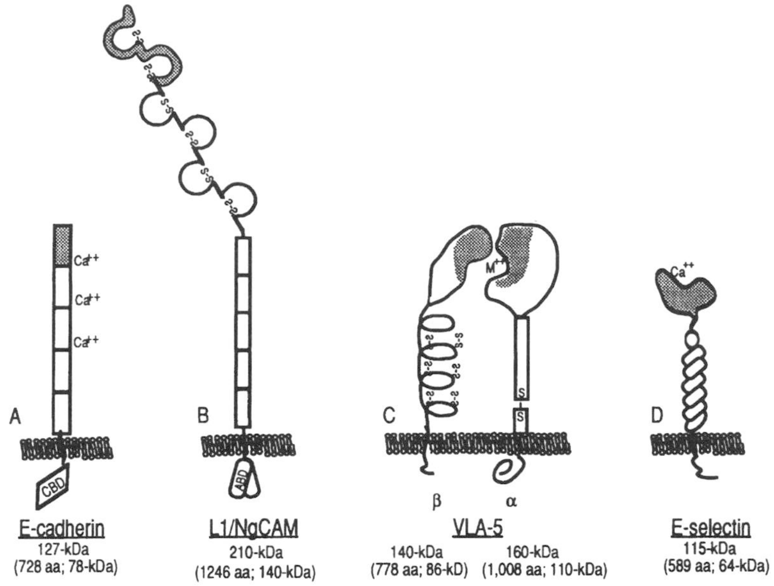

Figure 2.2. Schematic diagrams of representatives of the four major families of cell adhesion molecules. In each case the amino-terminus is located extracellularly (top of figure). The molecular mass of each glycoprotein is listed together with the size of the core polypeptide chain in amino acids and the predicted molecular mass of the unglycosylated core polypeptide, for comparison. Binding domains are indicated by stippling. (A) The extracellular domain of *cadherins* typically contains three to five repeating domains of 110–115 amino acids each. The binding specificity of individual cadherins resides in the first domain; conserved calcium binding sites are typically situated near the carboxyl-terminus of the first three or four domains. A highly conserved ~150-amino-acid cytoplasmic segment contains a catenin binding domain (CBD). (B) *Ig-related CAMs* contain from one to six disulfide-bonded Ig folds beginning at the amino-terminus, followed in many cases by several ~100-amino-acid fibronectin type III repeats (represented as five rectangular boxes for L1/NgCAM). Ig-CAMs frequently have a distinct hinge region in their extracellular domain. Several Ig-related CAMs, including L1/NgCAM, have a conserved cytoplasmic ankyrin-binding domain (ABD); a few have no cytoplasmic or transmembrane domain and are attached to the membrane by phosphatidylinositol linkage. (C) *Integrins* are heterodimers: the larger α subunit contains Ca^{2+}/Mg^{2+} binding sites (M^{++}) while the β subunit contains a series of cysteine-rich repeats near the transmembrane domain. The amino-terminal portion of the two subunits interact to form a globular head which contains the ligand binding site. (D) *Selectins* contain a 120-amino-acid Ca^{2+}-dependent lectin domain at their amino-terminal end, followed by a 30-amino-acid EGF-like domain (circle) and a variable number of 60-amino-acid complement regulatory-like repeats (ovals).

P-cadherin.[47] Similar strategies involving the conserved amino-terminal domain will likely yield other members of the cadherin multigene family. A bigger challenge will be to characterize the role of each in membrane physiology.

2.3.2. Immunoglobulin (Ig)-Related CAMs

Most Ig-related CAMs mediate Ca^{2+}-independent adhesion, typically via homophilic or heterophilic binding to other Ig-related CAMs. The I-CAM subgroup represents a diversion from this theme by mediating divalent cation-dependent adhesion by interaction with components of the integrin CAM family. No standardized naming scheme has been adopted for the Ig-related family. Many are named according to the cell types initially thought to utilize that particular CAM. The unifying feature of Ig-related CAMs is the presence of one to six repeated immunoglobulin-like folds or domains each approximately 100 amino acids in length and containing a pair of cysteine residues 55–75 aa acids apart (Figure 2.2B). Three variations of the basic immunoglobulin domain have been described, and designated V, C1, and C2 after their original description in the variable versus constant region of immunoglobulins. Most CAM Ig domains are of the C2 variety. The immunoglobulin superfamily[48,49] appears to be an ancient family of cell surface receptor, adhesion, and recognition proteins which predates the evolution of the vertebrate immune system and traditional immunoglobulin molecules. Based on structural determinations of immunoglobulin Bence-Jones proteins,[50] the basic Ig domain is thought to form a compact 4-nm × 2.5-nm cylindrical structure stabilized by interactions of antiparallel β strands (Figure 2.3). This produces two β sheets each containing three to four β strands crosslinked by disulfide bonding between the pair of cysteines contained in each Ig fold. CAM Ig domains are thought to be arranged in a linear, unpaired, head-to-tail arrangement producing a flexible rod. Ligand binding functions of NCAM and ICAM-1 have been studied in molecular detail. Adhesion functions appear to directly involve the amino-terminal-most Ig folds.[36,51–53]

In addition to the defining Ig domain, many CAM members of this multigene family have multiple fibronectin type III (FNIII) repeats located between the amino-terminal Ig domains and the cell membrane attachment site. These FNIII domains are also found in the extracellular matrix glycoproteins fibronectin and tenascin as well as in a putative adhesion molecule KALIG-1 whose expression is altered in Kallmann's

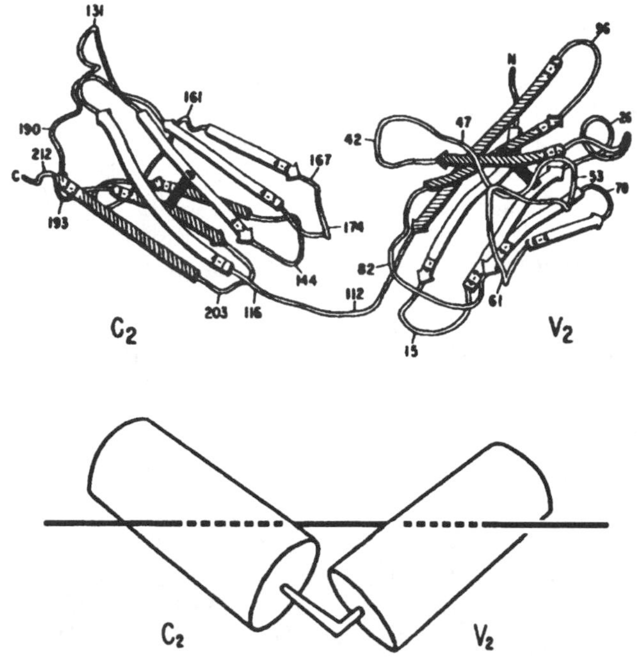

Figure 2.3. Schematic drawing of adjacent C and V domains within an immunoglobulin light chain polypeptide, as deduced from x-ray crystallography. (Top) Amino-to-carboxyl-terminus arrows are indicated for each 5- to 10-aa-long β strand. Four β strands form one β sheet in each fold (white arrows) while the second is formed by three β strands (hatched arrows). A solid black bar denotes the position of the disulfide bond linking the two β sheets of each Ig fold near their centers. Variation in the size of individual Ig domains is produced by addition of amino acid residues in the sequence connecting the two β sheets within a given Ig fold. Numbers indicate the position of specific amino acids within the crystal structure. (Bottom) Cylindrical schematic diagram illustrating the volume occupied by the two Ig domains. (From Ref. 50 with permission.)

syndrome (see Section 2.6.4). In addition, they have been found in the extracellular portion of membrane-associated protein tyrosine phosphatases.[54,55] The crystal structure has been determined for a tenascin FNIII domain.[56] Similar to the Ig fold, it is comprised of seven antiparallel β strands forming two β sheets. In the case of the tenascin FNIII repeat, the amino acid triplet Arg-Gly-Asp, thought to be directly involved in adhesion, was found exposed at a surface edge of this compact structure. If the Ig-related CAM FNIII repeats, like the Ig domains, are involved in intermolecular adhesion, their localization near the membrane attachment site suggests they function to produce lateral intermolecular binding of adjacent CAMs on the cell surface or CAM binding to other membrane proteins such as the recently identified FNIII-containing protein tyrosine phosphatases noted above.

A subset of the Ig-related CAMs (ABGP, neurofascin, NrCAM, NgCAM, and L1) contain a highly conserved 110-aa cytoplasmic domain which appears to confer binding of the cytoskeletal protein ankyrin.[57] This association suggests that adhesion mediated by these CAMs may be directly linked to, and perhaps modify, the cytoskeleton.

2.3.3. Integrins

Integrins are a family of adhesion receptors with widespread distributions on vertebrate cells. All cells express at least one and often several members of this adhesion family. Individual integrins were initially identified by biologists working on highly divergent problems including identification of lymphocyte surface proteins induced by antigen exposure, study of leukocyte–endothelial cell adhesion, identification of myoblast and fibroblast cell surface proteins involved in adhesion to extracellular matrix proteins, and study of platelet function and hemostasis. The variety of names initially given to individual members of the integrin family reflect this diverse heritage and include lymphocyte function antigen-1 (LFA-1), the subfamily of very late antigens (VLA-1 through VLA-6) appearing on the lymphocyte surface several days following lymphocyte activation, the Peyer's patch homing receptor, fibronectin, laminin, and vitronectin receptors, and platelet band IIb/IIIa. Comparison of immunologic cross-reactivity, study of the subunit structure and cDNA sequencing of these various receptors showed that they were in fact closely related but distinct receptors. The name *integrin* was initially proposed by Tamkun *et al.*[58] for the chick fibronectin receptor to denote its integral membrane nature and its role in maintaining the integrity of linkage between the cytoskeleton and extracellular matrix. The term has subsequently come to be used to refer to the entire family of receptors. Integrins all share a noncovalently associated αβ heterodimer subunit structure (Figure 2.2C), and are increasingly identified by their subunit composition; for example, $\alpha_L\beta_2$ for LFA-1, $\alpha_5\beta_1$ for VLA-5 (also formerly known as integrin or the fibronectin receptor), or $\alpha_{IIb}\beta_3$ for the platelet band IIb/IIIa. Eight different 90- to 110-kDa β subunits and fourteen different 120- to 180-kDa α subunits have been identified to date, forming some 15–20 well-characterized integrin heterodimers. Many α subunits are actually composed of two polypeptides originating from a single precursor polypeptide and bound by a cysteine–cysteine disulfide bond. All α subunits contain a seven-fold repeated aspartic acid-rich domain thought to be involved in divalent cation (Mg^{2+} and Ca^{2+}) binding, an essential feature of integrin receptor function. The β subunit contains a repeating structure consisting of 20% cysteine residues with multiple intramolecular disulfide bonds of unknown function.

The better-described integrins can be divided functionally into three groups (Table 2.2): the leukocyte adhesins involved in leukocyte cell–cell adhesion; the VLA family of proteins, most of which bind to one or more extracellular matrix components; and the cytoadhesins, a collection of integrins important for platelet function and hemostasis. The leukocyte adhesins share a common β_2 subunit, while the VLA family all incorporate a β_1 subunit which is constitutively expressed in nearly all cell types. Integrins all mediate adhesion via heterophilic binding to nonintegrin proteins. Most bind more than one ligand, and several ligands (including fibronectin, collagen, and laminin) are bound by more than one integrin. Some integrins mediate cell–cell adhesion by binding to Ig-related CAMs. A binding feature shared by the collagen, fibronectin, and vitronectin integrins is the presence of the amino acid triplet Arg-Gly-Asp (RGD) within the binding domain of their respective ligand proteins.[59,60] Core peptide sequences have similarly been identified within ligands for other integrins. Synthetic peptides containing these sequences have been used experimentally as probes of integrin function by competitively blocking integrin adhesion. The actual binding site is thought to be formed by the interaction of the N-terminal domains of the α and β subunits, while the cytosolic C-terminal domains are thought to interact with components of the cytoskeleton including talin[61] and α-actinin.[62]

2.3.4. Selectins

Selectins are the most recently discovered family of CAMs. They number three as of this writing and all contain an amino-terminal type C (Ca^{2+}-dependent) carbohydrate-binding lectin domain,[63] which represents their hallmark feature as well as the basis of the name *selectin.* The extracellular domain of selectins additionally contains an epidermal growth factor-like (EGF) domain and two to nine complement-binding protein-like repeats followed by a transmembrane domain and a short cytoplasmic domain (Figure 2.2). Individual selectins are named using a letter prefix indicating the cell type of their initial isolation (e.g., *l*eukocytes, *e*ndothelium, and *p*latelets). The three selectin genes are tightly linked on human chromosome 1, giving further evidence of their likely origin by local gene duplication. As elaborated in Section 2.5.2, L-selectin is constitutively expressed by subsets of circulating lymphocytes and mediates their adhesion to high endothelial venules of peripheral lymph nodes.[64–66] E-selectin is induced via transcriptional activation in vascular endothelial cells by inflammatory stimuli.[67,68] Endothelial cell E- and P-selectins are involved in leukocytes rolling, an early step in the binding of leukocytes to activated endothelium. P-selectin is stored in platelet α-granules and endothelial cell storage granules, and is mobilized within minutes to the cell surface following thrombotic activation of platelets or inflammatory cytokine activation of endothelia. Platelet P-selectin acts to promote the binding of monocytes and neutrophils to activated platelets, thereby incorporating these phagocytic cells into nascent thrombi as well as providing a mechanism to clear circulating activated platelets.[69] Selectin-mediated adhesion appears to involve recognition of specific sialyl Lewis^x-like glycoconjugates present on the cognate cell surface.[70–74]

2.4. CELL ADHESION MOLECULES DO MORE THAN BIND CELLS TOGETHER

As more is learned about these growing collections of CAMs, it is becoming increasingly clear that the term *cell adhesion molecule* refers more to an *in vitro* operational criterion than a definition of their role *in vivo.* Taken to the extreme, CAMs have been shown to act in certain circumstances as antiadhesins. For example, the neural cell adhesion molecule F11 displays several distinct binding activities. Adhesion to one ligand results in stimulation of axon growth and continued cell–cell contact while contact with a second ligand, the extracellular matrix component J1-160/180, leads to cell process withdrawal and ultimately the loss of cell adhesion.[75,76] A second example of a paradoxical antiadhesion role involves the posttranslational modification of NCAM by α-2,8-linked polysialic acid. This unusual form of sialic acid can act to prevent cell adhesion not only by NCAM but also by other CAMs (Section 2.5.3).

Cell contact activation of specific CAMs can lead to rearrangement of the cytoskeleton producing focal contact structures or leading to the generation of highly organized junctional complexes (Section 2.5.1). In other cases activation of CAMs plays a role in the mobilization of intracellular storage granules and endocytotic events (Section 2.6.1). Frequently these events are thought to involve tyrosine-specific protein phosphorylation events regulated by CAM activation (see Refs. 77–80 for examples). In other cases ligand binding by CAMs has been shown to result in activation of calcium channels raising intracellular free Ca^{2+},[81–84] stimulation of sodium/hydrogen exchangers leading to cytoplasmic alkalinization,[85,86] or activation of potassium channels leading to membrane hyperpolarization.[87] Modulation of key intracellular components (such as pH, free calcium levels, or membrane potential) by CAM binding can both trigger specific cellular responses and also provide a context within which cells react to other stimuli. One well-studied example of the latter is the cytokine-induced respiratory burst of neutrophils. Cytokines, such as tumor necrosis factor produced in response to tissue injury, trigger dramatic changes in neutrophils including reorganization of the cytoskeleton, release of proteases, and a prolonged respiratory burst. A key feature of this response is the requirement for the simultaneous ligand stimulation of neutrophil β_2 integrins[88,89] by virtue of neutrophil adhesion to other cells or to the extracellular matrix. This prerequisite for neutrophil adhesion in order to respond to activating signals is important in restricting the release of potentially toxic agents to the region of tissue damage by preventing soluble cytokines from activating neutrophils which pass through, but do not arrest, in areas of inflammation.

Thus, while contributing to the specificity and strength of cell–cell or cell–substratum adhesion, CAM binding also triggers changes in the cytoskeleton, generates second messengers, and leads to cell differentiation, proliferation, or migration. As roles such as these become better understood, labels such as *cell adhesion receptor* or *cell contact receptor* will increasingly replace the term *cell adhesion molecule* for this group of glycoproteins.

2.5. CELL ADHESION PROCESSES TYPICALLY INVOLVE MULTIPLE CAMS

2.5.1. Cell Junctional Complexes

A key feature of epithelial and endothelial cells layers is the presence of a specialized junctional complex marking the

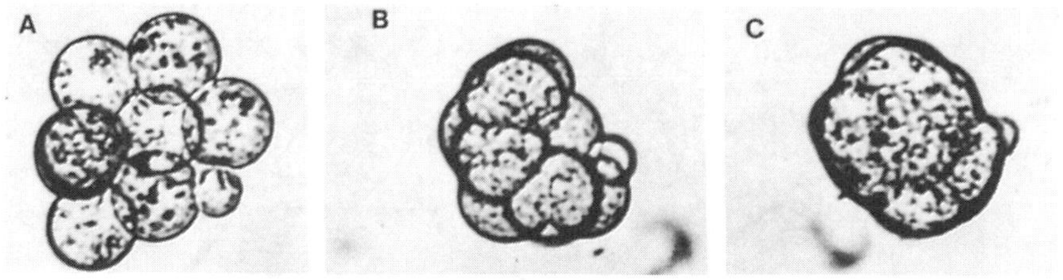

Figure 2.4. Compaction of the eight-cell morula. Photomicrographs of eight-cell mouse embryos with the zona pellucida removed obtained at 59, 62, and 65 hr postcoitus (panels A, B, and C, respectively). The dramatic compaction of the eight-cell embryo during this early developmental period is the result of increased cell–cell adhesion mediated by E-cadherin (uvomorulin), resulting in the development of organized intercellular junctions. Incubation of eight-cell embryos with antiuvomorulin antibodies prevents this compaction process and can cause the decompaction of fully compacted eight-cell embryos. The small cell at the lower right of each embryo is the polar body. (Adapted from Ref. 93 with permission.)

boundary between their apical and basolateral cell surfaces.[90] The junctional complex serves as a permeability barrier to paracellular movement of molecules across epithelial and endothelial cell layers, thus providing the opportunity for regulated vectorial transport of molecules between tissue compartments via transcellular pathways (for reviews see Refs. 91, 92). Experimental support for the involvement of E-cadherin (uvomorulin) in the development of epithelial junctions is now quite strong.

Specialized intercellular junctions first appear as early as the eight-cell embryo in mammalian development.[93] Formation of these junctions is the driving force for compaction of the preimplantation mouse embryo (Figure 2.4). Furthermore, it allows the generation of a polarized cell layer which is a prerequisite for the first cell differentiation step—establishment of the trophectoderm versus inner cell mass. Using the panspecific immunologic blockade approach, Jacob's laboratory identified a CAM that mediates compaction of the eight-cell embryo.[12] The name *uvomorulin* was coined for this glycoprotein to denote its involvement in the transition "between grape-like (Latin *uva*) and mulberry-like (Latin *morum,* whence morula) structures of cleavage-stage embryos."[94] The timing of morula compaction is not triggered by the onset of uvomorulin expression which occurs at the two-cell stage,[95] but rather by stimulation of uvomorulin interaction with cytoplasmic ligands by intracellular phosphorylation events.[96]

Subsequent electron microscopic studies have demonstrated the localization of E-cadherin at epithelial junctional complexes, specifically the zonula adherens component.[97] A more direct demonstration of the importance of E-cadherin in epithelial junction formation was provided by functional assays for membrane proteins involved in the formation of epithelial junctions *in vitro.* Gumbiner and Simons[98] screened randomly generated mAbs reactive with the epithelial cell surface for those that blocked the rapid re-formation of junctional complexes *in vitro.* The resulting mAb, rr1, was subsequently shown to react with uvomorulin (E-cadherin). Perhaps most convincingly, transfection of cultured cells with E-cadherin cDNAs results in the generation of adherens-type junctions, generation of epithelial monolayers, and polarization of the cell surface.[99,100]

Fully developed epithelial junctional complexes (Figure 2.5) consist of three distinct regions: the zonula occludens or tight junction, the zonula adherens or intermediate junction, and the desmosome. At least four different members of the cadherin family participate in the formation of mature junctional complexes. Uvomorulin (E-cadherin) has been localized to the intermediate junction.[97] The closely related N-cadherin (A-cadherin) is found at zonula adherens-like junctions in heart and optic lens.[101] More recently, purification and cloning of the principal transmembrane components of the desmosomal structure has led to the discoveries of the desmocollin and desmoglein cadherin subfamilies.[102] Transmembrane proteins of the apical tight junction have not been identified. One of the desmoglein-type cadherins (Dsg3) is the antigen recognized by autoantibodies in the potentially life-threatening skin disease, pemphigus vulgaris.[103] Antibodies against the pemphigus vulgaris antigen as well as antibodies directed at a second desmoglein, Dsg 1, both directly disrupt cell–cell adhesion.[104,105] The binding of actin bundles at the zonula adherens versus intermediate filaments at desmosomes reflects the disparate cytoplasmic domains of the E- and N-cadherins versus desmocollin and desmoglein subfamily of cadherins.

While E-cadherin only contributes to the zonula adherens component of the junctional complex, the ability of antiuvomorulin antibodies to block development of all three junctional components[106] indicates a central role for uvomorulin and the adherens junction in triggering formation of the complete junctional complex. Recent demonstrations that a principal tight junction cytoskeletal protein, ZO-1, colocalizes with E-, P-, and N-cadherin in nonepithelial cells but not in fully developed tight junctions[107] suggest the possibility that E-cadherin functions to trigger cytoskeleton-linked events involving catenins, actin bundles, and the tight junction protein ZO-1 which nucleates the organization of the tripartite junctional complex.

2.5.2. Leukocyte Adhesion

The highly effective surveillance capacity of the immune system requires that leukocytes be highly mobile, relatively nonadhesive cells during their transport through the vascula-

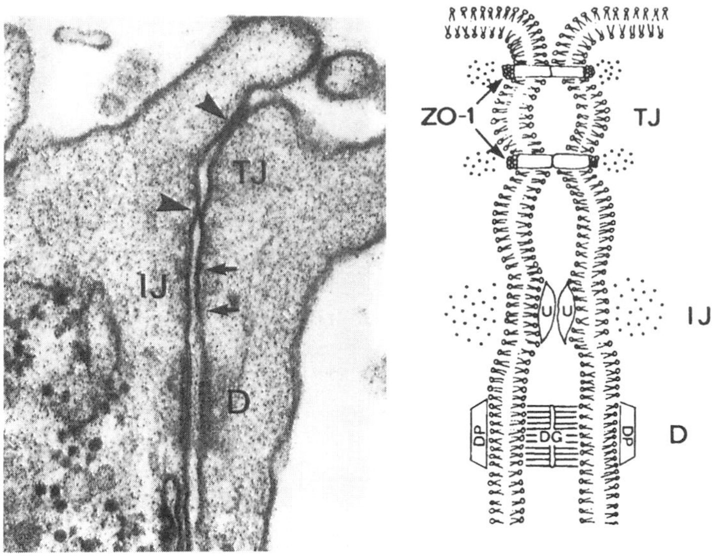

Figure 2.5. Components of the tripartite epithelial junctional complex. The schematic diagram (right), and electron micrograph (left) illustrate key features of a junctional complex between two rat pulmonary epithelial cells. The apical tight junction (TJ) features focal, close appositions (right-facing arrowheads in the electron micrograph) of adjacent cell membranes. The cytoskeletal protein ZO-1 is a cytoplasmic component of the tight junction. The adherens or intermediate junction (IJ) component typically has a slightly narrowed intercellular space containing scant electron-dense material comprised in part by uvomorulin (U). Intermediate junctions have an associated cytoplasmic fibrillar meshwork (bracketed by the two left-pointing arrows in the micrograph) corresponding to a circumferential ring of actomyosin. The third component of the junctional complex, the desmosome (D) is characterized by dense cytoplasmic plaques containing desmoplakins (DP), to which intermediate-sized filaments bind, and by a fine fibrillar, extracellular, intermembrane meshwork of desmoglea (DG) containing the desmoglein and desmocollin cadherins. Final magnification of the micrograph is approximately 73,000×. (Adapted from Ref. 91 with permission.)

ture, and yet be able to exit the circulation at specific sites to invade tissue elements and reversibly bind to potential antigen-presenting and target cells. For lymphocytes, this series of events repeats itself countless times as individual cells continuously recirculate from the vasculature into tissue spaces and back for weeks to years. Not surprisingly, regulated cell–cell adhesion plays a key role in these processes. Study of the specific molecules involved and their regulation has recently become a central focus of cellular immunology (for recent reviews see Refs. 108–111). What follows is a summary of some highlights from this work.

The vertebrate immune system has incorporated a rather complex pattern of recirculation of lymphocytes through subsets of body tissues in ways that foster the chance encounter of randomly generated T and B lymphocytes, each expressing one of millions of potential unique antigen-binding receptors, with their cognate antigens. This process, referred to as lymphocyte trafficking, recirculation, or homing, has been studied at a cellular level *in vivo* by syngeneic transfer of labeled subsets of circulating lymphocytes in rodents. When reinjected into host animals, donor lymphocytes tend to "home" back to the homologous site from which they were obtained. Key aspects of the pattern of lymphocyte recirculation (reviewed in Refs. 111, 112) are as follows. Newly generated lymphocytes enter the general circulation from their sites of development in the bone marrow and thymus. These naive lymphocytes then begin a continual recirculation between blood and the lymphatic system, exiting the circulation via specific post-capillary venules in organized lymphoid tissues including peripheral lymph nodes, the spleen, and Peyer's patches of the intestinal mucosa. These sites act as filters concentrating foreign antigen for presentation to recirculating lymphocytes. This recirculation of naive lymphocytes is not entirely random as subsets of lymphocytes show preference for recirculation through peripheral lymph nodes versus intestinal Peyer's patches. By avoiding recirculation through other body tissues, this selective trafficking of naive lymphocyte increases the opportunity for antigen stimulation, clonal expansion, and the development of memory and effector lymphocytes.

Following antigen stimulation the recirculation pattern of activated memory and effector lymphocytes is much different than that of naive cells. First, memory and effector lymphocytes acquire the ability to circulate between the vasculature and general body tissues as opposed to the selective homing of naive cells to lymph nodes and organized lymph tissues. Second, activated lymphocytes prefer to recirculate within those body tissue compartments whose lymphatic drainage returns via the lymphoid tissue in which antigen exposure first occurred, thereby focusing their circulation to those body tissues most likely to harbor primary sources of that particular antigen. Third, restrictions of lymphocyte recirculation are reduced at sites of active inflammation. These restricted patterns of lymphocyte recirculation enhance the potential both for the initial antigen encounter by naive lymphocytes and for the interaction of activated lymphocytes with tissue pathogens.

Progress in identification of adhesion molecules involved in this selective lymphocyte recirculation has centered on the role of lymphocyte adhesion to the blood vessel endothelium as the initial process in selective recirculation. In many sites (including lymph nodes, Peyer's patches, and articular synovium) specialized postcapillary venules composed of columnar and cuboidal endothelial cells—high endothelial venules (HEV)—act as preferred sites for lymphocyte extravasation. These specialized endothelial cells constitutively express lymphocyte adhesion factors termed *vascular addressins*. The initial identification of HEV addressins and their cognate lymphocyte homing receptors relied on a simple *in vitro* assay, first introduced by Stamler and Woodruff,[113] involving the static binding of test lymphocytes to cryostat sections of peripheral lymph nodes or Peyer's patches. Mirroring lymphocyte homing *in vivo*, lymphocytes selectively bind to cross sections of HEV present within these tissue slices, thus providing a functional assay to screen mAbs raised to lymphocyte and HEV cell surface components for those that block lymphocyte binding. Thus, antibodies have been developed to lymphocyte cell surface proteins (homing receptors) which block lymphocyte homing to peripheral lymph node but not to mucosal HEVs.[64,114] Antibodies to other lymphocyte cell surface components block homing to mucosal but not to peripheral lymph node HEVs.[115–117] Similarly, antibodies to endothelial proteins (vascular addressins) have been developed that selectively block lymphocyte homing to peripheral[118] or mucosal[119,120] HEVs.

The lymphocyte homing receptor for peripheral lymph nodes has been cloned[65,66] and is now known as L-selectin. Lymphocyte L-selectin binds to glycoconjugates defined by the peripheral lymph node vascular addressin (PNAd) mAb MECA-79 and selectively expressed by peripheral lymph node HEV.[73,118] Antibodies to Peyer's patch homing receptor(s) have led to the identification of both CD44[121,122] and the VLA-4 integrin[117] as intestinal lymphocyte homing receptors. CD44 has been shown to bind the mucosal HEV vascular addressin, a 58- to 66-kDa glycoprotein defined by mAb MECA-367.[123] The mucosal HEV ligand for VLA-4 has yet to be identified. Activation of naive lymphocytes yields memory and effector lymphocytes that express increased levels of several adhesion molecules including LFA-1, LFA-3, ICAM-1, VLA-4, VLA-5, VLA-6, as well as an additional tissue-specific homing receptor for skin-associated memory T cells. The cutaneous lymphocyte homing receptor, defined by the cutaneous lymphocyte antigen (CLA), is a glycoconjugate recognized by E-selectin present on the cell surfaces of inflamed cutaneous vascular endothelium.[124,125] Antibodies to these four homing receptors—L-selectin, CD44, VLA-4, and CLA—have been used to define distinct subsets of human T lymphocytes[126] supporting the idea that the differential expression of these homing receptors contributes to selective lymphocyte homing.

Study of adhesive interactions of a second class of leukocyte, the neutrophil, with inflamed endothelium indicates that the intricacy and hence potential for regulation of leukocyte–endothelial recognition are even more involved than that just described. *In vivo* microscopic examination of this process reveals that leukocytes begin interacting with postcapillary venule endothelium within minutes of injury to adjacent tissue. The initial interaction is a slow rolling of leukocytes along the vessel wall followed by their arrest, subsequent flattening, and emigration through the endothelium by extension of a pseudopod between endothelial cells. Studies of leukocyte adhesion to model endothelium *in vitro* as well as antibody perturbation studies of leukocyte interaction in real time with inflamed venules *in vivo*[127,128] have led to the hypothesis that, in inflamed tissues, leukocyte–endothelial cell adhesion is an interactive process involving at least four steps and multiple CAMs[129] (Figure 2.6). First, proinflammatory agents including tumor necrosis factor, interleukin 1, bacterial lipopolysaccharide, histamine, and thrombin induce the expression of at least four adhesion molecules—P-, E-selectin, ICAM-1, and VCAM-1—by endothelial cells.[68,130–134] Endothelial cell surface P-selectin levels are increased within minutes of activation via mobilization of P-selectin present in intracellular storage granules. Levels of endothelial cell surface E-selectin, ICAM-1, and VCAM-1 increase over a period of 2–4 hr and require new protein synthesis. Second, circulating leukocytes contacting the inflamed endothelium begin to slowly roll along its surface. Their passage is slowed because of weak adhesive interactions between the induced endothelial P- and E-selectins and carbohydrate ligands constitutively present on leukocytes, including those associated with L-selectin.[127,128,135,136] This nearly immediate induction of leukocyte rolling following endothelial activation is completely inhibited in genetically engineered P-selectin-deficient mice.[137] The absence of P-selectin results in a 2-hr delay of neutrophil recruitment into inflammatory sites, presumably reflecting the time required for the *de novo* synthesis of other endothelial addressins. Third, activation of rolling leukocytes by chemoattractants and activating factors presented by the activated endothelium leads to a posttranslational functional activation of preexisting leukocyte β_2-integrins Mac-1 and LFA and the β_1-integrin VLA-4.[138,139] In one system this functional activation involves a protein kinase C pathway leading to the formation of a low-molecular-weight lipid, termed integrin modulating factor (IMF-1), which reversibly activates β_2-integrin function.[140] The particular integrin involved and mechanism of activation appear to vary based on the specific activator signal and leukocyte pairing. Finally, arrest and stable binding of activated leukocytes to the inflamed endothelium, via interactions of activated leukocyte VLA-4 and β_2-integrins with both constitutive and induced endothelial cell VCAM-1, ICAM-1, and ICAM-2,[127,132,133,135,141] leads to the transendothelial migration of adherent leukocytes. The functional implication of this multistep process, involving the regulated expression and function of several adhesion molecule pairs by intercellular signals, is to provide a combinatorial mechanism allowing both flexibility and specificity for the regulation of leukocyte–endothelial interactions leading to transendothelial migration of subsets of leukocytes into specific tissue settings.

2.5.3. Adhesion and Neural Development

Our understanding of the mechanisms involved in the development and remodeling of neural connections that underlie

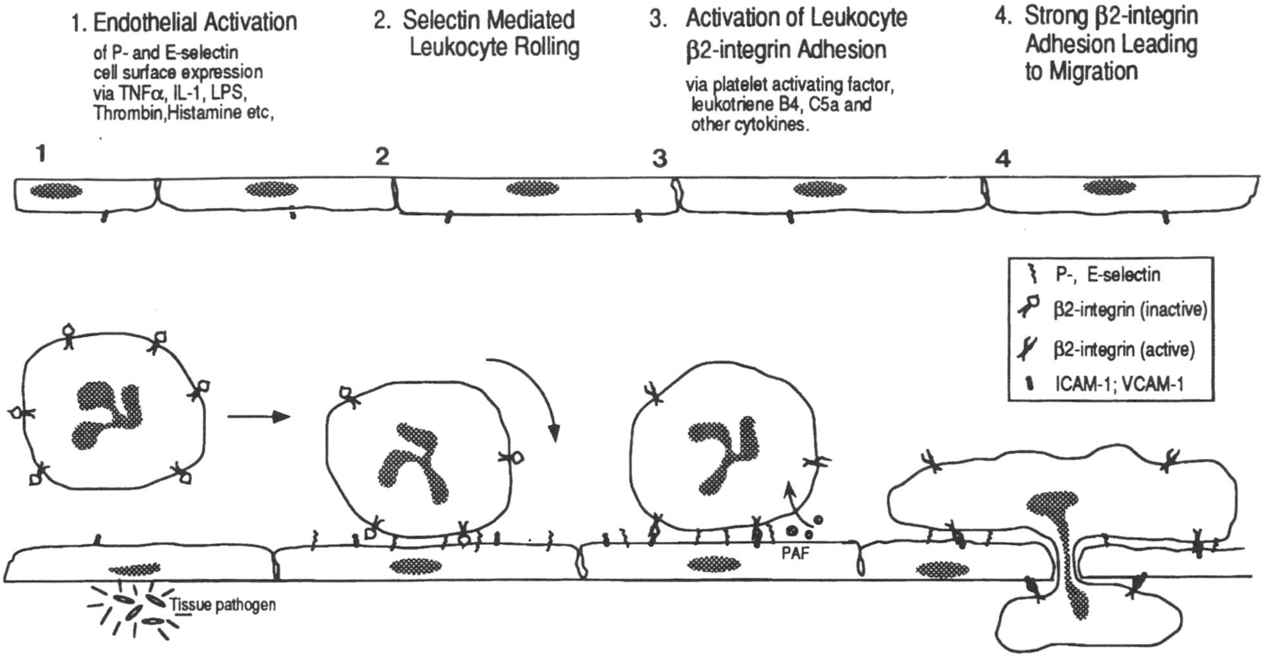

Figure 2.6. Cell–cell interactions leading to leukocyte binding and extravasation across postcapillary venule endothelia. Schematically illustrated are four key events in this process, together with some of the leukocyte and endothelial cell adhesion molecules. In the first step, inflammatory agents produced by tissue damage locally activate endothelial cells to express P- and E-selectin, as well as to increase their expression of ICAM-1 and VCAM-1. In step two, random leukocyte–endothelial cell contact leads to weak binding and slow leukocyte rolling along the endothelial surface as a result of adhesion mediated by constitutively expressed leukocyte glycoconjugates and P- and E-selectins on the activated endothelium. This enhanced leukocyte–endothelial contact is required for the efficient functional activation of preexisting β_2-integrins (shown in step three) by a variety of cytokines released from the activated endothelium. In step four, strong leukocyte–endothelial adhesion (via β_2-integrin–ICAM-1,-2,-3 and VCAM-1 interactions) acts in concert with further cytokine activation of leukocyte functions to produce transepithelial leukocyte migration.

the functional nervous system is in its infancy. Nonetheless, most workers in this field now agree that CAMs are important contributors to these processes. Over two dozen CAMs have been described in the nervous system, each with characteristic patterns of temporal and spatial expression. Figure 2.7 illustrates the distinct pattern of three Ig-related CAMs in the developing vertebrate spinal cord. The expression of individual CAMs by different subsets of neurons and even portions of neurons (e.g., axons versus dendrites and cell bodies) suggests they may serve as an important substrate contributing to the selective migration of cells, growth of axons, and synapse formation. Antibody blockade and more recently genetic perturbation experiments have been used to confirm the involvement of specific CAMs in neural cell migration and axon growth (see Refs. 20, 142–145 for specific examples). Adhesion molecules can be either (1) instructive, actively promoting decision-making turns by growth cone; (2) permissive, by providing a conducive pathway conducting later-growing axons along previously pioneered pathways; or (3) restrictive, by producing a tightly fasciculated (bundled) pattern of axon growth, minimizing the tendency for axon tips to explore their local environment. Once axons have reached their target, adhesion molecules may act to restrict further axon growth and to participate in the induction of nerve terminal differentiation. Finally, changes in the levels of adhesion molecules expressed by target cells in response to neural stimulation may serve as stimuli for the continual modification of synaptic connections by experience.

An interesting caveat of the regulation of neural adhesion is the role played by sialic acid modification of NCAM. NCAM, the most widespread and abundant neural CAM, is unique in serving as scaffold for the addition of polysialic acid—long homopolymers of α-2,8-linked neuraminic acid units.[146] This form of polysialic acid is unusual for vertebrates but has long been recognized as a principal component (known as colominic acid) of the outer glycocalyx of gram-negative bacteria including group B meningococcus, an important pathogen responsible for meningitis in children and infants. The difficulties that have been encountered in developing effective vaccines against this type of meningococcus appear to relate to the immunologic similarity of the bacterial capsular polysaccharide and NCAM polysialic acid. Indeed, stimulation of humoral immunity directed at the glycocalyx may in fact be detrimental, as cross-reactive autoimmune anti-NCAM polysialic acid antibodies have been detected in the serum of patients with group B meningitis[147] and may contribute to the development of permanent neurological sequelae following resolution of the acute infection. The extent of NCAM polysialic acid modification varies during development from over 30% (polysialic acid wt/NCAM wt) during embryonic and neonatal periods to less than 10% in most adult brain areas.[148,149] High levels of polysialic acid persist in adult brain

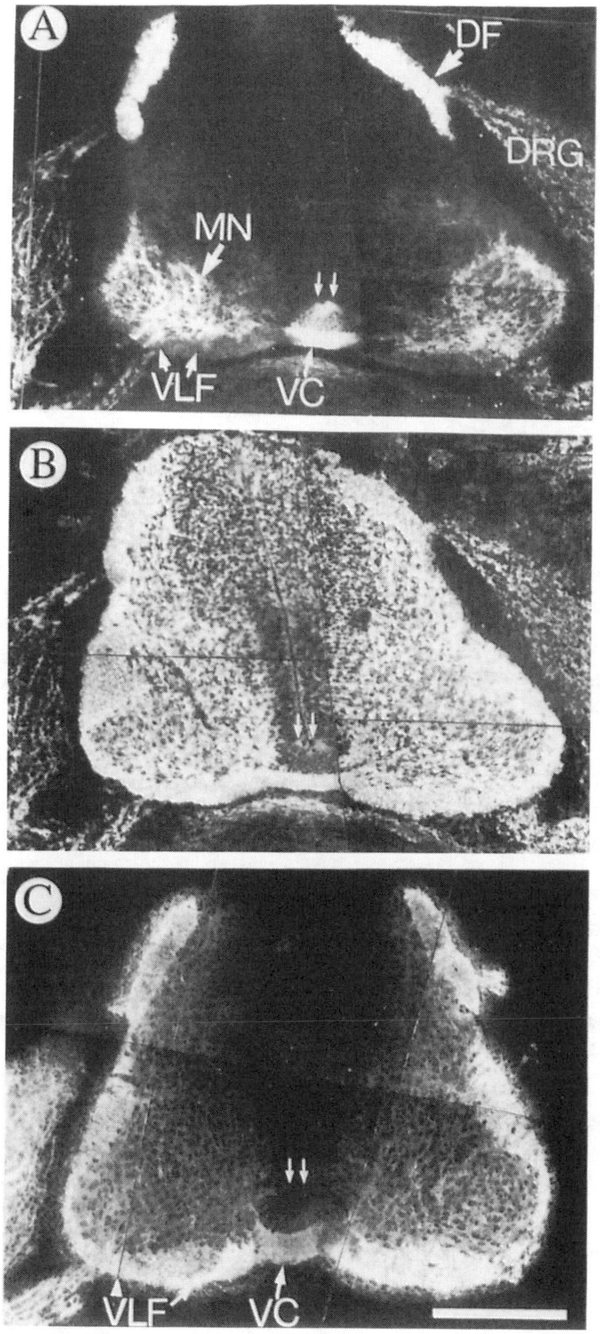

Figure 2.7. Comparative distributions of three Ig-related CAMs in the developing chick spinal cord. Photomontages of cryostat cross sections of a stage 28 chick embryo spinal cord and dorsal root ganglia stained with fluorescent antibodies to SC1/GRASP (panel A), NCAM (panel B), or NgCAM (panel C). Stained areas appear white or light gray. SC1/GRASP is localized to cell bodies and axons of spinal motor neurons (MN) and dorsal root ganglia (DRG) sensory neurons. The brightly stained SC1/GRASP-positive dorsal funiculus (DF) contains cross sections of DRG axons projecting to more rostral spinal cord segments. In contrast, other axon tracts such as the ventrolateral funiculus (VLF) which are stained by anti-NgCAM antibodies (C) do not express SC1/GRASP. NgCAM is expressed on most axons, but not to a significant degree on CNS neuron cell bodies or dendrites. In contrast to the restricted distributions of SC1/GRASP and NgCAM, NCAM is expressed at high levels by nearly all developing neurons and axons (B). Scale bar in panel C represents 200 μm. (From Ref. 160 with permission.)

regions which continue to have axon regeneration such as the olfactory bulb. NCAM polysialic acid imparts a severalfold increase in the hydrated volume occupied by the modified versus native glycoprotein, resulting in a steric hindrance of close cell–cell apposition.[150] As a result, high levels of NCAM polysialic acid typical of embryonic neurons can diminish membrane contact-dependent events. These include NCAM- and non-NCAM-mediated neural adhesion as well as CAM-independent contact-mediated transmembrane signaling events.[151] Thus, modulation of NCAM polysialic acid levels is thought to serve as a nonspecific mechanism to prevent strong neural cell adhesion at times when neural plasticity, rather than stability, is needed in order to allow growing neural processes to respond to subtle guidance cues.

Until relatively recently, the mechanism by which adhesion molecules selectively guide growing axons was thought to be a reflection of the relative strength of binding which they imparted. This view grew largely from the demonstration by Letourneau[152] that when given a choice of collagen, palladium-coated surfaces, or bare tissue culture plastic, the preferences demonstrated by growing axons correlated with strengths of growth cone–substrate binding. More recently, when axon growth rates versus growth cone adhesion were compared for three physiologic substrates—L1, N-cadherin, and laminin—little correlation was found between adhesive strength and stimulation of axon growth or fasciculation.[153] Indeed, it is becoming clear that perhaps the critical event produced as a result of adhesion molecule binding is the generation of intracellular second messengers which trigger cytoskeletal reorganization and growth cone activity.[83,84] Different adhesion molecules may utilize distinct but convergent intracellular signaling pathways. For example, in an *in vitro* system utilizing NCAM- versus N-cadherin-transfected fibroblasts as substrata, axon growth rates were linearly related to N-cadherin expression levels but showed a highly cooperative relationship to NCAM levels,[83a] suggesting the involvement of distinct signaling pathways. Thus, an updated hypothesis of how adhesion molecules regulate axon growth should consider both their adhesive quality as well as their ability to trigger, directly or indirectly, cytoskeletal changes within the growing axon. Modulation of second messenger levels and substrate phosphorylation by activation of traditional neurotransmitter and trophic factor receptors, as well as by contact-mediated activation of CAMs, provides a common pathway for the biochemical integration of diverse signals impinging on the growth cone.

In order to illustrate some of these process we will review what is currently known about the involvement of CAMs in spinal motor neuron axon growth and synaptogenesis in the peripheral nervous system. General aspects of this system in the chick hindlimb are illustrated in Figure 2.8A. Motor axons emerging from the developing spinal cord express at least four Ig-related CAMs (NCAM, NgCAM/L1, axonin-1, SC1/GRASP) and two cadherins (N- and T-cadherin) as well as several members of the β1-integrin family of extracellular matrix receptors. Several of these, including NgCAM/L1 and SC1/GRASP, promote axon fasciculation. Others—NCAM, the cadherins and integrins—subserve axon–

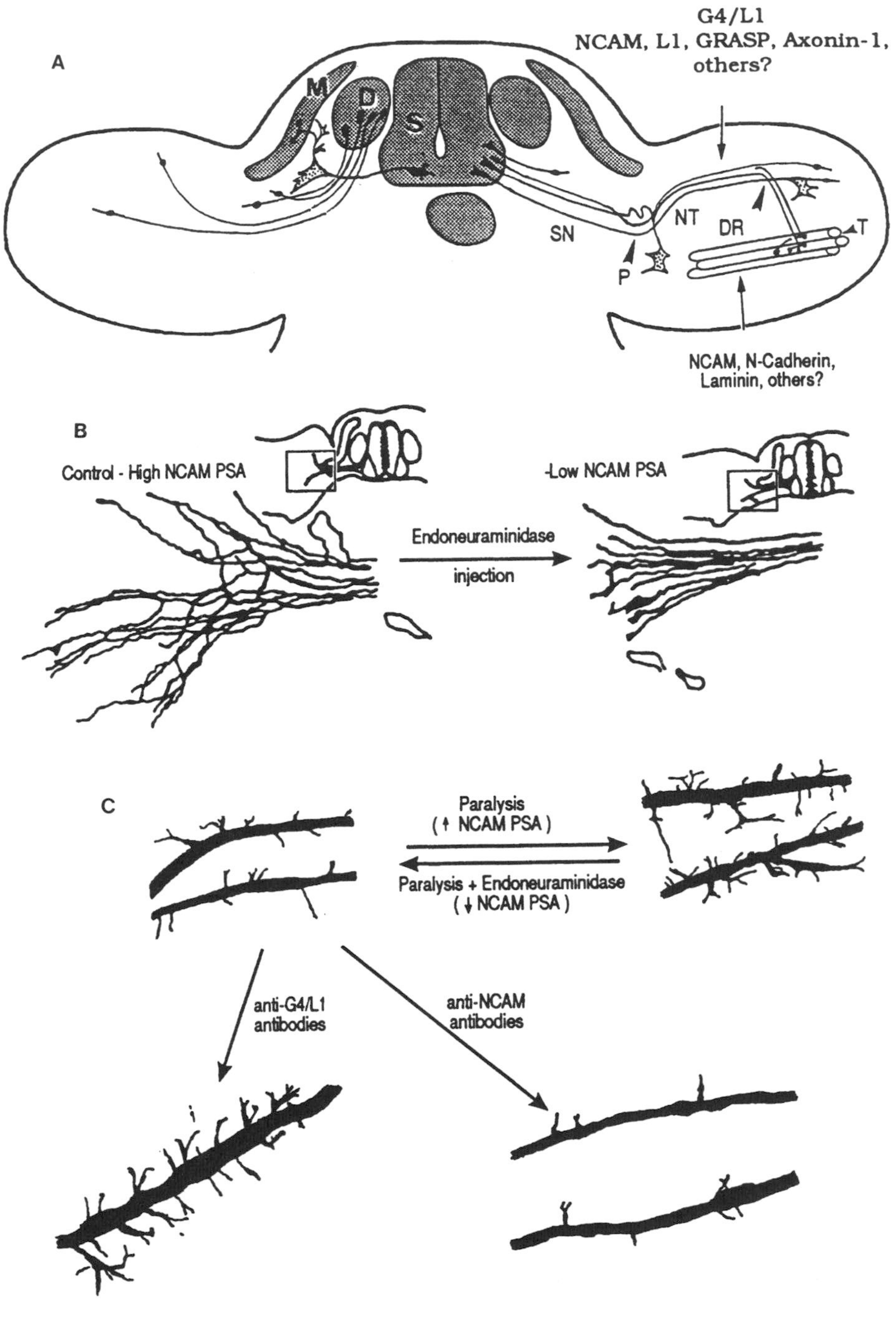

Figure 2.8. CAMs and selected aspects of neuromuscular development. (Panel A) Schematic summary of axon growth patterns and growth cone morphologies in the developing chick hindlimb. The trajectories of sensory axons arising from segmental dorsal root ganglion (D), and of myotomal (M) motor axons are illustrated on the left. Key features of motor axons innervating limb muscles are illustrated on the right. Motor axons exiting the spinal cord (S) via the ventral root merge with sensory axons in the dorsal root forming the mixed spinal nerve (SN); for simplicity these two axon populations are illustrated separately. On reaching the plexus region (P) at the limb base, some motor axons make sharp turns as they sort themselves into the crural and sciatic nerve trunks (NT). At characteristic decision regions (DR) near individual muscle masses, subsets of motor axons diverge from the main nerve trunk to form individual muscle nerves leading to their target (T) muscle. The size and morphology of growth cones are dependent on their position within the developing limb. Motor neuron growth cones are small and varicose when in the more fasciculated spinal nerve and distal nerve trunks. They are typically large and lamellipodial in the plexus region and muscle nerve decision regions. As indicated at the right, several adhesion molecules have been localized to growing motor axons and myotubes. (Panel B) Alteration of motor axon trajectories in the plexus region following experimental perturbation. Camera lucida tracings of motor axon profiles in stage 24 embryos 24 hr after injection of either saline (left) or endoneuraminidase (right). Endoneuraminidase removes polysialic acid (PSA) from NCAM, producing increased axon fasciculation and inhibiting widely divergent turns by growing axons. The resulting restriction of axon-sorting within the plexus region produces errors in the final pattern of neuromuscular connections. (Panel C) Modulation of muscle nerve side branch number by experimental perturbation of adhesion molecule function. Tracings of intramuscular nerve branching within stage 32 chick iliofibularis muscles. Individual axons within the intramuscular nerve are so numerous as not to be individually discernible. Dozens of myotubes surround and course parallel to the main nerve trunk; for clarity they are not shown. In comparison to saline-injected muscles (upper left), injection of antibodies to the axon–axon CAM L1/G4 (lower left) results in less tightly fasciculated intramuscular nerves and an increased number of side branches interacting with nearby myotubes. Embryos receiving intramuscular injections of anti-NCAM antibodies, which block both axon–axon and axon–myotube adhesion, have approximately one-half the normal number of intramuscular nerve side branches (lower right). Paralysis of embryos by *d*-tubocurare induces higher levels of NCAM PSA on motor axons, and results in axon defasciculation and an increased number of side branches. Removal of PSA by concurrent endoneuraminidase injections reverses the effect of paralysis on excessive side branch formation. The scale bar in panel C represents 400 μm for A and 50 μm for B and C. (Panels A–C adapted with permission from Refs. 154, 156, 161, 162, respectively.)

pathway and axon–target adhesion and recognition. NCAM additionally acts by virtue of polysialic acid modification as a global regulator of axon–axon contact.[151]

The initial growth of each spinal nerve is moderately well fasciculated until reaching the future nerve plexus region at the base of the limb. Once reaching the plexus region, motor axons become less tightly adherent; their growth cones can be seen to broaden and to make trajectory decisions producing relatively sharp turns.[154] The result of this process is to re-sort axons into several principal axon bundles prior to their invasion of the developing limb. After leaving the plexus region, motor axons once again assume a more tightly fasciculated pattern. This stereotyped defasciculation of the motor axon bundle at the limb base appears to result from an inhibition of close axon–axon adhesion produced by an increase in NCAM polysialic acid content at this developmental point rather than by a change in the relative amounts of individual CAMs.[155] This brief nonspecific decrease in axon–axon adhesion appears to be necessary in order to allow growing axons to respond to more subtle instructive re-sorting cues. Enzymatic removal of polysialic acid moieties at this critical time, via experimental injection of endoneuraminidase into the developing limb, results in reduced numbers of turning axons (Figure 2.8B), ultimately generating errors of axon growth manifest by the innervation of target muscles by inappropriate sets of motor axons.[155,156]

On arriving in the vicinity of individual muscle masses, subsets of axons diverge from the principal limb nerves to form individual muscle nerves. The factor responsible for the divergence of subsets of otherwise highly similar axons is completely unknown but is likely to involve soluble chemoattractants in addition to CAMs. With the exception of higher levels of NCAM polysialic acid on axons projecting to dorsal versus ventral muscle masses,[155] no adhesion molecules marking subsets of vertebrate motor axons have been identified to date. In contrast, studies of molecular specificities of invertebrate motor axons suggest they may indeed exist. Goodman's group has identified, using a rather tedious but powerful genetic enhancer trap technique, a homophilic adhesion molecule called connectin which is expressed by a subset of eight muscle fibers and the corresponding motor axons that innervate them within each abdominal body segment of *Drosophila*.[157]

Once properly delivered to the developing muscle mass, individual axons begin to splay out from the muscle nerve bundle to form close contacts with myotubes. The choice by growth cones to elongate along myotubes rather than along other axons appears to be controlled by the relative "adhesiveness" of these two alternative substrates. The expression of myotube NCAM and N-cadherin increase as axons invade muscle masses,[158,159] and the level of at least one homophilic axon–axon adhesion molecule, SC1/GRASP, declines during this period.[160] These temporal changes in CAM expression produce a relative increase in myotube attractiveness to growing axons, together with a decrease in axon–axon adhesion. Landmesser's group[161] has demonstrated by injection of antibodies preferentially blocking axon–axon versus axon–myotube interactions (Figure 2.8C), that a competition or comparison of "adhesive" qualities between neighboring motor axons versus the myotubes surface influences the generation of intramuscular nerve branches and ultimately of synapse formation. Blockade of axon–axon adhesion by anti-NgCAM/G4 antibodies resulted in a 70% increase in the number of axon side branches leaving the intramuscular nerve (Figure 2.8C, lower left tracing), while blockade of axon–myotube adhesion using anti-NCAM antibodies caused a 60% decrease in the number of axon side branches (Figure 2.8C, lower right tracing). This principle of comparison between potential growth substrates also appears to form the basis for an excess number of axon branches in chronically paralyzed embryos.[162] In this case treatment with curare causes an increase in axon NCAM polysialic acid content by an unknown mechanism, leading to defasciculation of intramuscular axon bundles and to the formation of double the normal number of axon side branches (Figure 2.8C, upper right tracing). Removal of NCAM polysialic acid by addition of endoneuraminidase reverses the effect of curare on the formation of excess intramuscular nerve branches.[162] The long-term effects of these rather subtle changes in the relative attractiveness of the myotube surface versus adjacent motor axons at this critical juncture are reflected by increased numbers of side branches, an increase in the number of synapses ultimately formed, and an increased number of motor neurons which survive into later stages of development.[163]

Following contact with their myotube target, developing axons induce the accumulation of focal deposits of synaptic adhesion molecules including a laminin-like molecule, S-laminin.[164] While S-laminin is highly adhesive for isolated neurons, more importantly it can arrest growth cones and may induce their differentiation into the specialized motor nerve terminal.

Completing the series of adhesive events leading to muscle fiber innervation is the activity-dependent downregulation of muscle fiber NCAM and N-cadherin.[159,165] Decreased levels of muscle fiber neural adhesion molecules are thought to contribute to the well-known refractoriness of normally innervated muscle fibers to accept implanted foreign nerves. In contrast, denervation or paralysis of adult muscle leads to the renewed expression of adhesion molecules including NCAM and N-cadherin[159,166] while injection of antibodies to NCAM can block motor nerve sprouting induced by muscle paralysis.[167] Together these findings suggest that dynamic feedback loops between pre- and postsynaptic cells involving changes in adhesion molecule expression driven by synaptic activity may play important roles in maintaining appropriate quantitative functional innervation of target cells and thereby provide a mechanism for the modification of synaptic size or number in response to changes in synaptic activity or experience. Additional experimental support for this notion comes from cellular and molecular studies of learning in the *Aplysia* model system. Kandel's group has reported that changes in the level of an NCAM-related cell adhesion molecule, apCAM, at specific synapses may be one of the earliest steps involved in long-term synaptic plasticity induced by behavioral training in this organism.[168,169] In this case a decrease in apCAM was triggered by synaptic activity and was suggested to destabilize synaptic contacts, promoting remodeling at those CNS synapses involved in producing specific learned behaviors.

2.6. DISORDERS OF CELL ADHESION IN DISEASE PROCESSES

2.6.1. CAM Involvement in the Intracellular Uptake of Viruses and Parasites

The spectrum of cells infected by a given virus is determined in part by the presence or absence of specific viral receptor proteins on the cell surface. Isolation and study of such receptors has led to the discovery that many microbes gain access to the host cell interior by pirating functions of host cell CAMs. CAMs have been shown to be involved in both initial binding events as well as transmembrane uptake of several intracellular pathogens.

Ig-related CAMs serve as receptors for several viruses. Examples include ICAM-1 which acts a receptor for the human rhinovirus,[170–172] the human immunodeficiency virus receptor CD4,[173,174] the murine hepatitis virus receptor BGP,[175,176] and the human poliovirus receptor, a previously undescribed Ig-related multigene family member.[177,178] The interaction of rhinovirus with ICAM-1 has been perhaps best described in molecular terms. Rhinoviruses, the most frequent cause of the common cold, are picornaviruses containing a single strand of RNA within an icosahedral shell composed of 60 protomers. There are over 100 distinct rhinovirus serotypes, with 90% of these utilizing the same cellular receptor, ICAM-1. Yet there is almost no cross-immunity between rhinovirus subtypes, suggesting the viral surface binding site is protected from humoral immune surveillance. The crystal structure of rhinovirus[179] revealed the presence of 12 "canyon"-like depressions on the capsid surface encircling each of the 12 capsid pentamer vertices. These narrow 25-Å-deep canyons, which taper from 30 Å at the rim to 12 Å at the base, were suggested to house the viral receptor binding site, as they would provide a site sterically protected from antibody binding (Figure 2.9A). This canyon hypothesis, developed by Rossmann and colleagues,[179,180] has become more secure with the recent characterization of the interaction between the host cellular receptor, ICAM-1, and rhinovirus particles. Correlation between structural models of the rhinovirus capsid and ICAM-1 with the results of site-directed ICAM-1 mutagenesis indicates that the unpaired first Ig domain of ICAM-1 fits nicely into the rhinovirus canyon. Furthermore, seven out of nine ICAM-1 site-specific mutations that decrease virus binding are located along the predicted ICAM-1–rhinovirus interface[181] (Figure 2.9B). Finally, recent cryoelectron microscopy of rhinovirus complexed with soluble ICAM-1 has provided visual support for the binding of ICAM-1 in the rhinovirus canyon.[182] These results supporting the canyon hypothesis for rhinovirus binding, together with the involvement of at least four different Ig-related CAMs as viral receptors, suggest that viral binding to the rodlike monomeric CAM proteins by way

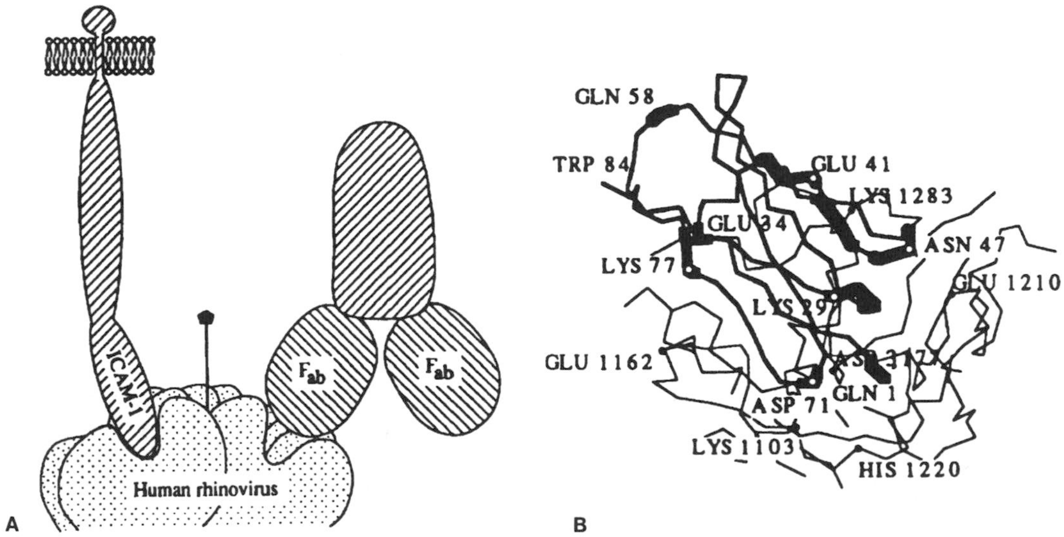

Figure 2.9. Binding of human rhinovirus to ICAM-1. (A) Schematic cross-section diagram of a small portion of the rhinovirus capsid surface illustrating a 12- to 15-Å-wide canyon which can accommodate the narrow ICAM-1 molecule. The wider antigen-binding region of immunoglobulins (Fab) contain paired Ig domains, and according to the canyon hypothesis, are sterically prevented from access into the recessed rhinovirus binding domain. (B) Alpha-carbon diagram of the amino-terminal Ig fold of ICAM-1 binding in a portion of the rhinovirus canyon. The virus interior is toward the bottom. The virus capsid protein backbone is shown in thin lines, ICAM-1 in medium lines, and regions of ICAM-1 in which site-directed mutations block binding are shown in heavy lines. Selected residues in each protein are indicated; open circles (ICAM-1) and solid circles (rhinovirus) represent predicted interacting amino acids. Seven of the nine mutation sites are predicted by structural studies to reside at ICAM-1/rhinovirus canyon interface sites. The exceptions are Gln-58 and Arg-49. (Adapted from Ref. 181 with permission.)

of recessed canyons or pits on the viral surface may be a particularly useful adaptation. An interesting caveat of the ICAM-1–rhinovirus pairing is that the low-level basal expression of ICAM-1 on epithelial and endothelial cell surfaces is converted to a robust, high level of expression by cytokines generated by nonspecific host defense reactions. Thus, early inflammatory nonspecific responses to the common cold enhance the local cellular spread of viral infection by promoting virus binding to cytokine-activated host epithelial cells.

Integrins have also been implicated in the binding and internalization of specific viral, bacterial, and fungal intracellular parasites. The VLA-2 ($\alpha_2\beta_1$) integrin receptor for collagen and laminin serves as a receptor for members of the human echovirus family.[183] This group of viruses are commonly involved in viral meningitis and fatal disseminated infections in the newborn. A second important human virus, adenovirus, interacts with the vitronectin integrins $\alpha_v\beta_3$ and $\alpha_v\beta_5$. In this case the integrin is not involved in initial viral binding but triggers viral uncoating and endocytosis by susceptible respiratory and gastrointestinal epithelial cells.[184]

Although the mechanism is not yet understood, it appears that under appropriate circumstances, microbial binding to integrin receptors triggers cytoskeletal events leading to phagocytosis. Unlike the immunoglobulin Fc receptor phagocytic pathway, this integrin pathway can occur in the absence of a destructive oxidative burst.[185] Instead, invading microbes are incorporated into specialized nondestructive phagosomes in which they can grow and divide protected from humoral immune surveillance. For example, *Mycobacterium avium-M. intracellulare,* one of the most common causes of disseminated bacterial infection in AIDS patients, utilizes the $\alpha_v\beta_3$ vitronectin receptor to gain entry into macrophages.[186] β1-integrins act as a receptor for invasin, a protein that promotes bacterial cell uptake by mammalian cells[187] while the enteric bacterium *Yersinia pseudotuberculosis* invades a variety of nonphagocytic cells by trickery of VLA integrins $\alpha_3\beta_1$, $\alpha_4\beta_1$, $\alpha_5\beta_1$, $\alpha_6\beta_1$.[187] Several other facultative intracellular macrophage pathogens, including the yeast form of *Histoplasma capsulatum,*[188] *Legionella pneumophilia,*[189] *Bordetella pertussis* bacteria,[190] and *Leishmania* parasites,[191] utilize the complement receptor 3, $\alpha_m\beta_2$ and $\alpha_x\beta_2$, integrins for cell binding and internalization.

2.6.2. Altered Cell Adhesion and Cancer

Tumor progression to cancer involves alterations in a large number of cellular processes including cell–cell adhesion. Reduced homotypic cell–cell adhesion is required for metastatic cells to separate from primary tumors while specific *de novo* heterotypic cell–cell adhesion is important for the efficient establishment of distant metastasis. The importance of cell adhesion in cancer has been reaffirmed by the identification of specific adhesion molecules linked to these processes (for a detailed discussion of this topic see Ref. 192).

In general there is a strong correlation between the invasiveness or metastatic potential of a given tumor and its state of differentiation. Poorly differentiated carcinomas have few close cell–cell contacts and are typically highly invasive with a poor prognosis. By maintaining close cell–cell contacts, CAMs both physically bind cells together and promote a variety of contact-mediated cell–cell exchanges required for normal development and differentiation. The vast majority of human cancers are epithelial tumors or carcinomas. Recognition that E-cadherin plays a pivotal role in promoting epithelial polarity and differentiation has focused attention on its role in epithelial tumor progression to cancer.[193] E-cadherin expression has been shown to be inversely correlated with invasiveness and metastatic activity for a variety of human carcinomas.[194–197] The functional significance of this correlation is strongly supported by E-cadherin gene transfer studies.[194,198] Transfection of highly invasive transformed carcinoma cell lines with E-cadherin cDNA expression vectors resulted in the loss of invasive properties, and in some cases production of differentiated tumors at the local site of tumor cell injection into syngeneic animals. In the converse experiment, transfection of noninvasive transformed cells with an antisense E-cadherin rDNA construct caused a downregulation of E-cadherin expression and the generation of an invasive phenotype.

The results of these gene transfer studies indicate that E-cadherin acts as a tumor suppressor, and suggest that some human cancers may result from the decreased expression of E-cadherin. In support of this notion, deletions involving the region of chromosome 16 containing the human E-cadherin gene have been reported for prostatic, breast, hepatocellular, and neuroectodermal cancers.[193] Deletion of one E-cadherin allele could unmask previously silent mutations in the second allele. A second, perhaps more frequent mechanism of E-cadherin downregulation appears to involve local epigenetic regulatory influences. Mareel *et al.*[199] have shown that tumor cell lines that stably express E-cadherin under controlled conditions *in vitro,* produce tumors with heterogeneous levels of E-cadherin expression when injected into nude mice. Levels of E-cadherin expression in individual tumor cells correlated with their degree of histologic differentiation. Subsequent *in vitro* passage of E-cadherin-negative, poorly differentiated tumors derived from these mice led to the reexpression of E-cadherin, demonstrating its reversible regulation in these tumor cells. Thus, local, dynamic epigenetic regulation of E-cadherin expression in tumor cells may be a focal turning point in the progression of epithelial dysplasia to cancer. Agents that promote the continued or re-expression of E-cadherin may slow or reverse the development of carcinomas.

Genetic identification of two other tumor suppressor genes suggests the involvement of additional CAMs in decreased cell adhesion leading to tumor development. Deletion of a gene, DCC (deleted in colon cancer), encoding a putative homotypic Ig-related CAM occurs in 70% of human colorectal cancers.[200] The DCC gene product contains four Ig domains and a single fibronectin type III domain and is expressed in a wide variety of cells including normal colonic mucosa. Details of how the loss of DCC-mediated cell–cell interaction is related to tumor progression in colon cancer remain to be determined. The third example of a CAM tumor suppressor gene has been provided by the cloning of the *fat* tumor suppressor gene in *Drosophila*.[201] Mutation of this gene, which encodes a novel cadherin, leads to the development of lethal, hyperplastic, tumorlike growth of larval imaginal discs.

In some cancers, disturbance of cell–cell adhesion may involve the expression of highly glycosylated CAMs which disrupt cell–cell contact. As noted earlier, NCAM containing high levels of polysialic acid can prevent close cell–cell apposition, and consequently cell contact-dependent signaling.[151] High levels of polysialic acid-rich NCAM have been shown to be characteristic of Wilms' tumor, small-cell lung carcinomas, and neuroblastomas[202–204] raising the possibility that it may contribute to altered growth characteristics. Similarly, overexpression of the highly glycosylated carcinoembryonic antigen (CEA) and its redistribution from apical to basolateral cell surfaces is thought to contribute to the disruption of normal epithelial structure and the metastatic progression of gastrointestinal tumors.[205]

In the case of blood-borne tumor cells, the efficiency of developing distant metastasis relates to the ability of tumor cells to bind to, and extravasate through, local blood vessel endothelium. For example, aberrant expression of a splice variant of the CD44 lymphocyte endothelial homing receptor[206] can confer metastatic potential to rat pancreatic tumor cells. Similar CD44 variants are expressed by many human carcinomas[207] and may be related to the metastatic potential of these cancers. Similarly, abnormally high levels of expression by gastrointestinal tumors of sialylated and fucosylated Lewis^x-like carbohydrate structures, which can act as ligands for endothelial cell E-selectin, may be important determinants in the ability of these tumor cells to bind to distant endothelial sites, initiating secondary metastasis.[208] Indeed, the characteristic spectrum of metastatic sites for specific primary tumors is thought to be determined in significant measure by selective adhesion between tumor cells and tissue-specific endothelial cells.[209] In parallel with the process of lymphocyte homing and the binding of leukocytes to select endothelia, specific, nonuniformly distributed endothelial cell adhesion receptors,[210,211] and the overexpression of their counterreceptors by specific tumor cells, have been suggested to be required for the efficient development of metastatic foci.

2.6.3. Disorders of Leukocyte Adhesion

Careful study of several patients referred for evaluation of recurrent life-threatening bacterial infections revealed defects in leukocyte adhesion and the absence of a 150-kDa granulocyte cell surface protein identical to the Mac-1 β subunit.[212–214] Subsequently, more than 50 patients with a similar molecular defect and immune deficiency have been identified. This group of patients represent a newly recognized disease, leukocyte adhesion deficiency type 1, caused by a congenital autosomal recessive defect in the β_1-integrin subunit common to the leukocyte adhesion molecules LFA-1 and Mac-1.[215,216] A range of defects have been identified in the β_1-integrin subunit gene as well as in its expression.[217–219] The hallmark functional deficiency seen in these patients is the inability of neutrophils and monocytes to migrate to sites of inflammation, resulting in recurrent severe bacterial infections. Patients fail to produce pus: infected tissues are devoid of neutrophils despite the presence of 5- to 20-fold higher than normal levels of circulating granulocytes. Patients with complete β_1-integrin subunit deficiency die in infancy of sepsis, while partial expression produces a milder disease with survival into adulthood. In addition to granulocyte migration deficits, phagocytes are unable to ingest serum-opsonized bacteria secondary to their inability to bind the iC3b fragment of complement via the Mac-1 (complement receptor 3) integrin. Despite similar decreased levels of expression of the β_1-integrin LFA-1 on lymphocytes in these patients, clinically there appears to be no impairment of lymphocyte migration. Redundant CD2: LFA-3-mediated lymphocyte–endothelial cell adhesion is able to offset the lack of β1-integrin lymphocyte adhesion. This unfortunate experiment of nature serves to graphically demonstrate and confirm the importance of the LFA-1 and Mac-1 adhesion molecules in granulocyte function. Study of leukocytes from these patients has been of great benefit in working out details of leukocyte–endothelial cell adhesion described in Section 2.5.2.

Recently, two unrelated children with severe recurrent bacterial infections have been identified who exhibit many features of patients with leukocyte adhesion deficiency type 1, but whose leukocytes express normal or even increased levels of β_1-integrin.[220] Neutrophils from these patients display severe deficits in endothelial adhesion and fail to migrate into sites of tissue infection. Consistent with their normal expression of β_1-integrin, neutrophils from these patients display potent opsonophagocytic and bactericidal activity. The adhesion defect in these patients is based on an inherited defect in fucosyltransferase activity. Granulocytes fail to express the sialyl-Lewis^x carbohydrate ligand recognized by endothelial E- and P-selectin, and thus are unable to participate in the initial weak adhesion rolling phase of leukocyte–endothelial cell interaction, a necessary prerequisite for the induction of integrin-mediated strong adhesion. Thus, identification of a second congenital form of leukocyte adhesion deficiency, type 2, confirms the experimental studies described earlier delineating a multistep adhesion cascade leading to leukocyte transendothelial migration (Section 2.5.2 and Figure 2.7). Further studies of leukocyte function in these two groups of patients will undoubtedly provide further clues to the intricacies of adhesion and leukocyte function and will hopefully suggest therapeutic interventions of benefit to these and other patients.

2.6.4. Disorders of Neural Development

Recent studies of genes involved in X-linked forms of two different neurological syndromes have implicated defective expression of adhesion molecules as their likely molecular basis. In one, X-linked congenital hydrocephalus, the genetic defect has been mapped to a region of the X chromosome containing the human equivalent of the L1 adhesion molecule. Northern blot analysis of brain mRNA from individuals with X-linked hydrocephalus shows unusual patterns of L1 mRNA splicing.[221] While further studies identifying the nature of the altered splicing and its effect on the protein product will be required to define the specific defect, the results to date suggest that the altered expression of this widespread, strongly adhesive neural cell adhesion molecule involved in both cellular migration and axon growth can have devastating effects on brain development.

The second X-linked developmental disorder, Kallmann's syndrome, is defined clinically by the presence of anosmia (loss of the sense of smell) together with hypogonadism.[222] The hypogonadism results from the complete absence of neurons containing gonadotropin-releasing hormone (GnRH) in the hypothalamic preoptic area, while the anosmia occurs secondary to the absence of an intact olfactory nerve and bulb. Both GnRH-containing neurons and primary olfactory neurons arise from the olfactory placode. Early in development primary olfactory neuron axons normally grow through the cribriform plate, forming the first cranial nerve invading the CNS and innervating mitral cells of the olfactory bulb. Developing GnRH neurons generated in the placode utilize the developing cranial nerve I as a pathway for their peripheral-to-central migration which is thought to be completed by the 13th to 15th week of human gestation. Postmortem study of a fetus at 19 weeks' gestation which carried the Kallmann's gene deletion (discussed below) showed the presence of intact GnRH and olfactory neurons in the olfactory placode, with axons of both cell types extending through the cribriform plate but failing to make contact with the central nervous system.[223] This developmental failure of placodal axons to complete their peripheral-to-central growth is the central pathophysiologic feature linking the two defining but otherwise disparate features of this syndrome. Study of the involved region of the X chromosome, Xp22.3, led to the identification of a novel gene, KALIG-1, encoding a putative cell adhesion molecule.[224,225] The predicted protein product has an N-terminal domain related to functional domains of protease and ATPase inhibitors, and more centrally, two fibronectin type III domains most closely related to those present in the neural cell adhesion molecules TAG-1, F11, and L1. Unlike these molecules, the KALIG-1 gene product does not contain Ig-like domains. Thus, while its structure in part resembles known axon growth-stimulating adhesion molecules, it does not appear to belong to any of the previously identified families of adhesion molecules. The chicken homologue of KALIG-1 has now been cloned, allowing developmental studies of its pattern of expression.[226] Chicken KALIG-1 mRNA expression is restricted to relatively limited groups of neurons. In normal animals, KALIG-1 is not expressed by olfactory placode neurons but rather by olfactory bulb mitral cells, the target of primary olfactory axons. This finding suggests the hypothesis that the absence of the KALIG-1 protein on the surface of mitral cells in Kallmann's patients prevents growing olfactory axons from forming stable adhesion to, or perhaps recognition of, their postsynaptic target.

In addition to its expression in the olfactory bulb, KALIG-1 mRNA is also found in Purkinje cells of the cerebellar cortex, in the oculomotor nucleus, and in the ventrolateral area of the anterior forebrain of developing chicks. In nonneural tissue the most prominent hybridization was to developing mesenchyme in mandibular arches and in the proximal portions of limb buds. Expression of KALIG-1 gene in nonolfactory areas may relate to the constellation of more variable features found in many Kallmann's patients. These include neurologic features such as mirror movements of the hands and feet, ocular motor abnormalities, and cerebellar dysfunction,[227] as well as nonneurologic manifestations such as developmental bone defects and in some patients unilateral renal aplasia.[222] As our understanding of adhesion and neurodevelopment continues to grow, additional examples of the involvement of adhesion molecules in neurologic and psychiatric disorders will undoubtedly unfold.

2.7. POTENTIAL THERAPEUTIC MODULATION OF CAM ACTIVITIES

Provision of exogenous CAMs associated with inert substrata has the potential of promoting cellular migration during axon regeneration and wound repair. Hydrogel implants containing axon growth-promoting adhesive molecules such as laminin, collagen, and fibronectin have the potential for promoting regeneration of axons through an area of damage.[228,229] The subsequent ability of regenerating CNS axons to make appropriate connections with their former target tissues will impact greatly on the potential benefit of such implants for spinal cord trauma patients. An analogous approach is being tested for the enhancement of dermal wound repair. Telioderm is a biologically active gel incorporating synthetic peptides containing RGD recognition sequences involved in integrin-mediated adhesion and locomotion of dermal fibroblasts and epithelial cells. This biological glue has shown utility in clinical trials for the treatment of chronic dermal ulcers and is currently awaiting final FDA approval.[230]

Inhibition of CAM function offers several therapeutic opportunities to augment the control of cancer metastasis, inflammation, and thrombosis. Blockade of blood-borne tumor cell adhesion to endothelial cells has the potential to reduce metastatic disease, particularly when administered prophylactically prior to surgical removal of a primary cancer. In animal models, administration of antibodies to a lung-specific endothelial CAM (Lu-ECAM-1) required for efficient seeding of melanoma cells produced a 90% reduction in the number of metastatic foci induced by intravenous injection of highly metastatic melanoma cells. Similarly, blockade of integrin-mediated tumor cell adhesion to extracellular matrix components can inhibit experimental metastasis formation *in vivo*.[231–233]

Control of inflammatory responses by CAM modulation is currently an area of active research. While both nonspecific and specific inflammatory responses are crucial for survival, unwanted or misdirected inflammation is an important clinical problem. Examples include chronic autoimmune disorders, transplant rejection, and acute inflammatory reactions such as cerebral edema in meningitis, adult respiratory distress syndrome, and ischemia-reperfusion injury. In the latter example, release of inflammatory cytokines generated during short periods of ischemia such as during a heart attack, stroke, hemorrhagic shock, or frostbite, triggers an inflammatory response which in many cases is more damaging than the ischemic event itself and ultimately leads to organ failure following an initial recovery. Corticosteroids, one of the most potent currently available anti-inflammatory agents, prevent the recruitment of leukocytes to inflammatory sites in part by blocking the induction of E-selectin and ICAM-1 expression by cytokine-activated endothelial cells.[234] As described in Section 2.5.2, E-selectin is involved in the initial adhesion and rolling phase of leukocyte–endothelial interaction, while ICAM-1 contributes importantly in the subsequent firm adhesion of leukocytes to inflamed endothelium. Other potent anti-inflammatory agents also reduce cell adhesion. The nonsteriodal class of anti-inflammatory agents reduce homotypic neutrophil adhesion,[235] while colchicine, an anti-inflammatory agent that has been used for hundreds of years, has recently been shown to reduce the expression of L-selectin on the surface of leukocytes and of I-CAM on endothelial cells.[236] The effects of corticosteroids, nonsteroidal anti-inflammatory agents, and colchicine are broad and varied; significant side effects frequently limit their more aggressive use. Attempts to develop more specific antiadhesion strategies to block unwanted inflammation have recently focused on the potential use of specific antibody-based blockade of leukocyte–endothelial cell adhesion. Antibodies to LFA-1, Mac-1, and ICAM-1 have been shown (1) to significantly reduce tissue injury and to increase survival in several animal studies of ischemia-reperfusion injury[237–239]; (2) to prevent asthmatic responses to antigen stimulation[240,241]; and (3) to reduce inflammatory tissue injury in experimental glomerulonephritis,[242] bacterial meningitis,[243] and endotoxin-induced uveitis.[244] Blockade of immune cell adhesion may also be useful in the induction of tolerance to grafted tissues. Combined short-term treatment with antibodies to ICAM-1 and LFA-1 have been shown to induce long-term survival of allogeneic cardiac transplants in rodents,[245,246] presumably both by limiting acute leukocyte–endothelial adhesion involved in nonspecific inflammation as well as by blocking T-cell activation events which are enhanced by LFA-1/ICAM-1-mediated lymphocyte–target cell adhesion. Similarly, short-term administration of antibodies to CD4 and CD8, Ig-related CAMs which strengthen the binding of lymphocytes to antigen-presenting and target cells, respectively, can induce a state of antigen-specific tolerance when coadministered with antigen or allograft.[247,248]

Platelet adhesion, a key process in hemostasis and thrombosis, is mediated in large part by the platelet integrin $\alpha_{IIb}\beta_3$. Reversible inhibition of integrin adhesion could provide an important addition to currently available antithrombotic therapy of cardiopulmonary bypass surgery, acute myocardial ischemia, and pulmonary emboli. A promising modality is the use of disintegrins, 5 to 9-kDa polypeptide components of certain snake venoms.[249] All disintegrins contain an RGD sequence and can block RGD-dependent integrin adhesion. A distinct advantage of disintegrins compared with antibody-based therapies is their reversible binding and rapid renal excretion. Thus, in dog models, bleeding times could be increased fourfold within 15 min of disintegrin administration, yet return to normal within 3 hr.[250] When used in combination with the currently used thrombolytic agent tPA (tissue-type plasminogen activator) in a canine model of coronary artery thrombosis, disintegrins produced a 70–80% decrease in the time to reperfusion.[251] Finally, disintegrins may be of great use in preventing damage and loss of platelets via integrin-mediated adhesion to fibrinogen-coated surfaces of extracorporeal membrane oxygenators of cardiopulmonary bypass equipment.[252] Platelet consumption and damage resulting from such extracorporeal adhesion contribute to prolonged postoperative bleeding following cardiac surgery.

Other potential therapeutic modulation of CAM activities include the use of soluble analogues of CAMs which have been pirated as viral receptors in order to block virus–cell adhesion. Genetically engineered soluble forms of ICAM-1 and CD4 have been shown to block the binding and uptake of rhinovirus and human immunodeficiency virus, respectively.[253–257] Unfortunately, clinical trials of soluble CD4 in AIDS patients have only been marginally successful in reducing levels of HIV antigen,[258] in part because of the need to maintain a high titer of the relatively low-affinity soluble CD4 ligand. Oligomerization and site-directed mutagenesis of soluble receptors may provide both agents with higher avidity[259] as well as ones that selectively disrupt viral but not cell–cell-mediated adhesion,[36,260] and may thus prove to have greater clinical utility.

2.8. SUMMARY

Cell–cell adhesion has long been recognized as an essential prerequisite and regulator of the development and functioning of animals. Only during the past 10 years has our understanding of cell adhesion on a molecular level begun to take clear shape. There are now scores of molecularly cloned, distinct CAMs, many of which comprise new gene families. Individual cells each express multiple CAMs. Distinct combinatorial expression of subsets of CAMs provides a basis for generating intricate selectivity in cell–cell adhesion. The details of such selectivity will certainly be tedious to tease out. Nonetheless, as illustrated by results from studies in the immune and nervous system, investigators are now making considerable progress in deciphering the molecular basis of selective cell–cell adhesion.

Several CAMs have been shown to interact with components of the cytoskeleton providing a direct linkage between extracellular binding events and the generation of cell–cell binding. In other cases the end result of CAM binding includes generation of intracellular second messengers and stimulation of protein phosphorylation events. The relatively recent realization that CAMs are much more than structural elements cementing cells together will likely motivate a change toward the perception of these molecules as cell contact receptors rather than simply adhesion molecules. Finally, given the central role of cell adhesion in the maintenance of animal tissues, it is not surprising that specific disturbances of cell adhesion are associated with the pathophysiology of disease states. Understanding the details of how CAMs regulate cell contact events in normal and diseased tissues will make possible the development of novel therapeutic treatments.

ACKNOWLEDGMENTS. The assistance of Robert Pilarski in preparation of the manuscript and of Mary Jane Spring in preparation of figures was much appreciated. Work in the author's laboratory has been sponsored by grants from the NIH (NS25264), the March of Dimes Birth Defects Foundation, and the University of Connecticut Research Foundation.

REFERENCES

1. Holtfreter, J. (1939). Gewebeaffinitat, ein Mittel der embryonalen Formbildung. Arch. Exp. Zellforsch. **23:**169–209, as translated in *Foundations in Experimental Embryology* (B. Willier and J. Oppenheimer, eds.), Prentice–Hall, Englewood Cliffs, NJ (1964).
2. Weiss, P. (1947). The problem of specificity in growth and development. *Yale J. Biol. Med.* **19:**235–278.
3. Steinberg, M. S. (1963). Reconstruction of tissues by dissociated cells. *Science* **141:**401–408.
4. Wilson, H. V. (1907). On some phenomena of coalescence and regeneration in sponges. *J. Exp. Zool.* **5:**245–258.
5. Moscona, A. A. (1962). Analysis of cell recombinations in experimental synthesis of tissues in vitro. *J. Cell. Comp. Physiol.* **60:**65–81.
6. Roth, S. (1968). Studies on intercellular adhesive selectivity. *Dev. Biol.* **18:**602–631.
7. Brackenbury, R., Thiery, J. P., Rutishauser, U., and Edelman, G. M. (1977). Adhesion among neural cells of the chick embryo. I. An immunological assay for molecules involved in cell–cell binding. *J. Biol. Chem.* **252:**6835–6840.
8. Thiery, J. P., Brackenbury, R., Rutischauser, U., and Edelman, G. M. (1977). Adhesion among neural cells of the chick embryo. II. Purification and characterization of a cell adhesion molecule from neural retina. *J. Biol. Chem.* **252:**6841–6845.
9. Kemler, R., Babinet, C., Eisen, H., and Jacob, F. (1977). Surface antigen in early differentiation. *Proc. Natl. Acad. Sci. USA* **74:** 4449–4452.
10. Grumet, M., and Edelman, G. M. (1984). Heterotypic binding between neuronal membrane vesicles and glial cells is mediated by a specific cell adhesion molecule. *J. Cell Biol.* **98:**1746–1756.
11. Bertolotti, R., Rutishauser, R., and Edelman, G. M. (1980). A cell surface molecule involved in aggregation of embryonic liver cells. *Proc. Natl. Acad. Sci. USA* **77:**4831–4835.
12. Hyafil, F., Morello, D., Babinet, C., and Jacob, F. (1980). A cell surface glycoprotein involved in the compaction of embryonal carcinoma cells and cleavage stage embryos. *Cell* **21:**927–934.
13. Damsky, C. H., Knudsen, K. A., Dorio, R. J., and Buck, C. A. (1981). Manipulation of cell–cell and cell–substratum interactions in mouse mammary tumor epithelial cells using broad spectrum antisera. *J. Cell Biol.* **89:**173–184.
14. Damsky, C. H., Richa, J., Solter, D., Knudsen, K., and Buck, C. A. (1983). Identification and purification of a cell surface glycoprotein mediating intercellular adhesion in embryonic and adult tissue. *Cell* **34:**455–466.
15. Knudsen, K. A., Rao, P. E., Damsky, C. H., and Buck, C. A. (1981). Membrane glycoproteins involved in cell–substratum adhesion. *Proc. Natl. Acad. Sci. USA* **78:**6071–6075.
16. Nagafuchi, A., Shirayoshi, Y., Okazaki, K., Yasuda, K., and Takeichi, M. (1987). Transformation of cell adhesion properties by exogenously introduced E-cadherin cDNA. *Nature* **329:**341–343.
17. Filbin, M. T., Walsh, F. S., Trapp, B. D., Pizzey, J. A., and Tennekoon, G. I. (1990). Role of myelin Po protein as a homophilic adhesion molecule. *Nature* **344:**871–872.
18. St. John, T., Meyer, J., Idzerda, R., and Gallatin, W. M. (1990). Expression of CD44 confers a new adhesive phenotype on transfected cells. *Cell* **60:**45–52.
19. Nose, A., Nagafuchi, A., and Takeichi, M. (1988). Expressed recombinant cadherins mediate cell sorting in model systems. *Cell* **54:** 993–1001.
20. Hynes, R. O., and Lander, A. D. (1992). Contact and adhesive specificities in the associations, migrations and targeting of cells and axons. *Cell* **68:**303–322.
21. Geiger, B., and Ayalon, O. (1992). Cadherins. *Annu. Rev. Cell Biol.* **8:**307–332.
22. Takeichi, M. (1990). Cadherins: A molecular family important in selective cell–cell adhesion. *Annu. Rev. Biochem.* **59:**237–252.
23. Edelman, G. M., and Crossin, K. L. (1991). Cell adhesion molecules: Implications for a molecular histology. *Annu. Rev. Biochem.* **60:**155–190.
24. Hynes, R. O. (1992). Integrins: Versatility, modulation, and signaling in cell adhesion. *Cell* **69:**11–25.
25. Lasky, L. A. (1992). Selectins: Interpreters of cell-specific carbohydrate information during inflammation. *Science* **258:**964–969.
26. Stoeckli, E. T., Kuhn, T. B., Duc, C. O., Ruegg, M. A., and Sonderegger, P. (1991). The axonally secreted protein axonin-1 is a potent substratum for neurite growth. *J. Cell Biol.* **112:**449–455.
27. Low, M. (1989). Glycosyl-phosphatidylinositol: A versatile anchor for cell surface proteins. *FASEB J.* **3:**1600–1608.
28. Stefanova, I., Horejsi, V., Ansotegui, I. J., Knapp, W., and Stockinger, H. (1991). GPI-anchored cell-surface molecules complexed to protein tyrosine kinases. *Science* **254:**1016–1019.
29. Owens, G. C., Edelman, G. M., and Cunningham, B. A. (1987). Organization of the neural cell adhesion molecule (N-CAM) gene: Alternative exon usage as the basis for different membrane-associated domains. *Proc. Natl. Acad. Sci. USA* **84:**294–298.
30. Small, S. J., Haines, S. L., and Akeson, R. A. (1988). Polypeptide variation in an NCAM extracellular immunoglobulin-like fold is developmentally regulated through alternative splicing. *Neuron* **1:**1007–1017.
31. Thompson, J., Dickson, G., Moore, S. E., Gower, H. J., Putt, W., Kenimer, J. G., Barton, C. H., and Walsh, F. S. (1989). Alternative splicing of the neural cell adhesion molecule gene generates variant extracellular domain structure in skeletal muscle and brain. *Genes Dev.* **3:**348–357.
32. Reyes, A. A., Schulte, S. V., Small, S., and Akeson, R. (1993). Distinct NCAM splicing events are differentially regulated during rat brain development. *Mol. Brain Res.* **17:**201–211.
33. Pollerberg, G. E., Schachner, M., and Davoust, J. (1986). Differentiation state-dependent surface mobilities of two forms of the neural cell adhesion molecule. *Nature* **324:**462–465.
34. Hall, A. K., and Rutishauser, U. (1987). Visualization of neural cell adhesion molecule by electron microscopy. *J. Cell Biol.* **104:** 1579–1586.
35. Becker, J. W., Erickson, H. P., Hoffman, S., Cunningham, B. A., and Edelman, G. M. (1989). Topology of cell adhesion molecules. *Proc. Natl. Acad. Sci. USA* **86:**1088–1092.
36. Staunton, D. E., Dustin, M. L., Erickson, H. P., and Springer, T. A. (1990). The arrangement of the immunoglobulin-like domains of ICAM-1 and the binding sites for LFA-1 and rhinovirus. *Cell* **61:** 243–254.
37. Nermut, M. V., Green, N. M., Eason, P., Yamada, S. S., and Yamada, K. M. (1988). Electron microscopy and structural model of human fibronectin receptor. *EMBO J.* **7:**4093–4099.
38. Ringwald, M., Schuh, R., Vestweber, D., Eistetter, H., Lottspeich, F., Engel, J., Dolz, R., Jahnig, F., Epplen, J., Mayer, S., Muller, C., and Kemler, R. (1987). The structure of cell adhesion molecule uvo-

morulin. Insights into the molecular mechanism of Ca-dependent cell adhesion. *EMBO J.* **6:**3647–3653.
39. Nagafuchi, A., Takeichi, M. (1988). Cell binding function of E-cadherin is regulated by the cytoplasmic domain. *EMBO J.* **7:** 3679–3684.
40. Ozawa, M., Baribault, H., and Kemler, R. (1989). The cytoplasmic domain of the cell adhesion molecule uvomorulin associates with three independent proteins structurally related in different species. *EMBO J.* **8:**1711–1717.
41. Ozawa, M., Ringwald, M., and Kemler, R. (1990). Uvomorulin–catenin complex formation is regulated by a specific domain in the cytoplasmic region of the cell adhesion molecule. *Proc. Natl. Acad. Sci. USA* **87:**4246–4250.
42. Nagafuchi, A., and Takeichi, M. (1989). Transmembrane control of cadherin-mediated cell adhesion: A 94 kDa protein functionally associated with a specific region of the cytoplasmic domain of E-cadherin. *Cell Regul.* **1:**37–44.
43. Hirano, S., Kimoto, N., Shimoyama, Y., Hirohashi, S., and Takeichi, M. (1992). Identification of a neural α-catenin as a key regulator of cadherin function and multicellular organization. *Cell* **70:**293–301.
44. Nose, A., Tsuji, K., and Takeichi, M. (1990). Localization of specificity determining sites in cadherin cell adhesion molecules. *Cell* **61:**147–155.
45. Blaschuk, O. W., Sullivan, R., David, S., and Pouliot, Y. (1990). Identification of a cadherin cell adhesion recognition sequence. *Dev. Biol.* **139:**227–229.
46. Ozawa, M., Engel, J., and Kemler, R. (1990). Single amino acid substitutions in one Ca binding site of uvomorulin abolish the adhesive function. *Cell* **63:**1033–1038.
47. Suzuki, S., Sano, K., and Tanihara, H. (1991). Diversity of the cadherin family: Evidence for eight new cadherins in nervous tissue. *Cell Regul* **2:**261–270.
48. Williams, A. F., and Barclay, A. N. (1988). The immunoglobulin superfamily-domains for cell surface recognition. *Annu. Rev. Immunol.* **6:**381–405.
49. Hunkapiller, T., and Hood, L. (1989). Diversity of the immunoglobulin gene superfamily. *Adv. Immunol.* **44:**1–63.
50. Edmundson, A. B., Ely, K. R., Abola, E. E., Schiffer, M., and Panagiotopoulos, N. (1975). Rotational allomerism and divergent evolution of domains in immunoglobulin light chains. *Biochemistry* **14:** 3953–3961.
51. Frelinger, A. L., III, and Rutishauser, U. (1986). Topography of N-CAM structural and functional determinants. II. Placement of monoclonal antibody epitopes. *J. Cell Biol.* **103:**1729–1737.
52. Cole, G. J., and Akeson, R. (1989). Identification of a heparin binding domain of the neural cell adhesion molecule NCAM using synthetic peptides. *Neuron* **2:**1157–1165.
53. Diamond, M. S., Staunton, D. E., Marlin, S. D., and Springer, T. A. (1991). Binding of the integrin Mac-1 (CD11b/CD18) to the third immunoglobulin-like domain of ICAM-1 (CD54) and its regulation by glycosylation. *Cell* **65:**961–971.
54. Tian, S. S., Tsoulfas, P., and Zinn, K. (1991). Three receptor-linked protein-tyrosine phosphatases are selectively expressed on central nervous system axons in the Drosophila embryo. *Cell* **67:**675–685.
55. Yang, X., Seow, K. T., Bahri, S. M., Oon, S. H., and Chia, W. (1991). Two Drosophila receptor-like tyrosine phosphatase genes are expressed in a subset of developing axons and pioneer neurons in the embryonic CNS. *Cell* **67:**661–673.
56. Leahy, D. J., Hendrickson, W. A., Aukhil, I., and Erickson, H. P. (1992). Structure of a fibronectin type III domain from tenascin phased by MAD analysis of the selenomethionyl protein. *Science* **258:**987–991.
57. Davis, J. Q., McLaughlin, T., and Bennett, V. (1993). Ankyrin-binding proteins related to nervous system cell adhesion molecules: Candidates to provide transmembrane and intercellular connections in adult brain. *J. Cell Biol.* **121:**121–133.
58. Tamkun, J. W., DeSimone, D. W., Fonda, D., Patel, R. S., Buck, C., Horwitz, A. F., and Hynes, R. O. (1986). Structure of integrin, a glycoprotein involved in the transmembrane linkage between fibronectin and actin. *Cell* **46:**271–282.
59. Pierschbacher, M. D., and Ruoslahti, E. (1984). Cell attachment activity of fibronectin can be duplicated by small synthetic fragments of the molecule. *Nature* **309:**30–33.
60. Ruoslahti, E., and Pierschbacher, M. D. (1987). New perspectives in cell adhesion: RGD and integrins. *Science* **238:**491–497.
61. Horwitz, A., Duggan, K., Buck, C., Beckerle, M. C., and Burridge, D. (1986). Interaction of plasma membrane fibronectin receptor with talin—a transmembrane linkage. *Nature* **320:**531.
62. Otey, C. A., Pavalko, F. M., and Burridge, K. (1990). An interaction between α-actinin and the β1 integrin subunit in vitro. *J. Cell Biol.* **111:**721–729.
63. Erbe, D. V., Watson, S. R., Presta, L. G., Wolitsky, B. A., Foxall, C., Brandley, B. K., and Lasky, L. A. (1993). P- and E-selectin use common sites for carbohydrate ligand recognition and cell adhesion. *J. Cell Biol.* **120:**1227–1235.
64. Gallatin, W. M., Weissman, I. L., and Butcher, E. C. (1983). A cell-surface molecule involved in organ-specific homing of lymphocytes. *Nature* **304:**30–34.
65. Siegelman, M. H., van de Rijn, M., and Weissman, I. L. (1989). Mouse lymph node homing receptor cDNA clone encodes a glycoprotein revealing tandem interaction domains. *Science* **243:**1165–1172.
66. Lasky, L. A., Singer, M. S., Yednock, T. A., Dowbenko, D., Fennie, C., Rodriguez, H., Nguyen, T., Stachel, S., and Rosen, S. D. (1989). Cloning of a lymphocyte homing receptor reveals a lectin domain. *Cell* **56:**1045–1055.
67. Bevilacqua, M. P., Pober, J. S., Mendrick, D. L., Cotran, R. S., and Gimbrone, M. A., Jr., (1987). Identification of an inducible endothelial–leukocyte adhesion molecule. *Proc. Natl. Acad. Sci. USA* **84:** 9238–9242.
68. Bevilacqua, M. P., Stengelin, S., Gimbrone, M. A., and Seed, B. (1989). Endothelial leukocyte adhesion molecule 1: An inducible receptor for neutrophils related to complement regulatory proteins and lectins. *Science* **243:**1160–1165.
69. Larsen, R., Celi, A., Gilbert, G. E., Furie, B. C., Erban, J. K., Bonfanti, R., Wagner, D. D., and Furie, B. (1989). PADGEM protein: A receptor that mediates the interaction of activated platelets with neutrophils and monocytes. *Cell* **59:**305–312.
70. Lowe, J. B., Stoolman, L. M., Nair, R. P., Larsen, R. D., Berhend, T. L., and Marks, R. M. (1990). ELAM-1-dependent cell adhesion to vascular endothelium determined by a transfected human fucosyltransferase cDNA. *Cell* **63:**475–484.
71. Foxall, C., Watson, S. R., Dowbenko, D., Fennie, C., Lasky, L. A., Kiso, M., Hasegawa, A., Asa, D., and Brandley, B. K. (1992). The three members of the selectin receptor family recognize a common carbohydrate epitope, the sialyl Lewis^x oligosaccharide. *J. Cell Biol.* **117:**895–902.
72. Moore, K. L., Stults, N. L., Diaz, S., Smith, D. F., Cummings, R. D., Varki, A., and McEver, R. P. (1992). Identification of a specific glycoprotein ligand for P-selectin (CD62) on myeloid cells. *J. Cell Biol.* **118:**445–456.
73. Lasky, L. A., Singer, M. S., Dowbenko, D., Imai, Y., Henzel, W. J., Grimley, C., Fennie, C., Gillett, N., Watson, S. R., and Rosen, S. D. (1992). An endothelial ligand for L-selectin is a novel mucin-like molecule. *Cell* **69:**927–938.
74. Levinovitz, A., Muhlhoff, J., Isenmann, S., and Vestweber, D. (1993). Identification of a glycoprotein ligand for E-selectin on mouse myeloid cells. *J. Cell Biol.* **121:**449–459.
75. Brummendorf, T., Hubert, M., Treubert, U., Leuschner, R., Tarnok, A., and Rathjen, F. G. (1993). The axonal recognition molecule F11

is a multifunctional protein: Specific domains mediate interactions with Ng-CAM and restrictin. *Neuron* **10:**711–727.

76. Pesheva, P., Gennarini, G., Goridis, C., and Schachner, M. (1993). The F3/11 cell adhesion molecule mediates the repulsion of neurons by the extracellular matrix glycoprotein J1-160/180. *Neuron* **10:** 69–82.
77. Ferrel, J. E., and Martin, G. S. (1989). Tyrosine-specific protein phosphorylation is regulated by glycoprotein IIb-IIa in platelets. *Proc. Natl. Acad. Sci. USA* **86:**2234–2238.
78. Atashi, J. R., Klinz, S. G., Ingraham, C. A., Matten, W. T., Schachner, M., and Maness, P. F. (1992). Neural cell adhesion molecules modulate tyrosine phosphorylation of tubulin in nerve growth cone membranes. *Neuron* **8:**831–842.
79. Burridge, K., Turner, C. E., and Romer, L. H. (1992). Tyrosine phosphorylation of paxillin and pp125FAK accompanies cell adhesion to extracellular matrix: A role in cytoskeletal assembly. *J. Cell Biol.* **119:**893–903.
80. Kornberg, L. J., Earp, H. S., Turner, C. E., Prockop, C., and Juliano, R. J. (1991). Signal transduction by integrins: Increased protein tyrosine phosphorylation caused by clustering of β1 integrins. *Proc. Natl. Acad. Sci. USA* **88:**8392–8396.
81. Schwartz, M. A. (1993). Spreading of human endothelial cells on fibronectin or vitronectin triggers elevation of intracellular free calcium. *J. Cell Biol.* **120:**1003–1010.
82. Jaconi, M. E. E., Theler, J. M., Schlegel, W., Appel, R. D., Wright, S. D., and Lew, P. D. (1991). Multiple elevations of cytosolic-free Ca^{2+} in human neutrophils: Initiation by adherence receptors of the integrin family. *J. Cell Biol.* **112:**1249–1257.
83. Doherty, P., Ashton, S. V., Moore, S. E., and Walsh, F. S. (1991). Morphoregulatory activities of NCAM and N-cadherin can be accounted for by G protein-dependent activation of L- and N-type neuronal Ca^{2+} channels. *Cell* **67:**21–33.
83a. Doherty, P., Rowett, L. H., Moore, S. E., Mann, D. A., and Walsh, F. S. (1991). Neurite outgrowth in response to transfected N-CAM and N-cadherin reveals fundamental differences in neuronal responsiveness to CAMs. *Neuron* **6:**247–258.
84. Williams, E. J., Doherty, P., Turner, G., Reid, R. A., Hemperly, J. J., and Walsh, F. S. (1992). Calcium influx into neurons can solely account for cell contact-dependent neurite outgrowth stimulated by transfected L1. *J. Cell Biol.* **119:**883–892.
85. Ingber, D. E., Prusty, D., Frangioni, J. V., Cragoe, E. J., Lechene, C., and Schwartz, M. A. (1990). Control of intracellular pH and growth by fibronectin in capillary endothelial cells. *J. Cell Biol.* **110:**1803–1811.
86. Schwartz, M. A., Lechene, C., and Ingber, D. E. (1991). Insoluble fibronectin activates the Na/H antiporter by clustering and immobilizing integrin α5β1, independent of cell shape. *Proc. Natl. Acad. Sci. USA* **88:**7849–7853.
87. Becchetti, A., Arcangeli, A., Riccarda del Bene, M., Olivotto, M., and Wanke, E. (1992). Response to fibronectin–integrin interaction in leukaemia cells: Delayed enhancing of a K^+ current. *Proc. R. Soc. London B Ser.* **248:**235–240.
88. Nathan, C., Srimal, S., Farber, C., Sanchez, E., Kabbash, L., Asch, A., Gailit, J., and Wright, S. D. (1989). Cytokine-induced respiratory burst of human neutrophils: Dependence on extracellular matrix proteins and CD11/CD18 integrins. *J. Cell Biol.* **109:**1341–1349.
89. Fuortes, M., Jin, W.-W., and Nathan, C. (1993). Adhesion-dependent protein tyrosine phosphorylation in neutrophils treated with tumor necrosis factor. *J. Cell Biol.* **120:**777–784.
90. Farquhar, M. G., and Palade, G. E. (1963). Junctional complexes in various epithelia. *J. Cell Biol.* **17:**375–412.
91. Schneeberger, E. E., and Lynch, R. D. (1992). Structure, function, and regulation of cellular tight junctions. *Am. J. Physiol.* **262:** L647–L661.
92. Gumbiner, B. (1987). The structure, biochemistry and assembly of epithelial tight junctions. *Am. J. Physiol.* **253:**C749–C758.
93. Ducibella, T., and Anderson, E. (1975). Cell shape and membrane changes in the eight-cell mouse embryo: Prerequisites for morphogenesis of the blastocyst. *Dev. Biol.* **47:**45–58.
94. Hyafil, F., Babinet, C., and Jacob, F. (1981). Cell–cell interactions in early embryogenesis: A molecular approach to the role of calcium. *Cell* **26:**447–454.
95. Vestweber, D., Gossler, A., Boller, K., and Kemler, R. (1987). Expression and distribution of cell adhesion uvomorulin in mouse preimplantation embryos. *Dev. Biol.* **124:**451–456.
96. Winkel, G. K., Ferguson, J. E., Takeichi, M., and Nuccitelli, R. (1990). Activation of protein kinase C triggers premature compaction in the four-cell stage mouse-embryo. *Dev. Biol.* **138:**1–15.
97. Boller, K., Vestweber, D., and Kemler, R. (1985). Cell–adhesion molecule uvomorulin is localized in the intermediate junctions of adult intestinal epithelial cells. *J. Cell Biol.* **100:**327–332.
98. Gumbiner, B., and Simons, D. (1986). A functional assay for proteins involved in establishing an epithelial occluding barrier: Identification of a uvomorulin-like polypeptide. *J. Cell Biol.* **102:**457–468.
99. Mege, R.-M., Matsuzaki, F., Gallin, W. J., Goldberg, J. I., Cunningham, B. A., and Edelman, G. M. (1988). Construction of epithelioid sheets by transfection of mouse sarcoma cells with cDNAs for chicken cell adhesion molecules. *Proc. Natl. Acad. Sci. USA* **85:** 7274–7278.
100. McNeill, H., Ozawa, M., Kemler, R., and Nelson, W. J. (1990). Novel function of the cell adhesion molecule uvomorulin as an inducer of cell surface polarity. *Cell* **62:**309–316.
101. Volk, T., and Geiger, B. (1986). A-CAM: A 135-kD receptor of intercellular adherens junctions. I. Immunoelectron microscopic localization and biochemical studies. *J. Cell Biol.* **103:**1441–1450.
102. Buxton, R. S., Cowin, P., Franke, W. W., Garrod, D. R., Green, K. J., King, I. A., Koch, P. J., Magee, A. I., Rees, D. A., Stanley, J. R., and Steinberg, M. S. (1993). Nomenclature of the desmosomal cadherins. *J. Cell Biol.* **121:**481–483.
103. Amagai, M., Klaus-Kovtun, V., and Stanley, J. R. (1991). Autoantibodies against a novel epithelial cadherin in pemphigus vulgaris, a disease of cell adhesion. *Cell* **67:**869–877.
104. Schiltz, J. R., and Michel, B. (1976). Production of epidermal acantholysis in normal human skin in vitro by the IgG fraction from pemphigus serum. *J. Invest. Dermatol.* **67:**254–260.
105. Rock, B., Labib, R. S., and Diaz, L. A. (1990). Monovalent Fab′ immunoglobulin fragments from endemic pemphigus foliaceus autoantibodies reproduce the human disease in neonatal Balb/c mice. *J. Clin. Invest.* **85:**296–299.
106. Gumbiner, B., Stevenson, B., and Grimaldi, A. (1988). The role of the cell adhesion molecule uvomorulin in the formation and maintenance of the epithelial junctional complex. *J. Cell Biol.* **107:**1575–1587.
107. Itoh, M., Nagafuchi, A., Yonemura, S., Kitani-Yasuda, T., Tsukita, S., and Tsukita, S. (1993). The 220-kD protein colocalizing with cadherins in non-epithelial cells is identical to ZO-1, a tight junction-associated protein in epithelial cells: cDNA cloning and immunoelectron microscopy. *J. Cell Biol.* **121:**491–502.
108. Shimizu, Y., Newman, W., Tanaka, Y., and Shaw, S. (1992). Lymphocyte interactions with endothelial cells. *Immunol. Today* **13:**106–112.
109. Zimmerman, G. A., Prescott, S. M., and McIntyre, T. M. (1992). Endothelial cell interactions with granulocytes: Tethering and signalling molecules. *Immunol. Today* **13:**93–100.
110. Dustin, M. L., and Springer, T. A. (1991). Role of lymphocyte adhesion receptors in transient interactions and cell locomotion. *Annu. Rev. Immunol.* **9:**27–66.
111. Picker, L. J., and Butcher, E. C. (1992). Physiological and molecular mechanisms of lymphocyte homing. *Annu. Rev. Immunol.* **10:** 561–591.
112. Woodruff, J. J., Clarke, L. M., and Chin, Y. H. (1987). Specific cell-adhesion mechanisms determining migration pathways of recirculating lymphocytes. *Annu. Rev. Immunol.* **5:**201–222.

113. Stamler, H. B., and Woodruff, J. J. (1976). Lymphocyte homing into lymph nodes: In vitro demonstration of the selective affinity of recirculating lymphocytes for high-endothelial venules. *J. Exp. Med.* **144:**828–833.
114. Rasmussen, R. A., Chin, Y.-H., Woodruff, J. J., and Easton, T. G. (1985). Lymphocyte recognition of lymph node high endothelium. VII. Cell surface proteins involved in adhesion defined by monoclonal anti-HEBF$_{LN}$(A.11) antibody. *J. Immunol.* **135:**19–24.
115. Chin, Y.-H., Rasmussen, R. A., Woodruff, J. J., and Easton T. G. (1986). A monoclonal anti-HEBF$_{PP}$ antibody with specificity for lymphocyte surface molecules mediating adhesion to Peyer's patch high endothelium of the rat. *J. Immunol.* **136:**2556–2561.
116. Jalkanen, S., Bargatze, R. F., de los Toyos, J., and Butcher, E. C. (1987). Lymphocyte recognition of high endothelium: Antibodies to distinct epitopes of an 85–95-kD glycoprotein antigen differentially inhibit lymphocyte binding to lymph node, mucosal, or synovial endothelial cells. *J. Cell Biol.* **105:**983–990.
117. Holzmann, B., McIntyre, B. W., and Weissman, I. L. (1989). Identification of a murine Peyer's patch-specific lymphocyte homing receptor as an integrin molecule with an α chain homologous to human VLA-4α. *Cell* **56:**37–46.
118. Streeter, P. R., Rouse, B. T. N., and Butcher, E. C. (1988). Immunohistologic and functional characterization of a vascular addressin involved in lymphocyte homing into peripheral lymph nodes. *J. Cell Biol.* **107:**1853–1862.
119. Streeter, P. R., Berg, E. L., Rouse, B. T. N., Bargatze, R. F., and Butcher, E. C. (1988). A tissue-specific endothelial cell molecule involved in lymphocyte homing. *Nature* **331:**41–46.
120. Nakache, M., Berg, E. L., Streeter, P. R., and Butcher, E. C. (1989). The mucosal vascular addressin is a tissue-specific endothelial cell adhesion molecule for circulating lymphocytes. *Nature* **337:**179–181.
121. Stamenkovic, I., Amiot, M., Pesando, J. M., and Seed, B. (1989). A lymphocyte molecule implicated in lymph node homing is a member of the cartilage link protein family. *Cell* **56:**1057–1062.
122. Goldstein, L. A., Zhou, D. F. H., Picker, L. J., Minty, C. N., Bargatze, R. F., Ding, J. F., and Butcher, E. C. (1989). A human lymphocyte homing receptor the Hermes antigen, is related to cartilage proteoglycan core and link proteins. *Cell* **56:**1063–1072.
123. Picker, L. J., Nakache, M., and Butcher, E. C. (1989). Monoclonal antibodies to human lymphocyte homing receptors define a novel class of adhesion molecules on diverse cell types. *J. Cell Biol.* **109:**927–937.
124. Berg, E. L., Yoshino, T., Rott, L. S., Robinson, M. K., Warnock, R. A., Kishimoto, T. K., Picker, L. J., and Butcher, E. C. (1991). The cutaneous lymphocyte antigen is a skin lymphocyte homing receptor for the vascular lectin endothelial cell–leukocyte adhesion molecule 1. *J. Exp. Med.* **174:**1461–1466.
125. Picker, L. J., Kishimoto, T. K., Smith, C. W., Warnock, R. A., and Butcher, E. C. (1991). ELAM-1 is an adhesion molecule for skin-homing T cells. *Nature* **349:**796–799.
126. Picker, L. J., Terstappen, L. W. M. M., Rott, L. S., Streeter, P. R., Stein, H., and Butcher, E. C. (1990). Differential expression of homing-associated adhesion molecules by T cell subsets in man. *J. Immunol.* **145:**3247–3255.
127. von Andrian, U. H., Chambers, J. D., McEvoy, L. M., Bargatze, R. F., Arfors, K. E., and Butcher, E. C. (1991). Two-step model of leukocyte–endothelial cell interaction in inflammation: Distinct roles for LECAM-1 and the leukocyte β_2 integrins in vivo. *Proc. Natl. Acad. Sci. USA* **88:**7538–7542.
128. von Andrian, U. H., Hansell, R., Chambers, J. D., Berger, E. M., Filho, I. T., Butcher, E. C., and Arfors, K. E. (1992). L-selectin function is required for β_2-integrin-mediated neutrophil adhesion at physiological shear rates in vivo. *Am. J. Physiol.* **263:**H1034–H1044.
129. Butcher, E. C. (1991). Leukocyte–endothelial cell recognition: Three (or more) steps to specificity and diversity. *Cell* **67:**1033–1036.
130. Sugama, Y., Tiruppathi, C., Janakidevi, K., Andersen, T. T., Fenton, J. W., II, and Malik, A. B., (1992). Thrombin-induced expression of endothelial P-selectin and intercellular adhesion molecule-1: A mechanism for stabilizing neutrophil adhesion. *J. Cell Biol.* **119:**935–944.
131. Lorant, D. E., Patel, K. D., McIntyre, T. M., McEver, R. P., Prescott, S. M., and Zimmerman, G. A. (1991). Coexpression of GMP-140 and PAF by endothelium stimulated by histamine or thrombin: A justacrine system for adhesion and activation of neutrophils. *J. Cell Biol.* **115:**223–234.
132. Shimizu, Y., Newman, W., Gopal, T. V., Horgan, K. J., Graber, N., Beall, L. D., van Seventer, G. A., and Shaw, S. (1991). Four molecular pathways of T cell adhesion to endothelial cells: Roles of LFA-1, VCAM-1, and ELAM-1 and changes in pathway hierarchy under different activation conditions. *J. Cell Biol.* **113:**1203–1212.
133. Dustin, M. L., and Springer, T. A. (1988). Lymphocyte function-associated antigen-1 (LFA-1) interaction with intercellular adhesion molecule-1 (ICAM-1) is one of at least three mechanisms for lymphocyte adhesion to cultured endothelial cells. *J. Cell Biol.* **107:**321–331.
134. Osborn, L., Hession, C., Tizard, R., Vassallo, C., Luhowskyj, S., Chi-Rosso, G., and Lobb, R. (1989). Direct expression cloning of vascular cell adhesion molecule 1, a cytokine-induced endothelial protein that binds to lymphocytes. *Cell* **59:**1203–1211.
135. Lawrence, M. B., and Springer, T. A. (1991). Leukocytes roll on a selectin at physiologic flow rates: Distinction from and prerequisite for adhesion through integrins. *Cell* **65:**859–873.
136. Picker, L. J., Warnock, R. A., Burns, A. R., Doerschuk, C. M., Berg, E. L., and Butcher, E. C. (1991). The neutrophil selectin LECAM-1 presents carbohydrate ligands to the vascular selectins ELAM-1 and GMP-140. *Cell* **66:**921–933.
137. Mayadas, T. N., Johnson, R. C., Rayburn, H., Hynes, R. O., and Wagner, D. D. (1993). Leukocyte rolling and extravasation are severely compromised in P selectin-deficient mice. *Cell* **74:**541–554.
138. Kishimoto, T. K., Jutila, M. A., Berg, E. L., and Butcher, E. C. (1989). Neutrophil Mac-1 and MEL-14 adhesion proteins inversely regulated by chemotatic factors. *Science* **245:**1238–1241.
139. Detmers, P. A., Lo, S. K., Olsen-Egbert, E., Walz, A., Baggiolini, M., and Cohn, Z. A. (1990). Neutrophil-activating protein 1/interleukin 8 stimulates the binding activity of the leukocyte adhesion receptor CD11b/CD18 on human neutrophils. *J. Exp. Med.* **171:**1155–1162.
140. Hermanowski-Vosatka, A., Van Strijp, J. A. G., Siggard, W. J., and Wright, S. D. (1992). Integrin modulating factor-1: A lipid that alters the function of leukocyte integrins. *Cell* **68:**341–352.
141. Elices, M. J., Osborn, L., Takada, Y., Crouse, C., Luhowskyj, S., Hemler, M. E., and Lobb, R. R. (1990). VCAM-1 on activated endothelium interacts with the leukocyte integrin VLA-4 at a site distinct from the VLA-4/fibronectin binding site. *Cell* **60:**577–584.
142. Covault, J. (1989). Molecular biology of cell adhesion in neural development. In *Molecular Neurobiology* (D. M. Glover and B. D. Hames, eds.), Oxford University Press, London, pp. 143–200.
143. Hortsch, M., and Goodman, C. S. (1991). Cell and substrate adhesion molecules in Drosophila. *Annu. Rev. Cell Biol.* **7:**505–557.
144. Reichardt, L. F., and Tomaselli, K. J. (1991). Extracellular matrix molecules and their receptors: Functions in neural development. *Annu. Rev. Neurosci.* **14:**531–570.
145. Rathjen, F. G., and Jessell, T. M. (1991). Glycoproteins that regulate the growth and guidance of vertebrate axons: Domains and dynamics of the immunoglobulin/fibronectin type III subfamily. *Semin. Neurosci.* **3:**297–307.
146. Finne, J., Finne, U., Deagostini-Bazin, H., and Goridis, C. (1983). Occurrence of α2-8 linked polysialosyl units in a neural cell adhesion molecule. *Biochem. Biophys. Res. Commun.* **112:**482–487.

147. Nedelec, J., Boucraut, J., Garnier, J. M., Bernard, D., and Rougon, G. (1990). Evidence for autoimmune antibodies directed against embryonic neural cell adhesion molecules (NCAM) in patients with group B meningitis. *J. Neuroimmunol.* **29:**49–56.
148. Chuong, C.-M., and Edelman, G. M. (1984). Alterations in neural cell adhesion molecules during development of different regions of the nervous system. *J. Neurosci.* **4:**2354–2368.
149. Sunshine, J., Balak, K., Rutishauser, R., and Jacobson, M. (1987). Changes in neural cell adhesion molecule (NCAM) structure during vertebrate neural development. *Proc. Natl. Acad. Sci. USA* **84:** 5986–5990.
150. Yang, P., Yin, X., and Rutishauser, U. (1992). Intercellular space is affected by the polysialic acid content of NCAM. *J. Cell Biol.* **116:**1487–1496.
151. Rutishauser, U., Acheson, A., Hall, A. K., Mann, D. M., and Sunshine, J. (1988). The neural cell adhesion molecule (NCAM) as a regulator of cell–cell interactions. *Science* **240:**53–57.
152. Letourneau, P. C. (1975). Cell-to-substratum adhesion and guidance of axonal elongation. *Dev. Biol.* **44:**92–101.
153. Lemmon, V., Burden, S. M., Payne, H. R., Elmslie, G. J., and Hlavin, M. L. (1992). Neurite growth on different substrates: Permissive versus instructive influences and the role of adhesive strength. *J. Neurosci.* **12:**818–826.
154. Tosney, K. W., and Landmesser, L. (1985). Growth cone morphology and trajectory in the lumbosacral region of the chick embryo. *J. Neurosci.* **5:**2345–2358.
155. Tang, J., Landmesser, L., and Rutishauser, U. (1992). Polysialic acid influences specific pathfinding by avian motoneurons. *Neuron* **8:** 1031–1044.
156. Tang, J. (1992). Role of polysialic acid in motoneuron pathfinding and naturally occurring cell death. Ph. D. thesis. University of Connecticut, Storrs.
157. Nose, A., Mahajan, V. B., and Goodman, C. S. (1992). Connectin: A homophilic cell adhesion molecule expressed on a subset of muscles and the motoneurons that innervate them in Drosophila. *Cell* **70:** 553–567.
158. Tosney, K. W., Watanabe, M., Landmesser, L., and Rutishauser, U. (1986). The distribution of NCAM in the chick hindlimb during axon outgrowth and synaptogenesis. *Dev. Biol.* **114:**437–452.
159. Hahn, C.-G., and Covault, J. (1992). Neural regulation of N-cadherin gene expression in developing and adult skeletal muscles. *J. Neurosci.* **12:**4677–4687.
160. El-Deeb, S., Thompson, S. C., and Covault, J. (1992). Characterization of a cell surface adhesion molecule expressed by a subset of developing chick neurons. *Dev. Biol.* **149:**213–227.
161. Landmesser, L., Dahm, L., Schultz, K., and Rutishauser, U. (1988). Distinct roles for adhesion molecules during innervation of embryonic chick muscle. *Dev. Biol.* **130:**645–670.
162. Landmesser, L., Dahm, L., Tang, J., and Rutishauser, U. (1990). Polysialic acid as a regulator of intramuscular nerve branching during embryonic development. *Neuron* **4:**655–667.
163. Tang, J., and Landmesser, L. (1993). Reduction of intramuscular nerve branching and synaptogenesis is correlated with decreased motoneuron survival. *J. Neurosci.* **13:**3095–3103.
164. Sanes, J. R., Hunter, D. D., Green, T. L., and Merlie, J. P. (1990). S-laminin. *Cold Spring Harbor Symp. Quant. Biol.* **55:**419–430.
165. Covault, J., and Sanes, J. R. (1986). Distribution of N-CAM in synaptic and extrasynaptic portions of developing and adult skeletal muscle. *J. Cell Biol.* **102:**716–730.
166. Covault, J., and Sanes, J. R. (1985). Neural cell adhesion molecule (N-CAM) accumulates in denervated and paralyzed skeletal muscles. *Proc. Natl. Acad. Sci. USA* **82:**4544–4548.
167. Booth, C. M., Kemplay, S. K., and Brown, M. C. (1990). An antibody to neural cell adhesion molecule impairs motor nerve terminal sprouting in a mouse muscle locally paralyzed with botulinum toxin. *Neuroscience,* **35:**85–91.
168. Mayford, M., Barzilai, A., Keller, F., Schacher, S., and Kandel, E. R. (1992). Modulation of an NCAM-related cell adhesion molecule with long-term synaptic plasticity in *Aplysia. Science* **256:**638–644.
169. Bailey, C. H., Chen, M., Keller, F., and Kandel, E. R. (1992). Serotonin-mediated endocytosis of apCAM: An early step of learning-related synaptic growth in *Aplysia. Science* **256:**645–649.
170. Greve, J. M., Davis, G., Meyer, A. M., Forte, C. P., Connolly Yost, S., Marlor, C. W., Kamarck, M. E., and McClelland, A. (1989). The major human rhinovirus receptor is ICAM-1. *Cell* **56:**839–847.
171. Stauton, D. E., Merluzzi, V. J., Rothlein, R., Barton, R., Marlin, S. D., and Springer, T. A. (1989). A cell adhesion molecule, ICAM-1, is the major surface receptor for rhinoviruses. *Cell* **56:**849–853.
172. Tomassini, J. E., Grahm, D., DeWitt, C. M., Lineberger, D. W., Rodkey, J. A., and Colonno, R. J. (1989). cDNA cloning reveals that the major group rhinovirus receptor on HeLa cells is intercellular adhesion molecule 1. *Proc. Natl. Acad. Sci. USA* **86:**4907–4911.
173. Maddon, P. J., Dalgleish, A. G., McDougal, J. S., Clapham, P. R., Weiss, R. A., and Axel, R. (1986). The T4 gene encodes the AIDS virus receptor and is expressed in the immune system and the brain. *Cell* **47:**333–348.
174. Sattentau, Q. J., and Weiss, R. A. (1988). The CD4 antigen: Physiological ligand and HIV receptor. *Cell* **52:**631–633.
175. Dveksler, G. S., Pensiero, M. N., Cardellichio, C. B., Williams, R. K., Jiang, G.-S., Holmes, K. V., and Dieffenbach, C. W. (1991). Cloning of the mouse hepatitis virus (MHV) receptor: Expression in human and hamster cell lines confers susceptibility to MHV. *J. Virol.* **65:**6881–6891.
176. Yokomori, K., and Lai, M. (1992). Mouse hepatitis virus utilizes two carcinoembryonic antigens as alternative receptors. *J. Virol.* **66:** 6194–6199.
177. Mendelsohn, C. L., Wimmer, E., and Racaniello, V. R. (1989). Cellular receptor for poliovirus: Molecular cloning, nucleotide sequence, and expression of a new member of the immunoglobulin superfamily. *Cell* **56:**855–865.
178. Koike, S., Taya, C., Kurata, T., Abe, S., Ise, I., Yonekawa, H., and Nomoto, A. (1991). Transgenic mice susceptible to poliovirus. *Proc. Natl. Acad. Sci. USA* **88:**951–955.
179. Rossmann, M. G., Arnold, E., Erickson, J. W., Frankenberger, E. A., Griffith, J. P., Hecht, H.-J., Johnson, J. E., Kamer, G., Luo, M., Mosser, A. G., Rueckert, R. R., Sherry, B., and Vriend, G. (1985). Structure of a human common cold virus and functional relationship to other picornaviruses. *Nature* **317:**145–153.
180. Rossmann, M. G. (1989). The canyon hypothesis. *J. Biol. Chem.* **264:**14587–14590.
181. Giranda, V. L., Chapman, M. S., and Rossmann, M. G. (1990). Modeling of the human intercellular adhesion molecule-1, the human rhinovirus major group receptor. *Proteins* **7:**227–233.
182. Olson, N. H., Kolatkar, P. R., Oliveira, M. A., Cheng, R. H., Greve, J. M., McClelland, A., Baker, T. S., and Rossmann, M. G. (1993). Structure of a human rhinovirus complexed with its receptor. *Proc. Natl. Acad. Sci. USA* **90:**507–511.
183. Bergelson, J. M., Shepley, M. P., Chan, B. M. C., Hemler, M. E., and Finberg, R. W. (1992). Identification of the integrin VLA-2 as a receptor for echovirus 1. *Science* **255:**1718–1720.
184. Wickham, T. J., Mathias, P., Cheresh, D. A., and Nemerow, G. R. (1993). Integrins $\alpha v\beta 3$ and $\alpha v\beta 5$ promote adenovirus internalization but not virus attachment. *Cell* **73:**309–319.
185. Wright, S. D., and Silverstein, S. C. (1983). Receptors for C3b and C3bi promote phagocytosis but not the release of toxic oxygen from human phagocytes. *J. Exp. Med.* **158:**2016–2023.
186. Rao, S. P., Ogata, K., Catanzaro, A. (1993). Mycobacterium avium-M, intracellulare binds to the integrin receptor $\alpha v\beta 3$ on human monocytes and monocyte-derived macrophages. *Infect. Immun.* **61:**663–670.
187. Isberg, R. R., and Leong, J. M. (1990). Multiple B1 chain integrins are receptors for invasin, a protein that promotes bacterial penetration into mammalian cells. *Cell* **60:**861–871.

188. Bullock, W. E., and Wright, S. D. (1987). Role of the adherence-promoting receptors, CR3, LFA-1, and p150,95, in binding of Histoplasma capsulatum by human macrophages. *J. Exp. Med.* **165:** 195–210.
189. Payne, N. R., and Horwitz, M. A. (1987). Phagocytosis of Legionella pneumophila is mediated by human monocyte complement receptors. *J. Exp. Med.* **166:**1377–1389.
190. Relman, D., Tuomanen, E., Falkow, S., Golenbock, D. T., Saukkonen, K., and Wright, S. D. (1990). Recognition of a bacterial adhesin by an integrin: Macrophage CR3 (amB2, CD11b/CD18) binds filamentous hemagglutinin of Bordetella pertussis. *Cell* **61:** 1375–1382.
191. Talamas-Rohana, P., Wright, S. D., Lennartz, M. R., and Russell, D. G. (1990). Lipophosphoglycan from Leishmania mexicana promastigotes binds to members of the CR3, p150,95 and LFA-1 family of leukocyte integrins. *J. Immunol.* **144:**4817–4824.
192. Albelda, S. M. (1993). Role of integrins and other cell adhesion molecules in tumor progression and metastasis. *Lab. Invest.* **68:**4–17.
193. Giroldi, L. A., and Schalken, J. A. (1993). Decreased expression of the intercellular adhesion molecule E-cadherin in prostate cancer: Biological significance and clinical implications. *Cancer Metast. Rev.* **12:**29–37.
194. Frixen, U. H., Behrens, J., Sachs, M., Eberle, G., Voss, B., Warda, A., Löchner, D., and Birchmeier, W. (1991). E-cadherin-mediated cell–cell adhesion prevents invasiveness of human carcinoma cells. *J. Cell Biol.* **113:**173–185.
195. Oka, H., Shiozaki, H., Kobayashi, K., Inoue, M., Tahara, H., Kobayashi, T., Takatsuka, Y. Matsuyoshi, N., Hirano, S., Takeichi, M., and Mori, T. (1993). Expression of E-cadherin cell adhesion molecules in human breast cancer tissues and its relationship to metastasis. *Cancer Res.* **53:**1696–1701.
196. Shiozaki, H., Tahara, H., Oka, H., Miyata, M., Kobayashi, K., Tamura, S., Iihara, K., Koki, Y., Hirano, M., Takeichi, M., and Mori, T. (1991). Expression of immuno-reactive E-cadherin adhesion molecules in human cancers. *Am. J. Pathol.* **139:**17–23.
197. Schipper, J. H., Frixen, E. H., Behrens, J., Unger, A., Jahnke, K., and Birchmeier, W. (1991). E-cadherin expression in squamous cell carcinomas of head and neck: Inverse correlation with tumor dedifferentiation and lymph node metastasis. *Cancer Res.* **51:**6328–6337.
198. Vleminckx, K., Vakaet, L., Jr., Mareel, M., Fiers, W., and Van Roy, F. (1991). Genetic manipulation of E-cadherin expression by epithelial tumor cells reveals an invasion suppressor role. *Cell* **66:**107–119.
199. Mareel, M. M., Behrens, J., Birchmeier, W., De Bruyne, G. K., Vleminckz, K., Hoogewijs, A., Fiers, W. C., and Van Roy, F. M. (1991). Down-regulation of E-cadherin expression in Madin Darby canine kidney (MDCK) cells inside tumors of nude mice. *Int. J. Cancer* **47:**922–928.
200. Fearon, E. R., Cho, K. R., Nigro, J. M., Kern, S. E., Simons, J. W., Ruppert, J. M., Hamilton, S. R., Preisinger, A. C., Thomas, G., Kinzler, K. W., and Vogelstein, B. (1990). Identification of a chromosome 18q gene that is altered in colorectal cancers. *Science* **247:**49–56.
201. Mahoney, P. A., Weber, U., Onofrechuk, P., Biessmann, H., Bryant, P. J., and Goodman, C. S. (1991). The fat tumor suppressor gene in Drosophila encodes a novel member of the cadherin gene superfamily. *Cell* **67:**853–868.
202. Roth, J., Zuber, C., Wagner, P., Blaha, I., Bitter-Suermann, D., and Heitz, P. U. (1988). Presence of the long chain form of polysialic acid of the neural cell adhesion molecule in Wilms' tumor: Identification of a cell adhesion molecule as an onco-developmental antigen and implications for tumor histogenesis. *Am. J. Pathol.* **133:**227.
203. Moolenaar, K. C. E. C., Muller, E. G., Schol, D. J., Figdor, C. G., Bock, E., Bitter-Suermann, D., and Michalides, R. J. A. M. (1990). Expression of neural cell adhesion molecule-related sialoglycoprotein in small cell lung cancer and neuroblastoma cell lines H69 and CHP-212. *Cancer Res.* **50:**1102–1106.
204. Livingstone, B., Jacobs, J. L., Glick, M. C., and Troy, F. A. (1989). Extended polysialic acid chains (on 755) in glycoproteins from human neuroblastoma cells. *J. Biol. Chem.* **263:**9443–9448.
205. Benchimol, S., Fuks, A., Jothy, S., Beauchemin, N., Shirota, K., and Stanners, C. P. (1989). Carcinoembryonic antigen, a human tumor marker, functions as an intercellular adhesion molecule. *Cell* **57:**327–334.
206. Günthert, U., Hofmann, M., Rudy, W., Reber, S., Zöller, M., Haubmann, I., Matzku, S., Wenzel, A., Ponta, H., and Herrlich, P. (1991). A new variant of glycoprotein CD44 confers metastatic potential to rat carcinoma cells. *Cell* **65:**13–24.
207. Hofmann, M., Rudy, W., Zöller, M., Tölg, C., Ponta, H., Herrlich, P., and Günthert, U. (1991). CD44 splice variants confer metastatic behavior in rats: Homologous sequences are expressed in human tumor cell lines. *Cancer Res.* **51:**5292–5297.
208. Itzkowitz, S. H., Yuan, M., Fukushi, Y., Palekar, A., Phelps, P. C., Shamsuddin, A. M., Trum, B. F., Hakomori, S.-I., and Kim, Y. S. (1986). Lewis^x- and sialylated Lewis^x-related antigen expression in human malignant and nonmalignant colonic tissues. *Cancer Res.* **46:**2627–2632.
209. Auerbach, R., Lu, W. C., Pardon, E., Gumkowski, F., Kaminska, G., and Kaminski, M. (1987). Specificity of adhesion between murine tumor cells and capillary endothelium: An in vitro correlate of preferential metastasis in vivo. *Cancer Res.* **47:**1492–1496.
210. Rice, G. E., and Bevilacqua, M. P. (1989). An inducible endothelial cell surface glycoprotein mediates melanoma adhesion. *Science* **246:** 1303–1306.
211. Zhu, D., Cheng, C.-F., and Pauli, B. U. (1991). Mediation of lung metastasis of murine melanomas by a lung-specific endothelial cell adhesion molecule. *Proc. Natl. Acad. Sci. USA* **88:**9568–9572.
212. Arnaout, M. A., Pitt, J., Cohen, H. J., Melamed, J., Rosen, F. S., and Colten, H. R. (1982). Deficiency of a granulocyte-membrane glycoprotein (gp150) in a boy with recurrent bacterial infections. *N. Engl. J. Med.* **306:**693–699.
213. Crowley, C. A., Curnette, J. T., Rosin, R. E., Andre-Schwartz, J., Klempner, M., Snyderman, R., Southwick, F. S., Stossel, T. P., and Babior, B. M. (1980). An inherited abnormality of neutrophil adhesion. Its genetic transmission and its association with a missing protein. *N. Engl. J. Med.* **302:**1163–1168.
214. Bowen, T. J., Ochs, H. D., and Altman, L. C. (1982). Severe recurrent bacterial infections associated with defective adherence and chemotaxis in two patients with neutrophils deficient in a cell-associated glycoprotein. *J. Pediatr.* **101:**932.
215. Anderson, D. C., Smith, C. W., and Springer, T. A. (1989). Leukocyte adhesion deficiency and other disorders of leukocyte motility. In *The Metabolic Basis of Inherited Disease,* 6th ed. (C. R. Scriver, A. L. Beaudet, W. S. Sly, and D. Valle, eds.) McGraw–Hill, New York, pp. 2751–2777.
216. Arnaout, M. A. (1990). Leukocyte adhesion molecules deficiency: Its structural basis, pathophysiology and implications for modulating the inflammatory response. *Immunol. Rev.* **114:**145–180.
217. Kishimoto, T. K., Hollander, N., Roberts, T. M., Anderson, D. C., and Springer, T. A. (1987). Heterogeneous mutations in the β subunit common to the LFA-1, Mac-1 and p150,95 glycoproteins cause lymphocyte adhesion deficiency. *Cell* **50:**193–202.
218. Kishimoto, T. K., Hollander, N., Roberts, T. M., Anderson, D. C., and Springer, T. A. (1989). Leukocyte adhesion deficiency: Aberrant splicing of a conserved integrin sequence causes a moderate deficiency phenotype. *J. Biol. Chem.* **264:**3588.
219. Arnaout, M. A., Dana, N., Gupta, S. K., Tenen, D., and Fathallah, D. (1990). Point mutations impairing cell surface expression of the common β subunit CD18 in a patient with Leu-CAM deficiency. *J. Clin. Invest.* **85:**977–981.

220. Etzioni, A., Frydman, M., Pollack, S., Avidor, I., Phillips, M. L., Paulson, J. C., and Gershoni-Baruch, R. (1992). Recurrent severe infections caused by a novel leukocyte adhesion deficiency. *N. Engl. J. Med.* **327:**1789–1792.
221. Rosenthal, A., Jouet, M., and Kenwrick, S. (1992). Aberrant splicing of neural cell adhesion molecule L1 mRNA in a family with X-linked hydrocephalus. *Nature Genet* **2:**107–112.
222. White, B. J., Rogol, A. D., Brown, K. S., Lieblich, J. M., and Rosen, S. W. (1983). The syndrome of anosmia with hypogonadotropic hypogonadism: A genetic study of 18 new families and a review. *Am. J. Med. Genet.* **15:**417–435.
223. Schwanzel-Fukuda, M., Bick, D., and Pfaff, D. W. (1989). Luteinizing hormone-releasing hormone (LHRH)-expressing cells do not migrate normally in an inherited hypogonadal (Kallmann) syndrome. *Mol. Brain Res.* **6:**311–326.
224. Franco, B., Guioli, S., Pragliola, A., Incerti, B., Bardoni, B., Tonlorenzi, R., Carrozzo, R., Maestrini, E., Pieretti, M., Taillon-Miller, P., Brown, C. J., Willard, H. F., Lawrence, C., Persico, M. G., Camerino, G., and Ballabio, A. (1991). A gene deleted in Kallmann's syndrome shares homology with neural cell adhesion and axonal path-finding molecules. *Nature* **353:**529–536.
225. Legouis, R., Hardelin, J.-P., Levilliers, J., Claverie, J.-M., Compain, S., Wunderle, V., Millasseau, P., Le Paslier, D., Cohen, D., Caterina, D., Bougueleret, L., de Waal, H. D.-V., Lutfalla, G., Weissenbach, J., and Petit, C. (1991). The candidate gene for the X-linked Kallmann syndrome encodes a protein related to adhesion molecules. *Cell* **67:** 423–435.
226. Rugarli, E. I., Lutz, B., Kuratani, S. C., Wawersik, S., Borsani, G., Ballabio, A., and Eichele, G. (1993). Expression pattern of the Kallmann syndrome gene in the olfactory system suggests a role in neuronal targeting. *Nature Genet.* **4:**19–26.
227. Schwankhaus, J. D., Currie, J., Jaffe, M. J., Rose, S. R., and Sherins, R. J. (1989). Neurologic findings in men with isolated hypogonadotropic hypogonadism. *Neurology* **39:**223–226.
228. Woerly, S., Marchand, R., and Lavallee, C. (1990). Intracerebral implantation of synthetic polymer/biopolymer matrix: A new perspective for brain repair. *Biomaterials* **11:**97–107.
229. Woerly, S., Maghami, G., Duncan, R., Subr, V., and Ulbrich, K. (1993). Synthetic polymer derivatives as substrata for neuronal adhesion and growth. *Brain Res. Bull.* **30:**423–432.
230. Travis, J. (1993). Biotech gets a grip on cell adhesion. *Science* **260:**906–908.
231. Vollmers, H. P., and Birchmeier, W. (1983). Monoclonal antibodies inhibit the adhesion of mouse B16 melanoma cells in vitro and block lung metastasis in vivo. *Proc. Natl. Acad. Sci. USA* **80:**3729–3733.
232. Humphries, M. J., Olden, K., and Yamada, K. M. (1986). A synthetic peptide from fibronectin inhibits experimental metastasis of murine melanoma cells. *Science* **233:**467–470.
233. Iwamoto, Y., Robey, F. A., Graf, J., Sasaki, M., Kleinman, H. K., Yamada, Y., and Martin, G. R. (1987). YIGSR, a synthetic laminin pentapeptide, inhibits experimental metastasis formation. *Science* **238:** 1132–1134.
234. Cronstein, B. N., Kimmel, S. C., Levin, R. I., Martiniuk, F., and Weissmann, G. (1992). A mechanism for the antiinflammatory effects of corticosteroids: The glucocorticoid receptor regulates leukocyte adhesion to endothelial cells and expression of endothelial–leukocyte adhesion molecule 1 and intercellular adhesion molecule 1. *Proc. Natl. Acad. Sci. USA* **89:**9991–9995.
235. Abramson, S. B., and Weissmann, G. (1989). The mechanisms of action of nonsteroidal antiinflammatory drugs. *Arthritis Rheum.* **32:**1–9.
236. Cronstein, B. N., and Weissmann, G. (1993). The adhesion molecules of inflammation. *Arthritis Rheum.* **36:**147–157.
237. Vedder, N. B., Winn, R. K., Rice, C. L., Chi, E. Y., Arfors, K.-E., and Harlan, J. M. (1988). A monoclonal antibody to the adherence-promoting leukocyte glycoprotein, CD18, reduces organ injury and improves survival from hemorrhagic shock and resuscitation in rabbits. *J. Clin. Invest.* **81:**939–944.
238. Simpson, P. J., Todd, R. F., III, Fantone, J. C., Mickelson, J. K., Griffin, J. D., and Lucchesi, B. R. (1988). Reduction of experimental canine myocardial reperfusion injury by a monoclonal antibody (anti-Mo1, anti-CD11b) that inhibits leukocyte adhesion. *J. Clin. Invest.* **81:**624–629.
239. Mileski, W. J., Winn, R. K., Vedder, N. B., Pohlman, T. H., Harlan, J. M., and Rice, C. L. (1990). Inhibition of CD18-dependent neutrophil adherence reduces organ injury after hemorrhagic shock in primates. *Surgery* **108:**206–212.
240. Gundel, R. H., Wegner, C. D., Torcellini, C. A., and Letts, L. G. (1992). The role of intercellular adhesion molecule-1 in chronic airway inflammation. *Clin. Exp. Allergy* **22:**569–575.
241. Wegner, C. D., Gundel, R. H., Reilly, P., Haynes, N., Letts, L. G., and Rothlein, R. (1990). Intercellular adhesion molecule-1 (ICAM-1) in the pathogenesis of asthma. *Science* **247:**456–459.
242. Kawasaki, K., Yaoita, E., Yamamoto, T., Tamatani, T., Miyasaka, M., and Kihara, I. (1993). Antibodies against intercellular adhesion molecule-1 and lymphocyte function-associated antigen-1 prevent glomerular injury in rat experimental crescentic glomerulonephritis. *J. Immunol.* **150:**1074–1083.
243. Tuomanen, E. I., Saukkonen, K., Sande, S., Cioffe, C., and Wright, S. D. (1989). Reduction of inflammation, tissue damage, and mortality in bacterial meningitis in rabbits treated with monoclonal antibodies against adhesion-promoting receptors of leukocytes. *J. Exp. Med.* **170:**959–968.
244. Whicup, S. M., DeBarge, L. R., Rosen, H., Nussenblatt, R. B., and Chan, C.-C. (1993). Monoclonal antibody against CD11b/CD18 inhibits endotoxin-induced uveitis. *Invest. Ophthalmol. Vis. Sci.* **34:** 673–681.
245. Isobe, M., Yagita, H., Okumura, K., and Ihara, A. (1992). Specific acceptance of cardiac allograft after treatment with antibodies to ICAM-1 and LFA-1. *Science* **255:**1125–1127.
246. Kameoka, H., Ishibashi, M., Tamatani, T., Takano, Y., Moutabarrik, A., Jiang, H., Kokado, Y., Takahara, S., Okuyama, A., Kinoshita, T., and Miyasaka, M. (1993). The immunosuppressive action of anti-CD18 monoclonal antibody in rat heterotopic heart allotransplantation. *Transplantation* **55:**665–667.
247. Qin, S. X., Cobbold, S., Benjamin, R., and Waldmann, H. (1989). Induction of classical transplantation tolerance in the adult. *J. Exp. Med.* **169:**779–794.
248. Hutchings, P., O'Reilly, L., Parish, N. M., Waldmann, H., and Cooke, A. (1992). The use of a non-depleting anti-CD4 monoclonal antibody to re-establish tolerance to βcells in NOD mice. *Eur. J. Immunol.* **22:**1913–1918.
249. Williams, J. A. (1992). Disintegrins: RGD-containing proteins which inhibit cell/matrix interactions (adhesion) and cell/cell interactions (aggregation) via the integrin receptors. *Pathol. Biol.* **40:** 813–821.
250. Shebuski, R. J., Ramjit, D. R., Bencen, G. H., and Polokoff, M. A. (1989). Characterisation and platelet inhibitory activity of Bitistain, a potent arginine-glycine-aspartic acid-containing peptide from the venom of the viper Bitis arietans. *J. Biol. Chem.* **264:** 21550–21556.
251. Yasuda, T., Gold, H. K., Leinbach, R. C., Yaoita, H., Fallon, J. T., Guerrero, L., Napier, M. A., Bunting, S., and Collen, D. (1991). Kistrin, a polypeptide platelet GPIIb-IIIa receptor antagonist, enhances and sustains coronary arterial thrombolysis with recombinant tissue-type plasminogen activator in a canine preparation. *Circulation* **83:**1038–1047.
252. Musical, J., Niewiarowski, S., Rucinski, B., Stewart, G., Cook, J., Williams, J. A., and Edmunds, L. H. (1990). Inhibition of platelet adhesion to surfaces of extracorporeal circuits by distintegrins. RGD-containing peptides from viper venoms. *Circulation* **82:**261–273.

253. Marlin, S. D., Staunton, D. E., Springer, T. A., Stratowa, C., Sommergruber, W., and Merluzzi, V. J. (1990). A soluble form of intercellular adhesion molecule-1 inhibits rhinovirus infection. *Nature* **344:**70–72.
254. Smith, D. H., Byrn, R. A., Masters, S. A., Gregory, T., Groopman, J. E., and Capon, D. J. (1987). Blocking of HIV-1 infectivity by a soluble, secreted form of the CD4 antigen. *Science* **238:**1704–1707.
255. Fisher, R. A., Bertonis, J. M., Meier, W., Johnson, V. A., Costopoulos, D. S., Liu, T., Tizard, R., Walker, B. D., Hirsh, M. S., Schooley, R. T., and Flavell, R. A. (1988). HIV infection is blocked in vitro by recombinant soluble CD4. *Nature* **331:**76–78.
256. Hussey, R. E., Richardson, N. E., Kowalski, M., Brown, N. R., Chang, H.-C., Siliciano, R. F., Dorfman, T., Walker, B., Sodroski, J., and Reinherz, E. L. (1988). A soluble CD4 protein selectively inhibits HIV replication and syncytium formation. *Nature* **331:**78–81.
257. Deen, K. C., McDougal, J. S., Inacker, R., Folena-Wasserman, G., Arthos, J., Rosenberg, J., Maddon, P. J., Axel, R., and Sweet, R. W. (1988). A soluble form of CD4 (T4) protein inhibits AIDS virus infection. *Nature* **331:**82–84.
258. Schooley, R. T., Merigan, T. C., Gaut, P., Hirsh, M. S., Holodniy, M., Flynn, T., Liu, S., Byington, R. E., Henochowicz, S., Gubish, E., Spriggs, D., Kufe, D., Schindler, J., Dawson, A., Thomas, D., Hanson, D. G., Letwin, B., Liu, T., Gulinello, J., Kennedy, S., and Fisher, R. (1990). Recombinant soluble CD4 therapy in patients with the acquired immunodeficiency syndrome (AIDS) and AIDS-related complex. *Ann. Intern. Med.* **112:**247–253.
259. Martin, S., Casasnovas, J. M., Staunton, D. E., and Springer, T. A. (1993). Efficient neutralization and disruption of rhinovirus by chimeric ICAM-1/immunoglobulin molecules. *J. Virol.* **67:**3561–3568.
260. Lamarre, D., Ashkenazi, A., Fleury, S., Smith, D. H., Sekaly, R.-P., and Capon, D. J. (1989). The MHC-binding and gp120-binding functions of CD4 are separable. *Science* **245:**743–746.
261. Crossin, K. L., Chuong, C. M., and Edelman, G. M. (1985). Expression sequences of cell adhesion molecules. *Proc. Natl. Acad. Sci. USA* **82:**6942–6946.
262. Napolitano, E. W., Venstrom, K., Wheeler, E. F., and Reichardt, L. F. (1991). Molecular cloning and characterization of B-cadherin, a novel chick cadherin. *J. Cell Biol.* **113:**893–905.
263. Koch, P. J., Goldschmidt, M. D., Walsh, M. J., Zimbelmann, R., Schmelz, M., and Franke, W. W. (1991). Amino acid sequence of bovine muzzle epithelial desmocollin derived from cloned cDNA: A novel subtype of desmosomal cadherins. *Differentiation* **47:**29–36.
264. Koch, P. J., Goldschmidt, M. D., Zimbelmann, R., Troyanovsky, R., and Franke, W. W. (1992). Complexity and expression patterns of the desmosomal cadherins. *Proc. Natl. Acad. Sci. USA* **89:**353–357.
265. Mechanic, S., Raynor, K., Hill, J. E., and Cowin, P. (1991). Desmocollins form a distinct subset of the cadherin family of cell adhesion molecules. *Proc. Natl. Acad. Sci. USA* **88:**4476–4480.
266. Troyanovsky, S. M., Eshkind, L. G., Troyanovsky, R. B., Leube, R. E., and Franke, W. W. (1993). Contributions of cytoplasmic domains of desmosomal cadherins to desmosome assembly and intermediate filament anchorage. *Cell* **72:**561–574.
267. Koch, P. J., Walsh, M. J., Schmelz, M., Goldschmidt, M. D., Zimbelmann, R., and Franke, W. W. (1990). Identification of desmoglein, a constitutive desmosomal glycoprotein, as a member of the cadherin family of cell adhesion molecules. *Eur. J. Cell Biol.* **53:**1–12.
268. Koch, P. J., Goldschmidt, M. D., Walsh, J. J., Zimbelmann, R., and Franke W. W. (1991). Complete amino acid sequence of the epidermal desmoglein precursos polypeptide and identification of a second type of desmoglein gene. *Eur. J. Cell Biol.* **55:**200–208.
269. Wheeler, G. N., Parker, A. E., Thomas, C. L., Ataliotis, P., Poynter, D., Arnemann, J., Rutman, A. J., Pidsley, S. C., Watt, F. M., Rees, D. A., Buxton, R. S., and Magee, A. I. (1991). Desmosomal glycoprotein DGI, a component of intercellular desmosome junctions, is related to the cadherin family of cell adhesion molecules. *Proc. Natl. Acad. Sci. USA* **88:**4796–4800.
270. Gallin, W. J., Sorkin, B. C., Edelman, G. M., and Cunningham, B. A. (1987). Sequence analysis of a cDNA clone encoding the liver cell adhesion molecule, L-CAM. *Proc. Natl. Acad. Sci. USA* **84:**2808–2812.
271. Donalies, M., Cramer, M., Ringwald, M., and Starzinski-Powitz, A. (1991). Expression of M-cadherin, a member of the cadherin multigene family, correlates with differentiation of skeletal muscle cells. *Proc. Natl. Acad. Sci. USA* **88:**8024–8028.
272. Hatta, K., Takagi, S., Fujisawa, H., and Takeichi, M. (1987). Spatial and temporal expression pattern of N-cadherin cell adhesion molecules correlated with morphogenetic processes of chicken embryos. *Dev. Biol.* **120:**215–227.
273. Hatta, K., Nose, A., Nagafuchi, A., and Takeichi, M. (1988). Cloning and expression of cDNA encoding a neural calcium-dependent cell adhesion molecule: Its identity in the cadherin gene family. *J. Cell Biol.* **106:**873–881.
274. Hatta, K., and Takeichi, M. (1986). Expression of N-cadherin adhesion molecules associated with early morphogenetic events in chick development. *Nature* **320:**447–449.
275. Nose, A., and Takeichi, M. (1986). A novel cadherin cell adhesion molecule: Its expression patterns associated with implantation and organogenesis of mouse embryos. *J. Cell Biol.* **103:**2649–2658.
276. Nose, A., Nagafuchi, A., and Takeichi, M. (1987). Isolation of placental cadherin cDNA: Identification of a novel gene family of cell–cell adhesion molecules. *EMBO J.* **6:**3655–3661.
277. Inuzuka, H., Miyatani, S., and Takeichi, M. (1991). R-Cadherin: A novel calcium-dependent cell–cell adhesion molecule expressed in the retina. *Neuron* **7:**69–79.
278. Ranscht, B., and Dours-Zimmermann, M. T. (1991). T-cadherin, a novel cadherin cell adhesion molecule in the nervous system lacks the conserved cytoplasmic region. *Neuron* **7:**391–402.
279. Ranscht, B., and Bronner-Fraser, M. (1991). T-cadherin expression alternates with migrating neural crest cells in the trunk of the avian embryo. *Development* **111:**15–22.
280. Hasler, T. H., Rader, C., Stoeckli, E. T., Zuellig, R. A., and Sonderegger, P. (1993). cDNA cloning, structural features, and eucaryotic expression of human TAG-1/axonin-1. *Eur. J. Biochem.* **211:**329–339.
281. Furley, A. J., Morton, S. B., Manalo, D., Karagogeos, D., Dodd, J., and Jessell, T. M. (1990). The axonal glycoprotein TAG-1 is an immunoglobulin superfamily member with neurite outgrowth-promoting activity. *Cell* **61:**157–170.
282. Zuellig, R. A., Rader, C., Schroeder, A., Kalousek, M. B., von Bohlenlund Halback, F., Osterwalder, T., Inan, C., Stoeckli, E. T., Affolter, H. U., Fritz, A., Hafen, E., and Sonderegger, P. (1992). The axonally secreted cell adhesion molecule, axonin-1: Primary structure, immunoglobulin- and fibronectin-type III-like domains, and glycosyl-phosphatidylinositol anchorage. *Eur. J. Biochem.* **204:**453–463.
283. Hinoda, Y., Neumaier, M., Hefta, S. A., Drzeniek, Z., Wagener, C., Shively, L., Hefta, L. J. F., Shively, J. E., and Paxton, R. J. (1988). Molecular cloning of a cDNA coding biliary glycoprotein I: Primary structure of a glycoprotein immunologically crossreactive with carcinoembryonic antigen. *Proc. Natl. Acad. Sci. USA* **85:**6959–6963.
284. Barnett, T. R., Kretschmer, A., Austen, D. A., Goebel, S. J., Hart, J. T., Elting, J. J., and Kamarck, M. E. (1989). Carcinoembryonic antigens: Alternative splicing accounts for the multiple mRNAs that code for novel members of the carcinoembryonic antigen family. *J. Cell Biol.* **108:**267–276.
285. Rojas, M., Fuks, A., and Stanners, C. (1990). Biliary glycoprotein, a member of the immunoglobulin supergene family, functions in vitro

as a Ca-dependent intercellular adhesion molecule. *Cell Growth Differ.* **1:**527–533.
286. Selvaraj, P., Plunkett, M. L., Dustin, M., Sanders, M. E., Shaw, S., and Springer, T. A. (1987). The T lymphocyte glycoprotein CD2 binds the cell surface ligand LFA-3. *Nature* **326:**400–403.
287. Driscoll, P. C., Cyster, J. G., Campbell, I. D., and Williams A. F. (1991). Structure of domain 1 of rat T lymphocyte CD2 antigen. *Nature* **353:**762–765.
288. Robey, E., and Axel, R. (1990). CD4: Collaborator in immune recognition and HIV infection. *Cell* **60:**697–700.
289. Doyle, C., and Strominger, J. L. (1987). Interaction between CD4 and class II MHC molecules mediates cell adhesion. *Nature* **330:** 256–259.
290. Norment, A. M., Salter, R. D., Parham, P., Engelhard, V. H., and Littman D. R. (1988). Cell–cell adhesion mediated by CD8 and MHC class I molecules. *Nature* **336:**79–81.
291. Beauchemin, N., Benchimol, S., Cournoyer, D., Fuks, A., and Stanners, C. P. (1987). Isolation and characterization of full-length functional cDNA clones for human carcinoembryonic antigen. *Mol. Cell. Biol.* **7:**3221–3230.
292. Oikawa, S., Imajo, S., Noguchi, T., Kosaki, G., and Nakazato, H. (1987). The carcinoembryonic antigen (CEA) contains multiple immunoglobulin-like domains. *Biochem. Biophys. Res. Commun.* **144:** 634–642.
293. Brummendorf, T., Wolff, J. M., Frank, R., and Rathjen, F. G. (1989). Neural cell recognition molecule F11: Homology with fibronectin type III and immunoglobulin type C domains. *Neuron* **2:**1351–1361.
294. Gennarini, G., Cibelli, G., Rougon, G., Mattei, M.-G., and Goridis, C. (1989). The mouse neuronal cell surface protein F3: A phosphatidylinositol-anchored member of the immunoglobulin superfamily related to chicken contactin. *J. Cell Biol.* **109:**775–788.
295. Ranscht, B. (1988). Sequence of contactin, a 130-kD glycoprotein concentrated in areas of interneuronal contact, defines a new member of the immunoglobulin supergene family in the nervous system. *J. Cell Biol.* **107:**1561–1573.
296. Rathjen, F. G., Wolff, J. M., and Chiquet-Ehrismann, R. (1991). Restrictin: A chick neural extracellular matrix protein involved in cell attachment co-purifies with the cell recognition molecule F11. *Development* **113:**151–164.
297. Stauton, D. E., Marlin, S. D., Stratowa, C., Dustin, M. L., and Springer, T. A. (1988). Primary structure of intracellular adhesion molecule 1 (ICAM-1) demonstrates interaction between members of the immunoglobulin and integrin supergene families. *Cell* **52:**925–933.
298. Dustin, M. L., Rothlein, R., Bhan, A. K., Dinarello, C. A., and Springer, T. A. (1986). Induction by IL-1 and interferon, tissue distribution, biochemistry and function of a natural adherence molecule (ICAM-1). *J. Immunol.* **137:**245–254.
299. Staunton, D. E., Dustin, M. L., and Springer, T. A. (1989). Functional cloning of ICAM-2, a cell adhesion ligand for LFA-1 homologous to ICAM-1. *Nature* **339:**61–64.
300. de Fougerolles, A. R., Stacker, S. A., Schwarting, R., and Springer, T. A. (1991). Characterization of ICAM-2 and evidence for a third counter-receptor. *J. Exp. Med.* **174:**253.
301. de Fougerolles, A. R., and Springer, T. A. (1992). Intercellular adhesion molecule 3, a third adhesion counter-receptor for lymphocyte function-associated molecule 1 on resting lymphocytes. *J. Exp. Med.* **175:**185–190.
302. de Fougerolles, A. R., Klickstein, L. B., and Springer, T. A. (1993). Cloning and expression of intercellular adhesion molecule 3 reveals strong homology to other immunoglobulin family counter-receptors for lymphocyte function-associated antigen 1. *J. Exp. Med.* **177:** 1187–1192.
303. Fawcett, J., Holness, C. L. L., Needham, L. A., Turley, H., Gatter, K. C., Mason, D. Y., and Simmons, D. L. (1992). Molecular cloning of ICAM-3, a third ligand for LFA-1, constitutively expressed on resting leukocytes. *Nature* **360:**481–484.
304. Vazeux, R., Hoffman, P. A., Tomita, J. K., Dickinson, E. S., Jasman, R. L., St. John, T., and Gallatin, W. M. (1992). Cloning and characterization of a new intercellular adhesion molecule ICAM-R. *Nature* **360:**485.
305. Linder, J., Rathjen, F. F., and Schachner, M. (1983). L1 mono- and poly-clonal antibodies modify cell migration in early postnatal mouse cerebellum. *Nature* **305:**427–430.
306. Rathjen, F. G., and Schachner, M. (1984). Immunocytological and biochemical characterization of a new neuronal cell surface component (L1 antigen) which is involved in cell adhesion. *EMBO J.* **3:**1–10.
307. Moos, M., Tacke, R., Scherer, H., Teplow, D., Furth, K., and Schachner, M. (1988). Neural adhesion molecule L1 as a member of the immunoglobulin superfamily with binding domains similar to fibronectin. *Nature* **334:**701–703.
308. Stallcup,W. B., and Beasley, L. (1985). Involvement of the nerve growth factor-inducible large external glycoprotein (NILE) in neurite fasciculation in primary cultures of rat brain. *Proc. Natl. Acad. Sci. USA* **82:**1276–1280.
309. Seed, B. (1987). An LFA-3 cDNA encodes a phospholipid-linked membrane protein homologous to its receptor, CD2. *Nature* **329:** 840–842.
310. Barbosa, J. A., Mentzer, S. J., Kamark, M. E., Hart, J., Strominger, J. L., Biro, P. A., and Burakoff, S. J. (1985). Somatic cell hybrid analysis of human lymphocyte function associated antigen (LFA-3): Gene mapping and role in CTL–target cell interactions. *I. C. S. U. Short Rep.* **2:**107–108.
311. Trapp, B. D., Quarles, R. H., and Suzuki, K. (1984). Immunocytochemical studies of quaking mice support a role for the myelin-associated glycoprotein in forming and maintaining the periaxonal space and periaxonal cytoplasmic collar of myelinating Schwann cells. *J. Cell Biol.* **99:**594–606.
312. Arquint, M., Roder, J., Chia, L., Down, J., Wilkinson, D., Bayley, H., Braun, P., and Dunn, R. (1987). Molecular cloning and primary structure of myelin-associated glycoprotein. *Proc. Natl. Acad. Sci. USA* **84:**600–604.
313. Salzer, J. L., Holmes, W. P., and Colman, D. R. (1987). The amino acid sequences of the myelin-associated glycoproteins: Homology to the immunoglobulin gene superfamily. *J. Cell Biol.* **104:** 957–965.
314. Poltorak, M., Sadoul, R., Keilhauer, G., Landa, C., Fahrig, T., and Schachner, M. (1987). Myelin-associated glycoprotein, a member of the L2/HNK-1 family of neural cell adhesion molecules, is involved in neuron–oligodendrocyte and oligodendrocyte–oligodendrocyte interaction. *J. Cell Biol.* **105:**1893–1899.
315. Fahrig, T., Landa, C., Pesheva, P., Kuhn, K., and Schachner, M. (1987). Characterization of binding properties of the myelin-associated glycoprotein to extracellular matrix constituents. *EMBO J.* **6:** 2875–2883.
316. Cunningham, B. A., Hemperly, J. J., Murray, B. A., Prediger, E. A., Brackenbury, R., and Edelman, G. M. (1987). Neural cell adhesion molecule: Structure, immunoglobulin-like domains, cell surface modulation, and alternative RNA splicing. *Science* **236:**799–806.
317. Volkmer, H., Hassel, B., Wolff, J. M., Frank, R., and Rathjen, F. G. (1992). Structure of the axonal surface recognition molecule neurofascin and its relationship to a neural subgroup of the immunoglobulin superfamily. *J. Cell Biol.* **118:**149–161.
318. Rathjen, F. G., Wolff, J. M., Chang, L., Bonhoeffer, F., and Raper, J. (1987). Neurofascin: A novel chick cell-surface glycoprotein involved in neurite–neurite interactions. *Cell* **51:**841–849.

319. Burgoon, M. P., Grumet, M., Mauro, V., Edelman, G. M., and Cunningham, B. A. (1991). Structure of the chicken neuron–glia cell adhesion molecule, Ng-CAM: Origin of the polypeptides and relation to the Ig superfamily. *J. Cell Biol.* **112:**1017–1029.
320. Grumet, M. Hoffman, S., and Edelman, G. M. (1984). Two antigenetically related neuronal cell adhesion molecules of different specificities mediate neuron–neuron and neuron–glia adhesion. *Proc. Natl. Acad. Sci. USA* **81:**267–271.
321. Grumet, M., Hoffman, S., Chuong, C.-M., and Edelman, G. M. (1984). Polypeptide components and binding functions of neuron–glia cell adhesion molecules. *Proc. Natl. Acad. Sci. USA* **81:**7989–7993.
322. Kuhn, T. B., Stoeckli, E. T., Rathjen, F. G., and Sonderegger, P. (1991). Neurite outgrowth on immobilized axonin-1 is mediated by a heterophilic interaction with L1(G4). *J. Cell Biol.* **115:**1113–1126.
323. Lemmon, V., Farr, K. L., and Lagenaur, C. (1989). L1-mediated axon outgrowth occurs via a homophilic binding mechanism. *Neuron* **2:**1597–1603.
324. Grumet, M., Mauro, V., Burgoon, M. P., Edelman, G. M., and Cunningham, B. A. (1991). Structure of a new nervous system glycoprotein, Nr-CAM, and its relationship to subgroups of neural cell adhesion molecules. *J. Cell Biol.* **113:**1399–1412.
325. Kayyem, J. F., Roman, J. M., de la Rosa, E. J., Schwarz, U., and Dreyer, W. J. (1992). Bravo/Nr-CAM is closely related to the cell adhesion molecules L1 and Ng-CAM and has a similar heterodimer structure. *J. Cell Biol.* **118:**1259–1270.
326. Lemke, G., and Axel, R. (1985). Isolation and sequence of a cDNA encoding the major structural protein of peripheral myelin. *Cell* **40:**501–508.
327. Newman, P. J., Berndt, M. C., Gorski, J., While, G. C., II, Lyman, S., Paddock, C., and Muller, W. A. (1990). PECAM-1 (CD31) cloning and relation to adhesion molecules of the immunoglobulin gene superfamily. *Science* **247:**1219–1222.
328. Albelda, S. M., Muller, W. A., Buck, C. A., and Newman, P. J. (1991). Molecular and cellular properties of PECAM-1 (endo-CAM/CD31): A novel vascular cell–cell adhesion molecule. *J. Cell Biol.* **114:**1059–1068.
329. Stockinger, H., Gadd, S. J., Eher, R., Majdic, O., Schreiber, W., Kasinrerk, W., Strass, B., Schnabl, E., and Knapp, W. (1990). Molecular characterization and functional analysis of the leukocyte surface protein CD31. *J. Immunol.* **145:**3889–3897.
330. Burns, F. R., von Kannen, S., Guy, L., Raper, J. A., Kamholz, J., and Chang, S. (1991). DM-GRASP, a novel Ig superfamily axonal surface protein that supports neurite extension. *Neuron* **7:**209–220.
331. Pourquie, O., Corbel, C., Le Caer, J.-P., Rossier, J., and Le Douarin, N. M. (1992). BEN, a surface glycoprotein of the immunoglobulin superfamily, is expressed in a variety of developing systems. *Proc. Natl. Acad. Sci. USA* **89:**5261–5265.
332. Tanaka, H., Matsui, T., Agata, A., Tomura, M., Kubota, I., McFarland, K. C., Kohr, B., Lee, A., Phillips, H. S., and Shelton, D. L. (1991). Molecular cloning and expression of a novel adhesion molecule, SC1. *Neuron* **7:**535–545.
333. Rice, G. E., Munro, J. M., and Bevilacqua, M. P. (1990). Inducible cell adhesion molecule 110 (INCAM-110) is an endothelial receptor for lymphocytes. A CD11/CD18-independent adhesion mechanism. *J. Exp. Med.* **171:**1369–1374.
334. Davignon, D., Martz, E., Reynolds, T., Kurzinger, K., and Springer, T. A. (1981). Monoclonal antibody to a novel leukocyte function-associated antigen (LFA-1): Mechanism of blocking of T lymphocyte-mediated killing and effects on other T and B lymphocyte functions. *J. Immunol.* **127:**590–595.
335. Kishimoto, T. K., O'Connor, K., Lee, A., Roberts, T. M., and Springer, T. A. (1987). Cloning of the β subunit of the lymphocyte adhesion proteins: Homology to an extracellular matrix receptor defines a novel supergene family. *Cell* **48:**681–690.
336. Marlin, S. D., and Springer, T. A. (1987). Purified intercellular adhesion molecule-1 (ICAM-1) is a ligand for lymphocyte function-associated antigen 1(LFA-1). *Cell* **51:**813–819.
337. Smith, C. W., Marlin, S. D., Rothlein, R., Toman, C., and Anderson, D. C. (1989). Cooperative interactions of LFA-1 and Mac-1 with intercellular adhesion molecule-1 in facilitating adherence and transendothelial migration of human neutrophils in vitro. *J. Clin. Invest.* **83:**2008–2017.
338. Dustin, M. L., and Springer, T. A. (1989). T cell receptor cross-linking transiently stimulates adhesiveness through LFA-1. *Nature* **341:**619–624.
339. Anderson, D. C., Miller, L. J., Schmalstieg, F. C., Rothlein, R., and Springer, T. A. (1986). Contributions of the Mac-1 glycoprotein family to adherence-dependent granulocyte functions: Structure–function assessments employing subunit-specific monoclonal antibodies. *J. Immunol.* **137:**15–27.
340. Lo, S. K., van Seventer, G., Levin, S. M., and Wright, S. D. (1989). Two leukocyte receptors (CD11a/CD18 and CD11b/CD18) mediate transient adhesion to endothelium by binding to different ligands. *J. Immunol.* **143:**3325–3329.
341. Diamond, M. S., Staunton, D. E., deFougerolles, A. R., Stacker, S. A., Garcia-Aguilar, J., Hibbs, M. L., and Springer, T. A. (1990). ICAM-1 (CD54): A counter-receptor for Mac-1 (CD11b/CD18). *J. Cell Biol.* **111:**3129–3139.
342. Keizer, F. D., Borst, J., Visser, W., Schwarting, R., deVries, J. E., and Figdor, C. G. (1987). Membrane glycoprotein p150,95 of human cytotoxic T cell clones is involved in conjugate formation with target cells. *J. Immunol.* **138:**3130–3136.
343. Keizer, G. D., te Velde, A. A., Schwarting, R., Figdor, C. G., and deVries, J. E. (1987). Role of p150,95 in adhesion, migration, chemotaxis and phagocytosis of human monocytes. *Eur. J. Immunol.* **17:** 1317–1322.
344. Ignatius, M. J., and Reichardt, L. F. (1988). Identification of a neuronal laminin receptor: An Mr 200k/120k integrin heterodimer that binds laminin in a divalent cation-dependent manner. *Neuron* **1:** 713–725.
345. Takada, Y., Wayner, E. A., Carter, W. G., and Hemler, M. E. (1988). Extracellular matrix receptors, ECMRii and ECMRI, for collagen and fibronectin correspond to VLA-2 and VLA-3 in the VLA family of heterodimers. *J. Cell. Biochem.* **37:**385–393.
346. Buck, C. A., Shea, E., Duggan, K., and Horwitz, A. F. (1986). Integrin (the CSAT antigen): Functionality requires oligomeric integrity. *J. Cell Biol.* **103:**2421–2428.
347. Takada, Y., Elices, M. J., Crouse, C., and Hemler, M. E. (1989). The primary structure of the alpha-4 subunit of VLA-4: Homology to other integrins and a possible cell–cell adhesion function. *EMBO J.* **8:**1361.
348. Guan, J.-L., and Hynes, R. O. (1990). Lymphoid cells recognize an alternatively spliced segment of fibronectin via the integrin receptor $\alpha 4\beta 1$. *Cell* **60:**53–61.
349. Hemler, M. E., Elices, M. J., Parker, C., and Takada, Y. (1990). Structure of the integrin VLA-4 and its cell–cell and cell–matrix adhesion functions. *Immunol. Rev.* **114:**45–65.
350. Hemler, M. E., Christina, H., and Schwarz, L. (1987). The VLA protein family. *J. Biol. Chem.* **262:**3300–3309.
351. Gailit, J., Pierschbacher, M., and Clark, R. A. F. (1993). Expression of functional $\alpha 4\beta 1$ integrin by human dermal fibroblasts. *J. Invest. Dermatol.* **100:**323–328.
352. Argraves, W. S., Suzuki, S., Arai, H., Thompson, K., Pierschbacher, M. D., and Ruoslahti, E. (1987). Amino acid sequence of the human fibronectin receptor. *J. Cell Biol.* **105:**1183–1190.
353. Werb, Z., Tremble, P. M., Behrendtsen, O., Crowley, E., and Damsky, C. H. (1989). Signal transduction through the fibronectin receptor induces collagenase and stromelysin gene expression. *J. Cell Biol.* **109:**877.

354. Gehlsen, K. R., Dillner, L., Engvall, E., and Ruoslahti, E. (1988). The human laminin receptor is a member of the integrin family of cell adhesion receptors. *Science* **241:**1228–1229.
355. Sonnenberg, A., Modderman, P. W., and Hogervorst, F. (1988). Laminin receptor on platelets is the integrin VLA-6. *Nature* **336:** 487–489.
356. Holzmann, B., and Weissman, I. L. (1989). Peyer's patch-specific lymphocyte homing receptors consist of a VLA-4-like α chain associated with either of two integrin β chains, one of which is novel. *EMBO J.* **8:**1735–1741.
357. Kilshaw, P. J., and Murant, S. J. (1991). Expression and regulation of $\beta_7(\beta_p)$ integrins on mouse lymphocytes: Relevance to the mucosal immune system. *Eur. J. Immunol.* **21:**2591–2597.
258. Erle, D. J., Reugg, C., Sheppard, D., and Pytela, R. (1991). Complete amino acid sequence of an integrin β subunit (β7) identified in leukocytes. *J. Biol. Chem.* **266:**11009–11016.
359. Ruegg, C., Postigo, A. A., Sikorski, E. E., Butcher, E. C., Pytela, R., and Erle, D. J. (1992). Role of $\alpha_4\beta_7/\alpha_4\beta_p$ integrin in lymphocyte adherence of fibronectin and VCAM-1 and in homotypic cell clustering. *J. Cell Biol.* **117:**179–190.
360. Stepp, M. A., Spurr-Michaud, S., Tisdale, A., Elwell, J., and Gipson, I. K. (1990). $\alpha_6\beta_4$ integrin heterodimer is a component of hemidesmosomes. *Proc. Natl. Acad. Sci. USA* **87:**8970–8974.
361. Sonnenberg, A., Calafat, J., Hanssen, H., Daams, H., van der Raaij-Helmer, L. M., Falcioni, R., Kennel, S. J., Aplin, J. D., Baker, J., Loizidou, M., and Garrod, D. (1991). Integrin β_6/α_4 complex is located in hemidesmosomes, suggesting a major role in epidermal cell–basement membrane adhesion. *J. Cell Biol.* **113:**907–917.
362. Kurpakus, M. A., Quaranta, V., and Jones, J. C. R. (1991). Surface relocation of alpha$_6$beta$_4$ integrins and assembly of hemidesmosomes in an in vitro model of wound healing. *J. Cell Biol.* **115:** 1737–1750.
363. Plow, E. F., McEver, R. P., Coller, B. S., Woods, V. L., and Marguerie, G. A. (1985). Related binding mechanisms for fibrinogen, fibronectin, von Willebrand factor, and thrombin-stimulated human platelets. *Blood* **66:**724–727.
364. Plow, E. F., Pierschbacher, M. D., Ruoslahti, E., Marguerie, G. A., and Ginsberg, M. H. (1985). The effect of Arg-Gly-Asp-containing peptides on fibrinogen and von Willebrand factor binding to platelets. *Proc. Natl. Acad. Sci. USA* **82:**8057–8061.
365. Pytela, R., Pierschbacher, M. D., Ginsberg, M. H., Plow, E. F., and Ruoslahti, E. (1986). Platelet membrane glycoprotein IIb/IIIa: Member of a family of Arg-Gly-Asp-specific adhesion receptors. *Science* **231:**1559–1562.
366. Suzuki, S., Argraves, W. S., Pytela, R., Arai, H., Krusius, T., Pierschbacher, M. D., and Ruoslahti E. (1986). cDNA and amino acid sequences of the cell adhesion protein receptor recognizing vitronectin reveal a transmembrane domain and homologies with other adhesion protein receptors. *Proc. Natl. Acad. Sci. USA* **83:**8614–8618.
367. Pytela, R., Pierschbacher, M. D., and Ruoslahti, E. (1985). A 125/115-kDa cell surface receptor specific for vitronectin interacts with the arginine-glycine-aspartic acid adhesion sequence derived from fibronectin. *Proc. Natl. Acad. Sci. USA* **82:**5766–5770.
368. Cheresh, D. A. (1987). Human endothelial cells synthesize and express an Arg-Gly-Asp-directed adhesion receptor involved in attachment ot fibrinogen and von Willebrand factor. *Proc. Natl. Acad. Sci. USA* **84:**6471–6475.
369. Charo, I. F., Bekeart, L. S., and Phillips, D. R. (1987). Platelet glycoprotein IIb-IIIa-like proteins mediate endothelial cell attachment to adhesive proteins and the extracellular matrix. *J. Biol. Chem.* **262:**9935–9938.
370. Johnston, G. I., Cook, R. G., and McEver, R. P. (1989). Cloning of GMP-140, a granule membrane protein of platelets and endothelium: Sequence similarity to proteins involved in cell adhesion and inflammation. *Cell* **56:**1033–1044.
371. Cheresh, D. A., Smith, J. W., Cooper, H. M., and Quaranta, V. (1989). A novel vitronectin receptor ($\alpha_v\beta_x$) is responsible for distinct adhesive properties of carcinoma cells. *Cell* **57:**59–69.

3

The Membrane-Associated Cytoskeleton and Exoskeleton

Robert J. Bloch

3.1. INTRODUCTION

The shape and organization of the plasma membrane are influenced not only by the lipid bilayer and its integral membrane proteins, but also by structures lying either just inside or just outside of the cell that provide a scaffolding for the membrane. Structures that interact closely with the extracellular face of the membrane may act as an "exoskeleton." Structures on the cytoplasmic face of the plasma membrane are generally considered to be part of the "cytoskeleton." A subset of the cytoskeleton that associates primarily or exclusively with the plasma membrane has been termed the "membrane skeleton." This chapter considers the biochemistry and morphology of the exoskeleton and of the membrane-associated cytoskeleton, how these two macromolecular complexes interact with the plasma membrane, and finally, how they interact with each other through receptors located in the plasma membrane. A major emphasis is placed on the role of these complexes in forming distinct membrane domains, enriched in particular integral membrane proteins with specific biological functions.

3.2. CYTOSKELETAL–MEMBRANE INTERACTIONS

The plasma membrane is intimately associated with the cytoskeleton. The combination of rigid but changeable intracellular struts—the cytoskeleton—with the more fluid cell surface—the plasmalemma—provides each cell with its unique shape and governs many aspects of cellular behavior. Each of the three major components of the cytoskeleton—actin microfilaments, microtubules, and intermediate filaments—has been reported to associate with the plasmalemma in one cell or another. Intermediate filaments, for example, associate with the cytoplasmic surface of the membrane at desmosomes, although the mechanism of this association is still poorly understood. Microtubules have also been reported to associate with the plasma membrane. Recently, for example, a peripheral membrane protein, termed "gephyrin," that binds to microtubules and to an integral membrane receptor for glycine, has been characterized.[1] These examples are relatively rare, however. Most of the interactions between the cytoskeleton and the plasma membrane appear to involve actin, usually organized as bundles of long microfilaments or as short filamentous oligomers. Typical of these two types of organization are the focal adhesion plaques seen in cells in tissue culture, and the membrane skeleton of the human erythrocyte. These and related structures are considered in detail below.

3.2.1. Membrane Skeleton

The classic example of a membrane skeleton, that of the human erythrocyte, has been especially amenable to study because it can be easily isolated in large amounts. Repeated lysis of erythrocytes yields pale, biconcave disks called "ghosts." The major peripheral membrane proteins of ghosts, visible in the electron microscope as a thin, fuzzy layer of filaments on the cytoplasmic face of the membrane,[2,3] can be extracted by altering the ionic strength. As these proteins dissociate, the plasma membrane vesiculates. If, instead, the lipid bilayer is dissolved in neutral detergents, the peripheral membrane proteins may retain the form of a biconcave disk.[4] These observations indicate that the distinctive shape of the erythrocyte is generated by the peripheral proteins of the membrane skeleton. They also provide procedures to separate the membrane skeleton, facilitating studies of its structure and chemistry.

3.2.1.1. Spectrin

The major protein of the erythrocyte membrane skeleton is "spectrin," so called because it was first isolated from red cell "ghosts." The properties of spectrin, and especially its flexibility and ability to form polymorphic networks, contribute more than any other single factor to the remarkable shape and plasticity of the erythrocyte.

Spectrin is a heterodimer of two large, rodlike subunits, α and β, that are similar in size and organization. The primary

Robert J. Bloch • Department of Physiology, University of Maryland School of Medicine, Baltimore, Maryland 21201.
Molecular Biology of Membrane Transport Disorders, edited by S.G. Schultz *et al.* Plenum Press, New York, 1996

sequence of each is dominated by tripical-helical domains, or "repeats" (α-spectrin has 22, β-spectrin, 17), containing an average of 106 amino acids each. Individual domains can fold independently to form protease-resistant cores,[5] of compact conformation,[5a,5b] that are joined by protease-sensitive linkers to form an extended structure. Both subunits have distinguishing features, however. The α subunit has two EF hands at its carboxyl-terminus, suggesting possible regulation by Ca^{2+},[6] whereas the β subunit contains an amino-terminal domain of ~16.5 kDa, homologous to the amino-terminal, actin-binding domain of α-actinin, that is involved in binding to actin[7] (see Section 3.2.1.2). The β subunit is also responsible for binding to the other peripheral membrane proteins involved in assembling the membrane skeleton and attaching it to the plasma membrane (see Section 3.2.1.4).

The spectrin subunits assemble to form heterodimers that in turn polymerize to form larger oligomers. To form heterodimers, α- and β-spectrin align in antiparallel fashion, so that the amino-terminus of one is adjacent to the carboxyl-terminus of its partner. Although interactions between the subunits occur at many sites, binding appears to be dominated by sites in only a few of the 106 amino acid repeats present in the α and β subunits. These sites, located in repeats 1–4 of β and 19–22 of α, nucleate the formation of the heterodimer.[8,8a] Once bound, the two subunits probably wrap around each other to form a supercoiled double helix,[9,10] the pitch of which can change depending on how much the molecule is compressed or extended.[9]

Its supercoiled, double-helical structure allows spectrin to undergo large changes in shape. Recent ultrastructural evidence suggests that spectrin is compressed *in situ,* and that full extension only occurs on its release from the erythrocyte membrane.[11–15] Purified spectrin heterodimers in solution measure ~100 nm in length, while structures likely to be heterodimers of spectrin measure only ~30 nm in the intact membrane skeleton.[15] When spectrin is released from the bilayer, the ~30 nm structures gradually expand.[9–11] The molecular basis for these changes in spectrin is still controversial, but it may be understood once the structure of the 106 amino acid repeats[5,5a,16] and the relationship between neighboring repeats[17,18] is determined. Whatever the explanation, it is likely that spectrin's ability to change shape contributes to the remarkable plasticity of the circulating erythrocyte.[19]

The ability of spectrin to polymerize to different extents *in situ* also contributes to its polymorphism. Purified spectrin is an αβ heterodimer or an $(\alpha\beta)_2$ tetramer formed by two such molecules that join "head to head." This association is through formation of an interchain, triple-helical structure composed of one helix from the amino-terminus of the α subunit and two helices from the 17th repeat of the β subunit.[18] Equilibrium between dimer and tetramer is strongly influenced by temperature, with dimers favored at 37°C and tetramers favored at 4°C.[20] The "head" of the molecule, defined as the end at which contact between heterodimers occurs, is the end containing the amino-terminus of the α subunit and the carboxyl-terminus of the β subunit.

Although oligomers larger than tetramers are not seen in purified preparations, they are present *in situ.* Cross-linking experiments performed on erythrocyte ghosts indicate that most spectrin is hexameric and octameric, and that some is present in even larger oligomers.[21,22] Spectrin oligomers have also been seen in ultrastructural studies.[10,23] It seems likely that the high local concentrations of spectrin that underlie the red cell membrane promote the formation of larger oligomers, and that the less favorable conditions following extraction promote their dissociation.[22] Thus, spectrin alone is polymorphic both because it can assume different lengths and because it can use different interchain interactions to oligomerize to different extents.

3.2.1.2. Spectrin–Actin Complex

Spectrin is not the only protein in the membrane skeleton of the human erythrocyte: another prominent, and polymorphic, protein in the skeleton is actin, specifically the "β" isoform. The isoelectric point of erythrocyte β-actin lies between those of skeletal muscle actin (α-actin) and that of smooth muscle, fibroblasts, and other cells (τ-actin). Erythrocyte β-actin is the product of a distinct actin gene[24–26] and can be distinguished from other isoactins with antibodies specific for its N-terminal sequence.[e.g., 27]

Unlike the long F-actin filaments seen in other cells, the actin in erythrocytes is assembled into short oligomers composed of approximately a dozen G-actin subunits.[28] The factors that limit actin polymerization in erythrocytes are still not fully understood, but two probably predominate. First, there are not enough actin subunits (4–5 × 10^5/cell) for many longer actin filaments to form. In addition, erythrocytes contain actin-associated proteins, including protein 4.9,[29] and erythroid forms of myosin[30] and tropomyosin,[31] that may assist in maintaining actin in short oligomers. An additional protein, termed "tropomodulin," has also been identified in erythrocytes.[32] Found at the ends of actin filaments in the contractile apparatus of skeletal muscle,[33] it may cap the ends of actin oligomers in the erythrocyte membrane skeleton by virtue of its ability to bind to the end of the tropomyosin molecule.[32]

The binding of spectrin to actin occurs at the "tail" of the spectrin heterodimer, through the actin binding domain in the amino-terminal portion of β-spectrin.[7] The affinity of erythrocyte spectrin for actin is very poor unless protein 4.1 is present. This protein, named for its relative position in dodecyl sulfate–polyacrylamide gel electrophoresis,[34] has a polypeptide chain molecular weight of ~68,000, determined by molecular cloning of the erythroid cDNA.[35] Protein 4.1 binds to the "tail" region of spectrin[36] at the actin-binding domain of the β subunit.[37,38] In the presence of actin, protein 4.1–spectrin complexes will bind actin to form a ternary complex. The binding sites for actin and spectrin are contained within an 8- to 10-kDa fragment in the middle of the 4.1 molecule. This region is subject to phosphorylation by cAMP-dependent protein kinase[39,40] and by a protein tyrosine kinase,[41] raising the possibility that actin–spectrin interactions may be subject to metabolic regulation. The interactions of protein 4.1 with actin and spectrin are also regulated by Ca^{2+}-calmodulin,[42–44] primarily via the binding of calmodulin to protein 4.1.[43] At micromolar concentrations of Ca^{2+}, formation of the ternary complex is inhibited. Thus, 4.1 is likely to play an important role in the stability, and potentially in the remodeling, of the spectrin–actin network.

Adducin is another protein of the erythrocyte skeleton that promotes actin–spectrin binding by forming a ternary complex.[45,46] Adducin is a heterodimer of two related subunits of M_r's of 97,000 and 103,000[47] and, like protein 4.1, it is subject to phosphorylation by protein kinases.[48] Unlike 4.1, however, adducin binds preferentially to actin and binding occurs in the absence of spectrin.[45] It shares sequence homology with the actin-binding domain of β-spectrin and α-actinin, which may provide a stable actin-binding site in the linear "tail" domain of the adducin molecule,[47] the domain responsible for promoting actin–spectrin binding.[48a] Adducin and protein 4.1 therefore contribute different primary "anchorage" sites to actin–spectrin interactions—adducin to actin and protein 4.1 to spectrin.

The erythrocyte skeleton contains only $\sim 3 \times 10^4$ copies of adducin, insufficient to place this protein at each of the 1–2 $\times 10^5$ actin–spectrin junctions present in each cell. By contrast, protein 4.1 is present at $\sim 2 \times 10^5$ copies/cell, sufficient to mediate most actin–spectrin binding in the erythrocyte skeleton. Protein 4.1 inhibits the binding of exogenous adducin.[49] It is not clear how adducin and protein 4.1 combine to mediate the formation of the "junctional complex" between actin and spectrin. It seems likely, however, that not all spectrin–actin junctions in the erythrocyte skeleton are identical. For example, a single adducin heterodimer anchors two spectrin heterodimers to a single site on the actin oligomer.[50] By contrast, the variable stoichiometry of protein 4.1 in reconstituted complexes suggests that this dimeric protein can anchor from one to four different spectrin molecules.[51] Perhaps the spectrin–actin junctions that form near the ends of the actin oligomer differ from those that form in the middle of the oligomer. This could account for the ability of protein 4.1, but not adducin, to cap the growth of actin filaments in the presence of spectrin.

3.2.1.3. Ultrastructure of the Membrane Skeleton

The product of the interactions among these proteins is a complex that, in electron microscopic studies of spread erythrocyte skeletons, appears as a set of long, thin filaments attaching to a central thicker filament. The dimensions, their association with ankyrin[10,14] (see Section 3.2.1.4a) as well as recent immunolabeling studies[10] identify the long, thin filaments as spectrin. Several criteria, including their dimensions and their association with spectrin, identify the shorter filaments as actin. The smaller molecules involved in joining actin and spectrin have not yet been visualized within this complex.

The association of actin oligomers with five to seven spectrin molecules,[10,12,14] which in turn can interact at their distal ends with other actin oligomers, produces an extended, two-dimensional, filamentous network. This network appears different *in situ* and in isolated preparations. In skeletons released from the lipid bilayer, actin oligomers and spectrin tetramers resemble the hubs and spokes in an interlocking set of irregular wheels. *In situ,* where spectrin oligomers are more common,[15,21,23] spectrin–spectrin and spectrin–actin junctions are difficult to distinguish from one another.[15] Furthermore, unlike its thinner profile in spread skeletons, spectrin *in situ* is approximately as thick as actin oligomers.[15] In addition, the number of filaments that come together to form the network, ~3.4 per intersection,[15] appears to be the same for both types of junctions. Thus, the oligomerization and compaction of spectrin, together with the junctions it forms with actin, give the intact membrane skeleton *in situ* the appearance of a meshwork made up of irregular triangles and rectangles (Figure 3.1). This picture of the membrane skeleton *in situ* has not yet been reconciled to the hexagonal pattern seen in spread skeletons.

3.2.1.4. Attachment of the Skeletal Network to the Membrane

In principle, the skeletal network of actin and spectrin could bind to the membrane by interacting with the hydropho-

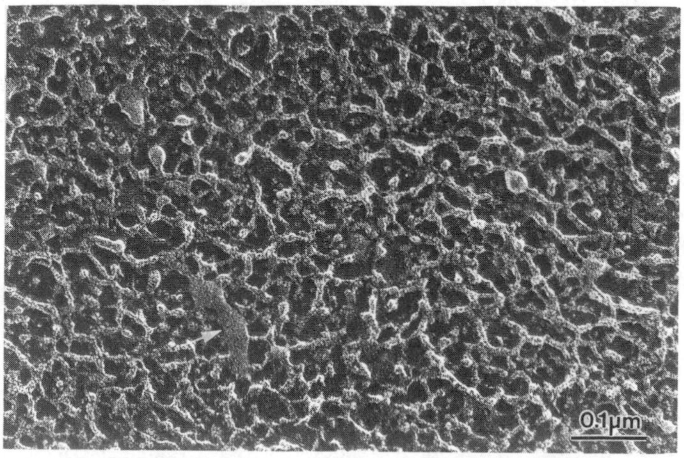

Figure 3.1. The membrane skeleton of the human erythrocyte. Human erythrocyte ghosts were fixed, attached to an adhesive coverslip, and sheared to expose the cytoplasmic face of the membrane. Samples were processed for ultrastructural studies by quick-freeze, deep-etch, rotary-replication. The membrane skeleton appears as an irregular network of short filaments. The amorphous structures (e.g., arrow) are small patches of ice left after etching. Adapted from Ursitti *et al.*[15] Scale bar = 0.1 μm.

bic core of the lipid bilayer, the phospholipid head groups, or the integral membrane proteins in the bilayer. Although both actin[52] and spectrin[53,54] can bind to lipid bilayers, these low-affinity interactions cannot account for the association of the spectrin–actin network with the erythrocyte membrane. This association is in fact mediated primarily by protein–protein interactions, the most important of which involves the peripheral membrane protein, ankyrin, and its integral membrane protein ligand, the anion exchanger.

3.2.1.4a. Ankyrin and the Anion Exchanger. Spectrin is bound to the membrane largely through ankyrin,[55] a peripheral protein that links spectrin to the anion exchanger (band 3 protein[56]; reviewed in Refs. 49,57). Ankyrin is almost as large a protein as β-spectrin, but, unlike spectrin, it is organized into three compact domains and assumes an almost globular conformation.[57] The amino-terminal domain, which consists of a series of 33 amino acid repeats, is ~90 kDa in mass and contains the binding site for the anion exchanger. A central, ~62-kDa portion of the molecule contains the binding site for β-spectrin within a 12-kDa region.[57a] Flanking sequences in this region modulate the affinity of this peptide for β-spectrin.[57a] The C-terminal, ~55-kDa domain has regulatory functions; it governs the affinity of ankyrin for the anion exchanger and for spectrin. Alternatively spliced variants in this region, in which 163 amino acids are missing, show higher affinity for spectrin[58] and an increased number of membrane binding sites.[58,59] The large number of such splice variants[60] suggests the possibility of considerable modulation of ankyrin interactions with spectrin and with its membrane ligands. These alternatively spliced products may also increase the potential number of ligands with which ankyrin can interact (see Sections 3.2.1.6a and 3.2.1.7).

Ankyrin binds with high affinity to repeat 15 of the β-spectrin subunit.[61] This places the ankyrin binding site closer to the "head" than to the "tail" of the spectrin heterodimer, a position confirmed by electron microscopic studies of the spectrin–ankyrin complex.[10,14] Stoichiometry[49] and ultrastructural studies[10,14] suggest that approximately one ankyrin molecule is bound for every spectrin heterotetramer in the erythrocyte membrane. Many potential sites on β-spectrin that could bind ankyrin therefore remain empty in the intact skeleton. Ankyrin also binds tightly to a portion of the anion exchanger, an integral membrane protein of the erythrocyte involved in Cl–bicarbonate exchange that projects into the cytoplasm from the erythrocyte membrane.[15,56,62] The portion of the anion exchanger that dominates binding to ankyrin is contained in a sequence of 13 amino acids (residues 174–186),[63] but other sequences are likely to be needed for high-affinity binding. Ankyrin bound to this region of the exchanger is able simultaneously to bind to spectrin and so anchor the spectrin–actin network to the membrane.

As a result of these interactions, a considerable fraction of the anion exchanger molecules bind to the membrane skeleton and become immobilized.[64,65] Protein determinations show that there are approximately 1×10^6 exchanger polypeptide chains per red cell, or 10 for every ankyrin monomer.[49] As the exchanger is likely to be tetrameric (see Ref. 15 for discussion) and as each tetramer is likely to interact with a single ankyrin molecule, one would expect only ~40% to be bound to the skeleton. In fact, ~60% of the anion exchangers appear immobile in the membrane.[65] This suggests that additional mechanisms contribute to preventing the lateral diffusion of the anion exchanger in the membrane (see Section 3.2.1.4b). Extraction of the membrane skeleton releases nearly all bound anion exchangers, however, suggesting that these additional mechanisms also involve the spectrin–actin skeleton.

3.2.1.4b. Other Cytoskeletal–Membrane Attachments in the Erythrocyte. Although the most important means of attaching the spectrin–actin network to the membrane involves ankyrin and its binding to the anion exchanger, the network is also bound to the membrane through interactions that do not require ankyrin. The most important of these alternative mechanisms involves protein 4.1.

Extraction of red cell ghosts with different aqueous solutions provided the first evidence for a stable association of 4.1 with the membrane, independent of spectrin and actin. In these experiments, 4.1 was found attached to the membrane when most of the actin and spectrin were removed by buffers of low ionic strength.[49] Subsequent studies identified the cytoplasmic tail of glycophorin C, another integral protein of the erythrocyte membrane, as a ligand for 4.1.[66] The portion of 4.1 responsible for this interaction lies in the carboxyl-terminal 30-kDa fragment, distinct from the region responsible for actin–spectrin binding.[67] Protein 4.1 has also been shown to bind to the anion exchanger directly,[68] suggesting an alternative mode of anchoring the spectrin–actin skeleton to the integral proteins of the bilayer. Additional interactions that link the membrane to its cytoskeleton, including direct, ankyrin-independent binding of spectrin to erythrocyte and brain membranes,[69,69a,69b] are under investigation in a number of laboratories.

3.2.1.5. Contributions to Erythrocyte Shape and Stability

There are, therefore, multiple interactions that can occur between the membrane skeleton and its anchor in the plasma membrane, as well as among the membrane skeletal proteins themselves. This tendency adds flexibility to the overall structure and increases the possibilities for selectively incorporating and differentially regulating its constituent proteins without affecting stability in general. Furthermore, it reduces the possibility of the structure being absent completely if only one component is defective.

The properties of the complex formed between the membrane skeleton and the plasma membrane can also account for the unusual shape and plasticity of the human erythrocyte. The erythrocyte is normally biconcave, but it is distorted as it circulates. The properties of the highly polymorphic spectrin–actin membrane skeleton appear to account for these reversible shape changes. A model of the membrane skeleton as a two-dimensional protein ionic gel linked to a lipid bilayer[19] predicts a biconcave discoid shape for the erythrocyte under physiological conditions of temperature, pH, and ionic strength. It further predicts that relatively small changes in these conditions will cause cells to become stomatocytes or echinocytes, as sometimes occurs in healthy red blood cells.

Defects in the membrane skeleton can also account for the unusual shapes of erythrocytes in some disease states, as well as for the instability of erythrocytes in many hemolytic anemias (reviewed in Ref. 70). Hereditary spherocytosis, for example, is associated with a deficiency in spectrin, and the severity of the disease is closely correlated with the extent of spectrin deficiency. In some cases, the primary deficiency may be in ankyrin,[e.g., 71] which can in turn lead to reduced binding of spectrin to the membrane of the diseased erythrocyte. Other hereditary diseases of the erythrocyte, such as elliptocytosis and pyropoikilocytosis, have also been traced to defects in the membrane skeleton, including protein 4.1.[e.g., 72] Several forms of these diseases are linked to mutations in the regions of spectrin that are involved in the formation of dimers or tetramers.[70] Additional mutations, related to the ankyrin-binding domain in repeat 15 of β-spectrin, are predicted.[61]

Knowledge of the detailed structure and chemistry of the erythrocyte skeleton has therefore proved to be important in understanding the molecular basis of many hereditary hemolytic anemias. Defects in other membrane skeletal proteins have also been linked to hereditary diseases in humans and other mammals (see Section 3.2.1.6c).

3.2.1.6. The Membrane Skeleton in Nonerythroid Cells

Although the membrane skeleton has been most extensively studied in the human erythrocyte, spectrin-rich structures associated with the plasma membrane have been found in many other types of eukaryotic cells, including slime mold, plants, and the lymphocytes, epithelia, neurons, and muscle cells of higher vertebrates. The general principles, outlined above for the erythrocyte, apply to spectrin-based membrane skeletons in other tissues, but there are important differences.

3.2.1.6a. Isoforms of Spectrin, Ankyrin, and Protein 4.1. Like many proteins expressed in eukaryotic cells, the spectrin and ankyrin of the human erythrocyte are members of larger gene families, the products of which are differentially expressed and alternatively spliced in a tissue-specific manner.

The form of spectrin commonly found in most tissues has been referred to as "fodrin," "tissue spectrin," spectrin$_G$, or spectrin II.[73] It is composed of α and β subunits that are highly homologous to the subunits of erythrocyte spectrin[74–76] (also called spectrin$_R$, or spectrin I[73]), and that perform many of the same functions. There are, however, structural differences that are probably related to the specialized functions of tissue spectrin. In particular, α-fodrin contains an SH3 domain in a nonhomologous 62-amino-acid sequence at the beginning of repeat 10,[77] and a putative Ca^{2+}-calmodulin-binding site, which probably lies between repeats 11 and 12.[78] The physiological significance of the SH3 domain remains obscure, but it may subject fodrin, and thus the membrane skeleton in nonerythroid cells, to control by the tyrosine kinase signaling pathway. Recent studies indicate that this domain can be mobilized to the cytoplasmic surface of the plasma membrane at the leading lamellae of motile cells,[78a] consistent with a role in modulating membrane structure even in the absence of a β subunit.

The binding of Ca^{2+}-calmodulin modulates the structure of fodrin, rendering it more susceptible to Ca^{2+}-dependent proteolysis,[79] which in turn alters actin–fodrin binding,[80] and making it less capable of binding to brain membranes.[81] It is of considerable interest that all of these effects involve regions of the fodrin molecule distant from the Ca^{2+}-calmodulin-binding sites. This suggests that the binding of Ca^{2+}-calmodulin to fodrin and subsequent proteolysis near the binding site cause significant changes in regions of the molecule many nanometers distant.

As one would expect for similar isoforms expressed in the same organism, the tissue distributions of fodrin and spectrin are quite distinctive. In mammals, for example, the expression of α-spectrin appears to be limited to erythropoietic cells (though a few exceptions have been reported,[e.g., 81a]), and α-fodrin is expressed preferentially in other tissues. β-Spectrin is expressed in many tissues, including brain and muscle, but in an alternatively spliced form that has a different C-terminal sequence than that found in the erythrocyte.[82–84] Alternative splicing has also been reported for α-fodrin[85] and is likely to be a common feature of this family of proteins[73] (see below). Interestingly, the alternatively spliced exon encodes the calmodulin-binding domain of α-fodrin.[85]

The presence of β-spectrin, but not α-spectrin, in most nonerythroid tissue suggests that if β-spectrin is associated with an α subunit at all, it must associate with α-fodrin. Recent evidence supports this idea,[84] and further suggests that, in muscle, some β-spectrin may associate with the sarcolemma in the absence of an associated α subunit[86] (Porter *et al.*, submitted for publication). Cells that express β-spectrin and α-fodrin also express β-fodrin, raising the possibility that mixed heterotetramers of spectrin and fodrin will form. Evidence for these structures has not yet been reported.[e.g., 87] In the brain, β-fodrin and β-spectrin appear to segregate into different neuronal domains,[88,89] suggesting that mixed heterotetramers may not be present, at least in mature neurons.

There are at least three different genes for ankyrin,[57] and these, too, are subject to alternative splicing to produce further variants.[90,91] In addition to the erythroid ankyrin gene, the alternative splicing of which was discussed briefly above, two other genes encode brain ankyrin (ankyrin$_B$)[92] and a form of ankyrin found primarily at the node of Ranvier (ankyrin$_{node}$ or ankyrin$_G$).[92a] An epithelial form has also been molecularly cloned[92b] and is homologous to ankyrin$_G$. Ankyrin$_B$ is made in at least two alternatively spliced forms, one of 220 kDa and the other of 440 kDa, formed by the addition of a large insert between the membrane- and spectrin-binding domains and the regulatory domain of the 220-kDa form. The 440-kDa form is especially prevalent in neonatal brain and is largely replaced by the smaller form of ankyrin$_B$ during postnatal development.[90] Ankyrin$_G$ or ankyrin 3, present at epithelia and at nodes of Ranvier, are also made in different alternatively spliced forms.[92a,92b] The physiological roles of these ankyrin variants are still poorly understood, but they almost certainly have an important role to play in the creation of distinctive membrane domains in the nervous system and in other tissues (see Section 3.2.1.7).

Perhaps the most diverse of the membrane skeletal proteins is protein 4.1. This protein has been found in many different tissues in a variety of alternatively spliced forms. In the chicken erythrocyte, for example, seven variants of 4.1 are found with molecular masses ranging from 77 to 175 kDa. All appear to be made by the same gene as a result of alternative RNA splicing.[93] A smaller, but still diverse, set of 4.1 variants is expressed in human erythroid and other mammalian cells.[94] The expression of these 4.1 variants is regulated developmentally and is tissue-specific.[95] Some of these alternatively spliced products may associate with the membrane without linking actin to spectrin, or associate with the cytoskeleton without binding to the membrane.[e.g., 67]

3.2.1.6b. Dystrophin and Related Proteins. Spectrin is the prototype of a family of membrane skeletal molecules, all of which contain a central rodlike domain consisting of largely α-helical repeats. In addition to spectrin and fodrin, other members of the family include α-actinin, dystrophin and dystrophin-related protein, or utrophin.[96,97]

Dystrophin was first identified as the protein missing from young boys with Duchenne muscular dystrophy. It is a large protein, 427 kDa in mass, that is encoded by an immense gene, greater than 2300 kb in length, located on the X chromosome and containing more than 70 exons. The size and complexity of this gene are probably related to the high frequency (1 in 3500 live male births) with which Duchenne muscular dystrophy is observed in the population.

Dystrophin is synthesized primarily by muscle tissue, although it is also present in different alternatively spliced forms in brain and other organs.[98,98a,98b] The full-length protein is organized into four different domains. Two domains, the amino-terminal actin-binding domain and a series of triple, α-helical repeats, are homologous to regions of β-spectrin. In the case of dystrophin, these repeats have an average of 109 amino acids each,[99] although recent experiments suggest that 117–119 amino acids may be required for proper folding and conformational stability.[99a] The amino-terminal domain has recently been shown to bind to actin.[100–102] The other two domains, C-terminal to the helical repeats, are a cysteine-rich region, homologous to a similar region of α-actinin, and a C-terminal domain, which is unique, being shared only by utrophin and an 87-kDa protein that is homologous to the cysteine-rich and C-terminal portion of dystrophin.[98] These C-terminal domains of dystrophin, which confer the ability to bind to the membrane, are also expressed in alternatively spliced polypeptides in tissues other than skeletal muscle, especially brain.[98,98a,98b] Indeed, dystrophin is expressed as many different alternatively spliced products,[98] the specific functions of which are still not known.

Like the spectrins, dystrophin is a peripheral membrane protein. Instead of binding to the membrane via ankyrin, however, it interacts with a distinctive complex of membrane-associated glycoproteins.[103,103a] This complex consists of three integral membrane-spanning glycoproteins with apparent molecular masses of 35, 43, and 50 kDa, a nonglycosylated integral membrane protein of 25 kDa, an extracellular glycoprotein of 156 kDa, termed "dystroglycan," and an intracellular protein of 59 kDa,[104] recently named "syntrophin."[105] Syntrophin binds to the C-terminal region of dystrophin,[106,106a,106b] specifically to an alternatively spliced exon.[106c] It also binds to itself, to form stable trimers.[106a] The glycoprotein complex also binds to dystrophin in its cysteine-rich, C-terminal region.[106d,106e] The 156-kDa dystroglycan component of the complex binds laminin[102,107] and so is likely to anchor the membrane-bound complex to the basal lamina (see Sections 3.3.4 and 3.3.5). The 156- and 43-kDa proteins are products of the same gene but are proteolytically cleaved following synthesis.[107]

Dystrophin *in vitro* forms dimers and larger molecules,[108,109] but its state *in situ* is not yet known. Although dystrophin may interact with spectrin or with utrophin, such mixed complexes have not yet been described. As a result, dystrophin's relationship to other membrane skeletal proteins is still poorly understood. In isolated preparations of acetylcholine receptor aggregates, studied by quick-freeze, deep-etch, rotary-replication electron microscopy, spectrin and dystrophin are present in the same membrane skeletal network and cannot be readily distinguished by their ultrastructural appearance.[110] Despite the fact that it binds to F-actin *in vitro*,[100–102] dystrophin in receptor aggregates is not closely associated with recognizable filaments of F-actin in these preparations.[110]

In adult striated muscle, dystrophin is present in a filamentous network at the cytoplasmic face of the sarcolemma,[111,112] in regions that are also rich in β-spectrin.[83] These regions, termed "costameres" because they appear as a series of "ribs" overlying the Z and M lines,[83,113,114] are also rich in ankyrin[115] and in several membrane cytoskeletal proteins typical of "focal adhesions," including vinculin, talin, and integrin[83,113,114,116,117] (see Section 3.3.3). Dystrophin binding to talin[118] may promote its ability to accumulate at focal adhesion plaques in some cells,[119] as well as in costameres.[83,120] In smooth muscle, however, dystrophin segregates from the proteins of focal adhesions.[121]

Utrophin, or dystrophin-related protein, is a more generally expressed, autosomal homologue of dystrophin[122–124] (much as fodrin is a more general type of spectrin). It is slightly smaller than dystrophin, with a mass of 397 kDa,[123] but like dystrophin, it is also made in several alternatively spliced forms.[124a] Nevertheless, it shares the same overall structure, including an amino-terminal actin-binding domain,[124b] the long, central rod region with average repeats of 109 amino acids, and the cysteine-rich, carboxyl-terminal region that binds syntrophin[106,106c] and the glycoprotein complex.[124c] Unlike dystrophin, which is expressed throughout the muscle fiber and is necessary for fiber integrity, utrophin is present in the mature myofiber only at the neuromuscular junction,[125] where it may be involved in stabilizing the postsynaptic accumulation of receptors for the neurotransmitter, acetylcholine.[126,127] The β2 isoform of syntrophin is similarly restricted to the neuromuscular junction in adult muscle.[127a] In immature muscle cells, however, the distribution of utrophin overlaps with that of dystrophin, for example in the membrane skeletal complex responsible for clustering acetylcholine receptors (R. Bloch, unpublished). In the *mdx* mouse, which lacks dystrophin, utrophin may be upregulated to subserve the same functions as dystrophin,[128] thus helping to stabilize the sarcolemma in regions where it is usually not found. In the central nervous system, dystrophin appears to be concentrated at synapses,[129] while utrophin is present primarily at astrocytic endfeet.[130]

Unlike its larger cousins, α-actinin is an exceptional member of this family, because it is primarily a cytoplasmic cytoskeletal protein, associated with actin in contractile filaments and stress fibers (reviewed in Refs. 96, 131). Like spectrin, α-actinin forms antiparallel dimers, but these are composed of two identical subunits. Also like spectrin, α-actinin has an N-terminal actin-binding domain, and a short span of four rodlike, but slightly larger, triple-helical repeats, with an average of 120 amino acids per repeat.[132,133] These are organized into compact, protease-resistant units resembling those of spectrin and dystrophin.[133a] Although α-actinin lacks a membrane-binding domain, a short sequence in α-actinin allows it to bind with high affinity to the cytoplasmic "tail" of the integral membrane protein, integrin,[134,135] perhaps accounting for its ability to concentrate at focal adhesion plaques in some cells[136] (reviewed in Ref. 137; see Section 3.3.3.1).

3.2.1.6c. The Membrane Skeleton and Muscular Dystrophies. As mentioned above, dystrophin is present in costameres. Extensive ultrastructural evidence indicates that costameres link the sarcolemma to the contractile apparatus and, simultaneously, to the extracellular matrix.[138–141] Considerable contractile force is transmitted from the contractile apparatus to the extracellular matrix and through the matrix to the tendon—a "parallel" pathway—rather than serially from sarcomere to sarcomere within each myofibril to the myotendinous junction.[138] By binding simultaneously to dystroglycan, which in turn binds laminin in the extracellular matrix (see Sections 3.3.1.1b, 3.3.4, and 3.3.5) and to actin emerging from the contractile apparatus,[102] dystrophin may establish a link between contractile and extracellular structures that is required for force transmission along this "parallel" pathway. Alternatively, dystrophin may stabilize the membrane during muscle contraction, much as spectrin stabilizes the erythrocyte membrane.

It is presumably the weakening of the sarcolemma when dystrophin is missing or mutated that causes myopathy.[83,102,141a] The absence of intact dystrophin in humans with Duchenne muscular dystrophy is caused by premature chain termination or deletions that result in the synthesis of defective proteins.[142] Becker's muscular dystrophy is usually less severe than Duchenne and is associated with less drastic changes in the dystrophin molecule.[142] In one unusual example, in which most of the central rodlike domain is missing because of a deletion between the otherwise intact N- and C-terminal domains, myopathy is mild.[143] In cases in which the rod domain is present but the membrane-binding domain is missing, myopathy is severe.[142,143a] Deletion of the amino-terminal, actin-binding domain also causes a more severe myopathy than in-frame deletions of the rod domain, though less severe than deletion of the C-terminal domain.[143b] This evidence suggests that the C-terminal, membrane-binding domain of dystrophin is most essential for its function.

The C-terminal region of dystrophin may also be involved in intra- or intercellular regulatory pathways, through its ability to bind nitric oxide synthetase (NOS). This enzyme is prevalent in skeletal muscle, where it is concentrated under the sarcolemma,[143c] presumably because of its ability to bind to dystrophin.[143d] It shares limited homology with syntrophin,[143d] suggesting that binding occurs near dystrophin's C-terminus. The concentration of NOS at the sarcolemma may regulate the activity of the muscle fiber, as indicated by the ability of NOS inhibitors to alter contractile strength.[143c] The ability of nitric oxide to permeate membranes and to alter blood flow through its action on capillary endothelial cells further suggests that NOS may help to regulate blood flow in skeletal muscle tissue. This in turn may affect the ability of muscle fibers to survive, especially in the absence of the enzyme that occurs when dystrophin is missing.[143d]

In some animals—in contrast to humans, in which it is almost invariably fatal—the absence of dystrophin may not cause severe myopathy. The mild myopathy seen in the *mdx* mouse is probably related to the reduced load experienced by muscles in the mouse, as increased load to muscles lacking dystrophin is associated with more severe myopathy.[144–146] This suggests that dystrophin is not essential for normal muscle function unless the sarcolemma is subjected to considerable stress.

The absence of dystrophin has additional consequences for the membrane skeleton, however. In addition to NOS,[143d] mentioned above, syntrophin is also substantially reduced in muscle fibers that lack dystrophin.[147,148] Syntrophin has recently been shown to exist in at least three distinct isoforms, synthesized by three different genes.[105,149,150] Although all three bind to the same site on dystrophin,[106c] the α1 isoform appears to be selectively depleted from sarcolemma when dystrophin is absent.[106a,127a]

Recent evidence implicates still another component of the dystrophin-associated glycoprotein complex in both human and animal myopathies. The 50-kDa membrane glycoprotein of the complex, termed "adhalin," is absent from the sarcolemma of humans with severe childhood autosomal recessive muscular dystrophy (SCARMD) and the dystrophy seen in cardiomyopathic hamsters.[151,151a] Despite these observations, recent evidence suggests that the levels of adhalin mRNA are unaffected in the hamster model.[151b] The loss of this protein in hamsters is, however, related to a selective weakening of the association of the 156-kDa extracellular protein with the complex in heart but not in skeletal muscle.[151] This suggests that the factors that regulate the stability of the dystrophin-associated glycoprotein complex differ in these two striated muscles, perhaps resulting in selective loss of individual components even when they are synthesized at normal levels. It seems likely that other human and animal myopathies will be explained by either mutation or loss of other dystrophin-associated proteins (see Ref. 151c for a recent review).

3.2.1.7. Spectrin-Based Membrane Domains

One consequence of anchoring the membrane skeleton to integral membrane proteins is the immobilization of those proteins. When this occurs in a limited region, rather than over the entire cell membrane as it does in the erythrocyte, it creates distinctive domains that can be highly enriched in particular integral membrane proteins. These plasmalemmal specializations are likely to serve important cellular functions, such as initiating postsynaptic potentials or action potentials in excitable cells.

The general organization of these spectrin-rich skeletons appears to be similar to those of the human erythrocyte.[97] In each cell that has been studied, a network of spectrin is bound to the membrane through the mediation of another peripheral membrane protein, usually a form of ankyrin, which itself is anchored to an integral membrane protein. The best examples of such structures are found in vertebrate muscle, brain, and kidney, but additional examples are likely to be found in other cell types as well.

The basolateral surfaces of kidney epithelial cells are enriched in cell adhesion molecules, required to form a transporting monolayer of cells, and in the Na^+,K^+-ATPase, the enzyme that creates a gradient of Na^+ and K^+ across the epithelial layer. Both the cell adhesion molecule, known variously as uvomorulin, E-cadherin, or LCAM, and the Na^+,K^+-ATPase bind to a spectrin-rich membrane skeleton.[152,153] The Na^+,K^+-ATPase binds through the mediation of ankyrin[154,155] and so fulfills the role that the anion exchanger plays in the human erythrocyte membrane: it serves as a membrane anchor for the membrane skeleton. (The Na^+,K+-ATPase shares no significant homology with the anion exchanger, however, and its binding to ankyrin utilizes sequences on that molecule distinct from those used in binding the anion exchanger of the erythrocyte.)[156] Simultaneously, the Na^+,K^+-ATPase becomes localized to the basolateral membrane; recent studies indicate that its localization there is related in part to its ability to interact with spectrin.[156a] It seems likely that this polarity is generated as a result of the adhesive interactions between neighboring epithelial cells, which in turn help to concentrate the membrane-skeletal proteins nearby.[157]

Similarly, in vertebrate brain, spectrin-rich structures bind to integral membrane proteins. The best-studied example is the voltage-gated Na^+ channel protein, which binds to spectrin through the mediation of ankyrin.[158,159] Voltage-gated Na^+ channels of myelinated nerves are concentrated at the nodes of Ranvier,[160] where spectrin and an unusual form of ankyrin, termed "ankyrin$_{node}$," are also enriched.[161] Ultrastructural studies[162] of these regions show small cytoplasmic filaments applied to the membrane at the node. This region is also associated with both intracellular and extracellular structures. The intracellular structures are microtubules and neurofilaments, consistent with the ability of the ankyrin and spectrin to bind tubulin and intermediate filament proteins, respectively.[163,164] Extracellular structures at the nodes are less well understood, but, like the association of the epithelial membrane skeleton with sites of cell adhesion, their presence suggests that factors acting on the external surface of the membrane can influence the local accumulation of membrane skeletal proteins. This idea has received support recently from the discovery in mammalian brain of a set of integral membrane proteins, structurally related to NCAM, that are capable of binding to ankyrin.[165] The importance of these interactions for neuronal integrity is underscored by the observation that a deficiency in the erythroid form of ankyrin in *nb/nb* mice is associated with degeneration of cerebellar Purkinje cells.[166]

The association of a spectrin-based membrane skeleton with adhesive and cytoskeletal structures is also seen in rat myotubes developing in culture. Aggregates of acetylcholine receptors form in these cells at sites just adjacent to sites of focal adhesions with the culture substrate,[167] and they are probably stabilized in part by the collection of proteins that associate with such adhesions (see Section 3.2.2). The cytoplasmic face of the membrane at receptor aggregates has an unusual form of β-spectrin, but, unlike most other membrane skeletons, it has no identifiable α subunit.[86] The β-spectrin is organized as a network of filaments that is very similar to that seen in the intact erythrocyte membrane[97]: filaments are ~ 30 nm long and ~ 6 nm wide, and make an average of 3.2 connections at each intersection of the network. Unlike the erythrocyte, however, the spectrin network appears to anchor long actin filaments,[110] thus establishing connections to deeper cytoplasmic structures analogous to those seen between the membrane and cytoskeletal elements at nodes of Ranvier.

Unlike the structures studied at the node of Ranvier and the basolateral membrane, acetylcholine receptor clusters on rat myotubes have not yet been shown to contain a form of ankyrin. This could reflect the diversity of this protein family and the absence until recently of antibodies capable of detecting its various members. Alternatively, interactions between the membrane skeleton and the membrane in this structure may be mediated by ankyrin-independent binding interactions. Examples of binding of the spectrins to membrane ligands other than ankyrin have been provided by several laboratories. These include a form of NCAM,[168] the cadherin-binding protein, α-catenin,[169] and as yet undefined ligands in erythrocyte and brain membranes.[69] Although the importance of these interactions is not yet understood, they do suggest that membrane skeletons can be assembled and used to create distinctive membrane domains in ways that may not be limited by interactions with the ankyrins.

These and other examples of spectrin-based membrane domains establish some of the common features of membrane skeletons as well as some of the features that distinguish one skeleton from another. The several common features include the presence of spectrin; its linkage to an integral membrane protein through stoichiometric amounts of some other peripheral membrane protein, usually ankyrin; and its formation of a filamentous network with intersections containing three to four filaments. The distinguishing features include participation of unusual isoforms of the membrane skeletal proteins, binding through ankyrin-dependent and -independent mechanisms, and association with different cytoskeletal and extracellular structures.

3.2.2. Focal Adhesions

Focal adhesions were first identified in cells in tissue culture as specialized regions of the plasma membrane that make very close contact with the substrate (membrane-to-substrate distance of 10–15 nm) and that have specialized ultrastructure and composition (reviewed in Ref. 137). Like the proteins of the membrane skeleton, the proteins of focal adhesions fall into several categories: integral membrane proteins, consisting primarily of one of the members of the integrin family; an ordered series of peripheral membrane proteins, including vinculin and talin, that bind to each other and to integrin; cytoskeletal structures, especially actin filaments, that are anchored to the membrane through the peripheral membrane proteins; and extracellular structures that are bound to integrin. Focal adhesions are found in nearly all cells that attach to substrates in tissue culture, and related structures are present in many cells *in vivo*.[137]

Membrane–Cytoskeletal Interactions at Focal Adhesions

The cytoplasmic face of the plasma membrane at focal adhesions appears to contain high levels of several peripheral membrane proteins, including vinculin, talin, paxillin, α-actinin, and actin (reviewed in Ref. 137). At the level of resolution afforded by the light microscope, vinculin, talin, and paxillin are specifically concentrated at focal adhesions. By contrast, α-actinin and actin are present in other regions of the cell as part of the cytoskeleton. The presence at focal adhesions of actin and α-actinin reflects at least in part the fact that the cytoplasmic surface of focal adhesions are important sites at which bundles of actin filaments ("stress fibers") anchor to the membrane (Figure 3.2). Biochemical studies have consequently focused on the proteins that are specifically associated with focal adhesions, and on their interactions with each other.

The first cytoskeletal protein shown, perhaps serendipitously, to be localized preferentially at focal adhesions was vinculin (reviewed in Ref. 170). Vinculin was first identified as a distinct polypeptide in column eluates obtained during the purification of α-actinin.[171,172] (In fact, vinculin binds to α-actinin with low affinity.[173]) Antibodies prepared to the purified protein labeled focal adhesions almost exclusively,[171,172] although considerable amounts of vinculin are also present in soluble form in the cytoplasm.[174] Subsequent molecular characterization of the protein confirmed a polypeptide mass of 117 kDa and the presence of several distinct regions capable

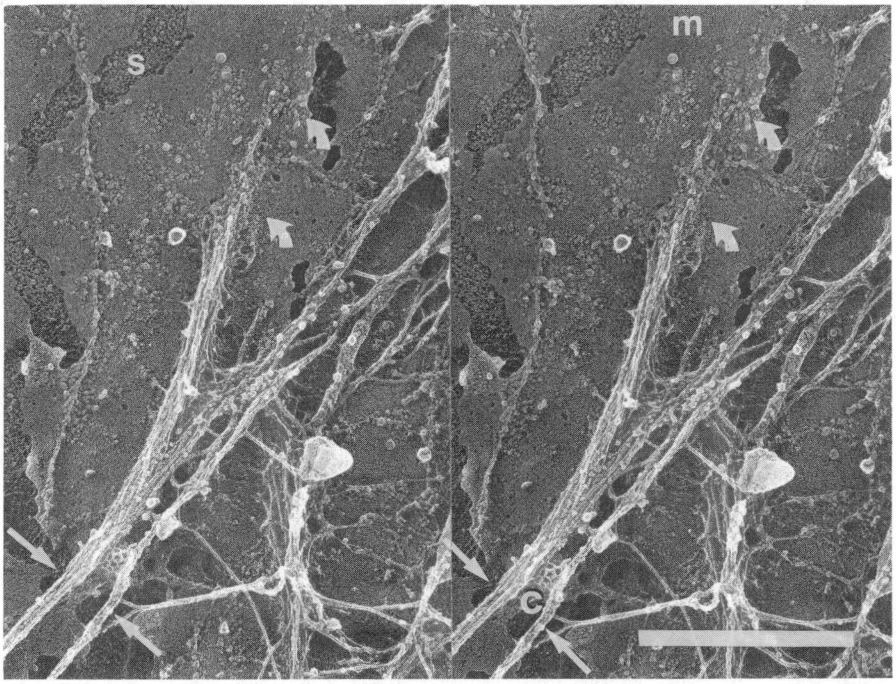

Figure 3.2. Cytoplasmic surface of the membrane at focal adhesions. Focal adhesions in cultured *Xenopus* cells were exposed by shearing with a stream of buffer. After fixation, samples were processed for ultrastructural studies by quick-freeze, deep-etch, rotary-replication. This stereo-pair shows two bundles of microfilaments (arrows), one of which approaches and associates with the membrane. Below and beyond this filament bundle are aggregates of particles that project from the membrane. Immunocytochemical studies reveal that these particles contain vinculin, talin, and integrin. Adapted from Samuelsson *et al.*[193] M, membrane; S, substrate; C, coated vesicle. Scale bar = 0.5 μm.

of interacting with proteins at focal adhesions.[175,176] Vinculin purified from chicken gizzard appears in electron micrographs as a globular protein with an extended tail.[177] The tail region, rich in proline, has binding sites for other vinculin molecules,[177] and for paxillin.[178] The globular region binds talin. An alternatively spliced variant of vinculin, termed "metavinculin," is formed by the insertion of an additional 68 amino acids between the globular and tail domains.[179] Unlike vinculin, which is expressed in a large number of cells, meta-vinculin appears to be expressed almost exclusively by smooth[180,181] and some striated[83] muscles, but its function is unknown.

Talin was first identified as a vinculin-binding protein[182] (reviewed in Ref. 183). Binding of vinculin to talin is of moderate affinity (10^{-8} M), and is governed by two stretches of amino acids in vinculin, located at the N-terminus and between residues 167 and 207.[182] Molecular cloning of talin revealed it to have a polypeptide mass of 270 kDa, arranged as a flexible rod.[184] The 47-kDa, N-terminal domain shares homology with protein 4.1 of the human erythrocyte, and with another membrane-associated protein, ezrin, isolated from intestinal brush border. The C-terminal region, too, may interact with membranes through integrin binding, as well as with the cytoskeleton, as both vinculin and integrin binding are localized to the C-terminal, 190-kDa fragment.[183,185]

Two other molecules that are selectively enriched at focal adhesions—paxillin and tensin—have only recently been identified and characterized. Paxillin is a 68-kDa protein that binds to the tail domain of vinculin[178] and is phosphorylated on tyrosines during cell attachment and spreading.[186] Tensin is a 170-kDa protein that, while concentrated at focal adhesions, is an avid actin binder. Indeed, actin polymerization and cosedimentation assays performed in the presence of different fragments of tensin show the protein to have three distinct actin-binding domains, two within the amino-terminal 463 amino acids that, when present in the tensin dimer, are each capable of cross-linking actin filaments, and one between amino acid residues 888 and 989 that may cap actin filaments.[187] More recent studies identify another actin binding site in this molecule.[187a] The native protein probably dimerizes at its carboxyl-terminus, forming an extended structure that can cross-link nearby actin filaments and slow or prevent their continued growth. No interactions between tensin and other proteins of the focal adhesion have yet been described, however. To account for its presence at focal adhesions, it has been proposed that the SH2 domain near the C-terminus of tensin[188] recognizes phosphotyrosine residues of other membrane-associated proteins (e.g., paxillin, vinculin, talin) and so promotes the association of actin filaments with the membrane.[187]

This brief discussion indicates that, just as in the erythrocyte membrane skeleton, multiple interactions occur among peripheral membrane proteins at focal adhesions that are likely to be important in organizing the membrane-associated cytoskeleton. In the case of the focal adhesion, tensin and probably α-actinin are likely to be the primary actin-binding proteins. These proteins probably stabilize the attachment of actin microfilaments to the membrane at focal adhesions through their interactions with paxillin and vinculin, as well as through direct binding of talin and α-actinin to integral membrane proteins at focal adhesions—especially integrins.

Integrins are a family of integral membrane proteins that mediate the binding of cells to extracellular matrix, and, in some cases, to other cells. Considerable evidence now indicates that some isoforms of integrin are concentrated at focal adhesions (see Section 3.3.3). The specific enrichment of proteins such as vinculin and talin with the membrane at focal adhesions therefore led to the suggestion that they may bind directly to integrin. Early experiments confirmed this suggestion. These experiments involved chromatography of integrin through a gel filtration matrix that contained a uniform concentration of vinculin or talin.[189] Earlier elution of the integrin from the matrix indicated a larger apparent molecular size, consistent with the formation of a complex. When this procedure was followed with vinculin, no shift in elution was seen. When incubated with talin, however, integrin eluted slightly earlier, and when incubated with a mixture of vinculin and talin, integrin eluted earlier still.[189] This suggested that integrin may bind preferentially to talin, and that it interacts with vinculin only after it binds to talin. As the apparent affinity of integrin for talin (10^{-6} M) was very low, these results have been difficult to reconcile with conditions *in situ*. This, together with studies of the src protein kinase[190,191] and other protein kinases (e.g., fak[192]), are consistent with the idea that the interactions of the actin cytoskeleton with integrins at focal adhesions are regulated by tyrosine phosphorylation (see Section 3.3.3.2). More recently, however, α-actinin was shown to bind to integrin with high affinity,[134,135] raising the possibility that integrin–cytoskeleton interactions might be mediated primarily by α-actinin rather than by talin.

These two contrasting possibilities have recently been addressed in ultrastructural studies of the cytoplasmic surface of focal adhesions.[193] In these studies, cells cultured on a collagen substrate were sheared with a stream of buffer to remove the cytoplasm, leaving tightly adherent structures remaining on the substrate. These were immediately fixed, and quick-frozen, freeze-dried, and coated with platinum/carbon to produce metal replicas of the cytoplasmic face of the adherent structures. In some experiments, samples were also labeled with antibodies to focal adhesion proteins, followed by immunogold particles. Focal adhesions in these samples usually contained bundles of actin filaments that approached the membrane at a shallow angle (Figure 3.2). Microfilaments could be seen to attach laterally to the membrane at many points, and especially at sites that contained aggregates of globular material. Immunogold labeling of these structures revealed that vinculin, talin, and β_1-integrin were present in substantial amounts in the globular aggregates. In contrast, α-actinin and filamin (another protein associated with actin microfilaments deeper in the cytoplasm) were only poorly labeled in these structures. Although this evidence indicates that α-actinin may not mediate the attachment of microfilaments to the membrane, more recent studies suggest that α-actinin is indeed present at the cytoplasmic face of focal adhesions, but is simply not accessible to antibodies when the cytoskeleton is intact.[193a]

3.3. INTERACTIONS OF THE EXOSKELETON WITH THE PLASMA MEMBRANE

The exoskeleton can be loosely defined as that portion of the extracellular matrix that interacts extensively and closely with the cell surface and that influences the local organization of the plasma membrane for protracted periods of time. Although extracellular matrices vary greatly from tissue to tissue and interact with the plasma membranes of cells in numerous ways, there are few matrices that act as an exoskeleton. The most obvious example of an exoskeleton, and the one discussed in detail here, is the basal lamina, found in close proximity to epithelial, endothelial, skeletal muscle, Schwann, and other cells. The interaction of fibronectin with the plasma membrane is also considered, in light of its key role in elucidating the structure and function of the integrin family of extracellular matrix receptors. Fibrous, cartilaginous, and more fluid extracellular matrices are not considered.

3.3.1. The Basal Lamina

Ultrastructural studies of cells surrounded by a basal lamina show three distinct compartments or layers, which together are sometimes referred to as a "basement membrane." The most electron-dense layer is the basal lamina, also known as the *lamina densa.* The basal lamina can vary greatly in thickness, but it usually lies between 25 and 50 nm above the lipid bilayer of the plasma membrane. Beyond the basal lamina lies the *lamina rara externa* or *lamina reticularis*—a region that is less electron dense than the basal lamina and that is composed of a loosely organized, often fibrous, extracellular matrix. Between the *lamina densa* and the cell membrane lies the *lamina rara interna* or *lamina lucida,* a region that is also much less electron dense than the basal lamina. This region contains regular structures that extend from the plasma membrane to attach to the basal lamina (e.g., in skeletal muscle[141]). When viewed from above, the basal lamina forms a sheath overlying large parts of the cell surface. This structure contributes significantly to the permeability properties of transporting epithelia and to the regenerative properties of skeletal muscle.

Basal laminae for many years resisted study at the biochemical level, in large part because methods for dissociating them into their component proteins were unavailable and also because there were few tissues that provided a large but homogeneous source. The introduction of a variety of chaotropic and denaturing agents, coupled with the discovery of tumors such as the Engelbreth–Holm–Swarm (EHS) sarcoma that synthesize large amounts of basal lamina, has led to rapid progress in this area since the early 1980s.

3.3.1.1. Proteins of the Basal Lamina

It is now clear that the biochemical composition of the basal lamina is rather simple. The EHS basal lamina is composed largely of four proteins: type IV collagen, a large, low-density heparan sulfate proteoglycan called "perlecan," laminin, and a laminin-binding protein termed "nidogen" or "entactin." Much is now known about the chemistry of these proteins, but how they become organized into a basal lamina is still poorly understood.

3.3.1.1a. Type IV Collagen. Type IV collagen is the major structural protein of the basal lamina. Unlike most other members of the collagen superfamily, type IV collagen does not form thick fibrils. Instead, it forms networks of thin fibers that, when stacked and combined with laminin, proteoglycan, and other proteins, creates a proteinaceous sheet. Like the other collagens, type IV collagen is assembled from monomeric units that themselves are composed of three polypeptide chains. At least five type IV collagen chains have been identified to date. The most widespread, designated $\alpha 1(IV)$ and $\alpha 2(IV)$, assemble to form $\alpha 1(IV)_2\alpha 2(IV)$ filaments typical of type IV collagen. Other chains are used for specialized basal laminae (e.g., at the neuromuscular junction; see Section 3.3.5).

The assembly of the type IV collagen subunits and their formation of triple helices occurs after the amino-termini of three chains come together to form a 7 S globule. The collagenous, triple-helical regions then "zip up," leaving only the small, globular NC1 regions at the carboxyl-terminus free. This monomeric unit, ~540 kDa in mass, is further stabilized by interchain disulfide and other covalent bonds. In electron micrographs, it appears as a flexible, "threadlike molecule,"[194] approximately 400 nm in length. End-to-end polymerization is promoted by the interaction of NC1 domains with single NC1 domains emerging from other triple helices. Formation of a network occurs when 7 S domains from four different triple helices associate. Side-to-side aggregation, so typical of the fibrous collagens, is inhibited in type IV collagen, probably because its regular triplet repeats containing glycine and proline are interrupted periodically by the introduction of other amino acids. These interruptions tend to occur at the same points along the polypeptide chains of the different subunits that assemble to form the monomeric units, suggesting that the noncollagenous sequences form distinctive domains within the triple-helical segment.

Based on the characteristics of the type IV collagen molecule and its tendency to polymerize end-to-end and not side-to-side, collagen would be expected to form a loose network of 800-nm filaments, with fourfold branching at the 7 S regions that come together to form each intersection. When type IV collagen is polymerized *in vitro,* or when type IV collagen networks are observed *in situ,* however, a much tighter network is seen.[194] This network contains thicker filaments that are only ~30 nm in length, and intersections with an irregular number of branch points. Type IV collagen therefore undergoes limited side-to-side associations, perhaps promoted by the short, nonhelical domains in the molecule. As a result of these additional interactions, the type IV collagen network shows $1–2 \times 10^3$ nm^2 openings, rather than the $\sim 6 \times 10^5$ nm^2 openings predicted for the looser network. This tighter mesh is more in keeping with the dense basal laminar structures of epithelia, for example, but it is not tight enough to account for the ability of such basal laminae to exclude many macromolecules.

Another level of organization, and thus increased density, may occur as networks of type IV collagen are stacked one on top of another to form a thick collagenous layer. Additional links between collagen filaments are also provided by laminin–nidogen, and by heparan sulfate proteoglycan (see Sections 3.3.1.1b and 3.3.1.1c). The presence of these large anionic molecules in the network would further reduce the mesh and permeability of the collagen network.

3.3.1.1b. Laminin and Nidogen. Laminin isolated from EHS sarcoma is a large (~800 kDa) cruciform molecule composed of three distinct chains, termed A (~400 kDa), B1 and B2 (~200 kDa each) (reviewed in Refs. 195,195a; for a revised nomenclature of the laminins and their subunits, see Ref. 195b). The ~75-nm long arm of the cross is formed by association of the three subunits, organized as an α-helical coiled coil, that extends to the ends of B1 and B2, but leaves a portion of A free. This carboxyl-terminal portion of the A chain folds into a series of five small, globular domains. The more internal globular units, together with the terminal region of the long arm, form a cell binding domain that interacts preferentially with one of the integrins (reviewed in Ref. 196; see Section 3.3.3). The last two globular domains in this region also bind heparan sulfate proteoglycan.[197] The three short (~35 nm) arms of the cross are formed by the individual subunits as they emerge from the long arm. Each has two or three globular domains. Several of these globular regions, located 10–15 nm in from the ends of each chain, are homologous to each other and to the cysteine-rich repeats found in epidermal growth factor (EGF). Although the functions of most of the EGF-like domains are still unknown, one may be involved in nidogen binding.[198] Laminin binds nidogen between the branch point of the cross and the EGF-like domain of the B2 chain.[199] Other globular regions, located at the ends of each chain, are involved in self-assembly of laminin and, for the B chains, in collagen binding. Integrin and heparan sulfate proteoglycan binding have also been localized to the amino-terminal region of the A chain, at the top of the cross.[199a]

Laminins from other tissues share many features with EHS laminin, but they may differ in their subunit composition. Some tissues, such as embryonic Schwann cells and skeletal myotubes, may synthesize laminin without an A chain. In such cases, the B1 and B2 chains assemble to form a helical structure with two branches. In other tissues, isoforms of the A chain, such as "merosin"[200] (also called "laminin Am," to distinguish it from the A chain of EHS laminin, known as Ae), and of the B1 chain, such as S-laminin[201] (or laminin B1s), and kalinin[202] (laminin B1k) are expressed. These tissues form laminin in the usual way, with one large and two small chains, and in various combinations. Recently, a small variant of the B2 chain has also been reported.[203]

Assembly of the laminin subunits is driven primarily by the formation of the α-helical coiled coil domain by the A, B1, and B2 chains.[203a,203b] This assembly is dependent on specific interactions that determine the particular combination of chains that form the trimer.[203a] Further interactions between assembled laminin molecules can result in the formation of oligomers and extended polymers.[194] This process is promoted by Ca^{2+} and heparin or related heparan sulfates.[204] Interactions between chains occur between globular ends of the short arms and between globular regions at the C-terminus of the A chain.[199a,205,206] Polymerization in solution occurs at concentrations in the micromolar range—concentrations that are likely to be met as laminin is secreted into narrow extracellular spaces. This suggests that laminin polymerization is physiological.[194] Recent studies[205] of the factors that retain laminin within native basal laminae indicate that only ~20% of the laminin is tightly bound to type IV collagen, whereas the remainder (~80%) can be solubilized under mild conditions by introducing peptide fragments that inhibit laminin oligomerization. It is therefore possible that laminin in the basal lamina forms distinct layers that interact only intermittently with type IV collagen, or that laminin can form a type of basal lamina without the participation of type IV collagen. Indeed, laminin is expressed without type IV collagen at the 8- to 16-cell stages of embryonic development.[207]

The interactions of laminin with collagen are probably promoted by two processes: direct but low-affinity binding of globular units at the ends of the short arms of laminin to collagen, and indirect but high-affinity binding through nidogen. Nidogen (also known as entactin) is a 148-kDa protein that binds tightly (K_d ~10^{-9} M) to laminin and is routinely found in a stoichiometric complex following laminin purification. It is composed of three globular domains, separated by rodlike spacers.[208,209] The C-terminal globular domain, G3, binds to laminin, while the globular domain in the middle of the molecule, G2, binds type IV collagen.[199,210] The G2 globular domain also binds to heparan sulfate proteoglycan[210] and to fibronectin.[211] Thus nidogen promotes laminin–collagen binding as well as providing an additional means through which laminin can interact with other components of the extracellular matrix.

3.3.1.1c. Heparan Sulfate Proteoglycan. Like laminin and type IV collagen, the low-density heparan sulfate proteoglycan of EHS sarcoma has been purified and characterized at the molecular level.[211] It is composed of a large (396 kDa, ~85 nm) polypeptide chain organized into five to seven globular domains, aligned like pearls on a string[212]—hence its name, "perlecan." Each polypeptide chain is covalently linked to two or three heparan sulfate side chains of 30–60 kDa in mass and 100–170 nm in length, depending on the tissue of origin. These attach to the N-terminal region of the protein. Perlecan binds to itself, to form dimers and trimers,[213] as well as to laminin–nidogen.[210,214] Much of the central and C-terminal region of the polypeptide is organized into Ig-like domains that most closely resemble those in neural cell adhesion molecule, suggesting that perlecan may bind to other extracellular and cell surface molecules via protein–protein interactions, as well as through its heparan sulfate chains. Indeed, recent results suggest that perlecan, like laminin and collagen type IV, can bind to receptors on the cell surface via an integrinlike receptor.[211]

3.3.2. Fibronectin

Fibronectin has been much studied since it was first shown to be related to growth control mechanisms in normal and transformed cells in culture[e.g., 215–217] (reviewed in Refs. 218–220). Fibronectin is synthesized in many alternatively spliced forms,[221–223] but all are products of a single gene.[224,225] The form of fibronectin most commonly associated with the cell surface is composed of two disulfide-linked polypeptide chains that are identical in all features but one, namely, one subunit contains an alternatively spliced domain involved in binding to the cell surface that is missing in the other. The rest of the molecule (~235 kDa per subunit) is constructed in modular form, with distinctive domains involved in binding to fibrin and heparin, and to collagen and gelatin.[226–229] Individual domains have recently been expressed in bacteria and crystallized, to facilitate structural studies.[229a,229b] In solution, the individual domains fold independently of one another, and the molecule is quite flexible. Fibronectin can, however, adopt an extended conformation, in which each subunit is 2–3 nm in diameter and 60–70 nm in length.[220] It is in this extended conformation that, through further polymerization and disulfide cross-linking, fibronectin forms fibrils on the cell surface. Indeed, polymerization is promoted by interaction with fibronectin receptors in the plasma membrane.[230]

These fibrils form distinctive complexes with the cell membrane. In embryonic fibroblasts, for example, the fibronectin fibrils seem to emanate from the same spot at the cell surface at which actin filaments insert intracellularly.[231] This structure, called a "fibronexus," constituted strong anatomical grounds for believing that fibronectin and actin were linked across the plasma membrane. The protein that was eventually shown to mediate this indirect linkage, the fibronectin receptor, is a member of the integrin family, mentioned above (see Section 3.2.2) as the site at which actin filaments become anchored on the cytoplasmic face of the plasma membrane at focal adhesions.

3.3.3. Integrins as Receptors for Extracellular Proteins

The integrins are a large family of proteins that mediate many of the interactions of a cell with surrounding structures, be they substrate, extracellular matrix, or other cells (reviewed in Refs. 232–235). All integrins are heterodimers of α and β subunits. The number of α- and β-integrin subunits now exceeds 20. Although there are a growing number of exceptions (reviewed in Ref. 234), a given α-integrin subunit tends to associate with only a single type of β subunit, but a given β subunit can interact with any of several α subunits, to produce integrins with distinctive specificities. Those integrins with β_2 subunits have so far been found only in the immune system, where they mediate cell–cell interactions. LFA-1, for example, mediates binding between T and B cells, and Mac-1, between T cells and macrophages. Other integrins that have been extensively studied contain the β_1 and β_3 subunits and are involved primarily in cell–substrate and cell–matrix interactions. I focus on these particular classes of integrins below.

The fibronectin receptor is the prototypic integrin and methods used to characterize its interactions with fibronectin have now become standard (reviewed in Ref. 236). After it was found that fibronectin bound specifically and saturably to intact cells and to preparations of cell membranes, the portion of the fibronectin molecule that was responsible for binding was identified. Fibronectin was partially degraded and the protease-resistant domains[226–229] were assayed for their ability to inhibit the binding of intact fibronectin. In early experiments, a single region, called the "cell binding domain," was identified and its amino acid sequence was determined. Overlapping oligopeptides covering this region were synthesized and one was shown to bind to the cell and to inhibit fibronectin binding.[237] This led to the discovery of a short stretch of amino acids—Arg-Gly-Asp, or RGD, in the single-letter code—that was essential for binding. The intact protein was used to prepare an affinity matrix to purify the putative receptor. Elution was effected by small peptides containing the RGD sequence.[238] The purified binding protein consisted of two subunits that were subsequently cloned and sequenced and later identified as a member of the "integrin" family of proteins containing a β_1 subunit.[238] Use of the same approach to purify the receptor for the serum protein, vitronectin, identified an RGD-binding protein homologous to the fibronectin receptor[239,240] but containing a β_3 subunit. The α subunits of these two receptors also differ.

As the purification of the fibronectin receptor was progressing, other groups were studying the surface proteins responsible for cell–substrate attachment. For example, monoclonal antibodies were identified that blocked the attachment of myoblasts and fibroblasts to the tissue culture substrate.[241] These antibodies were then used to isolate their antigens from cell membrane preparations. The precipitated polypeptides had properties that subsequently identified them as integrin subunits related to the fibronectin receptor.[242] Use of the antibodies in immunolocalization experiments revealed that their antigens were present at or near the focal adhesions made by the cells with the substrate.[243] Thus, the fibronectin receptor and closely related proteins bind to fibronectin and through that binding promote the adhesion of the cell to appropriate substrates. These same approaches have since been used to characterize a wide variety of receptors for extracellular matrix and basal laminar proteins.

The structure of integrin has been deduced from cDNA sequencing and from utrastructural studies of the purified protein.[236,244] Both subunits are integral membrane proteins with large extracellular domains that contain covalently bound carbohydrate, a single transmembrane domain, and a short cytoplasmic domain (except β_4, which has a large cytoplasmic domain[234]). The α subunits vary in mass from 130 kDa to 210 kDa; some are proteolytically cleaved into large and small (e.g., 30 kDa) fragments. The β subunits tend to be more uniform in size (110–140 kDa), but their mobility in dodecyl sulfate-gel electrophoresis decreases in the presence of reducing agents. This behavior is believed to be related to the presence of several intrachain disulfide bonds in the middle of the extracellular domain of the β chain that, on reduction, cause the molecule to assume a more extended conformation when denatured with dodecyl sulfate. The native heterodimer seems to resemble a kidney bean mounted on stilts that attach at each rounded end.[244] The bean-shaped portion sits approximately 12–14 nm off the membrane and measures 8 nm high and 12–15 nm wide. The stilts, corresponding to portions of the α and β subunits, are each ~2 nm wide and ~12–14 nm high. The molecule may extend as little as 2 nm into the cytoplasm.[244]

Integrins require divalent cations, usually Ca^{2+}, to be fully active.[e.g., 245] The Ca^{2+} binding sites are located in the middle of the extracellular domain of the α subunit, near the region of α involved in ligand binding. Binding of extracellular matrix proteins to integrin involves sites on both α- and β-integrin subunits, however, as shown in studies in which portions of the ligand were covalently linked to the active site. Labeling occurred in the middle of the divalent cation binding domain of α and just N-terminal to the cysteine-rich region of β.[227,246–248]

One of the striking features of many of the integrins is their ability to bind to several different extracellular ligands, sometimes with remarkably different affinities. If binding were only controlled by a simple sequence, such as RGD, it would be difficult to account for the specificity that different integrins show for their respective ligands, or that different ligands show for the same integrin. This can now be explained by at least two factors. First, different integrins that recognize RGD show preferences for the sequences flanking the RGD and for the conformation of the RGD-containing peptide. The role of conformation was investigated with linear and cyclic peptides.[249] The vitronectin receptor was found to have a higher affinity for the cyclic RGD peptide, whereas the fibronectin receptor preferred the linear conformation.[249] Second, cells bind and adhere to extracellular ligands through RGD-independent mechanisms. An interesting example of this was found with fibronectin. In addition to the cell binding domain containing RGD, a second region was found to which cells can bind even in the presence of saturating concentrations of RGD. On further analysis of this region, the specific sequence responsible for binding and adhesion was found within a 12-amino-acid oligopeptide, LHGPEILDVPST, with no RGD sequence. The integrin responsible for binding this sequence was subsequently shown to be a distinct integrin.[250] Similarly, integrins bind to other extracellular ligands at sites that do not contain RGD. Laminin, for example, binds a number of different integrins at several distinct sites, not all of which contain RGD.[196,251]

3.3.3.1. Regulation of Integrin Activity

Some differences in the affinity of integrins for their extracellular binding proteins cannot be explained by differences in the ligands or the ligand binding region of integrin, however. Platelets and endothelial cells express the same $\alpha_2\beta_1$ integrin, for example, but endothelial cells attach preferentially to a fragment of fibronectin, while platelets attach peferentially to collagen.[252] Thus, other factors must regulate the binding of integrin to its ligands.

One factor that may modulate the activity of integrin is the ability of integrin to bind to other molecules in the plasma membrane. Integrin in some cells is closely associated with other macromolecules, including glycolipid,[253,254] and proteins of 50 and 100 kDa.[255,256] The roles of these interactions in cell–substrate or cell–matrix adhesion are still poorly understood.

The activity of integrin will also be affected by its ability to cluster in the cell membrane. The apparent affinity of the cell for the ligand will increase as the number of integrins that can participate in binding increases. Thus, cells that are capable of mobilizing more integrins into areas involved in adhesion to matrix or substrate will adhere more tightly than cells that can mobilize fewer integrins. These processes are probably regulated by the ability of the cell to change shape (and so bring larger or smaller areas of the cell surface into contact with the substrate), the "fluidity" of the plasma membrane, and the extent to which the pool of integrin in the membrane is already associated with the cytoskeleton. Studies of the mobility and clustering of integrin[e.g., 257] have only just begun to address some of these issues. It is nevertheless clear that particular integrins accumulate in limited regions of the membrane (e.g., focal adhesions) in response to binding to their extracellular ligands, and that this process is both integrin- and ligand-specific.[258–261]

3.3.3.2. Signaling through the Integrins

Recent reports have suggested that a series of intracellular events akin to signaling by second messengers is triggered by the clustering of integrins in the plasma membrane (reviewed in Ref. 262). Cells that adhere to substrates coated with ligands for integrin, or that bind anti-integrin antibodies, accumulate integrins at specific sites.[243,258–261,263] A protein tyrosine kinase called fak, for "focal adhesion kinase" (reviewed in Ref. 263a), becomes localized on the cytoplasmic face of the membrane at these sites.[186,192,264] Its level of protein kinase activity also increases on adhesion.[265] Adhesion that does not involve integrins fails to induce the localization and activation of fak,[192,264] and blockade of protein tyrosine kinase activity in cells adhering through an integrin-mediated mechanism prevents subsequent oganization of cytoplasmic proteins, including talin and actin filaments.[186] Thus, fak or a closely related protein is likely to be involved in mediating the relationship between adhesion and cytoskeletal reorganization.

Fak was first identified as a substrate for pp60src,[266] another intracellular protein kinase that in some cells also accumulates at focal contacts.[190,267] Like src, fak is a soluble protein, but it has a molecular mass of 125 kDa, and, aside from its kinase domain, has no homology to any known protein.[264] This, together with the effects of tyrosine kinase inhibitors,[186] implicates a "tyrosine kinase cascade" as part of the response to ligand–integrin interactions. The downstream events presumably include activation of ras and intracellular protein kinases, and modulation of gene expression in the nucleus, as is the case with other tyrosine kinases. One example of an integrin-mediated change in gene expression involves extracellular proteases, especially collagenase and metalloproteinase,[267a,267b] both of which are upregulated by cell adhesion to fibronectin substrates. Several other adhesion molecules have also been implicated in signal transduction pathways, suggesting that adhesion and intracellular signaling may in general be closely linked.

Phosphorylation may also modulate the interactions of integrin with its various ligands. As mentioned above, phosphorylation of the intracellular domain of integrin can alter integrin's association with vinculin and talin,[191] favoring dissociation of the complex and thus destabilizing focal adhesion plaques. Similarly, intracellular events may modulate the ability of an integrin to bind to its extracellular ligand (reviewed in Ref. 268). In platelets, binding of fibrinogen to integrin, and the conformational changes that integrin undergoes on ligand binding, were augmented by hormonal activation of the platelets and subsequent protein phosphorylation.[269] This "inside-out" signaling, like integrin's association with the cytoskeleton, is dependent primarily on interactions mediated by the cytoplasmic portion of the β subunit.[269a]

Thus, the response of a cell to an extracellular ligand via its integrins may control not only its morphology, through the formation of focal adhesions, but also its biochemistry and even gene expression. Likewise, the biochemical state of the cell can regulate its ability to respond to extracellular matrix proteins via the integrins.

3.3.3.3. Signaling by Soluble Molecules Bound to the Exoskeleton

In addition to signaling via the integrins, the exoskeleton can signal indirectly, by sequestering, concentrating, or releasing soluble factors that can bind to receptors of their own at the cell membrane. The classic example of such a role for the exoskeleton was discovered with basic fibroblast growth factor (bFGF) (reviewed in Ref. 270). This hormone was found to bind via positively charged regions in its sequence to the polyanionic sites in the extracellular matrix provided by heparan sulfates. bFGF can be released from the matrix enzymatically and bind to its receptor in the cell membrane. In addition, bFGF and related hormones, such as vascular endothelial growth factor,[271] may be sequestered near the cell surface by binding to proteoglycans and so become available at elevated concentrations for interaction with their receptors. Indeed, these hormones appear to activate their membrane receptors more effectively when they are "presented" by nearby proteoglycans to which they are bound.[272] These multiple interactions between the basal lamina, hormones, and their surface receptors vastly expand the potential physiological role for the exoskeleton.

3.3.4. Other Extracellular Matrix Receptors

Although the integrins form an extensive and overlapping set of receptors for the extracellular matrix, they are not the only receptors at the cell surface for the proteins of the exoskeleton. Laminin alone can interact with several different classes of proteins, each of which may have distinct consequences for organization of the plasma membrane.

One set of proteins are lectins, i.e., proteins with specificity for the carbohydrate moieties of glycoproteins. A series of such molecules, with masses of ~13, ~33, and 67 kDa, have been characterized and shown to bind to several sites on laminin (reviewed in Ref. 196). This binding can generally be blocked by galactoside sugars. Binding on the long arm of the A chain can be blocked with a pentapeptide, IKVAV, while binding on the short arm of the B1 chain can be blocked by other oligopeptides, YIGSR and LGTIPG. All of these laminin-binding lectins are structurally related, as they contain sugar-binding domains that are antigenically similar. Their physiological role is still in question, however, as some of these proteins are found primarily in the cytoplasm, and one, the 14-kDa lectin, is translated on cytoplasmic ribosomes. It is conceivable, however, that these proteins, when released into the extracellular milieu (e.g., through shedding of large blebs, or after the death of a cell), could play an important role in binding laminin to molecules on the cell surface or to other proteins or glycoproteins in the extracellular matrix.

Another potential set of ligands for laminin is provided by membrane-intercalated heparan sulfate proteoglycans, such as syndecan[273] and α-dystroglycan.[107] α-Dystroglycan is one of the subunits of the membrane glycoprotein complex that anchors dystrophin to the cytoplasmic surface of the plasmalemma of skeletal muscle.[274] Of the five proteins in the complex, α-dystroglycan is the major extracellular component, with an apparent size in dodecyl sulfate gel electrophoresis of 120–200 kDa. Only ~56 kDa of the mass is protein, however; the remainder represents covalently bound carbohydrate, including heparan sulfate. A recent report indicates that the protein portion of α-dystroglycan is organized into two globular domains linked by a central region resembling mucin.[274a] α-Dystroglycan binds laminin in a Ca^{2+}-dependent manner that is inhibited by heparin and by antibodies that recognize its heparan sulfate side chains.[275] Binding is also inhibited by digestion with specific glycanases. It therefore seems likely that laminin in the muscle basal lamina can bind to the membrane through α-dystroglycan as well as through integrins.

The linkage of α-dystroglycan to laminin in the basal lamina is biologically essential, at least in skeletal muscle. Recent reports indicate that deletion or mutation of the merosin chain (Am), the major laminin A chain in the basal lamina of skeletal muscle, is linked to congenital muscular dystrophy in humans,[275a] and in the dy/dy mouse model for this disease.[275b] This is consistent with the fact that other muscular dystrophies are related to the absence of other proteins involved in linking the extracellular matrix with the internal, membrane skeletal structures present in striated muscle (see Section 3.2.1.6c).

The interaction between α-dystroglycan and laminin may also serve as a model for interactions between other laminin-like molecules in the extracellular matrix or in the membrane (e.g., neurexins[276]), and membrane-bound heparan sulfate proteoglycans. An example of considerable interest in cellular neurobiology, that of agrin binding to α-dystroglycan, is discussed briefly in Section 3.3.5.

3.3.5. Membrane Domains and the Exoskeleton

In addition to forming stable structures to which the plasma membrane can bind, the exoskeleton, and in particular the basal lamina, has morphogenetic properties in some tissues. Epithelial cells, for example, sit atop a basal lamina in a polar fashion, with their basal surfaces close to and their apical surfaces away from the lamina. If epithelial cells are brought into suspension and then replated onto an artificial basal lamina, they will recover their apical-to-basal polarity. The interaction between the plasma membrane and the extracellular substrate, therefore, contributes to the morphological specialization of the cell.[277]

A more striking example of the morphogenetic potential of the basal lamina is found in vertebrate skeletal muscle fibers. Most of the surface of myofibers is covered by a basement membrane containing the usual basal laminar constituents, described above. At the neuromuscular junction, the site where the muscle fiber is innervated by a motor neuron, the basal lamina is thicker and its composition changes. Junctional laminin is composed of S-laminin (B1s) and the A chain (Ae), while in extrajunctional regions the typical laminin is composed of B1e and M (or Am) chains.[278] The type IV collagen at the neuromuscular junction is also distinct, as it is composed of α3(IV) and α4(IV) subunits, rather than the usual α1(IV) and α2(IV) chains.[278] Acetylcholinesterase, the protein responsible for degrading the acetylcholine released by the motor neuron, is anchored specifically to the junctional basal lamina,[279] perhaps to the heparan sulfate proteoglycan.[280] This specialized basal lamina lies over a region of the muscle membrane that is almost paracrystalline in acetylcholine receptors,[281,282] and nearby regions of the nerve terminal that are specialized for neurotransmitter released. The postsynaptic cytoskeleton is highly specialized,[282a] consistent with the presence of extensive links between the basal lamina and the cytoplasm.

At least two components of the junctional basal lamina have potent morphogenetic properties that help to account for the organization of this synapse. In an initial series of ultrastructural studies, McMahan and his colleagues showed that when both nerve and muscle are damaged and the cellular debris is subsequently removed by macrophages, the only structure that remains in the junctional region is the junctional basal lamina.[283] If the nerve is permitted to regenerate without regeneration of the muscle, it will reinnervate the original junctional site and replace release sites for neurotransmitter at the appropriate presynaptic locations.[283] Subsequent studies by Merlie, Sanes, and colleagues showed that the reassociation of nerve and with the junctional exoskeleton is probably mediated by a special three-amino-acid sequence in S-laminin—LRE.[284] This sequence appears to induce growing motor neurons to stop growing.[284a] A strain of mice in which S-laminin has been deleted by homologous recombination shows significant changes in the organization of the nerve terminals, consistent with a role for this protein in presynaptic differentiation.[284b]

In a variation of the regeneration experiment, McMahan and his colleagues[285,286] allowed the muscle fiber to regenerate in the absence of reinnervation. In this case, the muscle fiber forms a new postsynaptic membrane immediately under the junctional basal lamina that is highly enriched in acetylcholine receptors. The active agent in the basal lamina that induces the accumulation of receptors in this, and only in this, region of the sarcolemma is agrin, a large (~220 kDa) extracellular glycoprotein[287–292] that is synthesized by the nerve, transported to the terminal, and deposited at the junctional region.[293–296] Agrin has since been identified as the major heparan sulfate proteoglycan in the brain.[296a] Recently, several groups have produced compelling evidence that α-dystroglycan is a low-affinity receptor for agrin that is present in high amounts on muscle fibers,[126,297–299] though it may not have the binding specificity expected of the physiological agrin receptor.[299]

Both agrin and S-laminin are likely to play important roles in synaptogenesis and regeneration of the neuromuscular junction, as well as elsewhere in the nervous system. Their remarkable activities at the regenerating neuromuscular junction illustrate how basal laminae can act as morphogens to induce highly specialized membrane domains in receptive cells.

3.4. CONCLUDING REMARKS

Our understanding of the organization of the plasma membrane has undergone considerable evolution since the "fluid mosaic model" of Singer and Nicholson[300] appeared in 1972. That model proposed that integral membrane proteins float freely in a layer of phospholipids. It is still valid, with one important caveat: that all of the constraints normally placed on the membrane lipids and proteins are first released. In the native membrane, these constraints can be prodigious, indeed. The structures discussed in this chapter are only some of the many that organize the plasma membrane. They operate by binding to integral membrane proteins at their intracellular and extracellular surfaces, by gathering them into concentrated patches, or membrane domains, that serve important functions in specialized cells, and by activating different receptors and associated proteins that alter the organization of the membrane and the biochemistry of the cell interior.

The importance of these functions is underscored by the complexity of these structures and the multiple interactions among their constituent proteins. In some cases, such as hereditary spherocytosis, these overlapping interactions save the organism when one component is defective, whereas in others, such as Duchenne muscular dystrophy, the component is so important that its loss or alteration can be lethal.

Modern cell biology and biochemistry have succeeded in identifying and characterizing many—though by no means all—of the proteins and glycoproteins of the membrane-associated cytoskeleton and the exoskeleton. The interactions among the proteins in these structures, the linkages that form between structures lying on the intracellular and extracellular faces of the plasma membrane, and the ways in which these structures control cellular fate and function are now under intensive study. The results of these studies will fill a crucial gap in our understanding of cellular physiology and contribute to our understanding of the cellular and molecular basis of disease.

REFERENCES

1. Kirsch, J., Langosch, D., Prior, P., Littauer, U. Z., Schmitt, B., and Betz, H. (1991). The 93-kDa glycine receptor-associated protein binds to tubulin. *J. Biol. Chem.* **266:**22242–22245.
2. Tilney, L. G., and Detmers, P. (1975). Actin in erythrocyte ghosts and its association with spectrin. *J. Cell Biol.* **66:**508–520.
3. Tsukita, S., and Ishikawa, H. (1980). Cytoskeletal network underlying the human erythrocyte membrane. *J. Cell Biol.* **85:**567–576.
4. Lange, Y., Hadesman, R. A., and Steck, T. L. (1982). Role of the reticulum in the stability and shape of the isolated human erythrocyte membrane. *J. Cell Biol.* **92:**714–721.
5. Winograd, E., Hume, D., and Branton, D. (1991). Phasing the conformational unit of spectrin. *Proc. Nat. Acad. Sci. USA* **88:** 10788–10791.

5a. Yan, Y., Winograd, E., Viel, A., Cronin, T., Harrison, S. C., and Branton, D. (1993). Crystal structure of the repetitive segments of spectrin. *Science* **262:**2027–2030.

5b. Speicher, D. W., and Ursitti, J. A. (1994). Conformation of a mammoth protein. *Curr. Biol.* **4:**154–157.

6. Lundberg, S., Lehto, V.-P., and Backman, L. (1992). Characterization of calcium binding to spectrins. *Biochemistry* **31:**5665–5671.
7. Karinch, A. M., Zimmer, W. E., and Goodman, S. R. (1990). The identification and sequence of the actin-binding domain of human red blood cell β-spectrin. *J. Biol. Chem.* **265:**11833–11840.
8. Speicher, D. W., Weglarz, L., and DeSilva, T. M. (1992). Properties of human red cell spectrin heterodimer (side-to-side) assembly and identification of an essential nucleation site. *J. Biol. Chem.* **267:** 14775–14782.

8a. Viel, A., and Branton, D. (1994). Interchain binding at the tail end of the Drosophila spectrin molecule. *Proc. Natl. Acad. Sci. USA* **91:**10839–10843.

9. McGough, A. M., and Josephs, R. (1990). On the structure of erythrocyte spectrin in partially expanded membrane skeletons. *Proc. Natl. Acad. Sci. USA* **87:**5208–5212.
10. Ursitti, J. A., and Wade, J. B. (1993). Ultrastructure and immunocytochemistry of the isolated human erythrocyte membrane skeleton. *Cell. Motil. Cytoskel.* **25:**30–42.
11. Shen, B. W., Josephs, R., and Steck, T. L. (1986). Ultrastructure of the intact skeleton of the human erythrocyte membrane. *J. Cell Biol.* **102:**997–1006.
12. Liu, S.-C., Derick, L. H., and Palek, J. (1987). Visualization of the hexagonal lattice in the erythrocyte membrane skeleton. *J. Cell Biol.* **104:**527–536.
13. Shotton, D. M., Burke, B. E., and Branton, D. (1979). The molecular structure of human erythrocyte spectrin: Biophysical and electron microscopic studies. *J. Mol. Biol.* **131:**303–329.
14. Byers, T. J., and Branton, D. (1985). Visualization of the protein associations in the erythrocyte membrane skeleton. *Proc. Natl. Acad. Sci. USA* **82:**6153–6157.
15. Ursitti, J. A., Pumplin, D. W., Wade, J. B., and Bloch, R. J. (1991). Ultrastructure of the human erythrocyte cytoskeleton and its attachment to the membrane. *Cell Motil. Cytoskel.* **19:**227–243.
16. Parry, D. A. D., Dixon, T. W., and Cohen, C. (1992). Analysis of the three-α-helix motif in the spectrin superfamily of proteins. *Biophys. J.* **61:**858–867.
17. Bloch, R. J., and Pumplin, D. W. (1992). A model of spectrin as a concertina in the erythrocyte membrane skeleton. *Trends Cell Biol.* **2:**186–189.
18. Kennedy, S. P., Weed, S. A., Forget, B. G., and Morrow, J. S. (1994). A partial structural repeat forms the heterodimer self-association site of all β-spectrins. *J. Biol. Chem.* **269:**11400–11408.
19. Elgsaeter, A., Stokke, B. T., Mikkelsen, A., and Branton, D. (1986). The molecular basis of erythrocyte shape. *Science* **234:**1217–1223.
20. Ralston, G. B., and Dunbar, J. C. (1979). Salt and temperature-dependent conformation changes in spectrin from human erythrocyte membranes. *Biochim. Biophys. Acta* **579:**20–30.
21. Morrow, J. S., Jr., Haigh, W. B., and Marchesi, V. T. (1981). Spectrin oligomers: A structural feature of the erythrocyte cytoskeleton. *J. Supramol. Struct. Cell Biochem.* **17:**275–287.
22. Morrow, J. S., and Marchesi, V. T. (1981). Self-assembly of spectrin oligomers in vitro: A basis for a dynamic cytoskeleton. *J. Cell Biol.* **88:**463–468.
23. Liu, S.-C., Windisch, P., Kim, S., and Palek, J. (1984). Oligomeric states of spectrin in normal erythrocyte membranes: Biochemical and electron microscopic studies. *Cell* **37:**587–594.
24. Vandekerckhove, J., and Weber, K. (1978). Mammalian cytoplasmic actins are the products of at least two genes and differ in primary structure in at least 25 identified positions from skeletal muscle actins. *Proc. Natl. Acad. Sci. USA* **75:**1106–1110.
25. Vandekerckhove, J., and Weber, K. (1978). At least six different actins are expressed in a higher mammal: An analysis based on the amino acid sequence of the amino-terminal tryptic peptide. *J. Mol. Biol.* **126:**783–802.
26. Gunning, P., Ponte, P., Okayama, H., Engel, J., Blau, H., and Kedes, L. (1983). Isolation and characterization of full-length cDNA clones for human α-, β- and τ-actin mRNA's: Skeletal but not cytoplasmic actins have an amino terminal cysteine that is subsequently removed. *Mol. Cell Biol.* **3:**787–795.
27. Gimona, M., Vandekerckhove, J., Goethals, M., Herzog, M., Lando, Z., and Small, J. V. (1994). β-Actin specific monoclonal antibody. *Cell Motil. Cytoskel.* **27:**108–116.
28. Pinder, J. C., and Gratzer, W. B. (1983). Structural and dynamic states of actin in the erythrocyte. *J. Cell Biol.* **96:**768–775.
29. Siegel, D. L., and Branton, D. (1985). Partial purification and characterization of an actin-bundling protein, band 4.9, from human erythrocytes. *J. Cell Biol.* **100:**775–785.
30. Fowler, V. M., Davis, J. Q., and Bennett, V. (1985). Human erythrocyte myosin: Identification and purification. *J. Cell Biol.* **100:**47–55.
31. Fowler, V. M., and Bennett, V. (1984). Erythrocyte membrane tropomyosin: Purification and properties. *J. Biol. Chem.* **259:** 5978–5989.
32. Fowler, V. M. (1990). Tropomodulin. A cytoskeletal protein that binds to the end of erythrocyte tropomyosin and inhibits tropomyosin binding to actin. *J. Cell Biol.* **111:**471–482.
33. Fowler, V. M., Sussman, M. A., Miller, P. G., Flucher, B. E., and Daniels, M. P. (1993). Tropomodulin is associated with the free (pointed) ends of the thin filaments in rat skeletal muscle. *J. Cell Biol.* **120:**411–420.

34. Fairbanks, G., Steck, T. L., and Wallach, D. F. H. (1971). Electrophoretic analysis of the major polypeptides of the human erythrocyte membrane. *Biochemistry* **10:**2606–2624.
35. Leto, T. L., and Marchesi, V. T. (1984). A structural model of human erythrocyte protein 4.1. *J. Biol. Chem.* **259:**4603–4608.
36. Tyler, J. M., Reinhardt, B. N., and Branton, D. (1980). Associations of erythrocyte membrane proteins: Binding of purified bands 2.1 and 4.1 to spectrin. *J. Biol. Chem.* **255:**7034–7039.
37. Cohen, C. M., and Langley, R. C., Jr. (1984). Functional characterization of human erythrocyte spectrin α and β chains: Association with actin and erythrocyte protein 4.1. *Biochemistry* **23:**4488–4495.
38. Coleman, T. R., Harris, A. S., Mische, S. M., Mooseker, M. S., and Morrow, J. S. (1987). β-Spectrin bestows protein 4.1 sensitivity on spectrin–actin interactions. *J. Cell Biol.* **104:**519–526.
39. Correas, I., Leto, T. L., Speicher, D. W., and Marchesi, V. T. (1986). Identification of the functional site of erythrocyte protein 4.1 involved in spectrin–actin associations. *J. Biol. Chem.* **261:**3310–3315.
40. Correas, I., Speicher, D. W., and Marchesi, V. T. (1986). Structure of the spectrin–actin binding site of erythrocyte protein 4.1. *J. Biol. Chem.* **261:**13362–13366.
41. Subrahmanyan, G., Bertics, P. J., and Anderson, R. A. (1991). Phosphorylation of protein 4.1 on tyrosine-418 modulates its function in vitro. *Proc. Natl. Acad. Sci. USA* **88:**5222–5226.
42. Fowler, V., and Taylor, D. L. (1980). Spectrin plus band 4.1 crosslink actin: Regulation by micromolar calcium. *J. Cell Biol.* **85:** 361–376.
43. Tanaka, T., Kadowaki, K., Lazarides, E., and Sobue, K. (1991). Ca^{2+}-dependent regulation of the spectrin/actin interaction by calmodulin and protein 4.1. *J. Biol. Chem.* **266:**1134–1140.
44. Anderson, J. P., and Morrow, J. S. (1987). The interaction of calmodulin with human erythrocyte spectrin: Inhibition of protein 4.1-stimulated actin binding. *J. Biol. Chem.* **262:**6365–6372.
45. Mische, S. M., Mooseker, M. S., and Morrow, J. S. (1987). Erythrocyte adducin: A calmodulin-regulated actin-binding protein that stimulates spectrin–actin binding. *J. Cell Biol.* **105:**2837–2845.
46. Gardner, K., and Bennett, V. (1986). A new erythrocyte membrane-associated protein with calmodulin binding activity: Identification and purification. *J. Biol. Chem.* **261:**1339–1348.
47. Joshi, R., Gilligan, D. M., Otto, E., McLaughlin, T., and Bennett, V. (1991). Primary structure and domain organization of human α and β adducin. *J. Cell Biol.* **115:**665–675.
48. Cohen, C. M., and Foley, S. F. (1986). Phorbol ester- and Ca^{2+}-dependent phosphorylation of human red cell membrane skeletal proteins. *J. Biol. Chem.* **261:**7701–7709.
48a. Hughes, C. A., and Bennett, V. (1995). Adducin: A physical model with implications for function in assembly of spectrin–actin complexes. *J. Biol. Chem.* **270:**18990–18996.
49. Bennett, V. (1990). Spectrin-based membrane skeleton: A multipotential adaptor between plasma membrane and cytoplasm. *Physiol. Rev.* **70:**1029–1065.
50. Gardner, K., and Bennett, V. (1987). Modulation of spectrin–actin assembly by erythrocyte adducin. *Nature* **328:**359–362.
51. Cohen, C. M., and Foley, S. F. (1984). Biochemical characterization of complex formation by human erythrocyte spectrin, protein 4.1 and actin. *Biochemistry* **23:**6091–6098.
52. St-Onge, D., and Gicquaud, C. (1990). Research on the mechanism of interaction between actin and membrane lipids. *Biochem. Biophys. Res. Commun.* **167:**40–47.
53. Williamson, P., Bateman, J., Kozarsky, K., Mattocks, K., Hermanowicz, N., Choe, H.-R., and Schlegel, R. A. (1982). Involvement of spectrin in the maintenance of phase-state asymmetry in the erythrocyte membrane. *Cell* **30:**725–733.
54. Cohen, A. M., Liu, S.-C., Derick, L. H., and Palek, J. (1986). Ultrastructural studies of the interaction of spectrin with phosphatidylserine liposomes. *Blood* **68:**920–926.
55. Bennett, V., and Stenbuck, P. J. (1979). Identification and partial purification of ankyrin, the high affinity membrane attachment site for human erythrocyte spectrin. *J. Biol. Chem.* **254:**2533–2541.
56. Bennett, V., and Stenbuck, P. J. (1979). The membrane attachment protein for spectrin is associated with band 3 in human erythrocyte membranes. *Nature* **280:**468–473.
57. Bennett V. (1992). Ankyrins. Adaptors between diverse plasma membrane proteins and the cytoplasm. *J. Biol. Chem.* **267:**8703–8706.
57a. Platt, O. S., Lux, S. E., and Falcone, J. F. (1993). A highly conserved region of human erythrocyte ankyrin contains the capacity to bind spectrin. *J. Biol. Chem.* **268:**24421–24426.
58. Hall, T. G., and Bennett, V. (1987). Regulatory domains of erythrocyte ankyrin. *J. Biol. Chem.* **262:**10537–10545.
59. Davis, L. H., and Bennett, V. (1990). Mapping the binding sites of human erythrocyte ankyrin for the anion exchanger and spectrin. *J. Biol. Chem.* **265:**10589–10596.
60. Gallagher, P. G., Tse, W. T., Scarpa, A. L., Lux, S. E., and Forget, B. G. (1992). Large number of alternatively spliced isoforms of the regulatory region of human erythrocyte ankyrin. *Trans. Assoc. Am. Physicians* **105:**268–277.
61. Kennedy, S. P., Warren, S. L., Forget, B. G., and Morrow, J. S. (1991). Ankyrin binds to the 15th repetitive unit of erythroid and nonerythroid β-spectrin. *J. Cell Biol.* **115:**267–277.
62. Bennett, V., and Stenbuck, P. J. (1980). Association between ankyrin and the cytoplasmic domain of band 3 isolated from the human erythrocyte membrane. *J. Biol. Chem.* **255:**6424–6432.
63. Davis, L., Lux, S. E., and Bennett, V. (1989). Mapping the ankyrin-binding site of the human erythrocyte anion exchanger. *J. Biol. Chem.* **264:**9655–9672.
64. Golan, D. E., and Veatch, W. (1980). Lateral mobility of band 3 in the human erythrocyte membrane studied by fluorescence photobleaching recovery: Evidence for control by cytoskeletal interactions. *Proc. Natl. Acad. Sci. USA* **77:**2537–2541.
65. Tsuji, A., and Ohnishi, S.-I. (1986). Restriction of the lateral motion of band 3 in the erythrocyte membrane by the cytoskeletal network: Dependence on spectrin association state. *Biochemistry* **25:** 6133–6139.
66. Anderson, R. A., and Lovrien, R. E. (1984). Glycophorin is linked by band 4.1 protein to the human erythrocyte membrane skeleton. *Nature* **307:**655–658.
67. Anderson, R. A., Correas, I., Mazzucco, C., Castle, J. D., and Marchesi, V. T. (1988). Tissue-specific analogues of erythrocyte protein 4.1 retain functional domains. *J. Cell. Biochem.* **37:**269–284.
68. Pasternack, G. R., Anderson, R. A., Leto, T. L., and Marchesi, V. T. (1985). Interactions between protein 4.1 and band 3: An alternative binding site for an element of the membrane skeleton. *J. Biol. Chem.* **260:**3676–3683.
69. Steiner, J. P., and Bennett, V. (1988). Ankyrin-independent membrane protein-binding sites for brain and erythrocyte spectrin. *J. Biol. Chem.* **263:**14417–14425.
69a. Lombardo, C. R., Weed, S. A., Kennedy, S. P., Forget, B. G., and Morrow, J. S. (1994). Beta II-spectrin (fodrin) and beta I sigma II spectrin (muscle) contain NH_2- and COOH-terminal membrane association domains (MAD1 and MAD2). *J. Biol. Chem.* **269:**29212–29219.
69b. Davis, L. H., and Bennett, V. (1994). Identification of two regions of β_G spectrin that bind to distinct sites in brain membranes. *J. Biol. Chem.* **269:**4409–4416.
70. Gallagher, P. G., and Forget, B. G. (1993). Spectrin genes in health and disease. *Semin. Hematol.* **30:**4–21.
71. Lux, S. E., Tse, W. T., Menninger, J. C., John, K. M., Harris, P., Shalev, O., Chilcote, R. R., Marchesi, S. L., Watkins, P. C., Bennett, V., McIntosh, S., Collins, F. S., Francke, U., Ward, D. C., and Forget, B. G. (1990). Hereditary spherocytosis associated with

deletion of human erythrocyte ankyrin gene on chromosome 8. *Nature* **345:**736–739.
72. Conboy, J., Mohandas, N., Tchernia, G., and Kan, Y. W. (1986). Molecular basis of hereditary elliptocytosis due to protein 4.1 deficiency. *N. Engl. J. Med.* **315:**680–685.
73. Winkelmann, J. C., and Forget, B. G. (1993). Erythroid and nonerythroid spectrins. *Blood* **81:**3173–3185.
74. Levine, J., and Willard, M. (1981). Fodrin: Axonally transported polypeptides associated with the internal periphery of many cells. *J. Cell Biol.* **90:**631–643.
75. Goodman, S. R., Zagon, I. S., and Kulikowski, R. R. (1981). Identification of a spectrin-like protein in nonerythroid cells. *Proc. Natl. Acad. Sci. USA* **78:**7570–7574.
76. Davis, J., and Bennett, V. (1983). Brain spectrin. Isolation of subunits and formation of hybrids with erythrocyte spectrin subunits. *J. Biol. Chem.* **258:**7757–7766.
77. Wasenius, V.-M., Saraste, M., Salven, P., Eramaa, M., Holm, L., and Lehto, V.-P. (1989). Primary structure of the brain α-spectrin. *J. Cell Biol.* **108:**79–93.
78. Harris, A. S., Croall, D. E., and Morrow, J. S. (1988). The calmodulin binding site in α-fodrin is near the calcium-dependent protease-I cleavage site. *J. Biol. Chem.* **263:**15754–15761.
78a. Merilainen, J., Palovuori, R., Sormunen, R., Wasneius, V.-M., and Lehto, V.-P. (1993). Binding of the α-fodrin SH3 domain to the leading lamellae of locomoting chicken fibroblasts. *J. Cell Sci.* **105:**647–654.
79. Harris, A. S., Croall, D. E., and Morrow, J. S. (1989). Calmodulin regulates fodrin susceptibility to cleavage by calcium-dependent protease I. *J. Biol. Chem.* **264:**17401–17408.
80. Harris, A. S., and Morrow, J. S. (1990). Calmodulin and calcium-dependent protease I coordinately regulate the interaction of fodrin with actin. *Proc. Natl. Acad. Sci. USA* **87:**3009–3013.
81. Steiner, J. P., Walke, H. T., Jr., and Bennett, V. (1989). Calcium/calmodulin inhibits direct binding of spectrin to synaptosomal membranes. *J. Biol. Chem.* **264:**2783–2791.
81a. Clark, M. B., Ma, Y., Bloom, M. L., Barker, J. E., Zagon, I. S., Zimmer, W. E., and Goodman, S. R. (1994). Brain erythroid α spectrin: Identification, compartmentalization and β spectrin associations. *Brain Res.* **663:**223–236.
82. Winkelmann, J. C., Costa, F. F., Linzie, B. L., and Forget, B. G. (1990). β-Spectrin in human skeletal muscle—Tissue specific differential processing of a 3′ β-spectrin pre-messenger RNA generates a β-spectrin isoform with a unique carboxyl terminus. *J. Biol. Chem.* **265:**20449–20454.
83. Porter, G. A., Dmytrenko, G. M., Winkelmann, J. C., and Bloch, R. J. (1991). Dystrophin colocalizes with β-spectrin in distinct subsarcolemmal domains in mammalian skeletal muscle. *J. Cell Biol.* **117:**997–1005.
84. Malchiodi-Albedi, F., Ceccarini, M., Winkelmann, J. C., Morrow, J. S., and Petrucci, T. C. (1993). The 270 kDa splice variant of erythrocyte β-spectrin (βIΣ2) segregates in vivo and in vitro to specific domains of cerebellar neurons. *J. Cell Sci.* **106:**67–78.
85. Moon, R. T., and McMahon, A. P. (1990). Generation of diversity in nonerythroid spectrins. Multiple polypeptides are predicted by sequence analysis of cDNAs encompassing the coding region of human non-erythroid α-spectrin. *J. Biol. Chem.* **265:**4427–4432.
86. Bloch, R. J., and Morrow, J. S. (1989). An unusual β-spectrin associated with clustered acetylcholine receptors. *J. Cell Biol.* **108:** 481–493.
87. Glenney, J., and Glenney, P. (1984). Co-expression of spectrin and fodrin in Friend erythroleukemic cells treated with DMSO. *Exp. Cell Res.* **152:**15–21.
88. Riederer, B. M., Zagon, I. S., and Goodman, S. R. (1986). Brain spectrin (240/235) and brain spectrin (240/235E): Two distinct spectrin subtypes with different locations within mammalian neural cells. *J. Cell Biol.* **102:**2088–2097.
89. Riederer, B. M., Lopresti, L. L., Krebs, K. E., Zagon, I. S., and Goodman, S. R. (1988). Brain spectrin (240/235) and brain spectrin (240/235E): Conservation of structure and location within mammalian neural tissue. *Brain Res. Bull.* **21:**607–616.
90. Kunimoto, M., Otto, E., and Bennett, V. (1991). A new 440-kD isoform is the major ankyrin in neonatal rat brain. *J. Cell Biol.* **115:** 1319–1331.
91. Birkenmeier, C. S., White, R. A., Peters, L. L., Barker, J. E., and Lux, S. E. (1993). Complex patterns of sequence variation and multiple 5′ and 3′ ends are found among transcripts of the erythroid ankyrin gene. *J. Biol. Chem.* **268:**9533–9540.
92. Otto, E., Kunimoto, M., McLaughlin, T., and Bennett, V. (1991). Isolation and characterization of cDNAs encoding human brain ankyrins reveal a family of alternatively spliced genes. *J. Cell Biol.* **114:**241–253.
92a. Kordeli, E., Lambert, S., and Bennett, V. (1995). Ankyrin_G: A new ankyrin gene with neural-specific isoforms localized at the axonal initial segment and node of Ranvier. *J. Biol. Chem.* **270:**2352–2359.
92b. Peters, L. L., John, K. M., Lu, F. M., Eicher, E. M., Higgins, A., Yialamas, M., Turtzo, L. C., Otsuka, A. J., and Lux, S. E. (1995). Ank3 (epithelial ankyrin), a widely distributed new member of the ankyrin gene family and the major ankyrin in kidney, is expressed in alternatively spliced forms, including forms that lack the repeat domain. *J. Cell Biol.* **130:**313–330.
93. Ngai, J., Stack, J. H., Moon, R. T., and Lazarides, E. (1987). Regulated expression of multiple chicken erythroid membrane skeletal protein 4.1 variants is governed by differential RNA processing and translational control. *Proc. Natl. Acad. Sci. USA* **84:**4432–4436.
94. Conboy, J. G., Chan, J., Mohandas, N., and Kan, Y. W. (1988). Multiple protein 4.1 isoforms produced by alternative splicing in human erythroid cells. *Proc. Natl. Acad. Sci. USA* **85:**9062–9065.
95. Tang, T. K., Leto, T. L., Correas, I., Alonso, M. A., Marchesi, V. T., and Benz, E. J., Jr. (1988). Selective expression of an erythroid-specific isoform of protein 4.1. *Proc. Natl. Acad. Sci. USA* **85:** 3713–3717.
96. Dhermy, D. (1991). The spectrin super-family. *Biol. Cell* **71:**249–254.
97. Pumplin, D. W., and Bloch, R. J. (1993). The membrane skeleton. *Trends Cell Biol.* **3:**113–117.
98. Ahn, A. H., and Kunkel, L. M. (1993). The structural and functional diversity of dystrophin. *Nature Genet.* 3:283–291.
98a. Schofield, J. N., Blake, D. J., Simmons, C., Morris, G. E., Tinsley, J. M., Davies, K. E., and Edwards, Y. H. (1994). Apo-dystrophin-1 and apo-dystrophin-2, products of the Duchenne muscular dystrophy locus: Expression during mouse embryogenesis and in cultured cell lines. *Hum. Mol. Genet.* **3:**1309–1316.
98b. Lidov, H. G., Selig, S., and Kunkel, L. M. (1995). Dp140: A novel 140 kDa CNS transcript from the dystrophin locus. *Hum. Mol. Genet.* **4:**329–335.
99. Koening, M., Monaco, A. P., and Kunkel, L. M. (1988). The complete sequence of dystrophin predicts a rod-shaped cytoskeletal protein. *Cell* **53:**219–228.
99a. Kahana, E., and Gratzer, W. B. (1995). Minimal folding unit of dystrophin rod domain. *Biochemistry* **34:**8110–8114.
100. Hemmings, L., Kuhlman, P. A., and Critchely, D. R. (1992). Analysis of the actin-binding domain of α-actinin by mutagenesis and demonstration that dystrophin contains a functionally homologous domain. *J. Cell Biol.* **116:**1369–1380.
101. Way, M., Pope, B., Cross, R. A., Kendrick-Jones, J., and Weeds, A. G. (1992). Expression of the N-terminal domain of dystrophin in E. coli and demonstration of binding to F-actin. *FEBS Lett.* **301:**243–245.

102. Ervasti, J. M., and Campbell, K. P. (1993). A role for the dystrophin–glycoprotein complex as a transmembrane linker between laminin and actin. *J. Cell Biol.* **122:**809–823.
103. Campbell, K. P., and Kahl, S. D. (1989). Association of dystrophin and an integral membrane glycoprotein. *Nature* **338:**259–262.
103a. Yoshida, M, and Ozawa, E. (1990). Glycoprotein complex anchoring dystrophin to sarcolemma. *J. Biochem.* **108:**748–752.
104. Ervasti, J. M., and Campbell, K. P. (1991). Membrane organization of the dystrophin–glycoprotein complex. *Cell* **66:**1121–1131.
105. Adams, M. E., Butler, M. H., Dwyer, T. M., Peters, M. F., Murnane, A. A., and Froehner, S. C. (1993). Two forms of mouse syntrophin, a 58 kd dystrophin-associated protein, differ in primary structure and tissue distribution. *Neuron* **11:**531–540.
106. Kramarcy, N. R., Vidal, A., Froehner, S. C., and Sealock, R. (1994). Association of utrophin and multiple dystrophin short forms with the mammalian Mr 58,000 dystrophin-associated protein (syntrophin). *J. Biol. Chem.* **269:**2870–2876.
106a. Yang, B., Jung, D., Rafael, J. A., Chamerlain, J. S., and Campbell, K. P. (1995). Identification of α-syntrophin binding to syntrophin triplet, dystrophin and utrophin. *J. Biol. Chem.* **270:**4975–4978.
106b. Suzuki, A., Yoshida, M., and Ozawa, E. (1995). Mammalian α1- and β1-syntrophin bind to the alternative splice-prone region of the dystrophin COOH terminus. *J. Cell Biol.* **128:**373–381.
106c. Ahn, A., and Kunkel, L. M. (1995). Syntrophin binds to an alternatively spliced exon of dystrophin. *J. Cell Biol.* **128:**363–371.
106d. Suzuki, A., Yoshida, M., Yamamoto, H., and Ozawa, E. (1992). Glycoprotein-binding site of dystrophin is confined to the cysteine-rich domain and the first half of the carboxy-terminal domain. *FEBS Lett.* **308:**154–160.
106e. Suzuki, A., Yoshida, M., Hayashi, K., Mizuno, Y., Hagiwara, Y., and Ozawa, E. (1994). Molecular organization at the glycoprotein-complex-binding site of dystrophin. Three dystrophin associated proteins bind directly to the carboxy-terminal portion of dystrophin. *Eur. J. Biochem.* **220:**283–292.
107. Ibraghimov-Beskrovnaya, O., Ervasti, J. M., Leveille, C. J., Slaughter, C. A., Sernett, S. W., and Campbell, K. P. (1992). Primary structure of dystrophin-associated glycoproteins linking dystrophin to the extracellular matrix. *Nature* **355:**696–702.
108. Pons, F., Augier, N., Heilig, R., Leger, J., Mornet, D., and Leger, J. J. (1990). Isolated dystrophin molecules as seen by electron microscopy. *Proc. Natl. Acad. Sci. USA* **87:**7851–7855.
109. Sato, O., Nonomura, Y., Kimura, S., and Maruyama, K. (1992). Molecular shape of dystrophin. *J. Biochem.* **112:**631–636.
110. Dmytrenko, G. M., Pumplin, D. W., and Bloch, R. J. (1993). Membrane and cytoskeletal localization of dystrophin in cultured rat myotubes. *J. Neurosci.* **13:**547–558.
111. Wakayama, Y. (1991). Dystrophin is localized to the plasma membrane of human skeletal muscle fibers by electron-microscopic cytochemical study. *Muscle Nerve* **14:**576–577.
112. Wakayama, Y., and Shibuya, S. (1991). Gold-labelled dystrophin molecule in muscle plasmalemma of mdx control mice as seen by electron microscopy of deep etching replica. *Acta Neuropathol.* **82:**178–184.
113. Pardo, J. V., Siliciano, J. D., and Craig, S. W. (1983). A vinculin-containing cortical lattice in skeletal muscle: Transverse lattice elements ("costameres") mark sites of attachment between myofibrils and sarcolemma. *Proc. Natl. Acad. Sci. USA* **80:**1008–1012.
114. Pardo, J. V., Siliciano, J. D., and Craig, S. W. (1983). Vinculin is a component of an extensive network of myofibril–sarcolemma attachment regions in cardiac muscle fibers. *J. Cell Biol.* **97:**1081–1088.
115. Nelson, W. J., and Lazarides, E. (1984). Goblin (ankyrin) in striated muscle: Identification of the potential membrane receptor for erythroid spectrin in muscle cells. *Proc. Natl. Acad. Sci. USA* **81:** 3292–3296.
116. Terracio, L., Gullberg, D., Rubin, K., Craig, S., and Borg, T. K. (1989). Expression of collagen adhesion proteins and their association with the cytoskeleton in cardiac myocytes. *Anat. Rec.* **223:**62–71.
117. Lakonishok, M., Muschler, J., and Horwitz, A. F. (1992). The α5β1 integrin associates with a dystrophin-containing lattice during muscle development. *Dev. Biol.* **152:**209–220.
118. Senter, L., Luise, M., Presotto, C., Betto, R., Teresi, A., Ceoldo, S., and Salviati, G. (1993). Interaction of dystrophin with cytoskeletal proteins: Binding to talin and actin. *Biochem. Biophys. Res. Commun.* **192:**899–904.
119. Kramarcy, N. R, and Sealock, R. (1990). Dystrophin as a focal adhesion protein. Colocalization with talin and the Mr 48,000 sarcolemmal protein in cultured Xenopus muscle. *FEBS Lett.* **274:**171–174.
120. Straub, V., Bittner, R. E., Leger, J. J., and Voit, T. (1992). Direct visualization of the dystrophin network on skeletal muscle fiber membrane. *J. Cell Biol.* **119:**1183–1191.
121. North, A. J., Galazkiewicz, B., Byers, T. J., Glenney, J. R., Jr., and Small, J. V. (1993). Complementary distributions of vinculin and dystrophin define two distinct sarcolemma domains in smooth muscle. *J. Cell Biol.* **120:**1159–1167.
122. Khurana, T. S., Hoffman, E. P., and Kunkel, L. M. (1990). Identification of a chromosome 6-encoded dystrophin-related protein. *J. Biol. Chem.* **265:**16717–16720.
123. Love, D. R., Hill, D. F., Dickson, G., Spurr, N. K., Byth, B. C., Marsden, R. F., Walsh, F. S., Edwards, Y. H., and Davies, K. E. (1989). An autosomal transcript in skeletal muscle with homology to dystrophin. *Nature* **339:**55–58.
124. Man, N. T., Thanh, L. T., Blake, D. J., Davies, K. E., and Morris, G. E. (1992). Utrophin, the autosomal homologue of dystrophin, is widely-expressed and membrane-associated in cultured cell lines. *FEBS Lett.* **313:**19–22.
124a. Man, N. T., Helliwell, T. R., Simmons, C., Winder, S. J., Kendrick-Jones, J., Davies, K. E., and Morris, G. E. (1995). Full-length and short forms of utrophin, the dystrophin-related protein. *FEBS Lett.* **358:**262–266.
124b. Winder, S. J., Hemmings, L., Maciver, S. K., Bolton, S. J., Simmons, C., Tinsley, J. M., Davies, K. E., Critchley, D. R., and Kendrick-Jones, J. (1995). Utrophin actin binding domain: Analysis of actin binding and cellular targeting. *J. Cell Sci.* **108:**63–71.
124c. Matsumura, K., Ervasti, J. M., Ohlendieck, K., Kahl, S. D., and Campbell, K. P. (1992). Association of dystrophin-related protein with dystrophin-associated proteins in mdx mouse muscle. *Nature* **360:**588–591.
125. Ohlendieck, K., Ervasti, J. M., Matsumura, K., Kahl, S. D., Leveille, C. J., and Campbell, K. P. (1991). Dystrophin-related protein is localized to neuromuscular junctions of skeletal muscle. *Neuron* **7:**499–508.
126. Campanelli, J. T., Roberds, S. L., Campbell, K. P., and Scheller, R. H. (1994). A role for dystrophin-associated glycoproteins and utrophin in agrin-induced AChR clustering. *Cell* **77:**663–674.
127. Phillips, W. D., Noakes, P. G., Roberds, S. L., Campbell, K. P., and Merlie, J. P. (1993). Clustering and immobilization of acetylcholine receptors by the 43-kD protein: A possible role for dystrophin-related protein. *J. Cell Biol.* **123:**729–740.
127a. Peters, M. F., Kramarcy, N. R., Sealock, R., and Froehner, S. C. (1944). β2-Syntrophin: Localization at the neuromuscular junction in skeletal muscle. *Neuroreport* **5:**1577–1580.
128. Pons, F., Augier, N., Leger, J. O. C., Robert, A., Tome, F. M. S., Fardeau, M., Voit, T., Nicholson, L. V. B., Mornet, D., and Leger, J. J. (1991). A homologue of dystrophin is expressed at the neuromuscular junctions of normal individuals and DMD patients, and of normal and mdx mice. *FEBS Lett.* **282:**161–165.
129. Lidov, H. G. W., Byers, T. J., Watkins, S. C., and Kunkel, L. M. (1990). Localization of dystrophin to postsynaptic regions of central nervous system cortical neurons. *Nature* **348:**725–728.

130. Khurana, T. S., Watkins, S. C., and Kunkel, L. M. (1992). The subcellular distribution of chromosome 6-encoded dystrophin-related protein in the brain. *J. Cell Biol.* **119:**357–366.
131. Matsudaira, P. (1991). Modular organization of actin crosslinking proteins. *Trends Biochem. Sci.* **17:**87–92.
132. Davison, M. D., and Critchley, D. R. (1988). α-Actinins and the DMD protein contain spectrin-like repeats. *Cell* **52:**159–160.
133. Davison, M.D., Baron, M. D., Critchley, D. R., and Wootton, J. C. (1989). Structural analysis of homologous repeated domains in α-actinin and spectrin. *Int. J. Biol. Macromol.* **11:**81–90.
133a. Gilmore, A. P., Parr, T., Patel, B., Gratzer, W. B., and Critchley, D. R. (1994). Analysis of the phasing of four spectrin-like repeats in α-actinin. *Eur. J. Biochem.* **225:**235–242.
134. Otey, C. A., Pavalko, F. M., and Burridge, K. (1990). An interaction between α-actinin and the β-1 integrin subunit in vitro. *J. Cell Biol.* **111:**721–729.
135. Otey, C. A., Vasquez, G. B., Burridge, K., and Erickson, B. W. (1993). Mapping of the α-actinin binding site within the beta 1 integrin cytoplasmic domain. *J. Biol. Chem.* **268:**21193–21197.
136. Lazarides, E., and Burridge, K. (1975). α-Actinin: Immunofluorescent localization of a muscle structural protein in nonmuscle cells. *Cell* **6:**289–298.
137. Burridge, K., Fath, K., Kelly, T., Nuckolls, G., and Turner, C. (1988). Focal adhesions: Transmembrane junctions between the extracellular matrix and the cytoskeleton. *Annu. Rev. Cell Biol.* **4:**487–525.
138. Street, S. F. (1983). Lateral transmission of tension in frog myofibers: A myofibrillar network and transverse cytoskeletal connections are possible transmitters. *J. Cell. Physiol.* **114:**346–364.
139. Pierobon-Bormioli, S. (1981). Transverse sarcomere filamentous systems: 'Z- and M-cables.' *J. Musc. Res. Cell Motil.* **2:**401–413.
140. Garamvölgi, N. (1965). Inter-Z bridges in the flight muscle of the bee. *J. Ultrastruct. Res.* **13:**435–443.
141. Shear, C. R., and Bloch, R. J. (1985). Vinculin in subsarcolemmal densities in chicken skeletal muscle: Localization and relationship to intracellular and extracellular structures. *J. Cell Biol.* **101:**240–256.
141a. Pasternak, C., Wong, S., and Elson, E. L. (1995). Mechanical function of dystrophin in muscle cells. *J. Cell Biol.* **128:**355–361.
142. Hoffman, E. P., Fischbeck, K. H., Brown, R. H., Johnson, M., Medori, R., Loike, J. D., Harris, J. B., Waterston, R., Brooke, M., Specht, L., Kupsky, W., Chamberlain, J., Caskey, C. T., Shapiro, F., and Kunkel, L. M. (1988). Characterization of dystrophin in muscle-biopsy specimens from patients with Duchenne's or Becker's muscular dystrophy. *N. Engl. J. Med.* **318:**1363–1368.
143. England, S. B., Nicholson, L. V. B., Johnson, M. A., Forrest, S. M., Love, D. R., Zubrzycka-Gaarn, E. E., Bulman, D. E., Harris, J. B., and Davies, K. E. (1990). Very mild muscular dystrophy associated with the deletion of 46% of dystrophin. *Nature* **343:** 180–182.
143a. Matsumura, K., Tome, F. M., Ionasescu, V., Ervasti, J. M., Anderson, R. D., Romero, N. B., Simon, D., Recan, D., Kaplan, J. C., Fardeau, M., *et al.* (1993). Deficiency of dystrophin-associated proteins in Duchenne muscular dystrophy patients lacking COOH-terminal domains of dystrophin. *J. Clin. Invest.* **92:**866–871.
143b. Matsumura, K., Burghes, A. H., Mora, M., Tome, F. M., Morandi, L., Cornello, F., Leturcq, F., Jeanpierre, M., Kaplan, J. C., Reinert, P., *et al.* (1994). Immunohistochemical analysis of dystrophin-associated proteins in Becker/Duchenne muscular dystrophy with huge in-frame deletions in the NH_2-terminal and rod domains of dystrophin. *J. Clin. Invest.* **93:**99–105.
143c. Kobzik, L., Reid, M. B., Bredt, D. S., and Stamler, J. S. (1994). Nitric oxide in skeletal muscle. *Nature* **372:**546–548.
143d. Brenman, J. E., Chao, D. S., Zia, H. H., Aldape, K., and Bredt, D. S. (1995). Nitric oxide synthase complexed with dystrophin and absent from muscle sarcolemma in Duchenne muscular dystrophy. *Cell* **82:**743–752.
144. Florence, J. M., Fox, P. T., Planer, G. J., and Brooke, M. H. (1985). Activity creatine kinase, and myoglobin in Duchenne muscular dystrophy: A clue to etiology? *Neurology* **35:**758–761.
145. Webster, C., Silberstein, L., Hays, A. P., and Blau, H. M. (1988). Fast muscle fibers are preferentiailly affected in Duchenne's muscular dystrophy. *Cell* **52:**503–513.
146. Weller, B., Karparti, G., and Carpenter, S. (1990). Dystrophin-deficient mdx muscle fibers are preferentially vulnerable to necrosis induced by experimental lengthening contractions. *J. Neuro. Sci.* **100:**9–13.
147. Froehner, S. C., Murnane, A. A., Tobler, M., Peng H. B., and Sealock, R. (1987). A postsynaptic M_r 58,000 (58K) protein concentrated at acetylcholine receptor-rich sites in Torpedo electroplaques and skeletal muscle. *J. Cell Biol.* **104:**1633–1646.
148. Ohlendieck, K., and Campbell, K. P. (1991). Dystrophin-associated proteins are greatly reduced in skeletal muscle from mdx mice. *J. Cell Biol.* **115:**1685–1694.
149. Ahn, A. H., Yoshida, M., Anderson, M. S., Feener, C. A., Selig, S., Hagiwara, Y., Ozawa, E., and Kunkel, L. M. (1994). Cloning of human basic A1, a distinct 59-kDa dystrophin-associated protein encoded on chromosome 8Q23-24. *Proc. Natl. Acad. Sci. USA* **91:** 4446–4450.
150. Yang, B., Ibraghimov-Beskrovnaya, O., Moomaw, C. R., Slaughter, C. A., and Campbell, K. P. (1994). Heterogeneity of the 59-kDa dystrophin-associated protein revealed by cDNA cloning and expression. *J. Biol. Chem.* **269:**6040–6044.
151. Matsumura, K., Tome, F. M. S., Collin, H., Azibi, K., Chaouch, M., Kaplan, J.-C., Fardeau, M., and Campbell, K. P. (1992). Deficiency of the 50K dystrophin-associated glycoprotein in severe childhood autosomal recessive muscular dystrophy. *Nature* **359:**320–322.
151a. Yamanouchi, Y., Mizuno, Y., Yamamoto, H., Takemitsu, M., Yoshida, M., Nonaka, U., and Ozawa, E. (1994). Selective defect in dystrophin-associated glycoproteins 50DAG (A2) and 35DAG (A4) in the dystrophic hamster: An animal model for severe childhood autosomal recessive muscular dystrophy (SCARMD). *Neuromusc. Disorders* **4:**49–54.
151b. Roberds, S. L., and Campbell, K. P. (1995). Adhalin mRNA and cDNA sequence are normal in the cardiomyopathic hamster. *FEBS Lett.* **364:**245–249.
151c. Campbell, K. P. (1995). Three muscular dystrophies: Loss of cytoskeleton–extracellular matrix linkage. *Cell* **80:**675–679.
152. Nelson, W. J., Shore, E. S., Wang, A. Z., and Hammerton, R. W. (1990). Identification of a membrane-cytoskeletal complex containing the cell adhesion molecule uvomorulin (E-cadherin), ankyrin, and fodrin in Madin-Darby canine kidney epithelial cells. *J. Cell Biol.* **110:**349–357.
153. Nelson, W. J., and Hammerton, R. W. (1989). A membrane-cytoskeletal complex containing Na,K-ATPase, ankyrin and fodrin in Madin-Darby canine kidney (MDCK) cells: Implications for the biogenesis of epithelial cell polarity. *J. Cell Biol.* **108:**893–902.
154. Nelson, W. J., and Veshnock, P. J. (1987). Ankyrin binding to (Na^+ + K^+) ATPase and implications for the organization of membrane domains in polarized cells. *Nature* **328:**533–536.
155. Morrow, J. S., Cianci, C. D., Ardito, T., Mann, A. S., and Kashgarian, M. (1989). Ankyrin links fodrin to the α subunit of Na,K-ATPase in Madin-Darby canine kidney cells and in intact renal tubule cells. *J. Cell Biol.* **108:**455–465.
156. Davis, J., Davis, L., and Bennett, V. (1989). Diversity in membrane binding sites of ankyrins. Brain ankyrin, erythrocyte ankyrin, and processed erythrocyte ankyrin associate with distinct sites in kidney microsomes. *J. Biol. Chem.* **264:**6417–6426.
156a. Hu, R.-J., Moorthy, S., and Bennett, V. (1995). Expression of functional domains of $beta_G$-spectrin disrupts epithelial morphology in cultured cells. *J. Cell Biol.* **128:**1069–1080.

157. Gumbiner, B. (1987). Structure, biochemistry, and assembly of epithelial tight junctions. *Am. J. Physiol.* **253:**C749–C758.
158. Srinivasan, Y., Elmer, L., Davis, J., Bennett, V., and Angelides, K. (1988). Ankyrin and spectrin associate with voltage-dependent sodium channels in brain. *Nature* **333:**177–180.
159. Srinivasan, Y., Lewallen, M., and Angelides, K. J. (1992). Mapping the binding site on ankyrin for the voltage-dependent sodium channel from brain. *J. Biol. Chem.* **267:**7483–7489.
160. Black, J. A., Friedman, B., Waxman, S. G., Elmer, L. W., and Angelides, K. J. (1989). Immuno-ultrastructural localization of sodium channels at nodes of Ranvier and perinodal astrocytes in rat optic nerve. *Proc. Roy. Soc. London Ser. B* **238:**39–51.
161. Kordeli, E., and Bennett, V. (1991). Distinct ankyrin isoforms at neuron cell bodies and nodes of Ranvier resolved using ankyrin-deficient mice. *J. Cell Biol.* **1145:**1243–1259.
162. Ichimura, T., and Ellisman, M. H. (1991). Three-dimensional fine structure of cytoskeletal–membrane interactions at nodes of Ranvier. *J. Neurocytol.* **20:**667–681.
163. Davis, J. Q., and Bennett, V. (1984). Brain ankyrin. A membrane-associated protein with binding sites for spectrin, tubulin, and the cytoplasmic domain of the erythrocyte anion channel. *J. Biol. Chem.* **259:**13550–13559.
164. Langley, R. C., Jr., and Cohen, C. M. (1986). Association of spectrin with desmin intermediate filaments. *J. Cell Biochem.* **30:**101–109.
165. Davis, J. Q., McLaughlin, T., and Bennett, V. (1993). Ankyrin-binding proteins related to nervous system cell adhesion molecules: Candidates to provide transmembrane and intercellular connections in adult brain. *J. Cell Biol.* **121:**121–133.
166. Peters, L. L., Birkenmeier, C. S., Bronson, R. T., White, R. A., Lux, S. E., Otto, E., Bennett, V., Higgins, A., and Barker, J. E. (1991). Purkinje cell degeneration associated with erythroid ankyrin deficiency in nb/nb mice. *J. Cell Biol.* **114:**1233–1241.
167. Bloch, R. J., and Geiger, B. (1980). The localization of acetylcholine receptor clusters in areas of cell–substrate contact in cultures of rat myotubes. *Cell* **21:**25–35.
168. Pollerberg, G. E., Burridge, K., Krebs, K. E., Goodman, S. R., and Schachner, M. (1987). The 180-kD component of the neural cell adhesion molecule N-CAM is involved in cell–cell contacts and cytoskeleton–membrane interactions. *Cell Tissue Res.* **250:**227–236.
169. Lombardo, C. R., Rimm, D. L., Kennedy, S. P., Forget, B. G., and Morrow, J. S. (1993). Ankyrin independent membrane sites for non-erythroid spectrin. *Mol. Biol. Cell* **4:**57.
170. Otto, J. J. (1990). Vinculin. *Cell Motil. Cytoskel.* **16:**1–6.
171. Geiger, B. (1979). A 130K protein from chicken gizzard: Its localization at the termini of microfilament bundles in cultured chicken cells. *Cell* **18:**193–205.
172. Feramisco, J. R., and Burridge, K. (1980). A rapid purification of α-actinin, filamin, and a 130,000-dalton protein from smooth muscle *J. Biol. Chem.* **255:**1194–1199.
173. Wachsstock, D. H., Wilkins, J. A., and Lin, S. (1987). Specific interaction of vinculin with α-actinin. *Biochem. Biophys. Res. Commun.* **146:**554–560.
174. Geiger, B. (1982) Microheterogeneity of avian and mammalian vinculin distinctive subcellular distribution of different isovinculins. *J. Mol. Biol.* **159:**685–701.
175. Price, G. J., Jones, P., Davison, M. D., Patel, B., Bendori, R., Geiger, B., and Critchley, D. R. (1989). Primary sequence and domain structure of chicken vinculin. *Biochem. J.* **259:**453–461.
176. Bendori, R., Salomon, D., and Geiger, B. (1989). Identification of two distinct functional domains on vinculin involved in its association with focal contacts. *J. Cell Biol.* **108:**2383–2393.
177. Molony, L., and Burridge, K. (1985). Molecular shape and self-association of vinculin and meta-vinculin. *J. Cell Biochem.* **29:** 31–36.
178. Turner, C. E., Glenney, J. R., and Burridge, K. (1990). Paxillin—a new vinculin-binding protein present in focal adhesions. *J. Cell Biol.* **111:**1059–1068.
179. Koteliansky, V. E., Ogryzko, E. P., Zhidkova, N. I., Weller, P. A., Critchley, D. R., Vancompernolle, K., Vandekerckhove, J., Strasser, P., Way, M., Gimona, M., and Small, J. V. (1992). An additional exon in the human vinculin gene specifically encodes meta-vinculin-specific difference peptide. Cross-species comparison reveals variable and conserved motifs in the meta-vinculin insert. *Eur. J. Biochem.* **204:**767–772.
180. Feramisco, J. R., Smart, J. E., Burridge, K., Helfman, D. M., and Thomas, G. P. (1982). Co-existence of vinculin and a vinculin-like protein of higher molecular weight in smooth muscle. *J. Biol. Chem.* **257:**11024–11031.
181. Glukhova, M. A., Kabakov, A. E., Belkin, A. M., Frid, M. G., Ornatsky, O. I., Zhidkova, N. I., and Koteliansky, V. E. (1986). Meta-vinculin distribution in adult human tissues and cultured cells. *FEBS Lett.* **207:**139–141.
182. Gilmore, A. P., Jackson, P., Waites, G. T., and Critchley, D. R. (1992). Further characterisation of the talin-binding site in the cytoskeletal protein vinculin. *J. Cell Sci.* **103:**719–731.
183. Beckerle, M. C., and Yeh, R. K. (1990). Talin: Role at sites of cell–substratum adhesion. *Cell Motil. Cytoskel.* **16:**7–13.
184. Molony, L., McCaslin, D., Abernethy, J., Paschal, B., and Burridge, K. (1987). Properties of talin from chicken gizzard smooth muscle. *J. Biol. Chem.* **282:**7790–7795.
185. Burridge, K., and Mangeat, P. (1984). An interaction between vinculin and talin. *Nature* **308:**744–746.
186. Burridge, K., Turner, C. E., and Romer, L. H. (1992). Tyrosine phosphorylation of paxillin and pp125FAK accompanies cell adhesion to extracellular matrix: A role in cytoskeletal assembly. *J. Cell Biol.* **119:**893–903.
187. Lo, S. H., Janmey, P. A., Hartwig, J. H., and Chen, L. B. (1994). Interactions of tensin with actin and identification of its three distinct actin-binding domains. *J. Cell Biol.* **125:**1067–1075.
187a. Chuang, J. Z., Lin, D. C., and Lin, S. (1995). Molecular cloning, expression, and mapping of the high affinity, actin-capping domain of chicken cardiac muscle. *J. Cell Biol.* **128:**1095–1109.
188. Davis, S., Lu, M. L., Lo, S. H., Lin, S., Butler, J. A., Druker, B. J., Roberts, T. M., An, Q., and Chen, L. B. (1991). Presence of an SH2 domain in the actin-binding protein tensin. *Science* **252:**712–715.
189. Horwitz, A. F., Duggan, K., Buck, C., Beckerle, M. C., and Burridge, K. (1986). Interaction of plasma membrane fibronectin with talin—A transmembrane linkage. *Nature* **320:**531–533.
190. Rohrschneider, L. R. (1980). Adhesion plaques of Rous sarcoma virus-transformed cells contain the src gene product. *Proc. Natl. Acad. Sci. USA* **77:**3514–3518.
191. Hirst, R., Horwitz, A., Buck, C., and Rohrschneider, L. (1986). Phosphorylation of the fibronectin receptor complex in cells transformed by oncogenes that encode tyrosine kinases. *Proc. Natl. Acad. Sci USA* **83:**6470–6474.
192. Kornberg, L., Earp, H. S., Parsons, J. T., Schaller, M., and Juliano, R. L. (1992). Cell adhesion or integrin clustering increases phosphorylation of a focal adhesion-associated tyrosine kinase. *J. Biol. Chem.* **267:**23439–23442.
193. Samuelsson, S. J., Luther, P. W., Pumplin, D. W., and Bloch, R. J. (1993). Structures linking microfilament bundles to the membrane at focal contacts. *J. Cell Biol.* **122:**485–496.
193a. Pavalko, F. M., Schneider, G., Burridge, K., and Lim, S. S. (1995). Immunodetection of α-actinin in focal adhesions is limited by antibody inaccessibility. *Exp. Cell Res.* **217:**534–540.
194. Yurchenco, P. D. (1990). Assembly of basement membranes. *Ann. N.Y. Acad. Sci.* **580:**195–213.
195. Engel, J. (1992). Laminins and other strange proteins. *Biochemistry* **31:**10643–10651.

195a. Timpl, R., and Brown, J. C. (1994). The laminins. *Matrix Biol.* **14:**275–281.
195b. Burgeson, R. E., Chiquet, M., Deutzmann, R., Ekblom, P., Engel, J., Kleinman, H., Martin, G. R., Meneguzzi, G., Paulsson, M., Sanes, J., *et al.* (1994). A new nomenclature for the laminins. *Matrix Biol.* **14:**209–211.
196. Mecham, R. P. (1991). Receptors for laminin on mammalian cells. *FASEB J.* **5:**2538–2546.
197. Yurchenco, P. D., Sung, U., Ward, M. D., Yamada, Y., and O'Rear, J. J. (1993). Recombinant laminin G domain mediates myoblast adhesion and heparin binding. *J. Biol. Chem.* **268:**8356–8365.
198. Mayer, U., Nischt, R., Poschl, E., Mann, K., Fukuda, K., Gerl, M., Yamada, Y., and Timpl, R. (1993). A single EGF-like motif of laminin is responsible for high affinity nidogen binding. *EMBO J.* **12:**1879–1885.
199. Gerl, M., Mann, K., Aumailley, M., and Timpl, R. (1991). Localization of a major nidogen-binding site to domain III of laminin B2 chain. *Eur. J. Biochem.* **202:**167–174.
199a. Colognato-Pyke, H., O'Rear, J. J., Yamada, Y., Carbonetto, S., Cheng, Y. S., and Yurchenco, P. D. (1995). Mapping of network-forming, heparin-binding, and $\alpha_1\beta_1$ integrin-recognition sites within the α-chain short arm of laminin-1. *J. Biol. Chem.* **270:** 9398–9406.
200. Lievo, I., and Engvall, E. (1988). A protein specific for basement membranes of Schwann cells, striated muscle, and trophoblast, is expressed late in nerve and muscle development. *Proc. Natl. Acad. Sci. USA* **85:**1544–1548.
201. Hunter, D. D., Shah, V., Merlie, J. P., and Sanes, J. R. (1989). A laminin-like adhesive protein concentrated in the synaptic cleft of the neuromuscular junction. *Nature* **338:**229–234.
202. Gerecke, D. R., Wagman, D. W., Champliaud, M.-F., and Burgeson, R. E. (1994). The complete primary structure for a novel laminin chain, the laminin B1k chain. *J. Biol. Chem.* **269:**11073–11080.
203. Vailly, J., Verrando, P., Champliaud, M. F., Gerecke, D., Wagman, D. W., Baudoin, C., Aberdam, D., Burgeson, R., Bauer, E., and Ortonne, J.-P. (1994). The 100-kDa chain of nicein/kalinin is a laminin B2 chain variant. *Eur. J. Biochem.* **219:**209–218.
203a. Utani, A., Nomizu, M., Timpl, R., Roller, P. P., and Yamada, Y. (1994). Laminin chain assembly. Specific sequences at the C terminus of the long arm are required for the formation of specific double- and triple-stranded coiled-coil structures. *J. Biol. Chem.* **269:**19167–19175.
203b. Kammerer, R. A., Antonsson, P., Schulthess, T., Fauser, C., and Engel, J. (1995). Selective chain recognition in the C-terminal α-helical coiled-coil region of laminin. *J. Mol. Biol.* **250:**64–73.
204. Yurchenco, P. D., Cheng, Y.-S., and Schittny, J. C. (1990). Heparin modulation of laminin polymerization. *J. Biol. Chem.* **265:**3981–3991.
205. Yurchenco, P. D., Cheng, Y.-S., and Colognato, H. (1992). Laminin forms an independent network in basement membranes. *J. Cell Biol.* **117:**1119–1133.
206. Schittny, J. C., and Yurchenco, P. D. (1990). Terminal short arm domains of basement membrane laminin are critical for its self-assembly. *J. Cell Biol.* **110:**825–832.
207. Leivo, I., Vaheri, A., Timpl, R., and Wartiovaara, J. (1980). Appearance and distribution of collagens and laminin in the early mouse embryo. *Dev. Biol.* **76:**100–114.
208. Fox, J. W., Mayer, U., Nischt, R., Aumailley, M., Reinhardt, D., Wiedemann, H., Mann, K., Timpl, R., Krieg, T., Engel, J., and Chu, M.-L. (1991). Recombinant nidogen consists of three globular domains and mediates binding of laminin to collagen type IV. *EMBO J.* **10:**3137–3146.
209. Nagayoshi, T., Sanborn, D., Hickok, N. J., Olsen, D. R., Fazio, M. J., Chu, M.-L., Knowlton, R., Mann, K., Deutzmann, R., Timpl, R., and Uitto, J. (1989). Human nidogen: Complete amino acid sequence and structural domains deduced from cDNAs, and evidence for polymorphism of the gene. *DNA* **8:**581–594.
210. Reinhardt, D., Mann, K., Nischt, R., Fox, J. W., Chu, M.-L., Krieg, T., and Timpl, R. (1993). Mapping of nidogen binding sites for collagen type IV, heparan sulfate proteoglycan, and zinc. *J. Biol. Chem.* **268:**10881–10887.
211. Noonan, D. M., Fulle, A., Valente, P., Cai, S., Horigan, E., Sasaki, M., Yamada, Y., and Hassell, J. R. (1991). The complete sequence of perlecan, a basement membrane heparan sulfate proteoglycan, reveals extensive similarity with laminin A chain, low density lipoprotein-receptor, and the neural cell adhesion molecule. *J. Biol. Chem.* **266:**22939–22947.
212. Laurie, G. W., Inoue, S., Bing, J. T., and Hassell, J. R. (1988). Visualization of the large heparan sulfate proteoglycan from basement membrane. *Am. J. Anat.* **181:**320–326.
213. Yurchenco, P. D., Cheng, Y.-S., and Ruben, G. C. (1987). Self-assembly of high molecular weight basement membrane heparan sulfate proteoglycan into dimers and oligomers. *J. Biol. Chem.* **262:**17668–17676.
214. Battaglia, C., Mayer, U., Aumailley, M., and Timpl, R. (1992). Basement-membrane heparan sulfate proteoglycan binds to laminin by its heparan sulfate chains and to nidogen by sites in the protein core. *Eur. J. Biochem.* **208:**359–366.
215. Ali, I. U., Mautner, V., Lanza, R., and Hynes, R. O. (1977). Restoration of normal morphology, adhesion and cytoskeleton in transformed cells by addition of a transformation-sensitive surface protein. *Cell* **11:**115–126.
216. Yamada, K. M., and Weston, J. A. (1974). Isolation of a major cell surface glycoprotein from fibroblasts. *Proc. Natl. Acad. Sci. USA* **71:**3492–3496.
217. Yamada, K. M., Yamada, S. S., and Pastan, I. (1976). Cell surface protein partially restores morphology, adhesiveness, and contact inhibition of movement to transformed fibroblasts. *Proc. Natl. Acad. Sci. USA* **73:**1217–1221.
218. Hynes, R. O., and Yamada, K. M. (1982). Fibronectins: Multifunctional modular glycoproteins. *J. Cell Biol.* **95:**369–377.
219. Hynes, R. O. (1990). *Fibronectins,* Springer-Verlag, Berlin.
220. Hynes, R. O. (1993). Fibronectins. In *Guidebook to the Extracellular Matrix and Adhesion Proteins* (T. Kreis and R. Vale, eds.), Oxford University Press, London, pp. 56–58.
221. Paul, J. I., Schwarzbauer, J. E., Tamkun, J. W., and Hynes, R. O. (1986). Cell-type-specific fibronectin subunits generated by alternative splicing. *J. Biol. Chem.* **261:**12258–12265.
222. Colombi, M., Barlati, S., Kornblihtt, A., Baralle, F. E., and Vaheri, A. (1986). A family of fibronectin mRNAs in human normal and transformed cells. *Biochim. Biophys. Acta* **868:**207–214.
223. Hayashi, M., and Yamada, K. M. (1981). Differences in domain structures between plasma and cellular fibronectins. *J. Biol. Chem.* **256:**11292–11300.
224. Tamkun, J. W., Schwarzbauer, J. E., and Hynes, R. O. (1984). A single rat fibronectin gene generates three different mRNAs by alternative splicing of a complex exon. *Proc. Natl. Acad. Sci. USA* **81:**5140–5144.
225. Hirano, H., Yamada, Y., Sullivan, M., DeCroimbrugghe, B., Pastan, I., and Yamada, K. M. (1983). Isolation of genomic DNA clones spanning the entire fibronectin gene. *Proc. Natl. Acad. Sci. USA* **80:**46–50.
226. Ruoslahti, E., Hayman, E. G., Kuusela, P., Shively, J. E., and Engvall, E. (1979). Isolation of a tryptic fragment containing the collagen-binding site of plasma fibronectin. *J. Biol. Chem.* **254:**6054–6059.
227. Hayashi, M., Schlesinger, D. H., Kennedy, D. W., and Yamada, K. M. (1980). Isolation and characterization of a heparin-binding domain of cellular fibronectin. *J. Biol. Chem.* **255:**10017–10020.

228. Wagner, D. D., and Hynes, R. O. (1980). Topological arrangement of the major structural features of fibronectin. *J. Biol. Chem.* **255:** 4304–4312.
229. Ruoslahti, E., Hayman, E. G., and Engvall, E. (1981). Alignment of biologically active domains in the fibronectin molecule. *J. Biol. Chem.* **256:**7277–7281.
229a. Dickinson, C. D., Gay, D. A., Parello, J., Ruoslahti, E., and Ely, K. R. (1994). Crystals of the cell-binding module of fibronectin obtained from a series of recombinant fragments differing in length. *J. Mol. Biol.* **238:**123–127.
229b. Dickinson, C. D., Veerapandian, B., Dai, X. P., Hamlin, R. C., Xuong, N. H., Ruoslahti, E., and Ely, K. R. (1994). Crystal structure of the tenth type III cell adhesion module of human fibronectin. *J. Mol. Biol.* **236:**1079–1092.
230. Wu, C., Bauer, J. S., Juliano, R. L., and McDonald, J. A. (1993). The α5β1 integrin fibronectin receptor, but not the α5 cytoplasmic domain, functions in an early and essential step in fibronectin matrix assembly. *J. Biol. Chem.* **268:**21883–21888.
231. Singer, I. I. (1979). The fibronexus: A transmembrane association of fibronectin-containing fibers and bundles of 5 nm microfilaments in hamster and human fibroblasts. *Cell* **16:**675–685.
232. Ruoslahti, E., and Pierschbacher, M. D. (1987). New perspectives in cell adhesion: RGD and integrins. *Science* **238:**491–497.
233. Hynes, R. O. (1987). Integrins: A family of cell surface receptors. *Cell* **48:**549–554.
234. Hynes, R. O. (1992). Integrins: Versatility, modulation, and signaling in cell adhesion. *Cell* **69:**11–25.
235. Albelda, S. M., and Buck, C. A. (1990). Integrins and other cell adhesion molecules. *FASEB J.* **4:**2868–2880.
236. Ruoslahti, E. (1988). Fibronectin and its receptor. *Annu. Rev. Biochem.* **57:**375–413.
237. Pierschbacher, M., Hayman, E. G., and Ruoslahti, E. (1983). Synthetic peptide with cell attachment activity of fibronectin. *Proc. Natl. Acad. Sci. USA* **80:**1224–1227.
238. Pytela, R., Pierschbacher, M. D., and Ruoslahti, E. (1985). Identification and isolation of a 140 kd cell surface glycoprotein with properties expected of a fibronectin receptor. *Cell* **40:**191–198.
239. Suzuki, S., Argraves, W. S., Pytela, R., Arai, H., Krusius, T., Pierschbacher, M. D., and Ruoslahti, E. (1986). cDNA and amino acid sequences of the cell adhesion protein receptor recognizing vitronectin reveal a transmembrane domain and homologies with other adhesion protein receptors. *Proc. Natl. Acad. Sci. USA* **83:**8614–8618.
240. Pytela, R., Pierschbacher, M. D., and Ruoslahti, E. (1985). A 125/115-kDa cell surface receptor specific for vitronectin interacts with the arginine-glycine-aspartic acid adhesion sequence derived from fibronectin. *Proc. Natl. Acad. Sci. USA* **82:**5766–5770.
241. Neff, N. T., Lowrey, C., Decker, C., Tovar, A., Damsky, C., Buck, C., and Horwitz, A. F. (1982). A monoclonal antibody detaches embryonic skeletal muscle from extracellular matrices. *J. Cell Biol.* **95:**654–666.
242. Horwitz, A. F., Duggan, K., Greggs, R., Decker, C., and Buck, C. (1985). The cell substrate attachment (CSAT) antigen has properties of a receptor for laminin and fibronectin. *J. Cell Biol.* **101:** 2134–2144.
243. Damsky, C. H., Knudsen, K. A., Bradley, D., Buck, C. A., and Horwitz, A. F. (1985). Distribution of the cell substratum attachment (CSAT) antigen on myogenic and fibroblastic cells in culture. *J. Cell Biol.* **100:**1528–1539.
244. Nermut, M. V., Green, N. M., Eason, P., Yamada, S. S., and Yamada, K. M. (1988). Electron microscopy and structural model of human fibronectin receptor. *EMBO J.* **7:**4093–4099.
245. Kirchhofer, D., Gailit, J., Ruoslahti, E., Grzesiak, J., and Pierschbacher, M. D. (1990). Cation-dependent changes in the binding specificity of the platelet receptor GPIIIb/IIIa. *J. Biol. Chem.* **265:** 18525–18530.
246. Smith, J. W., and Cheresh, D. A. (1990). Integrin (αvβ3)–ligand interaction: Identification of a heterodimeric RGD binding site on the vitronectin receptor. *J. Biol. Chem.* **265:**2168–2172.
247. D'Souza, S. E., Ginsberg, M. H., Burke, T. A., and Plow, E. F. (1990). The ligand binding site of the platelet integrin receptor GPIIb-IIIa is proximal to the second calcium binding domain of its α subunit. *J. Biol. Chem.* **265:**3440–3446.
248. Loftus, J. C., O'Toole, T. E., Plow, E. F., Glass, A., Frelinger, A. L., III, and Ginsberg, M. H. (1990). A β3 integrin mutation abolishes ligand binding and alters divalent cation-dependent conformation. *Science* **249:**915–918.
249. Pierschbacher, M. D., and Ruoslahti, E. (1987). Influence of stereochemistry of the sequence of Arg-Gly-Asp-Xaa on binding specificity in cell adhesion. *J. Biol. Chem.* **262:**17294–17298.
250. Mould, A. P., Wheldon, L. A., Komoriya, A., Wayner, E. A., Yamada, K. M., and Humphries, M. J. (1990). Affinity chromatographic isolation of the melanoma adhesion receptor for the IIICS region of fibronectin and its identification as the integrin α4β1. *J. Biol. Chem.* **265:**4020–4024.
251. Gehlsen, K. R., Dillner, L., Engvall, E., and Ruoslahti, E. (1988). The human laminin receptor is a member of the integrin family of cell adhesion receptors. *Science* **241:**1228–1229.
252. Kirchhofer, D., Languino, L. R., Ruoslahti, E., and Pierschbacher, M. D. (1990). α2β1 integrins from different cell types show different binding specificities. *J. Biol. Chem.* **265:**615–618.
253. Cheresh, D. A., and Klier, F. G. (1986). Asialoganglioside GD_2 distributes preferentially into substrate-associated microprocesses on human melanoma cells during their attachment to fibronectin. *J. Cell Biol.* **102:**1887–1897.
254. Yamada, K. M., Critchley, D. R., Fishman, P. H., and Moss, J. (1983). Exogenous gangliosides enhance the interaction of fibronectin with ganglioside-deficient cells. *Exp. Cell Res.* **143:** 295–302.
255. Brown, E., Hooper, L., Ho, T., and Gresham, H. (1990). Integrin-associated protein: A 50 kDa plasma membrane antigen physically and functionally associated with integrin. *J. Cell Biol.* **111:** 2785–2794.
256. Agraves, W. S., Dickerson, K., Burgess, W. H., and Ruoslahti, E. (1990). Fibulin, a novel protein that interacts with the fibronectin receptor β subunit cytoplasmic domain. *Cell* **58:**623–629.
257. Regen, C. M., and Horwitz, A. F. (1992). Dynamics of β1 integrin-mediated adhesive contacts in motile fibroblasts. *J. Cell Biol.* **119:** 1347–1359.
258. Singer, I. I., Scott, S., Kawka, D. W., Kazazis, D. M., Gailit, J., and Ruoslahti, E. (1988). Cell surface distribution of fibronectin and vitronectin receptors depends on substrate composition and extracellular matrix accumulation. *J. Cell Biol.* **106:**2171–2182.
259. LaFlamme, S. E., Akiyama, S. K., and Yamada, K. M. (1992). Regulation of fibronectin receptor distribution. *J. Cell Biol.* **117:** 437–447.
260. Wayner, E. A., Orlando, R. A., and Cheresh, D. A. (1991). Integrins αvβ3 and αvβ5 contribute to cell attachment to vitronectin but differentially distribute on the cell surface. *J. Cell Biol.* **113:**919–929.
261. Fath, K. R., Edgell, C.-J. S., and Burridge, K. (1989). The distribution of distinct integrins in focal contacts is determined by the substratum composition. *J. Cell Sci.* **92:**67–75.
262. Juliano, R. L., and Haskill, S. (1993). Signal transduction from the extracellular matrix. *J. Cell Biol.* **120:**577–585.
263. Chen, W.-T., Greve, J. M., Gottlieb, D. I., and Singer, S. J. (1985). Immunocytochemical localization of 140 kD cell adhesion molecules in cultured chicken fibroblasts, and in chicken smooth muscle and intestinal epithelial tissues. *J. Histochem. Cytochem.* **33:**576–586.
263a. Schaller, M. D., Parsons, J. T. (1994). Focal adhesion kinase and associated proteins. *Curr. Opin. Cell Biol.* **6:**705–710.

264. Schaller, M. D., Borgman, C. A., Cobb, B. S., Vines, R. R., Reynolds, A. B., and Parsons, J.T. (1992). pp125FAK, a structurally distinctive protein-tyrosine kinase associated with focal adhesions. *Proc. Natl. Acad. Sci. USA* **89:**5192–5196.

265. Lipfert, L., Haimovich, B., Schaller, M. D., Cobb, B. S., Parsons, J. T., and Brugge, J. S. (1992). Integrin-dependent phosphorylation and activation of the protein tyrosine kinase pp125FAK in platelets. *J. Cell Biol.* **119:**905–912.

266. Reynolds, A. B., Rosel, D. J., Kanner, S. B., and Parsons, J. T. (1989). Transformation-specific tyrosine phosphorylation of a novel cellular protein in chicken cells expressing oncogenic variants of the avian cellular src gene. *Mol. Cell. Biol.* **9:**629–638.

267. Nigg, E. A., Sefton, B. M., Hunter, T., Walter, G., and Singer, S. J. (1982). Immunofluorescent localization of the transforming protein of Rous sarcoma virus with antibodies against a synthetic src peptide. *Proc. Natl. Acad. Sci. USA* **79:**5322–5326.

267a. Huhtala, P., Humphries, M. J., McCarthy, J. B., Tremble, P. M., Werb, Z., and Damsky, C. H. (1995). Cooperative signaling by $\alpha_5\beta_1$ and $\alpha_4\beta_1$ integrins regulates metalloproteinase gene expression in fibroblasts adhering to fibronectin. *J. Cell Biol.* **129:**867–879.

267b. Tremble, P. M., Damsky, C. H., and Werb, Z. (1995). Components of nuclear signaling cascade that regulate collagenase gene expression in response to integrin-derived signals. *J. Cell Biol.* **129:** 1707–1720.

268. Ginsberg, M. H., Du, X., and Plow, E. F. (1992). Inside-out integrin signalling. *Curr. Opin. Cell Biol.* **4:**766–771.

269. Sims, P. J., Ginsberg, M. H., Plow, E. F., and Shattil, S. J. (1991). Effect of platelet activation on the conformation of the plasma membrane glycoprotein IIb-IIIa complex. *J. Biol. Chem.* **266:**7345–7352.

69a. Chen, Y.-P., O'Toole, T. E., Shipley, T., Forsyth, J., LaFlamme, S. E., Yamada, K. M., Shattil, S. J., and Ginsberg, M. H. (1994). "Inside-out" signal transduction inhibited by isolated integrin cytoplasmic domains. *J. Biol. Chem.* **269:**18307–18310.

270. Vlodavsky, I., Bar-Shavit, R., Ishai-Michaeli, R., Bashkin, P., and Fuks, Z. (1991). Extracellular sequestration and release of fibroblast growth factor: A regulatory mechanism? *Trends Biochem. Sci.* **16:**268–271.

271. Gitay-Cohen, H., Soker, S., Vlodavsky, I., and Neufeld, G. (1992). The binding of vascular endothelial growth factor to its receptors is dependent on cell surface-associated heparin-like molecules. *J. Biol. Chem.* **267:**6093–6098.

272. Yayon, A., Klagsbrun, M., Esko, J. D., Leder, P., and Ornitz, D. M. (1991). Cell surface, heparin-like molecules are required for binding of basic fibroblast growth factor to its high affinity receptor. *Cell* **64:**841–848.

273. Kokenyesi, R., and Bernfield, M., (1994). Core protein structure and sequence determine the site and presence of heparan sulfate and chondroitin sulfate on syndecan-1. *J. Biol. Chem.* **269:**12304–12309.

274. Ohlendieck, K., Ervasti, J. M., Snook, J. B., and Campbell, K. P. (1991). Dystrophin-glycoprotein complex is highly enriched in isolated skeletal muscle sarcolemma. *J. Cell Biol.* **112:**135–148.

274a. Brancaccio, A., Schulthess, T., Gesemann, M., and Engel, J. (1995). Electron microscopic evidence for a mucin-like region in chick muscle α-dystroglycan. *FEBS Lett.* **368:**139–142.

275. Hagiwara, Y., Yoshida, M., Nonaka, I., and Ozawa, E. (1989). Developmental expression of dystrophin on the plasma membrane of rat muscle cells. *Protoplasma* **151:**11–18.

275a. Tome, F. M., Evangelista, T., Leclerc, A., Sunada, Y., Manole, E., Estournet, B., Barois, A., Campbell, K. P., and Fardeau, M. (1994). Congenital muscular dystrophy with merosin deficiency. *C. R. Acad. Sci. Ser. C* **317:**351–357.

275b. Xu, H., Christmas, P., Wu, X.-R., Wewer, U. M., and Engvall, E. (1994). Defective muscle basement membrane and lack of M-laminin in the dystrophic dy/dy mouse. *Proc. Natl. Acad. Sci. USA* **91:**5572–5576.

276. Ushkaryov, Y. A., Petrenko, A. G., Geppert, M., and Sudhof, T. C., (1992). Neurexins: Synaptic cell surface proteins related to the α-latrotoxin receptor and laminin. *Science* **257:**50–56.

277. Sugrue, S. P., and Hay, E. D. (1981). Response of basal epithelial cell surface and cytoskeleton to solubilized extracellular matrix molecules. *J. Cell Biol.* **91:**45–54.

278. Sanes, J. R., Engvall, E., Butkowski, R., and Hunter, D. D. (1990). Molecular heterogeneity of basal laminae: Isoforms of laminin and collagen IV at the neuromuscular junction and elsewhere. *J. Cell Biol.* **111:**1685–1699.

279. McMahan, U. J., Sanes, J. R., and Marshall, L. M. (1978). Acetylcholinesterase is associated with the basal lamina at the neuromuscular junction. *Nature* **193:**281–282.

280. Brandan, E., Maldonado, M., Garrido, J., and Inestrosa, N. C. (1985). Anchorage of collagen-tailed acetylcholinesterase to the extracellular matrix is mediated by heparan sulfate proteoglycans. *J. Cell Biol.* **101:**985–992.

281. Fertuck, H. C., and Salpeter, M. M. (1976). Quantitation of junctional and extrajunctional acetylcholine receptors by electron microscope autoradiography after ^{125}I-α-bungarotoxin binding at mouse neuromuscular junctions. *J. Cell Biol.* **69:**144–158.

282. Heuser, J. E., and Salpeter, S. R. (1979). Organization of acetylcholine receptors in quick-frozen, deep-etched, and rotary-replicated Torpedo postsynaptic membrane. *J. Cell Biol.* **82:**150–173.

282a. Froehner, S. C. (1991). The sub-membrane machinery for nicotinic acetylcholine receptor clustering. *J. Cell Biol.* **114:**1–7.

283. Marshall, L. M., Sanes, J. R., and McMahan, U. J. (1977). Reinnervation of original synaptic sites on muscle fiber basement membrane after disruption of the muscle cells. *Proc. Natl. Acad. Sci. USA* **74:**3073–3077.

284. Hunter, D. D., Cashman, N., Morris-Valero, R., Bulock, J. W., Adams, S. P., and Sanes, J. R. (1991). An LRE (leucine-arginine-glutamate)-dependent mechanism for adhesion of neurons to S-laminin. *J. Neurosci.* **11:**3960–3971.

284a. Porter, B. E., Weis, J., and Sanes, J. R. (1995). A motoneuron-selective stop signal in the synaptic protein S-laminin. *Neuron* **14:** 549–559.

284b. Noakes, P. G., Gautam, M., Mudd, J., Sanes, J. R., and Merlie, J. P. (1995). Aberrant differentiation of neuromuscular junctions in mice lacking s-laminin/laminin β2. *Nature* **374:**258–262.

285. Burden, S. J., Sargent, P. B., and McMahan, U. J. (1979). Acetylcholine receptors in regenerating muscle accumulate at original synaptic sites in the absence of the nerve. *J. Cell Biol.* **82:**412–425.

286. McMahan, U. J., and Slater, C. R. (1984). The influence of basal lamina on the accumulation of acetylcholine receptors at synaptic sites in regenerating muscle. *J. Cell Biol.* **98:**1453–1473.

287. Ruegg, M. A., Tsim, K. W. K., Horton, S. E., Kroger, S., Escher, G., Gensch, E. M., and McMahan, U. J. (1992). The agrin gene codes for a family of basal lamina proteins that differ in function and distribution. *Neuron* **8:**691–699.

288. Tsim, K. W. K., Ruegg, M. A., Escher, G., Kroger, S., and McMahan, U. J. (1992). cDNA that encodes active agrin. *Neuron* **8:** 677–689.

289. Rupp, F., Payan, D. G., Magill-Solc, C., Cowan, D. M., and Scheller, R. H. (1991). Structure and expression of a rat agrin. *Neuron* **6:**811–823.

290. Ferns, M., Hoch, W., Campanelli, J. T., Rupp, F., Hall, Z. W., and Scheller, R. H. (1992). RNA splicing regulates agrin-mediated acetylcholine receptor clustering activity on cultured myotubes. *Neuron* **8:**1079–1086.

291. Hoch, W., Ferns, M., Campanelli, J. T., Hall, Z. W., and Scheller, R. H. (1993). Developmental regulation of highly active alternatively spliced forms of agrin. *Neuron* **11:**479–490.

292. Ferns, M. J., Campanelli, J. T., Hoch, W., Scheller, R. H., and Hall, Z. W. (1993). The ability of agrin to cluster AChRs depends on alternative splicing and on cell surface proteoglycans. *Neuron* **11:** 491–502.

293. Nitkin, R. M., Smith, M. A., Magill, C., Fallon, J. R., Yao, Y.-M. M., Wallace, B. G., and McMahan, U. J. (1987). Identification of agrin, a synaptic organizing protein from Torpedo electric organ. *J. Cell Biol.* **105:**2471–2478.

294. Magill-Solc, C., and McMahan, U. J. (1988). Motor neurons contain agrin-like molecules. *J. Cell Biol.* **107:**1825–1833.

295. Reist, N. E., Werle, M. J., and McMahan, U. J. (1992). Agrin released by motor neurons induces the aggregation of acetylcholine receptors at neuromuscular junctions. *Neuron,* **8:**865–868.

296. Cohen, M. W., and Godfrey, E. W. (1992). Early appearance of and neuronal contribution to agrin-like molecules at embryonic frog nerve–muscle synapses formed in culture. *J. Neurosci.* **12:**2982–2992.

296a. Tsen, G., Halfter, W., Kroger, S., and Cole, G. J. (1995). Agrin is a heparan sulfate proteoglycan. *J. Biol. Chem.* **270:**3392–3399.

297. Gee, S. H., Montanaro, F., Lindenbaum, M. H., and Carbonetto, S. (1994). Dystroglycan-α, a dystrophin-associated glycoprotein, is a functional agrin receptor. *Cell* **77:**675–686.

298. Bowe, M. A., Deyst, K. A., Leszyk, J. D., and Fallon, J. R. (1994). Identification and purification of an agrin receptor from Torpedo postsynaptic membranes: A heteromeric complex related to the dystroglycans. *Neuron* **12:**1173–1180.

299. Sugiyama, J., Bowen, D. C., and Hall, Z. W. (1994). Dystroglycan binds nerve and muscle agrin. *Neuron* **13:**1–20.

300. Singer, S. J., and Nicholson, G. L. (1972). The fluid mosaic model of the structure of cell membranes. *Science* **175:**720–731.

4

Methods of Reconstitution of Ion Channels

William P. Dubinsky and Otilia Mayorga-Wark

4.1. INTRODUCTION

The advent of single-channel technology has expanded the physiologist's ability to understand gross cellular and organellar function at the molecular level. At the same time that we are increasingly employing these microscopic techniques to characterize the molecular aspects of ion channels, we are better able to define their macroscopic behavior. Although the goal of many of these studies is to identify a specific ion channel with a specific cellular or organellar function, there are numerous problems and limitations in confirming this identity. Frequently, for example, there may be several distinct channel types present in the same membrane. The identification of a function with a specific ion channel thus depends on the characterization of the activity at the cellular and at the molecular, single-channel level. Several broad approaches are routinely applied to study the molecular aspects of specific ion channels. Patch clamp methodology identifies the channels in native membranes and provides the biophysical fingerprint as well as key regulatory features. These properties of the channel are subsequently used to compare and identify channel activities in an expression system or in a reconstituted membrane. A powerful technique for the molecular and electrophysiological characterization of an ion channel is to inject mRNA from a donor species into the *Xenopus* oocyte or other expression system. The oocyte expression system offers numerous advantages, particularly availability, ease of handling and injecting, and amenability to electrophysiological measurements.[1] One of the most powerful aspects of this technique is the ability to identify ion channels that may be expressed in the cells at very low abundance. The second approach, and the focus of this chapter, is to use resolution–re-constitution to purify and characterize ion channel activities. Regardless of the approach, initial studies in the native tissue define the various parameters that allow the correlation of the macroscopic currents with single-channel activities. Thus, individual properties such as voltage dependence, current-to-voltage relation, inhibitor sensitivity, activation, and regulation are combined to provide the complete channel profile. It is this profile that enables the final identification.

4.1.1. Resolution–Reconstitution Approaches to Study Ion Channels

The approach of resolution–reconstitution was introduced and developed by Racker and his associates as a means of dissecting and analyzing complex membrane phenomena. The principles are based on traditional biochemical rules for the elucidation of biochemical pathways. Substrates and products of a specific reaction or enzymatic pathway are identified and then used to establish a functional assay. The specific activity of the enzyme is determined at each stage of purification and provides an estimate of the degree of purification. If during the purification there is an abrupt loss of activity, then either there was some destructive event, or an essential component of the reaction sequence was removed. At this point fractions are recombined and the missing component identified by the restoration of activity. Racker's innovation was to extend these principles to membrane processes in which *there is a clear role of the two-compartment membrane system* such as electron transport, coupled phosphorylation, transporting ATPases, and other transport phenomena.[2–4] Some of these included traditional enzymatic activities; however, in the case of transport proteins such as ion channels, generally, there is no enzymatic activity in the classical sense of a substrate and a product, rather the catalyst simply lowers the activation energy for the transmembranous movement of an impermeant species. Purification of membrane proteins requires the disruption of the membrane structure, and usually a loss of the three-dimensional organization of the complex. Reconstitution of the biological activity necessitates the restoration of this three-dimensional structure in a membrane system in order to detect the translocation of solutes from one compartment to the other. Using this approach, one may then identify all of the necessary components of a complex, identify the role of each of the components in the overall process, begin to

William P. Dubinsky and Otilia Mayorga-Wark • Department of Integrative Biology, The University of Texas–Houston Medical School, Houston, Texas 77225.

Molecular Biology of Membrane Transport Disorders, edited by S.G. Schultz *et al.* Plenum Press, New York, 1996

Table 4.1. Purified and Reconstituted Ion Channels

	Refs.
Voltage-gated channels	
Na^+ channel	7–11
Ca^{2+} channel	12
K^+ channel	13–15
Ligand-gated channels	
Acetylcholine receptor	16–19
cGMP-dependent cation channel of rod outer segment	20,21
Ca^{2+} release channel of sarcoplasmic reticulum	22,23
K^+ channels	
K(ATP)	44,24
Ca^{2+}-dependent	25–28
Cl^- channels	29–33
CFTR	82,34
Gap Junctions	35

study structure–function relations, and assess the role of phospholipid in the process.

4.1.2. General Strategies for the Reconstitution of Ion Channels

A study using resolution–reconstitution would begin with the development of an assay for the biological activity. In the case of transport phenomena, this is usually the measurement of the flux of a solute into (or out of) a vesicular structure. These methods have been described and reviewed extensively.[5,6] Since the focus of this chapter is on reconstitution of ion channels, the biological activity is the movement of ions across a membrane in a manner that is consistent with channel-mediated fluxes. Parameters such as conductance, inhibitor sensitivity, ion selectivity, regulation, or other characteristics identified by studies of the native, intact tissue or cells would establish the criteria for an effective specific assay.

Numerous types of channels have been purified and reconstituted to assay for function. Examples of the purification

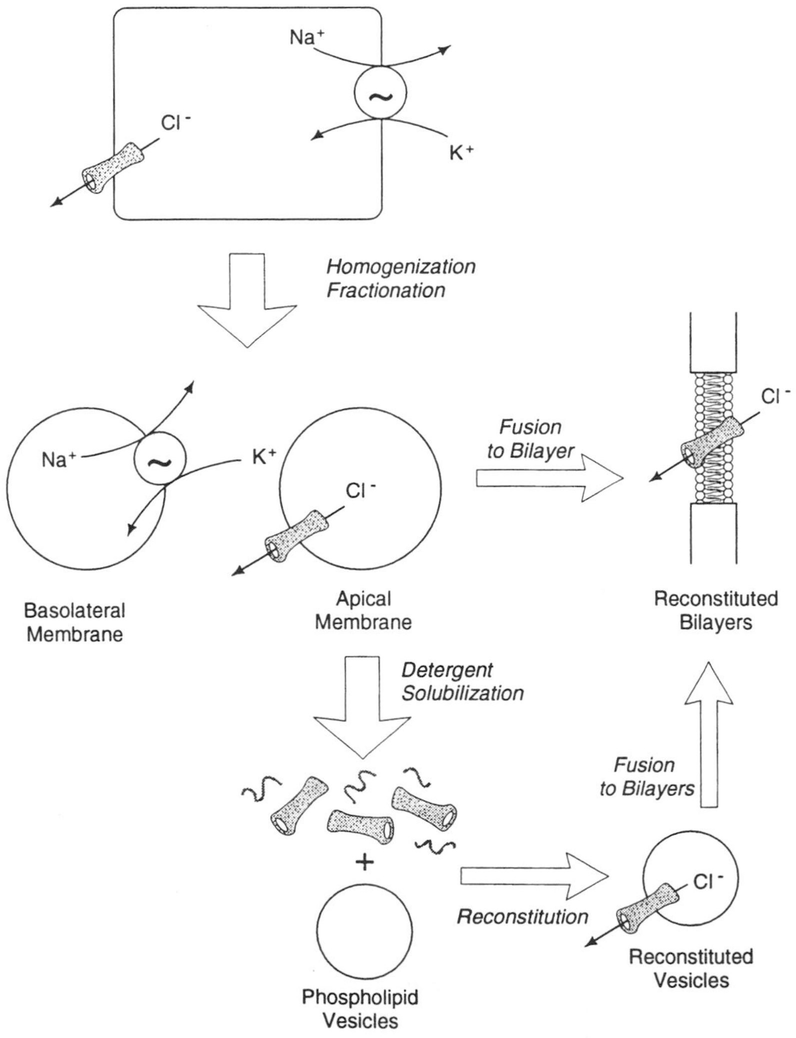

Figure 4.1. General strategies for the resolution and reconstitution of membrane transport proteins.

and assay of channel activities are compiled in Table 4.1. These examples illustrate a breadth of purification strategies using ligand affinity, biochemical and immunological techniques to isolate, purify, and identify ion channels. A general strategy for the purification of an ion channel would be to begin by identifying a tissue known to possess the activity of interest, can be handled and processed with relative ease, and is available in large quantity. An additional asset is that the tissue be composed mainly of the cell type possessing the channel of interest. As a first step, a crude homogenate of the tissue is prepared and the major membrane fractions separated by differential or gradient centrifugation or other techniques.[36–38] The membranes of interest are identified on the basis of specific marker enzymes such as the (Na+K)-ATPase for the basolateral membrane and alkaline phosphatase for the apical membrane of epithelial cells. Other specific markers have been identified for subcellular membranes.[39,40] Purification of the membrane is estimated by the increase in the specific activity of the marker enzyme activity in the fraction as compared to the crude homogenate. Alternatively, the ion channel activity itself may be used to assay for purification of the membrane population.[41,42] Two distinct methods are available for the study of ion channels. Vesicle-based assays usually determine net fluxes of ions into or out of the vesicles with radioisotopic flux measurements, fluorescence indicators, or ion-selective electrodes. These methods afford an assessment and characterization of simple kinetic parameters, inhibitor sensitivity, and activators of channel activity. More detailed analysis is obtained by reconstitution of ion channels into planar phospholipid bilayers for the electrophysiological characterization of single-channel properties. Either one of these approaches would be used to identify a specific ion channel activity in a membrane preparation. The membrane population exhibiting the channel activity would be purified and then serve as the starting material for the purification of the ion channel.

4.1.3. Vesicles versus Bilayers

The ability to measure single-channel activities in planar bilayers is a tremendous asset in the identification of specific ion channels in the sense that there is a clear advantage of performing a kinetic analysis on virtually a single molecule or molecular complex. However, since the analysis is only detecting a single molecule, there is no easy way of estimating the total number of transporters or its degree of purification in the preparation. Vesicle-based assays allow an estimate of the degree of purification on the basis of the specific activity of transport; however, the extremely high turnover rates of ion channels limit the utility of vesicles in the study of reconstituted ion channels. Modifications to the vesicle assay have been introduced that greatly extend the time course of equilibration, and thus permit a limited kinetic analysis and assessment of the effects of inhibitors or other pharmacologic agents on channel activity.[42] Another reason to reconstitute into phospholipid vesicles is to provide a vehicle for the delivery of a purified channel to a phospholipid bilayer for electrophysiological characterization.[43,44] Thus, a concerted approach to the purification of an ion channel would employ a vesicle-based assay and planar bilayers to complement each other. These different approaches are summarized in Figure 4.1.

4.2. CHARACTERIZATION AND ISOLATION OF AN EPITHELIAL K CHANNEL

4.2.1. Identifying the Ion Conductance in the Intact Tissue or Cell

In epithelial tissues there is a complex regulatory interaction that couples the activity of one cell surface, the apical, to the activity of the (Na+K)-ATPase and the K conductance of the basolateral membrane.[45] For example, in the intestine, on initiation of solute absorption there is a coordinated increase in Na extrusion from the cell at the basolateral surface which parallels the increase in absorptive activity. The phenomenon is readily demonstrated by analysis of intracellular recordings of the membrane potential and short-circuit currents in epithelial sheets of *Necturus* small intestine mounted in Ussing chambers.[46] The membrane potential was assessed by microelectrode impalements of the epithelial cells. Microelectrode analysis of the membrane potentials in these cells in intact epithelial sheets under short-circuited conditions demonstrates that there is an increase in the basolateral conductance to potassium after an increase in the absorptive activity of the tissue. In similar experiments addition of the venom from the scorpion *Leiurus quinquestriatas* (*LQV*) to the serosal bathing solution caused a decrease in the ratio of the apical to basolateral resistance (r^m/r^s) and a depolarization of the membrane potential which is consistent with the blockage of a basolateral K conductance.[38] Similarly, Ba^{2+} addition to the serosal solution also blocked the basolateral K conductance.[38] Increased Na transport occurs across the mucosal membrane and is accompanied by an increase in the K conductance and enhanced (Na+K)-ATPase in the basolateral membrane. Thus, there must be some mechanism coupling these events; however, the mechanism of coupling is completely unknown. What is clear from the data is that with increasing absorptive activity there is a parallel increase in the basolateral K conductance, and it is postulated that the increase in the K conductance in turn will increase the (Na+K)-ATPase via enhanced cycling of K across the basolateral membrane.[45] Further study of the basolateral K permeability was limited by accessibility to the basolateral surface.

4.2.2. Reconstitution of the *Necturus* Basolateral K Channel

4.2.2.1. Isolation of the Basolateral Membrane

Because of the geometry of the surface and basement membrane a patch clamp analysis of the basolateral membrane in the intact epithelia was impossible to perform. However, tissue was available in sufficient quantity to isolate specific cellular membranes and use resolution–reconstitution techniques to identify and characterize the ion channels in the basolateral membrane of these cells. Initial reconstitution experiments focused on the identification of K channels in planar phospholipid bilayers. An enriched basolateral membrane preparation was obtained from homogenates of intestinal scrapings by a series of differential and Percol gradient centrifugation steps.[38] Specific enzymatic activities were used as membrane markers to assess the degree of purification and separation of the major membrane fractions. The final gradient yielded the most highly enriched basolateral fraction [1.2 μmole hr^{-1} mg^{-1} versus 28 μmole hr^{-1} mg^{-1} (Na+K)-ATPase activity for the crude homogenate versus purified membranes, or 23-fold enrichment]. This highly enriched fraction was used in subsequent planar bilayer studies to identify and characterize the K permeability of these membranes. The *Necturus* basolateral membrane permeabilities were more fully characterized by incorporation of the basolateral membrane fraction into planar phospholipid bilayers for single-channel analysis. Only two different channel types have been observed in this membrane preparation.[38] An anion channel has been rarely observed and has not been characterized. A second, cation-selective channel of the K(ATP) type is frequently observed, has been characterized extensively, and has been partially purified.

4.2.2.2. Planar Bilayer Techniques

The technique of planar bilayers for the electrophysiological analysis of reconstituted membrane proteins was developed by Mueller and Rudin in studies aimed at identifying the molecular basis of membrane excitability.[47] The approach was then adapted to the study of individual ion channels as techniques were developed for the incorporation of membrane vesicles isolated from various cell types to planar phospholipid bilayers. Since then, there have been numerous modifications and applications of this powerful tool. Much of the technology has been reviewed,[48] and the present chapter will employ as an example the system[38,49] in use in our laboratories to demonstrate the principles of the technique. A convenient and workable bilayer apparatus has been described in detail by Alvarez.[50] A Lucite chamber is machined to have two compartments that are adjacent and opened to each other. One compartment accepts a small (3 ml) Delrin cup that has a 300-μm hole in the side. The cup is placed in the chamber with the aperture separating the two compartments. The two compartments are filled with 3 ml of buffered 5 mM KCl solution. The bilayer is formed by spreading a mixture of phospholipid consisting of phosphatidylethanolamine and phosphatidylserine (1:1) dissolved in decane (10 mg/ml) over the aperture with a small Teflon paddle. After thinning of the membrane, a salt gradient was established by bringing the *cis* chamber to 150 mM KCl through the addition of 3 M KCl. The salt gradient provides an osmotic driving force that facilitates fusion of the membrane vesicles to the planar bilayer. Fusion of the membranes containing the channel was achieved by the addition of an aliquot of membranes to the *cis* compartment and stirring until fusion occurred as evidenced by the appearance of channel activity. An alternative and more efficient fusion procedure was to directly apply the membranes to the bilayer with a pipette tip and a small amount of the vesicle suspension from the *trans* compartment.

Channel activity was monitored with a List EP7 amplifier (List Electronics DA-Eberstadt, FRG). The *cis* chamber was connected to the head stage of the amplifier through a Ag/AgCl electrode and an agar salt bridge equilibrated with 100 mM KCl. The *trans* solution was held at ground potential through the same electrode configuration. The analog output from the amplifier was passed through an 8-pole Bessel filter with a corner frequency of 0.5 kHz (Frequency Devices, Springfield, MA) and displayed on a Hitachi oscilloscope. The unfiltered signal was digitized using a PCM-2 analog-to-digital converter (Medical Systems, Greenvale, NY). The digitized signal was stored on video cassettes. The signals were analyzed by passing through an 8-pole Bessel filter and digitizing at 0.4-msec intervals with a Labmaster DMA Interface (Axon Instruments, Foster City, CA) and stored on a personal computer. Data were acquired and analyzed using the pClamp software package (Axon Instruments). By convention a positive current reflects the flow of cations from the *cis* to the *trans* compartment or the flow of anions in the reverse direction. The holding potential (V_m) is the electrical potential of the *cis* compartment with reference to that of the *trans* compartment. Since there is a likelihood that channels may incorporate randomly (i.e., "inside out" or "right side out"), an important consideration is the orientation when comparing to the intact cell. Thus, conventional electrophysiology would identify the *cis* compartment as the intracellular compartment and *trans* the bath, or reference (Figure 4.2).

4.2.3. Characterization of the Reconstituted Basolateral K Channel

4.2.3.1. Biophysical Properties

Characterization of the permeability begins with an analysis of the single-channel activity as a function of the holding potential, V_m. Recordings of single-channel currents under conditions of an asymmetric ionic gradient (150 mM KCl : 5 mM KCl, *cis* versus *trans*) are shown in Figure 4.3. Single-channel currents are plotted as a function of V_m to yield the current-to-voltage relation (*I–V* relation, Figure 4.4) for the channel. The relationship is linear illustrating the ohmic behavior of this channel. The conductance of the channel is readily determined as the slope of the *I–V* curve from Ohm's

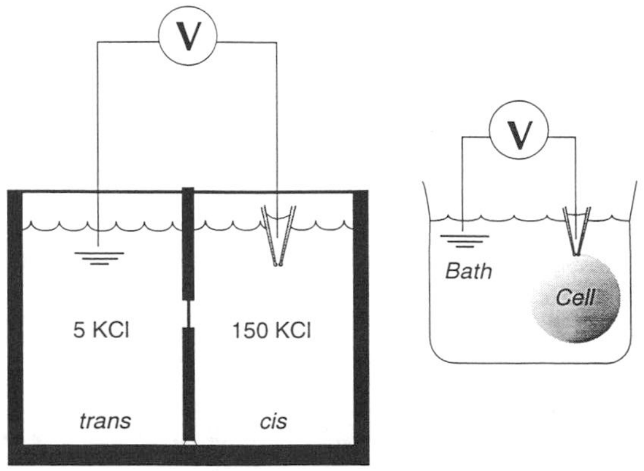

Figure 4.2. Electrophysiological configuration of planar bilayer studies as compared to conventional intracellular recording techniques.

law where $\Delta I = g\Delta E$ where g is the slope conductance. The conductance of the channel is 203 pS. The second property of the channel that can be determined from the I–V relation is the relative permeability of the channel to different ions. The holding potential at which the electrical potential is equal to the chemical potential of all permeable ionic species, or reversal potential, E_{rev}, is dependent on the relative permeabilities of the different ions. If the channel were perfectly K selective, then from the Nernst equation, E_K, the equilibrium potential for K, would be 88 mV. If the channel conducted other species, then the selectivity could be determined from the shift in the E_{rev} from the ideal. In the simple case of a single ionic pair, the relative permeability of K versus Cl can be calculated from the ionic activities,[53] according to the equation for a diffusion potential:

$$E_{rev} - \left(\frac{P_K - P_{cl}}{P_K + P_{cl}}\right) \times \frac{2.3RT}{F} \log\left(\frac{[KCl]_{trans}}{[KCl]_{cis}}\right)$$

Thus, from the E_{rev} determined from the data in Figure 4.4, the channel is very highly selective for K and imparts virtually no Cl permeability.

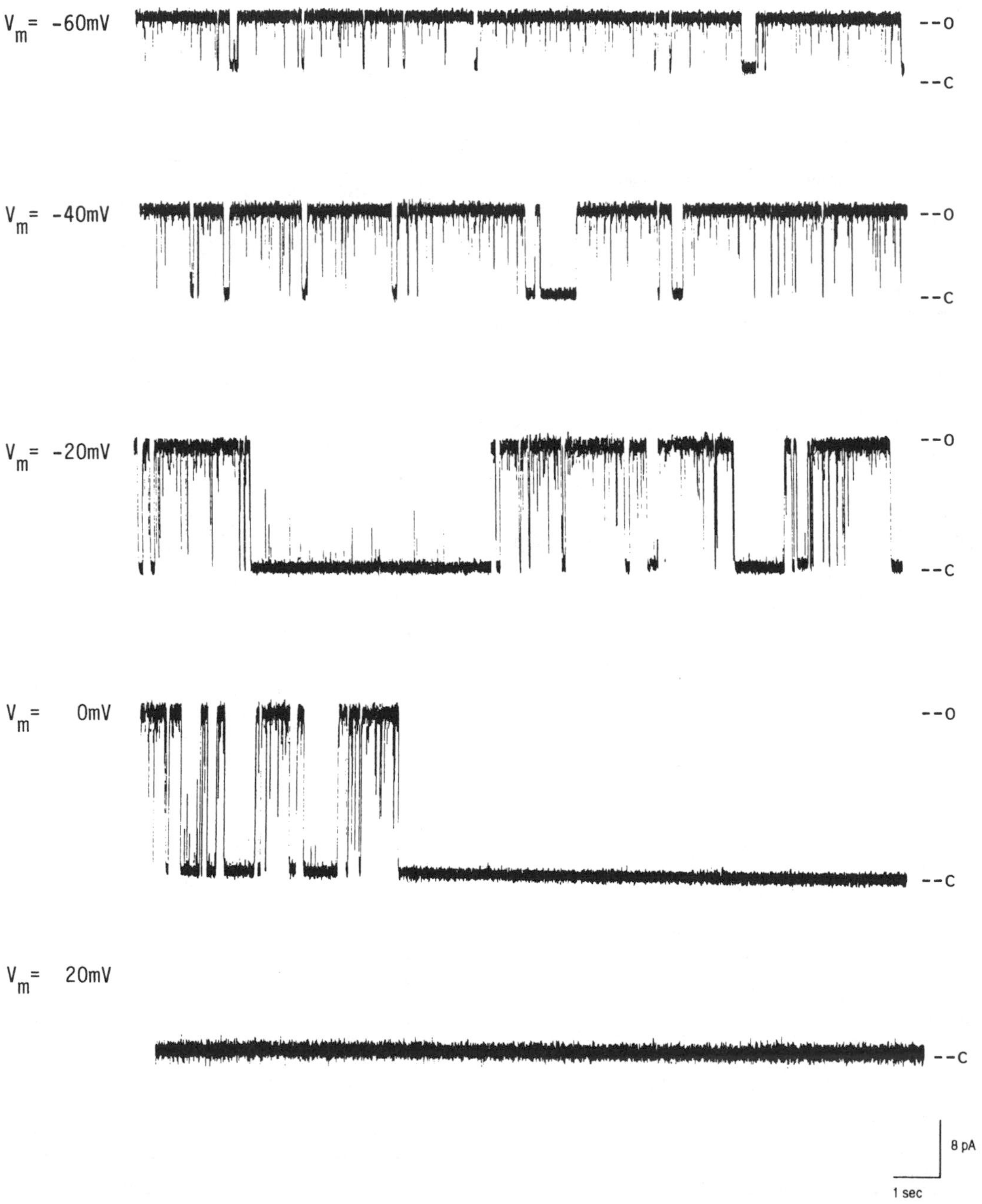

Figure 4.3. Single-channel recordings obtained from reconstituted *Necturus* basolateral membranes. The open and closed states are designated as "o" and "c," respectively. V_m, holding potential.

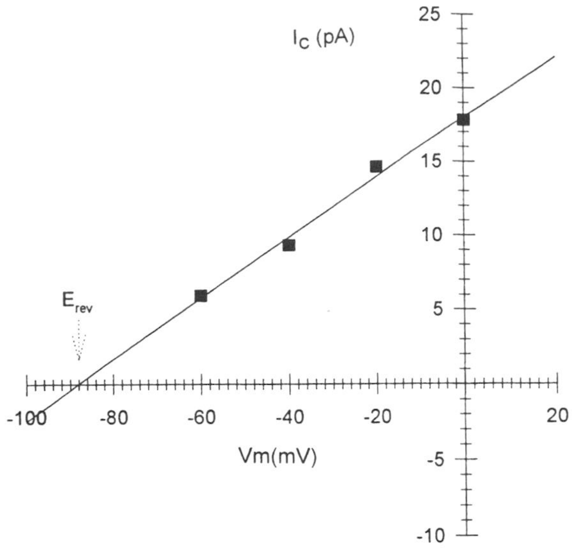

Figure 4.4. Relation of the current, *I*, to the holding potential, V_m, for the channel shown in Figure 4.3.

In a parallel series of experiments, the relative permeability of K versus Na was estimated by determining E_{rev} in the presence of asymmetric cations across the membrane (Figure 4.5). In this experiment the *trans* solution salt concentration was raised to the *cis* concentration with NaCl instead of KCl, and thus there is an asymmetric distribution of the cations, but not anions. Any shift in E_{rev} is therefore related to the change in chemical potentials across the membrane due to cations and proportional to their relative permeabilities. These relative permeabilities can be calculated from the constant field equation[51]:

$$E_{rev} = \frac{2.3RT}{F}\log\left(\frac{[K]_{trans} + \alpha[Na]_{trans}}{[K]_{cis} + \alpha[Na]_{cis}}\right)$$

where

$$\alpha = \frac{P_{Na}}{P_K}$$

In this equation the activities of the ions in solution are calculated from their concentrations.[52] Solving the expression yields a P_{Na}/P_K of approximately 0.09, indicating a good selectivity for K over Na.

Further examination of the recordings of channel activity presented in Figure 4.3 reveals another property of this channel. The channel activity is markedly reduced at positive values of V_m. This becomes apparent as a decrease in the amount of time that the channel spends in the open state and is estimated as the open time probability, P_o, which is the proportion of the time that the channel spends in the open state. Thus, the voltage dependence of the channel is illustrated in Figure 4.6 as the P_o versus V_m. What the data demonstrate is a very marked voltage dependence with a complete closure of the channel at positive potentials. In order to assess physiological function, a question remains as to the orientation of the channel in the planar bilayer. This was determined largely through the analysis of the inhibitor sensitivity of the reconstituted channel.

4.2.3.2. Inhibitors of the Channel

Electrophysiological studies on the intact tissue identified *LQV* and Ba^{2+} as effective blockers of the basolateral K

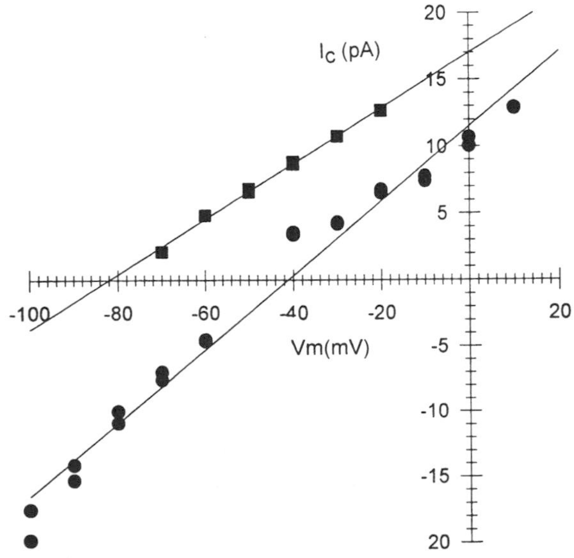

Figure 4.5. Determination of the ion selectivity of the channel for K versus Na. The square symbols represent the *I–V* relation in 150 mM KCl : 5 mM KCl *cis* : *trans*. The circles represent the *I–V* after the addition of 150 mM NaCl to the *trans* compartment.

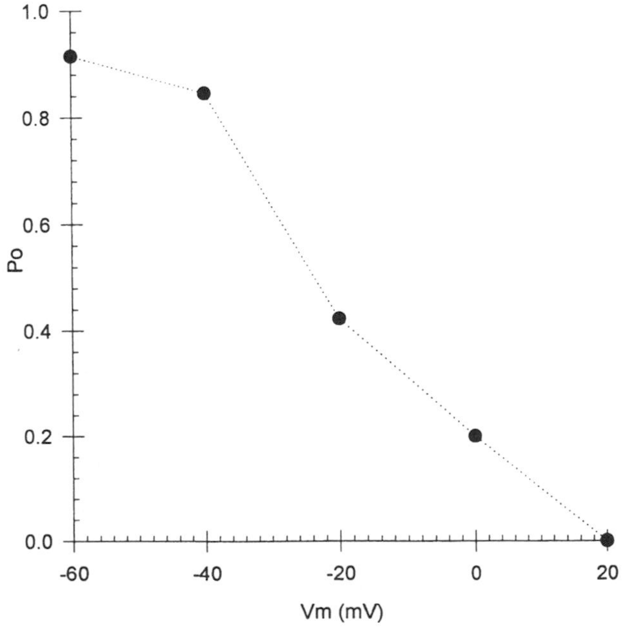

Figure 4.6. Relation of the open time probability, P_o, versus the holding potential, V_m, of the channel shown in Figure 4.3.

conductance when added to the serosal bathing solution. Thus, inhibition of single-channel activity was examined to identify specific blockers of the channel which in turn could be used to establish the orientation of the reconstituted channel, and to assist in establishing the identity of the channel. The venom from the scorpion *LQV* was previously shown to block the native K conductance when added to the basolateral membrane of the intact tissue. This venom was also an effective blocker of the channel in the reconstituted bilayer. In reconstitution experiments *LQV* was only effective as a channel blocker when added to the *trans* compartment and had no effect when added to the *cis*. Similarly, in the native epithelia, Ba^{2+} was found to block the K conductance from the basolateral surface. In the reconstituted planar bilayer, Ba^{2+} addition to the *trans* compartment at 5–10 mM markedly reduced the open time probability of the channel. Another well-characterized K channel blocker that acts at the inner pore of the channel near the inactivation site of inactivating K channels[53] is TEA^+. TEA^+ effects were assessed on channel activity and as with *LQV* there was a sidedness to the inhibition, only effective when added to the *cis* compartment. These results indicate there is a similarity to the basolateral conductance observed in the intact epithelial sheets, and, in addition, establish the orientation of the channel in the reconstituted bilayer. The sidedness of the inhibition is such that the vesicles must always fuse in a consistent manner. Referring to Figure 4.2, the inner, or cytoplasmic, face of the membrane faces the *cis* compartment, and the outer, or serosal, surface faces the *trans* compartment.

Charybdotoxin (CTX), a component of *LQV,* is a potent Ca^{2+}-activated K+ channel blocker with a K_i of less than 10 nM.[54,55] The most well-characterized component isolated from *LQV* is CTX. When purified CTX was tested on this channel it had no effect on channel activity, indicating that another component in the venom was the effective agent.[38] This observation helps to establish the identity of the channel when compared to other K channels and their response to different agents. The only other reported K channel in *Necturus* enterocytes is a Ca^{2+}-activated K channel. Patch clamp analysis of *Necturus* epithelia showed the presence of a Ca^{2+}-activated K channel in the basolateral membrane which is CTX sensitive.[56] The channel in the present studies is not Ca^{2+} dependent, and CTX insensitive indicating that it is distinct from the one identified in the patch clamp studies.

4.2.3.3. Structure–Function Relations

Structural homology with other K channels was examined in a series of experiments that utilized a synthetic analogue of the N-terminal 22 amino acids of the *Drosophila melanogaster Shaker B* K channel (ShB_{N22}). Hoshi *et al.*[57] and Zagotta *et al.*[58] demonstrated that the N-terminal 20 amino acids of the *Shaker* channel are responsible for the spontaneous, voltage-independent inactivation of the channel. According to the model, this peptide fragment is the inactivation component, or "ball" of the "ball and chain" model of K channel inactivation. The peptide folds into a globular structure that is connected to the channel through a peptide chain.[57–59] The globular head of the peptide interacts with a specific receptor region on the inner surface of the membrane. Further studies demonstrated that the synthetic analogue of this peptide fragment also inactivated the voltage-gated K channel from rat brain (RBK1) that has extensive homology with the *Shaker* K channel.[58] In studies of the reconstituted *Necturus* K channel,[60,61] addition of the peptide to the *trans* compartment was without effect on channel activity; however, *cis* addition resulted in the complete closure of the channel (Figure 4.7). The blockage of the channel was complete and irreversible. On the basis of site-directed mutagenesis of the native protein, Zagotta *et al.*[58] demonstrated two domains that were required for blocking activity. The ball has positively charged residues in positions 17–19. Trypsinization of the synthetic peptide which cleaves at these residues resulted in the loss of blocking activity in the reconstituted system. The Zagotta study also demonstrated the need for a hydrophobic residue in position 7. Several modifications of the peptide were synthesized and tested on the reconstituted *Necturus* channel (Figure 4.7). In the native peptide there is a leucine in position 7. Replacement of L7 with a glutamic acid residue abolished its ability to inactivate the channel. Replacement with another hydrophobic residue did not significantly alter activity. Replacement with the neutral glycine resulted in an increase in the flickering activity of the channel and a significant reduction in the P_o (Figure 4.7). The mechanism of interaction of the peptide with the channel was further examined in the reconstituted system.[60,61] An antibody against the synthetic peptide was used to immunoadsorb the peptide in an experiment. The channel was incorporated and recorded at −40 mV. The potential was raised to +20 mV, and the channel closed. The peptide, ShB_{N22}, was added to the *cis* chamber, incubated, and the anti-ShB_{N22} was added to the chamber. Immunoadsorption of the peptide resulted in the failure of the peptide to block the channel. In a parallel series of experiments, the reconstituted channel was initially treated with the well-characterized K-channel blocker, TEA. TEA alone resulted in complete blockage of the channel activity. However, the subsequent addition of ShB_{N22} to the chamber resulted in a reversal of the block and a significant increase in the mean open time of the channel. The result suggests that there is an interaction between the TEA and the peptide in the binding domain. Since TEA is a known pore blocker, the conclusion is that the peptide is a pore blocker that only interacts with the open channel. Taken together, these results are similar to what was observed with *Shaker B.* Thus, it may be surmised that there is significant structural homology in the region of the binding domain of this peptide on the *Necturus* K channel, a further indication that this may be a highly conserved region in a number of different types of K channels.

4.2.3.4. Regulation of the Channel

An important physiological regulator of the K channel is ATP. A series of experiments examining ATP and several of its analogues demonstrated that the channel itself is inhibited by physiological levels of ATP.[62] A necessary cofactor for the ATP-dependent inhibition is the presence of Mg^{2+} for inhibi-

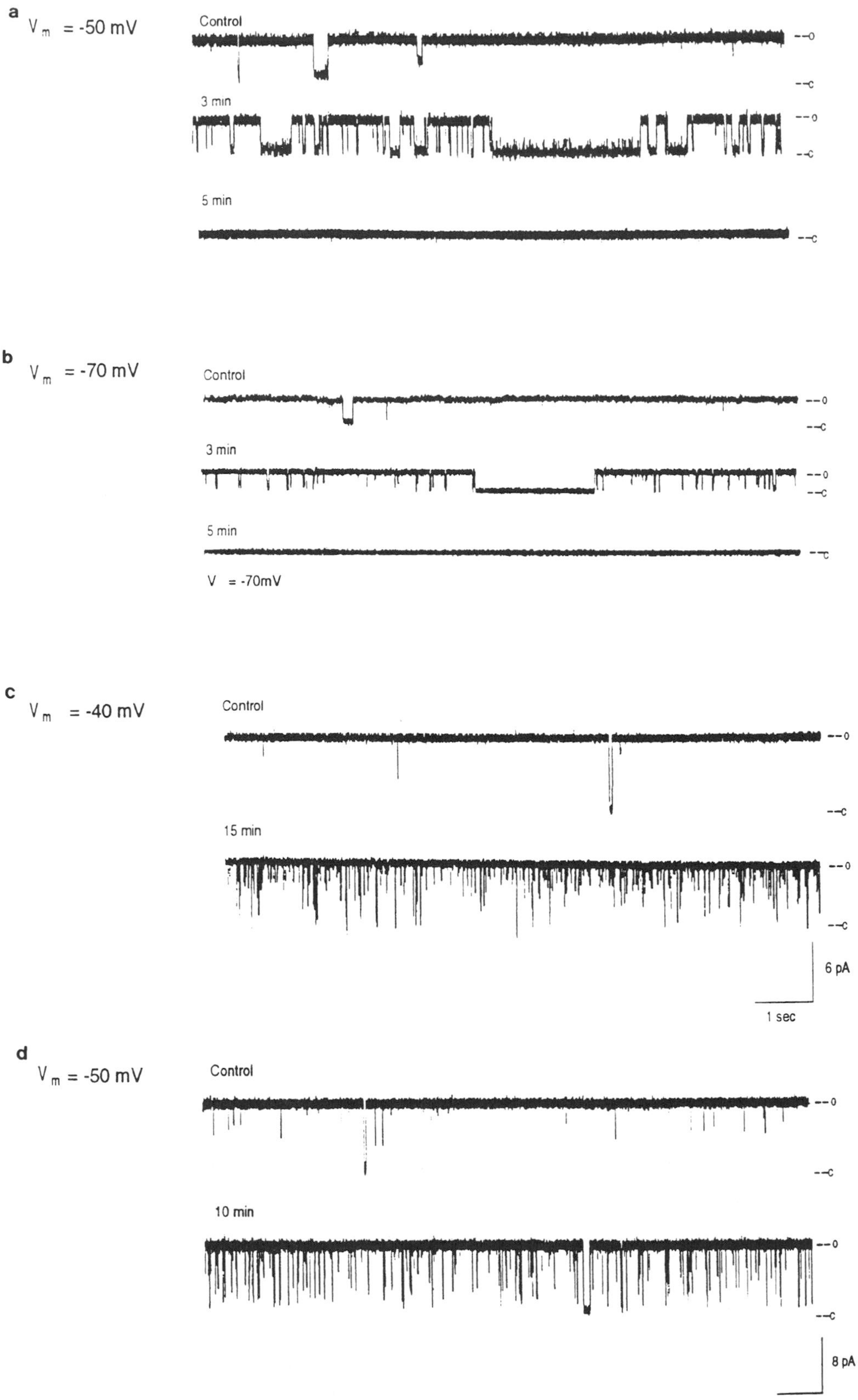

Figure 4.7 Effects of synthetic ShB_{N22} and its analogues on single-channel activities of the reconstituted K channel. (A) Control, ShB_{N22}; (B) L7F-ShB_{N22}; (C) L7K-ShB_{N22}; (D) L7G-ShB_{N22}. All peptides were added at a concentration of 50 μM to the *cis* compartment. The holding potentials are to the left of the recordings and the open and closed states are designated as "o" and "c," respectively. The calibration in panel C is the same for A and B. Note the different calibration for D.

tion to be observed. Addition of Na_2ATP to the *cis* compartment had no effect on channel activity up to a level of 25 mM. However, the subsequent addition of Mg^{2+} to the chamber led to a rapid inactivation of the K channel. Addition of Mg^{2+} alone had no significant effect on channel activity, although there was a slight decrease in the conductance of the channel. The addition of NaATP, or Mg^{2+} to the *trans* compartment had no effect on channel activity. Substitution of ATP with the nonhydrolyzable analogue of ATP, ATP-γ-S, resulted in a similar inactivation of the K channel only in the presence of Mg^{2+}. Finally, the addition of MgADP to the *cis* solution had no effect on channel activity. To the contrary, MgADP prevented the inhibitory effect of ATP, and was capable of reversing the MgATP block of the channel. On the basis of these data, it may be concluded that this channel is one of a class of channels designated K(ATP) which have been identified in a wide range of different tissues and organs.[63]

The identification of the channel as one of the K(ATP) types is further confirmed by analysis of the effects of the sulfonyl ureas on channel activity. These agents, long used as antihyperglycemics, are believed to exert their action by inactivation of K channel activity in pancreatic βcells leading to a cellular depolarization.[64] The shift in membrane potential activates voltage-sensitive Ca^{2+} channels allowing Ca^{2+} entry and stimulation of insulin secretion. Both tolbutamide and glibenclamide completely inactivated the reconstituted channel. The inactivation was only observed when the inhibitor was added to the *cis* chamber or cytoplasmic face of the channel. The physiological importance of these channels is that they provide a mechanism to couple the activity of the channel to the metabolic state of the cell via the ATP levels.

4.2.4. Purification of the Channel

The data obtained with the *Shaker* peptide suggested that there may be some structural homology among a variety of K channels. An IgG fraction was purified from an antiserum raised against the ShB_{N22} peptide.[61] Using the anti-ShB_{N22} we probed a blot of the *Necturus* basolateral membrane preparation.[44] The Western analysis revealed a single major protein of about 140 kDa, and two minor bands of 224 and 85 kDa, depending on the preparation.

4.2.4.1. Developing an Assay for Channel Proteins

A strategy for the purification of the immunostaining proteins was developed on the basis of the antibody binding properties.[44] Since the protein purification would depend on detergent solubilization of the membrane, a liposomal reconstitution system was deemed necessary to demonstrate solubilization of the transporter assay for the K channel, and as a vehicle to introduce the purified channel into the planar bilayers for single-channel characterization. The transport assay was used to detect channel activity in purified protein fractions. The assay was based on a procedure developed by Garty to measure channel-mediated transport in vesicles.[42] Initial experiments were carried out with a solubilized basolateral membrane fraction in order to characterize the assay as well as establish baseline values for the flux measurements. Vesicles were reconstituted by the freeze thaw sonication procedure of Kasahara and Hinkle.[65] An aliquot of the solubilized protein was mixed with preformed phospholipid vesicles, frozen in liquid N_2, thawed to room temperature, and sonicated briefly. The vesicles were prepared in a buffer that contained 100 mM K_2SO_4. Vesicles were passed through a cation exchange resin to remove extravesicular K, creating a large K_i–K_o gradient. Since the anion is impermeant, net flux of K is limited by electroneutrality. Vesicles are then added to a transport buffer that contains tracer $^{86}Rb^+$ to measure transport activity. Initial characterization included a time course of uptake, protein titration, and analysis of inhibitor sensitivity. Uptake of the isotope was linear with increasing protein concentrations and afforded roughly linear rates of uptake for the first several minutes of transport. The reconstituted transport activity was blocked by preincubation with TEA demonstrating inhibitor-sensitive ion fluxes.

4.2.4.2. Immunoisolation of Channel Proteins

The second step in the strategy was to demonstrate that the antibody was capable of immunoprecipitating the K channel. The basolateral membrane preparation was solubilized with 3-[(3-cholamidopropyl)-dimethylammonio]-1-propane sulfonate (CHAPS). The mixture was centrifuged and the soluble supernatant fraction treated with anti-ShB_{N22}. The extract was then treated with immunobeads and centrifuged to remove IgG's. Control experiments included immunobead exposure alone and treatment with the preimmune serum. The treated extract was then reconstituted into phospholipid vesicles and assayed for $^{86}Rb^+$ fluxes. There was no significant loss of activity in the preimmune, or immunobead-exposed extract compared to untreated controls. In contrast, the anti-ShB_{N22} IgG-treated extract showed a marked loss of transport activity, suggesting that the antibodies precipitated the channel, or at least a component of the functional channel. As a final stage in the purification, the purified anti-ShB_{N22} IgG fraction was coupled to CNBr-activated Sepharose beads in order to prepare an immunoaffinity resin to isolate the channel proteins. Basolateral membranes were solubilized as described above and incubated with the resin. The resin was then washed extensively to remove any unbound proteins. The column was eluted with pH 2.5 glycine buffer to remove the fraction of protein bound to the antibody. Aliquots of the eluant were reconstituted and assayed for transport activity. A fraction was obtained that showed a marked increase in the specific activity of Rb transport. Although we were unable to get an accurate protein concentration on the active fractions, uptake was demonstrated to be linear with increasing protein in control experiments.

4.2.4.3. Reconstitution of the Channel Protein in Planar Bilayers

Finally, the isolated channel was analyzed in planar bilayers for characterization and comparison to the native channel.[44] Proteoliposomes prepared from the immunoisolated protein were fused with the planar lipid bilayer following the standard protocol used for native membrane fractions. The biophysical properties of the purified channel were virtually indistinguishable from the native channel. The channel was blocked by the ShB_{N22} peptide at the same concentration as the native channel. On the basis of these comparisons, one may conclude that there is a significant enrichment of the voltage-gated K(ATP) channel in these studies. The specificity of the IgG purification suggests that there is some homology in the inhibitory domains of these two channels.

4.3. CHARACTERIZATION OF ION CHANNELS IN SUBCELLULAR MEMBRANES

Another useful application of resolution–reconstitution techniques is the study and characterization of ion channels in subcellular membranes such as the sarcoplasmic reticulum, endoplasmic reticulum, endocytotic vesicles, or partially purified plasma membrane fractions. These membranes are generally not amenable to analysis by patch clamp techniques because of their geometry, the nature of the membrane, and/or its localization within the cell. There are numerous techniques to isolate, characterize, and identify various membranes with regard to transport and enzymatic activities. There is also a wealth of reconstitution studies to serve as examples in the characterization of their ion channels as well as other transport proteins.[66–68]

4.3.1. Tracheal Apical Membrane Permeabilities

4.3.1.1. Ion Transport Studies

Ion transport studies in a highly enriched preparation of tracheal apical membrane vesicles identified both Na and Cl conductances.[36,39] The vesicles transported Na in an amiloride-sensitive fashion, and maximal rates of Na uptake were observed in the presence of permeant anions, including Cl. Similarly, $^{36}Cl^-$ uptake into the vesicles was stimulated by permeant cations, a K gradient in the presence of valinomycin, or an inwardly directed Na gradient. The Na-dependent uptake of Cl was inhibited by amiloride, suggesting parallel conductances for the two ions. Cl uptake was not inhibited by furosemide, suggesting that it was not mediated by the $NaKCl_2$ cotransport system. The radioisotopic technique is limited with certain isotopes because of the low available specific activities; thus, a more flexible flux analysis was developed to assess anion conductances in these membranes.[41,70] An anion-selective electrode was used to measure the efflux of the anion from preloaded membrane vesicles. The sensitivity of the assay is dependent on removing all extravesicular anions by passage through a Sephadex desalting column and elution into a reaction chamber fitted with the electrode. Maximal rates of efflux are achieved by making the vesicles in KCl media, and then assaying efflux in the presence of the K ionophore valinomycin. The valinomycin hyperpolarizes the membrane and allows net K efflux as counterion to the Cl moving out of the vesicle. In these experiments, Cl efflux from tracheal apical membrane vesicles was stimulated by the K conducting ionophore valinomycin, but was not supported by the electrically neutral nigericin. The electrodes also detect other halides and permitted the demonstration of a selectivity of the conductance for Cl over Br. This assay was convenient and efficient and was applicable to measure Cl fluxes in any vesicular system. Thus, it provided an assay to demonstrate the solubilization and reconstitution of the apical membrane Cl conductance. The reconstituted channel exhibited the same anion selectivity. Additional studies demonstrated the sensitivity of the conductance to DIDS, and was also labile to incubation at 37°C.

4.3.1.2. Planar Bilayer Studies of Tracheal Apical Membrane Ion Channels

In parallel studies, the same apical membrane vesicle preparation was reconstituted into planar phospholipid bilayers for single-channel analysis essentially as described above.[49] In all, five distinct channel types were identified, three anion channels and two cation channels. The most frequently observed channel (40 out of 56 observations) was a 71-pS channel that exhibited a linear *I–V* relation and an ion selectivity Cl > Br > F > I. The stilbene disulfonate, DIDS, produced a flickery block of open current at low concentrations (5–25 μM) and completely shut down the current at higher concentrations. The channel activity exhibited a rundown of activity with complete shutdown within 2–3 min following incorporation. Thus, there were a number of similarities between the net flux analysis and the single-channel characteristics. The ion selectivity of the reconstituted channel was comparable to that observed in the net flux studies in the native vesicles and reconstituted vesicles. The inhibitor sensitivity of the two preparations was comparable. Finally, both in the vesicle system and in the planar bilayers there was a clear lability or rundown of the channel resulting in a loss of activity. Taken together, these data suggest that the most frequently observed anion channel is the dominant membrane permeability mediating the Cl flux in the intact vesicle.

4.3.2. A Tracheal Subcellular Membrane Anion Channel

4.3.2.1. Biochemical Properties of the Membrane Fraction

During the course of an analysis of ion channels in other tracheal membrane fractions, a distinct subcellular membrane was identified that possessed an anion channel with biophysical properties characteristic of the plasma membrane anion channel described above. In these studies the ion channel and transport activity were used as a specific marker to assay for purification.[40,71] A purification scheme was developed that yielded a very highly enriched membrane fraction designated M_{PS}. Marker enzyme analyses failed to find any enrichment of the major membrane markers, including alkaline phosphatase (apical membrane) or (Na+K)-ATPase (basolateral membrane) in the M_{PS} preparation. Centrifugation on sucrose density gradients demonstrated a buoyant density of the membranes that is clearly distinct from either the apical or basolateral membranes. Analysis of the protein composition shows that several proteins are uniquely enriched in these membranes. One of these, a 200-kDa protein, was purified and used to raise an antiserum for immunological identification of the protein.[71] Of the various cellular fractions, trace amounts of the 200-kDa protein are found in most membrane fractions; however, there is a clear and dramatic enrichment in the M_{PS} fraction. The antiserum was used in immunolocalization studies of frozen sections of tracheal epithelia. The majority of, and densest, staining was in the region of the apical membranes of the cells; however, the staining was mostly present as punctate, or vesicular-like structures near the surface of the cells. These structures were visible at lower densities in the cytoplasm below the apical surface. Thus, the fraction designated M_{PS} appears to be a unique subcellular membrane which is distinct from the plasma membranes of these cells. Analysis of different tissues by Western blots shows that the 200-kDa protein is found in a number of different epithelial tissues, but not in muscle. Parallel studies on kidney cortex identified a membrane fraction that is virtually identical in all respects to that obtained from the trachea.[40,71]

4.3.2.2. Electrophysiology of the M_{PS} Anion Channel

The transport properties of M_{PS} have been characterized in the native membrane and in reconstituted channels. On the basis of net flux analysis there was approximately a 40-fold enrichment of the Cl conductance in M_{PS} compared to the crude homogenate, confirming that M_{PS} is a highly purified membrane preparation. Single-channel analysis was performed by fusion of M_{PS} with planar phospholipid bilayers. In the planar bilayer studies, only a single channel type has been observed in well over 100 observations. The channel is virtually identical to the channel that was the most frequently observed in the apical plasma membrane described above. The *I–V* relation of the tracheal and renal M_{PS} Cl channels are superimposable with that observed in the apical membrane. As with the apical membrane Cl channel, the ion selectivity sequence is Cl > Br > I. Its voltage dependence of open time probability and sensitivity to the stilbene disulfonate derivatives is indistinguishable from the apical Cl channel. The only property that differed was that the M_{PS} Cl channel did not exhibit the rundown observed with the plasma membrane channel. Once incorporated into the bilayer, the channel activity could be recorded for hours. A likely conclusion based on the combined results of biochemical, immunocytochemical, and electrophysiological data is that these are the same channel but present in two different locations. They exhibit an altered regulation, in that the apical channel exhibits a spontaneous rundown and can be activated by cAMP-dependent protein kinase, and the M_{PS} channel exhibits neither of these properties. This altered regulation suggests that there may be additional regulatory subunits involved in the apical function of this channel. Thus, the M_{PS} may be an intermediate stage in the processing or trafficking of the channel to the apical membrane. The localization of a high density of this membrane in the apical region suggests that it plays a role in the regulation of this membrane permeability or composition.

4.4. CONFIRMATION OF ION CHANNEL ACTIVITY

4.4.1. Cystic Fibrosis Transmembrane Conductance Regulator

Studies on epithelial tissues affected in cystic fibrosis (CF) have revealed a defect in the ionic permeability of the apical membrane. Several early studies indicated that the major defect was associated with Cl impermeability in epithelia. Analysis of the conductances of the reabsorptive duct of the sweat gland from normals and cystics clearly demonstated CF ducts to have very low Cl permeabilities relative to their Na permeability.[72] Thus, the characteristically high salt content of CF sweat was related to a failure to reabsorb NaCl because of decreased Cl permeability. Other studies on the secretory coil of the sweat gland also identified a decreased Cl permeability. In the case of the secretory coil, the gland failed to respond to β-adrenergic stimulation of Cl secretion.[73] Subsequent studies in airway epithelia from cystics also identified a failure to activate Cl channels in the apical membrane in response to normal secretagogues.[74–76] Thus, a model evolved that assumed that the defect in CF was likely related to the Cl channel or a protein involved in its activation.[77] On the discovery of the gene, and identification of the gene product, there was considerable debate as to the actual function of the protein. It bore the strongest resemblance to P-glycoprotein believed to be a transporting protein, but not necessarily an ion channel. The gene product was termed the cystic fibrosis transmembrane conductance regulator (CFTR). Complicating the interpretation of the structural data was the subsequent finding that the most common mutation of CFTR resulted in the failure to traffic the protein to the plasma membrane.[78] Processing of the mutant form was arrested and it accumulated in a subcellular compartment. These findings suggested that CFTR played a role in the regulation of membrane trafficking; thus, a likely mechanism of the defect would be the failure to recruit the appropriate anion channel to the apical cell surface. The functional defect in CF was not necessarily a defect in the channel *per se,* but merely the inability to get the appropriate channel to the membrane in response to the correct signals. Evidence supporting the ion channel model came from studies in which the defect in CF cells could be repaired by expression of wt CFTR.[79] Expression of engineered site-directed mutants demonstrated an altered ion selectivity.[80] Some ambiguities remained as to the role(s) of CFTR in Cl secretion. To address these questions, CFTR was purified from an insect cell line transfected to overproduce CFTR and reconstituted into planar phospholipid bilayers.[81] The reconstituted protein exhibited similar biophysical properties to the native channel as identified by patch clamp analysis. In addition, it demonstrated a similar profile of regulation as the native channel.

The results helped to demonstrate that CFTR has a complex role in cellular function contributing to both the trafficking and remodeling of cellular membranes, and mediating the Cl secretion.

4.4.2. Protein Replacement Therapy

A therapeutic role for reconstitution studies is projected from experiments testing the feasibility of protein replacement therapy in diseases in which there is a defective or missing protein. CFTR is not present or defective in affected individuals.[82] A novel approach to treatment of CF is to restore function through the addition of exogenous wild type protein.[83] Current studies are using a model cell line, Chinese hamster ovary cells (CHO) transfected with wt CFTR. CHO cells produce fully processed CFTR which is translocated to the plasma membrane. Functional CFTR was demonstrated by cAMP-dependent increase in membrane Cl permeability in the transfected cells. The functional CFTR was harvested by isolating the plasma membranes of the transfected CFTR-producing cells. The target cell line in these studies was a fibroblast cell line (Ha3b) that is devoid of cAMP-dependent Cl permeabilities and that expresses the viral fusion protein hemagglutinin (HA) of influenza virus. HA is expressed as a precursor in the target Ha3b cells and is activatable by trypsinization. Overall fusion of the CHO plasma membranes to Ha3b could be assessed by morphological changes in the target cell. Functional transfer of CFTR from the CHO-CFTR plasma membranes to the Ha3b cells was detected by assessing the cAMP-stimulated Cl permeability in target cells. Transfer of CFTR was only observed in the trypsin-activated Ha3b cells. If the HA was not activated, or if control CFTR-deficient CHO plasma membranes were used, there was no increase in the cAMP-dependent Cl permeability. There are clearly limitations to this approach at present, namely, low efficiency of expression and low efficiency of functional transfer of CFTR. However, future studies will begin to dissect the necessary components to effect successful transfer which should allow the construction of an artificial reconstituted vector to mediate more efficient transfer of functional protein.[83]

REFERENCES

1. Lester, H. (1988). Heterologous expression of excitability proteins: Route to more specific drugs? *Science* **241:**1057–1063.
2. Racker, E. (1977). Perspectives and limitations of resolution–reconstitution experiments. *J. Supramol. Struct.* **6:**215–228.
3. Miller, C., and Racker, E. (1979). Reconstitution of membrane transport functions. In *The Receptors* (R. D. O'Brien, ed.), Vol. 1, Plenum Press, New York, pp. 1–31.
4. Racker, E., Violand, B., O'Neal, S. A. M., and Telford, J. (1979). Reconstitution, a way of biochemical research; some new approaches to membrane bound enzymes. *Arch. Biochem. Biophys.* **198:**470–477.
5. Wright, E. M. (1984). Electrophysiology of plasma membrane vesicles. *Am. J. Physiol.* **246:**F363–F372.
6. Illsley, N. P., and Verkman, A. S. (1987). Membrane chloride transport measured using a chloride-sensitive fluorescent probe. *Biochemistry* **26:**1215–1219.
7. Tamkun, M. M., Talvenheimo, J. A., and Catterall, W. A. (1984). The sodium channel from rat brain. Reconstitution of neurotoxin-activated ion flux and scorpion toxin binding from purified components. *J. Biol. Chem.* **259:**1676–1688.
8. Barchi, R. L., Tanaka, J. C., and Furman, R. E. (1984). Molecular characteristics and functional reconstitution of muscle voltage-sensitive sodium channels. *J. Cell. Biochem.* **26:**135–146.
9. Hartshorne, R. P., Keller, B. U., Talvenheimo, J. A., Catterall, W. A., and Montal, M. (1986). Functional reconstitution of purified sodium channels from brain in planar lipid bilayers. *Ann. N.Y. Acad. Sci.* **479:**293–305.
10. Barhanin, J., Pauron, D., Lombet, A., Hanke, W., Boheim, G., and Lazdunski, M. (1984). New scorpion toxins with a very high affinity for Na^+ channels. Biochemical characterization and use for the purification of Na^+ channels. *J. Physiol. (Paris)* **79:**304–308.
11. Kraner, S. D., Tanaka, J. C., and Barchi, R. L. (1985). Purification and functional reconstitution of the voltage-sensitive sodium channel from rabbit T-tubular membranes. *J. Biol. Chem.* **260:**6341–6347.
12. Curtis, B. M., and Catterall, W. A. (1986). Reconstitution of the voltage-sensitive calcium channel purified from skeletal muscle transverse tubules. *Biochemistry* **25:**3077–3083.
13. Prestipino, G., Valdivia, H. H., Li'evano, A., Darszon, A., Ramirez, A. N., and Possani, L. D. (1989). Purification and reconstitution of potassium channel proteins from squid axon membranes. *FEBS Lett.* **250:**570–574.
14. Santacruz, T. L., Perozo, E., and Papazian, D. M. (1994). Purification and reconstitution of functional Shaker K^+ channels assayed with a light-driven voltage-control system. *Biochemistry* **33:**1295–1299.
15. Sun, T., Naini, A. A., and Miller, C. (1994). High-level expression and functional reconstitution of Shaker K^+ channels. *Biochemistry* **33:**9992–9999.
16. Huganir, R. L., Schell, M. A., and Racker, E. (1979). Reconstitution of the purified acetylcholine receptor from Torpedo californica. *FEBS Lett.* **108:**155–160.
17. Popot, J. L., Cartaud, J., and Changeux, J. P. (1981). Reconstitution of functional acetylcholine receptor. Incorporation into artificial lipid vesicles and pharmacology of the agonist-controlled permeability changes. *Eur. J. Biochem.* **118:**203–214.
18. Lindstrom, J., Anholt, R., and Einarson, B. (1980). Purification of acetylcholine receptors, reconstitution into lipid vesicles, and study of agonist-induced cation channel regulation. *J. Biol. Chem.* **255:**8340–8350.
19. Nelson, N., Anholt, R., Lindstrom, J., and Montal, M. (1980). Reconstitution of purified acetylcholine receptors with functional ion channels in planar lipid bilayers. *Proc. Natl. Acad. Sci. USA* **77:**3057–3061.
20. Matesic, D., and Liebman, P. A. (1987). cGMP-dependent cation channel of retinal rod outer segments. *Nature* **326:**600–603.
21. Wohlfart, P., Muller, H., and Cook, N. J. (1989). Lectin binding and enzymatic deglycosylation of the cGMP-gated channel from bovine rod photoreceptors. *J. Biol. Chem.* **264:**20934–20935.
22. Corondado, R., and Campbell, K. P. (1987). Purified ryanodine receptor from skeletal muscle sarcoplasmic reticulum is the Ca^{2+}-permeable pore of the calcium release channel. *J. Biol. Chem.* **262:**16636–16643.
23. Lai, F. A., Erickson, H. P., Rousseau, E., Liu, Q. Y., and Meissner, G. (1988). Purification and reconstitution of the calcium release channel from skeletal muscle. *Nature* **331:**315–319.
24. Paucek, P., Mironova, G., Mahdi, F., Beavis, A. D., Woldergiorgis, G., and Garlid, K. D. (1992). Reconstitution and partial purification of the glibenclamide-sensitive, ATP-dependent K^+ channel from rat liver and beef heart mitochondria. *J. Biol. Chem.* **267:**26062–26069.

25. Klaerke, D. A., Karlish, S. J., and Jorgensen, P. L. (1987). Reconstitution in phospholipid vesicles of calcium-activated potassium channel from outer renal medulla. *J. Membr. Biol.* **95:**105–112.
26. Klaerke, D. A., Petersen, J., and Jorgensen, P. L. (1987). Purification of Ca^{2+}-activated K^+ channel protein on calmodulin affinity columns after detergent solubilization of luminal membranes from outer renal medulla. *FEBS Lett.* **216:**211–216.
27. Liu, S., Dubinsky, W. P., Haddox, M. K., and Schultz, S. G. (1991). Reconstitution of isolated Ca^{2+}-activated K^+ channel proteins from basolateral membranes of rabbit colonocytes. *Am. J. Physiol.* **261:** C713–C717.
28. Garcia-Calvo, M., Knaus, H. G., McManus, O. B., Giangiacomo, K. M., Kaczorowski, G. J., and Garcia, M. L. (1994). Purification and reconstitution of the high-conductance, calcium-activated potassium channel from tracheal smooth muscle. *J. Biol. Chem.* **269:**676–682.
29. Akabas, M. H., Redhead, C., Edelman, A., Cragoe, E. A., Jr., and Al-Awqati, Q. (1989). Purification and reconstitution of chloride channels from kidney and trachea. *Science* **244:**1469–1472.
30. Schlesinger, P. H. (1990). Purification of a stilbene sensitive chloride channel and reconstitution of chloride conductivity into phospholipid vesicles. *Biochem. Biophys. Res. Commun.* **171:**920–925.
31. Ran, S., and Benos, D. J. (1991). Isolation and functional reconstitution of a 38-kDa chloride channel protein from bovine tracheal membranes. *J. Biol. Chem.* **266:**4782–4788.
32. Goldberg, A. F., and Miller, C. (1991). Solubilization and functional reconstitution of a chloride channel from Torpedo californica electroplax. *J. Membr. Biol.* **124:**199–206.
33. Finn, A. L., Gaido, M. L., and Dillard, M. (1992). Reconstitution and regulation of an epithelial chloride channel. *Mol. Cell Biochem.* **114**(1–2):21–26.
34. Jovov, B., Ismailov, I. I., and Benos, D. J. (1995). Cystic fibrosis transmembrane conductance regulator is required for protein kinase A activation of an outwardly rectified anion channel purified from bovine tracheal epithelia. *J. Biol. Chem.* **270:**1521–1528.
35. Donaldson, P. J., and Peracchia, C. (1991). Channel reconstitution in liposomes and planar bilayers with HPLC-purified MIP26 of bovine lens. *J. Membr. Biol.* **124:**21–32.
36. Langridge-Smith, J. E., Field, M., and Dubinsky, W. P. (1983). Isolation of transporting membrane vesicles from bovine tracheal epithelium. *Biochim. Biophys. Acta* **731:**318–328.
37. Haase, W., Schafer, A., Murer, H., and Kinne, R. (1978). Studies on the orientation of brush-border membrane vesicles. *Biochem. J.* **172:** 57–62.
38. Costantin, J., Alcalen, S., de Sousa-Otero, A., Dubinsky, W. P., and Schultz, S. G. (1989). Reconstitution of an inwardly rectifying potassium channel from the basolateral membranes of *Necturus* enterocytes into planar lipid bilayers. *Proc. Natl. Acad. Sci. USA* **86:**5212–5216.
39. Hochstadt, J., Quinlan, D. C., Rader, R. L., Chen-Chang, L., and Dowd, D. (1975). Use of isolated membrane vesicles in transport studies. In *Methods in Membrane Biology* (E. D. Korn, ed.), Plenum Press, New York, Vol. 5, pp. 117–162.
40. Preston, C. L., Calenzo, M. A., and Dubinsky, W. P. (1992). Isolation of a chloride channel-enriched membrane fraction from tracheal and renal epithelia. *Am. J. Physiol.* **263:**C879–C887.
41. Dubinsky, W. P., and Monti, L. B. (1986). Resolution of apical from basolateral membranes of shark rectal gland. *Am. J. Physiol.* **251:** C721–C726.
42. Garty, H., Rudy, B., and Karlish, S. J. (1983). A simple and sensitive procedure for measuring isotope fluxes through ion-specific channels in heterogeneous populations of membrane vesicles. *J. Biol. Chem.* **258:**13094–13099.
43. Woodbury, D. J., and Miller, C. (1990). Nystatin induced liposome fusion. *Biophys. J.* **58:**833–839.
44. Dubinsky, W. P., Mayorga-Wark, O., Garretson, L. T., and Schultz, S. G. (1993). Immunoisolation of a K+ channel from basolateral membranes of Necturus enterocytes. *Am. J. Physiol.* **1265:**C548–C555.
45. Schultz, S. G. (1981). Homocellular regulatory mechanisms in sodium transporting epithelia. Avoidance of extinction by "flush through." *Am. J. Physiol.* **241:**F579–F590.
46. Gunter-Smith, P., Grasset, E., and Schultz, S. G. (1982). Sodium-coupled amino acid and sugar transport by Necturus small intestine: An equivalent electrical circuit analysis of a rheogenic co-transport system. *J. Membr. Biol.* **66:**25–39.
47. Mueller, P., and Rudin, D. O. (1969). Bimolecular lipid membranes. Techniques of formation, study of electrical properties, and induction of ionic gating phenomenon. In *Laboratory Techniques in Membrane Biophysics* (H. Passow and R. Stampfli, eds.), Springer-Verlag, Berlin, pp. 141–156.
48. Miller, C., ed. (1986). *Ion Channel Reconstitution,* Plenum Press, New York.
49. Valdivia, H. H., Dubinsky, W. P., and Coronado, R. (1988). Reconstitution and phosphorylation of chloride channels from airway epithelial membranes. *Science* **242:**1441–1444.
50. Alvarez, O. (1986). How to set up a bilayer system. In *Ion Channel Reconstitution* (C. Miller, ed.), Plenum Press, New York, pp. 115–130.
51. Schultz, ed. (1980). *Basic Principles of Membrane Transport,* Cambridge University Press, London.
52. Robinson, R. A., and Stokes, R. H. (1959). *Electrolyte Solutions,* 2nd ed., Academic Press, New York.
53. Grissmer, S., and Cahalan, M. (1989). TEA^+ prevents inactivation while blocking open K^+ channels in human T lymphocytes. *Biophys. J.* **55:**203–206.
54. Latorre, R. (1986). The large calcium-activated potassium channel. In *Ion Channel Reconstitution* (C. Miller, ed.), Plenum Press, New York, pp. 451–467.
55. MacKinnon, R., Heginbotham, L., and Abramson, T. (1990). Mapping the receptor site for charybdotoxin, a pore blocking potassium channel inhibitor. *Neuron* **5:**767–771.
56. Sheppard, D. N., Giraldez, F., and Sepulveda, F. V. (1988). Kinetics of voltage- and Ca^{2+}-activation and Ba^{2+} blockade of a large conductance K^+ channel from *Necturus* enterocytes. *J. Membr. Biol.* **105:** 67–75.
57. Hoshi, T., Zagotta, W. N., and Aldrich, R. W. (1990). Biophysical and molecular mechanisms of *Shaker* potassium channel inactivation. *Science* **250:**533–538.
58. Zagotta, W. N., Hoshi, T., and Aldrich, R. W. (1990). Restoration of inactivation in mutants of *Shaker* potassium channels by a peptide derived from *ShB*. *Science* **250:**568–571.
59. Stuhmer, W., Conti, F., Suzuki, H., Wang, X. D., Noda, M., Kubo, H., and Numa, S. (1989). Structural parts involved in activation and inactivation of the sodium channel. *Nature* **339:**597–603.
60. Dubinsky, W. P., Mayorga-Wark, O., and Schultz, S. G. (1992). A peptide from *Drosophila* Shaker K^+ channel inhibits voltage-gated K^+ channels in basolateral membranes of *Necturus* enterocytes. *Proc. Natl. Acad. Sci. USA* **89:**1770–1774.
61. Mayorga-Wark, O., Costantin, J., Dubinsky, W. P., and Schultz, S. G. (1993). Effects of a *Shaker* K^+ channel peptide and trypsin on a K^+ channel in *Necturus* enterocytes. *Am. J. Physiol.* **265:**C541–C547.
62. Mayorga-Wark, O., Dubinsky, W. P., and Schultz, S. G. (1995). Reconstitution of a K(ATP) channel from basolateral membranes of *Necturus* enterocytes. *Am. J. Physiol.* **269:**C464–C471.
63. Ashcroft, S. J. H., and Ashcroft, F. M. (1990). Properties and function of ATP-sensitive K-channels. *Cell Signalling* **2:**197–214.
64. Edwards, G., and Weston, A. H. (1993). The pharmacology of ATP-sensitive potassium channels. *Annu. Rev. Pharmacol. Toxicol.* **33:** 597–637.

65. Kasahara, M., and Hinkle, P. C. (1977). Reconstitution and purification of the D-glucose transporter from human erythrocytes. *J. Biol. Chem.* **252:**7384–7390.
66. Miller, C., and Racker, E. (1976). Ca^{2+}-induced fusion of fragmented sarcoplasmic reticulum with artificial planar bilayers. *J. Membr. Biol.* **30:**283–300.
67. Reinhart, P. H., Chung, S., and Levitan, I. B. (1989). A family of calcium-dependent potassium channels from rat brain. *Neuron* **2:**1031–1041.
68. Ran, S., Fuller, C. M., Arrate, M. P., Latorre, R., and Benos, D. J. (1992). Functional reconstitution of a chloride channel protein from bovine trachea. *J. Biol. Chem.* **267:**20630–20637.
69. Langridge-Smith, J. E., Field, M., and Dubinsky, W. P. (1984). Cl^- transport in apical plasma membrane vesicles isolated from bovine tracheal epithelium. *Biochim. Biophys. Acta* **777:**84–92.
70. Dubinsky, W. P., and Monti, L. B. (1986). Solubilization and reconstitution of a chloride transporter from tracheal apical membrane. *Am. J. Physiol.* **251:**C713–C720.
71. Dubinsky, W. P., Preston, C. L., Calenzo, M. A., White, G. J., and Decker, E. R. (1992). Immunolocalization of chloride-transporting membrane vesicles in tracheal epithelial cells. *Am. J. Physiol.* **263:** C888–C895.
72. Quinton, P. M. (1983). Chloride impermeability in cystic fibrosis. *Nature* **301:**421–422.
73. Sato, K., and Sato, F. (1984). Defective β-adrenergic response of cystic fibrosis sweat glands in vivo and in vitro. *J. Clin. Invest.* **73:** 1763–1771.
74. Widdicombe, J. H. (1986). Cystic fibrosis and β-adrenergic response of airway epithelial cell cultures. *Am. J. Physiol.* **251:**R818–R822.
75. Welsh, M. J., Anderson, M. P., Rich, D. P., Berger, H. A., Denning, G. M., Ostedgaard, L. S., Sheppard, D. N., Cheng, S. H., Gregory, R. J., and Smith, A. E. (1992). Cystic fibrosis transmembrane conductance regulator: A chloride channel with novel regulation. *Neuron* **8:**821–829.
76. Frizzell, R. A. (1993). The molecular physiology of cystic fibrosis. *NIPS* **8:**117–120.
77. Riordan, J. R., Rommens, J. M., Karem, B.-S., Alon, N., Rozmahel, R., Grzelczak, Z., Zielenski, J., Lok, S., Plausic, N., Chou, J. L., Drumm, M. L., Iannuzzi, M. C., Collins, F. S., and Tsui, L. C. (1989). Identification of the cystic fibrosis gene: Cloning and characterization of complementary DNA. *Science* **245:**1066–1073.
78. Cheng, S. H., Gregory, R. J., Marshall, J., Paul, S., Souza, D. W., White, G. A., O'Riordan, C. R., and Smith, A. E. (1990). Defective intracellular transport and processing of CFTR is the molecular basis of most cystic fibrosis. *Cell* **63:**827–834.
79. Rich, D. P., Anderson, M. P., Gregory, R. J., Cheng, S. H., Paul, S., Jefferson, D. M., McCann, J. D., Klinger, K. W., Smith, A. E., and Welsh, M. J. (1990). Expression of cystic fibrosis transmembrane conductance regulator corrects defective chloride channel regulation in cystic fibrosis airway epithelial cells. *Nature* **347:**358–363.
80. Anderson, M. P., Gregory, R. J., Thompson, S., Souza, D. W., Paul, S., Mulligan, R. C., Smith, A. E., and Welsh, M. J. (1991). Demonstration that CFTR is a chloride channel by alteration of its anion selectivity. *Science* **253:**202–205.
81. Bear, C. E., Li, C., Kartner, N., Bridges, R. J., Jensen, T. J., Ramjeesingh, M., and Riordan, J. R. (1992). Purification and functional reconstitution of the cystic fibrosis transmembrane conductance regulator (CFTR). *Cell* **68:**809–818.
82. Welsh, M. J., and Smith, A. E. (1993). Molecular mechanisms of CFTR chloride channel dysfunction in cystic fibrosis. *Cell* **73:** 1251–1254.
83. Marshall, J., Fang, S., Ostedgaard, L. S., O'Riordan, C. R., Ferrara, D., Amara, J. F., Hoppe, H., IV, Scheule, R. K., Welsh, M. J., Smith, A. E., and Cheng, S. H. (1994). Stoichiometry of recombinant cystic fibrosis transmembrane conductance regulator in epithelial cells and its functional reconstitution into cells in vitro. *J. Biol. Chem.* **269:** 2987–2995.

5

Permeability and Conductance of Ion Channels
A Primer

David C. Dawson

5.1. INTRODUCTION

The purpose of this chapter is to provide an introduction to conductance and permeability, the two phenomenological parameters that are universally used to describe the process of ion permeation through channels. Both of these parameters contain information about how easy or difficult it is for an ion to enter a channel, move through it, and exit on the other side, but in some cases conductance and permeability can report different aspects of the permeation process. This difference arises because it is possible to use the term *permeability* in two different ways. Permeability can be used as a measure of how readily an ion will cross a membrane via an open channel. I will refer to this as the *apparent* permeability because it is a reflection of the nature of the interaction of a particular test ion with the channel, but can also include a contribution due to competition from other ions that can limit accessibility of the channel to the test ion. It is possible in principle, however, to circumvent these latter effects experimentally, and determine what I shall call the *intrinsic* permeability, that is, a measure of the interaction of the test ion with the channel in the absence of competition from other ions. This distinction will be important for understanding the significance of selectivity measurements that employ either *permeability ratios* estimated from reversal potentials or *conductance ratios* estimated from current–voltage relations. Channel conductance and conductance ratios are directly related to the apparent permeability whereas, under ideal conditions, reversal potentials measure the ratio of the intrinsic permeabilities. These two analyses can provide very different, but complementary, information about of the permeation process.

I will begin by developing operational definitions of conductance and permeability, the latter being based on a hypothetical tracer flux experiment conducted on a single channel. This leads to the development of a relation between permeability ratios and reversal potentials. In the next section the classical Nernst–Planck electrodiffusion model is used to explore the relation between conductance and permeability for channels in which permeant ions do not interact and to clarify the operational definition of conductance when the i–V plot is nonlinear. In the final section I introduce two of the modifications required to account for the real properties of ion channels, ion binding and ion–ion interactions.

5.2. MEASURABLE PROPERTIES OF ION CHANNELS: CONDUCTANCE AND PERMEABILITY

As a starting point it will be useful to adopt a standard physical configuration for a model channel that we will utilize for the remainder of this discussion. The channel is envisioned as embedded in a planar lipid bilayer (Figure 5.1) that separates two aqueous solutions, the composition of which we may vary at will. The bathing solutions are equipped with two pairs of electrodes, one pair for measuring the transmembrane electrical potential difference and the other for passing current across the membrane. In most cases it will be convenient to assume that the electrodes are connected to a voltage clamp device so that the transmembrane potential may be varied by passing current across the membrane. This configuration is the minimum required to determine the current–voltage relation for the channel, the starting point for any characterization of the conduction process.

5.2.1. Conductance: Ohm's Law

Any electrophysiological analysis of the conduction process must involve, in one way or another, the application of Ohm's law. We will see, however, that precisely how we apply Ohm's law will depend on the physical configuration of the channel, particularly as regards the composition of the bathing solutions, so as a point of departure we consider the bathing solutions to contain identical KCl concentrations so that the only driving force for ion flow is the transmembrane electrical potential, V_m, and current flow through the open channel is described by the equation

$$i = \gamma V_m \tag{5.1}$$

David C. Dawson • Department of Physiology, The University of Michigan Medical School, Ann Arbor, Michigan 48109.

Molecular Biology of Membrane Transport Disorders, edited by S.G. Schultz *et al.* Plenum Press, New York, 1996

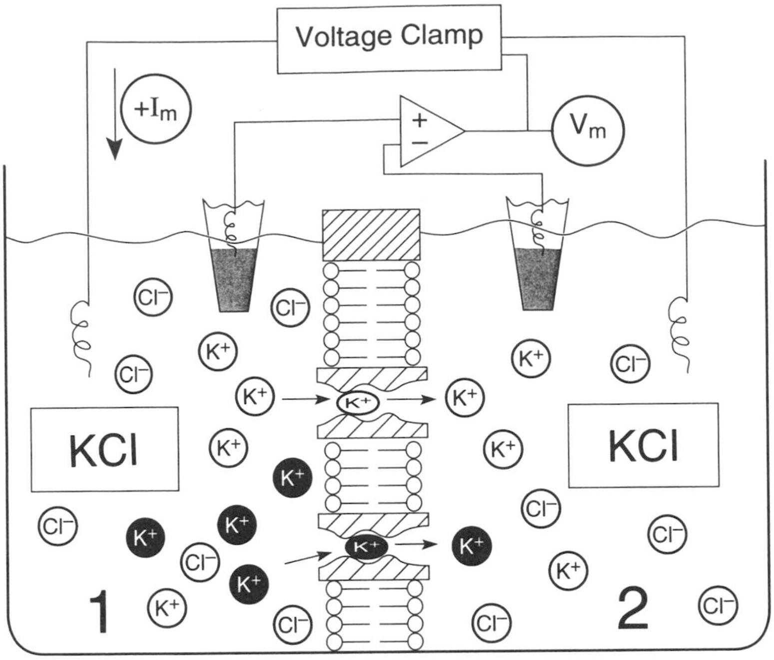

Figure 5.1. Diagrammatic representation of the physical setting in which conductance and permeability are measured. Shown is a lipid bilayer containing K^+-selective channels bathed by KCl solutions and connected through appropriate electrodes to a voltage clamp that can report membrane potential (V_m) and membrane current (I_m). Sign conventions are chosen such that a positive V_m will drive a positive I_m from side 1 to side 2. The measurement of single-channel current at different values of V_m defines the K^+ conductance of the channel. The black K^+ ions represent radioactively labeled tracer ions. Initially added to side 1, the rate of tracer flow through the channel into side 2 is used to operationally define the K^+ permeability of the channel.

where i is the total current flow through the channel in units of amperes (1 A = 1 coulomb/sec), V_m is the transmembrane electrical potential in volts, and γ is the single-channel conductance in units of siemens (1 S = 1 A/volt). In measurements of single-channel current the appropriate units are most likely to be a voltage in the range of 10–100 millivolts (1 mV = 10^{-3} V), associated with currents in the range of picoamperes (1 pA = 10^{-12} A) and single-channel conductances ranging from 5 to 500 picosiemens (pS). A transmembrane voltage of 100 mV, for example, applied across a channel of conductance 10 pS would result in a current of 1 pA.

Any practical application of Eq. (5.1) requires that we adopt sign conventions for voltage and current. The necessity for sign conventions is a result of the fact that instruments used for the measurement of voltage and current can be connected one of two ways (see Ref. 2 for discussion). Here we assume that our measuring apparatus is hooked up such that current flow from side 1 to side 2 is defined as positive, as is the outward current in a cellular preparation, and the sign of the membrane voltage will reflect the sign of side 1 (inside of cell) so that if V_m is positive, side 1 is positive.

The definition of γ is completely independent of the ion that is the actual charge carrier. In the present example, for instance, the channel current could be due to the movement of either K^+ or Cl^-, or both of these ions in any proportion. The operational definition of γ is based solely on the movement of charge, and this distinguishes conductance from *permeability,* for which the definition is more likely to be based on the behavior of a specific ion.

Current flow through the channel results from the movement of ions and it is useful at the outset to look at Ohm's law from the perspective of ionic flow. Let us assume, for example, that the channel is perfectly K^+ selective, and a current of 1 pA flows that is entirely due to the net flow of K^+ ions from side 1 to side 2. Faraday's constant (F = 96,500 C/equivalent) tells us the amount of charge associated with a mole of univalent ions, so that we can convert from units of charge flow (amperes) to units of mass flow (moles/time) and, using Avogadro's number ($N_A \sim 10^{23}$ ions/mole), convert to units of ions per second. For example, 1 pA = (10^{12} C/sec) (1 mole K^+/96,500 C)(10^{23} ions/mole) = $\sim 10^6$ K^+ ions/sec. Thus, a single-channel current of 1 pA, relatively small by household standards, represents the flow of a million ions per second. This high turnover rate is generally regarded as a signal feature of ion channels, a consequence of the fact that, unlike carriers or pumps, ion conduction through a pore is not coupled to conformational cycling of the protein.[2–4]

By measuring i at different values of V we obtain the current–voltage relation (i–V plot) for the channel, which is the basis for the operational definition of the single-channel conductance. The definition of γ is most unambiguous when the i–V plot can be described by a straight line, for which γ is the slope. Clearly, γ is a measure of how readily ions move through the channel and as such, is a concise summary of all of the physical processes involved in the entry of an ion into the channel from the aqueous solution, the translocation of the ion through the channel and the exit on the other side. One motivation for analyzing the physical basis for ion conduction is the desire to use ions as probes that, by virtue of their conduction rate, report the properties of the pore; properties that we are now in a position to modify, for example, by substituting one amino acid residue for another.

It is useful at this point to consider the form of Eq. (5.1). Most important is the fact that i is zero and when V is zero. This is an expression of the fact that the channel is a purely dissipative or permissive element; current only flows in the presence of an applied driving force, in this case a voltage. The form of Eq. (5.1) should not be taken to indicate that the current–voltage relations of channels are expected to be linear, however. As indicated in Figure 5.2, the current–voltage relation can be decidedly nonlinear even under symmetric ionic conditions, despite the fact that it is a thermodynamic necessity that the plot pass through the point $i = 0$, $V = 0$.

In the case of a nonlinear *i–V* relation the definition of γ becomes a bit more murky inasmuch as its value depends on the operating point, i.e., on the particular values of i and V at which the conductance is defined. We shall have more to say about the shapes of these *i–V* plots and the definition of γ later. For the time being we define γ as the slope of the *i–V* plot at any point. Having done this we must contend with the fact that γ can be strongly voltage dependent, and we shall see that there are a variety of ways in which this can come about. It is important, however, that this phenomenon be clearly distinguished from voltage-dependent *macroscopic* conductance that can arise from voltage-dependent channel *gating,* i.e., a change in the probability of finding the channel in the open state. Here we refer only to the conduction properties of an *open* channel, where we find that in some cases the same voltage will produce different currents depending on the direction of current flow. This is the definition of rectification.

5.2.2. Permeability: Tracer Flows

The term *permeability* is widely and somewhat loosely used to describe the leakiness of a membrane to some substance. Here we aim to begin with an unambiguous *operational* definition of ion permeability so that we will be well equipped to explore its physical significance. Inasmuch as conductance and ion permeability are both measures of the leakiness of a channel, we expect that they will be related; but we also find that these two parameters can provide different kinds of information about a channel that in some cases can even appear to be contradictory.

An operational definition of permeability is somewhat more difficult than that for single-channel conductance, because it must be based on an experiment that is not really possible for a single ion channel, the measurement of the flow of radioactively labeled ions. Despite this limitation, a definition based on a hypothetical tracer flow measurement has great heuristic value. It provides us with a basis for clarifying the relationship between permeability and conductance, and for developing an electrophysiological approach to the determination of permeability.

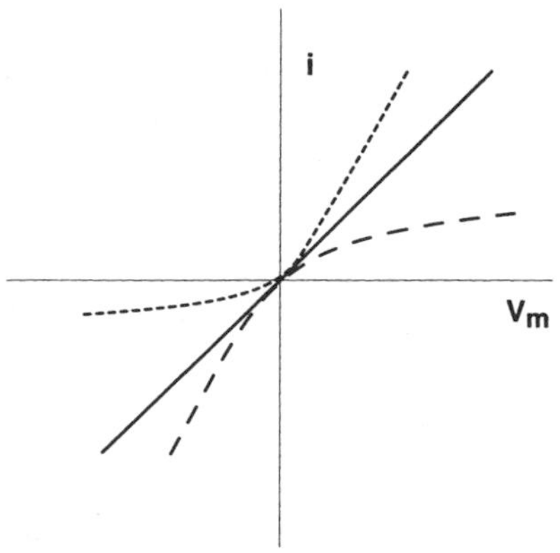

Figure 5.2. Three representative *i–V* plots for a K^+-selective channel bathed by identical salt solutions on either side. The *i–V* plots may be linear or may rectify in the outward or inward direction but all must pass through the point $i_K = 0$, $V_m = 0$.

Referring again to Figure 5.1 we define the permeability of the channel in terms of an experiment in which we add a "small amount" of radioactive K (i.e., ^{42}K) to side 1 and measure the appearance of radioactivity on side 2. The term "small amount" is used to call attention to the fact that the amount of ^{42}K required to accomplish this is so small on a molar basis that it is not expected to perturb the properties of the system. This is the most important difference between a conductance measurement and the determination of tracer flow. To measure γ, permeating ions must be reasonably abundant, otherwise there will be little or no current flow, no conductance. The radioactivity associated with the tracer ion, however, allows us in principle to determine its rate of movement in the virtual absence of permeant ions. The molar abundance of tracer ion required for this measurement is so small that we can imagine measuring K permeability in the absence of K ions.

To measure tracer flow we measure the appearance of radioactivity on the "cold" side (dn^*/dt) and define this as J_{K^*}, the *unidirectional* flow of labeled K where the units might be simply counts per minute (cpm) per second:

$$dn^*/dt = J_{K^*} = \text{cpm/sec} \tag{5.2}$$

The term *unidirectional* refers to the fact that the experiment is conducted under conditions such that the amount of tracer K that accumulates on the cold side is a negligible fraction of that which was initially added to side 1, i.e., this is the *initial rate* of tracer flow. This is, in part, why the measurement implied by this definition cannot actually be done. Technical problems aside, even at a flow rate of 10^6 ions/sec the initial molar flux of tracer, and hence the flow of radioactivity, through a *single* channel will be too small to detect even under ideal conditions. This does not reduce the heuristic value of the approach, however, which allows us to define a permeability on the basis of how readily radioactive ions move through the channel. Because the total flow rate will obviously depend on the amount of tracer added to side 1, we define the measured variable, the tracer *rate coefficient,* as the *ratio* of the flow rate to the concentration of counts on the hot side, i.e.,

$$\lambda_{K^*} = J_{K^*} / [K^*] \tag{5.3}$$

where J_{K^*} is the tracer flow in cpm per second, $[K^*]$ is the concentration of tracer in cpm per milliliter, and λ_{K^*} is the *rate coefficient* for ^{42}K flow through the channel.

The rate coefficient, defined in this way, will have units of cubic centimeters per second because, implicit in the definition of J_{K^*} and hence in λ_{K^*}, is a term proportional to the effective *area* of the conduction path. Thus, λ^* may be viewed as having the dimension of velocity (cm/sec) times area (cm^2).

λ_{K*} is a concise description of the overall permeation process. Like γ, it will depend on the details of the entry, translocation, and exit of the tracer ion from the channel. Note, however, that the definition of λ_{K*} can also include an intrinsic contribution of the membrane potential. If V_m is not zero, then tracer flow, J_{K*}, will be either enhanced or attenuated by the additional electrical driving force. This is not related to any voltage dependence of the properties of the channel, it is simply a reflection of the fact that the operational definition of λ^* does not permit us to factor out the membrane voltage. Thus, we find that the apparent K *permeability* of the channel, P_K, is most unamiguously defined as

$$P_K = (\lambda_{K*})_{V_m = 0} \tag{5.4}$$

the tracer rate coefficient at zero membrane voltage, where we expect it to reflect the physical interaction of the labeled K with the channel.

It was assumed in the foregoing discussion that our hypothetical experiment can be conducted using a channel that is always open. Otherwise, the rate of tracer flow would be attenuated by gating processes, i.e., the probability of finding the channel in the open state would be less than unity. This practical limitation need not limit our use of the tracer concept. We will find that the quantity most often of interest is the *ratio* of the permeabilities of two ions, say ^{42}K and ^{86}Rb, that we can envision obtaining in a double-label flux experiment so that the ratio would be independent of gating.

The operational definition of permeability is based on a determination of net *tracer* flow. In the condition chosen here the net flow of *unlabeled* K would be zero, because $V_m = 0$ and there is no K concentration gradient. In order to connect our definition of P_K with the behavior of the unlabeled ions that are contributing to K conductance, we introduce the assumption that the two isotopes experience the same interaction with the channel,[8] i.e.,

$$P_K = (\lambda_K)_{V_m = 0} = (\lambda_{K*})_{V_m = 0} \tag{5.5}$$

The relation of permeability to conductance will depend on the nature of the permeation process, but the operational definitions already suggest an important difference between the two parameters. Conductance is based on a *net* movement of a great many ions (ionic current) while permeability can be thought of as what is seen by a visiting test ion, the tracer, that is present in vanishingly small amounts on a molar basis. Conductance is going to depend on the local concentration, or abundance, of ions whereas permeability is more properly thought of as a property of a single ion. Thus, we expect to find that conductance will be related to the *product* of the permeability and the concentration of the permeating ion.

5.2.3. The Flux-Ratio Equation

The flux-ratio equation introduced by Ussing[5–7] provides a link between the concepts of permeability and conductance that is not dependent on a detailed model of the permeation process. The derivation of the flux-ratio criterion is based on recognition of the fact that *ratios* of rate coefficients provide information about the *energetics* of the permeation process[2,5–10] so that the flux ratio provides a link between the notions of a tracer flux and the electrochemical driving force that drives current flow. Here we imagine that we can measure the rate coefficients for tracer K flow through the channel in either direction, i.e., λ^*_{12} or λ^*_{21}. If the only driving force for tracer flow is the electrochemical gradient of tracer, i.e., the concentration gradient of tracer and the electrical potential, possible values for the ratio, $\lambda^*_{12}/\lambda^*_{21}$, are highly constrained.[8,9] The simplest illustration of this point is the condition $V_m = 0$ and $[K]_1 = [K]_2$, for which it must be true that

$$(\lambda^*_{12})_{V_m = 0} = (\lambda^*_{21})_{V_m = 0} \tag{5.6}$$

Clearly if this were not true, i.e., if $\lambda^*_{12} > \lambda^*_{21}$, it would suggest that the process involved some sort of energy-converting, pumplike mechanism that would be expected to produce a *net* flux of tracer (or unlabelled ion) in the absence of gradients of concentration or electrical potential. If there is no interaction between the tracer ions and unlabeled ions in the channel (see below) then this equality must hold *regardless* of the distribution or flow of unlabeled ions.[8,9]

Equation (5.6) can be generalized to nonzero values of V_m by noting that, in the absence of coupling between the flow of tracer and any other species, including the nonradioactive isotope of the tracer species, the ratio of the rate coefficients must be such that

$$\lambda^*_{12}/\lambda^*_{21} = \exp\,[zF/RT\,(V_1 - V_2)] \tag{5.7}$$

This result can be derived in a number of ways[5,8,9,16,17,22] and reflects the fact that at *non*zero values for V_m the *ratio* of the bidirectional rate coefficients is a measure of the total electrochemical driving force for tracer flow just as the ratio of the rates for a chemical reaction is a measure of the total free energy change that drives the reaction. We will find that this constraint applies to channels that are expected to be occupied by no more than one ion at a time, whereas multiple ion occupancy can give rise to coupling between the flows of tracer and abundant ions and a departure from Eq. (5.7).

The implications of this constraint for unidirectional fluxes of tracer or the unlabeled species can be seen by calculating the *flux ratio,* first introduced by Ussing[43] and given by

$$J^*_{12}/J^*_{21} = \lambda^*_{12}[K^*]1/\lambda^*_{21}[K^*]_2 = [K^*]_1/[K^*]_2 \exp[RT/zF(V_1 - V_2)]$$

which can be condensed to

$$J^*_{12} / J^*_{21} = \exp\,(\Delta\tilde{\mu}_{K*}/RT) \tag{5.8}$$

where $\Delta\mu_{K*}$ is the electrochemical potential difference for the tracer. Similarly, by virtue of the equality of the rate coefficients for the two isotopes we obtain for the nonradioactive, abundant isotope

$$J_{12}^{K}/J_{21}^{K} = (\lambda_{12}^{*}/\lambda_{21}^{*})\,[K]_1/[K]_2 = \exp[\Delta\tilde{\mu}_K/RT] \quad (5.9)$$

where $\Delta\mu_K$ is the difference in electrochemical potential for the abundant isotope. These results emphasize that, in the absence of coupled processes, the ratio of the unidirectional fluxes is a measure of the difference in electrochemical potential, the driving force for ion flow.

5.2.4. Conductance/Permeability Relation

A calculation of the *net* K flux expected when ions move independently results in an equation that leads directly to a relation between conductance and permeability. The net K flow is given by the difference between the two one-way flows,

$$J_{\text{net}}^{K} = J_{12}^{K} - J_{21}^{K}$$

substituting from Eq. (5.3) we obtain

$$J_{\text{net}}^{K} = \lambda_{12}^{K}\,[K]_1 - \lambda_{21}^{K}\,[K]_2 \quad (5.10)$$

Here we see the net K flow, and implicitly the K current, expressed as the difference between two unidirectional rates. Each unidirectional rate is the product of a K concentration and a rate coefficient, λ, that is a reflection of the interaction of the ion with the channel, but also depends on the membrane potential. This can be seen more clearly if we transform Eq. (5.10) into a form that is more akin to Ohm's law. Consider the current flow through the channel where $[K]_1 = [K]_2 = [K]_b$. Because $I_K = zFJ_{\text{net}}^{K}$ we have from Eq. (5.10)

$$I_K = zF[K]_b\,(\lambda_{12}^{K} - \lambda_{21}^{K})$$

and from Eq. (5.7)

$$\lambda_{12}^{K} / \lambda_{21}^{K} = e^{\zeta} \qquad \text{where } \zeta = zF(V_1 - V_2)/RT$$

which combine to yield

$$I_K = zF[K]_b\,\lambda_{21}^{K}\,(e^{\zeta} - 1)$$

For a K channel that obeys the Ussing equation the current is given by the product of the concentration, and $\lambda_{21}^{K}(V_m)$ and a second term in V_m that also reflects the dependence of λ_{12}^{K} and λ_{21}^{K} on the membrane potential.

If we assume for this channel that in the condition $[K]_1 = [K]_2$, the i–V relation will be linear over some arbitrarily small range around $V = 0$, then we can define the single-channel conductance by simply dividing I_K by $V_1 - V_2$ yielding

$$\gamma_K = I_K / (V_1 - V_2) = zF[K]_b\,\lambda_{21}^{K}\,(e^{\zeta} - 1)/(V_1 - V_2) \quad (5.11)$$

At this point we seem to have lost track of the permeability, $(\lambda_K^{*})_{V=0}$, because in Eq. (5.11), λ_{21}^{K} is a function of V_m. If we take the limit of γ_K as $V_1 - V_2$ approaches zero, however, we find the desired relation between conductance and permeability, i.e.,

$$(\gamma_K)_0 = \{(zF)^2 / RT\}(\lambda_{K*})_0\,[K]_b \quad (5.12)$$

where $(\gamma_K)_0$ is the conductance at $V_m = 0$ and $(\lambda_{K*})_0$ is the tracer rate coefficient at $V_m = 0$, i.e., the apparent permeability. This result confirms the intuitive expectation that single-channel conductance is proportional to the permeability of the channel and that the two parameters are related through the concentration. Clearly, high permeability will give rise to high conductance only if the permeant ion is sufficiently abundant.

5.2.5. Ion Gradients, Electromotive Forces, and Reversal Potential

5.2.5.1. Perfect Selectivity

We now consider the physical situation that is pertinent to most determinations of channel conduction properties, the measurement of single-channel currents as a function of V_m in the presence of a gradient of the permeant ion. Thus, we now modify the situation diagrammed in Figure 5.1 and impose a gradient of KCl across the membrane that, for the sake of concreteness, we will choose to be such that $[KCl]_1 = 10\,[KCl]_2$.

In the presence of a concentration gradient the single-channel i–V relation is expected to be altered in two ways. The voltage at which the current reverses in sign (the reversal potential) may not be zero; rather, it will be the voltage at which the total, electrochemical driving force for current flow becomes zero. The electrochemical driving force for an ideally K^+-selective channel is given by the difference in electrochemical potential, $\Delta\tilde{\mu}_K$, where

$$\Delta\tilde{\mu}_K = RT\ln([K]_1 / [K]_2) + zF\,(V_1 - V_2) \quad (5.13)$$

Converting to units of voltage by dividing by zF, we obtain the total driving force in electrical units,

$$\Delta\mu_K/zF = V_m - E_K \quad (5.14)$$

where $E_K = RT/zF\ln([K]_2 / [K]_1)$, the equilibrium potential for K^+, and V_m is equal to $V_1 - V_2$. If the channel is ideally selective for K^+, it is a thermodynamic necessity that the current reverse when $V_m = -E_K$, so that the reversal potential, E_r, is equal to E_K. The i–V relation is described by the more general form of Ohm's law,

$$i = \gamma(V_m - E_r) \quad (5.14a)$$

Because E_K is nonzero, ionic current will flow when $V_m = 0$. The current at $V_m = 0$ is of interest because here we can quite readily view the process of ion permeation in two ways, from the standpoint of Ohm's law or from the perspective of diffusional ion flow driven by a concentration gradient. In this condition Ohm's law yields

$$(i_K)_{V_m=0} = \gamma_K\,E_K \quad (5.15)$$

Equally appropriate, however, is the form suggested by Eq. (5.10) or an analysis of simple diffusion according to Fick's law[2,3,15,16,17,22] that yields for the net ion flow

$$(J_K)_{V_m=0} = P_K([K]_1 - [K]_2) \qquad (5.16)$$

where J_K is the total flow of K through the channel (moles/time), P_K is the apparent permeability, including a factor for the area of the channel, so that the current, I_K, is given by

$$(i_K)_{V_m=0} = zFJ_K = zFP_K([K]_1 - [K]_2) \qquad (5.17)$$

The comparison of Eqs. (5.15) and (5.17) is not particularly comforting. In the former i_K is given by the product of the conductance and the magnitude of the ionic emf, E_K, that may be thought of as the driving force for K flow due to the K concentration gradient expressed in electrical units. In the latter the current is seen as the product of the permeability and the difference in K concentrations. The Ohm's law version prompts a mechanical view of the conduction process. The flow, i_K, is driven by the force, E_K, and if E_K is zero the current (net ionic flow) vanishes. The second equation evokes a more statistical view of the same process that is more obvious if we expand Eq. (5.16) to yield

$$(J_K)_{V_m=0} = (zFP_K)[K]_1 - (zFP_K)[K]_2 \qquad (5.18)$$

Here the net flow of K is readily envisioned as representing the difference between two one-way processes, represented by J^K_{12} and J^K_{21}. Regardless of the direction in which it would traverse the channel, any single ion would experience the same resistance to flow; P_K would be the same. A *net* flow occurs only if there are more ions on one side than on the other. Here again we see the fundamental difference between conductance and permeability, the former pertaining to total net flow and the latter reflecting the experience of a single ion in the channel. Note that when V_m is nonzero, we must replace P in Eq. (5.18) with λ, the rate coefficient for K flow. In this formulation the effect of voltage on K flow is expressed as the dependence of the rate coefficient on membrane potential [see Eq. (5.11)].

5.2.5.2. Imperfect Selectivity

Consider now the behavior of a channel that excludes anions, but does not discriminate perfectly among cations, so that, for example, K and Na may both permeate. It is most convenient to examine the *i–V* plot under conditions such that the channel is bathed by two solutions in which the total cation content is identical, but opposing gradients of K and Na are present. In our model system, for example, let the solution on side 1 contain 100 mM KCl and 10 mM NaCl, and the solution on side 2 contain 10 mM KCl and 100 mM NaCl. This is referred to as the "bi-ionic" condition. A convenient point of reference is the condition $V_m = 0$ in which we expect opposing net flows of the two ions due to the opposing concentration gradients. In the case of equal gradients and if the ions do not "see" each other, the *net* charge flow, the current, through the channel will be determined by the relative permeabilities of the two ions. If $P_K = P_{Na}$, then the current at $V_m = 0$ will be zero, despite the fact that there are substantial net flows of K and Na in opposite directions.

The single-channel current for the condition $P_{Na} = P_K$ will be given by

$$i = \gamma(V_m)$$

as shown in Figure 5.3. Note that the *slope* of the *i–V* plot defines γ but does not betray the identities of the current carriers. The intercept, $i = 0$, $V_m = 0$, however, suggests to us that $P_K = P_{Na}$. Alternatively, we might say that the fact that the single-channel current reverses at $V_m = 0$ in the face of equal and opposite gradients of permeant ions is *prima facie* evidence that $P_{Na}/P_K = 1.0$.

The magnitude of i at $V_m = 0$ will be determined by the relative magnitudes of P_K and P_{Na}. As indicated in Figure 5.3, if $P_K > P_{Na}$, $(i)_{V_m=0}$ will be positive because $i_K > i_{Na}$, whereas if $P_K < P_{Na}$, $(i)_{V_m=0}$ will be negative. Thus, the value of the current at $V_m = 0$ is one measure of the *selectivity* of the channel. In addition, as suggested by the curves sketched in Figure 5.3, a change in the selectivity of the channel will also alter the point of zero current, the *reversal potential* (E_r). For a K-selective channel, E_r will be at negative values of V_m, and for a

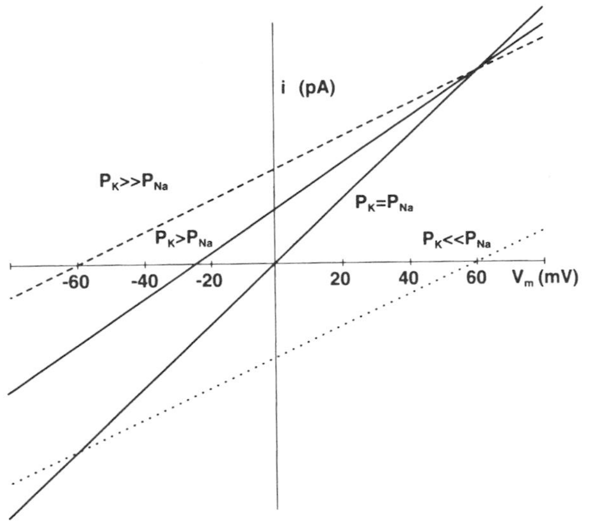

Figure 5.3. *i–V* plots demonstrating the effect of relative permeability (P_K/P_{Na}) on the reversal potential (E_r). The channel is exposed to equal and opposing gradients of K and Na such that $E_K = -60$ mV and $E_{Na} = 60$ mV. The plots are shown as being linear for simplicity and to emphasize that the *net* current will be the algebraic sum of the currents due to K and Na. The dashed and dotted lines represent the plots for perfect K selectivity ($P_K >> P_{Na}$) and perfect Na selectivity ($P_K << P_{Na}$), respectively. The plot for a nonselective channel ($P_K = P_{Na}$) is the sum of those for perfect K and Na selectivity. Hence, at $V_m = 0$ the flows of K and Na are equal and opposite so that $E_r = 0$. Note that at $V_m = -60$ the current through the nonselective channel will be carried exclusively by Na and at $V_m = 60$, exclusively by K. Moderate K selectivity ($P_K > P_{Na}$) shifts E_r to the left, moderate Na selectivity shifts E_r to the right (not shown).

Na-selective channel, E_r will be at positive values. The limiting values for E_r are E_K and E_{Na}, representing perfect K or Na selectivity, respectively. The position of E_r on the voltage axis between E_K and E_{Na} provides a quantitative measure of the selectivity of the channel, but the use of this approach requires a relation between the value of E_r and the relative permeability, P_K/P_{Na}.

5.2.5.3. Reversal Potentials and Permeability Ratios

The relation between E_r and ion permeability ratios can be arrived at by a number of routes, one being an explicit model for the conduction process based on simple diffusion that we will encounter in a subsequent section.[3,11,13,14,16,17,22] Here, instead, we use the approach suggested by Patlak[18] that relies only on the assumption that the movements of ions conform to Ussing's flux-ratio criterion. This derivation is instructive as regards the relation of permeability to tracer rate coefficients. For the sake of simplicity we allow the channel to be permeable only to Na and K, and define the net flux of each ion, i, at any value of V_m as the difference between the two unidirectional fluxes, i.e.,

$$J^i_{net} = J^i_{12} - J^i_{21} = \lambda^i_{12}\,[i]_1 - \lambda^i_{21}[i]_2 \qquad (5.19)$$

In addition, we require that the unidirectional fluxes for each ion conform to the Ussing flux ratio so that

$$\lambda^i_{12} / \lambda^i_{21} = \exp(zFV_m/RT) \qquad (5.20)$$

When $V_m = E_r$, the *net* ion flow through the channel, the sums of the flows of Na and K, will be zero. Summing equations like 5.19 for the net flows of Na and K and substituting from Eq. (5.20), we obtain the expression for E_r as

$$E_r = (V_m)_{i=0} = (RT/zF)\ln\frac{\lambda^{Na}_{21}\,[\mathrm{Na}]_2 + \lambda^{K}_{21}\,[\mathrm{K}]_2}{\lambda^{Na}_{21}\,[\mathrm{Na}]_1 + \lambda^{K}_{21}\,[\mathrm{K}]_1} \qquad (5.21)$$

This expression has the same *form* as that which can be obtained by solving the Nernst–Planck equations using the assumption of a constant electric field (see Refs. 3,11,17), but in the latter equation the rate coefficients are replaced by permeabilities. Equation (5.21), despite its interesting form, is of less apparent value because the rate coefficient (λ's) are functions of V_m and can only be equated with permeabilities in the condition $V_m = 0$.

On the other hand, rewriting Eq. (5.21) in terms of *ratios* of rate coefficients produces an expression that is a bit more user-friendly, i.e.,

$$E_r = (RT/zF)\ln\frac{[\mathrm{K}]_2 + (\lambda^{Na}_{21} / \lambda^{K}_{21})\,[\mathrm{Na}]_2}{[\mathrm{K}]_1 + (\lambda^{Na}_{21} / \lambda^{K}_{21})\,[\mathrm{Na}]_1} \qquad (5.22)$$

Here we see that the position of E_r on the voltage axis under bi-ionic conditions would be determined by the ratio, $\lambda^K_{21} / \lambda^{Na}_{21}$, measured *at that potential.* For conduction by simple electrodiffusion in the constant field condition it turns out that

$$\lambda^K_{21}(V) / \lambda^{Na}_{21}(V) = (\lambda^K_{21})_{V=0} / (\lambda^{Na}_{21})_{V=0} = P_K/P_{Na} \qquad (5.23)$$

If the voltage dependence of the two rate coefficients is the same, then the *ratio* of the two rate coefficients is independent of voltage and equal to what we have defined empirically as the ratio of the permeabilities. There is no guarantee, of course, that this equality will hold in general; but this relation provides a useful way of thinking about the physical significance of the permeability ratios gleaned from the determination of reversal potentials. There is another feature of the equality expressed in Eq. (5.23) that is more subtle, however. The parameters $(\lambda^{Na}_{21})_{V=0}$ and $(\lambda^K_{21})_{V=0}$ correspond to what I have termed the *apparent* channel permeabilities, but from a measurement of E_r we can derive only the *ratio* of these permeabilities. The determination of E_r is thus analogous to a "double label" tracer experiment in which the fluxes of ^{42}K and ^{22}Na are determined simultaneously, under identical conditions. As we shall see later, in this condition the ratio of the rate coefficients is equal to the ratio of the *intrinsic* channel permeabilities, P^0_K/P^0_{Na}. We have already acknowledged that an evaluation of this ratio by means of tracer flow is not generally possible for a single channel, but it is important to note that such a determination is perfectly feasible for an ensemble of channels; and like E_r, the *ratio* of the simultaneously determined tracer rate coefficients would be independent of the *gating* of the channels as long as gating is not greatly altered by the nature of the permeating ion. We will find later that the mental image of permeability as reflecting the behavior of a visiting tracer ion will provide a useful device for understanding the behavior of permeability for specific channel conduction mechanisms; in particular it will be useful for understanding why reversal potentials and conductances provide different information about the permeation process. It will be clear that the measurement of E_r can, in principle, provide information about ion permeability ratios that is impossible to obtain from measurements of single-channel conductance.

This derivation of the relation between permeability ratios and E_r has the advantage of emphasizing the fact that the *form* of Eq. (5.22) is very general and depends only on the assumption that the movements of ions through the channel are not coupled. The equation also makes intuitive sense. It resembles the Nernst equation except that the single ion concentration ratio is replaced by the ratio of the sum of [K] and [Na] with each value weighted by the permeability ratio. In the present example, if $\lambda^{Na}_{21}/\lambda^K_{21} = 1$, then Na is behaving as if it were K so that there is, effectively, no K gradient and $E_{r=0}$. If $\lambda^{Na}_{21}/\lambda^K_{21} < 1.0$, then it is as if the Na concentration behaves as a lesser concentration of K. The influence of K is weighted more heavily and E_r moves toward the limiting value of E_K. Diminished K selectivity is reflected by an increase in $\lambda^{Na}_{21}/\lambda^K_{21}$ such that the influence of [Na] on E_r is weighted more heavily and E_r moves to less negative values. Increasing Na selectivity $(\lambda^{Na}_{21}/\lambda^K_{21} > 1)$ means that the argument of the logarithm will be dominated by the terms in [Na] and E_r moves to positive values. E_r, as indicated in the derivation of Eq. (5.22), defines the point at which $J_K = -J_{Na}$. If E_r lies to the left of zero voltage, then this potential will decrease J_K and increase J_{Na}. The more negative the value of V_m required to equalize the fluxes, the more favored by the channel is K over Na.

Examining the effect of ion substitution on the reversal potential has a special place in the determination of ion channel conduction properties, and it is worth reflecting on the reasons why it is the method of choice for characterizing ion selectivity.[3,19] Perhaps most important is that E_r is determined by the point at which $i = 0$. Thus, the method can be applied to macroscopic as well as single-channel current measurements, because it is independent of channel *gating*; the zero point for both currents must be the same. In the macroscopic case, it is necessary that the current due to the ensemble of channels of interest be well defined, a criterion that is not always easily met for channels that lack distinctive voltage-dependence or toxin sensitivities. In this context it is useful to recall that the remarkable accuracy with which single-channel currents are determined is due to the fact that they are defined by the gating events that continuously reestablish the zero current and open channel current magnitudes. In both macroscopic and single-channel records, E_r also has the advantage of having well-defined limits, because for perfect selectivity E_r is defined on purely thermodynamic grounds as the potential at which the ion of interest is at equilibrium.

As we probe some of the mechanistic details of the conduction process it will be apparent that the interpretation of E_r in terms of permeability ratios depends on some assumptions about the conduction process that may not necessarily be met under all conditions. For example, in channels that exhibit a high degree of coupling between the fluxes of ions, the simple notion of permeability will break down because the flux of tracer ion will be affected by the flux of any other ion that can permeate the channel, even the nonradioactive isotope of the tracer species. Despite this essential limitation, the determination of changes in E_r remains the first approach to the characterization of an ion channel because it is operationally straightforward and has empirical value, if not always perfect analytical precision.

5.3. THE NERNST–PLANCK, CONSTANT FIELD CHANNEL

5.3.1. Conductance and Permeability

In this section we develop an electrodiffusional model for an ion conduction that will permit us to connect measurable parameters like single-channel conductance and permeability ratios to the underlying physics of the conduction process. Although simplified, it provides a basic intellectual framework for thinking about the ion permeation process, and is a standard against which the complexities of real life can be compared as required by experimental observations.

We will represent the conduction pathway as a right, cylindrical pore embedded in a dielectric matrix, a lipid bilayer. The approach to describing the conduction process will be that often referred to as "solubility-diffusion," as it is applied to analyzing the permeability of a lipid bilayer to nonelectrolytes.[2,20] This model is based on the notion that membranes are sufficiently thick that it is permissible to imagine ion permeation as consisting of two spatially separate processes, one the "partitioning" process that occurs at the membrane–solution interface (the mouth of the channel) and the second the intramembrane, translocation process. If the latter is assumed to be slow (rate-limiting), then the entry and exit of ions from the channel can be described by an equilibrium partitioning process and that is characterized by a partition coefficient. This approach has deficiencies, but its simplicity and generality of its predictions justify its use as a first approach.

The intramembrane ion translocation is described by the classic equation of electrodiffusion, the Nernst–Planck equation, where for an ion, i, we have

$$J_i = -(u_i C_i A)\, d\tilde{\mu}_i / dx \tag{5.24}$$

where J_i is the total ion flow at a point, x, in the channel, u_i is the absolute mobility of the ion in the channel, C_i is the concentration at point x, A is the area, and μ_i is the electrochemical potential.[2,3,11,15–17,19,21,22]

It is important to realize that the Nernst–Planck equation has a distinctly mechanical flavor; the local gradient of the electrochemical potential is seen as a "force" that produces the net flow of ions through an influence on their time-average velocity. This is an advantage in that it gives the equation a particularly palpable nature, but suffers from the limitation that the stochastic nature of the underlying movements of single ions are reflected in Eq. (5.18) is obscured. Thus, for the initial explication of the simple model the electrodiffusion approach serves admirably, but for more complicated models it must be supplemented by rate theory approaches that focus on the motion of individual ions.[2–4,19,21,23,24]

The approach is best illustrated by considering first a simple physical situation for which the solution is particularly straightforward. Let the membrane be bathed by identical salt solutions so that the only driving force for net ion flow is the membrane potential, V_m. Assume that the ionic mobility is the same along the length of the channel and that the ion concentration within the pore is uniform along the length of the channel. Equation (5.24) can then be integrated straight away[2] to yield the single-channel current for a K-selective channel, i_K,

$$i_K = (zF)^2 A\, [\mathrm{K}]_c\, u_K / l] V_m \tag{5.25}$$

where $[\mathrm{K}]_c$ is the K concentration *within* the channel, l is the length of the channel, and V_m is the membrane potential, $V_1 - V_2$.

To complete the model we assume that it is possible to relate the K concentration in the bathing solution, $[\mathrm{K}]_b$, to that in the channel by an equilibrium partition coefficient, β, where

$$\beta = [\mathrm{K}]_c / [\mathrm{K}]_b \tag{5.26}$$

so that the single-channel current is given by

$$i = \{(zF)^2 A\ \beta[\mathrm{K}]_b\ u_K / l\}\ V_m \tag{5.27}$$

The form of the equation suggests that the quantity in brackets represents the single-channel conductance, γ, where

$$\gamma = (zF)^2 A\ \beta[\mathrm{K}]_b\ u_K/l \tag{5.28}$$

The result makes physical sense in that it describes conductance, a measure of the rate of charge flow per unit voltage, as being proportional to two terms. One is the mobility, u_K, that represents the ease with which an ion can move in the channel. The second, $\beta[\mathrm{K}]_b$, is the ion concentration *within* the channel, perhaps best thought of as a measure of the time-average probability of finding an ion in the channel.[2]

The notion of ion permeability is not explicit in Eq. (5.28) but can be introduced by equating Eq. (5.28) with (5.12), the relation between the single-channel conductance at $V_m = 0$ when $[\mathrm{K}]_1 = [\mathrm{K}]_2$ and the tracer rate coefficient. This yields

$$[(zF)^2 / RT]\ [\mathrm{K}]_b\ (\lambda_{K^*})_0 = [(zF)^2 A\ [\mathrm{K}]_b\ \beta\ u_K/l \tag{5.29}$$

Clearly, for this simple channel the tracer rate coefficient at $V_m = 0$ (the permeability) is given by

$$P_K = (\lambda_{K^*})_0 = A\beta\ RT\ u_K/l \tag{5.30}$$

which can be rewritten in a somewhat more familiar form as

$$P_K = A\beta\ D_K/l \tag{5.31}$$

where $D_K = RTu_K$, the tracer or "self" diffusion coefficient for K in the channel.

Thus, P_K, operationally defined on the basis of a tracer rate coefficient, would be related to the product of the partition coefficient, β, and the intrachannel diffusion coefficient, D, so that the permeability ratio, say of Rb to K, would be given by

$$P_{Rb}/P_K = (\beta_{Rb}/\beta_K)(D_{Rb}/D_K) \tag{5.32}$$

where the first term, the ratio of the partition coefficients, is sometimes thought of as the *equilibrium selectivity,* in recognition of the fact that in this simple model β would determine the probability of the channel being occupied by an ion at equilibrium ($i = 0$). Similarly, the ratio of the diffusion coefficients is thought of as determining the *nonequilibrium* selectivity inasmuch as, in this mechanical model, D reflects restrictions to ion movement within the channel, due for example to steric constraints.

It is important to note that, for this simple ion channel, the ratios of single-channel conductance determined using identical concentrations of the two permeant ions would be identical to the ratios of the intrinsic channel permeabilities. That is to say, selectivity, defined using reversal potentials or conductance ratios, would be the same. This is a result of the fact that the solubility-diffusion model places no constraint on the occupancy of the channel; the probability of finding an ion in the channel simply increases linearly with the concentration of ions in the bath. For such a channel, γ would be a linear function of the concentration of the permeant ion [Eq. (5.28)]. In actual fact, however, γ is often found to be a saturable function of permeant ion concentration and for such channels we expect generally that ratios of intrinsic channel permeability obtained from reversal potentials will not be the same as conductance ratios that measure apparent permeability (see below and Refs. 3,19).

Two other features of the conductance of our model pore are worth noting. First, in the condition $[\mathrm{K}]_1 = [\mathrm{K}]_2 = [\mathrm{K}]_b$, γ is independent of V_m. This is because in the presence of symmetric solutions, the time-average probability that the channel contains an ion, expressed as $\beta\ [\mathrm{K}]_b$, is voltage-independent. If voltage is increased, ions traverse the channel and exit more rapidly, but they also enter more rapidly so the average occupancy is unchanged. The situation is quite different if a concentration gradient is imposed across the channel as we shall see below. Also important is the fact that the valence, z, and Faraday's constant, F, appear raised to the second power. This is because z and F enter into the model in two ways: in the definition of the force per ion *and* in the amount of current produced by the net flow of ions. A divalent ion experiences twice as much force as a univalent ion in the same electric field, but each ion also carries twice as much current. Thus, if the conductances for a Na-selective and Ca-selective Nernst–Planck channel were compared, even if the values for β and D were identical for the two ions, the single-channel conductance would be *fourfold* higher for the calcium channel.

5.3.2. Equilibrium Selectivity

The concept of equilibrium selectivity as expressed in the partition coefficient, β, leads naturally to a consideration of the physical interactions that would determine how an ion would partition between an aqueous solution and the channel interior. This view can be made explicit by expressing the partition coefficient in terms of the work required to transfer an ion from water to a "site" within the channel:

$$\beta_i = \exp\{-\Delta W_{w\text{-site}}/RT\} \tag{5.33}$$

where $\Delta W_{w\text{-site}}$ is the work required to transfer an ion from an aqueous solution (w) into the channel interior, and thus is a reflection of the net result of physical forces acting on the ion as it leaves water and associates with the channel interior. The total work can be thought of as the sum of two components: that resulting from ion–water interactions and that resulting from ion–channel (or ion–site) interactions so that

$$\Delta W_{w\text{-site}} = \Delta W_{ion,w} + \Delta W_{ion,site} \tag{5.34}$$

The term $\Delta W_{ion,w}$ represents the work required to remove an ion from the shell of highly polar water molecules that effectively screen its strong electric field. In order for an ion to enter the channel, this term must be balanced by some energetically favorable interaction of the ion with the channel interior, and it is generally presumed that this is brought about by the presence of polar or charged amino acid side chains that line the channel interior. The interaction of the ion with these groups is lumped into the term $\Delta W_{\text{ion,site}}$. This simple partitioning of the total energy term into ion–water and ion–site terms provides a thermodynamic framework for analyzing the equilibrium distribution of an ion between water and the channel. The affinity of the ion for the channel can be envisioned as depending on the relative size of these two terms.

The energy associated with the interaction of ions with water has been extensively characterized.[3,19,25–27] In contrast, the nature of the interaction with the channel lining is largely unknown and, hence, must be modeled on the basis of fairly general considerations having to do with the possible electrostatic character of the channel interior. An early attempt at this was the fixed charge membrane of Teorell[16] in which a matrix of fixed charges provided for counterion permeability and coion exclusion. The most successful model, however, is that developed by Eisenman[26–28] based on a comparison of the relative interactions of an ion with water and with charged or polar ligands within the channel. The model is remarkable for at least two reasons. First, it is based on only a few general assumptions, and nevertheless predicts virtually all of the known selectivity sequences for a variety of ion binding and translocation processes. Second, its success in predicting ion channel selectivity suggests that the simple notion of equilibrium selectivity is a very useful first approximation to understanding ion channel selectivity from the standpoint of permeability ratios.

5.3.2.1. Eisenman's Theory

The Eisenman approach to equilibrium selectivity is based on an electrostatic model of the forces that are expected to determine the equilibrium distribution of an ion between the aqueous solution and the channel interior.[3,19,26–28] The ion is represented as a rigid sphere that experiences coulombic interactions with water molecules in the solution and with the groups that form the channel lining. As both of these interactions are viewed as being dominated by coulombic forces, they depend on the apparent radius of the ion in question. The larger the radius of an ion, the weaker will be its interactions with either surrounding water molecules or the channel-lining ligands. Small ions, therefore, are most strongly hydrated and would also be expected to interact most strongly with charged or polar sites within the channel. The equilibrium partitioning of the ion into the pore is then a matter of the result of a tug-of-war between the channel sites and the aqueous solution. If the channel sites are characterized by a "high field strength," then the energy of the ion–channel interaction will be the dominant influence on β, and smaller ions will be more strongly attracted to the channel than larger ions. For alkali metal cations the partitioning selectivity sequence for high field strength is, therefore, predicted to be in the order of increasing ionic radius, i.e.,

$$\text{Li} > \text{Na} > \text{K} > \text{Rb} > \text{Cs} \qquad \text{high field strength}$$

in which Li is preferred over the larger Cs. In contrast, for a site of "low field strength," the energy of interaction will be dominated by the free energy of hydration so that the smaller the ion, the more likely it will be retained in the aqueous phase. The selectivity sequence, reflecting the ions most likely to enter the channel, will thus be in the order of *decreasing* ionic radius, i.e.,

$$\text{Cs} > \text{Rb} > \text{K} > \text{Na} > \text{Li} \qquad \text{low field strength}$$

What happens at intermediate field strength for which the energies of interaction with the site and water are comparable? Here the result is dependent on the shape of the curves relating interaction energy to ionic radius.[26] If the ion–water and ion–site have the same shape, then only two selectivity sequences are possible, the high field and the low field strength, because the curves never intersect. One of the energies is always larger (or smaller) than the other. If, however, the shapes of the two functions differ, then the two curves intersect and the point of intersection depends on the field strength of the site. Asymmetry between the ion-radius dependence of hydration energy and ion–site interaction means that a plot of the difference of the two energies, the determinant of equilibrium selectivity, will exhibit a maximum, and the position of the maximum will be field strength dependent. Using an electrostatic model it is possible to calculate, for example, the selectivity expected for a model carbonyl group.[28] One way to appreciate the effect of field strength is to calculate the net work for transferring ions from water to an anionic site, the field strength of which can be varied by varying its radius. The results of this calculation are summarized in Figure 5.4 in which the total work relative to that for cesium is calculated for each of the alkali metal cations and plotted versus the radius of the model site. By varying the field strength it is possible to generate 9 intermediate sequences to give a total of 11 as indicated in Figure 5.4. The high and low field strength extremes represent the order of decreasing or increasing atomic radius, respectively, but in the intermediate sequences, due to the crossing of the energy functions for $\Delta W_{\text{ion,w}}$ and $\Delta W_{\text{ion,site}}$, the overall selectivity is not in the order of size. In this view the molecular origin of equilibrium selectivity lies in an asymmetry in ion–water and ion–site interaction energies. The essential result of this asymmetry is that "ion–site interactions fall off as a function of cation size as a lower power of the cation radius than do the ion–water interaction energies."[26] At the molecular level the field strength would presumably be a reflection of the electronic structure of the ion binding site that could be comprised of a variety of individual ligands.

5.3.3. Nonequilibrium Selectivity

In the Nernst–Planck model, "nonequilibrium selectivity" is primarily a matter of the value of the ionic mobility within the channel. In the continuum view of ionic diffusion the mobility is taken to be a reflection of the "frictional" interactions of an ion with its surroundings. In aqueous solutions,

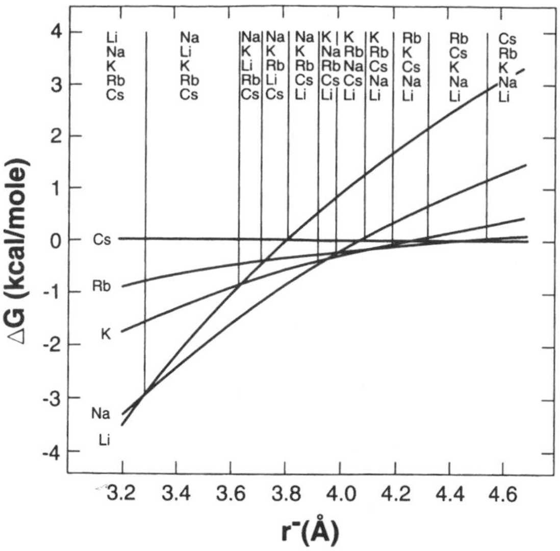

Figure 5.4. Diagram indicating the origin of ion selectivity according to the Eisenman model in which the work required to move an ion between an aqueous solution and the channel interior depends on the free energy associated with ion–water and ion–site interaction. (Redrawn from Ref. 51).

for example, the motion of ions is described fairly well by the Stokes–Einstein relation in which mobility is inversely proportional to the viscosity of the medium.[2,25] The temperature dependence of aqueous diffusion is predictable on the basis of the temperature coefficient of the viscosity of water. Although such a simple continuum view would seem to hold less promise for describing the behavior of ions in channels, this frictional view does evoke notions of steric constraints on ion flow that depend on the physical size of the permeation path.

An alternate view of the partitioning and translocation process is provided by a rate theory approach to diffusion in which the movement of the ion is envisioned as a series of discrete jumps or hopping events that take the ion from one side of the channel to the other.[3,14,23,24,29,30,46] In the rate theory approach, each of these hopping events is characterized by forward and reverse rate coefficients (see Figure 5.8). The magnitude of the individual rate coefficients is determined by the heights of energy barriers that must be traversed by the ion as it hops through the channel. The energy peaks are separated by energy minima or "wells" that indicate the energetically most favorable location of an ion in the channel. These energy wells, in some cases, can be identified with binding sites with which a permeant ion associates transiently as it moves through the channel. The selectivity of ion binding to such sites would seem to represent a natural application of Eisenman's theory. It is less obvious, however, how to apply this approach to the peak energy barriers that are traversed by permeant ions. Hille[19] has suggested that elements of Eisenman's theory can be applied in as much as a quasi-equilibrium, ion-site configuration can be envisioned as an intermediate in the ions' journey across the energy barrier. We will revisit this view of ion translocation in a later section.

5.3.4. *i–V* Relation for the Nernst–Planck Channel

In the presence of symmetric solutions, the *i–V* relation for the Nernst–Planck channel is a straight line passing through the origin. When an ion gradient is imposed, the situation is expected to be different in two ways. Generally, the reversal potential will be nonzero. Consider, for example, an ideally K-selective, Nernst–Planck channel such as that envisioned previously. If a 15:1 gradient of K is imposed, then the predicted shift in E_r will be about −70 mV according to Eq. (5.14). In addition, however, we find that this asymmetry in the bathing solutions produces a bending in the *i–V* relation as diagrammed in Figure 5.5. Shown there are *i–V* plots for three identical K-selective, Nernst–Planck channels, bathed by solutions containing either symmetric low K concentration, symmetric high K concentration, or a concentration gradient where $[K]_{hi}/[K]_{lo} = 15$. The slopes of the *i–V* relations for the two symmetric cases show the expected dependence of slope conductance on ion concentration. The increased conductance of the channel in the presence of high K is a reflection of the fact that the probability of finding an ion in the channel is greater, simply because ions are more abundant in the bathing solution. The K *permeability* of the channel, as defined by a tracer flow experiment, would be identical in either condition. In the presence of a K gradient, the *i–V* relation appears to be a combination of the two symmetric cases. At large positive voltages, when current is flowing from the high-concentration

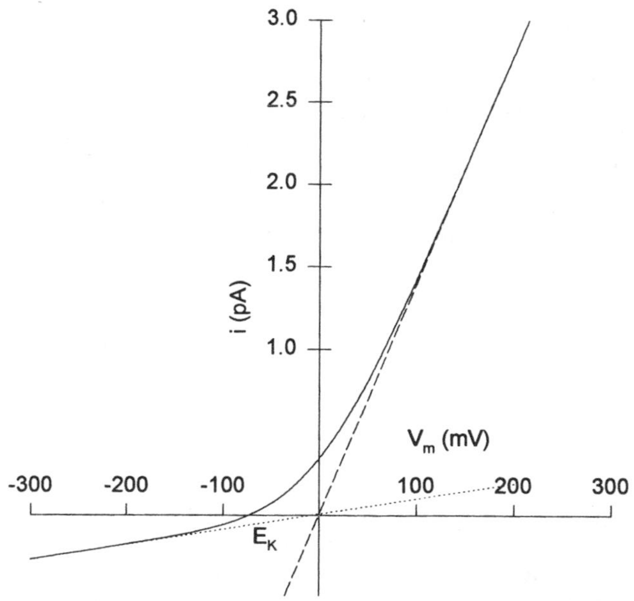

Figure 5.5. Predicted *i–V* plots for a Nernst–Planck, constant field channel. The channel is envisioned as a cylinder of aqueous solution of radius 2Å and length 50Å in which the diffusion coefficient for K is 10^{-5} cm²/sec and the partition coefficient (β) is unity. Currents were calculated using Eq. 5.37 for the condition $[K]_1 = 150$ mM and $[K]_2 = 10$ mM (solid line) to demonstrate Goldman rectification. The dotted and dashed lines, respectively, represent the predicted plots for symmetric low ($[K]_1 = [K]_2 = 10$) and high ($[K]_1 = [K]_2 = 150$ mM) K concentrations. The limiting single channel conductances predicted with symmetric solutions are about 0.9 pS (low k) and 14 pS (high K).

bath to the low-concentration bath, the slope of the i–V relation approaches the high-concentration limit, and at large negative potentials when current flows from low to high K concentration the conductance approaches the low-concentration limit. This comparison suggests that the bending of the i–V relation is due entirely to the imposition of the K gradient and is a reflection of a voltage-dependent change in the probability of finding a K ion in the channel. Driving current from the high-concentration bath tends to increase the probability of finding an ion in the channel whereas the opposite effect is achieved by driving current from the low-concentration bath. Another statement of this effect is that the concentration profile for K within the channel, $C_K(x)$, will be voltage dependent. It is important to bear in mind that rectification of currents through an ensemble of K channels can arise in a number of ways that we will not consider here, but are very important physiologically. One mechanism for rectification is voltage-dependent open probability, as exemplified by outwardly rectifying K channels.[3] Inward rectification, on the other hand, is due primarily to voltage-dependent block of outward K currents by intracellular magnesium ions and polyamines.[48] Finally, it seems likely that some channels exhibit an intrinsic rectification of ion flow that is apparent when the channel is bathed by identical solutions and may reflect the distribution of the barriers to ion flow within the channel.[14]

The bending of the i–V plot for a Nernst–Planck channel seen in the presence of a concentration gradient is often referred to as "Goldman rectification" in recognition of David Goldman's seminal paper describing this behavior using the Nernst–Planck equations.[11] In that paper Goldman presented an analytical solution to this problem that depends on an ad hoc assumption regarding the nature of the membrane potential profile (the electric field) within the membrane. Examination of the local form of the Nernst–Planck equation written for K reveals the necessity for this assumption. From Eq. (5.24), we have at any point in the channel

$$i_K = zFJ_K = -(zF)\,A\,(RTu_K)\,\frac{dC_K}{dx} - (zF)^2 A\,(u_K)\,C_K\,\frac{dV}{dx} \qquad (5.35)$$

where dC_K/dx and dV/dx are the local gradients of the concentration and electrical potential, respectively, and C_K should be written as $C_K(x)$ to emphasize that it is the K concentration at a point. A little fiddling will reveal that Eq. (5.35) cannot be integrated to yield a relation between i and the transmembrane gradients of concentration and voltage without introducing some assumption about the spatial profile of either [K] or V. Goldman's solution was to let the gradient of the potential (the field) be constant, i.e.,

$$-dV/dx = E = \text{constant} = V_m/l$$

where E is the electric field, V_m is the transmembrane potential, and l is the membrane thickness. Hence, the resulting relation between i, [K], and V_m is known as the constant field equation. It is also referred to as the Goldman–Hodgkin–Katz (GHK) equation, because the solution also appears in a quite accessible form in a paper by Hodgkin and Katz.[13] It turns out that the constant field assumption may have some justification based on consideration of the electrical properties of biological membranes, but I leave the reader to pursue this further.[47] Here I just present the results of the integration, the details of which can be found in the original papers or a variety of secondary sources.[11,13,17,22]

There is a subtlety about the integration that often escapes attention in a first pass, namely, that the result is restricted to the steady-state condition. This can be seen more clearly by rearranging Eq. (5.35) to a form that we could integrate under the condition of a constant field, i.e.,

$$dC_K/dx = -i_K/(zFAD)_K + (zFV_m/RTl)\,C_K \qquad (5.36)$$

where

$$dV/dx = V_m/l \quad \text{and} \quad D_K = RTu_K$$

Note that time does not appear explicitly as a variable in Eq. (5.36), despite the fact that, as argued above, we expect the shape of the concentration profile, $C_K(x)$, to be voltage-dependent. In order for the integration of Eq. (5.36) to make sense, the concentration profile must be that which pertains to the particular value of voltage, V_m. This will only be true in the stationary or steady-state condition, i.e., the result only pertains to the concentration profile that is obtained after all transients have ceased.[15] Integrating yields the well-known result in which we have incorporated the assumption of the equilibrium ionic distribution at the channel ends characterized by the partition coefficient, β:

$$i_K = (zF)\,P_K\,(V_m/\theta)\,\frac{[C_K^1 - C_K^2\exp(V_m/\theta)]}{[1 - \exp(V_m/\theta)]} \qquad (5.37)$$

where

$$\theta = RT/F \quad \text{and} \quad P_K = A\beta\,D/l$$

The best way to appreciate this result is to compare the limiting values of i_K at large positive or negative values of V_m. We let the K concentrations be in the ratio of 15:1 with $C_K(1) > C_K(2)$. For a large positive V_m, which would drive current from the high-concentration bath to the low-concentration bath, we have

$$i_K = \{(zF)^2/RT\}P_K C_K(1)\,V_m \qquad (5.38)$$

and for large negative V_m where current flows against the concentration gradient, we have likewise

$$i_K = \{(zF)^2/RT\}P_K C_K(2)\,V_m \qquad (5.39)$$

These two equations, in fact, describe the two straight lines plotted in Figure 5.5, and comparison with Eq. (5.27) recalls that the slope of each is the predicted value for the conductance of a channel bathed by symmetric solutions containing either high or low K. This comparison underlines the point that the rectification predicted by the Goldman constant field equation is completely attributable to a voltage-dependent change in the probability of finding an ion in the channel. At large positive potentials the slope of the *i–V* relation is maximal and the concentration of K *within* the channel approaches its maximum value given by $\beta C_K(1)$. At large negative potentials the concentration of K within the channel is a minimum given by $\beta C_K(2)$.

Another useful way of visualizing the origin of Goldman rectification is to plot the predicted concentration profiles for K within the channel at different values of V_m, as shown in Figure 5.6. The concentration profile is simply the distribution of ions within the pore expressed as a time-average concentration, which can be thought of as the probability of finding an ion at a particular point, *x*, in the channel. The most convenient reference point is $V_m = 0$. In this condition the single-channel current represents simply ion movement down a concentration gradient and, as expected for steady-state diffusion with D = constant, the concentration profile is linear.[2] As the voltage is increased in the positive direction, the profiles become increasingly convex, and it is clear that the average concentration within the channel is increasing toward a limit that is the concentration on side 1. Negative voltages produce the opposite effect and the concentration profile becomes increasingly concave as V_m approaches the low-concentration limit.

A consideration of the concentration profiles shows clearly that the shape of the Goldman *i–V* relation is a reflection of the fact that each value of V_m is associated with a different steady-state concentration profile, and, hence, a different conductance. This raises the question of how the channel gets from one steady state to the next. For example, if V_m is suddenly changed from zero to some positive value, what happens? At $t = 0^-$, just before the change in V_m, the concentration profile is linear and the concentration of ions at any

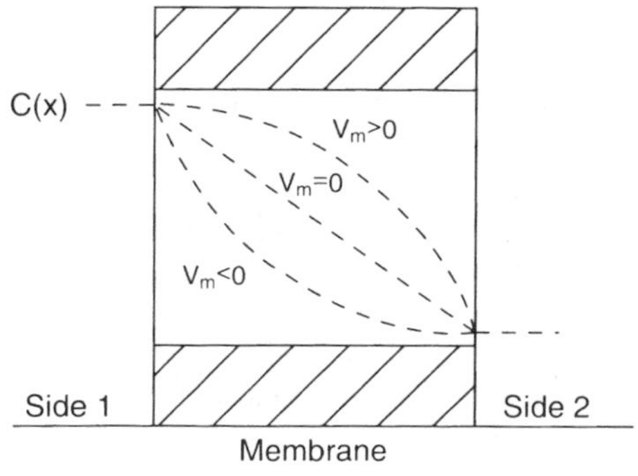

Figure 5.6. Predicted concentration profiles [$C(x)$]within a K^+-selective, Nernst–Planck channel in the presence of a transchannel K gradient. At $V_m = 0$, the concentration profile in linear as expected for simple diffusion. At $V_m > 0$, the concentration profile is concave upward and the time-average ion concentration within the channel is increased, leading to an increased single-channel conductance. If $V_m < 0$, the profile is concave downward, and the time-average ion concentration, and conductance are reduced.

point is not changing with time. On a global scale this time independence is a reflection of the fact that the rates of ion entry and exit from the channel are identical. At $t = 0^+$, just after the imposition of a positive V_m, the situation is quite different. Consider the movement of K as described by the Nernst–Planck equation, for ions that are just inside the channel on side one (high-K side) and just inside the channel on side 2 (low-K side). For ions in either location

$$i_K = zFAD_K\,(dC_K/dx) + \{(zF)^2/RT\}\,AD_KC_K\,(dV/dx) \qquad (5.40)$$

When the voltage is switched on, each ion experiences an additional force proportional to dV/dx. Note, however, that the effect of the membrane field on *total* ion flow is proportional to the local ion concentration, C_K. Because of the concentration gradient, the imposition of an electric field will result in a much greater increase in ion *entry* into the channel than in ion exit. Thus, at $t = 0^+$, just after V_m is switched on, the channel will begin to "gain ions" and this process will continue until the balance between entry and exit is restored. In this new steady state the concentration gradient near the high-concentration bath is "flattened out" and the reduction in dC_K/dx brings the influx and efflux into balance. The constant field equation describes the relation between the membrane voltage and the steady-state current that flows *after* this transient rearrangement of concentration profiles has ceased.

5.3.5. The Definition of Conductance: Slope vs. Chord

The slope of the steady-state *i–V* plot provides an operational definition of the *slope* conductance, γ_{slope}, given by

$$\gamma_{slope} = (\delta i/\delta V_m)_{t=\infty} \qquad (5.41)$$

written to emphasize that the plot only pertains to steady-state values of i at $t = \infty$. This definition, although operationally straightforward, is a little troubling, because it is not easily connected with the description of channel current provided by Ohm's law [Eq. (5.14a)]. For example, it was argued above that at large positive values of V_m the conductance for the constant field channel approaches a limiting value equal to that expected in the presence of symmetric high K concentration. If this interpretation is applied literally to Figure 5.5, it might be taken to suggest that at high values of V_m the apparent reversal potential (the intercept) would be zero. A moment's reflection reveals that the source of this problem is the fact that the Goldman equation applies *only* to the steady-state values of i attained at each value of V_m. Thus, if the voltage is stepped from zero, for example, to some positive value, the system starts off in one state (linear concentration profile) and ends up in another (convex concentration profile), so that the steady state *i–V* trajectory describes the path connecting a sequence of different states of the channel and the *slope* of this trajectory does not meet the criteria for an ohmic conductance.[15,20]

From the standpoint of Ohm's law, a more appropriate definition of the conductance is that illustrated in Figure 5.7

where the conductance at a point on the steady-state i–V trajectory is defined by the slope of a straight line connecting that point with the reversal potential, E_r. This conductance, typically referred to as the *chord* conductance, γ_{chord}, is defined on the basis of Ohm's law, i.e.,

$$\gamma_{chord} = i/(V_m - E_r) \tag{5.42}$$

According to this definition, the ohmic, voltage-dependent conductance of the constant field Nernst–Planck channel would be described by a series of chords connecting successive points on the steady-state i–V trajectory with E_r. It is useful to note for future reference that in the vicinity of the reversal potential the slope conductance and the chord conductance are equal.

The dashed i–V plot in Figure 5.2 provides a good example of the difference between the slope and chord conductances. Here the slope approaches zero in the lower left quadrant. Clearly, this does not imply that the channel conductance is zero; current flows at every value of V_m. The ohmic conductance of the channel can only be obtained by computing the chord conductance at each voltage. The chord conductance, therefore, is most appropriate for characterizing the conduction properties of the channel, but the slope conductance provides a useful measure of the shape of the i–V plot.

The chord conductance is also referred to as the "instantaneous conductance" because it describes the current that would flow "just after" a voltage perturbation at $t = 0^+$, before any rearrangement of ionic profiles has taken place. The analytical basis for this interpretation is found in a somewhat different approach to the i–V relations obtained by rewriting the Nernst–Planck equation as

$$i = [(zF)^2/RT]A\, D_K C_K \{(RT/zF)d \ln C_K/dx + dV/dx\} \tag{5.43}$$

that can be integrated in the form

$$i \int \frac{dx}{[(zF)^2/RT]AD_K C_K(x)} = \int d\{(RT/zF)\ln C_K(x) + V(x)\} \tag{5.44}$$

The right-hand side of Eq. (5.44) is a perfect integral that yields the difference between E_K and V_m. The integral on the left-hand side can be carried out only if $C_K(x)$ is known. Here again *time* does not appear as an explicit variable, but for the integral on $C(x)$ to make sense it must refer to $C(x)$ at some specified time. But that can be any time after the change in V; from $t = 0^+$, just after the change in V_m), to $t = \infty$ after all transients have ceased. Thus, to determine the steady-state value of the chord, or instantaneous, conductance, we must pick a steady-state operating point (i, V_m), determine the concentration profiles (by the Goldman analysis), and then integrate to get the chord conductance where γ is given by

$$\gamma = \left(\int \frac{dx}{[(zF)^2/RT]AD_K C_K(x)} \right)^{-1} \tag{5.45}$$

This value of γ will describe the slope of a line connecting the operating point i, V with $i = 0$, $V = E_r$ and would be operationally defined as

$$\gamma_{chord} = (\delta i/\delta V)_{t=0^+} \tag{5.46}$$

the "instantaneous" change in i produced by a change in V. The most straightforward integration of Eq. (5.45) is carried out at $V_m = 0$, because in this condition the concentration profile is linear. Integration yields for γ_{chord}

$$\gamma_{chord} = [(zF)^2/RT]AP_K\bar{C}_K \tag{5.47}$$

where $P_K = A\beta D/l$ and $\bar{C}_K$ is an "average concentration" given by

$$\bar{C}_K = [C_K(1) + C_K(2)] \ln[C_K(1)/C_K(2)]$$

This expression is reassuringly similar to that obtained (Eq. 5.28) in the condition $C_K(1) = C_K(2)$, the difference being the appearance of the average concentration, $\bar{C}_K$. At $V_m = 0$, therefore, the single-channel current would be given by

$$i_K = \gamma_K E_K \tag{5.48}$$

If the expressions for C_K and E_K are substituted into Eq. (5.48), the result is

$$i_K = zFP_K\{C_K(1) - C_K(2)\} \tag{5.49}$$

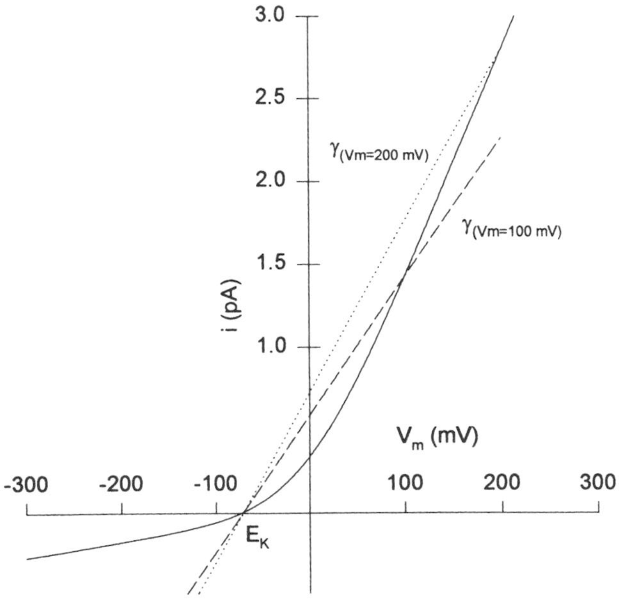

Figure 5.7. i–V plot for the Nernst–Planck, constant field channel described in Figure 5.5. The dotted and dashed lines show the trajectories describing ohmic chord conductances at 200 mV and 100 mV. The slopes of these lines predict chord conductances of 10.3 pS and 8.4 pS, respectively.

in which the ionic current at $V_m = 0$ is clearly depicted as the result of net diffusional ion flow driven by a concentration gradient.

Analysis of the constant field channel has led us to two forms of the expression for the single-channel current at $V_m = 0$ [Eqs. (5.48) and (5.49)]which we encountered earlier in the context of the operational definitions of conductance and permeability. The analytical solution of the constant field, Nernst–Planck channel shows that i_K can be expressed as either the product of a conductance and an ionic emf or the product of a permeability and a concentration difference. An example illustrates how these two equations represent two different views of the same process. Consider the behavior of our Nernst–Planck channel when exposed to two different ion concentration gradients chosen so that the concentration *ratios* differ but the concentration *difference* remains the same; for example, $C_K(1) = 50$ mM, $C_K(2) = 5$ mM and $C_K(1) = 100$ mM, $C_K(2) = 50$ mM. Equation (5.49) sees these two conditions as being identical. The single-channel currents will be the same because the channel permeability and the "driving force," ΔC_K, are identical in the two situations. Equation (5.48), on the other hand, sees these two sets of conditions as being quite different. The driving forces, E_K, are about 60 and 20 mV, respectively, but the conductance in the latter case is about threefold greater so that the product of γ_K and E_K yields identical single-channel currents in both cases. Despite being somewhat contrived, this is an excellent example of the way in which one's analytical "take" on a particular situation can influence the perception of the underlying physics.

5.4. CHANNELS THAT BIND IONS: RATE THEORY INTERPRETATIONS OF CONDUCTANCE AND PERMEABILITY

5.4.1. Beyond Nernst–Planck: Ion–Channel and Ion–Ion Interactions

The Nernst–Planck continuum approach to ion movement through channels is a direct extension of the analysis of solute movement in dilute, aqueous solution; and, as such, represents channel conduction as a funny sort of billiard game in which the balls interact with the channel but not with each other. Ions are seen as rigid, charged spheres that move about randomly, are acted on by electrostatic forces, and experience some vague frictional interaction with the walls of the channel that is reflected in the diffusion coefficient. The palpably mechanical nature of the basic equations lends itself to the development of an intuitive framework in which concepts such as conductance, permeability, selectivity, and rectification can be understood in terms of a set of basic electrophysiological measurements and the Nernst–Planck approach leads to a correct way of thinking about the interpretation of basic electrophysiological data.

The limitations in the Nernst–Planck approach are most clearly seen in several types of experimental results for which the simple continuum approach provides no physical basis. These are saturability in the relation between conductance and ion concentration, evidence for ion–ion interaction, blockade of channels by poorly permeating ions, and the observation that some channels rectify in the absence of an imposed ion concentration gradient.[3,4,14,21a,23,31,31a] These results can be interpreted to suggest that, in real channels, permeating ions can interact with the channel and with each other in ways that cannot be accounted for by simple electrodiffusion. All of these phenomena suggest that a realistic view of the permeation process must allow for the ion to associate transiently with a site or sites in the channel, so that the time the ion spends within the channel is increased over that expected on the basis of simple diffusion. A simple calculation[2] based on treating the channel interior as a very small equivalent volume of the surrounding aqueous bathing solution suggests that the probability of finding an ion within our Nernst–Planck channels if the $[K]_0$ were 100 mM would be about 0.003! Under "physiological conditions" the Nernst–Planck channel is rarely occupied so that it is no surprise that conductance is predicted to increase linearly with external bath ion concentration and that two ions would never be expected to be in the channel at the same time. The association of permeating ions with binding sites increases "the dwell time" of the ion within the channel, thus placing an upper limit on the degree to which single-channel conductance can be increased by increasing ion concentration in the bath. We will see that it is precisely this phenomenon that necessitates that we differentiate between the apparent and intrinsic permeability of the channel to ions.

5.4.2. Rate Theory and Ion Flows

The Nernst–Planck analysis can be modified to allow for ion binding and ion–ion interactions within a channel, but this approach is less attractive as an intuitive tool.[24,30] A more congenial approach is that provided by the theory of absolute reaction rates popularized by physical chemist Henry Eyring and, consequently, often referred to as "Eyring rate theory" or simply rate theory. In the rate theory approach ion translocation is envisioned as the result of ions hopping from one site to another and the *net* hopping rate is represented as the difference between two one-way, or unidirectional, hopping rates. In this way, rate theory recognizes the thermally driven, molecular motion that must underlie any ion flow process and, as such, lends itself to thinking about the goings-on inside of channels in terms of the fluxes of tracer ions.

For example, consider the net rate of ion translocation between two points, 1 and 2. The net hopping rate from site 1 to site 2 is defined as the difference between two one-way hopping rates so that the current, i, would be given by

$$i = zF\,(J_{12} - J_{21}) \tag{5.50}$$

where J_{12} and J_{21} are the one-way hopping rates in units of moles per second. Each of the one-way hopping rates can be expressed as the product of an intrinsic hopping rate, k_{12} or

k_{21}, and the probability of finding an ion at site 1 or site 2, respectively, so that

$$J_{12} = (n/N_A)P(1)\,k_{12}$$
$$J_{21} = (n/N_A)\,P(2)\,k_{21} \tag{5.51}$$

where k_{12} and k_{21} are the intrinsic hopping rates (in sec^{-1}), and $P(1)$ and $P(2)$ are, respectively, the time-average probabilities of finding an ion at site 1 or site 2. n represents the number of ions that is translocated in each hopping event (always one) and N_A is Avogadro's number so that the units are in moles per second rather than ions per second.

The aim of rate theory was to relate the magnitude of the intrinsic, thermally driven hopping rates, k_{12} and k_{21}, to the nature of the underlying process. Accordingly, the transport path was represented by a series of energy barriers such as those suggested in Figures 5.8, 5.9, and 5.10. The energy minima, or wells, between the peaks represent the most probable location of the ion, i.e. the presumed "sites." The translocation process is the net result of hopping from one well to the next over the intervening barrier. The height of the barrier conveys information about the likelihood of a jump from one site to the next. The maximum possible rate is taken to be the frequency of thermal vibration, given by kT/h, where k is Boltzmann's constant, T is absolute temperature, and h is Planck's constant so that kT/h is equal to about 6×10^{12} sec^{-1} at 25°C. The actual rate was then expressed as the product of kT/h and an exponential term determined by the height of the barrier, i.e.,

$$k_{12} = (kT/h)\exp(-\Delta G_{1P}/RT)$$
$$k_{21} = (kT/h)\exp(-\Delta G_{2P}/RT) \tag{5.52}$$

Where ΔG_{1P} and ΔG_{2P} are the magnitudes of the energy differences between site 1 and the peak of the energy barrier and site 2 and the peak energy, respectively. The ratio of the two rate coefficients is the equilibrium constant that would describe the equilibrium distribution of ions (zero flow) between the two sites, and is proportional to ΔG_{1P}–ΔG_{2P} or ΔG_{12}.

The possible physical origin of the peaks and wells is largely a matter of speculation, but it is useful in a general way to think of this profile as reflecting the nature of the local environment seen by a test ion within the channel; determined by the interactions of the ion with the residues lining the channel as well as the electrostatic forces that derive from the location of the channel in a lipid bilayer.[2,3,19,26,32–36]

The influence of the membrane electric field on the hopping process is included by adding to the argument of each exponential in equations 5.52 a term representing the fraction of the total membrane potential seen by the hopping ions (see 3, 12, 19, 23). For the purpose of this exposition, however, it will be sufficient to consider only the condition $V_m = 0$, and focus on the rate coefficients themselves, rather than on the values of ΔG (see Figure 5.8). This simplified analysis will give us a feel for how ion binding within the channel will affect the channel conductance, γ_i, the apparent permeability, P_i, and the intrinsic permeability P_i^o.

5.4.3. Single-Site, One-Ion Channel

5.4.3.1 Permeability of the One-Ion Channel

Here we consider the behavior of a channel that contains one "binding site" as indicated in Figure 5.8 and we stipulate that, due to electrostatic constraints, the channel can be occupied by only one ion at a time. The single-site, one-ion channel is represented by an energy profile consisting of two peaks and one well, and is thus characterized by four rate coefficients. The first step toward an expression for the permeability or conductance is a description of the association of the ion with the channel. This resembles a simple equilibrium binding problem with the small additional complication that the ion can enter or leave from either end of the pore. For the sake of simplicity, we consider the condition in which the transchannel voltage is zero and the concentration of the permeant ion, assumed to be K^+, is always the same on both sides.

The rate of entry of ions into an *empty* channel from side 1, J^e_{1m}, is governed by the intrinsic rate constant, k_{1m}. The one-way rate of entry into the channel is then given by the product of k_{1m} and the probability of finding an ion at a "site" just outside the channel from which it can enter the channel in one jump. This probability can be expressed as the product of the concentration of the ion in the bathing solution (assumed here to be K) and the magnitude of a small volume, v, from which one ion could enter the channel in one hop,[23] so that

$$P(1) = N^K_1 v \tag{5.53}$$

where N^K_1 is the number of K ions per unit volume of bathing solution. Note that v must have the units of volume per ion, i.e., that volume from which one ion could enter the channel in a single hop, so that $P(1)$ will be unitless as expected for a probability. We anticipate that v cannot exceed $(N_K)^{-1}$, the volume of solution in which the probability of finding an ion is unity, and generally will be less than this so that $P(1) \leq 1.0$. By analogy with Eq. (5.51) we have

$$J^e_{1m} = (n/N_A)(N^K_1 v)\,k_{1m} \tag{5.54}$$

where again n simply reminds us that there is only one ion jumping at a time. The ratio N^K_1/N_A is simply the molar concentration in the bath, $[K]_1$, so that

$$J^e_{1m} = nv\,[K]_1\,k_{1m} \tag{5.55}$$

This relation pertains only to an *empty* channel (an ion cannot enter an occupied channel) so in general we have

$$J_{1m} = nv\,[K]_1\,k_{1m}\,(1 - f_o) \tag{5.56}$$

where J_{1m} is the time-average molar flux into the channel and f_o is the probability that the channel is occupied.

The one-way *exit* rate from the channel into left-hand bath, J_{m1}, is the product of the intrinsic exit rate, k_{m1}, and the probability of finding an ion on the site. The latter quantity, however, can be identified with f_o, so that J_{m1} is given by

$$J_{m1} = (n/N_A) f_o k_{m1} \quad (5.57)$$

For steady ion flow through the channel, the rates of net ion entry and exit must be equal, i.e.,

$$J_{1m} - J_{m1} = J_{m2} - J_{2m} \quad (5.58)$$

Using an analogous expression for J_{m2} and J_{2m} and recalling that $[K]_1 = [K]_2 = [K]_b$, we have

$$\begin{aligned} &v[K]_b k_{1m} (1 - f_o) - (1/N_A) k_{m1} f_o = \\ &(1/N_A) k_{m2} f_o - v[K]_b k_{2m} (1 - f_o) \end{aligned} \quad (5.59)$$

where the common parameter, n, has dropped out. Rearranging yields

$$v(1 - f_o) [K]_b (k_{1m} + k_{2m}) = f_o (k_{m1} + k_{m2}) (1/N_A) \quad (5.60)$$

where the right-hand and left-hand sides of the equation represent the total on and off rates, respectively.

Solving Eq. 5.60, we obtain an expression for f_o,

$$f_o = [K]_b / ([K]_b + K_{1/2}) \quad (5.61)$$

where $K_{1/2} = (1/vN_A)(k_{m1} + k_{m2})/(k_{1m} + k_{2m})$ and the term $(vN_A)^{-1}$ ensures that the units of $K_{1/2}$ are moles per liter.

Not surprisingly, the expression for f_o has the form of a simple binding isotherm characterized by a $K_{1/2}$, or dissociation constant, that is proportional to the ratio of the sum of the off rates and the sum of the on rates. In this simple model with two equal barriers and one well, $K_{1/2}$ is determined by the difference in free energy between the bathing solution and the energy well, i.e., $K_{1/2} = (1/vN_A) \exp\{-\Delta Gw/RT\}$, where ΔG_w is the difference between the energy of the well and that of the bathing solution. The term $(vN_A)^{-1}$ has the units of concentration and because an exact value for v is not easily specified, for the purpose of calculations $(vN_A)^{-1}$ is usually set at 1 molar so that the calculated values of $K_{1/2}$ are with reference to a 1 molar standard state.[37] At low concentrations of permeant ion ($[K]_b << K_{1/2}$), the value of f_o (and the rate of ion flow) increases linearly with ion concentration, but at higher concentrations ($[K]_b >> K_{1/2}$) f_o saturates at a limiting value of 1.0. Similarily, the rate of ion flow through the channel will have an upper limit. As the concentration of the permeant ion is increased, the one-way exit rate (Eq. 5.57) will approach a maximum value determined by the intrinsic exit rate from a channel that is occupied 100% of the time.[2,3,8,37,38] The intrinsic maximum exit rate will be determined by ΔG_{wp}, the energy difference between the well and the peak of the energy barrier (Figure 5.8), and can also be thought of as being related to the inverse of the mean "dwell time" of a single ion in the channel.[2,38]

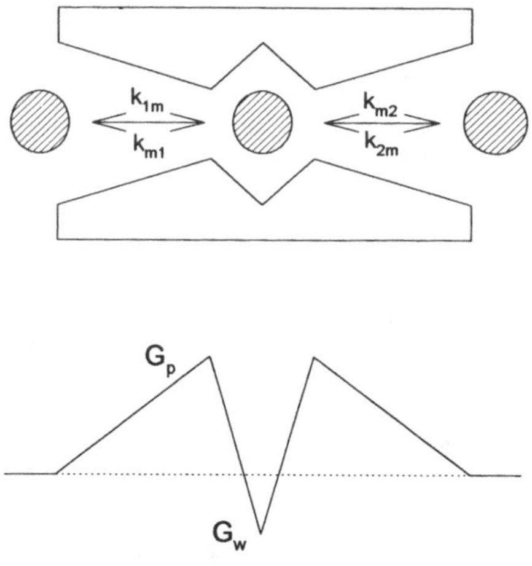

Figure 5.8. Cartoon depicting a channel that may contain, at most, one ion at a time. Entry and exit from the channel are described by the rate coefficients (*k*s) indicated. Shown below is a corresponding symmetric 2-barrier, 1-site (2B1S) energy profile. The entry rates (k_{1m} and k_2) are determined by the difference between the energy of the bath (G_b) and that of the peak (G_p), ΔG_{bp}. The exit rate coefficients and the maximum conductance are determined by the difference between the energy of the well (G_w) and that of the peak (G_p), ΔG_{wp}. The equilibrium distribution of ions and the $K_{1/2}$ for the conductance concentration relations is determined by the difference between the bath and well energies, ΔG_{bw}.

The apparent permeability of the single site, one ion channel is defined by a hypothetical single-channel tracer flow measurement, but the concentration-dependent loading of the channel introduces a phenomenon not encountered previously; namely, that permeant ions compete with each other for entry into the channel. We anticipate, therefore, that the rate of tracer flow will be strongly influenced by the concentration of the unlabelled ion; apparent permeability will be reduced as $[K]_b$ is increased. Because the rate coefficients for the flow of tracer and unlabelled ion are assumed to be equal, we proceed here by writing all of the expressions with reference to the unlabelled ion. The transchannel, unidirectional flow of abundant K, J_{12}, can be written as

$$J_{12} = J_{1m} J_{m2}/(J_{m1} + J_{m2}) \quad (5.62)$$

This equation expresses the fact that, in the steady state, the one-way rate of flow through the channel is equal to the rate of entry, J_{1m}, multiplied by the fraction of entering ions that exit toward side 2. Substituting using expressions similar to Eqs. 5.56 and 5.57 and dividing by $[K]_b$, yields the transchannel rate coefficient, λ^K_{12}, where

$$\lambda^K_{12} = J^K_{12}/[K]_b = nv (1 - f_o) [k_{1m} k_{m2}/(k_{m1} + k_{m2})] \quad (5.63)$$

Substituting for $(1 - f_o)$ from (Eq. 5.61) we obtain

$$\lambda^K_{12} = nv \{K_{1/2}/(K_{1/2} + [K]_b)\} k_{1m} k_{m2}/(k_{m1} + k_{m2}) = P_K \quad (5.64)$$

λ^K_{12}, which must be identical for tracer and abundant K, is the apparent permeability of the channel. The apparent permeability is seen to be the product of two terms, the right pertaining to the intrinsic hopping rate for an ion in the channel, and the left reporting on how likely it is that the channel will be empty and, hence, accessible to an entering ion. This latter term reflects the concentration-dependent loading of the channel and the competition between tracer and unlabelled ions for entry. As $[K]_b$ becomes large, λ^K_{12} approaches zero because the channel is occupied by unlabelled ions 100% of the time.

The maximum value of λ^K_{12}, seen when $[K]_b = 0$, is set by the intrinsic hopping rate through a channel and from equation 5.64 is given by

$$(\lambda^K_{12})_{max} = (nv)\, k_{1m}\, k_{m2} /(k_{m2} + k_{m1}) \tag{5.65}$$

where the term nv ensures that the units for λ^K_{12} are cubic centimeters per second. $(\lambda^K_{12})_{max}$ is the rate coefficient for tracer flow through a channel that is unoccupied virtually 100% of the time, and represents the intrinsic permeability of the channel, P°_K, uncomplicated by the competition between ions for entry into the channel. The intrinsic permeability is the rate of entry into an empty channel, k_{1m}, multiplied by the fraction of entering ions that leave on side 2. Thus, we see that equation 5.63 tells us that the apparent permeability of the channel is equal to the intrinsic permeability of the channel, multiplied by a factor that accounts for the fact that tracer may enter only unoccupied channels. P_K will equal P°_K only in the limit $[K]_b = 0$.

The importance of understanding the difference between the apparent and intrinsic permeabilities of the channel is well illustrated if we consider the effect of differences in well depth (binding affinity) and peak barrier height on each parameter. Consider, for example, two permeant ions, A and B. Let A and B see identical peak energy barriers, as indicated in Figure 5.9, but let B bind more tightly to the site so that the depth of the energy well is greater. The apparent permeabilities could be compared by bathing the channel with symmetric concentration of A or B and measuring the flow of labelled A or B across the membrane. Because the occupancy of the channel by A or B would be a function of concentration, the apparent permeabilities could be characterized only by making measurements at a number of different concentrations of A or B. At any identical concentration of A or B, we would find that $P_B < P_A$, because the higher affinity of B for the site would dictate that the probability of the channel being occupied by unlabelled ions would be greater in the presence of B than in the presence of A (Eq. 5.64).

The result would be quite different, however, if fluxes of A or B were determined in the absence of unlabelled ions. In the limit of zero ion concentration the flow of A or B would measure the intrinsic permeability, which in this example would be identical for A and B.[50] This follows from Eq. 5.65. The intrinsic permeability of the channel is the entry rate, k_{1m}, multiplied by the probability that on entering ion will exit on the opposite side, $k_{m2}/(k_{m1} + k_{m2})$. The higher affinity of B for the site in our 2B1S channel means that the well-to-peak energy barrier will be greater for B than A. Hence the values of k_{m2} and k_{m1} will both be reduced, but by identical amounts, so that the fraction, $k_{m2}/(k_{m1} + k_{m2})$, will be invariant. Thus the intrinsic permeability of the channel to A or B is determined by the peak barrier height, which, in this example, is identical for the two ions. The increased well depth for ion B is not reflected in the intrinsic channel permeability, so that in the limit of zero ion concentration, $P^\circ_A = P^\circ_B$. A moments' reflection will reveal that this result would also pertain if the flows of tracer A and B were determined simultaneously, in a double label experiment. Regardless of the ambient ion concentration, the ratio, $\lambda^A_{12}/\lambda^B_{12}$, would be equal to the ratio of the intrinsic permeabilities, in this case unity.

5.4.3.2. One-Ion Channel Obeys the Flux-Ratio Equation

The competition between tracer and unlabeled ions for the single site in the one-ion channel represents a strong interaction between permeating species that might, at first glance, be expected to produce some deviation from the Ussing flux-ratio equation that we typically associate with simple diffusion. That this is not the case can be gleaned readily by considering the *ratio* of rate coefficients measured from 1 to 2 or from 2 to 1 for a single-site, one-ion K channel.

At $V_m = 0$, for example, the tracer rate coefficients measured in either direction must be equal, regardless of the concentration of unlabeled K in either bath.[3] This follows from the fact that only one ion can occupy the channel at any time. An occupied channel is seen by tracer moving in either direction as unavailable; and an unoccupied channel is, likewise, equally available from either side. Thus, both rate coefficients will be strong functions of $[K]_b$, but the *ratio* will be independent of $[K]_b$. This result pertains in the presence or absence of a concentration gradient.

The fact that competition for a single site does not produce a deviation from the Ussing flux ratio is a reflection of the fact that this interaction provides no mechanism for the *coupling* of the two flows, i.e., the driving of one flow by the gradient of another species. In a multiion channel, however, flux coupling produces deviations from the flux-ratio equation and renders the concept of permeability a good deal more murky.

5.4.3.3. Conductance of the One-Ion Channel

If the rate coefficients for ion flows through the one-ion channel must conform to the Ussing flux-ratio equation, we can calculate the single-channel conductance for the condition $V_m = 0$ and $[K]_1 = [K]_2 = [K]_b$ using Eqs. (5.12) and (5.65):

$$(\gamma_K)_0 = [(zF)^2/RT]nv\, K_{1/2}\, [k_{1m}\, k_{m2}/(k_{m1} + k_{m2})] \{[K]_b/([K]_b + K_{1/2})\} \tag{5.66}$$

As expected from the foregoing analysis of channel occupancy and tracer flows, $(\gamma_K)_0$ is a saturable function of $[K]_b$, a reflection of the fact that the rate of ionic throughput at high concentrations of the permeant ion is limited by the rate at

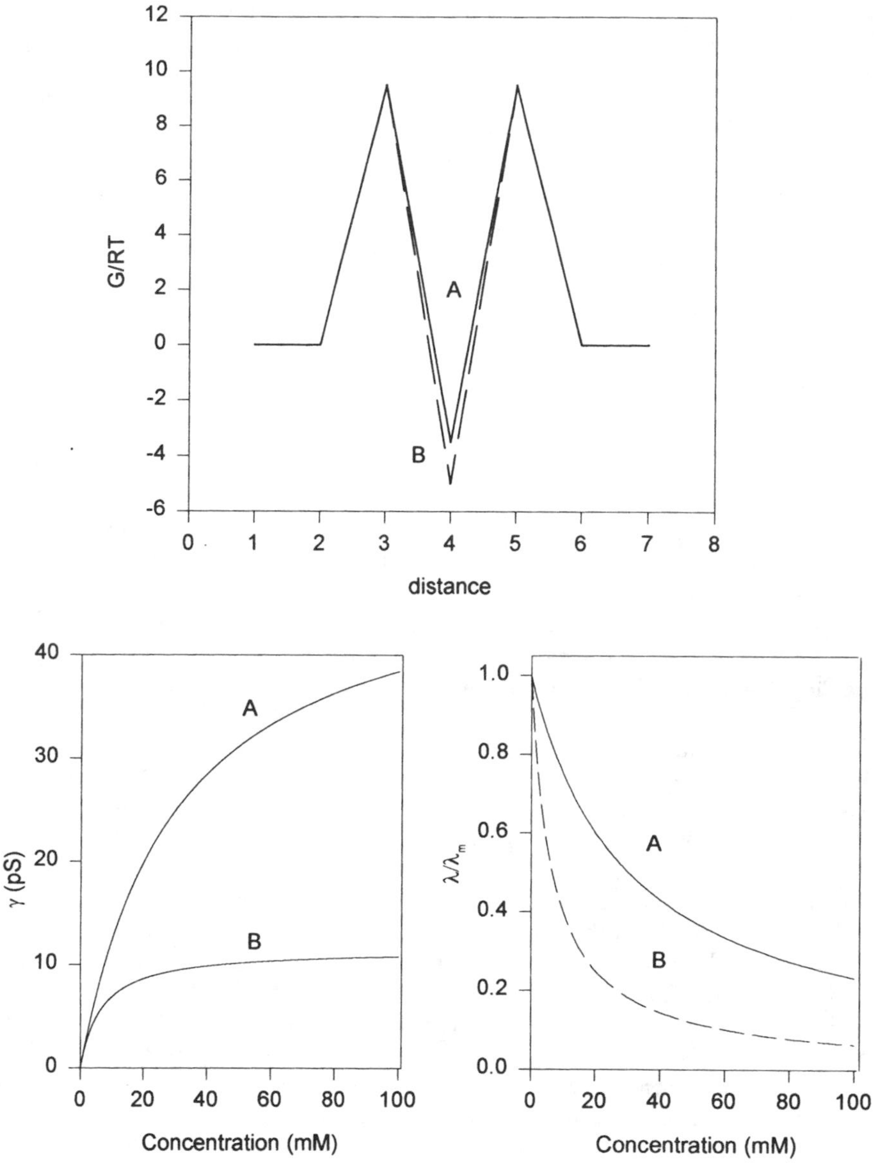

Figure 5.9. Predicted behavior for two ions, A and B, in a 2B-1S channel. Energy barrier profiles are shown for the two ions as multiples of *RT*. Peak energy barriers are symmetric and identical for A and B ($G_p = 9.5\ RT$). The well energies (G_w) differ for the two ions such that $\Delta G^{A}_{bw} = -3.5\ RT$ and $\Delta G^{B}_{bw} = -5.0\ RT$ so that B binds more tightly in the channel. These energies predict values for the $K_{1/2}$ (Eq. 5.61) and γ_{max} (Eq. 5.69) of $K^{A}_{1/2} = 30.2$ mM, $K^{B}_{1/2} = 6.7$ mM, $\gamma^{A}_{m} = 43.8$ pS, $\gamma^{B}_{m} = 9.8$ pS with respect to a 1 M standard state (see text). Plots in the lower part of the figure show the single channel conductances and relative tracer rate coefficients plotted for each ion as a function of bath concentrations of A or B.

which occupied channels unload, determined by k_{m2} (Figure 5.8).

The maximum conductance, seen when $[K]_b >> K_{1/2}$, is given by

$$(\gamma_K)_{max} = [(zF)^2/RT]nv\ K_{1/2}\ [K_{1m}\ k_{m2}/(k_{m1} + k_{m2})] \quad (5.67)$$

which can be simplified using the definition for $K_{1/2}$ to yield

$$(\gamma_K)_{max} = [(zF)^2/RT](n/N_A)\ [k_{1m}\ k_{m2}/(k_{1m} + k_{2m})] \quad (5.68)$$

Here $(\gamma_K)_{max}$ is seen to be proportional the the intrinsic exit rate, k_{m2}, multiplied by the fraction of exiting ions that entered on the opposite side.

For a symmetric, two-barrier model, however, $k_{1m} = k_{2m}$ so that

$$(\gamma_K)_{max} = [(zF)^2/RT](n/N_A)\ k_{m2}/2 \quad (5.69)$$

Equation (5.69) provides a means of estimating k_{m2}, a measure of the well-to-peak energy barrier that must be traversed by an ion leaving the channel.[37]

5.4.4. Selectivity: Conductance Ratios and Permeability Ratios

In this section we use the 2B 1S channel model to explore the interpretations of experiments designed to measure the ion selectivity of a channel. Generally this is done in one of two ways: by comparing the conductance of the channel in the presence of different ions, or by measuring the shift in the reversal potential resulting from the substitution of one ion for another. For example, with reference to Figure 5.1, conductance ratios could be measured by completely replacing K in both bathing solutions with other cations (e.g., Rb, Cs, Na, NH_4^+) and measuring the channel conductance in each condition. Alternatively, the K on side 2 could be substituted entirely or in part with another cation, the shift in E_r determined, and equation 5.22 could be used to calculate the ratio of the intrinsic permeabilities from the shift in E_r. For example, if the channel in Figure 5.1 were bathed initially by solutions in which K was the only cation and the K on side 2 only was substituted wholly or in part by Na the shift in the reversal potential, ΔE_r would be given by

$$\Delta E_r = \frac{RT}{zF} \ln \frac{[K]_2}{[K]_1} - \frac{RT}{zF} \ln \frac{[K]_2^* + (P_{Na}^{\circ} / P_K^{\circ})\,[Na]_2}{[K]_1}$$

which can be rearranged to yield:

$$\Delta E_r = \frac{RT}{zF} \ln \frac{[K]_2}{[K]_{2-}' + (P_{Na}^{\circ} / P_K^{\circ})\,[Na]_2}$$

and $[K]_2'$ is the new value of $[K]_2$ after substitution with $[Na]_2$

It is instructive to consider the expected results of applying these two measures of selectivity to our symmetric, 2B 1S channel, because we find that conductance ratios or permeability ratios (obtained from shifts in reversal potential) can provide very different but complementary information about the permeation pathway as seen by the visiting ions. To explore this point we again compare the permeation of ion A and ion B in our model channel. We will also assume that the interaction of the two ions with the channel differs in only one way: B binds more tightly to the single binding site so that $(K_{1/2})_A > (K_{1/2})_B$. On the basis of the foregoing analysis we can immediately predict that if we compared the values of γ measured at a series of symmetric concentrations of either A or B, the results would be as shown in Figure 5.9. Both γ_A and γ_B would be saturable functions of [A] or [B], respectively, but γ_B would saturate more rapidly with increasing ion concentration so that at most values of concentration, and particularly those in excess of $K_{1/2}$, $\gamma_A > \gamma_B$. The maximum attainable conductance would be greater for A because of the greater intrinsic rate of exit of A from the loaded channels. The ratio of the conductances determined in this way would be identical to the ratio of what we have called the "apparent permeabilities" of A or B determined by measuring, in separate experiments, transmembrane tracer fluxes of A or B under conditions of symmetric abundant A or B, respectively. Both P_A and P_B would be dependent on the concentration of the respective abundant ion (Figure 5.9), but at any particular identical concentration of A and B we would find that

$$\gamma_A/\gamma_B = P_A/P_B$$

As noted earlier a very different result is obtained if we measure the flows of tracer A or tracer B *simultaneously,* either in the absence or in the presence of unlabeled A or B. The ratio of rate coefficients determined in this way would be equal to the ratio of what we have called the "intrinsic" channel permeabilities for A and B, P_A^0/P_B^0. This ratio would be independent of the concentrations of A and B in the bathing solutions, because tracer A or tracer B can only enter and traverse unoccupied channels and their relative rates depend, therefore, only on the intrinsic permeability of the open channel for each ion. As discussed above, these intrinsic permeabilities are independent of well depth (binding affinity), hence, in the present example the ratio would be unity because the barrier heights were chosen to be the same. The equality of P_A^0 and P_B^0 This also tells us that the limiting ratio of γ_A and γ_B at very low concentrations of permeant ions would be unity, because the channels would be unoccupied most of the time. At ion concentrations less than $K_{1/2}$ the conductance ratios are expected to be highly concentration dependent, a fact that needs to be taken into account if this is used as an analytical tool.

What would the measurement of reversal potentials tell us about the permeation of A and B? We see from the outset that this is a very different measurement, if only because the measured variable is the shift in E_r produced by replacing, for example, most of ion A on one side of the membrane by ion B. This contrasts with the conductance determination that was done one ion at a time. Because the Ussing flux ratio is obeyed by the single-ion channel, we can legitimately use Eq. (5.22) to calculate ratios of tracer rate coefficients, which is equal to the ratio of the intrinsic permeabilities, P_A^0/P_B^0. Because both ions now experience exactly the same conditions within the channel, in contrast to the two separate experiments that led to the determination of γ_A and γ_B, the reversal potential assay is equivalent to conducting the *double label* flux experiment (alluded to earlier) in which the rate coefficients for A and B are determined simultaneously so that both tracer ions would experience exactly the same competition for the site from unlabeled A or B ions. Any change in the concentration of either ion is likely to alter the flow of tracer A or B, but both will change by an identical amount. Thus, in our one-ion channel the permeability ratios obtained from the shift in reversal potential will be independent of ion concentration and, if the voltage-dependence of λ_A and λ_B does not differ, then the ratio λ_A/λ_B will be equal to the ratio of the "intrinsic" channel permeabilities, which for our example is unity.

If we view permeant ions as probes of the channel's conduction pathway then we can now see how conductance ratios and intrinsic permeability ratios report on different aspects of the ion's experience in the pore. In a one-ion channel conductance ratios are very sensitive to ion binding. Conductance may be thought of as a measure of ionic "throughput" and the time-average rate is reduced if ions bind in the pore. Increased affinity for a binding site decreases the $K_{1/2}$ for the conductance–concentration relation and decreases the maximal conductance due to the increased barrier (ΔG_{wp}) for the exit of an ion from an occupied channel. The ratio of intrinsic permeabilities obtained from shifts in reversal potential, in contrast, is measured under the condition of zero current and is relatively insensitive to ion binding, but highly sensitive to the height of the barriers to ion permeation within the channel.

In a typical ion substitution experiment the contrast between the behavior of conductance and permeability can be quite striking. Imagine, for example, an experiment in which the channel depicted in Figure 5.9 is bathed by symmetric concentrations of ion A, so that the *i–V* relation would be characterized by a reversal potential of zero. If all of the A ion on one side were replaced by ion B, the reversal potential would not change because the peak heights of the energy barriers are identical for the two ions. The conductance, however, would be diminished because of the tighter binding of ion B in the channel. The reduced time-average rate of throughput of ion B produces an intrinsically lower conductance and also impedes the flow of ion A. In some Cl channels it has been found that thiocyanate ion (SCN^-) sees lower peak energy barriers than Cl^-, but also binds much more tightly in the channel, giving rise to the seemingly paradoxical result that reversal potential shifts seen with SCN^- substitution for Cl^- indicate that $P_{SCN}/P_{Cl} > 1$ while conductance ratios report that $\gamma_{SCN} < \gamma_{Cl}$. This behavior is diagnostic for ion binding in the channel.

5.4.5. Multiion Channels and the Interaction of Permeating Ions

In the previous section we explored some of the consequences for ion conduction of the transient association of permeating ions with a binding site in the pore, but the discussion was limited to channels that could be occupied by only one ion at a time. This constraint has the important consequence of ensuring that unidirectional fluxes through the channel will be as predicted by the Ussing flux-ratio equation. In a one-ion channel, permeant ions certainly influence one another in that an occupied channel cannot be entered by another ion, but the flows of the two ions are not coupled. That is to say, the gradient of one ion (say A) cannot produce the net flow of a second ion (B). It is precisely this type of energy conversion process that is excluded by the flux-ratio criterion.[8,9]

The term *multi-ion channel* refers to one in which two or more ions may reside simultaneously as indicated in Figure 5.10. Here the channel is depicted as containing two physically distinct binding sites although it has been suggested that multiple occupancy could also result from the binding of multiple ions by a single, “fuzzy” site.[49] Shown also in Figure 5.10 is corresponding energy barrier profile consisting of two energy wells (binding sites) and three intervening barriers. A moment’s reflection, however, reveals that such a barrier diagram may not be well-suited for analyzing phenomena associated with multiple ion occupancy, because the presence of one ion in the channel will alter the energy barriers for a second ion due to electrostatic repulsion.

A more congenial approach to multiion behavior is the state diagram illustrated in Figure 5.11 for a channel envisioned as having two binding sites, each of which may be occupied by one ion at a time. In the presence of a single permeant ion, this gives rise to four possible states, as shown, and the permeation process can be envisioned as the translation of an ion through these various states. The rate of permeation is dependent on the probability of finding the channel in its various states as determined by the rate coefficients indicated in the diagram, along with the ion concentrations in the bathing solutions and the transmembrane electrical potential.

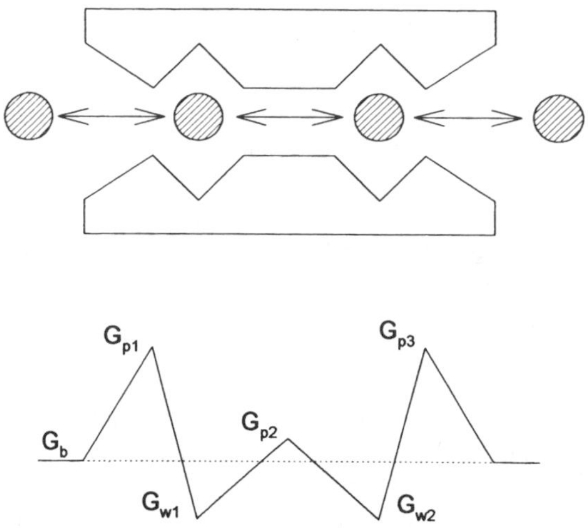

Figure 5.10. Cartoon describing a channel that can be occupied by two ions at a time. The three barrier, two site energy profile below describes the experience of a *single* ion in entering the channel moving from one site to the other and exiting on the other side.

It is most convenient to view the permeation process in terms of an ion that cycles through a specific series of occupancy states. In a two-site channel there are two such cycles, one involving single occupancy, the other double occupancy.[12] For example, in the case of a single permeant ion, an

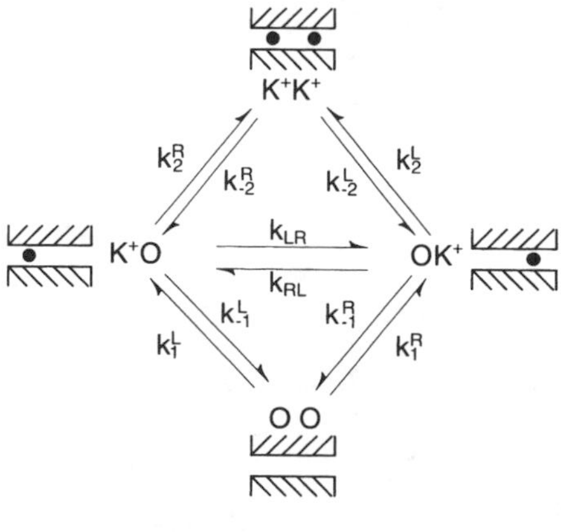

Figure 5.11. Diagram illustrating the possible states for a two-site K channel in the presence of a single permeant ion. The rate of formation of each state is determined by rate coefficients and the ambient ion concentrations. The rate of formation of KO, for example, is proportional to the product of k_1^L and the K concentration in the left-hand bath, $[K]_L$. The rate of formation of KK from KO is proportional to k_2^R and the K concentration in the right-hand bath, $[K]_R$. The coefficients k_{LR} and k_{RL} describe the rate of ion translocation from one site to another.

ion may enter an unoccupied channel, associating with site 1 and then translocate to site 2 and exit the channel on the other side—the result being one ion translocated and an unloaded channel open for another cycle. Alternatively, after an ion has entered from side 1 (site 1) and translocated to site 2, a second ion may enter the channel from side 1, generating the doubly occupied state. In the first ion jumps out on side 2, the net result is one ion translocated but the channel is left in the single occupancy state.

The predictions of this type of kinetic scheme have been examined in detail elsewhere.[3,39] Our goal here is less ambitious, namely, to get a feel for how multiple occupancy can give rise to certain anomalous results when the permeability ratios or conductance ratios are examined. We will see that the single most important feature is that the flows of two permeant ions can be coupled, and it will become apparent that this flux coupling makes it difficult or impossible to interpret parameters like permeability ratios in a way that is independent of models for the ion translocation process.

5.4.5.1. Coupling of Ion Flows

Figure 5.12 shows the state diagram for a two-site channel in the presence of two permeant ions, A and B. The channel may now exist in any one of nine possible states, but those of particular interest are the two in which the channel contains one of each ion: A(1), B(2) and B(1), A(2), where the numbers in parentheses refer to site 1 or site 2. The simplest example of the profound effect of multiple occupancy on ion translocation is that in which one ion is the radioactive isotope (tracer) for the other, e.g., abundant K and ^{42}K. In this case, it would be reasonable to presume that the rate coefficients that govern

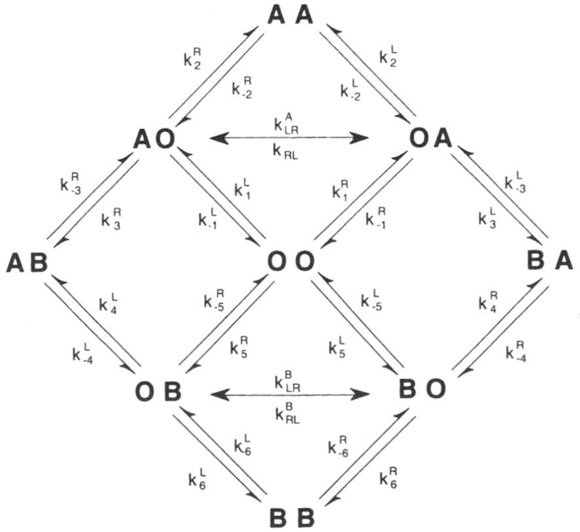

Figure 5.12. Diagram illustrating occupancy states for a two-ion channel in the presence of two permeant ions, A and B. In such a channel the definition of permeability is rendered ambiguous by the possible coupling between the flows of A and B that occurs due to the existence of multiple occupancy states AB and BA.

the interaction of any ion with any site will be identical for the two ions so that the only basis for anomalous behavior is multiple occupancy rather than differential competition for the sites in the channel.

Consider a hypothetical experiment in which we let $V_m = 0$, impose a gradient of unlabeled A to promote net flow of A across the channel, and allow tracer, B*, to be present on both sides of the membrane in identical concentrations. In this condition, there can be *net flow* of B* from side 1 to side 2 in the absence of an electrochemical potential gradient for B*, i.e., multiple occupancy provides a basis for coupling the free energy in the gradient of A to the flow of B.[1,9,10,39]

The basis for the coupling can be understood by imagining the sequence of events that would result in tracer translocation. Clearly, tracer ions could move in either direction via a "single occupancy cycle," and this flow would represent a measure of what we have called the "intrinsic permeability" of the two-ion channel. The rate coefficient would be identical in either direction, however, and this permeation mechanism would not give rise to any net flow of tracer. In keeping with our usual tracer definitions, we would assume that the *molar* abundance of the tracer is sufficiently low so as to render it very unlikely that we find a channel occupied by two tracer ions at the same time. The molar abundance of the unlabeled isotope dictates that most of the time the channel will be occupied, singly or doubly, by an unlabeled ion. The doubly occupied channel would clearly be unavailable to the tracer, thus reducing the apparent membrane permeability, but, not producing net flow of tracer. The basis for net flow is found in states involving one abundant ion and one tracer, i.e., A(1), B*(2) and B*(1), A(2). In the presence of a gradient of abundant A from side 1 to side 2, the state A(1), B*(2) will be favored so that tracer will be preferentially directed toward side 2. This simple example reveals a fundamental difficulty with the concept of permeability as applied to a multiion channel, namely, that the rate of movement of a tracer ion through the channel is now expected to be highly dependent on the distribution of the unlabeled species. The potential for complexity can be readily appreciated by considering the effect of changing the concentration gradient of A, for example, by increasing the concentration of side 1. The increased abundance of A will increase the probability that the channel will be occupied by two unlabeled ions and, thereby, reduce the availability of the channel to B*. On the other hand, increasing the gradient will *increase* the positive coupling effect and tend to *enhance* net flow of B*.

In the experimental configuration that is typically used to determine permeability ratios by measuring the shift in the reversal potential due to the addition of a substitute ion, the complexity of the multiion channel is compounded, because there are now opposing gradients of two permeant ions. Once again, we can obtain insight into the permeability ratios by imagining what would be seen if we measured the flow of two tracer ions, say ^{42}K and ^{86}Rb, simultaneously. As with the one-ion channel, there will be states of the multiion channel that preclude either tracer from entering and so would not affect the ratio of the tracer flows. Differences in the interaction of the two tracer ions with the channel, however, will predispose them to different degrees of positive coupling from *both* of the unlabeled ions! The permeability ratios, not surprisingly, are expected to be highly dependent on the concentrations of the permeant ions and on the membrane potential, so that although the concept of the permeability ratio is still useful as an empirical description, it loses the simple physical significance associated with ion–channel interaction in a Nernst–Planck or one-ion channel.

5.4.5.2. Anomalous Mole Fraction Effects

Multi-ion channels are predicted to exhibit concentration-dependent changes in reversal potential or conductance that are collectively referred to as "anomalous mole fraction effects."[3,12a,40,41] The term anomalous refers to the fact that as an ion is replaced on an equimolar basis by a substitute ion, the reversal potential and/or conductance can exhibit a minimum at an intermediate mole fraction of the substitute ion. For example, at low concentration a substitute ion that binds tightly in the channel might attenuate conductance by acting as a blocker. As the mole fraction of the substitute ion is increased, however, multiple occupancy by the substitute can lead to an *increase* in the conductance because ionic repulsion within the channel can facilitate ionic throughput. It has been suggested that multi-ion occupancy can permit a channel to utilize tight binding as a way of selecting for a specific ion because the ionic repulsion effects compensate for the retarding effects on ion binding on conductance. A striking example of the consequences of multi-ion occupancy is seen with calcium-selective channels. In the absence of calcium these channels are highly permeable to Na and K. Low concentrations of calcium (10^{-8}–10^{-5}M) block monovalent cation flow due to the tight binding of calcium to sites in the channel. The tight binding might be expected to severely attenuate calcium currents, but at higher calcium concentrations occupancy of the channel by more than one calcium ion and the resultant ionic repulsion effect speeds up the conduction process so that the currents are comparable to those seen with monovalent ions. The combination of high affinity for the binding sites and strong repulsion between the two occupying ions gives calcium an advantage over the univalent ions.[12a,40,41]

In the multiion channel, we see that the permeabilities and conductances can behave in a way that is not at first very intuitive, but reflects the fact that the nature of the interactions of individual ions with the channel proteins can differ far more than might have been predicted on the basis of the simple Nernst–Planck, or single site models. In the case of the calcium channel, the anomalous behavior that is attributed to double occupancy by calcium ions has clear physiological significance; these effects are the basis for the channel's selectivity for calcium, an ion that is 100-fold less abundant in physiological solutions. In other instances, however, such effects might at first be thought of as biophysical curiosities, of rather little significance. As we now approach channel structure in a more sophisticated way, however, it is apparent that it is precisely these subtleties of channel behavior that provide the basis for productive inquiry into the relationship between channel structure and function.

ACKNOWLEDGMENTS. I am grateful to Myrna Pancost for her work on the manuscript and to Monique Mansoura for comments on the text. The writing of this chapter was supported by the National Institutes of Health (DK 29786 and DK 45880), the Cystic Fibrosis Foundation, and the University of Michigan Gastrointestial Peptide Center.

REFERENCES

1. Kirk, K. L., and Dawson, D. C. (1983). Basolateral potassium channel in turtle colon. *J. Gen. Physiol.* **82:**297–313.
2. Dawson, D. C. (1991). Principles of Membrane Transport, *Handbook of Physiology, The Gastrointestinal System IV,* Physiological American Society, Bethesda, MD.
3. Hille, B. (1992). *Ionic Channels of Excitable Membranes,* 2nd ed., Sinauer, Sunderland, MA.
4. Moczydlowski, E. (1986). Single-channel enzymology. In *Ion Channel Reconstitution,* (C. Miller, ed.), Plenum Press, New York, pp. 75–113.
5. Ussing, H. H. (1949). The distinction by means of tracers between active transport and diffusion. *Acta Physiol. Scand.* **19:**43–56.
6. Ussing, H. H. (1952). Some aspects of the application of tracers in permeability studies. *Adv. Enzymol.* **13:**21–65.
7. Ussing, H. H. (1978). Interpretation of tracer fluxes. In *Membrane Transport in Biology,* (G. Giebisch, D. C. Testeson, and H. H. Ussing, eds.), Springer-Verlag, Berlin, pp. 115–140.
8. Dawson, D. C. (1977). Tracer flux ratios: A phenomenological approach. *J. Membr. Biol.* **31:**351–358.
9. Dawson, D. C. (1982). Thermodynamic aspects of radiotracer flow. In *Biological Transport of Radiotracers* (L. G. Colombetti, ed.), CRC Press, Boca Raton, pp. 79–95.
10. Finkelstein, A., and Rosenberg, P. A. (1979). Single-file transport: Implications for ion and water movement through gramicidin A channels. In *Membrane Transport Processes* (C. F. Stevens and R. W. Tsien, eds.), Raven Press, New York, Volume 3, pp. 73–88.
11. Goldman, D. E. (1943). Potential, impedance, and rectification in membranes. *J. Gen. Physiol.* **27:**37–60.
12. Finkelstein, A., and Andersen, O. S. (1981). The gramicidin A channel: A review of its permeability characteristics with special reference to the single-file aspect of transport. *J. Membr. Biol.* **59:**155–171.

12a. Hess, P., and Tsien, R. W. (1984). Mechanism of ion permeation through calcium channels. *Nature* **309:**453–456.

13. Hodgkin, A. L., and Katz, B. (1949). The effect of sodium ions on the electrical activity of the giant axon of the squid. *J. Physiol. (London)* **108:**37–77.
14. Jack, J. J. B., Noble, D., and Tsien, R. W. (1975). *Electric Current Flow in Excitable Cells,* Oxford University Press, (Clarendon), London.
15. Finkelstein, A., and Mauro, A. (1963). Equivalent circuits as related to ionic systems. *Biophys. J.* **3:**215–237.
16. Teorell, T. (1953). Transport processes and electrical phenomena in ionic membranes. *Prog. Biophys. Chem.* **3:**305–369.
17. Schultz, S. G. (1980). *Basic Principles of Membrane Transport,* Cambridge University Press, London.

18. Patlak, C. S. (1960). Derivation of an equation for the diffusion potential. *Nature* **188:**944–945.
19. Hille, B. (1975). Ionic selectivity of Na and K channels of nerve membranes. In *Membranes. Lipid Bilayers and Biological Membranes: Dynamic Properties* (G. Eisenman, ed.), Dekker, New York, pp. 255–323.
20. Finkelstein, A., and Mauro, A. (1977). Physical principles and formalisms of electrical excitability. In *Handbook of Physiology. The Nervous System. Cellular Biology of Neurons,* American Physiological Society, Bethesda, Section 1, Volume I, Part 1, pp. 161–213.
21. Dani, J. A., and Levitt, D. G. (1990). Diffusion and kinetic approaches to describe permeation in ionic channels. *J. Theor. Biol.* **146:**289–301.
21a. Hodgkin, A. L., and Fuxley, A. F. (1952). Currents carried by sodium and potassium ions through the membrane of the giant axon of Loligo. *J. Physiol. (London)* **116:**449–472.
22. Sten-Knudson, O. (1978). Passive transport processes. In *Membrane Transport in Biology* (G. Giebisch, D. C. Tosteson, and H. H. Ussing, eds.), Springer-Verlag, Berlin, Volume I, pp. 5–113.
23. Lauger, P. (1973). Ion transport through pores: A rate-theory analysis. *Biochim. Biophys. Acta* **311:**423–441.
24. Levitt, D. G. (1982). Comparison of Nernst–Planck and reaction-rate models for multiply occupied channels. *Biophys. J.* **37:**575–587.
25. Bockris, J. O., and Reddy, A. K. N. (1970). *Modern Electrochemistry,* Volume 1, Plenum Press, New York.
26. Eisenman, G., and Horn, R. (1983). Ionic selectivity revisited: The role of kinetic and equilibrium processes in ion permeation through channels. *J. Membr. Biol.* **76:**197–225.
27. Krasne, S. (1978). Ion selectivity in membrane permeation. In *Physiology of Membrane Disorders* (T. E. Andreoli, J. F. Hoffman, and D. D. Fanestil, eds.), Plenum Press, New York, pp. 217–241.
28. Krasne, S., and Eisenman, G. (1973). The molecular basis of ion selectivity. In *Membranes: Lipid Bilayers and Antibiotics* (G. Eisenman, ed.), Dekker, New York, Volume 2, pp. 277–328.
29. Lauger, P. (1979). Transport of noninteracting ions through channels. In *Membrane Transport Processes* (C. F. Stevens and R. W. Tsien, eds.), Raven Press, New York, Volume 3, pp. 17–27.
30. Levitt, D. G. (1986). Interpretation of biological ion channel flux data-reaction-rate versus continuum theory. *Annu. Rev. Biophys. Chem.* **15:**29–57.
31. Levitt, D. G. (1984). Kinetics of movement in narrow channels. In *Ion Channels: Molecular and Physiological Aspects. Current Topics in Membranes and Transport,* Volume 21 (W. D. Stein, ed.), Academic Press, New York.
31a. Pappone, P. A., and Cahalan, M. D. (1986). Ion permeation in cell membranes. In *Physiology of Membrane Disorders* (T. E. Andreoli, J. F. Hoffman, D. D. Fanestil, and S. G. Schultz, eds.), Plenum Press, New York, pp. 249–272.
32. Anderson, O. S., and Procopio, J. (1980). Ion movement through gramicidin A channels: On the importance of the aqueous diffusion resistance and ion–water interactions. *Acta Physiol. Scand. Suppl.* **481:** 27–35.
33. Eisenman, G., and Dani, J. A. (1987). An introduction to molecular architecture and permeability of ion channels. *Annu. Rev. Biophys. Chem.* **16:**205–226.
34. Jordan, P. C. (1993). *Interactions of Ions with Membrane Proteins,* CRC Press, Boca Raton.
35. Parsegian, V. A. (1969). Energy of an ion crossing a low dielectric membrane: Solutions to four relevant electrostatic problems. *Nature* **221:**844–846.
36. Parsegian, V. A. (1975). Ion–membrane interactions as structural forces. *Ann. N. Y. Acad. Sci.* **264:**161–174.
37. Coronado, R., Rosenberg, R., and Miller, C. (1980). Ionic selectivity, saturation, and block in an K^+-selective channel from sarcoplasmic reticulum. *J. Gen. Physiol.* **76:**425–446.
38. Latorre, R., and Miller, C. (1983). Conduction and selectivity in potassium channels. *J. Membr. Biol.* **71:**30.
39. Hille, B., and Schwarz, W. (1978). Potassium channels as multiion single-file pores. *J. Gen. Physiol.* **72:**409–442.
40. Almers, W., and McCleskey, E. W. (1984). Non-selective conductance in calcium channels of frog muscle: Calcium selectivity in a single-file pore. *J. Physiol. (London)* **353:**585–608.
41. Almers, W., McCleskey, E. W., and Palade, P. T. (1984). A non-selective cation conductance in frog muscle membrane blocked by micromolar external calcium ions. *J. Physiol. (London)* **353:**565–583.
42. Thompson, S. M. (1986). Relations between chord and slope conductances and equivalent electromotive forces. *Am. J. Physiol.* **250:** C333–C339.
43. Ussing, H. H. (1949). The distinction by means of tracers between active transport and diffusion. *Acta Physiol. Scand.* **19:**43–56.
44. Ussing, H. H. (1952). Some aspects of the application of tracers in permeability studies. *Adv. Enzymol.* **13:**21–65.
45. Ussing, H. H. (1978). Interpretation of tracer fluxes. In *Membrane Transport in Biology,* (G. Giebisch, D. C. Testeson, and H. H. Ussing, eds.), Springer-Verlag, Berlin, pp. 115–140.
46. Woodbury, J. W. (1971). Eyring rate theory model of the current–voltage relationships of ion channels in excitable membranes. In *Chemical Dynamics: Papers in Honor of Henry Eyring* (J. Hirschfelder, ed.), Wiley, New York, pp. 601–617.
47. Moore, W. J., (1972). Physical Chemistry, 4th Ed., Prentice-Hall, Englewood Cliffs, NJ.
48. Lopatin, A. N., and Makhina, E. N., and Nichols, C. G. (1995). The mechanism of inward rectification of potassium channels "Long pore plugging" by cytoplasmic polyamines. *J. Gen. Physiol.* **106:** 923–955.
49. Yang, J., Ellinor, P. T., Sather, W. A., Zhang, J. F., and Tsien, R. W. (1993). Molecular determinants of Ca 2+ selectivity and ion permeation in L-type Ca 2+ channels. *Nature* **366:**158–161.
50. Bezanilla, F., and Armstrong, C. M. (1972). Negative conductance caused by the entry of sodium and cesium ions into the potassium channels of squid axons. *J. Gen. Physiol.*
51. Junge, D. (1992). *Nerve and Muscle Excitations,* 3rd ed., Sinauer, Sunderland.

6

The Lactose Permease of *Escherichia coli*
An Update

H. Ronald Kaback

An unsolved basic biochemical phenomenon of critical importance is the general problem of energy transduction in biological membranes. Although the driving force for a variety of seemingly unrelated functions (e.g., secondary active transport, oxidative phosphorylation, rotation of the bacterial flagellar motor) is a bulk-phase, transmembrane electrochemical ion gradient, the molecular mechanism(s) by which free energy stored in such gradients is transduced into work or into chemical energy remains enigmatic. Nonetheless, gene sequencing and analyses of deduced amino acid sequences indicate that many biological machines involved in energy transduction, secondary transport proteins in particular,[1,2] fall into families encompassing proteins from archaebacteria to the mammalian central nervous system, thereby suggesting that the members may have common basic structural features and mechanisms of action. Moreover, certain of these proteins have been implicated in human disease (e.g., glucose/galactose malabsorption, certain forms of drug abuse, depression).

As postulated originally by Peter Mitchell[3,4] and demonstrated conclusively in bacterial membrane vesicles (see Refs. 5–7 for reviews), accumulation of a wide variety of solutes against a concentration gradient is driven by a proton electrochemical gradient $\Delta\bar{\mu}_{H^+}$; interior negative and/or alkaline). The work discussed here focuses on a specific secondary transport protein, the lactose or lac permease (also known as the lactose/H^+ symporter or lac carrier protein), as a paradigm. β-Galactoside accumulation in *Escherichia coli* is catalyzed by lac permease, a hydrophobic polytopic cytoplasmic membrane protein that carries out the coupled stoichiometric translocation of a β-galactoside with H^+ (i.e., β-galactoside/H^+ symport or cotransport). Physiologically, the permease utilizes free energy released from downhill translocation of H^+ to drive accumulation of β-galactosides against a concentration gradient (Figure 6.1A). In the absence of $\Delta\bar{\mu}_{H^+}$, lac permease catalyzes the converse reaction, utilizing free energy released from downhill translocation of β-galactosides to drive uphill translocation of H^+ with generation of a $\Delta\bar{\mu}_{H^+}$ the polarity of which depends on the direction of the substrate concentration gradient (Figure 6.1B,C). The essential question is how energy stored in the form of either $\Delta\bar{\mu}_{H^+}$ or a substrate concentration gradient is transduced to drive the coupled process.

Lac permease is encoded by the *lacY* gene, the second structural gene in the *lac* operon, and it has been cloned into a recombinant plasmid[8] and sequenced.[9] By combining overexpression of *lacY* with the use of a highly specific photoaffinity probe for the permease[10] and reconstitution of transport activity in artificial phospholipid vesicles (i.e., proteoliposomes),[11] the permease has been solubilized from the membrane, purified to homogeneity,[12–14] and shown to catalyze all of the translocation reactions typical of the β-galactoside transport system *in vivo* with comparable turnover numbers.[15,16] Therefore, a single gene product—the product of *lacY*—is solely responsible for all of the translocation reactions catalyzed by the β-galactoside transport system.

This chapter discusses selected observations with the lac permease at the molecular level, but it should be reemphasized that there are a huge number of proteins that catalyze similar reactions in virtually all biological membranes. Furthermore, it should be stated at the outset that structural information at high resolution is particularly difficult to obtain with hydrophobic membrane proteins.[17] The great majority of membrane proteins, lac permease in particular, have yet to be crystallized and it is becoming increasingly apparent that structural information is a prerequisite for mechanistic considerations.

6.1. LAC PERMEASE CONTAINS 12 TRANSMEMBRANE DOMAINS IN α-HELICAL CONFORMATION

Circular dichroic measurements on purified lac permease indicate that the protein is over 80% helical in conformation, an estimate consistent with the hydropathy profile of the permease which suggests that approximately 70% of its 417 amino acid residues are found in hydrophobic domains with a mean length of 24 ± 4 residues.[18] Based on these findings, a secondary structure was proposed in which the permease is

H. Ronald Kaback • Howard Hughes Medical Institute, Departments of Physiology and Microbiology and Molecular Genetics, Molecular Biology Institute, University of California, Los Angeles, Los Angeles, California 90024

Molecular Biology of Membrane Transport Disorders, edited by S.G. Schultz *et al.* Plenum Press, New York, 1996

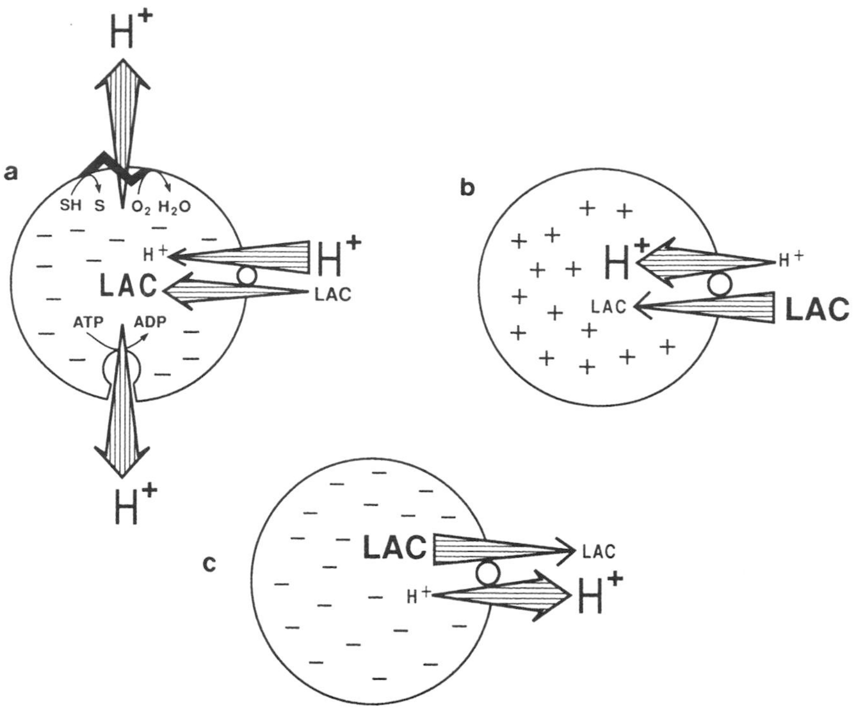

Figure 6.1. H^+/lactose symport in *E. coli.* (**a**) Lactose accumulation in response to $\Delta\bar{\mu}_{H^+}$ (interior negative and alkaline) generated either by respiration or by ATP hydrolysis. (**b**) Uphill H^+ transport in response to an inwardly directed lactose gradient. (**c**) Uphill H^+ transport in response to an outwardly directed lactose gradient.

composed of a hydrophilic N-terminus followed by 12 hydrophobic segments in α-helical conformation that traverse the membrane in zigzag fashion connected by hydrophilic domains (loops) with a 17-residue C-terminal hydrophilic tail (Figure 6.2). Support for the general features of the model and evidence that both the N- and C-termini, as well as the second and third cytoplasmic loops are exposed to the cytoplasmic face of the membrane were obtained subsequently from laser Raman[19] and Fourier transform infrared* spectroscopy, immunological studies,[20–28] limited proteolysis,[29,30] and chemical modifications.[31] However, none of these approaches differentiates between the 12-helix model and others containing 10 (Ref. 19) or 13 (Ref. 32) putative transmembrane helices.

Calamia and Manoil[33] have provided elegant and unequivocal support for the topological predictions of the 12-helix model by analyzing an extensive series of lac permease–alkaline phosphatase (*lacY–phoA*) chimeras. Alkaline phosphatase is synthesized as an inactive precursor in the cytoplasm of *E. coli* with a short signal sequence that directs its secretion into the periplasmic space where it dimerizes to form active enzyme. If the signal sequence is deleted, the enzyme remains in the cytoplasm in an inactive form. When alkaline phosphatase devoid of the signal sequence is fused to the C-termini of fragments of a cytoplasmic membrane protein, enzyme activity reflects the ability of the N-terminal portions of the chimeric polypeptides to translocate alkaline phosphatase to the outer surface of the membrane.[34] Alkaline phosphatase activity in cells independently expressing each of 36 *lacY–phoA* fusions exclusively favors the model of lac permease with 12 transmembrane domains.

*P. D. Roepe, K. Rothschild, and H. R. Kaback, unpublished information.

In addition, it was demonstrated[33] that approximately half of a transmembrane domain is needed to translocate alkaline phosphatase to the external surface of the membrane. Thus, the alkaline phosphatase activity of fusions engineered at every third amino acid residue in putative helices III and V (Figure 6.2) increases as a step function as the fusion junction proceeds from the 8th to the 11th residue of each of these transmembrane domains. Furthermore, when fusions are constructed at each amino acid residue in putative helices IX and X of the permease, the data are in excellent agreement with the model.* The implication of these experiments is that it may be possible to approximate the middle of transmembrane domains by means of serial alkaline phosphatase fusions. In addition, expression of contiguous, nonoverlapping permease fragments with discontinuities in either loops or transmembrane domains may provide a means of approximating the boundaries of transmembrane domains (see Section 6.8.2).

Purified lac permease reconstituted into proteoliposomes exhibits a notch or cleft on freeze-fracture electron microscopy,[35,36] an observation independently documented using completely different techniques.[37] The presence of a solvent-filled cleft in the permease has important implications with regard to the mechanism of β-galactoside/H^+ symport, as the barrier within the permease may be thinner than the full thickness of the membrane. Therefore, the number of amino acid residues in the protein directly involved in translocation may be fewer than required for substrates to traverse the entire thickness of the membrane. Furthermore, the existence of a solvent-filled cleft in the permease may present a caveat with

*M. L. Ujwal, E. Bibi, C. Manoil, and H. R. Kaback, manuscript in preparation.

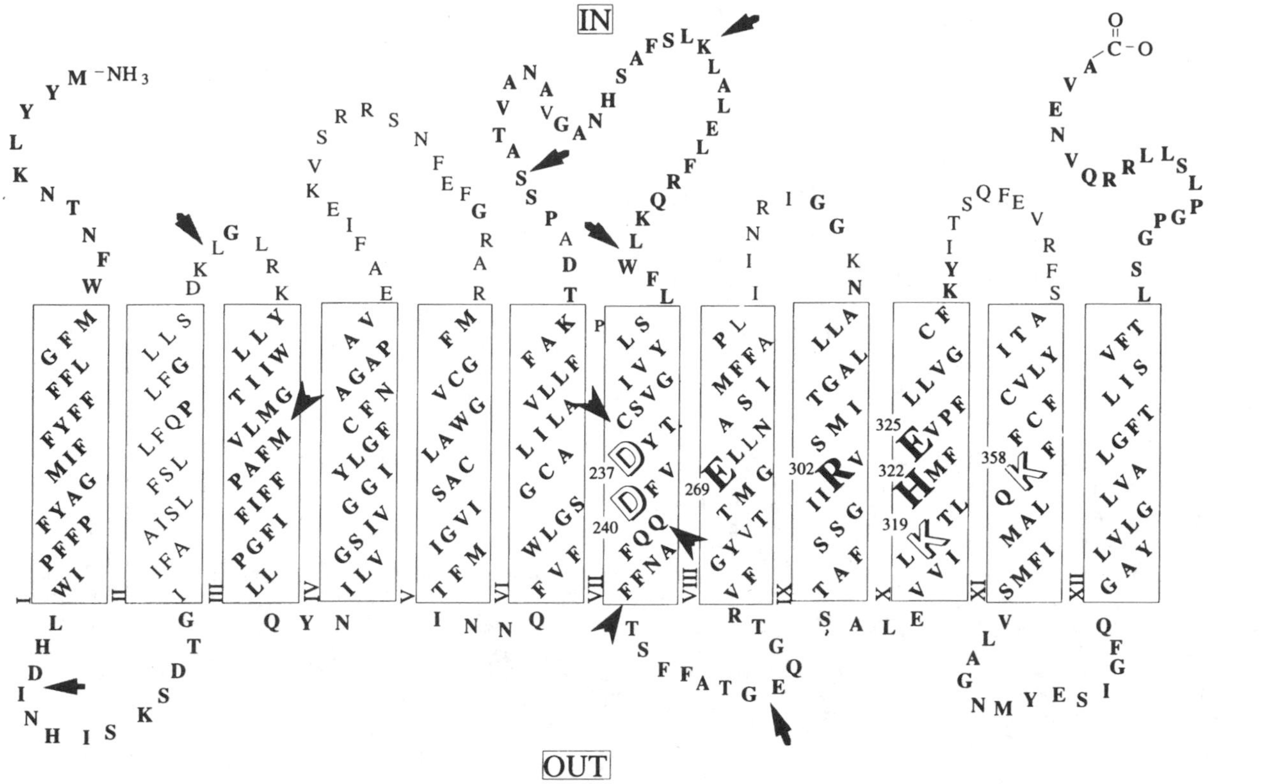

Figure 6.2. Secondary-structure model of lac permease based on the hydropathy profile of the protein. The single-letter amino acid code is used, and hydrophobic segments are shown in boxes as transmembrane α-helical domains connected by hydrophilic loops. Residues in normal type yield active permease when replaced with Cys; residues in solid block letters yield inactive permease when replaced with a number of different residues; residues in open block letters are charge-paired (Asp237–Lys358 and Asp240–Lys319); residues in light type have not been mutagenized. Permease with A177C is inactive, but other replacements have not been made; permease with Y336F is inactive, but other replacements have not been made. Grey arrowheads indicate active split permease constructs; black arrowheads signify inactive constructs.

respect to the topological placement of certain residues. That is, a residue located in the middle of a transmembrane domain lining such a cleft may be in a hydrophilic rather than a hydrophobic environment (see Section 6.4.1).

6.2. LAC PERMEASE IS FUNCTIONAL AS A MONOMER

One particularly difficult problem to resolve with hydrophobic membrane transport proteins is their functional oligomeric state. Although early evidence (summarized in Ref. 38) suggested that oligomerization might be important for activity, current data indicate that the permease is fully functional as a monomer. By studying eosinylmaleimide-labeled permease, Dornmair *et al.*[39] utilized phosphorescence anisotropy to demonstrate that purified, reconstituted lac permease exhibits a rotational diffusion consistent with a 46.5-kDa size particle, the molecular mass of the permease as determined from the deduced amino acid sequence,[9] and that the diffusion constant is not altered in the presence of $\Delta\bar{\mu}_{H^+}$. Costello *et al.*[36] reconstituted purified lac permease and cytochrome *o* into proteoliposomes under conditions in which both proteins are fully functional, and examined the preparations by freeze-fracture electron microscopy. In nonenergized proteoliposomes, the permease appears to reconstitute as a monomer based on: (1) the variation of intramembrane particle density with protein concentration; (2) the ratio of particles corresponding to each protein in proteoliposomes reconstituted with a known ratio of permease to oxidase; and (3) the dimensions of the particles observed in tantalum replicas. None of the parameters is altered in the presence of $\Delta\bar{\mu}_{H^+}$. Importantly, moreover, the initial rate of $\Delta\bar{\mu}_{H^+}$ driven lactose transport in proteoliposomes varies linearly with the ratio of lac permease to phospholipid over a range at which there is statistically between 0 and 12 molecules of permease per proteoliposome. If more than a single molecule of lac permease is required for active lactose transport, an exponential relationship would be expected, particularly at low protein:phospholipid ratios. At this point, it seemed to have been clearly demonstrated that lac permease is functional as a monomer.

However, Bibi are Kaback[38] then demonstrated that certain paired in-frame deletion mutants complement functionally. [The nomenclature of the constructs (N_xC_y) describes the number of putative transmembrane helices retained in the N-terminal (N_x) and C-terminal (C_y) portions of the permease before and after the deletion (Figure 6.3).] Although cells expressing each deletion individually are unable to catalyze active lactose accumulation, cells simultaneously expressing

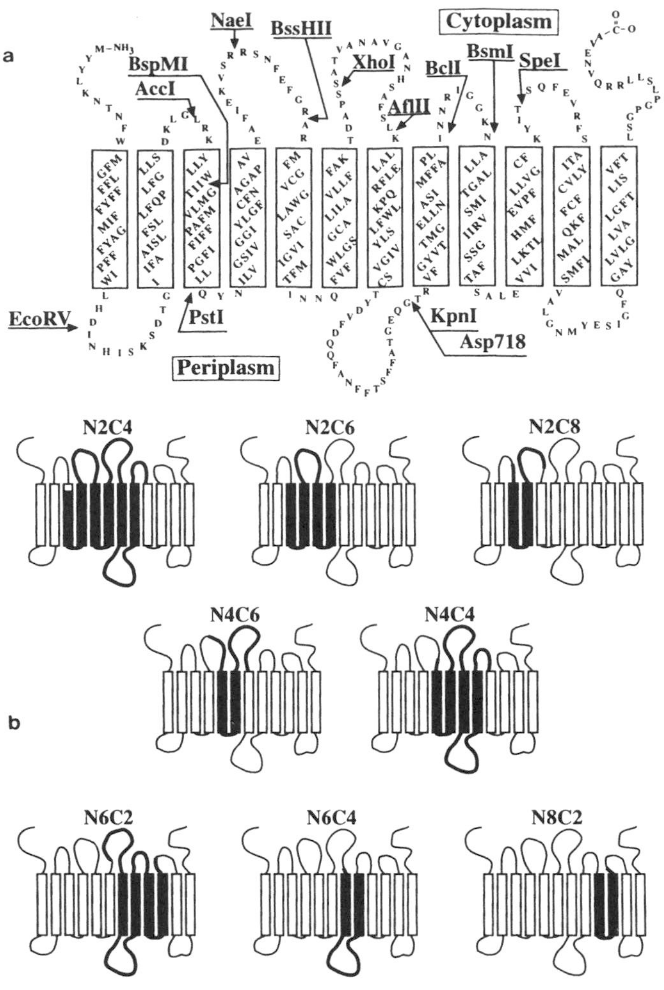

Figure 6.3. Secondary structure of lac permease (**a**) showing polypeptides encoded by the *lacY* deletion mutants described (b). The single-letter amino acid code is used, hydrophobic transmembrane helices are shown in boxes, and deleted regions are blackened. (**a**) Wild-type lac permease with restriction sites in the cassette *lacY* DNA indicated; (**b**) constructs in which putative transmembrane helices were deleted. A cassette *lac*Y gene (EMBL-X56095) containing the lac promoter/operator was cloned into plasmic pT7-5 and used for all *lac*Y gene manipulations. The cassette gene contains unique restriction sites in each segment of the gene encoding a putative loop (i.e., approximately every 100 bp). In most of the constructs, the plasmid was digested with appropriate restriction enzymes to remove the desired segment of the gene, treated with DNA polymerase (Klenow fragment) and ligated to itself. In one construct (N_2C_6), a linker was synthesized and inserted, and the cohesive ends were ligated. The fusion junctions of each construct were sequenced using the dideoxyoligonucleotide method. From Bibi and Kaback.[38]

N_2C_8 and N_8C_2 catalyze lactose transport 60% as well as cells expressing wild-type permease, while N_2C_6 and N_8C_2 or N_4C_6 and N_8C_2 exhibit diminished but significant transport activity. On the other hand, N_4C_6 and N_6C_4 or N_2C_6 and N_6C_2 exhibit only marginal activity, and the combinations N_4C_4–N_8C_2, N_2C_4–N_8C_2 or N_6C_4–N_8C_2 exhibit no activity whatsoever. Moreover, the following pairs of missense mutations or single amino acid deletions also exhibit no activity: P28S–H322K, E325C–H322K, E325C–K319L, E325C–R302L, or ΔW38–ΔH322. Importantly, it has been shown that complementation between N_2C_8/N_8C_2 occurs at the protein level and *not* the DNA level. Therefore, the ability to complement functionally is apparently a specific property of pairs of permease molecules containing relatively large deletions separated by at least two transmembrane hydrophobic domains, and it is not observed with pairs of missense mutations or point deletions.

One possible interpretation of the results is that specific interactions occur between transmembrane helices in wild-type permease and that disruption of these interactions by deletion leaves a "potential gap" in the structure that can be filled by interaction with another molecule containing the deleted segment. For instance, perhaps putative helix VIII has a high affinity for helix IX (Figure 6.2; note that the loop between putative transmembrane helices VIII and IX is relatively short) and poor affinity for helix XI. By this means, a permease molecule deleted of helices IX and X (e.g., N_8C_2) might "accept" these helices from a "donor" molecule deleted of helices III and IV (e.g., N_2C_8) and/or vice versa. However, *E. coli* T184 transformed with plasmids encoding P28S and N_8C_2 or H322K and N_2C_8 as potential donor–acceptor pairs do not complement functionally. Similarly, ΔH322 does not exhibit functional complementation with N_2C_8 nor does ΔW78 complement functionally with N_8C_2. Thus, the simplistic explanation does not appear to be the case.

Certain contiguous, nonoverlapping fragments of lac permease yield functionally active duplexes when expressed together (see Section 6.8.2). This phenomenon may be related to functional complementation between permease molecules containing large deletions in transmembrane domains. Thus, it is suggested that permease mutants containing missense mutations or point deletions, like the intact wild-type molecule, form relatively compact structures that are unable to form intermolecular complexes. On the other hand, molecules containing deletions in certain hydrophobic transmembrane domains (e.g., N_2C_8 and N_8C_2) may be in a more "relaxed" state and therefore able to interact to form functional dimers (i.e., *trans* complementation). In any case, the observation that certain pairs of deletion mutants can complement functionally rekindled concern regarding the oligomerization state of wild-type permease.

Recently, Sahin-Tóth and Kaback[40] engineered a fusion protein containing two lac permease molecules covalently linked in tandem (permease dimer). Permease dimer is inserted into the membrane in a functional state, and each half of the dimer exhibits equal activity. Thus, point mutations in either half of the *lacY* tandem repeat lead to 50% inactivation of transport. Furthermore, the activity of a permease dimer composed of wild-type permease and a mutant devoid of Cys is inactivated ca. 50% by *N*-ethylmaleimide (NEM). Clearly, therefore, the phenomenon of negative dominance is not observed; however, it is possible that the wild-type halves of the wild-type/mutant dimers might complement in *trans*. In order to test this caveat, a permease dimer was constructed that contains two different deletion mutants which complement when expressed as untethered molecules (N_2C_8 and N_8C_2). This construct does not catalyze lactose accumulation to any extent whatsoever, suggesting that permease dimers do not oligomerize in *trans*. Taken as a whole, the experiments clearly favor the conclusion that wild-type lac permease is functional as a monomer.

6.3. SITE-DIRECTED MUTAGENESIS

6.3.1. Few Amino-Acid Residues Are Essential for Activity

By using site-directed mutagenesis with wild-type permease or a functional mutant devoid of Cys residues (C-less permease; see Section 6.3.2),[41] individual amino acid residues in the permease that are essential mechanistically have been identified (Figure 6.2). Over 300 of the 417 residues in C-less permease have been mutagenized, mostly by Cys-scanning mutagenesis (see Section 6.3.3), and *remarkably,* less than a half-dozen residues have been identified thus far as being clearly essential for activity.[42–44]* Of the few mutants that do not catalyze active transport, most retain the ability to catalyze partial reactions or bind ligand. More specifically, none of the 8 Cys,[41,45–52] 6 TRP,[53] 12 Pro,[54,55] or 36 Gly residues† in the permease is obligatory for active transport. Only one of the 4 His residues[56–59] and possibly one of the 14 Tyr residues[60]‡ are important. On the other hand, Glu269 (helix VIII),[61,62] Arg302 (helix IX),[63,64] His322 (helix X),[56–59] and Glu325 (helix)[65,66] are essential for substrate accumulation and/or binding. Moreover, as will be discussed (see Section 6.6), differences in the properties of the mutants suggest that Arg302, His322, and Glu325 may function in a H^+ translocation pathway, although it is possible that the residues also form part of a coordination site for H_3O^+. In any event, very few residues appear to be mandatory for transport. Therefore, the chemistry involved in the mechanism must be relatively simple. Furthermore, the observation that the great majority of the residues in the permease can be mutagenized without abolishing function provides a strong indication that individual amino acid replacements do not cause gross conformational changes in the protein.

*In addition to helices, I, IX, X, and XI, Cys-scanning mutagenesis has been carried out with helices III (M. Sahin-Tóth, S. Frillingos, E. Bibi, A. Gonzalez, and H. R. Kaback, manuscript in preparation), V (C. Weizmann, M. Sahin-Tóth, and H. R. Kaback, unpublished information), and VII (S. Frillingos, M. Sahin-Tóth, and H. R. Kaback, manuscript in preparation) (see Figure 6.2).

†K. Jung, H. Jung, P. Colachurchio, and H. R. Kaback, manuscript in preparation.

‡Although Tyr—>114Phe replacements indicate that three Tyr residues are important for activity, Cys-scanning mutagenesis has revealed that Tyr26 and Tyr236 can be replaced with Cys with retention of significant activity.

6.3.2. Cys Residues and Construction of a Functional Permease Devoid of Cys Residues

Fox and Kennedy[67,68] demonstrated initially that lac permease is irreversibly inactivated by NEM and that protection is afforded by substrates such as β,D-galactopyranosyl 1-thio-β,D-galactopyranoside (TDG). On the basis of these findings, it was postulated[67] that a Cys residue is at or near the substrate-binding site of lac permease, and Beyreuther *et al.*[69] later showed that the substrate-protectable residue is Cys148 (Figure 6.2). In addition, the permease is reversibly inactivated by other sulfhydryl reagents like *p*-chloromercuribenzenesulfonate or by sulfhydryl oxidants such as diamide[70] or plumbagin,[71] and TDG blocks inactivation by these reagents as well.

In view of the importance attributed to sulfhydryl groups in lac permease (see Refs. 71–73 in addition), particularly Cys148, site-directed mutagenesis was used initially to replace Cys148 with Gly[45,47] or Ser.[48,49] Surprisingly, although Cys148 is required for substrate protection against alkylation by NEM, it is not important for lactose/H^+ symport. Subsequently, it was shown[6,46] that replacement of Cys154 with Gly leads to complete loss of transport activity although the permease binds the high-affinity ligand *p*-nitrophenyl-α,D-galactopyranoside (NPG) normally. Moreover, replacement of Cys 154 with Ser or Val yields permease with 10 or 30%, respectively, of the wild-type rate, indicating that although Cys154 is needed to full activity, it is not mandatory. Brooker and Wilson[50] then replaced Cys176 or 234 with Ser, and Menick *et al.*[51] replaced Cys117, 333 or 353, and 355 with Ser with little or no effect on activity. Therefore, out of a total of eight cysteinyl residues in the permease, only Cys154 appears to be relatively important for transport, but even this residue is not essential. Finally, experiments in which each of the Cys mutants was purified and reconstituted[52] indicate that sulfhydryl–disulfide interconversion does not play a role in regulation of permease activity.[71,73]

More recent studies[41] provide definitive support for the contention that Cys residues in lac permease do not play an essential role in the mechanism of lac permease. When Cys154 is replaced with Val and each of the other Cys residues in lac permease is replaced with Ser, "C-less" permease catalyzes active lactose transport at about 30% of the initial rate and at about 60% of the steady-state level of accumulation of wild-type permease. Moreover, active lactose transport in right-side-out vesicles containing C-less permease is not inactivated by

NEM, in dramatic contrast to vesicles containing wild-type permease.

Although Cys148 is not essential for permease activity, recent experiments[74,75] demonstrate that this residue is indeed located in a substrate binding site, as postulated originally.[67] Thus, Cys148 was replaced with hydrophobic (Ala, Val, Ile, Phe), hydrophilic (Ser, Thr), or charged (Asp, Lys) residues, and the properties of the mutants analyzed.[74] The results demonstrate that the size and polarity of the side chain at this position modify transport activity and substrate specificity. Small hydrophobic side chains (Ala, Val) generally increase the apparent affinity of the permease for substrate, while hydrophilic side chains (Ser, Thr, Asp) decrease apparent affinity and bulky or positively charged side chains (Phe, Lys) virtually abolish activity. In addition, hydrophilic substitutions (Ser, Thr, Asp) decrease the activity of the permease toward monosaccharides (thiomethyl-β,D-galactopyranoside and galactose) relative to disaccharides (lactose, TDG, and melibiose).

Site-directed fluorescence (see Section 6.4) was also used to study Cys148 and other residues in the vicinity.[75] In the absence of ligand, permease with a single Cys residue at position 148 reacts rapidly with 2-(4′-maleimidylanilino)-naphthalene-6-sulfonic acid (MIANS), a fluorophore whose quantum yield increases dramatically on reaction with a thiol, indicating that this residue is readily accessible to the probe. Various ligands of the permease block the reaction, and the concentration dependence is commensurate with the affinity of each ligand for the permease (i.e., TDG<<lactose<galactose), but neither sucrose nor glucose has any effect whatsoever. Thus, the specificity of the permease for substrate appears to be related exclusively to the asymmetry at the fourth carbon of the galactosyl portion of the substrate. Interestingly, labeling of Cys145 which is presumed to be one helical turn removed from Cys148 displays properties similar to those observed with Cys148 permease, but the effects of ligand are far less dramatic. On the other hand, permease with a single Cys residue at position 146 or 147 behaves in a completely different manner. Studies with iodide show that MIANS at position 145 or 148 is accessible to the collisional quencher, indicating that this face of helix V is solvent exposed, while MIANS at position 146 or 147 is not quenched by iodide in the presence or absence of ligand. Finally, iodide quenching of MIANS at position 145 is clearly diminished in the presence of TDG. Taken together with the findings discussed above, the results indicate that Cys148 is a likely component of a substrate binding site that interacts hydrophobically with the galactosyl portion of the substrate, but does not play an obligatory role. In addition, the observations indicate that residue 145 is in close proximity.

6.3.3. Cys-Scanning Mutagenesis

By using C-less permease,[41] more than 300 individual residues in the protein have been replaced with Cys. The approach is invaluable from a number of points of view. First, it allows identification of residues that are *not* essential. If a given residue is replaced with Cys and cells express the mutant and catalyze active transport, it is apparent that the residue is not important for insertion or activity. On the other hand, if cells expressing the mutant do not catalyze transport, multiple amino acid replacements must be made before it can be concluded that the residue is essential. It is also important to realize that Cys-scanning mutagenesis is carried out with C-less permease which already contains eight mutations and that residues which appear to be important in the C-less background may behave differently in the wild-type. For example, each of the 36 Gly residues in C-less permease was replaced recently with Cys*, and three of the mutants (G64C, G115C, and G147C) were found to be inactive. Each of the other mutants exhibits significant ability to catalyze lactose accumulation. When Gly in each of the three inactive Cys-replacement mutants is replaced with Ala, G115A or G147A is active, but G64A, G64V, or G64P permease is completely inactive. Remarkably, however, when the G64A mutation is placed in the wild-type background, highly significant activity is observed. Similarly, Leu76 in transmembrane domain III appears to be an important residue, as permease activity is markedly compromised when Leu76 is replaced with Cys in the C-less background. When the analogous mutation is placed in the wild-type background, however, the permease is active.† Clearly, therefore, a number of mutations must be made in both the C-less and wild-type backgrounds and more than a single transport substrate should be tested before it can be concluded that a given residue is essential. Permease with A177C or T348C is inactive in the C-less background, and permease with Y336P or Y336C is inactive (Figure 6.2), but other replacements have not been made nor have these mutations been placed in the wild-type background. For these reasons, it is not yet clear whether these residues are essential.

In addition to testing for activity *per se,* each active Cys-replacement mutant can be assayed for sensitivity to NEM, and the effect of ligand on NEM sensitivity can also be tested in order to determine whether Cys at a given position of the protein exhibits reduced or enhanced reactivity with the alkylating agent (see Section 6.4.2). Finally, it should be emphasized that each single-Cys replacement mutant represents a unique molecule that can be used for site-directed spectroscopic studies (see Sections 6.4 and 6.5).

6.4. HELIX PACKING IN THE C-TERMINAL HALF OF THE PERMEASE

6.4.1. Functional Interactions between Putative Intramembrane Charged Residues

In 1991, King *et al.*[76] found that lac permease mutants with neutral amino acid substitutions for Lys358 or Asp237 (Thr or Asn, respectively) do not catalyze active transport. Second-site suppressor mutations of K358T exhibit neutral amino acid replacements for Asp237 (Asn, Gly, or Tyr), while

*K. Jung, H. Jung, P. Colachurchio, and H. R. Kaback, manuscript in preparation.

†S. Frillingos and H. R. Kaback, unpublished information.

suppressors of D237N have Gln in place of Lys358. It was proposed that Asp237 and Lys358 interact via a salt bridge, thereby neutralizing each other. Presumably, neutral replacement of *either* charged residue individually causes a functional defect because of the remaining unpaired charge, while neutral substitutions for *both* residues do not inactivate because the unpaired charge is removed. Consequently, the secondary-structure model proposed for the permease[18] was altered to accommodate a putative salt bridge between Asp237 and Lys358 in the low dielectric of the membrane by moving Asp237 from the hydrophilic domain between helices VII and VIII into the middle of transmembrane helix VII.[76]

Using the *lacY* gene encoding C-less permease, putative intramembrane residues Asp237, Asp240, Glu269, Arg302, Lys319, His322, Glu325, and Lys358 were systematically replaced with Cys[77] (Figure 6.2). Individual replacement of any of these residues essentially abolishes lactose accumulation against a concentration gradient. Starting with the single-Cys mutants D237C and K358C, a double-Cys mutant was constructed containing Cys replacements for *both* Asp237 and Lys358 in the same molecule. D237C/K358C transports lactose at about half the rate of C-less permease to almost the same steady state. Remarkably, replacement of Asp237 *and* Lys358, respectively, with Ala and Cys or Cys and Ala or even *interchanging* Asp237 with Lys358 causes relatively little change in activity. Subsequently, the side chains of Asp237 and/or Lys358 were extended by replacement with Glu and/or Arg or by site-specific derivatization of single-Cys replacement mutants.[78,79] Iodoacetic acid was used to carboxymethylate Cys or methanethiosulfonate derivates[80] were used to attach negatively charged ethylsulfonate or positively charged ethylammonium groups. Replacement of Asp237 with Glu, carboxymethyl-Cys, or sulfonylethylthio-Cys yields active permease with Lys or Arg at position 358. Similarly, the permease tolerates replacement of Lys358 with Arg or ammoniumethylthio-Cys with Asp or Glu at position 237. Moreover, permease with Lys, Arg, or ammoniumethylthio-Cys in place of Asp237 is highly active when Lys358 is replaced with Asp or Glu, confirming the conclusion that the polarity of the charge interaction can be reversed without loss of activity and demonstrating that the distance between positions 237 and 358 can be extended by up to five bond-lengths.[79] Therefore, neither Asp237, Lys358, nor the interaction between these residues is important for permease activity, and Asp237 and Lys358 must interact in a highly "flexible" manner to form a salt bridge.

Lac permease mutants in the charge pair Asp237–Lys358 are inserted into the membrane at wild-type levels if the charge pair is maintained with either polarity. On the other hand, disruption of the interacting pair often causes a marked decrease in the amount of protein inserted into the membrane, suggesting a role for the salt bridge in permease following and/or stability.[79,80] Significantly, as opposed to certain C-terminal truncation mutants which are proteolyzed *after* insertion,[81–83] these permease mutants are probably degraded prior to or during insertion into the membrane, as the mutant proteins are inserted into the membrane in a stable form if they are overproduced at a high rate from the T_7 promoter.[79] In any case, the observations raise the possibility that Asp237 and Lys358 may interact in a folding intermediate, but not in the mature molecule. As discussed above, however, inactive single mutants with Cys in place of Asp237 or Lys358 regain full activity on carboxymethylation or treatment with methanethiosulfonate ethylammonium, respectively, which restores a negative charge at position 237 or a positive charge at position 358, and similar results are obtained when the charges are reversed. Therefore, it seems very likely that although neither residue nor the charge pair is important for activity, the interaction between Asp237 and Lys358 plays a role in folding/stability, and the residues maintain proximity in the mature permease. The results are also interesting because they suggest that there must be a step(s) between translation of the permease and its insertion into the membrane. Furthermore, the findings raise the possibility that the C-terminal half of the permease may be inserted posttranslationally into the membrane (i.e., helix VII with Asp240 may have to wait for helix XI with Lys358 before insertion can take place).

To test the possibility that other charged residues in transmembrane helices are neutralized by charge-pairing ("charge-pair neutralization"), 13 additional *double* mutants were constructed in which all possible interhelical combinations of negatively and positively charged residues were replaced pairwise with Cys.[77] Out of all of the combinations of double-Cys mutants, only D240C/K319C exhibits significant transport activity. However, the functional interaction between Asp240 and Lys319 is different phenomenologically from Asp237–Lys358. Thus, D240C/K319C catalyzes lactose transport at about half the rate of C-less to a steady-state level of accumulation that is about 25–30% of the control. Moreover, although significant activity is observed with the double-Ala mutant or with the two possible Ala–Cys combinations, interchanging Asp240 and Lys319 completely abolishes active transport. In addition, replacement of Asp240 with Glu abolishes lactose transport, and permease with carboxymethyl-Cys at position 240 is inactive when paired with Lys319, but exhibits significant activity with Arg319.[79] Sulfonylethylthio-Cys substitution for Asp also results in significant transport activity. Permease with Arg or ammoniumethylthio-Cys in place of Lys319 exhibits high activity with Asp240 as the negative counterion, but no lactose transport is observed when either of these modifications is paired with Glu240. Finally, mutations in Asp240-Lys319 do not effect insertion of the permease into the membrane. Therefore, although neither Asp240 nor Lys-319 *per se* is mandatory for active transport, the interaction between this pair of charged residues exhibits more stringent properties than Asp237–Lys358, and the polarity of the interaction appears to be important for activity.

Using a different experimental approach, Lee *et al.*[84] have also shown that there is an interaction between Asp240 and Lys319. These workers replaced Asp240 with Ala by site-directed mutagenesis and found little or no active sugar transport. Two second-site revertants were then isolated, one with Gln in place of Lys319 and the other with Val in place of Gly268. The double mutants exhibit little or no accumulation of sugar, but manifest significant rates of lactose entry down a concentration gradient. Although suppression of D240A by G268V was not explained (see Section 6.4.2), the properties of the double mutant D240A/K319Q are consistent with the observations of Sahin-Tóth *et al.*[77] and Sahin-Tóth and Kaback.[79] However, the suggestion that the Asp240–Lys319 interaction may play a direct role in H^+ translocation[84] is inconsistent with the observation that various combinations of Cys–Ala replacements at these positions yield permease that is able to concentrate lactose against a concentration gradient.[77]

The charge-pair neutralization approach is dependent on permease activity, and it should be stressed that the approach will not reveal charge-paired residues if they are essential for activity. In this context, it is noteworthy that double-Cys mutants involving residues suggested to be H-bonded and directly involved in lactose-coupled H^+ translocation and/or substrate recognition [i.e., Arg302, His322, and Glu325 (see Refs. 6,7), as well as Glu269 which has also been shown to be an important residue[61,62,85]] are defective with respect to active lactose transport. On the other hand, as discussed below (see Section 6.4.2), it is likely that certain pairs of these residues are in close proximity in the tertiary structure of the permease and probably interact.

Despite the indication that Asp237–Lys358 and Asp240–Lys319 may participate in salt bridges and are located in the middle of transmembrane domains, the evidence is indirect. Therefore, other approaches are required to determine the location of the residues relative to the plane of the membrane and to demonstrate directly that the pairs are in close proximity. C-less permease mutants containing double-Cys or paired Cys–Ala replacements are particularly useful in this respect. Efforts to estimate the accessibility of Cys residues at positions 237 and 358 with water- or lipid-soluble sulfhydryl reagents suggest that Cys residues at the two positions are accessible to both types of reagents, although the lipid-soluble reagents are relatively more effective.[78] Furthermore, sited-directed spin and fluorescent labeling of permease molecules with single Cys residues at position 237 or 240 indicate that these residues are in an amphipathic environment at a membrane–water interface.* However, functional complementation experiments[86] with contiguous, nonoverlapping permease fragments indicate that the approximate cytoplasmic and periplasmic boundaries of transmembrane domain VII are Leu222 and Gly254, respectively (Figure 6.2), and favor the interpretation that Asp237 and Asp240 are disposed toward the middle of helix VII. One possible explanation for the apparent amphiphilic nature of the protein at these positions is that the hydrophilic faces of helices VII, X, and XI may line the solvent-filled cleft within the permease.[35–37] In summary, Asp237–Lys358 and Asp240–Lys319 interact functionally, and it is likely that both pairs of residues are contained within transmembrane domains that are in close proximity. Thus, putative helix VII (Asp 237 and Asp240) probably neighbors helices X (Lys319) and XI (Lys358) in the tertiary structure of the permease (Figure 6.3).

*H. Jung, M. L. Ujwal, C. Altenbach, W. L. Hubbell, and H. R. Kaback, manuscript in preparation.

6.4.2. Site-Directed Excimer Fluorescence

Although attempts are being made to crystallize lac permease (see Section 6.7), development of alternative methods for obtaining structural information is essential. With this objective in mind, pairs of charged residues in putative transmembrane helices of C-less permease were replaced with Cys in order to provide specific sites for labeling with *N*-(1-pyrene)-maleimide (pyrene maleimide).[87] Pyrene maleimide was chosen as a fluorophore because two pyrene moieties can form an excited-state dimer (excimer) that exhibits a unique emission maximum at approximately 470 nm if the conjugated ring systems are within about 3.5 Å of each other and in the correct orientation[88] and because this fluorophore has been used previously to study proximity relationships between Cys residues (see Refs. 89–92 for examples).

Initially, Arg302 (putative helix IX), His322 (helix X), and Glu325 (helix X) were studied because these residues are important for activity and postulated to interact.[56–59,63,65,66,93] To test the proximity of the residues, the double-Cys mutants H322C/E325C, R302C/H322C, and R302C/E325C and the corresponding single-Cys mutants were constructed with the biotinylation domain from a *Klebsiella pneumoniae* oxalacetate decarboxylase[94] in the middle cytoplasmic loop, purified by avidin-affinity chromatography,[95] labeled with pyrene maleimide, reconstituted into proteoliposomes, and subjected to fluorescence studies.[87]

With the double mutant H322C/E325C, a typical pyrene excimer fluorescence emission band is observed at about 470 nm. The observation is consistent with the postulate that His322 and Glu325 are in a portion of the permease that is α-helical, since the residues are predicted to lie on the same face of an α-helix. Mutant R302C/E325C labeled with pyrene maleimide also exhibits strong excimer emission band. However, insignificant excimer fluorescence is observed with mutant R302C/H322C. As a whole, the data indicate that helix IX is in close proximity to helix X, but Arg302 appears to be close to Glu325 rather than His322 (see Ref. 63).

Since hydrophobic proteins have a strong tendency to aggregate, it is important to determine whether the excimer fluorescence observed with lac permease results from an *intramolecular* interaction within single molecules or from an *intermolecular* interaction between two permease molecules. The following experiments were performed to test these alternatives: (1) Each single-Cys mutants was analyzed, and no excimer fluorescence is observed at 470 nm. (2) Single-Cys mutants R302C and E325C were purified separately, labeled, mixed, and reconstituted into proteoliposomes. No excimer fluorescence is observed, and the fluorescence emission spectrum is identical to those obtained from the unmixed single-Cys mutants. (3) If excimer fluorescence results from an intermolecular interaction, intensity at 470 nm should be inversely related to lipid:protein ratio. Therefore, pyrene maleimide- labeled R302C/H322C permease was reconstituted at lipid:protein ratios of 128:1, 385:1, and 1000:1 (w/w). All three samples exhibit no excimer fluorescence. Based on these three control experiments, it seems highly likely that excimer fluorescence observed with E325C/H322C and R302C/

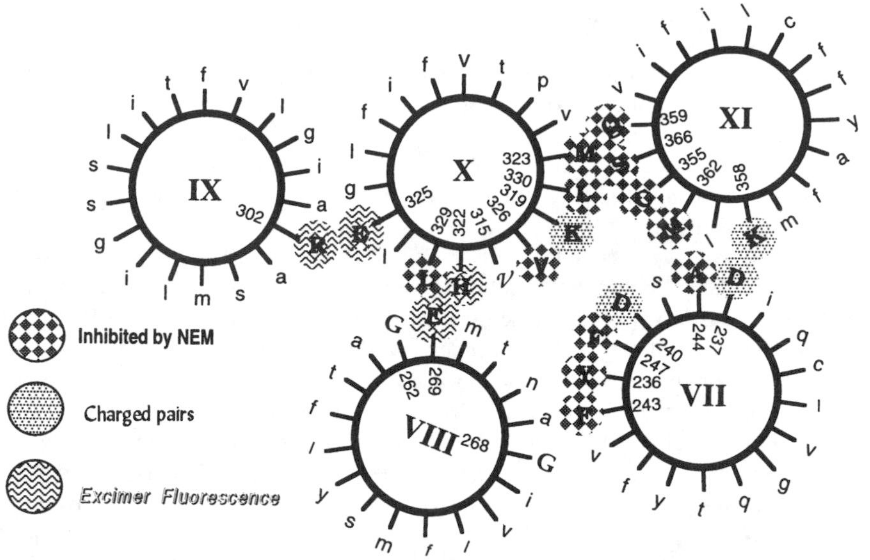

Figure 6.4. Helical wheel model of putative helices VII–XI in lac permease viewed from the periplasmic surface. In addition to the symbols given in the figure, G268 and G262 (helix VIII) which are sites of second-site suppressors for D240 and E325, respectively, are boldfaced, and V315 is highlighted.

E325C permeases results from intramolecular interactions between pyrene molecules attached to Cys residues within single molecules.

As discussed above, studies on second-site suppressor mutations[76,84] and site-directed mutagenesis studies on C-less permease[77–79] indicate that helix VII (Asp237 and Asp240) is close to helices X (Lys319) and XI (Lys358). Combining the results obtained from pyrene fluorescence with the previous studies leads to the model shown in Figure 6.4 where helices VII, IX, X, and XI are shown to interact via ion pairs between R302 and E325, K319 and D240, and K358 and D237. A side-view projection of the helices reveals that these residues are at approximately the same level with respect to depth in the membrane.[87]

In addition to neutral substitutions for Lys319, as mentioned above, a second-site suppressor mutant for Asp240 has been described[84] with Val in place of Gly268. Also, recent experiments* demonstrate that the phenotype of E325C is suppressed by substitution of Ser for Gly262. These observations suggest that helix VIII may interact with helices VII and X so as to bring Glu269 into proximity with His322 (Figure 6.4).

Regarding Glu269, by carrying out cassette mutagenesis on *lacY* DNA encoding putative helix VIII, Hinkle *et al.*[85] identified mutants that retain the ability to catalyze lactose accumulation. A strip of residues, largely on one side of helix VIII opposite Glu269, was shown to tolerate mutations with relatively little effect on activity, suggesting that this mutable strip of low information content is probably in contact with the membrane phospholipids. No active mutants in Glu269 were identified, however, and this residue was subjected to site-directed mutagenesis.[61,62] Permease with Cys or Gln in place of Glu269 is completely inactive, while E269D permease is completely detective with respect to lactose transport, but transporters TDG reasonably well. In addition, as noted previously, paired double mutants containing E269C and Cys replacements for each of the other charged residues in transmembrane domains are inactive.[77] Therefore, Glu269 plays an important role in transport, but the charge-pair neutralization approach gives no hint as to whether or not Glu269 interacts with another residue. However, additional second-site suppressor studies[96] indicate that Lys319 may also interact with Glu269, and it has been suggested that Lys319 (helix X) can interact with either Asp240 (helix VII) or Glu269 (helix VIII) as part of the transport mechanism.

To test the idea that Glu269 interacts with His322, the double-Cys mutant E269C/H322C was constructed.[87] The mutant and the corresponding single-Cys mutants containing biotin acceptor domains in the middle cytoplasmic loop were purified, labeled with pyrene maleimide, and reconstituted into proteoliposomes. Strikingly, pyrene maleimide-labeled E269C/H322C permease exhibits a distinct excimer, and importantly, no excimer is observed with pyrene maleimide-labeled E269C or H322C nor when the single-Cys mutants are labeled and mixed prior to reconstitution. The results provide a strong indication that Glu269 is close to His322 and imply that helix VIII is close to helix X (Figure 6.4). Moreover, experiments with permease containing His in place of Glu269 demonstrate that a divalent metal binding site is created in this mutant because of the presence of bis-His residues at positions 269 and 322.* However, no excimer fluorescence is observed with E269C/L319C permease labeled with pyrene maleimide in the presence or absence of TDG.† The lack of excimer formation in this double mutant in addition to the observation that Cys or Ala substitutions for *both* Lys319 and Asp240 do not abolish active lactose transport argue against an important mechanistic interaction between Lys319 and either Asp240 or Glu269. Nonetheless, the observation that certain mutations at Glu269 partially suppress the phenotype of Lys319 supports the contention that helices VIII and X are in close proximity.

*J. Wu and H. R. Kaback, unpublished information.

*K. Jung, J. Voss, and H. R. Kaback, unpublished information.
†K. Jung and H. R. Kaback, unpublished information.

Additional support for the model come from Cys-scanning mutagenesis of helix XI in C-less permease.[43] When each residue in helix XI (from Ala347 to Ser366) is replaced with Cys, most of the mutants exhibit highly significant activity, and the single-Cys mutants that are strongly inhibited by NEM fall on the same face of helix XI (Figure 6.4). Similarly, Cys-scanning mutagenesis of helix X[42] reveals that with the exception of Val331, the NEM-inhibitable Cys-replacement mutants are present on the same face of helix X as Val315, Lys319, His322, and Glu325, and the NEM-sensitive Cys-replacement mutants in helix VII also fall on the same face of the helix (Figure 6.4). Finally, although highly speculative, Baldwin[97] has put forward a model of the erythrocyte glucose facilitator in which the arrangement of the helices in the C-terminal half of the molecule is virtually identical to that discussed here for lac permease. If this is the case, it would suggest that these two transport proteins are likely to have a similar tertiary structure.

6.5. LIGAND BINDING INDUCES WIDESPREAD CONFORMATIONAL CHANGES

Excimer fluorescence can also be used to study dynamic aspects of permease folding.[98] The excimer observed with reconstituted pyrene maleimide-labeled R302C/E325C or E269C/H322C permease is markedly diminished by increasing concentrations of sodium dodecylsulfate (up to 0.6%), while excimer fluorescence with the H322C/E325C mutant is unaffected. Apparently, the detergent disrupts tertiary interactions within the permease with little effect on secondary structure. Consistently, the double mutants E269C/H322C and R302C/ E325C do not exhibit excimer fluorescence after labeling with pyrene maleimide in octyl-β,D-glucopyranoside, but do so after reconstitution into proteoliposomes.

Although individual replacement of each of the charged residues in transmembrane domains inactivates active lactose transport, some of the constructs exhibit ligand-induced conformational alterations.[98] Excimer fluorescence in proteoliposomes containing pyrene maleimide-labeled E269C/H322C permease is quenched by Tl^+, and the effect is strongly attenuated by permease substrates. It has also been shown[42] that C-less permease with a single Cys residue in place of Val315 [presumably the N-terminal residue in helix X (Figure 6.2) which is on the same face as His322 and E325 (Figure 6.4)] is inactivated by NEM much more rapidly in the presence of TDG or $\Delta\bar{\mu}_{H^+}$. Recently,[99] V315C permease containing the biotin acceptor domain in the middle cytoplasmic loop has been purified, and the kinetics of pyrene maleimide labeling has been studied in the presence and absence of TDG or $\Delta\bar{\mu}_{H^+}$. The studies confirm the observations described with right-side-out membrane vesicles and provide an exciting preliminary suggestion that ligand binding or $\Delta\bar{\mu}_{H^+}$ may cause the permease to assume the same conformation. In addition to V315C permease, single-Cys replacements for His322, Glu325* [helix X] or Glu269 [helix VIII],[98] as well as Gly147 [helix V],[75] Phe162 [helix V], Pro31 or Pro28† [helix I] exhibit increased reactivity in the presence of TDG. The results, albeit preliminary, suggest that ligand binding or $\Delta\bar{\mu}_{H^+}$ may induce widespread conformational changes in the permease.

*K. Jung and H. R. Kaback, unpublished information.
†J. Wu and H. R. Kaback, unpublished information.

6.6. LACTOSE-COUPLED H^+ TRANSLOCATION

The most provocative and controversial findings from site-directed mutagenesis studies on lac permease began when the four His residues in the protein were mutagenized.[56–59] Replacement of His35 and His39 (Figure 6.2, first periplasmic loop) with Arg or replacement of His205 (Figure 6.2, third cytoplasmic loop) with Arg, Asn, or Gln has no effect on active lactose transport, whereas replacement of His322 (putative helix X) with Arg, Asn, Gln, or Lys causes dramatic loss of activity. Conversely, a permease mutant with a single His at position 322 exhibits properties identical to wild-type permease. Although inactive with respect to lactose accumulation, H322R permease catalyzes downhill lactose influx at high substrate concentrations without concomitant H^+ translocation.

Efflux, exchange, and counterflow are useful for studying permease turnover because specific steps in the overall kinetic cycle can be delineated.[99–102] Permease with Arg, Asn, Gln, or Lys in place of His322 is markedly defective in all translocation reactions presumed to involve protonation or deprotonation (Figure 6.5). Furthermore, the primary kinetic effect of $\Delta\bar{\mu}_{H^+}$ (i.e., a decrease in apparent K_m for lactose[100,102]) is not observed. Interestingly, permease with Asn, Gln, or Lys in place of His322 catalyzes downhill efflux, as well as influx, but both processes occur without concomitant H^+ translocation.[58]

Since His322 may be directly involved in lactose-coupled H^+ translocation and this residue is located in putative helix X (Figure 6.2), attention focused on Glu325, which should be on the same face of helix X as His322 and may be ion-paired with this residue. In addition, structure/function studies on chymotrypsin[103] and other serine proteases have led to the notion that acidic amino acid residues may function with His as components of a charge-relay system, a type of mechanism that might conceivably be related to H^+ translocation. For these reasons, Glu325 was subjected to site-directed mutagenesis.[65,66] Permease with Ala, Gln, Val, His, Cys, or Trp in place of Glu325 catalyzes downhill influx of lactose without H^+ translocation, but does not catalyze either active transport or efflux. Remarkably, the rate of equilibrium ex-

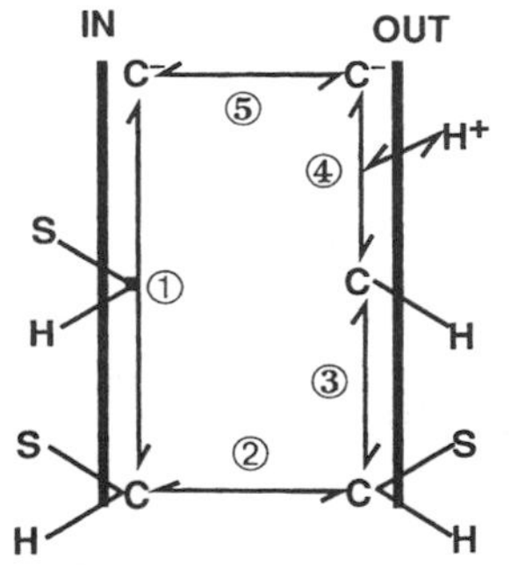

Figure 6.5. Schematic representation of reactions involved in lactose efflux, exchange, and counterflow. C represents lac permease; S is substrate (lactose). The order of substrate and H^+ binding at the inner surface of the membrane is not implied.

change with the altered permeases is at least as great as that observed with wild-type permease. Moreover, permease mutated at position 325 catalyzes counterflow at the same rate and to the same extent as wild-type permease, but the internal concentration of [^{14}C]lactose is maintained for a prolonged period because of the defect in efflux. It is also noteworthy that permease mutated at position 325 catalyzes counterflow three to four times better than wild-type permease when the external lactose concentration is below the apparent K_m.[66]

The results can be rationalized by the simple kinetic scheme shown in Figure 6.5. Efflux down a concentration gradient is thought to consist of a minimum of five steps: (1) binding of substrate and H^+ on the inner surface of the membrane (order unspecified); (2) translocation of the ternary complex to the outer surface; (3) release of substrate; (4) release of H^+; (5) return of the unloaded permease to the inner surface. Alternatively, exchange and counterflow with external lactose at saturating concentrations involve steps 1–3 only. Furthermore, release of H^+ appears to be rate-limiting for the overall cycle.[104,105]

All steps in the mechanism presumed to involve protonation or deprotonation appear to be blocked in the His322 mutants. Therefore, it seems reasonable to suggest that protonation of His322 may be involved in step 1. In contrast, replacement of Glu325 results in a permease that is defective in all steps involving net H^+ translocation but catalyzes exchange and counterflow normally. Clearly, therefore, permease mutated at position 325 is probably blocked in step 4 (i.e., it is unable to lose H^+).

Experiments in which Glu325 was replaced with Asp have yielded interesting results.[106] Permease with Asp325 is partially uncoupled and catalyzes lactose accumulation about 30% as well as wild-type permease.[66] The observation is not surprising, as the side chain containing the carboxylate is about 1.5 Å shorter in Asp relative to Glu. However, E325D permease catalyzes equilibrium exchange normally below pH 7.7, but as ambient pH is increased, exchange activity is progressively and reversibly inhibited with a midpoint at about pH 8.5, while equilibrium exchange with wild-type or E325A permease is unaffected by bulk-phase pH over a wide range. The findings indicate that translocation of the fully loaded permease does not tolerate a negative charge at position 325 and suggest that the carboxylate at 325 may undergo protonation and deprotonation during lactose/H^+ symport. The observation that equilibrium exchange with wild-type permease is insensitive to pH over the same range, is consistent with the notion that Glu325 is strongly H-bonded.

Replacement of Arg302 with Leu, His, or Lys (putative helix IX; Figure 6.1) yields permease with properties similar to those of H322R permease.[63] In marked contrast, replacement of most of the other residues with Cys in putative helices IX and X of C-less permease has no significant effect on active lactose transport, with the exceptions of Lys319 and Lys358 which interact with Asp240 and Asp237, respectively.[76–79,84] The findings highlight the unique importance of Arg302, His322, and Glu325. Molecular modeling of putative helices IX and X[63] suggested that the guanidino group in Arg302 may be sufficiently close to His322 to participate in H-bonding with the imidazole ring that, in turn, may be H-bonded to the carboxylate of Glu325. However, as discussed above, site-directed pyrene excimer fluorescence studies indicate that Arg302 interacts with Glu325, but not with His322 (Figure 6.4). Therefore, although the three residues may be involved in H^+ translocation, they cannot behave like the charge relay in the serine proteases.

As evidenced by binding studies with the high-affinity ligand NPG, permease mutated at position 325 has a K_d approximating that of wild-type permease.[66] The finding is consistent with the observation that counterflow, a process that exhibits an apparent K_m similar to that observed for active transport, is intact in the mutants, but is in marked contrast to findings with permeases mutated at Arg302 or His322 which exhibit markedly decreased affinities.[58,63] Therefore, it is tempting to speculate that the pathways for H^+ and lactose may overlap [i.e., that Arg302 and His322 may also be components of the substrate binding site in addition to being involved in H^+ translocation (see Refs. 107,108 in addition)].

If Arg302, Glu325, and His322 function in H^+ translocation, the polarity, distance, and orientation between the residues should be critical.[93] The importance of polarity between His322 and Glu325 was studied by interchanging the residues, and the modified permease is inactive in all modes of translocation. The effect of distance and/or orientation between His322 and Glu325 was investigated by interchanging Glu325 with Val326, thereby moving the carboxylate one residue around putative helix X. The resulting permease molecule is also completely inactive, and control mutations indicate that a Glu residue at position 326 inactivates the permease. The wild-type orientation between His and Glu was then restored by further mutation to introduce a His residue into position 323 or by interchanging Met323 with His322. The resulting permease molecules contain the wild-type His/Glu orientation, but the putative His/Glu ion-pair is rotated about the helical axis by 100° relative to Arg302 in putative helix IX. Both mutants are inactive with respect to all modes of translocation. The results support the contention that the polarity between His322 and Glu325 and the geometric relationships between Glu269, Arg302, Glu325, and His322 are critical for activity. In addition, they suggest that perturbation of the putative His322/Glu325 ion-pair alone is insufficient to account for inactivation (i.e., Glu322/His325 should remain ion-paired) and are consistent with a possible role for His322 and Glu325 H^+ translocation.

Other experiments question the notion that His322 is obligatory for lactose-coupled H^+ translocation. From studies on permease mutants with Tyr or Phe in place of His322, King and Wilson[109–111] conclude that sugar-dependent H^+ transport is observed, albeit with low efficiency, that melibiose efflux, in particular, remains coupled to H^+ translocation in the mutants, and that reactions involving exchange are limiting for lactose but not melibiose efflux suggesting that slow exchange is substrate-specific in the His322 mutants. In addition, Brooker[112] has shown that a permease double mutant with Val in place of Ala177 *and* Asn in palce of His322 catalyzes lactose-dependent H^+ influx with a stoichiometry close to unity. Taken at face value, the results are difficult to reconcile with the contention that His322 is obligatory for lactose-driven H^+ translocation. However, it should be emphasized that H^+ may be able to interact with the π electrons of the aromatic rings of Phe or Tyr and that *all* of the His322 mutants isolated thus far are *grossly defective* with regard to sugar accumulation. Thus, whatever its precise role in the mechanism, a His residue at position 322 is very important for β-galactoside accumulation against a concentration gradient.

On a broader level, H^+- and Na^+-coupled symport are conceptually and thermodynamically analogous, but it is unclear whether the two types of transport occur by the same general mechanism. Boyer[113] has suggested that H_3O^+, rather than H^+, may be the symported species. Appropriately placed N or O atoms in symporters like lac permease could provide cation-binding domains akin to those in the crown ethers or cryptates, both of which form coordination complexes with Na^+ and H_3O^+. In this context, however, each translocation reaction catalyzed by lac permease in which H^+ transfer is rate-limiting exhibits a significant D_2O effect[105] which is not expected if coordination with H_3O^+ is the rate-limiting step in translocation (i.e., the mass of D_3O^+ is only 3/19 greater than H_3O^+). Furthermore, efforts to drive lactose/Na^+ symport at alkaline pH values have been negative. Finally, it is difficult to explain the behavior of the Glu325 mutants with this type of model. In summary, therefore, although the contrast between H^+- and Na^+-coupled symport is of singular importance, the mechanistic relationship between the two is presently unclear.

6.7. A NOVEL APPROACH TO CRYSTALLIZATION OF HYDROPHOBIC MEMBRANE PROTEINS

A major thrust of this laboratory is aimed at 2-D and 3-D crystallization. Since hydrophobic membrane proteins like lac permease are especially difficult to crystallize in 3-D, Privé *et al.*[114] have devised a novel approach in which a fusion is constructed between the permease and a "carrier" protein. The carrier is a soluble, stable protein with its C- and N-termini close together in space at the surface of the protein, so that it can be introduced into an internal position of the permease without distorting either molecule. In this regard, McKenna *et al.*[115] demonstrated that all but three of the hydrophilic domains in the permease can be disrupted by the insertion of two or six contiguous His residues without abolishing activity. The carrier is chosen with convenient spectral properties, making the fusion protein considerably easier to characterize than the native molecule. A chimeric protein with *E. coli* cytochrome b_{562} fused into the middle cytoplasmic loop of lac permease and six His residues attached to the C-terminus has been constructed, expressed, and purified to a high state of purity by nickel chelate and ion-exchange chromatography. The chimera exhibits transport activity comparable to wild-type lac permease and has a visible absorption spectrum and a redox potential that are identical to cytochrome b_{562}. The chimera has a higher proportional polar surface area than wild-type permease and should have better possibilities of forming the strong, directional intermolecular contacts required of a crystal lattice.

6.8. INSERTION OF PERMEASE INTO THE MEMBRANE

6.8.1. Membrane Insertion and Stability

Stochaj *et al.*[116,117] demonstrated that sequences within the N-terminal 170 amino acid residues of lac permease may be important for insertion. Moreover, a truncated permease containing only the N-terminal 50 amino acid residues is inserted into the membrane, and it was proposed that this region contains an internal "start transfer" sequence resulting in the insertion of the N-terminus as a "helical hairpin."[118,119]

With respect to the C-terminus, the 17-amino-acid C-terminal hydrophilic tail is not involved in insertion of the permease into the membrane, its stability, or its ability to catalyze transport. On the other hand, a 4-amino-acid sequence at the end of the last putative transmembrane helix (...VFTL...; Figure 6.2) is critical for stability and hence activity once the protein is inserted into the membrane.[81,82] When stop codons (TAA) are placed sequentially at amino acid codons 396–401, permease truncated at residue 396 or 397 is completely defective with respect to lactose transport, while molecules truncated at residues 398, 399, 400, and 401, respectively, exhibit 15–25, 30–40, 40–45, and 70–100% of wild-type activity. As judged by pulse-chase experiments with [^{35}S]methionine, wild-type permease or permease truncated at residue 401 is stable, while permease molecules truncated at residue 400, 399, 398, 397, or 396 are degraded at increasingly rapid rates. Finally, replacement of residues 397–400 with LeuLeuLeuLeu or AlaAlaAlaAla yields a stable, fully functional permease, while replacement with GlyProGlyPro yields an unstable molecule with minimal transport activity.[83] The results indicate that the last turn of putative helix XII is important for proper folding and protection against proteolytic degradation.

The overall topology of polytopic membrane proteins like lac permease is thought to result from either the oriented insertion of the N-terminal α-helical domain followed by passive, serpentine insertion of subsequent helices[120–122] or from the function of independent topogenic determinants dispersed throughout the molecules.[123–125] In order to test these alternatives, even or odd numbers of putative transmembrane domains in lac permease were deleted, and the effect of the deletions on insertion, stability, and the orientation of the C-terminus with respect to the plane of the membrane were examined (Figure 6.6).[126] The strategy is that deletion of odd numbers of transmembrane domains might be expected to alter the position of the C-terminus relative to the plane of the membrane, while deletion of even numbers of transmembrane domains would not be expected to do so. As demonstrated, so long as the first N-terminal and the last four C-terminal putative α-helical domains are retained, stable polypeptides are inserted into the membrane, even when an odd number of helical domains is deleted. Moreover, even when an odd number of helices is deleted, the C-terminus remains on the cytoplasmic surface of the membrane, as judged by the activity of

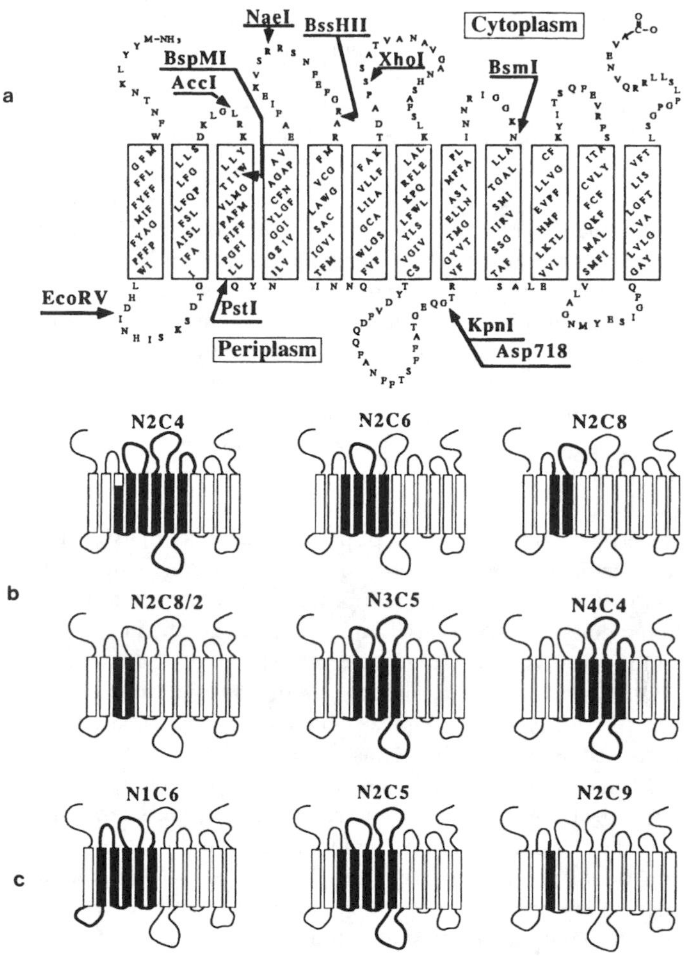

Figure 6.6. Secondary structure of lac permease (a) showing polypeptides encoded by the *lacY* deletion constructs described in panels b and c. The single-letter amino acid code is used, hydrophobic transmembrane helices are shown in boxes, and deleted regions are solid boxes and thick lines. (**a**) Wild-type lac permease with restriction sites in the cassette *lacY* DNA indicated. (**b**) Constructs in which even numbers of putative transmembrane helices were deleted. (**c**) Constructs in which odd numbers of putative transmembrane helices were deleted. From Bibi and Kaback.[126]

C-terminal lac permease–alkaline phosphatase fusions. Interestingly, although none of the deletions catalyzes lactose accumulation against a concentration gradient, permease molecules devoid of even or odd numbers of putative transmembrane helices retain a specific pathway for "downhill" lactose translocation. One construct, in particular, which is devoid of putative helices II–V exhibits about 80% of the downhill transport activity of intact permease, suggesting that the pathway for lactose translocation may be largely contained within the last six transmembrane domains. The results indicate that relatively short C-terminal domains in the permease contain topological information sufficient for insertion in the native orientation regardless of the orientation of the N-terminus. This conclusion has been confirmed and extended by Calamia and Manoil[127] who demonstrated recently that many individual membrane-spanning domains of lac permease act as independent export signals for attached alkaline phosphatase.

6.8.2. Expression of Lac Permease in Fragments as a Probe for Membrane-Spanning Domains

Bibi and Kaback[38] demonstrated that cells expressing N_6 and C_6 fragments of the permease catalyze significant lactose accumulation, while cells expressing either half independently are devoid of activity. Intact permease is completely absent from the membrane of cells expressing fragments either individually or together. Thus, transport activity must result from association between independently synthesized portions of lac permease. When the gene fragments are expressed individually, the N-terminal portion of the permease is observed sporadically and the C-terminal portion is not observed; when expressed together, the N- and C-terminal moieties of the permease are found in the membrane. Clearly, the N- or C-terminal halves are proteolyzed when synthesized independently, and association between the two complementing polypeptides leads to a more stable, catalytically active complex.

Coexpression of independently cloned fragments of *lacY* encoding N_2 and C_{10},[128] N_1 and C_{11}, or N_7 and C_5[86] also form stable molecules in the membrane which interact to form functional permease (Figure 6.2). Thus, *lacY* gene fragments encoding contiguous, nonoverlapping peptides with discontinuities in either cytoplasmic or periplasmic loops are able to complement functionally. On the other hand, Zen *et al.*[86] have demonstrated that peptide fragments with discontinuities in transmembrane domains are unable to form functional complexes, implying that the "split permease approach" may be useful for approximating helical boundaries. Based on this notion, a series of contiguous, nonoverlapping permease fragments with discontinuities at various positions in loop 6 (cytoplasmic), putative helix VII, and loop 7 (periplasmic) were coexpressed in order to approximate the boundaries of transmembrane domain VII.[86] Contiguous fragments with a discontinuity between Leu222 and Trp223 (loop 6) or Gly254 and Glu255 (loop 7) are functional, but fragments with discontinuities between Cys234 and Thr235, Gln241, and Gln242, or Phe247 and Thr248 are inactive (Figure 6.2). Therefore, it is likely that Leu222 and Gly254 are located in hydrophilic loops 6 and 7, respectively, while Cys234, Gln241, and Phe-247 are probably within transmembrane domain VII. These and other considerations (see Section 6.4.1) are consistent with a secondary-structure model of lactose permease in which Asp237 and Asp240 are contained within domain VII rather than loop 7, as predicted by hydropathy profiling.[76–79,86]

Regarding the contention that contiguous permease fragments with discontinuities in transmembrane domains do not exhibit functional complementation, evidence has been presented[81–83] that the last turn of putative helix XII must be intact for the permease to insert into the membrane in a stable form. It seems likely, therefore, that incomplete transmembrane domains are not inserted into the membrane in a stable form. However, it should be emphasized that proteolytic degradation cannot explain the lack of transport activity observed with permease duplexes containing discontinuities in transmembrane domains. Thus, an immunoreactive C-terminal fragment is observed with $N_{2.5}/C_{9.5}$. Moreover, permease deleted of the first 22 amino acid residues (the N-terminus and half of the first transmembrane domain) is stable and functional when expressed at a high rate.[129] Therefore, it seems reasonable to suggest that the lack of transport activity observed with duplexes split in transmembrane domains may be related to an alteration in the transfer of conformational information from one side of a transmembrane domain to the other.

6.8.3. Membrane Insertion May Involve Multiple Mechanisms

The demonstration that polypeptides corresponding to N_1 and C_{11} form a relatively stable, functional complex argues against the notion that the N-terminus of the permease inserts into the membrane as a helical hairpin. On the other hand, Zen *et al.** (see Ref. 95) have shown that insertion of the biotin-acceptor domain into the second or the fourth periplasmic loops of the permease (between helices III and IV or helices VII and VIII, respectively; Figure 6.2) blocks insertion of transmembrane helices III and IV or VII and VIII without altering the insertion of the remainder of the protein, suggesting that these portions of the permease may be inserted as a helical hairpin. It is also important that Dunten *et al.*[78] showed that disruption of the salt bridge between Asp237 and Lys358 causes the permease to be inserted into the membrane much less efficiently (see Section 6.4.1) which raises the possibility that the C-terminal half of the polypeptide may be inserted posttranslationally. Finally, as discussed above, the first 22 amino acid residues in the permease which represent the N-terminal hydrophilic domain and the first half of putative helix I are not important for activity, but enhance the efficiency of insertion into the membrane.[129]

*K. H. Zen, T. G. Consler, and H. R. Kaback, manuscript in preparation.

6.9. SUMMARY AND CONCLUDING REMARKS

The lac permease of *E. coli* is providing a paradigm for secondary active transport proteins that transduce the free energy stored in electrochemical ion gradients into work in the form of a concentration gradient. This hydrophobic, polytopic, cytoplasmic membrane protein catalyzes the coupled, stoichiometric translocation of β-galactosides and H^+, and it has been solubilized, purified, reconstituted into artificial phospholipid vesicles, and shown to be solely responsible for β-galactoside transport as a monomer. The *lacY* gene which encodes the permease has been cloned and sequenced, and based on spectroscopic analyses of the purified protein and hydropathy profiling of its amino acid sequence, a secondary structure has been proposed in which the protein has 12 transmembrane domains in α-helical configuration that traverse the membrane in zigzag fashion connected by hydrophilic loops with the N- and C-termini on the cytoplasmic face of the membrane. Unequivocal support for the topological predictions of the 12-helix model has been obtained by analyzing a large number of the lac permease–alkaline phosphatase (*lacY–phoA*) fusions. Extensive use of site-directed and Cys-scanning mutagenesis indicates that very few residues in the permease are directly involved in the transport mechanism. Interestingly, however, Cys148 which is not essential for activity is located in a binding site and probably interacts weakly and hydrophobically with the galactosyl moiety of the substrate.

Second-site suppressor analysis and site-directed mutagenesis with chemical modification have provided evidence that helix VII is probably close to helices X and XI in the tertiary structure of the permease. Experiments in which paired Cys replacements in C-less permease were labeled with pyrene, a fluorophore that exhibits excimer fluorescence when two of the unconjugated ring systems are in close approximation, indicate that His322 and Glu325 are located in an α-helical region of the permease and that helix IX is close to helix X. Based on certain second-site suppressor mutants, it was suggested that helix VIII (Glu269) is close to helix X (His322), and site-directed pyrene maleimide labeling experiments provide strong support for this postulate. Taken together, the findings lead to a model describing helix packing in the C-terminal half of the permease. Other findings indicate that widespread conformational changes in the permease result from either ligand binding or imposition of $\Delta\bar{\mu}_{H^+}$.

Since many membrane transporters appear to have similar secondary structures based on hydropathy profiling, it seems likely that the basic tertiary structure and mechanism of action of these proteins have been conserved throughout evolution. Therefore, studies on bacterial transport systems which are considerably easier to manipulate than their eukaryotic counterparts have important relevance to transporters in higher-order systems, particularly with respect to the development of new approaches to structure–function relationships. Although it is now possible to manipulate membrane proteins to an extent that was unimaginable only a few years ago, it is unlikely that transport mechanisms can be defined on a molecular level without information about tertiary structure. In addition to structure, however, dynamic information is required at high resolution, and as suggested by some of the experiments discussed here, site-directed spectroscopy will be particularly useful in this regard.

Remarkably few amino acid residues appear to be critically involved in the transport mechanism. On the other hand, certain active Cys-replacement mutants are altered by alkylation, and these mutants cluster on helical faces so as to line a cavity in the C-terminal half of the permease. The results suggest that surface contours within the permease may be important. This surmise coupled with the indication that few residues are essential to the mechanism is encouraging in that it suggests the possibility that a relatively low-resolution structure (i.e., helix packing) plus localization of the critical residues might begin to provide important mechanistic insights.

NOTE ADDED IN PROOF

For a more recent review, the reader is referred to Kaback, H. R. (1996) The lactose permease of *Escherichia coli:* Past, Present and Future. In *Handbook of Biological Physics:* Transport Processes in Eukaryotic and Prokargotic Organisms (W. N. Konings and H. R. Kaback, eds.), Elsevier, Amsterdam, in press.

REFERENCES

1. Henderson, P. J. (1990). Proton-linked sugar transport systems in bacteria. *J. Bioenerg. Biomembr.* **22:**525–569.
2. Marger, M. D., and Saier, M. H., Jr. (1993). A major superfamily of transmembrane facilitators that catalyze unipolar symptort and antiport. *Trends Biochem. Sci.* **18:**13–20.
3. Mitchell, P. (1963). Molecule, group and electron transport through natural membranes. *Biochem. Soc. Symp.* **22:**142–168.
4. Mitchell, P. (1968). Chemiosmotic Coupling and Energy Transduction, *Glyňn Research Ltd, Bodmin, England.*
5. Kaback, H. R. (1986). In *Physiology of Membrane Disorders,* (T. E. Andreoli, J. F. Hoffman, D. D. Fanestil, and S. G. Schultz, eds.) Plenum, New York p. 387.
6. Kaback, H. R. (1989). Molecular biology of active transport: From membranes to molecules to mechanism, *Harvey Lect.* **83:**77–103.
7. Kaback, H. R. (1992). In and out and up and down with the lactose permease of *Escherichia coli. Int. Rev. Cytol.* **137A:**97–125.
8. Teather, R. M. Müller-Hill, B., Abrutsch, U., Aichele, G., and Overath, P. (1978). Amplification of the lactose carrier protein in *Escherichia coli* using a plasmid vector. *Molec. Gen. Genet.* **159:**239–48.
9. Buchel, D. E., Gronenborn, B., and Muller-Hill, B. (1980). Sequence of the lactose permease gene, *Nature* **283:**541–545.
10. Kaczorowski, G. J., Leblanc, G., and Kaback, H. R. (1980). Specific labeling of the *lac* carrier protein in membrane vesicles of *Escherichia coli* by a photoaffinity reagent. *Proc. Natl. Acad. Sci. USA* **77:**6319–23.
11. Newman, M. J., and Wilson, T. H. (1980). Solubilization and reconstitution of the lactose transport system from *Escherichia coli. J. Biol. Chem.* **255:**10583–10586.
12. Newman, M. J., Foster, D. L., Wilson, T. H., and Kaback, H. R. (1981). Purification and reconstitution of functional lactose carrier from *Escherichia coli. J. Biol. Chem.* **256:** 11804–11808.
13. Foster, D. L., Garcia, M. L., Newman, M. J., Patel, L., and Kaback, H. R. (1982). Lactose-proton symport by purified *lac* carrier protein. *Biochemistry* **21:**5634–5638.
14. Viitanen, P., Newman, M. J., Foster, D. L., Wilson, T. H., and Kaback, H. R. (1986). Purification, reconstitution, and characterization of the *lac* permease of *Escherichia coli,* Methods Enzymol. **125:**429–452.
15. Viitanen, P., Garcia, M. L., and Kaback, H. R. (1984). Purified reconstituted lac carrier protein from *Escherichia coli* is fully functional. *Proc. Natl. Acad. Sci. USA* **81:**1629–1633.
16. Matsushita, K., Patel, L., Gennis, R. B., and Kaback, H. R. (1983). Reconstitution of active transport in proteoliposomes containing cytochrome o oxidase and *lac* carrier protein purified from *Escherichia coli. Proc. Natl. Acad. Sci. USA* **80:**4889–4893.
17. Deisenhofer, J., and Michel, H. (1991). High-resolution structures of photosynthetic reaction centers. *Annu. Rev. Biophys. Biophys. Chem.* **20:**247–266.
18. Foster, D. L., Boublik, M., and Kaback, H. R. (1983). Structure of the *lac* carrier protein of *Escherichia coli. J. Biol. Chem.* **258:**31–34.
19. Vogel, H., Wright, J. K., and Jahnig, F. (1985). The structure of the lactose permease derived from Raman spectroscopy and prediction methods. *EMBO J.* **4:** 3625–3631.
20. Carrasco, N., Tahara, S. M., Patel, L., Goldkorn, T., and Kaback, H. R. (1982). Preparation, characterization, and properties of monoclonal antibodies against the *lac* carrier protein from *Escherichia coli. Proc. Natl. Acad. Sci. USA* **79:**6894–6898.
21. Seckler, R., Wright, J. K., and Overath, P. (1983). Peptide-specific antibody locates the COOH terminus of the lactose carrier of *Escherichia coli* on the cytoplasmic side of the plasma membrane. *J. Biol. Chem.* **258:**10817–10820.
22. Carrasco, N., Viitanen, P., Herzlinger, D., and Kaback, H. R. (1984). Monoclonal antibodies against the *lac* carrier protein from *Escherichia coli.* 1. Functional studies. *Biochemistry* **23:**3681–3687.
23. Herzlinger, D., Viitanen, P., Carrasco, N., and Kaback, H. R. (1984). Monoclonal antibodies against the *lac* carrier protein from *Escherichia coli.* 2. Binding studies with membrane vesicles and proteoliposomes reconstituted with purified *lac* carrier protein. *Biochemistry* **23:**3688–3693.
24. Carrasco, N., Herzlinger, D., Mitchell, R., DeChiara, S., Danho, W., Gabriel, T. F., and Kaback, H. R. (1984). Intramolecular dislocation of the COOH terminus of the *lac* carrier protein in reconstituted proteoliposomes. *Proc. Natl. Acad. Sci. USA* **81:**4672–4676.
25. Seckler, R., and Wright, J. K. (1984). Sidedness of native membrane vesicles of *Escherichia coli* and orientation of the reconstituted lactose/H+ carrier. *Eur. J. Biochem.* **142:**269–279.
26. Herzlinger, D., Carrasco, N., and Kaback, H. R. (1985). Functional and immunochemical characterization of a mutant of *Escherichia coli* energy uncoupled for lactose transport. *Biochemistry* **24:**221– 229.
27. Danho, W., Makofske, R., Humiec, F., Gabriel, T. F., Carrasco, N., and Kaback, H. R. (1985). Use of site-directed polyclonal antibodies as immunotopological probes for the *lac* permease of *Escherichia coli.* In *Peptides: Structure and Function* (C. M. Deber, V. J. Hruby, K. D. Kopple, eds.), Pierce Chemical Co., Rockford, IL, pp. 59–62.
28. Seckler, R., Moroy, T., Wright, J. K., and Overath, P. (1986). Anti-peptide antibodies and proteases as structural probes for the lactose/H+ transporter of *Escherichia coli:* a loop around amino acid residue 130 faces the cytoplasmic side of the membrane. *Biochemistry* **25:** 2403–2409.

29. Goldkorn, T., Rimon, G., and Kaback, H. R. (1983). Topology of the *lac* carrier protein in the membrane of *Escherichia coli. Proc. Natl. Acad. Sci. USA* **80:**3322–3326.
30. Stochaj, U., Bieseler, B., and Ehring, R. (1986). Limited proteolysis of lactose permease from *Escherichia coli. Eur. J. Biochem.* **158:**423–428.
31. Page, M. G., and Rosenbusch, J. P. (1988). Topography of lactose permease from *Escherichia coli. J. Biol. Chem.* **263:**15906–15914.
32. Bieseler, B., Prinz, H., and Beyreuther, K. (1985). Topological studies of lactose permease of *Escherichia coli* by protein sequence analysis. *Ann. N Y Acad. Sci.* **456:**309–325.
33. Calamia, J., and Manoil, C. (1990). *Lac* permease of *Escherichia coli:* topology and sequence elements promoting membrane insertion. *Proc. Natl. Acad. Sci. USA* **87:**4937–4941.
34. Manoil, C., and Beckwith, J. (1986). A genetics approach to analyzing membrane protein topology. *Science* **233:**1403–1408.
35. Costello, M. J., Escaig, J., Matsushita, K., Viitanen, P. V., Menick, D. R., and Kaback, H. R. (1987). Purified *lac* permease and cytochrome o oxidase are functional as monomers. *J. Biol. Chem.* **262:** 17072–17082.
36. Costello, M. J., Viitanen, P., Carrasco, N., Foster, D. L., and Kaback, H. R. (1954). Morphology of proteoliposomes reconstituted with purified *lac* carrier protein from *Escherichia coli. J. Biol. Chem.* **259:**15579–15586.
37. Li, J., Tooth, P. (1987). Size and shape of the *Escherichia coli* lactose permease measured in filamentous arrays. *Biochemistry* **26:**4816–4823.
38. Bibi, E., and Kaback, H. R. (1990). In vivo expression of the *lacY* gene in two segments leads to functional *lac* permease. *Proc. Natl. Acad. Sci. USA* **87:**4325–4329.
39. Dornmair, K., Corin, A. F., Wright, J. K., and Jahnig, F. (1985). The size of the lactose permease derived from rotational diffusion measurements. *EMBO J.* **4:**3633–3638.
40. Sahin-Tóth, M., Lawrence, M. C., and Kaback, H. R. (1994). Properties of permease dimer, a fusion protein containing two lactose permease molecules from *Escherichia coli. Proc. Natl. Acad. Sci. USA* **91:**5421–5425.
41. van Iwaarden, P. R., Pastore, J. C., Konings, W. N., and Kaback, H. R. (1991). Construction of a functional lactose permease devoid of cysteine residues. *Biochemistry* **30:**9595–9600.
42. Sahin-Tóth, M., and Kaback, H. R. (1993). Cysteine scanning mutagenesis of putative transmembrane helices IX and X in the lactose permease of *Escherichia coli. Protein Sci.* **2:**1024–1033.
43. Dunten, R. L., Sahin-Toth, M., and Kaback, H. R. (1993). Cysteine scanning mutagenesis of putative helix XI in the lactose permease of *Escherichia coli. Biochemistry* **32:**12644–12650.
44. Sahin-Tóth, M., Persson, B., Schweiger, J., Cohan, M., and Kaback, H. R. (1994). Cysteine scanning mutagenesis of the N-terminal 32 amino acid residues in the lactose permease of *Escherichia coli. Protein Sci.* **3:**240–247.
45. Trumble, W. R., Viitanen, P. V., Sarkar, H. K., Poonian, M. S., and Kaback, H. R. (1984). Site-directed mutagenesis of cys148 in the *lac* carrier protein of *Escherichia coli. Biochem. Biophys. Res. Commun.* **119:**860–867.
46. Menick, D. R., Sarkar, H. K., Poonian, M. S., and Kaback, H. R. (1985). Cys154 is important for *lac* permease activity in *Escherichia coli. Biochem. Biophys. Res. Commun.* **132:**162–170.
47. Viitanen, P. V., Menick, D. R., Sarkar, H. K., Trumble, W. R., and Kaback, H. R. (1985). Site-directed mutagenesis of cysteine-148 in the *lac* permease of *Escherichia coli:* effect on transport, binding, and sulfhydryl inactivation. *Biochemistry* **24:**7628–7635.
48. Neuhaus, J.-M., Soppa, J., Wright, J. K., Riede, I., Bocklage, H., Frank, R., and Overath, P. (1988). Properties of a mutant lactose carrier of Escherichia coli with a Cys148–Ser148 substitution. *FEBS Lett.* **185:**83–88.
49. Sarkar, H. K., Menick, D. R., Viitanen, P. V., Poonian, M. S., and Kaback, H. R. (1986). Site-specific mutagenesis of cysteine 148 to serine in the *lac* permease of *Escherichia coli. J. Biol. Chem.* **261:** 8914–8918.
50. Brooker, R. J., and Wilson, T. H. (1986). Site-specific alteration of cysteine 176 and cysteine 234 in the lactose carrier of *Escherichia coli. J. Biol. Chem.* **261:**11765–11771.
51. Menick, D. R., Lee, J. A., Brooker, R. J., Wilson, T. H., and Kaback, H. R. (1987). Role of cysteine residues in the *lac* permease of *Escherichia coli. Biochemistry* **26:**1132–1136.
52. van Iwaarden, P. R., Driessen, A. J., Menick, D. R., Kaback, H. R., and Konings, W. N. (1991). Characterization of purified, reconstituted site-directed cysteine mutants of the lactose permease of *Escherichia coli. J. Biol. Chem.* **266:**15688–15692.
53. Menezes, M. E., Roepe, P. D., and Kaback, H. R. (1990). Design of a membrane transport protein for fluorescence spectroscopy. *Proc. Natl. Acad. Sci. USA* **87:**1638–1642.
54. Lolkema, J. S., Puttner, I. B., and Kaback, H. R. (1988). Site-directed mutagenesis of Pro327 in the *lac* permease of *Escherichia coli. Biochemistry* **27:**8307–8310.
55. Consler, T. G., Tsolas, O., and Kaback, H. R. (1991). Role of proline residues in the structure and function of a membrane transport protein. *Biochemistry* **30:**1291–1298.
56. Padan, E., Sarkar, H. K., Viitanen, P. V., Poonian, M. S., and Kaback, H. R. (1985). Site-specific mutagenesis of histidine residues in the *lac* permease of *Escherichia coli. Proc. Natl. Acad. Sci. USA* **82:**6765–6768.
57. Püttner, I. B., Sarkar, H. K., Poonian, M. S., and Kaback, H. R. (1986). *Lac* permease of *Escherichia coli:* His-205 and His-322 play different roles in lactose/H^+ symport. *Biochemistry* **25:**4483–4485.
58. Püttner, I. B., and Kaback, H. R. (1988). *Lac* permease of *Escherichia coli* containing a single histidine residue is fully functional. *Proc. Natl. Acad. Sci. USA* **85:**1467–1471.
59. Püttner, I. B., Sarkar, H. B., Padan, E., Lolkema, J. S., and Kaback, H. R. (1989). Characterization of site-directed mutants in the *lac* permease of *Escherichia coli:* I. Replacement of histidine residues. *Biochemistry* **28:**2525–2533.
60. Roepe, P. D., and Kaback, H. R. (1989). Site-directed mutagenesis of tyrosine residues in the *lac* permease of *Escherichia coli. Biochemistry* **28:**6127–6132.
61. Ujwal, M. L., Sahin-Toth, M., Persson, B., and Kaback, H. R. (1994). Role of glutamate-269 in the lactose permease of *Escherichia coli. Mol. Membr. Biol.* **1:**9–16.
62. Franco, P. J., and Brooker, R. J. (1994). Functional roles of Glu-269 and Glu-325 within the lactose permease of *Escherichia coli. J. Biol. Chem.* **269:**7379–7386.
63. Menick, D. R., Carrasco, N., Antes, L. M., Patel, L., and Kaback, H. R. (1987). *Lac* permease of *Escherichia coli:* Arginine-302 as a component of the postulated proton relay. *Biochemistry* **26:**6638–6644.
64. Matzke, E. A., Stephenson, L. J., and Brooker, R. J. (1992). Functional role of arginine 302 within the lactose permease of *Escherichia coli. J. Biol. Chem.* **267:**19095–19100.
65. Carrasco, N., Antes, L. M., Poonian, M. S., and Kaback, H. R. (1986). *Lac* permease of *Escherichia coli:* histidine-322 and glutamic acid-325 may be components of a charge-relay system. *Biochemistry* **25:**4486–4488.
66. Carrasco, N., Puttner, I. B., Antes, L. M., Lee, J. A., Larigan, J. D., Lolkema, J. S., Roepe, P. D., and Kaback, H. R. (1989). Characterization of site-directed mutants in the *lac* permease of *Escheq*2539.
67. Fox, C. F., and Kennedy, E. P. (1965). Specific labeling and partial purification of the M protein, a component of the β-galactoside transport system of *Escherichia coli. Proc. Natl. Acad. Sci. USA* **51:**891.

68. Kennedy, E. P., Rumley, M. K., and Armstrong, J. B. (1974). Direct measurement of the binding of labeled sugars to the lactose permease M protein. *J. Biol. Chem.* **249:**33–37.
69. Beyreuther, K., Bieseler, B., Ehring, R., and Müller-Hill, B. (1981). Identification of internal residues of lactose permease of *Escherichia coli* by radiolabel sequences of peptide mixtures. In *Methods in Protein Sequence Analysis* (M. Elzinga, ed.), Humana, Clifton, NY, pp. 139–148.
70. Kaback, H. R., and Patel, L. (1978). The role of functional sulfhydryl groups in active transport in *Escherichia coli* membrane vesicles. *Biochemistry* **17:**1640–1646.
71. Konings, W. N., and Robillard, G. T. (1982). Physical mechanism for regulation of proton solute symport in Escherichia coli, *Proc. Natl. Acad. Sci. USA* **79:**5480–5484.
72. Kaback, H. R., and Barnes, E. M., Jr. (1971). Mechanisms of active transport in isolated membrane vesicles. II. The mechanism of energy coupling between D-lactic dehydrogenase and β-galactoside transport in membrane preparations from *Escherichia coli. J. Biol. Chem.* **246:**5523–5531.
73. Robillard, G. T., and Konings, W. N. (1982). A hypothesis for the role of dithiol-disulfide interchange in solute transport and energy-transducing processes. *Eur. J. Biochem.* **127:**597–604.
74. Jung H., Jung, K., and Kaback, H. R. (1994). Cysteine 148 in the lactose permease of *Escherichia coli* is a component of a substrate binding site. I. Site-directed mutagenesis studies. *Biochemistry* **33:** 12160–12165.
75. Wu, J., and Kaback, H. R. (1994). Cysteine 148 in the lactose permease of *Escherichia coli* is a component of a substrate binding site. 2. Site-directed fluroescence studies. *Biochemistry* **33:**12166–12171.
76. King, S. C., Hansen, C. L., and Wilson, T. H. (1991). The interaction between aspartic acid 237 and lysine 358 in the lactose carrier of *Escherichia coli. Biochem. Biophys. Acta.* **1062:**177–186.
77. Sahin-Tóth, M., Dunten, R. L., Gonzalez, A., and Kaback, H. R. (1992). Functional interactions between putative intramembrane charged residues in the lactose permease of *Escherichia coli. Proc. Natl. Acad. Sci. USA* **89:**10547–10551.
78. Dunten, R. L., Sahin-Toth, M., and Kaback, H. R. (1993). Role of the charge pair formed by aspartic acid 237 and lysine 358 in the lactose permease of *Escherichia coli. Biochemistry* **32:**3139–3145.
79. Sahin-Tóth, M., and Kaback, H. R. (1993). Properties of interacting aspartic acid and lysine residues in the lactose permease of *Escherichia coli. Biochemistry* **32:**10027–10035.
80. Akabas, M. H., Stauffer, D. A., Xu, M., and Karlin, A. (1992). Acetylcholine receptor channel structure probed in cysteine-substitution mutants. *Science* **258:**307–310.
81. Roepe, P. D., Zbar, R. I., Sarkar, H. K., and Kaback, H. R. (1989). A five-residue sequence near the carboxyl terminus of the polytopic membrane protein *lac* permease is required for stability within the membrane. *Proc. Natl. Acad. Sci. USA* **86:**3992–3996.
82. McKenna, E., Hardy, D., Pastore, J. C., and Kaback, H. R. (1991). Sequential truncation of the lactose permease over a three-amino acid sequence near the carboxyl terminus leads to progressive loss of activity and stability. *Proc. Natl. Acad. Sci. USA* **88:**2969–2973.
83. McKenna, E., Hardy, D., and Kaback, H. R. (1992). Evidence that the final turn of the last transmembrane helix in the lactose permease is required for folding. *J. Biol. Chem.* **267:**6471–6474.
84. Lee, J. L., Hwang, P. P., Hansen, C., and Wilson, T. H. (1992). Possible salt bridges between transmembrane α-helices of the lactose carrier of *Escherichia coli. J. Biol. Chem.* **267:**20758–20764.
85. Hinkle, P. C., Hinkle, P. V., and Kaback, H. R. (1990). Information content of amino acid residues in putative helix VII of *lac* permease from *Escherichia coli. Biochemistry* **29:**10989–10994.
86. Zen, K. H., McKenna, E., Bibi, E., Hardy, D., and Kaback, H. R. (1994). Expression of lactose permease in contiguous fragments as a probe for membrane-spanning domains. *Biochemistry* **33:**8198–8206.
87. Jung, K., Jung, H., Wu, J., Privé, G. G., and Kaback, H. R. (1993). Use of site-directed fluorescence labeling to study proximity relationships in the lactose permease of *Escherichia coli. Biochemistry* **32:**12273–12278.
88. Kinnunen, P. K. J., Koiv, A., and Mustonen, P. (1993). Pyrene-labelled lipids as fluorescent probes in studies on biomembranes and membrane models. In *Fluorescence Spectroscopy,* (O. S. Wolfbeis, ed.), Springer-Verlag, New York, pp. 159–71.
89. Betcher-Lange, S. L., and Lehrer, S. S. (1978). Pyrene excimer fluorescence in rabbit skeletal alphaalphatropomyosin labeled with N-(1-pyrene)maleimide: A probe of sulfhydryl proximity and local chain separation. *J. Biol. Chem.* **253:**3757–3760.
90. Ishii, Y., and Lehrer, S. S. (1987). Fluorescence probe studies of the state of tropomyosin in reconstituted muscle thin filaments. *Biochemistry* **26:**4922–4925.
91. Ludi, H., and Hasselbach, W. (1987). Excimer formation pyrenemaleimide-labeled sarcoplasmic reticulum ATPase. *Biophys. J.* **51:**513–515.
92. Sen, A. C., and Chakrabarti, B. (1990). Proximity of sulfhydryl groups in lens proteins. Excimer fluorescence of pyrene-labeled crystallins. *J. Biol. Chem.* **265:**14277–14284.
93. Lee, J. A., Püttner, I. B., and Kaback, H. R. (1989). Effect of distance and orientation between arginine-302, histidine-322, and glutamate-325 on the activity of *lac* permease from *Escherichia coli. Biochemistry* **28:**2540–2544.
94. Cronan, J. E., Jr. (1990). Biotination of proteins in vivo. A post-translational modification to label, purify, and study proteins. *J. Biol. Chem.* **265:**10327–10333.
95. Consler, T. G., Persson, B. L., Jung, H., Zen, K. H., Jung, K., Prive, G. G., Verner, G. E., and Kaback, H. R. (1993). Properties and purification of an active biotinylated lactose permease from *Escherichia coli. Proc. Natl. Acad. Sci. USA* **90:**6964.
96. Lee, J. I., Hwang, P. P., and Wilson, T. H. (1993). Lysine 319 interacts with both glutamic acid 269 and aspartic acid 240 in the lactose carrier of Escherichia coli. *J. Biol. Chem.* **268:**20007–20015.
97. Baldwin, S. A. (1993). Mammalian passive glucose transporters: members of an ubiquitous family of active and passive transport proteins. *Biochim. Biophys. Acta.* **1154:**17–49.
98. Jung, K., Jung, H., and Kaback, H. R. (1994). Dynamics of lactose permease of *Escherichia coli* determined by site-directed fluroescence labeling. *Biochemistry* **33:**3980–3985.
99. Jung, H., Jung, K., and Kaback, H. R. (1994). A conformational change in the lactose permease of *Escherichia coli* is induced by ligand binding or membrane potential. *Protein Sci.* **3:**1052–1057.
100. Kaczorowski, G. J., and Kaback, H. R. (1979). Mechanism of lactose translocation in membrane vesicles from *Escherichia coli.* 1. Effect of pH on efflux, exchange, and counterflow. *Biochemistry* **18:**3691–3697.
101. Kaczorowski, G. J., Robertson, D. E., and Kaback, H. R. (1979). Mechanism of lactose translocation in membrane vesicles from *Escherichia coli.* 2. Effect of imposed delta psi, delta pH, and Delta mu H+. *Biochemistry* **18:**3697–3704.
102. Robertson, D. E., Kaczorowski, G. J., Garcia, M. L., and Kaback, H. R. (1980). Active transport in membrane vesicles from *Escherichia coli:* the electrochemical proton gradient alters the distribution of the *lac* carrier between two different kinetic states. *Biochemistry* **19:**5692–5702.
103. Blow, D. M., Birktoft, J. J., and Hartley, B. S. (1969). Role of a buried acid group in the mechanism of action of chymotrypsin. *Nature* **221:**337–340.
104. Garcia, M. L., Viitanen, P., Foster, D. L., and Kaback, H. R. (1983). Mechanism of lactose translocation in proteoliposomes reconstituted with *lac* carrier protein purified from *Escherichia coli.* 1. Ef-

fect of pH and imposed membrane potential on efflux, exchange, and counterflow. *Biochemistry* **22:**2524–2531.

105. Viitanen, P., Garcia, M. L., Foster, D. L., Kaczorowski, G. J., and Kaback, H. R. (1983). Mechanism of lactose translocation in proteoliposomes reconstituted with *lac* carrier protein purified from *Escherichia coli.* 2. Deuterium solvent isotope effects. *Biochemistry* **22:**2531–2536.
106. Kaback, H. R. (1990). *Lac* permease of *Escherichia coli:* on the path of the proton. *Phil. Trans. Roy Soc. Lond. B Biol. Sci.* **326:**425–436.
107. Collins, J. C., Permuth, S. F., and Brooker, R. J. (1989). Isolation and characterization of lactose permease mutants with an enhanced recognition of maltose and diminished recognition of cellobiose. *J. Biol. Chem.* **264:** 14698–14703.
108. Franco, P. J., Eelkema, J. A., and Brooker, R. J. (1989). Isolation and characterization of thiodigalactoside-resistant mutants of the lactose permease which possess an enhanced recognition for maltose. *J. Biol. Chem.* **264:**15988–15992.
109. King, S. C., and Wilson, T. H. (1989). Galactoside-dependent proton transport by mutants of the Escherichia coli lactose carrier. Replacement of histidine 322 by tyrosine or phenylalanine, *J. Biol. Chem.* **264:**7390–7394.
110. King, S. C., and Wilson, T. H. (1989). Galactoside-dependent proton transport by mutants of the Escherichia coli lactose carrier: substitution of tyrosine for histidine-322 and of leucine for serine-306. *Biochim. Biophys. Acta.* **982:**253–264.
111. King, S. C., and Wilson, T. H. (1990). Sensitivity of efflux-driven carrier turnover to external pH in mutants of the *Escherichia coli* lactose carrier that have tyrosine or phenylalanine substituted for histidine-322. A comparison of lactose and melibiose. *J. Biol. Chem.* **265:**3153–3160.
112. Brooker, R. J. (1990). Characterization of the double mutant, Val-177/Asn-322, of the lactose permease. *J. Biol. Chem.* **265:**4155–4160.
113. Boyer, P. D. (1988). Bioenergetic coupling to protonmotive force: should we be considering hydronium ion coordination and not group protonation? *Trends Biochem. Sci.* **13:**5–7.
114. Privé, G. G., Verner, G. E., Weitzman, C., Zen, K. H., Eisenberg, D., and Kaback, H. R. (1994). Fusion proteins as tools for crystallization: the lactose permease of *Escherichia coli. Acta Cryst.* 375–379.
115. McKenna, E., Hardy, D., and Kaback, H. R. (1992). Insertional mutagenesis of hydrophilic domains in the lactose permease of *Escherichia coli. Proc. Natl. Acad. Sci. USA* **89:**11954–11958.
116. Stochaj, U., and Ehring, R. (1987). The N-terminal region of *Escherichia coli* lactose permease mediates membrane contact of the nascent polypeptide chain. *Eur. J. Biochem.* **163:**653–658.
117. Stochaj, U., Fritz, H. J., Heibach, C., Markgraft, M., von Schaewen, A., Sonnewald, U., and Ehring, R. (1988). Truncated forms of *Escherichia coli* lactose permease: models for study of biosynthesis and membrane insertion. *J. Bacteriol.* **170:**2639–2645.
118. von Heijne, G., and Blomberg, C. (1979). Trans-membrane translocation of proteins. The direct transfer model. *Eur. J. Biochem.* **97:** 175–181.
119. Engelman, D. M., and Steitz, T. A. (1981). The spontaneous insertion of proteins into and across membranes: the helical hairpin hypothesis. *Cell* **23:**411–422.
120. Wickner, W. T., and Lodish, H. F. (1985). Multiple mechanisms of protein insertion into and across membranes. *Science* **230:**400–407.
121. Rapoport, T. A. (1986). Protein translocation across and integration into membranes. *CRC Crit. Rev. Biochem.* **20:**73–137.
122. Singer, S. J., Maher, P. A., and Yaffe, M. P. (1987). On the transfer of integral proteins into membranes. *Proc. Natl. Acad. Sci. USA* **84:** 1960.
123. Blobel, G. (1980). Intracellular protein topogenesis. *Proc. Natl. Acad. Sci. USA* **77:**1496–1500.
124. Popot, J. L., and de Vitry, C. (1990). On the microassembly of integral membrane proteins. *Annu. Rev. Biophys. Biophys. Chem.* **19:**369–403.
125. Popot, J. L., and Engelman, D. M. (1990). Membrane Protein Models: Possibilities and Probabilities. In *Protein Form and Function,* (R. A. Bradshaw, and M. Purton, eds.), Elsevier, Cambridge, pp. 147–151.
126. Bibi, E., Verner, G., Chang, C. Y., and Kaback, H. R. (1991). Organization and stability of a polytopic membrane protein: deletion analysis of the lactose permease of *Escherichia coli. Proc. Natl. Acad. Sci. USA* **88:**7271–7275.
127. Calamia, J., and Manoil, C. (1992). Membrane protein spanning segments as export signals. *J. Mol. Biol.* **224:**539–543.
128. Wrubel, W., Stochaj, U., Sonnewald, U., Theres, C., and Ehring, R. (1990). Reconstitution of an active lactose carrier in vivo by simultaneous synthesis of two complementary protein fragments. *J. Bacteriol.* **172:**5374–5381.
129. Bibi, E., Stearns, S. M., and Kaback, H. R. (1992). The N-terminal 22 amino acid residues in the lactose permease of *Escherichia coli* are not obligatory for membrane insertion or transport activity. *Proc. Natl. Acad. Sci. USA* **89:**3180–3184.

7

Structure and Function of Voltage-Gated Ion Channels

William A. Catterall

7.1. INTRODUCTION

The voltage-sensitive ion channels are responsible for generation of conducted action potentials in excitable cells and for a wide range of regulatory events in nonexcitable cells. On depolarization, permeability to sodium, calcium, or potassium increases dramatically over a period of 0.5 to hundreds of milliseconds and then decreases to the baseline level over a period of 2 msec to seconds. This biphasic behavior is described in terms of two experimentally separable processes that control ion channel function: activation, which controls the rate and voltage dependence of the ion permeability increase following depolarization, and inactivation, which controls the rate and voltage dependence of the subsequent return of ion permeability to the resting level during a maintained depolarization. These channels can therefore exist in three functionally distinct states or groups of states: resting, active, and inactivated. Both resting and inactivated states are nonconducting, but channels that have been inactivated by prolonged depolarization are refractory unless the cell is repolarized to allow them to return to the resting state. In addition, to the regulation of ion channel activity on the millisecond time scale by voltage-dependent gating processes, the activity of these channels is also regulated on longer time scales by neurotransmitters and hormones acting through G proteins and second-messenger-activated protein phosphorylation. These slower modulatory processes are critical elements in cellular regulation and play an essential role in control of physiological functions such as contraction, secretion, and neurotransmitter release. This chapter describes the proteins that comprise the voltage-gated ion channels and focuses on common themes that have been revealed by studies of their structure and function.

7.2. ION CHANNEL SUBUNITS

7.2.1. Structures of Sodium Channel Subunits

The purified sodium channel from rat brain consists of three polypeptides: α of 260 kDa, β1 of 36 kDa, and β2 of 33 kDa[1] (Figure 7.1). The β2 subunit is covalently attached to the α subunit by disulfide bonds while the β1 subunit is associated noncovalently. The subunits are present in a 1:1:1 stoichiometry and the sum of their molecular weights (329,000) agrees closely with the oligomeric molecular weight of the solubilized sodium channel. Antibodies against either the β1 or β2 subunits immunoprecipitate nearly all adult brain sodium channels indicating that they all have a heterotrimeric structure.[2,3] Sodium channels from eel electroplax contain only a single α subunit while sodium channels from rat skeletal muscle have α and β1 subunits.[4,5]

Availability of purified and functionally characterized sodium channel preparations provided the necessary starting material for identification of the genes encoding the sodium channel subunits and determination of their primary structures. Oligonucleotides encoding short segments of the electric eel electroplax sodium channel and antibodies directed against it were used to isolate cDNAs encoding the entire polypeptide from expression libraries of electroplax mRNA.[6] The deduced amino acid sequence revealed a protein of nearly 2000 amino acid residues with four internally homologous domains, each containing multiple potential α-helical transmembrane segments (Figure 7.2). The wealth of information contained in this deduced primary structure has revolutionized research on voltage-gated ion channels.

The cDNAs encoding the electroplax sodium channel were used to isolate cDNAs encoding three distinct, but highly homologous, rat brain sodium channels (Types I, II, and III[7,8]). cDNAs encoding the alternatively spliced Type IIA sodium channel were isolated independently by screening expression libraries with antibodies against the rat brain sodium channel α subunit.[9–11] These sodium channels have a close structural relationship. In general, the similarity in amino acid sequence is greatest in the homologous domains from transmembrane segment S1 through S6 while the intracellular connecting loops are not highly conserved.

The primary structures of sodium channel β1 subunits have been determined only recently.[12] The β1 subunit cloned from rat brain is a small protein of 218 amino acids (22,821 Da) with a substantial extracellular domain having four poten-

William A. Catterall • Department of Pharmacology, University of Washington, Seattle, Washington 98195.
Molecular Biology of Membrane Transport Disorders, edited by S.G. Schultz *et al.* Plenum Press, New York, 1996

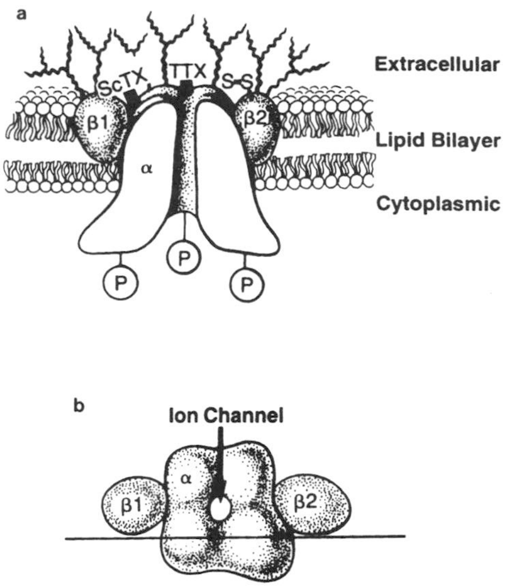

Figure 7.1. Subunit structure of the brain sodium channel. (A) A view of a cross section of a hypothetical sodium channel consisting of a single transmembrane α subunit of 260 kDa in association with a β1 subunit of 36 kDa and a β2 subunit of 33 kDa. The β1 subunit is associated noncovalently while the β2 subunit is linked through disulfide bonds. All three subunits are heavily glycosylated on their extracellular surfaces and the αsubunit has receptor site for α-scorpion toxins (ScTX) and tetrodotoxin (TTX). The intracellular surface of the α subunit is phosphorylated by multiple protein kinases (P). (B) A view of the sodium channel from the extracellular side illustrating the formation of the transmembrane pore in the center of the α subunit.

tial sites of *N*-linked glycosylation, a single α-helical membrane-spanning segment, and a very small intracellular domain. Distinct β1 subunits may form specific associations with different α subunits and contribute to the diversity of sodium channel structure and function.

7.2.2. Functional Expression of Sodium Channel Subunits

α-Subunit mRNAs isolated from rat brain by specific hybrid selection with Type IIa cDNAs[9] and RNAs transcribed from cloned cDNAs encoding α subunits of rat brain sodium channels,[10,11,13,14] rat skeletal muscle sodium channels,[15] and rat heart sodium channels[16,17] are sufficient to direct the synthesis of functional sodium channels when injected into *Xenopus* oocytes. These results establish that the protein structures necessary for voltage-dependent gating and ion conductance are contained within the α subunit itself.

Although α subunits alone are sufficient to encode functional sodium channels, their properties are not normal. Inactivation is slow relative to that observed in intact neurons and its voltage dependence is shifted to more positive membrane potentials. Coexpression of low-molecular-weight RNA from brain can accelerate inactivation, shift its voltage dependence to more negative membrane potentials, and increase the level of expressed sodium current.[10] These results suggested that the low-molecular-weight β1 or β2 subunits may modulate functional expression of the α subunit. Coexpression of RNA transcribed from cloned β1 subunits directly demonstrates this modulation.[12] Coexpression of β1 subunits in *Xenopus* oocytes accelerates the decay of the sodium current 5-fold, shifts the voltage dependence of sodium channel inactivation 20 mV in the negative direction, and increases the level of sodium current 2.5-fold. Evidently, β1 subunits are essential for normal functional expression of rat brain sodium channels.

Sodium channel α subunits can also be functionally expressed in mammalian cells in culture. Stable lines of Chinese hamster ovary (CHO) cells expressing the Type IIA sodium channel α subunits generate sodium currents with normal time course and voltage dependence, even though there is no evidence that these cells express an endogenous β1 subunit to form a complex with the transfected α subunit.[18,19] Evidently, β1 subunits do not have as important a functional impact when the α subunit is expressed in the genetic background of a mammalian somatic cell. The α subunits expressed in CHO cells have normal pharmacological properties as well. They have high-affinity receptor sites for saxitoxin and tetrodotoxin and are inhibited by low concentrations of tetrodotoxin. The voltage dependence of their activation is shifted in the negative direction and they are persistently activated by veratridine in a stimulus-dependent manner. Their inactivation is slowed by α-scorpion toxins. In addition, they are inhibited in a strongly frequency- and voltage-dependent manner by local anesthetic, antiarrhythmic, and anticonvulsant drugs.[20] Thus, the receptor sites for all of these diverse pharmacological agents are located on the α subunits.

7.2.3. Structures of Calcium Channel Subunits

Voltage-gated calcium channels mediate calcium influx into cells in response to depolarization of the plasma membrane. They are responsible for initiation of excitation–contraction and excitation–secretion coupling, and the calcium that enters cells through this pathway is important in regulation of protein phosphorylation, gene transcription, and other intracellular events. Four physiological classes of voltage-gated calcium channels have been defined based on electrophysiological and pharmacological properties.[21] Members of all four physiological classes of calcium channels are expressed in neurons. L-type, dihydropyridine-sensitive calcium channels mediate long-lasting calcium currents and are the most abundant calcium channels in muscle tissues where they initiate excitation–contraction coupling. The transverse tubule membrane of skeletal muscle has an unusually high density of voltage-gated calcium channels allowing these channels to be purified in good yield. Studies of the biochemical properties and functional reconstitution have therefore focused on skeletal muscle calcium channels.[21,22] Figure 7.3A illustrates a model of the subunit structure proposed for the skeletal muscle calcium channel based on biochemical studies.[22,23] A central transmembrane α1 subunit (175 kDa) is associated with four other polypeptides. The disulfide-linked α2 (143 kDa)

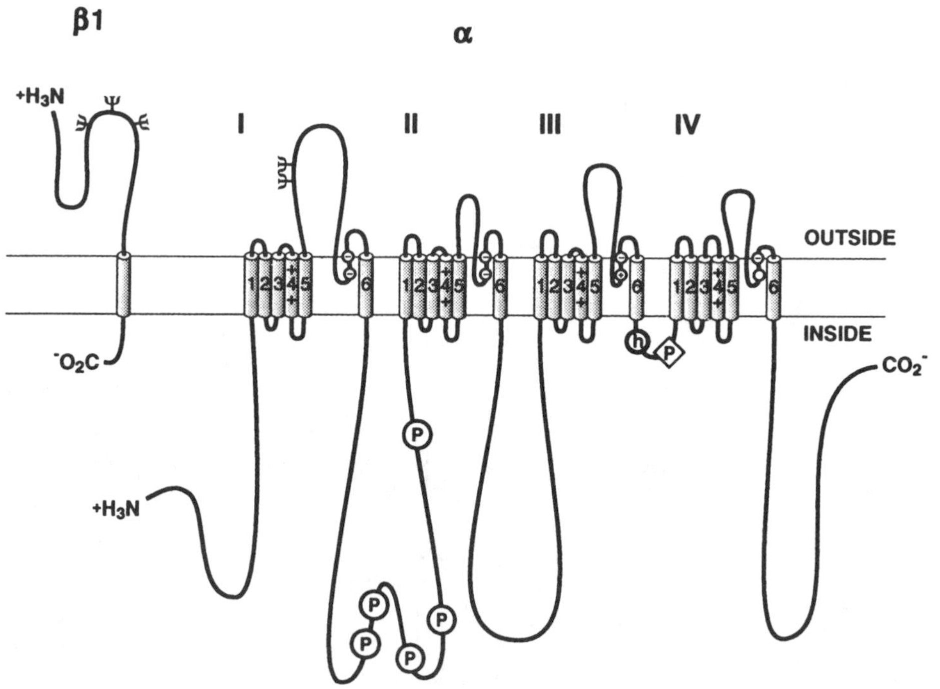

Figure 7.2. Primary structures of the α and β1 subunits of the sodium channel. The bold line represents the polypeptide chains of the αand β1 subunits with the length of each segment approximately proportional to its true length in the rat brain sodium channel. Cylinders represent probable transmembrane α helices. Other probable membrane-associated segments are drawn as loops in extended conformation like the remainder of the sequence. Sites of experimentally demonstrated glycosylation (Ψ), cAMP-dependent phosphorylation (circled P), protein kinase C phosphorylation (boxed P), amino acid residues required for tetrodotoxin binding (small circles with +, −, or open fields depict positively charged, negatively charged, or neutral residues), and amino acid residues that form the inactivation particle (circled h).

and δ (27 kDa) subunits are membrane-associated glycoproteins as is the noncovalently associated γ (30 kDa) subunit. The hydrophilic β subunit (50 kDa) is intracellularly disposed. The cDNA sequence of the α1 subunit predicts a protein of 1873 amino acids, whose structure is similar to the sodium channel α subunit.[24] Two size forms of this subunit are present in skeletal muscle: a major form of about 1700 amino acids (190 kDa) and a full-length form of 212 kDa.[25,26]

A combination of protein chemistry and cDNA cloning experiments have defined the primary structures of the other four subunits.[27–30] The α1, β, and γ subunits are encoded by separate genes. The α2 and δ subunits are encoded by the same gene and are disulfide-linked and proteolytically cleaved in posttranslational processing reactions. Figure 3B illustrates the primary structures of these proteins with their predicted α-helical transmembrane segments. The points of interaction among these subunits are not known, but their specific association as a complex is supported by copurification and by coimmunoprecipitation with antibodies directed against α1, α2, β, and γ subunits.[22–24]

7.2.4. Functional Expression of Calcium Channels

Although the calcium channel is an oligomeric structure, the α1 subunit alone is able to function autonomously in some respects. Complementary DNAs or the corresponding mRNAs encoding skeletal muscle or cardiac α1 subunits can direct expression of functional voltage-gated calcium channels in recipient mammalian cells or *Xenopus* oocytes, respectively,[31,32] and can restore both calcium currents and excitation–contraction coupling in myocytes from mice having the *muscular dysgenesis* mutation which disrupts the endogenous α1 gene.[33] What are the roles of the other subunits of the calcium channel complex? Coexpression studies reveal important modulatory effects of these other subunits on the level of expression and the functional properties of α1 subunits.[34–37]

While expression of α1 subunits in *Xenopus* oocytes or L cells yields voltage-activated calcium currents, the level of current and the number of high-affinity binding sites for dihydropyridine calcium channel blockers are very low. Coexpression of α2δ subunits increases the level of calcium currents significantly for cardiac α1 in *Xenopus* oocytes and more modestly for skeletal muscle α1 in L cells. β has even more substantial effects on expression of cardiac α1 in oocytes and also increases calcium channel expression in mammalian cells. In contrast, coexpression of γ subunits has little effect on the level of calcium current or DHP binding sites. Overall, these results argue that formation of an oligomeric calcium channel complex greatly improves functional expression of α1 subunits.

Activation of cardiac or skeletal muscle α1 subunits expressed alone is slow and requires a stronger depolarizing stimulus than the corresponding native channels. Coexpression with β subunits strongly affects these channel gating characteristics. The time required for channel activation to reach peak calcium current is markedly reduced, and the voltage dependence of activation is shifted toward more negative membrane potentials. α2δ and γ subunits have only modest effects. Like activation, the inactivation of α1 subunits expressed alone is slow and stronger depolarizing stimuli are required to inactivate the expressed channels than native channels. Coexpression of β subunits markedly accelerates channel inactivation, and coexpression of α subunits shifts the voltage dependence of inactivation toward more negative

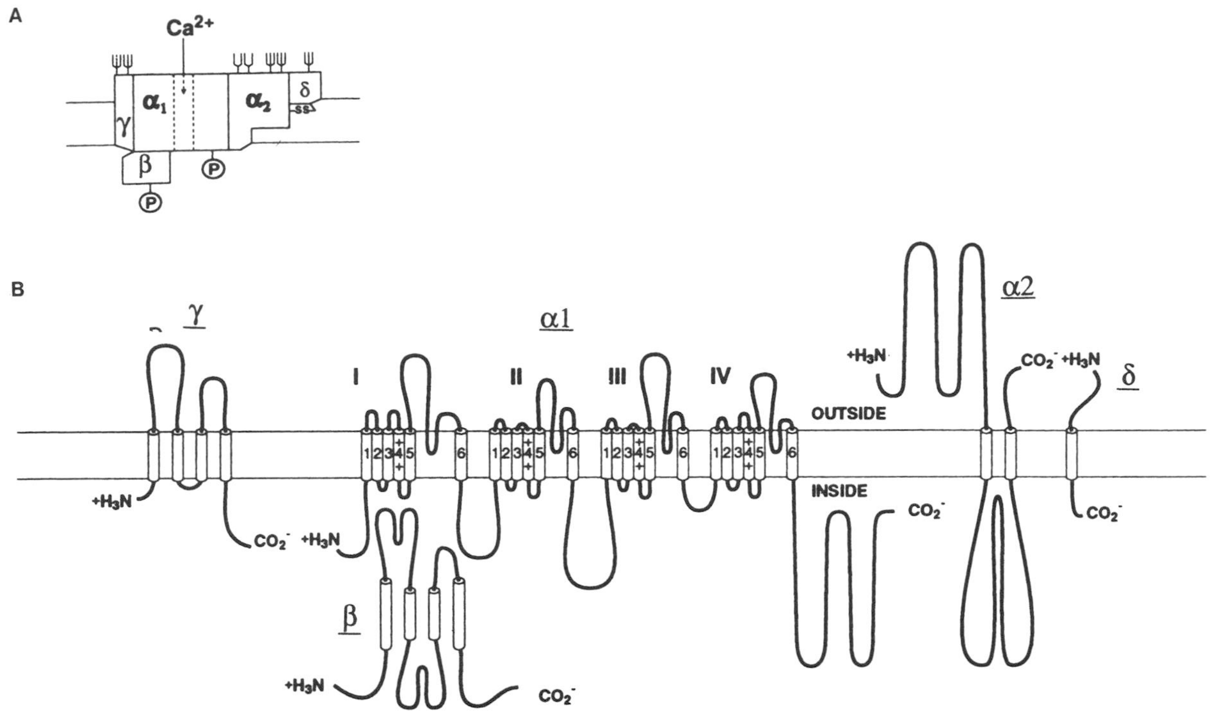

Figure 7.3. Subunit structure of skeletal muscle calcium channels. (A) A model of the subunit structure of the skeletal muscle Ca^{2+} channel derived from biochemical experiments. P, sites of cAMP-dependent protein phosphorylation; Ψ, sites of *N*-linked glycosylation. (B) Transmembrane folding models of the Ca^{2+} channel subunits derived from primary structure determination and analysis.

membrane potentials. Overall, these results show decisively that efficient expression of calcium channels with normal physiological properties is greatly enhanced by coexpression of all five of the subunits identified as components of the oligomeric calcium channel complex.

Multiple calcium subtypes have been described with different physiological and functional properties. Do all calcium channel subtypes have a similar functional subunit structure? Several recent results suggest that they may. L-type, dihydropyridine-sensitive calcium channels in mammalian brain have α1-, α2δ-, and β-like subunits.[38] N-type, ω-conotoxin-sensitive calcium channels in mammalian brain also have α1-, α2-, and β-like subunits.[39] The expression of brain calcium channel α1 subunits is dramatically enhanced by coexpression with α2δ and β subunits.[40–42] While still fragmentary, these results suggest that the complex oligomeric structure of the skeletal muscle L-type calcium channel may be recapitulated in most calcium channel subtypes.

7.2.5. Structures of Potassium Channel Subunits

Voltage-gated K^+ channels in mammalian brain are functionally diverse.[43,44] They can be classified into two major groups based on physiological properties: delayed rectifiers which activate slowly on membrane depolarization and either inactivate slowly or do not inactivate and A-type K^+ channels which are fast activating and inactivating. The molecular structure of the voltage-gated K^+ channels was first revealed by molecular cloning of the gene encoding the *Shaker* mutation in *Drosophila*.[43,44] A-type K^+ channels and delayed rectifier K^+ channels in *Drosophila* and vertebrates have principal subunits whose polypeptide backbones are 60 to 70 kDa and are homologous in structure to a single domain of the α or α1 subunits of Na^+ or Ca^{2+} channels[45,46] (Figure 7.4). Several lines of evidence now indicate that, like Na^+ and Ca^{2+} channels, they also contain one or more modulatory subunits as components of their oligomeric structure *in situ*.

Some brain K^+ channels contain a receptor for dendrotoxins (DTX), a family of neurotoxins isolated from the venom of the black mamba snake, *Dendroaspis polyepsis*.[46] These basic polypeptide toxins inhibit A-type K^+ channels, resulting in the facilitation of neurotransmitter release and epileptiform activity. The DTX binding site is the receptor for three additional peptide ligands: mast cell degranulating peptide, β-bungarotoxin, and charybdotoxin. These toxins have been successfully used as molecular probes to identify and purify components of A-type and delayed rectifier K^+ channels from mammalian brain. Purification of DTX-binding proteins from detergent-solubilized rat or bovine brain membranes revealed a noncovalently associated glycoprotein complex containing polypeptides of 74–80 kDa in association with smaller polypeptides of 42, 38, and 35 kDa when analyzed by SDS–PAGE.[47–49] The purified preparation contained receptor sites for mast cell degranulating peptide and β-bungarotoxin as well as DTX. The 80-kDa DTX-binding proteins from rat and bovine brain represent a family of pharmacologically and

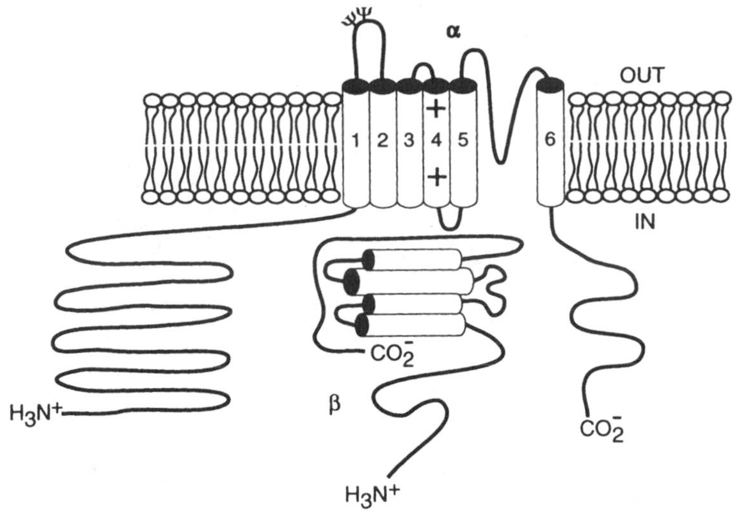

Figure 7.4. Subunit structure of a brain potassium channel. A model of the subunit structure of dendrotoxin-sensitive potassium channels from brain. Symbols are defined as in Figures 7.2 and 7.3.

structurally related glycoproteins, which are products of the mammalian homologues of the *Drosophila Shaker* gene.[49–53] Neuraminidase treatment reduced its apparent molecular mass to 70 kDa, and subsequent treatment with endoglycosidase F further reduced this to 65 kDa, indicating that the 80-kDa subunit was a sialylated membrane protein that was exposed to the extracellular surface. In contrast, treatment of the 38-kDa polypeptide with neuraminidase or endoglycosidase F had no effect on its mobility on SDS–PAGE, indicating that this component most likely did not contain *N*-linked sugars.[51,52]

Antibodies against a fusion protein generated from a cDNA were used to identify a delayed rectifier K^+ channel polypeptide from rat brain.[54] Similar to A-type K^+ channels identified via their sensitivity to DTX, this putative delayed rectifier K^+ channel also contained a low-molecular-weight subunit. Antibodies against it specifically immunoprecipitated a complex of 130- and 38-kDa polypeptides. Because the antibodies recognized the 130-kDa polypeptide exclusively and coprecipitated the 38-kDa subunit, it was concluded that these two polypeptides were immunologically distinct proteins in a heterooligomeric complex with a 1:1 stoichiometry. Thus, both delayed rectifier and A-type neuronal K^+ channels may have β subunits.

Oligonucleotides based on the amino acid sequence derived from the 38-kDa subunit of the DTX-sensitive K^+ channel were used to clone this subunit by a combination of polymerase chain reaction and library screening.[55] The resulting cDNAs encoded a new 38-kDa protein designated the β subunit of the K^+ channel which has no transmembrane segments or glycosylation sites (Figure 7.4). The β subunit of the K^+ channel has no amino acid sequence homology with any of the subunits of sodium channels, but it resembles the β subunit of calcium channels in primary and secondary structure.

7.2.6. Functional Expression of Potassium Channel Subunits

Potassium channel α subunits are fully functional when expressed in *Xenopus* oocytes or in mammalian cells.[45,46] They form homomeric and heteromeric tetramers of noncovalently associated subunits. Individual channels have a wide range of activation and inactivation rates as observed for K^+ channels in different cell types. As for sodium and calcium channels, the functional properties of K^+ channels are modified by coexpression of the β subunit.[55] Inactivation of the channel is accelerated and the voltage dependence of activation and inactivation is shifted to more negative membrane potentials. Thus, the auxiliary subunits of all three classes of voltage-gated ion channels serve as modulators of channel activity.

7.3. STRUCTURE AND FUNCTION OF THE VOLTAGE-GATED ION CHANNELS

Purification, molecular cloning, and determination of the primary structures of the principal subunits of sodium, calcium, and potassium channels have provided a molecular template for probing the relationship between their structure and function. The structure of the principal subunits of each of these channels is based on the same motif (Figure 7.5A): four homologous transmembrane domains which contain six probable transmembrane α helices and surround a central ion pore.[56–58] Analyses of the structure/function relationships of these channels have revealed conserved structural motifs that are responsible for voltage-dependent activation, ion conductance, and inactivation.

7.3.1. Voltage-Dependent Activation

Activation of the voltage-gated ion channels is thought to result from a voltage-driven conformational change which opens a transmembrane pore through the protein. Depolarization of the membrane exerts an electrical force on voltage sensors which contain the gating charges of the channel located within the transmembrane electrical field. These gating charges are likely to be charged amino acid residues located in trans-

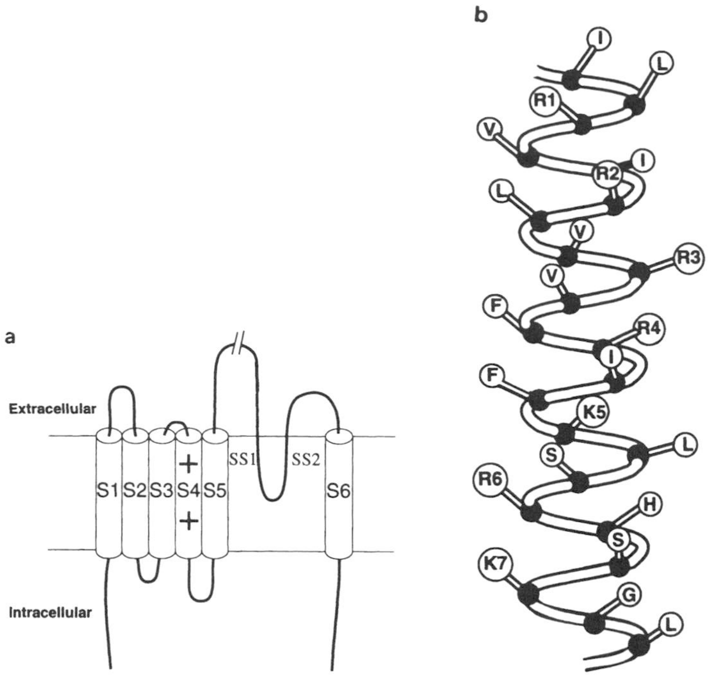

Figure 7.5. Common structural motif of the voltage-gated ion channels. (A) Transmembrane folding diagram of a homologous domain of a voltage-gated ion channel. Alpha-helical transmembrane segments S1 through S6 are represented as cylinders while the short membrane-associated segments SS1 and SS2 and other connecting segments are indicated as bold lines. (B) The S4 segment of the *Shaker* potassium channel is illustrated in a ball-and-stick α-helical representation. Amino acids are indicated in single-letter code and the positively charged amino acids in every third position are numbered from the extracellular end of the helix.

membrane or membrane-associated segments of the protein. The movement of the gating charges of the sodium channel through the membrane driven by depolarization has been directly measured as an outward gating current.[59] The gating current for sodium channels was estimated to correspond to the movement of approximately six charges all the way across the membrane; movement of a larger number of charges across a fraction of the membrane electric field would be equivalent. More detailed analyses of potassium channel gating currents suggest even larger gating charge movements. Identification of the voltage sensors and gating charges of the voltage-gated ion channels is the first critical step toward understanding the molecular basis of voltage-dependent activation.

Inspection and analysis of the primary structure of the sodium channel α subunit, the first member of the voltage-gated ion channel gene family, led to the prediction that the fourth transmembrane segment in each domain (the S4 segment) might serve as the voltage sensor.[56,57,60] These segments contain repeated motifs of a positively charged residue followed by two hydrophobic residues. For comparison among different channels, these positive charges have been numbered sequentially from the extracellular end of the segment (Figure 7.5B). The prediction that these positive charges serve as gating charges has been tested by mutagenesis and expression studies of both sodium and potassium channels. In each case, the gating charge has been inferred from measurements of the steepness of voltage dependence of channel activation at low levels of activation when this provides an indirect estimate of gating charge.

Neutralization of the four positively charged residues in the S4 segment of domain I of the sodium channel by site-directed mutagenesis has major effects on the voltage dependence of activation.[61] Neutralization of the arginine residue in position 1 (R1) had little effect on the steepness of sodium channel activation, but neutralization of the positively charged residues in positions 2 through 4 in this S4 segment reduced the apparent gating charge by 0.9 to 1.8 charges. Combined neutralization of multiple charged residues and mutations of positive charges to negative charges causes progressively increasing reduction of gating charge, but the reduction of apparent gating charge is less than proportional to the expected reduction in total charge of the S4 segment. In addition, most of the mutations also caused shifts of the voltage dependence of activation to more positive or more negative membrane potentials.

Neutralization of the positively charged residues in the S4 segments of potassium channels also caused reduction in apparent gating charge and shifts of the voltage dependence of activation.[62–64] For the *Shaker* potassium channel of *Drosophila,* neutralization of R1 (Figure 7.5B) causes an unexpectedly large reduction in apparent gating charge while charge reversal has little additional effect.[63,64] Most mutations at this site caused a positive shift in the voltage dependence of gating. In contrast, neutralization and charge reversal at R2 and K7 caused changes in apparent gating charge that were closely correlated with the changes in charge caused by the mutations. Most mutations which reduce positive charge at R2 caused a negative shift in the voltage dependence of activation while those at K7 caused a positive shift. Overall, the studies of S4 segments in potassium channel gating support the conclusion that the positively charged residues in these segments are indeed gating charges involved in the voltage sensors of the ion channels. The individual residues appear to contribute differentially to the overall apparent gating charge, and their size and chemical properties other than charge also have important influences on gating.

If the S4 segments must move through the protein structure during the process of activation, the size and shape of the hydrophobic residues in these segments should also have an important influence on voltage-dependent activation. Mutation of a single leucine residue to phenylalanine in an S4 segment of a sodium channel shifts the voltage dependence of gating 20 mV.[65] Similarly, mutation of several hydrophobic residues in the S4 segments of potassium channels also causes dramatic shifts (up to 80 mV) in the voltage dependence of activation.[66,67] In contrast, mutations of several hydrophobic residues in other transmembrane segments did not have major effects on activation.[67] These results, together with the effects of charge neutralization mutations, provide strong support for identification of the S4 segments as the voltage sensors of the voltage-gated ion channels and for identification of the positively charged residues within them as the gating charges of the channel.

Because the voltage-gated ion channels are tetramers of homologous domains, it is important to consider whether the domains interact cooperatively during the process of activation. For the sodium and calcium channels, it is difficult to distinguish cooperativity from domain-specific differences. In contrast, in the homotetrameric potassium channels, cooperativity can be distinguished by comparison of wild-type channels with mutants in which only a single domain has been mutated to shift its voltage dependence of activation to more negative membrane potential. Applying this kind of analysis of potassium channels engineered to have four covalently connected domains with a single mutant domain in different positions in the tetramer demonstrated marked positive cooperativity in the activation process.[68] The results were fit best by a model in which a cooperative conformational change in all four domains is required in order for activation to occur efficiently and the transition to the activated state was favored sixfold by association with a mutant subunit having a negative voltage dependence of activation. These results complicate the quantitative interpretation of previous experiments which neutralize gating charges in the S4 segments. Evidently, large changes in the slope of voltage-dependent activation can result from changes in cooperativity as well as actual changes in gating charge. Thus, the reduction in the slope of voltage-dependent activation due to charge neutralization cannot easily be equated with the direct contribution of the charged residue to overall gating charge movement. Direct estimation of gating charge movement by measurement of gating currents will be required to clearly interpret the effects of these mutations.

The mechanism by which the S4 segments serve as voltage sensors is not known. The "sliding helix" or "helical screw" models[56,60] proposed that the entire S4 helix moves across the membrane along a spiral path exchanging ion pair partners between its positively charged residues and fixed negative charges in surrounding transmembrane segments. This model implies a large (but unknown) energy barrier for breaking and remaking numerous ion pairs within the protein structure and suggests an approximate equivalence of gating charge movement among the different charged residues in the S4 helices. Because neutralization of individual charged residues has very different effects on the voltage dependence of channel activation, it is unlikely that this simple model can be correct in detail. A more complex "propagating helix" model proposes that the S4 transmembrane segments undergo an α helix–β sheet transition which propagates outward to move charged residues across the membrane.[60] This model also has no direct experimental support, but it has the potential to accommodate at least some of the differences observed among individual residues because not all charged residues in the S4 segment are proposed to move the same distance across the membrane. Thus, although the S4 segments are clearly implicated as the voltage sensors of the voltage-gated ion channels, the mechanism through which they initiate activation of the channels remains unknown.

7.3.2. Ion Conductance

Essentially all models for the structure of the voltage-gated ion channels include a transmembrane pore in the center of a square array of homologous transmembrane domains. Each domain would contribute one-fourth of the wall of the pore. Identification of the segments that line the transmembrane pore and define the conductance and ion selectivity of the channels is of great interest and importance. A number of toxins, drugs, and inorganic cations are blockers of the voltage-gated ion channels. In several cases, detailed biophysical analysis of their mechanism of action indicates that these molecules enter and bind within the transmembrane pores of the channels and compete with permeant ions for occupancy of the pore.[69] These channel blockers therefore serve as molecular markers and specific probes of the pore region of the ion channels. Amino acid residues that form the extracellular and intracellular mouths of the transmembrane pores have been identified by their interaction with pore-blocking drugs and toxins.

7.3.2.1. The Extracellular Mouth of the Pore

Tetrodotoxin and saxitoxin are thought to block sodium channels by binding with high affinity to the extracellular mouth of the pore.[69] Their binding is so specific that they were used as a marker in the initial purification of sodium channels from excitable cell membranes.[1,4,5] Block of their binding by protonation or covalent modification of carboxyl residues led to the model that these cationic toxins bind to a ring of carboxyl residues at the extracellular mouth of the pore.[69] These residues have now been identified by site-directed mutagenesis. Glu387 in rat brain sodium channel II was neutralized by mutagenesis to glutamine and expressed in *Xenopus* oocytes to analyze the functional properties of the mutant channel.[70] The affinity for tetrodotoxin was reduced over 10,000-fold. This amino acid residue is located in segment SS2 (Figure 7.6) on the extracellular side of the S6 transmembrane segment in domain I of the sodium channel. Subsequent exten-

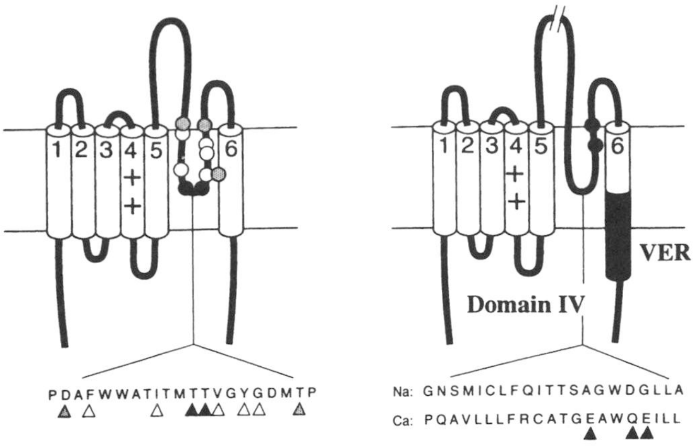

Figure 7.6. Amino acid residues required for pore formation: (left) potassium channels; (right) sodium and calcium channels. The positions of amino acid residues required for ion conductance and selectivity (open circles), high-affinity binding of pore blockers (shaded circles), or both (filled circles) and the peptide segment covalently labeled by the phenylalkylamine verapamil (shaded bar) are illustrated. The amino acid sequences of the SS1/SS2 region of the *Drosophila Shaker* potassium channel, the Type II brain sodium channel, and the L-type skeletal muscle calcium channel are presented below each drawing. The arrowheads below each sequence illustrate the positions at which residues required for ion conductance and selectivity (open symbols), high-affinity binding of pore blockers (shaded symbols), or both (filled symbols) are located.

sion of their analysis identified acidic amino acid residues in the same position as Glu387 in each domain which were all required for high-affinity tetrodotoxin binding.[71] These residues are therefore likely to surround the extracellular opening of the pore and contribute to a receptor site for tetrodotoxin. In addition to this ring of carboxyl residues, a second ring of amino acids located three residues on the amino-terminal side of these is also required for tetrodotoxin binding (Figure 7.6). These are acidic amino acids in domains I and II, basic in domain III, and neutral in domain IV. If this region is in α-helical conformation, this second set of residues required for tetrodotoxin binding would fall on the same side of the helix and form a second inner ring of residues at the opening of the pore.

Cardiac sodium channels bind tetrodotoxin with 200-fold lower affinity than brain or skeletal muscle sodium channels but retain all of the eight residues described above that are required for high-affinity binding. However, at position 385, two residues toward the N-terminal from Glu387 in the brain sodium channel II sequence, there is a change of a tyrosine or phenylalanine in the brain and skeletal muscle channels to cysteine in the cardiac sodium channel. Mutation of this residue from Cys to Phe or Tyr causes an increase of 200-fold in the cardiac sodium channel and the converse mutation causes a loss of affinity of 200-fold in the brain or skeletal muscle channel.[72–74] Thus, it is likely that this residue also contributes in an essential way to the tetrodotoxin receptor site. Cadmium is a high-affinity blocker of cardiac sodium channels but not of brain or skeletal muscle sodium channels. Substitution of this critical cysteine in the skeletal muscle or brain sodium channel confers high-affinity block by cadmium on these channels.[72,73] Analysis of the voltage dependence of cadmium block suggests that this ion passes 20% of the way through the membrane electrical field in reaching its binding site formed by this cysteine residue.[73] Thus, this residue may be approximately 20% of the way through the electrical field within the pore of the sodium channel.

The outer mouth of the potassium channel has been mapped in a similar way. The polypeptide charybdotoxin is an extracellular blocker of potassium channels which binds in the outer mouth of the pore. Identification of amino acids that contribute to binding of charybdotoxin reveals glutamic acid, aspartic acid, and threonine residues which are required for high-affinity binding.[75,76] The required residues cluster on both sides of the SS1–SS2 region as illustrated in Figure 7.6, and the residues closest to these short segments are most important for charybdotoxin binding. Conversion of any of these residues to a positively charged amino acid increases the K_d for charybdotoxin binding more than 300-fold, suggesting that positively charged amino acids in the toxin may normally interact with the negatively charged and hydroxylic residues in these positions in the wild-type channel.

Tetraethylammonium ions also block potassium channels from the extracellular side. Analysis of their affinity for block of the same family of potassium channel mutants reveals that residues on both sides of SS1 and SS2 are required, with the amino acid residue in position 449 on the carboxyl-terminal side of SS2 being dominant.[77] Tyrosine or phenylalanine in this position confers high-affinity block. Carboxyl residues in this position give intermediate affinity, and positively charged residues prevent tetraethylammonium binding completely.[77] Phenylalanine residues in this position in all four subunits of a potassium channel can participate in binding tetraethylammonium ion, suggesting that the four phenyl rings coordinate a single tetraethylammonium molecule through cation–π orbital interactions.[78,79] These residues are similar in position in the amino acid sequence to those that are required for tetrodotoxin binding to sodium channels. Thus, it seems likely that tetraethylammonium ions in potassium channels and tetrodotoxin in sodium channels occupy similar receptor regions at the extracellular mouth of the pore when they block the channels.

7.3.2.2. The Intracellular Mouth of the Pore

Local anesthetics and related antiarrhythmic drugs are thought to bind to a receptor site on the sodium channel which is accessible only from the intracellular side of the membrane and is more accessible when the sodium channel is open.[69] Similarly, the phenylalkylamine class of calcium channel antagonists are characterized as intracellular open channel blockers, and tetraithylammonium and related monoalkyltrimethylammonium derivatives can block potassium channels from the intracellular side of the membrane when the channel is open[69]. Both biochemical and molecular biological approaches have been used to probe the peptide segments of the principal subunits of the voltage-gated ion channels which interact with these intracellular pore blockers.

Verapamil and related phenylalkylamine calcium channel antagonist drugs are the highest-affinity ligands among the diverse intracellular pore blockers. Desmethoxyverapamil and its photoreactive azido derivative ludopamil have K_d's for equilibrium binding to purified calcium channels in the range of 30 nM and therefore can be used as highly specific binding probes of their receptor site in the intracellular mouth of the calcium channel. Covalent labeling of purified calcium channels with ludopamil results in incorporation into the $\alpha 1$ subunit only.[80] The site of covalent labeling was located by extensive proteolytic cleavage of the labeled $\alpha 1$ subunit followed by identification of the photolabeled fragments by immunoprecipitation with site-directed antipeptide antibodies. All of the covalent label recovered was incorporated into a peptide fragment containing the S6 segment of domain IV of the $\alpha 1$ subunit and several amino acid residues at the intracellular end of this transmembrane segment. Since phenylalkylamines act only from the inside of the cell, it was concluded that the intracellular end of transmembrane segment IVS6 and the adjacent intracellular residues form part of the receptor site for phenylalkylamines and therefore part of the intracellular mouth of the calcium channel.[80,81]

Analyses of mutations that alter block of potassium channels from the intracellular side by tetraalkylammonium ions point to both the amino acid residues between SS1 and SS2 and the intracellular end of the S6 segment as components of the intracellular mouth of the pore. Mutation of a critical threonine residue in the sequence between SS1 and SS2 to the closely related amino acid serine is sufficient to increase the K_d for block of potassium channels by intracellular tetraethylammonium ion 10-fold.[82] Tetraethylammonium must traverse only 15% of the membrane electrical field in reaching this binding site from the intracellular solution. Thus, it is likely that this threonine residue forms part of a binding site for tetraethylammonium ions at the intracellular mouth of the potassium channel. Alkyltriethylammonium ions with long carbon chains in the alkyl group also require this threonine residue for high-affinity binding and block. In addition, mutation of a threonine residue near the middle of the S6 segment to a hydrophobic residue increases the affinity for C_8 and C_{10} alkyltriethylammonium ions.[83] A more hydrophobic residue in this position gives higher affinity for the alkyltriethylammonium ions, and the effect is greater for larger alkyl substituents consistent with hydrophobic interactions between residues at this position and the alkyl group of the substituted tetraalkylammonium ion. Thus, in potassium channels as well as calcium channels, the S6 segments contribute to formation of the intracellular mouth of the pore and to binding of hydrophobic pore-blocking drugs.

7.3.2.3. Ion Conductance and Selectivity

Consistent with the idea that the amino acid residues that are required for binding of pore blockers are also required for interaction with permeant ions, changes in these residues have dramatic effects on ion conductance and selectivity. A clear demonstration of the close relationship between the amino acid residues that determine pore blocking properties and those that determine ion conductance and selectivity came from studies of chimeric potassium channels in which the SS1/SS2 region is transferred between channel types which differ in both single-channel conductance and affinity for tetraethylammonium ion at intracellular and extracellular sites.[84] Such chimeric channels have the ion conductance, affinity for extracellular tetraethylammonium ion, and affinity for intracellular tetraethylammonium ion specified by the SS1/SS2 region with little effect of the remainder of the channel structure. Small changes in individual amino acids within this region also have dramatic effects on ion selectivity (Figure 7.6). Changes of threonine to serine or phenylalanine to serine increase the conductance of Rb^+ and NH_4^+.[85] Coordinate changes of leucine to valine and valine to isoleucine in two potentially interacting positions in the deep pore region are responsible for the differences in conductance and binding affinity for intracellular tetraethylammonium ion between chimeric channels which differ in the SS1/SS2 region.[86] Deletion of two residues (Tyr445, Gly446) from the SS2 region of the *Shaker* voltage-gated potassium channel to yield a sequence similar in length to the corresponding region of the distantly related cyclic nucleotide-gated ion channels causes loss of potassium selectivity and increased channel block by divalent cations which are characteristic of the cyclic nucleotide-gated ion channels.[87] These results indicate that even small alterations in the amino acid residues in the SS1/SS2 region have crucial effects on ion conductance and selectivity, supporting the conclusion that these residues form part of the lining of the transmembrane pore and interact directly with permeant ions. Moreover, changes in amino acid sequence in this region can also determine ion selectivity properties of different families of potassium channels. It is surprising that so many of the amino acid residues in this region are hydrophobic and that hydrophobic residues are critical determinants of ion selectivity. It remains to be determined how these residues interact with permeant ions to allow rapid and selective ion conductance.

As for potassium channels, changes in the amino acid residues in the SS1/SS2 region that are important for binding of the pore blocker tetrodotoxin are also critical determinants of sodium channel ion conductance and selectivity. Single-channel conductance values for mutations that neutralize single charges among the six negatively charged amino acid residues that are important for tetrodotoxin binding also reduce single-channel conductance, in some cases to as little as

10% of wild-type levels.[71] Sodium and calcium channels have similar overall structures, but strikingly different ion selectivity. Sodium ions are essentially impermeant through calcium channels in the presence of calcium, but are rapidly permeant in the absence of divalent cations.[69] This property is thought to arise from high-affinity binding of calcium ions to two sites in the ion conductance pathway which blocks sodium ion entry and allows rapid calcium conductance. Calcium ions are less than 10% as permeable as sodium ions through the sodium channel. Remarkably, mutation of only two amino acid residues in the sodium channel is sufficient to confer calcium channel-like permeability properties.[88] Mutation of Lys1422 and Ala1714 to negatively charged glutamate residues not only altered tetrodotoxin binding but also caused a dramatic change in the ion selectivity of the sodium channel from sodium-selective to calcium-selective. In addition, these changes created a high-affinity site for calcium binding and block of monovalent ion conductance through the sodium channel, as has been previously described for calcium channels. Thus, in the mutant sodium channel with two additional negative charges near the extracellular mouth of the putative pore region, monovalent cation conductance is high in the absence of calcium. At calcium concentrations in the 10 μM range, monovalent cation conductance is strongly inhibited by high-affinity calcium binding. As calcium concentrations are increased, calcium conductance is preferred over sodium conductance. These results mirror the ion conductance properties of calcium channels and indicate that a key structural determinant of the ion selectivity difference between calcium and sodium channels is specified by the negatively charged amino acid residues at the mouth of the putative pore-forming region.

7.3.3. Inactivation

7.3.3.1. Fast Inactivation of Sodium Channels

Fast inactivation of the sodium channel acquires most of its voltage dependence from coupling to voltage-dependent activation, and the inactivation process can be specifically prevented by treatment of the intracellular surface of the sodium channel with proteolytic enzymes.[59] These results led to the proposal of an autoinhibitory, "ball-and-chain" model for sodium channel inactivation in which an inactivation particle tethered on the intracellular surface of the sodium channel (the ball) diffuses to a receptor site in the intracellular mouth of the pore, binds, and blocks the pore during the process of inactivation.[59] This model predicts that an inactivation gate on the intracellular surface of the sodium channel may be responsible for its rapid inactivation.

The sodium channel segments that are required for fast inactivation have been identified by use of a panel of site-directed antipeptide antibodies against peptides corresponding to short (15 to 20 residue) segments of the α subunit. These antipeptide antibodies were applied to the intracellular surface of the sodium channel from the recording pipette in whole-cell voltage clamp experiments or from the bathing solution in single-channel recording experiments in excised, inside-out membrane patches.[89,90] In both cases, only one antibody, directed against the short intracellular segment connecting homologous domains III and IV (Figure 7.7), inhibited sodium channel inactivation. Inhibition of fast inactivation of antibody-modified sodium channels in membrane patches was complete. The binding and effect of the antibody were voltage-dependent. At negative membrane potentials where sodium channels are not inactivated, the antibody bound rapidly and inhibited channel inactivation; at more positive membrane potentials where the sodium channel is inactivated, antibody binding and action were greatly slowed or prevented. Based on these results, it was proposed that the segment that this antibody recognizes is directly involved in the conformational change leading to channel inactivation. During this conformational change, this inactivation gating segment was proposed to fold into the channel structure, serve as the inactivation gate by occluding the transmembrane pore, and become inaccessible to antibody binding (Figure 7.7).[89,90]

A similar model is supported by site-directed mutagenesis experiments.[61] Expression of the sodium channel α subunit in *Xenopus* oocytes as two pieces corresponding to the first three domains and the fourth domain results in channels that activate normally but inactivate 20-fold more slowly than normal. The physiological characteristics of these cut channels are similar to those of sodium channels with inactivation blocked by the site-directed antibody. In contrast, sodium channel α subunits cut between domains II and III have normal functional properties. These two independent approaches using site-directed antibodies and cut mutations provide strong support for identification of the short intracellular segment connecting domains III and IV as an inactivation gating loop.

The inactivation gating loop contains highly conserved clusters of positively charged and hydrophobic amino acid residues. Neutralization of the positively charged amino acid residues in the inactivation gating loop of the sodium channel by site-directed mutagenesis does not have a profound effect on channel inactivation,[91,92] although neutralization of the cluster of positively charged residues at the amino-terminal end of the loop does slow inactivation and shift the voltage dependence of both activation and inactivation.[92] In contrast, deletion of the 10-amino-acid segment at the amino-terminal end of the loop completely blocks fast sodium channel inactivation.[92] Scans of the hydrophobic amino acid residues in this segment of the inactivation gating loop by mutation to the hydrophilic, but uncharged residue glutamine show that mutation of the three-residue cluster IFM to glutamine completely blocks fast sodium channel inactivation.[93] The single phenylalanine in the center of this cluster at position 1489 in sodium channel II is the critical residue (Figure 7.7). Conversion of this residue to glutamine is sufficient by itself to nearly completely prevent fast channel inactivation. Mutation of the adjacent isoleucine and methionine residues to glutamine also has substantial effects. Substitution of glutamine for isoleucine slows inactivation twofold and makes inactivation incomplete leaving 10% sustained current at the end of long depolariza-

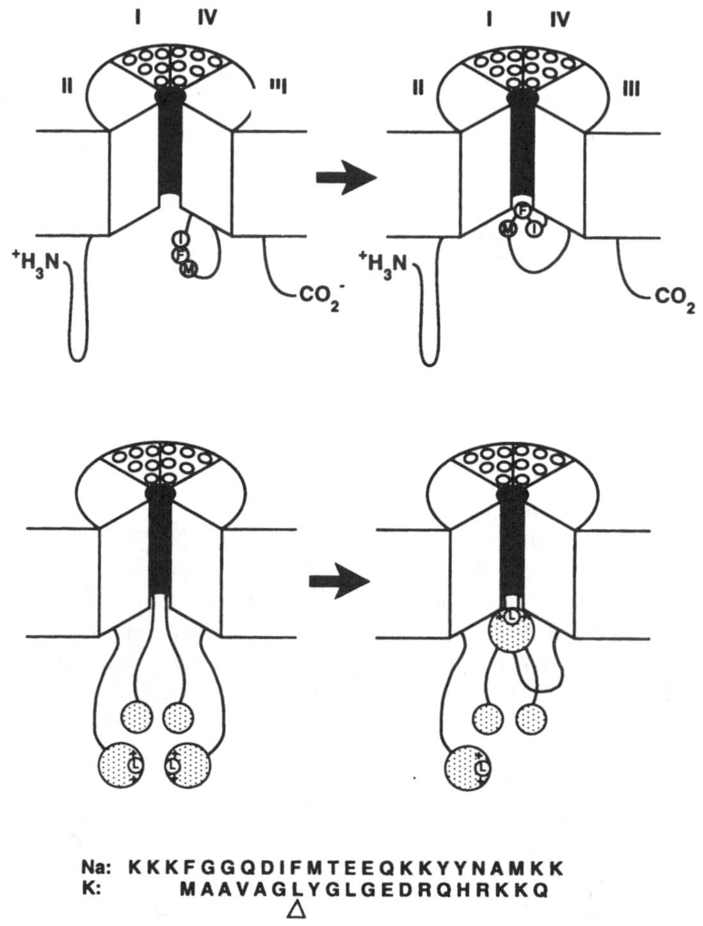

Figure 7.7. Mechanisms of inactivation of sodium and potassium channels. The hinged-lid mechanism of sodium channel inactivation (top) and the ball-and-chain mechanism of potassium channel inactivation (bottom) are illustrated. The intracellular loop connecting domains III and IV of the sodium channel is depicted as forming a hinged lid. The critical residues Leu7 and Phe1489 are shown as occluding the intracellular mouth of the pores in each case. The amino acid sequences of the inactivation particle region of each channel are illustrated below.

tions. Substitution of glutamine for methionine slows inactivation threefold. The interaction of Phe1489 with the receptor of the inactivation gating particle is likely to be hydrophobic because there is a close correlation between the hydrophobicity of the residue at that position and the extent of fast sodium channel inactivation.[94] On the basis of these results, it has been proposed that these residues serve as the inactivation gating particle entering the intracellular mouth of the transmembrane pore of the sodium channel and blocking it during channel inactivation. The intracellular loop between domains III and IV therefore serves as an inactivation gate and closes the transmembrane pore of the sodium channel from the intracellular side of the membrane.

7.3.3.2. Inactivation of Potassium Channels

In contrast to these results with sodium channels, the inactivation particle responsible for rapid inactivation of potassium channels by occlusion of the intracellular mouth of the pore is located at the N-terminus of the polypeptide.[95] Removal of the N-terminus by deletion mutagenesis prevents fast inactivation. Mutations of single positively charged amino acids in that region slow inactivation and mutation of Leu7 to a hydrophilic amino acid prevents fast inactivation almost completely.[95] Free peptides with amino acid sequences modeled on the inactivation particle restore fast, voltage-dependent inactivation to mutant potassium channels whose N-termini have been deleted.[96] These results fit the expectations of the "ball-and-chain" model of ion channel inactivation originally proposed for sodium channels by Armstrong and Bezanilla.[59] The N-terminal segment of the potassium channels is envisioned as an inactivation particle tethered on the end of a chain of approximately 200 amino acids (Figure 7.7). The positively charged and hydrophobic residues in the ball are thought to interact with an inactivation receptor at the intracellular mouth of the channel through a process of restricted diffusion and binding. This mechanistic model predicts that shortening the chain of amino acids connecting the ball to the rest of the channel should accelerate inactivation while elongating the chain should slow inactivation and also that extracellular permeant ions should oppose inactivation as they diffuse through the pore and compete with the inactivation particle for its binding site. These effects were in fact observed for expressed potassium channels. Fifty-residue segments deleted or inserted in the "chain" between the N-terminal inactivation particle and the membrane accelerate or slow inactivation.[97] External potassium ions also slow inactivation and speed recovery from inactivation as expected if inactivation cannot occur while the pore is occupied by permeant ions.[97] These results support the "ball-and-chain" mechanism as a valid model of potassium channel inactivation.

The "receptor" occupied by the N-terminal inactivation particle as it occludes the mouth of the potassium channel may include amino acid residues in the short intracellular loop connecting transmembrane segments S4 and S5. Mutation of charged residues in this loop to neutral, but hydrophilic ones and mutations of hydrophobic residues to alanine markedly reduced fast potassium channel inactivation.[98] Mutations in this loop also reduce conductance of the potassium channels. Thus, this short intracellular loop may contribute to formation of the intracellular mouth of the pore, along with sequences in the intracellular end of the S6 segment and the SS1/SS2 region as described above, and may serve to form a receptor for the N-terminal inactivation particle.

In addition to the rapid inactivation of potassium channels mediated by the N-terminal inactivation particle (N-type inactivation), potassium channels with deleted N-termini inactivate by another mechanism which requires specific amino acid residues in the extracellular end of the S6 segment (C-type inactivation).[99] N-type and C-type inactivation occur through distinct, but interactive, pathways; N-type inactivation accelerates C-type inactivation. Mutations of the valine residue at position 463 in the *Shaker* potassium channel can change the rate of C-type inactivation nearly 100-fold. C-type inactivation can also be strongly affected by the binding of permeant ions in the extracellular mouth of the pore and by mutations in the SS1/SS2 region of the channel sequence. Evidently, it represents a conformational change at the extracellular mouth of the pore which either closes the pore or alters an ion binding site that is essential for effective permeation.

7.3.3.3. Comparison of the Mechanisms of Inactivation of Sodium and Potassium Channels

The processes of N-type potassium channel inactivation and fast sodium channel inactivation are analogous in many respects. Both lead to rapid inactivation of the channels via occlusion of the intracellular mouth of the pore. In addition, sodium channels have a less well characterized slow inactivation mechanism that has some features in common with C-type potassium channel inactivation. It is unaffected by preventing fast inactivation by protease treatment[100] or by blocking the movement of the inactivation gate with specific antibodies against it[90] and therefore is likely to be an independent gating process. It will be important to determine whether these processes are also similar at the molecular level.

Molecular analysis of the mechanism of fast inactivation of sodium and potassium channels has revealed both similarities and differences. In both N-type potassium channel inactivation and fast sodium channel inactivation, an intracellular inactivation particle is thought to interact with the intracellular mouth of the pore and occlude it during inactivation. In both cases, hydrophobic amino acids are essential components of the inactivation particle. However, positively charged amino acids in the inactivation particle are required for normal N-type inactivation of potassium channels but not for fast inactivation of sodium channels. The amino acid sequences of the inactivation particles of sodium channels and potassium channels do not share any sequence homology (Figure 7.7). Moreover, the structures of the inactivation gate regions of the two channels are strikingly different. The potassium channel structure fits the "ball-and-chain" model of an inactivation particle tethered on a long chain that reaches its receptor by restricted diffusion. In contrast, the inactivation gate loop of the sodium channel places the inactivation particle within 15 residues of the membrane in a segment that is likely to be highly structured. It is unlikely to function as a loosely tethered ball and chain because mutations that place cuts in the inactivation gate segment, potentially allowing more facile movement of a ball and chain, slow inactivation instead,[61] and deletions that place the inactivation particle closer to the membrane, potentially allowing more rapid inactivation via a ball-and-chain mechanism, also slow the inactivation process.[92] In contrast to the potassium channel, the sodium channel inactivation particle may reach its receptor site as integral part of a stepwise conformational change which opens the pore, reveals a receptor site that can bind the inactivation particle, and finally inactivates the channel.

Although the processes of inactivation of sodium channels and potassium channels are distinct, it is likely that the sodium channel inactivation process evolved from potassium channel inactivation.[69] Does the potassium channel inactivation particle receptor recognize its likely descendant, the sodium channel inactivation particle? Insertion of the sodium channel inactivation gate at the N-terminal of a noninactivating potassium channel restores fast inactivation and the IFM sequence is required for restoration of the inactivation process.[101] Thus, the potassium channel does indeed recognize the sodium channel inactivation particle, even though there is no amino acid sequence identity between sodium and potassium channel inactivation particles. These distinct sequences must form inactivation particles having similar three-dimensional structures.

7.3.3.4. A Hinged-Lid Model of Sodium Channel Fast Inactivation

The structure of the sodium channel inactivation gate does not fit the expectations of the "ball-and-chain" model of inactivation very well. The inactivation gate is a short structured loop of the channel which places the inactivation particle close to the mouth of the transmembrane pore. The inactivation gate loop resembles more closely the "hinged lids" of allosteric enzymes which are rigid peptide loops that fold over enzyme active sites and control substrate access. Conformational changes induced by the binding of allosteric ligands move the hinged lid away from the active site and allow substrate access and catalytic activity. In analogy to the allosteric enzymes, it has been proposed that the sodium channel inactivation gate functions as a hinged lid which pivots to place the inactivation particle in a position to bind to the intracellular mouth of the transmembrane pore of the sodium channel (Figure 7.7).[93] The three-dimensional structures of some hinged lids of allosteric enzymes which are known from x-ray crystallographic and two-dimensional NMR studies provide a valuable structural model for design of further experiments to define the mechanism of sodium channel inactivation in more detail.

7.4. NATURALLY OCCURRING MUTATIONS IN HUMAN SODIUM CHANNEL GENES

7.4.1. Hyperkalemic Periodic Paralysis

An exciting new development in the biology of sodium channels is the discovery of mutations that cause human muscle disease. Hyperkalemic periodic paralysis is an autosomal dominant disorder characterized by episodic muscle weakness associated with mild elevation of serum potassium.[102] Electrophysiological recordings of acutely dissociated muscle fibers from patients with hyperkalemic periodic paralysis show persistent depolarization of the fibers and an abnormal, noninactivating sodium conductance which is activated above -60 mV and blocked by tetrodotoxin.[103,104] Whole-cell voltage clamp in normal physiological solution does not reveal any major abnormality,[105] but single-channel analysis shows that elevated external potassium causes episodes of repetitive single-channel openings with low probability.[106] This behavior is sufficient to cause a sustained sodium current during episodes of late channel openings.

The genetic locus responsible for hyperkalemic periodic paralysis is tightly linked to the skeletal muscle sodium channel $\mu 1$ α subunit gene on human chromosome 17.[107,108] Analysis of the nucleotide sequence of sodium channel genes reveals that multiple defective sodium channel alleles are responsible for the disease in different kindreds[109,110] (Figure 7.8). These missense mutations include substitution of valine for methionine near the intracellular end of transmembrane segment IVS6 and substitution of methionine for threonine at the intracellular end of transmembrane segment IIS5. Evidently, these mutations can cause subtle changes in the inactivation gating properties of sodium channels, even though they are located at a substantial distance from the inactivation gating segment in the primary structure of the sodium channel α subunit.

7.4.2. Paramyotonia Congenita

Paramyotonia congenita is a related disorder of periodic muscle paralysis induced by cold exposure and by exercise in the cold.[102] Changes in serum potassium are not generally correlated with periods of muscle weakness. Paramyotonia congenita maps to the same muscle sodium channel gene locus on chromosome 17 as hyperkalemic periodic paralysis.[111] Analysis of the DNA sequence of mutations in different kindreds also reveals multiple alleles. Two mutations cause substitutions of histidine or cysteine for arginine at the extracellular end of transmembrane segment IVS4, a voltage-sensing segment of the sodium channel[111] (Figure 7.8). Two distinct mutations cause substitutions of methionine for threonine and valine for glycine in the intracellular loop between domains III and IV, the inactivation gating segment of the sodium channel[112] (Figure 7.8).

7.4.3. Molecular Mechanisms of the Periodic Paralyses

The genetic results establish the skeletal muscle sodium channel as the locus of the genetic defects causing most of the inherited periodic paralyses of human skeletal muscle. While the molecular mechanisms by which these mutations cause sodium channel dysfunction remain unknown, comparison of the location of the mutations with the known functional regions of the sodium channel allows some suggestions for molecular mechanisms. These mutations all seem likely to alter the process of sodium channel inactivation, as has been demonstrated directly for hyperkalemic periodic paralysis.[106]

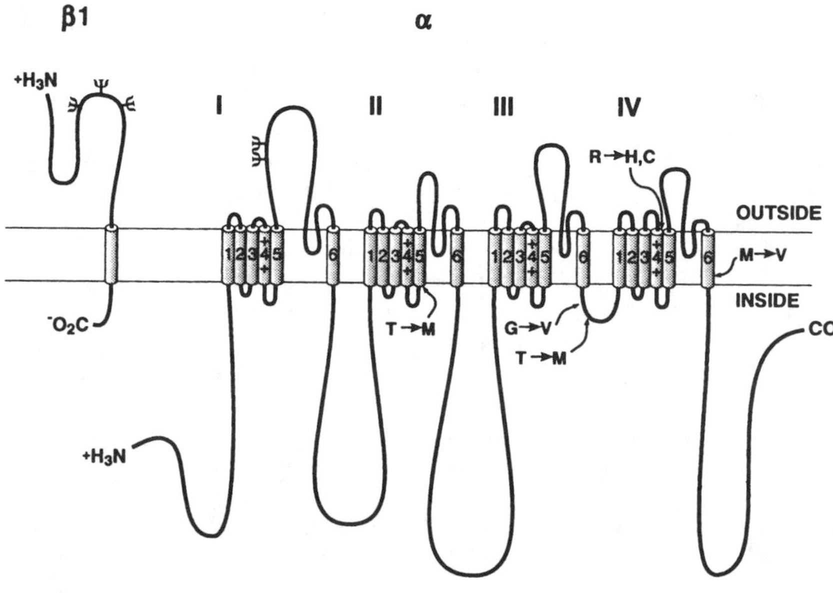

Figure 7.8. Sites of mutation in the periodic paralyses. Transmembrane folding model of the skeletal muscle sodium channel is illustrated with mutations that give rise to hyperkalemic periodic paralysis and paramyotonia congenita and their approximate locations in the amino acid sequence indicated. Individual mutations are described in the text.

Two paramyotonia congenita mutations are located within the inactivation gating loop near residues that are essential for inactivation and are likely to form the inactivation gate itself. Therefore, these mutations likely interfere with inactivation gating directly by altering the structure of the inactivation gate itself and impairing its closure. The two paramyotonia congenita mutations at the extracellular end of segment IVS6 neutralize a potential gating charge in the voltage-sensing S4 segment. If the S4 segment in domain IV plays a key role in coupling of activation to inactivation, a mutation at this site would be expected to modify the voltage dependence and time course of closure of the inactivation gate. The mutations that cause hyperkalemic periodic paralysis are surprisingly on the intracellular end of hydrophobic transmembrane segments. The mechanism by which extracellular potassium may influence channel function at these positions is unclear. However, the positions of these mutations at the intracellular ends of S5 and S6 segments suggest that they may form part of the "receptor" for the inactivation gate as it closes. Mutations at the intracellular end of an S5 segment in the *Shaker* potassium channel inhibit channel inactivation in a manner that is interpreted as modification of the inactivation gate receptor in a ball-and-chain inactivation mechanism.[98] Sequences at the intracellular end of segment IVS6 of the calcium channel are important for binding of phenylalkylamines which are intracellular open channel blockers and enhance channel inactivation.[113] These regions of the sodium channel may also surround the intracellular mouth of the sodium channel and serve as an intracellular receptor for closure of the inactivation gate. Structural changes in these sites are likely to affect inactivation by altering this receptor site structure.

7.5. CONCLUSION

Remarkable progress has be made in the past several years toward definition of the structural elements that are responsible for the basic functions of the voltage-gated ion channels. Beyond the new insights into the structure and function of the individual channels that have been gained, a striking commonality of functional design has emerged that allows a range of channel gating and permeability properties to be derived from subtle variations on a common structural theme. Moreover, the emerging molecular insights into ion channel function have allowed determination of the molecular basis for the period paralyses, a group of rare, autosomal dominant genetic diseases. Future investigations using the presently available methods of protein chemistry, mutagenesis, functional expression, and molecular modeling should delineate the primary structural basis for the essential functions of the ion channels and for the functional differences between different channel types and between wild-type and mutant sodium channels at ever higher resolution. However, a mechanistic understanding of these processes in terms of protein structure will require determination of the three-dimensional structure of a member of the ion channel family. While much effort is directed toward that goal presently, its attainment is likely to require many years of hard work.

REFERENCES

1. Catterall, W. A. (1986). Molecular properties of voltage-sensitive sodium channels. *Annu. Rev. Biochem.* **55:**953–985.
2. Wollner, D. A., Messner, D. J., and Catterall, W. A. (1987). Beta 2 subunits of sodium channels from vertebrate brain. Studies with subunit-specific antibodies. *J. Biol. Chem.* **262:**14709–14715.
3. McHugh-Sutkowski, E., and Catterall, W. A. (1990). β1 subunits of sodium channels. Studies with subunit-specific antibodies. *J. Biol. Chem.* **265:**12393–12399.
4. Agnew, W. S. (1984). Voltage-regulated sodium channel molecules. *Annu. Rev. Physiol.* **46:**517–530.
5. Barchi, R. L. (1988). Probing the molecular structure of the voltage-dependent sodium channel. *Annu. Rev. Neurosci.* **11:**455–495.
6. Noda, M., Shimizu, S., Tanabe, T., Takai, T., Kayano, T., Ideka, T., Takahashi, H., Nakayama, H., Kanaoka, Y., Minamino, N., Kangawa, K., Matsuo, H., Raftery, M., Hirose, T., Inayama, S., Hayashida, H., Miyata, T., and Numa, S. (1984). Primary structure of *Electrophorus electricus* sodium channel deduced from cDNA sequence. *Nature* **312:**121–127.
7. Noda, M., Ikeda, T., Kayano, T., Suzuki, H., Takeshima, H., Kurasaki, M., Takahashi, H., and Numa, S. (1986). Existence of distinct sodium channel messenger RNAs in rat brain. *Nature* **320:**188–192.
8. Kayano, T., Noda, M., Flockerzi, V., Takahashi, H., and Numa, S. (1988). Primary structure of rat brain sodium channel III deduced from the cDNA sequence. *FEBS Lett.* **228:**187–194.
9. Goldin, A. L., Snutch, T. P., Lubbert, H., Dowsett, A., Marshgall, J., Auld, V., Downey, W., Fritz, L. C., Lester, H. A., Dunn, R., Catterall, W. A., and Davidson, N. (1986). Messenger RNA coding for only the α subunit of the rat brain Na channel is sufficient for expression of functional channels in *Xenopus* oocytes. *Proc. Natl. Acad. Sci. USA* **83:**7503–7507.
10. Auld, V. J., Goldin, A. L., Krafte, D. S., Marshall, J., Dunn, J. M., Catterall, W. A., Lester, H. A., Davidson, N., and Dunn, R. J. (1988). A rat brain Na^+ channel alpha subunit with novel gating properties. *Neuron* **1:**449–461.
11. Auld, V. J., Goldin, A. L., Krafte, D. S., Catterall, W. A., Lester, H. A., Davidson, N., and Dunn, R. J. (1990). A neutral amino acid change in segment IIS4 dramatically alters the gating properties of the voltage-dependent sodium channel. *Proc. Natl. Acad. Sci. USA* **87:**323–327.
12. Isom, L. L., De Jongh, K. S., Patton, D. E., Reber, B. F. X., Offord, J., Charbonneau, H., Walsh, K., Goldin, A. L., and Catterall, W. A. (1992). Primary structure and functional expression of the β1 subunit of the rat brain sodium channel. *Science* **256:**839–842.
13. Noda, M., Ikeda, T., Suzuki, T., Takeshima, H., Takahashi, T., Kuno, M., and Numa, S. (1986). Expression of functional sodium channels from cloned cDNA. *Nature* **322:**826–828.
14. Suzuki, H., Beckh, S., Kubo, H., Yahagi, N., Ishida, H., Kayano, T., Noda, M., and Numa, S. (1988). Functional expression of cloned cDNA encoding sodium channel III. *FEBS Lett.* **228:**195–200.
15. Trimmer, J. S., Coopersmith, S. S., Tomiko, S. A., Zhou, J. Y., Crean, S. M., Boyle, M. B., Kallen, R. G., Sheng, Z. H., Barchi, R. L., Sigworth, F. J., Goodman, R. H., Agnew, W. S., and Mandel, G. (1989). Primary structure and functional expression of a mammalian skeletal muscle channel. *Neuron* **3:**33–49.
16. Cribbs, L. L., Satin, J., Fozzard, H. A., and Rogart, R. B. (1990). Functional expression of the rat heart I Na^+ channel isoform. Demonstration of properties characteristic of native cardiac Na^+ channels. *FEBS Lett.* **275:**195–200.

17. White, M. M., Chen, L., Kleinfield, R., Kallen, R. G., and Barchi, R. L. (1991). SkM2, a Na^+ channel cDNA clone from denervated skeletal muscle, encodes a tetrodotoxin-insensitive Na^+ channel. *Mol. Pharmacol.* **39:**604–608.
18. West, J. W., Scheuer, T., Maechler, L., and Catterall, W. A. (1992). Efficient expression of rat brain type IIA Na^+ channel α subunits in a somatic cell line. *Neuron* **8:**59–70.
19. Ragsdale, D., Scheuer, T., and Catterall, W. A. (1991). Frequency and voltage-dependent inhibition of type IIA Na^+ channels, expressed in a mammalian cell line, by local anesthetic, antiarrhythmic, and anticonvulsant drugs. *Mol. Pharmacol.* **40:**756–765.
20. Tsien, R., Elinor, P. T., and Horne, W. A. (1991). Molecular diversity of voltage-dependent calcium channels. *Trends Neurosci* **12:**349–354.
21. Campbell, K. P., Leung, A. T., and Sharp, A. H. (1988). The biochemistry and molecular biology of the dihydropyridine-sensitive calcium channel. *Trends Neurosci* **11:**425–430.
22. Catterall, W. A., Seager, M. J., and Takahashi, M. (1988). Molecular properties of dihydropyridine-sensitive calcium channels in skeletal muscle. *J. Biol. Chem.* **263:**3535–3538.
23. Catterall, W. A. (1991). Functional subunit composition of voltage-gated calcium channels. *Science* **253:**1499–1500.
24. Tanabe, T., Takeshima, H., Mikami, A., Flockerzi, V., Takahashi, H., Kangawa, K., Kojima, M., Matsuo, H., Hirose, T., and Numa, S. (1987). Primary structure of the receptor for calcium channel blockers from skeletal muscle. *Nature* **328:**313–318.
25. De Jongh, K. S., Merrick, D. K., and Catterall, W. A. (1989). Subunits of purified calcium channels: A 212-kDa form of α1 and partial amino acid sequence of a phosphorylation site of an independent β subunit. *Proc. Natl. Acad. Sci. USA* **86:**8585–8589.
26. De Jongh, K. S., Warner, C., Colvin, A. A., and Catterall, W. A. (1991). Characterization of the two size forms of the α1 subunit of skeletal muscle L-type calcium channels. *Proc. Natl. Acad. Sci. USA* **88:**10778–10782.
27. Ellis, S. B., William, M. E., Ways, N. R., Brenner, R., Sharp, A. H., Leung, A. T., Campbell, K. P., McKenna, E., Koch, W. J., Hui, A., Schwartz, A., and Harpold, M. M. (1988). Sequence and expression of mRNAs encoding the alpha 1 and alpha 2 subunits of a DHP-sensitive calcium channel. *Science* **241:**1661–1664.
28. Ruth, P., Rohrkasten, A., Biel, M., Bosse, E., Regulla, S., Meyer, H. E., Flockerzi, V., and Hofmann, F. (1989). Primary structure of the beta subunit of the DHP-sensitive calcium channel from skeletal muscle. *Science* **245:**1115–1118.
29. Jay, S. D., Ellis, S. B., McCue, A. F., Williams, M. E., Vedvick, T. S., Harpold, M. M., and Campbell, K. P. (1990). Primary structure of the gamma subunit of the DHP-sensitive calcium channel from skeletal muscle. *Science* **248:**490–492.
30. De Jongh, K. S., Warner, C., and Catterall, W. A. (1990). Subunits of purified calcium channels: α2 and δ are encoded by the same gene. *J. Biol. Chem.* **265:**14738–14741.
31. Mikami, A., Imoto, K., Tanabe, T., Niidome, T., Mori, Y., Takeshima, H., Narumiya, S., and Numa S. (1989). Primary structure and functional expression of the cardiac dihydropyridine-sensitive calcium channel. *Nature* **340:**230–233.
32. Perez-Reyes, E., Kim, H. S., Lacerda, A. E., Horne, W., Wei, X. Y., Rampe, D., Campbell, K. P., Brown, A. M., and Birnbaumer, L. (1989). Induction of calcium currents by the expression of the alpha 1-subunit of the dihydropyridine receptor from skeletal muscle. *Nature* **340:**233–236.
33. Tanabe, T., Beam, K. G., Powell, J. A., and Numa, S. (1988). Restoration of excitation–contraction coupling and slow calcium current in dysgenic muscle by dihydropyridine receptor complementary DNA. *Nature* **336:**134–139.
34. Singer, D., Biel, M., Lotan, I., Flockerzi, V., Hofmann, F., and Dascal, N. (1991). The roles of the subunits in the function of the calcium channel. *Science* **253:**1553–1557.
35. Varadi, G., Lory, P., Schultz, D., and Schwartz, A. (1991). Acceleration of activation and inactivation by the beta subunit of the skeletal muscle calcium channel. *Nature* **352:**159.
36. Lacerda, A. E., Kim, H. S., Ruth, P., Perez-Reyes, E., Flockerzi, V., Hofmann, F., Birnbaumer, L.., Brown, A. M. (1992). Normalization of current kinetics by interaction between the α1 and β subunits of the skeletal muscle dihydropyridine-sensitive Ca^{2+} channel. *Nature* **352:**527–530.
37. Wei, X., Perez-Reyes, E., Lacerda, A. E., Schuster, G., Brown, A. M., and Birnbaumer, L. (1991). Heterologous regulation of the cardiac Ca^{2+} channel α1 subunit by skeletal muscle β and γ subunits. Implications for the structure of cardiac L-type Ca^{2+} channels. *J. Biol. Chem.* **266:**21943–21947.
38. Ahlijanian, M. K., Westenbroek, R. E., and Catterall, W. A. (1990). Subunit structure and localization of dihydropyridine-sensitive calcium channels in mammalian brain, spinal cord and retina. *Neuron* **4:**819–832.
39. McEnery, M. W., Snowman, A. M., Sharp, A. H., Adams, M. G., and Snyder, S. H. (1992). Purified ω-conotoxin GVIA receptor of rat brain resembles a dihydropyridine-sensitive L-type calcium channel. *Proc. Natl. Acad. Sci. USA* **88:**11095–11099.
40. Mori, Y., Friedrich, T., Kim, M. S., Mikami, A., Kanaki, J., Ruth, P., Bosse, E., Hofmann, F., Flockerzi, V., Furuichi, T., Mikoshiba, K., Imoto, K., Tanabe, T., and Numa, S. (1991). Primary structure and functional expression from complementary DNA of a brain calcium channel. *Nature* **350:**398–402.
41. Williams, M. E., Feldman, D. H., McCue, A. F., Brenner, R., Velicelebi, G., Ellis, S. B., and Harpold, M. M. (1992). Structure and functional expression of α1, α2 and β subunits of a novel human neuronal calcium channel subtype. *Neuron* **8:**71–84.
42. Williams, M. E., Brust, P. F., Feldman, D. H., Patthi, S., Simerson, S., Maroufi, A., McCue, A. F., Velicelebi, G., Ellis, S. B., and Harpold, M. M. (1992). Structure and functional expression of an ω-conotoxin-sensitive human N-type calcium channel. *Science* **257:**389–395.
43. Papazian, D. M., Schwarz, T. L., Tempel, B. L., Jan, Y. N., and Jan, L. Y. (1987). Cloning of genomic and complementary DNA from *Shaker,* a putative potassium channel gene from *Drosophila. Science* **237:**749–753.
44. Tempel, B. L., Papazian, D. M., Schwarz, T. L., Jan, Y. N., and Jan, L. Y. (1987). Sequence of a probable potassium channel component encoded at Shaker locus of Drosophila. *Science* **237:**770–775.
45. Jan, L. Y., and Jan, Y. N. (1989). Voltage-sensitive ion channels. *Cell* **56:**13–25.
46. Pongs, O. (1992). Molecular biology of voltage-dependent potassium channels. *Physiol. Rev.* **72:**S69–S88.
47. Rehm, H., and Lazdunski, M. (1988). Purification and subunit structure of a putative K^+ channel protein identified by its binding properties for dendrotoxin 1. *Proc. Natl. Acad. Sci. USA* **85:**4919–4923.
48. Parcej, D. N., and Dolly, J. O. (1989). Dendrotoxin acceptor from bovine synaptic plasma membranes. Binding properties, purification and subunit composition of a putative constituent of certain voltage-activated K^+ channels. *Biochem. J.* **257:**899–903.
49. Newitt, R. A., Houamed, K. M., Rehm, H., and Tempe, B. L. (1991). Potassium channels and epilepsy: Evidence that the epileptogenic toxin, dendrotoxin, binds to potassium channel proteins. *Genetic Strategies in Epilepsy Research (Epilepsy Res. Suppl.)* **4:**263–273.
50. Rehm, H., and Lazdunski, M. (1988). Existence of different populations of the dendrotoxin 1 binding protein associated with neuronal K^+ channels. *Biochem. Biophys. Res. Commun.* **153:**231–240.
51. Rehm, H., Pelzer, S., Cochet, C., Chambaz, E., Tempel, B. L., Trauwein, W., Pelzer, D., and Lazdunski, M. (1989). Dendrotoxin-binding brain membrane protein displays a K^+ channel activity that is stimulated by both cAMP-dependent and endogenous phosphorylation. *Biochemistry* **28:**6455–6460.
52. Scott, V. E. S., Parcej, D. N., Keen, J. N., Findlay, J. B. C., and

Dolly, J. O. (1990). α-dendrotoxin acceptor from bovine brain is a K^+ channel protein. Evidence from the N-terminal sequence of its larger subunit. *J. Biol. Chem.* **265:**20094–20097.
53. Rehm, H., Newitt, R. A., and Tempel, B. L. (1988). Immunological evidence for a relationship between Dendrotoxin binding protein and the mammalian *Shaker* K^+ channel. *FEBS Lett.* **249:**224–228.
54. Trimmer, J. S., (1991). Immunological identification and characterization of a delayed rectifier K^+ channel polypeptide in rat brain. *Proc. Natl. Acad. Sci. USA* **88:**10764–10768.
55. Scott, V. E. S., Rettig, J., Parcej, D. N., Keen, J. N., Findlay, J. B. C., Pongs, O., and Dolly, J. O. (1993). Primary structure of a β subunit of α-dendrotoxin-sensitive K^+ channels from bovine brain. *Proc. Natl. Acad. Sci. USA* **91:**1637–1641.
56. Catterall, W. A. (1988). Structure and function of voltage-sensitive ion channels. *Science* **242:**50–61.
57. Numa, S. (1989). A molecular view of neurotransmitter receptors and ionic channels. *Harvey Lect.* **83:**121–165.
58. Jan, L. Y., and Jan, Y. N. (1992). Structural elements involved in specific K^+ channel functions. *Annu. Rev. Physiol.* **54:**537–555.
59. Armstrong, C. M. (1991). Sodium channels and gating currents. *Physiol. Rev.* **61:**644–682.
60. Guy, H. R., and Conti, F. (1990). Pursuing the structure and function of voltage-gated channels. *Trends Neurosci.* **13:**201–206.
61. Stühmer, W., Conti, F., Suzuki, H., Wang, X., Noda, M., Yahadi, N., Kubo, H., and Numa, S. (1989). Structural parts involved in activation and inactivation of the sodium channel. *Nature* **339:**597–603.
62. Papazian, D. M., Timpe, L. C., Jan, Y. N., and Jan, L. Y. (1991). Alteration of voltage-dependence of *Shaker* potassium channel by mutations in the S4 sequence. *Nature* **349:**305–310.
63. Liman, E. R., and Hess, P. (1991). Voltage-sensing residues in the S4 region of a mammalian K^+ channel. *Nature* **353:**752–756.
64. Logothetis, D. E., Movahedi, S., Satler, C., Lindpainter, K., Bisbas, D., and Nadal-Ginard, B. (1992). Incremental reductions of positive charge within the S4 region of a voltage-gated K^+ channel result in corresponding decreases in gating charge. *Neuron* **8:**531–540.
65. Auld, V. J., Goldin, A. L., Krafte, D. S., Catterall, W. A., Lester, H. A., Davidson, N., and Dunn, R. J. (1990). A neutral amino acid change in segment IIS4 dramatically alters the gating properties of the voltage-dependent sodium channel. *Proc. Natl. Acad. Sci. USA* **87:**323–327.
66. McCormack, K., Tanouye, M. A., Iverson, L. E., Lin, J.-W., Ramaswami, M., McCormack, T., Campanelli, J. T., Mathew, M. K., and Rudy, B. (1991). A role for hydrophobic residues in the voltage-dependent gating of *Shaker* K^+ channels. *Proc. Natl. Acad. Sci. USA* **88:**2931–2935.
67. Lopez, G. A., Jan, Y. N., and Jan, L. Y. (1991). Hydrophobic substitution mutations in the S4 sequence alter voltage-dependent gating in *Shaker* K^+ channels. *Neuron* **7:**327–336.
68. Tytgat, J., and Hess, P. (1992). Evidence for cooperative interactions in potassium channel gating. *Nature* **359:**420–423.
69. Hille, B. (1992). *Ion Channels of Excitable Membranes,* Sinauer Associates Inc., Sunderland, MA.
70. Noda, M., Suzuki, H., Numa, S., Stuhmer, W. (1990). A single point mutation confers tetrodotoxin and saxitoxin insensitivity on the sodium channel II. *FEBS Lett.* **259:**213–216.
71. Terlau, H., Heinemann, S. H., Stühmer, W., Pusch, M., Conti, F., Imoto, K., and Numa, S. (1991). Mapping the site of block by tetrodotoxin and saxitoxin of sodium channel II. *FEBS Lett.* **293:** 93–96.
72. Satin, J., Kyle, J. W., Chen, M., Bell, P., Cribbs, L. L., Fozzard, H. A., and Rogart, R. B. (1992). A point mutation of TTX-resistant cardiac Na channels confers three properties of TTX-sensitive Na channels. *Science* **256:**1202–1205.
73. Backx, P. H., Yue, D. T., Lawrence, J. H., Marban, E., and Tomaselli, G. F. (1992). Molecular localization of an ion-binding site within the pore of mammalian sodium channels. *Science* **257:** 248–251.
74. Heinemann, S. H., Terlau, H., and Imoto, K. (1992). Molecular basis for pharmacology differences between brain and cardiac sodium channels. *Pflueger's Arch.* **422:**90–92.
75. MacKinnon, R., and Miller C. (1989). Mutant potassium channels with altered binding of charybdotoxin, a pore-blocking peptide inhibitor. *Science* **245:**1382–1385.
76. MacKinnon, R., Heginbotham, L., and Abramson, T. (1990). Mapping the receptor site for charybdotoxin, a pore-blocking potassium channel inhibitor. *Neuron* **5:**767–771.
77. MacKinnon, R., and Yellen, G. (1990). Mutations affecting TEA blockade and ion permeation in voltage-activated K^+ channels. *Science* **250:**276–279.
78. Heginbotham, L., and MacKinnon, R. (1992). The aromatic binding site for tetraethylammonium ion on potassium channels. *Neuron* **8:**483–491.
79. Kavanaugh, M. P., Varnum, M. D., Osborne, P. B., Christie, M. J., Busch, A. G., Adelman, (1991). Interaction between tetraethylammonium and amino acid residues in the pore of cloned voltage-dependent potassium channels. *J. Biol. Chem.* **266:**7583–7587.
80. Striessnig, J., Glossman, H., and Catterall, W. A. (1990). Identification of a phenylalkylamine binding region within the α1 subunit of skeletal muscle Ca^{2+} channels. *Proc. Natl. Acad. Sci. USA* **87:**9108–9112.
81. Catterall, W. A., and Striessnig, J. (1992). Receptor sites for Ca^{2+} channel antagonists. *Trends Pharmacol. Sci.* **13:**256–262.
82. Yellen, G., Jurman, M. E., Abramson, T., and MacKinnon, R. (1991). Mutations affecting internal TEA blockade identify the probable pore-forming region of a K^+ channel. *Science* **251:**939–942.
83. Choi, K. L., Mossman, C., Aubé, J., and Yellen, G. (1993). The internal quaternary ammonium receptor site of *Shaker* potassium channels. *Neuron* **10:**533–541.
84. Hartmann, H. A., Kirsch, G. E., Drewe, J. A., Taglalatela, M., Joho, R. H., and Brown, A. M. Exchange of conduction pathways between two related K^+ channels. *Science* **251:**942–944.
85. Yool, A. J., and Schwarz, T. L. (1991). Alteration of ionic selectivity of a K^+ channel by mutation of the H5 region. *Nature* **349:**700–704.
86. Kirsch, G. E., Drewe, J. A., Hartmann, H. A., Taglialatela, M., deBiasi, M., Brown, A. M., and Joho, R. H. (1992). Differences beween the deep pores of K^+ channels determined by an interacting pair of nonpolar amino acids. *Neuron* **8:**499–505.
87. Heginbotham, L., Abramson, A., and MacKinnon, R. (1992). A functional connection between the pores of distantly related ion channels as revealed by mutant K^+ channels. *Science* **258:**1152–1155.
88. Heinemann, S. H., Terlau, H., Stühmer, W., Imoto, K., and Numa, S. (1992). Calcium channel characteristics conferred on the sodium channel by single mutations. *Nature* **356:**441–443.
89. Vassilev, P. M., Scheuer, T., and Catterall, W. A. (1988). Identification of an intracellular peptide segment involved in sodium channel inactivation. *Science* **241:**1658–1661.
90. Vassilev, P., Scheuer, T., and Catterall, W. A. (1989). Inhibition of inactivation of single sodium channels by a site-directed antibody. *Proc. Natl. Acad. Sci. USA* **86:**8147–8151.
91. Moorman, J. R., Kirsch, G. E., Brown, A. M., and Joho, R. H. (1990). Changes in sodium channel gating produced by point mutations in a cytoplasmic linker. *Science* **250:**688–691.
92. Patton, D. E., West, J. W., Catterall, W. A., and Goldin, A. L. (1992). Amino acid residues required for fast sodium channel inactivation. Charge neutralizations and deletions in the III–IV linker. *Proc. Natl. Acad. Sci. USA* **89:**10905–10909.
93. West, J. W., Patton, D. E., Scheuer, T., Wang, Y., Goldin, A. L., and Catterall, W. A. (1992). A cluster of hydrophobic amino acid residues required for fast Na^+ channel inactivation. *Proc. Natl. Acad. Sci. USA* **89:**10910–10914.

94. Scheuer, T., West, J. W., Wang, Y. L., and Catterall, W. A. (1993). Effects of amino acid hydrophobicity at position 1489 on sodium channel inactivation. *Biophys. J.* **64:**A88.
95. Hoshi, T., Zagotta, W. N., and Aldrich, R. W. (1990). Biophysical and molecular mechanisms of *Shaker* potassium channel inactivation. *Science* **250:**533–538.
96. Zagotta, W. N., Hoshi, T., and Aldrich, R. W. (1990). Restoration of inactivation in mutants of *Shaker* potassium channels by a peptide derived from ShB. *Science* **250:**568–571.
97. Demo, S. D., and Yellen, G. (1991). The inactivation gate of the *Shaker* K^+ channel behaves like an open-channel blocker. *Neuron* **7:**743–753.
98. Isacoff, E. Y., Jan, Y. N., and Jan, L. Y. (1991). Putative receptor for the cytoplasmic inactivation gate in the *Shaker* K^+ channel. *Nature* **353:**86–90.
99. Hoshi, T., Zagotta, W. N., and Aldrich, R. W. (1991). Two types of inactivation in *Shaker* K^+ channels: Effects of alterations in the carboxy-terminal region. *Neuron* **7:**547–556.
100. Rudy, B. (1978). Slow inactivation of the sodium conductance in squid giant axons. Pronase resistance. *J. Physiol. (London)* **283:** 1–21.
101. Patton, D. E., West, J. W., Catterall, W. A., and Goldin, A. L. (1993). A peptide segment critical for sodium channel inactivation functions as an inactivation gate in a potassium channel. *Neuron* **11:**1–20.
102. Rüdel, R., and Ricker, K. (1985). The primary periodic paralyses. *Trends Neurosci.* **8:**467–470.
103. Lehmann-Horn, F., Rüdel, R., Ricker, K., Lorkovic, H., Dengler, R., and Hopf, H. C. (1983). Two cases of adynamia episodica hereditaria: In vitro investigation of the muscle cell membrane and contraction parameters. *Muscle Nerve* **6:**113–121.
104. Lehamnn-Horn, F., Rüdel, R., and Ricker, K. (1987). Membrane defects of paramyotonia congenita (Eulenberg). *Muscle Nerve* **10:** 633–641.
105. Rüdel, R., Ruppersberg, J. P., and Spittelmeister, W. (1989). Abnormalities of the fast sodium current in myotonic dystrophy, recessive generalized myotonia, and adynamica episodica. *Muscle Nerve* **12:** 281–287.
106. Cannon, S. C., Brown, R. H., and Corey, D. P. (1991). A sodium channel defect in hyperkalemic periodic paralysis: Potassium-induced failure of inactivation. *Neuron* **6:**619–626.
107. Fontaine, B., Khurana, T. S., Hoffman, G. A., Bruns, G. A. P., Haines, J. L., Trofatter, J. A., Hanson, M. P., Rich, J., McFarlane, H., Yasek, D. M., Romano, D., Gusella, J. F., and Brown, R. H., Jr. (1990). Hyperkalemic periodic paralysis and the adult muscle sodium channel α subunit gene. *Science* **250:**1000–1002.
108. Ptacek, L. J., Tyler, F., Trimmer, J. S., Agnew, W. S., and Leppert, M. (1991). Analysis in a large hyperkalemic periodic paralysis pedigree supports tight linkage to a sodium channel locus. *Am. J. Hum. Genet.* **49:**378–382.
109. Ptacek, L. J., George, A. L., Griggs, R. C., et al. (1991). Identification of a mutation in the gene causing hyperkalemic periodic paralysis. *Cell* **67:**1021–1027.
110. Rojas, C. V., Wang, J., Schwartz, S., Hoffman, E. P., Powell, B. R., and Brown, R. H., Jr. (1991). A Met-to-Val mutation in the skeletal muscle sodium channel α subunit in hyperkalemic periodic paralysis. *Nature* **354:**387–389.
111. Ptacek, L. J., George, A. L., Barchi, R. L., Griggs, R. C., Riggs, J. E., Robertson, M., and Leppert, M. F. (1992). Mutations in an S4 segment of the adult skeletal muscle sodium channel cause paramyotonia congenita. *Neuron,* **8:**891–897.
112. McClatchey, A. I., Van Den Bergh, P., Pericak-Vance, M. A., Raskind, W., Verellen, C., McKenna-Yasek, D., Rao, K., Haines, J. L., Bird, T., Brown, Jr., R. H., and Gusella, J. F. (1992). Temperature-sensitive mutations in the III–IV cytoplasmic loop region of the skeletal muscle sodium channel gene in paramyotonia congenita. *Cell* **68:**769–774.
113. Striessnig, J., Glossmann, H., and Catterall, W. A. (1990). Identification of a phenylalkylamine binding region within the $\alpha 1$ subunit of skeletal muscle calcium channels. *Proc. Natl. Acad. Sci. USA* **87:** 9108–9112.

8

Nicotinic Receptors in the Central Nervous System

C. K. Ifune and Joe Henry Steinbach

8.1 INTRODUCTION

This chapter will review the structure, pharmacology, and physiology of neuronal nicotinic receptors, then consider the evidence that dysfunction of nicotinic receptors at interneuronal synapses is associated with any particular disorder. Only studies of receptors in vertebrates will be considered.

Nicotinic receptors at the skeletal neuromuscular junction underlie rapid excitatory transmission. Many studies of muscle-type nicotinic receptors (AChR) have provided a wealth of quantitative functional data,[1] and we have a reasonably quantitative picture of synaptic transmission at this synapse.[2–5] The cleft concentration of nerve-released ACh reaches a peak greater than 100 μM in less than 100 μsec. Binding to AChR occurs rapidly (driven by this high concentration), and the peak of the postsynaptic current occurs in less than 1 msec after the presynaptic action potential. Any free ACh in the cleft is rapidly hydrolyzed by acetylcholinesterase so the cleft concentration of ACh falls from its peak within another few hundred microseconds. Hence, the decay of the postsynaptic current is largely determined by the duration of the open-channel bursts produced by AChR which bound ACh during the first 100 μsec of transmission. The conductance increase produced during transmission is so large that essentially every presynaptic action potential results in a muscle action potential, so the neuromuscular synapse is an extremely fast and effective excitatory synapse.

Nicotinic receptors also underlie fast excitatory transmission at interneuronal synapses in the periphery. Although there are less quantitative data, transmission at autonomic synapses is qualitatively quite similar to neuromuscular transmission.[6]

In the central nervous system, however, fast excitatory transmission is carried out by excitatory amino acids and the family of glutamate ionotropic receptors. A central question, therefore, exists regarding the role of nicotinic receptors in the CNS, as discussed below (see Section 8.5).

Neuronal nicotinic receptors in the CNS have been demonstrated physiologically (from responses to nicotine and other specific activators), biochemically (as high-affinity binding sites for nicotine or other specific ligands), immunochemically (as epitopes for specific antisera), and from *in situ* hybridization of probes to nucleic acid sequences. In terms of overall cholinergic systems, nicotinic receptors are less prevalent than muscarinic receptors in the CNS, less than one-tenth the number,[7–9] although the pharmacological diversity of nicotinic receptors has made it difficult to obtain accurate total numbers.

8.2. STRUCTURE: HOW MANY KINDS OF NICOTINIC RECEPTORS?

Cloning of nucleic acid sequences, beginning with the muscle nicotinic receptor expressed in electric fish electrocytes, has identified a gene family of transmitter-gated membrane channels. The family includes the muscle nicotinic receptor subunits, neuronal nicotinic receptor subunits, neuronal 5-hydroxytryptamine-3 receptor subunit, glycine receptor subunits, and γ-aminobutyric acid-A receptor subunits.[10,11] These transmitter-gated membrane channels are clearly distinct from the muscarinic receptors which belong to the family of G-protein-coupled transmitter receptors (see Chapter 17), which includes muscarinic acetylcholine receptors.

8.2.1. Primary Amino Acid Sequence and Membrane Disposition

When the primary amino acid sequences of subunits in this family are aligned, there is an overall pattern of residues shown in Figure 8.1. Each subunit is thought to cross the cell membrane several times (Figure 8.2a), and the complete receptor is formed by side-to-side association of five subunits in a doughnut (Figure 8.2b,c). There is still active research about the disposition of subunits with respect to the membrane and

C. K. Ifune • Washington University School of Medicine, St. Louis, Missouri 63110. **Joe Henry Steinbach** • Department of Anesthesiology, Washington University School of Medicine, St. Louis, Missouri 63110.

Molecular Biology of Membrane Transport Disorders, edited by S.G. Schultz *et al.* Plenum Press, New York, 1996

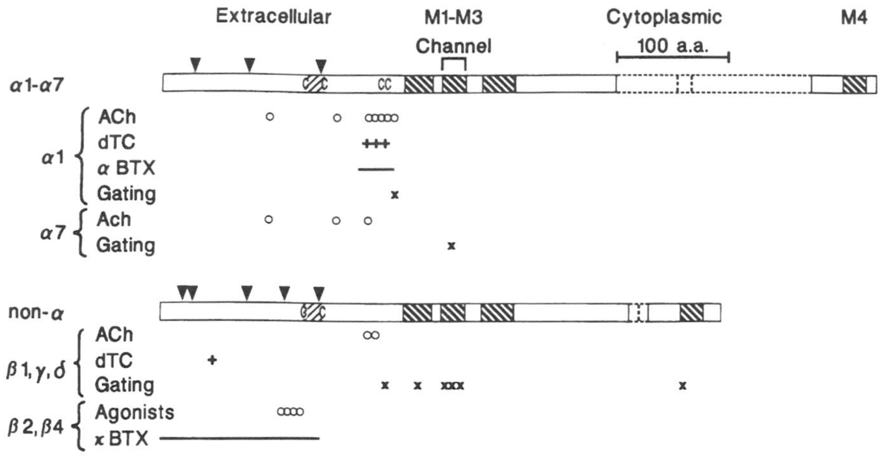

Figure 8.1. Schematic drawing of the primary amino acid sequences for the α (top) and non-α (bottom) subunits of nicotinic receptors. There are four regions: an N-terminal, hydrophilic stretch which lies on the extracellular side of the membrane (left end). This region contains sites for *N*-linked glycosylation (arrowheads) and residues involved in forming binding sites for agonists (circles, "ACh," "Agonists"), competitive antagonists (pluses, "dTC"), and snake neurotoxins (lines, "αBTX," "κBTX"). There is then a section with three closely spaced regions of greater hydrophobicity, the membrane-spanning segments M1, M2, and M3. Residues in the M2 segment (bracketed) form the lining of the ion channel, and also the binding site for channel blocking drugs. M3 is followed by a variable-length hydrophilic region (indicated by dashed lines), which lies on the cytoplasmic side of the membrane. The region contains consensus sequences for phosphorylation. Finally, there is a fourth membrane-spanning segment followed by short extracellular tail. Residues throughout the primary sequence have been shown to have some effects on channel gating or desensitization (crosses, "Gating"). (See text and Refs. 1, 15, 16.)

the arrangement of the subunits in the receptor.[12,13] Recent reviews of structural studies include Refs. 14–16.

8.2.2. The Ion Channel

All nicotinic receptors studied to date have channels which are cation selective, but discriminate poorly among cations. Apparently, the major determinant in permeability for cations is the size of the ion.[17–19] From the results of a variety of biophysical approaches, a model has been developed for the structure of the open AChR channel which features wide vestibules at each end, tapering to a short narrow region (see Figure 8.2c).[20–22] The narrow region is thought to be about 0.65 nm square,[18,19] based on the sizes of permeant ions, and is thought to contain a single saturable site for binding permeant ions.[21,22]

Structural studies indicate that the ion channel is formed at the central axis of the receptor, with portions of the channel

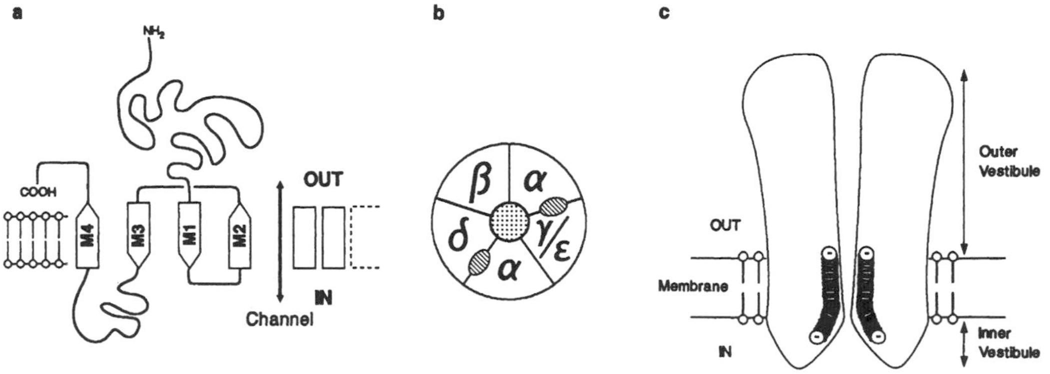

Figure 8.2. Three cartoons of receptor structure.
(a) The likely disposition of a single subunit across the cell membrane is diagrammed (see also Figure 8.1). The arrows on regions M1–M4 indicate the postulated N-terminal-to-C-terminal direction of the segment as it crosses the membrane. The ion channel is indicated by the double-headed arrow, next to the M2 segment.
(b) The complete AChR is formed by a doughnutlike association of five subunits, shown here in a view from the extracellular side. For the muscle receptor, the binding sites for ACh are formed at the interfaces between the α1 subunit and adjacent δ and γ/ε subunits (oval regions). The ion channel lies in the middle of the pentamer (hatched).
(c) The pentamer protrudes on both sides of the membrane, as shown in this side view of a cross section. There is a larger vestibule on the external surface, a smaller one on the internal surface. The channel lining is formed by uncharged amino acid residues in the M2 segment (indicated by the coils), but there are rings of negatively charged residues at either end (minus signs). The channel itself has a short narrow portion, with the greatest constriction probably near the cytoplasmic end. (See the text and, for panels a and b, Refs. 15 and 16; panel c, Refs. 13 and 20.)

lining contributed by each of the subunits. Selectivity for cations is thought to arise from two sources: a negative charge in the vestibules which would serve to enhance the local cation concentration,[20] and the presence of particular amino acid residues in the M2 region of the sequence.[23,24] Evidence to support the model in which M2 forms part of the channel lining includes studies in which the alteration of three residues in and near the M2 region into the residues found in corresponding region of the $GABA_A$ and glycine receptors converted the normally cation-selective channel formed by chicken nicotinic α7 subunits into an anion-selective channel.[24]

8.2.3. The Quarternary Structure Is Pentameric

The quaternary structure of the receptors formed by subunits in this family is most likely a pentameric doughnut of subunits (Figure 8.2b). Again, this is most clearly documented for the muscle nicotinic receptor.[15,16] However, data for a chicken neuronal receptor expressed in *Xenopus* oocytes are also consistent with a pentameric structure (see Section 8.2.5).

The finding that the receptor is a pentamer raises the question of subunit stoichiometry. The complexity of the situation can be exemplified by a brief discussion of the assembly of muscle nicotinic receptors. Five subunits have been identified as contributing to muscle nicotinic receptors (see Figure 8.2b). Three are apparently found in all receptors: α1, β1, and δ. There are two copies of α1 and one each of β1 and δ. The fifth subunit is either the ε subunit (in adult tissues) or the γ subunit (in developing or injured tissues).[25] A number of studies have supported the idea that the binding sites for ACh and for competitive antagonists lie at the interfaces between the α1 and δ subunits and the α1 and γ/ε subunits.[26–29] This implies that the δ and γ/ε subunits occupy equivalent positions with respect to the α1 subunit, as shown in Figure 8.2b. Assembly of functional AChR composed of only two or three subunits can also occur,[30–35] although apparently less efficiently. There is no indication that muscle receptors assemble in the absence of α1, nor that α1 subunits assemble by themselves. There is evidence that AChR assembly might proceed initially by formation of α1–γ and α1–δ pairs, followed by overall assembly with β1.[36,37] Likely, the preferred subunit composition is favored by asymmetric faces on the β1 subunit which complement faces on the appropriate partner. In summary, the muscle AChR is a pentamer that normally contains four different subunits, but the γ and ε isoforms readily swap and functional pentamer containing only three subunits can be formed.

8.2.4. Neuronal Nicotinic Receptor Subunits: Primary Sequences

Two major genera of neuronal nicotinic subunits have been identified. A set of subunits which have sequence similarity to the muscle α1 subunit have been numbered α2 through α8.[38–45] These subunits all have a pair of neighboring cysteine residues, which have been shown to be part of the ACh-binding site of the muscle α1 subunit (Figure 8.1; see Section 8.3.3.2). Members of a second genus have been named either non-α1 through non-α3 (for chicken sequences[42,44]) or β2 through β4 (for rat[46–48]). The decision to use a β designation was based on the observation that the first rat sequence cloned, β2 could result in functional AChR when coexpressed with muscle α1, γ, and δ subunits,[46] although sequence analysis indicates that the neuronal β subunits are not closely related to the muscle β1 subunit. For brevity we will use the terms β2, β3, and β4 for both the mammalian subunits and chicken subunits.

Within the neuronal α genus, two subgenera have been separated, based on pharmacological criteria. The α7 and α8 subunits[41,43] were first identified as subunits which would bind the snake neurotoxin, α-bungarotoxin (αBTX[45]), which binds tightly to the muscle AChR. Receptors containing the α2 through α5 subunits do not bind α-neurotoxins.

8.2.5. Neuronal Nicotinic Receptor Subunits: Assembly by Expression

All of the neuronal subunits have been tested for expression in *Xenopus* oocytes, and assayed by the expression of a functional AChR. Hence, not only assembly but also generation of a detectable response must occur.

The α7 subunit forms functional AChR when expressed without other subunits,[41,49] although to date no expression of α8 has been reported. When α7 has been expressed with non-α subunits, the only functional receptors have properties of α7 homooligomers,[41,49] suggesting that it does not coassemble with other known subunits. The rat[39] but not chick[50] α4 subunit also can form a functional AChR, although much less efficiently.

The α2, α3, and α4 subunits will assemble with either the β2 or β4 subunits when expressed in pairs (see Table 8.1). The α5[40] or β3[47] subunit has not formed a functional AChR in any of the pairwise combinations tested. When injected with muscle receptor subunits, the rat β2[46] and β4[48] subunits will substitute for the muscle β1 subunit. None of the α2, α3, α4, α5, or α7 subunits can substitute for the α1 subunit, when injected with β1, γ, and δ subunit sequences (J. Patrick, personal communication).

These studies suggest that the α7/α8 subunits form one assembly class, and the α2/α3/α4 subunits form a second class which coassembles with β2/β4. The rules for assembly are clearly broad enough to generate a wide variety of neuronal nicotinic receptors.

Two complementary studies of chicken subunits expressed in *Xenopus* oocytes indicate that one form of the receptor has composition $(\alpha4)_2(\beta2)_3$. In one, site-directed mutagenesis of a residue in the channel-lining region produced AChR of altered channel conductance; analysis of the prevalence of different single-channel conductance levels provided results consistent with the idea that two α4 and three β2 subunits contribute to the channel.[51] In the second, quantitative analysis of the incorporation of radiolabeled amino acids indicated that the ratio of subunits in the surface AChR was 2:2.92, α4:β2.[52]

The simplicity of this picture is complicated by another expression study, however. When the rat α2/β2 pair is ex-

Table 8.1. Pharmacological Properties of Native and Expressed Nicotinic Receptors[a]

Source	αBTX	κBTX	Agonist	Log–log slope	ACh EC_{50}(μM)	Subunits	References
			Part 1: Receptor in oocytes				
Chick							
α3β2	−	+	N.D.	1.4	5.2		100
α4β2	−	−	N=A=D	1.5	0.3–0.8		50, 101
α7	+	+	N>C>>A	1.4	115		49, 251
Rat							
α2β2	−	+[b]	N>D=A>C	1.2	N.D.		38, 85, 97, 252
α2β4	−	+[b]	C>N>A>D				85, 97
α3β2	−	+	D=A>N>C	1.4–1.5	354		39, 43, 97, 120
α3β4	−	+[b]	C>N=A=D	1.8	30		43, 49, 97, 120
α4β2	−	+	N=A>D>C	1.1–1.4	N.D.		43, 97
α4β4	N.D.	+[b]	C>N>A>D	N.D.	N.D.		97, 120
α7	+	N.D.	N>C>D>A	N.D.	N.D.		41
			Part 2: Native receptor				
Chick							
Ciliary ganglion	−	+	N.D.	1.7	100	α3, α4, α5, α7 β2, β4	84, 134, 253–255
Sympathetic ganglion	−	+	N.D.	N.D.	30	α3, α4, α5, α7 β2, β4	65, 256
Rat							
Hippocampus							
"Fast" current	+	+	N>D>C>A	1.3	129	α2, α3, α4, α7	41, 82, 83, 257
"Slow" current	−	−	A>N	1.5	65	β2	
Interpeduncular nucleus							
Presynaptic afferents	−	−	N>C>A>D	N.D.	100	N.D.	178, 258
Cell soma	−	−	N.D.	N.D.	28	α2, α3, α4, α5 β2	
Medial habenula	−	−	D>N>C>A	N.D.	60–80	α3, α4 β2, β3, β4	48, 115, 193, 257
Cardiac ganglion	−	+	N.D.	N.D.	N.D.	N.D.	19, 259
Retinal ganglion	−	+	N.D.	N.D.	N.D.	α3, α4 β2	165, 257
Sympathetic ganglion	−	+	N.D.	N.D.	45	N.D.	151, 260, 261
PC12 cells	−	+	C>N>A	N.D.	40	α3, α5, α7 β2, β3 β4	46, 80, 262, 263

[a]Column 1 gives the source of the receptors and the subunits injected into *Xenopus* oocytes in Part 1. Columns 2 and 3 show the sensitivity of the various receptors to αBTX and κBTX: − indicates that the receptor is not blocked and +, that it is blocked by the toxin. Column 4 lists the rank order of agonist sensitivity (N = nicotine, A = ACh, C = cytisine, and D = DMPP). Column 5 gives the log–log slopes of the dose–response relationship for ACh. For the receptors expressed in oocytes, the slope was determined at low agonist concentrations so that the effects of receptor desensitization were minimized. Column 6 gives the EC_{50}'s for ACh. For the native receptors, column 7 lists the subunit mRNAs that have been found to be expressed by the various cells or the cell type in Part 2. "N.D." indicates that the data are not available. Data on single-channel conductances are not listed, as determinations were made in varied ion concentrations, which make comparisons problematical.

[b]Block of these receptors is characterized by quick onset and recovery and is kinetically distinct from the block by κBTX of α3β2 receptors.

pressed in *Xenopus* oocytes, channels of two conductance classes are observed.[53] If the ratio of mRNA for the two subunits is altered, the prevalence of the classes is altered, which was interpreted to mean that the number of copies of the two subunits in a single AChR could be altered by the prevalence of the subunits. This observation is only to be expected from consideration of the symmetry of the subunit relationships in a circular pentamer. Clearly, both the α and the β subunit must have complementary surfaces on both sides for a pentamer to be possible.

Another piece of evidence indicating the presence of more than one ACh-binding subunit is that the response to ACh increases more than linearly with ACh concentration (see Table 8.1).

Overall, the expression studies indicate that neuronal AChR are pentamers, that the functional AChR studied to date have more than one α subunit, that not all subunits will coassemble (at least in pairs), and that the subunit stoichiometry can differ even for the same pair of subunits.

8.2.6. Neuronal Nicotinic Receptor Subunits: Assembly *in Vivo*

Biochemical studies have examined the composition of two of the major CNS neuronal nicotinic receptors. These are the high-affinity nicotine-binding sites in the brain and sites which bind the snake neurotoxin, BTX.[54–56] The nicotine-binding sites have been purified by immunoprecipitation and the subunits separated by electrophoresis. Amino-terminal sequences of peptides isolated from the two major protein bands were found to correspond to the α4[57] and β2 subunits.[58] Similarly, the high-affinity nicotine-binding sites from rat brain have been examined by immunoprecipitation with subunit-specific antisera; only antisera to α4 and β2 were able to precipitate the sites (sera to α2, α3, α5, α7, β3, and β4 subunits were tested[59]). These studies indicate that α4 and β2 subunits are major components of the high-affinity nicotine-binding sites.

A recent study of chicken brain nicotinic receptors used specific immunoprecipitation followed by immunoblotting of subunits separated by gel electrophoresis.[60] These results confirm that α4 is a major component for the total population of CNS nicotinic receptors, and that α3 and α5 both occur in a smaller proportion of the receptors. More surprisingly, the data clearly show that a small proportion of the receptors contain both α4 and α5, or α3 and α5 subunits. It is not known whether the failure of α5 to produce a functional receptor in expression studies means that it does not assemble by itself with β subunits, or whether the assembled receptor is simply nonfunctional, but the immunoprecipitation studies strongly indicate that it is a part of some brain nicotinic receptors. The β2 subunit was also detected in the precipitated receptors.

CNS αBTX-binding sites have been purified and characterized from chicken retina and brain. These sites contain either α7 alone, α8 alone, or both α7 and α8 subunits.[61] Other subunits have not been identified in these studies, although earlier work had suggested that there might be as many as four subunits distinguishable on the basis of electrophoretic migration.[62,63] The possibility of additional subunits is also suggested by the finding that αBTX-binding sites isolated from chicken brain have different pharmacological properties than homooligomeric α7 receptors isolated after expression in oocytes.[64] None of the immunoprecipitation studies have provided any evidence that the α7 or α8 subunits form detectable numbers of complexes with β2, α3, α4, or α5 subunits.[59,60]

Two studies of peripheral ganglionic neurons in the chicken indicate some additional features of assembly. Chicken sympathetic ganglion neurons express message for α3, α4, α5, α7, β2, and β4.[65] Exposure of cultured neurons to an antisense sequence to α3 greatly reduces the overall sensitivity of the cells to ACh (by 50–90%), suggesting that the α3 subunit occurs in many functional AChR. The pharmacology of the remaining response changes so that αBTX now blocks most of the response, although in control cells the α-toxin has no effect. This observation suggests that some of the functional AChR contain α7, after reduction of the amount of α3 subunit available. Exposure to α7 antisense sequences alone has little effect, so the assembly and/or functional role of α7 subunits in the control situation is obscure. Other studies have shown that chick sympathetic neurons express more than one functional type of AChR, and that the cells can apparently segregate the different types on the surface of the cell body.[66]

Chicken ciliary ganglion neurons express both an αBTX-binding site and nicotinic receptors; the receptors are localized immediately under synaptic terminals whereas the α-toxin binding sites are excluded from the postsynaptic region.[67–69] The αBTX sites include the α7 subunit and no other identified subunits, whereas the postsynaptic receptors contain at least the α3, β4, and α5 subunits, but not α7.[70] These data demonstrate that cells can assemble and separately localize different types of nicotinic receptors, selecting some subunits from the total complement synthesized by the cells.

These studies of AChR expressed *in vivo* suggest that there are preferred compositions for nicotinic receptors. The studies in peripheral ganglia indicate that a single cell can assemble receptors of different composition, although some implications of the results are not clear. What are the mechanisms and limitations for preferentially assembling receptors from particular subunits? The subunit composition has major effects on the physiological and pharmacological properties of the AChR (as will be described below), so rules for assembly may play a major role in shaping the phenotype generated from a given set of expressed subunit genes. One important observation from the studies of ganglionic neurons is that cells are able to separately localize AChR of different composition.

8.2.7. Receptor Structure: A Summary

Neuronal nicotinic receptors are pentameric, contain more than one α subunit (which can be different), and can be differentially localized on the surface of neurons. Although a major form of the receptor (the brain high-affinity nicotinebinding component) appears to have a preferred subunit stoichiometry, it is clear that subunit stoichiometry can be affected by subunit protein levels in expression systems. It is possible that the α7 and α8 subunits only contribute to the αBTX-binding proteins, not to other receptor types.

8.3. PHARMACOLOGY OF NEURONAL NICOTINIC RECEPTORS

Nicotinic receptors are pharmacologically defined as a class by activation by acetylcholine and nicotine with insensitivity to muscarine. However, quantitative analysis of agonist and antagonist actions shows differences between receptors on different cells or receptors composed of different subunits. Because of the heterogeneous nature of neuronal nicotinic receptors, there is no simple set of pharmacological probes that can reliably distinguish between classes of nicotinic receptors. In addition, some "nicotinic" blocking agents can show higher-affinity interactions with other transmitter-gated channels of the gene family (see Section 8.3.2). Nicotinic pharmacology has recently been reviewed in Refs. 71 and 72.

This section begins with some fundamentals of drug action at nicotinic receptors, reviews general aspects of nicotinic pharmacology, and discusses the relationship between receptor structure and pharmacology.

8.3.1. Sites of Drug Action

Studies at the neuromuscular junction and peripheral ganglionic synapses have distinguished three sites of action for pharmacological agents. The first is the ACh-binding site. Nicotinic agonists bind to this site and this results in channel activation. The basic scheme for channel activation was proposed a number of years ago[1,73] for the neuromuscular junction. There are two binding sites on a receptor, both of which must be occupied by agonist before the channel has a high probability of opening (see Figure 8.3a). Competitive antagonists interact with the ACh-binding site, but the interaction does not result in channel opening.

The second site is thought to be located in the ion channel, and to be revealed when the channel is opened by an agonist. When a drug binds to this site, ion movement through the channel is prevented (see Figure 8.3b). In the initial formulation of this model for drug action, channels were prevented from closing until the drug left its site.[3,74] However, analysis of blocking action suggests that some of the drugs show more complicated activities such as blocking when the channel is closed[75] or becoming trapped within the channel when the channel closes.[76] "Open channel blockers" are a heterogeneous group of molecules. At the muscle receptor, they include nicotinic agonists, such as ACh and nicotine, and many other molecules that contain a nitrogen and a relatively hydrophobic portion.[14,77] Channel blocking drugs show a noncompetitive block of ACh-elicited currents.

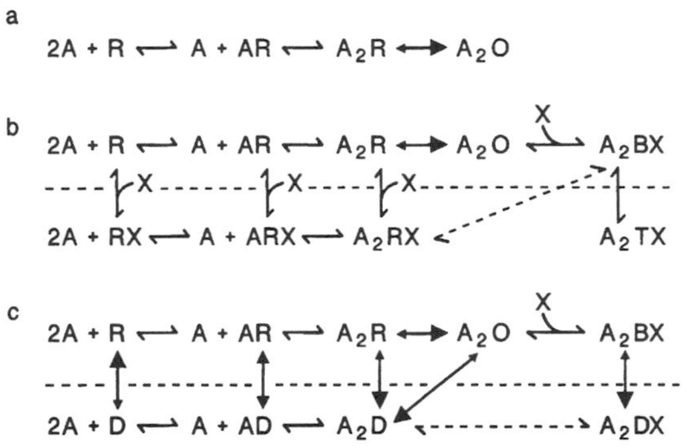

Figure 8.3. Schematic diagrams of some basic features of nicotinic receptor pharmacology.
(a) Activation by agonist occurs as a result of the binding of two agonist molecules (A) to a receptor with a closed channel (R), followed by channel opening (the receptor with an open channel is indicated by O). A receptor with two agonist molecules bound (A_2R) is very likely to have an open channel (A_2O), as indicated by the larger arrowhead toward A_2O. Competitive antagonists of activation bind to the same sites as agonist but do not produce channel opening.
(b) Open channel block occurs when a blocking drug (X) binds to an open channel (A_2O) to form a blocked (nonconducting) complex (indicated by A_2BX). A drug which acts by strict open channel block will have no effect until channels are opened by an agonist. The forms of the receptor below the dashed line indicate that drugs may bind to receptors with closed channels, or blocked channels may close. If an open channel blocker is trapped on its site by channels closure (A_2TX), agonist (and channel opening) may be required before its antagonism can be relieved. A drug which binds to closed channels can act before agonist is applied. These channel blocking mechanisms produce a noncompetitive antagonism of ion flux, without preventing binding of compounds to the agonist binding site.
(c) Desensitization of the receptor probably occurs from all states shown in parts (a) and (b)—closed channel states, the open channel state, and the blocked channel state. The desensitized receptor is indicated by D. The sizes of the arrowheads on the vertical lines indicate the overall equilibrium; for example, the R (closed) state is greatly favored over the D (desensitized) state. On the other hand, the A_2D state is greatly favored over the A_2R and A_2O states.

The third site of action involves receptor desensitization and is much more poorly defined. Nicotinic receptors, in general, desensitize in the continued presence of agonist (see Figure 8.3c, Section 8.4). Desensitization results in a diminished response, and an increased binding affinity for agonists (as first predicted by Katz and Thesleff[73]). As a group, nicotinic agonists desensitize nicotinic receptors. Other compounds, however, can block responses by acting as noncompetitive channel blockers and also enhance agonist affinity without interacting at the ACh-binding site.[14]

Many nicotinic drugs show several modes of action. For example, *d*-tubocurarine (dTC) is both a competitive inhibitor and a channel blocker at the muscle nicotinic receptor, although its action on neuromuscular transmission is essentially exclusively related to competitive inhibition.[78] At ganglionic synapses dTC is a weaker competitive inhibitor and its antagonism is largely the result of channel block.[6] Hence, even though a drug may be a functional antagonist for several types of nicotinic receptors, its mechanism of action may differ. Actions of agonists may also reflect multiple effects: nicotine is a relatively weaker agonist and a stronger channel blocker than ACh for the muscle-type receptors found in *Torpedo* electroplax.[79] Despite the limitations the countereffects on agonists and antagonists have been used to define classes of nicotinic receptors. As will be discussed in the following section, agonists and antagonists can be used to distinguish and define the diversity of neuronal receptors.

8.3.2. Nicotinic Agonists and Antagonists

Nicotinic agonists include such compounds as acetylcholine, nicotine, cytisine, 1,1-dimethyl-1,4-phenylpiperazinium (DMPP), suberyldicholine, and carbamylcholine. The efficacy and potency of these compounds as agonists differ depending on the nicotinic receptor examined. For example, muscle receptors are distinguished by their higher sensitivity to acetylcholine and suberyldicholine compared to neuronal receptors.[80] On the other hand, nicotinic and cytisine are more potent agonists of some neuronal receptors than muscle receptors.[80,81] There is also heterogeneity in response to agonists within classes of nicotinic receptors. Neurons from different areas of the brain have been shown to display different rank orders of agonist potency (see Table 8.1).

Two components of the venom from the elapid snake *Bungarus multicinctus,* αBTX and κBTX (or neuronal BTX), have played a prominent role in defining different classes of nicotinic receptors. αBTX binds with high affinity to muscle-type receptors and acts as a virtually irreversible competitive antagonist. In contrast, it does not inhibit the function of ganglionic receptors, and most brain receptors that have been examined are not functionally blocked by αBTX (Table 8.1). An exception is one of the ACh-activated currents in rat hippocampal cells[82,83] (see Table 8.1). αBTX has also been used to purify αBTX-binding proteins from brain (Section 8.2.6).

κBTX,[84] in contrast to αBTX, blocks ganglionic and some brain receptors (see Table 8.1). It also blocks muscle receptors, but the block is kinetically distinguishable from the block of ganglionic receptor. The block of muscle receptors is characterized by rapid binding and rapid dissociation.[85] In contrast, the binding to and unbinding from most ganglionic receptors is slow.

There are other compounds that can be used to distinguish receptor types. For example, hexamethonium and mecamylamine are more potent inhibitors of ganglionic receptors than muscle receptors.[80,86] Neosurugatoxin blocks brain and ganglionic receptors, but block of muscle receptors requires concentrations that are two orders of magnitude higher than those needed to block the neuronal receptors.[87] Brain αBTX-binding sites can be distinguished from muscle receptors by methyllycaconitine which have a higher affinity for brain αBTX sites.[88] Finally, there are compounds such as dihydro-β-erythroidine which have little effect on ganglionic receptors, but can potently inhibit the function of the high-affinity nicotine receptor in brain.[80]

There is an important caveat to be considered when antagonists are used in the studies of receptors. That is, several drugs demonstrate surprising cross-reactivities. Drugs thought to be selective for other receptors can have profound effects on nicotinic receptors. As mentioned above, the rat α7 nicotinic receptor is sensitive to strychnine,[41] a potent antagonist of the glycine receptor. In addition, bicuculline, known as a selective $GABA_A$ antagonist, also blocks responses to nicotinic agonists in pig pituitary cells.[89] Many noncompetitive inhibitors of the NMDA receptor, such as MK-801[90] and phencyclidine,[91] can also block nicotinic receptors. The converse is also true. dTC is a very potent blocker of some $5\text{-}HT_3$ receptors[92,93] and can also block $GABA_A$ responses in some cells.[94,95] Similarly, another nicotinic antagonist, mecamylamine, can block NMDA receptors.[96]

8.3.3. Receptor Structure and Pharmacological Properties

8.3.3.1. Receptor Subunits Expressed in Oocytes

The expression of receptors in *Xenopus* oocytes after the injection of particular combinations of subunit mRNAs has been a valuable tool in assaying the specificity of various pharmacological agents. One general conclusion is that pharmacological properties are determined by all of the subunits expressed. For neuronal receptors, both agonist and antagonist sensitivities are dependent on the α and β subunit expressed.[97–100] Results obtained with homologous subunits from the chick and rat brain are in general agreement, although some quantitative differences are apparent.

Rat subunits have been expressed in pairs (α2β2, α3β2, α4β2, α2β4, α3β4, and α4β4) and responsiveness to the agonists ACh, DMPP, nicotine, and cytisine compared.[97] Receptors containing the β4 subunit are more sensitive to cytisine than those containing the β2 subunit. Interestingly, the lower responsiveness associated with the β2 subunit appears to result from an enhanced noncompetitive inhibition by cytisine. For receptors containing the β2 subunit, the effect of the different α subunits can be seen; the α2β2 receptor is more sensitive to nicotine than ACh, the α4β2 is about equally sensitive and the α3β2 receptor is more sensitive to ACh than to nicotine. Muscle receptors are less responsive to nicotine than to ACh.[79,97]

Block by κBTX is profound and long-lasting for the rat α3β2 combination, while the α4β2 and α4β4 combinations are less strongly affected. All other combinations of rat subunits show a rapidly developing and reversing block by κBTX.[85] Chicken α4β2 receptors are not blocked by κBTX,[101] while the sensitivity of other chicken subunit combinations has not been reported.

Receptors composed of the α7 subunit have rather different pharmacological properties. The sensitivity to ACh is lower than to nicotine and cytisine, which are about equally effective as agonists.[41,102] The response is blocked by αBTX, dTC, and dHβE,[41,49,102] and, at least with rat α7, is very sensitive to strychnine (IC_{50} comparable to dTC[41]). Chick α7 is also blocked by κBTX albeit at higher concentrations compared to αBTX.[43]

Based on these observations, a high efficacy of cytisine relative to ACh would suggest the presence of the β4 subunit, slowly reversing block by κBTX would suggest α3β2 subunits, and block by αBTX and strychnine would suggest α7.

8.3.3.2. Sites of Drug Binding

Several regions of the primary sequence appear to be important in interactions with drugs, channel gating, and ion permeation, as summarized in Figure 8.1. This work began with labeling and microsequencing studies of the muscle-type receptors expressed in *Torpedo* electroplax (reviewed in Refs. 14–16). It has been complemented by studies of the properties of subunits with mutated residues. The sequencing studies have helped to determine which residues are physically close to the binding sites, while the mutation studies have confirmed that specific residues are involved in particular functions.

Residues in and near the M2 region of all of the subunits are involved in determining ion permeation properties,[23,24] suggesting that this region lines at least part of the ion channel. Residues in this region are also covalently labeled by noncompetitive channel blocking drugs,[14,16] and alterations of the residues in this region change the affinity of channel blocking drugs.[103–105]

The binding sites for agonists and competitive antagonists are formed by the N-terminal extracellular regions. An adjacent pair of cysteines at positions 192 and 193 in the muscle α1 subunit were the first residues shown to be near the ACh-binding site.[106] Subsequent work has shown that the binding site for ACh is formed from residues distributed over the N-terminal region of the α1 subunit (residues Y93, W149, Y190, C192, C193, Y198, and D200: labeling studies reviewed in Refs. 15 and 16, mutations in Refs. 107 and 108, and homologous residues in the α7 subunit.[109] The ACh-binding site also involves sites on the δ (and presumably γ) subunit (D180, E189, and D214[29,110]). The binding site of the competitive inhibitor dTC also resides on the α1, γ, and δ subunits.[26,28,111] Binding of αBTX to the α1 subunit is affected by residues between 180 and 200 of the α1 subunits.[112,113] Some studies of neuronal nicotinic AChR have extended these observations by indicating regions in the β subunit which are also important for drug binding to receptors composed of rat α3 and β2 or β4 subunits. Receptors containing α3β2 are blocked in a very slowly reversible fashion by κBTX, whereas α3β4 receptors are blocked in a rapidly reversible way. The 121 amino acids at the N-terminal end of the β2 subunit confer slow reversibility.[85] Similarly, the rat α3β2 receptor is more strongly activated by cytisine and tetramethylammonium than α3β4. The different sensitivities were largely determined by residues between 105 to 111 (cytisine) and 111 to 116 (tetramethylammonium).[99] These data indicate that the binding site for agonists and competitive antagonists is formed from residues distributed in the N-terminal regions of both α and non-α subunits. Presumably, the residues are brought into proximity by folding of the peptide chain and association between subunits.

Residues which affect gating of muscle-type AChR have been identified by mutations in several regions (M1, M2, and M4) of several subunits (reviewed in Refs. 1 and 16). One very interesting observation is the finding that conversion of a leucine to threonine in the middle of the M2 region of the α7 subunit appears to produce a desensitized receptor which has an open channel.[114] This observation implies that the channel has at least two "gates," one for closing and one for desensitization. Some nicotinic antagonists (dTC, dHβE) for the wild-type α7 receptor cause channel opening for these altered receptors, which suggests that nicotinic antagonists may stabilize a desensitized state.[102] Actually, direct effects on ligand binding can be very difficult to separate from alterations of receptor state transitions, and there is evidence that some mutations in the ACh-binding site affect gating as well as binding.[29,108,110]

The structural studies of nicotinic receptor subunits show that widely separated residues, even on separate subunits, contribute to the binding sites for agonists and competitive antagonists. A major part of the channel lining, and the site for many channel blocking drugs, is formed by a contiguous stretch of residues including the M2 region. Channel gating, on the other hand, has been harder to localize, and residues affecting gating are widely distributed.

8.3.4. Correspondence between Nicotinic Receptors on Cells and Expressed Subunits

It has been pointed out on several occasions[11,72,82,115] that the nicotinic responses seen on cells do not correspond exactly to physiological or pharmacological responses of subunit combinations expressed in oocytes. The lack of correspondence may reflect the structural data suggesting that some naturally occurring receptors include more than a single α and β subunit (see Section 8.2.6). Alternatively, native receptors may contain subunits that have yet to be identified.

8.3.5. Summary of Pharmacology

Neuronal nicotinic receptors are pharmacologically defined as being sensitive to ACh and nicotine. However, nACh receptors are heterogeneous; receptors from different sources display varying sensitivity to agonists and antagonists. Studies with expressed receptors suggest that both the α and β subunits are involved in determining the pharmacological characteristics of the receptor.

8.4. DESENSITIZATION, MODULATION, AND ENDOGENOUS LIGANDS

8.4.1. Desensitization

Desensitization is a phenomenon in which the response to a nicotinic agonist wanes in the continued presence of the agonist. All of the nicotinic receptors studied show some amount of desensitization. Katz and Thesleff[73] analyzed the phenomenon at the neuromuscular junction and proposed that the desensitized form of the receptor has a higher affinity for agonist. Subsequent binding studies have confirmed this prediction. For muscle receptors, the affinity of the desensitized receptor for ACh varies from about 10 to about 500 nM, depending on the source,[116] while binding to the nondesensitized receptor has an apparent dissociation constant greater than 10 μM.[1] Some nicotinic receptors from brain display a greatly increased affinity after prolonged exposure to ACh.[54,117] In contrast, receptor desensitization does not appear to induce much change in affinity in some ganglionic receptors[117] or αBTX-binding sites from brain.[55] It is interesting that some receptors from brain show a much higher affinity for nicotine when desensitized than do muscle-type receptors.[117]

Desensitization of muscle receptors shows at least two phases of development (fast and slow), the rates of which increase with increasing agonist concentration.[3,118,119] Some muscle AChR, at least when studied in membrane fragments, show a measurable resting level of desensitization in the absence of any agonist (about 10%[116]). On the whole, desensitization is the least well characterized aspect of nicotinic receptor function.

While desensitization is characterized by a reversible loss of response in the continued presence of agonist, there is a possibly related phenomenon, often called "run down," in which activity disappears irreversibly over a few minutes even in the absence of agonist. This is a particular problem after excision of a membrane patch. Every study of neuronal nicotinic receptors has found that activity is lost within minutes of patch excision. Other membrane channels (e.g., Ca^{2+} channels, $GABA_A$ receptors) may also show rundown, which can sometimes be reduced by inclusion of Ca^{2+} buffers or ATP in the intracellular recording solution, but these treatments have not proven effective with nicotinic receptors. Another phenomenon which resembles desensitization is slowly developing and reversing open channel block (see below).

The rate and extent of desensitization depend on the subunits expressed in oocytes. In studies of rat subunits, the α3β4 receptor desensitized more rapidly and completely than the α3β2 receptor,[120] suggesting a role for the β subunit. Conversely, chicken α3β2 receptors desensitized more completely than α4β2 over a 10-sec exposure to agonist,[100] indicating the role of the α subunit. (This observation is somewhat surprising, since the high-affinity nicotine binding is thought to be α4β2 receptor; see Section 8.2.6.)

One unusual aspect of desensitization is the finding that some neuronal receptors show essentially no desensitization if ACh is applied to a cell held at a positive membrane potential (e.g., rat sympathetic neurons[121]), whereas others desensitize at all potentials tested (e.g., rat PC12 cells[122]). Chicken α4β2 receptors expressed in oocytes showed no desensitization at +40 mV, while α3β2 receptors desensitized strongly.[100] Surprisingly, when the N-terminal extracellular regions of the α subunits were switched by making chimeric subunits joined immediately before the M1 region, both chimeric subunits produced receptors which did not desensitize at +40 mV. This finding suggests that some interaction between domains affects the voltage sensitivity of desensitization.

All nicotinic responses recorded from neurons show some degree of desensitization. Different components in the same cell can desensitize at widely different rates; e.g., the two ACh-elicited current in rat hippocampal neurons.[82,83] The role of desensitization in physiological processes is unknown, and none of these studies have attempted to model the concentration transients likely to be present at CNS nicotinic synapses. The amount of tonic activity in the presence of a constant concentration of agonist is not known accurately, although it might be of physiological relevance (see Section 8.5.4).

Measured values for the concentration of ACh in the cerebrospinal fluid range from 2 to 500 nM, with a median around 100 nM.[123] It seems likely that the concentration at a cholinergic synapse is lower, since acetylcholinesterase activity is high, but these concentrations are comparable to the concentrations of radiolabeled ACh used in assaying high-affinity ACh-binding in brain homogenates (5 to 50 nM).[124]

Several studies of muscle AChR support the idea that increased phosphorylation of the receptor results in increased steady-state desensitization, without a change in the rate of development.[125] The role of phosphorylation of nicotinic receptors is not completely clear, however. Many of the drugs used in altering levels of phosphorylation (e.g., forskolin, IBMX, and cyclic nucleotide derivatives) also independently cause a slowly developing channel block which resembles desensitization.[126–128] Also, when the cytoplasmic surface of excised patches is exposed to kinases or phosphatases, no effects on muscle AChR are seen.[129,130] Further work will be required to clarify the effects of phosphorylation on nicotinic receptor function, and any relationship to desensitization.

8.4.2. Modulation of Neuronal Receptor Function

Developmental changes in neuronal nicotinic receptors have been reported,[11,131] but will not be considered here. Shorter-term alterations have been reported in which function of an existing population of receptors on neurons is apparently altered. Substance P increases the rate of decline of ACh-elicited current in chicken sympathetic ganglion neurons.[132] This effect of substance P is not mediated by direct binding to the AChR, because the activity in a cell-attached patch decreases even when substance P is applied to the cell outside the patch. Since activators of protein kinase C have similar effects,[133] it may be that substance P acts indirectly by stimulating phosphorylation of the nicotinic receptor.

Treatments which increase the intracellular levels of cAMP also cause an increase in the number of functional nicotinic receptors on chicken ciliary ganglion neurons, as does the inclusion of cAMP and ATP in the intracellular recording solution.[134] This is not accompanied by an increase in the total number of AChR on the cell surface, suggesting that nonfunctional surface receptors are converted into functional ones. Internal dialysis of rat PC12 cells with ATP, CTP, or ATP-γ-S did not increase the nicotinic responses of these cells.[122]

It was originally thought that PKA stimulation could increase nicotinic responses of bovine chromaffin cells,[135] but subsequent reexamination has determined that stimulation of PKA activity has no rapid effect on the responses of chromaffin cells.[136] Indeed, treatment with neuropeptide Y reduces nicotine evoked currents in these cells.[137] Inhibition by neuropeptide Y was only seen if the intracellular solution contained GTP, and was blocked by GDP-β-S or pertussis toxin pretreatment, suggesting that the action of neuropeptide Y involves a GTP-binding protein (see Chapter 17). Neuropeptide Y also inhibits adenylate cyclase activity in these cells in a pertussis toxin-sensitive fashion.[138]

Thymopoietin has been studied as a possible endogenous modulator of neuronal nicotinic receptors. Recent reexamination has indicated that the ability of thymopoietin preparations to inhibit the binding of αBTX may reflect the presence of contaminating snake α-neurotoxin, and its ability to enhance desensitization may reflect the presence of phospholipase activity.[139]

8.4.3. Endogenous Ligands: Peptides, Steroids, and Ca^{2+}

Several peptides show effects on neuronal nicotinic receptor function which apparently result from peptide binding to site(s) on the AChR. Substance P increases the rate at which the nicotinic response declines in the presence of agonist for bovine chromaffin cells[140] and rat pheochromocytoma cells.[141] For these cells, it has been proposed that substance P binds to the nicotinic receptor and either blocks channels which then desensitize, or directly increases the rate of desensitization.[140,141] In addition, substance P analogues which are agonists or antagonists at the substance P receptor have similar effects on ion fluxes through nicotinic receptors on PC12 cells.[142] Atrial natriuretic peptide also enhances decay of nicotinic currents on bovine chromaffin cells, apparently as a direct result of binding to the nicotinic receptor.[143]

Progesterone reversibly blocks neuronal nicotinic receptors.[144] The block is greater at higher doses of ACh, as seen with open channel blockers, but develops in the absence of ACh. It is possible that progesterone could act at the channel blocking site, reaching it by a hydrophobic pathway through the membrane.

Increases in extracellular calcium can potentiate the response to agonist of some neuronal nicotinic receptors.[145,146]

There is a two- to threefold increase in the agonist-elicited current as the extracellular Ca^{2+} is raised from less than 1 mM to 10 mM, with a half-maximal increase at approximately 2–3 mM Ca^{2+}. It is thought that this potentiation is mediated by Ca^{2+} binding to an external site on the receptor. The concentration of free Ca^{2+} in the CSF is approximately 1.4 mM and it decreases to as low as 0.5 mM after periods of high neuronal activity.[147,148] Hence, some reduction in nicotinic responsiveness might occur during intense activity.

8.4.4. Summary of Desensitization and Modulation

Nicotinic receptors undergo a reversible decrease in activity in the maintained presence of agonist (desensitize). Some studies support the idea that phosphorylation of the receptor increases the extent of desensitization. For some neuronal nicotinic receptors it may also be possible to change the number of activatable receptors in a population. A number of endogenous agents have been shown to alter receptor function after they have bound to presently undefined sites on the receptor itself. These results indicate that cellular processes and endogenous ligands can alter the response generated by a population of receptors, but the physiological consequences on neuronal function have yet to be determined.

8.5. PHYSIOLOGY: WHAT DO NICOTINIC RECEPTORS IN THE CNS DO?

The cholinergic innervation of brain is widespread but fairly diffuse. Cells with high levels of choline acetyltransferase activity are found organized in two major groups,[149] although there also appear to be cholinergic interneurons in several brain areas. These groups of cells each span several nuclei in the forebrain or brain stem, and project to a variety of locations. However, there are less than 10% as many nicotinic receptors as muscarinic receptors in the brain, overall,[7–9] so the pattern of cholinergic neurons does not directly address the distribution or importance of nicotinic networks.

Establishment of the anatomical connections in the cholinergic system and better definition of the compositions of the nicotinic receptors in different regions will be required to fully understand the physiological properties of a nicotinic cholinergic network. Unfortunately, at the moment the major deficit in our understanding of brain nicotinic functions lies in the paucity of physiological data.

8.5.1. Neuronal Nicotinic Receptor Function

The mean length of time a muscle nicotinic receptor stays active at low ACh concentration is very close to the time constant for the decay of the endplate current at the neuromuscular junction,[150] and so the receptor gating properties play a major role in determining the time course of the postsynaptic response. The mean burst durations measured for neuronal nicotinic receptors vary from 1 msec to 40 msec, with most in the range of 5 to 10 msec at room temperature. On this basis, neuronal receptors might be associated with relatively brief synaptic currents.

Most studies of activation of neuronal AChR can be interpreted using the same basic scheme as for muscle receptors (see Figure 8.3). Current evoked by agonist increases more than linearly with agonist concentration at low doses, indicating that most receptors with open channels have more than one bound ACh molecule (Table 8.1). One basic parameter is the probability of being open for a channel with 2 ACh molecules bound. This probability is greater than 0.9 for muscle AChR,[1] and is greater than 0.5 for neuronal nicotinic receptors.[104,151,152] Activation of most types of neuronal nicotinic receptors appears to occur over a similar concentration range of ACh as muscle receptors. Using relatively slow applications, the concentration of ACh which gives a half-maximal current is in the range of 1 to 100 μM (see Table 8.1), which is comparable to values for muscle receptors.[1] Muscle nicotinic receptors open very rapidly: after binding steps are complete, channels open within 50 μsec [rates greater than 20,000 sec^{-1} (Ref. 1)]. The channel opening rate may be slower for some neuronal nicotinic receptors [longer than 1 msec; rates less than 1000 sec^{-1} (Refs. 152, 153)], although the receptors on rat superior cervical ganglion cells may be more similar to muscle.[151] However, much more work is required on the receptor types expressed in the brain to obtain more quantitative data.

In sum, although there are quantitative differences in functional properties of nicotinic receptors, in a qualitative sense they appear quite similar. There is no clear indication that the nicotinic receptors found in the brain differ in major functional respects from their better-characterized relatives in skeletal muscle.

8.5.2. Postsynaptic Nicotinic Responses in the CNS

Cells from many brain regions show responses to applied nicotinic agonists.[154–156] However, nicotinic antagonists do not block excitatory synaptic transmission to these cells. The only known example of a fast nicotinic postsynaptic response in the vertebrate CNS is the synapse between motoneuron collaterals and the Renshaw cells in the spinal cord.[157,158]

There have been several instances in which early reports had suggested that synapses might be nicotinic; e.g., in the interpeduncular nucleus,[159] the neostriatum,[160] or the supraoptic nucleus.[161] In each of these cases, subsequent work has shown that the rapid excitation is probably glutamatergic (IPN,[159] striate cortex,[162] SON[163]).

One example in which postsynaptic nicotinic receptors probably play a physiological role is the mammalian retina.[164] Retinal ganglion cells show nicotinic responses,[165] and it has been proposed that nicotinic receptors receive input from cholinergic amacrine cells involved in establishment of an excitatory input driving direction selectivity.[166] No evoked nicotinic postsynaptic events have been recorded.

Every nicotinic receptor channel studied in vertebrates carries a net inward current at physiological ion gradients and resting membrane potential. However, nicotine activates a potassium-selective conductance in neurons in the rat dorsolateral septal nucleus,[167] resulting in hyperpolarization. The response is blocked by κBTX and mecamylamine, but not αBTX or dTC, suggesting that α3 but not α7 subunits might be involved. This K^+ conductance is also blocked by intracellular injection of Ca^{2+}-chelating agents and by extracellular application of the bee venom toxin apamin. Therefore, it seems likely that the K^+ current is carried by Ca^{2+}-activated voltage-independent K^+ channels. It may be that Ca^{2+} entering through a nicotinic receptor channel activates this current, as has been reported for bullfrog sympathetic ganglion neurons.[168] It is not known whether this response can be elicited by synaptic transmission.

The overall result of the physiological studies of nicotinic responses on central neurons is that cells from a variety of regions show responses on the cell body, but that evidence for the involvement of postsynaptic nicotinic receptors in synaptic transmission is lacking. The "reason" for the existence of nicotinic receptors on neuronal cell bodies is still not clear.

8.5.3. Nicotinic Agents and Release: Presynaptic Receptors

8.5.3.1. Anatomical Evidence for Presynaptic Receptors

Three lines of research have converged to indicate the possible importance of presynaptically located nicotinic receptors. The first line is the results of lesions, which have given several instances in which nicotinic sites decrease greatly in target tissue after destruction of the projecting neurons. Some clear examples are in the mammalian visual system,[8,169,170] the medial habenular projections to the interpeduncular nucleus[171] and substantia nigral projections to neostriatum[172,173] in rat and projections from the lateral spiriform nucleus to the optic tectum in the chicken.[174]

8.5.3.2. Physiological Evidence for Presynaptic Receptors

The second line of evidence is physiological. Neurons in the lateral geniculate nucleus (LGN) of the cat (but not the guinea pig[175]) are depolarized by nicotinic agonists, although there is no indication that transmission in the LGN is nicotinic. However, there is evidence that the synaptic terminals of LGN axons in the cat visual cortex have functional nicotinic receptors.[8] Nicotine enhances and mecamylamine reduces the extracellularly recorded responses of cortical cells to light. Furthermore, lesion of the LGN markedly reduced the number of high-affinity nicotine-binding sites in visual cortex to which lesioned areas projected.[8] These observations led to the conclusion that the nicotinic receptors were located presynaptically on the LGN terminals, and that nicotinic activation modulates release of the actual transmitter.

The rat interpeduncular nucleus receives cholinergic innervation from the medial habenular nucleus. Neurons in the medial habenular nucleus also are excited by nicotine.[176] Brown *et al.*[59] showed that cells in the rat interpeduncular nucleus were excited by nicotine, but that the synaptic transmitter was most likely an excitatory amino acid, and that nicotinic blocking agents had no effect on transmission. However, application of nicotinic agonists reduced the amplitude of extracellular recorded action potentials in unmyelinated terminal axons in the interpeduncular nucleus, and the reduction could be attenuated by nicotinic blockers.[177] Mulle *et al.*[178] found that the nicotinic responses of habenular neurons, interpeduncular neurons, and the afferent axons in the interpeduncular nucleus have different sensitivities to blockers. This observation suggests that a different type of nicotinic receptor is present on the axon than on habenular neurons. The decrease in compound action potential amplitude probably reflects inactivation of Na^+ channels by a steady depolarization of the axon terminals, since a raised external K^+ concentration has a similar effect.[177] No physiological role for these presynaptic nicotinic receptors has been shown.

Finally, intracellular records from postsynaptic cells have shown that nicotinic activation can increase the frequency of spontaneously occurring synaptic currents. These include GABAergic currents in slices from chicken lateral spiriform nucleus,[179] chicken ventral lateral geniculate nucleus,[180] and in the rat medial habenula.[181] The enhanced rate of spontaneous release is seen with dissociated cells from the medial habenula, indicating that the action must be on detached synaptic boutons.[181] In the lateral spiriform nucleus and medial habenula the enhanced release is blocked by tetrodotoxin, suggesting that nicotinic activation results in enhanced sodium-dependent action potential frequency (Section 8.5.4.3). However, in the chicken ventral lateral geniculate the increase is not blocked by tetrodotoxin.

8.5.3.3. Biochemical Assays of Transmitter Release

The third line of evidence involves biochemical analyses of transmitter release. There is now reasonable evidence that motor nerve terminals have nicotinic receptors, which are likely to be involved in positive feedback (that is, activation leads to enhanced release), although the studies have not succeeded in defining the precise nature of the effects.[182,183]

There is stronger evidence for nicotinic enhancement of transmitter release in the brain (reviewed in Refs. 184 and 185). We will concentrate on the most extensively studied system, nicotinic enhancement of dopamine release from rat striate cortex.[186]

The most recent studies have used synaptosomes, to obviate actions at areas away from the terminal itself. Synaptosomes are closed, membrane-delimited structures prepared by homogenization and centrifugation of brain. The presynaptic terminal is recovered as a sack, whereas the postsynaptic elements are usually ruptured and recovered as membrane fragments attached to the terminal. Hence, release of transmitter or release of sequestered ions occurs from presynaptic terminals, while the binding of compounds may occur to either pre- or postsynaptic structures.

The release of radioactively labeled dopamine taken up by synaptosomes prepared from rat striate cortex is increased by nicotinic activators.[187–189] The enhanced release requires calcium ions in the external medium. It is not blocked by tetrodotoxin, so probably does not require activation of voltage-dependent sodium channels.[187] The nicotine-induced increase in release from striata excised as a block from rat brains is reduced by ω-conotoxin (50 nM toxin reduced release to 30%), which suggests that N-type voltage-gated Ca^{2+} channels are involved.[190] This has not yet been studied in synaptosomes.

A recent study has compared two aspects of nicotinic receptors in mouse striatal synaptosomes, enhancement of dopamine release and block of binding of radiolabeled nicotine,[189] and a second study compared nicotinic binding and Rb^+ efflux in mouse midbrain synaptosomes.[81] The comparison of a number of different nicotinic agonists suggests that the rubidium efflux is elicited by activation of the nicotinic receptor corresponding to the high-affinity nicotine-binding site.[81] However, stimulation of dopamine release appears to result from action at a different receptor.[189] A clear distinction is that κBTX had no effect on nicotine binding, yet effectively blocked nicotine-enhanced dopamine release. αBTX had no effect on any of the measures of nicotinic receptor function.

The reported presynaptic nicotinic actions involve an increase in transmitter release. A possible exception is the decrease in terminal action potentials seen in the interpeduncular nucleus, although a lower level of depolarization would likely result in increased spontaneous activity.

8.5.4. Physiology: What Might Nicotinic Receptors Do?

Two consequences of nicotinic receptor activation are likely to be important: depolarization and Ca^{2+} influx. For postsynaptic receptors, depolarization and gating of voltage-activated conductances are major, rapid effects, although Ca^{2+} transients may be important for longer-term effects. For presynaptic receptors, the ultimate effector for transmitter release is most likely the cytoplasmic Ca^{2+} concentration.

8.5.4.1. Ca^{2+} Entry through Nicotinic Receptor Channels

Takeuchi[191] first demonstrated that Ca^{2+} can permeate muscle AChR channels. Decker and Dani[192] estimated that under normal ionic conditions about 2% of the total inward current through muscle AChR channels is carried by Ca^{2+}. Subsequent studies of nicotinic receptors on rat habenular neurons,[193] sympathetic ganglion neurons,[194] and bovine chromaffin cells[195] have found that 2–10% of the current is likely to be carried by Ca^{2+}. The permeability ratio of Ca^{2+} to Cs^+ is 0.2 for muscle AChR channels.[17,145] The P_{Ca}/P_{Na} is greater than 1 for channels on bovine chromaffin cells,[145] rat PC12 cells,[196] and rat sympathetic ganglion neurons.[194] Rat α7 homooligomers expressed in *Xenopus* oocytes have a very high P_{Ca}/P_{Na}, 20 (Ref. 41), even higher than that of the NMDA receptor channel.[197,198] The receptors of rat parasympathetic ganglion neurons,[199] rat medial habenula neurons,[193] and rat α4/β2, α2/β2, and α4β4 subunits expressed in oocytes[41,145] also have substantial permeability to Ca^{2+}.

A number of studies using fluorescent indicators have shown that the intracellular Ca^{2+} is increased by activation of nicotinic receptors, even when activation of Ca^{2+} channels is blocked or controlled.[193–195]

One interesting study found that chick ciliary ganglion cells show a nicotine-induced increase in intracellular Ca^{2+} which can be blocked by αBTX.[200] Surprisingly, there was no detectable effect of αBTX on nicotine-induced membrane currents, suggesting that the αBTX-sensitive ion flux was very small. The possibility exists that the αBTX-sensitive AChR on these cells produce very little membrane conductance, and that any current is largely carried by Ca^{2+} ions.

8.5.4.2. Postsynaptic Receptors: Effects Mediated by Ca^{2+} Influx

No consequences of postsynaptic Ca^{2+} influx through nicotinic receptor channels have been demonstrated using physiological stimuli. However, the K^+ current activated by nicotinic agonists in neurons may be a consequence of Ca^{2+} entry.[167,168] Furthermore, stimulation by 30 μM nicotine in medial habenular neurons has been shown to both elicit a Cl^- current and reduce GABA responses.[193] The possibility exists, as well, for additional effects. For example, Ca^{2+} entry through various channels has been implicated in induction of long-term potentiation and depression, and it has been suggested that αBTX-sensitive nicotinic receptors on hippocampal neurons are involved in long-term regulation of neuronal properties.[201]

8.5.4.3. Presynaptic Depolarization

A synaptic bouton is a small structure. The resting electrical properties of the terminal axon and bouton are not known, but we will assume that the specific membrane resistance is 10^4 ohm-cm^2 and that the cytoplasm and CSF have a resistivity of 100 ohm-cm. Assuming that the bouton is a sphere 1 μm in diameter, its input resistance is 3×10^{11} ohm. If the terminal axon is a semi-infinite cylinder 0.4 μm in diameter surrounded by an extracellular space 20 nm wide, then the input resistance of the process is 6×10^9 ohm. The input resistance at the bouton is, therefore, about 6×10^9 ohm and the input conductance is about 2×10^{-10} siemens (equivalent to 4 open channels of 50 pS conductance). For this "typical" bouton, a single channel contributing 2 pA of inward current (one nicotinic receptor channel open at about −65 mV) would depolarize the bouton by about 10 mV.

8.5.4.4. Presynaptic Receptors: Effects on Release

We assume that enhanced transmitter release reflects an increased Ca^{2+} concentration in the nerve terminal. How much membrane current is needed? A Ca^{2+} current of 1 pA flowing for 1 msec carries a net influx of 3000 ions. This would raise the average intrabouton Ca^{2+} concentration to 10 μM. If only 2% of the total current is carried by Ca^{2+}, one channel open for 5 msec and carrying 1 pA of current would allow the entry of about 300 Ca^{2+} ions, raising the intrabouton Ca^{2+} concentration to about 1 μM. This is a high concentration for the cytoplasm, in which Ca^{2+} is normally less than 100 nM, but the concentration needed for rapid evoked transmitter release may be higher.[202,203]

The Ca^{2+} influx could (1) result directly from Ca^{2+} entry through nicotinic receptor channels. Alternatively, the depolarization could (2) open voltage-gated Ca^{2+} channels, or less directly (3) open Na^+ channels which then cause further depolarization and the opening of voltage-gated Ca^{2+} channels. In case 3, the Na^+ channel blocker, tetrodotoxin, should prevent enhanced release. This observation has been made for two cases of increased GABA release, but not for dopamine release in the striatum (see Sections 8.5.3.2 and 8.5.3.3). In case 2, tetrodotoxin should have no effect but Ca^{2+}-channel blockers should be effective. This seems to be the case for dopamine release (Section 8.5.3.3). Release induced by direct Ca^{2+} entry through the nicotinic receptor channel (case 1) would be blocked only by nicotinic receptor channel blockers. It is not known which drugs block the enhanced release in the chicken ventral lateral geniculate.

None of the studies performed to date have indicated that direct Ca^{2+} entry is important for enhanced release in the CNS, so it is possible that the Ca^{2+} influx through the AChR channels is too small to directly increase release. However, longer-term Ca^{2+}-dependent processes may modulate release. For example, a small but long-lasting Ca^{2+} increase is postulated to underlie facilitation of release at the neuromuscular junction,[204] and Ca^{2+}-activated kinase activity can affect release at the squid giant synapse.[205]

8.5.4.5. How Many Receptors for a Presynaptic Effect?

How many receptors might it take to generate an effect? There can only be speculation at present. First, we will make some simplifying assumptions about receptor activation: assume that a doubly liganded receptor has probability 1.0 of having an open channel (i.e., that ACh is very effective as an agonist). Further, assume that the number of channels open in the absence of desensitization is described by the equation $Y = NC^2/(1 + C^2)$, where C is the ratio of [ACh] to the EC_{50} and N is the total number of receptors. Rearranging, the total number of receptors needed to have Y open channels at a given [ACh] is $N = Y(1 + C^2)/C^2$. Clearly, the required number of receptors depends very strongly on the values for [ACh] and the EC_{50}, neither of which is known.

We will assume that the desired effect is to have one channel open on average, and that the ACh concentration is 100 nM, the average value for CSF.[123] For an EC_{50} of 10 μM, it would take $N = 10{,}000$ AChR on a bouton. This number corresponds to a density of 3000 AChR/μm² on the "typical" bouton (surface area 3.1 μm²), which seems ridiculously high. On the other hand, a value for the EC_{50} of 0.5 μM (as reported for the chicken α4β2 receptor[100] would require only about 25 AChR. Alternatively, if the [ACh] reached the EC_{50}, there would have to be only 2 AChR on the bouton to have an average of 1 open channel at steady state.

For a tonic response, the balance between steady desensitization and activation must also be included. Katz and Miledi[206] first demonstrated a steady activation of AChR at the neuromuscular junction when acetylcholinesterase activity is blocked. This "window current" implies that there are concentrations of ACh which produce appreciable steady activation even in the presence of some equilibrium level of desensitization. Lipton[207] has made similar observations for cultured rat retinal ganglion cells, and the data presented in Revah *et al.*[114] suggest that chick α7 AChR in oocytes would show a steady response at some ACh concentrations. Studies of other nicotinic receptors have not provided clear results in this context. Since desensitization is a complex phenomenon which can develop very slowly and at low agonist concentrations, it is not clear whether the residual response after a 10-sec exposure is relevant to any physiological condition. However, since even a very low level of presynaptic nicotinic receptor activity could have marked effects, the possible existence of tonic activation must be considered.

8.5.5. Summary of Physiology

The functional properties of neuronal nicotinic receptors qualitatively resemble those of muscle nicotinic receptors: the receptor has a relatively low affinity for ACh, a relatively brief burst duration, and shows desensitization. However, there is no evidence that nicotinic receptors mediate rapid excitatory synaptic transmission in the CNS, as they do in the periphery. Most results suggest that presynaptically located nicotinic receptors are of greater importance than postsynaptic receptors in the function of the CNS. Activation of presynaptic receptors results in an enhancement of transmitter release. There can be appreciable Ca^{2+} influx through nicotinic receptor channels, but the available data have not supported the idea that this direct entry affects transmitter release. Because of the small dimensions of the presynaptic terminal, very low levels of receptor activity could have large effects on membrane potential and intercellular Ca^{2+} levels. There can be tonic activation even at steady-state levels of desensitization, which indicates that both phasic and tonic patterns of activity should be considered.

Nicotinic receptor activation can produce appreciable changes in postsynaptic intracellular Ca^{2+}, and some data indicate that receptor-mediated Ca^{2+} fluxes might be significant even in cases in which membrane depolarization is minor. An increase of intracellular Ca^{2+} produced by applied nicotinic agonists has been shown to affect other channels postsynaptically. However, no physiological stimuli have been found that produce these actions.

8.6. DISORDERS OF NEURONAL NICOTINIC RECEPTOR FUNCTION

The major known disorder of neuronal nicotinic receptor function is nicotine dependence. To the best of our knowledge no involvement of peripheral neuronal nicotinic receptors occurs in myasthenia gravis (see Chapter 28), and there are no autoimmune disorders of neuronal nicotinic receptors.

8.6.1. Nicotine Dependence

Smokers find it very difficult to quit.[208] This is surprising, because the effects of nicotine in normal use, and the symptoms of nicotine withdrawal, are relatively subtle compared to other drugs which induce dependence. It is now clear, however, that abstaining from taking tobacco products does result in identifiable withdrawal symptoms which nicotine relieves.[208,209] It is not clear whether nicotine consumption *per se* has significant health consequences, but nicotine dependence when satisfied by consumption of tobacco products is a major personal health factor. The central questions about nicotine actions in the CNS are what effects nicotine has, how those effects are mediated, and what the nature of nicotine dependence is. These questions have not been answered, and we will only indicate some of the directions research is taking.

Large doses of nicotine cause a variety of effects (including vomiting, hallucinations, and coma) which form the basis for the use of nicotine in some shamanistic rituals,[210] but habitual users avoid such treatments. The earliest reports from European explorers encountering indigenous peoples of the New World indicate that consumption of nicotine-containing products was widespread throughout the Americas and occurred in social as well as ceremonial settings.[210–212] Europeans rapidly accepted the practice, and tobacco use was globally distributed within 200 years. In 1571 Pena and de l'Obel published these comments about tobacco smoking from crew members returning from the New World: ". . . they say their hunger and thirst are allayed, their strength restored, and their spirits refreshed. And they declare also that their brain is lulled with a pleasing drunkenness . . ." (quoted in Ref. 212).

These subjective impressions have persisted in smokers' accounts, and an extensive literature has developed of attempts to identify and quantify nicotine effects (reviewed recently in Refs. 208, 209, 213). The most robust effects appear to be an increase in "attention," and a possible enhancement of recall. We will not try to summarize this literature here, and only note that none of the observations have implicated any particular cellular or molecular process in mediating the effects of nicotine.

Assuming that nicotine acts at CNS nicotinic receptors, what aspect of nicotinic receptor function might be affected? The nicotinic antagonist mecamylamine crosses the blood–brain barrier, and has relatively small peripheral action. Effects of administered mecamylamine suggest that smokers seek activation of nicotinic receptors. That is, mecamylamine treatment increases the rate of nicotine intake from cigarettes,[214,215] reduces the subjective reactions to nicotine,[216,217] and reduces the ability of smoking to alleviate craving for cigarettes.[218] Further, mecamylamine has opposite effects to nicotine on some tests of attention,[219] and can block the effects of nicotine.[220] These observations indicate that receptor activation underlies the effects of nicotine, including the acute effects of nicotine, the normalization of cognitive measures by smoking or nicotine administration in acutely deprived smokers, and the reduction of craving for cigarettes.

During inhalation of a cigarette containing 1 to 2 mg nicotine, the plasma nicotine concentration rises very rapidly to about 350 nM, then falls rapidly at first (half-time about 10 min) followed by a slower fall (half-time about 120 min).[221] A smoker consuming a pack a day has repeated peaks, and a gradually increasing trough level during the morning, until the trough reaches about 200 nM.[221–223] Cigar or pipe smokers (who do not inhale) and consumers of smokeless tobacco products do not show peaks in nicotine levels, but develop similar steady concentrations.[222,223] Brain concentrations are higher; the ratio of brain to blood concentrations ranges from about 1.5 to 5.5 in mammals,[224–226] and one study comparing CSF to plasma levels obtained a ratio of 1.3 (Jellinek, mentioned in Rand).[227] Assuming that the brain concentration is twice the plasma level, the peak concentration would be about 700 nM and the trough 400 nM. The peak level would activate about 1% of AChR which have an EC_{50} of 7 μM. On the other hand, 50% of receptors with an EC_{50} of 500 nM (as reported for chicken $\alpha4\beta2$)[100] would be active. However, if the site of action is the high-affinity nicotine-binding site, then the plateau concentration is much greater than the apparent K_d (about 10 nM[54]). The high-affinity binding, in turn, is likely to reflect desensitized receptors, so essentially all of these receptors would be inactive.

The mechanism for nicotine dependence is not known, although it has been suggested that it might be related to stimulation of the mesolimbic dopaminergic system.[228] One effect of chronic nicotine administration is well documented: there is an increase in the number of high-affinity nicotine-binding sites,[229] including an increase in postmortem human brain of smokers compared to age-matched nonsmokers.[230] The increase seen in rat brain reflects an increase in the amount of $\alpha4\beta2$ receptors,[59] although there is no increase in the amount of mRNA for any nicotinic receptor subunit.[231] The suggestion has been made that the increase might be a compensatory consequence of nicotine-induced desensitization of these receptors,[186] to provide a constant number of nondesensitized receptors. The occurrence of withdrawal symptoms is perhaps surprising in this case, since it might be thought that the additional receptors would recover from desensitization and that physiological levels of ACh might then provide a supranormal stimulation. Possibly, tonic activation is the desired stimulus, while physiological patterns of activity might provide phasic stimuli. The resolution of some of these questions may lie in clearer definition of the multiplicity of receptor types and locations, as it is likely that global measurements lump apples and oranges together. Also, better data on the activation and desensitization properties of receptor types may provide some insight.

In sum, it appears that receptor activation by nicotine underlies both the acute actions of nicotine and the alleviation of nicotine withdrawal symptoms. Additional tests with more agonists and antagonists would be valuable, given the imprecision of the pharmacology and complexity of the intact CNS, but desensitization-induced reduction of nicotinic effects appears unlikely as a mechanism. This statement begs the question, however, of the role of the high-affinity nicotine-binding sites. It is very likely that these receptors will be in the high-affinity state at brain nicotine levels reached by smokers, and the increase in number indicates that there is an effect of chronic nicotine on these AChR. One possibility is that these receptors are not physiologically relevant. Another is that desensitization is not complete, since even a small amount of activation would have significant effects on a terminal.

8.6.2. Alzheimer's Disease

Alzheimer's disease is a dementia which is characterized by various transmitter deficits as well as by its histopathology (amyloid plaques and neurofibrillary tangles).[232] It has been observed that the brains of individuals with Alzheimer's disease display a profound decrease in nicotinic binding sites,[233] particularly in the frontal[234] and temporal cortices[234] and the hippocampus,[235] when compared to age-matched controls. The loss of nicotinic binding sites is not global; there is no significant decrease in subcortical regions such as the striatum and thalamus.[9] The loss of cortical nicotinic sites is thought to be related to the loss of presynaptic nicotinic receptors, since the nucleus basalis of Meynert provides the major cholinergic innervation to the cortex and these neurons degenerate in Alzheimer's disease. Consistent with this hypothesis is the accompanying loss of cortical presynaptic markers for cholinergic neurons such as choline acetyltransferase[235]; however, see Ref. 233.

The relationship of the decrease of nicotinic receptors to the deficits seen in individuals with Alzheimer's disease is not known. A possible connection is suggested by the finding that administration of nicotine to patients with Alzheimer's disease can significantly reduce some of the deficits experienced by these patients.[236,237] The most promising treatment for Alzheimer's disease currently being tested is an anticholinesterase (Tacrine; 4-aminoacridine). This drug can reduce symptoms in about 30% of patients, but it is not clear why it is not more generally effective, nor what cellular effects might be mediating the improvements.[238,239]

8.6.3. Parkinson's Disease

Parkinson's disease is a progressive CNS disorder which results from the loss of dopaminergic neurons of the substantia nigra. It is characterized by poverty of movement, muscular rigidity, and resting tremor. In addition to the loss of dopaminergic neurons, a reduction in the number of cortical and hippocampal nicotinic binding sites is seen.[240]

A relationship between Parkinson's disease and neuronal nicotinic receptors has been suggested from the apparent protective effect of cigarette smoking on the risk of Parkinson's disease. A number of studies have found an inverse relationship between cigarette smoking and the risk of disease (for review, see Ref. 241), in which the risk was reduced by 20% to 70% for smokers. There are three hypotheses for how smoking might produce a protective effect. Nicotine could enhance dopamine release, and hence delay the onset of symptoms.[241] The second hypothesis is that chronic exposure to nicotine might reduce dopaminergic activity, and hence result in lower production of oxidation products and hence less damage to neurons.[242] Finally, a protective effect could be produced by increased levels of carbon monoxide. The carbon monoxide might produce a reducing environment which would also reduce the amount of free radicals resulting from catecholamine oxidation.[243] It is important to mention that there is debate about whether the protective effect is real. The age of onset of Parkinson's disease in smokers is younger than in the general population,[244] and there is no apparent dose–effect relationship between smoking and the reduction in risk.[244,245] These observations have led to the proposal that the protective effect is illusory, and can be explained by early mortality among cigarette smokers.[244,246]

8.6.4. Schizophrenia

A possible role of neuronal nicotinic receptors in schizophrenia is suggested from the observation that smoking is more prevalent in patients with schizophrenia (88%) than in the general population.[208,247,248] It is unlikely that cigarette smoking, and by implication nicotine, is playing a causal role in schizophrenia, but the possibility exists that patients may be self-medicating. A connection has been hypothesized based on observations that schizophrenics show reduced paired depression of auditory evoked potentials (the P50 wave[249]), and that nicotine can restore normal patterns of paired pulse depression in relatives of schizophrenics.[250]

8.6.5. Summary of Disorders of Neuronal Nicotinic Receptors

It is surprising, given the global popularity of tobacco consumption, that there are so few clear insights into the role of nicotinic receptors in normal human performance. Possibly the popularity reflects the mild consequences of moderate nicotine consumption, and underscores the relatively subtle role of the nicotinic system in brain function. No disorders have been conclusively, or even strongly, linked to dysfunction of nicotinic receptors.

8.7. OVERALL SUMMARY

Nicotinic receptors in the CNS have been characterized as sites for ligands, as polypeptides, as mRNA species, and as membrane conductances. They are distributed widely in the brain, but the overall prevalence is less than one-tenth that of muscarinic receptors. They act as ligand-gated nonspecific cation channels, which can carry a significant Ca^{2+} flux. Their basic properties resemble the nicotinic receptor found at the neuromuscular junction, but there is no evidence that postsynaptic nicotinic receptors mediate excitatory synaptic transmission in the brain. Indeed, it is likely that presynaptically localized nicotinic receptors are more physiologically relevant than are postsynaptically localized receptors. The relatively high Ca^{2+} permeability of some types of brain nicotinic receptor channels suggests that Ca^{2+} fluxes, rather than depolarization, might underlie some consequences of nicotinic receptor activation. The overall role of nicotinic receptors in brain function is unclear; studies in humans and animals suggest that nicotinic receptors are involved in attention and possibly memory and mood. The known disorder which can reliably be ascribed to brain nicotinic receptors is nicotine dependence. However, the basis for development of dependence is not known.

ACKNOWLEDGMENTS. We thank E. Albuquerque, D. Berg, V. Chiapinelli, A. Collins, G. Hatton, J. Lindstrom, C. Lingle, L. McMahon, D. Parkinson, J. Patrick, and C. Zorumski for advice, comments, or unpublished manuscripts. J.H.S. is supported in part by NIH grants R01 NS-22356 and P01 GM-47969.

REFERENCES

1. Lingle, C., Maconochie, D., and Steinbach, J. H. (1992). Activation of skeletal muscle nicotinic acetylcholine receptors. *J. Membr. Biol.* **126:**195–217.
2. Colquhoun, D. (1986). On the principles of postsynaptic action of neuromuscular blocking agents. In *Handbook of Experimental Pharmacology* (D. A. Kharkevich, ed.), Springer-Verlag, Berlin, Vol. 79, pp. 59–113.
3. Adams, P. R. (1987). Transmitter action at endplate membrane. In *The Vertebrate Neuromuscular Junction* (E. M. Salpeter, ed.), Alan R. Liss, New York, pp. 317–359.
4. Salpeter, M. M. (1987). Vertebrate neuromuscular junctions: General morphology, molecular organization, and functional consequences. In *The Vertebrate Neuromuscular Junction* (M. M. Salpeter, ed.), Alan R. Liss, New York, pp. 1–54.
5. Bartol, T. M., Jr., Land, B. R., Salpeter, E. E., and Salpeter, M. M. (1991). Monte Carlo simulation of miniature endplate current generation in the vertebrate neuromuscular junction. *Biophys. J.* **59:** 1290–1307.
6. Ascher, P., Large, W. A., and Rang, H. P. (1979). Studies on the mechanism of action of acetylcholine antagonists on rat parasympathetic ganglion cells. *Eur. J. Physiol.* **295:**139–170.
7. Marks, M. J., and Collins, A. C. (1982). Characterization of nicotinic binding in mouse brain and comparison with the binding of α-bungarotoxin and quinuclidinyl benzilate. *Mol. Pharmacol.* **22:** 554–564.
8. Parkinson, D., Kratz, K. E., and Daw, N. W. (1988). Evidence for a nicotinic component to the actions of acetylcholine in cat visual cortex. *Exp. Brain Res.* **73:**553–568.
9. Aubert, I., Araujo, D. M., Cecyre, D., Robitaille, Y., Gauthier, S., and Quirion, R. (1992). Comparative alterations of nicotinic and muscarinic binding sites in Alzheimer's and Parkinson's diseases. *J. Neurochem.* **58:**529–541.
10. Cockcroft, V. B., Osguthorpe, D. J., Barnard, E. A., Friday, A. E., and Lunt, G. G. (1990). Ligand-gated ion channels. *Mol. Neurobiol.* **4:**129–169.
11. Sargent, P. B. (1993). The diversity of neuronal nicotinic acetylcholine receptors. *Annu. Rev. Neurosci.* **16:**403–443.
12. Pedersen, S. E., Bridgman, P. C., Sharp, S. D., and Cohen, J. B. (1990). Identification of a cytoplasmic region of the *Torpedo* nicotinic acetylcholine receptor α-subunit by epitope mapping. *J. Biol. Chem.* **265:**569–581.
13. Unwin, N. (1993). Nicotinic acetylcholine receptor at 9 Å resolution. *J. Mol. Biol.* **229:**1101–1124.
14. Changeux, J.-P. (1990). Functional architecture and dynamics of the nicotinic acetylcholine receptor: An allosteric ligand-gated channel. In *Fidia Research Foundation Neuroscience Award Lectures,* Raven Press, New York, Vol. 4, pp. 21–168.
15. Galzi, J.-L., Revah, F., Bessis, A., and Changeux, J.-P. (1991). Functional architecture of the nicotinic acetylcholine receptor: From electric organ to brain. *Annu. Rev. Pharmacol. Toxicol.* **31:**37–72.
16. Karlin, A. (1993). Structure of nicotinic acetylcholine receptors. *Curr. Opin. Neurobiol.* **3:**299–309.
17. Adams, D. J., Dwyer, T. M., and Hille, B. (1980). The permeability of endplate channels to monovalent and divalent metal cations. *J. Gen. Physiol.* **75:**493–510.
18. Cohen, B. N., Labarca, C., Davidson, N., and Lester, H. A. (1992). Mutations in M2 alter the selectivity of the mouse nicotinic acetylcholine receptor for organic and alkali metal cations. *J. Gen. Physiol.* **100:**373–400.
19. Dwyer, T. M., Adams, D. J., and Hille, B. (1975). The permeability of the endplate channel to organic cations in frog muscle. *J. Gen. Physiol.* **75:**469–492.
20. Dani, J. A. (1986). Ion-channel entrances influence permeation: Net charge, size, shape, and binding considerations. *Biophys. J.* **49:** 607–618.
21. Dani, J. A. (1989). Open channel structure and ion binding sites of the nicotinic acetylcholine receptor channel. *J. Neurosci.* **9:**884–892.
22. Dani, J. A., and Eisenman, G. (1987). Monovalent and divalent cation permeation in acetylcholine receptor channels: Ion transport related to structure. *J. Gen. Physiol.* **89:**959–983.
23. Imoto, K., Busch, C., Sakmann, B., Mishina, M., Konno, T., Nakai, J., Hideaki, B., Mori, Y., Fukuda, K., and Numa, S. (1988). Rings of negatively charged amino acids determine the acetylcholine receptor channel conductance. *Nature* **335:**645–648.
24. Galzi, J.-L., Devillers-Thiery, A,. Hussy, N., Bertrand, S., Changeux, J.-P., and Bertrand, D. (1992). Mutations in the channel domain of a neuronal nicotinic receptor convert ion selectivity from cationic to anionic. *Nature* **359:**500–504.
25. Mishina, M., Takai, T., Imoto, K., Noda, M., Takahashi, T., Numa, S., Methfessel, C., and Sakmann, B. (1986). Molecular distinction between fetal and adult forms of muscle acetylcholine receptor. *Nature* **321:**406–410.
26. Blount, P., and Merlie, J. P. (1989). Molecular basis of the two nonequivalent ligand binding sites of the muscle nicotinic acetylcholine receptor. *Neuron* **3:**349–357.
27. Sine, S. M., and Claudio, T. (1991). γ- and δ-subunits regulate the affinity and the cooperativity of ligand binding to the acetylcholine receptor. *J. Biol. Chem.* **266:**19369–19377.
28. Pedersen, S. E., and Cohen, J. B. (1990). *d*-Tubocurarine binding sites are located at α-γ and α-δ subunit interfaces of the nicotinic acetylcholine receptor. *Proc. Natl. Acad. Sci. USA* **87:**2785–2789.
29. Czajkowski, C., Kaufmann, C., and Karlin, A. (1993). Negatively charged amino acid residues in the nicotinic receptor δ subunit that contribute to the binding of acetylcholine. *Proc. Natl. Acad. Sci. USA* **90:**6285–6289.
30. Camacho, P., Liu, Y., Mandel, G., and Brehm, P. (1993). The epsilon subunit confers fast channel gating on multiple classes of acetylcholine receptors. *J. Neurosci.* **13:**605–613.
31. Golino, M. D., and Hamill, O. P. (1992). Subunit requirements for *Torpedo* AChR channel expression: A specific role for the δ-subunit in voltage-dependent gating. *J. Membr. Biol.* **129:**297–309.
32. Liu, Y., and Brehm, P. (1993). Expression of subunit-omitted mouse nicotinic acetylcholine receptors in *Xenopus Laevis* oocytes. *J. Physiol. (London)* **470:**349–363.

33. Charnet, P., Labarca, C., and Lester, H. A. (1992). Structure of the γ-less nicotinic acetylcholine receptor: Learning from omission. *Mol. Pharmacol.* **41:**708–717.
34. Jackson, M. B., Imoto, K., Mishina, M., Konno, T., Numa, S., and Sakmann, B. (1990). Spontaneous and agonist-induced openings of an acetylcholine receptor channel composed of bovine muscle α, β, and δ-subunits. *Pfluegers Arch.* **417:**129–135.
35. Kullberg, R., Owens, J. L., Camacho, P., Mandel, G., and Brehm, P. (1990). Multiple conductance classes of mouse nicotinic acetylcholine receptors expressed in *Xenopus* oocytes. *Proc. Natl. Acad. Sci. USA* **87:**2067–2071.
36. Blount, P., Smith, M. M., and Merlie, J. P. (1990). Assembly intermediates of the mouse muscle nicotinic acetylcholine receptor in stably transfected fibroblasts. *J. Cell Biol.* **111:**2601–2611.
37. Gu, Y., Forsayeth, J. R., Verrall, S., Yu, X. M., and Hall, Z. W. (1991). Assembly of the mammalian muscle acetylcholine receptor in transfected COS cells. *J. Cell Biol.* **114:**799–807.
38. Wada, K., Ballivet, M., Boulter, J., Connolly, J., Wada, E., Deneris, E. S., Swanson, L. W., Heinemann, S., and Patrick, J. (1988). Functional expression of a new pharmacological subtype of brain nicotinic acetylcholine receptor. *Science* **240:**330–334.
39. Boulter, J., Connolly, J., Deneris, E., Goldman, D., Heinemann, S., and Patrick, J. (1987). Functional expression of two neuronal nicotinic acetylcholine receptors from cDNA clones identifies a gene family. *Proc. Natl. Acad. Sci. USA* **84:**7763–7767.
40. Boulter, J., O'Shea-Greenfield, A., Duvoisin, R. M., Connolly, J. G., Wada, E., Jensen, A., Gardner, P. D., Ballivet, M., Deneris, E. S., McKinnon, D., Heinemann, S., and Patrick, J. (1990). α3, α5, and β4: Three members of the rat neuronal nicotinic acetylcholine receptor-related gene family form a gene cluster. *J. Biol. Chem.* **265:**4472–4482.
41. Seguela, P., Wadiche, J., Dineley-Miller, K., Dani, J. A., and Patrick, J. W. (1993). Molecular cloning, functional properties, and distribution of rat brain α_7: A nicotinic cation channel highly permeable to calcium. *J. Neurosci.* **13:**596–604.
42. Nef, P., Oneyser, C., Alliod, C., Couturier, S., and Ballivet, M. (1988). Genes expressed in the brain define three distinct neuronal nicotinic acetylcholine receptors. *EMBO J.* **7:**595–601.
43. Couturier, S., Bertrand, D., Matter, J.-M., Hernandez, M.-C., Bertrand, S., Millar, N., Valera, S., Barkas, T., and Ballivet, M. (1990). A neuronal nicotinic acetylcholine receptor subunit (α7) is developmentally regulated and forms a homo-oligomeric channel blocked by α-BTX. *Neuron* **5:**847–856.
44. Couturier, S., Erkman, L., Valera, S., Rungger, D., Bertrand, S., Boulter, J., Ballivet, M., and Bertrand, D. (1990). α5, α3, and non-α3: Three clustered avian genes encoding neuronal nicotinic acetylcholine receptor-related subunits. *J. Biol. Chem.* **265:**17560–17567.
45. Schoepfer, R., Conroy, W. G., Whiting, P., Gore, M., and Lindstrom, J. (1990). Brain α-bungarotoxin binding protein cDNAs and MAbs reveal subtypes of this branch of the ligand-gated ion channel gene family. *Neuron* **5:**35–48.
46. Deneris, E. S., Connolly, J., Boulter, J., Wada, E., Wada, K., Swanson, L. W., Patrick, J., and Heinemann, S. (1988). Primary structure and expression of β2: A novel subunit of neuronal nicotinic acetylcholine receptors. *Neuron* **1:**45–54.
47. Deneris, E. S., Boulter, J., Swanson, L. W., Patrick, J., and Heinemann, S. (1989). β3: A new member of nicotinic acetylcholine receptor gene family is expressed in brain. *J. Biol. Chem.* **264:** 6268–6272.
48. Duvoisin, R. M., Deneris, E. S., Patrick, J., and Heinemann, S. (1989). The functional diversity of the neuronal nicotinic acetylcholine receptors is increased by a novel subunit: β4. *Neuron* **3:**487–496.
49. Bertrand, D., Bertrand, S., and Ballivet, M. (1992). Pharmacological properties of the homomeric α7 receptor. *Neurosci. Lett.* **146:**87–90.
50. Ballivet, M., Nef, P., Couturier, S., Rungger, D., Bader, C. R., Bertrand, D., and Cooper, E. (1988). Electrophysiology of a chick neuronal nicotinic acetylcholine receptor expressed in *Xenopus* oocytes after cDNA injection. *Neuron* **1:**847–852.
51. Cooper, E., Couturier, S., and Ballivet, M. (1991). Pentameric structure and subunit stoichiometry of a neuronal nicotinic acetylcholine receptor. *Nature* **350:**235–238.
52. Anand, R., Conroy, W. G., Schoepfer, R., Whiting, P., and Lindstrom, J. (1991). Neuronal nicotinic acetylcholine receptors expressed in *Xenopus* oocytes have a pentameric quarternary structure. *J. Biol. Chem.* **266:**11192–11198.
53. Papke, R. L., Boulter, J., Patrick, J., and Heinemann, S. (1989). Single-channel currents of rat neuronal nicotinic acetylcholine receptors expressed in *Xenopus* oocytes. *Neuron* **3:**589–596.
54. Wonnacott, S. (1987). Brain nicotine binding sites. *Hum. Toxicol.* **6:**343–353.
55. Wonnacott, S. (1986). α-bungarotoxin binds to low-affinity nicotine binding sites in rat brain. *J. Neurochem.* **47:**1706–1712.
56. Lindstrom, J., Schoepfer, R., and Whiting, P. (1987). Molecular studies of the neuronal nicotinic acetylcholine receptor family. *Mol. Neurobiol.* **1:**281–337.
57. Whiting, P., Esch, F., Shimasaki, S., and Lindstrom, J. (1987). Neuronal nicotinic acetylcholine receptor β-subunit is coded for by the cDNA clone α_4. *FEBS Lett.* **219:**459–463.
58. Schoepfer, R., Whiting, P., Esch, F., Blacher, R., Shimasaki, S., and Lindstrom, J. (1988). cDNA clones coding for the structural subunit of a chicken brain nicotinic acetylcholine receptor. *Neuron* **1:**241–248.
59. Flores, C. M., Rogers, S. W., Pabreza, L. A., Wolfe, B. B., Kellar, K. J. (1991). A subtype of nicotinic cholinergic receptor in rat brain is composed of α4 and β2 subunits and is up-regulated by chronic nicotine treatment. *Mol. Pharmacol.* **41:**31–37.
60. Conroy, W. G., Vernallis, A. B., and Berg, D. K. (1992). The α5 gene product assembles with multiple acetylcholine receptor subunits to form distinctive receptor subtypes in brain. *Neuron* **9:**679–691.
61. Keyser, K. T., Britto, L. R. G., Schoepfer, R., Whiting, P., Cooper, J., Conroy, W., Brozozowska-Prechtl, A., Karten, H. J., and Lindstrom, J. (1993). Three subtypes of α-bungarotoxin-sensitive nicotinic acetylcholine receptors are expressed in chick retina. *J. Neurosci.* **13:**442–454.
62. Conti-Tronconi, B. M., Dunn, S. M. J., Barnard, E. A., Dolly, J. O., Lai, F. A., Ray, N., and Raftery, M. A. (1985). Brain and muscle nicotinic acetylcholine receptors are different but homologous proteins. *Proc. Natl. Acad. Sci. USA* **82:**5208–5212.
63. Whiting, P., and Lindstrom, J. (1987). Purification and characterization of a nicotinic acetylcholine receptor from rat brain. *Proc. Natl. Acad. Sci. USA* **84:**595–599.
64. Anand, R., Peng, X., and Lindstrom, J. (1993). Homomeric and native α7 acetylcholine receptors exhibit remarkably similar but non-identical pharmacological properties, suggesting that the native receptor is a heteromeric protein complex. *FEBS Lett.* **327:** 241–246.
65. Listerud, M., Brussaard, A. B., Devay, P., Colman, D. R., and Role, L. W. (1991). Functional contribution of neuronal AChR subunits revealed by antisense oligonucleotides. *Science* **254:**1518–1521.
66. Moss, B. L., and Role, L. W. (1993). Enhanced ACh sensitivity is accompanied by changes in ACh receptor channel properties and segregation of ACh receptor subtypes on sympathetic neurons during innervation *in vivo*. *J. Neurosci.* **13:**13–28.
67. Jacob, M. H., Berg, D. K., and Lindstrom, J. M. (1984). Shared antigenic determinant between the *Electrophorus* acetylcholine receptor and a synaptic component on chicken ciliary ganglion neurons. *Proc. Natl. Acad. Sci. USA* **81:**3223–3227.
68. Jacob, M. H., and Berg, D. K. (1983). The ultrastructural localization of α-bungarotoxin binding sites in relation to synapses on chick ciliary ganglion neurons. *J. Neurosci.* **3:**260–271.

69. Loring, R. H., and Zigmond, R. E. (1987). Ultrastructural distribution of ^{125}I-toxin F binding sites on chick ciliary neurons: Synaptic localization of a toxin that blocks ganglionic nicotinic receptors. *J. Neurosci.* **7:**2153–2162.
70. Vernallis, A. B., Conroy, W. G., and Berg, D. K. (1993). Neurons assemble acetylcholine receptors with as many as three kinds of subunits while maintaining subunit segregation among receptor subtypes. *Neuron* **10:**451–464.
71. Chiappinelli, V. A. (1993). Neurotoxins acting on acetylcholine receptors. In *Natural and Synthetic Neurotoxins* (A. L. Harvey, ed.), Academic Press, London, pp. 66–128.
72. Role, L. W. (1992). Diversity in primary structure and function of neuronal nicotinic acetylcholine receptor channels. *Curr. Opin. Neurobiol.* **2:**254–262.
73. Katz, B., and Thesleff, S. (1957). A study of the 'desensitization' produced by acetylcholine at the motor end-plate. *J. Physiol. (London)* **138:**63–80.
74. Neher, E., and Steinbach, J. H. (1978). Local anaesthetics transiently block currents through single acetylcholine-receptor channels. *J. Physiol. (London)* **277:***153–176.*
75. Adams, P. R. (1977). Voltage pump analysis of procaine action at frog end-plate. *Eur. J. Physiol.* **268:**291–318.
76. Neely, A., and Lingle, C. J. (1986). Trapping of an open-channel blocker at the frog neuromuscular acetylcholine channel. *Biophys. J.* **50:**981–986.
77. Lambert, J. J., Durant, N. N., and Henderson, E. G. (1983). Drug-induced modification of ionic conductance at the neuromuscular junction. *Annu. Rev. Pharmacol. Toxicol.* **23:**505–539.
78. Colquhoun, D., Dreyer, F., and Sheridan, R. E. (1979). The actions of tubocurarine at the frog neuromuscular junction. *J. Physiol. (London)* **293:**247–284.
79. Tonner, P. H., Wood, S. C., and Miller, K. W. (1992). Can nicotine self-inhibition account for its low efficacy at the nicotinic acetylcholine receptor from *Torpedo? Mol. Pharmacol.* **42:**890–897.
80. Lukas, R. J. (1989). Pharmacological distinctions between functional nicotinic acetylcholine receptors on the PC12 rat pheochromocytoma and the TE671 human medulloblastoma. *J. Pharmacol. Exp. Ther.* **251:**175–182.
81. Marks, M. J., Farnham, D. A., Grady, S. R., and Collins, A. C. (1993). Nicotine receptor function determined by stimulation of rubidium efflux from mouse brain synaptosomes. *J. Pharmacol. Exp. Ther.* **264:**542–554.
82. Alkondon, M., and Albuquerque, E. X. (1993). Diversity of nicotinic acetylcholine receptors in rat hippocampal neurons. I. Pharmacological and functional evidence for distinct structural subtypes. *J. Pharmacol. Exp. Ther.* **265:**1455.
83. Zorumski, C. F., Thio, L. L., Isenberg, K. E., and Clifford, D. B. (1992). Nicotinic acetylcholine currents in cultured postnatal rat hippocampal neurons. *Mol. Pharmacol.* **41:**931–936.
84. Ravdin, P. M., and Berg, D. K. (1979). Inhibition of neuronal acetylcholine sensitivity by α-toxins from Bungarus multicintus venom. *Proc. Natl. Acad. Sci. USA* **76:**2072–2076.
85. Parke, R. L., Duvoisin, R. M., and Heinemann, S. F. (1993). The amino terminal half of the nicotinic β-subunit extracellular domain regulates the kinetics of inhibition by neuronal bungarotoxin. *Proc. R. Soc. London Ser. B.* **252:**141–148.
86. Paton, W. D. M., and Zaimis, E. J. (1949). The pharmacological actions of polymethylene bistrimethylammonium salts. *Br. J. Pharmacol.* **4:**381–400.
87. Hong, S. J., Tsuji, K., and Chang, C. C. (1992). Inhibition of neurosurugatoxin and ω-conotoxin of acetylcholine release and muscle and neuronal nicotinic receptors in mouse neuromuscular junction. *Neuroscience* **48:**727–735.
88. Ward, J. M., Cockcroft, V. B., Lunt, G. G., Smillie, F. S., and Wonnacott, S. (1990). Methyllycaconitine: A selective probe for neuronal α-bungarotoxin binding sites. *FEBS Lett.* **270:**45–48.
89. Zhang, Z.-W., and Feltz, P. (1991). Bicuculline blocks nicotinic acetylcholine response in isolated intermediate lobe cells of the pig. *Br. J. Pharmacol.* **102:**19–22.
90. Amador, M., and Dani, J. A. (1991). MK-801 inhibition of nicotinic acetylcholine receptor channels. *Synapse* **7:**207–215.
91. Aguayo, L. G., and Albuquerque, E. X. (1986). Effects of phencyclidine and its analogs on the end-plate current of the neuromuscular junction. *J. Pharmacol. Exp. Ther.* **239:**15–24.
92. Yakel, J. L., and Jackson, M. B. (1988). 5-HT_3 receptors mediate rapid responses in cultured hippocampus and a clonal cell line. *Neuron* **1:**615–621.
93. Peters, J. A., Hales, T. G., and Lambert, J. J. (1988). Divalent cations modulate 5-HT_3 receptor-induced currents in N1E-115 neuroblastoma cells. *Eur. J. Pharmacol.* **151:**491–495.
94. Siebler, M., Koller, H., Schmalenbach, C., and Muller, H. W. (1988). GABA activated chloride currents in cultured rat hippocampal and septal region neurons can be inhibited by curare and atropine. *Neurosci. Lett.* **93:**220–224.
95. Lebeda, F. J., Hablitz, J. J., and Johnston, D. (1982). Antagonism of GABA-mediated responses by *d*-tubocurarine in hippocampal neurons. *J. Neurophysiol.* **48:**622–632.
96. O'Dell, T. J., and Christensen, B. N. (1988). Mecamylamine is a selective non-competitive antagonist of N-methyl-D-aspartate- and aspartate-induced currents in horizontal cells dissociated from the catfish retina. *Neurosci. Lett.* **94:**93–98.
97. Luetje, W., and Patrick, J. (1991). Both α- and β-subunits contribute to the agonist sensitivity of neuronal nicotinic acetylcholine receptors. *J. Neurosci.* **11:**837–845.
98. Luetje, C. W., Wada, K., Rogers, S., Abramson, S. N., Tsuji, K., Heinemann, S., and Patrick, J. (1990). Neurotoxins distinguish between different neuronal nicotinic acetylcholine receptor subunit combinations. *J. Neurochem.* **55:**632–640.
99. Figl, A., Cohen, B. N., Quick, M. W., Davidson, N., and Lester, H. A. (1992). Regions of β4-β2 subunit chimeras that contribute to the agonist selectivity of neuronal nicotinic receptors. *FEBS Lett.* **308:**245–248.
100. Gross, A., Ballivet, M., Rungger, D., and Bertrand, D. (1991). Neuronal nicotinic acetylcholine receptors expressed in *Xenopus* oocytes: Role of the αsubunit in agonist sensitivity and desensitization. *Pfluegers Arch.* **419:**545–551.
101. Bertrand, D., Ballivet, M., and Rungger, D. (1990). Activation and blocking of neuronal nicotinic acetylcholine receptor reconstituted in *Xenopus* oocytes. *Proc. Natl. Acad. Sci. USA* **87:**1993–1997.
102. Bertrand, D., Devillers-Thiery, A., Revah, F., Galzi, J.-L., Hussy, N., Mulle, C., Bertrand, S., Ballivet, M., and Changeux, J.-P. (1992). Unconventional pharmacology of a neuronal nicotinic receptor mutated in the channel domain. *Proc. Natl. Acad. Sci. USA* **89:** 1261–1265.
103. Leonard, R. J., Labarca, C. G., Charnet, P., Davidson, N., and Lester, H. A. (1988). Evidence that the M2 membrane-spanning region lines the ion channel pore of the nicotinic receptor. *Science* **242:** 1578–1581.
104. Charnet, P., Cohen, B., Davidson, N., Lester, H. A., and Pilar, G. (1992). Pharmacological and kinetic properties of α4β2 neuronal nicotinic acetylcholine receptors expressed in *Xenopus* ooxytes. *J. Physiol. (London)* **450:**375–394.
105. Revah, F., Galzi, J.-L,. Giraudat, J., Haumont, P.-Y., Lederer, F., and Changeux, J.-P. (1990). The noncompetitive blocker [^{3}H] chlorpromazine labels three amino acids of the acetylcholine receptor α subunit: Implications for the α-helical organization of regions MII and for the structure of the ion channel. *Proc. Natl. Acad. Sci. USA* **87:**4675–4679.
106. Kao, P. N., Dwork, A. J., Kaldany, R.-R. J., Silver, M. L., Wideman, J., Stein, S., and Karlin, A. (1984). Identification of the α subunit half-cystine specifically labeled by an affinity reagent for the acetylcholine receptor binding site. *J. Biol. Chem.* **259:**11662–11665.

107. Tomaselli, G. F., McLaughlin, J. T., Jurman, M., Hawrot, E., and Yellen, G. (1991). Site-directed mutagenesis alters agonist sensitivity of the nicotinic acetylcholine receptor. *Biophys. J.* **60:**721–729.
108. O'Leary, M. E., and White, M. M. (1992). Mutational analysis of ligand-induced activation of the *Torpedo* acetylcholine receptor. *J. Biol. Chem.* **267:**8360–8365.
109. Galzi, J.-L., Bertrand, D., Devillers-Thiery, A., Revah, F., Bertrand, S., and Changeux, J.-P. (1991). Functional significance of aromatic amino acids from three peptide loops of the α7 neuronal nicotinic receptor site investigated by site-directed mutagenesis. *FEBS Lett.* **294:**198–202.
110. Czajkowski, C., and Karlin, A. (1991). Agonist binding site of *Torpedo* electric tissue nicotinic acetylcholine receptor. *J. Biol. Chem.* **266:**22603–22612.
111. Chiara, D. C., and Cohen, J. B. (1992). Identification of amino acids contributing to high and low affinity D-tubocurarine (dTC) sites on the *Torpedo* nicotinic acetylcholine receptor (nAChR) subunits. *Biophys. J.* **61:**A106.
112. Barchan, D., Kachalsky, S., Neumann, D., Vogel, Z., Ovadia, M., Kochva, E., and Fuchs, S. (1992). How the mongoose can fight the snake: The binding site of the mongoose acetylcholine receptor. *Proc. Natl. Acad. Sci. USA* **89:**7717–7721.
113. Tzartos, S. J., and Remoundos, M. S. (1990). Fine localization of the major α-bungarotoxin binding site to residues α189–195 of the *Torpedo* acetylcholine receptor. *J. Biol. Chem.* **265:**21462–21467.
114. Revah, F., Bertrand, D., Galzi, J.-L., Devillers-Thiery, A., Mulle, C., Hussy, N., Bertrand, S., Ballivet, M., and Changeux, J.-P. (1991). Mutations in the channel domain alter desensitization of a neuronal nicotinic receptor. *Nature* **353:**846–849.
115. Mulle, C., and Changeux, J.-P. (1990). A novel type of nicotinic receptor in the rat central nervous system characterized by patch-clamp techniques. *J. Neurosci.* **10:**169–175.
116. Karlin, A. (1980). Molecular properties of nicotinic acetylcholine receptors. In *Cell Surface Reviews* (C. W. Cotman, G. Poste, and G. L. Nicolson, eds.), North Holland, Amsterdam, pp. 192–260.
117. Lukas, R. J. (1990). Heterogeneity of high-affinity nicotinic [^{3}H]acetylcholine binding sites. *J. Pharmacol. Exp. Ther.* **253:**51–57.
118. Feltz, A., and Trautmann, A. (1982). Desensitization at the frog neuromuscular junction: A biphasic response. *J. Physiol. (London)* **322:** 257–272.
119. Cachelin, A. B., and Colquhoun, D. (1988). Desensitization of the acetylcholine receptor of frog end-plates measured in a vaseline-gap voltage clamp. *J. Physiol. (London)* **415:**159–188.
120. Cachelin, A. B., and Jaggi, R. (1991). β subunits determine the time course of desensitization in rat α3 neuronal nicotinic acetylcholine receptors. *Pfluegers Arch.* **419:**579–582.
121. Mathie, A., Colquhoun, D., and Cull-Candy, G. (1990). Rectification of currents activated by nicotinic acetylcholine receptors in rat sympathetic ganglion neurones. *J. Physiol. (London)* **427:**625–655.
122. Ifune, C. K., and Steinbach, J. H. (1993). Modulation of acetylcholine-elicited currents in clonal rat phaeochromocytoma (PC12) cells by internal polyphosphates. *J. Physiol. (London)* **463:**431–447.
123. Giacobini, E. (1986). Brain acetylcholine—A view from the cerebrospinal fluid (CSF). *Neurobiol. Aging* **7:**392–395.
124. Whitehouse, P. J., Martino, A. M., Antuono, P. G., Lowenstein, P. R., Coyle, J. T., Price, D. L., and Kellar, K. J. (1986). Nicotinic acetylcholine binding sites in Alzheimer's disease. *Brain Res.* **371:**146–151.
125. Miles, K., and Huganir, R. L. (1988). Regulation of nicotinic acetylcholine receptors by protein phosphorylation. *Mol. Neurobiol.* **2:** 91–124.
126. Wagoner, P. K., and Pallotta, B. S. (1988). Modulation of acetylcholine receptor desensitization by forskolin is independent of cAMP. *Science* **240:**1655–1657.
127. Reuhl, T. O. K., Amador, M., Moorman, J. R., Pinkham, J., and Dani, J. A. (1992). Nicotinic acetylcholine receptors are directly affected by agents used to study protein phosphorylation. *J. Neurophysiol.* **68:**407–416.
128. Aylwin, M. L,. and White, M. M. (1992). Forskolin acts as a noncompetitive inhibitor of nicotinic acetylcholine receptors. *Mol. Pharmacol.* **41:**908–913.
129. Covarrubias, M., and Steinbach, J. H. (1990). Excision of membrane patches reduces the mean open time of nicotinic acetylcholine receptors. *Plfuegers Arch.* **416:**385–392.
130. Siara, J., Ruppersberg, J. P., and Rudel, R. (1990). Human nicotinic acetylcholine receptor: The influence of second messengers on activation and desensitization. *Pfluegers Arch.* **415:**701–706.
131. Schuetze, S. M., and Role, L. W. (1987). Developmental regulation of nicotinic acetylcholine receptors. *Annu. Rev. Neurosci.* **10:** 403–457.
132. Simmons, L. K., Schuetze, S. M., and Role, L. W. (1990). Substance P modulates single-channel properties of neuronal nicotinic acetylcholine receptors. *Neuron* **2:**393–403.
133. Downing, J. E. G., and Role, L. W. (1987). Activators of protein kinase C enhance acetylcholine receptor desensitization in sympathetic ganglion neurons. *Proc. Natl. Acad. Sci. USA* **84:**7739–7743.
134. Margiotta, J. F., Berg, D. K., and Dionne, V. E. (1987). Cyclic AMP regulates the proportion of functional acetylcholine receptors on chicken ciliary ganglion neurons. *Proc. Natl. Acad. Sci. USA* **84:** 8155–8159.
135. Higgins, L. S., and Berg, D. K. (1988). Cyclic AMP-dependent mechanism regulates acetylcholine receptor function on bovine adrenal chromaffin cells and discriminates between new and old receptors. *J. Cell Biol.* **107:**1157–1165.
136. Dubin, A. E., Rathouz, M. M., Mapp, K. S., and Berg, D. K. (1992). Cyclic AMP and the nicotinic response of bovine adrenal chromaffin cells. *Brain Res.* **586:**344–347.
137. Norenberg, W., Illes, P., and Takeda, K. (1991). Neuropeptide Y inhibits nicotinic cholinergic currents but not voltage-dependent calcium currents in bovine chromaffin cells. *Pfluegers Arch.* **418:** 346–352.
138. Zhu, J., Li, W., Toews, M. L., and Hexum, T. D. (1992). Neuropeptide Y inhibits forskolin-stimulated adenylate cyclase in bovine adrenal chromaffin cells via a pertussis toxin-sensitive process. *J. Pharmacol. Exp. Ther.* **263:**1479–1486.
139. Quik, M., Cook, R. G., Revah, F., Changeux, J.-P., and Patrick, J. (1993). Presence of α-cobratoxin and phospholipase A_2 activity in thymopoietin preparations. *Mol. Pharmacol.* **44:**368–369.
140. Clapham, D. E., and Neher, E. (1984). Substance P reduces acetylcholine-induced currents in isolated bovine chromaffin cells. *J. Physiol. (London)* **347:**255–277.
141. Boyd, N. D., and Leeman, S. E. (1987). Multiple actions of substance P that regulate the functional properties of acetylcholine receptors of clonal rat PC12 cells. *J. Physiol. (London)* **389:**69–97.
142. Eardley, D., and McGee, R. (1985). Both substance P agonists and antagonists inhibit ion conductance through nicotinic acetylcholine receptors on PC12 cells. *Eur. J. Pharmacol.* **114:**101–104.
143. Bormann, J., Flugge, G., and Fuchs, E. (1989). Effect of atrial natriuretic factor (ANF) on nicotinic acetylcholine receptor channels in bovine chromaffin cells. *Pfluegers Arch.* **414:**11–14.
144. Valera, S., Ballivet, M., and Bertrand, D. (1992). Progesterone modulates a neuronal nicotinic acetylcholine receptor. *Proc. Natl. Acad. Sci. USA* **89:**9949–9953.
145. Vernino, S., Amador, M., Luetje, C. W., Patrick, J., and Dani, J. A. (1992). Calcium modulation and high calcium permeability of neuronal nicotinic acetylcholine receptors. *Neuron* **8:**127–134.
146. Mulle, C., Lena, C., and Changeux, J.-P. (1992). Potentiation of nicotinic receptor response by external calcium in rat central neurons. *Neuron* **8:**937–945.
147. Benninger, C., Kadis, J., and Prince, D. A. (1980). Extracellular calcium and potassium changes in hippocampal slices. *Brain Res.* **187:**165–182.

148. Pumain, R., and Heinemann, U. (1985). Stimulus- and amino acid-induced calcium and potassium changes in rat neocortex. *Neurophysiology* **53:**1–16.
149. Semba, K., and Fibiger, H. C. (1989). Organization of central cholinergic systems. *Prog. Brain Res.* **79:**37–63.
150. Anderson, C. R., and Stevens, C. F. (1973). Voltage-clamp analysis of acetylcholine produced end-plate current fluctuations at frog neuromuscular junction. *J. Physiol. (London)* **235:**655–691.
151. Mathie, A., Cull-Candy, S. G., and Colquhoun, D. (1991). Conductance and kinetic properties of single nicotinic acetylcholine receptor channels in rat sympathetic neurones. *J. Physiol. (London)* **439:** 717–750.
152. Papke, R. L., and Heinemann, S. F. (1991). The role of the β_4-subunit in determining the kinetic properties of rat neuronal nicotinic acetylcholine α_3-receptors. *J. Physiol. (London)* **440:**95–112.
153. Maconochie, D. J., and Knight, D. E. (1992). A study of the bovine adrenal chromaffin nicotinic receptor using patch clamp and concentration-jump techniques. *J. Physiol. (London)* **454:**129–153.
154. Nicoll, R. A., Malenka, R. C., and Kauer, J. A. (1990). Functional comparison of neurotransmitter receptor subtypes in mammalian central nervous system. *Physiol. Rev.* **70:**514–566.
155. Egan, T. M. (1989). Single cell studies of the actions of agonists and antagonists on nicotinic receptors of the central nervous system. *Prog. Brain Res.* **79:**73–83.
156. Clarke, P. B. S. (1990). The central pharmacology of nicotine: Electrophysiological approaches. In *Nicotine Psychopharmacology* (S. Wonnacott, N. A. H. Russell, and I. P. Stolerman, eds.), Oxford University Press, London, pp. 158–193.
157. Eccles, J. C., Eccles, R. M., and Fatt, P. (1956). Pharmacological investigations on a central synapse operated by acetylcholine. *Eur. J. Physiol.* **131:**154–169.
158. King, K. T., and Ryal, R. W. (1981). A re-evaluation of acetylcholine receptors on feline Renshaw cells. *Br. J. Pharmacol.* **73:** 455–460.
159. Brown, D. A., Docherty, R. J., and Halliwell, J. V. (1983). Chemical transmission in the rat interpeduncular nucleus *in vitro. J. Physiol. (London)* **341:**655–670.
160. Misgeld, B., Weiler, M. H., and Bak, I. J. (1980). Intrinsic cholinergic excitation in the rat neostriatum: Nicotinic and muscarinic receptors. *Ex. Brain Res.* **39:**401–409.
161. Hatton, G. I., Ho, Y. W., and Mason, W. T. (1983). Synaptic activation of phasic bursting in rat supraoptic nucleus neurones recorded in hypothalamic slices. *J. Physiol. (London)* **345:**297–317.
162. Cordingley, G. E., and Weight, F. F. (1986). Non-cholinergic synaptic excitation in neostriatum: Pharmacological evidence for mediation by a glutamate-like transmitter. *Br. J. Pharmacol.* **88:**847–856.
163. Gribkoff, V. K., and Dudek, E. (1990). Effects of excitatory amino acid antagonists on synaptic responses of supraoptic neurons in slices of rat hypothalamus. *J. Neurophysiol.* **63:**60–71.
164. Daw, N. W., Brunken, W. J., and Parkinson, D. (1989). The function of synaptic transmitters in the retina. *Annu. Rev. Neurosci.* **12:** 205–225.
165. Lipton, S. A., Aizenman, E., and Loring, R. H. (1987). Neuronal nicotinic acetylcholine responses in solitary mammalian retinal ganglion cells. *Pfluegers Arch.* **410:**37–43.
166. Ariel, M., and Daw, N. W. (1982). Pharmacological analysis of directionally sensitive rabbit retinal ganglion cells. *J. Physiol. (London)* **324:**161–185.
167. Wong, L. A., and Gallagher, J. P. (1991). Pharmacology of nicotinic receptor-mediated inhibition in rat dorsolateral septal neurones. *J. Physiol. (London)* **436:**325–346.
168. Tokimasa, T., and North, R. A. (1984). Calcium entry through acetylcholine-channels can activate potassium conductance in bullfrog sympathetic neurons. *Brain Res.* **295:**364–369.
169. Swanson, L. W., Simmons, D. M., Whiting, P. J., and Lindstrom, J. (1987). Immunohistochemical localization of neuronal nicotinic receptors in the rodent central nervous system. *J. Neurosci.* **7:** 3334–3342.
170. Prusky, G. T., Shaw, C., and Cynader, M. S. (1987). Nicotine receptors are located on lateral geniculate nucleus terminals in cat visual cortex. *Brain Res.* **412:**131–138.
171. Clarke, P. B. S., Hamill, G. S. Nadi, N. S., Jacobowitz, D. M., and Pert, A. (1986). ^{3}H-nicotine- and ^{125}I-α-bungarotoxin-labeled nicotinic receptors in the interpeduncular nucleus of rats. II. Effects of habenular deafferentation. *J. Comp. Neurol.* **251:**407–413.
172. Schwartz, R. D., Lehmann, J., and Kellar, K. J. (1984). Presynaptic nicotinic cholinergic receptors labeled by [^{3}H]acetylcholine on catecholamine and serotonin axons in brain. *J. Neurochem.* **42:**1495–1498.
173. Clarke, P. B. S., and Pert, A. (1985). Autoradiographic evidence for nicotine receptors on nigrostriatal and mesolimbic dopaminergic neurons. *Brain Res.* **348:**355–358.
174. Swanson, L. W., Lindstrom, J., Tzartos, S., Schmued, L. C., O'Leary, D. D. M., and Cowan, W. M. (1983). Immunohistochemical localization of monoclonal antibodies to the nicotinic acetylcholine receptor in chick midbrain. *Proc. Natl. Acad. Sci. USA* **80:**4532–4536.
175. McCormick, D. A., and Prince, D. A. (1987). Actions of acetylcholine in the guinea-pig and cat medial and lateral geniculate nuclei, *in vitro. J. Physiol. (London)* **392:**147–165.
176. McCormick, D. A., and Prince, D. A. (1987). Acetylcholine causes rapid nicotinic excitation in the medial habenular nucleus of guinea pig, *in vitro. J. Neurosci.* **7:**742–752.
177. Brown, D. A., Docherty, R. J., and Halliwell, J. V. (1984). The action of cholinomimetic substances on impulse conduction in the habenulointerpeduncular pathway of the rat *in vitro. J. Physiol. (London)* **353:**101–109.
178. Mulle, C., Vidal, C., Benoit, P., and Changeux, J.-P. (1991). Existence of different subtypes of nicotinic acetylcholine receptors in the rat habenulo-interpeduncular system. *J. Neurosci.* **11:**2588–2597.
179. McMahon, L. L., Yoon, K.-W., and Chiappinelli, V. A. (1994). Nicotinic receptor activation facilitates GABAergic neurotransmission in the avian lateral spiriform nucleus. *Neuroscience* **59:**689–698.
180. McMahon, L. L., Yoon, K.-W., and Chiappinelli, V. A. (1993). Electrophysiological evidence for presynaptic nicotinic receptors in the avian ventral lateral geniculate nucleus. *J. Neurophysiol.* **71:**826–829.
181. Lena, C., Changeux, J.-P., and Mulle, C. (1993). Evidence for "preterminal" nicotinic receptors on GABAergic axons in the rat interpeduncular nucleus. *J. Neurosci.* **13:**2680–2688.
182. Bowman, W. C., Marshall, I. J., Gibb, A. J., and Harborne, A. J. (1988). Feedback control of transmitter release at the neuromuscular junction. *Trends Pharmacol. Sci.* **9:**16–20.
183. Wessler, I. (1989). Control of transmitter release from the motor nerve by presynaptic nicotinic and muscarinic autoreceptors. *Trends Pharmacol. Sci.* **10:**110–114.
184. Chesselet, M.-F. (1984). Presynaptic regulation of neurotransmitter release in the brain: Facts and hypothesis. *Neuroscience* **12:** 347–375.
185. Rowell, P. P. (1987). Current concepts on the effects of nicotine on neurotransmitter release in the central nervous system. In *Tobacco Smoking and Nicotine* (W. R. Martin, G. R. Van Loon, E. T. Iwamoto, and L. Davis, eds.), Plenum Press, New York, pp. 191–208.
186. Wonnacott, S., Drasdo, A., Sanderson, E., and Rowell, P. (1990). Presynaptic nicotinic receptors and the modulation of transmitter release. *Ciba Found. Symp.* **152:**87–101.
187. Rapier, C., Lunt, G. G., Wonnacott, S. (1988). Stereoselective nicotine-induced release of dopamine from striatal synaptosomes: Concentration dependence and repetitive stimulation. *J. Neurochem.* **50:**1123–1130.

188. Rapier, C., Lunt, G. G., Wonnacott, S. (1990). Nicotinic modulation of [^{3}H]dopamine release from striatal synaptosomes: Pharmacological characterisation. *J. Neurochem.* **54:**937–945.
189. Grady, S., Marks, M. J., Wonnacott, S., and Collins, A. C. (1992). Characterization of nicotinic receptor-mediated [^{3}H]dopamine release from synaptosomes prepared from mouse striatum. *J. Neurochem.* **59:**848–856.
190. Harsing, L. G. J., Serchen, H., Vizi, S. E., and Lajtha, A. (1992). N-type calcium channels are involved in the dopamine releasing effect of nicotine. *Neurochem. Res.* **17:**729–734.
191. Takeuchi, N. (1963). Effects of calcium on the conductance change of the end-plate membrane during the action of transmitter. *Eur. J. Physiol.* **167:**141–155.
192. Decker, E. R., and Dani, J. A. (1990). Calcium permeability of the nicotinic acetylcholine receptor: The single-channel calcium influx is significant. *J. Neurosci.* **10:**3413–3420.
193. Mulle, C., Choquet, D., Korn, H., and Changeux, J.-P., (1992). Calcium influx through nicotine receptor in rat central neurons: Its relevance to cellular regulation. *Neuron* **8:**135–143.
194. Trouslard, J., Marsh, S. J., and Brown, D. A. (1993). Calcium entry through nicotinic receptor channels and calcium channels in cultured rat superior cervical ganglion cells. *J. Physiol. (London)* **468:**53–71.
195. Zhou, Z., and Neher, E. (1993). Calcium permeability of nicotinic ACh-receptors in bovine chromaffin cells. *Soc. Neurosci. Abstr.* **19:**280.
196. Sands, S. B., and Barish, M. E. (1991). Calcium permeability of neuronal nicotinic acetylcholine receptor channels in PC12 cells. *Brain Res.* **560:**38–42.
197. Mayer, M., and Westbrook, G. (1987). Permeation and block of N-methyl-D-aspartic acid receptor channels by divalent cations in mouse cultured central neurons. *Eur. J. Physiol.* **394:**501–527.
198. Inno, M., Ozawa, S., and Tsuzuki, K. (1990). Permeation of calcium through excitatory amino acid receptor channels in cultured rat hippocampal neurones. *Eur. J. Physiol.* **424:**151–165.
199. Adams, D. J., and Nutter, T. J. (1992). Calcium permeability and modulation of nicotinic acetylcholine receptor-channels in rat parasympathetic neurons. *J. Physiol. (Paris)* **86:**67–76.
200. Vijayaraghavan, S., Pugh, P. C., Zhang, Z.-W., Rathouz, M. M., and Berg, D. K. (1992). Nicotinic receptors that bind α-bungarotoxin on neurons raise intracellular free Ca^{2+}. *Neuron* **8:**353–362.
201. Freedman, R., Wetmore, C., Stromberg, I., Leonard, S., and Olson, L. (1993). α-Bungarotoxin binding to hippocampal interneurons: Immunocytochemical characterization and effects on growth factor expression. *J. Neurosci.* **13:**1965–1975.
202. Augustine, G. J., and Neher, E. (1992). Neuronal Ca^{2+} signalling takes the local route. *Curr. Opin. Neurobiol.* **2:**302–307.
203. Llinas, R., Sugimori, M., and Silver, R. B. (1992). Microdomains of high calcium concentration in a presynaptic terminal. *Science* **256:** 677–679.
204. Katz, B., and Miledi, R. (1968). The role of calcium in neuromuscular facilitation. *J. Physiol. (London)* **195:**481–492.
205. Llinas, R., Gruner, J. A., and Sugimori, M. (1991). Regulation by synapsin I and Ca^{2+}-calmodulin-dependent protein kinase II of the transmitter release in squid giant synapse. *Eur. J. Physiol.* **436:**257–282.
206. Katz, B., and Miledi, R. (1977). Transmitter leakage from motor nerve endings. *Proc. R. Soc. London Ser. B* **196:**59–72.
207. Lipton, S. A. (1988). Spontaneous release of acetylcholine affects the physiological nicotinic responses of rat retinal ganglion cells in culture. *J. Neurosci.* **8:**3857–3868.
208. Surgeon General. (1988). Nicotine Addiction—The Health Consequences of Smoking, U.S. Department of Health and Human Services, Rockville, MD.
209. Anon. (1992). Special issue on nicotine. *Psychopharmacology* **108:**393–526.
210. Wilberg, J. (1987). *Tobacco and Shamanism in South America,* Yale University Press, New Haven, CT.
211. Fairholt, F. W. (1859). *Tobacco: Its History and Associations: An Account of the Plant and its Manufacture; with its Modes of use in all Ages and Countries,* Chapman & Hall, London.
212. Mackenzie, C. (1957). *Sublime Tobacco,* Chatto & Windus, London.
213. Anon. (1991). From directions in tobacco research. *Br. J. Addict.* **86:**483–666.
214. Stolerman, I. P., Goldfarb, T., Fink, R., and Jarvik, M. E. (1973). Influencing cigarette smoking with nicotine antagonists. *Psychopharmacologia,* **28:**247–259.
215. Pomerleau, C. S., Pomerleau, O. F., and Majchrzak, M. J. (1987). Mecamylamine pretreatment increases subsequent nicotine self-administration as indicated by changes in plasma nicotine level. *Psychopharmacology* **91:**391–393.
216. Nemeth-Coslett, R., Henningfield, J. E., O'Keeffe, M. K., and Griffiths, R. R. (1986). Effects of mecamylamine on human cigarette smoking and subjective ratings. *Psychopharmacology* **88:**420–425.
217. Pickworth, W. B., Herning, R. I., and Henningfield, J. E. (1988). Mecamylamine reduces some EEG effects of nicotine chewing gum in humans. *Pharmacol. Biochem. Behav.* **30:**149–153.
218. Rose, J. E., Sampson, A., Levin, E. D., and Henningfield, J. E. (1989). Mecamylamine increases nicotine preference and attenuates nicotine discrimination. *Pharmacol. Biochem. Behav.* **32:**933–938.
219. Newhouse, P. A., Potter, A., Corwin, J., and Lenox, R. (1992). Acute nicotinic blockade produces cognitive impairment in normal humans. *Psychopharmacology* **108:**480–484.
220. Elrod, K., Buccafusco, J. J., and Jackson, W. J. (1988). Nicotine enhances delayed matching-to-sample performance by primates. *Life Sci.* **43:**277–287.
221. Feyerabend, C., Ings, R. M., and Russell, M. A. H. (1985). Nicotine pharmacokinetics and its application to intake from smoking. *Br. Clin. Pharm. Ther.* **19:**239–247.
222. Russell, M. A. H. (1987). Nicotine intake and its regulation by smokers. In *Tobacco Smoking and Nicotine* (W. R. Martin, G. R. Van Loon, E. T. Iwamoto, and L. Davis, eds.), Plenum Press, New York, pp. 25–50.
223. Benowitz, N. L., Porchet, H., and Jacob, P. I. I. I. (1989). Nicotine dependence and tolerance in man: Pharmacokinetic and pharmacodynamic investigations. *Prog. Brain Res.* **79:**279–287.
224. Plowchalk, D. R., Andersen, M. E., and Debethizy, D. (1992). A physiologically based pharmacokinetic model for nicotine disposition in the Sprague–Dawley rat. *Toxicol. Appl. Pharmacol.* **116:** 177–188.
225. Stalhandske, T. (1970). Effects of increased liver metabolism of nicotine on its uptake, elimination and toxicity in mice. *Acta Physiol. Scand.* **80:**222–234.
226. Nordberg, A., Hartvig, P., Lundqvist, H., Antoni, G., Ulin, J., and Langstrom, B. (1989). Uptake and regional distribution of (+)-(R)- and (−)-(S)-N-[methyl-^{11}C]-nicotine in the brains of rhesus monkey: An attempt to study nicotinic receptors *in vivo. J. Neural Transm.* **1:**195–205.
227. Rand, M. J. (1989). Neuropharmacological effects of nicotine in relation to cholinergic mechanisms. *Prog. Brain Res.* **79:**3–11.
228. Clarke, P. B. S. (1990). Mesolimbic dopamine activation—the key to nicotine reinforcement? *Ciba Found. Symp.* **152:**153–168.
229. Wonnacott, S. (1990). The paradox of nicotinic acetylcholine receptor upregulation by nicotine. *Trends Pharmacol. Sci.* **11:**216–219.
230. Benwell, M. E. M., Balfour, D. J. K., and Anderson, J. M. (1988). Evidence that tobacco smoking increases the density of (−)-[^{3}H]nicotine binding sites in human brain. *J. Neurochem.* **50:** 1243–1247.
231. Marks, M. J., Pauly, J. R., Gross, D., Deneris, E. S., Hermans-Borgmeyer, I., Heinemann, S. F., and Collins, A. C. (1992). Nicotine binding and nicotinic receptor subunit RNA after chronic nicotine treatment. *J. Neurosci.* **12:**2765–2784.

232. Gottfries, C. G. (1985). Alzheimer's disease and senile dementia: Biochemical characteristics and aspects of treatment. *Psychopharmacology* **86:**245–252.
233. Flynn, D. D., and Mash, D. C. (1986). Characterization of L-[^{3}H]nicotine binding in human cerebral cortex: Comparison between Alzheimer's disease and the normal. *J. Neurochem.* **47:**1948–1954.
234. Nordberg, A., Nilsson-Hakansson, L., Adem, A., Hardy, J., Alfuzoss, I., Lai, Z., Herrera-Marschitz, M., and Winblad, B., (1989). The role of nicotinic receptors in the pathophysiology of Alzheimer's disease. *Prog. Brain Res.* **79:**353–362.
235. Perry, E. K., Perry, R. H., Smith, C. J., Dick, D. J., Candy, J. M., Edwardson, J. A., Fairbairn, A., and Blessed, G. (1987). Nicotinic receptor abnormalities in Alzheimer's and Parkinson's diseases. *J. Neurol. Neurosurg. Psychiatry* **50:**806–809.
236. Jones, G. M. M., Sahakian, B. J., Levy, D. R., Warburton, D. M., and Gray, J. A. (1992). Effects of acute subcutaneous nicotine on attention, information processing and short-term memory in Alzheimer's disease. *Psychopharmacology* **108:**485–494.
237. Newhouse, P. A., Sunderland, T., Narang, P. K., Mellow, A. M., Fertig, J. B., Lawlor, B. A., and Murphy, D. (1990). Neuroendocrine, physiologic, and behavioral responses following intravenous nicotine in nonsmoking healthy volunteers and in patients with Alzheimer's disease. *Psychoneuroendocrinology,* **15:**471–484.
238. Farlow, M., Gracon, S. I., Hershey, L. A., Lewis, K. W., Sadowsky, C. H., and Dolan-Ureno, J. (1992). A controlled trial of Tacrine in Alzheimer's disease. *Am. Med. Assoc.* **268:**2523–2529.
239. Davis, K. L., Thal, L. J., Gamzu, E. R., Davis, C. S., Woolson, R. F., Gracon, S. I., Drachman, D. A., Schneider, L. S., Whitehouse, P. J., Hoover, T. M., Morris, J. C., Kawas, C. H., Knopman, D. S., Earl, N. L., Kumar, V., and Doody, R. S. (1992). A double-blind, placebo-controlled multicenter study of Tacrine for Alzheimer's disease. *N. Engl. J. Med.* **327:**1253–1259.
240. Lange, K. W., Wells, F. R., Jenner, P., and Marsden, C. D. (1993). Altered muscarinic and nicotinic receptor densities in cortical and subcortical brain regions in Parkinson's disease. *J. Neurochem.* **60:**197–203.
241. Baron, J. A. (1986). Cigarette smoking and Parkinson's disease. *Neurology* **36:**1490–1496.
242. Kirch, D. G., Alho, A. M., and Wyatt, R. J. (1988). Hypothesis: A nicotine–dopamine interaction linking smoking with Parkinson's disease. *Cell. Mol. Neurobiol.* **8:**285–291.
243. Calne, D. B., and Langston, J. W. (1983). Aetiology of Parkinson's disease. *Lancet* **2:**1257–1459.
244. Rajput, A. H., Offord, K. P., Beard, C. M., and Kurland, L. T. (1987). A case–control study of smoking habits, dementia, and other illnesses in idiopathic Parkinson's disease. *Neurology* **37:**226–232.
245. Golbe, L. I., Cody, R. A., and Duvoisin, R. C. (1986). Smoking and Parkinson's disease: Search for a dose–response relationship. *Arch. Neurol.* **43:**774–778.
246. Riggs, J. E. (1992). Cigarette smoking and Parkinson's disease: The illusion of a neuroprotective effect. *Clin. Neuropharmacol.* **15:** 88–99.
247. Hughes, J. R., Hatsukami, D. K., Mitchell, J. E., and Dahlgren, L. A. (1986). Prevalence of smoking among psychiatric outpatients. *Am. J. Psychiatry* **143:**993–997.
248. O'Farrell, T. J., Connors, G. J., and Upper, D. (1986). Prevalence of smoking among psychiatric outpatients. *Addict. Behav.* **8:**329–333.
249. Freedman, R., Waldo, M., Bickford-Wimer, P., and Nagamoto, H. (1991). Elementary neuronal dysfunctions in schizophrenia. *Schizophrenia Res.* **4:**233–243.
250. Adler, L. E., Hoffer, L. J., Griffith, J., Waldo, M. C., and Freedman, R. (1992). Normalization by nicotine of deficient auditory sensory gating in the relatives of schizophrenics. *Soc. Biol. Psychiatry* **32:**607–616.
251. Ogden, D. C., and Colquhoun, D. (1985). Ion channel block by acetylcholine, carbachol and suberyldicholine at the frog neuromuscular junction. *Proc. R. Soc. London Ser. B* **225:**329–355.
252. Connolly, J., Boulter, J., and Heinemann, S. (1992). α4-β2 and other nicotinic acetylcholine receptor subtypes as targets of psychoactive and addictive drugs. *Br. J. Pharmacol.* **105:**657–666.
253. Margiotta, J. F., Berb, D. K., and Dionne, V. E. (1987). The properties and regulation of functional acetylcholine receptors on chick ciliary ganglion neurons. *J. Neurosci.* **7:**3612–3622.
254. Benson, J. (1988). Pharmacology of a locust thoracic ganglion somal nicotinic acetylcholine receptor. In *NATO ASI Series V H25: Nicotinic Acetylcholine Receptors in the Nervous System* (F. Clementi, C. Gotti, and E. Sher, eds.), Springer-Verlag, Berlin, pp. 227–240.
255. Vernallis, A. B., Conroy, W. G., Corriveau, R. A., Halvorsen, S. W., and Berg, D. K. (1991). AChR gene products in chick ciliary gangliary transcripts subunits and receptor subtypes. *Soc. Neurosci. Abstr.* **17:**12.
256. Role, L. W. (1988). Neural regulation of acetylcholine sensitivity in embryonic sympathetic neurons. *Proc. Natl. Acad. Sci. USA* **85:** 2825–2829.
257. Wada, E., Wada, K., Boulter, J., Deneris, E., Heinemann, S., Patrick, J., and Swanson, L. W. (1989). Distribution of α2, α3, α4, and β2 neuronal nicotinic receptor subunit mRNAs in the central nervous system: A hybridization histochemical study in the rat. *J. Comp. Neurol.* **284:**314–335.
258. Higgins, L. S., and Berg, D. K. (1987). Immunological identification of a nicotinic acetylcholine receptor on bovine chromaffin cells. *J. Neurosci.* **7:**1792–1798.
259. Fieber, L. A., and Adams, D. J. (1991). Acetylcholine-evoked currents in cultured neurones dissociated from rat parasympathetic cardiac ganglia. *J. Physiol. (London)* **434:**215–237.
260. Mathie, A., Cull-Candy, S. G., and Colquhoun, D. (1987). Single-channel and whole-cell currents evoked by acetylcholine in dissociated sympathetic neurons of the rat. *Proc. R. Soc. London Ser. B* **232:**239–248.
261. Sah, D. W. Y., Loring, R. H., and Zigmond, R. E. (1987). Long-term blockade by toxin F of nicotinic synaptic potentials in cultured sympathetic neurons. *Neuroscience* **20:**867–874.
262. Patrick, J., and Stallcup, W. B. (1977). Immunological distinction between acetylcholine receptor and the α-bungarotoxin-binding component on sympathetic neurons. *Proc. Natl. Acad. Sci. USA* **74:** 4689–4692.
263. Roger, S. W., Mandelzys, A., Deneris, E. S., Cooper, E., and Heinemann, S. (1992). The expression of nicotinic acetylcholine receptors by PC12 cells treated with NGF. *J. Neurosci.* **12:**4611–4623.

9

GABA-, Glycine-, and Glutamate-Gated Channels and Their Possible Involvement in Neurological and Psychiatric Illness

Mark G. Darlison and Robert J. Harvey

9.1. GENERAL INTRODUCTION AND SCOPE OF THE CHAPTER

Rapid chemical communication between cells in the vertebrate central nervous system is mediated by ligand-gated ion-channel receptors (also called ionotropic receptors), which are multisubunit complexes, that each contain an ion-selective channel. In response to an appropriate signal, neurotransmitter is released from the storage vesicles in the presynaptic terminal of a neuron into the synaptic cleft; the signaling molecules then diffuse across this intercellular compartment and bind to receptors located in the membrane of the postsynaptic cell. The binding of neurotransmitter to an ionotropic receptor results, via an unknown mechanism that is assumed to involve a conformational change in the protein, in the opening of the ion channel and the flux of either cations or anions. This rapid neurotransmission, which occurs on the millisecond time scale, is terminated by the closure of the channel as a result of either agonist dissociation, or receptor desensitization, and either transmitter reuptake or hydrolysis.

In vertebrate species, the amino acids γ-aminobutyric acid (GABA) and glycine are the predominant inhibitory neurotransmitters. The ligand-gated ion channels that are the receptors for these molecules are the GABA type A ($GABA_A$) receptors, which are the major class of inhibitory receptors in brain, and glycine receptors which are most abundant in spinal cord; both of these receptor types possess a chloride-selective channel. In contrast, rapid excitatory neurotransmission is largely mediated by the binding of L-glutamate to ionotropic receptors which have been pharmacologically classified according to their agonist affinities as either α-amino-3-hydroxy-5-methyl-4-isoxazole propionate (AMPA)-selective, kainate-selective, or *N*-methyl-D-aspartate (NMDA)-selective (see, for example, Ref. 1). Each of these receptors contains a channel that is permeable only to cations (sodium and potassium and, in some cases, calcium). In addition, separate cation-selective channels exist for the neurotransmitters acetylcholine (nicotinic acetylcholine receptors, which occur in brain as well as muscle[2]) and serotonin [5-hydroxytryptamine type 3 ($5HT_3$) receptors[3]], and probably also for adenosine 5′-triphosphate.[4,5]

During the last 7 years or so, tremendous advances have been made in the elucidation of the sequences of the polypeptides that associate to form either GABA-, glycine-, or glutamate-gated channels, and this has resulted from the application of the powerful techniques of molecular biology. In addition, much has been learned about the spatial and temporal patterns of expression, in brain and spinal cord, of the various genes that encode the different receptor subunits. From this information, and from the results of expression studies in heterologous systems, it has generally been concluded that subtypes of $GABA_A$, glycine, and glutamate receptors must exist *in vivo*, and these presumably play subtly different physiological roles in neuronal inhibition and excitability, and perhaps other processes. Since the genes that encode these receptors are expressed in nervous tissue, it is evident that a defect in any one of them could result in some form of neurological or psychiatric illness. Indeed, a dysfunction of $GABA_A$ receptors and of ionotropic glutamate receptors has been implicated in the etiology of, for example, various types of epilepsy.[6–8]

Much has already been written on the isolation of complementary DNAs (cDNAs) for ligand-gated ion-channel subunits, the sequences of the encoded polypeptides, the patterns of expression of the corresponding genes, and the pharmacological properties of receptors that have been reconstituted by the expression of either cloned cDNAs (in transfected mammalian cells) or *in vitro*-transcribed RNAs (in *Xenopus laevis* oocytes). For information on these aspects of $GABA_A$, glycine, and glutamate receptors, the authors recommend sev-

Mark G. Darlison and Robert J. Harvey • Institute for Cell Biochemistry and Clinical Neurobiology, University Hospital Eppendorf, University of Hamburg, 20246 Hamburg, Germany.

Molecular Biology of Membrane Transport Disorders, edited by S.G. Schultz et al. Plenum Press, New York, 1996

eral reviews.[9–14] In comparison, relatively little attention has been paid to the genes that encode the various receptor subunits, and the likelihood that defects in either the sequences or expression of these underlie various forms of mental illness. In this chapter, we will outline our current knowledge of the number and types of polypeptides that constitute GABA-, glycine-, and glutamate-gated channels, and discuss the structures and human chromosomal locations of the corresponding genes and the possible involvement of some of these in certain neurological and psychiatric disorders.

9.2. GABA-, GLYCINE-, AND GLUTAMATE-GATED ION CHANNELS

9.2.1. $GABA_A$ Receptors and Their Polypeptides

The isolation,[15] in 1987, of the first $GABA_A$ receptor cDNAs revealed that subunits of this protein are related in both structure and sequence to nicotinic acetylcholine receptor (AChR) and glycine receptor polypeptides. Although the quaternary structure of $GABA_A$ receptors has not yet been determined, it is assumed by analogy with AChRs[16] that they are pentamers. cDNA cloning experiments, which initially took advantage of chemically determined peptide sequences[15] and which subsequently involved screening libraries for homologous sequences, have revealed the existence of 13 mammalian $GABA_A$ receptor genes. The encoded products have been classified on the basis of sequence similarity as either α, β, γ, or δ subunits. Hydropathic analysis predicts that each type of polypeptide possesses four membrane-spanning domains (M1 to M4). Since each cloned cDNA also encodes an amino-terminal signal peptide, the mature subunits are assumed to have a long (220 to 240 amino acid) amino-terminal extracellular domain, a short loop between M1 and M2 and between M2 and M3, a large intracellular segment between M3 and M4, and a short carboxy-terminus that is extracellular.

In mammals, six isoforms of the α subunit, and three each of the β and γ subunits are predicted to exist; in contrast, only one form of the δ subunit is thought to occur. In addition, two $GABA_A$ receptor-like polypeptides, named ρ1 and ρ2, have been identified by cDNA cloning from a human retinal library (see Ref. 17). Although these exhibit between 31 and 43% identity to $GABA_A$ receptor subunits, they are probably components of so-called $GABA_C$ receptors which, unlike $GABA_A$ receptors, are bicuculline insensitive.[18] Polypeptides within a particular $GABA_A$ receptor subunit class (e.g., α1 and α2) typically display ~60 to ~80% sequence identity, while polypeptides from different subunit classes (e.g., α1 and γ2) usually share ~30 to ~40% identity. Homologues of many of the mammalian $GABA_A$ receptor subunits appear to occur, also, in avian species (see, for example, Ref. 19); however, two additional receptor subunits (named β4 and γ4) have been identified by cDNA cloning in the chicken.[20,21] Although the polypeptide compositions of many of the *in vivo* $GABA_A$ receptor subtypes are unknown, most are thought to contain at least one α, one β, and either one γ or one δ subunit.[9]

Variants of three different $GABA_A$ receptor polypeptides have also been recognized by the isolation of cDNAs that correspond to alternatively spliced primary gene transcripts. The γ2 subunit exists (in the human, bovine, and rat,[22] the mouse,[23] and the chicken[24]) in two forms, named γ2S and γ2L, that differ by the absence or presence, respectively, of eight amino acid residues in the large intracellular loop between M3 and M4; this extra sequence is encoded by a separate 24-nucleotide exon. Interestingly, the insertion contains a consensus protein kinase C (PKC) phosphorylation site[25] which has been shown to be a substrate for this enzyme.[22,26] It has been reported by Wafford and co-workers[27] that, when reconstituted in *Xenopus* oocytes, the GABA-induced responses of receptors containing the γ2L subunit (but not those containing, instead, the γ2S subunit), together with an α and a β subunit, can be potentiated by ethanol. Furthermore, this potentiation appears to require the PKC site that is present in the long form of the γ2 subunit.[28] However, it should be pointed out that Sigel and colleagues[29] have recently reported that they are unable to reproduce this ethanol effect on expressed $GABA_A$ receptors with the same subunit combination used by Wafford *et al.*[27] The only difference between the two series of experiments appears to be that Wafford *et al.* expressed mouse cDNAs while Sigel *et al.* used rat clones. This situation is further complicated by the recent observation[30] that ethanol can inhibit the activity of PKC.

The two other $GABA_A$ receptor polypeptides which occur in more than one form are the β2 and β4 subunits. Two variants of the chicken β2 subunit, named β2S and β2L, differ by the absence or presence, respectively, of 17 amino acids in the large intracellular loop region.[31] Like the situation for the γ2 polypeptide, the insertion in the β2L subunit is encoded by a separate exon and contains a consensus sequence for phosphorylation[25] by PKC (see Figure 9.1); the serine residue that is assumed to be phosphorylated by PKC is also a target for modification by casein kinase I.[31] Surprisingly, alternative splicing of the β2-subunit primary gene transcript does not seem to occur in either the bovine or the rat[31]; in these two species, at least, only the β2S subunit appears to be present. Two forms of the chicken β4 subunit, named β4 and β4′, which either lack or contain an additional four amino acids in the loop between M3 and M4 also exist. These variants result from the alternate choice of one of two 5′-donor splice sites.[20]

9.2.2. Glycine Receptors and Their Polypeptides

The purification, in 1982, of glycine receptors from rat spinal cord, using an aminostrychnine-agarose affinity column, revealed[32] a complex that was comprised of subunits of 48,000, 58,000, and 93,000 Da. Several years later, chemically determined peptide sequences were used to isolate a cDNA for the strychnine-binding 48,000-Da component,[33] which is now known as the α1 subunit. This revealed, as noted earlier, that glycine receptor polypeptides are related in both structure and sequence to $GABA_A$ receptor and AChR polypeptides. Currently, the sequences of three different α subunits[33–35] and one β subunit[36] have been deduced by mo-

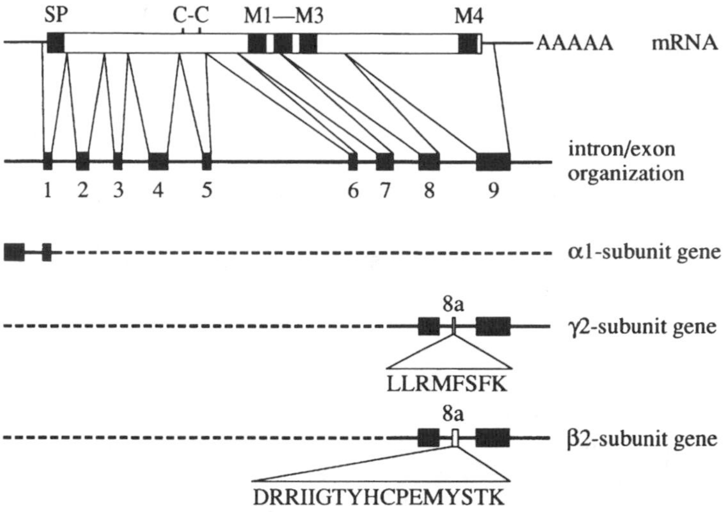

Figure 9.1. $GABA_A$ receptor gene organization. The figure (not drawn to scale) shows the intron/exon structure of a typical $GABA_A$ receptor subunit gene,[68–70] in which the coding region is specified by nine exons (numbered), under a cartoon of a $GABA_A$ receptor mRNA. Exons and introns are indicated by black boxes and lines, respectively. Displayed below this are representations of parts of a vertebrate α1-subunit gene, a vertebrate γ2-subunit gene, and the chicken β2-subunit gene. Note that an intron, which splits the 5′-untranslated region, is present in the chicken and human α1-subunit genes (Ref. 74, and A. N. Bateson and M. G. Darlison, unpublished results); an intron is also found, in a similar position, in the human β3-subunit gene[75] (see text). The amino acid sequences of the insertions found in the γ2L and β2L polypeptides, that are each encoded by a separate exon (named[31] exon 8a; indicated by an open box), which contain consensus sequences for serine phosphorylation, are shown in single-letter code. AAAAA, polyadenylic acid tail; C-C, dicysteine loop; M1–M4, the four proposed membrane-spanning domains; SP, signal peptide.

lecular cloning techniques; in addition, the existence of a fourth α subunit has been reported.[37] The sequences of the three fully characterized α subunits display between 81 and 86% identity to one another, but each is only between 51 and 53% identical to the β subunit. cDNA cloning studies have shown[38] that the 93,000-Da polypeptide, which has been named gephyrin, has no sequence similarity to ligand-gated ion-channel subunits. Recent data[39] suggest that the function of gephyrin is to cluster glycine receptors at synapses through an interaction with cytoskeletal components such as tubulin. Interestingly, the gephyrin primary gene transcript is alternatively spliced to yield at least five mRNAs[38] that are differentially distributed in brain.[40]

Since gephyrin is not considered to be an integral glycine receptor component, glycine-gated channels are thought to consist of only α and β subunits. Since four different α subunits are known to exist, and since glycine receptors have been shown by chemical cross-linking experiments[41] to be pentameric like the AChR, it seems likely that subtypes of this receptor exist *in vivo*. In support of this notion, *in situ* hybridization studies have revealed[37] that the α-subunit mRNAs have different and restricted distributions in the mammalian brain and spinal cord; in contrast, the β-subunit transcript is abundant in many areas of the brain. Since the *in situ* hybridization data also reveal that the β-subunit mRNA is found in certain brain regions (e.g., several cell layers of the olfactory bulb and the cerebellum, and two thalamic nuclei) that lack α1-, α2-, and α3-subunit transcripts,[37] the intriguing possibility that the β subunit either forms part of another receptor or fulfills some other function must be considered. Lastly, the α2 polypeptide has been shown[42] to be present in spinal cord only during early postnatal development; it is replaced during ontogeny by the α1 subunit.

Glycine receptor polypeptide variants have also been identified, in the rat, by cDNA cloning. Two forms of the α1 subunit, which differ in length by eight amino acids,[43] arise by the choice of one of two 3′-acceptor splice sites; the longer form contains an additional serine residue that might serve as a phosphorylation target for cyclic nucleotide-dependent protein kinases. Alternative splicing of one of two homologous 68-nucleotide exons, in the rat glycine receptor α2-subunit gene, which encode a small part of the large presumed extracellular domain, yields two further variants (named α2A and α2B) that differ in sequence by two amino acids.[44] In addition, a rat α2-subunit variant (α2*) has been described.[45] While the human and the rat α1 and α2 subunits bind strychnine,[34,43,44] the rat α2* polypeptide does not.[45] This pharmacological disparity is due to a single amino acid difference (a glutamic acid residue in the α2*-subunit sequence for a glycine in the normal α2 subunit) in the large amino-terminal presumed extracellular domain.[45]

9.2.3. Glutamate Receptors and Their Polypeptides

The first isolation of a cDNA for an ionotropic glutamate receptor subunit was reported in 1989 by the laboratory of Stephen Heinemann.[46] This was obtained by the expression of *in vitro*-transcribed RNAs, synthesized from pools of clones from a rat brain cDNA library, in *Xenopus* oocytes. The encoded polypeptide was found to be much greater in size (namely, ~100,000 Da) than other ligand-gated ion-channel receptor subunits (which are all ~50,000 Da). However, hydropathic analysis of the sequence of this glutamate receptor subunit (now referred to as either GluR1 or GluR-A) suggested that it contained four membrane-spanning segments (TM1 to TM4)[46] like acetylcholine, $GABA_A$, and glycine receptor subunits and, therefore, had a similar topology. Initially, there was some disagreement over the exact positions of these membrane-spanning domains; the locations subsequently proposed by Keinänen and co-workers[47] are now generally accepted. It should be noted, however, that this assumed topology of ionotropic glutamate receptor subunits has recently been called into question[48] through the finding that the carboxy-terminal portion (i.e., after TM4) of an NMDA receptor polypeptide can be phosphorylated and is, thus, probably intracellular. The GluR1 subunit was initially thought to be part of a kainate-selective channel.[46] In fact, the GluR1 polypeptide is now known to form cationic channels that are selectively gated by AMPA.[47]

Through the use of the polymerase chain reaction, a very powerful technique for the isolation of sequence-related cDNAs, followed by cDNA library screening, a family of glutamate receptor polypeptides was identified (for reviews, see Refs. 12–14). These are known as GluR1 to GluR4 (or GluR-A to GluR-D), which form AMPA-selective channels, GluR5 which forms domoate- and kainate-selective channels,[49] GluR6 which assembles to produce channels that respond preferentially to kainate,[50] GluR7, KA-1 and KA-2 that do not form functional homo-oligomeric channels but which yield nanomolar-affinity binding sites for kainate when the corresponding cDNAs are transfected into mammalian cells,[51–53] and δ1 and δ2 which exhibit between 25 and 32% identity to other non-NMDA glutamate receptor subunits but for which the ligand specificity is unknown.[54] As for $GABA_A$ and glycine receptor polypeptides (discussed earlier), each of the glutamate receptor subunits is encoded by a separate gene.

Variants of glutamate receptor polypeptides arise by the alternate splicing of primary gene transcripts and by RNA editing. For example, two forms (called "flip" and "flop") of each of the GluR1 to GluR4 subunits are generated by the use of one of two sequence-related 115-bp exons that encode part of the large loop between TM3 and TM4.[55] Channels formed by these two types of variant differ in the pharmacological and kinetic properties of the currents that are evoked by either glutamate or AMPA, but not those elicited by kainate. *In situ* hybridization localization of the mRNAs encoding the "flip" and "flop" forms of each of the GluR1 to GluR4 subunits, in rat brain, reveals distinct patterns of expression; this is particularly apparent in the CA1 and CA3 fields of the hippocampus. Furthermore, significant differences are found in the amounts of the alternatively spliced transcripts during brain development, with the "flop" mRNAs being present at only low levels prior to postnatal day 8 and then increasing up to postnatal day 14.[56] In contrast, the "flip" mRNA levels remain essentially invariant during postnatal development.

Several of the glutamate receptor subunit mRNAs are also subject to RNA editing, a process in which the sequence of an mRNA is altered by an as yet unknown mechanism. Editing occurs in RNA sequences that specify TM1 and TM2.[57,58] Particularly striking is the editing of one codon from CAG (in the GluR2-subunit gene) to CGG (in the translated mRNA). This appears to occur in 100% of cases and results in the presence of an arginine residue in the proposed pore-lining TM2 domain instead of a glutamine[57]; the corresponding position in GluR1, GluR3, and GluR4 subunits is always occupied by a glutamine. Comparison of the properties of "edited" and "unedited" (obtained by *in vitro* mutagenesis) GluR2 subunits revealed that channels formed by polypeptides that each possess a glutamine residue in TM2 are permeable to calcium (as well as sodium and potassium) ions, while those that include an arginine-containing GluR2 subunit exhibit only low calcium permeability.[59] In addition, differences are observed in the current–voltage relationships of receptors that either contain or lack a GluR2 subunit.[59,60]

Recently, cDNA clones that encode NMDA receptor subunits have been obtained; the first of these was isolated by expression cloning in *Xenopus* oocytes.[61] The encoded polypeptide (called NMDAR1 or ζ1) displays between 26 and 29% identity to other ionotropic glutamate receptor subunits. Subsequently, the isolation of four other cDNAs, which specify polypeptides (named NR2A to NR2D, or ε1 to ε4) which exhibit between 42 and 56% identity to one another but much lower identity (between 26 and 28%) to the NMDAR1 subunit, was reported.[62–65] Each of these five polypeptides is encoded by a separate gene. Since the NMDAR1-subunit gene is expressed widely throughout the brain,[61] whereas the NR2A- to NR2D-subunit genes have different and more restricted distributions,[62,65] it is likely that native NMDA receptors comprise the NMDAR1 subunit complexed with at least one of the NR2A to NR2D subunits. Lastly, in common with other ligand-gated ion-channel receptors, alternative splicing of NMDA receptor primary gene transcripts occurs. To date, nine different NMDAR1-subunit mRNAs have been detected.[66,67] These transcripts arise from the alternate use of one exon that encodes part of the amino-terminal, proposed extracellular domain and two exons that specify parts of the carboxy-terminus (i.e., after TM4). A subset of the encoded NMDAR1-polypeptide variants can be distinguished by their sensitivity to the potentiating effect, on agonist-induced currents, of submicromolar concentrations of zinc.[67]

9.3. THE GENES FOR GABA-, GLYCINE-, AND GLUTAMATE-GATED ION CHANNELS

9.3.1. $GABA_A$ and Glycine Receptor Genes

Very little has been published, to date, on the structural organization of GABA-, glycine-, and glutamate-gated ion-channel genes. Part of the reason for this may be that many of these genes are large (see later). The greatest amount of information on this topic has come from studies on $GABA_A$ receptors. The positions of the introns that split the coding regions of three $GABA_A$ receptor genes, namely, those for the human β1, chicken β4 and mouse δ subunits,[68–70] have been determined. These studies have revealed that each polypeptide is encoded by nine exons, and that the relative positions of the intervening introns are strongly conserved (see Figure 9.1). The most 5′ of these introns interrupts the DNA sequence that specifies the end of the signal peptide and the start of the mature polypeptide. The sequence encoding the amino-terminal, presumed extracellular domain of the mature polypeptide is split by five introns; interestingly, the dicysteine loop, which is found in all acetylcholine, $GABA_A$, glycine, and $5HT_3$ receptor subunits, is encoded by a single exon. The remaining two introns in $GABA_A$ receptor genes interrupt the DNA sequence that codes for the M2 domain, and that encoding the large intracellular loop between M3 and M4. The position of the latter appears to be the most variable, and this is probably related to the fact that the size of the long intracellular loop varies considerably among different subunits. These data strongly suggest that $GABA_A$ receptor subunit genes evolved from a common ancestral gene.

As noted earlier, two forms of the $GABA_A$ receptor γ2 subunit occur in all vertebrates studied, and two forms of the β2 subunit have been found in the chicken. The two variants of the γ2 and β2 polypeptides arise from the presence of an additional exon (named exon 8a; see Figure 9.1 and Ref. 31), in the corresponding genes, that is located between exons 8 and 9, which when used gives rise to the longer form of the subunit (i.e., γ2L and β2L). These results indicate that the primordial gene contained a minimum of ten (and not nine) exons, since the insertion of exon 8a at the same relative position into the γ2- and β2-subunit genes seems less likely than the loss of this exon from some of the descendants of the ancestral $GABA_A$ receptor gene. Lastly, many $GABA_A$ receptor genes are large: both the human β1-subunit gene[68] and the chicken β4-subunit gene[69] are at least 65 kb, the human β3-subunit gene[71] is ~250 kb (in part because the third intron is 150 kb long), and the human α5-subunit gene[71] is ~70 kb. In contrast, the murine δ-subunit gene[70] is only 13 kb in size.

The complete structure of a glycine receptor gene has yet to be described. It is not, therefore, possible to draw conclusions about whether, based on intron/exon organizations, the genes for subunits of $GABA_A$ and glycine receptors (which both contain chloride-selective channels) arose from a common ancestor. However, Sommer and co-workers[70] have noted that the DNA sequence coding for the second putative membrane-spanning domain of the glycine receptor α1 subunit is not interrupted by an intron. The significance of this is unclear; however, it should be noted that the M2 segment of the *Drosophila* $GABA_A$ receptor-like *Rdl* subunit is encoded by a single exon.[72] In addition, although an intron splits the sequence encoding the M2 domain of a molluscan (*Lymnaea stagnalis*) $GABA_A$ receptor-like ζ subunit,[73] this is located 39 bp 3′ of that found in vertebrate genes. Note also that at least one additional intron occurs in the chicken and human $GABA_A$ receptor α1-subunit genes (Ref. 74, and A. N. Bateson and M. G. Darlison, unpublished results) and in the human β3-subunit gene.[75] In the two α1-subunit genes, the 5′-untranslated region is split by an intron. In the β3-subunit gene, there are two alternate first exons (named exon 1 and exon 1a) that encode two different signal peptides. Interestingly, a strong promoter element has been shown[75] to be located within the intron between exons 1 and 1a in the human β3-subunit gene. The presence of an extra intron, in a similar position, in the $GABA_A$ receptor α1- and β3-subunit genes suggests that this feature was present in the ancestral gene and that it occurs in other extant $GABA_A$ receptor genes.

9.3.2. Glutamate Receptor Genes

The complete structural organization has been described for only one ionotropic glutamate receptor gene, namely, that encoding the NMDAR1 subunit.[67] Although this gene comprises 22 exons, some of which (as mentioned earlier) are alternatively spliced to generate NMDAR1-subunit variants, it spans only ~25 kb. In addition, the positions of several introns within the GluR1- to GluR6-subunit genes have been reported.[55,57,58] From the limited amount of data that are available, it is evident that the intron/exon organizations of AMPA-selective and kainate-selective receptor subunit genes are not identical, although they may be alike. However, based on amino acid sequence similarity, it is probable that the genes encoding the different glutamate receptor polypeptides arose by duplication of a common ancestor. What seems clear is that the origins of $GABA_A$ (and glycine) receptor genes and glutamate receptor genes are different.

9.4. CHROMOSOMAL LOCATIONS OF THE GENES THAT ENCODE GABA-, GLYCINE-, AND GLUTAMATE-GATED ION-CHANNEL POLYPEPTIDES

9.4.1. The Mapping of Human $GABA_A$ Receptor Genes

The human chromosomal locations of most $GABA_A$ receptor subunit genes are known (Table 9.1). These have been determined by a variety of methods, including the hybridization of probes to sorted chromosome DNA "spot blots,"[76] *in situ* hybridization to metaphase chromosomes,[76,77] mapping in panels of hybrid cell-lines,[78–80] and genetic mapping using the Centre d'Etude du Polymorphisme Humain (CEPH) panel of reference families.[80–82] The data reveal a clustering of genes, which might have been predicted given the likelihood that they arose by duplication of a common ancestral gene. For example, the α1-, α6-, and γ2-subunit genes are located on 5q32–q35 (Figure 9.2).[76,79,80] The α1- and γ2-subunit genes lie very close to one another; at least parts of these two genes are present on an ~450-kb human yeast artificial chromosome (YAC) clone.[78] The α2- and β1-subunit genes have been mapped[76] to 4p12–p13, the γ1-subunit gene has been localized[78] to 4p14–q21.1, and the α4-subunit gene has been mapped (M. E. S. Bailey, A. A. Hicks, B. P. Riley, W. Kamphuis, R. J. Harvey, M. G. Darlison, and K. J. Johnson, unpublished results) to the same chromosome. It is, therefore, plausible that these four genes are also clustered.

The α5- and β3-subunit genes are both located on chromosome 15q11–q13 (Figure 9.2), oriented in a "head-to-head" manner with an intergenic distance of less than 100 kb.[71,77] In addition, the γ3-subunit gene has recently been shown[83] to reside on an ~900-kb YAC contig together with the α5- and β3-subunit genes. This is consistent with the finding[84] that all of the γ3- and α5-subunit genes and the 5′ end of the β3-subunit gene, as well as part of the *p* gene, are deleted in the pink-eyed cleft-palate (p^{cp}) mouse mutant. Of the other human $GABA_A$ receptor genes, that for the α3 subunit has been localized[76] to Xq28 while that for the δ subunit lies on the short arm of chromosome 1[70]; the two $GABA_A$ receptor-like ρ-subunit genes map[17] to 6q14–q21 (see Table 9.1). The isolation and development of highly informative polymorphisms for several of the human $GABA_A$ receptor genes has also been reported (see

Table 9.1. Human Chromosomal Locations of GABA-Gated Ion-Channel Receptor Genes

Subunit	Gene name	Location	Polymorphism	References
		$GABA_A$ receptor polypeptides		
α1	GABRA1	5q32–q35	yes	76, 79, 82
α2	GABRA2	4p12–p13	yes[a]	76
α3	GABRA3	Xq28	yes	76, 85
α4	GABRA4	4[a]	yes[a]	—
α5	GABRA5	15q11–q13	yes	87
α6	GABRA6	5q32–q35	yes	80
β1	GABRB1	4p12–p13	yes	76, 81, 86
β2	GABRB2	N.D.[b]	no	—
β3	GABRB3	15q11–q13	yes	77, 88
γ1	GABRG1	4p14–q21.1	no	78
γ2	GABRG2	5q31.1–q33.2	yes[a]	78, 79
γ3	GABRG3	15q11–q13	no	83
δ	GABRD1	1p	no	70
		$GABA_C$ receptor polypeptides		
ρ1	GABRR1	6q14–q21	no	17
ρ2	GABRR2	6q14–q21	no	17

[a]M. E. S. Bailey, A. A. Hicks, B. P. Riley, W. Kamphuis, E. J. Feldman, R. J. Harvey, K. J. Johnson, and M. G. Darlison, unpublished results.
[b]N.D., chromosomal location not yet determined.

Refs. 80–82, 85–88, and Table 9.1), thus permitting the possible involvement of these genes in various neurological and psychiatric disorders to be tested.

9.4.2. $GABA_A$ Receptor Genes and Neurological/Psychiatric Illness

As mentioned earlier, the human $GABA_A$ receptor α3-subunit gene has been shown[76] to reside in Xq28. This localization generated a certain amount of interest since this region of the X chromosome had previously been suggested (see, for example, Refs. 89, 90) to contain a locus that predisposes to bipolar affective disorder (more commonly known as manic depression or manic depressive illness). The most convincing evidence in support of this hypothesis came from studies on several multigeneration Israeli families in which linkage to the X-linked loci for color blindness and glucose-6-phosphate dehydrogenase deficiency was described.[89] Consistent with this was a report of linkage of the coagulation factor IX locus at Xq27 to manic depression in Belgian pedigrees.[90] To test whether a defect of the $GABA_A$ receptor α3 subunit was involved in the etiology of this condition, this laboratory developed a highly informative polymorphism for the corresponding gene[85] and typed this (in collaboration with the groups of Conrad Gilliam and Miron Baron) in the Israeli pedigrees, which were concurrently reevaluated and extended by the inclusion of some new individuals, and (in collaboration with the groups of Christine van Broeckhoven and Julien Mendlewicz) in the originally identified Belgian pedigrees and additional Belgian families that appeared to segregate an X-linked form of the disorder. In both cases, linkage to other markers in the region of Xq28 was also tested. The results obtained with the Israeli families showed[91] greatly diminished support for linkage of a locus for manic depression to Xq28. However, it was not possible to exclude the involvement of the $GABA_A$ receptor α3-subunit gene in the pathogenesis of this disorder in one of the pedigrees. In the study on Belgian families, it was possible to exclude the α3-subunit gene, and a

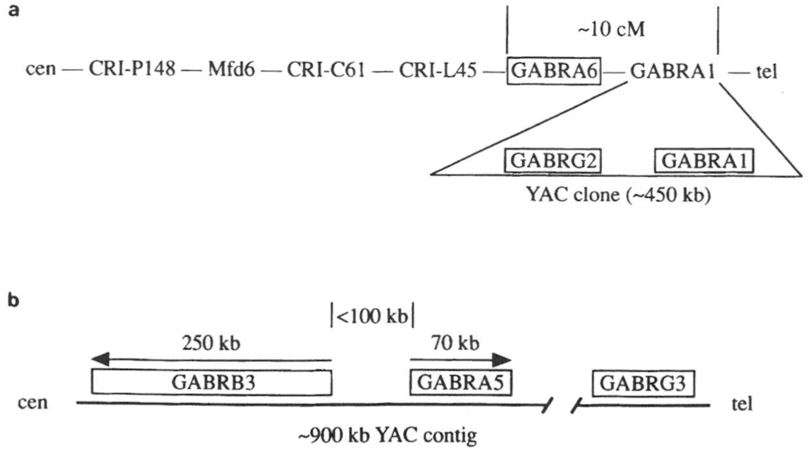

Figure 9.2. Clustering of human $GABA_A$ receptor genes. (a) Part of chromosome 5q32–q35 showing the positions of the α1- (GABRA1), α6- (GABRA6), and γ2- (GABRG2) subunit genes relative to several polymorphic markers (for details, see Ref. 80). Note that at least parts of GABRA1 and GABRG2 reside on the same ~450-kb YAC clone,[78] and that the gene order of GABRA1 and GABRG2 with respect to GABRA6 is unknown. (b) Part of chromosome 15q11–q13 showing the relationship of the α5- (GABRA5), β3- (GABRB3), and γ3- (GABRG3) subunit genes. Note that the three genes are located within a ~900-kb YAC contig,[83] but that the exact distance between GABRA5 and GABRG3 is unknown. Arrows indicate the directions of gene transcription (see Ref. 71). cen, centromere; cM, centiMorgan; tel, telomere.

distance of 7.5 centiMorgans either side of it.[92] Also, in an association study of patients with bipolar affective disorder, and age- and sex-matched controls,[93] no evidence was found for the involvement of either the $GABA_A$ receptor α1- or α3-subunit genes.

Localization of the human $GABA_A$ receptor α5- and β3-subunit genes to chromosome 15q11–q13 (see Refs. 77, 87) also proved of interest since deletions of this region of the genome cause two diseases, which are believed to involve genomic imprinting (i.e., the differential expression of one or more genes on either paternally or maternally inherited chromosomes), namely, the Angelman and Prader-Willi syndromes. Although at least part of the β3-subunit gene was found to be deleted in certain patients affected with these illnesses,[77] this locus has been excluded as the cause of Angelman syndrome.[94] Since the $GABA_A$ receptor α5-, β3-, and γ3-subunit genes map telomeric to the critical disease regions,[83,95] they are unlikely to be involved in the pathogenesis of either Angelman or Prader-Willi syndromes.

The assignment[76,79,80,82] of the human α1-, α6-, and γ2-subunit genes to chromosome 5q32–q35 resulted in their consideration as candidates in the neurological illness hyperekplexia (also known as startle disease[96]). This is a motor disorder of central nervous system origin that is characterized by sustained muscular hypertonia in infancy and by an exaggerated startle response to sudden acoustic or tactile stimuli. Interestingly, patients with this illness respond dramatically to treatment with the benzodiazepine compound clonazepam, a positive allosteric modulator of the $GABA_A$ receptor. However, the human α1- and γ2-subunit genes have been excluded as candidates in hyperekplexia,[96] and the homologous genes have been excluded[97] as the cause of *spasmodic* (*spd*), a mouse mutant that is characterized by fine motor tremor, leg clasping and stiffness, which maps to a part of the genome that exhibits synteny with human chromosome 5q21–q31. In fact, hyperekplexia has recently been shown to be the result of mutation of another ligand-gated ion-channel gene (see later).

The only mutation thus far detected in a $GABA_A$ receptor gene is that found within the α6-subunit gene of a strain of selectively outbred alcohol-nontolerant (ANT) rats that are highly susceptible to impairment of postural reflexes by the administration of certain benzodiazepine agonists such as diazepam.[98] In these animals, the cerebellum lacks diazepam-insensitive high-affinity binding sites for the benzodiazepine compound Ro15-4513; the presence of this feature is characteristic of receptors that contain the α6 subunit.[99] A point mutation in the α6-subunit gene results in the substitution of an arginine residue, in wild-type animals, for a glutamine residue in ANT rats.[98] When this mutation is introduced into the normal α6 subunit, the binding of Ro15-4513, to receptors containing the mutant polypeptide, is diazepam sensitive.

9.4.3. The Mapping of Human Glycine Receptor Genes

As far as the authors are aware, only two glycine receptor genes have been localized in the human genome, namely, those encoding the α1 and α2 subunits. It was originally reported[100] that the human α1-subunit gene mapped to Xp21.2–p22.1. However, it was subsequently shown[34] that the gene that lies in this segment is, in fact, that coding for the α2 subunit, and not that for the α1 subunit which is now known to be located on chromosome 5q31.3 (Ref. 101). The localization of the mouse α2-subunit gene to the X chromosome,[102] in a region syntenic with human Xp22.1–p22.3, supports the chromosomal assignment of the homologous human gene. Lastly, restriction fragment length polymorphisms have been reported for the human α2-subunit gene.[100]

9.4.4. Glycine Receptor Genes and Neurological/Psychiatric Illness

Although, as mentioned earlier, patients suffering from hyperekplexia respond dramatically to treatment with the $GABA_A$ receptor modulator clonazepam,[96] this inherited neurological disorder is now known to be the result of a point mutation in the glycine receptor α1-subunit gene.[103] This change, which occurs at the same position in four different families, results in the substitution of a positively charged arginine residue (in unaffected individuals), at the carboxy-terminal end of the second putative membrane-spanning domain, for either a glutamine or a leucine residue (in affected individuals). Such a mutation is predicted to have a significant influence on the channel activity of glycine receptors that contain the α1 subunit. The beneficial effect to hyperekplexic patients of clonazepam, which does not bind to glycine receptors, is likely to be due to the augmentation of GABAergic inhibition in the brain stem and spinal cord, which presumably compensates for the loss or impairment of glycinergic neurotransmission.

A glycine receptor deficit has also been reported in two neurological diseases in animals. The mouse mutant *spastic* suffers from a motor disorder characterized by tremor, episodic spasms, and a disturbed righting response, which only becomes apparent 2 weeks after birth. Biochemical studies have shown that the number of glycine receptors that contain the (adult) α1 subunit, in the nervous system of the *spastic* mouse, is drastically reduced compared with that found in wild-type animals.[104,105] However, the transcript that encodes the α1 subunit appears to be present at normal levels in the mutant.[105] A glycine receptor deficiency also occurs in inherited myoclonus in Poll Hereford calves.[106] These animals exhibit a severe reduction in the number of strychnine-binding sites in brain stem and spinal cord when compared to healthy controls. To date, the underlying genetic defects in these two disorders have not been established.

9.4.5. The Mapping of Human Glutamate Receptor Genes

Many of the genes that encode ionotropic glutamate receptor polypeptides have recently been mapped to human chromosomes using *in situ* hybridization and/or hybrid cell-line panels (Table 9.2). The locations of the genes for the GluR1 to GluR4 subunits are 5q33 (Refs. 107–109), 4q32–q33 (Refs. 108, 109), Xq25–q26 (Ref. 109), and 11q22–q23 (Ref.

Table 9.2. Human Chromosomal Locations of Glutamate-Gated Ion-Channel Receptor Genes

Subunit	Location	References
GluR1	5q33	107–109
GluR2	4q32–q33	108, 109
GluR3	Xq25–q26	109
GluR4	11q22–q23	109
GluR5	21q21.1–q22.1	110, 111
GluR6	6	111
GluR7	1p32–p34	111
KA-1	N.D.	—
KA-2	N.D.	—
δ1	N.D.	—
δ2	N.D.	—
NMDAR1	9q34.3	112–114
NR2A	16p13	114
NR2B	N.D.	—
NR2C	17q25	114
NR2D	N.D.	—

[a]Note that, at the time of writing, no official gene names have been attributed to ionotropic glutamate receptor genes. N.D., chromosomal location not yet determined.

109), respectively. The gene encoding the GluR5 subunit has been mapped to 21q21.1–q22.1 (Refs. 110, 111), and that specifying the GluR7 subunit has been reported[111] to be located in 1p32–p34. In addition, the GluR6-subunit gene is most probably situated on chromosome 6 (see Ref. 111). The locations of three NMDA receptor genes have been established, namely, those for the NMDAR1, NR2A, and NR2C subunits which are located on chromosomes 9q34.3 (Refs. 112–114), 16p13 (Ref. 114), and 17q25 (Ref. 114), respectively. Thus, unlike the situation for $GABA_A$ receptor genes, ionotropic glutamate receptor genes are not clustered within the human genome. This suggests that glutamate receptor genes either arose earlier in evolution and/or evolved faster, and were subsequently distributed throughout the genome, than $GABA_A$ receptor genes. This is consistent with the logical assumption that rapid excitatory neurotransmission appeared in animals before rapid inhibitory neurotransmission. To date, no polymorphisms have been described for ionotropic glutamate receptor genes.

9.4.6. Glutamate Receptor Genes and Neurological/Psychiatric Illness

The chromosomal locations of certain of the ionotropic glutamate receptor genes correspond to regions of the human genome that have previously been implicated in the etiology of various neurological and psychiatric disorders. For example, the GluR3-subunit gene maps to an area of the X chromosome that contains the loci for three disorders, including oculocerebralrenal syndrome of Lowe, two of which are characterized by mental deficits.[109] The gene for the GluR4 subunit is located in 11q22–q23, a region that contains a gene that predisposes to ataxia telangiectasia, a translocation breakpoint in a single pedigree with bipolar affective disorder, and another translocation that is associated with schizophrenia, schizoaffective disorder, and major depression in multiple pedigrees.[109] The location of the GluR5-subunit gene, close to the locus for familial amyotrophic lateral sclerosis (also known as Lou Gehrig's disease), which is a neurodegenerative motorneuron disorder, implicated it in the pathogenesis of this disease.[110,111] However, two recombinants were subsequently found between the GluR5-subunit gene and the gene that predisposes to amyotrophic lateral sclerosis,[111] which suggested that these two loci were distinct. Recently, this motorneuron disorder has been shown to result from mutations within a gene[115] that encodes a cytosolic copper/zinc-binding superoxide dismutase. Lastly, the human NMDAR1-polypeptide gene maps close to the region containing the disease genes for idiopathic torsion dystonia and tuberous sclerosis.[112,113]

9.5. CONCLUDING REMARKS

To date, only one neurological or psychiatric disorder of humans has been shown to result from mutation of either a $GABA_A$, glycine, or ionotropic glutamate receptor subunit gene, namely, hyperekplexia which is caused by changes in the glycine receptor α1-subunit gene. However, only a very small number of such conditions have been mapped in the human genome. Clearly, as discussed, neurological illness can result from an alteration in either the sequence or the expression of a ligand-gated ion-channel receptor gene. Furthermore, defects in GABAergic and glutamatergic neurotransmission are still considered by many to be the possible cause of neurological disorders such as epilepsy (see, for example, Refs. 6–8). We consider it important to map the human chromosomal locations of all brain-expressed genes since, even if a particular candidate gene is subsequently found not to be involved in a given disorder, knowledge of the locations of genes of defined function will be of value (as markers) to geneticists studying other inherited diseases. In the long term, understanding the molecular basis of a particular neurological or psychiatric illness is likely to be of great value in the design of a new generation of drugs that can alleviate the distressing symptoms that affect so many individuals.

NOTE ADDED IN PROOF

Since this manuscript was submitted for publication, many important developments have taken place in the areas reviewed in this chapter. Unfortunately, because of publication schedules, it was not possible to up-date this contribution prior to type-setting. However, we wish to draw attention to a recently-discovered error, namely, that the dinucleotide repeat polymorphism reported to be associated with the human $GABA_A$ receptor α6-subunit gene (GABRA6; Ref. 80) is, in fact, located within the $GABA_A$ receptor γ2-subunit gene (GABRG2; M.E.S. Bailey, B.E. Albrecht, K.J. Johnson, and M.G. Darlison, unpublished observations). Thus, the intergenic distance between GABRA6 and the $GABA_A$ receptor α1-subunit gene (GABRA1), shown in Figure 9.2a, is not known. Note, however, that GABRA6 does map to human chromosomal segment 5q32–q33 (A.S. Wilcox, R. Berry, N.A.R. Walter, P.J. Whiting, and J.M. Sikela, manuscript submitted for publication).

ACKNOWLEDGMENTS. The authors apologize to all those whose work, for reasons of space, has not been cited. Some of the studies described here, on the mapping of human $GABA_A$ receptor subunit genes, could not have been performed without the fruitful collaboration and expertise of Professor Keith J. Johnson and co-workers at Charing Cross and Westminster Medical School, in London. M.G.D. gratefully acknowledges support from the Deutsche Forschungsgemeinschaft (SFB232/B7) and the Dr. Hans Ritz und Liselotte Ritz Stiftung, Hamburg, for research in the area covered by this chapter.

REFERENCES

1. Barnard, E. A., and Henley, J. M. (1990). The non-NMDA receptors: Types, protein structure and molecular biology. *Trends Pharmacol. Sci.* **11**:500–507.
2. See elsewhere in this volume.
3. Maricq, A. V., Peterson, A. S., Brake, A. J., *et al.* (1991). Primary structure and functional expression of the $5HT_3$ receptor, a serotonin-gated ion channel. *Science* **254**:432–437.
4. Edwards, F. A., Gibb, A. J., and Colquhoun, D. (1992). ATP receptor-mediated synaptic currents in the central nervous system. *Nature* **359**:144–147.
5. Evans, R. J., Derkach, V., and Surprenant, A. (1992). ATP mediates fast synaptic transmission in mammalian neurons. *Nature* **357**: 503–505.
6. Chugani, H. T., and Olsen, R. W. (1986). Benzodiazepine/GABA receptor binding in vitro and in vivo in analysis of clinical disorders. In *Benzodiazepine/GABA Receptors and Chloride Channels: Structural and Functional Properties.* (R. W. Olsen and J. C. Venter, eds.), Alan R. Liss, New York, pp. 315–335.
7. Olsen, R. W., Bureau, M., Houser, C. R., *et al.* (1992). GABA/benzodiazepine receptors in human focal epilepsy. In *Neurotransmitters in Epilepsy* (Epilepsy Res. Suppl. 8) (G. Avanzini, J. Engel, Jr., R. Fariello, *et al.*, eds.), Elsevier, Amsterdam, pp. 383–391.
8. Choi, D. W. (1988). Glutamate neurotoxicity and diseases of the nervous system. *Neuron* **1**:623–634.
9. Burt, D. R., and Kamatchi, G. L. (1991). $GABA_A$ receptor subtypes: From pharmacology to molecular biology. *FASEB J.* **5**:2916–2923.
10. Wisden, W., and Seeburg, P. H. (1992). $GABA_A$ receptor channels: From subunits to functional entities. *Curr. Opin. Neurobiol.* **2**: 263–269.
11. Betz, H. (1991). Glycine receptors: Heterogeneous and widespread in the mammalian brain. *Trends Neurosci.* **14**:458–461.
12. Nakanishi, S. (1992). Molecular diversity of glutamate receptors and implications for brain function. *Science* **258**:597–603.
13. Seeburg, P. H. (1993). The molecular biology of mammalian glutamate receptor channels. *Trends Neurosci.* **16**:359–365.
14. Wisden, W., and Seeburg, P. H. (1993). Mammalian ionotropic glutamate receptors. *Curr. Opin. Neurobiol.* **3**:291–298.
15. Schofield, P. R., Darlison, M. G., Fujita, N., *et al.* (1987). Sequence and functional expression of the $GABA_A$ receptor shows a ligand-gated receptor superfamily. *Nature* **328**:221–227.
16. Raftery, M. A., Hunkapiller, M. W., Strader, C. D., *et al.* (1980). Acetylcholine receptor: Complex of homologous subunits. *Science* **208**:1454–1457.
17. Cutting, G. R., Curristin, S., Zoghbi, H., *et al.* (1992). Identification of a putative γ-aminobutyric acid (GABA) receptor subunit rho_2 cDNA and colocalization of the genes encoding rho_2 (GABRR2) and rho_1 (GABRR1) to human chromosome 6q14–q21 and mouse chromosome 4. *Genomics* **12**:801–806.
18. Shimada, S., Cutting, G., and Uhl, G. R. (1992). γ-Aminobutyric acid A or C receptor? γ-Aminobutyric acid ρ_1 receptor RNA induces bicuculline-, barbiturate-, and benzodiazepine-insensitive γ-aminobutyric acid responses in *Xenopus* oocytes. *Mol. Pharmacol.* **41**:683–687.
19. Glencorse, T. A., Darlison, M. G., Barnard, E. A., *et al.* (1993). Sequence and novel distribution of the chicken homologue of the mammalian γ-aminobutyric $acid_A$ receptor γ1 subunit. *J. Neurochem.* **61**:2294–2302.
20. Bateson, A. N., Lasham, A., and Darlison, M. G. (1991). γ-Aminobutyric $acid_A$ receptor heterogeneity is increased by alternative splicing of a novel β-subunit gene transcript. *J. Neurochem.* **56**:1437–1440.
21. Harvey, R. J., Kim, H.-C., and Darlison, M. G. (1993). Molecular cloning reveals the existence of a fourth γ subunit of the vertebrate brain $GABA_A$ receptor. *FEBS Lett.* **331**:211–216.
22. Whiting, P., McKernan, R. M., and Iversen, L. L. (1990). Another mechanism for creating diversity in γ-aminobutyrate type A receptors: RNA splicing directs expression of two forms of γ2 subunit, one of which contains a protein kinase C phosphorylation site. *Proc. Natl. Acad. Sci. USA* **87**:9966–9970.
23. Kofuji, P., Wang, J. B., Moss, S. J., *et al.* (1991). Generation of two forms of the γ-aminobutyric $acid_A$ receptor γ_2-subunit in mice by alternative splicing. *J. Neurochem.* **56**:713–715.
24. Glencorse, T. A., Bateson, A. N., and Darlison, M. G. (1992). Differential localization of two alternatively spliced $GABA_A$ receptor γ2-subunit mRNAs in the chick brain. *Eur. J. Neurosci.* **4**:271–277.
25. Woodgett, J. R., Gould, K. L., and Hunter, T. (1986). Substrate specificity of protein kinase C. Use of synthetic peptides corresponding to physiological sites as probes for substrate recognition requirements. *Eur. J. Biochem.* **161**:177–184.
26. Moss, S. J., Doherty, C. A., and Huganir, R. L. (1992). Identification of the cAMP-dependent protein kinase and protein kinase C phosphorylation sites within the major intracellular domains of the β_1, γ_2S, and γ_2L subunits of the γ-aminobutyric acid type A receptor. *J. Biol. Chem.* **267**:14470–14476.
27. Wafford, K. A., Burnett, D. M., Leidenheimer, N. J., *et al.* (1991). Ethanol sensitivity of the $GABA_A$ receptor expressed in Xenopus oocytes requires 8 amino acids contained in the γ2L subunit. *Neuron* **7**:27–33.
28. Wafford, K. A., and Whiting, P. J. (1992). Ethanol potentiation of $GABA_A$ receptors requires phosphorylation of the alternatively spliced variant of the γ2 subunit. *FEBS Lett.* **313**:113–117.
29. Sigel, E., Baur, R., and Malherbe, P. (1993). Recombinant $GABA_A$ receptor function and ethanol. *FEBS Lett.* **324**:140–142.
30. Slater, S. J., Cox, K. J. A., Lombardi, J. V., *et al.* (1993). Inhibition of protein kinase C by alcohols and anaesthetics. *Nature* **364**: 82–84.
31. Harvey, R. J., Chinchetru, M. A., and Darlison, M. G. (1994). Alternative splicing of a 51-nucleotide exon that encodes a putative protein kinase C phosphorylation site generates two forms of the chicken γ-aminobutyric $acid_A$ receptor β2 subunit. *J. Neurochem.* **62**:10–16.
32. Pfeiffer, F., Graham, D., and Betz, H. (1982). Purification by affinity chromatography of the glycine receptor of rat spinal cord. *J. Biol. Chem.* **257**:9389–9393.
33. Grenningloh, G., Rienitz, A., Schmitt, B., *et al.* (1987). The strychnine-binding subunit of the glycine receptor shows homology with nicotinic acetylcholine receptors. *Nature* **328**:215–220.

34. Grenningloh, G., Schmieden, V., Schofield, P. R., *et al.* (1990). Alpha subunit variants of the human glycine receptor: Primary structures, functional expression and chromosomal localization of the corresponding genes. *EMBO J.* **9:**771–776.
35. Kuhse, J., Schmieden, V., and Betz, H. (1990). Identification and functional expression of a novel ligand binding subunit of the inhibitory glycine receptor. *J. Biol. Chem.* **265:**22317–22320.
36. Grenningloh, G., Pribilla, I., Prior, P., *et al.* (1990). Cloning and expression of the 58 kd β subunit of the inhibitory glycine receptor. *Neuron* **4:**963–970.
37. Malosio, M.-L., Marquèze-Pouey, B., Kuhse, J., *et al.* (1991). Widespread expression of glycine receptor subunit mRNAs in the adult and developing rat brain. *EMBO J.* **10:**2401–2409.
38. Prior, P., Schmitt, B., Grenningloh, G., *et al.* (1992). Primary structure and alternative splice variants of gephyrin, a putative glycine receptor–tubulin linker protein. *Neuron* **8:**1161–1170.
39. Kirsch, J., Wolters, I., Triller, A., *et al.* (1993). Gephyrin antisense oligonucleotides prevent glycine receptor clustering in spinal neurons. *Nature* **366:**745–748.
40. Kirsch, J., Malosio, M.-L., Wolters, I., *et al.* (1993). Distribution of gephyrin transcripts in the adult and developing rat brain. *Eur. J. Neurosci.* **5:**1109–1117.
41. Langosch, D., Thomas, L., and Betz, H. (1988). Conserved quaternary structure of ligand-gated ion channels: The postsynaptic glycine receptor is a pentamer. *Proc. Natl. Acad. Sci. USA* **85:**7394–7398.
42. Becker, C.-M., Hoch, W., and Betz, H. (1988). Glycine receptor heterogeneity in rat spinal cord during postnatal development. *EMBO J.* **7:**3717–3726.
43. Malosio, M.-L., Grenningloh, G., Kuhse, J., *et al.* (1991). Alternative splicing generates two variants of the α_1 subunit of the inhibitory glycine receptor. *J. Biol. Chem.* **266:**2048–2053.
44. Kuhse, J., Kuryatov, A., Maulet, Y., *et al.* (1991). Alternative splicing generates two isoforms of the α2 subunit of the inhibitory glycine receptor. *FEBS Lett.* **283:**73–77.
45. Kuhse, J., Schmieden, V., and Betz, H. (1990). A single amino acid exchange alters the pharmacology of neonatal rat glycine receptor subunit. *Neuron* **5:**867–873.
46. Hollmann, M., O'Shea-Greenfield, A., Rogers, S. W., *et al.* (1989). Cloning by functional expression of a member of the glutamate receptor family. *Nature* **342:**643–648.
47. Keinänen, K., Wisden, W., Sommer, B., *et al.* (1990). A family of AMPA-selective glutamate receptors. *Science* **249:**556–560.
48. Tingley, W. G., Roche, K. W., Thompson, A. K., *et al.* (1993). Regulation of NMDA receptor phosphorylation by alternative splicing of the C-terminal domain. *Nature* **364:**70–73.
49. Sommer, B., Burnashev, N., Verdoorn, T. A., *et al.* (1992). A glutamate receptor channel with high affinity for domoate and kainate. *EMBO J.* **11:**1651–1656.
50. Egebjerg, J., Bettler, B., Hermans-Borgmeyer, I., *et al.* (1991). Cloning of a cDNA for a glutamate receptor subunit activated by kainate but not AMPA. *Nature* **351:**745–748.
51. Bettler, B., Egebjerg, J., Sharma, G., *et al.* (1992). Cloning of a putative glutamate receptor: A low affinity kainate-binding subunit. *Neuron* **8:**257–265.
52. Werner, P., Voigt, M., Keinänen, K. *et al.* (1991). Cloning of a putative high-affinity kainate receptor expressed predominantly in hippocampal CA3 cells. *Nature* **351:**742–744.
53. Herb, A., Burnashev, N., Werner, P., *et al.* (1992). The KA-2 subunit of excitatory amino acid receptors shows widespread expression in brain and forms ion channels with distantly related subunits. *Neuron* **8:**775–785.
54. Lomeli, H., Sprengel, R., Laurie, D. J., *et al.* (1993). The rat delta-1 and delta-2 subunits extend the excitatory amino acid receptor family. *FEBS Lett.* **315:**318–322.
55. Sommer, B., Keinänen, K., Verdoorn, T. A., *et al.* (1990). Flip and flop: A cell-specific functional switch in glutamate-operated channels of the CNS. *Science* **249:**1580–1585.
56. Monyer, H., Seeburg, P. H., and Wisden, W. (1991). Glutamate-operated channels: Developmentally early and mature forms arise by alternative splicing. *Neuron* **6:**799–810.
57. Sommer, B., Köhler, M., Sprengel, R., *et al.* (1991). RNA editing in brain controls a determinant of ion flow in glutamate-gated channels. *Cell* **67:**11–19.
58. Köhler, M., Burnashev, N., Sakmann, B., *et al.* (1993). Determinants of Ca^{2+} permeability in both TM1 and TM2 of high affinity kainate receptor channels: Diversity by RNA editing. *Neuron* **10:** 491–500.
59. Hume, R. I., Dingledine, R., and Heinemann, S. F. (1991). Identification of a site in glutamate receptor subunits that controls calcium permeability. *Science* **258:**1028–1031.
60. Verdoorn, T. A., Burnashev, N., Monyer, H., *et al.* (1991). Structural determinants of ion flow through recombinant glutamate receptor channels. *Science* **252:**1715–1718.
61. Moriyoshi, K., Masu, M., Ishii, T., *et al.* (1991). Molecular cloning and characterization of the rat NMDA receptor. *Nature* **354:**31–37.
62. Monyer, H., Sprengel, R., Schoepfer, R., *et al.* (1992). Heteromeric NMDA receptors: Molecular and functional distinction of subtypes. *Science* **256:**1217–1221.
63. Meguro, H., Mori, H., Araki, K., *et al.* (1992). Functional characterization of a heteromeric NMDA receptor channel expressed from cloned cDNAs. *Nature* **357:**70–74.
64. Ikeda, K., Nagasawa, M., Mori, H., *et al.* (1992). Cloning and expression of the ε4 subunit of the NMDA receptor channel. *FEBS Lett.* **313:**34–38.
65. Ishii, T., Moriyoshi, K., Sugihara, H., *et al.* (1993). Molecular characterization of the family of the *N*-methyl-D-aspartate receptor subunits. *J. Biol. Chem.* **268:**2836–2843.
66. Sugihara, H., Moriyoshi, K., Ishii, T., *et al.* (1992). Structures and properties of seven isoforms of the NMDA receptor generated by alternative splicing. *Biochem. Biophys. Res. Commun.* **185:**826–832.
67. Hollmann, M., Boulter, J., Maron, C., *et al.* (1993). Zinc potentiates agonist-induced currents at certain splice variants of the NMDA receptor. *Neuron* **10:**943–954.
68. Kirkness, E. F., Kusiak, J. W., Fleming, J. T., *et al.* (1991). Isolation, characterization, and localization of human genomic DNA encoding the β1 subunit of the $GABA_A$ receptor (GABRB1). *Genomics* **10:**985–995.
69. Lasham, A., Vreugdenhil, E., Bateson, A. N., *et al.* (1991). Conserved organization of γ-aminobutyric $acid_A$ receptor genes: Cloning and analysis of the chicken β4-subunit gene. *J. Neurochem.* **57:** 352–355.
70. Sommer, B., Poustka, A., Spurr, N. K., *et al.* (1990). The murine $GABA_A$ receptor δ-subunit gene: Structure and assignment to human chromosome 1. *DNA Cell Biol.* **9:**561–568.
71. Sinnett, D., Wagstaff, J., Glatt, K., *et al.* (1993). High-resolution mapping of the γ-aminobutyric acid receptor subunit β3 and α5 gene cluster on chromosome 15q11–q13, and localization of breakpoints in two Angelman syndrome patients. *Am. J. Hum. Genet.* **52:**1216–1229.
72. ffrench-Constant, R. H., and Rocheleau, T. (1992). *Drosophila* cyclodiene resistance gene shows conserved genomic organization with vertebrate γ-aminobutyric $acid_A$ receptors. *J. Neurochem.* **59:**1562–1565.
73. Hutton, M. L., Harvey, R. J., Earley, F. G. P., *et al.* (1993). A novel invertebrate $GABA_A$ receptor-like polypeptide: Sequence and pattern of gene expression. *FEBS Lett.* **326:**112–116.
74. Ultsch, A., Bateson, A. N., and Darlison, M. G. (1990). Isolation

and characterization of the 5′end of the chicken γ-aminobutyric $acid_A$ receptor α1-subunit gene. *Biochem. Soc. Trans.* **18:**437–438.
75. Kirkness, E. F., and Fraser, C. M. (1993). A strong promoter element is located between alternative exons of a gene encoding the human γ-aminobutyric acid-type A receptor β3 subunit (GABRB3). *J. Biol. Chem.* **268:**4420–4428.
76. Buckle, V. J., Fujita, N., Ryder-Cook, A. S., *et al.* (1989). Chromosomal localization of $GABA_A$ receptor subunit genes: Relationship to human genetic disease. *Neuron* **3:**647–654.
77. Wagstaff, J., Knoll, J. H. M., Fleming, J., *et al.* (1991). Localization of the gene encoding the $GABA_A$ receptor β3 subunit to the Angelman/Prader-Willi region of human chromosome 15. *Am. J. Hum. Genet.* **49:**330–337.
78. Wilcox, A. S., Warrington, J. A., Gardiner, K., *et al.* (1992). Human chromosomal localization of genes encoding the γ1 and γ2 subunits of the γ-aminobutyric acid receptor indicates that members of this gene family are often clustered in the genome. *Proc. Natl. Acad. Sci. USA* **89:**5857–5861.
79. Warrington, J. A., Bailey, S. K., Armstrong, E., *et al.* (1992). A radiation hybrid map of 18 growth factor, growth factor receptor, hormone receptor, or neurotransmitter receptor genes on the distal region of the long arm of chromosome 5. *Genomics* **13:**803–808.
80. Hicks, A. A., Bailey, M. E. S., Riley, B. P., *et al.* (1994). Further evidence for clustering of human $GABA_A$ receptor subunit genes: Localization of the α_6-subunit gene (GABRA6) to distal chromosome 5q by linkage analysis. *Genomics* **20:**285–288.
81. Dean, M., Lucas-Derse, S., Bolos, A., *et al.* (1991). Genetic mapping of the β1 GABA receptor gene to human chromosome 4, using a tetranucleotide repeat polymorphism. *Am. J. Hum. Genet.* **49:**621–626.
82. Johnson, K. J., Sander, T., Hicks, A. A., *et al.* (1992). Confirmation of the localization of the human $GABA_A$ receptor α1-subunit gene (GABRA1) to distal 5q by linkage analysis. *Genomics* **14:**745–748.
83. Phillips, R. L., Rogan, P. K., Culiat, C. T., *et al.* (1993). A YAC contig spanning 4 genes in distal human chromosome 15q11–q13, mapping of the human *GABRG3* gene, and effect of homozygous deletion of three $GABA_A$ receptor genes in mouse. *Am. J. Hum. Genet.* **53**(Suppl.):A1345.
84. Nakatsu, Y., Tyndale, R. F., DeLorey, T. M., *et al.* (1993). A cluster of three $GABA_A$ receptor subunit genes is deleted in a neurological mutant of the mouse *p* locus. *Nature* **364:**448–450.
85. Hicks, A. A., Johnson, K. J., Barnard, E. A., *et al.* (1991). Dinucleotide repeat polymorphism in the human X-linked $GABA_A$ receptor α3-subunit gene. *Nucleic Acids Res.* **19:**4016.
86. Polymeropoulos, M. H., Rath, D. S., Xiao, H., *et al.* (1991). Dinucleotide repeat polymorphism at the human β1 subunit of the $GABA_A$ receptor gene (GABRB1). *Nucleic Acids Res.* **19:**6345.
87. Glatt, K. A., Sinnett, D., and Lalande, M. (1992). Dinucleotide repeat polymorphism at the $GABA_A$ receptor α5 (GABRA5) locus at chromosome 15q11–q13. *Hum. Mol. Genet.* **1:**348.
88. Mutirangura, A., Ledbetter, S. A., Kuwano, A., *et al.* (1992). Dinucleotide repeat polymorphism at the $GABA_A$ receptor β3 (GABRB3) locus in the Angelman/Prader-Willi region (AS/PWS) of chromosome 15. *Hum. Mol. Genet.* **1:**67.
89. Baron, M., Risch, N., Hamburger, R., *et al.* (1987). Genetic linkage between X-chromosome markers and bipolar affective illness. *Nature* **326:**289–292.
90. Mendlewicz, J., Simon, P., Sevy, S., *et al.* (1987). Polymorphic DNA marker on X chromosome and manic depression. *Lancet* **i:**1230–1232.
91. Baron, M., Freimer, N. F., Risch, N., *et al.* (1993). Diminished support for linkage between manic depressive illness and X-chromosome markers in three Israeli pedigrees. *Nature Genet.* **3:**49–55.
92. Van Broeckhoven, C., De bruyn, A., Raeymaekers, P., *et al.* (1991). Exclusion of manic depressive illness from the chromosomal regions Xq27–q28 and 11p15. *Cytogenet. Cell. Genet.* **58:**1973–1974.
93. Walsh, C., Hicks, A. A., Sham, P., *et al.* (1992). $GABA_A$ receptor subunit genes as candidate genes for bipolar affective disorder—an association analysis. *Psychiatr. Genet.* **2:**239–247.
94. Reis, A., Kunze, J., Ladanyi, L., *et al.* (1993). Exclusion of the $GABA_A$-receptor β3 subunit gene as the Angelman's syndrome gene. *Lancet* **341:**122–123.
95. Wagstaff, J., Shugart, Y. Y., and Lalande, M. (1993). Linkage analysis in familial Angelman syndrome. *Am. J. Hum. Genet.* **53:** 105–112.
96. Ryan, S. G., Dixon, M. J., Nigro, M. A., *et al.* (1992). Genetic and radiation hybrid mapping of the hyperekplexia region on chromosome 5q. *Am. J. Hum. Genet.* **51:**1334–1343.
97. Buckwalter, M. S., Testa, C. M., Noebels, J. L., *et al.* (1993). Genetic mapping and evaluation of candidate genes for Spasmodic, a neurological mouse mutation with abnormal startle response. *Genomics* **17:**279–286.
98. Korpi, E. R., Kleingoor, C., Kettenmann, H., *et al.* (1993). Benzodiazepine-induced motor impairment linked to point mutation in cerebellar $GABA_A$ receptor. *Nature* **361:**356–359.
99. Lüddens, H., Pritchett, D. B., Köhler, M., *et al.* (1990). Cerebellar $GABA_A$ receptor selective for a behavioural alcohol antagonist. *Nature* **346:**648–651.
100. Siddique, T., Phillips, K., Betz, H., *et al.* (1989). RFLPs of the gene for the human glycine receptor on the X-chromosome. *Nucleic Acids Res.* **17:**1785.
101. Hung, W.-Y., Betz, H., Hu, P., *et al.* (1993). Human glycine receptor alpha 1 subunit is localized at 5q 31.3. *Am. J. Hum. Genet.* **53**(Suppl.):A1784.
102. Derry, J. M. J., and Barnard, P. J. (1991). Mapping of the glycine receptor α2-subunit gene and the $GABA_A$ α3-subunit gene on the mouse X chromosome. *Genomics* **10:**593–597.
103. Shiang, R., Ryan, S. G., Zhu, Y.-Z., *et al.* (1993). Mutations in the α_1 subunit of the inhibitory glycine receptor cause the dominant neurologic disorder, hyperekplexia. *Nature Genet.* **5:**351–358.
104. Becker, C.-M. (1990). Disorders of the inhibitory glycine receptor: The *spastic* mouse. *FASEB J.* **4:**2767–2774.
105. Becker, C.-M., Schmieden, V., Tarroni, P., *et al.* (1992). Isoform-selective deficit of glycine receptors in the mouse mutant *spastic*. *Neuron* **8:**283–289.
106. Gundlach, A. L. (1990). Disorder of the inhibitory glycine receptor: Inherited myoclonus in Poll Hereford calves. *FASEB J.* **4:** 2761–2766.
107. Puckett, C., Gomez, C. M., Korenberg, J. R., *et al.* (1991). Molecular cloning and chromosomal localization of one of the human glutamate receptor genes. *Proc. Natl. Acad. Sci. USA* **88:**7557–7561.
108. Sun, W., Ferrer-Montiel, A. V., Schinder, A. F., *et al.* (1992). Molecular cloning, chromosomal mapping, and functional expression of human brain glutamate receptors. *Proc. Natl. Acad. Sci. USA* **89:**1443–1447.
109. McNamara, J. O., Eubanks, J. H., McPherson, J. D., *et al.* (1992). Chromosomal localization of human glutamate receptor genes. *J. Neurosci.* **12:**2555–2562.
110. Eubanks, J. H., Puranam, R. S., Kleckner, N. W., *et al.* (1993). The gene encoding the glutamate receptor subunit GluR5 is located on human chromosome 21q21.1–22.1 in the vicinity of the gene for familial amyotrophic lateral sclerosis. *Proc. Natl. Acad. Sci. USA* **90:**178–182.
111. Gregor, P., Reeves, R. H., Jabs, E. W., *et al.* (1993) Chromosomal localization of glutamate receptor genes: Relationship to familial amyotrophic lateral sclerosis and other neurological disorders of mice and humans. *Proc. Natl. Acad. Sci. USA* **90:**3053–3057.

112. Collins, C., Duff, C., Duncan, A. M. V., *et al.* (1993). Mapping of the human NMDA receptor subunit (NMDAR1) and the proposed NMDA receptor glutamate-binding subunit (NMDARA1) to chromosomes 9q34.3 and chromosome 8, respectively. *Genomics* **17:** 237–239.
113. Karp, S. J., Masu, M., Eki, T., *et al.* (1993). Molecular cloning and chromosomal localization of the key subunit of the human *N*-methyl-D-aspartate receptor. *J. Biol. Chem.* **268:**3728–3733.
114. Takano, H., Onodera, O., Tanaka, H., *et al.* (1993). Chromosomal localization of the ε1, ε3 and ζ1 subunit genes of the human NMDA receptor channel. *Biochem. Biophys. Res. Commun.* **197:** 922–926.
115. Rosen, D. R., Siddique, T., Patterson, D., *et al.* (1993). Mutations in Cu/Zn superoxide dismutase gene are associated with familial amyotrophic lateral sclerosis. *Nature* **362:**59–62.

10

The Genetic and Physiological Basis of Malignant Hyperthermia

David H. MacLennan, Michael S. Phillips, and Yilin Zhang

10.1. INTRODUCTION

A goal of research in basic biological science is to develop understanding of normal processes that will be relevant to the understanding of disease processes. This goal has been realized to a large extent for the inherited neuromuscular abnormality malignant hyperthermia (MH), which is manifested most commonly in humans as anesthetic-induced muscle contracture, accompanied by high fever. The syndrome became widely recognized in the mid-1950s, following the introduction of halothane and succinylcholine as the most widely used combination of anesthetic and muscle relaxant. In certain families, the combination of inhalational anesthetics with a depolarizing muscle relaxant was fatal to individuals genetically predisposed to MH.[1]

During the same period, studies of the mechanisms controlling muscle contraction revealed that the interaction between actin and myosin was regulated by Ca^{2+} and that Ca^{2+} regulation was mediated through the Ca^{2+} binding protein troponin.[2,3] Moreover, muscle Ca^{2+} concentrations, in turn, were shown to be regulated by a unique muscle membrane system, the sarcoplasmic reticulum. Ca^{2+} stored in the membrane system is released to the sarcoplasm through a Ca^{2+} release channel to initiate muscle contraction, while Ca^{2+} pumps transport Ca^{2+} from the sarcoplasm to the luminal space of the membrane, thereby initiating muscle relaxation.[4] Ca^{2+} is stored in the lumen of the membrane, in association with the Ca^{2+} binding protein calsequestrin.[5] Ca^{2+} was also shown to play a role in the control of glycolysis in muscle through its activation of phosphorylase kinase.[6] From such studies, it became apparent that a defect in Ca^{2+} regulation, leading to the continued presence of Ca^{2+} within the sarcoplasm, could induce muscle contracture, extensive glycolysis, and enhanced mitochondrial oxidation of glycolytic end products, leading to the high turnover of ATP which could be responsible for the elevated temperatures associated with MH episodes.[6–9]

Defects in Ca^{2+} regulation could result from defects in the Ca^{2+} pump, responsible for Ca^{2+} uptake, or from defects in the Ca^{2+} release channel. No primary defect in Ca^{2+} uptake could be confirmed, even in the presence of halothane.[10,11] With the development of new assays for Ca^{2+} release,[9,12–14] it became possible to observe defects in Ca^{2+} release associated with MH muscle.[15–19]

The Ca^{2+} release channel was identified through its high-affinity binding of the modulator ryanodine, hence the name "ryanodine receptor," and purified.[20–22] Soon thereafter, cDNA encoding the Ca^{2+} release channel was cloned.[23,24] Comparative studies were then possible both for the protein[18,19,25–28] and for the DNA encoding it,[29–34] making it possible to demonstrate the association of MH with defects in the *RYR1* gene and the Ca^{2+} release channel expressed from it.

In this chapter we review studies that have led to the identification of defects in the Ca^{2+} release channel gene that are associated with both porcine and human MH and with central core disease. We also describe our present understanding of the physiological basis for MH and draw attention to the heterogeneity that exists in the genotype leading to this group of abnormalities and in the resulting phenotypes. Several excellent, comprehensive reviews of MH[35–37] and shorter reviews concerning the genetic basis of MH have been written in recent years.[38–42]

10.2. MALIGNANT HYPERTHERMIA

10.2.1. Human Malignant Hyperthermia

MH defines a clinical syndrome in which genetically susceptible individuals respond to the administration of potent inhalational anesthetics and depolarizing skeletal muscle relaxants with high fever and skeletal muscle rigidity.[35] These symptoms are accompanied by hypermetabolism, leading to hyperventilation, hypoxia, and lactic acidosis. Tachycardia, arrhythmia, and unstable blood pressure are also associated with an MH reaction and these probably result from cellular damage, which brings about electrolyte imbalance and eleva-

David H. MacLennan, Michael S. Phillips, and Yilin Zhang • Banting and Best Department of Medical Research, University of Toronto, C.H. Best Institute, Toronto, Ontario M5G 1L6, Canada.

Molecular Biology of Membrane Transport Disorders, edited by S.G. Schultz *et al.* Plenum Press, New York, 1996

tion in the serum and urine levels of muscle enzymes and myoglobin. If therapy is not initiated immediately, the patient may die within minutes from ventricular fibrillation, within hours from pulmonary edema or coagulopathy, or within days from neurological damage or obstructive renal failure, resulting largely from the release of muscle proteins into the circulation.

In a modern operating room setting, many patients at risk will have been identified in advance of anesthesia by knowledge of their kinship to individuals who have had an MH reaction. In these cases, anesthetic routines are varied to include nontriggering anesthetics, thereby preventing the onset of an MH episode. During the course of anesthesia for most patients, regardless of predicted MH status, heart rate, body temperature, and end tidal CO_2 production are monitored constantly. An increase in any of these will lead to an evaluation which may establish the diagnosis of MH. If MH is diagnosed, then the anesthesiologist stops administering the MH-triggering anesthetics, hyperventilates the patient with 100% oxygen, and administers the clinical antidote, dantrolene. These practices have lowered the death rate from MH episodes from over 80% to less than 7% in recent years. Neurological or kidney damage, however, still contribute to the morbidity resulting from MH episodes, since damage to muscle cells, leading to leakage of myoplasmic contents, occurs early in the reaction.[35]

Human MH was first defined as a genetic disease in the early 1960s.[1,43] In the first family to be studied, 10 members had died from the administration of ethyl chloride and ether as the anesthetic agent. The inheritance pattern suggested that a dominant gene was responsible for the MH abnormality. MH reactions do not occur in every case of administration of anesthetics to susceptible individuals, however. Indeed, about half of MH reactions occur in susceptible individuals who have previously had uneventful general anesthetics.[35] Conditions of anesthesia and the condition of the patient at the time of anesthesia can influence the occurrence of an MH reaction. Because of the incomplete penetrance of the gene, the difficulty in defining mild reactions, and the caution and care now taken by anesthesiologists, it is difficult to determine the actual incidence of MH susceptibility genes in the general population. Analysis of North American and European data available to Britt and Kalow[44] suggested that the incidence of recognized MH reactions is about 1 in 15,000 anesthetics in children and about 1 in 50,000 to 1 in 100,000 anesthetics in adults. These figures were supported in a study by Ording,[45] who found the incidence of MH episodes to be about 1 in 65,000 anesthetics in Denmark. For the reasons outlined above, this is probably an underestimate of the true incidence of MH susceptibility.

10.2.2. Diagnostic Tests for Malignant Hyperthermia

The health of those who inherit an MH susceptibility gene is seldom impaired. However, MH episodes may occur as a consequence of other inherited muscle diseases with more deleterious phenotypes such as central core disease[46] King–Denborough syndrome[47,48]; Duchenne muscular dystrophy[49–52] and possibly other myopathies.[53–55] Thus, a major goal of MH research has been to identify MH-susceptible individuals prior to the administration of anesthetics. If MH susceptibility is known, the use of alternate anesthetics and nondepolarizing muscle relaxants can circumvent the triggering of an MH reaction. Accordingly, *in vitro* diagnostic tests were developed in the early 1970s.[7]

The *in vitro* halothane caffeine contracture test (CHCT) was developed on the premise that the muscle from MH-susceptible (MHS) individuals might be more sensitive to agents inducing contraction and, therefore, might contract in the presence of lower amounts of either caffeine[7] or halothane[56] than the muscle from normal individuals. Two variants of the test, the North American test protocol[57] and the European test protocol,[58] have been standardized.

In the North American test, a freshly and cleanly excised muscle fascicle is secured by clamp or silk suture to a plastic electrode frame and immersed in Krebs–Ringer's solution at 37°C and pH 7.4. The other end of the muscle is then attached to a force displacement transducer and isometric tension is recorded with a polygraph. Initial resting tension is 1 to 2 g, but the muscle is stimulated every 5 sec to ensure its viability. After stable twitch and baseline tensions are achieved, caffeine is added directly to the bath in incremental doses from 0.5 to 32 mM. Measurement is made of the contracture achieved 4 min after the addition of each dose of caffeine, expressed as grams of tension. The most sensitive and specific measurement is of grams tension induced at 2 mM caffeine. MH susceptibility is associated with an increase in tension of 0.2 g or more at 2 mM caffeine. An alternative measurement is of the dose of caffeine required to raise the resting tension by 1 g. This dose is termed the caffeine specific concentration (CSC). If 1 g of increased tension is achieved with 4 mM caffeine or less, the patient is considered to be MHS. The amplitude of contractions induced by 3% halothane is measured in separate muscle strips placed simultaneously in separate baths. MHS individuals are defined as those with muscle strips that produce greater than 0.5 g to 0.7 g of tension in response to 3% halothane, depending on the laboratory.

In the European protocol, halothane is added to the solution bathing a single muscle fascicle at 0.5, 1.0, 1.5, 2.0, and 3.0% by volume. A threshold of 0.2 g contracture seen at 2% (0.44 mM) halothane or less is considered to indicate MH susceptibility. Caffeine is also added to a second muscle fascicle at concentrations of 0.5, 1.0, 1.5, 2.0, 3.0, and 4.0 mM or until a threshold of 0.2 g contracture is obtained. A threshold of contracture tension of 0.2 g or more, observed at 2 mM caffeine or less, is considered to indicate MH susceptibility. At the termination of addition of either caffeine or halothane, 32 mM caffeine is added to ensure that the muscle is still viable. In the European protocol, only those whose muscles respond abnormally to both caffeine and halothane are considered MHS, while those who react abnormally with one, but not the other, are considered to be MH equivocal (MHE).

The CHCT is a valuable clinical test.[59] Tests of this sort can be evaluated for their clinical effectiveness by determination of their sensitivity and specificity. The sensitivity of a test is defined as the percentage of positive test results in the abnormal population and is calculated from the formula: 100 × [true positives/(true positives + false negatives)]. The specificity of a test is defined as the percentage of negative test results in the absence of the abnormal gene and is calculated from the formula: 100 × [true negatives/(true negatives + false positives)]. Since failure to detect MH susceptibility can result in a serious or fatal outcome, sensitivity approaching 100% is more important for a clinical diagnosis than is specificity.

The North American CHCT currently achieves 92 to 95% sensitivity for two-, three-, and four-component tests and 75 to 53% specificity for the same tests.[60] By decreasing the contracture cutoff point for 3% halothane from > 0.7 g to > 0.5 g and increasing the cutoff point for 2 mM caffeine from > 0.2 g to > 0.3 g, the resulting two-component CHCT would approach 100% sensitivity and 78% specificity. As a clinical test, the CHCT would then assure that appropriate anesthetics would be administered to all of those patients who are MHS, while those diagnosed as normal would be treated with normal anesthetic routines in a cost-effective manner.

Problems with the CHCT are that it is invasive and expensive to perform and, as defined above under specificity, has a significant false-positive error rate[61] and a very small false-negative error rate.[62] While the error rate does not pose any significant problem for clinicians, inaccurate diagnosis creates enormous difficulties for geneticists attempting to link the inheritance of MH susceptibility to inheritance of a specific allele.

The validity of the premise on which the *in vitro* CHCT is based has been approached indirectly in a study by Kalow *et al.*[63] They plotted the concentration of caffeine (CSC) or of halothane plus caffeine (HCSC) that induced contracture thresholds in 1192 individuals subjected to the CHCT in Toronto over a 20-year period. Had there been a consistent contracture response of MHS individuals to specific concentrations of caffeine and had normal individuals responded consistently to specific higher levels of caffeine, then these two discrete populations of MHS and normal individuals should have been clustered in the plot as two distinct peaks. In fact, the plot showed a continuum of several peaks and troughs.

There are limitations in the study by Kalow *et al.*[63] in that all of the data were from one testing center, that the tests were carried out over a period of several years, allowing minor modifications in protocols, and that the CSC and HCSC tests are the least specific components of the North American CHCT. Thus, it is not reasonable to extrapolate these findings to those of other diagnostic centers or to the 3% halothane and 2 mM caffeine contracture tests. Nevertheless, the results of the study by Kalow *et al.*[63] graphically illustrate the fact that contracture is a multifactorial event, resulting from the interactions of dozens of gene products, each of which might influence the end contracture response to a specific concentration of halothane or of caffeine. Thus, while there is a rational basis for the postulate that *in vitro* contracture responses to halothane or caffeine can distinguish normal from MH muscle, the heterogeneity that exists in the population may make it difficult to determine precise thresholds for contracture and for drug concentration that will differentiate all individuals in the population at the level of accuracy required for genetic linkage analysis.

In view of previous limitations in accuracy of using the CHCT for MH diagnosis, researchers at the North American Malignant Hyperthermia Registry have spent several years clinically defining the MHS population[64] and then optimizing the North American CHCT protocols.[60,61] These studies are helping to resolve the problems concerning differentiation of MHS and normal individuals in the population.

An example of the problems of using the North American test response as the basis for MH status was presented in studies in which DNA linkage tests were compared with CHCT.[65] In a Canadian family many members were diagnosed as MHS and no linkage to a potential causal gene, *RYR1*, could be established. In order for MacKenzie *et al.*[65] to demonstrate linkage of MH susceptibility to the *RYR1* gene, they had to increase the contracture cutoff point for 2 mM caffeine from 0.2 g to 1.0 g tension. By doing this, all clearly MHS individuals were included in the MHS category and linkage of MH susceptibility to the *RYR1* gene with a lod score (log of the odds favoring linkage) of 3.6 at a recombination frequency of 0.0 was established for this family. While this increase in the cutoff point might be valid for the specific tests carried out on the specific family, it is clearly unwarranted from a clinical standpoint. Data from the North American MH Registry show that a cutoff point of 0.2 g tension for 2 mM caffeine has a sensitivity of 79% and a specificity of 79%, but a cutpoint of 1.0 g has a sensitivity of only 50%. This would lead to an unacceptable 50% false-negative rate for the diagnosis of a potentially life-threatening abnormality.

Other *in vitro* tests for MH have been developed. In recent studies, contractures induced by ryanodine have been evaluated for their ability to discriminate between MHS and MHN individuals.[66,67] The experimental protocol is similar to that used for the CHCT and the results are consistent with those of other components of the CHCT.

A considerable effort has been made to distinguish between MHS and MHN individuals on the basis of cytosolic free Ca^{2+} concentration in lymphocytes, both before and after addition of halothane.[68,69] A rational basis for these studies would be that the gene that is defective in Ca^{2+} metabolism, thereby bringing about MH episodes, might also be expressed in lymphocytes. Indeed, it appeared from these studies that defects might be present in both cellular and intracellular membranes. Defects in the Ca^{2+} release channel of skeletal muscle sarcoplasmic reticulum, encoded by the *RYR1* gene, have been implicated in some MH families.[31–34,70] This gene, however, seems to be expressed almost exclusively in skeletal muscle.[71–73] Thus, it is not clear that tests carried out on lymphocytes would be a good indicator of defects in *RYR1* in those families in which such defects are causal of MH. From a practical point of view, there is overlap between MH and normal individuals in this test, limiting its usefulness.[74]

The level of resting Ca^{2+} in the muscle of MH individuals has been investigated by two techniques, the direct use of implanted Ca^{2+} electrodes and the use of Fura 2 to measure intracellular Ca^{2+} concentration. Lopez-Padrino[75] used Ca^{2+} electrodes to demonstrate that the intracellular resting Ca^{2+} concentration from biopsies of MH patient is nearly fourfold higher than that in control patients. He postulated that this imbalance in resting intracellular Ca^{2+} homeostasis is a direct consequence of alteration of the intracellular mechanisms controlling myoplasmic Ca^{2+}. Iaizzo *et al.*[76] were unable to measure any difference in the resting Ca^{2+} concentrations in muscle biopsies from MH and normal swine using Fura 2 as a Ca^{2+} indicator. Thus, the question of resting Ca^{2+} levels is controversial. Since differences may arise from the differences in the techniques used to measure intracellular Ca^{2+}, it is not clear that measurement of intracellular Ca^{2+} homeostasis can form the basis for MH diagnostic tests.

Phosphorus nuclear magnetic resonance spectroscopy has also been tested as a means of discriminating between MHS and MHN individuals.[77,78] In this test, changes in ATP, phosphocreatine, inorganic phosphate, and acidity are measured noninvasively in whole muscles. In variants of this test, the arm or leg is placed within a magnet and muscles are flexed against a weight that is slowly increased. Concentrations of ATP, phosphocreatine, inorganic phosphate, and acid production are measured before, during, and after exercise. If differences in energy utilization can ultimately be established between MHN and MHS individuals, this might form the basis for a noninvasive diagnostic test for MH.

10.2.3. Porcine Malignant Hyperthermia

In the early 1960's it became evident that pigs, which were being used in experimental surgery and, therefore, were being administered halothane and succinylcholine, were susceptible to MH reactions.[79,80] The same course of events that was well known in human MH was observed in swine. Symptoms included high fever, skeletal muscle rigidity, cyanosis, hyperventilation, hypoxia, lactic acidosis, and death. A connection was soon made between the response of certain swine to anesthetics and the fact that individuals among herds of lean, heavily muscled swine were susceptible to fatal episodes of fever, skeletal muscle rigidity, and hyperventilation. This syndrome, referred to as the porcine stress syndrome (PSS), is brought on by various forms of stress, including overheating, exercise, mating, or transportation to market. It occurs in about 1 in 12 animals that are homozygous for the genetic abnormality, but rarely in carriers.

Stress-induced deaths in swine occur predominantly with homozygous MH animals, illustrating the recessive nature of the genetics of stress-induced deaths. Nevertheless, the presumed heterozygous MH genotype does lead to enhanced meat production,[81,82] to intermediate sensitivity of the Ca^{2+} release channel to ligand-induced gating,[17] and to intermediate contracture responses to halothane or to halothane plus succinylcholine.[83–86] The genetic background in which the porcine MH gene is expressed, however, might influence the heterozygote response.[87]

A serious problem in the pork industry is the incidence of pale, soft, exudative (PSE) meat that is found in large segments of the carcasses of many animals.[88] A high incidence of PSE meat has long been associated with a high incidence of the PSS gene. The deterioration in meat quality leading to the PSE label is not an "all or nothing" phenomenon. Meat is graded visually and mechanically for several qualities, with a central numerical ranking being reserved for a perfect cut of meat. PSE meat is ranked at lower numerical values, according to its severity, while "dry firm and dark" cuts of meat receive a derogatory ranking higher than the perfect cut. These cuts of meat, which may account for the entire carcass or only small parts thereof, are devalued, leading to considerable economic loss.

When it became possible to identify MH heterozygotes and homozygotes through a blood test,[89] it was possible to relate the incidence of PSE to the presence of the MH gene defect.[90–92] In an initial study, random samples of Canadian swine showed a 15% incidence of heterozygotes and a 0.6% incidence of homozygotes. In a second study, 913 loins were selected on the cutting line, characterized for PSE and genotyped. In the normal meat category, 11% of the loins were from MH heterozygotes (*N/n* genotype) while 0.6% were from MH homozygotes (*n/n* genotype). By comparison, 28.7% of the PSE class loins were from *N/n* pigs and 3.6% were from *n/n* pigs. Since 67.7% of the PSE meat came from normal *(N/N)* animals, it is apparent that elimination of the PSS gene alone will not eradicate the PSE problem. The proportion of PSE meat is also determined by preslaughter management of pigs. The quality of meat obtained from heterozygous animals will benefit the most from improved preslaughter management practices.[92]

When the MH (PSS) gene was recognized as contributing to economic loss, both through mortality and through devalued meat products, attempts were made to remove the gene from swine populations. This turned out to be very difficult for two reasons. First, there was no satisfactory way of identifying heterozygous carriers of the MH gene, although they could be identified retrospectively from genetic trials. Second, it became apparent that the MH gene was contributing to lean body mass[81,82] and, in selecting breeding stock for such characteristics as large ham conformation, large loin eye area, and excessive leanness, breeders were inadvertently selecting for the MH gene. Although it is not known in which breed the MH gene arose, its advantage for lean meat production was sufficiently obvious to ensure its dissemination throughout the world among lean, heavily muscled breeds of swine. Although the deleterious effects of the gene were recognized as early as 1953,[93] attempts to eliminate the gene from breeding stocks were frustrated by the continued selection for desirable meat characteristics and by the inability of breeders to identify heterozygous carriers. As a result, the incidence of the MH gene was stabilized in most lean, heavily muscled breeds of swine.

10.3. THE PHYSIOLOGICAL AND GENETIC BASIS FOR MALIGNANT HYPERTHERMIA

10.3.1. The Ca^{2+} Release Channel

Research on the regulation of excitation/contraction coupling and glycogen metabolism, initiated in the early 1960s, demonstrated that Ca^{2+} plays a major regulatory role in muscle. Studies with isolated sarcoplasmic reticulum[94–96] showed that the sarcoplasmic reticulum is capable of pumping Ca^{2+} to its interior through a Ca^{2+} pump and releasing it under appropriate conditions.[13,97]

Since MH reactions occur in a physiological context, it is essential to the understanding of the basis for MH to understand how Ca^{2+} release and reuptake are regulated *in vivo*. The regulatory cascade for Ca^{2+} release into skeletal muscle begins with depolarization of nerve and muscle membranes. Depolarization courses through the transverse tubule, an invagination of the plasma membrane into the interior of skeletal muscle cells. Polarization and depolarization are not properties of the sarcoplasmic reticulum membrane, since it is relatively permeable to monovalent ions.[98] Depolarization of the transverse tubular membrane is followed by release of Ca^{2+} from the sarcoplasmic reticulum through a Ca^{2+} release channel located at the junctional face of the sarcoplasmic reticulum, the region of the sarcoplasmic reticulum that abuts and associates with the transverse tubule. Ca^{2+} released to the sarcoplasm is then transported back to the lumen of the sarcoplasmic reticulum by a Ca^{2+} pump located throughout all regions of the sarcoplasmic reticulum, with the exception of the junctional face membrane. Ca^{2+} is stored in association with an acidic, luminal calcium-binding protein, calsequestrin,[5] localized in the junctional terminal cisternae.[99] Its localization may be the result of an initial nucleating event brought about by its association with a basic, transmembrane protein, triadin,[100–102] located exclusively at the junctional face (Figure 10.1).

Studies of Ca^{2+} release channels showed that they are localized in heavy sarcoplasmic reticulum vesicles representing the terminal cisternae of the membrane system.[20,97] The development of rapid filtration assays to measure the kinetics of Ca^{2+} release from heavy terminal cisternal vesicles permitted evaluation of the regulatory effects of specific physiological and pharmacological ligands.[13,103–105] More precise characterization was possible when single Ca^{2+} release channel proteins were incorporated into planar lipid bilayers, where they were shown to form ligand-gated channels with a conductance greater than 100 pS in 50 mM Ca^{2+}.[14] Ca^{2+} release is activated by micromolar Ca^{2+}, millimolar ATP, and millimolar caffeine and inhibited by millimolar Mg^{2+}.[9,13,14,105–107] The channel is completely inhibited by ruthenium red and it is locked in an open subconductance state by

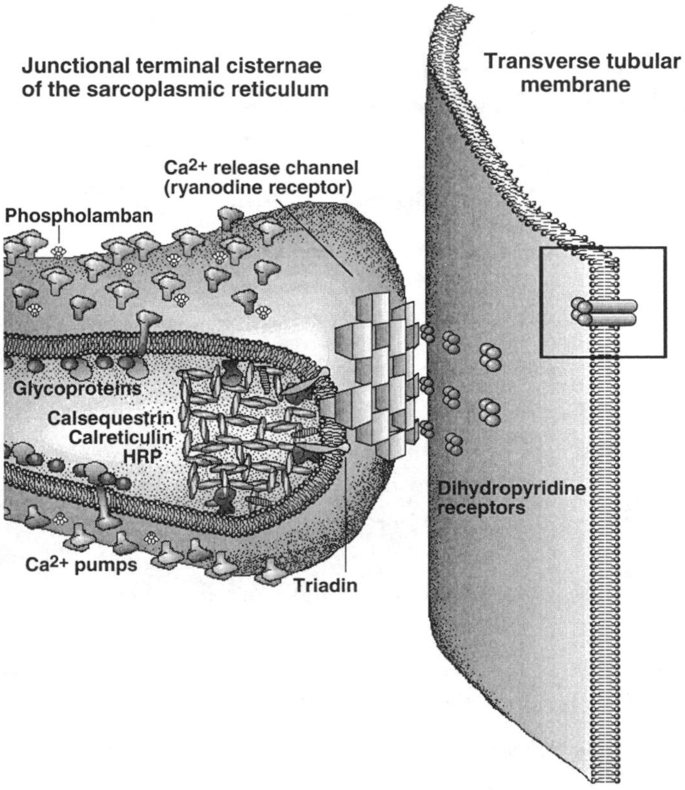

Figure 10.1. Proposed arrangement of proteins in the sarcoplasmic reticulum and transverse tubular membranes. A junctional terminal cistern and its contiguous longitudinal sarcoplasmic reticulum are shown abutting the transverse tubular membrane. Dihydropyridine receptors in the transverse tubular membrane (here shown peeled away from their normal association with the sarcoplasmic reticulum) are illustrated as transmembrane tetrad complexes (boxed), physically apposed to every other calcium release channel (ryanodine receptor). The ryanodine receptor is visualized as a square pyramidal structure in the junctional face of the terminal cistern. An aggregate of elongated calsequestrin molecules is shown within the lumen of the terminal cistern, anchored to triadin, a transmembrane protein in the junctional face membrane. The Ca^{2+}-ATPase is the major protein within the longitudinal membrane of sarcoplasmic reticulum. A pentamer of phospholamban molecules is shown extending from the cytoplasmic face of the longitudinal sarcoplasmic reticulum. Also present in the lumen are calreticulin (a shaded oblong), HRP (shown as a more globular bilobed membrane-associated structure), and the two major glycoproteins, gp53 and gp160 (sarcalumenin), which are shown attached to the membrane and also to each other. (Adapted from Ref. 134 with copyright permission from Raven Press.)

nanomolar ryanodine.[14,108] At higher concentrations, ryanodine closes the channel. Micromolar calmodulin partially inhibits the Ca^{2+} release channel, apparently by direct protein–protein interactions.[103]

Studies of the association between the sarcoplasmic reticulum and the transverse tubules (Figure 10.1) led to the isolation of triads[109,110] and terminal cisternae.[111,112] Morphological studies of the junctional face membranes from these preparations showed them to contain an offset checkerboard pattern of square structures averaging 240 Å across, extending 120 Å from the membrane and having a center-to-center distance of 400 Å. These membrane preparations retained Ca^{2+} poorly. Ruthenium red improved Ca^{2+} retention, while ryanodine, at very low concentrations, enhanced Ca^{2+} release.[113] Since ryanodine modulated Ca^{2+} release with such high affinity, it was used as a ligand for the identification and purification of the Ca^{2+} release channel.[20–22] The purified ryanodine receptor retained the morphology of the junctional face membrane structures and exhibited Ca^{2+} release channel activities identical to those associated with the native channel in planar lipid bilayers.[108,114,115] Thus, the ryanodine receptor and the Ca^{2+} release channel are the same proteins. The Ca^{2+} release channel is a homotetramer of 564,000-Da subunits (Figure 10.2). It has fourfold symmetry and the channel, which originates in a small, transmembrane base plate, branches into four radial channels which empty at peripheral vestibules on both top and bottom surfaces of the huge, cytoplasmic domain that projects from the membrane.[115–117]

Ca^{2+} release channels have been cloned from skeletal *(RYR1)*,[23,24,29] cardiac *(RYR2)*,[71,118] and nonmuscle sources *(RYR3)*,[119,120] The proteins contain from 4872 to 5037 amino acids with masses between 550,000 and 564,000 Da. The number of transmembrane sequences in ryanodine receptors is controversial. Takeshima *et al.*[23] predicted 4 transmembrane sequences, while Zorzato *et al.*[24] predicted 12 (Figure 10.3). Each of the three ryanodine receptor isoforms has four strongly hydrophobic sequences which are well conserved and are almost certainly transmembrane.

The proteins have four repeated sequences in two tandem pairs that are well conserved, but no function has been assigned to these repeat sequences. Residues 2809 and 2843 in cardiac and skeletal isoforms, respectively, are the major phosphorylation sites in these proteins.[121,122] A predicted ATP-binding sequence of structure GXGXXG begins at residue 2652. Since the region between residues 2800 and 3050 also contains predicted calmodulin-binding sites, sequence 2600–3050 has been proposed to be a modulator binding domain.[71] This domain might be extended upstream to residue 2434 (Figure 10.3) where a regulatory mutation giving rise to central core disease and MH has been located.[33] ATP, Ca^{2+}, and calmodulin binding sites were proposed by Takeshima *et al.*[23] to cluster around the proposed transmembrane sequences in the COOH-terminal end of the protein and to encompass the region between residues 4253 and 4499, forming a second predicted modulator binding region. There is overall homology between ryanodine receptors and IP_3 receptors (Figure 10.3), although the IP_3 receptor is only half as long.[123] In addition, two short sequences including residues 3886–3896 and residues 3965–3973 (skeletal isoform) are fully conserved in all ryanodine and IP_3 receptors and may have an essential function.[120]

10.3.2. Excitation–Contraction Coupling

Studies carried out over the past two decades have highlighted the fact that depolarization of the transverse tubule leads to the opening of the Ca^{2+} release channel in the sarcoplasmic reticulum membrane.[124] During depolarization,

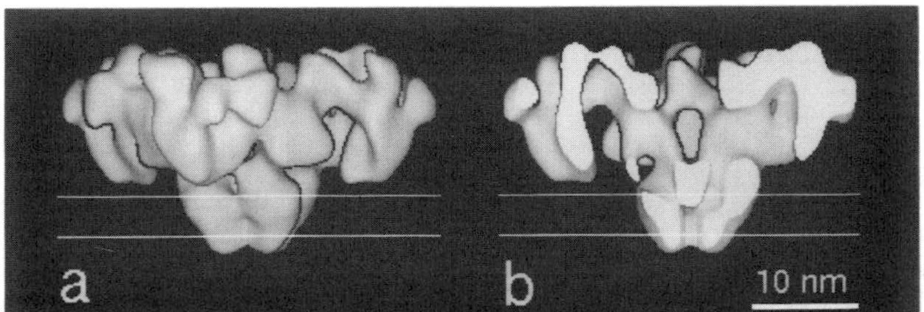

Figure 10.2. Overall surface structure (a) and internal structure (b) of the Ca^{2+} release channel. The three dimensional reconstruction, determined from cryo-electron microscopic images, is presented as a surface representation in side view (a), or cut in half along one orthogonal direction (b). The Ca^{2+} release channel is a homotetramer and has fourfold symmetry. In (a), the transmembrane domain appears as a pillar supporting a platform representing the lace-like cytoplasmic domain. In cross section, the transmembrane domain is seen as a plugged channel. The location of the plug may determine channel opening or closing. (Adapted from Radermacher, M., *et al., J. Cell Biol.* **127,** 411, 1994, copyright permission of the Rockefeller University Press).

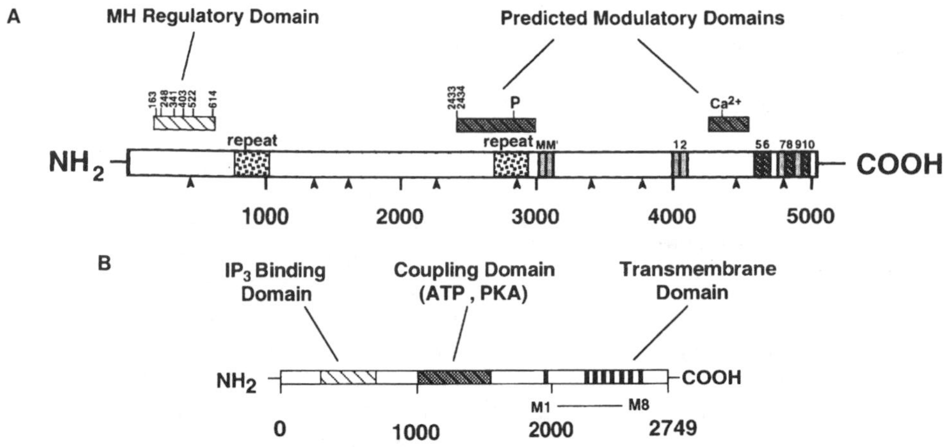

Figure 10.3. (A) A model of the structure of the Ca^{2+} release channel, based on its linear sequence. The transmembrane domain is made up of 4 to 10 transmembrane sequences, of which 4 to 8 are predicted to be located in the COOH-terminal fifth of the protein and 2 are predicted to be more central. Predicted transmembrane sequences 3 and 4[24] are deleted because they are not conserved among ryanodine receptors.[120] There are two tandem repeat sequences (dotted rectangles), but their function is unknown. Nine sites of proteolytic cleavage (arrowheads), perhaps bracketing functional domains, are indicated. Regulatory domains (hatched rectangles) are those predicted to lie between residues 2400 and 3000 by Otsu *et al.*[71] and Zhang *et al.*,[33] where MH or CCD mutations at 2433 and 2434 are located, and between residues 4250 and 4500 by Takeshima *et al.*[23] A third regulatory domain is proposed to lie near the NH_2-terminal end of the protein, since this is the region in which MH or CCD mutations at 163-614 are clustered and in which IP_3 binds to a homologous Ca^{2+} release channel, the IP_3 receptor.

(B) A model of the IP_3 receptor is included for comparative purposes. The IP_3 binding site is homologous to the MH regulatory region in the ryanodine receptor.

movement of a fixed charge in the transverse tubular membrane has been observed to precede Ca^{2+} release[125] and has been proposed to be a prerequisite for Ca^{2+} release.[126] Since dihydropyridine (DHP) blocks both charge movement and Ca^{2+} release, the fixed charge movement has been proposed to occur within the DHP receptor, a slow Ca^{2+} channel that is concentrated in transverse tubular membranes. Indeed, there is a structural feature of the α-1 subunit of the DHP receptor that is consistent with charge movement. The fourth transmembrane sequence in each of the four repeat sequences of the DHP receptor is basic, repeating Lys or Arg residues in these sequences being separated by three hydrophobic residues.[127] If each of these basic sequences were to slide transversely within the bilayer in response to depolarization, a fixed charge movement would be observed. From these observations, it has been postulated that physical movements in the DHP receptor are critical to excitation–contraction coupling in skeletal muscle.[126]

The most important, unsolved problem in excitation–contraction coupling is to understand how the charge movement and/or potential Ca^{2+} channel activity of the DHP receptor opens the Ca^{2+} release channel in the sarcoplasmic reticulum membrane, thereby initiating muscle contraction. The DHP receptor can function both as a slow Ca^{2+} channel and as a voltage sensor.[72,124,128] All evidence suggests that the slow Ca^{2+} channel function of the DHP receptor is not critical for excitation–contraction coupling in skeletal muscle. For example, repeated contraction can be initiated in a Ca^{2+}-free medium, demonstrating that extracellular Ca^{2+} is not required for regulation of contraction. Moreover, the kinetics of opening of slow Ca^{2+} channels are not synchronized with the speed of excitation–contraction coupling. Accordingly, Ca^{2+}-induced Ca^{2+} release, involving the influx of extracellular Ca^{2+} through the DHP receptor, is unlikely to be the initial trigger for Ca^{2+} release in skeletal muscle. As an alternative to Ca^{2+}-induced Ca^{2+}-release, it has been proposed that a physical interaction between DHP and ryanodine receptors occurs in conjunction with the charge movement that accompanies depolarization.[125,126]

The triad junction (Figure 10.1) is formed by the close apposition of channels, originating in the sarcoplasmic reticulum, with other proteins originating in the transverse tubule.[129] Morphological studies suggest that the protein in the transverse tubule that interacts with the ryanodine receptor is the DHP receptor. Block *et al.*[130] showed that the DHP receptors are positioned in the transverse tubule membrane in a pattern that would permit their physical interaction with alternate ryanodine receptors (Figure 10.1). Biochemical studies[72] and studies of coexpression[131] have so far failed to demonstrate a physical interaction between the DHP and ryanodine receptor proteins. It is possible that interactions have not been found because a third component is involved in the interaction between the two proteins. Kim *et al.*[100] proposed that triadin, a 95-kDa membrane-bound, junctional face protein, might play such a role on the basis of its binding to both DHP and ryanodine receptors in protein blots. Knudson *et al.*[101,102] cloned cDNA encoding triadin and, on the basis of its deduced amino acid sequence, predicted that only a short segment of the protein would be located in the cytoplasm. The rest would consist of a transmembrane sequence and a long, basic, luminal sequence that would be ideal for interaction with the acidic residues in calsequestrin.[132] Thus, its properties were not consistent with its role as a mediator between DHP and ryanodine receptors. The mechanism of interaction between DHP receptors and the Ca^{2+} release channel is still not clearly understood.

10.3.3. The Role of the Ca^{2+} Release Channel in Malignant Hyperthermia

The release of Ca^{2+} is the end result of a cascade of events including depolarization of nerve, muscle, and transverse tubular membranes; charge movement associated with the slow Ca^{2+} channel of the transverse tubular membrane;[133] and opening of the Ca^{2+} release channel. Many of the proteins involved in this cascade are well characterized,[134] but others, undoubtedly, are yet to be identified. Ca^{2+} pumps and exchangers in the plasma membrane and carriers in the mitochondrial membrane are also regulated by Ca^{2+} and contribute to Ca^{2+} regulation within muscle cells.[135]

Abnormalities in regulation of the intracellular concentrations of Ca^{2+} that lead to MH might result from mutations in genes encoding the Ca^{2+} pump, the Ca^{2+} release channel, or other proteins that participate in the cascade of excitation–contraction coupling. Abnormalities in the Ca^{2+} pump were ruled out in biochemical studies,[10,11,36] but higher rates of Ca^{2+}-induced Ca^{2+} release, particularly at low levels of inducing Ca^{2+} have been observed in membrane vesicle preparations from both human[15] and porcine[16,136,137] muscle. Closing of single porcine MH channels at high Ca^{2+} concentrations was shown to be inhibited.[18,28] In comparable studies of humans, Ca^{2+} release channels with abnormally greater caffeine sensitivity were detected in MH individuals.[19] In sarcoplasmic reticulum from swine with MH, ryanodine binding, which is dependent on the open state of the Ca^{2+} release channel, is enhanced.[25] Digestion with trypsin revealed an alteration in the amino acid sequence of the Ca^{2+} release channel in MH animals.[26] Thus, from these physiological studies, the Ca^{2+} release channel was implicated as a potential causal factor for MH.

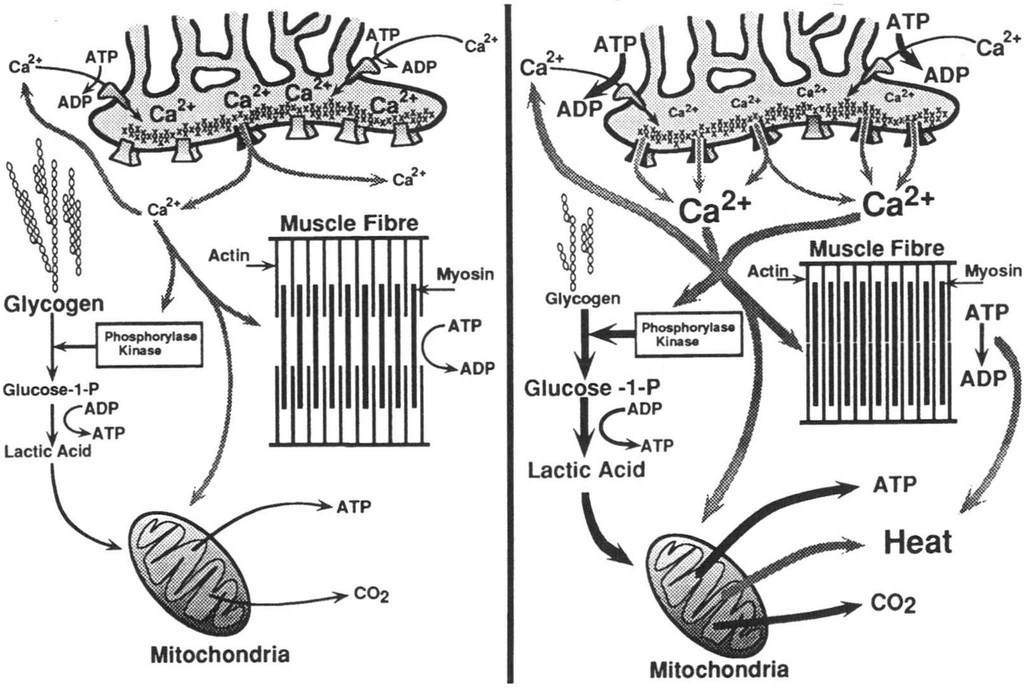

Figure 10.4. A proposed mechanism for induction of malignant hyperthermia via abnormalities in the Ca^{2+} release channel of skeletal muscle sarcoplasmic reticulum. Muscle contraction, glycolysis, and mitochondrial function are regulated by cytoplasmic Ca^{2+} concentrations. In a normal relaxation–contraction cycle (left), Ca^{2+} is pumped into the sarcoplasmic reticulum by a Ca^{2+}-ATPase to initiate relaxation, stored within the lumen in association with calsequestrin, and released through a Ca^{2+} release channel to initiate contraction. Glycolytic and aerobic metabolism proceed only rapidly enough to maintain the energy balance of the cell. The Ca^{2+} release channel can be regulated by Ca^{2+} itself, ATP, Mg^{2+}, and calmodulin, as well as by its association with the dihydropyridine receptor, and, even when stimulated, has a relatively short open time. The abnormal malignant hyperthermia Ca^{2+} release channel (right) may be sensitive to lower concentrations of stimulators of opening, it releases Ca^{2+} at enhanced rates and does not close readily. In the presence of anesthetic agents, the abnormal channel floods the cell with Ca^{2+} and overpowers the Ca^{2+} pump in its attempts to lower cytoplasmic Ca^{2+}. Chronically high levels of Ca^{2+} stimulate muscle contracture and glycolysis, accounting for rigidity and the generation of lactic acid. Damage to cell membranes and imbalances of ion transport resulting from lactic acid production and lowered ATP concentration can account for the life-threatening systemic problems that appear during a malignant hyperthermia episode. Heat is generated by the hydrolysis of ATP by the contracted muscle and the Ca^{2+} pumps. Regeneration of ATP through aerobic metabolism accounts for CO_2 generation and enhanced O_2 uptake. (Adapted from Ref. 39 with permission, ©AAAS.)

A defect in the Ca^{2+} release channel giving rise to abnormal Ca^{2+} regulation within skeletal muscle could account for all of the symptoms of MH (Figure 10.4). If the Ca^{2+} release channels had longer open times in the presence of anesthetic agents, intracellular Ca^{2+} might be chronically elevated, resulting in muscle contracture and activation of the first steps in glycogenolysis through activation of phosphorylase kinase.[6] Muscle contracture and the pumping of cytoplasmic Ca^{2+} to the lumen of the sarcoplasmic reticulum would consume large amounts of ATP, generating heat. The ADP formed would stimulate glycolysis and the mitochondrial oxidation of pyruvate derived from glucose. These hypermetabolic responses would lead to depletion of ATP, glycogen, and oxygen, to the production of excess lactic acid, CO_2, and heat, and, ultimately, to the disruption of cellular and extracellular ion balance.

10.3.4. Dantrolene and Ryanodine

Dantrolene sodium is the antidote for MH reactions.[35,138,139] Its mechanism of action, however, is still not understood. It is absorbed slowly and completely when administered orally and is used to relieve muscle spasm.[140] As an antidote for acute MH reactions, however, it is administered intravenously, following reconstitution of the appropriate solution from a lyophilized preparation of dantrolene, mannitol, and sodium hydrochloride. The fact that dantrolene is a universal antidote to MH reactions makes it important to understand its site of action. If its site of action is the Ca^{2+} release channel of the sarcoplasmic reticulum, then this focuses attention on the *RYR1* gene as the primary gene responsible for MH.

Dantrolene appears to have no significant effects on cardiac or smooth muscle, but it attenuates contraction in neurally stimulated skeletal muscle.[141,142] Dantrolene does not alter the nondepolarizing neuromuscular blocking action of *d*-tubocurarine or the anticholinesterase action of edrophonium on twitch tension. Therefore, it must act at a point distal to the motor nerve.[143] Doses of dantrolene that reduce the twitch response of directly stimulated skeletal muscle have no effect on either resting or action potentials of the muscle membrane. Dantrolene also does not alter total membrane capacitance or membrane resistance.[143] It does not influence the resting influx of $^{45}Ca^{2+}$ from extracellular spaces,[144] but does inhibit K^+-induced contraction. Studies utilizing microelectrodes show that dantrolene significantly lowers the cytoplasmic Ca^{2+} concentration of MH muscle.[145] These studies suggest that dantrolene acts at the level of Ca^{2+} release through the Ca^{2+} release channel or at the level of signal transduction between the transverse tubule and the Ca^{2+} release channel. Ohnishi *et al.*[146] showed that dantrolene altered the caffeine- and halothane-induced increment of Ca^{2+} release from the sarcoplasmic reticulum, although several workers were unable to demonstrate significant effects of dantrolene on Ca^{2+} release.[139,147,148] Ohta *et al.*[149] were, however, able to demonstrate inhibition by dantrolene of Ca^{2+} release from guinea pig sarcoplasmic reticulum incubated at 37°C but not at 25°C.

In very recent experiments, Nelson and Lin[150] were able to demonstrate effects of dantrolene on single Ca^{2+} release channel function. They found that dantrolene first activates and then inhibits the Ca^{2+} release channel. Thus, part of the difficulty in establishing a role for dantrolene as an inhibitor of the Ca^{2+} release channel may have arisen from the complexity of the interaction of dantrolene with the channel protein. Similar problems plagued early studies of ryanodine binding to the channel protein. These problems were only overcome when it was understood that ryanodine binds only to open channels so that binding must be carried out under conditions where Ca^{2+} release channels are predominantly open.[151] Thus, evidence to date, although not definitive, suggests that the Ca^{2+} release channel is the primary site of action of dantrolene.

Just as studies of the mechanism of action of dantrolene may provide insight into the gene product causing MH, so studies with ryanodine might also provide important clues to identification of the causal factor. It is of interest that, in *in vitro* contracture tests, ryanodine-induced contractures appear to have high specificity and selectivity.[66,67] Ryanodine binds selectively and with very high affinity only to the Ca^{2+} release channel of the sarcoplasmic reticulum.[113,151] Thus, if mutations affecting ryanodine association with the Ca^{2+} release channel were, in fact, the cause of MH susceptibility, then a ryanodine contracture test should discriminate between MH mutant and normal Ca^{2+} release channels. Accordingly, if a ryanodine contracture test should prove accurate for all MH testing, then the genetic basis for MH might logically be assigned entirely to the *RYR1* gene. Since ryanodine binds with highest affinity to open Ca^{2+} release channels,[113,151] then a further corollary might be that all MH mutations should be defined as those that prolong the open state of the Ca^{2+} release channel.

10.3.5. Linkage of the *RYR1* Gene to Malignant Hyperthermia in Humans and Swine

In early studies of porcine MH, Andersen and Jensen[152] demonstrated linkage between inheritance of MH and polymorphisms in the gene encoding glucose phosphate isomerase *(GPI)*. Later studies[153,154] established a linkage group for the porcine *HAL* gene, the designation of the MH gene giving rise to halothane sensitivity, *GPI,* and the gene for 6-phosphogluconate dehydrogenase *(PGD),* localized near the centromere of pig chromosome 6.[155–157] The syntenic region around the human *GPI* locus was known to be on the long arm of chromosome 19,[158] making this a candidate region for human MH localization. Cloning of the human skeletal muscle ryanodine receptor *(RYR1)* cDNA[24] led to the localization of *RYR1* to human chromosome 19q13.1, in the same region as human *GPI.*[159]

Cloning of human RYR1 also led to the discovery of several restriction fragment length polymorphisms (RFLPs) in the human *RYR1* gene.[160] In a study of linkage between inheritance of MH and one or more *RYR1* polymorphisms and polymorphisms in flanking markers, cosegregation with *RYR1* markers was found in 23 meioses in nine families, leading to a lod score of 4.2 favoring linkage with a recombinant fraction of 0.0. The probability of linkage of more than 10,000 to 1 identified *RYR1* as a candidate gene for MH in humans.[158] In an independent study of linkage of human MH to a series of chromosome 19q markers, the MH locus was assigned to the region of human chromosome 19 where *RYR1* was localized.[161]

10.3.6. Malignant Hyperthermia Mutations in Swine

Linkage of *RYR1* to MH provided the incentive to begin parallel searches in both swine and humans for sequence differences in the *RYR1* gene between MH and normal individuals.[29–32] In a comparison of the cDNA sequences of MH (Pietrain) and normal (Yorkshire) pigs, 18 nucleotide polymorphisms were found, but only one of these altered the amino acid sequence.[29] The substitution of T for C at nucleotide 1843 leads to the substitution of Cys for Arg at amino acid residue 615 (Table 10.1). Since this mutation resulted in the loss of a *Hin*PI restriction endonuclease site and the gain of an *Hgi*AI site, it was possible to analyze the mutation either by restriction endonuclease digestion or by differential oligonucleotide probe analyses. Initial studies showed an association between inheritance of the mutation and inheritance of MH in some 80 animals from five different breeds. To establish tight linkage, studies were carried out on backcrosses between British Landrace heterozygous animals of the *N/n* genotype and homozygous MH animals of the *n/n* genotype.[30] In this study, 376 animals were tested, including 338 representing informative meioses. Phenotypic diagnoses were based on the halothane challenge test and confirmed by *GPI* and *PGD* haplotype analysis. Cosegregation of the phenotype with the Cys for Arg^{615} substitution was complete, leading to a lod score favoring linkage of 102 with a recombinant fraction of 0.0.

The fact that the identical mutation appeared in five lean, heavily muscled pig breeds led to the possibility that the mutation arose in a founder animal. Haplotype and genotype analysis using three markers covering over 100 kb within the *RYR1* gene showed the inheritance of a $Hin\mathrm{PI}^{-/-}Ban\mathrm{II}^{+/+}Rsa\mathrm{I}^{-/-}$ genotype in every *n/n* animal examined and the potential for the inheritance of this haplotype in every *N/n* animal. By contrast, the $Hin\mathrm{PI}^{+/+}Ban\mathrm{II}^{-/-}Rsa\mathrm{I}^{+/+}$ genotype was found in normal animals in all breeds examined. Association of a specific haplotype with the MH phenotype suggests that the disease did originate in a founder animal and was selected for in breeding stock.

The disease gene seems to have been selected because it contributes 2 to 3% to dressed carcass weight and contributes to leanness and heavy muscling.[81,82] A hypersensitive Ca^{2+} release channel could give rise to spontaneous muscle contraction and the continual toning of the muscle may lead to muscle hypertrophy.[39] The utilization of ATP for spontaneous contraction would limit the deposition of fat. A similar situation has arisen where selection for heavily muscled individuals among quarter horses has led to selection of a defect in the sodium channel (SCN4A) gene giving rise to equine hyperkalemic periodic paralysis.[162] These animals may be heavily muscled because of spontaneous contractions induced by the sodium channel defect.

10.3.7. Malignant Hyperthermia Mutations in Humans

The demonstration of linkage between MH and a substitution of Cys for Arg^{615} in the *RYR1* gene in swine led to a search for the corresponding mutation in human MH families. The equivalent mutation, Cys for Arg^{614}, was found in a single family of five members in which the mutation segregated with MH. The mutation results in the loss of an *Rsa*I restriction endonuclease site (Table 10.1), making it easy to diagnose by PCR amplification and cleavage.[89] This mutation has been found in several MH families worldwide.[163] In most families it segregates with individuals who have been diagnosed by CHCT as MHS,[31,70] but in other families it does not.[163,164]

Table 10.1. *RYR1* Mutations Associated with MH or CCD[a]

Amino acid substitution	Nucleotide substitution	Detection	Association	Reference
Cys for Arg^{163}	T for C487	Loss of *Bst* UI	MH, CCD	34
Arg for Gly^{248}	A for G742	Allele-specific PCR	MH	32
Arg for Gly^{341}	A for G1021	SSCP	MH	166
Met for Ile^{403}	G for C1209	Loss of *Mbo*I	MH, CCD	34
Ser for Tyr^{522}	C for A1565	SSCP	MH, CCD	—
Cys for Arg^{614}	T for C1840	Loss of *Rsa*I	MH	31
Cys for Arg^{615} (pig)	T for C1843 gain of *Hgi*AI	Loss of *Hin*PI,	MH, PSS	29
Arg for Gly^{2433}	A for G7297	gain of *Dde*I	MH	167, 168
His for Arg^{2434}	A for G7301	Loss of *Hga*I	MH, CCD	33

[a] Key: MH, malignant hyperthermia; CCD, central core disease; PSS, porcine stress syndrome; PCR, polymerase chain reaction; SSCP, single strand conformational polymorphism; *Bst*VI, *Mbo*I, *Rsa*I, *Hin*PI, *Dde*I, and *Hga*I are restriction endonucleases.

The assessment of the Arg^{614}-to-Cys mutation as causative of MH is based on strong genetic and biochemical evidence. A lod score of 102 favoring linkage at θ max = 0.0 in swine, combined with the crossing of a species barrier between swine and humans provides the genetic evidence, while the demonstration of a measurable defect in closing of the Ca^{2+} release channel[19,28] provides the biochemical evidence. The Arg^{614}-to-Cys mutation has not been found in studies of normal chromosomes which now probably exceed more than 1000. Thus, when linkage of this mutation to MH in humans cannot be demonstrated, it is possible that the cause is inaccurate diagnosis of MH through the CHCT.[165] This problem of potential inaccuracy in the CHCT will make it very difficult to prove the causal nature of other MH mutations through linkage of inheritance of MH and mutant *RYR1* alleles.

In sequencing of a full-length *RYR1* cDNA from an MH proband, the mutation of Gly^{248} to Arg was found to segregate with MH in the proband, her brother, who was also MHS, and her mother, who was not diagnosed.[32] This mutation has not been found in analysis of 155 individuals from more than 100 unrelated Canadian families who had been diagnosed as MHS.[33] Nevertheless, it is a candidate mutation for MH in man. Other amino acid substitutions noted in the proband were Cys for Arg^{470}, Glu for Pro^{1785}, and Cys for Gly^{2057}. None of these substitutions were shared between the proband and her MHS brother. The latter two substitutions appear to be very common polymorphisms, not associated with MH.

The mutation of Gly^{341} to Arg has been found to segregate with MH susceptibility in about 10% of MH families studied.[166] In the seven families in which the mutation was found, six MHE individuals were noted. Of these, five did not contain the Gly^{341}-to-Arg mutation, suggesting that the majority of patients diagnosed as MHE by the European CHCT may not be susceptible to MH from a genetic viewpoint.

The mutation of Gly^{2433} to Arg has been found in 8 MH families, representing about 4% of those families screened for the mutation.[167,168] The mutation was linked to MH in four European families, but in only two of four Canadian MH families investigated. In one small Canadian family, a case could be made that two MH mutations might be present in the same family. An exceptionally strong CHCT result was found in one sibling carrying the Gly^{2433} to Arg mutation, while the other sibling, who did not carry the mutation, had a positive, but not exceptional CHCT result. In the second discordant family, one false negative and two false positive CHCT results would have to be invoked to achieve linkage.

The search for additional MH mutations is hampered by the size of the cDNA and the gene from which it is derived. Sequencing of cDNA, or the combination of sequencing with other methods for detecting polymorphisms prior to sequencing[32–34] is fully feasible, but is laborious and expensive. Sequencing from genomic DNA is feasible, since all 106 exon/intron boundaries have been defined.[169] Data from phage cloning and from the isolation of YAC and cosmid clones containing the gene[170] showed it to be about 205 kbp, but more refined measurement have shown it to be about 161 kbp.[169] Thus, while the 15,000-bp cDNA encoding the 564,000-Da protein sequence is one of the longest known, the gene is not exceptionally large.

10.3.8. Searching for a Second MHS Locus

On the basis of linkage of MH to chromosome 19 and to *RYR1* mutations, *RYR1* is established as a very strong candidate gene for MH, but it is not necessarily the only candidate gene for MH. There are cases where diagnosis appears to be accurate and where no linkage between *RYR1* and MH can be discerned.[163,164,171–173] In at least some of these cases, both false-positive and false-negative diagnoses would have to be invoked to prove linkage. The finding of genetic heterogeneity is perhaps not surprising in light of reports that patients with other muscle diseases are subject to MH episodes or to positive CHCT.[46–55,174,175] Thus, individuals with central core disease, myotonia congenita, myotonia fluctuans, myotonia dystrophica, limb-girdle muscular dystrophy, Brody's disease, or Duchenne or Becker's muscular dystrophy have apparently had MH reactions or have been diagnosed as MHS by CHCT. Such a reaction could be postulated to result from a normal Ca^{2+} release channel which is triggered to open by a rise in Ca^{2+} in a muscle cell in which poor Ca^{2+} regulation exists as a result of defects in the Ca^{2+} pump, calsequestrin, or DHP receptor subunits, for example, or from increased membrane permeability related to secondary causes.[173] Thus, other defective proteins leading to poor Ca^{2+} regulation within the cell may eventually be shown to give rise to other forms of MH susceptibility. However, the fact that in some laboratories as few as 35–50% of MH families[42] can be linked to chromosome 19q13.1, suggests that a second unique MH gene locus must be considered.

Levitt *et al.*[176] and Olckers *et al.*[177] presented evidence for a second MHS locus on chromosome 17q21. There are three potential candidate genes on chromosome 17, the β and the γ subunits of the DHP receptor[178] and a sodium channel gene, SCN4A, which has been linked to hyperkalemic periodic paralysis.[179] Attempts to confirm the localization of a second MHS locus on chromosome 17 were unsuccessful.[178] The $\alpha 1$ subunit of the DHP receptor, localized to chromosome 1, has also been excluded as a candidate gene for MH susceptibility.[180] Further studies suggest that MH susceptibility may segregate with a SCN4A mutation, which gives rise to both myotonia fluctuans and masseter muscle rigidity.[174,175]

MH has been linked to chromosome 7q, with a lod score of less than 3 in a single family.[181] The presence of the gene encoding the $\alpha 2/\delta$ subunit of the dihydropyridine receptor (CACNL2A) on chromosome 7q21-q22 establishes it as a possible candidate gene for MH.

In an effort to determine if any other MHS loci exist, a consortium of scientists forming the genetics section of the European Malignant Hyperthermia Group screened some of their large, non-chromosome 19 linked families against several hundred polymorphic microsatellite markers developed by Généthon, that cover the entire human genome. In this study they linked MH to a locus on chromosome 3q13.1 with a lod score of 3.22 in a single family.[182] This high lod score suggests that true heterogeneity exists for MH. This chromosome 3-linked MH family offers an excellent opportunity for the identification of an additional causal gene for MH.

10.4. CENTRAL CORE DISEASE

Central core disease (CCD) is a rare, nonprogressive myopathy characterized by hypotonia and proximal muscle weakness and presenting in infancy.[183] Additional variable clinical features include pes cavus, kyphoscoliosis, foot deformities, congenital hip dislocation, and joint contractures.[184–187] Although symptoms may be severe, up to 40% of patients demonstrating central cores may be clinically normal.[187] Diagnosis is made on the basis of the lack of oxidative enzyme activity in central regions of skeletal muscle cells,[188] observed on histological examination of biopsies. Electron microscopic analysis shows disintegration of the contractile apparatus ranging from blurring and streaming of the Z lines to total loss of myofibrillar structure.[189,190] The sarcoplasmic reticulum and transverse tubular systems are greatly increased in content and are, in general, less well structured.[191] Mitochondria are depleted in the cores, but may be enriched around the surfaces of the cores. Genetic analysis indicates that the disorder is inherited as an autosomal dominant trait with variable penetrance.[189,190,192,193]

An important feature of CCD is its close association with susceptibility to MH.[46,53,187,193,194] This association led investigators to establish linkage between CCD and markers in the long arm of chromosome 19 in large Australian[195] and European[196] CCD pedigrees. A lod score of 11.8 favoring linkage with a recombinant fraction of 0.0, using markers within the *RYR1* gene, was also established in the Australian family.[197]

Analysis of *RYR1* cDNA sequences in several CCD families has led to the discovery of four mutations that are linked to CCD and/or MH (Table 10.1). Zhang *et al.*[33] linked the substitution of Arg^{2434} with His to CCD in a Canadian family, obtaining a lod score of 4.8 favoring linkage with a recombinant fraction of 0.0. This mutation was found in only a single pedigree of more than a dozen examined. Quane *et al.*[34] found that the substitution of Met for Ile^{403} and the substitution of Cys for Arg^{163} in CCD families were linked to inheritance of either MH or CCD in these families. Quane *et al.*[198] also linked the mutation of Tyr^{522}-to-Ser to CCD. The Arg^{163}-to-Cys mutation gave rise to invariant MH susceptibility, but to variable formation of central cores. The Ile^{403}-to-Met and the Tyr^{522}-to-Ser mutations were found in single CCD pedigrees.

It is of interest that six out of eight MH or CCD mutations lie between amino acids 163 and 614 (Figure 10.3), in a region homologous to the IP_3 binding region in the IP_3 receptor.[199,200] Such a cluster of mutations would suggest that this region of the molecule forms a regulatory domain in the Ca^{2+} release channel. Studies of tryptic digestion[201] suggest that proteolytic cleavage sites occur at about residues 500 and 1400, the first site lying in the midst of the MH mutation cluster (Figure 10.3). These observations may indicate that two interacting regulatory domains lie within the first 1400 residues of the Ca^{2+} release channel. The two additional MH mutations (amino acids 2433 and 2434) lie at the beginning of a sequence predicted earlier[71] to be a regulatory sequence of about 600 amino acids (Figure 10.3). A second interesting feature of MH mutations discovered so far is the fact that six of the eight involve either loss or gain of an Arg residue. This suggests that positive changes within the domain are critical to regulatory function.

The discovery of mutations in *RYR1* potentially causal of CCD and MH raises interesting questions concerning the pathophysiology of these diseases. It is clear that, in both cases, Ca^{2+} regulation is imbalanced, leading to the contracture and hypermetabolism that characterize MH. In some cases, however, the altered Ca^{2+} regulation leads to disorganization of the contractile proteins in the central core, a proliferation of sarcoplasmic reticulum and transverse tubules, and a loss of functional mitochondria. It is possible that these structural alterations arise developmentally and are the result of disorganization imposed on the developing fiber by alterations in the physical properties of the Ca^{2+} release channel, which may play a critical role in the organization of the sarcotubular membrane system.

Another possibility is that these alterations occur subsequent to the initial formation of a fully functional muscle fiber and are the result of physiological adaptation to functional alterations in the Ca^{2+} release channel that lead to elevated Ca^{2+} levels within the myofibril. Myofibrils regulate Ca^{2+} through at least four systems: plasma membrane Ca^{2+} pumps (PMCAs); sarco (endo)plasmic reticulum Ca^{2+} pumps (SERCAs); Na^+/Ca^{2+} exchangers; and mitochondria. Ca^{2+} pumps and Na^+/Ca^{2+} exchangers in the plasma membrane can remove Ca^{2+} from the muscle cell (Figure 10.5). Of these two systems, the Ca^{2+} pump has the higher affinity for Ca^{2+}, while the Na^+/Ca^{2+} exchanger is more active with elevated levels of intracellular Ca^{2+}. The sarcoplasmic reticulum is the major regulator of Ca^{2+} within the muscle cell, removing it from the cytoplasm, storing it, and releasing it again to initiate muscle contraction. If Ca^{2+} concentrations are elevated, the mitochondria can transport Ca^{2+} to matrix spaces, thereby protecting the cell from Ca^{2+}-induced damage. If the Ca^{2+} release channel were to release excessive amounts of Ca^{2+} within the muscle cell, then the Na^+/Ca^{2+} exchanger and mitochondria might play a more important role in Ca^{2+} regulation in a CCD cell than in a normal cell. Extrusion of excess Ca^{2+} from the cell might, itself, have deleterious effects on the skeletal muscle cell, in which a constant amount of Ca^{2+} is normally cycled internally, without any requirement for extracellular Ca^{2+}.

Pumps and exchangers in the plasma membrane might be more effective in protecting the periphery of the cell than the interior of the cell where the full burden of regulation of excess Ca^{2+} would fall on the sarcoplasmic reticulum and mitochondria. It is possible that mitochondria, which have a high capacity for Ca^{2+} uptake and would, undoubtedly, participate in removal of excess Ca^{2+} from central areas of the cell, might destroy themselves in an effort to protect the cell from Ca^{2+}-induced necrosis.[199] Loss of mitochondria from the center of the cell would, in turn, lead to lower ATP synthesis and might

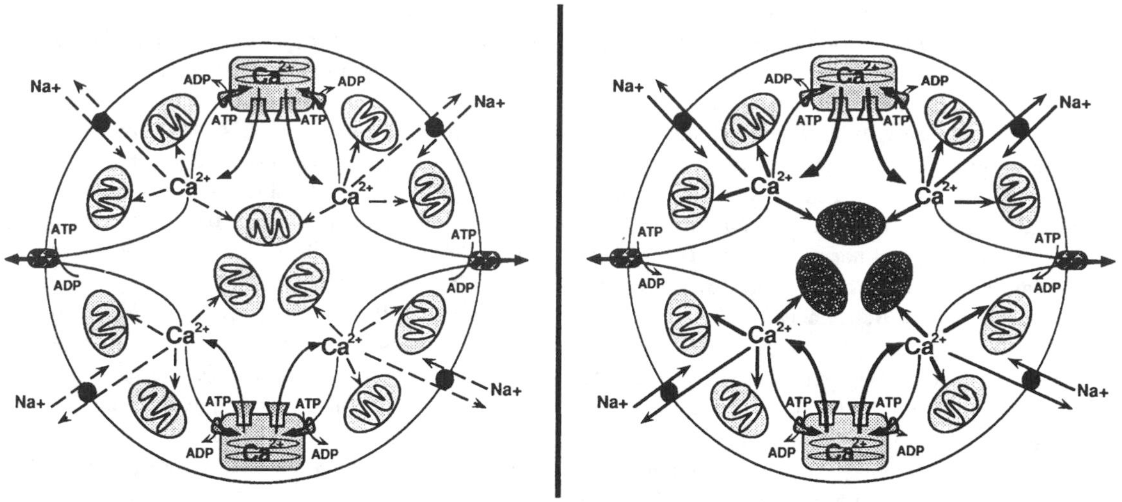

Figure 10.5. A proposed mechanism for the development of central cores resulting from defects in the ryanodine receptor. In normal muscle fibers, Ca^{2+} released from the sarcoplasmic reticulum can be regulated by four systems including the organellar sarcoplasmic reticulum and mitochondria and the plasma membrane Ca^{2+} pumps and Na^+/Ca^{2+} exchangers. Under normal circumstances, the bulk of the Ca^{2+} is cycled only through the sarcoplasmic reticulum. If enhanced Ca^{2+} release occurred spontaneously, in CCD muscle, the additional Ca^{2+} regulatory systems might be co-opted to regulate Ca^{2+}. The plasma membrane exchangers and pumps would be effective in regulating Ca^{2+} near the periphery, but, in the core, mitochondria may be forced to bear this load, destroying themselves in the process. The degeneration of mitochondria could lead to the degeneration of a central, possibly compartmented, core.

be an underlying cause of the disorganization of the central core. Another possible cause of the disorganization of myofibrils in the central core might be differential contraction of fibrils in the core, where Ca^{2+} concentrations remain high relative to fibrils in the periphery, where Ca^{2+} concentrations might be better regulated. This differential contraction could lead to blurring and streaming of the fibrils and the two lines. This disorganization could be a primary cause of muscle weakness and atrophy. The profusion of sarcoplasmic reticulum and transverse tubules might be induced at the gene level by high local Ca^{2+} concentrations.

A problem with this postulate is that the cores are clearly delineated, which would not be consistent with the view that the core might arise from an intracellular gradient of Ca^{2+} which would be elevated in the core relative to the periphery of the cell. Perhaps there are functional "compartments" within the myofibril, of which we are currently unaware, but which are delineated by their different abilities to deal with excess Ca^{2+} in fibrils from CCD patients. Central cores in skeletal muscle have also been observed in patients with familial hypertrophic cardiomyopathy.[203] This disease has been linked to defects in the β myosin heavy chain. Thus, it is clear that there can be different routes to the formation of central cores, making the etiology of central core formation a fascinating research topic.

It is of interest that the Arg^{615}-to-Cys mutation is associated with muscle hypertrophy in swine[30,39] while the Arg^{2434}-to-His, Tyr^{522}-to-Ser, Ile^{403}-to-Met, and Arg^{163}-to-Cys mutations are associated with variable degrees of muscle atrophy, metabolically inert cores, and proximal muscle weakness. If all of these mutations led to poorly regulated Ca^{2+} release into the muscle cell, they could trigger spontaneous muscle contractions. Such spontaneous contractions could lead to the muscle hypertrophy observed in swine. In this case, the system of pumps and exchangers in the plasma membrane and organelle systems of mitochondria and sarcoplasmic reticulum within the cell could remove excess Ca^{2+} from the sarcoplasm without deleterious effects on the muscle cell. As we have outlined above, however, the CCD mutations might be more severe, leading to damage to the interior of the cell and to loss of mitochondrial function and structural abnormalities in the central core. These, in turn, could lead to disruption of core fibrils and to muscle weakness and atrophy. Thus, mutations in *RYR1* can lead to a spectrum of pathophysiological responses ranging from muscle hypertrophy to muscle atrophy.

10.5. KING–DENBOROUGH SYNDROME

Anesthetic-induced malignant hyperthermia was reported by King and Denborough[47] to occur in children with particular congenital abnormalities such as short stature, scoliosis, pectus deformity, delay in motor development, ptosis, low-set ears, anti-mongolian slanted eyes, and cryptorchidism.[204] This syndrome is inherited as an autosomal dominant trait. It is not yet known whether the syndrome is linked to chromosome 19 and whether it could result from defects in *RYR1*.

10.6. FUTURE DIRECTIONS

Research over the past quarter century has led to a very good understanding of the physiological and genetic basis for MH in human and in swine. All of the important goals of MH research have not yet been realized, however, and these must be pursued. Perhaps the most important immediate goal is to define all of the genes and all of the mutations in those genes that are causal of human MH. This will provide a firm basis for diagnosis of MH susceptibility for a large fraction of those families in which MH is inherited. A second important goal is to utilize the MH gene in ways that will be of most benefit to the pork industry. A third goal will be to understand the functional consequences of the structural alterations in the Ca^{2+} release channel that lead to its involvement in MH episodes. A fourth goal will be to understand, at the physiological level, how MH relates to additional abnormalities in muscle cells, such as the formation of central cores, and how other myopathies can underlie anesthetic-induced malignant hyperthermia.[53]

The search for a second MH gene locus has been described. The key to success will be the accuracy of the CHCT which has provided most of the available data on inheritance of the abnormality. While the CHCT is a valuable clinical test, evidence is accumulating that it is inadequate for all genetic studies. Since it may be critical to the discovery of a second or third genetic locus to be able to diagnose inheritance of MH accurately, more research into MH diagnosis is clearly warranted. On the other hand, continued investigation of linkage of *RYR1* mutations to MH may permit investigators to identify probable false-positive and false-negative diagnoses.[65,163–165] Such studies may indicate that more MH susceptibility is associated with defects in *RYR1* than the 30 to 50% that can now be demonstrated.

In the search for MH mutations in the *RYR1* gene to date, mutations have been discovered, either through sequencing of full-length cDNAs from human MH patients or through PCR amplification of segments of *RYR1* cDNA, or genomic DNA, followed by analysis of single-strand conformational polymorphisms (SSCPs).[205] The identification of all 106 exons and their flanking sequences now makes it possible to sequence *RYR1* either from cDNA or from genomic DNA, following PCR amplification of exons. Recent advances in automated sequencing may make it feasible to carry out rapid sequencing of *RYR1* exons cDNAs from any MH proband. Thus, the rate of identification of MH mutations may be dramatically increased.

Once most of the MH mutations are known, it should be possible to follow inheritance of these mutations and to designate those members of MH families who are MHS. This will ensure that MHS individuals will receive a regimen of "safe" anesthetics, while normal family members can utilize more conventional anesthetics. There will always be a caveat in interpreting such diagnostic tests, since new mutations or the presence of an unsuspected second mutation within a family could give rise to false-negative diagnoses. The fact that there will be a large number of independent mutations in *RYR1,* and possibly in other MHS genes, makes it apparent that diagnosis of MH susceptibility through simple DNA-based tests applied at random will be very difficult. Thus, it is unlikely that population screening for MH susceptibility will be feasible in the near future.

For the swine industry, the outlook is more straightforward. With a single refined, restriction endonuclease test for the disease,[89] muscle or blood samples as small as 50 μl can be used to obtain rapid, accurate diagnosis. Accordingly, it is feasible to identify and remove the MH gene from the porcine population within a very short time, if that should prove desirable.[206] On the other hand, since the gene is beneficial in terms of food conversion, leanness, and muscularity, breeding programs utilizing *n/n* × *N/N* lines may be set up that will produce *N/n* slaughter animals.[207] Research is currently under way to evaluate the usefulness of the gene to the pork industry. The finding that two-thirds of PSE comes from normal animals[91] makes research into preslaughter management of all swine a high priority.[92]

The investigation of structure–function relationships in the Ca^{2+} release channel is just beginning.[208] Very little is known about the structural and functional domains of the protein and how they affect gating of the channel. An important concept is emerging from studies of MH mutations showing that these are clustered between residues 163 and 614 in a domain that corresponds to the IP_3 binding domain of the IP_3 receptor. These mutations frequently involve the basic residue, arginine. Thus, studies of the region in which MH mutations are clustered may lead to the recognition of an important regulatory domain in the Ca^{2+} release channel.

The question of how central cores form in muscle from a mutation that also causes MH raises the important question of how chronic variation in Ca^{2+} levels in a muscle cell can affect its physiology. The development of animal models to investigate these processes will be an important component of future research on MH and central core disease. The question of how other myopathies, such as the King–Denborough syndrome and Duchenne muscular dystrophy, can lead to Ca^{2+} imbalances that lead to MH episodes also deserves investigation.

In this chapter we have outlined how basic science has illuminated our understanding of the cause of an abnormality that is important both to human medicine and to the agricultural industry and provided the basis for its genetic dissection and diagnosis. Understanding of the genetic basis for the abnormality will ultimately provide not only a more accurate and less invasive means of diagnosis for human carriers of the defective gene(s) but will also be of tremendous economic benefit to the pork industry.

Acknowledgments. We thank our many colleagues for advice and discussion in the preparation of this review. We also thank Drs. Marilyn Larach, Tommie McCarthy, and Cheryl Greenberg for critical review of the manuscript. Research grants to D.H.M., supporting original work from our laboratory, were from the Medical Research Council of Research Council of Canada (MRCC), the Muscular Dystrophy Association of Canada (MDAC), the Heart and Stroke Foundation of Ontario (HSFO), and the Canadian Genetic Diseases Network of Centers of Excellence. M.S.P. was a predoctoral fellow of the HSFO and of the MRCC. Y.Z. was a postdoctoral fellow of the MDAC/MRCC.

REFERENCES

1. Denborough, M. A., and Lovell, R. R. H. (1960). Anaesthetic deaths in a family. *Lancet* **2:**45.

2. Ebashi, S. (1963). Third component participating in the superprecipitation of "natural actomyosin." *Nature* **200:**1010.
3. Zot, A. S., and Potter, J. D. (1987). Structural aspects of troponin–tropomyosin regulation of skeletal muscle contraction. *Annu. Rev. Biophys. Biophys. Chem.* **16:**535–559.
4. Ebashi, S., Endo, M., and Ohtsuki, I. (1969). Control of muscle contraction. *Q. Rev. Biophys.* **2:**351–384.
5. MacLennan, D. H., and Wong, P. T. S. (1971). Isolation of a calcium-sequestering protein from sarcoplasmic reticulum. *Proc. Natl. Acad. Sci. USA* **68:**1231–1235.
6. Brostrom, C. O., Hunkeler, F. L., and Krebs, E. G. (1971). The regulation of skeletal muscle phosphorylase kinase by Ca^{2+}. *J. Biol. Chem.* **246:**1961–1967.
7. Kalow, W., Britt, B. A., and Terreau, M. E. (1970). Metabolic error of muscle metabolism after recovery from malignant hyperthermia. *Lancet* **2:**895–898.
8. Berman, M. C., Harrison, G. G., Bull, A. B., and Kench, J. E. (eds.) (1970). Changes underlying halothane-induced malignant hyperthermia in Landrace pigs. *Nature* **225**:653–655.
9. Endo, M. (1977). Calcium release from the sarcoplasmic reticulum. *Physiol. Rev.* **57:**71–108.
10. O'Brien, P. J. (1986). Porcine malignant hyperthermia susceptibility: Increased calcium sequestering activity of skeletal muscle sarcoplasmic reticulum. *Can. J. Vet. Res.* **50:**329–337.
11. Nelson, T. E. (1988). SR function in malignant hyperthermia. *Cell* **9:**257–265.
12. Ohnishi, S. T. (1979). Calcium-induced calcium release from fragmented sarcoplasmic reticulum. *J. Biochem.* **86:**1147–1150.
13. Meissner, G. (1984). Adenine nucleotide stimulation of Ca^{2+} induced Ca^{2+} release in sarcoplasmic reticulum. *J. Biol. Chem.* **259:** 2365–2374.
14. Smith, J. S., Coronado, R., and Meissner, G. (1985). Sarcoplasmic reticulum contains adenine nucleotide activated calcium channels. *Nature* **316:**446–449.
15. Endo, M., Yagi, S., Ishizuka, T., Horiuti, K., Koga, Y., and Amaha, K. (1983). Changes in the Ca-induced Ca release mechanism in sarcoplasmic reticulum from a patient with malignant hyperthermia. *Biomed. Res.* **4:**83–92.
16. Ohnishi, S. T., Taylor, S., and Gronert, G. A. (1983). Calcium-induced Ca^{2+} release from sarcoplasmic reticulum of pigs susceptible to malignant hyperthermia. The effects of halothane and dantrolene. *FEBS Lett.* **161:**103–107.
17. O'Brien, P. J. (1986). Porcine malignant hyperthermia susceptibility: Hypersensitive calcium-release mechanism of skeletal muscle sarcoplasmic reticulum. *Can. J. Vet. Res.* **50:**318–328.
18. Fill, M., Coronado, R., Mickelson, J. R., Vilven, J., Ma, J., Jacobson, B. A., and Louis, C. F. (1990). Abnormal ryanodine receptor channels in malignant hyperthermia. *Biophys. J.* **50:**471–475.
19. Fill, M., Stefani, E., and Nelson, T. E. (1991). Abnormal human sarcoplasmic reticulum Ca^{2+} release channels in malignant hyperthermia skeletal muscle. *Biophys. J.* **59:**1085–1090.
20. Inui, M., Saito, A., and Fleischer, S. (1987). Purification of the ryanodine receptor and identity with feet structures of junctional terminal cisternae of sarcoplasmic reticulum from fast skeletal muscle. *J. Biol. Chem.* **262:**1740–1747.
21. Lai, F. A., Erickson, H., Block, B. A., and Meissner, G. (1987). Evidence for a junctional feet–ryanodine receptor complex from sarcoplasmic reticulum. *Biochem. Biophys. Res. Commun.* **143:**704–709.
22. Campbell, K. P., Knudson, C. M., Imagawa, T., Leung, A. T., Sutko, J. L., Kahl, S. D., Raab, C. R., and Madson, L. (1987). Identification and characterization of the high affinity [^{3}H]ryanodine receptor of the junctional sarcoplasmic reticulum Ca^{2+} release channel. *J. Biol. Chem.* **262:**6460–6463.
23. Takeshima, H., Nishimura, S., Matsumoto, T., Ishida, H., Kangawa, K., Minamino, N., Matsuo, H., Ueda, M., Hanoka, M., Hirose, T., and Numa, S. (1989). Primary structure and expression from complementary DNA of skeletal muscle ryanodine receptor. *Nature* **339:**439–445.
24. Zorzato, F., Fujii, J., Otsu, K., Phillips, M. S., Green, N. M., Lai, F. A., Meissner, G., and MacLennan, D. H. (1990). Molecular cloning of cDNA encoding human and rabbit forms of the Ca^{2+} release channel (ryanodine receptor) of skeletal muscle sarcoplasmic reticulum. *J. Biol. Chem.* **265:**2244–2256.
25. Mickelson, J. R., Gallant, E. M., Litterer, L. A., Johnson, K. M., Rempel, W. E. and Louis, C. F. (1988). Abnormal sarcoplasmic reticulum ryanodine receptor in malignant hyperthermia. *J. Biol. Chem.* **263:**9310–9315.
26. Knudson, C. M., Mickelson, J. R., Louis, C. F., and Campbell, K. P. (1990). Distinct immunopeptide maps of the sarcoplasmic reticulum Ca^{2+} release channel in malignant hyperthermia. *J. Biol. Chem.* **265:**2421–2424.
27. Carrier, L., Villaz, M., and Dupont, Y. (1991). Abnormal rapid Ca^{2+} release from sarcoplasmic reticulum of malignant hyperthermia susceptible pigs. *Biochim. Biophys. Acta* **1064:**175–183.
28. Shomer, N. H., Louis, C. F., Fill, M., Litterer, L. A., and Mickelson, J. R. (1993). Reconstitution of abnormalities in the malignant hyerthermia-susceptible pig ryanodine receptor. *Am. Physiol. Soc.* **264:**C125–C135.
29. Fujii, J., Otsu, K., Zorzato, F., deLeon, S., Khanna, V. K., Weiler, J., O'Brien, P. J., and MacLennan, D. H. (1991). Identification of a mutation in porcine ryanodine receptor associated with malignant hyperthermia. *Science* **253:**448–451.
30. Otsu, K., Khanna, V. K., Archibald, A. L., and MacLennan, D. H. (1991). Co-segregation of porcine malignant hyperthermia and a probable causal mutation in the skeletal muscle ryanodine receptor gene in backcross families. *Genomics* **11:**744–750.
31. Gillard, E. F., Otsu, K., Fujii, J., Khanna, V. K., deLeon, S., Derdemezi, J., Britt, B. A., Duff, C. L., Worton, R. G., and MacLennan, D. H. (1991). A substitution of cysteine for arginine-614 in the ryanodine receptor is potentially causative of human malignant hyperthermia. *Genomics* **11:**751–755.
32. Gillard, E. F., Otsu, K., Fujii, J., Duff, C. L., deLeon, S., Khanna, V. K., Britt, B. A., Worton, R. G., and MacLennan, D. H. (1992). Polymorphisms and deduced amino acid substitutions in the coding sequence of the ryanodine receptor (RYR1) gene in individuals with malignant hyperthermia. *Genomics* **13:**1247–1254.
33. Zhang, Y., Chen, H. S., Khanna, V. K., de Leon, S., Phillips, M. S., Schappert, K., Britt, B. A., Brownell, A. K. W., and MacLennan, D. H. (1993). A mutation in the human ryanodine receptor gene associated with central core disease. *Nature Genet.* **5:**46–50.
34. Quane, K. A., Healy, J. M. S., Keating, K. E., Manning, B. M., Couch, F. J., Palamucci, L. M., Douguzzi, C., Fiagerlund, T. H., Berg, K., Ording, H., Bendixen, D., Mortier, W., Linz, V., Muller, C. R., and McCarthy, T. V. (1993). Mutations in the ryanodine receptor gene in central core disease and malignant hyperthermia. *Nature Genet.* **5:**51–55.
35. Britt, B. A. (1991). Malignant hyperthermia: A review. In *Thermoregulation: Pathology, Pharmacology and Therapy* (E. Schonbaum and P. Lomax, eds.), Pergamon Press, New York, pp. 179–292.
36. O'Brien, P. J. (1987). Etiopathogenetic defect of malignant hyperthermia: Hypersensitive calcium-release channel of skeletal muscle sarcoplasmic reticulum. *Vet. Res. Commun.* **11:***527–559.*
37. Ohnishi, S. T., and Ohnishi, T. (eds.) (1993). *Malignant Hyperthermia: A Genetic Membrane Disease,* CRC Press, Boca Raton.
38. MacLennan, D. H. (1990). Molecular tools to elucidate problems in excitation contraction coupling. *Biophys. J.* **58:**1355–1365.
39. MacLennan, D. H., and Phillips, M. S. (1992). Malignant hyperthermia. *Science* **256:**789–794.
40. Levitt, R. C., Meyers, D., Fletcher, J. E., and Rosenberg, H. (1991). Molecular genetics and malignant hyperthermia. *Anesthesiology* **75:** 1–3.

41. Levitt, R. C. (1992). Prospects for the diagnosis of malignant hyperthermia susceptibility using molecular genetic approaches. *Anesthesiology* **76:**1039–1048.
42. Ball, S. P., and Johnson, K. J. (1993). The genetics of malignant hyperthermia. *J. Med. Genet.* **30:** 89–93.
43. Denborough, M. A., Forster, J. F. A., and Lovell, R. R. H. (1962). Anaesthetic deaths in a family. *Br. J. Anaesth.* **34:**395–396.
44. Britt, B. A., and Kalow, W. (1970). Malignant hyperthermia: A statistical review. *Can. Anaesth. Soc. J.* **17:**293–315.
45. Ording, H. (1985). Incidence of malignant hyperthermia in Denmark. *Anesth. Analg.* **64:**700–704.
46. Denborough, M. A., Dennett, X., and Anderson, R. M. (1973). Central-core disease and malignant hyperpyrexia. *Br. Med. J.* **1:**272–273.
47. King, J. O., and Denborough, M. A. (1973). Anaesthetic-induced malignant hyperpyrexia in children. ***J. Pediatr.*** 83:37–40.
48. Isaacs, H., and Badenhorst, M. E. (1992). Dominantly inherited malignant hyperthermia (MH) in the King–Denborough syndrome. *Muscle Nerve* **15:**740–742.
49. Brownell, A. K. W., Paasuke, R. T., Elash, A., Fowlow, S. B., Seagram, C. G. F., Diewold, A. J., and Friesen, C. (1983). Malignant hyperthermia in Duchenne muscular dystrophy. *Anesthesiology* **58:** 180–182.
50. Kefler, H. M., Singer, W. D., and Reynolds, R. N. (1983). Malignant hyperthermia in a child with Duchenne muscular dystrophy. *Pediatrics* **71:**118–119.
51. Sethna, N. F., and Rockoff, M. A. (1986). Cardiac arrest following inhalation induction of anaesthesia in a child with Duchenne's muscular dystrophy. *Can. Anaesth. Soc. J.* **33:**799–802.
52. Delphin, E., Jackson, D., and Rothstein, P. (1987). Use of succinylcholine during elective pediatric anesthesia should be reevaluated. *Anesth. Analg.* **66:**1190–1192.
53. Brownell, A. K. W. (1988). Malignant hyperthermia: Relationship to other diseases. *Br. J. Anaesth.* **60:**303–308.
54. Heiman-Patterson, T., Rosenberg, H., Fletcher, J. E., Tahmoush, A. J. (1988). Malignant hyperthermia in myotonia congenita. Halothane–caffeine contracture testing in neuromuscular disease. *Muscle Nerve* **11:**453–457.
55. Karpati, G., Charuk, J., Carpenter, S., Jablecki, C., and Holland, P. (1986). Myopathy caused by a deficiency of Ca^{2+}-adenosine triphosphatase in sarcoplasmic reticulum (Brody's disease). *Ann. Neurol.* **20:**38–49.
56. Ellis, F. R., and Harriman, D. G. F. (1973). A new screening test for susceptibility to malignant hyperpyrexia. *Br. J. Anaesth.* **45:**638.
57. Larach, M. G., for The North American Malignant Hyperthermia Group. (1989). Standardization of the caffeine halothane muscle contracture test. *Anesth. Analg.* **69:**511–515.
58. European MH Group. (1984). Malignant hyperpyrexia: A protocol for the investigation of malignant hyperthermia (MH) susceptibility. *Br. J. Anaesth.* **56:**1267–1269.
59. Larach, M. G. (1993). Should we use muscle biopsy to diagnose malignant hyperthermia susceptibility. *Anesthesiology* **79:**1–4.
60. Larach, M. G., Landis, J. R., Shirk, B. S., and Diaz, M. (1992). Prediction of malignant hyperthermia susceptibility in man: Improving sensitivity of the caffeine halothane contracture test. *Anesthesiology* **77:**A1052.
61. Larach, M. G., Landis, J. R., Bunn, J. S., and Diaz, M. (1992). Prediction of malignant hyperthermia susceptibility in low-risk subjects; An epidemiologic investigation of caffeine halothane contracture responses. *Anesthesiology* **76:**16–27.
62. Isaacs, H., and Badenhorst, M. (1993). False-negative results with muscle caffeine halothane contracture testing for malignant hyperthermia. *Anesthesiology* **79:**5–9.
63. Kalow, W., Sharer, S., and Britt, B. (1991). Pharmacogenetics of caffeine and caffeine–halothane contractures in biopsies of human skeletal muscle. *Pharmacogenetics* **I:** 126–135.
64. Larach, M. G., Localio, A. R., Allen, G. C., Denborough, M. A., Ellis, F. R., Gronert, G. A., Kaplan, R. F., Muldoon, S. M., Nelson, T. E., Ording, H., Rosenberg, H., Waud, B. E., and Wedel, D. J. (1994). A clinical grading scale to predict malignant hyperthermia susceptibility. *Anesthesiology* **80:**771–779.
65. MacKenzie, A. E., Allen, G., Lahey, D., Crossan, M. L., Nolen, K., Mettler, G., Worton, R. G., MacLennan, D. H., and Korneluk, R. G. (1991). A comparison of the caffeine halothane muscle contracture test with the molecular genetic diagnosis of malignant hyperthermia. *Anesthesiology* **75:**4–8.
66. Hopkins, P. M., Ellis, F. R., and Halsall, P. J. (1991). Ryanodine contracture: A potentially specific in vitro diagnostic test for malignant hyperthermia. *Br. J. Anaesth.* **66:**611–613.
67. Lenzen, C., Roewer, N., Wappler, F., Scholz, J., Kahl, J., Blank, M., Rumberger, E., and Schulte, J. (1993). Accelerated contractures after administration of ryanodine to skeletal muscle of malignant hyperthermia susceptible patients. *Br. J. Anaesth.* **71:**242–246.
68. Klip, A., Britt, B. A., Elliott, M. E., Pegg, W., Frodis, W., and Scott, E. (1987). Anaesthetic-induced increase in ionised calcium in blood mononuclear cells from malignant hyperthermia patients. *Lancet* **I:**463–466.
69. Klip, A., Ramlal, T., Walker, D., Britt, B. A., and Elliott, M. E. (1987). Selective increase in cytoplasmic calcium by anesthetic in lymphocytes from malignant hyperthermia-susceptible pigs. *Anesth. Analg.* **66:**381–385.
70. Hogan, K., Couch, F., and Powers, P. A. (1992). A cysteine-for-arginine substitution (R614C) in the human skeletal muscle calcium release channel cosegregates with malignant hyperthermia. *Anesth. Analg.* **75:**441–448.
71. Otsu, K., Willard, H. F., Khanna, V. K., Zarzato, F., Green, N. M., and MacLennan, D. H. (1990). Molecular cloning of cDNA encoding the Ca^{2+} release channel (ryanodine receptor) of rabbit cardiac muscle sarcoplasmic reticulum. *J. Biol. Chem.* **265:**13472–13483.
72. McPherson, P. S., and Campbell, K. P. (1993). The ryanodine receptor/Ca^{2+} release channel. *J. Biol. Chem.* **268:**13765–13768.
73. Sorrentino, V., and Volpe, P. (1993). Ryanodine receptors: How many, where and why? *Trends Pharmacol. Sci.* March, pp. 98–103.
74. Ording, H., Foder, B., and Scharff, O. (1990). Cytosolic free calcium concentrations in lymphocytes from malignant hyperthermia susceptible patients. *Br. J. Anaesth.* **64:**341–345.
75. Lopez-Padrino, J. R. (1993). Free calcium concentration in skeletal muscle of malignant hyperthermia susceptible subjects: Effects of ryanodine. In *Malignant Hyperthermia: A Genetic Membrane Disease* (S. T. Ohnishi and T. Ohnishi, eds.), CRC Press, Boca Raton, pp. 133–150.
76. Iaizzo, P., Klein, W., and Lehman-Horn, F. (1988). Fura 2 detected myoplasmic calcium and its correlation with contracture force in skeletal muscle from normal and malignant hyperthermia susceptible pigs. *Pfluegers Arch* **411:**648–653.
77. Webster, D. W., Thompson, R. T., Gravelle, D. R., Laschuk, M. J., and Driedger, A. A. (1990). Metabolic response to exercise in malignant hyperthermia sensitive patients measured by ^{31}P magnetic resonance spectroscopy. *Magn. Reson. Med.* **15:**81–89.
78. Payen, J., Bosson, J., Bourdon, L., Jacquout, C., Le Bas, J., Steiglitz, P., and Benahid, A. (1993). Improved noninvasive diagnostic testing for malignant hyperthermia susceptibility from a combination of metabolites determined *in vivo* with ^{31}P-magnetic resonance spectroscopy. *Anesthesiology* **78:**848–855.
79. Hall, L. W., Woolf, N., Bradley, J. W., and Jolly, D. W. (1966). Unusual reaction to suxanethonium chloride. *Br. Mol. J.* **2:**1305.
80. Harrison, G. G. (1979). Porcine malignant hyperthermia. *Int. Anesthesiol. Clin.* **17:**25–62.
81. Simpson, S. P., and Webb, A. J. (1989). Growth and carcass performance of British Landrace pigs heterozygous at the halothane locus. *Anim. Prod.* **49:**503–509.

82. Webb, A. J., and Simpson, S. P. (1986). Performance of British Landrace pigs selected for high and low incidence of halothane sensitivity. 2. Growth and carcass traits. *Anim. Prod.* **43:**493–503.
83. Webb, A. J., Imlah, P., and Carden, A. E. (1986). Succinylcholine and halothane as a field test for the heterozygote at the halothane locus in pigs. *Anim. Prod.* **42:**275–279.
84. Seeler, D. S., McDonell, W. N., and Basrur, P. K. (1983). Halothane and halothane/succinylcholine induced malignant hyperthermia (porcine stress syndrome) in a population of Ontario boars. *Can. J. Comp. Med.* **47:**284–290.
85. Gallant, E. M., Mickelson, J. R., Roggow, B. D., Donaldson, S. K., Louis, C. F., and Rempel, W. E. (1989). Halothane-sensitivity gene and muscle contractile properties in malignant hyperthermia. *Am. Physiol. Soc.* C781–C786.
86. Nelson, T. E., Flewellen, E. H., and Gloyna, D. F. (1983). Spectrum of susceptibility to malignant hyperthermia—Diagnostic dilemma. *Int. Anesth. Res. Soc.* **62:**545–552.
87. Fletcher, J. E., Calvo, P. A., and Rosenberg, H. (1993). Phenotypes associated with malignant hyperthermia susceptibility in swine genotyped as homozygous or heterozygous for the ryanodine receptor mutation. *Br. J. Anaesth.* **71:**410–417.
88. Eikelenboom, G., and Minkema, D. (1974). Prediction of pale, soft, exudative muscle with a non-lethal test for halothane induced porcine malignant hyperthermia syndrome. *Neth. J. Vet. Sci.* **99:** 421–426.
89. Otsu, K., Phillips, M. S., Khanna, V. K., deLeon, S., and MacLennan, D. (1992). Refinement of diagnostic assays for a probable causal mutation for porcine and human malignant hyperthermia. *Genomics* **13:**835–837.
90. Pommier, S. A., Houde, A., Rousseau, F., and Savoie, Y. (1992). The effect of malignant hyerthermia as determined by a restriction endonuclease assay on carcass characteristics of commercial crossbred pigs. *Can. J. Anim. Sci.* **72:**973–976.
91. Pommier, S. A., and Houde, A. (1993). Effect of the genotype for malignant hyperthermia as determined by a restriction endonuclease assay on the quality characteristics of commercial pork loins. *J. Anim. Sci.* **71:**420–425.
92. Fortin, A., Pommier, S. A., and Houde, A. (1993). PSE in pork: The relationship between genotype as determined by the restriction endonuclease assay, and the environment. *Proceedings of the 38th International Congress of Meat Science and Technology* **2:**173–176.
93. Ludvigsen, J. (1953). Muscular degeneration in hogs. *15th International Veterinary Congress,* Stockholm **1:**602–606.
94. Marsh, B. B. (1951). A factor modifying muscle fiber syneresis. *Nature* **167:**1065–1067.
95. Hasselbach, W., and Makinose, M. (1961). Die calciumpumpe der "Erschlaffungsgrana" des muskels und ihr abhängigkeit von der ATP-spaltung. *Biochem. Z.* **333:**518–528.
96. Ebashi, S., and Lipman, F. (1962). Adenosine triphosphate-linked concentration of calcium ions in a particulate fraction of rabbit muscle. *J. Cell Biol.* **14:**389–400.
97. Weber, A., and Herz, R. (1968). The relationship between caffeine contracture of intact muscle and the effect of caffeine on reticulum. *J. Gen. Physiol.* **52:**750–759.
98. McKinley, D., and Meissner, G. (1978). Evidence for a K^+, Na^+ permeable channel in sarcoplasmic reticulum. *J. Membr. Biol.* **44:** 159–186.
99. Meissner, G. (1975). Isolation and characterization of two types of sarcoplasmic reticulum vesicles. *Biochim. Biophys. Acta* **389:**51–68.
100. Kim, K. C., Caswell, A. H., Talvenheimo, J. A., and Brandt, N. R. (1990). Isolation of a terminal cisternae protein which may link the dihydropyridine receptor to the junctional foot protein in skeletal muscle. *Biochemistry* **29:**9281–9289.
101. Knudson, C. M., Stang, K. J., and Jorgensen, A. O. (1993). Biochemical characterization and ultrastructural localization of a major junctional sarcoplasmic reticulum glycoprotein (Triadin). *J. Biol. Chem.* **268:**12637–12645.
102. Knudson, C. M., Stang, K. K., Moomaw, C. R., Slaughter, C. A., and Campbell, K. P. (1993). Primary structure and topological analysis of a skeletal muscle-specific junctional sarcoplasmic reticulum glycoprotein (Triadin). *J. Biol. Chem.* **268:**12646–12654.
103. Meissner, G. (1986). Evidence of a role for calmodulin in the regulation of calcium release from skeletal muscle sarcoplasmic reticulum. *Biochemistry* **25:**244–251.
104. Meissner, G. (1986). Ryanodine activation and inhibition of the Ca^{2+} release channel of sarcoplasmic reticulum. *J. Biol. Chem.* **261:** 6300–6306.
105. Meissner, G., Darling, E., and Eveleth, J. (1986). Kinetics of rapid Ca^{2+} release by sarcoplasmic reticulum. Effects of Ca^{2+}, Mg^{2+}, and adenine nucleotides. *Biochemistry* **25:**236–244.
106. Miyamoto, H., and Racker, E. (1982). Mechanism of calcium release from skeletal sarcoplasmic reticulum. *J. Membr. Biol.* **66:** 193–201.
107. Morii, H., and Tonomura, Y. (1983). The gating behavior of a channel for Ca^{2+}-induced Ca^{2+} release in fragmented sarcoplasmic reticulum. *J. Biochem.* **93:**1271–1285.
108. Smith, J. S., Imagawa, T., Ma, J., Fill, M., Campbell, K. P., and Coronado, R. (1988). Purified ryanodine receptor from rabbit skeletal muscle is the Ca^{2+} release channel of sarcoplasmic reticulum. *J. Gen. Physiol.* **92:**1–26.
109. Mitchell, R. D., Palade, P., and Fleischer, S. (1983). Purification of morphologically intact triad structures from skeletal muscle. *J. Cell Biol.* **96:**1008–1016.
110. Mitchell, R. D., Saito, A., Palade, P., and Fleischer, S. (1983). Morphology of isolated triads. *J. Cell Biol.* **96:**1017–1029.
111. Costello, B., Chadwick, C., Saito, A., Maurer, A., and Fleischer, S. (1986). Characterization of the junctional face membrane from terminal cisternae of sarcoplasmic reticulum. *J. Cell Biol.* **103:**741–753.
112. Saito, A., Seiler, S., Chu, A., and Fleischer, S. (1984). Preparation and morphology of sarcoplasmic reticulum terminal cisternae from rabbit skeletal muscle. *J. Cell Biol.* **99:**875–885.
113. Fleischer, S., Ogunbunmi, E. M., Dixon, M. C., and Fleer, E. A. M. (1985). Localization of Ca^{2+} release channels with ryanodine in junctional terminal cisternae of sarcoplasmic reticulum of fast skeletal muscle. *Proc. Natl. Acad. Sci. USA* **82:**7256–7259.
114. Hymel, L., Inui, M., Fleischer, S., and Schindler, H. G. (1988). Purified ryanodine receptor of skeletal muscle sarcoplasmic reticulum forms Ca^{2+}-activated oligomeric Ca^{2+} channels in planar bilayers. *Proc. Natl. Acad. Sci. USA* **85:**441–445.
115. Lai, F. A., Erickson, H. P., Rousseau, E., Liu, Q.-Y., and Meissner, G. (1988). Purification and reconstitution of the calcium release channel from skeletal muscle. *Nature* **331:**315–319.
116. Radermacher, M., Rao, R., Grassucci, R., Frank, A., Timerman, S., Fleischer, S., and Wagenknecht, T. (1994). Cryo-electron microscopy and three dimensional reconstruction of the calcium release channel/ryanodine receptor from skeletal muscle. *J. Cell Biol.* **127:**411–423.
117. Wagenknecht, T., Grassucci, R., Frank, J., Saito, A., Invi, M., and Fleischer, S. (1989). Three-dimensional architecture of the calcium channel/foot structure of sarcoplasmic reticulum. *Nature* **338:**167–170.
118. Nakai, J., Imagawa, T., Hakamata, Y., Shigekawa, M., Takeshima, M., and Numa, S. (1990). Primary structure and functional expression from cDNA of the cardiac ryanodine receptor/calcium release channel. *FEBS* **271:**169–177.
119. Giannini, G., Clementi, E., Ceci, R., Marziali, G., and Sorrentino, V. (1992). Expression of a ryanodine receptor-Ca^{2+} channel that is regulated by TGF-β. *Science* **257:**91–94.
120. Hakamata, Y., Nakai, J., Takeshima, H., and Imota K. (1992). Primary structure and distribution of a novel ryanodine receptor/cal-

cium release channel from rabbit brain. *Fed. Eur. Biochem. Soc.* **312:**229–235.
121. Suko, J., Maurer-Fogy, I., Plank, B., Bertel, O., Wyskovsky, W., Hohenegger, M., and Hellmann, G. (1993). Phosphorylation of serine 2843 in ryanodine receptor-calcium release channel of skeletal muscle by cAMP-, cGMP- and CaM-dependent protein kinase. *Biochim. Biophys. Acta* **1175:**193–206.
122. Witcher, D. R., Kovacs, R. J., Schulman, H., Cefali, D. C., and Jones, L. R. (1991). Unique phosphorylation site on the cardiac ryanodine receptor regulates calcium channel activity. *J. Biol. Chem.* **266:**11144–11152.
123. Miyawaki, A., Furuichi, T., Ryou, Y., Yoshikawa, S., Nakagawa, T., Saitoh, T., and Mikoshiba, K. (1991). Structure-function relationships of the mouse inositol 1, 4, 5-triphosphate receptor. *Proc. Natl. Acad. Sci. USA* **88:**4911–4915.
124. Catterall, W. A. (1991). Excitation–contraction coupling in vertebrate skeletal muscle: A tale of two calcium channels. *Cell* **64:**871–874.
125. Schneider, M. F., and Chandler, W. K. (1973). Voltage dependent charge movement in skeletal muscle: A possible step in excitation–contraction coupling. *Nature* **242:**747–751.
126. Rios, E., and Pizzaro, G. (1991). Voltage sensor of excitation–contraction coupling in skeletal muscle. *Physiol. Rev.* **71:**849–908.
127. Tanabe, T., Takeshima, H., Mikami, A., Flockerzi, V., Takahashi, H., Kangawa, K., Kojima, M., Matsuo, H., Hirose, T., and Numa, S. (1987). Primary structure of the receptors or calcium channel blockers from skeletal muscle. *Nature* **328:**313–318.
128. Tanabe, T., Beam, K. G., Powell, J. A., and Numa, S. (1988). Restoration of excitation–contraction coupling and slow calcium current in dysgenic muscle by dihydropyridine receptor complementary DNA. *Nature* **336:**134–139.
129. Franzini-Armstrong, C. (1970). Studies of the triad. I. Structure of the junction in frog twitch fibers. *J. Cell Biol.* **47:**488–499.
130. Block, B. A., Imagawa, T., Campbell, K. P., and Tranzini-Armstrong, C. (1988). Structural evidence for direct interaction between the molecular junction in skeletal muscle. *J. Cell Biol.* **107:** 2587–2600.
131. Takekura, H., Takeshima, H., Nishimura, S., Takahashi, M., Tanabe, T., Numa, S., Flockerzi, V., Hoffman, F., and Franzini-Armstrong, C. (1993). Coexpression of ryanodine and dihydropyridine receptors is not sufficient to form a junction. *Biophys. J.* **64:**A153.
132. Fliegel, L., Ohnishi, M., Carpenter, M. R., Khanna, V. K., Reithmeier, R. A. F., and MacLennan, D. H. (1987). Amino acid sequence of rabbit fast-twitch skeletal muscle calsequestrin deduced from cDNA and peptide sequencing. *Proc. Natl. Acad. Sci. USA* **84:** 1167–1171.
133. Fleischer, S., and Inui, M. (1989). Biochemistry and biophysics of excitation–contraction coupling. *Annu. Rev. Biophys. Chem.* **18:** 333–364.
134. Lytton, J., and MacLennan, D. H. (1992). Sarcoplasmic reticulum. In *Heart and Cardiovascular System: Scientific Foundations,* 2nd Ed. (H. A. Fozzard, E. Haber, R. B. Jennings, A. M., Katz, and H. E. Morgan, eds.), Raven Press, New York, Vol. 2, pp. 1203–1222.
135. Carafoli, E. (1987). Intracellular calcium homeostasis. *Annu. Rev. Biochem.* **56:**395–433.
136. Nelson, T. E. (1983). Abnormality in calcium release from skeletal sarcoplasmic reticulum of pigs susceptible to malignant hyperthermia. *J. Clin. Invest.* **72:**862–870.
137. Kim, D. H., Sreter, F. A., Ohnishi, S. T., Ryan, J. F., Roberts, J., Allen, P. D., Meszaros, L. G., Antoniu, B., and Jikemoto, N. (1984). Kinetic studies of Ca^{2+} release from sarcoplasmic reticulum of normal and malignant hyperthermia susceptible pig muscles. *Biochim. Biophys. Acta* **775:**320–327.
138. Harrison, G. G. (1975). Control of the malignant hyperpyrexic syndrome in MHS swine by dantrolene sodium. *Br. J. Anaesth.* **47:** 62–65.
139. Britt, B. A. (1984). Dantrolene. *Can. Anaesth. Soc. J.* **31:**61–75.
140. Chyatte, S. B., Birdsong, J. H., and Robertson, D. L. (1973). Dantrolene sodium in athetoid cerebral palsy. *Arch. Phys. Med. Rehabil.* **54:**365–368.
141. Ellis, K. O., and Carpenter, J. F. (1972). Studies on the mechanism of action of dantrolene sodium, a skeletal muscle relaxant. *Naunyn Schmiedebergs Arch. Pharmacol.* **275:**83–85.
142. Ellis, K. O., Castellion, A. W., Honkomp, L. J., Wessels, F. L., Carpenter, J. F., and Halliday, R. P. (1973). Dantrolene, a direct acting skeletal muscle relaxant. *J. Pharm. Sci.* **62:**948–951.
143. Ellis, K. O., and Bryant, S. H. (1972). Excitation–contraction uncoupling in skeletal muscle by dantrolene sodium. *Naunyn Schmiedebergs Arch. Pharmacol.* **274:**107–109.
144. Desmedt, J. E., and Hainaut, K. (1977). Inhibition of the intracellular release of calcium by dantrolene in barnacle giant muscle fibers. *J. Physiol. (London)* **265:**565.
145. Lopez, J. R., Allen, P., Alamo, L., Ryan, J. F., Jones, D. E., and Sreter, F. (1987). Dantrolene prevents the malignant hyperthermic syndrome by reducing free intracellular calcium concentration in skeletal muscle of susceptible swine. *Cell Calcium* **8:**385–396.
146. Ohnishi, S. T., Waring, A. J., Fang, S.-R. G., Horiuchi, K., Flick, J. L., Sadanaga, K. K., and Ohnishi, T. (1986). Abnormal membrane properties of the sarcoplasmic reticulum of pigs susceptible to malignant hyperthermia: Modes of action of halothane, caffeine, dantrolene, and two other drugs. *Arch. Biochem. Biophys.* **247:**294–301.
147. Nelson, T. E. (1984). Dantrolene does not block calcium pulse-induced calcium release from a putative calcium channel in sarcoplasmic reticulum from malignant hyperthermia and normal pig muscle. *FEBS Lett.* **167:**123–126.
148. White, M. D., Collins, J. G., and Denborough, M. A. (1983). The effect of dantrolene on skeletal-muscle sarcoplasmic-reticulum function in malignant hyperpyrexia in pigs. *Biochem. J.* **212:**399–405.
149. Ohta, T., Ito, S., and Ohga, A. (1990). Inhibitory action of dantrolene on Ca^{2+}-induced Ca release from sarcoplasmic reticulum in guinea pig skeletal muscle. *Eur. J. Pharmacol.* **178:**11–19.
150. Nelson, T. E., and Lin, M. (1993). Dantrolene activates, then blocks the ryanodine-sensitive calcium release channel in a planar lipid bilayer. *Biophys. J.* **64:**A380.
151. Pessah, I. N., Francini, A. O., Scales, D. J., Waterhouse, A. L., and Casida, J. E. (1986). Calcium–ryanodine receptor complex: Solubilization and partial characterization from skeletal muscle junctional sarcoplasmic reticulum vesicles. *J. Biol. Chem.* **261:**8643–8648.
152. Andersen, E., and Jensen, P. (1977). Close linkage established between the HAL locus for halothane sensitivity and the PHI (phosphohexose isomerase) locus in pigs of the Danish Landrace breed. *Nord. Vet. Med.***29:**502–504.
153. Gahne, B., and Juneja, R. K. (1985). Prediction of the halothane (Hal) genotypes of pigs by deducing Hal, Phi, Po2, Pgd haplotypes of parents and offspring: Results from a large-scale practice in Swedish breed. *Anim. Blood Groups Biochem. Genet.* **16:**265–283.
154. Archibald, A. L., and Imlah, P. (1985). The halothane sensitivity locus and its linkage relationships. *Anim. Blood Groups Biochem. Genet.* **16:**253–263.
155. Davies, W., Harbitz, I., Fries, R., Stranzinger, G., and Hauge, J. G. (1988). Porcine malignant hyperthermia carrier detection and chromosomal assignment using a linked probe. *Anim. Genet.* **19:**203–212.
156. Chowdhary, B. P., Harbitz, I., Makinen, A., Davies, W., and Gustavsson, I. (1989). Localization of the glucose phosphate isomerase gene to p12–q21 segment of chromosome 6 in pig by in situ hybridization. *Hereditas* **111:**73–78.
157. Harbitz, I., Chowdhary, B., Thomsen, P., Davies, W., Kaufman, V., Kran, S., Gustavsson, I., Christensen, K., and Hauge, J. (1990). Assignment of the porcine calcium release channel gene, a candidate

for the malignant hyperthermia locus, to the 6p11–q21 segment of chromosome 6. *Genomics* **9:**243–248.

158. Lusis, A. J., Heinzmann, C., Sparkes, R. S., Scott, J., Knott, T. J., Geller, R., Sparkes, M. C., and Mahandas, T. (1986). Regional mapping of human chromosome 19: Organization of genes for plasma lipid transport (APOC1, -C2 and -E and LDLR) and the genes C3, PEPD and GPI. *Proc. Natl. Acad. Sci. USA* **83:**3929–3933.
159. MacKenzie, A. E., Korneluk, R. G., Zorzato, F., Fujii, J., Phillips, M. S., Iles, D., Wieringa, B., LeBlond, S., Bailly, J., Willard, H. F., Duff, C. L., Worton, R. G., and MacLennan, D. H. (1990). The human ryanodine receptor gene: Its mapping to 19q13.1, placement in a chromosome 19 linkage group and exclusion as the gene causing myotonic dystrophy. *Am. J. Hum. Genet.* **46:**1082–1089.
160. MacLennan, D. H., Duff, C., Zorzato, F., Fujii, J., Phillips, M. S., Korneluk, R. G., Frodis, W., Britt, B. A., and Worton, R. G. (1990). Ryanodine receptor gene is a candidate for predisposition to malignant hyperthermia. *Nature* **343:**559–561.
161. McCarthy, T. V., Healy, J. M. S., Heffron, J. J. A., Lehane, M., Deufel, T., Lehmann-Horn, F., Faralli, M., and Johnson, K. (1990). Localization of the malignant hyperthermia susceptibility locus to human chromosome 19q12–13.2. *Nature* **343:**562–564.
162. Rudolph, J. A., Spier, S. J., Byrns, G., Rojas, C. V., Bernoco, D., and Hoffman, E. P. (1992). Periodic paralysis in quarter horses: A sodium channel mutation disseminated by selective breeding. *Nature Genet.* **2:**144–147.
163. Deufel, T., Sudbrak, R., Feist, Y., Rubsam, B., DuChesne, I., Schafer, K.-L., Roewer, N., Grimm, T., Lehmann-Horn, F., Hartung, E. J., and Muller, C. R. (1995). Discordance in a malignant hyperthermia pedigree between in vitro contractive-test phenotypes and haplotypes for the MHS1 region on chromosome 19q12-13.2 comprising the C1840T transition in the *RYR1* gene. *Am. J. Hum. Genet.* **56:**1334–1342.
164. Serfas, K. D., Bose, D., Patel, L., Wrogemann, K., Phillips, M. S., MacLennan, D. H., and Greenberg, C. R. (1996). Comparison of the segregation of the *RYR1* C1840 T mutation with segregation of the caffeine/halothane contractive test results for malignant hyperthermia susceptibility in a large Manitoba mennonite family. *Anesthesiology* **84:**322–329.
165. MacLennan, D. H. (1995). Discordance between phenotype and genotype in malignant hyperthermia. *Curr. Opinions Neurol.* **8:**397–401.
166. Quane, K. A., Keating, K. E., Manning, B. M., *et al.* (1994). Detection of a novel common mutation in the ryanodine receptor gene in malignant hyperthermia: Implications for diagnosis and heterogeneity studies. *Hum. Mol. Genet.,* **3:**471–476.
167. Quane, K. A., Keating, K. E., Manning, B. M., Healy, J. M. S., Monsieurs, K., Heffron, J. J. A., Lehane, M., Heytens, L., Krivosic-Horber, R., Adnet, P., Ellis, F. R., Monnier, N., Lumardi, J., and McCarthy, T. V. (1994). Detection of a novel common mutation in the ryanodine receptor gene in malignant hyperthermia: Implications for diagnosis and heterogeneity studies. *Hum. Mol. Gen.* **3:**471–476.
168. Phillips, M. S., Khanna, V. K., de Leon, S., Frodis, W., Britt, B. A., and MacLennan, D. H. (1994). The substitution of Arg for Gly2433 in the human skeletal muscle ryanodine receptor is associated with malignant hyperthermia. *Hum. Mol. Genet.* **3:**2181–2186.
169. Philips, M. S., Fujii, J., Khanna, V. K., de Leon, S., Yokobata, K., deJong, P. J., and MacLennan, D. H. (1996). The structural organization of the human skeletal muscle ryanodine receptor *(RYR1)* gene. *Genomics* (in press).
170. Rouquier, S., Giorgi, D., Trask, B., Bergmann, A., Phillips, M. S., MacLennan, D. H., and deJong, P. (1993). A cosmid and yeast artificial chromosome contig containing the complete ryanodine receptor *RYR1* gene. *Genomics* **17:**330–340.
171. Levitt, R. C., Nouri, N., Jedlicka, A. E., McKusick, V. A., Marks, A. R., Shutack, J. G., Fletcher, J. E., Rosenberg, H., and Meyers, D. A. (1991). Evidence for genetic heterogeneity in malignant hyperthermia susceptibility. *Genomics* **11:**543–547.
172. Deufel, T., Golla, A., Iles, D., Meindl, A., Meitinger, T., Schindelhauer, D., DeVries, A., Pongratz, D., MacLennan, D. H., Johnson, K. J., Lehmann-Horn, F. (1992). Evidence for genetic heterogeneity of malignant hyperthermia susceptibility. *Am. J. Hum. Genet.* **50:** 1151–1161.
173. Fagerlund, T., Islander, G., Ranklev, E., Harbitz, I., Hauge, J. G., Mekleby, E., and Berg, K. (1991). Genetic recombination between malignant hyperthermia and calcium release channel in skeletal muscle. *Clin. Genet.* **41:**270–272.
174. Vita, G. M., Olckers, A., Jedicka, A. E., George, A. L., Heiman-Patterson, T., Rosenberg, H., and Fletcher, J. E. (1995). Masseter muscle rigidity associated with glycine[1306]-to-alanine mutation in the adult muscle sodium channel α-subunit gene. *Clin. Invest.* **82:**1097–1103.
175. Iaizzo, P. A., and Lehmann-Horn, F. (1995). Anesthetic complications in muscle disorders. *Anesthesiology* **82:**1093–1096.
176. Levitt, R. C., Olckers, A., Meyers, S., Fletcher, J. E., Rosenberg, H., Isaac, H., and Meyers, D. A. (1992). Evidence for the localization of a malignant hyperthermia susceptibility locus (MHS2) to human chromosome 17q. *Genomics* **14:**562–566.
177. Olckers, A., Meyers, D. A., Meyers, S., Taylor, E. W., Fletcher, J. E., Rosenberg, H., Isaacs, H., and Levitt, R. D. (1992). Adult muscle sodium channel α-subunit is a gene candidate for malignant hyperthermia susceptibility. *Genomics* **14:**829–831.
178. Iles, D. E., Segers, B., Sengers, R. C. A., Monsieurs, K., Heytens, L., Halsall, P. J., Hopkins, P. M., Ellis, F. R., Hall-Curran, J. L., Stuart, A. D., and Wieringa, B. (1993). Genetic mapping of the β1 and γ subunits of the human skeletal muscle L-type voltage dependent calcium channel on chromosome 17q and exclusion as a candidate gene for malignant hyperthermia susceptibility. *Hum. Mol. Genet.* **2:**863–868.
179. Fontaine, B., Khurana, T. S., Hoffman, E. P., Bruns, G. A. P., Haines, J. L., Trofatter, J. A., Hanson, M. P., Rich, J., Mc Farlane, H., Yasik, D. M., Romano, D., Gusella, J. F., and Brown, R. H. (1990). Hyperkalemic periodic paralysis and the adult muscle sodium channel α-subunit gene. *Science* **250:**1000–1002.
180. Sudbrak, R., Golla, A., Powers, P., Gregg, R., Duchesne, I., Lehmann-Horn, F., and Deufel, T. (1993). Exclusion of malignant hyperthermia susceptibility (MHS) from a putative MHS2 locus on chromosome 17q and of the α1, β1, γ subunits of the dihydropyrimidine receptor calcium channel as candidates for the molecular defect. *Hum. Mol. Genet.* **2:**857–862.
181. Iles, D. E., Lehmann-Horn, F., Scherer, S. W., Tsui, L. C., Weghuis, D. O., Suijkerbuijk, R. F., Heytens, L., Mikala, G., Schwartz, A., Ellis, F. R., Stewart, A. D., and Wieringa, B. (1994). Localization of the gene encoding the α2/δ-subunits of the L-type voltage-dependent calcium channel to chromosome 7q and analysis of the segregation of flanking markers in malignant hyperthermia susceptible families. *Hum. Mol. Genet.* **3:**969–975.
182. Sudbrak, R., Procaccio, V., Klausnitzer, M., Curran, J. L., Monsieurs, K., Van Broeckenhoven, C., Ellis, R., Heyetens, L., Hartung, E. J., Kozak-Ribbens, G., Heilinger, D., Weissenbach, J., Lehman-Horn, F., Mueller, C. R., Deufel, T., Stewart, A. D., and Lunardi, J. (1995). Mapping of a further malignant hyperthermia susceptibility locus to chromosome 3q13.1. *Am. J. Hum. Genet.* **56:**684–691.
183. Shy, G. M., and Magee, K. R. (1956). A new congenital non-progressive myopathy. *Brain* **79:**610–621.
184. Patterson, V. H., Hill, T. R., Fletcher, P. J., and Heron, J. R. (1979). Central core disease: Clinical and pathological evidence within a family. *Brain* **102:**581–594.
185. Ramsey, P. L., and Hensinger, R. N. (1975). Congenital dislocation of the hip associated with central core disease. *J. Bone Jt. Surg.* **57A:**648–651.

186. Dubowitz, V., and Platts, M. (1965). Central core disease of muscle with focal wasting. *J. Neurol. Neurosurg. Psychiatry* **28:**432–437.
187. Shuaib, A., Paasuke, R. T., and Brownell, K. W. (1987). Central core disease: Clinical features in 13 patients. *Medicine* **66:**389–396.
188. Dubowitz, V., and Pearse, A. G. E. (1960). Oxidative enzymes and phosphorylase in central-core disease of muscle. *Lancet* 23–24.
189. Isaacs, H., Heffron, J. J. A., and Badenhorst, M. (1975). Central core disease. A correlated genetic, histochemical, ultramicroscopic and biochemical study. *J. Neurol. Psychiatry* **38:**1177–1186.
190. Byrne, E., Blumbergs, P. C., and Hallpike, J. F. (1982). Central core disease. Study of a family of five affected generations. *J. Neurol. Sci.* **53:**77–83.
191. Hayashi, K., Miller, R. G., and Brownell, A. K. W. (1989). Central core disease: Ultrastructure of the sarcoplasmic reticulum and T-tubules. *Muscle Nerve* **12:**95–102.
192. Dubowitz, V., and Roy, S. (1970). Central core disease of muscle: Clinical histochemical and electron microscopic studies of an affected mother and child. *Brain* **93:**133–146.
193. Eng, G. D., Epstein, B. S., Engel, W. K., McKay, D. W., and McKay, R. (1978). Malignant hyperthermia and central core disease in a child with congenital dislocating hips. *Arch. Neurol.* **35:**189–197.
194. Frank, J. P., Harati, Y., Butler, I. J., Nelson, T. E., and Scott, C. I. (1980). Central core disease and malignant hyperthermia syndrome. *Ann. Neurol.* **7:**11–17.
195. Haan, E. A., Freemantle, C. J., McCure, J. A., Friend, K. L., and Mulley, J. C. (1990). Assignment of the gene for central core disease to chromosome 19. *Hum. Genet.* **86:**187–190.
196. Kausch, K., Lehmann-Horn, F., Janka, M., Wieringa, B., Grimm, T., and Müller, C. R. (1991). Evidence for linkage of the central core disease locus to the proximal long arm of human chromosome 19. *Genomics* **10:**765–769.
197. Mulley, J. C., Kozman, H. M., Phillips, H. A., Gedeon, A. K., McCure, J. A., Iles, D. E., Gregg, R. G., Hogan, K., Couch, F. J., Weber, J. L., MacLennan, D. H., and Haan, E. A. (1993). Refined genetic localization for central core disease. *Am. J. Hum. Genet.* **52:** 398–405.
198. Quane, K. A., Keating, K. E., Healy, J. M. S., Manning, B. M., Krivosic-Horber, R., Krivosic, I., Monnier, N., Lunardi, J., and McCarthy, T. V. (1994). Mutation screening of the *RYR1* gene in malignant hyperthermia: Detection of a novel Tyr to Ser mutation in a pedigree with associated central cores. *Genomics* **23:**236–239.
199. Mignery, G. A., and Südhof, T. C. (1990). The ligand binding site and transduction mechanism in the inositol-1,4,5-triphosphate receptor. *EMBO J.* **9:**3893–3898.
200. Miyawaki, A., Furuichi, T., Ryou, Y., Yoshikawa, S., Nakagawa, T., Scutoh, T., and Mikoshiba, K. (1991). Structure–function relationships of the mouse inositol 1,4,5-trisphosphate receptor. *Proc. Natl. Acad. Sci. USA* **88:**4911–4915.
201. Chen, S. R. W., Airey, J. A., and MacLennan, D. H. (1993). Positioning of major tryptic fragments in the Ca^{2+} release channel (ryanodine receptor) of rabbit skeletal muscle sarcoplasmic reticulum. *J. Biol. Chem.* **268:**22642–22649.
202. Wrogemann, K., and Pena, S. D. J. (1976). Mitochondrial calcium overload: A general mechanism of cell necrosis in muscle diseases. *Lancet* **1:**672–673.
203. Fananapazir, L., Dalakas, M. C., Cyran, F., Cohn, G., and Epstein, N. D. (1993). Missense mutations in the β-myosin heavy-chain gene cause central core disease in hypertrophic cardiomyopathy. *Proc. Natl. Acad. Sci. USA* **90:**3993–3997.
204. Heiman-Patterson, T. D., Rosenberg, H. R., Binning, C. P. S., and Tahmoush, A. J. (1986). King–Denborough syndrome: Contracture testing and literature review. *Pediatr. Neurol.* **2:**175–177.
205. Orita, M., Suzuki, Y., Sekiya, T., and Hayashi, K. (1989). Rapid and sensitive detection of point mutations and DNA polymorphisms using the polymerase chain reaction. *Genomics* **5:**874–879.
206. Dickson, D. (1993). DNA testing helps British bring better pig to market. *Nature* **362:**688.
207. Vansickle, J. (1993). Report: Stress gene, National Hog Farmer, November 15, pp. 22–31.
208. MacLennan, D. H., and Chen, S. R. W. (1993). The role of the calcium release channel of skeletal muscle sarcoplasmic reticulum in malignant hyperthermia. *Ann. N.Y. Acad. Sci* **707:**294–304.

11

Stretch-Activated Ion Channels

Henry Sackin

11.1. INTRODUCTION

Mechanosensitive or stretch-activated (SA) channels respond to membrane stress by changes in open probability. These channels exist in auditory cells, stretch receptors, muscle spindles, vascular endothelium, and other neurosensory tissues where their physiological function seems readily apparent. It is less obvious why *nonexcitable* cells, such as those of blood and epithelial tissues, need channels that respond to mechanical stimuli. Clearly, all cells must cope with the dual problems of volume regulation and electrolyte homeostasis. Since the primary function of epithelia is salt and water transport, these cells face both extracellular and intracellular osmotic challenges. For example, Na-transporting epithelia in the intestine and kidney must accommodate significant variations in net solute uptake without suffering destructive changes in cell volume, caused by slight discrepancies between influx and efflux.[1]

Since volume is a physical property of the cell, it can only be sensed physically, or mechanically, although cell swelling may dilute the concentration of certain impermeants that could conceivably function as intracellular volume sensors, a volume controller that relies on chemical sensors is inherently unstable because chemical concentration is affected by multiple factors within the cell. A reliable feedback system for short-term volume regulation ultimately requires some type of *mechanical* sensor to convey information about cell size. Similarly, regulation of cell growth may also require specific mechanotransducers that detect physical changes in cell size and shape. Conversely, the abnormal growth of cancer cells could involve a breakdown of such a mechanotransduction system.

Early studies implied that mechanosensitive channels belonged to a fairly homogeneous family.[2–5] Their ion selectivity, single-channel conductance, kinetics, and stretch sensitivity were surprising similar, despite diverse tissues of origin. This led to the concept of a primitive, mechanosensitive channel type that persisted through the course of evolution. Perhaps it started by conferring the important property of volume regulation to its one-celled host, and was then retained in higher animals as a multipurpose channel, serving as sensory transducer in one tissue and volume regulator in another.

Although the concept of a single SA channel with multiple functions was appealing from a theoretical point of view, it soon became apparent that SA channels were sufficiently different to defy description by a single model. Nonetheless, one can still define a family of mechanosensitive channels, with common properties, that are clearly distinguishable from the many channel types that are not stretch-activated. In an attempt to group SA channels into logical categories, I will stress a functional rather than a phylogenetic scheme, with emphasis on volume regulation and electrolyte homeostasis. Next, I will consider the properties that characterize SA channels, emphasizing the various hypotheses for electromechanical transduction. Finally, I will address the important question of how SA channels might be involved in the physiological regulation of transport and volume. This treatment of SA channels is not intended to be all-inclusive and the reader is referred to reviews that deal more extensively with the growing family of mechanosensitive channels.[6–8]

11.2. FUNCTIONAL CLASSIFICATION OF SA CHANNELS

11.2.1. SA Cation Channels

The original family of mechanosensitive, cation-selective, ion channels was discovered (probably by accident), in chick skeletal muscle[3] and embryonic *Xenopus* muscle,[2] while applying pipette suction during seal formation. An early record of mechanosensitive channel activity is illustrated in Figure 11.1, taken from a patch-clamp study on muscle from larval *Xenopus*.[2] Clearly, the application of suction via the patch pipette increases channel activity without significant alterations in current amplitude or conductance, indicating that mechanical forces affect channel gating and not channel conductance.

Subsequently, SA cation channels were identified in choroid plexus,[9] corneal epithelium,[10] neuroblastoma cells,[11] osteoblasts,[12] Ehrlich ascites tumor cells,[13] opossum kidney cells,[14] *Xenopus* oocytes,[15] aortic endothelial cells,[16] amphibian smooth muscle cells,[17] frog diluting segments,[18] basolateral

Henry Sackin • Department of Physiology and Biophysics, Cornell University Medical College, New York, New York 10021.
Molecular Biology of Membrane Transport Disorders, edited by S.G. Schultz *et al.* Plenum Press, New York, 1996

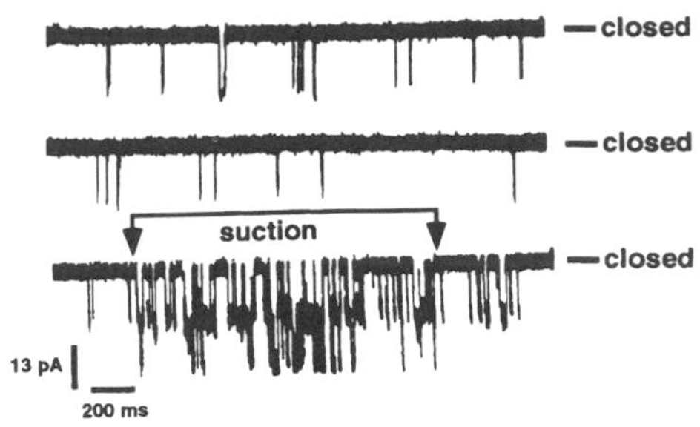

Figure 11.1. Early record of mechanosensitive channel activity. Channels were recorded from a cell-attached patch on myotomal muscle of *Xenopus laevis* tadpoles in the absence of exogenous ACh (seal >100 GΩ). The top two panels indicate openings that are apparently spontaneous in origin. Gentle suction was applied in the interval bracketed by the arrows. This resulted in a large increase in the frequency of opening of this channel type. [Reproduced from Figure 9 of Brehm *et al., J. Physiol. (London)* **350:**631–648, 1984.]

membrane of frog proximal tubule,[19] and apical membrane of *Necturus* renal proximal tubule.[20]

The channels were consistently cation selective, sometimes with a slight preference for K over Na; hence, the nomenclature SA-*cat* will be used. Their single-channel conductance ranged between 25 and 35 pS (in normal bathing solutions). The trivalent lanthanide, gadolinium (Gd), was found to completely block the channel at low concentrations.[15,20] Although this blocker is a useful tool for identifying SA channels and relating them to macroscopic currents (see Section 11.4), it is not 100% specific for the SA-cat channel since it also blocks some calcium channels[21] and endplate channels.[15] Fortunately, this is not a serious drawback for epithelial studies.

The significant calcium permeability of the SA-cat channel raised the possibility that Ca might function as a second messenger for translating mechanical stress into regulation of ion transport. This would be particularly appropriate for volume regulation. An example is the choroid plexus response to hypotonic shock, where opening of SA-cat channels presumably increases cytosolic Ca sufficiently to activate Ca-dependent maxi K channels.[9] This type of coupling seemed to establish a clear function for the SA-cat channel as a regulator of cytosolic calcium.

11.2.1.1. Cell Volume Regulation

The SA-cat channel was subsequently implicated as a mediator for two types of Ca-dependent K channels, both of which were activated by hypotonic shock: (1) a small-conductance (15 pS) Ca-dependent K channel in opossum kidney cells[22] and (2) a large-conductance (150 pS) K channel in cultured MTAL cells.[23] Stretch and osmotic activation of both of these channels appeared to depend on extracellular Ca. The presumption was that mechanical deformations of the membrane were turning on SA-cat channels and increasing cytosolic Ca which then gated the Ca-dependent channels.

Actually, the details of this cascade may be more complex than originally proposed. Studies on both opossum kidney cells and cultured TALH indicate that release of Ca from internal stores may be as important as extracellular Ca for the hypotonic RVD response.[24] Fluorescence measurements of cytosolic Ca in rabbit proximal tubule during cell swelling suggested that the changes in free intracellular Ca during swelling were too small to activate Ca-dependent maxi K channels.[25] It is still possible that SA-cat channels could establish unstirred layers of elevated Ca concentration adjacent to the interior of the cell membrane, and that these (micromolar?) levels of Ca could gate Ca-dependent channels. In any case, calcium appears to be important for normal RVD in a number of epithelial preparations.[26–28] In experiments on isolated cells, waves of Ca were produced by mechanically prodding heart cells (possessing SA-cat channels) with a smooth pipette.[29] Presumably, this calcium is extracellular in origin and enters via SA-cat channels. Final evaluation of the *SA-cat* calcium-influx hypothesis must await higher-resolution techniques for calcium imaging in volume-regulating cells.

Figure 11.2 illustrates what may be a subgroup of the SA-cat family.[14] These channels are seen in the opossum kidney cell line, and resemble the SA-cat channel in conductance

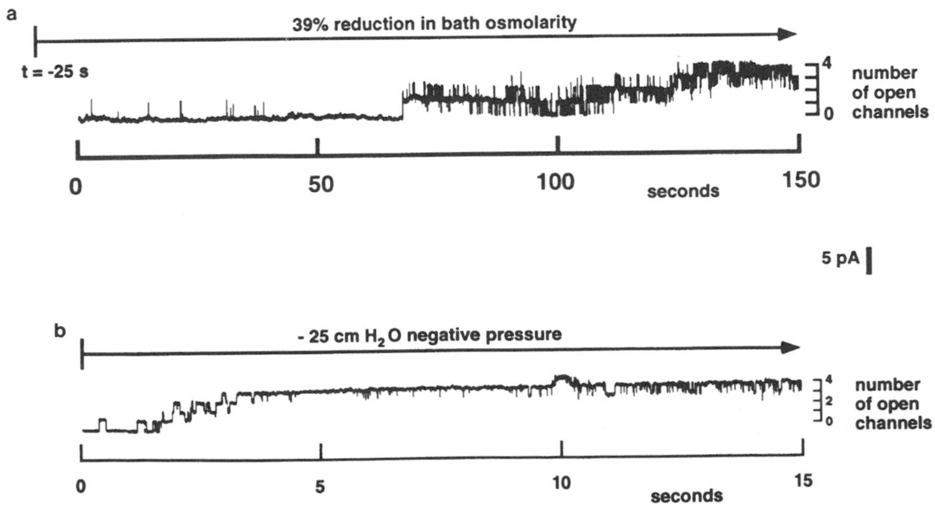

Figure 11.2. Activation of a channel population in opossum kidney cells by hyposmotic shock (a) and pipette suction (b). The channels are similar to the SA-cat channel. (Redrawn from Figure 4 of Ubl *et al., J. Membr. Biol.* **104:**223–232, 1988.)

(22 pS) and permeability to divalents like Ca. However, they differ from most SA-cat channels by displaying a significant permeability to anions like Cl. The exact nature of anion permeation through this channel is unclear. Furthermore, it appears that the cation conductance of the channel is affected by the presence of chloride.

As indicated in Figure 11.2, these opossum kidney channels are activated equally well by either pipette suction or hypotonic shock. This makes them possible mediators of the renal RVD response, provided they allow significant amounts of Ca to enter the cell during swelling. A small-conductance channel (15 pS), which is gated by calcium,[22] may be functionally coupled to this SA channel since it is active during hyposmotic shock. In contrast to maxi K channels, the 15-pS Ca-dependent K channel activates at lower levels of intracellular Ca (at normal membrane potential), making it a better candidate than the maxi K for eliminating K from swollen cells. Another case of functional coupling between ion-selective channels and the SA-cat channel seems to occur in Ehrlich ascites tumor cells where cell swelling activates independent K and Cl fluxes via channels that are not themselves mechanosensitive.[13]

An obvious problem with the SA-cat channel and volume regulation is that cation-selective channels also allow Na to enter the cell down its electrochemical gradient. The influx of osmotically active Na (and anion) would offset the volume regulatory effect of Ca-dependent K efflux. One way around this problem would be for the cell to use Ca as a biological amplifier, so that a relatively small influx of Ca would gate a large efflux of K. The steep Ca-dependence of the maxi K channel makes this a possible candidate for an amplification scheme. However, such a volume control system would also require Na levels in the cell to be under tight regulation by an independent system. This could be tested experimentally by assessing the dependence of RVD on the Na pump. In fact, renal tubules, whose basement membranes have been removed by collagenase, lose the ability to volume regulate if the Na pump is blocked by ouabain.[30] However, if the basement membrane and the cell cytoskeleton remain intact, these tubules can still volume regulate after pump inhibition.[31] Hence, there may be redundant systems for RVD, some of which seem to be independent of Na transport.

Calcium permeation through SA-cat channels may also regulate chloride currents essential for hypotonic volume regulation. The prevalence of SA-cat channels in *Xenopus* oocytes (>10^6 per oocyte) suggests that these channels might play a role during ovulation into pond water. Given the low osmotic strength of this environment, some type of volume regulatory process would be advantageous for species survival. The finding of several types of Ca-activated chloride currents in *Xenopus* oocytes[32–34] raises the possibility that oocyte volume regulation may be mediated by swelling-induced Ca permeation through SA channels.

Consistent with this hypothesis was the observation that removal of external Ca prevented chloride currents associated with normal RVD in *Xenopus* oocytes.[32] Although single channels were not recorded in these experiments, the implication was that SA-cat channels were responsible for the transient increase in cytosolic Ca required to activate the Ca-dependent chloride channels essential for RVD. In contrast to this result, two-electrode voltage-clamp experiments suggest that hyposmotic solutions may directly activate oocyte Cl currents without an increase in inward Ca flux, although activation did require the presence of Ca in the bath.[33] Since macroscopic Cl current increased twofold in these experiments, it is always possible that these Cl currents masked small changes in Ca current occurring via stretch-sensitive channels.

Recent experiments in our laboratory indicate a temporal correlation between oocyte swelling and increases in the open probability of single Sa-cat channels. Since the P_0 of these channels was effectively zero in the absence of pipette suction or cell swelling, any volume-induced increase in SA-cat activity would provide a pathway for Ca influx into the oocyte. In cell-attached patches on five separate *Xenopus* oocytes, a 50% reduction in bath osmolarity increased single-channel P_0 from 0 to 0.06 ± 0.01 within 6 min. Although correlations of this kind do not prove that Ca influx is essential for volume regulation, they do establish that cell swelling can produce sufficient membrane tension to activate endogenous SA-cat channels that are permeable to Ca.

11.2.1.2. Morphogenesis

In contrast to the oocytes of freshwater animals, the oocytes of ocean-dwelling ascidians (like the sea squirt, *Boltenia villosa*) would not be subjected to the hypotonic stress of freshwater ovulation. The existence of the same SA-cat channels in oocytes of both freshwater *Xenopus* and saltwater ascidians suggests that mechanosensitive channels may have functions besides volume regulation.[35] Specifically, mechanosensitive channels may be important in the cytokinetic events associated with morphogenesis and embryonic development. The dramatic changes in cell shape that occur during cleavage cycles of the early embryo would produce sufficient increases in membrane tension to induce Ca influx via these SA-cat channels. This influx of Ca coupled with release of Ca from intracellular stores could trigger active contraction of cytoskeletal elements essential for the morphogenetic movements associated with gastrulation. For example, there is evidence that Ca entry via SA-cat channels stimulates a propagated SA contraction that causes neural tube closure in developing oocytes.[36]

The relatively high density of SA channels would not only favor significant influx of Ca during cytokinesis but might also depolarize the oocyte sufficiently to allow further entry of Na and Ca via voltage-gated channels.[35] Since the SA-cat channel persists from prefertilization through the early stages of morphogenesis, it may help to orchestrate the complex pattern of appearance and disappearance of Na, Ca, and K currents during embryogenesis.

Although the SA channel in fish embryos is distinctly K selective, it might also contribute to morphogenesis in a manner similar to the Sa-cat channel of ascidians. The membrane stresses and cytoskeletal rearrangements that occur during cleavage of fish embryos appear to activate a transient, outward K current that then turns off during the period of reduced membrane tension between cleavages.[37] This results in a cyclic hyperpolarization that can trigger other events in embryonic development.

11.2.1.3. Capillary Endothelium and Atrial Cells

The SA-cat channel may also be involved in mechanotransduction of shear stress forces in blood vessels. Studies on cultured endothelial cells from pig aorta reveal a cation-selective channel with a single-channel conductance that varies between 39 pS in physiological saline solutions and 19 pS in solutions where calcium is the primary current-carrying ion.[16] Under these conditions Ca fluxes appear as unitary currents of about 1 to 2 pA that are reversibly activated by 15 mm Hg pipette suction. The kinetics of this vascular endothelial channel are similar to those of the SA-cat channel found in chick skeletal muscle and *Xenopus* oocyte. The channels are normally quiescent in the absence of pipette suction, but application of negative pressure to the pipette increases P_0 and decreases the duration of the longer closed state. This raises the possibility that changes in blood pressure and shear stress associated with flow over the capillary endothelium might be sufficient to turn on SA-cat channels. The relatively high selectivity of these channels ($P_{Ca}/P_{Na} = 6$)[16] could permit sufficient entry of Ca to stimulate synthesis and release of endothelial relaxing factor and subsequent vascular vasodilation.

Similar SA-cat channels have been reported in cells isolated from pig coronary arteries where single-channel conductance varied from 36 pS in normal-K solutions to 11 pS in high-Ca solutions.[38] Smooth muscle cells isolated from the stomach muscularis of the toad *Bufo marinus* exhibit a similar SA-cat channel whose single-channel conductance ranges from 60 pS in normal-K solutions to 19 pS in high-calcium solutions.[17] It is not clear whether the conductance differences in channels from these three preparations are significant enough to suggest more than one class of SA-cat channel in smooth muscle.

A physiological role for these SA-cat channels is suggested by results of experiments on primary cultures of human umbilical endothelial cells grown on a deformable silicon membrane. In these studies, cytosolic Ca was measured fluorometrically while the cell layer was subjected to a bending stress (Figure 11.3).[39] The observed increase in cytosolic Ca was consistent with activation of calcium-permeable Sa-cat channels. Although single channels could not be recorded in this apparatus, independent experiments demonstrated the existence of SA-cat channels in this preparation.[39] Furthermore, the increase in Ca was abolished either by removal of extracellular Ca or by external application of 10 μM gadolinium, a known inhibitor of SA-cat channels.[15] In contrast to the block by gadolinium, nifedipine had no effect on the stretch-induced increase in cell calcium. Even assuming that gadolinium is not 100% specific for SA-cat channels, the lack of nifedipine effect argues against influx of Ca via L-type voltage-dependent Ca channels. An interesting aspect of the mechanosensitive response in these endothelial cells is that cell Ca increases in two phases: first, a Ca influx via SA-cat channels, and second a Ca-induced Ca release from endoplasmic reticulum.[39]

In addition to the bending stress depicted in Figure 11.3, vascular smooth muscle cells have been individually stretched by the technique illustrated in Figure 11.4.[38] This procedure allowed direct recording of macroscopic currents while cells were subjected to forces similar to what they would experience during changes in vascular pressure or flow. Results indicate that longitudinal cell stretch (Figure 11.4) reversibly increased inward current in a manner consistent with stimulation of microscopic SA-cat channels, known to be present at this membrane.[38] The increases in macroscopic inward current illustrated in Figure 11.4 were insensitive to nifedipine and were associated with increases in cytosolic Ca whose origin appeared to be extracellular.[40] Hence, SA-cat

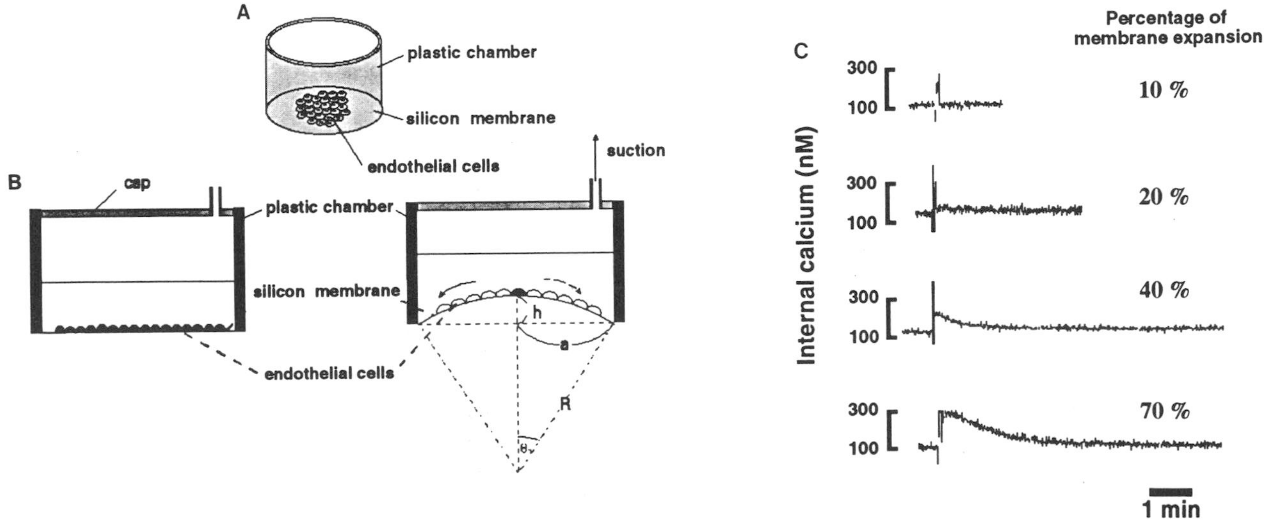

Figure 11.3. (A) Overview of the apparatus used to stretch a layer of endothelial cells grown on an elastic silicon membrane that is attached to the bottom of a plastic tube. (B) Side view of the apparatus. Suction was applied for 3 sec through the tubing. Percent expansions of membrane area were calculated from the parameters: *h, a, R,* and θ_a. (C) Elevation of intracellular Ca concentration versus the percentage of membrane expansion. (Reproduced from Naruse and Sokabe, *Am. J. Physiol.* **264:**C1037–C1044, 1993.)

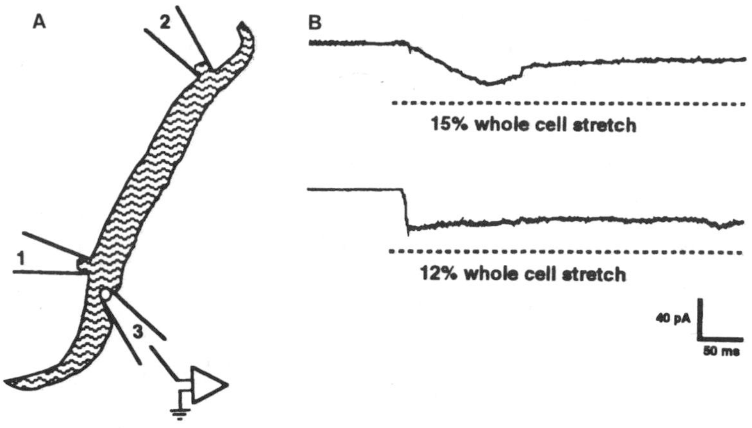

Figure 11.4. Technique for longitudinal stretching of isolated vascular smooth muscle cells. (A) Cells were stretched by using pipette #1 to stabilize one end of the cell, and pulling the cell with pipette #2. Pipette #3 was used for recording. (B) Under voltage clamp conditions, cell stretching elicited inward currents that were sustained for the duration of the stretch. Sometimes these currents appeared to be biphasic, as indicated in the top trace. Both records were obtained with the cell potential clamped at −80 mV. (Reproduced from Davis *et al., Am. J. Physiol.* **262:**C1083–C1088, 1992.)

channels may constitute the initial transduction step in the myogenic autoregulatory response of vascular smooth muscle either by allowing a significant influx of Ca or by depolarizing the cell sufficiently to activate voltage-gated Ca channels.

Similarity in the effects of cell swelling and pipette suction seen in capillary endothelial cells may not extend to cardiac tissues. Hyposmotic challenge to rat atrial cells[41] consistently activated a 36-pS SA-cat channel (140 mM K solutions), similar to the cation-selective channel reported for chick skeletal muscle[3] and *Xenopus* oocytes.[15] This swelling-induced increase in channel activity was also associated with a small but significant increase in cytosolic Ca, measured ratiometrically with Fura-2.[41] However, application of pipette suction affected the open probability of this channel in a peculiar manner. In cell-attached patches on preswollen cells, application of negative pipette pressure sometimes increased and sometimes decreased open probability. In one case, negative pressure even decreased P_0, while positive pressure, applied to the same patch, increased P_0.[41] It is difficult to construct a simple model for phenomena of this kind since one would expect that both positive and negative pipette pressures would exert similar increases in membrane tension.

11.2.1.4. Osteoblasts, Lens, and Plant Cells

Osteoblasts at the surface of bone matrix are instrumental in synthesizing bone matrix proteins. Clonal UMR-106 cells derived from rat osteogenic sarcoma possessed an 18-pS SA channel capable of conducting barium and calcium into the cell.[12] This channel was voltage insensitive, but its susceptibility to membrane tension suggested a role in volume regulation or as a stimulus for bone metabolism in response to mechanical stress. Furthermore, the finding of three types of SA channels in human osteoblast osteosarcoma G292 cells[42] suggests the paradigm where mechanical stress activates a cation-selective channel allowing a sufficient influx of Ca into the cell to gate Ca-dependent K channels. Although the role of Ca-dependent K channels in bone growth is still unclear, one possibility is that an increase in K channel activity offsets the cell depolarization produced during stress-induced elevations of cyclic AMP.[43]

Recent experiments on the osteosarcoma cell line UMR-106 imply a physiological role for mechanosensitive channels in the transduction of mechanical stress and bone growth. Cells subjected to chronic strain exhibited a lower threshold for SA-cat channel activation (Duncan, personal communication). This would support the notion that long-term information on mechanical stress can be translated into specific signals for bone development.

A mechanosensitive channel has also been seen in amphibian lens preparations.[10] Its ion selectivity is similar to SA-cat but it has a slightly higher single-channel conductance, 50 pS, compared to the 20–30 pS of other SA-cat's. Since it is the only channel in the lens that appears to be stretch sensitive, it is distinguished by the term *CAT-50.* This channel could play a role in cataract formation since any increase in lens pressure would activate CAT-50's and allow abnormal amounts of sodium, calcium, and water to enter the cell. The influx of Na would also depolarize the lens, producing a greater driving force for K exit. The resulting increase in Na and decrease in K is observed in many types of cataracts.

An interesting variety of SA-cat channels is found in plant and fungus cells. Opening of SA ion channels may play a crucial role in the geotrophic response of plants by selectively permitting Ca entry into cells responsible for root orientation. The finding that micromolar concentrations of the specific inhibitor gadolinium block this trophic response implies that the SA channels of plant cells bear a remarkable similarity to the SA-cat channels of animal cells.[44]

An SA-cat channel of much larger conductance has been reported in membrane patches of the fungus *Uromyces.*[45] This channel is cation selective, permeable to divalent ions like Ca, blocked by low concentrations of gadolinium, but has a single-channel conductance of 600 pS, more than 20 times the conductance of most SA-cat channels in animal cells. This mechanosensitive channel does not seem to be involved in volume regulation but does seem to be essential for allowing the fungus to gain entry into wheat and bean plants. SA plant channels seem to be more involved in transducing topographical and geotropic signals than in maintaining cell homeostasis. In higher plants, mechanosensitive channels may transduce mechanical (i.e., gravitational) signals into elevations in cytoplasmic Ca, thereby causing membrane kinases to selectively phosphorylate specific transporters of gravitropic hormones.[46,47]

11.2.1.5. SA-cat Channel Caveats

The presence of SA-cat channels does not necessarily require that they serve a physiological function. A number of recent reports have suggested that mechanosensitive channels may be activated by cell swelling, but that this activation is not necessarily a cause of the volume regulation *per se.* For example, the mechanism for maxi K channel involvement in RVD is still somewhat unclear. Ca-dependent maxi K channels exhibit a substantial increase in P_0 following exposure of rat CCD cells to hyposmotic media.[48] This response seems to be dependent on extracellular Ca, and would be consistent with swelling activation of Ca-permeable SA-cat channels. In cultured rabbit proximal tubule cells a 25% hypotonic bath elevated cytosolic Ca from 96 to 468 nM and dramatically activated Ca-dependent maxi K channels.[49,50] In these experiments, it was estimated that 84% of the Ca increase arose from G-protein-mediated Ca influx. The rest of the rise in Ca during swelling presumably occurred via $IP_{3,4}$ release of Ca from intracellular stores. In support of this hypothesis are experiments with atrial tissues which indicate that atrial stretch increases levels of IP_3.[51]

In rabbit cortical collecting tubule cultures, hypotonic solutions also increase activity of a maxi K channel at the apical membrane of principal cells.[52] This response can be mimicked by arachidonic acid or PGE_2, which presumably release Ca from intracellular stores. The lag between hypotonic exposure and channel activation suggests that intracellular intermediates (e.g., eicosanoids), rather than SA-cat channels, may be producing the rise in cytosolic Ca. Precedents for such a stretch-induced increase in prostaglandin synthesis come from studies in vascular endothelium.[53] Ca-dependent maxi K channels were also implicated in the RVD after ouabain-induced swelling of perfused rabbit CCTs. Again the results could be mimicked by exogenous PGE_2, suggesting that volume regulation via maxi K channels may not require direct influx of Ca from the bath.[54]

Finally, as discussed in the next section, a class of maxi K channels observed in intercalated cells of freshly dissected rabbit CCTs can be turned on by changes in membrane tension, independent of local calcium concentrations.

11.2.2. SA Potassium Channels

In addition to the large *SA-cat* family of mechanosensitive channels, there are smaller families of mechanosensitive channels that are less widely distributed among different species but are more selective for particular ions. SA channels displaying a predominant selectivity for K will be labeled *SAK's* and can be grouped into two broad classes based on their sensitivity to Ca.

11.2.2.1. Ca-Insensitive SAKs

A flickery K-selective SA channel with two open and three closed states has been reported in molluskan heart cells.[55] The spontaneous activity of this ventricular SAK, in the absence of applied stretch, raises the possibility that this channel could contribute to the normal resting potential. Application of pipette suction up to 25 mm Hg dramatically increased the P_0 of this channel without altering its selectivity or its 33-pS single-channel conductance. The kinetics and density of distribution of this channel are similar to the less selective *SA-cat* channel described above. However, the characteristic K selectivity of this SA channel makes it a better candidate for mediating RVD than the SA-cat channel. Since hypotonic volume regulation depends on an efflux of cellular cations, any channel with a significant selectivity for K over Na would be useful in restoring cell volume.

In many cells, short-term volume regulation after exposure to hypotonic media is accompanied by efflux of solutes and water out of the cell.[8] In epithelia, this volume regulatory phase (or RVD) seems to occur via a barium-sensitive K efflux,[56,57] together with a chloride flux,[58] where K and Cl ions leave the cell via separate channels.[59] Although K channels definitely seem to be involved in short-term epithelial volume regulation, there is no unified theory for how this occurs. In mammalian renal cells, K channels seem to be regulated indirectly by Ca, pH, and possibly ATP levels. In contrast, many of the K-selective channels found in amphibian cells are directly activated by stretch.

Two-types of SAK channels were found at the basolateral membrane of *Necturus* proximal tubule: (1) a short-open-time flickery SAK with conductance of about 45 pS and mean open time on the order of 1.5 msec,[60,61] and (2) a longer-open-time SAK with conductance of about 30 pS and mean open time between 40 and 50 msec.[61,62] In addition to being activated by pipette suction, these K channels were also sensitive to osmotic swelling. Hyposmotic solutions increased P_0, without changes in conductance or channel selectivity, in both K-depolarized cells[63] as well as cells maintained in Na-Ringer solutions (Figure 11.5).[61] In the latter case, measurements of electrical potential during hyposmotic shock indicated that the accompanying changes in membrane potential were not large enough to affect the relatively weak voltage dependence of this channel. The similarity in the kinetic response to both swelling and pipette suction[63] suggests that cell swelling activates these SAK channels via mechanical forces rather than by release of some intracellular mediator. The role of chemical signaling during RVD may be more important in mammalian cells than in amphibian cells.

11.2.2.2. Ca-Sensitive SAKs

There are now a number of preparations in which what appear to be maxi K channels also display a direct sensitivity to stretch. The channels retain their characteristic steep dependence on voltage and Ca sensitivity but also respond directly

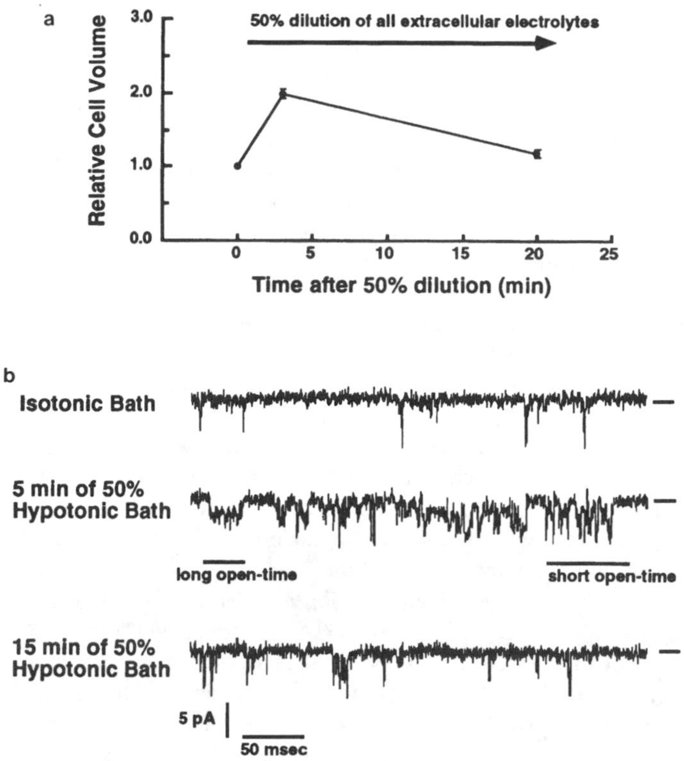

Figure 11.5. Effect of bath dilution on cell volume and channel activity. All experiments on isolated, polarized *Necturus* proximal cells, suspended by patch pipettes in flowing solution. (a) Relative cell volume as a function of time after a 50% dilution of extracellular electrolytes ($n = 10$ animals). (b) Basolateral single-channel currents during one experiment. *Isotonic bath:* sodium chloride Ringer solution; *50% dilution:* half sodium chloride Ringer; *pipette:* potassium chloride Ringer. Transpatch potential-0mV. Downward deflections from the closed state (horizontal bar) denote inward (K) current. (Reproduced from Filipovic and Sackin, *Am. J. Physiol.* **262:**F857–F870, 1992.)

to increases in membrane tension, even when Ca concentrations are tightly controlled. Maxi K channels have been identified at the apical membrane of rat and rabbit CCT that react independently to the three variables: voltage, cytosolic Ca, membrane stretch.[64] Surprisingly, the density of maxi K channels on intercalated cells (IC) was much higher than on principal cells (PC). The example of Figure 11.6 is taken from an excised (inside-out) patch on an intercalated cell of rabbit CCT where the calcium at the cytoplasmic side of the patch was buffered to 10^{-7} M, and there was no Ca added to the NaCl pipette solution.[64]

Successive application of negative pipette pressure dramatically increased the mean number of open channels in this excised patch (Figure 11.6). Upward deflections denote outward K movement from a 140 mM KCl cytoplasmic-side solution into the Na-containing pipette. Although these channels are sensitive to elevations in cytosolic-side Ca, it is unlikely that the observed increase in P_0 is the result of a change in calcium. With excised patches formed on steeply tapered pipettes, the patch remains close to the pipette tip so that the Ca at the inside face of the membrane should be effectively buffered by BAPTA in the external bath.

Although there appeared to be no obvious effect of apical hypotonicity in these cells, experiments on everted rat CCTs indicate that reducing the osmolarity at the basolateral side of the cell activates an apical, maxi K channel that might be involved in RVD.[64a] However, it is not known whether these channels responded to increases in cell volume via changes in membrane tension or via some intracellular mediator.

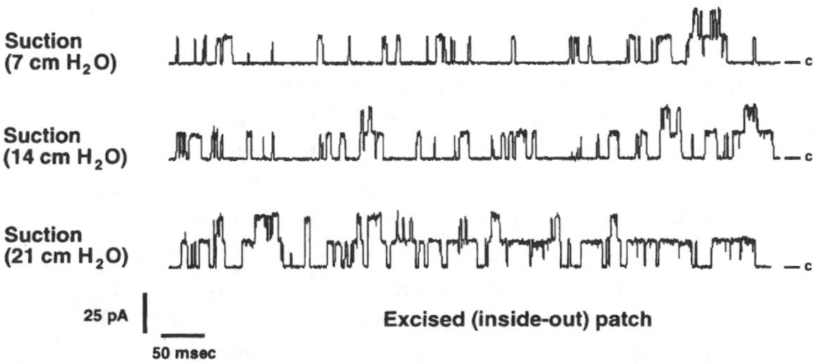

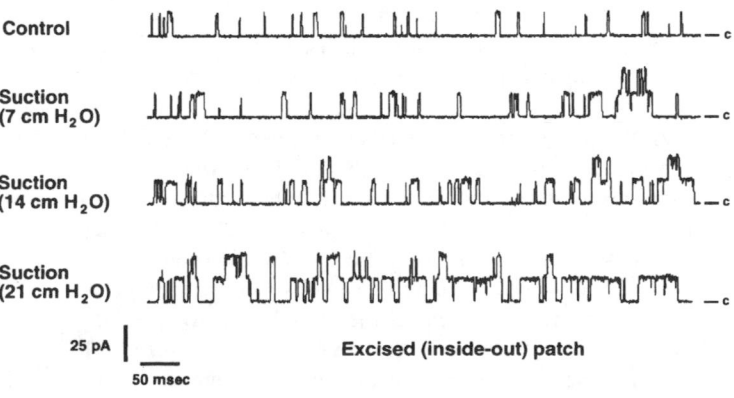

Figure 11.6. Effect of pipette suction on single-channel currents in an excised (inside-out) patch on an intercalated cell from rabbit cortical collecting tubule. Pipette contained 50 mM NaCl, but no added Ca. Bath (cytosolic side) contained 140 mM KCl with Ca maintained at pCa = 7 with BAPTA. Holding potential was −40 mV, interior side of the membrane positive. (Reproduced from Pacha *et al.*, *Am. J. Physiol.* **261:**F696–F705, 1991.)

In addition to possibly being involved in hypotonic RVD, the stretch-sensitive class of maxi K channels could be important for flow-dependent K secretion in distal tubule and CCT. Since increases in flow rate (distal delivery) could increase CCT intraluminal pressure by as much as 6 or 7 mm Hg, SA maxi K channels, at the apical membrane of intercalated (and principal) cells, might contribute to the flow-dependent K secretion observed in this nephron segment.

Ca-dependent maxi K channels that are directly activated by changes in membrane tension (independent of Ca) do not seem to be an anomaly of certain peculiar preparations. In addition to the above findings in CCT, types of mechanosensitive K(Ca) channels have also been reported in osteoblast cell lines[43] and pulmonary smooth muscle.[65] In the pulmonary smooth muscle experiments, four distinct types of stimuli were effective at increasing the P_0 of this large-conductance K channel: cytosolic Ca, cell depolarization, membrane stretch, and exogenous fatty acids, where the effect of stretch does not seem to be mediated by calcium. This was elegantly demonstrated by applying pipette suction to a series of excised (inside-out) patches with 5 mM EGTA and no added Ca on both sides of the patch. It is interesting that these channels could also be activated by fatty acids via a mechanism that did not seem to involve (1) formation of biologically active metabolites, (2) phosphorylation, or (3) Ca. This raises the possibility that increases in membrane tension could stimulate a membrane-bound phospholipase to release endogenous fatty acids from the membrane which then activate the channel. This mechanism, and its reverse, are discussed further in Section 11.3.2.4.

11.2.3. Anion Channels

Mechanosensitive anion channels are not as prevalent as SA-cat or SAK channels. With the exception of chloride channels in Ehrlich ascites tumor cells, most mechanosensitive anion channels *(SA-an)* are large (>300 pS) and often subject to regulation by specific chemical mediators. For example, a mechanosensitive anion channel found in *E. coli* spheroblasts has a conductance between 650 and 970 pS[66] and is readily activated by amphipathic compounds via a mechanism that may (or may not) be distinct from its mechanosensitive activation.[67] A mechanosensitive anion channel of about 100 pS has been observed in tobacco protoplasts that is consistently stretch-activated in both cell-attached and excised patches.[68] It is not known whether this channel responds to changes in osmolarity.

A 305-pS mechanosensitive anion-selective channel has also been described in a cell line (RCCT-28A), which has the phenotype of α-intercalated renal CCT cells. Both pipette suction and hyposmotic solutions activated this channel, which suggests that it might be involved in hyposmotic volume regulation.[69] In addition, exposing these cells to dihydrocytochalasin B dramatically increased the stretch-sensitivity of the channel.[70] This effect is similar to what has been seen with SA-cat channels.[3] However, an important difference is that the *SA-an* channel is activated not only by stretch/swelling, but also by an inositol lipid cascade involving: a membrane-bound phospholipase C, diacylglycerol, protein kinase C, and a G protein.[71] The net result is the opening of a significant pathway for Cl efflux during RVD. Direct stretch-activation of protein kinase C has also been reported for rat skeletal muscle.[72]

In Ehrlich ascites tumor cells, cell swelling activates a 23-pS conductance channel with an 11-fold preference for chloride over potassium.[73] Exposure to 10 mM glycine produced a slow increase in the volume of Ehrlich ascites tumor cells which was the result of accumulation of osmotically active Na and glycine within the cells. Activation of chloride channels in cell-attached patches occurred at the same time as the increase in cell volume. These Cl channels could also be activated by reducing the osmolarity of the external solution. This suggests that the effect of glycine on channel activity may be mediated by mechanical forces associated with cell swelling.

Hoffmann has reported a volume-sensitive, small-conductance (7 pS) chloride channel in Ehrlich ascites tumor cells, which can be activated both by swelling and by increases in cytosolic calcium.[8,13] Interestingly, these Cl channels cannot be directly activated by pipette suction. This suggests that they may function in concert with SA-cat channels that are also present in ascites tumor cells. In this scenario, stretch or swelling would activate SA-cat channels, thereby allowing sufficient Ca to enter the cell and gate the small-conductance Cl channels, permitting exit of this anion during hyposmotic RVD. This would be similar to the virtual coupling between the SA-cat channel and the maxi K channel which was discussed in the previous section.

Since many Cl channels are activated by cAMP, it is interesting that cAMP production is sometimes modulated by membrane tension.[53] This may constitute a novel "indirect" form of anion channel stretch-activation. In fact, both mechanical deformation and hyposmotic swelling increase the production of cAMP in S49 mouse lymphoma cells.[74] The possibility that membrane tension activates adenylate cyclase in much the same way that it activates certain ion channels is reinforced by how closely the secondary and tertiary structure of adenylate cyclase resembles the multiple membrane-spanning domains of many ion channels.[75] Furthermore, disruption of actin microfilaments by cytochalasin B increased cAMP content of S49 cells[74]; however, cells that were pretreated with cytochalasins exhibited no further increase in cAMP after osmotic swelling.[53] Hence, an intact cytoskeleton may be required for mechanostransduction systems of this type. Presumably, mechanical forces are focused via cytoskeletal filaments onto the membrane-bound adenylate cyclase, perturbing its structure in a way that increases catalytic activity.

11.2.4. SA Nonselective Channels and Stretch-Inactivated Channels

Truly nonselective channels *(SA-non)* which exhibit little discrimination between anions and cations are unusual. Perhaps the best example is the 36-pS nonselective channel in yeast plasma membrane which has been studied at both the whole-cell and single-channel level.[76] Although this yeast channel is blocked by 10 μM gadolinium, its inability to select between cations and anions, and its kinetic response to suction, distinguish it from the SA-cat family, and even from the slightly nonselective channel of opossum kidney cells. Presumably, the physiological function of SA-non and SA-cat channels are similar, namely, to elevate intracellular divalents (Ca) during changes in cell volume or membrane tension.

Stretch-inactivated channels (SI) are still relatively rare. However, they have now been identified in dystrophic muscle from *mdx* mice,[77] toad gastric smooth muscle,[78] astrocytes,[79] snail neurons,[80] and atrial myocytes.[81] Their presence in dystrophic muscle is particularly intriguing since normal muscle contains a high proportion of SA channels and almost no SI channels (2%), whereas dystrophic muscle contains a much higher proportion of SI channels. The significant P_0 and Ca permeability of SI channels in dystrophic muscle could account for the elevated Ca seen in *mdx* myotubes since the SI channels would be open at rest and allow large quantities of Ca to enter the cells.[82]

Aside from obvious differences in their response to membrane stretch, SI and SA channels from the same tissue have remarkably similar single-channel conductances, Ca permeability, and sensitivity to extracellular Mg.[78] Although SI channels in dystrophic tissues could explain their high intracellular Ca, the function of SI channels in normal muscle is more obscure. One possibility is that cation entry through these "normally open" SI channels may depolarize the resting potential of smooth muscle enough to activate voltage-dependent Ca channels whose function is to maintain the basal tension of smooth muscle.

Along these lines, the coexistence of both SA and SI channels in the same preparation would function as a "notch filter," where the effective P_0 of both channels reaches a minimum within a narrow region of membrane tension. For example, in snail neuron, K-selective SA and SI channels together produce a minimum K cell permeability in a region of intermediate membrane tension.[80] Within this region, the cell is sufficiently depolarized to render voltage-gated Ca channels hyperexcitable. This might have important consequences for neuronal growth cone motility.

A final implication of SI channels is that their very existence refutes the notion that mechanosensitivity is just a subtle form of channel enlargement. The fact that SI channels shut down during increases in membrane tension proves that stretch is a true gating process, rather than a mechanically induced leakage current.

11.3. THE MECHANICS OF STRETCH-ACTIVATION

11.3.1. Properties Common to SA Channels

11.3.1.1. Increase in Open Probability

Despite profound differences in kinetics, ion selectivity, and stretch sensitivity, there are a surprising number of properties common to the different classes of SA channels. In all SA channels studied to date, mechanical stimuli induce an increase in open probability, without significant changes in either single-channel conductance or ion selectivity. This preservation of selectivity and conductance reinforces the notion that membrane stretch gates the channel. At high levels of suction "breakdown currents" do occur, but these can be readily distinguished from discrete openings that characterize channel currents.

11.3.1.2. A Membrane-Specific Phenomenon

Stretch-activation is a membrane (or membrane-cytoskeletal) phenomenon. It does not seem to involve soluble cytosolic messengers or insertion of preexisting cytoplasmic channels. The principal evidence for this comes from the repeated observation that mechanosensitive channels can be stretch-activated in cell-free excised patches. This effectively rules out mediators such as Ca or ATP, provided the excised patch is formed near the tip of the pipette. Nonetheless, a variable amount of cytoskeleton often remains attached to an excised patch, and the stimulus for an increase in P_0 may be more complicated than simply membrane tension.

Stretch-activation in excised patches also rules out any mechanism dependent on insertion of channels from the cytoplasm. This has been proposed to explain osmotically induced changes in urinary bladder conductance.[83] Rapid reversibility of channel activation is another argument against channel insertion as an explanation of mechanosensitivity. Many SA channels reversibly activate on a time scale that is much faster than membrane fusion processes. Finally, SA channels can be turned on by both positive and negative pipette pressures, even though it is technically more difficult to stabilize the seal during application of positive pressure. This finding rules out the possibility that pipette suction is somehow mechanically drawing channels or channel components into the patch. If this were the case, positive and negative pipette pressures should not have the same effect on P_0.

11.3.1.3. A Common Kinetic Scheme

In addition to gross similarities in their general mode of activation, mechanosensitive channels also display subtle similarities in the kinetics of their activation. Most SA channels activate in bursts of flickery openings, although the extent of flicker seems to be greater in SAKs than SA-cats. Figures 11.7 and 11.8 illustrate the effect of suction on K-selective and cation-selective SA channels, respectively. Figure 11.7 is taken from a study on molluskan heart cells,[55] and Figure 11.8 was obtained on chick skeletal muscle.[3] Despite differences in conductance, selectivity, and kinetics, suction appears to increase the frequency of bursts to the open state, in both types of channels.

Some of these qualitative features of SA kinetics were described in models proposed by Sachs *et al.* (Figure 11.9A). In this scheme there is one open state, with a mean open time that is relatively unaffected by stretch.[3] Three closed states are arranged in a linear fashion, where only the rate constant governing transitions out of the longest closed state is stretch sen-

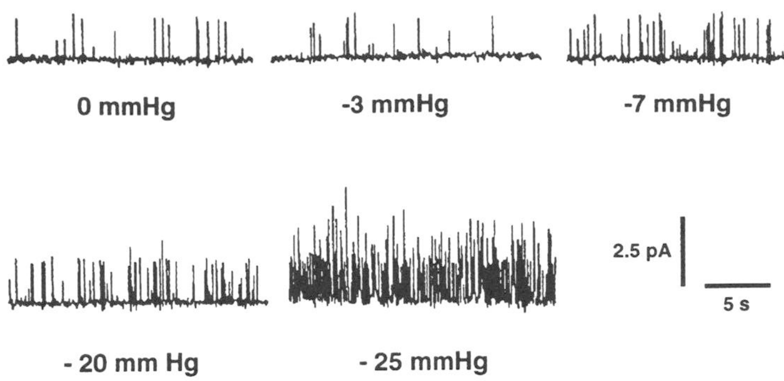

Figure 11.7. Effect of pipette suction on K channel activity in molluskan heart cells. Cell-attached patch with 30 mM K in the pipette and a pipette holding potential of +40 mV relative to bath. Higher negative pressures produced increased flickering to the open state, with no apparent change in mean open time. (Reproduced from Sigurdson *et al., J. Exp. Biol.* **127:** 191–209, 1987.)

sitive. This model (Figure 11.9A) is consistent with data on SA channels from molluskan heart cells[55] and chick skeletal muscle.[3] Mechanosensitive channels from amphibian proximal tubule have a similar kinetic scheme (Figure 11.9B), except that only two distinct closed states can be resolved.[63] As with the first model, the predominant effect of pipette suction is to increase the rate constant (k_{21}) governing transitions out of the longest closed state.

11.3.1.4. Alternative Kinetic Schemes

The linear kinetic scheme with a single open state and contiguous closed states is based to some extent on exclusion of other possible models. For example, Models B and C (of Figure 11.9) were compared using single-channel data from patches on the basolateral membranes of *Necturus* proximal tubules and equations developed by Colquhoun and Hawkes.[84] In Model B (contiguous closed states) both suction and hypotonic swelling significantly increased the rate constant k_{21} governing transitions between the interburst closed state and the flickery closed state (Table 11.1). Pipette suction also produced a small increase in the rate constant k_{13} compared to control; however, the change in k_{13} was not significant during hypotonic swelling. Even if stretch-activation involves more than one rate constant, the direction of the changes are consistent with the general appearance of SA currents during stretch.

If the same data are fit to the model of Figure 11.9C with a single open state flanked by two closed states, the predicted changes in rate constants are counterintuitive to the process of stretch-activation (Table 11.2). Although suction and swelling still increase k_{21}, which governs transitions between the interburst state and the open state, suction now increases k_{12} by 37%. Since k_{12} is the rate constant for transitions between the open state and the longer closed state, it is hard to reconcile an increase in k_{12} with the observation that suction always *decreases* the interburst closed time of the channel. This does not prove that Figure 11.9C is *incorrect,* but comparisons of this type can provide an intuitive basis for selecting one particular kinetic scheme over another.

11.3.1.5. Sigmoidal Dependence of Open Probability

Early models for mechanosensitive gating relied on a strict application of Hooke's law to the elastic elements of the membrane with the result that the free energy of gating was dependent on the square of applied tension.[3,6,85] This model predicted a sigmoidal dependence similar to that of Eq. (11.1), where K is a pressure-independent constant, θ is the stretch sensitivity of the membrane, and T is the applied tension. Although there is a good empirical basis for Eq. (11.1) and it is usually possible to obtain a "good" fit to this equation, the underlying assumptions of a T^2 dependence are incomplete.[86]

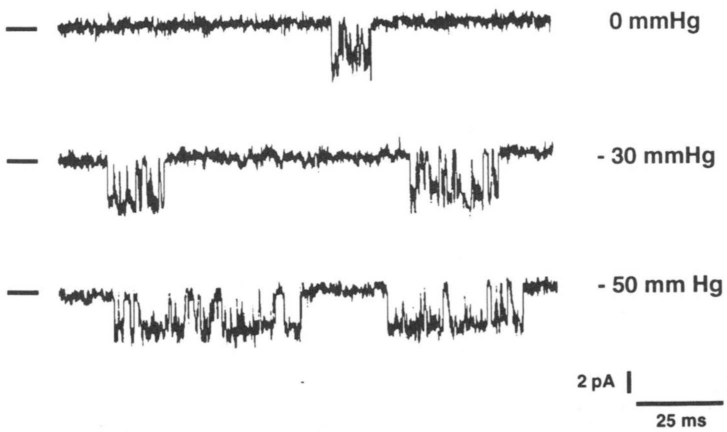

Figure 11.8. Effect of pipette suction on cation-selective channel in embryonic chick skeletal muscle. Inward currents (down deflections) from an excised (inside-out) patch with 150 mM K in the pipette and normal saline in the bath. Pipette potential was +50 mV with respect to the bath. [Reproduced from Figure 1 of Guharay and Sachs, *J. Physiol. (London)* **352:**685–701, 1984.]

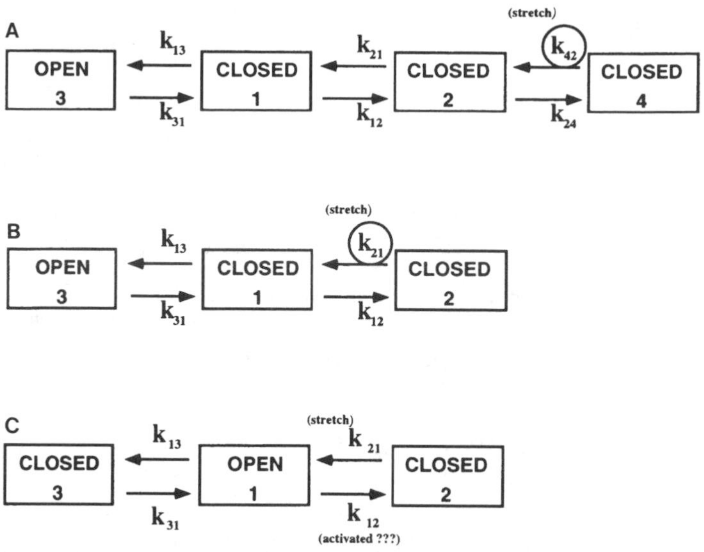

Figure 11.9. Possible kinetic schemes for stretch-activated channels. Model A was originally proposed for SA channels in chick skeletal muscle cells.[3] The mechanosensitive rate constant governs transitions out of the longest closed state. Model B is a variation of this, appropriate for amphibian preparations, in which the mechanosensitive rate constant governs transitions between the interburst closed state (#2) and the flickery closed state (#1). In Model C, the open state is surrounded by two closed states. Rate constants (Table 11.2) derived from model C are counterintuitive to the qualitative behavior of currents during suction.

$$P_0 = \frac{P_{max}}{1 + K \bullet \exp[-\theta \bullet T^2]} \quad (11.1)$$

A better representation for open probability can be derived by assuming that the free energy available for gating is linearly[87] (rather than quadratically) related to applied tension T. Open probability, P_0, will still be a sigmoidal function of tension, but the curve will have a slightly different shape and obey the equation:

$$P_0 = \frac{P_{max}}{1 + k_{eq} \bullet \exp\left[\frac{-T\Delta A - q \bullet V}{k\,T_k}\right]} \quad (11.2)$$

where k_{eq} is the equilibrium constant for channel opening in the absence of stretch, ΔA is the increase in channel cross-sectional area during opening (ΔA>0), q is the total gating charge (in coulombs) moving across the potential field during channel opening (q>0), k is the Boltzmann constant, and T_k is the absolute temperature in degrees Kelvin.

In general, P_0 can be thought of as a sigmoidal function of tension with a dependence on T, T^2, and voltage V. If the detailed dynamic parameters of the channel are unknown (as is often the case), P_0 can be expressed in terms of four independent parameters (k_{eq}, a, b, c):

$$P_0 = \frac{P_{max}}{1 + k_{eq} \bullet \exp[-a \bullet T^2 - b \bullet T - c \bullet V]} \quad (11.3)$$

Without prior information about the values of these parameters, a standard fitting algorithm will result in many, equally good fits to a given data set, all of which will be sigmoidal functions of applied tension. For example, Figure 11.10 illustrates data for the SA-cat channel of the crayfish stretch receptor.[85] When P_0 is plotted as a function of pipette suction (in cm H_2O), the data can be fit slightly better by an equation that assumes that the free energy is linearly related to tension [dashed line in Figure 11.10, fit by Eq. (11.2)] than by an equation that assumes that free energy varies as the square of the tension [solid line in Figure 11.10, fit by Eq. (11.1)]. However, the difference in "goodness" of the fit is relatively minor. Slightly more scatter in the data would make it impossible to distinguish the two models on an empirical basis. The pressure dependence of the SAK channel of *Necturus* proxi-

Table 11.1 Rate Constants for Linear Kinetic Model with Two Contiguous Closed States[a]

	Open time rate constant $(msec)^{-1}$	Closed time rate constants $(msec)^{-1}$		
	k_{31}	k_{13}	k_{12}	k_{21}
Control	1.3 ± 0.2	0.19 ± 0.04	0.31 ± 0.04	0.06 ± 0.004
Suction (−6 cm H_2O)	1.2 ± 0.05	0.29 ± 0.06*	0.22 ± 0.04	0.14 ± 0.02
Cell swelling (95 mOsm)	1.1 ± 0.1	0.23 ± 0.05	0.17 ± 0.02[b]	0.20 ± 0.03

[a]Data for this table were obtained on K-selective SAKs at the basolateral membrane of *Necturus* proximal tubules. Rate constants were determined from Model B (Figure 11.9), using equations developed by Colquhoun and Hawkes (*Proc. R. Soc. London Ser. B* **211**:205–235, 1981).
[b]Significantly different from control values.

Table 11.2. Rate Constants for Linear Kinetic Model with Open State Surroundedby Two Closed States[a]

	Open time rate constants (msec)$^{-1}$			Closed time rate constants (msec)$^{-1}$	
	k_{13}	k_{12}	$k_{13}+k_{12}$	k_{21}	k_{31}
Control	0.91 ± 0.2	0.38 ± 0.04	1.3 ± 0.2	0.02 ± 0.003	0.54 ± 0.07
Suction (-6 cm H_2O)	0.68 ± 0.1	0.52 ± 0.04	1.2 ± 0.1	0.07 ± 0.01[b]	0.59 ± 0.10
Cell swelling (95 mOsm)	0.77 ± 0.2	0.36 ± 0.06	1.1 ± 0.2	0.07 ± 0.01	0.54 ± 0.06

[a]Data for this table were obtained on K-selective SAKs at the basolateral membrane of *Necturus* proximal tubules. Rate constants were determined from Model C (Figure 11.9), using equations developed by Colquhoun and Hawkes (*Proc. R. Soc. London Ser. B* **211**:205–235, 1981).
[b]Significantly different from control values.

mal tubule is illustrated in Figure 11.11. Both the quadratic and linear dependence models fit the data equally well.

The sigmoidal curve in Figure 11.11 was fit using Eq. (11.2) with an effective stretch sensitivity of 0.62 (cm $H_2O)^{-1}$ or 0.84 (mm Hg)$^{-1}$. This value is between two and three times larger than the stretch sensitivity of the crayfish stretch receptor (Figure 11.10), and is consistent with the half-maximal suction being much less for the *Necturus* SAK than for the crayfish SA-cat channel.

11.3.1.6. Adaptation

Several varieties of SA channels exhibit a time-dependent decrease in P_0 on exposure to a constant or repeated pipette suction. This adaptation can be distinguished from inactivation since the channel remains sensitive to mechanical stimuli, but with a reduced stretch-sensitivity. Adaptation has been reported in *Xenopus* oocyte,[88] yeast,[76] and higher plants.[46] It is independent of Ca and is evident at both the single-channel[88] and whole-cell level.[89] Adaptation occurs at negative cell potentials, and is virtually abolished when the cell membrane potential is clamped to a positive value. Since adaptation, but not mechanosensitivity, is easily abolished after strong suction, the mechanism of adaptation may involve

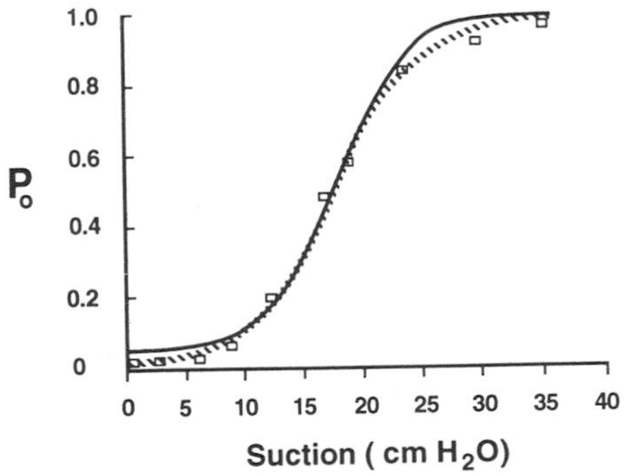

Figure 11.10. Effect of pipette suction on crayfish abdominal stretch receptor neuron. Single-channel open probability P_o is sigmoidally dependent on pipette suction, in which the exponential term is either a quadratic function of tension [solid line, Eq. (11.1)] or a linear function of tension [dashed line, Eq. (11.2)]. (Reproduced from Erxleben, *J. Gen. Physiol.* **94**:1071–1083, 1989.)

membrane-cytoskeletal interactions that can be decoupled by mechanical stress.[88]

11.3.2. Theories of Mechanosensitive Channel Activation

11.3.2.1. Tension versus Pressure

Early studies of mechanosensitive channels often contained tacit assumptions about the equivalence of hydrostatic pressure and membrane tension. In fact, these two forces are orthogonal to each other, with pressure being exerted perpendicular to the plane of the membrane and tension being a force *in* the plane of the membrane. In thin-walled spheres, hydrostatic pressure (*P*) and membrane tension (*T*) are related according to Laplace's law, where d_c is the diameter of the cell or sphere:

$$T = \frac{P \bullet d_c}{4} \tag{11.4}$$

The issue of pressure versus tension was addressed in whole-cell clamps on yeast spheroblasts of various sizes.[76] As indicated by the data of Figure 11.12, plots of mechanosensitive current versus tension were independent of spheroblast diameter, whereas plots of current versus pressure exhibited a clear dependence on diameter. These studies indicate that membrane *tension* and not pressure is the relevant parameter that controls SA channel activity.

11.3.2.2. Swelling and Membrane Tension

The discovery of SA channels in nonexcitable cells immediately raised the possibility that these channels were somehow involved in cell volume regulation. The theoretical basis for the relation between cell volume changes and membrane tension is deceptively simple, and depends primarily on application of Laplace's law [Eq. (11.4)] to an ideally elastic system. However, the assumptions of the model are less obvious, and reflect our basic lack of information. For example, it is usually assumed that increases in cell volume are associated with significant changes in membrane tension. However, most epithelial cells possess multiple membrane infoldings that may change shape during swelling without producing a real

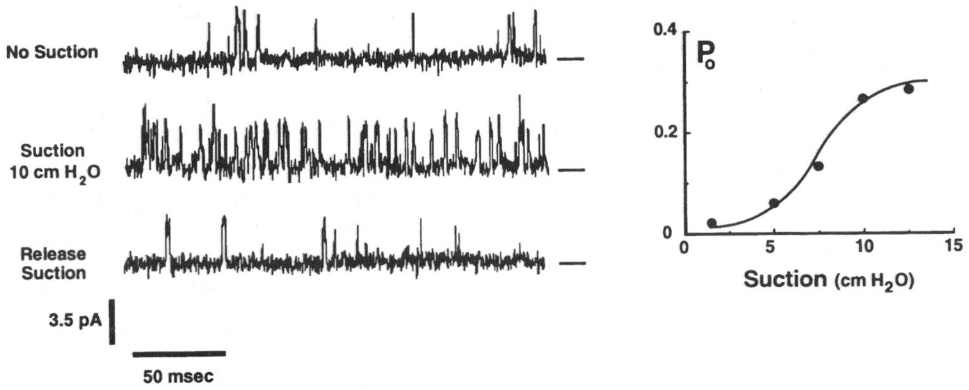

Figure 11.11. Effect of pipette suction on the short-open-time, stretch-activated K channel at the basolateral membrane of *Necturus* proximal tubule. Cell-attached patch on an isolated tubule. Basement membrane was manually removed without the use of enzymes. *Pipette:* Na Ringer; *bath:* K Ringer, zero applied potential. The open probability P_o is a sigmoidal function of applied tension. Curve was fit to Eq. (11.2) of the text using an effective stretch sensitivity of 0.6 $(\text{cm } H_2O)^{-1}$.

change in membrane tension. Since we cannot visualize dynamic changes in cell surface morphology during swelling of living cells, we can only guess at the changes in membrane structure that are occurring.

11.3.2.2a. Spherical Cell Model. Nonetheless, it is instructive to examine the simplest model consisting of a spherical cell with no membrane infolding. In this idealized case, the membrane tension on the patch (T_p) that is associated with a specific increase in cell surface area ΔA can be described by a two-dimensional form of Hooke's law[3]:

$$T_p = K_A(\Delta A/A)_p \tag{11.5}$$

where $(\Delta A/A)_p$ is the fractional increase in membrane patch area and K_A is the area elasticity coefficient. The latter parameter can be estimated from the measured hydrostatic pressure and the associated increase in relative area that produces membrane breakdown (i.e., lysis).

If the relative increase in surface area associated with *cell* swelling is $(\Delta A/A)_c$, then the membrane tension in a normal (unclamped) cell is T_c where

$$T_c = K_A(\Delta A/A)_c \tag{11.6}$$

The elasticity coefficient K_A is determined empirically, and $(\Delta A/A)_c$ is geometrically related to the relative change in cell volume $\Delta V/V$, according to

$$\left(\frac{\Delta A}{A}\right)_c = \left(1 + \frac{\Delta V}{V}\right)^{2/3} - 1 \tag{11.7}$$

which follows directly from the equations for surface area and volume of an ideal sphere.

Combining Eqs. (11.6) and (11.7), the cell tension associated with a particular relative increase in cell volume $(\Delta V/V)$ is

$$T_c = K_A \bullet \left[\left(1 + \frac{\Delta V}{V}\right)^{2/3} - 1\right] \tag{11.8}$$

Even though Eq. (11.8) represents the cell as an ideal sphere, we can still use it to estimate the increase in cell volume that would generate sufficient membrane tension, T_c, to activate SA channels in an unclamped cell. If a 30-μm-diameter amphibian cell is hypotonically swollen by 1% (i.e., $\Delta V/V = 0.01$), the relative volume factor in Eq. (11.8) becomes

$$\left[\left(1 + \frac{\Delta V}{V}\right)^{2/3} - 1\right] = (1.01)^{2/3} - 1 = 0.007 \tag{11.9}$$

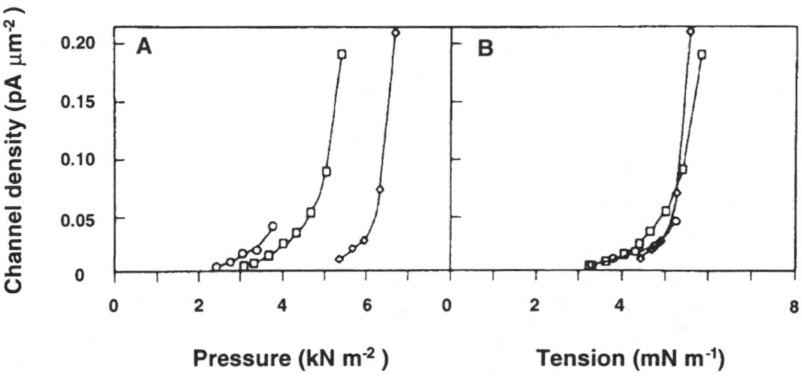

Figure 11.12. Current density through mechanosensitive channels of three cells of different sizes plotted against the applied pressure or tension. (A) Pressure dependence of SA channels in three whole-cell recordings. Cell diameters were 5.7 μm (○), 4.3 μm (□), and 3.3 μm (◇). (B) Tension dependence of SA channels in the same three cells. Tension was calculated from Laplace's law. Applied voltage = −60 mV. (Reproduced from Gustin *et al., Science* **242**:762–765, 1988.)

The elasticity coefficient K_A for the "unclamped" cell can be estimated by patching the cell and noting the maximum suction that can be sustained by the membrane. In general, amphibian proximal tubule cells can withstand 50 cm H_2O in a cell-attached patch of about 0.5-μm diameter. Assuming that cell lysis occurs after a 3% increase in real area,[90,91] the K_A for the whole cell is about 12 dyn/cm.

This elasticity coefficient can then be used to estimate the membrane tension, T_c, that would occur in an "unclamped" cell that was subjected to a 1% increase in cell volume. From Eqs. (11.8) and (11.9), the membrane tension produced by the increase in cell volume is

$$T_c = 12.2 \text{ dyn/cm} \cdot 0.007 = 0.09 \text{ dyn/cm} \tag{11.10}$$

It turns out that this tension is sufficient to activate at least some classes of SA channels. For example, SAK channels at the basolateral membrane of amphibian proximal tubule can be activated by 6 cm H_2O negative pressure in patches of 0.5-μm diameter. Applying Laplace's law to a hemispherical patch, 6 cm H_2O negative pressure is equivalent to a cell-attached patch tension of

$$T = \frac{P \cdot d}{4} = \frac{6 \cdot 980 \text{ dyn/cm}^2 \cdot 0.5 \cdot 10^{-4} \text{ cm}}{4} \\ = 0.07 \text{ dyn/cm} \tag{11.11}$$

Hence, the cell membrane tension (T_c) of 0.09 dyn/cm that is generated by only a 1% increase in cell volume [Eq. (11.10)] could be sufficient to turn on SA channels in this preparation.

11.3.2.2b. Mismatch Problem between Cell and Pipette. If this is the case, why is it usually necessary to swell cells by about 50% in order to detect significant increases in P_0? This apparent paradox arises from the mismatch between the diameter of the cell-attached patch and the diameter of the cell, and is illustrated schematically in Figure 11.13. If it were possible to form gigaohm seals on half the cell (pipette B, Figure 11.13), a 1% increase in cell volume *would* produce sufficient tension in this "super patch" to activate mechanosensitive channels.

In practice, pipette tips larger than 1 μm fail to form stable high-resistance seals on most cells. An important exception to this is the giant patches that have been formed on cardiac myocytes.[92] It is not clear that the giant patch technique can be applied to epithelial cells; however, it is certainly worth considering since it would favorably increase the ratio of patch to cell diameter (d_p/d_c). When single-channel currents are recorded from patches whose area is small compared to the cell surface area, the mechanical constraints at the tight seal reduce the tension in the cell-attached patch (T_p) in proportion to the ratio d_p/d_c. This situation is expressed by Eq. (11.12), which follows directly from Laplace's law and the requirement that both the unclamped region of the cell and the section of membrane within the pipette must experience the same hydrostatic pressure *(P)*.

$$P = \frac{T_c}{d_c} = \frac{T_p}{d_p} \tag{11.12}$$

Combining Eqs. (11.8) and (11.12), one can compute the actual tension that a cell-attached patch experiences from a 60% increase in cell volume ($\Delta V/V$):

$$T_p = \frac{d_p}{d_c} K_A \cdot \left[\left(1 + \frac{\Delta V}{V}\right)^{2/3} - 1\right] \\ = \frac{0.5}{30} \cdot 12.2 \text{ dyn/cm } [(1 + 0.6)^{2/3} - 1] \\ = 0.08 \text{ dyn/cm} \tag{11.13}$$

In other words, a 60% increase in relative cell area is required to produce sufficient tension (>0.07 dyn/cm) in a cell-attached patch to turn on SA channels to the same extent as a direct application of 6 cm H_2O suction [see Eq. (11.11)].

11.3.2.2c. Spheroblasts and Vesicles. Although the use of giant patches[92] would establish a more favorable ratio of patch area to cell area, the details of single-channel activation would be difficult to discern in giant patches. Further clarification of the role of SA channels in volume regulation could be obtained by studying vesicles whose diameter was comparable to the tip diameter of standard patch pipettes. Mechanical expansion of yeast spheroblasts produced an important correlation between SA single-channel currents and whole-cell currents[76,93]; however, osmotic perturbations were not investigated in this study.

Even though mechanical stretching is analogous to osmotic swelling, the exact nature of the forces may not be completely equivalent. In this regard, we have attempted to record osmotic effects on single-channel activity in 2- to 3-μm-diameter vesicles formed by excision of cell-attached patches (con-

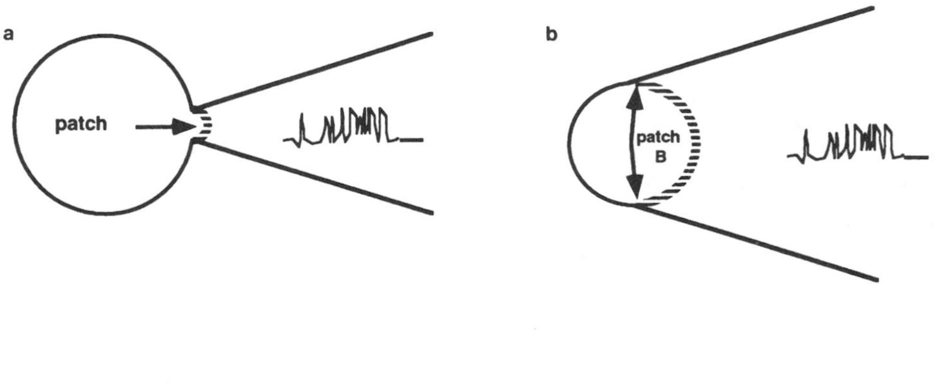

Figure 11.13. Cell-attached patches are mechanically as well as electrically clamped. (a) Small patch pipettes require a large change in volume to produce sufficient tension in a cell-attached patch to activate mechanosensitive channels. (b) If the cell-attached patch were equal to the diameter of the cell, SA channel activity could be recorded with relatively small changes in cell volume.

taining SA-cat channels) from *Xenopus* oocytes. The presence of a vesicle within the pipette tip (as opposed to an inside-out patch) was verified by low-angle, high-resolution imaging of the pipette. Vesicles were also confirmed by the characteristic rounding of single-channel currents, which reflects the capacitance of a closed structure.[94] Mechanosensitive channel activity could be elicited from these vesicles either by pipette suction or by a reduction in external osmolarity. Within 5 min after a 50% decrease in bath osmolarity, there was visible movement of the inner membrane of the vesicle and an associated increase in P_0 from 0 to 0.07 ± 02 (n=5).

11.3.2.3. Nonideal Membrane Mechanics

Although the above calculations are instructive, it is a considerable simplification to assume that the shape of the membrane at the tip of the patch pipette is an ideal hemisphere. The spherical portion of the patch often occurs quite far from the pipette tip so that the patch assumes an "omega shape" with a long straight section of membrane in close contact with the glass wall of the pipette.[95] Presumably, this straight section is the site of the gigaohm seal. Light microscopic observations[95] as well as electron microscopy[96] reveal cytoplasmic and cytoskeletal material within the patch pipette (Figure 11.14). This material often persists even after the patch is excised.

In some cases, suction completely distorted the shape of the membrane, whereas in other cases, suction simply decreased the radius of curvature of the domed portion of membrane spanning the inside of the pipette, while the straight section of membrane, attached to the glass, remained unchanged.[97] Hence, some of the basic concepts of ideal hemispherical patches considered in the previous section could still be applied to the spherical portion of membrane spanning the pipette. However, the presence of a viscous cytoplasm within the pipette and different estimates of K_A resulting from the omega shape of the patch could alter some of the predictions about how cell swelling affects membrane tension. Furthermore, the observation that the patches of membrane held in long-tipped pipettes are stretched flat at zero applied pressure[97] would mean that unstretched patches have a radius of curvature similar to that of the cell. In this case, both the cell and the patch could experience the same tensions during osmotic stress. Consequently, in certain cases, small changes in cell volume could produce measurable channel activity in cell-attached patches.

Difficulties in reproducing the stretch sensitivity of a particular channel type may be partially related to an inability to control patch geometry. Variability in the "omega" shape of a patch makes it difficult to reproduce specific membrane tensions with the same pipette pressures.

High-resolution video microscopy indicates that pipette suction proportionally increases both apparent patch area and patch capacitance so that the *specific capacitance* of the patch remains relatively unchanged.[95] Hence, patch area is increased at nearly constant membrane thickness. Consequently, the increase in dome area (Figure 11.14) does not arise from membrane thinning but from an influx of lipid recruited from along the walls of the patch pipette. Release of suction causes retraction of the patch, probably arising from elastic restoring forces of the underlying cytoskeleton.

Despite movement of lipid along the wall into the dome region, the membrane itself remains rigidly attached to the glass in both excised and cell-attached patches.[95] This suggests that the membrane–glass attachment involves specific proteins that bind to the glass, while surface glycoproteins provide a high-resistance "sucrose gap" between the outer membrane leaflet and the glass pipette. In such a model, suction could produce real increases in patch area, via lipid flow, without pulling new channel proteins into the patch.

11.3.2.4. Lipid-Soluble Mediators and Membrane Deformation

The process of stretch-activation may mimic normally occurring physiological regulators. For example, endogenous fatty acids seem to augment the open probability of certain SA channels in a fashion similar to membrane stretch. In Ehrlich ascites tumor cells, arachidonic acid metabolites such as the leukotrienes activate a K conductance pathway that seems to be essential for normal volume regulation.[98] It is also becoming apparent that fatty acids can activate channels directly by processes that do not depend on the arachidonic acid pathways. For example, recent studies on vascular smooth muscle

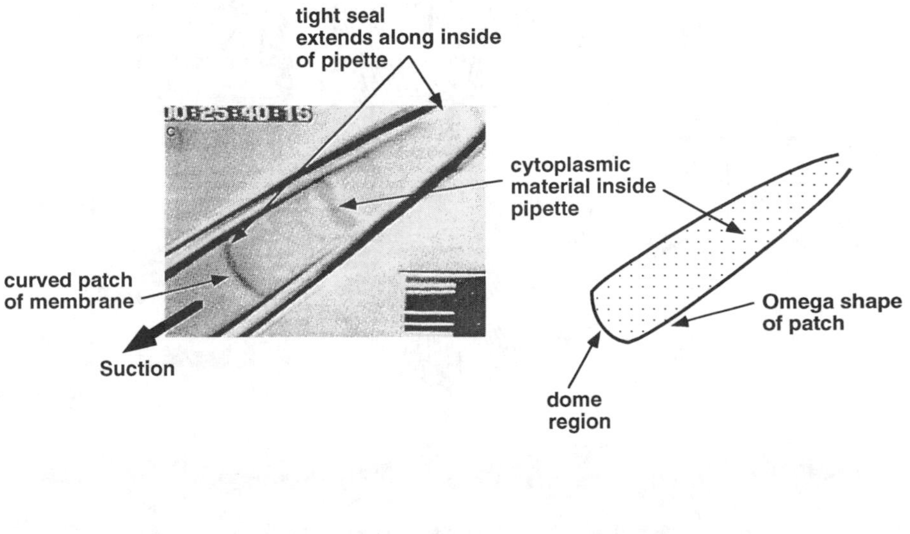

Figure 11.14. Cell-attached patch on cultured chick skeletal muscle cell. Applied pressure on the patch was −10 mmHg. Note that membrane curvature (dome region) can occur up to 10 μm from the tip of the pipette. (Reproduced from Sokabe *et al.*, *Biophys. J.* **59:**722–728, 1991.)

cells have demonstrated a large-conductance K channel that is independently regulated by four types of stimuli: calcium, voltage, stretch, and free fatty acids.[65] The demonstration of mechanosensitive channels with multiple agonists reinforces the possibility that membrane deformation might be an important clue to a more general process of channel gating. Dual activation by both free fatty acids and stretch was also observed for a lower-conductance K channel, isolated from smooth muscle cells of toad stomach.[17,99,100]

One hypothesis for fatty acid mediation of mechanotransduction is that physical deformation of the patch activates membrane-bound phospholipases. These release fatty acids from the bilayer which could then directly gate the channel protein.[101] The mechanism may involve binding of the fatty acid to the channel, via a process similar to fatty acid activation of purified protein kinase C.[102–104] It is equally possible that endogenous fatty acids, released by stretch or cell swelling, indirectly affect P_0 by altering either membrane fluidity or the lipid environment of the channel.[101]

A clue to the process of fatty acid channel activation may be the finding that lipid-soluble amphipaths activate mechanosensitive channels of *E. coli,* by a mechanism that appears to alter the curvature of the bilayer.[67] Following a model originally proposed by Sheetz and Singer to explain drug–erythrocyte interactions,[105] Martinac *et al.* suggested that amphipathic compounds (molecules having both hydrophobic and hydrophilic characteristics) would preferentially insert into the inner or outer leaflet of the bilayer. This would expand one monolayer relative to the other and produce a microscopic bending of the bilayer that transmits tension to mechanosensitive receptors on the channel protein or cytoskeleton (Figure 11.15). In this model, hydrophobic *cations* would insert into the relatively negatively charged inner leaflet causing a "cup-shaped" bending (as seen in red cells, Figure 11.15). Conversely, hydrophobic *anions* would insert into the less negative outer leaflet, expanding it and crenating a local region of the cell membrane (also seen in red cells, Figure 11.15).

Interestingly, hydrophobic cations and anions were equally effective at increasing the P_0 of mechanosensitive channels.[67] In the simplest model (Figure 11.15), amphipathic increases in bilayer tension could distort channel proteins in a way that increases their open probability. Alternatively, the changes in bilayer curvature could be communicated to the cytoskeletal network, which may then gate the channel. Both of these mechanisms are consistent with the experimental observation that most SA channels can be activated by both positive and negative pipette pressures.

Hence, we are left with two alternative theories to explain fatty acid activation of mechanosensitive channels. In the first model, the primary process is fatty acid activation of the channel. Pipette suction, swelling, and membrane stretch all act to release endogenous fatty acids from the bilayer which then move to their active site on the channel. In the second (or bilayer couple) model, a microscopic change in membrane tension is the primary process that gates the channel. Exogenous free fatty acids mimic stretch by preferentially inserting into the inner or outer leaflet of the bilayer and thereby mechanically deforming the membrane. Although this is an intriguing hypothesis, the time course of fatty acid activation of SA channels in gastric smooth muscle cells[99] and pulmonary vascular smooth muscle[65] is much faster than what is seen with amphipathic channel activators in *E. coli.*[67] This implies that the bilayer couple model may not be a universal mechanism for the stretch-activation process.

11.3.2.5. The Cytoskeleton and Stretch-Activation

Given the intimate association between channels and the cytoskeleton,[106,107] it would not be too surprising if suction forces were coupled to SA channels via the cytoskeletal network. Early studies by Guharay and Sachs found dramatic increases in the stretch sensitivity (θ) of mechanosensitive channels after treatment with a class of cytochalasins, known to disrupt actin filaments.[3] At the time, these results were interpreted as indicating a network of cytoskeletal strands funneling membrane strain energy into widely spaced nodal points (Figure 11.16).[5] The increased sensitivity to stretch after cytochalasin would then be consistent with disruption of the elastic elements in parallel with the channel since this would increase the effective stress on the individual channel proteins.

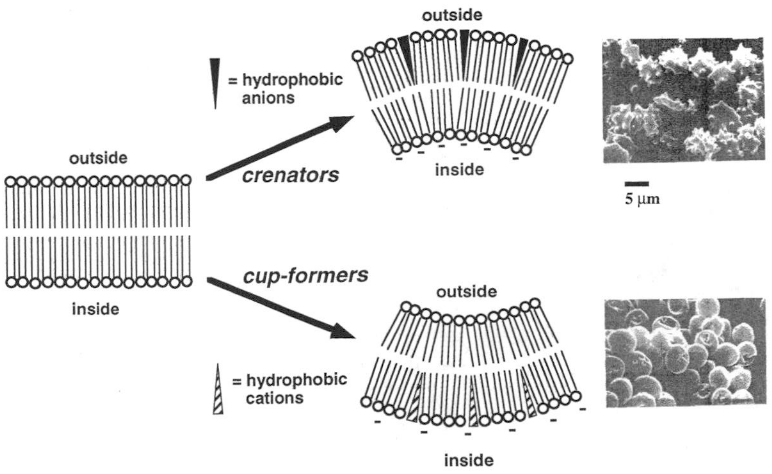

Figure 11.15. The bilayer couple model. The binding of amphipathic cations into the more negatively charged inner leaflet causes it to expand relative to the extracellular side, thereby bending the bilayer and locally distorting the membrane into a cup shape. Conversely, amphipathic anions would bind to the less negatively charged outer leaflet, expanding it relative to the inside monolayer and crenating local regions of the cell. (Reproduced from Sheetz and Singer, *Proc. Natl. Acad. Sci. USA* **71:**4457–4461, 1974.)

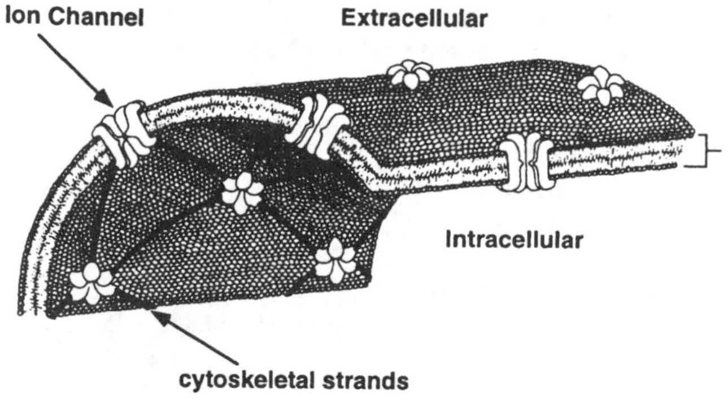

Figure 11.16. Diagram of the hypothetical linkage between ion channels and submembranous cytoskeleton. (Reproduced from Sachs, *Fed. Proc.* **46:**12–16, 1987.)

More recent studies in which elastic constants (K_A) of individual membrane patches were determined from direct video imaging indicated no effect of cytochalasins on K_A, and agents that disrupted tubulin and actin did not block SA channel activity.[95] This finding does not completely exclude the model of Figure 11.16, but it does imply that the elastic elements are not actin based, although they could consist of spectrin. Alternatively, it may be that the model of Figure 11.16 is too simple a representation of how the cytoskeletal network actually interacts with membrane channels.

The proximity of an actin-based cytoskeleton to Na channels in A6 cells[108] suggests that the cytoskeletal network may be essential for the proper regulation of at least some types of channels. This structural finding was reinforced by observations that anything that increased the relative amount of short actin filaments (cytochalasin D, exogenous action + ATP, or gelsolin) increased Na channel open probability in A6 cells.[108] The relationship between SA channels and cytoskeleton is indirectly strengthened by work with *E. coli* mutants lacking Braun's lipoprotein, which normally links the bacterial cell wall with the stretch-sensitive membrane.[109] Bacteria lacking this lipoprotein exhibit major alterations in their mechanosensitive channels. The possibility of producing selective mutations in cytoskeletal elements of *E. coli* makes the study of bacterial SA channels a very promising avenue of exploration.

There are well-known links between the cytoskeleton and volume regulation in a variety of tissues. Disruption of cytoskeletal elements by cytochalasin B inhibits volume regulation in *Necturus* gallbladder,[110] rabbit proximal tubule,[31] PC12 cells,[111] and Ehrlich ascites tumor cells.[112] This establishes a tentative link between the cytoskeleton and membrane ion channels, at least in those cases where hypotonic volume regulation is known to depend on electrolyte efflux through channels. In more recent experiments, cytochalasin B not only disrupted volume regulation but also activated mechanosensitive channels and decreased cellular content of K and Cl.[8] One possibility is that microfilaments are somehow involved in sensing changes in cell volume and transmitting this information two specific ion channels. Disruption of this network could have two simultaneous effects: (1) a loss of ability to transmit volume or stretch information to the ion channels and (2) an increase in stretch sensitivity to the point where the channel spends much more time in the open state.

Another line of evidence supporting cytoskeletal–channel interactions comes from ultrastructural observations of PC12 cells after hyposmotic swelling.[111] During hyposmotic shock, homogeneously distributed microfilaments (but not microtubules) become concentrated around the nucleus with radial extensions to the cell periphery.[111] These cytoskeletal changes take place on a time scale coincident with the time required for RVD in normal PC12 cells.

In summary, the association between SA channels and the cytoskeleton is intriguing, but unclear at this time. Disruption of cytoskeletal elements by cytochalasins has prevented volume regulation in some *but not all* of the tissues where it was examined. Morphological changes in the cytoskeleton during swelling and volume regulation suggest that these elements may function as volume transducers, but experiments with fixed cells cannot resolve whether cytoskeletal changes precede or follow volume regulation.

11.4. DO MECHANOSENSITIVE CHANNELS MEDIATE CELLULAR CURRENTS?

11.4.1. SA-cat, SA-non Channels and Macroscopic Currents

It has recently been proposed that the phenomenon of stretch-activation may be dependent on the manner in which the high-resistance gigaohm seal is formed.[113] This hypothesis arose from experiments that showed no correlation between activation of single mechanosensitive channels and changes in macroscopic currents.[113] Morris and Horn generalized this hypothesis to suggest that the process of gigaseal formation artificially sensitizes channels to mechanical stimuli. Although this issue is still in dispute, it has provided an important stimulus for reexamining mechanosensitive phenomena using a variety of tools.

A number of investigators have begun to compare single-channel and whole-cell currents in the same preparation. Some of these results constitute important counterexamples to the Morris–Horn hypothesis. In one case, Gustin reported that inflating yeast spheroblasts with positive pressure elicited macroscopic currents of a magnitude that were consistent with the expected total current from individual SA channels in these spheroblasts.[93] Furthermore, the ion selectivity and sensitivity of these channels to inhibitors such as gadolinium were identical for individual single channels and macroscopic currents. This seems to clearly establish (at least for yeast) that SA ion channels underlie the whole-cell currents produced by a mechanical deformation.

Experiments in other preparations also support the notion that SA channels form the basis for whole-cell currents. Stretched smooth muscle cells exhibited a net inward current that could be attributed to microscopic SA channels.[38] In liver cells, hyposmotic stress activated Ca-permeable, SA channels concomitant with an increase in cytosolic calcium.[114] Hyposmotically stressed intestinal cells (I407 line) manifested large gadolinium-sensitive inward currents that were correlated with Ca influx,[115] as did *Xenopus* myocytes.[116]

In nystatin-permeabilized,[117] whole-cell recordings from isolated opossum kidney cells, stepwise reductions in bath osmolarity simultaneously increased cell size, inward current, and total conductance.[118] However, these whole-cell currents were not completely blocked by gadolinium, although the cells possessed SA channels that were readily activated by either pipette suction or hyposmotic stress.[8] Since the opossum SA channels are different from most SA-cat channels (by virtue of their anion permeability), they also may not be directly blockable by gadolinium. Alternatively, hyposmotic swelling may have more than simply a mechanical effect on SA channels. For example, there is some evidence that swelling alters the cytoskeleton in a way that activates a membrane-bound phospholipase C. The IP_3 produced from this process would mobilize Ca from intracellular stores, and the diacylglycerol produced would elevate protein kinase C, which could, in turn, stimulate specific ion channels.[71] Diacylglycerol could also function as a precursor in arachidonic acid metabolism. Some of the end products of this pathway, such as the leukotrienes, may also gate specific K or Cl channels essential for RVD.

11.4.2. SAK Channels and Macroscopic Currents

The absence of highly specific blockers for the different classes of SA channels makes it difficult to attribute macroscopic currents to SA channels with absolute certainty. Gustin's experiments may be the most convincing (to date) that a direct relationship exists between mechanosensitive channels and volume- or stress-induced macroscopic currents.[93]

In a somewhat different approach, we have sought to circumvent the absence of specific blockers for SAK channels by working with the amphibian proximal tubule, where all of the demonstrable K conductance seems to be mechanosensitive. In this preparation, a relatively unspecific blocker like barium can be used to assess the amount of macroscopic current directly attributable to potassium channels.

In these experiments, isolated frog proximal tubule cells were voltage clamped and the effects of glucose (Glc) and phenylalanine (Phe) on cell volume and whole-cell current were examined. Consistent with previous studies, both Glc and Phe were cotransported with Na into proximal tubule cells. The principal new finding was that cotransport of Glc and Phe with Na increased both cell volume and barium-sensitive current on about the same time scale.[119] In 11 conventional whole-cell clamps, isosmotic addition of 40 mM glucose to the bathing solution increased cell volume by 23 ± 4% and increased barium-sensitive (i.e., K) conductance by 40 ± 10%. Similar results were obtained for isosmotic addition of Phe, and for addition of cotransported, but nonmetabolized, Glc analogues.

Figure 11.17 illustrates the current–voltage relations for one of the glucose experiments.[119] Only the region of negative cell potential (left of the ordinate) is meaningful because of a voltage-dependent barium block at positive cell potentials (bending over of the curves to the right of the ordinate). The linear portion of the current–voltage relation (negative potentials) depicts both inward and outward currents in the range of physiological membrane potentials. The dashed line denotes the control state (absence of transportable substrates), and the solid line denotes the *I–V* relation after isosmotic replacement of NMDG-methanesulfonate by an equivalent amount of D-Glc. Although the Glc-induced increase in barium-sensitive conductance was relatively modest (33% in this case), each cell was used as its own control and the differences were highly significant ($p < 0.005$). The swelling-associated change in conductance also occurred without significant shifts in the barium-sensitive reversal potential, indicating that addition of Glc and Phe do not alter channel selectivity.

In another series of experiments, *Necturus* proximal tubule cells were electrically uncoupled with 0.6 mM octanol and then subjected to an abrupt change in bath osmolarity while recording whole-cell currents. As illustrated in Figure 11.18, the decrease in bath osmolarity at constant ionic strength (by removal of sucrose) caused a significant increase in the barium-sensitive whole-cell conductance. Since barium-sensitive SAK channels have been identified at the basolateral membrane of these cells,[60,63,120,121] the increase in *macroscopic* barium-sensitive conductance can be tentatively attributed to osmotic activation of these stretch channels. In the region of positive cell potentials, any effect of osmotic activation is obscured by the voltage dependence of the barium block, similar to Figure 11.17.

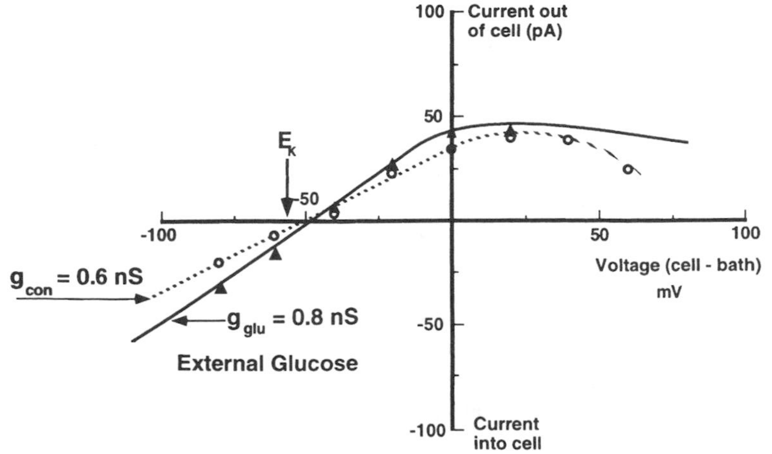

Figure 11.17. Effect of glucose on barium-sensitive current in isolated frog proximal tubule cells. The *control I–V* relation (dashed line) was calculated as the barium-sensitive current in the absence of external substrates. The solid line labeled *external glucose* was calculated as the barium-sensitive current after isosmotic addition of 40 mM external glucose. Each cell served as its own control. The terms g_{con} and g_{glu} refer respectively to the linear portions of the barium-sensitive current in the absence and presence of glucose. In all cases, the pipette contained high K Ringer. (Reproduced from Cemerikic and Sackin, *Am. J. Physiol.* **264:**F697–F714, 1993)

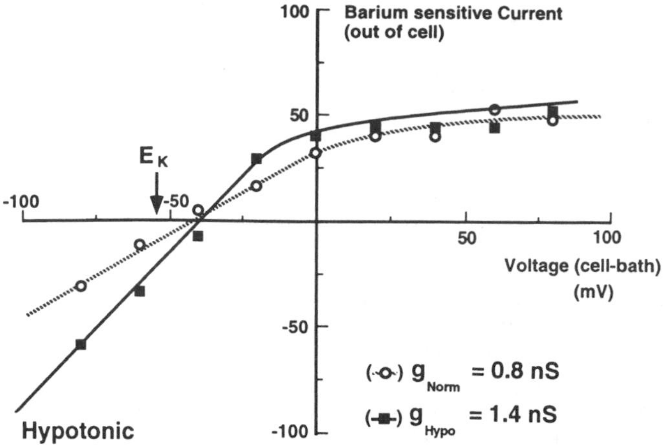

Figure 11.18. Effect of hypotonic bath solutions on whole-cell current from "uncoupled" *Necturus* proximal tubule cells. Cells were uncoupled with 0.6 mM octanol solutions. Electrical isolation was verified by capacitance measurement and lack of cell-to-cell communication. Pipette contained 64 mM K-methanesulfonate and bath contained 6.4 mM k-methanesulfonate ± sucrose. Barium-sensitive currents were measured under isotonic (dashed line) and hypotonic (solid line) conditions. The decrease in bath osmolarity was produced by removal of 100 mOsmolar sucrose.

A priori, the temporal association between cell swelling and increased barium-sensitive conductance does not prove that the increase in conductance is caused by mechanosensitive channels. However, one can propose the following argument. Since all K channels observed in amphibian proximal tubule cells are stretch-activated and barium sensitive, all barium-sensitive macroscopic currents can be attributed to SAKs. Use of the conventional whole-cell technique in these experiments rather than the permeabilized patch procedure[117] effectively washes out diffusible intracellular mediators that might link substrate transport to increased K conductance (Figure 11.19). Therefore, the substrate-induced increase in G_K probably arises from a swelling-associated deformation of the submembrane cytoskeleton or a direct change in membrane tension. In this regard, mechanosensitive K channels would be the microscopic correlates of the macroscopic change in G_K.

In some of the experiments, cells were specifically loaded with high levels of ATP to exclude this substance as a possible mediator. Therefore, the observed increase in macroscopic barium-sensitive (K) conductance must be mediated by a "membrane-linked" mechanism that does not depend on subtle changes in cytosolic pH, or intracellular metabolites (Figure 11.19). This mechanism could involve changes in P_0 produced by the physical forces of membrane deformation[86,122] [see Eq. (11.3)], or by subtle changes in the cytoskeleton that communicates cell size information to specific ion channels. There is ample evidence that even excised patches contain viable cytoskeletal elements. In any case, the above results strongly suggest that SAKs underlie the increases in macroscopic K conductance that accompany Glc and Phe cotransport. This does not exclude other parallel processes that may also increase K conductance in a way that does not depend on SA channels. The whole-cell experiments described above were designed to maximize cell–pipette equilibration and, therefore, provide no information about chemically mediated conductance signals.

11.4.3. Concluding Remarks

The Morris–Horn hypothesis[113] that the patch-clamp technique hypersensitizes channels to stretch raises a more general concern, namely, is it ever possible to design a totally noninvasive experiment? Intracellular electrodes, cell-attached patches, excised patches, whole-cell clamps, and nystatin-permeabilized whole-cell clamps all perturb the cell to some degree from its normal state; this is in addition to the very process of cell isolation that traumatically wrenches a cell away from its neighbors. Clearly, gigaseal formation is not without its hazards to the cell. In video-micrographs, the formation of a cell-attached patch is often accompanied by much initial distortion of the membrane and possible modification of the cytoskeleton.[95,97]

There are already indications that the amount of pipette suction required to form the initial seal can affect the kinetic properties of the channels being studied. Certain channels that appear to be linked to the cytoskeletal network exhibit adap-

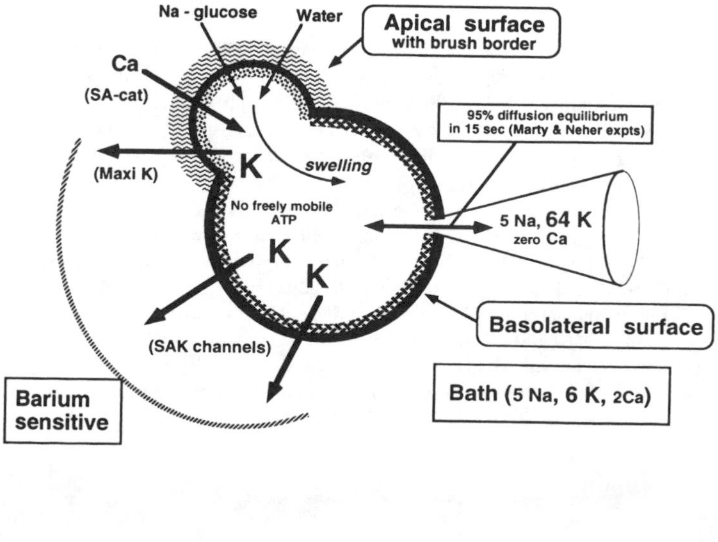

Figure 11.19. Hypothetical model of the substrate-induced increase in K conductance observed in isolated, proximal tubule cells during voltage clamp. Stippled regions within the cell denote possible unstirred layers where osmolarity transiently exceeds that of the bulk cytosol. The cross-hatched region denotes a submembranous cytoskeletal network that may be altered by cell swelling. *SA-cat* refers to stretch-activated, cation-selective channels found at the apical membrane. *Maxi K* refers to Ca-dependent, high-conductance K channels at the apical surface. *SAK* refers to stretch-activated K channels. (Reproduced from Cemerikic and Sackin, *Am. J. Physiol.* **264:**F697–F714, 1993)

tive responses to suction that are only seen with "gentle" sealing.[88] Seals formed with more vigorous suction abolish the adaptation and possibly increase the stretch sensitivity of the channel. Hence, it is conceivable that SA currents are not always seen in the whole-cell mode because these currents would adapt (shut down) quickly, whereas cell-attached patches formed with enough suction to destroy adaptation would remain sensitive to membrane tension.

Ultimately, one is left with a sort of "biological uncertainty principle" where the process of measurement itself disturbs the system (even when dealing on a scale much larger than the wavelength of light). It may be inevitable that the method of forming gigaseals, which are necessary to observe single channels, increases the stretch-sensitivity of those channels. Even if this turns out to be the case, the pessimism of Morris and Horn is unjustified.[113] Much useful information and insight can still be gleaned from discovering how the patching process itself alters channel gating. As more data become available on mechanosensitive whole-cell currents and more selective blockers are utilized, the function of SA channels will undoubtedly be clarified.

ACKNOWLEDGMENTS. The author gratefully acknowledges the assistance and advice of Professor Larry G. Palmer (Cornell University Medical College) and Professor Fred Sachs (Department of Biophysics, SUNY Buffalo). This work was supported in part by a grant (R01-DK38596) from the National Institute of Diabetes and Digestive and Kidney Diseases.

REFERENCES

1. Schultz, S. G. (1981). Homocellular regulatory mechanisms in sodium-transporting epithelia: Avoidance of extinction by "flush-through." *Am. J. Physiol.* **241**:F579–F590.
2. Brehm, P., Kullberg, R., and Moody-Corbett, F. (1984). Properties of non-junctional acetylcholine receptor channels in innervated muscle of *Xenopus* larvae. *J. Physiol. (London)* **350:**631–648.
3. Guharay, F., and Sachs, F. (1984). Stretch-activated single ion channel currents in tissue-cultured embryonic chick skeletal muscle. *J. Physiol. (London)* **352:**685–701.
4. Sachs, F. (1986). Biophysics of mechanoreception. *Membr. Biochem.* **6:**173–195.
5. Sachs, F. (1987). Baroreceptor mechanisms at the cellular level. *Fed. Proc.* **46:**12–16.
6. Morris, C. E. (1990). Mechanosensitive ion channels. *J. Membr. Biol.* **113:**93–107.
7. Sachs, F. (1989). Ion channels as mechanical transducers. In *Cell Shape: Determinants, Regulation and Regulatory Role* (F. Bonner and W. Stein, eds.), Academic Press, New York, pp. 63–92.
8. Hoffmann, E. K., and Kolb, H.-A. (1991). Mechanisms of activation of regulatory volume responses after cell swelling. In *Comparative and Environmental Physiology,* Vol. 9 (R. Gilles, E. K. Hoffmann, and L. Bolis, eds.), Springer-Verlag, Berlin, pp. 140–177.
9. Christensen, O. (1987). Mediation of cell volume regulation by Ca influx through stretch-activated channels. *Nature* **330:**66–68.
10. Cooper, K. E., Tang, J. M., Rae, J. L., Eisenberg, R. S. (1986). A cation channel in frog lens epithelia responsive to pressure and calcium. *J. Membr. Biol.* **93:**259–269.
11. Falke, L. C., and Misler, S. (1989). Activity of ion channels during volume regulation by clonal N1E115 neuroblastoma cells. *Proc. Natl. Acad. Sci. USA* **86:**3919–3923.
12. Duncan, R., and Misler, S. (1989). Voltage-activated and stretch-activated Ba^{2+} conducting channels in an osteoblast-like cell line (UMR 106). *FEBS Lett.* **251:**17–21.
13. Christensen, O., and Hoffmann, E. K. (1992). Cell swelling activates K and Cl channels as well as non-selective stretch-activated cation channels in Ehrlich ascites tumor cells. *J. Membr. Biol.* **129:** 13–36.
14. Ubl, J., Murer, H., and Kolb, H. A. (1988). Ion channels activated by osmotic and mechanical stress in membranes of opossum kidney cells. *J. Membr. Biol.* **104:**223–232.
15. Yang, X., and Sachs, F. (1989). Block of stretch-activated ion channels in *Xenopus* oocytes by gadolinium and calcium ions. *Science* **243:**1068–1071.
16. Lansman, J. B. (1987). Single stretch-activated ion channels in vascular endothelial cells as mechanotransducers. *Nature* **325:**811–813.
17. Kirber, M. T., Walsh, J. V., and Singer, J. J. (1988). Stretch-activated ion channels in smooth muscle: A mechanism for the initiation of stretch-induced contraction. *Pfluegers Arch.* **412:**339–345.
18. Hurst, A. M., and Hunter, M. (1990). Stretch-activated channels in single early distal tubule cells of the frog. *J. Physiol. (London)* **430:** 13–24.
19. Hunter, M. (1990). Stretch-activated channels in the basolateral membrane of single proximal cells of frog kidney. *Pfluegers Arch.* **416:**448–453.
20. Filipovic, D., and Sackin, H. (1991). A calcium-permeable stretch-activated cation channel in renal proximal tubule. *Am. J. Physiol.* **260:**F119–F129.
21. Lansman, J. B. (1990). Blockage of current through single calcium channels by trivalent lanthanide cations. Effect of ionic radius on the rates of ion entry and exit. *J. Gen. Physiol.* **95:**679–696.
22. Ubl, J., Murer, H., and Kolb, H.-A. (1988). Hypotonic shock evokes opening of Ca-activated K channels in opossum kidney cells. *Pfluegers Arch.* **412:**551–553.
23. Taniguchi, J., and Guggino, W. B. (1989). Membrane stretch: A physiological stimulator of Ca-activated K channels in thick ascending limb. *Am. J. Physiol.* **257:**F347–F352.
24. Montrose-Rafizadeh, C., and Guggino, W. B. (1991). Role of intracellular calcium in volume regulation by medullary thick ascending limb cells. *Am. J. Physiol.* **260:**402–409.
25. Beck, J. S., Breton, S., Laprade, R., and Giebisch, G. (1991). Volume regulation and intracellular calcium in the rabbit proximal convoluted tubule. *Am. J. Physiol.* **260:**F861–F867.
26. Okada, Y., and Hazama, A. (1989). Volume-regulatory ion channels in epithelial cells. *News Physiol. Sci.* **4:**238–242.
27. Hazama, A., and Okada, Y. (1988). Ca sensitivity of volume-regulatory K and Cl channels. *J. Physiol. (London)* **402:**687–702.
28. Wong, S. M., DaBell, M. C., and Chase, H. (1990). Cell swelling increases intracellular free [Ca] in cultured toad bladder cells. *Am. J. Physiol.* **258:**F292–F296.
29. Sigurdson, W. S., Ruknudin, A., and Sachs, F. (1992). Calcium imaging of mechanically induced fluxes in tissue-cultured chick heart: Role of stretch-activated ion channels. *Am. J. Physiol.* **262:** H1110–H1115.
30. Linshaw, M. A., and Grantham, J. J. (1980). Effect of collagenase and ouabain on renal cell volume in hypotonic media. *Am. J. Physiol.* **238:**F491–F498.
31. Linshaw, M., Fogel, C. A., Downey, G. P., Koo, E. W. Y., Gotlieb, A. (1992). Role of cytoskeleton in volume regulation of rabbit proximal tubule in dilute medium. *Am. J. Physiol.* **262:**F144–F150.
32. Barish, M. E. (1983). A transient calcium-dependent chloride current in the immature *Xenopus* oocyte. *J. Physiol. (London)* **342:** 309–325.
33. Chen, J. G., Chen, Y., Kempson, S. A., and Yu, L. (1993). Hypotonicity potentiates chloride currents in *Xenopus* oocytes. *Biophys. J.* **64:**A389.

34. Boton, R., Dascal, N., Gillo, B., and Lass, Y. (1989). Two calcium-activated chloride conductances in *Xenopus laevis* oocytes permeabilized with the ionophore A23187. *J. Physiol. (London)* **408:** 511–534.
35. Moody, W. J., and Bosma, M. M. (1989). A nonselective cation channel activated by membrane deformation in oocytes of the ascidian *Boltenia villosa. J. Membr. Biol.* **107:**179–188.
36. Odell, G. M., Oster, G., Alberch, P., and Burnside, B. (1981). The mechanical basis of morphogenesis. I. Epithelial folding and invagination. *Dev. Biol.* **85:**446–462.
37. Medina, I. R., and Bregestovski, P. D. (1988). Stretch-activated ion channels modulate the resting membrane potential during early embryogenesis. *Proc. R. Soc. London Ser. B* **235:**95–102.
38. Davis, M. J., Donovitz, J. A., and Hood, J. D. (1992). Stretch-activated single-channel and whole cell currents in vascular smooth muscle cells. *Am. J. Physiol.* **262:**C1083–C1088.
39. Naruse, K., and Sokabe, M. (1993). Involvement of stretch-activated ion channels in Ca^{2+} mobilization to mechanical stretch in endothelial cells. *Am. J. Physiol.* **264:**C1037–C1044.
40. Davis, J. M., Donovitz, J. A., Zawieja, D. C., and Meininger, G. A. (1990). Whole-cell currents and intracellular calcium changes elicited by longitudinal stretch of single vascular smooth muscle cells. *FASEB J.* **4:**A844.
41. Kim, D., and Fu, C. (1993). Activation of a nonselective cation channel by swelling in atrial cells. *J. Membr. Biol.* **135:**27–37.
42. Davidson, R., Tatakis, D., and Auerbach, A. (1990). Multiple forms of mechanosensitive ion channels in osteoblast-like cells. *Pfluegers Arch.* **416:**646–651.
43. Davidson, R. (1993). Membrane stretch activates a high conductance K channel in G292 osteoblastic-like cells. *J. Membr. Biol.* **131:**81–92.
44. Millet, B., and Pickard, B. G. (1988). Gadolinium ion is an inhibitor suitable for testing the putative roles of stretch-activated ion channels in geotropism and thigmotropism. *Biophys. J.* **53:**155a.
45. Zhou, X., Stumpf, M., Hoch, H., and Kung, C. (1991). A mechanosensitive channel in whole cells and in membrane patches of the fungus *Uromyces. Science* **253:**1415–1417.
46. Pickard, B. G., and Ding, J. P. (1993). The mechanosensory calcium-selective ion channel: Key component of a plasmalemmal control centre? *Aust. J. Plant Physiol.* **20:**439–459.
47. Pickard, B. G., and Ding, J. P. (1992). Gravity sensing in higher plants. In *Advances in Comparative and Environmental Physiology* (F. Ito, ed.), Springer-Verlag, Berlin, pp. 81–110.
48. Hirsch, J., Leipziger, J., Fröbe, U., and Schlatter, E. (1993). Regulation and possible physiological role of the Ca-dependent K channel of cortical collecting ducts of the rat. *Pfluegers Arch* **442:**492–498.
49. Suzuki, M., Kawahara, K., Ogawa, A., Morita, T., Kawaguchi, Y., Kurihara, S., and Sakai, O. (1990). Ca rises via G protein during regulatory volume decrease in rabbit proximal tubule cells. *Am. J. Physiol.* **258:**F690–F696.
50. Kawahara, K., Ogawa, A., and Suzuki, M. (1991). Hyposmotic activation of Ca-activated K channels in cultured rabbit kidney proximal tubule cells. *Am. J. Physiol.* **260:**F27–F33.
51. Harsdorf, R. V., Lang, R., Fullerton, M., Smith, A., and Woodcock, E. A. (1988). Right atrial dilation increases inositol-(1,4,5) triphosphate accumulation. *FEBS Lett.* **233:**201–205.
52. Ling, B., Webster, C., and Eaton, D. (1992). Eicosanoids modulate apical Ca-dependent K channels in cultured rabbit principal cells. *Am. J. Physiol.* **263:**F116–F126.
53. Watson, P. (1991). Function follows form: Generation of intracellular signals by cell deformation. *FASEB J.* **5:**2013–2019.
54. Strange, K. (1990). Volume regulation following Na pump inhibition in CCT principal cells: Apical K loss. *Am. J. Physiol.* **258:** F732–F740.
55. Sigurdson, W. J., Morris, C. E., Brezden, B. L., and Gardner, D. R. (1987). Stretch activation of a K channel in molluscan heart cells. *J. Exp. Biol.* **127:**191–209.
56. Welling, P. A., Linshaw, M. A., and Sullivan, L. W. (1985). Effect of barium on cell volume regulation in rabbit proximal straight tubules. *Am. J. Physiol.* **249:**F20–F27.
57. Macleod, R. J., and Hamilton, J. R. (1991). Separate K and Cl transport pathways are activated for regulatory volume decrease in jejunal villus cells. *Am. J. Physiol.* **260:**G405–G415.
58. Welling, P. A., and Linshaw, M. A. (1988). Importance of anion in hypotonic volume regulation of rabbit proximal straight tubule. *Am. J. Physiol.* **255:**F853–F860.
59. Welling, P. A., and O'Neil, R. G. (1990). Cell swelling activates basolateral Cl and K conductances in rabbit proximal tubule. *Am. J. Physiol.* **258:**F951–F962.
60. Sackin, H. (1987). Stretch-activated potassium channels in renal proximal tubule. *Am. J. Physiol.* **253:**F1253–F1262.
61. Filipovic, D., and Sackin, H. (1992). Stretch and volume activated channels in isolated proximal tubule cells. *Am. J. Physiol.* **262:** F857–F870.
62. Kawahara, K. (1990). A stretch-activated K channel in the basolateral membrane of *Xenopus* kidney proximal tubule cells. *Pfluegers Arch.* **415:**624–629.
63. Sackin, H. (1989). A stretch-activated potassium channel sensitive to cell volume. *Proc. Natl. Acad. Sci. USA* **86:**1731–1735.
64. Pacha, J., Frindt, G., Sackin, H., and Palmer, L. (1991). Apical maxi K channels in intercalated cells of CCT. *Am. J. Physiol.* **261:** F696–F705.
64a. Stoner, L. C., and Morley, G. E. (1995). Effect of basolateral or apical hyposmolarity on apical maxi K channels of everted rat collecting tubule. *Am. J. Physiol.* **268:** F569–F580.
65. Kirber, M. T., Ordway, R. W., Clapp, L. H., Walsh, J. V., and Singer, J. J. (1991). Both membrane stretch and fatty acids directly activate large conductance Ca-activated K channels in vascular smooth muscle cells. *FEBS Lett.* **297:**24–28.
66. Martinac, B., Buechner, M., Delcour, A., Adler, J., and Kung, C. (1987). Pressure-sensitive ion channel in *Escherichia coli. Proc. Natl. Acad. Sci. USA* **84:**2297–2301.
67. Martinac, B., Adler, J., and Kung C. (1990). Mechanosensitive ion channels of *E. coli* activated by amphipaths. *Nature* **348:**261–263.
68. Falke, L. C., Edwards, K., Pickard, B., and Misler, S. (1988). A stretch-activated anion channel in tobacco protoplasts. *FEBS Lett.* **237:**141–144.
69. Stanton, B. A., Dietl, P., and Schwiebert, E. (1990). Cell volume regulation in the cortical collecting duct: Stretch activated Cl channels. *J. Am. Soc. Nephrol.* **1:**692 (abstract).
70. Stanton, B. A., Mills, J. A., and Schwiebert, E. M. (1991). Role of the cytoskeleton in regulatory volume decrease in cortical collecting duct cells in culture. *J. Am. Soc. Nephrol.* **2:**751 (abstract).
71. Schwiebert, E. M., Karlson, K., Friedman, P. A., Dietl, P., Spielman, W. S., and Stanton, B. (1992). Adenosine regulates a chloride channel via protein kinase C and a G protein in a rabbit cortical collecting duct cell line. *J. Clin. Invest.* **89:**834–841.
72. Richter, E. A., Cleland, P. J., Rattigan, S., and Clark, M. G. (1987). Contraction-associated translocation of protein kinase C in rat skeletal muscle. *FEBS Lett.* **217:**232–236.
73. Hudson, R. L., and Schultz, S. G. (1988). Sodium-coupled glycine uptake by Ehrlich ascites tumor cells results in an increase in cell volume and plasma membrane channel activities. *Proc. Natl. Acad. Sci. USA* **85:**279–283.
74. Watson, P. A. (1989). Accumulation of cAMP and calcium in S49 mouse lymphoma cells following hyposmotic swelling. *J. Biol. Chem.* **264:**14735–14740.
75. Krupinski, J., Coussen, F., Bakalyar, H., Tang, W., Feinstein, P. G., Orth, K., Slaughter, C., Reed, R., and Gilman, A. (1989). Adenylyl cyclase amino acid sequence: Possible channel- or transport-like structure. *Science* **244:**1558–1564.
76. Gustin, M. C., Zhou, X., Martinac, B., and Kung, C. (1988). A mechanosensitive ion channel in the yeast plasma membrane. *Science* **242:**762–765.

77. Franco, A., and Lansman, J. B. (1990). Calcium entry through stretch-inactivated ion channels in *mdx* myotubes. *Nature* **344:** 670–673.
78. Hisada, T., Walsh, J. V., and Singer, J. (1993). Stretch-inactivated cationic channels in single smooth muscle cells. *Pfluegers Arch.* **422:**393–396.
79. Ding, J. P., Bowman, C. L., Sokabe, M., and Sachs, F. (1989). Mechanical transduction in glial cells: SACs and SICs. *Biophys. J.* **55:**244a.
80. Morris, C. E., and Sigurdson, W. J. (1989). Stretch-inactivated ion channels coexist with stretch-activated ion channels. *Science* **243:** 807–809.
81. Wagoner, D. R. V. (1991). Mechanosensitive ion channels in atrial myocytes. *Biophys. J.* **59:**546a.
82. Fong, P., Turner, P. R., Denetclaw, W. F., and Steinhardt, R. (1990). Increased activity of calcium leak channels in myotubes of Duchenne human and *mdx* mouse origin. *Science* **250:**673–676.
83. Lewis, S. A., and Clausen, C. (1991). Urinary proteases degrade epithelial sodium channels. *J. Membr. Biol.* **122:**77–88.
84. Colquhoun, D., and Hawkes, A. G. (1981). On the stochastic properties of single ion channels. *Proc. R. Soc. London Ser. B* **211:**205–235.
85. Erxleben, C. (1989). Stretch-activated current through single ion chanels in the abdominal stretch receptor organ of the crayfish. *J. Gen. Physiol.* **94:**1071–1083.
86. Sachs, F., and Lecar, H. (1991). Stochastic models for mechanical transduction. *Biophys. J.* **59:**1143–1145.
87. Howard, J., Roberts, W. M., and Hudspeth, A. J. (1988). Mechanoelectrical transduction by hair cells. *Annu. Rev. Biophys. Biophys. Chem.* **17:**99–124.
88. Hamill, O. P., and McBride, D. W. (1992). Rapid adaptation of single mechanosensitive channels in *Xenopus* oocytes. *Proc. Natl. Acad. Sci. USA* **89:**7462–7466.
89. Gustin, M. C. (1992). Mechanosensitive ion channels in yeast. Mechanisms of activation and adaptation. In *Advances in Comparative and Environmental Physiology,* Vol. 10 (F. Ito, ed.), Springer-Verlag, Berlin, pp. 19–35.
90. Kwok, R., and Evans, E. (1981). Thermoelasticity of large lecithin bilayer vesicles. *Biophys. J.* **35:**637–652.
91. Evans, E. A., Waugh, R., Melnik, L. (1976). Elastic area compressibility modulus of red cell membrane. *Biophys. J.* **16:**585–595.
92. Collins, A., Somlyo, A. V., and Hilgemann, D. W. (1992). The giant cardiac membrane patch method: Stimulation of outward Na-Ca exchange current by MgATP. *J. Physiol (London)* **454:**27–57.
93. Gustin, M. C., Sachs, F., Sigurdson, W., Ruknudin, A., Bowman, C., and Morris, C. (1991). Single-channel mechanosensitive currents. *Science* **253:**800–802.
94. Hamill, O. P., Marty, A., Neher, E., Sakmann, B., and Sigworth, F. J. (1981). Improved patch-clamp techniques for high resolution current recording from cells and cell-free membrane patches. *Pfluegers Arch.* **391:**85–100.
95. Sokabe, M., Sachs, F., and Jing, Z. (1991). Quantitative video microscopy of patch clamped membranes: Stress, strain, capacitance, and stretch activation. *Biophys. J.* **59:**722–728.
96. Ruknudin, A., Song, M. J., and Sachs, F. (1989). The ultrastructure of patch-clamped membranes: A study using high voltage electron microscopy. *Biophys. J.* **112:**125–134.
97. Sokabe, M., and Sachs, F. (1990). The structure and dynamics of patch-clamped membranes: A study using differential interference contrast light microscopy. *J. Cell Biol.* **111:**599–606.
98. Lambert, I. H., Hoffmann, E. K., and Christensen, P. (1987). Role of prostaglandins and leukotrienes in volume regulation by Ehrlich ascites tumor cells. *J. Membr. Biol.* **98:**247–256.
99. Ordway, R. W., Walsh, J. V., and Singer, J. J. (1989). Arachidonic acid and other fatty acids directly activate potassium channels in smooth muscle cells. *Science* **244:**1176–1178.
100. Ordway, R. W., Petrou, S., Kirber, M. T., Walsh, J. V., and Singer, J. J. (1992). Two distinct mechanisms of ion channel activation by membrane stretch: Evidence that endogenous fatty acids mediate stretch activation of K channels. *Biophys. J.* **61:**A391 (abstract).
101. Ordway, R. W., Singer, J. J., and J. V. W. Jr. (1991). Direct regulation of ion channels by fatty acids. *Trends Neurol. Sci.* **14:**96–100.
102. Seifert, R., Schächtele, C., Roenthal, W., and Schultz, G. (1988). Activation of protein kinase C by cis- and trans-fatty acids and its potentiation by diacylglycerol. *Biochem. Biophys. Res. Commun.* **154:** 20–26.
103. McPhail, L. C., and Snyderman, R. (1984). A potential second messenger role for unsaturated fatty acids: Activation of Ca-dependent protein kinase. *Science* **244:**622–625.
104. Morimoto, Y. M. (1988). Activation of protein kinase C by fatty acids and its dependency on Ca and phospholipid. *Cell Struct. Funct.* **13:**45–49.
105. Sheetz, M. P., and Singer, S. J. (1974). Biological membranes as bilayer couples. A molecular mechanism of drug–erythrocyte interactions. *Proc. Natl. Acad. Sci. USA* **71:**4457–4461.
106. Branton, D. (1981). Membrane cytoskeletal interactions. *Cold Spring Harbor Symp. Quant. Biol.* **46:**1–5.
107. Bennett, V. (1985). The membrane skeleton of human erythrocytes and its implications for more complex cells. *Annu. Rev. Biochem.* **54:**273–304.
108. Cantiello, H. F., Stow, J. L., Prat, A. G., and Ausiello, D. A. (1991). Actin filaments regulate Na channel activity. *Am. J. Physiol.* **261:** C882–C888.
109. Kubalski, A., Martinac, B., Adler, J., and Kung, C. (1991). Altered properties of the mechanosensitive ion channel in a lipoprotein mutant of *E. coli. Biophys. J.* **59:**455a.
110. Foskett, J. K., and Spring, K. R. (1985). Involvement of calcium and cytoskeleton in gallbladder epithelial cell volume regulation. *Am. J. Physiol.* **248:**C27–C36.
111. Cornet, M., Delpire, E., and Gilles, R. (1987). Study of microfilaments network during volume regulation process of cultured PC 12 cells. *Pfluegers Arch* **410:**223–225.
112. Cornet, M., Lambert, I. H., and Hoffmann, E. K. (1992). Relation between cytoskeleton, hypo-osmotic treatment and volume regulation in Ehrlich escites tumor cells. *J. Membr. Biol.* **131:**55–66.
113. Morris, C. E., and Horn, R. (1991). Failure to elicit neuronal macroscopic mechanosensitive currents anticipated by single-channel studies. *Science* **251:**1246–1249.
114. Bear, C. E. (1990). A nonselective cation channel in rat liver cells is activated by membrane stretch. *Am. J. Physiol.* **258:**C421–C428.
115. Okada, Y., Hazma, A., and Yuan, W. L. (1990). Stretch-induced activation of Ca permeable ion channels is involved in the volume regulation of hypotonically swollen epithelial cells. *Neurosci. Res* **12:** S5–S13.
116. Sachs, F. (1991). Mechanical transduction by membrane ion channels: A mini review. *Mol. Cell. Biochem.* **104:**57–60.
117. Horn, D. (1988). Muscarinic activation of ionic currents measured by a new whole-cell recording method. *J. Gen. Physiol.* **92:**145– 159.
118. Ubl, J., Murer, H., and Kolb, H.-A. (1989). Simultaneous recording of cell volume, membrane current and membrane potential: Effect of hypotonic shock. *Pfluegers Arch.* **415:**381–383.
119. Cemerikic, D., and Sackin, H. (1993). Substrate activation of mechanosensitive, whole-cell currents in renal proximal tubule. *Am. J. Physiol.* **264:**F697–F714.
120. Sackin, H., and Palmer, L. G. (1987). Basolateral potassium channels in renal proximal tubule. *Am. J. Physiol.* **253:**F476–F487.
121. Kawahara, K., Hunter, M., and Giebisch, G. (1987). Potassium channels in *Necturus* proximal tubule. *Am. J. Physiol.* **253:**F488–F494.
122. Sachs, F. (1992). Stretch-sensitive ion channels: An update. In *Sensory Transduction: Soc. of Gen. Physiol. 45th Ann. Symp.* (D. Corey and S. Roper, ed.), Rockefeller University Press, New York, pp. 242–260.

12

Cation Transport ATPases

Douglas M. Fambrough and Giuseppe Inesi

12.1. INTRODUCTION

The P-type ATPases comprise a large superfamily of cation transporters that are able to utilize the energy of ATP to transport ions against their electrochemical gradients across cell membranes. In the process of ion transport, the γ-phosphate of the ATP is cleaved and the phosphate becomes covalently linked transiently to an aspartyl carboxyl group in the ATPase molecule. It has been proposed that the ion transport process also involves cycling of the ATPase molecule between two sets of conformational states, termed E1 and E2, and hence, these ion transporters are sometimes referred to as E1E2 ATPases.

Most of the P-type ATPases that occur in vertebrates contain a large subunit of about 100 to 130 kDa (Figure 12.1). This subunit is the only known subunit of the calcium-transporting P-type ATPases, while the Na^+,K^+- and H^+,K^+-ATPases have this kind of "catalytic" subunit (α in Figure 12.1) plus a smaller glycoprotein subunit (β) as well. The best-described bacterial member of the P-type ATPase superfamily, the Kdp-ATPase,[1] is comprised of three subunits, one of which (B in Figure 12.1) shares a fairly high level of amino acid sequence homology with the ATP-binding domain and the phosphorylation site of eukaryotic P-type ATPases (Figure 12.1). The recently discovered human copper transport ATPases may be multisubunit proteins, but only one subunit, which apparently has only four to six membrane spans, has been identified.[2]

The mode of ion transport and the protein structure of the P-type ATPases distinguish them from two other classes of ion transport ATPases, the V-type[3] and the F1F0-type.[4] These latter two classes have much more complex subunit structures, do not form a phosphorylated intermediate in the transport cycle, and are exclusively involved with the transport of protons. The F1F0 ATPases are also called ATP synthases, because they generally run in reverse, coupling the flow of protons down an electrochemical gradient across the membrane to the synthesis of ATP from ADP and inorganic phosphate. They are involved in ATP synthesis in mitochondria and chloroplasts and in bacteria. In vertebrates, the F1F0 ATPase occurs in the mitochondrial inner membrane and mediates most of the synthesis of ATP that occurs during aerobic respiration. The V-type ATPases, which take their name from the fact that they occur in the vacuolar membranes of plants and fungi, share structural similarities with the F1F0-type ATPases, and they are involved with ATP-dependent proton transport. In vertebrates, V-type ATPases occur in several organellar membranes, including those of the lysosome and synaptic vesicle. V-type ATPases are involved in generating the low pH characteristic of the lumen of lysosomes and in generating the pH gradient that is used as an energy source for the transport of neurotransmitters into synaptic vesicles. A V-type ATPase is also involved along with a P-type ATPase (an H^+, K^+-ATPase) in acid secretion in the kidney.[5]

In this chapter we will touch on the major aspects of all of the P-type ATPases known to occur in humans. The greatest emphasis is placed on the SERCA-type Ca^{2+}-ATPases, because they are the best understood in terms of structure and function and because they embody characteristics common to all of the P-type ATPases, such as general mechanism of ion transport and basic structure.

12.2. EVOLUTION OF THE MAJOR FAMILIES OF P-TYPE ATPASES

The molecular evolution of the P-type ATPases has recently been studied by computer-based comparisons of amino acid sequences. The most comprehensive analyses[6,7] compare nearly 50 sequences, derived from bacteria, fungi, plants, and animals. Figure 12.2 shows the type of phylogenetic trees these analyses have yielded. The length of each branch is drawn proportional to evolutionary distance. There is no attempt to identify the order in which different branches of the tree diverge from a common origin; in other words, this is not a rooted tree. What the figure shows is that all of the P-type ATPases can be treated as having diverged from a common ancestor of unknown ion selectivity very early in evolution. Over evolution, molecules with the same ion selectivity, such as those that transport potassium (the Na^+,K^+ and H^+,K^+-ATPases), have remained more similar to one another than to pumps with a different ion selectivity, for example, calcium. The pattern formed by the tree shows five clusters, or fami-

Douglas M. Fambrough • Department of Biology, The Johns Hopkins University, Baltimore, Maryland 21218 **Giuseppe Inesi** • Department of Biological Chemistry, The University of Maryland School of Medicine, Baltimore, Maryland 21201.

Molecular Biology of Membrane Transport Disorders, edited by S.G. Schultz et al. Plenum Press, New York, 1996

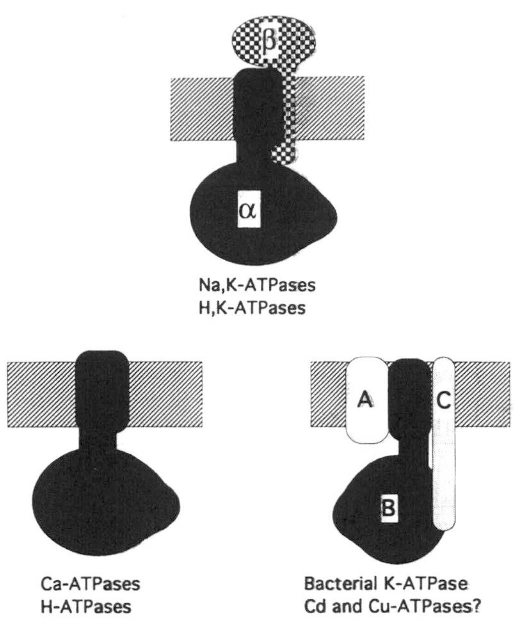

Figure 12.1. Subunit structures of P-type ATPases. The horizontal bars represent the lipid bilayer of the membrane, cytosolic face is down. Each P-type ATPase contains a large subunit (black) that spans the lipid bilayer multiple times. In all cases this subunit includes a large globular head outside the membrane; in published diagrams this head is represented either below or above the plane of the membrane, in all cases indicating cytosolic exposure. The Na^+,K^+- and H^+,K^+-ATPases have a smaller subunit, β, that spans the plasma membrane bilayer a single time and bears several N-linked oligosaccharides on its extracellular domain. In some tissues a small γ subunit, thought to span the bilayer a single time, associates with the Na^+,K^+-ATPase. The organellar and plasma membrane Ca^{2+}-ATPases consist of a single subunit. However, in some tissues the SERCA-type CA^{2+}-ATPases are associated with the small, pentameric integral membrane protein, phospholamban, that is involved in regulation of Ca^{2+} transport. One bacterial Mg^{2+}-importing ATPase closely resembles eukaryotic Ca^{2+}-ATPases.[182,183] The bacterial Kdp-ATPase is comprised of three subunits, the B subunit showing good homology with other P-type ATPases. Other bacterial ATPases that transport heavy metals have a B-like subunit that is most similar to the Kdp-ATPase, but their full subunit compositions are unknown. Human copper transport ATPases most closely resemble the bacterial P-type ATPases, but their subunit compositions are not known.

lies, of P-type ATPases. The Na^+,K^+- and H^+,K^+-ATPases form one cluster at the top. The SERCA-type Ca^{2+}-ATPases form a second cluster at the upper left, with a third cluster, the plasma membrane Ca^{2+}-ATPases, as a distant offshoot below. The fourth is a loosely related group of P-type ATPases that occur both in bacteria and in eukaryotes and are specialized for transport of a variety of different cations, including cadmium and copper. The fifth cluster is the H^+-ATPases of fungi and plants,[8,9] to the right in Figure 12.2. Mammals, including humans, have representatives of four of the five families of P-type ATPases, lacking only the plant- and fungal-type H^+-ATPases.

Interestingly, molecular-phylogenetic trees for the P-type ATPases like that in Figure 12.2 cannot be constructed for the region of the molecules C-terminal to the ATP-binding domain. This is because the sequences for different families are so different that it has not been possible to align the sequences with each other. Either the sequences have diverged so much over evolution that their residual similarities are too few to recognize, or this portion of the molecules might not have a common evolutionary origin. Each subfamily might have arisen by combination of the common N-terminal and central regions with different genes that evolved to encode the C-terminal portions.

Although the similarities between families are easily recognized, the levels of amino acid identity between families are fairly low. For example, the SERCA-type Ca^{2+}-ATPases share only about 30% amino acid sequence identity with the plasma membrane Ca^{2+}-ATPases, and only 25% amino acid sequence identity with Na^+,K^+-ATPase α subunits (see Song and Fambrough[10]). Therefore, in general, antibodies to one P-type ATPase are unlikely to cross-react with any other, and encoding DNAs for one type do not cross-hybridize sufficiently to identify genes or mRNAs encoding another. To put it more positively, most immunological reagents and nucleic acid probes can be used under conditions where they are entirely selective, and conventional procedures for probing protein blots, RNA blots, and DNA blots are completely adequate to distinguish each P-type ATPase.

Isoforms

Many of the genes in vertebrate organisms are members of multigene families. The proteins encoded by the genes of a multigene family carry out similar or identical functions and are called isoforms. Most or all of the major P-type ATPases in vertebrates are encoded by multigene families. The vertebrate SERCA-type Ca^{2+}-ATPases are encoded by at least four genes. In addition, transcripts of the SERCA genes are subjected to alternative splicing, resulting in additional variations at the protein level, mainly in the C-terminal region. Similarly, for the plasmalemmal Ca^{2+}-ATPases, four genes have been described so far in mammals, and alternative splicing has also been found. For the Na^+,K^+-ATPase there are three well-characterized genes encoding the α subunit and at least two genes encoding the β subunit, and additional genes may yet be discovered.[11] Likewise, there are several genes encoding H^+,K^+-ATPase α subunits, and presumably several genes encode H^+,K^+-ATPase β subunits as well. Alternative splicing has not been found to generate any functional Na^+,K^+- or H^+,K^+-ATPase forms. However, a splice variant of the Na^+,K^+-ATPases α1 subunit has been found in smooth muscle. This form is a highly truncated α subunit, called α1-T, that results when an intron near the middle of the coding region is not spliced out of the transcript, and translation of this normally intronic sequence leads to a premature stop-translation codon.[12]

Most of the P-type ATPase multigene families that exist in vertebrate organisms probably arose by gene duplications within the vertebrate lineage, for no isoform sets homologous to vertebrate isoforms have been found in invertebrate organisms. For example, the two isoforms of the Na^+,K^+-ATPase α subunit in the brine shrimp appear to have arisen indepen-

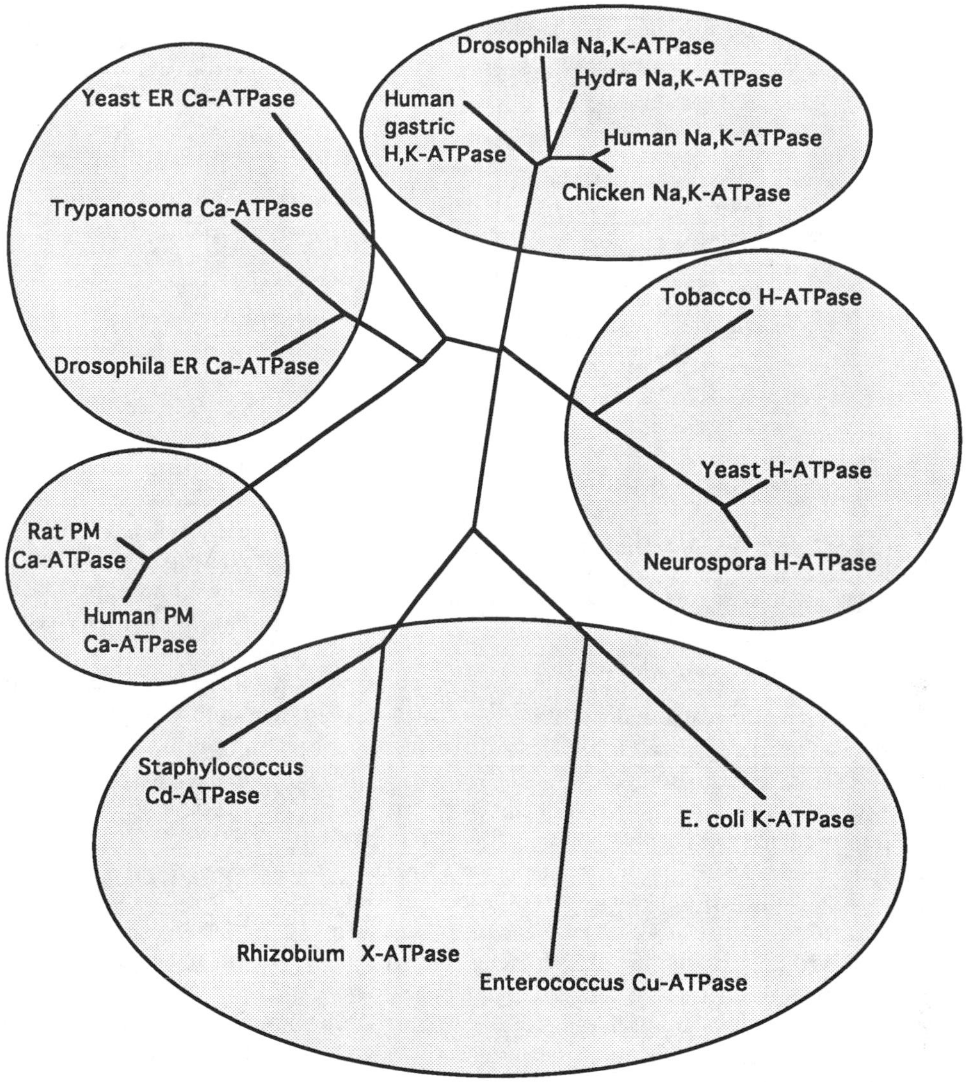

Figure 12.2. Unrooted phylogenetic tree of P-type ATPases. This tree depicts the phylogenetic relationshps determined by maximum parsimony calculations based on the aligned amino acid sequences that precede the large cytosolic domain of 44 P-type ATPases. The lengths of the lines connecting the ATPases are proportional to predicted evolutionary distances. The ion selectivity of the bacterial ATPase, *Rhizobium* X-ATPase, is unknown, Adapted from Fagan and Saier.[6]

dently of the origin of the three well-known vertebrate α-subunit isoforms.

The biological function of the isoforms is not clear. There are two major ideas.[13] First, the isoforms of any particular P-type ATPase may be functionally equivalent, and multiple genes exist to facilitate the regulation of developmental and cell-type-specific expression by providing a set of specific promoters. Perhaps this is the case for the α subunits of the Na^+,K^+-ATPase. It has been found that the α2 isoform, which is the principal form expressed in skeletal muscle, fails to upregulate under conditions of low potassium, while the isoforms expressed in most other tissues are upregulated.[14–16] In times of potassium deficiency, this difference in promoter function should facilitate the loss of potassium stores from skeletal muscle, making this potassium available to other tissues, in keeping with the function of skeletal muscle as a reservoir for various metabolites, including amino acids and glucose. Second, the isoforms may be functionally distinct, each isoform evolved to function optimally in specific cell types or developmental stages. In this regard, it has been suggested that the Na^+,K^+-ATPase α-subunit isoforms may differ in their affinities for sodium, so that different isoformic combinations may mediate sodium transport under different physiological conditions.[17,18] In addition, the isoforms may differ in their responses to regulatory signals such as phosphorylation.[19–21]

The two hypotheses about the roles of isoforms of P-type ATPase are difficult to test rigorously. The ultimate test may involve transgenic mice in which the transcribed regions of the ATPase genes are exchanged. If the proteins are functionally equivalent, then such exchanges should have no effect on survival or function.

12.3. SUBCELLULAR DISTRIBUTIONS OF P-TYPE ATPASE

Each P-type ATPase resides in a defined subset of cellular membranes, as diagrammed in Figure 12.3. The SERCA-type Ca^{2+}-ATPases take their name from their occurrence in

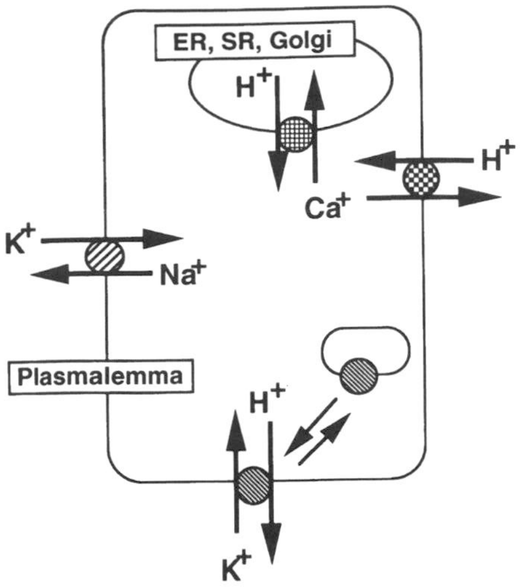

Figure 12.3. Subcellular distribution of P-type ATPases in mammalian cells. The figure schematically depicts the locations of vertebrate P-type ATPases in their usual cellular membranes and indicates the directions of ion transport.

sarcoplasmic reticulum and endoplasmic reticulum.[22] There is also a SERCA-like form of Ca^{2+}-ATPases in the Golgi apparatus of yeast.[23] A homologue of this Golgi-type Ca^{2+}-ATPase may have been discovered in mammals,[24] but its organellar location has not been determined. Since some of the intracellular Ca^{2+}-ATPases may not reside in the SR or ER, we will refer to them as "SERCA-type Ca^{2+}-ATPases" instead of simply "SERCAs." The Na^+,K^+-ATPase and the plasma membrane Ca^{2+}-ATPase reside in the plasma membrane. The gastric H^+,K^+-ATPase functions as a plasma membrane protein, but its activity is regulated by shuttling of the H^+,K^+-ATPase between the plasma membrane and subsurface vesicles, described later in this chapter. The Cu^{2+} transport ATPases are presumably plasma membrane proteins, but their cellular locations have not been rigorously determined.

In polarized epithelia, the Na^+,K^+-ATPase is restricted to a part of the plasma membrane. In most such epithelia the Na^+,K^+-ATPase is a basolateral membrane protein. However, in the choroid plexus and in the pigmented epithelium of the retina, the Na^+,K^+-ATPase is an apical membrane protein.

How cells regulate the distribution of their membrane proteins is a major focus of current studies in cell biology, and, indeed, some of the current research focuses on the mechanisms through which different P-type ATPases become and remain localized. A lively controversy has arisen in the Na^+,K^+-ATPase field. One body of evidence suggests that the basolateral distribution of the Na^+,K^+-ATPase results from interactions between ATPase molecules and ankyrin, a component of the membrane cytoskeleton.[25] This interaction is thought to stabilize the Na^+,K^+-ATPase, resulting in a very long lifetime. Molecules sent to the apical membrane after biosynthesis, according to this evidence, are rapidly endocytosed and degraded. Other evidence suggests that polarized epithelial cells have mechanisms that target newly synthesized Na^+,K^+-ATPases molecules directly to the correct, generally basolateral, membrane.[26] As with many controversies in biology, it may well turn out that both mechanisms, selective targeting and selective stabilization, operate in parallel to robustly regulate Na^+, K^+-ATPase distribution.

12.4. CA^{2+} TRANSPORT ATPASES

Ca^{2+} plays an essential role in regulation of cytosolic enzymes and functions, serving as a second messenger in a variety of signal transduction mechanisms.[27] Cytosolic Ca^{2+} is normally maintained below the micromolar level by intervention of Ca^{2+} transport ATPase, either bound to the plasma membrane and operating an outward-direct Ca^{2+} pump, or bound to intracellular membranes and sustaining active transport of Ca^{2+} from the cytosol to intracellular organelles ("Ca^{2+} stores"), as diagrammed in Figure 12.3. In turn, activation of various signal transduction mechanisms increases cytosolic Ca^{2+} by means of passive diffusion of Ca^{2+} from extracellular media or intracellular stores through specific channels into the cytosol. Following inactivation of the transduction signals, the transport ATPases are then involved in lowering again the cytosolic Ca^{2+} concentration. Na^+–Ca^{2+} exchange proteins also contribute to this function, whereby free energy derived from a preexisting Na^+ gradient is utilized for outward Ca^{2+} transport.

12.4.1. Intracellular Ca^{2+} Transport ATPases

Intracellular Ca^{2+} stores play an important role in regulation of the cytosolic Ca^{2+} concentration by releasing stored Ca^{2+} and rendering possible its function of second messenger.[28] In turn, the intracellular stores are filled by active transport of cytosolic Ca^{2+} through the intervention of membrane-bound ATPases. The first of these enzymes to be discovered was the Ca^{2+} transport ATPase of sarcoplasmic reticulum, which was obtained as a microsomal fraction of muscle homogenates producing relaxation of myofibrils by sequestration of Ca^{2+}.[29,30] Subsequently, cloning of cDNAs encoding homologous ATPases demonstrated the presence of several isoforms not only in SR, but also in the endoplasmic reticulum of a wide variety of cells.[22,31–35] The intracellular Ca^{2+}-ATPase isoforms display quite similar functional behavior, although some differences have been noted in their turnover number and in their Ca^{2+} concentration dependence.[36] We will discuss here these enzymes in terms of protein structure and topology, catalytic and transport functions, and mechanisms of regulation and inhibition.

12.4.2. Protein Structure

Much of the information available on structure and function of intracellular Ca^{2+} transport ATPases derives from studies of the sarcoplasmic reticulum (SR) ATPase. This is due to the abundant and convenient source of protein provided by vesicular fragments of SR, which can be obtained by differential centrifugation of muscle homogenates. The main protein component of the isolated SR vesicles is the Ca^{2+} transport ATPase, which is densely spaced within the plane of the membrane. The isolated vesicles retain a homogeneous membrane orientation, with the cytosolic surface of the SR membrane facing the outer medium.

The main protein component of the isolated SR vesicles, the Ca^{2+} transport ATPase, can be visualized by electron microscopy of negatively stained SR vesicles. The EM micrographs show outer granules corresponding to single ATPase monomers.[37,38] The outer granules are connected by stalks to corresponding intramembranous particles which can be demonstrated in freeze-fracture preparations.[39] X-ray and electron diffraction patterns[40–43] have confirmed the asymmetric ATPase disposition within the membrane and yielded an image that includes a large lobe projecting from the cytosolic side of the membrane, a stalk connecting the outer lobe to a region penetrating the membrane bilayer, and only a much smaller mass emerging from the luminal side of the membrane (Figure 12.4).

The Ca^{2+}-ATPase is composed of a single polypeptide chain of about 1000 amino acids whose sequence was determined by a combination of amino acid[44] and cDNA sequencing.[33,34] A hydropathy plot of the amino acid sequence reveals ten hydrophobic regions which are likely to be transmembrane helical segments,[34] except in one splice form of the SERCA-2 isoform expressed primarily in brain,[45] where there is an extra membrane span near the C-terminus. A model predicting that the ten helices cross the membrane as five hairpins is supported by the reactivity of relevant epitopes to monoclonal antibodies.[46–50] This model (Figure 12.5) leaves both N- and C-terminal residues on the cytosolic side of the membrane, as well as two large loops that fold to form the large head and stalk of the functional protein (Figure 12.4). The luminal loops of the five hairpins are predicted to be very short, consistent with the protein shape revealed by physical methods.

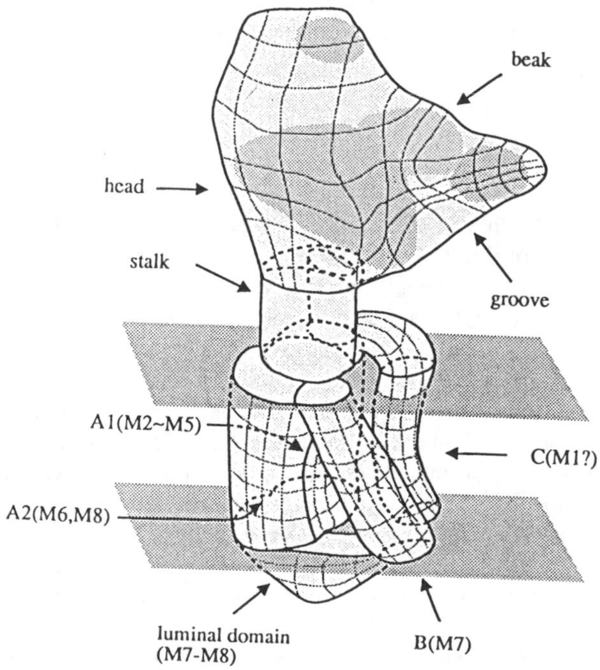

Figure 12.4 Overall three-dimensional structure of the Ca^{2+}-ATPase. A large segment (over 50% of the total protein) protrudes on the cytosolic side of the membrane with a narrow stalk, 16 Å long and 20 Å wide, connected to a large oblong head, 65 × 40 × 50 Å. Clustered helical segments cross the membrane. Very little of the protein extends out of the luminal side of the membrane. From Toyoshima *et al.*[140]

12.4.3. Catalytic and Transport Cycle

The cation transport ATPases share common features with respect to the mechanism of ATP utilization and cation transport, including formation of a phosphorylated intermediate by transfer of ATP terminal phosphate to an aspartyl residue at the catalytic site. For this reason they are referred to as "P-type ATPases." Here we will use the SR Ca^{2+}-ATPase as an example, since the natural abundance of this enzyme has permitted the most convenient experimentation.

If ATP is added to a suspension of SR vesicles in the presence of Ca^{2+} and Mg^{2+}, medium Ca^{2+} is taken up and transported into the lumen of the vesicles while ATP is utilized by the ATPase. The use of native SR vesicles presents a distinctive advantage in equilibrium and pre-steady-state experiments designed to characterize the partial reactions of a single transport cycle, since the high density of ATPase units yields measurable concentrations of enzyme intermediates.[51] On the other hand, observation of steady-state transport activity is favored by conditions permitting the luminal Ca^{2+} concentration to remain low for a longer time. This can be accomplished by sequestration of luminal Ca^{2+} with anions such as oxalate,[52] or simply by reconstituting proteoliposomal vesicles with an excess of exogenous phospholipids thereby increasing up to 50- to 100-fold the luminal volume per ATPase molecule.[53]

It was found by the complementary use of these experimental systems that the ATPase is activated by high-affinity and cooperative binding of 2 moles of medium Ca^{2+} per mole of enzyme, in exchange for 2 H^+. When ATP is added to enzyme preincubated with Ca^{2+}, a phosphorylated intermediate is rapidly formed by transfer of the ATP terminal phosphate to an aspartyl residue (Asp351) of the enzyme.[54,55] In parallel with enzyme phosphorylation, and before hydrolytic cleavage of P_i, the bound Ca^{2+} is internalized by the enzyme and then released into the lumen of the vesicles against a concentration gradient (Figure 12.6[56,57]). The step intervening between internalization and release contributes significantly to the rate limitation, and corresponds to an occluded state of the bound Ca^{2+}.[58,59] The catalytic and transport cycle is finally completed by hydrolytic cleavage of the phosphorylated enzyme intermediate. The cycle is reversible, as demonstrated by ATP synthesis associated with efflux of luminal Ca^{2+} from the vesicles,[60] and by the occurrence of ATP–P_i exchange.[61,62]

Under optimal conditions, utilization of 1 mole of ATP, including intermediate enzyme phosphorylation, is accompanied by transport of 2 moles of medium Ca^{2+} into the lumen of

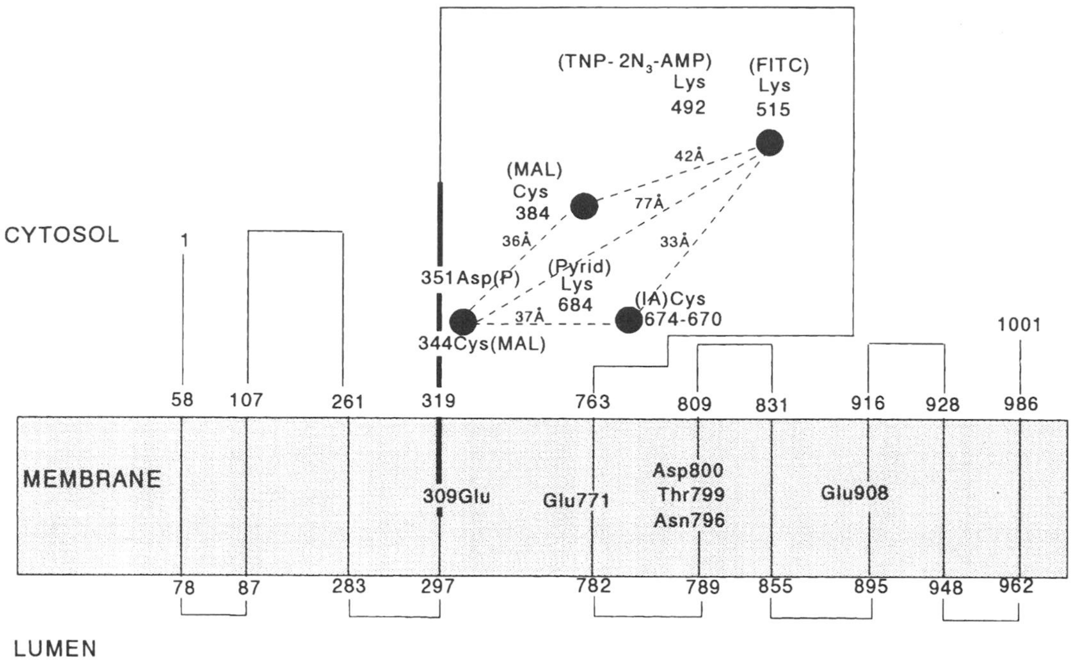

Figure 12.5. Topology of the Ca^{2+}-ATPase in the sarcoplasmic reticulum membrnae. The diagram is based on the analysis of MacLennan *et al.*[34] and derivatization of various residues as reviewed by Bigelow and Inesi.[72] The cytosolic head piece is composed of two major segments. The larger, residues 329 to 743, contains the catalytic site Asp351, which is phosphorylated during catalytic turnover, and the ATP binding site. The smaller cytosolic loop, residues 107 to 261, is likely to participate also in the folded structure of the cytosolic head. Ten helical segments cross the membrane, and the six residues essential for Ca^{2+} binding are shown within the membrane in helices 4, 5, 6, and 8. Specific points of reference are shown in the extramembranous region, including cysteine residues that are reactive to fluorescent maleimide (MAL) or iodoacetamine (IA) derivatives, as well as the site of FITC derivatization. The points of reference corresponding to the related chromophores are shown as black dots, in a trigonal arrangement as suggested by spectroscopic studies. Lys492 (derivatized by TNP-N_3-AMP), Lys684 (derivatized by ATP pyridoxal), and Asp351 (undergoing phosphorylation on utilization of ATP) are shown near the appropriate points of reference. These reference points denote the ATP binding and phosphorylation domain. The thick line denotes a consensus sequence linking the phosphorylation and cation binding domains in several cation transport ATPases. Note that distance between phosphorylation site and Ca^{2+} binding site is approximately 50 Å.

the vesicles and ejection of 2 moles of luminal H^+ into the medium. The simplest scheme for the corresponding reaction sequence is shown in Figure 12.7.

It is noteworthy that the ATPase reaction cycle is likely to include isomeric transitions as consequences of ligand binding and catalytic events. In fact, transport against a concentration gradient requires that the Ca^{2+} sites undergo a reduction in affinity and a change in vectorial orientation on utilization of ATP. This change has been explained[63] by assuming that the enzyme resides normally in two interconverting states. In the first state (E1) the Ca^{2+} sites are open to the cytosol and bind medium (i.e., cytosolic) Ca^{2+} with high affinity. In the second state (E2) the Ca^{2+} sites are converted to low affinity and luminal exposure, thereby permitting dissociation of bound Ca^{2+} in the presence of a luminal concentration higher than the cytosolic concentration. Enzyme phosphorylation by ATP would then shift the equilibrium in favor of E2 states, so that the catalytic cycle shown in Figure 12.7 would include a transition from E-P1 to E-P2 on phosphorylation, and a transition from E2 to E1 on hydrolytic cleavage of P_i. Additionall isomeric transitions, such as those involved in the cooperative binding of Ca^{2+}[64] and in the kinetic effect of ATP as a ligand,[65] have been described. It is then possible that the catalytic and transport cycle includes a series of conformational adjustments of the enzyme, in parallel with ligand binding and chemical reactions.

It should also be pointed out that diagrams portraying a sequence of rigidly coupled reactions may be misleading. For instance, the scheme of Figure 12.7 implies that Ca^{2+} transport and ATP utilization are unavoidably coupled with a stoichiometric ratio of two. In fact, a ratio of nearly two is only observed when the concentration of Ca^{2+} transported into the lumen of the vesicles is very low. The ratio decreases progressively as the luminal Ca^{2+} increases, indicating that alternate pathways of phosphoenzyme cleavage, without net release of Ca^{2+} into the lumen, are possible.[66] In principle, a protein in equilibrium with a series of ligands will yield all possible bound species and pertinent isomerization states, according to the related equilibrium constants and concentration of ligands.[67] If the protein is an enzyme, flux through branched pathways will occur depending on the concentration of intermediate species and the related rate constants for their transformations. This explains the variable stoichiometric efficiencies observed even in the absence of passive leak.

Although the stoichiometry and cation specificity may be different in other ATPases, it is apparent that the main features of catalysis and transport are similar to those described for intracellular Ca^{2+}-ATPase.

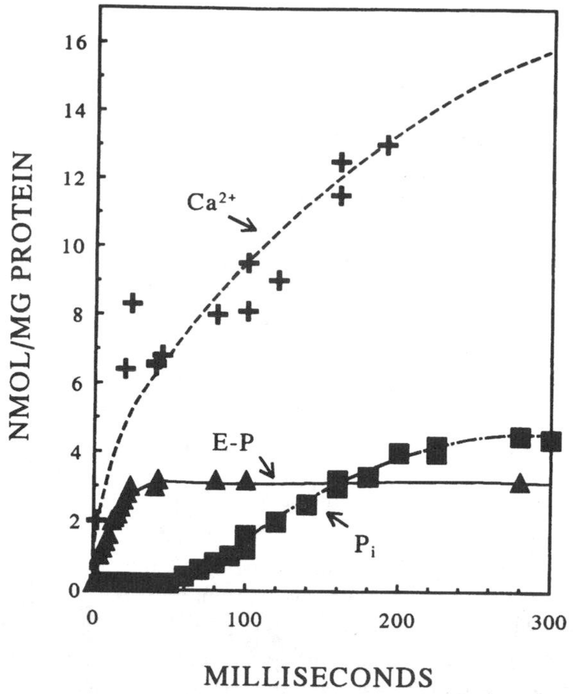

Figure 12.6. Pre-steady-state events in the catalytic and transport cycle of the Ca^{2+}-ATPase. Formation of phosphorylated intermediate, internalization of bound Ca^{2+}, and hydrolytic cleavage of the phosphorylated intermediate (i.e., P_i production) following addition of ATP to sarcoplasmic reticulum vesicles preincubated with Ca^{2+}. Time resolution was obtained with rapid kinetic methods. From Inesi.[184]

12.4.4. The Energetics of the Catalytic and Transport Cycle

If we consider the partial reactions comprising a productive cycle of the Ca^{2+}-ATPase, we can determine their kinetic and equilibrium constants by appropriate experimental measurements and analysis.[68] A list of the partial reactions, and their equilibrium constants measured under conditions permitting temperature (25°C) and pH (7.0) to remain constant, is as follows:

1. $H_2E + 2Ca^{2+}_{out} \longleftrightarrow E{\bullet}Ca_2 + 2H^+_{out}$ $\quad (3 \times 10^{12}\ M^{-2})$
2. $E{\bullet}Ca_2 + ATP \longleftrightarrow ATP{\bullet}E{\bullet}Ca_2$ $\quad (1 \times 10^{5}\ M^{-1})$
3. $ATP{\bullet}E{\bullet}Ca_2 \longleftrightarrow ADP{\bullet}E{-}P{\bullet}Ca_2$ $\quad (0.3)$
4. $ADP{\bullet}E{-}P{\bullet}Ca_2 \longleftrightarrow E{-}P{\bullet}Ca_2 + ADP$ $\quad (7 \times 10^{-4}\ M)$
5. $E{-}P{\bullet}Ca_2 + 2H^+_{in} \longleftrightarrow H_2E{-}P + 2Ca^{2+}_{in}$ $\quad (3 \times 10^{-6}\ M^2)$
6. $E{-}P \longleftrightarrow E{\bullet}P_i$ $\quad (1)$
7. $E{\bullet}P_i \longleftrightarrow E + P_i$ $\quad (1 \times 10^{-2}\ M)$

In the reaction sequence given above, the initial high-affinity acquisition (reaction 1) of two Ca^{2+} from the outer medium activates the enzyme, permitting utilization of ATP and formation of a phosphorylated intermediate (reactions 2–4). In turn, enzyme phosphorylation destabilizes and changes the vectorial orientation of the bound Ca^{2+}, thereby increasing the probability of its dissociation into the lumen of the vesicles (reaction 5). Note that the equilibrium constant for the enzyme phosphorylation by ATP is nearly 1, indicating that the free energy of ATP is conserved by the enzyme, and utilized to change the Ca^{2+} binding characteristics. Finally, the phosphoenzyme undergoes hydrolytic cleavage and releases P_i (reactions 6 and 7) before entering another cycle.

All isomeric transitions occurring within the cycle are implicitly coupled with the chemical reactions subjected to experimental measurement, and their influence is reflected by the equilibrium constants given above. In fact, the standard free energies ($-RT\ln K$) of the partial reactions add up to the standard free energy of ATP hydrolysis (γ-phosphate), as expected. The chemical potential of ATP does not manifest itself in the phosphoryl transfer or hydrolytic cleavage reactions (K_4 and K_6 are near 1), but rather in the drastic reduction of the enzyme affinity for Ca^{2+}. We can then write that, under standard conditions,

$$\Delta G = RT\ \ln\ (K_a^{CaEP}/K_a^{CaE})$$

per n number of ions transported per cycle. K_a^{CaE} (reaction 1) and K_a^{CaEP} (reversal of reaction 5) are the association constants of the enzyme for Ca^{2+} in the ground state and following activation by ATP. Correction of the standard free energies for the actual concentrations of substrates and products (i.e., ATP, ADP, P_i, Ca^{2+}_{out}, Ca^{2+}_{in}) yields net free energy changes permitting net forward or reverse fluxes as observed under any particular experimental condition.

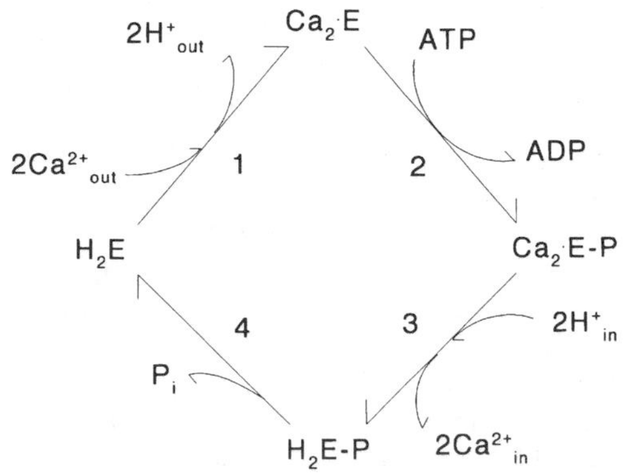

Figure 12.7. Simple reaction cycle for the Ca^{2+}-ATPase. The diagram shows Ca^{2+} binding required for enzyme activation, ATP utilization, inward displacement of bound Ca^{2+}, and hydrolytic cleavage of the phosphoenzyme intermediate. In reality, the catalytic and transport cycle is likely to include iosmeric transitions and branched pathways as explained in the text.

12.4.5. Topology of Functional Domains within the ATPase Structure

Identification of the aspartyl residue (Asp351) undergoing phosphorylation was the first lead to localization of the phosphorylation site.[69,70] Furthermore, evidence derived from chemical derivatization and site-directed mutagenesis indicates that the entire catalytic domain, comprising phosphorylation and ATP binding sites, resides within the cytosolic globular region of the ATPase. Analogies with nucleotide handling kinases, whose crystal structures are known, suggest that three-dimensional folding of the ATPase cytosolic region includes two domains connected by a hinge that permits approximation of bound nucleotide to the phosphoryl transfer site on enzyme activation.[71]

Approximate distances separating fluorescent labels placed as spectroscopic reference points within the folded structure of the cytosolic region were obtained from resonance energy transfer. Consideration of these distances suggests a trigonal arrangement for approximation of various peptide segments within the cytosolic region, and their interaction with adenine, ribose, and phosphate moieties of bound ATP. Diffraction studies[42] and triangulation of spectroscopic distances[72] indicate that the cytosolic region containing the catalytic domain is an elongated globular structure of approximate dimensions 40 × 50 × 65 Å, connected to the membrane-bound region by a stalk 16 Å long and 28 Å wide. Estimate of the distance between a Lys515 label (likely at the upper end of the cytosolic region) and the membrane surface is approximately 60 Å. Most importantly, the distance between the phosphorylation site (Asp351) and the membrane surface is approximately 50 Å (Figure 12.5).

While the phosphorylation and ATP binding domains reside within the cytosolic region of the ATPase, the Ca^{2+} binding domain is located within the membrane-bound region of the enzyme, as indicated by experiments involving site-directed mutagenesis,[73] chemical derivatization,[74] and chimeric recombination of Ca- and Na^+,K^+-ATPases.[75] Six amino acyl residues (Glu309, Glu771, Asn796, Thr799, Asp800, and Glu908) within membrane-spanning segments 4, 5, 6, and 8 of the ATPase polypeptide chain have been identified as residues participating in Ca^{2+} complexation.[73] Molecular modeling indicates that the four helical segments 4, 5, 6, and 8 can be clustered within the membrane to form a narrow channel, and rotated to optimize participation of at least five of the six residues mentioned above in direct complexation of two Ca^{2+}.

12.4.6. Long-Range Linkage between Phosphorylation and Cation Binding Sites

A fundamental question remains the mechanism by which Ca^{2+} binding is coupled to catalytic site activation and, conversely, enzyme phosphorylation by ATP is coupled to translocation of bound Ca^{2+}. The first question to be considered is whether the transmission of forces between chemical groups undergoing catalytic transfer and cations undergoing vectorial translocation is direct and short range, or indirect and long range. However, as explained above, the experimental evidence indicates that the phosphorylation site in the cytosolic region of the enzyme and the Ca^{2+} binding site at the center of the membrane bilayer are separated by a total distance of approximately 50 Å (Figure 12.7), thereby requiring a long-range intramolecular linkage.[68]

It is of interest, in this regard, that the sequence intervening between residues 309 and 351 is highly conserved not only in Ca^{2+}-ATPase isoforms, but also in other cation transport ATPases. This is by far the most extensive region of amino acid sequence homology between the Ca^{2+}-ATPase and the other cation ATPases. This homologous segment is of particular interest as it begins within transmembrane helix 4 which contains the residue Glu309 critical for Ca^{2+} binding, and extends in the C-terminal direction to Asp351 which is the residue undergoing phosphorylation. While the entire protein may be involved in the mechanism of energy transduction, it is likely that the homologous segment intervening between phosphorylation and Ca^{2+} binding sites plays a specific role in long-range functional linkage, serving as a transmitter of structural perturbations from one site to the other. In fact, recent work shows that the Ca^{2+}-ATPase function is highly sensitive to single mutations within this segment.[76]

The conformational changes implied by the long-range linkage of phosphorylation site and Ca^{2+} site are likely to involve tertiary structure and/or segmental reorientation, with minimal alterations of secondary structure.[77,78] A key structural feature of the Ca^{2+} binding site is the presence of residues originating from four different helices which are separated by long segments of primary sequence, but approximated in the three-dimensional protein folding. Cooperative binding of two Ca^{2+} then produces strong stabilization of the approximated residues. It also renders possible enzyme phosphorylation by ATP, indicating that the stabilizing effect is transmitted to the catalytic site. Consistent with the occurrence of long-range conformational effects, Ca^{2+} binding is followed by fluorescence changes involving tryptophanyl residues near the membrane-bound region and extended to a Lys515 label at the far end of the cytosolic region (reviewed by Bigelow and Inesi[72]). Conversely, Ca^{2+} dissociation destabilizes the protein and renders possible the phosphorylation reaction with P_i,[79] indicating that even the destabilizing effect is transmitted to the catalytic site. The extent of this destabilizing effect can be appreciated considering that, in the absence of P_i, Ca^{2+} dissociation facilitates denaturation of the ATPase.[80]

In the presence of Ca^{2+}, destabilization of the binding structure requires input of energy by utilization of ATP at the phosphorylation site, and transmission of structural perturbations by the same long-range linkage involved in the effect of Ca^{2+} binding on catalytic activation. Rotation of channel helices and displacement of residues participating in Ca^{2+} binding are expected to destabilize the bound cation and change its vectorial exposure. Displacement of the acidic side chains participating in complexation is likely to be responsible for the observed[81] rise in their pK, facilitating the exchange of Ca^{2+} for H^+. Changes in polarization energies favoring ion charge interaction with the channel may contribute to vectorial displacement of the bound cation.

It is of interest that, although no large effects of ligand binding or phosphorylation can be demonstrated on the secondary structure of the enzyme, global effects involving the entire protein have been observed. For instance, changes in the spatial relationship of ATPase molecules with each other can be demonstrated by spectroscopic methods following formation of the phosphorylated enzyme intermediate. Furthermore, a tendency of the ATPase to form ordered dimer ribbons within the plane of the membrane depends on Ca^{2+} dissociation and interaction with P_i or decavanadate.[82,83] Finally, changes in diffraction patterns following ATP utilization have been attributed to effects on the shape of the enzyme and its relationship with the membrane bilayer.[84]

12.4.7. Regulation of Intracellular Ca^{2+} Transport

For the intracellular Ca^{2+} transport ATPases, it was originally found in experiments with SR isolated from cardiac muscle that transport activity is stimulated by kinase dependent phosphorylation of phospholamban.[85–88] Phospholamban is a homopentamer protein which is formed by five 6-kDa subunits, which are thought to have their hydrophobic carboxyl-terminus inserted as helical segments in the lipid bilayer of the SR membrane.[89–91] Phospholamban is expressed predominantly in cardiac and smooth muscle, and it copurifies with the cardiac Ca^{2+}-ATPase.[92] The unphosphorylated form of phospholamban suppresses the activity of the Ca^{2+}-ATPase, while phosphorylation causes phospholamban to dissociate from the Ca^{2+}-ATPase, increasing the rate of Ca^{2+} transport two- to threefold. The physiological result of disinhibition of the Ca^{2+}-ATPase is that calcium ions are sequestered more rapidly in the SR, allowing faster relaxation of muscle. Decreasing the relaxation phase of the cardiac muscle contraction cycle makes possible more rapid heartbeat. One of the kinases that phosphorylates phospholamban is protein kinase A, which is activated in response to adrenalin, a hormone that results in accelerated heartbeat. Thus, the phosphorylation of phospholamban provides a prominent pathway for the cardiac inotropic effect of adrenergic stimulation.[93] The importance of phospholamban in regulation of cardiac contractile activity has recently been shown by gene knock-out experiments.[94]

The ATPase regions undergoing functionally relevant interaction with phospholamban have been identified with segments between amino acids 336 and 412 within the catalytic domain, and between amino acids 467 and 762 within the nucleotide binding region.[95,96] Of the various Ca^{2+}-ATPase isoforms, SERCA-1 and SERCA-2 are sensitive to phospholamban, while SERCA-3 is not, because of amino acid sequence divergence within the interacting regions.

12.4.8. Thapsigargin, a Specific Inhibitor of Intracellular Ca^{2+}-ATPases

A specific inhibitor of intracellular Ca^{2+}-ATPases has been recognized only recently. Thapsigargin is a sesquiterpene lactone isolated from root extracts of *Thapsia garganica*[97] which was found to inhibit the endoplasmic[98] and sarcoplasmic[99] reticulum Ca^{2+} pumps. It was then established that thapsigargin is a highly specific inhibitor of various intracellular Ca^{2+}-ATPases expressed in transfected COS1 cells, including SERCA-1, SERCA-2A, and SERCA-2B.[32,100,101] The inhibitory effect is produced by amounts of thapsigargin stoichiometrically equivalent to the ATPase present in the reaction mixture, with a dissociation constant lower than nanomolar. The inhibition involves Ca^{2+} transport as well as Ca^{2+}-dependent ATPase activity.

Studies of the mechanism of inhibition reveal that both the Ca^{2+} binding and phosphorylation functions are affected by interaction of the enzyme with 1 mole of thapsigargin. Ca^{2+} binding (reaction 1 in the sequence given above) measured under equilibrium conditions in the absence of ATP is found to be inhibited by thapsigargin. On the other hand, enzyme phosphorylation with P_i (reverse of reactions 6 and 7 in the sequence given above) measured under equilibrium conditions in the absence of Ca^{2+} is also found to be inhibited by thapsigargin.[102] These experiments demonstrate that two partial reactions occurring in distant domains within the enzyme molecule are inhibited by thapsigargin under conditions permitting their occurrence independent of each other (i.e., not requiring enzyme cycling). It is then apparent that thapsigargin is a global inhibitor inasmuch as its effect is transmitted over a long distance through the enzyme structure.[103] In fact, this long-range effect determines, under appropriate conditions, whether the ATPase does or does not form bidimensional crystalline arrays within the plane of the membrane.[104]

12.4.9. Disorders Involving Intracellular Ca^{2+}-ATPases

From the physiological standpoint, Ca^{2+} transport ATPases are required to lower the cytosolic Ca^{2+} concentration following Ca^{2+} transient rises originating from various membrane signals. For instance, they are involved in lowering the sarcoplasmic Ca^{2+} concentration in order to allow relaxation of contractile tension in muscle fibers. Alternations in expression of SR Ca^{2+}-ATPase and phospholamban have been noted in animal models of cardiac hypertrophy and dilatation, as well as in human dilated cardiomyopathy.[105–109] In fact, the failing heart exhibits abnormal Ca^{2+} handling properties including a prolongation of the Ca^{2+} transient and a 50% reduction of Ca^{2+} uptake into SR fractions.[110] Molecular studies have shown that mRNA encoding the SERCA-2 are dramatically decreased in human failing heart of patients with hypertension, dilated cardiomyopathy, and ischemia.[108,111] The role of SERCA ATPase isoforms in cardiovascular physiology and pathology was recently reviewed by Lompre *et al.*[112]

With regard to skeletal muscle, it was reported that a myopathy known as Brody's disease is, in fact, caused by a deficiency of Ca^{2+}-ATPase in SR,[113] but the genetic basis of the disorder has not been defined.

12.4.10. The Plasmalemmal Ca^{2+} Transport ATPases

It was first noted by Schatzmann[114] that Ca^{2+} is extruded from erythrocytes through an active transport mechanism which is modulated by calmodulin.[115,116] The ATPase responsible for Ca^{2+} extrusion is a plasma membrane intrinsic protein which, although present in relatively small amounts, can be purified by calmodulin affinity chromatography.[117] Functional studies have been carried out with erythrocyte plasma membrane ghosts, and with reconstituted proteoliposomal vesicles containing purified ATPase. In both preparations the orientation of the enzyme is inverted, and Ca^{2+} is actively transported into the lumen of the vesicles. ATP utilization by the plasma membrane ATPase includes Ca^{2+}- and Mg^{2+}-dependent formation of a phosphorylated intermediate,[118–120] and results in electrogenic transport of 1 Ca^{2+} per enzyme cycle in exchange for 1 H^+.[81,121]

Cloning of genes and cDNA encoding the plasma membrane Ca^{2+}-ATPase in rat and human tissues has revealed various enzyme isoforms and their sequences, deriving from human or rat libraries.[122–127] Additional heterogeneity is provided by RNA splicing of single gene products.[128] The molecular mass of the plasma membrane Ca^{2+}-ATPase is approximately 130 kDa, somewhat larger than is typical for catalytic subunits of P-type ATPases, mostly the result of segments involved in calmodulin binding at the carboxyl-terminal.

Models for the topological distribution of the plasma membrane Ca^{2+}-ATPase include ten transmembrane helical segments, minimal protein mass on the extracellular side of the membrane, and a large portion protruding from the cytosolic side of the membrane. This is the same topology now predicted for most of the catalytic subunits of P-type ATPases. It was originally reported by Gopinath and Vincenzi[115] and Jarrett and Penniston[129] that the plasmalemmal Ca^{2+} pump is activated by calmodulin. The calmodulin effect is manifested by an increase of the Ca^{2+} concentration required for activation of the pump[130,131] and by higher velocities of transport.[121,132] Calmodulin binds stoichiometrically to the plasmalemmal ATPase, interacting with the carboxyl-terminal segment of the enzyme,[133] which is exposed to the cytosol. In turn, the calmodulin binding segment of the ATPase interacts with other segments which are involved in catalysis.[134]

It is apparent that the kinetics of catalysis are interfered with by interaction of the calmodulin binding segment with the catalytic domain of the plasmalemmal ATPase. This interference can be relieved by calmodulin binding to the carboxyl-terminal segment of the plasmalemmal ATPase in the presence of sufficient Ca^{2+} to activate calmodulin (Figure 12.8).

12.5. NA^+,K^+-ATPASES, THE SODIUM PUMPS

The Na^+,K^+-ATPase, or sodium pump, plays a fundamental role in animal cell physiology. It is present in every animal cell and has been estimated to consume up to 70% of cellular ATP. By exchanging internal sodium ions for external potassium ions, the sodium pump participates in maintenance of an internal ionic milieu appropriate for intracellular processes, and, at the same time, compensates for passive permeability characteristics of the cell membrane that otherwise would lead to osmotic imbalance and cell lysis. The ion gradients established by its action serve as energy sources in the coupled transport of other ions, sugars, and amino acids. The transmembrane sodium and potassium gradients built by the action of the sodium pump are also essential for cell functions that involve transmembrane voltage changes, such as the classical excitable cell functions of nerve and muscle and a great variety of "nonclassical' cellular functions, such as regulation of insulin secretion by pancreatic islet cells.

One curious aspect of the Na^+,K^+-ATPase is that the β-subunit isoform, β2, which is expressed particularly in the nervous system,[135] has been shown to be identical to a cell adhesion molecule called AMOG (adhesion molecule on glia).[136] AMOG had previously been characterized with monoclonal antibodies, shown to be active in *in vitro* cell adhesion assays between certain CNS neurons and glia, and correlated with cell interactions in the developing brain of rats. Recently an AMOG gene knockout mouse was produced.[137] The mice displayed grave problems in CNS cellular physiology consistent with inadequate Na^+,K^+-ATPase function in the brain, and the mice died a few weeks after birth. While this gene

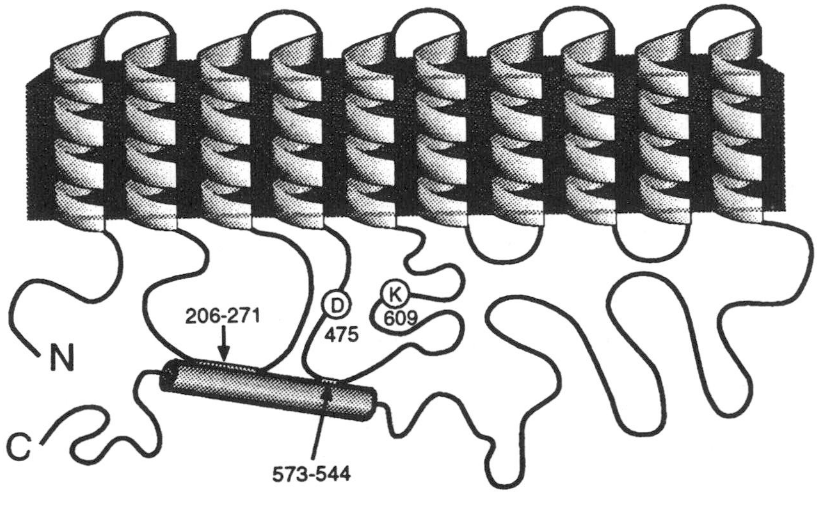

Figure 12.8. Topology of the Ca^{2+}-ATPase in the plasmalemmal membrane. In analogy with the sarcoplasmic reticulum ATPase, the plasmalemmal ATPase has ten helical segments crossing the membrane, and a large cytosolic head piece which is composed of two major segments. Regulation by calmodulin occurs through binding of calmodulin with the ATPase segment symbolized as a shaded cylinder. In turn, this cylinder interacts with two distinct "receptor" sites within the two peptide segments composing the head piece. D475 is the aspartate undergoing phosphorylation on utilization of ATP. K609 is likely to be in close proximity of the ATP binding site, as revealed by selective labeling. From Falchetto *et al.*[134]

knockout experiment showed the essential role of the Na^+,K^+-ATPase β2 subunit in mice, it did not reveal a unique, necessary role for AMOG in prenatal or neonatal brain development. On the other hand, the experiment did not exclude the possibility that AMOG does function in cellular interactions in brain development but that its function is redundant to the function of other cell adhesion molecules.

Its central importance in physiology, its relative abundance in a variety of cell types, and its simple quaternary structure have made the sodium pump a popular focus for studies of membrane protein structure, function, and regulation. The potential involvement of the sodium pump in major human disorders, such as hypertension, gives added impetus to such studies.

12.5.1. Structure

An α–β heterodimer is the minimum functional unit for Na^+,K^+-ATPase activity (see review by Mercer[138]). A possible third subunit, γ, M_r about 6 kDa, has been found in a few tissues,[139] but functional studies on pumps expressed *in vitro* have failed to show any effect of this subunit. The locations of the ion binding sites are not known, but are thought to involve primarily the membrane-spanning regions. α-Subunit topology has been a serious issue that appears to be nearing resolution in favor of a ten-membrane span structure (see Figure 12.1), although the precise locations of some membrane spans remain uncertain.[11] Low-resolution images of the sodium pump and much better images of the Ca^{2+}-ATPase[140] (Figure 12.4), obtained from two-dimensional crystals, show the overall shape of the molecules to be similar. The shape of the β subunit of the Na^+,K^+-ATPase, which should show up as a major difference between Na^+,K^+- and Ca^{2+}-ATPase structure, remains poorly defined.

12.5.2. Mechanism of Cation Transport

The sodium pump is the prototypic cation transport ATPase and has been the object of intensive study as an enzyme. A rather large set of partial reactions has been demonstrated, including steps in which each type of ion becomes sequestered in a nonexchangeable "occluded" state during its transport. A variety of techniques have been used to demonstrate that the transport process involves large conformational changes. Correlations between partial functions and molecular structure are just now being obtained through the study of mutant and chimeric pumps expressed in oocytes, yeast, and tissue-cultured cells. Basically, the reaction/transport cycle is like that of the Ca^{2+}-ATPase described above, obviously with a change in ion selectivity.[11,141]

In the transport steps bringing ions from the cytosol across the plasma membrane into extracellular space, the sodium pump binds three sodium ions. The apparent K_m for cytosolic sodium is about 10–30 mM, depending on the Na^+,K^+-ATPase isoform, and there is high affinity for ATP. After ATP hydrolysis and a shift of the phosphorylated enzyme to the E-2 conformation, sodium ions dissociate from the enzyme. This step has been determined to be the principal voltage-sensitive step in the transport cycle.[142,143] Once the three sodium ions have dissociated, the enzyme is available for binding two extracellular potassium ions, with an apparent K_m in the 0.5–1.5 mM range. This affinity appears to be influenced by the particular β-subunit isoform. The conformation of the enzyme that binds extracellular potassium ions is also the conformation that has a high affinity for ouabain and related cardiac glycosides such as digitalis, which act as inhibitors of the transport cycle.

On binding potassium, the enzyme enters a conformation in which the potassium ions are occluded within the complex. Reversion of the enzyme to the E-1 form and release of the two potassium ions and the phosphate into the cytosol are facilitated by low-affinity ATP binding.

Under normal physiological conditions the 3 for 2 exchange of Na^+ and K^+ ions is obligatory and the transport process carries a net charge per transport cycle. Thus, the pump creates membrane current, and hence it is called "electrogenic." Normally, pump current does not contribute substantially to the membrane potential—no more than 1 or 2 mV. However, when sodium and potassium electrochemical gradients have been run down by rapid, repetitive action potentials in electrically excitable cells, the sodium pump may contribute substantially to the resting potential and thereby help to relieve depolarization-induced ion channel inactivation, promoting excitable cell function.

12.5.3. Na^+,K^+-ATPase Regulation

The sodium pump is under exquisite control. There are interesting questions about the mechanisms of regulation of the sodium pump for each cell type that has been examined. Examples of regulation at virtually every level have been described (Figure 12.9). These include regulation of (1) transcription of sodium pump genes, (2) translation of sodium pump mRNAs, (3) assembly of α and β subunits, (4) distribution of sodium pump molecules in the plasma membrane, (5) interactions with cytoskeletal proteins, (6) endocytosis and exocytosis of sodium pump molecules, and (7) degradation. In addition, there is evidence that cation transport by the sodium pump is depressed by mechanisms involving arachidonic acid and related eicosanoids, by protein kinases, and by endogenous ouabain-like substances. For each level of regulation, important questions about molecular mechanisms remain to be answered. In most cases, answers should shed light more broadly on the basic molecular processes themselves.

Probably the two most well-known instances of Na^+,K^+-ATPase regulation are (1) upregulation in response to demand for ion transport and (2) upregulation in the distal nephron in response to aldosterone. Increased demand for ion transport can occur if the Na^+,K^+-ATPase is partially inhibited by cardiac glycosides or by decreased extracellular potassium ion concentration, or if permeability of the plasma membrane to sodium is increased (reviewed by Pressley[144]). In each of

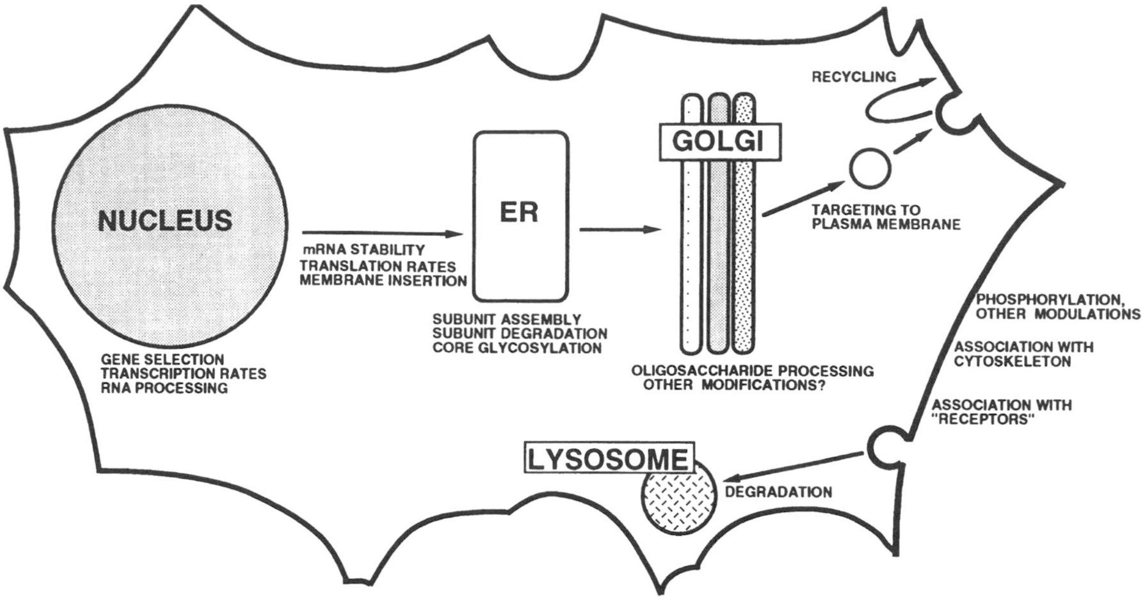

Figure 12.9. Mechanisms of cellular regulation of the Na^+,K^+-ATPase. The figure symbolizes a cell with some of its major organelles. Arrows indicate the biosynthetic pathway and pathways of endocytosis and exocytosis. Types of regulatory processes are indicated along the pathways. Adapted from Fambrough *et al.*[172]

these cases, intracellular sodium ion concentration will increase. Since intracellular sodium concentration is approximately at the K_m for transport by the sodium pump, small increases will lead to accelerated rates of ion transport. In addition, the increased intracellular sodium may serve in some unknown way as a signal for upregulation of Na^+,K^+-ATPase expression. In some instances, this upregulation has been shown to operate at the level of gene expression.[145] While the accelerated transport is an instantaneous effect, the upregulatory response is relatively slow, generally occurring over hours.

Regulation of the sodium pump by aldosterone is part of the mechanism for regulation of sodium ion homeostasis in the body. Aldosterone is released from the adrenal cortex in response to angiotensin II as part of the renin–angiotensin-related hormonal control of blood volume. In response to aldosterone, the distal nephron more avidly recaptures sodium from the filtrate. The response involves increased sodium channels in the apical plasma membrane as well as enhanced sodium pump activity in the basolateral membrane, resulting in vectorial transport of sodium from the filtrate back through the renal epithelium to the blood. Part of the aldosterone response is the activation of gene expression for the Na^+,K^+-ATPase $\alpha 1$ and $\beta 1$ subunits, resulting in enhanced sodium pump biosynthesis.[146–149]

Another interesting regulatory response of the Na^+,K^+-ATPase is involved in the response of skeletal muscle to insulin. In this complex response, acute insulin treatment results in recruitment of Na^+,K^+-ATPase molecules (containing the $\alpha 2$ subunit) from intracellular vesicular stores to the plasma membrane,[150] increasing surface Na^+,K^+-ATPase severalfold.

12.5.4. Involvement of the Na^+,K^+-ATPase in Disease

The sodium pump is of considerable medical importance as the receptor for digitalis and related cardiac glycosides used traditionally to strengthen and steady the heartbeat. Recently, endogenous digitalis-like substances that inhibit the Na^+,K^+-ATPase have come under intense study.[151] Elevated levels in humans correlate strongly with hypertension.[140,152,153] Chronic treatment of rats with low doses of ouabain also produces hypertension.[154] The postulated link to hypertension involves a reduction in Na^+,K^+-ATPase activity in vascular smooth muscle because of the circulating inhibitor, leading to elevated cytosolic sodium, and to a change in the balance of sodium–calcium exchange and consequently an elevation of cytosolic calcium. The increased cytosolic calcium ion concentration causes an increase in vascular muscle tone. Vascular muscle tone is also heightened by hyperactivity of autonomic innervation that is a consequence of depolarization of autonomic neurons following partial inhibition of neuronal sodium pumps. These findings make the sodium pump a focal point for studies of hypertension.

While human sodium pumps are encoded by at least three α- and two β-subunit genes, in no case, as yet, has a human genetic disorder been shown to result from mutation in a sodium pump gene. However, mutations in sodium pump genes in *Drosophila* and *Caenorhabditis* illustrate the subtlety of phenotypic consequences of sodium pump mutations. The fly mutants[155] show deficiencies in phototaxis, lethargic behavior, and excitation-induced paralysis; the worm mutant (unpublished observations of W. Davis and L. Avery) was first isolated in a screen for eating disorders.

Given the importance of the sodium pump in so many basic cell functions, it is not surprising that the sodium pump is involved in a wide variety of disease states. Alterations in sodium pump function have been linked not only to hypertension, but also to diabetes,[156] epilepsy,[157] mood disorders,[158,159] and polycystic kidney disease (PKD). In PKD, an autosomal dominant genetic disease that is among the most prevalent human genetic disorders, the sodium pump is misdirected to the apical membrane of kidney tubule cells as they form cysts. In this location the pump contributes to the enlargement of cysts and deterioration of kidney function. Approximately 10% of patients on dialysis have PKD. The primary genetic defect in PKD involves a recently cloned gene of unknown function.[160] However, the link between the genetic defect and cyst formation may well operate through a change in β-subunit gene expression, leading to mistargeting of the sodium pump to the apical membrane.[161]

12.6. H^+,K^+-ATPASE

Function of the gastric H^+,K^+-ATPase involves an impressive reorganization of cell membranes.[162] In resting parietal cells the H^+,K^+-ATPase is highly concentrated in the membranes of tubulovesicular structures that lie beneath the apical plasma membrane. Acid secretion in the stomach is triggered when secretagogues, such as histamine, acetylcholine, or gastrin, stimulate the fusion of the tubulovesicular membrane system with the apical plasma membrane to form the secretory canaliculi. Thereupon, the H^+,K^+-ATPase mediates export of protons in exchange for extracellular potassium ions.[163] Chloride secretion is accomplished in parallel, involving a basolateral chloride/bicarbonate anion exchanger and apical chloride channels.

The first H^+,K^+-ATPase to be characterized at the molecular level is the one that mediates acid secretion by gastric parietal cells.[164–168] cDNA clones encoding other H^+,K^+-ATPase α-subunit isoforms have been obtained from libraries representing several tissues, including colon[169] and toad urinary bladder.[170] As the phylogenetic tree in Figure 12.2 shows, the H^+,K^+-ATPase α subunits are in the same subfamily as the Na^+,K^+-ATPase α subunits. In fact, there is 63% amino acid sequence identity between gastric H^+,K^+-ATPase and the Na^+,K^+-ATPase αsubunits. The H^+,K^+-ATPase, like the Na^+,K^+-ATPase, also has a second subunit, the β subunit. H^+,K^+- and Na^+,K^+-ATPase β subunits share the same structural motifs, including three disulfide bridges and several *N*-glycosylation sites, and they have about 35% amino acid identity. Expression experiments in which H^+,K^+-ATPase β subunits are coexpressed with Na^+,K^+-ATPase α subunit have revealed that cross-assembly can occur,[171,172] and the hybrid α–β molecules are functional cation transport ATPases.[173,174] Evidence indicates that the ion selectivity goes with the α subunit, although the β subunit influences potassium ion affinity.

The H^+,K^+-ATPases are inhibited by omeprazole and related benzimidazole derivatives and by Schering compound SCH 28080 and related pyridyl imidazoles.[175] Recently, an H^+,K^+-ATPase that is inhibited by both SCH 28080 and by ouabain has been discovered in toad urinary bladder,[170] attesting to the close relationship between the Na^+,K^+ and the H^+,K^+-ATPases.

12.6.1. Involvement in Autoimmune Gastritis and Pernicious Anemia

Pernicious anemia, a vitamin B_{12} deficiency disease, is the end stage of autoimmune gastritis, a disease that affects predominantly people of Northern European origin and is characterized by the presence of circulating antibodies to parietal cell antigens, submucosal lymphocytic infiltration, and, in pernicious anemia, the loss of both parietal cells and pepsinogen-secreting chief cells from the gastric pits (see review by Gleeson and Toh[176]). The α and β subunits of the gastric H^+,K^+-ATPase appear to be the major autoimmune targets in the disease. A mouse model of autoimmune gastritis has been produced by thymectomy at postnatal day 2–4.[177] In greater than 50% of Balb/c *nu/+* mice, this leads within 3 months to autoimmune gastritis, with lesions similar to those seen in humans, and among monoclonal antibodies derived from such mice are antibodies directed at the α or β subunit of the H^+,K^+-ATPase.[176]

Passive transfer experiments have established that T cells are required for transfer of the disorder from mice with autoimmune gastritis induced by neonatal thymectomy. A hypothetical mechanism for generation of the mucosal lesions is T-cell-mediated attack on parietal cells that display on their basolateral surfaces peptides derived from H^+,K^+-ATPase molecules, in association with MHC class I molecules. Autoimmune gastritis leads indirectly to pernicious anemia because the gastric parietal cells are the source of intrinsic factor, which is required for dietary absorption of vitamin B_{12}.

12.7. CU^{2+}-TRANSPORTING ATPASES IN MENKES DISEASE AND WILSON DISEASE

Menkes disease is a rare X-linked genetic disorder, characterized by a defect in copper ion export from many cell types, including intestinal cells but not hepatocytes. Dietary copper enters intestinal cells but is not transported across the intestinal epithelium into the body, resulting in severe copper deficiency. Recently the gene thought to be responsible for this disorder was cloned and characterized.[178] The gene encodes a P-type ATPase that falls within the family that includes bacterial Cd^{2+}- and Cu^{2+}-transporting ATPases. Based on this phylogenetic relationship and the fact that the disease involves copper metabolism, it has been proposed that the gene codes for a Cu^{2+}-transporting ATPase.[2,179] Another genetic disorder involving copper metabolism is Wilson disease, which maps to chromosome 13. In Wilson disease copper accumulates in liver rather than being incorporated into ceruloplasmin and secreted into the bile. Copper accumulates to toxic levels in the liver and also in the brain and kidney, leading to liver failure and to neurological deterioration. The gene defect in Wilson disease has been cloned and shown to have 60% amino acid sequence similarity with the Menkes gene.[180,181] It is proposed that the two P-type ATPases play comparable copper-exporting roles but in different cell types.

The two putative copper-transporting ATPases are the first eukaryotic P-type ATPases found that appear to transport heavy metals. Bacterial homologues include copper-, cadmium-, and mercury-exporting ATPases. All of these molecules have potential heavy metal binding amino acid sequence motifs in their N-terminal cytosolic domain. These motifs, up to six repeats, have a consensus sequence GMTCXXC, where the CXXC motif is also found in other proteins that are known to bind copper, including the metallothionines. These ATPases also have the characteristic ATP binding domain and phosphorylation site of P-type ATPases, but they appear to have fewer membrane spans than the ten thought to occur in other eukaryotic P-type ATPases.

12.8. SUMMARY

There are many different P-type ATPases in mammals, representing most of the families of P-type ATPases found in all kinds of organisms. All of these share structural features and carry out homologous transport processes, although they differ widely in ion selectivity. Each type of ATPase occurs in specific cellular locations, and each is regulated by a variety of hormonal and other physiological signals. These P-type ATPases play fundamental roles in cells, setting the sodium and potassium ion gradients across plasma membrane, participating in the regulation of cytosolic calcium ion levels, mediating acid secretion in some tissues, and involved in the transport of some heavy metal cations, such as copper. P-type ATPases are implicated in a variety of disorders, including heart failure, hypertension, pernicious anemia, and genetic defects in copper metabolism. Given their central roles in cell physiology and their normally tight regulation, it is not surprising that the P-type ATPases often play roles in disease states even when they are clearly not the primary defect.

REFERENCES

1. Hesse, J. E., Wieczorek, L., Altendorf, K., Reicin, A. S., Dorus, E., and Epstein, W. (1984). Sequence homology between two membrane transport ATPases, the Kdp-ATPase of *Escherichia coli* and the Ca^{2+}-ATPase of sarcoplasmic reticulum. *Proc. Natl. Acad. Sci. USA* **81:**4746–4750.
2. Bull, P. C., and Cox, D. W. (1994). Wilson disease and Menkes disease: New handles on heavy-metal transport. *Trends Genet.* **10:** 246–252.
3. Harvey, W. R. (1992). Physiology of V-ATPases. *J. Exp. Biol.* **172:** 1–17.
4. Pedersen, P. L., and Amzel, L. M. (1993). ATP synthases. Structure, reaction center, mechanisms, and regulation of one of nature's most unique machines. *J. Biol. Chem.* **268:**9937–9940.
5. Wingo, C. S., and Cain, B. D. (1993). The renal H-K-ATPase: Physiological significance and role in potassium homeostasis. *Annu. Rev. Physiol.* **55:**323–347.
6. Fagan, M. J., and Saier, M. H. (1994). P-type ATPases of eukaryotes and bacteria: Sequence analyses and construction of phylogenetic trees. *J. Mol. Evol.* **38:**57–99.
7. Saier, M. H. (1994). Computer-aided analysis of transport protein sequences: Gleaning evidence concerning function, structure, biogenesis, and evolution. *Microbiol. Rev.* **58:**71–93.
8. Addison, R. (1986). Primary structure of the *Neurospora* plasma membrane H^+-ATPase deduced from the gene sequence. Homology to Na^+/K^+-, Ca^{2+}-, and K^+-ATPases. *J. Biol. Chem.* **261:**14896–14901.
9. Rao, R., and Slayman, C. W. (1992). Mutagenesis of the yeast plasma membrane H^+-ATPase, a novel expression system. *Biophys. J.* **62:**235–242.
10. Song, Y., and Fambrough, D. M. (1994). Molecular evolution of the calcium-transporting ATPases analyzed by the maximum parsimony method. In *Molecular Evolution of Physiological Processes* (D.M. Fambrough, ed.), Rockefeller University Press, New York, pp. 271–283.
11. Horisberger, J.-D. (1994). *The Na,K-ATPase: Structure–Function Relationship,* R. G. Landes Co, Austin, TX.
12. Medford, R. M., Hyman, R., Ahmad, M., Allen, J. C., Pressley, T. A., Allen, P. D., and Nadal-Ginard, B. (1991). Vascular smooth muscle expresses a truncated Na^+,K^+-ATPase α-1 subunit isoform. *J. Biol. Chem.* **266:**18308–18312.
13. Sweadner, K. J. (1989). Isozymes of the Na^+/K^+-ATPase. *Biochim. Biophys. Acta* **988:**185–220.
14. McDonough, A. A., Azuma, K. K., Hensley, C. B., and Magyar, C. E. (1994). Physiological relevance of the $\alpha2$ isoform of Na,K-ATPase in muscle and heart. In *The Sodium Pump* (E. Bamberg and W. Schoner, eds.), Steinkopff, Darmstadt, pp. 170–180.
15. McDonough, A. A., Azuma, K. K., Lescale-Mathys, L., Tang, M. J., Nakhoul, F., Hensley, C. B., and Komatsu, Y. (1992). Physiological rationale for multiple sodium pump isoforms—Differential regulation of $\alpha1$ and $\alpha2$ by ionic stimuli. *Ann. N.Y. Acad. Sci.* **671:** 156–169.
16. Clausen, T. (1990). Significance of Na,K-ATPase pump regulation in skeletal muscle. *News Physiol. Sci.* **5:**148–151.
17. Thirien, A. G., Ball, W. J., and Blostein, R. (1995). Tissue-specific versus isoform-specific differences in cation activation kinetics of the Na,K-ATPase. *Biophys. J.* **68:**A307.
18. Jewell, E. A., and Lingrel, J. B. (1991). Comparison of the substrate dependence properties of the rat Na,K-ATPase $\alpha1$, $\alpha2$, and $\alpha3$ isoforms expressed in HeLa cells. *J. Biol. Chem.* **266:**16925–16930.
19. Bertorello, A. M., Aperia, A., Walaas, S. I., Niarn, A. C., and Greengard, P. (1991). Phosphorylation of the catalytic subunit of Na^+,K^+-ATPase inhibits the activity of the enzyme. *Proc. Natl. Acad. Sci. USA* **88:**11359–11362.
20. Aperia, A., Ibarra, F., Svensson, L.-B., Klee, C., and Greengard, P. (1992). Calcineurin mediates α-adrenergic stimulation of Na^+,K^+-ATPase activity in renal tubule cells. *Proc. Natl. Acad. Sci. USA* **89:** 7394–7397.
21. Feschienko, M. S., and Sweadner, K. J. (1995). Structural basis for species specificity in phosphorylation of Na,K-ATPase by protein kinase C. *Biophys. J.* **68:**A308.
22. Burk, S. E., Lytton, J., MacLennan, D. H., and Shull, G. E. (1989). cDNA cloning, functional expression, and mRNA tissue distribution of a third organellar Ca^{2+} pump. *J. Biol. Chem.* **264:**18561–18568.
23. Randolph, H. K., Antebi, A., Fink, G. R., Buckley, C. M., Dorman, T., Levitre, J., Davidow, L. S., Mao, J., and Moir, D. T. (1989). The yeast secretory pathway is perturbed by mutations in PMR1, a member of Ca^{2+} ATPase family. *Cell* **58:**133–145.
24. Shull, G. E., Clarke, D. M., and Gunteski-Hamblin, A. M. (1992).

cDNA cloning of possible mammalian homologs of the yeast secretory pathway Ca(2+)-transporting ATPase. *Ann. N.Y. Acad. Sci.* **671:**70–80.
25. Hammerton, R. W., Krzeminski, K. A., Mays, R. W., Ryan, T. A., Wollner, D. A., and Nelson, W. J. (1991). Mechanism for regulating cell surface distribution of Na^+,K^+-ATPase in polarized epithelial cells. *Science* **254:**847–850.
26. Zurzolo, C., Rodriguez-Boulan, E., Gottardi, C. J., Caplan, M. J., Krzeminski, K. A. S., Hammerton, R. W., Mays, R. W., Ryan, T. A., Wolner, D. A., and Nelson, W. J. (1993). Delivery of Na^+,K^+-ATPase in polarized epithelial cells. [Technical Comments]. *Science* **260:** 550–556.
27. Carafoli, E. (1987). Intracellular calcium homeostasis. *Annu. Rev. Biochem.* **56:**395–433.
28. Berridge, M. J. (1993). Inositol trisphosphate and calcium signalling. *Nature* **361:**315–325.
29. Hasselbach, W., and Makinose, M. (1961). Die Calciumpumpe der "Erschlaffungsgrana" des Muskels and ihre Abhangigkeit von der ATP-Spaltung. *Biochem. Z.* **333:**518–528.
30. Ebashi, S., and Lippman, F. (1962). Adenosine triphosphate-linked concentration of calcium ions in a particular fraction of rabbit muscle. *J. Cell Biol.* **14:**389–400.
31. Gunteski-Hamblin, A. M., Greeb, J., and Shull, G. E. (1988). A novel Ca^{2+} pump expressed in brain, kidney, and stomach is encoded by an alternative transcript of the slow-twitch muscle sarcoplasmic reticulum Ca-ATPase gene. Identification of cDNAs encoding Ca^{2+} and other cation-transporting ATPases using an oligonucleotide probe derived from the ATP-binding site. *J. Biol. Chem.* **263:** 15032–15040.
32. Campbell, A. M., Kessler, P. D., Sagara, Y., Inesi, G., and Fambrough, D. M. (1991). Nucleotide sequences of avian cardiac and brain SR/ER Ca^{2+}-ATPases and functional comparison with fast twitch Ca^{2+}-ATPase: Calcium affinities and inhibitor effects. *J. Biol. Chem.* **266:**16050–16055.
33. Karin, N. J., Kaprielian, Z., and Fambrough, D. M. (1989). Expression of avian Ca^{2+}-ATPase in cultured mouse myogenic cells. *Mol. Cell. Biol.* **9:**1978–1986.
34. MacLennan, D. H., Brandl, C. J., Korczak, B., and Green, N. M. (1985). Amino-acid sequence of a Ca^{2+} +Mg^{2+} dependent ATPase from rabbit muscle sarcoplasmic reticulum, deduced from its complementary DNA sequence. *Nature* **316:**696–700.
35. Lytton, J., and MacLennan, D. H. (1988). Molecular cloning of cDNAs from human kidney coding for two alternatively spliced products of the cardiac Ca^{2+}-ATPase gene. *J. Biol. Chem.***263:**15024–15031.
36. Lytton, J., Westin, M., Burk, S. E., Shull, G. E., and MacLennan, D. H. (1992). Functional comparisons between isoforms of the sarcoplasmic or endoplasmic reticulum family of calcium pumps. *J. Biol. Chem.* **267:**14483–14489.
37. Inesi, G., and Asai, H. (1968). Trypsin digestion of fragmented sarcoplasmic reticulum. *Arch. Biochem. Biophys.* **126:**469–477.
38. Scales, D. J., and Insei, G. (1976). Assembly of ATPase protein in sarcoplasmic reticulum membranes. *Biophys. J.* **16:**735–751.
39. Deamer, D., and Baskin, T. (1969). Ultrastructure of sarcoplasmic reticulum preparations. *J. Cell Biol.* **42:**296–307.
40. Dupont, Y., Harrison, S., and Hasselbach W. (1973). Molecular organization in the sarcoplasmic reticulum membrane studied by x-ray diffraction. *Nature* **244:**554–558.
41. Herbette, L. G., DeFoor, P., Fleischer, S., Pascolini, D., Scarpa, A., and Blasie, J. K. (1985). The separate profile structures of the calcium pump proteins and the phospholipid bilayer within isolated sarcoplasmic reticulum membranes determined by x-ray and neutron diffraction. *Biochim. Biophys. Acta* **817:**103–122.
42. Stokes, D. L., and Green, N. M. (1990). Structure of Ca-ATPase: Electron microscopy of frozen-hydrated crystals at 6 Å resolution in projection. *J. Mol. Biol.* **213:**529–538.
43. Taylor, K. A., Dux, L., and Martonosi, A. (1986). Three-dimensional reconstruction of negatively stained crystals of the Ca^{2+}-ATPase from muscle sarcoplasmic reticulum. *J. Mol. Biol.* **187:**417–427.
44. Allen, G., Trinnaman, B. J., Green, N. M. (1980). The primary structure of the calcium ion-transporting adenosine triphosphatase protein of rabbit skeletal sarcoplasmic reticulum. Peptides derived from digestion with cyanogen bromide, and the sequences of three long extramembranous segments. *Biochem. J.* **187:**591–616.
45. Campbell, A. M., Kessler, P. D., and Fambrough, D. M. (1992). The alternative carboxyl termini of avian cardiac and brain sarcoplasmic reticulum/endoplasmic reticulum Ca^{2+}-ATPases are on opposite sites of the membrane. *J. Biol. Chem.* **267:**9321–9325.
46. Matthews, A. M., Sharma, R. P., Lee, A. G., and East, J. M. (1990). Transmembranous organization of (Ca^{2+} +Mg^{2+})-ATPase from sarcoplasmic reticulum. Evidence or lumenal location of residues 877–888. *J. Biol. Chem.* **265:**18737–18740.
47. Clarke, D. M., Loo, T. W., and MacLennan, D. H. (1990). The epitope for monoclonal antibody A20 (amino acids 870–890) is located on the luminal surface of the Ca-ATPase of the sarcoplasmic reticulum. *J. Biol. Chem.* **265:**17405–17408.
48. Matthews, A. M., Tunwell, R. E., Sharma, R. P., and Lee, A. G. (1992). Definition of surface-exposed and trans-membrane regions of the (Ca^{2+} + Mg^{2+})-ATPase of sarcoplasmic reticulum using anti-peptide antibodies. *Biochem. J.* **286:**567–580.
49. Matthews, A. M., Colyer, J., Mata, M., Green, N. M., Sharma, R. P., Lee, A. G., and East, J. M. (1989). Evidence for the cytoplasmic location of the N- and C-terminal segments of sarcoplasmic reticulum (Ca^{2+} + Mg^{2+})-ATPase. *Biochem. Biophys. Res. Commun.* **161:**683–688.
50. Reithmeier, R. A. F., and MacLennan, D. H. (1981). The NH_2 terminus of the Ca^{2+} + Mg^{2+})-adenosine triphosphatase is located on the cytoplasmic surface of the sarcoplasmic reticulum membrane. *J. Biol. Chem.* **256:**5957–5960.
51. Inesi, G., Kurzmack, M., Coan, C., and Lewis, D. (1980). Kinetic and equilibrium characterization of an energy-transducing enzyme and its partial reactions. *Methods Enzymol.* **157:**154–190.
52. Hasselbach, W. (1964). Relaxing factor and the relaxation of muscle. *Prog. Biophys. Biophys. Chem.* **14:**169–222.
53. Levy, D., Seigneuret, M., Bluzat, A., and Rogaud, J. L. (1990). Evidence for proton countertransport by the sarcoplasmic reticulum Ca^{2+}-ATPase during calcium transport in reconstituted proteoliposomes with low ionic permeability. *J. Biol. Chem.* **265:**19524–19534.
54. Makinose, M. (1969). The phosphorylation on the membrane protein of the sarcoplasmic vesicles during active calcium transport. *Eur. J. Biochem.* **10:**74–82.
55. Yamamoto, T., and Tonomura, Y. (1968). Reaction mechanism of the Ca^{++}-dependent ATPase of sarcoplasmic reticulum from skeletal muscle. II. Intermediate formation of phosphoryl protein. *J. Biochem. (Tokyo)* **64:**137–145.
56. Inesi, G., Kurzmack, M., and Verjovski-Almeida, S. (1978). ATP phosphorylation and calcium ion translocation in the transient state of sarcoplasmic reticulum activity. *Ann. N.Y. Acad. Sci.* **307:**224–227.
57. Sudima, M., and Tonomura, Y. (1974). Reaction mechanism of the Ca^{2+} dependent ATPase of sarcoplasmic reticulum from skeletal muscle. Direct evidence for Ca^{2+} translocation coupled with formation of a phosphorylated intermediate. *J. Biochem. (Tokyo)* **75:** 283–297.
58. Takakuwa, Y., and Kanazawa, T. (1979). Slow transition of phosphoenzyme from ADP-sensitive to ADP-insensitive forms in solubilized Ca^{2+}+Mg^{2+}-ATPase of sarcoplasmic reticulum. Evidence for retarded dissociation of Ca^{2+} from the phosphoenzyme. *Biochem. Biophys. Res. Commun.* **88:**1209–1216.
59. Dupont, Y. (1980). Occlusion of divalent cations in the phosphorylated calcium pump of sarcoplasmic reticulum. *Eur. J. Biochem.* **109:**231–238.

60. Bargolie, B., Hasselbach, W., and Makinose, M. (1971). Activation of calcium efflux by ADP and inorganic phosphate. *FEBS Lett.* **12:**267–268.
61. De Meis, L., and Carvalho, M. G. C. (1974). Role of the Ca^{2+} concentration gradient in the ATP–Pi exchange catalyzed by sarcoplasmic reticulum. *Biochemistry* **13:**5032–5038.
62. Makinose, M., and Hasselbach, W. (1971). ATP synthesis by the reverse of the sarcoplasmic reticulum calcium pump. *FEBS Lett.* **121:**271–272.
63. De Meis, L., and Vianna, A. (1979). Energy interconversion by the Ca^{2+}-dependent ATPase of the sarcoplasmic reticulum. *Annu. Rev. Biochem.* **48:**275–292.
64. Inesi, G., Kurzmack, M., Coan, C., and Lewis, D. (1980). Kinetic and equilibrium characterization of an energy-transducing enzyme and its partial reactions. *Methods Enzymol.* **157:**154–190.
65. Jencks, W. P. (1989). How does the calcium pump pump calcium? *J. Biol. Chem.* **264:**18855–18858.
66. Yu, S., and Inesi, G. (1995). Variable stoichiometric efficiency of Ca^{2+} and Sr^{2+} transport by the sarcoplasmic reticulum ATPase. *J. Biol. Chem.* **270:**4361–4367.
67. Hill, T. (1991). *Statistical Thermodynamics,* Addison–Wesley, Reading, MA.
68. Inesi, G., Lewis, D., Nikic, D., and Kirtley, M. E. (1992). Long range intramolecular linked functions in the calcium transport ATPase. In *Advances in Enzymology* (A. Meister, ed.), Wiley, New York, pp. 185–215.
69. Degani, C., and Boyer, P. D. (1973). A borohydride reduction method for characterization of the acyl phosphate linkage in proteins and its application to sarcoplasmic reticulum adenosine triphosphatase. *J. Biol. Chem.* **248:**8222–8226.
70. Bastide, F., Meissner, G., Fleischer, S., and Post, R. L. (1973). Similarity of the active site of phosphorylation of the ATPase for transport of sodium and potassium ions in kidney to that for transport of calcium ion in sarcoplasmic reticulum of muscle. *J. Biol. Chem.* **248:**8385–8391.
71. Taylor, W. R., and Green, N. M. (1989). The predicted secondary structures of the nucleotide-binding sites of six cation-transporting ATPases lead to a probable tertiary fold. *Eur. J. Biochem.* **179:**241–248.
72. Bigelow, D. J., and Inesi, G. (1992). Contributions of chemical derivatization and spectroscopic studies to the characterization of the Ca^{2+} transport ATPase of sarcoplasmic reticulum. *Biochim. Biophys. Acta* **1113:**323–338.
73. Clarke, D. M., Loo, T. W., Inesi, G., and MacLennan, D. H. (1989). Location of high affinity Ca^{2+}-binding sites within the predicted transmembrane domain of the sarcoplasmic reticulum Ca^{2+}-ATPase. *Nature* **339:**476–478.
74. Sumbilla, C., Cantilinia, T., Collins, J. H., Malak, H., Lakowicz, J. R., and Inesi, G. (1991). Structural perturbation of the transmembrane region interferes with calcium binding by the Ca^{2+} transport ATPase. *J. Biol. Chem.* **266:**12682–12689.
75. Sumbilla, C., Lu, L., Inesi, G., Ishii, T., Takeyasu, K., Feng, Y., and Fambrough, D. M. (1993). Ca^{2+} dependent and thapsigargin-inhibited phosphorylation of Na^+,K^+-ATPase catalytic domain following chimeric recombination with the Ca^{2+}-ATPase. *J. Biol. Chem.* **268:** 21185–21192.
76. Zhang, Z., Sumbilla, C., Lewis, D., and Insei, G. (1993). High sensitivity to site directed mutagenesis of the peptide segment connecting phosphorylation and Ca^{2+} binding domains in the Ca^{2+} transport ATPase. *FEBS Lett.* **335:**261–264.
77. Nakamoto, R. K., and Inesi, G. (1986). Retention of ellipticity between enzymatic states of the Ca^{2+}-ATPase of sarcoplasmic reticulum. *FEBS Lett.* **194:**258–262.
78. Girardet, J.-L., and Dupont, Y. (1992). Ellipticity changes of the sarcoplasmic reticulum Ca^{2+}-ATPase induced by cation binding and phosphorylation. *FEBS Lett.* **296:**103–106.
79. Matsuda, H., and De Meis, L. (1973). Phosphorylation of the sarcoplasmic reticulum membrane by orthophosphate. Inhitibion by calcium ions. *Biochemistry* **12:**4581–4585.
80. Duggan, P. F., and Martonosi, A. N. (1970). Sarcoplasmic reticulum. IX. The permeability of sarcoplasmic reticulum membranes. *J. Gen. Physiol.* **56:**147–167.
81. Yu, S., Hao, L., and Inesi, G. (1994). A pK change of acidic residues contributes to cation countertransport in the Ca-ATPase of sarcoplasmic reticulum. Role of H^+ in Ca^{2+}-ATPase countertransport. *J. Biol. Chem.* **269:**16656–16661.
82. Dux, L., and Martonosi, A. N. (1983). Two dimensional arrays of proteins in sarcoplasmic reticulum and purified Ca^{2+}-ATPase vesicles treated with vanadate. *J. Biol. Chem.* **258:**2599–2603.
83. Dux, L., and Martonosi, A. N. (1983). The regulation of ATPase–ATPase interactions in sarcoplasmic reticulum membrane. I. The effects of Ca^{2+}, ATP, and inorganic phosphate. *J. Biol. Chem.* **258:** 11896–11902.
84. Blasie, J. K., Pascolini, D., Asturias, L. G., Herbette, L. G., Pierce, D., and Scarpa, A. (1990). Large scale structural changes in the sarcoplasmic reticulum ATPase appear essential for calcium transport. *Biophys. J.* **58:**687–693.
85. Kirchberger, M. A., Tada, M., and Katz, A. M. (1974). Adenosine 3′:5′-monophosphate-dependent protein kinase-catalyzed phosphorylation reaction and its relationship to calcium transport in cardiac sarcoplasmic reticulum. *J. Biol. Chem.* **249:**6166–6173.
86. Le Peuch, C. J., Haiech, J., and Demaille, J. G. (1979). Concerted regulation of cardiac sarcoplasmic reticulum calcium transport by cyclic adenosine monophosphate dependent and calcium-calmodulin-dependent phosphorylations. *Biochemistry* **18:**5150–5157.
87. Movsesian, M. A., Nishikawa, M., and Adelstein, R. S. (1984). Phosphorylation of phospholamban by calcium-activated, phospholipid-dependent protein kinase. Stimulation of cardiac sarcoplasmic reticulum calcium uptake. *J. Biol. Chem.* **259:**8029–8032.
88. Tada, M., Kirchberger, M. A., Repke, D. I., and Katz, A. M. (1974). The stimulation of calcium transport in cardiac sarcoplasmic reticulum by adenosine 3′:5′-monophosphate-dependent protein kinase. *J. Biol. Chem.* **249:**6174–6180.
89. Jones, L., Wegener, A. D., and Simmerman, H. K. (1988). Purification of phospholamban from canine cardiac sarcoplasmic reticulum vesicles by use of sulfhydryl group affinity chromatography. *Methods Enzymol.* **157:**360–369.
90. Tada, M., Kadoma, M., Inui, M., and Fujii, J. (1988). Regulation of Ca^{2+} pump from cardiac sarcoplasmic reticulum. *Methods Enzymol.* **157:**107–154.
91. Arkin, I. T., Adams, P. D., Mackenzie, K. R., Lemmon, M. A., Brunger, A. T., and Engleman, D. M. (1994). Structural organization of the pentameric transmembrane α-helices of phospholamban, a cardiac ion channel. *EMBO J.* **13:**4757–4764.
92. Wegener, A. D., and Jones L. R. (1984). Phosphorylation-induced mobility shift in phospholamban in sodium dodecyl sulfate–polyacrylamide gels. Evidence for a protein structure consisting of multiple identical phosphorylatable subunits. *J. Biol. Chem.* **259:**1834–1841.
93. Kirchberger, M. A., and Tada, M. (1976). Effects of adenosine 3′:5′-monophosphate-dependent protein kinase on sarcoplasmic reticulum isolated from cardiac and slow and fast contracting skeletal muscles. *J. Biol. Chem.* **251:**725–729.
94. Luo, W., Grupp, I. L., Harrer, J., Ponniah, S., Grupp, G., Duffy, J. J., Doetschman, T., and Kranias, E. G., (1994). Targeted ablation of the phospholamban gene is associated with markedly enhanced myocardial contractility and loss of beta-agonist stimulation. *Circ. Res.* **75:**401–409.
95. James, P., Inui, M., Tada, M., Chiesi, M., and Carafoli, E. (1989). Nature and site of phospholamban regulation of the Ca^{2+} pump of sarcoplasmic reticulum. *Nature* **342:**90–92.
96. Toyofuku, T., Kurzydlowski, K., Tada, M., and MacLennan, D. H.

(1993). Identification of regions in the Ca^{2+}-ATPase of sarcoplasmic reticulum that affect functional association with phospholamban. *J. Biol. Chem.* **268:**2809–2815.
97. Rasmussen, U., Christensen, S. B., and Sandberg, F. (1978). Thapsigargin and thapsigargicin, two new histamine liberators from *Thapsia garganica. Acta Pharm. Suec.* **15:**133–140.
98. Thastrup, O., Cullen, P. J., Drobak, B. K., Hanley, M. R., and Dawson, A. P. (1990). Thapsigargin, a tumor promoter, discharges intracellular Ca^{2+} stores by specific inhibition of the endoplasmic reticulum Ca-ATPase. *Proc. Natl. Acad. Sci. USA* **87:**2466–2470.
99. Sagara, Y., and Inesi, G. (1991). Inhibition of the sarcoplasmic reticulum Ca^{2+} transport ATPase by thapsigargin at subnanomolar concentrations. *J. Biol. Chem.* **266:**13503–13506.
100. Kijima, Y., Ogunbunmi, E., and Fleischer, S. (1991). Drug action of thapsigargin on the Ca^{2+} pump protein of sarcoplasmic reticulum. *J. Biol. Chem.* **266:**22912–22918.
101. Lytton, J., Westin, M., and Hanley, M. R. (1991). Thapsigargin inhibits the sarcoplasmic or endoplasmic reticulum Ca-ATPase family of calcium pumps. *J. Biol. Chem.* **266:**17067–17071.
102. Sagara, Y., Fernandez-Belda, F., De Meis, L., and Inesi, G. (1992). Characterization of the inhibition of intracellular Ca^{2+} transport ATPases by thapsigargin. *J. Biol. Chem.* **267:**12606–12613.
103. Inesi, G., and Sagara, Y. (1994). Specific inhibitors of intracellular Ca^{2+} transport ATPases. *J. Membr. Biol.* **141:**1–6.
104. Sagara, Y., Wade, J., and Inesi, G. (1992). A conformational mechanism for formation of a dead-end complex by the sarcoplasmic reticulum ATPase with thapsigargin. *J. Biol. Chem.* **267:**1286–1292.
105. Whitmer, J. T., Kumar, P., and Solaro, R. J. (1988). Calcium transport properties of cardiac sarcoplasmic reticulum from cardiomyopathic Syrian hamsters (Bio 53.58 and 14.6): Evidence for a quantitative defect in dilated myopathic hearts not evident in hypertrophic hearts. *Circ. Res.* **62:**81–85.
106. Takahashi, T., Allen, P. D., and Izumo, S. (1992). Expression of A-, B-, and C-type natriuretic peptide genes in failing and developing human ventricles: Correlation with expression of the Ca^{2+}-ATPase gene. *Circ. Res.* **71:**9–17.
107. Nagai, R., Zarain-Herzberg, A., Brandl, C. J., Fujii, J., Tada, M., MacLennan, D. H., Alpert, N. R., and Periasamy, M. (1989). Regulation of myocardial Ca^{2+}-ATPase and phospholamban mRNA expression in response to pressure overload and thyroid hormone. *Proc. Natl. Acad. Sci. USA* **86:**2966–2970.
108. Mercadier, J. J., Lompre, A. M., Duc, P., Boheler, K. R., Fraysse, J. B., Wisnewsky, C., Allen P. D., Komajda, M., and Schwartz, K. (1990). Altered sarcoplasmic reticulum Ca^{2+}-ATPase gene expression in the human ventricle during end-stage heart failure. *J. Clin. Invest.* **85:**305–309.
109. Feldman, A. M., Weinberg, E. O., Ray, P. E., and Lorell, B. H. (1993). Selective changes in cardiac gene expression during compensated hypertrophy and the transition to cardiac decompensation in rats with chronic aortic banding. *Circ. Res.* **73:**184–192.
110. Limas, C. J. (1978). Calcium transport ATPase of cardiac sarcoplasmic reticulum in experimental hyperthyroidism. *Am. J. Physiol.* **235:**H745–H751.
111. Arai, M., Alpert, N. R., MacLennan, D. H., Barton, P., and Periasamy, M. (1993). Alterations in sarcoplasmic reticulum gene expression in human heart failure. A possible mechanism for alterations in systolic and diastolic properties of the failing myocardium. *Circ. Res.* **72:**463–469.
112. Lompre, A. M., Anger, M., and Levitsky, D. (1994). Sarco(endo) plasmic reticulum calcium pumps in the cardiovascular system: Function and gene expression. *J. Mol. Cell. Cardiol.* **26:**1109–1132.
113. Karpati, G., Charuk, J., Carpenter, S., Jablecki, C., and Holland, P. (1986). Myopathy caused by a deficiency of Ca^{2+}-adenosine triphosphatase in sarcoplasmic reticulum (Brody's disease). *Anal. Neurol.* **20:**38–49.
114. Schatzmann, H. J. (1973). Dependence on calcium concentration and stoichiometry of the calcium pump in human red cells. *J. Physiol. (London)* **235:**551–569.
115. Gopinath, R. M., and Vincenzi, F. F. (1977). Phosphodiesterase protein activator mimics red blood cell cytoplasmic activator of (Ca^{2+}–Mg^{2+})-ATPase. *Biochem. Biophys. Res. Commun.* **77:**1203–1209.
116. Jarrett, H. W., and Penniston, J. T. (1977). Partial purification of the Ca^{2+}–Mg^{2+} ATPase activator from human erythrocytes: Its similarity to the activator of 3′:5′-cyclic nucleotide phosphodiesterase. *Biochem. Biophys. Res. Commun.* **77:**1210–1216.
117. Niggli, V., Penniston, J. T., and Carafoli, E. (1979). Purification of the (Ca^{2+} +Mg^{2+})-ATPase from human erythrocyte membranes using a calmodulin affinity column. *J. Biol. Chem.* **254:**9955–9958.
118. Rega, A. F., and Garrahan, P. J. (1975). Calcium ion-dependent phosphorylation of human erythrocyte membranes. *J. Membr. Biol.* **22:**313–327.
119. Rega, A. F., and Garrahan, P. J. (1975). Calcium ion-dependent dephosphorylation of the Ca^{2+}-ATPase of human red-cells by ADP. *Biochim. Biophys. Acta* **507:**182–184.
120. Katz, S., and Blostein, R. (1975). Ca^{2+}-stimulated membrane phosphorylation and ATPase activity of the human erythrocyte. *Biochim. Biophys. Acta* **389:**314–324.
121. Niggli, V., Adunyah, E. S., Penniston, J. T., and Carafoli, E. (1981). Purified (Ca^{2+} +Mg^{2+})-ATPase of the erythrocyte membrane: Reconstitution and effect of calmodulin and phospholipids. *J. Biol. Chem.* **256:**395–401.
122. Verma, A. K., Filoteo, A. G., Stanford, D. R., Wieben, E. D., Penniston, J. T., Strehler, E. E., Fischer, R., et al. (1988). Complete primary structure of a human plasma membrane Ca^{2+} pump. *J. Biol. Chem.* **263:**14152–14159.
123. Shull, M. M., and Greeb, J. (1988). Molecular cloning of two isoforms of the plasma membrane Ca^{2+}-transporting ATPase from rat brain. Structural and functional domains exhibit similarity to Na^{+},K^{+}- and other cation transport ATPases. *J. Biol. Chem.* **263:** 8646–8657.
124. Strehler, E. E., James, P., Fischer, R., Heim, R., Vorherr, T., Filoteo, A. G., Penniston, J. T., and Carafoli, E. (1990). Peptide sequence analysis and molecular cloning reveal two calcium pump isoforms in the human erythrocyte membrane. *J. Biol. Chem.* **265:**2835–2842.
125. Keeton, T. P., Burk, S. E., and Shull, G. E. (1993). Alternative splicing of exons encoding the calmodulin-binding domains and C termini of plasma membrane Ca^{2+}-ATPase isoforms 1, 2, 3, and 4. *J. Biol. Chem.* **268:**2740–2748.
126. Burk, S. E., and Shull, G. E. (1992). Structure of the rat plasma membrane Ca(2+)-ATPase isoform 3 gene and characterization of alternative splicing and transription products. Skeletal muscle-specific splicing results in a plasma membrane Ca(2+)-ATPase with a novel calmodulin-binding domain. *J. Biol. Chem.* **267:**19683–19690.
127. Brandt, P., Neve, R. L., Kammescheidt, A., Rhoads, R. E., and Vanaman, T. C. (1992). Analysis of the tissue-specific distribution of mRNAs encoding the plasma membrane calcium pumping ATPases and characterization of an alternatively spliced form of PMCA4 at the cDNA and genomic levels. *J. Biol. Chem.* **267:**4376–4385.
128. Strehler, E. E. (1991). Recent advances in the molecular characterization of plasma membrane Ca^{2+} pumps. *J. Membr. Biol.* **120:**1–15.
129. Jarrett, H. W., and Penniston, J. T. (1978). Purification of the Ca^{2+}-stimulated ATPase activator from human erythrocytes: Its membership in the class of Ca^{2+}-binding modular proteins. *J. Biol. Chem.* **253:**4676–4682.
130. Graf, E., and Penniston, J. T. (1981). Equimolar interaction between calmodulin and the Ca^{2+} ATPase from human erythrocyte membranes. *Arch. Biochem. Biophys.* **210:**257–262.
131. Foder, B., and Scharff, O. (1981). Decrease of apparent calmodulin

affinity of erythrocyte (Ca^{2+} +Mg^{2+})-ATPase at low Ca^{2+} concentrations. *Biochim. Biophys. Acta* **649:**367–376.

132. Larsen, F. L., and Vincenzi, F. F. (1979). Concerted regulation of cardiac sarcoplasmic reticulum calcium transport by cyclic adenosine monophosphate dependent and calcium-calmodulin-dependent phosphorylations. *Science* **204:**306–309.
133. James, P., Maeda, M., Fischer, R., Verma, A. K., Krebs, J., Penniston, J. T., and Carafoli, E. (1988). Identification and primary structure of a calmodulin binding domain of the Ca^{2+} pump of human erythrocytes. *J. Biol. Chem.* **263:**2905–2910.
134. Falchetto, R., Vorherr, T., and Carafoli, E. (1992). The calmodulin-binding site of the plasma membrane Ca^{2+} pump interacts with the transduction domain of the enzyme. *Protein Sci.* **1:**1613–1621.
135. Martin-Vasallo, P., Dackowski, W., Emanuel, J. R., and Levenson, R. (1989). Identification of a putative isoform of the Na,K-ATPase β subunit. *J. Biol. Chem.* **264:**4613–4618.
136. Gloor, S., Antonicek, H., Sweadner, K. J., Pagliusi, S., Frank, R., Moos, M., and Schachner, M. (1990). The adhesion molecule on glia (AMOG) is a homologue of the β subunit of the Na,K-ATPase. *J. Cell Biol.* **110:**165–174.
137. Magyar, J. P., Bartsch, V., Wang, Z.-Q., Howells, N., Aguzzi, A., Wagner, E. F., and Schachner, M. (1994). Degeneration of neural cells in the central nervous system of mice deficient in the gene for the adhesion molecule on glia, the β2 subunit of murine Na,K-ATPase. *J. Cell Biol.* **127:**835–845.
138. Mercer, R. W. (1993). Structure of the Na,K-ATPase. *Int. Rev. Cytol.* **137C:**139–168.
139. Mercer, R. W., Biemesderfer, D., Bliss, D. P., Collins, J. H., and Forbush, B. (1993). Molecular cloning and immunological characterization of the gamma polypeptide, a small protein associated with the Na,K-ATPase. *J. Cell Biol.* **121:**579–586.
140. Toyoshima, C., Sasabe, H., and Stokes, D. L. (1993). Three-dimensional cryo-electron microscopy of the calcium ion pump in the sarcoplasmic reticulum membrane. *Nature* **362:**469–471.
141. Läuger, P. (1991). *Electrogenic Ion Pumps,* Sinauer, Sunderland, MA.
142. Gadsby, D., Rakowski, R. F., and De Weer, P. (1993). Extracellular access to the Na,K pump: Pathway similar to ion channel. *Science* **260:**100–103.
143. Hilgemann, D. W. (1994). Channel-like function of the Na,K pump probed at microsecond resolution in giant membrane patches. *Science* **263:**1429–1432.
144. Pressley, T. A. (1988). Ion concentration-dependent regulation of Na,K-pump abundance. *J. Membr. Biol.* **105:**187–195.
145. Taormino, J. P., and Fambrough, D. M. (1990). Pre-translational regulation of the (Na^{+} + K^{+})-ATPase in response to demand for ion transport in cultured chicken skeletal muscle. *J. Biol. Chem.* **265:** 4116–4123.
146. Geering, K., Girardet, M., Bron, C., Kraehenbuhl, J.-P., and Rossier, B. C. (1982). Hormonal regulation of (Na^{+},K^{+})-ATPase biosynthesis in the toad bladder. *J. Biol. Chem.* **257:**10338–10343.
147. Rossier, B. C., and Palmer, L. G. (1992). Mechanisms of aldosterone action on sodium and potassium transport. In *The Kidney: Physiology and Pathophysiology,* 2nd ed. (D.W. Seldin and G. Giebisch, eds.), Raven Press, New York, pp. 1373–1409.
148. Verrey, F. (1990). Regulation of gene expression by aldosterone in tight epithelia. *Semin. Nephrol.* **10:**410–420.
149. Oguchi, A., Ikeda, U., Kanbe, T., Tsuruya, Y., Yamamoto, K., Kawakami, K., Medford, R. M., and Shimada, K., (1993). Regulation of Na-K-ATPase gene expression by adlosterone in vascular smooth muscle cells. *Am. J. Physiol.* **265:**H1167–H1172.
150. Marette, A., Krisher, J., Lavoie, L., Ackerley, C., Carpenter, J.-L., and Klip, A. (1993). Insulin increases the Na^{+}–K^{+}-ATPase α2-subunit in the surface of rat skeletal muscle: Morphological evidence. *Am. J. Physiol.* **265:**C1716–C1722.
151. Blaustein, M. P. (1993). Physiological effects of endogenous ouabain: Control of intracellular Ca^{2+} stores and cell responsiveness. *Am. J. Physiol.* **264:**C1367–C1387.
152. Hamlyn, J. M., and Manunta, P. (1992). Ouabain, digitalis-like factors and hypertension. *J. Hypertens.* **10**(Suppl. 7):S99–S111.
153. Hamilton, B. P., Manunta, J., Laredo, B. S., Hamilton, J. H., and Hamlyn, J. M. (1994). The new adrenal steroid hormone ouabain. *Curr. Opin. Endocrinol. Diabetes* **1994:**123–131.
154. Yuan, C. M., Manunta, P., Hamlyn, J. M., Chen, S., Bohen, E., Yeun, J., Haddy, F. J., and Pamnani, M. B. (1993). Long-term ouabain administration produces hypertension in rats. *Hypertension* **22:**178–187.
155. Schubiger, M., Feng, Y., Fambrough, D. M., and Palka, J. (1994). A mutation of the *Drosophila* sodium pump α-subunit gene results in bang-sensitive paralysis. *Neuron* **12:**373–381.
156. Wald, H., Scherzer, P., and Rasch, R. (1991). Renal tubular Na^{+}–K^{+}-ATPase in diabetes mellitus: Relationship to metabolic abnormality. *Am. J. Physiol.* **265:**E96–E101.
157. Grisar, T., Guillaume, D., and Delgado-Escueta, A. V. (1992). Contribution of Na^{+},K^{+}-ATPase to focal epilepsy: A brief review. *Epilepsy Res.* **12:**141–149.
158. El Mallakh, R. S. (1983). The Na,K-ATPase hypothesis for manic-depression. II. The mechanism of action of lithium. *Med. Hypotheses* **12:**269–282.
159. Reddy, P. L., Khanna, S., and Subhash, M. N. (1992). Erythrocyte membrane sodium-potassium adenosine triphosphatase activity in affective disorders. *J. Neural Transm.* **89:**209–218.
160. European Polycystic Kidney Disease Consortium. (1994). The polycystic kidney disease 1 gene encodes a 14 kb. transcript and lies within a duplicated region on chromosome 16. *Cell* **77:**881–894.
161. Wilson, P. D., Fatti, L., and Burrow, C. R. (1993). Expression of the beta-2 isoform of Na-K-ATPase during human renal development and in polycystic kidney disease (PKD). *Mol. Biol. Cell* **4:**34a.
162. Black, J. A., Forte, T. M., and Forte, J. G. (1980). Structure of oxyntic cell membranes during conditions of rest and secretion of HCl as revealed by freeze-fracture. *Anat. Rec.* **196:**163–172.
163. Rabon, E. C., and Reuben, M. A. (1990). The mechanism and structure of gastric H,K-ATPase. *Annu. Rev. Physiol.* **52:**321–344.
164. Shull, G. E., and Lingrel, J. B. (1986). Molecular cloning of the rat stomach (H^{+}+ K^{+})-ATPase. *J. Biol. Chem.* **261:**16788–16791.
165. Newman, P. R., Greeb, J., Keeton, T. P., Reyes, A. A., and Shull, G. E. (1990). Structure of the human gastric H,K-ATPase gene and comparison of the 5′-flanking sequences of the human and rat genes. *DNA Cell Biol.* **9:**749–762.
166. Newman, P. R., and Shull, G. E. (1991). Rat gastric H,K-ATPase β-subunit gene: Intron/exon organization, identification of multiple transcription initiation sites, and analysis of the 5′-flanking region. *Genomics* **11:**252–262.
167. Maeda, M., Ishizaki, J., and Futai, M. (1988). cDNA cloning and sequence determination of pig gastric (H^{+}+K^{+})-ATPase. *Biochem. Biophys. Res. Commun.* **157:**203–209.
168. Maeda, M., Oshima, K.-I., Tamura, S., Kaya, S., Mahmood, S., Reuben, M. A., Lasater, L. S., Sachs, G., and Futai, M. (1991). The rat H^{+}/K^{+}-ATPase βsubunit gene and recognition of its control region by gastric DNA binding protein. *J. Biol. Chem.* **266:**21584–21588.
169. Crowson, M. S., and Shull, G. E. (1992). Isolation and characterization of a cDNA encoding the putative distal colon H^{+},K^{+}-ATPase. Similarity of deduced amino acid sequence to gastric H^{+},K^{+}-ATPase and Na^{+},K^{+}-ATPase and mRNA expression in distal colon, kidney and uterus. *J. Biol. Chem.* **267:**13740–13748.
170. Jassier, F., Horisberger, J.-D., Geering, K., and Rossier, B. C. (1993). Mechanisms of urinary K^{+} and H^{+} excretion: Primary structure and functional expression of a novel H,K-ATPase. *J. Cell Biol.* **123:**1421–1429.

171. Lemas, M. V., Yu, H. Y., Takeyasu, K., Kone, B., and Fambrough, D. M. (1994). Assembly of Na,K-ATPase α-subunit isoforms with Na^+,K-ATPase β-subunit isoforms and H,K-ATPase β-subunit. *J. Biol. Chem.* **269:**18651–18655.
172. Fambrough, D. M., Wolitzky, B. A., Taormino, J. P., Tamkun, M. M., Takeyasu, K., Somerville, D., Renaud, K. J., and Lemas, M. V. (1991). A cell biologist's perspective on sites of Na,K-ATPase regulation. In *The Sodium Pump: Structure, Mechanism, and Regulation* (J.H. Kaplan and P. DeWeer, eds.), Rockefeller University Press, New York, pp. 17–30.
173. Jaisser, F., Canessa, C. M., Horisberger, J.-D., and Rossier, B. C. (1992). Primary sequence and functional expression of a novel ouabain-resistant Na,K-ATPase: The β subunit modulates potassium activation of the Na,K-pump. *J. Biol. Chem.* **267:**16895–16903.
174. Eakle, K. A., Kabalin, M. A., Wang, S.-G., and Farley, R. A. (1994). The influence of the β subunit structure on the stability of Na^+/K^+-ATPase complexes and interaction with K^+. *J. Biol. Chem.* **269:** 6550–6557.
175. Pope, A. J., and Sachs, G. (1992). Reversible inhibitors of the gastric (H^+/K^+)-ATPase as both potential therapeutic agents and probes of pump function. *Biochem. Soc. Trans.* **20:**566–572.
176. Gleeson, P. A., and Toh, B.-H. (1991). Molecular targets in pernicious anaemia. *Immunol. Today* **12:**233–238.
177. Kojima, A., and Prehn, R. T. (1981). Genetic susceptibility to post-thymectomy autoimmune diseases in mice. *Immunogenetics* **14:** 15–27.
178. Vulpe, C., Levinson, B., Whitney, S., Packman, S., and Gitschier, J. (1993). Isolation of a candidate gene for Menkes disease and evidence that it encodes a copper-transporting ATPase. *Nature Genet.* **3:**7–13.
179. Silver, S., Nucifora, G., and Phung, L. T. (1993). Human Menkes X-chromosome disease and the staphylococcal cadmium-resistance ATPase: A remarkable similarity in protein sequences. *Mol. Microbiol.* **10:**7–12.
180. Bull, P. C., Thomas, G. R., Rommens, J. M., Forbes, J. R., and Cox, D. W. (1993). The Wilson disease gene is a putative copper transporting P-type ATPase similar to the Menkes gene. *Nature Genet.* **5:**327–337.
181. Tanzi, R. E., Petrukhin, K., Chernov, I., Pellequer, J. L., Wasco, W., Ross, B., Romano, D. M., Parano, E., Pavone, L., Brzustowicz, L. M., Devoto, M., Peppercorn, J., Bush, A. I., Sternlieb, I., Pirasty, M., Gusella, J. F., Evgrafav, O., Penchaszadeh, G. K., Honig, B., Edelman, I. S., Soares, M. B., Scheinberg, I. H., and Gilliam, T. C. (1993). The Wilson disease gene is a copper transporting ATPase with homology to the Menkes disease gene. *Nature Genet.* **5:**344–350.
182. Snavely, M. D., Miller, C. G., and Maguire, M. E. (1991). The *mgtB* Mg^{2+} transport locus of *Salmonella typhimurium* encodes a P-type ATPase. *J. Biol. Chem.* **266:**815–823.
183. Smith, D. L., Tao, T., and Maguire, M. E. (1994). Membrane topology of a P-type ATPase: The MgtB magnesium transport protein of *Salmonella typhimurium. J. Biol. Chem.* **268:**22469–22479.
184. Inesi, G. (1985). Mechanism of calcium transport. *Annu. Rev. Physiol.* **47:**573–601.

13

Multidrug Resistance Transporter

Michael M. Gottesman, Suresh V. Ambudkar, Marilyn M. Cornwell, Ira Pastan, and Ursula A. Germann

13.1. INTRODUCTION

The mechanisms by which cells evade the lethal effects of cytotoxic drugs have been the subject of intense investigation by cell and molecular biologists for more than 20 years. Much of this work was stimulated by the initial success of cancer chemotherapy in disseminated childhood cancers such as leukemias, neuroblastoma, and sarcomas, in germ-cell tumors such as choriocarcinoma and testicular cancer, and in the responsiveness of other cancers such as lymphomas, breast, and ovarian cancers. These clinical successes suggested that metastatic cancer could be cured with chemotherapy, yet in many cases promising remissions were followed by regrowth of drug-resistant cancers. The possibility of cure, coupled with the hope that a defined set of reversible resistance mechanism could be delineated, led to the intense interest in studies of drug resistance.

Because of the complexity of studying drug resistance in intact animals or humans, *in vitro* model systems were developed based on cultured cancer cells (reviewed in Ref. 1). In general, drug-sensitive cancer cells from rodents or humans were used as a starting point, and cells were selected for growth in cytotoxic anticancer drugs. Early observations[2] suggested that for certain classes of natural product anticancer drugs, selection for resistance to one such agent led to cross-resistance to all members of the group. This phenomenon became known as multidrug resistance (MDR), and its discovery engendered considerable excitement since the patterns of resistance and the ability to develop resistance to multiple agents in a single step mimicked observations made on human cancers in patients.[3] Although other patterns of MDR affecting different classes of drugs (such as nucleotide analogues, *cis*-platinum, antimetabolites, and alkylating agents) have since been found in cultured cells, the "classical" MDR pattern, including most natural product anticancer drugs, remains the best studied and perhaps most clinically relevant. Thus, "classical" MDR is the subject of this review.

13.1.1. Physiology of Multidrug Resistance

Our studies have focused on human KB carcinoma cells (a subclone of HeLa, a cervical adenocarcinoma) selected either in doxorubicin (adriamycin), vinblastine, or colchicine. Selection in single steps results in cross-resistance at each step to these drugs as well as many natural products (reviewed in Ref. 1) including other anthracyclines or related agents (daunorubicin, idarubicin, mitoxantrone), *Vinca* alkaloids (vincristine), epipodophyllotoxins [VP-16 (etoposide) and VM-26 (teniposide)], and other potent anticancer drugs (e.g., actinomycin D, mitomycin C, and topotecan). In addition to these anticancer drugs, other cytotoxic agents including protein synthesis inhibitors (emetine), DNA intercalating agents (ethidium bromide), and toxic peptides (valinomycin, gramicidin D) are part of the pattern of cross-resistance. These drugs share hydrophobic and amphipathic properties, but have no other obvious chemical features in common. Most, but not all, are positively charged, but none are anionic.

Resistance appears to be related in essentially all cases to reduced accumulation of drug within cells as a result of the activity of an energy-dependent efflux pump.[4,5] In studies in which Chinese hamster ovary (CHO) cells were selected as populations to very high levels of drug resistance in multiple steps, the resistant cells contained within their plasma membrane a high-molecular-weight phospho-glycoprotein,[6] which has since been shown to be the product of the *mdr* gene responsible for MDR.[7]

13.1.2. Cloning of the MDR Genes

The cloning of the human *MDR*1 gene, and its rodent homologues, was achieved using cultured cells selected to very high levels of MDR. In these cells, the *MDR*1 gene is frequently amplified, and amplified DNA segments were screened for their ability to hybridize to an mRNA whose expression was proportional to levels of drug resistance in vari-

Michael M. Gottesman • Laboratory of Cell Biology, National Cancer Institute, National Institutes of Health, Bethesda, Maryland 20892. **Suresh V. Ambudkar** • Department of Medicine, The Johns Hopkins University School of Medicine, Baltimore, Maryland 21218. **Marilyn M. Cornwell** • Fred Hutchinson Cancer Research Center, Seattle, Washington 98195. **Ira Pastan** • Laboratory of Molecular Biology, National Cancer Institute, National Institutes of Health, Bethesda, Maryland 20892. **Ursula A. Germann** • Vertex Pharmaceuticals Incorporated, Cambridge, Massachusetts 02139.

Molecular Biology of Membrane Transport Disorders, edited by S.G. Schultz et al. Plenum Press, New York, 1996

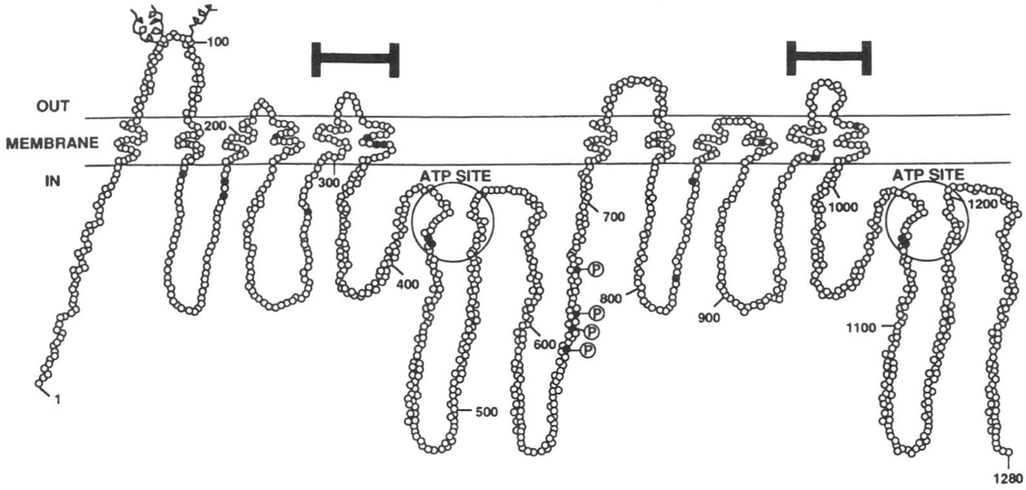

Figure 13.1. Two-dimensional model of human P-glycoprotein based on hydropathy analysis of the amino acid sequence.[16] The ATP sites (nucleotide-binding folds) are circled and the putative N-linked glycosylation sites are indicated by wiggly lines. The bars above the model show the regions labeled with photoaffinity analogues. Darkened circles represent amino acid residues, except for those in the ATP sites, in which mutations have been shown to affect drug transport specificity. The mutations in the amino acid residues in the 6th and 11th transmembrane domains were originally described in Chinese hamster and mouse Pgp, respectively. Serine residues at positions 661, 667, 671, and 683 have been shown *in vitro* to be phosphorylated either by protein kinase C or by protein kinase A.[149,150] (Adapted from Ref. 17.)

ous cell lines.[8,9] Isolation of a full-length cDNA for the human and mouse *mdr* genes was achieved using this general approach, and these cDNAs were soon shown to confer the complete pattern of MDR on cells into which they were transfected or retrovirally transduced.[10–13] *mdr* cDNAs were also isolated from MDR hamster cells using cloning technique based on differential expression of *mdr* RNAs[14] and by cloning the cDNA that encoded P-glycoprotein detected using a P-glycoprotein monoclonal antibody.[15]

Based on the amino acid sequence predicted by the nucleotide sequence of the full-length functional cDNA for the human *MDR*1 gene, a putative primary structure for P-glycoprotein was postulated.[16] This structure (reviewed in Refs. 1, 17), illustrated schematically in Figure 13.1, predicts 12 transmembrane segments, two potential ATP binding/utilization sites, and three extracellular glycosylation sites. Although the basic features of this structure have been confirmed, it is possible that some of the transmembrane segments are actually extracellular, and that a structure with 8 or 10 transmembrane segments can exist in cellular membranes[18–20] (reviewed in Refs. 21, 22). The conserved ATP binding regions and the presence of multiple hydrophobic transmembrane segments define the *mdr* gene products as members of the ATP binding cassette (ABC) family of transporters.

13.2. THE ABC SUPERFAMILY OF TRANSPORTERS

As already noted, P-glycoproteins belong to an ancient family of membrane transport proteins known as ABC transporters[23] or traffic ATPases.[24] Organisms ranging from archebacteria, bacteria, yeast, plants, and insects to mammals including humans produce ABC transporters to serve a great variety of biological functions, such as the uptake of nutrients, the extrusion of noxious compounds, the secretion of toxins, the transport of ions and peptides, or cell signaling (for reviews see Refs. 25–27). To date, over 50 ABC transporters have been identified that are localized at the cell surface or within intracellular membranes (e.g., peroxisomes, endoplasmic reticulum) depending on their physiological role.[27] Human members of this family include the *MDR*1 multidrug transporter,[16] which is the subject of this review, the highly homologous *MDR*2 gene product that may act as a phospholipid flippase or translocase,[28,29] the MDR-associated MRP protein,[30] the cystic fibrosis transmembrane conductance regulator (CFTR) that is mutated in patients with cystic fibrosis,[31] the 70-kDa peroxisomal membrane protein PMP70 which in a defective form may be responsible for a fatal cerebro-hepato-renal dysfunction known as Zellweger syndrome,[32,33] the product of the adrenoleukodystrophy gene ALDP,[34,35] and the TAP1 and TAP2 gene products that have been implicated in the transport of peptides for major histocompatibility class I antigen presentation.[36]

Two basic structural domains are common to all ABC transporters, namely, a hydrophobic transmembrane region that harbors multiple (most often six) putative membrane-spanning segments, and a hydrophilic ATP-binding fold. As shown in Figure 13.2, functionally active ABC transporters typically consist of two transmembrane domains and two nucleotide binding domains (reviewed by Higgins[27]) that may represent part of a single polypeptide chain (e.g., *MDR*1[16] and *MDR*2[37,38] P-glycoprotein, MRP protein,[30] CFTR gene product,[31] or the *Saccharomyces cerevisiae* STE 6 **a**-mating factor transport protein[39]). Alternatively, these four domains may

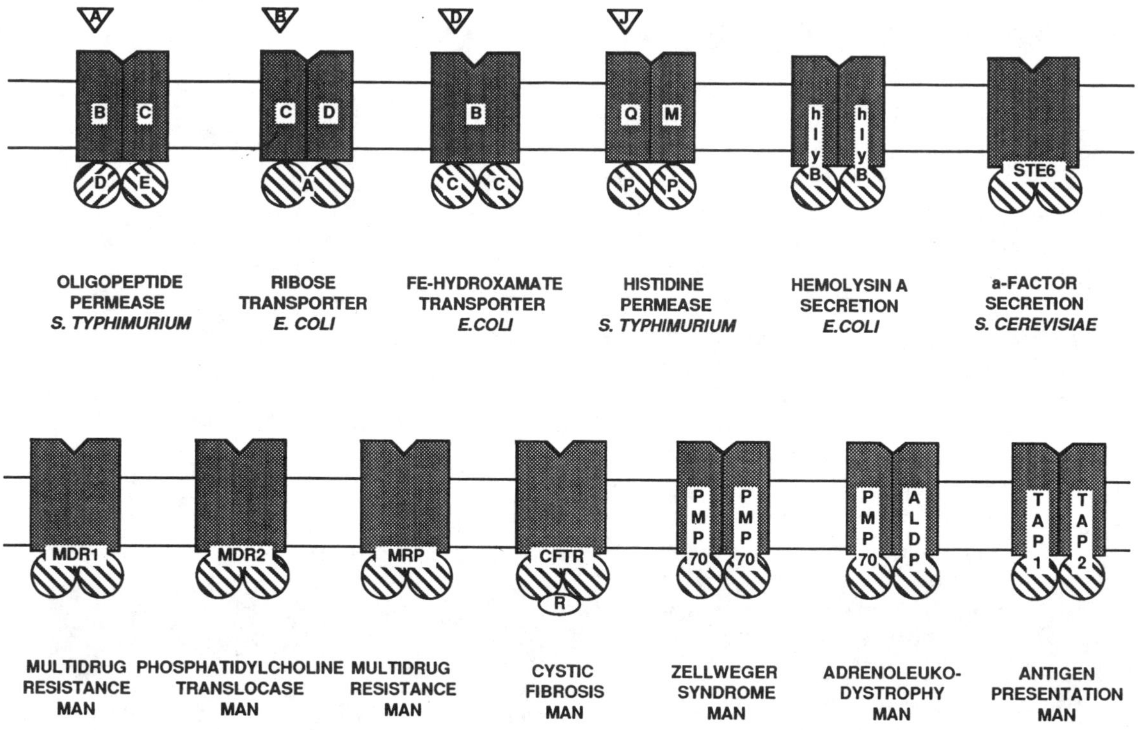

Figure 13.2. Domain organization of ATP binding cassette (ABC) transporters consisting of two highly hydrophobic transmembrane regions (shaded) and two nucleotide binding folds (shaded). Domains may represent part of a single polypeptide chain, or be derived from different subunits that function as homodimers, heterodimers, or in oligomeric complexes. Some transporters have additional domains (unshaded) that do not represent part of the core transmembrane translocation mechanism. (Adapted from Ref. 27.)

be organized within distinct subunits that are believed to cooperate as homodimers (e.g., PMP70,[32,33] bacterial hemolysin B[40]), heterodimers (e.g., TAP1 and TAP2,[36] PMP70 and ALDP[35]), or in oligomeric complexes (e.g., *Salmonella typhimurium* histidine permease[41] and oligopeptide permease,[42] *Escherichia coli* ribose transporter,[43] and Fe-hydroxamate transporter[44,45]). In most prokaryotic traffic ATPases, the transmembrane domains and the nucleotide binding folds are encoded by separate open reading frames within an operon. In eukaryotic ABC transporters, a single gene usually codes for both the membrane-integral part and the nucleotide binding fold.

All members of the ABC superfamily share extensive sequence and structural similarity and, thus, may operate by similar mechanisms. The ATP binding folds as the regions of highest homology presumably energize translocation of substrates across the membrane through the use of ATP. Four amino acid sequence motifs within the ATP binding domains of ABC transporters are highly conserved. These include the two "Walker A" and "Walker B" core consensus motifs that appear to be directly involved in ATP binding and are also shared by many nucleotide binding proteins outside of the ABC transporter superfamily.[46] In contrast, two further short stretches including a "center region" located approximately in the middle of the loop between the two Walker motifs and the so-called "linker peptide" immediately preceding the "Walker motif B" are characteristic for traffic ATPases only.[47] The overall amino acid sequence identity among ABC transporters is approximately 30% in the nucleotide binding folds, but much lower within the membrane-associated domains. Among a few ABC transporters the loop regions that connect putative membrane-spanning regions exhibit similarities in length and primary structure.[48] The putative membrane-spanning regions, however, show almost no resemblance among various ABC transporters. Besides anchoring the protein to the appropriate biological membrane, the membrane-integral regions may, therefore, be involved in directing the individual substrate specificity for the vectorial transport function, as will be discussed for P-glycoprotein in Section 13.4.2.

13.3. GENERAL MECHANISMS OF ACTION

The widespread occurrence of ABC transporters, and the extreme conservation of the ATP binding cassettes, suggests that these transporters may share a mechanism of action; or at least that they transduce the energy of ATP in similar ways for their transport functions. Thus, it has been of interest to attempt to understand the mechanism of action of the multidrug transport protein as a representative of this fascinating and growing group of transporters. In addition, information about mechanism of action could yield useful insights into ways of circumventing MDR in human cancers.

The extremely broad range of substrates recognized by P-glycoprotein suggests an unusual mechanism of substrate recognition. In addition to the long list of cytotoxic drugs enumerated in Section 13.1.1, an even longer list of noncytotoxic agents appears to interact with P-glycoprotein. These interactions were first detected owing to the ability of many hydrophobic, amphipathic drugs to "reverse" drug resistance by inhibiting the multidrug transporter. Dozens of pharmaceuticals, including the calcium channel blocker verapamil, the antihypertensive reserpine, and the antiarrhythmic and antimalarial enantiomers quinidine and quinine have been shown to be substrates for P-glycoprotein and to inhibit its pump function by saturating its ability to pump drugs out of cells (for review, see Ref. 1). Thus, any model of pump–substrate interaction must take into account this broad substrate recognition.

Furthermore, as promiscuous in substrate choice as is P-glycoprotein, there is also ample evidence that minor changes in substrate structure[49] and single amino acid substitutions (see Section 13.4.2) can produce dramatic alterations in substrate specificity. In addition, specific sites in P-glycoprotein are involved in interaction with substrate (see Section 13.4.1). How can this seeming paradox be reconciled? One class of explanations is that these drugs all interact with a common carrier, such as a normal component of the lipid bilayer, and that it is the carrier bound to drug that is recognized as a substrate. A second possibility is that it is the physical properties of the substrates (i.e., their size, shape, charge, and overall hydrophobicity) that allow them to interact with a "groove" or "pore" in P-glycoprotein, or to be transported by the transporter as passive participants in a bulk flow process.

Several lines of evidence suggest that the drugs themselves are recognized within the lipid bilayer: (1) Drug substrates are concentrated in membranes because of their hydrophobicity; (2) energy transfer,[50] confocal fluorescence,[51] and photoactivation experiments[52] suggest that in cells with active transporters the concentration of substrate drugs in the plasma membrane is reduced; (3) kinetic studies indicate that for some substrates the rate of influx of drugs into the cytoplasm may be reduced, consistent with removal of drug as it is crossing the plasma membrane[53]; (4) studies with acetoxy methyl esters of fluorescent indicator drugs (e.g., Fura-2) which are hydrolyzed into nonsubstrate anions when they enter the cytoplasm, indicate that they are P-glycoprotein substrates while crossing the plasma membrane in their neutral state[54]; and (5) recent evidence based on genetic knockouts of the mouse *mdr*2 gene, a close homologue of the multidrug transporter, suggests that this gene product "pumps" phospholipids including phosphatidylcholine,[28] a function compatible with removal of substrates directly from the membrane. This model of P-glycoprotein as an energy-dependent pump that detects and ejects drugs present in the lipid bilayer is illustrated in cartoon form in Figure 13.3.

We can only speculate about how the energy of ATP is harnessed for the detection and extrusion of drugs from the lipid bilayer. Several kinds of models have been proposed, none of which is completely satisfactory, and none of which is supported by definitive experimental evidence. P-glycoprotein may be viewed as a flippase, which, in analogy to the phospholipid flippases, takes drugs in the inner leaflet of the plasma membrane bilayer and "flips" them to the outer leaflet, or into the extracellular space.[55] A different model views P-glycoprotein as an "osmotic engine" which draws amphipathic drugs out of this membrane as a result of active transport of a cation, such as H^+, or an anion, such as a chloride.[53] Recent evidence that P-glycoprotein may be, or may activate, a chloride channel raises the possibility that anion transport or an anion channel could be involved some way in its mechanism of action.[56,57]

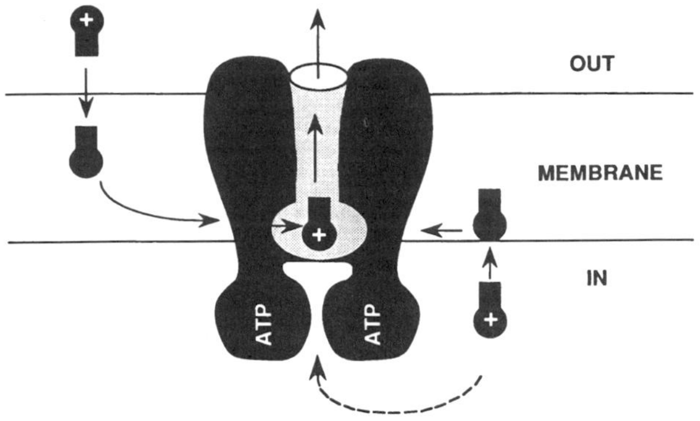

Figure 13.3. Speculative model for mechanism of action of P-glycoprotein as a scavenger of hydrophobic, amphipathic plasma membrane-associated compounds and also as a pump for efflux of hydrophobic compounds from the cytosol. (Reproduced from Ref. 1.)

The studies described in the remainder of this review on the biochemistry of P-glycoprotein are aimed at providing a better database for generation of new hypotheses about the mechanism of action of P-glycoprotein and other members of the ABC superfamily.

13.4. SUBSTRATE–TRANSPORTER INTERACTIONS

Two general approaches have been taken to define the residues involved in interactions between the multidrug transporter and its substrates. The first is a direct biochemical approach in which substrates that are also photoaffinity labels are used as probes to bind covalently to regions of the transporter that interact with substrate. The second approach involves the use of molecular genetic analysis to define residues and domains within the transporter that affect drug specificity and ability to transport drugs.

13.4.1. Photoaffinity Labeling

After the initial demonstration that P-glycoprotein bound vinblastine,[58] the first photoaffinity labeling of P-glycoprotein used an analogue of vinblastine ($[^{125}I]$-NASV) that had been derivatized with a nitrene group to allow covalent association with P-glycoprotein after exposure to UV light.[59,60] Other P-glycoprotein substrates including $[^3H]$azidopine,$[^{125}I]$dideoxyforskolin, and $[^{125}I]$iodoarylazidoprazosin prove to be easier to use either because they are commercially available or because they are more easily synthesized.[61–65] Studies in which photoaffinity-labeled P-glycoprotein was degraded with a variety of proteases and/or cyanogen bromide defined two major binding regions in the transporter comprising: (1) the 5th and 6th transmembrane domains, the extracytoplasmic loop between them, and a cytoplasmic segment immediately following the 6th transmembrane region, and (2) the homologous region around the 11th and 12th transmembrane regions.

The finding that a variety of structurally different photoaffinity labels identified similar regions of P-glycoprotein suggests that these regions are important in drug–transporter interactions. In addition, other substrates of P-glycoprotein, such as vinblastine, are effective inhibitors of this photoaffinity labeling, and seem to inhibit labeling of both sites with approximately equal ability.[62] Thus, a simple hypothesis is that there is one major drug interaction site which is comprised of residues from the N- and C-terminal halves of P-glycoprotein. Note that there is no evidence that this domain defines a drug binding site, but rather a region of the transporter through which drugs must pass during the transport process. Thus, studies that suggest more than one drug binding site[66] are not inconsistent with these photoaffinity labeling results.

13.4.2. Mutants

A structure–function analysis of spontaneously occurring and genetically engineered P-glycoprotein mutants has facilitated the identification of protein regions and amino acid residues within the multidrug transporter that may participate in substrate recognition, binding, or release. Hybrids obtained by exchanging homologous regions between the mouse *mdr*1 and *mdr*2,[67] or the mouse *mdr*1 and *mdr*3 gene products[68] have confirmed results from photoaffinity studies that the transmembrane domains in both homologous halves of P-glycoprotein may be critical determinants of drug interactions. Results with human *MDR*1/*MDR*2 chimeras[69] and deletions involving the second to fourth transmembrane domains[70] which are nonfunctional suggest that these transmembrane regions are essential for proper folding or function of the transporter.

Several point mutations that are scattered throughout the primary structure, but predicted to reside in close proximity or within membrane-spanning segments, modulate the cross-resistance pattern conferred by the multidrug transporter (Figure 13.1). These include a Gly185-to-Val185 substitution near the putative transmembrane 3 in the first intracytoplasmic loop that causes an increase in resistance to colchicine and etoposide, with a concomitant decrease in resistance to *Vinca* alkaloids and actinomycin D.[71–73,158] The introduction of an additional mutation by exchanging Asn183 with Ser near the Val substitution of Gly185 causes a relative increase in resistance to actinomycin D and vinblastine, without decreasing the relative colchicine resistance.[69] Replacement by Ala of 2 out of 13 Pro residues that reside within or close to predicted transmembrane segments (Pro223 within the predicted transmembrane domain 4 and Pro866 at an analogous position within predicted transmembrane domain 10) results in a P-glycoprotein mutant with a reduced capacity to confer resistance to colchicine, adriamycin, and actinomycin D, while the ability to confer resistance to vinblastine is barely affected.[74] An increased resistance to actinomycin D is conferred by a multidrug transporter mutant carrying both an Ala substitution for Gly338 and a Pro substitution for Ala339 in the predicted transmembrane domain 6.[75] Apparently, the Ala339 Pro mutation appears to be largely responsible for this preferential substrate specificity (P. W. Melera, personal communication).

Site-directed mutagenesis of 2 out of 31 Phe residues in predicted transmembrane regions also alters the drug resistance profile conferred by the appropriate mutant multidrug transporter.[76] Substitution of Phe335 within the predicted transmembrane domain 6 with either Ala or Ser reduces the capacity to confer resistance to colchicine or adriamycin. Substitution of the same Phe335 with Tyr has no effect, whereas an intermediate effect can be observed for a Leu substitution. Similarly, substitution of Phe978 with Ala or Ser at an analogous position within the putative transmembrane domain 12 markedly lowers the relative resistance to vinblastine and actinomycin D. Furthermore, the capacity to confer resistance to colchicine or adriamycin is almost completely abrogated. Again, the drug resistance profile conferred by a P-glycoprotein mutant carrying a Tyr at position 978 is comparable to that conferred by wild-type multidrug transporter, whereas a Leu residue at this position causes an intermediate effect.

Within the putative transmembrane domain 11, a single substitution of a Ser residue with Phe (at position 941 in the mouse *mdr*1, or at position 939 in the mouse *mdr*3 gene product) lowers the capacity to confer colchicine and adriamycin resistance drastically, but does not affect the relative vinblastine resistance.[77] Interestingly, this Ser-to-Phe mutation also modulates the MDR reversal activity of various agents including verapamil, progesterone, and cyclosporin A,[78,79] and interferes with the labeling by photoaffinity analogues.[78] Substitution of this Ser residue in either the *mdr*1 or *mdr*3 gene product with Ala or Cys has minor effects on the relative drug substrate specificity.[80] In contrast, nonconservative substitutions with Thr, Tyr, Trp, or Asp alter the drug resistance profile considerably with respect to colchicine, adriamycin, and actinomycin D, but not vinblastine. A P-glycoprotein mutant carrying Thr939/941 confers decreased resistance to colchicine and adriamycin, whereas a Tyr, Trp, or Asp mutant is basically incapable of conferring resistance to these drugs. With respect to the capacity to confer resistance to actinomycin D, Tyr and Trp substitutions cause opposite effects depending on the P-glycoprotein backbone (increasing relative resistance in *mdr*1 versus decreasing relative resistance in *mdr*3).

Taken together, these findings provide further evidence that the transmembrane domains in both halves of P-glycoprotein make unique contributions in determining the substrate specificity of the drug transport function. Amino acids that are widely separated from each other within the primary structure may act as key determinants for drug interactions. To date, however, it is unclear whether these amino acids are directly involved in drug–substrate interactions. Alternatively, they may participate in drug transport through allosteric effects. Interactions with different drugs (e.g., vinblastine or colchicine) appear to require distinct subsets of amino acids which only partially overlap. The described mutational analysis of *mdr* genes indicates that the cross-resistance pattern accompanying the MDR phenotype may be modulated by genetic variants. It remains to be determined whether clinically significant mutations occur spontaneously that affect functionally important areas of P-glycoprotein and, thus, allow cells to attain and/or alter a pattern of drug resistance.

13.5. ATP BINDING/UTILIZATION

13.5.1. Mutational Analysis

P-glycoprotein as a member of the ABC superfamily of transporters contains two highly homologous ATP binding/utilization domains that transduce the energy of ATP to support drug transport. The amino- and carboxy-terminal ATP binding loops of the human multidrug transporter are approximately 64% identical in their primary structure. An even higher degree of amino acid sequence identity is shared by the ATP binding folds in the amino-terminal half (and carboxy-terminal half, respectively) of different P-glycoproteins (e.g., 88% identity of the amino-terminal ATP binding domains of human *MDR*1 and *MDR*2 gene products). The regions of highest amino acid sequence identity encompass the "Walker motifs A and B"[46] ["GQT(V/L)ALVG(N/S)SGCGKST" and "RALVR(N/Q)P(K/H)ILLLDE" within the two halves of the *MDR*1 gene product], the "center region"[47] ["(T/C)(T/S) IAENI(R/A)YG(R/D)(E/N)" within the two halves of the *MDR*1 gene product], and the "linker peptide"[47] ["(R/K)G (A/T)QLSGGQKQRIAIA" within the two halves of the *MDR*1 gene product]. It is quite likely that all of these conserved polypeptide regions assume some important role during the drug transport process, e.g., the "Walker motifs" contribute to nucleotide binding.

A molecular genetic approach involving site-directed mutagenesis has demonstrated that both ATP binding folds of P-glycoprotein are crucial for its drug transport activity. Introduction of single discrete mutations within the "Walker motif A" in either the amino- or the carboxy-terminal half (e.g., the substitution of the conserved Gly with Ala, or the substitution of the conserved Lys with Arg) yield mutant mouse *mdr*1 gene products that are incapable of conferring the MDR phenotype to drug-sensitive cells.[81] Surprisingly, ATP binding does not seem to be impaired in such mutants of the multidrug transporter, since they are still photoaffinity labeled by 8-azido-ATP, even when carrying substitutions in both halves of the molecule. Thus, a step(s) subsequent to ATP binding may be affected in these mutants, possibly hydrolysis of ATP, release of ADP, or secondary conformational changes induced on ATP binding, hydrolysis, or release.[81] The observation that a single point mutation in one ATP binding domain is sufficient to abrogate the multidrug transporter function of P-glycoprotein may also indicate cooperative action between the two ATP binding domains.[81]

In either half of the human *MDR*1 gene product, the introduction of a more radical mutation of the conserved Lys residue by replacing it with a Met residue reduces ATP binding (B. Schott, B. M. Morse, and I. B. Roninson, personal communication). Surprisingly, low-level resistance to vinblastine can be observed for cells that have been selected for expression of a neomycin resistance gene after transfection with a plasmid carrying both this selectable marker and the mutated *MDR*1 cDNA under the control of different promoters. Subsequent selection of these transfectants for increased resistance to vinblastine is accompanied by enhanced expression of these mutant forms of P-glycoprotein (B. Schott, B. M. Morse, and I. B. Roninson, personal communication).

Preliminary data have been obtained for P-glycoprotein mutants carrying an amino-terminal type of ATP binding fold in both halves of the molecule, or a carboxy-terminal type, respectively (U. A. Germann, P. Wu, C. Hrycyna, T. Licht, S. J. Currier, I. Aksentijevich, I. Pastan, and M. M. Gottesman, unpublished data). A P-glycoprotein mutant carrying two carboxy-terminal ATP binding folds does not appear to be translocated to the cell surface and, therefore, is unable to confer drug resistance to drug-sensitive cells on transfection. In contrast, a multidrug transporter mutant with two amino-terminal ATP binding folds is expressed within the plasma membrane and retains multidrug transporter activity. Thus, the amino-terminal binding fold can structurally and functionally substitute for the carboxy-terminal one, but not vice versa. In this respect, it is interesting to note that the amino-terminal half of P-glycoprotein, but not the carboxy-terminal half, has the capacity to independently catalyze ATP hydrolysis.[82,83]

An analysis involving hybrid mouse *mdr1*/*mdr*2 gene products has revealed that the appropriate *mdr*2 ATP binding domains in either half of P-glycoprotein complement the drug transport activity of the *mdr*1 gene product.[67] Similar transfection experiments were performed with human *MDR*1/*MDR*2 chimeras to confirm this finding (U. A. Germann, P. Wu, C. Hrycyna, T. Licht, S. J. Currier, I. Aksentijevich, I. Pastan, and M. M. Gottesman, unpublished data). The human *MDR*2/*MDR*3[37,38] and the mouse *mdr*2[13,84] gene products have so far not been directly associated with the drug resistance phenomenon[85] despite a high degree (approximately 60%) of sequence identity with the human *MDR*1 and mouse *mdr*1 and *mdr*3 gene products, respectively. Again the ATP binding folds represent the regions of highest homology (around 90%). The studies involving ATP binding domain exchanges suggest that within the mdr family structurally and functionally important residues are conserved in these regions. In contrast, *MDR*1 ATP binding folds are not interchangeable with CFTR ATP binding folds that are approximately 30% identical (U. A. Germann, P. Wu, C. Hrycyna, T. Licht, S. J. Currier, I. Aksentijevich, I. Pastan, and M. M. Gottesman, unpublished data). P-glycoprotein mutants in which either the amino- or the carboxy-terminal *MDR*1 ATP binding domain is replaced by the appropriate CFTR ATP binding domain are not functional. These chimeras appear to be unstable, presumably because of the lack of determinants for proper folding.

13.5.2. Biochemistry

Studies in the 1960s shortly after identification of the MDR phenomenon, indicated that treatment with azide resulted in accumulation of anticancer drugs in multidrug-resistant cells, suggesting the involvement of an ATP-driven process.[4] In addition, molecular biological studies including the mutational analysis of conserved residues in the ATP binding sites (see Section 13.5.1) further strengthened this view. However, the P-glycoprotein-associated ATPase activity has only recently been biochemically characterized. Crude membranes of Sf9 (insect) cells infected with baculovirus-*MDR*1 exhibit high levels of vinblastine- or verapamil-stimulated ATPase activity (3–5 μmole/min per mg P-glycoprotein[86]). Although some of the recombinant P-glycoprotein is associated with the plasma membranes of Sf9 cells, the majority of the protein is retained in intracellular vesicles, which can be separated from other membranes by centrifugation of the postnuclear lysate at 5000*g*. Both the plasma- and intracellular-membrane fractions exhibit similar drug-stimulated ATPase activity. However, the octylglucoside-solubilized protein from these membranes elutes at different salt concentrations during anion exchange chromatography (S. V. Ambudkar, U. A. Germann, I. Pastan, and M. M. Gottesman, unpublished data). This heterogeneity may be related to differences in posttranslational modifications of P-glycoprotein and also in lipid composition of these membranes. The ATPase activity of P-glycoprotein in plasma membranes isolated from highly drug-resistant Chinese hamster ovary cell lines has been characterized by other workers.[87–90] These cells were selected for resistance to colchicine at very high concentrations (5 to 30 μg/ml) resulting in P-glycoprotein levels from 5 to 30% by weight of total plasma membrane protein, and this might result from overexpression of various *mdr* genes (i.e., *mdr*1, *mdr*3, and *mdr*2[91]). Thus, the observed ATPase activities may be related to P-glycoproteins encoded by one or more than one *mdr* gene.

We have partially purified P-glycoprotein from human multidrug-resistant carcinoma KB-V1 cells that have been shown to overexpress only *MDR*1 (P. V. Schoenlein and M. M. Gottesman, unpublished data) and reconstituted it into artificial membrane vesicles prepared from a mixture of lipid containing *E. coli* bulk phospholipid, phosphatidylcholine, phosphatidylserine, and cholesterol.[92] P-glycoprotein is reconstituted into proteoliposomes predominantly with inside-out orientation (>90%) and the ATPase activity is stimulated only by drugs or other agents that are known to be its substrates (Figure 13.4). Camptothecin, the hydrophobic anticancer drug that is not a substrate for P-glycoprotein, does not stimulate ATPase activity. Although the basal ATPase activity of P-glycoprotein is preserved in the detergent solution, the stimulation by the cytotoxic drug-substrates (vinblastine) or the reversing agents (verapamil) is not (Figure 13.5). The lack of effect of these agents may be related to an interaction of hydrophobic detergent with the drug-substrate binding site(s) on the P-glycoprotein. Alternatively, the conformation of soluble P-glycoprotein may not be suitable for substrate-induced activation. The requirement of a membrane environment for the substrate-induced stimulation has also been reported for other members of the superfamily of ABC transporters[93,94] and it is quite possible that this property will turn out to be one of the diagnostic features of these transporters.

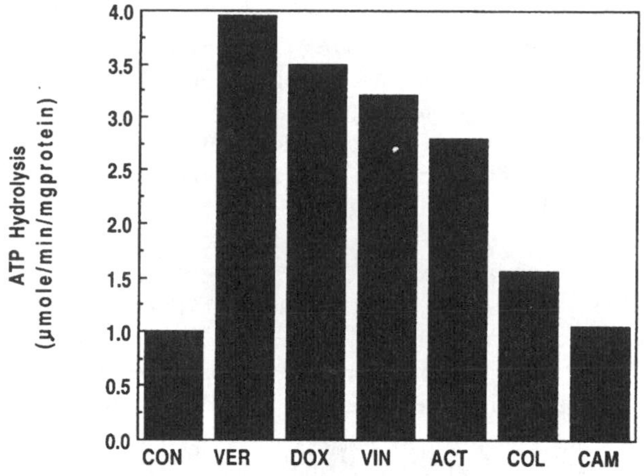

Figure 13.4. Effect of various agents on ATPase activity of reconstituted P-glycoprotein. The partially purified human P-glycoprotein was reconstituted into proteoliposomes by detergent-dilution procedure and the ATPase activity in the presence and absence of 100 μM vanadate was measured.[92] CON, control (dimethylsulfoxide treated); VER, 100 μM verapamil; DOX, 20 μM doxorubicin; VIN, 20 μM vinblastine; ACT, 20 μM actinomycin D; COL, 100 μM colchicine; and CAM, 100 μM camptothecin. Only the vanadate-sensitive activities are shown. [Adapted from Ref. 92 and unpublished data of authors (S. V. Ambudkar, I. Lelong, J. Zhang, I. Pastan, and M. M. Gottesman).]

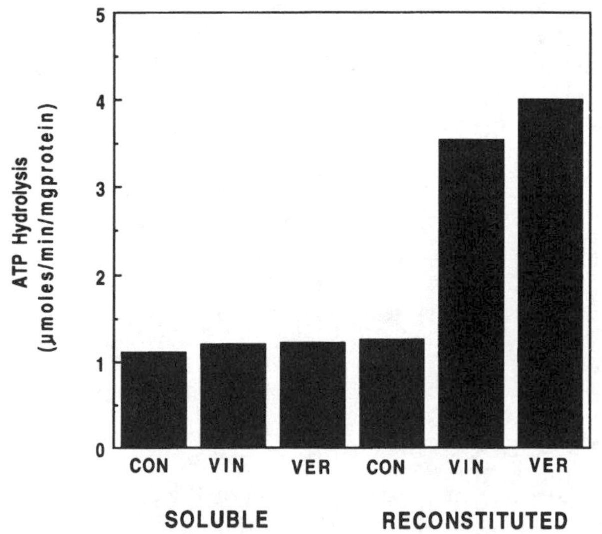

Figure 13.5. Drug-stimulated ATPase activity of soluble and reconstituted P-glycoprotein. The ATPase activity of P-glycoprotein in octylglucoside solution and in proteoliposomes was measured in the presence and absence of 100 μM vanadate. Other details as described in the legend to Figure 13.4. (Adapted from Ref. 92.)

The interaction of substrate-drugs with reconstituted P-glycoprotein leads to an increase in the maximal velocity of ATPase activity without affecting the apparent K_m for ATP (300 μM[92]). The vinblastine-stimulated ATPase activity of purified P-glycoprotein in proteoliposomes ranges from 5 to 25 μmole/min per mg protein, depending on the method of reconstitution, confirming our earlier work with the partially purified protein showing that this is a high-capacity pump similar to ion-transporting ATPases (S. V. Ambudkar, J. Zhang, I. Lelong, G. Park, C. O. Cardarelli, I. Pastan, and M. M. Gottesman, unpublished data). Several cytotoxic drugs, such as vinblastine, doxorubicin, daunomycin, and actinomycin D, stimulate the ATPase activity in a dose-dependent manner.[86–90,92] Similarly, hydrophobic acetoxymethyl ester derivatives of calcium and pH indicators, Fura-2 and BCECF [2′,7′-bis(2-carboxyethyl)-5(6)-carboxyfluorescein], respectively, which are substrates for P-glycoprotein, also stimulate ATPase activity.[54] It should be noted, however, that colchicine is only marginally effective as a stimulator of P-glycoprotein-associated ATPase activity even in Chinese hamster cells, which are selected in high concentrations of this drug (Figure 13.4 and Refs. 86–90). The reason for this discrepancy is not known. In addition to ATP, 2′-dATP, 8-azido-ATP, and 2-azido-ATP are also good substrates for P-glycoprotein-ATPase.[89]

Vanadate is a potent inhibitor of ATPase activity,[86–90,92] as well as the drug transport activity.[95] The mechanism of vanadate inhibition of Pgp is not clear. Sulfhydryl reagents such as *N*-ethylmaleimide also inhibit the P-glycoprotein-ATPase activity and Mg-ATP protects from such inactivation.[89] This suggests that the sulfhydryl groups may be located within the catalytic site. The analysis of primary structure indicates that the cysteine residue in the Walker A region of both ATP binding domains is conserved in all mammalian P-glycoproteins encoded by either *MDR*1 or *MDR*2 genes and also in some of the other ABC transporters.[16,23,91,96] It is possible that N- and C-terminal ATP domains are linked by a disulfide bond. Directed mutagenesis of these cysteines will help to elucidate their role in P-glycoprotein function. Although it is known that the ATP-binding domain alone (Figure 13.1) is sufficient for ATP binding, it is not clear whether other parts of the molecule, such as the transmembrane region(s), are also required for the hydrolysis of ATP. The availability of large quantities of pure ATP-binding domains by overexpression in heterologous expression systems may be helpful to resolve some of these issues.

13.6. REGULATION OF *MDR* GENE EXPRESSION

Because increased expression of P-glycoprotein-related genes results in biologically interesting and clinically important phenotypes in human tumors and some pathogenic organisms, the regulation of expression of the *mdr* genes is under intense investigation. Two complementary approaches have been used to investigate regulation of *mdr* gene expression. The first involves defining cellular conditions under which the level of expression of the various *mdr* genes are modulated. The second involves defining the *cis*-acting DNA sequences and *trans*-acting factors that influence basal and regulated gene expression.

13.6.1. Expression in Normal Tissues and Cultured Cells

Studies of *mdr* gene regulation have focused primarily on the rodent and human *mdr* genes. The first evidence that expression of the *mdr* genes was regulated came from studies of expression in normal tissues. The human and rodent genes corresponding to *MDR*1 and *MDR*2 exhibit overlapping but distinct patterns of expression.[97,98] The human *MDR*1 gene is expressed at relatively low levels in most normal human tissues, but relatively high levels of *MDR*1 RNA have been observed in adrenal, kidney, colon, and liver, as well as brain and testis.[99,100] Recent data indicate that *MDR*1 is also expressed in normal human bone marrow progenitor ($CD34^+$) cells and peripheral blood lymphocytes.[101–103] The distinct patterns of expression suggest that specific *mdr* gene regulatory sequences confer tissue specificity; however, enhancers for specific tissues have not been characterized.

Recent evidence suggests that the *mdr* genes respond to developmental and differentiation signals. Induction of mouse *mdr*1b gene expression has been observed in the secretory epithelium of the mouse uterus during pregnancy.[104,105] This effect is thought to be mediated through promoter sequences within the *mdr*1b first exon (+1 to +98).[106] While the human *MDR*1 and mouse *mdr*1b promoters share some sequence similarity within this region, progesterone responsiveness of the *MDR*1 promoter has not been reported. Increased *MDR*1 RNA levels have been observed after treatment of human neuroblastoma cells[107] and colon carcinoma cells[108] with differentiating agents such as retinoic acid, sodium butyrate, dimethylsulfoxide and dimethylformamide. Further, Chaudhary, Mechetner, and Roninson have shown that treatment of some hematopoietic cell lines with the protein kinase C (PKC) activators TPA and diacylglycerol (DAG) results in increased steady-state levels of *MDR*1 RNA.[103] TPA- and DAG-induced *MDR*1 expression in K562 cells was blocked by the PKC inhibitor staurosporine, implicating PKC in the pathway of activation in these cells. Under the differentiating conditions tested,[103,106–109] increases in *MDR*1 RNA were reflected in increased P-glycoprotein levels. Finally, recent studies have shown that TPA-induced increases in steady-state levels of *MDR*1 RNA in K562 cells are correlated with activation of the *MDR*1 promoter in TPA-treated K562 cells (McCoy and Cornwell, unpublished; and as discussed below), indicating that increased transcription is a component of TPA-induced activation of the *MDR*1 gene in these cells.

One implication of these data is that the *MDR* genes are transcriptionally regulated by cellular gene products that regulate normal growth control processes. Evidence supporting this hypothesis comes from studies showing that genes known to be involved in cell proliferation can influence *MDR*1 gene expression. The activated proto-oncogenes v-*ras* and v-*raf* have been shown to increase steady-state *mdr* RNA levels in transfected rat cells[110] and the proto-oncogene c-Ha-*ras* has been shown to activate the *MDR*1 promoter in transfected NIH3T3 cells.[111] Further, a mutant form of the cell cycle regulatory gene p53 has been shown to activate the *MDR*1 promoter in transfected NIH3T3 cells.[111] Additional evidence that genes involved in cellular growth control signals can regulate activity of the *MDR*1 promoter was provided by studies of quiescent NIH3T3 and human IMR-90 lung fibroblasts stimulated to enter the cell cycle by the addition of serum or mitogens. The results demonstrated (1) that serum- and mitogen-activation of the *MDR*1 promoter in transfected NIH3T3 cells is mediated by Raf-1 kinase[112] and (2) that the endogenous *MDR*1 gene can be stimulated by addition of serum to quiescent human cells (Stommel and Cornwell, unpublished data). Taken together, these data suggest that the human *MDR*1 gene is transcriptionally regulated through normal cellular processes involved in cell growth and differentiation.

Of considerable interest is whether carcinogens and chemotherapeutic agents directly activate *mdr* gene expression. There is evidence that the rodent *mdr* genes are activated by treatment of cells with carcinogens[113–115] and chemotherapeutic agents.[116] Two recent reports suggest that steady-state levels of the human *MDR*1 gene can be increased by drugs that act as P-glycoprotein antagonists[117] and some chemotherapeutic agents.[103] However, the mechanism of activation by these agents is not known. Heat shock and sodium arsenite treatment of cells have been shown to increase *MDR*1 RNA levels in a human kidney cell line[118] and the heat shock response depends on the integrity of heat shock-like elements within the *MDR*1 promoter,[119] suggesting that the heat shock and sodium arsenite response is controlled at least in part at the level of transcription. These results suggest that the cellular stress response may influence *mdr* gene expression in some cells.

13.6.2. Functional Studies of *MDR* Regulatory Sequences

While there are data suggesting that *mdr* gene expression is regulated by both transcriptional and posttranscriptional controls,[106,111,112,115–117] transcriptional control has been the focus of most molecular analysis of *MDR*1 regulation. To identify DNA sequences that regulate *mdr* transcription, DNA flanking the human and rodent structural sequences has been cloned and partially characterized.[120] Analysis of the 5′-flanking sequence has shown that human *MDR*1 and rodent *mdr*1a transcripts can be generated from two promoters in some cells.[121,122] The sequences defining the distal human and rodent promoters, which in the human gene are approximately 10 to 20 kilobases upstream of the stuctural gene, are not well characterized. It is not known whether distal *MDR*1 promoter transcripts encode functional protein. Hsu *et al.*[122] reported that transcript initiation from the *mdr*1a distal promoter correlated with a 70 to 85% decrease in the ratio of *mdr*1a protein to transcripts at steady state, suggesting that distal promoter transcripts are not translated into protein. In most normal tissues and most tumors, *MDR*1 and *mdr*1b transcripts are generated from proximal promoter sequences.[122,123]

Deletion analysis of the proximal human *MDR*1 and rodent *mdr*1a and *mdr*1b promoters indicates that basal transcription is conferred by sequences within approximately 200 base pairs centered about the transcription initiation site.[123–127] Within the core promoter sequences, the human *MDR*1 and rodent *mdr* proximal promoters share putative regulatory motifs, such as consensus binding sites for the transcriptional activators AP1 and SP1.[122,123,128,129] However, there exist clear distinctions in the sequence and organization of potential regulatory motifs in the different rodent and human promoters. For example, the mouse *mdr*1a, *mdr*1b, and rat *mdr*1b promoters all contain putative TATA-boxes, whereas the human *MDR*1 gene does not.[121,122,129] There are only a few studies in which the importance of the many putative regulatory sites have been examined and, as expected, the data suggest that there are dramatic differences in the influence of specific regulatory motifs depending on the cell type in which the human or rodent promoter function is measured.[112,116,122,130–132] The differences between promoter sequences among the rodent genes, as well as between rodent and human genes, may account for the differences in activation observed in response to the various agents described above.

A schematic summary of the human *MDR*1 core promoter is shown in Figure 13.6. Only those sites that have been demonstrated by deletion and mutation analysis to alter promoter activity and transcription factor binding are shown. Three protein binding regions upstream of the initiation site influence promoter activity. The first, a GC-rich region between about -120 and -100, has been identified as a transcriptional repressor binding site in drug-resistant human erythroleukemia K562/ADR cells[133] and drug-resistant KB-8-5 carcinoma cells.[127] The putative repressor proteins from the two cell types have different binding characteristics and their relationship is not known. Second, a consensus binding site for the NF-Y family of proteins between -70 and -80 has been shown to influence the basal level of *MDR*1 transcription,[132] although it is not yet clear which of the NF-Y family members regulate the *MDR* promoters through this site. Third, an over-

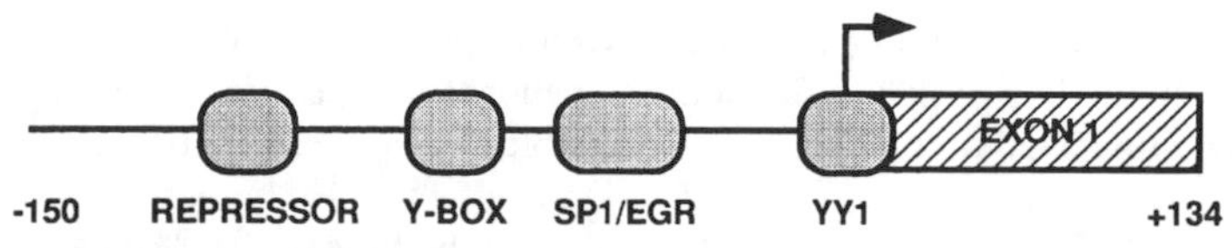

Figure 13.6 Schematic diagram of functionally important transcription factor binding sites in the *MDR*1 promoter. Core promoter sequences (solid line) are indicated relative to the transcription initiation site at +1 (arrow). The first noncoding exon (exon 1, hatched box) is shown from +1 to +134. Transcription factor binding sites (ovals) are indicated below the diagram.

lapping binding site for SP1 and members of the EGR (early growth response) family of transcription factors has been shown to influence basal promoter activity.[127] While purified SP1 and EGR-1 bind this site, thus far only SP1 has been shown directly to activate the promoter through binding to this GC-rich region.[127] However, mutations that inhibit EGR binding but not SP1 binding to this site inhibit TPA-activation of the *MDR*1 promoter in K562 cells (McCoy *et al.*, unpublished data), suggesting that integrity of and binding to the EGR site is functionally important for *MDR*1 regulation in some cells. The stimulatory effect of mutant forms of p53 also appears to be mediated through sequences or factors interacting with the basal promoter and/or transcription initiation regions (Chin *et al.*, unpublished). Finally, the 13-base-pair region surrounding the human *MDR*1 initiation site has been shown to specify accurate *MDR*1 initiation *in vitro*[134] and *in vivo*[135] and these *MDR*1 sequences specifically bind the initiator protein YY1 (Cornwell, unpublished data).

Our understanding of the transcriptional regulation of *MDR* genes is far from complete. The mechanistic details of how NF-Y, SP1 (and/or EGR) interact with their respective binding sites in the *MDR*1 promoter and whether the interactions are independent or cooperative remain to be elucidated. Furthermore, only a few preliminary studies of how *MDR* transcription is activated at the molecular level have been reported. As discussed above, the sodium arsenite response of the *MDR*1 promoter has been reported to require tandem repeats of heat-shock responsive elements in the *MDR*1 promoter.[120] Second, deletion analysis localized progesterone responsive sequences of the *mdr*1b promoter to the first exon of this gene,[106] although detailed analysis of this region has not been reported. Finally, activation of transcription is dependent on *MDR*1 core promoter sequences in at least two cases: (1) activation in response to v-raf, and to serum stimulation in quiescent NIH3T3 cells (Means and Cornwell, unpublished) and (2) activation in response to TPA stimulation in K562 cells (McCoy and Cornwell, unpublished). In both cases, activation is conferred by *MDR*1 promoter sequences including only the SP1/EGR binding site and the INR.

Although the molecular characterization of *MDR* gene transcription is incomplete, taken together the data suggest (1) that regulation of transcription is an important component of the expression of these genes and (2) that transcription is regulated by cellular proteins involved in cell growth, differentiation, and homeostasis.

13.7. POSTTRANSLATIONAL MODIFICATIONS

As already noted, when P-glycoprotein was first discovered, it was described as a phospho-glycoprotein.[6] The role of these posttranslational modifications has been the subject of some controversy, which will be reviewed in this section.

13.7.1. Glycosylation

There are ten potential glycosylation sites in human P-glycoprotein as determined by the presence of Asn-X-Ser or Asn-X-Thr residues in the primary amino acid sequence. Existing evidence suggests that only the three potential sites in the first extracytoplasmic loop are glycosylated (see Figure 13.1). This evidence is as follows: (1) Photoaffinity-labeled P-glycoprotein fragments from the amino-terminal half of P-glycoprotein migrate more slowly than predicted, and their migration increases as predicted after treatment with endoglycosidase F;[61] (2) purified P-glycoprotein fragments, during chromatography on wheat germ agglutinin column, behave as if only the amino-terminal fragments are glycosylated[92] (S. Ambudkar, I. Pastan, and M. M. Gottesman, unpublished data); (3) deletion of the three glycosylation sites in the first extracytoplasmic loop produces a P-glycoprotein that migrates as expected for a completely unglycosylated protein.[136] Some alternative models of P-glycoprotein structure predict that one or more of the potential internal glycosylation sites might be extracellular, and hence subject to glycosylation.[18] Hence, these results showing glycosylation of the sites in the first extracytoplasmic loops only support the structural model of P-glycoprotein presented in Figure 13.1, although they do not prove it since the presence of an extracellular glycosylation site does not guarantee that it will be glycosylated.

Several lines of evidence argue against a major role for glycosylation in affecting the transport function of P-glycoprotein: (1) Growth of MDR cells in tunicamycin has no effect on drug resistance[137]; (2) MDR cells can be readily selected from among lectin-resistant CHO mutants which are incapable of carrying out Asn-linked glycosylation[138]; and (3) mutants lacking one or all three Asn-linked glycosylation sites in the first extracytoplasmic domain of P-glycoprotein are capable of conferring MDR when expressed in drug-sensitive cells.[136] It is worth noting, however, that mutants lacking all three glycosylation sites are somewhat less efficient in conferring drug resistance, consistent with a possible role for glycosylation in helping to target P-glycoprotein to the cell surface, or in stabilizing protein against proteolytic digestion once it is on the cell surface.[136]

13.7.2. Phosphorylation

There is considerable evidence to suggest that protein kinases may be involved in the expression and function of P-glycoprotein and MDR. Several groups have reported an increase in kinase activities or increased phosphorylation of P-glycoprotein in multidrug-resistant cells.[139–143] Elevated levels of PKC are associated with increased MDR of some but not all multidrug-resistant cells.[140,141,144] The transfection of a plasmid expressing functional PKC-α, but not PKC-γ, is associated with an increase in MDR in BC-19 cells.[145,146] This indicates that the regulation of P-glycoprotein may be dependent on specific isoform(s) of PKC. However, it is possible that in these transfected cells PKC-α may have an indirect effect on P-glycoprotein, such as by increasing mRNA.[147] Recently, Bates *et al.*[148] have demonstrated that the inhibition of PKC with calphostin C or staurosporine or prolonged treatment with the phorbol ester TPA led to decreased phosphorylation of P-glycoprotein and these treatments also affected the action of certain antagonists, but a direct effect of phosphorylation has not been shown.

The specific serine residues in human P-glycoprotein phosphorylated *in vitro* by PKC have been identified.[149] The serine residues 661, 667, and 671 are phosphorylated by PKC whereas serine 683 is phosphorylated by only PKA. These residues are clustered in the linker region located between the two halves of P-glycoprotein (see Figure 13.1). Similarly, in mouse *mdr*1b gene-encoded P-glycoprotein, serine 669 and serine 681 are phosphorylated by PKC and PKA, respectively.[150] This region (amino acids 629 to 687) encoded by exon 16 contains 38% charged residues and has been called a "mini R domain" as it is analogous to the R domain of the CFTR protein.[31] The R domain of CFTR contains several PKA and PKC consensus phosphorylation sites and this region is implicated in the regulation of CFTR function.[151] Recently, we have purified a novel 55 to 60-kDa membrane-associated kinase from human multidrug-resistant KB-V1 cells that predominantly phosphorylates P-glycoprotein. The kinase can function in the presence of Mg^{2+} or Mn^{2+} and it is inhibited by >1 mM calcium. This kinase, referred to as "V-1 kinase," does not phosphorylate histone, or PKC or PKA peptide-substrates, and it is inhibited only by staurosporine (S. V. Ambudkar, G. Park, C. O. Cardarelli, I. Pastan, and M. M. Gottesman, unpublished data). The analysis of the amino acid sequence of the purified V-1 kinase will facilitate the determination of its identity. Recent evidence indicates that substitution of serine residues 661, 667, 671, 675, and 683 with either alanine or aspartate has no effect on the ability of P-glycoprotein to confer drug resistance.[159]

13.8. RELEVANCE OF MOLECULAR BIOLOGY TO HUMAN DISEASE

The insights revealed by molecular analysis of P-glycoprotein promise to improve the treatment of cancer and may be useful in the treatment of other diseases as well. As the substrate and inhibitor interaction sites on P-glycoprotein are defined, it should be possible to predict what molecules will be the best inhibitors of P-glycoprotein. Such information will be invaluable for the design of drugs to reverse MDR in human cancer. Current information suggests that approximately 50% of human cancers will express the *MDR*1 gene at physiologically significant levels at some time during the development of the cancer, and the treatment of most of these tumors would benefit from potent agents to reverse MDR due to P-glycoprotein (reviewed in Gottesman[152]). Similarly, more information about the mechanism of action of the transporter, especially the mode of energy transduction, would allow targeting of new drugs to inhibit the P-glycoprotein pump.

The ability to use *MDR*1 vectors to confer MDR on sensitive cells has suggested that *MDR*1 vectors might be extremely useful as dominant selectable markers for gene therapy. Bone marrow cells in mice can be protected from the toxic effects of chemotherapy by expression of the human *MDR*1 gene[153–157] suggesting that such an approach might prove useful to protect human bone marrow from the toxic effects of chemotherapy. Knowledge about mutations that alter substrate and inhibitor specificity (see Section 13.4.2) will facilitate development of "designer" *MDR*1 vectors that specifically protect bone marrow and other tissues against certain potent anticancer drugs, or allow resistance to MDR "reversing agents," a desirable property if the cancer being treated expresses a wild-type *MDR*1 gene. *MDR*1 vectors expressing other genes, which are otherwise nonselectable, could be used to select for cells *in vivo* that have been transduced with genes whose expression is defective owing to inborn errors of metabolism.

Finally, knowledge gleaned from molecular studies of the multidrug transporter will be useful in understanding the growing family of ABC transporters, many of which are essential for normal cellular physiology, and which, when defective, contribute to a growing list of human diseases.

ACKNOWLEDGMENTS. S.V.A. was supported by NIH Grant 1R55 CA57154-01 and American Cancer Society Grant BE-157. M.M.C. was supported by PHS Grants RO1 CA51728 and RO1 CA63419.

REFERENCES

1. Gottesman, M. M., and Pastan, I. (1993). Biochemistry of multidrug resistance mediated by the multidrug transporter. *Annu. Rev. Biochem.* **62:**385–427.
2. Danø, K. (1972). Cross resistance between vinca alkaloids and anthracyclines in Ehrlich ascites tumor *in vivo. Biochim. Biophys. Acta* **323:**466–483.
3. Ling, V., and Thompson, L. H. (1974). Reduced permeability in CHO cells as a mechanism of resistance to colchicine. *J. Cell. Physiol.* **83:**103–116.
4. Danø, K. (1973). Active outward transport of daunomycin in resistant Ehrlich ascites tumor cells. *Biochim. Biophys. Acta* **323:**466–483.
5. Fojo, A., Akiyama, S.-i., Gottesman, M. M., and Pastan, I. (1985). Reduced drug accumulation in multiply drug-resistant human KB carcinoma cell lines. *Cancer Res.* **45:**3002–3007.
6. Juliano, R. L., and Ling, V. (1976). A surface glycoprotein modulating drug permeability in Chinese hamster ovary cell mutants. *Biochim. Biophys. Acta* **455:**152–162.
7. Ueda, K., Cornwell, M. M., Gottesman, M. M., Pastan, I., Roninson, I. B., Ling, V., Riordan, J. R. (1986). The *mdr*l gene, responsible for multidrug-resistance, codes for P-glycoprotein. *Biochem. Biophys. Res. Commun.* **141:**956–962.
8. Shen, D.-W., Fojo, A. T., Chin, J. E., Roninson, I. B., Richert, N., Pastan, I., and Gottesman, M. M. (1986). Human multidrug resistant cell lines: Increased *mdr*l expression can precede gene amplification. *Science* **232:**643–645.
9. Roninson, I. B., Chin, J. E., Choi, K., *et al.* (1986). Isolation of human *mdr* DNA sequences amplified in multidrug-resistant KB carcinoma cells. *Proc. Natl. Acad. Sci. USA* **83:**4538–4552.
10. Ueda, K., Cardarelli, C., Gottesman, M. M., *et al.* (1987). Expression of a full-length cDNA for the human "*MDR*1" (P-glycoprotein) gene confers multidrug resistance to colchicine, doxorubicin, and vinblastine. *Proc. Natl. Acad. Sci. USA* **84:**3004–3008.
11. Pastan, I., Gottesman, M. M., Ueda, K., *et al.* (1988). A retrovirus carrying an *MDR*1 cDNA confers multidrug resistance and polarized expression of P-glycoprotein in MDCK cells. *Proc. Natl. Acad. Sci. USA* **85:**4486–4490.

12. Gros, P., Ben Neriah, Y., Croop, J., *et al.* (1986). Isolation and characterization of a complementary DNA that confers multidrug resistance. *Nature* **323**:728–731.
13. Gros, P., Raymond, M., Bell, J., *et al.* (1988). Cloning and characterization of a second member of the mouse *mdr* gene family. *Mol. Cell. Biol.* **8**:2770–2778.
14. Scotto, K. W., Biedler, J. L., and Melera, P. W. (1986). The differential amplification and expression of genes associated with multidrug-resistance in mammalian cells. *Science* **232**:751–755.
15. Gerlach, J. H., Endicott, J. A., Juranka, P. F., et al. (1986). Homology between P-glycoprotein and a bacterial haemolysin transport protein suggests a model for multidrug resistance. *Nature* **324**: 485–489.
16. Chen, C.-J., Chin, J. E., Ueda, K., Clark, D., Pastan, I., Gottesman, M. M., and Roninson, I. B. (1986). Internal duplication and homology with bacterial transport proteins in the *mdr*1 (P-glycoprotein) gene from multidrug-resistant human cells. *Cell* **47**:381–389.
17. Ambudkar, S. V., Pastan, I., and Gottesman, M. M. (1995). Cellular and biochemical aspects of multidrug resistance. In *Drug Transport in Antimicrobial and Anticancer Chemotherapy* (N.H. Georgapapadakou, ed.), Dekker, New York, pp. 525–547.
18. Zhang, J.-T., and Ling, V. (1991). Study of membrane orientation and glycosylated extracellular loops of mouse P-glycoprotein by *in vitro* translation. *J. Biol. Chem.* **266**:18224–18232.
19. Zhang, J. T., Duthie, M., and Ling, V. (1993). Membrane topology of the N-terminal half of the hamster P-glycoprotein molecule. *J. Biol. Chem.* **268**:15101–15110.
20. Skach, W. R., Calayag, M. C., and Lingappa, V. R. (1993). Evidence for an alternate model of human P-glycoprotein structure and biogenesis. *J. Biol. Chem.* **268**:6903–6908.
21. Germann, U. A., Pastan, I., and Gottesman, M. M. (1993). P-glycoproteins: Mediators of multidrug resistance. *Semin. Cell Biol.* **4**: 63–76.
22. Germann, U. A. (1993). Molecular analysis of the multidrug transporter. *Cytotechnology* **12**:33–62.
23. Hyde, S. C., Emsley, P., Hartshorn, M. J., *et al.* (1990). Structural model of ATP-binding proteins associated with cystic fibrosis, multidrug resistance, and bacterial transport. *Nature* **346**:362–365.
24. Mimura, C. S., Holbrook, S. R., and Ames, G. F.-L. (1991). Structural model of the nucleotide-binding conserved component of periplasmic permeases. *Proc. Natl. Acad. Sci. USA* **88**:84–88.
25. Ames, G. F.-L., Mimura, C. S., Holbrook, S. R., *et al.* (1992). Traffic ATPases: A superfamily of transport proteins operating from *Escherichia coli* to humans. *Adv. Enzymol.* **65**:1–47.
26. Ames, G. F.-L., and Lecar, H. (1992). ATP-dependent bacterial transporters and cystic fibrosis: Analogy between channels and transporters. *FASEB J.* **6**:2660–2666.
27. Higgins, C. F. (1992). ABC transporters—from microorganisms to man. *Annu. Rev. Cell Biol.* **8**:67–113.
28. Smit, J. J. M., Schinkel, A. H., Oude Elferink, R. P. J., *et al.* (1993). Homozygous disruption of the murine *mdr*2 P-glycoprotein gene leads to a complete absence of phospholipid from bile and to liver disease. *Cell* **75**:451–462.
29. Ruetz, S., and Gros, P. (1994). Phosphatidylcholine translocase: A physiological role for the *mdr*2 gene. *Cell* **77**:1071–1081.
30. Cole, S. P. C., Bhardwaj, G., Gerlach, J. H., *et al.* (1992). Overexpression of a transporter gene in a multidrug-resistant human lung cancer cell line. *Science* **258**:1650–1654.
31. Riordan, J. R., Rommens, J. M., Kerem, B.-s., *et al.* (1989). Identification of the cystic fibrosis gene: Cloning and characterization of complementary DNA. *Science* **245**:1066–1073.
32. Kamijo, K., Taketani, S., Yokota, S., *et al.* (1989). The 70-kDa peroxisomal membrane protein is a member of the Mdr (P-glycoprotein)-related ATP binding superfamily. *J. Biol. Chem.* **265**:4534–4540.
33. Gärtner, J., Moser, H., and Valle, D. (1992). Mutations in the 70K peroxisomal membrane protein gene in Zellweger syndrome. *Nature Genet* **1**:16–23.
34. Mosser, J., Douar, A., Sarde, C., *et al.* (1993). Putative X-linked adrenoleukodystrophy gene shares unexpected homology with ABC transporters. *Nature* **361**:726–730.
35. Valle, D., and Gärtner, J. (1993). Penetrating the peroxisome. *Nature* **361**:682–683.
36. Monaco, J. J. (1992). Major histocompatibility complex-linked transport proteins and antigen processing. *Immunol. Res.* **11**:125–132.
37. Van der Bliek, A. M., Baas, F., Ten Houte de Lange, T., *et al.* (1987). The human *mdr*3 gene encodes a novel P-glycoprotein homologue and gives rise to alternatively spliced mRNAs in liver. *EMBO J.* **6**:3325–3331.
38. Van der Bliek, A. M., Kooiman, P. M., Schneider, C., *et al.* (1988). Sequence of *mdr*3 cDNA, encoding a human P-glycoprotein. *Gene* **71**:401–411.
39. McGrath, J. P., and Varshavsky, A. (1989). The yeast *STE6* gene encodes a homologue of the mammalian multidrug resistance P-glycoprotein. *Nature* **340**:400–404.
40. Hess, J., Wels, W., Vogel, M., *et al.* (1986). Nucleotide sequence of a plasmid-encoded haemolysin determinant and its comparison with a corresponding chromosomal haemolysin sequence. *FEMS Microbiol. Lett.* **34**:1–11.
41. Higgins, C. F., Haag, P. D., Nikaido, K., *et al.* (1982). Complete nucleotide sequence and identification of membrane components of the histidine transport operon of *S. typhimurium. Nature* **298**: 723–727.
42. Hiles, I. D., Gallagher, M. P., Jamieson, D., *et al.* (1987). Molecular characterization of the oligopeptide permease of *Salmonella typhimurium. J. Mol. Biol.* **195**:125–142.
43. Bell, A. W., Buckel, S. D., Groarke, J. M., *et al.* (1986). The nucleotide sequence of the *rbsD, rbsA* and *rbsC* genes of *Escherichia coli. J. Biol. Chem.* **261**:7652–7658.
44. Kostler, W., and Braun, V. (1986). Iron hydroxamate transport of *Escherichia coli:* Nucleotide sequence of the *fhuB* gene and identification of the protein. *Mol. Gen. Genet.* **204**:435–442.
45. Coulton, J. W., Mason, P., and Allatt, D. (1987). *fhuC* and *fhuD* genes for iron(III)-ferrichrome transport into *Escherichia coli* K-12. *J. Bacteriol.* **169**:3844–3849.
46. Walker, J. E., Saraste, M., Runswick, M. J., *et al.* (1982). Distantly related sequences in the α- and β-subunits of ATP synthase, myosin, kinases and other ATP-requiring enzymes and a common nucleotide binding fold. *EMBO J.* **1**:945–951.
47. Shyamala, V., Baichwald, V., Beall, E., *et al.* (1991). Structure–function analysis of the histidine permease and comparison with cystic fibrosis mutations. *J. Biol. Chem.* **266**:18714–18719.
48. Manavalan, P., Smith, A. E., and Mcpherson, J. M. (1993). Sequence and structural homology among membrane-associated domains of CFTR and certain transporter proteins. *J. Protein Chem.* **12**:279–290.
49. Tang-Wai, D. F., Brossi, A., Arnold, L. D., *et al.* (1993). The nitrogen of the acetamido group of colchicine modulates P-glycoprotein-mediated multidrug resistance. *Biochemistry* **32**:6470–6476.
50. Kessel, D. (1989). Exploring multidrug resistance by using rhodamine 123. *Cancer Commun.* **1**:145–149.
51. Weaver, J. L., Ine, P. S., Aszalos, A., *et al.* (1991). Laser scanning and confocal microscopy of daunorubicin, doxorubicin and rhodamine 123 in multidrug-resistant cells. *Exp. Cell Res.* **196**:323–329.
52. Raviv, Y., Pollard, H. B., Bruggemann, E. P., *et al.* (1990). Photosensitized labeling of a functional multidrug transporter in living drug-resistant tumor cells. *J. Biol. Chem.* **265**:3975–3980.
53. Stein, W. D., Cardarelli, C. O., Pastan, I., *et al.* (1994). Kinetic evidence suggesting that the multidrug transporter differentially han-

dles influx and efflux of its substrates. *Mol. Pharmacol.* **45:**763–772.
54. Homolya, L., Hollo, Z., Germann, U. A., *et al.* (1993). Fluorescent cellular indicators are extruded by the multidrug resistance protein. *J. Biol. Chem.* **268:**21493–21496.
55. Higgins, C. F., and Gottesman, M. M. (1992). Is the multidrug transporter a flippase? *Trends Pharmacol. Sci.* **17:**18–21.
56. Gill, D. R., Hyde, S. C., Higgins, C. F., *et al.* (1992). Separation of drug transport and chloride channel functions of the human multidrug resistance P-glycoprotein. *Cell* **71:**23–32.
57. Valverde, M. A., Diáz, M., Sepúlveda, F. V., *et al.* (1992). Volume-regulated chloride channels associated with the human multidrug resistance P-glycoprotein. *Nature* **355:**830–833.
58. Cornwell, M. M., Gottesman, M. M., and Pastan, I. (1986). Increased vinblastine binding to membrane vesicles from multidrug resistant KB cells. *J. Biol. Chem.* **262:**7921–7928.
59. Cornwell, M. M., Safa, A. R., Felsted, R. L., *et al.* (1986). Membrane vesicles from multidrug-resistant human cancer cells contain a specific 150–170kDa protein detected by photoaffinity labeling. *Proc. Natl. Acad. Sci. USA* **83:**3847–3850.
60. Safa, A. R., Glover, C. J., Meyets, M. B., *et al.* (1986). Vinblastine photoaffinity labeling of a high-molecular-weight surface membrane glycoprotein specific for multidrug-resistant cells. *J. Biol. Chem.* **261:**6137–6140.
61. Bruggemann, E. P., Germann, U. A., Gottesman, M. M., *et al.* (1989). Two different regions of P-glycoprotein are photoaffinity labeled by azidopine. *J. Biol. Chem.* **264:**15483–15488.
62. Bruggemann, E. P., Currier, S. J., Gottesman, M. M., *et al.* (1992). Characterization of the azidopine and vinblastine binding site of P-glycoprotein. *J. Biol. Chem.* **267:**21020–21026.
63. Greenberger, L. M., Lisanti, C. J., Silva, J. T., *et al.* (1991). Domain mapping of the photoaffinity drug-binding sites in P-glycoprotein encoded mouse *mdr*1b. *J. Biol. Chem.* **266:**20744–20751.
64. Morris, D. I., Speicher, L. A., Ruoho, A. E., *et al.* (1991). Interaction of forskolin with the P-glycoprotein multidrug transporter. *Biochemistry* **30:**8371–8379.
65. Greenberger, L. M. (1993). Major photoaffinity drug labeling sites for iodoarylazidoprazosin in P-glycoprotein are within, or immediately C-terminal to, transmembrane domain-6 and domain-12. *J. Biol. Chem.* **268:**11417–11425.
66. Tamai, I., and Safa, A. R. (1991). Azidopine noncompetitively interacts with vinblastine and cyclosporin A binding to P-glycoprotein in multidrug resistant cells. *J. Biol. Chem.* **266:**16796–16800.
67. Buschman, E., and Gros, P. (1991). Functional analysis of chimeric genes obtained by exchanging homologous domains of the mouse *mdr*1 and *mdr*2 genes. *Mol. Cell. Biol.* **11:**595–603.
68. Dhir, R., and Gros, P. (1992). Functional analysis of chimeric proteins constructed by exchanging homologous domains of two P-glycoproteins conferring distinct drug resistance profiles. *Biochemistry* **31:**6103–6110.
69. Currier, S. J., Kane, S. E., Willingham, M. C., *et al.* (1992). Identification of residues in the first cytoplasmic loop of P-glycoprotein involved in the function of chimeric human *MDR*1–*MDR*2 transporters. *J. Biol. Chem.* **267:**25153–25159.
70. Currier, S. J., Ueda, K., Willingham, M. C., *et al.* (1989). Deletion and insertion mutants of the multidrug transporter. *J. Biol. Chem.* **264:**14376–14381.
71. Choi, K., Chen, C.-J., Kriegler, M., *et al.* (1989). An altered pattern of cross-resistance in multidrug-resistant human cells results from spontaneous mutations in the *mdr*1 (P-glycoprotein) gene. *Cell* **53:** 519–529.
72. Kioka, N., Tsubota, J., Kakehi, Y., *et al.* (1989). P-glycoprotein gene (*MDR*1) cDNA from human adrenal: Normal P-glycoprotein carries Gly^{185} with an altered pattern of multidrug resistance. *Biochem. Biophys. Res. Commun.* **162:**224–231.
73. Safa, A. R., Stern, R. K., Choi, K., *et al.* (1990). Molecular basis of preferential resistance to colchicine in multidrug-resistant human cells conferred by Gly to Val-185 substitution in P-glycoprotein. *Proc. Natl. Acad. Sci. USA* **87:**7225–7229.
74. Loo, T. W., and Clarke, D. M. (1993). Functional consequences of proline mutations in the predicted transmembrane domain of P-glycoprotein. *J. Biol. Chem.* **268:**3143–3149.
75. Devine, S. E., Ling, V., and Melera, P. W. (1992). Amino acid substitutions in the 6th transmembrane domain of P-glycoprotein alter multidrug resistance. *Proc. Natl. Acad. Sci. USA* **89:**4564–4568.
76. Loo, T. W., and Clarke, D. M. (1993). Functional consequences of phenylalanine mutations in the predicted transmembrane domain of P-glycoprotein. *J. Biol. Chem.* **268:**19965–19972.
77. Gros, P., Dhir, R., Croop, J., *et al.* (1991). A single amino acid substitution strongly modulates the activity and substrate specificity of the mouse *mdr*1 and *mdr*3 drug efflux pumps. *Proc. Natl. Acad. Sci. USA* **88:**7289–7293.
78. Kajiji, S., Talbot, F., Grizzuti, K., *et al.* (1993). Functional analysis of P-glycoprotein mutants identifies predicted transmembrane domain-11 as a putative drug binding site. *Biochemistry* **32:**4185–4194.
79. Kajiji, S., Dreslin, J. A., Grizzuti, K., *et al.* (1994). Structurally distinct MDR modulators show specific patterns of reversal against P-glycoproteins bearing unique mutations at serine 939/941. *Biochemistry* **33:**5041–5048.
80. Dhir, R., Grizzuti, K., Kajiji, S., *et al.* (1993). Modulatory effects on substrate specificity of independent mutations at the serine (939/941) position in predicted transmembrane domain-11 of P-glycoproteins. *Biochemistry* **32:**9492–9499.
81. Azzaria, M., Schurr, E., and Gros, P. (1989). Discrete mutations introduced in the predicted nucleotide-binding sites of the *mdr*1 gene abolish its ability to confer multidrug resistance. *Mol. Cell. Biol.* **9:**5289–5297.
82. Shimabuku, A. M., Saeki, T., Ueda, K., *et al.* (1991). Production of a site specifically cleavable P-glycoprotein-β-galactosidase fusion protein. *Agric. Biol. Chem.* **55:**1075–1080.
83. Shimabuku, A. M., Nishimoto, T., Ueda, K., *et al.* (1992). P-glycoprotein—ATP hydrolysis by the N-terminal nucleotide-binding domain. *J. Biol. Chem.* **267:**4308–4311.
84. Buschman, E., Arceci, R. J., Croop, J. M., *et al.* (1992). *mdr*2 encodes P-glycoprotein expressed in the bile canalicular membrane as determined by isoform-specific antibodies. *J. Biol. Chem.* **267:** 18093–18099.
85. Schinkel, A. H., Roelofs, M. E. M., and Borst, P. (1991). Characterization of the human *MDR*3 P-glycoprotein and its recognition by P-glycoprotein-specific monoclonal antibodies. *Cancer Res.* **51:** 2628–2635.
86. Sarkadi, B., Price, E. M., Boucher, R. C., *et al.* (1992). Expression of the human multidrug resistance cDNA in insect cells generates a high activity drug-stimulated membrane ATPase. *J. Biol. Chem.* **267:**4854–4858.
87. Doige, C. A., Yu, X. H., and Sharom, F. J. (1992). ATPase activity of partially purified P-glycoprotein from multidrug-resistant Chinese hamster ovary cells. *Biochim. Biophys. Acta* **1109:**149–160.
88. Al-Shawi, M. K., and Senior, A. E. (1993). Characterization of the adenosine triphosphatase activity of Chinese hamster P-glycoprotein. *J. Biol. Chem.* **268:**4197–4206.
89. Al-Shawi, M. K., Urbatsch, I. L., and Senior A. E. (1994). Covalent inhibitors of P-glycoprotein ATPase activity. *J. Biol. Chem.* **269:** 8986–8992.
90. Shapiro, A. B., and Ling, V. (1994). ATPase activity of purified and reconstituted P-glycoprotein from Chinese hamster ovary cells. *J. Biol. Chem.* **269:**3745–3754.
91. Ng, W. F., Sarangi, F., Zastawny, R. L., *et al.* (1989). Identification of members of the P-glycoprotein multigene family. *Mol. Cell. Biol.* **9:**1224–1232.

92. Ambudkar, S. V., Lelong, I. H., Zhang, J. P., *et al.* (1992). Partial purification and reconstitution of the human multidrug-resistance pump—Characterization of the drug-stimulatable ATP hydrolysis. *Proc. Natl. Acad. Sci. USA* **89:**8472–8476.
93. Davidson, A. L., Shuman, H. A., and Nikaido, H. (1992). Mechanism of maltose transport in *Escherichia coli:* Transmembrane signaling by periplasmic binding proteins. *Proc. Natl. Acad. Sci. USA* **89:**2360–2364.
94. Bishop, L., Agbyani, R., Ambudkar, S. V., *et al.* (1989). Reconstitution of a bacterial periplasmic permease in proteoliposomes and demonstration of ATP hydrolysis concomitant with transport. *Proc. Natl. Acad. Sci. USA* **86:**6953–6957.
95. Horio, M., Gottesman, M. M., and Pastan, I. (1988). ATP-dependent transport of vinblastine in vesicles from human multidrug-resistant cells. *Proc. Natl. Acad. Sci. USA* **85:**3580–3584.
96. Gros, P., Croop, J., and Housman, D. E. (1986). Mammalian multidrug resistance gene: Complete cDNA sequence indicates strong homology to bacterial transport proteins. *Cell* **47:**371–380.
97. Croop, J. M., Raymond, M., Haber, D., *et al.* (1989). The three mouse multidrug resistance *(mdr)* genes are expressed in a tissue-specific manner in normal mouse tissues. *Mol. Cell. Biol.* **9:** 1346–1350.
98. Chin, J. E., Chen, C.-J., Kriegler, M., *et al.* (1989). Structure and expression of the human MDR (P-glycoprotein) gene family. *Mol. Cell. Biol.* **9:**3808–3820.
99. Cardon-Cardo, C., O'Brien, J. P., Boccia, C., *et al.* (1990). Expression of multidrug resistance gene product (P-glycoprotein) in human normal and tumor tissues. *J. Histochem. Cytochem.* **38:**1277–1287.
100. Gottesman, M. M., Willingham, M. C., Theibaut, F., *et al.* (1991). Expression of the *MDR*1 gene in normal human tissues. In *Molecular and Cellular Biology of Multidrug Resistance in Tumors* (I. B. Roninson, ed.), Plenum Press, New York, pp. 279–289.
101. Chaudhary, P. M., and Roninson, I. B. (1991). Expression and activity of P-glycoprotein, a multidrug efflux pump, in human hematopoietic stem cells. *Cell* **66:**85–94.
102. Drach, D., Zhao, S. R., Drach, J., *et al.* (1992). Subpopulations of normal peripheral blood and bone marrow cells express a functional multidrug resistant phenotype. *Blood* **80:**2729–2734.
103. Chaudhary, P. M., Mechetner, E. B., and Roninson, I. B. (1992). Expression and activity of the multidrug resistance P-glycoprotein in human peripheral blood lymphocytes. *Blood* **80:**2735–2739.
104. Arceci, R. J., Croop, J. M., Horwitz, S. B., *et al.* (1988). The gene encoding multidrug resistance is induced and expressed at high levels during pregnancy in the secretory epithelium of the uterus. *Proc. Natl. Acad. Sci. USA* **85:**4350–4354.
105. Arceci, R. J., Baas, F., Raponi, R., *et al.* (1990). Multidrug resistance gene expression is controlled by steroid hormones in the secretory epithelium of the uterus. *Mol. Reprod. Dev.* **25:**101–109.
106. Piekarz, R. L., Cohen, D., and Horwitz, S. B. (1993). Progesterone regulates the murine multidrug resistance *mdr*1b gene. *J. Biol. Chem.* **268:**7613–7616.
107. Bates, S. E., Mickley, L. A., Chen, Y.-N., *et al.* (1989). Expression of a drug resistance gene in human neuroblastoma cell lines. *Mol. Cell. Biol.* **9:**4337–4344.
108. Mickley, L. A., Bates, S. E., Richert, N. D., *et al.* (1989). Modulation of the expression of a multidrug resistance gene by differentiating agents. *J. Biol. Chem.* **264:**18031–18040.
109. Frommel, T. O., Coon, J. S., Tsuruo, T., *et al.* (1993). Variable effects of sodium butyrate on the expression and function of the *MDR*1 (P-glycoprotein) gene in colon carcinoma cell lines. *Int. J. Cancer* **55:**297–302.
110. Burt, R. K., Garfield, S., Johnson, K., *et al.* (1988). Transformation of rat liver epithelial cells with v-Ha-ras or v-raf causes expression of *MDR*1, glutathione-S-transferase-P and increased resistance to cytotoxic chemicals. *Carcinogenesis* **9:**2329–2332.
111. Chin, K.-V., Ueda, K., Pastan, I., *et al.* (1992). Modulation of the activity of the promoter of the human *MDR*1 gene by Ras and p53. *Science* **255:**459–462.
112. Cornwell, M. M., and Smith, D. E. (1993). A signal transduction pathway for activation of the *mdr*1 promoter involves the proto-oncogene c-raf kinase. *J. Biol. Chem.* **268:**15347–15350.
113. Thorgeirsson, S. S., Huber, B. E., Sorrell, S., *et al.* (1987). Expression of the multidrug-resistance gene in hepatocarcinogenesis and regenerating rat liver. *Science* **236:**1120–1122.
114. Fairchild, C., Ivy, S., Rushmore, T., *et al.* (1987). Carcinogen-induced mdr overexpression is associated with xenobiotic resistance in rat preneoplastic nodules and hepatocellular carcinomas. *Proc. Natl. Acad. Sci. USA* **84:**7701–7705.
115. Gant, T. W., Silverman, J. A., Bisgaard, H. C., *et al.* (1991). Regulation of 2-AAF and methylcholanthrene-mediated induction of multidrug resistance and cytochrome P450IA gene family expression in primary hepatocyte cultures and rat liver. *Mol. Carcinogenesis* **4:** 499–509.
116. Chin, K.-V., Chauhan, S., Pastan, I., *et al.* (1990). Regulation of *mdr* gene expression in acute response to cytotoxic insults in rodent cells. *Cell Growth Differ.* **1:**361–365.
117. Herzog, C. E., Tsokos, M., Bates, S. E., *et al.* (1993). Increased mdr-1/P-glycoprotein expression after treatment of human colon carcinoma cells with P-glycoprotein antagonists. *J. Biol. Chem.* **268:** 2946–2952.
118. Chin, K.-V., Tanaka, S., Darlington, G., *et al.* (1990). Heat shock and arsenite increase expression of the multidrug resistance (*MDR*1) gene in human renal carcinoma cells. *J. Biol. Chem.* **265:**221–226.
119. Kioka, N., Yamano, Y., Komano, T., *et al.* (1992). Heat-shock responsive elements in the induction of the multidrug resistance gene (*MDR*1). *FEBS Lett.* **301:**37–40.
120. Chin, K.-V., Pastan, I., and Gottesman, M. M. (1993). Function and regulation of the human multidrug resistance gene. *Adv. Cancer Res.* **60:**157–180.
121. Ueda, K., Clark, D. P., Chen, C.-j., *et al.* (1987). The human multidrug-resistance (*mdr*1) gene: cDNA cloning and transcription initiation. *J. Biol. Chem.* **262:**505–508.
122. Hsu, S. I., Lothstein, L., and Horwitz, S. B. (1989). Differential overexpression of three *mdr* gene family members in multidrug-resistant J774.2 mouse cells. *J. Biol. Chem.* **264:**12053–12062.
123. Ueda, K., Pastan, I., and Gottesman, M. M. (1987). Isolation and sequence of the promoter region of the human multidrug-resistance (P-glycoprotein) gene. *J. Biol. Chem.* **262:**17432–17436.
124. Raymond, M., and Gros, P. (1990). Cell-specific activity of *cis*-acting regulatory elements in the promoter of the mouse multidrug resistance gene mdr1. *Mol. Cell. Biol.* **10:**6036–6040.
125. Cohen, D., Piekarz, R. I., Hsu, S. I., *et al.* (1991). Structural and functional analysis of the mouse mdr1b gene promoter. *J. Biol. Chem.* **266:**2239–2244.
126. Madden, M. J., Morrow, C. S., Nakagawa, M., *et al.* (1993). Identification of 5′and 3′sequences involved in the regulation of transcription of the human *mdr*l gene *in vivo. J. Biol. Chem.* **268:**8290–8297.
127. Cornwell, M. M., and Smith, D. E. (1993). SP1 activates the MDR1 promoter through one of two distinct G-rich regions that modulate promoter activity. *J. Biol. Chem.* **268:**19505–19511.
128. Zastawny, R. L., and Ling, V. (1993). Structural and functional analysis of 5′ flanking and intron-1 sequences of the hamster P-glycoprotein pgp1 and pgp2 genes. *Biochim. Biophys. Acta* **1173:**303–313.
129. Silverman, J. A., Raunio, H., Gant, T. W., *et al.* (1991). Cloning and characterization of a member of the rat multidrug resistance *(mdr)* gene family. *Gene* **106:**229–236.
130. Ikeguchi, M., Teeter, L., Eckersberg, T., *et al.* (1991). Structural and functional analyses of the promoter of the murine multidrug resistance gene *mdr3/mdr*1a reveal a negative element containing the AP1-site. *DNA Cell Biol.* **10:**639–649.

131. Teeter, L. D., Eckersberg, T., Tsai, Y., *et al.* (1991). Analysis of the Chinese hamster P-glycoprotein/multidrug resistance gene pgp1 reveals that the AP-1 site is essential for full promoter activity. *Cell Growth Differ.* **2:**429–437.
132. Goldsmith, M. E., Madden, M. J., Morrow, C. S., *et al.* (1993). A y-box consensus sequence is required for basal expression of the human multidrug resistance (*mdr*1) gene. *J. Biol. Chem.* **268:**5856–5860.
133. Ogura, M., Takatori, T., and Tsuruo, T. (1992). Purification and characterization of NF-R1 that regulates the expression of the human multidrug resistance (MDR1) gene. *Nucleic Acids Res.* **20:**5811–5817.
134. Cornwell, M. M. (1990). The human multidrug-resistance (MDR1) gene: Sequences upstream and downstream of the initiation site influence transcription. *Cell Growth Differ.* **1:**607–615.
135. Van Groenigen, M., Valentijn, L. J., and Baas, F. (1993). Identification of a functional initiator sequence in the human *MDR*1 promoter. *Biochim. Biophys. Acta* **1172:**138–146.
136. Schinkel, A. H., Kemp, S., Dolle, M., *et al.* (1993). N-glycosylation and deletion mutants of the human *MDR*1 P-glycoprotein. *J. Biol. Chem.* **268:**7474–7481.
137. Beck, W. T., and Cirtain, M. (1982). Continued expression of *Vinca* alkaloid resistance by CCRF-CEM cells after treatment with tunicamycin or pronase. *Cancer Res.* **42:**184–189.
138. Ling, V., Kartner, N., Sudo, T., *et al.* (1983). The multidrug resistance phenotype in Chinese hamster ovary cells. *Cancer Treat. Rep.* **67:**869–874.
139. Center, M. S. (1985). Mechanisms regulating cell resistance to adriamycin: Evidence that drug accumulation in resistant cells is modulated by phosphorylation of a plasma membrane glycoprotein. *Biochem. Pharmacol.* **34:**1471–1476.
140. Mellado, W., and Horwitz, S. B. (1987). Phosphorylation of the multidrug resistance associated glycoprotein. *Biochemistry* **26:**6900–6904.
141. Hamada, H., Hagiwara, K.-I., Nakajima, T., *et al.* (1987). Phosphorylation of the Mr 170,000 to 180,000 glycoprotein specific to multidrug-resistant tumor cells: Effects of verapamil, trifluoperazine, and phorbol esters. *Cancer Res.* **47:**2860–2865.
142. Fine, R. L., Patel, J., and Chabner, B. A. (1988). Phorbol esters induce multidrug resistance in human breast cancer cells. *Proc. Natl. Acad. Sci. USA* **85:**582–586.
143. Epand, R. M., and Stafford, A. R. (1993). Protein kinases and multidrug resistance. *Cancer J.* **6:**154–158.
144. Chambers, T. C., McAvoy, E. M., Jacobs, J. W., *et al.* (1990). Protein kinase C phosphorylates P-glycoprotein in multidrug resistant human KB carcinoma cells. *J. Biol. Chem.* **265:**7679–7686.
145. Ahmad, S., Trepel, J. B., Ohno, S., *et al.* (1992). Role of protein kinase-C in the modulation of multidrug resistance—Expression of the atypical gamma-isoform of protein kinase-C does not confer increased resistance to doxorubicin. *Mol. Pharmacol.* **42:**1004–1009.
146. Ahmad, S., and Glazer, R. I. (1993). Expression of the antisense cDNA for protein kinase-C-alpha attenuates resistance in doxorubicin-resistant MCF-7 breast carcinoma cells. *Mol. Pharmacol.* **43:** 858–862.
147. Chaudhary, P. M., and Roninson, I. B. (1992). Activation of *MDR*1 (P-glycoprotein) gene expression in human cells by protein kinase-C agonists. *Oncol. Res.* **4:**281–290.
148. Bates, S. E., Lee, J. S., Dickstein, B., *et al.* (1993). Differential modulation of P-glycoprotein transport by protein kinase inhibition. *Biochemistry* **32:**9156–9164.
149. Chambers, T. C., Pohl, J., Raynor, R. L., *et al.* (1993). Identification of specific sites in human P-glycoprotein phosphorylated by protein kinase-C. *J. Biol. Chem.* **268:**4592–4595.
150. Orr, G. A., Han, E. K.-H., Browne, P. C., *et al.* (1993). Identification of the major phosphorylation domain of murine *mdr*1b P-glycoprotein. *J. Biol. Chem.* **268:**25054–25062.
151. Riordan, J. R. (1993). The cystic fibrosis transmembrane conductance regulator. *Annu. Rev. Physiol.* **55:**609–630.
152. Gottesman, M. M. (1993). How cancer cells evade chemotherapy—Sixteenth Richard and Linda Rosenthal Foundation Award Lecture. *Cancer Res.* **53:**747–754.
153. Galski, H., Sullivan, M., Willingham, M. C., *et al.* (1989). Expression of a human multidrug-resistance cDNA(*MDR*1) in the bone marrow of transgenic mice: Resistance to daunomycin-induced leukopenia. *Mol. Cell. Biol.* **9:**4357–4363.
154. Mickisch, G., Merlino, G. T., Galski, H., *et al.* (1991). Transgenic mice that express the human multidrug resistance gene in bone marrow enable a rapid identification of agents that reverse drug resistance. *Proc. Natl. Acad. Sci. USA* **88:**547–551.
155. Mickisch, G. H., Aksentijevich, I., Schoenlein, P. V., *et al.* (1992). Transplantation of bone marrow cells from transgenic mice expressing the human *MDR*1 gene results in long-term protection against the myelosuppressive effect of chemotherapy in mice. *Blood* **79:** 1–7.
156. Sorrentino, B. P., Brandt, S. J., Bodine, D., *et al.* (1992). Retroviral transfer of the human *MDR*1 gene permits selection of drug resistant bone marrow cells *in vivo*. *Science* **257:**99–103.
157. Podda, S., Ward, M., Himelstein, A., *et al.* (1992). Transfer and expression of the human multiple drug resistance gene into live mice. *Proc. Natl. Acad. Sci. USA* **89:**9676–9680.
158. Cardarelli, C. O., Aksentijevich, I., Pastan I., *et al.* (1995). Differential effects of P-glycoprotein inhibitors on NIH3T3 cells transfected with wild-type (G185) or mutant (V185) multidrug transporters. *Cancer Res.* **55:**1086–1091.
159. Germann, U. A., Chambers, T. C., Ambudkar, S. V., *et al.* (1996). Characterization of phosphorylation-defective mutants of human P-glycoprotein expressed in mammalian cells. *J. Bio. Chem.* **271:** 1708–1716.

14

Molecular Studies of Members of the Mammalian Na^+/H^+ Exchanger Gene Family

Mark Donowitz, Susan A. Levine, C. H. Chris Yun, Steven R. Brant, Samir Nath, Jeannie Yip, Sandra Hoogerwerf, Jacques Pouysségur, and Chung-Ming Tse

14.1. INTRODUCTION

The brief history of the contribution of molecular biologic studies to the understanding of the Na^+/H^+ exchanger gene family is not unlike the history of studies of other transport proteins. Many years of results from physiologic and biochemical studies provided the background to allow strategies for the molecular recognition of an initial member of the Na^+/H^+ exchanger gene family. This was followed by recognition of the existence of a gene family, which even now is only partially defined. Rapid advances followed concerning location, regulation, and structure/function relationships, all of which have served to extend the previous physiologic studies. Current studies involve "torturing" the specific transport proteins by deletion and point mutation and creation of chimeric constructs to further explore structure/function studies. These are descriptive studies that are attempting to gain clues as to how the proteins carry out transport and are regulated. However, they fall short of defining how the proteins *work,* which presumably will follow from crystallagraphic techniques, although no mammalian transport protein has yet yielded the required information using any approach or combination of approaches.

Na^+/H^+ exchange was initially proposed as a mechanism for renal acidification in 1949.[1] Twenty-seven years later, Na^+/H^+ exchange was demonstrated in renal brush border membrane vesicles by Murer, Hopfer, and Kinne.[2] In 1989, the first mammalian Na^+/H^+ exchanger was cloned by Sardet, Franchi, and Pouysségur.[3] The existence of a gene family of mammalian Na^+/H^+ exchangers (the cloned Na^+/H^+ exchangers are referred to as NHEs) was reported by Tse, Pouysségur, Donowitz, and co-workers in 1991,[4] and epithelial-specific isoforms were identified by Tse, Pouysségur, Donowitz, and co-workers in 1991, 1992, and 1993[5–7] and separately by Orlowski, Shull, and co-workers in 1992 and 1993.[8,9]

This review will concentrate on defining the members of the Na^+/H^+ exchanger gene family, structure/function relationships which have largely evolved from use of the cloned proteins, and studies of regulation of members of this gene family (for recent reviews see Refs. 10–12).

Na^+/H^+ exchangers are integral plasma membrane proteins that catalyze the electroneutral exchange of extracellular Na^+/H^+ for intracellular H^+ with a stoichiometry of one for one. A consistent feature of Na^+/H^+ exchangers is their allosteric activation by intracellular protons which are presumed to interact at a "modifier" site that is separate from the sites involved in Na^+ and H^+ transport. Na^+/H^+ exchangers have multiple functions, including pH homeostasis, volume regulation, cell proliferation, and transcellular Na^+ absorption (reviewed in Ref. 13). In no cell is it the only mechanism for any one of these functions. For instance, multiple mechanisms of pH homeostasis are present in most eukaryotic cells including Cl^-/HCO_3^- exchange, $NaHCO_3^-$ cotransport, Na^+-dependent Cl^-/HCO_3^- exchange, and multiple mechanisms of H^+ extrusion (reviewed in Ref. 14), including H^+,K^+-ATPase pumps.

The existence of multiple isoforms of mammalian Na^+/H^+ exchangers had been predicted on the basis of: (1) while all Na^+/H^+ exchangers are inhibited by the diuretic amiloride, they have widely different sensitivities to inhibition by amiloride from cell type to cell type and even between Na^+/H^+ exchangers on different plasma membrane domains (apical versus basolateral in polarized epithelial cells) in the same cell (for a recent review see Clark and Limbird[15]); (2) protein kinases have different effects in regulating Na^+/H^+ exchangers depending not only on cell type, but also on different plasma membrane domains in the same cell (the latter has been reviewed recently[11]); (3) while it has been documented that regulation of Na^+/H^+ exchangers can occur by a mechanism that shifts the pK value for intracellular H^+ of the exchangers, protein kinase regulation of some Na^+/H^+ exchanger isoforms involves a mechanism that changes the V_{max} of Na^+/H^+ exchange; such a change in V_{max} may or may not be accompanied by a change in pH dependence on intracellular H^{+}[16–18];

Mark Donowitz, Susan A. Levine, C. H. Chris Yun, Steven R. Brant, Samir Nath, Jeannie Yip, Sandra Hoogerwerf, and Chung-Ming Tse • Departments of Medicine and Physiology, GI Unit, The Johns Hopkins University School of Medicine, Baltimore, Maryland 21205. **Jacques Pouysségur** • Department of Biochemistry, University of Nice, Nice, France.

Molecular Biology of Membrane Transport Disorders, edited by S.G. Schultz *et al.* Plenum Press, New York, 1996

(4) Na^+/H^+ exchangers have been shown to have multiple physiologic roles, making it difficult to understand how a single transport protein could carry out so many functions; (5) by genomic Southern blot analysis, we demonstrated that the housekeeping Na^+/H^+ exchanger (NHE1) cDNA can hybridize to other closely related but not identical genes under low-stringency hybridization and washing conditions.[4] To date, four members of the mammalian NHE gene family have been recognized molecularly and expressed in mammalian expression systems, which has allowed their characterization as to general topology, amiloride sensitivity, cellular localization, kinetic parameters for Na^+ and H^+ transport, and regulation of transport function. A fifth isoform has been partially cloned,[80] as has a NHE3-related pseudogene.[81]

14.2. MOLECULAR IDENTIFICATION OF NA^+/H^+ EXCHANGER GENE FAMILY

Molecular identification of the mammalian Na^+/H^+ exchanger was pioneered by Sardet, Pouysségur, and co-workers who used genetic complementation[3,19] with fibroblast cell lines that they had selected to lack all endogenous Na^+/H^+ exchangers (the Chinese hamster lung fibroblast-derived cell line PS120 and the mouse fibroblast-derived cell line LAP1[20,21]) to clone a human cDNA that encoded an amiloride-sensitive, growth factor-activated Na^+/H^+ exchanger. Since then, four additional members of this mammalian gene family have been identified by our group and by Orlowski and Shull using screening of mammalian cDNA libraries under conditions of varying stringency.[8,9] We have named the members of the Na^+/H^+ exchange gene family in the order of their molecular identification as NHE1 (first cloned NHE), etc. NHE1 is the Na^+/H^+ exchanger isoform initially cloned by Pouysségur *et al.*[3] It has been cloned from human, rabbit, rat, pig (LLC-PK_1 cells) and Chinese hamster and contains 815–820 amino acids (species variation).[3,4,8,22] NHE2 has been cloned from rabbit and rat and contains 809 and 813 amino acids, respectively.[6,9] NHE3 has been cloned from rat, rabbit, and human and has 831, 832, and 832 amino acids, respectively.[7,8,79] NHE4 has been cloned from rat and has 717 amino acids.[8] The corresponding predicted sizes of NHE1, NHE2, NHE3, and NHE4 based on amino acid composition as predicted from cDNAs without considering glycosylation are ~91, ~91, ~93, and ~81 kDa, respectively. BNHE is an additional vertebrate isoform cloned from trout red blood cells.[25]

All NHEs exhibit about 50% amino acid identity with respect to each other. NHE2 and NHE4 are most closely related with 60% amino acid identity and NHE3 is the least related isoform[11] (Figure 14.1). There is great amino acid conservation among species of the individual NHE isoforms with NHE1 and NHE3 having at least 89% identity among the species cloned to date. NHE1 and B-NHE1 are the isoforms that resemble each other most in the C-terminal portion.[25]

As predicted from NHE primary structure, with differences being present throughout the entire amino acid sequence, NHEs are separate gene products. The NHE1 gene has been localized to human chromosome 1 p35–p36.1 by *in situ* hybridization[26] and the NHE3 gene has been physically mapped to the distal portion of chromosome 5p 15.3 where it is the most telomeric marker yet identified.[27] There are as yet no examples of alternately spliced Na^+/H^+ exchanger isoforms for NHE1, NHE3, and NHE4. A rat NHE2 isoform that lacks the first two membrane-spanning domains has recently been reported.[28,29] Whether this represents an alternately spliced NHE isoform or a cloning artifact remains to be determined.

14.3. TISSUE AND CELLULAR DISTRIBUTION OF NA^+/H^+ EXCHANGER MESSAGE AND PROTEIN

Based on Northern analysis and ribonuclease protection assays, NHE1 message is present in nearly all mammalian cells.[3,4,6–9,11] The only mammalian cells studied in which NHE1 message was not identified are the OK cells (opossum renal proximal tubule cell line) and rat proximal tubule cortical segments S_1 and S_2.[30] These cells are known to lack functional basolateral membrane Na^+/H^+ exchangers. Using NHE1 antibody, NHE1 has been found in plasma membrane of fibroblasts and A431 epidermoid cells.[19]

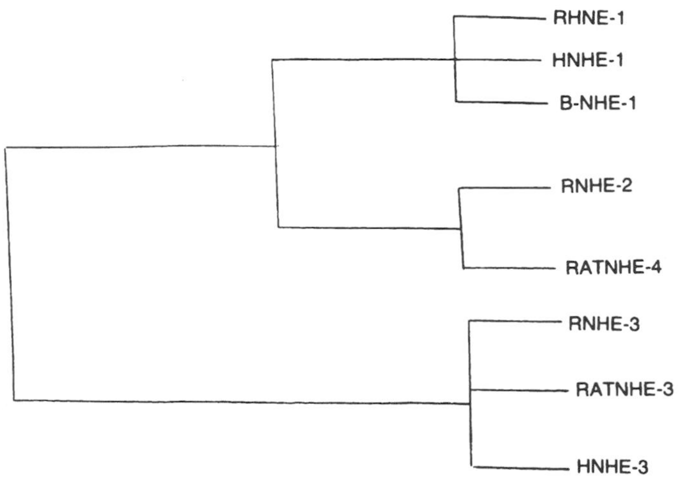

Figure 14.1. Relationship among members of the eukaryotic Na^+/H^+ exchanger gene family cloned to date based on amino acid identify using PC gene subprogram Clustal dendogram (R, rabbit; B, trout red blood cell; H, human).

At the message level, the other mammalian NHE isoforms are more restricted in tissue distribution (Table 14.1). NHE2, NHE3, and NHE4 are predominantly expressed in stomach, intestine, and kidney and thus are epithelial cell-specific isoform Na^+/H^+ exchangers.[6–8] NHE2 message is present in large amounts in stomach, uterus, kidney, intestine, adrenal gland and much less in trachea and skeletal muscle.[6,8] The message is most expressed in the kidney medulla exceeding that in the kidney cortex. In the gastrointestinal tract, the ascending colon has most message followed by descending colon > jejunum > ileum > duodenum. NHE3 message is found exclusively in kidney, intestine, and stomach.[7,8] Most message is present in the kidney cortex, exceeding that in the medulla. The area of second most message is rabbit ascending colon which is approximately equal to ileum > jejunum. NHE3 message is not present in the duodenum or descending colon. Thus, the presence of NHE3 message in rabbit gastrointestinal tract colocalizes with the expression of the neutral NaCl absorptive process.[31] In human intestine, NHE3 message relative to NHE1 message is more abundant in ileum > proximal colon > duodenum, jejunum = distal colon.[32] Of note, human distal colon does possess neutral NaCl absorption, and thus NHE3 is only present in tissues with this transport process.

NHE4 message is present in greatest amount in the rat stomach (maximum in gastric antrum) > proximal small intestine = cecum = proximal colon with much smaller amounts in the uterus, brain, kidney, and skeletal muscle.[8]

Tissue localization studies of members of the NHE gene family are more preliminary. Again in preliminary studies, Soleimani reported that NHE4 was present in the basolateral membrane of rat cortex and inner and outer medullary collection ducts.[33] NHE1 is found on the basolateral surface of several epithelia. In the rabbit ileum, NHE1 is restricted to the basolateral membrane of both the villus and the crypt epithelial cells, but appears to be diffusely present in the plasma membrane of goblet cells.[4] In addition, it is restricted to the basolateral membrane of the human colon cancer Cl secretory cell line, Caco-2, under certain growth conditions[34] and to the basolateral membrane of the porcine renal epithelial cell line, LLC-PK1.[22] In rabbit kidney, NHE1 is present on the basolateral membrane of proximal tubule cells, distal convoluted tubules, thick ascending limb of Henle's loop, and the collecting duct but is absent from glomeruli, the thin descending limb of Henle's loop, intercalated cells of the connecting tubules and collecting ducts, and nonepithelial cells of the renal cortex and medulla.[35] Using an anti-NHE3 antibody, Aronson *et al.* showed that NHE3 is restricted to the brush border of renal proximal tubule cells,[36] while Soleimani and coworkers found NHE3 in the rat kidney cortex proximal tubule brush border and thick ascending limb of Henle's loop.[33] Immunocytochemical studies showed that NHE3 and NHE2 are present in the brush border but not basolateral membranes of villus cells (small intestine) and surface cells (colon). They also are present to a lesser extent in the brush border of crypt cells, but not in goblet cells of human, rabbit, dog, and rat jejunum, ileum, and ascending colon, and human and rat (but not rabbit) descending colon.[82] In all these areas the distribution of NHE3 parallels the presence of the neutral NaCl absorptive process.

Table 14.1. Tissue Distribution of the mRNAs of the Cloned Isoforms NHE1, NHE2, NHE3, and NHE4 as Determined by Northern Analysis and RNase Protection Assays

Isoform	Tissue distribution
NHE1	Ubiquitous except rat renal proximal tubule cells of the S_1 and S_2 cortical segments and rabbit renal glomeruli, thin descending limb of Henle's loop, and collecting ducts
NHE2	Rabbit[6]: Kidney medulla>ascending colon > kidney cortex > adrenal gland > descending colon > jejunum > ileum > duodenum > trachea = skeletal muscle Rat[9]: Small intestine = colon = stomach > > skeletal muscle > brain > kidney > testis > uterus > heart > lung
NHE3	Rabbit[7]: Kidney cortex > ileum = ascending colon > kidney medulla > jejunum > stomach (absent descending colon, duodenum) Rat[8]: Large intestine > small intestine > kidney > stomach Human[79]: Kidney > > small intestine > > testes > ovary > colon = prostate > thymus > leukocytes = brain > spleen > placenta
NHE4	Rat[8]: Stomach > small and large intestine > > uterus = brain = kidney = skeletal muscle

14.4. TOPOLOGY OF THE MAMMALIAN NHE GENE FAMILY: TWO FUNCTIONAL DOMAINS

The primary structure (amino acids) and the predicted secondary structure (hydrophobicity profile) of all four identified mammalian Na^+/H^+ exchanger gene family members are similar. Figure 14.2 depicts the predicted general membrane topology of the NHEs, based on the hydrophobicity plots using the method of Engelman or Kyte and Doolittle. They are predicted to have two structural domains: an approximately 500-amino-acid N-terminus which contains 10–12 membrane-spanning domains and an approximately 300-amino-acid cytoplasmic C-terminus. Antibody studies confirm the NHE membrane topology that the C-terminus for NHE1 and NHE2 is intracellular,[15,37] based on the requirement for membrane permeabilization to visualize the epitope.

The most highly conserved portions of the molecule among the identified isoforms are the membrane-spanning domains. The latter exhibit 50–60% amino acid identify among isoforms. Of the membrane-spanning domains, 5A and 5B are most conserved, being almost identical among the isoforms. This suggests that these transmembrane domains are likely to be vital for structural and/or functional aspects of Na^+/H^+ exchange. Each protein is predicted to contain a cleavable signal peptide at the N-terminus and the first membrane-spanning domain appears to be cleaved off in the intact protein.[74] A single putative N-linked glycosylation consensus sequence is present in extracytoplasmic loop D in all isoforms.[3,4,6,8] However, this site does not appear to be glycosylated (see below). The areas least related among the Na^+/H^+ exchanger isoforms

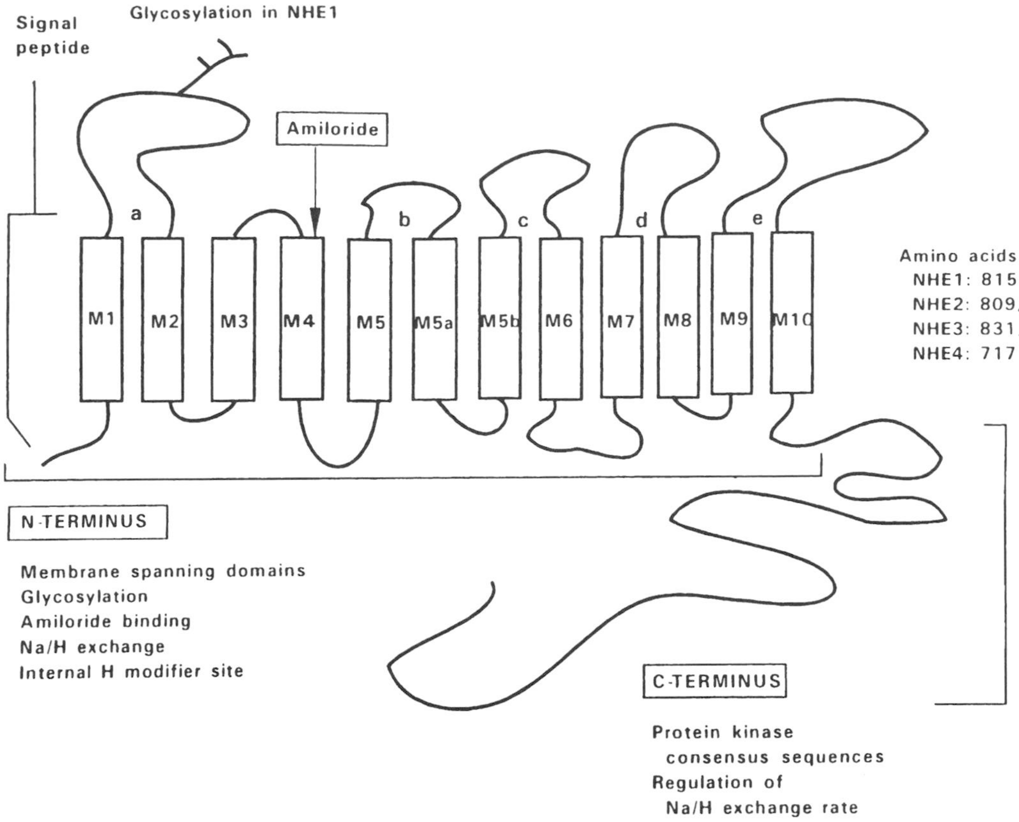

Figure 14.2. Topology of mammalian Na^+/H^+ exchangers as predicted from their hydrophobicity plots and as determined by biochemical studies; the branched line represents the N-linked glycosylation site in NHE1.

include: (1) the first membrane-spanning domain (the signal peptide) and extracellular loop A and (2) the intracytoplasmic C-terminal domain. The C-terminus for all isoforms contains multiple putative protein kinase consensus sequences which are extremely variable among members of the gene family but are more conserved for each isoform among species. The C-terminus is only approximately 30% conserved among isoforms, except for NHE1 and BHNE1 which are 48% identical.[25]

Structure/function studies using NHE1, NHE2, and NHE3 have also suggested that each NHE molecule can be divided into two functional domains, i.e., the transporter domain and the regulatory domain. The transport domain is the N-terminal membrane-spanning domain, which in NHE1 is predicted to consist of amino acids 1–499,[38] in NHE2 1–479,[6] and in NHE3 1–455.[9,11] This N-terminal membrane-spanning domain retains the ability to (1) be inserted into the plasma membrane; (2) catalyze amiloride-sensitive Na^+/H^+ exchange activity, although at a much slower rate than the wild-type exchangers; (3) be activated by intracellular H^+, i.e., has an H^+-modifier site; (4) be inhibited by amiloride.

The regulatory domain is the long cytoplasmic C-terminus which mediates growth factor regulation of NHEs. Wakabayashi *et al.*[38] constructed a set of deletion mutants within the cytoplasmic C-terminus of human NHE1 and stably expressed the truncated cDNAs in PS120 fibroblasts. It was found that deletion of the cytoplasmic C-terminus of NHE1, in particular the region from amino acids 567 to 635, completely abolished the growth factor activation.

14.5. GLYCOSYLATION

NHE1 is an N-linked glycoprotein. While NHE1 contains three consensus sequences for N-linked glycosylation, with asparagines (nxs/t) at residues 75, 370, and 410. It is only the first site that is glycosylated.[74,83] Deglycosylation by neuraminidase and endoglycosidase F reduced the size of NHE1 from 110 kDa to 90 kDa.[19,40] However, NHE1 is resistant to endoglycosidase H digestion, suggesting the carbohydrate is a high-mannose complex type. The combination of endoglycosidase F and neuraminidase yields a more complex deglycosylation of NHE1 than endoglycosidase F alone.[40] NHE2 consists of 85- and 75-kDa proteins.[82] NHE2 has one potential N-linked glycosylation sequence located on the putative extracellular loop D[6,9] and this is conserved in all cloned Na^+/H^+ exchangers. However, NHE3 and NHE2 are resistant to PNGase F and Endo H digestion. NHE2 is *O*-glycosylated, the first example of a transport protein that is not *N*-glycosylated or linked to an *N*-glycosylated protein.[84] Thus, NHE3 and NHE2 are not *N*-glycosylated.[37,83,84] When expressed in fibroblasts, NHE3 is not glycosylated.[83] It is not known whether NHE4 is glycosylated. It is also not known if the NHE glycosylation status is dependent on the cells expressing the Na^+/H^+ exchanger, especially epithelial cells.

The functional significance of *N*-glycosylation in NHE1 is not known. Deglycosylation of NHE1 with endoglycosidase F in placental brush border membrane vesicles had no effect on its Na^+/H^+ exchange activity.[40] On the other hand, it was reported that deglycosylation of rat renal brush border membranes with endoglycosidase F reduced the V_{max} of the Na^+/H^+ exchanger by 50% without causing a change in the apparent $K_m(Na^+)$.[41] Of note, this renal brush border membrane Na^+/H^+ exchanger is likely to be NHE3. However, the method of study deglycosylated many renal brush border proteins and it is not known if the change in Na^+/H^+ exchange is related to a change in the exchanger or in some other protein.

14.6. OLIGOMERIZATION

All NHE isoforms appear to consist of a single type of subunit based on complementation of Na^+/H^+ exchange activity by a single cDNA in an exchanger-deficient cell[3,4,19,23,42,43] including PS120 cells, LAP cells, and Chinese hamster ovary (CHO) cells. However, there is substantial evidence that NHEs operate as an oligomeric structure: (1) Kiniella *et al.* studied pre-steady-state Na^+ kinetics in renal brush border membrane vesicles.[44] They found that, although the steady-state Na^+ uptake into renal brush border membrane vesicles obeyed Michaelis–Menten kinetics, the pre-steady-state Na^+ kinetics exhibited positive cooperativity. Therefore, they concluded that the change in multiplicity of Na^+ transport sites that occurs in the transition between the pre-steady-state and the steady-state Na^+ uptake in renal brush border membrane vesicles reflects the oligomeric nature of the Na^+/H^+ exchanger which operates in a "flip-flop" mechanism. (2) Beliveau *et al.*[45] estimated by radiation inactivation analysis that the renal brush border Na^+/H^+ exchanger has a molecular size *in situ* 321 kDa. Recently, several groups demonstrated that NHE3 is a renal brush border Na^+/H^+ exchanger with a size of ~85 kDa.[36,82] If NHE3 is the only exchanger isoform present in the renal brush border membrane, this result suggests that the renal brush border Na^+/H^+ exchanger (NHE3) exists as a tetramer. (3) In contrast, treatment of NHE1 in PS120 with disuccinimidylsuberate (DSS) shifted the size of NHE1 from 100 kDa to 200 kDa.[46] This suggests that NHE1 exists as a dimer, and dimerization of NHE1 seems to be thio-sensitive.[47] Although this evidence suggests that the NHEs, at least NHE1 and NHE3, exist as multimers (dimer or tetramer), it remains to be determined whether the functional unit or NHE is a monomer or an oligomer.

14.7. AMILORIDE BINDING DOMAIN

Clark and Limbird reviewed the amiloride sensitivity of Na^+/H^+ exchangers in different tissues and cells.[15] While all Na^+/H^+ exchangers are inhibited by amiloride and are more sensitive to 5-amino-substituted amiloride analogues than to amiloride itself, the different Na^+/H^+ exchanger isoforms differ greatly as to sensitivity based on cell type. In polarized cells, the apical and basolateral membrane Na^+/H^+ exchangers are differently sensitive to amiloride with greater sensitivity in the basolateral Na^+/H^+ exchanger.

The porcine kidney LLC-PK1 cell line provides a good illustration of this feature since this cell line was demonstrated to display pharmacologically distinguishable Na^+/H^+ exchange activities on its apical and basolateral surfaces when grown on permeable filters to promote polarization.[48] The apical Na^+/H^+ exchange activity has an apparent inhibitory constant ($K_{0.5}$) for the amiloride analogue EIPA (ethylisopropylamiloride) of 10 μM compared to a value of 20 nM for its basolateral counterpart. The $K_{0.5}$ value for amiloride for the Na^+/H^+ exchanger present in nonepithelial cells or in the basolateral membrane of polarized epithelial cells has been reported to be in the low micromolar range; this value is one to two orders of magnitude greater for the Na^+/H^+ exchanger on the apical membrane of polarized epithelial cells from kidney and intestine (reviewed in Ref. 15).

Table 14.2 summarizes the amiloride sensitivity of the recently cloned Na^+/H^+ exchangers (NHE1, NHE2, and NHE3) expressed in either PS120 cells or CHO cells. For all three expressed exchangers, Na^+/H^+ exchange[42,43] is entirely inhibited by amiloride and its 5-amino-substituted amiloride analogues. NHE1 and NHE2 are equally sensitive to amiloride[6,49] while NHE3 is an amiloride-resistant isoform. NHE1 is sensitive to ethylisopropylamiloride while NHE2 and NHE3 are resistant, with NHE3 being more resistant than NHE2. The amiloride-related diuretic HOE694 has the greatest differential sensitivity among the NHEs (Table 14.2).[85]

Amiloride and its 5-amino-substituted derivatives inhibit the Na^+/H^+ exchanger by competing with Na^+ on its external binding site (see Ref. 50 for review). The amiloride binding site is located in the N-terminal domain of the exchanger since NHE1, after removal of the entire C-terminal domain, remains sensitive to amiloride.[41]

Table 14.2. Effect of Amiloride and Its Analogues on NHE1, NHE2, and NHE3[a]

	Inhibition constants (K_i) (μM)		
Inhibitor	NHE1	NHE2	NHE3
Amiloride	1[49]	1[49]	39[49]
	3[53]	1.4[55]	100[43]
	1.6[43]		
Ethylisopropyl-amiloride	0.02[43,49,53]	1[49]	8[49]
	0.05[23]	0.08[55]	2.4[43]
Methylpropyl-amiloride	0.05[53]		
Dimethyl-amiloride	0.02[43]	0.25[55]	14[43]
Benzamil	120[43]	320[55]	100[43]
HOE694	0.15[85]	5[85]	650[85]

[a]Values are taken from references shown in superscripts.

The putative fourth membrane-spanning domain has been shown to be part of the amiloride binding domain. This has been reviewed[11] and thus will be briefly discussed here. Counillon, Franchi, and Pouysségur selected mutant cells expressing an exchanger resistant to amiloride or its analogues,[51] and subsequently cloned and sequenced the exchanger cDNA; they identified a one-base-pair mutation: leucine to phenylalanine at position 167 (hamster amino acid sequence) which is responsible for the decreased affinity for 5-amino-substituted amiloride observed in the selected cells[52] (Table 14.3). It is noteworthy that this residue is located within a highly conserved part of the fourth putative transmembrane domain of the antiporter. However, NHE2, NHE3, and NHE4 differ subtly from NHE1 in this area. Counillon and colleagues[53] made a series of site-directed mutants of NHE1 in the fourth putative transmembrane domain based on this information, to mimic NHE2, NHE3, NHE4, and the mutant NHE. They showed that the NHE1 mutants, F165Y (mimicking NHE4) and L167F (mimicking the mutant NHE), are 30- and 5-fold more resistant to amiloride, respectively, when compared with the wild-type NHE1; and suggested that Phe 165 and Leu 167 are important for amiloride inhibition of Na^+/H^+ exchange.

Although NHE2 has the same amiloride sensitivity as NHE1, it is 20- to 50-fold more resistant to EIPA than NHE1.[6] NHE1 has a Phe 168 in the fourth transmembrane domain, which corresponds to Tyr 144 in NHE2. Therefore, our group made a site-directed mutant of NHE2, which mimics NHE1 (the mutant Y144F).[49] Interestingly, this mutant behaved as the wild-type NHE2, being resistant to EIPA. Conversely, it was found that the NHE1 mutant, F168Y which mimics NHE2, had the same sensitivity to EIPA as the wild-type NHE1.[53] Thus, we additionally conclude that there are other amiloride binding domains elsewhere in the NHE. The affinity for external Na^+ is not significantly changed by any of these mutants in the fourth transmembrane domain.[49] This supports that the amiloride binding site and the external Na^+ binding site are not identical, as was previously suggested based on functional studies of native fibroblast Na^+/H^+ exchangers.[51]

Table 14.3. Comparison of Amino Acid Composition of NHE1 and NHE2 in the Area of the Putative Fourth Membrane-Spanning Domain Which Appears to Be Involved in Determining Sensitivity of the Na^+/H^+ Exchanger Isoform to Amiloride and 5′-Amino-Substituted Amiloride

cDNA	Amino acid sequence	IC_{50} (μM) Amiloride	MPA	EIPA
NHE1	V F F L F L	3	0.05	
NHE1 → AR300	V F F *F* F L	15	1.5	
NHE1 → NHE2	V F F L *Y* L	3	0.05	
NHE1 → NHE3	V F F *F* *Y* L	15	1	
NHE1 → NHE4	V *Y* F L *Y* L	100	1	
NHE2	V F F L Y L	1		0.5
NHE2 → AR300	V F F *F* *F* L	10		5
NHE2 → NHE1	V F F L *F* L	1		0.3
NHE2 → NHE3	V F F *F* Y L	4		10

[a]Point mutations were made in NHE1 (top) and NHE2 (bottom) to mimic the amino acid composition in comparable areas in NHE1, 2, 3, 4, and the amiloride-sensitive mutant AR300. Italic amino acids show those mutated in NHE1 or NHE2 to the amino acid present in the isoform shown at the right of the arrow. Modified from Ref. 11.

14.8. KINETICS

The kinetics of Na^+/H^+ exchange in cells and tissues has been previously described.[13] We will focus here on the kinetics of the cloned mammalian Na^+/H^+ exchangers expressed in Na^+/H^+ exchanger-deficient cells (Table 14.4).

14.8.1. Na^+ Kinetics

NHE1, NHE2, and NHE3 have been studied when stably expressed in antiporter-deficient cells.[42,43] The expression system used for these functional studies has primarily been the PS120 fibroblast cell line. NHE1 has also been expressed in LAP cells and CHO cells[43] and NHE2 and NHE3 were also expressed in CHO cells.[9,43] The cloned exchangers show similar kinetic characteristics when undergoing Na^+-dependent pH recovery following an acid load,[42] although the exchangers differ in their response to growth factors and phorbol esters[42,43,54] (discussed below). The kinetics for external Na^+ follows a classical Michaelis–Menten model with a K_m of 4.7–18 mM and a Hill coefficient of 1 (Table 14.4),[42] suggesting that there is a single binding site for external Na^+. When rabbit NHEs were expressed in PS120 cells, we found that all isoforms had similar $K_m(Na^+)$ (15–18 mM) (Table 14.4) and there was no significant difference among the isoforms. However, when rat NHEs were expressed in CHO cells,[43,86] Orlowski found that $K_m(Na^+)$ was lower (5–10 mM) for NHE1 and NHE3 and higher (50 mM) for NHE2 (Table 14.3). The reason why different $K_m(Na^+)$ values were observed among rat NHE isoforms and the same $K_m(Na^+)$ was found in all rabbit

Table 14.4. Kinetics of Cloned Mammalian Na^+/H^+ Exchangers Expressed in Na^+/H^+ Exchanger-Deficient Cells Shown in Parentheses[a]

	NHE1	NHE2	NHE3
$K_m(Na^+)$ (mM)	15 (PS120)[42]	18 (PS120)[42]	17 (PS120)[42]
	4.7 (CHO)[43]	50 (CHO)[86]	10 (CHO)[43]
pK $[H^+]$			
−serum	6.20 (PS120)[42]	6.45 (PS120)[42]	6.45 (PS120)[42]
+ serum	6.41 (PS120)[42]	6.38 (PS120)[42]	6.44 (PS120)[41]
	6.70 (PS120)[38]		
	6.75 (CHO)[43]		6.45 (CHO)[43]

[a]Values are taken from references shown in superscripts.

NHE isoforms is not known. It may be related to species variation and/or different assay systems for Na^+ kinetics (spectrofluorometry using BCECF versus $^{22}Na^+$ uptake for rabbit and rat NHEs, respectively) or differences in the expression systems.

External Li^+ and H^+ compete with external Na^+ to inhibit NHE1 and NHE3. In contrast, external K^+ inhibits Na^+ uptake by NHE1 but has no effect on NHE2 and NHE3.[43,55]

14.8.2. H^+ Kinetics

The kinetics with respect to internal H^+ are also similar for the three exchangers with all three deviating from the hyperbolic response expected with Michaelis–Menten kinetics and all having a Hill coefficient of 2–3.[42] The data describing the effect of intracellular H^+ best fit an allosteric model with at least two independent binding sites for H^+.[56] In addition to the internal H^+ transport site, there is thought to be an internal modifier site for intracellular H^+, which can regulate the activity of the exchanger.[56] The pK values for intracellular H^+ [$pK(H_i^+)$]are similar among different isoforms of NHE when measured by the spectrofluorometric method using the pH-sensitive dye, BCECF, under conditions in which cells expressing NHE1 are not serum deprived (Table 14.4). Since NHE is regulated by growth factors by increasing its sensitivity to intracellular H^+, the $pK(H_i^+)$ value for NHE1 will decrease (i.e., lower affinity) when the cells are serum starved, as has been reported. When Orlowski assayed the H kinetics of NHE1 and NHE3 expressed in CHO cells, it was found that the affinity of NHE1 was twofold higher than that of NHE3, with $pK(H_i^+)$ values of 6.75 and 6.45, respectively (Table 14.4).[43]

All NHEs exhibit an internal modifier site. There has been some question whether NHEs in intestinal cells exhibit the modifier. However, most studies which failed to demonstrate the modifier site used a limited number of intravesicular pHs in plasma membrane vesicle uptake studies, while other studies using plasma membrane vesicles showed the presence of the modifier site, making it likely that this site is present.

14.8.3. Steady-State pH_i and Initial Rate of Na^+/H^+ Exchange

It has been suggested that the NHEs, in particular NHE1, have a "set point."[38] However, it is not known whether at the "set point" pH, the Na^+/H^+ exchangers are totally "turned off" or are operating at a minimal rate that just balances the rate of intracellular acid production. It has previously been reported by Pollock *et al.* that in OK cells, when Na^+/H^+ exchange activity was inhibited by amiloride, the steady-state pH_i fell, and when Na^+/H^+ exchange activity was increased by monensin (a Na^+/H^+ exchanging ionophore), the steady-state pH_i increased.[57] Thus, they concluded that at the "set point" pH, the Na^+/H^+ exchanger is kinetically at equilibrium. We recently also demonstrated[54] that the Na^+/H^+ exchanger is not turned off at the "set point" pH_i. As shown in Figure 14.3, inhibition of NHE3 by amiloride at the "set point" caused intracellular acidification. Further, when we stably expressed NHE2 in PS120 cells, we obtained clones with sixfold differences in Na^+/H^+ exchange rate. The "set point" pH_i reached by the NHE2 clones was related to the initial rate of Na^+/H^+ exchange.[42] If cells exhibited a high Na^+/H^+ exchange range at the beginning of pH recovery, they reached equilibrium at a relatively high pH_i. Conversely, cells with a slower initial H^+ efflux rate reached "set point" at a lower pH_i (Levine, Donowitz, and Tse, unpublished results). This further supports that at the steady-state pH_i, Na^+/H^+ exchange activity and other pH regulatory mechanisms within the cells interact to maintain intracellular pH within a physiological range and the Na^+/H^+ exchangers are not "off," but are balancing intracellular acidification processes.

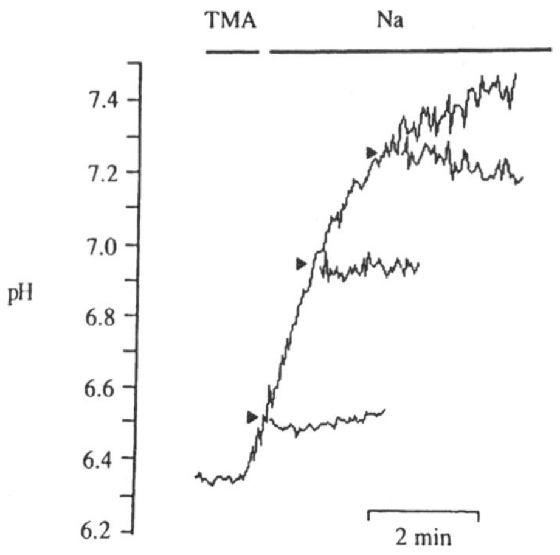

Figure 14.3. Effect of amiloride at "set point" pH of NHE3/PS120 cells. NHE3/PS120 cells loaded with BCECF were studied spectrofluorometrically to measure intracellular pH (pH_i). This composite tracing from four representative experiments shows the effect of 3 mM amiloride added during the Na^+-dependent recovery from an acid load at pH 6.50, 6.90, and 7.25 (arrowheads indicate amiloride addition), compared with untreated control. Na^+/H^+ exchange was inhibited 80% at 6.50 and 93% at 6.90 in these experiments. Amiloride addition at pH 7.25 caused a decrease in pH_i, presumably because of unmasking of an underlying acidification process present at high pH_i. This suggests that NHE3 is not turned off at the "set point." Reproduced from Ref. 54.

14.9. REGULATION

Different mechanisms of regulation of Na^+/H^+ exchange exist. Regulation has been described by protein kinases, growth factors, response to changes in cell volume and to cell spreading. These different regulatory mechanisms occur with differing time courses: for instance, protein kinases and serum cause changes in Na^+/H^+ exchange rate which starts within seconds of exposure; stimulation by changes in osmolarity has a delay of less than a minute; while some hormones including glucocorticoids take hours before changes in Na^+/H^+ exchange rate can be detected. In addition, the nature of the cell which contains a specific NHE isoform influences the regulation by protein kinases of that exchanger.

14.9.1. Short-Term Regulation

14.9.1.1. Regulation of Cloned Na^+/H^+ Exchangers by Growth Factors/Protein Kinases

When the three cloned Na^+/H^+ exchangers (NHE1, NHE2, and NHE3) were stably expressed in the same cell line (PS120), they differed in their response to growth factors.[42,54] Thus, there are intrinsic differences among NHE proteins that allow them to respond differently to growth factors. Table 14.5 summarizes the effects of various agents on the rate of Na^+/H^+ exchange by NHE1, NHE2, and NHE3 stably expressed in PS120 cells. Fetal bovine serum and FGF stimulate all three cloned exchangers. In contrast, the phorbol ester, phorbol myristate acetate (PMA), stimulates NHE1 and NHE2 but inhibits NHE3. Hyperosmolarity stimulates NHE1, but inhibits NHE2 and NHE3.

cAMP does not have any effect on NHE1, NHE2, or NHE3 expressed in PS120 fibroblasts. However, cAMP can effect Na^+/H^+ exchange in PS120 cells. The B-NHE isoform isolated from trout red cells has two clustered cAMP-dependent protein kinase consensus sites located in the cytoplasmic domain.[25] B-NHE in trout red blood cells is activated by an increase in cAMP.[58] When expressed in PS120 fibroblasts, B-NHE was also stimulated by cAMP.[25] The deletion or point mutation of the two protein kinase A consensus sites of B-NHE strongly reduced its response to cAMP.[25] Replacing the human NHE1 cytoplasmic tail by the B-NHE corresponding tail conferred cAMP-sensitivity to the chimeric NHE1 isoform. These results reinforce the notion that the cytoplasmic domain mediates and dictates the nature of the hormonal response.

The cell specificity of growth factor/protein kinase regulation of the NHEs has recently become clearer. For instance, there is a difference in regulation of the NHEs by protein kinase A based on which cell the exchangers are expressed in. In PS120 cells and in the human colon cancer cell line Caco-2, cAMP has no effect on any NHE, but in AP-1 cells (derived from Chinese hamster ovary cells) cAMP stimulates NHE1 and NHE2, but inhibits NHE3.[87] The explanation for these differences is not yet understood, but attention is being paid to differences in associated regulatory proteins present in the different cell lines. It is of note that cAMP increased NHE3 phosphorylation when expressed in AP-1 cells.[91]

The biochemical mechanism by which protein kinases regulate Na^+/H^+ exchangers has been studied in greatest detail by Pouysségur *et al.* with NHE1.[19,59] They demonstrated that NHE1 is stimulated by thrombin and EGF, with thrombin acting via phosphatidylinositol turnover, while EGF acts by affecting tyrosine phosphorylation.[59] NHE1 is a phosphoprotein[19,59] and the amount of phosphorylation on serine of the same set of specific phosphopeptides in NHE1 changes with exposure to EGF and thrombin in parallel with the changes in intracellular pH, likely the result of a change in Na^+/H^+ exchange rate. Also the phosphatase 1 and 2A inhibitor okadaic acid, which would be expected to increase phosphorylation on the assumption that phosphorylation is present under basal conditions, increases basal phosphorylation of NHE1 and also increases the basal rate of Na^+/H^+ exchange. Okadaic acid adds to the stimulatory effects of EGF and thrombin on both phosphorylation of NHE1 and rate of Na^+/H^+ exchange.[59] Nonetheless, it is unknown if changes in the phosphorylation of the Na^+/H^+ exchanger or in an associated protein lead to changes in the Na^+/H^+ exchange rate.

Phosphopeptide mapping demonstrated that NHE1 shows changes in phosphorylation in response to both thrombin and EGF but that these changes occur only on serine residues with no tyrosine phosphorylation identified,[19] and the same set of phosphopeptides are changed by both agents. This indicates that an intermediate kinase is involved in EGF regulation. This has been postulated to involve MAP kinase (mitogen-activated protein kinase).[19] In addition, it has been postulated that NHE1 may be more directly regulated by a kinase associated with the exchanger (NHE1 kinase),[59] perhaps similar to the regulation of the β-adrenergic receptor by a receptor-related kinase.

Concerning mechanisms of regulation, the C-terminus of NHEs is required for growth factor/protein kinase regulation. For NHE1, Pouysségur *et al.* proposed that phosphorylation of the C-terminus is required to allow interaction of the C-terminus with the H^+-modifier site, since depleting intracellular ATP alters the H^+-modifier site functionally.[38] However, the domain on the C-terminus of NHE1 which is phosphorylated by protein kinases and growth factors is different from the domain that interacts with the H^+-modifier site. Importantly, Pouysségur identified a domain in the C-terminus of approximately 70 amino acids (amino acids 567–635) which had to be

Table 14.5. Effect of Various Agents on Rate of Na^+/H^+ Exchange in the Three Transfected PS120 Cell Lines, with Kinetic Parameters Shown in Parentheses[42]

Agent	NHE1	NHE2	NHE3
FBS (10%)	Stimulation (K')	Stimulation (V_{max})	Stimulation (V_{max})
FGF (10 ng/ml)	Stimulation (K')	Stimulation (V_{max})	Stimulation (V_{max})
Thrombin (1 U/ml)	Not done	Stimulation (V_{max})	Stimulation (V_{max})
PMA (1 uM)	Stimulation (K')	Stimulation (V_{max})	Inhibition (V_{max})
8-Br-cAMP (0.5 mM)	No effect	No effect	No effect
8-Br-cGMP (0.5 mM)	No effect	No effect	No effect
Hyperosmolarity	Stimulation (K')	Inhibition (V_{max})	Inhibition (V_{max})
Okadaic acid	Stimulation (K')	Stimulation (V_{max})	Stimulation (V_{max})
Elevated Ca^{2+}	Stimulation (K')	No effect	No effect
W13	No effect	Stimulation (V_{max})	Stimulation (V_{max})

present to allow any regulation by protein kinases of NHE1, indicating that this is the part of the exchanger that affects the internal "modifier site" in the N-terminus. This is not the part of NHE1 that is phosphorylated. In fact, all major phosphorylation sites are located in the cytoplasmic tail between amino acids 636 and 815. Deletion of the 636–816 region led to a 50% decrease in kinase regulation of NHE1, which established the existence of at least one additional mechanism not involving direct phosphorylation of the Na^+/H^+ exchanger. This has led to the concept that phosphorylation-independent regulation of the NHEs occurs and is assumed to involve associated regulatory proteins. The first of these recognized is calmodulin (see below).

There are significant differences in the protein kinase regulation of NHE3 compared to NHE1, when both isoforms are expressed in the PS120 fibroblast cell line. While all protein kinase regulation of NHE1 is stimulatory, NHE3 can be both up- and downregulated; this is one of the defining characteristics of the intestinal brush border neutral NaCl absorptive process, of which NHE3 is thought to be part. For NHE3, serum fibroblast growth factor and thrombin stimulate, while phorbol esters inhibits Na^+/H^+ exchange. Protein kinase regulation of NHE1 is only by an increase in sensitivity of intracellular H^+. In contrast to NHE1, NHE3 regulation by growth factors and protein kinases occurs only by a change in the V_{max}, with no change in sensitivity to intracellular H^+ (Figure 14.4).

Furthermore, while all protein kinases appear to act on NHE1 via a common intermediate, this does not appear to be true for NHE3. A series of COOH-terminal deletion mutants of NHE3 was used to identify specific regions of the COOH-terminus, which are responsible for growth factor/protein kinase regulation.[93] These studies showed that there were two regulatory domains, one stimulatory (aa 475–585) and one inhibitory (aa 585–832) (Fig. 14.5). Both of these domains were made up of subdomains which acted independently. The stimulatory domain contained serum, FGF, and okadaic acid responsive elements, while the inhibitory domain contained a protein kinase C and calmodulin responsive elements. The presence of discrete subdomains for individual growth factors and protein kinases is a major difference from NHE1. Whether each of these domains is separately phosphorylated by protein kinases or acts by associated regulatory proteins is not known.

While studies with NHE2 are more preliminary, all growth factor/protein kinase regulation involves changes in V_{max} without any accompanying changes in sensitivity to intracellular H^+. In addition, all protein kinase regulation demonstrated has been stimulatory, including the effect of phorbol esters.[42]

The first recognized NHE-associated regulatory protein is calmodulin. This has been best studied for NHE1 expressed in PS120 fibroblasts by Wakabayashi and his colleagues.[94,95] They showed that calmodulin binds to and stimulates NHE1 by two sites, one of high affinity and one of lower affinity. The former is immediately C-terminal (aa 636–656) of the domain which interacts with the H^+-modifier site. Under basal Ca^{2+} conditions, the high affinity site inhibits NHE1, presumably by binding to the H^+ modifier site; while elevating intracellular Ca^{2+} removes the inhibition to stimulate NHE1. NHE3 also is affected by calmodulin. Under basal conditions, calmodulin inhibits NHE3.[93] Removing the calmodulin effect stimulates NHE3; the calmodulin inhibition is exerted by both calmodulin kinase II dependent and independent mechanisms, with the calmodulin kinase independent mechanism predominant. The calmodulin effect on NHE3 is exerted C-terminal of aa 756.

14.9.1.2. Regulation of Cloned Na^+/H^+ Exchangers by ATP Depletion

Another way to study the role of phosphorylation in regulation of Na^+/H^+ exchange is via studying the effects of ATP depletion. In all three cloned exchangers, cellular ATP depletion eliminates regulation of Na^+/H^+ exchange by growth factors.[42] In addition, cellular ATP depletion also changes the kinetics of Na^+/H^+ exchange (Figure 14.6) ATP depletion reduces the cooperative nature of Na^+/H^+ exchange with the Hill coefficient decreased from 2 to 1.2, which is not significantly different from 1 and fits Michaelis–Menten kinetics.[42] This observation suggests that: (1) the H^+-modifier site of the Na^+/H^+ exchangers requires phosphorylation or another ATP-dependent process to function; (2) under ATP-depleted conditions, the affinity of the H^+ substrate site can be distinguished from the H^+-modifier site since only one H^+ binding site is acting (Hill coefficient 1) and that must be the substrate transport site since Na^+/H^+ exchange still occurs. Therefore, it can be concluded that NHE1 has lower affinity for intracellular H^+ at the substrate site than NHE2 and NHE3 which have similar affinity (Figure 14.6). Further, ATP depletion decreases the V_{max} of all three cloned exchangers suggesting that basal phosphorylation of the NHE and/or regulatory protein(s) associated with NHEs is required for exchange activity. Pouysségur *et al.*[38] reported that cellular ATP depletion results in a decrease in the affinity of NHE1 for intracellular H^+ but there is still cooperativity of intracellular H^+. Their observation that phosphorylation or another ATP-dependent process is not required for the function of the H^+-modifier site is in contradiction with our studies.

14.9.1.3. Regulation of Na^+/H^+ Exchange by Cell Shrinkage

Osmotic shrinkage activates Na^+/H^+ exchange in a variety of cell types such as lymphocytes[60] and red blood cells.[61] Mechanistically, cell shrinkage activates Na^+/H^+ exchange by increasing its sensitivity to intracellular H^+, an effect similar to that of growth factor activation of NHE1.[60] Recently, Grinstein *et al.* showed that NHE1 is involved in cellular volume regulation.[62] They showed that NHE1, stably transfected in PS120 cells, responds to hyperosmolarity and causes cellular alkalinization. Although this process is ATP dependent, the hyperosmolar activation of NHE1 changes neither the amount of phosphorylation of NHE1 nor the pattern of the NHE1

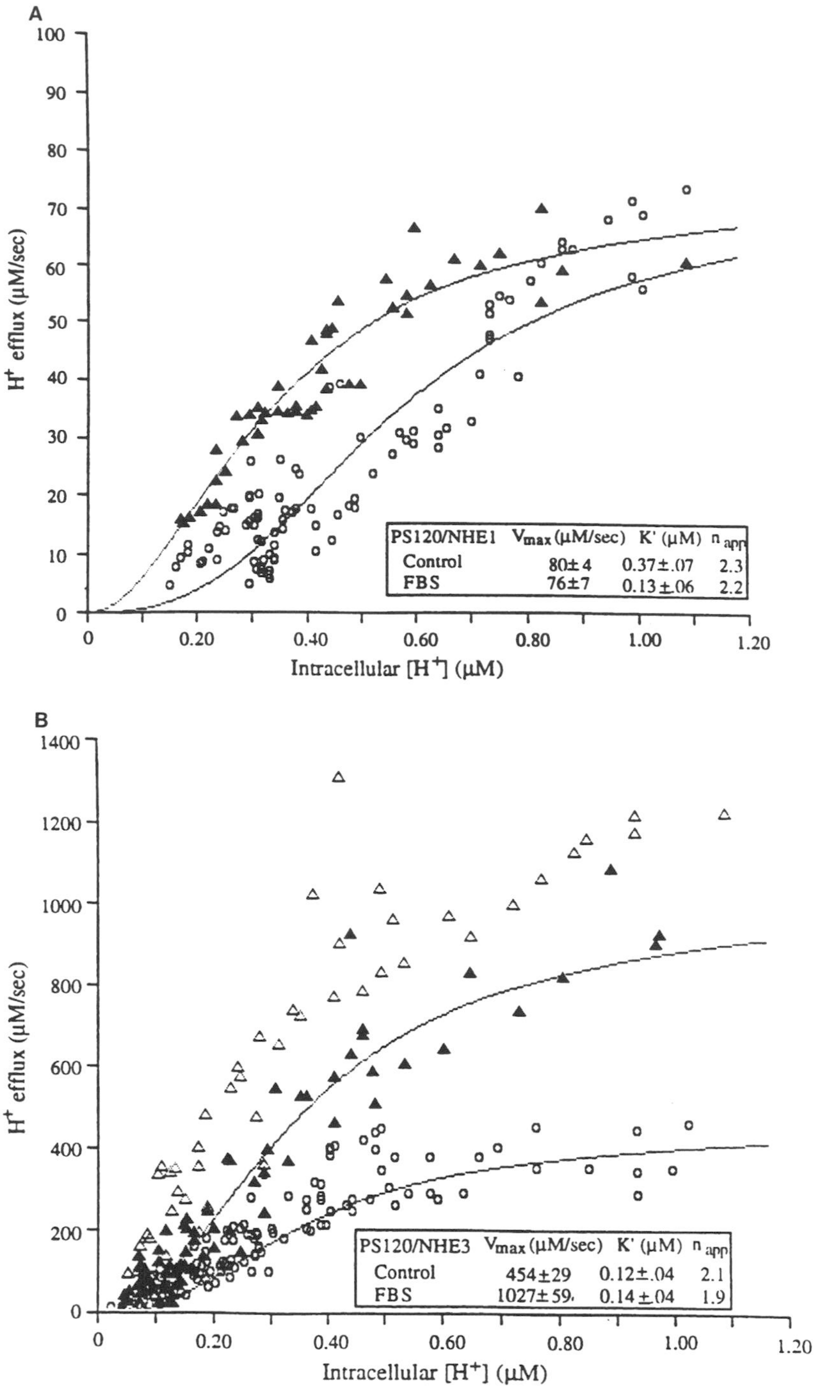

Figure 14.4. Effect of serum on Na^+/H^+ exchange rate of NHE1 and NHE3, stably expressed in PS120 fibroblasts. FBS stimulates Na^+/H^+ exchange rate in PS120/NHE1 and NHE3 cells when added at the beginning of the Na^+-dependent pH recovery. Control cells (○) were acidified with an NH_4Cl prepulse and allowed to recover in Na^+ medium; treated cells (▲) were similarly acidified, then perfused with Na^+ medium containing 10% FBS. (A) for PS120/NHE1 cells, the stimulation in exchanger activity was not reflected in an elevated V_{max}; rather, a decrease in $K'H^+_1$ was seen. In contrast, for PS120/NHE3 cells, (B) there was an increase in V_{max} with addition of FBS. When PS120/NHE3 cells (▲) were incubated with the C kinase inhibitor H7 (65 μM) for 10 min prior to addition of Na^+ medium with 10% FBS, there was a greater stimulation of exchanger activity. Incubation with H7 alone did not change the exchanger activity compared with control cells (data not shown). Reproduced from Ref. 42.

phosphopeptide map. Thus, they proposed that osmotic activation of NHE1 does not require phosphorylation of the Na^+/H^+ exchanger and that this mechanism of regulation of NHE1 is distinct from the growth factor regulation which is phosphorylation dependent.[62] However, it is not yet known how the cell shrinkage signal is "transmitted" to the Na^+/H^+ exchanger. Until recently, NHE4 had not been shown to carry out Na^+/H^+ exchange when expressed in mammalian cells. Based on its location in the most hypertonic part of the kidney,[33] in preliminary studies, Bookstein, Chang, and co-workers showed that it could be activated by hyperosmolarity.[63]

Other NHEs are also regulated by hyperosmolarity. When stably expressed in PS120 fibroblasts, NHE2 and NHE3 were inhibited by hyperosmolarity.[89] In contrast, when stably expressed in AP-1 cells, NHE2 was stimulated and NHE3 inhibited by hyperosmolarity.[90] The mechanism for hyperosmolar regulation of the NHEs is unknown as is what explains differences in NHE2 regulation based on cell type of expression.

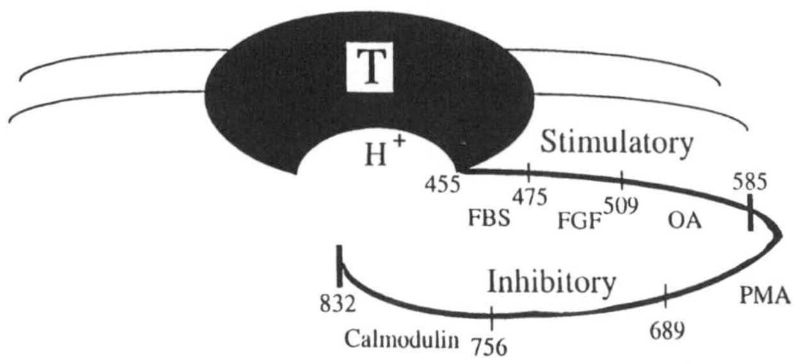

Figure 14.5. Stimulatory and inhibitory regions of cytoplasmic COOH-terminus of NHE3. Cytoplasmic tail of NHE3 contains separate sites for stimulation by fetal bovine serum (FBS), fibroblast growth factor (FGF), and okadaic acid (OA) and for inhibition by phorbol myristate acetate (PMA) and calmodulin.[93] (Reproduced from ref. 93).

14.9.1.4. Regulation of Na^+/H^+ Exchange by Cell Spreading

Schwartz, Lechene *et al.*[64,65] showed that cells grown on nonadhesive polymer, polyHEMA, are rounded and not able to grow while cells grown on plastic spread out and are able to proliferate. It has been proposed that an alkaline intracellular pH is a signal for cell proliferation and that well-spread cells have a higher intracellular pH (0.13–0.2 pH unit higher) than rounded cells because of the activation of Na^+/H^+ exchange by cell spreading. This increase in pH_i closely parallels increased DNA synthesis, thus suggesting a role for Na^+/H^+ exchange in cellular proliferation. The effect of cell spreading on Na^+/H^+ exchange occurs within minutes and is reversible and mediated by extracellular matrix proteins such as fibronectin and vitronectin and their integrin receptors.[66] The downstream intracellular signaling pathways that mediate the effect of extracellular matrix proteins and receptors on pH_i are not well-defined but might involve the activation of protein kinase C which in turn activates Na^+/H^+ exchange.[67] NHE1 but not NHE3 expressed in PS120 fibroblasts is activated within minutes by cell spreading (Lechene and Pouysségur, unpublished observations).

14.9.2. Chimeric Na^+/H^+ Exchangers

Studies were done to determine (i) which domain, the transmembrane N-terminus or cytoplasmic C-terminus, plays the critical role in determining the V_{max} *vs.* $K'(H^+_i)$ regulatory effect of the Na^+/H^+ isoforms, and (ii) if the second messenger regulation of the exchangers can be modified by swapping the cytoplasmic C-termini, which are thought to be solely responsible for the protein kinase regulation. Chimeric Na^+/H^+ exchangers were constructed by exchanging the N- and C-termini among three cloned rabbit Na^+/H^+ exchangers (NHE1–3).[88,92] All the chimeric cDNAs resulted in functional Na^+/H^+ exchangers when expressed in PS120 fibroblasts with the activities comparable to the wild-type exchangers.

The protein kinase regulation of the chimeric exchangers was studied when they were stably expressed in P120 fibroblasts. FBS was studied as the most potent stimulator of NHEs. Chimera M3C1 (membrane domain of NHE3 and cytoplasmic domain of NHE1) responded to FBS with an increase in V_{max} [no change in $K'(H^+_i)$ or Hill coefficient], while M1C2 also responded to FBS but with an increase in affinity for intracellular H^+ and no change in V_{max} of the Hill coefficient. It seems that the NHE contributing the N-terminus determined the kinetics of regulation [NHE3 regulated by V_{max} mechanism, NHE1 regulated by $K'(H^+_i)$]. These results suggest that the previously reported V_{max} *vs.* $K'(H^+_i)$ effect of growth factor/protein kinase NHE regulation originates from the membrane-bound N-termini of the exchangers.

Phorbol myristate acetate is known to stimulate NHE1 by increasing the affinity for H^+, and NHE2 by increasing V_{max}. In contrast, NHE3 is inhibited by PMA through a change in V_{max}. PMA did not affect M1C3, M1C2, M2C1, and M3C1, all of which consist of one domain from the housekeeping isoform NHE1 and the other domain from an epithelial isoform. On the other hand, PMA resulted in changes in the transport activities of M3C2 and M2C3. V_{max} of M3C2 was increased by PMA without any changes in $K'(H^+_i)$ and n_{app}. Of note, M3C2 and M2C3 consist of a membrane-bound N-terminus of an epithelial exchanger and a cytoplasmic C-terminus of another epithelial exchanger. The stimulation of M3C2 and inhibition of M2C3 by PMA followed the effect of PMA on NHE2 and NHE3, respectively.

The response of the chimeric exchangers to FGF was similar in that stimulation only occurred with M3C2 and M2C3, which are made up of epithelial isoforms, with an increase in V_{max}, whereas the kinetic properties of M2C1, M1C2, M1C3, and M3C1 were not affected by FGF.

The above observation suggests that the regulation of the chimeras by PMA and FGF is dictated by the C-terminus, but this requires a specific interaction between the membrane domain and the cytoplasmic domain of a Na^+/H^+ exchanger. The lack of regulation of M1C3, M1C2, M2C1 and M3C1 by PMA and FGF is due to the inability of the N- and C-termini of the chimeric proteins to interact. On the other hand, M2C3 and M3C2 are regulated by PMA and FGF because there is a common structural element between C2 and C3, and also between M2 and M3, which is involved in V_{max} regulation of these epithelial Na^+/H^+ exchangers. Similarly, the previously reported cAMP-dependent stimulation of a chimera with the N-terminus of NHE1 and the cytoplasmic C-terminal domain of βNHE is due to a specific interaction between the membrane domain of NHE1 and the cytoplasmic domain of βNHE, which allows a $K'(H^+_i)$ regulation of Na^+/H^+ exchange.[88]

14.9.3. Effect of Cell Type on the NHE Plasma Membrane Location and Protein Kinase Regulation

There is evidence that the cell type and membrane location (apical versus basolateral membrane in epithelial cells) can influence second messenger regulation of Na^+/H^+ exchange in addition to the nature of the Na^+/H^+ exchanger isoform. This cell specificity is demonstrated by analysis of the

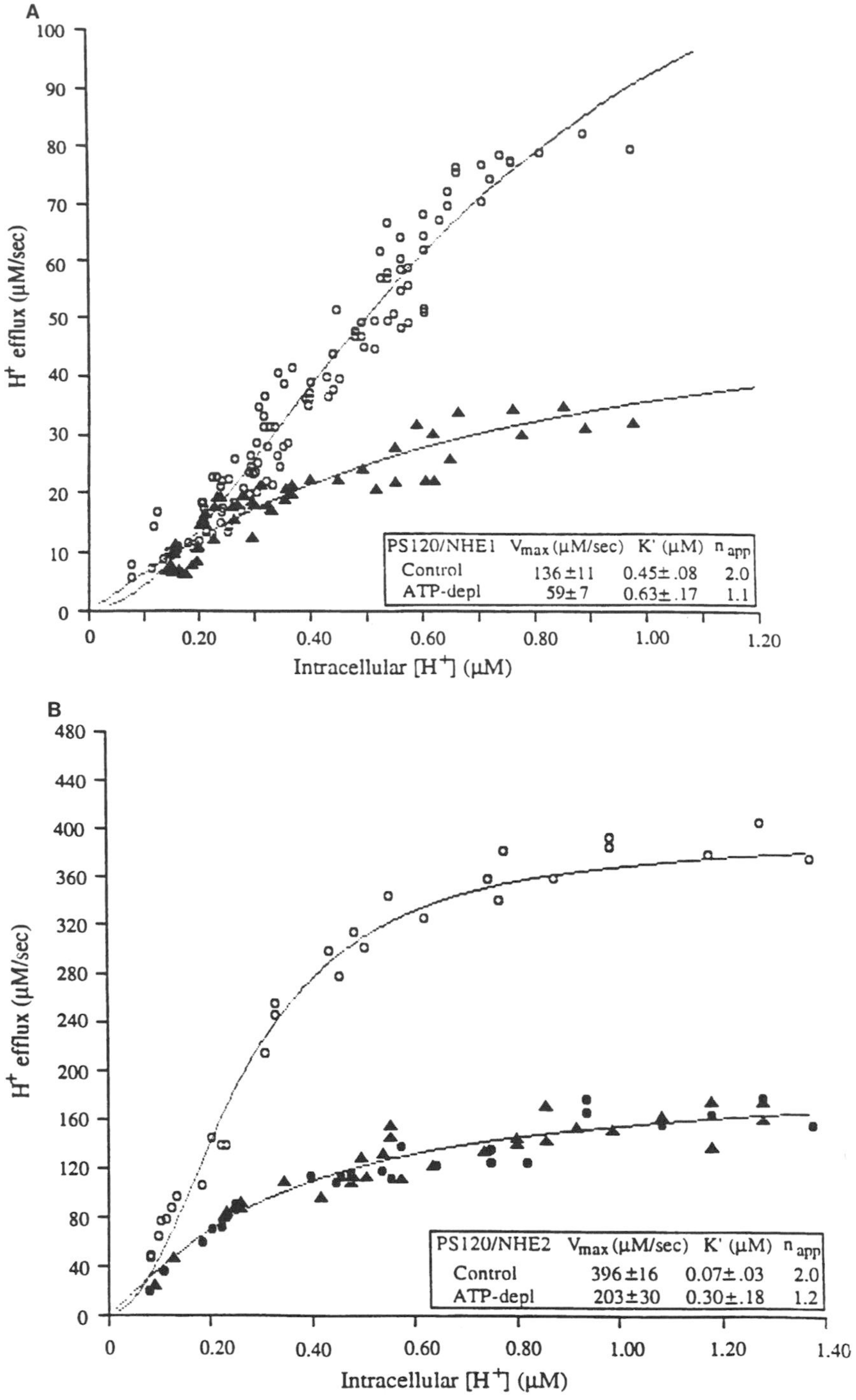

Figure 14.6. ATP depletion of NHE1-, NHE2-, and NHE3-transfected cells lines causes a marked decrease in Na^+/H^+ exchanger activity. This decrease in activity was characterized both by a decrease in V_{max} and K' estimates and by a decrease in the apparent Hill coefficient. Cells were incubated with 2 μg/ml oligomycin and 5 mM 2-deoxy-D-glucose for 30 min at 37°C during dye loading. Shown here are data from five experiments in which control (○) or ATP-depleted (▲) cells were loaded and allowed to recover in Na^+ medium. (A) PS120/NHE1 cells showed a 57% decrease in V_{max} and a 40% increase in K'. (B) PS120/NHE2 cells showed a 49% decrease in V_{max} and a 329% increase in K'. Addition of 10% fetal bovine serum to ATP-depleted cells (●) during pH recovery did not alter V_{max} or K'. (C) PS120/NHE3 cells showed a 68% decrease in V_{max} and a 142% increase in K'. These data are from five or more experiments for each condition in the three cell lines. Reproduced from Ref. 42.

effects of C kinase on specific NHE isoforms. PMA stimulates cloned NHE1 expressed in PS120 cells[4,42] but has no effect on the cloned NHE1 expressed in LAP cells.[68] PMA inhibits endogenous NHE1 in human A431 cells[68] but has no effect on endogenous NHE1 in granulocytic HL-60 cells.[69] In polarized Caco-2 cells, NHE1 is on the basolateral membrane but this NHE1 is not regulated acutely by serum, phorbol esters, or growth factors,[30] whereas NHE1 on the basolateral membrane of other epithelial cells is regulated by second messengers (for review, see Ref. 11). Further, OK cells have an endogenous amiloride-resistant Na^+/H^+ exchanger on the apical membrane but lack any basolateral membrane Na^+/H^+ exchanger. When cloned NHE1 is stably expressed in OK cells, it is expressed partially in the basolateral membrane. The transfected NHE1 on the basolateral membrane of OK cells is inhibited by protein kinase C but is stimulated by protein kinase A.[70] In PS120 cells transfected with NHE1, NHE1 is stimulated by protein kinase C, while protein kinase A has no effect. Also of note is that NHE1 does not have any cAMP-dependent protein kinase consensus sequence, suggesting that an intermediate is in-

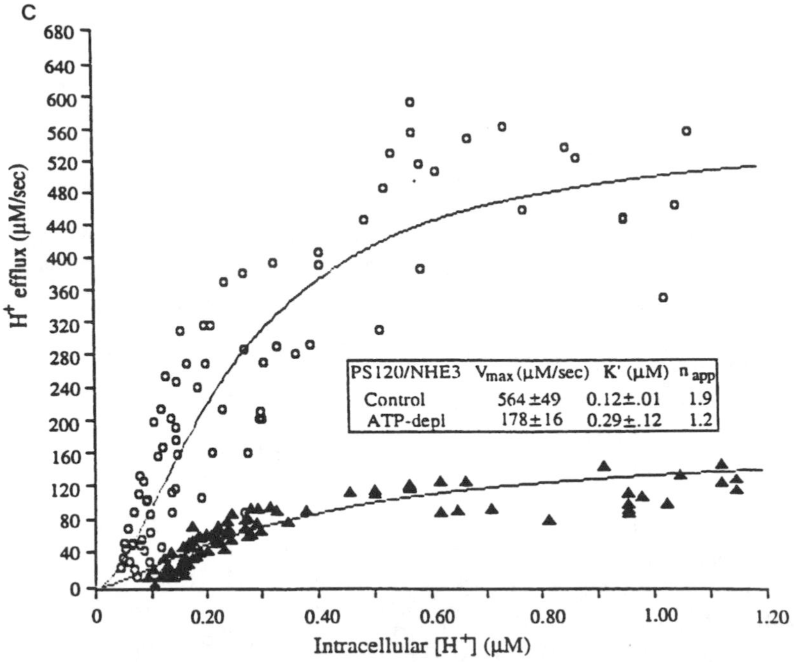

Figure 14.6. *Continued*

volved in the regulation. These studies show that the specific cell type contributes to the nature of regulation by kinases of a specific isoform of NHE. In addition, they raise the possibility that in addition to the nature of the NHE isoforms, NHEs are closely associated with other "NHE specific kinase(s) or regulatory protein(s)" which adds to the diversity in regulation of Na^+/H^+ exchangers in different cell types and even in different cellular domains in the same cells (polarized cells).

14.9.4. Long-Term Regulation

14.9.4.1. Effect of Glucocorticoids on Ileal Brush Border Na^+/H^+ Exchange

It had previously been shown that glucocorticoids stimulate intestinal water and NaCl absorption, an effect occurring in hours.[71] Methylprednisolone stimulates rabbit ileal neutral NaCl absorption, whereas induction of glucocorticoid deficiency with aminoglutethimide inhibits NaCl absorption.[72] Studies were done to determine whether the mechanism of these longer-term effects involved stimulation of ileal villus cell brush border Na^+/H^+ exchange and if glucocorticoid regulation potentially involved a change in the amount of message of any NHE isoform (transcriptional regulation or effect on mRNA stability).[73] Rabbits treated with methylprednisolone for 24 and 72 hr had increased ileal brush border Na^+/H^+ exchange by ~100%, whereas aminogluthethimide treatment led to a 50% decrease in Na^+/H^+ exchange. Quantitation of message of NHE1, NHE2, and NHE3 showed that methylprednisolone stimulated NHE3 mRNA level by four- to sixfold. In contrast, messages for NHE1 and NHE2 were not affected by methylprednisolone. These results demonstrated that glucocorticoids regulate ileal Na^+ uptake by an effect on the brush border Na^+/H^+ exchanger and that basal Na^+/H^+ exchange is glucocorticoid dependent. This effect occurs either by an effect on transcription and/or by a change in mRNA stability. These results are analogous to the earlier findings of Freiberg *et al.*[74] that the glucocorticoid dexamethasone, but not the mineralocorticoid aldosterone, increases rat proximal tubule brush border Na^+/H^+ exchange since an increase only in message for NHE3 parallels the increase in brush border Na^+/H^+ exchanger. This study further suggests that NHE3 is the brush border Na^+/H^+ exchanger isoform involved in ileal NaCl absorption.

14.9.4.2. Chronic Acidosis and Na^+/H^+ Exchange

Chronic acidosis (metabolic acidosis and respiratory acidosis) causes an increase in Na^+/H^+ exchange activity in renal proximal tubule brush border membranes and cultured renal cell lines. With the advance in cloning and identification of Na^+/H^+ exchangers, NHE cDNA probes, in particular to NHE1, have been used to study the relationship between the increase in Na^+/H^+ exchange activity by chronic acidosis and the change in NHE message level. Studies in a renal epithelial cell line (MCT) have found that both metabolic and respiratory acidosis increase Na^+/H^+ exchanger activity with a parallel increase in amount of NHE1 mRNA.[75] On the other hand, in an animal model, metabolic acidosis, but not respiratory acidosis, increases the amount of NHE1 mRNA in renal proximal tubules.[76] The discrepancy of this observation between the animal model and the cell culture model is not understood. Apparently, this is a specific response of renal proximal tubule cells to chronic acidosis as cultured human foreskin fibroblasts do not have increased but rather have decreased

NHE1 message with metabolic acidosis.[75] This increase in the Na^+/H^+ exchange of renal cells in response to chronic acidosis is not an immediate response and is regulated at the level of transcription and/or posttranscriptionally, for instance, by an increase in the NHE1 mRNA transcription and/or its stability, and requires protein synthesis. As recently reviewed by Krapf and Alpern,[14,77] the mechanism that mediates increased Na^+/H^+ exchanger mRNA expression in response to acidosis appears to involve protein kinase C.

14.10. NHE1 IS THE HOUSEKEEPING ISOFORM NA^+/H^+ EXCHANGER AND NHE3 IS THE BRUSH BORDER NA^+/H^+ EXCHANGER OF INTESTINAL AND RENAL EPITHELIAL CELLS

By Northern blot analysis and by immunocytochemistry, NHE1 has been found ubiquitously in nearly all cells.[4,35] It is sensitive to amiloride inhibition, activated by cell shrinkage[62] and cell spreading, and stimulated by growth factors and protein kinases, in particular protein kinase C. These characteristics support that NHE1 is the "housekeeping isoform" Na^+/H^+ exchanger involved at least in regulation of intracellular pH and volume and perhaps cell proliferation.

NHE3 is resistant to amiloride and its message is restricted to intestine, kidney, and stomach.[7,54] The expression of NHE3 message in rabbit and human gastrointestinal tract colocalizes with the expression of the neutral NaCl absorptive process. When expressed in PS120 cells, NHE3 is the only isoform exchanger that is inhibited by protein kinase C. Glucocorticoids stimulate ileal brush border Na^+/H^+ exchange. NHE3 message, but not NHE1 or NHE2 message, increases with glucocorticoid stimulation of Na^+/H^+ exchange.[73] Further, anti-NHE3 antibodies have shown that NHE3 is the brush border Na^+/H^+ exchanger of kidney proximal tubule cells, and of multiple small intestinal and colonic absorptive cells (excluding rabbit descending colon). These pharmacological, functional, and immunolocalization studies suggest that NHE3 is the major brush border Na^+/H^+ exchanger of intestinal and renal epithelial cells and is involved in cellular Na^+ absorption. While NHE2 is also an intestinal brush border Na^+/H^+ exchanger, its function is not established, although it may play a backup role under conditions of increased Na^+ absorption.

14.11. CONCLUDING REMARKS AND FUTURE STUDIES

With the identification of the Na^+/H^+ exchanger gene family, there has been an advance in understanding of the molecular properties, physiology, and structure/function relationships of Na^+/H^+ exchangers. Five NHEs have been identified to date. However, additional members of the mammalian gene family almost certainly remain to be identified. At the least, these include a hippocampal isoform which is totally amiloride resistant,[78] and perhaps additional renal/intestinal forms.

Why are there so many Na^+/H^+ exchanger isoforms? As discussed in the Introduction, it is predicted that each Na^+/H^+ exchanger assumes a specialized function. NHE1 has been shown to be involved in regulation of cellular volume and cellular proliferation, in addition to pH homeostasis. NHE3 is a brush border Na^+/H^+ exchanger in intestinal and renal epithelial cells and as such is probably fine-tuned for Na^+ absorption. NHE4 appears to be the renal isoform that responds to hyperosmolarity. The physiological role of NHE2, in addition to that of a Na^+/H^+ exchanger to regulate intracellular pH, is not clear. It will be interesting to determine whether NHE2, NHE3, and other isoform Na^+/H^+ exchangers can perform specialized functions like regulation of cellular volume and cellular proliferation.

There are several major areas relating to structure/function studies in which more insight will be added rapidly in the near future: (1) How are Na^+/H^+ exchangers regulated by external signals, e.g. growth factors, cell shrinkage, and cell spreading? Phosphorylation of the NHEs is not the only mechanism that regulates NHE activity; what are the other mechanisms? Are all other isoform NHEs phosphoproteins? Are NHEs associated with other regulatory protein(s) that are cell type specific and even cellular domain specific within the same cell type? (2) How are NHEs directed to specific domains in epithelial cells? Preliminary experiments in Pouysségur's and in our laboratory suggest that NHE3 is directed exclusively to the apical membrane of epithelial cells like Caco-2 and MDCK cells and NHE1 is directed predominantly, but not exclusively, to the basolateral membrane in these two cell lines.[6,54] (3) What are the crucial amino acids involved in Na^+/H^+ transport, the H^+-modifier site function, and the regulatory effect of the C-terminus on rate of Na^+/H^+ exchange? Truncation and point mutations, chimera production, and random mutagenesis all are likely to contribute to this area, while we wait for crystal structure to reveal details of the reaction mechanisms.

REFERENCES

1. Pitts, R. F., Ayer, J. L., and Schiess, W. A. (1949). The renal regulation of acid–base balance in man. III. The regulation and excretion of bicarbonate. *J. Clin. Invest.* **28:**35–44.
2. Murer, H., Hopfer, U., and Kinne, R. (1989). Sodium, proton antiport in brush border membranes isolated from rat small intestine and kidney. *Biochem. J.* **154:**597–602.
3. Sardet, C., Franchi, A., and Pouysségur, J. (1989). Molecular cloning, primary structure and expression of the human growth factor-activatable Na^+/H^+ antiporter. *Cell* **56:**271–280.
4. Tse, C. M., Ma, A. I., Yang, V. W., Watson, A. J. M., Potter, J., Sardet, C., Pouysségur, J., and Donowitz, M. (1991). Molecular cloning of cDNA encoding the rabbit ileal villus cell basolateral membrane Na^+/H^+ exchanger. *Embo J.* **10:**1957–1967.
5. Watson, A. J. M., Tse, C. M., Ma, A. I., Pouysségur, J., and Donowitz, M. (1991). Cloning and functional expression of a sec-

ond novel rabbit ileal villus epithelial cell Na^+/H^+ exchanger (NHE2). *Gastroenterology* **100:**258.

6. Tse, C. M., Levine, S. A., Yun, C. H. C., Montrose, M. H., Little, P. J., Pouysségur, J., and Donowitz, M. (1993). Cloning and expression of a rabbit cDNA encoding a serum-activated ethylisopropyl amiloride resistant epithelial Na^+/H^+ exchanger isoform (NHE-2). *J. Biol. Chem.* **268:**11917–11924.
7. Tse, C. M., Brant, S. R., Walker, S., Pouysségur, J., and Donowitz, M. (1992). Cloning and sequencing a rabbit cDNA encoding an intestinal and kidney-specific Na^+/H^+ exchanger isoform (NHE-3). *J. Biol. Chem.* **267:**9340–9346.
8. Orlowski, J., Kandasamy, R. A., and Shull, G. E. (1992). Molecular cloning of putative members of the Na^+/H^+ exchanger gene family. *J. Biol. Chem.* **267:**9332–9339.
9. Wang, Z., Orlowski, J., and Shull, G. E. (1993). Primary structure and functional expression of a novel gastrointestinal isoform of the rat Na^+/H^+ exchanger. *J. Biol. Chem.* **268:**11925–11928.
10. Counillon, L., and Pouysségur, J. (1993). Structure, function and regulation of vertebrate Na^+/H^+ exchangers (NHE). *Curr. Opin. Nephrol. Hypertension* **2:**708–714.
11. Tse, C., Levine, S., Yun, C., Bryant, S., Counillon, L., Pouysségur, J., and Donowitz, M. (1993). Structure/function studies of the epithelial isoforms of the mammalian Na^+/H^+ exchanger gene family. *J. Membr. Biol.* **135:**93–108.
12. Yun, C. H. C., Tse, C.-M., Nath, S. K., Levine, S. A., Brant, S. A., and Donowitz, M. (1995). Mammalian Na^+/H^+ exchanger gene family. *Am. J. Physiol.* **269:**G1–G11.
13. Grinstein, S., Rotin, D., and Marson, M. J. (1989). Na^+/H^+ exchanger and growth factor-induced cytosolic pH change. Role in cellular proliferation. *Biochim. Biophys. Acta* **988:**73–91.
14. Krapf, R., and Alpern, R. J. (1993). Cell pH and transepithelial H/HCO_3 transport in renal proximal tubule. *J. Membr. Biol.* **131:** 1–10.
15. Clark, J. D., and Limbird, L. L. (1991). Na^+/H^+ exchanger subtypes: A predictive review. *Am. J. Physiol.* **261:**G945–G953.
16. Helme-Kolb, C., Montrose, M. H., and Murer, H. (1990). Parathyroid hormone regulation of Na^+/H^+ exchange in opossum kidney cells. Polarity and mechanisms. *Pfluegers Arch.* **416:**615–623.
17. Vigne, P., Frelin, C., and Lazdunski, M. (1985). The Na^+/H^+ antiporter is activated by serum and phorbol ester in proliferating myoblasts but not in differentiated myotubes. Properties of the activation process. *J. Biol. Chem.* **260:**8008–8013.
18. Miller, R. T., and Pollock, A. S. (1987). Modification of the internal pH sensitivity of the Na^+/H^+ antiporter by parathyroid hormone in a cultured renal cell line. *J. Biol. Chem.* **262:**9115–9120.
19. Sardet, C., Counillon, L., Franchi, A., and Pouysségur, J. (1990). Growth factors induce phosphorylation of the Na^+/H^+ antiporter, a glycoprotein of 110 kD. *Science* **247:**723–726.
20. Pouysségur, C., Sardet, C., Franchi, A., L'Allemain, G., and Paris, S. (1984). A specific mutation abolishing Na^+/H^+ antiport activity in hamster fibroblasts precludes growth at neutral and acidic pH. *Proc. Natl. Acad. Sci. USA* **81:**4833–4837.
21. Franchi, A., Perucca-Lostanten, D., and Pouysségur, J. (1986). Functional expression of a human Na^+/H^+ antiporter gene transfected into antiporter-deficient mouse L cells. *Proc. Natl. Acad. Sci. USA* **83:**9388–9392.
22. Reilly, R. F., Hildebrandt, F., Biemesderfer, D., Sardet, C., Pouysségur, J., Aronson, P. S., Slayman, C. N., and Igarashi, P. (1991). cDNA cloning and immunolocalization of a Na^+/H^+ exchange in LLC-PK_1, renal epithelial cells. *Am. J. Physiol.* **261:**F1088–F1094.
23. Takaichi, K., Wang, D., Blakovets, D. F., and Warnock, D. G. (1992). Cloning, sequencing, and expression of Na^+/H^+ antiporter cDNAs from human tissues. *Am. J. Physiol.* **262:**C1069–C1076.
24. Counillon, L., and Pouysségur, J. (1993). Nucleotide sequence of the Chinese hamster Na^+/H^+ exchanger NHE1. *Biochim. Biophys. Acta* **1172:**343–345.
25. Borgese, F., Sardet, C., Cappadoro, M., Pouysségur, J., and Motais, R. (1992). Cloning and expressing a cAMP-activated Na^+/H^+ exchanger: Evidence that the cytoplasmic domain mediates hormonal regulation. *Proc. Natl. Acad. Sci. USA* **89:**6768–6769.
26. Mattei, M. G., Sardet, C., Franchi, A., and Pouysségur, J. (1988). The human amiloride-sensitive Na^+/H^+ antiporters: Localization to chromosome 1 by in situ hybridization. *Cytogenet. Cell. Genet.* **48:** 6–8.
27. Brant, S. R., Bernstein, M., Wasmuth, J. J., Taylor, E. W., McPherson, M. D., Li, X., Walker, S. A., Pouysségur, J., Donowitz, M., Tse, C. M., and Jabs, E. W. (1993). Physical and genetic mapping of a human apical epithelial Na^+/H^+ exchanger (NHE3) isoform to chromosome 5p15.3. *Genomics* **15:**668–672.
28. Collins, J. F., Honda, T., Knobel, S., Bulus, N. M., Conary, J., Dubois, R., and Ghishan, F. K. (1993). Molecular cloning, sequencing, tissue distribution, and functional expression of a Na^+/H^+ exchanger (NHE2). *Proc. Natl. Acad. Sci. USA* **90:**3938–3942.
29. Honda, T., Knobel, S. M., Bulus, N. M., and Ghishan, F. K. (1993). Kinetic characterization of a stable expressed novel Na^+/H^+ exchanger (NHE2). *Biochim. Biophys. Acta* **1150:**199–202.
30. Krapf, G., and Solioz, M. (1991). Na^+/H^+ antiporter in RNA expression in single nephron segments of rat kidney cortex. *J. Clin. Invest.* **88:**783–788.
31. Donowitz, M., and Welsh, M. J. (1987). Regulation of mammalian small intestinal electrolytic secretion. In *Physiology of the Gastrointestinal Tract* (L. R. Johnson, ed.), Raven Press, New York, pp. 1351–1388.
32. Syed, I., Rao, D. D., Bavishi, D., Dahdal, R. Y., Harig, J. M., Ramaswamy, K., and Dudeja, P. K. (1994). Regional distribution of Na/H exchanger isoforms, NHE1 and NHE3. *Gastroenterology,* in press.
33. Depauli, A. M., Chang, E. B., Bookstein, C., Rao, M. C., Musch, M., Bizal, G. L., Evan, A., and Soleimani, M. (1993). *J. Am. Soc. Nephrol.* **4:**836 (abstract).
34. Watson, A. J. M., Levine, S., Donowitz, M., and Montrose, M. H. (1991). Kinetics and regulation of polarized Na^+/H^+ exchanger from Caco-2 cells, a human intestinal cell line. *Am. J. Physiol.* **261:** G229– G238.
35. Biemesderfer, D., Reilly, R., Exner, M., and Igarasgu, P. (1992). Immunocytochemical characterization of Na^+/H^+ exchanger isoform NHE1 in rabbit kidney. *Am. J. Physiol.* **263:**F833–F840.
36. Biemesderfer, D., Pizzonia, J., Abu-Alfa, A., Exner, M., Reilly, R., Isarashi, R., and Aronson, D. S. (1993). NHE3: A Na/H exchange isoform of renal brush border. *Am. J. Physiol.* **965:**F736–F742.
37. Tse, C. M., Levine, S., Yun, C. H., Khurana, S., and Donowitz, M. (1994). Na^+/H^+ exchanger-2 is an *O*-linked but not an N-linked sialoglycoprotein. *Biochemistry* **33:**12954–12961.
38. Wakabayashi, S., Fafournoux, P., Sardet, C., and Pouysségur, J. (1992). The Na^+/H^+ antiporter cytoplasmic domain mediates growth factor signals and controls H^+-sensing. *Proc. Natl. Acad. Sci. USA* **88:**2424–2428.
39. Tse, C. M., Levine, S., Yun, C., Montrose, M., and Donowitz, M. (1992). The cytoplasmic domain of the epithelial specific Na^+/H^+ exchanger isoform (NHE3) mediates inhibition by phorbol myristate acetate (PMA) but not stimulation by serum. *Gastroenterology* **104:**A285.
40. Haworth, R. S., Frohloch, O., and Fliegel, L. (1993). Multiple carbohydrate moieties on the Na^+/H^+ exchanger. *Biochem. J.* **289:**637–640.
41. Yusufi, A. N. K., Szczeparska-Konkel, M., and Dousa, T. P. (1988). Role of N-linked oligosaccharides in the transport activity of the

Na^+/H^+ antiporter in rat renal brush border membrane. *J. Biol. Chem.* **263:**13683–13691.

42. Levine, S., Montrose, M., Tse, C. M., and Donowitz, M. (1993). Kinetic and regulation of three cloned mammalian Na^+/H^+ exchangers stably expressed in a fibroblast cell line. *J. Biol. Chem.* **268:**25527–25535.
43. Orlowski, J. (1993). Heterologous expression and functional properties of amiloride high affinity (NHE1) and low affinity (NHE3) isoforms of the rat Na^+/H^+ exchanger. *J. Biol. Chem.* **268:**16369–16377.
44. Kinsella, J., Heller, P., and Froehlich, J. P. (1993). Sodium dependence of the Na^+/H^+ exchanger in the pre-steady state: Implications for the exchanger mechanism. *J. Biol. Chem.* **268:**3184–3193.
45. Beliveau, R., Demeule, M., and Potier, M. (1988). Molecular size of the Na^+/H^+ antiport in renal brush border membranes, as estimated by radiation inactivation. *Biochem. Biophys. Res. Commun.* **152:** 484–489.
46. Wakabayashi, S,. Sardet, C., Fafournoux, P., Counillon, L., Meloche, S., Pages, G., and Pouysségur, J. (1992). Structure function of the growth factor-activatable Na^+/H^+ exchanger (NHE1). *Rev. Physiol. Biochem. Pharmacol.* **119:**157–186.
47. Fliegel, L., Haworth, R. S., and Dyck, J. R. B. (1993). Characterization of the placental brush border membrane Na^+/H^+ exchanger: Identification of thio-dependent transitions in apparent molecular size. *Biochem. J.* **289:**101–107.
48. Haggerty, J. G., Agarwal, N., Reilly, R. F., Adelberg, E. A., and Slayman, C. W. (1988). Pharmacologically different Na/H antiporters on the apical and basolateral surfaces of cultured porcine kidney cells (LLC-PK1). *Proc. Natl. Acad. Sci. USA* **85:**6797–6801.
49. Yun, C. H. C., Little, P. J., Nath, S. K., Levine, S. A., Pouysségur, J., Tse, C. M., and Donowitz, M. (1993). Leu143 in the putative fourth membrane spanning domain is critical for amiloride inhibition of an epithelial Na^+/H^+ exchanger isoform (NHE2). *Biochem. Biophys. Res. Commun.* **193:**532–539.
50. Benos, D. J. (1988). Amiloride: Chemistry, kinetics, and structure–activity relationships. In *Na^+/H^+ Exchange* (S. Grinstein, ed.), CRC Press, Boca Raton, FL, pp. 121–136.
51. Franchi, A., Cragoe, J., and Pouysségur, J. (1986). Isolation and properties of fibroblast mutants overexpressing an altered Na^+/H^+ antiporter. *J. Biol. Chem.* **261:**14614–14620.
52. Counillon, L., Franchi, A., and Pouysségur, J. (1993). A point mutation of the Na^+/H^+ exchanger gene (NHE1) confers amiloride resistance upon chronic acidosis. *Proc. Natl. Acad. Sci. USA* **90:**4508–4512.
53. Counillon, L., Franchi, A., and Pouysségur, J. (1993). A point mutation of the Na^+/H^+ exchanger gene (NHE1) and amplification of the mutated allele confer amiloride-resistance upon chronic acidosis. *Proc. Natl. Acad. Sci. USA* **90:**4508–4512.
54. Tse, C. M., Levine, S. A., Yun, C. H., Brant, S. R., Pouysségur, J., and Donowitz, M. (1993). Functional characteristics of a cloned epithelial Na^+/H^+ exchanger (NHE3): Resistant to amiloride and inhibition of protein kinase C. *Proc. Natl. Acad. Sci. USA* **90:**9110–9114.
55. Yu, F. H., Shull, G. E., and Orlowski, J. (1993). Functional properties of the rat Na/H exchanger NHE2 isoform expressed in Na/H exchanger-deficient Chinese hamster ovary cells. *J. Biol. Chem.* **268:** 25536–25541.
56. Aronson, P. S., Nee, J., and Suhm, N. A. (1982). Modifier role of internal H^+ inactivating the Na^+/H^+ exchange in renal microvillus membrane vesicles. *Nature* **299:**161–163.
57. Pollock, A. S., Warnock, D. G., and Strewler, G. J. (1986). Parathyroid hormone inhibition of Na^+/H^+ antiporter activity in a cultured renal cell line. *Am. J. Physiol.* **250:**F217–F225.
58. Motais, R., and Garcia-Romeu, F. (1988). Effects of catecholamines and cyclic nucleotides on Na^+/H^+ exchange. In *Na^+/H^+ Exchange* (S. Grinstein, ed.), CRC Press, Boca Raton, FL, pp. 255–270.
59. Sardet, C., Fafournoux, P., and Pouysségur, J. (1991). Thrombin, epidermal growth factor, and okadaic acid activate the Na^+/H^+ exchanger, NHE1, by phosphorylating a set of common sites. *J. Biol. Chem.* **266:**19166–19171.
60. Grinstein, S., Cohen, S., Goetz, J. D., Rothstein, A., Mellors, A., and Gelfand, E. W. (1986). Activation of the Na^+/H^+ antiport by changes in cell volume and by phorbol ester; possible role of protein kinase. *Curr. Top. Membr. Transp.* **26:**115–136.
61. Cala, P. M. (1986). Volume sensitive alkali metal-H transport in Amphiuma red blood cells. *Curr. Top. Membr. Transp.* **26:**79–100.
62. Grinstein, S., Woodside, M., Sardet, C., Pouysségur, J., and Rotin, D. (1992). Activation of the Na^+/H^+ antiporter during cell volume regulation. Evidence for a phosphorylation-independent mechanism. *J. Biol. Chem.* **267:**23823–23828.
63. Bookstein, C., Musch, M. W., Depauli, D., Xie, Y., Hornung, S., Rao, M. C., Villereal, M., and Chang, E. B. (1994). Characterization of an unusual Na/H exchange isoform in NHE1 of inner renal medullary collecting tubules which is activated under hyperosmolar stress in transfected NHE-deficient Chinese hamster lung fibroblasts. *Gastroenterology,* in press.
64. Schwartz, M. A., Cragoe, E. J., and Lechene, C. P. (1990). pH regulation in spread cells and round cells. *J. Biol. Chem.* **265:**1327–1332.
65. Schwartz, M. A., Both, G., and Lechene, C. (1989). Effect of cell spreading on cytoplasmic pH in normal and transformed fibroblasts. *Proc. Natl. Acad. Sci. USA* **86:**4525–4529.
66. Schwartz, M. A., Lechene, C., and Ingber, D. E. (1991). Insoluble fibronectin activates the Na^+/H^+ antiporter by clustering and immobilizing integrin $\alpha 5\beta 1$, independent of cell shape. *Proc. Natl. Acad. Sci. USA* **88:**7849–7853.
67. Schwartz, M. A., and Lechene, C. (1992). Adhesion is required for protein kinase C-dependent activation of the Na^+/H^+ antiporter by platelet-derived growth factor. *Proc. Natl. Acad. Sci. USA* **89:**6138–6141.
68. Takaichi, K., Balkovetz, D. F., Meier, E. V., and Warnock, D. G. (1993). Cytosolic pH sensitivity of an expressed human NHE1 Na^+/H^+ exchanger. *Am. J. Physiol.* **264:**C944–C950.
69. Rao, G. N., Sardet, C., Pouysségur, J., and Berk, B. C. (1993). Phosphorylation of Na^+/H^+ antiporter is not stimulated by phorbol ester and acidification in granulocytic HL-60 cells. *Am. J. Physiol.* **264:**C1278–C1284.
70. Helmle-Kolb, C., Counillon, L., Roux, D., Pouysségur, J., Mrkic, B., and Murer, H. (1993). Na^+/H^+ exchange activates in NHE1-transfected OK-cells: Cell polarity and regulation. *Pfluegers Arch.* **424:** 377–384.
71. Charney, A. N., Kinsey, M. D., Meyers, L,. Giannella, R. A., and Gots, R. (1978). Na-K activated adenosine triphosphatase and intestinal electrolyte transport—effect of adrenal steroids. *J. Clin. Invest.* **56:**653–660.
72. Sellin, J. H., and Field, M. (1981). Physiologic and pharmacologic effects of glucocorticoids on ion transport across rabbit ileal mucosa in vitro. *J. Clin. Invest.* **67:**770–778.
73. Yun, C. H., Gurubhagavatula, S., Levine, S. A., Montgomery, J. M., Brant, S. R., Cohen, M. E., Pouysségur, J., Tse, C. M., and Donowitz, M. (1993). Glucocorticoid stimulation of ileal Na^+ absorptive cell brush border Na^+/H^+ exchange and association with an increase in message for NHE-3, an epithelial isoform Na^+/H^+ exchanger. *J. Biol. Chem.* **268:**206–211.
74. Counillon, L., Pouysségur, J., and Reithmeier, R. A. (1994). The Na^+/H^+ exchanger NHE-1 possesses N- and O-linked glycosylation restricted to the first N-terminal extracellular domain. *Biochemistry* **33:**10463–10469.

74a. Freiberg, J. M., Kinsella, J., and Sacktor, B. (1982). Glucocorticoids increase the Na^+/H^+ exchange and decrease the Na^+ gradient-dependent phosphate-uptake systems in renal brush border membrane vesicles. *Proc. Natl. Acad. Sci. USA* **79:**4932–4936.

75. Moe, O. W., Miller, R. T., Horie, S., Cano, A., Preisig, P. A., and Alpern, R. J. (1991). Differential regulation of Na+/H+ antiporter by acid in renal epithelial cells and fibroblasts. *J. Clin. Invest.* **88:** 1703–1708.
76. Krapf, R., Pearce, D., Lynch, C., Xi, X. P., Reudelhuber, T. L., Pouysségur, J., and Rector, F. C. (1991). Expression of rat renal Na+/H+ antiporter mRNA levels in response to respiratory and metabolic acidosis. *J. Clin. Invest.* **87:**747–751.
77. Alpern, R. J., Yamaji, Y., Cano, A., Horie, S., Miller, R. T., Moe, O., and Preisig, P. A. (1993). Chronic regulation of the Na+/H+ antiporter. *J. Lab. Clin. Med.* **122:**137–140.
78. Raley-Susman, K. M., Cragoe, E. J., Jr., Sapolsky, R. M., and Kopito, R. R. (1991). Regulation of intracellular pH in cultured hippocampal neurons by an amiloride-insensitive Na+/H+ exchanger. *J. Biol. Chem.* **266:**2739–2745.
79. Brant, S. R., Yun, C. H. C., Donowitz, M., and Tse, C.-M. (1995). Cloning, tissue distribution and functional analysis of the human Na+/H+ exchanger isoform, NHE3. *Am. J. Physiol.* **269:**C198–C206.
80. Klanke, C., Su, Y. R., Callen, D. F., Wang, Z., Meneton, P., Baird, N., Kandasamy, R. A., Orlowki, J., Otterud, B. E., Leppert, M., Shull, G. E., and Menon, A. (1995). Molecular cloning and physical and genetic mapping of a novel Na+/H+ exchanger (NHE5/SCL 9A5) to chromosome 16q22.1. *Genomics* **25:**615–622.
81. Kokke, F. T. M., Elsawy, T., Bengtsson, I., Wasmuth, J. J., Jabs, E. W., Tse, C.-M., Donowitz, M., and Brant, S. R. (1996). A NHE3-related pseudogene is on human chromosome 10; the functional gene maps to 5p15.3. *Mammalian Genome* **7:** in press.
82. Hoogerwerf, W. A., Tsao, S. C., Devuyst, O., Levine, S. A., Yung, C. H. C., Yip, J. W., Cohen, M. E., Wilson, P. T., Lazenby, A., Montgomery, J., Tse, C.-M., and Donowitz, M. (1996). NHE2 and NHE3 are human and rabbit brush border proteins. *Am. J. Physiol.,* **270:** 629–641.
83. Counillon, L., Pouysségur, J., and Reithmeier, R. A. F. (1994). The Na+/H+ exchanger NHE1 possesses N- and O-linked glycosylation restricted to the first N-terminal extracellular domain. *Biochemistry* **33:**10463–10469.
84. Tse, C. M., Levine, S. A., Yun, C. H. C., Khurana, S., and Donowitz, M. (1994). The Na+/H+ exchanger-2 (NHE2) is an O-linked but not a N-linked sialoglycoprotein. *Biochemistry* **33:**12954–12961.
85. Counillon, L., Scholz, W., Lang, H. J., and Pouysségur, J. (1993). Pharmacological characterization of stably tranfected Na+/H+ antiporter isoforms using amiloride analogs and a new inhibitor exhibiting anti-ischemic properties. *Am. Soc. Pharmacol. Exp. Ther.* **44:**1041–1045.
86. Yu, F. H., Shull, G., and Orlowski, J. (1993). Functional properties of the rat Na+/H+ exchanger NHE2 isoform expressed in Na+/H+ exchanger deficient Chinese hamster ovary cells. *J. Biol. Chem.* **268:**25536–25541.
87. Kandasamy, R. A., Yu, F. H., Harris, R., Boucher, A., Hanrahan, J. W., and Orlowski, J. (1995). Plasma membrane Na+/H+ exchanger isoforms (NHE -1, -2, and -3) are differentially responsive to second messenger agonists of the protein kinase A and C pathways. *J. Biol. Chem.* **270:**29209–29216.
88. Borgese, F., Sardet, C., Cappadoro, M., Pouysségur, J., and Motais, R. (1992). Cloning and expression of a cAMP-activatable Na+/H+ exchanger. Evidence that the cytoplasmic domain mediates hormonal regulation. *Proc. Natl. Acad. Sci. USA* **89:**6765–6769.
89. Nath, S. K., Hang, C. Y. H., Levine, S. A., Yun, C. H. C., Montrose, M. H., Donowitz, M., and Tse, C.-M. (1996). Hyperosmolarity inhibits the cloned epithelial specific Na/H exchanger isoforms, NHE2 and NHE3: An effect opposite to that of the housekeeping isoform NHE1. *Am. J. Physiol.,* in press.
90. Kapus, A., Grinstein, S., Wasan, S., Kandasamy, R., and Orlowski, J. (1995). Functional characterization of three isoforms of the Na+/H+ exchanger stably expressed in Chinese hamster ovary cells. ATP dependence, osmotic sensitivity, and role in cell proliferation. *J. Biol. Chem.* **269:**23544–23552.
91. Moe, O. W., Amemiya, M., and Yamaji, Y. (1995). Activation of protein kinase A acutely inhibits and phosphorylates Na/H exchanger NHE3. *J. Clin. Invest.* **96:**2187–2194.
92. Yun, C. H. C., Tse, C.-M., and Donowitz, M. (1995). Chimeric Na+/H+ exchangers: An epithelial N-terminal domain requires an epithelial cytoplastic C-terminal domain for regulation by protein kinases. *Proc. Natl. Acad. Sci. USA* **92:**10723–10727.
93. Levine, S. A., Nath, S. K., Yun, C. H. C., Yip, J. W., Montrose, M., Donowitz, M., and Tse, C.-M. (1995). Separate C-terminal domains of the epithelial specific brush border Na+/H+ exchanger isoform NHE3 are involved in stimulation and inhibition by protein kinases/growth factors. *J. Biol. Chem.* **270:**13716–13725.
94. Wakabayashi, S., Bertrand, B., Ikeda, T., Pouysségur, J., and Shigekawa, J. (1994). Mutation of calmodulin-binding site renders the Na+/H+ exchanger (NHE1) highly H+-sensitive and Ca^{2+} regulation defective. *J. Biol. Chem.* **269:**13710–13715.
95. Bertrand, B., Wakabayashi, S., Ikeda, T., Pouysségur, J., and Shigekawa, M. (1994). The Na+/H+ exchanger isoform 1 (NHE1) is a novel member of the calmodulin-binding proteins. Identification and characterization of the calmodulin-binding sites. *J. Biol. Chem.* **269:** 13703–13709.

15

Mitochondrial Transport Processes

Ronald S. Kaplan

15.1. INTRODUCTION

The chief functions of mitochondria are to supply the cell with ATP made via oxidative phosphorylation, to regulate the level of reduced pyridine nucleotides in the cytoplasm, to supply the cytoplasm with sufficient carbon precursor for fatty acid and sterol biosyntheses, and to catalyze certain of the reactions involved in gluconeogenesis and ureogenesis. In order for mitochondria to perform these processes, a high-magnitude flux of numerous metabolites must occur across their membranes. Movement of metabolites, as large as 6–8 kDa, occurs across the outer mitochondrial membrane primarily via diffusion through a channel protein known as the voltage-dependent anion channel.[1,2] In contrast to the broad specificity of the outer membrane channel, the inner mitochondrial membrane maintains a highly selective permeability resulting from the existence of specific membrane transport proteins. To date, at least 13 major inner membrane metabolite transporters have been identified and extensively characterized (see Table 15.1). During the last 25 years, these transporters have been studied in isolated mitochondria and more recently many have been investigated at the molecular level using purified, reconstituted transport proteins.

The purpose of this chapter is to give an overview of the current state of knowledge about the major metabolite transporters with an emphasis on progress made at the molecular and genetic levels during the last 5 years. The reader is referred to excellent earlier reviews for additional information on transporter function either in isolated mitochondria[3–9] or at the molecular level,[10–14] as well as for in-depth information on specific transport proteins (e.g., ADP/ATP carrier,[15,16] phosphate carrier,[17,18] citrate carrier,[19] and the uncoupling protein[20,21]). It should be noted that information on the uncoupling protein is included throughout this chapter since, based on its structural and functional characteristics, it clearly can be assigned membership in the same family as several of the metabolite carriers. Finally, in order to cover the extensive literature on the major metabolite transporters in sufficient depth, this review will not include information on the mitochondrial channel proteins or the small cation transporters. For information on these subjects the reader is directed to several recent reviews[22–25] and original articles.[26–29]

15.2. IDENTIFICATION, METABOLIC SIGNIFICANCE, AND ORGAN DISTRIBUTION OF THE MAJOR MITOCHONDRIAL TRANSPORTERS

Table 15.1 lists the physiological substrates and the metabolic significance of the 13 major metabolite transporters that have been well characterized to date plus the uncoupling protein. The functions of most of these transporters are associated with metabolic pathways that control cellular bioenergetics. However, since mitochondria also carry out other essential processes (e.g., DNA replication and transcription, protein synthesis) and since the inner membrane maintains a tightly controlled permeability, it is apparent that mitochondria likely contain a large number of metabolite transporters in addition to those listed in Table 15.1. Experimental evidence for some of these has recently been obtained. For example, data exist indicating the presence of uptake systems for choline,[30] coenzyme A,[31] polyamines,[32] thiamine pyrophosphate,[33] glutathione,[34] *N*-acetylglutamate,[35] and deoxyguanosine.[36] Clearly, the discovery and molecular characterization of additional transport proteins that are metabolically important represents an essential area for future research.

The major mitochondrial transporters do not merely serve as passive conduits allowing metabolite flux through the inner membrane, but instead may regulate substrate supply to a given metabolic pathway and possibly control flux through the pathway. Although the complexities of the interrelationships between metabolic pathways and associated transport processes have led to difficulties in clearly establishing the extent to which a given transporter modulates flux through a pathway,[8] the bulk of the experimental evidence supports (under certain metabolic conditions) a rate-controlling role for the pyruvate transporter on hepatic gluconeogenesis,[37–40] for the ADP/ATP translocase on respiration,[8,41–43] and for the glutamine transporter on urea synthesis in the liver and kidney.[44,45]

Ronald S. Kaplan • Department of Pharmacology, College of Medicine, University of South Alabama, Mobile, Alabama 36688.
Molecular Biology of Membrane Transport Disorders, edited by S.G. Schultz *et al.* Plenum Press, New York, 1996

Table 15.1. Major Mitochondrial Metabolite Transporters[a]

Transporter name	Physiological substrates	Metabolic importance
ADP/ATP	ADP, ATP	Oxidative phosphorylation
Phosphate	Phosphate	Oxidative phosphorylation
ATP-Mg/phosphate	ATP-Mg, phosphate	Matrix adenine nucleotide content
Citrate	Citrate, isocitrate, *cis*-aconitate, malate, phosphoenolpyruvate, succinate	Fatty acid and sterol biosyntheses, gluconeogenesis, isocitrate-α-ketoglutarate shuttle
α-Ketoglutarate	α-Ketoglutarate, malate, succinate, oxaloacetate	Malate/aspartate shuttle, gluconeogenesis, isocitrate-α-ketoglutarate shuttle, nitrogen metabolism
Dicarboxylate	Phosphate, malate, succinate, oxaloacetate, sulfate, sulfite	Gluconeogenesis, urea synthesis, sulfur metabolism
Pyruvate	Pyruvate, other monocarboxylates, ketone bodies	Citric acid cycle, gluconeogenesis
Carnitine	Carnitine, acylcarnitines	Fatty acid oxidation
Aspartate/glutamate	Aspartate, glutamate, cysteine-sulfinate	Malate/aspartate shuttle, gluconeogenesis, urea synthesis, cysteine catabolism
Glutamate	Glutamate	Urea synthesis
Ornithine/citrulline	Ornithine, citrulline, lysine, arginine	Urea synthesis
Glutamine	Glutamine	Glutamine breakdown
Branched chain α-keto acid	Branched chain α-keto acids	Branched chain amino acid catabolism
Uncoupling protein	H^+, halide anions	Thermogenesis

[a]Data from Refs. 7, 8, 12, 13, 189.

The expression of specific mitochondrial transport proteins is organ-specific and appears to reflect the metabolic needs of a given type of cell.[4,6,8,13,20,46–48] Thus, the ADP/ATP and phosphate carriers, which are required for ATP synthesis, are present in mitochondria from all cell types. Also, the pyruvate, carnitine, oxoglutarate, aspartate/glutamate, and the branched chain α-keto acid transporters, which are required for the transport of substrates and reducing equivalents into the mitochondrial matrix for oxidative phosphorylation, are present in most types of mitochondria. In contrast, other carriers which participate in more specialized metabolic functions, display a more restricted expression. Transporters in this category include the citrate, dicarboxylate, and glutamate carriers which are expressed mainly in liver, and the ornithine/citrulline, glutamine, and ATP-Mg/P_i transporters which are found mainly in liver and kidney. The uncoupling protein, which is uniquely expressed in brown fat mitochondria, represents an excellent example of tissue specificity. It should be noted, however, that since molecular probes are not yet available for many of the major metabolite transporters, and since the function of certain of the transporters has only been assayed in mitochondria from several different organs, it is possible that the distribution of some of the carriers may in fact be more widespread than present data would suggest.

15.3. MOLECULAR PROPERTIES OF MITOCHONDRIAL TRANSPORT PROTEINS

15.3.1. Purification and Functional Reconstitution

As a result of extensive efforts during the last 15 years, many of the major mitochondrial metabolite transporters have now been purified in reconstitutively active form (Table 15.2). This progress has enabled the study of transporter structure and function at the molecular level. The strategies utilized for the purification and functional reconstitution of these transporters have recently been reviewed extensively[11] and thus this section will focus on highlights and new developments. Most purification procedures begin with the extraction of mitochondria (or mitoplasts) with a nonionic detergent, often in the presence of exogenous phospholipid, followed by chromatography on hydroxylapatite. The resulting eluate is typically highly enriched with respect to mitochondrial transporters. Final purification is then achieved via chromatography on a variety of resins (e.g., DEAE, SH-affinity resins, a variety of dye resins, silica, and Celite). The presence of a given transporter in a fraction is usually detected by functional reconstitution in a liposomal system in which inhibitor-sensitive substrate-specific transport is measured. The state of purity is assessed by the combination of SDS–PAGE and functional reconstitution. With respect to the latter, it must be noted that many variables can affect the observed reconstituted specific activity (e.g., liposome composition and internal size, detergent concentration, the presence of interacting proteins in a given fraction, the proportion of the isolated protein that is in a functional conformation) and thus the magnitude of this value must be interpreted with great caution.[11] As indicated in Table 15.2, most of the purified transporters display apparent molecular mass values in the range of 28–34 kDa with two important exceptions which are discussed below. Also, as depicted in Table 15.2, the yield of the different transporters varies greatly presumably reflecting a combination of the abundance of a given carrier within the inner membrane and the effectiveness of a given purification procedure. With certain transporter preparations, the low yield has led to

Table 15.2. Purification of Mitochondrial Metabolite Transport Proteins from Mammalian Tissues[a]

Carrier	Source	Apparent mol. wt. (kDa)[b]	Detection method	Reconstituted transport activity (μmole/min/mg)[c]	Protein yield (mg)	Refs.
ADP/ATP	Bovine heart	30	Inhibitor binding	0.017[d]	2.5	92, 93
	Bovine heart	30	Reconstitution	0.5	—[e]	238
Phosphate	Rat liver	33, 35[f]	NEM-bind., reconstitution	22.6	0.1	104, 239
	Rat liver	34	NEM-bind., antibody inhib.	—	—	240
	Pig heart	33	NEM-bind., reconstitution	26.0	—	241, 242
	Bovine heart	34	NEM-bind., reconstitution	—	5.9	243, 244
	Bovine heart	33	Reconstitution	90	—	113, 245
Citrate	Rat liver	30	Reconstitution	2.0	0.002	52, 246
	Rat liver	32.5	Reconstitution	1.9	0.013	51, 19
	Rat liver	38–39	Reconstitution	1.6	0.008	50
	Bovine liver	37–38	Reconstitution	0.4	0.035	49
α-Ketoglutarate	Pig heart	31.5	Reconstitution	2.3	0.016	247
	Bovine heart	31.5	Reconstitution	22.2	—	248
	Bovine heart	31.5	Reconstitution	0.2	0.021	249
	Bovine liver	31.5–32	Reconstitution	28.5	0.03	250
	Rat liver	32.5	Reconstitution	0.3	0.006	251
Dicarboxylate	Rat liver	28	Reconstitution	6.0	—	251, 252, 252a
	Bovine liver	28	Reconstitution	1.0	—	253
	Bovine heart	34,36[g]	Reconstitution	0.5	0.004	254
Pyruvate	Bovine heart	34	Reconstitution	114	0.024	249, 255
	Rat liver	34[h]	Reconstitution	232	—	255
	Rat brain	34[h]	Reconstitution	202	—	255
Carnitine	Rat liver	32.5	Reconstitution	1.7	—	256, 257
	Rat brain	≥4 bands	Reconstitution[i]	0.01	—	258
Aspartate/glutamate	Bovine heart	31.5	Reconstitution	0.4	0.002	259
Ornithine/citrulline	Rat liver	33.5	Reconstitution	0.4	—	260
Branched chain α-keto acid	Rat heart	4 bands[j]	Reconstitution	2.4	0.028	46
Uncoupling protein	Hamster brown fat[k]	32	GDP binding	0.016[l]	8–12	88, 261
	Hamster brown fat	32	Reconstitution	2–4	0.28	262, 263

[a]Data included represent only highly purified and/or active preparations that have been obtained from a given source, unless otherwise indicated.
[b]Based on SDS–PAGE analysis.
[c]All specific transport activity values have been normalized to a per minute basis. Only the highest value reported for a given preparation has been tabulated.
[d]The purified ADP/ATP translocase displayed a carboxyatractyloside binding stoichiometry of 17.4 nmole/mg protein.
[e]—, not reported.
[f]The phosphate transporter was purified in two forms (based on SDS–PAGE analysis) consisting of either a single 33-kDa band (Form I) or 33-kDa + 35-kDa bands (Form II) both of which were catalytically active. Forms I and II could be converted into a single 35-kDa band by alkylation with NEM.
[g]This preparation contains two main protein bands. However, it has not yet been determined which band represents the dicarboxylate transporter.
[h]The pyruvate transporter preparations from both rat liver and rat brain contain several minor protein contaminants.
[i]This is a partially purified preparation. Reconstituted carnitine transport activity correlates with the appearance of a 33-kDa protein.
[j]Reconstituted branched chain α-keto acid transport activity is associated with 39-kDa and/or 41-kDa proteins.
[k]Data for brown fat mitochondrial uncoupling protein purified from a variety of species are presented in Refs. 88, 261, 264.
[l]The purified uncoupling protein displayed a GDP binding stoichiometry of 16.2 nmole/mg protein.

difficulty in obtaining sufficient material for amino acid sequencing (and subsequent isolation and sequencing of transporter cDNAs) and thus has hindered progress at the structural level.

Recently, Azzi's group has reported the isolation of a citrate transport protein from both bovine[49] and rat liver[50] that has a molecular mass of ~38 kDa. This value is somewhat higher than the values observed for other mitochondrial transporters and contrasts with the rat liver citrate transporter purified by both our group[51] and Palmieri's group[52] which displayed lower molecular mass values (i.e., 32.5 and 30 kDa, respectively). Peptide mapping studies have shown that the 38-kDa protein is not structurally related to the lower-molecular-mass protein.[53] Determination of the amino acid sequence of the 32.5-kDa protein[54] indicates that it contains the structural characteristics that typify inner mitochondrial membrane transporters (see Section 15.3.2). In contrast, the amino acid sequence of the Azzi preparation does not display significant homology to the other mitochondrial carriers sequenced to date. It will be of great interest to determine whether the high-molecular-mass form does in fact represent a second inner mitochondrial membrane citrate transport protein. With respect to this point, a concern centers around the fact that Azzi's group has employed mitochondria isolated by differential centrifugation as the starting material, and that such preparations are typically contaminated with considerable amounts of lysosomes, peroxisomes, and endoplasmic reticulum.[55] Thus, one can envision that the 38-kDa citrate transport protein may have arisen from a membrane other than the mitochondrial inner membrane, thereby explaining the atypical molecular mass value. Clearly, such a finding would be quite interesting in its own right. To resolve this confusion, investigations are needed to determine whether the 38-kDa protein can be purified using either mitoplasts or purified inner membrane as the starting material. While the Azzi group has recently shown that the 38 kDa band can, in fact, be isolated from mitoplasts,[53] neither the details of the mitoplast preparation nor the transport capability of the isolated protein were presented. Finally, a second transporter which appears to display a higher than usual molecular mass value is the branched chain α-keto acid transporter that has recently been partially purified from rat heart by Hutson and Hall.[46] Transport activity was associated with a 39-kDa and/or a 41-kDa protein. Most interestingly, several pieces of evidence suggest that the transport activity may be catalyzed by the mitochondrial branched chain aminotransferase enzyme, thus rendering this a bifunctional protein.

In summary, impressive progress has been made in the purification and functional reconstitution of many of the mitochondrial metabolite transporters. This progress has provided the foundation for a wide range of studies (which are described in the sections to follow) that have resulted in (1) a determination of the total amino acid sequence of two of the transporters by direct protein sequencing; (2) acquisition of partial amino acid sequence information thereby permitting the design of oligonucleotide probes that allowed the successful isolation and sequencing of transporter cDNAs; (3) the development of antibody probes for examination of transporter expression and transmembrane topology; and (4) a rigorous examination of transport mechanism at the molecular level. Finally, with minor modification, these purification and functional reconstitution procedures have enabled an examination of the effects of site-directed mutations on transporter function.

15.3.2. Primary Structure

Through a combination of direct amino acid sequencing and/or cDNA sequencing, the complete primary structure has been determined for five mitochondrial transport proteins of known function: the ADP/ATP translocase,[56] the phosphate,[10,57–60] the oxoglutarate,[61] and the citrate[54] carriers, and the uncoupling protein.[62] These carriers have been sequenced from a variety of sources (e.g., ADP/ATP translocase, 18 sources; phosphate carrier, 5 sources; α-ketoglutarate carrier, 3 sources; citrate carrier, 2 sources; uncoupler protein, 6 sources; see Refs. 14, 54, 63, 64, 134b for a review of these sequences). Table 15.3 shows the characteristics of one representative sequence for each of the five mitochondrial transport proteins of known function that have been sequenced to date. The length of the amino acid sequences of the mature (i.e., processed) form of the transporters fall in the narrow range of 297 to 314 residues with a calculated molecular mass range of 32.6 to 34.7 kDa. The mitochondrial transporters are rather polar for integral membrane proteins and each displays a net positive charge. It is of interest to note that the citrate[54] and phosphate[60] transporters possess presequences of 13 and 44 residues, respectively, which likely function as targeting sequences that are cleaved during the import of these transporters to the inner membrane. In contrast, the ADP/ATP[56,65–67] and oxoglutarate[61] carriers, as well as the uncoupling protein[68,69] do not have cleavable presequences. Presumably, the latter group of transporters are targeted to the inner membrane via information residing within the sequence of the mature protein.

Further analysis of the amino acid sequences of the five mitochondrial transporters included in Table 15.3 reveals several interesting features. First, comparison between these sequences indicates approximately 25% identity and, allowing for conservative substitutions, approximately 50% similarity. Despite these relatively low homology values, dot matrix analysis (see Figure 15.1 and Refs. 10,12,14,15,54,59–62,70) indicates that each of the transporter sequences displays stretches of homologous amino acids that are displaced approximately 100 and 200 residues, respectively, from the main diagonal. This pattern indicates that each transporter displays a tripartite structure with three homologous 100-amino-acid domains. Furthermore, among the five different transporters all of these domains show homology with respect to each other. A more fine-tuned analysis of the homology between the 100-residue repetitive segments is obtained by alignments such as that depicted in Figure 15.2. A consensus sequence can be derived that contains 33 positions at which residues are either identical or conservatively substituted in at least 10 of

TABLE 15.3. Comparison of the Molecular Characteristics of the Five Mitochondrial Transport Proteins of Known Function that Have Been Sequenced to Date[a]

Feature	ADP/ATP translocase	Citrate carrier	α-Ketoglutarate carrier	Phosphate carrier	Uncoupling protein
Molecular mass (kDa)	32.8	32.6	34.2	34.7	33.2
Number of amino acids	297	298	314	312	306
Polarity[b] (%)	37.7	40.3	38.2	35.9	41.8
Isoelectric point	10.7	10.6	10.7	9.7	9.6
Net charge[c]	+18	+17	+15	+11	+9

[a]Parameters were calculated from published sequences for the bovine heart ADP/ATP translocase,[56] the rat liver citrate transport protein,[54] the bovine heart α-ketoglutarate carrier,[61] the rat liver phosphate carrier,[60] and the hamster brown fat uncoupling protein[62] using the GCG PeptideSort program. Modified from Kaplan *et al.*[54] with permission from *J. Biol. Chem.*

[b]Polarity is defined as the sum of the residue mole percentages of the polar amino acids Asp, Asn, Glu, Gln, Lys, Ser, Arg, Thr, and His as previously detailed.[265]

[c]Net charge was calculated taking into account whether or not a given protein was modified at its amino-terminus.

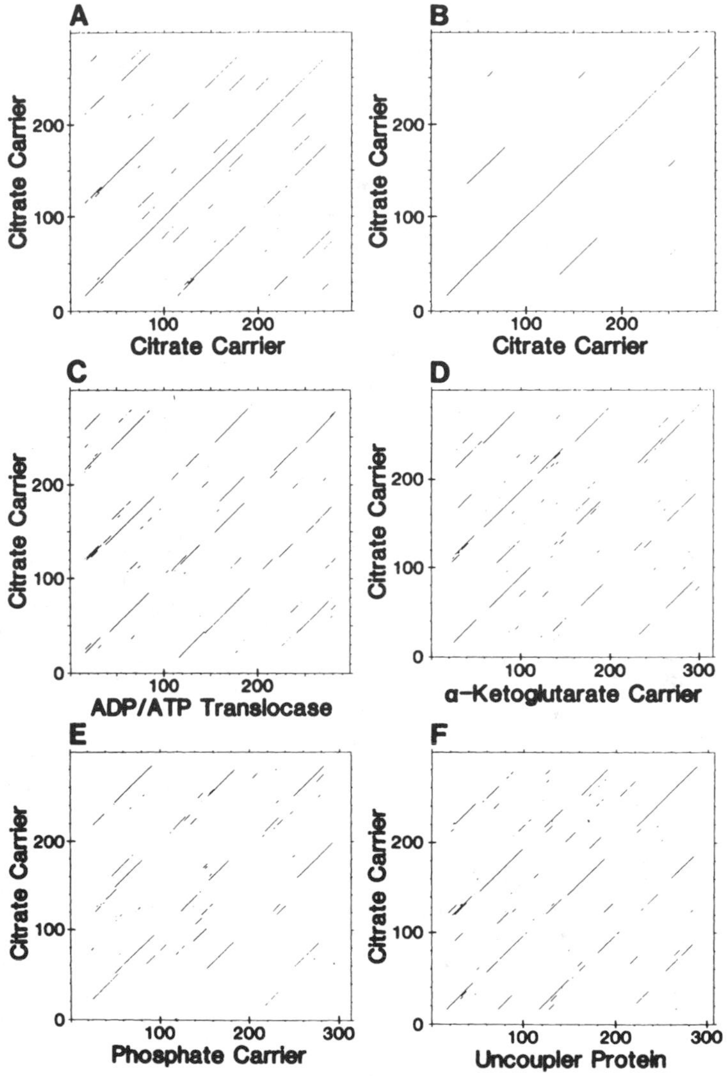

Figure 15.1. DotPlot comparisons of the amino acid sequence of the citrate carrier against itself and against the sequences of other mitochondrial carriers. The citrate carrier sequence was compared with itself (panels A and B); the beef heart ADP/ATP translocase (panel C); the beef heart α-ketoglutarate carrier (panel D); the rat liver phosphate carrier (panel E); and the hamster brown fat uncoupling protein (panel F). The sequences for the carriers listed on the *X*-axes in panels C–F were obtained from Refs. 54,56,60–62. Comparisons were made with the GCG programs using a window of 30 and a stringency of 13, except for panel B in which a window of 30 and a stringency of 18 were employed. Reproduced from Kaplan *et al.*[54] with permission from *J. Biol. Chem.*

the 15 sequence domains. Additionally, the transporters display the signature sequence Pro-X-(Asp/Glu)-X-(Val/Ile/Ala)-(Lys/Arg)-X-(Arg/Lys/Gln)-(Leu/Met/Phe/Ile) which repeats two to three times in each of the mitochondrial transporter sequences. Based in part on these sequence similarities, it has been proposed that a superfamily of mitochondrial transport proteins exists that is derived from a common ancestral gene encoding approximately 100 amino acids.[10,12,14,15,59,70]

Recently, numerous other sequences have been obtained that encode proteins of unknown function but can nonetheless be tentatively assigned to the mitochondrial superfamily based on the criteria that they display (1) a tripartite structure consisting of repeats of 100-amino-acid segments, (2) the appropriate signature sequence, and (3) the predicted presence of six membrane-spanning segments (two per 100-amino-acid repeat; see Section 15.3.3). These sequences include numerous sequences from the yeast *Saccharomyces cerevisiae* genome,[14,71,72] from the genome of the nematode worm *Caenorhabditis elegans,*[14,73] from the human and bovine genomes,[74,75] as well as a sequence present in the hypotrychous ciliated protozoan *Oxytricha fallax.*[76] Presently, over 100 sequences have been assigned to the mitrochondrial transporter superfamily.[14,54,63,64]

15.3.3. Membrane Topology

In order to obtain insight into the participation of specific structural domains within the mitochondrial carriers in the transport mechanism, it is necessary to obtain high-resolution information on the three-dimensional structure of these proteins. In the absence of x-ray crystallographic data, a variety of other approaches have been utilized to obtain initial information on the secondary structure of the mitochondrial carriers within the inner membrane. These approaches include hydropathy analysis, as well as experiments involving the use of sequence-specific antibodies, site-specific proteolysis, and chemical modification.

Hydropathy analysis has led to a prediction of six putative α-helical membrane-spanning domains present within each of the five transporters that have been sequenced to

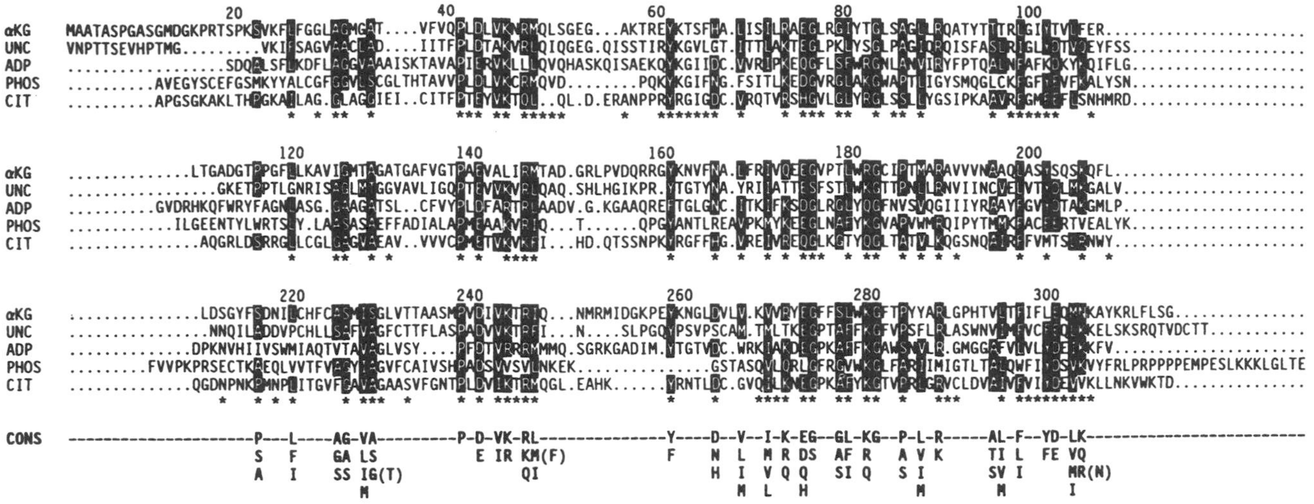

Figure 15.2. Alignment of the citrate transporter repetitive sequence domains with those present in other mitochondrial transporters. The α-ketoglutarate carrier,[61] uncoupling protein,[62] ADP/ATP translocase,[56] and phosphate carrier[60] were each divided into three sequence domains as described by Runswick *et al.*[61] On analysis of the citrate transport protein sequence, it was similarly divided into three sequence elements. The resulting 15 sequence domains were then compared using the GCG PileUp program and the alignment is depicted. A consensus sequence was derived based on the criterion that a given residue(s) was identical or conservatively substituted in at least 10 out of the 15 sequence domains. Conserved residues were assigned based on the MDM_{78} table[266] and are depicted as white letters on a black background, as well as on the consensus line. The vertical order of conserved residues in the consensus sequence reflects the frequency of occurrence. Residues in parentheses represent alternative substitutions with respect to the adjacent residue. Positions indicated by an asterisk indicate residues that are conserved in at least four out of five sequences within a given domain. Numbers refer to the α-ketoglutarate carrier sequence. Reproduced from Kaplan *et al.*[54] with permission from *J. Biol. Chem.*

date.[10,12,14,15,54,59–62,70] This type of analysis has permitted the development of initial models for transporter topography. Figure 15.3 shows the proposed topography of the citrate carrier[54] and is similar to models that have been proposed for the other carriers.[10,12,14,15,59–62,70] It is noteworthy that the transporter appears to be asymmetrically distributed within the inner membrane since the three interconnecting helical loops exposed to one side of the membrane (i.e., loops A, C, and E) are considerably larger than the two loops facing the other side (i.e., loops B and D). Additionally, the sequences of the three large hydrophilic loops are conserved to a much greater extent than are the sequences of the two smaller loops (i.e., 43 versus 15% sequence conservation, respectively, in at least four out of five transporter sequences). It is tempting to speculate that the two smaller loops may play a role in substrate recognition thereby differentiating one transporter from another.[12,61] Finally, the amino- and carboxy-termini are predicted to reside on the same side of the membrane.

As mentioned above, the hypothetical topography model for the citrate transport protein that is shown in Figure 15.3 can be viewed as a representative example of a general model for the transmembrane folding of the mitochondrial metabolite carriers. As indicated by others,[10,12,13] an important aspect of the general model is that it takes into account the threefold repeating 100-amino-acid sequence domain which is thought to contain two membrane-spanning α-helices per repeat. The general topography model has been extensively probed with several transporters via the use of sequence-specific antibodies, proteases, and chemical modification. For example, immunological studies carried out with the tricarboxylate transporter indicate that the amino and carboxy terminii are exposed to the cytoplasmic side of the inner membrane, thus supporting an even number of transmembrane segments as depicted in the above model.[77] Studies with the phosphate carrier[77a,78] clearly indicate that both the amino- and carboxy-termini face the cytosol. Other experiments[78] showed that cleavage of the transporter with Arg-endoprotease occurs at Arg140 and/or Arg152 only in inside-out submitochondrial particles but not in right-side-out freeze-thawed mitochondria, thereby indicating a matrix location for the residues in loop C between helices III and IV (see Figure 15.3). Similar experiments with Lys-endoprotease[79] indicate that loops B and D containing Lys96 and Lys198 (or Lys203) face the cytosol, as does Lys288 which is located near the carboxy-terminus. Thus, compartments A and B in the general model can be defined as representing the matrix and intermembrane spaces, respectively. With the ADP/ATP carrier, immunological studies[80] have verified the cytosolic location of the amino-terminus, but the orientation of the carboxy-terminus is more uncertain probably related to the fact that it has a short exposed tail which in the native membrane may be insufficiently accessible to antibodies. Additionally, enzymatic cleavage of the transporter in inside-out submitochondrial particles with a variety of endoproteases indicates that hydrophilic loops A, C, and E are accessible to the matrix.[80,81] With regard to the α-ketoglutarate carrier, results from immunological and proteolytic cleavage studies support an even number of transmembrane segments in the monomer, with the amino- and carboxy-terminal regions facing the cytosol.[81a] With respect to the uncoupling protein, localization of a trypsin cleavage site

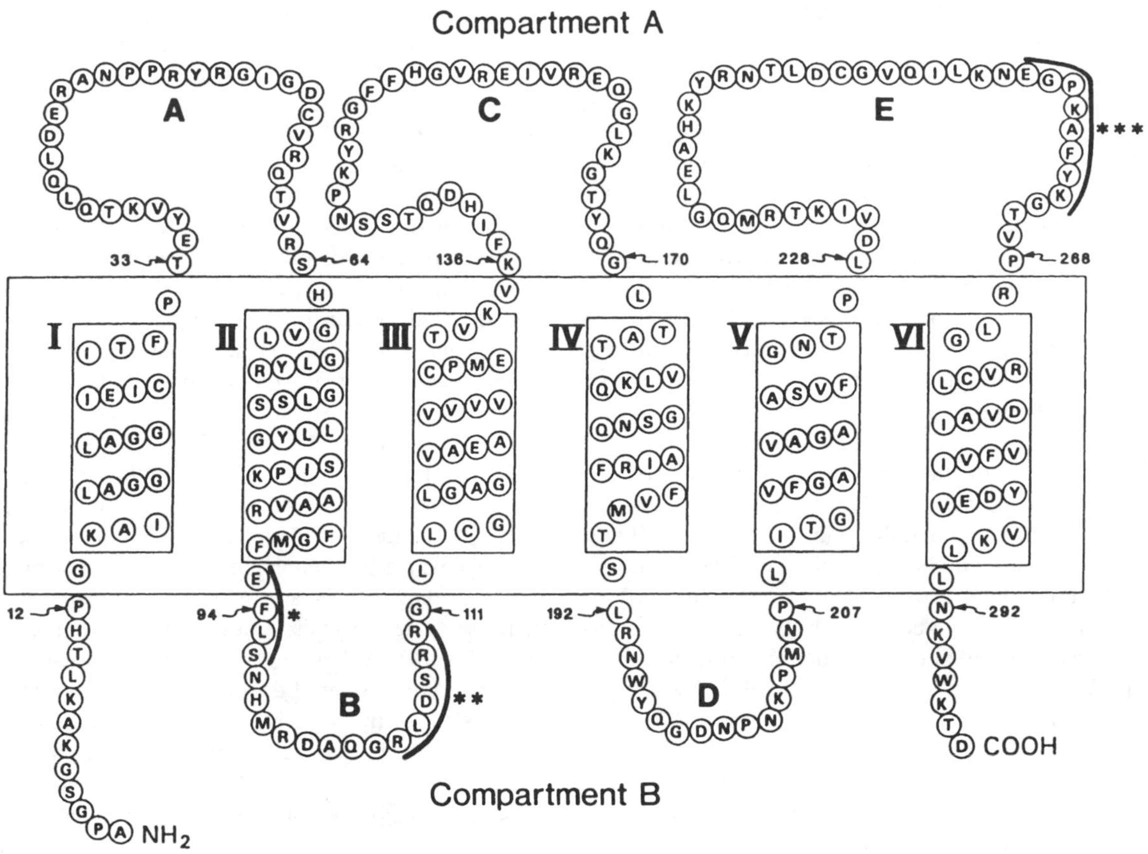

Figure 15.3. Proposed model for the membrane topography of the mitochondrial citrate transport protein based on its hydropathy profile. The model depicts six putative transmembrane α-helical segments (I–VI) which are enclosed in boxes and are connected by five hydrophilic loops (A–E). Numbers refer to the location of a given residue within the citrate transport protein primary sequence. Based on data obtained with the citrate, phosphate, and ADP/ATP carriers, as well as with the uncoupling protein, compartments A and B have been indentified as the matrix and intermembrane space, respectively. Single and double asterisks denote phosphorylation consensus sequences[200] for casein kinase I and protein kinase C, respectively. Triple asterisk denotes sequence that is highly conserved (i.e., six out of eight residues) with respect to the first zinc finger of nuclear receptors.[267] Modified from Kaplan *et al.*[54] with permission from *J. Biol. Chem.*

in isolated mitochondria to Lys292,[82] as well as intermolecular cross-linking at Cys304 with the impermeant reagent Cu^{2+}-phenanthroline,[83] clearly establish a cytosolic location for the carboxy-terminus. In an elegant series of experiments, Miroux *et al.*[84,85] constructed a library that expressed short subsequences of the uncoupler protein fused to the malE periplasmic protein of *E. coli.* The fusion proteins were used to select sequence-specific antibodies from anti-uncoupler protein serum which were then used to probe the uncoupler protein topology within the inner membrane. This approach has enabled a determination of the orientations of the amino-terminal extremities of membrane-spanning α-helices I–IV and VI, and has provided solid support for the model shown in Figure 15.3.

In summary, evidence obtained through the use of multiple approaches indicates that the general topology model of Figure 15.3 is likely to be correct for the uncoupler protein, the ADP/ATP translocase, and the phosphate carrier. It will be of interest to determine whether the validity of this model extends to other mitochondrial carriers such as the α-ketoglutarate and citrate transporters. This would be expected based on the similar sequence properties (i.e., tripartite structure, hydropathy profile) that characterize the different transporters, as well as the initial topology investigations.

15.3.4. Dimeric Nature

An important issue related to both the membrane topography and the molecular mechanism of the mitochondrial transporters concerns whether the transporters function as monomers, dimers, or as larger oligomers. As recently pointed out,[86,87] a large number of transporters from both prokaryotes and eukaryotes are thought to consist of a monomer with 12 α-helical transmembrane segments. It has been hypothesized that the monomers themselves have substructure and often have homology, such that they may in fact act as a functional dimer.[86] Thus, the proposal that the mitochondrial transporters, which possess 6 membrane-spanning segments per monomer, act as functional dimers would enable a uniform structural design to provide the physical basis for a wide variety of transport processes. Experimental evidence verifies that several of the mitochondrial transporters do in fact function as dimers. For example, in both the isolated state as well as in intact mitochondria, cross-linking of the uncoupling protein by Cu^{2+}-phenanthroline oxidation results in the formation of a dimer via the formation of a disulfide bond.[83] With the isolated protein, the cross-linking was found to be independent of the uncoupling protein concentration but could be prevented by SDS, thus providing solid evidence that cross-linking occurs between two monomers of a native dimer. Other evidence in support of the dimer model includes a nucleotide binding stoichiometry of one nucleotide per two uncoupling protein monomers[88] and hydrodynamic studies with the isolated protein.[89] Similarly, with the ADP/ATP transporter, the combination of hydrodynamic[90] and cross-linking[91] studies, as well as an inhibitor binding stoichiometry of one inhibitor molecule bound (i.e., carboxyatractyloside) per translocase dimer[92,93] have led to the conclusion that this transporter is also a dimer. Finally, cross-linking studies[79] using Cu^{2+}-phenanthroline support a dimeric model for the α-ketoglutarate carrier. Thus far, little data are available concerning the oligomeric structure of the phosphate and citrate transporters. In summary, the data obtained to date with three of the mitochondrial carriers indicate that they likely function as homodimers. Consequently, the functional unit of the mitochondrial transporters appears to consist of 12 membrane-spanning regions, and so these carriers may operate utilizing structural and mechanistic features that are similar to those inherent in transporters from other membranes.

15.3.5. Kinetic Mechanism

The mechanism by which the mitochondrial transport proteins catalyze substrate exchange has been investigated in detail.[79,94–99] Studies have been conducted with reconstituted, purified transporters in which two-reactant initial velocity measurements were made where both the external and internal substrate concentrations were varied within a single experiment.[79,94–98] Subsequent Lineweaver–Burk plots resulted in a series of straight lines that intersected at a common point on the abscissa. This intersection pattern (in contrast to a parallel pattern that would have been observed for a Ping-Pong mechanism) is characteristic of a sequential mechanism in which both countersubstrate molecules must form a ternary complex with the transport protein prior to the translocation event.[79,94–98] In other words, a second substrate (i.e., transport substrate present on the far side of the membrane) must bind before the first substrate (i.e., transport substrate present on the near side of the membrane) is released. Furthermore, the finding of a common point of intersection on the abscissa in double reciprocal plots in concert with the results obtained from secondary plots (see Refs. 79, 94–99 for additional details) suggests (1) a random order of substrate binding and (2) that the binding of the first substrate does not alter the apparent affinity of the transporter for the second substrate on the opposite side of the membrane. This type of sequential mechanism (i.e., a rapid-equilibrium random mechanism) has been demonstrated with the reconstituted aspartate/glutamate,[94] oxoglutarate,[79,97] dicarboxylate,[95] and citrate[96] carriers, and in intact mitochondria with the aspartate/glutamate,[98] oxoglutarate,[99,100] and ADP/ATP transporters,[99] although there is some controversy with respect to the latter transporter (see Refs. 13 and 15 for further details). Moreover, the sequential mechanism has important structural consequences since it requires the presence of a total of two binding sites each located at opposite sides of the transport protein which are accessible and occupied by substrate.[13] Since the transporters are functional dimers, it is conceivable that each monomer could provide one binding site. (Note, however, that with the dicarboxylate carrier the situation is somewhat more complex in that two separate binding sites are thought to exist at each side of the membrane for a total of four binding sites.[95]). Finally, it is important to consider that in contrast to this view, Klingenberg[15,101] maintains that the ADP/ATP translocase functions via a "single binding center gated pore" mechanism in which a single binding site is accessible at different stages in the transport cycle to substrate at one side of the membrane and then the other. Although this model is consistent with and supported by substantial data (see Ref. 15 for a review), it is inconsistent with the observed sequential kinetics (i.e., a Ping-Pong kinetic mechanism would be expected).[13] A resolution of this controversy will require additional insight into structure/function correlations at the molecular level.

In summary, the available evidence indicates that several mitochondrial transporters display an identical kinetic reaction mechanism (i.e., a sequential mechanism) and thus can be grouped into a single family based on functional as well as structural considerations. It is tempting to speculate that the transporters may function via similar molecular/chemical mechanisms that originate from the participation of common structural motifs.

15.3.6. The Molecular/Chemical Bases for Transporter Function

The ultimate goal of much transporter research is an elucidation at the molecular and chemical levels of the mechanism of transporter function. This would include a detailed understanding of the residues that comprise the substrate binding site, the translocation pathway within a given transporter, and the nature of the transporter's significant conformations. Since none of the mitochondrial transporters have been crystallized (although efforts directed towards this goal are currently underway), x-ray diffraction analysis has not been possible, and thus high-resolution three-dimensional structural information has not been obtained. Nonetheless, progress toward a partial understanding of the location of essential active site residues has been made. The approaches used to date have involved chemical modification of the transporters either with group-directed reagents or with affinity labels (i.e., substrate or inhibitor analogues) followed, in some cases, by localization of the modified residues within the primary sequence. Most recently, site-directed mutagenesis utilizing both yeast and *E. coli* expression systems has been used to probe residues of potential interest. These studies will be discussed in the sections to follow.

15.3.6.1. Identification of Essential Types of Residues with Protein Labeling Agents

Most of the transporters listed in Table 15.1 have been probed with group-selective protein labeling agents in intact mitochondria and/or in reconstituted systems. Generally, these transporters contain essential cysteine groups,[51,102–110] as well as essential lysine[51,104,109,111–113] and/or arginine[49,51,114] residues. Several of the carriers have also been shown to contain essential histidines.[51,104,107,109,112] Such results are not unexpected since one can readily envision (1) the participation of positively charged residues in the binding of negatively charged substrates and (2) the involvement of a histidine residue(s) in proton transport. Moreover, cysteine residues appear to play an important structural role in a variety of transport proteins. Based on the criterion that substrate protection against the inhibition caused by a given reagent constitutes one piece of evidence that the modified residue may be localized near or within the substrate binding site,[115] the binding site composition has been partially determined for several of the transporters (e.g., citrate carrier: lysine,[116] arginine[114] residues; oxoglutarate carrier: cysteine[105] residues; glutamate/aspartate carrier: lysine,[112] cysteine,[109] and possibly histidine[112] residues; ADP/ATP carrier: arginine residues[117,118]; and carnitine carrier: cysteine residues.[118a]).

15.3.6.2. Localization of Substrate and Inhibitor Binding Sites

Several groups have begun to map substrate and inhibitor binding sites to specific residues within the primary sequence of a given transporter. For example, with the ADP/ATP translocase from beef heart mitochondria, Boulay and Vignais[102] have determined that *N*-ethylmaleimide inhibited transport by binding to Cys56. This group[119] has also conducted studies using radioactive azido derivatives of atractyloside, an impermeant inhibitor of the translocase. Photolabeling with these derivatives resulted in specific labeling within the Cys159-to-Met200 domain although the actual residues that were labeled remain unidentified. Studies utilizing 2-azido-ADP,[120] a nontransportable photoactivatable substrate analogue that binds to the cytoplasmic face of the translocase and inhibits transport by competing with externally added ADP, indicated labeling at Lys162, Lys165, Ile183, Val254, and Lys259. With the exception of Ile183 (which is most likely located in helix IV), the labeled residues are located in hydrophilic loops C and E (see Figure 15.3), respectively. The fact that the labeled residues are present in domains thought to face the mitochondrial matrix (based on a variety of topology studies discussed in Section 15.3.3) would appear to contradict the finding that 2-azido-ADP competes for the binding of cytoplasmic substrate. The discrepancy has been explained by the postulation that the hydrophilic loops are accessible to opposite sides of the membrane during different stages in the transport cycle and perhaps form a portion of the translocation pathway.[12,15] This possibility might be especially likely if the hydrophilic loops can actually embed within the inner membrane to form a nucleotide binding pocket[16,121] (rather than protrude into the matrix as depicted in Figure 15.3). In yeast mitochondria, labeling of the translocase with the substrate analogue 2-azido-ATP has been localized to the Gly172-to-Met210 domain.[122] This segment contains Lys179 and -182 which are homologous to the Lys162 and -165 present in the bovine translocase. Finally, based on experiments using pyridoxal phosphate as an impermeant covalent probe for translocase lysine residues in both mitochondria and inverted submitochondrial particles,[111] it has been concluded that Lys22, -106, -162, and -165 likely line the hydrophilic translocation pathway, with Lys42 and 48 forming a portion of the entrance to this pathway from the matrix side.[15,111] In combination, the above data (1) provide firm support for the postulation that Lys162 and -165 are important components of the translocase substrate binding site and (2) begin to define the substrate translocation pathway through the translocase.

Similar studies have been conducted with the uncoupler protein utilizing 8-azido-ATP,[123] 2-azido-ATP,[124] and 3′-*O*-(5-fluoro-2,4-dinitrophenyl)adenosine 5′-triphosphate,[124] and indicate likely labeling at Thr260, Thr263, and Cys253, respectively, thus localizing a portion of the nucleotide binding site to hydrophilic domain E (Figure 15.3) which faces the matrix. It is thought that ATP interacts via the 2-adenine position with Thr263 and via the ribose 3′hydroxyl group with Cys253.[124] Since in intact mitochondria the uncoupler protein interacts with nucleotides only from the cytosolic side, it is proposed that the labeled hydrophilic domain is partially embedded within the membrane where it forms a nucleotide binding pocket which is accessible from both the cytosol and the matrix.[123,124]

In summary, it is apparent that the chemical labeling approach has enabled important initial progress in delineating potential roles of certain structural elements in the molecular mechanism(s) of the ADP/ATP translocase and the uncoupler protein. However, little additional information is currently available on mapping the substrate binding sites and/or the translocation pathways present within the other mitochondrial transporters. Clearly, this represents an important area for future study.

15.3.6.3. Expression of Mutated Transporters in Yeast

An extremely exciting recent development has been the use of yeast to express mutant forms of mitochondrial transporters in order to probe more precisely structure–function relationships. To date, the ADP/ATP,[125,126] phosphate,[127] pyruvate,[128] and citrate[129] transporters have been purified to varying extents from yeast mitochondria. The corresponding genes for the ADP/ATP,[130–133] phosphate,[134] and citrate[134a,134b] carriers have been identified and sequenced. With respect to the ADP/ATP transporter, the yeast *S. cerevisiae* contains three genes that encode three different transport proteins (AAC1, AAC2, AAC3). At the amino acid level, the three transporters display 75–93% identity. Disruption of the AAC1 gene has revealed that it is not required for growth on the nonfermentable carbon source glycerol.[132] In contrast, disruption of the AAC2 gene, which is expressed under normal growth conditions, renders the yeast unable to grow on glycerol.[135] Interestingly, on disruption of both the AAC1 and the AAC2 genes, yeast were still viable on a fermentable carbon source, thus leading to the postulation that another route for ADP/ATP transport exists other than via AAC1 and AAC2.[135] As predicted, a third ADP/ATP transporter gene has been discovered which is expressed almost exclusively under anaerobic conditions.[133] A recent determination of the molecular basis for the pet9 mutation, in which a defective ADP/ATP transporter is expressed, has revealed that it consists of a single G-to-A transition which results in an Arg-to-His change at position 96.[135]

Recently, site-directed mutagenesis has been applied to AAC2.[63,135,135a] Utilizing the failure of yeast to grow on glycerol as a phenotypic indicator of translocase function, four site-directed mutants at Arg96 failed to restore growth on glycerol, suggesting that transporter function is impaired in each mutant and that Arg96 is important in the transport mechanism.[135] In fact, single mutations of the four charged amino acids that are located within the intramembranous segments (i.e., Lys38, Arg-96, -204, and -294) caused a significant decrease in function as did mutations in the matrix-facing arginine cluster (Arg252, -253, -254).[63,135a] Importantly, reconstitution studies indicate that certain of the mutations differentially altered the transport of ATP versus ADP, thereby changing the efficiency of the transporter.[135a] In a related idea, it has been suggested that the arginine cluster functions to control the diffusion of negatively charged substrate into the translocation site.[136] It is noteworthy that none of the mutated cysteine residues (i.e., Cys73, -244, -271) were required for growth on glycerol or for reconstituted exchange activity, thereby indicating that these residues are not essential for function and that the NEM sensitivity of the transporter arises from steric effects rather than the direct participation of the modified residues in the translocation mechanism.[63,135a] Of great interest is the recent approach taken by Nelson and Douglas[137] in which mutants of AAC2 that could not grow on glycerol were selected for spontaneous suppressors. Following single Ile mutations at the Arg252–254 cluster, revertant mutations that occurred at sites other than the Arg cluster were analyzed. Most of these revertant mutations mapped to a side of the membrane opposite that of the matrix-facing arginine cluster, at positions that were very near the membrane surface. It has been proposed that the locations of many of these revertant mutations either comprise portions of the membrane channel and/or are involved in helix–helix contacts near the cytosolic side of the membrane. This approach has led to the development of detailed, experimentally testable models for the structure of the AAC transmembrane channel and holds great promise for unraveling complex, long-range structure–function relationships within the mitochondrial transporters.

Other investigators have demonstrated that in contrast to the ADP/ATP transporter, the yeast phosphate and citrate transport proteins are probably encoded by single genes.[134,134b] Site-directed mutagenesis of the phosphate carrier has indicated that replacement of Thr43 with a Cys renders the otherwise insensitive transporter highly sensitive to NEM.[138] Thus, by analogy, the Cys42 that is present in the beef heart transporter is likely responsible for that transporter's acute NEM sensitivity.[127] Also, the observation that the Thr43-to-Cys mutation in the yeast transporter caused a substantial decline in the observed transport activity, suggests that Thr43 occupies an important position within the yeast phosphate carrier structure.[138] Other mutagenesis experiments have indicated that an intermolecular disulfide bond formed at Cys28 is likely responsible for the oxygen-induced formation of a homodimer which is associated with an inhibition of transporter function.[139] In an important related development, Casteilla *et al.*[140] have achieved the stable expression of functional rat uncoupling protein in Chinese hamster ovary cells. The protein was targeted to the mitochondria in these cells where it functioned normally. The expression of the uncoupler protein in a heterologous cell system should now permit site-directed mutagenesis to be used to unravel the structure–function relationships within this transporter.

15.3.6.4. Overexpression of Mitochondrial Transporters

In order to obtain high-resolution, three-dimensional structural information for the mitochondrial transporters, it is necessary to obtain quality crystals of these carriers that are suitable for x-ray crystallographic analysis. While there are several difficulties in attaining this goal, the first major impediment has been obtaining a sufficient quantity of purified transport protein to permit crystallization trials. Ferreira and Pedersen[141] have been able to overexpress in *E. coli* truncated forms of the phosphate carrier (fused to a portion of the ATP synthase α-subunit protein) in which 20% or more of the transporter's carboxy-terminal sequence, corresponding to at least one of the six membrane-spanning α-helices, is removed. However, the native protein could not be expressed in this system. In an important recent development, Murdza-Inglis *et al.*[142] have achieved high-level expression (i.e., 70–100 μg/mg mitochondrial protein) of the uncoupler protein in yeast mitochondria. Purification of the heterologously expressed transporter yielded 500–700 μg of protein which on reconstitution exhibited GDP-sensitive Cl^- and H^+ transport activities. Recently, Fiermonte *et al.*[143] reported a major breakthrough wherein they expressed high levels of the bovine heart mitochondrial oxoglutarate carrier in *E. coli* (i.e., 10–15 mg of oxoglutarate carrier were obtained per liter of culture). The transporter accumulated in inclusion bodies which were then disaggregated with the detergent *N*-dodecanoylsarcosine, thereby allowing the subsequent reconstitution of transporter function in a liposomal system. The authors showed that the expressed protein contained the same amino-terminal sequence (except for the absence of a modified amino-terminus) as the native transporter, reacted with antiserum raised against the bovine heart transporter, and on reconstitution displayed a similar substrate specificity and inhibitor sensitivity as the native transporter. Utilizing a similar approach Kaplan *et al.*[134b] and Xu *et al.*[143a] have overexpressed the tricarboxylate transporter from both yeast and rat liver mitochondria, respectively. Approximately 25 mg of the yeast mitochondrial tricarboxylate transporter at a purity of 75–80% and 90 mg of the rat liver transporter at a purity of greater than 90% were obtained per liter of starting *E. coli* culture. In both cases the overexpressed transporter was reconstitutively active when incorporated into liposomes following extraction from an isolated inclusion body fraction with sarkosyl. This general approach has also enabled the overexpression of the phosphate carrier.[143b] Thus, this system appears to have broad applicability for overexpression of a variety of mitochondrial transporters from different sources and has allowed, for the first time, investigators to obtain abundant quantities of both wild-type and mutated[143b,143c] forms of these transport proteins, thereby enabling a variety of structure/function studies.

15.4. MITOCHONDRIAL TRANSPORTER GENES

15.4.1. The Number of Genes Encoding a Given Transporter

The first evidence for the existence of multiple transporter isoforms was obtained at the immunological level. Results from several groups indicated the presence of tissue-specific isoforms of the ADP/ATP translocase,[144,145] as well as the phosphate[146] and α-ketoglutarate[147] carriers. With respect to the ADP/ATP translocase, these observations were subsequently confirmed at the genetic level where it has been shown that the human genome contains at least three different genes for the translocase[65,67,148–151] as does the yeast *S. cerevisiae*.[130–133] Two genes have been identified in the bovine,[66] rat,[64] and *Zea mays*[152,153] genomes. With respect to other mitochondrial transporters, Southern blot analysis indicates the possible existence of multiple genes (or pseudogenes) for the rat phosphate[60] and citrate[54] transporters. In contrast, single genes have been found to encode the yeast phosphate and citrate transporters,[134,134b] the bovine α-ketoglutarate carrier,[61] and the uncoupling protein from mouse,[154] rat,[155] and human.[156] With regard to the α-ketoglutarate carrier, the discrepancy between the immunological data (suggesting multiple isoforms)[147] and the genetic data (indicating a single gene)[61] may possibly have arisen from a tissue-specific posttranslational modification of this carrier and/or the presence of more distantly related genes that code for isoforms but were not identified at the stringencies employed in the Southern blot and PCR analyses.[61] Finally, it is important to note that to date, multiple isoforms that display different functional properties have not been identified or isolated for any transporter from mammalian species.[13] It will be of great interest to determine whether *functionally distinct* isoforms (which may be differentially expressed) in fact exist for a given transporter.

15.4.2. Gene Structure

Table 15.4 summarizes characteristics of selected mitochondrial transporter genes that have been sequenced from higher eukaryotic cells. To date, genomic sequences have been published for the ADP/ATP translocase,[64,67,148,151] the α-

Table 15.4. Characteristics of Selected Genes Encoding Mitochondrial Transporters in Higher Eukaryotic Organisms

Transporter	Source	Length[a] (kb)	Exons	Coding sequence identity[b] (%)	Protein sequence identity (%)	5′ regulatory elements[c]	Refs.
ADP/ATP T1 (ANT-1)[d]	Human	4.0	4	—	—	A, B, C, D, E, F	67, 148
ADP/ATP T2 (ANT-3)	Human	5.9[e]	4	78	88	C, D, G, H	67
ADP/ATP T3 (ANT-2)	Human	2.8	4	77	89	A, B, D, E*, H, I, J*, K*	151
α-Ketoglutarate	Human	2.5[e]	8	—	—	L	157, 158
	Bovine	2.4[e]	6	93	97	L	157, 158
Uncoupling protein	Human	≈9[e]	6	—	—	A, G, M*	156
	Rat	8.4	6	79	79	A, B, G, M*, N*, O*, P*	159

[a]Denotes the length of DNA over which the exons are spread.
[b]Both nucleotide coding sequence identity and protein amino acid sequence identity values are calculated as a comparison to the first sequence listed for a given transporter.
[c]Regulatory element abbreviations: A, TATA box; B, CCAAT box; C, CpG-rich island; D, SP1 binding sites present within the first intron; E, SV40 enhancer element; F, a 13-nucleotide pair sequence and an overlapping 8-nucleotide pair sequence which are present in the ADP/ATP T1 translocase and the ATP synthase β gene and have been referred to as OXBOX and REBOX, respectively; G, G+C-rich region; H, SP1 binding sites; I, CCACT sequence (i.e., a modified CCAAT box); J, OCT-1 binding site motif; K, AP-2 binding site motif; L, the 5′ upstream sequence has not been determined and therefore the regulatory elements have not yet been characterized; M, a consensus cAMP regulatory element; N, a 13-bp fat-specific element; O, a glucocorticoid regulatory element; P, a thyroid regulatory element; asterisk denotes a sequence that is similar but not identical to a given regulatory element.
[d]Parentheses contain an alternative nomenclature which is also widely used.
[e]This length represents a minimal value because of the fact that the transcriptional start site was not determined and thus the 5′ extremity of the first exon is not known.

ketoglutarate carrier,[157,158] and the uncoupling protein.[156,159,160] Interestingly, with each of these genes, the introns mostly interrupt sequences that encode regions that are either near or within the extramembranous hydrophilic loops (i.e., loops A–E in Figure 15.3) rather than the α-helical transmembrane domains.[67,158,160] It is thought that the location of these introns may reflect evolution of the transporter genes which included two tandem duplications of an ancestral gene encoding a domain of about 100 amino acids.[67,158] Thus, the introns may have originated from these earlier fusion events. Interestingly, each of the above genes contains introns that immediately follow the first and precede the sixth transmembrane domains, thus giving rise to the speculation that the 100-amino-acid repeat may itself have arisen via duplication of a single transmembrane segment.[67,158] However, as previously noted,[67,158] this hypothesis is not supported by sequence comparisons. Finally, both the ADP/ATP translocase T1 gene[148] and the uncoupler protein gene[156] have been localized to chromosome 4 in humans, providing further suggestion of their common origin.[156]

Elucidation of the molecular genetic regulatory mechanisms that control mitochondrial transporter gene expression represents an area of great interest. Since this expression is both tissue-specific as well as environmentally and hormonally sensitive, as genomic sequences have become available investigators have begun to examine the 5′ regulatory regions for clues concerning potential regulatory motifs. For example, it is interesting to note that while the three ADP/ATP translocase structural genes have similar features, their 5′-flanking regulatory regions are quite distinct, a property that is thought to influence their expression (see Section 15.4.3). Thus, the T1 gene displays features of a typical eukaryotic promoter including the TATA and CCAAT motifs immediately upstream of the transcription initiation site, SV40 enhancer elements farther upstream, and a CpG island near its 5′end.[67,148] In contrast, the T2 gene displays characteristics suggestive of a housekeeping gene that is expressed in all tissues (i.e., it lacked both the TATA and CCAAT boxes, but contained many potential SP1 binding sites as well as a G+C-rich promoter).[67] The T3 gene displays features intermediate between the other two genes in that it contains a TATA motif but not a nearby CCAAT motif.[151] However, it does contain a CCACT sequence which may function as a CCAAT box.[151,161] Also, it contains several SP1 binding sites, and sequences resembling the OCT-1 binding site motif, the AP-2 binding site motif, and the SV40 enhancer core.[151] All three translocase genes contain SP1 binding sites within the first intron. Such sites have been implicated in regulating transcription of the pro-α1(I) collagen gene.[151,162] Recently, the T1 gene has been shown to contain two overlapping sequence elements in the promoter region, namely, a 13-bp OXBOX[163] and an 8-bp REBOX.[164] The OXBOX is a muscle-specific positive control element for the transcription of genes involved in oxidative phosphorylation. Its binding factors appear to be present only in myogenic cell lines. The REBOX is a redox-sensitive (possibly negative) control element which binds sequence-specific factors that are present in most cell types. Interestingly, the binding of REBOX factors is sensitive to NADH, pH, dithiothreitol, and thyroxine, suggesting that it may modulate gene expression in response to both environmental (e.g., cellular energy requirements) and hormonal changes. OXBOX and REBOX factors do not bind concurrently. A type of push–pull model has been proposed in which OXBOX factors activate expression and REBOX factors repress expression.[164]

With respect to the uncoupler protein, the 5′ regulatory region of the human gene contains a TATA box which is preceded by a G+C-rich region.[156] Comparison with the rat upstream sequence indicates an identical sequence at position −569 in the human gene[156] and −565 in the rat gene[159] which is 85% homologous to the consensus CRE sequence, thereby suggesting a role for cAMP in the regulation of uncoupler protein gene expression in humans and rats.[156] Additionally, in the rat gene, regions partially homologous to a fat-specific element, as well as to glucocorticoid- and thyroid-regulatory elements were observed.[159] Recently, in a series of elegant transfection experiments, Cassard-Doulcier *et al.*[165] have shown that most of the *cis*-acting regulatory elements which control uncoupler protein gene expression (including both tissue-specific and β-adrenergic response elements) were present within 4.5 kb of the 5′ region that flanked the uncoupler structural gene. Deletion analysis of the uncoupler promoter region in transfected cells indicated that the minimal region that exhibited both promoter activity and tissue specificity is located between −157 and −57 bp. A 211-bp activator element was located between −2494 and −2283 bp and an inhibitory element was possibly located between −400 and −157 bp. Recent studies focusing on the *in vitro* interactions between nuclear proteins and the uncoupling protein gene promoter have revealed the existence of several putative transactivating factors including Ets 1, the retinoid X receptor, the thyroid hormone receptor, and a CACCC box-binding protein.[165a]

In summary, the genomic sequences of the structural genes that encode three of the mitochondrial transport proteins have been obtained, and with two of these (i.e., the genes encoding the ADP/ATP translocase and the uncoupling protein) solid progress has been made in the identification of the 5′ regulatory elements. To date, no information is available regarding the structure of other mitochondrial transporter genes. Clearly, this represents an exciting area for future research.

15.4.3. Differential Expression of Transporter Genes

As mentioned earlier (Section 15.4.1), evidence for the organ-specific expression of ADP/ATP translocase isoforms was first provided at the immunological level, where Schultheiss and Klingenberg[144,145] demonstrated differences in the antigenic determinants in translocase from bovine heart, kidney, and liver mitochondria. At the genetic level, it was found that two genes (i.e., T1 and T2) encode the bovine translocase which differ in their organ-specific expression.[66]

For example, T1 mRNA was present mainly in the heart whereas T2 mRNA was found mainly in the intestine and kidney.[66] In the human, Li *et al.*[148] demonstrated that T1 transcripts were present at high levels in the heart and skeletal muscle, and at very low levels in liver, kidney, and brain. In contrast, the T3 transcript was present in all five tissues. Lunardi *et al.*[166] have shown that the three isoforms are differentially expressed during human muscle development. For example, T1 mRNA is detectable only in the adult muscle, whereas T3 mRNA is present at high levels in growing myoblasts and then progressively decreases during development, and T2 mRNAs are present at high levels in both myoblasts and myotubes before subsequently decreasing in amount in the mature adult muscle. Lunardi and Attardi[167] have shown that in HeLa cells, levels of the T2 and T3 mRNAs are differentially affected by growth conditions while T1 mRNA was not detectable. For example, the T2 mRNA level remains constant during both the exponential and stationary phases, whereas the T3 mRNA level decreases progressively during the second half of the exponential phase and the stationary phase.[167] These quantitative changes reflected gene expression rather than message stability. Exposure of cells to agents that induce cessation of cell proliferation and differentiation caused a decrease in both T2 and T3 mRNA transcripts. Incubation with dinitrophenol, an uncoupler of oxidative phosphorylation, also caused a decrease in the mRNA transcript levels which in this instance could be accounted for by a decrease in message stability. Thus, in combination, these results indicate that the expression of multiple translocase genes is responsive to cell physiological conditions by changes in the rate of synthesis or stability of their mRNAs.[167] In a related finding, Battini *et al.*[150] have shown that the steady-state level of T3 mRNA is responsive to growth factors. Thus, the steady-state T3 mRNA level increased when quiescent BALB/c/3T3 fibroblast cells were stimulated by serum, platelet-derived growth factor, or epidermal growth factor, but not by platelet-poor plasma or insulin.

In contrast to the ubiquitous expression observed with the ADP/ATP translocase protein, the uncoupling protein (and its mRNA) is uniquely expressed in brown adipocytes where it confers on these cells the ability to dissipate oxidation energy as heat. It has been shown that environmental stimuli such as exposure of rats to cold, as well as nutritional changes such as starvation followed by refeeding, induce the synthesis of both the uncoupler protein and its mRNA.[20,21,154,168–170] Nuclear run-on experiments have indicated that cold induction causes a tenfold increase in the rate of uncoupler protein gene transcription.[21,168] Moreover, treatment of rats with epinephrine[171] or β-adrenoceptor agonist[168] also cause increases in uncoupling protein gene transcription. Little information is currently available concerning the differential expression of other mitochondrial transporter genes.

In summary, with regard to the mitochondrial transporter genes, it is our view that the characterization of the (1) structural genes for the other important metabolite carriers, (2) tissue-specific, environmentally sensitive genetic regulatory elements that modulate the expression of transporter genes, and (3) expression of functionally distinct isoforms for a given transport protein represent important challenges that will likely be met in the near future.

15.5. HORMONAL REGULATION OF TRANSPORTER FUNCTION

It is well established that thyroid hormone plays an important role in the regulation of hepatic bioenergetics. It influences mitochondrial respiration, volume, cytochrome content, and membrane lipid composition, as well as the rate of mitochondrial protein and RNA syntheses (see Refs. 172–174 for reviews). The long-term effect (days) of thyroid hormone is mediated via binding to a nuclear receptor which then regulates the transcription of a variety of genes. The short-term effect (minutes to hours) occurs via an extranuclear pathway which is not well defined and presumably involves either binding to a plasma membrane receptor followed by generation of an intracellular signal, or possibly by a direct interaction with intracellular proteins. With respect to the transporters, hyperthyroidism has been shown to increase the function of the phosphate,[175] pyruvate,[176] citrate,[177] and ADP/ATP[178] transporters. Furthermore, the activities of the ADP/ATP[179,180,180a] and the pyruvate[181] transporters have been found to decrease in hypothyroid animals. Inhibitor binding studies have consistently shown that the amount of a given transporter does not appear to change in response to thyroid hormone and thus does not account for the observed activity changes.[175–177] However, it has been well documented[175–177] that in parallel to the observed changes in transporter function, hyperthyroidism causes increases in the amount of negatively charged mitochondrial phospholipid such as cardiolipin and phosphatidylserine, as well as in the fatty acid 20:4/18:2 ratio. This finding has led to the suggestion that the alteration in transporter function arises from hormone-mediated changes in the membrane lipid composition which alters the fluidity of the transporters' microenvironment, thereby affecting transporter function.[175–177] Finally, an observation of related interest is that following the injection of thyroid hormone, the mRNA level for the growth-activated T3 isoform of the ADP/ATP transporter is increased 13-fold.[182] However, neither the amount nor the activity of the transporter was measured and thus it is unclear if these changes influenced the amount of expressed protein.

With respect to the uncoupling protein, a series of experiments carried out in rat brown adipose tissue have indicated that its gene is under the dual control of norepinephrine and triiodothyronine.[171] The primary signal appears to be generated by norepinephrine which increases uncoupling protein mRNA by stimulating the transcription rate of the corresponding gene. In addition, triiodothyronine amplifies the transcriptional response to norepinephrine by a factor of four- to fivefold. Studies carried out with adipocytes in culture indicated that both norepinephrine (an α_1-, β-agonist) and isoprenaline (a pure β-agonist) selectively increase uncoupling protein synthesis, thus indicating that β-adrenergic receptors are primarily involved.[183] Elevation of intracellular cAMP also stimulated uncoupling protein synthesis. With respect to the effects of other hormones, investigations by Burcelin *et al.*[184] showed that streptozotocin-induced diabetes caused a decrease in rat brown adipose tissue uncoupling protein and its mRNA, which was prevented by insulin infusion into the diabetic animals. When the hyperglycemia associated with the diabetes was prevented by infusing phlorizin into diabetic rats, the levels of both the uncoupling protein and its corresponding mRNA were further reduced. Thus, the uncoupling protein gene appears to be both insulin- and glucose-dependent.

Studies[185] involving the pyruvate transporter have indicated that a 30-min *in vivo* exposure to epinephrine results in an increase in transport activity which does not arise from an alteration in either the membrane potential or the pH gradient. The increase in pyruvate transporter function may be partly responsible for the observed epinephrine-mediated increase in hepatic gluconeogenesis.[185,186] Investigations involving the *in vivo* administration of glucagon showed a stimulation in pyruvate transporter function which was characterized by an increase in the apparent V_{max},[187] which likely arose from a glucagon-induced increase in the pH gradient across the mitochondrial inner membrane.[185,187] Finally, studies carried out by Aprille's group[188,189] have shown that the ATP-Mg/P_i carrier is activated by Ca^{2+} and that the increase in mitochondrial adenine nucleotide content that occurs after the administration of glucagon, vasopressin, or epinephrine likely arises from a Ca^{2+}-induced increase in ATP-Mg/P_i transport activity.

Related to the topic of hormonal regulation of transporter function is the idea (about which there is little available information) that the regulatory mechanism may involve phosphorylation–dephosphorylation reactions. We believe this possibility is of particular interest since (1) other metabolically important proteins within mitochondria are regulated by phosphorylation (e.g., carnitine palmitoyltransferase I,[190] pyruvate dehydrogenase,[191] branched chain α-ketoacid dehydrogenase[192]); (2) several different protein kinases and phosphatases are present within mitochondria;[191–197] (3) incubation of mitochondria with [^{32}P]-ATP results in the phosphorylation of unidentified proteins in the range 30–40 kDa[198,199]; and (4) inspection of the sequences of the citrate (see Figure 15.3)[54] and phosphate[60] transporters, as well as the uncoupling protein[62] reveals the presence of phosphorylation consensus sequences specific for several protein kinases.[200] Accordingly, we think research into this possibility is needed and may represent a fruitful avenue of investigation.

15.6. ALTERATION OF MITOCHONDRIAL TRANSPORTER FUNCTION IN DISEASE, DEVELOPMENT, AND AGING

Mitochondrial diseases can be classified according to several different schemes.[201,202] For example, one scheme[201] divides mitochondrial disease into inherited versus acquired conditions. The inherited conditions consist of those that affect nuclear DNA, mitochondrial DNA, or intergenomic signaling mechanisms. Since all mitochondrial transporters are encoded by nuclear DNA, any inherited disease involving a transporter would probably involve defects in the nuclear genome, although alterations in intergenomic signaling events that affect transporter function cannot be ruled out. According to a second classification scheme,[202] mitochondrial dysfunction can be considered as either a primary or a secondary disease. Primary diseases arise from a genetic defect in a mitochondrial enzyme or transporter. The defect may also occur in a protein involved in the expression (transcription, translation, posttranslational modification, or import) of a given component. All other diseases causing mitochondrial dysfunction are considered to be secondary. To date, most of the diseases that have been reported to involve the transporters would be classified as acquired, secondary conditions. Alteration of transporter function in aging is also placed in this category. However, it is important to note that with even the most well researched of the carriers, the experimental tools required to determine the molecular basis for observed altered function (e.g., isoform-specific antibodies and cDNAs) have only become available within the last several years. In fact, for many of the transporters these tools are still not available. Thus, transporter function in most of the conditions discussed below has not yet been investigated at the molecular level, and therefore the conclusions reached need to be interpreted cautiously. Clearly, this represents an exciting area for future research.

15.6.1. Heart Disease, Primary Biliary Cirrhosis, and Other Mitochondrial Myopathies

The function of the ADP/ATP translocase is altered in several different disease states. For example, during myocardial ischemia the translocase is severely inhibited, possibly as a result of a depletion of endogenous nucleotides.[203–205] Borutaite *et al.*[206] have shown that, based on inhibitor binding studies, 45 min of ischemia causes a decrease in the amount of active translocase protein and an increase in the control coefficient of the translocase in state 3 respiration. They concluded that the transporter is one of the most important steps in the regulation of oxidative phosphorylation during the development of ischemic heart injury. Duan and Karmazyn[205] showed that following reperfusion of the ischemic heart a significant relationship exists between the oxidative phosphorylation rate and ADP/ATP translocase function, as well as between translocase function and post-reperfusion contractile recovery. They concluded that depression of the activity of the translocase may be an important mechanism involved in cardiac injury during acute ischemia and reperfusion. In a series of papers of great interest, Schmid's group[207,208] has noted that exposure of rat myocardial membranes to free-radical-generating systems resulted in lipid peroxidation and an apparent formation of adducts between membrane proteins and the peroxidized phospholipid. The most prominent effect was observed with the ADP/ATP translocase which, based on SDS–PAGE, was found to increase in molecular mass by approximately 1 kDa at low oxidant levels. This increase was then followed by the complete disappearance of the transporter from the protein profile following either prolonged or more drastic incubation conditions. Since it is widely believed that ischemia-reperfusion injury involves membrane damage which is mediated by oxygen free radicals, these authors suggest that free-radical-mediated alteration of the translocase may play a role in such injury.

Patients with dilated cardiomyopathy, an autoimmune disease with a suspected viral etiology, display circulating autoantibodies directed against the ADP/ATP translocase.[209] These autoantibodies are organ specific since they exhibit significantly less binding to liver mitochondria than to heart mitochondria. Moreover, autoantibodies obtained from about 60% of the patients were able to inhibit adenine nucleotide exchange *in vitro.*[210] An evaluation of whether the antitranslocase antibodies altered translocase function *in vivo* was performed as follows.[210] Guinea pigs were injected with isolated translocase in order to raise the desired antibodies. With those animals that produced specific antibodies capable of inhibiting transport *in vitro,* cytosolic and mitochondrial adenine nucleotide levels were measured following the freeze-clamping of perfused hearts. The results obtained indicated that relative to control hearts, the hearts from the immunized animals displayed a decrease in both ATP level and the ATP/ADP ratio in the cytosol, and an increase in both the ATP level and the ATP/ADP ratio in the mitochondria, thereby indicating that an inhibition of the translocase had occurred in the immunized animals. These authors[210] suggested that the autoantibodies may contribute to the pathophysiology observed in dilated cardiomyopathy by causing an imbalance between energy delivery and demand within cardiac myocytes. Additionally, Kuhl *et al.*[211] have identified two proteins on the cell surface of cardiac myocytes that bind the antitranslocase antibodies and may serve as receptors for receptor-mediated endocytosis of the autoantibodies, thus enabling their penetration into the cell. Finally, in a finding of related interest, Schultheiss *et al.*[212,213] have discovered that specific antitranslocase antibodies are also present in patients with primary biliary cirrhosis. The antibodies were organ specific (i.e., reaction with liver > heart > kidney) and in about half of the patients were able to inhibit translocase activity in isolated mitochondria. It is unclear whether transporter function is altered *in vivo.*

With respect to the alteration of transporter function in other mitochondrial myopathies, Western blot analysis has revealed that a patient with a mitochondrial myopathy, presenting with lactic acidosis, displayed a fourfold decrease in the concentration of the ADP/ATP translocase in muscle tissue.[214] It is thought that the decrease in the quantity of translocase was the primary cause of this disease. Additionally, with respect to other mitochondrial myopathies, studies have indicated defects in the malate–aspartate shuttle[215] and possibly in the carnitine transporter.[216] In fact, recent studies have documented several cases of carnitine translocase deficiency which presented with prominent cardiac (and other) abnormalities shortly after birth.[216a]

15.6.2. Cancer

There are a number of reports documenting altered mitochondrial transporter function in different types of tumors. For example, the V_{max} of the pyruvate transporter is found to decrease in Ehrlich ascites tumor cells and in two different Morris hepatomas.[217,218] A change (i.e., increase) in the K_m value was observed with one of the hepatomas. With respect to the ADP/ATP translocase, observed alterations vary depending on the type of tumor examined. For example, Senior *et al.*[219] found a severalfold increase in the V_{max} and decrease in the K_m for ADP uptake into mitochondria isolated from a rat mammary tumor relative to values observed with mitochondria from normal rat mammary gland. In contrast, Kolarov *et al.*[220] observed the opposite results with mitochondria from a Zajedal hepatoma as well as from an Ehrlich ascites carcinoma. Studies with the citrate transporter[221] indicated that at 10°C citrate transport kinetics were unchanged in mitochondria from a rapidly growing Morris hepatoma, relative to control mitochondria. However, as the incubation temperature was increased, the rate of increase in citrate transport in the tumor mitochondria was greater than in the normal mitochondria. Extrapolation of the data to 37°C indicated that at physiological temperatures, transport in the tumor mitochondria would likely occur twofold faster than in normal control mitochondria. These studies also demonstrated that comparisons of transport activity measurements carried out at the reduced temperatures that are typically utilized to facilitate accurate quantification (e.g., 4–10°C) do not necessarily prove valid at higher physiological temperatures. This is especially true if the disease in question alters the lipid composition of the mitochondrial inner membrane which may in turn change the membrane fluidity and/or the lipid phase transition temperature.[221] Two other points should be noted. First, the altered transporter functions that have been observed in tumors likely represent secondary changes rather than primary ones, but may nonetheless be important for the maintenance of the aberrant intermediary metabolism that characterizes many tumors. Second, at the time these studies were conducted, molecular probes were not available for most of the carriers. Thus, it would be of interest to carry out analogous studies in which changes in transport activities were correlated with protein and mRNA levels, as well as the rate of transcription of transporter mRNAs. It is likely that important information regarding the molecular mechanisms by which the transporters are regulated might emerge from such studies.

15.6.3. Diabetes Mellitus

Early work in this area provided suggestive evidence for alterations in mitochondrial transporter function in diabetes mellitus. For example, Titheradge and Coore[185] observed that administration of streptozotocin to rats resulted in increased liver mitochondrial pyruvate transport activity. In contrast,

Kielducka *et al.*[222] reported a decreased pyruvate transport capacity in alloxan-induced diabetic animals. Interestingly, a preincubation with ketone bodies restored the diabetic V_{max} value to control levels. They suggested that these changes arose from a decreased pH gradient across the diabetic inner mitochondrial membrane and possibly a change in the intrinsic molecular properties of the pyruvate carrier. Rahman *et al.*[223] reported that insulin treatment of rat epididymal adipose tissue caused an increase in the rate of pyruvate transport into subsequently isolated mitochondria. With respect to other transporters, diabetes was found to cause an increase in the K_m of the citrate transporter[224] and a decrease in the activity of the ADP/ATP translocase.[225] These changes were attributed to increased levels of long-chain acyl-CoA esters in diabetic liver mitochondria. Thus, in many of these early studies, changes in transporter function were thought to be consequences that were secondary to diabetes-induced alterations in other parameters.

During the last several years, a major focus of our laboratory has been the use of recently developed molecular approaches in order to clearly determine (1) the effect of streptozotocin-induced type 1 diabetes (i.e., insulin-dependent diabetes mellitus; juvenile diabetes) on mitochondrial transporter function and (2) the molecular basis for diabetes-induced alterations in transporter function. The underlying hypothesis for our studies[226] was that the liver mitochondrial transporters are regulated in coordination with the enzymes of metabolic pathways to which they either supply substrate or remove product. Based on this hypothesis, and the demonstration by other investigators of specific alterations in a variety of hepatic metabolic pathways in diabetes, we predicted that the functional levels of the liver mitochondrial: (1) pyruvate and dicarboxylate transporters would be increased; (2) citrate transporter would be decreased; and (3) phosphate transporter would be unchanged in diabetes (see Figure 15.4 and Ref. 226 for a review of the metabolic rationale supporting these predictions). Our experimental strategy consisted of the extraction (via previously optimized procedures) of each of the above transporters from mitochondria prepared from control versus diabetic animals, followed by the functional reconstitution of each transporter in well-defined liposomal systems. This approach offered important advantages relative to the more classical approach (which would involve conducting transport measurements with intact isolated mitochondria) since it allowed us to place transporters from the two groups of animals in an identical lipid environment, with the same intra- and extra-liposomal substrate concentrations, in the absence or presence of an identical pH gradient and/or potential difference across the liposomal bilayer. Thus, transport measured under these conditions enabled a direct examination of the effect of diabetes on the intrinsic functional properties of a given transport protein *per se,* separate from other (possibly artifactual) variables that might indirectly influence the observed transport rates. To our knowledge, this study represented the first use of this approach to probe mitochondrial transporter function in diabetes. Our results[226] indicated that diabetes caused an increase in the level of extractable, reconstitutable total pyruvate and dicarboxylate transporter activities (i.e., 104 and 69%, respectively; diabetic versus control; $p < 0.001$), a decrease in the citrate transporter activity (i.e., 45%; $p < 0.001$), and no change in the phosphate transporter activity (i.e., 14% increase; $p > 0.10$). Similar findings were noted with respect to the specific transport activity values. Thus, each of our predictions was proven correct, thereby suggesting the validity of the underlying hypothesis. Additionally, the finding that diabetes differentially affected transporter function in a predictable manner suggested that the changes were specific and did not reflect a more generalized,

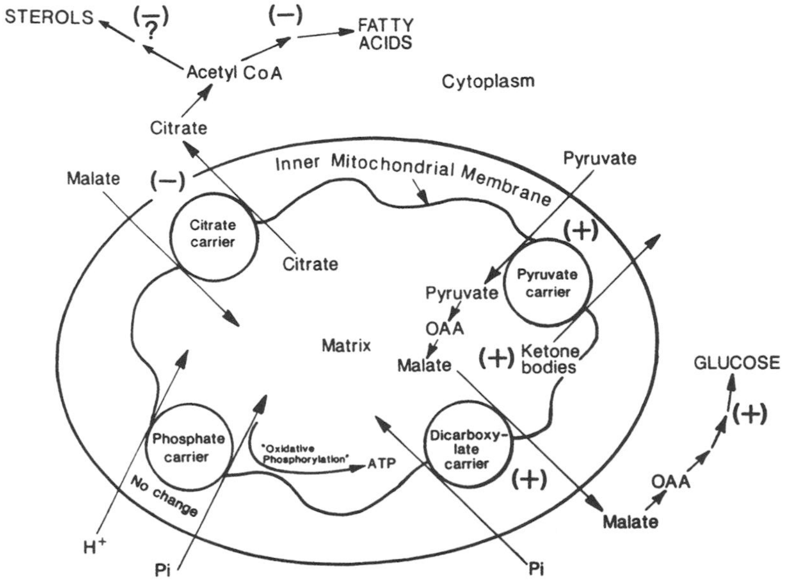

Figure 15.4. Importance of mitochondrial anion transporters in the altered hepatic intermediary metabolism that characterizes experimental diabetes. This figure summarizes the effect of experimental diabetes on (1) the levels of functional mitochondrial anion transporters that have been characterized in the author's laboratory[226,227] and (2) the levels of associated metabolic pathways to which the mitochondrial transporters either supply substrate or remove product as documented by other investigators (see Ref. 226 for a detailed reference list). Increases are denoted as (+) and decreases as (–). "Ketone bodies" refers both to the levels of ketone bodies and to the rate of fatty acid oxidation. "?" indicates conflicting data as to whether or not diabetes in fact causes a decrease in sterol biosynthesis. Reproduced from Kaplan *et al.*[226] with permission from *Arch. Biochem. Biophys.*

nonspecific alteration in mitochondrial function. In a subsequent study,[227] we showed that treatment of diabetic animals with insulin resulted in a reversal of the alteration in the functional levels of the pyruvate, dicarboxylate, and citrate transporters. In combination, the above findings provided the first clear-cut evidence that (1) diabetes causes changes in the functional levels of the mitochondrial anion transporters and (2) these changes arise as a consequence of the insulin deficiency that characterizes this disease. In a final series of studies,[228] we observed that in an animal model for type 2 diabetes (i.e., non-insulin-dependent diabetes mellitus; adult-onset diabetes), in which the level of insulin was not profoundly deficient, the functional levels of the pyruvate, dicarboxylate, and citrate transporters were unaltered, thus lending further support to our contention that insulin plays an important role in the regulation of these transporters. Finally, it is important to note that it is our belief that the ability to chemically induce a pathology that gives rise to specific changes in mitochondrial transporter function, which can be reversed by insulin therapy, provides an excellent physiologically/pathologically relevant model system for elucidating the molecular events involved in the regulation of transporter function.

15.6.4. Other Diseases

The syndrome of hyperornithinemia, hyperammonemia, and homocitrullinuria is an inherited metabolic disorder that is characterized by elevated blood ornithine levels, liver and renal tubular dysfunction, mental retardation, and sometimes seizure. The primary defect in this disorder resides in the transport of ornithine across the mitochondrial inner membrane.[229,230] A drastic reduction in ornithine transport is observed with liver mitochondria obtained from these patients,[229] while a partial reduction in transport is observed with skin fibroblast mitochondria.[230] Another syndrome, familial hyperlysinemia, is associated with mental and motor retardation, and has been proposed to arise from a defect in lysine transport through both the plasma membrane and the inner mitochondrial membrane.[231]

15.6.5. Development and Aging

Recent work by Schonfeld *et al.*[232] utilized quantitative inhibitor binding studies to show that the amount of ADP/ATP translocase in isolated liver mitochondria increased 2-fold to the adult level during the first 2 days after birth. This increase is thought to be partly responsible for the increase in state 3 respiration that occurs during this time. Hutson and Hall[46] have shown that mitochondria from fetal liver display an 85-fold higher level of functional branched chain α-keto acid transporter compared with mitochondria from adult liver. With respect to a subsequent age-dependent alteration of transporter function, numerous studies have shown that the aging of adult animals causes a decrease in the function of the ADP/ATP,[233] pyruvate,[234] phosphate,[235] and acylcarnitine–carnitine[236] transporters. Typically, these changes involve decreases in the apparent V_{max} values with little change in the K_m. Despite this finding, inhibitor binding studies[233–235] indicate that the abundance of a given transport protein within the inner membrane does not significantly change with aging. Related to these findings is the observation that the lipid composition is markedly altered in mitochondria from aged animals such that the cholesterol content is increased, whereas the phospholipid level including the cardiolipin level is decreased.[233–235] It has been proposed[233–236] that the age-related decreases in transporter function arise from (1) the altered lipid milieu surrounding a given transporter within the inner membrane and possibly (2) a decreased intramitochondrial substrate concentration. It should be noted that to date, these studies have not been extended to the molecular level. It would be of interest to determine whether a different complement of transporter isoforms, or their mRNA transcripts are expressed in senescent animals versus their juvenile counterparts.

15.7. SUMMARY

It is the aim of this review to document recent advances in our knowledge of diverse aspects of the mitochondrial metabolite transporters. During the last 5–10 years, impressive strides have been made in this field. These include (1) the purification and functional reconstitution of most of the known major metabolite transporters; (2) the completion of detailed kinetic analyses and the resulting determination of the kinetic reaction mechanism for about half of the different purified transport proteins; (3) a determination of the primary structures for five of the transporters; (4) a partial elucidation of the membrane topography and portions of the substrate binding sites for several of the transporters; (5) a determination of the genomic sequence of the structural genes encoding three of the transporters and the initial analyses of upstream regulatory elements for two of the carriers; (6) the first uses of molecular probes to elucidate transporter (i.e., the ADP/ATP translocase and the uncoupling protein) regulation in physiological and pathological states.

Despite these impressive advances, it is fair to say that we still do not have a clear idea as to how any of these transporters function at the molecular level to translocate substrate across the inner membrane. Clearly, considerably more structural information will be required in order to elucidate the precise roles of specific structural domains within the translocation mechanism. It is the opinion of this author that two recent advances represent important breakthroughs that will likely enable substantial additional progress to be made during the next several years: (1) the demonstration that wild-type and mutated forms of the ADP/ATP translocase and the phosphate transporters can be expressed in functional form in yeast mitochondria (see Section 15.3.6.3) and (2) the successful overexpression and recovery of large amounts of functional wild type and site-directed mutated forms of mitochondrial transport proteins in *E. coli* (see Section 15.3.6.4).

With respect to the first development, the usefulness of this system has already been demonstrated with two mitochondrial transporters and it is likely that development of the appropriate yeast hosts for the expression of other mitochondrial carriers will be accomplished in the near future. Expression in yeast will permit rigorous analysis of the effects of site-directed mutations, random mutations, as well as both intragenic and extragenic second site revertants on transporter function in intact mitochondria as well as following purification and functional reconstitution. It should be noted that we are in agreement with the assessment of Kramer and Palmieri[13] that the success of the various types of mutagenesis studies will require a rigorous functional analysis of the resulting mutants. Regarding the second development, the *E. coli* expression system provides, for the first time, sufficient quantities of the mitochondrial transporters to enable crystallization trials, with the hope that in the not too distant future, high-resolution three-dimensional structural information will be obtainable via x-ray crystallography. The *E. coli* system also enables the abundant expression of site-directed mutated forms of these transporters.

We would like to note several other promising avenues for future studies. First, at the genomic level, we think that it is important to determine the sequences of the various transporter genes followed by an identification of the 5′ regulatory elements. Specifically, what are the genetic regulatory elements that enable tissue-specific, hormonally- and/or environmentally-sensitive transporter expression? Second, it will be of great interest to use recently developed molecular probes to elucidate the mechanism(s) that gives rise to the documented alterations in transporter function in certain diseases (e.g., diabetes mellitus). Third, we think that characterization of the interaction of mitochondrial transporters with other proteins will likely prove important in understanding transporter function and regulation. Srere's group[237] has already demonstrated the usefulness of the yeast system in this context. Finally, it is important to point out that as alluded to earlier (Section 15.2), there are a number of mitochondrial transporters that have been only minimally characterized, and it is likely that transporters exist that have yet to be discovered, but will nonetheless prove essential to mitochondrial function. Thus, we advocate the continued search for various mitochondrial transport activities and the subsequent extension of these studies to the molecular level. Such information will prove valuable to our understanding of the full spectrum of molecular mechanisms that mitochondria utilize to regulate metabolite flow across the inner membrane.

Acknowledgments. The author wishes to thank June A. Mayor who has been a major contributor to work described from the author's laboratory. Many thanks also go to collaborators Drs. David O. Wood and Glenn L. Wilson who have participated greatly in our molecular biology and diabetes investigations, respectively, and to Talina Boone and Barbara Shartava for carefully proofreading the manuscript. Studies carried out in the author's laboratory, which are cited in this review, were supported by grants from the National Science Foundation (MCB-9219387) and the National Institutes of Health (DK44993 and GM-38785).

REFERENCES

1. Mannella, C. A., Forte, M., and Colombini, M. (1992). Toward the molecular structure of the mitochondrial channel, VDAC. *J. Bioenerg. Biomembr.* **24:**7–19.
2. De Pinto, V., and Palmieri, F. (1992). Transmembrane arrangement of mitochondrial porin or voltage-dependent anion channel (VDAC). *J. Bioenerg. Biomembr.* **24:**21–26.
3. Chappell, J. B. (1968). Systems used for the transport of substrates into mitochondria. *Br. Med. Bull.* **24:**150–157.
4. LaNoue, K. F., and Schoolwerth, A. C. (1979). Metabolite transport in mitochondria. *Annu. Rev. Biochem.* **48:**871–922.
5. Klingenberg, M. (1979). Overview on mitochondrial metabolite transport systems. *Methods Enzymol.* **56:**245–252.
6. Bryla, J. (1980). Inhibitors of mitochondrial anion transport. *Pharmacol. Ther.* **10:**351–397.
7. Meijer, A. J., and Van Dam, K. (1981). Mitochondrial ion transport. In *Membrane Transport* (S. L. Bonting and J. J. H. H. M. de Pont, eds.), Elsevier, Amsterdam, pp. 235–256.
8. LaNoue, K. F., and Schoolwerth, A. C. (1984). Metabolite transport in mammalian mitochondria. In *Bioenergetics* (L. Ernster, ed.), Elsevier, Amsterdam, Volume 9, pp. 221–268.
9. Schoolwerth, A. C., and LaNoue, K. F. (1985). Transport of metabolic substrates in renal mitochondria. *Annu. Rev. Physiol.* **47:** 143–171.
10. Aquila, H., Link, T. A., and Klingenberg, M. (1987). Solute carriers involved in energy transfer of mitochondria form a homologous protein family. *FEBS Lett.* **212:**1–9.
11. Kramer, R., and Palmieri, F. (1989). Molecular aspects of isolated and reconstituted carrier proteins from animal mitochondria. *Biochim. Biophys. Acta* **974:**1–23.
12. Walker, J. E. (1992). The mitochondrial transporter family. *Curr. Opin. Struct. Biol.* **2:**519–526.
13. Kramer, R., and Palmieri, F. (1992). Metabolite carriers in mitochondria. In *Molecular Mechanisms in Bioenergetics* (L. Ernster, ed.), Elsevier, Amsterdam, pp. 359–384.
14. Walker, J. E., and Runswick, M. J. (1993). The mitochondrial transport protein superfamily. *J. Bioenerg. Biomembr.* **25:**435–446.
15. Klingenberg, M. (1989). Molecular aspects of the adenine nucleotide carrier from mitochondria. *Arch. Biochem. Biophys.* **270:** 1–14.
16. Klingenberg, M. (1993). Mitochondrial carrier family: ADP/ATP carrier as a carrier paradigm. In *Molecular Biology and Function of Carrier Proteins* (L. Reuss, J. M. Russell, Jr., and M. L. Jennings, eds.), The Rockefeller University Press, New York, pp. 201–212.
17. Pedersen, P. L., and Wehrle, J. P. (1982). Phosphate transport processes of animal cells. In *Membranes and Transport* (A. N. Martonosi, ed.), New York, Plenum Press, Volume 1, pp. 645–663.
18. Wohlrab, H. (1986). Molecular aspects of inorganic phosphate transport in mitochondria. *Biochim. Biophys. Acta.* **853:**115–134.
19. Kaplan, R. S., and Mayor, J. A. (1993). Structure, function and regulation of the tricarboxylate transport protein from rat liver mitochondria. *J. Bioenerg. Biomembr.* **25:**503–514.
20. Klaus, S., Casteilla, L., Bouillaud, F., and Ricquier, D. (1991). The uncoupling protein UCP: A membraneous mitochondrial ion carrier exclusively expressed in brown adipose tissue. *Int. J. Biochem.* **23:**791–801.

21. Ricquier, D., Casteilla, L., and Bouillaud, F. (1991). Molecular studies of the uncoupling protein. *FASEB J.* **5:**2237–2242.
22. Beavis, A. D. (1992). Properties of the inner membrane anion channel in intact mitochondria. *J. Bioenerg. Biomembr.* **24:**77–90.
23. Kinnally, K. W., Antonenko, Y. N., and Zorov, D. B. (1992). Modulation of inner mitochondrial membrane channel activity. *J. Bioenerg. Biomembr.* **24:**99–110.
24. Diwan, J. J. (1987). Mitochondrial transport of K^+ and Mg^{2+}. *Biochim. Biophys. Acta* **895:**155–165.
25. Gunter, T. E., and Pfeiffer, D. R. (1990). Mechanisms by which mitochondria transport calcium. *Am. J. Physiol.* **258:**C755–C786.
26. Li, X., Hegazy, M. G., Mahdi, F., Jezek, P., Lane, R. D., and Garlid, K. D. (1990). Purification of a reconstitutively active K^+/H^+ antiporter from rat liver mitochondria. *J. Biol. Chem.* **265:**15316–15322.
27. Garlid, K. D., Shariat-Madar, Z., Nath, S., and Jezek, P. (1991). Reconstitution and partial purification of the Na^+-selective Na^+/H^+ antiporter of beef heart mitochondria. *J. Biol. Chem.* **266:**6518–6523.
28. Li, W., Shariat-Madar, Z., Powers, M., Sun, X., Lane, R. D., and Garlid, K. D. (1992). Reconstitution, identification, purification, and immunological characterization of the 110-kDa Na^+/Ca^{2+} antiporter from beef heart mitochondria. *J. Biol. Chem.* **267:** 17983–17989.
29. Paucek, P., Mironova, G., Mahdi, F., Beavis, A. D., Woldegiorgis, G., and Garlid, K. D. (1992). Reconstitution and partial purification of the glibenclamide-sensitive, ATP-dependent K^+ channel from rat liver and beef heart mitochondria. *J. Biol. Chem.* **267:** 26062–26069.
30. Porter, R. K., Scott, J. M., and Brand, M. D. (1992). Choline transport into rat liver mitochondria. Characterization and kinetics of a specific transporter. *J. Biol. Chem.* **267:**14637–14646.
31. Tahiliani, A. G., and Neely, J. R. (1987). A transport system for coenzyme A in isolated rat heart mitochondria. *J. Biol. Chem.* **262:**11607–11610.
32. Toninello, A., Via, L. D., Siliprandi, D., and Garlid, K. D. (1992). Evidence that spermine, spermidine, and putrescine are transported electrophoretically in mitochondria by a specific polyamine uniporter. *J. Biol. Chem.* **267:**18393–18397.
33. Barile, M., Passarella, S., and Quagliariello, E. (1990). Thiamine pyrophosphate uptake into isolated rat liver mitochondria. *Arch. Biochem. Biophys.* **280:**352–357.
34. Kurosawa, K., Hayashi, N., Sato, N., Kamada, T., and Tagawa, K. (1990). Transport of glutathione across the mitochondrial membranes. *Biochem. Biophys. Res. Commun.* **167:**367–372.
35. Meijer, A. J., Van Woerkom, G. M., Wanders, R. J. A., and Lof, C. (1982). Transport of N-acetylglutamate in rat-liver mitochondria. *Eur. J. Biochem.* **124:**325–330.
36. Watkins, L. F., and Lewis, R. A. (1987). The metabolism of deoxyguanosine in mitochondria. Characterization of the uptake process. *Mol. Cell. Biochem.* **77:**71–77.
37. Thomas, A. P., and Halestrap, A. P. (1981). The role of mitochondrial pyruvate transport in the stimulation by glucagon and phenylephrine of gluconeogenesis from L-lactate in isolated rat hepatocytes. *Biochem. J.* **198:**551–564.
38. Rognstad, R. (1983). The role of mitochondrial pyruvate transport in the control of lactate gluconeogenesis. *Int. J. Biochem.* **15:** 1417–1421.
39. Patel, T. B., Barron, L. L., and Olson, M. S. (1984). The stimulation of hepatic gluconeogenesis by acetoacetate precursors. A role for the monocarboxylate translocator. *J. Biol. Chem.* **259:**7525–7531.
40. Martin-Requero, A., Ayuso, M. S., and Parrilla, R. (1986). Rate-limiting steps for hepatic gluconeogenesis. Mechanism of oxamate inhibition of mitochondrial pyruvate metabolism. *J. Biol. Chem.* **261:**13973–13978.
41. Gellerich, F. N., Bohnensack, R., and Kunz, W. (1983). Control of mitochondrial respiration. The contribution of the adenine nucleotide translocator depends on the ATP- and ADP-consuming enzymes. *Biochim. Biophys. Acta* **722:**381–391.
42. Westerhoff, H. V., Plomp, P. J. A. M., Groen, A. K., Wanders, R. J. A., Bode, J. A., and Van Dam, K. (1987). On the origin of the limited control of mitochondrial respiration by the adenine nucleotide translocator. *Arch. Biochem. Biophys.* **257:**154–169.
43. Bohnensack, R., Gellerich, F. N., Schild, L., and Kunz, W. (1990). The function of the adenine nucleotide translocator in the control of oxidative phosphorylation. *Biochim. Biophys. Acta* **1018:**182–184.
44. Simpson, D. P. (1975). Glutamine transport in dog kidney mitochondria. A new control mechanism in acidosis. *Med. Clin. North Am.* **59:**555–567.
45. Lenzen, C., Soboll, S., Sies, H., and Haussinger, D. (1987). pH control of hepatic glutamine degradation. Role of transport. *Eur. J. Biochem.* **166:**483–488.
46. Hutson, S. M., and Hall, T. R. (1993). Identification of the mitochondrial branched chain aminotransferase as a branched α-keto acid transport protein. *J. Biol. Chem.* **268:**3084–3091.
47. Nosek, M. T., and Aprille, J. R. (1992). ATP-Mg/P_i carrier activity in rat liver mitochondria. *Arch. Biochem. Biophys.* **296:**691–697.
48. Hagen, T., Joyal, J. L., Henke, W., and Aprille, J. R. (1993). Net adenine nucleotide transport in rat kidney mitochondria. *Arch. Biochem. Biophys.* **303:**195–207.
49. Claeys, D., and Azzi, A. (1989). Tricarboxylate carrier of bovine liver mitochondria. Purification and reconstitution. *J. Biol. Chem.* **264:**14627–14630.
50. Glerum, D. M., Claeys, D., Mertens, W., and Azzi, A. (1990). The tricarboxylate carrier from rat liver mitochondria. Purification, reconstitution and kinetic characterization. *Eur. J. Biochem.* **194:** 681–684.
51. Kaplan, R. S., Mayor, J. A., Johnston, N., and Oliveira, D. L. (1990). Purification and characterization of the reconstitutively active tricarboxylate transporter from rat liver mitochondria. *J. Biol. Chem.* **265:**13379–13385.
52. Bisaccia, F., De Palma, A., and Palmieri, F. (1989). Identification and purification of the tricarboxylate carrier from rat liver mitochondria. *Biochim. Biophys. Acta* **977:**171–176.
53. Azzi, A., Glerum, M., Koller, R., Mertens, W., and Spycher, S. (1993). The mitochondrial tricarboxylate carrier. *J. Bioenerg. Biomembr.* **25:**515–524.
54. Kaplan, R. S., Mayor, J. A., and Wood, D. O. (1993). The mitochondrial tricarboxylate transport protein. cDNA cloning, primary structure, and comparison with other mitochondrial transport proteins. *J. Biol. Chem.* **268:**13682–13690.
55. Pedersen, P. L., Greenawalt, J. W., Reynafarje, B., Hullihen, J., Decker, G. L., Soper, J. W., and Bustamente, E. (1978). Preparation and characterization of mitochondria and submitochondrial particles of rat liver and liver-derived tissues. *Methods Cell Biol.* **20:** 411–481.
56. Aquila, H., Misra, D., Eulitz, M., and Klingenberg, M. (1982). Complete amino acid sequence of the ADP/ATP carrier from beef heart mitochondria. *Hoppe-Seyler's Z. Physiol. Chem.* **363:**345–349.
57. Kolbe, H. V. J., and Wohlrab, H. (1985). Sequence of the N-terminal formic acid fragment and location of the N-ethylmaleimide-binding site of the phosphate transport protein from beef heart mitochondria. *J. Biol. Chem.* **260:**15899–15906.
58. Walker, J. E., Fearnley, I. M., and Blows, R. A. (1986). A rapid solid-phase protein microsequencer. *Biochem. J.* **237:**73–84.

59. Runswick, M. J., Powell, S. J., Nyren, P., and Walker, J. E. (1987). Sequence of the bovine mitochondrial phosphate carrier protein: Structural relationship to ADP/ATP translocase and the brown fat mitochondria uncoupling protein. *EMBO J.* **6:**1367–1373.
60. Ferriera, G. C., Pratt, R. D., and Pedersen, P. L. (1989). Energy linked anion transport. Cloning, sequencing, and characterization of a full length cDNA encoding the rat liver mitochondrial proton/phosphate symporter. *J. Biol. Chem.* **264:**15628–15633.
61. Runswick, M. J., Walker, J. E., Bisaccia, F., Iacobazzi, V., and Palmieri, F. (1990). Sequence of the bovine 2-oxoglutarate/malate carrier protein: Structural relationship to other mitochondrial transport proteins. *Biochemistry* **29:**11033–11040.
62. Aquila, H., Link, T. A., and Klingenberg, M. (1985). The uncoupling protein from brown fat mitochondria is related to the mitochondrial ADP/ATP carrier. Analysis of sequence homologies and of folding of the protein in the membrane. *EMBO J.* **4:**2369–2376.
63. Nelson, D. R., Lawson, J. E., Klingenberg, M., and Douglas, M. G. (1993). Site-directed mutagenesis of the yeast mitochondrial ADP/ATP translocator. Six arginines and one lysine are essential. *J. Mol. Biol.* **230:**1159–1170.
64. Shinohara, Y., Kamida, M., Yamazaki, N., and Terada, H. (1993). Isolation and characterization of cDNA clones and a genomic clone encoding rat mitochondrial adenine nucleotide translocator. *Biochim. Biophys. Acta* **1152:**192–196.
65. Neckelmann, N., Li, K., Wade, R. P., Shuster, R., and Wallace, D. C. (1987). cDNA sequence of a human skeletal muscle ADP/ATP translocator: Lack of a leader peptide, divergence from a fibroblast translocator cDNA, and coevolution with mitochondrial DNA genes. *Proc. Natl. Acad. Sci. USA* **84:**7580–7584.
66. Powell, S. J., Medd, S. M., Runswick, M. J., and Walker, J. E. (1989). Two bovine genes for mitochondrial ADP/ATP translocase expressed differently in various tissues. *Biochemistry* **28:**866–873.
67. Cozens, A. L., Runswick, M. J., and Walker, J. E. (1989). DNA sequences of two expressed nuclear genes for human mitochondrial ADP/ATP translocase. *J. Mol. Biol.* **206:**261–280.
68. Bouillaud, F., Weissenbach, J., and Ricquier, D. (1986). Complete cDNA-derived amino acid sequence of rat brown fat uncoupling protein. *J. Biol. Chem.* **261:**1487–1490.
69. Ridley, R. G., Patel, H. V., Gerber, G. E., Morton, R. C., and Freeman, K. B. (1986). Complete nucleotide and derived amino acid sequence of cDNA encoding the mitochondrial uncoupling protein of fat brown adipose tissue: Lack of a mitochondrial targeting presequence. *Nucleic Acids Res.* **14:**4025–4035.
70. Saraste, M., and Walker, J. E. (1982). Internal sequence repeats and the path of polypeptide in mitochondrial ADP/ATP translocase. *FEBS Lett.* **144:**250–254.
71. Colleaux, L., Richard, G.-F., Thierry, A., and Dujon, B. (1992). Sequence of a segment of yeast chromosome XI identifies a new mitochondrial carrier, a new member of the G protein family, and a protein with the PAAKK motif of the H1 histones. *Yeast* **8:**325–336.
72. Weisenberger, G., Link, T. A., Von Ahsen, U., Waldherr, M., and Schweyen, R. J. (1991). MRS3 and MRS4, two suppressors of mtRNA splicing defects in yeast, are new members of the mitochondrial carrier family. *J. Mol. Biol.* **217:**23–37.
73. Waterston, R., Martin, C., Craxton, M., Huynh, C., Coulson, A., Hillier, L., Durbin, R., Green, P., Skownkeen, R., Halloran, N., Metzstein, M., Hawkins, T., Wilson, R., Berks, M., Du, Z., Thomas, K., Thierry-Mieg, J., and Sulston, J. (1992). A survey of expressed genes in *Caenorhabditis elegans. Nature Genet.* **1:**114–123.
74. Zarrilli, R., Oates, E. L., McBride, O. W., Lerman, M. I., Chan, J. Y., Santisteban, P., Ursini, M. V., Notkins, A. L., and Kohn, L. D. (1989). Sequence and chromosomal assignment of a novel cDNA identified by immunoscreening of a thyroid expression library: Similarity to a family of mitochondrial solute carrier proteins. *Mol. Endocrinol.* **3:**1498–1508.
75. Fiermonte, G., Runswick, M. J., Walker, J. E., and Palmieri, F. (1992). Sequence and pattern of expression of a bovine homologue of a human mitochondrial transport protein associated with Grave's disease. *DNA Sequence* **3:**71–78.
76. Williams, K. R., and Herrick, G. (1991). Expression of the gene encoded by a family of macronuclear chromosomes generated by alternative DNA processing in *Oxytricha fallax. Nucleic Acids Res.* **19:**4717–4724.
77. Capobianco, L., Bisaccia, F., Michel, A., Sluse, F. E., and Palmieri, F. (1995). The N- and C-termini of the tricarboxylate carrier are exposed to the cytoplasmic side of the inner mitochondrial membrane. *FEBS Lett.* **357:**297–300.
77a. Ferreira, G. C., Pratt, R. D., and Pedersen, P. L. (1990). Mitochondrial proton/phosphate transporter. An antibody directed against the COOH terminus and proteolytic cleavage experiments provides new insights about its membrane topology. *J. Biol. Chem.* **265:**21202–21206.
78. Capobianco, L., Brandolin, G., and Palmieri, F. (1991). Transmembrane topography of the mitochondrial phosphate carrier explored by peptide-specific antibodies and enzymatic digestion. *Biochemistry* **30:**4963–4969.
79. Palmieri, F., Bisaccia, F., Capobianco, L., Dolce, V., Iacobazzi, V., Indiveri, C., and Zara, V. (1992). Structural and functional properties of two mitochondrial transport proteins: The phosphate carrier and the oxoglutarate carrier. In *Molecular Mechanisms of Transport* (E. Quagliariello and F. Palmieri, eds.), Elsevier, Amsterdam, pp. 151–158.
80. Brandolin, G., Boulay, F., Dalbon, P., and Vignais, P. V. (1989). Orientation of the N-terminal region of the membrane-bound ADP/ATP carrier protein explored by antipeptide antibodies and an arginine-specific endoprotease. Evidence that the accessibility of the N-terminal residues depends on the conformational state of the carrier. *Biochemistry* **28:**1093–1100.
81. Marty, I., Brandolin, G., Gagnon, J., Brasseur, R., and Vignais, P. V. (1992). Topography of the membrane-bound ADP/ATP carrier assessed by enzymatic proteolysis. *Biochemistry* **31:**4058–4065.
81a. Bisaccia, F., Capobianco, L., Brandolin, G., and Palmieri, F. (1994). Transmembrane topography of the mitochondrial oxoglutarate carrier assessed by peptide-specific antibodies and enzymatic cleavage. *Biochemistry* **33:**3705–3713.
82. Eckerskorn, C., and Klingenberg, M. (1987). In the uncoupling protein from brown adipose tissue the C-terminus protrudes to the c-side of the membrane as shown by tryptic cleavage. *FEBS Lett.* **226:**166–170.
83. Klingenberg, M., and Appel, M. (1989). The uncoupling protein dimer can form a disulfide cross-link between the mobile C-terminal SH groups. *Eur. J. Biochem.* **180:**123–131.
84. Miroux, B., Frossard, V., Raimbault, S., Ricquier, D., and Bouillaud, F. (1993). The topology of the brown adipose tissue mitochondrial uncoupling protein determined with antibodies against its antigenic sites revealed by a library of fusion proteins. *EMBO J.* **12:**3739–3745.
85. Miroux, B., Casteilla, L., Klaus, S., Raimbault, S., Grandin, S., Clement, J. M., Ricquier, D., and Bouillaud, F. (1992). Antibodies selected from whole antiserum by fusion proteins as tools for the study of the topology of mitochondrial membrane proteins. Evidence that the N-terminal extremity of the sixth α-helix of the uncoupling protein is facing the matrix. *J. Biol. Chem.* **267:**13603–13609.
86. Maloney, P. C. (1990). A consensus structure for membrane transport. *Res. Microbiol.* **141:**374–383.
87. Griffith, J. K., Baker, M. E., Rouch, D. A., Page, M. G. P., Skurray, R. A., Paulsen, I. T., Chater, K. F., Baldwin, S. A., and Henderson,

P. J. F. (1992). Membrane transport proteins: Implications of sequence comparisons. *Curr. Opin. Cell Biol.* **4:**684–695.

88. Lin, C.-S., and Klingenberg, M. (1982). Characteristics of the isolated purine nucleotide binding protein from brown fat mitochondria. *Biochemistry* **21:**2950–2956.
89. Lin, C. S., Hackenberg, H., and Klingenberg, M. (1980). The uncoupling protein from brown adipose tissue mitochondria is a dimer. A hydrodynamic study. *FEBS Lett.* **113:**304–306.
90. Hackenberg, H., and Klingenberg, M. (1980). Molecular weight and hydrodynamic parameters of the adenosine 5′-diphosphate-adenosine 5′-triphosphate carrier in Triton X-100. *Biochemistry* **19:**548–555.
91. Klingenberg, M., Hackenberg, H., Eisenreich, G., and Mayer, I. (1979). The interaction of detergents with the ADP,ATP carrier from mitochondria. In *Function and Molecular Aspects of Biomembrane Transport* (E. Quagliariello, F. Palmieri, S. Papa, and M. Klingenberg, eds.), Elsevier, Amsterdam, pp. 291–303.
92. Riccio, P., Aquila, H., and Klingenberg, M. (1975). Purification of the carboxy-atractylate binding protein from mitochondria. *FEBS Lett.* **56:**133–138.
93. Klingenberg, M., Riccio, P., and Aquila, H. (1978). Isolation of the ADP,ATP carrier as the carboxyatractylate–protein complex from mitochondria. *Biochim. Biophys. Acta* **503:**193–210.
94. Dierks, T., Riemer, E., and Kramer, R. (1988). Reaction mechanism of the reconstituted aspartate/glutamate carrier from bovine heart mitochondria. *Biochim. Biophys. Acta* **943:**231–244.
95. Indiveri, C., Prezioso, G., Dierks, T., Kramer, R., and Palmieri, F. (1993). Kinetic characterization of the reconstituted dicarboxylate carrier from mitochondria: A four-binding-site sequential transport system. *Biochim. Biophys. Acta* **1143:**310–318.
96. Bisaccia, F., De Palma, A., Dierks, T., Kramer, R., and Palmieri, F. (1993). Reaction mechanism of the reconstituted tricarboxylate carrier from rat liver mitochondria. *Biochim. Biophys. Acta* **1142:** 139–145.
97. Palmieri, F., Bisaccia, F., Capobianco, L., Iacobazzi, V., Indiveri, C., and Zara, V. (1990). Structural and functional properties of mitochondrial anion carriers. *Biochim. Biophys. Acta* **1018:**147–150.
98. Sluse, F. E., Evens, A., Dierks, T., Duychaerts, C., Sluse-Goffart, C. M., and Kramer, R. (1991). Kinetic study of the aspartate/glutamate carrier in intact rat heart mitochondria and comparison with a reconstituted system. *Biochim. Biophys. Acta* **1058:**329–338.
99. Sluse, F. E., Sluse-Goffart, C. M., and Duyckaerts, C. (1989). Kinetic mechanisms of the adenylic and the oxoglutaric carriers: A comparison. In *Anion Carriers of Mitochondrial Membranes* (A. Azzi, K. A. Nalecz, M. J. Nalecz, and L. Wojtczak, eds.), Springer-Verlag, Berlin, pp. 183–195.
100. Indiveri, C., Dierks, T., Kramer, R., and Palmieri, F. (1991). Reaction mechanism of the reconstituted oxoglutarate carrier from bovine heart mitochondria. *Eur. J. Biochem.* **198:**339–347.
101. Klingenberg, M. (1992). Structure-function of the ADP/ATP carrier. *Biochem. Soc. Trans.* **20:**547–550.
102. Boulay, F., and Vignais, P. V. (1984). Localization of the N-ethylmaleimide reactive cysteine in the beef heart mitochondrial ADP/ATP carrier protein. *Biochemistry* **23:**4807–4812.
103. Kaplan, R. S., and Pedersen, P. L. (1985). Isolation and reconstitution of the n-butylmalonate-sensitive dicarboxylate transporter from rat liver mitochondria. *J. Biol. Chem.* **260:**10293–10298.
104. Kaplan, R. S., Pratt, R. D., and Pedersen, P. L. (1986). Purification and characterization of the reconstitutively active phosphate transporter from rat liver mitochondria. *J. Biol. Chem.* **261:**12767–12773.
105. Zara, V., and Palmieri, F. (1988). Inhibition and labelling of the mitochondrial 2-oxoglutarate carrier by eosin-5-maleimide. *FEBS Lett.* **236:**493–496.
106. Jezek, P., and Drahota, Z. (1989). Sulfhydryl groups of the uncoupling protein of brown adipose tissue mitochondria. Distinction between sulfhydryl groups of the H^+ channel and the nucleotide binding site. *Eur. J. Biochem.* **183:**89–95.
107. Hutson, S. M., Roten, S., and Kaplan, R. S. (1990). Solubilization and functional reconstitution of the branched-chain α-keto acid transporter from rat heart mitochondria. *Proc. Natl. Acad. Sci. USA* **87:**1028–1031.
108. Nalecz, K. A., Muller, M., Zambrowicz, E. B., Wojtczak, L., and Azzi, A. (1990). Significance and redox state of SH groups in pyruvate carrier isolated from bovine heart mitochondria. *Biochim. Biophys. Acta.* **1016:**272–279.
109. Stappen, R., Dierks, T., Broer, A., and Kramer, R. (1992). Probing the active site of the reconstituted aspartate/glutamate carrier from mitochondria. Structure/function relationship involving one lysine and two cysteine residues. *Eur. J. Biochem.* **210:**269–277.
110. Indiveri, C., Tonazzi, A., Dierks, T., Kramer, R., and Palmieri, F. (1992). The mitochondrial carnitine carrier: Characterization of SH-groups relevant for its transport function. *Biochim. Biophys. Acta* **1140:**53–58.
111. Bogner, W., Aquila, H., and Klingenberg, M. (1986). The transmembrane arrangement of the ADP/ATP carrier as elucidated by the lysine reagent pyridoxal 5-phosphate. *Eur. J. Biochem.* **161:** 611–620.
112. Dierks, T., Stappen, R., Salentin, A., and Kramer, R. (1992). Probing the active site of the reconstituted aspartate/glutamate carrier from bovine heart mitochondria: Carbodiimide-catalyzed acylation of a functional lysine residue. *Biochim. Biophys. Acta* **1103:**13–24.
113. Genchi, G., Petrone, G., De Palma, A., Cambria, A., and Palmieri, F. (1988). Interaction of phenylisothiocyanates with the mitochondrial phosphate carrier. I. Covalent modification and inhibition of phosphate transport. *Biochim. Biophys. Acta* **936:**413–420.
114. Stipani, I., Zara, V., Zaki, L., Prezioso, G., and Palmieri, F. (1986). Inhibition of the mitochondrial tricarboxylate carrier by arginine-specific reagents. *FEBS Lett.* **205:**282–286.
115. Plapp, B. V. (1982). Application of affinity labeling for studying structure and function of enzymes. *Methods Enzymol.* **87:**469–499.
116. Gremse, D. A., Dean, B., and Kaplan, R. S. (1995). Effect of pyridoxal 5′-phosphate on the function of the purified mitochondrial tricarboxylate transport protein. *Arch. Biochem. Biophys.* **316:**215–219.
117. Block, M. R., Lauquin, G. J. M., and Vignais, P. V. (1981). Chemical modifications of atractyloside and bongkrekic acid binding sites of the mitochondrial adenine nucleotide carrier. Are there distinct binding sites? *Biochemistry* **20:**2692–2699.
118. Vignais, P. V., and Lunardi, J. (1985). Chemical probes of the mitochondrial ATP synthesis and translocation. *Annu. Rev. Biochem.* **54:**977–1014.

118a. Indiveri, C., Tonazzi, A., Giangregorio, N., and Palmieri, F. (1995). Probing the active site of the reconstituted carnitine carrier from rat liver mitochondria with sulfhydryl reagents: A cysteine residue is localized in or near the substrate binding site. *Eur. J. Biochem.* **228:**271–278.

119. Boulay, F., Lauquin, G. J. M., Tsugita, A., and Vignais, P. V. (1983). Photolabeling approach to the study of the topography of the atractyloside binding site in mitochondrial adenosine 5′-diphosphate/adenosine 5′-triphosphate carrier protein. *Biochemistry* **22:**477–484.
120. Dalbon, P., Brandolin, G., Boulay, F., Hoppe, J., and Vignais, P. V. (1988). Mapping of the nucleotide-binding sites in the ADP/ATP carrier of beef heart mitochondria by photolabeling with 2-azido[α-^{32}P]adenosine diphosphate. *Biochemistry* **27:**5141–5149.
121. Vignais, P. V., Brandolin, G., Boulay, F., Dalbon, P., Block, M. R., and Gauche, I. (1989). Recent developments in the study of the conformational states and the nucleotide binding sites of the

ADP/ATP carrier. In *Anion Carriers of Mitochondrial Membranes* (A. Azzi, K. A. Nalecz, M. J. Nalecz, and L. Wojtczak, eds.), Springer-Verlag, Berlin, pp. 133–146.

122. Mayinger, P., Winkler, E., and Klingenberg, M. (1989). The ADP/ATP carrier from yeast (AAC-2) is uniquely suited for the assignment of the binding center by photoaffinity labeling. *FEBS Lett.* **244:**421–426.
123. Winkler, E., and Klingenberg, M. (1992). Photoaffinity labeling of the nucleotide-binding site of the uncoupling protein from hamster brown adipose tissue. *Eur. J. Biochem.* **203:**295–304.
124. Mayinger, P., and Klingenberg, M. (1992). Labeling of two different regions of the nucleotide binding site of the uncoupling protein from brown adipose tissue mitochondria with two ATP analogs. *Biochemistry* **31:**10536–10543.
125. Knirsch, M., Gawaz, M. P., and Klingenberg, M. (1989). The isolation and reconstitution of the ADP/ATP carrier from wild-type *Saccharomyces cerevisiae.* Identification of primarily one type (AAC-2). *FEBS Lett.* **244:**427–432.
126. Brandolin, G., Le Saux, A., Trezeguet, V., Vignais, P. V., and Lauquin, G. J.-M. (1993). Biochemical characterization of the isolated Anc2 adenine nucleotide carrier from *Saccharomyces cerevisiae* mitochondria. *Biochem. Biophys. Res. Commun.* **192:**143–150.
127. Guerin, B., Bukusoglu, C., Rakotomanana, F., and Wohlrab, H. (1990). Mitochondrial phosphate transport. N-ethylmaleimide insensitivity correlates with absence of beef heart-like Cys^{42} from the *Saccharomyces cerevisiae* phosphate transport protein. *J. Biol. Chem.* **265:**19736–19741.
128. Nalecz, M. J., Nalecz, K. A., and Azzi, A. (1991). Purification and functional characterization of the pyruvate (monocarboxylate) carrier from baker's yeast mitochondria (*Saccharomyces cerevisiae*). *Biochim. Biophys. Acta* **1079:**87–95.
129. Persson, L.-O., and Srere, P. A. (1992). Purification of the mitochondrial citrate transporter in yeast. *Biochem. Biophys. Res. Commun.* **183:**70–76.
130. O'Malley, K., Pratt, P., Robertson, J., Lilly, M., and Douglas, M. G. (1982). Selection of the nuclear gene for the mitochondrial adenine nucleotide translocator by genetic complementation of the op_1 mutation in yeast. *J. Biol. Chem.* **257:**2097–2103.
131. Adrian, G. S., McCammon, M. T., Montgomery, D. L., and Douglas, M. G. (1986). Sequences required for delivery and localization of the ADP/ATP translocator to the mitochondrial inner membrane. *Mol. Cell. Biol.* **6:**626–634.
132. Lawson, J. E., and Douglas, M. G. (1988). Separate genes encode functionally equivalent ADP/ATP carrier proteins in *Saccharomyces cerevisiae.* Isolation and analysis of AAC2. *J. Biol. Chem.* **263:**14812–14818.
133. Kolarov, J., Kolarova, N., and Nelson, N. (1990). A third ADP/ATP translocator gene in yeast. *J. Biol. Chem.* **265:**12711–12716.
134. Phelps, A., Schobert, C. T., and Wohlrab, H. (1991). Cloning and characterization of the mitochondrial phosphate transport protein gene from the yeast *Saccharomyces cerevisiae. Biochemistry* **30:** 248–252.

134a. Holmstrom, K., Brandt, T., and Kallesoe, T. (1994). The sequence of a 32420 bp segment located on the right arm of chromosome II from *Saccharomyces cerevisiae. Yeast* **10:**S47–S62.

134b. Kaplan, R. S., Mayor, J. A., Gremse, D. A., and Wood, D. O. (1995). High level expression and characterization of the mitochondrial citrate transport protein from the yeast *Saccharomyces cerevisiae. J. Biol. Chem.* **270:**4108–4114.

135. Lawson, J. E., Gawaz, M., Klingenberg, M., and Douglas, M. G. (1990). Structure–function studies of adenine nucleotide transport in mitochondria. I. Construction and genetic analysis of yeast mutants encoding the ADP/ATP carrier protein of mitochondria. *J. Biol. Chem.* **265:**14195–14201.

135a. Klingenberg, M., and Nelson, D. R. (1994). Structure-function relationships of the ADP/ATP carrier. *Biochim. Biophys. Acta* **1187:** 241–244.

136. Klingenberg, M., Gawaz, M., Douglas, M. G., and Lawson, J. E. (1992). Mutagenized ADP/ATP carrier from *Saccharomyces.* In *Molecular Mechanisms of Transport* (E. Quagliariello and F. Palmieri, eds.), Elsevier, Amsterdam, pp. 187–195.
137. Nelson, D. R., and Douglas, M. G. (1993). Function-based mapping of the yeast mitochondrial ADP/ATP translocator by selection for second site revertants. *J. Mol. Biol.* **230:**1171–1182.
138. Phelps, A., and Wohlrab, H. (1991). Mitochondrial phosphate transport. The *Saccharomyces cerevisiae* (threonine 43 to cysteine) mutant protein explicitly identifies transport with genomic sequence. *J. Biol. Chem.* **266:**19882–19885.
139. Phelps, A., and Wohlrab, H. (1993). Cys28 of the mitochondrial phosphate transport protein is responsible for the inhibition of transport by oxygen. *FASEB J.* **7:**A1107.
140. Casteilla, L., Blondel, O., Klaus, S., Raimbault, S., Diolez, P., Moreau, F., Bouillaud, F., and Ricquier, D. (1990). Stable expression of functional mitochondrial uncoupling protein in Chinese hamster ovary cells. *Proc. Natl. Acad. Sci. USA* **87:**5124–5128.
141. Ferreira, G. C., and Pedersen, P. L. (1992). Overexpression of higher eukaryotic membrane proteins in bacteria. Novel insights obtained with the liver mitochondrial proton/phosphate symporter. *J. Biol. Chem.* **267:**5460–5466.
142. Murdza-Inglis, D. L., Patel, H. V., Freeman, K. B., Jezek, P., Orosz, D. E., and Garlid, K. D. (1991). Functional reconstitution of rat uncoupling protein following its high level expression in yeast. *J. Biol. Chem.* **260:**11871–11875.
143. Fiermonte, G., Walker, J. E., and Palmieri, F. (1993). Abundant bacterial expression and reconstitution of an intrinsic membrane-transport protein from bovine mitochondria. *Biochem. J.* **294:**293–299.

143a. Xu, Y., Mayor, J. A., Gremse, D., Wood, D. O., and Kaplan, R. S. (1995). High-yield bacterial expression, purification, and functional reconstitution of the tricarboxylate transport protein from rat liver mitochondria. *Biochem. Biophys. Res. Commun.* **207:**783–789.

143b. Wohlrab, H., and Briggs, C. (1994). Yeast mitochondrial phosphate transport protein expressed in *Escherichia coli.* Site-directed mutations at threonine-43 and at a similar location in the second tandem repeat (isoleucine-141). *Biochemistry* **33:**9371–9375.

143c. Xu, Y., Gremse, D. A., Mayor, J. A., and Kaplan, R. S. (1996). Functional consequences of mutating two cysteine residues in the yeast mitochondrial citrate transport protein. *Biophys. J.,* **70:**A414.

144. Schultheiss, H. P., and Klingenberg, M. (1984). Immunochemical characterization of the adenine nucleotide translocator. Organ specificity and conformation specificity. *Eur. J. Biochem.* **143:**599–605.
145. Schultheiss, H.-P., and Klingenberg, M. (1985). Immunoelectrophoretic characterization of the ADP/ATP carrier from heart, kidney, and liver. *Arch. Biochem. Biophys.* **239:**273–279.
146. Rasmussen, U. B., and Wohlrab, H. (1986). Conserved structural domains among species and tissue-specific differences in the mitochondrial phosphate-transport protein and the ADP/ATP carrier. *Biochim. Biophys. Acta* **852:**306–314.
147. Zara, V., De Benedittis, R., Ragan, C. I., and Palmieri, F. (1990). Immunological characterization of the mitochondrial 2-oxoglutarate carrier from liver and heart: Organ specificity. *FEBS Lett.* **263:**295–298.
148. Li, K., Warner, C. K., Hodge, J. A., Minoshima, S., Kudoh, J., Fukuyama, R., Maekawa, M., Shimizu, Y., Shimizu, N., and Wallace, D. C. (1989). A human muscle adenine nucleotide translocator gene has four exons, is located on chromosome 4, and is differentially expressed. *J. Biol. Chem.* **264:**13998–14004.

149. Houldsworth, J., and Attardi, G. (1988). Two distinct genes for ADP/ATP translocase are expressed at the mRNA level in adult human liver. *Proc. Natl. Acad. Sci. USA* **85:**377–381.
150. Battini, R., Ferrari, S., Kaczmarek, L., Calabretta, B., Chen, S.-T., and Baserga, R. (1987). Molecular cloning of a cDNA for a human ADP/ATP carrier which is growth-regulated. *J. Biol. Chem.* **262:** 4355–4359.
151. Ku, D.-H., Kagan, J., Chen, S.-T., Chang, C.-D., Baserga, R., and Wurzel, J. (1990). The human fibroblast adenine nucleotide translocator gene. Molecular cloning and sequence. *J. Biol. Chem.* **265:**16060–16063.
152. Baker, A., and Leaver, C. J. (1985). Isolation and sequence analysis of a cDNA encoding the ATP/ADP translocator of *Zea mays L. Nucleic Acids Res.* **13:**5857–5867.
153. Bathgate, B., Baker, A., and Leaver, C. J. (1989). Two genes encode the adenine nucleotide translocator of maize mitochondria. Isolation, characterisation and expression of the structural genes. *Eur. J. Biochem.* **183:**303–310.
154. Jacobsson, A., Stadler, U., Glotzer, M. A., and Kozak, L. P. (1985). Mitochondrial uncoupling protein from mouse brown fat. Molecular cloning, genetic mapping, and mRNA expression. *J. Biol. Chem.* **260:**16250–16254.
155. Bouillaud, F., Ricquier, D., Thibault, J., and Weissenback, J. (1985). Molecular approach to thermogenesis in brown adipose tissue: cDNA cloning of the mitochondrial uncoupling protein. *Proc. Natl. Acad. Sci. USA* **82:**445–448.
156. Cassard, A.-M., Bouillaud, F., Mattei, M.-G., Hentz, E., Raimbault, S., Thomas, M., and Ricquier, D. (1990). Human uncoupling protein gene: Structure, comparison with rat gene, and assignment to the long arm of chromosome 4. *J. Cell. Biochem.* **43:**255–264.
157. Palmieri, F., Bisaccia, F., Iacobazzi, V., Indiveri, C., and Zara, V. (1992). Mitochondrial substrate carriers. *Biochim. Biophys. Acta* **1101:**223–227.
158. Iacobazzi, V., Palmieri, F., Runswick, M. J., and Walker, J. E. (1992). Sequences of the human and bovine genes for the mitochondrial 2-oxoglutarate carrier. *DNA Sequence* **3:**79–88.
159. Bouillaud, F., Raimbault, S., and Ricquier, D. (1988). The gene for rat uncoupling protein: Complete sequence, structure of primary transcript and evolutionary relationship between exons. *Biochem. Biophys. Res. Commun.* **157:**783–792.
160. Kozak, L. P., Britton, J. H., Kozak, U. C., and Wells, J. M. (1988). The mitochondrial uncoupling protein gene. Correlation of exon structure to transmembrane domains. *J. Biol. Chem.* **263:**12274–12277.
161. Chodosh, L. A., Baldwin, A. S., Carthew, R. W., and Sharp, P. A. (1988). Human CCAAT-binding proteins have heterologous subunits. *Cell* **53:**11–24.
162. Rossouw, C. M. S., Vergeer, W. P., du Plooy, S. J., Bernard, M. P., Ramirez, F., and de Wet, W. J. (1987). DNA sequences in the first intron of the human pro-α1(I) collagen gene enhance transcription. *J. Biol. Chem.* **262:**15151–15157.
163. Li, K., Hodge, J. A., and Wallace, D. C. (1990). OXBOX, a positive transcriptional element of the heart-skeletal muscle ADP/ATP translocator gene. *J. Biol. Chem.* **265:**20585–20588.
164. Chung, A. B., Stepien, G., Haraguchi, Y., Li, K., and Wallace, D. C. (1992). Transcriptional control of nuclear genes for the mitochondrial muscle ADP/ATP translocator and the ATP synthase β subunit. Multiple factors interact with the OXBOX/REBOX promoter sequences. *J. Biol. Chem.* **267:**21154–21161.
165. Cassard-Doulcier, A.-M., Gelly, C., Fox, N., Schrementi, J., Raimbault, S., Klaus, S., Forest, C., Bouillaud, F., and Ricquier, D. (1993). Tissue-specific and β-adrenergic regulation of the mitochondrial uncoupling protein gene: Control by cis-acting elements in the 5′-flanking region. *Mol. Endocrinol.* **7:**497–506.
165a. Cassard-Doulcier, A.-M., Larose, M., Matamala, J. C., Champigny, O., Bouillaud, F., and Ricquier, D. (1994). *In vitro* interactions between nuclear proteins and uncoupling protein gene promoter reveal several putative transactivating factors including Ets1, retinoid X receptor, thyroid hormone receptor, and a CACCC box-binding protein. *J. Biol. Chem.* **269:**24335–24342.
166. Lunardi, J., Hurko, O., Engel, W. K., and Attardi, G. (1992). The multiple ADP/ATP translocase genes are differentially expressed during human muscle development. *J. Biol. Chem.* **267:**15267–15270.
167. Lunardi, J., and Attardi, G. (1991). Differential regulation of expression of the multiple ADP/ATP translocase genes in human cells. *J. Biol. Chem.* **266:**16534–16540.
168. Ricquier, D., Bouillaud, F., Toumelin, P., Mory, G., Bazin, R., Arch, J., and Penicaud, L. (1986). Expression of uncoupling protein mRNA in thermogenic or weakly thermogenic brown adipose tissue. Evidence for a rapid β-adrenoreceptor-mediated and transcriptionally regulated step during activation of thermogenesis. *J. Biol. Chem.* **261:**13905–13910.
169. Ricquier, D., and Bouillaud, F. (1986). The brown adipose tissue mitochondrial uncoupling protein. In *Brown Adipose Tissue* (P. Trayhurn and D. G. Nicholls, eds.), Edward Arnold, Baltimore, pp. 86–104.
170. Champigny, O., and Ricquier, D. (1990). Effects of fasting and refeeding on the level of uncoupling protein mRNA in rat brown adipose tissue: Evidence for diet-induced and cold-induced responses. *J. Nutr.* **120:**1730–1736.
171. Bianco, A. C., Sheng, X., and Silva, J. E. (1988). Triiodothyronine amplifies norepinephrine stimulation of uncoupling protein gene transcription by a mechanism not requiring protein synthesis. *J. Biol. Chem.* **263:**18168–18175.
172. Soboll, S. (1993). Thyroid hormone action on mitochondrial energy transfer. *Biochim. Biophys. Acta* **1144:**1–16.
173. Nelson, B. D. (1990). Thyroid hormone regulation of mitochondrial function. Comments on the mechanism of signal transduction. *Biochim. Biophys. Acta* **1018:**275–277.
174. Soboll, S. (1993). Long-term and short-term changes in mitochondrial parameters by thyroid hormones. *Biochem. Soc. Trans.* **21:**799–803.
175. Paradies, G., and Ruggiero, F. M. (1990). Stimulation of phosphate transport in rat-liver mitochondria by thyroid hormones. *Biochim. Biophys. Acta* **1019:**133–136.
176. Paradies, G., and Ruggiero, F. M. (1988). Effect of hyperthyroidism on the transport of pyruvate in rat-heart mitochondria. *Biochim. Biophys. Acta* **935:**79–86.
177. Paradies, G., and Ruggiero, F. M. (1990). Enhanced activity of the tricarboxylate carrier and modification of the lipids in hepatic mitochondria from hyperthyroid rats. *Arch. Biochem. Biophys.* **278:** 425–430.
178. Babior, B. M., Creagan, S., Ingbar, S. H., and Kipnes, R. S. (1973). Stimulation of mitochondrial adenosine diphosphate uptake by thyroid hormones. *Proc. Natl. Acad. Sci. USA* **70:**98–102.
179. Mowbray, J., and Corrigall, J. (1984). Short-term control of mitochondrial adenine nucleotide translocator by thyroid hormone. *Eur. J. Biochem.* **139:**95–99.
180. Mak, I. T., Shrago, E., and Elson, C. E. (1983). Effect of thyroidectomy on the kinetics of ADP–ATP translocation in liver mitochondria. *Arch. Biochem. Biophys.* **226:**317–323.
180a. Sterling, K., and Brenner, M. A. (1995). Thyroid hormone action: effect of triiodothyronine on mitochondrial adenine nucleotide translocase *in vivo* and *in vitro*. *Metabolism* **44:**193–199.
181. Paradies, G., and Ruggiero, F. M. (1989). Decreased activity of the pyruvate translocator and changes in the lipid composition in heart mitochondria from hypothyroid rats. *Arch. Biochem. Biophys.* **269:**595–602.
182. Luciakova, K., and Nelson, B. D. (1992). Transcript levels for nuclear-encoded mammalian mitochondrial respiratory-chain components are regulated by thyroid hormone in an uncoordinated fashion. *Eur. J. Biochem.* **207:**247–251.

183. Kopecky, J., Baudysova, M., Zanotti, F., Janikova, D., Pavelka, S., and Houstek, J. (1990). Synthesis of mitochondrial uncoupling protein in brown adipocytes differentiated in cell culture. *J. Biol. Chem.* **265:**22204–22209.
184. Burcelin, R., Kande, J., Ricquier, D., and Girard, J. (1993). Changes in uncoupling protein and GLUT4 glucose transporter expressions in interscapular brown adipose tissue of diabetic rats: Relative roles of hyperglycaemia and hypoinsulinaemia. *Biochem. J.* **291:**109–113.
185. Titheradge, M. A., and Coore, H. G. (1976). Hormonal regulation of liver mitochondrial pyruvate carrier in relation to gluconeogenesis and lipogenesis. *FEBS Lett.* **71:**73–78.
186. Garrison, J. C., and Haynes, R. C., Jr. (1975). The hormonal control of gluconeogenesis by regulation of mitochondrial pyruvate carboxylation in isolated rat liver cells. *J. Biol. Chem.* **250:**2769–2777.
187. Halestrap, A. P. (1978). Stimulation of pyruvate transport in metabolizing mitochondria through changes in the transmembrane pH gradient induced by glucagon treatment of rats. *Biochem. J.* **172:** 389–398.
188. Nosek, M. T., Dransfield, D. T., and Aprille, J. R. (1990). Calcium stimulates ATP-Mg/P_i carrier activity in rat liver mitochondria. *J. Biol. Chem.* **265:**8444–8450.
189. Aprille, J. R. (1988). Regulation of the mitochondrial adenine nucleotide pool size in liver: Mechanism and metabolic role. *FASEB J.* **2:**2547–2556.
190. Harano, Y., Kashiwagi, A., Kojima, H., Suzuki, M., Hashimoto, T., and Shigeta, Y. (1985). Phosphorylation of carnitine palmitoyltransferase and activation by glucagon in isolated rat hepatocytes. *FEBS Lett.* **188:**267–272.
191. Randle, P. J., Sugden, P. H., Kerbey, A. L., Radcliffe, P. M., and Hutson, N. J. (1978). Regulation of pyruvate oxidation and the conservation of glucose. *Biochem. Soc. Symp.* **43:**47–67.
192. Bradford, A. P., and Yeaman, S. J. (1986). Mitochondrial protein kinases and phosphatases. *Adv. Prot. Phosphatases* **3:**73–106.
193. Reed, L. J., and Damuni, Z. (1987). Mitochondrial protein phosphatases. *Adv. Prot. Phosphatases* **4:**59–76.
194. Schwoch, G., Trinczek, B., and Bode, C. (1990). Localization of catalytic and regulatory subunits of cyclic AMP-dependent protein kinases in mitochondria from various rat tissues. *Biochem. J.* **270:** 181–188.
195. Muller, G., and Bandlow, W. (1987). Protein phosphorylation in yeast mitochondria: cAMP-dependence, submitochondrial localization and substrates of mitochondrial protein kinases. *Yeast* **3:** 161–174.
196. Henriksson, T., and Jergil, B. (1979). Protein kinase activity and endogenous phosphorylation in subfractions of rat liver mitochondria. *Biochim. Biophys. Acta* **588:**380–391.
197. Vardanis, A. (1977). Protein kinase activity at the inner membrane of mammalian mitochondria. *J. Biol. Chem.* **252:**807–813.
198. Ferrari, S., Moret, V., and Siliprandi, N. (1990). Protein phosphorylation in rat liver mitochondria. *Mol. Cell. Biochem.* **97:**9–16.
199. Technikova-Dobrova, Z., Sardanelli, A. M., and Papa, S. (1993). Phosphorylation of mitochondrial proteins in bovine heart. Characterization of kinases and substrates. *FEBS Lett.* **322:**51–55.
200. Kemp, B. E., and Pearson, R. B. (1990). Protein kinase recognition sequence motifs. *Trends Biochem. Sci.* **15:**342–346.
201. De Vivo, D. C. (1993). The expanding clinical spectrum of mitochondrial diseases. *Brain Dev.* **15:**1–22.
202. Scholte, H. R. (1988). The biochemical basis of mitochondrial diseases. *J. Bioenerg. Biomembr.* **20:**161–191.
203. LaNoue, K. F., Watts, J. A., and Koch, C. D. (1981). Adenine nucleotide transport during cardiac ischemia. *Am. J. Physiol.* **241:** H663–H671.
204. Regitz, V., Paulson, D. J., Hodach, R. J., Little, S. E., Schaper, W., and Shug, A. L. (1984). Mitochondrial damage during myocardial ischemia. *Basic Res. Cardiol.* **79:**207–217.
205. Duan, J., and Karmazyn, M. (1989). Relationship between oxidative phosphorylation and adenine nucleotide translocase activity of two populations of cardiac mitochondria and mechanical recovery of ischemic hearts following reperfusion. *Can. J. Physiol. Pharmacol.* **67:**704–709.
206. Borutaite, V., Mildaziene, V., Katiliute, Z., Kholodenko, B., and Toleikis, A. (1993). The function of ATP/ADP translocator in the regulation of mitochondrial respiration during development of heart ischemic injury. *Biochim. Biophys. Acta* **1142:**175–180.
207. Parinandi, N. L., Zwizinski, C. W., and Schmid, H. H. O. (1991). Free radical-induced alterations of myocardial membrane proteins. *Arch. Biochem. Biophys.* **289:**118–123.
208. Zwizinski, C. W., and Schmid, H. H. O. (1992). Peroxidative damage to cardiac mitochondria: Identification and purification of modified adenine nucleotide translocase. *Arch. Biochem. Biophys.* **294:**178–183.
209. Schultheiss, H.-P., Schwimmbeck, P., Bolte, H.-D., and Klingenberg, M. (1985). The antigenic characteristics and the significance of the adenine nucleotide translocator as a major autoantigen to antimitochondrial antibodies in dilated cardiomyopathy. In *Advances in Myocardiology* (N. S. Dhalla and D. J. Hearse, eds.), Plenum Press, New York, pp. 311–327.
210. Schultheiss, H.-P., Schulze, K., Kuhl, U., Ulrich, G., and Klingenberg, M. (1986). The ADP/ATP carrier as a mitochondrial auto-antigen: Facts and perspectives. *Ann. N.Y. Acad. Sci.* **488:**44–63.
211. Kuhl, U., Ulrich, G., and Schultheiss, H.-P. (1987). Cross-reactivity of antibodies to the ADP/ATP translocator of the inner mitochondrial membrane with the cell surface of cardiac myocytes. *Eur. Heart J.* **8:(Suppl. J)**:219–222.
212. Schultheiss, H.-P., Berg, P., and Klingenberg, M. (1983). The mitochondrial adenine nucleotide translocator is an antigen in primary biliary cirrhosis. *Clin. Exp. Immunol.* **54:**648–654.
213. Schultheiss, H.-P., Berg, P. A., and Klingenberg, M. (1984). Inhibition of the adenine nucleotide translocator by organ specific autoantibodies in primary biliary cirrhosis. *Clin. Exp. Immunol.* **58:** 596–602.
214. Bakker, H. D., Scholte, H. R., Bogert, C. V. D., Ruitenbeek, W., Jeneson, J. A. L., Wanders, R. J. A., Abeling, N. G. G. M., Dorland, B., Sengers, R. C. A., and Gennip, A. H. V. (1993). Deficiency of the adenine nucleotide translocator in muscle of a patient with myopathy and lactic acidosis: A new mitochondrial defect. *Pediatr. Res.* **33:**412–417.
215. Hayes, D. J., Taylor, D. J., Hilton-Jones, D., Arnold, D. L., Bone, P. J., and Radda, G. K. (1986). A new metabolic myopathy: A malate-aspartate-shuttle defect. *Biochem. Soc. Trans.* **14:**1208–1209.
216. Scholte, H. R., Luyt-Houwen, I. E. M., Blom, W., Busch, H. F. M., De Jonge, P. C., De Visser, M., Huijmans, J. G. M., Jennekens, F. G. I., Mooy, P. D., Przyrembel, H., Schutgens, R. B. H., Vaandrager-Verduin, M. H. M., and Van Coster, R. N. A. (1986). Defects in mitochondrial beta oxidation. *Ann. N.Y. Acad. Sci.* **488:**511–512.
216a. Pande, S. V., and Murthy, M. S. R. (1994). Carnitine-acylcarnitine translocase deficiency: Implications in human pathology. *Biochim. Biophys. Acta* **1226:**269–276.
217. Paradies, G., Capuano, F., Palombini, G., Galeotti, T., and Papa, S. (1983). Transport of pyruvate in mitochondria from different tumor cells. *Cancer Res.* **43:**5068–5071.
218. Eboli, M. L., Paradies, G., Galeotti, T., and Papa, S. (1977). Pyruvate transport in tumour-cell mitochondria. *Biochim. Biophys. Acta* **460:**183–187.
219. Senior, A. E., McGowan, S. E., and Hilf, R. (1975). A comparative study of inner membrane enzymes and transport systems in mito-

chondria from R3230AC mammary tumor and normal rat mammary gland. *Cancer Res.* **35:**2061–2067.

220. Kolarov, J., Kuzela, S., Krempasky, V., and Ujhazy, V. (1973). Some properties of coupled hepatoma mitochondria exhibiting uncoupler-insensitive ATPase activity. *Biochem. Biophys. Res. Commun.* **55:**1173–1179.

221. Kaplan, R. S., Morris, H. P., and Coleman, P. S. (1982). Kinetic characteristics of citrate influx and efflux with mitochondria from Morris hepatomas 3924A and 16. *Cancer Res.* **42:**4399–4407.

222. Kielducka, A., Paradies, G., and Papa, S. (1981). A comparative study of the transport of pyruvate in liver mitochondria from normal diabetic rats. *J. Bioenerg. Biomembr.* **13:**123–132.

223. Rahman, R. O'Rourke, F., and Jungas, R. L. (1983). Effects of insulin on CO_2 fixation in adipose tissue. Evidence for regulation of pyruvate transport. *J. Biol. Chem.* **258:**483–490.

224. Cheema-Dhadli, S., and Halperin, M. L. (1973). The role of the mitochondrial citrate transporter in the regulation of fatty acid synthesis: Effect of fasting and diabetes. *Can. J. Biochem.* **51:**1542–1544.

225. Lerner, E., Shug, A. L., Elson, C., and Shrago, E. (1972). Reversible inhibition of adenine nucleotide translocation by long chain fatty acyl coenzyme A esters in liver mitochondria of diabetic and hibernating animals. *J. Biol. Chem.* **247:**1513–1519.

226. Kaplan, R. S., Oliveira, D. L., and Wilson, G. L. (1990). Streptozotocin-induced alterations in the levels of functional mitochondrial anion transport proteins. *Arch. Biochem. Biophys.* **280:**181–191.

227. Kaplan, R. S., Mayor, J. A., Blackwell, R., Maughon, R. H., and Wilson, G. L. (1991). The effect of insulin supplementation on diabetes-induced alterations in the extractable levels of functional mitochondrial anion transport proteins. *Arch. Biochem. Biophys.* **287:**305–311.

228. Kaplan, R. S., Mayor, J. A., Blackwell, R., Wilson, G. L., and Schaffer,S. W. (1991). Functional levels of mitochondrial anion transport proteins in non-insulin-dependent diabetes mellitus. *Mol. Cell. Biochem.* **107:**79–86.

229. Inoue, I., Saheki, T., Kayanuma, K., Uono, M., Nakajima, M., Takeshita, K., Koike, R., Yuasa, T., Miyatake, T., and Sakoda, K. (1988). Biochemical analysis of decreased ornithine transport activity in the liver mitochondria from patients with hyperornithinemia, hyperammonemia and homocitrullinuria. *Biochim. Biophys. Acta* **964:**90–95.

230. Hommes, F. A., Ho, C. K., Roesel, R. A., and Coryell, M. E. (1982). Decreased transport of ornithine across the inner mitochondrial membrane as a cause of hyperornithinaemia. *J. Inher. Metab. Dis.* **5:**41–47.

231. Oyanagi, K., Aoyama, T., Tsuchiyama, A., Nakao, T., Uetsuji, N., Wagatsuma, K., and Tsugawa, S. (1986). A new type of hyperlysinaemia due to a transport defect of lysine into mitochondria. *J. Inher. Metab. Dis.* **9:**313–316.

232. Schonfeld, P., Fritz, S., Halangk, W., and Bohnensack, R. (1993). Increase in the adenine nucleotide translocase protein contributes to the perinatal maturation of respiration in rat liver mitochondria. *Biochim. Biophys. Acta* **1144:**353–358.

233. Nohl, H., and Kramer, R. (1980). Molecular basis of age-dependent changes in the activity of adenine nucleotide translocase. *Mech. Ageing Dev.* **14:**137–144.

234. Paradies, G., and Ruggiero, F. M. (1990). Age-related changes in the activity of the pyruvate carrier and in the lipid composition in rat-heart mitochondria. *Biochim. Biophys. Acta* **1016:**207–212.

235. Paradies, G., and Ruggiero, F. M. (1991). Effect of aging on the activity of the phosphate carrier and on the lipid composition in rat liver mitochondria. *Arch. Biochem. Biophys.* **284:**332–337.

236. Hansford, R. G. (1978). Lipid oxidation by heart mitochondria from young adult and senescent rats. *Biochem. J.* **170:**285–295.

237. Grigorenko, E. V., Small, W. C., Persson, L.-O., and Srere, P. A. (1990). Citrate synthase 1 interacts with the citrate transporter of yeast mitochondria. *J. Mol. Recog.* **3:**215–219.

238. Kramer, R., and Klingenberg, M. (1977). Reconstitution of adenine nucleotide transport with purified ADP, ATP-carrier protein. *FEBS Lett.* **82:**363–367.

239. Kaplan, R. S., Pratt, R. D., and Pedersen, P. L. (1989). Purification and reconstitution of the phosphate transporter from rat liver mitochondria. *Methods. Enzymol.* **173:**732–745.

240. Gibb, G. M., Reid, G. P., and Lindsay, J. G. (1986). Purification and characterization of the phosphate/hydroxyl ion antiport protein from rat liver mitochondria. *Biochem. J.* **238:**543–551.

241. De Pinto, V., Tommasino, M., Palmieri, F., and Kadenback, B. (1982). Purification of the active mitochondrial phosphate carrier by affinity chromatography with an organomercurial agarose column. *FEBS Lett.* **148:**103–106.

242. Bisaccia, F., and Palmieri, F. (1984). Specific elution from hydroxylapatite of the mitochondrial phosphate carrier by cardiolipin. *Biochim. Biophys. Acta* **766:**386–394.

243. Kolbe, H. V. J., Costello, D., Wong, A., Lu, R. C., and Wohlrab, H. (1984). Mitochondrial phosphate transport. Large scale isolation and characterization of the phosphate transport protein from beef heart mitochondria. *J. Biol. Chem.* **259:**9115–9120.

244. Wohlrab, H., Kolbe, H. V. J., and Collins, A. (1986). Isolation and reconstitution of the phosphate transport protein from mitochondria. *Methods Enzymol.* **125:**697–705.

245. Stappen, R., and Kramer, R. (1993). Functional properties of the reconstituted phosphate carrier from bovine heart mitochondria: Evidence for asymmetric orientation and characterization of three different transport modes. *Biochim. Biophys. Acta* **1149:**40–48.

246. Bisaccia, F., De Palma, A., Prezioso, G., and Palmieri, F. (1990). Kinetic characterization of the reconstituted tricarboxylate carrier from rat liver mitochondria. *Biochim. Biophys. Acta* **1019:**250–256.

247. Bisaccia, F., Indiveri, C., and Palmieri, F. (1985). Purification of reconstitutively active α-oxoglutarate carrier from pig heart mitochondria. *Biochim. Biophys. Acta* **810:**362–369.

248. Indiveri, C., Palmieri, F., Bisaccia, F., and Kramer, R. (1987). Kinetics of the reconstituted 2-oxoglutarate carrier from bovine mitochondria. *Biochim. Biophys. Acta* **890:**310–318.

249. Bolli, R., Nalecz, K. A., and Azzi, A. (1989). Monocarboxylate and α-ketoglutarate carriers from bovine heart mitochondria. Purification by affinity chromatography on immobilized 2-cyano-4-hydroxycinnamate. *J. Biol. Chem.* **264:**18024–18030.

250. Claeys, D., Muller, M., and Azzi, A. (1989). Purification and reconstitution of the 2-oxoglutarate carrier from bovine heart and liver mitochondria. In *Anion Carriers of Mitochondrial Membranes* (A. Azzi, K. A. Nalecz, M. J. Nalecz, and L. Wojtczak, eds.), Springer-Verlag, Berlin, pp. 17–34.

251. Bisaccia, F., Indiveri, C., and Palmieri, F. (1988). Purification and reconstitution of two anion carriers from rat liver mitochondria: The dicarboxylate and the 2-oxoglutarate carrier. *Biochim. Biophys. Acta* **933:**229–240.

252. Indiveri, C., Capobianco, L., Kramer, R., and Palmieri, F. (1989). Kinetics of the reconstituted dicarboxylate carrier from rat liver mitochondria. *Biochim. Biophys. Acta* **977:**187–193.

252a. Lancar-Benba, J., Foucher, B., and Saint-Macary, M. (1994). Purification of the rat-liver mitochondrial dicarboxylate carrier by affinity chromatography on immobilized malate dehydrogenase. *Biochim. Biophys. Acta* **1190:**213–216.

253. Nalcec, M. J., Szewczyk, A., Broger, C., Wojtczak, L., and Azzi, A. (1989). Isolation and functional reconstitution of the dicarboxylate carrier from bovine liver mitochondria. In *Anion Carriers of Mitochondrial Membranes* (A. Azzi, K. A. Nalecz, M. J. Nalecz, and L. Wojtczak, eds.), Springer-Verlag, Berlin, pp. 71–85.

254. Szewczyk, A., Nalecz, M. J., Broger, C., Wojtczak, L., and Azzi, A. (1987). Purification by affinity chromatography of the dicarboxylate carrier from bovine heart mitochondria. *Biochim. Biophys. Acta* **894:**252–260.
255. Nalecz, K. A., Kaminska, J., Nalecz, M. J., and Azzi, A. (1992). The activity of pyruvate carrier in a reconstituted system: Substrate specificity and inhibitor sensitivity. *Arch. Biochem. Biophys.* **297:**162–168.
256. Indiveri, C., Tonazzi, A., and Palmieri, F. (1990). Identification and purification of the carnitine carrier from rat liver mitochondria. *Biochim. Biophys. Acta* **1020:**81–86.
257. Indiveri, C., Tonazzi, A., Prezioso, G., and Palmieri, F. (1991). Kinetic characterization of the reconstituted carnitine carrier from rat liver mitochondria. *Biochim. Biophys. Acta* **1065:**231–238.
258. Kaminska, J., Nalecz, K. A., Azzi, A., and Nalecz, M. J. (1993). Purification of carnitine carrier from rat brain mitochondria. *Biochem. Mol. Biol. Int.* **29:**999–1007.
259. Bisaccia, F., De Palma, A., and Palmieri, F. (1992). Identification and purification of the aspartate/glutamate carrier from bovine heart mitochondria. *Biochim. Biophys. Acta* **1106:**291–296.
260. Indiveri, C., Tonazzi, A., and Palmieri, F. (1992). Identification and purification of the ornithine/citrulline carrier from rat liver mitochondria. *Eur. J. Biochem.* **207:**449–454.
261. Lin, C. S., and Klingenberg, M. (1980). Isolation of the uncoupling protein from brown adipose tissue mitochondria. *FEBS Lett.* **113:** 299–303.
262. Klingenberg, M., and Winkler, E. (1985). The reconstituted isolated uncoupling protein is a membrane potential driven H^+ translocator. *EMBO J.* **4:**3087–3092.
263. Klingenberg, M., and Winkler, E. (1986). Reconstitution of an H^+ translocator, the "uncoupling protein" from brown adipose tissue mitochondria, in phospholipid vesicles. *Methods Enzymol.* **127:** 772–779.
264. Ricquier, D., Lin, C.-S., and Klingenberg, M. (1982). Isolation of the GDP binding protein from brown adipose tissue mitochondria of several animals and amino acid composition study in rat. *Biochem. Biophys. Res. Commun.* **106:**582–589.
265. Capaldi, R. A., and Vanderkooi, G. (1972). The low polarity of many membrane proteins. *Proc. Natl. Acad. Sci. USA* **69:**930–932.
266. Gribskov, M., and Devereux, J. (1991). *Sequence Analysis Primer,* Stockton, New York, p. 233.
267. Bouillaud, F., Casteilla, L., and Ricquier, D. (1992). A conserved domain in mitochondrial transporters is homologous to a zinc-finger knuckle of nuclear hormone receptors. *Mol. Biol. Evol.* **9:** 970–975.

16

Receptor-Mediated Endocytosis

Victoria P. Knutson, Patricia V. Donnelly, Maria M. Lopez-Reyes, and Yvonne L. O. Balba

16.1. INTRODUCTION

In order to sustain normal metabolic processes, the cell must recruit from its external environment an array of ions, molecules, and macromolecules. Small molecules, such as ions, water, and monosaccharides, can gain entry into the cytosol of the cell through water-filled channels. However, molecules larger than approximately 1000 Da cannot pass through these channels. Therefore, if diffusion through plasma membrane pores or channels were the only method of entry into the cell cytoplasm, larger molecules such as polypeptides, polysaccharides, and other nutrients could not be utilized by the cell for normal metabolic processes. In addition, these larger molecules are often present in the extracellular environment of the cell at much lower concentrations than ions or monosaccharides. Therefore, in order to overcome problems associated with internalizing large and scarce molecules, the cell has developed the process of receptor-mediated endocytosis. The extracellular macromolecules bind to specific, high-affinity receptors on the cell surface, thereby trapping them and concentrating these macromolecules at the plasma membrane. The receptor-bound ligands then are internalized into the cell in membrane-bound vesicles, endosomes. The internalized ligands are sorted in and transported through a series of morphologically distinct endosomal compartments before ultimate delivery to the lysosomes. The hydrolytic enzymes resident in the lysosomes "digest" the macromolecules liberating the component amino acids, triglycerides, and so forth, for use as cellular nutrients. The purpose of this review is to provide an overview of the process of endocytosis and its relevance to medicine. This review covers not only endocytotic processes by which nutrients are "captured" and delivered to the cell, but also the role of endocytosis in signal transduction, antigen presentation, neurotransmitter uptake, and the entry of viral and bacterial toxins and pathogens into cells.

16.2 RECEPTOR-MEDIATED ENDOCYTOSIS: GENERAL PROCESSES

The process of receptor-mediated endocytosis was initially defined through studies with the low-density lipoprotein (LDL) receptor, and the uptake of LDL by its receptor serves as the prototype for the process.[1] The general aspects of the process are schematized in Figure 16.1.

16.2.1. Plasma Membrane Processes

16.2.1.1. Binding at the Plasma Membrane

The extracellular ligand binds to the receptor at the cell plasma membrane. The receptor is specific for the ligand, and the binding occurs with high affinity in a time-, temperature-, and concentration-dependent manner, which is saturable at high ligand concentrations.[2] Since the ligands internalized via receptor-mediated endocytosis are present in relatively low concentrations in the extracellular milieu, the binding to the receptor concentrates the ligand at the plasma membrane, leading to an increased efficiency of uptake of the ligand into the cell.[1,3]

16.2.1.2. Coated Pit Formation

The receptors then migrate into coated pits.[4] The electron-dense coat of the coated pits is formed from a multisubunit protein, clathrin, that polymerizes into a three-armed structure called a triskelien. Current data suggest that a cytosolic adaptor protein binds to the cytosolic side of the plasma membrane and then acts as the nucleation site for the generation of the clathrin lattices that make up the coated pit.[5–7] Nucleation of the triskeliens into the lattice structure that forms the coated pits is initiated by the binding of adaptor molecules, specifically the α subunit of adaptin molecule AP-2, to specific

Victoria P. Knutson, Patricia V. Donnelly, Maria M. Lopez-Reyes, and Yvonne L. O. Balba • Department of Pharmacology, University of Texas–Houston Medical School, Houston, Texas 77225.

Molecular Biology of Membrane Transport Disorders, edited by S.G. Schultz *et al.* Plenum Press, New York, 1996

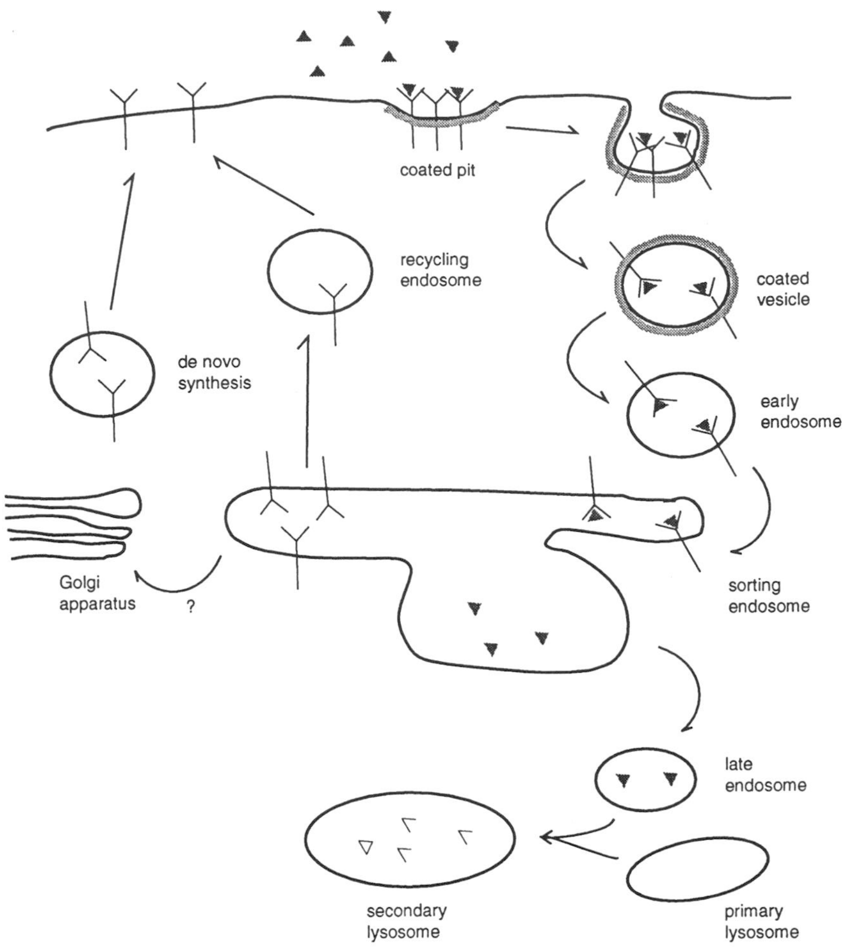

Figure 16.1. Trafficking of receptor and ligand in receptor-mediated endocytosis. A schematic of the general processes involved in receptor-mediated endocytosis, as exemplified by the LDL receptor. Modifications in this pathway are both receptor- and ligand-specific, and these changes are detailed in the text.
Y, receptor; ▲, intact ligand; △, ∠, degraded ligand; ▒, clathrin coat.

sites on the cytosolic side of the plasma membrane.[8] This binding requires the aggregation of AP-2 molecules. Regulation of AP-2 aggregation is achieved by inositol phosphates (IPs), suggesting that hormones and growth factors that induce IP production regulate coated pit formation.[8] To date, the high-affinity binding site on the plasma membrane for AP-2 has not been identified.[9]

Once the coated pits have formed, the plasma membrane receptors migrate into the pits,[10] and a weak association between the amino-terminal domain of a subunit of AP-2[11] and the cytoplasmic tail region of the receptor maintains the receptor within the coated pit.[8] The domain within the cytoplasmic tail region of the receptor that interacts with AP-2 has been characterized as a tyrosine-containing sequence, NPXY (where X is any amino acid), which has the secondary structure of a tight turn.[12,13] Interestingly, however, the lack of this internalization motif does not result in the active exclusion of a protein from coated pits. Instead, a mechanism has been proposed whereby the cytoplasmic tail of the receptor interacts with elements of the cortical cytoskeleton in a regulated manner, resulting in the active exclusion of receptors from coated pits.[14,15]

Following the clustering of receptors into the coated pits, the pit invaginates and buds off of the plasma membrane to form coated vesicles. This process requires ATP, calcium, other unidentified cytosolic factors, and the clathrin lattice.[16] In addition, low-molecular-weight GTP-binding proteins also participate in this process.[17,18] The clathrin coat surrounding the vesicle then disassembles, permitting the coat proteins to be reutilized in another cycle of pit formation.[19] The resulting coatless vesicle is termed an endosome.

16.2.2. Endosomes

16.2.2.1. Acidification

Within 5 min after endocytosis, the lumen of the endosome becomes mildly acidic. Studies utilizing pH-sensitive fluorescent dyes[20,21] or monitoring pH-induced conformational changes in viral proteins[22] have demonstrated that while the lumen of the clathrin-coated vesicle is neutral, the early endosome has a luminal pH of approximately 6.[22] Within 8 to 10 min after endocytosis, in the late endosome, the pH declines further to approximately 5.[22] This late endosome mediates delivery of the endosomal contents to the lysosome.

The acidic pH in the lumen of the endosome is regulated and maintained by at least two activities: an ATP-dependent proton pump, the V (vacuolar)-ATPase, and Na^+/K^+-ATPase.

Protons are pumped into the lumen of the endosome by the V-ATPase.[23] This macromolecular complex is composed of two structural domains (a transmembrane domain and a cytosolic domain), and each domain is composed of multiple subunits. The aggregate mass of the complex is approximately 750 kDa. The ATP-dependent acidification mediated by the V-ATPase is electrogenic, resulting in an interior-positive membrane potential.[24] Once attained, this interior-positive membrane potential opposes further acidification of the lumen, thereby limiting the degree of acidification. In addition to the V-ATPase, the early endosomes contain a Na^+/K^+-ATPase activity that pumps Na^+ into the lumen of the endosome. This activity also generates an interior-positive membrane potential, further limiting the degree of acidification of this endosomal compartment. The Na^+/K^+-ATPase activity is found in the early endosomes but not in the late endosomes. As noted above, the pH of the lumen of the early endosome is approximately 6, while that of the late endosome is 5. Modulation of the luminal pH of the endosomes may be achieved by the activation of the Na^+/K^+-ATPase in the early endosome, and subsequent inactivation of the pump in the late endosome.[25] However, how the Na^+/K^+-ATPase activity is regulated is currently unknown.

Regulation of endosomal pH may also be achieved by modulating the activity of the V-ATPase. Recent studies of the V-ATPase in mutant yeast cells have demonstrated that the appropriate assembly of the multiple subunits of the V-ATPase complex or the targeting of the complex to the lysosomal membrane requires the presence of specific subunits of both the cytosolic and membrane domains of the complex.[26,27] In addition, a 25-kDa protein has been identified whose presence is required for assembly or targeting of the ATPase. Interestingly, this 25-kDa protein is not itself a component subunit of the V-ATPase.[28] Therefore, it is possible that the V-ATPase complex is associated with the membranes of the endosomes, but association and dissociation of critical subunits of the complex regulate the activity of the V-ATPase in specific endosomal compartments.

As the endosomal vesicles mature, their intraluminal pH becomes progressively more acidic. This drop in pH serves multiple purposes. In the early endosomes, the slightly acidic environment serves to facilitate the dissociation of ligands from receptors. It has been proposed[29] that whether a receptor recycles back to the plasma membrane or is, instead, routed to and degraded in the lysosomes is determined by the pH sensitivity of the association between ligand and receptor. If the ligand dissociates from the receptor at the mildly acidic conditions of the early endosome, the receptor will recycle quickly between early endosomes and the plasma membrane. If the ligand–receptor complex requires the more acidic pH of late endosomes (pH 5–5.5) for dissociation, receptor recycling will be slower, as occurs, for instance, with the mannose-6-phosphate receptor.[30] If the ligand–receptor complex is resistant to dissociation by the acidic conditions of the late endosome, both ligand and receptor will be degraded in the lysosomes, as is found to occur with the Fc receptor–IgG complex.

While this model presents an intriguing functional rationale for the graded decrease in pH between early endosomes and lysosomes, many systems do not conform to this model. An example is the insulin receptor. Insulin is remarkably resistant to discharge from the insulin receptor by acid. Dissociation in what is likely to be an early endosomal compartment is facilitated by the presence of a specific insulin-degrading proteinase.[31,32] While data suggest that insulin is discharged from the receptor before it is a substrate for the insulinase, the data do not entirely rule out the possibility that insulin in complex with the receptor is a substrate for the insulinase.[32] Therefore, in contrast to the model suggested above, the pH sensitivity of the insulin–receptor complex is not predictive of insulin receptor recycling: the insulin–receptor complex is acid resistant, yet the receptor recycles.

Also in conflict with the pH-sensitivity model is the observation that receptor recycling can be a metabolically regulated process. On chronic (6 hr) stimulation of cultured cells with insulin, the rate of insulin receptor recycling slows compared to cells stimulated with insulin for 30 min.[33] On acute stimulation with insulin, the insulin receptor primarily recycles, but on chronic stimulation with insulin, the insulin receptor is primarily degraded. Since it is unlikely that the pH sensitivity of the insulin–receptor complex is modified by chronic stimulation with insulin, these data suggest that pH sensitivity of the ligand–receptor complex alone does not govern whether a receptor recycles or is degraded. Other cellular activities and factors, which may, in fact, be receptor- or ligand-specific, govern these processes.

16.2.2.2. Proteolysis

The regulation of proteolytic activity may be one such process that governs recycling versus degradation. A significant level of acid-dependent proteolysis occurs within the endosomes prior to fusion with lysosomes. Incubation of macrophages with mannose-BSA, a ligand for the mannose receptor, for less than 10 min results in the proteolysis of the ligand in a nonlysosomal compartment.[34,35] This proteolysis is inhibited by the protease inhibitors pepstatin A and leupeptin,[36] and requires acidification by an ATP-dependent process.[34]

Several proteases have been identified within endosomal vesicles, the best characterized to date being cathepsin D.[18,37] While the bulk of this enzyme is localized in lysosomes, considerable amounts are also found in the endosomes, with no detectable activity at the plasma membrane.[35,38] This interesting subcellular localization has led to the suggestion that newly synthesized cathepsin D is routed from the Golgi to the endosomes. Delivery to the lysosomes is thereafter mediated by the fusion of the endosomes with primary lysosomes. Cathepsin B activity has also been localized to early endosomes, with detectable activity observed less than 2 min after endocytosis of ligand.[39] Like cathepsin D, cathepsin B cannot be detected on the plasma membrane, and requires an acidic pH for activity. Therefore, both cathepsins D and B may utilize endosomes as shuttles for the delivery of the enzymes from the Golgi to the lysosomes.

Proteases that demonstrate ligand specificity are also expressed in endosomes. A protease has been identified that hydrolyzes insulin in an early endosomal compartment.[31,32] This enzyme is specific for insulin: it does not degrade epidermal growth factor or prolactin. Unlike the findings on the subcellular localization of the cathepsins, this insulin protease is found not only in the early endosomes, but also in the plasma membrane and in the cytosol. Localization in the plasma membrane suggests that the protease is posttranslationally inserted into the plasma membrane and then colocalizes into the endosomes with the ligand–receptor complex on endocytosis. In the acidic environment of the endosomes, the insulin protease is active. However, the functional significance of the insulin protease in the cytosol is currently unexplained.

Degradation of the toxin ricin has provided insight into the role of decreasing luminal pH in regulation of proteolysis as endosomes mature from early to late species.[40] Both cathepsins B and D have been identified in early endosomes. However, cathepsin B is active between pH 4.5 and 7, the pH of the early endosomes, while cathepsin D is active at below pH 5.5, the pH of the late endosomes and lysosomes. These data suggest that cathepsin B proteolyzes ricin in an early endosome. While cathepsin D is present in the early endosome, the local pH renders it inactive. As the endosome matures into a late endosome with the concomitant decrease in pH, cathepsin D is then activated, leading to further degradation of ricin. Similar studies have been performed on the degradation of epidermal growth factor (EGF).[41] These studies demonstrated a sequential proteolytic processing of EGF. Within the early endosome, a carboxypeptidase B activity initially cleaves the EGF molecule. On maturation to a late endosome, with the corresponding decline in pH, a serine protease is activated, resulting in the further cleavage of EGF.

Proteolytic activity expressed within the endosomes is directed not only at the ligands, but also at the corresponding receptors. An acidification-dependent thiol protease has been identified in an endosomal compartment which cleaves the insulin receptor.[42] The endosomal compartment wherein insulin receptor proteolysis occurs is physically distinct from the early endosome where insulin degradation occurs (V. P. Knutson, manuscript in review). The nerve growth factor receptor has also been shown to be proteolyzed within a nonlysosomal acidified vesicle.[43]

Therefore, the steps in the maturation of endosomes from early to late species provide loci for regulatory control. The maturation process permits differential expression of proteases (and potentially other regulatory factors) within different endosomes, and the activities of the expressed proteases are regulated by the decreasing pH within the maturing endosomes.

Defects in the ability to acidify endosomes, or mutations leading to changes in the proteolytic clip site of proteins degraded in endosomes could lead to disease states. Mutations in the primary sequence of arylsulfatase B have been shown to lead to the lysosomal storage disease Maroteaux–Lamy syndrome. In the normal, nondisease state, arylsulfatase B is synthesized as a precursor protein that moves through the Golgi into the late endosomes. Within the late endosomes, limited proteolysis converts the precursor form to the mature form of the enzyme. The mutant form of the enzyme, which contains a glycine-to-valine substitution, is extremely susceptible to catabolic proteolysis by an enzyme present in an endosomal compartment intermediate between the Golgi and the late endosome. Therefore, this conservative substitution in the primary sequence of arylsulfatase B makes the enzyme a better substrate for a resident endosomal protease, resulting in the loss of arylsulfatase activity and a concomitant phenotype of an intermediate-severity mucopolysaccharidosis.[44]

Endocytosis has ramifications in cellular processes as diverse as nutrient uptake, immune function, and signal transduction. The rest of this chapter will address some of these processes with emphasis on how abnormalities in these endocytotic events lead to human disease.

16.3. ENDOCYTOSIS AND NUTRIENT UPTAKE

16.3.1. The Low-Density Lipoprotein Receptor

The LDL receptor binds to a protein component (apolipoprotein B-100) of the cholesterol-rich LDL particle. LDL binding at the plasma membrane and endocytosis of the complex leads to the lysosomal degradation of LDL and the corresponding release of cholesterol into the cell. In the process of providing cholesterol to the cells for membrane synthesis, steroidogenesis, and bile acid synthesis, endocytosis of LDL results in a decrease in the blood levels of cholesterol.

The endocytosis cycle of the LDL receptor is schematized in Figure 16.1. While the LDL particle is degraded in the lysosomes, the LDL receptor is spared from degradation and, after discharge of LDL, is recycled back to the plasma membrane for subsequent cycles of endocytosis.[1] The level of the expression of the LDL receptor is regulated by the level of cholesterol in the cell.[45] At high intracellular concentrations of cholesterol, the level of LDL receptor mRNA and LDL receptor protein is decreased: at low levels of intracellular cholesterol, receptor expression is elevated. This feedback system permits the cell to fine-tune the amount of LDL, and thereby the amount of cholesterol that is taken up into the cell.

Mutations in the LDL receptor that lead to a reduced ability of the cells to take up LDL result in the disease familial hypercholesterolemia (FH), an autosomal dominant condition. In homozygous individuals, these mutations lead to serum cholesterol levels that are 2.5- to 6-fold higher than normal, and result in the development of and death from coronary artery disease at an early age.[46] To date, approximately 150 natural mutations in the LDL receptor have been described.[1,45] These mutations have permitted the mapping of functional domains of the LDL receptor protein, and the identification of regions of the receptor that are required for LDL binding, receptor clustering in clathrin coated pits, and receptor endocytosis.[46] Abnormalities in any of these functions of the receptor lead to hypercholesterolemia.

Mutations in the ligand, apolipoprotein B-100, that result in a reduced affinity of ligand for receptor also lead to hypercholesterolemia.[47] This defect, termed familial defective apolipoprotein B-100 (FDB), is the result of a single base change in the gene coding for apolipoprotein B-100, substituting a glutamine for an arginine residue. With this mutation, the reduced affinity of the LDL particle for the receptor results in decreased uptake of LDL by the receptor with concomitant elevations of plasma cholesterol, and premature coronary artery disease in affected individuals.

Therefore, with the LDL receptor, defects in cholesterol uptake can be mediated on two fronts: mutations in the LDL receptor, which decreases the receptor's ability to bind and/or internalize LDL, and mutations in the ligand, apolipoprotein B-100, that decrease the affinity of ligand for receptor. Both types of mutations result in a decreased ability of the cell to internalize cholesterol.

16.3.2. The Transferrin Receptor

Iron is taken up into all cells, but this uptake is most pronounced and most studied in liver and erythroid cells. Inside the erythroid cell, iron is utilized in the biosynthesis of heme. In all cells, iron is required for the biosynthesis of other iron-containing proteins. In the liver, iron accumulates and is stored in dense bodies termed ferritin deposits. One pathway facilitating the cellular uptake of iron is mediated by the transferrin receptor (TfR). In erythroid cells, the exclusive mechanism by which iron enters the cells is by means of the TfR.[48] However, in liver, which experiences locally high iron concentrations via the portal circulation, considerable amounts of iron also enter the cells by fluid phase endocytosis and another poorly defined nonspecific pathway.[49]

The endocytotic pathway by which the TfR mediates the cellular uptake of iron is slightly different from the LDL receptor pathway.[50] Within the serum, iron binds to a large glycoprotein, transferrin. At the physiological pH of blood, the iron-loaded ferrotransferrin binds to the TfR on the plasma membrane of the cell. Following endocytosis, the endosome acidifies to a pH of approximately 5.1. Although this pH favors the release of iron from the transferrin, the association of the apotransferrin with the receptor remains intact. An iron transporter presumably exists in the membrane of the endosome facilitating the movement of iron from the lumen of the endosome into the cytosol of the cell. The apotransferrin–transferrin receptor complex then recycles back to the plasma membrane of the cell. At the neutral pH of the interstitial fluid, the binding of apotransferrin to the receptor is disfavored, and the apotransferrin dissociates from the receptor. The apotransferrin can then again bind iron, and the recycled TfR can rebind ferrotransferrin. In this system, the pH sensitivities of the binding constants of iron for transferrin and transferrin for the receptor are beautifully exploited in the endosomal acidification processes: at neutral pH, the binding of iron to transferrin is favored, and the binding of ferrotransferrin to the receptor is also favored. In addition, at neutral pH, the discharge of apotransferrin from the receptor is favored. However, at more acidic pH, the binding of apotransferrin to the receptor is favored, but the discharge of iron from ferrotransferrin is also favored. This cycle optimally delivers iron to the cell, but protects both transferrin and the TfR from lysosomal degradation, thereby permitting reutilization of both molecules in subsequent cycles of iron binding and endocytosis. This endocytosis process of the TfR is very rapid, with a single cycle complete in 3–4 min.[51]

In a manner parallel to that described above with the LDL receptor, the level of intracellular iron regulates the level of expression of the TfR. In liver cells, a high intracellular level of iron leads to decreased levels of TfR mRNA, and subsequent decreased expression of TfR protein on the plasma membrane of the cell.[52,53] This negative feedback regulation between iron and the TfR is illustrated in the iron overload state of hemochromatosis, where antibody staining of hepatocytes from patients with varying degrees of iron overload demonstrates the inverse relationship between cellular iron levels and TfR expression. Under conditions of severe iron overload, no TfR expression can be detected.[54] Correspondingly, when intracellular iron levels are low, elevated expression of hepatocyte TfR is demonstrated. Consistent with what has been demonstrated in hepatocytes, in erythroid cells, when blood levels of iron decrease, an elevated level of TfR mRNA is detected, and the level of TfR protein is increased. Interestingly, however, this elevated level of expression of receptor protein is not seen on the plasma membrane of the cell: only the intracellular, endosomal population of TfR is elevated.[55] How this tight control of the subcellular expression of the TfR is regulated, and its physiological significance in the erythroid cell, are currently unknown. Potentially, the larger intracellular pool of TfR in the iron-depleted erythroid cell may, by mass action, accelerate the recycling pathway of the receptor, increase the flux of the receptor through the endocytotic cycle, and thereby increase the amount of iron delivered to the cell.

In view of the sensitive feedback loop between intracellular iron levels and TfR expression, it has been proposed that the level of expression of the TfR on blood cells could be utilized as a diagnostic tool in iron deficiency states.[56] In particular, mild forms of tissue iron deficiency could be detected. In addition, quantitation of the TfR would permit a differential diagnosis of anemias associated with iron deficiency, as opposed to chronic disease-related anemias. Also, pregnancy-related iron deficiencies could be readily detected and subsequently treated.

The TfR also plays an important role in delivering iron across specialized epithelial cell barriers. Because of the charge on iron, simple diffusion of iron across the highly impermeant endothelial cells comprising the blood–brain barrier is unlikely. Recent studies utilizing microvessel endothelial cells from brain indicate that TfR on the apical (blood) side of the polarized cells bind and internalize iron, and that a fraction of this iron is then vectorially transported to the basolateral (brain) side of the cell barrier.[57] These studies suggest a role of endocytosis in the regulated delivery of iron to the brain.

16.4. "SCAVENGER" RECEPTOR SYSTEMS

A variety of macromolecular compounds are found in the blood that are either end products of catabolic processes or the cellular components of lysed cells. These compounds are too large for the kidney to clear from the circulation, and hence a number of endocytotic processes exist to remove these large molecules from the circulation for processing to smaller compounds that can be utilized subsequently by the cell itself, or secreted from the cell. Two general classes of scavenger systems are discussed here: the carbohydrate binding receptors and the α_2-macroglobulin/lipoprotein receptor-related protein (LPR) receptor. From a general perspective, the endocytotic itinerary of these ligand/receptor systems is much like the LDL receptor. That is, the complexes are internalized, the ligand and receptor are sequestered, and the receptor recycles back to the plasma membrane of the cell while the ligand is degraded.

16.4.1. The Carbohydrate Binding Receptors

As suggested by their general classification, the carbohydrate binding receptors have high affinity for specific carbohydrate moieties. Hence, they are classified as mammalian lectins, with characteristics much like their plant counterparts.[58] These receptors are multisubunit membrane proteins in which each subunit contains a single transmembrane domain and a short cytoplasmic domain. The extracellular, carboxy-terminal domain contains the carbohydrate recognition domain, a highly conserved domain characteristic of all lectins. These receptors undergo constitutive clustering and endocytosis and a large percentage of the total cellular receptor is found not in the plasma membrane, but in an intracellular location.[59]

16.4.1.1. The Asialoglycoprotein Receptor

The asialoglycoprotein (ASGP) receptor binds glycoproteins that have been subjected to neuraminidase activity, resulting in the removal of terminal sialic acids. Galactose and to some extent *N*-acetylgalactosamine then become the terminal carbohydrate residues, and these residues direct the binding of the desialylated glycoprotein to this receptor. Because of the specificity of the ASGP receptor for galactose, this receptor is also referred to as the galactose (Gal) receptor. The binding of glycoproteins to the Gal receptor is calcium dependent and the affinity of binding is determined by the number of terminal galactose residues that are present on the glycoprotein ligand. A protein with a complex carbohydrate chain terminating in a single galactose residue binds to the receptor with a K_d of 10^{-3} M, whereas a carbohydrate chain terminating in three galactose residues (a triantennary structure) binds with a K_d of 5×10^{-9} M.[62] Three different subunits of the Gal receptor have been identified, but determination of the number of subunits and their relative stoichiometry within the native Gal receptor have remained elusive.[63]

Like the LDL receptor, the level of the ASGP receptor in the plasma membrane is subject to negative regulation by its ligand: on chronic exposure to ligand the ASGP receptor level decreases.[60] Because the affinity of the ASGP receptor for ligand and calcium is greatly decreased on exposure to the acidic pH of the early endosome,[61] the ASGP receptor is commonly utilized as a marker protein for early endocytotic compartments.

The Gal receptor form which resides only in hepatocytes[30,63] has been proposed to function in the removal of glycoproteins from the serum once they have been desialylated by neuraminidases in the blood. This function has been amplified recently by Weigel[63] into a general scheme called the "galactosyl homeostasis hypothesis." Its underlying premise is that in normal tissues, the maintenance of cell–cell interactions, in addition to cell–matrix and matrix–matrix interactions, relies on the binding of galactose glycoconjugates to cognate lectins and/or receptors. If the level of galactose/*N*-acetylgalactosamine glycoconjugates in the interstitial fluid becomes high, it will lead to disruption of the normal cellular organization in the differentiated tissue. Any galactose and *N*-acetylgalactosamine glycoconjugates that may be present in the serum as a result of disease, inflammation, or cell death will be removed by the Gal receptor, maintaining the glycoconjugate concentration in the serum at the normal level of approximately 10 nM and thereby maintaining tissue organization.

The Gal receptor has also been implicated in the removal of chylomicron remnants from the serum. It has been demonstrated that on incubation of HepG2 cells with antibody against the Gal receptor, the ability of the hepatoma cells to remove chylomicron remnants from the culture media was inhibited.[64] This finding was consistent with the earlier observation that apolipoprotein E, the major protein component of chylomicrons, is secreted as a sialylated glycoprotein but is subsequently desialylated in the plasma.[65] It therefore appears that the apo E component of the chylomicron remnants binds to the Gal receptor and the remnant is removed from the serum by receptor-mediated endocytosis.

A number of studies have implicated the Gal receptor in the uptake of IgA from the serum. In rodents and rabbits, serum polymeric IgA binds to a specific hepatocyte membrane receptor, secretory component, and the complex is then translocated across the liver cell to the bile canaliculus. The IgA is then released into the bile with a portion of secretory component, and the resulting secretory IgA plays a major role in mucosal immune defense.[66] In humans, however, the secretory component pathway is ineffective in the delivery of IgA to the mucosal surface. However, the Gal receptor has been shown to mediate the endocytosis of human IgA. This endocytotic process requires calcium and is inhibited by desialylated glycoproteins, indicating specificity for the hepatic Gal receptor.[67,68] However, it is not clear if the IgA molecule is ultimately routed to the bile. Therefore, Gal receptor-mediated uptake of IgA may represent a mechanism to remove IgA from the serum under pathological conditions where IgA concentrations increase to high levels, but not for the delivery of IgA to the bile.

A Gal receptor has also been identified in resident peritoneal macrophages.[30] While homologous to the Gal receptor on hepatocytes, the macrophage receptor is nevertheless distinct. The function of this receptor in macrophages has been proposed to be in the phagocytosis of aged red blood cells.[69,70]

16.4.1.2. The Mannose Receptor

The mannose receptor is also a scavenger receptor which specifically recognizes ligands containing terminal mannose and fucose residues. This receptor is localized to the endothelial cells of the liver and macrophage cells (both resident and elicited) from a wide variety of sources.[71] It functions in both the endocytosis of soluble ligands and phagocytosis. The location of these receptors on cells of the reticuloendothelial system makes them well-suited for the clearance of mannose-glycoconjugates from either the bloodstream or sites of inflammation. A number of natural soluble ligands have been identified for this receptor. These include lysosomal hydrolases, which may be present in the plasma as a result of tissue damage and/or inflammation,[72] tissue plasminogen activator,[73] and amylase.

While the ligands for the Gal receptor and LDL receptor are degraded on delivery to the lysosomes, degradation of the ligands of the mannose receptor begins within the endosomal compartment.[74] Acidification within the endosome activates a cathepsin D-like activity that initiates ligand degradation. It is not clear if the final stages of degradation occur on delivery to the lysosome. Recent studies have focused on the role of the macrophage mannose receptor in phagocytosis. Specific binding to the receptor on macrophages results in the efficient engulfment of bacteria, yeast, and other pathogenic organisms, and recent data have led to the suggestion that the macrophage mannose receptor may be the first line of defense against infection by *Pneumocystis carinii*.[75] This receptor has been cloned and when transfected into the nonmacrophage COS cell line, the COS cells could bind and ingest microbes,[75] indicating that expression of the mannose receptor is sufficient for phagocytosis. Further use of this clone of the mannose receptor in the production of chimeric receptors between the mannose receptor and the high-affinity Fc receptor demonstrated that both the transmembrane domain and the cytoplasmic domain of the mannose receptor are essential in mediating phagocytosis.[76] Currently, it is unclear if the same mannose receptor protein is mediating both the endocytosis of soluble ligands and the phagocytosis of large particles, or if, as described above for the Gal receptor, different receptor species exist for these distinct functions.

Recently, data have been reported that indicate that modulators of host defense can act by modifying the level of expression of the macrophage mannose receptor. Treatment of a macrophage cell line with interferon-gamma for time periods as short as 4–8 hr resulted in decreased levels of mannose receptor mRNA, while an overnight incubation reduced mRNA levels to an undetectable level.[77] HIV infection has been documented to decrease the ability of the macrophage mannose receptor to function in the phagocytosis of *P. carinii*.[78]

16.4.2. The α_2-Macroglobulin/Lipoprotein Receptor-Related Protein (LRP) Receptor

The function of this receptor is very diverse, primarily as a consequence of the multifunctional characteristics and diversity of its ligands.

α_2-Macroglobulin (α_2M) is a large multisubunit protein of the family that includes complement proteins C3 and C4. It comprises up to 10% of the total protein in serum.[79] Each subunit of α_2M contains a stretch of 25 amino acid residues that can act as a substrate for nearly any known proteinase. On cleavage by the proteinase, a conformational change is induced in α_2M that traps the proteinase. While the active site of the proteinase is not inactivated, the proteinase is effectively sequestered away from any substrates. In this way, α_2M protects from both endogenous proteinase activities and proteinases produced by pathogenic organisms such as *Trypanosoma cruzi*[80] and *Pseudomonas aeruginosa*.[81] As a result of these characteristics, α_2M has been characterized as a "pan proteinase inhibitor." The α_2M loaded with the trapped proteinases then binds to its specific receptor. Although this receptor is found on a number of different cell types, such as fibroblasts, adipocytes, and astrocytes, the majority of uptake of ligand is found in macrophages and hepatocytes.[79]

The pH dependence of binding of α_2M to its receptor demonstrates a very sharp profile, such that dissociation of ligand from receptor occurs at pH 6.8.[82] Therefore, dissociation of α_2M from its receptor occurs early in the endocytotic pathway, with ultimate degradation of α_2M and associated proteinases in the lysosome. However, not all proteinases trapped by α_2M are degraded. In particular, the proteinase from *Serratia marcescens* is trapped by α_2M in the plasma, but on endocytosis, the change in endosomal pH reactivates the enzyme with resulting cytotoxic activity.[83] Therefore, this cytotoxic proteinase depends on α_2M for its cell-killing activity. The α_2M receptor has also been implicated recently in mediating infection by the common cold virus. The minor-group human rhinovirus has been shown to specifically bind to the α_2M receptor, and this binding leads to the production of infectious virus.[84] This report did not ascertain the mechanism by which the α_2M receptor delivered the virus particles into the cell. However, endosomal acidification may play a role in this process.

α_2M also mediates the endocytosis of peptides and cytokines.[79] The interactions between α_2M and the cytokines have dissociation constants in the nanomolar range,[85,86] similar to the high-affinity interactions of hormones with their receptors. The α_2M molecule then acts to deliver the cytokines, such as nerve growth factor,[86] to the interior of the cell where it can act on signal transduction cascades. The mechanisms by which the cytokines escape from the endosome and interact with the cytosolic components of the signal transduction cascades are not yet defined. More recent data suggest that there may be two distinct α_2M receptors, one of which is itself a signaling receptor. These studies demonstrated that the binding of α_2M-methylamine to macrophages results in changes in intracellular calcium levels, IPs, and cAMP,[87] and that these effects are mediated through coupling of this receptor to a G-protein.[87]

The α_2M receptor not only binds the α_2M molecule, but also specifically interacts with other molecules, mediating their endocytosis. These molecules include: exotoxin A of *Pseudomonas aeruginosa*,[88] with resulting cytotoxic effects; urokinase plasminogen activator,[89,90] resulting in the clearance and ultimate degradation of the molecule; apolipoprotein E.[91]

The role of the Gal receptor in clearing chylomicron remnants from the serum through its interaction with apolipoprotein E has been well-documented and discussed above. Remnant clearance is also mediated by the LRP receptor. Recently, it has been found that the LRP receptor, which specifically binds to apoE and facilitates the clearance of chylomicron remnants from the circulation, is the same protein molecule as the α_2M receptor.[92] In fact, it appears that the binding site for α_2M and apoE are either identical or share overlapping domains, since α_2M is a competitive inhibitor of apoE-mediated uptake of chylomicron remnants.[93] The relative roles of the Gal receptor and the α_2M/LRP receptor in chylomicron remnant uptake are currently unknown.

By *in situ* hybridization and immunocytochemistry, the α_2M/LRP receptor has been localized in both early and advanced atherosclerotic lesions in smooth muscle cells.[94] The LDL receptor, however, was absent from the lesion area. These findings suggest a role of the α_2M/LRP receptor in the development of atherosclerotic lesions.

The α_2M receptor has also been found to be a constituent of the amyloid plaques found in Alzheimer's disease.[95] The enzyme transglutaminase has been found to catalyze the generation of covalently cross-linked polymers of β-amyloid protein, a major constituent of senile plaques, and the α_2M receptor.[95] This finding poses the question of whether the elevated expression of α_2M receptor in the brain predisposes an individual to Alzheimer's disease.

16.5. ENDOCYTOSIS IN IMMUNE FUNCTION: ANTIGEN PRESENTATION THROUGH THE MHC CLASS II MOLECULE

The immune response is composed of an antigen-specific recognition system mediated by both T and B lymphocytes. The antigens act as ligands that bind to specific receptors on the T and B cells. The antigens recognized by these receptors are fragments of proteins (designated epitopes) that are complexed with an endogenous or "self protein." The complex (peptide fragment plus self protein) is denoted the antigen-presenting molecule, and the cells that express these molecules on their plasma membrane are denoted antigen-presenting cells. The antigen-presenting molecule is encoded by the genes of the major histocompatibility complex (MHC), and the two classes of MHC molecules are denoted class I and class II. In general, MHC class II molecules bind peptides that are derived from extracellular proteins. These extracellular proteins, derived from invasive organisms such as bacteria, virus, fungus, or parasites, are internalized via the endocytotic pathway and are degraded within the endosomes to generate the antigenic peptides. The assembly of self protein and antigenic peptide also occurs within the endosomal apparatus. Hence, the endosomal pathway plays a pivotal role in immune function.

The self protein component of the class II molecule is composed of two transmembrane glycoprotein subunits, designated α and β. When loaded with peptide, which binds in a groove between the two subunits,[96] the class II molecule is expressed on the plasma membrane of B cells, some activated T cells, and some macrophage cells.[96] The peptide that binds in the intersubunit groove ranges in size from 12 to 25 amino acid residues.[97,98] It is the endocytotic machinery of the lymphocyte that generates the antigenic peptides and loads them onto the newly synthesized α and β subunits of the class II molecule.

The α and β chains of the class II molecule are synthesized and glycosylated in the endoplasmic reticulum. From there, the molecule is targeted to the endocytotic apparatus. This targeting is achieved by the multifunctional protein invariant chain.[99] Invariant chain complexes with the class II molecule and acts as a chaperone. Recent data indicate that two dileucine motifs in the cytoplasmic tail of invariant chain target the invariant chain–class II complex to the endosomal apparatus.[100] In the absence of invariant chain, or if a modification is made in the dileucine motifs, the class II molecules are routed by default to the plasma membrane.[100] In addition, invariant chain is envisioned to form a cap on the peptide binding groove in the class II molecule, thereby inhibiting the binding of peptide to the αβ dimer before delivery to the endosomal compartment.[101] Once within the acidic environment of the endosome, the invariant chain is proteolyzed. Under *in vitro* conditions, invariant chain can be released from class II molecules by the action of the lysosomal proteinase cathepsin B, and this is a likely mechanism of release of invariant chain under *in vivo* conditions.[102] The acidic environment of the endosome is a requirement for ultimate antigen presentation. When acidification processes are inhibited through the use of ammonia or chloroquine, no antigen presentation can be demonstrated by class II molecules.[103,104] Therefore, acidification serves at least two purposes in MHC class II assembly: removal of the invariant chain from the class II molecule, and degradation of the antigenic proteins to peptides.

The peptides that are ultimately complexed to the αβ dimer of the class II molecule are derived from extracellular proteins. These extracellular proteins are initially present in the serum complexed to antibodies, and these antibody–protein complexes then bind to specific receptors (slg or Fc receptors) on the surface of the antigen-presenting cell. By receptor-mediated endocytosis, these complexes gain entry into the endosomal apparatus of the cell. Peptide generation in the endosomal apparatus requires at least two steps: unfolding of the protein through the reduction of disulfide bonds,[104] and partial degradation of the protein.[105] Some proteins are more resistant to proteolysis than others, and currently, it is not clear how some peptides are protected from the more active proteolytic action required by other proteins. Interestingly, binding of peptides to the class II molecule has been shown to protect the peptide from additional degradation,[106] and this finding may provide insight into how the antigen-presenting cell controls differential sensitivity to proteinase activity.

Recent studies have focused on the identification of the endosomal compartment in which the peptide is loaded onto the class II molecule. A novel endosomal compartment has been identified where this loading occurs.[107–109] This compartment, termed the "compartment for peptide loading" (CPL), is biochemically and morphologically distinct from other endosomal compartments and lysosomes, as evidenced by the absence of markers unique to the endosomes or lysosomes. It has been proposed[110] that the newly synthesized $\alpha\beta$–invariant chain complexes move through the Golgi to an early endosomal compartment where proteolysis of the invariant chain begins. From there, the complexes move into the CPL where degradation of invariant chain goes to completion, leaving the $\alpha\beta$ subunits free of invariant chain, but in an aggregated state. This aggregated state causes retention of the $\alpha\beta$ subunits in the CPL, and prevents the routing of the peptide-free molecules to the plasma membrane.[111] Nevertheless, in an aggregated state, the subunits are in a conformation capable of binding peptide. Antigenic proteins are routed through early and possibly late endosomes to the CPL, undergoing progressive proteolysis. In the CPL, the peptides are loaded onto aggregated $\alpha\beta$ subunits. When loaded with peptide, the aggregated class II molecules disaggregate,[111] facilitating their transport to the plasma membrane. An additional class II-like molecule, DM, has been shown to be required for class II–peptide assembly,[112,113] and these proteins may be instrumental in the transport of class II–peptide complexes from the CPL to the plasma membrane.

Defects in any component of the lymphocyte endocytotic apparatus could lead to profound impairment of immune function.

Once at the plasma membrane, the class II molecule binds to CD4, a protein component of the T-cell receptor (TCR).[114] This binding initiates a signal transduction cascade that ultimately culminates in events such as lymphokine secretion, cellular proliferation and differentiation.[115] CD4 has been shown to be a high-affinity receptor for type 1 human immunodeficiency virus (HIV-I).[116] Binding of HIV-I, or a glycoprotein component of the virus, gpl20, inhibits the interaction of the MHC class II molecule with CD4.[117] This inhibition of binding and the subsequent depression of the immune response provide a molecular basis for the compromised immune function demonstrated in patients infected with HIV.

16.6. ENDOCYTOSIS IN NEUROTRANSMISSION

Neurotransmitters are amino acids, or derivatives thereof, that are synthesized and packaged into vesicles in the presynaptic terminals. On arrival of an action potential at the nerve terminal, the synaptic vesicles fuse with the presynaptic membrane and release their contents into the synaptic cleft. To terminate the effect of the neurotransmitter on the postsynaptic cell, the neurotransmitters can either be degraded within the cleft, or they can be reinternalized into the presynaptic cell and repackaged into synaptic vesicles for another round of exocytosis. Many of the processes related to the repackaging of the neurotransmitters into synaptic vesicles involve elements of the endocytotic apparatus.

The reutilization of neurotransmitter occurs through two discrete steps: (1) transport of the neurotransmitter across the plasma membrane of the presynaptic cell into the cell cytosol and (2) uptake of the neurotransmitter from the cytosol into the lumen of the synaptic vesicle.

The transport of neurotransmitters from the synaptic cleft into the cell occurs at the plasma membrane and is driven by sodium, potassium, and chloride gradients generated by the Na^+/K^+-ATPase. Neurotransmitter-specific carrier proteins in the plasma membrane are coupled to sodium and/or chloride cotransport to facilitate the delivery of the neurotransmitters to the cytosol.[118] For the transport of some neurotransmitters into the cytosol, such as glutamate and serotonin, the antiport of potassium is required.[119]

Once in the cytoplasmic milieu, the neurotransmitters are actively sequestered into the synaptic vesicles by neurotransmitter-specific transporter proteins localized in the membrane of the vesicle.[120] The driving force for this uptake is provided by the electrochemical proton gradient generated by the vesicle membrane-associated V-ATPase. The V-ATPase of the synaptic vesicle is similar in composition and activity to the ATP-dependent proton pumps found in endosomes and lysosomes.[118,121,122] Different neurotransmitters require the potential difference across the vesicle membrane, the pH gradient, or both as the driving force for the uptake process.[119] Interestingly, the neurotransmitter transport molecules found in the plasma membrane demonstrate no homology to the transporters found in the membrane of the synaptic vesicle.[118]

An understanding of the biogenesis and recycling of the synaptic vesicle has been the subject of much recent scrutiny. Recent data suggest that the proteins unique to the synaptic vesicle are synthesized in the endoplasmic reticulum, routed through the Golgi apparatus, and then delivered to the plasma membrane of the cell in a constitutive manner.[122,123] Constitutive endocytosis of these proteins places them in the endocytotic apparatus, where the proteins destined for the synaptic vesicles are sorted from other endocytosed proteins. Following sorting, budding from the larger endosomal vesicles results in the production of synaptic vesicles.[124] On activation of the V-ATPase, cytosolic neurotransmitters are pumped into the lumen of the vesicles, as described above. These loaded vesicles are then ready for fusion with the plasma membrane. After fusion, the proteins in the synaptic vesicles are recycled. The manner in which this recycling occurs is under debate. Data have been described in which the synaptic vesicles are derived directly from the plasma membrane by a direct budding process, with little, if any, membrane intermixing between plasma membrane and synaptic vesicle membrane.[125,126] Other investigators, however, have demonstrated that the regeneration of synaptic vesicles requires an additional round of sorting, involving an endosomal intermediate.[121,127,128] A clearer picture of these processes should be forthcoming soon.

The endocytosis process does appear to involve clathrin-coated pits and vesicles.[128] Several variants of the coat components, such as AP1, AP2,[129,130] and clathrin light chains,[131] have been found to be specific to cells of neuronal origin.[132] This specialization of the clathrin-associated proteins may be a reflection of the high degree of specialization of neuronal cells in membrane trafficking. Immunochemical studies utilizing antibodies directed against proteins found specifically in the mature synaptic vesicles have demonstrated that the same proteins are also found in the coated vesicles.[120] Interestingly, the stoichiometry of these vesicle-specific proteins is the same in the coated vesicles as is found in the mature synaptic vesicle.[120] This suggests that as the vesicle-specific proteins are routed from the plasma membrane and through the endosomal apparatus, there is very little "dilution" of the associated core of vesicle proteins by nonvesicle proteins either in the plasma membrane or in the coated vesicle. It is possible that the synaptic vesicle proteins cluster together during routing, and that they may be sorted in the endosomal apparatus as a whole, rather than as individual proteins. These findings, in fact, support the model whereby the synaptic vesicles are formed by direct budding from the plasma membrane with no endosomal intermediates.

The formation of endosome-like compartments as an intermediate in the recycling pathway has been suggested, although its involvement is still not firmly established. Under various experimental conditions, such as intense nerve stimulation, large endocytotic intermediates appear near the nerve terminal.[127,133] These structures vary in size and form, but eventually give rise to synaptic vesicles.[127] Recycling synaptic vesicles were found to be adjacent to these large endosomal intermediates,[133] and this compartment gradually disappeared as newly formed synaptic vesicles increased in number.[127,134] Acidification has been documented to occur within the lumen of the synaptic vesicle protein-sorting endosomes.[135] However, the functional significance of acidification in this sorting endosome has not been established.

Synaptic vesicles contain characteristic membrane proteins that have been utilized extensively as specific markers. Through monitoring these unique proteins, a decrease has been observed in the synaptic vesicle pool in individuals affected with Alzheimer's disease.[136] The defect leading to decreased synaptic vesicle number has not, however, been described.

Recently, serum from patients with certain autoimmune diseases was shown to immunologically react with membrane proteins unique to synaptic vesicles. Synaptotagmin, a vesicle protein implicated in the docking and fusion of the vesicle with the plasma membrane,[121,122] was identified as the putative antigenic protein in Lambert–Eaton myasthenic syndrome (LEMS). This autoimmune disease is a presynaptic disorder of the neuromuscular junction and is characterized by muscle weakness.[137] It is thought that the autoantibody binds to synaptotagmin-associated calcium channels, thereby blocking the influx of calcium and consequently affecting the release of acetylcholine into the synaptic cleft.[138] Another vesicle-associated membrane protein, amphiphysin,[139] was also linked as the autoantigen in cancer patients with Stiffman syndrome, a central nervous system autoimmune disease characterized by muscle rigidity.[140]

Thus, neurotransmitter packaging and reutilization are absolutely dependent on processes occurring in the endosomal apparatus.

16.7. ENDOCYTOSIS IN SIGNAL TRANSDUCTION: THE RECEPTOR-LINKED TYROSINE KINASES

In the process of signal transduction, extracellular molecules interact with a cell to effect a change in intracellular processes. Membrane receptors play a central role in this process. The binding of peptide hormones and other signaling ligands to their specific receptors is the first step in the signal transduction cascade. This binding event is often followed by endocytosis and the subsequent processing of the ligand–receptor complex. The signal-transducing receptors about which the most is known are the receptor-linked tyrosine kinases (RTKs), which include the epidermal growth factor (EGF) receptor, platelet-derived growth factor (PDGF) receptor, and the insulin receptor.[141] For these receptors, the binding of ligand to the extracellular domain of the receptors induces the clustering of the ligand–receptor complexes over clathrin-coated pits. The clustering then leads to the endocytosis of the complexes. Interestingly, endocytosis of these signal-transducing receptors occurs primarily when the receptors are loaded with ligand.[142] This differs significantly from the other receptor systems described above which undergo endocytosis on a constitutive basis, whether loaded with ligand or not. The loading and clustering of the ligand-bound receptors also stimulate both the autophosphorylation of the receptors on tyrosine residues and the activation of the intrinsic tyrosine kinase activity of the receptors.[143] While the binding of ligand occurs on the extracellular domain of the receptor molecules, the autophosphorylation sites are exclusively located in the cytoplasmic domains of the receptors. The tyrosine kinase activity and autophosphorylation of these receptors is intimately linked with their respective abilities to induce signal transduction cascades.[143–145]

Endocytosis and receptor-linked signal transduction are intimately linked. Endocytosis serves three discrete functions with respect to signal transduction by RTKs: (1) sequestration and degradation of ligand within the endosomes, (2) dephosphorylation of the receptor within the endosomes, and (3) signal transduction by the receptor within the endosomes.

16.7.1. Ligand Sequestration and Degradation

The binding of ligand and subsequent internalization of the ligand–receptor complexes is followed by acidification within the early endosomal compartments.[146] As has been found for LDL and a large number of other ligand–receptor systems, this acidification promotes the dissociation of ligand from the RTKs. While some ligand–receptor complexes, such as EGF and the EGF receptor, are jointly routed to the lysosomes where both are degraded,[147] neither insulin nor the insulin receptor is degraded within the lysosomes. Insulin-degrading activity has been characterized within the acidic endosomal compartment, and this enzymatic activity appears to degrade the insulin molecule to its component amino acids.[31,148] With short-term stimulation of the cell with insulin, the majority of the insulin receptor recycles back to the plasma membrane.[33] In effect, endocytosis of the ligand–RTK complex serves the purpose of clearing the stimulating ligand from the extracellular environment, thereby attenuating the signal.

With extended incubation with ligand, the insulin receptor level of the cell is downregulated[149,150] and this downregulation is mediated through endocytosis.[151] Insulin receptor downregulation has been found to occur by degradation of - the receptor by a thiol protease which requires an acid pH for optimal activity.[42] Interestingly, the degradation of insulin receptor, like degradation of insulin, occurs in an endosomal compartment.[151] It is not yet clear how chronic insulin stimulation of the cells induces this thiol protease activity. Two possibilities exist: The protease could exist in a preformed endosomal compartment, and chronic stimulation of the cells with insulin results in the alternate routing of the receptor from the recycling endosome into this degradative endosome. Alternately, chronic insulin treatment could lead to the biosynthetic insertion and/or activation of the protease in the endosomal compartment normally traversed by the insulin receptor. Which, if either, of these possibilities is true is not yet known. In addition, it is not yet clear whether proteolysis of insulin and the insulin receptor occurs in the same or different endosomal compartments.

16.7.2. Dephosphorylation of the Receptor

As indicated above, the insulin receptor, unlike the EGF receptor, is recycled back to the plasma membrane after endocytosis. In order for the recycled insulin receptor to mediate additional rounds of signal transduction, the tyrosine kinase activity of the receptor must be inactivated. This inactivation occurs through the dephosphorylation of the insulin receptor within the endosomal compartment.[142] Several lines of evidence indicate that dephosphorylation of the insulin receptor occurs in the endosomal apparatus. The laboratories of Posner and Bergeron[152] isolated both a plasma membrane and an endosomal fraction from insulin-stimulated liver cells. They demonstrated that there is less tyrosine phosphorylation of the insulin receptor localized in the endosomes than receptor localized in the plasma membrane. Other studies have utilized antiphosphotyrosine antibodies and demonstrated that the insulin receptor that recycles back into the plasma membrane of the cell is devoid of tyrosine phosphate.[153] Further studies indicate that there is a defined sequence of receptor dephosphorylation.[154] Phosphotyrosine phosphatases that act on the insulin receptor have been identified in both membrane-bound and cytosolic fractions.[154] In addition, a phosphatase activity that acts on both the insulin and EGF receptors has been localized within endosomal membranes.[155] There is, therefore, ample evidence to support the model that the endosomal routing of RTKs functions in the deactivation of their respective tyrosine kinase activities. This deactivation permits the recycling of a receptor molecule which, on rebinding ligand, is ready to become reactivated for the transduction of signal.

16.7.3. Endosomal RTKs as Signal Transducers

The underlying assumption of many investigators is that signal transduction is mediated exclusively by the RTK localized within the plasma membrane of the cell. In 1980, it was proposed by Posner *et al.* that receptors localized within the endosomal compartments may act as mediators of signaling.[156] As the RTKs are internalized into the cell, their tyrosine kinase domains, which mediate signal transduction, maintain an orientation of continued exposure to cytosolic factors. Therefore, as the receptors move through the endosomal apparatus, they are exposed to factors, such as kinases and phosphatases, that may be inaccessible to the receptors localized within the plasma membrane. These interactions may facilitate the propagation of signaling pathways that are unavailable to the receptor localized within the plasma membrane.

In support of this hypothesis, it has been demonstrated that the autophosphorylation activity of the insulin receptor changes as the receptor moves from the plasma membrane to the endosomes.[152] Our own data indicate that on evaluation of the ability of the endosomal insulin-activated insulin receptor to phosphorylate specific domains of its immediate downstream substrate, insulin receptor substrate-1, changes in both K_m and V_{max} values are observed (Wang and Knutson, manuscript in preparation).

Studies evaluating the role of endocytosis in RTK signal transduction have been considerably hampered by the interesting finding that tyrosine kinase-deficient RTKs do not undergo ligand-induced endocytosis.[157–159] In studies with the insulin receptor, kinase defective receptors were shown to be defective in endocytosis and also incapable of mediating glucose uptake, endogenous substrate phosphorylation, glycogen synthesis, or thymidine incorporation into DNA.[160–162] While these studies have suggested that tyrosine autophosphorylation is necessary for the endocytosis of the RTKs, it has made it difficult to dissect the role of the endosomal receptor in signal transduction.

The finding that tyrosine autophosphorylation is necessary for endocytosis of the RTKs is clearly a distinct requirement compared to the LDL receptor. The consensus sequence for endocytosis found in the LDL receptor, NPXY, has been identified in the cytosolic domains of 11 other intrinsic membrane proteins, including the insulin and EGF receptors.[142] It is possible that this consensus sequence confers constitutive endocytosis properties on proteins. This view is substantiated by the finding with the EGF receptor that on rendering it tyrosine kinase deficient through a point mutation in the ATP-binding domain, the EGF receptor is incapable of EGF-induced endocytosis, but nevertheless can undergo endocytosis in a constitutive, ligand-independent manner.[163] It has therefore been suggested that the autophosphorylation of the RTKs results in sorting signals to rout receptors into and through a ligand-induced endosomal pathway that is distinct from the constitutive pathway.[164]

In view of the role of the RTKs in mediating signals that are involved in growth,[165,166] it can be anticipated that defects in the endosomal routing of the RTKs might lead to either amplified or depressed growth. This suggestion is substantiated by work employing an engineered mutant of the EGF receptor defective in endocytosis. Cells transfected with this construct demonstrated reduced serum requirements for growth, loss of contact inhibition of growth, and enhanced EGF-dependent morphological transformation.[167] However, natural, disease-inducing modifications in endosomal trafficking of RTKs have been difficult to identify. A possible exception to this may be found with the insulin receptor. We had previously demonstrated that chronic stimulation of cells with insulin results in the proteolytic cleavage of the insulin receptor within the endosomes of the cell, resulting in the generation of a cytosolic fragment of the insulin receptor containing the signal transduction domain of the receptor.[42] We have demonstrated that this fragment of the insulin receptor is also found in tissues of hyperinsulinemic, diabetic rats, and that it stoichiometrically interacts with the intact insulin receptor and inhibits the autophosphorylation of the insulin receptor.[168] Since inhibition of the generation of the fragment decreases the diabetic symptoms of the animals, we propose that the proteolytic processing of the insulin receptor in the endosomes directly causes the insulin-resistant state associated with adult-onset diabetes mellitus. It remains to be shown that this inhibitor of receptor autophosphorylation also inhibits the endocytosis of the receptor.

Thus, endocytosis of signal-transducing receptors serves to not only attenuate the cellular signal through ligand degradation and receptor deactivation, but may also provide the subcellular site appropriate for coupling to effector molecules, facilitating signal transduction cascades.

16.8. SUMMARY

The processes of endocytosis serve a myriad of functions in cells, tissues, and intact organisms. In addition to the delivery of nutrients to the cell, endocytosis functions in the clearance of molecules and microbes from the circulation, the presentation of antigen for intact immune function, efficient neurotransmission, and the precise regulation of signal transduction. In fact, it appears that all cellular functions, both general and cell-specific, are impacted by endocytosis. Many human diseases are the direct result of defects, either genetic or environmentally induced, in endocytotic processes. However, in view of the breadth of functions subserved by endocytosis, it is remarkable that more disease states cannot be directly attributed to defective endocytosis. This is probably a reflection of our superficial knowledge of the molecular events surrounding endocytosis. As our knowledge of endocytosis increases, its role in human disease will probably become more evident. Hopefully, understanding these molecular mechanisms will also improve our ability to treat and cure these diseases.

REFERENCES

1. Goldstein, J. L., Brown, M. S., Anderson, R. G. W., Russell, D. W., and Schneider, W. J. Receptor-mediated endocytosis: Concepts emerging from the LDL receptor system. *Annu. Rev. Cell Biol.* **1:** 1–39.
2. Levitzki, A. (1985). Receptors. In *Endocytosis* (I. Pastan and M. C. Willingham, eds.), Plenum Press, New York, pp. 45–68.
3. Anderson, R. G. W. (1991). Molecular motors that shape endocytic membrane. In *Intracellular Trafficking of Proteins* (C. J. Steer and J. A. Hanover, eds.), Cambridge University Press, London, pp. 13–47.
4. Pearse, B. M. F., and Robinson, M. S. (1990). Clathrin, adaptors, and sorting. *Annu. Rev. Cell Biol.* **6:**151–171.
5. Hansen, S. H., Sandvig, K., and Van Deurs, B. (1993). Clathrin and HA2 adaptors: Effects of potassium depletion, hypertonic medium, and cytosol acidification. *J. Cell Biol.* **121:**61–72.
6. Mahaffey, D. T., Moore, M. S., Brodsky, F. M., and Anderson, R. G. W. (1989). Coat proteins isolated from clathrin coated vesicles can assemble into coated pits. *J. Cell Biol.* **108:**1615–1624.
7. Moore, M. S., Mahaffey, D. T., Brodsky, F. M., and Anderson, R. G. W. (1989). Assembly of clathrin-coated pits onto purified plasma membranes. *Science* **263:**558–563.
8. Chang, M.-P., Mallet, W. G., Mostov, K. E., and Brodsky, F. M. (1993). Adaptor self-aggregation, adaptor–receptor recognition and binding of alpha-adaptin subunits to the plasma membrane contribute to recruitment of adaptor (AP2) components to clathrin-coated pits. *EMBO J.* **12:**2169–2180.
9. Peeler, J. S., Donzell, W. C., and Anderson, R. G. W. (1993). The appendage domain of the AP-2 subunit is not required for assembly or invagination of clathrin-coated pits. *J. Cell Biol.* **120:**47–54.
10. Heuser, J. E., and Anderson, R. G. W. (1988). Hypertonic media inhibit receptor-mediated endocytosis by blocking clathrin-coated pit formation. *J. Cell Biol.* **108:**389–400.
11. Sorkin, A., and Carpenter, G. (1993). Interaction of activated EGF receptors with coated pit adaptins. *Science* **261:**612–615.
12. Collawn, J. F., Stangel, M., Kuhn, L. A., Esekogwu, V., Jing, S., Trowbridge, I. S., and Tainer, J. A. (1990). Transferrin receptor in-

ternalization sequence YXRF implicates a tight turn as the structural recognition motif for endocytosis. *Cell* **63:**1061–1072.

13. Trowbridge, I. S. (1991). Endocytosis and signals for internalisation. *Curr. Opin. Cell Biol.* **3:**634–641.
14. Watts, C., and Marsh, M. (1992). Endocytosis: What goes in and how? *J. Cell Sci.* **103:**1–8.
15. Vallee, R. B. (1992). Dynamin: Motor protein or regulatory GTPase. *J. Muscle Res. Cell. Motil.* **13:**493–496.
16. Lin, H. C., Moore, M. S., Sanan, D. A., and Anderson, R. G. W. (1991). Reconstitution of coated pit budding from plasma membranes. *J. Cell Biol.* **114:**881–891.
17. Carter, L. L., Redelmeier, T. E., Woollenweber, L. A., and Schmid, S. L. (1993). Multiple GTP-binding proteins participate in clathrin-coated vesicle-mediated endocytosis. *J. Cell Biol.* **120:**37–45.
18. Blum, J. S., Diaz, R., Mayorga, L. S., and Stahl, P. D. (1993). Reconstitution of endosomal transport and proteolysis. In *Subcellular Biochemistry. Endocytic Components: Identification and Characterization* (J. J. M. Bergeron and J. R. Harris, eds.), Plenum Press, New York, pp. 69–93.
19. Brown, M. S., Anderson, R. G. W., and Goldstein, J. L. (1983). Recycling receptors: The round-trip itinerary of migrant membrane proteins. *Cell* **32:**663–667.
20. Maxfield, F. R. (1985). Acidification of endocytic vesicles and lysosomes. In *Endocytosis* (I. Pastan and M. C. Willingham, eds.), Plenum Press, New York, pp. 235–257.
21. Salzman, N. H., and Maxfield, F. R. (1993). Quantitative fluorescence techniques for the characterization of endocytosis in intact cells. In *Subcellular Biochemistry. Endocytic Components: Identification and Characterization* (J. J. M. Bergeron and J. R. Harris, eds.), Plenum Press, New York, pp. 95–123.
22. Schmid, S., Fuchs, R., Kielian, M., Helenius, A., and Mellman, I. (1989). Acidification of endosome subpopulations in wild-type Chinese hamster ovary cells and temperature-sensitive acidification-defective mutants. *J. Cell Biol.* **108:**1291–1300.
23. Forgac, M. (1992). Structure, function and regulation of the coated vesicle V-ATPase. *J. Exp. Biol.* **172:**155–169.
24. Fuchs, R., Male, P., and Mellman, I. (1989). Acidification and ion permeabilities of highly purified rat liver endosomes. *J. Biol. Chem.* **264:**221–230.
25. Fuchs, R., Schmid, S., and Mellman, I. (1989). A possible role for Na^+,K^+-ATPase in regulating ATP-dependent endosome acidification. *Proc. Natl. Acad. Sci. USA* **86:**539–543.
26. Kane, P. M., Kuehn, M. C., Howald-Stevenson, I., and Stevens, T. H. (1992). Assembly and targeting of peripheral and integral membrane subunits of the yeast vacuolar (H(+)-ATPase. *J. Biol. Chem.* **267:**447–454.
27. Ho, M. N., Hill, K. J., Lindorfer, M. A., and Stevens, T. H. (1993). Isolation of vacuolar membrane H(+)-ATPase-deficient yeast mutants; the VMA5 and VMA4 genes are essential for assembly and activity of the vacuolar H(+)-ATPase. *J. Biol. Chem.* **268:**221–227.
28. Hirata, R., Umemoto, N., Ho, M. N., Ohya, Y., Stevens, T. H., and Anraku, Y. (1993). VMA12 is essential for assembly of the vacuolar H(+)-ATPase subunits onto the vacuolar membrane in Saccharomyces cerevisiae. *J. Biol. Chem.* **268:**961–967.
29. Mellman, I. (1992). The importance of being acid: The role of acidification in intracellular membrane traffic. *J. Exp. Biol.* **172:**39–45.
30. Geffen, I., and Spiess, M. (1992). Asialoglycoprotein receptor. *Int. Rev. Cytol.* **137B:**181–219.
31. Hamel, F. G., Posner, B. I., Bergeron, J. J. M., Frank, B. H., and Duckworth, W. C. (1988). Isolation of insulin degradation products from endosomes derived from intact rat liver. *J. Biol. Chem.* **263:** 6703–6708.
32. Doherty, J.-J., II, Kay, D. G., Lai, W. H., Posner, B. I., and Bergeron, J. J. M. (1990). Selective degradation of insulin within rat liver endosomes. *J. Cell Biol.* **110:**35–42.
33. Knutson, V. P. (1992). Ligand-independent internalization and recycling of the insulin receptor. Effects of chronic treatment of 3T3-C2 fibroblasts with insulin and dexamethasone. *J. Biol. Chem.* **267:** 931–937.
34. Diment, S., and Stahl, P. (1985). Macrophage endosomes contain proteases which degrade endocytosed protein ligands. *J. Biol. Chem.* **260:**15311–15317.
35. Rodman, J. S., Levy, M. A., Diment, S., and Stahl, P. D. Immunolocalization of endosomal cathepsin D in rabbit alveolar macrophages. *J. Leuk. Biol.* **48:**116–122.
36. Blum, J. S., Fiani, M. L., and Stahl, P. D. (1989). Characterization of neutral and acidic proteases in endosomal vesicles. *J. Cell Biol.* **109:**188a (Abstract).
37. Casciola-Rosen, L. A., Renfrew, C. A., and Hubbard, A. L. (1992). Lumenal labeling of rat hepatocyte endocytic compartments: Distribution of several acid hydrolases and membrane receptors. *J. Biol. Chem.* **267:**11856–11864.
38. Rodman, J. S., Levy, M. A., Diment, S., and Stahl, P. D. (1990). Immunolocalization of endosomal cathepsin D in rabbit alveolar macrophages. *J. Leuk. Biol.* **48:**116–122.
39. Roederer, M., Bowser, R., and Murphy, R. F. (1987). Kinetics and temperature dependence of exposure of endocytosed material to proteolytic enzymes and low pH; evidence for a maturation model for the formation of lysosomes. *J. Cell. Physiol.* **131:**200–209.
40. Blum, J. S., Fiani, M. L., and Stahl, P. D. (1992). Proteolytic cleavage of ricin A chain in endosomal vesicles. Evidence for the action of endosomal proteases at both neutral and acidic pH. *J. Biol. Chem.* **266:**22091–22095.
41. Renfrew, C. A., and Hubbard, A. L. (1991). Sequential processing of epidermal growth factor in early and late endosomes of rat liver. *J. Biol. Chem.* **266:**4348–4356.
42. Knutson, V. P. (1991). Proteolytic processing of the insulin receptor β subunit is associated with insulin-induced receptor down-regulation. *J. Biol. Chem.* **266:**15656–15662.
43. Zupan, A. A., and Johnson, E. M., Jr. (1991). Evidence for endocytosis-dependent proteolysis in the generation of soluble truncated nerve growth factor receptors by A875 human melanoma cells. *J. Biol. Chem.* **266:**15384–15390.
44. Wicker, G., Prill, V., Brooks, D., Gibson, G., Hopwood, J., Von Figura, K., and Peters, C. (1992). Mucopolysaccharidosis VI (Maroteaux–Iamy syndrome). An intermediate clinical phenotype caused by substitution of valine for glycine at position 137 of arylsulfatase B. *J. Biol. Chem.* **266:**21386–21391.
45. Hobbs, H. H., Brown, M. A., and Goldstein, J. L. (1992). Molecular genetics of the LDL receptor gene in familial hypercholesterolemia. *Hum. Mutat.* **1:**445–466.
46. Grossman, M., and Wilson, J. M. (1992). Frontiers in gene therapy: LDL receptor replacement for hypercholesterolemia. *J. Lab. Clin. Med.* **119:**457–460.
47. Tybjaerg-Hansen, A., and Humphries, S. E. (1992). Familial defective apolipoprotein B-100: A single mutation that causes hypercholesterolemia and premature coronary artery disease. *Atherosclerosis* **96:**91–107.
48. Morgan, E. H., and Baker, E. (1988). Role of transferrin receptors and endocytosis in iron uptake by hepatic and erythroid cells. *Ann N.Y. Acad. Sci.* **526:**65–82.
49. Bonkovsky, H. L. (1991). Iron and the liver. *Am. J. Med. Sci.* **301:** 32–43.
50. Seligman, P. A., Klausner, R. D., and Huebers, H. A. (1987). Molecular mechanisms of iron metabolism. In *The Molecular Basis of Blood Diseases* (G. Stamatoyannopoulos, A. W. Nienhuis, P. Leder, et al., eds.), Saunders, Philadelphia, pp. 219–244.
51. Iacopetta, B. J., and Morgan, E. H. (1983). The kinetics of transferrin endocytosis and iron uptake from transferrin in rabbit reticulocytes. *J. Biol. Chem.* **258:**9108–9115.

52. Hentze, M. W., Rouault, T. A., Harford, J. B., and Klausner, R. D. (1989). Oxidation–reduction and the molecular mechanism of a regulatory RNA–protein interaction. *Science* **244:**357–359.
53. Rouault, T. A., Hentze, M. W., Haile, D. J., Klausner, R. D., and Harford, J. B. (1989). The RNA-binding protein that interacts with the iron-responsive elements found in transferrin receptor and ferritin mRNA's: Isolation and partial characterization of the human protein. *Abstracts of IXth International Conference on Proteins of Iron Transport and Storage*, Brisbane, p. 19.
54. Sciot, R., Paterson, A. C., Van den Oord, J. J., and Desmet, V. J. (1987). Lack of hepatic transferrin receptor expression in hemochromatosis. *Hepatology* **7:**831–837.
55. Abe, Y., Muta, K., Mishimura, J., and Nawata, H. (1992). Regulation of transferrin receptors by iron in human erythroblasts. *Am. J. Hematol.* **40:**270–275.
56. Thorstensen, K., and Romslo, I. (1993). The transferrin receptor: Its diagnostic value and its potential as therapeutic target. *Scand. J. Clin. Lab. Invest.* **215:**113–120.
57. Raub, T. J., and Newton, C. R. (1991). Recycling kinetics and transcytosis of transferrin in primary cultures of bovine brain microvessel endothelial cells. *J. Cell. Physiol.* **149:**141–151.
58. Drickamer, K., and Taylor, M. E. (1993). Biology of animal lectins. *Annu. Rev. Cell Biol.* **9:**237–264.
59. Geffen, I., Wessels, H. P., Roth, J., Shia, M. A., and Spiess, M. (1989). Endocytosis and recycling of subunit H1 of the asialoglycoprotein receptor is independent of oligomerization with H2. *EMBO J.* **8:**2855–2862.
60. Steer, C. J., Weiss, P., Huber, B. E., Wirth, P. J., Thorgeirsson, S. S., and Ashwell, G. (1987). Ligand-induced modulation of the hepatic receptor for asialoglycoproteins in the human hepatoblastoma cell line, Hep G2. *J. Biol. Chem.* **262:**17524–17529.
61. Breitfeld, P. P., Simmons, C. F., Strous, G. J. A. M., Geuze, H. J., and Schwartz, A. L. (1985). Cell biology of the asialoglycoprotein receptor system: A model of receptor-mediated endocytosis. *Int. Rev. Cytol.* **97:**47–95.
62. Lee, Y. C., Townsend, R. R., Hardy, M. R., Lonngren, J., Arnarp, J., Haraldsson, M. and Lonn, H. (1983). Binding of synthetic oligosaccharides to the hepatic Gal/GalNAc lectin. Dependence on fine structural features. *J. Biol. Chem.* **258:**199–202.
63. Weigel, P. H. (1993). Endocytosis and function of the hepatic asialoglycoprotein receptor. *Subcell. Biochem.* **19:**125–161.
64. Windler, E., Greeve, J., Levkau, B., Kolb-Bachofen, V., Daerr, W., and Greten, H. (1991). The human asialoglycoprotein receptor is a possible binding site for low-density lipoproteins and chylomicron remnants. *Biochem. J.* **276:**79–87.
65. Zannis, V. I., McPherson, J., Goldberger, G., Karathanasis, S. K., and Breslow, J. L. (1984). Synthesis, intracellular processing, and signal peptide of human apolipoprotein E. *J. Biol. Chem.* **259:** 5495–5499.
66. Brown, W. R., and Kloppel, T. M. (1989). The liver and IgA: Immunological, cell biological and clinical implications. *Hepatology* **9:**763–784.
67. Daniels, C. K., Schmucker, D. L., and Jones, A. L. (1989). Hepatic asialoglycoprotein receptor-mediated binding of human polymeric immunoglobulin A. *Hepatology* **9:**229–234.
68. Mestecky, J., Moldoveanu, Z., Tomana, M., Epps, J. M., Thorpe, S. R., Phillips, J. O., and Kulhavy, R. (1989). The role of the liver in catabolism of mouse and human IgA. *Immunol. Invest.* **18:**313–324.
69. Aminoff, D. (1988). The role of sialoglycoconjugates in the aging and sequestration of red blood cells from circulation. *Blood Cells* **14:**229–247.
70. Schlepper-Schafer, J., and Kolb-Bachofen, V. (1988). Red cell aging results in a change of cell surface carbohydrate epitopes allowing for recognition by galactose-specific receptors of rat liver macrophages. *Blood Cells* **14:**259–269.
71. Pontow, S. E., Kery, V., and Stahl, P. D. (1992). Mannose receptor. *Int. Rev. Cytol.* **137B:**221–244.
72. Shepherd, V. L., and Hoidal, J. R. (1990). Clearance of neutrophil-derived myeloperoxidase by the macrophage mannose receptor. *Am. J. Respir. Cell. Mol. Biol.* **2:**335–340.
73. Otter, M., Kuiper, J., van Berkel, T. J., and Rijken, D. C. (1992). Mechanisms of tissue-type plasminogen activator (tPA) clearance by the liver. *Ann. N.Y. Acad. Sci.* **667:**431–442.
74. Rijnboutt, S., Kal, A. J., Geuze, H. J., Aerts, A., and Strous, G. J. (1991). Mannose 6-phosphate-independent targeting of cathepsin D to lysosomes in HepG2 cells. *J. Biol. Chem.* **266:**23586–23592.
75. Ezekowitz, R. A., Williams, D. J., Koziel, H., (1991). Uptake of Pneumocystis carinii mediated by the macrophage mannose receptor. *Nature* **351:**155–158.
76. Kruskal, B. A., Sastry, K., Warner, A. B., Mathieu, C. E., and Ezekowitz, R. A. (1992). Phagocytic chimeric receptors require both transmembrane and cytoplasmic domains from the mannose receptor. *J. Exp. Med.* **176:**1673–1680.
77. Harris, M., Super, M., Rits, M., Chang, G., and Ezekowitz, R. A. (1992). Characterization of the murine macrophage mannose receptor: Demonstration that the downregulation of receptor expression mediated by interferon-gamma occurs at the level of transcription. *Blood* **80:**2363–2373.
78. Koziel, H., Kruskal, B. A., Ezekowitz, R. A., and Rose, R. M. (1993). HIV impairs alveolar macrophage mannose receptor function against Pneumocystis carinii. *Chest* **103:**111S–112S.
79. Borth, W. (1992). α_2-Macroglobulin, a multifunctional binding protein with targeting characteristics. *FASEB J.* **6:**3345–3353.
80. Araujojorge, T. C., Lage, M. J. F., Rivera, M. T., Carlier, Z., and Van-Leuven, F. (1992). Trypanosoma cruzi-enhanced alpha-macroglobulin levels correlate with the resistance of BALB/cj mice to acute infection. *Parasitol. Res.* **78:**215–221.
81. Miyagawa, S., Nishino, N., Kamata, R., Okamura, R., and Maeda, H. (1991). Effects of proteinase inhibitors on growth of Serratia marcescens and Pseudomonas aeruginosa. *Microb. Pathol.* **11:**137–141.
82. Yamashiro, D. J., Borden, L. A., and Maxfield, F. R. (1989). Kinetics of α_2-macroglobulin endocytosis and degradation in mutant and wild-type Chinese hamster ovary cells. *J. Cell. Physiol.* **139:**377–382.
83. Maeda, H., Molla, A., Oda, T., and Katsuki, T. (1987). Internalization of serratial protease into cells as an enzyme–inhibitor complex with α_2-macroglobulin and regeneration of protease activity and cytotoxicity. *J. Biol. Chem.* **262:**10946–10950.
84. Hofer, F., Gruenberger, M., Kowalski, H., Machat, H., Huettinger, M., Kuechler, E., and Blass, D. (1994). Members of the low density lipoprotein receptor family mediate cell entry of a minor-group common cold virus. *Proc. Natl. Acad. Sci. USA* **91:**1839–1842.
85. Gonias, S. J. (1992). α_2-Macroglobulin: A protein at the interface of fibrinolysis and cellular growth regulation. *Exp. Hematol.* **20:**302–311.
86. Koo, P. H., and Stach, R. W. (1989). Interaction of nerve growth factor with murine α-macroglobulin. *J. Neurosci. Res.* **22:**247–261.
87. Misra, U. K., Chu, C. T., Gawdi, G., and Pizzo, S. V. (1994). Evidence for a second alpha 2-macroglobulin receptor. *J. Biol. Chem.* **269:**12541–12547.
88. Kounnas, M. Z., Morris, R. E., Thompson, M. R., FitzGerald, D. J., Strickland, D. K., and Saelinger, C. B. (1992). The α_2-macroglobulin receptor/low density lipoprotein receptor-related protein binds and internalizes *Pseudomonas* exotoxin A. *J. Biol. Chem.* **267:** 12420–12423.
89. Nykjoer, A., Peterson, M. C., Moller, M., Jensen, P. H., Moestrup, S. K., Holtet, T. L., Etyerodt, M., Thogersen, J. C., Munch, M., Andreasen, P. A., and Gliemann, J. (1992). Purified α_2-macroglobulin receptor/LDL receptor-related protein binds urokinase plasminogen

activator inhibitor type 2 complex. Evidence that the α_2-macroglobulin receptor mediates cellular degradation of urokinase receptor-bound complexes. *J. Biol. Chem.* **267:**14543–14546.

90. Andreasen, P. A., Sottrup-Jensen, L., Kjoller, L., Nykjaer, A., Moestrup, S. K., Peterson, C. M., and Gliemann, J. (1994). Receptor-mediated endocytosis of plasminogen activators and activator/inhibitor complexes. *FEBS Lett.* **338:**239–245.
91. Cooper, A. D. (1992). Hepatic clearance of plasma chylomicron remnants. *Semin. Liver Dis.* **12:**386–396.
92. Strickland, K. D., Ashcom, J. D., and Williams S. (1990). Sequence identity between the α_2-macroglobulin receptor and low density lipoprotein receptor-related protein suggests that this molecule is a multifunctional receptor. *J. Biol. Chem.* **265:**17401–17404.
93. Hussain, M. M., Maxfield, F. R., and Mas-Oliva, J. (1991). Clearance of chylomicron remnants by the low density lipoprotein receptor-related protein/α_2-macroglobulin receptor. *J. Biol. Chem.* **266:** 13936–13940.
94. Luoma, J., Hiltunen, T., Sarkioja, T., Moestrup, S. K., Gliemann, J., Kodama, T., Nikkari, T., and Yla-Herttuala, S. (1994). Expression of alpha 2-macroglobulin receptor/low density lipoprotein receptor-related protein and scavenger receptor in human atherosclerotic lesions. *J. Clin. Invest.* **93:**2014–2021.
95. Rasmussen, L. K., Sorensen, E. S., Petersen, T. E., Gliemann, J., and Jensen, P. H. (1994). Identification of glutamine and lysine residues in Alzheimer amyloid beta A4 peptide responsible for transglutaminase-catalyzed homopolymerization and cross-linking to alpha 2 M receptor. *FEBS Lett.* **338:**161–166.
96. Barber, L. D., and Parham, P. (1993). Peptide binding to major histocompatibility complex molecules. *Annu. Ref. Cell Biol.* **9:**163–206.
97. Chicz, R. M., Urban, R. G., Lane, W. S., Gorga, J. C., and Stern, L. J. (1992). Predominant naturally processed peptides bound to HLA-DR1 are derived from MHC-related molecules and are heterogeneous in size. *Nature* **358:**764–768.
98. Nelson, C. A., Roof, R. W., McCourt, D. W., and Unanue, E. R. (1992). Identification of the naturally processed form of hen egg white lysozyme bound to the murine histocompatibility complex class II molecule I-A^k. *Proc. Natl. Acad. Sci. USA* **89:**7380–7383.
99. Brodsky, F. M. (1990). The invariant dating service. *Nature* **348:** 581–582.
100. Pieters, J., Bakke, O., and Dobberstein, B. (1993). The MHC class II-associated invariant chain contains two endosomal targeting signals within its cytoplasmic tail. *J. Cell Sci.* **106:**831–846.
101. Roche, P. A., and Cresswell, P. (1990). Invariant chain association with HLA-DR molecules inhibits immunogenic peptide binding. *Nature* **345:**615–618.
102. Roche, P. A., and Cresswell, P. (1991). Proteolysis of the class II-associated invariant chain generates a peptide binding site in intracellular HLA-DR molecules. *Proc. Natl. Acad. Sci. USA* **88:**3150–3154.
103. Ziegler, H. K., and Unanue, E. R. (1982). Decrease in macrophage antigen catabolism caused by ammonia and chloroquine is associated with inhibition of antigen presentation of T cells. *Proc. Natl. Acad. Sci. USA* **79:**175–178.
104. Collins, D. S., Unanue, E. R., and Harding, C. V. (1991). Reduction of disulphide bonds within lysosomes is a key step in antigen processing. *J. Immunol.* **147:**4054–4059.
105. Vidard, L., Rock, K. L., and Benacerraf, B. (1991). The generation of immunogenic peptides can be selectively increased or decreased by proteolytic enzyme inhibitors. *J. Immunol.* **147:**1786–1791.
106. Mouritsen, S., Meldal, M., Werdelin, O., Stryhn Hansen, A., and Buus, S. (1992). MHC molecules protect T cell epitopes against proteolytic destruction. *J. Immunol.* **149:**1987–1993.
107. Amigorena, S., Drake, J. R., Webster, P., and Mellman, I. (1994). Transient accumulation of new class II MHC molecules in a novel endocytic compartment in B lymphocytes. *Nature* **369:**113–120.
108. Tulp, A., Verwoerd, D., Dobberstein, B., Ploegh, H. L., and Pieters, J. (1994). Isolation and characterization of the intracellular MHC class II compartment. *Nature* **369:**120–126.
109. West, M. A., Lucocq, J. M., and Watts, C. (1994). Antigen processing and class II MHC peptide-loading compartments in human B-lymphoblastoid cells. *Nature* **369:**147–151.
110. Schmid, S. L., and Jackson, M. R. (1994). Making class II presentable. *Nature* **369:**103–104.
111. Germain, R. N., and Rinker, A. G. J. (1993). Peptide binding inhibits protein aggregation of invariant-chain free class II dimers and promotes surface expression of occupied molecules. *Nature* **363:**725–728.
112. Morris, P., Shaman, J., Attaya, M., Amaya, M., Goodman, S., Bergman, C., Monco, J. J., and Mellins, E. (1994). An essential role for HLA-DM in antigen presentation by class II major histocompatibility molecules. *Nature* **368:**551–553.
113. Fling, S. P., Arp, B., and Pious, D. (1994). HLA-DMA and -DMB genes are both required for MHC class II/peptide complex formation in antigen-presenting cells. *Nature* **368:**554–558.
114. Klatzmann, D., Champagne, E., Chamaret, S., Gruest, J., Guetard, D., Hercend, T., Gluckman, J. C., and Montagnier, L. (1984). T-lymphocyte T4 molecule behaves as the receptor for human retrovirus LAV. *Nature* **312:**767–768.
115. De Rossi, A., Franchini, G., Aldovini, A., Del Mistro, A., Chieco-Bianchi, L., Gallo, R. C., and Wong-Staal, F. (1986). Differential response to the cytopathic effects of human T-cell lymphotrophic virus type III (HTLV-III) superinfection in $T4^+$ (helper) and $T8^+$ (suppressor) T-cell clones transformed by HTLV-I. *Proc. Natl. Acad. Sci. USA* **83**:4297–4301.
116. Folks, T., Kelly, J., Benn, S., Kinter, A., Justement, J., Gold, J., Redfield, R., Sell, K. W., and Fauci, A. S. (1986). Susceptibility of normal human lymphocytes to infection with HTLV-III/LAV. *J. Immunol.* **136:**4049–4053.
117. Lasky, L. A., Nakamura, G., Smith, D. H., Fennie, C., Shimasaki, C., Patzer, E., Berman, P., Gregory, T., and Capon, D. J. (1987). Delineation of a region of the human immunodeficiency virus type 1 gp120 glycoprotein critical for interaction with the CD4 receptor. *Cell* **50:**975–985.
118. Rudnick, G., and Clark, J. (1993). From synapse to vesicle: The reuptake and storage of biogenic amine neurotransmitter. *Biochim. Biophys. Acta* **1144:**249–263.
119. McMahon, A. T., and Nicholls, D. G. (1991). The bioenergetics of neurotransmitter release. *Biochim. Biophys. Acta* **1059:**243–264.
120. Maycox, P. R., Link, E., Reetz, A., Morris, S. A., and Jahn, R. (1992). Clathrin-coated vesicles in nervous tissue are involved primarily in synaptic vesicle recycling. *J. Cell Biol.* **118:**1379–1388.
121. Sudhof, T. C., and Jahn, R. (1991). Proteins of synaptic vesicles involved in exocytosis and membrane recycling. *Neuron* **6:**665–677.
122. Kelly, R. B. (1993). Storage and release of neurotransmitters. *Cell* (Suppl.) **72:**43–53.
123. Regnier-Vigoroux, A., and Huttner, W. B. (1994). Biogenesis of small synaptic vesicles and synaptic-like microvesicles. *Neurochem. Res.* **18:**59–64.
124. Linstedt, A. D., and Kelly, R. B. (1991). Synaptophysin is sorted from endocytotic markers in neuroendocrine PC12 cells but not transfected fibroblasts. *Neuron* **7:**309–317.
125. Ceccarelli, B., Hurlbut, W. P., and Mauro, A. (1973). Turnover of transmitter and synaptic vesicles at the frog neuromuscular junctions. *J. Cell Biol.* **57:**499–524.
126. Valtorta, F., Fesce, R., Grohovaz, F., Haimann, C., Hurlbut, W. P., Lezzi, H., Torri-Tarelli, F., Villa, A., and Ceccarelli, B. (1990). Neurotransmitter release and synaptic vesicle recycling. *Neuroscience* **35:**477–489.
127. Miller, T. M., and Heuser, J. E. (1984). Endocytosis of synaptic vesicle membrane at the frog neuromuscular junction. *J. Cell Biol.* **98:** 685–698.

128. Heuser, J. (1989). The role of coated vesicles in recycling of synaptic vesicle membrane. *Cell Biol. Int. Rep.* **13:**1063–1076.
129. Robinson, M. S. (1989). Cloning of cDNAs encoding two related 100-kD coated vesicle proteins (α-adaptins). *J. Cell Biol.* **108:**833–842.
130. Ponnambalam, S., Robinson, M. S., Jackson, A. P., Peiper, L., and Parham, P. (1993). Conservation and diversity in families of coated vesicle adaptins. *J. Biol. Chem.* **265:**4814–4820.
131. Jackson, A. P., Sewo, H. F., Holmes, N., Drickamer, K., and Parhem, P. (1987). Clathrin light chains contain brain-specific insertion sequences and a region of homology with intermediate filaments. *Nature* **326:**154–159.
132. Pearse, B. M., and Robinson, M. S. (1990). Clathrin, adaptors and sorting. *Annu. Rev. Cell Biol.* **6:**151–171.
133. Haimann, C., Torri-Tarelli, F., Fesce, R., and Ceccarelli, B. (1985). Measurement of quantal secretion induced by ouabain and its correlation with depletion of synaptic vesicles. *J. Cell Biol.* **101:** 1953–1965.
134. Bennett, M. V. L., Model, P. G., and Highstein, S. M. (1976). Stimulation-induced depletion of vesicles, fatigue of transmission and recovery processes at a vertebrate central synapse. *Cold Spring Harbor Symp. Quant. Biol.* **40:**25–36.
135. Sulzer, D., and Holtzman, E. (1989). Acidification and endosome-like compartments in the presynaptic terminals of frog retinal photoreceptors. *J. Neurocytol.* **18:**529–540.
136. Lassman, H., Weiler, R., Fischer, P., Bancher, C., Jellinger, K., Floor, E., Danielczyk, W., Seitelberger, F., and Winkler, H. (1992). Synaptic pathology in Alzheimer's disease: Immunological data for markers of synaptic and large dense core vesicles. *Neuroscience* **46:**1–8.
137. Lisak, R. P. (1994). *Handbook of Myasthenia Gravis and Myasthenic Syndromes,* Dekker, New York, pp. 81–102.
138. Leveque, C., Hoshino, T., David, P., Shoji-Kasai, Y., Leys, K., Omori, A., Lang, B., El Far, O., Sato, K., Martin, M. N., Newsom-Davis, J., Takahashi, M., and Seagar, M. (1992). The synaptic vesicle protein synaptotagmin associates with calcium channels and is a putative Lambert–Eaton myasthenic syndrome antigen. *Proc. Natl. Acad. Sci. USA* **89:**3625–3629.
139. Lichte, B., Veh, R. W., Meyer, H. E., and Kilimann, M. W. (1992). Amphiphysin, a novel protein associated with synaptic vesicles. *EMBO J.* **11:**2521.
140. De Camilli, P., Thomas, A., Cofiel, R., Folli, F., Lichte, B., Piccolo, G., Meinck, H., Austoni, M., Fassetta, G., Bottazzo, G., Bates, D., Cartlidge, N., Solimena, M., and Kilimann, M. W. (1993). The synaptic vesicle-associated protein amphyphysin is the 128 kD autoantigen of Stiff-Man Syndrome with breast cancer. *J. Exp. Med.* **178:**2219–2223.
141. Cadena, D. H., and Gill, G. N. (1992). Receptor tyrosine kinases. *FASEB J.* **6:**2332–2337.
142. Kahn, M. N., Lai, W. H., Burgess, J. W., Posner, B. I., and Bergeron, J. J. M. (1993). Potential role of endosomes in transmembrane signaling. In *Subcellular Biochemistry. Endocytic Components: Identification and Characterization* (J. J. M. Bergeron and J. R. Harris, eds.), Plenum Press, New York, pp. 223–254.
143. Fantl, W. J., Johnson, D. E., and Williams, L. T. (1993). Signalling by receptor tyrosine kinases. *Annu. Rev. Biochem.* **62:**453–481.
144. Ullrich, A., and Schlessinger, J. (1990). Signal transduction by receptors with tyrosine kinase activity. *Cell* **61:**203–212.
145. Pazin, M. J., and Williams, L. T. (1992). Triggering signaling cascades by receptor tyrosine kinases. *Trends Biochem. Sci.* **17:**374–378.
146. Knutson, V. P., Ronnett, G. V., and Lane, M. D. (1985). The effects of cycloheximide and chloroquine on insulin receptor metabolism. Differential effects on receptor recycling and inactivation and insulin degradation. *J. Biol. Chem.* **260:**14180–14188.
147. Lai, W. H., Cameron, P. H., Wada, I., Doherty, J. J., Posner, B. I., and Bergeron, J. J. M. (1989). Ligand mediated internalization, recycling, and down regulation of the epidermal growth factor receptor in vivo. *J. Cell Biol.* **109:**2741–2749.
148. Duckworth, W. C., Hamel, F. G., Peavy, D. E., Liepnieks, J. J., Ryan, M. P., Hermodson, M. A., and Frank, B. H. (1988). Degradation products of insulin generated by hepatocytes and by insulin protease. *J. Biol. Chem.* **263:**1826–1833.
149. Knutson, V. P., Ronnett, G. V., and Lane, M. D. (1982). Control of insulin receptor level in 3T3 cells: Effect of insulin-induced down-regulation and dexamethasone-induced up-regulation on rate of receptor inactivation. *Proc. Natl. Acad. Sci. USA* **79:**2822–2826.
150. Ronnett, G. V., Knutson, V. P., and Lane, M. D. (1982). Insulin-induced down-regulation of insulin receptors in 3T3-L1 adipocytes. Altered rate of receptor inactivation. *J. Biol. Chem.* **257:**4285–4291.
151. Knutson, V. P., Ronnett, G. V., and Lane, M. D. (1983). Rapid, reversible internalization of cell surface insulin receptors. Correlation with insulin-induced down-regulation. *J. Biol. Chem.* **258:**12139–12142.
152. Burgess, J. W., Wada, I., Ling, N., Kahn, M. N., Bergeron, J. J. M., and Posner, B. I. (1992). Decrease in β-subunit phosphotyrosine correlates with internalization and activation of the insulin receptor kinase. *J. Biol. Chem.* **267:**10077–10086.
153. Backer, J. M., Kahn, C. R., and White, M. F. (1989). Tyrosine phosphorylation of the insulin receptor during insulin-stimulated internalization in rat hepatoma cells. *J. Biol. Chem.* **264:**1694–1701.
154. King, M. J., and Sale, G. J. (1990). Dephosphorylation of insulin-receptor autophosphorylation sites by particulate and soluble phosphotyrosyl-protein phosphatases. *Biochem. J.* **266:**251–259.
155. Faure, R., Baquiran, G., Bergeron, J. J. M., and Posner, B. I. (1992). The dephosphorylation of insulin and epidermal growth factor receptors: Role of endosome-associated phosphotyrosine phosphatase(s). *J. Biol. Chem.* **267:**11215–11221.
156. Posner, B. I., Patel, B., Verma, A. K., and Bergeron, J. J. M. (1980). Uptake of insulin by plasmalemma and Golgi subcellular fractions of rat liver. *J. Biol. Chem.* **255:**735–741.
157. Sorkin, A., Waters, C., Overholser, K. A., and Carpenter, G. (1991). Multiple autophosphorylation site mutations of the epidermal growth factor receptor. Analysis of kinase activity and endocytosis. *J. Biol. Chem.* **266:**8355–8362.
158. Felder, S., LaVin, J., Ullrich, A., and Schlessinger, J. (1992). Kinetics of binding, endocytosis and recycling of EGF receptor mutants. *J. Cell Biol.* **117:**203–212.
159. Carpentier, J.-L., Paccaud, J.-P., Backer, J., Gilbert, A., Orci, L., and Kahn, C. R. (1993). Two steps of insulin internalization depend on different domains of the β subunit. *J. Cell Biol.* **122:**1243–1252.
160. McClain, D. A., Maegawa, H., Lee, J., Dull, T. J., Ullrich, A., and Olefsky, J. M. (1987). A mutant insulin receptor with defective tyrosine kinase displays no biologic activity and does not undergo endocytosis. *J. Biol. Chem.* **262:**14663–14671.
161. Chou, C. K., Dull, T. J., Russell, D. S., Cherzi, R., Lebwohl, D., Ullrich, A., and Rosen, O. M. (1987). Human insulin receptor mutated at the ATP-binding site lack protein tyrosine kinase activity and fail to mediate postreceptor effects of insulin. *J. Biol. Chem.* **262:** 1842–1847.
162. Ebina, Y., Araki, E., Taira, M., Shimada, R., Mori, M., Craik, C. S., Siddle, K., Pierce, S. B., Roth, R. A., and Rutter, W. J. (1987). Replacement of lysine residue 1030 in the putative ATP-binding region of the insulin receptor abolishes insulin- and antibody-stimulated glucose uptake and receptor kinase activity. *Proc. Natl. Acad. Sci. USA* **84:**704–708.
163. Honegger, A. M., Dull, T. J., Felder, S., Van Obberghen, E., Bellot, F., Szapary, D., Schmidt, A., Ullrich, A., and Schlessinger, J. (1987).

Point mutation at the ATP binding site of EGF receptor abolishes protein-tyrosine kinase activity and alters cellular routing. *Cell* **51:** 199–209.

164. Felder, S., Miller, K., Moehren, G., Ullrich, A., Schlessinger, J., and Hopkins, C. R. (1990). Kinase activity controls the sorting of the epidermal growth factor receptor within the multivesicular body. *Cell* **61:**623–634.

165. Iwashita, S., and Kobayashi, M. (1992). Signal transduction system for growth factor receptors associated with tyrosine kinase activity: Epidermal growth factor receptor signalling and its regulation. *Cell. Signal.* **4:**123–132.

166. Crouch, M. F., and Hendry, I. A. (1993). Growth factor second messenger systems: Oncogenes and the heterotrimeric GTP-binding protein connection. *Med. Res. Rev.* **13:**105–123.

167. Chen, W. S., Lazar, C. S., Lund, K. A., Welsh, J. B., Chang, C.-P., Walton, G. M., Oer, C. J., Wiley, H. S., Gill, G. N., and Rosenfeld, M. G. (1989). Functional independence of the epidermal growth factor receptor from a domain required for ligand-induced internalization and calcium regulation. *Cell* **59:**33–43.

168. Knutson, V. P., Donnelly, P. V., Balba, Y., et al. (1995). Insulin resistance mediated by a proteolytic fragment of the insulin receptor. *J. Biol. Chem.* **270:**24972–24981.

17

Signal Transduction by G Protein-Coupled Receptors

Mariel Birnbaumer and Lutz Birnbaumer

17.1. INTRODUCTION

The primary structure of most of the components involved in G protein-mediated signal transduction are now well known. They include a large family of transmembrane receptors, a large family of heterotrimeric ($\alpha\beta\gamma$) G proteins activated by GTP under the influence of receptors, and a series of molecularly unrelated effectors that are regulated by G protein α, G protein $\beta\gamma$ dimer, or both G protein α and $\beta\gamma$ dimers (Table 17.1). The list of extracellular compounds depending on G proteins for their signaling includes hormones, neurotransmitters, auto- and paracrine factors. The abundance and diversity of the signaling molecules illustrate the central role of G protein-mediated signal transduction in cell regulation and body homeostasis.

The cellular effector functions regulated by the activated G proteins include, foremost, adenylyl cyclase (AC) and the type Cβ phospholipase (PLCβ), which are the enzymes responsible for the production of the classical second messengers, cAMP—by AC—and inositol 1,4,5-trisphosphate (IP3) and diacylglycerol (DAG)—by PLCβ. Other effectors that are expressed on some but not all cells are also targets of G protein regulation. These include visual phosphodiesterase (PDE), Ca^{2+} channels, and a variety of K^+ channels. Yet there may be more (Table 17.2).

Specificity and selectivity parameters define which G protein is activated by which receptor, and which effector is activated by which G protein. These specificity and selectivity parameters are stringent enough to avoid undesired cross talk between distinct G protein-regulated pathways, for example, AC versus PLC, but they are not absolute. Because of this it is possible that under extreme conditions the activation of one G protein-mediated signaling pathway, say activation of AC, may be accompanied by that of another, e.g., PLC. This gives rise to responses that may vary in interesting ways, depending on receptor density and receptor occupancy.

Here we will summarize the relevant aspects of the structural and functional features of the "receptor —> G protein —> effector" axis (Figure 17.1). We shall begin with the G proteins which transduce the conformational change of the receptor induced by agonist into a regulated effector function. This will be followed by a discussion of structural features of the receptors. The chapter will finish with a short description of molecular diversity within the two main effector functions: ACs and PLCβs and their responses to G protein subunits. An overview of the historical development of this field was published and will not be reiterated here.[1] As a "precursor" review to this one we suggest Birnbaumer *et al.*[2]

17.2. G PROTEIN-COUPLED RECEPTORS

Receptors that act through G proteins are encoded in a large structurally related superfamily of molecules, referred to variously as *G protein-coupled receptors,* which relates to their mode of action, or as *seven-transmembrane-spanning or heptahelical receptors,* which relates to their common structure. The arrangement of these seven transmembrane segments (TMs) in relation to each other was initially inferred from the structural homology of rhodopsin to bacteriorhodopsin, for which it has been possible to obtain a three-dimensional structure by electron microscopy with a resolution between 3.5 and 7 Å (Figure 17.2A). The resolved image showed the TMs to be perpendicular to the plane of the membrane enclosing a central space (Figure 17.2B).[3,4] A similar picture was obtained for rhodopsin on the basis of electron diffraction projection maps derived from two-dimensional crystals that formed on reconstitution of the purified protein into phospholipid vesicles,[5] which then allowed the construction of a detailed amino acid sequence comparison map of the TMs of G protein-coupled receptors.[6] Figure 17.2C shows an idealized view from the bottom or intracellular side of a G protein-coupled receptor, based on these observations. In this view αhelices have clockwise connectivity as they progress from the intra- to the extracellular face of the membrane. Figure 17.3 shows a

Mariel Birnbaumer and Lutz Birnbaumer • Departments of Anesthesiology and Biological Chemistry, School of Medicine, and the Molecular Biology Institute, University of California at Los Angeles, Los Angeles, California 90095.

Molecular Biology of Membrane Transport Disorders, edited by S.G. Schultz *et al.* Plenum Press, New York, 1996

Table 17.1. Molecular Elements of the G Protein Transmembrane Signal Transduction Path

A. > 200 receptors

- For neurotransmitters and autacoids
  - Biogenic amines
    - catecholamines
      - adrenaline
      - noradrenaline
      - dopamine
    - serotonin (5HT)
    - histamine
    - melatonin
  - Nonbiogenic amines
    - γ-aminobutyric acid (GABA)
    - glutamate
    - acetylcholine
    - cannabinoid: anandamide
    - purines
      - adenosine
      - ATP
- For opioid peptides
  - β-endorphin
  - leu- and met-enkephalins
  - dynorphin
  - neo-dynorphin
- For pituitary hormones
  - adrenocorticotropin (ACTH)
  - glycoprotein hormones
    - luteinizing hormone (LH)
    - thyrotropin (TSH)
    - follicle-stimulating hormone (FSH)
  - melanocyte-stimulating hormones (α-, β-, & γ-MSH)
  - vasopressin
  - oxytocin
- For pituitary release factors
  - thyrotropin release hormone (TRH)
  - corticotropin release factor (CRF)
  - somatotropin release factor (GRF)
  - gonadotropin release hormone (GnRH)
  - pituitary adenylate cyclase activating peptide (PACAP)
  - somatostatin (SST, SRIF)
- For kinins
  - tachykinins or neurokinins
    - Substance P
    - Substance K
    - Neurokinin B
  - bradykinin
  - angiotensin II
  - endothelin
- For gastrointestinal peptides
  - cholecystokinin/pancreozymin (CCK)
  - gastrin
  - secretin
  - glucagon
  - glucagon-like peptide 1
  - glucose-dependent insulinotropic peptide (GIP)
  - gastrin releasing peptide (GRP)
  - gastrin releasing peptide (GRP)
  - peptide tyrosine-tyrosineamide (PYY)
- For neuropeptides
  - neuropeptide Y (NPY)
  - vasoactive intestinal peptide (VIP)
  - peptide histidine-methionineamide (PHM)
- For endocrine and paracrine hormones
  - parathyroid hormone (PTH)
  - calcitonin
  - calcitonin gene-related product (CGRP)
  - amylin
  - pancreatic polypeptide (PP)
  - pancreastatin
  - galanin
- For chemokines and chemotactic agents
  - interleukin 8
  - formyl-peptides
  - C5a
  - C3a
  - platelet-activating factor (PAF)
  - thrombin
- For arachidonic acid metabolites
  - prostaglandins and prostacyclin
  - thromboxanes
  - leukotrienes
- For sensory inputs
  - light: opsins (dim, red, green, blue)
  - odors
  - taste (sweet, bitter)
- For divalent cations
  - Ca^{2+}

B. G proteins: combinations of α and βγ

- α subunits
  - α-s
  - α-olf
  - α-q
  - α-11
  - α-14
  - α-15/16
  - α-12
  - α-13
  - α-o1(α-oA)
  - α-o2 (α-oB)
  - α-i1
  - α-i2
  - α-i3
  - α-z
  - α-t(rod)—transducin
  - α-t(cone)
  - α-gust—gustducin
- βγ dimers
  - β1 through β5
  - γ1 through γ7

C. Response elements (effectors)

- Adenylyl cyclases—8 genes
- Phospholipases (PLCβ)—4 genes
- Ion channels
  - K channels
    - inward rectifier—1 gene

Table 17.1. Continued

ATP-sensitive—1 gene
Ca^{2+}-activated—>1 gene
Ca channels
cardiac—C type
skeletal muscle—S type
presynaptic—types:
N, L, ?others
Others
plasma membrane Ca-pump
?glucose transporter
?Na/H exchanger
D. Examples of molecular diversity of cloned receptors
Catecholamine receptors
3 β-adrenergic (β1, β2, and β3)
3 α_1-adrenergic (1A, 1B, and 1C)
3 α_2-adrenergic (C10, C2, C4)
5 dopaminergic (D1 through D5)
Serotonin receptors
> 10: 5 type 1: 5HT-1A, -1B (also 1Dβ), -1D (also 1Dα), -1E, -1F
3 type 2: 5HT-2A (former 2), -2B (former 2F), -2C (former 1C)
other: 5HT-4, 5HT-5, 5HT-7
Acetylcholine
5 muscarinic (M1 through M5)
Peptide hormones
5 somatostatin
3 vasopressin (VP-1a, VP-1b, VP-2)
3 opioid: μ, δ, κ
3 MSH
2 NPY/PYY (Y1, Y2)
2 VIP (type I and II)
2 endothelin (ET-a, ET-b)
2 angiotensin II (type I and II)
2 bradykinin (BK1, BK2)
Odorant
> 200

scheme in which the receptor has been flattened to simplify description of its primary features.

In addition to the seven transmembrane domains with an extracellular N-terminal domain, these features include three intra- and three extracellular loops, and an intracellular C-terminal tail connected by the transmembrane domains. Most receptors, but not all, have a cysteine about 13 amino acids from the end of TM-VII that is palmitoylated, providing for what is assumed to be another membrane anchor and a fourth intracellular loop. In some cases there is no C-terminal cysteine, in others there is a pair of vicinal cysteines (e.g., rhodopsin, type 2 vasopressin receptor). While removal by mutagenesis of the palmitoylation site does not affect function in some receptors (rhodopsin, α_2-adrenergic), it impairs function in others. For example, a β_2-adrenergic receptor mutant (Cys341 to Gly) is hyperphosphorylated and largely uncoupled from the G_s/adenylyl cyclase system. Most, but also not all, have a cysteine in each of the first and second extracellular loops, which in rhodopsin, and by inference in other receptors, may form a disulfide bridge that contributes to the stability of the seven transmembrane structure. A changeable number of variably located *N*-glycosylation sites have been found in the N-terminal extension or in one of the extracellular loops. While not required for function once assembled, for some receptors removal of the glycosylation sites impairs their proper assembly and/or membrane insertion, resulting in defective cell surface expression.

Little is known about the folding and membrane insertion process except for the fact that the receptor can be constructed from two independently folding domains, one encompassing an N-terminal half (N-terminus to third intracellular loop) and a C-terminal half (from third intracellular loop to the end). Once assembled, intra-transmembrane domain forces and possibly disulfide bridges maintain the tertiary structure, and it is possible to fragment the molecule by proteolytic digestion without loss of function.

Most G protein-coupled receptors have been cloned by now, giving a wealth of information about the conserved and divergent parts of the receptors. While the transmembrane core, extracellular loops, and first and second intracellular loops are relatively similar in size forming a similarly sized family of proteins of ca. 350 amino acids, their N-termini, third intracellular loop, and C-termini vary very much in length. The longest N-termini appear to be the glycoprotein hormone-binding N-termini of the receptors for LH, FSH, and TSH with ca. 350 amino acids. Hormone specificity of these receptors lies in their N-termini. Depending on the length, the N-termini either contain a signal sequence (LH, FSH/TSH receptors; glucagon, PTH) or do not (neurotransmitter receptors, rhodopsin, α-thrombin receptor). Table 17.3 presents an amino acid sequence alignment for a selected few of these receptors.

17.2.1. Subfamilies

Analysis of sequence alignments has defined the existence of at least three distinct subfamilies. The first subfamily has as its structural signature a DRY motif and encompasses the vast majority of the greater than 200 G protein-coupled receptors. They have about 12 highly conserved amino acids in their TMs: 3–4 in TM-II, 2 in TM-VI, and 2–3 in TM-VII, and, at the interface between the membrane and the cytoplasm, at the end of TM-III, a so-called DRY consensus motif (Asp-Arg-Tyr), of which the R is absolutely conserved in all receptors of this subfamily, the D is mostly D, but can also be E (rhodopsin) and the Y, while present in about 100 members, varies in the others, the most frequently found replacement being F, H, and C. An amino acid sequence alignment of a selected few of the receptors belonging to the class with DRY motif can be found in Table 17.3. DRY-motif-containing receptors do not show a G protein preference, or preference for neurotransmitter (small ligand) versus peptide hormone (large ligand).

Table 17.2. Examples of Mammalian Effector Systems Regulated by G Proteins

Effector class (structural/functional characteristics)	Comments (molecular diversity—types/subtypes)
Adenylyl cyclases (ACs) 12 transmembrane helices/domains (TMs)	8 types (I–VIII) All stimulated by $G_s\alpha$, except 40-kDa sperm AC; some stimulated by Ca^{2+}/CaM (e.g., types I and III) Two or more are inhibited by $G_i\alpha$'s (e.g., types V, VI, and II) Effect of $G\beta\gamma$ may be: inhibition (type I) stimulation, conditional on $G_s\alpha$-stimulation (types II, IV) no effect (e.g., types III, V, VI)
cGMP-phosphodiesterase (cG-PDE) Peripheral membrane protein: $\alpha\beta\gamma_2$	$G_t\alpha$ (α_T) interacts with and relieves the inhibitory action of γ_{PDE} on activity of α β complex
PIP_2-specific phospholipase C's (PLCβ's) Peripheral membrane protein, single subunit enzyme, part of family of PIP_2-specific PLC's: β,γ,δ	4 isoforms: β1, β2, β3, β4 PLCβ_1: stimulated by α of G_q, G_{11}, G_{14}, and (less) G_{16}; expressed ubiquitously PLCβ_2: stimulated by $\beta\gamma$ and (less) by $G_{16}\alpha$, as well as by high concentrations of other G_q-type α's; expressed predominantly in blood-borne cells PLCβ_3: sensitive to both $\beta\gamma$ and some G_q-type α's PLCβ_4: αvs. $\beta\gamma$ sensitivity not known
Inwardly rectifying K^+ channel(s) Transmembrane protein, structure: GIRK1: 2 TMs	Several subtypes (?), e.g., 40-pS cardiac muscarinic K^+ channel; 50-pS GH cell acetylcholine and somatostatin-stimulated inward rectifier Stimulated in heart by three $G_i\alpha$'s, not by $G_o\alpha$ One or more may be stimulated by $\beta\gamma$ and/or arachidonic acid metabolite(s)
Ca^{2+} channels Voltage-gated; variable sensitivity to: DHP's, ω-CTx-GVIA, ω-CTx-MVIIC, ω-Aga-IVA Transmembrane, 24 TMs in α_1 Alternative splicing of α_1 Six α_1 genes known: α1A, α1B, α1C, α1D, α1E, α1S β isoforms (β_1–β_4) General subunit structure $(\alpha_2\delta)\alpha_1\beta[\pm\gamma]$, subtype defined by type of α_1, presence of γ proven only in skeletal muscle	> 6 subtypes: Type S: skeletal muscle, L-current [inhibited (inh.) by DHP] Type C: heart/smooth muscle/neurons/endocrine cells, L-current, inh. by DHP Type B: neurons, N-current, inh. by ω-CTx-GVIA Type D: neurons/endocrine cells, L-current, inh. by DHP Type A: neurons, Q-current, inh. by ω-CTx-MVIIC Type E: neurons, R (residual)-current Type P: P-current, inh. by ω-Aga-IVA Stimulation by $G_s\alpha$: proven for type S (skeletal muscle) and type C (cardiac) Inhibition by $G_{o1}\alpha$ and $G_{o2}\alpha$ mediated pathways seen in whole cell recordings, direct regulation by $G\alpha$ not yet proven in a cell-free system, L- and N-type Ca^{2+} channels are regulated in this way Stimulation by either a $G_i\alpha$ or a $G_i\beta\gamma$ triggered pathway, e.g., PTX-sensitive stimulation of Ca^{2+} currents in Y1 adrenal cells by angiotensin II and in GH3 cells by TRH
ATP-sensitive K^+ channel(s) (Inhibited by sulfonylureas (SUs), Transmembrane protein, structure/subunit composition unknown	Several subtypes by pharmacology and affinity labeling of SU receptor site Stimulated by three $G_i\alpha$'s, not by $G_s\alpha$ or $G_o\alpha$ in inside-out membrane patches
Ca^{2+}-dependent K^+ channels Charybdotoxin sensitive, Transmembrane proteins, 6 TMs; cloned mSlo K_{Ca} channel: 1200 aa; purified tracheal channel: 62-kDa α and 31-kDa β; subunit structure: $\alpha_4\beta_4$?	Several subtypes based on electrophysiology One type stimulated by $G_s\alpha$: proven for pig coronary artery and rat uterine smooth muscle channels in lipid bilayers

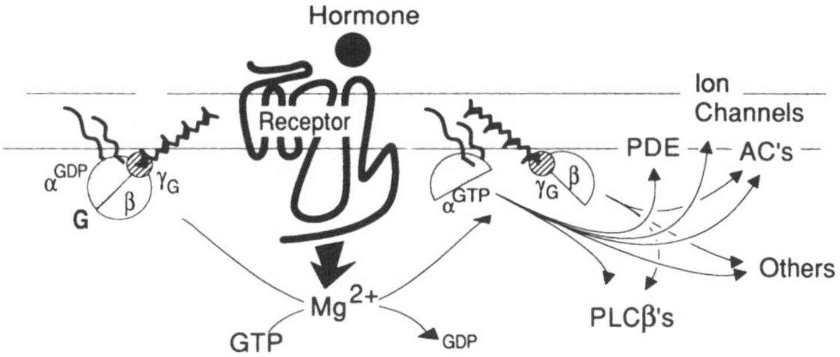

Figure 17.1. The transduction of the signal emitted by the hormone-occupied receptor by a G protein involves the activation of the G protein by GTP and concomitant dissociation of the protein into two signaling molecules, α-GTP and βγ. These then regulate diverse effector functions: ACs, adenylyl cyclases; PLCβs, β-type phospholipase Cs; PDE, visual phosphodiesterase; ion channels; others, yet to be defined functions. Heavy wavy lines emerging from α subunits denote myristoylation and palmitoylation.

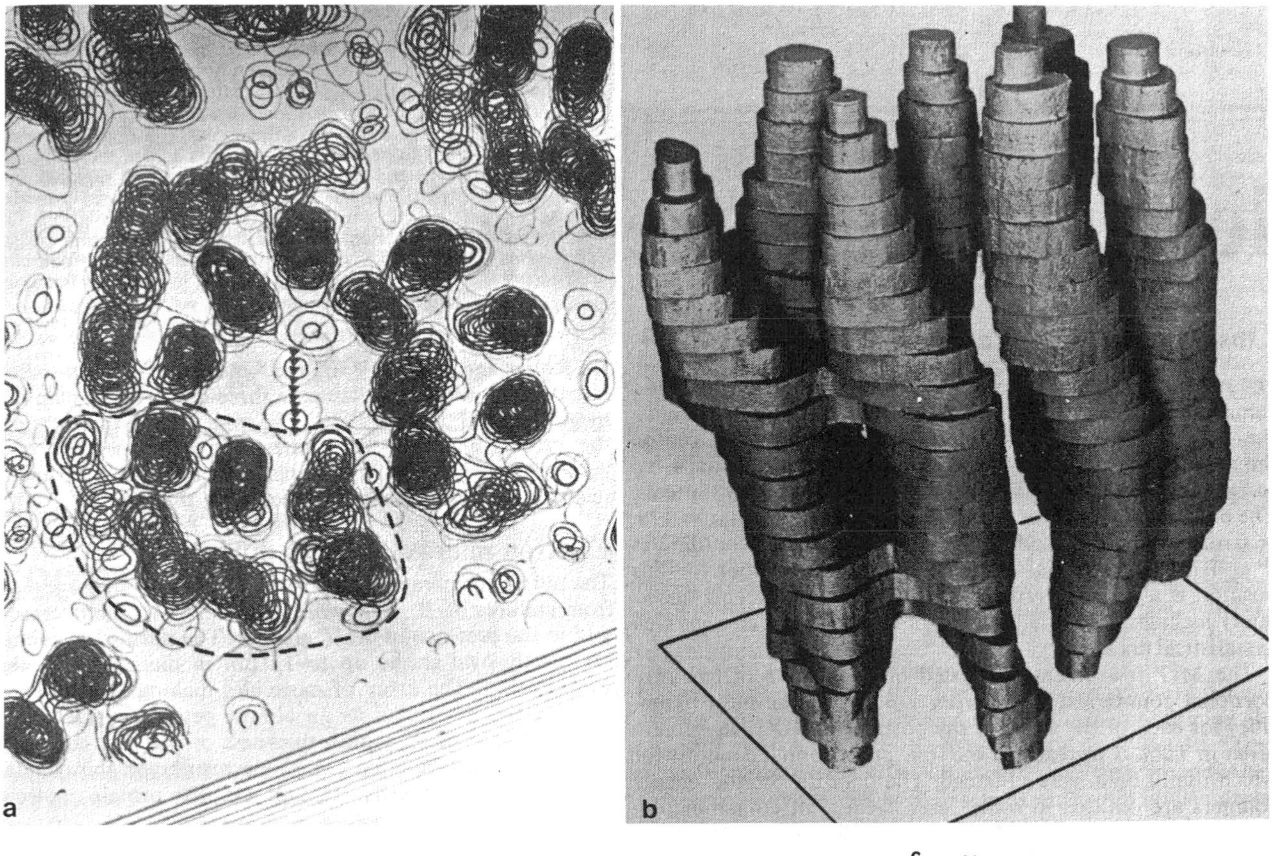

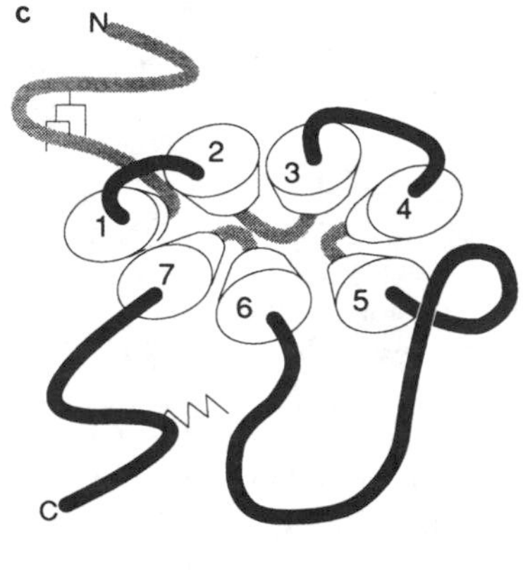

Figure 17.2. Basis for the three-dimensional structure assumed for G protein-coupled receptors. (a) Electron density map of the bacteriorhodopsin in purple membrane of *Halobacterium halobium* showing three molecules each with seven transmembrane regions, grouped around a threefold axis. The probable boundary of one of them is indicated by the broken lines. (b) Model of a single bacteriorhodopsin molecule in the purple membrane (7-Å resolution) showing seven closely packed α-helical segments (TMs) that span the plasma membrane in a roughly perpendicular fashion (from Ref. 3). (c) Model of "view" from the cell's inside of a typical G protein-coupled receptor based on the bacteriorhodopsin model and multiple amino acid sequence alignments. Gray, extracellular; black, intracellular.

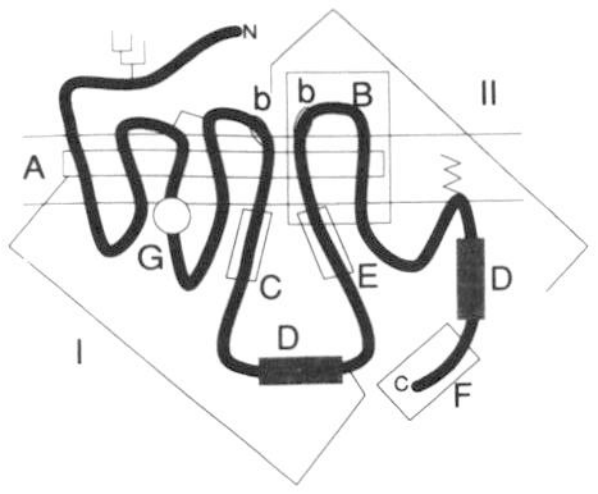

Figure 17.3. Structural elements of a G protein-coupled receptor and their functional import. I, II, A through G and other are of the same hierachy. (I, II) The tertiary structure and relative positioning of the transmembrane segments depends not only on connecting extra- and intracellular loops, but also on interactions among transmembrane segments. Coexpression of an N-terminal half (N-terminus plus TMs I–V plus part of the third intracellular loop) and a C-terminal half leads to an active receptor[189]; proteolytic digestion of purified receptor protein leads to a limit digest formed of all seven TMs[190]; mutations that interfere with TM-I–TM-VII interaction prevent receptor folding.[191]

(A) Small agonists (retinal catecholamines) interact with the intramembrane region of the receptor to trigger a response. Larger agonists are also assumed to interact with the transmembrane domain to activate the receptor, but their overall binding involves more than the transmembrane domains. This has been proven for glycoprotein hormones and is likely to apply also to intermediate-size peptides such as vasopressin, kinins, and angiotensin. The N-terminal domain of the LH/CG receptor binds hCG with the same affinity as the entire receptor.[192]

(B) Sites on transmembrane regions VI and VII define specificities for antagonists in biogenic amine receptors. A chimera that is β_2AR for TMs I–V and α_2AR for TMs VI and VII (plus C-tail) activates adenylyl cyclase (β_2 effect) and is blocked by yohimbine (α_2AR blocker[189]; a single amino acid change in TM VII of the 5HT$_{1B}$ changes the affinity for an antagonist (pindolol) by three orders of magnitude.[193] (b) Antagonist binding may involve more than B domain [nonpeptide NK1 antagonist interacts with epitopes that are on top of both TM-VI and TM-VII, even though neither epitope is required for NK1 (substance P) binding].[194]

(C) N-terminal segment of third intracellular loop is involved in defining specificity of G protein interaction. Replacement of 17 amino acids of the N-terminal end of intracellular loop 3 of the PLC-activating M3 muscarinic receptor, with the cognate 16-amino-acid stretch of the adenylyl cyclase inhibiting M2 muscarinic receptor, resulted in loss of PLC stimulation and acquisition of adenylyl cyclase inhibitory activity to a level that was 25% of control.[18,195]

(D) Ser-Thr-rich regions in either or both the third intracellular loop and the C-terminal tail are substrates for G protein-coupled receptor kinases (GRKs). These sites are exposed on activation of receptor by agonist and initiate a desensitization cascade. Except for rhodopsin kinase which is anchored to the membrane through a C-terminal polyisoprene and phosphorylates light-activated rhodopsin without requiring other proteins, the other GRKs appear to phosphorylate HRs only when presented in the context of an HR–$\beta\gamma$ complex[196–201] (reviewed in Ref. 202).

(E) Potential protein–protein interaction site for receptor–Gαinteraction. Point mutations in the C-terminal end of the third intracellular loop may lead to constitutive (agonist independent) activation of receptors as shown through artificial mutations (for the α_1-[23,24] α_2-,[26] and β_2-[25] adrenergic receptors and found in two natural mutations of the TSH receptor.[174] Peptides derived from this region can activate purified G proteins *in vitro*.[21,29,40,203]

(F) Addition of GTP to membranes lowers affinity of receptor for glucagon,[49] catecholamines,[50,51] and carbachol,[52,53] but increases affinity for prostaglandin in the case of platelet prostaglandin receptor,[204] and has no effect on the vasopressin–V2R interaction (M. Birnbaumer, unpublished). In the case of the cloned EP3-prostaglandin receptor, which mediates inhibition of adenylyl cyclase in adipose tissue, there are two splice variants that structurally differ in their C-terminus. For one variant, GTP addition lowers agonist affinity, for the other it increases it. The C-terminal tail may thus contribute to interaction with G protein.[205]

(G) A "DRY" motif (consensus D/e-R-Y/f/h/c, of which the R is invariant) is found in all 200 (or thereabout) G protein-coupled receptors, including yeast STE2. Exceptions: receptors for the glucagon-related peptides [glucagon, GLP-1, GIP, secretin, VIP (types 1 and 2), GRF, PACAP], CRF, PTH, and calcitonin, which constitute a subfamily of heptahelical G protein-coupled receptors. The DRY motif appears to be important for coupling to G protein. Mutation of its R in the VP2 vasopressin receptor results in loss of coupling to G$_s$ without changes in agonist binding or loss sequestration in response to hormone binding[31] (M. Birnbaumer, unpublished).

(Other) N-termini tend to be glycosylated. Extracellular loops 1 and 2 may be linked by disulfide bridges,[206] and C-terminal tails tend to exhibit at 13–15 amino acids from TM-VII one or two vicinal palmitoylated cysteines. Removal of the cysteine impaired function in the β-adrenergic receptor,[207,208] but had no functional effect on rhodopsin[206] or the α_2-adrenergic receptor.[209]

To the second subfamily belong the receptors for the glucagon-related peptides and for CRF, PTH, and calcitonin. Receptors of this class share little sequence homology with the DRY-motif-containing G protein-coupled receptors, and they all stimulate adenylyl cyclase, i.e., activate G$_s$. Because of the presence of conserved cysteines, they may have extracellular loops 1 and 3 connected by a disulfide bridge. An alignment of several of their sequences is presented in Table 17.4. It is of interest that seven of the ten evolutionarily related receptors bind the seven also evolutionarily related glucagon-related peptide hormones. Although this may represent an example of coevolution of interacting elements, the evolutionary paths followed from ancestral genes to the present are not obvious. The N-termini of all members of this family (ca. 120 amino acids in length), with the apparent exception of one (gastric inhibitory peptide, Table 17.1), have a hydrophobic sequence encoding a signal peptide, followed by a stretch of amino acids that, like the transmembrane cores, are evolutionarily related.

To the third subfamily belong the four metabotropic glutamate receptors and the Ca^{2+} sensing receptor. Because of the small number of members, this does not distinguish itself for any special character other than having a very long extracellular N-terminal domain, in spite of the modest size of the ligands—Glu and Ca^{2+}—and to have the general feature of (putatively) traversing the plasma membrane seven times, and to be functionally able to activate a G protein.

The following three major functions are encoded into the primary structure of a G protein-coupled receptor:

1. A binding site that when occupied changes conformation in a manner that can be sensed by a G protein
2. The ability to form a complex with a specific G protein, which alters in some but not all cases the affinity parameters of the receptor–agonist interaction
3. The susceptibilty to be desensitized by a desensitization machinery that senses the receptor's occupation by agonist

For small ligands the binding site is fully located within the plane of the membrane involving residues of the transmembrane domains (Figure 17.3). Receptor activation involves

Table 17.3 Sequence Comparison among Various DRY-Motif-Containing G Protein-Coupled Receptors

a. TSH receptor vs. LH receptor. Number of identities in transmembrane core: 224

```
tsh   M--RPADLLQLVLLLDLPRDLGGMGCSSPPCECHQEEDFRVTCKDIQRIPSLPPSTQTLKLIETHLRTIPSHAFSNLPNISRIYVSIDVTLQQLESHSFY   98
lh    MGRRVPALRQLLVLAMLVLKQSQL-HSPELSGSRCPEPCDCAPDGALRCPGPRAGLARLSLTYLPVKVIPSQAFRGLNEVVKIEISQSDSLERIEANAFD   99
      *  *   * **  *  *         *          *          * *       * *      *** **  *     *   *     *   *   *

tsh   NLSKVTHIEIRNTRNLTYIDPDALKELPLLKFLGIFNTGLKMFPDLTKVYSTDIFFILEITDNPYMTSIPVNAFQGLCNETLTLKLYNNGFTSVQGYAFN  198
lh    NLLNLSEILIQNTKNLLYIEPGAFTNLPRLKYLSICNTGIRTLPDVSKISSSEFNFILEICDNLYITTIPGNAFQGMNNESITLKLYGNGFEEVQSHAFN  199
      **       * * ** ** ** * *    ** ** * * ***      **  *  *     ***** ** * * ** *****  **  ***** ***  ** ***

tsh   GTKLDAVYLNKNKYLTVIDKDAFGGVYSGPSLLDVSQTSVTALPSKGLEHLKELIARNTWTLKKLPLSLSFLHLTRADLSYPSHCCAFKNQKKIRGILES  298
lh    GTTLISLELKENIYLEKMHSGTFQGA-TGPSILDVSSTKLQALPSHGLESIQTLIATSSYSLKTLPSREKFTSLLVATLTYPSHCCAFRNLPKKEQNFSF  288
      ** *     *  * **        * *   *** **** *   **** ***     ***      ** **     *   *   * * ******** *  *

tsh   LMCNESSMQSLRQRKSVNALNSPLHQEYEENLGDSIVGYKEKSKFQDTHNNAHYYVFFEEQEDEIIGFGQELKNPQEETLQAF 380  hum-tsh: M31774
lh    SIFENFSKQC-------------------------------ESTVREANNETLYSAIFEENE-----------------LSGW 333  mur-LH:  M81310
          * *                                   *       *    *   *** *                 *

                                                     ==========I===========       ==========II=========
tsh   381  D---SHYDYTICGDSEDMVCTPKSDEFNPCEDIMGYKFLRIVVWFVSLLALLGNVFVLLILLTSHYKLNVPRFLMCNLAFADFCMGMYLLLIA  472
lh    334  D---YDYDFC---SPKTLQCTPEPDAFNPCEDIMGYAFLRVLIWLINILAIFGNLTVLFVLLTSRYKLTVPRFLMCNLSFADFCMGLYLLLIA  421
           *     **            ***  * ********** ***        ******  **  **** *** *** ***** ******* ******

==                           ==========III=========                      ===========IV===========
tsh   SVDLYTHSEYYNHAIDWQTGPGCNTAGFFTVFASELSVYTLTVITLERWYAITFAMRLDRKIRLRHACAIMVGGWVCCFLLALLPLVGISSYAKVSICLP  572
lh    SVDSQTKGQYYNHAIDWQTGSGCSAAGFFTVFASELSVYTLTVITLERWHTITYAVQLDQKLRLRHAIPIMLGGWIFSTLMATLPLVGVSSYMKVSICLP  511
      ***  *    ***********  **  ************************  ** *   ** * *****  ** ***      * * ***** *** *******

                ==========V===========                      ==========VI=============          ==========
tsh   MDTETPLALAYIVFVLTLNIVAFVIVCCCHVKIYITVRNPQYNPGDKDTKIAKRMAVLIFTDFICMAPISFYALSAILNKPLITVSNSKILLVLFYPLNS  671
lh    MDVESTLSQVYILSILLLNAVAFVVICACYVRIYFAVQNPELTAPNKDTKIAKKMAILIFTDFTCMAPISFFAISAAFKVPLITVTNSKVLLVLFYPVNS  620
      ** *  *    **    * ** ****  * * * **   * **          ******* ** ****** ******* * **     ***** *** ******* **

=VII=======
tsh   CANPFLYAIFTKAFQRDVFILLSKFGICKRQAQAYRGQRVPPKNSTDIQVQKVTHDMRQGLHNMEDVYELIENSHLTPKKQGQISEEYMQTVL@  764
lh    CANPFLYAVFTKAFQRDFFLLLSRFGCCKHRAELYRRKEFSA----------CTFNSKNGFPRSSKPSQAALKLSIVHCQQPTPPRVLIQ@     700
      ******** ******** * *** ** **   *  **                        *      *                  *        *
```

b. ACTH receptor vs. MSH receptor. Number of identities in transmembrane core: 115

```
                                                ===========I============        =============II=======
acth                   MKHIINSYENINNTARN--NSDCPRVVLPEEIFFTISIVGVLENLIVLLAVFKNKNLQAPMYFFICSLAISDMLGSLYKILE   80
msh           MSTQEPQKSLLGSLNSNATSHLGL-ATNQSEPWCLYVSIPDGLFLSLGLVSLVENVLVVIAITKNRNLHSPMYYFICCLALSDLMVSVSIVLE   92
                                         * *     *   *   *   *     *  ***  *   *  ** **  *** *** ** **   *    **

=======                          ==========III==========                        =========IV=========  =======
acth  NILIIILRNMGYLKPRGSFETTADDIIDSLFVLSLLGSIFSLSVIAADRYITIFHALRYHSIVTMRRTVVVLTVIWTFCTGTGITMVIFSHHVPTVITFTS  180
msh   TTIILLLEVGILVARVALVQQLDNLIDVLICGSMVSSLCFLGIIAIDRYISIFYALRYHSIVTLPRARRAVVGIWMVSIVSSTLFITYYKHTAVLLCLVT  192
         * *   * * *         *  ** *    *    *    *  ** **** ** *********  *          **                 *
```

(Continued)

Table 17.3 *Continued*

```
        ====V===========                                  ============VI===========                  ==========VII===
acth    LFPLMLVFILCLYVHMFLLARSHTRKISTLPRANM--------KGAITLTLTILLGVFIFCWAPFVLHVLLMTFCPSNPYCACYMSLFQVNGMLIMCNAVID  272
msh     FFLAMLALMAILYAHMFTRACQHVQGIAQLHKRRRSIRQGFCLKGAATLTLTILLGIFFLCWGPFFLHLLLIVLCPQHPTCSCIFKNFNLFLLLIVLSSTVD  292
          *  **      ** ***  *  *   *  *             *** ******* *  ** ** ** **   **  * * *    *    **      *

        ======
acth    PFIYAFRSPELRDAFKKMIFCSRYW@    297  acth   X65633   human ACTH
msh     PLIYAFRSQELRMTLKEVLLCS--W@    315  msh    X65635   mouse MSH
        * ****** ***    *    **  *
```

c. β_2 Adrenoceptor vs. LH, type 2 vasopressin, α_1-adrenergic, α_2-adrenergic, light (rho), ACTH, and α-thrombin receptors. Number of identical residues are listed at the end of the sequences.

```
                                                            ======= ======I== =========         =   =======II===
beta2              MGQPGNGSAFLLAP-NRSHAPDHDVTQQRDEVWVVGMGIVMS-WIVLAIVFG-NVLVITAIAKFERLQ--TVTNY---FITSLACADLVM   82
lh       334    D---YDYDFC----SPKT-LQCTPEPDAFNPCEDIMGYAFLRVLILWINILAIFG-NLTVLFVLLTSRYKL--TVPRF---LMCNLSFADFCM  412
                                    *                  *     ** *  *               **            *  **  *

vp2               MLMASTTSAVPGHPSLPS-LPSNSSQERPLDTRDPLLARAELALLS-IVFVAVALS-NGLVLAALARRGRRGHWAPIHV---FIGHLCLADLAV   88
                                *    *          **             *     *     * **  * *   *            **  *  ***

alpha1  MNPDLDTGHNTSAPAQWGELKDANFTGPNQTSSNSTLPQLDVTRAISVGLVLGAFILFAIV-G-NILVILSVACNRHLR--TPTNY---FIVNLAIADLLL   94
                                                   * *      * *     *  *** * * ***          *   * ***    **  ** ***

alpha2             MGSLQPDAGNASWNGT-EAPGGGARATPYSLQVTLT--LVCLAG-LLMLLTVFG-NVLVIIAVFTSRALK--APQNL---FLVSLASADILV   82
                                                       *  *  ***  ***** *        *             *  *** **

rho               MNGTEGPNFYVPFS-NKTGVVRSPFEAPQYYLAEPWQFSMLAAYM-FLLIMLGFPINFLTLYVTVQHKKLR--TPLNY---ILLNLAVADLFM   86
                                                  * *   *              * *            *   *  **        ** *** *

acth                              MKHIINSYENINNTARN---NSDCPRVVLPEEIFFTISIVGVLE-NLIVLLAVFKNKNLQ--APMYF---FICSLAISDMLG   73
                                                                         *    *    *    *    *    **        ** ***  *

α-thr   MGPRRLLLVAACFSLCGPLLSARTRARRPESKATNATLDPR ↓ SFLLRNPNDKYEPFWEDEE 60
α-thr   61  -KNESGLTEYRLVSINKSSPLQKQLPAFISEDASGYLTSSWLTLFVP-S-VYTGVFVVS-LPLNIMAIVVFILKM--KVKKPAVVYMLHLATADVLF  151
                                              *                  *      *     * * **   *         *        ** **
```

```
        ======== ======                    =========III====  ======                           ==========IV=========
beta2   GLAVVPFG-AAHILMKMW-------TFGNFWCEFWTSIDVLCVTASIET--LCVIAVDRYFAITSPFKYQSLLTKNKARVIILMVWIVSGLTSFLPIQMH  172
lh      GLYLLLIA-SVDSQTKGQYYNHAIDWQTGSGCSAAGFFTVFASELSVYT--LTVITLERWHTITYAVQLDQKLRLRHAIPIMLGGWIFSTLMATLPL---  506
        **         *                     *          *      *   *  * **    *   **          *   *  * *  ** *  *   **

vp2     ALFQVLPQ-LAWKATDRF-------RGPDALCRAVKYLQMVGMYASSYM--ILAMTLDRHRAICRPM----LAYRHGSGAHWNRPVLVAWAFSLLLSLPQ  174
          *  *    *                     *                *             ** **  *       *                     *   * *

alpha1  SFTVLPFS-ATLEVLGYW-------VLGRIFCDIWAAVDVLCCTASILS--LCAISIDRYIGVRYSLQYPTLVTRRKAILALLSVWVLSTVISIGPL-LG  183
           * **  *       *          *   *   *   **** ****    ** *  ***             *   *  *  **      * **  *   *  *

alpha2  ATLVIPFS-LANEVMGYW-------YFGKTWCEIYLALDVLFCTSSIVH--LCAISLDRYWSITQAIEYNLKRTPRRIKAIIITCWVISAVISFPPL---  169
           * **  *  *  *         **  ***       ***  * **    ** *  ***  **       *       *       **   *   *   ** *

rho     VFGGFTTT-LYTSLHGYF-------VFGPTGCNLEGFFATLGGEIALWS--LVVLAIERYVVVCKPMSNFRFGENHAIMGVAFTWVMALACAAPPLVGWS  176
                      *              **   *        *        * * *  **      *

acth    SLYKILEN-ILIIILRNMGYLK-PRGSFETTADDIIDSLFVLSLLGSIFS--LSVIAADRYITIFHALRYHSIVTMRRTVVVLTVIWTFCTGTGITMVIFS  169
        *        **  *          *         *  **       **      * *** ***  *       *  *  *         *       *      *

α-thr   V-SVLPFKISYYFSGSDW-------QFGSELCRFVTAAFYCNMYASILL--MTVISIDRFLAVVYPMQSLSWRTLGRASFTCLAIWA-LAIAGVVPLVLK  240
           * **               *          *    *  *  *         ***       **  **   *      *      *   *   *      *  *              *
```

```
              ==                                       ==========V============
beta2   WYRATHQEAINCYANETCCD-------FFTNQAYAIASSIVSFYVPLVIMVFVYSRVFQEAKRQLQKIDKSEGR-------------------------  239
lh      VGVSSYMKVSICLPMDV--E-------STLSQVYILSILLLNAVAFVVICACYVRIYFAVQNPELTAPNKDT---------------------------  569
                     *                * *        **         *    *    *

vp2     LFIFAQRNVEGGSGVTDCWA------------CFAEPWGRRTYVTWIALMVFVAPTLGIAACQVLIFREIHASLVP------------------------  238
                         *              *               ****     * *

alpha1  WKEPAPNDDKECGVTE--------------EPFYALFSSLGSFYIPLAVILVMYCRVYIVAKRTTKNLEAGVMKEMSNSKELTLRIHSKN----------  259
        *           *   *            **  **  *** **      * **   ***

alpha2  ISIEKKGGGGGPQPAEPRCE-------INDQKWYVISSCIGSFFAPCLIMILVYVRIYQIAKRRTRVPPSRRGPDAVAAPPGGTERRPNGLGPERSAGPG  262
                     *  *            * * * * **  *   **   ** *   * ***             *

rho     RYIPEGMQCSCGIDYYTPHE-------ETNNESFVIYMFVVHFIIPLIVIFFCYGQLV------------------------------------------  210
         *              *             *    *    * *  **    * *

acth    HH-----------------------------VPTVITFTSLFPLMLVFILCLYVHMFLLARSHTRKIST-------------------------------  209
                                                *  **      *   *   *     **

α-thr   --EQTIQVP--GLNITTCHDVLNETLLEGYYAYYFSAFSAVFFFVPLIIS------------------------------------------------  286
           * *      ** *              *       * * *** *

beta2   --------------------------------------------------------------------------------------FHVQNLSQVEQDG  252
lh      -----------------------------------------------------------------------------------------------------  569

vp2     --------------------------------------------------------------------------------------GPSERPGGRRRGR  252

alpha1  --------------------------------------------------------------------------------------FHEDTLSSTKAKG  272
                                                                                                **   **     *

alpha2  GAEAEPLPTQLNGAPGEPAPAGPRDTDALDLEESSSSDHAERPPGPRRPERGPRGKGKAARSQVKPGDSLRGAGRGRRGSGRRLQGRGRSASGLPRRRAG  362
                                                                                                            *

rho     --------------------------------------------------------------------------------------FTVKEAAAQQQES  240
                                                                                                * *         *

acth    -----------------------------------------------------------------------------------------------------  209

α-thr   ---------------------------------------------------------------------------------------------TVCYVS  292

                               ====== ====VI=============            ========VII==========
beta2   RTGHGLRRSSKF-CLKEHKALKTLGIIMG-TFTLCWLPFFIVNIVHVIQDNL----IR-KEVYILLNWIGYVNSGFNPLIY-CRSPDFRIAFQELLCL--  342
lh      -----------------KIAKKMAILIF-TDFTCMAPISFFAISAAFKVPL----ITVTNSKVLLVLFYPVNSCANPFLY----AVFTKAFQRDFFL--  640
                          *  *   *    *   *   *    *        *     *       **    ***  **  *      * ***    *

vp2     RTG-SPGEGAHV-SAAVAKTVRMTLVIVV-VYVLCWAPFFLVQLWAAWDPEA----PLEGAPFVLLMLLASLNSCTNPWIY---ASFSSSVSSELRS----  338
        ***                *        *     *** *** *                 **      **  **  **                **

alpha1  HNPRSSIAVKLFKFSREKKAAKTLGIVVG-MFILCWLPFFIALPLGSLFSTL----KPPDAVFKVVFWLGYFNSCLNPIIYPCSSKEFKRAFMRILGCQC  358
                    *     * **  **  **   *   * ********          *          *     * ** **   ** ** * *   *   **   *

alpha2  AGGQNLEKRFTF-VLAEKRFTFVLAVVIG-VFVVCWFPFFFTYTLTAVGCSV----PRT--LFKFFFWFGYCNSSLNPVIYTIFNHDFRRAFKKILCR--  443
         *  *     *  *  *       *      *  *   ** ***                 *         * ** **   ** **    *** **   **

rho     ATTQ-----------KAEKEVTRMVIIMVIAFLICWLPYAGVAFYIFTHQG----SDFGPIFMTIPAFFAKTSAVYNPVIYIMMNKQF------------  313
         *             *  *      ***   *  ****    *                                      ** **       *

acth    -----LPRANM-------KGAITLTLTILLG-VFIFCWAPFVLHVLLMTFCPSNPY-CACYMSLFQVNGMLIMCNAVIDPFIYAFRSPELRDAFK-------  288
              * *              *   ** *   *   *   ** **                                  *    * **    **   * *

α-thr   IIRCLSSSAVAN-RSKKSRALFLSAAVFC-IFIICFGPTNVLLIAHYSFLSHTSTTEAAYFAYLLCVCVSSISSCIDPLIY-YYASSECQRYVYSILCC-  388
                         **             *    *    *       * *                  * *          *     ****
```

(Continued)

Table 17.3 *Continued*

```
beta2   ---------------------------------RRSSLKAYGNGYSSN----GNTGEQSGYHVEQEKE--------------------------------  373
lh      ---------------------------------LLSR-------FGCC----KHRAEL--YRRKEFSA--------------------------------  662
                                           *                    *   *
vp2     ----------------------------------------------------------------------------------------------------  338
alpha1  RSGRRRRRRRRLGACAYTYRPWTRGGSLERSQSRKDSLDDSGSCMSGSQRTLPSASPSPGYLGRGAQPPLELCAYPEWKSGALLSLPEPPGRRGRLDSGP  467
                                          *  **   *   *             **
alpha2  ----------------------------------------------------------------------------------------------------  443
rho     ----------------------------------------------------------------------------------------------------  313
acth    ----------------------------------------------------------------------------------------------------  288
α-thr   ----------------------------------------------------------------------------------------------------  388

beta2   ---NKLLCE-DLPGTEDFVGHQGTVPSDNIDSQGRNCSTNDSLL@      413
lh      ---CTFNSK-NGFPRSSKPSQAALKLSIVHCQQPT--PPRVLIQ@      700
                                  *      *
vp2     -----LLCC-ARGRTPPSLGPQDESCTTASSSLAK-----DTSS@      371
             ***      *    * *         *        *
alpha1  LFTFKLLGEPESPGTEGDASNGGCDATTDLANGQPGFKSNMPLAPGHF@  515
            *** *   ****      *                *  *
alpha2  -------------------------------------GDRKRIV@      450
rho     ---RNCMVT-TLCCGKNPLGDDEASTTVS-----KTETSQVAPA@      348
                   *       *
acth    -----------------------------------KMIFCSRYW@      297
α-thr   KESSDPSSYN---SSGQLMASKMDTCSSNLN----NSIYKKLLT@      425
                                   * *      *      *
```

Abbr.	Accession #	Receptor	aa hemology to beta2
beta2	J02960	human β2AR	100
lh	M81310	murine LH/CG	70
vp2	Z11687	human type-2 AVP	70
alpha1	J04084	hamster α_{1B}AR	127
alpha2	M18415	human $\alpha_{2\text{-}C10}$AR	109
rho	M12689	bovine rhodopsin	50
acth	X65633	human ACTH	61
α-thr	M62424	human α-thrombin	67

d. GRF receptor vs. β_2-adrenoceptor. Number of identities in transmembrane core: 43

```
h-grf   MDRRMWGAHVFCVLSPLPTVLGHMHPECDFITQLREDESACLQAAEEMPNTTLGCPATWDGLLCWPTAGSGEWVTLPCPDFFSHFSSESGAVKRDCTITG  100

                                    ***********I************                    **********II***********
h-grf   WSEPFPPYPVACPVPLELLAEEESYFSTVKIIYTVGHSISIVALFVAITILVALRRLHC---PRNYVHTQLFTTFILKAGAVFLKDAALFHSDDTDHCSF  197
                                         **  *      **    *          *    *  * *    *  *    *
beta2    MGQPGNGSAFLLAPNRSHAPDHDVTQQRDEVWVVGMGIVMSWIVLAIVFGNVLVITAIAKFERLQTVTNYFITSLACADLVMGLAVVPFGA--------  91
                                          ===========I==============                    ==========II=========

              *********III**********                                              **********IV**********
h-grf   STVLCKVSVAASHFATMTNFSWLLAEAVYLNCLLASTSP------------------------SSRRAFWWLVLAGWGLPVLFTGTWVSCKLAFED-IAC  272
                       *  **         *  *                                   *    *   **                  * *
beta2   -------AHILMKMWTFGNFWCEFWTSIDVLCVTASIETLCVIAVDRYFAITSPFKYQSLLTKNKARVIILMVWIVSGLTSFLPIQMHWYRATHQEAINC  184
                           =========III=========                      ===========IV========

                  ************V***********                                                          *****
h-grf   ---------WDLDDTSPYWWIIKGPIVLSVGVNFGLFLNIIRILVRK--------------------------------LEPAQGSLHTQSQYWRLSKST  340
                                *     *   *    *                                                           *
beta2   CYANETCCDFFTNQAYAIASSIVSFYVPLVIMVFVYSRVFQEAKRQLQKIDKSEGRFHVQNLSQVEQDGRTGHGLRRSSKFCLKEHKALKTLGIIMGTFT  284
                  ==========V============                                                         =========

****VI**************             *********VII*********
h-grf   LFLIPLFGIHYIIFNFLPDNAGLGIRLPLELGLGSFQ-GFIVAILYCFLNQEVRTEISRKWHGHDPELLPAWRTRAKWTTPSRSAAKVLTSMC@        423
        *    * * *  *        **          *    *    **   * **
beta2   LCWLPFF-IVNIV-HVIQDNLIRKEVYILLNWIGYVNSGFNPLI-YCRSPDFRIAFQELLCLRRSSLKAYGNGYSSNGNTGEQSGYHVEQEKENKLLCED  380
        =VI=============        =========VII=============

beta2   LPGTEDFVGHQGTVPSDNIDSQGRNCSTNDSLL@     413     Accession #'s: h-grfR: L09237; beta2-AR: J02960
```

Table 17.4 Sequence Alignment of Receptors for Glucagon-Related Peptides, PTH, and Calcitonin

```
h-ct                                                   MQFSGEKISGQRDLQKSKMRFTFTSRCL   28
h-pth                            MGTARIAPGLALLLCCPVLSSAYALVDADDVMTKEEQIFLLHRAQ   45
r-secr                                                                       MLST    4

h-ct    ALFLLLNHPTPILPAFSNQTYPTIEPKPFLYVVGRKKMMDAQYKCYD-RMQQLPAYQGEGPYCNRTWDGW-   97
h-pth   AQCEKRLKEVLQRPASIMESDKGWTSASTSGKPRKDKASGKLYPESEEDKEAPTGSRYRGRPCLPEWDHI-  115
r-gluc    MLLTQLHCPYLLLLLVVLSCLPK-APSAQVMDFLFEKWKLYSDQ-CHHNLSLLPPPTELVCNRTFDKY-   66
r-glp1   MAVTPSLLRLALLLLGAVGRAGPRPQGATVSLSETVQKWREYRHQCQRFLTEAPLLATLGFCNRTFDDY-   69
r-secr   MRPRLSLLLLRLLLLTKAAHTVGVPPRLCDVRRVLLEERAHCLQQLSKEKKGALGPETASGCEGLWDNM-   73
r-pacap            MARVLQLSLTALLLPVAIAMHSDCIFKKEQAMCLERIQRANDLMGLNESSPCPGMWDNI-   60
h-grf           MDRRMWGAHVFCVLSPLPTVLGHMHPECDFITQLREDESACLQAAEEMPNTTLGCPATWDGL-   62
                                                                           +

h-ct    LCWDDTP---AGVLSYQFCPDYFPD--FD-PS----------------------EKVTKYCDEKGVWFKH   139
h-pth   LCWPLGA---PGEVVAVPCPDYIYD--FN-HK----------------------GHAYRRCDRNGSWELV   157
r-gluc  SCWPDTP---PNTTANISCPWYLPWYHKV-QH----------------------RLVFKRCGPDGQWV-R   109
r-glp1  ACWPDGP---PGSFVNVSCPWYLPWASSV-LQ----------------------GHVYRFCTAEGIWLHK   113
r-secr  SCWPSSA---PARTVEVQCPKFL-LMLSN-KN----------------------GSLFRNCTQDG-----   111
r-pacap TCWKPAQ---VGEMVLVSCPEVF-RIFNPDQVWMTETIGDSGFADSNSLEITDMGVVGRNCTEDG-----   121
h-grf   LCWPTAG---SGEWVTLPCPDFF-SHFSS-ES----------------------GAVKRDCTITG-----   100
         *                **                                        *   *
                                             ============I=============
h-ct    PENNRT-WSNYTMCN-AF-TPEKLKN-A--Y-VLYYLAI---VGHSLSIFTLVISLGIFVFFRKLTTIFP   199
h-pth   PGHNRT-WANYSECV-KF-LTNETRE-R--E-VFDRLGMIYTVGYSVSLASLTVAVLILAYFR-------   213
r-gluc  GPRGQS-WRDASQCQMDDDEIEVQKGVA--K-MYSSYQVMYTVGYSLSLGALLLALVILLGLR-------   168
r-glp1  DNSSLP-WRDLSECE-ES-KQGERNSPE--E-QLLSLYIIYTVGYALSFSALVIASAILVSFR-------   170
r-secr  WSETFP-RPDLACGV-NI-NNSFNERRH--A-YLLKLKVMYTVGYSSSLAMLLVALSILCSFR-------   168
r-pacap WSEPFP-HYFDACGF-DD-YEPESG--DQDY-YYLSVKALYTVGYSTSLATLTTAMVILCRFR-------   178
h-grf   WSEPFP-PYPVACPV-PL-ELLAEE--E--S-YFSTVKIIYTVGHSISIVALFVAITILVALR-------   155
                                                   *   *   *          *

                         ===========II========
h-ct    LNWKYRKALSLGCQRVTLHKNMFLTYILNSMIIIIHLVEVVPNGEL-----------------------V   246
h-pth   ---------RLHCTRNYIHMHLFLSFMLRAVSIFVKDAVLYSGATLDEAERLTEEELRAIAQAPPPPATA   274
r-gluc  ---------KLHCTRNYIHGNLFASFVLKAGSVLVIDWLLKTRYSQKIGDDLSVSVWL-----------S   218
r-glp1  ---------HLHCTRNYIHLNLFASFILRALSVFIKDAALKWMYS-TAAQQHQWDGLL-----------S   219
r-secr  ---------RLHCTRNYIHMHLFVSFILRALSNFIKDAVLFS----------SDDVTY-----------C   208
r-pacap ---------KLHCTRNFIHMNLFVSFMLRAISVFIKDWILYA----------EQDSSH-----------C   218
h-grf   ---------RLHCPRNYVHTQLFTTFILKAGAVFLKDAALFH----------SDDTDH-----------C   195
                    *              *
                =======III===============              ============IV========
h-ct    RRDPVSCKILHFFHQYMMACNYFWMLCEGIYLHTLIVVAVFTEKQRLRWYYLLGWGFPLVPTTIHAITRA   316
h-pth   AAGYAGCRVAVTFFLYFLATNYYWILVEGLYLHSLIFMAFFSEKKYLWGFTVFGWGLPAVFVAVWVSVRA   349
r-gluc  DGAVAGCRVATVIMQYGIIANYCWLLVEGVYLYSLLSITTFSEKSFFSLYLCIGWGSPLLFVIPWVVVKC   288
r-glp1  YQDSLGCRLVFLLMQYCVAANYYWLLVEGVYLYTLLAFSVFSEQRIFKLYLSIGWGVPLLFVIPWGIVKY   289
r-secr  DAHKVGCKLVMIFFQYCIMANYAWLLVEGLYLHTLLAISFFSERKYLQAFVLLGWGSPAIFVALWAITRH   278
r-pacap FVSTVECKAVMVFFHYCVVSNYFWLFIEGLYLFTLLVETFFPERRYFYWYTIIGWGTPTVCVTVWAVLRL   288
h-grf   SFSTVLCKVSVAASHFATMTNFSWLLAEAVYLNCLLASTSPSSRRAFWWLVLAGWGLPVLFTGTWVSCKL   265
              *             *  *   *  **                      *** *
                         ===========V=============
h-ct    VYFNDN---CW-LSV--ETHLLYIIHGPVMAALVVNFFFLLNIVRVLVTKMRETHE--AESHMY------   372
h-pth   TLANTG---CW-DLS--SGNKKWIIQVPILASIVLNFILFINIVRVLATKLRETNA-GRCDTRQQY----   403
r-gluc  LFENVQ---CWTSND--NMGFWWILRIPVLLAILINFFIFVRIIHLLVAKLRAHQM--HYADYK------   345
r-glp1  LYEDEG---CWTRNS--NMNYWLIIRLPILFAIGVNFLVFIRVICIVIAKLKANLM--CKTDIK------   346
r-secr  FLENTG---CWDINA--NASVWWVIRGPVILSILINFIFFINILRILMRKLRTQETRGSETNHY------   337
r-pacap YFDDAG---CWDMND--STALWWVIKGPVVGSIMVNFVLFIGIIIILVQKLQSPDMGGNESSIYLTNLRL   353
h-grf   AFEDIA---CWDLDD--TSPYWWIIKGPIVLSVGVNFGLFLNIIRILVRKLEPAQGSLHTQSQY------   328
                 **            *           **            *
                                                             ===========VI====
h-ct    --------------------------------------------------LKAVKATMILVPLLGIQFVV   392
h-pth   --------------------------------------------------RKLLKSTLVLMPLFGVHYIV   423
r-gluc  --------------------------------------------------FRLARSTLTLIPLLGVHEVV   365
r-glp1  --------------------------------------------------CRLAKSTLTLIPLLGTHEVI   366
r-secr  --------------------------------------------------KRLAKSTLLLIPLFGIHYIV   357
r-pacap RVPKKTREDPLPVPSDQHSPPFLSCVQKCYCKPQRAQQHSCKMSELSTITLRLARSTLLLIPLFGIHYTV   423
h-grf   --------------------------------------------------WRLSKSTLFLIPLFGIHYII   344
                                                                *  * **
        ====                ==========VII===========
h-ct    FPWRPS-NK-ML--GKIYDYVMHS---LIHFQGFFVATIYCFCNNEVQTTVKRQWAQFKIQWNQRWGRRP   455
h-pth   FMATPY-TE-VS--GTLWQVQMHYEMLFNSFQGFFVAIIYCFCNGEVQAEIKKSWSRWTLALDFKRKARS   487
r-gluc  FAFVTDEHA-QGTLRSTKLFFDLF---FSSFQGLLVAVLYCFLNKEVQAELLRRWRRWQEGKALQEERMA   431
r-glp1  FAFVMDEHA-RGTLRFVKLFTELS---FTSFQGFMVAVLYCFVNNEVQMEFRKSWERWRLERLNIQRDS-   431
r-secr  FAFSPEDAM-E-----VQLFFELA---LGSFQGLVVAVLYCFLNGEVQLEVQKKWRQWHLQEFPLRPVA-   417
r-pacap FAFSPENVSKR-----ERLVFELG---LGSFQGFVVAVLYCFLNGEVQAEIKRKWRSWKVNRYFTMDFK-   484
h-grf   FNFLPDNAGLG-----IRLPLELG---LGSFQGFIVAILYCFLNQEVRTEISRKW---HGHDPELLPA--   401
        *                             ***  *   *** * **
```

Table 17.4 *Continued*

h-ct	SNRSARAAAAAAAEAGDIPIYICHQEPNEPANNQGEESAEIIPLNIIEQESSA@	508	end
h-pth	GSSSYSYGPMVSHTSVTNVGPRVGLGLPLSPRLLPTATTNGHPQLPGHAKPGTPALETLETTPPAMAAPK	559	
r-gluc	SSHGSHMAPAGTCHGDPCEKLQLMSAGSSSGTGCEPSAKTSLASSLPRLADSPT@	485	end
r-glp1	-SMK-P-----LKCPTSSVSSGATVGSSVYAATCQNSCS@	463	end
r-secr	-FNN-S-----FSNATNGPTHSTKASTEQSRSIPRASII@	449	end
r-pacap	-HRHPS-----LASSGVNGGTQLSILSK-SSSQLRMSSLPADNLAT@	523	end
h-grf	-----------WRTRAKWTTPSRSAAKVLTSMC@	423	end
h-pth	DDGFLNGSCSGLDEEASGPERPPALLQEEWETVM@	593	end

ct, calcitonin; grf, growth hormone-releasing factor; pth, parathyroid hormone; glp1, glucagon-related peptide-1; pacap, pituitary adenylated cyclase-activating peptide. **GenBank receptor accession numbers:** human CT: L00587; human PTH: L04308; rat glucagon: L04796; rat GLP-1: S44970; rat secretin: X59132; human GRF: L09237; rat PACAP: Z23722. +, *little* mouse mutation of GRF receptor.

conformational changes that alter the relationship among the transmembrane domains, which cause acquisition of high affinity for one or more subtypes of G proteins. While for larger ligands the primary ligand binding function resides in their N-termini and various aspects of the extracellular loops, it may be assumed on the basis of structural similarity of the transmembrane cores and certain aspects of the intracellular appendages, that the "activating" function of the ligand is still exerted at the level of the transmembrane aspect of the receptor.

Extensive mutational analysis of the β_2-adrenergic receptor has led to the identification of the major residues in the binding of the activating catecholamine[7] (reviewed in Ref. 8; Figure 17.4). Two cases of special interest are rhodopsin and the α-thrombin receptors. Both are "preassociated" with their ligand and receptor activation is in fact the consequence of ligand activation. In the case of rhodopsin, the ligand, 11-*cis*-retinal, is bound through a Schiff base to Lys296 of TM-VII. The ligand is activated on light absorption which induces photoisomerization to all-*trans*-retinal followed by deprotonation of the Schiff base and relocation of the proton possibly to neutralize Glu113 (Figure 17.4). In the case of α-thrombin receptor, the N-terminus contains a latent receptor ligand and is a substrate for thrombin. The proteolytic action of thrombin results in the removal of the first 41 amino acids of the receptor and "exposure" of the ligand. The new N-terminus now constitutes a tethered ligand that activates the receptor by presumably curling onto itself and entering into a binding pocket formed by the TMs of the receptor.[9]

Screening of amino acid sequence alignments identified conserved aspartic acids in biogenic amine binding receptors. Using the adrenergic receptors as the main test subjects, these aspartates have been subjected to analysis through removal by site-directed mutagenesis (reviewed in Refs. 7, 8, 10). These studies identified β_2-adrenergic receptor Asp113 in TM-II, and its cognate in other receptors, as a counterion for the binding of the amino-group of catecholamines and acetylcholine. They also identified β_2-adrenoceptor Asp79 (80 in the D_2 dopamine receptor and 383 in the receptor for luteinizing hormone) as responsible for regulation of binding affinity by sodium (decreased for agonists and increased for antagonists) with consequences on coupling to G proteins that varied from nondetectable—LH receptor—to selective for one versus another type of G protein—α_2-adrenergic receptor—to mere changes in agonist concentrations required for half-maximal activation—β_2-adrenergic receptor (Table 17.5). Of these possibly the most interesting is the finding that for the α_2-adrenergic receptor the Asp79-to-Asn mutation led to loss of K^+ channel regulation, an effect of G_i, with unaltered inhibition of adenylyl cyclase, an effect of G_i or G_o, and unaltered inhibition of Ca^{2+} currents, an effect of G_o.[11] Opposing regulatory effects of antagonist and agonist binding by Na^+ were an early finding for opioid receptors, before they were known to be G protein-coupled receptors.[12] The sodium-induced state was then referred to as the antagonist state to differentiate it from an agonist state. The functional role of this regulation is still unclear. Mutations of the third conserved aspartate, which is the D of the DRY motif, uncouples the receptor, while increasing agonist affinity (Table 17.5).

The G protein-coupled receptors have recognition elements for both the G protein α subunit and the G protein $\beta\gamma$ dimer, whereas α's will not interact with receptors in the absence of $\beta\gamma$'s,[13–15] but $\beta\gamma$'s interact with a receptor in the absence of an α subunit.[16,17] Current thoughts as to the events that precede and follow interaction of receptor with the G protein, and the effect that interaction with a G protein may have on the affinity of the receptor for agonists will be discussed below in conjunction with mechanism of G protein activation.

Efforts are being made to locate the regions of a receptor responsible for G protein activation, measured either as stimulation of GTPase, as stimulation of GTPγS binding, or as activation of the G protein response pathway in whole cells. Attention has focused on the third intracellular loop. This loop is highly variable in length and composition, and this made it initially a candidate for conferring G protein specificity to a receptor. Tests of this hypothesis have borne interesting fruits. Exchange of a 17-amino-acid sequence of the N-terminal end of the third intracellular loop of M3 and M2 muscarinic receptors switched substantially the receptor G protein specificity of these receptors.[18] Removal of a similarly located sequence from the β_2-adrenergic receptor uncoupled it from G protein interaction.[19] A β_2-adrenergic peptide comprising N-terminal sequences required for function is able to stimulate GTP hydrolysis of G_s.[20] Also the C-terminal end of intracellular loop 3 is important, for β_2-adrenergic receptor peptides comprising this end of the loop are also able to stimulate GTP hydrolysis and GTPγS binding.[20,21] Through the use of peptides derived from receptor sequences, it has been suggested that other non-

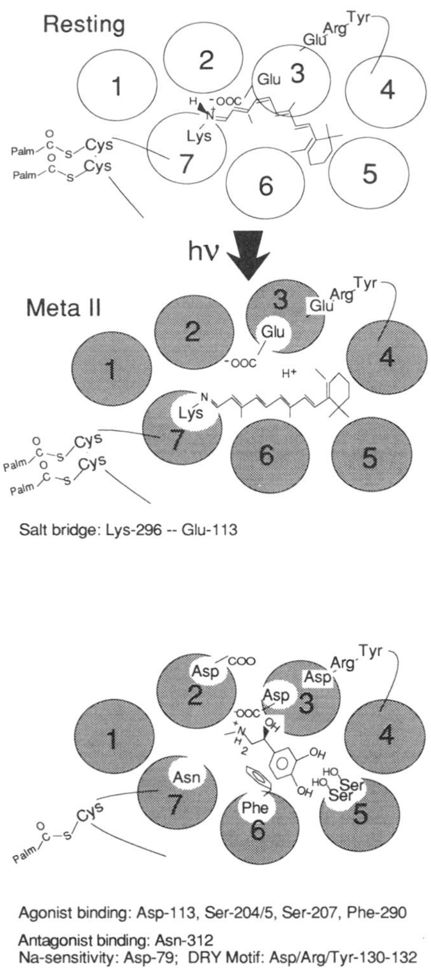

Figure 17.4. Proposed intramembrane, inter-TM domain location of 11-*cis*-retinal, the light-sensitive ligand of rhodopsin (upper), and of norepinephrine, the agonist of the β-adrenergic receptor (lower). (Upper) The drawings depict rhodopsin with protonated Schiff-based bonded to Lys296 of TM-VII of rhodopsin, as it is thought to be in the nonexcited resting state, and all-*trans*-retinal bonded to the same Lys296 but now unprotonated as it is assumed to be in the fully activated Meta-II. The location of the relocated proton has not been determined. This form of rhodopsin has a relatively long life, allowing it to sequentially catalyze activation of many transducin molecules before it is deactivated (after Refs. 210–212). Although the natural photoreceptor has the chromophore covalently attached through the Schiff base, this is not an absolute requirement. [Lys296→Gly]opsin binds the *n*-propylamine Schiff base of 11-*cis*-retinal and is activated by light[213] (Table 17.6). (Lower) Mutational studies that monitored the effect of structural changes on ligand affinity suggest that binding of agonist—a catecholamine—and activation of the β-adrenergic receptor involve Asp113 of TM-III, acting as a counterion to the cationic amino group of the ligand, Ser204 and 207 of TM-V, creating hydrogen bonds with the two -OH of the catechol, and Phe290 of TM-VI, that may stabilize the aromatic group of the catecholamine through π–π bonding of the aromatic rings (after Ref. 7). Asn312 in TM-VII has been shown to be intimately involved in conferring high-affinity binding to antagonist of the pindolol type, not only in this,[214] but also in the type-1B/Dβ serotonin receptor.[193]

heptahelical receptors may also activate G proteins.[21,22] Similarities in amino acid composition of the C-terminal ends of the intracellular loops of the α_1- and β_2-adrenergic receptors led Lefkowitz and collaborators to explore chimeras in which the α_1 receptor was made β_2-like and vice versa. This led to identification of α_1-, β_2-, and α_2-adrenergic receptor mutants with constitutively enhanced agonist-independent activity.[23–26]

The potential role(s) of the third intracellular loop in the coupling process between receptor and G protein was also investigated by transient expression of the entire loop or N- and C-terminal 27-amino-acid-long portions of the loop.[27] Using as template α_{1B}-adrenergic receptor sequences the loop or loop fragments had no effect by themselves on IP3 production—the result of activating one of the G proteins that couple the α_{1B}-receptor to phosphoinositide hydrolysis. In contrast, coexpression of the full α_{1B}-receptor with its full third intracellular loop resulted in inhibition of the PLC activating effect of the receptor. This was mimicked partially by the N-terminal segment of the loop but not the C-terminal segment. Expression of the third intracellular loop of the D_{1A}-dopamine receptor had no effect on α_{1B}-receptor action, but inhibited the effect of the D_{1A}-receptor on PLC. The α_{1B}-loop also has a small effect on M1-muscarinic receptor-stimulated PLC activation. Increasing the concentrations of full-length receptors expressed in cells tended to overcome the inhibitory effect of the loop. These data are consistent with the idea that the loop acted by interfering with protein:protein interaction and hence that the third intracellular loop may physically contact the G protein.

17.2.2. Naturally Occurring Receptor Mutations

Insight into structure–function relations have also come from the study of genetic diseases that were shown to be the result of mutations of G protein-coupled receptors (summarized in Table 17.5). Two of these receptors have given a wealth of information because they give nonlethal phenotypes: rhodopsin and the VP2-vasopressin receptor.

17.2.2.1. Rhodopsin Mutations

Mutations of the rhodopsin receptor cause retinitis pigmentosa, a group of dominant autosomal diseases in which malfunction of rhodopsin leads to progressive retinal degeneration. In some cases this may occur because the mutant opsin is not properly processed causing damage to the protein processing and vesicle trafficking components of the rod cells[28]; in other cases, the reasons for the retinal degeneration are not clear. Of interest within the context of the present discussion is the opsin Lys296-to-Glu mutant. A study of this mutant opsin, expressed in COS cell membranes that were purified and then reconstituted with transducin, showed it to have constitutive activity, independent of light or retinal addition. An artificial mutation Glu113 to Gln, based on the fact that Glu113 is the counterion to the Lys296/retinal Schiff base, was also found to have constitutive activity in the absence of retinal.

Table 17.5. Mutant Forms of G Protein-Coupled Receptors

Receptor	Site	Mutation	Comments (receptor properties, phenotype, etc.)	Reference
A. Naturally occurring				
ACTHR (human)	TM III	S120R	Inactive—Hereditary familial glucocorticoid deficiency; autosomal recessive	Tsigos *et al.*[167]
	3i loop	R201Stop	Inactive—Hereditary familial glucocorticoid deficiency; autosomal recessive	Tsigos *et al.*[167]
	TM II	S74I	Inactive—Familial glucocorticoid deficiency; autosomal recessive	Clark *et al.*[168]
CaSensingR (human)	3i loop	R796W	Fails to elicit response to Ca^{2+} in *Xenopus* oocytes after cRNA injection—cause not determined. Familial hypocalciuric hypercalcemia (FHH) (heterozygous)	Pollak *et al.*[169]
	N-term	R198E	FHH—not functionally expressed	Pollak *et al.*[169]
	N-term	E298K	FHH; homozygous: neonatal severe hyperparathyroidism (NSHPT)—not functionally expressed	Pollack *et al.*[169]
GRFR (mouse)	N-term	D60G	*lit/lit* mouse: hypoplastic anterior pituitary: lack of GRF action	Lin *et al.*[170]
LHR (human)	TM VI	D578G	Constitutive activation, partial and stimulable by LH Hereditary autosomal dominant male precocious puberty	Shenker *et al.*[35]
	TM VI	M571I	Same as D578G	Kremer *et al.*[171]
MSHR (mouse)	1i loop	S69L	Constitutive activation, hyperresponsive to MSH Phenotype: dominant extension of black: E^{tob} (tobacco coat)	Robbins *et al.*[33]
	TM II	E92K	Constitutive activation, unresponsive to MSH Phenotype: dominant extension of black: $E^{so,3J}$ (somber coat)	Robbins *et al.*[33]
	1e loop	L98P	Constitutive activation Phenotype: dominant extension of black: E^{so} (somber)	Robbins *et al.*[33]
	2e loop	H183frshft	Inactive (frameshift) Phenotype: nonextension of black: *e* (yellow coat)	Robbins *et al.*[33]
Opsin (human)	TM VII	K296E	Constitutive activation—retinitis pigmentosa	Robinson *et al.*[29]
	TM VII	A292E	Constitutive activation—stationary night blindness	Dryja *et al.*[172]
	TM II	G90D	Constitutive activation—stationary night blindness	Rao *et al.*[173]
	N-term	V20G	Impedes processing and damages cell causing autosomal dominant retinitis pigmentosa	Nash *et al.*[28]
	N-term	P23H	Same as V20G	Nash *et al.*[28]
	N-term	P27L	Same as V20G	Nash *et al.*[28]
TSHR (human)	3i loop	D619G	Constitutive activity—thyroid adenoma	Parma *et al.*[174]
	3i loop	A623I	Constitutive activity—thryoid adenoma	Parma *et al.*[174]
TSHR (mouse)	TM IV	P556L	*hyt/hyt* mouse: hypothyroid mouse—loss of receptor expression on cell surface	Stein *et al.*[175]
V2R (human)	3i loop	G246frshft	Inactive—CNDI (Q5 allele)	Rosenthal *et al.*[176]
	Dry motif	R137H	Inactive—CNDI (Q2 allele)	Rosenthal *et al.*[31]
	1e loop	R113W	Reduced affinity for ligand, reduced expression on cell surface and reduced coupling—CNDI (Q3 allele)	Birnbaumer *et al.*[32]
B. Man-made				
Expansion of natural mutations:				
Opsin (human)	TM VII	K296G	Constitutive activation, suppressed by *n*-propylamine Schiff base of 11-*cis*-retinal giving light-sensitive receptor	Robinson *et al.*[29]
	TM VII	K296A	Constitutive activation, suppressed by *n*-ethylamine Schiff base of 11-*cis*-retinal giving light-ensitive sreceptor	Robinson *et al.*[29]

(Continued)

Table 17.5. *Continued*

Receptor	Site	Mutation	Comments (receptor properties, phenotype, etc.)	Reference
	TM III	E113Q (counterion)	Constitutive activation, suppressed by 11-*cis*-retinal giving light-sensitive receptor at pH 6.7 instead of pH 7.5	Robinson *et al.*[29]
LHR (rat)	TM VI	D578N	Unchanged activity: nature of mutation matters	Ji and Ji[36]
TSH (human)	3i loop	A623K	No effect	Kosugi *et al.*[177]
		D623E	No effect	Kosugi *et al.*[177]
Conserved Asp of transmembrane domain II:				
α_2-AR	TM II	D79N	Unchanged ligand binding, but loss of regulation by Na	Horstman *et al.*[178]
			Signaling: inhibition of AC and Ca^{2+} currents unaltered but stimulation of K^+ currents (inward (rectifier) severely impaired	Surprenant *et al.*[11]
β_2-AR	TM II	D79A	Increase in K_d and K_{act} for Iso (10×); antagonist binding unaffected; G_s activated to 50% of control	Strader *et al.*[179]
		D79N	Normal low-affinity agonist binding, no GTP-sensitive high-affinity binding, increased K_{act} to cause G_s activation to ca. 15% of control	Chung *et al.*[180]
D_2R	TM II	D80A D80E	Loss of Na^+-induced increase in antagonist binding and of Na^+-induced decrease of agonist binding	Neve *et al.*[181]
LHR (rat)	TM II	D383N	Loss of Na^+-induced decrease in hormone binding	Quintana *et al.*[182]
Constitutive activations due to mutation of C-terminal end of third intracellular loop:				
α_{1B}-AR	3i loop	triple: R288K, K290H, and A293L	Constitutive activation, lower K_d's for agonist	Cotecchia *et al.*[23]
	3i loop	A293 any AA	Constitutive activation, graded, lower K_d for agonist	Kjelsberg *et al.*[24]
α_{2C10}-AR	3i loop	T348F,A,E, C, or K	Constitutive activation, graded (K is best), lower K_d's for agonist	Ren *et al.*[26]
β_2-AR	3i loop	quadruple: L266S, H269K, & L272A	Constitutive activation	Samama et al.[25]
Opsin (human)	ERY motif	E134R, R135RE	Binds retinal but not transducin	Franke *et al.*[183,184]
		R135Q	Binds retinal but does not activate G_t	Franke *et al.*[184]
		R135L	Loss of retinal binding	Franke *et al.*[184]
		R135W	Loss of retinal binding	Franke *et al.*[184]
		E134Q	Binds retinal and activates transducin at 150% of control efficacy	Franke *et al.*[184]
β2-AR	DRY motif	D130N	Normal antagonist binding, increased high-affinity agonist binding, altered GTP shift, no G_s activation	Fraser *et al.*[185]

TM: transmembrane domain; i loop, intracellular loop; e loop, extracellular loop; N-term, N-terminal; frshft, frameshift. A, Ala; C, Cys; D, Asp; E, Glu; F, Phe; G, Gly; H, His; K, Lys; L, Leu; M, Met; N, Asn; P, Pro; Q, Gln; R, Arg; S, Ser; T, Thr; V, Val; W, Trp; Y, Tyr.

The intrinsic ligand-independent activity can be suppressed by addition of 11-*cis*-retinal. This novel light-sensitive receptor differs from the natural in that suppression of activity requires a lower pH to aid in the protonation of the Schiff base.[29] Analysis of the man-made Lys296-to-Ala and Lys296-to-Gly mutants, which are also constitutively active, showed that their activity could be suppressed by alkylamine Schiff bases of 11-*cis*-retinal and reactivated by light. This indicated that, mechanistically, the attachment of retinal to opsin is a convenience designed by nature to maximize the light-capturing efficiency, without being an essential part of the response mechanism.[29] Two other mutations, Ala292 to Glu in TM-VII and Gly90 to Asp in TM-II, also lead to constitutive activity in the absence of retinal and to visual defects. Taken together, ligand-independent activity of the apoprotein is obtained any time the Lys296 → Glu113 salt bridge is interfered with, and activation appears to involve a shift in the position of TM-III away from TM-VII (Figure 17.4, Table 17.6).

17.2.2.2. VP2-Vasopressin Receptor Mutations

A second large group of mutations has been identified in the type 2 vasopressin receptor gene. This receptor was cloned in 1991, located to the X chromosome, and soon proven to be the cause of many if not all cases of X-linked congenital nephrogenic diabetes insipidus.[30–32] This disease manifests itself mainly through the inability of the kidney to respond to the antidiuretic effect of vasopressin [also known as antidiuretic hormone (ADH)]; patients cannot concentrate urine. Newborns become dehydrated, which results in reduced growth, severe mental retardation, and if untreated—by the administration of water—leads to death. Water is the only treatment required.

Thirty-eight independent mutations of the VP2 gene have been identified at this time (Figure 17.5). Eleven result in frameshifts with codon changes and subsequent premature protein truncations, seven create a stop codon and also cause premature protein termination. All others are single amino acid changes of which ten occur in the predicted transmembrane domains, seven in the second extracellular domain, and two just prior to the beginning of TM-III. One, Arg137 to His (Q2 allele), is cytosolic just after TM-III at the center of the DRY motif—which is DRH for wild-type VP2. As mentioned, this mutation completely uncouples the receptor from G_s, while other agonist-dependent functions (binding and receptor sequestration) remain unaltered[31] (Figure 17.6, Table 17.5). This identifies the DRH as important in G protein–receptor coupling. A second, Arg113 to Trp (Q3 allele), just prior to TM-III, is located in the putative first extracellular loop of the VP2R next to a frequently conserved Cys thought to interact via a disulfide bridge with a Cys of the second extracellular loop. This mutation lowers the affinity for AVP by a factor of 20, interferes with its maturation/transport to the cell surface by a factor of 5 or so, and diminishes the efficacy of the receptor so that at equal receptor density it requires a threefold higher concentration of AVP

Table 17.6. Effects of Opsin Mutations that Interrupt the Lys296/Glu113 Salt Bridge

I. Apoprotein (opsin)			
wt	Lys^{296}-NH_3^+—$^-$OOC-Glu^{113}		Inactive
K296E	Glu^{296}	$^-$OOC-Glu^{113}	Active in the dark; retinitis pigmentosa
K296G	Gly^{296}	$^-$OOC-Glu^{113}	Active in the dark
E113Q	Lys^{296}-NH_3^+	NH_2CO-Gln^{113}	Active in the dark
A292E	Glu^{292}-COO^- Lys^{296}-NH_3^+	$^-$OOC-Glu^{113}	Active in the dark; night blindness
G90D	$^-$OOC-Asp^{90} Lys^{296}-NH_3^+	$^-$OOC-Glu^{113}	Active in the dark; night blindness
II. Holoprotein (rhodopsin)			
wt	Lys^{296}-NH^+—$^-$OOC-Glu^{113} | 11-*cis*-retinal (Schiff base)		Inactive in the dark
	Lys^{296}-N H$^+$ | all-*trans*-retinal (deprotonated Schiff base)	$^-$OOC-Glu^{113}	Active (Meta II, ε_{max} = 380 nm)
K296G	Propyl | $Gly^{296}NH^+$—$^-$OOC-Glu^{113} | 11-*cis*-retinal (Schiff base)		Inactive in the dark, activated by light
E113Q	Lys^{296}-NH^+ NH_2CO-Gln^{113} | 11-*cis*-retinal (Schiff base)		Inactive in the dark, activated by light at pH 6.7

Lys296 and Ala292 are in TM-VII separated by one helical turn; Glu113 is in TM-III; Gly90 is in TM-II. Glu292 and Asp90 are each able to establish a salt bridge with Lys296 disrupting the Lys296/Glu113 salt bridge. For references see Table 17.5.

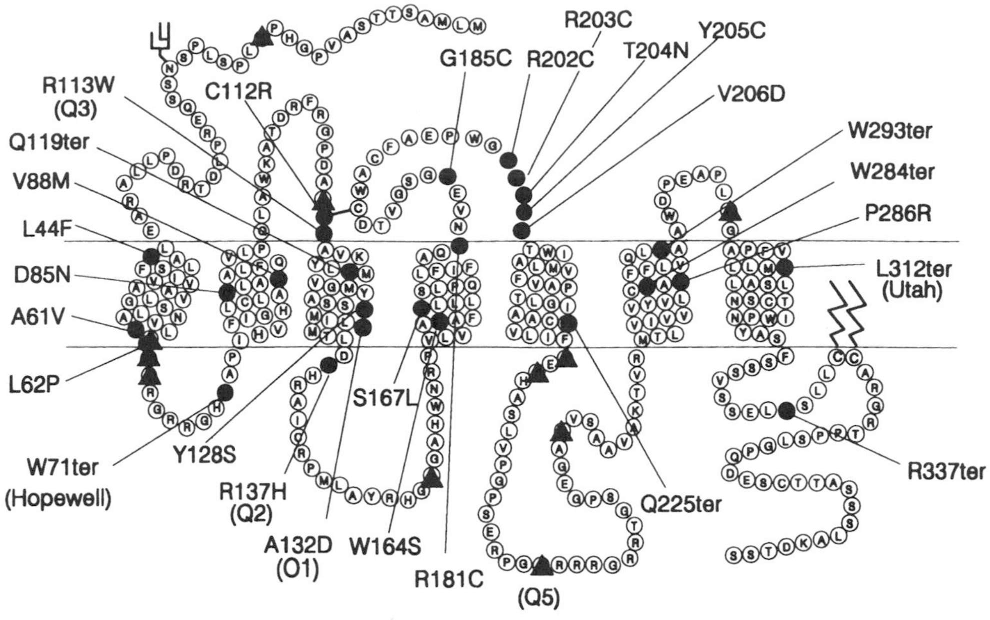

Figure 17.5. Physical map of VP2-vasopressin receptor mutants found in patients suffering from congenital nephrogenic diabetes insipidus. From Refs. 31, 32, 176, 215–220.

for adenylyl cyclase stimulation than the wild-type receptor.[32] Being extracellular, the effect of the mutation to alter binding affinity and interfere with the maturation of the protein is not unexpected, but the reduction in signal transduction is. This suggests that TM-III is structurally involved in the transmembrane transmission of the hormone binding signal. It is to be expected that further analysis of other amino acid changes will reveal new aspects of structure–function relations in the VP2 receptor that might be extrapolated to other receptors as well.

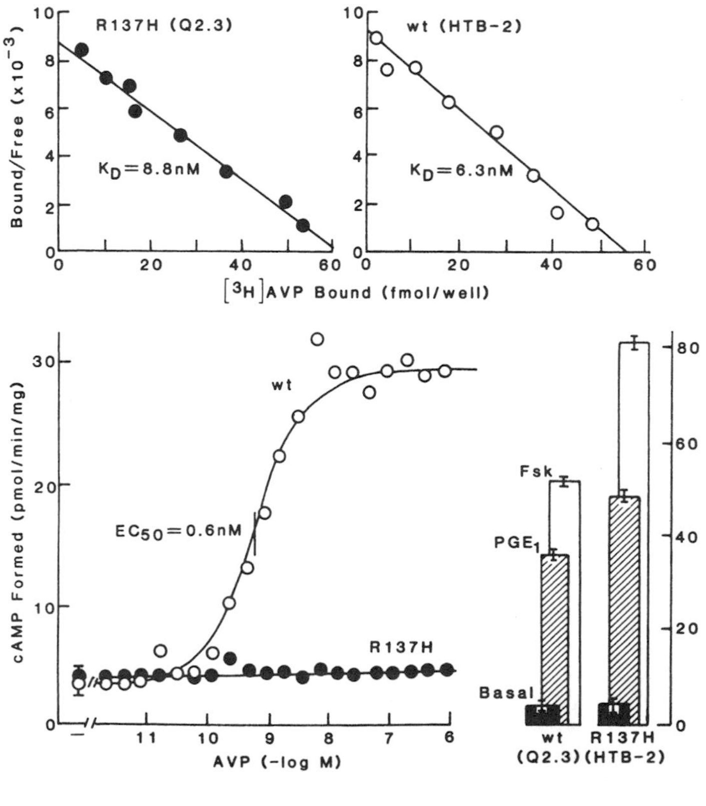

Figure 17.6. Loss of G protein activating capacity is the molecular basis of the congenital diabetes insipidus in male members of Quebec family Q2 who carry the Arg137-to-His mutant of the VP2 receptor gene. Transfection of the mutated receptor cDNA into L cells results in normal expression of the protein as seen by Scatchard analysis of binding sites, with normal affinity for the hormone. However, the mutant receptor fails to catalyze activation of G_s by GTP as seen by the failure of vasopressin-mediated stimulation of adenylyl cyclase. (Adapted from Ref. 31.)

17.2.2.3. The Melanocyte-Stimulating Hormone (MSH) Receptor as a Genetic Model

An interesting model that has potential for additional genetic analysis emerged from the study of MSH receptor mutants. MSH regulates melanin synthesis in melanocytes of the hair follicle cells and skin. The MSH receptor is coupled to adenylyl cyclase through G_s, and its stimulation by the intermediary pituitary lobe hormone MSH elevates cAMP, which in turn regulates the transcription of the enzyme tyrosinase. Tyrosinase is the rate-limiting step in the synthesis of melanins, of which melanocytes make two kinds: phaeomelanin and eumelanin. Hair made from follicular cells synthesizing phaeomelanin is yellow/red in color, hair made from follicular cells synthesizing excess eumelanin over phaeomelanin is brown/black. At low tyrosinase activity the levels of dihydroxyphenylalanine (DOPA) made from tyrosine are low and the default pathway is synthesis of DOPA-quinone followed by cysteinyl-DOPA which is incorporated into melanin to give phaeomelanin. On excess production of DOPA, as results after maximal stimulation of the MSH/G_s system, excess DOPA is converted to DOPA-chrome, which when incorporated into melanin gives black eumelanin (summarized in Figure 17.7). One of the MSH receptor genes, of which there are at least three, has been identified as the gene product of the *extension* locus. This is the gene that controls the degree to which dark color extends into the red/yellow background of hair[33] (reviewed in Ref. 34).

As shown in Figure 17.7C, hair color is also controlled by another genetic locus, the *agouti* locus. The gene product of *agouti* is a protein with the characteristics of a secreted polypeptide complete with signal sequence and processing signals. The agouti signal is made in paracrine fashion by cells that surround hair follicles, and although its mechanism of action has not been definitively established, it is most likely a competitive inhibitor of MSH action on the MSH receptor.

Inactivation of the *extension* locus is autosomal recessive, activation of the *extension* locus is dominant, inactivation of the *agouti* locus is recessive, and activation of the *agouti* locus can have variable effects depending on the strength of the normal *extension* locus. Three dominant murine *extension* mutations have been characterized as MSH receptor mutants, and the biochemical consequence was determined for two[33] (Figure 17.7D). One, $E^{so,3J}$ (sombre), causes the receptor to be constitutively activated, giving 50% of the activity obtained with wild-type receptor after MSH stimulation, but unresponsive to further stimulation by MSH; the second, E^{tob} (tobacco), causes the receptor to have some constitutive activity and to be hyperstimulated by MSH, as a result of what appears to be an increased efficacy. This mutation, Ser69 to Leu, affects the first intracellular loop; the activating $E^{so,3J}$ mutation, Glu92 to Lys, is close to the end of TM-II. The *extension* locus could be a powerful model to search for receptor mutants using coat color as initial selection marker. Especially parameters such as intrinsic activity (efficacy) of a receptor, which confers at equal receptor abundance an enhanced or decreased capacity of the receptor to stimulate G_s, should be amenable to study in this way.

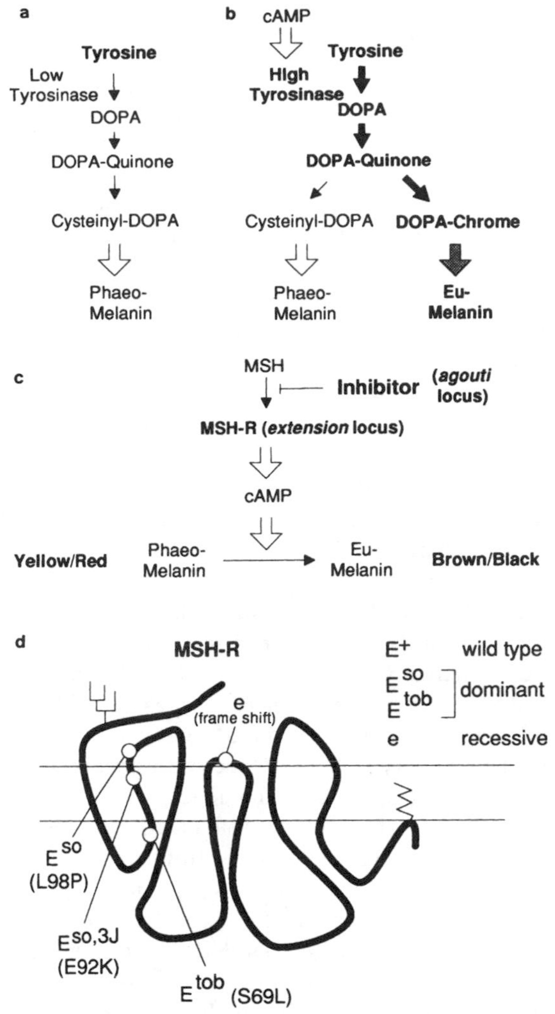

Figure 17.7. Summary of regulation of synthesis in melanocytes of phaeomelanin (a), of phaeomelanin plus eumelanin after MSH receptor stimulation (b), of the interactions of the *extension* (MSH receptor) and *agouti* loci (c), and of murine MSH receptor mutants responsible for the *e*, E^{so}, $E^{so,3J}$, and E^{tob} alleles (d). For details see text and Ref. 33.

Naturally occurring mutations that have been found in several other receptors are summarized in Table 17.5. It is interesting to note that several of them are of the activating type and that two of these, found in the TSH receptor, are related to a change in amino acid composition of the C-terminal end of the third intracellular loop, where Lefkowitz and collaborators had previously identified the potential for such functional activity. One, causing constitutive activation of the LH/hCG receptor, changes an Asp in TM-VI to Gly.[35] Although it has been speculated that activation may be the result of disruption of ion-pairing with a counterion from another TM, the structural cause for activation must be another since mutating the same Asp to Asn instead of Gly did not activate the receptor.[36]

Taken together, studies of naturally occurring receptor variants and mutants, as well as man-made mutants indicate a participation of all of the intracellular appendages in their interaction with G proteins.

17.2.3. G Protein Activation by Non-Heptahelical Receptors

In 1989, Nishimoto and collaborators demonstrated functional coupling of Man-6P/IGF-II receptor to G_{i2} in reconstituted phospholipid vesicles. GTPγS decreased the affinity of the receptor for IGF-II by a factor of close to 8 and IGF-II stimulated GTPγS binding in a PTX-sensitive manner.[37] This last effect was mimicked by a Man-6P/IGF-II receptor peptide comprising a 14-amino-acid sequence just inside the plasma membrane.[38] The peptide mimicked kinetically "classical" G protein-coupled receptors in that it acted to reduce the concentration at which Mg^{2+} promotes nucleotide exchange.[38] Residue substitution studies suggested the signaling motif to be between 14 and 20 amino acids long, to have two basic residues at the N-terminal side, and to contain the C-terminal submotif B-B-X-B or B-B-X-X-B, where B is a basic amino acid and X is any amino acid (Figure 17.8). Sequences with these general characteristics can be found in the second and third intracellular loops of several heptahelical receptors, including the C-terminal sequence of the human β_2-adrenergic receptor. A test of the β_2-adrenergic receptor sequence found it able to stimulate GTPγS binding to G_s.[21] Further studies identified a G_i/G_o activating sequence in the C-terminal end of the third intracellular loop of the M4-muscarinic receptor[39] and a variant but very potent G_i/G_o activating sequence at the comparable position of the α_2C10-adrenergic receptor, having at its C-terminus a B-B-X-X-F, instead of B-B-X-X-B.[40]

These rather tantalizing studies (reviewed in Ref. 22) provide for a structural basis for the PTX-sensitivity of some of the effects of non-heptahelical receptors, such as those of the already mentioned Man-6P/IGF-II receptor[37] and of the transforming activity of the type II TGFβ receptor in NIH-3T3 cells. This latter receptor has a "Nishimoto couplone" in the middle of its cytosolic kinase domain.[41] Also the Alzheimer β/A_4-amyloid precursor protein (APP) has a Nishimoto motif, and, on examination, APP was found to form a complex with G_o, the most abundant neuronal G protein, in a Mg^{2+}-dependent and GTPγS-sensitive manner. Mutations of the motif interfered with complex formation.[42,cf. 22]

These types of studies are not only interesting in their own right; they also provide a focus for further studies on structural aspects of the mechanics by which heptahelical receptors activate a G protein. As a word of caution, it must be mentioned that the determination of both the actual structural pattern of a receptor that activates the G protein and its specificity will not be easy. Other substances mimic receptors, just as Nishimito couplones do. These include mastoparan and polylysine.[43,44]

17.3. THE G PROTEINS

Three separate aspects of heterotrimeric G proteins are important:

1. The existence of a basic regulatory cycle involving a GTPase-dependent unidirectional subunit dissociation–reassociation cycle by which hormone receptor drives the activation by GTP and the consequential dissociation into an active GTP·αcomplex plus a βγ dimer, and the GTPase activity, which, by converting α·GTP into α·GDP, drives the deactivation of the α subunit with concomitant acquisition of high affinity for βγ dimer (Figure 17.9)
2. The existence of a rather large degree of molecular diversity among all three of the G protein subunits, of which the different α subunits can be grouped into structurally and functionally related subgroups
3. The fact that both α's and βγ dimers are signaling molecules, i.e., have the potential of regulating effector functions, and that single α's and βγ's are able each to regulate more than a single effector function

The regulatory cycle ensures that G protein-mediated processes are rapidly reversible and dependent on a second-by-second basis on agonist-driven receptor activity. The cycle is relatively slow and appears to involve separation of the activated form of the G protein from the receptor, so that during the lifetime of an activated G protein a single receptor has the ability to activate several G protein molecules. Thus, occupancy of one receptor may lead to activation of not one but several effector molecules. G protein-mediated signaling not

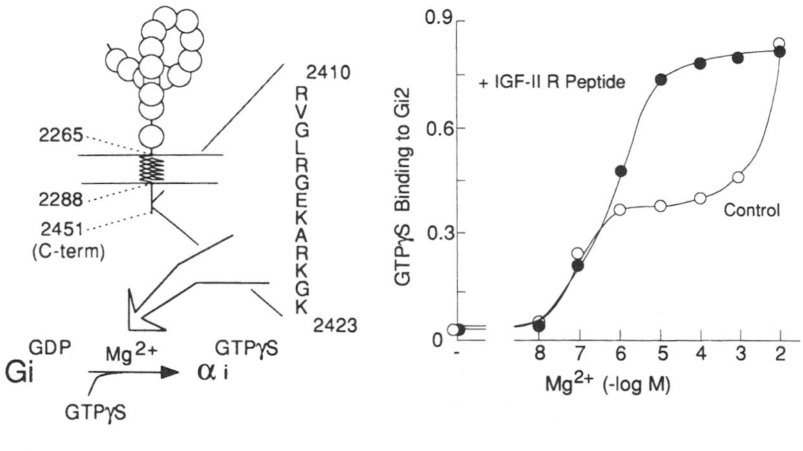

Figure 17.8. Summary of mode of action of the "Nishimoto couplone" of the cation-independent mannose-6-phosphate/insulin-like growth factor II (Man-6P/IGF-II) receptor as seen on addition of the couplone peptide shown in the left panel to phospholipid vesicles reconstituted with G_{i2}. Note the effect of the couplone to left-shift the concentration of Mg^{2+} at which GDP/GTPγS exchange at G_{i2} is obtained.

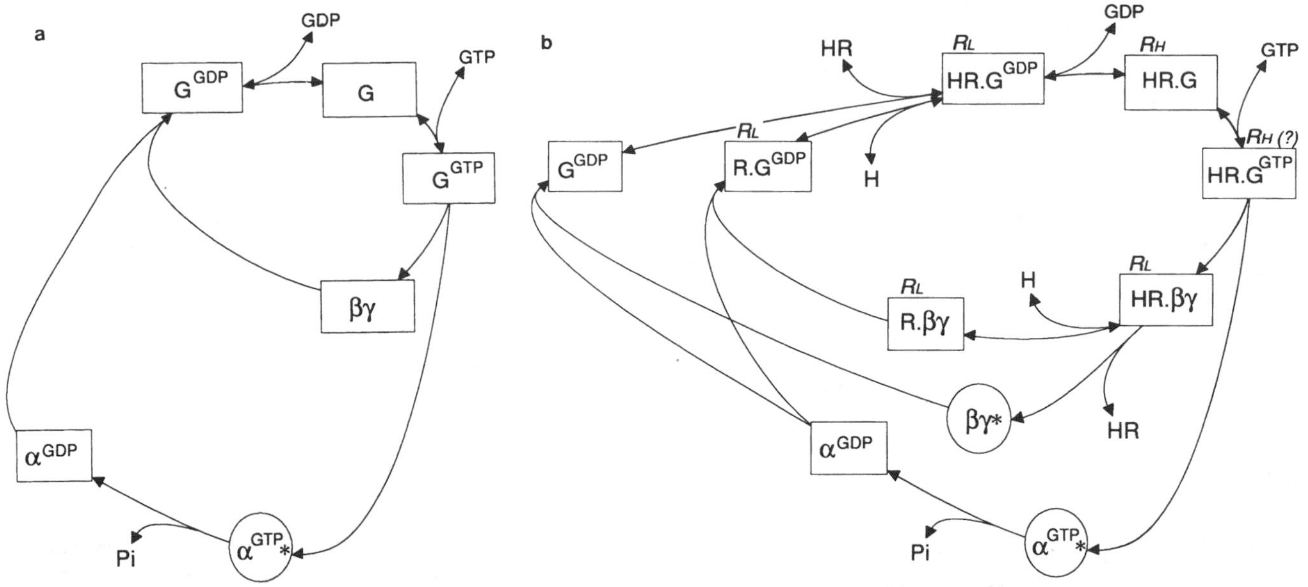

Figure 17.9. Basic regulatory cycles of a G protein involving both GTP-induced activation and subunit dissociation and GTPase-dependent inactivation and subunit reassociation. Panel A shows this cycle as it may occur in the absence of receptor. Panel B shows receptor–G protein interactions within the context of a G protein regulatory cycle incorporating various possible fates of the HR·βγ complex. Although only free α-GTP and free βγ are assumed to be competent to regulate an effector, this has not been tested. Note that α-GTP may form without or with formation of free βγ, and that HR·βγ is a postulated substrate of G protein-coupled receptor kinases. A tentative assignment of which forms of receptor–G protein complexes would exhibit high and low affinity for agonist is also shown, assuming that the receptor is a β-adrenergic receptor.

only involves the transduction of the input signal, receptor occupancy by ligand, into modulation of effector activity altering the intracellular level of a second messenger, but also the amplification of the input signal whereby occupation of a single receptor may activate several G proteins, and each G protein, as a result of generation of both an α signal and a βγ signal, may modulate not one but two effector molecules.

Molecular diversity of G protein subunits provides both for a certain degree of redundancy, most clearly shown in a recent knockout experiment, and for the potential ability of a given cell to alter its primary or secondary response patterns by up- or downregulating selected isoforms of α's, β's, and γ's.

The potential ability of multiple actions of a single G protein in turn provides for a potentially large variation in the response pattern that may be elicited in a given cell by activation of a given G protein. The particular response pattern elicited depends on the type of G proteins and the type of effector functions that the cell is expressing.

17.3.1. The Regulatory Cycle—Mechanism of Action of a Receptor

The overall effect of a receptor is to accelerate the activation of a G protein by guanine nucleotide. The experiment in Figure 17.10 illustrates what appears to be the molecular basis for this effect of the receptor, i.e., the reduction in the requirement for Mg^{2+} of G protein activation by guanine nucleotide, from supraphysiologic (app. K_m ca. 15–20 mM) to well below that of normal prevailing intracellular levels (app. K_m ca. 10 μM). The effect of Mg^{2+} in turn is to promote nucleotide exchange, GTP or GTPγS for GDP. Only two proteins—receptor and G protein reconstituted in phospholipid vesicles—can mimic this phenomenology in full. The mechanism by which the increase in affinity of G protein for Mg^{2+} occurs is not well understood, but, rather than the more commonly considered nucleotide exchange reaction, it is the likely underlying mechanism of receptor-mediated activation of a G protein.

Activation of a purified G protein by a nonhydrolyzable GTP analogue [GMP-P(NH)P or GTPγS] leads, as it does in membranes, to its persistent, quasi-irreversible activation. Hydrodynamic analysis of the G protein before and after activation showed that activation is associated not only with stable binding of the guanine nucleotide, but also with a decrease in molecular weight via a dissociation reaction of the type

$$\alpha^{GDP}\cdot\beta\gamma + GTP\gamma S \rightarrow \alpha^{GTP\gamma S} + \beta\gamma + GDP$$

In the presence of GTP, the reversal reaction involves GTP hydrolysis and reassociation of α with βγ to give the starting αGDP·βγ complex according to the reaction sequence

$$\alpha^{GTP} \rightarrow P_i + \alpha^{GDP} + \beta\gamma \rightarrow \alpha^{GDP}\cdot\beta\gamma$$

Taken together, these reactions allow for the postulation of the minimal regulatory cycle of the type shown in Figure 17.9A.

Several individual rate constants that make up this cycle have been assessed with native[e.g.,45] (reviewed in Ref. 46) and recombinant[e.g., 47,48] G proteins. From these measurements it is

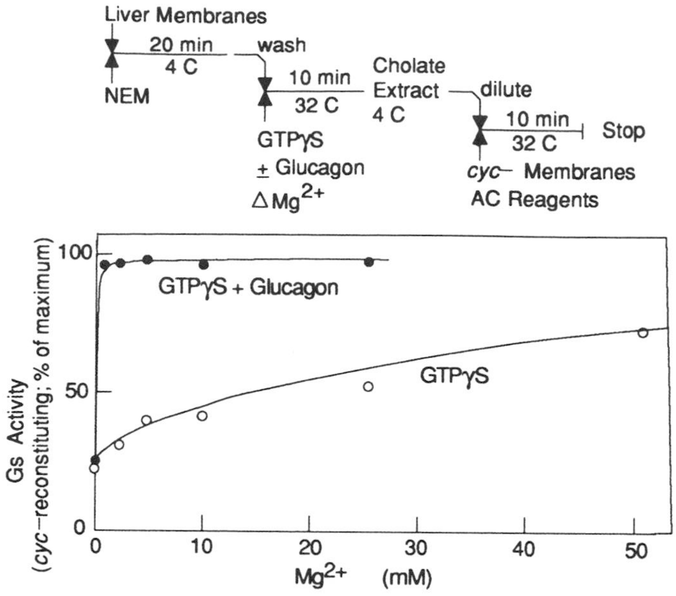

Figure 17.10. Key characteristic of receptor-mediated stimulation of G protein activation. Receptor reduces the concentration of Mg^{2+} required for activation of the G protein by GTP or a GTP analogue. (Adapted from Ref. 221.)

now clear that in the absence of Mg^{2+} $\beta\gamma$ inhibits GDP dissociation from α, but in the presence of Mg^{2+} it destabilizes the α-GDP complex and promotes GTP:GDP exchange, and that it is this activity of $\beta\gamma$ dimer that receptors stimulate by reducing the Mg^{2+} requirement.

The interaction of G protein with receptor can be assessed not only by studying changes in nucleotide binding and hydrolysis, but, because of a reciprocal effect of G protein on receptor, by determining the agonist affinity state of the receptor (Figure 17.11). The formation of this complex requires an agonist, and can generally be detected because it stabilizes the receptor in a conformation that, often but not always, has higher affinity for agonist (R_h) than what is observed in the absence of the G protein (R_l).[49–53] In support of the assumption that for receptors such as the β-adrenergic receptor, the high-affinity form of the receptor represents G·R, purified G protein-free receptor shows the same low affinity for agonist as seen after saturation of membranes with guanine nucleotide, and acquires high affinity for agonist when reconstituted with purified G protein. Since either a guanine nucleoside diphosphate (GDP or GDPβS) or a guanine nucleotide triphosphate (GTP or one of its nonhydrolyzable analogues) promotes low-affinity binding [e.g., 49] high-affinity binding can be taken largely as the measure of the receptor complexed with nucleotide free heterotrimeric G protein (HR·G), poised for association with either GDP or GTP, thus allowing "free" exchange to occur. In one case, GDP, the complex relaxes without activation of G. In the other case, GTP, the complex advances along a not totally understood path to new forms of HR·G. Because of a conformational change in α induced by MgGTP binding the α-$\beta\gamma$ interaction is reduced. This promotes dissociation of the HR·G(Mg)·GTP(Mg) complex into the activated α·GTP·Mg plus HR·$\beta\gamma$·Mg.

Possible fates of the HR·$\beta\gamma$ complex and the cycling of $\beta\gamma$ (or HR·$\beta\gamma$) to restore the $\alpha\beta\gamma$ trimer (or HR·$\alpha\beta\gamma$) are shown in Figure 17.9B. One path leads to formation of $\beta\gamma$ dimer plus HR complex, the other lets H dissociate before $\beta\gamma$ does, leading to formation of an R·$\beta\gamma$ complex. There is no information as to the role of Mg^{2+} in determining the type of path followed, but whether or not $\beta\gamma$ forms appears to be important from the signaling point of view because there are G protein signaling pathways that are mediated by the dimers

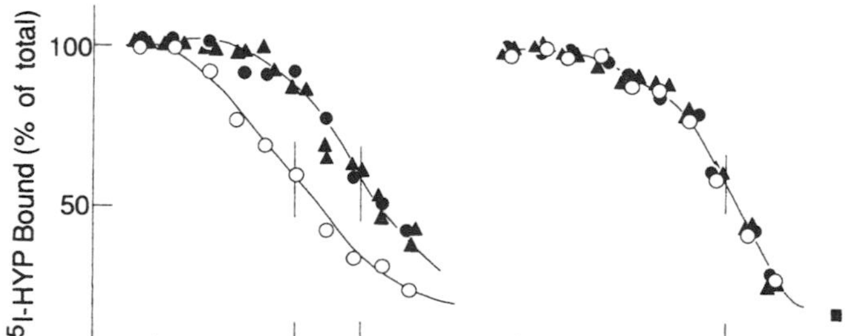

Figure 17.11. Guanine nucleotide and Mg^{2+} regulation of affinity of receptor for agonist reflects the interaction of G proteins with receptor. Regulation of receptor affinity for agonist is dependent on G_s, as seen from a comparison of agonist displacement curves obtained with G_s-containing (left) and G_s-deficient (right) S49 cell membranes, and on Mg^{2+} in the medium. (From Ref. 222.)

rather than the α subunits. α subunits activate adenylyl cyclases (for molecular diversity see below), a type of Ca^{2+} dependent K^+ channel, K^+ channels of the inwardly rectifying and ATP-sensitive type, skeletal and cardiac voltage-dependent Ca^{2+} channels, visual phosphodiesterase, various isoforms of PLCβ, and others. βγ dimers, on the other hand, stimulate certain forms of PLCβ as well as certain types of adenylyl cyclase, and also the inwardly rectifying K^+ channel and one of the Ca^{2+}-activated K^+ channels. While in the case of α subunits it appears to be clear that it is the activated free α subunit that is responsible for receptor signaling, it has not been shown for example whether only free βγ dimers or whether βγ·HR and βγ·R complexes can also modulate effector functions.

17.3.2. βγ Dimers as Potential Activators of α Subunits by Transphosphorylation

A surprising and not as yet much recognized finding is that βγ dimers can activate the G_s-adenylyl cyclase system by a mechanism involving α subunit activation. Activation occurs if βγ dimers are prepared from heterotrimeric G protein by activation with GTPγS, but not if prepared using GMP-P(NH)P as the dissociating agent. By the use of ^{35}S-labeled GTPγS and [γ-^{32}P]-GTP, it was now shown that the β subunit can become phosphorylated, and that the phosphorylated βγ can transfer its phosphate to membranes under conditions that lead to activation of adenylyl cyclase, or onto GDP, but not ADP, GMP, or GDPβS. The overall reaction is that of a nucleoside diphosphate kinase and the acceptor is very likely α_s-bound GDP. Phosphorylation of βγ requires micromolar concentrations of Mg^{2+}. Chemical properties of the bound phosphate are such that the most likely accept or amino acid is histidine and treatment of βγ with a histidine-modifying reagent inhibited its phosphorylation.[54,55] This then presents an alternate possibility for receptors to catalyze G protein reactivation:

$$\mathrm{HR}\cdot\alpha^{\mathrm{GDP}}\cdot\beta\gamma + \mathrm{GTP} \rightarrow [\mathrm{HR}\cdot\alpha^{\mathrm{GDP}}\cdot\beta\gamma{\sim}\mathrm{P} \rightarrow \mathrm{HR}\cdot\alpha^{\mathrm{GTP}}\cdot\beta\gamma] \rightarrow \mathrm{HR}\cdot\beta\gamma + \alpha^{\mathrm{GTP}}$$

It is clear that many if not all the arguments made above for how receptors promote activation of a G protein are applicable here as well. The reaction may proceed only slowly in the absence of receptors, receptors could accelerate the reaction by lowering requirement for Mg^{2+}, and so on. One might be tempted to suggest that βγ-catalyzed activation of α-GDP to α-GTP by direct transfer of phosphate may be the preferred mode of activation of G proteins. Yet, the fact that adenylyl cyclase, and hence G_s, is activated also with the nonphosphorylating analogue GMP-P(NH)P, and that this reaction is stimulated by receptor, would indicate that phosphorylation of bound GDP cannot be the whole story. The question of reagent purity—i.e., whether GMP-P(NH)P preparations are really free of GTP—becomes of paramount importance to interpret correctly experiments done with βγ dimers.

17.3.3. G Protein Structure

17.3.3.1. Molecular Diversity of G Protein Subunits

Purified unactivated G proteins have molecular weights in the range of 100,000, being composed of three subunits α, β, and γ, in decreasing order of size. Biochemical, chemical, and eventually molecular cloning of mRNAs showed that each of the subunits exhibits molecular diversity. This diversity is large for α's and γ's and moderate for β's. All are subject to posttranslational modifications, notably lipidations. Without being transmembrane in character, G proteins nevertheless localize to membranes within the cell, being found not only in plasma membranes where they are clearly engaged in signal transduction but also in Golgi and endosome membranes where their role(s) is(are) not yet clear.

The α subunits constitute both the most numerous and the most complex set. There are 16 nonallelic mammalian α subunit genes known. At 354 to 380 amino acids, they are similar but on the average only 47% homologous (the least: 36%, α_{olf} versus α_{12}; the most: 88%, α_q versus α_{11} or α_{i1} versus α_{i2}; the most frequent homology among them: 40–45%).

From a functional viewpoint, four α subunits are sense related: two for vision (rod and cone transducins), one for olfaction (α_{olf}), and one for taste (α_{gust} or gustducin). The others are α_s (four splice variants of uncertain functional difference), three α_i's, α_o (two splice variants), α_z, α_q, α_{11}, α_{12}, α_{13}, α_{14}, and α_{15}(mouse)/α_{16}(human). An amino acid sequence alignment of the principal G protein α subunits and a tree that relates them phylogenetically are shown in Table 17.7.

In addition, it is known that there are five nonallelic β genes (four cDNA sequences published), and seven nonallelic γ genes (five cDNAs published) (Table 17.8). From a structural view, the β subunits (each of 340 amino acids) are the least variant of the three G protein subunits, being 78 and 86% homologous. They belong to the type of proteins with WD-40 motifs (highlighted in the inset of Table 17.8). WD-40 motifs divide the β's into eight blocks of which the first has a hypervariable region (β25–40) and the remaining seven have either a complete WD-40 motif (blocks 2, 4, 5, 6, and 8) or only the second (B) half of the motif (blocks 3 and 7). Homologues have been found in nonmammalian organisms as well as in yeast where the β subunit is known as STE4. The WD-40 motif is intriguing, in that it has been found in a variety of apparently unrelated proteins that include transcription factors, a component of yeast cytoskeleton, a product of a chicken MHC locus, and a gene responsible for a form of hereditary lissencephaly (see references in Refs. 56 and 57). Like the phosphotyrosine-binding SH2 domains and polyproline-rich motifs recognizing SH3 domains of nonreceptor tyrosine kinases and the GRB2/sem5 adaptor protein, the WD-40 motif may encode a protein–protein interacting function for which the partner has yet to be identified. Another possibility is that they contribute to a metal binding site, such as appears to be required for the β subunit function to stimulate nucleotide exchange.

Table 17.7 Sequence Alignment of All Major Classes of Mammalian G Protein α Subunits

```
                                 (palmitoyl)
GenBank #                  .         |  .         .         .         .         .
X04408                             MGCLGNS---KT-EDQRNEEKAQREANKKIEKQLQKDKQVYRATHR   42    hum-gs
M26718                        (mry)MGCLGNSS--KTAEDQGVDEKERREANKKIEKQLQKERLAYKATHR   44    rat-golf
M27543                        (mry)MGCTL-----------SAEDKAAVERSKMIDRNLREDGEKAAKEVK   35    hum-gi3
M36777                        (mry)MGCTL-----------SAEERAALERSKAIEKNLKEDGISAAKDVK   35    mus-go1
K03253                        (mry)MGAGA-----------SAEEK----HSRELEKKLKEDAEKDARTVK   31    bov-gtrod
D90150                        (mry)MGCRQ-----------SSEEKEAARRSRRIDRHLRSESQRQRREIK   35    rat-gz
M55412                          MTLESIMACCL-----------SEEAKEARRINDEIERHVRRDKRDARRELK   41    mus-gq
M63359                   MSGVVRTLSRCLLPAEAGARERRAGAARDAERRARRRSRDIDALLARERRAVRRLVK   57    mus-g12
M63904                          MARSLTWRCCPW---------CLTEDEKAAARVDQEINRILLEQKKQDRGELK   44    hum-g16
                                                          *       h-+

    .         .         .         .         .         .         .
LLLLGAGESGKSTIVKQMRILHVNGFNGEGGEEDPQAARSNSDGEKATKVQDIKNNLKEAIETIVAAMSN  112    hum-gs
LLLLGAGESGKSTIVKQMRILHVNGFNPE---------------EKKQKILDIRKNVKDALVTIISAMST   99    rat-golf
LLLLGAGESGKSTIVKQMKIIHEDGYSED---------------ECKQYKVVVYSNTIQSIIAIIRAMGR   90    hum-gi3
LLLLGAGESGKSTIVKQMKIIHEDGFSGE---------------DVKQYKPVVYSNTIQSLAAIVRAMDT   90    mus-go1
LLLLGAGESGKSTIVKQMKIIHQDGYSLE---------------ECLEFIAIIYGNTLQSILAIVRAMTT   86    bov-gtrod
LLLLGTSNSGKSTIVKQMKIIHSGGFNLE---------------ACKEYKPLIIYNAIDSLTRIIRALAA   90    rat-gz
LLLLGTGESGKSTFIKQMRIIHGSGYSDE---------------DKRGFTKLVYQNIFTAMQAMIRAMDT   96    mus-gq
ILLLGAGESGKSTFLKQMRIIHGREFDQK---------------ALLEFRDTIFDNILKGSRVLVDARDK  112    mus-g12
LLLLGPGESGKSTFIKQMRIIHGAGYSEE---------------ERKGFRPLVYQNIFVSMRAMIEAMER   99    hum-g16
 ****   *****  *** * *                                   *          *

    .         .         .         .         .         .         .
LVPPVELANPE--NQFRVDYILSVMNVPDFDFPPEFYEHAKALWEDEGVRACYERSNEYQLIDCAQYFLD  180    hum-gs
IIPPVPLANPE--NQFRSDYIKSIAPITDFEYSQEFFDHVKKLWDDEGVKACFERSNEYQLIDCAQYFLE  167    rat-golf
LKIDFGEAARA--DDARQLFVLAGSAE-EGVMTPELAGVIKRLWRDGGVQACFSRSREYQLNDSASYYLN  157    hum-gi3
LGVEYGDKERK--TDSKMVCDVVSRMEDTEPFSAELLSAMMRLWGDSGIQECFNRSREYQLNDSAKYYLD  158    mus-go1
LNIQYGDSARQ--DDARKLMHMADTIE-EGTMPKEMSDIIQRLWKDSGIQACFDRASEYQLNDSAGYYLS  153    bov-gtrod
LRIDFHNPDRA--YDAVQLFALTGPAESKGEITPELLGVMRRLWADPGAQACFSRSSEYHLEDNAAYYLN  158    rat-gz
LKIPYKYEHNK--A--HAQLVREVDVEKVSAFENPYVDAIKSLWNDPGIQECYDRRREYQLSDSTKYYLN  162    mus-gq
LGIPWQHSENEKHGMFLMAFENKAGLPVEPATFQLYVPALSALWRDSGIREAFSRRSEFQLGESVKYFLD  182    mus-g12
LQIPFSRPESK--H--HASLVMSQDPYKVTTFEKRYAAAMQWLWRDAGIRACYERRREFHLLDSAVYYLS  165    hum-g16
                                              ** * *   +   *  *  *   * *

    .         .         .         .         .         .         .
KIDVIKQADYVPSDQDLLRCRVLTSGIFETKFQVDKVNFHMFDVGGQRDERRKWIQCFNDVTAIIFVVAS  250    hum-gs
RIDSVSLVDYTPTDQDLLRCRVLTSGIFETRFQVDKVNFHMFDVGGQRDERRKWIQCFNDVTAIIYVAAC  237    rat-golf
DLDRISQSNYIPTQQDVLRTRVKTTGIVETHFTFKDLYFKMFDVGGQRSERKKWIHCFEGVTAIIFCVAL  227    hum-gi3
SLDRIGAGDYQPTEQDILRTRVKTTGIVETHFTFKNLHFRLFDVGGQRSERKKWIHCFEDVTAIIFCVAL  228    mus-go1
DLERLVTPGYVPTEQDVLRSRVKTTGIIETQFSFKDLNFRMFDVGGQRSERKKWIHCFEGVTCIIFIAAL  223    bov-gtrod
DLERIAAADYIPTVEDILRSRDMTTGIVENKFTFKELTFKMVDVGGQRSERKKWIHCFEGVTAIIFCVEL  228    rat-gz
DLDRVADPSYLPTQQDVLRVRVPTTGIIEYPFDLQSVIFRMVDVGGQRSERRKWIHCFENVTSIMFLVAL  232    mus-gq
NLDRIGQLNYFPSKQDILLARKATKGIVEHDFVIKKIPFKMVDVGGQRSQRQKWFQCFDGITSILFMVSS  252    mus-g12
HLERITEEGYVPTAQDVLRSRMPTTGINEYCFSVQKTNLRIVDVGGQKSERKKWIHCFENVIALIYLASL  235    hum-g16
           *   * *  *  * ** *  *           *****   *  * o*

    .         .         .         .         .         .         .
SSYNMVIREDNQTNRLQEALNLFKSIWNNRWLRTISVILFLNKQDLLAEKVLAGKSKIEDYFPEFARYT-  319    hum-gs
SSYNMVIREDNNTNRLRESLDLFESIWNNRWLRTISIILFLNKQDMLAEKVLAGKSKIEDYFPEYANYT-  306    rat-golf
SDYDLVLAEDEEMNRMHESMKLFDSICNNKWFTETSIILFLNKKDLFEEKI--KRSPLTICYPEYTGSN-  294    hum-gi3
SGYDQVLHEDETTNRMHESLMLFDSICNNKFFIDTSIILFLNKKDLFGEKI--KKSPLTICFPEYPGSN-  295    mus-go1
SAYDMVLVEDDEVNRMHESLHLFNSICNHRYFATTSIVLFLNKKDVFSEKI--KKAHLSICFPDYNGPN-  290    bov-gtrod
SAYDLKLYEDNQTSRMAESLRLFDSICNNNWFINTSLILFLNKKDLLAEKI--RRIPLTICFPEYKGQN-  295    rat-gz
SEYDQVLVESDNENRMEESKALFRTIITYPWFQNSSVILFLNKKDLLEEKI--MYSHLVDYFPEYDGPQR  300    mus-gq
SEYDQVLMEDRRTNRLVESMNIFETIVNNKLFFNVSIILFLNKMDLLVEKV--KSVSIKKHFPDFKGDPH  320    mus-g12
SEYDQCLEENNQENRMKESLALFGTILELPWFKSTSVILFLNKTDILEEKI--PTSHLATYFPSFQGPKQ  303    hum-g16
*      *    *  *    * *         *  ***** *    *

    .         .         .         .         .         .         .
TPEDATPEPGEDPRVTRAKYFIRDEFLRISTASGD-----------GRHYCYPHFTCAVDTENIRRVFND  378    hum-gs
VPEDATPDAGEDPKVTRAKFFIRDLFLRISTATGD-----------GKHYCYPHFTCAVDTENIRRVFND  365    rat-golf
TYEEAAA-------------YIQCQFEDLNRRKDT-----------KE--IYTHFTCATDTKNVQFVFDA  338    hum-gi3
TYEDAAA-------------YIQTQFESKNRS-PN-----------KE--IYCHMTCATDTNNIQVVFDA  338    mus-go1
TYEDAGN-------------YIKVQFLELNMRRDV-----------KE--IYSHMTCATDTQNVKFVFDA  334    bov-gtrod
TYEEAAV-------------YIQRQFEDLNRNKET-----------KE--IYSHFTCATDTSNIQFVFDA  339    rat-gz
DAQAARE-------------FILKMFVDLNPDSD------------KI--IYSHFTCATDTENIRFVFAA  343    mus-gq
RLEDVQR-------------YLVQCFDRKRRNRS------------KP--LFHHFTTAIDTENIRFVFHA  363    mus-g12
DAEAAKR-------------FILDMYTRMYTGCVDGPEGSKKGARSRR--LFSHYTCATDTQNIRKVFKD  358    hum-g16
                                                      * *+*    *   **

CRDIIQRMHLRQYELL  394  hum-gs
CRDIIQRMHLKQYELL  381  rat-golf
VTDVIIKNNLKECGLY  354  hum-gi3
VTDIIIANNLRGCGLY  354  mus-go1
VTDIIIKENLKDCGLF  350  bov-gtrod
VTDVIIQNNLKYIGLC  355  rat-gz
VKDTILQLNLKEYNLV  359  mus-gq
VKDTILQENLKDIMLQ  379  mus-g12
VRDSVLARYLDEINLL  374  hum-g16
   *    *    *
```

Note: there are 16 nonallelic mammalian genes known: four are sense related: two transducins, one olfaction, and one taste (gustducin), the remainder are α_s (four splice variants), three α_i's, α_0 (two splice variants,), α_z, α_q, α_{11}, α_{12}, α_{13}, α_{14}, and α_{15}. α_{16} is the human homolgue of mouse α_{15}.

*, absolutely conserved amino acid in all mammalian α subunits (including α_{o2}, α_{i1}, α_{i2},$\alpha_{t\text{-cone}}$, α_{11}, α_{13}, α_{14}, and α_{gust} (gustducin). **o**, Cys conserved in all α's. +, cysteine conserved in all except α_{12} and α_{13}. **h-+**, amino acids thought to contact βγ dimer. Cys108 is unique to α_o and reactive to NEM, and does not participate in cross-linking to β.[186,187]

Table 17.8 Amino Acid Sequence Alignment of G Protein β and γ Subunits

a. β subunits

```
β Loligo forbesi (Lf)                                                      MT---EA----T--------E----A--T--AMA-A-VE-V   41
β Drosophila melanogaster (Dm)                                              MN---S------S---A-------AC-TS-L-AATSLE-I   40
β Caenorhabditis elegans (Ce)                                               ---------------S---E---SAN-T--ATVAS-LE-I   40
β1                                                                          MSELDQLRQEAEQLKNQIRDARKACADATLSQITNNIDPV   40
β2                                                                          ----E---------R---------CG-S--T---AGL---   40
β3                                                                  (myr?)  -G-MEQ---------K--A-----C--V--AELVSGLEV-   40
β4                                   .     .     .     .     .     .     .  .----E---------R---Q-----CN----V---S-M-S-   40
STE4          MAAHQMDSITYSNNVTQQYIQPQSLQDISAVEEEIQNKIEAA---SK--HAQINKAKHKIQD-S-F-MA-KVTSL   75
                                                                            * *  *********  ** ****** * **          *

                                                                          1:     A                            B
βLf                                                                     __------------------------AS---N-----------V--G   85
βDm                                                                     __-------------------------N---N-----------V---   84
βCe                                                                     __------------------------AS---N-----------V---   84
β1                                                                      __GRIQMRTRRTLRGHLAKIYAMHWGTDSRLLVSASQDGKLIIWDS   84
β2                                                                      __--------------------------------------------   84
β3                                                                      __--V---------------------A---K-----------V---   84
β4                                                                      __-------------------------Y------------------   84
STE4                                                                    TKNK-NLKPNIV-K--NN--SDFR-SR--KRIL------FML----  121
                                                                          ** ******************** ** ************ ***

                                                                                                          2:    B
βLf                                                                        -------------------C----------C------C-----  127
βDm                                                                        H------S-----------C------S---C-----MC---N-  126
βCe                                                                        -------------------C------SF--C------C-----  126
β1                                                                         YTTNKVHAIPLRSSWVMTCAYAPSGNYVACGGLDNICSIYSL  126
β2                                                                         -------------------C-------F--C------C---S-  126
β3                                                                         -------------------C-------F--C-----MC-----  126
β4                                                                         -----M-------------C----------C------C-----  126
STE4                                                                       ASGL-QN----D-Q--LSC-IS--STL--SA--N-NCT--RV  163
                                                                           ***** ******************* ******** **** *

                                                                             3:      A                              B
βLf                                                                       _-------------P-------CC--I-----------M-C-----  172
βDm                                                                       _-------------P--G----CC--------------MSCG----  171
βCe                                                                       _-------------P-------CC--------------M-C-----  171
β1                                                                        _KTREGNVRVSRELAGHTGYLSCCRFLDDNQIVTSSGDTTCALWDI  171
β2                                                                        _-------------P-------CC--------I-------C-----  171
β3                                                                        _-S-----K-----SA------CC------N---------C-----  171
β4                                                                        _-----D---------------CC-----G--I-------C-----  171
STE4                                                                      S-ENRVAQNVASIFK---C-I-DIE-T-NAH-L-A---M-C-----  209
                                                                          * *** * *****  *************  * *************

                                                                              4:      A                                  B
βLf                                                                     ---N-I-S-G-N------------M-T_______-----C------F-I   214
βDm                                                                     ---L-V-S-L-------A-----QCK-_______-----C--------I   213
βCe                                                                     -----C-A-------------S--F-T_______-I---C--------I   313
β1                                                                      ETGQQTTTFTGHTGDVMSLSLAPDTRL_______FVSGACDASAKLWDV   213
β2                                                                      -----VG-A--S------------G-T_______-----C---I-----   213
β3                                                                      ----K-V-V------C---AVS--FN-_______-I---C---------   213
β4                                                                      ------------S--------S--LKT_______-----C---S----I   213
STE4                                                                    PKAKRVREYSD-L---LA-AIPEEPN-ENSSNT-A-CGS-GYTYI--S   257
                                                                        ****    * **  ** ***   **         * ******* ****

                                                                            5:      A                             B
βLf                                                                         -D-ICK-------------TY----F----------C----I   266
βDm                                                                         -E-VCK---P---------VT-----Q----------C----I   265
βCe                                                                         -D--CK---P---------VA---S--R---------C----I   265
β1                                                                          RQGMCRQTFTGHESDINAICFFPNGNAFATGSDDATCRLFDL   265
β2                                                                          -DS-C----I---------VA-----Y--T-------C-----   265
β3                                                                          -E-TC--------------C-----E-IC------SC-----   265
β4                                                                          -D--C--S----I-----VS---S-Y-----------C-----   265
STE4                                                                        ESPSAV-S-YVND-----LR--KD-MSIVA---NGAINMY--   298
                                                                            *   *** * ** *****  *** * *  ****** ******

                                                                                                      6:    B
βLf           -----IGM-----_____________________________________________--C-----A---------G------C----V   300
βDm           ------AM-----_____________________________________________--C-----A-----------------C----T   299
βCe           ------AM-----_____________________________________________--C-----A--------F--------C----S   299
β1            RADQELMTYSHDN_____________________________________________IICGITSVSFSKSGRLLLAGYDDFNCNVWDA   299
β2            ------LM-----_____________________________________________--C-----A--R--------------C-I---   299
β3            ------ICF--ES_____________________________________________--C-----A--L-----F--------C----S   299
β4            ------LLLY---_____________________________________________--C-----A-----------------CS----   299
STE4          -S-CSIA-F-LFRGYEERTPTPTYMAANMEYNTAQSPQTLKSTSSSYLDNQ-VV-LD--A----MYSC-T-IGCV---V   378
              ******    *                                                ******** ** ***** ********  **

                                                                             7:     A                          B
βLf                                                                          L-QE------------C----E-----------------   341
βDm                                                                          M--E-S-I---------C----EN-------------RV--   340
βCe                                                                          MRQE-------------C----E-----C------------   340
β1                                                                           LKADRAGVLAGHDNRVSCLGVTDDGMAVATGSWDSFLKIWN   340
β2                                                                           M-G--------------C-----------------------   340
β3                                                                           M-SE-V-I-S-------C----A------------------   340
β4                                                                           --GG-S-----------C-------------------R---   340
STE4                                                                         --GEIV-K-E--GG--TGVRSSP--L--C------TM---SPGYQ  423
                                                                              *  * * * ************ ************** ***
```

```
              A                          B
LxGHxxxIxxΦxδ --- ΦΦSGGxDxxΦxIWDδ
F        L           TAA N  C LFN
         V            S       VY

           WD-40 Motif
Φ, hydrophobic; δ, not charged
```

	GenBank Accession #
βLf	X56757
βDm	J04083
βCe	X17497
β1	M13236
β2	M16538
β3	M31328
β4	M87286
STE4	—

(Continued)

Table 17.8 *Continued*

b. γ subunits

```
γ2             MASNNT_______________ASIA___QARKLVE___QLKMEANIDRIKVSKAAADLMAYCEA_     43
γ3         MKGETPV-S-_______________M--G___----M--___---I--SLC------------T-CD-_     48
γ5               MSGS_______________S-V-___AMK-V-Q___--RL--GLN-V---Q-----KQFCLQ_     41
γ7               MSA-_______________NN--___-------___--RI--G-E-------SSE--S-C-Q_     41
γ1          MPVINIEDL_______________TEKD___KLKME-D___---K-VTLE-ML---CCEEFRD-V-E_     42
STE18      MTSVQN-PRLQQPQEQQQQQQLSLKIKQLKLKRIN-LNNK-RK-LSRE--TA-NCCLTIIN-TSNT     77
                                          *    **  *    *  **

γ2     HAKEDPLLTPVP_ASENPFR__EKKFF________________CAIL(gg)    71
γ3     --C----I----_T------__-----________________C-L-(gg)    75
γ5     N-QH-----G-S_S-T----__PQ-_V________________CSF-(gg)    68
γ7     --RN----VG--_------K__D--P_________________CI--(gg)    68
γ1     RSG----VKGI-_EDK---K___L-GG________________CV-S(f)     73
STE18  KDYTL-E-WGY-V-GS-HFIEG_L-NAQKNSQMSNSNSVCCTLM(f)     110
          ***         ***     *                 *
```

-, amino acid identical to those in β1 or γ2; _, gap; **C**, all cysteines are highlighted; ∗, amino acids conserved in all mammalian β's or γ's. Cys25 of β1 can be cross-linked to Cys36 or Cys37 of γ_1 (Ref. 188); ∗, conserved in all; ∗, conserved in mammalian sequences only; under- and over-lined boldfaced sequences in STE4 are important for activation of the STE20 kinase. Box in **a**: consensus sequences of WD-40 motif. The five complete (A–B) and two partial (B) WD-40 of G protein β subunits are highlighted.

The γ subunits, 68–75 amino acids long, are only 28 to 43% homologous and constitute a set of the subunits that is much more heterogeneous than α or β subunits (Table 17.8).

Combinatorial analysis gives the possibility of forming hundreds of distinct G proteins. However, not all of these genes are expressed in any single cell, so that the actual cellular G protein complexity is somewhat less staggering, but still impressive.

The interaction of β's with γ's to form dimers may involve a coil–coil interaction with the participation of a large proportion of γ and the N-terminus of β.[58] Given the rather large sequence variability of γ and the fact that the different β's show sequence variability in their N-termini, it is thus not surprising that not all of the β's combine with all of the γ's. For example, on expression in cells or *in vitro,* β2 did not dimerize with γ1, and β3 did not dimerize with either γ1 or γ2.[59–61] An antisense signaling interference assay showed pairing of β3 with γ4.[62,63]

Coexpression of γ, β1(1–129), and β1(130–340) leads to formation of a stable βγ dimer.[58] β1(1–129) encompasses the N-terminal variable domain, the first (complete) WD-40 motif and the first incomplete WD-40 motif (Table 17.8), and β1 (130–340) encompases the remainder of the β subunit, including at its N-terminus a sequence found in yeast STE4 to be important for interaction with its effector, the protein kinase STE20. This suggests the existence of at least two independently folding structural domains in β.

17.3.3.2. Lipid Modification of G Protein Subunits—Membrane Attachment and Function in Signaling

17.3.3.2a. Myristoylation. Some G protein α subunits have been shown to be myristoylated at the N-termini. These include the two α_t's, the three α_i's, the two α_o's, α_{gust}, and α_z, which are the α subunits that have the MGXXXS consensus myristoylation signal. Since the N-termini of all subunits of purified G proteins are blocked, it stands to reason that nonmyristoylated α subunits are nevertheless modified at their N-termini. For transducin, it has been shown that the myristic acid at position 2 may be replaced by other lipids.

Mutations of α subunits that prevent myristoylation (e.g., G2A) prevent their localization to the membrane.[64,65] In one *in vitro* reconstitution assay that measures inhibition of adenylyl cyclase by recombinant $G_i\alpha$, only the myristoylated form of $G_i\alpha$ was found to be active[66] (Codina and Birnbaumer, unpublished). Myristoylated α's exhibit a markedly enhanced affinity for βγ dimers.[67] It remains to be seen whether membrane localization is driven by high affinity for βγ dimers, or whether the myristic acid contributes to membrane localization by serving as a lipophilic anchor. It is likely that both factors contribute to the membrane localization of G_α's.

17.3.3.2b. Palmitoylation. In contrast to myristoylation, which affects only a few α subunits, most if not all α's appear to be palmitoylated at Cys3, or another cysteine located near the N-terminus.[68–70] The function of this posttranslational modification appears to be to contribute further to the membrane localization of Gα's, as shown for α_s and α_q. Two cysteines (Cys9 and 10) are palmitoylated in α_q and mutation of both to Ser (α_qC9S,C10S) not only delocalizes α_q but also interferes with its capacity to be activated by receptor or to activate phosphoinositide breakdown, even if it is activated by the Arg183-to-Cys mutation (R183C) (see below). Receptor-mediated stimulation of phosphoinositide turnover by α_qC9S,C10S can be restored by attachment of the myristoylated N-terminus of α_t, α_t(1–9), to nonpalmitoylated α_q(16–rest). For α_s, removal of the palmitoylation site by the α_sC3S mutation produces loss of membrane localization accompanied by loss of receptor-mediated activation of adenylyl cyclase, but results only in a minor impairment in its action if it is constitutively activated by an R → C mutation. In contrast to the result with the α_t/α_q chimera, construction of the myristoylated, nonpalmitoylated α_t(1–9)α_s(17–rest) chimera restores membrane localization but confers an intrinsic activity not present in the wild-type α's without restoring receptor-mediated activation of adenylyl cyclase.[70]

Also in contrast to myristoylation, which is permanent, palmitoylation is a reversible modification that varies with the metabolic or regulatory state of cells. It may thus be that cells contain "active" and "inactive" pools of G proteins, that are regulated by palmitoylation. Changes in palmitoylation of $G_s\alpha$ after stimulation of cells through the G_s pathway have been observed.[71]

17.3.3.2c. Polyisoprenylation. Like α subunits, β and γ subunits are also blocked at their N-termini with an as yet unknown blocking group (Codina and Birnbaumer, unpublished). However, no specific posttranslational lipidations have thus far been described for β's. γ subunits, on the other hand, have at their C-termini a CAAX consensus polyisoprenylation signal, and are polyisoprenylated.[72–74] Full processing of γ's involves both the polyisoprenylation of the Cys at −4 followed by cleavage of the three terminal amino acids and carboxymethylation of the polyisoprenylated cysteine. γ1, which is expressed almost exclusively in the retinal cells (γ-transducin), is farnesylated (C15); the remainder of the known γ's are geranylgeranylated (C20). While polyisoprenylation is not necessary for association with β's, it was found essential for interaction with adenylyl cyclase[59] and it increases affinity for α subunits. Thus, both myristoylation and polyisoprenylation contribute to the high-affinity interaction between inactive GDP-liganded α and βγ dimers.

17.3.4. Patterns of G Protein Subunit Expression

Of the genes listed above, α_s, α_{i2}, and α_{11} appear to be expressed in all cells. Most cells also express α_q and either α_{i1} or α_{i3} (functional homologues of α_{11} and α_{i2}, respectively), and one or both the α_{12} and α_{13} genes. Thus, all cells express eight "ubiquitous" α subunits.

Expression of α_o, α_{14}, α_{15}, and α_z is not ubiquitous but also not exclusive to single cell types, as appear to be α_t's and α_{gust}. Their products are found in groups of cells or tissues that often but not always have common embryonic origin. α_{olf}, which was originally thought to be expressed exclusively in cells of the olfactory neuroepithelium, has a somewhat wider but nevertheless still restricted expression, being found also in basal ganglia of the central nervous system, in pancreatic islets, testis, lung, and liver. α_{olf} comigrates with the short form(s) of α_s, as shown by immunoblotting, and interpretation of effects of receptors, originally attributed by default to activation of a single G_s, may have to be reconsidered because of the presence of two "G_s" with differing efficiencies in both receptor coupling and effector activation.

α_o's (α_{o1} and α_{o2}) are preferentially expressed in cells derived from the neural crest and in endocrine cells (pituitary, pancreatic β cell) as well as in other selected cell types such as cardiac myocytes. α_{14} is expressed primarily in stromal and epithelial cells, and α_{15} is expressed in many but not all cells with hematopoietic lineage. α_z is found primarily in neurons and platelets, and in small quantities also in red blood cells and other cell types. It follows that out of the repertoire of 16 α subunit genes, a standard cell expresses between 9 and 10.

On a comparative basis, it is difficult to properly measure the relative amounts of the different α subunits in a cell. The difficulty arises from the large difference in sensitivity of the antibodies available for quantification (mostly antipeptide) and from the heterologous nature of the standards (recombinant proteins) that are used to quantify the measurement of the proteins (SDS-denatured membrane-associated proteins). There is, however, a consensus, derived not only from immunoblotting but also from the yields with which individual proteins are purified, as to which are more abundant and which are less so. Thus, levels of α_s, now sum of α_s plus α_{olf}, in nonolfactory cells are on the low side, which for the sake of this discussion can be set at a value of 1.0. In contrast, α_{olf} is a major protein in cells of the olfactory neuroepithelium where it accumulates in cilia to very high levels (to ca. 500 times standard α_s levels), and where it appears to act to transduce the input of any one of over 200 olfactory receptors. Like α_s, the levels of expression of α_q and α_{11} are also low, differing from those of α_s by at most a factor of two. The same appears to apply to α_{14}, $\alpha_{15/16}$, α_{12}, and α_{13}. In contrast, the levels of expression of α_{i2} appear to be on the average at least 5 times those of α_s, α_{i1}, and α_{i3} (of which at least one is coexpressed with α_{i2} at about half the abundance of α_{i2}). In neutrophils, and possibly also in other white blood cells, levels of α_{i2} and α_{i3} are higher than in other cells, probably 10 to 20 times those of α_s. In neurons, α_o may be 50 to 100 times more abundant than α_s, contributing to up to 1% of total membrane protein. Highest of all appears to be transducin (α_{tr}) in rod cell disks, where it accumulates to 5% of total membrane protein.

Although the tissue distribution of β and γ has been less well studied, it is clear that most cells express at least two β and two γ genes and more likely three of each. Together with the standard α subunit repertoire, this makes for 40 to 90 distinct G proteins engaged in transducing receptor signals into modulated effector activities, if, as one assumes, all α subunits have the ability to combine with any combination of βγ dimers.

17.3.5. The α Subunit: Structure and Function

The α subunits have been the object of extensive study in recent years. Secondary and tertiary structures have been inferred on the basis of sequence homology with other regulatory GTPases that had been crystallized, notably bacterial elongation factor EF-TU and p21-ras (Ras), the effect on activity of directed mutagenesis of selected amino acids and N-terminal truncations, and blockade or mimicry of G protein regulation by receptors with receptor peptides. These studies have led to a general picture in which regions of the α chain have been identified that are involved in GTP binding and hydrolysis, in recognition of receptor, in interaction with effector, and in interaction with $\beta\gamma$ dimer (reviewed in Refs. 75 and 76).

The receptor interaction with C-termini of α subunits was deduced primarily from the finding that pertussis toxin (PTX) uncouples receptors from G proteins (reviewed in Ref. 77) by ADP-ribosylating the cysteine at -4 from the C-terminus,[78] and that the uncoupled phenotype of the UNC allele of S49 lymphoma cells was caused by the mutation of the arginine at -6 from the C-terminus to proline.[79] A recent switch in receptor specificity caused by the switching of merely three amino acids in the C-terminus[80] and the finding that a synthetic 20-amino-acid polypeptide encoding transducin 309–328 mimics transducin in stabilizing photoactivated rhodopsin in its Meta II state[81] support the notion that the C-terminal section of α subunits is involved in receptor–G protein interaction.

The N-terminal end (first 21 amino acids) on the other hand participates in association with $\beta\gamma$ subunits. This interaction involves various aspects: lipidation of the γ subunit of $\beta\gamma$ and at least two subdomains of the N-terminus, one close to the N-terminus proper and another at about 20 amino acids from it (Refs. 82, 83 and references therein, 84).

Mutational analysis of α_s showed that determinants that define effector interaction, in this case adenylyl cyclase, are in the last third of the molecule.[85] Consistent with this, α_s peptides derived from regions around 50 amino acids from the C-terminus can stimulate adenylyl cyclase, and a similar peptide from α_t interacts with γ_{PDE} and leads to stimulation of the catalytic activity of $\alpha\beta_{PDE}$.[86]

Activation by GTP—experimentally by GTPγS—causes a fundamental conformational change which for the α subunits of heterotrimeric G proteins involves a hinge region that is part of the so-called G3 or switch-II region of Ras and contains the DVGGQ motif. Specifically, Gly (226 in α_s) is necessary for activation by GTP. Its replacement in α_s with Ala to give α_s G226A (also H21a or reverse-UNC) results in inhibition of activation by GTPγS, without blocking the binding of GTPγS or the guanine nucleotide-induced regulation of the affinity of β_2-adrenergic receptor for agonist of the type shown in Figure 17.11.

17.3.5.1. Natural and Man-Made Activating Mutations of α Subunits

Two activating mutations merit comment on this point. One, Q→L, is a mutation in which the conserved glutamine of the DVGGQ motif in α_s (Q227) is changed to L. In Ras this mutation is oncogenic and inhibits its GTPase activity. In α_s, it also inhibits GTPase causing persistent activation in the absence of receptor stimulation. This mutation was found in pituitary and thyroid adenomas and given the oncogenic name *gsp*[87] (Figure 17.12). The other is a mutation in which the arginine of the RVXT motif, α_s R201 and α_{i2} R179, is changed to either histidine, cysteine, or serine, R→H/C/S. In α_s and transducin α this arginine is the site of ADP-ribosylation by cholera toxin, a modification that inactivates the GTPase activity of these α subunits, causing their receptor-independent activation by GTP.[88–91] Like Q→L mutations, α_s mutants R→C, R→H, and R→S were found in growth-hormone-secreting tumors and thyroid adenomas, and also in patients suffering from McCune–Albright's syndrome, a mosaic endocrinopathy (Figure 17.12). The α_{i2} mutants R→H and R→C were found in adrenal cortical carcinomas and ovarian granulosa and theca cell tumors (Figure 17.12). Expression of mutationally activated α_{i2}, referred to as *gip2,* in Rat-1 cells leads to MAP kinase activation and induces their transformation to an oncogenic state.[92,93]

G_s and G_{i2} are not the only G protein α subunits with oncogenic potential, for also the Q→L forms of α_q, α_o, and α_{12} induce transformed states.[94–96] Even though it would thus appear that each of the major classes of α subunits may be potential oncogenes, i.e., proto-oncogenes, this concept has recently been challenged by the demonstration that mutationally activated α_s, rather than potentiating, suppresses Ras-induced transformation,[97] and that colonic epithelium of an α_{i2}-knockout mouse tends to develop adenocarcinomas.[98]

An inactivating mutation of α_s found in two male patients with a combined pseudohypoparathyroidism/male precocious puberty syndrome is also of interest (Figure 17.12). In most cases of PHP1a, the syndrome is related to a loss of function by mutation in which one of the α_s alleles is inactive. Patients of this type are hypoparathyroid, hypothyroid, and have elevated serum calcium, among other symptoms. As presented in Table 17.5, patients with male precocious puberty (also testotoxicosis) can result as a consequence of a gain of function: a mutation in the LH receptor gene causing it to be constitutively activated in the absence of LH. As a result, males begin secreting testosterone at an early age—2 to 3 years. Development of the female's estrogen-secreting capacity in response to LH requires prior priming of the follicles by FSH. Since this does not happen until puberty, females with constitutively activated LH receptor do not show prepubertal signs as males do.[35] Two male patients, from independent families, have been identified with the seemingly paradoxical combination of PHP1a/testotoxicosis, one thought to be related to loss of α_s function, the other to stimulation of α_s at the wrong time. The reason in both patients is a single point mutation causing the change of Ala[366] to Ser in α_s (Figure 17.12).[99] Ala[366] lies in the G5 region of this GTP binding protein (see Figure 17.13). Molecular analysis of recombinant α_s[A366S] showed it to be a temperature-sensitive protein stable at 32°C but not at 37°C, and, at the same time, to have lost high affinity for GDP and hence to be activated by GTP without requiring $\beta\gamma$/receptor-mediated exchange activity. This accounts for the dominant gain of function in the testis—LH-independent testosterone production—where the mutant α_s is functional, and for the also dominant loss of function in the remainder of the body where the mutant α_s inactivates as it is made.

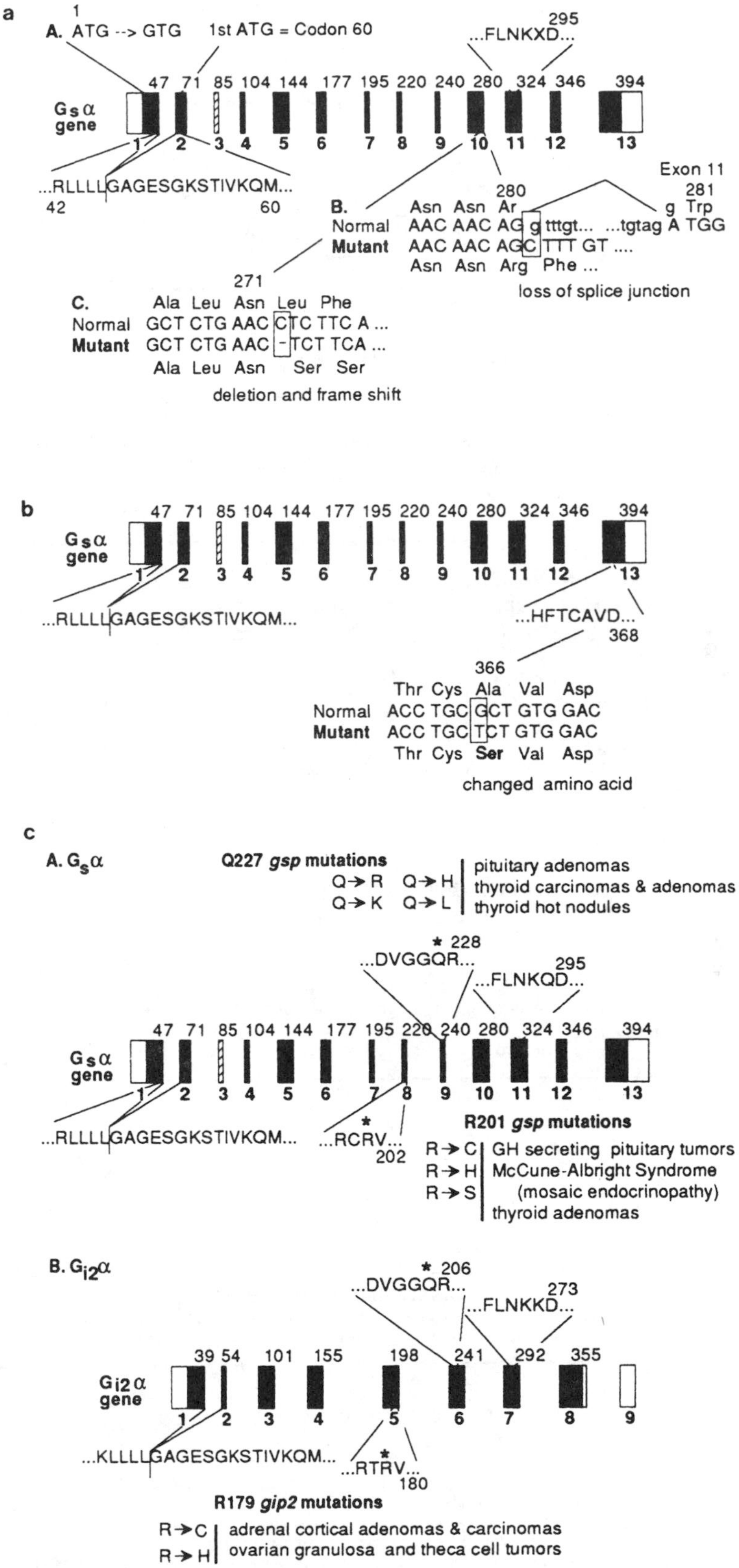

Figure 17.12. G_s-inactivating and -activating mutations found in Albright's hereditary osteodystrophy [pseudohypoparathyroidism Ia (PHP1a)], McCune–Albright's syndrome or polyostotic fibrous dysplasia and in combined PHP1a and testotoxicosis. From Refs. 99, 223–227.

```
                                                                              S43
                                                                    <-β1->   <---α1---->
bov-αtr   (mry )MGAGA___________SAEEKHSRELEKKLKEDAEKDARTVKLLLLGAGESGKSTIVKQMKII    52
                bbbbbbbbbbbbbbbbbbbbbbbbbbbbbbbbbbb                       PPPPPP
                                                                            m
                                                                        ***G1***
xl-αs           M---------------V---T-------------------------------------------    63
hum-αs        (p3)MGCLGNSKTEDQRNEEKAQREANKKIEKQLQKDKQVYRATHRLLLLGAGESGKSTIVKQMRIL    63
mus-αq    (p9)MTLESIMACCL_____SE-A-EARRI-DE-ERHVRR--RDA-RELK-----T-------FI-----I    62
mus-αi2   (myr1)M--TV________SA=D-AAA-RS-M-D-N-RE-GEKAAREVK------------------K-I    56
_____________________________________________________________________________________

          <------------------------------------------------------------- I-1 ----
          58
       L1 <------------αA------------>           <---αB---->              <----
bov-αtr HQDGYSLEECLEFIAIIYGNTLQSILAIVRAMTTLNIQYGDSARQDDARKLMHMADTIE_EGTMPKEMSD  121

          V: +   ++          +          +  +  +      +        +    +++ ++  + +
xl-αs   ------A---KI--------I--------T--G--S-----V-------I----NLP-YK--E-S-----  133
hum-αs  HVNGFNGEEKATKVQDIKNNLKEAIETIVAAMSNLVPPVELANPENQFRVDYILSVMNVPDFDFPPEFYE  133
mus-αq  -GS-YSD-D-RGFTKLVYQ-IFT-MQAMIR--DT-KI-YKYEH__-KAHAQLVREVDVE-VSA-ENPYVD  130
mus-αi2 -ED-YSE--CRQYRAVVYS-TIQS-LA--K--G--QIDFADPQRADDA-QLFA-SCAAEEQGMLPEDLSG  126
_____________________________________________________________________________________

        ------------------------------------------------------>
                               146                         171 174
        -αC---> <-αD->           <---αE--->              <-αF-> _L2     <-β2-->   <-
bov-αtr IIQRLWKDSGIQACFDRASEYQLNDSAGYYLSDLERLVTPGYVPTEQDVLRSRVKTTGIIETQFSFKDLN  191
                                gg  g                          g   g P  P
                               K266                                ss   m
         +  + +                                                ***G2***
xl-αs   -T-T--Q---------------------------IV--N--T---------CTX----------------  203
hum-αs  HAKALWEDEGVRACYERSNEYQLIDCAQYFLDKIDVIKQADYVPSDQDLLRCRVLTSGIFETKFQVDKVN  203
mus-αq  AI-S--N-P-IQE--D-RR----S-STK-Y-NDL-R-ADPS-L-TQ--V--V--P-T--I-YP-DLQS-I  200
mus-αi2 VIRR--A-H--Q--FG--R----N-S-A-Y-NDLER-A-S--I-TQ--V--T--K-T--V--H-TFKDLH  196
_____________________________________________________________________________________

             Switch II                     <---- I-2 ---->
             203
        -β3-->    <--α2--->     <-β4-->               <----α3----->          <β5
bov-αrt FRMFDVGGQRSERKKWIHCFEGVTCIIFIAALSAYDMVLVEDDEVNRMHESLHLFNSICNHRYFATTSIV  261
            P          eeeee                 eeeeeeeeeeeeeeeeeeeeeeeeeeee
                    343NLKDCGL349
          **G3**     I: +  ++                 II: ++          III: +  + + ++
xl-αs   -----------------------------------------H---------------------------  273
hum-αs  FHMFDVGGQRDERRKWIQCFNDVTAIIFVVASSSYNMVIREDNQTNRLQEALNLFKSIWNNRWLRTISVI  273
mus-αq  -R-V------S------H--EN--S-M-L--L-E-DQ-LV-SDNE--ME-SKA--RT-ITYP-FQNS---  270
mus-αi2 -K--------S--K---H--EG------C--L-A-DL-LA--EEM--MH-SMK--D--C--K-FTDT-I-  266
_____________________________________________________________________________________

                    <------- I-3 --------->                          <--I-4-->
          266
        β5-> <-αG-->                     <-------------α4---------->
bov-αtr LFLNKKDVFSEKI__KKAHLSICFPDYNGPN_TYEDAGN______________YIKVQFLELNMRRDV__KE  314
          gg g                          eeeeeeeeeeeeeeeeeeeeeeeeeeeeeeeeeeeeee
          D146                                                           rrrrrr
        **G4**                                                  IV: ++++++++
xl-αs   -----------N------------------D-----V-----------------------------------  343
hum-αs  LFLNKQDLLAEKVLAGKSKIEDYFPEFARYTTPEDATPEPGEDPRVTRAKYFIRDEFLRISTASGDGRHY  343
mus-αq  -----K---E--I__MY-HLV-----YDGPQRDAQ-AR-______________--LKM-VDLN_PDS-__KI  323
mus-αi2 -----K--FE--I__TQ-SLTIC---YTGANKYDE-AS______________Y-QSK-EDLNKRKDT__KE  319
_____________________________________________________________________________________

         <β6->     <-------α5------->
bov-αtr IYSHMTCATDTQNVKFVFDAVTDIIIKENLKDCGLF  350
           ggg                           PTX
        rrrrrrrrrrrrr             rrrrrr|rrr
                                  212EGVT215
           *G5*
xl-αs   ------------------------------------  379
hum-αs  CYPHFTCAVDTENIRRVFNDCRDIIQRMHLRQYELL  379
mus-αq  I-S-----T------F--AAVK-T-LQLN-KE-N-V  359
mus-αi2 I-T-----T--K-VQF--DAVT-V-IKNN-KDCG-F  355
```

Figure 17.13. Delineation of amino acid sequences of α subunits involved in GTP binding and hydrolysis, in interaction with receptor, and in interaction with effector as deduced from analysis of the crystal structure of bovine human α_t, from analysis of α_s mutations, and from effects of synthetic α subunit peptides.*Features of crystal structure.* Heterotrimeric G protein α's are formed of two distinct domains: a GTPase domain and a ca. 115-amino-acid helical domain, connected by two linkers L1 and L2. The alignment analyzes the primary α subunit sequence in terms of homology to the sequence of the smaller p21 Ha-*ras*. Since G α's are longer, this gives rise to inserts 1 (I-1) through 4 (I-4), of which I-1 constitutes a structurally separate *helical domain.* α helices and β pleats of sequences homologous to those of Ras are numbered sequentially α1 through α5 and β1 through β6. α helices arising from, or due to, the inserts are denotated as αA through αG. Amino acids contacting the guanine base, ribose, and phosphates are subscripted with **g, s,** and **p,** respectively. Amino acids involved in coordinating the Mg are subscripted with **m.** GTP hydrolysis is proposed to involve a general base attack of the γ-phosphoryl group by Glu203, and both Arg174 and Lys42. Sequences subscripted *eee* are implicated in the interaction with effector on the basis of mimicry of PDE activation (boldfaced amino acids) or simple binding to PDEγ.[76,86,228] Sequences subscripted *rrr* are implicated in containing receptor (rhodopsin) on the basis of the ability of a peptide out to rhodopsin and to stabilize the active Meta II state of rhodopsin.[81] In agreement with this, ADP-ribosylation of Cys at −4 from C termini of G protein α subunits, impedes receptor-mediated activation. The C-terminus makes van der Waals contacts with α2/β4 loop, which is part of a conformational GDP/GTP switch (Switch II region). Trp in the Switch II/α2 region served to monitor Mg-induced conformational changes in α_o–GTP complexes on hydrolysis to GDP.[45,229,230] *bbb,* sequences that are known to be required for interaction with βγ dimers (e.g., Ref. 83 and references therein).

Alignment with sequences of other α subunits and the structure–function inferences obtained from mutations and chimeras. The sequences of *Xenopus laevis* α_s, murine α_q, and murine α_{i2} are compared to that human α_s (splice variant: short, with Ser). –, amino acid identity; __, gap. Amino acids in bold in hum-α_s have been shown to confer α_s activity to chimeras. Those comprising regions I–III plus the extended sequence that includes region IV confer α_s activity to an α_i[1–212/α_s][220–341]/α_i[318–355].[85] Amino acids in α_s superscripted by + in regions I through IV are essential for α_s activity of an α_i[1–212]/α_s[220–379] chimera. Amino acids in hum-α_s subscripted in region V by + are necessary for α_s activity of an xl-α_s. Xl-α_s does not activate mammalian adenylyl cyclase but a ^{l}-α_s[1–35]/hum-α_s[35–157]/xl-α_s[158–379] does (Antonielli, Olate, Birnbaumer, Allende, and Allende, unpublished). α_s[320–341] stimulates adenylyl cyclase (H. Hamm, personal communication). *G1* through *G5* are amino acid sequences that had been inferred from EF-Tu and Ras structures to be involved in binding and hydrolyzing GTP. Actual points of contact are shown below the sequence of α_t.

17.3.5.2. The Three-Dimensional Structure of the G Protein α Subunit

The 2.2-Å crystal structure of transducin α ($\alpha_{t\text{-rod}}$) has been solved. Though it is missing the first 25 amino acids, it represents a landmark that allows for the first time a comparison between the real structure of a GTP-liganded G protein α subunit and the inferred, ras/EF-TU-based structure. By and large there have been no major surprises. The GTP-binding domain of α_t is indeed similar to that of Ras and EF-TU. What the crystal provided is the location of the region that represents the major difference in amino acid composition between Ras and α_t. This sequence is a 120-amino-acid insert plus connecting linkers, lies between the GAGES G1 and the DVQQ G2 motifs, and constitutes a helical domain. Its presence is responsible for major functional differences between the small and the large GTPases. The crystal shows that this helical domain forms a lid over the GTP-binding site of the GTPase domain, and that its linker at the C-terminal end (linker 2) contains R174, the cognate of α_s R201 and α_{i2} R179. The lid structure of the helical domain explains the much slower nucleotide exchange rates of G protein α subunits compared to other regulatory GTPases. The arginine, which has no parallel in other types of GTPases, turned out to be at the center of the GTP-hydrolysis mechanism, together with a lysine next to the G1 motif (K42) and a general base that catalyzes the nucleophilic attack required for the separation of the γ-phosphate. Figure 17.13 shows the comparative sequence alignments illustrating structure–function relationships as they are currently understood. Figure 17.14 shows a model of the transducin α-GTPγS complex. In terms of effector regulation, the picture that is be emerging is that loops connecting α helices and β pleats in the C-terminal part of α subunits may constitute the effector-regulatory domains. Substantiation of this concept—which will surely be tested in many laboratories—would imply that effector regulation involves not one but several independent points of contact (summarized in Ref. 76).

17.3.5.3. G Protein-Based Human Diseases

Human diseases based on G protein dysfunction are summarized in Table 17.9. While most are caused by genetic mutations and of rare occurrence, two, cholera and whooping cough, are infectious diseases that afflict a large number of people. The molecular basis of cholera is the ADP-ribosylation of the α subunit of G_s at position Arg201 catalyzed by the ADP-ribosyltransferase activity of the A-subunit of the toxin of *Vibrio cholera* [cholera toxin (CTX)] according to the reaction

$$G_s + NAD^+ \xrightarrow[\text{(ARF, GTP)}]{\text{CTX}} G_s(\text{ADP-ribosyl-Arg}^{201}\text{-}\alpha_s) + \text{nicotinamide}$$

Like the Arg201-to-Cys or -His mutation, ADP-ribosylation of this Arg causes inhibition of GTPase activity and spontaneous βγ/receptor-independent, activation by GTP. This stimulates adenylyl cyclase and elevates cAMP in intestinal epithelial cells, which are the first cells of the body exposed to the toxin. cAMP in turn causes secretion of water into the intestinal lumen, diarrhea, dehydration, and, if unchecked, death by dehydration.[88,100,101]

The clinical symptoms of whooping cough are almost exclusively caused by the ADP-ribosylation of α subunits of G_i/G_o proteins at a Cys located at −4 from the carboxy-

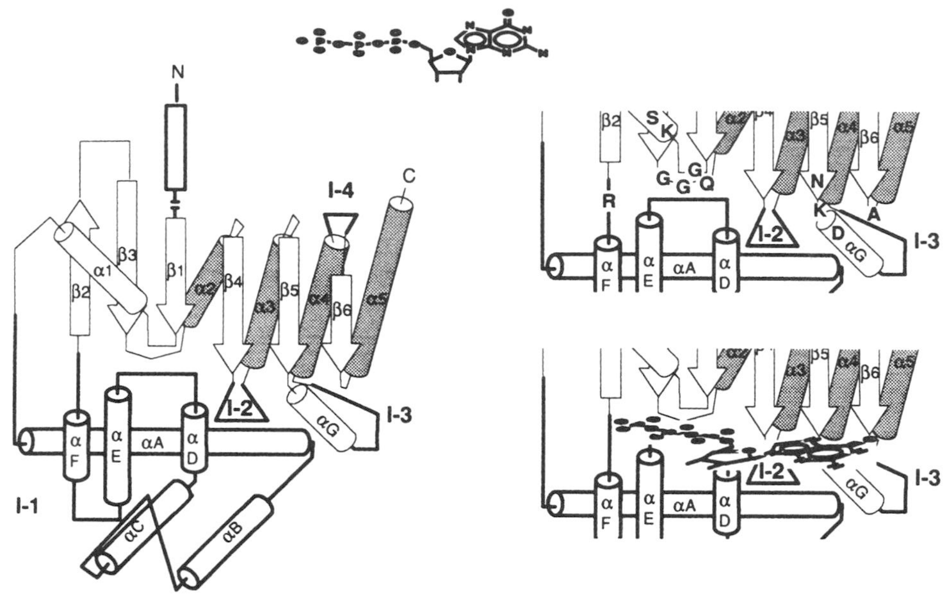

Figure 17.14. Ribbon model of the GTPγS–transducin derived from its crystal structure. In this model the guanine nucleotide (top) should be envisioned as being bound along the horizontal cleft delimited by the GTPase domain (thin lines) and the I1 helical domain (thick lines) as shown in the lower right panel. Several amino acids involved in binding of GTP and defined by mutational analysis are highlighted in the upper right panel. G in β1–α1 loop, Gly38, cognate to Ras Gly12; K-S in α1, Lys42,Ser43, involved in GTP hydrolysis and Mg^{2+} coordination; R in linker 2, Arg174 CTX substrate involved in GTP hydrolysis and cognate to α_s Arg201 and α_{i2} Arg179; GGQ in β3–α2 loop includes Gln200, cognate to α_s Gln227 and Ras Gln61, and Gly199, cognate to α_s Gly226, site of H21a or reverse UNC mutation that impedes activation by GTP but leaves unaffected GTP regulation of agonist binding; NKXD (X not shown) in β5–αG, residues involved in guanine ring recognition and highly conserved for all α subunits; A in β6–α5 loop, Ala322, cognate to α_s Ala366. Regions defined by mutations or effects of peptides to be involved in effector regulation and receptor interactions are all confined to the α2 through C-terminus of the molecule (*eeee* and *rrrr* in Figure 17.13). Orientation of the N-terminus is unknown. Based on the three-dimensional structure of α_t of Noel, Hamm, and Sigler.[76]

Table 17.9. G Protein-Based Diseases[a]

Disease	Molecular basis
Due to G protein activation	
Cholera	GTPase inhibition due to ADP-ribosylation of α_s Arg179 by cholera toxin (substrate: G_s)
McCune–Albright's syndrome (polyostotic fibrous dysplasia)	Dominant somatic mutation of α_s Arg179 codon in early embryonic life with resultant loss of GTPase activity
Various endocrine adenomas and possibly also some carcinomas[b]	Clonal expansion of primary somatic mutation of either Arg201 or Gln227 of α_s or Arg179 of α_{i2}
Due to G protein inactivation	
Whooping cough	Loss of G protein regulation by receptor(s) (uncoupling) due to ADP-ribosylation by pertussis toxin (PTX) of cysteine at −4 from C-terminus of PTX-sensitive G protein(s): α_{i1}, α_{i2}, α_{i3}, α_{o1}, and/or α_{o2}. Substrate(s): αβγtrimers
Pseudohypoparathyroidism type Ia (Albright's hereditary osteodystrophy)	Autosomal dominant due to mutations that cause loss of functional α_s
Due to tissue-specific activation and inactivation	
Van Dop's combined testotoxicosis and pseudohypoparathyroidism	Dominant gain of function in testis and dominant loss of function in remainder of body due to $G_s\alpha$Ala366-to-Ser mutation

[a]For references see text.
[b]Unequivocal cause–effect relation still needs to be established.

terminus. The reaction is catalyzed by pertussis toxin (PTX), the main exotoxin of *Bordetella pertussis,* according to the reaction

$$G_i/G_o + NAD^+ \xrightarrow[\text{(ATP, GTP)}]{\text{PTX}} G_i/G_o(\text{ADP-ribosyl-Cys}^{351/352}\text{-}\alpha_i/\alpha_o) + \text{nicotinamide}$$

G_i/G_o proteins ADP-ribosylated at this Cys are sterically hindered from interacting with receptors and hence not able to fulfill their signal transducing functions, leading to loss of pre- and postsynaptic effects of "inhibitory" receptors such as inhibition of adenylyl cyclase, inhibition of Ca^{2+} channels, and stimulation of inwardly rectifying and ATP-sensitive K^+ channels. It is of interest that CTX and PTX have similar architectural composition.

CTX is an $\alpha_1\beta_5$ heterohexamer formed of an A subunit and five B subunits that comprise a pentamer referred to as the B-oligomer. PTX is an $\alpha\beta\gamma_2\delta\varepsilon$ heterohexamer formed of an S_1 subunit (the ADP-ribosyltransferase proper) and a pentameric oligomer formed of four B-like subunits: one S_2, two S_3, one S_4, and one S_5. In each case the role of the pentamer is merely to bind to specific (toxin) receptors on target cells (GM1 gangliosides for CTX, unknown for PTX). A and S_1 subunits are then internalized by a not-well-understood mechanism and activated following reduction of a critical disulfide bond by intracellular glutathione. Activation of PTX is then completed by association with ATP.[77,101] CTX activity on the other hand requires the participation of a GTP-activated "ADP-ribosylation factor" (ARF), a low-molecular-weight (GTPase that physiologically plays a central role in intracellular vesicle trafficking.[103]

On using [^{32}P] NAD^+, the reaction products of CTX- and PTX-catalyzed ADP-ribosyl alkylations contain [^{32}P]-ADP-ribose. This has been used extensively to label *in vitro* the respective susceptible G protein α subunits. After SDS–PAGE and autoradiography, these are then easily identified as discrete bands.

17.4. FUNCTIONAL CORRELATES TO MOLECULAR DIVERSITY IN G PROTEINS AND EFFECTORS

17.4.1. Multiple Effectors for a Single G Protein

Activated α subunits are regulators of effectors. The structural subdivision that emerged from the phylogenetic analysis of α subunits (Figure 17.15) correlates rather well with their known and suspected functional subdivision. Activated forms of the closely related α_s and α_{olf} both stimulate adenylyl cyclases; those of α_q, α_{11}, α_{14}, and $\alpha_{15/16}$ stimulate PI-specific PLCs of the β type and the three α_i's, the type 1 α_o splice variant, and α_z inhibit adenylyl cyclase. Among the three sensory PTX sensitive α's, α_{tr}, α_{tc}, and α_{gust}, the mechanism of action of α_{tr}—activation of visual PDE through interaction with inhibitory PDEγ subunits—is well understood;

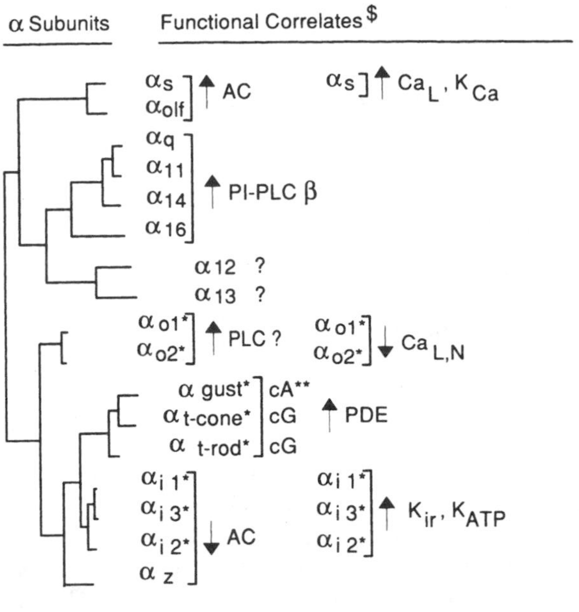

Figure 17.15. Summary of known effects of G proteins on cellular effector functions. (Adapted from Birnbaumer.[104])

that of α_{tc} is assumed to be similar, and that of α_{gust}, which is involved in perception of sweet and bitter tastes, is still in need of confirmation at the biochemical level. The effector complements of α_{12} and α_{13}, a subbranch of the α_q family, are still under debate and the subject of intense research. A similar situation applies to α_o's. While one is able to inhibit adenylyl cyclase in a reconstitution assay, the other is not. Both have been clearly shown able to mediate the inhibition of L- and N-type Ca^{2+} channels in mammalian cells and in molluskan synapses, and stimulation of a PLC in *Xenopus* oocytes, but for neither has the effect been shown to be direct. These two α subunits, each of 355 amino acids, are encoded by the same gene and differ in that the open reading frame of α_{o1} uses exons 7 and 8 that differ from exons 7′ and 8′ used to build α_{o2} As a result α_o's differ in 26 of their last 106 amino acids in a region into which both receptor and effector specificities are encoded.

For α_s and α_i's it has been shown that they can regulate more than a single effector system. In addition to stimulating adenylyl cyclase, α_s can stimulate Ca channels of the L-type and a subclass of smooth muscle Ca^{2+}-activated voltage-dependent K channels (K_{Ca} channels). These are functions that are expressed only in specific tissues and cell types so that responses to receptors that activate G_s may vary from tissue to tissue. For example, Ca channel stimulation by α_s occurs in tissues such as skeletal muscle and heart where it potentiates the stimulatory effect of the cAMP-PKA system, but not in liver or endothelial cells that lack voltage-gated Ca channels. Stimulation of the K_{Ca} channel by α_s occurs in a tissue such as coronary smooth muscle, where it potentiates the relaxing effect of the cAMP-PKA system, but not in atrial cells of the heart that lack K_{Ca} channels. Likewise, α_{i1}, α_{i2}, and α_{i3}, which inhibit adenylyl cyclases, can also activate at least two classes of K^+ channels, the "muscarinic"-type inwardly rectifying K^+

channel found primarily in cardiac atrial cells and in neuroendocrine cells, and the ATP-sensitive K^+ channel found in cardiac ventricle cells and pancreatic islet cells. These channels are also referred to as G protein-gated K channels.

The fact that single α subunits may stimulate or inhibit more than one effector is relevant in terms of the ultimate elucidation of the roles that each G protein may play in a cell. The complexity of the responses of a cell to G protein stimulation is likely to be large.

17.4.2. Signaling through $\beta\gamma$ Dimers

While signaling through α subunits is the most common mechanism by which receptors activate or inhibit effector functions—e.g., stimulation and inhibition of adenylyl cyclase by α_s and α_i's, stimulation of the β subtype of C-type phospholipases (PLCβs)—G proteins also signal through their $\beta\gamma$ dimers. These responses require both a respectable level or concentration of G protein, and expression of the adequate $\beta\gamma$-responsive effector system (reviewed in Ref. 104).

17.4.3. Molecular and Functional Diversity of Adenylyl Cyclases

Eight nonallelic mammalian adenylyl cyclase (AC) genes have been identified. ACs are true transmembrane proteins, traversing the membrane 12 times, beginning with a cytosolic N-terminus and ending with a cytosolic C-terminus. Most intracellular loops are predicted to be short, but the fourth intracellular loop and the C-terminal are quite long. N-termini are variable in length (e.g., short for type I AC). This separates the first six transmembrane segments from the second six transmembrane segments. ACs are thus constituted of two similarly organized halves each with six transmembrane domains or segments and a rather long C-terminus. Catalysis depends on the cooperative interaction of both halves and is carried out by the segments that are common to all ACs (reviewed in Ref. 105).

Figure 17.16 incorporates in cartoon form these structural features and summarizes the rather surprising differences in regulatory features of the various AC subtypes. All ACs have in common that they are stimulated by the activated form of α_s and by the diterpene drug forskolin, but they differ strikingly in their responses to Ca^{2+} and calcium/calmodulin (Ca/CaM). In addition, some but not all ACs are effectors for $\beta\gamma$ dimers.

Two, type I and type III, are stimulated by Ca/CaM. Two, type V and VI, are inhibited by micromolar Ca^{2+}. One, type I, is inhibited by $\beta\gamma$ dimers; in contrast, two others, type II and type IV, are stimulated by $\beta\gamma$ dimers. While inhibition of type I is independent of other regulatory input(s), i.e., affects basal, α_s-, forskolin-, and Ca/CaM-stimulated activity stimulation of type II and IV ACs only happens if the cyclase is simultaneously stimulated by α_s, i.e., $\beta\gamma$'s act to potentiate an existing stimulatory input. There is a close phylogenetic relation of type VIII AC to type I AC, found on the basis of knowledge of about 60% of the molecule. It will be interesting to see whether on examination of the properties of the complete enzyme encoded in the type VIII gene this relationship will hold up at the functional level by showing stimulation by Ca/CaM and inhibition by $\beta\gamma$ dimers.

Studies are being carried out on the tissue distribution of the various ACs. The picture that is emerging is complex. Type I and VIII cyclases appear to be of primarily neuronal expression. Type III, which is very highly expressed in olfactory neuroepithelium and which like α_{olf} was originally thought to be restricted to this cell type, is, however, expressed quite ubiquitously, as are all other types studied (V, VI, II, and IV). It is worth mentioning that in one study in which expression of AC subtypes was tested for by analyzing RNA by RT-PCR, a single cell type was found to express five out of six ACs tested for (Codina and Birnbaumer, unpublished).

Inhibition of Adenylyl Cyclase

The first indications that guanine nucleotides are involved in inhibitory regulation of AC were published in 1973[106,107] and the first purifications of the PTX-sensitive putative G_i proteins were published in 1983.[108,109] All indications were that ACs should be direct targets of α_i-mediated inhibition of activity, as indicated by the presence of inhibitory regulation in α_s-negative S49 cells,[110,111] and the lack of an effect

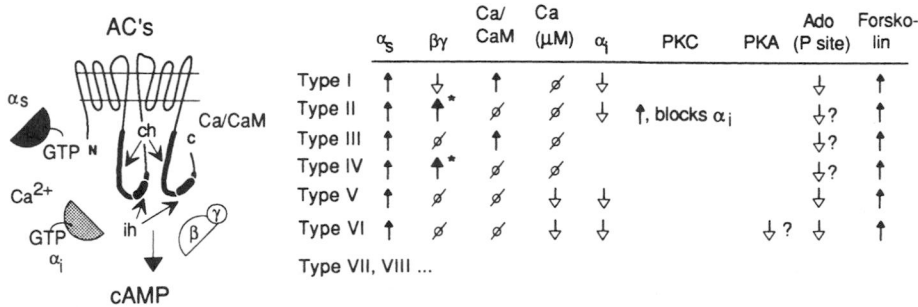

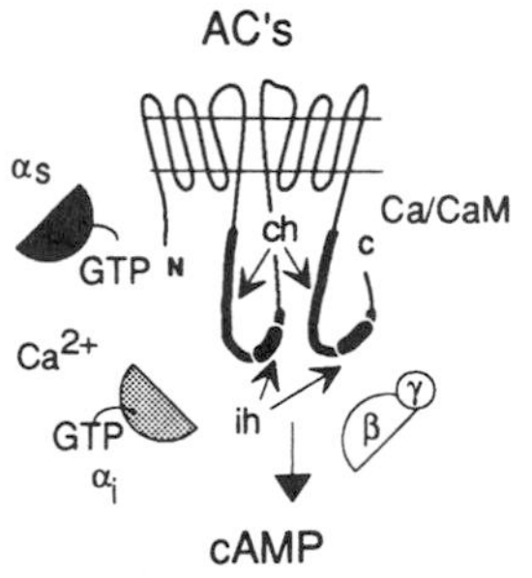

	α_s	$\beta\gamma$	Ca/CaM	Ca (µM)	α_i	PKC	PKA	Ado (P site)	Forskolin
Type I	↑	↓	↑	ø	↓			↓	↑
Type II	↑	↑*	ø	ø	↓	↑, blocks α_i		↓?	↑
Type III	↑	ø	↑	ø				↓?	↑
Type IV	↑	↑*	ø	ø				↓?	↑
Type V	↑	ø	ø	↓	↓			↓	↑
Type VI	↑	ø	ø	↓	↓		↓ ?	↓	↑
Type VII, VIII ...									

Figure 17.16. Transmembrane model of adenylyl cyclases (ACs) as originally proposed by Krupinski *et al.*[231] and supported by hydrophobicity and similarity analysis, and summary of main regulatory features (adapted from Ref. 232). *ch (common homology),* regions of ACs with conserved amino acids among all; *ih (internal homology),* regions where sequences in the first half are similar to sequences in the second half of the molecule. Cytoplasmic regions of high homology among the different ACs are assumed to form the catalytic core of the enzyme.

of cholera toxin, known to reduce affinity of α_s for $\beta\gamma$, on inhibitory regulation of AC.[112] Nevertheless, the *in vitro* reconstitution of inhibition of AC by α_i subunits was not achieved until 1993[66] (Codina and Birnbaumer, unpublished) (Figure 17.17). The inhibition is noncompetitive with respect to activation of α_s, indicating that the ACs should have independent sites for interacting with α_s and α_i. Although not as marked as with $\beta\gamma$ dimers which act at the 100 through 1000 nM level, the concentrations of recombinant α_i's required for half-maximal inhibition (5–10 nM; Figure 17.17A) are about 100-fold higher than those required for half-maximal stimulation with recombinant α_s under the same conditions. As seen with natural membranes and in transient expression assays[113,114] (Codina and Birnbaumer, unpublished), the extent of inhibition varies with AC subtype. Inhibition appears to be 100% when operating on AC-VI but less so when operating on AC-II. Moreover, AC-II inhibition by α_i is suppressed by phorbol esters, presumably through PKC-mediated phosphorylation. In agreement with differential sensitivity of AC subtypes to inhibition by α_i, a test of the effect of α_{i3} on AC in different membranes, each expressing a different complement of AC subtypes, shows degrees of inhibition that vary with the membrane tested (Figure 17.17B).

Regulation by Phosphorylation

Although studies on regulation of ACs by kinases are still in their infancy, the initial results point to a complex and again type-specific response pattern. Type II AC is stimulated by PKC, while types V and VI appear to be inhibited by PKA.

17.4.4. Phospholipase C

Known to be a target of G protein regulation,[115,116] phosphoinositide (PI)-specific phospholipase C (PI-PLC) turned out to be a complex family of related proteins. The structures of PI-PLCs were deduced from combined biochemical, genetic, and molecular biology studies[117–121] (reviewed in Ref. 122). PI-PLCs are structurally subclassified into β, γ, and δ, and for each of them there are at least three closely related subtypes, of which only the β subclass is a target of G protein regulation. Four mammalian members of the type β PLCs are known, as well as one *Xenopus laevis* homologue and two *Drosophila melanogaster* homologues, norpA and PLC 21.

The amino acid alignment of β, γ, and δ PLCs reveals two stretches of highly conserved sequences, X and Y, which presumably constitute the enzymatic core, and divergent N- and C-termini as well as diverging X-Y linkers. PLCγs have the longest X-Y linkers, containing two *src* homology 2 (SH2) domains and one *src* homology 3 (SH3) domain. They participate in the intracellular transduction of signaling pathways that involve protein tyrosine kinase (PTK) activities. SH2 domains bind autophosphorylated PTKs and are phosphorylated by them. This results in PLCγ activation, formation of diacylglycerol (DAG) and inositol trisphosphate (IP3), and the consequential activation of protein kinases of the C type (PKCs) and Ca^{2+} mobilization.

In view of the fact that both PKC and Ca^{2+}, alone or in combination with calmodulin (CaM), modulate ACs in a type-specific manner, PLCγ activation may or may not be accompanied by cAMP changes. Type δ PLCs have both short X-Y linkers and short N- and C-termini. They lack SH2 and SH3 domains and little is known at this point about factors or stimuli that regulate their activity.

All four mammalian type β PLCs are regulated by G protein subunits (summarized in Figure 17.18). Their X-Y linker segments lack SH2 and SH3 domains and are short; in contrast, they have long C-termini when compared to γ and δ PLC C-termini. Mammalian type β PLCs are structurally more related to drosophila's norpA PLC than to γ- and δ-type mammalian PLCs, and this was the first indication that β-type PLCs were likely to be the G protein-sensitive PLCs. NorpA (no receptor potential A) flies are blind because of disruption of the insects' rhodopsin signaling pathway, which in insects is PLC- rather than PDE-dependent.[123,124]

PLC-β regulation by G proteins is being worked out in reconstitution studies[125–137] (summarized in Figure 17.18).

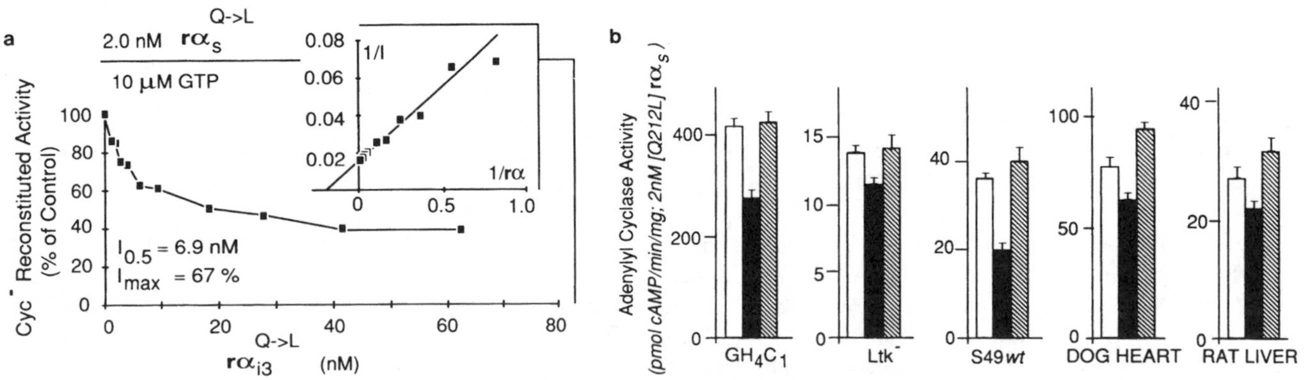

Figure 17.17. Characteristics of α_i-mediated inhibition of adenylyl cyclases. (a) Dependence on concentration of α_i. In the experiment shown, ca. 5 nM α_{i3}(Q→L) caused 50% of maximal inhibition which in several experiments averaged 60%. Assays were in the presence of GTP. Similar data in terms of potency and extent of inhibition were obtained with wild-type α_{i3} in the presence of GTPγS. (b) Extent of inhibition depends on the origin of membranes presumably due to expression of different adenylyl cyclase subtypes. (Unpublished data from Codina and Birnbaumer.)

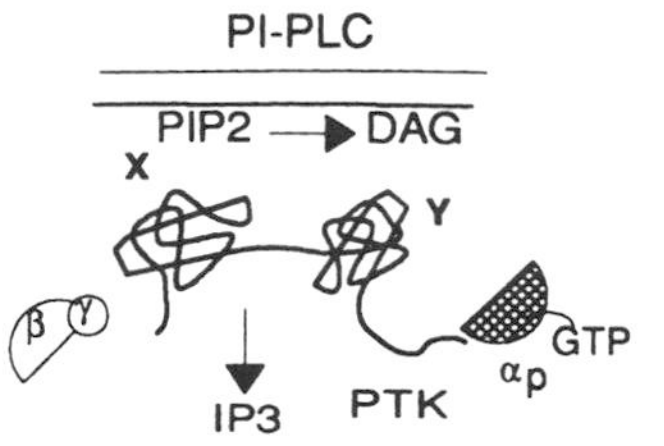

Type	*src* Homology Domains	Stimulated by
PI-PLC β1	None	αq, α11, α14 >> α16
PI-PLC β2	None	βγ >> α16 >> α11
PI-PLC β3	None	βγ >>> αq, α11
PI-PLC β4	None	α and/or βγ
PI-PLC γ	SH2 & SH3	PTK's
PI-PLC δ	None	?

Figure 17.18. Summary of some of the structural and regulatory features of phosphoinositide specific C-type phospholipases (PI-PLCs). Based on Rhee and Choi[122] and references listed in the text. Amino acid alignment of PI-PLCs revealed two regions of high homology termed X and Y, which are thought to contribute to the enzymatic function of the proteins. N-termini, X-Y linkers, and C-termini of β, γ, and δ PLCs show low homology, if any, and vary in length. SH2 and SH3 domains are found in the X-Y linker of type γ PLCs. Only type β PLCs are regulated by α subunits (α_q, α_{11}, α_{14}, and α_{16}) due to interaction with the C-terminal extension. βγ's may interact either with the X-Y linker or the N-terminal extension or both of these regions of type β PLCs. Type γ PLCs are substrates of and activated by tyrosine kinases.

The picture that is emerging from these studies is that PLC-γ's show differential sensitivities to α and βγ dimers. Studies have been reported with PLC-β1, -β2, and -β3; PLC-β4 was cloned later and little is known about its regulation by G protein subunits. The α_q/α_{11} subunits have been ranked for effectiveness to stimulate PLC activity in one laboratory as PLCβ1≥ PLCβ3>>PLCβ2, and the response to βγ as PLCβ3>PLCβ2>> PLCβ1, with PLCβ4 being possibly unresponsive. For PLCβ2 the rank order of responses appears to be $\alpha_{16}>\alpha_q/\alpha_{11}$. Studies of this type are ongoing and ranking of reactivities may change depending on the type of assay, e.g., *in vitro* reconstitution versus overexpression in COS or HEK-293 cells versus mere analysis of responses of normal cells. Regardless of the exact final outcome, it is clear that, as is the case with ACs, the responses of cells to the activation of a given G protein by a receptor may vary in intensity and complexity depending on the complement of PLCs that is expressed in these cells.

C-terminal truncation of PLCβ results in loss of stimulation by α subunit without loss of response to βγ dimers (133, 138; Schnabel and Gierschik, personal communication). This indicates the existence of separate response domains for the two effects.

17.5. SUBUNIT CONCENTRATIONS THAT CAUSE HALF-MAXIMAL EFFECTS—ROLE OF GAP ACTIVITY OF THE EFFECTOR

As is the case for AC regulation by βγ dimers, that of PLC by βγ dimers also requires high levels of this protein, 100–1000 μM, as compared to the much lower concentrations required to obtain regulation with α subunits (compare Refs. 128 and 129 versus 139), raising the question of physiological relevance of the βγ regulation. Two arguments support a physiological role for βγ dimers in PLC regulation: one relates to the cell type in which βγ regulation is proposed to be relevant: the neutrophil and the neutrophil-like HL-60 cell; the other relates to the finding that PLCβ is a GTPase-activating protein (GAP) for α_q.[140]

In neutrophils and HL-60 cells, PLC activation by the formyl-Met-Leu-Phe (fMLP) receptor is blocked by PTX[141–143] and hence not likely to be mediated by any of the PTX-insensitive G proteins of the G_q class that stimulate β-type PLCs.[127,131,144,145] On the other hand, fMLP stimulates cholera toxin-mediated ADP-ribosylation of the HL-60's two main PTX substrates α_{i2} and α_{i3}, which supports the idea that in this cell G_i's rather than the G_q class of G protein(s) may signal PLC activation.[146,147] Both the levels of G_i proteins (>100 pmole/mg protein) and the fMLP receptor density (>100,000/cell) are very high in HL-60 cells, making it plausible that βγ dimers rather than α's mediate PLC activation in this cell, in spite of the high concentration required.

The finding that PLCβ1 stimulates the intrinsic GTPase activity of α_q indicates that reconstitutions with α subunits that have been persistently activated with nonhydrolyzable GTP analogues may be left shifted with respect to what would be obtained with GTP-activated and hence GTPase-sensitive α subunits.[66,148,149] As analyzed by the simplified model of Taussig *et al.*,[66] the EC_{50} is a composite value that incorporates the affinity of the active form—ratio of on and off rate constants—and the rate at which an α subunit inactivates due to GTP hydrolysis, so that the EC_{50} or $K_{act} = (k_{off} + k_{GTPase})/k_{on}$. For GTPase-stimulating effectors, it follows that EC_{50} values for effector regulation—stimulatory or inhibitory—by α subunits activated by GTP are right-shifted with respect to EC_{50} values for α's activated with nonhydrolyzable GTP analogues or AlF_4^- by a factor given by the k_{GTPase}/k_{on} ratio. This ratio has been reasoned to be on the order of 100 or more,[66] thus bringing the effective concentration of α subunit needed to modulate PLC into the range of the concentrations required for stimulation of PLC by βγ dimers.

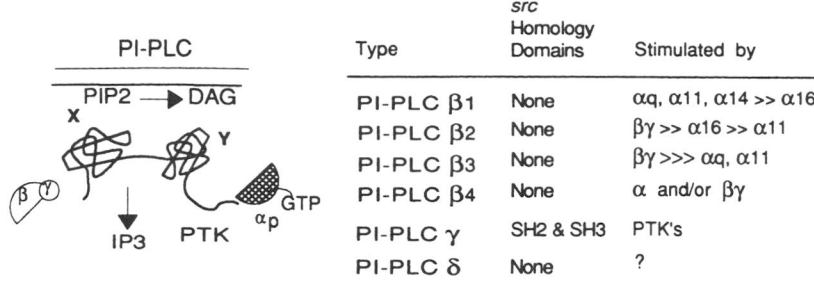

Since βγ dimers are shared by α subunits to form G proteins, and apparently α subunits do not discriminate in a major way between βγ dimer subtypes, βγ signaling is the most likely basis for cross talk between AC and PLC signaling pathways (reviewed in Ref. 150).

17.6. SPECIFICITY OF RECEPTORS FOR G PROTEINS—DUAL COUPLING PHENOMENA

Analysis of electrophysiological responses to agonists in cells previously injected with antisense nucleotide, and studies of cells transiently expressing receptors and G protein subunits indicate that receptors select a G protein not only on the basis of its α subunit but also on the basis of its β and γ subunits. The specificity rules emerging from these studies are by no means simple and easy to understand. For example, by measuring agonist-induced inhibition of voltage-activated Ca^{2+} currents in pituitary cells previously injected with subunit specific oligonucleotides, it was shown that the pituitary somatostatin (SST) receptor interacts with a G_o of composition $\alpha_{o2}\beta_1\gamma_4$ but not with any G_o having α_{o1}, β_3, or γ_3 as its component, and that the M3-muscarinic receptor in these cells interacts selectively with a G_o of composition $\alpha_{o1}\beta_1\gamma_3$, and not with a G_o having either α_{o2}, β_2, or γ_4 as its component.[62,63,151] Yet, the subunit selectivity is not absolute. The M3 and SST receptors inhibit adenylyl cyclase and activate inwardly rectifying K^+ channels, which are effects that are mediated by G_i rather than G_o protein(s).[127,152]

Results obtained via transient expression of various types of receptors in COS cells alone and in combination with G protein subunits are consonant with those obtained by suppression of G protein subunit synthesis. For example, expression of C5a receptors in COS cells has no effect on phosphoinositide hydrolysis unless they are coexpressed with α_{16}. α_q or α_{11} were unable to substitute for α_{16}. Expression of platelet-activating factor (PAF) receptor, in contrast, mediates PTX-insensitive stimulation of the COS cell's PLCβ(s), occurring through G_q and/or G_{11}.[153] Expression of interleukin 8 (IL-8) receptors leads in COS cells to stimulation of PLC if coexpressed with α_{14}, α_{15}, or α_{16}, but not with α_q or α_{11}, and this stimulation is PTX-insensitive. Expression of IL-8 receptors with $\beta\gamma$-sensitive PLCβ2 on the other hand leads to PTX-sensitive phosphoinositide hydrolysis, indicating interaction with non-α_{14} or non-$\alpha_{15/16}$ type G protein that is of the G_i-type.[154] Thus, IL-8 receptors interact with PTX-insensitive G_{16} and G_{14}, with PTX-sensitive G_i(s), but do not interact with G_q or G_{11}.

The first unequivocal report of dual coupling of a receptor was that of Ashkenazi *et al.*[155] who noted that the cloned M2 acetylcholine receptor expressed in CHO cells inhibited cAMP accumulation and stimulated PIP_2 breakdown. Curiously, while AC inhibition occurred at low concentrations of agonist (EC_{50} ca. 1 μM) and was unaffected by receptor density over a range of 150,000 to 2,500,000 sites per cell, PLC stimulation occurred at much higher concentrations, and increased in degree and required progressively lower concentrations of agonist to reach 50% of maximum as receptor abundance increased. Several other cloned receptors that mediate inhibition of AC were subsequently shown also to stimulate cellular PIP-PLC activity, including the serotonin 5HT-1A receptor,[156] the histamine H1 receptor,[157] and the D2 dopamine receptor.[158]

One hallmark of PLC stimulation by $\beta\gamma$ dimers is the already mentioned fact that it requires high nanomolar levels of $\beta\gamma$, which does not appear to be the case for stimulation by the α_q's. The reasons for low PLC sensitivity to $\beta\gamma$'s could be twofold. One is that the $\beta\gamma$ signal is teleologically meant to be secondary to an α signal and this is accomplished by a low affinity of the effector for the $\beta\gamma$ dimer. This would explain why inhibitory regulation of AC by the M2 receptor, here assumed to be an α_i signal, was fully developed at low receptor density, while PLC stimulation, here assumed to be a $\beta\gamma$ signal, became more and more intense as receptor density increased.[155] The second is that the low sensitivity to $\beta\gamma$ dimers in reconstitution assays is artifactual and but a reflection of a low relative abundance of an active $\beta\gamma$ subtype among $\beta\gamma$'s that are either inactive or active only at very high concentrations. The results obtained with the transfected M2 receptor could then be explained by assuming that the receptor selected G_i proteins with $\beta\gamma$ complement(s) of low intrinsic PLC stimulating activity or that perhaps the high selectivity of the receptor for the $\beta\gamma$ complement associated with α is not absolute and lost at high occupancy and massive overexpression.

Several cloned receptors that activate the G_s/AC system have also been found to have the potential to stimulate PLC. Included in this group are the TSH and LH receptors,[159,160] and the calcitonin (CT) and parathyroid hormone (PTH) receptors.[161,162] As was the case for the M2 receptor, the dose–response relations for PLC stimulation of AC stimulating receptors are also up-shifted (by factors of 20 to 50) with respect to AC stimulation[159–161] (Figure 17.19). None of the effects is PTX sensitive. As was the case for M2 receptors, PLC stimulation by cyclase stimulatory receptors can be interpreted either in terms of coupling to G_s plus coupling to a G_p, i.e., a G_q class of G proteins, or in terms of generation of G_s-derived $\beta\gamma$. G_s is a low-abundance G protein, so that the high hormone requirement for PLC stimulation through $\beta\gamma$ would have to come about either because of a naturally lower sensitivity of PLC to $\beta\gamma$'s (e.g., high pM–low nM) relative to AC's exquisitely high sensitivity to α_s (EC_{50} ca. 10 pM[163,164]) or because of the existence of activation cycles that while generating active α_s in each round generate active $\beta\gamma$ in only a fraction of the rounds (Figure 17.9). If, however, dual coupling is related to activation of both a G_s and a G_p, the effect of activated α_p's can be expected to be right-shifted with respect to α_s because PLCs appear to have GAP activity while ACs do not. Modeling of a hormone-sensitive guanine nucleotide-activated AC with GTPase activity has shown that an increase in GTPase activity causes a right shift in the EC_{50} values of an activating hormone.[148,149]

A very curious situation exists with α_2-adrenergic receptors. At low density of expression or at low agonist concentration, they inhibit AC activity via G_i, and at high levels of expression or high agonist concentration, they stimulate AC via G_s[165] as well as PLC via G_q. Since the delineation of many

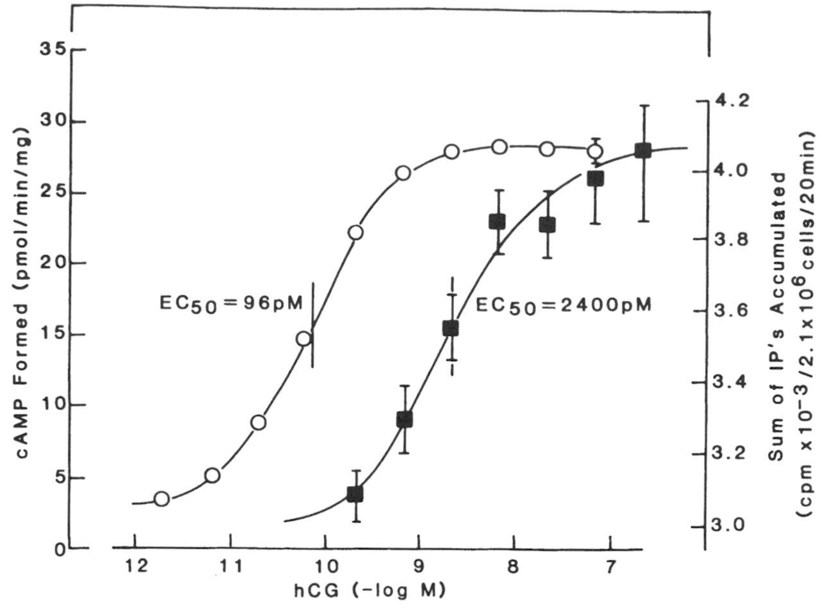

Figure 17.19. Example of dual action of a receptor. The cloned murine luteinizing receptor (LHR) expressed in L cells is shown to activate both the adenylyl cyclase system and phospholipase C (from Ref. 160). While the intervention of G_s in the activation of adenylyl cyclase has been established, the identity of the G protein or G protein subunit mediating the PLC stimulation is unclear. One possibility is that AC is stimulated by α_s and PLC is stimulated by $\beta\gamma$ of G_s; another is that LHR interacts with G_s to activate AC and with a G_p to activate PLC via an activated α_p (i.e., α_q, α_{11}, α_{14}, or α_{16}). The right-shifted dose–response curve for PLC stimulation could arise either because only a fraction of the G_s cycles generate $\beta\gamma$'s (see Figure 17.18) or because the GAP activity of PLC reduces the effective concentration of PLC activating α in comparison to that of the AC activating α (based on calculations published in Refs. 148 and 149).

of these interaction pathways involves overexpression of the interacting components, some of them may never occur under normal circumstances and their acceptance in physiological terms may have to be reviewed. On the other hand, failure of establishing an interaction under conditions of overexpression and proper controls is probably a very stringent and meaningful test.

In vitro reconstitution assays point to a role for γ in specifying with which heterotrimer a receptor interacts. This was shown for the interaction of rhodopsin with transducin reconstituted from α_t-GTP and recombinant $\beta_1\gamma_1$, $\beta_1\gamma_2$, or $\beta_1\gamma_3$. It was found that only $\beta_1\gamma_1$ binds to light-activated rhodopsin.[166]

17.7. CONCLUSION

The broad strokes of G protein-mediated signal transduction as well as many of the molecular players are now well known: hormone binds to receptor, receptor changes conformation—which constitutes the first response to agonist—and promotes activation of a G protein by GTP (Figures 17.1 and 17.20). However, the details of these interactions are for the most part nebulous. One would like to know which aspect of a receptor interacts with which subunit of a G protein and then with which aspect of the G protein subunit. While it is clear that receptors "look" at the carboxyl-terminal of α subunits to decide with which G protein to interact, it is not known whether other important points of contact exist on α's and which the points of contact on β's and γ's are. We have pointed out that $\beta\gamma$, on binding Mg^{2+}, acquires the function of a nucleotide exchanger, and raised the possibility that receptors may be acting merely by aiding in the activation of $\beta\gamma$ by Mg^{2+}. Implicit, of course, is also that $\beta\gamma$'s have one or more specific site(s) for Mg^{2+} and we have raised the possibility that the WD40 motif may be an ion binding motif.

A comparison of receptor sequences reveals three subfamilies—and possibly more if receptors from lower eukaryotic species are taken into consideration. Yet the three appear to operate functionally in the same way. For example, the shift

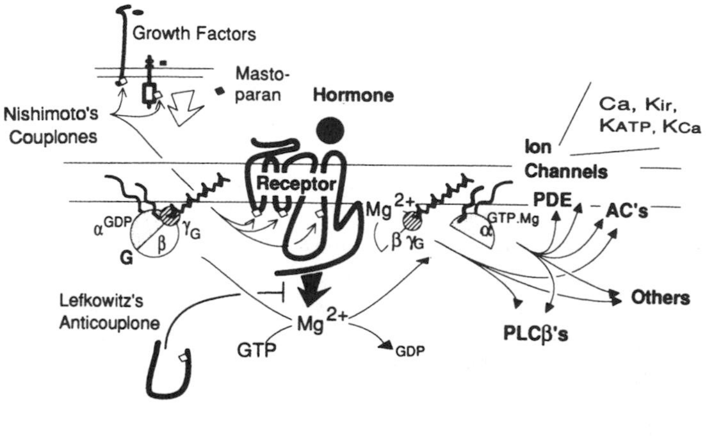

Figure 17.20. Signal transduction by G proteins incorporating possible locations of Mg^{2+} binding sites, the existence of Nishimoto couplones, and the paradoxical finding that expression of the third intracellular loop containing a Nishimoto couplone and bearing the site responsible for conferring constitutive activity of receptors, acts as an anticouplone.

in agonist binding affinity by GTP was discovered studying the interaction of glucagon with its receptor—a non-DRY-motif-containing receptor—and the same shift in affinity for agonists is also found with receptors containing the DRY motif. Even though the motif is absent from two of the three receptors subfamilies, mutation of the R results in complete loss of interaction with G protein without an effect on agonist binding, except for the fact that only low-affinity binding is seen because of lack of interaction with G protein. For the glucagon or PTH receptor, one would like to know what—or where—the functional equivalent is of the DRY motif. Sequence alignments do not provide clarification in this regard (Table 17.3d).

Receptors have what can be called "intrinsic activity" or efficacy. In its simplest terms this is the number of G protein molecules that it can activate in the presence of GTP per unit time. This then translates into number of effector molecules affected per unit time by a single receptor and hence the x-fold stimulation measured with one receptor versus another receptor. Mutations are being found that affect intrinsic activity, some natural and others man-made. One would like to know the kinetic parameter or parameters responsible for altered intrinsic activity. Possibilities include rate of nucleotide exchange and rate of dissociation from one G protein and association to another, dissociation or not from $\beta\gamma$ in the course of an α subunit activation cycle.

Intrinsic activity of DRY-type receptors have been elevated markedly by mutation, primarily at the C-terminal end of the third intracellular loop, but in other places as well (first intracellular loop, sixth transmembrane domain). One would like to know whether equivalent mutations can be made or will be found in receptors of the glucagon-responsive family. An understanding of the role of the N- and C-termini of the third intracellular loop in normal receptor functioning should be helpful, but data with respect to this are difficult to interpret. One would like to know why the expression of a segment of a receptor in a cell interferes with receptor–G protein coupling, but addition of a polypeptide derived from this segment promotes activation of a purified G protein by GTP (Figure 17.19).

Also in terms of cell physiology there are important questions that need clarification. We now know that a receptor may signal to more than one effector using two distinct α's by interacting with two G proteins, or by using α and $\beta\gamma$ from the same G protein. Some of the effector systems have the capacity of responding to both the α and the $\beta\gamma$, showing synergistic effects. One would like to know the physiologic relevance of these findings. Such dually responsive effectors may act as signal integrators being modulated by costimulation of distinct α subunit pathways. One interesting aspect is that $\beta\gamma$ only operates at high concentration, much higher than α's, notwithstanding the GTPase activating argument. One would like to have better measurements of both potencies of α and $\beta\gamma$ subunits and of the GTPase activating effect of the effectors. Convincing evidence has thus far been obtained only for the GAP activity of PLCβ1. Do all PLCs have GAP activity? Do any of the ACs have GAP activity?

The role that one may assign to $\beta\gamma$'s as signaling molecules depends on the levels of activated G protein. For circulating hormones, full receptor occupancy does not occur. It may thus be that at any time only a small proportion of the target G protein is activated, yielding submaximal amounts of α's and because of the higher concentration requirements, to ineffective amounts of $\beta\gamma$'s. On the other hand, for postsynaptic neurotransmitter receptors, full occupancy and high local receptor density is the rule, so that in this case one may expect signaling through both α and $\beta\gamma$. $\beta\gamma$'s also appear to be the signaling arm of receptors in nonneuronal cells such as neutrophils. Here it is clear that PLC (β2) activation has to be via $\beta\gamma$'s because the effect of ligands such as fMLP is blocked by PTX. Neutrophils contain no PTX substrate other than G_i's and α_i's do not stimulate PLCs.

One would like to know the subcellular distribution and relative concentrations of G proteins and effectors, not only in general, but especially in cases where $\beta\gamma$ mediation is suspected. The concentration argument raised above may be irrelevant, for example, if receptor, G protein, and effector exist as a preassembled complex. One would like to know whether receptors exist as monomers, dimers or oligomers. One would also like to know to what extent signaling through G proteins occurs within macromolecular complexes in which one could find several receptors (or the same of mixed type), a set of G proteins (also different kinds), and relevant effectors—a *transducesome* (pronounced transdusome). Signal transduction, cross talk between signaling pathways, and signal integration would occur locally under conditions where the concentrations required in reconstituted systems have no relevance.

Other questions relate to the effector systems. Some, like AC and PLCβ, are primary, expressed in all cells and shown unequivocally in reconstituted systems to be regulated by one or the other of the two signaling arms of G_s, G_q, etc. Ion channels on the other hand are responses restricted to cells expressing the particular channel of interest. At the moment of this writing, G protein regulation of ion channels has not yet been accomplished by reconstitution from purified components, and in some cases laboratories cannot agree on the results. In several cases the channel in question is regulated not only by G protein but also by phosphorylation, making it difficult to establish whether the effect of G protein activation is caused by a bonafide G protein–effector interaction or the result of a feedback regulation via a second messenger activated protein kinase or via the activation of a phosphatase. One would therefore like to see purified channel proteins reconstituted with purified G proteins akin to what was done to demonstrate the interaction of receptor and G proteins, and this under conditions where the state of phosphorylation is known.

Finally, one would like to know all of the biochemical functions of each of the G protein α subunits and of each of the $\beta\gamma$ dimers. At the time of this writing, G_{12} and G_{13}, neither of which is PTX-sensitive, are still orphans. But it may be that all others are partially orphans because while we know of one or two effectors, there may be more.

ACKNOWLEGMENTS. We thank Dr. Heidi Hamm for making the crystal structure of α_t available to use before publication. We also thank Dr. Juan Codina for providing us continued encouragement and constructive criticisms. This work was supported by NIH grants DK-19318 and HL-45198 to L.B. and DK-41244 to M.B. We acknowledge Baylor College of Medicine's Molecular Biology Computation Resource (MBCR) that made it easy for us to have access to the molecular biology computation softwares, Eugene and GCG Suite, which were needed to handle, analyze, and print the sequences presented in the figures.

REFERENCES

1. Birnbaumer, L. (1990). Transduction of receptor signal into modulation of effector activity by G proteins: The first 20 years or so. . . *FASEB J.* **4:**3178–3188.
2. Birnbaumer, L., Codina, J., Mattera, R., Yatani, A., Scherer, N. M., Toro, M.-J., and Brown, A. M. (1987). Signal transduction by G proteins. *Kidney Int.* **32**(Suppl. 23):S14–S37.
3. Henderson, R., and Unwin, P. N. T. (1975). Three-dimensional model of purple membrane obtained by electron microscopy. *Nature* **257:**28–32.
4. Henderson, R., Baldwin, J. M., Ceska, T. A., Zemlin, F., Beckmann, E., and Downing, K. H. (1990). Model for the structure of bacteriorhodopsin based in high-resolution electron cryomicroscopy. *J. Mol. Biol.* **213:**899–929.
5. Schertler, G. F. X., Villa, C., and Henderson, R. (1993). Projection structure of rhodopsin. *Nature* **362:**770–772.
6. Baldwin, M. J. (1993). The probable agreement of the helices in G-protein coupled receptors. *EMBO J.* **12:**1693–1703.
7. Dixon, R. A. F., Sigal, I. S., and Strader, G. D. (1988). Structure function analysis of the β-adrenergic receptor. *Cold Spring Harbor Symp. Quant. Biol.* **53:**487–497.
8. Savarese, T. M., and Fraser, C. M. (1992). *In vitro* mutagenesis and the search for structure–function relationships among G protein-coupled receptors. *Biochem. J.* **283:**1–19.
9. Vu, T.-K. H., Hung, D. T., Wheaton, V. I., and Coughlin, S. R. (1991). Molecular cloning of a functional thrombin receptor reveals a novel proteolytic mechanism of receptor activation. *Cell* **64:**1057–1068.
10. Ostrowski, J., Kjelsberg, M. A., Caron, M. G., and Lefkowitz, R. J. (1992). Mutagenesis of the β2-adrenergic receptor: How structure elucidates function. *Annu. Rev. Pharmacol. Toxicol.* **32:**167–183.
11. Surprenant, A., Horstman, D., Akbarali, H., and Limbird, L. E. (1992). A point mutation of cloned α-adrenoceptor 1 that blocks coupling to potassium but not calcium currents. *Science* **257:**977–980.
12. Pasternak, G. W., and Snyder, S. H. (1975). Identification of novel high affinity opiate receptor binding in rat brain. *Nature* **253:**563–565.
13. Kanaho, Y., Tsai, S.-C., Adamik, R., Hewlett, E. L., Moss, J., and Vaughan, M. (1984). Rhodopsin-enhanced GTPase activity of the inhibitory GTP-binding protein of adenylate cyclase. *J. Biol. Chem.* **259:**7378–7381.
14. Florio, V. A., and Sternweis, P. C. (1985). Reconstitution of resolved muscarinic cholinergic receptors with purified GTP-binding proteins. *J. Biol. Chem.* **260:**3477–3483.
15. Florio, V. A., and Sternweis, P. C. (1989). Mechanism of muscarinic receptor action on G_o in reconstituted phospholipid vesicles. *J. Biol. Chem.* **264:**3909–3915.
16. Phillips, W. J., and Cerione, R. A. (1992). Rhodopsin/transducin interactions. I. Characterization of the binding of the transducin–βγ subunit complex to rhodopsin using fluorescence spectroscopy. *J. Biol. Chem.* **267:**17032–17039.
17. Phillips, W. J., Wong, S. C., and Cerione, R. A. (1992). Rhodopsin/transducin interactions. II. Influence of the transducin–βγ subunit complex on the coupling of the transducin-α subunit to rhodopsin. *J. Biol. Chem.* **267:**17040–17046.
18. Wess, J., Bonner, T. I., Derje, F., and Brann, M. R. (1990). Delineation of muscarinic receptor domains conferring selectivity of coupling to guanine nucleotide-binding proteins and second messengers. *Mol. Pharmacol.* **38:**517–523.
19. Cheung, A. H., Sigal, I. S., Dixon, R. A. F., and Strader, C. D. (1989). Agonist-promoted sequestration of the β2-adrenergic receptor requires regions involved in functional coupling with G_s. *Mol. Pharmacol.* **34:**132–138.
20. Cheung, A. H., Huang, R.-R. C., Graziano, M. P., and Strader, C. D. (1991). Specific activation of G_s by synthetic peptides corresponding to an intracellular loop of the β-adrenergic receptor. *FEBS Lett.* **279:**277–280.
21. Okamoto, T., Murayama, Y., Hayashi, Y., Inagaki, M., Ogata, E., and Nishimoto, I. (1991). Identification of a G_s activator region of the β2-adrenergic receptor that is autoregulated via protein kinase A-dependent phosphorylation. *Cell* **67:**723–730.
22. Nishimoto, I. (1993). The IGF-II receptor system: A G protein-linked mechanism. *Mol. Reprod. Dev.* **35:**398–407.
23. Cotecchia, S., Exum, S., Caron, M. G., and Lefkowitz, R. J. (1990). Regions of the α_1-adrenergic receptor involved in coupling to phosphatidylinositol hydrolysis and enhanced sensitivity of biological function. *Proc. Natl. Acad. Sci. USA* **87:**2896–2900.
24. Kjelsberg, M. A., Cotecchia, S., Ostrowski, J., Caron, M. G., and Lefkowitz, R. J. (1992). Constitutive activation of the α_{1B}-adrenergic receptor by all amino acid substitutions at a single site. *J. Biol. Chem.* **267:**1430–1433.
25. Samama, P., Cotecchia, S., Costa, T., and Lefkowitz, R. J. (1993). A mutation induced activated state of the β_2-adrenergic receptor. Extending the ternary complex model. *J. Biol. Chem.* **268:**4625–4636.
26. Ren, Q., Kurose, H., Lefkowitz, R. J., and Cotecchia, S. (1993). Constitutively active mutants of the α_2-adrenergic receptor. *J. Biol. Chem.* **268:**16483–16487.
27. Luttrell, L. M, Ostrowski, J., Cotecchia, S., Kendall, H., and Lefkowitz, R. J. (1993). Antagonism of catecholamine receptor signaling by expression of cytoplasmic domains of the receptor. *Science* **259:**1453–1457.
28. Nash, M. I., Hollyfield, J. G., Al-Ubaidi, M. R., and Baehr, W. (1993). Simulation of autosomal dominant retinitis pigmentosa in transgenic mice expressing a mutated murine opsin gene. *Proc. Natl. Acad. Sci. USA* **90:**5499–5503.
29. Robinson, P. R., Cohen, G. B., Zhukovsky, E. A., and Oprian, D. D. (1992). Constitutively active mutants of rhodopsin. *Neuron* **9:**719–725.
30. Birnbaumer, M., Seibold, A., Gilbert, S., Ishido, M., Barberis, C., Antaramian, A., Brabet, P., and Rosenthal, W. (1992). Molecular cloning of the human antidiuretic hormone receptor. *Nature* **357:**333–335.
31. Rosenthal, W., Antaramian, A., Gilbert, S., and Birnbaumer, M. (1993). Nephrogenic diabetes insipidus: A V2 vasopressin receptor unable to stimulate adenylyl cyclase. *J. Biol. Chem.* **268:**13030–13033.
32. Birnbaumer, M., Gilbert, S., and Rosenthal, W. (1994). An extracellular NDI mutation of the vasopressin receptor reduces cell surface expression, affinity for ligand and coupling to the G_s/adenylyl cyclase system. *Mol. Endocrinol.* **8:**886–894.
33. Robbins, L. S., Nadeau, J. H., Johnson, K. R., Kelly, M. A., Roselli-Rehfuss, L., Baack, E., Mountjoy, K. G., and Cone, R. D. (1993). Pigmentation phenotypes of variant extension locus alleles result from point mutations that alter MSH receptor function. *Cell* **72:**827–834.

34. Jackson, I. J. (1993). More colour than meets the eye. *Curr. Biol.* **3:**510–521.
35. Shenker, A., Laue, L., Kosugi, S., Merendino, J. J., Minegishi, T., and Cutler, G. B., Jr. (1993). A constitutively activating mutation of the luteinizing hormone receptor in familial male precocious puberty. *Nature* **365:**652–654.
36. Ji, I., and Ji, T. H. (1991). Asp^{383} in the second transmembrane domain of the lutropin receptor is important for high affinity hormone binding and cAMP production. *J. Biol. Chem.* **266:**14953–14957.
37. Nishimoto, I., Murayama, Y., Katada, T., Ui, M., and Ogata, E. (1989). Possible direct linkage of insulin-like growth factor-II receptor with guanine nucleotide-binding proteins. *J. Biol. Chem.* **264:**14029–14038.
38. Okamoto, T., Katada, T., Murayama, Y., Ui, M., Ogata, E., and Nishimoto, I. (1990). A simple structure encodes G protein-activating function of the IGF-II/mannose 6-phosphate receptor. *Cell* **62:** 709–717.
39. Okamoto, T., and Nishimoto, I. (1992). Detection of G protein activator regions in the M4 subtype muscarinic cholinergic and α_2-adrenergic receptors based upon characteristics of primary structure. *J. Biol. Chem.* **267:**8342–8346.
40. Ikezu, T., Okamoto, T., Ogata, E., and Nishimoto, I. (1992). Amino acids 356–372 constitute a G_i-activator sequence of the α_2-adrenergic receptor and have a phe substitute in the G protein-activator sequence motif. *FEBS Lett.* **311:**29–32.
41. Kataoka, R., Sherlock, J., and Lanier, S. M. (1993). Signaling events initiated by transforming growth factor-β_1 that require $G_{i\alpha 1}$. *J. Biol. Chem.* **263:**19851–19857.
42. Nishimoto, I., Okamoto, T., Matsuura, Y., Takahashi, S., Okamoto, T., Murayama, Y., and Ogata, E. (1993). Alzheimer amyloid protein precursor complexes with brain GTP-binding protein G_o. *Nature* **362:**75–79.
43. Ross, E. M., Wong, S. K. F., Rubenstein, R. C., and Higashijima, T. (1988). Functional domains in the β-adrenergic receptor. *Cold Spring Harbor Symp. Quant. Biol.* **53:**499–506.
44. Antonelli, M., Olate, J., Graf, R., Allende, C. C., and Allende, J. E. (1992). Differential stimulation of the GTPase activity of G-proteins by polylysine. *Biochem. Pharmcol.* **44:**547–551.
45. Higashijima, T., Ferguson, K. M., Sternweis, P. C., Smigel, M. D., and Gilman, A. G. (1987). Effects of Mg^{2+} and the *beta-gamma* subunit complex on the interactions of guanine nucleotides with G proteins. *J. Biol. Chem.* **262:**762–766.
46. Casey, P. J., and Gilman, A. G. (1988). G protein involvement in receptor–effector coupling. *J. Biol. Chem.* **263:**2577–2580.
47. Graziano, M. P., Freissmuch, M., and Gilman, A. G. (1989). Expression of $G_{s\alpha}$ in *Escherichia coli.* Purification and properties of two forms of the protein. *J. Biol. Chem.* **264:**409–418.
48. Lee, E., Taussig, R., and Gilman, A. G. (1992). The G226AA mutant of $G_s\alpha$ highlights the requirement for dissociation of G protein subunits. *J. Biol. Chem.* **267:**1212–1218.
49. Rodbell, M., Krans, H. M. J., Pohl, S. L., and Birnbaumer, L. (1971). The glucagon-sensitive adenyl cyclase system in plasma membranes of rat liver. IV. Binding of glucagon: Effect of guanyl nucleotides. *J. Biol. Chem.* **246:**1872–1876.
50. Maguire, M. E., Van Arsdale, P. M., and Gilman, A. G. (1976). An agonist-specific effect of guanine nucleotides on binding to the beta adrenergic receptor. *Mol. Pharmacol.* **12:**335–339.
51. Lefkowitz, R. J., Mullikan, D., and Caron, M. G. (1976). Regulation of β-adrenergic receptors by guanyl-5′-yl imidodiphosphate and other purine nucleotides. *J. Biol. Chem.* **251:**4686–4692.
52. Berrie, C. P., Birdsall, N. J. M., Burgen, A. S. V., and Hulme, E. C. (1979). Guanine nucleotides modulate muscarinic receptor binding in the heart. *Biochem. Biophys. Res. Commun.* **87:**1000–1005.
53. Rosenberger, L. B., Roeske, W. R., and Yamamura, H. I. (1979). The regulation of muscarinic cholinergic receptors by guanine nucleotides in cardiac tissue. *Eur. J. Pharmacol.* **56:**179–180.
54. Wieland, T., Hunzan, M., and Jakobs, K. H. (1992). Stimulation and inhibition of human platelet adenylcyclase by thiophosphorylated transducin βγ-subunits. *J. Biol. Chem.* **267:**20791–20797.
55. Wieland, T., Nurnbarg, B., Ulibarr, I., Kaldenberg-Stasch, S., Schultz, G., and Jakobs, K. H. (1993). Guanine nucleotide-specific phosphate transfer by guanine nucleotide-binding regulatory protein β subunits. Characterization of the phosphorylated amino acid. *J. Biol. Chem.* **268:**18111–18118.
56. van der Voorn, L., and Ploegh, H. L. (1992). The WD-40 repeat. *FEBS Lett.* **307:**131–134.
57. Reiner, O., Carrozo, R., Shen, Y., Wehnert, M., Faustinella, F., Dobyns, W. B., Caskey, C. T., and Ledbetter, D. H. (1993). Isolation of a Miller–Dieker lissencephaly gene containing G protein β-subunit-like repeats. *Nature* **346:**717–721.
58. Garritsen, A., van Galen, P. J. M., and Simonds, W. F. (1993). The N-terminal coiled-coil domain of β is essential for γ association. A model for G-protein βγ subunit interaction. *Proc. Natl. Acad. Sci. USA* **90:**7706–7710.
59. Iñiguez-Lluhi, J. A., Simon, M. I., Robishaw, J. D., and Gilman, A. G. (1992). G protein βγ subunits synthesized in Sf9 cells. Functional characterization and the significance of prenylation of γ. *J. Biol. Chem.* **267:**23409–23417.
60. Pronin, A. N., and Gautham, N. (1992). Interaction between G protein β and γ subunit types is selective. *Proc. Natl. Acad. Sci. USA* **89:**6220–6224.
61. Schmidt, C. J., Thomas, T. C., Levine, M. A., and Neer, E. J. (1992). Specificity of G protein β and γ subunit interactions. *J. Biol. Chem.* **267:**13807–13810.
62. Kleuss, C., Scherübl, H., Hescheler, J., Schultz, G., and Wittig, B. (1992). Different β-subunits determine G protein interaction with transmembrane receptors. *Nature* **358:**424–426.
63. Kleuss, C., Scherübl, H., Hescheler, J., Schultz, G., and Wittig, B. (1993). Selectivity in signal transduction determined by γ subunits of heterotrimeric G proteins. *Science* **259:**832–834.
64. Mumby, S. M., Heukeroth, R. O., Gordon, J. E., and Gilman, A. G. (1990). G-protein α-subunit expression, myristoylation, and membrane association in COS cells. *Proc. Natl. Acad. Sci. USA* **87:**728–732.
65. Jones, R. L. Z., Simonds, W. F., Merendino, J. J., Jr., Brann, M. R., and Spiegel, A. M. (1990). Myristoylation of an inhibitory GTP-binding protein α subunit is essential for its membrane attachment. *Proc. Natl. Acad. Sci. USA* **87:**568–572.
66. Taussig, R., Iñiguez-Lluhi, J. A., and Gilman, A. G. (1993). Inhibition of adenylyl cyclase by $G_{i\alpha}$. *Science* **261:**218–221.
67. Linder, M. E., Pang, I. H., Duronio, R. J., Gordon, J. I., Sternweis, P. C., and Gilman, A. G. (1991). Lipid modification of G protein subunits. Myristoylation of $G_o\alpha$ increases its affinity for βγ. *J. Biol. Chem.* **266:**4654–4659.
68. Linder, M. E., Middleton, P., Hepler, J. R., Taussig, R., Gilman, A. G., and Mumby, S. M. (1993). Lipid modifications of G proteins: α subunits are palmitoylated. *Proc. Natl. Acad. Sci. USA* **90:**3675–3679.
69. Parenti, M., Vigano, M. A., Newman, C. M. H., Milligan, G., and Magee, A. I. (1993). A novel N-terminal motif for palmitoylation of G-protein α subunits. *Biochem. J.* **291:**349–353.
70. Wedegaertner, P. B., Chu, D. H., Wilson, P. T., Levis, M. J., and Bourne, H. R. (1993). Palmitoylation is required for signaling function and membrane attachment of $G_q\alpha$ and $G_s\alpha$. *J. Biol. Chem.* **268:**25001–25008.
71. Degtyarev, M. Y., Spiegel, A. M., and Jones, T. L. Z. (1993). Increased palmitoylation of the G_s protein α subunit after activation by the β-adrenergic receptor or cholera toxin. *J. Biol. Chem.* **268:** 23769–23772.
72. Mumby, S. M., Casey, P. J., Gilman, A. G., Gutowski, S., and Sternweis, P. C. (1990). G protein γ subunits contain a 20-carbon isoprenoid. *Proc. Natl. Acad. Sci. USA* **87:**5873–5877.

73. Simonds, W. F., Butrynski, J. E., Gautman, N., Unison, C. G., and Spiegel, A. M. (1991). G-protein βγ dimers. Membrane targetting requires subunit co-expression and intact γ CAAX domain. *J. Biol. Chem.* **266:**5363–5366.
74. Sanford, J., Codina, J., and Birnbaumer, L. (1991). γ-subunits of G proteins, but not their α- or β-subunits, are polyisoprenylated. Studies on post-translational modifications using *in vitro* translation with rabbit reticulocyte lysates. *J. Biol. Chem.* **266:**9570–9579.
75. Conklin, B. R., and Bourne, H. R. (1993). Structural elements of Gα subunits that interact with Gβγ, receptors, and effectors. *Cell* **73:** 631–641.
76. Noel, J. P., Hamm, H. E., and Sigler, P. B. (1993). The 2.2 Åcrystal structure of transducin α-GTPγS. *Nature* **366:**654–663.
77. Ui, M. (1984). Islet-activating protein, pertussis toxin: A probe for functions of the inhibitory guanine nucleotide regulatory component of adenylate cyclase. *Trends Pharmacol. Sci.* **5:**277–279.
78. West, R. E., Jr., Moss, J., Vaughan, M., Liu, T., and Liu, T.-Y. (1985). Pertussis toxin-catalyzed ADP-ribosylation of transducin. Cysteine 347 is the ADP-ribose acceptor site. *J. Biol. Chem.* **260:**14428–14430.
79. Masters, S. B., Sullivan, K. A., Beiderman, B., Lopez, N. G., Ramachandran, J., and Bourne, H. R. (1988). Carboxy terminal domain of G_s-*alpha* specifies coupling of receptors to stimulation of adenylyl cyclase. *Science* **241:**448–451.
80. Conklin, B. R., Farfel, Z., Lustig, K. D., Julius, D., and Bourne, H. R. (1993). Substitution of three amino acids switches receptor specificity. *Nature* **363:**274–276.
81. Dratz, E. A., Furstenau, J. E., Lanbert, C. G., Thireault, D. L., Rarick, H., Schepers, T., Pakhlevaniants, S., and Hamm, H. E. (1993). NMR structure of a receptor-bound G protein peptide. *Nature* **363:**276–281.
82. Navon, S. E., and Fung, B. K.-K. (1987). Characterization of transducin from bovine retinal rod outer segments. Participation of the amino-terminal region of T_α in subunit interaction. *J. Biol. Chem.* **262:**15746–15751.
83. Graf, R., Mattera, R., Codina, J., Estes, M. K., and Birnbaumer, L. (1992). A truncated recombinant α subunit of G_{i3} with a reduced affinity for βγ dimers and altered GTPγS binding. *J. Biol. Chem.* **267:**24307–24314.
84. Slepak, V. Z., Wilkie, T. M., and Simon, M. I. (1993). Mutational analysis of G protein α subunit $G_o\alpha$ expressed in *Escherichia coli. J. Biol. Chem.* **268:**1414–1423.
85. Berlot, C. H., and Bourne, H. R. (1992). Identification of effector-activating residues of $G_{s\alpha}$. *Cell* **68:**911–922.
86. Rarick, H. M., Artemyev, N. O., and Hamm, H. E. (1992). A site on rod G protein α subunit that mediates effector activation. *Science* **256:**1031–1033.
87. Lyons, J., Landis, C. A., Harsh, G., Vallar, L., Grünewald, K., Feichtinger, H., Duh, Q.-Y., Clark, O. H., Kawasaki, E., Bourne, H. R., and McCormick, F. (1990). Two G protein oncogenes in human endocrine tumors. *Science* **249:**655–659.
88. Cassel, D., and Selinger, Z. (1977). Mechanism of adenylate cyclase activation by cholera toxin: Inhibition of GTP hydrolysis at the regulatory site. *Proc. Natl. Acad. Sci. USA* **74:**3307–3311.
89. Abood, M. E., Hurley, J. B., Pappone, M.-C., Bourne, H. R., and Stryer, L. (1982). Functional homology between signal-coupling proteins. Cholera toxin inactivates the GTPase activity of transducin. *J. Biol. Chem.* **257:**10540–10543.
90. Van Dop, C., Tsubokawa, M., Bourne, H. R., and Ramachandran, J. (1984). Amino acid sequence of retinal transducin at the site ADP-ribosylated by cholera toxin. *J. Biol. Chem.* **259:**696–699.
91. Graziano, M. P., and Gilman, A. G. (1990). Synthesis in *Escherichia coli* of GTPase-deficient mutants of $G_s\alpha$. *J. Biol. Chem.* **264:**15475–15482.
92. Gupta, S. K., Gallego, C., Johnson, G., and Heasley, L. E. (1992). MAP kinase is constitutively activated in *gip2* and *src* transformed Rat-1a fibroblasts. *J. Biol. Chem.* **267:**7987–7990.
93. Gupta, S. K., Gallego, C., and Johnson, G. L. (1992). Mitogenic pathways regulated by G protein oncogenes. *Mol. Biol. Cell* **3:**123–128.
94. De Vivo, M., Chen, J., Codina, J., and Iyengar, R. (1992). Enhanced phospholipase C stimulation and transformation in NIH-3T3 cells expressing Q209L G_q-α subunits. *J. Biol. Chem.* **267:**18263–18266.
95. Kroll, S. D., Chen, J., De Vivo, M., Carthy, D. J., Buku, A., Premont, R. T., and Iyengar, R. (1992). The Q205L G_o-α subunit expressed in NIH-3T3 cells induces transformation. *J. Biol. Chem.* **267:**23182–23188.
96. Jiang, H., Wu, D., and Simon, M. I. (1993). The transforming activity of activated Gα12. *FEBS Lett.* **330:**319–322.
97. Chen, J. C., and Iyengar, R. (1994). Suppression of *ras*-induced transformation of NIH-3T3 cells by activated α_s. *Science* **263:**1278–1281.
98. Rudolph, U., Finegold, M. J., Rich, S. S., Harriman, G. R., Srinivasan, Y., Brabet, P., Bradley, A., and Birnbaumer, L. (1995). G_{i2}-deficient mice develop inflammatory bowel disease and adenocarcinomas of the colon. *Nature Genetics* **10:**143–150.
99. Iiri, T., Herzmark, P., Nakamoto, J. M., van Dop, C., and Bourne, H. R. (1994). Rapid GDP release from Gs alpha in patients with gain and loss of endocrine function. *Nature* **371:**164–168.
100. Gill, D. M. (1974). The enzymatic nature of cholera toxin. In *Proceedings of the 10th Joint Conference, US–Japan Cooperative Medical Science Program, Cholera Panel* (H. Fukumi and M. Chashi, eds.), National Institute of Health, Tokyo, pp. 119–128.
101. Cassel, D., and Pfeuffer, T. (1978). Mechanism of cholera toxin action: Covalent modification of the guanyl nucleotide-binding protein of the adenylate cyclase system. *Proc. Natl. Acad. Sci. USA* **75:** 2669–2673.
102. Mattera, R., Codina, J., Sekura, R. D., and Birnbaumer, L. (1986). The interaction of nucleotides with pertussis toxin. Direct evidence for a nucleotide binding site on the toxin regulating the rate of ADP-ribosylation of N_i, the inhibitory regulatory component of adenylyl cyclase. *J. Biol. Chem.* **261:**11173–11179.
103. Balch, W. E. (1990). Small GTP-binding proteins in vesicular transport. *Trends Biochem. Sci.* **15:**473–477.
104. Birnbaumer, L. (1992). Receptor-to-effector signaling through G proteins: Roles for βγ dimers as well as for α subunits. *Cell* **71:**1069–1072.
105. Tang, W. J., and Gilman, A. G. (1991). Type specific regulation of adenylyl cyclase by G protein βγ subunits. *Science* **254:**1500–1503.
106. Harwood, J. P., Löw, H., and Rodbell, M. (1973). Stimulatory and inhibitory effects of guanyl nucleotides on fat cell adenylate cyclase. *J. Biol. Chem.* **248:**6239–6245.
107. Birnbaumer, L. (1973). Hormone-sensitive adenylyl cyclases: Useful models for studying hormone receptor functions in cell-free systems. *Biochim. Biophys. Acta (Reviews on Biomembranes)* **300:** 129–158.
108. Bokoch, G. M., Katada, T., Northup, J. K., Hewlett, E. L., and Gilman, A. G. (1983). Identification of the predominant substrate for ADP-ribosylation by islet activating protein. *J. Biol. Chem.* **258:** 2071–2075.
109. Codina, J., Hildebrandt, J. D., Iyengar, R., Birnbaumer, L., Sekura, R. D., and Manclark, C. R. (1983). Pertussis toxin substrate, the putative N_i of adenylyl cyclases, is an *alpha-beta* heterodimer regulated by guanine nucleotide and magnesium. *Proc. Natl. Acad. Sci. USA* **80:**4276–4280.
110. Hildebrandt, J. D., Hanoune, J., and Birnbaumer, L. (1982). Guanine nucleotide inhibition of *cyc*$^-$s49 mouse lymphoma cell membrane adenylyl cyclase. *J. Biol. Chem.* **257:**14723–14725.
111. Hildebrandt, J. D., Codina, J., and Birnbaumer, L. (1984). Interaction of the stimulatory and inhibitory regulatory proteins of the adenylyl cyclase system with the catalytic component of *cyc*$^-$ s49 cell membranes. *J. Biol. Chem.* **259:**13178–13185.

112. Toro, M.-J., Montoya, E., and Birnbaumer, L. (1987). Inhibitory regulation of adenylyl cyclases. Evidence against a role for $\beta\gamma$ complexes of G proteins as mediators of G_i-dependent hormonal effects. *Mol. Endocrinol.* **1**:669–676.
113. Jacobowitz, O., Chen, J., Premont, R. T., and Iyengar, R. (1993). Stimulation of specific types of G_s-stimulated adenylyl cyclases by phorbol ether treatment. *J. Biol. Chem.* **268**:3829–3832.
114. Chen, J. C., and Iyengar, R. (1993). Inhibition of cloned adenylyl cyclases by mutant-activated G_i-α and specific suppression of type 2 adenylyl cyclase inhibition by phorbol ester treatment. *J. Biol. Chem.* **268**:12253–12256.
115. Litosch, I., Wallis, C., and Fain, J. N. (1985). 5-Hydroxytryptamine stimulates inositol phosphate production in a cell-free system from blowfly salivary glands. Evidence for a role of GTP in coupling receptor activation to phosphoinositide breakdown. *J. Biol. Chem.* **260**:5464–5471.
116. Cockroft, S., and Gomperts, B. D. (1985). Role of guanine nucleotide binding protein in the activation of polyphosphoinositide phosphodiesterase. *Nature* **314**:534–536.
117. Suh, P.-G., Ryu, S. H., Moon, K. H., Suh, H. W., and Rhee, S. G. (1988). Cloning and sequence of multiple forms of phospholipase C. *Cell* **54**:161–169.
118. Katan, M., Kriz, R. W., Totty, N., Philp, R., Meldrum, E., Aldape, R. A., Knopf, J. L., and Parker, P. J. (1988). Determination of the primary structure of PLC-154 demonstrates diversity of phosphoinositide-specific phospholipase C activities. *Cell* **54**:171–177.
119. Cooper, C. L., Morris, A. J., and Harden, T. K. (1989). Guanine nucleotide-sensitive interaction of a radiolabeled agonist with a phospholipase C-linked P_{2y}-purinergic receptor. *J. Biol. Chem.* **264**: 6202–6206.
120. Boyer, J. L., Downes, C. P., and Harden, T. K. (1989). Kinetics of activation of phospholipase C by P_{2y} purinergic receptor agonists and guanine nucleotides. *J. Biol. Chem.* **264**:884–890.
121. Bloomquist, B. T., Shortridge, R. D., Schneuwly, S., Perdew, M., Montell, C., Steller, H., Rubin, G., and Pak, W. L. (1988). Isolation of a putative phospholipase C gene of *Drosophila, norpA,* and its role in phototransduction. *Cell* **54**:723–733.
122. Rhee, S. G., and Choi, K. D. (1992). Regulation of inositol phospholipid-specific phospholipase C isozymes. *J. Biol. Chem.* **267**:12393–12398.
123. Selinger, Z., and Minke, B. (1988). Inositol lipid cascade of vision studied in mutant flies. *Cold Spring Harbor Symp. Quant. Biol.* **53**:333–341.
124. Selinger, Z., Dora, Y. N., and Minke, B. (1993). Mechanisms and genetics of photoreceptor desensitization in *Drosophila* flies. *Biochim. Biophys. Acta* **1179**:283–299.
125. Waldo, G. L., Boyer, J. L., Morris, A. J., and Harden, T. K. (1991). Purification of an AlF^-_4 and G-protein $\beta\gamma$-subunit-regulated phospholipase C-activating protein. *J. Biol. Chem.* **266**:14217–14255.
126. Park, D., Jhon, D. Y., Kritz, R., Knopf, J., and Rhee, S. G. (1992). Cloning, expression and G_q-independent activation of phospholipase C-β2. *J. Biol. Chem.* **267**:16048–16055.
127. Lee, C. H., Park, D., Wu, D., Rhee, S. G., and Simon, M. I. (1992). Members of the G_q alpha subunit gene family activate phospholipase C-beta isozymes. *J. Biol. Chem.* **267**:16044–16047.
128. Camps, M., Hou, C., Sidiropoulos, D., Stock, J. B., Jakobs, K. H., and Gierschik, P. (1992). Stimulation of phospholipase C by guanine-nucleotide-binding protein $\beta\gamma$ subunits. *Eur. J. Biochem.* **206**: 821–831.
129. Camps, M., Carozzi, A., Schnabel, P., Scheer, A., Parker, P. J., and Gierschik, P. (1992). Isozyme-selective stimulation of phospholipase C-β_2 by G protein $\beta\gamma$-subunits. *Nature* **360**:684–686.
130. Katz, A., Wu, D., and Simon, M. I. (1992). $\beta\gamma$ subunits of the heterotrimeric G protein activate the β2 isoform of phospholipase C. *Nature* **360**:686–689.
131. Wu, D., Lee, C. H., Rhee, S. G., and Simon, M. I. (1992). Activation of phospholipase C by the α subunit of the G_q and G_{11} protein in transfected COS-7 cells. *J. Biol. Chem.* **267**:1811–1817.
132. Wu, D., Katz, A., and Simon, M. I. (1993). Activation of phospholipase C β_2 by the α and $\beta\gamma$ subunits of trimeric GTP-binding protein. *Proc. Natl. Acad. Sci. USA* **90**:5297–5301.
133. Wu, D., Jiang, H., Katz, A., and Simon, M. I. (1993). Identification of critical regions on phospholipase C-β1 required for activation by G-proteins. *J. Biol. Chem.* **268**:3704–3709.
134. Blank, J. L., Brattain, K. A., and Exton, J. H. (1992). Activation of cytosolic phosphoinositide phospholipase C by G-protein $\beta\gamma$ subunits. *J. Biol. Chem.* **267**:23069–23075.
135. Hepler, J. R., Kozasa, T., Smrcka, A. V., Simon, M. I., Rhee, S. G., Sternweis, P. C., and Gilman, A. G. (1993). Purification from Sf9 cells and characterization of recombinant $G_{q\alpha}$ and $G_{11\alpha}$. Activation of purified phospholipase C isozymes by Gα subunits. *J. Biol. Chem.* **268**:14367–14375.
136. Carozzi, A., Camps, M., Gierschik, P., and Parker, P. J. (1993). Activation of phosphatidylinositol lipid-specific phospholipase C-β3 by G-protein $\beta\gamma$ subunits. *FEBS Lett.* **315**:340–342.
137. Schnabel, P., Schreck, R., Schiller, D. L., Camps, M., and Gierschik, P. (1993). Stimulation of phospholipase C by a mutationally activated G protein α16 subunit. *Biochem. Biophys. Res. Commun.* **188**: 1018–1023.
138. Park, D., Jhon, D. Y., Lee, C. W., Ryu, S., and Rhee, S. G. (1993). Removal of the carboxy-terminal region of phospholipase C-β1 by calpain abolishes activation by $G_q\alpha$. *J. Biol. Chem.* **268**:3710–3714.
139. Berstein, G., Blank, J. L., Smrcka, A. V., Higashijima, T., Sternweis, P. C., Exton, J. H., and Ross, E. M. (1992). Reconstitution of agonist-stimulated phosphatidylinositol 4,5-tris-phosphate hydrolysis using purified M1 muscarinic receptor. $G_{q/11}$ and phospholipase C-β1. *J. Biol. Chem.* **267**:8081–8088.
140. Berstein, G., Blank, J. L., Jhon, D. Y., Exton, J. H., Rhee, S. G., and Ross, E. M. (1992). Phospholipase C-β1 is a GTPase-activating protein for $G_{q/11}$, its physiologic regulator. *Cell* **70**:411–418.
141. Okajima, F., and Ui, M. (1984). ADP-ribosylation of the specific membrane protein by islet-activating protein, pertussis toxin, associated with inhibition of a chemotactic peptide-induced arachidonate release in neutrophils. A possible role of the toxin substrate in Ca^{2+}-mobilizing biosignaling. *J. Biol. Chem.* **259**:13863–13871.
142. Ohta, H., Okajima, F., and Ui, M. (1985). Inhibition by islet-activating protein of a chemotactic peptide-induced early breakdown of inositol phospholipids and Ca^{2+} mobilization in guinea pig neutrophils. *J. Biol. Chem.* **260**:15771–15780.
143. Kikuchi, A., Kozawa, O., Kaibuchi, K., Katada, T., Ui, M., and Takai, Y. (1986). Direct evidence for involvement of a guanine nucleotide-binding protein in chemotactic peptide-stimulated formation of inositol bisphosphate and trisphosphate in differential human leukemic (HL-60) cells. Reconstitution with G_i or G_o of the plasma membranes ADP-ribosylated by pertussis toxin. *J. Biol. Chem.* **261**: 11558–11562.
144. Taylor, S. J., Chae, H. Z., Rhee, S. G., and Exton, J. H. (1991). Activation of the β1 isozyme of phospholipase C by α subunits of the G_q class of G proteins. *Nature* **350**:516–518.
145. Smrcka, A. V., Helper, J. R., Brown, K. O., and Sternweis, P. C. (1991). Regulation of polyphosphoinositide-specific phospholipase C activity by purified G_q. *Science* **251**:804–807.
146. Gierschik, P., and Jakobs, K. H. (1987). Receptor mediated ADP-ribosylation of a phospholipase C-stimulating g protein. *FEBS Lett.* **224**:219–223.
147. Gierschik, P., Sidiropoulos, D., and Jakobs, K. H. (1989). Two distinct G_i-proteins mediate formyl peptide receptor signal transduction in human leukemia (HL-60) cells. *J. Biol. Chem.* **264**:21470–21473.

148. Birnbaumer, L., Bearer, C. F., and Iyengar, R. (1980). A two-state model of an enzyme with an allosteric regulatory site capable of metabolizing the regulatory ligand: Simplified mathematical treatments of transient and steady state kinetics of an activator and its competitive inhibition as applied to adenylyl cyclases. *J. Biol. Chem.* **255:**3552–3557.
149. Iyengar, R., Abramowitz, J., Bordelon-Riser, M. E., and Birnbaumer, L. (1980). Hormone receptor-mediated stimulation of adenylyl cyclase systems. Nucleotide effects and analysis in terms of a two-state model for the basic receptor-affected enzyme. *J. Biol. Chem.* **255:**3558–3564.
150. Pieroni, J. P., Jakobowitz, O., Chen, J., and Iyengar, R. (1993). Signal recognition and integration by G_s-stimulated adenylyl cyclases. *Curr. Opin. Neurobiol.* **3:**345–351.
151. Kleuss, C., Hescheler, J., Ewel, C., Rosenthal, W., Schultz, G., and Wittig, B. (1991). Assignment of G-protein subtypes to specific receptors inducing inhibition of calcium currents. *Nature* **353:**43–48.
152. Yatani, A., Codina, J., Sekura, R. D., Birnbaumer, L., and Brown, A. M. (1987). Reconstitution of somatostatin and muscarinic receptor mediated stimulation of K^+ channels by isolated G_k protein in clonal rat anterior pituitary cell membranes. *Mol. Endocrinol.* **1:** 283–289.
153. Amatruda, T. T., III, Gerard, N. P., Gerard, C., and Simon, M. I. (1993). Specific interactions of chemoattractant factor receptors with G-proteins. *J. Biol. Chem.* **268:**10139–10144.
154. Wu, D., LaRosa, G. J., and Simon, M. I. (1993). G protein-coupled signal transduction pathways for interleukin-8. *Science* **261:**101–103.
155. Ashkenazi, A., Winslow, J. W., Peralta, E. G., Peterson, G. L., Schimerlik, M. I., Capon, D. J., and Ramachandran, J. (1987). An M2 muscarinic receptor subtype coupled to both adenylyl cyclase and phosphoinositide turnover. *Science* **238:**672–675.
156. Fargin, A., Raymond, J. R., Regen, J. W., Cotecchia, S., Lefkowitz, R. J., and Caron, M. G. (1989). Effector coupling mechanisms of the cloned 5-HT1A receptor. *J. Biol. Chem.* **264:**14848–14852.
157. Raymond, J. R., Albers, F. J., Middleton, J. P., Lefkowtiz, R. J., Caron, M. G., Obeid, L. M., and Denis, V. W. (1991). 5-HT_{1A} and histamine H_1 receptors in HeLa cells stimulate phosphoinositide hydrolysis and phosphate uptake via distinct G protein pools. *J. Biol. Chem.* **266:**372–379.
158. Vallar, L., Muca, C., Magni, M., Albert, P., Bunzow, J., Meldolesi, J., and Civelli, O. (1990). Differential coupling of dopaminergic D_2 receptors expressed in different cell types. Stimulation of phosphatidylinositol 4,5-bisphosphate hydrolysis in *Ltk*-fibroblasts, hyperpolarization, and cytosolic free Ca^{2+} concentration decrease in GH_4C_1 cells. *J. Biol. Chem.* **265:**10320–10326.
159. VanSande, J., Raspe, E., Perret, J., Lejeune, C., Manhaut, C., Vassart, G., and Dumont, J. E. (1990). Thyrotropin activates both the cAMP and the PIP_2 cascade in CHO cells expressing the human cDNA of the TSH receptor. *Mol. Cell. Endocrinol.* **74:**R1–R6.
160. Gudermann, T., Birnbaumer, M., and Birnbaumer, L. (1992). Evidence for dual coupling of the murine LH receptor to adenylyl cyclase and phosphoinositide breakdown/Ca^{2+} mobilization. Studies with the cloned murine LH receptor expressed in L cells. *J. Biol. Chem.* **267:**4479–4488.
161. Chabre, O., Conklin, B. R., Lin, H. Y., Lodish, H. F., Wilson, E., Ives, H. E., Catanzariti, L., Hemmings, B. A., and Bourne, H. R. (1992). A recombinant calcitonin receptor independently stimulates 3′,5′-cyclic adenosine monophosphate and Ca^{2+}/inositol phosphate signaling pathways. *Mol. Endocrinol.* **6:**551–556.
162. Abou-Sambra, A. B., Jüpmer, H., Force, T., Freeman, M. W., Kong, X. F., Schipani, E., Urena, P., Richards, J., Bonventre, J. V., Potts, J. T., Jr., Kronenberg, H. M., and Segre, G. V. (1992). Expression cloning of a common receptor for parathyroid hormone and parathyroid hormone-related peptide from rat osteoblast-like cells: A single receptor stimulates intracellular accumulation of both cAMP and inositol trisphosphates and increases intracellular free calcium. *Proc. Natl. Acad. Sci. USA* **89:**2732–2736.
163. Northup, J. K., Smigel, M. D., Sternweis, P. C., and Gilman, A. G. (1983). The subunits of the stimulatory regulatory component of adenylate cyclase. Resolution of the activated 45,000-dalton (alpha) subunit. *J. Biol. Chem.* **258:**11369–11376.
164. Codina, J., Hildebrandt, J. D., Sekura, R. D., Birnbaumer, M., Bryan, J., Manclark, C. R., Iyengar, R., and Birnbaumer, L. (1984). N_s and N_i, the stimulatory and inhibitory regulatory components of adenylyl cyclases. Purification of the human erythrocyte proteins without the use of activating regulatory ligands. *J. Biol. Chem.* **259:**5871–5886.
165. Eason, M. G., Kurose, H., Holt, B. D., Raymond, J. R., and Liggett, S. B. (1992). Simultaneous coupling of α_2-adrenergic receptors to two G-proteins with opposing effects. Subtype selective coupling of α_2C10, α_2C4 and α_2C2 adrenergic receptors to G_i and G_s. *J. Biol. Chem.* **267:**15795–15801.
166. Kisselev, O., and Gautam, N. (1993). Specific interaction with rhodopsin is dependent on the γ subunit type in a G protein. *J. Biol. Chem.* **268:**24519–24522.
167. Tsigos, C., Arai, K., Hung, W., and Chrousos, G. P. (1993). Hereditary isolated glucocorticoid deficiency is associated with abnormalities of the adrenocorticotropin receptor. *J. Clin. Invest.* **92:**2458–2461.
168. Clark, A. J. L., McCloughlin, L., and Grossman, A. (1993). Familial glucocorticoid deficiency associated with a point mutation in the adrenocorticotropin receptor. *Lancet* **341:**461–462.
169. Pollak, M. R., Brown, E. M., Chou, Y.-H. W., Marx, S. J., Steinman, B., Levi, T., Seidman, C. E., and Seidman, J. D. (1993). Mutations in the human Ca^{2+} sensing receptor gene cause familial hypocalciuric hypercalcemia and neonatal severe hyperparathyroidism. *Cell* **75:** 1297–1303.
170. Lin, S.-L., C. R., Gukovsky, I., Lusis, A. J., Sawchenko, P. E., and Rosenfeld, M. G. (1993). Molecular basis of the *little* mouse phenotype and implications for cell type-specific growth. *Nature* **364:** 208–213.
171. Kremer, H., Mariman, E., Otten, B. J., Moll, G. W., Jr., Stoelinga, G. B. A., Wit, J. M., Jansen, M., Drop, S. L., Faas, B., Ropers, H.-H., and Brunner, H. G. (1993). Cosegregation of missense mutations of the luteinizing hormone receptor gene with familial male-limited precocious puberty. *Hum. Mol. Genet.* **2:**1779–1783.
172. Dryja, T. P., Berson, E. L., Rao, V. R., and Oprian, D. D. (1993). Heterozygous missense mutation in the rhodopsin gene as a cause of congenital stationary night blindness. *Nature Genet.* **4:**280–283.
173. Rao, V. R., Cohen, G. B., and Oprian, D. D. (1994). Rhodopsin mutation G90D and a molecular mechanism for congenital night blindness. *Nature* **367:**639–642.
174. Parma, J., Duprez, L., Van Sande, J., Cochaux, P., Gervy, C., Mockel, J., Dumont, J., and Vassart, G. (1993). Somatic mutations in the thyrotropin receptor gene cause hyperfunctioning thyroid adenomas. *Nature* **365:**649–651.
175. Stein, S. A., Oats, E. L., Hall, C. R., Grumbles, R. M., Fernandez, L. M., Taylor, N. A., Puett, D., and Gin, S. (1994). Identification of a pointmutation in the thyrotropin receptor of the *hyt/hyt* hypothyroid mouse. *Mol. Endocrinol.* **8:**129–138.
176. Rosenthal, W., Seibold, A., Antaramian, A., Lonergan, M., Arthus, M.-F., Hendy, G. N., Birnbaumer, M., and Bichet, D. G. (1992). Molecular identification of the gene responsible for congenital nephrogenic diabetes insipidus. *Nature* **359:**233–235.
177. Kosugi, S., Okajima, F., Ban, T., Hidaka, A., Shenker, A., and Kohn, L. D. (1992). Mutation of alanine 623 in the third cytoplasmic loop of the rat thyrotropin receptor results in a loss in the phosphoinositide but not cAMP signal induced by TSH and receptor autoantibodies. *J. Biol. Chem.* **267:**24153–24156.

178. Horstman, D., Brandon, S., Wilson, A. L., Guyer, C. A., Cragoe, E. J., Jr., and Limbird, L. E. (1990). An aspartate conserved among G-protein receptors confers allosteric regulation of α_2-adrenergic receptors by sodium. *J. Biol. Chem.* **265:**21590–21595.
179. Strader, C. D., Sigal, I. S., Candelone, M. R., Rands, E., Gill, W. S., and Dixon, R. A. F. (1988). Conserved aspartic acid residues 79 and 113 of the β-adrenergic receptor have different roles in receptor function. *J. Biol. Chem.* **263:**10267–10271.
180. Chung, P. Z,. Wang, C. D., Potter, P. C., Venter, J. C., and Fraser, C. M. (1988). Site-directed mutagenesis and continuous expression of human β-adrenergic receptor. Identification of a conserved aspartate residue involved in agonist binding and receptor activation. *J. Biol. Chem.* **163:**4052–4055.
181. Neve, K. A., Cox, B. A., Henningsen, R. A., Spanoyannis, A., and Neve, R. L. (1991). Pivotal role for asparate-80 in the regulation of dopamine D2 receptor affinity for drugs and inhibition of adenylyl cyclase. *Mol. Pharmacol.* **39:**733–739.
182. Quintana, J., Wang, H., and Ascoli, M. (1993). The regulation of the binding affinity of the luteinizing hormone/choriogonadotropin receptor by sodium ions is mediated by a highly conserved aspartate located in the second transmembrane domain of G-protein coupled receptors. *Mol. Endocrinol.* **7:**767–775.
183. Franke, R. R., Koenig, B., Sakmar, T. P., Khorana, H. G., and Hofmann, K. P. (1990). Rhodopsin mutants that bind but fail to activate transducin. *Science* **250:**123–125.
184. Franke, R. R., Sakmar, T. P., Graham, R. M., and Khorana, H. G. (1992). Structure and function in rhodopsin. Studies of the interaction between the rhodopsin cytoplasmic domain and transducin. *J. Biol. Chem.* **267:**14767–14774.
185. Fraser, C. M., Chung, F. Z., Wang, C. D., and Venter, J. C. (1988). Site directed mutagenesis of human β-adrenergic receptors: Substitution of aspartic acid-130 by asparagine produces a receptor with high affinity agonist binding that is uncoupled from adenylate cyclase. *Proc. Natl. Acad. Sci. USA* **85:**5478–5482.
186. Winslow, J. W., Van Amsterdam, J. R., and Neer, E. J. (1986). Conformations of the α_{39}, α_{41} and βγ components of brain guanine nucleotide-binding proteins: Analysis by limited proteolysis. *J. Biol. Chem.* **261:**14428–14430.
187. Yi, F., Denker, B. M., and Neer, E. J. (1991). Structural and functional studies of cross-linked G_o protein subunits. *J. Biol. Chem.* **266:**3900–3906.
188. Bubis, J., and Khorana, H. G. (1990). Sites of interaction in the complex between β- and γ-subunits of transducin. *J. Biol. Chem.* **265:** 12995–12999.
189. Kobilka, B. K., Kobilka, T. S., Daniel, K., Regan, J. W., Caron, M. G., and Lefkowitz, R. J. (1988). Chimeric α_2/β_2-adrenergic receptors: Delineation of domains involved in effector coupling and ligand binding specificity. *Science* **240:**1310–1316.
190. Rubenstein, R. C., Wong, S. K. P., and Ross E. M. (1987). The hydrophobic tryptic core of the β-adrenergic receptor retains G_s regulatory activity in response to agonists and thiols. *J. Biol. Chem.* **262:**16655–16662.
191. Suryanarayana, S., von Zastrow, M., and Kobilka, B. K. (1992). Identification of intramolecular interactions in adrenergic receptors. *J. Biol. Chem.* **267:**21991–21994.
192. Tsai-Morris, C. H., Buczko, E., Wang, W., and Dufau, M. L. (1990). Intronic nature of the rat luteinizing hormone receptor gene defines a soluble receptor subspecies with hormone binding activity. *J. Biol. Chem.* **265:**19385–19388.
193. Osenberg, D., Marsters, S. A., O'Dowd, B. F., Jin, H., Havlik, S., Peroutka, S. J., and Ashkenazi, A. (1992). A single amino acid difference confers major pharmacological variation between human and rodent 5HT-1B receptors. *Nature* **360:**161–163.
194. Gether, U., Johansen, T. E., Snider, M. R., Lowe, J. A., Nakanishi, S., and Schwartz, T. W. (1993). Different binding epitopes in the NK1 receptor for substance P and a non-peptide antagonist. *Nature* **362:**345–348.
195. Blüml, K., Mutschler, E., and Wess, J. (1994). Identification of an intracellular tyrosine residue critical for muscarinic receptor-mediated stimulation of phosphatidylinositol hydrolysis. *J. Biol. Chem.* **269:**402–405.
196. Benovic, J. L., Kühn, H., Weyand, I., Codina, J., Caron, M. G., and Lefkowitz, R. J. (1987). Functional desensitization of the isolated β-adrenergic receptor by the β-adrenergic receptor kinase: Potential role of an analog of the retinal protein arrestin (48k protein). *Proc. Natl. Acad. Sci. USA* **84:**8879–8882.
197. Benovic, J. L., De Blasi, A., Stone, W. C., Caron, M. G., and Lefkowitz, R. J. (1989). β-Adrenergic receptor kinase: Primary structure delineates a multigene family. *Science* **246:**235–240.
198. Lohse, M. J., Andexinger, S., Pitcher, J., Trukawinski, S., Codina, J., Faure, J.-P., Caron, M. G., and Lefkowitz, R. J. (1992). Receptor-specific desensitization with purified proteins. Kinase dependence and receptor specificity of β-arrestin and arrestin in the β_2-adrenergic receptor and rhodopsin systems. *J. Biol. Chem.* **267:**8558–8564.
199. Haga, K., and Haga, T. (1992). Activation by G protein βγsubunits of agonist- or light-dependent phosphorylation of muscarinic acetylcholine receptors and rhodopsin. *J. Biol. Chem.* **267:**2222–2227.
200. Kameyama, K., Haga, K., Haga, T., Kontani, K., Katada, T., and Fukada, Y. (1993). Activation by G protein βγ subunits of β-adrenergic and muscarinic receptor kinase. *J. Biol. Chem.* **268:**7753–7758.
201. Pitcher, J. A., Inglese, J., Higgins, J. B., Arriza, J. L., Casey, P. J., Kim, C., Benovic, J. L., Kwatra, M. M., Caron, M. G., and Lefkowitz, R. J. (1992). Role of βγ subunits of G proteins in targeting of the β-adrenergic receptor kinase to membrane-bound receptors. *Science* **257:**1264–1267.
202. Inglese, J., Freedman, N. J., Koch, W. J., and Lefkowitz, R. J. (1993). Structure and mechanism of the G protein coupled receptor kinases. *J. Biol. Chem.* **268:**23735–23738.
203. Dalman, H. M., and Neubig, R. R. (1991). Two peptides from the α_{2A}-adrenergic receptor alter G protein coupling by distinct mechanisms. *J. Biol. Chem.* **266:**11025–11029.
204. Grandt, R., Aktories, K., and Jakobs, K. H. (1982). Guanine nucleotides and monovalent cations increase agonist affinity of prostaglandin E_2 receptors in hamster platelets. *Mol. Pharmacol.* **22:**320–326.
205. Sugimoto, Y., Negishi, M., Hayashi, Y., Namba, T., Honda, A., Watabe, A., Hirata, M., Narumiya, S., and Ichikawa, A. (1993). Two isoforms of EP3 receptor with different carboxyl terminal domains. Identical ligand binding properties and different coupling properties with G_i proteins. *J. Biol. Chem.* **268:**2712–2718.
206. Karnik, S. S., Sakmar, T. P., Chen, H. B., and Khorana, H. G. (1988). Cysteine residues 110 and 187 are essential for the formation of correct structure of bovine rhodopsin. *Proc. Natl. Acad. Sci. USA* **85:** 8459–8463.
207. O'Dowd, B. F., Hnatowitch, M., Caron, M. G., Lefkowitz, R. J., and Bouvier, M. (1989). Palmitoylation of the human beta 2-adrenergic receptor: Mutation of Cys-341 in the carboxyl tail leads to an uncoupled non-palmitoylated form of the receptor. *J. Biol. Chem.* **264:** 7564–7569.
208. Moffett, S. W., Mouillac, B., Bonin, H., and Bouvier, M. (1993). Altered phosphorylation and desensitization patterns of a human β_2-adrenergic receptor lacking the palmitoylated Cys341. *EMBO J.* **12:** 349–356.
209. Kennedy, M. E., and Limbird, L. E. (1993). Mutations of the α_{2A}-adrenergic receptor that eliminate detectable palmitoylation do not perturb receptor–G protein coupling. *J. Biol. Chem.* **268:**8003–8011.
210. Chabre, M. (1985). Trigger and amplification mechanisms in visual phototransduction. *Annu. Rev. Biophys. Biophys. Chem.* **14:**331–347.
211. Stryer, I. (1988). Molecular basis for visual excitation. *Cold Spring Harbor Symp. Quant. Biol.* **53:**283–294.

212. Braiman, M., Bubis, J., Doi, T., Chen, H.-B., Flitsch, S. L., Franke, R. R., Giles-Gonzalez, M. A., Graham, R. M., Karnik, S. S., Khorana, G. G., Knox, B. E., Kebs, M. P., Marti, T., Mogi, T., Nakayama, T., Oprian, D. D., Puckett, K. L., Sakmar, T. P., Stern, L. J., Subramanian, S., and Thompson, D. A. (1988). Studies on light transduction by bacteriorhodopsin and rhodopsin. *Cold Spring Harbor Symp. Quant. Biol.* **53:**355–364.
213. Zhukovsky, E. A., Robinson, P. R., and Oprian, D. D. (1991). Transducin activation by rhodopsin without a covalent bond to the 11-*cis*-retinal chromophore. *Science* **251:**558–559.
214. Link, R., Daunt, D., Barsh, G., Chruscinski, A., and Kobilka, B. (1993). Cloning of two genes encoding α2-adrenergic receptor subtypes and identification of a single amino acid in the mouse α2-C10 homolog responsible for an interspecies variation in antagonist binding. *Mol. Pharm.* **42:**16–27.
215. Bichet, D. G., Arthus, M.-F., Lonergan, M., Hendy, G. N., Paradis, A. J., Fujiwara, T. M., Morgan, K., Gregory, M. C., Rosenthal, W., Antaramian, A., and Birnbaumer, M. (1993). X-linked nephrogenic diabetes insipidus mutations in North America and the Hopewell hypothesis. *J. Clin. Invest.* **92:**1262–1268.
216. Holzman, E. J., Harris, H. W. H., Kolakowski, L. F., Guay-Woodford, L. M., Botelho, B., and Aussiello, D. A. (1993). A molecular defect in vasopressin V2-receptor gene causing nephrogenic diabetes insipidus. *N. Engl. J. Med.* **328:**1534–1537.
217. Knoers, N. V. A. M., Verdijk, M., Monnens, L. A. H., van den Ouweland, A. M. W., and van Oost, B. A. (1993). Inheritance of mutations in the vasopressin V2 receptor gene in 15 Dutch families with congenital nephrogenic diabetes insipidus. In *Vasopressin* (P. Gross, D. Richter, and G. L. Robertson, eds.), J. Libbey, Eurotext, Proceedings of the IV International Vasopressin Conference, May, 1993, Berlin, pp. 571–572.
218. Merendino, J. J., Spiegel, A. M., Crawford, J. D., O'Carroll, A.-M., Brownstein, M. J., and Lolait, S. J. (1993). A mutation in the vasopressin V2 receptor gene in a kindred with X-linked nephrogenic diabetes insipidus. *N. Engl. J. Med.* **328:**1538–1541.
219. Pan, Y., Metzenberg, A., Das, S., and Gitschier, J. (1992). Mutations in the V2 vasopressin receptor gene are associated with X-linked nephrogenic diabetes insipidus. *Nature Genet.* **2:**103–106.
220. van den Ouweland, A. M. W., Dreesen, J. C. F. M., Verdijk, M., Knoers, N. V. A. M., Monnens, L. A. H., Rocchi, M., and van Oost, B. A. (1992). Mutations in the vasopressin type 2 receptor gene AVPR2 associated with nephrogenic diabetes insipidus. *Nature Genet.* **2:** 99–102.
221. Iyengar, R., and Birnbaumer, L. (1982). Hormone receptor modulates the regulatory component of adenylyl cyclases by reducing its requirement for Mg ion and enhancing its extent of activation by guanine nucleotides. *Proc. Natl. Acad. Sci. USA* **79:**5179–5183.
222. Birnbaumer, L., Abramowitz, J., and Brown. A. M. (1990). Signal transduction by G proteins. *Biochim. Biophys. Acta (Reviews in Biomembranes)* **1031:**163–224.
223. Levine, M. A., Ahn, T. G., Klupp, S. F., Kaufman, K. D., Smallwood, P. M., Bourne, H. R., Sullivan, K. A., and Van Dop, C. (1988). Genetic deficiency of the α subunit of the guanine nucleotide-binding protein G_s as the molecular basis for Albright's hereditary osteodystrophy. *Proc. Natl. Acad. Sci. USA* **85:**617–621.
224. Patten, J. I., Johns, D. R., Valle, D., Eil, C., Gruppuso, P. A., Steele, G., Smallwood, P. M., and Levine, M. A. (1990). Mutation in the gene encoding the stimulatory G protein of adenylate cyclase in Albright's hereditary osteodystrophy. *N. Engl. J. Med.* **322:**1412–1419.
225. Weinstein, L. S., Gejman, P. V., Friedman, E., Kadowaki, T., Collins, R. M., Gershon, E. S., and Spiegel, A. M. (1990). Mutations of the G_s α-subunit gene in Albright hereditary osteodystrophy detected by denaturing gradient gel electrophoresis. *Proc. Natl. Acad. Sci. USA* **87:**8287–8290.
226. Weinstein, L. S., Shenker, A., Gejman, P. V., Merino, M. J., Friedman, E., and Spiegel, A. M. (1990). Activating mutations of the stimulatory G protein in the McCune–Albright syndrome. *N. Engl. J. Med.* **325:**1688–1695.
227. Schwindinger, W. F., Francomano, C. A., and Levine, M. A. (1992). Identification of a mutation in the gene encoding the α subunit of the stimulatory G proteins of adenylyl cyclase in McCune–Albright syndrome. *Proc. Natl. Acad. Sci. USA* **89:**5151–5156.
228. Artemyev, N. O., Mills, J. S., Thornburg, K. R., Knapp, D. R., Schey, K. L., and Hamm, H. (1993). A site on transducin α-subunit of interaction with the polycationic region of cGMP phosphodiesterase inhibitory subunit. *J. Biol. Chem.* **268:**23611–23615.
229. Higashijima, T., Ferguson, K. M., Sternweis, P. C., Ross, E. M., Smigel, M. D., and Gilman, A. G. (1987). The effect of activating ligands on the intrinsic fluorescence of guanine nucleotide-binding regulatory proteins. *J. Biol. Chem.* **262:**752–756.
230. Higashijima, T., Ferguson, K. M., Smigel, M. D., and Gilman, A. G. (1987). The effect of GTP and MG^{2+} on the GTPase activity and the fluorescent properties of G_o. *J. Biol. Chem.* **262:**757–761.
231. Krupinski, J., Coussen, F., Bakalyar, H. A., Tang, W.-J., Feinstein, P. G., Orth, K., Slaughter, C., Reed, R. R., and Gilman, A. G. (1991). Adenylyl cyclase amino acid sequence: Possible channel- or transporter-like structure. *Science* **244:**1558–1564.
232. Iyengar, R. (1993). Molecular and functional diversity of mammalian G_s-stimulated adenylyl cyclases. *FASEB J.* **7:**768–775.

18

Egg Membranes during Fertilization

Laurinda A. Jaffe

18.1. INTRODUCTION

In most eggs, the outer few micrometers of the egg, often called the "cortex," includes three distinct membrane systems: the plasma membrane and its associated glycocalyx, a set of membrane-bound cortical granules that in most species will undergo exocytosis at fertilization, and the endoplasmic reticulum. In the interior of the egg are found membranes of the endoplasmic reticulum including the nuclear envelope, as well as mitochondria, yolk, and other organelles. This chapter focuses on the egg plasma membrane and endoplasmic reticulum, and their roles in transduction of signals from sperm. Functions of the sperm membrane are discussed in relation to sperm–egg plasma membrane fusion and the activation of the egg by the sperm; other aspects of sperm membrane function, including chemotaxis toward eggs and the acrosome reaction in response to contact with egg, have been reviewed recently.[1] This chapter emphasizes studies on sea urchins, starfish, frogs, and mammals; studies of the many other organisms that have contributed to understanding of the processes of fertilization, such as the marine worm *Urechis,* are discussed less extensively.

18.2. MEMBRANE FUSION EVENTS AT THE EGG PLASMA MEMBRANE

At fertilization, the plasma membranes of the sperm and egg fuse, forming a single continuous membrane around the zygote cytoplasm. In many marine organisms, such as sea urchins as well as the hemichordate *Saccoglossus* and the annelid *Hydroides* (the two species where fusion was first described, by Colwin and Colwin), the fusion occurs near the tip of a long membrane-bound extension of the sperm, the acrosomal process.[2–4] In mammals, fusion occurs on the equatorial and/or posterior parts of the head of the sperm (see Ref. 5). In some species, the sperm can fuse with any part of the egg membrane, but in other species, only particular regions can undergo fusion (see Ref. 4). Fusion is restricted to the "animal" half of a frog egg (the animal pole is defined as the position where polar bodies are produced in meiosis).[6] Likewise, fusion is restricted to a small patch at the animal pole of a hydrozoan egg,[7] and occurs preferentially in the animal half of an ascidian egg.[8] Fusion can occur everywhere except for the animal pole region of a mammalian egg.[9]

How sperm–egg fusion is mediated is unknown, but recent evidence indicates that the process may be similar to viral fusion with a host cell.[5,10] Fertilin, a plasma membrane protein from mammalian sperm (originally known as PH-30, for posterior head-30) appears to be involved in the sperm–egg fusion process, since antibodies against this protein, or peptides derived from it, inhibit sperm–egg fusion.[11,12] Recently, fertilin has been cloned and sequenced, and it has been found to contain a hydrophobic region with similarity to the hydrophobic regions of viral fusion proteins.[10,12a] It is proposed that, like viral fusion proteins, fertilin may mediate fusion of the sperm with the egg by inserting this hydrophobic peptide into the lipid bilayer of the egg plasma membrane.

It is likely that some step in the sperm–egg fusion process involves a specific egg component as well, since hamster and mouse oocyte membranes acquire the ability to fuse with sperm during oogenesis, and lose it after fertilization,[13,13a] and since trypsin or pronase treatment of mouse eggs makes them incapable of fusing with sperm (see Ref. 14). Species specificity of sperm–egg fusion also argues that complementary sperm and egg molecules are involved (see Refs. 4, 14). The fertilin molecule of mammalian sperm contains a "disintegrin"-like domain that binds to an egg plasma membrane integrin.[10,12,12a,15,15a] This may bring the two membranes into the close contact that then allows insertion of the hydrophobic peptide leading to fusion. A synthetic peptide corresponding to this sequence causes fusion of liposomes,[15b] as do proteins isolated from the acrosomal contents of sea urchin[16,17] and abalone[18] sperm.

In many species the fusion of sperm and egg plasma membranes is followed by a fusion of cortical granule membranes with the egg plasma membrane. This exocytosis serves to establish a permanent block to polyspermy by modifying the surface coats of the egg (see Refs. 19, 20). In sea urchins, starfish, fish, and frogs, the surface coat of the egg is caused to lift from the egg plasma membrane, apparently because proteases secreted from the cortical granules cleave the connec-

Laurinda A. Jaffe • Department of Physiology, University of Connecticut Health Center, Farmington, Connecticut 06032.
Molecular Biology of Membrane Transport Disorders, edited by S.G. Schultz *et al.* Plenum Press, New York, 1996

tions between the two structures. Other secreted proteins harden and thicken the surface coat so that it becomes impermeable to sperm. In many mammals, cortical granule exocytosis also establishes a permanent polyspermy block, but here the surface coat (zona pellucida) is already elevated before fertilization, and exocytosis serves to modify the zona, preventing subsequent sperm binding (see Ref. 21).

The fusion of cortical granule membranes with the egg plasma membrane is caused by the increase in intracellular free calcium that occurs at fertilization (see Section 18.5). Artificially raising intracellular Ca in an unfertilized egg causes exocytosis, while maintaining intracellular Ca at a low level by introduction of EGTA or BAPTA prevents exocytosis.[22–25] Like the calcium release, the exocytosis occurs in a wave starting at the site of sperm fusion.[26,26a] In the sea urchin egg, the wave of exocytosis takes about 30 sec to occur and results in an approximate doubling of the egg plasma membrane surface area. This has been determined both by electron microscopic measurements[27] and by measurement of an increase in membrane capacitance.[28]

Exocytosis in sea urchin eggs has been investigated using a "cortex preparation," consisting of the isolated plasma membrane and adhering cortical granules and endoplasmic reticulum[29,30]; exocytosis in isolated cortices can be stimulated by raising Ca^{2+}.[29,31] In such preparations, an antibody against calmodulin inhibits calcium-induced exocytosis.[32] Recent studies have suggested a role for protein phosphorylation in the pathway by which Ca stimulates exocytosis,[33,34] and have indicated that protein components on the secretory granule membrane may be sufficient to cause fusion.[35,36] However, as in other cells, this process is not fully understood. Recent progress in identifying proteins that may be involved in Ca-mediated exocytosis at neuronal synapses and in other somatic cells[37] offers leads for future studies in eggs.

Exocytosis of cortical granules is followed by a period of endocytosis, which has been detected by either morphological or capacitance measurements, in sea urchin,[26,38] frog,[39,40] fish,[41] and mammalian[42] eggs. It is possible that the successive addition and subtraction of membrane to the egg surface may change its composition, which could be important in the activation of the egg to begin development. However, in sea urchin eggs in which the exocytosis is suppressed by high hydrostatic pressure (8000 psi), cleavage and development to gastrula occur normally, if measures are taken to avoid polyspermy.[43,44]

18.3. ELECTRICAL REGULATION OF SPERM–EGG FUSION AND ENTRY OF THE SPERM NUCLEUS INTO THE EGG CYTOPLASM

In sea urchins and starfish, many other marine animals, frogs, and the seaweed *Fucus,* the egg's membrane depolarizes at fertilization,[45–47] and this fertilization potential provides a fast electrical block to polyspermy.[47,48] The evidence that the fertilization potential prevents multiple sperm entries is that artificially depolarizing the egg's membrane by passing current inhibits sperm entry, and holding the egg's membrane potential negative induces polyspermy.[49,50] Several observations indicate that entry of multiple sperm is blocked at the step of sperm–egg fusion. Electron microscopy of eggs and sperm fixed during the fertilization potential shows multiple sperm that have contacted but have not fused with the egg plasma membrane.[51] Similarly, a fluorescent dye does not transfer from the cytoplasm of the egg to the cytoplasm of other sperm that are attached to the egg surface, in gametes fixed during the period of the electrical polyspermy block.[52] The absence of a capacitance increase during fertilization at positive potentials also supports the conclusion that sperm–egg fusion is blocked.[53] The probability of sperm–egg fusion is a graded function of the egg's membrane potential, becoming less probable as the voltage becomes more positive.[50]

In some species, the egg's membrane hyperpolarizes at fertilization. In crabs, the hyperpolarizing fertilization potential provides an electrical block to polyspermy; in this case sperm entry is inhibited when the egg's membrane potential is clamped at the *negative* level attained during the fertilization potential.[54] In fish[55] and mammals,[56–58] the hyperpolarization at fertilization does not block sperm entry.

The mechanisms by which the egg's membrane potential affects sperm fusion are only partly understood. In the marine worm *Urechis,* ion substitution and ion flux studies have shown that the block of sperm–egg fusion by positive potentials is related to the potential change itself and not to secondary effects on ion fluxes.[50] In a series of cross-fertilization studies involving marine and amphibian species, it has been found that the voltage-dependence of sperm–egg fusion depends on a property of the sperm rather than the egg.[59–61] These experiments have determined the voltage-dependence of sperm–egg fusion when the gametes come from different species that have different voltage sensitivity of fertilization. Some species, such as most salamanders, show voltage-independent fertilization, and among those species where fertilization is voltage-dependent, the magnitude of the voltage required for inhibition varies. Although fertilization between different species only rarely occurs in nature, it can sometimes occur readily under experimental conditions. In such crosses, it is always the voltage-dependent properties of the sperm species that determine the voltage-dependence of the sperm–egg interaction. For example, sperm from a voltage-independent salamander species will fuse with frog eggs clamped at +10 mV, even though frog sperm will be blocked at this potential.[60] Conversely, a cross between a voltage-dependent sperm species and a voltage-independent egg species is voltage-dependent.[61] These observations argue against the hypothesis that the voltage-dependence of fertilization results from a voltage-dependent component (e.g., receptor) in the egg membrane. They support the hypothesis that fertilization is voltage-dependent because the sperm membrane contains a charged component (fusion peptide?) that must insert in the egg membrane to initiate fusion. Species without voltage-dependent fertilization might have uncharged fusion peptides.

Not only sperm–egg fusion, but also the stimulation by the sperm of egg activation responses, such as cortical vesicle exocytosis and ion channel opening, shows voltage-dependence. Stimulation of these activation responses sometimes shows the same dependence on membrane potential as does sperm–egg fusion, but there are examples where the voltage-dependence differs for different fertilization events.[50,62–64]

In sea urchins, the fertilization potential not only serves as a block to polyspermy, but also functions to allow the nucleus of the fused sperm to move into the egg cytoplasm.[53,62,65] Normal fertilization involves a shift in potential from about –80 mV to about +20 mV. Sperm–egg fusion occurs normally at –80 mV, but entry of the sperm nucleus into the egg cytoplasm does not always proceed if the egg's membrane potential is prevented from shifting in a positive direction. The voltage must be –25 mV or more positive in order for sperm entry to occur in all eggs. At +20 mV, sperm–egg fusion cannot occur, but once fusion has occurred at –80 mV, entry of the sperm nucleus can proceed at +20 mV. Thus, normal fertilization of sea urchins depends on the fertilization potential in two ways, first to prevent the fusion of multiple sperm with the egg, and second to allow the nucleus of the one sperm that fuses to enter the egg cytoplasm. Recent evidence suggests that the failure of entry of the sperm nucleus at negative potentials is related to the effect of the voltage clamp on calcium influx.[66,67]

18.4. IONIC MECHANISMS OF THE FERTILIZATION POTENTIAL

The ion channels that open to produce the fertilization potential can be sodium or cation selective (most marine species[46,68,69]), chloride selective (frogs[70,71]), or potassium selective (fish,[41] frogs,[71] mammals,[56] crabs[54]). In most of the cases that have been studied, the channels are present but closed in the membrane of the unfertilized egg, and are caused to open by the rise in intracellular Ca at fertilization. Some eggs also have voltage-gated Na and/or Ca channels that amplify the egg's response to fertilization (see Ref. 46, 47). The conclusion that most of the channels open at fertilization are gated by Ca is based on observations in frog[72] and hamster[73] eggs that the Ca rise and potential change begin simultaneously, and on experimental injection of Ca or Ca buffers.[24,56,69,76] Buffers that maintain Ca at submicromolar levels suppress the opening of channels at fertilization, while buffers that raise Ca cause an opening of ion channels like that occurring at fertilization. In addition, the opening of channels occurs in a wave that crosses the egg surface with kinetics similar to the wave of increased free Ca.[71,75,75] As Ca falls back to a low level, the Ca-activated channels close, and the potential returns to its original level. In sea urchin eggs, the opening of K channels also contributes to the termination of the fertilization potential (see Ref. 46). The duration of the fertilization potential varies from several minutes to over an hour, depending on the species.

One exception to the generalization that Ca opens the channels responsible for the fertilization potential is found in the early phase of the fertilization potential in sea urchin. Voltage clamp experiments have shown that the initial conductance increase responsible for the fertilization potential occurs about 20–40 sec before a Ca rise is detected with fura-2.[77] Consistently, injection of 1 mM EGTA or 5 mM BAPTA does not prevent the initial conductance increase.[77] In contrast, the conductance increase underlying later phases of the fertilization potential is prevented by injection of EGTA or BAPTA.[77] These observations indicate that the initial conductance increase is not related to a rise in intracellular Ca.

One possible explanation of the initial phase of the fertilization potential in sea urchin eggs is that the initial conductance increase is caused by channels of the sperm membrane that become incorporated in the zygote membrane after fusion. This idea is supported by measurements showing that increases in conductance and capacitance in the patch of egg membrane at the site of sperm fusion occur simultaneously.[53] Electron microscopic and dye transfer observations indicate a delay of 5–10 sec between the rise of the fertilization potential and sperm–egg fusion, but this could be related to the methods of measurement.[52,78,78a] In any case, it is possible that fusion may occur simultaneously with the conductance increase, but may not cause it. One observation that cannot be simply explained by the idea that the channels that cause the initial phase of the fertilization potential are introduced from the sperm membrane is that the magnitude of the conductance seen when a sperm fuses with an immature oocyte is about half that when a sperm fuses with a mature egg.[79]

18.5. INTRACELLULAR FREE CALCIUM INCREASE AT FERTILIZATION

Fertilization causes a rise in intracellular free calcium; this was discovered in 1977 by Ridgway *et al.*[80] in fish and by Steinhardt *et al.*[81] in sea urchins, and subsequently has been seen in all species examined (see Refs. 82–85). The rise in Ca is usually related primarily to Ca release from intracellular stores, although Ca influx from the extracellular medium, through voltage-gated or possibly other Ca channels, is important in some species. Ca levels in unfertilized eggs are $\sim$100 nM; they rise to $\geq$1 μM at fertilization. The Ca rise occurs in a wave that starts from the point of contact and fusion of the successful sperm, and passes through the egg in $\sim$10 sec to 5 min, depending on the size of the egg (fish,[86] sea urchin,[87] frog,[72] hamster[88]). Ca then returns to resting levels over a period of minutes. In mammals and ascidians, there are several successive Ca transients, although the first rise is the largest (see Ref. 89).

The best established source of intracellular Ca release at fertilization is IP_3-induced Ca release from the endoplasmic reticulum. IP_3-mediated Ca release at fertilization is indicated by several observations. (1) IP_3 levels and phosphatidylinositol lipid turnover increase rapidly at fertilization (sea urchin,[90–93] frog[94]). (2) Injecting IP_3 into unfertilized eggs (sea urchin,[95,96] starfish,[97] frog,[98] hamster[99]; see also Ref. 84), or applying IP_3 to the endoplasmic reticulum of isolated sea urchin egg cortices,[100] causes Ca release. Injected IP_3 is effective at a concentration similar to the binding constant of the mammalian IP_3 receptor (~30 nM).[101] (3) Injection of an antibody against the IP_3 receptor blocks Ca release at fertilization (hamster[88]). Experiments with heparin, an inhibitor of IP_3-induced Ca release, also support a role for IP_3 in Ca release at fertilization, but because of heparin's nonspecificity and its competitive mode of action, these experiments must be interpreted cautiously (see Refs. 72, 102). In sea urchin eggs, ~100–600 μg/ml heparin causes a fourfold increase in the latent period between the rise of the fertilization potential and the rapid rise in intracellular Ca, slows the rate of wave propagation about twofold, and in some cases reduces the amplitude of the Ca rise somewhat.[103–106] 500–1000 μg/ml heparin or another IP_3 receptor inhibitor, pentosan polysulfate, slows the onset of Ca release and reduces its amplitude; in some experiments, Ca release is completely inhibited.[106a] In frog eggs, ~300 μg/ml heparin inhibits Ca release at fertilization.[72,106] In hamster eggs, 200–800 μg/ml heparin only partially inhibits Ca release at fertilization,[107] in contrast to the complete inhibition obtained with the IP_3 receptor antibody.[88]

A protein related to the mammalian ryanodine receptor is present in sea urchin eggs, as identified by an antibody against the skeletal muscle ryanodine receptor (~565 kDa) which recognizes an ~380-kDa protein localized in the egg cortex.[108,109] Extracellular application or intracellular injection of ryanodine can cause Ca release in sea urchin[104,105,110–112] and mouse[113] eggs. However, >50 μM ryanodine must be injected, while the binding constant of ryanodine for its receptor in muscle is ~10 nM[114]; even with injections of 200 μM ryanodine, the Ca release is variable and the average amplitude is less than at fertilization.[104] Injection of 50 μM ryanodine in frog eggs[72] or 25 μM ryanodine in hamster eggs[88] does not cause Ca release. In sea urchin eggs, it has been proposed that IP_3 receptors and ryanodine receptors both function in Ca release at fertilization.[106]

ADP ribose has also been found to release Ca in sea urchin eggs, and it has been proposed that this may be by way of a ryanodine receptor.[105,106,111,115,116] This, however, remains to be established, and it is also unknown whether there is a rise in ADP ribose at fertilization. Evidence that ADP ribose may function as a second messenger at fertilization is that coinjection of an antagonist of ADP ribose (8-amino-ADP ribose) and heparin inhibits egg activation at fertilization in sea urchins.[116] Although this result is suggestive, it will be important to test if injection of 8-amino-cADP ribose inhibits Ca release in response to injection of IP_3. cGMP has also been found to cause Ca release in sea urchin eggs,[117,118] although at a concentration ~500 × higher than that required for half-maximal activation of cGMP-dependent kinase.[119] It has been proposed that cGMP causes Ca release by stimulating the production of cADP ribose, but the reported cGMP stimulation of β-NAD^+ metabolism[118] is too slow to support this hypothesis. Whether cGMP rises at fertilization, and whether it functions in Ca release at fertilization, are unknown.

The return of the free Ca concentration to a low level after fertilization appears to result from resequestration of Ca into the endoplasmic reticulum, since injection of IP_3 into sea urchin eggs at 20 min after fertilization causes Ca release comparable to that at fertilization.[120] However, uptake into mitochondria may also occur, since the mitochondrial uncoupler FCCP (carbonyl cyanide-*p*-trifluormethoxy phenylhydrazone) causes Ca release when applied to sea urchin eggs after but not before fertilization.[121] In addition, measurements of the Ca content of sea urchin eggs before and after fertilization suggest the possibility that some Ca may be pumped out of the fertilized egg through its plasma membrane.[122,123]

As discussed above, the rise in intracellular Ca at fertilization causes cortical granule exocytosis and ion channel opening, and both of these events contribute to polyspermy prevention. Based on experimental manipulation of cytosolic Ca levels, it has been concluded that the rise in Ca also appears to be at least in part responsible for many subsequent developmental events: the resumption of meiosis in frog and mouse eggs,[24,25] the decondensation of sperm chromatin and pronuclear formation and migration in frog eggs,[24] as well as increases in NAD kinase activity, oxygen consumption, and DNA and protein synthesis in sea urchin eggs (see Ref. 83).

18.6. OTHER CHANGES IN EGG MEMBRANE TRANSPORT AT FERTILIZATION

In sea urchin eggs, the intracellular pH rises gradually from ~6.8 to ~7.2 during the first ~4 min after fertilization.[124,125] The pH change is accompanied by hydrogen ion efflux across the egg plasma membrane[126] and is dependent on extracellular Na, suggesting the involvement of a Na/H exchanger in the plasma membrane.[127] There is evidence that the pH change is stimulated either by a Ca/calmodulin-dependent pathway and/or by a protein kinase C-dependent pathway that is activated by the production of diacylglycerol at fertilization.[125,128] pH increases at fertilization also occur in frog eggs (0.3 pH unit)[129] and *Urechis* eggs (0.3 pH unit).[130] In *Urechis* eggs, the pH change is essential for the stimulation of germinal vesicle breakdown and the establishment of a permanent polyspermy block (see Ref. 130). The function of the pH change in sea urchin and frog eggs is not well understood (see Refs. 82,83,131).

Na-dependent transport of nucleosides and amino acids across the sea urchin egg plasma membrane also increases dramatically within 3 min of fertilization (see Ref. 132). The uptake of dissolved organic material from sea water appears to be important for embryonic growth.[133] The signal transduction pathways leading to the activation of these transport mechanisms are unknown.

As mentioned earlier, the sea urchin egg becomes more permeable to potassium at the termination of the fertilization potential; potassium permeability then remains high (see Ref. 46). The sea urchin egg plasma membrane also becomes more permeable to chloride[134] and phosphate (see Ref. 132) after fertilization, but the mechanism and significance of these increases are unknown.

18.7. SIGNAL TRANSDUCTION DURING SPERM INTERACTION WITH THE EGG PLASMA MEMBRANE

Current ideas about the "first messenger" by which a sperm activates an egg to begin development can be divided into hypotheses that propose that the signal results from contact of the sperm and egg membranes, and those that propose that it results from fusion of the membranes (see Refs. 84, 135–137). Contact hypotheses propose that there is a receptor for sperm in the egg plasma membrane that is linked to egg activation by pathways like those involved in activation of somatic cells of neurotransmitters, hormones, growth factors, antigens, or extracellular matrix components. Fusion hypotheses propose that the fusion of the sperm and egg membranes delivers an activating substance from the sperm cytoplasm to the egg cytoplasm (e.g., Ca, IP_3, cADP ribose, or cGMP), or from the sperm membrane to the egg membrane (e.g., a Ca channel, a G-protein, a protein that could activate a G-protein, a protein tyrosine kinase, or guanylate cyclase). At present, it is not possible to say which if either or both of these mechanisms operate at fertilization, and whether or not all species use the same mechanism.

Capacitance measurements[53] and morphological measurements[52,78] indicate that sperm–egg fusion in sea urchins occurs rapidly enough to be considered as a first messenger. By these means, fusion has been detected within 0–10 sec after the rise of the fertilization potential, which in turn occurs within <1–3 sec of sperm contact with the egg.[62,78] In favor of a fusion-mediated mechanism are experiments in sea urchins[138,139] and mammals[140,141,141a,141b] in which sperm extracts or whole sperm have been injected into eggs and shown to cause egg activation. Such experiments do not, however, truly mimic the delivery of sperm cytoplasm into the egg cytoplasm at fertilization, since an extract of sperm has a composition different from the free cytoplasm of an intact sperm. Organelle-bound substances, such as proteases in the acrosomal vesicle, are not under normal circumstances introduced into the free cytoplasm of the egg. Furthermore, control experiments have suggested the possibility of injection artifacts in the experiments with sea urchin eggs.[138]

A contact-mediated mechanism is consistent with data showing that externally applied proteases[142,143,143a] and lectins[144,145] can in certain cases cause egg activation. In particular, trypsin and chymotrypsin activate starfish eggs even in the absence of extracellular calcium, indicating that an extracellularly exposed egg protein is coupled to intracellular Ca release.[143a] This protein may be a receptor that transduces a signal from the sperm to initiate egg activation at fertilization. Also supporting a contact-mediated mechanism is evidence that a surface protein exposed during the acrosome reaction of sperm of the marine worm *Urechis* can cause egg activation.[63,146] Some caution is needed since detergents were used to isolate the protein; however, a small synthetic peptide corresponding to a six-amino-acid sequence of the protein can also cause egg activation.[147]

Recently, a series of experiments have addressed the question of whether somatic cell receptors, if introduced into an egg membrane, can initiate events like those occurring at fertilization. Positive results have been obtained with two G-protein-linked receptors (serotonin 1c and muscarinic m1) introduced into frog,[148,149] starfish,[102,150] and mouse[151,152] eggs, as well as two tyrosine kinase receptors (EGF and PDGF/FGF receptors) introduced into frog[153] and starfish[102] eggs, respectively. The G-protein-linked and tyrosine kinase receptors appear to be equivalently effective; both can elicit early events such as a calcium rise and exocytosis as well as later events such as DNA synthesis and cleavage. These experiments indicate that at least two signal transduction pathways are present in eggs (G-protein-linked and tyrosine kinase-linked). Whether the egg contains an endogenous receptor coupled to either or both of these pathways is unknown.

There is evidence, however, that sperm activate protein tyrosine kinases in eggs at fertilization. Several proteins in sea urchin eggs are phosphorylated on tyrosine by 15–60 sec after insemination.[154,155] Recently, a 350-kDa transmembrane glycoprotein of sea urchin eggs, which has been identified as a sperm receptor because it binds to the acrosomal region of sperm and because an antibody against it blocks fertilization,[156–159] has been found to be phosphorylated on tyrosine within 5 sec after fertilization.[160] This sperm receptor does not itself have sequences resembling a protein tyrosine kinase,[158] but it could conceivably activate a protein tyrosine kinase, that could phosphorylate PLC-γ (like $p56^{lck}$ or $p59^{fynT}$ in lymphocytes).[161–163] A possible candidate is a 220-kDa protein tyrosine kinase in the sea urchin egg cortex that is activated during the first few minutes post-insemination.[164] A study in which protein tyrosine kinase inhibitors were applied to sea urchin eggs indicates that Ca release may occur without protein tyrosine phosphorylation, but the occurrence of polyspermy in these experiments suggests that the inhibitors may slow Ca release.[164a] That the 350-kDa sperm receptor may function in egg activation is suggested by the observation that a small fraction (2–20%) of eggs were activated when an antibody against the receptor was applied to their surfaces.[157]

Whether G-proteins are activated at fertilization is unknown. However, in hamster eggs, injection of GDP-β-S inhibits the Ca increase at fertilization, but not the Ca increase in response to subsequent injection of IP_3.[99] GDP-β-S does not prevent Ca release at fertilization of sea urchin eggs, although it is unknown whether GDP-β-S may have an inhibitory effect on the amplitude of the Ca response or its latency.[103] If a G-protein is involved, it is of the pertussis toxin-insensitive class, since PTX does not inhibit egg activation in starfish,[165] frog,[149] or mouse.[151]

One additional suggestion that egg activation at fertilization might be mediated by contact of the sperm with an egg membrane receptor comes from recent studies showing that binding of sperm to mammalian eggs is mediated by an integrin.[15a] Integrins can be coupled to activation events in other cells.[166]

It may be that multiple membrane molecules function in the process of egg activation by the sperm, as occurs in activation of lymphocytes by the T-cell receptor complex.[167] The contact area between the sperm and egg membranes is large in molecular terms; the acrosomal process of sea urchin sperm is ~0.1 μm in diameter, and sites on mammalian sperm that contact the egg are even larger (see References in Section 18.2).

18.8. STRUCTURAL CHANGES IN THE EGG'S ENDOPLASMIC RETICULUM AT FERTILIZATION

As described above, the ER is very probably the principal intracellular source of Ca at fertilization. The ER in sea urchin and starfish eggs has been labeled by injection of an oil drop saturated with the long-chain dicarbocyanine dye DiI and observed by confocal microscopy.[168–170] Recently, the ER has also been labeled by injection of RNA encoding a green fluorescent protein (GFP) targetted to the ER lumen.[170a] These studies show that the ER is one continuous membrane throughout the egg cytoplasm, including a tubular network in the cortex, an array of large irregularly oriented cisternae in the interior, and the nuclear envelope. The cortical ER tubules surround the cortical granules, and the interior cisternae (also known as lamellae or sheets) surround the numerous yolk platelets.

At fertilization, the ER is altered such that the ER sheets become smaller. This change occurs in a wave corresponding well with the Ca wave, and it continues for 1–2 min. Then, starting at a few minutes postfertilization, the ER structure returns gradually to the prefertilization appearance; the recovery is complete by about 10 min after fertilization. Accompanying these changes in the form of the interior ER cisternae, the cortical ER also undergoes a transient disruption (see also Ref. 171). The time course of the changes in ER structure is similar to the time course of the rise and fall of intracellular free Ca, and artificially raising Ca by injection of IP_3 causes similar structural changes.[170] These observations suggest that the Ca rise and fall in the cytosol, and/or release and uptake of Ca in the ER lumen, may cause the structural changes in the ER.

Attempts to characterize these changes in ER structure precisely have not succeeded, because the dimensions of the ER are close to the resolution limit of the light microscope, and because the ER is moving. However, observations of the rate of spreading of DiI in eggs before and during fertilization, as well as observations of fluorescence recovery after photobleaching of GFP and DiI, indicate that the ER becomes transiently discontinuous in the first few minutes after fertilization.[169,170,170a] Presumably this results from membrane fusion.

At the same time as the ER throughout the egg is breaking down and then re-forming, the nuclear envelope of the sperm that has just entered the egg cytoplasm also breaks down and then re-forms (see Ref. 172). This is necessary because the sperm chromatin decondenses and consequently the sperm nucleus enlarges. Since the nuclear envelope is part of the ER, it may be that the general ER breakdown and sperm nuclear envelope breakdown are closely related. Interestingly, the envelope surrounding the egg nucleus does not break down during this period.

Before fertilization and throughout the period of breakdown and re-formation, the sea urchin egg's ER is uniformly distributed throughout the cytoplasm. By 6 min, however, a radial array of microtubules begins to assemble around the centrosome brought into the egg by the sperm, and the ER membrane moves to accumulate within the microtubule array.[168,173] As this "sperm aster," consisting of microtubules, ER, and the male pronucleus and centriole, enlarges, it moves toward the center of the egg. During this time the microtubules are lengthening, and come to contact the female pronucleus. Like the ER throughout the egg, the female pronuclear envelope is drawn to the center of the microtubule array where it contacts and fuses with the male pronuclear envelope at ~20 min after fertilization. By this time, the ER is highly concentrated in the center of the egg and remains there even after the microtubule array disassembles, throughout the ensuing period of rapid cell division.[168,171,174]

The functional importance of the transient breakdown of ER structure at fertilization and the subsequent movement of the ER to the egg center are unknown. Speculations include the possibility that the breakdown of the ER partitions facilitates entry of the sperm into the egg cytoplasm, and that the cellwide ER breakdown and re-formation are a mechanism that ensures that wherever the sperm nucleus enters the egg cytoplasm, its envelope will be stimulated to break down and re-form. A possible role in the activation of the egg's metabolism and the initiation of the cell cycle at fertilization might also be imagined as functions for these dramatic changes in ER structure. This question as well as the generality of ER breakdown at fertilization in other species remain to be examined.

18.9. DEVELOPMENT OF EGG MEMBRANES DURING OOCYTE MATURATION

In response to hormonal stimuli, a fully grown immature oocyte becomes a mature fertilizable egg (see Ref. 175). Immature oocytes of many species, such as starfish,[176] frogs,[177] and mammals[13,13a] bind sperm and undergo fusion with the sperm plasma membrane. However, the ability to undergo Ca release develops during oocyte maturation in starfish[97] and hamsters.[178] Indirect evidence indicates that Ca release mechanisms develop during frog oocyte maturation as well.[179] Maturation of the oocyte membranes in preparation for undergoing Ca release at fertilization involves an increase in the ability to release Ca in response to IP_3.[97,178,180] Most of the IP_3-sensitive Ca stores are already present in the immature oocytes,[97,178,180] but more IP_3 is required to obtain equivalent Ca release.

Maturation-inducing hormones cause changes in ER membrane structure that may be related to the increase in sensitivity to sperm- or IP_3-induced Ca release. During frog oocyte maturation, the amount of cortical ER membrane increases; this membrane is derived from the annulate lamellae of the immature oocyte subcortex.[181–183] The ER cisternae come to surround the cortical granules, in an appropriate position for releasing Ca to cause exocytosis. In mouse oocyte maturation[184] and in late stages of oogenesis in sea urchin oocytes,[185] there is a similar increase in the amount of cortical ER. In both mouse and hamster oocytes, maturation causes the appearance of a highly organized array of cortical ER clusters.[185a,185b] In starfish oocytes, the form of the ER in the cell interior changes during maturation, to include membranes in the shape of incomplete spherical shells.[170]

The ability to undergo cortical granule exocytosis at fertilization also develops during oocyte maturation (starfish,[97,176] frog and toad,[177,181,186] mouse[187]). This is related in large part to the development of the ability to release Ca. In starfish but not frog, there also appears to be an increase in the Ca sensitivity of exocytosis.[97,179] The cortical granules are in place adjacent to the plasma membrane in the fully grown immature oocyte (starfish,[188] frog,[181–183] mammal[189]), but the transduction mechanisms that allow exocytosis to be stimulated are not yet developed.

The oocyte plasma membrane also matures in preparation for fertilization. A common feature among sea urchins, starfish, and frogs is a decrease in membrane conductance [sea urchin,[79] starfish and frogs (see Ref. 190)]. Since these species prevent polyspermy by a change in the egg's membrane potential, the conductance decrease serves the function of allowing the conductance increases that occur at fertilization to have a larger effect on the membrane potential (see urchin,[79] starfish,[191,192] frog and toad[193,194]). The decrease in membrane conductance is sometimes, but not always, accompanied by a decrease in the surface area of the oocyte plasma membrane [starfish (see Ref. 190), frog[195]]. Oocyte maturation in frog is accompanied by a dramatic increase in expression of the α_q family of G-proteins, which function in stimulation of phospholipase C.[195a] This may be significant in preparing the oocyte to release Ca at fertilization.

Fertilization before the completion of oocyte maturation results in polyspermy (and failure of embryogenesis), because ER and plasma membrane transport mechanisms are not fully developed. The occurrence of polyspermy in human *in vitro* fertilization, if eggs are inseminated before they are fully mature,[196] might be related to such "membrane transport disorders," particularly the presence of less cortical ER and less release of calcium from the ER.

18.10. SUMMARY

Fertilization involves a complex sequence of membrane fusion events: fusion between the sperm and egg plasma membranes, exocytosis, endocytosis, fusion of ER membranes to cause the ER to become transiently discontinuous and to allow enlargement of the male pronucleus, and finally fusion of the male and female pronuclear envelopes to allow combination of the chromosomes. Closely interrelated with the fusion events are membrane signaling and ion transport events. By an incompletely understood pathway, the sperm causes Ca to be released from the ER, and the Ca in turn opens ion channels in the plasma membrane, causing a potential change that in many species is essential for polyspermy prevention. The Ca rise also causes exocytosis, and other developmental events. In addition, fertilization causes the egg's pH to rise and the activity of many other membrane transport processes to increase. The egg prepares for these once-in-a-lifetime events of fertilization during the period of oocyte maturation, tuning up its calcium release and other machinery.

ACKNOWLEDGMENTS. I thank Diana Myles, Mark Terasaki, Nick Cross, David Carroll, Fraser Shilling, and Chris Gallo for their valuable comments on the manuscript, and the NIH and NSF for their financial support.

REFERENCES

1. Ward, C. R., and Kopf, G. S. (1993). Molecular events mediating sperm activation. *Dev. Biol.* **158:**9–34.
2. Colwin, L. H., and Colwin, A. L. (1967). Membrane fusion in relation to sperm–egg association. In *Fertilization* (C. B. Metz and A. Monroy, eds.), Academic Press, New York, Volume 1, pp. 295–367.
3. Summers, R. G., and Hylander, B. L. (1974). An ultrastructural analysis of early fertilization in the sand dollar, *Echinarachnius parma. Cell Tissue Res.* **150:**343–368.
4. Yanagimachi, R. (1988). Sperm–egg fusion. *Curr. Top. Membr. Transp.* **32:**3–43.
5. Myles, D. G. (1993). Molecular mechanisms of sperm–egg membrane binding and fusion in mammals. *Dev. Biol.* **158:**35–45.
6. Elinson, R. P. (1975). Site of sperm entry and a cortical contraction associated with egg activation in the frog *Rana pipiens. Dev. Biol.* **47:**257–268.
7. Freeman, G., and Miller, R. L. (1982). Hydrozoan eggs can only be fertilized at the site of polar body formation. *Dev. Biol.* **94:**142–152.
8. Speksnijder, J. E., Jaffe, L. F., and Sardet, C. (1989). Polarity of sperm entry in the ascidian egg. *Dev. Biol.* **133:**180–184.
9. Ebensperger, C., and Barros, C. (1984). Changes at the hamster oocyte surface from the germinal vesicle stage to ovulation. *Gamete Res.* **9:**387–397.
10. Blobel, C. P., Wolfsberg, T. G., Turck, C. W., Myles, D. G., Primakoff, P., and White, J. M. (1992). A potential fusion peptide and an integrin ligand domain in a protein active in sperm–egg fusion. *Nature* **356:**248–252.
11. Primakoff, P., Hyatt, H., and Tredick-Kline, J. (1987). Identification and purification of a sperm surface protein with a potential role in sperm–egg membrane fusion. *J. Cell Biol.* **104:**141–149.

12. Myles, D. G., Kimmel, L. H., Blobel, C. P., White, J. M., and Primakoff, P. (1994). Identification of a binding site in the disintegrin domain of fertilin required for sperm–egg fusion. *Proc. Natl. Acad. Sci. USA* **91:**4195–4198.
12a. Wolfsberg, T. G., Straight, P. D., Gerena, R. L., Huovila, A. P. J., Primakoff, P., Myles, D. G., and White, J. M. (1995). ADAM, a widely distributed and developmentally regulated gene family encoding membrane proteins with a disintegrin and metalloprotease domain. *Dev. Biol.* **169:**378–383.
13. Zuccotti, M., Yanagimachi, R., and Yanagimachi, H. (1991). The ability of hamster oolemma to fuse with spermatozoa: Its acquisition during oogenesis and loss after fertilization. *Development* **112:**143–152.
13a. Zuccotti, M., Piccinelli, A., Marziliano, N., Mascheretti, S., and Redi, C. A. (1994). Development and loss of the ability of mouse oolemma to fuse with spermatozoa. *Zygote* **2:**333–339.
14. Ponce, R. H., Yanagimachi, R., Urch, U. A., Yamagata, T., and Ito, M. (1993). Retention of hamster oolemma fusibility after various enzyme treatments: A search for the molecules involved in sperm–egg fusion. *Zygote* **1:**163–171.
15. Tarone, G., Russo, M. A., Hirsch, E., Odorisio, T., Altruda, F., Silengo, L., and Siracusa, G. (1993). Expression of β1 integrin complexes on the surface of unfertilized mouse oocyte. *Development* **117:**1369–1375.
15a. Almeida, E. A. C., Huovila, A. P. J., Sutherland, A. E., Stephens, L. E., Calarco, P. G., Shaw, L. M., Mercurio, A. M., Sonnenberg, A., Primakoff, P., Myles, D. G., and White, J. M. (1995). Mouse egg integrin a6b1 functions as a sperm receptor. *Cell* **81:**1095–1104.
15b. Muga, A., Neugebauer, W., Hirama, T., and Surewicz, W. K. (1994). Membrane interactive and conformational properties of the putative fusion peptide of PH-30, a protein active in sperm–egg fusion. *Biochemistry* **33:**4444–4448.
16. Glabe, C. G. (1985). Interaction of the sperm adhesive protein, bindin, with phospholipid vesicles. II. Bindin induces the fusion of mixed-phase vesicles that contain phosphatidylcholine and phosphatidylserine in vitro. *J. Cell Biol.* **100:**800–806.
17. Miraglia, S. J., Craddock, A. G., and Glabe, C. G. (1993). The sea urchin acrosomal adhesive protein, bindin, induces the rapid fusion of sphingomyelin-cholesterol vesicles. *Mol. Biol. Cell* **4:**205a.
18. Hong, K., and Vacquier, V. D. (1986). Fusion of liposomes induced by a cationic protein from the acrosome granule of abalone spermatozoa. *Biochemistry* **25:**543–549.
19. Schuel, H. (1985). Functions of egg cortical granules. In *Biology of Fertilization* (C. B. Metz and A. Monroy, eds.), Academic Press, Orlando, Volume 3, pp. 1–43.
20. Jaffe, L. A., and Gould, M. (1985). Polyspermy-preventing mechanisms. In *Biology of Fertilization* (C. B. Metz and A. Monroy, eds.), Academic Press, Orlando, Volume 3, pp. 223–250.
21. Wassarman, P. M. (1993). Mammalian eggs, sperm and fertilisation: Dissimilar cells with a common goal. *Semin. Dev. Biol.* **4:** 189–197.
22. Zucker, R. S., and Steinhardt, R. A. (1978). Prevention of the cortical reaction in fertilized sea urchin eggs by injection of calcium-chelating ligands. *Biochim. Biophys. Acta* **541:**459–466.
23. Hamaguchi, Y., and Hiramoto, Y. (1981). Activation of sea urchin eggs by microinjection of calcium buffers. *Exp. Cell Res.* **134:** 171–179.
24. Kline, D. (1988). Calcium-dependent events at fertilization of the frog egg: Injection of a calcium buffer blocks ion channel opening, exocytosis, and formation of pronuclei. *Dev. Biol.* **126:**346–361.
25. Kline, D., and Kline, J. T. (1992). Repetitive calcium transients and the role of calcium in exocytosis and cell cycle activation in the mouse egg. *Dev. Biol.* **149:**80–89.
26. Chandler, D. E., and Heuser, J. (1979). Membrane fusion during secretion: Cortical granule exocytosis in sea urchin eggs as studied by quick-freezing and freeze-fracture. *J. Cell Biol.* **83:**91–108.
26a. Terasaki, M. (1995). Visualization of exocytosis during sea urchin egg fertilization using confocal microscopy. *J. Cell Sci.* **108:**2293–2300.
27. Schroeder, T. E. (1979). Surface area change at fertilization: Resorption of the mosaic membrane. *Dev. Biol.* **70:**306–326.
28. Jaffe, L. A., Hagiwara, S., and Kado, R. T. (1978). The time course of critical vesicle fusion in sea urchin eggs observed as membrane capacitance changes. *Dev. Biol.* **67:**243–248.
29. Vacquier, V. D. (1975). The isolation of intact cortical granules from sea urchin eggs: Calcium ions trigger granule discharge. *Dev. Biol.* **43:**62–74.
30. Terasaki, M., Henson, J., Begg., D., Kaminer, B., and Sardet, C. (1991). Characterization of sea urchin egg endoplasmic reticulum in cortical preparations. *Dev. Biol.* **148:**398–401.
31. Whitaker, M. J., and Baker, P. F. (1983). Calcium-dependent exocytosis in an *in vitro* secretory granule plasma membrane preparation from sea urchin eggs and the effects of some inhibitors of cytoskeletal function. *Proc. R. Soc. London Ser. B* **218:**397–413.
32. Steinhardt, R. A., and Alderton, J. M. (1982). Calmodulin confers calcium sensitivity on secretory exocytosis. *Nature* **295:**154–155.
33. Bement, W. M., and Capco, D. G. (1990). Protein kinase C acts downstream of calcium at entry into the first mitotic interphase of *Xenopus laevis*. *Cell. Regul.* **1:**315–326.
34. Whalley, T., Crossley, I., and Whitaker, M. (1991). Phosphoprotein inhibition of calcium-stimulated exocytosis in sea urchin eggs. *J. Cell Biol.* **113:**769–778.
35. Vogel, S. S., and Zimmerberg, J. (1992). Proteins on exocytic vesicles mediate calcium-triggered fusion. *Proc. Natl. Acad. Sci. USA* **89:**4749–4753.
36. Vogel, S. S., Chernomordik, L. V., and Zimmerberg, J. (1992). Calcium-triggered fusion of exocytotic granules requires proteins in only one membrane. *J. Biol. Chem.* **267:**25640–25643.
37. Sudhof, T. C. (1995). The synaptic vesicle cycle: a cascade of protein-protein interactions. *Nature* **375:**645–653.
38. Fisher, G. W., and Rebhun, L. I. (1983). Sea urchin egg cortical granule exocytosis is followed by a burst of membrane retrieval via uptake into coated vesicles. *Dev. Biol.* **99:**456–472.
39. Jaffe, L. A., and Schlichter, L. C. (1985). Fertilization-induced ionic conductances in eggs of the frog, *Rana pipiens*. *J. Physiol. (London)* **358:**299–319.
40. Bernardini, G., Ferraguti, M., and Peres, A. (1986). The decrease of *Xenopus* egg membrane capacity during activation might be due to endocytosis. *Gamete Res.* **14:**123–127.
41. Nuccitelli, R. (1980). The electrical changes accompanying fertilization and cortical vesicle secretion in the medaka egg. *Dev. Biol.* **76:**483–498.
42. Kline, D., and Stewart-Savage, J. (1994). The timing of cortical granule fusion, contents dispersal and endocytosis during fertilization of the hamster egg: An electrophysiological and histochemical study. *Dev. Biol.* **162:**277–287.
43. Chase, D. G. (1967). Inhibition of the cortical reaction with high hydrostatic pressure and its effects on the fertilization and early development of sea urchin eggs. Thesis, University of Washington.
44. Fisher, G. W., Summers, R. G., and Rebhun, L. I. (1985). Analysis of sea urchin egg cortical transformation in the absence of cortical granule exocytosis. *Dev. Biol.* **109:**489–503.
45. Tyler, A., Monroy, A., Kao, C., Y., and Grundfest, H. (1956). Membrane potential and resistance of the starfish egg before and after fertilization. *Biol. Bull.* **111:**153–177.
46. Hagiwara, S., and Jaffe, L. A. (1979). Electrical properties of egg cell membranes. *Annu. Rev. Biophys. Bioeng.* **8:**385–416.

47. Kline, D. (1991). Electrical characteristics of oocytes and eggs. *Curr. Top. Membr.* **39:**89–120.
48. Jaffe, L. A., and Cross, N. L. (1986). Electrical regulation of sperm–egg fusion. *Annu. Rev. Physiol.* **48:**191–200.
49. Jaffe, L. A. (1976). Fast block to polyspermy in sea urchin eggs is electrically mediated. *Nature* **261:**68–71.
50. Gould-Somero, M., Jaffe, L. A., and Holland, L. Z. (1979). Electrically mediated fast polyspermy block in eggs of the marine worm, *Urechis caupo. J. Cell Biol.* **82:**426–440.
51. Paul, M., and Gould-Somero, M. (1976). Evidence for a polyspermy block at the level of sperm–egg plasma membrane fusion in *Urechis caupo. J. Exp. Zool.* **196:**105–112.
52. Hinkley, R. E., Wright, B. D., and Lynn, J. W. (1986). Rapid visual detection of sperm–egg fusion using the DNA-specific fluorochrome Hoechst 33342. *Dev. Biol.* **118:**148–154.
53. McCulloh, D. H., and Chambers, E. L. (1992). Fusion of membranes during fertilization. Increases of the sea urchin egg's membrane capacitance and membrane conductance at the site of contact with the sperm. *J. Gen. Physiol.* **99:**137–175.
54. Goudeau, H., and Goudeau, M. (1989). A long-lasting electrically mediated block, due to the egg membrane hyperpolarization at fertilization, ensures physiological monospermy in eggs of the crab *Maia squinado. Dev. Biol.* **133:**348–360.
55. Nuccitelli, R. (1980). The fertilization potential is not necessary for the block to polyspermy or the activation of development in the medaka egg. *Dev. Biol.* **76:**499–504.
56. Miyazaki, S., and Igusa, Y. (1982). Ca-mediated activation of a K current at fertilization of golden hamster eggs. *Proc. Natl. Acad. Sci. USA* **79:**931–935.
57. Jaffe, L. A., Sharp, A. P., and Wolf, D. P. (1983). Absence of an electrical polyspermy block in the mouse. *Dev. Biol.* **96:**317–323.
58. McCulloh, D. H., Rexroad, C. E., and Levitan, H. (1983). Insemination of rabbit eggs is associated with slow depolarization and repetitive diphasic membrane potentials. *Dev. Biol.* **95:**372–377.
59. Jaffe, L. A., Gould-Somero, M., and Holland, L. Z. (1982). Studies of the mechanism of the electrical polyspermy block using voltage clamp during cross-species fertilization. *J. Cell Biol.* **92:**616–621.
60. Jaffe, L. A., Cross, N. L., and Picheral, B. (1983). Studies of the voltage-dependent polyspermy block using cross-species fertilization of amphibians. *Dev. Biol.* **98:**319–326.
61. Iwao, Y., and Jaffe, L. A. (1989). Evidence that the voltage-dependent component in the fertilization process is contributed by the sperm. *Dev. Biol.* **134:**446–451.
62. Lynn, J. W., and Chambers, E. L. (1984). Voltage clamp studies of fertilization in sea urchin eggs. *Dev. Biol.* **102:**98–109.
63. Gould, M., and Stephano, J. L. (1987). Electrical responses of eggs to acrosomal protein similar to those induced by sperm. *Science* **235:**1654–1656.
64. Kobayashi, W., Baba, Y., Shimozawa, T., and Yamamoto, T. S. (1994). The fertilization potential provides a fast block to polyspermy in lamprey eggs. *Dev. Biol.* **161:**552–562.
65. Lynn, J. W., McCulloh, D. H., and Chambers, E. L. (1988). Voltage clamp studies of fertilization in sea urchin eggs. II. Current patterns in relation to sperm entry, nonentry, and activation. *Dev. Biol.* **128:**305–323.
66. McCulloh, D. H., Ivonnet, P. I., and Chambers, E. L. (1989). Blockers of calcium influx promote sperm entry in sea urchin eggs at clamped negative membrane potentials. *J. Cell Biol.* **109:**126a.
67. McCulloh, D. H., Ivonnet, P. I., and Chambers, E. L. (1990). Microinjection of a Ca^{2+} chelator, EGTA or BAPTA, promotes sperm entry in sea urchin eggs clamped at negative membrane potentials. *J. Cell Biol.* **111:**113a.
68. Gould-Somero, M. (1981). Localized gating of egg Na^+ channels by sperm. *Nature* **291:**254–256.
69. Kline, D., Jaffe, L. A., and Kado, R. T. (1986). A calcium-activated sodium conductance contributes to the fertilization potential in the egg of the nemertean worm *Cerebratulus lacteus. Dev. Biol.* **117:** 84–193.
70. Cross, N. L., and Elinson, R. P. (1980). A fast block to polyspermy in frogs mediated by changes in the membrane potential. *Dev. Biol.* **75:**187–198.
71. Jaffe, L. A., Kado, R. T., and Muncy, L. (1985). Propagating potassium and chloride conductances during activation and fertilization of the egg of the frog, *Rana pipiens. J. Physiol. (London)* **368:** 227–242.
72. Nuccitelli, R., Yim, D. L., and Smart, T. (1993). The sperm-induced Ca^{2+} wave following fertilization of the *Xenopus* egg requires the production of Ins(1,4,5)P_3. *Dev. Biol.* **158:**200–212.
73. Igusa, Y., and Miyazaki, S. (1986). Periodic increase of cytoplasmic free calcium in fertilized hamster eggs measured with calcium-sensitive electrodes. *J. Physiol. (London)* **377:**193–205.
74. Cross, N. L. (1981). Initiation of the activation potential by an increase in intracellular calcium in eggs of the frog, *Rana pipiens. Dev. Biol.* **85:**380–384.
75. Kline, D., and Nuccitelli, R. (1985). The wave of activation current in the *Xenopus* egg. *Dev. Biol.* **111:**471–487.
76. McCulloh, D. H., and Chambers, E. L. (1991). A localized zone of increased conductance progresses over the surface of the sea urchin egg during fertilization. *J. Gen. Physiol.* **97:**579–604.
77. Swann, K., McCulloh, D. H., McDougall, A., Chambers, E. L., and Whitaker, M. (1992). Sperm-induced currents at fertilization in sea urchin eggs injected with EGTA and neomycin. *Dev. Biol.* **151:** 552–563.
78. Longo, F. J., Lynn, J. W., McCulloh, D. H., and Chambers, E. L. (1986). Correlative ultrastructural and electrophysiological studies of sperm–egg interactions of the sea urchin, *Lytechinus variegatus. Dev. Biol.* **118:**155–166.
78a. Longo, F. J., Cook, S., McCulloh, D. H., Ivonnet, P. I., and Chambers, E. L. (1994). Stages leading to and following fusion of sperm and egg plasma membranes. *Zygote* **2:**317–331.
79. McCulloh, D. H., Lynn, J. W., and Chambers, E. L. (1987). Membrane depolarization facilitates sperm entry, large fertilization cone formation, and prolonged current responses in sea urchin oocytes. *Dev. Biol.* **124:**177–190.
80. Ridgway, E. B., Gilkey, J. C., and Jaffe, L. F. (1977). Free calcium increases explosively in activating medaka eggs. *Proc. Natl. Acad. Sci. USA* **74:**623–627.
81. Steinhardt, R., Zucker, R., and Schatten, G. (1977). Intracellular calcium release at fertilization in the sea urchin egg. *Dev. Biol.* **58:** 185–196.
82. Jaffe, L. F. (1985). The role of calcium explosions, waves, and pulses in activating eggs. In *Biology of Fertilization* (C. B. Metz and A. Monroy, eds.), Academic Press, Orlando, Volume 3, pp. 127–165.
83. Whitaker, M. J., and Steinhardt, R. A. (1985). Ionic signalling in the sea urchin egg at fertilization. In *Biology of Fertilization* (C. B. Metz and A. Monroy, eds.), Academic Press, Orlando, Volume 3, pp. 167–221.
84. Nuccitelli, R. (1991). How do sperm activate eggs? *Curr. Top. Dev. Biol.* **25:**1–16.
85. Deguchi, R., and Osanai, K. (1994). Repetitive intracellular Ca^{2+} increases at fertilization and the role of Ca^{2+} in meiosis reinitiation from the first metaphase in oocytes of marine bivalves. *Dev. Biol.* **163:**162–174.
86. Gilkey, J. C., Jaffe, L. F., Ridgway, E. B., and Reynolds, G. T. (1978). A free calcium wave transverses the activating egg of the medaka, *Oryzias latipes. J. Cell Biol.* **76:**448–466.
87. Hafner, M., Petzelt, C., Nobiling, R., Pauley, J. B., Kramp, D., and Schatten, G. (1988). Wave of free calcium at fertilization in the sea

urchin egg visualized with fura-2. *Cell Motil. Cytoskel.* **9**:271–277.
88. Miyazaki, S.-I., Yuzaki, M., Nakada, K., Shirakawa, H., Nakanishi, S., Nakade, S., and Mikoshiba, K. (1992). Block of Ca^{2+} wave and Ca^{2+} oscillation by antibody to the inositol 1,4,5-trisphosphate receptor in fertilized hamster eggs. *Science* **257**:251–255.
89. Miyazaki, S., Shirakawa, H., Nakada, K., and Honda, Y. (1993). Essential role of the inositol 1,4,5-trisphosphate receptor/Ca^{2+} release channel in Ca^{2+} waves and Ca^{2+} oscillations at fertilization of mammalian eggs. *Dev. Biol.* **158**:62–78.
90. Turner, P. R., Sheetz, M. P., and Jaffe, L. A. (1984). Fertilization increases the polyphosphoinositide content of sea urchin eggs. *Nature* **310**:414–415.
91. Kamel, L. C., Bailey, J., Schoenbaum, L., and Kinsey, W. (1985). Phosphatidylinositol metabolism during fertilization in the sea urchin egg. *Lipids* **20**:350–356.
92. Ciapa, B., and Whitaker, M. (1986). Two phases of inositol polyphosphate and diacylglycerol production at fertilisation. *FEBS Lett.* **195**:347–351.
93. Ciapa, B., Borg, B., and Whitaker, M. (1992). Polyphosphoinositide metabolism during the fertilization wave in sea urchin eggs. *Development* **115**:187–195.
94. Stith, B. J., Goalstone, M., Silva, S., and Jaynes, C. (1993). Inositol 1,4,5-trisphosphate mass changes from fertilization through first cleavage in *Xenopus laevis*. *Mol. Biol. Cell* **4**:435–443.
95. Whitaker, M., and Irvine, R. F. (1984). Inositol 1,4,5-trisphosphate microinjection activates sea urchin eggs. *Nature* **312**:636–639.
96. Swann, K., and Whitaker, M. (1986). The part played by inositol trisphosphate and calcium in the propagation of the fertilization wave in sea urchin eggs. *J. Cell Biol.* **103**:2333–2342.
97. Chiba, K., Kado, R. T., and Jaffe, L. A. (1990). Development of calcium release mechanisms during starfish oocyte maturation. *Dev. Biol.* **140**:300–306.
98. Busa, W. B., Ferguson, J. E., Joseph, S. K., Williamson, J. R., and Nuccitelli, R. (1985). Activation of frog (*Xenopus laevis*) eggs by inositol trisphosphate. I. Characterization of Ca^{2+} release from intracellular stores. *J. Cell Biol.* **101**:677–682.
99. Miyazaki, S. (1988). Inositol 1,4,5-trisphosphate-induced calcium release and guanine nucleotide-binding protein-mediated periodic calcium rises in golden hamster eggs. *J. Cell Biol.* **106**:345–353.
100. Terasaki, M., and Sardet, C. (1991). Demonstration of calcium uptake and release by sea urchin egg cortical endoplasmic reticulum. *J. Cell Biol.* **115**:1031–1037.
101. Mignery, G. A., Johnston, P. A., and Sudhof, T. C. (1992). Mechanism of Ca^{2+} inhibition of inositol 1,4,5-trisphosphate ($InsP_3$) binding to the cerebellar $InsP_3$ receptor. *J. Biol. Chem.* **267**:7450–7455.
102. Shilling, F. M., Carroll, D. J., Muslin, A. J., Escobedo, J. A., Williams, L. T., and Jaffe, L. A. (1994). Evidence for both tyrosine kinase and G-protein coupled pathways leading to starfish egg activation. *Dev. Biol.* **162**:590–599.
103. Crossley, I., Whalley, T., and Whitaker, M. (1991). Guanosine 5′-thiotriphosphate may stimulate phosphoinositide messenger production in sea urchin eggs by a different route than the fertilizing sperm. *Cell Regul.* **2**:121–133.
104. Buck, W. R., Rakow, T. L., and Shen, S. S. (1992). Synergistic release of calcium in sea urchin eggs by caffeine and ryanodine. *Exp. Cell Res.* **202**:59–66.
105. Shen, S. S., and Buck, W. R. (1993). Sources of calcium in sea urchin eggs during the fertilization response. *Dev. Biol.* **157**:157–169.
106. Galione, A., McDougall, A., Busa, W. B., Willmott, N., Gillot, I., and Whitaker, M. (1993). Redundant mechanisms of calcium-induced calcium release underlying calcium waves during fertilization of sea urchin eggs. *Science* **261**:348–352.
106a. Mohri, T., and Ivonnet, P. I., Chambers, E. L. (1995). Effect on sperm-induced activation current and increase of cytosolic Ca^{2+} of agents that modify the mobilization of $[Ca^{2+}]_i$ I. Heparin and pentosan polysulfate. *Dev. Biol.* **172**:139–157.
107. Miyazaki, S., Nakada, K., and Shirakawa, H. (1993). Signal transduction of gamete interaction and intracellular calcium release mechanism at fertilization of mammalian eggs. In *Biology of the Germ Line. In Animals and Man* (H. Mohri, M. Takahashi, and C. Tachi, eds.), Japan Scientific Societies Press, Tokyo, pp. 125–143.
108. McPherson, S. M., McPherson, P. S., Mathews, L., Campbell, K. P., and Congo, F. J. (1992). Cortical localization of a calcium release channel in sea urchin eggs. *J. Cell Biol.* **116**:1111–1121.
109. McPherson, P. S., and Campbell, K. P. (1993). The ryanodine receptor/Ca^{2+} release channel. *J. Biol. Chem.* **268**:13765–13768.
110. Fujiwara, A., Taguchi, K., and Yasumasu, I. (1990). Fertilization membrane formation in sea urchin eggs induced by drugs known to cause Ca^{2+} release from isolated sarcoplasmic reticulum. *Dev. Growth Differ.* **32**:303–314.
111. Galione, A., Lee, H. C., and Busa, W. B. (1991). Ca^{2+}-induced Ca^{2+} release in sea urchin egg homogenates: Modulation by cyclic ADP-ribose. *Science* **253**:1143–1146.
112. Sardet, C., Gillot, I., Ruscher, A., Payan, P., Girard, J.-P., and de Renzis, G. (1992). Ryanodine activates sea urchin eggs. *Dev. Growth Differ.* **34**:37–42.
113. Swann, K. (1992). Different triggers for calcium oscillations in mouse eggs involve a ryanodine-sensitive calcium store. *Biochem. J.* **287**:79–84.
114. Fleischer, S., and Inui, M. (1989). Biochemistry and biophysics of excitation–contraction coupling. *Annu. Rev. Biophys. Biophys. Chem.* **18**:333–364.
115. Dargie, P. J., Agre, M. C., and Lee, H. C. (1990). Comparison of Ca^{2+} mobilizing activities of cyclic ADP-ribose and inositol trisphosphate. *Cell Regul.* **1**:279–290.
116. Lee, H. C., Aarhus, R., and Walseth, T. F. (1993). Calcium mobilization by dual receptors during fertilization of sea urchin eggs. *Science* **261**:352–355.
117. Whalley, T., McDougall, A., Crossley, I., Swann, K., and Whitaker, M. (1992). Internal calcium release and activation of sea urchin eggs by cGMP are independent of the phosphoinositide signaling pathway. *Mol. Biol. Cell* **3**:373–383.
118. Galione, A., White, A., Willmott, N., Turner, M., Potter, B. V. L., and Watson, S. P. (1993). cGMP mobilizes intracellular Ca^{2+} in sea urchin eggs by stimulating cyclic ADP-ribose synthesis. *Nature* **365**:456–459.
119. Lincoln, T. M., and Corbin, J. D. (1983). Characterization and biological role of the cGMP-dependent protein kinase. *Adv. Cyclic Nucleotide Res.* **15**:139–192.
120. Twigg, J., Patel, R., and Whitaker, M. (1988). Translational control of $InsP_3$-induced chromatin condensation during the early cell cycles of sea urchin embryos. *Nature* **332**:366–369.
121. Eisen, A., and Reynolds, G. T. (1985). Source and sinks for the calcium released during fertilization of single sea urchin eggs. *J. Cell Biol.* **100**:1522–1527.
122. Azarnia, R., and Chamber, E. L. (1976). The role of divalent cations in activation of the sea urchin egg. I. Effect of fertilization on divalent cation content. *J. Exp. Zool.* **198**:65–78.
123. Gillot, I., Ciapa, B., and Sardet, C. (1991). The calcium content of cortical granules and the loss of calcium from sea urchin eggs at fertilization. *Dev. Biol.* **146**:396–405.
124. Shen, S. S., and Steinhardt, R. A. (1978). Direct measurement of intracellular pH during metabolic derepression of the sea urchin egg. *Nature* **272**:253–254.
125. Shen, S. S., and Buck, W. R. (1990). A synthetic peptide of the pseudosubstrate domain of protein kinase C blocks cytoplasmic al-

kalinization during activation of the sea urchin egg. *Dev. Biol.* **140:**272–280.

126. Holland, L. Z., and Gould-Somero, M. (1983). Fertilization acid of sea urchin eggs: Evidence that it is H^+, not CO_2. *Dev. Biol.* **92:** 549–552.
127. Shen, S. S., and Steinhardt, R. A. (1979). Intracellular pH and the sodium requirement at fertilisation. *Nature* **282:**87–89.
128. Shen, S. S. (1989). Na^+-H^+ antiport during fertilization of the sea urchin egg is blocked by W-7 but is insensitive to K252a and H-7. *Biochem. Biophys. Res. Commun.* **161:**1100–1108.
129. Webb, D. J., and Nuccitelli, R. (1981). Direct measurement of intracellular pH changes in *Xenopus* eggs at fertilization and cleavage. *J. Cell Biol.* **91:**562–567.
130. Gould, M. C., and Stephano, J. L. (1993). Nuclear and cytoplasmic pH increase at fertilization in *Urechis caupo. Dev. Biol.* **159:** 608–617.
131. Epel, D. (1989). Arousal of activity in sea urchin eggs at fertilization. In *Cell Biology of Fertilization,* (G. Schatten and H. Schatten eds.), Academic Press, New York, pp. 361–385.
132. Schneider, E. G. (1985). Activation of Na^+-dependent transport at fertilization in the sea urchin: Requirements of both an early event associated with exocytosis and a later event involving increased energy metabolism. *Dev. Biol.* **108:**152–163.
133. Schilling, F. M., and Manahan, D. T. (1990). Energetics of early development for the sea urchins *Strongylocentrotus purpuratus* and *Lytechinus pictus* and the crustacean *Artemia* sp. *Mar. Biol.* **106:**119–127.
134. Christen, R., Sardet, C., and Lallier, R. (1979). Chloride permeability of sea urchin eggs. *Cell Biol. Int. Rep.* **3:**121–128.
135. Jaffe, L. F. (1991). The path of calcium in cytosolic calcium oscillations: A unifying hypothesis. *Proc. Natl. Acad. Sci. USA* **88:** 9883–9887.
136. Whitaker, M., and Swann, K. (1993). Lighting the fuse at fertilization. *Development* **117:**1–12.
137. Kline, D. (1993). Cell signalling and regulation of exocytosis at fertilization of the egg. In *Signal Transduction during Biomembrane Fusion* (D. H. O'Day, ed.), Academic Press, San Diego, pp. 75–102.
138. Ehrenstein, G., Dale, B., and DeFelice, L. J. (1984). A soluble fraction of sperm triggers cortical granule exocytosis in sea urchin eggs. *Biophys. J.* **45:**23a.
139. Dale, B., DeFelice, L. J., and Ehrenstein, G. (1985). Injection of a soluble sperm fraction into sea-urchin eggs triggers the cortical reaction. *Experientia* **41:**1068–1070.
140. Stice, S. L., and Robl, J. M. (1990). Activation of mammalian oocytes by a factor obtained from rabbit sperm. *Mol. Reprod. Dev.* **25:**272–280.
141. Swann, K. (1990). A cytosolic sperm factor stimulates repetitive calcium increases and mimics fertilization in hamster eggs. *Development* **110:**1295–1302.
141a. Kimura, Y., and Yanagimachi, R. (1995). Intracytoplasmic sperm injection in the mouse. *Biol. Reprod.* **52:**709–720.
141b. Kimura, Y., and Yanagimachi, R. (1995). Mouse oocytes injected with testicular spermatozoa or round spermatids can develop into normal offspring. *Development* **121:**2397–2405.
142. Steinhardt, R. A., Lundin, L., and Mazia, D. (1971). Bioelectric responses of the echinoderm egg to fertilization. *Proc. Natl. Acad. Sci. USA* **68:**2426–2430.
143. Jaffe, L. A., Gould-Somero, M., and Holland, L. (1979). Ionic mechanism of the fertilization potential of the marine worm, *Urechis caupo* (Echiura). *J. Gen. Physiol.* **73:**469–492.
143a. Carroll, D. J., and Jaffe, L. A. (1995). Proteases stimulate fertilization-like responses in starfish eggs. *Dev. Biol.* **170:**690–700.
144. Zalokar, M. (1980). Activation of ascidian eggs with lectins. *Dev. Biol.* **79:**232–237.
145. Speksnijder, J. E., Sardet, C., and Jaffe, L. F. (1990). The activation wave of calcium in the ascidian egg and its role in ooplasmic segregation. *J. Cell Biol.* **110:**1589–1598.
146. Gould, M., Stephano, J. L., and Holland, L. Z. (1986). Isolation of protein from *Urechis* sperm acrosomal granules that binds sperm to eggs and initiates development. *Dev. Biol.* **117:**306–318.
147. Gould, M. C., and Stephano, J. L. (1991). Peptides from sperm acrosomal protein that initiate egg development. *Dev. Biol.* **146:** 509–518.
148. Kline, D., Simoncini, L., Mandel, G., Mave, R., Kado, R. T., and Jaffe, L. A. (1988). Fertilization events induced by neurotransmitters after injection of mRNA in *Xenopus* eggs. *Science* **241:** 464–467.
149. Kline, D., Kopf, G. S., Muncy, L. F., and Jaffe, L. A. (1991). Evidence for the involvement of a pertussis toxin-insensitive G-protein in egg activation of the frog, *Xenopus laevis. Dev. Biol.* **143:**218–229.
150. Shilling, F., Mandel, G., and Jaffe, L. A. (1990). Activation by serotonin of starfish eggs expressing the rat serotonin 1c receptor. *Cell Regul.* **1:**465–469.
151. Williams, C. J., Schultz, R. M., and Kopf, G. S. (1992). Role of G proteins in mouse egg activation: Stimulatory effects of acetylcholine on the ZP2 to $ZP2_f$ conversion and pronuclear formation in eggs expressing a functional m1 muscarinic receptor. *Dev. Biol.* **151:**228–296.
152. Moore, G. D., Kopf, G. S., and Schultz, R. M. (1993). Complete mouse egg activation in the absence of sperm by stimulation of an exogenous G protein-coupled receptor. *Dev. Biol.* **159:**669–678.
153. Yim, D. L., Opresko, L. K., Wiley, H. S., *et al.* (1994). Highly polarized EGF receptor tyrosine kinase activity initiates egg activation in *Xenopus. Dev. Biol.* **162:**41–55.
154. Peaucellier, G., Veno, P. A., and Kinsey, W. H. (1988). Protein tyrosine phosphorylation in response to fertilization. *J. Biol. Chem.* **263:**13806–13811.
155. Ciapa, B., and Epel, D. (1991). A rapid change in phosphorylation on tyrosine accompanies fertilization of sea urchin eggs. *FEBS Lett.* **295:**167–170.
156. Foltz, K. R., and Lennarz, W. J. (1990). Purification and characterization of an extracellular fragment of the sea urchin egg receptor for sperm. *J. Cell Biol.* **111:**2951–2959.
157. Foltz, K. R., and Lennarz, W. J. (1992). Identification of the sea urchin egg receptor for sperm using an antiserum raised against a fragment of its extracellular domain. *J. Cell Biol.* **116:**647–658.
158. Fotlz, K. R., Partin, J. S., and Lennarz, W. J. (1993). Sea urchin egg receptor for sperm: Sequence similarity of binding domain and hsp70. *Science* **259:**1421–1425.
159. Ohlendieck, K., Dhume, S. T., Partin, J. S., and Lennarz, W. J. (1993). The sea urchin egg receptor for sperm: Isolation and characterization of the intact, biologically active receptor. *J. Cell Biol.* **122:**887–895.
160. Abassi, Y. A., and Foltz, K. R. (1994). Tyrosine phosphorylation of the egg receptor for sperm at fertilization. *Dev. Biol.* **164:**430–443.
161. Veillette, A., and Davidson, D. (1992). Src-related protein tyrosine kinases and T-cell receptor signalling. *Trends Genet.* **8:**61–66.
162. Straus, D. B., and Weiss, A. (1992). Genetic evidence for the involvement of the *lck* tyrosine kinase in signal transduction through the T cell antigen receptor. *Cell* **70:**585–593.
163. Hall, C. G., Sancho, J., and Terhost, C. (1993). Reconstitution of T cell receptor ζ-mediated calcium mobilization in nonlymphoid cells. *Science* **261:**915–918.
164. Moore, K. L., and Kinsey, W. H. (1994). Identification of an abl-related protein tyrosine kinase in the cortex of the sea urchin egg: possible role at fertilization. *Dev. Bio.* **164:**444–455.

164a. Moore, K. L., and Kinsey, W. H. (1995). Effects of protein tyrosine kinase inhibitors on egg activation and fertilization-dependent protein tyrosine kinase activity. *Dev. Biol.* **168:**1–10.
165. Shilling, F., Chiba, K., Hoshi, M., *et al.* (1989). Pertussis toxin inhibits 1-methyladenine-induced maturation in starfish oocytes. *Dev. Biol.* **133:**605–608.
166. Schwartz, M. A. (1992). Transmembrane signalling by integrins. *Trends Cell Biol.* **2:**304–308.
167. Janeway, C. A. (1992). The T cell receptor as a multicomponent signalling machine: CD4/CD8 coreceptors and CD45 in T cell activation. *Annu. Rev. Immunol.* **10:**645–674.
168. Terasaki, M., and Jaffe, L. A. (1991). Organization of the sea urchin egg endoplasmic reticulum and its reorganization at fertilization. *J. Cell Biol.* **114:**929–940.
169. Jaffe, L. A., and Terasaki, M. (1993). Structural changes of the endoplasmic reticulum of sea urchin eggs during fertilization. *Dev. Biol.* **156:**566–573.
170. Jaffe, L. A., and Terasaki, M. (1994). Structural changes in the endoplasmic reticulum of starfish oocytes during meiotic maturation and fertilization. *Dev. Biol.* **164:**579–587.
170a. Terasaki, M., Jaffe, L. A., Hunnicutt, G. R., and Hammer, J. A. Evidence for ER fragmentation during fertilization by photobleaching of a green fluorescent protein targetted to the ER lumen. (Unpublished results).
171. Henson, J. H., Begg, D. A., Beaulieu, S. M., Fishkind, D. J., Bonder, E. M., Terasaki, M., Lebeche, D., and Kaminer, B. (1989). A calsequestrin-like protein in the endoplasmic reticulum of the sea urchin: Localization and dynamics in the egg and first cell cycle embryo. *J. Cell Biol.* **109:**149–161.
172. Longo, F. J. (1985). Pronuclear events during fertilization. In *Biology of Fertilization* (C. B. Metz and A. Monroy, eds.), Academic Press, Orlando, Volume 3, pp. 251–298.
173. Longo, F. J., and Anderson, E. (1968). The fine structure of pronuclear development and fusion in the sea urchin, *Arbacia punctulata. J. Cell Biol.* **39:**339–368.
174. Harris, P. (1975). The role of membranes in the organization of the mitotic apparatus. *Exp. Cell Res.* **94:**409–425.
175. Masui, Y., Clarke, H. J. (1979). Oocyte maturation. *Int. Rev. Cytol.* **57:**185–282.
176. Fol, H. (1877). Sur le commencement de l'henogénie chez divers animaux. *Arch. Zool. Exp. Gen.* **6:**145–169.
177. Elinson, R. P. (1977). Fertilization of immature frog eggs: Cleavage and development following subsequent activation. *J. Embryol. Exp. Morphol.* **37:**187–201.
178. Fujiwara, T., Nakada, K., Shirakawa, H., and Miyazaki, M. (1993). Development of inositol trisphosphate-induced calcium release mechanism during maturation of hamster oocytes. *Dev. Biol.* **156:**69–79.
179. Goldenberg, M., and Elinson, R. P. (1980). Animal/vegetal differences in cortical granule exocytosis during activation of the frog egg. *Dev. Growth Differ.* **22:**345–356.
180. Mehlmann, L. M., and Kline, D. (1994). Regulation of intracellular calcium in the mouse egg: calcium release in response to sperm or inositol trisphosphate is enhanced after meiotic maturation. *Biol. Reprod.* **51:**1088–1098.
181. Charbonneau, M., Grey, R. D. (1984). The onset of activation responsiveness during maturation coincides with the formation of the cortical endoplasmic reticulum in oocytes of *Xenopus laevis. Dev. Biol.* **102:**90–97.
182. Campanella, C., Andreuccetti, P., Taddei, C., and Talevi, R. (1984). The modifications of cortical endoplasmic reticulum during in vitro maturation of *Xenopus laevis* oocytes and its involvement in cortical granule exocytosis. *J. Exp. Zool.* **229:**283–293.
183. Larabell, C. A., and Chandler, D. E. (1988). Freeze-fracture analysis of structural reorganization during meiotic maturation in oocytes of *Xenopus laevis. Cell Tissue Res.* **251:**129–136.
184. Ducibella, T., Rangarajan, S., and Anderson, E. (1988). The development of mouse oocyte cortical reaction competence is accompanied by major changes in cortical vesicles and not cortical granule depth. *Dev. Biol.* **130:**789–792.
185. Henson, J. H., Beaulieu, S. M., Kaminer, B., and Begg, D. A. (1990). Differentiation of a calsequestrin-containing endoplasmic reticulum during sea urchin oogensis. *Dev. Biol.* **142:**255–269.
185a. Mehlmann, L. M., Terasaki, M., Jaffe, L. A., and Kline, D. (1995). Reorganization of the endoplasmic reticulum during meiotic maturation of the mouse oocyte. *Dev. Biol.* **170:**607–615.
185b. Shiraishi, K., Okada, A., Shirakawa, H., Nakanishi, S., Mikoshiba, K., and Miyazaki, S. (1995). Developmental changes in the distribution of the endoplasmic reticulum and inositol 1,4,5-trisphosphate receptors and the spatial pattern of Ca^{2+} release during maturation of hamster oocytes. *Dev. Biol.* **170:**594–606.
186. Iwao, Y. (1982). Differential emergence of cortical granule breakdown and electrophysiological responses during meiotic maturation of toad oocytes. *Dev. Growth Differ.* **24:**467–477.
187. Ducibella, T., Kurasawa, S., Duffy, P., Kopf, G. S., and Schultz, R. M. (1993). Regulation of the polyspermy block in the mouse egg: Maturation-dependent differences in cortical granule exocytosis and zona pellucida modifications induced by inositol 1,4,5-trisphosphate and an activator of protein kinase C. *Biol. Reprod.* **48:**1251–1257.
188. Longo, F. J., So, F., and Schuetz, A. W. (1982). Meiotic maturation and the cortical granule reaction in starfish eggs. *Biol. Bull.* **163:** 465–476.
189. Ducibella, T., Anderson, E., Albertini, D. F., Aaalberg, J., and Rangarajan, S. (1988). Quantitative studies of changes in cortical granule number and distribution in the mouse oocyte during meiotic maturation. *Dev. Biol.* **130:**184–197.
190. Simoncini, L., and Moody, W. J. (1990). Changes in voltage-dependent currents and membrane area during maturation of starfish oocytes: Species differences and similarities. *Dev. Biol.* **138:** 194–201.
191. Miyazaki, S., and Hirai, S. (1979). Fast polyspermy block and activation potential: Correlated changes during oocyte maturation of a starfish. *Dev. Biol.* **70:**327–340.
192. Miyazaki, S. (1979). Fast polyspermy block and activation potential: Electrophysiological bases for their changes during oocyte maturation of a starfish. *Dev. Biol.* **70:**341–354.
193. Schlichter, L. C., and Elinson, R. P. (1981). Electrical responses of immature and mature *Rana pipiens* oocytes to sperm and other activating stimuli. *Dev. Biol.* **83:**33–41.
194. Iwao, Y. (1987). The spike component of the fertilization potential in the toad, *Bufo japonicus:* Changes during meiotic maturation and absence during cross-fertilization. *Dev. Biol.* **123:**559–565.
195. Kado, R. T., Marcher, K., and Ozon, R. (1981). Electrical membrane properties of the *Xenopus laevis* oocyte during progesterone-induced meiotic maturation. *Dev. Biol.* **84:**471–476.
195a. Gallo, C. J., Jones, T. L. Z., Battey, J. F., Aragay, A. M., and Jaffe, L. A. (1995). Expression of alpha q family G-proteins during oocyte maturation and early development of *Xenopus laevis. Mol. Biol. Cell* **6:**90a.
196. Flood, J. T., Chillik, C. F., van Uem, J. F. H. M., Iritani, A., and Hodgen, G. D. (1990). Ooplasmic transfusion: Prophase germinal vesicle oocytes made developmentally competent by microinjection of metaphase II egg cytoplasm. *Fertil. Steril.* **53:**1049–1054.

19

Cell Volume Regulation

John R. Sachs

19.1. INTRODUCTION

The basic strategy of all cells—enclosing macromolecules, many of which are polyelectrolytes, within a space surrounded by a phospholipid bilayer membrane suspended in a solution of ions and small organic solutes—poses a daunting threat to their survival. Phospholipid membranes are permeable to water and to small ions.[1] If the external solution is free of macromolecules and ions to which the membrane is impermeable, the enclosed macromolecules produce an unbalanced oncotic pressure which draws water into the enclosed compartment, and the polyelectrolytes produce a transmembrane potential which accumulates ions together with water. In the absence of anything else, no equilibrium is possible; as long as the concentration of impermeable macromolecules and polyelectrolytes is greater within the space enclosed by the phospholipid membrane than in the external solution, the flow of water and ions into the enclosed compartment will continue until the membrane bursts. Many of the properties of biological membranes have developed in response to the challenge of overcoming this problem.

A priori, cells might solve their problem in several ways. By becoming impermeable to water, cells would eliminate the possibility of volume changes in response to osmotic forces. However, no measures are known for reducing the water permeability of phospholipid membranes other than a reduction of temperature (Arrhenius activation energy > 10 kcal/mole). In fact, the membranes of many cells (e.g., erythrocytes, mammalian epithelial cells, mammalian crystalline lens) contain water channel proteins which greatly increase their permeability to water.[2,3] By developing a structure which imparts tensile strength to the phospholipid bilayer, which has very little tensile strength *per se,* cells might develop a hydrostatic pressure to counteract the osmotic pressure generated by intracellular macromolecules. Plant cells have adopted the strategy of enclosing the phospholipid bilayer in a rigid cell wall which is able to resist a considerable hydrostatic pressure, but at the sacrifice of mobility. It has been suggested that the cytoskeleton of animal cells might restrict swelling in response to osmotic shock,[4–6] or that organization of cells in tissues might provide mechanical support for the phospholipid membrane through contact with other cells or with extracellular structures such as the basement membrane, and therefore permit the development of a hydrostatic pressure. Although such mechanisms might modulate the swelling which occurs when cells are exposed to hypotonic solutions, they do not seem capable of completely accounting for the observed response, and it is hard to see how they could account for cell shrinkage on exposure to hypertonic solutions. Red blood cells, which have an extensive cytoskeleton attached to the cell membrane, are unable to resist a hydrostatic pressure of more than a few centimeters H_2O.[7] Finally, cells might defend themselves against the unbalanced osmotic forces attributable to intracellular macromolecules by becoming impermeable to one or more solutes in the extracellular solution.

Although for many years impermeability to extracellular cations was the accepted explanation for the ability of animal cells to maintain their volume, by 1941 many pieces of evidence had accumulated which made it clear that cells are not impermeable to the major extracellular cations. Muscle cells were shown to lose K and gain Na during stimulation[8]; liver and muscle of rats lost K and gained Na when the animals were fed K-deficient diets[9]; and muscles bathed in K-free solutions lost K and gained Na, but when reimmersed in solutions which contained K, regained K and lost Na.[10] When radioactive isotopes of Na and K became available, both *in vivo* and *in vitro* experiments clearly showed that extracellular cations readily exchange with intracellular cations.[11,12] In 1941, Dean[13] pointed out that these findings require the presence of a sodium pump in the cell membrane, and by 1946 the existence of a cation pump was widely accepted.[14]

The concept of an energy-driven Na pump was thus suggested by the need to balance the passive leak movements of Na and K in order to maintain intracellular concentrations of these ions constant.[15] Wu[16] supposed that the presence of polyvalent macromolecules within the cell could be balanced if the extracellular solution contains at least one osmolyte to which the cell membrane is impermeable: the double Donnan equilibrium. The concentration of macromolecules in the extracellular solution is low compared to their intracellular concentration, but impermeability to small osmolytes will suffice to support the equilibrium. Wilson[17] and Leaf[18] realized that it

John R. Sachs • Department of Medicine, State University of New York at Stony Brook, Stony Brook, New York 11794.

Molecular Biology of Membrane Transport Disorders, edited by S.G. Schultz *et al.* Plenum Press, New York, 1996

is not necessary that the membrane be physically impermeable; by balancing the passive movement of Na into the cell drawn by the osmotic pressure and transmembrane potential generated by the intracellular negatively charged macromolecules with a net outward movement of Na through the pump supported by the hydrolysis of ATP, the membrane is rendered functionally impermeable to Na. The pump-leak hypothesis of volume regulation was elaborated by Ussing[19] and Tosteson and Hoffman[20] who pointed out that the major contribution of the Na pump is to maintain the asymmetric distribution of Na and K across the cell membrane. As a result of the asymmetry, it is possible for the leaks of the two species to be balanced in such a way that total intracellular ionic content, and therefore cell volume, remains constant. The theory has been extensively evaluated with preparations of tissues and cells and several reviews have appeared in which the correspondence of the theory with reality has been examined in detail.[21,22]

The pump-leak hypothesis was formulated to account for steady-state volume maintenance in an unchanging isotonic environment, and the leaks were envisioned as diffusion-mediated processes which do not involve specific membrane structures. It soon became evident that most of the transmembrane movements of ions are carried out by specific transport proteins and are under physiological control. Many of the transport processes are difficult to study in isovolemic cells because of their small magnitude, but it was found that exposure of the cells to anisosmotic solutions increased the activity of the transporters in specific ways. Early work in this area was done by Kregenow who showed that swelling of duck red blood cells in hypotonic solutions increased the efflux of K and Cl with return of the cells toward their original volume,[23] and shrinkage of cells in a hypertonic solution increased the activity of an Na,K,2Cl cotransporter which increased the cell content of K and Cl with resultant swelling of the cells toward their original volume.[24] These findings led to the concept of a volume regulatory mechanism illustrated in Figure 19.1. When cells are exposed to hypotonic solutions, they first swell rapidly as water enters the cell in response to the osmotic gradient, and then return toward their original volume as intracellular osmolytes are lost along with water; this response to swelling has been termed *regulatory volume decrease* (RVD),[25] and is a property of most eukaryotes. The intracellular osmolytes most commonly lost by cells of higher vertebrates during acute swelling experiments are K along with an anion, usually Cl, but sometimes HCO_3. In some cases intracellular organic osmolytes are also lost. Swelling after cell shrinkage in acute experiments [regulatory volume increase (RVI)] occurs less commonly in the cells of higher vertebrates, and when it occurs is mediated by the activation of transporters which accumulate Na and Cl within the cell. The accumulated Na is then exchanged for K by the Na pump. Although RVI is not commonly seen when cells are transferred from isotonic to hypertonic solutions, many cells demonstrate a phenomenon termed *RVI after RVD.* In this case, when cells are transferred from isotonic to hypotonic solutions, they first swell, and then return toward their original volume by losing water and intracellular electrolytes. If the cells are returned to the isotonic solution, they immediately shrink, but then undergo RVI and swell toward their original volume. The transport processes which account for RVI are the same in the two circumstances.

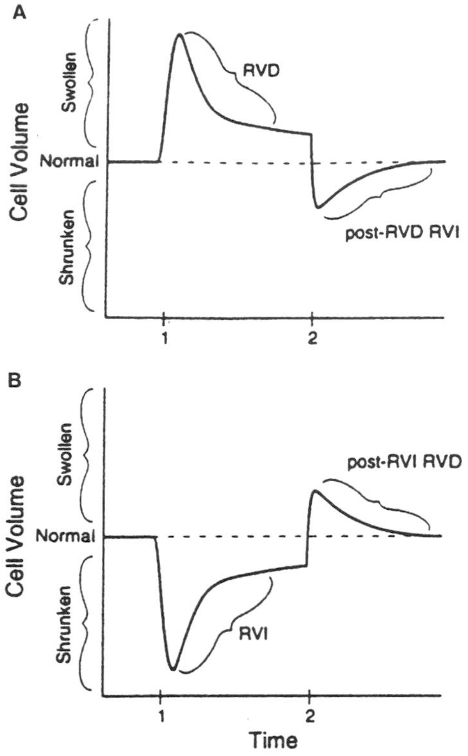

Figure 19.1. Cell responses to swelling and shrinking. (A) Cells exposed to hypotonic solutions at first swell because of water uptake, and then shrink toward their original volume as they lose solute. If the cells are returned to isotonic solutions, they rapidly shirnk as they lose water, and then gradually return toward their isotonic volume (RVI after RVD). (B) The inverse of the events shown in panel A. While most cells display RVI after RVD, many do not undergo RVI unless preceded by RVD. Taken from McCarty and O'Neill[66] and reprinted with permission.

Although studies of short-term adjustment of cell volume after drastic perturbation have focused attention on mechanisms which alter the cell content of inorganic ions, studies of volume regulation by cells from organisms which inhabit marine environments, and of cells from higher vertebrates over longer periods of time, have emphasized the importance of changes in the cell content of organic compounds in regulating cell volume. Cells of euryhaline invertebrates encounter environments in which osmolality changes drastically,[26] and marine vertebrates which do not regulate the osmolality of their extracellular fluid must cope with an environment in which the osmolality is about 1100 mOsm/kg H_2O, nearly four times that of mammalian extracellular fluid.[27] Inorganic cations and anions at high concentrations are known to modify the structure and function of proteins—to "destabilize" proteins—by a combination of effects including interaction with protein ligands, disruption of salt bridges, modification of the structure of water, and others. At osmolalities not much greater than those found in sea water, NaCl and KCl solutions are frequently used for removing extrinsic proteins by interfering with electrostatic interactions in protocols for the purification of membrane proteins. A large fraction of the osmotically active solutes of eukaryotes exposed to high and varying osmolalities is made up of small organic solutes. Among the solutes used are sugars (e.g., mannose, sucrose), polyols (e.g., sorbitol, inositol, mannitol), amino acids (e.g., glutamate, glutamine, alanine, proline) and substances derived from them (taurine), methylamines (e.g., betaine, glycerophosphorylcholine, trimethylamine oxide), and urea.[28] All of these (except urea) are believed to be "compatible" solutes in that, even at very high concentrations, they do not disturb protein structure or function, i.e., they do not destabilize proteins. In fact, many of these solutes, especially the amino acids and amines, are referred to as "compensatory" solutes since they seem to be able to counteract the destabilizing effects of small ions and urea. Urea at high concentrations is known to destablize proteins by breaking hydrogen bonds and perhaps by altering the structure of water. Trimethylamine oxide at a concentration ratio of about 2 urea : 1 trimethylamine oxide is known to counteract the destabilizing effect of urea. In the cells of elasmobranchs, urea is present at a concentration of 0.4 to 0.5 M, and the cells also contain trimethylamine oxide at the concentration appropriate to counteract the destabilizing effect of urea.[29]

In cells in which organic solutes make up a considerable proportion of intracellular osmolytes, loss or gain of these solutes is frequently used as a means of maintaining cell volume in response to an osmotic challenge.[27] Loss or gain of organic solutes is usually a relatively long-term process compared to the rapid response of ion transport pathways to changes in cell volume. Until recently, little attention has been paid to the role of organic solutes in volume regulation in the cells of higher vertebrates. It is clear, however, that renal medullary cells, which can be subjected to osmolalities approaching those in a marine environment, make use of organic osmolytes for both acute and long-term volume regulation.[30,31] Moreover, it seems likely that cells of many tissues, and especially those of neural tissues, make use of amino acids and other organic solutes to regulate their cytoplasmic solute content over longer time scales.[32]

Theoretically, the control of cell volume must involve a sensor, a signal transduction pathway, and a receptor; one or more of these functions could be carried out by the same structure. Although the strategy adopted by cells for coping with volume changes is readily outlined, the precise mechanisms involved are discouragingly diverse, and, perhaps because of the importance of volume control for cell survival, at least in the case of cell swelling, frequently redundant. A great deal has been learned about the characteristics of the transport mechanisms involved in the movement of solutes in response to volume changes, and the possibility that second messenger pathways are involved in the transmission of the signal of cell volume change from the sensor to the transporter has been thoroughly evaluated. These studies are sometimes complicated since the transporters responsible for volume regulation are frequently used for other purposes, and are often under endocrine control. Currently, much of the interest in this field centers on the nature of the sensor, and this is the aspect of cell volume control about which the least is known, although there is no shortage of ingenious theories.

This chapter will deal with some aspects of cell volume control, primarily as demonstrated by the cells of higher vertebrates, but no claim is made that the treatment is inclusive. Frequently, reference will be made to a relevant review rather than to an original publication in order to attempt to keep the bibliography manageable. In recent years a number of excellent general reviews of cell volume regulation have appeared (Refs. 33–37) and recently two volumes have been published containing comprehensive reviews of all aspects of this subject (Refs. 38, 39).

19.2. PROPERTIES OF CELL WATER AND OSMOTIC PHENOMENA

Cell volume regulation is, in reality, balancing the influx and efflux of water across cell membranes. It is customary in reviews of this subject to recount the experiment of Rev. Stephen Hales in 1733. Hales injected a large volume of water into an animal's circulation and observed a pronounced swelling of liver, kidneys, and other organs which was reversed by injection of a salt solution.[32] Variations of this experiment have been repeated under more controlled circumstances with essentially similar results.[22] The experiment suggests that movements of water are determined by the relative concentration of osmolytes in cytoplasm and in the extracellular solution.

19.2.1. Osmotic Equilibrium

Most cells are permeable to water since the lipid bilayer, which is the major component of the cell membrane, is permeable to water (although the partition coefficient of water in phospholipid bilayers is low, the concentration of water in aqueous solutions is very high). The permeability of cells to water can be largely accounted for by the permeability of the bilayer.[1] Movement of water across most cell membranes is rapid and it has been evident for some time that not enough energy is available from metabolism to maintain a concentration gradient of water (but see below). It is, therefore, generally believed that the chemical potential of water inside the cell is equal to the chemical potential of water outside the cell:

$$\mu_w^i = \mu_w^o \tag{19.1}$$

where μ_w^i is the chemical potential of cell water and μ_w^o is the chemical potential of extracellular water. In this section some of the evidence for this belief will be recounted.

For any species, μ_j is given by

$$\mu_j = \mu_{j,s} + RT\ln N_j + p\bar{V}_j + z_jF\Psi \tag{19.2}$$

where $\mu_{j,s}$ is the standard chemical potential of species j, N_j the mole fraction of species j, $\bar{V}_j$ the partial molar volume of species j, z_j the charge of species j, p the pressure, Ψ the electrical potential, and R, T, and F have their usual meaning. For two solutions separated by a semipermeable membrane (e.g., a cell suspended in a solution), Eq. (19.1) can be expanded to

$$\mu^{i}_{w,s} + RT\ln N^{i}_{w} + p_i \bar{V}_w = \mu^{o}_{w,s}\ RT\ln N^{o}_{w} + p_o \bar{V}_w \qquad (19.3)$$

since z_w is zero. If it is assumed that $\mu_{w,s}$ is the same on the two sides of the membrane,

$$RT\ln (N^{i}_{w}/N^{o}_{w}) = (p_o - p_i)\bar{V}_w = \Delta p \bar{V}_w \qquad (19.4)$$

Δp is the hydrostatic pressure which must be applied to the more concentrated solution in order to prevent the flow of water from the more dilute solution, and is usually referred to as the osmotic pressure difference, $\Delta\Pi$. If the solution on one side of the membrane is pure water,

$$RT\ln (N_w) = \Delta p \bar{V}_w \qquad (19.5)$$

and it can be shown that[40]

$$\Pi = RT\Sigma C_j \qquad (19.6)$$

where C_j is the molar concentration of solute j. This is the Boyle–van't Hoff law.

Equation (19.6) requires modification since the osmotic behavior of even dilute solutions is not ideal; the modification is made by inserting in Eq. (19.4) a correction factor, ϕ, the osmotic coefficient:

$$\Pi = RT\Sigma\phi_j C_j \qquad (19.7)$$

For small electrolytes ϕ is generally somewhat greater than 1, and for electrolytes somewhat less than 1 because of the interaction of positive and negative particles. For macromolecules ϕ is greater than 1 because of the anomalously high entropy of mixing which results when the volume of the solute is very large in relation to the volume of water.[41]

If water is at equilibrium across the membrane, and if there is no hydrostatic pressure or other source of pressure on either side of the membrane,

$$\Pi^{i} = \Pi^{o} = \Pi \qquad (19.8)$$

Equation (19.7) can be written

$$\Pi V_w = RT\Sigma\phi_j Q_j \qquad (19.9)$$

where $\phi_j Q_j = \phi_j C_j V_w$, and Q_j and V_w are the amounts of solute and water associated with a given volume of cells in isotonic solution. If $\Sigma\phi_j Q_j$ does not change when cells are transferred to a solution of different osmolality (i.e., the cells do not lose solute), the right-hand side of Eq. (19.9) is constant. Therefore, if V_o is the cell volume in isotonic solution of osmolality Π_o, and V the volume at osmolality Π, then

$$\Pi V = \Pi_o V_o \qquad (19.10)$$

and if b is the nonwater fraction of cell volume,

$$\Pi(V - b) = \Pi_o(V_o - b) \qquad (19.11)$$

which can be written:

$$V = (\Pi_o/\Pi)(V_o - b) + b \qquad (19.12)$$

If cell volume is plotted as a function of the reciprocal of Π, the result should be a straight line with slope $\Pi_o(V_o - b)$ and intercept b, which can be independently determined. This relation has been extensively evaluated using red blood cells, in which the absence of intracellular structures or compartments simplifies evaluation. In general, the behavior observed differs from that predicted by Eq. (19.12), and the deviation is greater at high osmolalities than at low. An attractive explanation for this behavior has been advanced.[42,43] The osmotic coefficient of hemoglobin, ϕ_H, is not constant, but increases as the concentration of hemoglobin increases.[44] Although ΣQ_j [Eq. (19.9)] is constant, $\Sigma\phi_j Q_j$ is not, but increases as the cell shrinks and the concentration of hemoglobin increases. It might be expected, then, that cell shrinkage as Π increases will be less than that predicted by Eq. (19.12), and since the increase in ϕ_H becomes greater at higher hemoglobin concentrations, the deviation should be greater at high osmolalities, as is observed. The osmotic behavior of red cells whose ionic composition has been modified by nystatin can be accounted for by a model which takes account of the anomalous osmotic behavior of hemoglobin.[45] This may not, however, be the complete explanation. Garay-Bobo and Solomon[46] found that the charge on hemoglobin (z_H) varies with hemoglobin concentration; as the concentration of hemoglobin increased, z_H decreased. As a result, the cells gained Cl (ΣQ_j was not constant), and this accounts for some of the deviation of the observed data from Eq. (19.12). That the variation of the behavior of intact red cells from simple osmotic theory can be attributed to the anomalous behavior of hemoglobin is supported by the observation that ghosts which have lost most of their hemoglobin behave as predicted by Eq (19.12).[47]

19.2.2. Osmotic Equilibrium in More Complicated Cells

Although simple cells like red blood cells can be shown to behave in accordance with the Boyle–vant Hoff law, questions have been raised about the behavior of more complex cells in which the cytoplasm is both crowded with intracellular structures and compartmentalized. Clegg[48] has reviewed evidence which suggests that, in complex cells, cytoplasm is far from homogeneous, but is highly structured. Perhaps the most striking support for this view is an experiment reported by Kempner and Miller[49] who centrifuged intact, living *Euglena* cells at 100,000*g* for 1 hr. The cytoplasm within the cells was highly stratified into distinct layers or zones into which the intracellular structures were segregated. Most of the soluble phase separated into the top layer. Analysis showed that the soluble layer contained small molecules and electrolytes, but very few macromolecules which were, however, present in the supernatant when the cells were fractionated by standard biochemical methods. It was concluded that macromolecules were not free in solution, but were bound to the cytoskeleton and to cell membranes.

A further suggestion that cell water may not have the physical properties of bulk water comes from the results of experiments in which the forces between phospholipid bilayers[50] or mica surfaces[51] were measured. In each case the results indicated that the surfaces influenced the properties of adjacent water to a distance as great as 50 Å from the surface. Although the effect was not directly shown to alter the osmotic properties of solutions, the possibility cannot be excluded. On the other hand, nuclear magnetic resonance studies indicate that only 5% of intracellular water has properties different from bulk water.[52]

The relevant question for the prediction of the effect of concentration changes on cell volume is whether or not the cell cytoplasm is in osmotic equilibrium with the extracellular solution. Early attempts to compare some colligative property of cell water, most often freezing point depression, with the same property in the extracellular solution bathing the cells yielded values for cell water activity which were lower for the cells than for the extracellular solution.[22] These studies were complicated by the difficulty of separating intracellular solution from extracellular solution during the measurement, and by the accumulation of osmotically active metabolites as a result of the cessation of cellular metabolism during preparation of the specimens. Maffly and Leaf[53] rapidly froze tissues and extracellular fluid in liquid nitrogen, and then pulverized the specimens while frozen. The samples were then suspended in chilled silicone and their temperature monitored as the samples were slowly melted. The melting curves of the pulverized tissue and the extracellular fluid were superimposable, indicating that no osmotic gradients existed. Most of the available evidence indicates, as suggested by the experiment of Rev. Hales, that, as with red cells, cells of more complicated tissues are in osmotic equilibrium with their extracellular environment.

19.2.3. Water Transport

Since net flow of water across semipermeable membranes occurs if the chemical potential of water differs on the two sides of the membrane, the driving force for water flow is taken as $\Delta\mu_w/\bar{V}_w$ where $\Delta\mu_w$ is the difference in μ_w across the membrane. The chemical potential of water is [Eq. (19.3)]

$$\mu_w = \mu_{w,s} + RT\ln N_w + p\bar{V}_w \tag{19.13}$$

or

$$\mu_w/\bar{V}_w = \mu_{w,s}/\bar{V}_w + RT/\bar{V}_w \ln N_w + p \tag{19.14}$$

and, with Eqs. (19.5) and (19.6)

$$\mu_w/\bar{V}_w = \mu_{w,s}/\bar{V}_w - RT\,\Sigma C_j + p \tag{19.15}$$

where C_j is the molar concentration of solute j. If $\mu_{w,s}$ is the same on the two sides of the membrane, the difference between μ_w/N_w across the membrane is

$$\Delta\mu_w/\bar{V}_w = \Delta p - RT\,\Sigma\Delta C_j \tag{19.16}$$

and

$$J_v^w = L_p(\Delta p - RT\,\Sigma\Delta C_j) \tag{19.17}$$

where L_p is the hydraulic conductivity of the membrane and J_v^w is the flow of water in units of volume. The equation assumes that the membrane is totally impermeable to the solutes. If this is not the case, solute will flow across the membrane and J_v will not equal J_v^w. In order to account for this, a correction factor, σ, the Staverman reflection coefficient, is introduced. The reflection coefficient takes account of the extent to which solute is "reflected" from the membrane and varies from 1 for a solute to which the membrane is completely impermeable to 0 for a solute to which the membrane is totally permeable. Introducing σ and ϕ, the osmotic coefficient [Eq. (19.17)], to account for the nonideal behavior of solutes, Eq. (19.17) becomes

$$J_v^w = L_p(\Delta p - RT\,\Sigma\sigma_j\Delta\phi_j C_j) = L_p(\Delta p - \sigma\Delta\Pi) \tag{19.18}$$

In order for J_v^w to be zero (equilibrium), it is necessary that L_p be zero (which probably is not possible for a cell enclosed in a phospholipid membrane), or that $\Delta p = \sigma\Delta\Pi$, i.e., as stated above, an osmotic (concentration) difference across the membrane must be balanced by a difference in hydrostatic pressure. J_v^w is not usually given in terms of membrane area, but as net flux across the total membrane since, in general, it is frequently difficult to form a reliable estimate of cell surface area because of complicated geometry, membrane infoldings, and the like. However, surface area is an important determinant of the rate of cell swelling; the greater the surface area available for water flow, the greater will be J_v^w for any given transmembrane hydrostatic pressure or concentration difference.

If it is true that animal cell membranes are not able to develop any signification tension, and therefore are not able to support a transmembrane hydrostatic pressure difference, water flow, and therefore volume change, depends only on the transmembrane difference in the concentrations of solutes. Equilibrium conditions can be disturbed in two ways (Figure 19.2). Cells in equilibrium with an isotonic solution can be transferred to a hypotonic or hypertonic solution, in which case water flows across the membrane until the concentration of osmolytes on the two sides of the membrane is equal or, in the case of a hypotonic solution, the cell lyses. Although much of what is known about the response of cells to acute changes in their volume comes from experiments in which cells are

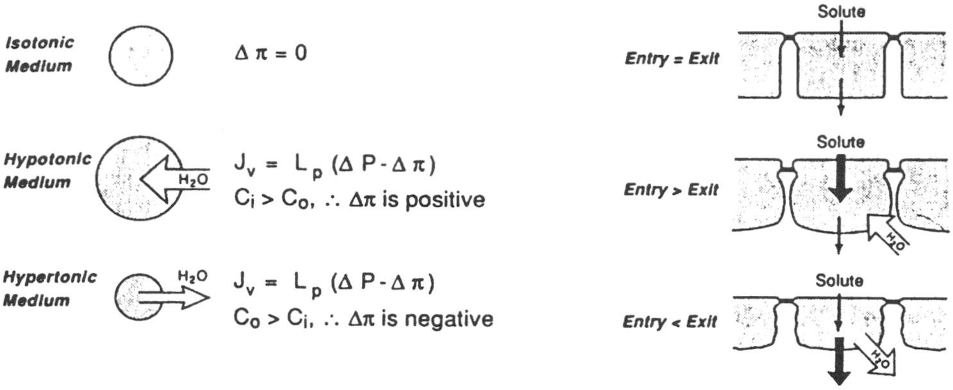

Figure 19.2. Water flow in cells as a result of changes in intracellular and extracellular solute concentration. (A) Anisosmotic volume changes. In hypotonic solutions, intracellular solute concentration is greater than the extracellular so that $\Delta\pi$ is positive and water flows into the cell (ΔP is approximately zero). In hypertonic solutions, extracellular solute concentration is greater than intracellular, $\Delta\pi$ is negative, and water flows out of the cell. (B) Isosmotic volume changes. Solute uptake (e.g., ions, amino acids, glucose) increases intracellular solute concentration and therefore water flows into the cell. The figure represents an epithelial cell, but similar phenomena occur in many cells, epithelial and nonepithelial. Taken from McCarty and O'Neill[66] and reprinted with permission.

transferred to anisosmotic solutions, it is often pointed out that, except for cells which reside in or pass through the renal medulla or papilla, cells in higher vertebrates, in which the composition of the extracellular solution is closely regulated, rarely encounter such situations. Anisosmotic volume change is mainly a problem for euryhaline invertebrates[54] and some marine vertebrates.[27] In higher vertebrates, the concentration of cell osmolytes changes mainly as a result of alteration in the uptake or release of solutes, or of changes in intracellular metabolism. Operation of the Na–glucose cotransporter increases the intracellular content of both of these solutes, and uptake of amino acids by Na-coupled mechanisms also increases the cell content of osmolytes. Change from glycogen synthesis to glycogenolysis in response to hormonal stimuli rapidly increases the content of osmotically active metabolic intermediates in hepatocytes. Large transcellular movements of ions occur in secreting and absorbing epithelia. Changes in intracellular solute content as a result of these processes change cell volume and activate the mechanisms which restore cell volume to normal. In higher vertebrates, volume regulation occurs, for the most part, under isotonic conditions.

Although it seems clear that most cells are not able to support a considerable transmembrane hydrostatic pressure, it has been suggested that the cytoskeleton, which is linked to the cell membrane by bonds with integral membrane proteins, may be able to modify the rate of cell swelling in hypotonic solutions, or reduce the tension which must develop in the cell membrane to resist a hydrostatic pressure.[6] In thin-walled spheres, Laplace's law relates membrane tension, T, to hydrostatic pressure, p:

$$T = \frac{pd}{4} \tag{19.19}$$

where d is the diameter of the sphere. The tension which develops in the membrane of a spherical cell at any given transmembrane hydrostatic pressure is proportional to the cell diameter. By reducing the radius of curvature, the attachment of the cytoskeleton to the cell membrane may reduce the tension which develops at any given hydrostatic pressure. For instance, the cytoskeleton of the human red blood cell membrane is attached to the membrane at about 2×10^5 sites.[55] If the cytoskeletal attachments divide the red cell membrane into subdomains, a rough estimate indicates that the diameter of each subdomain should be about 200 times less than the diameter of the spherical red cell. Nevertheless, the red cell membrane is unable to withstand a hydrostatic pressure greater than 2.3 mm H_2O[7] (estimated from measurements of membrane deformability) which corresponds to an osmotic pressure difference of 0.01 mOsm/kg H_2O or of 65 mm H_2O[55a] (calculated from the rehemolysis of resealed ghosts with varied hemoglobin concentrations), which corresponds to an osmotic pressure difference of 0.28 mOsm/kg H_2O. Perhaps the red cell cytoskeleton is anomalous since it is believed that the cytoskeletal attachment sites rapidly rearrange in response to cell deformation.[56] Neutrophils, which have a thicker cortical cytoskeleton than red blood cells, can support a hydrostatic pressure of 2.3 cm H_2O (0.1 mOsm/kg H_2O[57]), and rat renal proximal tubule cells can support a hydrostatic pressure of at least 20 cm H_2O (about 1 mOsm/kg H_2O).[58] It is not likely that the cytoskeleton is able to confer on the phospholipid membrane the ability to resist a considerable transmembrane hydrostatic pressure difference. In many cells, swelling leads to the rapid disruption of the actin cytoskeleton[6]; it may be that, as in red cell membranes, deformation leads to rapid rearrangement of cytoskeletal–membrane bonds. In the axon of the worm *Mercierella enigmata,* however, there is some evidence for an intracellular cytoskeletal structure which may resist osmotic swelling. This species can survive abrupt changes in blood osmolality between 80 and 2300 mOsm/kg H_2O. Close-packed hemidesmosome-like structures are associated with the axon membrane and connected to a cytoskeleton of neurofilaments. This structure may reduce the radius of curvature of the axon membrane, and oppose membrane distention by providing mechanical support for the membrane.[59]

Although most cell membranes are freely permeable to water, some cells swell slowly when transferred to hypotonic solutions—e.g., *Xenopus* oocytes[2]—and some cell membranes are much less permeable than others—apical membranes of many epithelial cells are, in the absence of antidiuretic hormone, much less permeable to water than the basolateral membranes of the same cells.[60] It is possible that variations in the phospholipid composition of the cell membrane may account for some of the variability in cell membrane permeability to water. Some cells—*Xenopus* oocytes—are surrounded by an extracellular matrix which may limit water flow,[61] but this cannot be the whole explanation since expression of the water channel CHIP28 in oocytes greatly increases their rate of swelling in hypotonic solutions.[2] The rate of cell swelling depends, of course, on the ratio of surface area to cell volume, and this may in part account for the slow rate of swelling of large cells such as oocytes in hypotonic solutions. In the absence of vasopressin, the apical membranes of many epithelial cells such as distal tubule cells and amphibian bladder are impermeable to water. Administration of vasopressin increases the concentration of cAMP and causes a population of vesicles which are located just beneath the apical membrane and which contain water channels to fuse with the cell membrane.[60] Water permeability increases both because of the insertion of channels into the membrane and the increase in membrane surface area.

Ito *et al.*[5] have explored the possibility that the cortical actin cytoskeleton may be capable of restricting the transport of water through cells. In *in vitro* experiments, they found that an actin network formed from F actin in the presence of actin-binding protein, which increases the density of interfilament cross-links and the viscosity of F actin networks, prevented water flow across an osmotic pressure difference of at least 65 cm H_2O, the highest value tested, which is equivalent to a concentration difference of about 3 mOsm/kg H_2O. It is perhaps of interest that in responsive epithelial cells, fusion of vesicles with the cell membrane in response to exposure to vasopressin is accompanied by depolymerization of the actin cytoskeleton, specifically at the apical membrane.[62]

Although the anatomical characteristics of the cell membrane may modify the transport of water across the membrane when the transmembrane osmolality differences are small, it seems clear that at sufficiently high $\Delta\Pi$, water flow will observe, at least qualitatively, the principles outlined above with consequent effects on cell volume.

19.3. VOLUME-REGULATED INORGANIC ION TRANSPORT PATHWAYS

Only a few ion transport processes are involved in the response of cells to changes in their volume; they are shown in Figure 19.3. For the most part, the ionic response to RVD is

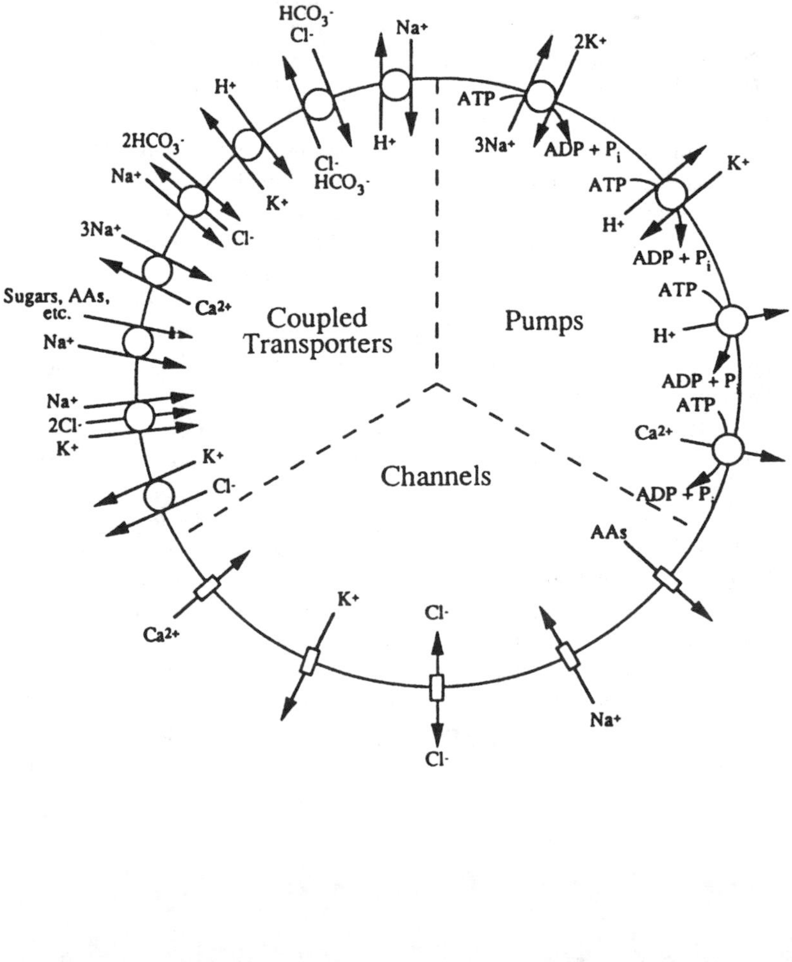

Figure 19.3. The transporters involved in cell volume regulation. The arrows show the usual direction of net flow, but many of the transporters (Na,K,2Cl cotransporter; K–Cl cotransporter; K channels) operate in either direction depending on the electrochemical potential gradients of the relevant species. Taken from Hallows and Knauf[39] and reprinted with permission.

the loss of KCl. KCl loss from swollen cells was first described in amphibian epithelial cells.[63] In most cells, K and Cl are lost through separate conductive pathways (channels), but in red blood cells from a variety of species, and in some other cells, K and Cl are lost together by means of an electroneutral K–Cl cotransporter. In either case, K loss is a passive process and does not require metabolic energy; the energy is supplied by the Na,K pump which maintains a much higher concentration of K inside the cell than outside. Red blood cells from two species, dog and *Amphiuma,* respond to swelling in unique ways which are discussed below.[64]

Na uptake during RVI and RVI after RVD takes place by means of an electroneutral Na–H exchanger, in parallel with the HCO_3–Cl exchanger, by means of an electroneutral Na,K,2Cl cotransporter, or by means of an Na-Cl cotransporter. In each case the direction of Na movement is determined by the electrochemical potential of the involved ions across the cell membrane, but the velocity of transport is determined by cell size. Accumulated Na is exchanged for external K by the Na,K pump so that the net result is the accumulation of KCl within the shrunken cells.

Some of the characteristics of these volume-sensitive transport processes are described next.

19.3.1. Swelling-Activated K and Cl Channels

Activation of conductive pathways for K and Cl by cell swelling was first described in mouse Ehrlich ascites tumor cells and in human peripheral lymphocytes.[36,64] Similar findings have now been described using many cells including platelets, astrocytes, granulocytes, some epithelial cells, and a variety of cultured cells. Swelling-stimulated K channels have been described in epithelial cells from many sources, enterocytes, and hepatocytes. A more complete list of the distribution of these pathways can be found in the review by Lang *et al.*[65] The characteristics of the process have been most extensively evaluated in ascites tumor cells and peripheral lymphocytes.

Movement of an ion across a cell membrane is driven by its electrochemical potential gradient. The electrochemical potential of K inside most cells is higher than the electrochemical potential of K in the outside solution so that, if the membrane permeability to K is increased, K will be lost from the cell. K loss hyperpolarizes the membrane and increases the transmembrane electrochemical potential gradient for Cl and other anions so that anion loss will also increase. On the other hand, loss of anions depolarizes the membrane and increases the driving force for K exit. In most cells, including ascites tumor cells and peripheral lymphocytes, the resting membrane potential is close to the equilibrium potential for K. As a result, activation of K channels in these cells would not be as efficient in promoting KCl loss as activation of anion channels.[65] In fact, it has been found that swelling of ascites tumor cells leads to an approximately 60-fold increase in Cl permeability, but only a 2-fold change in K permeability.[64] Similarly, response of peripheral lymphocytes to an increase in cell volume is characterized by preferential activation of Cl channels, and KCl loss in swollen cells is limited by the smaller increase in K permeability; addition of an ionophore such as valinomycin or gramicidin to swollen cells increases the rate of KCl loss because of the increase in K permeability, and also increases the rate at which the cells return to the original volume.[36] On the other hand, the resting membrane potential of hepatocytes is much closer to the Cl than to the K equilibrium potential so that the driving force for K exit is much higher than the driving force for Cl exit. Swelling of hepatocytes results in an increase in K permeability which, by hyperpolarizing the membrane, increases the driving force for Cl exit, and increases the rate of KCl efflux.[65]

In many cells, including peripheral lymphocytes and ascites tumor cells, swelling activation of KCl loss is not much affected by removal of Ca from the external solution as long as intracellular Ca stores are not modified.[66] Depletion of the intracellular Ca stores of lymphocytes or ascites tumor cells reduces, but does not abolish, the response to swelling in Ca-free solutions, but the response is restored to normal if Ca is added to the external solution. In lymphocytes, the major effect of Ca depletion is to reduce swelling-activated K conductance rather than Cl conductance. Patch clamp studies demonstrate the presence of Ca-activated K channels in these cells with electrophysiological properties similar to those of the swelling-activated K channels. Increased K permeability in response to cell swelling is abolished by Ba and quinine, inhibitors of Ca-activated K channels. Although it seems possible that Ca-activated K channels are responsible for K loss in swollen cells, the persistence of RVD even in Ca-depleted cells complicates interpretation. Treatment of ascites tumor cells with the divalent cation ionophore A23187 in the presence of Ca results in cell shrinkage and increases in K and Cl conductance even in isotonic solutions; in the case of ionophore-activated shrinkage, however, activation of K conductance is much greater than activation of Cl conductance, but the reverse is true in RVD activated by cell swelling.[64] RVD of rabbit renal proximal straight tubules is abolished when the external Ca concentration is reduced to low levels even in cells in which internal Ca stores are not disturbed.[66]

Figure 19.4 shows an experiment of Hoffmann *et al.*[67] in which ascites tumor cells were exposed to quinine and then swollen. The quinine-inhibited cells do not return toward their original volume as do the control cells. On the other hand, addition of the cation ionophore gramicidin to the quinine-inhibited, swollen cells results in their return toward their original volume at a rate even more rapid than that seen with the control cells. The experiment shows that, even if the increased permeability to K produced by cell swelling is blocked, Cl permeability increases and the increased permeability is uncovered when increased K permeability is restored by the ionophore; this is part of the evidence that leads to the conclusion that the swelling-activated increases in the permeabilities of K and Cl occur through separate pathways. In Figure 19.4, cells treated with gramicidin shrank faster than did the control cells since the cells respond to swelling by increasing their Cl permeability more than their K permeability.

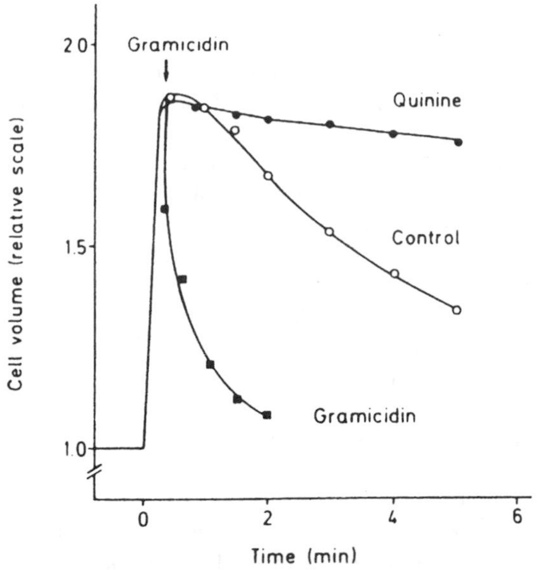

Figure 19.4. The characteristics of RVD in Ehrlich ascites tumor cells. The cells were swollen by exposure to a hypotonic solution. The control cells returned toward their original volume. Quinine, which blocks K channels, aborted RVD. When gramicidin (a cation ionophore) was added to the quinine-treated cells, return toward original volume was even faster than with the control cells, showing that Cl permeability was not rate limiting. Taken from Hoffmann *et al.*[67] and reprinted with permission.

The volume-sensitive channel is relatively specific for K; although Rb is also transported, Na permeability does not increase in response to cell swelling. In fact, in ascites tumor cells, it decreases. The direction of KCl movement in swollen lymphocytes depends on the electrochemical potential difference for the two ions across the membrane; if lymphocytes are swollen in KCl solutions, they swell further rather than shrink.[36]

Since their first discovery in chick skeletal muscle,[65] channels which are activated (increased open probability) by membrane stretch or other mechanical stimulation have been described in a wide variety of cells.[69,70] The stretch-activated channels belong to several classes determined by their ion selectivity. Some channels (SAcats) are selective for cations, but do not discriminate between Na, K, and divalent cations such as Ca. Other channels (SAKs) are selective for K, and still others are selective for anions. The presence of stretch-activated channels in cells which are not excitable or subject to mechanical stimulation suggests the possibility that they may be involved in cell volume regulation. The channels do not respond to transmembrane hydrostatic pressure differences but to membrane tension although the two forces are related by Laplace's law. This has been clearly shown in an experiment with yeast spheroplasts of different sizes. According to Laplace's law [Eq. (19.19)], the same transmembrane hydrostatic pressure will result in different membrane tensions in spheroplasts of differing diameters. Using the spheroplasts, it was found that plots of mechanosensitive current as a function of membrane tension were independent of spheroplast diameter, but plots of current against pressure varied with diameter.[71] It can be concluded that stretch-activated channels respond to membrane tension.

There have now been a number of reports of SAKs which are activated both by membrane stretch and by increases in cell volume, and the characteristics of the activated current are the same in the two circumstances.[69] To date, most of the observations have been made with amphibian renal epithelia. Activation of SAKs requires a fairly high membrane tension, and doubt has been expressed about whether hydrostatic pressure differences sufficient to generate such tension could be supported by cell membranes. Moreover, in the experiments in which the effect of cell swelling on the activity of SAKs is evaluated, cells are frequently swollen by 100% or more, which raises questions about the relevance of the findings to physiological conditions. However, Sackin[72] pointed out that, in a patch clamp experiment, the patched membrane is partially isolated from the tension which develops in the rest of the cell membrane when cells are swollen. Calculations indicate that, if a patch of the diameter of the whole cell were possible, a cell volume increase of as little as 1% would develop sufficient membrane tension to activate SAKs.

The chloride channel in ascites tumor cells has been demonstrated in intact cells in which volume was increased by adding glycine to the external solution with resultant increase in osmolyte uptake through the operation of the Na–amino acid cotransporter.[73] In ascites tumor cells, Cl permeability is activated when the divalent cation ionophore A23187 is added to the external solution containing Ca, and in cell-attached patches the same permeability is activated either by Ca or by increased cell volume. In membrane patches, however, the Cl channel is not activated by Ca so that the role of Ca in activation of permeability in intact cells must be indirect.[64] The channel is inhibited by the Cl channel blockers dephenylamine-2-carboxylate and by indocrinone. In lymphocytes, activation of Cl permeability exhibits a volume threshold, and spontaneously inactivates in a time-dependent manner even if cell volume is maintained elevated. Unlike the ascites tumor Cl channel, the lymphocyte channel does not seem to be Ca-dependent; Ca-depletion of lymphocytes does not inhibit swelling activation of Cl permeability. Ca activates a Cl permeability in lymphocytes, but the magnitude of the increase is less than that caused by cell swelling.[36] The Cl channels of both ascites tumor cells and lymphocytes are relatively nonspecific for small inorganic anions, but do not transport large organic anions. In mouse renal proximal tubule cells and MDCK cells[65] (but not rabbit renal proximal tubule cells[66]), HCO_3 rather than Cl accompanies K during RVD.

Stretch- and volume-activated Cl channels have now been found in a variety of epithelial cells, in cultured cells, and in *Xenopus* oocytes.[74] The swelling-activated channels in oocytes are Ca- and voltage-independent. The anion selectivity of this channel (Br >Cl >I) is somewhat different from that of voltage-activated Cl channels and the cystic fibrosis transport regulator (CFTR)-mediated, cAMP-activated Cl channel.

Overexpression of P-glycoprotein, the product of the multidrug resistance gene, in NIH3T3 fibroblasts leads to the appearance of volume-activated Cl currents in the cells.[75]

19.3.2. K–Cl Cotransport

For cells with low Cl concentrations, K–Cl cotransport is a relatively inefficient means of losing KCl since the driving force for KCl loss by coupled cotransport is proportional to $K_c \times Cl_c$ where K_c and Cl_c are the intracellular concentrations of K and Cl. Moreover, in such cells the maximal rate of KCl loss by means of a cotransporter will be less than the maximal rate of loss through conductive pathways. In most red blood cells, both K and Cl are present at relatively high concentration, and the red blood cells of a variety of species including duck, sheep, human, rabbit, pig, and teleost fishes[76,77] demonstrate K–Cl cotransport either in immature cells or in both mature and immature cells. Sheep are genetically dimorphic with respect to the regulation of red cell cation concentration. Cells with low K and high Na concentration (LK) have a much lower Na,K pump rate than cells with high K and low Na concentration (HK), and K–Cl cotransport can be demonstrated in mature LK cells but not HK cells.[76] Dog red cells have a low intracellular K concentration, but it is possible to demonstrate K–Cl cotransport in these cells provided intracellular K concentration is increased at the expense of Na.[36] K–Cl cotransport in red cells is activated by cell swelling; swelling of young rabbit red cells by 25% results in a tenfold increase in K–Cl activity,[78] and 10% swelling of LK sheep red cells increases cotransport activity severalfold.[76] Activation of K–Cl cotransport by cell swelling is followed by a decrease in cell size in duck red cells,[23] pig reticulocytes,[77] both HK and LK sheep reticulocytes,[76] and young human cells.[79] Swelling-activated K–Cl cotransport has been demonstrated in ascites tumor cells, but only if intracellular Ca is depleted, in which case conductive loss of KCl is diminished, or if the measurements are made at low pH.[64] Reports that K–Cl cotransport is activated by swelling of amphibian gallbladder cells,[37] partly based on the sensitivity of the fluxes to the inhibitor bumetanide, have subsequently been challenged since K loss persists when Cl is replaced by NO_3 or SCN.[80] K–Cl cotransport has been reported in *Necturus* choroid plexus epithelial cells[81] and in several renal epithelia.[76]

K–Cl cotransport is relatively specific for K and Cl. Rb replaces K, but Na does not. Br is a somewhat better substrate than Cl, but SCN, I, acetate, CH_3SO_4, PO_4, SO_4, and sulfonate will not substitute for Cl; in fact, NO_3 inhibits cotransport when added to a solution with fixed Cl concentration.[82] Inhibition of K transport on replacement of Cl with NO_3 is frequently taken as evidence for K–Cl cotransport. In addition to evidence that cotransport is specific for K and Cl (or a few substitutes), unequivocal confirmation of the presence of cotransport requires (1) demonstration that the fluxes of K and Cl through the cotransporter have a fixed, integral stoichiometry and (2) demonstration that the downhill movement of one of the participants through the cotransporter is able to drive the simultaneous uphill (against an electrochemical potential gradient) movement of the other. Because of the very high permeability of red cell membranes to Cl, it is very difficult to measure the stoichiometry of K and Cl movements carried out by the cotransporter. Lytle and McManus[83] showed that a sizeable fraction of Cl efflux from DIDS-treated duck red cells is inhibited by bumetanide and dependent on the presence of intracellular K, suggesting a fixed stoichiometry. If the fixed K : Cl stoichiometry is 1 : 1, then one would expect K–Cl cotransport to be electroneutral. K–Cl cotransport in duck red cells is electroneutral,[84] and so is K–Cl cotransport in human red cells since membrane potential does not change when K–Cl cotransport is activated by cell swelling.[85] Moreover, alteration of transmembrane potential over a wide range by the use of a Na ionophore and varied transmembrane Na gradients did not alter the magnitude of swelling-activated K uptake (K–Cl cotransport). In choroid plexus epithelial cells, the flux of K equals the flux of Cl over a wide range of experimental conditions.[81] Brugnara *et al.*,[86] using human, LK sheep, and rabbit red cells, showed that if both band 3-mediated anion exchange and carbonic anhydrase are inhibited, then clamping of the transmembrane potential by the use of anions more permeable than Cl (NO_3 and SCN) permitted the demonstration of the movement of K against an electrochemical potential gradient driven by an oppositely oriented electrochemical potential gradient for Cl. Uphill movement of K driven by a downhill movement of Cl and uphill movement of Cl driven by a downhill movement of K has been shown in choroid plexus epithelial cells.[81]

Several studies have been reported in which the steady-state kinetics of K–Cl cotransport were evaluated with a view toward establishing the reaction mechanism. The studies are hampered by the low rates of transport and the high values for $K_{1/2}$ for both K and Cl on the two sides of the membrane. The studies are also hampered by the need to replace the cotransported ions with substitutes which do not themselves interact with the cotransporter. Although this is relatively easy for K replacement, it is not always clear that the common substitutes for Cl are inert. In swollen human red cells the cotransporter is kinetically asymmetric: the $K_{1/2}$ for K at the outside is much greater than the $K_{1/2}$ for K at the inside, and V_M for K influx is greater than V_M for K efflux when the measurements are made at constant high Cl concentrations inside and outside.[86] But, in swollen LK sheep red cells, there is little difference between the $K_{1/2}$ for K at the outside (measured with Rb), and the $K_{1/2}$ for K at the inside (measured with K), and V_M for efflux was greater than V_M for influx.[87] In sheep red cells, influx is maximal when both internal K and Cl are absent, and addition of either intracellular K or Cl inhibits influx; further addition of the second ion is without effect. On the other hand, extracellular Cl alone, but not K alone, inhibits efflux, and addition of K in the presence of Cl further inhibits. From the results of these transinhibition experiments, and from the results of steady-state experiments in which the effect on V_M and $K_{1/2}$ of variation of the concentration of the two substrates on either side of the membrane was measured, Delpire and Lauf[87] concluded that, at the inside, binding of K and Cl is random, but at the outside, binding is ordered with Cl adding before K. In swollen human red cells, however, when measurements are made at high constant Cl both inside and outside, inside K is without effect on K influx, and outside K stimulates K efflux.[88] One conclusion to be drawn from these experiments is that there is very little K–K exchange through the K–Cl cotransporter, and that swelling increases transport in both directions. The net flow of KCl will be determined by the transmembrane electrochemical potential gradient of both K and Cl. Practically, the result confirms the validity of the common practice of measurement of K influx, which is easier and more reproducible than the measurement of K efflux, in evaluating the effect of swelling on K–Cl cotransport rate. Swelling of LK sheep red cells reduces $K_{1/2}$ for external K, and increases V_M for K influx.[89,90]

Investigation of swelling-activated K–Cl cotransport has been seriously handicapped by the lack of a specific inhibitor. The loop diuretics bumetanide and furosemide, which inhibit Na,K,2Cl cotransport at relatively low concentration, can also inhibit K–Cl cotransport, but only at millimolar concentrations. Inhibition is promoted by external K.[76,77] Inhibitors of anion exchange mediated by band 3—DIDS, H_2DIDS, and SITS—have also been shown to inhibit K–Cl cotransport and inhibition requires external Cl and is promoted by external K.[77] Vitoux *et al.*[91] found that [(dihydroxyindenyl)oxy] alkanoic acid inhibits K–Cl cotransport in human red cells with half-maximal inhibition at 10^{-5} M and little effect on Na,K,2Cl cotransport, but the inhibitor is not readily available. Two analogues of bumetanide have been reported to inhibit swelling-activated K–Cl cotransport in human red cells with half-maximal inhibition in the range of 10^{-4} M, and one of them has no effect on Na,K,2Cl cotransport.[92] In LK sheep red cells, an antibody raised by immunizing HK sheep with LK sheep red cells significantly inhibits K–Cl cotransport.[76,77] Caution must be exerted in identifying K–Cl cotransport from the effects of relatively nonspecific inhibitors. Mention was made above that bumetanide inhibition of KCl loss from swollen amphibian gallbladder cells confused the assignment of the fluxes to conductive pathways or cotransport,[80] and it has recently been reported that quinine, believed to be a fairly specific inhibitor of Ca-activated K channels, also inhibits K–Cl cotransport in LK sheep red cells.[93]

K–Cl cotransport is frequently difficult to demonstrate in mature red cells (except for LK sheep red cells), but is present in young red cells separated by gradient density centrifugation.[76,77] It has been suggested that the presence of volume-sensitive cotransport in young red cells is related to the transition of reticulocytes to mature cells during which cell volume decreases. Swelling-activated cotransport is also prominent in resealed ghosts prepared from human red cells even if the cells are prepared from mature cells in which K–Cl cotransport is hard to detect.[76,77] K–Cl cotransport is also prominent in the red cells of individuals homozygous for sickle-cell hemoglobin or hemoglobin C.[77] Although it is possible that increased cotransport in Hgb S cells is related to the relatively young age of the cells since hemolysis is a prominent feature of the disease, hemolysis is not significant in Hgb C disease. Olivieri *et al.*[94] have shown that a number of mutations of the locus that is mutated in Hgb S and Hgb C result in mature red cells in which K–Cl cotransport can be demonstrated. Treatment of mature red cells of a variety of species in which K–Cl cotransport cannot be demonstrated with *N*-ethylmaleimide unmasks latent cotransport activity by a mechanism which is at present unclear.[77]

Mg and other divalent cations have long been known to inhibit K–Cl activity in red cells.[77] The effect of Mg is complex and in part related to the effect of ATP. In red cell ghosts, ATP stimulates cotransport activity. When ghosts are prepared to contain ATP in the complete absence of Mg, K–Cl cotransport is no greater than it is in ghosts which do not contain ATP. When low concentrations of Mg are added to the ghosts which contain ATP, cotransport is markedly stimulated presumably because MgATP is formed.[95] There is evidence that in ghosts, stimulation of K–Cl cotransport activity by ATP results from the phosphorylation of a tyrosine group somewhere in the membrane since cotransport in ATP-free ghosts is higher when the ghosts contain vanadate, an inhibitor of tyrosine phosphatases, and stimulation of MgATP is prevented by genistein, a tyrosine kinase inhibitor.[95] At higher concentrations, Mg inhibits cotransport activity both in ghosts which contain ATP and in ATP-free ghosts as it does in intact cells. Some hint of the mechanism by which Mg inhibits cotransport activity may be provided by the observation that a number of soluble polycations and cationic amphiphiles also inhibit cotransport activity in ghosts.[96] Since Mg and the soluble polycations shield negative membrane surface charge, and since cationic amphiphiles insert into cell membranes and neutralize negative charges, it is possible that all of the agents exert their inhibitory effects by reducing the negative surface charge of the interior of the cell membrane. How such neutralization might affect cotransport rate is not known.

It was stated above that osmotic equilibrium across cell membranes cannot be maintained by active transport of water. However, many epithelia—gallbladder, intestine, renal proximal tubule, choroid plexus—are able to transport water against a considerable transepithelial osmotic gradient. The phenomenon has been explained by supposing there is a subcellular compartment—the lateral intracellular spaces—bounded by entrance and exit barriers with different permeability characteristics to ions and water so that active transport of ions accompanied by an osmotically obligated flow of water across the entrance barrier develops a hydrostatic pressure within the compartment which drives hydraulic flow of the solution across the exit barrier.[97] Recently, however, evidence has been presented that water movements against osmotic gradients in amphibian choroid plexus are phenomenologically coupled to the cotransport of K and Cl.[81] Some of the evidence is shown in Figure 19.5. In these experiments, intracellular K and Cl concentrations were estimated by means of intracellular electrodes and cell volume was estimated by means of intracellular electrodes sensitive to choline concentration; cells were loaded with choline before the start of the experiment. Addition of the indicated substances were made to the solution bathing the ventricular side of the epithelium. When ventricular osmolality was increased by the addition of mannitol, the

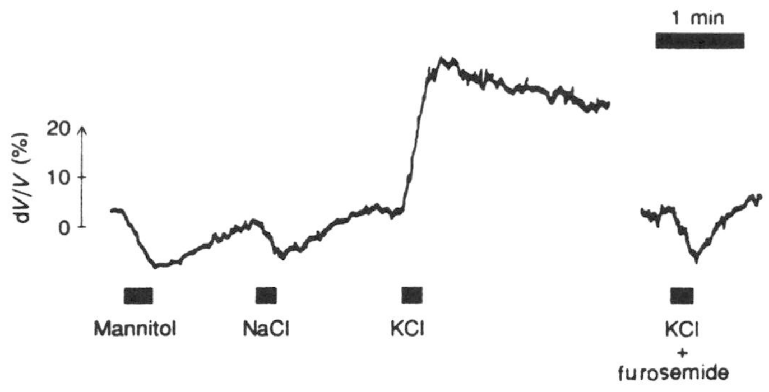

Figure 19.5. Volume changes of choroid plexus epithelial cells after exposure of the ventricular surface to hypertonic solutions of several solutes. Exposure to 100 mM mannitol or 50 mM NaCl caused cell shrinkage, as expected. Exposure to 50 mM KCl, however, resulted in immediate cell swelling which was abolished by furosemide. Taken from Zeuthen[81] and reprinted with permission.

cells, as expected, shrank. By varying the extracellular concentration of mannitol, it was possible to estimate the hydraulic permeability coefficient of the membrane. The cells behaved in a similar way when ventricular osmolality was increased by adding NaCl. When extracellular osmolality was increased by adding KCl, however, the cells swelled, and the increase in volume occurred before there was a significant increase in intracellular osmolality. Estimates of the hydraulic permeability coefficient in the presence of extracellular KCl were about twice as high as the estimates made with extracellular mannitol. Exposure of the cells to the loop diuretic furosemide or bumetanide reversed the volume increase when KCl was added to the ventricular solution. The characteristics of the procces satisfied the phenomenological criteria for cotransport. In all cases K flow, Cl flow, and H_2O flow were in the ratio 1K : 1Cl : 500 H_2O, so that the transported fluid had a fixed concentration of about 110 mM. Flow of K or water required Cl, and flow of Cl and water required K. Under a variety of circumstances the downhill flow of one of the participants was shown to energize the uphill flow of the other two. On the basis of these and other findings it was proposed that the cotransporter is an "osmotic engine"[81,98] and that some space within the cotransporter serves the same purpose as the lateral intracellular space in the Curran–MacIntosh model.[97]

19.3.3. K–H Exchange Coupled with Cl–HCO_3 Exchange

Amphiuma red cells respond to hypotonic swelling by losting KCl as do other red cells. However, the loss seems to be accomplished by a unique mechanism[99]: exchange of intracellular K for extracellular H, and parallel exchange of intracellular Cl for extracellular HCO_3 by means of the band 3 anion exchanger. Accumulated intracellular H and HCO_3 form H_2O and CO_2 which is lost from the cell, and the net result is KCl loss. Evidence for the mechanism includes the observation that the transmembrane potential does not change during K loss, and that variation of the membrane potential by manipulation of the ratio of intracellular K concentration to extracellular K concentration in the presence of valinomycin does not modify the magnitude of swelling-stimulated K loss. The pH of the extracellular solution increased during RVD.[33] Thermodynamic distinction between K–Cl cotransport and the K–H exchange mechanism cannot be made since the driving force for K–H exchange in the presence of vigorous Cl–HCO_3 exchange does not differ from the driving force for K–Cl cotransport.

The response of *Amphiuma* red cells to cell swelling is reproduced by increasing intracellular Ca concentration, and there is an increase in intracellular Ca concentration during RVD; calmodulin antagonists inhibit volume-stimulated K–H exchange.[33,36] The primary response to swelling of these cells must be an increase in intracellular Ca; how this is accomplished is not known.

Lew and Bookchin[100] suggested that the experimental findings with *Amphiuma* red cells might be explained if a Ca-dependent K channel is activated by swelling, and the membrane has a high conductance to H. Change in membrane potential would drive the accumulation of H producing the apparent K–H exchange. The observation that membrane potential does not change during RVD casts doubt on this mechanism.

19.3.4. Na–Ca Exchange

Mature red cells of dogs and other carnivores do not have Na,K pumps. As a result, the intracellular concentrations of Na and K differ little from their concentrations in the extracellular solution, and the cells are not able to maintain their cell volume, or respond to changes in cell volume, by regulating the downhill movement of Na and K. Dog red cells do, however, have a membrane-bound Na–Ca exchanger, and it has been shown that, both *in vivo* and *in vitro,* dog red cells maintain their steady-state volume by exchanging intracellular Na for extracellular Ca, and then pumping Ca out by means of the membrane Ca pump, a mechanism which has also been described in the red cells of the closely-related ferret[101a]; energy for outward movement of Na against its gradient is supplied by the downhill inward movement of Ca.[36,101] When dog red cells are swollen, Ca influx and Na efflux through the Na–Ca exchanger increase so that cell Na content and therefore cell volume decrease. Although swelling-activated K–Cl exchange in mature dog red cells under physiological conditions will not result in cell shrinkage because there is little difference in K concentration between the cells and plasma, swelling-activated K–Cl cotransport activity in these cells can be demonstrated if intracellular K is increased at the expense of Na.

19.3.5. Na,K,2Cl Cotransport

Early measurements of ouabain-insensitive Na and K movements in red cells suggested that the Na and K fluxes are coupled. It was later shown that the coupled movements require the presence of Cl; although Br can replace Cl, many other anions—e.g., I, NO_3, SCN, CH_3SO_4—are unable to support the coupled movements. This is the same anion selectivity as that of K–Cl cotransport. The loop diuretics furosemide and bumetanide, which inhibit K–Cl cotransport, also inhibit the coupled movements of Na and K, but at much lower concentrations.[101] Geck *et al.*[102] demonstrated with thermodynamic rigor that Na, K, and Cl movements in Ehrlich ascites tumor cells represent Na,K,2Cl cotransport. The cotransporter has now been found in many cells including red cells from a variety of species, cultured cells, secretory and absorptive epithelial cells, squid axons, and others, and seems to be ubiquitous (although it is not found in LK sheep red cells). Cotransport of Na,K,2Cl has been found to be electroneutral whenever the measurement has been made, but there is some doubt remaining about the stoichiometry. If the stoichiometry of Na and K through the cotransporter is 1 : 1, then electroneutrality requires simultaneous transport of two Cl. Stoichiometries such as 2Na : 1K : 3Cl, which would also result in electroneutral cotransport, have been suggested.[103] Such a stoichiometry implies a sigmoid relation between cotransport rate and Na concentration, but, when measured, the relation has usually been found to be hyperbolic. The cotransporter from shark rectal gland has now been cloned[104] and, using probes obtained from the shark cDNA, three isoforms of a closely related cotransporter have been identified in mammalian kidney[105]; the three isoforms have specific and different distribution within the kidney. Variation in the characteristics of cotransport in different cells may be attributable to the presence of different isoforms. In many cells, including those from renal tissues, Na–Cl cotransport has been described which has been difficult to distinguish from Na,K,2Cl cotransport, but the recent report of the cloning of an Na–Cl cotransporter[106] different from the Na,K,2Cl cotransporter[107] seems to confirm the reality of Na–Cl cotransport.

Na,K,2Cl cotransport kinetics have been studied in detail in duck red cells, and the results have been found to be consistent with a model proposed by Lytle and McManus and shown in Figure 19.6.[108] The model states that the addition of ions is ordered on both sides of the membrane, and that the ion which adds first at one side of the membrane is released first at the other side. Thus, in the model, Na adds first at the outside and is released first at the inside. The model assumes that the transport sites move from one side of the membrane to the other only when they are occupied by ions. The model predicts the observed dependence of transport rate on the concentration of ions at the two sides of the membrane. The cotransport of duck red cells is, under appropriate conditions, able to carry out bumetanide-sensitive exchanges of Na and K. The model predicts that K–K exchange requires Cl outside and Na and Cl inside, and Na–Na exchange requires K and Cl inside. These predictions have been verified experimentally. Although the model explains in detail the characteristics of Na,K,2Cl cotransport in duck red cells in which it has been studied in most detail, the model, at least in its unmodified version, has been difficult to reconcile with findings from some other preparations.[101]

It was early shown that the shrinkage of duck red cells in hyperosmotic solutions activates Na,K,2Cl cotransport and, if the extracellular K concentration is elevated, ions move into the cells and they return toward their volume in isotonic solutions.[24] Activation of Na,K,2Cl cotransport by cell shrinkage has now been demonstrated in a variety of cells including ascites tumor cells, rabbit renal medullary cells, squid axons, and retinal pigment epithelium.[65] However, RVI cannot be demonstrated in many cells in which cotransport is activated by hyperosmotic shrinkage, although RVI after RVD frequently can. The driving force for Na,K,2Cl cotransport is

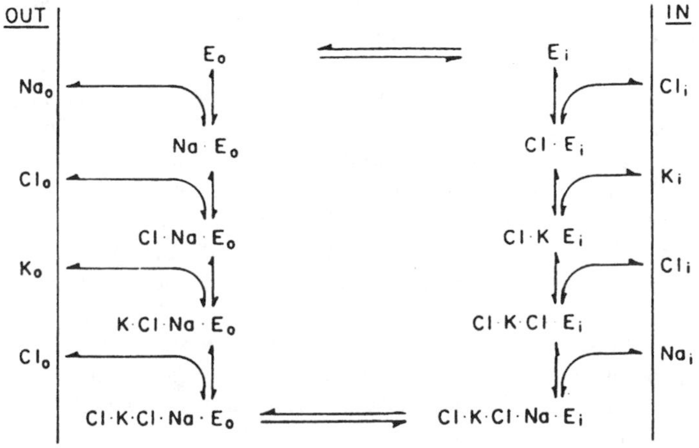

Figure 19.6. Glide symmetry model for Na,K,2Cl cotransport proposed by McManus and his collaborators. The ions are released at the inside in the same order that they add at the outside. Taken from Parker and Dunham[101] and reprinted with permission.

proportional to the transmembrane difference in the chemical potential of the involved ions (since cotransport is electroneutral, membrane potential will not affect the direction of net movement). In ascites tumor cells, shrinkage of cells in hyperosmotic solutions activates Na,K,2Cl cotransport, but the shrunken cells do not regain their original volume. If the cells are first swollen and allowed to undergo RVD, and then shrunk by resuspending them in an isotonic solution, the cells activate cotransport and return toward their original volume in isotonic solutions.[64] When cells shrink in hypertonic solutions, the intracellular concentrations of Na, K, and Cl increase in proportion to the increase in extracellular concentrations. In cells which have lost KCl during RVD, however, the intracellular concentrations of these ions relative to their extracellular concentrations when the cells are resuspended in isotonic solutions are less than in untreated cells suspended in isotonic solutions. In ascites tumor cells, uptake of ions during RVI after RVD and restoration of original cell volume correlates with an increase during swelling of cell Cl concentration, and the equilibrium volume is linearly related to cell Cl concentration.[109] Uptake of Cl by the cells is carried out by the Na,K,2Cl cotransporter, and cell Cl concentration in these cells is an important determinant of equilibrium cell volume.

In renal medullary cells, uptake of Na is believed to be carried out by Na–Cl cotransport in isotonic solutions, but by Na,K,2Cl cotransport in hypertonic solutions. Na–Cl cotransport has been difficult to distinguish from Na,K,2Cl cotransport, and it has been proposed that antidiuretic hormone changes the Na–Cl cotransporter to an Na,K,2Cl cotransporter.[110] The availability of a primary structure for the two cotransporters may permit resolution of the question of how such a change is accomplished.

19.3.6. Na–H Exchange

Na–H exchange has now been described in a great variety of cells including red cells of *Amphiuma,* dog, rabbit, and trout, in osteoclasts and many cultured cells, gallbladder cells, and many epithelial cells.[65] Na–H exchange is, as expected, electroneutral, and the stoichiometry is believed to be 1:1 since the relation between exchange rate and Na concentration is hyperbolic. The relation between exchange rate and internal H ion concentration is, however, markedly sigmoid, and this, together with some other observations, has led to the conclusion that, in addition to serving as a substrate, H acts at an inside modulator site to increase exchange activity.[111] The exchange is bidirectional, and the direction of net flow of Na depends on the transmembrane difference in the chemical potentials of Na and H. As expected, external H is a competitive (with Na) inhibitor of exchange. In many cells, Li can replace Na. Na–H exchange is inhibited by amiloride and, less effectively, by quinidine.

Stimulation of Na–H exchange by cell shrinkage with subsequent RVI was first described in *Amphiuma* red cells by Siebens and Kregenow[112] and by Cala.[113] Cell shrinkage has since been found to activate Na–H exchange in many cells.[36,37] In lymphocytes, cell shrinkage activates exchange by increasing the affinity of the external modulator site for H.[111] In contrast to the Na,K,2Cl cotransporter, the driving force for Na–H exchange greatly favors the accumulation of intracellular Na at the expense of intracellular H. In most cells, carbonic anhydrase forms H and HCO_3 from carbonic acid, and the HCO_3 remaining after the loss of H is exchanged for extracellular Cl by means of the band 3 anion exchanger. Na is then exchanged for K by the Na,K pump, and the net result of the operation of the Na–H exchanger is the accumulation of intracellular KCl and osmotically obligated water and an increase in cell volume. Although the energetics of the Na–H exchanger for increasing cell volume is much more favorable than the energetics of the Na,K,2Cl cotransporter, RVI after hyperosmotic shrinkage mediated by activation of Na–H exchange is rarely seen, although RVI and RVD is not uncommon. A reasonable case can be made that this observation is made in cells which rely on Na,K,2Cl cotransport since the intracellular concentrations of K and Cl are relatively low when cells which have undergone RVD are resuspended in isotonic solutions, but it is difficult to see how a similar explanation can be used for cells which rely on Na–H exchange.

Four isoforms of the Na–H exchanger have been identified in mammalian tissues. The most common and most intensively studied of these, NEH-1, is found in many cells and is present in the basolateral membranes of many epithelial cells. It is activated by a decrease in intracellular pH, by cell shrinkage, and by many hormones; its main physiological function may be regulation of intracellular pH. Na–H exchange is activated when cultured cells are exposed to mitogens. Operation of the exchanger results in intracellular alkalinization, which may be necessary for cell division, and accumulation of intracellular KCl and water, which is necessary for the increase in total cell volume after cell division.

A recently cloned isoform, NEH-4, when transfected into fibroblasts deficient in Na–H exchange, confers shrinkage-activated, amiloride-sensitive Na–H exchange on the cells. At high osmolality (420 mOsm/kg H_2O and greater), activation of NEH-4 was greater than activation of NEH-1. NEH-4 is expressed in renal inner medullary and papillary cells, which are exposed to environments in which high osmolality prevails.[114]

19.4. ROLE OF ORGANIC OSMOLYTES IN CELL VOLUME REGULATION

Although many of the immediate responses of cells to sudden changes in their volume involve the exchange of inorganic ions between the cytoplasm and the external solution, some of the immediate responses and many of the long-term responses involve the accumulation or release of organic osmolytes. Control of cell volume by regulation of cell organic osmolyte content has some obvious advantages. As mentioned above, inorganic ions at high concentration destabilize proteins, but the organic osmolytes most commonly involved in volume regulation—e.g., amino acids, methylamines, polyols, sugars—not only do not perturb protein structure, but in fact can counteract the destabilizing effects of high inorganic ion concentration. In cells in which membrane potential is primarily a K diffusion potential, changes in intracellular K concentration in response to cell volume changes will modify resting membrane potential and perhaps impair intercellular communication by means of action potentials. It is therefore not surprising that cells in which organic osmolytes are important for volume regulation (or at least cells in which such mechanisms are most often studied) are those which are exposed to the widest ranges of extracellular osmolality changes—marine invertebrates which inhabit intratidal environments, marine vertebrates, vertebrate inner medullary or papillary renal cells, and neural cells. It is, however, clear that cells from other tissues utilize organic osmolytes to regulate their volume to varying extents.

Inner medullary and papillary renal cells from vertebrate kidney respond to increased extracellular osmolality in part by increasing the intracellular concentrations of sorbitol, myoinositol, betaine, glycerophosphorylcholine, and taurine. Sorbitol is formed from glucose by the enzyme aldose reductase. Shortly after cells which have been maintained in an isotonic solution are exposed to a hypertonic solution, the cell content of aldose reductase begins to increase and continues to increase over the course of several days.[115] Increased aldose reductase content results from an increased rate of transcription of the aldose reductase gene rather than from increased translation or an increase in the life span of mRNA for aldose reductase or of the enzyme.[115] If the cells are returned to an isotonic solution, further synthesis of aldose reductase is reduced, but the intracellular content of the enzyme remains high for a number of days, and the intracellular concentration of sorbitol depends on the rate of sorbitol exit. Increase in myoinositol content results from increased uptake of the substance from the extracellular solution by a Na-dependent cotransporter, and increase in cell betaine and taurine results from increased uptake by means of Na- and Cl-dependent cotransporters. In each case, V_M for cotransport increases over the course of 24 hr, but $K_{1/2}$ for the transported species is little changed. Transcription of the genes for the Na–myoinositol cotransporter and for the Na–Cl–betaine cotransporter increases in hypertonically shrunken cells, and this increases cell mRNA content for the two cotransporters, and presumably increases the rate of cotransporter synthesis.[116]

The mechanism by which transription of the genes for aldose reductase and for the cotransporters is increased in shrunken cells is under active investigation. The process seems to be similar to that utilized by bacteria (*E. coli* and *S. typhimurium*) in responding to changes in external osmolality. Exposure of the bacteria to hypertonic solutions increases transcription of the kdp ABC operon which encodes a high-affinity K transporter (Kdp), and of the gene pro U which encodes a high-affinity betaine transporter. In hypertonically shrunken cells, Kdp synthesis is increased first, and increased transcription of the kdp ABC operon appears to be activated by decreased turgor. Increased activity of Kdp increases intracellular K concentration, and increased transcription of pro U begins only after the increase in the concentration of intracellular K.[117] There is evidence that in isotonic solutions, pro U transcription is diminished by a protein repressor. Inhibition by the repressor is relieved at a K concentration of about 250 mM by a direct interaction of K with the repressor. The concentration of K at which repression is relieved is in the range of the concentration at which K destabilizes proteins and other macromolecules.

Increased transcription of the gene for aldose reductase seems to be related to increased intracellular concentration of K, or of the sum of the Na and K concentration. Direct evidence that transcription of the gene for the betaine cotransporter in renal medullary cells is controlled by the Na + K concentration has been obtained in experiments in which luciferase reporter constructs containing the promoter region of the Na–betaine cotransporter gene were transfected in MDCK cells. Exposure of the transfected cells to hypertonic solutions induced luciferase activity.[118]

Exposure to hypotonic solutions leads to the rapid release of organic osmolytes. Hypotonic swelling of ascites tumor cells results in the release of taurine,[64] hypotonically swollen astrocytes lose glutamate, aspartate, and taurine,[119] and hypotonic swelling of inner medullary and papillary renal cells increases the efflux of sorbitol, betaine, myoinositol, and other organic osmolytes.[37] Sorbitol permeability of renal cells is low in isotonic solutions, but when cells are swollen in hypotonic solutions, sorbitol permeability increases within 10 sec. The increase in permeability increases with cell size, and may reach 80 times the baseline permeability in isotonic solutions.[37] Swelling induced increase in the sorbitol permeability of renal collecting duct cells requires external Ca, and there is evidence that sorbitol permeability is increased by the fusion of submembrane vesicles containing sorbitol channels with the cell membrane.[31]

Hypotonic swelling of skate red cells results in greatly increased efflux of taurine. On the basis of the sensitivity of the taurine efflux to the band 3 inhibitors DIDS, NAP-taurine, and niflumic acid, it was suggested that taurine efflux takes place by way of the band 3 anion exchanger.[120] Hypotonic swelling of flounder red cells, however, increased the permeability of the cells to taurine, glucose, and uridine.[121] The kinetic characteristics of the uptake of all three osmolytes were identical, and uptake was not saturated even at very high concentrations. Transport was inhibited by DIDS, niflumic acid, furosemide, and several Cl channel blockers. Transport of the three osmolytes was thought to take place through a channel rather than by means of transporters. Hypotonic swelling of C6 glioma cells causes the loss of inositol and taurine; characteristics of the efflux suggest that transport of the two substrates takes place by the same mechanism, and efflux of both was inhibited by the same Cl channel blockers. Evidence was presented that the cells displayed a volume-activated Cl conductance.[122] It may be that the efflux of a variety of hydrophilic osmolytes takes place through a swelling-activated Cl channel. It may be of interest that the P-glycoprotein, product of the multidrug resistance gene, which transports a variety of hydrophobic compounds including cytotoxic drugs out of cells, behaves as a swelling-activated Cl channel.[75]

Motais *et al.*[123] examined the response of flounder red cells to swelling. Hypotonic swelling of the cells by suspension in solutions in which Na concentration was reduced with or without urea to make up the osmotic deficit (the red cells are permeable to urea) resulted in K loss which was independent of the presence of Cl, and loss of taurine and choline. Isotonic cell swelling produced by hormonal activation of Na–H exchange or by replacement of part of the extracellular NaCl with NH_4Cl (the cells are permeable to NH_3) results in K loss which is Cl-dependent, and very little loss of taurine and choline. The responses were too rapid to be mediated by means of increased protein synthesis. Flounder red cells are able to adjust their response to swelling; swelling at low ionic strength activates organic osmolyte loss and K loss by a Cl-independent pathway, but swelling at physiological ionic strength activates Cl-dependent K loss, and little loss of organic solutes.

19.5. ISOVOLUMETRIC VOLUME REGULATION AND VOLUME REGULATION IN ISOTONIC SOLUTIONS

The experiments of Motais *et al.*[123] suggest that the response to cell swelling depends on the circumstances under which cell swelling occurs. In order to optimize the changes which occur in response to cell volume change, many experiments involve the sudden exposure of cells to solutions with osmolalities considerably different from that of physiological extracellular fluid. Grantham and his colleagues have reported a series of experiments in which the volume of proximal tubules was measured as the osmolality of the extracellular solution was gradually altered.

Proximal tubules exposed to sudden changes in extracellular osmolality from 293 mOsm/kg H_2O to 240, 190, or 140 mOsm/kg H_2O first swell as nearly perfect osmometers, and then undergo RVD and return toward their original volume over a 5-min period. If the tubules are exposed to solutions in which the osmolality of the extracellular solution is reduced at the rate of 1.5 mOsm/kg H_2O per min, they do not change volume until the extracellular osmolality reaches 167 mOsm/kg H_2O.[124] At lower osmolalities, the tubules gradually swell. Constant volume was maintained when osmolality was reduced at a rate of 27 mOsm/kg H_2O per min or less, but the tubules swelled when the rate of change was 42 mOsm/kg H_2O per min.

When exposed to hypertonic solutions, the tubules shrank, but did not return to isotonic volume over a 10-min period unless the extracellular solution contained 0.5 mM butyrate (along with alanine, citrate, lactate, acetate, and glucose) in which case sudden shrinkage of the cells in 400 mOsm/kg H_2O solutions was followed by RVI toward the isotonic volume.[125] If the extracellular solution did not contain butyrate, proximal tubules exposed to an increase in extracellular osmolality at a rate of 1.5 mOsm/kg H_2O per min maintained their volume up to an osmolality of 361 mOsm/kg H_2O, but shrank when the rate of change was 3 mOsm/kg H_2O per min. In solutions containing butyrate, however, the tubules maintained constant volume up to 450 mOsm/kg H_2O when the rate of change of extracellular osmolality was 2 mOsm/kg H_2O per min.

The change in content of cell solute depended on the protocol by which extracellular osmolality was altered. Sudden hypotonic swelling led to the loss of K, but little K was lost during gradual swelling.[126] In each case, the loss of inorganic ions did not account for the total loss of solute predicted from the volume measurements: inorganic ions accounted for 60–70% of the required solute loss during RVD after sudden osmolality change, but only 30% of the solute loss needed to maintain volume when osmolality was gradually reduced. Organic osmolytes may have made up the remainder of the lost solute. The inorganic ions Na, K, and Cl accounted for only 21% of the osmotically active solute taken up during RVI after sudden exposure to hypertonic solution, but 52% of the solute taken up by tubules exposed to gradual increase in osmolality. Most of the inorganic cation increase was accounted for by Na, and evidence was presented that the uptake occurred by means of parallel Na–H and Cl–HCO_3 exchange.[127]

The maintenance of proximal tubule volume during gradual change in osmolality was termed *isovolumetric regulation.*[124] Less than 3% change in cell volume occurred as osmolality was altered over a range of 167 to 450 mOsm/kg H_2O.

In vivo, vertebrate cells are rarely exposed to anisosmotic solutions. Changes in cell volume most often result from increases or decreases in cell solute content. Accumulation and loss of glucose, amino acids, and other metabolites and absorption and secretion of NaCl by epithelia may result in rapid changes in cell solute content. The extent to which the volume regulatory pathways described above are involved in the response to such changes has been of great interest.

Jejunal villous enterocytes respond to hypotonic swelling by activating separate conductive pathways for K and Cl and undergo RVD. When the cells are exposed to glucose or alanine in the presence of Na, the cells absorb the metabolites by means of Na-dependent cotransporters and swell, reaching peak volumes within 5 min. The cells then undergo RVD and return to their original volume within 1 to 3 min. K and Cl channel blockers inhibit RVD. The response to hypotonic swelling is not identical to the response to metabolite absorption. The protein kinase inhibitor H7 has no effect on RVD after hypotonic swelling, but inhibits RVD after alanine absorption. RVD after hypotonic swelling absolutely requires extracellular Ca, but RVD after alanine uptake occurs in calcium-free solutions, but does require intracellular Ca. RVD in response to hypotonic swelling and RVD secondary to osmolyte uptake utilize the same transport mechanisms, but the activation pathways differ.[128]

When hypoosmotically swollen, hepatic cells undergo RVD both *in vivo* and *in vitro* by activation of conductive pathways for K and Cl. *In vivo,* hypertonically shrunken liver cells undergo RVI by activating Na–H exchange, but *in vitro* RV1 occurs only after RVD. At physiological concentrations, extracellular glutamine and other amino acids which are accumulated by means of Na-coupled cotransporters cause hepatic cell swelling which is followed by RVD by means of KCl loss. RVI and RVD result in the approach of cell volume toward, but not to, its original volume. Swelling induced either by exposure to hypotonic solutions or by accumulation of amino acids reduces glycolysis, increases glycogen synthesis, and increases protein synthesis, and cell shrinkage either caused by exposure to hypertonic solutions or for other reasons has opposite effects—glycogenolysis and protein catabolism. Insulin, which has metabolic effects similar to cell swelling, increases hepatic cell size by activating Na–H exchange and Na,K,2Cl cotransport, and glucagon, which has effects similar to cell shrinkage, decreases cell size by activating K channels.[129] Glycogenolysis may be regulated by cell Cl concentration, which falls during RVD.[130]

During solute absorption or secretion, transcellular movement of osmolytes increases enormously. It is clear that the activity of the transporters in the apical and basolateral cell membranes must be closely coordinated if the cell is to avoid swelling or shrinking,[131] a phenomenon which has been called *transcellular cross-talk.*[132] It is tempting to believe that such coordination might be mediated by responses to changes in cell volume. For instance, for an epithelial cell absorbing Na, entrance of Na across the apical membrane increases intracellular Na content, and Na is exported by the Na,K pump at the basolateral membrane in exchange for K. Increase in cell volume should increase basolateral membrane K and Cl conductance so that KCl is lost and the cell maintains its original volume. While this is a satisfying scenario, it is far from certain that it is correct. There are many aspects of transmembrane coordination that are hard to explain: e.g., inhibition of Na,K pump activity at the basolateral membrane inhibits Na transport through the apical membrane; increased apical Na entry increases basolateral Na,K pump activity even though cell Na does not increase. The activity of most of the transporters involved in cell volume regulation is controlled by protein kinases and phosphatases (see below) and other signaling mechanisms. Moreover, it is troubling that calculation of the maximal solute flow through proximal tubules during isovolumetric regulation amounts to only 15% of the maximal solute flow during maximal NaCl reabsorption.[124]

19.6. SENSORS AND SIGNAL TRANSDUCTION

Up to this point, there has been little discussion about how volume is sensed, currently the most active area of investigation in this field. It seemed appropriate to assemble the information available from many different systems in a single section. In no instance is it known for sure how a cell senses volume, or how the signal of volume change is transmitted to the effector, although there is no dearth of hypotheses, some of which enjoy enthusiastic support. It should be borne in mind that many of the experiments described below involve large changes in cell volume although, under physiological conditions, a volume change of only a few percent is enough to elicit a volume regulatory response as in the experiments of Lohr and Grantham.[124] It should also be borne in mind, as suggested by the experiments of Motais *et al.,*[123] that even in the same cell sensors with different set points, or differing signals from a single sensor at differing volumes, may elicit different responses from the same transporter.

19.6.1. The Set Point

The set point for a transporter is the cell volume at which the transporter is activated or deactivated; e.g., the volume of a red cell just below that at which K–Cl cotransport is activated. It would be inefficient if shrinkage-activated transporters and swelling-activated transporters did not have the same set point since, in that event, energy would be wasted in countering the effects of the inappropriate response. But even if the set point is the same for shrinkage- and swelling-activated processes, it is not necessarily the case that they share the same sensor or signal transduction mechanism.

Evidence that something common to the sensor-transduction pathways for shrinkage- and swelling-activated processes exists in red cells comes from the experiments of Starke and McManus[133] who measured swelling-activated K–Cl cotransport and shrinkage-activated Na,K,2Cl cotransport in duck red cells as a function of cell volume. Measurements were made in cells in which intracellular Mg^{2+} concentration was varied. At each Mg^{2+}, a volume was found at which both K–Cl cotransport and Na,K,2Cl cotransport were minimal: the set point for the two cotransporters was the same. The common set point at low Mg^{2+} was lower than the common set point at high Mg^{2+} concentration. Since Mg^{2+} changed the set point of both shrinkage- and swelling-activated cotransport, the likely conclusion is that there is a common component in the two sensor-transduction pathways which is modified by Mg^{2+}. At the time the experiments were done, it was thought possible that Mg^{2+} converted Na,K,2Cl cotransporters to K–Cl cotransporters, but it was soon shown that dog red cells, which respond to shrinkage by increasing Na–H exchange and to swelling by increasing K–Cl cotransport, increase the common set point for the two processes when Mg^{2+} is increased. Since the two transporters are clearly different, there can be no possibility that one of the transporters is converted to the other by Mg^{2+}.[134] Na,K,2Cl cotransport increases as Mg^{2+} increases in human red cells,[135] and Mg^{2+}, as stated above, decreases K–Cl cotransport in a variety of red cells. It is possible that variation of the set point as Mg^{2+} increases in a general phenomenon, at least in red cells.

It is not known how Mg^{2+} alters the set point. There may be a clue in the observation that a variety of large hydrophilic polycations and cationic amphiphiles inhibit K–Cl cotransport in human red cell ghosts.[96] Mg^{2+} and soluble cations can be expected to shield negative charges at the interface between the cell membrane and the intracellular solution, and the amphiphilic cations, which insert in phospholipid bilayers, can be expected to neutralize negative charges at the membrane surface. Neutralization of membrane surface charges may be responsible for the alteration in the set point by Mg^{2+}.

Parker and his colleagues[134] have reported a number of other maneuvers which alter the common set point for Na–H exchange and K–Cl cotransport in dog red cells. Intracellular Li, like increased Mg^{2+} concentration, increases the common set point, and SCN, urea, and formamide decrease the set point. Treatment of duck red cells with catecholamines decreases the common set point for Na,K,2Cl cotransport and K–Cl cotransport.[136] Further evidence for a common element in the shrinkage and swelling signal transduction pathways in dog red cells comes from experiments in which the operational state of the transporters was fixed with glutaraldehyde or with *N*-phenylmaleimide.[134] If swollen cells were treated with these agents, K–Cl cotransport could not be turned off nor could Na–H exchange be turned on by cell shrinkage; if shrunken cells were treated, Na–H exchange could not be turned off nor could K–Cl cotransport be turned on by cell swelling.

Finally, Figure 19.7 shows the results of experiments in which the set point of resealed dog red cell ghosts was compared with the set point of intact cells.[137] In ghosts, the set point is reduced, but it is increased if hemoglobin or a mixture of hemoglobin and albumin is included when the ghosts are resealed. The result suggests that the set point in this case varies with intracellular protein concentration.

19.6.2. Macromolecular Crowding

Red cells have a high concentration (~7.5 mM) of hemoglobin, which is present in free solution. When concentrated protein solutions are diluted or further concentrated, macromolecules dissolved in the solution exhibit nonlinear thermodynamic behavior as a result of a phenomenon called *macromolecular crowding*.[138] The principle can be described with reference to Figure 19.8. Macromolecules in solution physically occupy a finite volume which is not available for solution of other molecules. The volume from which a small molecule is excluded is not much greater than the total volume physically occupied by protein molecules since small molecules are able to occupy most of the intermolecular volume. On the other hand, macromolecules are excluded from a much larger volume since they are too big to fit into the volume between protein molecules. Since a macromolecule has access to only a fraction of the solvent volume, its actual concentration is higher than the concentration which would be calculated from the total amount of macromolecule and the total amount of solvent, and since the activity of a substance depends on its concentration, the activity of a macromolecule in a concentrated protein solution is higher than the activity estimated from its calculated concentration. The activity is the thermodynamic parameter of interest in calculating ligand occupation of a receptor, enzyme activity as a function of enzyme concentration, and so forth. Further concentration of a concentrated protein solution increases the volume of the intermolecular spaces from which macromolecules are excluded more rapidly than the volume which is excluded by physical occupation by protein molecules, and the activity of the excluded macromolecule is an exponential function of protein concentration. The extent to which the activity of a macromolecule is increased by macromolecular crowding depends on its size: the larger the macromolecule, the greater the increase in its activity.

Macromolecular crowding is an attractive mechanism for accounting for the effect of cell volume on transport processes since it is reasonable to believe that, at least in red cells, intracellular protein concentration varies with cell volume. The nonlinear variation of the activity of macromolecules with varying protein concentration provides the amplification seen when the effect of volume changes on transport rate is measured: small changes in volume result in large changes in transport rate. The concept is supported by the observation that the set point of dog red cells varies with intracellular protein concentration.[139]

There is, however, a difficulty even with red cells.[82,95] Resealed ghosts can be prepared from human red cells which contain very little protein. Ghosts can be prepared so that swollen and shrunken ghosts suspended in the same extracellular solution have the same concentrations of intracellular solutes. The ghosts exhibit swelling-stimulated K–Cl cotrans-

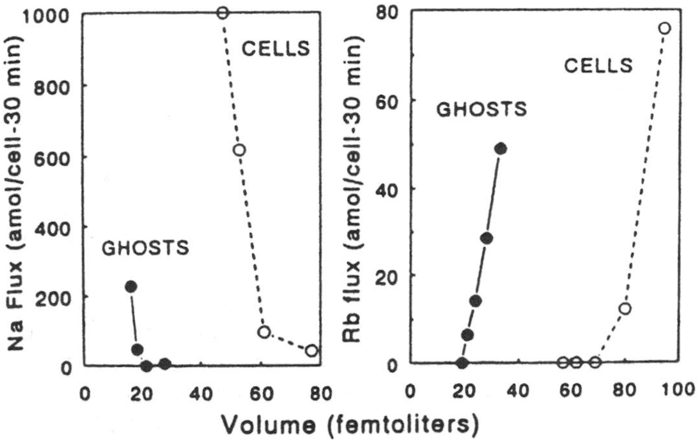

Figure 19.7. The set point. K–Cl cotransport (right) and Na–H exchange (left) are minimal at about the same cell volume. K–Cl cotransport increases with cell swelling and Na–H exchange is activated by cell shrinkage. Identical set points for K–Cl cotransport and Na–H exchange are found with intact cells and with resealed ghosts prepared from them. Taken from Parker[137] and reprinted with permission.

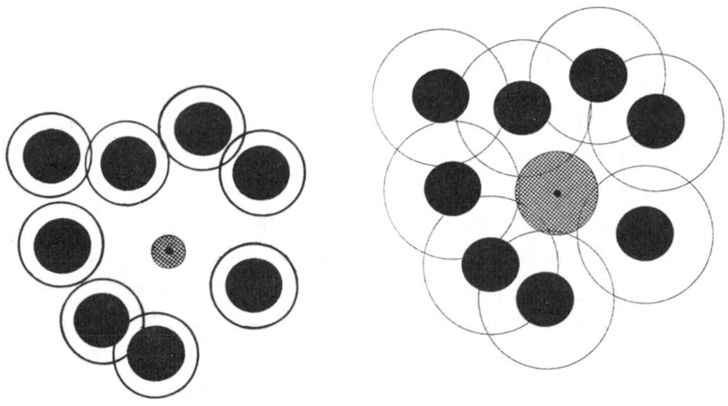

Figure 19.8. Macromolecular crowding. Probe molecules (hatched) in a solution of background macromolecules (solid). The circles surrounding the background molecules are drawn at a distance of one radius of the probe molecule from the surface of the background molecule. Thus, the center of the probe molecule cannot occupy any of the area within the circles. It is apparent that the large probe molecule is excluded from a much larger area than the small probe molecule, and that the area from which the probe molecule is excluded considerably exceeds the area physically occupied by the background molecules. Taken from Minton *et al.*[138] and reprinted with permission.

port with characteristics similar to those seen in intact red cells. The results lead to the inescapable conclusion that in the ghosts K–Cl cotransport is increased by ghost swelling even though there is no change in the concentration of any intracellular solute. The results have been criticized since the K–Cl cotransport rate in shrunken ghosts is greater than the cotransport rate in shrunken red cells.[137] Since the set point in the hemoglobin-free human ghosts is likely to be reduced just as it is in dog red cell ghosts,[137] it is possible that the human ghosts were not shrunk enough to completely turn off cotransport; a second explanation for the large residual cotransport rate is considered below. At any rate, the absolute cotransport rate in swollen, protein-free ghosts is high, and the rate is reduced significantly when the ghosts are shrunk.

Macromolecular crowding is proposed to change the activity of one or more regulators which directly influence the rate of volume-sensitive transporters.[140] The candidate for the role of regulator is discussed below. In this formulation, the state of crowding of intracellular macromolecules is the volume sensor, and the activity of the regulator is the signal transducer. An alternative formulation of the macromolecular crowding experiments is that the regulator changes the response of the sensor to cell volume and that this accounts for the change in the set point; the regulator does not itself alter the activity of the volume-sensitive transporters. An attractive candidate for the role of regulator in this formulation is hemoglobin whose activity is altered by macromolecular crowding; it was mentioned above that Hgb S and Hgb C have marked effects on K–Cl cotransport rate in human red cells. This formulation is consistent with the findings in ghosts of dog red cells[137] and in ghosts of human red cells.[82,95]

While it is likely that protein of red cells is in free solution, the state of protein in more complicated cells is uncertain, and the relevance of macromolecular crowding to volume regulation in complex cells is similarly uncertain.[141]

19.6.3. Phosphorylation–Dephosphorylation Events

Jennings and Al-Rohil[142] measured the time course of activation of K–Cl cotransport in rabbit red cells after sudden hypotonic swelling, and determined that there was a delay before the steady-state cotransport rate was reached. A similar delay has been reported with red cells of a number of other species including duck, dog, pig, and LK sheep. On the other hand, when swollen rabbit red cells were suddenly shrunk, deactivation of cotransport occurred without measurable delay. An analysis based on relaxation kinetics (which state that, for a reversible bimolecular process, the rate constant for relaxation from one state to the other is the sum of the forward and reverse rate constants) was applied, and it was concluded that activation of K–Cl cotransport by cell swelling resulted from a decease in the rate of transformation of active to inactive cotransporters (Figure 19.9). Subsequently, it was shown that activation of cotransport by cell swelling was prevented by the serine-threonine phosphatase inhibitor okadaic acid, and that submaximal concentrations of the inhibitor increased the delay in activation.[143,144] Similar results have been obtained with LK sheep red cells[145] and dog red cells.[146] It was con-

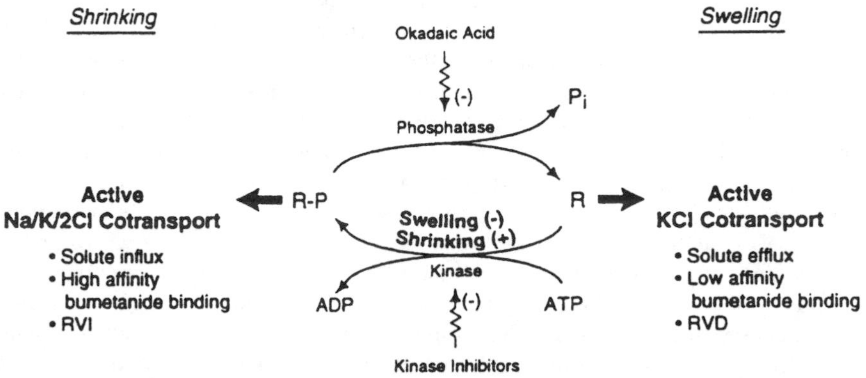

Figure 19.9. Regulation of the response of the Na,K,2Cl cotransporter and K–Cl cotransporter in response to cell swelling by phosphorylation and dephosphorylation of some common but unidentified component of the signal transduction pathway. Taken from McCarty and O'Neill[66] and reprinted with permission.

cluded that activation of K–Cl cotransport by cell swelling was caused by inhibition of a volume-sensitive serine-threonine kinase. In dog red cells, activation of Na–H exchange by cell shrinkage proceeds without delay, but there is a delay in the deactivation of the exchange when the cells are reswollen. Okadaic acid increased Na–H exchange activity so that it was concluded that activation of Na–H exchange by cell shrinkage results from activation of the volume-sensitive kinase.[146] The results led to the proposal of a simple model for the coordinated control of shrinkage-activated and swelling-activated transporters by cell volume (Figure 19.9): volume regulates the state of phosphorylation of serine-threonine groups; when the groups are phosphorylated, shrinkage-activated transporters function, and when the groups are dephosphorylated, swelling-activated transporters function.[147,148] If a single kinase were responsible for control of both swelling- and shrinkage-activated cotransporters, it would not be possible to obtain a volume at which both transporters are quiescent. In duck red cells, okadaic acid increases shrinkage-activated Na,K,2Cl cotransport and inhibits swelling-activated K–Cl cotransport. However, the protein kinase inhibitor K252a, which stimulates Na,K,2Cl cotransport in these cells, has no effect on K–Cl cotransport or its activation by hypotonic swelling.[149] The K–Cl cotransporter has not been identified, so that it is not possible to examine the effect of cell volume on its phosphorylation state. The Na,K,2Cl cotransporter of shark rectal gland is phosphorylated by cell shrinkage,[150] but the Na–H exchange of lymphocytes, whose activity is increased by cell shrinkage, is not.[151]

The remarkable ability of the simple application of relaxation kinetics to account for the observed findings is impressive. Consideration of the model proposed in Figure 19.9 shows that the rate constants in the kinetic analysis do not represent the forward and reverse steps of simple monomolecular transformation: the forward rate constant might include steps in which a phosphatase binds to a receptor, dephosphorylation occurs, the phosphatase is released, and each of these steps might be reversed. The lag in the activation and deactivation curves might be caused by the presence of some other rate-limiting event in the sensor–signal transduction pathway.

Once more, the results with resealed ghosts of human red cells are not easily reconciled with the model.[95] Activation of K–Cl cotransport by ghost swelling proceeds without a measurable delay, as does deactivation by ghost shrinkage. When preparing resealed ghosts, it is necessary to incubate them for a period in order to reseal them. When the resealing period is 1 hr, and okadaic acid is added at the beginning of the resealing period, swelling-activated K–Cl cotransport is nearly completely abolished. If okadaic acid is added at the end of the resealing period, little inhibition is seen. Addition of okadaic acid midway through the resealing period results in intermediate levels of inhibition. Apparently the ghosts have lost their kinase activity (this may in part account for the relatively high cotransport activity in ghosts which are not swollen) but retain some phosphatase activity, and during the resealing period dephosphorylation proceeds nearly to completion. Even though dephosphorylation is complete, the ghosts respond to swelling by increasing their K–Cl cotransport rate. Finally, ATP-free ghosts demonstrate swelling-activated K–Cl cotransport which is decreased when the ghosts are reshrunk. Ghosts incubated while swollen, and then reshrunk, deactivate cotransport; since the ghosts are ATP-free, the deactivation cannot result from phosphorylation. While there is little doubt that phosphorylation and dephosphorylation events modulate K–Cl cotransport, activation and deactivation in response to changes in ghost size occur even when phosphorylation is impossible.

Similar results have been obtained with LK sheep red cells.[145] Depletion of Mg^{2+} (which eliminates MgATP, the substrate for protein kinases) activates cotransport rate, but when the cells are swollen, further activation is seen. In the ghost experiments, it seems likely that kinase is lost during hemolysis, and in the red cell experiments, kinase activity is eliminated by removing substrate.

The results of the two studies indicate that there are two regulators of cotransport activity, one mediated by phosphorylation and one not. A possibility worth considering is that swelling induces a conformational change in the cotransporter which increases cotransport activity and increases the accessibility of a serine-threonine group to dephosphorylation, which further increases cotransport rate, and shrinkage induces a conformational change which decreases K–Cl cotransport activity and increases the accessibility of the serine-threonine group to phosphorylation, which further reduces cotransport rate. Such a mechanism would provide an amplification mechanism consistent with the marked sensitivity of cotransport rate to small volume changes. Serine phosphorylation dependent on transporter conformation has been reported with Na,K pump preparations.[152]

The kinase–phosphatase mechanism has been proposed as the point at which macromolecular crowding regulates cotransport rate; it is proposed that, when cells shrink, macromolecular crowding increases the activity of the kinase and so turns off K–Cl cotransport.[139] In order for this to be the case, if the enzymes are both soluble, the kinase must be larger than the phosphatase: if the enzymes are approximately the same size, their activity would be increased equally with no effect on the net phosphorylation level. In human red cells, the molecular mass of the catalytic unit of PP-1 phosphatase, the phosphatase relevant for cotransport activation, is 36,000 Da.[153] The identity of the relevant kinase is not known, but the casein kinase of human red cells has a molecular mass of 34,000 Da.[154] It is also possible that the phosphatase is bound to the membrane so that the effect of crowding is exerted only on the kinase. There is some evidence for this from the experiments with resealed ghosts since okadaic acid-sensitive dephosphorylation occurs in the absence of cytoplasm, but the rate is low compared to that in intact red cells, and the amount of membrane-bound phosphatase may be small.

Crowding markedly increases the activity of soluble macromolecules, but crowding also reduces their diffusion coefficients, and the magnitude of the diffusion coefficient reduction is comparable to the increase in activity. Diffusion coefficients enter into the rate constants for binding reactions, contributing to both the on and off rate constants, so they may have little effect on the equilibrium binding constant. However, the mechanism shown in Figure 19.9 is more complex than a simple bimolecular binding reaction. It involves the binding and debinding of two different macromolecules in two different branches. Evaluation of the effect of macromolecular crowding on the phosphorylation reaction involves the evaluation of the activities of two different enzymes, whose identity and state in the cell are uncertain, and on their diffusion coefficients, and evaluation of the effect of these parameters on the equilibrium level of phosphorylation. In view of these uncertainties, and the uncertainty about the role of phosphorylation–dephosphorylation events in the transduction pathway between sensor and transporter, it might be appropriate to reserve judgment about the role of macromolecular crowding.

Parker[137] demonstrated that urea alters the set point of dog red cells and suggested that it alters the activity of hemoglobin. Urea activates K–Cl cotransport in both LK sheep[155] and human red cells,[156] and its effect has been attributed to inhibition of the kinase. The concentration of urea was high, and urea is known to destabilize proteins, usually unfolding them and so increasing their size. Dunham[155] showed that the effect of urea is reduced if the cells are shrunk, and suggests that concentration of hemoglobin by cell shrinkage counteracts the effect of urea, which reduces molecular crowding. However, it is known that macromolecular crowding protects against the destabilizing effect of monovalent cations on enzymes, and it might also protect against the destabilizing effect of urea.

In human red cell ghosts, ATP increases K–Cl cotransport rate, as it does in other red cells. The effects of the inhibitors vanadate and genistein suggest that cotransport is increased by the phosphorylation of a tyrosine group.[95] It has been suggested that the effect of ATP results from stimulation of the activity of the phosphatase by phosphorylation.[144] However, in red cell ghosts, okadaic acid does not inhibit cotransport if added at the end of the resealing incubation whether or not the ghosts contain ATP so that dephosphorylation in either case is complete. Since ATP still increases cotransport rate, it does not seem possible that the ATP stimulation results from an increase in the activity of a phosphatase.

19.6.4. Signal Transduction

Many studies have addressed the question of whether or not any of the common signal transduction pathways are involved in transmitting information about cell volume from the volume sensor to the volume-sensitive transporters. These efforts have been complicated since the transporters which respond to cell volume changes are also used for other physiological functions and are therefore subject to controls which do not involve volume regulation.[157] For instance, cAMP activates Na,K,2Cl cotransport in duck red cells and so does cell shrinkage, but shrinkage does not increase cAMP content[158]; both cAMP and cell shrinkage phosphorylate the Na,K,2Cl cotransporter of shark rectal gland cells.[150] Growth promoters increase Na–H exchange and increase phosphorylation of the cotransporter, and cell shrinkage also increases exchange rate, but does not increase cotransporter phosphorylation.[151] In view of the multiple routes to transporter activation, it is frequently difficult to evaluate the role of signal transduction pathways in the control of cell volume.

If signal transduction pathways are involved in the regulation of volume-sensitive transporters, some attention must be given to the question of how the pathways are activated. With the discovery of stretch-activated ion channels, attractive candidates for such activators are the stretch-activated ion channels themselves, or enzymes in signal transduction pathways which are activated by a structural change in the membrane caused by cell volume change. It was mentioned above that stretch-activated K channels and Cl channels may mediate RVD directly in some epithelial cells. Soon after stretch-activated cation channels were described, Christensen[159] provided evidence that in choroid plexus epithelial cells, swelling activates SAcat channels which indiscriminately permit the passage of monovalent and divalent cations. Influx of Ca^{2+} through the channels increases the activity of K channels and Cl channels by calmodulin-dependent processes with resulting RVD. The role of Ca^{2+} in the activation of RVD has been extensively studied, and uncertainties persist.[66,160] Nevertheless, in some cells removal of Ca^{2+} from the outside solution aborts the ionic responses to swelling, and for these cells the mechanism proposed by Christensen is an attractive possibility.

In other cells, RVD is not affected by removal of Ca^{2+} from the extracellular solution, but extensive depletion of intracellular Ca^{2+} prevents RVD. In these cells, some mechanism must be sought to explain the release of Ca^{2+} from intracellular stores. In Ehrlich ascites tumor cells, swelling is associated with a decrease in membrane-bound phosphatidylinositol diphosphate and the appearance of inositol triphosphate and inositol tetrakisphosphate which could account for Ca^{2+} release.[64] The results are consistent with the view that some membrane change induced by swelling activates phospholipase C, but since phospholipase C is itself activated by Ca^{2+}, the alternative possibility that intracellular Ca^{2+} is increased by some other means with consequent activation of phospholipase C cannot be discounted. Many of the enzymes which initiate signal transduction pathways are associated with cell membranes and it has been suggested that alterations in the physical characteristics of the membranes may modulate their activity. Krupinski *et al.*[161] have pointed out that the structure of membrane-bound adenyl cyclase is similar to the structures of membrane transporters; adenyl cyclase activity is altered by mechanical forces such as stretch and by changes in volume.[162,163] Mechanical manipulation alters phosphatidylinositol turnover[164,165] and the association of protein kinase C with the cell membrane.[166]

Stretch applied to membranes activates channels, and it is possible that it can activate enzymes in signal transduction pathways. It is hard to see how epithelial cells, with extensive membrane infoldings, could develop much membrane tension after volume changes of a few percent. Red cells, which are biconcave disks, would not be expected to develop membrane tension until they become spheres which requires a near doubling of their volume. Stretch-activated channels respond to either positive or negative pressure applied to the pipette; the membrane bends in opposite directions according to the pressure applied. Sealed inside-out membrane vesicles prepared from LK sheep red cells activate K–Cl cotransport when swollen, a procedure which bends the membrane in a direction opposite to the direction it bends when intact red cells are swollen.[167] The change in the membrane generated by swelling or shrinkage may be more subtle than the development of tension. Either positively or negatively charged hydrophobic compounds activate stretch-activated channels in *E. coli*,[168] perhaps by a mechanism which bends the phospholipid bilayer, in opposite directions for the positively and negatively charged compounds. In ghosts of human red cells, both soluble polycations and cationic amphiphiles, which insert in phospholipid membranes, inhibit volume-sensitive K–Cl cotransport[96]; both classes of compounds are expected to neutralize negative charges at membrane surfaces. Watson[163] has discussed the possible role of electrostatic forces in volume regulatory phenomena.

A great deal of information has accumulated about the role of second messengers in signal transduction pathways in cell volume regulation. Unfortunately, no unifying pattern has emerged: the same messenger cascade frequently has different effects in different cells. The issue has been intensively examined in Ehrlich ascites tumor cells, and fairly complete information is available.[64] These cells respond to swelling by activating K and Cl conductance pathways whether or not Ca^{2+} is present in the outside solution, but not if the cells are depleted of Ca (but Ca-depleted cells do activate K–Cl cotransport). Activation of K and Cl channels is prevented by calmodulin antagonists. When Ca^{2+} is present in the outside solution, it is likely that it enters through stretch-activated cation channels. Application of Ca^{2+} to patched membranes activates K channels, but not Cl channels. Volume increase decreases the membrane content of phosphatidylinositol diphosphate and increases the appearance in the cytoplasm of inositol triphosphate and inositol tetrakisphosphate which may mobilize intracellular Ca^{2+}. Both K conductance and Cl conductance are increased by leukotriene LTD4, and synthesis of this compound is increased by cell swelling; inhibitors of leukotriene synthesis block RVD. Leukotriene synthesis depends on the release of arachidonic acid from membrane phospholipids by phospholipase A_2 which is activated by Ca^{2+} in a calmodulin-dependent manner; calmodulin antagonists block leukotriene synthesis and RVD. LTD4 seems to be essential for activation of Cl channels, and increases K channel activity.

In contrast to the complex situation with ascites tumor cells, there is evidence that no second messenger is involved in the activation of K–Cl cotransport in swollen resealed ghosts of human red cells.[95]

19.6.5. Cytoskeleton

An obvious candidate for a role in volume sensing and signal transduction is the cytoskeleton which is bound to the plasma membrane. Three possible mechanisms for cytoskeletal involvement have been suggested.[64] Microfilaments may take part in the fusion of cytoplasmic vesicles containing transporters with the plasma membrane in response to changes in cell size. The mechanism has been advanced as an explanation for the inhibition by cytochalasins of RVD after swelling of hepatocytes[169] and gallbladder epithelial cells,[170] to account for the appearance of sorbitol transporters in the cell membrane of renal tubule cells after hypertonic shrinkage,[31] and the fusion of cytoplasmic vesicles containing water channels with the apical membrane of epithelial cells after exposure to antidiuretic hormone.[60] Second, it has been suggested[171] that the actin cytoskeleton transmits tension developed by membrane stretch to stretch-activated channels. Finally, it is possible that some component of the actin cytoskeleton released during cell volume change is utilized as a messenger to transmit the signal of cell volume change to membrane transporters.

Actin microfilaments are disrupted after hypotonic swelling in many cells including PC12 cells,[172] ascites tumor cells,[173] shark rectal gland cells,[174] and HL60 cells.[175] Hallows *et al.*[175] have provided an analysis of the expected effects of cell volume change on the state of actin polymerization. With cytoplasmic dilution after hypoosmotic swelling, some depolymerization of F actin to G actin is expected since the intracellular concentration of G actin decreases, and the depolymerization reaction is monomolecular while the polymerization reaction is bimolecular. Similarly, hypertonic shrinkage is expected to promote polymerization. They point out, however, that the steady-state concentration of free G actin is low and that most intracellular G actin is combined with a variety of actin binding proteins. Using reasonable values for the binding constants and steady-state concentrations involved, they conclude that, if the change in polymerization state of F actin with cell volume change is accounted for by mass action considerations, there must be a modest change in the average K_D of the actin binding proteins or in the critical concentration of G actin or both. The authors did not consider the possible effects of macromolecular crowding on actin polymerization. For instance, in the binding of a G actin molecule with an F actin molecule, one substrate (F actin) and the product (F actin) are not much different in size, and the effect of macromolecular crowding on the activity of the two species will not be much different. The polymerization reactions therefore depend primarily on the activity of G actin which may be markedly affected by crowding. Effects of crowding on actin binding proteins add a further level of complication to the analysis. It is not inconceivable that the relatively small changes in the K_D values of the binding reactions and the critical concentration of G actin with cell volume which were calculated as necessary if the variation in the state of polymerization of F actin in HL60 cells with cell volume is accounted for by mass action considerations are related solely to crowding and concentration effects.

Much of the evidence for the role of the cytoskeleton in cell volume regulation comes from observations of the effect of cytochalasins, which disrupt the actin cytoskeleton, on

RVD after hypotonic swelling. Cytochalasins inhibit RVD in hepatocytes,[169] gallbladder epithelial cells,[170] isolated axons of the euryhaline crab *Carcinus maenas,*[4] and ascites tumor cells.[173] However, dihydrocytochalasin B does not prevent RVD in HL60 cells even though it causes considerable disruption of the actin cytoskeleton. Cytochalasins bind to the barbed, proliferating ends of actin filaments and therefore reduce the average filament size, just as cell swelling does. It is puzzling that disruption of the actin cytoskeleton by cytochalasins inhibits RVD, while disruption by hypoosmotic swelling does not.

There are some other inconsistencies in the observations which relate cytoskeletal disruption to the regulation of RVD. In HL60 cells, cytoskeletal swelling occurs with a marked delay after cell swelling, but RVD begins almost immediately,[175] and a similar temporal dissociation between the two events occurs in ascites tumor cells.[173] In ascites tumor cells, cytoskeletal disruption does not occur if swelling is carried out in Ca^{2+}-free solutions, but RVD occurs as long as cellular Ca stores are not depleted.

There are, however, some findings which, while they have not led to any detailed mechanism, do suggest that the cytoskeleton, or at least actin, may be involved in the signal transduction pathway between the volume sensor and the transporters.

1. Melanoma cells which are deficient in actin binding protein have a high resting K conductance and do not activate K permeability when swollen, nor do they activate RVD.[176] Cells which express actin binding protein have a lower resting K conductance which is activated when the cells swell, and undergo RVD toward their isotonic volume. Expression of the gene for actin binding protein in deficient cells "rescues" the cells and converts their behavior toward that of the nondeficient cells.
2. A cytosolic chloride conductance regulatory protein (pI_{Cln}) has been found in significant concentration in a number of cells including MDCK cells, rat cardiac myocytes, and *Xenopus* oocytes.[177] When overexpressed in *Xenopus* oocytes, this protein resulted in an increase in a chloride conductance which had characteristics similar to a chloride conductance activated by swelling of the oocytes. A monoclonal antibody directed against pI_{Cln}, when injected into *Xenopus* oocytes, abolished the volume sensitivity of the chloride conductance. pI_{Cln} binds to actin (among other cytosolic proteins), and the authors suggest that pI_{Cln} may transduce the signal of cell swelling by connecting actin to a membrane channel.
3. In a cell line derived from rabbit cortical collecting duct, hypotonic swelling activated a DIDS-sensitive Cl channel, and activation of the channel coincided temporally with activation of RVD.[6] In whole cells, dihydrocytochalasin B activated chloride conductance, and swelling did not further stimulate it. Phalloidin, which stabilizes actin filaments, prevented the activation of chloride conductance by swelling. Dihydrocytochalasin B, when applied to the cytoplasmic face of inside-out patches, activated a chloride conductance which was also activated by membrane stretch, and phalloidin prevented stretch activation of the channel. Cell swelling was associated with depolymerization and reorganization of the actin cytoskeleton, and so was treatment with dihydrocytochalasin B. In this study no findings were inconsistent with the hypothesis that hypotonic swelling depolymerizes the actin cytoskeleton, and that the depolymerization is associated with RVD.

Most of the studies which have addressed the relationship of cytoskeletal–membrane interaction to the activation of volume-sensitive transporters have concentrated on the behavior of actin. Jennings and Schulz[78] examined the effect of changes in cell morphology on K–Cl cotransport using rabbit red cells. They incubated the cells with hydrophobic agents which inserted into the outer leaflet of the bimolecular phospholipid membrane and converted the cells into echinocytes, or into the inner leaflet with the formation of stomatocytes. In neither case did the profound change in cell shape alter K–Cl cotransport or its response to cell volume change.

19.7. SUMMARY

Until a little more than two decades ago, little attention was paid to volume regulatory mechanisms, especially in vertebrate cells where they seemed to be superfluous. Moreover, at that time most of the transport mechanisms which cells use to respond to volume changes were completely unknown. At present, fairly complete information is available about the transport pathways which cells use to adjust their volumes, but it is not known why different cells use different mechanisms, or why the same cell uses different mechanisms under different circumstances.

Moreover, not much is known for sure about the volume sensor, or the signal transduction pathway. All of the commonly known cell signaling pathways have been evaluated as candidates for these roles, and convincing evidence for some of these mechanisms has been obtained in various preparations, but to date no mechanism has been shown to operate universally. It would not be surprising if many different means of monitoring cell volume have been developed over evolutionary time, but neither would it be surprising if there is some underlying mechanism waiting to be recognized.

Recent molecular biological studies of two volume-sensitive transporters have provided what may be a provocative clue. The chloride channel CIC-2 is regulated by voltage and by cell volume. When mRNA for the channel was expressed in *Xenopus* oocytes, a chloride conductance activated by cell swelling was found. A region in the cytoplasmic N-terminus of the channel was found to be essential for volume sensitivity; if the region was deleted, the channel was active at all cell volumes. The essential region could be transplanted to different cytoplasmic regions of the channel with restoration of volume sensitivity.[178] Two of the volume-sensitive isoforms of the Na–H exchanger (NHE1 and NHE2) when expressed in exchanger-deficient fibroblasts mediate Na–H exchange which is inhibited in hypotonic solutions. Truncation of the C-terminal cytoplasmic region of the cotransporter by 125 amino acids had no effect on exchange or its volume sensitivity, but removal of the next 54 amino acids produced an exchanger which was active at all cell volumes with no inhibition in hypotonic solutions.[179] The studies raise the possibility that the transporters are not activated by swelling or shrinking, but rather are deactivated at normal cell volume, and suggest the important possibility that regulation of the transporters may result from direct interaction of some portion of their cytoplasmic regions with a component of the membrane; no secondary messengers may be involved.

Although mechanisms of volume regulation are often supported enthusiastically based on convincing studies with a single system, it seems too early to commit to any one mechanism. "The unfacts, did we possess them, are too imprecisely few to warrant our certitude. . . ."[180]

ACKNOWLEDGMENTS. This work was supported in part by a grant from the American Heart Association, NY State Affiliate. The expert assistance of Ms. Shirley Murray in preparing the manuscript was indispensable.

REFERENCES

1. Finkelstein, A. (1987). *Water Movement through Lipid Bilayers, Pores and Plasma Membranes: Theory and Reality.* Wiley–Interscience, New York.
2. Preston, G. M., Carroll, T. P., Guggino, W. B., and Agre, P. (1992). Appearance of water channels in *Xenopus* oocytes expressing red cell CHIP28 protein. *Science* **256:**385–387.
3. Chrispeels, M. J., and Agre, P. (1994). Aquaporins: Water channel proteins of plant and animal cells. *Trends Biochem. Sci.* **19:** 421–425.
4. Gilles, R., Delpire, E., Duchene, C., Cornet, M., and Péquex, A. (1986). The effect of cytochalasin B on the volume regulation response of isolated axons of the green crab *Carcinus maenas,* submitted to hypo-osmotic media. *Comp. Biochem. Physiol.* **86A:** 523–525.
5. Ito, T., Suzuki, A., and Stossel, T. P. (1992). Regulation of water flow by actin-binding protein-induced actin gelation. *Biophys. J.* **61:**1301–1305.
6. Mills, J. W., Schwerbert, E. M., and Stanton, B. A. (1994). The cytoskeleton and cell volume regulation. In *Molecular Physiology of Cell Volume Regulation* (K. Strange, ed.), CRC Press, Boca Raton, pp. 241–258.
7. Rand, R. P., and Burton, A. L. (1964). Mechanical properties of the red cell membrane: I. Membrane stiffness and intracellular pressure. *Biophys. J.* **4:**115–135.
8. Fenn, W. O., and Cobb, D. M. (1936). Electrolyte changes in muscle during activity. *Am. J. Physiol.* **115:**345–356.
9. Heppel, L. A. (1939). The electrolytes of muscle and liver in potassium-depleted rats. *Am. J. Physiol.* **127:**385–392.
10. Steinback, H. B. (1940). Sodium and potassium in frog muscle. *J. Biol. Chem.* **133:**695–701.
11. Noonan, T. R., Fenn, W. O., and Haege, L. (1940). The distribution of injected radioactive potassium in rats. *Am. J. Physiol.* **132:**474–488.
12. Heppel, L. A. (1940). The diffusion of radioactive sodium into the muscles of potassium-deprived rats. *Am. J. Physiol.* **128:**449–454.
13. Dean, R. B. (1941). Theories of electrolyte equilibrium in muscle. *Biol. Symp.* **3:**331–348.
14. Krogh, A. (1946). The active and passive exchanges of inorganic ions through the surfaces of living cells and through living membranes generally. *Proc. R. Soc. London Ser. B.* **133:**140–200.
15. Davson, H. (1994). The permeability of the erythrocyte to cations. *Cold Spring Harbor Symp. Quant. Biol.* **8:**255–268.
16. Wu. H. (1926). Note on Donnan equilibrium and osmotic pressure relationship between the cells and the serum. *J. Biol. Chem.* **70:**203–205.
17. Wilson, T. H. (1954). Ionic permeability and osmotic swelling of cells. *Science* **120:**104–105.
18. Leaf, A. (1956). On the mechanism of fluid exchange of tissues in vitro. *Biochem. J.* **62:**241–248.
19. Ussing, H. H. (1960). Active and passive transport of the alkali metal ions. In *The Alkali Metal Ions in Biology* (H. H. Ussing, F. Kruhøffer, J. Hess Thaysen, and N. A. Thorn, eds.), Springer-Verlag, Berlin, pp. 45–143.
20. Tosteson, D. C., and Hoffman, J. F. (1960). Regulation of cell volume by active cation transport in high and low potassium sheep red cells. *J. Gen. Physiol.* **44:**169–194.
21. MacKnight, A. D. C. (1987). Volume maintenance in isosmotic conditions. *Curr. Top. Membr. Transp.* **30:**3–43.
22. MacKnight, A. D. C., and Leaf, A. (1977). Regulation of cellular volume. *Physiol. Rev.* **57:**510–573.
23. Kregenow, F. M. (1971). The response of duck erythrocytes to nonhemolytic hypotonic media. *J. Gen. Physiol.* **58:**372–394.
24. Kregenow, F. M. (1971). The response of duck erythrocytes to hypertonic media. *J. Gen. Physiol.* **58:**396–412.
25. Kregenow, F. M. (1981). Osmoregulatory salt transporting mechanisms: Control of cell volume in anisotonic media. *Annu. Rev. Physiol.* **43:**493–505.
26. Gilles, R. (1988). Comparative aspects of cell osmoregulation and volume control. *Renal Physiol. Biochem.* **3–5:**277–288.
27. Goldstein, L., and Kleinzeller, A. (1987). Cell volume regulation in lower vertebrates. *Curr. Top. Membr. Transp.* **30:**181–203.
28. Chamberlin, M. E., and Strange, K. (1989). Anisosmotic cell volume regulation: A comparative view. *Am. J. Physiol.* **257:**C159–C173.
29. Yancey, P. H., and Somero, G. N. (1979). Counteraction of urea destabilization of protein structure by methylamine osmoregulatory compounds of elasmobranch fishes. *Biochem. J.* **183:**317–323.
30. Spring, K. R., and Siebens, A. W. (1988). Solute transport and epithelial cell volume regulation. *Comp. Biochem. Physiol.* **90A:** 557–560.
31. Kinne, R. K. H., Czekay, R.-P., Grunewald, J. M., Mooren, F. C., and Kinne-Saffron, E. (1993). Hypotonicity-evoked release of organic osmolytes from distal renal cells: Systems, signals and sidedness. *Renal Physiol. Biochem.* **16:**66–78.
32. MacKnight, A. D. C., Grantham, J., and Leaf, A. (1993). Physiologic responses to changes in extracellular osmolality. In *Clinical Disturbances of Water Metabolism* (D. W. Seldin and G. Giebisch, eds.), Raven Press, New York, pp. 31–49.
33. Siebens, A. W. (1985). Cellular volume control. In *The Kidney: Physiology and Pathophysiology* (D. W. Seldin and G. Giebisch, eds.), Raven Press, New York, pp. 91–115.
34. Eveloff, J. L., and Warnock, D. G. (1987). Activation of ion transport systems during cell volume regulation. *Am. J. Physiol.* **252:** F1–F10.

35. Hoffmann, E. K., and Simonsen, L. O. (1989). Membrane mechanisms in volume and pH regulation in vertebrate cells. *Physiol. Rev.* **69:**315–382.
36. Sarkadi, B., and Parker, J. C. (1991). Activation of ion transport pathways by changes in cell volume. *Biochim. Biophys. Acta* **1071:** 407–427.
37. Spring, K. R., and Hoffmann, E. K. (1992). Cellular volume control. In *The Kidney: Physiology and Pathophysiology* (D. W. Seldin and G. Giebisch, eds.), Raven Press, New York, pp. 147–169.
38. Lang, F., and Haüssinger, D. (eds.) (1993). *Interaction of Cell Volume and Cell Function,* Springer-Verlag, Berlin.
39. Strange, K. (ed.) (1994). *Cellular and Molecular Physiology of Cell Volume Regulation,* CRC Press, Boca Raton.
40. Sachs, J. R., Knauf, P. A., and Dunham, P. B. (1975). Transport through red cell membranes. In *The Red Blood Cell,* 2nd ed. (D. M. Surgenor, ed.), Academic Press, New York, pp. 613–703.
41. Dick, D. A. T. (1965). *Cell Water,* Butterworths, London.
42. Dick, D. A. T., and Lowenstein, L. M. (1958). Osmotic equilibria in human erythrocytes studied by immersion refractometry. *Proc. R. Soc. London Ser. B* **148:**241–256.
43. McConaghey, P. D., and Maizels, M. (1961). The osmotic coefficients of hemoglobin in red cells under varying conditions. *J. Physiol. (London)* **155:**28–45.
44. Adair, G. S. (1928). A theory of partial osmotic pressures and membrane equilibria, with special reference to the application of Dalton's Law to hemoglobin solutions in the presence of salts. *Proc. R. Soc. London Ser. A* **120:**573–611.
45. Freedman, J. C., and Hoffman, J. F. (1979). Ionic and osmotic equilibria of human red blood cells treated with nystatin. *J. Gen. Physiol.* **74:**157–185.
46. Garay-Bobo, C. M., and Solomon, A. K. (1971). Hemoglobin charge dependence on hemoglobin concentration in vitro. *J. Gen. Physiol.* **57:**283–289.
47. Kwant, W. O., and Seeman, P. (1970). The erythrocyte ghost is a perfect osmometer. *J. Gen. Physiol.* **55:**208–219.
48. Clegg, J. S. (1984). Properties and metabolism of the aqueous cytoplasm and its boundaries. *Am. J. Physiol.* **246:**R133–R151.
49. Kempner, E. S., and Miller, J. H. (1968). The molecular biology of *Euglena gracilis.* IV. Cellular stratification by centrifuging. *Exp. Cell Res.* **51:**141–149.
50. LeNeveu, D. M., Rand, R. P., and Parsegian, V. A. (1976). Measurements of forces between lecithin bilayers. *Nature* **259:**601–603.
51. Israelachvilli, J. N., and Adams, G. E. (1976). Direct measurements of long range forces between two mica surfaces in aqueous KNO_3 solutions. *Nature* **262:**774–776.
52. Schporer, M., and Civan, M. M. (1977). The state of water and alkali cations within intracellular fluids: The contribution of NMR spectroscopy. *Curr. Top. Membr. Transp.* **9:**1–36.
53. Maffly, R. H., and Leaf, A. (1959). The potential of water in mammalian tissues. *J. Gen. Physiol.* **42:**1257–1275.
54. Gilles, R. (1987). Volume regulation in cells of euryhaline invertebrates. *Curr. Top. Membr. Transp.* **30:**205–247.
55. Bennett, V. (1990). Spectrin-based membrane skeleton: A multipotent adaptor between plasma membrane and cytoplasm. *Physiol. Rev.* **70:**1029–1065.
55a. Hoffman, J. F. (1958). Physiological characteristics of human red blood cell ghosts. *J. Gen. Physiol.* **42:**9–28.
56. Fischer, T. M. (1992). Is the surface area of the red cell membrane skeleton locally conserved? *Biophys. J.* **61:**298–305.
57. Zhelev, D. V., Needham, D., and Hochmuth, R. M. (1994). Role of the membrane cortex in neutrophil deformation in small pipets. *Biophys. J.* **67:**695–705.
58. Doctor, R. B., Zhelev, D., and Mandel, L. J. (1994). Description of plasma membrane–cytoskeletal interactions in anoxic rat renal proximal tubules. *J. Am. Soc. Nephrol.* **5:**896.
59. Skaer, H. L., Treherne, J. E., Benson, J. A., and Moreton, R. B. (1978). Axonal adaptations to osmotic and ionic stress in an invertebrate osmoconformer (Mercierella enigmata Fauvel) I. Ultrastructural and electrophysiological observations in axon accessibility. *J. Exp. Biol.* **76:**191–204.
60. Hays, R. M., Condulis, J., Gao, Y., Simon, H., Ding, G., and Franki, N. (1993). The effect of vasopressin on the cytoskeleton of the epithelial cell. *Pediatr. Nephrol.* **7:**672–679.
61. Kelly, S. M., and Macklem, P. T. (1991). Direct measurement of intracellular pressure. *Am. J. Phyiol.* **260:**C652–C657.
62. Simon, H., Gao, Y., Franki, N., and Hays, R. M. (1993). Vasopressin depolymerizes F-actin in rat inner medullary collecting duct. *Am. J. Physiol.* **265:**C757–C762.
63. MacRobbie, E. A. C., and Ussing, H. H. (1961). Osmotic behavior of the epithelial cells of frog skin. *Acta Physiol. Scand.* **53:**348–365.
64. Hoffmann, E. K., Simonsen, L. O., and Lambert, I. H. (1993). Cell volume regulation: Intracellular transmission. *Adv. Comp. Environ. Physiol.* **14:**187–248.
65. Lang, F., Ritter, M., Völk, H., and Häussinger, D. (1993). Cell volume regulatory mechanisms—An overview. *Adv. Comp. Environ. Physiol.* **14:**1–31.
66. McCarty, N. A., and O'Neill, R. G. (1992). Calcium signaling in cell volume regulation. *Physiol. Rev.* **72:**1037–1061.
67. Hoffmann, E. K., Lambert, I. H., and Simonsen, L. O. (1986). Separate Ca^{2+}-activated K^+ and Cl^- transport pathways in Ehrlich ascites tumor cells. *J. Membr. Biol.* **91:**227–244.
68. Guharay, F., and Sachs, F. (1984). Stretch-activated single ion channel currents in tissue-cultured embryonic chick skeletal muscle. *J. Physiol. (London)* **352:**685–701.
69. Sackin, H. (1994). Stretch-activated ion channels. In *Molecular Physiology of Cell Volume Regulation* (K. Strange, ed.), CRC Press, Boca Raton, pp. 215–240.
70. Morris, C. F. (1990). Mechanosensitive ion channels. *J. Membr. Biol.* **113:**93–107.
71. Gustin, M. C., Zhou, X., Martinoc, B., and Kung, C. (1988). A mechanosensitive ion channel in the yeast plasma membrane. *Science* **242:**762–765.
72. Sackin, H. (1989). A stretch-activated potassium channel sensitive to cell volume. *Proc. Natl. Acad. Sci. USA* **86:**1731–1735.
73. Hudson, L., and Schultz, S. G. (1988). Sodium-coupled glycine uptake by Ehrlich ascites tumor cells results in an increase in cell volume and plasma membrane channel activity. *Proc. Natl. Acad. Sci. USA* **85:**279–288.
74. Ackerman, M. J., Wickman, K. D., and Clapham, D. E. (1994). Hypotonicity activates a native chloride current in *Xenopus* oocytes. *J. Gen. Physiol.* **103:**153–179.
75. Valverde, M. A., Diaz, M., Sepulveda, F. P., Gill, D. R., Hyde, S. C., and Higgins, C. F. (1992). Volume-regulated chloride channels associated with the multidrug-resistance P-glycoprotein. *Nature* **355:**830–833.
76. Dunham, P. B. (1990). K,Cl cotransport in mammalian erythrocytes. In *Regulation of Potassium Transport Across Biological Membranes* (L. Reuss, J. M. Russell, and G. Szabo, eds.), University of Texas Press, Austin, pp. 331–360.
77. Lauf, P. K., Bauer, J., Adragna, N. C., Fujise, H., Zode-Oppen, A. M. M., Ryn, K. H., and Delpire, E. (1992). Erythrocyte K–Cl cotransport. Properties and regulation. *Am. J. Physiol.* **263:**C917–C932.
78. Jennings, M. L., and Schulz, R. K. (1990). Swelling-activated KCl cotransport in rabbit red cells: Flux is determined mainly by cell volume rather than shape. *Am J. Physiol.* **259:**C960–C967.
79. O'Neill, W. C. (1989). Cl-dependent K transport in a pure population of volume-regulating human erythrocytes. *Am. J. Physiol.* **256:** C858–C864.

80. Furlong, T. J., and Spring, K. R. (1990). Mechanisms underlying volume regulatory decrease by *Necturus* gallbladder epithelium. *Am. J. Physiol.* **258:**C1016–C1024.
81. Zeuthen, T. (1994). Cotransport of K^+,Cl^- and H_2O by membrane proteins from choroid plexus epithelium of *Necturus maculosus. J. Physiol. (London)* **478:**203–219.
82. Sachs, J. R. (1988). Volume-sensitive K influx in human red cell ghosts. *J. Gen. Physiol.* **92:**685–711.
83. Lytle, C., and McManus, T. J. (1987). Effect of loop diuretics and stilbene derivatives on swelling induced K–Cl cotransport. *J. Gen. Physiol.* **90:**28a.
84. McManus, T. J., Haas, M., Starke, L. C., and Lytle, C. Y. (1985). The duck red cell model of volume-sensitive chloride-dependent cation transport. *Ann. N.Y. Acad. Sci.* **456:**183–186.
85. Kaji, D. M. (1993). Effect of membrane potential on K–Cl transport in human erythrocytes. *Am. J. Physiol.* **264:**C376–C382.
86. Brugnara, C., Van Ha, T., and Tosteson, D. C. (1989). Role of chloride in potassium transport through a K–Cl cotransport system in human red blood cells. *Am. J. Physiol.* **256:**C994–C1003.
87. Delpire, E., and Lauf, P. K. (1991). Kinetics of Cl-dependent K fluxes in hyposmotically swollen LK sheep erythrocytes. *J. Gen. Physiol.* **97:**173–193.
88. Kaji, D. M. (1989). Kinetics of volume-sensitive K transport in human erythrocytes: Evidence for asymmetry. *Am. J. Physiol.* **256:** C1214–C1223.
89. Dunham, P. B., and Ellory, J. C. (1981). Passive transport in low potassium sheep red cells: Dependence on cell volume and chloride. *J. Physiol. (London)* **318:**511–530.
90. Bergh, C., Kelly, S. J., and Dunham, P. B. (1990). K–Cl cotransport in LK sheep erythrocytes: Kinetics of stimulation by cell swelling. *J. Membr. Biol.* **117:**177–188.
91. Vitoux, D., Olivieri, O., Garay, R. P., Cragoe, E. J., Galacteros, E., and Beuzzard, Y. (1989). Inhibition of K^+ efflux and dehydration of sickle cells by [(dihydroxyoindenyl)oxy] alkanoic acid: An inhibitor of the K^+Cl^--cotransport system. *Proc. Natl. Acad. Sci. USA* **86:**4273–4276.
92. Ellory, J. C., Hall, A. C., Ody, S. O., Englert, H. C., Mauia, D., and Lang, H.-J. (1990). Selective inhibitors of KCl cotransport in human red cells. *FEBS Lett.* **262:**215–218.
93. Adragna, N. C., and Lauf, P. K. (1994). Quinine and quinidine inhibit and reveal heterogeneity of K–Cl cotransport in low K sheep erythrocytes. *J. Membr. Biol.* **142:**195–207.
94. Olivieri, O., Vitoux, D., Galacteros, F., Bachir, D., Blouquit, Y., Beuzzard, Y., and Brugnara, C. (1992). Hemoglobin variant and activity of (K^+Cl^-) cotransport system in human erythrocytes. *Blood* **79:**793–797.
95. Sachs, J. R., and Martin, D. W. (1993). The role of ATP in swelling-stimulated K–Cl cotransport in human red cell ghosts. Phosphorylation–dephosphorylation events are not in the signal transduction pathway. *J. Gen. Physiol.* **102:**551–573.
96. Sachs, J. R. (1994). Soluble polycations and cationic amphiphiles inhibit volume-sensitive K–Cl cotransport in human red cell ghosts. *Am. J. Physiol.* **266:**C997–C1005.
97. Curran, P. F., and MacIntosh, J. R. (1962). A model system for biological water transport. *Nature* **193:**347–348.
98. Zeuthen, T., and Stein, W. D. (1994). Cotransport of salt and water in membrane proteins: Membrane proteins as osmotic engines. *J. Membr. Biol.* **137:**179–195.
99. Cala, P. M. (1980). Volume regulation by *Amphiuma* red blood cells: The membrane potential and its implications regarding the nature of the ion flux pathways. *J. Gen. Physiol.* **76:**683–708.
100. Lew, V. L., and Bookchin, R. M. (1986). Volume, pH, and ion-content regulation in human red cells: Analysis of transient behavior with an integrated model. *J. Membr. Biol.* **92:**57–74.
101. Parker, J. C., and Dunham, P. B. (1989). Passive cation transport. In *Red Blood Cell Membranes. Structure, Function, Clinical Implications* (P. Agre and J. C. Parker, eds.), Dekker, New York, pp. 507–561.
101a. Milanick, M. A., and Hoffman, J. F. (1986). Ion transport and volume regulation in red blood cells. *Annal. N.Y. Acad. Sci.* **488:** 174–186.
102. Geck, P., Pietrzyk, C., Burkhardt, B.-C., Pfeiffer, B., and Heinz, E. (1980). Electrically silent cotransport of Na,K and Cl in Ehrlich cells. *Biochim. Biophys. Acta* **600:**432–447.
103. Russell, J. M. (1983). Cation-coupled chloride influx in squid axon. Role of potassium and stoichiometry of the transport process. *J. Gen. Physiol.* **81:**909–925.
104. Xu, J.-C., Lytle, C., Zhu, T. T., Payne, J. A., Benz, E., Jr., and Forbush, B., III. (1994). Molecular cloning and functional expression of the bumetanide-sensitive Na–K–Cl cotransporter. *Proc. Natl. Acad. Sci. USA* **91:**2201–2205.
105. Payne, J., and Forbush, B., III. (1994). Alternatively spliced isoforms of the putative renal Na–K–Cl cotransporter are differently distributed within the rabbit kidney. *Proc. Natl. Acad. Sci. USA* **91:**4544–4548.
106. Gamba, G., Saltzberg, S. N., Lombardi, M., Mujanoshita, A., Lytton, J., Hediger, M. A., Brenner, B. M., and Hebert, S. C. (1993). Primary structure and functional expression of a cDNA encoding the thiazide-sensitive, electroneutral sodium-chloride cotransporter. *Proc. Natl. Acad. Sci. USA* **90:**2749–2753.
107. Gamba, G., Mujanoshita, A. M., Lombardi, M., Lytton, J., Lee, W. S., Hediger, M. A., and Hebert, S. C. (1994). Molecular cloning, primary structure, and characterization of two members of the mammalian electroneutral sodium-(potassium)-chloride cotransporter family expressed in kidney. *J. Biol. Chem.* **269:**17713–17722.
108. Lytle, C. Y., and McManus, T. J. (1986). A minimal kinetic model of [Na + K + 2Cl] cotransport with ordered binding and glide symmetry. *J. Gen. Physiol.* **88:**36a.
109. Levinson, C. (1990). Regulatory volume increase in Ehrlich ascites tumor cells. *Biochim. Biophys. Acta* **1021:**1–8.
110. Sun, A. M., Saltzberg, S. N., Deheri, D., and Hebert, S. C. (1990). Mechanisms of cell volume regulation by the mouse medullary thick ascending limb of Henle. *Kidney Int.* **38:**1019–1029.
111. Grinstein, S., Goetz, J. D., Cohen, S., Furuya, W., Rothstein, A., and Gelfand, E. W. (1985). Mechanism of regulatory volume increase in osmotically shrunken lymphocytes. *Mol. Physiol.* **8:**185–198.
112. Siebens, A. W., and Kregenow, F. M. (1985). Volume-regulatory responses of Amphiuma red cells in anisotonic media. The effect of amiloride. *J. Gen. Physiol.* **86:**527–564.
113. Cala, P. M. (1985). Volume regulation by *Amphiuma* red blood cells: Characteristics of volume-sensitive K/H and Na/H exchange. *Mol. Physiol.* **8:**199–214.
114. Bookstein, C., Musch, M. W., De Paoli, A., Zie, Y., Villereal, M., Rao, M. C., and Chang, E. B. (1944). A unique sodium–hydrogen exchange isoform (NHE-4) of the inner medulla of the rat kidney is induced by hyperosmolality. *J. Biol. Chem.* **269:**29704–29709.
115. Garcia-Perez, A., and Ferraris, J. D. (1994). Aldose reductase gene expression and osmoregulation in mammalian renal cells. In *Molecular Physiology of Cell Volume Regulation* (K. Strange, ed.), CRC Press, Boca Raton, pp. 373–381.
116. Handler, J. S., and Moo Kwon, H. (1993). Regulation of renal cell organic osmolyte transport by tonicity. *Am. J. Physiol.* **265:** C1449–C1455.
117. Higgins, C. F., Cairney, J., Stirling, D. A., Sutherland, L., and Booth, I. R. (1987). Osmotic regulation of gene expression: Ionic strength as an intracellular signal? *Trends Biochem. Sci.* **12:**339–344.
118. Tabenaka, M., Preston, A. S., Moo Kwan, H., and Handler, J. S. (1994). The tonicity-sensitive element that mediates increased

transcription of the betaine transporter gene in response to hypertonic stress. *J. Biol. Chem.* **269:**29379–29387.

119. Kimelberg, H. K., O'Connor, E. R., and Kettenmann, H. (1993). Effects of swelling on glial cell function. *Adv. Comp. Environ. Physiol.* **14:**157–185.
120. Goldstein, L., and Brill, S. R. (1991). Volume-activated taurine efflux from skate erythrocytes: Possible band 3 involvement. *Am. J. Physiol.* **260:**R1014–R1020.
121. Kirk, K., Ellory, J. C., and Young, J. D. (1992). Transport of organic substrates via a volume-activated channel. *J. Biol. Chem.* **267:**23475–23478.
122. Jackson, P. S., and Strange, K. (1995). Volume-sensitive anion channels mediate swelling-activated inositol and taurine efflux. *Am. J. Physiol.* **265:**C1489–C1500.
123. Motais, R., Guizouarn, H., and Garcia-Romeu, F. (1991). Red cell volume regulation: The pivotal role of ionic strength in controlling swelling-dependent transport systems. *Biochim. Biophys. Acta* **1075:** 169–180.
124. Lohr, J. W., and Grantham, J. J. (1986). Isovolumetric regulation of isolated S_2 proximal tubules in anisotonic media. *J. Clin. Invest.* **78:**1165–1172.
125. Rome, L., Grantham, J., Savin, V., Lohr, J., and Lechene, C. (1989). Proximal tubule volume regulation in hyperosmotic media: Intracellular K^+,Na^+, and Cl^-. *Am. J. Physiol.* **257:**C1093–C1100.
126. Rome, L., Lechene, C., and Grantham, J. J. (1990). Proximal tubule volume regulation in hypo-osmotic media: Intracellular K^+,Na^+, and Cl^-. *Am. Soc. Nephrol.* **1:**211–218.
127. Lohr, J. W., Sullivan, L. P., Cragoe, E. J., and Grantham, J. J. (1989). Volume regulation determinants in isolated proximal tubules in hypertonic medium. *Am. J. Physiol.* **256:**F622–F631.
128. MacLeod, R. J. (1994). How an epithelial cell swells is a determinant of signal pathways that activate RVD. In *Molecular Mechanisms of Cell Volume Regulation* (K. Strange, ed.), CRC Press, Boca Raton, pp. 191–200.
129. Häussinger, D., Gerok, W., and Lang, F. (1993). Cell volume and hepatic metabolism. *Adv. Comp. Environ. Physiol.* **14:**33–65.
130. Meijer, A. J., Baquet, A., Gustafson, L., van Woerkum, G. M., and Hue, L. (1992). Mechanism of activation of liver glycogen synthase by swelling. *J. Biol. Chem.* **267:**5823–5828.
131. Schultz, S. G. (1981). Homocellular regulatory mechanisms in sodium-transporting epithelia: Avoidance of extinction by "flush-through." *Am. J. Physiol.* **241:**F579–F590.
132. Diamond, J. M. (1982). Transcellular cross-talk between epithelial cell membranes. *Nature* **300:**683–685.
133. Starke, L. J., and McManus, T. J. (1990). Intracellular free magnesium determines the volume-regulatory set point in duck red cells. *FASEB J.* **4:**A818.
134. Parker, J. C. (1994). Coordinated regulation of volume-activated transport pathways. In *Molecular Mechanisms of Cell Volume Regulation* (K. Strange, ed.), CRC Press, Boca Raton, pp. 311–321.
135. Mairbäurl, H., and Hoffman, J. F. (1992). Internal magnesium, 2,3-diphosphoglycerate, and the regulation of the steady-state volume of human red blood cells by the Na/K/2Cl cotransport system. *J. Gen. Physiol.* **99:**721–746.
136. Haas, M., and McManus, T. J. (1985). Effect of norepinephrine on swelling-induced potassium transport in duck red cells. Evidence against a volume-regulatory decrease under physiological conditions. *J. Gen. Physiol.* **85:**649–667.
137. Parker, J. C. (1993). In defense of cell volume? *Am. J. Physiol.* **265:**C1191–C1200.
138. Minton, A. P. (1994). Influence of macromolecular crowding on intracellular association reactions: Possible role in volume regulation. In *Molecular Physiology of Cell Volume Regulation* (K. Strange, ed.), CRC Press, Boca Raton, pp. 181–190.
139. Minton, A. P., Colclasure, G. C., and Parker, J. C. (1992). Model for the role of macromolecular crowding in regulation of cellular volume. *Proc. Natl. Acad. Sci. USA* **89:**10504–10506.
140. Parker, J. C., Dunham, P. B., and Minton, A. P. (1995). Effects of ionic strength on the regulation of Na/H exchange and K-Cl cotransport in dog red blood cells. *J. Gen. Physiol.* **105:**677–700.
141. Garner, M. M., and Burg, M. B. (1994). Macromolecular crowding and confinement in cells exposed to hypertonicity. *Am. J. Physiol.* **266:**C877–C892.
142. Jennings, M. L., and Al-Rohil, N. (1990). Kinetics of activation and inactivation of swelling-stimulated K^+/Cl^- transport. The volume-sensitive parameter is the rate constant for inactivation. *J. Gen. Physiol.* **95:**1021–1040.
143. Jennings, M. L., and Schulz, R. K. (1991). Okadaic acid inhibition of KCl cotransport. Evidence that protein dephosphorylation is necessary for activation of transport by either cell swelling or N-ethylmaleimide. *J. Gen. Physiol.* **97:**799–818.
144. Kaji, D. M., and Tsukitani, Y. (1991). Role of protein phosphatase in activation of KCl cotransport in human erythrocytes. *Am. J. Physiol.* **260:**C178–C182.
145. Dunham, P. B., Klimczak, J., and Logue, P. J. (1993). Swelling-activation and K–Cl cotransport in LK sheep erythrocytes—a three state process. *J. Gen. Physiol.* **101:**733–766.
146. Parker, J. C., Colclasure, C. G., and McManus, T. J. (1991). Coordinated regulation of shrinkage-induced Na/H exchange and swelling-induced [K–Cl] cotransport in dog red cells. Further evidence from activation kinetics and phosphatase inhibition. *J. Gen. Physiol.* **98:**869–880.
147. Cossins, A. R. (1991). A sense of cell size. *Nature* **352:**667–668.
148. Grinstein, S., Furuya, W., and Bianchini, L. (1992). Protein kinases, phosphatases, and the control of cell volume. *News Physiol. Sci.* **7:**232–237.
149. Palfrey, H. C. (1994). Protein phosphorylation control in the activity of volume-sensitive transport systems. In *Molecular Physiology of Cell Volume Regulation* (K. Strange, ed.), CRC Press, Boca Raton, pp. 201–214.
150. Lytle, C. Y., and Forbush, B., III. (1992). The Na–K–Cl cotransport protein of shark rectal gland. II. Regulation by direct phosphorylation. *J. Biol. Chem.* **267:**25438–25443.
151. Grinstein, S., Woodside, M., Sardet, C., Pouyssegur, J., and Rotin, D. (1992). Activation of the Na^+/H^+ antiporter during cell volume regulation. Evidence for a phosphorylation-independent mechanism. *J. Biol. Chem.* **267:**23823–23828.
152. Feschenko, M. S., and Sweadner, K. J. (1994). Conformation-dependent phosphorylation of Na,K-ATPase by protein kinase A and protein kinase C. *J. Biol. Chem.* **269:**30436–30444.
153. Kiener, P. A., Carroll, D., Roth, B. J., and Westhead, E. W. (1987). Purification and characterization of a high molecular weight type phosphoprotein phosphatase from the human erythrocyte. *J. Biol. Chem.* **262:**2016–2024.
154. Tao, M., Conway, R., and Cheta, S. (1986). Purification and characterization of a membrane-bound protein kinase from human erythrocytes. *J. Biol. Chem.* **255:**2563–2568.
155. Dunham, P. B. (1995). Effects of urea on volume-sensitive K–Cl cotransport in LK sheep red blood cells—Evidence for two different signals of swelling. *Am. J. Physiol.* **268:**C1026–C1032.
156. Kaji, D. M., and Gusson, C. (1995). Urea activation of K : Cl transport in human erythrocytes. *Am. J. Physiol.* **268:**C1018–C1025.
157. Bianchini, L., and Grinstein, S. (1993). Regulation of volume-modulating ion transport systems by growth promoters. *Adv. Comp. Environ. Physiol.* **14:**247–277.
158. Pewitt, E. B., Hegde, R. S., Haas, M., and Palfrey, H. C. (1990). The regulation of Na/K/2Cl cotransport and bumetanide binding in

avian erythrocytes by protein phosphorylation and dephosphorylation. Effects of kinase inhibitors and okadaic acid. *J. Biol. Chem.* **265:**20747–20756.

159. Christensen, O. (1987). Mediation of cell volume regulation by Ca^{2+} influx through stretch-activated channels. *Nature* **330:**66–68.
160. Foskett, J. K. (1994). The role of calcium in the control of volume-regulatory transport pathways. In *Molecular Physiology of Cell Volume Regulation* (K. Strange, ed.), CRC Press, Boca Raton, pp. 259–277.
161. Krupinski, J., Coussen, F., Bakalyar, H. A., Tang, W.-J., Feinstein, P. G., Orth, K., Slaughter, C., Reed, R. R., and Gilman, A. G. (1989). Adenyl cyclase amino acid sequence: Possible channel- or transporter-like structure. *Science* **244:**1558–1569.
162. Mills, A., Leitsou, G., Rabban, J., Sumpio, B., and Ginretz, H. (1990). Mechanosensitive adenylate cyclase activity in coronary vascular smooth muscle cells. *Biochem. Biophys. Res. Commun.* **171:**143–147.
163. Watson, P. A. (1991). Function follows form: Generation of intracellular signals by cell deformation. *FASEB J.* **5:**2013–2019.
164. von Harsdorf, L., Lang, R., Fullerton M., Smith, A. I., and Woodcock, E. (1988). Right atrial dilatation increases inositol-(1,4,5) triphosphate accumulation. Implications for control of atrial natriuretic peptide release. *FEBS Lett.* **233:**201–205.
165. Yoo, J., Ellis, R., Morgan, K. G., and Hai, C.-M. (1994). Mechanosensitive modulation of myosin phosphorylation and phosphatidylinositol turnover in smooth muscle. *Am. J. Physiol.* **267:** C1657–C1665.
166. Richter, E. A., Cleland, P. J. F., Rattigan, S., and Clark, M. G. (1987). Contraction-associated translocation of protein kinase C in rat skeletal muscle. *FEBS Lett.* **213:**232–236.
167. Kracke, G. R., and Dunham, P. B. (1990). Volume-sensitive K–Cl cotransport in inside-out vesicles made from erythrocyte membranes from sheep of low-K phenotype. *Proc. Natl. Acad. Sci. USA* **87:**8575–8579.
168. Martinac, B., Adler, J., and Kung, C. (1990). Mechanosensitive ion channels in *E. coli* activated by amphipaths. *Nature* **348:**261–263.
169. von Rossum, G. D. V., and Russo, M. A. (1981). Ouabain-resistant mechanism of volume control and the ultrastructural organization of liver slices recovering from swelling in vitro. *J. Membr. Biol.* **59:**191–209.
170. Foskett, J. K., and Spring, K. R. (1985). Involvement of calcium and cytoskeleton in gall bladder epithelial cell volume regulation. *Am. J. Physiol.* **248:**C27–C36.
171. Sachs, F. (1987). Baroreceptor mechanisms at the cellular level. *Fed. Proc.* **46:**12–16.
172. Cornet, M., Delpire, E., and Gilles, R. (1987). Study of microfilaments network during regulation process of cultured PC12 cells. *Eur. J. Physiol.* **410:**223–225.
173. Cornet, M., Lambert, I. H., and Hoffmann, E. K. (1993). Relation between cytoskeleton, hypoosmotic treatment and volume regulation in Ehrlich ascites tumor cells. *J. Membr. Biol.* **131:**55–66.
174. Ziyadeh, F. N., Mills, J. W., and Kleinzeller, A. (1992). Hypotonicity and cell volume regulation in shark rectal gland: Role of organic osmolytes and F-actin. *Am. J. Physiol.* **262:**F468–F479.
175. Hallows, K. R., Packman, C. H., and Knauf, P. A. (1981). Acute cell volume changes in anisotonic media affect F-actin content of HL-60 cells. *Am. J. Physiol.* **261:**C1154–C1161.
176. Cantiello, H. F., Prat, A. G., Bonventure, J. V., Cunningham, C. C., Hartwig, J. H., and Ausiello, D. A. (1993). Actin-binding protein contributes to cell volume regulatory ion channel activation in melanoma cells. *J. Biol. Chem.* **268:**4596–4599.
177. Krapivinsky, G. B., Ackerman, M. J., Gordon, E. A., Krapivinsky, L. B., and Clapham, D. E. (1994). Molecular characterization of a swelling-induced chloride conductance regulatory protein, pl_{Cln}. *Cell* **76:**439–448.
178. Gründer, S., Thiemann, A., Pusch, M., and Jentsch, T. J. (1992). Regions involved in the opening of CIC-2 chloride channel by voltage and cell volume. *Nature* **360:**759–762.
179. Kapus, A., Bianchini, L., Pouyssegur, J., Orkowski, J., and Grinstein, S. (1995). Osmotic responsiveness of three isoforms of the Na^+/H^+ exchanger. *Biophys. J.* **68:**A309.
180. Joyce, J. (1976). *Finnegans Wake,* Penguin Books, New York, p. 57.

20

Regulation of Cell pH

Orson W. Moe and Robert J. Alpern

20.1. INTRODUCTION

All cells regulate their pH within a relatively narrow normal range. This subserves a number of functions. First, many cell proteins are regulated by pH, and have defined pH optima. In addition, there is evidence that changes in cell pH participate in cell signaling events related to growth and hormonal effects. H/HCO_3 transporters and cell pH also play a pivotal role in cell volume regulation. In this chapter we will briefly discuss general concepts of cell pH regulation. We will then concentrate on more recent advances in the field, and issues that are of importance at the present time.

20.2. ACID–BASE CHEMISTRY

Below is shown an association/dissociation reaction for an acid/base pair:

$$H^+B \rightleftharpoons H^+ + B$$

In this reaction, H^+B is an acid, defined as a proton donor, and B is a base, defined as a proton acceptor. The dissociation constant for this reaction is defined by

$$K = \frac{[H^+][B]}{[H^+B]} \quad (20.1)$$

The logarithmic transformation of this equation is the Henderson–Hasselbalch equation:

$$pH = pK + \log([B]/[H^+B]) \quad (20.2)$$

where pX is the negative logarithm of X.

Thus, acids and bases always exist in pairs; each acid has a conjugate base and each base has a conjugate acid. From a practical point of view, in biology the term *acid* is used for proton donors with a pK < 7.4, and the term *base* is used for proton acceptors with a pK > 7.4. Thus, at physiologic pH, acids tend to donate their protons, and bases tend to accept protons.

The distinction between strong and weak acids and bases is based on their dissociation constants. Strong acids such as HCI and H_2SO_4 have dissociation constants that are considerably greater than 1, and thus pKs that are less than 0. Given the fact that hydrogen ion concentrations in physiologic solutions are generally between 10^{-7} and 10^{-8} M, one can calculate that the ratio of conjugate base to acid will be greater than 10^7. Similarly, strong bases have very small dissociation constants, such that at physiologic pH most of the total base has accepted a proton and is in the conjugate acid form. Weak acids and weak bases have dissociation constants that are closer to 10^{-7}, and thus are less predisposed to dissociate or associate, respectively, at physiologic pH. For instance, most weak acids have dissociation constants around 10^{-4} M. Thus, at pH 7, 1/1000 of the total acid will be in the undissociated form.

20.3. MEASUREMENT OF INTRACELLULAR pH

Our understanding of cell pH regulation has recently exploded as a result of the general availability of methods for its measurement. Four general approaches have been used for the measurement of intracellular pH. We will briefly discuss these. The reader is referred to Ref. 1 for an in-depth discussion.

20.3.1. Distribution of Weak Acids and Bases

The use of weak acids or bases to measure cell pH relies on the fact that the uncharged weak acid or base is more permeable across cell membranes than their conjugate pair. In this setting, an equilibrium will be achieved where the concentrations of the weak acid or base are equal in and out of the cell. The relative concentrations of the conjugate pair inside and outside the cell are then determined by intracellular and extracellular pH. Thus, if one knows extracellular fluid composition and measures intracellular weak acid or base (and its conjugate pair) concentration, cell pH can be calculated.

Orson W. Moe • Department of Internal Medicine, The University of Texas Southwestern Medical Center, Dallas, Texas 75235 and Veterans Administration Medical Center, Dallas, Texas 75216. **Robert J. Alpern** • Department of Internal Medicine, The University of Texas Southwestern Medical Center, Dallas, Texas 75235

Molecular Biology of Membrane Transport Disorders, edited by S.G. Schultz *et al.* Plenum Press, New York, 1996

In order for this approach to work, the permeability of the uncharged weak acid or base must be substantially greater than that of the ionized conjugated pair. Any significant fluxes of the ionized moiety will prevent equilibration of the uncharged form. It is also important that the concentration of the weak acid or base be low. Most of the weak acid or base that enters the cell will donate or accept a proton. If this flux is substantial, the measurement will perturb cell pH. This is generally not a problem with the use of isotopically labeled weak acids. The most important problem relates to the movement of weak acids and bases into intracellular compartments. Thus, alkaline intracellular compartments such as mitochondria will trap weak acids leading to an erroneously high measurement of cell pH, and acid intracellular compartments such as endosomes, Golgi, endoplasmic reticulum, and lysosomes will trap weak bases leading to an erroneously low estimate of cell pH. Accurate measurements of cell pH by this approach require accurate measurement of intracellular isotope, and accurate correction for contamination of samples by extracellular fluid. This is generally done by including a marker for extracellular fluid volume such as labeled sucrose. A weakness of this approach is that one cannot continuously follow cell pH. Each sample can be used for only one measurement at one time point. The approach also requires the use of a considerable amount of tissue, and cannot be applied to single cells. Despite these problems, this approach has been widely used with the weak acid 5,5′-dimethyl-2,4-oxazolidine-dione (DMO). Measurements have been found to agree within 0.1 to 0.2 pH unit with measurements made by other techniques.

20.3.2. Magnetic Resonance Spectroscopy

This approach relies on measurement of the inorganic phosphate peak using ^{31}P spectroscopy. This peak shifts in a predictable manner in response to changes in pH caused by shifts in the relative abundance of $H_2PO_4^-$ and HPO_4^{2-}. The approach offers the advantage of continuous measurements, and the unique advantage of being able to measure cell pH in intact animals without the need for extensive surgery. It is limited by the cost of the equipment and the requirement for large amounts of tissue. Thus, it has been very difficult to apply this approach to cultured cells. In addition, the approach relies on the assumption that most of the inorganic phosphate is located intracellularly. This may be a considerable problem in certain tissues. One can also utilize pH-dependent shifts of phosphorus-containing metabolic intermediates which are more definitely intracellular. Although this technique allows continuous measurements, its time resolution is quite inferior to those of the methods described below.

20.3.3. pH-Sensitive Microelectrodes

This technique likely represents the "gold standard" for measurement of cell pH in individual cells. The most accurate pH-sensitive electrodes utilize pH-sensitive glass. However, in small cells this requires a recessed tip electrode to ensure that pH-sensitive glass is exposed only to intracellular fluid. More recently, a resin pH sensor has come into more widespread use because the electrodes are more easily made. A pH-sensitive microelectrode requires the simultaneous measurement of cell voltage. Thus, one must either use a double-barreled microelectrode, or place two electrodes in the same cell. In the case of small cells, a compromise is to place the two electrodes in different cells. This technique offers the advantage that one measures cell pH and cell voltage, and thus can examine whether changes in pH are secondary to changes in voltage. The major disadvantage is the great difficulty in making and utilizing microelectrodes that can work in small vertebrate cells.

20.3.4. pH-Sensitive Dyes

The recent explosion in our understanding of cell pH regulation is probably most attributable to the development of pH-sensitive absorbance and fluorescent dyes. Of these two approaches, that of fluorescence to measure cell pH has been used more often because measurements can be made with less complicated, less expensive equipment. Although the equipment for fluorescence measurement of cell pH is not inexpensive, fluorometers are generally widely available to investigators. Measurement of pH in individual cells can be performed with this technology utilizing epifluorescence microscopes with or without imaging techniques. Although a number of pH-sensitive fluorescent dyes have been developed, the most useful have been those based on fluorescein. Fluorescein has the highest quantum yield of any fluorophore and is naturally pH sensitive. The most widely used derivative, (2′,7′)-bis(carboxyethyl)-(5,6)-carboxyfluorescein (BCECF), has been modified to shift its pK to 7.0, and to include four or five negative charges which keep it trapped within the cell.[2] BCECF is synthesized as an acetoxymethyl ester which is uncharged and cell permeant. When perfused around cells, it diffuses into cells where cytoplasmic esterases cleave off the acetoxymethyl group, yielding a charged compound that is trapped in the cell.

The basis for the fluorescence measurement of cell pH is given in Figure 20.1,[3] which shows an excitation spectrum for BCECF. Emission is measured at 530 nm, while excitation wavelength is varied. It can be seen that at 436 nm (the isosbestic point), acidic and alkaline forms of BCECF are equally fluorescent, and the magnitude of the fluorescence is pH-independent. Peak fluorescence occurs with excitation at approximately 504 nm, where the alkaline form of the dye is substantially more fluorescent than the acidic form. Thus, fluorescence at around 500 nm excitation is dependent on the pH of the ambient fluid and the concentration of the dye, while fluorescence at 436 nm is dependent only on the concentration of the dye. The ratio of fluorescence intensity at these two wavelengths is independent of dye concentration and pH is the only determinant. While one is always concerned about fluorescence artifacts related to changes in intracellular composition or entrance of the dye into intracellular vesicular compartments, these have generally not been a problem. The fluorescence signal is usually calibrated by setting cell pH using nigericin in the presence of high-potassium solutions.[4]

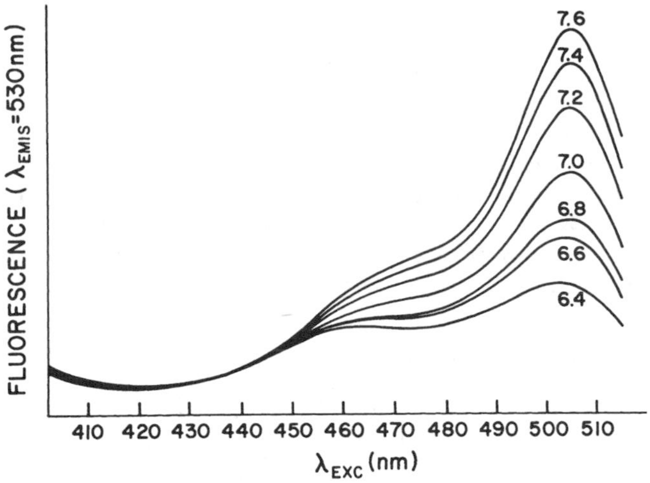

Figure 20.1. Excitation spectrum of BCECF. (From Ref. 3.)

This technique allows the measurement of cell pH in single cells or parts of single cells to an extremely high level of sensitivity. The only requirement is that the cells can be placed in a cuvette or studied with an epifluorescence microscope. The major drawback is the inability to simultaneously measure cell voltage. However, some investigators have used voltage-sensitive dyes in an attempt to compensate for this.

20.4. CELL pH REGULATION

20.4.1. Physicochemical Buffering

The buffer capacity (β) of any solution is defined as the amount of strong acid or base that one must add to a solution to achieve a change of 1 pH unit. This is quantitated as

$$\beta = d(\text{base or acid})/d\text{pH} \qquad (20.3)$$

Thus, if one adds an acid or base load to the cell and measures the change in cell pH achieved, one can calculate the buffer capacity, which will be in units of millimoles per liter per pH unit.

It is important to realize that buffer capacity is unique among the processes that regulate cell pH in that it has no effect on the steady-state cell pH. Thus, in the steady state, cell pH is that pH that is associated with equal rates of acid and base addition to the cell. Since physicochemical buffers neither donate nor accept hydrogen ions in the steady state, they have no effect on this value. If there is a sudden change in the rates of net acid or base addition to the cell, a new steady-state pH will be achieved. The buffer capacity then will determine the rate at which the new steady state is achieved.

Physicochemical buffers are generally considered to be of two types. Closed buffers are those whose total concentration in the cell is fixed, such that changes in cell pH lead to reciprocal changes in the quantities of the acidic and basic moieties, but do not change the total buffer concentration. Open buffers, on the other hand, vary in total concentration based on pH, because one of the buffer moieties is cell permeant. The most important open buffer system is the CO_2/HCO_3 buffer system. Because CO_2 is highly permeable across cell membranes, the intracellular CO_2 concentration is fixed and equal to that of extracellular fluid. The concentration of carbonic acid, the true acid in this acid–base system, is merely a function of the concentrations of CO_2 and water, and thus is also equal in intracellular and extracellular fluid. Thus, given the reaction

$$H^+ + HCO_3^- \rightleftharpoons H_2CO_3$$

the concentration of HCO_3 varies inversely with the concentration of H ions.

The interested reader is referred to Ref. 1 for derivations of the magnitude of buffer capacity for open and closed buffer systems. The important point is that open and closed buffers behave very differently. The buffer capacity for a closed buffer is maximal when pH equals pK and at this pH is equal to 0.58 times the concentration of total buffer. At pHs above or below pK, buffer capacity decreases. At pHs greater than 1 pH unit away from pK, buffering power is minimal.

The buffering power of open buffers, on the other hand, is related to the total amount of buffer. Thus,

$$\beta_{open} = 2.3 \times [\text{impermeant species}] \qquad (20.4)$$

The higher the concentration of the cell-impermeant species, the higher the buffer capacity is. For CO_2/HCO_3 this means that buffer capacity increases as cell pH increases, in spite of the fact that cell pH moves further away from the pK of H_2CO_3 dissociation.

To measure buffer capacity, one needs to add acid or alkali to cells acutely, and to measure the cell pH change. Ideally, this is performed by injecting a strong acid or base into the cell. In most cases, however, cell size prevents the use of this approach. More commonly, one provides acid or base loads by the sudden addition or removal of uncharged weak acids or bases. Thus, if one adds a substantial amount of a weak acid (HA) to a cell, HA rapidly enters the cell as a result of its high permeability across the cell membrane. Because cell pH is at least 2–3 pH units above the pK of the weak acid, most of the weak acid dissociates to form hydrogen plus the conjugate base. The net result is an acid load in the cell. At equilibrium, which is attained instantaneously, one can calculate the concentration of intracellular conjugate base from the concentration of intracellular weak acid (assumed to equal extracellular concentration) and the measured cell pH. The acid load per liter of cells is then equal to the concentration of conjugate base. Using the measured cell pH change, the amount of acid added to the cell, and Eq. (20.3), one can then calculate the buffer capacity. A similar approach can be used on removal of the weak acid or with a weak base such as ammonia. The major requirement for this approach is that the uncharged weak acid or base have a considerably higher permeability than the conjugate acid or base, such that transmembrane fluxes of conjugate acid or base are negligible. Any flux of the conjugate acid or base would lead to an addition of this moiety to the cell that does not correspond to addition of H or OH.

A major problem with the measurement of buffer capacity is that processes other than physicochemical buffering can defend cell pH relatively rapidly, decreasing the cell pH change and leading to an overestimation of buffer capacity. Processes that fall into this category include cell membrane transport, metabolism, and organellar transport. Changes in membrane transport and metabolism ideally should not be confused with physicochemical buffering, as they provide a continued source of net acid or alkali addition and will contribute to an altered steady-state cell pH. Changes in organellar transport, however, may justifiably belong under the category of cell buffer capacity, as this process would represent a one-time release of H or OH into the cytoplasm. There are presently no data indicating whether organellar H transport is regulated by changes in cytoplasmic pH, and thus contributes to buffer capacity. If changes in metabolism or membrane transport are slower than neutral acid or base diffusion, they will not interfere with the measurement of buffer capacity. However, rapid changes can interfere. One approach to minimizing the contribution from slower processes is to extrapolate cell pH back to the time of net acid or base addition or removal. It is also advisable when measuring buffer capacity to inhibit all important membrane transporters when possible.

The major use for the buffer capacity is to calculate transporter activity from measured rates of change in cell pH. When transporter rate suddenly changes, the rate of change in cell pH is a function of transporter activity and cell buffer capacity:

$$dpH_i/dt(\text{pH/min}) = J_H(\text{mmole/liter} \bullet \text{min})/\beta(\text{mmole/liter} \bullet \text{pH}) \qquad (20.5)$$

In studying regulation of transporters, one therefore needs to be certain that differences in dpH_i/dt are attributable to differences in transporter rate rather than to changes in buffer capacity. In practice, the authors are not aware of any condition in which buffer capacity of the cell has been shown to be regulated. One point worth emphasizing is that the buffer capacity of the cell varies markedly as a function of pH. Therefore, if one is to calculate rates of hydrogen transport from rates of change in cell pH, it is important that the buffer capacity used is measured at approximately the same cell pH.

Most measurements suggest that the CO_2/HCO_3 buffer system is the dominant component of buffer capacity in vertebrate cells. In addition, closed buffers include cell proteins, phosphate, and a number of phosphorylated and carboxylated compounds. One important physiologic point is whether closed buffers are mobile or immobile. Their ability to buffer cell pH changes is unaffected by their mobility. However, the rate of acid and base diffusion across cells is markedly dependent on buffer mobility. Because H and OH ions are present in low concentrations ($< 10^{-7}$ M), the amount of diffusion of these moieties is limited even though the diffusion coefficient for H is high. Thus, typically acid–base diffusion across solutions occurs as diffusion of buffers that exist in millimolar concentrations.[5,6] If closed buffers are immobile, then acid–base diffusion within the cytoplasm will be dependent on the CO_2/HCO_3 buffer system. This would mean that carbonic anhydrase, which catalyzes the reaction $OH^- + CO_2 \rightleftharpoons HCO_3^-$, would be required to maintain acid–base diffusion. The ability of acids and bases to diffuse across the cell is particularly important in epithelia that accomplish transepithelial H/HCO_3 transport.

20.4.2. Plasma Membrane Transporters

In this section, we will review the most important H/HCO_3 transporters that have been identified in plasma membranes. Figure 20.2 provides a summary of these transport mechanisms, divided according to whether they effect acid efflux from or acid entry into the cell. We will describe briefly the basic characteristics of the transporters, as well as functions that they have been shown to subserve. This will be followed by a discussion of the molecular mechanisms responsible for transport and of other issues of importance at present.

20.4.2.1. Na/H Antiporter

The Na/H antiporter is likely the most ubiquitous of all H transport mechanisms in vertebrate cells. It functions as an electroneutral 1:1 exchanger of Na and H ions. Because cell Na concentrations are maintained low by the Na,K-ATPase, there is an inward Na gradient which can drive H extrusion from the cell. The Na/H antiporter is inhibited by amiloride, and the potency of inhibition is increased by hydrophobic 5-amino group substitutions of amiloride.[7,8]

Assignment of a function to the Na/H antiporter generally involves demonstrating inhibition by amiloride or Na removal. As with most inhibitors, amiloride is somewhat nonspecific, and inhibits many Na transport processes including the Na,K-ATPase and Na channels, and other processes such

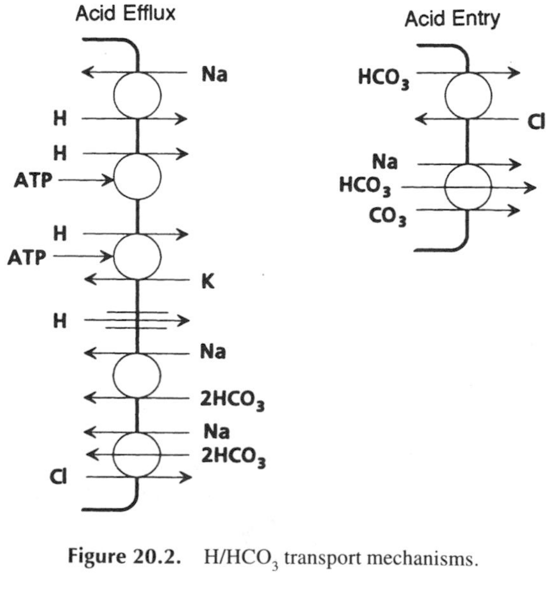

Figure 20.2. H/HCO_3 transport mechanisms.

as metabolism.[9] While Na removal appears at first glance to be more specific, it can have many effects on cells such as depolarization. In the renal outer medullary collecting duct, inner stripe, Na removal was found to raise intracellular Ca concentration by release of Ca from intracellular stores and Ca entry into cells.[10] This increase in cell Ca then inhibited the H pump. The net result was that Na removal inhibited H transport, independent of any role of an Na/H antiporter. Thus, great care must be taken in inferring a role for Na/H antiport in a physiologic process.

The Na/H antiporter subserves a number of functions. First of all, it appears to be the major transporter responsible for defense of cell pH against an acid load in vertebrate cells. This activity is enhanced by the fact that decreases in cell pH increase the activity of the transporter (see below). Na/H antiporters are also involved in the maintenance of cell volume following cell shrinkage (see Chapter 19). These housekeeping functions are believed to be mediated by a ubiquitous Na/H antiporter that is sensitive to low concentrations of amiloride and amiloride analogues.[11] In many epithelia, Na/H antiporters effect transepithelial transport. Alone they mediate $NaHCO_3$ transport; in parallel with Cl/base exchangers, Na/H antiporters mediate NaCl transport.[12] This function is mediated by an Na/H antiporter that is more resistant to amiloride and its analogues.[11] Lastly, increases in Na/H antiporter activity have been observed following treatment of cells with growth factors or hormones, and in cells from patients with hypertension. The importance of these findings is discussed below.

The gene for the Na/H antiport was first cloned by Sardet *et al.* using complementation,[13] and subsequently three other isoforms have been cloned from mammalian cDNA libraries by homology.[14–17] The genes encode proteins of predicted size 80–93 kDa. Amino acid homology between the isoforms ranges from 36 to 57%. All of these isoforms have a similar hydropathy profile which predicts a hydrophobic N-terminal region containing 10–12 transmembrane domains. The C-terminal third forms a large hydrophilic cytoplasmic loop. All of the isoforms contain at least one conserved potential N-linked glycosylation site, and all contain consensus phosphorylation sites. By sequence analysis and the study of deletion mutants, the functions of the domains have been determined. Deletion of the C-terminal cytoplasmic domain eliminates growth factor regulation of the Na/H antiporter, but the protein is able to effect Na/H exchange, and is activated by decreases in cell pH.[18] Thus, the N-terminal hydrophobic domain is believed to mediate transport, while the C-terminal cytoplasmic domain serves more of a regulatory function.

The roles of the four NHE isoforms are presently under investigation. NHE-1 is believed to encode the ubiquitous amiloride-sensitive housekeeping Na/H antiporter. Apical membrane H transport effecting transepithelial H transport in the intestine and renal proximal tubule is presently believed to be mediated at least in part by NHE-3. This is based on predominant NHE-3 mRNA expression in small intestine and renal cortex, and localization of NHE-3 protein to the proximal tubule apical membrane by immunocytochemistry.[14,15,19] Expression studies have shown that NHE-3 is amiloride-resistant.[20] In addition, OKP cells, which possess an amiloride-resistant Na/H antiporter with many similarities to the apical membrane Na/H antiporter, express NHE-3 mRNA.[21] It remains possible that other isoforms also contribute to this process. NHE-2 and -4 are expressed predominantly in stomach and intestine.[14,16,17,22] In rabbit, NHE-2 is expressed to a large extent in kidney, while in rat it is expressed in kidney to a lesser extent.[16,17,22]

20.4.2.2. ATP-Driven Proton Pumps

There are three major classes of ATP-driven cation pumps. One of these is the phosphorylated class of ATP-driven cation pumps referred to as ElE2 or P-type ATPases, which have a phosphorylated intermediate. P-type ATPases are reviewed extensively in Chapter 12. These pumps generally function to exchange cations but can also effect unidirectional cation transport. Pumps of this type include the Na,K-ATPase which exchanges Na for K, the Ca-ATPase which exchanges Ca for H, the H,K-ATPase which exchanges H for K, and the H-translocating ATPase of *Neurospora crassa* and yeast. With regard to plasma membrane H transport in higher eukaryotes, the H,K-ATPase plays an important role in certain epithelia. The H,K-ATPases mediate electroneutral exchange of H for K, utilizing ATP to drive uphill transport. As is characteristic of all members of this family, ATP phosphorylates the transporter, forming a high-energy β-aspartyl phosphate intermediate.[23] The H,K-ATPase mediates H secretion by the gastric parietal cell, and is thus responsible for the stomach's acidity.[23] In addition, an H,K-ATPase is present in the colon[24] and likely in the collecting duct of the kidney.[25]

The rat cDNA for the gastric H,K-ATPase catalytic or α subunit was first cloned based on homology to sequences in the Na,K-ATPase.[26] The protein is predicted to have a molecular mass of 114 kDa. The H,K-ATPase shares 63% amino acid homology with the Na,K-ATPase, 24% amino acid homology with the Ca-ATPase, 14% homology with the H-transporting ATPase of *N. crassa,* and 17% homology with the yeast H-transporting ATPase.[23,26] The regions of greatest homology occur around the phosphorylation site (Asp385), a proposed transmembrane energy transduction region, and a putative ATP binding domain.[26] By hydrophobicity, the protein is predicted to possess 6–10 transmembrane domains. Using tryptic digestion and monoclonal antibodies, a model with 8 transmembrane domains and cytoplasmic N- and C-termini has been derived.[27]

Similar to Na,K-ATPase, the H,K-ATPase has been found to possess a β subunit. The bovine β subunit was cloned by screening a bovine abomassum λgt11 cDNA library with a monoclonal antibody raised against a putative β subunit.[28] The rat β subunit was then cloned by PCR from a gastric cDNA library. The clones predict a protein of 34 kDa. Hydropathy analysis indicates a single transmembrane domain, similar to the β1 and β2 subunits of the Na,K-ATPase. The predicted amino acid sequence of the rat H,K-ATPase β subunit is 41% identical to the rat Na,K-ATPase β2 subunit, and 35% identical to the β1 subunit. The presumed C-terminal extracytoplasmic domain contains six cysteine residues which are highly conserved between the β subunits, and seven potential N-linked glycosylation sites, two of which are highly conserved relative to β1 and β2. By Western blot, digestion with N-glycanase F reduces the apparent molecular mass from 60–80 kDa to 32 kDa, suggesting considerable N-linked glycosylation. By Northern blot, H,K-ATPase β subunit mRNA is expressed only in stomach.

The role of the β subunit has not been determined. Similar to the Na,K-ATPase, all catalytic functions are performed by the α subunit. In transfected cos-1 cells, β subunits are transported to the cell surface in the absence of α subunits, but α subunits require β subunits for cell surface expression.[29] By generating chimeras between the gastric H,K-ATPase and Na,K-ATPase α subunits, it was found that the C-terminal half of the α subunit determines which β subunit is utilized for assembly.[29] This is of particular interest because Na,K-ATPases are uniformly basolateral while H,K-ATPases are apically located in epithelia. The sequences responsible for polarized distribution of H,K-ATPase subunits have been studied in LLC-PK1 cells which are polarized and express an endogenous basolateral membrane Na,K-ATPase, but no H,K-ATPase.[30] Transfection of these cells with the HK β subunit leads to apical membrane expression, while transfection with the α subunit alone leads to no surface expression. When cells are transfected with both HK α and β, both are expressed in the apical membrane. When cells are transfected with a chimera containing the N-terminal half of the H,K-ATPase α subunit and the C-terminal half of the Na,K-ATPase α subunit, which binds to the Na,K-ATPase β subunit,[29] the chimeric α subunit and the NaK β subunit are expressed on the apical membrane.[30] Thus, these studies suggest that both the α and β subunits of the H,K-ATPase encode independent signals that specify apical membrane localization.

A putative colonic H,K-ATPase catalytic α subunit was cloned by homology from a rat colonic cDNA library.[31] The cDNA encodes a protein of predicted size 115 kDa. Hydropathy analysis is similar to that of other members of the family. Amino acid homology is 63% with the gastric H,K-ATPase and 63% with Na,K-ATPase isoforms. Homology among Na,K-ATPase isoforms is much greater. By Northern blot, the putative colonic H,K-ATPase α subunit was found to be expressed at high abundance in distal colon, and at lower abundance in kidney and uterus.[31] By *in situ* hybridization, mRNA expression of this putative H,K-ATPase was found in surface epithelial cells of the rat distal colon, and between muscular layers of the myometrium (possibly in nerve cells).[32] It should be noted that the putative colonic H,K-ATPase has not yet been expressed, and thus has not been proven to be an H,K-ATPase. The above-described homology suggests that it is a K-transporting ATPase, while it is not sufficiently homologous to the Na,K-ATPases to likely be a member of that family. There are a number of similarities between the colonic cDNA and the gastric H,K-ATPase in the N-terminal region.[31] However, the major evidence that the colonic cDNA encodes an H,K-ATPase is the correlation between expression and physiology.[24,31]

It is likely that an H,K-ATPase mediates H secretion and K absorption in the renal collecting duct, based on inhibition of these process by SCH 28080 and omeprazole, inhibitors of the gastric H,K-ATPase.[25,33–35] However, the molecular basis of this transport remains unclear. Antibodies against the gastric H,K-ATPase label intercalated cells (acid-transporting cells) of the collecting duct.[36] Utilizing RT-PCR, expression of the gastric H,K-ATPase α subunit in kidney was detected by one group.[37] A number of groups have found expression of the putative colonic H,K-ATPase α subunit in kidney by Northern blot[31] and by RT-PCR.[37,38] One group was unable to detect expression of the colonic isoform in kidney by Northern blot, RNase protection, and *in situ* hybridization.[32] α and β subunits have recently been cloned from *Bufo marinus* toad bladder, which when coexpressed in *Xenopus laevis* oocytes, effect Rb uptake and H efflux.[39] These mRNAs are also expressed in kidney and eye.[39] Thus, while the renal collecting duct likely possesses a plasma membrane H,K-ATPase, the specific isoform expressed has not been resolved.

A second class of ATP-driven cation pumps is that of the F_0F_1 H-ATPases. These H pumps are present in bacteria, chloroplasts, and mitochondria. In these locations, the pump actually functions as an ATP-synthesizing proton channel. Because these pumps have not been found in the plasma membrane of eukaryotes, they will not be discussed further.

The third class of ATP-driven cation transporters are the vacuolar ATPases.[40] These ATP-driven H pumps have been found to transport H into a number of intracellular vesicular compartments such as endosomes, secretory granules, lysosomes, Golgi, clathrin-coated vesicles, and multivesicular bodies. The vacuolar and the F_0F_1 H-ATPases are both multisubunit and are believed to share a common evolutionary origin. Each possesses a number of membrane-spanning subunits, one of which binds to and is inhibited by DCCD.[40] They also possess a number of subunits that are not integral membrane polypeptides and function in catalysis. The best studied vacuolar H pump is that of the clathrin-coated vesicle. It contains nine subunits of 116, 70, 58, 40, 38, 34, 33, 19, and 17 kDa.[40,41] The 70-, 58-, 40-, and 33-kDa polypeptides are required for ATP hydrolysis.[42] The 70-kDa subunit is the catalytic subunit that is phosphorylated by ATP. The 58-kDa subunit is also phosphorylated by ATP, but this likely plays a regulatory role. The 17-kDa subunit is bound by DCCD and likely forms the transmembrane H channel.[43] All three of these subunits are present in multiple copies. The 70-, 58-, and 33-kDa subunits have now been cloned.[44–46] The vacuolar H-ATPase is ubiquitous, present in all cells in intracellular vesicular compartments. By immunocytochemistry, the vacuolar H pump has been demonstrated on the plasma membranes of renal proximal tubule and collecting duct cells.[47] A kidney-specific isoform of the 58-kDa subunit has now been cloned.[48] This isoform is only expressed in H-transporting intercalated cells of the renal collecting duct.

Definition of the functional role of these proton pumps in intact cells requires a specific assay which is not available. Plasma membrane H pump activity can be measured as transepithelial H secretion or as recovery of cell pH from an acid load. However, because the only required substrates for the H pump are H and ATP, it is difficult to manipulate H pump activity in a manner that is specific for the vacuolar H pump. While recovery of cell pH from an acid load in the absence of Na has been used as an assay for the vacuolar H pump, it is now clear that the H,K-ATPase and H leak pathways can participate in this assay (see below). Thus, the specificity of these assays has relied on the use of inhibitors. Vacuolar H-ATPases are inhibited by NEM and NBD Cl, but neither of these inhibitors is specific enough to provide convincing results. More recently, bafilomycin A_1 has been found to be a relatively specific inhibitor of these pumps.[49] Vacuolar ATPases are very sensitive to this inhibitor, P-type ATPases are moderately sensitive, and F_0F_1-ATPases are unaffected.[49,50] However, bafilomycin A_1 has been shown to inhibit Cl channels.[51] In addition, because intravesicular H transport relies on vacuolar H pumps, inhibitory effects of bafilomycin A_1 may be secondary to effects on these vesicles, rather than on the plasma membrane H-transporting mechanism.

The H,K-ATPase can also function to mediate transepithelial H secretion and as a Na-independent mechanism to defend cell pH from an acid load. The H,K-ATPase can be distinguished from the vacuolar H pump by its requirement for extracellular potassium. However, in intact cells the intracellular concentration of K is high and small amounts of leaked K may be sufficient to drive the H,K-ATPase. Omeprazole has been utilized as an inhibitor of the H,K-ATPase. However, omeprazole can also inhibit vacuolar H-ATPases, albeit at higher concentrations.[50] SCH 28080 [2-methyl-8-(phenylmethoxy)imidazo(1,2*a*)pyridine-3-acetonitrile] inhibits the H,K-ATPase by competing with K.[52] This may be the most specific inhibitor available. However, in the turtle urinary bladder, electrogenic, K-independent, bafilomycin A_1-sensitive H secretion is inhibited by SCH 28080.[53] Lastly, vanadate, an inhibitor of all P-ATPases, can be used to inhibit H,K-ATPases. However, vanadate inhibits many processes. It has recently been reported to inhibit the osteoclast vacuolar H-ATPase.[54] Thus, it is with great caution that one should infer the roles of H-ATPases in physiologic processes. The relative roles of vacuolar and P-ATPases in collecting duct H transport are not resolved.

20.4.2.3. Proton Conductance

The existence of a large plasma membrane proton conductance has been generally felt to be unlikely. This is based largely on the fact that such a conductance would provide a large acid load to the cell, requiring considerable energy input to extrude protons and maintain a normal cell pH. In addition, because H ions are present in physiologic solutions at concentrations many orders of magnitude lower than those of other cations such as Na and K, a H-selective channel would have to possess an unusually high selectivity.

In spite of these limitations, there is now clear evidence for the existence of a proton conductance in certain cell types. The interested reader is referred to a recent in-depth review of these channels.[55] In general, the cells that possess a H conductance are cells that undergo substantial depolarization at a time when cell pH markedly decreases. Once again, because there are no cotransported species, it is difficult to prove that a H flux is mediated by a H conductance if one merely measures cell pH. The most convincing data have been provided using electrophysiologic methods. A H conductance has now been demonstrated in snail neurons, *Amphiuma* oocytes, type II alveolar cells, and peritoneal macrophages.[55–59] The conductance is activated by cell depolarization, decreases in cell pH, and increases in extracellular pH. These regulatory influences ensure that when the channel is open it will function as a H efflux mechanism. Thus, in a resting cell with a voltage of −60 mV and a cell pH of 7.0, a proton conductance would cause entry of H ions. In this setting the H conductance is small. Cell depolarization and cell acidification increase the proton conductance and create a driving force that favors H efflux. Thus, when an action potential occurs in a neuron, cell pH decreases. Similarly, activated macrophages undergo a cell depolarization accompanied by a decrease in cell pH. The cell depolarization along with the decrease in cell pH increase the proton conductance and provide a driving force for protons to leave the cell.

It is not presently clear whether all cells possess this conductance, or whether this conductance is confined to cells that undergo considerable depolarization. This becomes of much methodologic importance because a frequent assay used for detection of ATP-driven H pumps is the recovery of cell pH from an acid load in the absence of Na (see above). Because Na removal may be expected to cause a substantial cell depolarization, it is possible that a H conductance could contribute to increases in cell pH in this setting. Thus, once again one relies on inhibitors to distinguish the mechanism of the cell pH recovery. H conductances studied thus far have been found to be inhibited by cadmium and zinc.[55,59]

The nature of the H conductance is not clear. The conductance has an unprecedented selectivity for H ions over other ions ($\geq 10^6$:1). It also has an extremely large permeability to H of approximately 10^{-1} cm/sec.[55] Given these characteristics, along with the regulation of the H conductance, it seems unlikely to be mediated by diffusion across a lipid bilayer. At present it is not clear whether the H conductance represents a channel, or an uncoupled electrogenic carrier.

20.4.2.4. Cl/HCO_3 Exchanger

The Cl/HCO_3 exchanger has been most extensively studied in the red blood cell where it is estimated to comprise approximately 25% of total membrane protein, and has an identifiable band on SDS–PAGE (band 3). In the red blood cell, the band 3 Cl/HCO_3 exchanger functions in both directions as an electroneutral exchanger of Cl and HCO_3. It can also transport other anions, but at varying rates. It is inhibited by disulfonic stilbenes such as SITS and DIDS. Most other cells also possess Cl/HCO_3 exchangers. In these cells the exchanger tends to run in the base efflux direction driven by the low cell Cl concentration. In addition, as described below these Cl/HCO_3 exchangers are activated by increases in cell pH which allow them to run at rapid rates in the presence of a high cell pH, and to shut off at low cell pH.[30,60–63] Thus, in most cells Cl/HCO_3 exchange serves as the major mechanism for defense against an alkaline pH challenge. The RBC band 3 Cl/HCO_3 exchanger is somewhat unique in that it is not activated by increases in cell pH,[64,65] and in fact is not regulated except by changes in driving force. Increases in cell $[HCO_3]$in peripheral blood vessels cause the transporter to run in a HCO_3 efflux mode, and decreases in cell $[HCO_3]$ in the pulmonary circulation cause the transporter to run in the HCO_3 influx mode. In certain cell types such as the gastric parietal cell and the renal collecting duct intercalated cell, Cl/HCO_3 exchange functions as a base efflux mechanism mediating transepithelial transport.

The cDNA encoding the red cell Cl/HCO_3 exchanger has been cloned by screening a murine spleen λgt11 cDNA library with a polyclonal antibody.[66] It encodes a protein of predicted size 103 kDa with an N-terminal hydrophilic cytoplasmic region and a C-terminal hydrophobic region that contains up to 14 membrane-spanning domains. The C-terminal part of the protein mediates transport and binding to disulfonic stilbenes, while the N-terminal cytosolic domain is responsible for cytoskeletal interactions.[67,68] A series of cDNAs have now been cloned encoding a family of anion exchange proteins. The three isoforms are termed AE1, AE2, and AE3. AE1 encodes the red blood cell Cl/HCO_3 exchanger. A truncated form of AE1, deficient in the first three exons of AE1, mediates basolateral membrane Cl/HCO_3 exchange in the acid-secreting renal collecting duct intercalated cell.[69–71] This transporter functions as a HCO_3 efflux mechanism in these cells. Both the renal and red cell Cl/HCO_3 exchanger are encoded by the same gene, but the renal exchanger utilizes a transcriptional initiation site in the middle of the third intron (with respect to the red blood cell mRNA).[72]

AE2 and AE3 were cloned from cDNA libraries by hybridization at low stringency.[73–76] The sequences of AE2 and AE3 are homologous to AE1, and can be divided into three regions. Sequence similarity is highest in the hydrophobic C-terminal domain, where there is 65–70% amino acid identity between the three isoforms. AE2 and AE3 encode proteins that are somewhat larger than AE1, the difference resulting from a 250-amino-acid extension at the N-terminus. This region, which is lacking in AE1, is similar between AE2 and AE3, containing a large abundance of acidic amino acid residues flanked by proline/glycine-rich clusters. This region also contains a striking histidine-rich sequence. The remaining N-terminal cytoplasmic domain shows some similarity between the three isoforms, albeit less than that present in the C-terminal hydrophobic domain. Hydropathy profiles of all three isoforms are very similar. By Northern blot, all three genes have been shown to encode transcripts of multiple sizes.[74] AE1 is expressed in greatest abundance in spleen (likely representing red blood cell precursors) and kidney. With prolonged exposures, transcripts can be found in numerous tissues. AE2 is expressed in highest abundance in stomach, large intestine, lung, and uterus, but is more ubiquitous. AE2 is a likely candidate for the above-described ubiquitous Cl/HCO_3 exchanger. AE3 mRNA is found in highest abundance in heart and brain, but once again with long exposures can be seen in multiple tissues. Given the pH sensitivity of the non-RBC Cl/HCO_3 exchanger, it is tempting to speculate that this function is encoded by the N-terminal 250 amino acids unique to AE2 and AE3, and possibly by the histidine-rich sequence.

20.4.2.5. Na/HCO_3 Cotransport

Electrogenic Na/HCO_3 cotransport was first described in the renal proximal tubule.[3,77,78] In general, these transporters possess the following characteristics: (1) they are dependent on the presence of CO_2/HCO_3, and can be driven in either direction by HCO_3 gradients; (2) they are dependent on the presence of sodium and can be driven in either direction by sodium gradients; (3) they are electrogenic and voltage-sensitive, carrying a net negative charge; and (4) they are inhibited by disulfonic stilbenes such as SITS and DIDS.

In the renal proximal tubule the Na/HCO_3 transporter has been found on the basolateral membrane and has been shown to likely effect most of base exit mediating transepithelial transport.[3,77,78] This transport has a stoichiometry of 1 Na:3 HCO_3:2 negative charges.[78,79] Based on the ability of SO^{2-}_3 to drive the transporter, it has been concluded that the actual substrates transported are 1 Na:1 HCO_3:1 CO_3 (thermodynamically equivalent to 1 Na:3 HCO_3.[80] Given a cell voltage of −50 to −70 mV in the proximal tubule, this transporter will function as an acid loader or base efflux mechanism. Based on these results, it was initially postulated that the transporter could provide a base efflux mechanism in a number of epithelia that mediate transepithelial H/HCO_3 transport. Because of the requirement for cell Na, it could only function at a significant rate as an efflux mechanism in cells that possess a Na entry mechanism functioning at a significant rate. In the proximal tubule the Na entry mechanism is provided by the apical membrane Na/H antiporter.

Subsequent studies have found an electrogenic Na/HCO_3 cotransport mechanism in many epithelial and nonepithelial cells including corneal endothelial cells, retinal pigment epithelial cells, gastric parietal cells, hepatocytes, bile duct epithelial cells, neuroglial cells, ciliary muscle cells, thick ascending limb cells, and in the outer stripe of the outer medullary collecting duct.[81–85] While the transport mechanism identified in these cell types is similar in many respects to that found in the proximal tubule, a few notable differences have been found. In the renal proximal tubule, inhibition of the Na/HCO_3 cotransporter by disulfonic stilbenes leads to cell alkalinization.[3] However, in retinal pigment epithelial cells, inhibition of the transporter by disulfonic stilbenes lead to cell acidification.[81] In addition, whereas in the renal proximal tubule inhibition of the transporter leads to cell hyperpolarization, in retinal pigment epithelia transporter inhibition leads to cell depolarization.[78,81] These results suggest that while in renal proximal tubule cells the transporter functions as a base efflux, acid loading mechanism across the basolateral membrane, in retinal pigment epithelial cells where the transporter is on the apical membrane it functions as a base entry step.

In retinal pigment epithelium, the effect of DIDS on cell voltage depended on cell voltage, demonstrating that the transporter could function in either direction.[81] The voltage at which DIDS had no effect on cell voltage, the "reversal potential," was used to determine the stoichiometry of the transporter which was 2 HCO_3:1 Na. Similar results were obtained in cultured BSC-1 monkey kidney epithelial cells,[86] and ciliary muscle cells,[83] where the transporter runs base into cells under resting conditions. In BSC-1 cells, kinetic data are consistent with 1:1 Na/CO_3 cotransport.[87] In addition, while hepatocytes transport HCO_3 into bile canaliculi, the Na/HCO_3 transporter was found on the apical membrane, where it was postulated to function as a base influx mechanism.[88] Lastly, gastric parietal cells have been demonstrated to possess a basolateral membrane Na/HCO_3 cotransporter. While once again this could function as a base efflux mechanism mediating gastric acid secretion, it was noted that the activity of the transporter decreases on stimulation of acid secretion.[89] In cardiac Purkinje fibers, a DIDS-sensitive Na/HCO_3 transporter has been found that appears to be electroneutral, suggesting a 1:1 stoichiometry.[90]

In proximal tubule cells, the Na/HCO_3 cotransporter functions as a base efflux step in response to an alkaline load, but also functions as a base entry step in response to an acid load.[91] The ability to function as a base entry step is not inconsistent with a 1:3 stoichiometry, given the presence of an acid load. Similarly, in other cell types, the demonstration that a Na/HCO_3 transporter functions as a base loader in the presence of an acid load should not be inferred to indicate that the transporter functions similarly at resting cell pH. Based on the response of cell pH and voltage to SITS, it has been found in *Necturus* proximal convoluted tubule that the Na/HCO_3 transporter functions as a base efflux mechanism under control conditions, but as a base entry mechanism with isohydric CO_2 increases.[92] By measuring intracellular Na, pH, and voltage, it was calculated that base efflux functioned with a 1:3 Na:HCO_3 stoichiometry, but that base influx required a 1:2 stoichiometry. Thus, in a single cell Na/HCO_3 transporters are present that function in opposite directions with different stoichiometries.

Based on the above, we would postulate that the electrogenic Na/HCO_3 cotransport mechanism exists in at least two forms. One form has a stoichiometry of 1 Na:3 HCO_3 and is likely a Na/HCO_3/CO_3 cotransporter. This transport mechanism functions as a base efflux mechanism, but can in certain circumstances in the presence of cell depolarization or cell acidity function in reverse. A second form of the transporter has a stoichiometry of 1 Na:2HCO_3, and may represent a Na/CO_3 cotransport mechanism. This likely functions as a base entry mechanism and may contribute to transepithelial transport in cells such as hepatocytes and to the cell pH defense against an acid load. At present the electrogenic Na/HCO_3 transporters have not been purified or cloned. Based on experiences with other genes, if there are indeed differential stoichiometries of the transporter, this will likely be related to distinct isoforms, splice variants, or different transcription initiation sites.

20.4.2.6. Na(HCO_3)$_2$/Cl Exchange

In invertebrate cells such as squid giant axon, snail neuron, giant barnacle muscle fiber, and crayfish neuron, the major mechanism for defense against an intracellular acid load is the Na(HCO_3)$_2$/Cl exchanger. The interested reader is referred to an in-depth review of these transporters.[1] The transporter carries base into the cell, is electroneutral, and is dependent on extracellular Na and HCO_3 and intracellular Cl. The stoichiometry is 1 Cl leaving:1 Na entering:2 HCO_3 or its equivalent entering the cell.[93] In squid axon, the transporter appears to exchange Cl for the Na carbonate ion pair.[94] However, in barnacle muscle the kinetic data are not consistent with such a transport mechanism.[95] The transporter is inhibited by SITS and DIDS. The gene encoding this transporter has not been cloned.

Although this transport mechanism appears to be of less importance in vertebrate cells, this may in many cases relate to the fact that it has not been looked for. Because vertebrate cells possess a Na/H antiporter that can be studied in the absence of CO_2/HCO_3, there has been a tendency for most studies to be performed in this setting. Thus, many of these cells could also possess a Na(HCO_3)$_2$/Cl exchanger if studied in the presence of CO_2/HCO_3. Indeed, studies have demonstrated this transporter to be present in fibroblasts, mesangial cells, esophageal mucosa, thyrocytes, and in the renal proximal tubule.[96–100] In the renal proximal tubule this transporter has been postulated to function as a Cl exit mechanism effecting transepithelial Cl transport. It should be noted, however, that a DIDS-sensitive electrogenic Na/HCO_3 cotransporter functioning in parallel with a Cl conductance can appear as a DIDS-sensitive electroneutral Na(HCO_3)$_2$/Cl exchanger if only cell pH is measured.

20.4.3. Regulation of H Transporter Activity

H transporters regulate cell pH, as well as a number of physiologic processes. As such, it is important for these transporters to themselves be precisely regulated. One method of regulation is related to substrate concentration. Thus, the rates of acid-loading transporters will be stimulated by increases in cell pH, and the rates of acid-extruding transporters will be stimulated by decreases in cell pH. In addition, a number of more sensitive regulatory mechanisms have been identified.

20.4.3.1. Allosteric Regulation

Allosteric regulation refers to the regulation of protein function by a substance that may or may not be a substrate, but that regulates protein function by binding to a nonsubstrate site. Thus, decreases in cell pH can stimulate acid-extruding transporters to a greater extent than may be expected for H interacting as a substrate. This was first shown for the Na/H antiporter using plasma membrane vesicles from renal proximal tubule.[101] Increases in intravesicular H ion concentration were found to accelerate Na uptake to a greater extent than could be attributed to interactions of H with the transport site. In addition, increases in intravesicular H ion concentration accelerated the rate of ^{22}Na efflux mediated by the Na/H antiporter. This effect is opposite in direction to that which would be expected for an effect of H as a substrate. Subsequent studies in intact cells have also demonstrated an exquisite pH sensitivity of the Na/H antiporter, with decreases in cell pH markedly increasing activity. It should be noted that in intact cells it is difficult to be certain whether one is dealing with allosteric activation, or with a more complicated sequence of events such as acid regulation of phosphorylation or trafficking. Decreases in cell pH also have been demonstrated to increase the activity of the $Na(HCO_3)_2$/Cl exchanger in barnacle muscle.[102] This effect cannot be attributed to effects on substrate levels, in that rates of Na efflux or influx are stimulated by decreases in cell pH. As described above, decreases in cell pH also increase the H conductance.[55–59]

Increases in cell pH have been demonstrated to allosterically activate the Cl/HCO_3 exchanger.[60–63] This has been shown convincingly in ileal brush border membrane vesicles, where increases in cell pH increase the rate of Cl–Cl exchange.[63] This is likely of considerable physiologic importance, in that the Cl/HCO_3 exchanger functions to defend cell pH against an alkaline load. It would be detrimental to many cells for the transporter to function at normal or acidic cell pHs as it would provide a continuous acid load. Thus, the system will function better given a higher sensitivity to cell pH, which likely is accomplished through allosteric regulation. As noted above, a major exception to this is the red blood cell band 3 Cl/HCO_3 exchanger which needs to function at acid and alkaline cell pH, and is not regulated allosterically by cell pH.[64,65]

20.4.3.2. Posttranslational Modification

Transporter activities can also be modified by posttranslational modification. A number of H/HCO_3 transporters have been shown to change their activity in response to activation of kinase pathways. Specific protein phosphorylation has only been shown in a few circumstances. Na/H antiporter activity has been shown to be increased by protein kinase C activation, and by activation of tyrosine kinase pathways. Both of these forms of activation are associated with phosphorylation of the antiporter.[103,104] In addition, phosphopeptide mapping suggests that the amino acids phosphorylated are similar in both cases.[104]

20.4.3.3. Membrane Trafficking

H transporter activity can be regulated by exocytotic insertion of membrane containing H transporters. This was first demonstrated in the gastric parietal cell, where it was shown that activators of gastric H secretion lead to exocytotic insertion of membrane containing H,K-ATPases.[105] Similarly, in the renal tubule and the turtle urinary bladder, decreases in cell pH lead to exocytotis.[106,107] These membranes contain H pumps, and thus exocytotic insertion may increase the capacity for H transport.

20.4.3.4. Chronic Regulation

In addition to acute regulation of H transporter activity, chronic regulation is also of importance. Chronic decreases in extracellular and intracellular pH lead to chronic increases in the activities of the renal proximal tubule Na/H antiporter and Na/HCO_3 cotransporter.[108–112] Kinetically, there is a change in the V_{max} of both transporters, with no change in the affinity for Na. Similar effects have been found with chronic K depletion and chronic increases in glomerular filtration rate, both of which are associated with increases in renal net acid excretion.[113,114]

The effect of chronic acidosis *in vivo* can be reproduced *in vitro*. Incubation of quiescent confluent primary cultures of rabbit proximal tubule cells in acid media for 48 hr leads to an increase in Na/H antiporter activity which is dependent on protein synthesis.[115] This represents a memory effect in that it persists after cells are removed from acid media. At present the proximal tubule is believed to express at least two Na/H antiporter isoforms, an amiloride-sensitive basolateral membrane Na/H antiporter encoded by NHE-1 and an amiloride-resistant apical membrane Na/H antiporter encoded by NHE-3.[11,19,116] Utilizing cell lines that express each of these Na/H antiporters, it has been shown that acid incubation increases activities of both Na/H antiporters, and increases abundance of mRNAs corresponding to both isoforms[117] (Amemiya, Moe, Yamaji, and Alpern, unpublished observation). In the renal proximal tubule these adaptations appear to be part of a generalized response to chronic acidosis which includes increases in activities and mRNA abundance of a number of ammoniagenic enzymes, an increase in the activity of the Na/citrate cotransporter, and hypertrophy.[12]

Recent studies have addressed the signaling mechanisms activated by acidosis. Exposure of MCT cells, an SV40-transformed renal proximal tubule cell line, to acid media leads to transcriptional increases in the abundance of c-fos, c-jun, junB, and egr-1 mRNA, a number of immediate early genes which themselves function as transcription factors.[118] The increased expression of these genes is blocked by the tyrosine kinase inhibitor, herbimycin A, and is unaffected by protein kinase C inhibition.[119] In addition, acid incubation leads to increased tyrosine phosphorylation of a number of proteins. Lastly, acid incubation has been found to activate c-src, a membrane-bound nonreceptor tyrosine kinase.[120] It is still not clear what protein is directly activated by decreases in pH, and initiates the signaling cascade described above (the pH sensor; see below). In addition, the interrelations between tyrosine kinase pathways, immediate early genes, and acid-induced adaptations remain to be defined.

20.4.3.5. Concept of a pH Sensor

A biologic homeostatic system in general has functional components that sense a parameter and compare it to an optimal set point, and components that correct deviations from the set point. Numerous mechanisms have been described for rectifying deviations from normal cell pH but little is known about the pH sensor. Mechanisms of pH sensing are ubiquitous as all cells are capable of regulating their intracellular pH. In certain highly differentiated cells such as renal epithelia and specialized neurons in the peripheral and central respiratory chemoreceptors, pH sensing is even more critical as it invokes responses that involve not only correction of cell pH but regulation of whole organism acid–base balance. In addition to pH homeostasis, any biologic effect triggered by a change in pH implies that the downstream target molecules are likely to have inherent pH-sensing capabilities.

Individual proteins that can be allosterically modified by intracellular pH have "built-in" pH-sensing mechanisms. This type of mechanism is convenient in that it combines the sensor and effector in the same molecule. This design is teleologically appropriate in situations where the only response to cell pH perturbation is cell pH correction by one or a few transporters. Alternatively, a pH sensor may trigger numerous effectors in concert. Such a scheme circumvents the need for each protein to evolve an independent sensing mechanism and is logical for cells such as renal epithelia where a perturbation in pH activates a complex set of biologic responses.

Of all of the amino acid side chains that can form acid–base pairs, only the imidazole group of histidine has a pK_a between 6 and 7 in solution at physiologic temperatures. The role of imidazole in ventilatory chemoreceptor control has received support from indirect experimental data using the imidazole reactive agent diethylpyrocarbonate (DEPC). Direct administration of DEPC to the ventilatory center results in hypoventilation and blunted ventilatory response to high CO_2 and low pH, while hypoxic drive remains intact.[121] The search for the exact pH-sensing cell and the pH-sensing protein in the respiratory center has been elusive. The interested reader is referred to a review[122] addressing the theoretical role of histidine in ventilatory control.

There are examples of pH-sensitive proteins with histidine residues that are critical for their pH responses. In the mold *Dictyostelium discoideum,* cell motility associated with chemotaxis and endocytosis involves rapid rearrangement of its actin-based cytoskeleton in response to external stimuli. Hisactophilin is a submembranous intracellular protein that senses small changes in cytoplasmic pH in its immediate microenvironment, binds to actin, and induces actin polymerization at pH values below 7.0.[123] Hisactophilin can thus transduce any extracellular signal that causes intracellular acidification, to induce actin polymerization. Hisactophilin contains 31 histidines out of 118 amino acids[123] and the three-dimensional structure predicts that 90% of the histidines are located in loops and turns at the surface of the molecule rendering all of the imidazole groups accessible to protonation.[124] When an extracellular ligand such as a chemotactic factor binds to a surface receptor and triggers a regional cytoplasmic acidification, hisactophilin acquires multiple protons, binds to actin as a polycationic protein akin to polylysine, and induces actin polymerization and cytoskeletal motility in the immediate vicinity of the ligand receptor complex.

The activity of nhaA, an *E. coli* Na/H exchanger isoform, is highly pH-dependent displaying a 1000-fold activation when cellular or extracellular pH is increased from 7 to 8.[125] This property enables nhaA to extrude Na at rapid rates even at high ambient pH when H driving forces are minimal. Since the same pH sensitivity is observed with the purified protein in proteoliposomes (2000-fold activation),[126] the nhaA protein must be furnished with an intrinsic pH sensor. NhaA is a transmembrane protein that contains only 8 histidines out of 450 amino acids. A single point mutation of His226 situated at a hydrophobic–hydrophilic junction to Arg results in dramatic changes in the pH sensitivity of nhaA.[127] While the normal protein increases its activity over 1000-fold from pH 7 to 8 (half max at 7.5) and retains the high activity above pH 8, the H226R mutant is activated to a similar degree but over a different pH range from 6.5 to 7.5 (half max at 7) and is quickly inactivated above pH 8.[127] Although the pH profile is much altered, the mutant protein is still capable of sensing pH. The structure of this protein has not been resolved so that the three-dimensional location of H226 is not known.

20.4.4. Integrated Cell pH Regulation

Cell pH is thus regulated by a number of processes functioning in parallel. Cell pH in the steady state will be that pH at which the rates of all acid-loading processes are equal to the rates of alkali-loading processes. Given this, one can envision two scenarios. In one case, which we will refer to as the "H quiescent cell," membrane transporters will have minimal activity and H-extruding mechanisms such as the Na/H antiporter will function at a minimal rate sufficient to balance H formed from metabolism and H entering the cells through H leaks of relatively small magnitude. A perturbation that leads to a change in cell pH will then activate mechanisms that tend to return cell pH toward normal values. Such activation could be brought about by changes in substrate H concentration, but more likely are mediated by allosteric regulation as discussed above. The new result is that in the resting state there is minimal energy consumption attributed to H transport.

We refer to the second scenario as the "H active cell." Here, acid and base loaders are functioning in the resting cell. The net result may be similar to the "H quiescent cell" with regard to cell pH, in that cell pH is again defined as that pH that causes acid- and alkali-loading processes to occur at equal rates. However, high rates of acid and alkali loading at resting cell pH lead to high rates of energy consumption. In a nonpolar cell, such energy consumption would be teleologically undesirable. However, in a polarized epithelium that effects transepithelial H/HCO_3 transport, such a system would be advantageous. Thus, if all of the acid-loading mechanisms were localized to one membrane and all of the alkali-loading mechanisms were localized to the other membrane, these transport mechanisms would result in transepithelial H secretion. In epithelia that effect transepithelial H/HCO_3 transport, membrane transporters must be active at resting values of cell pH, while in nonpolar cells it is more energetically favorable for these transporters to operate in the quiescent mode and to be activated when needed.

Differences in transporter activity at resting cell pH may be secondary to differential regulation in cells. Thus, a H extrusion mechanism may be inactive in fibroblasts but may be active in the renal proximal tubule because of a difference in lipid membrane milieu, posttranslational modification, or associated proteins. Conversely, differences in activity at resting cell pH values may be related to different molecular isoforms of the transporters. For instance, the ubiquitous Na/H exchanger that is relatively inactive at resting cell pH values is likely encoded by the NHE-1 isoform, while the proximal tubule Na/H antiporter which must function at rapid rates at resting cell pH values is likely encoded by the NHE-3 isoform.[13–15] The red blood cell Cl/HCO_3 exchanger encoded by the AE1 isoform functions bidirectionally and is active at resting cell pH.[64–66] In H-secreting cells of the mammalian collecting duct, the Cl/HCO_3 exchanger needs to function at baseline and in the HCO_3 efflux direction and is encoded by a variant of AE1 that is lacking the first three exons.[69,70] Many other cells have a Cl/HCO_3 exchanger which is inactive at resting cell pH, and is activated at alkaline cell pH.[60–62] These latter Cl/HCO_3 exchangers are likely encoded by AE2 or AE3.[73,74,76] As noted above, the Na/HCO_3 cotransporter seems to have a variable stoichiometry associated with functioning as an acid or alkali loader in different cells. The determination of molecular mechanisms responsible for this difference will await the cloning of this transporter.

Polarized distribution of pH-sensitive acid-loading and acid-extruding transporters provides a mechanism for simultaneous regulation of transepithelial transport and intracellular pH. While certain epithelia must effect transepithelial H transport, they must still defend cell pH. Thus, regulatory activation of an apical membrane H extrusion mechanism leads to an increase in cell pH, which will secondarily activate basolateral membrane base efflux. The increased rate of basolateral base efflux increases transepithelial transport and serves to minimize any change in cell pH. Transepithelial transport can be regulated with even lesser changes in cell pH, if regulation at the molecular level occurs at both membranes. Indeed, this has been found in the renal proximal tubule. Here, acidosis, K deficiency, angiotensin II, and increases in GFR have been found to lead to parallel increases in the activities of the apical membrane Na/H antiporter and the basolateral membrane $Na/HCO_3/CO_3$ cotransporter.[108,109,113,114,128] Similarly, protein kinase A, protein kinase C, and calcium/calmodulin-dependent protein kinase have all been found to regulate the apical Na/H antiporter and the basolateral $Na/HCO_3/CO_3$ cotransporter similarly in the proximal tubule.[129–132]

20.5. REGULATION OF CELL FUNCTION BY pH

It is likely that many cell functions are regulated by cell pH. Below we will discuss a few of these functions that are likely of importance and have received considerable attention.

20.5.1. Role of Cell pH in Growth Regulation

A number of growth factors have been shown to induce intracellular alkalinization as a result of activation of the plasma membrane Na/H exchanger and a rise in cell pH has been axiomatically accepted as a signal for cell growth. However, whether intracellular alkalinization is important for cell growth and exactly how intracellular alkalinization contributes to the growth process are not understood.

When the plasma membrane Na/H exchanger is blocked with amiloride analogues, there is a tight inverse correlation between antiporter blockade and growth factor-induced DNA synthesis in the absence of CO_2/HCO_3.[7] Also, in the absence of CO_2/HCO_3, mutant cells lacking the Na/H antiporter do not alkalinize and do not grow in response to growth factors.[133,134] When external pH is raised to increase cell pH, normal growth can be restored in these antiporter null mutants.[134] When CO_2/HCO_3 is present in the medium, cell growth is normal in these antiporter null cells, suggesting that HCO_3 transporters can replace the loss of Na/H antiporter function.[134] Thus, it appears that cell pH rather than the Na/H antiporter *per se* is key. Cytoplasmic alkalinization seems to have at least a permissive role in cell growth.

When nontransformed fibroblasts are constitutively alkalinized by expression of a yeast pump, these cells develop tumorigenic characteristics.[135] Maturation of *Xenopus* eggs can be artificially induced by weak bases in the absence of a growth factor.[136] The latter two observations suggest that cytoplasmic alkalinization is not just permissive but sufficient for cell growth. However, these findings are not universally true and nonmitogen-related alkalinization is not usually mitogenic.

The importance of cell pH in cell growth is not totally resolved. While some cells alkalinize on growth factor stimulation in the presence of HCO_3,[137] others do not.[138] When the growth factor vasopressin is added to mesangial cells in the presence of HCO_3, cell growth is associated with an intracellular acidification rather than alkalinization.[138] This is a result of activation of the Cl/HCO_3 exchanger to a greater extent than the Na/H exchanger and $Na(HCO_3)_2/Cl$ exchanger.[138] These studies suggest that either cell pH is not related to cell growth or the action of cell pH is highly cell-dependent. Since stimulation of both the Na/H exchanger and the $Na(HCO_3)_2/Cl$ exchanger increases Na entry, cell Na rather than cell pH may actually be the signal for cell growth.

20.5.2. Interaction of Cell pH and Cell Ca

Both intracellular Ca and intracellular H are present in the cytoplasm at nanomolar concentrations and play major roles in a number of cellular processes. The interaction between cell pH and Ca transients is highly cell-dependent and for a given cell multiple mechanisms likely prevail. In general the approach to analyzing the relationship between agonist-induced cell Ca and pH transients has been to examine whether Ca or pH transients induced by ionophores or permeant acids (or bases) can produce the other transient. Alternatively, buffers and inhibitors of transport are used to block one of the agonist-induced transients to see if the other is also abolished. This discussion will highlight some of the mechanisms of interaction that have been studied and their potential biologic relevance.

20.5.2.1. Ca/H Exchange

The most direct way for cell Ca and pH to influence each other is via Ca/H exchange. Although there is no molecular evidence for a simple Ca/H exchanger, there is evidence that the Ca-ATPase in plasma membrane and endoplasmic reticulum functions as a Ca:*n*H ($n \geq 2$) exchanger.[139–141] This mode of exchange can explain increases in Ca_i secondary to cytoplasmic acidification.

Another potential mechanism of Ca/pH coupling equivalent to Ca/H exchange is direct displacement of one by the other from binding sites on intracellular proteins. A transient rise in cell Ca will lead to occupancy of anionic Ca-binding sites on proteins resulting in release of H and a fall in cell pH.

20.5.2.2. Parallel Coupled Transporters: Na/H and 3Na/Ca Exchangers

This system has been studied in MDCK cells.[142] Under resting conditions these exchangers extrude H or Ca, driven by the low cell Na and the interior negative cell voltage. An acute increase in cell Ca or H, changing the rate of one transporter, can increase intracellular Na concentration and secondarily modify the rate of the other transporter. In this system, cell Na is the key intermediate and cell pH, Ca, and Na can regulate each other. Based on indirect evidence, this exchange coupling has been proposed as a mechanism by which high plasma pCO_2 stimulates the type 1 glomus cells in the carotid chemoreceptor,[143] CO_2 acidifies these specialized neurons and causes a cell Ca transient which triggers neurotransmitter release and excites the closely apposed afferent nerve terminals to signal an increase in ventilation.

20.5.2.3. Indirect Mechanisms of Ca/pH Coupling

In addition to direct coupling of Ca and H transport, Ca and cell pH can modulate each other in a variety of ways. H-sensitive Ca channels have been described in the endoplasmic reticulum (ER) and the plasma membrane. These are distinct from the Ca/H exchange process described above as they are ATP-independent and pH appears to only affect Ca flux from one side.[144,145] It is unclear at present whether cell pH is a link between receptor activation and Ca channel opening. Inhibition of the Ca-ATPase by low pH has also been described in isolated membranes.[146,147] This can be a direct effect of H on the ATPase or it may be secondary to changes in Ca-calmodulin activation of the Ca-ATPase as the binding reaction of Ca to calmodulin is itself pH-dependent.[148] Lowering of extracellular pH has been shown to induce inositol polyphosphate formation[149] and pH-sensitive IP_3-mediated Ca release has been suggested in some cells.[149,150] Conversely, cell Ca can indirectly affect H transporters. Ca-calmodulin kinase regulates H transport,[129,130] and cell Ca has been implicated in exocytotic insertion of H transporters in renal epithelia.[10,151]

20.5.2.4. Ca/pH Effects on Common Target Proteins and Physiologic Processes

There are many situations where cell Ca and pH act in conjunction to modify target proteins. This may result from allosteric regulation of Ca-regulated enzymes by H (or vice versa), or Ca and H may compete for similar binding sites on target proteins. Examples include the inhibition of Ca-calmodulin interaction by low pH,[148] the inhibition of Ca-activated maxi K channels by low pH,[152] and the synergistic and interdependent effects of cell Ca and pH on gap junction conductance.[153] Spermatozoa of externally fertilizing species initiate their motility immediately on changes in osmolality as the sperm traverse from the reproductive organ to the spawning ground. Motility is physiologically triggered by osmotically stimulated rises in cell pH and Ca.[154] Full motility can be artificially triggered by ammonium salts and Ca ionophore in isotonic states. Intracellular alkalosis has been shown to enhance myofilament responsiveness to Ca and increase cardiac contractility.[155] Both α- and β-adrenergic stimulation function as positive inotropes by increasing cell Ca transients during contractions, but α-adrenergic stimulation has the additional effect of activating the Na/H antiporter leading to cell alkalinization and increased myofilament responsiveness to cell Ca.[156]

20.5.3. Role of pH in Receptor-Mediated Endocytosis

Receptor-mediated endocytosis refers to the internalization of ligand–receptor complexes into endocytotic vesicles (endosomes) after binding of ligands to their cell surface receptors.[157] The role of vesicular acidification in the processing of ligand–receptor complexes has been long established. Recently, cytoplasmic pH has also been shown to contribute to this process. Although this chapter does not deal with vesicular acidification, the role of cytoplasmic and vesicular pH will be addressed together as they function in concert to regulate endocytosis and recycling. Two examples will be discussed, the transferrin receptor complex which is recycled and the epidermal growth factor (EGF) receptor complex which is degraded postendocytosis.

Plasma diferric-transferrin (2Fe:1 transferrin) is rapidly endocytosed on binding to its receptor and the endosome is acidified to a pH of 5.0 to 6.4.[158] At this pH the ferric ions dissociate from transferrin and enter the cytoplasm leaving apotransferrin on the receptor.[158,159] Apotransferrin is stably bound to the receptor at low pH and will not dissociate at pH > 3.[158] In contrast, in lysosome-destined endosomes such as those containing LDL and insulin receptors, the ligands rapidly disengage from their receptors at low pH.[160] When studied simultaneously in the same cell with lysosome-destined endosomes, transferrin-containing endosomes tend to consistently have a slightly higher pH.[161,162] The slightly higher pH in these transferrin-containing endosomes and the acid-stability of the apotransferrin–receptor complex may be crucial to chaperone the vesicle to escape a hydrolytic fate. On shedding its Fe, the apotransferrin–receptor complex rapidly returns to the plasma membrane.[163] Acid intravesicular pH facilitates this recycling as permeant weak bases slow down reinsertion.[163] On exposure to extracellular pH (> 7.0), apotransferrin dissociates very rapidly and recycles back to the extracellular space.[158,163] This dissociation is not caused by a change in receptor affinity as the K_d differs only 2-fold,[158] but rather there is a 28-fold increase in k_{off} and a 15-fold increase in k_{on} at pH > 7.0.[158]

Recent studies have shown that in addition to intravesicular pH, cytoplasmic pH also regulates transferrin receptor processing. Cytoplasmic acidification blocks the internalization and recycling of transferrin along with fluid-phase markers while transferrin surface binding is unaltered.[164] This effect is totally reversible when the cytoplasmic pH recovers to normal. The importance of the cell pH-sensitivity of Fe uptake is unclear but based on the changes in uptake of fluid-phase markers and clathrin morphology, cell pH may play a general role in endocytosis. At cell pH approaching 6.8, clathrin forms large aggregates that inhibit pinching off of vesicles from the plasma membrane.[165,166] This effect may reflect the acid-induced stabilization of clathrin polymers previously observed *in vitro*.[167] In addition to its effects on clathrin, cell pH also likely regulates intravesicular acidification to some extent. In starfish oocytes, the formation of acidic granules during fertilization is dependent on cytoplasmic alkalinization secondary to Na/H antiporter activation. Fertilization in the absence of extracellular Na abolishes the formation of acidic granules.[168] Lastly, the transport of solutes across isolated endosomal vesicles depends on not just the intravesicular pH but also the transmembrane H gradient.[169]

Within minutes after ligand binding, the EGF receptor clusters laterally and is endocytosed via coated pits to form microvesicular endosomes.[170–172] The intravesicular pH of these endosomes immediately drops to about 5.0 to 6.0.[161,173] Both EGF and its receptor remain intact in this acid environment as these endosomes are prelysosomal and do not contain hydrolases. Despite an intravesicular pH < 6.0, less than 30% of EGF is dissociated.[161] It appears that a fraction of the EGF–receptor complex persists in endosomes for up to 2–3 hr and still possesses kinase activity on its cytoplasmic side.[174] This translocation step has been postulated as a mechanism whereby the EGF–receptor complex is shuttled to substrates that are otherwise inaccessible to EGF-induced phosphorylation from the plasma membrane.[174] Although this functional role is still controversial, maneuvers that prolong the life span of internalized EGF receptors by inhibiting fusion with lysosomes lead to enhanced cellular DNA synthesis,[175] a finding consistent with the mitogenic potential of internalized EGF–receptor complexes. Although vesicular acidification clearly plays a role in the eventual degradation of EGF by lysosomal acid-activated hydrolases,[176] the purpose of the rapid and persistent acidification in the prelysosomal stage is unclear. Acidification likely has a role in vesicle trafficking as dissipation of acid intravesicular pH by weak bases disrupts ligand-independent receptor recycling.[177] Finally, if the EGF–receptor complex is translocated to activate signaling pathways distant from the cell surface, intravesicular acidification may allosterically modify the function of the receptor complex. At least for unbound EGF receptors on the cell surface, conformational transitions in response to low extracellular pH have been shown.[178]

Like the transferrin receptor, cytoplasmic pH also regulates EGF–receptor processing. EGF internalization but not surface binding was suppressed by cytoplasmic acidification in one study,[165] but stimulated in another study using different cells and methods.[179] Potassium depletion which leads to intracellular acidosis in whole animals[180,181] also decreases EGF endocytosis[182] and the mitogenic action of EGF[183] in fibroblasts. It is presently unclear how cytoplasmic pH affects EGF internalization, but these findings suggest that plasma membrane H equivalent transporters can also be potential regulators of EGF internalization.

REFERENCES

1. Roos, A., and Boron, W. F. (1981). Intracellular pH. *Physiol. Rev.* **61**:297–434.
2. Moolenaar, W. H., Tsien, R. Y., van der Saag, P. T., and de Laat, S. W. (1983). Na/H exchange and cytoplasmic pH in the action of growth factors in human fibroblasts. *Nature* **304**:645–648.
3. Alpern, R. J. (1985). Mechanism of basolateral membrane H/OH/HCO_3 transport in the rat proximal convoluted tubule. *J. Gen. Physiol.* **86**:613–636.
4. Thomas, J. A., Buchsbaum, R. N., Simniak, A., and Racker, E. (1979). Intracellular pH measurements in Ehrlich ascites tumor cells utilizing spectroscopic probes generated in situ. *Biochemistry* **18:** 2210–2218.
5. Gros, G., and Moll, W. (1974). Facilitated diffusion of CO_2 across albumin solutions. *J. Gen. Physiol.* **64:**356–371.
6. Gros, G., Moll, W., Hoppe, H., and Gros, H. (1976). Proton transport by phosphate diffusion—A mechanism of facilitated CO_2 transfer. *J. Gen. Physiol.* **67:**773–790.
7. L'Allemain, G., Franchi, A., Cragoe, E., Jr., and Pouyssegur, J. (1984). Blockade of the Na/H antiport abolishes growth factor-induced DNA synthesis in fibroblasts. *J. Biol. Chem.* **259:**4313–4319.
8. Kleyman, T. R., and Cragoe, E. J., Jr. (1988). Amiloride and its analogs as tools in the study of ion transport. *J. Membr. Biol.* **105:** 1–21.
9. Soltoff, S. P., Cragoe, E. J., Jr., and Mandel, L. J. (1986). Amiloride analogues inhibit proximal tubule metabolism. *Am. J. Physiol.* **250:** C744–C747.

10. Hays, S. R., and Alpern, R. J. (1991). Inhibition of Na-independent H pump by Na-induced changes in cell Ca^{2+}. *J. Gen. Physiol.* **98:** 791–813.
11. Haggerty, J. G., Agarwal, N., Reilly, R. F., Adelberg, E. A., and Slayman, C. W. (1988). Pharmacologically different Na/H antiporters on the apical and basolateral surfaces of cultured porcine kidney cells (LLC-PK_1). *Proc. Natl. Acad. Sci. USA* **85:**6797–6801.
12. Alpern, R. J. (1990). Cell mechanisms of proximal tubule acidification. *Physiol. Rev.* **70:**79–114.
13. Sardet, C., Franchi, A., and Pouyssegur, J. (1989). Molecular cloning, primary structure, and expression of the human growth factor-activatable Na/H antiporter. *Cell* **56:**271–280.
14. Orlowski, J., Kandasamy, R. A., and Shull, G. E. (1992). Molecular cloning of putative members of the Na/H exchanger gene family. *J. Biol. Chem.* **267:**9331–9339.
15. Tse, C. M., Brant, S. R., Walker, M. S., Pouyssegur, J., and Donowitz, M. (1992). Cloning and sequencing of a rabbit cDNA encoding an intestinal and kidney-specific Na/H exchanger isoform (NHE-3). *J. Biol. Chem.* **267:**9340–9346.
16. Tse, C. M., Levine, S. A., Yun, C. H. C., Montrose, M. H., Little, P. J., Pouyssegur, J., and Donowitz, M. (1993). Cloning and expression of a rabbit cDNA encoding a serum-activated ethylisopropyl-amiloride-resistant epithelial Na/H exchanger isoform (NHE-2). *J. Biol. Chem.* **268:**11917–11924.
17. Wang, Z., Orlowski, J., and Shull, G. E. (1993). Primary structure and functional expression of a novel gastrointestinal isoform of the rat Na/H exchanger. *J. Biol. Chem.* **268:**11925–11928.
18. Wakabayashi, S., Fafournoux, P., Sardet, C., and Pouyssegur, J. (1992). The Na/H antiporter cytoplasmic domain mediates growth factor signals and controls "H-sensing." *Proc. Natl. Acad. Sci. USA* **89:** 2424–2428.
19. Biemesderfer, D., Pizzonia, J., Exner, M., Reilly, R., Igarashi, P., and Aronson, P. S. (1993). NHE3: A Na/H exchanger isoform of the renal brush border. *Am. J. Physiol.* **265:**F736–F742.
20. Counillon, L., and Pouyssegur, J. (1993). Molecular biology and hormonal regulation of vertebrate Na/H exchanger isoforms. In *Molecular Biology and Function of Carrier Proteins* (L. Reuss, J. M. Russell, Jr., and M. L. Jennings, eds.) Rockefeller University Press, New York, pp. 169–185.
21. Amemiya, M., Moe, O. W., Yamaji, Y., Cano, A., Baum, M., and Alpern, R. J. (1993). NHE-3 mRNA is expressed in OKP cells and is regulated by glucocorticoids. *J. Am. Soc. Nephrol.* **4:**830.
22. Collins, J. F., Honda, T., Knobel, S., Bulus, N. M., Conary, J., DuBois, R., and Ghishan, F. K. (1993). Molecular cloning, sequencing, tissue distribution, and functional expression of a Na/H exchanger (NHE-2). *Proc. Natl. Acad. Sci. USA* **90:**3938–3942.
23. Rabon, E. D., and Reuben, M. A. (1990). The mechanism and structure of the gastric H,K-ATPase. *Annu. Rev. Physiol.* **52:**321–344.
24. Kaunitz, J. D., and Sachs, G. (1986). Identification of a vanadate-sensitive potassium-dependent proton pump from rabbit colon. *J. Biol. Chem.* **261:**14005–14010.
25. Wingo, C. S. (1989). Active proton secretion and potassium absorption in the rabbit outer medullary collecting duct: Functional evidence for H/K-ATPase. *J. Clin. Invest.* **84:**361–365.
26. Shull, G. E., and Lingrel, J. B. (1986). Molecular cloning of the rat stomach (H + K)-ATPase. *J. Biol. Chem.* **261:**16788–16791.
27. Sachs, G., Besancon, M., Shin, J. M., Mercier, F., Munson, K., and Hersey, S. (1992). Structural aspects of the gastric H,K-ATPase. *J. Bioenerg. Biomembr.* **24:**301–308.
28. Canfield, V. A., Okamoto, C. T., Chow, D., Dorfman, J., Gros, P., Forte, J. G., and Levenson, R. (1990). Cloning of the H,K-ATPase β subunit. *J. Biol. Chem.* **265:**19878–19884.
29. Gottardi, C. J., and Caplan, M. J. (1993). Molecular requirements for the cell-surface expression of multisubunit ion-transporting ATPases. *J. Biol. Chem.* **268:**14342–14347.
30. Jennings, M. L. (1992). Cellular anion transport. In *The Kidney* (D. W. Seldin and G. Giebisch, eds.), Raven Press, New York, pp. 113–145.
31. Crowson, M. S., and Shull, G. E. (1992). Isolation and characterization of a cDNA encoding the putative distal colon H,K-ATPase. *J. Biol. Chem.* **267:**13740–13748.
32. Jaisser, F., Coutry, N., Farman, N., Binder, H. J., and Rossier, B. C. (1993). A putative H-K-ATPase is selectively expressed in surface epithelial cells of rat distal colon. *Am. J. Physiol.* **265:**C1080–C1089.
33. Gifford, J. D., Rome, L., and Galla, J. H. (1992). H-K-ATPase activity in rat collecting duct segments. *Am. J. Physiol.* **262:**F692–F695.
34. Silver, R. B., and Frindt, G. (1993). Functional identification of H-K-ATPase in intercalated cells of cortical collecting tubule. *Am. J. Physiol.* **264:**F259–F266.
35. Okusa, M. D., Unwin, R. J., Velazquez, H., Giebisch, G., and Wright, F. S. (1992). Active potassium absorption by the renal distal tubule. *Am. J. Physiol.* **262:**F488–F493.
36. Wingo, C. S., Madsen, K. M., Smolka, A., and Tisher, C. C. (1990). H-K-ATPase immunoreactivity in cortical and outer medullary collecting duct. *Kidney Int.* **38:**985–990.
37. Kraut, J. A., Starr, F., Birmingham, S., Sachs, G., and Reuben, M. A. (1993). Expression of colonic and gastric H,K-ATPase mRNA in rat kidney. *J. Am. Soc. Nephrol.* **4:**870.
38. Tsuchiya, K., Giebisch, G., and Welling, P. A. (1993). Molecular characterization and distribution of H/K ATPase catalytic subunit gene products in the kidney. *J. Am. Soc. Nephrol.* **4:**881.
39. Jaisser, F., Horisberger, J., Geering, K., and Rossier, B. (1993). Molecular cloning and functional expression of a toad bladder H,K-ATPase. *J. Am. Soc. Nephrol.* **4:**868.
40. Forgac, M. (1989). Structure and function of vacuolar class of ATP-driven proton pumps. *Physiol. Rev.* **69:**765–796.
41. Stone, D. K., and Xie, X. S. (1988). Proton translocating ATPases: Issues in structure and function. *Kidney Int.* **33:**767–774.
42. Xie, X. S., and Stone, D. K. (1988). Partial resolution and reconstitution of the subunits of the clathrin-coated vesicle proton ATPase responsible for Ca^{2+} activated ATP hydrolysis. *J. Biol. Chem.* **263:** 9859–9867.
43. Sun, S. Z., Xie, X. S., and Stone, D. K. (1987). Isolation and reconstitution of the dicyclohexylcarbodiimide-sensitive proton pore of the clathrin-coated vesicle proton translocating complex. *J. Biol. Chem.* **262:**14790–14794.
44. Sudhof, T. C., Fried, V. A., Stone, D. K., Johnston, P. A., and Xie, X. S. (1989). The human endomembrane proton pump strongly resembles the archaebacter ATP synthetase. *Proc. Natl. Acad. Sci. USA* **86:**6067–6071.
45. Hirsch, S., Strauss, A., Masood, K., Lee, S., Sukhatme, V., and Gluck, S. (1988). Isolation and sequence of a cDNA clone encoding the 31 kDa subunit of bovine kidney vacuolar H ATPase. *Proc. Natl. Acad. Sci. USA* **85:**3004–3008.
46. Marushack, M. M., Lee, B. S., Masood, K., and Gluck, S. (1992). cDNA sequence and tissue expression of bovine vacuolar H-ATPase M_r 70,000 subunit. *Am. J. Physiol.* **263:**F171–F174.
47. Brown, D., Hirsch, S., and Gluck, S. (1988). Localization of a proton-pumping ATPase in rat kidney. *J. Clin. Invest.* **82:**2114–2126.
48. Nelson, R. D., Guo, X. L., Masood, K., Brown, D., Kalkbrenner, M., and Gluck, S. (1992). Selectively amplified expression of an isoform of the vacuolar H-ATPase 56-kilodalton subunit in renal intercalated cells. *Proc. Natl. Acad. Sci. USA* **89:**3541–3545.
49. Bowman, E. J., Siebers, A., and Altendorf, K. (1988). Bafilomycins: A class of inhibitors of membrane ATPases from microorganisms, animal cells, and plant cells. *Proc. Natl. Acad. Sci. USA* **85:**7972–7976.
50. Mattsson, J. P., Vaananen, K., Wallmark, B., and Lorentzon, P. (1991). Omeprazole and bafilomycin, two proton pump inhibitors:

Differentiation of their effects on gastric, kidney and bone H-translocating ATPases. *Biochim. Biophys. Acta* **1064:**261–268.

51. Pappas, C. A., and Koeppen, B. M. (1992). Electrophysiological properties of cultured outer medullary collecting duct cells. *Am. J. Physiol.* **263:**F1004–F1010.
52. Wallmark, B., Briving, C., Fryklund, J., Munson, K., Jackson, R., Mendlein, J., Rabon, E., and Sachs, G. (1987). Inhibition of gastric H,K-ATPase and acid secretion by SCH 28080, a substituted pyridyl(1,2a)imidazole. *J. Biol. Chem.* **262:**2077–2084.
53. Kohn, O. F., Mitchell, P. P., and Steinmetz, P. R. (1993). Sch-28080 inhibits bafilomycin-sensitive H secretion in turtle bladder independently of luminal [K]. *Am. J. Physiol.* **265:**F174–F179.
54. Chatterjee, D., Chakraborty, M., Leit, M., Neff, L., Jamsa-Kellokumpu, S., Fuchs, R., and Baron, R. (1992). Sensitivity to vanadate and isoforms of subunits A and B distinguish the osteoclast proton pump from other vacuolar H ATPases. *Proc. Natl. Acad. Sci. USA* **89:**6257–6261.
55. Lukacs, G. L., Kapus, A., Nanda, A., Romanek, R., and Grinstein, S. (1993). Proton conductance of the plasma membrane: Properties, regulation, and functional role. *Am. J. Physiol.* **265:**C3–C14.
56. Byerly, L., Meech, R., and Moody, W., Jr. (1984). Rapidly activating hydrogen ion currents in perfused neurones of the snail, *Lymnaea stagnalis*. *J. Physiol. (London)* **351:**199–216.
57. Barish, M. E., and Baud, C. (1984). A voltage-gated hydrogen ion current in the oocyte membrane of the axolotl, *Ambystoma*. *J. Physiol. (London)* **352:**243–263.
58. DeCoursey, T. E. (1991). Hydrogen ion currents in rat alveolar epithelial cells. *Biophys. J.* **60:**1243–1253.
59. Kapus, A., Romanek, R., Qu, A. Y., Rotstein, O. D., and Grinstein, S. (1993). A pH-sensitive and voltage-dependent proton conductance in the plasma membrane of macrophages. *J. Gen. Physiol.* **102:**729–760.
60. Olsnes, S., Tonnessen, T. I., Ludt, J., and Sandvig, K. (1987). Effect of intracellular pH on the rate of chloride uptake and efflux in different mammalian cell lines. *Biochemistry* **26:**2778–2785.
61. Olsnes, S., Tonnessen, T. I., and Sandvig, K. (1986). pH-regulated anion antiport in nucleated mammalian cells. *J. Cell Biol.* **102:**967–971.
62. Simchowitz, L., and Davis, A. O. (1991). Internal alkalinization by reversal of anion exchange in human neutrophils: Regulation of transport by pH. *Am. J. Physiol.* **260:**C132–C142.
63. Mugharbil, A., Knickelbein, R. G., Aronson, P. S., and Dobbins, J. W. (1990). Rabbit ileal brush-border membrane Cl-HCO_3 exchanger is activated by an internal pH-sensitive modifier site. *Am. J. Physiol.* **259:**G666–G670.
64. Funder, J., and Wieth, J. O. (1976). Chloride transport in human erythrocytes and ghosts: A quantitative comparison. *J. Physiol. (London)* **262:**679–698.
65. Gunn, R. B., Dalmark, M., Tosteson, D. C., and Wieth, J. O. (1973). Characteristics of chloride transport in human red blood cells. *J. Gen. Physiol.* **61:**185–206.
66. Kopito, R. R., and Lodish, H. F. (1985). Primary structure and transmembrane orientation of the murine anion exchange protein. *Nature* **316:**234–238.
67. Jennings, M. L. (1984). Oligomeric structure and the anion transport function of human erythrocyte band 3 protein. *J. Membr. Biol.* **80:**105–117.
68. Jennings, M. L. (1992). Anion transport properties. In *The Kidney* (D. W. Seldin and G. Giebisch, eds.), Raven Press, New York, pp. 503–535.
69. Brosius, F. C., III, Alper, S. L., Garcia, A. M., and Lodish, H. F. (1993). The major kidney band 3 gene transcript predicts an aminoterminal truncated band 3 polypeptide. *J. Biol. Chem.* **264:**7784–7787.
70. Kudrycki, K. E., and Shull, G. E. (1989). Primary structure of the rat kidney band 3 anion exchange protein deduced from a cDNA. *J. Biol. Chem.* **264:**8185–8192.
71. Wagner, S., Vogen, R., Lietzke, R., Koob, R., and Drenckhahn, D. (1987). Immunochemical characterization of a band 3-like anion exchanger in collecting duct of human kidney. *Am. J. Physiol.* **253:**F213–F221.
72. Kudrycki, K. E., and Shull, G. E. (1993). Rat kidney band 3 Cl/HCO_3 exchanger mRNA is transcribed from an alternative promoter. *Am. J. Physiol.* **264:**F540–F547.
73. Alper, S. L., Kopito, R. R., Libresco, S. M., and Lodish, H. F. (1988). Cloning and characterization of a murine band 3-related cDNA from kidney and from a lymphoid cell line. *J. Biol. Chem.* **263:**17092–17099.
74. Kudrycki, K. E., Newman, P. R., and Shull, G. E. (1990). cDNA cloning and tissue distribution of mRNAs for two proteins that are related to the band 3 Cl/HCO_3 exchanger. *J. Biol. Chem.* **265:**462–671.
75. Demuth, D. R., Showe, L. C., Ballantine, M., Palumbo, A., Fraser, P. J., Cioe, L., Rovera, G., and Curtis, P. J. (1986). Cloning and structural characterization of a human non-erythroid band 3-like protein. *EMBO J.* **5:**1205–1214.
76. Kopito, R. R., Lee, B. S., Simmons, D. M., Lindsey, A. E., Morgans, C. W., and Schneider, K. (1989). Regulation of intracellular pH by a neuronal homolog of the erythrocyte anion exchanger. *Cell* **59:**927–937.
77. Boron, W. F., and Boulpaep, E. L. (1983). Intracellular pH regulation in the renal proximal tubule of the salamander: Basolateral HCO_3 transport. *J. Gen. Physiol.* **81:**53–94.
78. Yoshitomi, K., Burckhardt, B. C., and Fromter, E. (1985). Rheogenic sodium–bicarbonate cotransport in the peritubular cell membrane of rat renal proximal tubule. *Pfluegers Arch.* **405:**360–366.
79. Soleimani, M., Grassl, S. M., and Aronson, P. S. (1987). Stoichiometry of Na–HCO_3 cotransport in basolateral membrane vesicles isolated from rabbit renal cortex. *J. Clin. Invest.* **79:**1276–1280.
80. Soleimani, M., and Aronson, P. S. (1989). Ionic mechanism of Na:HCO_3 cotransport in renal basolateral membrane vesicles (BLMV). *J. Biol. Chem.* **264:**18302–18308.
81. Hughes, B. A., Adorante, J. S., Miller, S. S., and Lin, H. (1989). Apical electrogenic $NaHCO_3$ cotransport. A mechanism for HCO_3 absorption across the retinal pigment epithelium. *J. Gen. Physiol.* **94:**125–150.
82. Alvaro, D., Cho, W. K., Mennone, A., and Boyer, J. L. (1993). Effect of secretin on intracellular pH regulation in isolated rat bile duct epithelial cells. *J. Clin. Invest.* **92:**1314–1325.
83. Stahl, F., Lepple-Wienhues, A., Koch, M., and Wiederholt, M. (1992). Na-dependent HCO_3 transport and Na/H exchange regulate pH_i in human ciliary muscle cells. *J. Membr. Biol.* **127:**215–225.
84. Krapf, R. (1988). Basolateral membrane H/OH/HCO_3 transport in the rat cortical thick ascending limb. *J. Clin. Invest.* **82:**234–241.
85. Jentsch, T. J., Stahlknecht, T. R., Hollwede, H., Fischer, D. G., Keller, S. K., and Wiederholt, M. (1985). A bicarbonate-dependent process inhibitable by disulfonic stilbenes and a Na/H exchange mediate ^{22}Na uptake into cultured bovine corneal endothelium. *J. Biol. Chem.* **260:**795–801.
86. Jentsch, T. J., Matthes, H., Keller, S. K., and Wiederholt, M. (1986). Electrical properties of sodium bicarbonate symport in kidney epithelial cells (BSC-1). *Am. J. Physiol.* **251:**F954–F968.
87. Jentsch, T. J., Schwartz, P., Schill, B. S., Langner, B., Lepple, A. P., Keller, S. K., and Wiederholt, M. (1986). Kinetic properties of the sodium bicarbonate (carbonate) symport in monkey kidney epithelial cells (BSC-1). *J. Biol. Chem.* **261:**10673–10679.
88. Renner, E. L., Lake, J. R., Scharschmidt, B. F., Zimmerli, B., and Meier, P. J. (1989). Rat hepatocytes exhibit basolateral Na/HCO_3 cotransport. *J. Clin. Invest.* **83:**1225–1235.

89. Curci, S., Debellis, L., and Fromter, E. (1987). Evidence for rheogenic sodium bicarbonate cotransport in the basolateral membrane of oxyntic cells of frog gastric fundus. *Pfluegers Arch.* **408:**497–504.
90. Dart, C., and Vaughan-Jones, R. D. (1992). Na–HCO_3 symport in the sheep cardiac Purkinje fibre. *J. Physiol. (London)* **451:**365–385.
91. Krapf, R., Berry, C. A., Alpern, R. J., and Rector, F. C., Jr. (1988). Regulation of cell pH by ambient bicarbonate, carbon dioxide tension, and pH in the rabbit proximal convoluted tubule. *J. Clin. Invest.* **81:**381–389.
92. Planelles, G., Thomas, S. R., and Anagnostopoulos, T. (1993). Change of apparent stoichiometry of proximal-tubule Na–HCO_3 cotransport upon experimental reversal of its orientation. *Proc. Natl. Acad. Sci. USA* **90:**7406–7410.
93. Boron, W. F., and Russell, J. M. (1983). Stoichiometry and ion dependencies of the intracellular-pH-regulating mechanism in squid giant axons. *J. Gen. Physiol.* **81:**373–399.
94. Boron, W. F. (1985). Intracellular pH-regulating mechanism of the squid axon. Relation between the external Na and HCO_3 dependences. *J. Gen. Physiol.* **85:**325–345.
95. Boron, W. F., McCormick, W. C., and Roos, A. (1981). pH regulation in barnacle muscle fibers: Dependence on extracellular sodium and bicarbonate. *Am. J. Physiol.* **240:**C80–C89.
96. Alpern, R. J., and Chambers, M. (1987). Basolateral membrane Cl/HCO_3 exchange in the rat proximal convoluted tubule. *J. Gen. Physiol.* **89:**581–598.
97. Sasaki, S., and Yoshiyama, N. (1988). Interaction of chloride and bicarbonate transport across the basolateral membrane of rabbit proximal straight tubule. *J. Clin. Invest.* **81:**1004–1011.
98. Guggino, W. B., London, R., Boulpaep, E. L., and Giebisch, G. (1983). Chloride transport across the basolateral cell membrane of the *Necturus* proximal tubule: Dependence on bicarbonate and sodium. *J. Membr. Biol.* **71:**227–240.
99. L'Allemain, G., Paris, S., and Pouyssegur, J. (1985). Role of a Na-dependent Cl/HCO_3 exchange in regulation of intracellular pH in fibroblasts. *J. Biol. Chem.* **260:**4877–4883.
100. Kayser, L., Hoyer, P. E., Perrild, H., Wood, A. M., and Robertson, W. R. (1992). Intracellular pH regulation in human thyrocytes: Evidence of both Na/H exchange and Na-dependent Cl/HCO_3 exchange. *J. Endocrinol.* **135:**391–401.
101. Aronson, P. S., Nee, J., and Suhm, M. A. (1982). Modifier role of internal H in activating the Na–H exchanger in renal microvillus membrane vesicles. *Nature* **299:**161–163.
102. Russell, J. M., Boron, W. F., and Brodwick, M. S. (1983). Intracellular pH and Na fluxes in barnacle muscle with evidence for reversal of the ionic mechanism of intracellular pH regulation. *J. Gen. Physiol.* **82:**47–78.
103. Sardet, C., Counillon, L., Franchi, A., and Pouyssegur, J. (1990). Growth factors induce phosphorylation of the Na/H antiporter, a glycoprotein of 110 kD. *Science* **247:**723–726.
104. Sardet, C., Fafournoux, P., and Pouyssegur, J. (1991). α-Thrombin, epidermal growth factor, and okadaic acid activate the Na/H exchanger, NHE-1, by phosphorylating a set of common sites. *J. Biol. Chem.* **266:**19166–19171.
105. Forte, T. M., Machen, T. E., and Forte, J. G. (1975). Ultrastructural and physiological changes in piglet oxyntic cells during histamine stimulation and metabolic inhibition. *Gastroenterology* **69:**1208–1222.
106. Gluck, S., Cannon, C., and Al-Awqati, Q. (1982). Exocytosis regulates urinary acidification in turtle bladder by rapid insertion of H pumps into the luminal membrane. *Proc. Natl. Acad. Sci. USA* **79:**4327–4331.
107. Schwartz, G. J., and Al-Awqati, Q. (1985). Carbon dioxide causes exocytosis of vesicles containing H pumps in isolated perfused proximal and collecting tubules. *J. Clin. Invest.* **75:**1638–1644.
108. Akiba, T., Rocco, V. K., and Warnock, D. G. (1987). Parallel adaptation of the rabbit renal cortical sodium/proton antiporter and sodium/bicarbonate cotransporter in metabolic acidosis and alkalosis. *J. Clin. Invest.* **80:**308–315.
109. Preisig, P. A., and Alpern, R. J. (1988). Chronic metabolic acidosis causes an adaptation in the apical membrane Na/H antiporter and basolateral membrane $Na(HCO_3)_3$ symporter in the rat proximal convoluted tubule. *J. Clin. Invest.* **82:**1445–1453.
110. Cohn, D. E., Klahr, S., and Hammerman, M. R. (1983). Metabolic acidosis and parathyroidectomy increase Na/H exchange in brush border vesicles. *Am. J. Physiol.* **245:**F217–F222.
111. Tsai, C. J., Ives, H. E., Alpern, R. J., Yee, V. J., Warnock, D. G., and Rector, F. C., Jr. (1984). Increased V_{max} for Na/H antiporter activity in proximal tubule brush border vesicles from rabbits with metabolic acidosis. *Am. J. Physiol.* **147:**F339–F343.
112. Kinsella, J. L., Cujkit, T., and Sactor, B. (1984). Na–H exchange activity in renal brush border membrane vesicles in response to metabolic acidosis: The role of glucocorticoids. *Proc. Natl. Acad. Sci. USA* **81:**630–634.
113. Soleimani, M., Bergman, J. A., Hosford, M. A., and McKinney, T. D. (1990). Potassium depletion increases luminal Na/H exchange and basolateral Na:CO_3:HCO_3 cotransport in rat renal cortex. *J. Clin. Invest.* **86:**1076–1083.
114. Preisig, P. A., and Alpern, R. J. (1991). Increased Na/H antiporter and Na/3HCO_3 symporter activities in chronic hyperfiltration. *J. Gen. Physiol.* **97:**195–217.
115. Horie, S., Moe, O., Tejedor, A., and Alpern, R. J. (1990). Preincubation in acid medium increases Na/H antiporter activity in cultured renal proximal tubule cells. *Proc. Natl. Acad. Sci. USA* **87:**4742–4745.
116. Biemesderfer, D., Reilly, R. F., Exner, M., Igarashi, P., and Aronson, P. S. (1992). Immunocytochemical characterization of Na–H exchanger isoform NHE-1 in rabbit kidney. *Am. J. Physiol.* **263:** F833–F840.
117. Moe, O. W., Miller, R. T., Horie, S., Cano, A., Preisig, P. A., and Alpern, R. J. (1991). Differential regulation of Na/H antiporter by acid in renal epithelial cells and fibroblasts. *J. Clin. Invest.* **88:** 1703–1708.
118. Horie, S., Moe, O., Yamaji, Y., Cano, A., Miller, R. T., and Alpern, R. J. (1992). Role of protein kinase C and transcription factor AP-1 in the acid-induced increase in Na/H antiporter activity. *Proc. Natl. Acad. Sci. USA* **89:**5236–5240.
119. Yamaji, Y., Moe, O. W., Miller, R. T., and Alpern, R. J. (1994). Acid activation of immediate early genes. *J. Clin. Invest.* **94:**1297–1303.
120. Yamaji, Y., Moe, O. W., Miller, R. T., and Alpern, R. J. (1993). Acidosis activates a src-related tyrosine kinase in renal cells. *J. Am. Soc. Nephrol.* **4:**506.
121. Nattie, E. E. (1988). Diethyl pyrocarbonate inhibits rostral ventrolateral medullary H sensitivity. *J. Appl. Physiol.* **64:**1600–1609.
122. Reeves, R. B. (1972). An imidazole alphastat hypothesis for vertebrate acid–base regulation: Tissue carbon dioxide content and body temperature in bullfrogs. *Respir. Physiol.* **14:**219–236.
123. Scheel, J., Ziegelbauer, K., Kupke, T., Humbel, B. M., Noegel, A. A., Gerisch, G., and Schleicher, M. (1989). Hisactophilin, a histidine-rich actin-binding protein from *Dictyostelium discoideum.* *J. Biol. Chem.* **264:**2832–2839.
124. Habazettl, J., Gondol, D., Wiltscheck, R., Otlewski, J., Schleicher, M., and Holak, T. A. (1992). Structure of hisactophilin is similar to interleukin-1β and fibroblast growth factor. *Nature* **359:**855–858.
125. Padan, E., Maisler, N., Taglicht, D., Karpel, R., and Schuldiner, S. (1989). Deletion of *ant* in *Escherichia coli* reveals its function in adaptation to high salinity and an alternative Na/H antiporter system(s). *J. Biol. Chem.* **264:**20297–20302.

126. Taglicht, D., Padan, E., and Schuldiner, S. (1991). Overproduction and purification of a functional Na/H antiporter coded by *nhaA* (*ant* from *Escherichia coli*). J. Biol. Chem. **266:**11289–11294.
127. Gerchman, Y., Olami, Y., Rimon, A., Taglicht, D., Schuldiner, S., and Padan, E. (1993). Histidine-226 is part of the pH sensor of nhaA, a Na/H antiporter in *Escherichia coli. Proc. Natl. Acad. Sci. USA* **90:**1212–1216.
128. Geibel, J., Giebisch, G., and Boron, W. F. (1990). Angiotensin II stimulates both Na–H exchange and Na/HCO_3 cotransport in the rabbit proximal tubule. *Proc. Natl. Acad. Sci. USA* **87:**7917–7920.
129. Ruiz, O. S., and Arruda, J. A. L. (1992). Regulation of the renal Na–HCO_3 cotransporter by cAMP and Ca-dependent protein kinases. *Am. J. Physiol.* **262:**F560–F565.
130. Weinman, E. J., Dubinsky, W. P., Fisher, K., Steplock, D., Dinh, Q., Chang, L., and Shenolikar, S. (1988). Regulation of reconstituted renal Na/H exchanger by calcium-dependent protein kinases. *J. Membr. Biol.* **103:**237–244.
131. Weinman, E. J., Dubinsky, W. P., and Shenolikar, S. (1988). Reconstitution of cAMP-dependent protein kinase regulated renal Na–H exchanger. *J. Membr. Biol.* **101:**11–18.
132. Weinman, E. J., and Shenolikar, S. (1986). Protein kinase C activates the renal apical membrane Na/H exchanger. *J. Membr. Biol.* **93:**133–139.
133. L'Allemain, G., Paris, S., and Pouyssegur, J. (1984). Growth factor action and intracellular pH regulation in fibroblasts. *J. Biol. Chem.* **259:**5809–5815.
134. Pouyssegur, J., Sardet, C., Franchi, A., L'Allemain, G., and Paris, S. (1984). A specific mutation abolishing Na/H antiport activity in hamster fibroblasts precludes growth at neutral and acidic pH. *Proc. Natl. Acad. Sci. USA* **81:**4833–4837.
135. Perona, R., and Serrano, R. (1988). Increased pH and tumorigenicity of fibroblasts expressing a yeast proton pump. *Nature* **334:**438–440.
136. Houle, C., and Wasserman, W. (1982). Intracellular pH plays a role in regulating protein synthesis in Xenopus oocytes. *Dev. Biol.* **97:** 302–312.
137. Schwartz, M. A., and Lechene, C. (1992). Adhesion is required for protein kinase C-dependent activation of the Na^+/H^+ antiporter by platelet-derived growth factor. *Proc. Natl. Acad. Sci. USA* **89:** 6138–6141.
138. Ganz, M. B., Boyarsky, G., Sterzel, R. B., and Boron, W. F. (1989). Arginine vasopressin enhances pH_i regulation in the presence of HCO_3 by stimulating three acid–base transport systems. *Nature* **337:**648–651.
139. Valant, P. A., and Haynes, D. H. (1993). The Ca^{2+}-extruding ATPase of the human platelet creates and responds to cytoplasmic pH changes, consistent with a 2 Ca^{2+}/nH exchange mechanism. *J. Membr. Biol.* **136:**215–230.
140. Smallwood, J. I., Waisman, D. M., Lafreniere, D., and Rasmussen, H. (1983). Evidence that the erythrocyte calcium pump catalyzes a Ca^{2+}:*n*H exchange. *J. Biol. Chem.* **258:**11092–11097.
141. Dixon, D. A., and Haynes, D. A. (1989). Ca pumping ATPase of cardiac sarcolemma is sensitive to membrane potential produced by K and Cl gradients but requires a source of counter-transportable H. *J. Membr. Biol.* **112:**169–183.
142. Borle, A. B., and Bender, C. (1991). Effects of pH on Ca_i^{2+}, Na_i, and pH_i of MDCK cells: Na–Ca^{2+} and Na–H antiporter interactions. *Am. J. Physiol.* **261:**C482–C489.
143. Kazemi, H., and Hitzig, B. M. (1992). Central chemical control of ventilation and acid–base balance in the kidney. In: *Physiology and Pathophysiology* (S. Giebirca, ed.) Raven Press, New York, pp. 2627–2644.
144. Muallem, S., Pandol, S. J., and Beeker, T. G. (1989). Modulation of agonist-activated calcium influx by extracellular pH in rat pancreatic acini. *Am. J. Physiol.* **257:**G917–G924.
145. Tsunoda, Y., Matsuno, K., and Tashiro, Y. (1991). Cytosolic acidification leads to Ca^{2+} mobilization from intracellular stores in single and populational parietal cells and platelets. *Exp. Cell Res.* **193:**356–363.
146. Kraus-Friedman, N., Biber, J., Murer, H., and Carafoli, E. (1982). Calcium uptake in isolated hepatic plasma-membrane vesicles. *Eur. J. Biochem.* **129:**7–12.
147. Kribben, A., Tyrakowski, T., and Schulz, I. (1983). Characterization of Mg-dependent Ca transport in cat pancreatic microsomes. *Am. J. Physiol.* **244:**G480–G490.
148. Iida, S., and Potter, J. D. (1986). Calcium binding to calmodulin: Cooperativity of the calcium binding sites. *J. Biochem.* **99:**1765–1772.
149. Smith, J. B., Dwyer, S. D., and Smith, L. (1989). Lowering extracellular pH evokes inositol polyphosphate formation and calcium mobilization. *J. Biol. Chem.* **264:**8723–8728.
150. Brass, L. F., and Joseph, S. K. (1988). A role for inositol triphosphate in intracellular Ca mobilization and granule secretion in platelets. *J. Biol. Chem.* **260:**15172–15179.
151. van Adelsberg, J., and Al-Awqati, Q. (1986). Regulation of cell pH by Ca-mediated exocytotic insertion of H-ATPase. *J. Cell Biol.* **102:**1638–1645.
152. Klaerke, D. A., Weiner, H., Zeuthen, T., and Jorgensen, P. L. (1993). Ca^{2+} activation and pH dependence of a maxi K channel from rabbit distal colon epithelium. *J. Membr. Biol.* **136:**9–21.
153. White, R. L., Doeller, J. E., Verselis, V. K., and Wittenberg, B. A. (1990). Gap junctional conductance between pairs of ventricular myocytes is modulated synergistically by H and Ca. *J. Gen. Physiol.* **95:**1061–1075.
154. Oda, S., and Morisawa, M. (1993). Rises of intracellular Ca^{2+} and pH mediate the initiation of sperm motility by hyperosmolality in marine teleosts. *Cell Motil. Cytoskel.* **25:**171–178.
155. Bountra, C., and Vaughan-Jones, R. D. (1989). Effect of intracellular and extracellular pH on contraction in isolated, mammalian cardiac muscle. *J. Physiol. (London)* **418:**163–187.
156. Gambassi, G., Spurgeon, H. A., Lakatta, E. G., Blank, P. S., and Capogrossi, M. C. (1992). Different effects of α- and β-adrenergic stimulation on cytosolic pH and myofilament responsiveness to Ca^{2+} in cardiac myocytes. *Circ. Res.* **71:**870–882.
157. Goldstein, J. L., Anderson, R. G. W., and Brown, M. S. (1979). Coated pits, coated vesicles, and receptor-mediated endocytosis. *Nature* **279:**679–685.
158. Klausner, R. D., Ashwell, G., van Renswoude, J., Harford, J. B., and Bridges, K. R. (1983). Binding of apotransferrin to K562 cells: Explanation of the transferrin cycle. *Proc. Natl. Acad. Sci. USA* **80:** 2263–2266.
159. Dautry-Varsat, A., Ciechanover, A., and Lodish, H. F. (1983). pH and the recycling of transferrin during receptor-mediated endocytosis. *Proc. Natl. Acad. Sci. USA* **80:**2258–2262.
160. Geuze, H. J., Slot, J. W., Strous, G. J. A. M., Lodish, H. F., and Schwartz, A. J. (1983). Intracellular site of asialoglycoprotein receptor–ligand uncoupling: Double-label immunoelectron microscopy during receptor-mediated endocytosis. *Cell* **32:**277–287.
161. Sorkin, A. D., Teslenko, L. V., and Nikolsky, N. N. (1988). The endocytosis of epidermal growth factor in A431 cells: A pH of microenvironment and the dynamics of receptor complex dissociation. *Exp. Cell Res.* **175:**192–205.
162. Yamashiro, D. J., Tycko, B., Fluss, S. R., and Maxfield, F. R. (1984). Segregation of transferrin to a mildly acidic (pH 6.5) para-Golgi compartment in the recycling pathway. *Cell* **37:**789–800.
163. Ciechanover, A., Schwartz, A. L., Dautry-Varsat, A., and Lodish, H. F. (1983). Kinetics of internalization and recycling of transferrin and the transferrin receptor in a human hepatoma cell line. *J. Biol. Chem.* **258:**9681–9689.
164. Cosson, P., Curtis, I. D., Pouyssegur, J., Griffiths, G., and Davoust, J. (1989). Low cytoplasmic pH inhibits endocytosis and transport

from the trans-Golgi network to the cell surface. *J. Cell Biol.* **108:** 377–387.
165. Sandvig, K., Olsnes, S., Petersen, O., and van Deurs, B. (1987). Acidification of the cytosol inhibits endocytosis from coated pits. *J. Cell Biol.* **105:**679–689.
166. Davoust, J., Gruenberg, J., and Howell, K. E. (1987). Two threshold values of low pH block endocytosis at different stages. *EMBO J.* **6:** 601–3609.
167. Unanue, E., Ungewickell, E., and Branton, D. (1981). The binding of clathrin triskelions to membranes from coated vesicles. *Cell* **26:**439–446.
168. Lee, H., and Epel, D. (1983). Changes in intracellular acidic compartments in sea urchin eggs after activation. *Dev. Biol.* **98:**446–454.
169. Anderson, R. G. W., and Orci, L. (1988). A view of acidic intracellular compartments. *J. Cell Biol.* **106:**539–543.
170. Pastan, I. H., and Willingham, M. C. (1981). Receptor-mediated endocytosis of hormones in cultured cells. *Annu. Rev. Physiol.* **43:**239–250.
171. Schlessinger, J., Schreiber, A. B., Levi, A., Lax, R., Libermann, T., and Yarden, Y. (1983). Regulation of cell proliferation by epidermal growth factor. *CRC Crit. Rev. Biochem.* **14:**93–111.
172. Hanover, J. A., Willingham, M. C., and Pastan, I. (1984). Kinetics of transit of transferrin and epidermal growth factor through clathrin-coated membranes. *Cell* **39:**283–293.
173. Tycko, B., and Maxfield, F. R. (1982). Rapid acidification of endocytic vesicles containing α_2-macroglobulin. *Cell* **28:**643–651.
174. Cohen, S., and Fava, R. A. (1985). Internalization of functional epidermal growth factor:receptor/kinase complexes in A-431 cells. *J. Biol. Chem.* **260:**12351–12358.
175. Friedkin, M., Legg, A., and Rozengurt, E. (1979). Antitubulin agents enhance the stimulation of DNA synthesis by polypeptide growth factors in 3T3 mouse fibroblasts. *Proc. Natl. Acad. Sci. USA* **76:** 3909–3912.
176. Gorden, P., Carpentier, J.-L., Cohen, S., and Orci, L. (1978). Epidermal growth factor: Morphological demonstration of binding, internalization, and lysosomal association in human fibroblasts. *Proc. Natl. Acad. Sci. USA* **75:**5025–5029.
177. Gonzalez-Noriega, A., Grubb, J. H., Talkad, V., and Sly, W. S. (1980). Chloroquine inhibits lysosomal enzyme pinocytosis and enhances lysosomal enzyme secretion by impairing receptor recycling. *J. Cell Biol.* **85:**839–852.
178. DiPaola, M., and Maxfield, F. R. (1984). Conformational changes in the receptors for epidermal growth factor and asialoglycoprotiens induced by the mildly acidic pH found in endocytic vesicles. *J. Biol. Chem.* **259:**9163–9171.
179. Hwang, J., Pouyssegur, J., Willingham, M. C., and Pastan, I. (1986). The role of intracellular pH in ligand internalization. *J. Cell. Physiol.* **128:**18–22.
180. Jones, B., and Simpson, D. P. (1983). Influence of alterations in acid–base conditions on intracellular pH of intact renal cortex. *Renal Physiol.* **6:**28–35.
181. Adam, W. R., Koretsky, A. P., and Weiner, M. W. (1986). ^{31}P NMR in vivo measurement of renal intracellular pH: Effects of acidosis and K depletion in rats. *Am. J. Physiol.* **251:**F904–F910.
182. Larkin, J. M., Brown, M. S., Goldstein, J. L., and Anderson, R. G. W. (1983). Depletion of intracellular potassium arrests coated pit formation and receptor-mediated endocytosis in fibroblasts. *Cell* **33:**273–285.
183. Lopez-Rivas, A., Adelberg, E. A., and Rozengurt, E. (1982). Intracellular K and the mitogenic response of 3T3 cells to peptide factors in serum-free medium. *Proc. Natl. Acad. Sci. USA* **79:**6275–6279.

21

Regulation of Intracellular Free Calcium

William B. Busa

21.1. OVERVIEW AND HISTORICAL PERSPECTIVE

It can be argued that the calcium ion is the oldest recognized signal transduction molecule, whether one dates the birth of this recognition with Ringer's 1883 studies[1] on the requirement for extracellular Ca^{2+} in myocardial contraction, or with Heilbrunn's much later proposal[2] that Ca^{2+} is the physiological inducer of muscle contraction. Perhaps because of this early focus on calcium's role in contractility in particular, and in the function of classically "excitable" cells generally (such as Katz and Miledi's seminal studies on the role of Ca^{2+} in neurotransmitter release[3]), the first 100 years of cell calcium research was dominated by efforts to understand how transient changes in intracellular free Ca^{2+} activity ($[Ca^{2+}]_i$) are achieved and transduced in but two very specialized cell types—muscle and neuron. That century of effort generated an enormous body of data concerning the fundamentals of $[Ca^{2+}]_i$ regulation (such as the realization that intracellular vesicular stores play an active role,[4,5] the isolation and ongoing characterization of the major protein components of these stores,[6] the discovery and characterization of the voltage-gated Ca^{2+} channels of the plasma membrane,[7] and the understanding that ubiquitous Ca^{2+}-binding proteins such as calmodulin both transduce and modulate $[Ca^{2+}]_i$ changes[8]), but it was largely biased toward a view of transient $[Ca^{2+}]_i$ changes as the norm in excitable cells and the *exception* in most other cell types, so much so that the concept of Ca^{2+} homeostasis, as opposed to Ca^{2+} signaling, long dominated thinking with regard to $[Ca^{2+}]_i$ regulation in nonexcitable cells.

That mind-set, already beginning to give way by 1970,[9] was finally and fatally challenged by two separate lines of work that reached fruition at the opening of the 1980s. One was the introduction, by Tsien and colleagues,[10,11] of fluorescent and cell-permeant indicators of free calcium ion concentration (Figure 21.1). Prior to the introduction of these probes, the vast majority of $[Ca^{2+}]_i$ regulation studies either relied on observation of the *consequences* of $[Ca^{2+}]_i$ changes (such as myofibril contraction) to indirectly report those changes, or else employed either metallochromic indicators, natural Ca^{2+}-sensitive photoproteins, or Ca^{2+}-selective microelectrodes to detect such changes directly (always with considerable technical difficulty, and with frustratingly limited applicability). It is probably not excessive to liken the change wrought by the introduction of Tsien's indicators (which can be employed, with due caution, in nearly any cell type, and which provide both visual and quantitative information concerning $[Ca^{2+}]_i$ changes with very high spatial and temporal resolution) to that which a blind man, suddenly gaining vision, would experience on realizing that the world around him was infinitely richer and more dynamic than the tip of his cane had previously reported. A flood of research ensued, demonstrating that $[Ca^{2+}]_i$ is an extremely dynamic parameter in nearly all cell types, and challenging the old view of Ca^{2+} homeostasis to account for these widespread dynamics. To a not insignificant extent, this challenge has still not been met, although considerable progress (outlined below) is being made. The other line of work that was to usher in the second century of cell calcium studies, and that would provide much of the answer to the question raised above, was that of Michell, Berridge, Irvine, and colleagues (see Ref. 12 for review), whose identification and characterization of the polyphosphoinositide signal transduction pathway (Figure 21.2) provided a crucial "missing link" between the first step in many calcium signaling events (binding of a ligand to a plasma membrane receptor) and the last (elevation of $[Ca^{2+}]_i$).

Perhaps it should not be surprising that $[Ca^{2+}]_i$ displays such richness of behavior. Other second messengers (e.g., cAMP) are, relatively speaking, impoverished in mechanisms regulating their intracellular levels: one enzyme generates the messenger, and another degrades it. The mechanisms in contention for control of $[Ca^{2+}]_i$ are, in contrast, extremely diverse (Figure 21.3): plasma membrane channels gated by either voltage, extracellular ligands or intracellular second messengers, plasma membrane Ca^{2+} pumps, organelles as diverse as the endoplasmic reticulum, the mitochondria, and the nuclear envelope, intracellular buffering molecules, and at least three second messengers. Many (perhaps most) of these $[Ca^{2+}]_i$-modulating mechanisms are themselves Ca^{2+}-sensitive (some displaying positive feedback, others negative) and all seem to be modulated by other second messenger pathways, as well. Most or all of these factors coexist in the single cell.

William B. Busa • Department of Biology, The Johns Hopkins University, Baltimore, Maryland 21218.
Molecular Biology of Membrane Transport Disorders, edited by S.G. Schultz *et al.* Plenum Press, New York, 1996

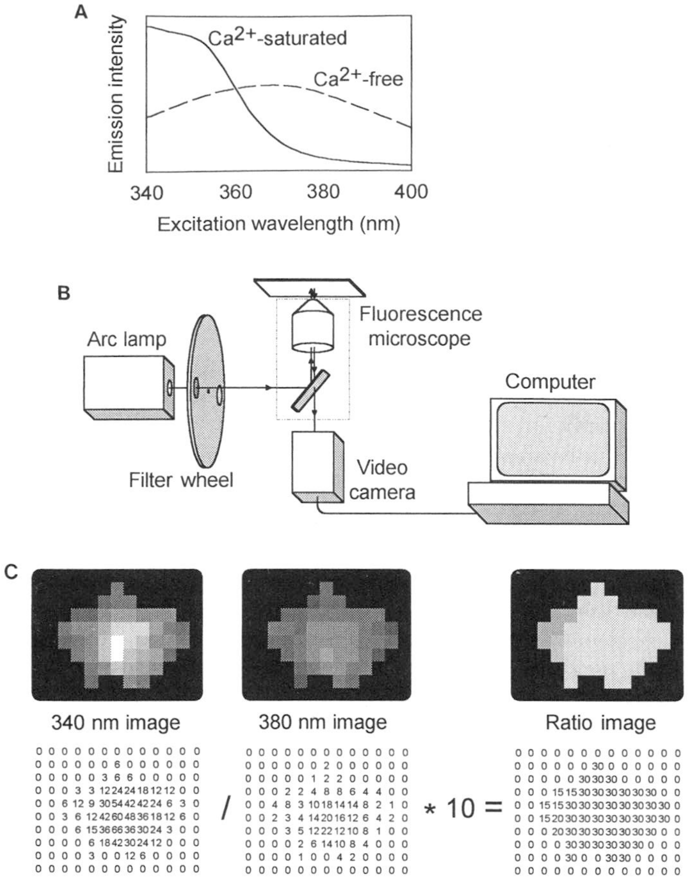

Figure 21.1. $[Ca^{2+}]_i$ visualization and quantification with fluorescent reporter dyes. (A) The excitation spectra of Ca^{2+}-free and Ca^{2+}-bound fura-2. On binding Ca^{2+}, fura-2's excitation spectrum undergoes a pronounced blueshift. At any $[Ca^{2+}]$ near fura-2's K_D, some fraction of the dye molecules will be Ca^{2+}-bound and some will be Ca^{2+}-free; sequential excitation at two appropriate wavelengths (e.g., 340 and 380 nm) permits determination of the fraction in each of these states. Knowledge of the dye's K_D then permits calculation of $[Ca^{2+}]_i$. (B) A typical arrangement for ratio imaging employing fura-2. A filter wheel bearing 340- and 380-nm interference filters allows sequential excitation of the specimen on an epifluorescence microscope. The fluorescence image is collected by a video camera (often, an intensified CCD camera) and passed to a computer. (C) Calculation of ratio images representing the distribution of $[Ca^{2+}]_i$ values. Each sequential (340 or 380 nm) video image received by the computer is digitized into an array of values representing the intensity of the image at each position (a single location in the image is a *pixel*). The digitized value of each pixel in the 340-nm image is then divided by the value of the corresponding pixel in the 380-nm image, the result is multiplied by an arbitrary constant, and the result is displayed in the *ratio image,* where image intensity at each point in the cell is directly proportional to $[Ca^{2+}]_i$ at that location. (In this example, $[Ca^{2+}]_i$ is homogeneous throughout most of the cell, but is lower at the left-hand edge.) Usually, the monochrome images are converted to *pseudocolor,* in which each gray level is assigned a different color, facilitating visual analysis of the results.

In Sections 21.2 and 21.3, I briefly review the fundamentals of $[Ca^{2+}]_i$ regulation: the physicochemical constraints of free Ca^{2+} concentration within the cell, the organelles employed in $[Ca^{2+}]_i$ homeostasis, and their molecular constituents. All of these topics have been the subjects of recent excellent reviews, to which the reader is directed for detailed insight; my goal here is merely to introduce all of the major players individually, noting some of their more relevant characteristics. Section 21.4 then introduces the emerging view of $[Ca^{2+}]_i$ in nonexcitable cells, with special attention to its spatial and temporal dynamics, and explores the extent to which we can account for these given current and emerging insights. Not addressed in this review is the intriguing question of the physiological *roles* of these dynamic responses in nonexcitable cells. This issue is (and will surely continue to be) a—if not *the*—dominant theme of the second century of cell calcium research.

21.2. PHYSICAL CONSTRAINTS ON $[Ca^{2+}]_i$

21.2.1. The Electrochemical Gradient for Ca^{2+} across the Plasma Membrane

As argued elegantly by Kretsinger,[13] choices made early on in the evolution of life, such as the general reliance on phosphorylated compounds as the structural components, intermediary metabolites, and chemical energy stores of life, made it necessary that cells should maintain very low internal free Ca^{2+} concentrations, owing to the very low solubility of calcium phosphates. As a consequence, whereas extracellular calcium levels are quite high (10 mM total calcium in seawater and 3 mM in plasma, about half of which is in the free, ionized form), basal $[Ca^{2+}]_i$ is routinely maintained on the order of 50–300 nM, i.e., a 10,000-fold concentration gradient of Ca^{2+} across the PM is a fact of life for all cells. In combination with the cell's negative-inside membrane potential, the electrochemical gradient is thus on the order of 8–9 kcal/mole.

21.2.2. Physicochemical Buffering

Obviously, the steep electrochemical gradient between the cytosol and the extracellular space presents a challenge for Ca^{2+} homeostasis. For example, the human erythrocyte membrane possesses the lowest observed permeability to Ca^{2+} of any cell yet studied, displaying a passive, unidirectional influx of just 0.4 fmole/cm^2 per sec.[14] Low as this leakage rate is, it would elevate basal $[Ca^{2+}]_i$ by tenfold in less than a minute if unopposed by buffering, extrusion, or sequestration mechanisms. Eukaryotic cells rely on all three of these to maintain low resting $[Ca^{2+}]_i$ levels. Here we consider only the first, physicochemical buffering related to Ca^{2+}-binding molecules in the cytosol. Physiological buffering by the energy-

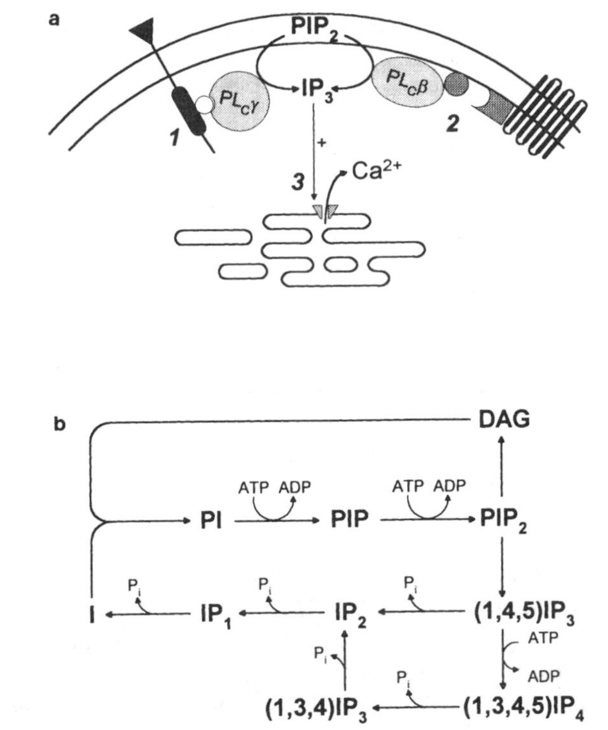

Figure 21.2. The polyphosphoinositide cycle signal transduction pathway. (A) On binding their ligands, numerous tyrosine kinase-linked receptors (*1*) or G protein-coupled receptors (*2*) activate different phospholipase C isoforms. This enzyme hydrolyzes the minor plasma membrane lipid, phosphatidylinositol 4,5-bisphosphate (PIP_2), releasing into the cytosol the second messenger molecule, inositol 1,4,5-trisphosphate (IP_3). The endoplasmic reticulum (and perhaps other organelles, as well) bear IP_3-gated Ca^{2+} channels (*3*), which release Ca^{2+} from the lumen of the ER into the cytosol, thus elevating $[Ca^{2+}]_i$. (B) Hydrolysis, phosphorylation, and dephosphorylation events produce numerous inositol polyphosphates (only some of which are shown here), but only two—1,4,5-IP_3 and 1,3,4,5-IP_4—are known to play regulatory roles. Additionally, PIP_2 hydrolysis produces the lipid-soluble second messenger, diacylglycerol (DAG), the endogenous activator of protein kinase C.

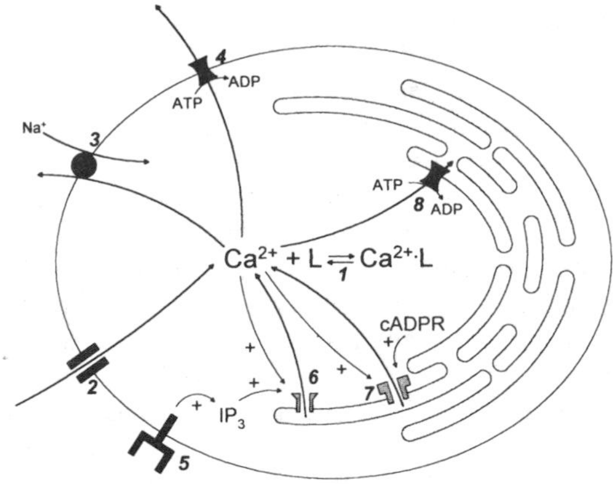

Figure 21.3. Most (if not all) cells possess multiple $[Ca^{2+}]_i$ regulatory mechanisms; at any one point in time and space, $[Ca^{2+}]_i$ will be determined by the *balance* of their activities. (*1*) High-affinity Ca^{2+}-binding proteins such as calmodulin buffer $[Ca^{2+}]_i$ changes. (*2*) Voltage-, ligand-, and second messenger-gated channels, as well as constitutive leak channels, support influx of Ca^{2+} down its electrochemical gradient. (*3*) The Na^+/Ca^{2+} antiporter supports efflux of Ca^{2+} from the cell, powered by the electrochemical gradient for Na^+ across the plasma membrane. (*4*) The Ca^{2+}-ATPase (Ca^{2+} pump) extrudes Ca^{2+} from the cell, powered by ATP hydrolysis. (*5*) Plasma membrane receptors activate the PI cycle (Figure 21.2), producing the Ca^{2+}-mobilizing second messenger, IP_3, which triggers efflux of Ca^{2+} from the endoplasmic reticulum via the IP_3 receptor (*6*). (*7*) Another Ca^{2+}-mobilizing second messenger, cyclic ADP-ribose (cADPR), triggers Ca^{2+} efflux from the ER via ryanodine receptors. (*8*) The lumen of the ER is recharged with Ca^{2+} via a family of pumps distinct from those of the plasma membrane.

requiring processes of extrusion or sequestration is discussed in Sections 21.3.2 and 21.3.3.

It is worthwhile to point out, as have many authors, that physicochemical buffering plays only a quantitatively minor role in Ca^{2+} homeostasis. In order to maintain a steady state, all Ca^{2+} entering the cell must ultimately be exported again; chemical buffers are readily exhausted, whereas extrusion mechanisms are not. But to end the discussion here is to fail to appreciate what is probably the most important role of the cell's chemical Ca^{2+} buffers: that of *shaping* $[Ca^{2+}]_i$ transients in both time and space. In the temporal domain, chemical buffering obviously serves to slow $d[Ca^{2+}]_i/dt$ (for a fixed rate of Ca^{2+} flux into or out of a compartment), and this may, in part, explain the common occurrence of high levels of Ca^{2+}-buffering proteins such as parvalbumin in neurons with high firing rates. Perhaps more importantly, as several authors have appreciated, the *spatial extent* of $[Ca^{2+}]_i$ transients or standing gradients will be radically altered by chemical buffering, the precise effect achieved depending on factors such as the kinetic constants, local concentrations, and mobilities of the buffering agent(s) involved.[15–17] This presents the potential (which, it seems, is realized in many cell types) for spatially restricting $[Ca^{2+}]_i$ transients within the otherwise continuous cytosol, allowing this single second messenger to *differentially* regulate numerous cell processes, provided their subcellular locations are discrete.

Cells possess a large number of calcium-binding small molecules, including metabolic acids, amino acids, nucleotides, and phospholipids, but most have pK_D values for Ca^{2+} that are too low (ranging from 1.5 to 4) and possess too little selectivity for Ca^{2+} over Mg^{2+} (the latter's free concentration in the cytosol is 10,000 times that of the former) to render them efficient buffers at pCa 7.[18] Some exceptions include the anionic phospholipids, particularly cardiolipin and the polyphosphoinositides, which have often been suggested to bind considerable amounts of cell calcium. But the major physicochemical Ca^{2+} buffers of the cytosol appear to be the EF-hand-containing proteins,[19] of which troponin C[20] and calmodulin[21] are the prototypes. As do these two, most if not all members of this increasingly large family of proteins probably play regulatory roles by conferring Ca^{2+}-sensitivity on the proteins with which they interact. Nevertheless, to the extent that they occur in substantial concentrations in the cytosol, they also function as Ca^{2+} buffers.

Various recent reviews have summarized the high-affinity Ca^{2+}-binding proteins of cells,[22–24] successfully conveying the impression that the cytosol is a highly buffered medium, but a quantitative analysis has been difficult to construct for a variety of technical reasons. The most useful work in this regard is that of Krinks *et al.*,[25] who employed the $^{45}Ca^{2+}$ overlay method to identify and quantify the major buffer proteins of squid axoplasm. The dominant components were found to be calmodulin (2.5% of total protein, or 120 μM, presenting 480 μM Ca^{2+}-binding sites) and an unidentified 17-kDa protein of similar abundance and calcium-binding affinity. Krinks *et al.* calculated the total high-affinity Ca^{2+}-binding capacity of axoplasmic proteins to be about 1 mM. Even this value will be something of an underestimate,[26] but the conclusion is nonetheless clear: the cytosol is rich in high-affinity Ca^{2+}-binding proteins. As a practical consequence, large $[Ca^{2+}]_i$ changes can only be achieved by very substantial fluxes of Ca^{2+} across membranes. If, as a first approximation, we take the apparent dissociation constant of the axoplasmic buffering proteins to be approximately equal to resting $[Ca^{2+}]_i$ (i.e., approximately 0.1 μM), then the *buffering power* (β) of the cytosol, defined as

$$\beta = d[\mathrm{Ca}]_T / d\mathrm{pCa}_F$$

[where $d[\mathrm{Ca}]_T$ is the change in *total* Ca^{2+} (bound + free) required to achieve a *free* $[Ca^{2+}]_i$ change of one pCa unit ($d\mathrm{pCa}_F$)] would be about 0.6 mM. Thus, for example, 400 μM Ca^{2+} must enter the cytosol in order to achieve a 400 nM increase in $[Ca^{2+}]_i$ from 100 to 500 nM (and an equal amount must be extruded from the cytosol during recovery from this transient). This might make evolution's choice of Ca^{2+} as a widely employed second messenger seem energetically very inefficient, inasmuch as 1000 times more ATP must be consumed during recovery than would be required if an unbuffered ion were employed instead. This is not the case, however, because of the very low basal $[Ca^{2+}]_i$ maintained by the cell (which, at least in the case of the mammalian erythrocyte, is estimated to require < 1% of glycolytically produced ATP to maintain[27]). As we have just seen, a fivefold increase in $[Ca^{2+}]_i$ above the basal value requires a transmembrane flux of 0.4 mmole Ca^{2+}/liter cytosol. For an *un*buffered ion with a basal free concentration of 1 mM, a fivefold increase in concentration would require a flux ten times as large. Thus, highly buffered ions can be employed efficiently in signaling, provided their basal concentrations are maintained at very low levels.

21.3. CALCIUM FLUXES ACROSS MEMBRANES

Unlike nearly all of the other chemical messengers employed by cells, Ca^{2+} can be neither synthesized nor degraded; thus, its fluxes into and out of the cell are of central importance. This is true even of those processes that are proximally regulated by Ca^{2+} efflux from and sequestration into intracellular organelles, since these are ultimately charged with Ca^{2+} entering the cell from the extracellular pool. Additionally, calcium fluxes across the PM are of paramount interest from a therapeutic point of view, since these are the fluxes that are most readily accessible to pharmacological manipulation. The clinical application of Ca^{2+} channel blockers in a wide range of disorders has been reviewed recently.[28]

The wide range of unique characteristics possessed by the proteins that mediate Ca^{2+} fluxes across the PM lend astonishing virtuosity to Ca^{2+}-based signaling. PM Ca^{2+} channels can give rise to either global or highly localized $[Ca^{2+}]_i$ changes, and they can even influence $[Ca^{2+}]_i$ by means that do not require their action as ion channels *per se.* When physically separated in the PM from the pumps or exchangers that extrude Ca^{2+}, their activity can give rise to a *gradient* of $[Ca^{2+}]_i$ across the cytoplasm, as well as to transcellular Ca^{2+} currents.[29] There is even evidence that the synchronized action of PM Ca^{2+} pumps and channels can maintain the concentration of free Ca^{2+} immediately subjacent to the PM at a value quite different from that of the bulk cytosol.[30] A growing body of evidence indicates that cells employ all of these tricks to lend specificity and precision to Ca^{2+}-mediated cell regulation. This brief review cannot do justice to the enormous literature on this topic. Instead, each section below focuses on one protein which here represents its class, in order to provide a broad overview of the range of strategies available to the cell. The examples chosen are often molecules from muscle fiber or neuron, but this reflects only the research emphasis of the last few decades, *not* the actual tissue distribution of the mechanisms they support. Indeed, molecules belonging to each class are found in most or all eukaryotic cells, although their molecular and physiological characterization in nonexcitable cells is still in its infancy.

21.3.1. Plasma Membrane Ca^{2+} Channels

Transient, passive influx of Ca^{2+} down its electrochemical gradient is observed in essentially all cell types in response to such diverse stimuli as PM depolarization, the binding of Ca^{2+}-mobilizing hormones or growth factors to their PM receptors, neurotransmitter binding, and cell–cell interactions and mechanical forces. This phenomenon regulates, in whole or in part, such important physiological processes as neurotransmission, muscle contraction, secretion, egg activation, cell motility, ciliary beating, cell growth, gene expression, and memory. The major classes of PM channels mediating these fluxes may be grouped into at least four categories based on their gating mechanisms (*voltage-, ligand-,* and *second messenger*-activated channels, as well as constitutively active *leak* channels).

21.3.1.1. Voltage-Activated Ca^{2+} Channels (VACCs)

VACCs have been the subject of intensive study for many years, and their characteristics have been thoroughly reviewed from many perspectives.[31–35] According to their electrophysiological and pharmacological characteristics, they are grouped into four categories: L-, T-, N-, and P-type channels, of which the L type are by far the best characterized.

L-type channels are the predominant VACCs of the skeletal muscle PM. They are composed of five polypeptide subunits, of which the α1 transmembrane subunit forms both the ion-permeable channel across the bilayer and the voltage-sensing element that gates the channel's activity.[36] They display the highest single-channel conductance of all VACCs, open in response to comparatively large PM depolarizations (to about −10 mV), inactivate only very slowly and are inhibited by dihydropyridines (and are therefore sometimes referred to as "dihydropyridine receptors"). Site-directed mutagenesis studies have begun to reveal the molecular details of the means by which these channels achieve their high Ca^{2+} selectivity and flux rates.[37]

Skeletal muscle contraction is triggered by a PM depolarization that commences at the neuromuscular junction and is propagated along PM invaginations known as the transverse tubules (T-tubules). T-tubule depolarization, in turn, triggers release of Ca^{2+} from the endoplasmic reticulum [termed *sarcoplasmic reticulum* (SR) in muscle cells], a Ca^{2+}-sequestering and -releasing organelle (see Section 21.3.3) whose bounding membrane is distinct from, but forms close appositions with, the T-tubule membrane (Figure 21.4). The means by which T-tubule depolarization elicits Ca^{2+} release from the SR (termed *excitation–contraction coupling*) is one of the enduring mysteries of physiology, but has begun to yield to molecular investigation (see Ref. 38 for review). L-type channels in the T-tubule membrane open during PM depolarization and support a modest flux of Ca^{2+} into the cytosol. However, this influx of Ca^{2+} is not required to trigger Ca^{2+} release from skeletal muscle SR, as demonstrated by the ability of muscle fibers to contract in nominally Ca^{2+}-free medium,[39] in the presence of certain Ca^{2+} channel blockers,[40] or when PM voltage is clamped at values that reverse the electrochemical gradient of Ca^{2+} across the membrane.[41] Nevertheless, the T-tubules' L-type channels are required, even if the Ca^{2+} they conduct is not: skeletal muscle fibers from mice bearing the *muscular dysgenesis* mutation are defective in excitation–contraction coupling because of the absence of L-type α1 subunits in these cells,[42] and normal coupling is restored when the cells express wild-type α1 subunits encoded by a microinjected expression vector.[43] The answer to this puzzle, as first suggested in brief by Chandler *et al.*,[44] appears to be that the α1 subunit is in physical contact (either directly or via a second protein), through one of its intracellular loops,[45] with the Ca^{2+} release channel of the SR (the so-called *ryanodine receptor;* see Section 21.3.3.3). Voltage-induced conformational changes in α1, resulting from transmembrane movements of charged amino acid residues in four of its membrane-spanning domains, appear to activate the ranodine receptor via an allosteric mechanism.[38]

Unlike skeletal muscle, cardiac muscle cells do require an influx of Ca^{2+} across the PM, largely via L-type channels, for excitation–contraction coupling, as demonstrated by the requirement for an inward-directed electrochemical gradient of this ion across the PM.[46] This absolute dependence on Ca^{2+} influx, coupled with observations that a cytosolic $[Ca^{2+}]_i$ increase can itself trigger Ca^{2+} release from cardiac muscle SR, has suggested a very different model of excitation–contraction coupling in this tissue.[47,48] In this model, Ca^{2+} influx via voltage-activated L channels is thought to raise $[Ca^{2+}]_i$ sufficiently to trigger *calcium-induced calcium release* (CICR) from the ryanodine receptor Ca^{2+} channels of the SR, and it is this latter flux that raises $[Ca^{2+}]_i$ sufficiently to trigger contraction. CICR (now thought to function in a variety of cell types) is discussed in Section 21.3.3.3.

Although VACCs are routinely studied in classically "excitable" cells (skeletal, cardiac and smooth muscle, neuronal and secretory cells), they are also present in many other cell types, such as fibroblasts,[49] megakaryocytes,[50] osteoblasts,[51] and glial cells.[52] Attesting to their important role in cell function, the inappropriate function of VACCs has been implicated in several disorders, including muscular dysgenesis (see above) and the destruction of pancreatic β cells in diabetes.[53]

21.3.1.2. Ligand-Gated Ca^{2+} Channels (LGCCs)

Several extracellular ligand-gated ion channels not normally thought of as Ca^{2+} channels are, nonetheless, sufficiently nonselective that some fraction of the ionic currents they support under physiological conditions will be carried by Ca^{2+}. The nicotinic acetylcholine receptor is one example; the ratio of its permeabilities to Ca^{2+} and Na^+ (P_{Ca}/P_{Na}) is about 0.2.[54] Because of the much higher extracellular concentration of the more permeant ion, however, the effect of this Ca^{2+} permeability on $[Ca^{2+}]_i$ will be very minor. In contrast, the *N*-methyl-D-aspartate (NMDA) receptor, a member of the glutamate neuronal receptor family, displays a P_{Ca}/P_{Na} of ~5.[55] This recently cloned[56,57] family of receptors displays complex behavior, being modulated by Mg^{2+}, glycine, Zn^{2+} polyamines, and membrane potential (see Ref. 58 for review). The effect of

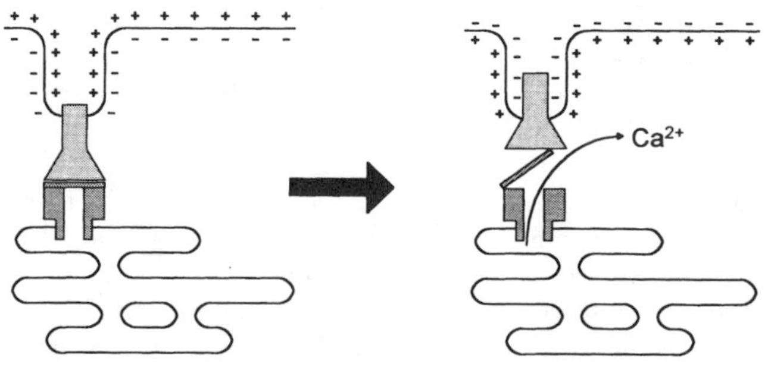

Figure 21.4. The current view of the mechanism of excitation–contraction coupling in skeletal muscle. Dihydropyridine (DHP) receptors in the plasma membrane are thought to physically associate (perhaps via accessory proteins) with the ryanodine receptor Ca^{2+} channel of the sarcoplasmic reticulum. Depolarization of the plasma membrane is sensed by charged amino acid residues of the DHP receptor, altering its conformation and/or position in the membrane. This change gates the ryanodine receptor.

Mg^{2+} is particularly important. When the PM's electrical potential (E_m) is at its resting level (about −70 mV), the ion channel is blocked by extracellular Mg^{2+}, even in the presence of the receptor's physiological ligand, glutamate. The Mg^{2+} block is relieved by PM depolarization, but ligand binding is still required to open the Ca^{2+} channel.[59] In other words, at least *two* inputs—ligand binding and membrane depolarization—must be present *simultaneously* to elicit Ca^{2+} influx through the NMDA receptor. This property of the receptor is thought to play a role in *long-term potentiation,* a selective sensitization of those synapses on a neuron that are most frequently stimulated, which may underlie some forms of learning and memory.[60] Postsynaptic regions stimulated at high frequency will experience localized $[Ca^{2+}]_i$ increases through the activation of NMDA receptors, whereas less frequently stimulated synapses *on the same neuron* will not. This, in turn, is thought to lead to spatially restricted activation of Ca^{2+}-sensitive enzymes such as protein kinase C, Ca^{2+}/calmodulin-dependent protein kinase II, and nitric oxide synthase.[61–64] Although details concerning the molecular bases of long-term potentiation are controversial, the occurrence of localized, NMDA receptor-mediated $[Ca^{2+}]_i$ increases specific to repetitively stimulated synapses has been documented elegantly in high-resolution Ca^{2+} imaging studies by Regehr and Tank.[65]

Although all of the well-established LGCCs occur in excitable cells, evidence is accumulating for their existence in other cell types as well. A cation conductance displaying a P_{Ca}/P_{Na} of 3.3, which is gated by extracellular ATP, was detected in smooth muscle cells by Benham and Tsien.[66] Other lines of evidence suggest that a similar ATP-gated channel (now referred to as the P_{2Z} purinergic receptor) exists in parotid acinar cells,[67] macrophages,[68] and other cell types, but progress in its study has been hampered by the lack of specific antagonists. Only one member of the P_2 receptor family has been cloned,[69] and it appears to be a G protein-linked metabotropic receptor; the ionotropic members of this family remain at large.

21.3.1.3. Modulation of Ca^{2+} Channel Activity by Second Messengers and G Proteins

An ever more obvious characteristic of signal transduction pathways generally is their marked interconnection. Indeed, it is nearer the truth to speak of signal transduction *networks* rather than pathways, since the numerous functional interconnections between the Ca^{2+}, cAMP, protein kinase (and other) signaling pathways tend to cause each to respond when any one is modulated, just as plucking one strand of a taut net sets the whole to vibrating. In this regard, PM Ca^{2+} channels are no exceptions.

The purified L-type VACC from muscle is a substrate for several protein kinases *in vitro,* including cAMP-dependent protein kinase (PK_A). PK_A-mediated phosphorylation both prolongs the open-channel lifetime and shortens the interval between openings at positive membrane potentials,[70,71] and this is likely the basis of the stimulation of whole-cell Ca^{2+} currents observed in β-adrenergic stimulated cells.[72] Additionally, the activated α subunit of G_S (the G protein that couples β-adrenergic receptor occupation to activation of adenylyl cyclase to produce cAMP) appears to be able to interact directly with and stimulate this same channel, even under nonphosphorylating conditions.[71] Indeed, G protein modulation of Ca^{2+} channel activity is now a widely observed phenomenon. In both neurons and endocrine cells, G protein-mediated *inhibition* of VACCs is thought to be involved in receptor-mediated inhibition of secretion (see Ref. 73 for review), and in nonexcitable cells GTP-sensitive single-channel Ca^{2+} currents have been observed,[74] although the latter remain to be molecularly characterized.

21.3.1.4. Second Messenger-Operated Channels (SMOCs)

A variety of hormones and growth factors elicit $[Ca^{2+}]_i$ increases by activating the polyphosphoinositide (PI) cycle, a signaling pathway generating a cytosolic second messenger that triggers Ca^{2+} release from the endoplasmic reticulum (ER) (Figure 21.2). Following the rapid, transient $[Ca^{2+}]_i$ increase resulting from release from the ER (Section 3.3), a prolonged plateau phase elevation of $[Ca^{2+}]_i$ is often observed, and this latter phase has been shown to depend on influx of Ca^{2+} across the PM (for review see Ref. 75). The channels responsible for this flux have not been identified, nor are their gating mechanisms understood. Patch clamp recordings of the Ca^{2+} influx in mast cells[76] and frog oocytes[77] reveal the unknown channel's pronounced selectivity for Ca^{2+} over Ba^{2+}, Sr^{2+}, or Mn^{2+}, a preference not displayed by any known Ca^{2+} channels. Other channels must also be involved in hormone-responsive Ca^{2+} influx, since in many cells external Mn^{2+} can substitute for Ca^{2+} (see Ref. 78 for review). A wide variety of agents have been suggested to mediate hormone-responsive Ca^{2+} influx. Most intriguing are the numerous studies suggesting that this flux is triggered by depletion of ER Ca^{2+} stores (see Ref. 79 for review). Although a large body of data now supports this view, it suffers from a lack of molecular detail: how the Ca^{2+} content of the ER lumen modulates PM Ca^{2+} channel activity remains unclear. Novel (and as yet undefined) diffusible second messengers have recently been implicated,[77,80] but evidence for these remains circumstantial, and some of the putative messenger's characteristics[80] seem very difficult to rationalize at present.

21.3.1.5. Constitutively Active Leak Channels

It is likely that basal $[Ca^{2+}]_i$ is determined by the balance of two factors: steady-state Ca^{2+} leakage into the cell (down its electrochemical gradient) and extrusion by active processes (Sections 21.3.2 and 21.3.3). Ca^{2+} channels displaying constitutive activity in unstimulated cells have been detected in smooth, skeletal, and cardiac muscle membranes,[81–84] and seem likely to occur in many other cells, but they remain uncharacterized. These channels are overactive in muscle fibers from muscular dystrophy patients and *mdx* mice,[84] giving rise to elevated basal $[Ca^{2+}]_i$ levels which appear to be at least partly responsible for the excessive levels of protein degradation observed in these cells.[85] Intriguingly, the Ca^{2+} channel blocker, diltiazem, dramatically reduces both muscle wasting and mortality in dystrophic hamsters.[86]

21.3.2. Plasma Membrane Ca^{2+} Exchangers and Pumps

As already discussed (Section 21.2.2), over the long term all of the Ca^{2+} entering the cell must be exported again in order to avoid saturation of the cell's chemical and physiological buffering systems. Cells employ two distinct means of extruding Ca^{2+}: a Na^+/Ca^{2+} antiporter that does not directly require ATP but derives the energy required for transport of Ca^{2+} up its steep electrochemical gradient by coupling this to the influx of Na^+ down its oppositely directed gradient, and a Ca^{2+}-ATPase, a primary active transporter powered by the hydrolysis of ATP.

21.3.2.1. Na^+/Ca^{2+} Antiporter

The Na^+/Ca^{2+} antiporter constitutes a very-high-capacity extrusion mechanism abundant in retinal rod and cardiac muscle PM (two cell types faced with very large Ca^{2+} influxes during their normal function), and is also found (at much lower abundance) in smooth and skeletal muscle, intestinal and renal epithelia, neurons, neutrophils, and secretory cells. In some cell types (e.g., frog oocytes) it is entirely undetectable. Biochemical studies of Na^+/Ca^{2+} exchange (reviewed in Ref. 87) reveal a stoichiometry of 3 Na^+ per calcium ion extruded (making the antiporter an electrogenic pump) and a K_M for cytosolic Ca^{2+} on the order of 0.5–2 μM (but this value is highly dependent on the ionic milieu, the presence of ATP, and the membrane potential). The fully active antiporter supports enormous transmembrane fluxes in cells where it is abundant—up to 50 pmole/cm^2 per sec in retinal rods.

The cloned canine cardiac Na^+/Ca^{2+} antiporter is a single polypeptide with a deduced size of 108 kDa possessing 12 putative transmembrane domains, and shares no extensive sequence homology with any other known protein.[88] The bovine retinal rod antiporter shares almost no sequence homology with the cardiac isoform, although its membrane topology appears to be similar.[89] No details of the regulation of either of these molecules are revealed by their primary structure,[90] but the antiporter appears to be regulated by allosteric binding of Ca^{2+} (see Ref. 91 for review). Thus, the often-expressed view that the Na^+/Ca^{2+} antiporter provides a "backup" mechanism for dealing with large $[Ca^{2+}]_i$ transients (as well as the corollary of this—that is otherwise contributes little to the dynamics of $[Ca^{2+}]_i$) cannot yet be refuted. Specific inhibitors of the antiporter are not available; they would be valuable tools in tests of this notion.

21.3.2.2. Ca^{2+}-ATPase

Unlike the Na^+/Ca^{2+} antiporter, the Ca^{2+}-ATPase ("Ca^{2+} pump") of the PM appears to be nearly ubiquitous in its distribution among eukaryotic cell types (citations in Ref. 92), and is thus viewed as the workhorse of cellular Ca^{2+} extrusion. It has been the subject of several excellent recent reviews.[92,93] Alterations in Ca^{2+} pump activity have been observed in spontaneously hypertensive rat strains and in humans suffering from essential hypertension, as well as in patients with sickle-cell disease (works cited in Ref. 92).

At least four genes code for PM Ca^{2+} pumps in mammals[93]; their gene products are termed PMCA1–PMCA4 and range in size from 127 to 136 kDa. Secondary structure predictions suggest a molecule with ten transmembrane domains, placing about 80% of the protein in the cytoplasmic space—a pattern corresponding with other P-type ATPases.[94] Alternative splicing products have been detected for all members of this family of genes. For PMCA1 (apparently the most nearly ubiquitously expressed member of the family), three of four possible splice variants possess a phosphorylation site for PK_A,[95] and PMCA1 transcripts encoding the PK_A-sensitive isoform range in abundance from 55% to 100% of all PMCA1 transcripts in tissues ranging from brain to liver, respectively.[96] This presents a possible means by which cells can modulate the degree of cross talk between the cAMP and Ca^{2+} signaling systems, since PK_A-mediated phosphorylation of the Ca^{2+} pump increases both its velocity and its affinity for Ca^{2+}.[97]

Certainly one of the most important means of regulating the PM Ca^{2+} pump involves its regulation by Ca^{2+}/calmodulin. High-affinity binding to Ca^{2+}/CaM served as the basis for the first purifications of the pump protein, and both chemical cross-linking studies and sequencing identify a CaM binding site in the C-terminal region. Ca^{2+}/CaM, binding increases both the pump's velocity and its Ca^{2+} affinity, but much more so in the absence of acid phospholipids than in their presence.[98] Curiously, the enzyme from hepatocytes (which has not yet been cloned) is unique in its insensitivity to CaM,[99] an observation that has not yet been satisfactorily explained from a physiological point of view.

CaM sensitivity presents an obvious means by which the activity of Ca^{2+}-mediating signaling processes can regulate Ca^{2+} extrusion from the cell (in this case, a $[Ca^{2+}]_i$ increase will stimulate Ca^{2+} extrusion). Additional observations suggest (albeit in a very speculative fashion, requiring further study) that other sequelae of Ca^{2+}-mediated signaling might modulate Ca^{2+} pump activity as well. It has often been observed that the activity of Ca^{2+} pumps, reconstituted in synthetic liposomes, is a sensitive function of the mole fraction of anionic lipids present in the bilayer. The polyphosphoinositides, substrates of the PI cycle signal transduction pathway (see Section 21.3.3.2), are the most potent phospholipid activators of the Ca^{2+} pump—in the presence of CaM, the free calcium concentration required for half-maximal stimulation of the pump decreases from about 200 nM to 60 nM as the mole fraction of phosphatidylinositol-4,5-bisphosphate (PIP_2) increases from 0 to 5%.[100] Because PI cycle-mediated Ca^{2+} signaling transiently depresses PM PIP_2 levels, the pump's lipid dependence might provide a means by which the earliest phase of a PI cycle-mediated $[Ca^{2+}]_i$ increase is prevented from being blunted by immediate stimulation of the pump. As well, the inositol phosphate products of PIP_2 hydrolysis have been observed to inhibit the pump, although in some details the data from these studies conflict (perhaps because of differences between the pump isoforms under study?).[101,102] Simultaneous optical measurements of both $[Ca^{2+}]_i$ and net Ca^{2+} efflux from single pancreatic acinar cells suspended in microdrops do reveal that hormone-evoked $[Ca^{2+}]_i$ increases often precede detectable Ca^{2+} efflux,[103] but it is unclear whether this study, which had other aims, possessed sufficient temporal resolution to definitively address this issue.

21.3.3. Intracellular Ca^{2+} Pumps and Channels

Although the extracellular medium represents an essentially infinite source (and sink) of Ca^{2+}, this pool is far removed from the heart of the cell, to which $[Ca^{2+}]_i$ transients employed in cell signaling must often penetrate. The very high buffer power of the cytoplasm seems likely to render Ca^{2+} signaling solely via influx across the PM inefficient, at least where *rapid, cellwide* responses are required, and most particularly in larger cells, where surface-to-volume ratios are low. Perhaps for these reasons, eukaryotic cells have evolved intracellular, membrane-bounded stores of Ca^{2+} that can be mobilized rapidly by unique second messengers and that, as we shall see in Section 21.4, make possible exceedingly complex temporal and spatial behaviors of $[Ca^{2+}]_i$. The growing awareness throughout the 1970s that a number of physiological responses in a very wide array of cell types involved $[Ca^{2+}]_i$ increases that were largely independent of extracellular Ca^{2+}, and that were not accompanied by PM depolarization, gave rise to a heightened focus on mechanisms involved in the mobilization of these intracellular stores of sequestered Ca^{2+}. Whereas much has been learned in the last decade, the discussion below will indicate that our understanding of this important and very widely employed mechanism is still very incomplete, although progress continues to be made at a rapid pace.

21.3.3.1. Sarco/endoplasmic Reticulum Ca^{2+}-ATPase

In order for the sarco/endoplasmic reticulum to serve as a mobilizable store of Ca^{2+} for signaling purposes, it must first be charged with Ca^{2+} by a transporter. Indeed, the first Ca^{2+} pumps cloned and sequenced were not those of the PM (Section 21.3.2.2) but those of the SR.[104] These are the products of at least three genes, and are termed SERCA1, SERCA2, and SERCA3. SERCA1 is expressed exclusively in fast-twitch skeletal muscles. Two developmentally regulated alternative splice forms are known, but no functional differences between these have yet been identified.[105] One splice variant of SERCA2 (SERCA2a) is expressed in slow-twitch and cardiac muscle, while SERCA2b is ubiquitous in the ER of most cells; the latter displays a lower turnover rate than that of SERCA1 or 2a.[106] SERCA3 is also widely expressed in nonmuscle cells, and is distinguished from the other members of this family by its lower affinity for Ca^{2+} and very different pH/activity profile,[106] as well as by its possesion of a potential PK_A phosphorylation site.[107]

Although the membrane topology of the SERCAs is similar to that of PM Ca^{2+} pumps (ten transmembrane domains with a large portion of the protein in the cytoplasm), the SERCAs are no more similar to the PM Ca^{2+} pumps than they are to other P-type ATPases. In particular, their regulation differs substantially from that of the PM pumps. SERCAs are not regulated directly by CaM. Instead, SERCA1 and 2 can bind phospholamban, an integral SR membrane protein and substrate for both cAMP- and Ca^{2+}/CaM-dependent protein kinases.[108] Phospholamban binding inhibits the pump, and only unphosphorylated phospholamban binds.[109] Thus, via CaM $[Ca^{2+}]_i$ increases can also activate SERCA1 and 2, albeit indirectly. However, phospholamban is not universally expressed in cell types expressing SERCA2,[110] whose regulation can therefore be expected to involve other mechanisms as well. A very useful tool in studies of intracellular Ca^{2+} pools is the plant toxin thapsigargin, a potent and specific inhibitor of all SERCAs that has no effect on PM Ca^{2+} pumps.[111]

21.3.3.2. Inositol 1,4,5-Trisphosphate Receptor (IP_3R)

Michell[112] first recognized that a variety of neurotransmitters, hormones, and growth factors that were thought to elicit mobilization of intracellular Ca^{2+} on binding to their cell surface receptors also stimulated the metabolism of PM inositol phospholipids (a phenomenon originally observed by Hokin and Hokin[113]), and suggested that hormone-stimulated phosphoinositide metabolism produced a second messenger mediating Ca^{2+} release from intracellular stores. The particular lipid species involved was soon thereafter found to be a quantitatively minor phosphorylated derivative of phosphatidylinositol, PIP_2.[114] Berridge proposed that IP_3, the water-soluble product of phospholipase C-mediated PIP_2 hydrolysis, was the second messenger involved, documented its rapid production on hormonal stimulation of blowfly salivary glands,[115] and he and his collaborators showed that authentic IP_3 potently triggers Ca^{2+} release from nonmitochondrial stores in permeabilized pancreatic acinar cells.[116] The decade since these discoveries has seen dramatic advances on two major fronts: first, in identifying an array of cell physiological events mediated in part or whole by PI cycle activation and IP_3-mediated Ca^{2+} release from the ER (see Refs. 75 and 117 for reviews), and, second, in identifying, cloning, and characterizing the proteins involved in the PI cycle and in IP_3-mediated Ca^{2+} mobilization. Pathological conditions involving IP_3-mediated Ca^{2+} release have begun to surface with the discovery that Lowe's syndrome, an X-linked disorder, arises from a chromosomal translocation within a gene encoding a possible isoform of IP_3 5-phosphatase.[118]

Observations that mammalian cerebellum is particularly rich in high-affinity IP_3 binding sites,[119] and that heparin binds these same sites avidly but reversibly,[120] made possible the purification of the IP_3R,[121] followed by its reconstitution as a functional IP_3-gated Ca^{2+} channel in vesicles.[122] Subsequently, an IP_3R was cloned and sequenced from both rat[123] and mouse,[124] followed by the cloning of three additional isoforms[125,126]; these are now termed IP_3R-I through -IV, and appear to be the products of different genes. As judged by PCR analysis, various tissues express different combinations of these isoforms, and a single cell type may express multiple isoforms.[126] Additionally, several alternative splicing variants of IP_3R-I are detected, differing in their ligand-binding and putative regulatory domains; within the early postnatal cerebellum they demonstrate differing patterns of developmental regulation.[127]

Molecular and biochemical studies have elucidated many of the characteristics of the IP_3R (for review and citations see Ref. 128). The monomer is a protein of about 2700 amino acids, but the functional receptor in the ER membrane is a homotetramer. Each monomer possesses six transmembrane domains at its C-terminus (forming the ion channel),[128a] a ligand-binding domain of about 400 amino acid residues at its N-terminus, and a hydrophilic, cytoplasmic domain of over 1000 amino acid residues connecting these two functional domains; the latter is often suggested to be the site of modulatory influences such as phosphorylation and ATP-binding (see below), and is usually referred to as the regulatory domain. IP_3Rs purified from various sources display apparent K_Ds for IP_3 on the order of 10 to 100 nM.

Although it is likely that the IP_3R is richly endowed with potential modulation mechanisms, much of the work in this field has thus far failed to distinguish between direct effects of a modulator on the IP_3R itself rather than, e.g., on the Ca^{2+} pump or other ER proteins, and the physiological significance of some of the suggested modulations is less than obvious. Additionally, the heterogeneity of receptor isoforms present in any tissue has complicated interpretation of many studies. For example, mammalian cerebellar IP_3R is phosphorylated *in vitro* by cAMP-, Ca^{2+}/CaM-, and Ca^{2+}/phospholipid-dependent protein kinases[129,130] (PK_A, CaMK, and PK_C, respectively; note that IP_3R-II lacks the putative PK_A phosphorylation sites[126,131]). It has proven tempting to speculate concerning the roles of IP_3R phosphorylation in providing cross talk between the PI cycle and cAMP signaling pathways and feedback regulation of IP_3-mediated Ca^{2+} release, but conflicting results from various labs have clouded this issue: phosphorylation-mediated enhancement[132,133] or inhibition[130,134] of IP_3-mediated Ca^{2+} release has variously been observed in crude systems composed of many proteins, making generalization difficult (neither PK_A- nor PK_C-mediated phosphorylation has a detectable effect on the receptor's affinity for IP_3[130,131a,132]). Specifically addressing this issue, Nakade *et al.*[131a] immunoaffinity-purified IP_3R-I, reconstituted it in lipid vesicles, and demonstrated a pronounced increase in Ca^{2+} flux in response to PK_A-mediated receptor phosphorylation. The effect on other isoforms of the enzyme remains to be investigated. The IP_3R also appears to possess the ability to autophosphorylate on serine residues, albeit only slowly and to substoichiometric levels.[135] The inverse dependence of both rate and stoichiometry of autophosphorylation on temperature raises the possibility that "autophosphorylation" may actually reflect the formation of a phosphoenzyme intermediate, although this itself would be surprising for an ion channel.

Speculation has also attended the discovery of one or more ATP-binding sites on the IP_3R. Studies with purified receptors reconstituted into vesicles[136] or planar lipid bilayers[137] demonstrate that IP_3-triggered Ca^{2+} flux through the channel is enhanced by ATP at concentrations below 10 μM, and is somewhat inhibited at higher concentrations. Here, too, physiological significance is unclear, as it is an open question whether intracellular ATP levels (normally millimolar) ever drop low enough, even locally, for ATP-mediated modulation to come into play.

The IP_3R's affinity for its ligand is negatively modulated by Ca^{2+} itself,[138] but previous suggestions that this occurs via an associated protein termed *calmedin*[139] now appear to be incorrect.[140] ^{45}Ca gel-overlay assays suggest that the IP_3R binds calcium, but with what affinity is not known. In a number of crude preparations (permeabilized cells, microsomes, or total microsomal proteins incorporated in planar lipid bilayers) a biphasic dependence of IP_3-mediated Ca^{2+} release on extravesicular $[Ca^{2+}]$ has been observed, such that $[Ca^{2+}]$ less than ~1 μM promotes IP_3-mediated Ca^{2+} release whereas higher concentrations are inhibitory (see Ref. 140 for citations). Because of the crude nature of all of these preparations, this effect cannot yet be attributed to the IP_3R itself (although it should be possible to test this by employing purified IP_3R functionally reconstituted in liposomes). Nevertheless, for the overall *process* of Ca^{2+} release via the IP_3R, it is clear that positive feedback is observed at $[Ca^{2+}]_i$ just above basal values, changing to negative feedback at levels approaching those observed in stimulated cells, and this is very likely to be of physiological importance.

Another modulator likely to be of some importance is the proton, H^+. Binding of IP_3 to its receptor is markedly dependent on pH[138] within the physiological range of intracellular pH values (~6–8), as expected for a ligand bearing multiple acidic groups with pKs near physiological pH. The EC_{50} for IP_3-mediated Ca^{2+} release from crude microsomes is thus quite pH-dependent.[141] The interdependence of intracellular pH and $[Ca^{2+}]_i$ in intact cells has been widely observed and is likely related to a number of factors (see Ref. 142 for review); this may well be one of them.

A recent study[143] demonstrates that the membrane cytoskeletal protein, ankyrin, binds with high affinity to an isoform of the IP_3R in T-lymphoma cells and potently inhibits both IP_3 binding and IP_3-mediated Ca^{2+} release. Although it remains to be seen whether these proteins associate *in vivo,* such observations certainly raise the interesting possibility of IP_3R modulation via protein–protein interactions.

Perhaps the most puzzling characteristic of IP_3R modulation is the phenomenon termed *quantal release.*[144] Normally, in the absence of desensitization or down regulation mechanisms, a ligand-gated ion channel might be expected to adopt a high open-probability conformation on binding of ligand, and to remain in that state until ligand is dissociated. Macroscopically, this would mean that the same total amount of Ca^{2+} release should be achieved by either high or low [IP_3] (within the effective range), albeit at different rates (since at higher concentrations more receptors will be occupied). This is, however, not observed. Purified IP_3R reconstituted into vesicles support IP_3-mediated $^{45}Ca^{2+}$ influx, leading to intravesicular ^{45}Ca contents that plateau rapidly.[145] Surprisingly, both the initial rate of ^{45}Ca accumulation and the *plateau value* achieved are direct functions of [IP_3], and successive increments in [IP_3] yield successively higher plateaus. Analogous behavior has been observed in crude microsomes, single cells, and cell populations: the total amount of Ca^{2+} released is a function of the concentration of IP_3. This could arise from multiple isoforms of IP_3R (with differing affinities for ligand) in various compartments [ER, calciosomes (see below) and the nuclear envelope in cells, or distinct vesicles in homogenates and reconstituted liposomes]. Alternatively, even a homogeneous population of receptors could demonstrate such behavior if, as recently proposed, the luminal surface of the receptor bears a binding site for Ca^{2+} capable of modulating the receptor's affinity for IP_3.[146] IP_3-mediated Ca^{2+} release would then be a function of luminal Ca^{2+} concentration ($[Ca^{2+}]_L$); at high $[Ca^{2+}]_L$, low concentrations of IP_3 would initiate release, which would in turn lower $[Ca^{2+}]_L$, thus partially desensitizing the receptor and requiring higher concentrations of IP_3 to evoke additional release. Circumstantial support exists for both of these proposed mechanisms, although neither alone can explain all of the experimental data. For example, in the reconstituted receptor studies described above,[145] $[Ca^{2+}]_L$ should have been constant throughout the experiments, but quantal release was still observed; receptor heterogeneity might thus explain these results, but modulation of the receptor by $[Ca^{2+}]_L$ cannot. Whereas modulation of IP_3-evoked Ca^{2+} release by $[Ca^{2+}]_L$ has been inferred in some studies,[147,148] others have failed to observe it.[149,150] Perhaps one aspect of IP_3R heterogeneity concerns its modulation by $[Ca^{2+}]_L$? The systems in which these issues are usually studied (permeabilized cells or crude microsomes) are probably too unwieldy to permit resolution of this issue, not least because $[Ca^{2+}]_L$ can neither be controlled precisely nor measured directly; single-channel recording from receptors in membrane patches or planar lipid bilayers is called for. Whatever its mechanism(s) may prove to be, however, the phenomenon of quantal release may help to explain some of the more puzzling behaviors of $[Ca^{2+}]_i$ in intact cells (see below).

The IP_3R appears unique among Ca^{2+} channels in the breadth of its subcellular distribution. In keeping with its presumed role in mediating mobilization of Ca^{2+} from the ER, immunogold studies document its concentration on intracellular tubular structures plausibly interpreted to be rough and smooth ER, with a notably patchy distribution.[124,151] Observations of PI cycle activity in the nucleus[154] and the apparent involvement of IP_3-mediated Ca^{2+} release in nuclear envelope assembly[155] suggest that nuclear IP_3R may have physiological roles. Although immunolabeling does not usually reveal IP_3R on the plasma membrane, more sensitive techniques such as patch-clamp electrophysiological recording and cell surface radioiodination reveal IP_3R on the PM of numerous cell types, including lymphocytes,[156,157] neurons,[158] and mast cells.[159]

Although the immuno-EM data cited above clearly define the ER as a major source of IP_3R, several cell fractionation studies have noted the failure of a significant fraction of IP_3R to copurify with ER vesicles (as judged by biochemical markers),[160,161] leading to the suggestion that a unique vesicular organelle, the "calciosome," may be another site of IP_3 action. Recent reevaluations of this issue[162,163] make it unclear whether the calciosome is indeed an organelle physically distinct from the ER, or whether calciosomes are simply spatially restricted specializations of the ER, perhaps even cyclically budding off of and rejoining the ER. Evidence from a variety of sources suggests just such a dynamic differentiation of the ER.[164] Further, the calciosome is not devoted solely to IP_3-mediated Ca^{2+} release, as first supposed, as both cell fractionation and immuno-EM studies now report colocalization of both IP_3R and a second important Ca^{2+} release channel, the ryanodine receptor (see below), in this fraction.[165,166] These issues reflect the current lack of clarity concerning the precise localization and number of (and interactions between) mobilizable Ca^{2+} stores in cells.

21.3.3.3. Ryanodine Receptor (RyR)

First identified as the Ca^{2+} release channel of SR (see Ref. 167 for review), the ryanodine receptor derives its name from its high-affinity binding of the plant alkaloid, ryanodine, which locks the channel in its open state. Although this compound proved crucial in the identification, purification, and cloning of the RyR, it has long been unclear whether physiological ligands also exist. Indeed, elucidation of the molecular mechanisms of excitation–contraction coupling in skeletal and cardiac muscle (see Section 21.3.1.1) initially seemed to suggest that the RyR was not a receptor in the classical sense of the term; rather, it seemed to be gated only by interaction with the voltage-sensing dihydropyridine receptor (in skeletal muscle) or by Ca^{2+} itself (in cardiac muscle). However, the recent discovery that the cardiac (and other) isoforms of the RyR are expressed in a wide range of tissues, combined with the serendipitous discovery of a cellular metabolite that potently gates the channel, have led to a new view of the RyR as a true receptor-gated Ca^{2+} channel, which (like the related IP_3 receptor) is involved in second messenger-mediated Ca^{2+} mobilization (see below).

RyRs have been implicated in at least two diseases. *Malignant hyperthermia* (MH) is a cryptic condition, revealed only when sufferers are administered volatile anesthetics, whereupon muscular rigidity, hypermetabolism, and fever result and can be swiftly fatal. In swine, MH is caused by a point mutation in the skeletal muscle RyR gene. In humans the disease is genetically heterogeneous, but in at least some lineages appears to be linked to an RyR mutation (see Ref. 168 for review). Additionally, recent insights into RyR gating in non-muscle cells suggest an indirect role for the RyR in *streptozotocin-induced diabetes mellitus* (see below).

The cloned RyRs from skeletal[169] and cardiac[170] muscle (now referred to as RyR-I and -II, respectively) are single polypeptides of M_r 565,000 and about 5000 amino acids in length, and are 66% identical. In both their deduced amino acid sequences and their presumed native conformations they bear a striking resemblence to the IP_3R, featuring a very large N-terminal cytoplasmic domain (the "foot region," thought to be the domain coupling the type I RyR to the dihydropyridine receptor in skeletal muscle), a short, cytoplasmic C-terminal domain, and (as proposed by various authors) 4 to 12 transmembrane domains toward the C-terminal tenth of the molecule, forming the ion channel. Electron microscopical and binding studies suggest that, like the IP_3R, the protein forms a homotetramer in the SR membrane. Muscle is the only tissue known to express RyR-I whereas RyR-II is expressed in heart, brain, stomach, and possibly other tissues. Repeated observations of [^{3}H]ryanodine binding or ryanodine-induced $[Ca^{2+}]_i$ transients in a wide array of cell types suggest an even broader tissue distribution of RyRs, and a third (RyR-III) has recently been partially cloned[171] and found to be expressed nearly ubiquitously, albeit at levels 100 times lower than those of the skeletal muscle and heart isoforms (see Ref. 172 for review). Intriguingly, RyR-III expression is induced by transforming growth factor β in mink lung epithelial cells.[171] A full-length cDNA clone from rabbit brain[173] may also encode RyR-III.

The search for the physiological triggers of Ca^{2+} release from the SR (i.e., through the RyR) has occupied generations of physiologists. Section 21.3.1.1 discussed the current view, regarding skeletal muscle, that the dihydropyridine receptor functions as a voltage sensor in the PM and itself gates the type I RyR. A wealth of experimental data (reviewed in Ref. 174) suggests, however, that a completely different control mechanism applies in cardiac muscle. Here, it is thought, the opening of PM Ca^{2+} channels permits a Ca^{2+} influx insufficient in itself to trigger myofilament contraction, but sufficient to trigger additional Ca^{2+} to be released from the SR in a process termed *Ca^{2+}-induced Ca^{2+} release* (CICR). Indeed, single-channel recordings from purified RyRs in planar lipid bilayers reveal activation of the cardiac isoform at much lower Ca^{2+} concentrations than those required for activation of the skeletal muscle isoform,[175] indicating that CICR is an inherent property of RyR-II, whereas RyR-I is a much less sensitive Ca^{2+}-gated Ca^{2+} release channel. How CICR might function in supporting Ca^{2+} waves and oscillations in intact cells is discussed in Section 21.4. A long-troubling problem with the view of the RyR-II as a Ca^{2+}-gated Ca^{2+} channel is that CICR is a positive feedback mechanism, and should therefore give rise to all-or-none responses, whereas *in vivo* CICR is a graded phenomenon. Recent single-channel recordings appear to resolve this problem by demonstrating that the receptor fairly rapidly adapts to elevated $[Ca^{2+}]$, becoming desensitized to this level of Ca^{2+} while retaining the ability to respond to further $[Ca^{2+}]$ increases.[176] The molecular mechanism of this adaptation is unclear; in particular, it is unknown whether adaptation is an intrinsic property of the RyR itself, or of a tightly associated protein such as the FK506 binding protein.[177] Other negative modulators of RyR activity might also be involved, such as phosphorylation by Ca^{2+}/CaM-dependent protein kinase[178,179] or binding of sphingosine.[180] Additionally, CICR is modulated according to the Ca^{2+} content of the SR lumen (see below).

Very recent studies have demonstrated that Ca^{2+} is not the only physiologically relevant activator of RyR-II. A recently discovered metabolite of NAD, cyclic ADP-ribose (cADPR), is at least as potent as IP_3 in triggering Ca^{2+} mobilization in preparations as diverse as sea urchin eggs,[181] rat pituitary cells,[182] and pancreatic β cells.[183] A pharmacological analysis of Ca^{2+} release from sea urchin egg microsomes revealed a striking relationship between cADPR-induced Ca^{2+} release and ryanodine- or caffeine-mediated release via RyRs: both are inhibited by procaine and ruthenium red (IP_3-mediated release is not), and microsomes desensitized to cADPR are also cross-desensitized to ryanodine and caffeine (and vice versa), but not to IP_3.[184] Additionally, very low concentrations of cADPR (30–40 nM) potentiate CICR.[185] The suggestion that cADPR might be a ligand for the RyR[184] was later confirmed by single-channel recording from cardiac RyRs.[186] Skeletal muscle RyRs, in contrast, do not respond to cADPR,[184,186] and it remains to be seen whether type III receptors are responsive. The first potentially physiological role discerned for cADPR-mediated Ca^{2+} release via RyRs (other than in egg activation; see below) is in glucose-stimulated insulin secretion from pancreatic β cells,[183] and it seems likely that many more roles for cADPR as a second messenger involved in Ca^{2+} mobilization will soon be defined. Recent observations that cGMP stimulates cADPR production[187] highlight a possible interconnection between the cyclic nucleotide and nitric oxide signaling pathways and RyR function.

21.3.3.4. Functional Interactions between Ca^{2+} Release Channels and ER Luminal Proteins

Mediation of excitation–contraction coupling via interaction between RyR-I and the dihydropyridine receptor (see above) establishes that protein–protein interactions can play an important role in RyR regulation. This principle appears to extend, as well, to interactions between RyRs and luminal proteins of the ER/SR. One such protein, *calsequestrin* (CS; for review and citations see Ref. 188), binds Ca^{2+} with high capacity (~40 mole/mole) but low affinity ($K_d \approx 1$ mM), and is thought to serve as an intra-SR Ca^{2+} buffer. SR regions richest in RyR-I are also enriched in CS, which is bound to the membrane in these regions. Whereas CICR is triggered by Ca^{2+} from the cytoplasmic side of RyR-I (see, e.g., Ref. 176), its sensitivity to cytoplasmic $[Ca^{2+}]$ is modulated by the *luminal* Ca^{2+} content,[189,190] and CS has been implicated as the sensor conveying luminal $[Ca^{2+}]$ sensitivity on the RyR.[189] A physical interaction between the RyR and CS, with functional consequences, is also suggested by observations that the RyR-activating compound, caffeine, triggers a rapid, transient *intravesicular* $[Ca^{2+}]$ increase prior to the onset of Ca^{2+} release from SR vesicles.[191] Even in unsealed SR membranes, caffeine elicits Ca^{2+} release from associated CS. In total, all of these observations suggest a functional and reciprocal interaction between CS and RyR-I, wherein the Ca^{2+} saturation state of CS alters the RyR's sensitivity to cytosolic [Ca^{2+}], and ligand binding to the RyR alters the conformation (and, thus, the Ca^{2+} binding affinity) of CS. Although a substantial body of evidence now supports the existence of such an interaction in skeletal muscle SR, it is not yet possible to extend this to the broader field of other excitable and nonexcitable cells generally, since it is unknown whether non-type-I RyR isoforms can also participate, and the same is true for the numerous isoforms of CS that have been characterized,[192,193] nor has a luminal [Ca^{2+}] dependence of CICR been so clearly established for nonmuscle cells. Nevertheless, indirect evidence from permeabilized hepatocytes[194] is in keeping with the proposal[195] that ER luminal [Ca^{2+}] modulates IP_3 receptor responsiveness, suggesting that luminal [Ca^{2+}] may be of general significance as a regulatory parameter. More direct studies of this issue are warranted since, as we shall see in Section 21.4, many of the current models of $[Ca^{2+}]_i$ wave and $[Ca^{2+}]_i$ oscillation generation in nonexcitable cells invoke this as a fundamental principle.

21.3.3.5. IP_3R Also Support Ca^{2+}-Induced Ca^{2+} Release

Berridge and co-workers[194] first suggested that spontaneous Ca^{2+} releases from permeabilized hepatocytes represent a form of CICR via the IP_3R, based on the ability of inhibitors of this receptor (heparin, high [ATP] or 2,3-bisphosphoglycerate) to block the response. However, hepatocytes appear to possess RyRs as well,[172,196] complicating the analysis of CICR in these cells. In contrast, *Xenopus* eggs and oocytes have no RyRs, as demonstrated by [^{3}H]ryanodine binding studies and Western blotting,[197] and they do not respond to microinjected cADPR or ryanodine (W. B. Busa, unpublished observations), but they possess IP_3R and respond to microinjection of IP_3 with dramatic $[Ca^{2+}]_i$ transients.[198,199] Microsomes from these cells nonetheless display CICR, and this is blocked by the IP_3R antagonist, heparin, and potentiated by subthreshold concentrations of IP_3.[200] Thus, CICR—once thought to be an exclusive feature of RyRs—is displayed by IP_3Rs as well. The available evidence suggests that CICR via the IP_3R requires basal IP_3 levels, and thus probably represents a Ca^{2+}-mediated sensitization of the receptor to low levels of ligand. The same may also be true for at least some isoforms of RyR, since subthreshold levels of cADPR potentiate CICR in sea urchin egg microsomes.[185]

21.4. Ca^{2+} DYNAMICS IN "NONEXCITABLE" CELLS

21.4.1. $[Ca^{2+}]_i$ Waves

In the late 1970s, Jaffe and colleagues[201] first employed microscopic imaging of microinjected aequorin to reveal that fusion of a fertilizing sperm with a medaka fish egg triggers a large, transient $[Ca^{2+}]_i$ increase that initiates at the sperm entry site and propagates without decrement across the egg at a rate of about 10 μm/sec. By the mid-1980s, this phenomenon of propagating $[Ca^{2+}]_i$ waves in both vertebrate and invertebrate zygotes had proven to be quite general, being observed in the fertilizing eggs of fish, sea urchins, frogs, and mammals (Ref. 202 and works cited therein), and was found to be independent of extracellular Ca^{2+}, thus representing a propagated wavelike release of Ca^{2+} from intracellular stores. Fertilization presents rather unique requirements, from a cell-regulatory point of view, in that information regarding a highly localized event (fusion of the very small spermatozoan at one point on the surface of the comparatively enormous egg cell) must be rapidly communicated throughout a large mass of cytoplasm, and so this propagated Ca^{2+} wave seemed an ideal mechanism for regulating the developmental activation of the zygote. Surprisingly, however, recent studies have documented that a variety of somatic cell types (including myocytes,[203] adrenal chromaffin cells,[204] astrocytes,[205] pancreatic acinar cells,[206] gonadotropes,[207] and hepatocytes[208]) also display propagated, wavelike $[Ca^{2+}]_i$ transients, often in response to *global* stimuli such as hormones. Thus, actively propagated intracellular Ca^{2+} waves are not an unusual response exhibited by only a single, uniquely challenged cell type (the egg), but appear to be quite general among eukaryotic cells. Nevertheless, for a variety of reasons the activating egg has proven to be the system of choice in studies aimed at elucidating the mechanism of wave propagation. Its large size facilitates both microinjection of test agents and spatial resolution of the wave, and (thus far) only in eggs is the physiological relevance of the wave and the stimulus required to elicit it well documented. The following discussion thus focuses on insights derived from studies of activating eggs; it seems likely that lessons learned in this cell system will prove applicable in studies of somatic cell $[Ca^{2+}]_i$ waves as well, but this remains to be seen.

The mechanism of Ca^{2+} wave propagation remains a matter of active debate (Figure 21.5). Common to all discussions of the issue is the recognition that, just as in the propagated electrical depolarization of the axonal plasmalemma, a *positive feedback* mechanism is required to propagate the response. CICR is an obvious candidate, and was the mechanism originally proposed to account for the medaka egg wave.[150] Later, after the discovery of IP_3-mediated Ca^{2+} release from the ER, a "two-pool" model was proposed for the fertilization-induced wave,[198] in which the wave is *initiated* when sperm–egg fusion triggers a localized production of IP_3 (which elicits a localized $[Ca^{2+}]_i$ mobilization from the IP_3-responsive store), and is subsequently *propagated* via CICR alone. An alternative model,[209] based on the observation that PIP_2 hydrolysis (and, thus, IP_3 production) is stimulated by micromolar free Ca^{2+} levels, envisioned the same initiation mechanism as the two-pool model, but proposed that propagation was the result of an autocatalytic cycle of Ca^{2+}-induced IP_3 production and IP_3-triggered Ca^{2+} release (the autocatalytic propagation model). These two models (and variations theron) have subsequently been advanced by several other investigators (most notably Berridge[210] and Stryer[211]) to account for Ca^{2+} waves and oscillations in somatic cells as well.

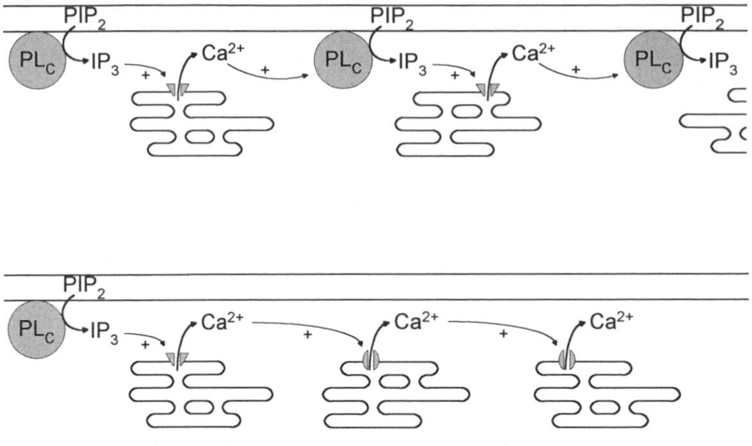

Figure 21.5. Two major models of the mechanism(s) of $[Ca^{2+}]_i$ waves. (A) The "autocatalytic propagation" model. Local elevation of $[Ca^{2+}]_i$ stimulates phospholipase C to produce IP_3, triggering Ca^{2+} release from nearby ER stores, which stimulates additional phospholipase C, and so on. (B) The "two-pool" model. The wave is initiated by local production of IP_3, triggering release from IP_3-sensitive stores. The resulting local elevation of $[Ca^{2+}]_i$ then triggers Ca^{2+}-induced Ca^{2+} release from adjoining stores, which further propagates the wave. The name of this model is actually a misnomer. It is now known that IP_3 receptors themselves can support Ca^{2+}-induced Ca^{2+} release; thus, two distinct Ca^{2+} pools are not a strict requirement in this model.

It has proven extremely difficult to unambiguously distinguish experimentally between these alternative mechanisms of Ca^{2+} wave propagation, not least because we lack highly specific pharmacological inhibitors of PIP_2 hydrolysis and CICR, but also, perhaps, because these models are not mutually exclusive; various cell types might employ either—or *both*—mechanisms. This latter point is emphasized by the observation that the Ca^{2+} wave propagation rate in acetylcholine-stimulated pancreatic acinar cells is influenced *both* by ryanodine (which should modulate CICR via RyRs) and by agonist concentration (which should influence cytosolic IP_3 levels).[206]

Recently, a comparative study[200] of the two cell types responsible for setting this debate in motion—the frog and sea urchin eggs—has yielded a plausible and satisfying resolution. Sea urchin egg microsomes display both IP_3- and cADPR-mediated Ca^{2+} release mechanisms, via IP_3Rs and RyRs, and either receptor type can support CICR when the other is inhibited (by heparin or ruthenium red, respectively). However, CICR is not observed when both receptor types are inhibited simultaneously. Similarly, in the intact sea urchin egg undergoing fertilization, neither heprin alone nor ruthenium red alone blocks Ca^{2+} wave propagation, whereas in combination they completely abrogate the wave (similar results were obtained when 8-amino-cADPR, an inhibitor of the RyR likely to be much more specific than ruthenium red, was employed[212]). In contrast, *Xenopus* egg microsomes display only an IP_3-mediated Ca^{2+} release mechanism (this egg lacks detectable ryanodine receptors[197]), and in these microsomes heparin alone is sufficient to block CICR; as well, heparin alone blocks the fertilization-induced wave in the intact frog egg. Thus, two types of eggs, both displaying similar Ca^{2+} waves, employ distinct but related propagation strategies. In the urchin, redundant CICR mechanisms are employed, either of which is alone sufficient to progpagate the wave. The frog, in contrast, relies exclusively on CICR via the IP_3R. Importantly, CICR via either the IP_3R or the RyR is dependent on (or, at least, is sensitized by) IP_3 or cADPR, respectively, just as IP_3- and cADPR-mediated Ca^{2+} releases are modulated by intra- and extraluminal $[Ca^{2+}]$ (see above), blurring the once seemingly solid distinction between Ca^{2+}- and ligand-induced Ca^{2+} releases. There are two lessons to be learned from these observations. First, even seemingly very similar cells can employ molecularly distinct strategies for Ca^{2+} wave propagation. Second, ligand- (IP_3 or cADPR) and Ca^{2+}-induced Ca^{2+} releases are two faces of the same coin, not mutually exclusive mechanisms. Whether or not propagated IP_3 production accompanies the Ca^{2+} wave (as the strict autocatalytic propagation model would predict) remains to be unambiguously established. However, microinjection of *Xenopus* oocytes with anti-PIP_2 antibodies (which inhibit PIP_2 hydrolysis[213]) reduces the peak cortical $[Ca^{2+}]_i$ observed during the wave but does not impede its propagation,[214] suggesting that autocatalytic IP_3 production *accompanies* wave propagation (through the cortex, at least) but is not *required* for propagation.

21.4.2. $[Ca^{2+}]_i$ Oscillations

As several authors[210,211,215,216] have pointed out, Ca^{2+} waves and Ca^{2+} oscillations (or, more properly, trains of spikes) are closely related phenomena, and thus probably rely on similar mechanisms. Again, the *Xenopus* oocyte has produced some of the clearest evidence: when the immature (ovarian) oocyte is artifically driven into oscillations (either by stimulation of exogenous hormone receptors[217] or by injection of nonhydrolyzable analogues of GTP or IP_3[199]) these oscillations sometimes take the form of trains of circular or spiral waves propagated from local "hot spots" throughout the cytoplasm. Oscillations have received more attention, both experimental and theoretical, than have waves, perhaps because as one- rather than four-dimensional responses they are more easily studied, and they have been detected in a very wide variety of cells (see Ref. 216 for review). The "two-pool" and "autocatalytic" models (discussed above with regard to wave propagation) have also been invoked to account for Ca^{2+} oscillations (Figure 21.6), as have two additional models involving negative rather than positive feedback (see Refs. 216 and 218 for reviews and citations). In the first of these negative feedback models, protein kinase C (stimulated by diacylglycerol produced via PI cycle activity) inhibits further receptor-medi-

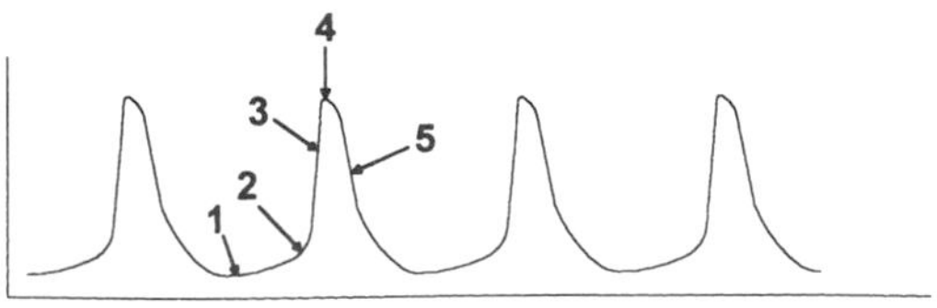

1	2	3	4	5	Ref.
Influx of Ca_e recharges pool(s)	Influx of Ca_e triggers CICR	Regenerative CICR	Pool(s) empty	Reuptake into pool(s) and extrusion	210
$[IP_3]$ increases slowly at low $[Ca]_i$	Increasing positive feedback between $[Ca]_i$ increase and IP_3 production		High $[Ca]_i$ inhibits IP_3R	Reuptake and extrusion; $[IP_3]$ declines	211

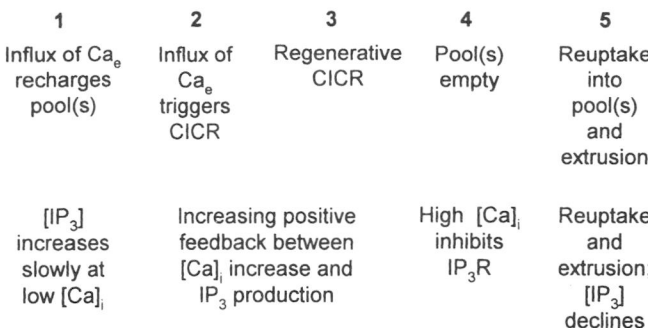

Figure 21.6. Two major models of the mechanism(s) of $[Ca^{2+}]_i$ oscillation. The first (Ref. 210) proposes that rounds of Ca^{2+}-induced Ca^{2+} release from the ER, followed by its recharging, underlie oscillations. The second (Ref. 211) proposes that positive feedback of $[Ca^{2+}]_i$ on IP_3 production, combined with negative feedback of $[Ca^{2+}]_i$ on IP_3 receptor-mediated Ca^{2+} release, produces oscillations.

ated diacylglycerol and IP_3 production, giving rise to oscillatory production of these second messengers. Conceptually, this is similar (although the sign of the feedback is opposite) to the "autocatalytic" model, in that a second messenger feeds back on the very earliest stage of second messenger production; it thus drives oscillations in second messenger production. In the second negative feedback model, elevated $[Ca^{2+}]_i$ inhibits additional IP_3-mediated Ca^{2+} release through the IP_3 receptor. This, again, bears some similarity to the "two-pool" model, because here the downstream response (Ca^{2+} release) feeds back on itself rather than on upstream components of the signaling pathway.

What is one to make of the plethora of models advanced to account for $[Ca^{2+}]_i$ oscillations? All are fairly firmly based on the established characteristics of the macromolecules involved. All can cite experimental data in their support, just as all face other data that seem to contradict them. Two points seem important to bear in mind. First, as Tsien and Tsien[218] have noted, many (though not all) examples of $[Ca^{2+}]_i$ oscillations in nonexcitable cells derive from cells under conditions far from physiological. It is certainly to be expected that complex pathways can be driven into oscillation when pushed beyond their design limits—and can be so pushed by many different means. Thus, the physiological relevance of $[Ca^{2+}]_i$ oscillations in some experimental systems is, at best, unclear, and in others (Ref. 217, as but one example) is clearly nonexistent. This is not to say that such studies themselves are irrelevant. They are not, because they provide important information regarding the repertoire of responses of which the Ca^{2+} signaling system is potentially capable, and (it is hoped) stimulate the search for similar responses under physiologically relevant conditions. The second, and perhaps most important, point to bear in mind is that the multiplicity of $[Ca^{2+}]_i$-regulating mechanisms possessed by eukaryotic cells generally (multiple and variously gated plasma membrane channels, numerous buffers, exchangers, pumps, and compartments, and at least two independent pathways for generating Ca^{2+}-mobilizing second messengers) enable distinct cell types to pick and choose among a variety of mechanisms for shaping, timing, and delimiting $[Ca^{2+}]_i$ signals in the manner most appropriate to their unique physiological requirements. It is thus less than fruitful to argue the relative merits of various models generally; a model can only be meaningful with respect to a particular cell type undergoing a specific—and, ideally, *physiological*—response. Viewed from this perspective, it seems obvious that different cells will often employ different mechanisms to achieve otherwise similar goals, just as we have seen that the sea urchin and frog eggs adopt distinct but related means to achieve the activation wave at fertilization. Versatility lies at the very heart of the Ca^{2+} signaling system.

As are Ca^{2+} waves, Ca^{2+} oscillations are perhaps best understood in fertilized eggs. Here, as with Ca^{2+} waves, the physiological relevance of oscillations is demonstrated,[219] and the ease of microinjection of these cells permits studies that are not yet possible in most somatic cell types. Fish, frog, and sea urchin eggs do not display oscillations, but mammalian eggs do.[219–222] In the hamster egg, the first two or three transients propagate as waves from the sperm attachment site, whereas subsequent transients, repeated at intervals of 2 to 4 min for an hour or more, occur synchronously over the entire egg.[220] Several lines of evidence (reviewed in Ref. 223) strongly suggest that these periodic Ca^{2+} mobilizations represent CICR occurring via the IP_3R, and that both cytosolic and ER luminal calcium concentration changes play roles in their regulation. Oscillations occur even during continuous injection of IP_3,[224] ruling out the involvement of either negative-feedback or autocatalytic mechanisms of repetitive bursts of IP_3 production. Thus, the propagation of Ca^{2+} waves and the generation of Ca^{2+} oscillations appear to involve the same mechanism in eggs. However, the fact that the initial Ca^{2+} wave is followed by repetitive transients in mammalian eggs, but not in those of fish, frogs, or urchins, indicates that there are aspects of this signaling system that remain to be elucidated.

21.5. CONCLUSION

Compared with the other familiar second messenger system, cAMP, the Ca^{2+} signaling system displays an astonishing degree of virtuosity. Its inherent complexity, offering the potential for achieving a single end (alteration of $[Ca^{2+}]_i$) by a number of molecularly distinct means, has presented an enormous challenge to researchers. As this review has shown, we now possess a fairly detailed (although certainly not complete) understanding of most (perhaps all?) of the major molecules involved in cell calcium regulation: we understand a good deal concerning their structures, their kinetic parameters, their modulation, their tissue and subcellular distributions, and their interactions. Nevertheless, it is not yet possible to predict successfully from first principles the spatiotemporal form of a $[Ca^{2+}]_i$ response prior to its direct observation; the study of cell calcium regulation remains largely an empirical science. Thus, a major task facing the next century of cell calcium research will be to achieve an *integrated* view of $[Ca^{2+}]_i$ regulation—in effect, to reconstitute the living cell. It is also important to recognize, however, that this same degree of complexity that currently impedes our achievement of a global view of $[Ca^{2+}]_i$ regulation also presents important possibilities for rather selective pharmacological interventions. In this regard, the current widespread clinical application of Ca^{2+} channel blockers probably represents merely the tip of the iceberg. Lithium, the drug of choice in the management of bipolar manic depression, is widely (although not universally) thought to act by moderating PI cycle overactivity[225]; if this proves true, then more selective and less toxic $[Ca^{2+}]_i$ modulators may find a role in the treatment of other psychiatric disorders as well. The increasingly apparent roles of Ca^{2+} in ischemia/reperfusion injury and in excitotoxicity[226] similarly provide motivation for the development of more effective Ca^{2+} antagonists. More speculatively, the recent implication of cADPR in lymphocyte proliferation and the identification of a lymphocyte marker antigen as a cADPR-generating enzyme[227] may prove to open new avenues in clinical immunology. Given the involvement of Ca^{2+} in nearly every aspect of the lives (and deaths) of cells, completing the elucidation of the complexities of intracellular calcium regulation during the next 100 years of cell calcium research should prove to be of as much—if not more—interest clinically as academically.

REFERENCES

1. Ringer, S. (1883). *J. Physiol. (London)* **4:**29–42.
2. Heilbrunn, L. V. (1940). The action of calcium on muscle protoplasm. *Physiol. Zool.* **13:**88–94.
3. Katz, B., and Miledi, R. (1965). The effect of calcium on acetylcholine release from motor nerve terminals. *Proc. R. Soc. London Ser. B* **161:**496–503.
4. Kumagi, H., Ebashi, S., and Takeda, F. (1955). Essential relaxing factor in muscle other than myokinase and creatine phosphokinase. *Nature* **176:**166.
5. Constantin, L. L., Franzini-Armstrong, C., and Podolsky, R. J. (1965). Localisation of calcium-accumulating structures in striated muscle fibres. *Science* **147:**158–159.
6. MacLennan, D. H., and Campbell, K. P. (1979). Structure, function and biosynthesis of sarcoplasmic reticulum proteins. *Trends Biochem. Sci.* **4:**148–151.
7. Hagiwara, S., and Byerly, L. (1981). Calcium channel. *Annu. Rev. Neurosci.* **4:**69–125.
8. Means, A. R., Tash, J. S., and Chafouleas, J. G. (1982). Physiological implications of the presence, distribution, and regulation of calmodulin in eukaryotic cells. *Physiol. Rev.* **62:**1–39.
9. Rasmussen, H. (1970). Cell communication, calcium ion, and cyclic nucleotide monophosphate. *Science* **170:**404–412.
10. Tsien, R. Y. (1980). New calcium indicators and buffers with high selectivity against magnesium and protons: Design, synthesis, and properties of prototype structures. *Biochemistry* **19:**2396–2404.
11. Grynkiewicz, G., Poenie, M., and Tsien, R. Y. (1985). A new genenration of Ca^{2+} indicators with greatly improved fluorescence properties. *J. Biol. Chem.* **260:**3440–3450.
12. Hokin, L. E. (1985). Receptors and phosphoinositide-generated second messengers. *Annu. Rev. Biochem.* **54:**205–235.
13. Kretsinger, R. H. (1990). Why cells must export calcium. In *Intracellular Calcium Regulation* (F. Bronner, ed.), Wiley–Liss, New York, pp. 439–457.
14. McNamara, M. K., and Wiley, J. S. (1986). Passive permeability of human red blood cells to calcium. *Am. J. Physiol.* **250:**C26–C31.
15. Simon, S. M., and Llinás, R. R. (1985). Compartmentalization of the submembrane calcium activity during calcium influx and its significance in transmitter release. *Biophys. J.* **48:**485–498.
16. Fogelson, A. L., and Zucker, R. S. (1985). Presynaptic calcium diffusion from various arrays of single channels: Implications for transmitter release and synaptic facilitation. *Biophys. J.* **48:**1003–1017.
17. Speksnijder, J. E., Miller, A. L., Weisenseel, M. H., Chen, T.-H., and Jaffe, L. F. (1989). Calcium buffer injections block fucoid egg development by facilitating calcium diffusion. *Proc. Natl. Acad. Sci. USA* **86:**6607–6611.
18. Campbell, A. K. (1983). *Intracellular Calcium: Its Universal Role as Regulator,* Wiley, New York.
19. Kretsinger, R. H. (1987). Calcium coordination and the calmodulin fold: Divergent versus convergent evolution. *Cold Spring Harbor Symp. Quant. Biol.* **52:**499–510.
20. Wnuk, W. (1988). Calcium binding to troponin C and the regulation of muscle contraction: A comparative approach. In *Calcium and Calcium Binding Proteins, Molecular and Functional Aspects* (C. Gerday, L. Bolis, and R. Gilles, eds.), Springer-Verlag, Berlin, pp. 44–68.
21. Persechini, A., Moncrief, N. D., and Kretsinger, R. H. (1989). The EF-hand family of calcium-modulated proteins. *Trends Neurosci.* **12:**462–467.
22. Carafoli, E. (1987). Intracellular calcium homeostasis. *Annu. Rev. Biochem.* **56:**395–433.
23. Heizmann, C. W., and Hunziker, W. (1990). Intracellular calcium-binding molecules. In *Intracellular Calcium Regulation* (F. Bronner, ed.), Liss, New York, pp. 211–248.
24. Blaustein, M. P. (1988). Calcium transport and buffering in neurons. *Trends Neurosci.* **11:**438–443.
25. Krinks, M. H., Klee, C. B., Pant, H. C., and Gainer, H. (1988). Identification and quantification of calcium-binding proteins in squid axoplasm. *J. Neurosci.* **8:**2172–2182.
26. Maruyama, K., Mikawa, T., and Ebashi, S. (1984). Detection of calcium binding proteins by ^{45}Ca autoradiography on nitrocellu-

lose membrane after sodium dodecyl sulfate gel electrophoresis. *J. Biochem.* **95:**511–519.
27. Ferreira, H. G., and Lew, V. L. (1975). Ca transport and Ca pump reversal in human red blood cells. *J. Physiol. (London)* **252:**86P–87P.
28. Hurwitz, L., Partridge, L. D., and Leach, J. K. (eds.) (1991). *Calcium Channels: Their Properties, Functions, Regulation, and Clinical Relevance,* CRC Press, Boca Raton.
29. Nuccitelli, R. (1988). Ionic currents in morphogenesis. *Experientia* **44:**657–666.
30. Rasmussen, H., and Rasmussen, J. E. (1990). Calcium as intracellular messenger: From simplicity to complexity. *Curr. Top. Cell. Regul.* **31:**1–109.
31. Tsien, R. W. (1983). Calcium channels in excitable cell membranes. *Annu. Rev. Physiol.* **45:**341–358.
32. Tsien, R. W., Lipscombe, D., Madison, D. V., Bley, K. R., and Fox, A. P. (1988). Multiple types of neuronal calcium channels and their selective modulation. *Trends Neurosci.* **11:**431–438.
33. Bean, B. P. (1989). Classes of calcium channels in vertebrate cells. *Annu. Rev. Physiol.* **51:**367–384.
34. Hess, P. (1990). Calcium channels in vertebrate cells. *Annu. Rev. Neurosci.* **13:**337–356.
35. Miller, R. J., and Fox, A. P. (1990). Voltage-sensitive calcium channels. In *Intracellular Calcium Regulation* (F. Bronner, ed.), Liss, New York, pp. 97–138.
36. Perez-Reyes, E., Kim, H. S., Lacerda, A. E., *et al.* (1989). Induction of calcium currents by the expression of the alpha 1-subunit of the dihydropyridine receptor from skeletal muscle. *Nature* **340:** 233–236.
37. Yang, J., Ellinor, P. T., Sather, W. A., *et al.* (1993). Molecular determinants of Ca^{2+} selectivity and ion permeation in L-type Ca^{2+} channels. *Nature* **366:**158–161.
38. Catterall, W. A. (1991). Excitation–contraction coupling in vertebrate skeletal muscle: A tale of two calcium channels. *Cell* **64:** 871–874.
39. Armstrong, C. M., Bezanilla, F. M., and Horowicz, P. (1972). Twitches in the presence of ethylene glycol bis (β-aminoethyl ether)-N-N′-tetraacetic acid. *Biochim. Biophys. Acta* **267:**605–608.
40. Gonzales-Serratos, H., Valle-Aguilera, R., Lathrop, D. A., *et al.* (1982). Slow inward calcium currents have no obvious role in muscle excitation–contraction coupling. *Nature* **298:**292–294.
41. Brum, G., Rios, E., and Stefani E. (1988). Effects of extracellular calcium on calcium movements of excitation–contraction coupling in frog skeletal muscle fibres. *J. Physiol. (London)* **398:**441–473.
42. Knudson, C. M., Chaudhauri, N., Sharp, A. H., *et al.* (1989). Specific absence of the alpha 1 subunit of the dihydropyridine receptor in mice with muscular dysgenesis. *J. Biol. Chem.* **264:**1345–1348.
43. Tanabe, T., Beam, K. G., Powell, J. A., *et al.* (1988). Restoration of excitation–contraction coupling and slow calcium current in dysgenic muscle by dihydropyridine receptor complementary DNA. *Nature* **336:**134–139.
44. Chandler, W. K., Rakowski, R. F., and Schneider, M. F. (1976). A non-linear voltage dependent charge movement in frog skeletal muscle. *J. Physiol. (London)* **254:**245–283.
45. Tanabe, T., Beam, K. G., Adams, B. A., *et al.* (1990). Regions of the skeletal muscle dihydropyridine receptor critical for excitation–contraction coupling. *Nature* **346:**567–569.
46. Rich, T. L., Langer, G. A., and Klassen, M. G. (1988). Two components of coupling calcium in single ventricular cells of rabbits and rats. *Am. J. Physiol.* **254:**H937–H946.
47. Fabiato, A., and Fabiato, F. (1975). Contractions induced by a calcium-triggered release of calcium from the sarcoplasmic reticulum of single skinned cardiac cells. *J. Physiol. (London)* **249:**469–495.
48. Fabiato, A., and Fabiato, F. (1979). Calcium and cardiac excitation–contraction coupling. *Annu. Rev. Physiol.* **41:**473–484.
49. Chen, C., Corbley, M. J., Roberts, T. M., *et al.* (1988). Voltage-sensitive calcium channels in normal and transformed 3T3 fibroblasts. *Science* **239:**1024–1026.
50. Kawa, K. (1990). Voltage-gated calcium and potassium currents in megakaryocytes dissociated from guinea-pig bone-marrow. *J. Physiol. (London)* **431:**187–206.
51. Chesnoy-Marchais, D., and Fritsch, J. (1988). Voltage-gated sodium and calcium currents in rat osteoblasts. *J. Physiol. (London)* **398:**291–311.
52. Barres, B. A., Chun, L. L. Y., and Corey, D. P. (1988). Ion channel expression by white matter glia. I. Type 2 astrocytes and oligodendrocytes. *Glia* **1:**10–30.
53. Junti-Berggren, L., Larsson, O., Rorsman, P., *et al.* (1993). Increased activity of L-type Ca^{2+} channels exposed to serum from patients with type I diabetes. *Science* **261:**86–90.
54. Edwards, C. (1982). The selectivity of ion channels in nerve and muscle. *Neuroscience* **7:**1335–1366.
55. MacDermott, A. B., Mayer, M. L., Westbrook, G. L., *et al.* (1986). NMDA-receptor activation increases cytoplasmic calcium concentration in cultured spinal cord neurones. *Nature* **321:**519–522.
56. Moriyoshi, K., Masu, M., Ishii, T., *et al.* (1991). Molecular cloning and characterization of the rat NMDA receptor. *Nature* **354:**31–37.
57. Kutsuwada, T., Kashiwabuchi, N., Mori, H., *et al.* (1992). Molecular diversity of the NMDA receptor channel. *Nature* **358:**36–41.
58. Monaghan, D. T., Bridges, R. J., and Cotman, C. W. (1989). The excitatory amino acid receptors: Their classes, pharmacology, and distinct properties in the function of the central nervous system. *Annu. Rev. Pharmacol. Toxicol.* **29:**365–402.
59. Mayer, M. L., Westbrook, G. L., and Guthrie, P. B. (1984). Voltage-dependent block by Mg^{2+} of NMDA responses in spinal cord neurones. *Nature* **309:**261–263.
60. Bliss, T. V. P., and Lynch, M. A. (1988). Long-term potentiation of synaptic transmission in the hippocampus—Properties and mechanisms. *Neurol. Neurobiol.* **34:**3–72.
61. Muller, D., Buchs, P. A., Stoppini, L., *et al.* (1991). Long-term potentiation, protein kinase C, and glutamate receptors. *Mol. Neurobiol.* **5:**277–288.
62. Massicotte, G., and Baudry, M. (1991). Triggers and substrates of hippocampal synaptic plasticity. *Neurosci. Biobehav. Rev.* **15:** 415–423.
63. Zhuo, M., Small, S. A., Kandel, E. R., *et al.* (1993). Nitric oxide and carbon monoxide produce activity-dependent long-term synaptic enhancement in hippocampus. *Science* **260:**1946–1950.
64. Colbran, R. J. (1992). Regulation and role of brain calcium/calmodulin-dependent protein kinase II. *Neurochem. Int.* **21:**469–497.
65. Regehr, W. G., and Tank, D. W. (1992). Calcium concentration dynamics produced by synaptic activation of CA 1 hippocampal pyramidal cells. *J. Neurosci.* **12:**4202–4223.
66. Benham, C. D., and Tsien, R. W. (1987). A novel receptor-operated Ca^{2+}-permeable chanel activated by ATP in smooth muscle. *Nature* **328:**275–278.
67. Soltoff, S. P., McMillian, M. K., and Talamo, B. R. (1992). ATP activates a cation-permeable pathway in rat parotid acinar cells. *Am. J. Physiol.* **262:**C934–C940.
68. Buisman, H. P., Steinberg, T. H., Fischbarg, J., *et al.* (1988). Extracellular ATP induces a large nonselective conductance in macrophage plasma membranes. *Proc. Natl. Acad. Sci. USA* **85:**7988–7992.
69. Lustig, K. D., Shiau, A. K., Brake, A. J., *et al.* (1993). Expression cloning of an ATP receptor from mouse neuroblastoma cells. *Proc. Natl. Acad. Sci. USA* **90:**5113–5117.
70. Flockerzi, V., Oeken, H.-J., Hofmann, F., *et al.* (1986). Purified dihydropyridine-binding site from skeletal muscle t-tubules is a functional calcium channel. *Nature* **323:**66–68.

71. Yatani, A., Imoto, Y., Codina, J., *et al.* (1988). The stimulatory G protein of adenylyl cyclase, G_s, also stimulates dihydropyridine-sensitive Ca^{2+} channels. *J. Biol. Chem.* **263:**9887–9895.
72. Kameyama, M., Hescheler, J., Hofmann, F., *et al.* (1986). Modulation of Ca current during the phosphorylation cycle in the guinea pig heart. *Pfluegers Arch.* **407:**123–128.
73. Schultz, G., Rosenthal, W., Hescheler, J., and Trautwein, W. (1990). Role of G proteins in calcium channel modulation. *Annu. Rev. Physiol.* **52:**275–292.
74. Mozhayeva, G. N., Naumov, A. P., and Kuryshev, Y. A. (1991). Variety of Ca^{2+}-permeable channels in human carcinoma A431 cells. *J. Membr. Biol.* **124:**113–126.
75. Berridge, M. J., and Irvine, R. F. (1989). Inositol phosphates and cell signalling. *Nature* **341:**197–205.
76. Hoth, M., and Penner, R. (1992). Depletion of intracellular calcium stores activates a calcium current in mast cells. *Nature* **355:**353–356.
77. Parekh, A. B., Terlau, H., and Stuhmer, W. (1993). Depletion of $InsP_3$ stores activates a Ca^{2+} and K^+ current by means of a phosphatase and a diffusible second messenger. *Nature* **364:**814–818.
78. Rink, T. J. (1990). Receptor-mediated calcium entry. *FEBS Lett.* **268:**381–385.
79. Menniti, F. S., Bird, G. S. J., Glennon, M. C., *et al.* (1992). Inositol polyphosphates and calcium signaling. *Mol. Cell. Neurosci.* **3:** 1–10.
80. Randriamampita, C., and Tsien, R. Y. (1993). Emptying of intracellular Ca^{2+} stores releases a novel small messenger that stimulates Ca^{2+} influx. *Nature* **364:**809–814.
81. Benham, C. D., and Tsien, R. W. (1987). Calcium-permeable channels in vascular smooth muscle: Voltage-gated, ligand-gated and receptor-operated channels. *Symp. Soc. Gen. Physiol.* **42:**45–64.
82. Rosenberg, R. L., Hess, P., and Tsien, R. W. (1988). Cardiac calcium channels in planar lipid bilayers. *J. Gen. Physiol.* **92:**27–54.
83. Coulombe, A., Lefevre, A., Baro, I., *et al.* (1989). Barium- and calcium-permeable channels open at negative membrane potentials in rat ventricular myocytes. *J. Membr. Biol.* **111:**57–67.
84. Fong, P., Turner, P. R., Denetclaw, W. F., *et al.* (1990). Increased activity of calcium leak channels in myotubes of Duchenne human and *mdx* mouse origin. *Science* **250:**673–676.
85. Turner, P. R., Westwood, T., Regen, C. M., *et al.* (1988). Increased protein degradation results from elevated free calcium levels found in muscle from *mdx* mice. *Nature* **335:**735–738.
86. Johnson, P. L., and Bhattacharya, S. K. (1993). Regulation of membrane-mediated chronic muscle degeneration in dystrophic hamsters by calcium-channel blockers: Diltiazem, nifedipine and verapamil. *J. Neurol. Sci.* **115:**76–90.
87. Reeves, J. P. (1990). Sodium–calcium exchange. In *Intracellular Calcium Regulation* (F. Bronner, ed.), Liss, New York, pp. 305–347.
88. Nicoll, D. A., Longoni, S., and Philipson, K. D. (1990). Molecular cloning and functional expression of the cardiac sarcolemmal Na^+–Ca^{2+} exchanger. *Science* **250:**562–565.
89. Reilander, H., Achilles, A., Friedel, U., *et al.* (1992). Primary structure and functional expression of the Na/Ca,K-exchanger from bovine rod photoreceptors. *EMBO J.* **11:**1689–1695.
90. Reeves, J. P. (1992). Molecular aspects of sodium–calcium exchange. *Arch. Biochem. Biophys.* **292:**329–334.
91. Dipolo, R., and Beauge, L. (1991). Regulation of Na–Ca exchange. An overview. *Ann. N.Y. Acad. Sci.* **639:**100–111.
92. Carafoli, E. (1991). Calcium pump of the plasma membrane. *Physiol. Rev.* **71:**129–153.
93. Grover, A. K., and Khan, I. (1992). Calcium pump isoforms: Diversity, selectivity and plasticity. *Cell Calcium* **13:**9–17.
94. Verma, A. K., Filoteo, A. G., Stanford, D. R., *et al.* (1988). Complete primary structure of a human plasma membrane Ca^{2+} pump. *J. Biol. Chem.* **263:**14152–14159.
95. Strehler, E., Strehler-Page, M., Vogel, G., *et al.* (1989). mRNAs for PM calcium pump isoforms differing in their regulatory domain are generated by alternative splicing that involves two internal donor sites in a single exon. *Proc. Natl. Acad. Sci. USA* **86:** 6908–6912.
96. Khan, I., and Gover, A. (1991). Expression of cyclic-nucleotide-sensitive and -insensitive isoforms of the plasma membrane Ca^{2+} pump in smooth muscle and other tissues. *Biochem. J.* **277:** 345–349.
97. James, P. H., Pruschy, M., Vorherr, T. E., *et al.* (1989). Primary structure of the cAMP-dependent phosphorylation site of the plasma membrane calcium pump. *Biochemistry* **28:**4253–4258.
98. Niggli, V., Adunyah, E. S., Penniston, J. T., *et al.* (1981). Purified (Ca^{2+}-Mg^{2+})-ATPase of the erythrocyte membrane. Reconstitution and effect of calmodulin and phospholipids. *J. Biol. Chem.* **256:** 395–401.
99. Kessler, F., Bennardini, F., Bachs, O., *et al.* (1990). Partial purification and characterization of the Ca^{2+}-pumping ATPase of the liver plasma membrane. *J. Biol. Chem.* **265:**16012–16019.
100. Choquette, D., Hakim, G., Filoteo, A. G., *et al.* (1984). Regulation of plasma membrane Ca^{2+} ATPases by lipids of the phosphatidylinositol cycle. *Biochem. Biophys. Res. Commun.* **125:**908–915.
101. Fraser, C. L., and Sarnacki, P. (1992). Regulation of plasma membrane-bound Ca^{2+}-ATPase pump by inositol phosphates in rat brain. *Am. J. Physiol.* **262:**F411–F416.
102. Davis, F. B., Davis, P. J., Lawrence, W. D., *et al.* (1991). Specific inositol phosphates inhibit basal and calmodulin-stimulated Ca^{2+}-ATPase activity in human erythrocyte membranes in vitro and inhibit binding of calmodulin to membranes. *FASEB J.* **5:**2992–2995.
103. Tepikin, A. V., Voronina, S. G., Gallacher, D. V., *et al.* (1992). Pulsatile Ca^{2+} extrusion from single pancreatic acinar cells during receptor-activated cytosolic Ca^{2+} spiking. *J. Biol. Chem.* **267:** 14073–14076.
104. MacLennan, D. H., Brandl, C. J., Korczak, B., *et al.* (1985). Amino-acid sequence of a Ca^{2+}+Mg^{2+}-dependent ATPase from rabbit muscle sarcoplasmic reticulum, deduced from its complementary DNA sequence. *Nature* **316:**696–700.
105. Maruyama, K., and MacLennan, D. H. (1988). Mutation of aspartic acid-351, lysine-352, and lysine-515 alters the Ca^{2+} transport activity of the Ca^{2+}-ATPase expressed in COS-1 cells. *Proc. Natl. Acad. Sci. USA* **85:**3314–3318.
106. Lytton, J., Westlin, M., Burk, S. E., *et al.* (1992). Functional comparisons between isoforms of the sarcoplasmic or endoplasmic reticulum family of calcium pumps. *J. Biol. Chem.* **267:**14483–14489.
107. Burk, S., Lytton, J., MacLennan, D. H., *et al.* (1989). cDNA cloning, functional expression, and mRNA distribution of a third organellar Ca-pump. *J. Biol. Chem.* **264:**18561–18568.
108. Wegener, A. D., Simmerman, K. B., Liepnieks, J., *et al.* (1986). Proteolytic cleavage of phospholamban purified from canine cardiac sarcoplasmic reticulum vesicles. *J. Biol. Chem.* **261:**5154–5159.
109. James, P., Inui, M., Tada, M., *et al.* (1989). Nature and site of phospholamban regulation of the Ca^{2+} pump of sarcoplasmic reticulum. *Nature* **342:**90–92.
110. Eggermont, J. A., Wuytack, F., Verbist, J., *et al.* (1990). Expression of endoplasmic-reticulum Ca^{2+} pump isoforms and of phospholamban in pig smooth-muscle tissues. *Biochem. J.* **271:**649–653.
111. Inesi, G., and Sagara, Y. (1992). Thapsigargin, a high affinity and global inhibitor of intracellular Ca^{2+} transport ATPases. *Arch. Biochem. Biophys.* **298:**313–317.
112. Michell, R. H. (1975). Inositol phospholipids and cell surface receptor function. *Biochim. Biophys. Acta* **415:**81–147.
113. Hokin, M. R., and Hokin, L. E. (1954). Effects of acetylcholine on phospholipides in the pancreas. *J. Biol. Chem.* **209:**549–558.

114. Adbel-Latif, A. A., Akhatar, R., and Hawthorne, J. N. (1977). Acetylcholine increases the breakdown of triphosphoinositide of rabbit iris muscle prelabelled with [^{32}P]phosphate. *Biochem. J.* **162:**61–73.
115. Berridge, M. J. (1983). Rapid accumulation of inositol trisphosphate reveals that agonists hydrolyse polyphosphoinositides instead of phosphatidylinositol. *Biochem. J.* **212:**849–858.
116. Streb, H., Irvine, R. F., Berridge, M. J., *et al.* (1983). Release of Ca^{2+} from nonmitochondrial intracellular store in pancreatic acinar cells by inositol-1,4,5-trisphosphate. *Nature* **306:**67–69.
117. Berridge, M. J. (1993). Inositol trisphosphate and calcium signalling. *Nature* **361:**315–325.
118. Attree, O., Olivos, I. M., Okabe, I., *et al.* (1992). The Lowe's oculocerebrorenal syndrome gene encodes a protein highly homologous to inositol polyphosphate-5-phosphatase. *Nature* **358:**239–242.
119. Worley, P. F., Baraban, J. M., Colvin, J. S., *et al.* (1987). Inositol trisphosphate receptor localization in brain: Variable stoichiometry with protein kinase C. *Nature* **325:**159–161.
120. Worley, P. F., Baraban, J. M., Supattapone, S., *et al.* (1987). Characterization of inositol trisphosphate receptor binding in brain. Regulation by pH and calcium. *J. Biol. Chem.* **262:**12132–12136.
121. Supattapone, S., Worley, P.F., Baraban, J. M., *et al.* (1988). Solubilization, purification, and characterization of an inositol trisphosphate receptor. *J. Biol. Chem.* **263:**1530–1534.
122. Ferris, C. D., Huganir, R. L., Supattapone, S., *et al.* (1989). Purified inositol 1,4,5-trisphosphate receptor mediates calcium flux in reconstituted lipid vesicles. *Nature* **342:**87–89.
123. Furuichi, T., Yoshikawa, S., Miyawaki, A., *et al.* (1989). Primary structure and functional expression of the inositol 1,4,5-trisphosphate-binding protein P400. *Nature* **342:**32–38.
124. Mignery, G. A., Sudhof, T. C., Takei, K., *et al.* (1989). Putative receptor for inositol 1,4,5-trisphosphate similar to ryanodine receptor. *Nature* **342:**192–195.
125. Sudhof, T. C., Newton, C. L., Archer, B. T., *et al.* (1991). Structure of a novel $InsP_3$ receptor. *EMBO J.* **10:**3199–3206.
126. Ross, C. A., Danoff, S. K., Schell, M. J., *et al.* (1992). Three additional inositol 1,4,5-trisphosphate receptors: Molecular cloning and differential localization in brain and peripheral tissues. *Proc. Natl. Acad. Sci. USA* **89:**4265–4269.
127. Nakagawa, T., Okano, H., Furuichi, T., *et al.* (1991). The subtypes of the mouse inositol 1,4,5-trisphosphate receptor are expressed in a tissue-specific and developmentally specific manner. *Proc. Natl. Acad. Sci. USA* **88:**6244–6248.
128. Ferris, C. D., and Snyder, S. H. (1992). Inositol 1,4,5-trisphosphate-activated calcium channels. *Annu. Rev. Physiol.* **54:**469–488.
128a. Michikawa, T., Hamanaka, H., Otsu, H., *et al.* (1994). Transmembrane topology and sites of *N*-glycosylation of inositol 1,4,5-trisphosphate receptor. *J. Biol. Chem.* **269:**9184–9189.
129. Ferris, C. D., Huganir, R. L., Bredt, D. S., *et al.* (1991). Inositol trisphosphate receptor: Phosphorylation by protein kinase C and calcium calmodulin-dependent protein kinases in reconstituted lipid vesicles. *Proc. Natl. Acad. Sci. USA* **88:**2232–2235.
130. Supattapone, S., Danoff, S. K., Theibert, A., *et al.* (1988). Cyclic AMP-dependent phosphorylation of a brain inositol trisphosphate receptor decreases its release of calcium. *Proc. Natl. Acad. Sci. USA* **85:**8747–8750.
131. Danoff, S. K., Ferris, C. D., Donath, C., *et al.* (1991). Inositol 1,4,5-trisphosphate receptors: Distinct neuronal and nonneuronal forms derived by alternative splicing differ in phosphorylation. *Proc. Natl. Acad. Sci. USA* **88:**2951–2955.
131a. Nakade, S., Rhee, S. K., Hamanaka, H., *et al.* (1994). Cyclic AMP-dependent phosphorylation of an immunoaffinity-purified homotetrameric inositol 1,4,5-trisphosphate receptor (type I) increases Ca^{2+} flux in reconstituted lipid vesicles. *J. Biol. Chem.* **269:**6735–6742.
132. Matter, N., Ritz, M.-F., Freyermuth, S., *et al.* (1993). Stimulation of nuclear protein kinase C leads to phosphorylation of nuclear inositol 1,4,5-trisphosphate receptor and accelerated calcium release by inositol 1,4,5-trisphosphate from isolated rat liver nuclei. *J. Biol. Chem.* **268:**732–736.
133. Burgess, G. M., Bird, G. St. J., Obie, J. F., *et al.* (1991). The mechanism for synergism between phospholipase C- and adenylylcyclase-linked hormones in liver. *J. Biol. Chem.* **266:**4772–4781.
134. Volpe, P., and Anderson-Lang, B. H. (1990). Regulation of inositol 1,4,5-trisphosphate-induced Ca^{2+} release. II. Effect of cAMP-dependent protein kinase. *Am. J. Physiol.* **258:**C1086–C1091.
135. Ferris, C. D., Cameron, A. M., Bredt, D. S., *et al.* (1992). Autophosphorylation of inositol 1,4,5-trisphosphate receptors. *J. Biol. Chem.* **267:**7036–7041.
136. Ferris, C. D., Huganir, R. L., and Snyder, S. H. (1990). Calcium flux mediated by purified inositol 1,4,5-trisphosphate receptor in reconstituted lipid vesicles is allosterically regulated by adenine nucleotides. *Proc. Natl. Acad. Sci. USA* **87:**2147–2151.
137. Maeda, N., Kawasaki, T., Nakade, S., *et al.* (1991). Structural and functional characterization of inositol 1,4,5-trisphosphate receptor channel from mouse cerebellum. *J. Biol. Chem.* **266:**1109–1116.
138. Worley, P. F., Baraban, J. M., Supattapone, S., *et al.* (1987). Characterization of inositol trisphosphate receptor binding in brain. Regulation by pH and calcium. *J. Biol. Chem.* **262:**12132.
139. Danoff, S. K., Supattapone, S., and Snyder, S. H. (1988). Characterization of a membrane protein from brain mediating the inhibition of inositol 1,4,5-trisphosphate receptor binding by calcium. *Biochem. J.* **254:**701–705.
140. Mignery, G. A., Johnston, P. A., and Südhof, T. C. (1992). Mechanism of Ca^{2+} inhibition of inositol 1,4,5-trisphosphate ($InsP_3$) binding to the cerebellar $InsP_3$ receptor. *J. Biol. Chem.* **267:**7450–7455.
141. Joseph, S. K., Rice, H. L., and Williamson, J. R. (1989). The effect of external calcium and pH on inositol trisphosphate-mediated calcium release from cerebellum microsomal fractions. *Biochem. J.* **258:**261–265.
142. Busa, W. B., and Nuccitelli, R. (1984). Metabolic regulation via intracellular pH. *Am. J. Physiol.* **246:**R409–R438.
143. Bourguignon, L. Y. W., Jin, H., Iida, N., *et al.* (1993). The involvement of ankyrin in the regulation of inositol 1,4,5-trisphosphate receptor-mediated internal Ca^{2+} release from Ca^{2+} storage vesicles in mouse T-lymphoma cells. *J. Biol. Chem.* **268:**7290–7297.
144. Muallem, S., Pandol, S. J., and Beeker, T. G. (1989). Hormone-evoked calcium release from intracellular stores is a quantal process. *J. Biol. Chem.* **264:**205–212.
145. Ferris, C. D., Cameron, A. M., Huganir, R. L., *et al.* (1992). Quantal calcium release by purified reconstituted inositol 1,4,5-trisphosphate receptors. *Nature* **356:**350–352.
146. Irvine, R. F. (1990). 'Quantal' Ca^{2+} release and the control of Ca^{2+} entry by inositol phosphates—A possible mechanism. *FEBS Lett.* **263:**5–9.
147. Missiaen, L., De Smedt, H., Droogmans, G., *et al.* (1992). Ca^{2+} release induced by inositol 1,4,5-trisphosphate is a steady-state phenomenon controlled by luminal Ca^{2+} in permeabilized cells. *Nature* **357:**599–602.
148. Missiaen, L., Taylor, C. W., Berridge, M. J. (1992). Luminal Ca^{2+} promoting spontaneous Ca^{2+} release from inositol trisphosphate-sensitive stores in rat hepatocytes. *J. Physiol. (London)* **455:**623–640.
149. Sayers, L. G., Brown, G. R., Michell, R. H., *et al.* (1993). The effects of thimerosal on calcium uptake and inositol 1,4,5-trisphosphate-induced calcium release in cerebellar microsomes. *Biochem. J.* **289:**883–887.

150. Shuttleworth, T. H. (1992). Ca^{2+} release from inositol trisphosphate-sensitive stores is not modulated by intraluminal [Ca^{2+}]. *J. Biol. Chem.* **267:**3573–3576.
151. Satoh, T., Ross, C. A., Villa, A., *et al.* (1990). The inositol 1,4,5-trisphosphate receptor in cerebellar Purkinje cells: Quantitative immunogold labeling reveals concentration in an ER subcompartment. *J. Cell Biol.* **111:**615–624.
152. Ross, C. A., Meldolesi, J., Milner, T. A., *et al.* (1989). Inositol 1,4,5-trisphosphate receptor localized to endoplasmic reticulum in cerebellar Purkinje neurons. *Nature* **339:**468–470.
153. Kume, S., Muto, A., Aruga, J., *et al.* (1993). The Xenopus IP_3 receptor: Structure, function, and localization in oocytes and eggs. *Cell* **73:**555–570.
154. Divecha, N., Banfic, H., and Irvine, R. F. (1991). The polyphosphoinositide cycle exists in the nuclei of Swiss 3T3 cells under the control of a receptor (for IGF-I) in the plasma membrane, and stimulation of the cycle increases nuclear diacylglycerol and apparently induces translocation of protein kinase C to the nucleus. *EMBO J.* **10:**3207–3214.
155. Sullivan, K. M. C., Busa, W. B., and Wilson, K. L. (1993). Calcium mobilization is required for nuclear vesicle fusion in vitro: Implications for membrane traffic and IP_3 receptor function. *Cell* **73:** 1411–1422.
156. Kuno, M., and Gardener, P. (1987). Ion channels activated by inositol 1,4,5-trisphosphate in plasma membrane of human T-lymphocytes. *Nature* **326:**301–304.
157. Khan, A. A., Steiner, J. P., Klein, M. G., *et al.* (1992). IP_3 receptor: Localization to plasma membrane of T cells and cocapping with the T cell receptor. *Science* **257:**815–818.
158. Fadool, D. A., and Ache, B. W. (1992). Plasma membrane inositol 1,4,5-trisphosphate-activated channels mediate signal transduction in lobster olfactory receptor neurons. *Neuron* **9:**907–918.
159. Penner, R., Matthews, G., and Neher, E. (1988). Regulation of calcium influx by second messengers in rat mast cells. *Nature* **334:**499–504.
160. Volpe, P., Krause, K. H., Hashimoto, S., *et al.* (1988). "Calciosome," a cytoplasmic organelle: The inositol 1,4,5-trisphosphate-sensitive Ca^{2+} store of nonmuscle cells?. *Proc. Natl. Acad. Sci. USA* **85:**1091–1095.
161. Lew, D. P. (1989). Receptor signalling and intracellular calcium in neutrophil activation. *Eur. J. Clin. Invest.* **19:**338–346.
162. Volpe, P., Pozzan, T., and Meldolisi, J. (1990). Rapidly exchanging Ca^{2+} stores of non-muscle cells. *Semin. Cell Biol.* **1:**297–304.
163. Villa, A., Podini, P., Clegg, D. O., *et al.* (1991). Intracellular Ca^{2+} stores in chicken Purkinje neurons: differential distribution of the low affinity–high capacity Ca^{2+} binding protein, calsequestrin, of Ca^{2+} ATPase and of the ER lumenal protein, Bip. *J. Cell Biol.* **113:**779–791.
164. Sitia, R., and Meldolesi, J. (1992). Endoplasmic reticulum: A dynamic patchwork of specialized subregions. *Mol. Biol. Cell* **3:** 1067–1072.
165. Walton, P. D., Airey, J. A., Sutko, J. L., *et al.* (1991). Ryanodine and inositol trisphosphate receptors coexist in avian cerebellar Purkinje neurons. *J. Cell Biol.* **113:**1145–1157.
166. Volpe, P., Villa, A., Damiani, E., *et al.* (1991). Heterogeneity of microsomal Ca^{2+} stores in chicken Purkinje neurons. *EMBO J.* **10:** 3183–3189.
167. Fill, M., and Coronado, R. (1988). Ryanodine receptor channel of sarcoplasmic reticulum. *Trends Neurosci.* **11:**453–457.
168. MacLennan, D. H., and Phillips, M. S. (1992). Malignant hyperthermia. *Science* **256:**789–794.
169. Takeshima, H., Nishimura, S., Matsumoto, T., *et al.* (1989). Primary structure and expression from complementary DNA of skeletal muscle ryanodine receptor. *Nature* **339:**439–445.
170. Otsu, K., Willard, H. F., Khanna, V. K., *et al.* (1990). Molecular cloning of cDNA encoding the Ca^{2+} release channel (ryanodine receptor) of rabbit cardiac muscle sarcoplasmic reticulum. *J. Biol. Chem.* **265:**13472–13483.
171. Giannini, G., Clementi, E., Ceci, R., *et al.* (1992). Expression of a ryanodine receptor-Ca^{2+} channel that is regulated by TGF-β. *Science* **257:**91–94.
172. Sorrentino, V., and Volpe, P. (1993). Ryanodine receptors: How many, where and why? *Trends Pharmacol. Sci.* **14:**98–103.
173. Hakamata, Y., Nakai, J., Takeshima, H., *et al.* (1992). Primary structure and distribution of a novel ryanodine receptor/calcium release channel from rabbit brain. *FEBS Lett.* **312:**229–235.
174. Fabiato, A. (1989). Appraisal of the physiological relevance of two hypotheses for the mechanism of calcium release from the mammalian cardiac sarcoplasmic reticulum: Calcium-induced release versus charge-coupled release. *Mol. Cell. Biochem.* **89:**135–140.
175. Hymel, L., Schindler, H., Inui, M., *et al.* (1988). Reconstitution of purified cardiac muscle calcium release channel (ryanodine receptor) in planar bilayers. *Biochem. Biophys. Res. Commun.* **152:**308–314.
176. Gyorke, S., and Fill, M. (1993). Ryanodine receptor adaptation: Control mechanism of Ca^{2+}-induced Ca^{2+} release in heart. *Science* **260:**807–809.
177. Jayaraman, T., Brillantest, A.-M., Timerman, A. P., *et al.* (1992). FK506 binding protein associated with the calcium release channel (ryanodine receptor). *J. Biol. Chem.* **267:**9474–9477.
178. Takasago, T., Imagawa, T., Furukawa, K., *et al.* (1991). Regulation of the cardiac ryanodine receptor by protein kinase-dependent phosphorylation. *J. Biochem.* **109:**163–170.
179. Wang, J., and Best, P. M. (1992). Inactivation of the sarcoplasmic reticulum calcium channel by protein kinase. *Nature* **359:**739–741.
180. Sabbadini, R. A., Betto, R., Teresi, A., *et al.* (1992). The effects of sphingosine on sarcoplasmic reticulum membrane calcium release. *J. Biol. Chem.* **267:**15475–15484.
181. Clapper, D. L., Walseth, T. F., Dargie, P. J., *et al.* (1987). Pyridine nucleotide metabolites stimulate calcium release from sea urchin egg microsomes desensitized to inositol trisphosphate. *J. Biol. Chem.* **262:**9561–9568.
182. Koshiyama, H., Lee, H. C., and Tashjian, A. H., Jr. (1991). Novel mechanism of intracellular calcium release in pituitary cells. *J. Biol. Chem.* **266:**16985–16988.
183. Takasawa, S., Nata, K., Yonekura, H., *et al.* (1993). Cyclic ADP-ribose in insulin secretion from pancreatic β cells. *Science* **259:**370–373.
184. Galione, A., Lee, H. C., and Busa, W. B. (1991). Ca^{2+}-induced Ca^{2+} release in sea urchin egg homogenates: Modulation by cyclic ADP-ribose. *Science* **253:**1143–1146.
185. Lee, H. C. (1993). Potentiation of calcium- and caffeine-induced calcium release by cyclic ADP-ribose. *J. Biol. Chem.* **268:**293–299.
186. Meszaros, L. G., Bak, J., and Chu, A. (1993). Cyclic ADP-ribose as an endogenous regulator of the non-skeletal type ryanodine receptor Ca^{2+} channel. *Nature* **364:**76–79.
187. Galione, A., White, A., Willmott, N., *et al.* (1993). cGMP mobilizes intracellular Ca^{2+} in sea urchin eggs by stimulating cyclic ADP-ribose synthesis. *Nature* **365:**456–459.
188. Campbell, K. P. (1986). Protein components and their roles in sarcoplasmic reticulum function. In *Sarcoplasmic Reticulum in Muscle Physiology* (M. L. Entman and W. B. Van Winkle, eds.), CRC Press, Boca Raton, Volume 1, pp. 65–99.
189. Ikemoto, N., Ronjat, M., Mészáros, L., *et al.* (1989). Postulated role of calsequestrin in the regulation of calcium release from sarcoplasmic reticulum. *Biochemistry* **28:**6764–6771.
190. Gilchrist, J. S. C., Belcastro, A. N., and Katz, S. (1992). Intraluminal Ca^{2+} dependence of Ca^{2+} and ryanodine-mediated regulation of skeletal muscle sarcoplasmic reticulum Ca^{2+} release. *J. Biol. Chem.* **267:**20850–20856.

191. Ikemoto, N., Antoniu, B., Kang, J.-J., *et al.* (1991). Intravesicular calcium transient during calcium release from sarcoplasmic reticulum. *Biochemistry* **30:**5230–5237.
192. Biral, D., Volpe, P., Damiani, E., *et al.* (1992). Coexistence of two calsequestrin isoforms in rabbit slow-twitch skeletal muscle fibers. *FEBS Lett.* **299:**175–178.
193. Lebeche, D., and Kaminer, B. (1992). Characterization of a calsequestrin-like protein from sea-urchin eggs. *Biochem. J.* **287:**741–747.
194. Missiaen, L., Taylor, C. W., and Berridge, M. J. (1991). Spontaneous calcium release from inositol trisphosphate-sensitive calcium stores. *Nature* **352:**241–244.
195. Irvine, R. (1990). 'Quantal' Ca^{2+} release and the control of Ca^{2+} entry by inositol phosphates—a possible mechanism. *FEBS Lett.* **263:**5–9.
196. Benzotte, R. B., Pereira, B., and Higham, S. (1991). Effects of ryanodine on calcium sequestration in the rat liver. *Biochem. Pharmacol.* **42:**1799–1803.
197. Parys, J. B., Sernett, S. W., Delisle, S., *et al.* (1992). Isolation, characterization, and localization of the inositol 1,4,5-trisphosphate receptor protein in *Xenopus laevis* oocytes. *J. Biol. Chem.* **267:**18776–18782.
198. Busa, W. B., Ferguson, J. E., Joseph, S. K., *et al.* (1985). Activation of frog (*Xenopus laevis*) eggs by inositol trisphosphate. I. Characterization of Ca^{2+} release from intracellular stores. *J. Cell Biol.* **101:**677–682.
199. Lechleiter, J. D., and Clapham, D. E. (1992). Molecular mechanisms of intracellular calcium excitability in X. laevis oocytes. *Cell* **69:**283–294.
200. Galione, A., McDougall, A., Busa, W. B., *et al.* (1993). Redundant mechanisms of calcium-induced calcium release underlying calcium waves during fertilization of sea urchin eggs. *Science* **261:**348–352.
201. Gilkey, J. C., Jaffe, L. F., Ridgway, E. B., *et al.* (1978). A free calcium wave traverses the activating egg of the medaka, *Oryzias latipes*. *J. Cell Biol.* **76:**448–466.
202. Busa, W. B., and Nuccitelli, R. (1985). An elevated free cytosolic Ca^{2+} wave follows fertilization in eggs of the frog, *Xenopus laevis*. *J. Cell Biol.* **100:**1325–1329.
203. Wier, W. G., Cannell, M. B., Berlin, J. R., *et al.* (1987). Cellular and subcellular heterogeneity of $[Ca^{2+}]_i$ in single heart cells revealed by fura-2. *Science* **235:**325–328.
204. O'Sullivan, A. J., Cheek, T. R., Moreton, R. B., *et al.* (1989). Localization and heterogeneity of agonist-induced changes in cytosolic calcium concentration in single bovine adrenal chromaffin cells from video imaging of fura-2. *EMBO J.* **8:**401–411.
205. Cornell-Bell, A. H., Finkbeiner, S. M., Cooper, M. S., *et al.* (1990). Glutamate induces calcium waves in cultured astrocytes: Long-range glial signaling. *Science* **247:**470–473.
206. Nathanson, M. H., Padfield, P. J., O'Sullivan, A. J., *et al.* (1992). Mechanism of Ca^{2+} wave propagation in pancreatic acinar cells. *J. Biol. Chem.* **267:**18118–18121.
207. Rawlings, S. R., Berry, D. J., and Leong, D. A. (1991). Evidence for localized calcium mobilization and influx in single rat gonadotropes. *J. Biol. Chem.* **266:**22755–22760.
208. Rooney, T. A., Sass, E. J., and Thomas, A. P. (1990). Agonist-induced cytosolic calcium oscillations originate from a specific locus in single hepatocytes. *J. Biol. Chem.* **265:**10792–10796.
209. Whitaker, M. J., and Irvine, R. F. (1984). Inositol 1,4,5-trisphosphate microinjection activates sea urchin eggs. *Nature* **312:**636–639.
210. Berridge, M. J. (1990). Calcium oscillations. *J. Biol. Chem.* **265:** 9583–9586.
211. Meyer, T., and Stryer, L. (1991). Calcium spiking. *Annu. Rev. Biophys. Biophys. Chem.* **20:**153–174.
212. Lee, H. C., Aarhus, R., and Walseth, T. F. (1993). Calcium mobilization by dual receptors during fertilization of sea urchin eggs. *Science* **261:**352–355.
213. Han, J. K., Fukami, K., and Nuccitelli, R. (1992). Reducing inositol lipid hydrolysis, $Ins(1,4,5)P_3$ receptor availability, or Ca^{2+} gradients lengthens the duration of the cell cycle in Xenopus laevis blastomeres. *J. Cell Biol.* **116:**147–156.
214. Larabell, C., and Nuccitelli, R. (1992). Inositol lipid hydrolysis contributes to the Ca^{2+} wave in the activating egg of Xenopus laevis. *Dev. Biol.* **153:**347–355.
215. Jaffe, L. F. (1991). The path of calcium in cytosolic calcium oscillations: A unifying hypothesis. *Proc. Natl. Acad. Sci. USA* **88:** 9883–9887.
216. Berridge, M. J., and Galione, A. (1988). Cytosolic calcium oscillators. *FASEB J.* **2:**3074–3082.
217. Lechleiter, J., Girard, S., Peralta, E., *et al.* (1991). Spiral calcium wave propagation and annihilation in Xenopus laevis oocytes. *Science* **252:**123–126.
218. Tsien, R. W., and Tsien, R. Y. (1990). Calcium channels, stores, and oscillations. *Annu. Rev. Cell Biol.* **6:**715–760.
219. Kline, D., and Kline, J. T. (1992). Repetitive calcium transients and the role of calcium in exocytosis and cell cycle activation in the mouse egg. *Dev. Biol.* **149:**80–89.
220. Miyazaki, S., Hashimoto, N., Yoshimoto, Y., *et al.* (1986). Temporal and spatial dynamics of the periodic increase in intracellular free calcium at fertilization of golden hamster eggs. *Dev. Biol.* **118:**259–267.
221. Sun, F. Z., Hoyland, J., Haung, X., *et al.* (1992). A comparison of intracellular changes in porcine eggs after fertilization and electroactivation. *Development* **115:**947–956.
222. Fissore, R. A., Dobrinksy, J. R., Balise, J. J., *et al.* (1992). Patterns of intracellular Ca^{2+} concentrations in fertilized bovine eggs. *Biol. Reprod.* **47:**960–969.
223. Miyazaki, S., Shirakawa, H., Nakada, K., *et al.* (1993). Essential role of the inositol 1,4,5-trisphosphate receptor/Ca^{2+} release channel in Ca^{2+} waves and Ca^{2+} oscillations at fertilization of mammalian eggs. *Dev. Biol.* **158:**62–78.
224. Swann, K., Igusa, Y., and Miyazaki, S. (1989). Evidence for an inhibitory effect of protein kinase C on G-protein-mediated repetitive calcium transients in hamster eggs. *EMBO J.* **8:**3711–3718.
225. Berridge, M. J., Downes, C. P., and Hanley, M. R. (1989). Neural and developmental actions of lithium: A unifying hypothesis. *Cell* **59:**411–419.
226. Mattson, M. P., Rydel, R. E., Lieberburg, I., *et al.* (1993). Altered calcium signaling and neuronal injury: Stroke and Alzheimer's disease as examples. *Ann. N.Y. Acad. Sci.* **679:**1–21.
227. Howard, M., Grimaldi, J. C., Bazan, J. F., *et al.* (1993). Formation and hydrolysis of cyclic ADP-ribose catalyzed by lymphocyte antigen CD38. *Science* **262:**1056–1059.

22

Sodium Transport by Epithelial Cells

Lawrence G. Palmer

22.1. INTRODUCTION

The Koefoed-Johnsen and Ussing model for Na transport by the frog skin is now more than 35 years old. One of the purposes of this chapter is to pay homage to this simple but brilliant insight into how epithelial cells work. In addition, the model will be updated in terms of the transporters involved in the Na reabsorption process. As we shall see, this is not really a refinement of the model, since most of these transporters are precisely those postulated in the original paper.[1] We do have, however, a great deal more information on their functional and molecular properties. Another goal of this chapter will be to review the regulation of the individual components of the system, and of the system as a whole. A final topic is the question of which ions are transported along with Na or in exchange for Na to preserve electroneutrality, and what mechanisms are involved in this transport.

22.1.1. The Koefoed-Johnsen and Ussing Model

The Koefoed-Johnsen and Ussing (KJ-U) model involves three essential transporters (Figure 22.1). Two of these confer ion permeability on the plasma membranes; they always facilitate ion flow in the direction predicted by an electrochemical activity gradient. In modern language these transporters are called channels. The apical plasma membrane (outer membrane of the frog skin, mucosal membrane of the urinary bladder or GI tract, luminal membrane of the renal tubule) contains Na-selective channels and is selectively permeable for Na over K. The basolateral membrane (inner membrane of the skin, serosal membrane of the bladder or GI tract, contraluminal membrane of the renal tubule) contains K channels and is selectively permeable for K over Na. This difference in the ion selectivity of the two membranes was the essential experimental finding underlying the model. The third component is an active transport system or ion pump that moves Na out of the cell and K into the cell. Both of these fluxes are in directions opposite to those predicted from electrochemical activities and are driven by metabolic energy. The Na/K exchange pump is assumed to be exclusively in the inner membrane. Thus, Na entering the cells across the outer membrane is pumped out across the inner, resulting in the net absorption of the ion. K pumped into the cell is recycled across the inner membrane, so that there will be little or no net movement of this ion.

22.1.2. Applicability of the Model

One of the strengths of the KJ-U model is its wide applicability to a variety of epithelia. The three transporters described above are present in most if not all vertebrate epithelia that absorb Na and that have high-resistance tight junctions between the cells. A high resistance is usually defined arbitrarily as one of 100 Ωcm^2 or more.[2] The amphibian skin is the prototype of these epithelia. Also included in this category are distal nephron segments of the kidney (distal tubule, collecting duct) and urinary bladder, the colon, the airways, and the ducts of salivary and sweat glands.[3–5] Although all of these organs make use of the basic KJ-U scheme, they differ considerably with respect to other ion transport systems. These other systems will determine in large part whether Na^+ is reabsorbed along with Cl^- (or HCO^-_3) or in exchange for another cation such as K^+ or H^+. This will be discussed in more detail below.

In addition to these epithelia in which the most literal form of the KJ-U system is used, there are others that reabsorb Na using variations on this basic theme. These are mostly the low-resistance or "leaky" epithelia such as the proximal renal tubule, the small intestine, and the gallbladder. Here the major variations are in the mechanisms of transport of Na across the outer membrane.[6,7] Instead of a channel-mediated conduction of Na^+ ions, Na^+ crosses the membrane by an obligatory exchange for H^+, or by an obligatory coupling with the influx of other inorganic ions (Cl^-, PO^-_4) or organic solutes (sugars, amino acids, bile salts). The essential steps at the basolateral membrane are the same: extrusion of Na in exchange for K by the energy-dependent pump and recycling of K through channels.

Thus, the key features of the model, which are preserved in many epithelia, are (1) the asymmetric distribution of transporters to the apical and basolateral membranes, (2) the coupling of transport to metabolic energy through the Na/K pump

Lawrence G. Palmer • Department of Physiology and Biophysics, Cornell University Medical College, New York, New York 10021.
Molecular Biology of Membrane Transport Disorders, edited by S.G. Schultz *et al.* Plenum Press, New York, 1996

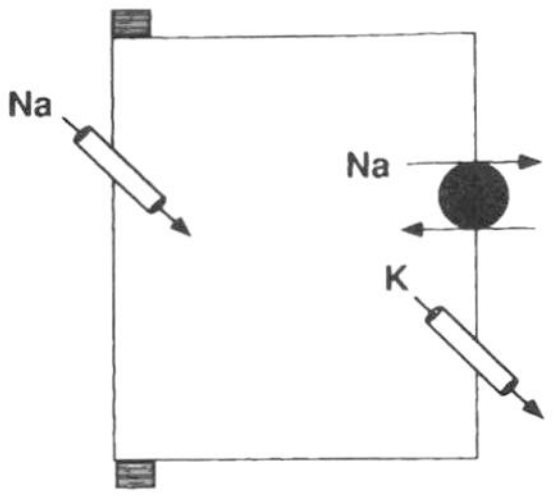

Figure 22.1 Basic model of a Na-transporting epithelium after Koefoed-Johnsen and Ussing. The essential features of the model are the Na-selective channels in the apical membrane, and the Na,K pump and K-selective channels in the basolateral membrane.

or ATPase in the basolateral membrane, (3) the movement of Na into the cell through specific channels or transporters, driven by the low electrochemical activity for Na in the cytoplasm, and (4) the recycling of K through ion channels in the basolateral membrane. In the next section the basic system will be considered in more detail.

22.2. THE COMPONENTS OF THE SYSTEM

22.2.1. The Na/K Pump

The nature and mechanism of transport by the Na/K pump has been the subject of a number of recent reviews[8–12] and is described in detail in another chapter of this volume (see Chapter 12). In this chapter, only those aspects of the pump that pertain particularly to epithelial transport will be examined.

22.2.1.1. Pump Proteins

The basis for the active transport of Na and K across the basolateral membrane of epithelial cells is the Na/K-ATPase enzyme. The enzymatic activity was first described in 1957 by J. C. Skou; the corresponding protein was first purified from the kidney by Jorgensen.[10,11] It was found to be composed of two subunits, termed α and β, which are sufficient for both enzyme activity and ATP-dependent transport of Na and K. The structure and function of these isoforms is discussed elsewhere in this volume (Chapter 12). Molecular cloning of the subunits documented the presence of a number of isoforms of the α subunit which had been suspected on the basis of inhibitor sensitivity and other differences between the enzymes from different tissues.[12]

The Na/K-ATPase in renal epithelia is predominantly the α_1 isoform, which is indeed sometimes called the kidney isoform. Some studies, however, have indicated the presence of other isoforms.[12] The α_1 is generally the isoform least sensitive to inhibition by cardiotonic steroids such as ouabain. It is possible that this may have physiological significance in that sensitivity to endogenous ouabainlike compounds may permit differential regulation of the enzyme in different tissues.[13] Hormonal regulation of the isoforms can also be different.[12]

22.2.1.2. Polarized Distribution of Pumps

An essential element of transepithelial Na reabsorption is the asymmetrical distribution of pumps to the basolateral membrane. Indeed, the Na/K-ATPase protein is restricted to the basolateral surface of most epithelial cells.[14]

The basis for this polarization is controversial.[15] Caplan *et al.*[16] reported that all of the newly synthesized pumps in a cultured kidney cell line (MDCK) appeared directly at the basolateral surface of the cell. This suggests that the pumps are targeted to the correct surface from the Golgi apparatus, which is the site of final processing of the protein. This implies the existence of a specific sorting signal somewhere within the protein, which directs it to the proper compartment within the cell, although the nature of this signal is unknown. On the other hand, Hammerton *et al.*[17] found that pumps could be delivered equally well to both apical and basolateral membranes. However, those in the basolateral membrane remained in the membrane for much longer times. This group proposed that the polarized distribution of the pump was attributable, at least in part, to its stabilization at the basolateral surface by interaction of the protein with cytoskeleton. In this view the polarization of the cytoskeleton determines the steady-state distribution of this membrane protein. It is unclear to what extent the differences in these findings reflect technical problems or different strategies by different cells for maintaining epithelial polarity.[15]

22.2.1.3. Stoichiometry of the Transport Reaction

The consensus view of the transport reaction catalyzed by the pump is that under reasonably physiological conditions the stoichiometry is 3Na:2K:1ATP.[8] This is thought to represent the normal reaction cycle of the pump. The pump can also mediate Na–Na exchange and K–K exchange. These operations, which do not require ATP hydrolysis, occur at low rates and will not normally affect stoichiometry measurements. They are revealed when the normal cycle is interrupted, as in the complete absence of Na or K or when the ADP/ATP ratio is high.

The stoichiometry of the normal cycle has been confirmed in some experiments on intact epithelia. In turtle colon, net fluxes of Na and K through the pump were found in a ratio of approximately 3:2.[18] In the rabbit urinary bladder, the kinetics of the pump were studied after removal of the apical membrane permeability barrier using the pore-forming antibiotic nystatin.[19] Activation of the pump by intracellular Na and extracellular K could best be described by cooperative binding of 3 Na and 2 K ions. A similar approach was used to study pump properties in the toad urinary bladder.[20] Here the ratio of short-circuit current (net charge transported) to net Na transport through the pump was approximately 1:3, again consistent with a 3Na:2K stoichiometry. Measurements of changes in intracellular Na and extracellular K in proximal tubular suspensions after activation of the pump indicated a 3:2 ratio over a more than fivefold range of flux values.[21]

Other studies have indicated that the pump stoichiometry can be variable and different from 3:2. In frog skin, the ratio of short-circuit current (net Na transport) to ouabain-sensitive K influx across the basolateral membrane of the frog skin was found to increase with increasing transport rates and could be as high as 9:1.[22] Similar elevations in the apparent pump stoichiometry during stimulation of transport were also observed in amphibian proximal tubule.[23] In the rabbit CCT, Na transport rates were compared with the pump current across the basolateral membrane estimated using an equivalent circuit model.[24] Stoichiometries were 3:2 with normal animals and 3:1 when the animals were treated with a mineralocorticoid to elevate the transport rate. So far these deviations from the expected stoichiometry of the pump remain unexplained.

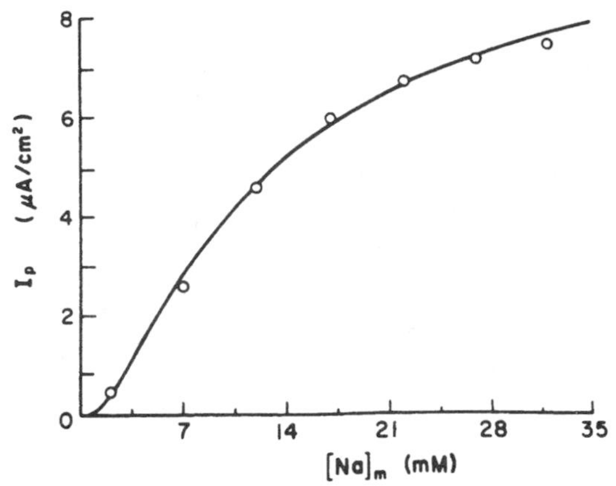

Figure 22.2 Activation of the Na,K pump in turtle colon. The apical membrane was permeabilized with nystatin and the various concentrations of Na were added to the mucosal medium. The increase in current was attributed to the operation of the electrogenic Na pump. The data were fit according to Eq. (22.2) with K_{Na} = 4mM, I_{max} = 11 $\mu A/cm^2$, and n = 3. Reproduced from *The Journal of General Physiology,* Volume 82, pp. 315–329 by copyright permission of The Rockefeller University Press.

22.2.1.4. Activation by Ions

The relationship between pump activity and intracellular Na activity is important in epithelial transport because it determines how well the pump will respond to changes in the rate of Na entry into the cell across the apical membrane. Studies of the pure enzyme indicated that when the K concentration is high, the ATPase is half-activated at a concentration of 37 mM Na.[10] Similar results were obtained in membrane preparations of RBC (24 mM[25]), kidney (~70 mM[26]), and brain (~65 mM[27]). The Na concentration required for half-maximal activation increases with increasing K concentration, indicating a competitive interaction between intracellular K and Na.

In intact cells the $K_{1/2}$ value for activation of active Na transport by intracellular Na is somewhat lower. Typical values in a variety of tissues range from 10 to 20 mM.[8] Numbers for this parameter obtained in Na-transporting epithelia also fall within this range. In nystatin-permeabilized tissues, $K_{1/2}$ was estimated to be 14 mM in rabbit urinary bladder,[19] 15 mM in turtle colon[28] and 15–20 mM in toad urinary bladder.[20] These lower values of $K_{1/2}$ for Na are close to the physiological range of intracellular Na concentrations of 10–20 mM. This implies that the pump runs at about half speed under physiological conditions.

The shape of the activation curve is sigmoid rather than hyperbolic, presumably because of the requirement for binding of three Na ions for activation (Figure 22.2). This can be analyzed in terms of a cooperative interaction:

$$J = J_{max}/[1 + (K_{Na}'/Na_i)^3] \qquad (22.1)$$

or by assuming three independent sites:

$$J = J_{max}/[1 + (K_{Na}'/Na_i)]^3 \qquad (22.2)$$

where Na_i is the intracellular Na concentration, J_{max} is the maximum pump flux, and K_{Na}' is the apparent binding constant for Na_i.[19] In the first case, $K_{1/2} = K_{Na}'$. In the second case, $K_{1/2} = 0.26 \times K_{Na}'$. In both cases the relationship between pump rate and Na_i is steeper than that predicted for a simple one-site Michaelis–Menten-type activation curve in the concentration range at or below $K_{1/2}$. This is especially true of the cooperative activation curve. For example, for $K_{Na} = K_{1/2}$ = 15 mM, and Na_i of 10 mM, 50% below K_{Na}, the pump rate will be only 23% of maximum. This provides a strong ability of the pump to respond to changes in cell Na by increasing its turnover rate. This will be considered again below.

Activation of the pump by extracellular K is half-maximal at around 1–2 mM K both in purified enzyme[10] and in intact cells.[8] Although the pump may not be completely saturated at normal external K concentrations, changes in extracellular K in the physiological range will not have strong effects on the turnover rate.

22.2.2. Na Channels

The major route for Na entry across the apical membrane of high-resistance, reabsorbing epithelia is through Na-selective channels. A hallmark of these channels is their sensitivity to the pyrazine diuretic amiloride. Several types of Na channels have been observed.[4] However, the predominant type that is expressed in most of these tissues is the high-selectivity, low-conductance channel designated Na(5). The properties of this channel will be reviewed below. For further information a number of recent reviews may be consulted.[3,29–32]

22.2.2.1. Selectivity

The ion selectivity of Na channels has been most extensively studied in frog skin and toad urinary bladder (for review and references see Ref. 33). The conductance of the channels to K is less than 1% that to Na, and neither Rb nor Cs passes through the pore to a measurable extent. Of the alkali metal cations, only Na and Li have appreciable conductivities; that to Li is 30 to 50% higher than that to Na. There is also evidence that protons can pass through the channel. A variety of small organic cations were also examined, but none of these could be shown to be permeant. The permeability of water through this pore also appears to be small.

Altogether, these findings suggest that the epithelial Na channel is much more selective than most channels that conduct Na, including voltage-gated Na channels. The ability to discriminate between Na and K is comparable to that of highly selective K channels.[34] The most likely explanation is that unlike the voltage-gated Na channels the epithelial Na channel conducts a completely dehydrated Na ion through a region of the pore that is just wide enough to accept Na and Li but too narrow for K, Rb, etc. Thus, the channel can select ions by size. In addition to this molecular sieving, there may also be selective interactions of ions with the walls of the pore.

22.2.2.2. Conductance

Patch-clamp studies of several Na-reabsorbing epithelia including rat CCT, cultured A6 cells, and toad urinary bladder have revealed the single-channel properties of the epithelial Na channels.[9,30] The single-channel conductance when inward current is carried by Na at room temperature is 4 to 5 pS (Figure 22.3). This is in good agreement with earlier measurements with noise analysis where single-channel currents were estimated from blocker-induced noise in a number of high-resistance epithelia.[35] The conductance is increased to about 9 pS at 37°C and to 7 pS when Li is the conducted ion. The conductance is about 13 pS when Li is conducted at 37°C. Thus, the Q_{10} for conductance through the channel is about 1.5. This is somewhat higher than that for simple diffusion through water (1.3)[34] but smaller than that expected for an enzymatic process.

The conductance of the channel saturates with increasing concentrations of either Na or Li. In the rat CCT, the single-channel currents at constant voltage increase with an apparent K_m for Na of 25 mM.[36] The apparent K_m for Li is 50 mM, suggesting that Li interacts with the pore in a qualitatively similar but quantitatively different way than does Na. These measurements are consistent with earlier findings indicating that the Na transport rate by several model epithelia is a saturable function of external or mucosal Na concentration, with apparent K_m values of 10 to 30 mM.[3] At least part of this limit on the rate of transport can therefore be attributed to the saturation of individual channels. Other mechanisms may also contribute and these are discussed below.

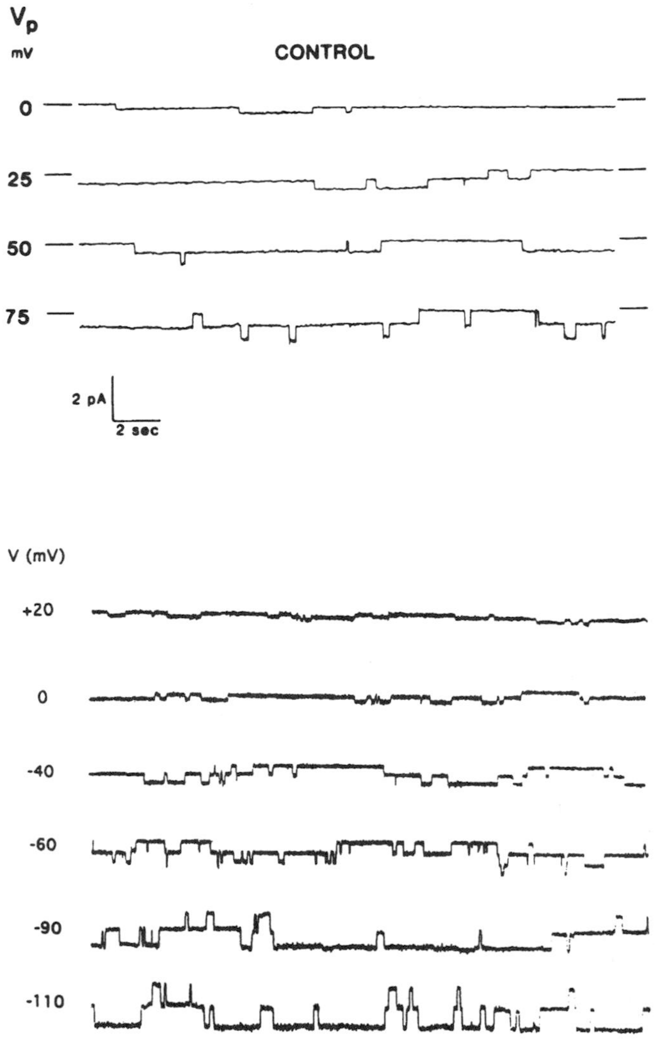

Figure 22.3 Patch-clamp records of epithelial Na channels. (Top) From the rat CCT (from Palmer and Frindt, 1986[40]). (Bottom) From oocytes injected with RNA obtained from the Na channel subunit clones α, β, and γrENaC. Reproduced with permission from *Nature*, Volume 367, pp. 463–467. Copyright 1994 Macmillan Magazines Limited.[52]

This saturation phenomenon implies the interaction of Na ions with sites within the pore. This is not surprising given the apparent narrow pore dimensions as discussed above. In some channels, ions can bind to several sites within the pore and move in single-file fashion across it.[34] One of the consequences of this can be a flux-ratio exponent (n') greater than one:

$$M_{12}/M_{21} = [(C_1/C_2)\exp(zFV/RT)]^{n'} \qquad (22.3)$$

where M_{12} and M_{21} are fluxes from compartment 1 to 2 and 2 to 1, respectively, C_1 and C_2 the concentrations of permeant ions in compartments 1 and 2, respectively, and V the voltage difference between the two compartments. Theoretical values of n' are 1 for a pore where at most one ion is present at a time, less than 1 for an obligatory ion exchange mechanism, and greater than 1 for a multiion pore. Estimates of n' in frog skin and toad bladder were close to 1, consistent with occupancy by a single ion.[32] This does not rule out the existence of multiple binding sites for the ions, as discussed below.

22.2.2.3. Channel Kinetics

Patch-clamp studies have also indicated that Na channels normally spend roughly equal times in open and closed states. Measurements of open probability under control conditions can vary considerably, but tend to average from 0.3 to 0.5.[36–38] The open probability is determined by the ratio of the rate constants switching between open and closed states:

$$\text{open} \underset{k_-}{\overset{k_+}{\longleftrightarrow}} \text{closed}$$

where k_+ is the rate of transition from closed to open, and k_- from open to closed. Although at least in some instances more than one open and/or closed state can be detected from kinetic analysis,[39] in most cases average mean lifetimes or rate constants are computed. These rates tend to be slow. For example, in the rat CCT k_+ and k_- are on the order of 0.3 sec^{-1} at room temperature.[40] This corresponds to mean open and closed times of 3 to 4 sec. These slow transition rates make it difficult to obtain sufficient kinetic data from a single patch to analyze the kinetics more extensively. At 37°C the rate constants increase to 2 to 7 sec^{-7}.[41] Thus, the Q_{10} for channel gating is approximately 5. This indicates that activation energies for gating are much higher than for channel conduction, as has been found for channels in general.[34] In contrast to the Na channels from muscle and nerve, the open probability of the epithelial Na channels is weakly increased by *hyper*polarization of the membrane.[39,41]

22.2.2.4. Pharmacology

The most important pharmacological tool for the study of Na channels is amiloride.[42–44] This diuretic consists of a pyrazine ring with a guanidinium side chain. Both of these structures are essential for high-affinity block of the channel, with apparent K_i values of around 0.1 μM. The action of this drug is rapid and easily reversible when applied to the outer surface of the cell, suggesting a receptor site on the external side of the membrane. The interaction with the channel can be described by a simple bimolecular reaction with rate constants of about 10–20 sec^{-1} μM^{-1} for the on rate and 2–4 sec^{-1} for the off rate. Many different analogues of amiloride have been synthesized and tested.[43] The analogues that are best able to block Na channels are those in which the guanidinium side chain is derivatized with a hydrophobic group such as benzene (benzamil) or phenol (phenamil).

The block of Na channels by amiloride is voltage dependent; cell-negative potentials facilitate inhibition.[32] This can be most easily explained if the positively charged guanidinium moiety penetrates into the pore so that it senses a portion of the electric field across the membrane. Consistent with this idea is the finding that guanidinium itself can block the channel with a similar voltage dependence, although the affinity of guanidinium for the channel is several orders of magnitude lower. This difference is presumably related to specific interactions of the pyrazine ring with the channel. Further evidence is reviewed elsewhere.[42]

There is evidence that amiloride can interact with both the open and closed states of the channel.[45] Both the on- and off-rate constants for the interaction with the closed state are lower than for the open state; the state that is both closed and blocked can last for many seconds. According to the pore-plugging model for amiloride action, this would indicate that the outer mouth of the pore is still open and accessible to the drug even when the conduction pathway through the channel is closed.

22.2.2.5. Channel Model

A quasiphysical kinetic model of ion conduction through the Na channel of the toad bladder has been proposed.[46] This model is based on the interactions of the permeant ions (Na and Li) and impermeant blocking ions (including both small cations such as K and larger ones such as amiloride). According to the model the pore contains at least three sites at which cations will interact. One is a binding site in the outer mouth of the pore that is accessible to all cations with a diameter of 5 Å or less, including the guanidinium portion of amiloride. About 15% of the electric field is sensed at this site. Another binding site, at which about 30% of the field is sensed, is accessible to small cations, including the monovalents K^+, Rb^+, and Cs^+ and the divalents Ca^{2+}, Mg^{2+}, and Sr^{2+}. Occupancy of one of the two sites excludes occupancy of the other in the model, so that only one ion is in the channel at a time, consistent with the flux ratio data. Beyond that site the pore narrows to a diameter that will allow only Na, Li, and H to pass through.

The model was used to estimate rate constants for entry of permeant and blocking ions and for exit of the permeant ions. An essential feature of the system that can be accounted for by the model is the finding that the apparent K_m for the single-channel currents is independent of the membrane voltage.

This implies that hyperpolarization of the apical membrane will have equal effects to accelerate Na entry into and Na exit from the channel, such that the occupancy of the channel remains constant. Consistent with this idea is the finding that the ability of Na to compete with amiloride for the outer binding site is also independent of voltage.

22.2.2.6. Channel Diversity

As discussed above, other epithelial Na channels have been described, although they are less well characterized.[4] One of these is a channel of 9–10 pS observed in the apical membrane of A6 cells cultured on solid supports rather than the more physiological permeable supports. This channel has a lower selectivity for Na over K (3–4:1). It appears to be regulated by the submembrane cytoskeleton.[47] Another is a 28-pS channel regulated by cGMP[48] which is found in cells cultured from the mammalian medullary collecting duct. This channel does not discriminate between Na and K. Both of these channel types are blocked by amiloride at concentrations similar to those that affect the highly selective channels. Yet another type of Na-selective channel with a conductance of 21 pS was found in cultured nasal epithelial cells.[49]

22.2.2.7. Molecular Identification of Na Channels

The epithelial Na channel from the rat colon has been identified and sequenced using the expression cloning approach.[50–52] A single subunit, called αrENac, induces amiloride-sensitive Na conductance in *Xenopus* oocytes. This clone has a predicted protein product of 78 kDa. It has sections of hydrophobic amino acids that could be embedded in the membrane, consistent with the hypothesis that this protein itself is a channel. Hydrophobicity plots suggest that the protein spans the membrane twice, with short cytoplasmic portions at the N- and C-terminals and most of the mass in the extracellular space.[52] This topology is supported by experimental evidence.[53,54]

Although expression of this clone alone leads to the formation of channels in oocytes, the total conductance induced was less than that which could be obtained by injected total mRNA from rat colon.[51] This implied the existence of other subunits which could enhance expression. Indeed, two additional subunits were identified by complementation expression cloning.[52] These clones, termed βrENac and γrENac, have predicted molecular masses and structures similar to those of the α subunit. These subunits did not themselves induce channel activity in the oocytes. However, coexpression of the three subunits resulted in Na channel activities that were 50-fold higher than those obtained with the α subunit alone.

The similarity of the three subunits led to the hypothesis that they are symmetrically arranged in the membrane, perhaps around a common pore. Such a structure would be similar to that of the nicotinic ACh receptor.[34] The minimal structure would have one of each of the subunits, but the actual number of subunits required and the stoichiometry are unknown.

The functional properties of the channels formed by these clones in oocytes are very similar to those of the Na channels in native epithelia[52] (Figure 22.3). They have a similar selectivity pattern, with Li preferred over Na and no measurable K conductance. Currents saturate as a function of extracellular Na or Li, with apparent K_m values somewhat higher for Li than for Na. The single-channel conductances of 4–5 pS and the opening and closing kinetics are slow. The open probability is around 0.5 and is increased slightly by membrane hyperpolarization.

These results suggest that these three subunits are sufficient to form the basic channel structure and to reproduce the essential permeation and gating characteristics. However, since the channels have been studied after expressed in oocytes, rather than as purified, reconstituted proteins, the possible contribution of components of the oocyte in forming the channels cannot be ruled out.

22.2.2.8. Biochemistry of the Channel

Candidate proteins for the Na channel have been identified using a photoactivatable amiloride analogue, which binds covalently in the presence of UV light, to label proteins in bovine renal medulla.[31] After solubilization and purification using standard biochemical procedures, a large 700-kDa complex was obtained. After treatment with a reducing agent, the complex could be separated into at least six peptides of different molecular masses (315, 150, 95, 70, 55, and 40 kDa). Amiloride binding was localized to the 150-kDa subunit.

Polyclonal antibodies against the protein preferentially mark the apical membrane of Na-transporting epithelial cells, consistent with the identification of the protein with the Na channel. In addition, incorporation of the protein into planar lipid bilayers resulted in the reconstitution of channels which were Na-selective and amiloride-sensitive.[55] There were small-conductance (3–4 pS) channels with short open times (20–40 msec) and larger-conductance (16 pS) channels with longer open times (>100 msec). Functional reconstitution of a partially separated preparation of this protein, containing the 150- and 40-kDa subunits, was also achieved.[56] These channels had a low conductance (3–10 pS) and slow kinetics, and the mean open time was reduced by 10^{-7} M amiloride. Selectivity was not assessed.

The relationship between this protein and that of the cloned channel is unknown. The molecular mass of the amiloride-binding and probable pore-forming subunit (150 kDa) is nearly twice that of αrENac. To what extent glycosylation could account for the difference has not been examined; αrENac has four possible *N*-linked glycosylation sites which could increase the weight of the mature protein considerably. Alternatively the 700-kDa protein could comprise another type of amiloride-sensitive channel.[4]

22.2.3. K Channels

The third essential component of the KJ-U system is the K conductance of the basolateral membrane, which is necessary to recycle the K that enters the cells through the Na/K pump and also serves to maintain an intracellular-negative membrane potential. The characteristics of this transporter have been more difficult to study than those of the apical Na channel because of the difficulty of gaining access to the basolateral membrane. Some properties of the basolateral membrane can be inferred from microelectrode studies of intact cells, but these measurements are complicated by the presence of apical membrane and paracellular conductance pathways. The basolateral membrane channels have also been studied more directly in several ways. First, elimination of the apical membrane resistance with ionophores allows the basolateral membrane conductance to be studied using extracellular electrodes. Second, treatment of the basal surface of some epithelial cells with collagenase has permitted the direct recording of channels using patch-clamp techniques. Third, isolated cells that maintain basolateral membrane properties have been examined. From this combination of approaches a picture of the characteristics of the major basolateral K channels has emerged.

22.2.3.1. Conductance and Inward Rectification

Hyperpolarization of the basolateral membrane of the frog skin is associated with an increase in membrane conductance.[57] Since the extracellular K concentration is always smaller than that of the cell, this "inward rectification" is opposite to what would be predicted from the concentration gradient by applying the constant-field equation, and therefore also qualifies as "anomalous" rectification. This phenomenon is also observed in the plasma membrane of a number of cells such as the skeletal muscle at rest and the starfish egg.[34] The finding has been confirmed in a number of epithelia using different techniques.[58–61]

This property seems odd given that the function of this conductance is to recycle K across the membrane, i.e., to conduct outward current. It should be pointed out, however, that the rectification observed in most epithelial K channels is not absolute, as it is in muscle and some other tissues such as starfish eggs.[34] Outward conductance is two to three times smaller than inward conductance, but it is not zero. In this respect the K conductance more resembles that of moderately inwardly rectifying channels such as the ATP-dependent channels in the heart or pancreas.[62] Similarities between these channels and K channels in the *apical* membrane of renal epithelia will be discussed further below.

Patch-clamp studies of the single-channel properties of basolateral K channels showed that inward rectification was a property of the single-channel conductance rather than an effect of channel gating (Figure 22.4). Patches with symmetrical

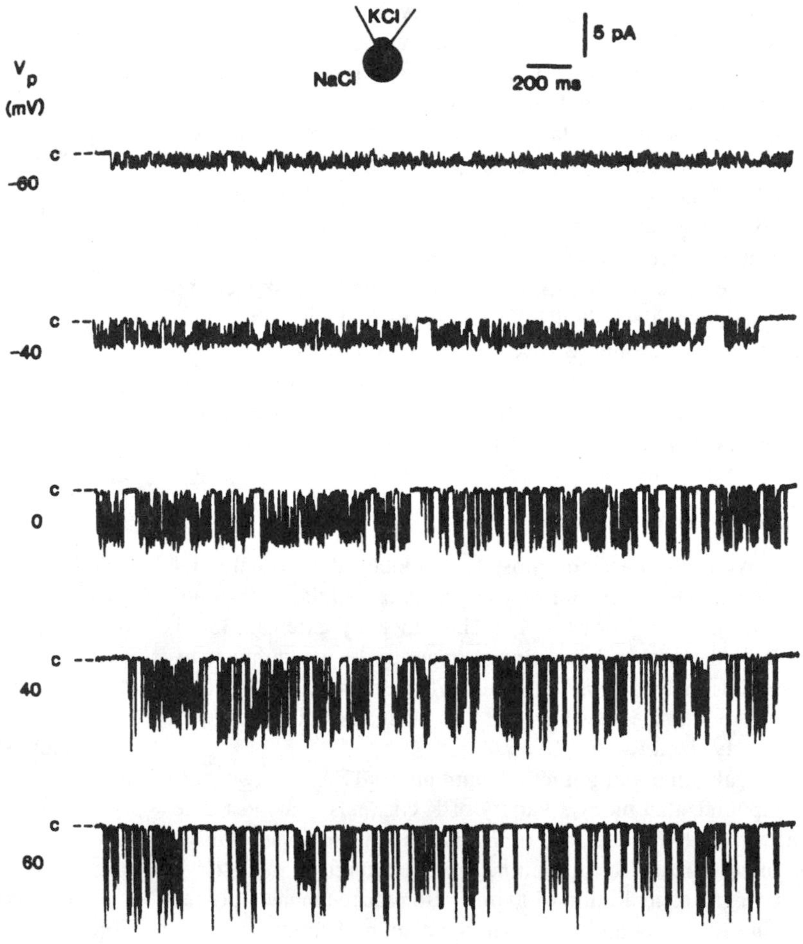

Figure 22.4 Patch-clamp records of basolateral K channels from frog skin. Reproduced from *The Journal of General Physiology,* Volume 98, pp. 131–161 by copyright permission of The Rockefeller University Press.[60]

K concentrations had a slope conductance of 44 pS at the reversal potential. The conductance increased at hyperpolarizing voltages and decreased at depolarizing voltages.[60] Patch-clamp studies of intact frog skin epithelia also revealed the presence of inwardly rectifying channels in both cell-attached and excised patches.[61] As in other inwardly rectifying K channels,[63] the rectification appears to reflect a voltage-dependent block by intracellular Mg^{2+}, as removal of this ion from the solution bathing the cytoplasmic surface of the patch resulted in a nearly linear *i–V* relationship.[61]

Another KJ-U epithelium in which basolateral membrane channels have been examined at the single-channel level is the rat CCT. Two types of K channels were observed which differed mainly in their single-channel conductances: 67 and 145 pS for inward currents under conditions similar to those used to measure g in the frog skin cells except that the temperature was 37°C.[64] Rectification was not examined.

Relatively little information is available on the selectivity of these channels. In turtle colon the basal conductance appears to conduct Rb nearly as well as K. Swelling the cells activated a different K channel which was highly selective for K over Rb.[65] The Rb conductance through the frog skin inward rectifier was about two-thirds that of K.[61]

22.2.3.2. Kinetics

Although, like many K channels in excitable cells, the basolateral K channel in frog skin has an open probability that increases with depolarization, the channel is active at the resting membrane potential. This is consistent with a role for this channel in the steady-state conductance of the membrane. Curiously, in studies of intact skin epithelium, the voltage dependence of the open probability was reversed when the major cation in the patch pipette, i.e., outside the channel, was K.[61] This phenomenon has not been explained.

The gating of the K channels in isolated frog skin epithelial cells was described by a single open state, with a mean open time of around 5 msec and two closed states, the predominant brief closed state having a mean lifetime of around 2 msec.[60] Membrane depolarization increased the mean open time and decreased the mean closed time.

22.2.3.3. Pharmacology

As is the case for most K channels, the basolateral K channels in epithelia are blocked by extracellular Ba^{2+} in millimolar concentrations.[57–61,64] The block by external Ba^{2+} is voltage dependent, becoming stronger as the membrane is hyperpolarized.[59,60] This is presumably because the Ba^{2+} enters the pore and binds to a site that is normally occupied by the similarly sized K^+ ion.

Quinidine and tetraethylammonium (TEA) are two other compounds that block a variety of K channels.[34] In most cases the resting K conductance of the basolateral membrane was not strongly affected by quinidine,[59,61,65] although in one study a very high concentration (500 mM) did reduce the inwardly rectifying K current and decreased the open probability of inward rectifier channels.[60] TEA was also ineffective at blocking basolateral K conductances in most studies,[60,65] although it was a potent blocker of channels of the rat CCT in outside-out patches.[64]

As mentioned above, the conductive properties of the basolateral K channels are similar to those of the ATP-sensitive channels in heart and pancreas;[62] in both cases the conductance is intermediate in magnitude (30–60 pS) and is moderately inwardly rectifying. The sulfonylureas such as tolbutamide and glybenclamide are potent blockers of the ATP-sensitive channels. The basolateral K conductance of A6 cells is also inhibited by these drugs, with apparent K_i values of 29 and 0.11 μM, respectively.[66] This is consistent with the idea that the basolateral K channels and the metabolically regulated K channels are related.

22.2.3.4. Molecular Identification

Basolateral K channels have not been positively identified at the protein or molecular level. However, several K channels in the category of inwardly rectifying channels have been cloned and sequenced. These include channels that behave as absolute inward rectifiers[67] as well as moderately inward rectifying channels from the kidney.[68,69] In the latter case, they are thought to represent the apical K conductance of the epithelial cells.

All of these channels clearly belong to the same family, based on both their sequence homology and their general structure. All have molecular masses of around 40 kDa. They appear to have long cytoplasmic C- and N-terminal ends, two clear membrane-spanning segments. The extracellular loop between these two segments includes a shorter hydrophobic region which may form the conducting pore.

In addition to inward rectification, these channels share with the basolateral K channels the properties of intermediate single-channel conductance, sensitivity to external Ba but not TEA, high open probability at the resting potential with mild activation by depolarization, and a high Rb conductance. It therefore seems likely that the basolateral K channels will be members of this gene family.

22.2.3.5. Other Basolateral K Conductances

In addition to the resting K conductance, the basolateral membrane of KJ-U epithelia also contains a second type of K channel which is activated when the cell volume increases. This pathway was first identified using nystatin-permeabilized turtle colon epithelia,[65] and was distinguished from the resting conductance channels by sensitivity to quinidine, poor conductance to Rb, and insensitivity to the muscarinic-cholinergic agonist carbachol. A swelling-activated K conductance in A6 cells was also found to be blocked by quinidine and was outwardly rectifying instead of inwardly rectifying.[59] Patch-clamp recordings indicate that this channel has a single-channel conductance of 17 pS in cell-attached patches under physiological conditions.[70]

Yet another K conductance was identified in the turtle colon when the apical membrane was permeabilized with digitonin and the cell Ca^{2+} increased.[71] This conductance, unlike the resting conductance, was sensitive to quinidine. Unlike the swelling-activated pathway, the conductance to Rb was high. Thus, this channel type, which has not been identified at the single-channel level, appears to be distinct from the other two.

Thus, it appears that the basolateral K conductance pathway of these epithelial cells is complex. At least three distinct pathways or channel types are involved. The relative conductance of the three pathways will depend on the physiological state of the cells.

22.2.4. Rate-Determining Steps

It has long been the consensus that entry of Na into the cell across the apical membrane is the rate-determining step in the Na transport process.[72] This idea is based on a number of observations. Addition of pore-forming antibiotics to the apical side of epithelia to increase the Na permeability of the membrane also stimulates the Na transport rate. Conversely, hormones known to increase Na transport have been shown to stimulate apical Na channels (see below). Finally, measurements of intracellular Na^+ in the cytoplasm of frog skin epithelial cells using a variety of techniques have indicated that the concentration is 15–20 mM under normal physiological conditions.[73] At this concentration the turnover rate Na pump is likely to be only about 50% of maximal values (see above).

More direct assessments of the degree of saturation of the Na pump can be made by comparing transport rates and maximal pump rates in the same tissues. In the rabbit urinary bladder, which transports Na by the KJ-U mechanism but at very low rates, the pump capacity was estimated by abolishing the apical membrane resistance with nystatin.[19] Maximal pump rates measured electrically were 20–30 μA/μF. Assuming a stoichiometry of 3Na:2K, this gives a maximal Na flux of 60–90 μA/μF. This flux is well above the rate of Na transport measured as the short-circuit current of 1–2 μA/μF, indicating that apical entry is definitely rate-limiting in this epithelium. Similar results were obtained in the toad urinary bladder, which transports Na at a moderate rate.[20]

In the mammalian CCT, however, the analysis gives different results.[74] Maximal transport rates in isolated perfused rat CCT were around 3000–9000 pmole/hr•mm, much higher than in toad or rabbit urinary bladder. The maximal pump rates, measured as Na/K-ATPase activity, were 1000–2000 pmole ATP/mm•hr or 3000–6000 pmole Na/mm•hr. Thus, the pump must have been working at near-maximal levels to achieve these transport rates. Obviously there is some uncertainty in that the measurements were made on different tubules in different labs. The general idea was confirmed, however, using whole-cell recording techniques on rat CCT cells.[74] In CCTs from aldosterone-treated rats the maximal Na pump current was 150 pA/cell (Na efflux of 450 pA/cell), whereas the rate of Na entry into the cell under physiological conditions was estimated to be around 300 pA/cell. In this epithelium, at least, it seems likely that the Na pump can be rate-limiting for Na transport under some conditions.

22.2.5. Conclusions

The three essential components of the KJ-U system have been examined in detail. The functional properties of apical Na and basolateral K conductances have been defined, and single-channel characteristics have been described. The molecular entities comprising these components are being identified. Na/K pump protein has been purified, and cDNAs encoding both the pump and the Na channels have been cloned.

22.3. REGULATION OF Na TRANSPORT

Na transport in tight epithelia is highly regulated. The regulatory systems that have been most extensively studied can be divided into two major categories. The first is hormonal regulation. Here transport rates are modulated up or down according to the needs of the *organism* to either retain or excrete salt. Changes in plasma volume or osmolarity stimulate, either directly or indirectly, various endocrine organs including the adrenal cortex, the posterior pituitary, and the heart. These organs subsequently secrete hormones that affect epithelial Na reabsorption. This type of regulation contributes to salt and water balance or *homeostasis of the organism.* The second type of regulation is intrinsic to the epithelial cells themselves. In this case the cells respond to changes in solute influx or efflux in a manner that tends to minimize changes in cell volume or ion composition. These responses may involve mechanisms known as feedback inhibition or membrane transporter cross talk. Alterations in the activity of one transporter result in compensatory effects on the same transporter (feedback inhibition) or adjustments in the activity of other transporters (cross talk) in order to maintain the cell composition. This type of regulation contributes to *homeostasis of the cell.*

22.3.1. Hormonal Regulation

22.3.1.1. Aldosterone

Certainly the most universal hormonal system that affects Na transport is that of the adrenal corticosteroid aldosterone. This hormone regulates transport in most high-resistance, Na-reabsorbing epithelia of vertebrates.[5,75,76] Aldosterone is involved in a homeostatic response mechanism for maintaining constant plasma volume. The adrenal cortex secretes aldosterone in response to angiotensin II, which in turn is controlled by plasma volume through the release of renin from juxtaglomerular cells.[77] A loss of plasma volume will lead to a cascade of events culminating in increases in circulating aldosterone and retention of Na.

The effects of the hormone can be extremely powerful; in some systems such as the rat colon and CCT, or the hen copradeum, transport rates can vary from undetectably low activity to a very high activity depending on the levels of circulating aldosterone.[5] Most of the effects of the hormone are mediated through intracellular steroid hormone receptors and appear to require the synthesis of new proteins.

Despite the central importance of aldosterone to the regulation of transport and salt and water balance, the mechanisms underlying the effects of the hormone have not been fully elucidated. This is related in part to the complexity of the response of the epithelial cell to the steroid. As described below, the hormone can exert its effects on at least three different time scales.

22.3.1.1a. Rapid Effects. Although the "classical" effects of aldosterone in stimulating epithelial transport have a lag time of at least 1 hr, during which synthesis of new mRNA and protein molecules presumably takes place, in some epithelia more rapid changes have been documented. One clear case of such a response is in cells of the amphibian diluting segment. The effect of the hormone is to increase cytoplasmic pH through activation of Na–H exchange.[78] The alkalinization may in turn activate K channels. This response is measurable 20 min after addition of the hormone and is virtually complete within 40 min. Similar effects were observed in the intact frog skin.[79] Basolateral K channel activity was increased 20 min after addition of aldosterone to the bathing medium. This was believed to be the result of an activation of a Na–H exchanger, as the effect was abolished when the exchanger was blocked by amiloride.

In the mammalian colon, Na reabsorption can also be increased rapidly by *glucocorticoids*.[80] This effect is observed 30 min after application of hormone *in vivo* and appears to be mediated by a glucocorticoid rather than mineralocorticoid receptor. Na transport was increased without an increase in the electrical potential across the tissue, suggesting that an electroneutral transporter such as an apical Na–H exchanger was activated, rather than Na channels.

All of these effects are consistent with the activation of Na–H exchange by a nongenomic pathway. Such a mechanism would not require the synthesis of new proteins. Further support for this concept comes from work on human mononuclear leukocytes.[81] In these cells Na/H antiport could be stimulated within minutes by physiological concentrations of aldosterone. The stimulation was not affected by inhibitors of protein synthesis, suggesting that the classical gene-induction pathway is not involved. It was postulated that membrane receptors for the steroid may mediate the response.

Although this regulatory pathway has been documented in a number of different systems, the classic KJ-U system of *electrogenic* Na reabsorption has not been shown to be affected by this rapid-response mechanism. In the classical frog skin model, rapid activation of K channels has been documented but this has not been shown to lead to an increase in Na transport as measured by short-circuit current. In the mammalian colon, Na transport can be modulated over this rapid time scale but this is an electroneutral system that does not involve channel-mediated Na entry.

22.3.1.1b. Short-Term Effects. The classical short-term effects of aldosterone typically are observed within 1–3 hr after injection of aldosterone into an animal or addition of hormone to an *in vitro* epithelial preparation.[75] Studies of amphibian systems such as the toad urinary bladder have shown that the increase in Na transport over this time period is mediated by mineralocorticoid receptors and depends on protein synthesis, although the identification or exact nature of the proteins synthesized in response to the hormone remains unclear.[5,75]

Effects on Na Channels. There is abundant evidence that the apical Na permeability, or the activity of the apical Na channels, is increased during this phase.[5,75] This raises the question of whether the channels themselves might be synthesized *de novo*. There is considerable indirect evidence suggesting that this is not the case.[76] Vesicles isolated from toad bladder after treatment for 3 hr with aldosterone do not retain the increased membrane Na permeability. Furthermore, mRNA from toad bladder treated for 3 hr with aldosterone does not appear to be enriched in mRNA coding for the channels according to a bioassay using the *Xenopus* oocyte expression system.[82] Finally, short-term incubation with aldosterone was shown to increase the mean open time and open probability of Na channels in A6 cells.[83] These data are consistent with the idea that preexisting channels are activated by the hormone.

One candidate for the activation mechanism involves protein methylation. Addition of a methyl donor was shown to increase the Na permeability of membranes isolated from aldosterone-depleted but not aldosterone-treated A6 cells.[84] Another possible mediator of this effect is cytoplasmic alkalinization. Increased pH has been shown to activate Na channels in intact frog skin[85] and in excised patches from rat CCT.[86] The cell pH in frog skin epithelium was shown to increase 2 hr after addition of aldosterone.[87]

Although *de novo* synthesis of the channels themselves may not account for the early effects of aldosterone, the sensitivity of the response to inhibitors of protein and RNA synthesis strongly suggests that some new protein must be made. It is not yet clear what sort of protein this might be. One protein whose biosynthesis is increased during this phase is comprised of a set of 65- to 70-kDa peptides which appears to be expressed at the apical surface of the cell.[88] The function of this protein is unknown.

Effects on Na Pumps. Although the apical Na permeability clearly increases after aldosterone treatment in all epithelia studied, the intracellular Na concentration rises relatively little.[20,89] This suggests that the Na pump is also upregulated under these conditions. In A6 cells, the number of ouabain-binding sites on the basolateral membrane increases in response to aldosterone, consistent with an increased number of glycoside-inhibitable pump units.[90]

Measurements of Na/K-ATPase in toad bladder showed a minimal increase in response to aldosterone.[91] In the rat CCT, however, there was a doubling of enzyme activity within 3 hr after treatment of the tubules *in vitro* with the steroid.[92] This effect apparently requires protein synthesis since it is blocked by protein synthesis inhibitors. As in the case of the channels, new pump proteins are probably not synthesized during this period. Increased rate of biosynthesis of the pump was observed after 6–18 hr of hormone exposure, but not after 3 hr.[93]

Therefore, this effect could not account for the earlier rise in pump activity or ouabain-binding sites.

There is evidence that the increased Na/K-ATPase activity might be related at least in part to an increase in intracellular Na secondary to an increased apical Na permeability. *In vivo* experiments showed that the increased enzyme activity in the CCT after acute (1.5 to 3 hr) treatment with aldosterone was abolished when rabbits were pretreated with amiloride to prevent an increase in Na influx. On the other hand, the increased enzyme activity observed over the same period *in vitro* was still observed in the presence of amiloride.[92] Recent experiments have indicated that aldosterone and increased intracellular Na may have synergistic effects on the acute stimulation of the pump.[94,95]

In summary, aldosterone can increase Na transport by KJ-U epithelia within a few hours of administration. The stimulation involves activation of Na channels and, perhaps to a lesser extent, activation of Na pumps. The available evidence so far suggests that the effects involve the synthesis of proteins that modulate the activity of preexisting transporters, or move them from intracellular stores to the plasma membrane.

22.3.1.1c. Long-Term Effects. Long-term exposure to high levels of aldosterone *in vivo* increases the rate of Na transport measured *in vitro* in renal and intestinal epithelia.[5] These chronic effects may involve mechanisms that are different from those observed in acute experiments.

Effects on Na Channels. Apical Na permeability is increased during long-term aldosterone treatment. This has been assessed using both macroscopic measurements of apical membrane conductance[96–98] as well as patch-clamp analysis of the density of conducting Na channels[99] (Figure 22.4). Unlike the short-term exposure, however, the apparent abundance of channel mRNA increases in the *Xenopus* oocyte assay, at least in coloncytes.[82] In addition, 18-hr treatment of toad bladders with aldosterone leads to an increase in the permeability of vesicles isolated from the treated tissues.[100] These findings are consistent with the idea of an increased biosynthesis of channel mRNA and protein. An increased abundance of mRNAs coding for both β and γ (but not α) rENac was observed in response to aldosterone in the rat colon.[101] At the protein level, all three subunits were dependent on the steroid status. On the other hand, in the rat kidney the subunits were constitutively expressed, suggesting that regulation was posttranscriptional in this tissue.

Effects on Na Pumps. The Na/K pump is also stimulated during chronic mineralocorticoid treatment *in vivo*. The stimulation can be measured either as an increase in enzyme activity[102–104] or as an increase in the electrical current generated by the pump.[74] In addition, the basolateral membrane area is greatly increased under these conditions.[105] Thus, the stimulation may not be specific for the pump but rather may involve many basolateral membrane enzymes and transporters.

Whether these late effects on the basolateral transport properties are direct effects of the steroid or secondary consequences of an increased apical Na permeability has been controversial. In amphibian cells the rate of Na/K-ATPase gene transcription was increased before a measurable increase in the rate of Na transport.[106] This implies that the effects of aldosterone on the biosynthesis of the pump are direct. On the other hand, chronic increases in endogenous aldosterone secretion in response to a low-Na diet lead to transient increases in Na/K-ATPase activity in the CCT.[107] Measurements of (ouabain-sensitive) pump current in rat CCT had similar results; pump current was increased by exogenous aldosterone treatment but not by a low-Na diet, even though aldosterone levels were much higher in the latter condition.[74]

One interpretation of these findings is that high intracellular Na, resulting from increased Na permeability, is necessary for the upregulation of the pump. Since activation of mRNA synthesis does not appear to require increased cell Na, it is possible that the effects of the intracellular ionic composition are exerted at a later stage of pump biosynthesis or insertion into the membrane. In animals on a low-Na diet, Na permeability is high but delivery of Na to the distal tubule may be low, so that relatively little Na enters the cells. When the aldosterone-induced Na permeability was prevented by amiloride, the increased Na/K-ATPase activity in response to infusion of aldosterone into the animal over 24 hr was abolished.[108] Similarly, stimulation of pump current by chronic aldosterone infusion was greatly diminished when amiloride was also infused.[74] These findings are consistent with the idea that a rise in intracellular Na accounts at least in part for the increased number of pumps. A number of studies have shown that increased cell Na can lead to increased pump biosynthesis in cultured cell systems.[109] An alternative interpretation was suggested by observations that animals on a low-Na diet became hypothyroid.[110] Normalization of thyroid levels led to an increased Na/K-ATPase activity in the rabbit colon. This implies that the lack of stimulation of the pump during Na restriction is an indirect effect of reduced thyroid hormone.

Although the mechanisms involved are controversial, there is now general agreement that both the short-term and the long-term action of aldosterone involve coordinated increases in apical membrane permeability and in basolateral pump activity. This allows an increased transepithelial transport rate without significant changes in intracellular Na concentration or content.[96] The cells appear to put a high priority on preserving constant cytoplasmic ion concentrations. This theme will be discussed further below.

22.3.1.2. Antidiuretic Hormone

In some, but not all, epithelia that operate via the KJ-U system, Na transport is stimulated by antidiuretic hormone (ADH; vasopressin) or its amphibian counterpart vasotocin. The main physiological effect of these hormones is to promote water retention through regulation of epithelial water channels.[111] However, the peptide can also be released under conditions of a large decrease in plasma volume, such as in hemorrhage.[112] In these circumstances ADH will promote fluid retention through activation of Na transport. The regulation of both Na and water transport systems is mediated by V_2 receptors through increases in the concentration of cAMP within the cell.[111]

Apical Na channels are stimulated by this pathway. This has been demonstrated using a number of different techniques including noise analysis[113,114] and more recently that of single-channel analysis using the patch-clamp approach.[37] In the latter study, treatment of A6 cells with ADH increased the density of conducting channels on the apical cell membrane (Figure 22.5). The conductance and open probability of the channels were not changed. Stimulation of the channels could not be demonstrated when the hormone was added during recording from cell-attached patches. This indicates that formation of the gigaohm seal somehow protected channels within the patch from the effects of the hormone. Similar results were obtained in the rat CCT.[115]

Two ideas have been put forward to explain the effects of ADH and cAMP on the channels. The most straightforward involves the activation of channels in the membrane by phosphorylation of the channel proteins. Recently this hypothesis was supported by the observation of channels incorporated into artificial lipid bilayers after purification of the channel-containing complex described above.[116] These channels could be opened by the application of the catalytic subunit of PKA, indicating that phosphorylation of the protein could affect channel activity. Unlike the action of cAMP on intact A6 cells, the effects of PKA were apparently to increase the open probability of the channels, rather than the number of active channels. PKA-dependent phosphorylation of the channel was demonstrated on the 315-kDa peptide. As discussed above, this peptide is probably not an essential part of the pore itself.

A second possibility is that channels are inserted into the membrane by vesicle fusion in response to the hormone. There is good evidence that this scheme accounts for the increased apical membrane water permeability in target epithelia.[117] Several findings are consistent with this interpretation, including the observations that PKA could not increase Na permeability in vesicles isolated from the toad bladder,[118] and that in the intact toad bladder ADH-stimulated channels were inaccessible to trypsin before stimulation with hormone.[119] The idea might also explain the inability to activate channels in cell-attached patches; it is possible that the plasma membrane may be mechanically separated from the channel-containing vesicles during formation of the patch. However, the evidence for the vesicle-fusion hypothesis is based primarily on negative results and will require confirmation using more direct techniques.

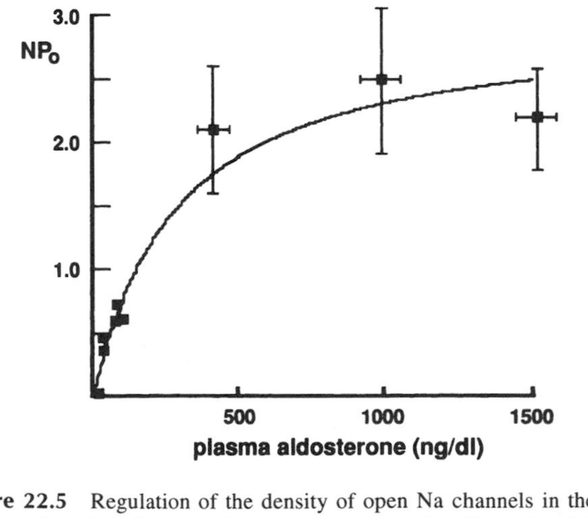

Figure 22.5 Regulation of the density of open Na channels in the rat CCT as a function of plasma aldosterone concentration. Reproduced from *The Journal of General Physiology,* Volume 102, pp. 25–42 by copyright permission of The Rockefeller University Press.[99]

22.3.1.3. Atrial Natriuretic Factor (ANF)

ANF is released from the heart in response to plasma volume *expansion*.[120] Signaling through this system is therefore opposite to that of aldosterone, and the peptide can inhibit Na reabsorption by a number of epithelia. These include epithelia that reabsorb NaCl through a tightly coupled, electroneutral system, presumably involving an apical membrane cotransporter. Two examples of this are the flounder intestine and the rat renal CCT, where the electroneutral cotransport system operates in parallel with that of the KJ-U mechanism.[120,121]

In addition, there is evidence that ANF can downregulate the KJ-U system in the medullary collecting tubule of the rat.[121] The peptide reduces both amiloride-sensitive oxygen consumption and amiloride-sensitive Na influx into cells from the renal medulla. The downregulation of Na channels has been shown more directly using patch-clamp methods in cells cultured from rat inner medulla.[122,123] Channels in cell-attached patches are inhibited by application of the hormone. Furthermore, channels in excised patches are deactivated by cGMP, the second messenger for ANF, via two different mechanisms. One involves a direct effect of the nucleotide; the other is mediated by a G-protein.

The Na channels in IMCD cells which are regulated by ANF and cGMP have very different properties from those in the CCT which have been the major focus of this review. The cGMP-sensitive channels have a higher single-channel conductance (28 pS) and a lower selectivity for Na over K (~1:1). This difference may explain the finding that Na channels in the CCT are not downregulated either by ANF or by cGMP.[120,124]

22.3.1.4. Insulin

Insulin is another hormone that can stimulate Na transport, at least in the amphibian models including frog skin, toad urinary bladder, and A6 cells.[125–127] The physiological importance of this natriferic effect is not obvious. Insulin is not known to be secreted in response to an electrolyte or fluid volume imbalance. It is therefore uncertain whether this is a homeostatic response. Recent work with the toad bladder has suggested that the receptor that mediates the insulin response is actually a receptor for insulinlike growth factor.[128] It is possible that the pathway is involved in growth or development.

The effect of insulin involves a stimulation of apical Na permeability.[126] This effect has been investigated in detail using patch-clamp techniques with A6 cells.[129] Insulin increases the open probability of a subset of channels. The intracellular events mediating the insulin response remain unknown. In frog skin there is evidence that insulin and PKC stimulate Na channels through a common pathway.[130] However, as discussed below, in other epithelia PKC appears to have inhibitory effects.

22.3.1.5. Osmolality

Na transport by the KJ-U system is quite sensitive to the osmolality of the bathing medium. Decreased osmolality of the internal or serosal bathing medium stimulates transport, while increased osmolality decreases transport. This finding was original made by Ussing[131] and has been confirmed in toad bladder[132] and most recently in A6 cells.[133] This response does not involve a hormonal signal. It would, however, serve directly to maintain osmotic balance in the whole organism. Decreased osmolality would be compensated by an enhanced reabsorption of Na. The mechanisms involved are not well understood.

22.3.2. Cellular Homeostasis

In the category of regulatory mechanisms for maintaining *cellular homeostasis* in Na-transporting epithelia, three types of responses have been considered in depth. The first is volume regulation, usually defined as a response to an altered osmolarity that tends to counteract an initial swelling or shrinkage of the cell. The second is called feedback regulation, in which a cell responds to changes in Na influx or efflux in ways to maintain constant intracellular Na. The third is called pump-leak balance, in which changes in pumping rate are accompanied by changes in passive permeability, particularly to K, to preserve cell volume K concentrations.

22.3.2.1. Cell Volume Regulation

This topic will be considered only briefly, since it is a main thesis of Chapter 23 in this volume. Epithelial cells of the frog skin swell in response to a decrease in osmolarity but then shrink back toward control levels.[134] This response is thought to involve a loss of cell Cl. Normally, cell Cl is above its electrochemical equilibrium level as a result of the operation of a Na/K/2Cl cotransporter on the basolateral membrane. Cl is normally prevented from moving out of the cell by the low Cl permeability of this membrane. Cell swelling may activate a latent Cl conductance, allowing net Cl efflux from the cell.[135,136] The channels involved in this response have not been identified with certainty. However, an outwardly rectifying Cl-selective channel has been observed in isolated frog skin cells.[60] The channels had a single-channel conductance of 48 pS at 0 mV with symmetrical Cl, and were activated by membrane depolarization. It is not known if this channel is regulated by cell volume.

This mechanism may not be applicable to all Na-transporting epithelia. In the rabbit CCT, principal cells undergo a regulatory volume decrease in response to hypotonic cell swelling, but this response apparently does not require an increase in either Cl or K permeability.[137] It was suggested that loss of organic ions may underlie the phenomenon. On the other hand, recent evidence using the patch-clamp technique[138] has indicated that an apical membrane K channel can be opened during cell swelling. This would lead to a cell volume decrease assuming that K is above its electrochemical equilibrium. These channels can be activated by increases in cell Ca, suggesting that Ca may be a mediator of the swelling response.

22.3.2.2. Feedback Inhibition of Na Channels

MacRobbie and Ussing[134] noted that inhibition of the Na pump of the frog skin appeared to diminish the apical Na permeability as assessed from swelling experiments. Similar results were obtained in principal cells of the rabbit CCT.[139] This suggests another type of cellular homeostatic mechanism: the cells protect themselves against substantial changes in the intracellular Na concentration. A number of different types of protocols have confirmed this general notion.[140,141] This phenomenon has been documented most directly in A6 cells and rat CCT, where changes in the Na entry rate across the apical membrane with amiloride or reduced mucosal Na stimulated apical Na channels which were protected from the perturbation by the patch-clamp pipette.[38,41] As discussed below, and diagrammed in Figure 22.6, several different types of cellular mechanisms may be involved.

Voltage. Cell voltage is perhaps the simplest of the proposed mediators of feedback control of Na channels. Although the gating of epithelial Na channels is not dramatically voltage dependent, the open probability is measurably increased by membrane hyperpolarization.[36] An increase in Na entry into the cell or a decrease in Na exit from the cell would both tend to depolarize the apical cell membrane and decrease channel activity. This mechanism has been postulated to account at least in part for alterations of channel open probabil-

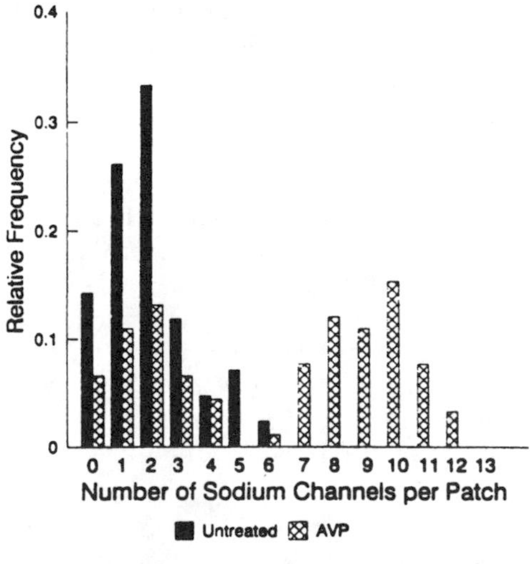

Figure 22.6 Regulation of the density of Na channels in A6 cells by ADH. Reproduced with permission from *The American Journal of Physiology*, Volume 260, pp. C1071–C1084.[37]

ity in response to modulation of Na entry rates in the rat CCT.[41]

Cell Ca^{2+}. Cell Ca has been suggested to play a key role in mediating feedback responses.[142] The idea is that intracellular Ca changes in parallel with changes in intracellular Na via the operation of a Na–Ca exchange mechanism in the basolateral membrane. Since the Ca gradient depends at least in part on the Na gradient across the membrane, increases in cell Na will reduce the gradients for both ions. Indeed, evidence for such a mechanism has been found in mammalian CCT.[143–145]

The mechanisms underlying the effects of Ca^{2+} are not yet resolved. A direct inhibition of channels by Ca was observed in vesicles from toad bladder but not in excised patches from the CCT.[3,140] Indirect, presumably enzyme-mediated effects of Ca were observed both in the vesicle preparation and in cell-attached patches of the CCT. When the CCT was treated with ouabain, similar to the original MacRobbie and Ussing experiment, inhibition of Na channels was observed in cell-attached patches.[144] This inhibition was well correlated with changes in intracellular Ca. The mechanisms involved are not known. One candidate is the Ca-dependent protein kinase (PKC). Inhibitors of PKC abolished the feedback response of A6 cells to low Na[38] as well as the feedback response of the CCT to ouabain.[146]

Cell pH. Another plausible mediator of feedback inhibition is cell pH. In the frog skin, both the apical Na channels and the basolateral K channels were found to be pH sensitive; acidification reduced both conductances.[85] Inhibition of the Na pump with ouabain decreased cell pH, probably by increasing intracellular Na. The cell pH in the frog skin is controlled in large part by a Na–H exchanger, such that extrusion of protons from the cell depends on the Na gradient, similar to the efflux of Ca^{2+} ions.[87] The acidification of the cell was sufficient to account for the inhibition of the apical Na conductance. The pH sensitivity of Na channels has also been observed in excised patches of the CCT.[140] In this epithelium, however, no change in cell pH was detected after ouabain treatment.[144] These findings indicate that variations in cell Ca and cell pH may be different in different species and under different conditions. The relative importance of these two factors in mediating the negative feedback response may be similarly variable.

Cell Metabolism. There is also evidence that the Na channels may be metabolically regulated. In the toad bladder, inhibition of metabolism results in a reversible inhibition of Na permeability.[140] If a similar response takes place during metabolic stress induced by high transport rates, this would serve as a feedback mechanism for matching the Na entry into the cell with the capability of the cell to provide energy for pumping the ion out again. Such a mechanism has been proposed to modulate the response of the cells to ADH or cAMP.[115]

22.3.2.3. Pump-Leak Balance

The idea that membrane K conductance should be regulated to match the activity of the Na/K pump in the basolateral membrane was put forward by Schultz.[147] The basic premise is that increases in the rate of pumping of Na ions out of the cell lead to parallel increases in K influx through the Na/K pump. In epithelia whose primary transport task is to reabsorb Na, most or all of this K will have to be recycled across the basolateral membrane. An increase in the recycling could be accomplished by an increase in the driving force for K to leave the cell through either a depolarization of the membrane or an increase in intracellular K concentration. It may be important, however, to keep these parameters constant, or at least to regulate them independently of the transport rate. If the K conductance were proportional to the rate of pumping, the recycling rate could increase at a constant driving force.

So far the concept has been examined most thoroughly in low-resistance, high-transport-rate epithelia such as the small intestine and proximal tubule. There is clear evidence, for example, that the basolateral membrane conductance of amphibian small intestine increases in parallel with the rate of Na entry through solute-coupled pathways.[148] However, the principle involved may apply also to high-resistance epithelia using the KJ-U mechanism.

Several mechanisms for mediating this "cross talk" between membrane transporters have been proposed and are summarized below. They are also illustrated in Figure 22.7.

Cell Volume. In a number of cells, swelling can trigger an activation of K channels which leads to the efflux of K and subsequent shrinking of the cells (see Chapter 11). Increased Na entry is expected to produce cell swelling via the accompanying influx of Cl^- and water. If the mechanism regulating basolateral K channels can sense this change in volume, this would provide a simple means of coupling the rates of Na transport and K recycling. K channels that are activated directly by membrane stretch or cell swelling have been described in amphibian proximal tubule (see Chapter 11). So far no comparable stretch-sensitive channels have been observed on the basolateral membrane of high-resistance epithelia.

Cell Ca^{2+}. Cell Ca^{2+} provides an attractive possibility for coupling Na entry with K conductance.[149] As discussed above, increases in cell Na concentration can lead to parallel increases in cell Ca through changes in the driving force for Na–Ca exchange. In addition, cell swelling has also been shown to lead to increases in cell Ca^{2+} in tight epithelia.[150] If the basolateral membrane contains Ca-activated K channels,

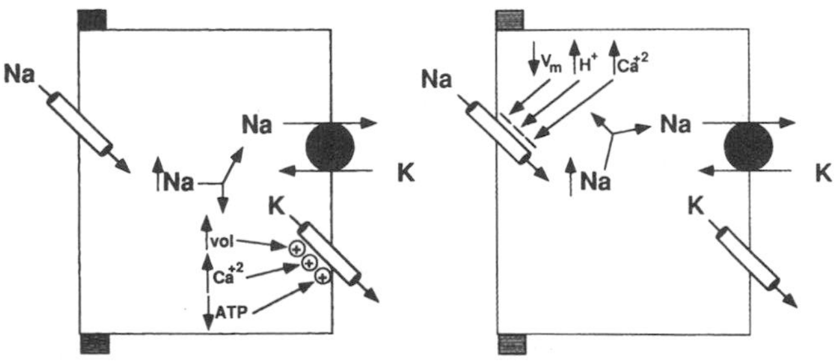

Figure 22.7 (A) Feedback regulation of Na channels. (B) Parallel pump-leak regulation at the basolateral membrane.

these channels could provide the necessary increase in basolateral K conductance when Na entry is stimulated.

The clearest evidence for an important Ca-activated basolateral K conductance is seen in the turtle colon.[71] Permeabilization of the apical membrane with digitonin revealed a basolateral K conductance that is closed at normal, low intracellular Ca^{2+} concentrations but that can be activated by Ca^{2+} in the micromolar range. Ca-activated channels have also been found in the basolateral membrane of the frog skin.[79] High-conductance, Ca- and voltage-activated K channels have also been observed in the *apical* membrane of the mammalian CCT.[151] Activation of these channels would also mediate K efflux and a balance of cell K and volume, although the net result would in this case be a transepithelial *secretion* rather than a recycling of K.

ATP. As discussed above, basolateral K channels in several KJ-U-type epithelia are sensitive to high levels of ATP, at least in excised patches. This could provide another means of coupling Na transport and K recycling rates. When the Na/K pump is operating at high rates, ATP levels may fall, particularly in the vicinity of the basolateral cell membrane where the ATPase is operating. If K channel activity is suppressed by physiological levels of ATP, then such a fall in ATP would activate the K conductance whenever the transport rate is high.

Strong evidence for a link between the Na pump and basolateral K channels was obtained in proximal tubule cells. Application of ouabain to block the pump resulted in the closure of K channels measured in cell-attached patches.[152] These channels were also shown to be ATP-sensitive in excised patches. Furthermore, the basolateral K conductance measured with microelectrode techniques was shown to increase when Na-coupled solute entry was enhanced, and this response was diminished when the cells were loaded with ATP.[153] However, in these experiments ATP loading also depressed the increase in apical conductance normally associated with electrogenic Na–substrate entry. This may indicate a complicated role for ATP in pump-leak regulation.

There is also evidence for regulation of basolateral K channels by cellular ATP in KJ-U epithelia. In A6 cells, the basolateral K conductance can be reduced by addition of glucose plus insulin, which increases ATP.[66] The conductance in these cells is also sensitive to sulfonylureas, which are known to block ATP-sensitive channels in the pancreas and heart. Inhibition by ATP of inwardly rectifying K channels from the basolateral membrane of frog skin has also been demonstrated.[61]

22.3.3. Conclusions

All three components of the KJ-U system are affected during regulation of Na transport. The rate-limiting step for transport is usually Na entry into the cell through apical Na channels. These channels seem to be upregulated in all cases where transport is increased. In many instances, however, the Na/K pump is also stimulated, allowing it to keep up with the increased Na entry rate without an increase in intracellular Na concentration. When this does not occur, the Na channels may be downregulated through feedback control mechanisms. To what extent the upregulation of the pump is dependent on the increased Na permeability, or is accomplished through independent mechanisms, is controversial. In any case, stimulation of the pump is in turn often accompanied by an increased basolateral K conductance (pump-leak parallelism). This is thought to be mediated by factors such as cell swelling or changes in intracellular Ca^{2+} or ATP.

22.4. COUNTERION MOVEMENTS IN Na-TRANSPORTING EPITHELIA

The model of Koefoed-Johnsen and Ussing explains the movement of Na across the epithelium but does not specify the route or mechanism of transport of other ions that must accompany the transport of Na. Except under short-circuited conditions, the absorption of each Na^+ ion must be balanced by the absorption of an anion or secretion of a cation to maintain electroneutrality. Subsequent work has identified several different counterions that cross various epithelia through different pathways. To a large extent the nature of the ions transported with or in exchange for Na depends on the particular functions of the epithelium as well as the particular needs of the organism. They can be different under different physiological conditions. Some of these pathways will be discussed below and are illustrated in Figure 22.8.

22.4.1. Reabsorption of Cl^-

When the active transport of Na^+ is neutralized by Cl^- movements, the net effect is the reabsorption of NaCl. Because the Na transport system generates a transepithelial potential difference in which the serosal side is positive with respect to the mucosa, the movement of Cl^- can in principle be passive, i.e., driven by the electrical potential. This is not necessarily the case, however, as discussed below. Three basically different routes of Cl movement have been postulated: between the cells (paracellular route), through the cells that transport Na (principal or granular cells), and through cells different from the major Na-transporting cells (intercalated or mitochondria-rich cells).

22.4.1.1. Paracellular Transport

Although the paracellular pathway is an important route of solute reabsorption in low-resistance epithelia, the extent of ion movement through this pathway in high-resistance epithelia is difficult to determine. Because the paracellular conductance is low—often much lower than that of the transcellular paths—it is difficult to separate from experimentally induced conductance or "edge damage."[154,155] In the frog skin, properties of the paracellular pathway have been estimated by inhibiting the conductive transporters in the cell membranes using amiloride (for the Na channels) and Cu (which apparently inhibits Cl channels). From such data it has been concluded that the normal paracellular conductance is around 75

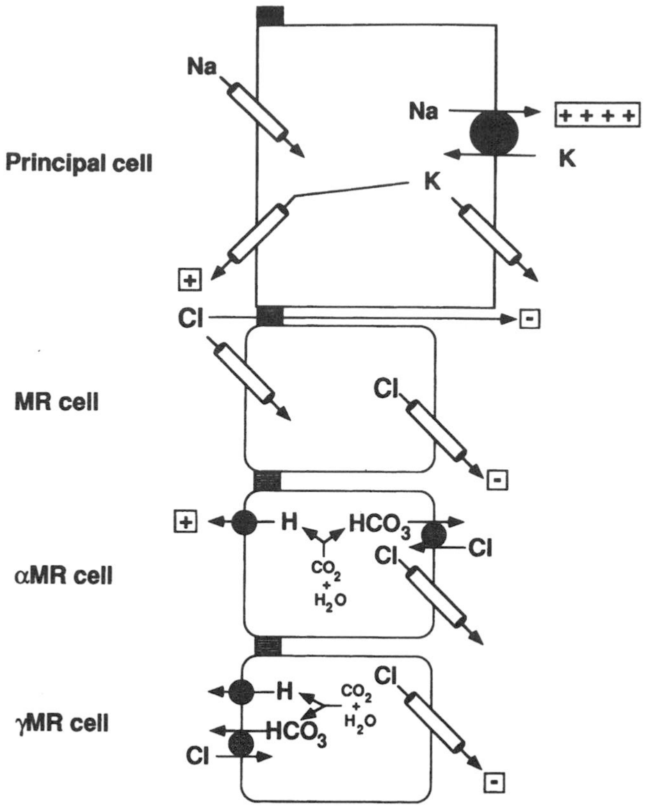

Figure 22.8. Ion movements balancing the reabsorption of Na in tight epithelia.

kΩcm^2, with a Cl permeability on the order of 1×10^{-8} cm/sec.[156] With Ringer's solution on both sides of the epithelium, and a transepithelial potential of 75 mV, this corresponds to a net influx of Cl^- of around 1 μA/cm^2 or 35 nmole/hr•cm^2. Since net Na fluxes measured under these conditions are around 700 nmole/hr•cm^2,[157] this would indicate that a rather small portion of the active transport of Na can be accompanied by Cl^- reabsorption through the paracellular pathway. When the mucosal solution is more dilute, corresponding more closely to pond water, the uptake of Cl^- by this mechanism will be even less. Experimental observations on frog skin under open-circuited conditions showed that Cl^- influx was considerably lower than Na^+ influx, despite the much more favorable electrical driving force for Cl^- uptake.[157]

In mammalian KJ-U epithelia, such as colon or CCT, the overall transepithelial resistance is much lower than in frog skin or toad urinary bladder. The paracellular resistance of these tissues has also been estimated from the transepithelial resistance after inhibiting the known cell membrane conductances with amiloride and Ba (to block K channels). In the rabbit CCT, values for the remaining resistance range from 100 to 1000 Ωcm^2.[158] Again, it is difficult to know to what extent the lower values of paracellular resistance might reflect experimentally induced conductances through damaged parts of the tissue. At the lowest values the paracellular pathway becomes a major route for Cl transport. When the urine NaCl concentration is isotonic, calculations indicate that as much as half the net Na reabsorption can be balanced by paracellular Cl reabsorption. On the other hand, this percentage becomes much lower if the higher values of paracellular resistance are used.[158]

It is energetically inefficient to reabsorb Cl^- through the paracellular pathway. Assuming that the permeabilities to Na^+ and Cl^- are about equal,[2] and that the mucosal fluid is isotonic NaCl, then for each Cl^- that is reabsorbed a Na^+ ion will be secreted. That means that to reabsorb 1 molecule of NaCl, 2 Na^+ ions will have to be actively transported. The process becomes even less efficient if the mucosal solution is hypotonic, since the backflux of Na^+ will be larger than the forward flux of Cl^- under these conditions.

22.4.1.2. Transcellular Cl Conductance

The Cl permeability of the major Na-transporting cell types in most KJ-U epithelia is very low.[156] Thus, Cl movement through these cells is probably not of quantitative importance in the balancing of Na fluxes. One exception to this rule is the reabsorptive cell of the human sweat duct.[159] Here the apical membrane contains both amiloride-sensitive Na^+ channels and Cl^- channels. These Cl^- channels are defective in cystic fibrosis. The basolateral membrane also has Cl^- channels although these have not been further identified. In this case Cl^- reabsorption can be driven by the transepithelial voltage established by the Na^+ transport system. This system is more efficient than that of the paracellular pathway, since there will be no backflux of Na^+ through the Cl^- selective channels.

In the amphibian skin another conductive path for Cl has been identified in the minority of mitochondrion-rich cells.[156] These channels are strongly voltage-dependent. Under short-circuited conditions, when the transepithelial potential is zero, the conductance through this pathway is very low. When the mucosal potential becomes negative, corresponding to the physiological situation, the conductance becomes activated. The activation process is slow compared to that of excitable cells, requiring minutes to become complete. Studies of local current densities using vibrating probes as well as optical measurements of cell volume changes in response to voltage perturbations have localized this conductance to the mitochondrion-rich cells.[156] The currents through this pathway can be large—on the order of 100 μA/cm^2 tissue—and could account for substantial Cl^- movements when the transepithelial potential is large enough to activate it. Again the pathway is selective for Cl^- over Na^+, making it energetically efficient. Because of the voltage dependence of the conductance, it will open when the Na transport system is activated. Thus far, similar channels have not been identified in the analogous intercalated cells of the mammalian kidney.

22.4.1.3. Active Cl Reabsorption

Under some conditions active Cl reabsorption across the frog skin has been observed.[160,161] Recently a model accounting for this transport has been suggested.[162] According to this scheme, the primary driving force for this transport is the ATP-dependent translocation of protons across the apical membrane of MR cells. The protons for this pump come primarily from the hydration of CO_2; the HCO_3^- formed by this reaction can leave the cell through an anion exchanger driving Cl^- into the cell across either the apical or basolateral membrane. When the exchanger is on the *apical* membrane, the efflux of HCO_3^- neutralizes the secretion of protons. The net result is the uptake of Cl^-, which can occur against an electrochemical activity gradient. Cl^- can then exit the cells through channels in the basolateral membrane. The overall reaction is then the transepithelial absorption of Cl^-, which can balance the uptake of Na^+ through the principal cells via the KJ-U mechanism. This is also an energetically expensive transport mechanism, as hydroysis of one ATP is required for each Cl^- ion absorbed, in addition to that needed for Na^+ uptake. It may be useful, however, for the uptake of NaCl from very dilute solutions, such as a freshwater pond.[162]

22.4.2. Proton Secretion

If, in the intercalated cell described above, the $Cl–HCO_3$ exchanger is on the *basolateral* membrane, the net transport reaction driven by the proton pump will be the secretion of a proton into the apical solution. This is the conventional transport scheme for acid secretion by the turtle urinary bladder and the mammalian CCT. The proton pumps are localized at the apical membrane of the α- or a-type intercalated cells.[163] The pumps themselves are of the V or vacuolar type.[164]

This system will be electrically coupled with the KJ-U system. The operation of one system will aid the other by providing a favorable electrical driving force. Under these circumstances Na^+ is absorbed in exchange for H^+ rather than as NaCl. There is evidence that when the amphibian skin reabsorbs Na from very dilute solutions, the uptake is primarily in exchange for H^+ and is therefore coupled to the proton secretory system.[165,166]

In other epithelia, the contribution of proton secretion to the neutralization of Na reabsorption may depend on the acid–base status of the animal, i.e., on the need to acidify the urine. The proton pumps are highly regulated, as studied most extensively in the turtle urinary bladder and the mammalian CCT.[163,164] The pumps are activated acutely by CO_2 in respiratory acidosis. In addition, the ability to secrete protons is also enhanced by chronic acidosis.

22.4.3. K Secretion

A third basic way of neutralizing Na^+ reabsorption is by secreting K^+ ions. As in the case with Cl^- absorption, the secretion of K^+ could in principle be through either paracellular or transcellular pathways. However, the low concentration of K^+ relative to Na^+ in the interstitial fluid would imply that the net movement of K^+ through paracellular spaces would be much smaller than the backflux of Na^+. Thus, the relatively nonselective nature of this pathway makes it poorly suited as a route for K^+ secretion.

On the other hand, the cells contain high K concentrations, so that the fluxes of K from cell to mucosal fluid can be large. Furthermore, since the Na pump moves two K^+ ions into the cell for every three Na^+ ions moved out, efflux of all of the pumped K^+ across the apical membrane would neutralize two-thirds of the active Na^+ transport. This assumes that no K^+ is recycled across the basolateral membrane. To the extent that K^+ is recycled, the fraction will be reduced.

K secretion through such a mechanism requires the presence of an apical K channel or conductance. In the amphibian KJ-U systems, the apical membrane K conductance is very low under most conditions. Changes in mucosal K concentration have little effect on the transepithelial voltage or conductance or on the apical membrane voltage or resistance divider ratio as measured with microelectrodes.[1,167,168] However, channels that can conduct both K^+ and Na^+ can be activated either by hyperpolarizing the epithelium[169] or by reducing the mucosal Ca^{2+} concentration.[170,171] The pathway has been tentatively identified at the single-channel level in toad bladder using patch-clamp techniques.[172] These channels may serve to secrete K^+ and to absorb Na^+ under conditions when the transepithelial voltage is high and/or the mucosal Ca^{2+} is low. It is not known to what extent they do so *in vivo*.

In the renal CCT, K secretion appears to be quantitatively more important in balancing Na movements. In isolated perfused rabbit CCT, the measured values of net K secretion are one-third to one-half those of net Na reabsorption.[173] Somewhat lower ratios (one-seventh to one-third) were found in the rat CCT.[174] In these cells the apical membrane K conductance is actually larger than the Na conductance.[97,98,175] The smaller net fluxes for K^+ relative to Na^+ are the result of the smaller electrochemical driving forces for K efflux across the apical membrane compared with Na influx.

The K channels involved in this K secretory process have been identified using the patch-clamp technique in the rat CCT.[151] Like the basolateral K channels discussed above, they are mildly inwardly rectifying, and have a single-channel conductance of 15–30 pS depending on whether inward or outward currents are measured. The open probability is normally quite high (>0.9) and is weakly voltage dependent. The pharmacological signature of the channels is a voltage-dependent block by millimolar concentrations of external Ba^{2+}, and an insensitivity to external TEA^+.

The density of these "SK" channels in the CCT will also vary with the physiological state, especially the K balance, of the animal.[151] Rats on a high-K diet, which need to secrete more K in the urine, have a higher density of conducting channels. The SK channels are also upregulated by phosphorylation with PKA and downregulated by phosphorylation with PKC and by high concentrations of ATP.

Based on the biophysical characteristics of these channels they are probably from a family of K-selective channels cloned from the kidney.[68,69] These channels, called ROMK1 and ROMK2, have structures that are quite different from voltage-gated channels, but are closely related to the inward rectifier channels that have also recently been cloned from macrophage and heart muscle.[67–69,176] The predicted channel proteins are small—around 40 kDa—and probably span the membrane twice. The number of subunits required to form a conducting unit is unknown.

22.4.4. Conclusions

Many epithelia use the basic KJ-U system for reabsorbing Na. The ions transported with or in exchange for Na to balance charge vary considerably from tissue to tissue and for various experimental conditions. These balancing ionic movements can be largely Cl^- reabsorption (sweat duct), or largely H^+ secretion (frog skin in dilute solutions). In some cases K^+ secretion is important (mammalian CCT). Indeed, the organism can make use of this versatility to maintain the balance of acid or of K^+ as the need arises.

22.5 SUMMARY

Since the original publication in 1958 by Koefoed-Johnsen and Ussing, many details of the mechanism underlying Na reabsorption by epithelia like the frog skin have come to light. This understanding has been driven in large part by technological innovations, particularly in the fields of electrophysiology and molecular biology. Some further work on the individual transporters will be necessary. Basolateral K channels, for example, still need to be identified at the molecular level. Additional directions for future studies will center on biochemistry, such as the nature of modification of Na channels during stimulation and inhibition of Na transport, and systems physiology, including the identification of hormones and cytoplasmic factors that regulate transport *in vivo.* In all cases the future and well as the past research will rest on the firm foundation of the KJ-U model.

REFERENCES

1. Koefoed-Johnsen, V., and Ussing, H. H. (1958). On the nature of the frog skin potential. *Acta Physiol. Scand.* **42:**298–308.
2. Palmer, L. G., and Sackin, H. (1992). Electrophysiological analysis of transepithelial transport. In *The Kidney: Physiology and Pathophysiology* (D. W. Seldin and G. Giebisch, eds.), Raven Press, New York, pp. 361–405.
3. Garty, H., and Benos, D. J. (1988). Characteristics and regulatory mechanisms of the amiloride-blockable Na^+ channel. *Physiol. Rev.* **68:**309–373.
4. Palmer, L. G. (1992). Epithelial Na channels: Function and diversity. *Annu. Rev. Physiol.* **54:**51–66.
5. Rossier, B. C., and Palmer, L. G. (1992). Mechanisms of aldosterone action on sodium and potassium transport in *The Kidney: Physiology and Pathophysiology* (D. W. Seldin and G. Giebisch, eds.), Raven Press, New York, Volume 1, pp. 1373–1409.
6. Burckhardt, G., and Greger, R. (1992). Principles of electrolyte transport across plasma membranes of renal tubular cells. In *Handbook of Physiology Section 8: Renal Physiology* (E. E. Windhager, ed.), Oxford University Press, London, pp. 639–657.
7. Burckhardt, G., and Kinne, R. K. H. (1992). Transport proteins: Cotransporters and countertransporters. In *The Kidney: Physiology and Pathophysiology* (D. W. Seldin and G. Giebisch, eds.), Raven Press, New York, pp. 537–586.
8. De Weer, P. (1992). Cellular sodium–potassium transport. In *The Kidney: Physiology and Pathophysiology* (D. W. Seldin and G. Giebisch, eds.), Raven Press, New York, pp. 93–112.
9. Horisberger, J.-D., Lemas, V., Kraehenbuhl, J. P., and Rossier, B. C. (1991). Structure–function relationship of Na,K-ATPase. *Annu. Rev. Physiol.* **53:**565–584.
10. Jorgensen, P. L. (1982). Mechanism of the Na,K-ion pump. Protein structure and conformations of the pure Na,K-ATPase. *Biochim. Biophys.* **694:**27–68.
11. Jorgensen, P. L. (1986). Structure, function and regulation of Na,K-ATPase in the kidney. *Kidney Int.* **29:**10–20.
12. Sweadner, K. J. (1989). Isozymes of the Na^+/K^+-ATPase. *Biochim. Biophys. Acta* **988:**185–220.
13. Tymiak, A. A., Norman, J. A., Bolgar, M., DiDonato, G. C., Lee, H., Parker, W. L., Lo, L.-C., Berova, N., Nakanishi, K., Haber, E., and Haupert, G. T. J. (1993). Physicochemical characterization of a ouabain isomer isolated from bovine hypothalamus. *Proc. Nat. Acad. Sci. USA* **90:**8189–8193.
14. Ernst, S. A., and Mills, J. W. (1980). Autoradiographic localization of tritiated ouabain-sensitive sites in ion transporting epithelia. *J. Histochem. Cytochem.* **28:**72–77.
15. Zurzolo, C., Rodriguez-Boulan, E., Gottardi, C. J., Caplan, M., Siemers, K. A., Wilson, R., Mays, R. W., Ryan, T. A., Wollner, D. A., and Nelson, W. J. (1993). Delivery of Na^+,K^+-ATPase in polarized epithelial cells (Technical Comment). *Science* **260:**550–556.
16. Caplan, M. J., Anderson, H. C., Palade, G. E., and Jamieson, J. D. (1986). Intracellular sorting and polarized cell surface delivery of Na,K-ATPase, an endogenous component of MDCK cell basolateral plasma membranes. *Cell* **46:**623–631.
17. Hammerton, R. W., Krzeminski, K. A., Mays, R. W., Wollner, D. A., and Nelson, W. J. (1991). Mechanism for regulating cell surface distribution of Na^+,K^+-ATPase in polarized epithelial cells. *Science* **254:**847–853.
18. Kirk, K. L., Halm, D. R., and Dawson, D. C. (1980). Active sodium transport by turtle colon via an electrogenic Na-K exchange pump. *Nature* **287:**237–239.
19. Lewis, S. A., and Wills, N. K. (1983). Apical membrane permeability and kinetic properties of the sodium pump in rabbit urinary bladder. *J. Physiol. (London)* **341:**169–184.
20. Palmer, L. G., and Speez, N. (1986). Stimulation of apical Na permeability and basolateral Na pump of toad urinary bladder by aldosterone. *Am. J. Physiol.* **250:**F273–F281.
21. Avison, M. J., Gullans, S. R., Ogino, T., Giebisch, G., and Shulman, R. G. (1987). Measurement of Na^+-K^+ coupling ratio of Na^+-K^+-ATPase in rabbit proximal tubules. *Am. J. Physiol.* **253:**C126–C136.
22. Cox, T. C., and Helman, S. I. (1986). Na^+ and K^+ transport at basolateral membranes of epithelial cells. I. Stoichiometry of the Na,K-ATPase. *J. Gen. Physiol.* **87:**467–483.
23. Sackin, H., and Boulpaep, E. L. (1983). Rheogenic transport in the renal proximal tubule. *J. Gen. Physiol.* **82:**819–851.
24. Sansom, S. C., and O'Neil, R. G. (1986). Effects of mineralocorticoids on transport properties of cortical collecting duct basolateral membrane. *Am. J. Physiol.* **251:**F743–F757.
25. Post, R. L., Merritt, C. R., Kinsolving, C. R., and Albright, C. D. (1960). Membrane adenosine triphosphatase as a participant in the active transport of sodium and potassium in the human erythrocyte. *J. Biol. Chem.* **235:**1796–1802.
26. Green, A. L., and Taylor, C. B. (1964). Kinetics of ($Na^+ + K^+$)-stimulated ATPase of rabbit kidney microsones. *Biochem. Biophys. Res. Commun.* **14:**118–123.
27. Lindenmayer, G. E., Schwartz, A., and Thompson, H. K. (1974). A kinetic description for sodium and potassium effects on ($Na^+ + K^+$)-

adenosine triphosphatase: A model for a two-nonequivalent site potassium activation and analysis of multiequivalent site models for sodium activation. *J. Physiol. (London)* **236:**1–28.
28. Halm, D. R., and Dawson, D. C. (1983). Cation activation of the basolateral sodium–potassium pump in turtle colon. *J. Gen. Physiol.* **82:**315–329.
29. Rossier, B. C., Canessa, C. M., Schild, L., and Horisberger, J.-D. (1994). Epithelial sodium channels. *Curr. Opin. Nephrol. Hypertens.,* **3:**487–496.
30. Eaton, D. C., and Hamilton, K. L. (1988). The amiloride-blockable sodium channel of epithelial tissue. In *Ion Channels* (T. Narahashi, ed.), Plenum Press, New York, Volume 1, pp. 151–182.
31. Smith, P. R., and Benos, D. J. (1991). Epithelial Na^+ channels. *Annu. Rev. Physiol.* **53:**509–530.
32. Palmer, L. G. (1991). The epithelial Na channel: Inferences about the nature of the conducting pore. *Comments Mol. Cell. Biophys.* **7:**259–283.
33. Palmer, L. G. (1987). Ion selectivity of epithelial Na channels. *J. Memb. Biol.* **96:**97–106.
34. Hille, B. (1992). *Ionic Channels of Excitable Membranes,* Sinauer, Sunderland, MA.
35. Lindemann, B. (1984). Fluctuation analysis of sodium channels in epithelia. *Annu. Rev. Physiol.* **46:**497–515.
36. Palmer, L. G., and Frindt, G. (1988). Conductance and gating of epithelial Na channels from rat cortical collecting tubules. Effects of luminal Na and Li. *J. Gen. Physiol.* **92:**121–138.
37. Marunaka, Y., and Eaton, D. C. (1991). Effects of vasopressin and cAMP on single amiloride-blockable Na channels. *Am. J. Physiol.* **260:**C1071–C1084.
38. Ling, B. N., and Eaton, D. C. (1989). Effects of luminal Na^+ on single Na^+ channels in A6 cells, a regulatory role for protein kinase C. *Am. J. Physiol.* **256:**F1094–F1103.
39. Palmer, L. G., and Frindt, G. (1996). Gating of Na channels in the rat cortical collecting tubule: Effects of voltage and membrane stretch. *J. Gen. Physiol.,* in press.
40. Palmer, L. G., and Frindt, G. (1986). Amiloride-sensitive Na channels from the apical membrane of the rat cortical collecting tubule. *Proc. Nat. Acad. Sci. USA* **83:**2767–2770.
41. Frindt, G., Silver, R. B., Windhager, E. E., and Palmer, L. G. (1993). Feedback inhibition of Na channels in rat CCT. II. Effects of inhibition of Na entry. *Am. J. Physiol.* **264:**F565–F574.
42. Palmer, L. G., and Kleyman, T. R. (1995). Potassium-sparing diuretics: Amiloride. In *Handbook of Experimental Pharmacology, Diuretics* (E. Mutschler and R. Greger, eds), Springer-Verlag, Berlin.
43. Kleyman, T. R., and Cragoe, E. J. J. (1988). Amiloride and its analogs as tools in the study of ion transport. *J. Memb. Biol.* **105:**1–21.
44. Benos, D. J. (1982). Amiloride: A molecular probe of sodium transport in tissues and cells. *Am. J. Physiol.* **242:**C131–C145.
45. Eaton, D. C., and Marunaka, Y. (1990). Ion channel fluctuations: "Noise" and single-channel measurements. *Curr. Top. Membr. Transp.* **37:**61–113.
46. Palmer, L. G., and Andersen, O. S. (1989). Interactions of amiloride and small monovalent cations with the epithelial sodium channel. Inferences about the nature of the channel pore. *Biophys. J.* **55:** 779–787.
47. Cantiello, H. F., Stow, J., Prat, A. G., and Ausiello, D. A. (1991). Actin filaments control epithelial Na^+ channel activity. *Am. J. Physiol.* **261:**C882–C888.
48. Light, D. B., Corbin, J. D., and Stanton, B. A. (1990). Amiloride-sensitive cation channel in apical membrane of inner medullary collecting duct. *Am. J. Physiol.* **255:**F278–F286.
49. Chinet, T. C., Fulton, J. M., Yankaskas, J. R., Coucher, R. C., and Stutts, M. J. (1993). Sodium-permeable channels in the apical membrane of human nasal epithelial cells. *Am. J. Physiol.* **265:**C1050–C1060.
50. Lingueglia, E., Voilley, N., Waldmann, R., Lazdunski, M., and Barbry, P. (1993). Expression cloning of an epithelial amiloride-sensitive Na^+ channel. *FEBS Lett* **318:**95–99.
51. Canessa, C. M., Horisberger, J.-D., and Rossier, B. C. (1993). Epithelial sodium channel related to proteins involved in neurodegeneration. *Nature* **361:**467–470.
52. Canessa, C. M., Schild, L., Buell, G., Thorens, B., Gautschi, Y., Horisberger, J.-D., and Rossier, B. C. (1994). The amiloride-sensitive epithelial sodium channel is made of three homologous subunits. *Nature* **367:**463–467.
53. Renard, S., Lingueglia, E., Voilley, N., Lazdunski, M., and Barbry, P. (1994). Biochemical analysis of the membrane topology of the amiloride-sensitive Na^+ channel. *J. Biol. Chem.* **269:**12981–12986.
54. Canessa, C. M., Mérillat, A.-M., and Rossier, B. C. (1994). Membrane topology of the epithelial sodium channel. *Am. J. Physiol.* **267:**C1682–C1690.
55. Oh, Y., and Benos, D. J. (1993). Single-channel characteristics of a purified bovine renal amiloride-sensitive Na^+ channel in planar lipid bilayers. *Am. J. Physiol.* **264:**C1489–C1499.
56. Sariban-Sohraby, S., Abramov, M., and Fisher, R. S. (1992). Single-channel behavior of a purified epithelial Na^+ channel subunit that binds amiloride. *Am. J. Physiol.* **263:**C1111–C1117.
57. Nagel, W. (1985). Basolateral membrane ionic conductance in frog skin. *Pfluegers Arch.* **405:**S39–S43.
58. Horisberger, J.-D., and Giebisch, G. (1988). Voltage dependence of the basolateral membrane conductance in the *Amphiuma* collecting tubule. *J. Membr. Biol.* **105:**257–263.
59. Broillet, M.-C., and Horisberger, J. D. (1993). Basolateral membrane potassium conductance of A6 cells. *J. Membr. Biol.* **124:**1–12.
60. Garcia-Diaz, J. F. (1991). Whole-cell and single channel K^+ and Cl^- currents in epithelial cells of frog skin. *J. Gen. Physiol.* **98:**131–161.
61. Urbach, V., Van Kerkhove, E., and Harvey, B. J. (1994). Inward-rectifier potassium channels in basolateral membranes of frog skin epithelium. *J. Gen. Physiol.* **103:**583–604.
62. Ashcroft, F. M. (1988). Adenosine 5′-triphosphate-sensitive potassium channels. *Annu. Rev. Neurosci.* **11:**97–118.
63. Matsuda, H., Saigusa, A., and Irisawa, H. (1987). Ohmic conductance through the inwardly rectifying K channel and blocking by internal Mg^{2+}. *Nature* **325:**156–159.
64. Hirsch, J., and Schlatter, E. (1993). K^+ channels in the basolateral membrane of rat cortical collecting duct. *Pfluegers Arch.* **424:**470–477.
65. Germann, W. J., Ernst, S. A., and Dawson, D. C. (1986). Resting and osmotically induced basolateral K conductances in turtle colon. *J. Gen. Physiol.* **88:**253–274.
66. Broillet, M.-C., and Horisberger, J.-D. (1993). Tolbutamide-sensitive potassium conductance in the basolateral membrane of A6 cells. *J. Membr. Biol.* **134:**181–188.
67. Kubo, Y., Baldwin, T. J., Jan, Y. N., and Jan, L. Y. (1993). Primary structure and functional expression of a mouse inward rectifier potassium channel. *Nature* **362:**127–133.
68. Zhou, H., Tate, S. S., and Palmer, L. G. (1994). Primary structure and functional properties of an epithelial K channel. *Am. J. Physiol.* **266:**C809–C824.
69. Ho, K. H., Nichols, C. G., Lederer, W. J., Lytton, J., Vassilev, P. M., Kanazirska, M. V., and Hebert, S. C. (1993). Cloning and expression of an inwardly rectifying ATP-regulated potassium channel. *Nature* **362:**31–37.
70. Richards, N. W., and Dawson, D. C. (1986). Single potassium channels blocked by lidocaine and quinidine in isolated turtle colon epithelial cells. *Am. J. Physiol.* **251:**C85–C89.
71. Chang, D., and Dawson, D. C. (1988). Digitonin-permeabilized colonic cell layers. *J. Gen. Physiol.* **92:**281–306.
72. Sharp, G. W. G., and Leaf, A. (1966). Mechanism of action of aldosterone. *Physiol. Rev.* **46:**593–633.

73. Civan, M. M., and Shporer, M. (1989). Physical state of cell sodium. *Curr. Top. Membr. Transp.* **34:**1–19.
74. Palmer, L. G., Antonian, L., and Frindt, G. (1993). Regulation of the Na-K pump of the rat cortical collecting tubule by aldosterone. *J. Gen. Physiol.* **102:**43–57.
75. Garty, H. (1986). Mechanisms of aldosterone action in tight epithelia (topical review). *J. Membr. Biol.* **90:**193–205.
76. Garty, H. (1992). Regulation of Na^+ permeability by aldosterone. *Semin. Nephrol.* **12:**24–29.
77. Laragh, J. (1992). The renin system and the renal regulation of blood pressure. In *The Kidney: Physiology and Pathophysiology, Second Edition* (D. W. Seldin and G. Giebisch, eds.), Raven Press, New York, pp. 1411–1453.
78. Oberleithner, H., Weigt, M., Westphale, H.-J., and Wang, W. (1987). Aldosterone activates Na^+-H^+ exchange and raises cytoplasmic pH in target cells of the amphibian kidney. *Proc. Nat. Acad. Sci. USA* **84:**1464–1468.
79. Harvey, B. J., and Urbach, V. (1993). Regulation of ion and water transport by hydrogen ions in high resistance epithelia. In *Advances in Environmental and Comparative Physiology 22. Acid–Base Regulation, Ion Transfer and Metabolism,* Springer-Verlag, Berlin, pp. 153–183.
80. Bastl, C. P., Schulman, G., and Cragoe, E. J. J. (1989). Low-dose glucocorticoids stimulate electroneutral NaCl absorption in rat colon. *Am. J. Physiol.* **257:**F1027–F1038.
81. Wehling, M., Käsmayr, J., and Theisen, K. (1991). Rapid effects of mineralocorticoids on sodium–proton exchanger: Genomic or nongenomic pathway? *Am. J. Physiol.* **260:**E719–E726.
82. Asher, C., Eren, R., Kahn, L., Yeger, O., and Garty, H. (1992). Expression of the amiloride-blockable Na^+ channel by RNA from control versus aldosterone-stimulated tissue. *J. Biol. Chem.* **267:**16061–16065.
83. Kemendy, A. E., Kleyman, T. R., and Eaton, D. C. (1992). Aldosterone alters the open probability of amiloride-blockable sodium channels in A6 epithelia. *Am. J. Physiol.* **263:**C825–C837.
84. Sariban-Sohraby, S., Burg, M., Wiesmann, W. P., Chiang, P. K., and Johnson, J. P. (1984). Methylation increases sodium transport into A6 apical membrane vesicles: Possible mode of aldosterone action. *Science* **225:**745–746.
85. Harvey, B. J., Thomas, S. R., and Ehrenfeld, J. (1988). Intracellular pH controls cell membrane Na^+ and K^+ conductances and transport in frog skin epithelium. *J. Gen. Physiol.* **92:**767–791.
86. Palmer, L. G., and Frindt, G. (1987). Effects of cell Ca and pH on Na channels from rat cortical collecting tubule. *Am. J. Physiol.* **253:** F333–F339.
87. Harvey, B. J., and Ehrenfeld, J. (1988). Role of Na^+/H^+ exchange in the control of intracellular pH and cell membrane conductances in frog skin epithelium. *J. Gen. Physiol.* **92:**793–810.
88. Szerlip, H. M., Weisberg, L., Clayman, M., Neilson, E., Wade, J. B., and Cox, M. (1989). Aldosterone-induced proteins: Purification and localization of GP65,70. *Am. J. Physiol.* **256:**C865–C872.
89. Rick, R., Spancken, G., and Dörge, A. (1988). Differential effects of aldosterone and ADH on intracellular electrolytes in the toad urinary bladder epithelium. *J. Membr. Biol.* **101:**275–282.
90. Pellanda, A. M., Gaeggler, H.-P., Horisberger, J.-D., and Rossier, B. C. (1992). Sodium-independent effect of aldosterone on initial rate of ouabain binding in A6 cells. *Am. J. Physiol.* **260:**C899–C906.
91. Park, C.-S., and Edelman, I. S. (1984). Effect of aldosterone on abundance and phosphorylation kinetics of Na-K-ATPase of toad urinary bladder. *Am. J. Physiol.* **246:**F509–F516.
92. Barlet-Bas, C., Khadouri, C., Marsy, S., and Doucet, A. (1988). Sodium-independent in vitro induction of Na^+-K^+ ATPase by aldosterone in renal target cells: Permissive effect of triiodothyronine. *Proc. Nat. Acad. Sci. USA* **85:**1701–1711.
93. Geering, K., Girardet, M., Bron, C., Kraehenbuhl, J.-P., and Rossier, B. C. (1982). Hormonal regulation of (Na^+,K^+)-ATPase biosynthesis in the toad bladder. Effect of aldosterone and 3,5,3′-triiodo-l-thyronine. *J. Biol. Chem.* **257:**10338–10343.
94. Barlet-Bas, C., Khadouri, C., Marsy, S., and Doucet, A. (1990). Enhanced intracellular sodium concentration in kidney cells recruits a latent pool of Na-K-ATPase whose size is modulated by corticosteroids. *J. Biol. Chem.* **265:**7799–7809.
95. Blot-Chabaud, M., Wanstok, F., Bonvalet, J. P., and Farman, N. (1990). Cell sodium-induced recruitment of Na^+-K^+ ATPase pumps in rabbit cortical collecting tubules is aldosterone dependent. *J. Biol. Chem.* **265:**11676–11681.
96. Turnheim, K., Hudson, R. L., and Schultz, S. G. (1987). Cell Na activities and transcellular Na absorption by descending colon from normal and Na-deprived rabbits. *Pfluegers Arch.* **410:**279–283.
97. Sansom, S. C., and O'Neil, R. G. (1985). Mineralocorticoid regulation of apical cell membrane Na^+ and K^+ transport of the cortical collecting duct. *Am. J. Physiol.* **248:**F858–F868.
98. Schlatter, E., and Schafer, J. A. (1987). Electrophysiological studies in principal cells of rat cortical collecting tubules. *Pfluegers Arch.* **409:**81–92.
99. Pácha, J., Frindt, G., Antonian, L., Silver, R., and Palmer, L. G. (1993). Regulation of Na channels of the rat cortical collecting tubule by aldosterone. *J. Gen. Physiol.* **102:**25–42.
100. Asher, C., and Garty, H. (1988). Aldosterone increases the apical Na^+ permeability of toad bladder by two different mechanisms. *Proc. Nat. Acad. Sci. USA* **85:**7413–7417.
101. Renard, S., Viollet, N., Bassilana, F., Lazdunski, M., and Barbry, P. (1995). Localization and regulation by steroids of the α, β and γ subunits of the amiloride-sensitive Na^+ channel in colon, lung and kidney. *Pfluegers Arch.* **430:**299–307.
102. Doucet, A., and Barlet-Bas, C. (1989). Involvement of Na^+,K^+-ATPase in antinatriuretic action of mineralocorticoids in mammalian kidney. *Curr. Top. Membr. Transp.* **34:**185–208.
103. Katz, A. I. (1990). Corticosteroid regulation of NaK-ATPase along the mammalian nephron. *Semin. Nephrol.* **10:**388–399.
104. Marver, D. (1992). Regulation of Na^+,K^+-ATPase by aldosterone. *Semin. Nephrol.* **12:**56–61.
105. Wade, J. B., Stanton, B. A., Field, M. J., Kashgarian, M., and Giebisch, G. (1990). Morphological and physiological responses to aldosterone: Time course and sodium dependence. *Am. J. Physiol.* **259:**F88–F94.
106. Verrey, F., Kraehenbuhl, J.-P., and Rossier, B. C. (1989). Aldosterone induces a rapid increase in the rate of Na,K-ATPase gene transcription in cultured kidney cells. *Mol. Endocrinol.* **3:**1369–1376.
107. O'Neil, R. G., and Hayhurst, R. A. (1985). Sodium-dependent modulation of the renal Na-K-ATPase: Influence of mineralocorticoids on the cortical collecting duct. *J. Membr. Biol.* **85:**169–179.
108. Hayhurst, R. A., and O'Neil, R. G. (1988). Time-dependent actions of aldosterone and amiloride on Na^+K^+ ATPase of cortical collecting duct. *Am. J. Physiol.* **254:**F689–F696.
109. Pressley, T. A. (1992). Ionic regulation of Na^+,K^+-ATPase expression. *Semin. Nephrol.* **12:**67–71.
110. Wiener, H., Nielsen, J. M., Klaerke, D. A., and Jørgensen, P. L. (1993). Aldosterone and thyroid hormone modulation of α1-, β1-mRNA and Na,K-pump sites in rabbit distal colon epithelium. Evidence for a novel mechanism of escape from the effect of hyperaldosteronism. *J. Membr. Biol.* **133:**203–211.
111. Skorecki, K. L. (1992). Molecular mechanisms of vasopressin action in the kidney. In *Handbook of Physiology Section 8: Renal Physiology* (E. E. Windhager, ed.), Oxford University Press, London, pp. 1185–1218.
112. Robertson, G. L. (1992). Regulation of vasopressin secretion. In *The Kidney: Physiology and Pathophysiology. Second Edition* (D. W. Seldin and G. Giebisch, eds.), Raven Press, New York, pp.1596–1613.
113. Helman, S. I., Cox, T. C., and Van Driessche, W. (1983). Hormonal control of apical membrane Na transport in epithelia. Studies with fluctuation analysis. *J. Gen. Physiol.* **82:**201–220.

114. Li, H.-Y., Palmer, L. G., Edelman, I. S., and Lindemann, B. (1982). The role of Na-channel density in the natriferic response of the toad urinary bladder to an antidiuretic hormone. *J. Membr. Biol.* **64:**77–89.
115. Frindt, G., Silver, R. B., Windhager, E. E., and Palmer, L. G. (1995). Feedback regulation of Na channels in rat CCT. III. Response to cAMP. *Am. J. Physiol.* **268:**F480–F489.
116. Oh, Y., Smith, P. R., Bradford, A. L., Keeton, D., and Benos, D. J. (1993). Regulation by phosphorylation of purified epithelial Na^+ channels in planar lipid bilayers. *Am. J. Physiol.* **265:**C85–C91.
117. Wade, J. B. (1986). Role of membrane fusion in hormonal regulation of epithelial transport. *Annu. Rev. Physiol.* **48:**213–223.
118. Lester, D. S., Sher, C. A., and Garty, H. (1988). Characterization of cAMP-induced activation of epithelial sodium channels. *Am. J. Physiol.* **254:**C802–C808.
119. Garty, H., and Edelman, I. S. (1983). Amiloride-sensitive trypsinization of apical sodium channels. Analysis of hormonal regulation of sodium transport in toad bladder. *J. Gen. Physiol.* **81:**785–803.
120. Atlas, S. A., and Maack, T. (1992). Atrial natriuretic factor. In *Handbook of Physiology Section 8: Renal Physiology* (E. E. Windhager, eds.), Oxford University Press, London, pp. 1577–1673.
121. Ballerman, B. J., and Zeidel, M. L. (1992). Atrial natriuretic hormone. In *The Kidney: Physiology and Pathophysiology. Second Edition* (D. W. Seldin and G. Giebisch, eds.), Raven Press, New York, pp. 1843–1884.
122. Light, D. B., Corbin, J. D., and Stanton, B. A. (1990). Dual ion-channel regulation by cyclic GMP and cyclic GMP-dependent protein kinase. *Nature* **344:**336–339.
123. Light, D. B., Schwiebert, E. M., Karlson, K. H., and Stanton, B. A. (1989). Atrial natriuretic peptide inhibits a cation channel in renal inner medullary collecting duct cells. *Science* **243:**383–385.
124. Das, S., Garepapaghi, M., and Palmer, L. G. (1991). Stimulation by cGMP of apical Na channels and cation channels in toad urinary bladder. *Am. J. Physiol.* **260:**C234–C241.
125. Siegel, B., and Civan, M. M. (1976). Aldosterone and insulin effects on driving force of Na^+ pump in toad bladder. *Am. J. Physiol.* **230:** 1603–1608.
126. Schoen, H. F., and Erlij, D. (1987). Insulin action on electrophysiological properties of apical and basolateral membrane of frog skin. *Am. J. Physiol.* **252:**C411–C417.
127. Fidelman, M. L., May, J. M., Biber, T. U. L., and Watlington, C. O. (1982). Insulin stimulation of Na^+ transport and glucose metabolism in cultured kidney cells. *Am. J. Physiol.* **242:**C121–C123.
128. Blazer-Yost, B., Cox, M., and Furlanetto, R. (1989). Insulin and IGFI receptor-mediated Na^+ transport in toad urinary bladders. *Am. J. Physiol.* **257:**C612–C620.
129. Marunaka, Y., Hagiwara, N., and Tohda, H. (1993). Insulin activates single amiloride-blockable Na channels in a distal nephron cell line (A6)*c251E. *Am. J. Physiol.* **263:** F392–F400.
130. Civan, M. M., Peterson-Yantorno, K., and O'Brien, T. G. (1988). Insulin and phorbol ester stimulate conductive Na^+ transport through a common pathway. *Proc. Nat. Acad. Sci. USA* **85:**963–967.
131. Ussing, H. H. (1965). Relationship between osmotic reactions and active sodium transport in frog skin epithelium. *Acta Physiol. Scand.* **63:**141–155.
132. Lipton, P. (1972). Effect of changes in osmolarity on sodium transport across toad bladder. *Am. J. Physiol.* **222:**821–828.
133. Wills, N. K., Millinoff, L. P., and Crowe, W. E. (1991). Na^+ channel activity in cultured renal (A6) epithelium: Regulation by solution osmolarity. *J. Membr. Biol.* **121:**79–90.
134. MacRobbie, E. A. C., and Ussing, H. H. (1961). Osmotic behavior of the epithelial cells of frog skin. *Acta Physiol. Scand.* **53:**348–365.
135. Strieter, J., Stephenson, J. L., Palmer, L. G., and Weinstein, A. W. (1990). Volume-activated chloride permeability can mediate cell volume regulation in a mathematical model of a tight epithelium. *J. Gen. Physiol.* **96:**319–344.
136. Ussing, H. H. (1982). Volume regulation of frog skin epithelium. *Acta Physiol. Scand.* **114:**363–369.
137. Strange, K. (1988). RVD in principal and intercalated cells of rabbit cortical collecting tubule. *Am. J. Physiol.* **255:**C612–C621.
138. Hirsch, J., Leipzinger, J., Fröbe, U., and Schlatter, E. (1993). Regulation and possible physiological role of the Ca^{2+}-dependent K^+ channel of cortical collecting ducts of the rat. *Pfluegers Arch.* **422:**492–498.
139. Strange, L. (1989). Ouabain-induced cell swelling in rabbit cortical collecting tubule: NaCl transport by principal cells. *J. Membr. Biol.* **107:**249–261.
140. Palmer, L. G., Frindt, G., Silver, R. B., and Strieter, J. (1989). Feedback regulation of epithelial sodium channels. *Curr. Top. Membr. Transp.* **34:**45–60.
141. Turnheim, K. (1991). Intrinsic regulation of apical sodium entry in epithelia. *Physiol. Rev.* **71:**429–445.
142. Taylor, A., and Windhager, E. E. (1979). Possible role of cytosolic calcium and Na–Ca exchange in regulation of transepithelial sodium transport. *Am. J. Physiol.* **236:**F505–F512.
143. Breyer, M. D. (1990). Regulation of water and salt transport in collecting duct through calcium-dependent signaling mechanisms. *Am. J. Physiol.* **260:**F1–F11.
144. Silver, R. B., Frindt, G., Windhager, E. E., and Palmer, L. G. (1993). Feedback regulation of Na channels in rat CCT. I. Effects of inhibition of the Na pump. *Am. J. Physiol.* **264:**F557–F564.
145. Taniguchi, S., Marchetti, J., and Morel, F. (1989). Na/Ca exchangers in collecting cells of rat kidney. A single tubule fura-2 study. *Plfuegers Arch.* **415:**191–197.
146. Frindt, G., Palmer, L. G., and Windhager, E. E. (1996). Feedback inhibition of Na channels in rat CCT. IV. Mediation by activation of PKC. *Am. J. Physiol.,* in press.
147. Schultz, S. G. (1981). Homeocellular regulatory mechanisms in sodium-transporting epithelia: Avoidance of extinction by "flush-through." *Am. J. Physiol.* **241:**F579–F590.
148. Schultz, S. G. (1989). Intracellular sodium activities and basolateral membrane potassium conductances of sodium-absorbing epithelial cells. *Curr. Top. Membr. Transp.* **34:**21–44.
149. Chase, H. S. (1984). Does calcium couple and apical and basolateral membrane permeabilities in epithelia? *Am. J. Physiol.* **247:**F869–F876.
150. Wong, S. M. E., DeBell, M. C., and Chase, H. S. J. (1990). Cell swelling increases intracellular free [Ca] in cultured toad bladder cells. *Am. J. Physiol.* **258:**F292–F296.
151. Wang, W., Sackin, H., and Giebisch, G. (1993). Renal potassium channels and their regulation. *Annu. Rev. Physiol.* **54:**81–96.
152. Hurst, A. M., Beck, J., Laprade, R., and Lapointe, J.-Y. (1993). Na pump inhibition downregulates an ATP-sensitive K channel in rabbit proximal tubule. *Am. J. Physiol.* **264:**F760–F764.
153. Tsuchiya, K., Wang, W., Giebisch, G., and Welling, P. A. (1992). ATP is a coupling modulator of parallel Na/K ATPase K channel activity in the renal proximal tubule. *Proc. Nat. Acad. Sci. USA* **89:** 6418–6422.
154. Helman, S. I., and Miller, D. A. (1971). In vitro techniques for avoiding edge damage in studies of the frog skin. *Science* **173:** 146–148.
155. Erlij, D. (1976). Basic electrical properties of tight epithelia determined with a simple method. *Pfluegers Arch.* **364:**91–93.
156. Larsen, E. H. (1991). Chloride transport by high resistance heterocellular epithelia. *Physiol. Rev.* **71:**235–283.
157. Ussing, H. H. (1949). The active ion transport through the isolated frog skin in the light of tracer studies. *Acta Physiol. Scand.* **17:**1–37.
158. Strieter, J., Stephenson, J. L., Giebisch, G., and Weinstein, A. M. (1992). A mathematical model of the rabbit cortical collecting tubule. *Am. J. Physiol.* **263:**F1063–F1075.
159. Quinton, P. M. (1990). Cystic fibrosis: A disease in electrolyte transport. *FASEB J.* **4:**2709–2717.

160. Zadunaisky, J. A., Candia, O. A., and Chiarandini, D. J. (1963). The origin of the short-circuit current in the isolated skin of the South American frog Leptodactyllus ocellatus. *J. Gen. Physiol.* **47:**393–402.
161. Levi, H, and Ussing, H. H. (1949). Resting potential and ion movement in the frog skin. *Nature* **164:**928.
162. Larsen, E. H., Willumsen, N. J., and Christoffersen, B. C. (1992). Role of proton pump of mitochondria-rich cells for active transport of chloride ions in toad skin epithelium. *J. Physiol. (London)* **450:** 203–216.
163. Steinmetz, P. R., and Kohn, O. F. (1992). Hydrogen ion transport in model epithelia. In *The Kidney: Physiology and Pathophysiology* (D. W. Seldin and G. Giebisch, eds.), Raven Press, New York, pp. 2563–2580.
164. Al-Awqati, Q., and Beauwens, R. (1992). Cellular mechanisms of H^+ and HCO^- transport in tight urinary epithelia. In *Handbook of Physiology Section 8: Renal Physiology* (E. E. Windhager, ed.), Oxford University Press, London, pp. 323–350.
165. Ehrenfeld, J., Lacoste, I., and Harvey, B. J. (1989). The key role of the mitochondria-rich cell in Na^+ and H^+ transport across frog skin epithelium. *Pfluegers Arch.* **414:**59–67.
166. Ehrenfeld, J., and Garcia-Romeu, F. (1977). Active hydrogen excretion and sodium absorption through isolated frog skin. *Am. J. Physiol.* **233:**F46–F54.
167. Palmer, L. G., Edelman, I. S., and Lindemann, B. (1980). Current–voltage analysis of apical sodium transport in toad urinary bladder: Effects of inhibitors of transport and metabolism. *J. Membr. Biol.* **57:**59–71.
168. Nagel, W. (1976). The intracellular electrical potential profile of the frog skin epithelium. *Pfluegers Arch.* **365:**135–143.
169. Palmer, L. G. (1986). Apical membrane K conductance in the toad urinary bladder. *J. Membr. Biol.* **92:**217–226.
170. Van Driessche, W., Aelvoet, I., and Erlij, D. (1987). Oxytocin and cAMP stimulate monovalent cation movements through a Ca^{2+}-sensitive, amiloride insensitive channel in the apical membrane of toad urinary bladder. *Proc. Nat. Acad. Sci. USA* **84:**313–317.
171. Van Driessche, W., and Zeiske, W. (1985). Ca^{2+}-sensitive, spontaneously fluctuating cation channels in the apical membrane of the adult frog skin epithelium. *Pfluegers Arch.* **405:**250–259.
172. Das, S., and Palmer, L. G. (1989). Extracellular Ca^{2+} controls outward rectification by apical cation channels in toad urinary bladder: Patch-clamp and whole-bladder studies. *J. Membr. Biol.* **107:**157–168.
173. Schwartz, G. J., and Burg, M. B. (1978). Mineralocorticoid effects on cation transport by cortical collecting tubules *in vitro. Am. J. Physiol.* **235:**F576–F585.
174. Tomita, K., Pisano, J. J., and Knepper, M. A. (1985). Control of sodium and potassium transport in the cortical collecting tubule of the rat. Effects of bradykinin, vasopressin, and deoxycorticosterone. *J. Clin. Invest.* **76:**132–136.
175. Koeppen, B. M., Biagi, B. A., and Giebisch, G. (1983). Intracellular microelectrode characterization of the rabbit cortical collecting duct. *Am. J. Physiol.* **244:**F35–F47.
176. Kubo, Y., Reuveny, E., Slesinger, P., Jan, Y. N., and Jan, L. Y. (1993). Primary structure and functional expression of a rat G-protein-coupled muscarinic potassium channel. *Nature* **364:**802–806.

23

Gastric Acid Secretion
The H,K-ATPase and Ulcer Disease

George Sachs, Jai Moo Shin, Krister Bamberg, and Christian Prinz

23.1. INTRODUCTION

Acid secretion by the parietal cell of the gastric mucosa depends on activation of the acid pump, the gastric H,K-ATPase. This enzyme is expressed in high quantities by the parietal cell. The site of acid secretion in the parietal cell is the secretory canaliculus. This specialized structure is considered to be an infolding of the apical plasma membrane. The active H,K-ATPase is present in the microvilli lining the secretory canaliculus. When inactive the pump is found in cytoplasmic tubules.

Gastric and duodenal ulcer, as well as reflux esophagitis depend on the presence of acid secretion. All clinical experience shows that without acid secretion, neither ulcers nor reflux esophagitis are found, and if acid secretion is inhibited these ulcers heal.

The acidity generated by the H,K-ATPase is remarkably high. In the canaliculus of the parietal cell, it is probably equivalent to 160 mM HCl, providing a pH of 0.8. Following secretion from the parietal cell into the lumen of the gastric gland and then into the lumen of the stomach, partial neutralization of the secreted HCl occurs, increasing the pH found in the stomach. In the absence of food, this pH may fall to about 1.0. Over a 24-hr period, with food, the average intragastric pH is about 1.4. Food itself acts as a buffer increasing the intragastric pH to about 4.0 immediately after meals. It appears that it is the intragastric pH, rather than the absolute rate of acid secretion that relates to the presence of esophagitis and duodenal ulcer disease. The situation is less clear with respect to gastric ulcer disease, since the wall of the stomach is at a lower pH than its contents. Hence, elevation of intragastric pH due to antisecretory drugs may be much more pronounced than elevation of wall pH due to antisecretory drugs.

However, the presence of acid alone is not sufficient for generation of lesions in the upper gastrointestinal tract. Eighty percent of the population has normal acid secretion but does not develop ulcer disease at any time. Other causes must be present. For esophagitis, acid reflux must be present, due to an inadequate lower esophageal sphincter. In duodenal ulcer, consensus is growing that there must be infection by *Helicobacter pylori.*[1] In gastric ulcer, it is now thought that either this infection must also be present or nonsteroidal anti-inflammatory drugs are being used.[2] Although perhaps 60% of the population above 60 years old has *H. pylori* infection, again at least 80% of those infected do not experience peptic ulcer disease. Perhaps, therefore, there are different strains of *H. pylori* of varying toxicity or there are variations in response to the presence of *H. pylori*. Low pH and high Hp are a prerequisite for the development of duodenal and non-drug-related gastric ulcers. *H. pylori* appears to play virtually no role in reflux esophagitis. Almost surely there will be exceptions to the *H. pylori* rule. For example, in the case of Zollinger–Ellison syndrome, acid secretion is so high that it can well damage the stomach on its own. Stress ulcers are also not related to *H. pylori* infection.

Therapy of acid-related diseases has focused on the reduction of acidity in the upper GI tract. Since *H. pylori* is also an essential factor, much effort is now being expended not only on reduction of acid by new types of pump inhibitors, but also in developing protocols for treating the *H. pylori* infection. Unfortunately, to date, there is no simple regimen that achieves the latter objective although there is increasing consensus that treatment of peptic ulcer disease should be accompanied by *H. pylori* eradication.

Reliable reduction of acid secretion was, until 1973, essentially the province of surgery. Then with the introduction of histamine 2 receptor antagonists,[3] reduction of acid secretion became amenable to medical therapy. The development of pump inhibitors of the substituted benzimidazole class[4] provided a more effective means of acid reduction. Pump inhibition alone is sufficient to heal the ulcer disease; it does not prevent recurrence after healing unless therapy is continued. It is now felt that prevention of recurrence requires removal of *H. pylori*.

Infection by *H. pylori* is also associated with gastritis. Prevention of acid secretion does not affect this gastritis,

George Sachs, Jai Moo Shin, Krister Bamberg, and Christian Prinz • Membrane Biology Laboratory, Department of Medicine, University of California, Los Angeles, and Wadsworth VA Medical Center, Los Angeles, California 90073.

Molecular Biology of Membrane Transport Disorders, edited by S.G. Schultz *et al.* Plenum Press, New York, 1996

which may predispose to gastric cancer.[5] Elimination of *H. pylori* treats the gastritis. The gastritis is thus largely a response to the bacterium, not to the presence of acid secretion.

Accordingly, peptic ulcer disease is now thought to be a pathological consequence of the simultaneous presence of acid and *H. pylori*. By concerted action, there is localized destruction of the epithelium of the affected organ. This must involve not only acid secretion, but also infective induction of an abnormal response of the organ to acid.

23.2. THE ACID BARRIER

Various hypotheses have been proposed to explain the resistance of the gastric and duodenal mucosae to the high acidity in the lumen of these organisms, lumped together by the term "gastric acid barrier." This idea is based on the intuition that ulcer disease would be an inevitable result of exposure of the apical membrane of the epithelia to the average diurnal pH of 1.4.

One explanation suggested for this "gastric acid barrier" was the presence of mucus secretion. It was suggested that in some way mucus prevented the back diffusion of acid.[6] However, mucus is a hydrophilic gel, and on its own cannot significantly restrict diffusion of protons. Recognition of this problem led to additional hypotheses. For example, it has been suggested that a lipid component exists within the gastric mucus that prevents proton back diffusion. However, against such a postulate is the finding that K^+ changes in Cl^--free solutions on the apical surface of amphibian mucosa result in a rapid change in transmucosal potential difference, in spite of the presence of mucus.[7] The rapid effect of K^+ changes also demonstrates the inherent inadequacy of the mucus component of the "gastric barrier." If K^+ can penetrate the mucus, so can H^+.

A second explanation was based on the observation that both stomach and duodenum secrete HCO_3^-.[8] Measurement of the rate of secretion of this neutralizing anion showed that maximally a rate equal to 10% of the rate of acid secretion could be achieved.[9] This may be sufficient to partially neutralize 1 mM HCl but cannot be sufficient to neutralize 100 mM HCl or even 10 mM HCl. The stomach on the average contains 50 mM HCl to give the mean diurnal pH of 1.4.

The source of HCO_3^- is the blood. At high rates of acid secretion, the HCO_3^- production by the parietal cell generates the "alkaline tide" which can also increase the source of HCO_3^-. Whether the majority of this anion appears in the gastric lumen via diffusion across the tight junction or via transcellular transport is not clear. In the duodenum, the tight junction is leakier than in the stomach, and there is perhaps sufficient HCO_3^- secretion to neutralize the acid entering the duodenum. In the stomach, simple HCO_3^- entry is insufficient to neutralize surface pH.

It was then considered that a mucus diffusion barrier in combination with bicarbonate secretion underneath the mucus layer provided the gastric acid barrier necessary for the prevention of ulcer disease, the unstirred layer concept. However, if the mucus layer provided little resistance to acid diffusion and if the secretory capacity for bicarbonate is relatively low, even a combination of these ideas still does not explain the ability of these epithelia to resist acid damage. Further, since *H. pylori* is not mucolytic, this idea does not account for the necessity of this organism for peptic ulcer disease.

Early data using model lipid membranes had suggested a remarkably high proton permeability for a phospholipid bilayer. However, it turned out that the proton permeability in these experiments was mostly related to contamination with protonophoric compounds such as weak acids or permeating protonated weak bases.[10] When these were excluded, the proton permeability of the bilayer became quite low, but still more than the permeability to sodium or potassium.[11] Proton gradients are the basis of bioenergetics. It makes little sense to create leaky barriers when the gradient is a necessity for ATP synthesis or nutrient transport. A pure phospholipid membrane without discontinuity has essentially no included water. The fatty acid backbone would have no means of allowing permeation of the hydronium ion, H_3O^+, any more than allowing permeation of Na^+ or K^+.

Thus, provided there are no transport pathways for H^+ present in the apical membrane of the gastric epithelium, high acidity on that surface will not necessarily result in cell damage and ensuing ulceration. This makes concepts such as the "gastric barrier" unnecessary.

Within the gastric band there is a mixture of peptic and parietal cells. At the neck of the gland there are progenitor cells and mucus cells. Obviously the peptic cell is exposed directly to the acid secreted by the parietal cell and cannot have protonophoric pathways present in its apical membrane if it is to survive. This must also be true of the progenitor cells and the neck cells. Further, the membrane enclosing the acid pump is made up of phospholipids of no remarkable composition. If this membrane were leaky to protons, at the very least the acid pump would be operating in an inefficient environment and probably the cytoplasm of the parietal cell would be exposed to local acid loads.[12]

Direct measurement of acid resistance of peptic cell monolayers has shown that these can survive a luminal pH as low as 2 *in vitro*[13] substantiating the idea that the apical membrane of this cell type is essentially proton impermeable. Thus, there does not have to be a barrier to acid above the epithelial cell itself for the cell to survive, if the peptic cell is an example.

In contrast to the apical membrane, all of these epithelial cells, including the parietal cell, have several proton transporters in their basal lateral membranes so that a low pH on that surface inevitably reduces intracellular pH.[14–16] Effective penetration of acid to that surface *in vivo* would result in acidic cell damage and death.

Functional separation of luminal and serosal surfaces of epithelia depends not only on the cell membranes but also on the tight junction between the cells. The gastric epithelium has a relatively "tight" tight junction providing significant but not infinite resistance to back diffusion of protons.[17] The tight junction between, for example, oxyntic cells can be visualized quite readily on electron microscopy using La^{3+} infiltration as shown in Figure 23.1A. The junction has a length of about

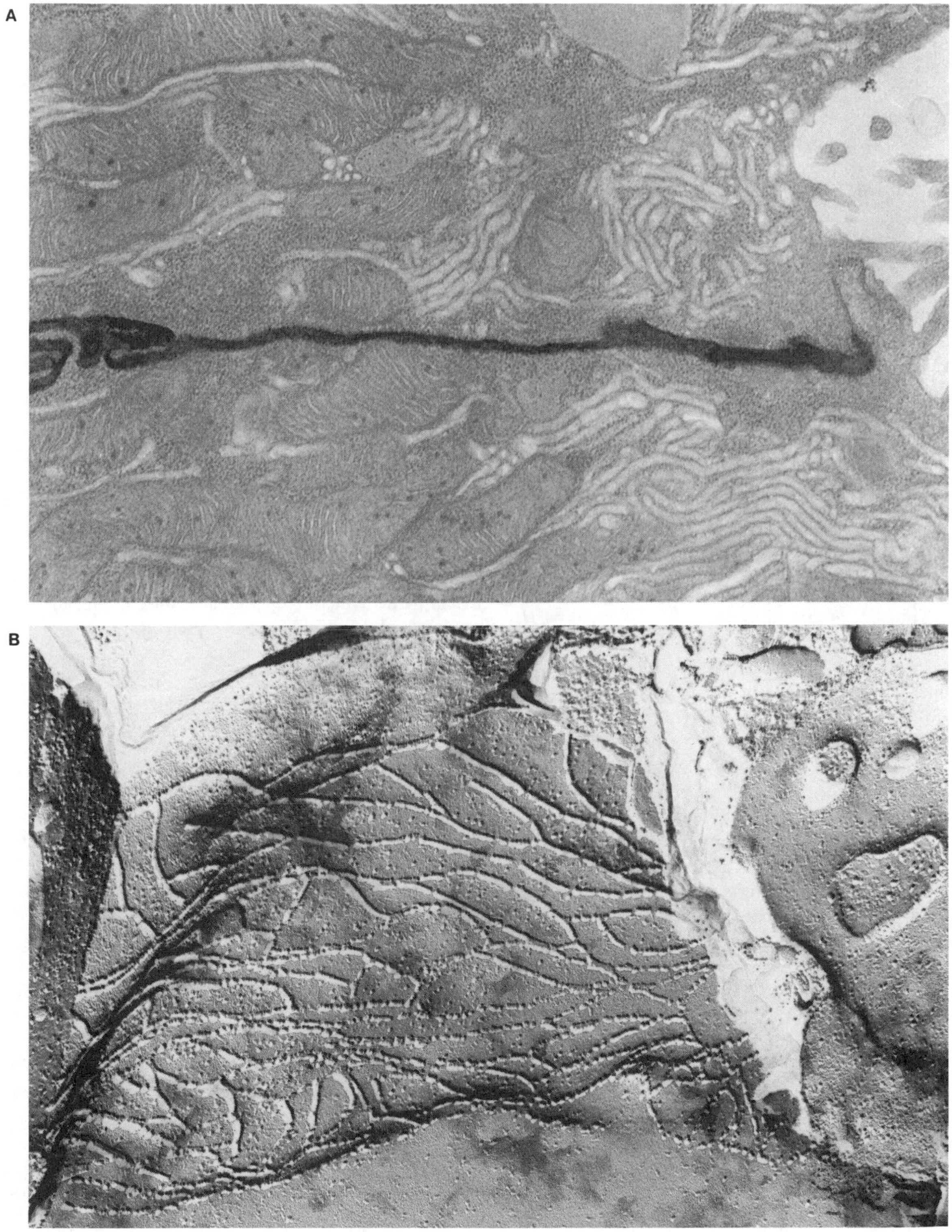

Figure 23.1. (A) An electron micrograph showing adjacent oxyntic cells of *Necturus* gastric mucosa, following lanthanum infiltration. The region of the tight junction can be seen as the narrow isthmus at the apex of the intercellular space. It is about 0.5 μm in length. (B) A freeze-fracture micrograph of a tight junction between gastric oxyntic cells showing the large number of strands (ca. 11) and particles, indicating that this is a "tight" tight junction. (Photographs courtesy of D. Anderson and H. F. Helander.)

0.5 μm. Freeze-fracture images, as presented in Figure 23.1B, show the presence of, on average, 11 strands of tight junction protein. Generally similar, but somewhat lower numbers are found for junctions between the surface cells (H. F. Helander, unpublished observations). The level of expression and functional assembly of the tight junction protein is a phenotypic property of epithelial cells. Evidently in the stomach, tight junction must provide a significant barrier to the back diffusion of protons.

The presence of acid is of course a necessity for acid back diffusion to occur. That *H. pylori* is also required suggests that somehow the presence of this organism facilitates acid back diffusion in an organ that normally does not allow this process to occur.

A pathological change in membrane protein targeting would be required to change the endogenous proton permeability of the apical membrane of peptic or surface epithelial cells. Addition of a protonophore such as acetyl salicylic acid to the membrane could also result in an abnormal proton permeability. Finally, disruption of tight junction would result in an increase in proton back diffusion across the epithelium.

It does not seem likely that *H. pylori* infection would result in a change in membrane protein structure or targeting. It further does not appear that *H. pylori* secretes a protonophoric substance.[18] The NH_3 produced by the urease secreted by *H. pylori* is not a protonophore since it penetrates as NH_3, not as NH_4^+. The penetration of NH_3 across the apical membrane will first alkalinize the cell because of intracellular formation of NH_4^+. Loss of NH_3 across the basal lateral membrane then tends to acidify the cell back to control values. A reasonably hypothesis, consistent with present data, is that *H. pylori* disrupts, or prevents adequate assembly of, tight junctions. This might be related to changes in intracellular pH induced by the production of NH_3. This disruption of the tight junctions will allow acid back diffusion as well as back diffusion of proteins secreted by the organism. Since these proteins, such as urease, are foreign, the immune response, visualized as gastritis, in initiated. This provides one scenario for the induction of gastritis.

Accordingly, esophagitis and peptic ulcer disease are a consequence of acid back diffusion across the epithelium and "oxynoptic" death of the epithelial cells. In the esophagus, acid reflux is all that is required, since these cells' apical surface or tight junctions do not provide adequate resistance to an acidity less than pH 4.[19] In stomach or duodenum, acid and *H. pylori* must be present. *H. pylori* is required to increase the rate of acid back diffusion across these tissues allowing insurmountable acid entry into the cells across their basal lateral surfaces. This idea is illustrated in the model of Figure 23.2.

In the case of duodenal ulcer disease, it seems likely that *H. pylori* survives on nests of ectopic parietal cells. As these cells secrete acid at the normal concentration of 160 mM HCl, disruption of their tight junctions will also result in unacceptably high levels of acid back diffusion.

Therefore, a discussion of the gastric H,K-ATPase in relation to the pathophysiology of membrane disorders must also now consider the presence and role of *H. pylori* in acid-related diseases. For example, does the acidic environment created by the ATPase generate a unique ecological niche for the growth of this spiral-shaped organism?

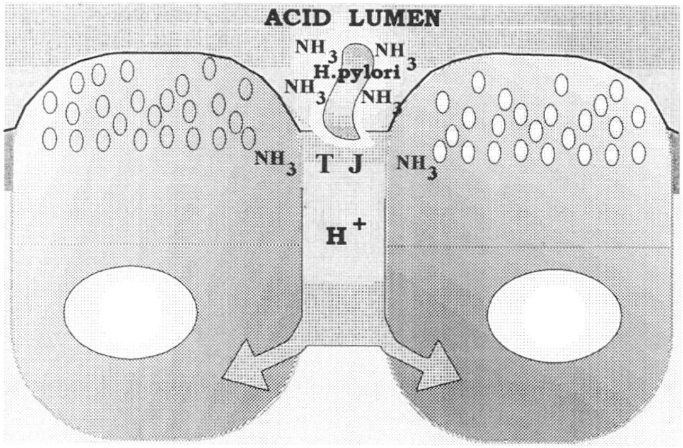

Figure 23.2. A model showing the possible mechanism of development of ulcer disease. A *H. pylori* organism is shown close to a tight junction between surface epithelial cells. The production of NH_3 by the bacterium alkalinizes the cell, with consequent increasing disruption of the tight junction. Acid back diffusion allows acidification of the intercellular space and entry of acid into the cell across the basal lateral membrane eventually causes oxynoptic cell death.

23.3. THE ACID PUMP

23.3.1. Stimulation of Acid Secretion

Acid secretion is a well-regulated process. It is present in the stomach of all vertebrates, and it seems that the same hormones or transmitters regulate parietal cell function. Also the same enzyme is responsible for the secretion of HCl.

There are three known stimulatory receptors on the parietal cell, the histamine 2 receptor,[20] the muscarinic M3 receptor,[21,22] and the gastrin/CCK-B receptor. Apparently the first two of these are capable of independent stimulation of parietal cell acid secretion. The role of the CCK-B receptor on the parietal cell is less clear, since the major secretory effect of gastrin is mediated by a CCK-B receptor-dependent stimulation of histamine release from the enterochromaffin-like (ECL) cell.[23] There is therefore no single common pathway for stimulation of the parietal cell.

Cholinergic stimulation of parietal cell acid secretion depends on vagal activity, initiated by central stimulation via hypothalamic centers.[24] Acetylcholine stimulates the M3 receptor on the parietal cell and also a muscarinic receptor on the ECL cell.[21–23] Peripheral stimulation of acid secretion depends on gastrin release from the G cells of the antrum stimulated by the presence of amino acids or peptides in the lumen of that part of the stomach.[25,26] Histamine release from the ECL cell is the major pathway initiated by gastrin and part of vagal stimulation also depends on histamine release from the ECL cell. Thus, histamine is often considered the major agonist in stimulation of gastric acid secretion. However, the role of vagal stimulation is also significant. Somatostatin, released from the D cell, inhibits gastrin release from G cells, histamine release from ECL cells, and also inhibits the Parietal cell directly (Ref. 24 and unpublished observations). Thus, the G, ECL, and D cells are a triumvirate of regulatory cells for parietal cell acid secretion.

Clinical and pH metric data using H2 receptor antagonists have shown good inhibition of nighttime acid secretion but poor inhibition of daytime acid secretion.[27] Thus, whereas nighttime secretion appears to depend virtually entirely on histamine release from the ECL cell, this is not so for secretion during the day. On the other hand, long-acting, "insurmountable" H2 receptor antagonists appear to control acid secretion as effectively as pump inhibitors.[28]

23.3.2. The Acid Pump and Stimulation

Whereas there is no common pathway for stimulation of acid secretion, the targets of both cAMP and elevated $[Ca]_i$, the second messengers affected by the H2 and M3 receptors, respectively, are protein kinases that activate the acid pump. This ATPase carries out the final step of acid secretion and therefore is the common target of all secretory stimuli. It has also become the target of the drugs that have been developed in the post-H2 era of acid secretory control.

The pump in the resting parietal cell is present in smooth membrane structures in the cytoplasm, namely, cytoplasmic tubules. On stimulation, the pump is seen in the microvilli of the secretory canaliculus.[29,30] Using an acid-activated cysteine reagent, the acid pump inhibitor omeprazole, it has been shown that activation of the pump occurs only in the canaliculus, as further demonstrated in Figure 23.3. As shown here, the initial site of labeling on stimulation as determined by electron microscopic autoradiography is in the canaliculus, not in the cytoplasmic tubules.[31] Since the binding of omeprazole depends on acid activation, only the canaliculus generates the acid necessary for both accumulation and acid activation.

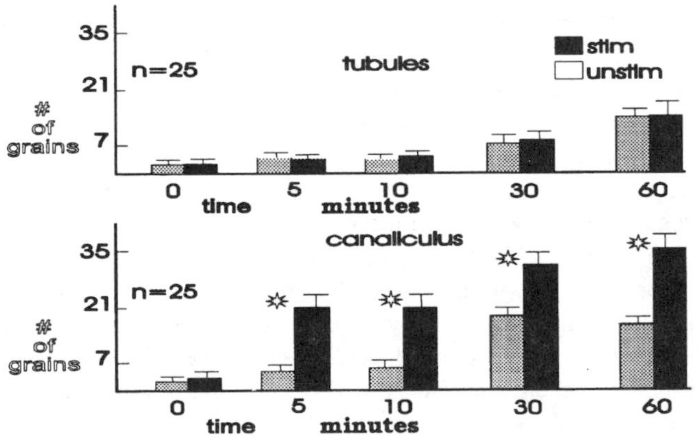

Figure 23.3. The distribution of [^{3}H]omeprazole counts over canaliculus and cytoplasm as a function of stimulation of a "resting" gastric gland preparation. It can be seen that the increase of counts occurs for the first 20 min only over the canaliculus. At later times, a redistribution to the cytoplasm is seen as the labeled pump cycles back into the cytoplasm. Omeprazole is an acid-activated pump inhibitor and will covalently label only acid-secreting pumps.

This change of location of the pump from cytoplasmic tubules to canalicular microvilli depends on a rearrangement of cytoskeletal proteins such as actin and ezrin.[32] It is not at all clear if this involves a fusion step or not. Given that the cytoplasmic location is tubular, it is unlikely that successive fusion steps are involved, in contrast to exocytosis in the parotid gland. Classical fusion proteins homologous to VAMPs, tagmins, NSF, and SNAPs have not yet been described in the parietal cell.

The gastric H,K-ATPase is an obligatory H-for-K exchange enzyme.[33] It therefore requires a supply of K to its extracytoplasmic surface. This does not occur in the tubular state of the pump, but only when the pump is present in the canalicular microvilli.[34] Accordingly, along with relocation of the pump there is also activation, in the canalicular membrane, of K and Cl conductances.[35] The relocation of the pump allows it to associate with these conductances, which are probably present in an inactive state in the canalicular membrane prior to stimulation.

It is not known whether there is modification of the pump protein as a function of stimulation. In the primary sequence of the α subunit, there is a single potential cAMP kinase phosphorylation site and several possible C kinase phosphorylation sites. Whether these are used is not known. Stimulation of acid secretion is related rather to cam II kinase than to C kinase.[36] In the β subunit there is a possible C kinase phosphorylation site present at position 11 on the cytoplasmic surface of this subunit. Perhaps one or more of these sites are subject to phosphorylation.

Inhibition of the pump will inhibit acid secretion, whatever the nature of the stimulus. Two types of inhibitors have been used in humans: the covalent substituted benzimidazoles (omeprazole, lansoprazole, pantoprazole, and E3810) and the K^+ competitive aryl quinolines (SK 574). These are more effective than H2 receptor antagonists or tolerated doses of the nonselective muscarinic antagonist, atropine, or the M1 selective antagonists, pirenzepine and telenzepine.

23.3.3. Coupling of Hydrolysis and Transport

The breakdown of ATP is scalar. Ion transport by the pump, protons outward and K inward, is vectorial in nature. This pump, like all others of its class (the EP class), couples the scalar and vectorial steps by conformational changes induced by a cycle of phosphorylation and dephosphorylation in the α subunit of the pump.

Phosphorylation is initiated after binding of the primary ion by binding of MgATP to the enzyme. The ion binding occurs on the cytoplasmic face of the pump. This results in the E_1–P conformational intermediate which has both Mg^{2+} and protons bound. Since the larger Na^+ can act as a surrogate for H^+, it is probable that the transported species is H_3O^+, the hydronium ion.[37] The pump is unlikely to contain a proton wire such as an array of OH or SH groups traversing the membrane. In this E_1–P form the pump is able to catalyze an

ADP/ATP exchange.[38] There is then, following phosphorylation, a rapid conversion to a form where the ion binding site is extracytoplasmic, the E_2–P form, with Mg^{2+} and hydronium ion (proton) still bound. However, ATP/ADP exchange is no longer catalyzed by this form of the pump. Hydronium ion is released into the luminal medium from the E_2–P form. This series of steps is electrogenic.[39]

Without luminal K^+ there is a slow conversion of the E_2–P form back to E_1. In the presence of luminal K^+ the potassium ion binds to the luminal side. The enzyme's subunit dephosphorylates and Mg^{2+} is released, and the E_2 form converts rapidly to the $E_1.K^+$ form. K^+ is then released back to the cytoplasm, allowing the cycle to begin again. This series of steps is also electrogenic.[40] The overall stoichiometry at neutral pH is 2 H transported per 2 K per ATP hydrolyzed.[41,42]

During the transport of K^+ inward (and presumably H^+ outward), the ion passes through an occluded form, where it is relatively inaccessible to both surfaces of the pump.[43] This occluded form is found even when the cytoplasmic domain has been cleaved from the enzyme and therefore occurs within the membrane domain of the enzyme.

These steps are illustrated in Figure 23.4. The arrival of molecular biology has allowed a better definition of the pathway taken by the ion across the membrane domain. Lacking a high-resolution three-dimensional structure, only the boundaries of the ion transport pathway can be determined by techniques that define the membrane-embedded portion of the pump and that define the amino acids "essential" for transport.

23.3.4. The Membrane Domain of the H,K-ATPase

The primary sequence of the α and β subunits of several species has been established either by cDNA or genomic sequencing.[e.g., 44–47] Hydropathic averaging of the β subunit predicts a single membrane-spanning subunit. When this method is applied to the α subunit, four membrane segments are predicted in the first one-third of the sequence, but four, five, or six segments are predicted in the C-terminal one-third of the enzyme. Experimental observations must be used to established the veracity of hydropathy.

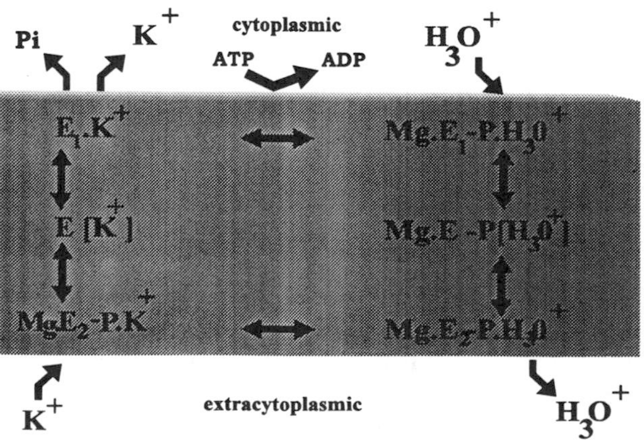

Figure 23.4. The catalytic cycle of the H,K-ATPase. The two major conformations are the E_1 form, with ion binding from the cytoplasm, and the E_2 form, where ion binding occurs from the extracytoplasmic face. The two forms interconvert as a function of phosphorylation and dephosphorylation.

The N-terminal part of the pump is cytoplasmic as shown by rapid tryptic cleavage at position Glu47 of the vesicular hog enzyme.[48] Iodination of the vesicular enzyme which terminates with two tyrosines has shown that the C-terminal end is also cytoplasmic. There is therefore an even number of membrane-spanning segments.[49]

Trypsinolysis of intact cytoplasmic-side-out H,K-ATPase vesicles followed by tricine gradient PAGE and N-terminal sequencing, provides direct evidence for four membrane-spanning segment pairs.[50] Similarly, covalent labeling with selected extracytoplasmic reagents such as the imidazopyridine [^{3}H]-MeDAZIP$^+$, and cysteine reagents omeprazole, lansoprazole, and pantoprazole also shows the presence of four pairs of membrane-spanning segments.[48,51–53]

The last pair of hydropathic segments is not revealed as membrane inserted by these methods. However, using *in vitro* translation, evidence for these has been obtained.[54] A protein with multiple membrane-spanning sequences usually does not have the cleavable signal sequence seen with secreted proteins. On the other hand, membrane insertion is driven by the presence of signal anchor sequences and stop transfer sequences. A signal anchor sequence allows the protein to insert into and cross the membrane. The protein continues to cross the membrane until a stop transfer sequence is encountered, at which point the nascent protein now remains on the cytoplasmic side of the membrane until a new signal anchor sequence appears.[55] This allows methods to be designed that test for the presence of signal anchor and stop transfer sequences within the primary amino acid sequence of the α subunit of the ATPase.

The principle of the method is illustrated in Figure 23.5. The idea is to generate fusion proteins that contain an N-terminal cytoplasmic domain and a C-terminal glycosylatable sequence which are separated by putative membrane-spanning segments. Translation is carried out in the presence of microsomes to detect glycosylation. A cDNA is constructed encoding the first 101 amino acids of the α subunit (M0 vector, to analyze signal anchor sequences) or the first 139 amino acids of the subunit (M1 vector to analyze stop transfer sequences), then a linker region into which various nucleotide sequences encoding membrane segments were inserted and then the last 177 amino acids of the β subunit to provide the glycosylation region. This cDNA is transcribed and translated using a coupled reticulocyte lysate in the absence and presence of microsomes and labeled methionine. The translation product is then separated on SDS gels, and autoradiography carried out.

If the sequence inserted into the linker region is read as a single or odd number of membrane-spanning segments, the product is glycosylated in the presence of microsomes and has an easily visualized higher molecular weight than when transcribed and translated without microsomes. Equally, if no membrane-inserted sequence is present, or if an even number is present, no glycosylation is seen even in the presence of

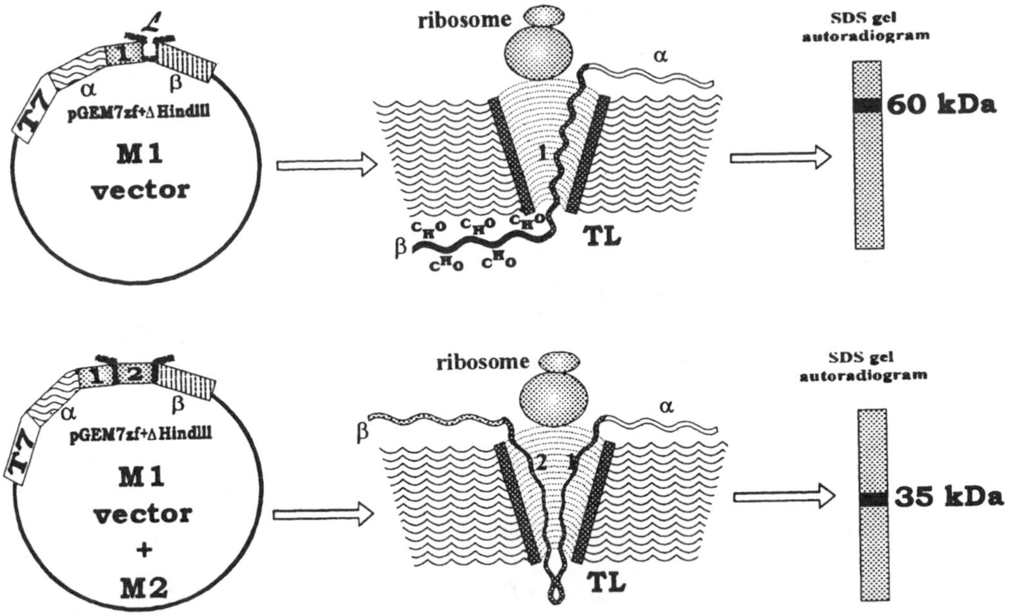

Figure 23.5. A diagram illustrating the method used for determining potential membrane-spanning segments using *in vitro* translation in the absence and presence of microsomes. A vector is constructed encoding the first 139 amino acids of the α subunit linked by a variable region to the last 177 amino acids of the β subunit. Into the variable region can be inserted any combination of putative membrane-spanning segments to result, when translated by a TNT reticulocyte lysate in the presence of microsomes, in membrane translocation of the β subunit with consequent glycosylation and a higher molecular weight on SDS gel autoradiography or in stop transfer with no glycosylation of the βC-terminal fragment and a lower relative molecular weight on SDS gel autoradiography.

microsomes. This is illustrated in Figure 23.5 for the M1 vector alone and for the M1 vector containing the M2 membrane segment.

As illustrated in Figure 23.6, it can be seen that the ninth hydrophobic sequence, H9, acts as a signal anchor sequence in the presence of microsomes. The tenth hydrophobic sequence, H10, acts as a stop transfer sequence. Thus, although these were not found by biochemical methods, they probably should be included in the membrane domain of the α subunit. It could still be that another translational event prevents H9 acting as a full signal anchor sequence, but what that might be is not clear.

Epitope mapping provided information as to the region of interaction between the α and β subunits. A monoclonal antibody was raised against intact rat parietal cells.[56] It was shown to recognize the H,K-ATPase and to label cells on their extracytoplasmic surface.[57]

Western analysis of intact rat enzyme showed that it was mainly the β subunit that was recognized, although the α subunit showed weaker reactivity. The hog β subunit was not recognized, and the presence of reducing agents abolished recognition of the β subunit of the rat. On inspection of the sequences of the β subunit of hog and rat, in the region of the second disulfide bridge, Cys161–Cys178, the hog had a nonconservative substitution Arg, Pro for Leu, Val in the rat. It seemed likely that this region contained the epitope of β subunit recognized by mAb 146.[58]

Western analysis of tryptic fragments of the α subunit showed that mAb 146 also recognized the M7/loop/M8 region of the enzyme.[58] The octamer walk technique in this region of the enzyme showed that it was the sequence in the α subunit between positions 873 and 877 that was also an epitope for mAb 146.[58] Expression using baculovirus in SF9 cells of the individual subunits also showed that both subunits were recognized by mAb 146. The α subunit was, however, only rec-

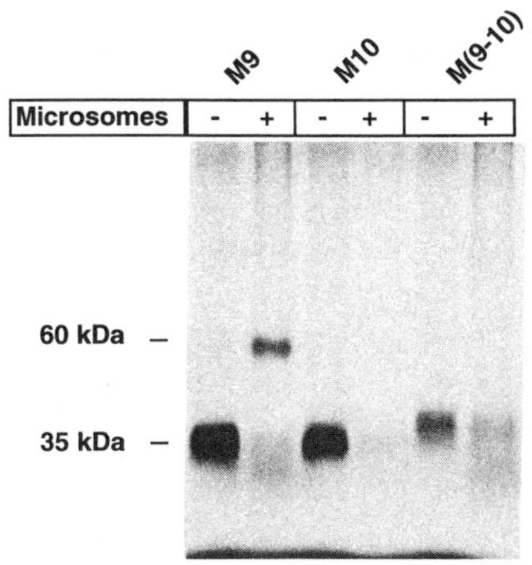

Figure 23.6. The result of an *in vitro* translation experiment where the sequences for H9, H10, and H9 + H10 were inserted into the linker region of the cDNA construct containing only the first 101 amino acids of the α subunit. It can be seen that H9 acts as a signal anchor sequence and H10 as a stop transfer sequence. Hence, they are able to act as a membrane-spanning segment pair.

ognized in Western blots but not in intact cells. This would be consistent with putative location of positions 873 and 877, at the extracytoplasmic border of M7.

The β subunit binds to lectins such as WGA with its extracytoplasmic surface.[59] After tryptic digestion and washing off the cytoplasmic fragments, the β subunit is bound to a WGA column following solubilization in nonionic detergent. Washing the column now removes the membrane segments not tightly bound to the β subunit. Washing with 0.1 N acetic acid now removes the β subunit along with any fragments of the α subunit retained by the β subunit. The sequence M7/loop/M8 is quantitatively retained and released when the β subunit is eluted. This result is consistent with the conclusions derived from the mAb 146 epitope mapping. If the C-terminal part of the α subunit remains intact in the sense that there is no cleavage after the N-terminal segment of M7, the M5/loop/M6 sector is also retained by the β subunit. These data suggest further that there is probably interaction within the α subunit between the M5/M6 region and H9/H10.[60] Thus, the β subunit binds M7/loop/M8 and in turn the M9/loop/M10 binds the M5/loop/M6 region.

Additional information has been obtained by mutagenesis of the β subunit of the Na,K-ATPase. Mutation of a few hydrophobic amino acids in the C-terminal end of the β subunit prevents association with the α subunit.[61] Perhaps it is the region of the β subunit that binds M7/loop/M8.

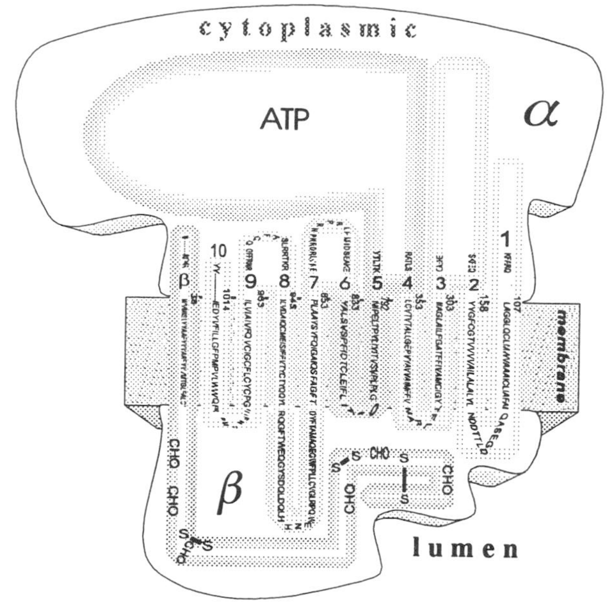

Figure 23.7. A two-dimensional representation of the membrane and extracytoplasmic domains of the αand β subunits of the H,K-ATPase. The α subunit is represented as having ten membrane segments, the β subunit as having a single membrane-spanning segment. One region of contact between the two subunits close to the membrane domain is shown to be in the vicinity of M7/M8.

23.3.5. The Ion Transport Pathway of the Gastric H,KATPase

From the above data, it is possible to define the membrane segments of the enzyme as illustrated in Figure 23.7. The location of hydrophilic amino acids (Asp, Glu, Ser, Thr, and Gln) is mainly in the M4, M5, M6, and M8 segments. The loop between M1 and M2 contains several carboxylic amino acids and a long extracytoplasmic loop exists between M7 and M8. On the assumption that ions cross the membrane domain by traversing a hydrophilic pathway containing hydrophilic amino acids (although an arrangement of aromatic amino acids could also be suitable for ion transport), the extracytoplasmic loop between M1 and M2, the conserved Glu or Asp in M4 and M5, M6 and M8 would be at least part of the boundaries of the ion transport pathway. Further information on this pathway can be obtained by discovering the sites of binding of extracytoplasmic reagents that inhibit ion transport by this enzyme.

Figure 23.8 illustrates a conceptual three-dimensional arrangement of the entry to the membrane domain of the H,K-ATPase, illustrating part of the region of the ion pathway discussed above.

23.3.6. Ion Transport Inhibitors

23.3.6.1. Substituted Benzimidazoles

Omeprazole was the first pump inhibitor introduced into clinical practice. It is a substituted benzimidazole and has a chemistry specialized for targeting to and reaction with the α subunit of the H,K-ATPase. The other benzimidazoles have a generally similar chemistry.

They are all weak bases of $pK_a = 4.0$. They will therefore accumulate effectively only in acidic compartments with a $pH \leq 4.0$. The only compartment in the body with a pH of this value or less is the secretory canaliculus of the active parietal cell. Since the pH of this compartment is about 1.0, these compounds accumulate about 1000-fold in this region. Weak bases of a higher pK_a will be able to accumulate in other acidic compartments such as those of lysosomes and neurosecretory granules. The pK_a of 4 is essential for specific targeting of the benzimidazoles.

The protonation occurs on the pyridine N. This is probably followed by an intramolecular transfer of the proton to one of the benzimidazole N residues. This deprotonation of the pyridine and protonation of the benzimidazole results in a large increase in the electrophilic reactivity of the C at the 2 position between the N's of the benzimidazole. This carbon then reacts with the pyridine N as the S═O group moves to the opposite N of the benzimidazole with loss of the O atom. In this way a tetracyclic sulfenamide is formed, with a positively charged pyridinium ring. This is the rate-limiting step of the inhibitory reaction with benzimidazoles.

Sulfenamides react rapidly with free cysteines in proteins to form a stable disulfide. Since these membrane-impermeant sulfenamides are formed virtually exclusively in the acid compartment contained within the secretory canaliculus of the parietal cell, they will react with the cysteines present on, or

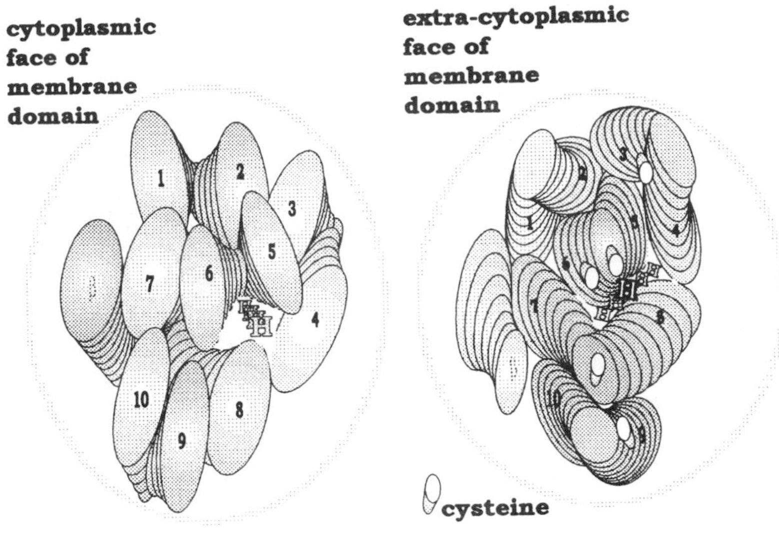

Figure 23.8. A conceptual three-dimensional model sectioning the pump on the cytoplasmic side of the membrane domain. The membrane segments are illustrated as pairs and the M5/M6 region is shown as being inserted into the pump protein, rather than into the lipid bilayer. Illustrated also is the possible region of ion transport, a region where there is a concentration of hydrophilic amino acids.

accessible from, the extracytoplasmic face of the H,K-ATPase. The product of reaction is a disulfide. The reaction pathway of omeprazole is shown in Figure 23.9.[62]

Considering the secondary structure of the enzyme as shown in Figure 23.7, of the 28 cysteines present in the α subunit, 4 or 5 would be predicted to be possibly accessible to the sulfenamide. The 6 extracytoplasmic cysteines of the β subunit are disulfide-linked and therefore do not react stably with the sulfenamide. The cysteines available to these sulfenamides are illustrated in the conceptual three-dimensional diagram of Figure 23.10, which shows a section through the outer surface of the membrane domain of the enzyme.

There is diagnostic selectivity between the different benzimidazoles studied in terms of the cysteines that react in a stable fashion. Omeprazole reacts with Cys813 (or 822) and Cys892. Lansoprazole reacts with Cys321, Cys813 (or 822), and Cys892. Pantoprazole reacts with Cys813 and Cys822.[51–53] All of these compounds inhibit ATPase activity and proton transport. The common site of action is at Cys813 (or 822). This

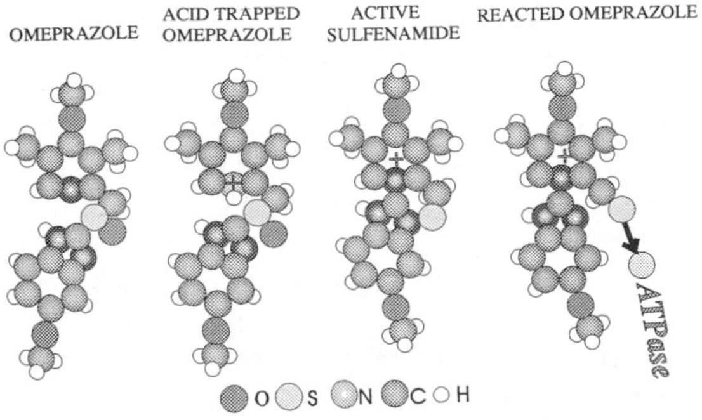

Figure 23.9. The mechanism of activation of omeprazole, a representative substituted benzimidazole. There is first acid compartment-dependent accumulation and then chemical conversion in an acid-catalyzed reaction to a tetracyclic cationic sulfenamide which reacts with accessible cysteines from the luminal surface of the H,K-ATPase.

would suggest that the loop between M5 and M6 (the putative location of Cys813) is on the transport pathway of the enzyme. This is further borne out by the finding that binding of these compounds results in proton leak across the membrane; the leak is reduced when a reducing agent such as tributyl phosphine is added to the inhibited enzyme and acid transport is also restored.[63] Mutation of the carboxylic acids in this region of the Na,K-ATPase results in inhibition of enzyme activity.[64]

The compounds differ in their rate of activation. Lansoprazole activates the quickest, pantoprazole the slowest. The compounds are also considerably stabilized by protein binding. When added to acid-transporting vesicles, inhibition of the ATPase and proton transport takes some considerable time, approximately correlated with the rate of activation of the different drugs. However, the rate of inactivation of transport is surprisingly slow, which might mean that these compounds bind first in the extracytoplasmic domain of the enzyme itself, presumably in protonated form. Formation of the pertinent sulfenamide then occurs within this domain. The sulfenamide formed in bulk solution does not react with the pump, at least not in the case of pantoprazole. The rate of formation of the sulfenamide within the extracytoplasmic membrane domain determines the cysteines that react. Lansoprazole reacts with the cysteines initially available, namely, Cys321, 813, and 892, and perhaps in this compound the sulfenamide being formed in free solution is responsible for inhibition. Omeprazole which converts somewhat more slowly to the sulfenamide reacts with Cys813 and 892. Pantoprazole which converts even more slowly to the sulfenamide has time to penetrate the vestibule of the enzyme and reacts therefore with both cysteines in the M5/loop/M6 sector, but not the more superficial cysteines at positions 321 and 892. This hypothesis suggests that it is not the free sulfenamide, at least for pantoprazole, that inhibits the enzyme, but rather the sulfenamide that is formed *in situ.* That binding of the protonated species may occur before reaction with the cysteines is sup-

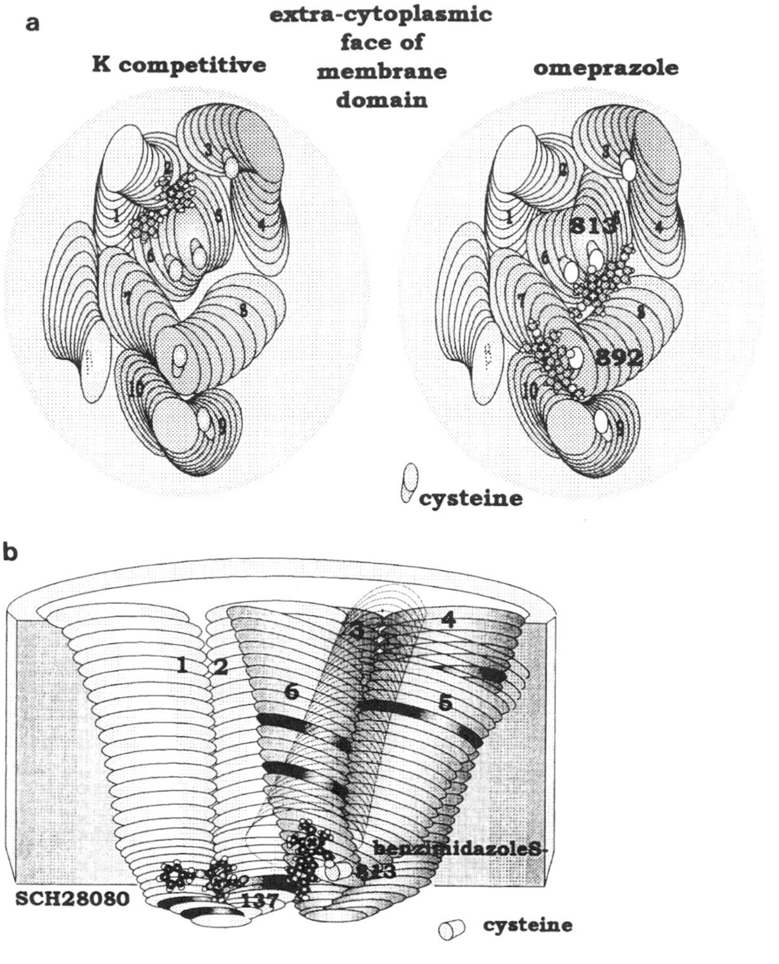

Figure 23.10. (a) A conceptual three-dimensional model of a a section of the membrane domain of the H,K-ATPase at the extracytoplasmic surface, showing the possible cysteines which could bind a sulfenamide formed in the extracytoplasmic space. (b) A conceptual two-dimensional model sectioning the membrane domain of the pump, showing the first six membrane-spanning segments, with the crucial binding site for the substituted benzimidazoles and for the K^+ competitive inhibitor, SCH 28080. Carboxylic acids in the M5, M6, and M4 segments are shaded dark. M1 through M4 are visualized as partially surrounding the shorter M5/M6 sector of the enzyme.

ported by the observation that K^+ competitive reagents prevent benzimidazole inhibition of acid transport by the enzyme, independently of their inhibition of the formation of an acid gradient. These compounds are known to bind in the region of the loop between the first and second membrane-spanning segments. If binding of a bulky substituent in this region prevents access of the benzimidazoles to their target cysteines, the extracytoplasmic domain must be a relatively compact structure.

23.3.6.2. K^+ Competitive Reagents

A variety of amines were shown to be K^+ competitive inhibitors of the H,K-ATPase.[65] The benzimidazoles as a class are not K^+ competitive. However, an imidazopyridine, SCH 28080, initially designed to mimic omeprazole, is a potent, K^+ competitive inhibitor of the enzyme.[66,67] The protonated species is the reactive form. This compound reacts only with the external face of the enzyme[67] and stabilizes the E_2–P form as well as the E_2 form.[67–69] An azido derivative of SCH 28080 ([^{3}H]-MeDAZIP$^+$) was synthesized as an *N*-methylated radioactive quaternary salt, a permanent cation, and shown to bind to the region between M1 and M2 after photolysis (Figure 23.11).[48] This region and the M3/M4 region of the Na,K-ATPase are thought to be responsible for the high-affinity binding of ouabain, a partially K^+ competitive inhibitor of the Na^+ pump.[70,71]

Other K^+ competitive reagents have been synthesized, in the search for antiulcer drugs of shorter duration of action as compared to the substituted benzimidazoles. An aryl quinoline has been tested in humans and shown to be an effective inhibitor of gastric acid secretion.[72] Figure 23.10B shows the sites of binding of the K^+ competitive reagents and the critical site for binding of the substituted benzimidazoles.

One of these aryl quinolines, MDPQ, is fluorescent (Figure 23.11). On binding to the H,K-ATPase, its fluorescence increases. Phosphorylation of the enzyme increases fluorescence even further. It appears that formation of the E_2– P.[I] form of the enzyme increased the hydrophobicity of the environment of MDPQ.[73] In contrast, formation of the $E_2.K^+$ form of the enzyme decreases the fluorescence of fluorescein isothiocyanate (FITC) bound to Lys517 in the large cytoplasmic loop between M4 and M5. Formation of the $E_1.Na^+$ form of the enzyme increases FITC fluorescence.[74] These fluorescence data show that the conformational changes in the enzyme that represent the two major forms E_1 and E_2 span the whole enzyme, in the cytoplasmic domain and in the extracytoplasmic domain, presumably transmitted by the membrane domain of the enzyme.

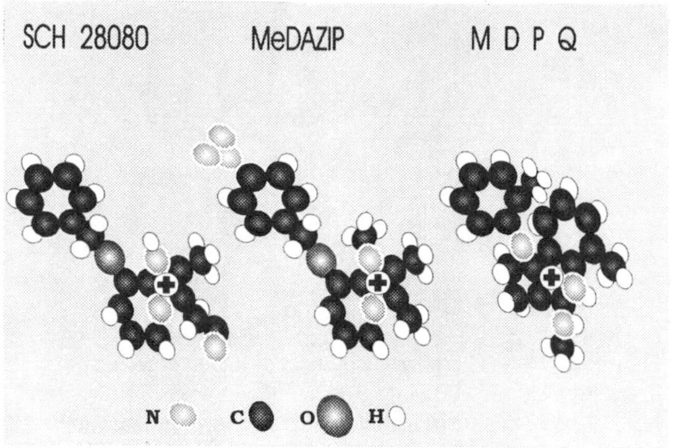

Figure 23.11. The chemical structure of the K^+ competitive inhibitors, SCH 28080, $MeDAZIP^+$, and MDPQ.

The reagents developed to inhibit the enzyme have proved useful as probes of structure–function. They have also proved to be improved therapeutic tools for treatment of acid-related diseases as compared to the H2 receptor antagonists.

23.3.7. Turnover of the Acid Pump

The acid pump in the rat appears to have a half-life of about 50–70 hr whether measured by the rate of inhibition induced by cycloheximide or by a pulse chase using ^{35}S-labeled methionine.[75,76] The half-time for recovery from omeprazole inhibition in the rat is about 30 hr, suggesting that omeprazole increases pump turnover.[75]

Elevation of serum gastrin as induced by omeprazole administration transiently increases mRNA levels in gastric mucosa.[77] Coadministration of an H2 antagonist blocks this transient effect, suggesting that it is activation of the H2 receptor that is elevating pump message.[78] In agreement with this, direct administration of histamine also elevates pump mRNA levels.[79] It therefore appears that activation of the H2 receptor increases gene expression of the pump.

In contrast to these acute effects, long-term administration of omeprazole in the rabbit appears to decrease pump protein levels[80,81] whereas long-term administration of the H2 antagonist, famotidine, to rabbits increases pump protein by the same amount.[81] These latter effects are likely related to changes in protein turnover and not gene expression.

Although the upstream gene sequence in the rat contains both cAMP and Ca responsive elements, this is not true of all species. The upstream sequences of the αand β subunit gene contain the nucleotide sequence GATAGC on either side of a TATA box, which is selectively recognized by gastric nuclear protein.[82–84] The regulation of pump levels is obviously precise and relates to the acid secretory demand. In part, this must be due to gene transcription regulation, in part related to changes in pump turnover. It is probably the latter that is changed with either the hypergastrinemia/histamine release caused by omeprazole or the H2 blockade induced by famotidine. The pump is retrieved under conditions of H2 blockade and is largely cytoplasmic; with omeprazole treatment the pump remains largely in the secretory canaliculus. Perhaps the anatomic location of the pump contributes to its rate of turnover.

23.3.8. Membrane Targeting of the Acid Pump

The parietal cell expresses the Na,K-ATPase on its basal lateral surface and the H,K-ATPase at the contralateral surface. The Na,K-ATPase is invariably expressed on the basal lateral surface of epithelial cells. Supposed exceptions are the cells of the choroid plexus and the nonpigmented epithelial cells of the ciliary. However, these cells may be considered inverted absorptive cells, the true basal lateral surface facing inward.

Both pumps are inserted into the endoplasmic reticulum cotranslationally. There is early association with the β subunit. Lacking a βsubunit, neither αsubunit can leave the ER.[85] In the presence of the β subunit of either pump, the α subunit reaches the plasma membrane of the oocyte. Hence, the β subunit is essential for plasma membrane targeting of the H,K-ATPase.[86] The β subunit is interchangeable between the two pumps.[87] The part of the pump that determines apical or basal lateral membrane selectivity has not been described at the time of writing this review. It seems evident that the pump moves selectively to its target site. It has been suggested that it is the stability of the pump in one or other membrane that determines its final cellular location, but the bilateral targeting of the Na,K- and H,K-ATPase argues for selective insertion rather than selective removal.

23.4. CLINICAL USE OF PUMP INHIBITORS

The benzimidazoles that are in current clinical use share the important property of being able at high doses to control acid secretion for 16 hr out of 24. Elevation of intragastric pH to 3 or thereabouts for a 2-week period appears optimal for healing of duodenal ulcer.[88] A required elevation to a pH of 4 suggests that more inhibition is necessary for treatment of esophageal disease.[89]

The covalent nature of the inhibition produces favorable pharmacokinetics. The half-life in plasma is short, about 60 min. However, the duration of action is prolonged, and is cumulative for about 3 days in humans.[90] A typical plasma concentration curve and duration of action is shown in Figure 23.12. The cumulative effect is a result of inhibition of only the active pumps in the cell (and not all are active at once) and resynthesis of the pump. Full acid secretory capacity is restored within 72 hr with a half-life of 30 hr. Therefore, treatment with omeprazole appears to increase pump turnover, with only a transient effect on synthesis. This explains the 25% lowering of pump levels seen in rabbit after 5 days' treatment.[81]

The effective inhibition of secretion during the day as well as at night on a once-a-day regimen is sufficient to improve healing rates and symptom relief in patients with duodenal and gastric ulcers and particularly in those with severe

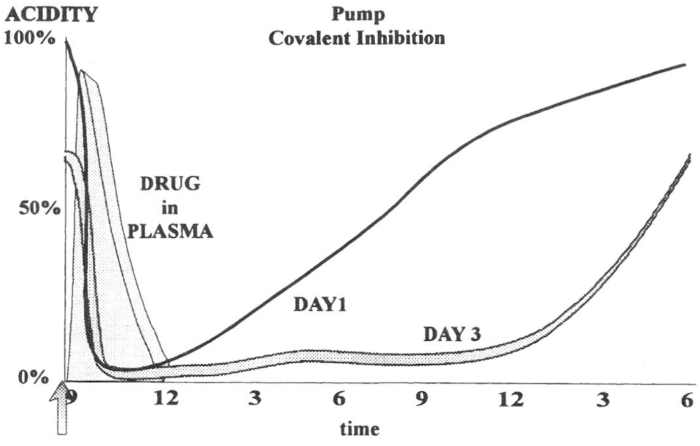

Figure 23.12. The pharmacokinetics of omeprazole inhibition of gastric acid secretion using once-a-day dosage. The compound has a plasma half-life of about 60 min. Acid secretion is rapidly inhibited after the first dose but returns rapidly as new pumps are activated. On the third day of dosing a steady state is reached between inactivation and resynthesis of the pump.

erosive esophagitis as compared to H2 receptor antagonists. However, these drugs heal the lesion, they do not cure the disease since there is generally recurrence after stopping the drug. The same phenomenon is found after stopping H2 receptor antagonists,[91] but in addition these compounds may induce acid rebound, related in part perhaps to the increased pump levels[81] and in part perhaps to upregulation of the H2 receptor or the muscarinic pathway. Tolerance to H2 antagonists is also found.[91] Treatment of duodenal and gastric ulcer in the future will almost surely involve additional steps to eradicate *H. pylori.*

23.5. *H. PYLORI* AND THE FUTURE OF PEPTIC ULCER TREATMENT

This organism has been mentioned as being an additional requirement for the development of duodenal and drug-unrelated gastric ulcer as well as responsible for gastritis. Single antibiotics such as amoxicillin can kill the bacterium at pH 7.0 *in vitro,* but are quite ineffective *in vivo.* Severe inhibition of acid secretion, so that mean diurnal pH $\geq$ 5.0, improves the effect of single antibiotics.[92] Evidently, *H. pylori* occupies a unique niche in a highly acidic environment. There has been much discussion as to the actual pH at the surface of the stomach. It is our strong conviction that the mucus layer does not provide a barrier to acid diffusion and that insufficient HCO_3^- secretion is present to allow neutralization of the acid at the epithelial surface when the pH of gastric contents drops below 3.0. Since average diurnal pH in humans is 1.4, it seems that *H. pylori* survives an acidic environment for much of its life.

A group of organisms survive and grow under acidic pH conditions in nature, the acidophiles. Most bacteria utilize the internal negative potential generated by their membrane redox pumps to produce an adequate protonmotive force for ATP synthesis across their F1/F0-ATPase. The acidophiles, as the external pH falls, develop an increasingly positive internal potential.[93] This maintains a constant protonmotive force and aids in prevention of acid back diffusion and oxynoptic cell death. The molecular biological basis for this has not been defined.

H. pylori secretes a Ni^{2+}-requiring urease.[94] Without urease the organism has difficulty in colonizing the stomach. The urease generates NH_3 from urea, $CO(NH_2)_2$, and this NH_3 is able to neutralize some of the acid in the environment, perhaps particularly in the periplasmic space. Intracellular urease will also generate intracellular buffer. An $NH^+{}_4$ transporter must be present for this to be a successful strategy for the organism's survival at acidic pH. It is of interest in this regard that Bi^{2+} is able to reduce the survival of *H. pylori.* Perhaps this cation acts as a surrogate for Ni^{2+} in an uptake system (probably an EP-type divalent cation-transporting ATPase), similar to other cation transporters found in bacteria such as the Cd^{2+}-ATPase preventing urease activity. The environment of *H. pylori* contains about 3 mM urea, and the $NH^+{}_4$ concentration in its environment must be quite high. However, at a pH of 1.4, it seems unlikely that elevation of pH to more than 3.0 or so can be achieved by this mechanism. On the other hand, NH_3 at the surface of the cells will tend to alkalinize the cell interior. Can this alkalinization alter the properties of the cell tight junctions? With infection, antibodies to urease are produced. Is this related to transcellular uptake or to diffusion of the urease across the disrupted tight junctions?

H. pylori must share some properties of acidophiles that confer the ability to resist a low environmental pH besides its urease secretion. Whether there is modification of the electrogenicity of redox pumps so that low potentials are generated, whether there is modification of the H^+ symporters so that they are gradient and not potential driven, whether the F1/F0-ATPase is used to export H^+ are questions that will be answered in the future.

It is not clear how *H. pylori* can disrupt the tight junction barrier. One hypothesis may be that the bacterium induces a mucosal inflammatory response of T and B lymphocytes which in turn stimulate the activity of monocytes and macrophages or the bacterium stimulates peripheral blood monocytes by antigenic exposure of certain lipopolysaccharides.[95] Monocytes can secrete a variety of inflammatory mediators, such as interleukins or tumor necrosis factor (TNF). Increased levels of interleukin 1, 2, and 6 and also of TNF have been found in patients with helicobacter-positive antral gastritis.[96] TNFα is a strong cytotoxin which might affect the structure of the biomembranes directly. Additionally, these cytotoxins might alter the structure of tight junctions or affect the function of other mucosal cells. The role of surfactant in resistance to acid is not clear.[97] It seems at least that it does not hinder the diffusion of K^+ since there are rapid responses of transmucosal potential with the addition of K^+ to the mucosal solution.[7] The secretion of phospholipases by *H. pylori,* if injurious, would relate more to cell membrane damage rather than to disruption of the surfactant. Enzymatic preparation of the gastric tissue (C. Prinz, personal communication) shows that isolated, single mucosal cells are obtained only in the presence of a proteolytic enzyme like trypsin. Therefore, it seems likely that the disruption of tight junctions may also demand the presence of a very strong proteolytic enzyme like pepsin, which has a pH optimum of 2. The common action of pepsin, acid exposure, and cytotoxic factors then would act together to destroy the tight junctions.

In the present, the knowledge that duodenal and drug-unrelated gastric ulcers are a consequence of both acidity and *H. pylori* has changed our understanding of the pathogenesis of these diseases. Treatment of ulcer disease will involve inhibition of acid secretion by pump inhibitors, coupled to a means of eradication of this acidophilic inhabitant of the gastric or duodenal mucosa. Prophylactic treatment will involve eradication of *H. pylori* probably using an as yet undiscovered monotherapy regimen.

ACKNOWLEDGMENT. This work was supported by USVA-SMI and NIH grants RO1 DK 40165 and 41301.

REFERENCES

1. Rauws, E. A. J., Langenberg, W., Houthoff, H. J., Zanen, H. C., and Tytgat, G. N. J. (1988). Campylobacter pyloridis-associated chronic active antral gastritis: A prospective study of its prevalence and the effects of antibacterial and antiulcer treatment. *Gastroenterology* **94:**33–40.
2. Howden, C. W., and Holt, S. (1991). Acid suppression as treatment for NSAID-related peptic ulcers. *Am. J. Gastroenterol.* **86:**1720–1722.
3. Black, J. W., Duncan, W. A. M., Emmett, J. C., Ganellin, C. R., Hesselbo, T., Parsons, M. E., and Wyllie, J. M. (1973). Metiamide—An orally active histamine H2-receptor antagonist. *Agents Actions* **3:** 133–134.
4. Fellenius, E., Berglindh, T., Sachs, G., Olbe, L., Elander, B., Sjostrand, S.-E., and Wallmark, B. (1981). Substituted benzimidazoles inhibit gastric acid secretion by blocking (H^++K^+)ATPase. *Nature* **290:** 159–161.
5. Correa, P. (1991). Is gastric carcinoma an infectious disease? *N. Engl. J. Med.* **325:**1170–1171.
6. Heatley, N. G. (1959). Mucosubstance as a barrier to diffusion. *Gastroenterology* **37:**313–318.
7. Schwartz, M., Chu, T. C., Carrasquer, G., Rehm, W. S., and Holloman, T. L. (1981). Origins of positive potential difference of frog gastric mucosa in Cl^--free solutions. *Am. J. Physiol.* **240:**G267–G273.
8. Flemstrom, G. (1987). Gastric and duodenal mucosal bicarbonate secretion. In *Physiology of the Gastrointestinal Tract*, 2nd ed. (L. R. Johnson, J. Christensen, M. J. Jackson, E. D. Jacobson, and J. H. Walsh, eds.), Raven Press, New York, pp. 1011–1029.
9. Engel, E., Peskoff, A., Kauffman, G. L., Jr., and Grossman, M. I. (1984). Analysis of hydrogen ion concentration in the gastric gel mucus layer. *Am. J. Physiol.* **247:**G321–G338.
10. Gutknecht, J. (1984). Proton/hydroxide conductance through lipid bilayer membranes. *J. Membr. Biol.* **82:**105–112.
11. Deamer, D. W. (1987). Proton permeation of lipid bilayers. *J. Bioenerg. Biomembr.* **19:**457–479.
12. Priver, N. A., Rabon, E. C., and Zeidel, M. L. (1993). Apical membrane of the gastric parietal cell: Water, proton, and nonelectrolyte permeabilities. *Biochemistry* **32:**2459–2468.
13. Ayalon, A., Sanders, M. J., Thomas, L. P., Amirian, D. A., and Soll, A. H. (1982). Electrical effects of histamine on monolayers formed in culture from enriched canine gastric chief cells. *Proc. Natl. Acad. Sci. USA* **79:**7009–7013.
14. Paradiso, A. M., Townsley, M. C., Wenzl, E., and Machen, T. E. (1989). Regulation of intracellular pH in resting and in stimulated parietal cells. *Am. J. Physiol.* **257:**C554–C561.
15. Muallem, S., Burnham, C., Blissard, D., Berglindh, T., and Sachs, G. (1985). Electrolyte transport across the baso lateral membrane of the parietal cell. *J. Biol. Chem.* **260:**6641–6653.
16. Negulescu, P. A., and Machen, T. E. (1990). Intracellular ion activities and membrane transport in parietal cells measured with fluorescent dyes. *Methods Enzymol.* **192:**38–81.
17. Spenney, J. G., Shoemaker, R. L., and Sachs, G. (1974). Microelectrode studies of fundic gastric mucosa: Cellular coupling and shunt conductance. *J. Membr. Biol.* **19:**105–128.
18. Halter, F., Hurlimann, S., and Inauen, W. (1992). Pathophysiology and clinical relevance of Helicobacter pylori. *Yale J. Biol. Med.* **65:** 625–638.
19. Orlando, R. C. (1991). Esophageal epithelial defense against acid injury. *J. Clin. Gastroenterol.* **13**(Suppl. 2):1–5.
20. Black, J. W., Duncan, W. A. M., Durant, C. J., Ganellin, C. R., and Parsons, M. E. (1972). Definition and antagonism of histamine H2-receptor. *Nature* **236:**321–326.
21. Pfeiffer, A., Rochlitz, H., Noelke, B., Tacke, R., Moser, U., Mutschler, E., Lambrecht, G., and Herawi, M. (1990). Muscarinic receptors mediating acid secretion in isolated rat gastric parietal cells are of M3 type. *Gastroenterology* **98:**218–222.
22. Kajimura, M., Reuben, M., and Sachs, G. (1992). The muscarinic receptor gene expressed in rabbit parietal cells is the M3 subtype. *Gastroenterology* **103:**870.
23. Prinz, C., Kajimura, M., Scott, D. R., Mercier, F., Helander, H. F., and Sachs, G. (1993). Histamine secretion from rat enterochromaffin-like cells. *Gastroenterology* **105:**459–461.
24. Tache, Y. (1987). Central nervous system regulation of acid secretion. In *Physiology of the Gastrointestinal Tract.* 2nd ed. (L. R. Johnson, ed.), Raven Press, New York, Volume 2, p. 911.
25. Walsh, J. H. (1988). Peptides as regulators of gastric acid secretion. *Annu. Rev. Physiol.* **50:**41–63.
26. Lichtenberger, L. M. (1982). Importance of food in the regulation of gastrin release and formation. *Am. J. Physiol.* **243:**G429.
27. Bertaccini, G., and Coruzzi, G. (1989). Control of gastric acid secretion by histamine H2 receptor antagonists and anticholinergics. *Pharmacol. Res.* **21:**339.
28. Feldman, M., and Burton, M. E. (1990). Histamine H2 receptor antagonist. *N. Engl. J. Med.* **323:**1672.
29. Helander, H. F., and Hirschowitz, B. I. (1972). Quantitative ultrastructural studies on gastric parietal cells. *Gastroenterology* **63:**951.
30. Gibert, A. J., and Hersey, S. J. (1982). Morphometric analysis of parietal cell membrane transformations in isolated gastric glands. *J. Membr. Biol.* **67:**113.

31. Scott, D. R., Helander, H. F., Hersey, S. J., and Sachs, G. (1993). The site of acid secretion in the mammalian parietal cell. *Biochim. Biophys. Acta* **1146:**73–80.
32. Yao, X., Thibodeau, A., and Forte, J. G. (1993). Ezrin–calpain I interacts in gastric parietal cells. *Am. J. Physiol.* **265:**C36.
33. Ganser, A., and Forte, J. G. (1973). K^+ stimulated ATPase in purified microsomes of bullfrog oxyntic cells. *Biochim. Biophys. Acta* **307:** 169–180.
34. Hersey, S. J., Perez, A., Matheravidathu, S., and Sachs, G. (1989). Gasric H^+,K^+-ATPase in situ: Evidence for compartmentalization. *Am. J. Physiol.* **257:**G539.
35. Wolosin, J. M., and Forte, J.. G. (1984). Stimulation of oxyntic cells triggers K^+ and Cl^- conductors in apical H,K-ATPase membrane. *Am. J. Physiol.* **246:**C537.
36. Mamiya, N., Goldenring, J. R., Tsunoda, Y., Modlin, I. M., Yasui, K., Usuda, N., Ishikawa, T., Natsume, A., and Hidaka, H. (1993). Inhibition of acid secretion in gastric parietal cells by the Ca^{2+}/calmodulin-dependent protein kinase II inhibitor KN-93. *Biochem. Biophys. Res. Commun.* **195:**608–615.
37. Polvani, C., Sachs, G., and Blostein, R. (1989). Sodium ions as substitutes for protons in the gastric H,K-ATPase. *J. Biol. Chem.* **264:** 17854–17859.
38. Wallmark, B., Stewart, H. B., Rabon, E., Saccomani, G., and Sachs, G. (1980). The catalytic cycle of gastric (H^++K^+)-ATPase. *J. Biol. Chem.* **255:**5313–5319.
39. Bamberg, E., Butt, H. J., Eisenrauch, A., and Fendler, K. (1993). Charge transport of ion pumps on lipid bilayer membranes. *Q. Rev. Biophys.* **26:**1–25.
40. Lorentzon, P., Sachs, G., and Wallmark, B. (1988). Inhibitory effects of cations on the gastric H^+,K^+-ATPase. A potential-sensitive step in the K^+ limb of the pump cycle. *J. Biol. Chem.* **263:**10705–10710.
41. Skrabiana, A. T., Dupont, J. J. H. M., and Bonting, S. L. (1984). H/ATP transport ratio of the K/H ATPase of pig gastric mucosa. *Biochim. Biophys. Acta* **774:**91–95.
42. Rabon, E. C., McFall, T. L., and Sachs, G. (1982). The gastric H,K-ATPase: H^+/ATP stoichiometry. *J. Biol. Chem.* **257:**6296–6299.
43. Rabon, E. C., Smillie, K., Seru, V., and Rabon, R. (1993). Rubidium occlusion within tryptic peptides of the H,K-ATPase. *J. Biol. Chem.* **268:**8012–8018.
44. Maeda, M., Ishizaki, J., and Futai, M. (1988). cDNA cloning and sequence determination of pig gastric H,K-ATPase. *Biochem. Biophys. Res. Commun.* **157:**203–209.
45. Shull, G. E., and Lingrel, J. B. (1986). Molecular cloning of the rat stomach H,K-ATPase. *J. Biol. Chem.* **261:**16788–16791.
46. Reuben, M. A., Lasater, L. S., and Sachs, G. (1990). Characterization of a β subunit of the gastric H/K-ATPase. *Proc. Natl. Acad. Sci. USA* **87:**6767–6771.
47. Toh, B.-H., Gleeson, P. A., Simpson, R. J., Moritz, R. L., Callaghan, J. M., Goldkorn, I., Jones, C. M., Martinelli, T. M., Mu, F. -T., Humphris, D. C., Pettitt, J. M., Mori, Y., Masuda, T., Sobieszczuk, P., Weinstock, J., Mantamadiotis, T., and Baldwin, G. S. (1990). The 60- to 90-kDa parietal cell autoantigen associated with autoimmune gastritis is a β subunit of the gastric H,K-ATPase(proton pump). *Proc. Natl. Acad. Sci. USA* **87:**6418–6422.
48. Munson, K. B., Gutierrez, C., Balaji, V. N., Ramnarayan, K., and Sachs, G. (1991). Identification of an extracytoplasmic region of H,K-ATPase labeled by a K^+-competitive photoaffinity inhibitor. *J. Biol. Chem.* **266:**18976–18988.
49. Scott, D., Munson, K., Modyanov, N., and Sachs, G. (1992). Determination of the sidedness of the C-terminal region of the gastric H,K-ATPase β subunit. *Biochim. Biophys. Acta* **1112:**246–250.
50. Besancon, B., Shin, J. M., Mercier, F., Munson, K., Rabon, E., Hersey, S., and Sachs, G. (1992). Chemomechanical coupling in the gastric H,K ATPase. *Acta Physiol. Scand.* **146:**77–88.
51. Besancon, M., Shin, J. M., Mercier, F., Munson, K., Miller, M., Hersey, S., and Sachs, G. (1993). Membrane topology and omeprazole labeling of the gastric H,K-adenosinetriphosphatase. *Biochemistry* **32:**2345–2355.
52. Sachs, G., Shin, J. M., Besancon, M., and Prinz, C. (1993). The continuing development of gastric acid pump inhibitors. *Alimen. Pharmacol. Ther.* **7:**4–12.
53. Shin, J. M., Besancon, M., Simon, A., and Sachs, G. (1993). The site of action of pantoprazole in the gastric H/K-ATPase. *Biochim. Biophys. Acta* **1148:**223–233.
54. Bamberg, K., and Sachs, G. (1993). Topology of the H,K-ATPase. *J. Cell. Biochem. Suppl.* **17C:**H148.
55. von Heijne, G. (1988). Transcending the impenetrable: How proteins come to terms with membranes. *Biochim. Biophys. Acta* **947:**307– 333.
56. Mercier, F., Reggio, H., Devilliers, G., Bataille, D., and Mangeat, P. (1989). A marker of acid-secreting membrane movement in rat gastric parietal cells. *Biol. Cell* **65:**7–20.
57. Mercier, F., Reggio, H., Devilliers, G., Bataille, D., and Mangeat, P. (1989). Membrane–cytoskeleton dynamics in rat parietal cells: Mobilization of actin and spectrin upon stimulation of gastric acid secretion. *J. Cell Biol.* **108:**441–453.
58. Mercier, F., Bayle, D., Besancon, M., Joys, T., Shin, J. M., Lewin, M. J. M., Prinz, C., Reuben, A. M., Soumarmon, A., Wong, H., Walsh, J. H., and Sachs, G. (1993). Antibody epitope mapping of the gastric H/K-ATPase. *Biochim. Biophys. Acta* **1149:**151–165.
59. Okamoto, C. T., Karpilow, J. M., Smolka, A., and Forte, J. G. (1990). Isolation and characterization of gastric microsomal glycoproteins. Evidence for a glycosylated β-subunit of the H/K-ATPase. *Biochim. Biophys. Acta* **1037:**360–372.
60. Shin, J. M., and Sachs, G. (1994). Identification of a region of the H,K-ATPase α subunit associated with the β subunit, *J. Biol. Chem.* **269:**8642–8646.
61. Beggah, A. T., Beguin, P., Juanin, P., Peitsch, M. C., and Geering, K. (1993). Hydrophobic C-terminal amino acids in the β-subunit are involved in assembly with the α-subunit of Na,K-ATPase. *Biochemistry* **32:**14117–14124.
62. Lindberg, P., Nordberg, P., Alminger, T., Brandstrom, A., and Wallmark, B. (1986). The mechanism of action of the gastric acid secretion inhibitor, omeprazole. *J. Med. Chem.* **29:**1327–1329.
63. Besancon, M., Simon, A., and Sachs, G. (1995). The critical cysteine in the gastric H,K ATPase for inhibition of acid secretion by proton pump inhibitors. *AGA* **A661:**2643 (abstract).
64. Jewell-Motz, E., and Lingrel, J. B. (1993). Site-directed mutagenesis of the Na,K-ATPase: Consequences of substitutions of negatively-charged amino acids localized in the transmembrane domains. *Biochemistry* **32:**13523–13530.
65. Kaminski, J. J., Wallmark, B., Briving, C., and Andersson, B. M. (1991). Antiulcer agents. 5. Inhibition of gastric H/K-ATPase by substituted imidazo[1,2a]pyridines and related analogues and its implication in modeling the high affinity potassium ion binding site of the gastric proton pump enzyme. *J. Med. Chem.* **34:**533–541.
66. Wallmark, B., Briving, C., Fryklund, J., Munson, K., Jackson, R., Mendlein, J., Rabon, E., and Sachs, G. (1987). Inhibition of gastric H, K-ATPase and acid secretion by SCH 28080, a substituted pyridyl [1,2a]imidazole. *J. Biol. Chem.* **262:**2077–2084.
67. Munson, K. B., and Sachs, G. (1988). Inactivation of H,K-ATPase by a K^+-competitive photoaffinity inhibitor. *Biochemistry* **27:**3932–3938.
68. Mendlein, J., and Sachs, G. (1990). Interaction of a K^+-competitive inhibitor, a substituted imidazo[1,2a]pyridine, with the phospho- and dephosphoenzyme forms of H,K-ATPase. *J. Biol. Chem.* **265:**5030–5036.
69. Keeling, D. J., Fallowfield, C., Lawrie, K. M. W., Saunders, D., Richardson, S., and Ife, R. J. (1989). Photoaffinity labelling of the lu-

menal potassium site of the gastric H,K ATPase. *J. Biol. Chem.* **264:** 5552–5558.

70. Canessa, C. M., Horisberger, J.-D., and Rossier, B. C. (1993). Mutation of a tyrosine in the H3–H4 ectodomain of Na,K-ATPase α subunit confers ouabain resistance. *J. Biol. Chem.* **268:**17722–17726.
71. Price, E. M., Rice, D. A., and Lingrel, J. B. (1990). Structure–function studies of Na,K-ATPase. Site-directed mutagenesis of the border residues from the H1–H2 extracellular domain of the α subunit. *J. Biol. Chem.* **265:**6638–6641.
72. Ife, R. J., Brown, T. H., Keeling, D. J., Leach, C. A., Meeson, M. L., Parsons, M. E., Reavill, D. R., Theobald, C. J., and Wiggall, K. J. (1992). Reversible inhibitors of the gastric H,K-ATPase. 3-substituted-4-(phenylamino)quinolines. *J. Med. Chem.* **35:**3413–3422.
73. Rabon, E., Sachs, G., Bassilian, S., Leach, C., and Keeling, D. (1991). A K^+-competitive fluorescent inhibitor of the H,K-ATPase. *J. Biol. Chem.* **266:**12395–12401.
74. Rabon, E., Bassilian, S., Sachs, G., and Karlish, J. D. (1990). Conformational transitions of the H,K-ATPase studied with sodium ions as surrogates for protons. *J. Biol. Chem.* **265:**19594–19599.
75. Im, W. B., Blakeman, D. P., and Davis, J. P. (1985). Irreversible inactivation of rat gastric H^+,K^+-ATPase in vivo by omeprazole. *Biochem. Biophys. Res. Commun.* **126:**78–82.
76. Gedda, K., Scott, D., Besancon, M., and Sachs, G. (1994). The half life of the gastric H,K ATPase. *Gastroenterology,* **109:**1134–1141.
77. Tari, A., Wu, V., Sumii, M., Sachs, G., and Walsh, J. H. (1991). Regulation of rat gastric H^+/K^+-ATPase α-subunit mRNA by omeprazole. *Biochim. Biophys. Acta* **1129:**49–56.
78. Tari, A., Yamamoto, G., Sumii, K., Sumii, M., Takehara, Y., Haruma, K., Kajiyama, G., Wu, H. V., Sachs, G., and Walsh, J. H. (1993). Role of histamine2 receptor in increased expression of rat gastric H^+,K^+-ATPase α-subunit induced by omeprazole. *Am. J. Physiol.* **265**(4 Pt. 1):G752–G758.
79. Tari, A., Yamamoto, G., Yonei, Y., Sumii, M., Sumii, K., Haruma, K., Kajiyama, G., Wu, V., Sachs, G., and Walsh, J. H. (1994). The effect of histamine on H,K ATPase α subunit expression. *Am. J. Physiol,* G444–G450.
80. Crothers, J. M., Chow, D. C., and Forte, J. G. (1993). Omeprazole decreases H^+,K^+-ATPase protein and increases permeability of oxyntic secretory membrane in rabbit. *Am. J. Physiol.* **265:**G231–G241.
81. Scott, D. R., Besancon, M., Sachs, G., and Helander, H. (1994). Effects of anti-secretory agents on parietal cell structure and H,K-ATPase levels in rabbit gastric mucosa in vivo. *Dig. Dis. Sci.* **39:**2118–2126.
82. Oshiman, K., Motojima, K., Mahmood, S., Shimada, S., Tamura, S., Maeda, M., and Futai, M. (1990). Control region and gastric specific transcription of the rat H,K ATPase α subunit gene. *FEBS Lett.* **281:** 250–254.
83. Campbell, V. W., and Yamada, T. (1989). Acid secretagogue-induced stimulation of gastric parietal cell gene expression. *J. Biol. Chem.* **264:**11381–11386.
84. Maeda, M., Oshiman, K., Tamura, S., Kaya, S., Mahmood, S., Reuben, M. A., Laster, L. S., Sachs, G., and Futai, M. (1991). The rat H,K-ATPase β-subunit gene and recognition of its control region by gastric DNA binding protein. *J. Biol. Chem.* **266:**21584–21588.
85. Ackermann, U., and Geering, K. (1990). Mutual dependence of Na,K-ATPase α- and β-subunits for correct posttranslational processing and intracellular transport. *FEBS Lett.* **269:**105–108.
86. Gottard, C. J., and Caplan, M. J. (1993). Molecular requirements for the cell-surface expression of multisubunit ion-transporting ATPases: Identification of protein domains that participate in Na,K-ATPase and H,K-ATPase. *J. Biol. Chem.* **268:**14342–14347.
87. Horisberger, J.-D., Jaunin, P., Reuben, M. A., Lasater, L. S., Chow, D. C., Forte, J. G., Sachs, G., Rossier, B. C., and Geering, K. (1991). The H,K-ATPase β-subunit can act as a surrogate for the β-subunit of Na,K-pumps. *J. Biol. Chem.* **266:**19131–19134.
88. Burget, D. W., Chiverton, S. G., and Hunt, R. H. (1990). Is there an optimal degree of acid suppression for healing of duodenal ulcers? A model of the relationship between ulcer healing and acid suppression. *Gastroenterology* **99:**345–351.
89. Bell, N. J. V., Burget, D., Howden, C. W., Wilkinson, J., and Hunt, R. H. (1992). Appropriate acid suppression for the management of gastroesophageal reflux disease. *Digestion* **51:**59.
90. Cederberg, C., Ekenved, G., Lind, T., and Olbe, L. (1985). Acid inhibitory characteristics of omeprazole in man. *Scand. J. Gastroenterol. Suppl.* **108:**105–112.
91. Feldman, M., and Burton, M. E. (1990). Histamine H2 receptor antagonists. Part two. *N. Engl. J. Med.* **323:**1749.
92. Bayerdorffer, E., Mannes, G. A., Sommer, A., Hochter, W., Weigart, J., Hatz, R., Lehn, N., Ruckdeschel, G., Dirschedl, P., and Stolte, M. (1992). High omeprazole treatment combined with amoxicillin eradicates Helicobacter pylori. *Eur. J. Gastroenterol. Hepatol.* **4:**697–702.
93. Pedersen, P. L., and Amzel, L. M. (1993). ATP synthases. Structure, reaction center, mechanism, and regulation of one of nature's most unique machines. *J. Biol. Chem.* **268:**9937–9940.
94. Dunn, B. E., Campbell, G. P., Perez-Perez, G. I., and Blaser, M. J. (1990). Purification and characterization of urease from Helicobacter pylori. *J. Biol. Chem.* **265:**9464–9469.
95. Birkholz, S., Knip, U., Nietzki, C., Adamek, R. J., and Opferkuch, W. (1993). Immunological activity of lipopolysaccharide of Helicobacter pylori on human peripheral mononuclear blood cells in comparison of lipopolysaccharides of other intestinal bacteria. *FEMS* **6:**317–324.
96. Crabtree, J. E., Shallcross, T. M., Heatley, R. V., and Wyatt, J. I. (1991). Mucosal tumor necrosis factor αand interleukin-6 in patients with helicobacter-pylori associated gastritis. *Gut* **32:**1473–1477.
97. Lichtenberger, L. M., Graziani, L., Dial, E. J., Butler, B. D., and Hills, B. A. (1983). Role of surface-active phospholipids in gastric cytoprotection. *Science* **219:**1327–1329.

24

Cell Death

Thomas J. Burke and Robert W. Schrier

24.1. INTRODUCTION

Alterations in membrane transport can be both the cause of and the result of cell injury and cell death. This chapter examines the causes of cell death that result from decreased ATP synthesis as occurs in anoxia or ischemia. Because some toxins also depress mitochondrial respiratory capacity, thereby reducing ATP production, several of these will also be considered. In both settings, at least part of the metabolic derangement in cell function that leads subsequently to cell death may be the primary result of the reduced ATP-mediated effects on membrane potential, altered ionic homeostasis, ion channel and enzyme phosphorylation, remodeling of transmembrane-spanning peptides including receptors and integrins, and retention and activity of antioxidants such as glutathione (GSH). The number of cell functions controlled by ATP content of cells is large.

Most of the focus of this chapter will be devoted to cell death in renal transporting epithelia after toxin or oxygen deprivation. Additionally, we will focus attention on increased cellular calcium burdens which may precede cell death. Cellular calcium can increase as a result of reduced ATP-mediated efflux, as a result of increased influx rate for Ca^{2+} from the extracellular fluid, or both. Therefore, there is a propensity to activate calcium-dependent phospholipases and proteases which then become the immediate causes of cell death. Despite such events, glycine and acidosis appear to delay or prevent cell death and their potential roles as cytoprotective agents will be discussed. Finally, despite these lethal events, it is clear that growth factors can provide for rapid tissue recovery after injury. The role of these growth factors in salvaging cells exposed to ischemic or noxious insults will also be considered.

24.2. ADENINE NUCLEOTIDE METABOLISM

Mitochondria, found in high density in the nephron segments engaged in transport, may be particularly sensitive to oxygen deprivation or toxins. The ability of renal epithelial cells to transport, reabsorb, or secrete solute at normal rates is compromised when tubular ATP levels decrease from physiologic values of 5–20 nmole/mg protein.[1–3] During hypoxic or toxic insults, ATP levels are not maintained. The absence of sufficient O_2 to act as an electron acceptor truncates ATP synthesis despite increases in ADP and PO_4. Additional factors that contribute to energy deficits include defects in mitochondrial ADP/ATP translocase, the transmitochondrial proton gradient, intramitochondrial and cytosolic pH, mitochondrial membrane potential, and the disturbed ion gradients.

One extremely important variable is the buffering capacity of mitochondria for extramitochondrial (cytosolic) calcium ions (Ca^{2+}). Mitochondria can accumulate Ca^{2+} or synthesize ATP but they are unable to do both simultaneously.[4] In hypoxia or during toxin injury, the decreased synthesis of ATP is accompanied by increased movement of Ca^{2+} from the extracellular fluid into the cytosol, partial buffering of this additional Ca^{2+} burden by mitochondria,[5] and therefore a further compromised ATP production rate. This effect is also seen in postischemic/hypoxic tissues when reflow to previously damaged cells presents a continuous burden of calcium which is accumulated by cells and this also attenuates ATP recovery.[6] Progressive accumulation of Ca^{2+} occurs in these tissues, and within their mitochondria until eventually mitochondrial Ca overload becomes so extensive that further buffering is not possible.[2,7]

Exogenous provision of ATP has been shown to delay or prevent cell injury related to hypoxic and other insults.[8] Magnesium ions (Mg^{2+}) must be added to the ATP to achieve the maximal protection. ATP is catabolized to adenosine which is then taken up by cells for subsequent ATP regeneration.[9,10] Inhibitors that prevent adenosine deamination are also protective.[11] The recovery of cell viability caused by ATP may be achieved, in part, via its effects on extruding Ca^{2+} and/or by enhancing mitochondrial Ca^{2+} buffering. ATP–$MgCl_2$ may also result in some degree of cellular acidosis[12] which also delays Ca^{2+} influx,[13] inactivates phospholipase A2,[14] and inhibits activation of calpain, a cytosolic Ca^{2+}-dependent neutral protease.[15] Extracellular acidosis is particularly beneficial even in the absence of ATP–$MgCl_2$ administration.

ATP depletion also alters the intracellular concentrations of other ions including Na^+, K^+, and Cl^-[5,16] and depresses signaling via reduced phosphorylation by kinases.[17] The multiple

Thomas J. Burke and Robert W. Schrier • Department of Medicine, University of Colorado Medical School, Denver, Colorado 80262.
Molecular Biology of Membrane Transport Disorders, edited by S.G. Schultz *et al.* Plenum Press, New York, 1996

effects of ATP depletion, therefore, provide opportunities to test additional therapeutic maneuvers since preventing ATP depletion itself is often not possible.

24.2.1. Cellular Consequences of ATP Depletion

The most evident secondary effect of ATP depletion is the loss of ionic homeostasis. Specifically, cellular K^+ content decreases and cellular Na^+, Cl^-, and Ca^{2+} increase, each moving down their respective concentration gradients. As a result of this reconfiguration of intracellular ion concentrations, membrane potential and intracellular pH buffering are disturbed.

24.2.1.1. Loss of Potassium

The decreased gradient of K^+ across the tubule membrane is accompanied by a reduced membrane potential.[18] Whether this loss of K^+ during hypoxia reflects simply normal outward conductivity, unaccompanied by energetic Na^+/K^+-ATPase-mediated K^+ transport, or whether there are additional events associated with ATP depletion such as changes in channel number, channel phosphorylation, or open time probability (P_o) is not certain. It is also likely that changes in K^+ permeability at apical and basolateral membranes occur over different times and with different kinetics. Furthermore, the K^+-selective pore of these channels differs substantially, being 8 amino acids in some species and 21 amino acids in others.[19,20] Some distinct insights into the precise events that regulate K^+ channel conductance are found elsewhere in this book (see Chapters 7 and 11).

The loss of K^+ and the coincidental onset of cell dysfunction has prompted studies evaluating the protection with tissue baths containing a high K^+ concentration. This delays cellular K^+ loss and prolongs tissue and organ viability during hypothermic, hypoxic storage as occurs during transplantation procedures. For example, liver, pancreas, kidney, heart, and lung removed from cadavers are flushed with cold solutions containing K^+ as the major cation. These maneuvers could simply preserve intracellular K^+ or there might be other accompanying biochemical changes that contribute to cytoprotection. The flux of Ca^{2+} into cells with decreased transmembrane K^+ gradients is very rapid.[21] These data suggest that the associated change in epithelial membrane potential may expose a latent voltage-dependent Ca^{2+} channel. Such channels have recently been demonstrated in rabbit proximal tubules and preliminary results show them to be sensitive to calcium channel blockers.[22]

K^+ channels are regulated, in part, by ATP (apical membrane) and by GTP binding proteins (basolateral).[23] GTP, like ATP, will decrease during oxygen deprivation. The potential for disturbing transmembrane K^+ gradients during oxygen deprivation or toxin administration is marked.

Thus, severe oxygen deprivation is accompanied by a decreased membrane potential which is a consequence, in part, of the change in transmembrane K^+ gradient. The decreased membrane potential, possibly acting through the voltage sensor component of transmembrane ion channel proteins, results in further increases in ion flux rates as noted by increases in P_o. At least one result of these changes is a progressive, time-related increase in tissue calcium levels. Calcium handling by hypoxic tissues will be discussed below.

24.2.1.2. Gain of Sodium

During severe hypoxia or ischemia, renal tubules gain Na^+.[16] This is a time-limited accumulation with most, if not all, of the increase being achieved by 20 min. Little further change in tissue Na^+ occurs between 20 and 60 min of ischemia. Dominguez and co-workers have examined the independent effects of altered transmembrane Na^+ gradients on Ca^{2+} influx into normoxic renal tubules.[24] By progressively lowering the media concentration of Na^+, Ca^{2+} efflux rate was reduced thereby effecting an increase in $[Ca^{2+}]_i$. Mitochondria may have buffered some of the increased calcium burden thereby reducing what otherwise would have been an even higher increase in $[Ca^{2+}]_i$. However, if ouabain is used to decrease the transmembrane Na^+ gradient, by increasing intracellular Na^+, $[Ca^{2+}]_i$ is not increased.[25] Both maneuvers are associated with increased influx rates for $^{45}Ca^{2+}$.[21,24] Because hypoxia reduces ATP levels, Ca^{2+}-ATPase-mediated efflux rate should be reduced even further when the transmembrane Na^+ gradient is reduced. Frindt *et al.* also showed a decrease in membrane potential and increased cytosolic Ca^{2+} activity when the transmembrane Na^+ gradient was reduced in renal tubules.[18]

In recent studies, using the message for the Na^+/Ca^{2+} exchanger and inhibitors of Na^+/Ca^{2+} exchange, Dominguez's group concluded that when the Na^+ gradient across rat tubules is decreased, the Na^+/Ca^{2+} exchanger working in the reverse or backward mode (i.e., Na^+ exit in exchange for Ca^{2+} entry) can also contribute to tissue and $[Ca^{2+}]_i$ overload.[26] Yu *et al.* have demonstrated the presence of cardiac-like Na^+/Ca^{2+} exchangers in renal tubules of the rat and these isoforms predominate in distal nephron segments.[27] The Na^+/Ca^{2+} exchanger is inhibitable by verapamil or other calcium channel blockers (CCB).[28]

Therefore, two conditions favoring accelerated entry rate for Ca^{2+} into tubules during oxygen deprivation or metabolic inhibition, occur simultaneously: namely, loss of cell K^+ and gain of cell Na^+. Because ATP for Ca^{2+}-mediated efflux is reduced, total tissue Ca^{2+} rises progressively. The absolute magnitude of the increase depends, in part, on Na^+/Ca^{2+} exchange.

Compelling evidence that altered ion gradients are important mediators of cell injury during hypoxia comes from studies by Weinberg and Humes.[9] Addition to hypoxic rabbit tubules of 200 μM boluses of ATP–$MgCl_2$ at 10-min intervals virtually totally prevented cell injury.[9] Since there was little O_2 to act as an electron acceptor, any adenosine formed was not used for new ATP synthesis. The effects of ATP were probably mediated by providing a surrogate source of energy for ion transport. Under these conditions Ca^{2+}-ATPase-mediated efflux and Na^+/K^+-ATPase activity were possibly maintained at nearer to normal levels and the magnitude of cellular calcium overload may, therefore, have been reduced.

24.2.1.3. Gain of Chloride

The movement of Cl^- down its concentration gradient into hypoxic cells has recently been appreciated as an additional cause of cell injury.[29] This recent focus on Cl^- came about in a unique way. Specifically, many studies have attempted to show the relevance of reactive oxygen species in mediating hypoxia–reoxygenation (H/R) or ischemia–reoxygenation injury. Provision of GSH attenuated H/R injury, but it was shown that glycine, a component of GSH which is one of the three amino acid components of GSH breakdown, was as protective as was GSH.[30] Since cell protection occurred even if intracellular GSH resynthesis was prevented,[31] additional studies were performed to determine the mechanism for this protective effect of glycine.

Because a glycine receptor in brain tissue gates chloride channels thereby inhibiting neuronal excitation, the search for a similar renal tubular glycine receptor was undertaken.[32,33] Strychnine, which in the brain blocks the effects of glycine at the receptor level, was examined for its potential effect in reversing glycine's protective effects. Surprisingly, strychnine was equally protective compared to glycine in both rat[34] and rabbit[35] tubules. Miller and Schnellmann showed that Cl^- flux into hypoxic renal tubules is attenuated by glycine or strychnine[35] and cytoprotection is observed under these conditions. Preliminary evidence from that group confirms the presence in rabbit tubules of a glycine–strychnine receptor.[33]

The protective effects of glycine/strychnine were also observed in the antimycin A model of ATP depletion and furthermore cell injury was attenuated if the extracellular NaCl concentration was reduced.[36] Finally, DIDS, a chloride channel antagonist, not known to alter cellular Ca^{2+} homeostasis, reduced the rate of Ca^{2+} influx and lactate dehydrogenase (LDH) release during metabolic inhibition.

However, replacement of NaCl by choline chloride also results in less cell injury in this model suggesting that Cl^- alone may not be sufficient to cause cell injury.

24.2.1.4. Calcium Ions

Many studies of cell injury implicate Ca^{2+} as a participating factor, although this is not a universal finding. Species or strain differences, tissue type, and other experimental conditions including whole animals, isolated organs, freshly isolated tissues, or primary or established cell lines may account for some of the disagreements.

Calcium flux rate into ATP-depleted cells proceeds at an accelerated rate[37] and concomitantly ATP-mediated Ca^{2+} efflux rate is reduced. Even if Ca^{2+} influx rate were not increased, the decrease in efflux rate should permit total tissue Ca^{2+} overload to occur. Calcium accumulation is a clear and unequivocal finding in renal tubule hypoxic injury, and was first demonstrated by Mandel's group[5] and by Weinberg.[1] Much of this calcium burden was found associated with mitochondria since ruthenium red could prevent total tissue Ca^{2+} overload during hypoxia.[5] In hypoxia, rabbit tubular mitochondria are apparently able to sequester most of this additional Ca^{2+} since increases in cytosolic Ca^{2+} are not observed.[38] The calcium-sequestering ability of mitochondria is probably related to the degree to which mitochondrial membrane potential is maintained and this in turn is related, in part, to the level of pO_2 under which hypoxic studies are performed. As total anoxia is approached, less Ca^{2+} is sequestered by mitochondria[5] and more is found in the cytosol.[39] Thus, during ATP depletion induced either by inhibitors of metabolism or by hypoxia or anoxia, Ca^{2+} influx rate can be shown to be increased. Depending on the species or the state of mitochondrial energization, some of this increased Ca^{2+} can be detected in the cytosol.

24.2.1.4a. Cytosolic Calcium. The evaluation of potential changes in cytosolic calcium as a prelude to cell death relies on the retention of fluorescent dyes in the cytosol. However, when cell injury and the damaged plasma membranes allow for leakage of both large molecules such as LDH and small dyes such as fura-2 from the cell, accurate measurement of ions with dyes becomes problematic.[40] Weinberg *et al.* have also described decreasing levels of intracellular fura-2 during hypoxia or metabolic inhibition.[41] Nevertheless, fura-2 can be a useful dye in measuring changes in Ca_i before overt membrane damage has occurred. We recently described an early, significant increase in Ca_i at 2 min of hypoxia.[40] Ca_i continued to rise for at least 10 min; thereafter, membrane damage detected by nuclear staining with propidium iodide (PI) became evident and the loss of fura-2 made accurate measurements of Ca_i impossible. Schwartz's group also was able to measure increases in Ca_i over 15 min in MDCK cells exposed to the inhibitors KCN and 2-deoxy-D-glucose.[42] A large part of the increase (~80%) was related to Ca influx from the media. About half of the residual increase (i.e., 10%) was related to release from TMB-8-blockable intracellular stores.

Rabbit tubules on the other hand do not demonstrate increased Ca_i in response to oxygen deprivation. Mandel's group showed that during hypoxia rabbit tubule mitochondria sequester Ca^{2+} avidly[5] and, in other studies, that rabbit tubule mitochondria retain Ca^{2+} much better than do rat tubule mitochondria.[38] These species' differences may involve other organs as well. LeMasters and his colleagues also find no increase in Ca_i in rabbit hepatocytes until these cells exhibit lethal injury.[43] Mitochondrial calcium buffering tends to delay or blunt the rise in cytosolic $[Ca^{2+}]_i$ and thus may be cytoprotective initially but then the eventual capacity of mitochondria to synthesize ATP during reoxygenation/reflow is compromised. Ruthenium red, which blocks mitochondrial Ca^{2+} overload during hypoxia, may also delay or prevent Ca^{2+} influx.[44] Whether or not Ca_i increases during oxygen deprivation is an important consideration since activation of phospholipases and proteolytic enzymes may be calcium dependent. Activation of these enzymes, after Ca_i has increased, could then cause cell damage, as assessed by phospholipid degradation, release of free fatty acids, LDH release, and/or PI uptake.

These activated enzymes could account for the altered cytoskeletal derangements, loss of membrane polarity, and separation of cells from their underlying matrix. Although some of the Ca^{2+} burden is partly buffered by mitochondria, as pO_2 falls to near zero or when cytochrome a/a_3 is reduced almost completely, mitochondrial Ca^{2+} buffering ceases.[5] Under these "anoxic" conditions $[Ca^{2+}]_i$ rises to higher levels than are seen in hypoxia. In modeling oxygen deprivation injury, complete anoxia is a singular insult whereas hypoxia may be mild, moderate, or severe. Depending on the experimental circumstances, therefore, "hypoxia" can be accompanied by severe tissue cytosolic and mitochondrial Ca^{2+} overload. When such changes are not seen, the relative degree of hypoxia must be considered as being partly responsible. Unfortunately, many studies fail to report the pO_2 level, which would help to characterize the severity of the hypoxic state and allow better interpretation of results.

24.2.1.4b. Calcium Channel Blockers. Various types of CCB have been shown to attenuate whole kidney or tubular injury related to oxygen deprivation. Often, however, the CCB concentrations used are quite high, which suggested indirectly that voltage-dependent Ca^{2+} channels are unlikely to be present in renal tubules. Recently, however, direct evidence for voltage-dependent, CCB-sensitive Ca^{2+} channels in rabbit proximal tubules has been reported.[22] More distal nephron segments, which also demonstrate voltage-dependent Ca^{2+} channels, are less injured by hypoxic conditions. However, in both proximal and distal nephron segments, depolarization results in increased P_o rather than increased number of Ca^{2+} channels.[22,45]

There are also other reported effects of CCB on cell viability that are demonstrable in the absence of altered Ca^{2+} influx. Verapamil inhibits Na^+ gradient-driven processes including Na^+/Ca^{2+} exchange in synaptosomes,[46] Na^+-induced Ca^{2+} release in mitochondria,[47] and decreases basolateral membrane potential in mouse proximal tubules.[48] We have demonstrated that verapamil treatment slows the decrease in cellular ATP concentration that accompanies hypoxia to rat proximal tubules, even in a calcium-free media.[49] To better understand this protective effect which seemed to involve mitochondrial function, we measured isolated rat renal cortical mitochondrial respiration and swelling without and with exposure to calcium. Verapamil and dibucaine both reduced the swelling and mitochondrial respiratory impairment induced by extramitochondrial calcium. Thus, verapamil may also prevent cell injury, in part, by not only preventing mitochondrial Ca uptake but also by diminishing mitochondrial respiratory dysfunction. Calcium clearly exerts damaging effects on mitochondria which may be mediated in part by activation of a Ca^{2+}-dependent phospholipase.[14] This would explain the protective effects of dibucaine seen by Malis and Bonventre,[50] by our group,[51] and by Lieberthal and his colleagues.[52]

Finally, although ATP depletion can elevate $[Ca^{2+}]_i$ which then can mediate cell injury in part by activating important cellular enzymes and by depressing mitochondrial ATP synthetic processes, an additional effect of diminished kinase activity may also explain certain aspects of cell injury in the absence of ATP.

24.2.1.5. Kinase Activity

Protein kinases including protein kinase C (PKC) and protein kinase A (PKA) are tightly regulated enzymes necessary for normal cell function including cell division, differentiation, and signaling. Phosphorylation by kinases is balanced by phosphatases so that signaling does not continue unabated. During ischemic or toxic insults, the lack of sufficient ATP would by expected to compromise these processes. Other kinases such as protein kinase G (PKG) use GTP, rather than ATP, for phosphorylation. PKG may not be able to phosphorylate K^+ channels during oxygen deprivation since GTP levels are also decreased.[23] PKG is an important regulator of basolateral K^+ permeability in renal collecting ducts. Moreover, with the recognition that many isoforms of kinases and phosphatases exist, the causes of cell death related to ATP depletion extend far beyond simple changes in ionic homeostasis.

24.2.1.5a. Protein Kinases. Most of the experimental studies examining PKC, PKA, and PKG during ischemic insults have used either brain or heart tissue.[53,54] Very little experimental work in this area concerning the kidney response during or following ischemia or toxins has been conducted. Part of the problem relates to the polarized nature of renal "transporting" epithelial cells, which express these kinases differently in (1) different membranes, i.e., basolateral and apical, and (2) in different nephron segments, i.e., proximal versus distal tubules. Brain and cardiac tissues do not have these constraints and are more easily studied. Thus, by necessity the following assessments of the role of changes in protein kinases in ischemic or toxic cell death in transporting epithelial cells will be quite general.

PKC activity is translocated from cytosolic to membrane sites during ischemia. The decrease in H-7-inhibitable cytosolic activity of 60–70% is paralleled by a twofold increase in particulate (membrane) activity after 30 min of ischemia.[55] With reflow and regeneration of ATP, the membrane would seem poised to phosphorylate important proteins needed to respond to growth factors that permit tissue remodeling, including proliferation and differentiation of damaged cells.

Reflow for 1 hr after ischemia is, in fact, accompanied by increased immunoreactivity and biochemical activity of alpha, delta, and epsilon isoenzymes of PKC. When these enzyme are inhibited with agents such as H-7 or staurosporine in spinal cord ischemia models, the animals become paraplegic with short periods of ischemia.[56] On the other hand, 1,2-oleoylacetylglycerol (OAG), an activator of PKC, modestly extends the ischemic time before significant paraplegia develops. By 48 hr of reflow, PKC levels in both the soluble and particulate fractions are markedly decreased. The unanswered question is whether this downregulation reflects, in part, a shutdown of growth factor activity. As discussed elsewhere in this book (Chapter 12), cells need not be exposed to ATP depletion to undergo toxic cell death. Glutamate neurotoxicity, for example, is associated with profound increases in Ca_i. The increased Ca_i is followed shortly thereafter by activation of PKC.[57] Even though Ca_i rises, cell death can be prevented by inhibitors of PKC (H-7 and calphostin C), but inhibition of PKA with HA-1004 does not abrogate cell death.[57] Clearly, and depending on the cell type or model of cell injury, inhibition of PKC can either delay or hasten injury.

In the brain, Ca^{2+}/calmodulin kinase activity (CaM-KII) is also reduced by ischemia.[58] CaM-KII falls to the lowest level of any cytosolic kinase yet measured; however, like PKC, there is an increase in membrane activity. Thus, as with PKC, cytosolic phosphorylation of proteins important in signal transduction of normal cells appears to be depressed by the ischemic insult. Perhaps this should not be surprising given the constraints for ATP regeneration, reestablishment of ion gradients, the need to remodel membranes and to return cytosolic and cell Ca^{2+} homeostasis to normal. All of these must be reestablished in ischemic/reoxygenated cells before kinase-related transport function can be normalized.

As noted above (Section 24.2.1.1), cells lose K^+ during oxygen deprivation and this ultimately can result in cell death. In rat renal cortical collecting duct cells, basolateral K^+ conductance is partly dependent on a cGMP-dependent protein kinase, namely, PKG.[23] Since anoxic conditions generally reduce kinase activity, the loss of K^+ could be related in part to depressed kinase regulation of K^+ channels. This observation suggests an additional point regarding cell salvage. If protein phosphorylation is depressed by the anoxic conditions, might agents that inhibit phosphatase activity prolong cell viability? Preliminary results form Mandel's laboratory suggest that such is the case. Calyculin-A, a phosphatase inhibitor, increased the rate of anoxic phosphorylation in rabbit kidney proximal tubules and lessened the release of LDH, which is a measure of cell death.[59]

Despite these encouraging results, it has also been reported that protracted inhibition of the serine/threonine protein phosphatases 1 and 2A with okadaic acid enhances neurotoxicity and this could be overcome by H-7 or by downregulation of PKG (i.e., 24-hr phorbol ester treatment).[60] Admittedly, these are different tissues and in the latter case the neurons were not rendered anoxic. The key observation appears to be the degree, and length of time in which protein phosphorylation is sustained. Cells exposed for too long to continued phosphorylation are susceptible to death, whereas a too abrupt truncation of protein phosphorylation can prevent cell survival. It is anticipated that novel studies, attempting to modulate cell death via regulation of kinases and phosphatases, will be soon forthcoming. These may reveal that the translocation of PKC, for example, from cytosol to membrane sites, prevents overphosphorylation of cytosolic proteins while at the same time, cytokine–growth factor responses, mediated via membrane receptors, are accentuated.

24.2.1.5b. Kinases/Growth Factors. Growth factors including epidermal growth factor (EGF) and others may play an important role both in salvaging sublethally injured cells and in hastening cell proliferation/differentiation. After renal ischemia, exogenous EGF treatment increased PKC and ribosomal protein and S6 kinase activity much more quickly than would normally occur.[61] Proliferation of renal tubules was also enhanced as evidenced by [^{3}H]thymidine incorporation. Thus, the phosphorylation of proteins, which may participate in renal cell repair or salvage, can be attributed directly to the stimulating effects of some growth factors.

Nerve growth factor (NGF), fibroblast growth factor (FGF), insulin and insulin-like growth factors (IGF) I and II all exert protective effects on cell viability in PC12 cells.[62] Elevated K^+ in the media also slows the rate of cell death (possibly via a delay in the rate of cellular Ca^{2+} overload). Importantly, these protective effects occur independent of RNA or protein synthesis, events that are likely to be compromised shortly after ischemic or toxic insults, in association with the decreased synthesis of ATP.

24.2.2. MgATP and Cell Protection

Several aspects of MgATP cytoprotection were discussed above. These beneficial effects include reestablishment of ionic homeostasis and thus less cellular calcium overload due to membrane depolarization or Na^+/Ca^{2+} exchange, enhanced ATP-mediated Ca^{2+} efflux, return of cellular glycine to preischemic levels, and recovery of kinase activity.

The need for Mg^{2+} has not been completely explained. One possible role for Mg^{2+} may be its effect maximizing Na^+/K^+-ATPase activity. In the presence of ATP, this cofactor would permit more complete or more rapid return of ionic homeostasis.

An additional relationship between cell injury and ATP depletion has recently been proposed. Siegel's group has reported that during severe ATP depletion, heat-shock protein (HSP) mRNA and heat-shock factor activation are rigorously induced.[63] It is thought that HSPs, and there are several isoforms, act as molecular chaperones that prevent proteolysis of structural or functional proteins during ischemia. During reflow, ATP would then allow the function of these proteins, which had been protected during ischemia by HSP, to begin anew.

Finally, many enzymes require energy-dependent dimerization or heterodimerization prior to being able to express full activity. Loss of ATP would negate this important organizational step and render these enzymes incapable of function; provision of Mg^{2+} and ATP would restore such activity. Thus, the loss of cell viability related to ATP depletion in affected cells is expressed at several structural and functional sites.

24.3. PHOSPHOLIPASE ACTIVATION

Phospholipase A_2 (PLA_2) activated by an increased Ca_i in hypoxia is an important potential mediator of cell injury. This activation occurs at a similar time as the marked release of free fatty acids, especially arachidonic acid, that characterizes ischemic injury.[64]

24.3.1. Calcium Dependent

Renal tubules demonstrate increased levels of unsaturated and saturated free fatty acids early during a hypoxic insult.[65] The release of arachidonic acid from its *sn*-2 position on glycerol phosphate suggests activation by PLA_2. Finally, PLA_2 activity and its movement from intercellular to extracellular sites is demonstrable during hypoxia.[66] Direct measurement of PLA_2 activity by our laboratory has shown the appearance during hypoxia of a low-molecular-weight form of PLA_2 (15 kDa) with a dependence on Ca^{2+} and a preference for phosphatidylethanolamine (PE).[66] A higher-molecular-weight form (100 kDa) is also observed in normoxic proximal tubules. The activity of this Ca^{2+}-dependent, PE-specific enzyme, however, declines during 25 min of hypoxia. An intermediate-size isoform (60 kDa) which is similar to cytosolic PLA_2 ($cPLA_2$) activity also decreased with hypoxia.

Thus, the increased PLA_2 activity during hypoxia appears to be related to the appearance of a low-molecularweight isoform concomitant with disappearance of the high-molecular-weight isoform. These determinations suggest that proteolysis has occurred possibly by one of the Ca^{2+}-dependent proteases (e.g., calpain). As mentioned above, the low-molecular-weight PLA_2 may reflect the existence of a catalytic unit that results from the loss of dimerization during ATP depletion. This subunit is not seen in normoxia, i.e., when Ca_i is low and cellular ATP is normal. Furthermore, the decreased activity of the 60-kDa isoform during hypoxia may be related to the loss of ATP-dependent phosphorylation. The balance between phosphorylation and phosphatase activity during hypoxia, in part, determines functional enzyme activity. These findings in isolated tubules are consistent with observations showing loss of $cPLA_2$ activity in the isolated perfused kidney during ischemia.[14]

24.3.2. Calcium Independent

Associated with ischemic or hypoxic insults is an increase in the release of free fatty acids which are normally esterified in phospholipids. As noted above, a substantial increase in extracellular (media) PLA_2 is observed during hypoxia.[66] This is not an unexpected finding given the release of other enzymes such as LDH. However, this extracellular form of PLA_2 is Ca^{2+} independent.[66] It is not known whether this extracellular PLA_2 is a normal form of PLA_2 or whether it represents a modification of the Ca^{2+}-dependent PLA_2. A Ca^{2+}-independent PLA_2 is also found during anoxia in rabbit tubules.[67]

Neither $cPLA_2$ activity nor the release of fatty acids is blocked by glycine, although it does prevent cell injury.[65,66] However, glycine treatment prevented the release of PLA_2 into the medium suggesting that extracellular PLA_2 may damage hypoxic cells by acting on the cell membrane at its external surface. High concentrations of exogenous PLA_2 exacerbate mild hypoxic injury and cause release of free fatty acids from normoxic renal tubules.[65,68] Glycine does not prevent injury in this latter model. Low concentrations of exogenous PLA_2 actually prevent hypoxic injury.[69]

The released, extracellular PLA_2 is sensitive to inhibition by dibucaine, more so than to mepacrine,[66] and both of these putative phospholipase inhibitors delay the appearance of lethal cell injury in hypoxic rat tubules.[51] Dibucaine attenuates damage to mitochondria exposed to Ca^{2+} and oxygen free radicals.[50]

A calcium-independent form of PLA_2 has been described in anoxic rabbit tubules.[67] This isoform is found in the plasma membrane of these tubules. As mentioned above, the rate of rise of Ca_i is less rapid in anoxic rabbit tubules (possibly because of avid mitochondrial sequestration or slower release of endogenous stores of Ca^{2+}) and LDH release occurs at a later time than in rat tubules. Thus, a species difference may account for why rat and rabbit tubules exhibit different forms of PLA_2, which are activated at different times and are located at different sites.

24.3.3. Inhibition of PLA_2

If PLA_2 activity is enhanced by oxygen deprivation or toxic insults, then inhibitors of PLA_2 should be beneficial. Clear evidence of PLA_2-mediated oxidant injury to mitochondria was demonstrated by Malis and Bonventre.[50] Dibucaine, one of several drugs with PLA_2 inhibitory activity, was partially protective. Since pO_2 levels are low, but not totally absent, in hypoxia, this mechanism for a mitochondrial site of injury may be important. The *in vitro* assay of PLA_2 activity in hypoxic tubule preparations shows concentration-dependent inhibition with dibucaine and mepacrine.[66] Intact rat tubules exposed to hypoxic insults are also protected for some time by mepacrine and dibucaine.[51] Cellular release of LDH and ^{45}Ca influx rates are lower during hypoxia in the presence of these inhibitors. Inhibition of PLA_2 activity, even for a short time, delays membrane damage, reduces the magnitude of free fatty acid release, and lowers LDH release to the media. Since CCB also delay the onset of cell injury, treatment of hypoxic tubules with both PLA_2 inhibitors and CCB simultaneously may exhibit additive effects.

24.3.4. pH Modulation

Another important variable modulating PLA_2 activity is intracellular pH (pH_i). PLA_2 has optimal activity at pH 8.0 but considerable activity is found at pH 7.4.[14] In various injury models, lowering pH to 7.0 or below eliminates PLA_2 activity,[14] reduces Ca^{2+} influx,[70] delays the onset of cell injury,[1,3,70] and prevents tissue Ca^{2+} overload during hypoxia.[1] The increase in $[Ca^{2+}]_i$ during hypoxia is, however, not reduced by acidotic conditions[71]; these data together suggest that lowering pH_i prevents mitochondrial Ca^{2+} overload and, despite a rise in Ca_i, PLA_2 is inactivated by the low pH_i. The mitochondrial PLA_2 studied by Malis and Bonventre[50] may also be inhibited by acidosis. Low pH_i will also reduce the activity of the neutral protease calpain. Anoxia but not hypoxia is associated with decreases in extracellular pH (pH_o) and presumably this reflects enhanced acid production by glycolysis.[1] When pO_2 is about 0.4 mm Hg, pH_o falls to 7.1.[72] Measurements of pH_i have been eased by the introduction of specific fluorescent dyes such as BCECF and SNARF-1. Since these dyes are retained only in cells with membrane integrity, the accuracy of pH_i measurements with hypoxia will depend on the integrity of membranes. Those cells that leak LDH are less likely to provide meaningful information about pH_i because of the decreasing intracellular concentration of the indicator dyes.

24.4. PROTEASES

Proteases are a group of cytosolic and membrane-bound (lysosomal) enzymes whose activities can be altered by changes in pH and/or calcium.[73] The lysosomal enzymes are compartmentalized to prevent inadvertent cellular digestion. However, when membrane damage occurs, as in hypoxia possibly related to cell swelling or phospholipase activation, these enzymes might gain access to the intracellular environment where additional cell injury could then occur. In addition, the hypoxia-induced increases in cytosolic calcium would favor activation of some of these proteases. Cytosolic enzymes such as calpain, a neutral cysteine protease, are also calcium dependent, and with the early increase in cytosolic free calcium that accompanies hypoxia,[40] calpain may be the first protease to be activated. Studies on the potential role of proteases in renal cell injury have only recently begun and these have been directed primarily at renal tubular injury due to cyclosporin.[74] As new, specific protease inhibitors are developed, the number of studies directed at examining the role of proteases in hypoxic and toxic cell injury is expected to increase. However, Gores and colleagues have recently shown that glycine treatment of hepatocytes, exposed to metabolic inhibitors, protects these cells by reducing the proteolysis caused by calpain.[75] Whether the activity of the native enzyme is reduced or whether the substrate proteins are protected from attack by glycine is as yet unclear. The latter may well occur since Kellerman *et al.* have shown early protection of the actin cytoskeleton with glycine treatment of $LLCPK_1$ cells treated with the cytotoxin ionomycin.[76]

24.4.1. Classification of Proteases

Extracellular and intracellular proteases are classified in part by the site of the peptide bond that they hydrolyze. Endopeptidases cleave polypeptide chains in their inner regions whereas exopeptidases act near the N- or C-terminal region. Oligopeptidases act on very small proteins (20–30 amino acids). The best studied proteases are extracellular and these include renin, pepsin, trypsin, and chymotrysin. However, the more interesting proteases, from a cell injury viewpoint, are the intracellular proteases.

The most prevalent and best characterized intracellular proteases are localized in lysosomes. These cysteine proteases are represented by the class called cathepsins of which there are several isoforms that are calcium independent. Other intracellular proteases, the calpains, are calcium-dependent, neutral proteases. Calpains, but not cathepsins, have been implicated in cyclosporin-mediated tubular injury[74] and, in preliminary reports, in hypoxic injury to human renal tubules in culture.[77] The well-recognized time-dependent increasing burden of tissue Ca^{2+} in hypoxia suggests that calpain may well become an important cytotoxic enzyme. As noted above, glycine apparently prevents the devastating effects of calpain in hepatocytes during metabolic inhibition. We have been unable to detect cathepsin activity in the media of hypoxic tubules nor is there any increase in cathepsin activity during hypoxia (unpublished results). Moreover, lucifer yellow which is accumulated by lysosomes does not leak into the cytosol during hypoxia (unpublished results). We have not examined the influence of longer periods of oxygen deprivation; with more extensive cell death, cathepsin activity may increase. At present, however, it appears that cathepsin activation is not an initiator of hypoxic cell injury.

24.4.2. Calpain

Calpain, the major extralysosomal, intracellular protease, is a heterodimer with two subunits, an 80-kDa catalytic unit and a 30-kDa regulatory subunit.[78] There are four domains in the 80-kDa subunit. Region II contains the protease activity and region IV is the calcium-binding site. This fourth region exists as a helix–loop–helix structure. Calcium binds to the loop when the proenzyme is activated. The 30-kDa subunit has two domains: region V is glycine-rich whereas region VI is very similar to region IV. Both regions IV and VI are structurally similar to calmodulin. Micromolar levels of calcium activate the calpain I isoform. The calpain II isoform is activated by millimolar levels of calcium. Thus, it is more likely that calpain I is involved in calcium-mediated cell injury caused by oxygen deprivation or toxins.

24.4.2.1. Activation of Calpain

Although calpain activation is partly dependent on micromolar levels of Ca^{2+}, an important role for phosphatidylinositol has also been demonstrated.[78] The membrane localization of calpain provides for the initiation of membrane cytoskeletal disassembly as well as for attack on phospholipase C, PKC, and other membrane proteins. The modified domains (I–IV) of the 80-kDa subunit and region VI of the 30-kDa subunit subsequently dissociate, or are proteolyzed, from the membrane and enter the cytosol to expand the targets of proteolysis. Clearly it becomes apparent that preventing a rise in cytosolic calcium will delay this injury process. Often, however, cell injury, related to increased Ca_i, has already occurred or is impossible to modify. The recent studies showing that glycine can effect cellular protection despite increases in Ca_i[40] suggest that Ca_i-induced cell injury caused, in part, by calpain activation may be prevented. Protection can also be demonstrated through use of specific calpain inhibitors.[79]

24.4.3. Protease Inhibitors

Protease inhibitors fall into two classes. The first, the oxiranes, act by covalent interaction between the cysteine thiol group (Cys108) on region II of the 80-kDa subunit of calpain and the electrophilic center of the inhibitor. The standard oxirane is E-64, an isolate from *Aspergillus japonicus*. However, E-64 is not specific for calpain because it also inhibits other cysteine proteases. The second class, the aldehydes, is represented by leupeptin. Leupeptin reversibly removes Ca^{2+} from calpain and other cysteine proteases; however, it also inactivates serine proteases. Both of these inhibitors are charged, thus limiting their access to the cell interior. Recently, E-64-d has become available and this orally active molecule easily enters cells. It has been used in clinical trials in Japan for patients with muscular dystrophisms. Finally, MDL 28170 (CBZ Val Phe H) and calpeptin are short hydrophobic, N-blocked dyspeptidyl aldehydes which enter cells by passive diffusion. They inhibit the activity of calpain, cathepsin B, and trypsin. The use of these inhibitors of proteases, as is the case for mepacrine and dibucaine for PLA_2 inhibition, is beginning to provide insights into the pathogenesis of cell injury. However, the importance of these enzymes will be completely resolved only when specific selective inhibitors are developed.

24.5. CELL POLARITY/CYTOSKELETON

The membrane cytoskeleton is a complex of proteins including, but not limited to, spectrin, gelsolin, microfilaments, microtubules, intermediate filaments, tektins, and two forms of actin [globular (G) and filamentous (F)]. Specific tissues often express additional and unique proteins. The cytoskeleton forms the basic cell structure/shape, maintains nuclear position and microvillar shape, determines cell polarity and the sorting of membrane enzymes in transporting epithelia, maintains cell–cell contact via gap junctions, and limits leakage of dissimilar fluids to or from urine and plasma by stabilizing the tight junction. Cytoskeletal structure also directs the location and positioning of cell surface receptors for ligands and second message cofactors such as G proteins. Finally, the integrity of the cytoskeleton is an important regulator of those transmembrane proteins that anchor cells to the underlying basement membrane. Different proteins and different components of the cytoskeleton are found in different regions of the same cell: apical, lateral, basolateral. Adhesion molecules that anchor cells to the basement membrane are disrupted rather late during ischemia whereas disruptions in cell–cell contact and polarity changes are early responses.[44,80]

24.5.1. Cell Polarity

Ischemic insults or ATP depletion states result in a loss of the vectorial location of enzymes that direct secretion, reabsorption, and transport. Molitoris *et al.* first described the translocation of Na^+/K^+-ATPase from its predominately basolateral location toward the apical surface of ischemic renal tubules *in situ*.[80] This observation provided an important clue as to why sodium reabsorption is reduced in postischemic acute renal failure since the novel relocation of the enzyme would favor extrusion of Na^+ into the luminal rather than interstitial fluid. Importantly, this relocation was preceded by a loss of cell–cell contact, especially at tight junctions.[44] Cytoskeletal integrity is therefore required for sorting of transport enzymes and other proteins including perhaps ion channels, to precise locations within a cell. It is well accepted that renal epithelial cells in culture do not polarize until confluence (cell–cell contact) is reached. A portion of this regulatory process may involve phosphorylation so that when ATP decreases, polarity changes associated with a disrupted cytoskeleton are expressed. The activation of calpain proteases, mentioned above, will result in degradation of many cellular proteins so that even if ATP was maintained at or near normal levels, proteolytic cell injury would still occur. The best example of such a process occurs when ionomycin raises cellular Ca^{2+} in normoxic, ATP-replete tubules; cytoskeletal changes and cell injury can be demonstrated quite promptly.[76]

24.5.2. Actin Cytoskeleton

Actin content of the cytoskeleton is found in two forms, G and F. G-actin is polymerized whereas F-actin is depolymerized. Fluorescence studies with rhodamine-phalloidin which binds to F-actin but not to G-actin revealed a time-dependent movement of F-actin both to the lumen (shed microvilli) and to the cell interior during renal ischemia.[81] Most of this loss occurs in the first 5 min of total renal ischemia, a time by which ATP depletion is virtually complete and Ca^{2+} has risen significantly. These changes are unique to renal proximal tubules since glomeruli and distal tubules do not demonstrate these changes. F-actin also decreases within 5 min of exposure of $LLCPK_1$ cells to 5 μM ionomycin,[76] but increases when ATP depletion is induced by antimycin A.[82]

Thus, it appears that the rise in Ca^{2+}, rather than the decrease in ATP, is responsible for decreasing cytoskeletal F-actin. The loss of F-actin may be partly responsible for the loss of membrane polarity of Na^+/K^+-ATPase during ischemia.

The cytoskeletal proteins villin, α-actinin, and calmodulin are associated with actin and provide some of the linkages between the plasma membrane and the microfilaments in microvilli. As F-actin redistributes into the cytosol during ischemia, these bridges become untethered at one end. MgATP treatment of intact rats with renal ischemia results in functional and morphologic cell protection and repletion of MgATP may restore F-actin to a normal distribution pattern. ATP may decrease Ca_i by promoting Ca efflux and/or by altering K^+ and Na^+ concentrations. In both cases, the effect of calcium to alter F-actin, possibly via calpain, would be eliminated.

It has recently been reported that the simple amino acid glycine conveys cytoskeletal protection. Glycine (2 mM) prevents decreases in F-actin in $LLCPK_1$ cells exposed to ionomycin[76] and 3 mM glycine protects hepatocytes exposed to KCN.[83] In renal tubules exposed to extensive hypoxia, 2 mM glycine prevents completely membrane damage as assessed by LDH release.[30] It would appear that increases in Ca_i activate a protease (calpain) which then disrupts the cytoskeleton, as measured by decreases in F-actin, so as to permit leakage of intracellular enzymes. Glycine protects cells from being injured, not by preventing the increase in Ca_i but at some subsequent step(s).[40] In ischemia, therefore, loss of glycine, down its concentration gradient, to the extracellular fluid may uncover a proteolytic mechanism that is activated by concomitant increases in Ca_i.

24.5.3. Cell Swelling/Vacuoles

Rat proximal tubules exposed to a zero phosphate medium (without glycine) develop cell injury, as assessed by LDH release and increased influx rates for Ca^{2+}, in concert with reduced ADP phosphorylation to ATP.[70] Morphologically, a marked degree of subapical vacuolization is seen. Glycine (2 mM) or acidosis (pH 6.7–6.9) completely prevented morphologic and membrane damage.[70] These vacuoles, the gaps in the submembrane cytoskeleton, may be the physical counterpart of F-actin loss. These gaps may be sites where LDH and PLA_2 release into the media can occur. Glycine, which prevents the escape of these large molecules and prevents the decrease in F-actin, virtually completely prevents the development of the vacuoles. Since glycine may prevent activation of calpain[83] despite the increase in Ca_i[40] and since acidosis would decrease the activity of both calpain and PLA_2, cytoprotection with glycine may occur at multiple sites. Clearly, glycine prevents the escape of large molecules, halts the decrease of F-actin, and prevents the formation of vacuoles.

Cells lose the ability to retain small as well as large molecules when cell swelling, related to oxygen deprivation and other insults that reduce cell ATP levels, occurs. This increased permeability can be reversed or largely prevented despite cytoskeletal derangements through use of impermeable solutes such as polyethylene glycol[84] and mannitol.[85] Amelioration of cell injury by preventing cell swelling was shown in *in vivo* renal ischemia in dogs[86] and in anoxic cell swelling.[87] Prevention of cell swelling potentially exerts two additional benefits to cell viability. First, mechanical distortion or stretch of normoxic cell membranes has been shown to increase the conductance of K^+. Thus, cell swelling during ischemia may accelerate K^+ loss and thus exacerbate the decrease in membrane potential and the increased influx rate of Ca^{2+}. Patch clamp studies have quantitated this increased K^+ conductance in response to dimpling of the surface of the cell with a micropipette or to suction.[88] Of interest is the observation that membrane perturbations in one cell are expressed as changed conductance in adjacent cells which receive some signal via gap junctions. Therefore, cells that undergo hypoxic swelling are likely to transmit this information to adjacent cells, at least until gap junctions are disrupted. Because altered K^+ conductance and membrane depolarization result in increased Ca^{2+} influx and decreased Ca^{2+} efflux, the message sent to adjacent cells may involve the disrupted regulation of Ca_i. It is clear that preventing cell swelling during ischemia/transplantation/organ retrieval may be beneficial via several mechanisms, one of which could be via altering the nature of information transmitted from one hypoxic, swollen cell to another.

24.6. ADHESION MOLECULES

Adhesion molecules are a group of transmembrane-spanning proteins that anchor cells to their extracellular matrix. The cytosolic anchor is at the level of the cytoskeleton. These molecules, the integrins, exhibit structural diversity that permits the recognition of specific attachment sites on extracellular collagen and other components of basement membranes. However, all appear to anchor to the B subunit of actin within the cell. The basement membrane domains may be more labile during hypoxia for two reasons: viable cells with proximal tubule characteristics have been detected in, and cultured from, the urine of patients with acute renal failure (ARF).[89] Presumably, these cells broke free from their basement membranes. In addition, it is clear that the tubular obstruction occurring in ARF may be related in part to these exfoliated cells lodged within the tubular lumen. Their close approximation to one another in the lumen of distal nephrons, which may reflect in part a low rate of urine flow and the slower passage time of urine in ARF, has recently been attributed to the binding of one cell to another.[90] These attachments are likely made between the free ends of the integrins. This dense cell-to-cell contact results in impenetrable obstructive material.

The integrins are a family of proteins with α as well as β subunits. Heterodimerization results in the formation of several integrins. Some integrins are specific to vascular tissues (i.e., $\alpha_2\beta_1$ is necessary for endothelial cell integrity[91]) or white blood cells or keratinocytes. Proximal tubules are enriched in the α_6 subunit which forms receptors for extracellular laminin. Other subunits bind fibronectin or collagen (see Ref. 90 for recent review).

Loss of ATP as occurs in various forms of cell injury disrupts actin integrity. Actin relocates toward the cell nucleus and little dispersed actin can be found near the plasma membrane.[81] These early changes in the cytoskeleton may contribute to the disorganization of cell–cell or cell–basement membrane contacts such that, on reperfusion, some of these cells become lodged in the tubular lumen. It is important to recognize that in spite of the fact that these cells are detached from their support, they are still viable.

Another important factor is the generation of reactive oxygen species during reflow. Normoxic, ATP-replete proximal cells exposed to oxidants exhibit altered levels of integrins and cell shape changes ensue. Again, this may be a specific response because distal cells or mesangial cells, which have different integrins, are rather insensitive to similar insults. Distal cells also express integrins at their apical surface thereby accounting in part for their ability to bind exfoliated proximal tubule cells which may exhibit decreased polarity of their specific integrins as occurs also for Na^+/K^+-ATPase. These viable cells relocated to more distal nephron sites have the capacity to synthesize extracellular matrix which would facilitate the obstructive process.[90] The denudation of the proximal tubule basement membrane seen after ischemia reflects this exfoliation.

Recovery from these forms of cell injury requires that new cells be replicated and eventually differentiated. The flat epithelium seen during recovery is probably a reflection of undifferentiated cells migrating along the basement membrane. Integrin formation, so as to provide proper and stable attachment of these new cells, must be one of the early processes that recover after cell injury. When cells have proliferated sufficiently to completely cover the denuded area, mitosis ceases and differentiation begins. Although decreased ATP and/or increased cellular calcium are contributors to the process of cytoskeletal disruption, it has recently been appreciated that phosphatase activity may also be important. When the phosphatase inhibitor calyculin-A is added to hypoxic rabbit proximal tubules, less cell injury occurs. The decreased phosphorylation of and by kinases cannot be altered during ATP depletion but the removal of important phosphate groups can be slowed by calyculin.[59] This observation points to the important role for ATP as a mediator of protein phosphorylation necessary for cell viability.

24.7. OXYGEN RADICALS

24.7.1. Intracellular Generation

Oxidants formed during prolonged hypoxia or during reperfusion after complete ischemia can severely damage renal tubules. These highly reactive oxygen species include superoxide anion (O_2^-), hydrogen peroxide (H_2O_2), hydroxyl radical ($\cdot OH$), and peroxynitrite ($ONOO\cdot$) and its metabolites. Specific loci within cells are particularly susceptible to local damage.

24.7.1.1. Mitochondrial Oxygen Free Radicals

Transport of electrons during oxidative phosphorylation is attended by continuous formation of small amounts of O_2^-. Primary detoxification of O_2^- occurs by redox coupling between GSH, GSSG, NADP, and NADPH and results initially in the formation of small amounts of H_2O_2 which are converted to less toxic alcohols and/or water. Mitochondrial GSH stores are retained longer than is cytosolic GSH during insults; loss of mitochondrial GSH correlates with the appearance of cell death (LDH release).

Malis and Bonventre exposed normal rat renal cortical mitochondria to the O_2^--generating system glucose–glucose oxidase[50] and very little mitochondrial dysfunction occurred. Exposure of the mitochondria to Ca^{2+} alone also caused only moderate mitochondrial respiratory impairment. However, addition of both Ca^{2+} and the oxidant-generating substrates caused profound mitochondrial respiratory damage which could largely be attenuated by the PLA_2 inhibitor dibucaine. The residual damage may have been related to calcium efflux from and recycling back into the mitochondria, an event that clearly damages this organelle in the presence of oxidants.[92]

During hypoxia in, or reoxygenation of, intact cells, all oxidant levels will increase and, in the presence of elevated tissue Ca^{2+}, result in mitochondrial membrane and functional damage. One postulated mechanism involves the release of unsaturated fatty acids, including arachidonic acid whose double bonds are susceptible to oxidant attack. Release of arachidonic acid during ischemia may also directly activate potassium channels thereby altering ionic homeostasis even before excessive oxidant formation begins during reflow.[93]

Normally, oxidant damage to mitochondrial membranes would be limited by the scavenging activities of GSH oxidoreductase enzymes and/or the dismutation by Mn^{2+} superoxide dismutase (SOD). However, during oxygen deprivation, increased O_2^- formation coupled with decreases in GSH and increased Ca^{2+}–mediated PLA_2 activity would result in a much increased H_2O_2 production. The partial protective effects observed with dibucaine could be attributed to PLA_2 inhibition although mitochondrial swelling and respiratory impairment related to Ca^{2+} recycling would still occur.[92] In addition, exogenous H_2O_2 in the presence of Ca^{2+} causes extensive cell death. We recently characterized a Ca^{2+}–mediated impairment of mitochondrial respiration that is temporally paralleled by mitochondrial swelling. These changes could be partially reversed by either verapamil or dibucaine and additive protective effects were observed (unpublished results).

24.7.1.2. Intracellular Oxidants

Although activated neutrophils and macrophages provide a source of extracellular oxidants that can injure nearby cells, important cellular damage resulting from toxins or oxygen deprivation may occur primarily via intracellular (tubular) generation of oxidants.

It is difficult to measure the effects of the formation of various oxidants within whole cells during oxygen deprivation because of the profound metabolic derangements. A much larger body of experimental evidence has been obtained from studies in which normal cells are exposed to exogenous peroxides, or superoxide anion-generating systems such as glucose–glucose oxidase or xanthine–xanthine oxidase. However, oxidants formed within renal tubules may be more physiologically relevant.

Paller has measured H_2O_2 formation in freshly isolated renal tubules after reoxygenation; surprisingly only modest increases were detected.[94] We have confirmed these findings.[95] We also measured H_2O_2 before reoxygenation, i.e., at the end of the induced hypoxia. Most of the increase in H_2O_2 in this study arose during hypoxia itself (pO_2 normal = 400 mm Hg; PO_2 hypoxia = 20–40 mm Hg). O_2^- measured during hypoxia and after reoxygenation was very low, which could explain the absence of appreciable increases in H_2O_2, the dismutation product of O_2^-. However, O_2^- is locally quite reactive, short-lived, and may therefore be technically difficult to measure. It is now apparent that O_2^- participates in other reactions, specifically in the formation of peroxynitrite (ONOO·).[96] Low values for O_2^-, therefore, could be the result of its disappearance as ONOO· is formed; this would also account, in part, for the relatively small increase in H_2O_2 in intact cells. The generation of ONOO· is speculative, since this reactive oxygen species itself has not yet been measured in hypoxia/reoxygenation. Peroxynitrite has a half-life of about 2–4 sec and is rapidly converted to other toxic metabolites, particularly in the presence of bicarbonate as would occur in hypoxia/anoxia/reperfusion *in vivo*.[97] Bicarbonate is often used in media in which freshly isolated tubules are studied.

The generation of peroxynitrite occurs when O_2^- and nitric oxide (NO) interact. NO is a by-product of L-arginine metabolism in renal tubules containing nitric oxide synthase (NOS). In rat renal tubules, addition of L-arginine worsens cell injury and this damage can be prevented by L-NAME, an inhibitor of NOS.[98] NOS has been detected in rat renal proximal tubules.[99] Exogenous NO donors such as sodium nitroprusside also increase the extent of hypoxic damage; this additional injury can be ameliorated by hemoglobin, a scavenger of NO.[98]

Finally, OH· radicals are generated from H_2O_2 in the presence of iron (Fenton reaction) and are a by-product of ONOO· degradation.[100] Renal tubules contain iron in heme proteins such as cytochrome a/a_3 and in Fe/sulfur flavin nucleotides which are components of electron transport proteins in mitochondria. Mitochondria therefore would appear to be likely targets of hydroxyl injury although, as mentioned above, mitochondrial damage appears to occur later than does plasma membrane injury. In short-term hypoxia (15 min), we have been unable to demonstrate increased hydroxyl radical generation in freshly isolated rat proximal tubules.[95] Zager *et al.* were also not able to detect OH· during 30 or 45 min of hypoxia.[101] Furthermore, the diversion of O_2^- to peroxynitrite could also reduce the subsequent generation of OH·, accounting for its absence in hypoxia/reoxygenation.[96] This would be particularly true during reoxygenation since return of O_2^- to normal levels should be associated with much larger increases in H_2O_2 as well as OH· radical formation.

24.7.2. Cyclooxygenases

Oxygen free radicals are also generated when prostanoids are formed from arachidonic acid. The previously mentioned increase in PLA_2, the well-documented generation of increased levels of arachidonate during hypoxia and reoxygenation, and the enhanced renal functional impairment after reflow if nonsteroidal antiinflammatory drugs are given suggest that the enhanced prostaglandin generation, despite being necessary to increase renal blood flow, may simultaneously be accompanied by increased production of reactive oxygen molecules. Protective vasodilators, such as CCB, may indirectly reduce cell injury by limiting the formation of arachidonate. This would decrease postischemic oxidant formation via the cyclooxygenase pathway.

24.7.3. Xanthine/Xanthine Oxidase Production of Oxygen Radicals

Protease cleavage of xanthine dehydrogenase during hypoxia yields xanthine oxidase. Xanthine levels are also increased during ischemia[6] so that cells are poised on reperfusion and the reintroduction of molecular oxygen, to generate substantial amounts of O_2^-. Extremely high concentrations of hypoxanthine and xanthine are present after 45 min of complete renal ischemia; by 1 hr of reflow these have decreased considerably.[6] Reflow is associated with disappearance of cellular xanthine, formation of oxygen free radicals, and importantly, elimination of xanthine as a potential source of ATP precursors. This may be related to losses from cells because xanthine oxidase levels in renal tubules may not be very high.[102] Treatment with allopurinol or oxypurinol does lessen the injury, presumably because of activated xanthine oxidase, in ischemic acute renal failure.[103,104] These inhibitors of xanthine oxidase would decrease xanthine losses, reduce oxidant formation, and permit more rapid ATP resynthesis. They would not, however, delay oxidant formation via cyclooxygenase activity, ONOO· generation, or mitochondrial free radical production. Thus, while certainly pallative, these inhibitors cannot completely prevent cell injury.

24.7.4. Extracellular Sources

In intact organs, cells such as neutrophils and macrophages are able to damage renal tubules. In addition, if red blood cell hemolysis occurs, iron from hemoglobin can exacerbate cell injury.[105] Finally, endothelial cells, lining capillaries that lie near renal tubules, can produce NO which could lead to tubular and nontubular oxidant damage via ONOO·.

24.7.4.1. Activated Neutrophils

It is believed that the physiologic response to ischemia or other forms of ATP depletion *in vivo,* is endothelial cell swelling and exposure of the underlying basement membrane to platelets and neutrophils. Platelet aggregation, release of neutrophil oxidants, and the presence of trapped erythrocytes containing heme iron provide a rich environment for continued oxidant injury especially in the outer medulla or medullary rays where tissue oxygen tension and blood flow are normally quite low.[106]

When primed or "activated" neutrophils are added to perfusate circulating through the isolated perfused kidney, functional injury is not seen; however, with only 15 min of ischemia, the addition of activated neutrophils induces renal failure.[107] Moreover, drug-induced depletion of neutrophils lessens functional injury. It is, therefore, likely that neutrophils activated during reflow and trapped in capillaries exert direct oxidant effects on nearby renal tubules and endothelial cells. Simultaneously, hemostasis at these sites prolongs local oxygen deprivation; renal tubules, including the proximal straight tubule and medullary thick ascending limb (MTAL), in such an environment, exhibit the most necrosis *in vivo*.

24.7.4.2. Iron/Hemoglobin

Iron released from intracellular heme proteins or from myoglobin or hemoglobin may contribute to the generation of OH· and thus to cell injury.[105] Catalytic iron (Fe^{2+}) increased more so in isolated tubules during hypoxia as compared to reoxygenation. Despite removal of Fe^{2+} activity with desferrioxamine (DFO) or phenanthroline, cytoprotection was not observed.[108] It was concluded that increased Fe^{2+} is a result, not a cause, of cell injury. DFO or phenanthroline cannot prevent deranged mitochondrial respiration or ATP depletion nor do they enhance recovery during reoxygenation. However, addition of exogenous H_2O_2 damages cultured renal epithelial cells and this form of damage can be attenuated by DFO or phenanthroline.[109]

Further studies have shown that both Fe^{2+} and Fe^{3+} result in increased malondialdehyde (MDA) formation, a sign of oxidant-mediated lipid peroxidation; only Fe^{2+} causes cytotoxicity as assessed by LDH release.[110] In addition, DMTU, benzoate, mannitol, or GSH, each of which either reduces the generation of OH· or scavenges OH·, is not cytoprotective. Both GSH and catalase prevent or limit formation of H_2O_2 and also decrease MDA formation. In models where DFO does exhibit cytoprotection, the mechanism may involve the scavenging of ONOO·.[100]

At present, it is well established that while Fe certainly enhances lipid peroxidation in renal tubules and cultured renal epithelia, this can be disassociated from cell injury. This mechanism of hypoxic cell injury appears to be related more so to H_2O_2 generation than to OH· formation. It must be noted, however, that a paradoxical protection against Fe^{2+}–mediated cell injury occurs when myoglobin is added.[112] It is possible that myoglobin, like hemoglobin, may trap NO thus preventing both formation of ONOO· and subsequent injury. The degree of cell injury in the presence of hemoglobin/myoglobin may be related to the balance between diminished ONOO· formation and release of iron from porphyrin rings.

24.7.4.3. Endothelial Cells/Platelets

Endothelial cells are now recognized as being the source of NO which, while beneficial via its vasodilatory action, may also incite injury under pathologic conditions. Normally during ischemia or anoxia, swollen or injured endothelial cells do not produce significant NO unless and until oxygen and possibly L-arginine is reintroduced. Since endothelial cells also produce endothelin, a potent vasoconstrictor, its effect to reduce blood flow at critical sites could further enhance or prolong endothelial swelling, limit oxygen delivery, and possibly modify the extent of reperfusion-mediated ONOO· formation.

In isolated renal vessels obtained from ischemic kidneys, NO production is maximal and cannot be regulated, for example, via acetylcholine infusions.[113] This observation explains why paradoxical renal vasoconstriction occurs after ischemia in response to reductions in renal perfusion pressure; the ability to generate NO is probably very high and cannot be increased further. However, high levels of NO production might well lead to enhanced ONOO· production by endothelial cells and therefore prolong oxidant injury. Finally, endothelial cells also produce a factor referred to as endothelial derived hyperpolarizing factor. It is not known whether the production of this substance is disturbed during reflow but any attenuation of its activity would delay the reestablishment of normal membrane potential and prolong voltage (potential)-dependent Ca^{2+} overload.

Platelets are activated in part by ischemia itself and by ADP released from cells with leaky membranes. Their subsequent aggregation within capillaries slows blood flow and delays recovery of both endothelial and adjacent tubular epithelial cells. Thus, especially in medulla which is relatively oxygen poor under the best of perfusion conditions, any further reduction in oxygen delivery caused by platelet aggregation results in prolonged hypoxia and thus in cell injury. The injury resulting from reduced O_2^- availability is best observed in the isolated perfused kidney, which does not contain platelets, using perfusates that contain no amino acids.[106] Ca^{2+} enhances MTAL tubular necrosis in a dose-dependent manner.[114] Verapamil attenuates this necrosis [115] as does acidosis[114]; glycine, added to the perfusate as the only source of an amino acid, totally prevents the tubular necrosis.[116]

Platelets may be activated in the postischemic kidney by release of platelet activating factor (PAF) derived from glomerular mesangial cells. Administration of PAF to intact kidneys reduces GFR and PAF antagonists reverses this effect. PAF may subsequently induce production of the potent vasoconstrictor TxA_2 in platelets since inhibitors of TxA_2 reduce functional or cellular damage induced by PAF infusion or ischemia. Two of the important stimuli for PAF production, namely, Ca^{2+} and oxidants, are increased by ischemia. *In vivo* renal ischemia can clearly cause several changes that initiate and prolong cell injury in and to renal tubules and vascular tissue.

24.7.5. Lipid Peroxidation

In intact animals, lipid peroxidation is detected by exhaled ethane, a β-scission product of lipid peroxides. Lipid peroxidation (or ethane), however, does not always correlate with cell injury. The rather late release of ethane or pentane, which occurs well after oxidant generation and membrane damage are detected, suggests that lipid peroxides are the result, not the cause, of cell death.

Enzymatic oxidation of fatty acids yields ATP in normoxic tissues. As ATP levels decrease, the enzymes responsible for β-oxidation of fatty acids become immeasurable and lipid peroxidation increases. This is an energetically unfavorable event but by diverting fatty acids to peroxides, any residual ATP, perhaps produced by glycolysis, will not be consumed via β-oxidation during early stages of reflow or reoxygenation. Drugs that inhibit lipid peroxidation prevent or attenuate ischemic acute renal failure in the rat. The mechanism may involve preservation of those unsaturated fatty acids released by PLA_2 activation, to be reutilized for membrane phospholipid resynthesis on reflow. Increases in MDA reflect the degree of lipid peroxidation but this does not indicate the sources of lipids that have been affected by oxidants. MDA products are increased by hypoxia or toxins, and by mixtures of Fe^{2+} and Fe^{3+}. Only Fe^{2+}, however, caused renal cell injury.[108] Whether certain substrates are preferentially affected by the redox state of iron is not known. However, the generalization that lipid peroxidation (i.e., MDA formation) always results in cell injury does not appear tenable. For example, despite marked increases in MDA formation associated with antimycin treatment, cell injury does not occur.[108] MDA levels can, however, be reduced by GSH addition or catalase treatment; these findings suggest that lipid peroxidation is an H_2O_2-directed process.

24.7.6. Oxygen Radical Scavengers

Renal tubules contain enzymes that detoxify oxygen radicals. Expression of some of these enzymes at high levels has been achieved in transgenic mice[117]; conversely, interbreeding programs have resulted in animals with very low levels of, for example, catalase.[118] These models are thus powerful tools for the study of oxidant injury. Basal levels of the oxygen radical scavengers are sufficient to defend against and protect cells from the small amounts of oxidants formed continuously by metabolism. In ischemia, however, SOD, catalase, glutathione peroxidase, and glutathione are reduced. The cells, therefore, are at risk from oxidant injury because even normal levels of oxidants would not be detoxified and the probability of enhanced oxidant formation during reflow makes cytotoxicity even worse. Most laboratories have shown that provision of extraneous antioxidants is beneficial to cell viability and organ function.

Paller and colleagues[103] treated intact rat kidneys with SOD, DMTU, or allopurinol prior to renal ischemia; each treatment lessened the severity of renal hemodynamic, tubular reabsorptive, and histologic injury in the ensuing 2 days of reflow. The degree of mitochondrial lipid peroxidation was also reduced. Catalase, perhaps because its large molecular weight which precludes filtration into the nephrons, was not protective.

Further studies in the pig kidney exposed first to warm ischemia, next to hypothermic pump perfusion, and then transplantation showed that organ function was markedly enhanced by SOD plus catalase treatment given just 3 min before reperfusion.[119] This result points clearly to the need for intact enzymes (at high concentrations) to be present at the time when an oxidant burst, associated with normothermic reperfusion, is initiated. Intracellular catalase undergoes proteolysis or its synthesis is impaired during reperfusion, thereby necessitating exogenous repletion. The proteolysis that degrades catalase also affects other enzymes present within catalase-containing peroxisomes. Specifically, acetyl CoA oxidase activity, responsible, in part, for maintaining oxidative phosphorylation, is decreased especially during reperfusion.[120] The decrease in β-oxidation of fatty acids, a process that provides substrate for energy production, is dependent on acetyl CoA oxidase and is therefore another marker of an ischemic insult.

Cisplatin treatment for malignancies also results in renal lipid peroxidation and nephrotoxicity. Sadzuka *et al.* showed that rat kidneys were particularly injured by cisplatin, which also effected marked decreases in catalase, glutathione peroxidase, glutathione-*S*-transferase, and SOD.[121] Since cisplatin itself does not cause lipid peroxidation, the reported increases in lipid peroxides must be related to the loss of the protective effects of these important antioxidants/scavengers. Losses of enzymes important in oxidant detoxification are also observed after ischemia.

Finally, because of the reduced antioxidant activity during ischemia and the proteolytic degradation of these detoxifying enzymes during reperfusion, greater reperfusion injury might be expected in the presence of high oxygen tension. When rats breathed 100% O_2 rather than room air during reperfusion, a further significant decrease in all enzymes was observed including catalase, SOD, and glutathione peroxidase.[122] Thus, oxidant formation during reperfusion must be considered to be a major mechanism responsible for cell injury and oxygen radical scavenger therapy thus has potential as an important cytoprotective maneuver.

GSH levels are reduced by oxygen deprivation. While Weinberg showed that provision of exogenous GSH could prevent hypoxic cell injury to isolated rabbit tubules, the protective effect was apparently mediated by glycine, one of the three amino acids that comprise GSH.[30,31] Intact rabbit kidneys also demonstrated less ischemic tubular injury when mannitol alone, mannitol plus SOD, or mannitol plus the xanthine oxidase inhibitor oxypurinol was administered.[123] Unlike studies in the rat, glomerular filtration rate was not improved. Thus, in different species the tubules appear to be primarily at risk for oxidant injury and they are amenable to antioxidant therapy with scavengers. A multiple therapy approach, using SOD, catalase, and nifedipine, also reduced vascular damage related to ischemia, as observed by protein leakage from capillaries in the rabbit kidney.[124] The individual treatments alone were not effective. This mixture of mannitol, SOD, and catalase also reduces postischemic proteinuria and urinary heparin sulfate loss. This result suggests that glomerular endothelial cells and/or the permeability barrier are also targets of ischemic oxidant injury.

The particular role of individual oxidants in causing cell injury has also received much attention. Linas *et al.*, using the isolated perfused rat kidney, showed that functional damage was likely induced by H_2O_2 or its by-product $OH\cdot$.[125] Importantly, GSH levels were reduced by H_2O_2 but not by elastase or collagenase which also lower cell ATP and cause cell death. Decreased tissue GSH then is an important marker of oxidant injury.

Kidneys reperfused after ischemia are at risk for interstitial infiltrates and neutrophil/macrophage/monocyte-associated injury. In a model of bivalent hapten immune complex injury, intact rats demonstrated partial protection, as assessed by proteinuria and/or monocytic infiltration, when SOD or catalase was administered. DMTU, a hydroxyl scavenger, however, was the most protective.[126] Thus, modest overexpression of oxidants by macrophages or neutrophils, which can usually be scavenged by normal renal tissues, became lethal when short-term ischemia inactivated tissue defensive enzymes such as GSH, SOD, and catalase.

24.8. MEMBRANE DAMAGE

24.8.1. The Permeability Barrier

The protein-phospholipid organization of the cell membrane provides several functions important for cell viability including receptor positioning, endo- and exocytosis, transport activity, and separation of internal cytosolic constituents from the extracellular media. Cells can become damaged so severely that large and small molecules move freely across this barrier.[127] One of the hallmarks of lethal cell injury is the release across damaged membranes of large, internal enzymes including the rather easily measured LDH. This is a presumptive sign of cell death for most tissues. Heart tissue damaged sufficiently to leak LDH also releases the enzyme creatine phosphokinase (CPK), whereas in liver tissue, SGOT and other hepatic specific enzymes are found in the extracellular compartment when this organ is damaged.

24.8.1.1. Large Molecules/LDH

LDH release becomes significant at 5–10 min of hypoxia at 37°C in most experimental preparations. However, in flow through chambers where extracellular constituents are continuously washed away, LDH release is not apparent until 15–20 min of hypoxia.[40] Therefore, perhaps another unmeasured enzyme which is released early and attacks cell membranes in static systems is washed away by the continuous perfusion systems. Because membranes are comprised of both phospholipids and protein, two possible candidates, i.e., phospholipases and proteases, should be considered as the injurious enzymes.

Glycine content of intact cells decreases rapidly with a hypoxic or anoxic insult.[30] Addition of glycine to the media of cells during hypoxia prevents cell membrane damage assessed by LDH release. In such conditions, LDH release is not increased over 60 min of severe hypoxia. Coincidentally, glycine also prevents the release to the media of a Ca^{2+}-independent form of PLA_2.[66] It is possible that this Ca^{2+}-independent PLA_2 attacks plasma membranes at the outer surface and thus permeabilizes the cell membrane to large molecules. A calcium-dependent substrate-specific $cPLA_2$ is activated by hypoxia and associated with this activation is a marked increase in renal tubule free fatty acid (FFA) release.[64,65] Others have also noted increased release of FFA during hypoxia or metabolic inhibition in rabbit tubules and cultured mouse tubule cells. Glycine, however, has no effect on the release of FFA.[65]

As noted previously, glycine appears to prevent the formation of increased G-actin in $LLCPK_1$ cells exposed to ionomycin. Since previous studies showed disruption of cytoskeletal actin homeostasis during complete ischemia, the protection against LDH and PLA_2 release from hypoxic cells treated with glycine may lie in its effects to neutralize protease activity. This speculation is consistent with the observation that despite glycine treatment, Ca_i still increases during hypoxia.[40] Since calpain is a calcium-activated neutral cysteine protease and studies have shown cytoprotection when this enzyme is inhibited, it may be that glycine interferes with an enzymatic attack of calpain (or other cysteine proteases) on the actin cytoskeleton. Preservation of the actin cytoskeleton, at a time when $cPLA_2$ activity (assessed by FFA release) and cytosolic Ca^{2+} changes are present, provides a possible explanation for the membrane protective effects of glycine in models of hypoxic injury.

The results of these basic studies are now being extended to experimental transplantation. Southards and Belzer's group is providing glycine both to perfusate used to flush the harvested liver, and to the recipient (dog or rat) in the posttransplant period. Preliminary results suggest that glycine may be organ sparing and cytoprotective under these conditions.[128]

24.8.1.2. Small Molecules

Often, experimental conditions are such that the standard measurements of LDH release are impractical. Thus, cell uptake of normally impermeable, small molecules such as trypan blue, nigrosin, or PI are used to measure the permeability of cell membranes. PI uptake is useful because it stains nuclei of injured cells and fluoresces at a different wavelength than fura-2 so that a determination can be made as to whether cytosolic calcium overload precedes, or follows, membrane injury.[40,127] PI staining also correlates very positively with LDH release. Thus, it appears that once membrane damage proceeds to a point where small molecules can enter, the permeability to larger molecules occurs almost simultaneously. Nigrosin dye penetrates damaged cells and stains the cytoplasm which can, as is the case with trypan blue, be visualized by light microscopic techniques.

Cell permeability changes can occur under experimental conditions that are not consistent with cell death.[129] Temperature-dependent cell swelling or metabolic inhibition led to increased trypan blue staining in cells that were subsequently found to be viable. PI staining of cell nuclei is becoming one of the standards for assessing the increased membrane permeability associated with cell death. Its use in several laboratories in a variety of models attests to its increasing acceptance as an important biological tool.

24.8.2. Fatty Acid Release

Cell death is accompanied by release of FFA. This is found in cell models in which it has been studied and the release is progressive, related either to more cell death or to progressive damage to the phospholipid pool in each cell. The latter would likely reflect late damage to intracellular membranes such as those surrounding the mitochondria, nucleus, lysosomes or those comprising the Golgi or endoplasmic reticulum.

Humes *et al.* have shown that a critical level of FFA must be reached before cell injury becomes apparent.[64] Hypoxic injury begins to become evident when very small increases in FFA occur. The early injury that is observed is correlated more so with FFA release, especially unsaturated FFA, than with accumulation of lipid by-products such as lysophospholipids. Importantly, addition of exogenous PLA_2 exacerbates markedly this modest injury.[64] As noted above, endogenous PLA_2 released into the media of hypoxic tubules was associated with severe cellular injury and both release of PLA_2 and injury could be prevented by incubation with glycine.[66] When PLA_2 is extracellular to hypoxic tissues, the injury process is greatly accelerated in time.[64–66] Binding of FFA with bovine serum albumin appears to arrest injury related to exogenous PLA_2 or during hypoxia. Therefore, FFA, in particular arachidonic acid, are injurious even to normal cells. Both albumin and glycine attenuate this damage. Mouse proximal tubules (MPT) in culture also release FFA in a hypoxic (low ATP) model induced by cyanide.[52] Curiously, neither MPT nor MDCK cells (used as a control in these studies) contain detectable amounts of arachidonic acid. This observation may help explain, in part, the resistance of distal tubules to hypoxic injury since the MDCK cell line exhibits properties more consistent with a distal, than a proximal, tubule origin.

24.9. APOPTOSIS

Apoptosis, or programmed cell death, is not a well-understood phenomenon but it is clearly developmentally and biologically important.[130] Since both increases in $[Ca^{2+}]_i$ and calpain have been implicated in apoptotic cell death, apoptosis may have some relevance to ischemic or toxic cell injury.[131] Apoptosis determines the truncation of neural growth once synaptic contacts are formed, and also terminates B- and T-cell proliferation once immunologic defenses have corrected an insult. Without such effects, uncontrolled cell proliferation would continue. The signals responsible for initiating programmed cell death are beginning to be recognized as are systems that keep this process in abeyance. If apoptosis is a contributor to ischemic or toxic cell death, the systems or responses that control apoptosis could be useful adjuncts to therapy. It must be emphasized, however, that morphologically and biochemically, apoptosis and necrosis have very few similarities. They may occur simultaneously in cells residing next to one another, but cells die from either necrosis or apoptosis. Apoptosis occurs in otherwise normal-appearing cells; mitochondria appear normal, membrane transport proteins remain polarized, ATP levels are normal, and leakage of enzymes to the extracellular space is not seen. There is a primary and absolute need for protein synthesis in order for apoptosis to occur; such protein synthesis is unlikely during ischemia.

Thus, apoptosis is initiated under conditions that are quite distinct from most of those that occur in necrosis. Characterization of apoptosis is usually accomplished by measurement of DNA "laddering"; this is usually preceded by a decrease in growth factor availability.

24.9.1. DNA "Laddering"

During apoptotic events, a Ca^{2+},Mg^{2+}-dependent endonuclease cleaves nuclear DNA into oligonucleoside fragments detectable on agarose gels.[132] This laddering pattern is demonstrable after 12 hr of reflow following renal ischemia and is even more apparent after 24 hr. Thus, apoptosis is an event that occurs during reflow, not during ischemia.[132] During reflow, the process of regeneration of renal tubule epithelium begins. The induction and expression of several different growth factors during regeneration, including transforming growth factor β (TGFβ), EGF, insulin-like growth factor, platelet-derived growth factor, and other cytokines participate in the recovery pattern of epithelium. At some point, however, the growth process ceases and the process of differentiation begins. In renal tissue exposed to lead nitrite, toxic renal hyperplasia accompanied by increased DNA synthesis (thymidine incorporation) and DNA content, increased mitosis, and increased organ weight reached a maximum at 2 days.[133] Thereafter, regression of renal size was accompanied by apoptotic events, mostly in proximal tubules; cell necrosis could not be demonstrated. Apoptosis was no longer measurable when kidneys reverted to their normal size. Thus, apoptosis is usually expressed as a mechanism to prevent uncontrolled growth as in regeneration of renal tubules. Based on studies with CNS tissue where apoptosis is initiated when growth factors are removed from cultured cells, it may be that the apoptotic laddering represents cessation of growth factor production or action as regeneration processes in individual nephrons becomes completed. Reflow is an important variable because continued ischemia for 24 hr is not associated with signs of apoptosis. Furthermore, only during reperfusion after renal ischemia is the mRNA for sulfated glycoprotein-2, a gene product of apoptotic cells, observed.[132]

24.9.2. Gene Expression

Apoptosis can be prevented or arrested in cells transfected with the gene Bcl-2.[134] In addition, some neoplastic tissues do not express Bcl-2.[135] Thus, uncontrolled growth is related to the absence of Bcl-2. Bcl-2 is found both in cytosol, usually at or near the plasma membrane, and in mitochondria.[136] These are two sites where the devastating effects of lethal cell injury are apparent. Growth factors such as those noted above keep mRNA for Bcl-2 repressed; when these are removed, mRNA for Bcl-2 increases markedly. Thus, the preservation of some cells while others proceed toward apoptotic death is apparently regulated by a signal that determines the activity of Bcl-2. How this regulation is achieved is unknown.

The onset of apoptosis suggests that an endonuclease is tightly regulated. It has been shown that transfection of Cos cells with the cDNA for rat parotid DNase I (which immunoprecipitates with nuclei) results in apoptosis.[137] Rat kidney, lymph node, and thymus contain an enzyme of identical size. It has been postulated that in these tissues, apoptosis is induced when this endonuclease penetrates the nuclear membrane.

Apoptosis is also a hallmark of proliferating tissues after x-irradiation.[138] Thus, apoptosis occurs in cells that are or were rapidly dividing such as those exposed to lead, ischemia/reperfusion, or irradiation. These findings suggest that cells with damaged DNA are the ones that will eventually become apoptotic, since rapid cell division is more likely to result in DNA damage. In mature, nondividing tissues, little evidence for ongoing apoptosis is seen, possibly because this process of removal of damaged cells has already been largely completed. Finally, some viruses have lytic potential in cells that are dividing but these same viruses exhibit persistent, productive infections in stable cells (i.e., postmitotic neurons). It was recently shown that Sindbis virus (SV) infection of baby hamster kidney (BHK) cells resulted in apoptotic cell death.[134] Only long-term persistent infection was seen in SV-infected AT-3 cells transfected with the Bcl-2 oncogene which should prevent apoptosis.[134] Viruses alter normal DNA replication and thus as with lead, ischemia, or x-ray-induced damage noted above, cells with damaged DNA are more likely to exhibit programed cell death, i.e., apoptosis. Bcl-2 may prevent injury to the genome. It shares structural similarity with ced-9, a gene of lower animals that prevents apoptosis. The recently reported structural homology between Bcl-2 and calpastatin is being explored as a possible link between calcium and apoptotic cell death.[131] Calpastatin inhibits calpain activity, alluded to earlier as a possible initiator of cell membrane damage.

Part of the protective effect of Bcl-2 may be related to its ability to maintain proliferating cell nuclear antigen (PCNA). This cell cycle-related protein is inhibited by ochratoxin in hamster kidney cells undergoing apoptosis.[139] PCNA is a necessary factor for DNA polymerase activity in dividing cells.

Recently, another gene, the testosterone-repressed prostate message-2 gene (and its RNA and protein products), has been implicated as an additional marker for apoptosis.[140] This gene is expressed differently in cells undergoing apoptosis from different causes. Therefore, it may be extremely useful to measure DNA laddering as well as other indices of apoptotic cell death in conjunction with Bcl-2 and other genes whose differential expression may indicate whether apoptosis is the result of Ca^{2+} overload, aging, or toxins. Apoptosis does not appear to be involved in the process of classical, necrotic cell death.

24.10. TUBULAR EPITHELIAL REGENERATION

Although understanding the mechanism of cell death is important in designing preventive therapy, often insults occur that are unpredictable or unpreventable. Development of processes that speed regeneration of injured tissue and salvage organ function is therefore an important research area. Although several approaches are being experimentally tested, we will focus on two: growth factors and heat-shock proteins (HSP).

24.10.1. Growth Factors

The interest in recovery of tissues in previously damaged kidneys was based on the observation that not all tubular epithelial cells were equally necrotic following organ ischemia or toxin administration. Thus, there appeared to be a population of tubular cells that could be induced to proliferate and differentiate so as to replace the lost, necrotic cells.[141] These "stem" cells are also present in cultures of renal epithelial cells which provide an additional experimental tool (in addition to intact kidneys or freshly isolated nephron segments) for examining the process of tubular cell regeneration.

Humes and Cieslinski[141] showed that provision of EGF, retinoic acid (RA), and TGFβ1 would induce differential growth of rabbit primary cultures of proximal tubule cells and that three-dimensional clusters reminiscent of intact tubules were formed. RA also induced laminin formation, thus supporting the view that extracellular matrix provides a critical component of cells' ability to proliferate, differentiate, and "recover" from a noxious stimulus. It is likely that signals that cause complete regeneration of renal tubules, *in vivo,* involve the interaction of growth factors as well as extracellular matrix proteins. This can be clearly demonstrated in cultured baby hamster kidney cells which only differentiate when grown on matrigel if either EGF or TGFβ2 is simultaneously present.[142]

One of the important *in vivo* signals appears to be triiodothyronine (T_3). It has been established that exogenous T_3 increases recovery from or lessens the severity of gentamicin and ischemic renal injury. T_3 now appears to provide at least part of this protection by increasing renal tubule EGF receptor gene expression.[143] Increased mitogenesis observed in these studies provides for the quicker recovery of injured tubules. Furthermore, newly formed renal cells are not yet differentiated and thus lack the full capacity for transport. In some renal cells, serosal EGF will limit sodium reabsorption from the lumen,[144] thereby possibly sparing ATP, normally consumed during transport, for replication and differentiation processes. EGF receptors are located on the basolateral surface of transporting renal epithelial cells including $LLCPK_1$, MDCK, and mouse proximal tubules.[145]

Thus, in the process of recovery from cell injury, basolateral EGF receptors appear to be important regardless of whether the cells are of proximal or distal origin and this is observed in cells derived from all species so far studied (dog, pig, and mouse).

EGF expression in renal tubules was associated with reduced hydraulic conductivity and a diuresis[144] possibly related to the finding that EGF receptors are abundant in TAL and early DCT cells.

Hepatocyte growth factor (HGF), important for liver regeneration, has recently been shown to participate in renal recovery from an ischemic insult or mercuric chloride administration.[146] HGF mRNA increases in the outer medulla by 6–12 hr after insult reaching a peak at 48 hr. HGF protein and mRNA appear to be expressed intrarenally by endothelial cells or macrophages. HGF, also produced by mesangial cells, may reach injured proximal nephrons and increase [^{3}H] thymidine incorporation and stimulate Na^+/K^+-ATPase. Finally, once cells reach confluence (or tubular regeneration is complete), HGF effects (i.e., DNA synthesis) are no longer demonstrable.[147]

24.10.2. Heat-Shock Proteins

HSP are induced in all organs or cells so far studied. These proteins with molecular masses at or near 70 kDa appear to act as "chaperones" to important cellular constituents, preserving for example the location and three-dimensional structure of proteins at risk for injury. Induction of HSP appears to be a temporizing measure to protect cellular constituents especially at times when DNA and RNA synthesis are reduced. Coincident with HSP induction and despite decreased total RNA synthesis, increased mRNA levels of immediate early response gene transcription factors such as Egr-1 and cfos occur.

HSP responses of renal tissues were increased at 3 hr following 15 min of renal ischemia.[148] Thus, transient ischemia and exposure to elevated temperature (42°C) induces a cytoprotective protein that may be involved in cell repair during reperfusion. Longer periods of ischemia (i.e., 45 min) result in a more rapid induction of HSP, i.e., at 15 min of reflow. By immunolocalization, HSP is found first at the apical border of proximal nephrons, whereas with reflow of 2–6 hr, a cytoplasmic location is observed.[63] Some HSP is found associated with lysosomes where they may act to prevent intracellular digestion from cathepsins. By 24 hr of reflow, no HSP is found at the apical domain. Repolarization of membrane transport proteins may be dependent on the induction of this molecular chaperone. Renal tissues are not unique in their responses to HSP induction. Studies with cultured myocytes have demonstrated that a latent pool of heat-shock factors bind to DNA in order to induce HSP formation during hypoxia.[149] ATP depletion of these cells with glucose deprivation alone was unable to induce HSP formation, whereas rotenone, a mitochondrial inhibitor, reduced ATP to lower concentrations and HSP induction occurred. Rotenone also lowers cellular pH and acidosis has been shown to delay the onset of cellular calcium overload as well as being cytoprotective. However, lowering pH_i by other maneuvers (unaccompanied by ATP depletion) does not result in HSP induction. Permeabilization of cells with nigericin at normal pH also resulted in reduced ATP stores and in HSP induction.[149] Therefore, it appears that induction of HSP occurs when some ATP-dependent mechanism is compromised during mitochondrial inhibition by hypoxia, ischemia, or rotenone. The pH effect and the need for mitochondrial respiratory depression suggest that transmitochondrial H+ gradient (uncoupled respiration) may play a role in the induction of this cytoprotective protein. Both more rapid ATP recovery and cytoprotection during HSP induction are consistent with a mitochondrial site of action, either direct or indirect, and underscore the need to continue efforts to bypass or reverse the devastating effects of ATP depletion.

24.11. SUMMARY

Renal injury is expressed in different kidney cells (mesangial, endothelial, interstitial, and epithelial) in different ways, at different times, and on different surfaces (basolateral and apical) during and following ischemic or toxic insults that result in water and solute transport defects. The initial event in most forms of cell injury is disturbed energy metabolism. This leads promptly to altered ionic homeostasis, cell swelling, and, shortly thereafter, to leakage of intracellular enzymes. Ca^{2+}-dependent and -independent enzymes appear to participate in the injury process by degrading phospholipids (phospholipases) and proteins (proteases). The simple amino acid, glycine, exhibits remarkable cytoprotective effects. It will be interesting to see if the preliminary results showing better organ and cell survival with glycine repletion to the recipient of a transplanted organ will be confirmed. If so, the problems associated with postischemic generation of oxidants, via iron-mediated reactions, may be less devastating.

Finally, ways to induce HSP and to take advantage of the cytoprotective effects of growth factors should evolve from studies currently in progress. Clearly, the key to cell survival is to find ways to preserve energy metabolism and the membrane cytoskeleton and its associated transmembrane proteins. These proteins anchor viable cells to their extracellular matrix, to one another, and provide the integrity of the microvilli, which are so critical for normal transport function. Because many of these processes are dependent on mitochondrial or nuclear genes, the integrity of genetic material is also critical to cell survival.

REFERENCES

1. Weinberg, J. M. (1985). Oxygen deprivation-induced injury to isolated rabbit kidney tubules. *J. Clin. Invest.* **76:**1193–1208.
2. Arnold, P. E., Van Putten, V. J., Lumlertgul, D., Burke, T. J., and Schrier, R. W. (1986). Adenine nucleotide metabolism and mitochondrial Ca^{2+} transport following renal ischemia. *Am. J. Physiol.* **250:**F357–F363.
3. Bonventre, J. V., and Cheung, J. Y. (1985). Effects of metabolic acidosis on viability of cells exposed to anoxia. *Am. J. Physiol.* **249:**C149–C159.
4. Rossi, C. S., and Lehninger, A. L. (1964). Stoichiometry of respiratory stimulation, accumulation of Ca^{2+} and phosphate, and oxidative phosphorylation in rat liver mitochondria. *J. Biol. Chem.* **239:** 3971–3980.
5. Takano, T., Soltoff, S. P., Murdaugh, S., and Mandel, L. J. (1985). Intracellular respiratory dysfunction and cell injury in short term anoxia of rabbit renal proximal tubules. *J. Clin. Invest.* **76:**2377–2384.
6. Arnold, P. E., Lumlertgul, D., Burke, T. J., and Schrier, R. W. (1985). In vitro versus in vivo mitochondrial calcium loading in ischemic acute renal failure. *Am. J. Physiol.* **248:**F845–F850.

7. Wilson, D. R., Arnold, P. E., Burke, T. J., and Schrier, R. W. (1984). Mitochondrial calcium accumulation and respiration in ischemic acute renal failure in the rat. *Kidney Int.* **25:**519–526.
8. Siegel, N. J., Glazier, W. B., Chaudry, I. H., Gaudio, K. M., Lytton, B., Baue, A. E., and Kashgarian, M. (1980). Enhanced recovery from acute renal failure by the post ischemic infusion of adenine nucleotides and magnesium chloride in rats. *Kidney Int.* **17:**338–349.
9. Weinberg, J. M., and Humes, H. D. (1986). Increases of cell ATP produced by exogenous adenine nucleotides in isolated rabbit kidney tubules. *Am. J. Physiol.* **250:**F720–F733.
10. Mandel, L. J., Takano, T., Soltoff, S. P., and Murdaugh, S. (1988). Mechanisms whereby exogenous adenine nucleotides improve rabbit renal proximal function during and after anoxia. *J. Clin. Invest.* **81:**1255–1264.
11. Stromski, M E., vanWaarde, A., Avison, M. J., Thulin, G., Gaudio, K. M., Kashgarian, M., Shulman, R. J., and Siegel, N. J. (1988). Metabolic and functional consequences of inhibiting adenosine deaminase during renal ischemia in rats. *J. Clin. Invest.* **82:**1694–1699.
12. Schrier, R. W., Arnold, P. E., Van Putten, V. J., and Burke, T. J. (1987). Cellular calcium in ischemic acute renal failure: Role of calcium entry blockers. *Kidney Int.* **32:**313.
13. Burnier, M., Van Putten, V. J., Schieppati, A., and Schrier, R. W. (1988). Effect of extracellular acidosis on ^{45}Ca uptake in isolated hypoxic proximal tubules. *Am. J. Physiol.* **254:**C839–C846.
14. Nakamura, H., Nemenoff, R. A., Gronich, J. H., and Bonventre, J. V. (1991). Subcellular characteristics of phospholipase A_2 activity in the rat kidney. Enhanced cytosolic, mitochondrial, and microsomal phospholipase A_2 enzymatic activity after renal ischemia and reperfusion. *J. Clin. Invest.* **87:**1810–1818.
15. Suzuki, K., Saido, T. C., and Hirai, S. (1992). Modulation of cellular signals by calpain. *Ann. N.Y. Acad. Sci.* **674:**218–227.
16. Mason, J., Beck, F., Dorge, A., Rick, R., and Thurau, K. (1981). Intracellular electrolyte composition following renal ischemia. *Kidney Int.* **20:**61–70.
17. Kinnula, V. L., and Hassinen, I. (1978). Metabolic adaptation to hypoxia. Redox state of the cellular NAD pools, phosphorylation state of the adenylate system and the (Na^+-K^+)-stimulated ATPase in rat liver. *Acta Physiol. Scand.* **104:**109–116.
18. Frindt, G., Lee, C. O., Yang, J. M., and Windhager, E. E. (1988). Potential role of cytoplasmic calcium ions in the regulation of sodium transport in renal tubules. *Miner. Electrolyte Metab.* **14:**40–47.
19. Yellen, G., Jurman, M. E., Abramson, T., and MacKinnon, R. (1991). Mutations affecting TEA blockade identify the probable pore forming region of a K^+ channel. *Science* **251:**939–942.
20. Hartmann, H. A., Kirsch, G. E., Drewe, J. A., Taglialatela, M., Joho, R. H., and Brown, A. M. (1991). Exchange of conduction pathways between two related K^+ channels. *Science* **251:**942–944.
21. Schrier, R. W., Conger, J. D., and Burke, T. J. (1991). Pathogenic role of calcium in renal cell injury. In *Nephrology* (M. Hatano, ed.), Springer-Verlag, Berlin, pp. 648–649.
22. Sanders, J. C. J., and Isaacson, L. C. (1990). Patch clamp study of Ca channels in isolated renal tubule segments. In *Calcium Transport and Intracellular Calcium Homeostasis* (D. Pansu and F. Bronner, eds.), Springer-Verlag, Berlin, pp. 27–34.
23. Kurachi, Y. (1989). Regulation of G-protein gated K^+ channels. *News Physiol. Sci.* **4:**158–161.
24. Dominguez, J. H., Rothrock, J. K., Macias, W. L., and Price, J. (1989). Na^+ electrochemical gradient and Na^+–Ca^{2+} exchange in rat proximal tubule. *Am. J. Physiol.* **257:**F531–F538.
25. Llibre, J., LaPointe, M. S., and Battle, D. C. (1988). Free cytosolic calcium in renal proximal tubules from the spontaneously hypertensive rat. *Hypertension* **12:**399–404.
26. Dominguez, J. H., Juhaszova, M., Kleinbocker, S. B., Hale, C. C., and Feister, H. A. (1992). Na^+–Ca^{2+} exchanger of rat proximal tubule: Gene expression and subcellular localization. *Am. J. Physiol.* **263:**F945–F950.
27. Yu, A. S. L., Hebert, S. C., Lee, S.-L., Brenner, B. M., and Lytton, J. (1992). Identification and localization of renal Na^+–Ca^{2+} exchange by polymerase chain reaction. *Am. J. Physiol.* **263:**F680–F685.
28. Ishikawa, S., and Saito, T. (1989). Enhancement by veratridine and 60 mM KCl of vasopressin-induced 3′,5′-cyclic adenosine monophosphate production and cellular free calcium concentration in rat renal papillary collecting tubule cells in culture. *Endocrinology* **124:**265–271.
29. Miller, G. W., and Schnellmann, R. G. (1993). Cytoprotection by inhibition of chloride channels: The mechanism of action of glycine and strychnine. *Life Sci.* **53:**1211–1215.
30. Weinberg, J. M., Davis, J. A., Abaruza, M., and Rajan, T. (1987). Cytoprotective effects of glycine and glutathione against hypoxic injury to renal tubules. *J. Clin. Invest.* **80:**1446–1454.
31. Mandel, L. J., Schnellmann, R. G., and Jacobs, W. R. (1990). Intracellular glutathione in the protection from anoxic injury in renal proximal tubules. *J. Clin. Invest.* **85:**316–324.
32. Heyman, S., Spokes, K., Rosen, S., and Epstein, F. H. (1992). Mechanism of glycine protection in hypoxic injury: Analogies with glycine receptor. *Kidney Int.* **42:**41–45.
33. Miller, G. W., and Schnellmann, R. G. (1993). The cytoprotective glycine/strychnine site on renal proximal tubules is related to the neuronal strychnine-sensitive glycine receptor [abstract]. *J. Am. Soc. Nephrol.* **4:**756.
34. Wetzels, J. F. M., Burke, T. J., and Schrier, R. W. (1992). Strychnine but not glycine increases cell potassium in isolated rat proximal tubules during reoxygenation after hypoxic injury [abstract]. *J. Am. Soc. Nephrol.* **3:**549.
35. Miller, G. W., and Schnellmann, R. G. (1993). Cytoprotection by inhibition of chloride channels: The mechanism of action of glycine and strychnine. *Life Sci.* **53:**1211–1215.
36. Aleo, M. D., and Schnellmann, R. G. (1992). The neurotoxins strychnine and bicuculline protect renal proximal tubules from mitochondrial inhibitor-induced cell death. *Life Sci.* **51:**1783–1787.
37. Almeida, A. R. P., Bunnachak, D., Burnier, M., Wetzels, J. F., Burke, T. J., and Schrier, R. W. (1992). Time-dependent protective effects of calcium channel blockers on anoxia- and hypoxia-induced proximal tubule injury. *J. Pharmacol. Exp. Ther.* **260:**526–532.
38. Jacobs, W. R., Sgamboli, M., Gomez, G., Vilaro, P., Higdon, M., Bell, P. D., and Mandel, L. J. (1991). Role of cytosolic Ca in renal tubule damage induced by anoxia. *Am. J. Physiol.* **260:**C545–C554.
39. Biscoe, T. J., and Duchen, M. R. (1990). Monitoring pO_2 by the carotid chemoreceptor. *News Physiol. Sci.* **5:**229–233.
40. Kribben, A., Wieder, E. D., Wetzels, J. F. M.,Yu, L., Gengaro, P. E., Burke, T. J., and Schrier, R. W. (1994). Evidence for a role of cytosolic free calcium in hypoxia-induced proximal tubule injury. *J. Clin. Invest.* **93:**1922–1929.
41. Weinberg, J. M., Davis, J. A., Roeser, N. F., and Venkatachalam, M. A. (1991). Role of increased cytosolic free calcium in the pathogenesis of rabbit proximal tubule cell injury and protection by glycine or acidosis. *J. Clin. Invest.* **87:**581–590.
42. McCoy, C. E., Selvaggio, A. M., Alexander, E. A., and Schwartz, J. H. (1988). Adenosine triphosphate depletion induces a rise in cytosolic free calcium in canine renal epithelial cells. *J. Clin. Invest.* **82:**1326–1332.
43. LeMasters, J. J., DiGuiseppi, J., Nieminen, A.-L., and Herman, B. (1987). Blebbing, free Ca^{2+} and mitochondrial membrane potential preceding cell death in hepatocytes. *Nature* **325:**78–81.
44. Molitoris, B. A., Falk, S. A., and Dahl, R. H. (1989). Ischemia-induced loss of epithelial polarity. Role of the tight junction. *J. Clin. Invest.* **84:**1334–1339.

45. Tan, S., and Lau, K. (1993). Patch-clamp evidence for calcium channels in apical membranes of rabbit kidney collecting tubules. *J. Clin. Invest.* **92:**2731–2736.
46. Erdreich, A., Spanier, R., and Rahaminoff, H. (1983). The inhibition of Na-dependent Ca uptake by verapamil in synaptic plasma membrane vesicles. *Eur. J. Pharmacol.* **90:**193–202.
47. Buss, W. C., Savage, D. D., Stepanek, J., Little, S. A., and McGuffe, L. J. (1988). Effect of calcium channel antagonists on calcium uptake and release by isolated rat cardiac mitochondria. *Eur. J. Pharmacol.* **152:**247–253.
48. Volkl, H., Greger, R., and Lang, F. (1987). Potassium conductance in straight proximal tubules of the mouse. Effect of barium, verapamil and quinidine. *Biochim. Biophys. Acta* **900:**275–281.
49. Wetzels, J. F. M., Yu, L., Wang, X., Kribben, A., Burke, T. J., and Schrier, R. W. (1993). Calcium modulation and cell injury in isolated rat proximal tubules. *J. Pharmacol. Exp. Ther.* **267:**176–180.
50. Malis, C. D., and Bonventre, J. V. (1986). Mechanisms of calcium potentiation of oxygen free radical injury to renal mitochondria. *J. Biol. Chem.* **261:**14201–14208.
51. Bunnachak, D., Almeida, A. R. P., Wetzels, J. F., Gengaro, P., Nemenoff, R. A., Burke, T. J., and Schrier, R. W. (1994). Time dependent phospholipase inhibition in protection against hypoxia-induced proximal tubule injury. *Am. J. Physiol.* **266:**F196–F201.
52. Sheridan, A. M., Schwartz, J. H., Kroshian, V. M., Tercyak, A. M., Laraia, J., Masino, S., and Lieberthal, W. (1993). Renal mouse proximal tubular cells are more susceptible than MDCK cells to chemical anoxia. *Am. J. Physiol.* **265:**F342–F350.
53. Ehrlich, Y. H., Snider, R. M., Kornecki, E., Garfield, M. G., and Lenox, R. H. (1988). Modulation of neuronal signal transduction systems by extracellular ATP. *J. Neurochem.* **50:**295–301.
54. Mery, P. F., Lohmann, S. M., Walter, U., and Fischmeister, R. (1991). Ca^{2+} current is regulated by cyclic GMP-dependent protein kinase in mammalian cardiac myocytes. *Proc. Natl. Acad. Sci. USA* **88:**1197–1201.
55. Prasad, M. R., and Jones, R. M. (1992). Enhanced membrane protein kinase C activity in myocardial ischemia. *Basic Res. Cardiol.* **87:**19–26.
56. Madden, K. P., Clark, W. M., Kochhar, A., and Zivin, J. A. (1991). Effect of protein kinase C modulation on outcome of experimental CNS ischemia. *Brain Res.* **547:**193–198.
57. Felipo, V., Minana, M. D., and Grisolia, S. (1993). Inhibitors of protein kinase C prevent the toxicity of glutamate in primary neuronal cultures. *Brain Res.* **604:**192–196.
58. Aronowski, J., Waxham, M. N., and Grotta, J. C. (1993). Neuronal protection and preservation of calcium/calmodulin dependent protein kinase II and protein kinase C activity by dextrorphan treatment in global ischemia. *J. Cereb. Blood Flow Metab.* **13:**550–557.
59. Kobryn, C., and Mandel, L. J. (1993). Effect of anoxia on protein phosphorylation in rabbit kidney proximal tubules [abstract]. *J. Am. Soc. Nephrol.* **4:**739.
60. Candeo, P., Favaron, M., Lengyel, I., Manev, R. M., Rimland, J. M., and Manev, H. (1992). Pathological phosphorylation causes neuronal death: Effect of okadaic acid in primary culture of cerebellar granule cells. *J. Neurochem.* **59:**558–561.
61. Alberli, P., Bardella, L., and Comolli, R. (1993). Ribosomal protein S6 kinase and protein kinase C activation by epidermal growth factor after temporary renal ischemia. *Nephron* **64:**296–302.
62. Ruckenstein, A., Rydel, R. E., and Greene, L. A. (1991). Multiple agents rescue PC12 cells from serum-free cell death by translation- and transcription-independent mechanisms. *J. Neurosci.* **11:**2552–2563.
63. Van Why, S. K., Hildebrandt, F., Ardito, T., Mann, A. S., Siegel, N. J., and Kashgarian, M. (1992). Induction and intercellular localization of HSP-72 after renal ischemia. *Am. J. Physiol.* **263:**F769–F775.
64. Humes, H. D., Nguyen, V. D., Cieslinski, D. A., and Messana, J. M. (1989). The role of free fatty acids in hypoxia-induced injury to renal proximal tubule cells. *Am. J. Physiol.* **256:**F688–F696.
65. Wetzels, J. F. M., Wang, X., Gengaro, P. E., Nemenoff, R. A., Burke, T. J., and Schrier, R. W. (1993). Glycine protection against hypoxic but not phospholipase A_2-induced injury in rat proximal tubules. *Am. J. Physiol.* **264:**F94–F99.
66. Choi, K. H., Edelstein, C. L., Gengaro, P., Schrier, R. W., and Nemenoff, R. A. (1995). Hypoxia induces changes in phospholipase A_2 in rat proximal tubules: Evidence for multiple forms. *Am. J. Physiol.,* **269:**F846–F853.
67. Portilla, D., Mandel, L. J., Bar-Sagi, D., and Millington, D. S. (1992). Anoxia induces phospholipase A_2 activation in rabbit renal proximal tubules. *Am. J. Physiol.* **263:**F354–F360.
68. Nguyen, V. D., Cieslinski, D. A., and Humes, H. D. (1988). Importance of adenosine triphosphate in phospholipase A_2-induced rabbit renal proximal tubule injury. *J. Clin. Invest.* **82:**1098–1105.
69. Zager, R. A., Schimpf, B. A., Gmur, D. J., *et al.* (1993). Phospholipase A_2 activity can protect renal cells from oxygen deprivation injury. *Proc. Natl. Acad. Sci. USA* **90:**8297–8301.
70. Almeida, A. R. P., Wetzels, J. F. M., Bunnachak, D., Burke, T. J., Chaimovite, C., Hammond, W. S., and Schrier, R. W. (1992). Acute phosphate depletion and in vitro rat proximal tubule injury: Protection by glycine and acidosis. *Kidney Int.* **41:**1494–1500.
71. Wieder, E. D., Kribben, A., Yu, L., Burke, T. J., and Schrier, R. W. (1993). Cytosolic calcium and intracellular pH in hypoxic rat proximal tubules in normal and acidotic conditions [abstract]. *J. Am. Soc. Nephrol.* **4:**746.
72. Borkan, S. C., and Schwartz, J. H. (1989). Role of oxygen free radical species in in vitro models of proximal tubular ischemia. *Am. J. Physiol.* **257:**F114–F125.
73. Barrett, A. J. (1992). Cellular proteolysis. An overview. *Ann. N.Y. Acad. Sci.* **647:**1–15.
74. Wilson, P. D., and Hartz, P. A. (1991). Mechanism of cyclosporine A toxicity in defined cells of renal tubule epithelia: A role for cysteine proteases. *Cell Biol. Int. Rep.* **15:**1243–1258.
75. Nichols, J. C., Bronk, S. F., and Gores, G. J. (1993). Glycine cytoprotection during anoxic hepatocyte injury is associated with inhibition of calpain protease activity [abstract]. *Am. Gastroenterol. Assoc. Meet.* p. 1017.
76. Kellerman, P. S., Murphy, J. L., and Burke, T. J. (1993). Glycine and alanine selectively inhibit ionomycin-induced actin depolymerization in renal proximal tubule cells [abstract]. *Clin. Res.* **41:**142A.
77. Wilson, P. D. (1993). Renal proteases in cellular injury. *XIIth Int. Congr. Nephrol.* p. 248.
78. Suzuki, K., Saido, T. C., and Hirari, S. (1992). Modulation of cellular signals by calpain. *Ann. N.Y. Acad. Sci.* **674:**218–227.
79. Ivy, G. O. (1992). Protease inhibition causes some manifestations of aging and Alzheimer's disease in rodent and primate brain. *Ann. N.Y. Acad. Sci.* **674:**89–102.
80. Molitoris, B. A., Wilson, P. D., and Schrier, R. W. (1985). Ischemia induces partial loss of surface membrane polarity and accumulation of putative calcium ionophores. *J. Clin. Invest.* **76:**2097–2105.
81. Kellerman, P. S., Clark, P. A., Holien, C. A., Linas, S. L., and Molitoris, B. A. (1990). Role of microfilaments in maintenance of proximal tubule structural and functional integrity. *Am. J. Physiol.* **259:**F279–F285.
82. Kellerman, P. S., Jones, G., and Burke, T. J. (1994). Glycine ameliorates cytoskeletal polymerization induced by metabolic inhibitors in proximal tubule cells. *Clin. Res.,* 221A.
83. Dickson, R. C., Bronk, S. F., and Gores, G. J. (1992). Glycine cytoprotection during lethal hepatocellular injury from adenosine triphosphate depletion. *Gastroenterology* **102:**2098–2107.

84. Flores, J., DiBona, D. R., Beck, C. H., and Leaf, A. (1972). The role of cell swelling in ischemic renal damage and the protective effect of hypertonic solute. *J. Clin. Invest.* **51:**118–126.
85. Schrier, R. W., Arnold, P. A., Gordon, J. A., and Burke, T. J. (1984). Protection of mitochondrial function by mannitol in ischemic acute renal failure. *Am. J. Physiol.* **247:**F365–F369.
86. Burke, T. J., Arnold, P. A., Gordon, J. A., Bulger, R. E., Dobyan, D. C., and Schrier, R. W. (1984). Protective effect of intrarenal calcium membrane blockers before or after renal ischemia. *J. Clin. Invest.* **74:**1830–1841.
87. Kreisberg, J. I., Mills, J. W., Jarrell, J. A., Rabito, C. A., and Leaf, A. (1980). Protection of cultured renal tubular epithelial cells from anoxic cell swelling. *Proc. Natl. Acad. Sci. USA* **77:**5445–5447.
88. Filipovic, D., and Sackin, H. (1991). A calcium-permeable stretch-activated cation channel in renal proximal tubules. *Am. J. Physiol.* **260:**F119–F129.
89. Racusen, L. C., Fivush, B. A., Li, Y.-L., Slatnick, I., and Solez, K. (1991). Dissociation of tubular cell detachment and tubular cell death in clinical and experimental acute tubular necrosis. *Lab. Invest.* **64:**546–556.
90. Goligorsky, M. S., Lieberthal, W., Racusen, L., and Simon, E. E. (1993). Integrin receptors in renal tubule epithelium: New insights into pathophysiology of acute renal failure. *Am. J. Physiol.* **264:**F1–F8.
91. Lampugnani, M. G., Resnati, M., Dejana, E., and Marchisio, P. C. (1991). The role of integrins in the maintenance of endothelial monolayer integrity. *J. Cell Biol.* **112:**479–490.
92. Richter, C., and Frei, B. (1988). Ca^{2+} release from mitochondria induced by prooxidants. *Free Radical Biol. Med.* **4:**365–375.
93. Ordway, R. W., Walsh, J. V., Jr., and Singer, J. J. (1989). Arachidonic acid and other fatty acids directly activate potassium channels in smooth muscle cells. *Science* **244:**1176–1179.
94. Paller, M. S. (1991). Hydrogen peroxide and ischemic renal injury: Effect of catalase inhibition. *Free Radical Biol. Med.* **10:**29–34.
95. Yu, L., Gengaro, P. E., Burke, T. J., and Schrier, R. W. (1992). Reactive oxygen species in renal tubular hypoxia/reoxygenation injury-role for hydrogen peroxide. *J. Am. Soc. Nephrol.* **3:**718 (abstract).
96. Radi, R., Beckman, J. S., Bush, K. M., and Freeman, B. A. (1991). Peroxynitrite-induced membrane lipid peroxidation: The cytotoxic potential of superoxide and nitric oxide. *Arch. Biochem. Biopyhys.* **288:**481–487.
97. Radi, R., Cosgrove, T.P., Beckman, J. S., and Freeman, B. A. (1993). Peroxynitrite-induced luminol chemiluminescence. *Biochem. J.* **290:** 51–57.
98. Yu, L., Gengaro, P. E., Niederberger, M., Burke, T. J., and Schrier, R. W. (1994). Nitric oxide: A mediator of rat tubular hypoxic/reoxygenation injury. *Proc. Natl. Acad. Sci. USA,* **91:**1691–1695.
99. Markewitz, B. A., Michael, J. R., and Kohan, D. E. (1993). Cytokine-induced expression of a nitric oxide synthase in rat renal tubule cells. *J. Clin. Invest.* **91:**2138–2143.
100. Beckman, J. S., Beckman, T. W., Chen, J., Marshall, P. A., and Freeman, B. A. (1990). Apparent hydroxyl radical production by peroxynitrite: Implications for endothelial injury from nitric oxide and superoxide. *Proc. Natl. Acad. Sci. USA* **87:**1620–1624.
101. Zager, R. A., Gmur, D. J., and Schrimpf, B. A. (1992). Evidence against increased hydroxyl radical production during oxygen deprivation–reoxygenation proximal tubule injury. *J. Am. Soc. Nephrol.* **2:**1627–1631.
102. Doctor, R. B., and Mandel, L. J. (1991). Minimal role of xanthine oxidase and oxygen free radicals in rat renal tubular reoxygenation injury. *J. Am. Soc. Nephrol.* **1:**959–969.
103. Paller, M. S., Hoidal, J. R., and Ferris, T. F. (1984). Oxygen free radicals in ischemic acute renal failure in the rat. *J. Clin. Invest.* **74:** 1156–1164.
104. Dillon, J. J., Grossman, S. H., and Finn, W. F. (1993). Effect of oxypurinol on renal reperfusion injury in the rat. *Renal Fail.* **15:** 37–45.
105. Paller, M. S. (1988). Hemoglobin- and myoglobin-induced acute renal failure in rats: Role of iron in nephrotoxicity. *Am. J. Physiol.* **255:**F539–F544.
106. Brezis, M., Rosen, S., Silva, P., and Epstein, F. H. (1984). Selective vulnerability of the medullary thick ascending limb to anoxia in the isolated perfused rat kidney. *J. Clin. Invest.* **73:**182–190.
107. Linas, S. L., Wittenbury, D., Parsons, P. E., and Repine, J. E. (1992). Mild renal ischemia activates primed neutrophils to cause acute renal failure. *Kidney Int.* **42:**610–616.
108. Zager, R. A., Schrimpf, B. A., Bredl, C. R., and Foerder, C. A. (1992). Increased proximal tubule cell catalytic iron content: A result, not a mediator, of hypoxia–reoxygenation injury. *J. Am°. Soc. Nephrol.* **3:**116–118.
109. Walker, P. D,. and Shah, S. V. (1991). Hydrogen peroxide cytotoxicity in LLC-PK1 cells: A role for iron. *Kidney Int.* **40:**891–898.
110. Zager, R. A., Schrimpf, B. A., and Bredl, C. R. (1993). Inorganic iron effects on in vitro hypoxic proximal renal tubular cell injury. *J. Clin. Invest.* **91:**702–708.
111. Beckman, J. S., Beckman, T. W., Chen, J., Marshall, P. A., and Freeman, B. A. (1990). Apparent hydroxyl radical production by peroxynitrite: Implications for endothelial injury from nitric oxide and superoxide. *Proc. Natl. Acad. Sci. USA* **87:**1620–1624.
112. Zager, R. A., and Foerder, C. A. (1992). Effects of inorganic iron and myoglobin on in vitro proximal tubule lipid peroxidation and cytotoxicity. *J. Clin. Invest.* **89:**989–995.
113. Robinette, J. B., and Conger, J. D. (1993). Nitric oxide synthase activity is increased—not decreased—in vasculature of acute renal failure kidneys [abstract]. *J. Am. Soc. Nephrol.* **4:**743.
114. Shanley, P. F., and Johnson, G. C. (1991). Calcium and acidosis in renal hypoxia. *Lab. Invest.* **65:**298–305.
115. Brezis, M., Shina, A., Kidroni, G., Epstein, F. H., and Rosen, S. (1988). Calcium and hypoxic injury in the renal medulla of the perfused rat kidney. *Kidney Int.* **34:**186–194.
116. Heyman, S., Spokes, K., Rosen, S., and Epstein, F. H. (1992). Mechanism of glycine protection in hypoxic injury: Analogies with glycine receptor. *Kidney Int.* **42:**41–45.
117. Shanley, P. F., White, C. W., Avraham, K. B., Groner, Y., and Burke, T. J. (1992). Use of transgenic animals to study disease models: Hyperoxic lung injury and ischemic acute renal failure in "high" SOD mice. *Renal Fail.* **14:**391–394.
118. Feinstein, R. N., Braun, J. T., and Howard, J. B. (1967). Acatalasemic and hypocatalasemic mouse mutants. II. Mutational variations in blood and solid tissue catalases. *Arch. Biochem. Biophys.* **120:**165–169.
119. Bosco, P. J., and Schweizer, R. T. (1988). Use of oxygen radical scavengers on autografted pig kidneys after warm ischemia and 48-hour perfusion preservation. *Arch. Surg.* **123:**601–604.
120. Gulati, S., Singh, A. K., Irazu, C., Orak, J., Rajagopalan, P. R., Fitts, C. T., and Singh, I. (1992). Ischemia–reperfusion injury: Biochemical alterations in peroxisomes of rat kidney. *Arch. Biochem. Biophys.* **295:**90–100.
121. Sadzuka, Y., Shoji, T., and Takino, Y. (1992). Effect of cisplatin on the activities of enzymes which protect against lipid peroxidation. *Biochem. Pharmacol.* **43:**1872–1875.
122. Sela, S., Shasha, S. M., Mashiach, E., Haj, M., Kristal, B., and Shkolnik, T. (1993). Effect of oxygen tension on activity of antioxidant enzymes and on renal function of the postischemic reperfused rat kidney. *Nephron* **63:**199–206.
123. Bratell, S., Haraldsson, G., Herlitz, H., Jonsson, O., Pettersson, S., Schersten, T., and Waldenstrom, J. (1990). Protective effects of pretreatment with superoxide dismutase, catalase and oxypurinol on

tubular damage caused by transient ischemia. *Acta Physiol. Scand.* **139:**417–425.

124. Bratell, S., Folmerz, P., Hansson, R., Jonsson, O., Lundstam, S., Pettersson, S., Rippe, B., and Schersten, T. (1988). Effects of oxygen free radical scavengers, xanthine oxidase inhibition and calcium entry-blockers on leakage of albumin after ischemia. An experimental study in rabbit kidneys. *Acta Physiol. Scand.* **134:**35–41.
125. Linas, S. L., Whittenburg, D., and Repine, J. E. (1987). O_2 metabolites cause reperfusion injury after short but not prolonged renal ischemia. *Am. J. Physiol.* **253:**F685–F691.
126. Omata, M. (1990). Role of reactive oxygen species (Ros) in model immune complex nephritis. *Nippon Jinzo Gakkai Shi* **32:**949–958.
127. Kribben, A., Wetzels, J. F. M., Wieder, E. D., Burke, T. J., and Schrier, R. W. (1993). New technique to assess hypoxia-induced cell injury in individual isolated renal tubules. *Kidney Int.* **43:**464–469.
128. den Butter, G., Lindell, S. L., Sumimotot, R., Schilling, M. K., Southard, J. H., and Belzer, F. O. (1993). Effect of glycine in dog and rat liver transplantation. *Transplantation* **56:**817–822.
129. Cook, J. A., and Mitchell, J. B. (1989). Viability measurements in mammalian cell systems. *Anal. Biochem.* **179:**1–7.
130. Gerschenson, L. E., and Rotello, R. J. (1992). Apoptosis: A different type of cell death. *FASEB J.* **6:**2450–2455.
131. Squier, M. K. T., Miller, A. C. K., Malkinson, A. M., and Cohen, J. J. (1994). Calpain activation in apoptosis. *J. Cell. Physiol.* **159:**229–237.
132. Schumer, M., Colombel, M. C., Sawczuk, I. S., Gobe, G., Connor, J., O'Toole, K. M., Olsson, C. A., Wise, G. J., and Buttyan, R. (1992). Morphologic, biochemical and molecular evidence of apoptosis during the reperfusion phase after brief periods of renal ischemia. *Am. J. Pathol.* **140:**831–838.
133. Ledda-Columbano, G. M., Columbano, A., Coni, P., Faa, G., and Pane, P. (1989). Cell deletion by apoptosis during regression of renal hyperplasia. *Am. J. Pathol.* **135:**657–662.
134. Levine, B., Huang, Q., Isaacs, J. T., Reed, J. C., Griffin, D. E., and Hardwick, J. M. (1993). Conversion of lytic to persistent alphavirus infection by the bcl-2 cellular oncogene. *Nature* **361:**739–742.
135. Richter, C. (1993). Pro-oxidants and mitochondrial Ca^{2+}: Their relationship to apoptosis and oncogenesis. *FEBS Lett.* **325:**104–107.
136. Hockenbery, D. M., Nunez, G., and Milliman, C. (1990). Bcl-2 is an inner mitochondrial membrane protein that blocks programmed cell death. *Nature* **348:**334–336.
137. Peitsch, M. C., Polzar, B., and Stephan, H. (1993). Characterization of the endogenous deoxyribonuclease involved in nuclear DNA degradation during apoptosis (programmed cell death). *EMBO J.* **12:**371–377.
138. Gobe, G. C., Axelsen, R. A., Harman, B. V., and Allan, D. J. (1988). Cell death by apoptosis following X-irradiation of the foetal and neonatal rat kidney. *Int. J. Radiat. Biol.* **54:**567–576.
139. Seegers, J. C., and Garlinski, P. J. (1993). Apoptosis-associated chromatin degradation underlies ochratoxin-A toxicity in kidney cells and lymphocytes [abstract]. *Proc. Annu. Meet. Am. Assoc. Cancer Res.* **34:**A235.
140. Pearse, M. J., O'Bryan, M., Fisicaro, N., Rogers, L., Murphy, B., and d'Apice, A. J. (1992). Differential expression of clusterin in inducible models of apoptosis. *Int. Immunol.* **11:**1225–1231.
141. Humes, H. D., and Cieslinski, D. A. (1992). Interaction between growth factors and retinoic acid in the induction of kidney tubulogenesis in tissue culture. *Exp. Cell Res.* **201:**8–15.
142. Taub, M., Wang, Y., Szczesny, T. M., and Kleinman, H. K. (1990). Epidermal growth factor or transforming growth factor alpha is required for kidney tubulogenesis in matrigel cultures of serum free media. *Proc. Natl. Acad. Sci. USA* **87:**4002–4006.
143. Humes, H. D., Cieslinski, D. A., Johnson, L. B., and Sanchez, I. O. (1992). Triiodothyronine enhances renal tubule cell replication by stimulating EGF receptor gene expression. *Am. J. Physiol.* **262:** F540–F545.
144. Vehaskari, V. M., Hering-Smith, K. S., Moskowitz, D. W., Weiner, I. D., and Hamm, L. L. (1989). Effect of epidermal growth factor on sodium transport in the cortical collecting duct. *Am. J. Physiol.* **256:**F803–F809.
145. Salido, E. C., Barajas, L., and Lechago, J. (1986). Immunocytochemical localization of epidermal growth factor in mouse kidney. *J. Histochem. Cytochem.* **34:**1155–1160.
146. Igawa, T., Matsumoto, K., Kanda, S., Saito, Y., and Nakamura, T. (1993). Hepatocyte growth factor may function as a renotropic factor for regeneration in rats with acute renal injury. *Am. J. Physiol.* **265:**F61–F69.
147. Igawa, T., Kanda, S., Kanetake, H., Saitoh, Y., Ichihara, A., Tomita, Y., and Nakamura, T. (1991). Hepatocyte growth factor is a potent mitogen for cultured rabbit tubular epithelial cells. *Biochem. Biophys. Res. Commun.* **174:**831–838.
148. Emami, A., Schwartz, J. H., and Borkan, S. C. (1991). Transient ischemia or heat stress induces a cytoprotectant protein in rat kidney. *Am. J. Physiol.* **260:**F479–F485.
149. Benjamin, I. J., Horie, S., Greenberg, M. L., Alpern, R. J., and Williams, R. S. (1992). Induction of stress proteins in cultured myogenic cells. Molecular signals for the activation of heat shock transcription factor during ischemia. *J. Clin. Invest.* **89:**1685–1689.

25

Genetic Variants of Erythrocytes

John C. Parker†

25.1. INTRODUCTION

This chapter discusses some genetic variants of the human red cell with clear-cut relationships to solute transport. These include band 7.2b deficiency, congenital increase in red cell pump sites, S and C hemoglobinopathies, congenital absence of carrier-mediated urea transport, and two mutations of the anion exchanger.

Because this volume focuses on transport, the present review does not discuss progress in the burgeoning field of red cell membrane structure. Recent work has revealed the molecular basis for many inherited red cell membrane abnormalities. Diseases that were once classified on the basis of red cell shape can now be understood at the level of DNA. Mutations involving alpha and beta spectrin, ankyrin, bands 3, 4.1, 4.2, and glycophorin C can give rise to a variety of alterations in the appearance and physical properties of red cells. In some of these conditions (e.g., hereditary spherocytosis) there are quantitative abnormalities in red cell sodium transport, as reviewed previously.[1,2] The basis of the altered ion movements is obscure, and it is not likely they contribute to the pathophysiology of shortened red cell survival. Inherited membrane skeletal defects have been discussed in two recent reviews.[3,4]

Previous versions of this paper contained discussions of red cell transport abnormalities in hypertension, diabetes mellitus, hyperthyroidism, congenital neuromuscular disorders, manic-depressive disease, cystic fibrosis, and other conditions.[1,2] These topics will be omitted here because they are not understood at the molecular level and the findings have not provided valuable insights into pathogenesis or therapy.[5–7]

25.2. BAND 7.2B DEFICIENCY (CONGENITAL HYDROCYTOSIS)

Since the original report by Zarkowsky *et al.*,[8] several people were found whose red cells have grossly abnormal contents of Na, K, and water. Some cells are overhydrated, called "hydrocytes,"[9] and others dry, called "dessiccytes"[9] or "xerocytes."[10] The term "hereditary stomatocytosis" often is used to apply to this group of conditions, but it is a misnomer: not all such cells have a stoma-like central pallor, and many that do have no abnormality of ion or water content. The evidence for inheritance, however, is convincing. Whereas an autosomal dominant pattern is found in most families, some show an autosomal recessive pattern.[10] Table 25.1 shows values for ion and water content in fresh red cells from such patients.

Both hydrocyte and xerocyte patients have a reduced red cell life span, with reticulocytosis. Density fractionation of the hydrocyte patients' blood shows that, unlike the normal situation, the reticulocytes are the densest cells. With time in the circulation they accumulate sodium and water. In dessicytosis or xerocytosis the cells lose potassium and water and become more and more dense with age.[10]

Most investigators of these disorders would agree that the red cells have normally functioning Na/K pumps, stimulated to an abnormally high degree of activity by the increased cell Na concentration and as a consequence of which there is an increased rate of glycolysis. The fundamental transport defect is thought to involve passive ion movements, which may be increased 2 to 40 times normal.[11] The roles of electrodiffusion and the known systems for co- and countertransport in mediating these high fluxes are not clear.

Mentzer *et al.*[12] published a remarkable experiment on the red cells of a patient with hydrocytosis (Table 25.2). During incubation *in vitro* with a bifunctional imidoester, the cells adjusted their ion and water content to normal. It was as if the cross-linking agent had selectively repaired the congenital red cell permeability defect and left other transport mechanisms functionally intact.[13]

Mentzer *et al.*[13] further found that red cell membranes from people with hydrocytosis showed on SDS–PAGE a decreased intensity of staining in the band 7 region (26–30 kDa), compared to controls. On further study, it was shown that red cell membranes from two unrelated patients with hydrocytosis lacked a 28-kDa component of band 7.[14] These findings were confirmed in other patients,[15] and the missing membrane component was dubbed band 7.2b. Further characterization of this protein showed it is not removed from the membrane by incu-

†Deceased November 22, 1993

John C. Parker • Department of Medicine, University of North Carolina, Clinical Research Center, Chapel Hill, North Carolina 27599.
Molecular Biology of Membrane Transport Disorders, edited by S.G. Schultz *et al.* Plenum Press, New York, 1996

Table 25.1. Data on Unfractionated, Freshly Drawn Red Cells from People with Abnormal Red Cell Ion and Water Content[a]

	Cell water (% wet wt)	Cell Na	Cell K
		(mmole/liter of cells)	
Normal	65.8	8	99
Hydrocytes	69.7	87	38
Dessic- or xerocytes	62.2	15	80

[a]Adapted from Wiley.[11]

bation in low ionic strength, 0.1 M NaOH, or EDTA. It is relatively insoluble in detergents, requiring 5% Triton X-100 for extraction from the membrane. Antibodies prepared from it recognized a 24-kDa polypeptide in red cell membranes from people, rats, cows, pigs, sheep, rabbits, chickens, and frogs in one study.[16] Another antibody reacted with human and rat red cell membranes, but not with comparable preparations from sheep, dogs, cats, cows, pigs, or chickens.[17] Red cell membrane preparations from people with hydrocytosis showed no immunoreactivity with antibodies to band 7.2b. Reactivity was found, however, in various nonerythroid cell lines. The band can be phosphorylated by cAMP kinase, and it is palmitylated but not glycosylated.[16,17] There are about 400,000 copies of it per red cell. The protein is accessible to proteolytic digestion from the inner but not the outer membrane surface.[16,18] Antigenic activity can be detected in Triton shell preparations made from human red cells, suggesting that protein 7.2b binds to membrane skeletal components.

Two groups cloned cDNAs for the band 7.2b polypeptide.[18,19] The deduced protein sequence has an alkaline isoelectric pH. A 29-residue hydrophobic segment near the N-terminus is likely embedded in the lipid bilayer, with the highly polar C-terminal segment facing the cytoplasm. Northern blots indicate mRNA coding for protein 7.2b in erythroid precursors plus a wide variety of nonerythroid tissues. The protein bears no homologies to known transporters.

Interestingly, mRNA for band 7.2b can be detected in reticulocytes of patients with hydrocytosis whose mature red cells are deficient in the polypeptide. This finding is in apparent conflict with the virtual absence of the protein in red cells from hemizygotes: perhaps the mutant band is unstable or leads to instability of oligomers composed of itself and copies of band 7.2b inherited from the normal parent.

Table 25.2. Effect of 6 hr *in Vitro* Incubation of Red Cells with 5 mM Dimethyl Adipimidate (DMA) versus a Sucrose Control[a]

	Cell water (% wet wt)	Cell Na	Cell K
		(mmole/liter cells)	
Before incubation	70.8	93.4	29.5
DMA incubation	62.7	3.7	102.8
Sucrose control	71.4	91.5	32.4

[a]From Ref. 12. Fresh red cells were washed and suspended in a plasmalike buffer with glucose and 1% albumin, in the presence of 5 mM DMA or 5 mM sucrose. After 6 hr incubation at 37°C, the cells were washed and analyzed for Na, K, and water content.

The mechanism of increased cation permeability in red cells lacking band 7.2b is not defined. Morle *et al.*[20] report the case of a young man with hemolytic anemia since birth. His circulating red cells were abnormally large and had a reduced concentration of hemoglobin. His red cell membranes had only 28% of the normal amount of band 7. Studies of Na and K transport led the authors to conclude that the patient's red cells were swollen because of a large increase in Na and K "leaks" and that as a consequence of the cells being swollen, K–Cl cotransport (defined as dihydroindenylocyalkanoic acid-sensitive K efflux) was maximally activated. The prominent swelling-induced K–Cl cotransport in this patient's red cells was undoubtedly related to their low mean age.[21] No confirmatory studies demonstrating chloride-dependence of the increased K flux were included in the report. The study by Olivieri *et al.*[22] of a patient whose red cells were stomatocytic, had reduced band 7, and yet had normal cations, lacks full characterization of the band 7 region and therefore may not be germane to the present topic.

There is increasing evidence that human red cells may have channels through which Na and K can move by electrodiffusion and which may be voltage-gated and responsive to acetylcholine.[23–27] Stewart *et al.*[18] suggest that the band 7.2b polypeptide may function as a physiological "plug" which limits the movement of cations through such a channel, in analogy with the model proposed for the *Shaker* K channel cloned from *Drosophila.*[28] Such a speculation may be unwarranted, however. There are 400,000 copies of band 7.2b per cell but only 300 channels.[27]

25.3. CONGENITAL INCREASE IN RED CELL PUMP SITES

DeLuise and Flier[29] described a morbidly obese woman whose red cells had a 10- to 20-fold increase in the number of Na/K pump sites. Her monocytes were not affected. Although one of her sons had a moderate increase in red cell Na/K pumps, the inheritance pattern was not clear. On further study, each of the propositus' red cell pumps had normal kinetic features. Red cells from the patient had an internal Na concentration of 2 mmole/liter cells, compared with a normal value of 8–10. The steady-state pump flux of these cells *in vivo* was normal, because although the number of pumps and therefore the maximum pump rate of the cells was increased, each pump worked at a much reduced fraction of its normal capacity due to the low cell Na concentration. When the patient's red cell Na was normalized *in vitro,* the pump flux greatly increased.[30] The biological basis of this abnormality is not clear, but a possibility mentioned by the authors is that it may involve a failure of the normal mechanism by which the very large number of pumps in the membrane of each red cell precursor becomes reduced to a level characteristic of mature, circulating erythrocytes. Such an explanation is reminiscent of reports in the dog red cell literature. Normal, circulating dog red cells lack Na/K pumps altogether, but their precursors in the marrow have pumps. In a rare breed of Japanese dogs, Na/K pumps are found in abundance in circulating red cells. The abnormality in such dogs may be a failure of the process by which Na/K pumps are depleted with red cell maturation.[31,32]

25.4. HEMOGLOBINOPATHIES

Recent developments have raised the hope that understanding a disease at the DNA level might clarify its pathogenesis. This promise has not been fulfilled in the case of S (β_6 Glu $\rightarrow$ Val) and C (β_6 Glu $\rightarrow$ Lys) hemoglobinopathies, which were among the first human illnesses described in molecular terms. To be sure, the single base mutations that cause these abnormalities give rise to altered properties of hemoglobin. In addition, however, there are consequences for red cell function that are indirect, not predictable *a priori* from the genetic defect, and responsible for much of the clinical picture. Hemoglobin S exerts important effects on the membrane which cause the cells to stick to endothelium and to have abnormal ion transport. Hemoglobin C causes red cells containing it to be deficient in ions and water and therefore poorly deformable in the microcirculation. These alterations play an important role in the vaso-occlusive events that underlie sickle-cell crises and in the pathogenesis of reduced red cell survival. This review focuses on membrane transport in S and C hemoglobinopathies. Comprehensive summaries of this topic have appeared recently,[33–35] to which the reader is referred for original references.

The subject of ions and water in red cells containing hemoglobins S and C is under investigation in several laboratories. Throughout the literature in this field the water content of red cells is regarded as being an inverse function of the mean corpuscular hemoglobin concentration (MCHC) and also of the cell density. This is because all are measured per mass or per volume of cells. For most of the present discussion, cell density can be regarded as proportional to MCHC, and a rise in cell density or MCHC can be taken to mean a fall in cell water content, and vice versa. The red cell water content (and MCHC and density) can change in response to anisotonicity or changes in content of inorganic ions or charged macromolecules.[36,37] However, there are situations in which it is important to recognize that two red cells of differing water content might have the same MCHC and density if they contained different amounts of hemoglobin. This consideration arises in relation to changes brought about by hydroxyurea therapy.

The MCHC of red cells in freshly drawn venous blood from sickle-cell patients is normal (Figure 25.1). However, there is an abnormally wide distribution of MCHC about the mean, with roughly equal numbers of abnormally well-hydrated and abnormally dehydrated cells.[38] The study of salt and water regulation in sickle cells is of interest because the abnormally dehydrated cell population is thought to play a role in obstruction of the microvasculature. In addition, small increases in cell water can greatly reduce the tendency of hemoglobin S to gel on deoxygenation.[39] Cell water is important

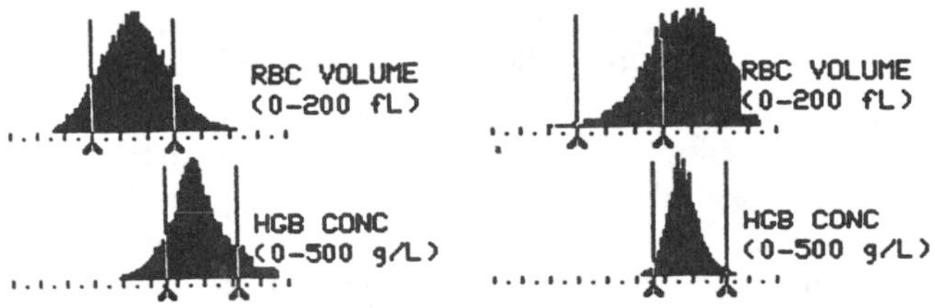

Figure 25.1. Technicon H1 data for a sickle-cell patient before (left) and during (right) treatment with hydroxyurea. Each graph is a histogram showing number of cells on the vertical axis and red cell volume (MCV) or hemoglobin concentration (MCHC) on the horizontal axis. The vertical lines on each plot represent the boundaries of normal: 60 to 120 fl for MCV, and 275 to 400 g/l red cells for MCHC. Note that hydroxyurea treatment raises the MCV from 85 to 130 fl. Although the MCHC is not changed by hydroxyurea, remaining at about 330 g/l, the distribution of red cell hemoglobin concentration is much narrower in the patient under treatment: both the highest and lowest density fractions disappear. (From Orringer *et al.*,[38] with permission.)

in hemoglobin C disease because fresh red cells from these patients are grossly dehydrated, cation depleted, and poorly deformable in the capillaries. Hemoglobin C does not gel on deoxygenation.[40]

Table 25.3 lists some of the major transport pathways of human red cells through which net cation movements can occur and potentially lead to the gain or loss of cell water. As indicated, several of these permeation routes are claimed to be abnormal in hemoglobin S and C disease. Some investigators hope that beneficial clinical effects might be obtained if one

Table 25.3. Major Routes of Net Cation Transport in Human Red Cells[a]

Transporter	Inhibitors	Affected in S and/or C hemoglobinopathies
The 3Na/2K pump[48]	Ouabain	Yes
Na–K–2Cl cotransport[49]	Bumetanide	No
K–Cl cotransport[51]	DIOA[50]	Yes
Na/H exchange	Amiloride	No
Ca-dependent K conductance[52]	Charybdotoxin[53,54]	Possibly
Nonspecific cation channels[27,52]	DIDS	Yes
$NaHCO_3$ ion pair movement via band 3[52]	DIDS	Possibly
An undefined, Na-dependent anion transport system expressed transiently during maturation of reticulocytes[43,47]		Possibly

[a]A mathematical model incorporating many of the features of these transporters was formulated by Lew *et al.*[47] and can be used to predict effects of various perturbations on cell water content.

or another of these transporters could be manipulated in such a way that the MCHC would decrease and the cells thereby rendered more deformable and less likely to sickle.

The mean red cell age in virtually all hemolytic anemia patients is less than in controls. Therefore, any claim that red cells in hemoglobin S or C disease have a distinctly abnormal mode of ion transport must be regarded with skepticism. Could this putative membrane lesion simply be a reflection of the younger red cell population? This is not always easy to resolve. It used to be thought that a good way to separate young from old red cells was via buoyant density centrifugation: the youngest cells would rise to the top and the oldest would sink to the bottom.[41] The validity of that principle for normal red cells is a matter of controversy: everyone agrees that reticulocytes are found in the lightest or least dense fraction, but there are differences of opinion regarding senescent cells. Some of the oldest cells are likely in the densest fraction, but others may undergo an agonal decrease in density.[41,42] In red cells from individuals homozygous for hemoglobins S and C, the situation is more complicated. Some hemoglobin S cells appear to undergo a rapid increase in density and become very dehydrated early in life,[43] while in blood from hemoglobin C homozygotes (and SC double heterozygotes) the reticulocytes are the densest cells.[44–46]

A pervasive problem in all work with red cells in general and hemoglobinopathic red cells in particular is the heterogeneity within a given blood sample. Phenomena described in unfractionated red cells may be much more comprehensible if studied on density-separated fractions, but heterogeneity can sometimes be revealed even among cells of the same density or within a population of reticulocytes, for example when cells in a given fraction are exposed to hypotonic media.[45]

With these caveats in mind, one can organize a discussion of ion transport in hemoglobinopathies according to the various pathways in human red cells that might possibly lead to net alterations in cell ion content (Table 25.3).

25.4.1. The Na/K Pump

There is evidence for impaired function of the Na/K pump in sickle cells, especially the very densest ones.[55] The lesion appears not to be related to a reduced number of pumps per cell but rather to pump regulation.[55–57] The Na/K pump has been implicated in red cell dehydration because it extrudes 3 Na ions for every 2 K taken up. As noted below, sickle cells undergo an increase in Na concentration during deoxygenation, the result of the opening up of passive permeability pathways for inorganic cations. The high cytosolic Na stimulates the pump and may, by virtue of the unequal stoichiometry mentioned above, result in net loss of cell solute.[58–60]

25.4.2. Na–K–2Cl Cotransport

This ubiquitous transporter is found in human red cells and can mediate net changes in cell ion content given appropriate driving forces. However, in physiological situations its substrates are at equilibrium,[47] and it is furthermore not very active in the red cells of black people, in whom hemoglobins S and C prevail.[33]

25.4.3. K–Cl Cotransport

Some investigators believe that K–Cl cotransport plays an important role in the pathogenesis of red cell dehydration in the hemoglobinopathies. K–Cl cotransport is an electroneutral, passive permeability route found in many red cell species.[51] It is highly regulated, being activated by cell swelling, cell magnesium depletion, cytoplasmic acidification, certain chaotropic agents such as thiocyanate[61] and urea,[62] oxidants,[63,64] and sulfhydryl reagents such as *N*-ethylmaleimide.[51] The pathway is downregulated by increases in cytosolic free magnesium, by okadaic acid, and by increases in cell lithium ion content.[65] In some nonhuman red cells, K–Cl cotransport responds to catecholamines and cyclic nucleotides. An inhibitor of the transporter itself is available.[50]

The activity of K–Cl cotransport is much greater in reticulocytes than mature red cells of hematologically normal people.[21,66,67] It is also greater in sickle and hemoglobin CC cells than in normal cells.[68–70] An important question is whether the increased K–Cl cotransport in these hemoglobinopathic red cells is related to their relative youth or whether it reflects an interaction between the mutant hemoglobin and the plasma membrane. In the case of sickle-cell disease the question seems unresolved. Even if K–Cl cotransport in hemoglobin SS cells were no more active than expected for a comparably young population of hemoglobin AA cells, triggering of this pathway by acidification could play a strong role in the pathogenesis of sickling. One can imagine that exposure of cells to the acid environment of an underperfused vascular bed might activate K–Cl cotransport, leading to salt and water loss from the cells, an increase in their MCHC, and an exponential tendency for their hemoglobin to gel on deoxygenation.[71] Incidentally, deoxygenation raises cytosolic free magnesium and reduces the activity of the K–Cl cotransporter.[33,66,72]

It is questionable, however, whether this red cell pathway ever functions *in vivo*. Conditions leading to its activation (hypotonicity, acidification, treatment with sulfhydryl reagents) would be lethal in the intact organism. The importance of K–Cl cotransport in the pathogenesis of crises and hemolysis in sickle-cell disease is therefore a matter of debate. In support of such a role, however, is the observation that the densest, most viscous fraction of sickle-cell blood contains some of the youngest cells. This would imply that some age-related ion transport system is instrumental in the genesis of vaso-occlusion.[43,73]

It is a different matter with hemoglobin C. In this condition the transporter is not only extremely active, perhaps out of proportion to the mean age of the red cell population, but the volume at which it becomes active is abnormally low. For present purposes the term "set point" is used to denote that cell volume (or density or water content) at which a transport pathway begins (or ceases) to function. The set point is usually determined by measuring flux through the pathway of interest as a function of cell volume. A swelling-activated pathway, such as K-Cl cotransport, would show no flux at volumes below the set point but would show an increasing flux as the cells were swollen above the set point. The dimensions of the set point are cell volume or water content or MCHC. When the water content of hemoglobin CC cells is raised (MCHC reduced) to levels found in normal red cells, K-Cl cotransport is activated and becomes downregulated only as the cells return to the dehydrated, cation-depleted state in which they normally circulate.[40,68,70]

In both SS and SC cells the brisk activation of K-Cl cotransport by cell swelling and acidification is counteracted by okadaic acid,[74] suggesting that the mutant hemoglobins exert their effects on the transporter via a kinase-phosphatase regulatory system.[75,76]

25.4.4. Na/H Exchange

Human red cells have the capacity to conduct Na/H exchange *in vitro* under conditions designed to stimulate that transporter maximally. It is unlikely that the experimental circumstances (acid pH) required to demonstrate Na/H exchange in human red cells could ever exist in the microcirculation of a living mammal. The exchanger as demonstrated *in vitro* is most active in the youngest red cells and diminishes with cell age at a slower rate than does K–Cl cotransport. The increased Na/H exchange found in sickle cells is probably explained by the young red cell population. It seems doubtful that this transporter plays a role in the pathogenesis of sickle-cell disease.[35,77]

25.4.5. Ca-Dependent K Conductance

It has been known for some time that sickle cells contain more calcium than normal and that sickling causes a net uptake of calcium. Most of the calcium is located in inside-out vesicles formed by endocytosis. Free calcium levels in sickle-cell cytoplasm are normal.[78–80]

Potassium channels with a conductance of about 20 pS can be reversibly activated in human red cells by raising the free calcium level at the cytosolic surface. The history and features of this so-called Gardos channel have been reviewed recently.[52] Intermittent elevations of free calcium during the life of a sickle cell might lead to potassium, chloride, and water depletion, not only via calcium-activated K channels but also because the opening of such channels would hyperpolarize the red cell, possibly causing an influx of protons, with resulting activation of the K–Cl cotransporter.[47]

The work of Murphy *et al.*[80] provides important evidence for such calcium transients in sickle cells. An impermeant reporter of calcium (FBAPTA) was endocytosed by sickle cells during an episode of deoxygenation, in the absence of external calcium. After reoxygenation and resickling in a medium containing calcium but no extracellular FBAPTA, a strong calcium signal was seen emanating from the previously endocytosed FBAPTA. The study was interpreted as showing that during sickling, some calcium must have entered the cells and been pumped into the vesicles. Such a transient increase in cytosolic free calcium could trigger a pulse of potassium loss and acidification and ultimately lead to cell dehydration,[43] as predicted by the model of Lew *et al.*[47]

The mechanism by which the sickle mutation causes calcium uptake is perhaps related to membrane distortion,[81] but this is a matter of conjecture.

Quite recently, Brugnara *et al.*[82] reported that K movements through the Ca-activated channel could be inhibited by clotrimazole at an IC_{50} in the nanomolar range. Clinical trials of this FDA-approved agent might clarify the importance of Ca-activated K channels in the pathogenesis of red cell dehydration in sickle-cell disease.

25.4.6. Nonspecific Cation Channels

Early observations of Tosteson *et al.*[83–85] showed that sickling caused a net, balanced exchange of Na and K across the membrane, although the mechanism of ion movements was not clear. Many investigators have confirmed the original observation.[86,87] Recent work has focused on the nature of the sickling-induced cation flux and the way in which the globin mutation might activate it.

The sickling-induced pathway is not inhibited by various identifying ligands such as ouabain or loop diuretics. It shows no selectivity among Na, K, Li, Rb, or Cs. Fluxes through it respond to imposed alterations in electrochemical driving force in a manner consistent with electrodiffusion.[35,88] It is not accessible to small inorganic nonelectrolytes such as mannitol.[89] It is inhibited by DIDS, but is ten times less sensitive to that reagent than is the anion exchanger, suggesting that the deoxygenation-induced nonspecific cation conductance in sickle cells is not mediated by band 3.[35,90]

Sickling also permeabilizes red cells to calcium[91] and magnesium[92] ions.

25.4.7. A Transiently Expressed Na-Dependent Anion Transporter

In their model for volume regulation, Lew *et al.*[47] postulate the existence of a Na-dependent anion entry pathway in reticulocytes that protects them from the dehydrative effects of the very active K-Cl cotransporter. For various reasons it seems unlikely that the pathway is conductive or that it is Na/H exchange coupled to Cl/HCO_3 exchange. The authors consider electroneutral [Na-Cl] cotransport[93] a possibility. Hemoglobin S might inactivate such a pathway and thereby contribute to the rapid dehydration of sickle-cell reticulocytes.

Another possibility, not considered in the model of Lew *et al.*,[47] is that Na gets smuggled into the cell in the guise of an anion: in any solution containing Na^+ and HCO_3^- there is a certain amount of the ion pair species $NaCO_3^-$, which can traverse the membrane via anion exchange on band 3.[52,94] One can utilize this pathway *in vitro* to change the ion and water content of red cells.[95] Deficiency of such a transport mechanism in sickle reticulocytes might lead to cell dehydration.

25.4.8. Relationships between the Mutant Globins and Membrane Transport

How might globin mutants interact with the membrane to alter its ion transport characteristics? Table 25.4 lists some suggested mechanisms. Cell age and heterogeneity may contribute to abnormal transport measurements in the hemoglobinopathies, but these factors are not directly related to effects of globin on membrane transporters. Most patients with sickle-cell disease have red cell inclusions because of functional hyposplenism, but hemoglobin AA homozygotes whose spleens have been removed manifest similar inclusions without red cell transport disorders.

An obvious difference between the hemoglobin variants under discussion is their charge. An anionic glutamyl side chain at the β_6 position in hemoglobin A is replaced by a neutral valine in hemoglobin S and a cationic lysine in hemoglobin C. Could this difference be the basis for the putative increase in K-Cl cotransport activity in cells containing the two mutant globins? The answer is no. Olivieri *et al.*[96] studied swelling-induced K–Cl cotransport in red cells containing a variety of mutant globins, some of which had amino acid substitutions identical to those of hemoglobins S and C but at loci remote from the 6 position on the βglobin chain. Others had charge substitutions near the β_6 locus. The conclusion from this study, which was appropriately controlled for cell age, is that the effect of the mutant globins on K–Cl cotransport was not related to the charge substitution *per se,* but rather to its location near the β_6 position.

There is evidence that hemoglobins S and C bind to the red cell membrane more avidly than does hemoglobin A,[97,98] but the binding site and the relation to transport function are not clear.

Of the transport abnormalities listed in Table 25.3, only the nonspecific cation leak seen in sickle cells is clearly related to deoxygenation. Membrane distortion is important in the genesis of this leak. It does not occur when cells are deoxygenated by gassing them with carbon monoxide, a ligand that simulates oxygen, keeps hemoglobin in the "oxy" conformation, and prevents sickling.[83] The degree of membrane distortion induced by the formation of deoxyhemoglobin S can be manipulated by varying the MCHC or water content of the cells *in vitro.* Dense, dehydrated SS cells on deoxygenation form multiple, independent hemoglobin nucleation sites, so that gelation of their contents does not lead to much membrane distortion.[39] Cells with a lower MCHC form aligned, long polymers that distort the membrane and in some cases lift the bilayer off the underlying spectrin–actin network.[99] Mohandas *et al.*[100] used solutions of varying tonicity to vary the MCHC of sickle cells and showed that the induction of the nonspecific cation leak could be inhibited by deoxygenating at high MCHC and accentuated by working at low MCHC, thus validating the concept that it is distortion of the membrane and not simply the presence of deoxyhemoglobin that leads to the permeability change. Induction of a reversible, balanced leak of Na and K in normal hemoglobin AA and oxygenated SS red cells can be achieved by subjecting them to shearing stress in a viscometer,[101,102] although the relationship of this type of deformation to that induced during sickling is not clear. These membrane distortion experiments do not explain the putative abnormalities in membrane transport associated with hemoglobin S that are not dependent on deoxygenation, nor do they address the basis of the altered set point volume for K–Cl cotransport in hemoglobin CC cells.

The membranes of sickle cells are especially vulnerable to oxidation, as evidenced by their content of lipid peroxides, oxidized thiol groups, and oxidized cytoskeletal components.[103–105] These lesions must be brought about by interaction of the mutant hemoglobin with oxygen. It is known that oxidative stress can greatly stimulate K–Cl cotransport in normal hemoglobin A red cells.[63,64] Perhaps this accounts for that portion of the K–Cl cotransport increase in sickle cells that cannot be simply attributed to the young cell population. Oxidation also makes red cells more vulnerable to shearing stress.[101,106]

Other sickle-cell membrane lesions include disruption of the membrane skeleton and its relationships with the lipid bilayer, plus loss of asymmetry of phospholipid distribution in the bilayer.[107,108] The relationship of these lesions to the transport abnormalities listed in Table 25.3 is not obvious.

A promising treatment for sickle-cell disease is hydroxyurea. Patients taking this drug produce red cells with an elevated fetal hemoglobin content. In one study average hemoglobin F levels in whole blood rose from 4 to 15% after 5 months of therapy.[38] Many individuals have had objective clinical responses to hydroxyurea. In some cases the benefit preceded the appearance of fetal hemoglobin and seemed more closely correlated with an increase in mean corpuscular volume. Red cells of hydroxyurea-treated sickle-cell patients are larger and more homogeneous. The very dense cells and the least dense cells no longer circulate (Figure 25.1). Whether these changes reflect the increased appearance of fetal hemoglobin or some other red cell membrane property is the subject of recent speculation.[38] It is remarkable that the increase in red cell volume (MCV) from 101 to 126 fl in patients treated with this drug was not associated with a change in MCHC or percent cell water (Figure 25.1): each cell contained more water,

Table 25.4 Consequences of Mutant Hemoglobinopathies that Might Give Rise to Differences in Ion Transport Measurements between Red Cells of Patients and Hemoglobin AA Controls

Altered cell property	Responsible hemoglobins
Younger red cell population	S and C
Heterogeneous red cell population	S and C
Failure of normal splenic processing of red cells	S
Changed isoelectric point of cell contents	S and C
Binding of globin to the membrane	S and C
Distortion of the membrane	S
Oxidation of cell membrane and cell contents	S
Disruption of the membrane skeleton	S
Disruption of the normal asymmetry of membrane phospholipids	S

more salt, and more protein. The transport features of these cells are under investigation.[38,109]

In conclusion, there are many differences in cation transport between normal and hemoglobinopathic red cells, some of which likely contribute to the clinical picture. In no case can a chain of causality be traced between the amino acid substitution and the alteration in ion permeability. Of the various transport abnormalities described, the ones most likely to be caused by the abnormal globin are the nonspecific conductive cation pathway triggered by sickling, the altered set point (see definition above) for K–Cl cotransport in red cells containing hemoglobin C, and the inferred failure of a transiently expressed sodium entry pathway in sickle reticulocytes. It would be important to discover whether the conductive pathway opened by sickling bears any relationship to the one that is unregulated in patients with band 7.2b deficiency. Another possibility is that the voltage-gated conductance described originally in red cells exposed to membrane depolarization is somehow involved in the sickling-induced leak.[24–27]

25.5. CONGENITAL ALTERATION OF UREA TRANSPORT

The Kidd blood group antigens, denoted Jk, are inherited as codominant alleles. People who are genotypically Jk^a/Jk^a, Jk^a/Jk^b, Jk^b/Jk^b, are phenotypically Jk(a+b−), Jk(a+b+), Jk(a−b+), respectively. A few people from around the Pacific basin lack Kidd antigens on their red cells. They are genotypically Jk_{null}/Jk_{null} and phenotypically Jk(a−b−).[110] Red cells from these individuals were noted to be unique in their capacity to resist lysis by 2 M urea.[111]

Fröhlich *et al.*[112] performed studies of urea, anion, water, and ethylene glycol transport in Jk(a−b−) red cells and found them normal in every respect except for a complete absence of the urea transporter characteristic of mammalian red cells.[113] Tracer fluxes of urea in the Jk(a−b−) cells were as little as 0.6% of normal. Urea moved across the membrane of such cells as if by simple nonionic diffusion through a bilayer, whereas in red cells containing Kidd antigens urea moves by saturable, facilitated diffusion. This discovery had several repercussions. It established the independence of urea and water transporters. It suggested experiments on urea transport in other tissues from Jk(a−b−) patients. Finally, it opened the way to test the hypothesis that the urea transporter and the Jk antigen are one and the same molecule: naturally occurring anti-Jk antibodies could be used to isolate a cDNA clone for the putative urea transporter from an expression library.[114]

The ability to concentrate urine is thought to depend in part on the presence of a facilitated diffusion system for urea transport in the kidney that has features in common with the red cell urea transporter. People with Jk(a−b−) red cells were found deficient in urinary concentrating ability to a degree consistent with a postulated deficiency in urea transport, but it was not clear whether the observations were related to slow equilibration of urea with red cells in the renal circulation or whether a renal tubular defect was also involved.[115]

Thus far the isolation of a cDNA for the Jk antigen/urea transporter has not been reported. Proof of the identity of the two molecules would depend on the ability to use the cDNA clone to express a facilitated diffusion system for urea in a cell system lacking one, as was recently done for the human red cell water channel.[116]

Another blood group variant leading to abnormal ion transport is the Rh_{null} phenotype. The mechanism of altered ion movements through the membrane of Rh_{null} cells is not known, and the participation of the protein bearing the Rh antigens in membrane transport is obscure. The subject was reviewed in the last edition of this book.[2]

25.6. MUTANTS OF BAND 3 THAT AFFECT ANION EXCHANGE

Band 3 is the principal integral membrane protein in red cells. There are 1.2 million copies of it per cell. It has a molecular mass of about 100 kDa and consists of a 911-amino-acid polypeptide chain that is glycosylated, palmitylated via a thioester bond, and has potential phosphorylation sites.[117]

The protein consists of two domains with distinct functions. The N-terminal portion (40 kDa) is relatively polar, faces the cytoplasm, and binds to proteins of the membrane skeleton, hemoglobin, and various enzymes. At least one phosphorylation site (Tyr8) may play a role in the control of glycolysis.[118] The C-terminal end (55 kDa) contains a relatively hydrophobic domain that is located in the lipid bilayer and subserves the function of anion exchange. Specifically, the red cell's ability to exchange chloride for bicarbonate across its membrane via band 3 makes hemoglobin available as a proton buffer during the changes in carbonic acid content that occur as blood circulates between the capillaries and the lungs. It was this transport function of the polypeptide that first led to its identification by Cabantchik and Rothstein,[119] who found that some disulfonic stilbene compounds bound covalently to a single externally facing red cell membrane protein and inhibited stoichiometrically the exchange of anions. The best of these reagents was 4,4′-diisothiocyano-2,2′-stilbene disulfonic acid (DIDS), and the DIDS binding site continues to be an important marker on the transmembranous, anion transporter portion of the band 3 molecule.

The primary structure of red cell band 3 has been inferred from cDNAs cloned from several species. Other cell types have analogous membrane transporters that mediate anion exchange. A recent review of the structure and function of band 3 has been published.[117]

Recently a group of individuals in Southeast Asia and Melanesia with hereditary ovalocytosis were found to have two structural mutations in red cell band 3.[120–122] One (Lys56 → Glu) is a common polymorphism found in 6–7% of random human blood samples, designated "band 3 Memphis." It is within the N-terminal cytoplasmic domain and has no known physiological or clinical consequences. The other is a deletion of amino acid residues 400–408, which in the normal polypeptide are predicted to be at the junction between the cytoplasmic and transmembrane domains. The mutant band 3 does not bind DIDS, and it lacks a polylactosaminyl oligosaccharide found on the normal gene product. Thus, deletion of 9 amino acids from the cytosolic domain of band 3 affects glycosylation and inhibitor binding on the exofacial portion of the mutant protein.[123] Heterozygotes with Southeastern Asian hereditary ovalocytosis have red cells that bind half the normal amount of DIDS per cell and transport anions at half the normal rate.[124] In addition, the cells are oval and more rigid than normal cells.[125] Heterozygosity for this red cell mutation is thought to confer resistance to malaria. Homozygosity is presumed incompatible with life.

A second variant of band 3 associated with altered transport is the so-called band 3-HT (HT denotes high transport rate), a point mutation in which the proline residue at position 868 is replaced by leucine. The substitution occurs at a site believed to be near the cytoplasmic boundary of the most C-terminal of the membrane-spanning segments of the protein.[126,127] The trait is presumed to be inherited as an autosomal recessive. The red cells of homozygotes have a maximum anion exchange rate of 1.5 to 2 times normal and a decreased number of high-affinity ankyrin binding sites. The number of H_2DIDS (4,4′-diisothiocyanato-1,2-diphenylethane-2,2′-disulfonic acid) binding sites is normal. The affected cells are acanthocytic or multispiculated in shape and have a normal survival *in vivo*. Homozygotes are not anemic, nor do they have an elevated reticulocyte count.

There are other band 3 mutations, some associated with spherocytosis or acanthocytosis, and some with abnormalities of the central nervous system.[117,128] Transport abnormalities in these variants have not been defined.

REFERENCES

1. Parker, J. C., Orringer, E. P., and McManus, T. J. (1978). Disorders of ion transport in red blood cells. In *Physiology of Membrane Disorders* (T. E. Andreoli, J. F. Hoffman, and D. D. Fanestil, eds.), Plenum Press, New York, pp. 773–800.
2. Parker, J. C., and Berkowitz, L. R. (1986). Genetic variants affecting the structure and function of the human red cell membrane. In *Pathology of Membrane Disorders* (T. E. Andreoli, J. F. Hoffman, D. D. Fanestil, and S. G. Schultz, eds.), Plenum Press, New York, pp. 785–813.
3. Palek, J., and Sahr, K. E. (1992). Mutations of the red blood cell membrane proteins: From clinical evaluation to detection of the underlying defect. *Blood* **80:**308–330.
4. Palek, J. (1993). Introduction: Red cell membrane proteins, their genes and mutations. *Semin. Hematol.* **30:**1–3.
5. Ives, H. E. (1989). Ion transport defects and hypertension. Where is the link? *Hypertension* **14:**590–597.
6. Lifton, R. P., Hunt, S. C., Williams, P. R., Pouyssegur, J., and Lalouel, J. M. (1991). Exclusion of the Na–H antiporter as a candidate gene in human essential hypertension. *Hypertension* **17:**8–14.
7. Rutherford, P. A., Thomas, T. H., and Wilkinson, R. (1992). Erythrocyte sodium–lithium countertransport: Clinically useful, pathophysiologically instructive, or just phenomenology? *Clin. Sci.* **82:**341–352.
8. Zarkowsky, H. S., Oski, F. A., Sha'afi, R., Shohet, S. B., and Nathan, D. G. (1968). Congenital hemolytic anemia with high sodium, low potassium red cells: 1. Studies of membrane permeability. *N. Engl. J. Med.* **278:**573–581.
9. Nathan, D. G., and Shohet, S. B. (1970). Erythrocyte transport defects and hemolytic anemia: "Hydrocytosis" and "dessiccytosis." *Semin. Hematol.* **7:**381–408.
10. Lande, W. M., and Mentzer, W. C. (1985). Haemolytic anaemia associated with increased cation permeability. *Clin. Haematol.* **14:** 89–103.
11. Wiley, J. S. (1977). Genetic abnormalities of cation transport in the human erythrocyte. In *Membrane Transport in Red Cells* (J. C. Ellory and V. L. Lew, eds.), Academic Press, New York, pp. 337–361.
12. Mentzer, W. C., Lubin, B. H., and Emmons, S. (1976). Correction of the permeability defect in hereditary stomatocytosis by dimethyl adipimidate. *N. Engl. J. Med.* **294:**1200–1204.
13. Mentzer, W. C., Lam, G. K. H., Lubin, B. H., Greenquist, A., Schrier, S. L., and Lande, W. (1978). Membrane effects of imidoesters in hereditary stomatocytosis. *J. Supramol. Struct.* **9:**275–288.
14. Lande, W. M., Thieman, P. V. W., and Mentzer, W. C. (1982). Missing band 7 membrane protein in two patients with high Na, low K erythrocytes. *J. Clin. Invest.* **70:**1273–1280.
15. Eber, S. W., Lande, W. M., Iarocci, T. A., Mentzer, W. C., Hohn, P., Wiley, J. S., and Schroter, W. (1989). Hereditary stomatocytosis: Consistent association with an integral membrane protein deficiency. *Br. J. Haematol.* **72:**452–455.
16. Hiebl-Dirschmied, C. M., Adolf, G. R., and Prohaska, R. (1991). Isolation and partial characterization of the human band 7 integral membrane protein. *Biochim. Biophys. Acta* **1065:**195–202.
17. Wang, D., Mentzer, W. C., Cameron, T., and Johnson, R. M. (1991). Purification of band 7.2b, a 31-kDa integral phosphoprotein absent in hereditary stomatocytosis. *J. Biol. Chem.* **266:**17826–17831.
18. Stewart, G. W., Hepworth-Jones, B. E., Keen, J.N., Dash, B. C. J., Argent, A. C., and Casimir, C. M. (1992). Isolation of cDNA coding for an ubiquitous membrane protein deficient in high Na, low K stomatocytic erythrocytes. *Blood* **79:**1593–1601.
19. Hiebl-Dirschmied, C. M., Entler, B., Glotzman, C., Maurer-Fogy, I., Stratowa, C., and Prohaska, R. (1991). Cloning and nucleotide sequence of cDNA encoding human erythrocyte band 7 integral membrane protein. *Biochim. Biophys. Acta* **1090:**123–124.
20. Morle, L., Pothier, B., Alloisio, N., Feo, C., Garay, R., Bost, M., and Delauney, J. (1989). Reduction of membrane band 7 and activation of volume stimulated [K–Cl]cotransport in a case of congenital stomatocytosis. *Br. J. Haematol.* **71:**141–146.
21. Hall, A. C., and Ellory, J. C. (1986). Evidence for the presence of volume-sensitive KCl transport in 'young' human red cells. *Biochim. Biophys. Acta* **858:**317–320.
22. Olivieri, O., Girelli, D., Vettore, L., Balercia, G., and Corrocher, R. (1992). A case of congenital dyserythropoietic anemia with stomatocytosis, reduced bands 7 & 8, and normal cation content. *Br. J. Haematol.* **80:**258–260.
23. Kracke, G. R., and Dunham, P. B. (1987). Effect of membrane potential in furosemide-inhibitable sodium influxes in human red blood cells. *J. Membr. Biol.* **98:**117–124.
24. Halperin, J. A., Brugnara, C., Tosteson, M. T., Van Ha, T., and Tosteson, D. C. (1989). Voltage-activated cation transport in human erythrocytes. *Am. J. Physiol.* **257:**C986–C996.
25. Halperin, J. A., Brugnara, C., Van Ha, T., and Tosteson, D. C. (1990). Voltage-activated cation permeability in high-potassium but not low-potassium red blood cells. *Am. J. Physiol.* **258:**C1169–C1172.
26. Christophersen, P., and Bennekou, P. (1991). Evidence of a voltage-gated, nonselective cation channel in the human red cell membrane. *Biochim. Biophys. Acta* **1065:**103–106.
27. Bennekou, P. (1993). The voltage-gated non-selective cation channel from human red cells is sensitive to acetylcholine. *Biochim. Biophys. Acta* **1147:**165–167.

28. Miller, C. (1991). 1990: Annus mirabilis of potassium channels. *Science* **252:**1092–1096.
29. DeLuise, M., and Flier, J. S. (1982). Functionally abnormal Na–K pump in erythrocytes of a morbidly obese patient. *J. Clin. Invest.* **69:**38–44.
30. Halperin, J. A., Brugnara, C., Kopin, A. S., Ingwall, J., and Tosteson, D. C. (1987). Properties of the Na–K pump in human red cells with increased number of pump sites. *J. Clin. Invest.* **80:**128–137.
31. Inaba, M., and Maeda, Y. (1986). Na,K-ATPase in dog red cells. Immunological identification and maturation-associated degradation by the proteolytic system. *J. Biol. Chem.* **261:**16099–16105.
32. Fujise, H., Yamada, I., Masuda, M., Miyazawa, Y., Ogawa, E., and Takahashi, R. (1991). Several cation transporters and volume regulation in high-K dog red blood cells. *Am. J. Physiol.* **260:**C589–C597.
33. Canessa, M. (1991). Red cell volume-related ion transport systems in hemoglobinopathies. *Hematol. Oncol. Clin. North Am.* **5:**495–516.
34. Moore, R. B. (1992). Pathophysiology of the sickle cell. In *Sickle Cell Disease. Pathophysiology, Diagnosis, and Management* (V. N. Mankad and R. B. Moore, eds.), Praeger Publishers, Westport, CT, pp. 44–104.
35. Joiner, C. H. (1993). Cation transport and volume regulation in sickle red blood cells. *Am. J. Physiol.* **264:**C251–C270.
36. Parker, J. C. (1971). Ouabain-insensitive effects of metabolism on ion and water content of dog red cells. *Am. J. Physiol.* **221:**338–342.
37. Lee, P., Kirk, R. G., and Hoffman, J. F. (1984). Interrelations among Na and K content, cell volume, and buoyant density in human red blood cell populations. *J. Membr. Biol.* **79:**119–126.
38. Orringer, E. P., Blythe, D. S. B., Johnson, A. E., Phillips, G., Dover, G., and Parker, J. C. (1991). Effects of hydroxyurea on hemoglobin F and water content in the red cells of dogs and of patients with sickle cell anemia. *Blood* **78:**212–216.
39. Eaton, W. A., and Hofrichter, J. (1990). Sickle cell hemoglobin polymerization. *Adv. Protein Chem.* **40:**63–279.
40. Murphy, J. R. (1968). Hemoglobin CC disease: Rheological properties of erythrocytes and abnormalities in cell water. *J Clin. Invest.* **47:**1483–1495.
41. Hoffman, J. F. (1958). On the relationship of certain erythrocyte characteristics to their physiological age. *J. Cell. Comp. Physiol.* **51:**415–423.
42. Piomelli, S., and Seaman, C. (1993). Mechanism of red cell aging: Relationship of cell density and cell age. *Am. J. Hematol.* **42:**46–52.
43. Bookchin, R. M., Ortiz, O. E., and Lew, V. L. (1991). Evidence for a direct reticulocyte origin of dense red cells in sickle cell anemia. *J. Clin. Invest.* **87:**113–124.
44. Fabry, M. E., Canessa, M., Romero, J., Lawrence, C., and Nagel, R. L. (1989). The unique density distribution of reticulocytes and ion transport properties in hemoglobin CC red cells. *Blood* **74:**261a.
45. Fabry, M. E., Romero, J. R., Buchanan, I. D., Suzuka, S. M., Stamatoyannopoulos, G., Nagel, R. L., and Canessa, M. (1991). Rapid increase in red blood cell density driven by [K–Cl]cotransport in a subset of sickle cell anemia reticulocytes and discocytes. *Blood* **78:**217–225.
46. Lawrence, C., Fabry, M. E., and Nagel, R. L. (1991). The unique red cell heterogeneity of SC disease: Crystal formation, dense reticulocytes, and unusual morphology. *Blood* **78:**2104–2112.
47. Lew, V. L., Freeman, C. J., Ortiz, O. E., and Bookchin, R. M. (1991). A mathematical model of the volume, pH, and ion content regulation in reticulocytes. Application to the pathophysiology of sickle cell dehydration. *J. Clin. Invest.* **87:**100–112.
48. Kaplan, J. (1989). Active transport of sodium and potassium. In *Red Blood Cell Membranes: Structure, Function, Clinical Implications* (P. Agre and J. C. Parker, eds.), Dekker, New York, pp. 455–480.
49. Haas, M. (1989). Properties and diversity of [Na-K-2Cl] cotransporters. *Annu. Rev. Physiol.* **51:**443–457.
50. Vitoux, D., Olivieri, O., Garay, R. P., Cragoe, E. J., Galacteros, F., and Beuzard, Y. (1989). Inhibition of K efflux and dehydration of sickle cells by [(dihydroindenyl)oxy] alkainoic acid, an inhibitor of the [K–Cl] cotransport system. *Proc. Natl. Acad. Sci. USA* **86:**4273–4276.
51. Lauf, P. K., Bauer, J., Adragna, N. C., Fujise, H., Zade-Oppen, A. M. M., Ryu, K. H., and Delpire, E. (1992). Erythrocyte [K–Cl] cotransport. Properties and regulation. *Am. J. Physiol.* **263:** C917–C932.
52. Parker, J. C., and Dunham, P. B. (1989). Passive cation movements. In *Red Blood Cell Membranes: Structure, Function, Clinical Implications* (P. Agre and J. C. Parker, eds.), Dekker, New York, pp. 507–561.
53. Brugnara, C., De Franceschi, L., and Alper, S. L. (1993). Ca-activated K transport in erythrocytes. Comparison of binding and transport inhibition by scorpion toxins. *J. Biol. Chem.* **268:**8760–8768.
54. Liu, D., and Hoffman, J. F. (1991). Interaction of charybdotoxin (ChTX) with Ca-activated K channels of human red blood cells. *Biophys. J.* **59:**639a.
55. Clark, M. R., Morrison, C. E., and Shohet, S. B. (1978). Monovalent cation transport in irreversibly sickled cells. *J. Clin. Invest.* **62:** 329–337.
56. Ortiz, O. E., Lew, V. L., and Bookchin, R. M. (1986). Calcium accumulated in sickle cell anemia red cells does not affect their potassium (^{86}Rb) flux components. *Blood* **67:**710–715.
57. Ortiz, O. E., Lew, V. L., and Bookchin, R. M. (1990). Deoxygenation permeabilizes sickle cell anemia red cells to magnesium and reverses its gradient in dense cells. *J. Physiol. (London)* **427:**211–226.
58. Clark, M. R., Guatelli, J. C., White, A. T., and Shohet, S. B. (1981). Study on the dehydrating effect of the red cell Na/K pump in nystatin-treated cells with varying Na and water contents. *Biochim. Biophys. Acta* **646:**422–432.
59. Izumo, H., Lear, S., Williams, M., Rosa, R., and Epstein, F. H. (1987). Sodium–potassium pump, ion fluxes, and cellular dehydration in sickle cell anemia. *J. Clin. Invest.* **79:**1621–1630.
60. Joiner, C. H., Platt, O. S., and Lux, S. E. (1986). Cation depletion by the sodium pump in red cells with pathological cation leaks: Sickle cells and xerocytes. *J. Clin. Invest.* **78:**1487–1496.
61. Parker, J. C., and Colclasure, G. C. (1992). Actions of thiocyanate and N-phenylmaleimide on volume-responsive Na and K transport in dog red cells. *Am. J. Physiol.* **262:**C418–C421.
62. Parker, J. C. (1993). Urea alters the set point for [K–Cl]cotransport, Na/H exchange, and Ca/Na exchange in dog red cells. *Am. J. Physiol.* **265:**C447–C452.
63. Orringer, E. P., and Parker, J. C. (1977). Selective increase of potassium permeability in red blood cells exposed to acetylphenylhydrazine. *Blood* **50:**1013–1020.
64. Orringer, E. P. (1984). Potassium leak pathways of the human erythrocyte. *Am. J. Hematol.* **16:**335–366.
65. Parker, J. C., McManus, T. J., Starke, L. C., and Gitelman, H. J. (1990). Co-ordinated regulation of Na/H exchange and [K–Cl] cotransport in dog red cells. *J. Gen. Physiol.* **96:**1141–1152.
66. Brugnara, C., and Tosteson, D. C. (1987). Cell volume, K transport, and cell density in human erythrocytes. *Am. J. Physiol.* **252:**C268–C276.
67. Canessa, M., Fabry, M. E., Blumenfeld, N., and Nagel, R. L. (1987). Volume-stimulated, Cl-dependent K efflux is highly expressed in young human red cells containing normal hemoglobin or HbS. *J. Membr. Biol.* **97:**97–105.
68. Brugnara, C., Kopin, A. S., Bunn, H. F., and Tosteson, D. C. (1985). Regulation of cation content and cell volume in erythrocytes from patients with homozygous hemoglobin C disease. *J. Clin. Invest.* **75:**1608–1617.
69. Brugnara, C., Bunn, H. F., and Tosteson, D. C. (1986). Regulation of erythrocyte cation and water content in sickle cell anemia. *Science* **232:**388–390.

70. Brugnara, C. (1989). Characteristics of volume- and chloride-dependent K transport in human erythrocytes homozygous for hemoglobin C. *J. Membr. Biol.* **111:**69–81.
71. Brugnara, C., Van Ha, T., and Tosteson, D. C. (1989). Acid pH induces formation of dense cells in sickle erythrocytes. *Blood* **74:** 487–495.
72. Canessa, M., Fabry, M .E., and Nagel, R. L. (1987). Deoxygenation inhibits the volume-stimulated, Cl-dependent K efflux in SS and young AA cells: A cytosolic Mg modulation. *Blood* **70:**1861–1866.
73. Bertles, J. F., and Milner, P. F. A. (1968). Irreversibly sickled erythrocytes. A consequence of the heterogeneous distribution of hemoglobin types in sickle cell anemia. *J. Clin. Invest.* **47:**1731–1741.
74. Orringer, E. P., Brockenbrough, J. S., Whitney, J. A., Glosson, P. S., and Parker, J. C. (1991). Okadaic acid inhibits activation of [K–Cl] cotransport in red cells containing hemoglobins S and C. *Am. J. Physiol.* **261:**C591–C593.
75. Kaji, D., and Tsukitani, Y. (1991). Role of protein phosphatase in activation of [K–Cl]cotransport in human erythrocytes. *Am. J. Physiol.* **260:**C178–C182.
76. Jennings, M. L., and Schulz, R. K. (1991). Okadaic acid inhibition of KCl cotransport. Evidence that protein dephosphorylation is necessary for activation of transport by either cell swelling or N-ethylmaleimide. *J. Gen. Physiol.* **97:**799–817.
77. Canessa, M., Fabry, M. E., Suzuka, S. M., Morgan, K., and Nagel, R. L. (1990). Na/H exchange is increased in sickle cell anemia and young normal red cells. *J. Membr. Biol.* **116:**107–115.
78. Eaton, J. W., Skelton, T., Swofford, M., Kolpin, C., and Jacob, H. S. (1973). Elevated erythrocyte calcium in sickle cell disease. *Nature* **246:**105.
79. Lew, V. L., Hockaday, A., Sepulveda, M. I., Somlyo, A. P., Somlyo, A. V., Ortiz, O. E., and Bookchin, R. M. (1985). Compartmentation of sickle-cell calcium in endocytic inside-out vesicles. *Nature* **315:** 586–588.
80. Murphy, E., Berkowitz, L. R., Orringer, E. P., Levy, S. A., Gabel, S. A., and London, R. E. (1987). Cytosolic free calcium levels in sickle red blood cells. *Blood* **69:**1469–1474.
81. Larsen, F. L., Katz, S., Roufogalis, B. D., and Brooks, D. E. (1981). Physiological shear stresses enhance the Ca permeability of human erythrocytes. *Nature* **294:**667–668.
82. Brugnara, C., De Franceschi, L., Armsby, C. C., and Alper, S. L. (1993). Inhibition of Ca-dependent K transport and cell dehydration by clotrimazole and other imidazole derivatives. Abstracts, 18th Annual Meeting of the National Sickle Cell Disease Program, Philadelphia.
83. Tosteson, D. C., Shea, E., and Darling, R. C. (1952). Potassium and sodium of red blood cells in sickle cell anemia. *J. Clin. Invest.* **31:**406–411.
84. Tosteson, D. C. (1955). The effects of sickling on ion transport. II. The effect of sickling on sodium and cesium transport. *J. Gen. Physiol.* **39:**53–67.
85. Tosteson, D. C., Carlson, E., and Dunham, E. T. (1955). The effects of sickling on ion transport. I. Effect of sickling on potassium transport. *J. Gen. Physiol.* **39:**31–63.
86. Berkowitz, L. R., and Orringer, E. P. (1985). Passive sodium and potassium movements in sickle cell erythrocytes. *Am. J. Physiol.* **249:**C208–C214.
87. Joiner, C. H., Dew, A., and Ge, D. L. (1988). Deoxygenation-induced fluxes in sickle cells. I. Relationship between net potassium efflux and net sodium influx. *Blood Cells* **13:**339–348.
88. Joiner, C. H., Morris, C. L., and Cooper, E. S. (1993). Deoxygenation-induced cation fluxes in sickle cells. III. Cation selectivity and responses to pH and membrane potential. *Am. J. Physiol.* **264:** C734–C744.
89. Clark, M. R., and Rossi, M. E. (1990). Permeability characteristics of deoxygenated sickle cells. *Blood* **76:**2139–2145.
90. Joiner, C. H. (1990). Deoxygenation-induced fluxes in sickle cells. II. Inhibition by stilbene disulfonates. *Blood* **76:**212–220.
91. Rhoda, M. D., Apova, M., Beuzard, Y., and Giraud, F. (1990). Ca permeability in deoxygenated sickle cells. *Blood* **75:**2453–2458.
92. Ortiz, O. E., Lew, V. L., and Bookchin, R. M. (1990). Deoxygenation permeabilizes sickle cell anemia red cells to magnesium and reverses its gradient in the dense cells. *J. Physiol. (London)* **427:** 211–226.
93. Gamba, G., Saltzberg, S. N., Lombardi, M., Miyanoshita, A., Lytton, J., Hediger, H. A., Brenner, B. M., and Hebert, S. C. (1993). Primary structure and functional expression of a cDNA encoding the thiazide-sensitive electroneutral sodium–chloride cotransporter. *Proc. Natl. Acad. Sci. USA* **90:**2749–2753.
94. Funder, J., Tosteson, D. C., and Wieth, J. O. (1978). Effects of bicarbonate on lithium transport in human red cells. *J. Gen. Physiol.* **71:**721–746.
95. Orringer, E. P., Roer, M. E. S., and Parker, J. C. (1980). Cell density profile as a measure of erythrocyte hydration. Therapeutic alteration of salt and water content in normal and SS red blood cells. *Blood Cells* **6:**345–353.
96. Olivieri, O., Vitoux, D., Galacteros, F., Bachir, D., Blouquit, Y., Beuzard, Y., and Brugnara, C. (1992). Hemoglobin variants and activity of the [K–Cl]cotransport system in human erythrocytes. *Blood* **79:**793–797.
97. Shaklai, N., Sharma, V. S., and Ranney, H. M. (1981). Interaction of sickle cell hemoglobin with erythrocyte membranes. *Proc. Natl. Acad. Sci. USA* **78:**65–68.
98. Reiss, G., Ranney, H. M., and Shaklai, N. (1982). Association of hemoglobin C with erythrocyte ghosts. *J. Clin. Invest.* **70:**946–952.
99. Liu, S., Derick, L. H., Zhai, S., and Palek, J. (1991). Uncoupling of the spectrin-based skeleton from the lipid bilayer in sickled red cells. *Science* **252:**574–576.
100. Mohandas, N., Rossi, M. E., and Clark, M. R. (1986). Association between morphologic distortion of sickle cells and deoxygenation-induced cation permeability increase. *Blood* **68:**450–454.
101. Ney, P. A., Christopher, M. M., and Hebbel, R. P. (1990). Synergistic effects of oxidation and deformation on erythrocyte monovalent cation leak. *Blood* **75:**1192–1198.
102. Sugihara, T. S., and Hebbel, R. P. (1992). Exaggerated cation leak from oxygenated sickle red blood cells during deformation: Evidence for a unique leak pathway. *Blood* **80:**2374–2378.
103. Jain, S. K., and Shohet, S. B. (1984). A novel phospholipid in irreversibly sickled cells: Evidence for in vivo peroxidative membrane damage in sickle cell disease. *Blood* **63:**362–367.
104. Rank, B. H., Carlson, J., and Hebbel, R. P. (1985). Abnormal redox status of membrane protein thiols in sickle erythrocytes. *J. Clin. Invest.* **75:**1531–1537.
105. Schwartz, R. S., Rybicki, A. C., Heath, R. H., and Lubin, B. H. (1987). Protein 4.1 in sickle erythrocytes. Evidence for oxidative damage. *J. Biol. Chem.* **262:**15666–15669.
106. Hebbel, R. P., and Mohandas, N. (1991). Reversible deformation-dependent erythrocyte cation leak. Extreme sensitivity conferred by minimal peroxidation. *Biophys. J.* **60:**712–715.
107. Chiu, D. T. Y., Lubin, B. H., and Shohet, S. B. (1979). Erythrocyte membrane lipid reorganization during the sickling process. *Br. J. Haematol.* **41:**223–224.
108. Blumenfeld, N., Sachowski, A., Galacteros, F., Beuzard, Y., and Devaux, P. F. (1991). Transmembrane mobility of phospholipids in sickle erythrocytes. Effect of deoxygenation on diffusion and asymmetry. *Blood* **77:**849–854.
109. Orringer, E. P., and Parker, J. C. (1992). Hydroxyurea and sickle cell disease. *Hematol. Pathol.* **6:**171–178.
110. Mollison, P. L. (1983). *Blood Transfusion in Clinical Medicine,* 7th ed., Blackwell, Oxford, p. 409.

111. Heaton, D. C., and McLaughlin, K. (1982). Jk(a−b−) red blood cells resist urea lysis. *Transfusion* **22:**70–71.
112. Fröhlich, O., Macey, R. I., Edwards-Moulds, J., Gargus, J. J., and Gunn, R. B. (1991). Urea transport deficiency in Jk(a−b−) erythrocytes. *Am. J. Physiol.* **260:**C778–C783.
113. Macey, R. I. (1984). Transport of water and urea in red blood cells. *Am. J. Physiol.* **246:**C195–C203.
114. Gargus, J. J., and Mitas, M. (1988). Physiological processes revealed through an analysis of inborn errors. *Am. J. Physiol.* **255:** F1047–F1058.
115. Sands, J. M., Gargas, J. J., Frohlich, O., Gunn, R. B., and Kokko, J. P. (1992). Urinary concentrating ability in patients with Jk(a−b−) blood type who lack carrier-mediated urea transport. *J. Am. Soc. Nephrol*. **2:**1689–1696.
116. Preston, G. M., Carroll, T. P., Guggino, W. B., and Agre, P. (1992). Appearance of water channels in Xenopus oocytes expressing red cell CHIP28 protein. *Science* **256:**385–387.
117. Tanner, M. J. A. (1993). Molecular and cellular biology of the erythrocyte anion exchanger (AE1). *Semin. Hematol.* **30:**34–57.
118. Harrison, M. L., Rathinavelu, P., Arese, P., Geahlen, Rp. L., and Low, P. S. (1991). Role of band 3 tyrosine phosphorylation in the regulation of erythrocyte glycolysis. *J. Biol. Chem.* **266:**4106–4111.
119. Cabantchik, Z. I., and Rothstein, A. (1974). Membrane proteins related to anion permeability of human red blood cells. I. Localization of disulfonic stilbene binding sites in proteins involved in permeation. *J. Membr. Biol.* **15:**207–226.
120. Liu, S. C., Zhai, S., Palek, J., Golan, D. E., Amato, D., Hassan, K., Nurse, G. T., Babona, D., Coetzer, T., and Jarolim, P. (1990). Molecular defect of the band 3 protein in Southeast Asian ovalocytosis. *N. Engl. J. Med.* **323:**1530–1538.
121. Tanner, M. J. A., Bruce, L., Martin, P. G., Rearden, D. M., and Jones, G. L. (1991). Melanesian hereditary ovalocytes have a deletion in red cell band 3. *Blood* **78:**2785–2786.
122. Jarolim, P., Palek, J., Amato, D., Hassan, K., Sapak, P., Nurse, G. T., Rubin, H. L., Zhai, S., Sahr, K. E., and Liu, S. C. (1991). Deletion in band 3 gene in malaria-resistant Southeast Asian ovalocytosis. *Proc. Natl. Acad. Sci. USA* **88:**11022–11026.
123. Sarabia, V. E., Casey, J. R., and Reithmeier, R. A. F. (1993). Molecular characterization of the band 3 protein from Southeastern Asian ovalocytosis. *J. Biol. Chem.* **268:**10676–10680.
124. Schofield, A. E., Reardon, D. M., and Tanner, M. J. A. (1992). Defective anion transport activity of the abnormal band 3 in hereditary ovalocytic red blood cells. *Nature* **355:**836–838.
125. Mohandas, N., Lie-injo, L. E., Friedman, M., and Mak, J. W. (1984). Rigid membranes of Malayan ovalocytes. A likely genetic barrier against malaria. *Blood* **63:**1385–1392.
126. Kay, M. M. B., Bosman, G. J. C. G. M., and Lawrence, C. (1988). Functional topography of band 3: Specific structural alteration linked to functional aberrations in human erythrocytes. *Proc. Natl. Acad. Sci. USA* **85:**492–496.
127. Bruce, L. J., Kay, M. M. B., Lawrence, C., and Tanner, M. J. (1993). Band 3HT, a human red-cell variant associated with acanthocytosis and increased anion transport, carries the mutation of Pro-868 →Leu in the membrane domain of band 3. *Biochem. J. 293:317–320.*
128. Kay, M. M. B. (1992). Senescent cell antigen and band 3 in aging and disease. In *Progress in Cell Research* (E. Bamberg and H. Passow, eds.), Elsevier, Amsterdam, Volume 2, pp. 245–250.

26

Disorders of Biliary Secretion

Piotr Zimniak and Roger Lester

26.1. INTRODUCTION

More than 0.5 liter of bile is produced per day in humans. The chemical composition of bile is complex. In addition to inorganic electrolytes, bile contains high (millimolar) concentrations of organic compounds, including glutathione, dipeptides, and amino acids (Table 26.1). While the above compounds are freely water-soluble, equally high concentrations of bilirubin, cholesterol, and phospholipids are present in bile. The latter material is kept in a micellar and/or vesicular form by another major biliary component, the bile acids. The physical chemistry of the ternary bile acid–cholesterol–phospholipid system is beyond the scope of the present chapter; it should be mentioned, however, that an imbalance in the secretion of these three components sets the stage for biliary stone formation. Finally, bile contains proteins, e.g., IgA. Except for the inorganic electrolytes, all of the compounds mentioned above are present in bile at concentrations that significantly exceed their concentrations in plasma. This indicates that biliary secretion is an active process. The latter conclusion is also consistent with the generally accepted model of bile formation. It is envisioned that the active transport of solutes into the canalicular space creates an osmotic force able to draw water, by a trans- or paracellular pathway, from blood into the canaliculus, leading to bile flow.

The physiological functions of bile formation and bile flow are manifold. Cholesterol can be excreted from the organism only through bile, either as the free compound or after metabolism of bile acids. Thus, bile formation is just as essential for maintaining cholesterol homeostasis as the intake and metabolism of cholesterol, aspects that have received much more attention. Bile acids, in addition to being end products of cholesterol catabolism, are essential for solubilization of both endogenous (cholesterol and lecithin) and dietary fat; the latter cannot be digested without the aid of bile acids. Finally, bile flow has important toxicological and pharmacological ramifications. The upper limit of molecular mass for renal excretion is less than 500 Da. Larger molecules, as well as a considerable fraction of smaller compounds of both endogenous and exogenous origin, are secreted into bile, often after preceding oxidative and/or conjugative metabolism by hepatic phase I and phase II detoxifying enzymes, respectively. For instance, the majority of bilirubin is excreted, in the form of a diglucuronide, by the biliary route, and disturbances in any step of this process result in hyperbilirubinemia and the associated clinical manifestations. Heavy metals may serve as an example of a toxic xenobiotic excreted into bile. Thus, an adequate bile flow not only is essential for liver metabolism, but is an integral part of the physiology of the organism as a whole.

The physiology and biochemistry of biliary secretion have continued to form the focus of intense investigation by numerous laboratories in recent years, and the available information and extent of understanding in many cases have reached the molecular level. The considerable effort devoted to the study of this process is related both to its intrinsic mechanistic interest, relevant to fundamental principles of bioenergetics and membrane biochemistry, and to the serious clinical consequences resulting from its disorders, e.g., gallstone formation or cholestasis. Numerous recent reviews cover various aspects of biliary secretion.[1–10] Therefore, the present chapter is not intended to provide an exhaustive coverage of the field. Rather, we will concentrate on an update summarizing the current views of the molecular transport processes relevant to and underlying biliary secretion. Furthermore, the importance of analyzing the pathophysiological consequences of blocks in specific transport processes for the understanding of the mechanisms and driving forces of biliary secretion will be emphasized, using the Dubin–Johnson syndrome and its animal models as an illustration.

26.2. BILIARY SECRETION

In spite of its more complex microanatomy, the liver is topologically equivalent to a secretory epithelium, and as such catalyzes vectorial transport of solutes. While in a typical epithelium the transport occurs across a layer of cells sealed by tight junctions, the transcellular transport in the liver is from either side of the cell plate, i.e., from the sinusoid, to spaces between the cells which are surrounded by the canalicular membranes of two adjoining hepatocytes (Figure 26.1); the

Piotr Zimniak • Departments of Internal Medicine and of Biochemistry and Molecular Biology, University of Arkansas for Medical Sciences, and McClellan VA Hospital, Little Rock, Arkansas 72205. **Roger Lester** • Department of Internal Medicine, University of Arkansas for Medical Sciences, and McClellan VA Hospital, Little Rock, Arkansas 72205.

Molecular Biology of Membrane Transport Disorders, edited by S.G. Schultz *et al.* Plenum Press, New York, 1996

Table 26.1 Approximate Composition of Human Hepatic Bile[7,8]

Component	Concentration (mM)
Inorganic electrolytes	
Na^+	130–160
K^+	3–7
Ca^{2+}	1.2–4.8
Mg^{2+}	1.4–3
Cl^-	100–130
HCO_3^-	17–55
Freely soluble organic compounds	
Glutathione (GSH)	3–5
Amino acids	1–2.5
Bile acids	2–45
Lipids (total concentrations)	
Phospholipids (mainly lecithin)	0.5–11
Cholesterol	2–8
Bilirubin and other bile pigments	0.3–3
Proteins	0.3–3 mg/ml

canalicular membrane corresponds to the apical pole of a typical epithelial cell. For substances taken up by the liver from the circulation, three consecutive transport events occur and have to be considered: sinusoidal uptake, transcellular transport, and canalicular secretion. Kinetic analysis of biliary transport has shown that in most cases the canalicular secretion is rate-limiting for the overall process.

The driving forces for biliary secretion across the canalicular membrane differ from compound to compound. For example, reduced glutathione is present in the cytoplasm of hepatocytes at a concentration that exceeds that in bile. Therefore, GSH can leave the cell by a facilitated diffusion process down its concentration gradient. Two distinct transport systems for GSH have been identified.[11–13] The transport process is augmented by the membrane potential (negative inside) across the canalicular membrane; indeed, the transporters are electrophoretic.[14] The membrane potential appears to determine the distribution of Cl^- across the canalicular membrane; a high-conductance Cl^- channel is present in the membrane. The significant Na^+ and perhaps also K^+ gradients, originally formed by the basolateral Na^+,K^+-ATPase (the canalicular membrane is devoid of this enzyme), are utilized for the reabsorption of a number of amino acids from bile. In the case of bile acids, the monomeric biliary concentration is significantly higher than that within the cell; an additional driving force is thus required for secretion. An electrophoretic transport system specific for bile acids has been identified in the canalicular membrane.[15] The transporter has been further characterized as a 100-kDa protein, and has been purified and functionally reconstituted.[16–18] It has been pointed out,[19] however, that the membrane potential of the canalicular membrane, approx. −35 mV, would be thermodynamically unable to sustain the observed concentration gradient of bile acids. According to the Nernst equation, a membrane potential of −35 mV is in equilibrium with an approx. 4-fold gradient of a monovalent ion, while the observed gradient is at least 100-fold. The pH difference between the cytoplasm and bile is minimal,[6] and no coupling of the canalicular bile acid transport to gradients of other ions, e.g., Na^+ or K^+, could be demonstrated. This apparent contradiction has been resolved by the discovery that part of the canalicular bile acid transport is directly driven by the free energy of ATP hydrolysis.[20–23]

The preceding brief discussion indicates that the transport processes that lead to bile formation are catalyzed by a variety of mechanistically dissimilar transporters. Ion channels, electrophoretic uniporters, symporters, and antiporters, and primary ATPases are represented.[6] In addition, vesicular transcellular transport has been demonstrated, at least for bile acids. For many of the processes the transport proteins have been identified and structurally characterized. Selected canalicular transporters are shown schematically in Figure 26.2. It is striking that no less than four distinct primary ATPases have been found in the canalicular membrane, even though not all of them have been assigned specific functions, and not all known ATP-dependent transport processes have been matched with a catalytic protein. The presence of active transporters is consistent with the role of biliary secretion not only in supplying bile constituents, but also in driving bile flow itself. The fact remains, however, that the canalicular membrane of the hepatocyte is devoid of ATPases typical for other plasma membranes, such as the Na^+,K^+-ATPase, H^+-ATPase, or Ca^{2+}-ATPase. Rather, the membrane appears to contain ATPases specific for the transport of organic substances. To

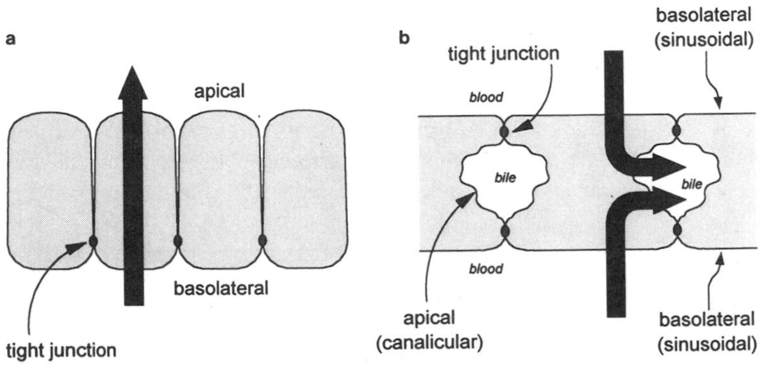

Figure 26.1. Comparison of the topology of a secretory epithelium (a) and a hepatocyte cell plate (b). The predominant direction of solute flow is shown with heavy arrows.

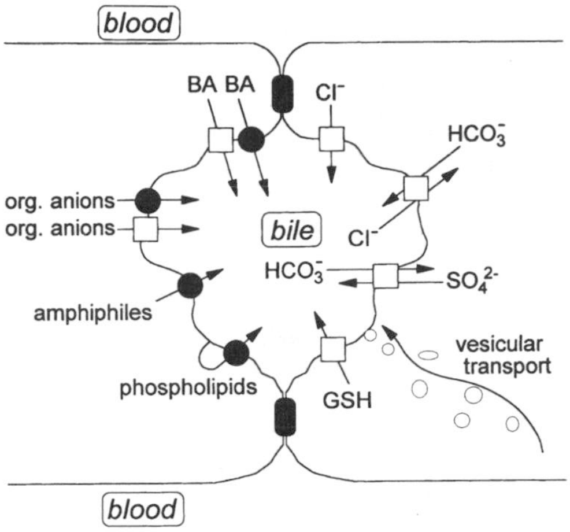

Figure 26.2. Schematic depiction of hepatic canalicular transport systems. ATP-dependent transporters are shown as closed circles, other transporters as open squares.

put this finding into a proper perspective, a short discussion of active transport processes is necessary.

26.3. ACTIVE TRANSPORT

The transport of solutes against their electrochemical gradient requires the expenditure of free energy. In biological systems, the free energy is originally derived from light or, in nonphotosynthesizing cells, from respiration (Figure 26.3). Both processes lead to the formation of an electrochemical proton gradient (proton-motive force, $\Delta\mu_{H}+$); in eukaryotic cells, this occurs at the membranes of specialized organelles, namely, chloroplasts and mitochondria. Both components of $\Delta\mu_{H}+$, i.e., the chemical concentration gradient of H^+ and the electrical membrane potential, can be used directly to drive transport of various solutes across the membrane of the above organelles. Even though $\Delta\mu_{H}+$ is delocalized along the surface of the membrane at which it was formed, e.g., the inner mitochondrial membrane, it is not available at other cellular membranes, most notably the cellular plasma membrane which catalyzes transport processes essential for the survival

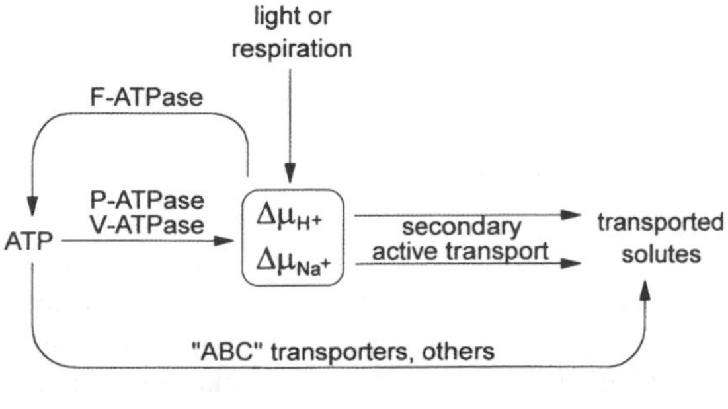

Figure 26.3. Cellular flow of free energy and its use for transport processes.

of the cell. Therefore, $\Delta\mu_{H}+$ is converted into a chemical intermediate, ATP, by a specialized ATPase (or, according to the physiological direction of its reaction, ATP synthetase). ATP is free to diffuse in the cytoplasm, and its free energy is converted back into an electrochemical cation gradient at membranes that need to be "energized" to carry out transport. In vacuolar membranes, which in mammalian cells include but are not limited to those of endosomes, lysosomes, and secretory vesicles, ATP is used to re-create $\Delta\mu_{H}+$ by V-ATPases.[24] V-ATPases are structurally related to the mitochondrial F-ATPase, although marked differences between the two classes exist (Figure 26.4). At the plasma membrane, ATP is used to create a Na^+ rather than H^+ electrochemical gradient. The Na^+,K^+-ATPase that is involved belongs to a distinct class, the P-ATPases, which differ from the F- and V-ATPases not only in their structures, but also in the reaction mechanism: P-ATPases go in their reaction cycle through a covalent aspartyl-phosphate high-energy intermediate absent in the F- and V-type enzymes (Figure 26.4). The electrochemical Na^+ gradient can then be used to drive transport processes.

In all of the above cases, the transport of organic compounds, e.g., amino acids or glucose into cells and end products of metabolism out of cells or organelles, is mediated by specific transport proteins. Energetically, these proteins can function in a variety of ways. In the simplest case, *facilitated diffusion,* the protein provides a pathway for the solute to cross the membrane, either by a channel mechanism or by a *cis*-binding/conformational change/*trans*-release reaction cycle. The solute then equilibrates across the membrane according to its concentration gradient. If an electrical membrane potential exists across the membrane and the solute is charged, the solute may accumulate against its concentration gradient until the free energy of the concentration difference reaches equilibrium with the membrane potential. Mechanistically, the transport protein acts as a *uniporter;* energetically, the process can be characterized as *secondary active transport.* Another example of a secondary active transport is the uptake of glucose or bile acids coupled with the uptake of Na^+ across the plasma membrane, i.e., a *symport* mechanism. The coupling is ensured by the transport protein which has to bind both the solute and Na^+ prior to the conformational change that leads to the release of both compounds on the other side of the membrane. If the Na^+ gradient is greater than that of the solute, the latter can be accumulated against its own concentration gradient. The mechanism of an *antiporter* is energetically similar, although the solute and the coupled ion move in opposite directions. In all cases, secondary active transporters make use of preexisting electrochemical gradients of ions, usually Na^+ or H^+. The formation of the latter gradients, e.g., Na^+ gradient by the Na^+,K^+-ATPase, requires the direct use of ATP and is termed *primary active transport.* The relationship between primary and secondary active transporters is schematically depicted in Figure 26.5.

The model described above, consisting of usually a single primary transporting ATPase creating an electrochemical gradient of an inorganic cation which then serves to drive a number of secondary transporters, has attained near-dogma

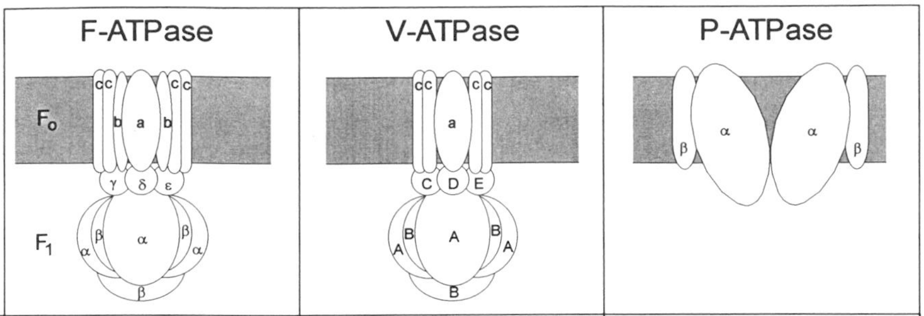

	F-ATPase		V-ATPase		P-ATPase	
total size	500 kDa		500 kDa		200-300 kDa	
subunits	α_3	55 kDa each	B_3	57 kDa each	α_2	110 kDa each
	β_3	50 kDa each	A_3	72 kDa each	(β_2	55 kDa each)
	γ	31 kDa	C	44 kDa		(β subunits are present
	δ	19 kDa	D	34 kDa		in Na^+,K^+-ATPase but
	ε	15 kDa	E	26 kDa		not in other P-ATPases)
	a	30 kDa	$a_?$	20 kDa each		
	b_2	17kDa each				
	$c_{10\text{-}12}$	8 kDa each	c_6	16 kDa each		
substrates	H^+, Na^+		H^+		H^+, Na^+, K^+, Ca^{2+}	
physiol. direction	ATP synthesis		ATP hydrolysis		ATP hydrolysis	
selected inhibitors	oligomycin, azide		NEM		vanadate	
phosphorylated intermediate	no		no		yes	

Figure 26.4. The three classes of ion-motive ATPases. The structures are highly schematized. Selected functional characteristics and typical subunit composition are shown.

status. The model, which has its roots in Mitchell's chemiosmotic theory, is elegant in its simplicity, and has proven to have considerable predictive power regarding new transport phenomena. Another factor that contributed greatly to the wide acceptance of the model was the considerable similarity of the primary ATPases. New ATPases that have been discovered could be always assigned to one of the few classes listed in Figure 26.4, even though they may have functions as dissimilar as gastric acid production, neurotransmitter release, or photosynthetic ATP production. In fact, a recent conference on ion-translocating ATPases has had its focus on the similarities of the enzymes.[25] As is frequently the case, however, the success of a paradigm does not necessarily mean that it is complete. Of the at least three and probably four transporting ATPases of the canalicular membrane of the hepatocyte, none belongs to the classes listed in Figure 26.4. Moreover, all of these ATPases transport directly organic molecules, thus circumventing the need for an intermediate electrochemical gradient of an inorganic ion. While these findings do not invalidate the "central dogma" of membrane transport, they add a new aspect to it.

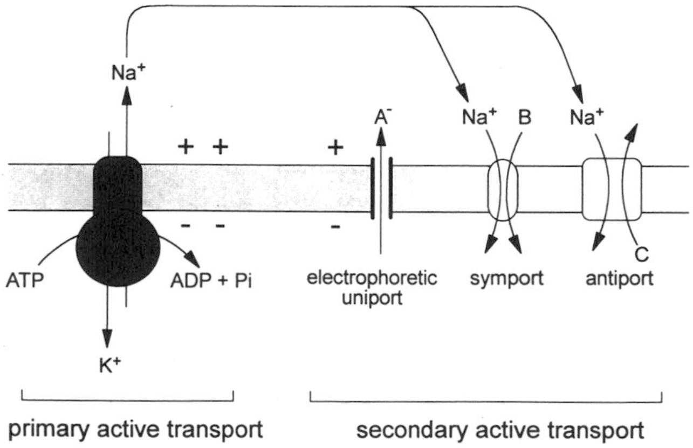

Figure 26.5. Energetic interdependence between a primary ion-motive ATPase and secondary active transporters that use the membrane potential and/or the ion gradient created by the ATPase for the transport of other solutes.

26.4. F-, V-, AND P-ATPASES

A brief description of the properties of the F-, V-, and P-type ATPases will be needed to contrast these enzymes, and their mode of action, with the organic compound-transporting primary ATPases present in the canalicular membrane of hepatocytes, and which will be discussed in the next section.

26.4.1. F-ATPases

F-type ATPases are found in eubacteria as well as in eukaryotic chloroplasts and mitochondria (Figure 26.6). These are the only ATPases that function physiologically in the direction of ATP synthesis, using the proton-motive force generated by photosynthesis or oxidation of organic substrates, and provide the majority of cellular ATP. F-ATPases consist of two segments, each containing multiple subunits. The F_1 segment is a peripheral membrane protein complex which can be removed from the membrane under retention of ATPase activity uncoupled from H^+ transport; the transmembranous, integral F_0 segment is responsible for H^+ transport but has no ATPase activity. The catalytic active sites are located on each of the three β subunits within the F_1 segment (compare Figure 26.4), with the α subunits having regulatory functions. The resulting three active sites of the enzyme do not function independently, even though a single, isolated αβ complex retains ATPase activity. The reason for the cooperativity between the three αβ moieties is thought to be the coupling to H^+ flow. Each of the αβ complexes can bind ADP and P_i, and the free energy of binding is used to form the covalent bond between these substrates, i.e., to synthesize ATP. However, the ATP remains tightly bound to the protein. The free energy of the flow of three protons through the F_0 segment is necessary to cause a conformational change in F_1 which ejects the ATP. The energy of H^+ translocation is transmitted to F_1 through the subunits linking F_0 and F_1, i.e., γ and δ. It is thought that, at any given time, only one of the three αβ pairs can be in contact with the γ subunit and thus be catalytically active; during that time, the other αβ pairs assume a conformation that promotes the binding of ADP and P_i and initiates another catalytic cycle.

Even though the mechanistic details of ATP synthesis by the F-ATPase, and especially of the coupling of H^+ flow with ATP synthesis, are not known, it is obvious that the enzyme is highly specific and adapted to its function. No ions other than H^+ support ATP synthesis by eukaryotic F-ATPases, and the system is tightly coupled, i.e., has an invariant H^+/ATP stoichiometry of 3.[26] An interesting variant of an F-type enzyme found in the bacterium *Propionigenium modestum* has the ability to use a Na^+ rather than H^+ gradient to drive ATP synthesis. The specificity for Na^+ resides in the F_0 segment, since a hybrid of *P. modestum* F_0 with *E. coli* F_1 can still use a Na^+ gradient.[27] Therefore, and despite the differences that exist between F-ATPases from various organisms, each enzyme is highly specific for its substrates. As will be shown later, this high substrate specificity is in sharp contrast with the properties of organic compound-transporting primary ATPases.

26.4.2. V-ATPases

The vacuolar V-ATPases are present in a variety of eukaryotic organelles that have an acidic lumen, e.g., Golgi, synaptic vesicles, chromaffin granules, clathrin-coated vesicles, and lysosomes, as well as yeast and plant vacuoles.[24,28] A V-ATPase is also present in the plasma membrane of acid-secreting cells in the kidney, mainly the proximal tubule, where it participates in the acidification of urine.[29] The A subunit of V-ATPases is homologous to the catalytic β subunit of F-ATPases,[30] as are the regulatory B and α subunits, respectively. The 16-kDa c subunit of the V_0 segment of V-ATPases (corresponding to F_0 of the F-type enzymes) probably evolved by gene duplication and fusion of the corresponding 8-kDa c subunit (proteolipid) of the F-ATPases.[24] It is thought that this difference in the H^+-transducing segments of the enzymes determines their opposite physiological modes of action, i.e., ATP synthesis for the F-ATPases and ATP hydrolysis for the V-ATPases.[31] In contrast to the F_0 segment from which F_1 has been stripped, V_0 after V_1 removal does not function as a proton channel.[32] In this context it is also interesting that the plasma membrane ATPase of Archaebacteria, which is structurally similar to the V-ATPases except that it uses an F-type proteolipid, catalyzes ATP synthesis rather than hydrolysis under physiological conditions.[33]

The structural homology between V- and F-type ATPases does not extend to subunits other than A, B, and c. For instance, the γ and C subunits of F- and V-ATPases, respectively, are structurally dissimilar although functionally equivalent, and therefore represent an example of convergent evolution. Since both enzymes are ancient and involved in processes fundamental for sustaining life, comparisons of V- and F-ATPases constitute a powerful tool in elucidating evolutionary relationships of Archaebacteria, eubacteria, and eukaryotes.[34]

It is assumed that the mechanism of action of V-ATPases is similar to that of F-ATPases. Both types of enzymes do not form a phosphoenzyme intermediate during the catalytic cycle, appear to have three pairs of subunits, in the case of the V-ATPases subunits A and B, which take turns in catalyzing

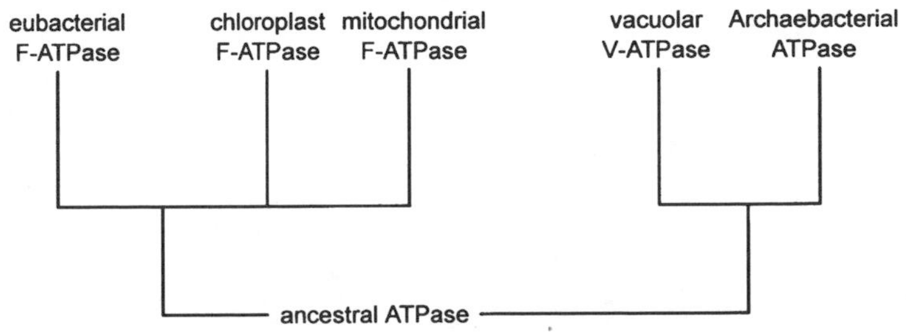

Figure 26.6. Phylogenetic relationships between the ion-motive ATPases.

ATP hydrolysis coupled to H^+ transport. V-ATPases are highly specific for H^+; no other ions are known to be transported by this type of enzyme.

26.4.3. P-ATPases

While the key subunits of the F- and V-ATPases are homologous and the evolutionary relationship of the two types of enzymes is undisputed, P-ATPases differ from them in both structure and mechanism of action. P-ATPases generally consist of a single, catalytically competent subunit, although the active form in the membrane is likely to be a dimer. An accessory subunit, named β, is present in the Na^+,K^+-ATPase, but probably has a structural rather than catalytic function; the β subunit also has been implicated to play a role in the assembly of the Na^+,K^+-ATPase in the membrane. Regulatory proteins are known, although they are not usually considered to be part of the enzyme. Phospholamban, a protein interacting with the 110-kDa Ca^{2+}-ATPase of heart muscle sarcoplasmic reticulum, may serve as an example. However, in the plasma membrane Ca^{2+}-ATPase, the regulatory calmodulin-binding domain is an integral part of the polypeptide chain, which increases in size to approx. 135 kDa. In this sense proteins such as phospholamban may in fact be counted as true subunits.

Among ion-motive ATPases, P-ATPases are most versatile in terms of substrate specificity. In addition to enzymes transporting Na^+/K^+, H^+/K^+, H^+, and the several ATPases specific for Ca^{2+}, a P-type enzyme transporting Mg^{2+} has been described recently,[35] and the list is likely to continue to grow. The mechanism of action of P-ATPases is clearly different from that of the F- and V-type enzymes. While in the latter the free energy released or consumed in ATP hydrolysis or synthesis is stored exclusively in the binding energy of the nucleotide to the protein or in the conformation of the enzyme, in P-ATPases an actual high-energy phosphoenzyme intermediate is present and can be characterized by chemical methods. In the phosphoenzyme, a specific aspartyl side chain is phosphorylated, i.e., a mixed anhydride is formed. Depending on the conformational state of the enzyme, the phosphate of the mixed anhydride can be either high-energy, i.e., in equilibrium with ATP, or low-energy, i.e., in equilibrium with inorganic phosphate. As was the case with the other types of ATPases, the free energy of these transitions is coupled through the enzyme protein with the flow of the transported ion(s). The interference of vanadate, a transition-state analogue of phosphate, with the phosphoenzyme makes vanadate a selective inhibitor that distinguishes between P-ATPases on one hand, and F- and V-ATPases on the other.

The reaction cycle of a P-ATPase, on the example of the Na^+,K^+-ATPase, is shown schematically in Figure 26.7. It illustrates an important principle applicable to many if not most transmembranous transport processes, namely, that of conformational changes of the enzyme that expose the binding site for the transported solute on alternating sides on the membrane. In the case of the Na^+,K^+-ATPase, the E_1 form of the enzyme has high affinity for Na^+, and has a Na^+ binding site accessible from the inside of the cell. The bound Na^+ is, however, exchangeable with the aqueous medium. Phosphorylation of the enzyme by ATP leads to the formation of a high-energy aspartyl phosphate and to closing of the ion binding site; the Na^+ is said to be occluded. The next event is a key step in the reaction cycle. The free energy stored in the aspartyl phosphate is used to effect a conformational change of the protein in which the ion binding site opens, but to the outside rather than the inside of the cell, and the affinity for Na^+ is reduced. This conformation of the enzyme is called E_2, and the aspartyl phosphate becomes low energy during the transition from E_1 to E_2. Na^+ leaves the enzyme and is replaced by K^+, for which the E_2 form of the ATPase has a high affinity. Dephosphorylation of the enzyme causes occlusion of K^+. A conformational change back to E_1 opens the ion binding site to the inside, the K^+ ions leave, and the enzyme can begin another catalytic cycle. Although the mechanistic details of ion transport are not fully understood for the Na^+,K^+-ATPase and,

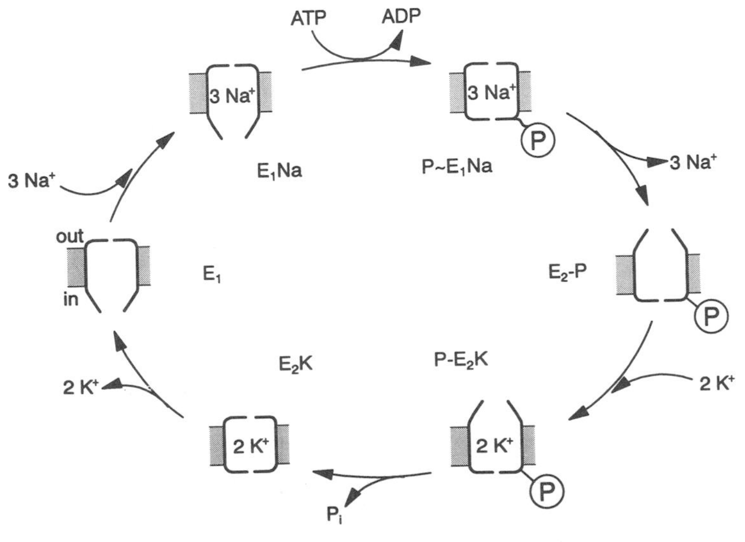

Figure 26.7. Reaction cycle of a P-ATPase on the example of the Na^+,K^+-ATPase. See text for details.

in fact, for any membrane transporter, the principle of conformational changes that modify the affinity of the transport protein for its solute, and make the binding site accessible from opposite sides of the membrane appears to be a useful model. Once more, specific binding of solutes, in the above case of Na^+ and K^+, to appropriate forms of the enzyme makes the progression through the catalytic cycle possible.

26.5. PRIMARY TRANSPORT OF ORGANIC COMPOUNDS

The classification of transporting ATPases into the F-, V-, and P-types, although extremely useful in organizing the wealth of available information, is incomplete. As mentioned previously, organic compounds can be transported directly by ATPases, and the enzymes involved in that process appear to differ in many respects from the classical ion-motive ATPases. Since all known transporting ATPases in the canalicular membrane of hepatocytes are of this type, the problem is of special significance for biliary secretion.

26.5.1. P-Glycoproteins

The P-glycoproteins were first identified in tumor cells that exhibit multidrug resistance.[36,37] It was observed that certain tumor cells express a glycosylated protein in their plasma membrane, and take up lesser amounts of various anticancer drugs, often structurally unrelated to each other. The latter phenomenon was attributed to a decreased permeability of the membrane, hence the name P-glycoprotein.[37] A conceptual difficulty arises from this model, since it is difficult to visualize how the presence of a protein in the membrane, even at a high concentration, could alter the membrane's bulk properties so as to decrease the permeability for some but not all solutes. Further studies revealed that the P-glycoproteins are in fact ATP-dependent pumps able to transport the drugs either from the core of the membrane or from the interior of the cell to the outside. The route may depend mainly on the hydrophobicity and therefore the partitioning of the drug between the aqueous phase and the hydrophobic core of the membrane. The possibility of expelling the drug from the plasma membrane, i.e., prior to its entry into the cell, explained the apparent decrease in membrane permeability. However, this finding led to a mechanistic question for which a satisfactory answer is still lacking: how can a single transporter recognize a variety of structurally unrelated compounds? Possible answers to this question will be discussed later, after a brief description of the structure and properties of P-glycoproteins and related proteins. At this point it is, however, important to note that the relative lack of selectivity of P-glycoproteins puts them in sharp contrast with the F-, V-, or P-type ion-motive ATPases which have clearly defined, narrow substrate specificities.

The P-glycoproteins are about 170 kDa in size, 30 kDa of which is in oligosaccharides concentrated in one region of the protein. The molecule consists of two similar but nonidentical halves, each of which probably contains six transmembranous segments and an ATP binding site close to the C-terminus. The ATP binding sites conform to the Walker motifs[38] found on many ATP-utilizing proteins. The topology of the protein has been worked out by a combination of different methods and is widely accepted (see Ref. 39 for a review), although alternate models have also been proposed.[40] The ATP binding sites are located on the cytoplasmic side of the membrane. In contrast, the regions responsible for drug binding are located on stretches of the sequence close to or within the membrane. This is consistent with the ability of the protein to extract its substrates from the core of the membrane.

At least two genes coding for distinct P-glycoproteins exist in humans (*MDR1* and *MDR2,* for multiple drug resistance), and three genes (*mdrla* or *mdr3, mdrlb,* and *mdr2*) in mice. Other species may have an even higher number of *mdr* genes. While the *MDR1* gene product is the drug-transporting P-glycoprotein, the *MDR2* gene product does not confer the multiple drug resistance phenotype on cells; rather, the latter protein acts as a phospholipid translocase ("flippase").[41] The *MDR1* P-glycoprotein is overexpressed in malignant cells as a result of gene amplification and/or increased transcription. Mutations in various parts of the protein occur spontaneously or have been introduced experimentally; some of these mutations cause changes in the specificity profile of the transporter.[39]

While the *MDR1* gene product is highly overexpressed in drug-resistant tumor cells, in some cases up to one-third of the total plasma membrane protein,[42] the P-glycoproteins are also present, at lower levels, in normal tissues. The *MDR1* gene product is localized predominantly in adrenal glands, kidneys, the intestine, and the blood–brain barrier.[43,44] The canalicular membrane of hepatocytes is the major site of localization for the *MDR2*-encoded protein[45,46] and, in mice, it also contains *mdr3* gene product. The latter is capable of drug transport, in which it resembles the *mdr1* but not the *mdr2* P-glycoprotein. While the substrate of the *mdr2* gene product appears to be phosphatidylcholine,[41] the natural substrates of the *mdr1*-type transporters remain unknown. Two possibilities have been entertained: a general detoxification role against xenobiotics, or a specific transporter role for as yet unidentified endogenous hydrophobic metabolites, perhaps steroids.[43]

The P-glycoproteins are part of a large family of proteins, mostly with transport function, termed ABC (ATP binding cassette) proteins[47] or traffic ATPases.[48] Approximately 40 members of this family are known, and are found both in bacteria and in eukaryotes. While the P-glycoproteins of eukaryotic plasma membranes consist of a single polypeptide chain containing two homologous domains, in intracellular organelles single-domain subunits exist but probably form functional dimers or higher oligomers.[39] In bacteria the transport complex consists of several subunits that may have fused together in higher organisms. A wide variety of compounds are transported by ABC proteins, including nutrients and proteinaceous toxins in bacteria, peptide mating factors in yeast, pigments in *Drosophila,* proteins and/or lipids in mammalian peroxisomes, and drugs in transformed cells but also in *Plas-*

modium parasites. In many cases, the similarity of transporters from widely different organisms is sufficient for functional replacement; for example, the human *MDR1* gene product is able to complement yeast mating deficiency resulting from a defect of the STE6 transporter for peptide mating factor a.[39]

Perhaps the most puzzling aspect of P-glycoproteins or, more generally, ABC transporters is their mechanism of action. No satisfactory explanation for the broad substrate specificity exists. The transported compounds share no obvious chemical or structural characteristics. This argues against a P-glycoprotein reaction mechanism similar to that of ion-motive ATPases or of secondary active transporters of organic molecules, where binding of the substrate to an enzymelike, specific binding site is an integral part of the transport process. Instead, the concept of a molecular "hydrophobic vacuum cleaner" has been proposed for P-glycoproteins.[39] P-glycoprotein substrates usually have an octanol : water partition coefficient close to 1,[39] i.e., they are amphiphilic; molecules that are either very hydrophilic or very hydrophobic are poor substrates. Consequently, the basic feature of the vacuum cleaner model is the existence of substrate entry gates within the core or close to the surface of the membrane, enabling the protein to intercept and expel molecules that partitioned into the membrane but have not yet reached the cytoplasm. The entry gates, of which there may be several for the recognition of substrates residing deeper or shallower in the membrane or entirely in the aqueous phase, would merge into a single channel, or barrel, of the transporter leading to the outside of the cell. The existence of multiple entry gates would explain the modulation of activity for specific substrates (or substrate groups) by amino acid substitutions in the protein; still, and unlike more traditional transporters, the specificity of the gates should be low, so that the proximity to the P-glycoprotein in the membrane would be the major determinant of what is transported.

A variation of the molecular vacuum cleaner concept is the "flippase" model.[39,49] In it, the P-glycoprotein would catalyze the transfer reaction of a substrate molecule from the inner to the outer leaflet of the plasma membrane, from which the compound could partition into the external aqueous phase. Mechanistically, the model could imply a moving part of the protein binding the substrate and then actually traversing the membrane in a rather major conformational change. However, a channel within the protein opening to the outer leaflet of the membrane rather than to the bulk aqueous phase could also function as a flippase, perhaps making the two models conceptual extremes on a continuum of possible mechanisms. In this context it is interesting to note that an ATP-dependent phospholipid flippase activity has been found,[50] and more recently that it has been determined that an *mdr2*-type P-glycoprotein catalyzes the transmembranous translocation of phosphatidylcholine in mouse liver canalicular membranes.[41]

The vacuum cleaner or flippase models do not address the problem of coupling of ATP hydrolysis with transport. Several possibilities have been advanced (reviewed in Ref. 39). The first is based on the finding that the P-glycoprotein, while in the membrane, has a basal, drug-independent ATPase activity in addition to substrate-stimulated ATPase,[42,51] i.e., has no defined stoichiometry of substrate molecules transported per ATP consumed. This again differentiates the P-glycoprotein from the more conventional ATPases, and perhaps also from the ABC-type bacterial permeases which are tightly coupled.[49] The bacterial enzymes, which often transport polar and/or charged solutes such as amino acids and sugars, may thus function in a way resembling the ion-motive ATPases. The relaxed coupling in the P-glycoproteins could, however, be interpreted in terms of constant conformational changes, taking place regardless of the presence or absence of substrate. Transport would occur when a substrate molecule happened to fall onto this continually moving "conveyor belt." In this model, the reaction cycle may be accelerated by, but not be strictly dependent on the presence of a substrate. Even though such model appears to constitute a radical departure from the mechanisms thought to apply to the more conventional transporting ATPases, it may in fact be a variation of a common theme, although perhaps an extreme one. Ion-motive ATPases, especially of the V type, show the phenomenon of slip, i.e., a partial decoupling of ion transport from ATPase activity,[52] which may be important for their function.[24] The slip could be more pronounced in P-glycoproteins, leading to the observed drug-independent ATPase activity. If so, the physiological significance of the slip remains to be explained; the rather low amounts of P-glycoproteins in normal tissues would, however, prevent this pump from becoming a major drain on the cell's energy resources.

An alternative possibility of coupling ATP hydrolysis with transport is based on recent research on the properties of P-glycoproteins. There is evidence that P-glycoproteins may be outward-directed H^+ pumps[39]; in addition, P-glycoproteins have cell volume-regulated Cl^- channel activity whose relationship to drug transport is kinetically complex.[39,53] In fact, the CFTR protein, which is a member of the ABC family phylogenetically somewhat removed from the P-glycoproteins,[54] may have lost transporter activity for amphiphiles but has retained Cl^- channel activity. The transport of H^+, and the following Cl^- movement, by P-glycoproteins could create an osmotic drag moving water from within the membrane to the outside; compounds present in the membrane but having, at the same time, sufficient solubility in water, i.e., amphiphiles, could thus be expelled from the cell, provided they could pass through the gates or filters mentioned above. The model[39,55] is attractive, but additional experimental evidence will need to be accumulated in its support.

The tantalizing finding that the *mdr1*-encoded P-glycoprotein[56] as well as the CFTR protein[57] are channels transporting ATP from the cytoplasm to the surrounding medium has important ramifications for various areas of biochemistry and physiology whose listing is beyond the scope of this chapter, and has caused considerable excitement.[58] One of the implications relevant to the present discussion is a proposed model for P-glycoproteins action in which the driving force for drug transport would be the electrochemical gradient of ATP.[56] Even though such a role for ATP is unconventional, the millimolar intracellular ATP concentration and a membrane potential of typically −50 mV make it thermodynamically feasible. At the same time, a high electrochemical ATP gradient in the presence of an unregulated[56] ATP conductance should lead to a considerable efflux of ATP. In fact, from the data presented in Ref. 56 it can be calculated that about 5% of the total cellular ATP is lost per minute through P-glycoprotein channels in normal CHO cells, and in cells overexpressing the P-glycoproteins almost half of the ATP leaves every minute. The released ATP is hydrolyzed by the ecto-ATPase (see next section), which rapidly establishes a steady state of the extracellular ATP concentration.[56] While the purine from the resulting ADP is recovered through the action of 5′-nucleotidase followed by a nucleoside uptake system, the energetic cost to the cell of maintaining this ATP cycling must be high. It appears unlikely that the process would occur for the benefit of expelling an occasional toxicant; it has been suggested that the transport of an as yet unidentified but essential amphiphilic metabolite is coupled to the outward ATP movement.

The hypothesis that P-glycoprotein-mediated drug transport is actually a symport with ATP leaves several important questions unanswered. The foremost problem is the incorporation into the model of ATP hydrolysis, which has been repeatedly shown to be necessary for drug efflux. Even though the ATP binding sites are involved in the ATP channel activity, ATP is not hydrolyzed in the process; the P-glycoprotein merely acts as an electrophoretic ATP channel.[56] Thus, it is possible that the P-glycoprotein is multifunctional,[53] with amphiphile transport, Cl^- channel, and ATP channel as essentially independent activities.

26.5.2. Bile Acid Transporters

The first bile acid transporter to be identified in canalicular plasma membranes of hepatocytes has been characterized as membrane potential-driven, and distinct from the basolateral Na^+-dependent activity.[15] Further work revealed that the canalicular transporter is a 100-kDa glycoprotein, which was subsequently purified and functionally reconstituted.[16–18] In a surprising development, three groups independently identified the existence of primary, ATP-dependent taurocholate transport in canalicular membranes of rat liver,[20–22] a result that has subsequently been shown to apply also to human liver.[23,59] The molecular weight, susceptibility to inhibitors, and recognition by antibodies are similar for the electrophoretic and ATP-driven bile acid transporters, and it has been suggested that the two activities are associated with the same transport protein.[60] However, subcellular fractionation studies have shown a different distribution of the two activities, and indicated that the membrane potential-dependent transporter found in canalicular plasma membrane fractions may in fact be the result of contamination with endoplasmic reticulum.[61] Moreover, the electrophoretic and ATP-dependent bile acid transport activities are differentially induced by phenobarbital.[62] Therefore, the notion of separate transport proteins catalyzing the two activities appears more likely at the present time, although more work will be needed to clarify the issue. Regardless of the multiplicity of canalicular bile acid transporters, it is generally agreed that this activity is not related to either the P-glycoproteins or the canalicular organic anion transporters which will be described in the next section.

The major ATP-hydrolyzing activity of the rat canalicular plasma membrane is the ecto-ATPase (or ectonucleotidase). The enzyme is unusual for an ATPase in having its active center for ATP hydrolysis located outside the cell.[63] The cloning of the cDNA coding for the rat liver ecto-ATPase has been reported,[64] and the deduced protein sequence was found to be homologous to the human biliary glycoprotein 1 and identical to the rat hepatocyte cell adhesion molecule C-CAM, which in turn show significant similarity to immunoglobulins with respect to domain structure. A model of the protein[63] based on its sequence[64] postulates a single membrane-spanning region, a 71-amino-acid-long cytoplasmic C-terminal

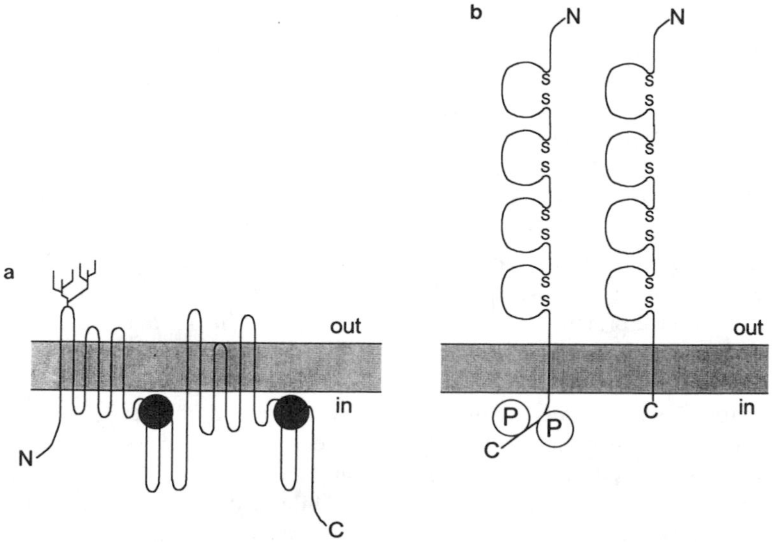

Figure 26.8. Hypothetical models of the structure of a P-glycoprotein (a) and the ecto-ATPase (b). Based on Refs. 55 and 63. See text for details.

domain carrying two potential tyrosine and one serine phosphorylation site, with the bulk of the protein including the N-terminus, two nucleotide binding sites, and multiple glycosylation sites located in the extracellular space. An immunoglobulin-like folding has been postulated for the extracellular domain based on the sequence homology with IgG molecules. The model of the ecto-ATPase/C-CAM is shown schematically in Figure 26.8B, along with a short form[65] in which the cytoplasmic part is truncated from 71 to 10 amino acid residues. The latter form, which may arise through alternative splicing of the mRNA[66] or could be coded for by a distinct gene,[67] no longer contains the phosphorylation sites. The functions of the protein are not fully understood. The ecto-ATPase can hydrolyze extracellular ATP (and other nucleotides),[68] e.g., the ATP released by P-glycoprotein- associated ATP channels.[56] In this capacity, the ecto-ATPase may serve to terminate ATP-triggered responses, and to enable the reuptake of nucleosides. In its C-CAM role, the protein mediates cell adhesion.[69] However, several workers failed to confirm the presence of ecto-ATPase activity in purified C-CAM.[70,71] Therefore, it could be argued that the ecto-ATPase and C-CAM functions should be ascribed to distinct proteins.

In a recent communication, the rat liver canalicular ecto-ATPase has been identified as a bile acid transporter.[72] This startling finding, based both on biochemical data and on transfection experiments in which ecto-ATPase/C-CAM expression conferred taurocholate transport activity on target COS cells, leaves a number of questions unanswered. The possible distinction between the ecto-ATPase and C-CAM proteins, and therefore the identity of the cDNA[64] used in the transfection experiments,[72] has already been mentioned. Even if these two functions are the result of a single protein, the data presented in Ref. 72 suggest that while bile acid transport is stimulated and perhaps even driven by ATP, the ecto-ATPase activity and bile acid transport are not necessarily coupled. For example, vanadate inhibited bile acid transport but not ecto-ATPase activity. The lack of coupling of taurocholate transport to ecto-ATPase activity has recently been confirmed by mutagenesis of the putative extracellular ATP binding site which abolished ecto-ATPase activity but not bile acid transport.[73] Somewhat surprisingly, it has also been shown that taurocholate transport, even by the mutated protein that lacks ecto-ATPase activity, still requires extracellular ATP which probably undergoes hydrolysis since it is not replaceable by the nonhydrolyzable analogue AMP-PNP.[73] Moreover, the same mutation that abrogated ecto-ATPase activity did not impair the photoaffinity labeling of the protein with ATPγS.[73] These results are difficult to reconcile, except by postulating that a second ATP binding and hydrolysis site exists on the protein, and that this site (rather than the one that has been mutated) is involved in bile acid transport. Moreover, it has recently been shown that taurocholate transport activity does not correlate with ecto-ATPase activity in rat liver canalicular membrane subfractions.[74] Clearly, further work will be necessary to establish the role and mechanism of the ecto-ATPase and/or C-CAM proteins in bile acid transport, and to determine whether these proteins may be identical with, or related to the ATP-driven canalicular bile acid transporter.

In spite of the problems mentioned previously, the possible role of the ecto-ATPase/C-CAM in bile acid transport raises a number of fundamental molecular and mechanistic questions. According to the model shown in Figure 26.8B, the protein has a single transmembranous region. This is in contrast with more conventional transport proteins, including the P-glycoprotein (Figure 26.8A), all of the ion-motive ATPases, and a variety of secondary transporters, which have several membrane-spanning domains thought to form the channel, or barrel, used in transport. While an assembly of several ecto-ATPase/C-CAM protein molecules could give rise to such a barrel, the bulky extracellular domain may not favor the formation of oligomeric structures because of steric crowding. More generally, the recent proliferation of reports describing membrane-bound enzymes proposed to be bifunctional and involved in transport is somewhat troubling, but also holds the promise of new, exciting insights. The ecto-ATPase has already been discussed. In another example, it has been suggested that the microsomal epoxide hydrolase acts as a bile acid transporter. If located in the basolateral plasma membrane of the hepatocyte, the transporter was Na^+-coupled.[75] In a different orientation and in the endoplasmic reticulum membrane, the same protein has been found to catalyze membrane potential-driven bile acid uptake.[76] It is striking that in all of the above cases the transported molecule, being amphiphilic, has a tendency to associate with membranes. One could speculate that integral membrane proteins undergoing conformational changes, e.g., in the course of a catalytic cycle or as a regulatory event, may transiently disturb the integrity of the membrane to a degree sufficient for "opportunistic" transport of molecules that are already concentrated at the membrane. The required movement of the protein within the membrane would be plausible for the enzyme, such as epoxide hydrolase, which undergoes a catalytic cycle. The ecto-ATPase is phosphorylated on its cytoplasmic domain. While the serine phosphorylation by protein kinase C has no effect on the extracellular ATP hydrolysis,[72] the ecto-ATPase is also a substrate for the insulin receptor tyrosine kinase, and tyrosine phosphorylation has been suggested to generate a regulatory signal transmitted to the external domain through a conformational change.[66]

Obviously, the model of a protein that undulates in the membrane for unrelated reasons but, in the process, catalyzes the transport of amphiphiles does not mean that the transport is an experimental artifact. The process may seem opportunistic, but so is evolution, and examples of bifunctional proteins abound. In fact, the model of a "wiggle transporter" could even accommodate unidirectional, active transport, provided the motion of the protein resembles peristalsis. In addition, it is our opinion that the distinction between a wiggle transporter and a classical one, e.g., an ion-motive ATPase, is not purely semantic. The first difference would be the lack, in the former, of tight coupling of transport with the driving force, i.e., the lack of a defined stoichiometry. The second consequence would be the absence of a defined substrate binding site. Instead, substrate–protein contact sites with limited specificity, such as those seen in P-glycoproteins, would suffice. Such relaxed specificity may explain the puzzling finding of a bile acid transporter in vacuolar membranes of plants which do not have bile acids.[77] Finally, amphiphilic compounds with a tendency to bind to membranes would be preferred over highly hydrophilic molecules. The above characteristics of a wiggle transporter should help to generate experimentally testable predictions needed to evaluate the model.

26.5.3. Transporters of Non-Bile Acid Organic Anions

Toxic endogenous metabolites and xenobiotics taken up by the liver or generated in hepatocytes usually undergo phase II (conjugative) detoxification reactions prior to secretion into bile. The increased hydrophilicity of the conjugates precludes nonmediated diffusion across the canalicular membrane. Instead, a specific transport event, termed phase III of detoxification,[78] is required. It is generally accepted that the transporters involved in phase III detoxification are distinct from both the *mdr1/mdr3* P-glycoproteins and the bile acid transporters, even though cross-inhibition of one transporter by substrates of another is observed in some cases; this inhibition is probably the result of nonproductive binding that does not lead to transport.[10]

The variety of substrates of the multispecific organic anion transporter (MOAT), and the lack of any obvious unifying structural characteristics, resemble the situation encountered with P-glycoproteins. A somewhat arbitrary selection of these substrates is shown in Figure 26.9. While most of the compounds carry two widely spaced negative charges, a feature proposed to be the criterion for recognition by the transporter,[79] exceptions to this rule exist, e.g., the singly charged glucuronides of 1-naphthol[80] or 4-nitrophenol.[81] The model substrate *S*-(2,4-dinitrophenyl)glutathione is often used experimentally. Cysteinyl leukotrienes, with K_m values in the submicromolar range,[82] have the highest affinity for the transporter, and bilirubin diglucuronide may well be the quantitatively predominant substrate in the liver.

The structural diversity of substrates of the organic anion transporter creates the same conceptual difficulty in understanding the reaction mechanism that was already discussed for P-glycoproteins. The problem is compounded by the considerable differences in hydrophobicity between the various substrates. While amphiphilic molecules, especially gluta-

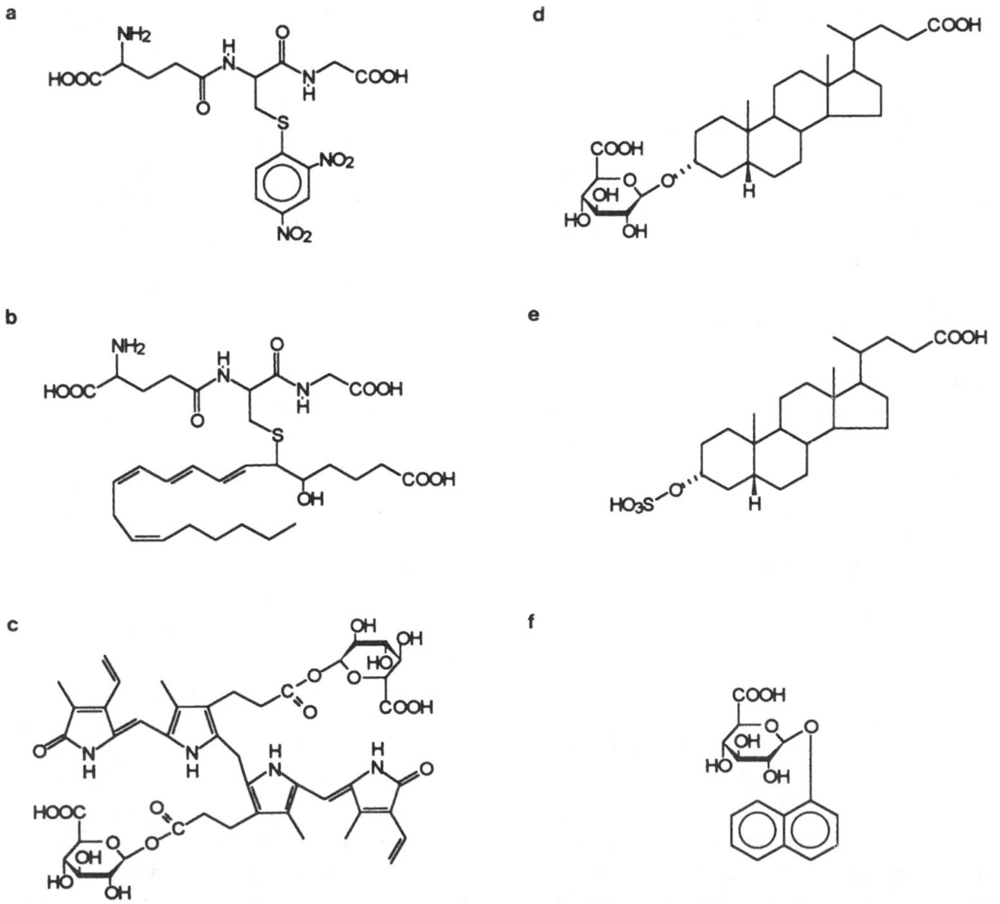

Figure 26.9. Selected substrates of the ATP-dependent canalicular organic anion transporter. a, *S*-(2,4-dinitrophenyl)glutathione; b, leukotriene LTC_4; c, bilirubin diglucuronide; d, 3-*O*-glucuronide of lithocholic acid; e, 3-*O*-sulfate of lithocholic acid; f, glucuronide of 1-naphthol.

thione *S*-substituted with long aliphatic chains, have a high affinity for the transporter,[78] at least some of the substrates, including *S*-(2,4-dinitrophenyl)glutathione, are highly polar and water-soluble. In light of the preceding discussion of transport processes, such compounds are likely to require specific and therefore selective binding sites on a transport protein. At present, it is difficult to reconcile this requirement with the lack of common structural elements among the substrates.[83]

Initial measurements of *S*-(2,4-dinitrophenyl)glutathione uptake into inside-out canalicular plasma membrane vesicles were interpreted in terms of an electrophoretic mechanism.[84] However, at approximately the same time it was demonstrated that *S*-(2,4-dinitrophenyl)glutathione stimulates ATP hydrolysis by rat liver plasma membranes,[85,86] a finding indicative of a transporting ATPase. Indeed, further work revealed that ATP-driven transport of *S*-(2,4-dinitrophenyl)glutathione and other organic anions occurs in the canalicular membrane.[82,87–91] The transport is unidirectional: in intact cells, it occurs from the inside to the outside, i.e., into bile. This has been experimentally verified by showing that only inside-out (as opposed to right-side-out) vesicles derived from the canalicular membrane are capable of ATP-driven uptake of organic anions.[92] The ATP-dependent and electrophoretic transport systems coexist in the same membrane, although the former is probably predominant.[93] The demonstration[93] that a spontaneous mutation affects only the ATP-dependent system (see below) strongly suggests that the two systems utilize distinct transport proteins.

The enzymology of the ATP-dependent organic anion transport has been characterized in considerable detail. Nucleotides other than ATP are poor substrates, and nonhydrolyzable ATP analogues do not support transport, indicating that ATP hydrolysis rather than binding is required.[92] Transport is inhibited by vanadate[92,94] and sulfhydryl reagents such as *N*-ethylmaleimide.[91] Neither reduced glutathione nor taurocholate competed with the transport of organic anions, but oxidized glutathione did.[91,92] Furthermore, the lack of competition by substrates of the P-glycoprotein[82,91,92] confirms that the latter is distinct from the organic anion transporter.

In an initially unrelated line of work, ATP-dependent transport of organic anions and ATPase activity stimulated by these substrates has been identified and characterized in human erythrocytes. The efflux of *S*-(2,4-dinitrophenyl)glutathione out of erythrocytes is energy-dependent.[95,96] The transporter has been shown to be a primary ATPase.[97,98] The protein responsible for the ATPase activity has been purified and characterized.[99–102] While it has been firmly established that this protein, which is 38 kDa in size in its denatured, monomeric form, is sufficient for the organic anion-dependent ATPase activity, at present it is not certain whether additional subunits may be necessary for transport.

In addition to erythrocytes, the 38-kDa organic anion transporter is present in other tissues, including the liver,[99,103,104] as demonstrated by Western blotting and purification. This raises the question of its relationship to the organic anion transport activity associated with the canalicular membranes of hepatocytes that has been described above. Recent evidence suggests that the 38-kDa system may be fairly ubiquitous, while the liver may have an additional organic anion transporter, perhaps necessary to cope with the greater metabolic needs of that organ, especially in terms of bilirubin glucuronide transport. This notion, which remains to be established in a more rigorous way, is based on the following observations.

1. A mutation that leads to a severe impairment of the liver canalicular ATP-dependent organic anion transport is known in rats, sheep, monkeys, and humans (reviewed in Ref. 10). This mutation, which is the subject of the final part of this chapter, affects the liver but not, or only insignificantly, organic anion transporters in erythrocytes[105] and in the intestine and kidney.[80]
2. Initial purification results indicate that, while a protein similar in size to the erythrocyte transporter is present in rat liver, the predominant species that can be isolated from this tissue has a monomer molecular mass of 85 to 90 kDa (Refs. 9, 10, 104, and unpublished results).
3. Differences in catalytic properties of the erythrocyte and hepatic ATP-dependent transporters include the fact that oxidized glutathione is a substrate for the latter (see above) but not the former.[106] Perhaps even more importantly, the erythrocyte transporter accepts substrates usually considered to be specific for *mdr1*-type P-glycoproteins, e.g., doxorubicin,[107] while the hepatic transporter is restricted to organic anions.[92] In agreement with the above, *S*-(2,4-dinitrophenyl)glutathione transport by the hepatic system is not inhibited by substrates of the P-glycoprotein (see above), while doxorubicin significantly inhibits *S*-(2,4-dinitrophenyl) glutathione efflux from erythrocytes.[108]

26.6. TRANSPORT DEFECTS

Because of the physiological importance of bile formation, defects in transporters essential for the process are probably lethal and are therefore not observed. Exceptions would be expected either for transporters that contribute to bile flow but are not absolutely indispensable for bile formation or bile function, or for transporters that occur in multiple forms having a similar function. For example, this rules out a viable mutant with a disrupted ATP-dependent bile acid transporter: especially in humans, bile flow is largely related to bile acid secretion, and bile acids are indispensable for bile function. Even though an electrophoretic bile acid transporter has been identified, it is not clear whether it is not merely an uncoupled mode of action of the ATP-driven transporter, and the available membrane potential would be in any case insufficient for the secretion of adequate amounts of bile acids. Both of the above conditions can, however, be satisfied for organic anion transport. Elimination of the bile acid-independent bile flow, a somewhat paradoxical consequence of the impairment of organic anion transport (see below), reduces but does not interrupt bile production. Functional redundancy resulting from the presence in liver of at least two and perhaps three organic anion transporters (ATP-dependent, membrane potential-dependent, and erythrocyte-type 38 kDa) would ensure the secretion of potentially toxic compounds, although perhaps at a reduced rate. As mentioned previously, a liver-specific impairment of the ATP-dependent organic anion transport has indeed been observed in at least two independent strains of rats, in sheep, and in humans. The clinical consequences of the mutation are mild, and in fact often go undiagnosed. The interest in this hereditary defect is therefore not aimed at finding a treatment; it is generally felt that treatment is unnecessary. Rather, the mutation is used as a way to investigate and understand the process of biliary secretion and, more generally, bile formation. Gaining such understanding is obviously invaluable from both the scientific and the clinical point of view.

26.7. DUBIN–JOHNSON SYNDROME

Dubin–Johnson syndrome is a form of familial chronic conjugated hyperbilirubinemia. The clinical syndrome has been examined exhaustively in recent reviews[10,109–111] and will be presented here in brief. Dubin–Johnson syndrome is inherited as an autosomal recessive, and is not detectable in the heterozygote by techniques presently available. The syndrome is benign; it exerts no known adverse effects on organ function and longevity is thought to be normal. No therapy is required. Its importance lies in two directions: It must be distinguished from forms of conjugated hyperbilirubinemia which are associated with clinically significant abnormal liver function. The examination of its pathogenesis has led to an improved understanding of the physiologic basis of bile secretion and the pathophysiology of jaundice.

The significance of the presence of *conjugated* (as opposed to an unconjugated) hyperbilirubinemia can be seen with a brief examination of bilirubin metabolism.[109,111] Bilirubin derives primarily from the degradation of the heme of hemoglobin in reticuloendothelial cells. Bilirubin is delivered in plasma bound to albumin and taken up by putative transporter-mediated processes at the sinusoidal surface of the liver cell. It diffuses to the endoplasmic reticulum (ER) bound to "ligandin" or possibly in association with vesicle membranes. It crosses the membrane of the ER possibly by diffusion or by transporter-mediated processes. At the luminal surface of the ER, bilirubin and its cosubstrate, uridine diphosphoglucuronic acid (UDPGA), interact with the active enzymatic site of bilirubin glucuronosyltransferase to form bilirubin glucuronides (mono- and diglucuronides). By ill-defined processes it then reaches the canalicular surface of the plasma membrane and is secreted into the canalicular lumen by the organic anion transporter/ATPase (MOAT) discussed above. Bilirubin glucuronides are sparingly absorbed from the gut and have a limited enterohepatic circulation.

From this brief overview it is possible to pinpoint the possible sites of disorders that lead to the development of unconjugated or conjugated hyperbilirubinemias. In short, an *unconjugated* hyperbilirubinemia might result from overproduction of bilirubin, failure of hepatic uptake of unconjugated bilirubin, or, theoretically, failure of transport of unconjugated bilirubin within the liver cell to the active enzymatic site for glucuronidation. Necessarily, if the defect in Dubin–Johnson syndrome is within the liver cell, it must be "after" those processes by which unconjugated bilirubin reaches and binds to the active enzymatic site of bilirubin glucuronosyltransferase, and "before" the entry of conjugated bilirubin into the canalicular lumen. The defect might be one of transport from the active enzymatic site to the canalicular membrane, but, as discussed above and below, the syndrome is now thought to be the result of a defect of bilirubin glucuronide transport across the canalicular membrane.

This transport defect leads to certain of the signs and measurable changes associated with the syndrome. The syndrome is characterized by the following findings: Chronic, variable conjugated hyperbilirubinemia is most commonly first noted between puberty and early adulthood.[112] The patient may or may not be overtly jaundiced. Total plasma bilirubin concentrations commonly range as high as 5 mg/100 ml, but may be higher.[10] The "direct-reacting" (conjugated) fraction is more than half of the total, but values vary. Typically, unlike acquired liver diseases, diglucuronide conjugates are in greater concentration in plasma than monoconjugates; diglucuronide conjugates also predominate in bile.[113] As would be expected with a chronic conjugated hyperbilirubinemia, a significant fraction of the bilirubin in plasma may be covalently linked to albumin, the result of acyl shift of the glucuronide and the formation of reactive intermediates.[114] Jaundice may increase and the syndrome may become overt during pregnancy and in patients receiving oral agents for birth control.[115] Estrogen may further inhibit the defective canalicular organic anion transport mechanism found in the patients. The incidence of Dubin–Johnson syndrome has been noted to be high in certain populations with high rates of intermarriage, e.g., in certain rural populations in Japan and among Iranian Jews.[109,116]

A number of vague symptoms have been reported in association with the syndrome (e.g., abdominal pain, nausea, weakness), but these complaints may not be causatively related to the syndrome, and rather may be nonspecific and the motive force behind the workup leading to establishing the diagnosis. Enlargement and tenderness of the liver have been described in some patients.

Conventional liver function tests other than the plasma bilirubin concentration are typically normal. An associated factor VII deficiency has been reported in Israeli patients, but the association is thought to be fortuitous and genetically distinct.[117] Opacification of the gallbladder by the dyes used in oral cholecystography is defective. Gallbladder visualization is possible using the 99mTc-labeled scanning agents for cholescintography.[118] The difference in results using the two techniques might reflect the incomplete nature of the defect in organic anion secretion, and/or minor methodological differences (e.g., oral versus intravenous administration of the test agent).

Striking increases are noted in the pigmentation of the liver which may be grossly black in appearance.[119] On examination with conventional light microscopy, large dark brown granules are noted in the hepatocytes of the centrilobular regions with spread to the periphery as the changes progress. Otherwise histology has been normal or characterized by minor nonspecific changes only. Electron microscopic examination of the liver shows that the pigment is deposited in lysosomes. The chemical nature of the pigment has been debated. It has been considered a form of lipofuscin, and, later, a form of melanin. More recently, evidence has been developed that the pigment consists of polymers of epinephrine metabolites.[120] The usually unstated assumption is that these accumulate in the hepatocyte as a result of the organic anion secretory defect that is the mechanism of pathogenesis of the syndrome. It is of interest that when radiolabeled epinephrine was injected intravenously into mutant rats with an organic anion secretory defect comparable to that of Dubin–Johnson syndrome (discussed below), the mutants retained more of the injected dose in liver lysosomes than the controls. Supplementation of the diet with tyrosine, tryptophan, and phenylalanine for 4 months resulted in the accumulation of lysosomal pigment in the mutants.[120] It has also been noted that the pigment decreased in a patient with Dubin–Johnson syndrome who had contracted hepatitis, the pigment then reappearing at a later date.

Other laboratory abnormalities are found in patients with Dubin–Johnson syndrome. BSP administrated intravenously (a previously commonly used liver function test now outmoded) may or may not be excessively retained in plasma at 45 min, but is consistently retained in excess at 90 min.[122–124] The glutathione conjugate of BSP is retained indicating normal uptake and metabolism, but defective secretion of the dye. Kinetic analyses of plasma disappearance of the injected BSP confirm this impression.[125] The initial limb of the plasma disappearance curve corresponding to hepatic uptake is normal, hepatic storage of the dye is normal, but hepatic excretion is markedly delayed, and the T_m (maximal excretory rate) drastically decreased. Again, these results are those expected to be associated with a defect in canalicular organic anion secretion.

One other abnormality is associated with the syndrome and has not been satisfactorily explained. Porphyrinogens, consisting of four pyrrole rings with side chains, connected by carbon bridges to form a tetrapyrrole ring, are normal intermediates of heme synthesis.[126] The porphyrinogens are traditionally classified according to the order of their side chains. Type III porphyrinogens are the isomers of the biochemical pathway leading to heme synthesis. Other isomers, notably type I, are also formed, but are metabolic "dead letters." The oxidation products of porphyrinogens, porphyrins, are found in bile (three-quarters) and urine (one-quarter).[127] Type III coproporphyrin is normally excreted predominantly in urine and to a lesser degree in bile. Type I coproporphyrin is predominantly excreted in bile. Thus, in normals, two-thirds of bile coproporphyrin is type I, while three-quarters of urine coproporphyrin is type III.

In Dubin–Johnson syndrome urine coproporphyrin excretion is altered.[128] Total excretion remains unchanged, but 80% of urine coproporphyrin is found to be type I. This finding is consistent with, and, in fact, has been used to establish the diagnosis of Dubin–Johnson syndrome. It differs from the findings in individuals with other liver diseases in whom total urine coproporphyrin may increase, presumably as the result of diversion of bile constituents to urine, but the proportion of type I isomers, while elevated, does not exceed 65%. This latter finding is what would be expected to occur as a result of diminished biliary secretion of normal bile constituents. The predominance of type I coproporphyrin in the urine of patients with Dubin–Johnson syndrome, therefore, does not appear to be the result of simple failure of coproporphyrin secretion into bile. It would seem that an alteration in the relative rates of type I and III excretion in bile and/or urine, or a change in rates of types I and III synthesis or degradation must occur. It is not clear why such changes should occur in association with an apparently specific genetic transporter defect. The alteration in coproporphyrin excretion in Dubin–Johnson syndrome has been confirmed fully, but no satisfactory explanation of the phenomenon has been advanced.

Finally, it should be mentioned that a similar, but distinct syndrome of benign familial chronic conjugated hyperbilirubinemia, Rotor's syndrome, has been partially characterized.[109,110] Like Dubin–Johnson syndrome, it is thought to be inherited as an autosomal recessive, and is commonly diagnosed in early adulthood. Unlike Dubin–Johnson syndrome, the liver is not pigmented, oral cholecystographic agents are secreted normally, and urinary coproporphyrin secretion resembles that of acquired liver disease. Liver biopsy analysis by light and electron microscopy demonstrates only nonspecific changes. Kinetic analyses of the plasma clearance of bilirubin and BSP suggest that hepatic organic anion uptake and storage are defective in Rotor's syndrome,[125] but the cellular mechanisms corresponding to these pathophysiologic phenomena have not been characterized. Patients with Rotor's syndrome may be difficult to distinguish from those with Dubin–Johnson syndrome with lesser degrees of hepatic pigmentation, but the pattern of urinary coproporphyrin excretion permits the diagnosis to be made.

26.8. ANIMAL MODELS

A syndrome similar to the Dubin–Johnson syndrome was first observed in mutant Corriedale sheep in the 1960s.[129] The abnormalities in the sheep closely resembled those in patients with the syndrome. It was characterized by a conjugated hyperbilirubinemia, abnormalities of organic anion secretion, and a pigmented liver in the absence of other major abnormalities of liver function and histology. Elucidation of the model was, however, limited by the cumbersome techniques necessary to maintain and study large animals, and by the limited understanding of hepatic transport physiology available during that era.

Because of model size and the elaboration of hepatic transport physiology in rats during the interval, the identification of a Dubin–Johnson-like rat model in the 1980s proved invaluable.[130] Mutant rats with congenital conjugated hyperbilirubinemia were identified by several groups and designated "TR−," "GY," and "EHBR."[10] Cross-breeding experiments established that the mutations were allelic, and thus, that the rats shared the same genetic defect.[120,131] A pattern of autosomal recessive inheritance was established. Serum total bilirubin was abnormal at birth, rose markedly during the first day of life, declined after the tenth day, and reached an elevated steady-state level thereafter.[130] The total of serum bilirubin in the mutants consisted almost exclusively of mono- and diconjugated bilirubin. This differs from that of Dubin–Johnson patients in whom a significant, though minor, fraction of the total is unconjugated bilirubin. A low-grade elevation of serum bile acids was present in the mutant rat, but not in Dubin–Johnson patients. The absence of hepatic pigment (in TR− and GY, but not EHBR mutants) was a still more striking difference. Histology by light and electron microscopy showed only minor abnormalities.

Basal bile bilirubin glucuronide secretion was diminished and a greater fraction of bile bilirubin was diconjugated than that found in normal rats.[130] Glutathione, normally present in rat bile in millimolar concentrations, was virtually absent from the bile of mutants.[131,132] "Bile acid-independent bile flow," which may be partially dependent on glutathione secretion, was markedly decreased.

The basic defect in mutants was ascribed to a failure of divalent organic anion secretion.[130,133,134] Bilirubin mono- and diglucuronides are divalent organic anions, and, as noted above, basal rates of secretion were decreased. It was clear that the defect was not absolute, since the bile of mutants did contain bilirubin glucuronides. Presumably, the maintenance of these lower than normal levels was at the cost of marked elevations of plasma conjugated bilirubin, which, in turn, was an indicator of the inefficiency of the excretory process in the mutant.

The biliary secretion of other organic anions, but not organic cations, was impaired.[130,134] The kinetic analysis of plasma BSP clearance showed results similar to those obtained in patients with Dubin–Johnson syndrome indicating impaired biliary secretion and reflux of BSP–glutathione conjugate into plasma.[10] The secretion of bile acid sulfates and glucuronides, both divalent organic anions, was markedly impaired when compared to conventional bile acid amino acid conjugates, which are monovalent anions.[79,131] As noted above, the transporter subserving the canalicular secretion of such compounds as bilirubin glucuronides and BSP (MOAT) appears primarily to utilize divalent organic anions as its substrates. The degree of impairment of organic anion secretion varied greatly with the substrate. Bilirubin diglucuronide and (synthetic) bilirubin ditaurine are both divalent organic anions with similar chemical structures. The impairment of bilirubin diglucuronide secretion, however, was severe, while the impairment of bilirubin ditaurine secretion was minor.[135] Similarly, the degree of impairment of lithocholate 3-*O*-glucuronide was much more severe than the impairment of the structurally similar cholate 3-*O*-glucuronide.[79] The secretion of the bioactive organic anion, cysteinyl leukotriene LTC_4, was grossly impaired.[134]

Abnormalities of urinary porphyrin excretion resembled those found in Dubin–Johnson patients.[130] Total coproporphyrin excretion was normal, but the absolute and relative percentage of type I isomer was increased. In normal rats, however, the urinary secretion of the type I isomer is only 5% and type I isomer excretion was increased to only 20% in the mutants.

Studies of the mutant defect in intact animals have been supplemented by studies of isolated perfused liver, isolated hepatocytes, and canalicular membranes obtained from mutants. These have confirmed and extended the observations above. Defective organic anion transport is demonstrable in isolated perfused liver and isolated liver cells, but while nonhepatic tissues including red cells, the kidney, and gut display transport activities for a similar array of organic anions, transport in mutant nonhepatic tissues is normal or minimally deranged.[80,105]

As noted above, the concentration of free glutathione in bile is markedly decreased. Defective biliary secretion of glutathione may, however, be an indirect effect of impairment of the putative organic anion transporter. Incubation of glutathione with liver plasma membranes from normal rats did not stimulate ATPase activity or inhibit transport of a substrate for the organic anion transporter.[85,87,90,91] This suggested that glutathione is secreted into bile by a separate, membrane potential-driven transporter. Evidence was developed to suggest that the glutathione transporter is inhibited in the mutant rat by the cellular accumulation of organic anions.[14] Indeed, in recent work, the canalicular glutathione transporter has been cloned, partially characterized, and definitively demonstrated to be distinct from the organic anion transporter.[13] It should soon be possible to define the relation between the defect in organic anion transport in mutant rats and the absence of glutathione in their bile.

In contrast to glutathione, glutathione conjugates, such as the synthetic conjugate, 2,4-dinitrophenylglutathione (DNP-SG), which has been studied extensively in experiments *in vitro* and *in vivo,* are defectively secreted in bile by mutants.[10]

As detailed above, the movement of organic anions from blood, across the liver cell, into bile involves complex mechanisms. While studies of whole animals, perfused liver, and isolated liver cells, leave open the possibility of a defect in many of the steps in this process, the site of the defect in the mutant was definitively established in studies in which vesicles of canalicular membrane isolated from normal and mutant liver were examined.[82] ATP-stimulated transmembrane transport of DNP-SG and of LTC_4 cysteinyl leukotriene was demonstrable in normal, but deficient in canalicular vesicles from mutant liver. These studies provided the basis for specifically characterizing the defect as one of canalicular transport of organic anions.

In summary, the defect in Dubin–Johnson syndrome and in the animal models appears to be similar or identical. While the syndrome has several signs, symptoms, laboratory findings, and pathological changes, the basic defect that accounts for most or all of these appears to be a failure of canalicular secretion of divalent organic anions. The study of animal models of Dubin–Johnson syndrome has clarified this concept and promises to form the basis of the ultimate definition of the defect at the molecular level.

26.9. POSSIBLE MOLECULAR BASIS OF DUBIN–JOHNSON SYNDROME AND SIMILAR CONJUGATED HYPERBILIRUBINEMIAS

The function of the multispecific organic anion transporter (MOAT) has been studied in considerable detail in a variety of experimental systems. These investigations included the physiologically important role of MOAT in affecting the canalicular secretion of reduced glutathione,[14] and thus in modulating the bile acid-independent fraction of bile formation. However, the protein responsible for the MOAT function has not been purified or otherwise characterized, which precluded elucidation of the molecular basis of the Dubin–Johnson syndrome and similar defects. Although a final answer to that question is still outstanding, considerable progress has been made in the last few years, and will be summarized in the final part of this chapter.

The work on the mechanisms of ATP-dependent organic anion transport proceeded on two fronts. Because of the availability of relatively abundant starting material for biochemical investigations, the characterization of the transporter from human erythrocytes is most advanced (reviewed in Ref. 9). Recently, MOAT from another extrahepatic tissue, the mastocytoma cells, has also been isolated.[136] The relevance of the information gained from these studies to the liver canalicular MOAT remains to be established since, as discussed previously, there is persuasive evidence that at least the predominant hepatic MOAT may be distinct from MOAT systems found in erythrocytes and other extrahepatic tissues. The recent purification, characterization, and functional reconstitution of the rat liver canalicular MOAT[137,138] appears to support that notion.

26.9.1. Extrahepatic MOAT

The organic anion transporters from human erythrocytes was the first such system to be identified[97,98] and purified.[99–101] Although the protein was initially characterized as having an apparent mass of 38 kDa, more recent results indicate that this may be a fragment of a larger parent molecule. Nevertheless, the purified 38-kDa protein was sufficient to catalyze ATPase activity stimulated by *S*-(2,4-dinitrophenyl)glutathione, bile acid 3-*O*-glucuronides, and other organic anions,[99–101] and to confer incremental transport activity, in this case of doxorubicin, upon erythrocyte membranes into which it was incorporated.[107] This indicates that the 38-kDa polypeptide, or perhaps its oligomers,[102] contain the core of the protein necessary for ATP-dependent transport of organic material.

The transport of *S*-(2,4-dinitrophenyl)glutathione and other conjugates of electrophilic compounds with glutathione, glucuronic acid, or sulfate has the characteristics of a phase III[78] detoxification process. A class of bioactive organic anions, the cysteinyl leukotrienes, is also transported by a MOAT system, and the K_m for these compounds is in the low micromolar range.[82,139] The mast cells constitute a site of active synthesis and secretion of cysteinyl leukotrienes. Consequently, a protein responsible for transport was identified in mastocytoma cells,[136] a cell line derived from mast cells. Differential photoaffinity labeling with a member of the cysteinyl leukotriene group revealed proteins of 190 kDa, 35 kDa, and small amounts of material close to 100 kDa, whereas 8-azido-ATP labeled mostly the 35-kDa protein. The 190-kDa material has been found to consist of a 140-kDa proteinaceous core and 50 kDa of N-linked oligosaccharides. The material did not cross-react with antibodies specific for the mdr-type P-glycoproteins, but may be related to a p-190, a protein that confers the mdr phenotype and belongs to the ABC family, but is distinct from the P-glycoproteins.[54] The nature of the 35-kDa protein remains unclear; it could conceivably represent a fragment similar to that found in human erythrocytes.

Organic anion transporters with properties differing from those mentioned above have also been described. In particular, a transporter specific to oxidized glutathione (GSSG) has been identified in human red blood cells.[140,141] The material, when purified by *S*-hexylglutathione affinity chromatography, was found to consist of nonidentical subunits of 65 and 85 kDa.

The variety of results obtained from attempts to purify the MOAT can be rationalized by a pronounced proteolytic lability of the protein. Depending on tissue and purification methodology, various fragments could be produced, some of which could retain at least partial activity. Alternatively, or perhaps concomitantly with that explanation, it appears likely that MOAT is in fact a family of proteins with overlapping substrate specificities. Many available functional data are best explained in this way, and the possibility is consistent with the situation found for essentially all detoxifying enzymes, e.g., cytochromes P450, UDP-glucuronosyltransferases, glutathione *S*-transferases. If so, the question whether the various MOAT isoforms are structurally related, or whether a number of distinct membrane proteins evolved to acquire the ability to transport, perhaps by the previously mentioned "wiggle" mechanism, will require a more thorough characterization than is available today.

26.9.2 Hepatic MOAT

A 38-kDa protein that is similar to the human erythrocyte material by structural, immunological, and functional criteria could be isolated from human or rat liver.[99,103,104] However, the liver has organic anion transport requirements that go beyond the relatively low-level detoxification needs of most tissues. For example, the concentration of bile pigments in bile is typically 2 mM; considering that 500 ml bile can be produced per day by humans, 1 mmole or almost a gram of material derived from heme degradation is excreted per day. Preliminary data indicate that these increased transport needs are met by an additional MOAT system specific for the liver.

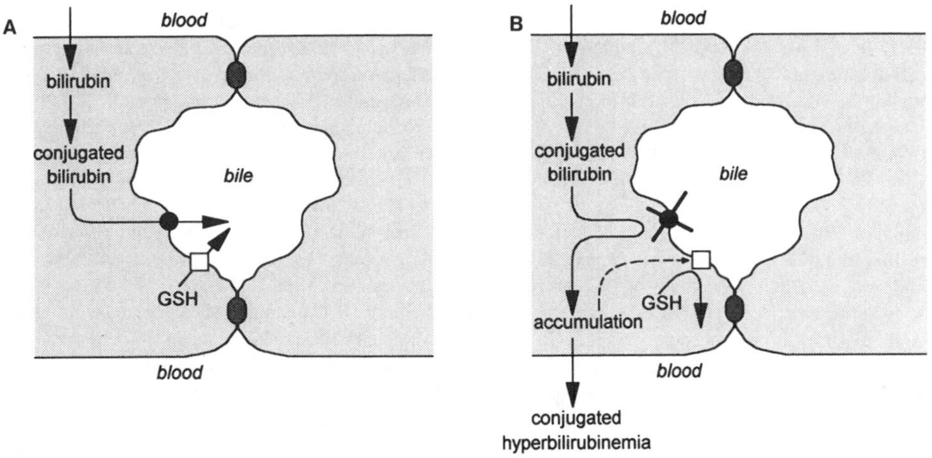

Figure 26.10. Schematic depiction of the hypothetical biochemical causes of the observed Dubin–Johnson phenotype. (A) Normal liver. Bilirubin is taken up from the circulation and undergoes intracellular conjugation with glucuronic acid. The conjugate is then transported across the canalicular membrane into bile by the MOAT (multispecific organic anion transport) system. Reduced glutathione is transported independently by an electrogenic, ATP-independent carrier. (B) Dubin–Johnson liver. The function of MOAT is impaired. This leads to an accumulation of conjugated bilirubin, which is eventually released into circulation leading to conjugated hyperbilirubinemia. At the same time, the intracellular conjugated bilirubin interferes with the GSH carrier, which leads to a depletion of biliary GSH and an abolishment of bile acid-independent bile flow.

By a combination of several affinity chromatography techniques, a 90-kDa glycoprotein could be isolated from rat liver plasma membranes.[137] The protein, which contained only about 5% N-linked sugars, under certain experimental conditions yielded fragments of 30 and 60 kDa, in agreement with the already mentioned possibility of proteolytic susceptibility. The 90-kDa protein had the partial activities expected of MOAT: it could be photoaffinity labeled by both *S*-(4-azido-2-nitrophenyl)glutathione,[142] an analogue of *S*-(2,4-dinitrophenyl) glutathione, and by 8-azido-ATP; it also bound TNP-ATP, a fluorescent probe derived from ATP. Furthermore, the purified 90-kDa protein had *S*-(2,4-dinitrophenyl) glutathione-dependent ATPase activity which was significantly stimulated by prior phosphorylation of the protein with protein kinase C. The latter finding is consistent with the previously reported modulation of MOAT activity by protein kinase C inhibitors and activators in intact cells.[143] The purified 90-kDa protein could be reconstituted into artificial proteoliposomes which then acquired the ability to take up *S*-(2,4-dinitrophenyl)glutathione in an ATP-dependent manner into their intravesicular lumen.[138] This proves that the 90-kDa protein is sufficient to carry out MOAT function, without the need for additional proteins or cofactors. Protein kinase C phosphorylation prior to reconstitution increased considerably the transport catalyzed by the vesicles. Taking into account the transport rates observed after reconstitution, it could be calculated that the amount of the 90-kDa protein present in the rat liver canalicular membrane is sufficient to account for the maximal biliary secretory rate of organic anions. It is at present not known whether the 90-kDa protein is a member of the ABC family.

26.9.3. Possible Primary Defects in the Dubin–Johnson Syndrome

As delineated previously, the available functional evidence indicates that the primary defect in the rat models of the Dubin–Johnson syndrome, and by extrapolation in the human syndrome as well, is in the transport of organic anions across the liver canalicular membrane into bile. Accumulation of conjugated bilirubin first in the liver and then in the circulation would constitute a primary consequence of the transport deficiency; the ensuing depletion of biliary glutathione and fall in the bile acid-independent bile flow are thought to be secondary effects. It has been proposed[14] that the substrates of the MOAT that accumulate in hepatocytes inhibit the transport of GSH from the intracellular side, and at the same time fail to stimulate GSH transport at the extracellular side. The transport of GSH itself is catalyzed by a separate ATP-independent transporter that has been characterized and cloned.[13] The above processes are schematically depicted in Figure 26.10A for the normal, and in Figure 26.10B for a Dubin–Johnson liver.

The primary defect in MOAT function could be related to the total absence of the MOAT protein, to the presence of a mutated and functionally impaired protein in the canalicular membrane, or to mistargeting or incorrect assembly of an otherwise intact protein. It has been previously demonstrated that the "erythrocyte-type" 38-kDa transporter is present and, at least on a gross level, is unchanged in the mutant GY rat.[104] Similar analyses for the 90-kDa protein have not been carried out. Final answers will have to await the cloning of the hepatic MOAT from normal and affected animals and humans.

REFERENCES

1. Arias, I. M., Che, M., Gatmaitan, Z., Leveille, C., Nishida, T., and St Pierre, M. (1993). The biology of the bile canaliculus, 1993. *Hepatology* **17**:318–329.
2. Boyer, J. L., Graf, J., and Meier, P. J. (1992). Hepatic transport systems regulating pH_i, cell volume, and bile secretion. *Annu. Rev. Physiol.* **54**:415–438.

3. Chowdhury, J. R., and Chowdhury, N. R. (1992). Bilirubin metabolism and its disorders. In *Biliary Diseases* (N. Kaplowitz, ed.), Williams & Wilkins, Baltimore, pp. 131–147.
4. Chowdhury, J. R., Wolkoff, A. W., and Arias, I. M. (1989). Hereditary jaundice and disorders of bilirubin metabolism. In *The Metabolic Basis of Inherited Disease,* 6th ed. (C. R. Scriver, A. L. Beaudet, W. S. Sly, and D. Valle, eds.), McGraw–Hill, New York, Volume 1, pp. 1367–1408.
5. Jansen, P. L. M., and Oude Elferink, R. P. J. (1993). Defective hepatic anion secretion in mutant TR-rats. In *Hepatic Transport and Bile Secretion: Physiology and Pathophysiology* (N. Tavoloni and P. D. Berk, eds.), Raven Press, New York, pp. 721–731.
6. Meier, P. (1993). Canalicular membrane transport processes. In *Hepatic Transport and Bile Secretion: Physiology and Pathophysiology* (N. Tavoloni and P. D. Berk, eds.), Raven Press, New York, pp. 587–596.
7. Erlinger, S. (1993). Secretion of bile. In *Diseases of the Liver,* 7th ed. (L. Schiff and E. R. Schiff, eds.), Lippincott, Philadelphia, Volume 1, pp. 85–107.
8. Scharschmidt, B. F. (1993). Bilirubin metabolism, bile formation, and gallbladder and bile duct function. In *Gastrointestinal Disease. Pathophysiology, Diagnosis, Treatment,* 5th ed. (M. H. Sleisenger and J. S. Fordtran, eds.), Saunders, Philadelphia, Volume 2, pp. 1730–1746.
9. Zimniak, P., and Awasthi, Y. C. (1993). ATP-dependent transport systems for organic anions. *Hepatology* **17:**330–339.
10. Zimniak, P. (1993). Dubin–Johnson and Rotor syndromes: Molecular basis and pathogenesis. *Semin. Liver Dis.* **13:**248–260.
11. Garcia-Ruiz, C., Fernandez-Checa, C. J., and Kaplowitz, N. (1992). Bidirectional mechanism of plasma membrane transport of reduced glutathione in intact rat hepatocytes and membrane vesicles. *J. Biol. Chem.* **267:**22256–22264.
12. Fernandez-Checa, J. C., Yi, J.-R., Garcia-Ruiz, C., Knezic, Z., Tahara, S. M., and Kaplowitz, N. (1993). Expression of rat liver reduced glutathione transport in *Xenopus laevis* oocytes. *J. Biol. Chem.* **268:**2324–2328.
13. Yi, J.-R., Lu, S., Fernandez-Checa, J., and Kaplowitz, N. (1994). Expression cloning of a rat hepatic reduced glutathione transporter with canalicular characteristics. *J. Clin. Invest.* **93:**1841–1845.
14. Fernandez-Checa, J. C., Takikawa, H., Horie, T., Ookhtens, M., and Kaplowitz, N. (1992). Canalicular transport of reduced glutathione in normal and mutant Eisai hyperbilirubinemic rats. *J. Biol. Chem.* **267:**1667–1673.
15. Inoue, M., Kinne, R., Tran, T., and Arias, I. M. (1984). Taurocholate transport by rat liver canalicular membrane vesicles. Evidence for the presence of an Na^+-independent transport system. *J. Clin. Invest.* **73:**659–663.
16. Ruetz, S., Fricker, G., Hugentobler, G., Winterhalter, K., Kurz, G., and Meier, P. J. (1987). Isolation and characterization of the putative canalicular bile salt transport system of rat liver. *J. Biol. Chem.* **262:**11324–11330.
17. Ruetz, S., Hugentobler, G., and Meier, P. J. (1988). Functional reconstitution of the canalicular bile salt transport system of rat liver. *Proc. Natl. Acad. Sci. USA* **85:**6147–6151.
18. Sippel, C. J., Ananthanarayanan, M., and Suchy, F. J. (1990). Isolation and characterization of the canalicular membrane bile acid transport protein of rat liver. *Am. J. Physiol.* **258:**G728–G737.
19. Anwer, M. S., and Hegner, D. (1982). Potassium is important for hepatic bile acid secretion. *Gastroenterology* **82:**1009 (abstr.).
20. Adachi, Y., Kobayashi, H., Kurumi, Y., Shouji, M., Kitano, M., and Yamamoto, T. (1991). ATP-dependent taurocholate transport by rat liver canalicular membrane vesicles. *Hepatology* **14:**655–659.
21. Muller, M., Ishikawa, T., Berger, U., Klunemann, C., Lucka, L., Schreyer, A., Kannicht, C., Reutter, W., Kurz, G., and Keppler, D. (1991). ATP-dependent transport of taurocholate across the hepatocyte canalicular membrane mediated by a 110-kDa glycoprotein binding ATP and bile salt. *J. Biol. Chem.* **226:**18920–18926.
22. Nishida, T., Gatmaitan, Z., Che, M., and Arias, I. M. (1991). Rat liver canalicular membrane vesicles contain an ATP-dependent bile acid transport system. *Proc. Natl. Acad. Sci. USA* **88:**6590–6594.
23. Wolters, H., Kuipers, F., and Vonk, R. J. (1992). ATP-dependent taurocholate transport in human liver plasma membrane vesicles. *Gastroenterology* **102:**A910 (abstr.).
24. Nelson, N. (1991). Structure and pharmacology of the proton-ATPases. *Trends Pharmacol. Sci.* **12:**71–75.
25. Scarpa, A., Carafoli, E., and Papa, S. (1992). *Ion-Motive ATPases: Structure, Function, and Regulation,* New York Academy of Sciences, New York.
26. Hinkle. P. C., Kumar, M. A., Resetar, A., and Harris, D. L. (1991). Mechanistic stoichiometry of mitochondrial oxidative phosphorylation. *Biochemistry* **30:**3576–3582.
27. Dimroth, P., Laubinger, W., Kluge, C., Kaim, G., Ludwig, W., and Schleifer, K. H. (1992). Sodium-translocating adenosinetriphosphatase of *Propionigenium modestum. Ann. N.Y. Acad. Sci.* **671:**311–322.
28. Nelson, N., and Taiz, L. (1989). The evolution of H^+ATPases. *Trends Biochem. Sci.* **14:**113–116.
29. Gluck, S. L. (1992). The structure and biochemistry of the vacuolar H^+ ATPase in proximal and distal urinary acidification. *J. Bioenerg. Biomembr.* **24:**351–359.
30. Zimniak, L., Dittrich, P., Gogarten, J. P., Kibak, H., and Taiz, L. (1988). The cDNA sequence of the 69-kDa subunit of the carrot vacuolar H^+-ATPase. Homology to the beta-chain of the F0F1-ATPases. *J. Biol. Chem.* **263:**9102–9112.
31. Supek, F., Supekova, L., Beltran, C., Nelson, H., and Nelson, N. (1992). Structure, function, and mutational analysis of V-ATPases. *Ann. N.Y. Acad. Sci.* **671:**284–292.
32. Beltran, C., and Nelson, N. (1992). The membrane sector of vacuolar H(+)-ATPase by itself is impermeable to protons. *Acta Physiol. Scand. (Suppl.)* **607:**41–47.
33. Schafer, G., and Meyering-Vos, M. (1992). The plasma membrane ATPase of Archaebacteria. A chimeric energy converter. *Ann. N.Y. Acad. Sci.* **671:**292–309.
34. Nelson, N. (1992). Evolution of organellar proton-ATPases. *Biochim. Biophys. Acta* **1100:**109–124.
35. Maguire, M. E. (1992). MgtA and MgtB: Prokaryotic P-type ATPases that mediate Mg^{2+} influx. *J. Bioenerg. Biomembr.* **24:**319– 328.
36. Biedler, J. L., Riehm, H., Peterson, R. H., and Spengler, B. A. (1975). Membrane-mediated drug resistance and phenotypic reversion to normal growth behavior of Chinese hamster cells. *J. Natl. Cancer Inst.* **55:**671–680.
37. Juliano, R. L., and Ling, V. (1976). A surface glycoprotein modulating drug permeability in Chinese hamster ovary cell mutants. *Biochim. Biophys. Acta* **455:**152–162.
38. Walker, J. E., Saraste, M., Runswick, M. J., and Gay, N. J. (1982). Distantly related sequences in the alpha- and beta-subunits of ATP synthase, myosin, kinases and other ATP-requiring enzymes and a common nucleotide binding fold. *EMBO J.* **1:**945–951.
39. Gottesman, M. M., and Pastan, I. (1993). Biochemistry of multidrug resistance mediated by the multidrug transporter. *Annu. Rev. Biochem.* **62:**385–427.
40. Skach, W. R., Calayag, M. C., and Lingappa, V. R. (1993). Evidence for an alternate model of human P-glycoprotein structure and biogenesis. *J. Biol. Chem.* **268:**6903–6908.
41. Smit, J. J. M., Schinkel, A. H., Oude Elferink, R. P. J., Groen, A. K., Wagenaar, E., van Deemter, L., Mol. C. A. A. M., Ottenhoff, R., van der Lugt, N. M. T., van Roon, M. A., van der Valk, M. A., Offerhaus, G. J. A., Berns, A. J. M., and Borst, P. (1993). Homozygous disruption of the murine mdr2 P-glycoprotein gene leads to a complete absence of phospholipid from bile and to liver disease. *Cell* **75:** 451–462.
42. Al-Shawi, M. K., and Senior, A. E. (1993). Characterization of the adenosine triphosphatase activity of Chinese hamster P-glycoprotein. *J. Biol. Chem.* **268:**4197–4206.

43. Gros, P., and Buschman, E. (1993). The mouse multidrug resistance gene family: Structural and functional analysis. *Int. Rev. Cytol.* **137C:**169–197.
44. Pastan, I., Willingham, M. C., and Gottesman, M. (1991). Molecular manipulations of the multidrug transporter: A new role for transgenic mice. *FASEB J.* **5:**2523–2528.
45. Thiebaut, F., Tsuruo, T., Hamada, H., Gottesman, M. M., Pastan, I., and Willingham, M. C. (1987). Cellular localization of the multidrug-resistance gene product P-glycoprotein in normal human tissues. *Proc. Natl. Acad. Sci. USA* **84:**7735–7738.
46. Buschman, E., Arceci, R. J., Croop, J. M., Che, M., Arias, I. M., Housman, D. E., and Gros, P. (1992). mdr2 encodes P-glycoprotein expressed in the bile canalicular membrane as determined by isoform-specific antibodies. *J. Biol. Chem.* **267:**18093–18099.
47. Hyde, S. C., Emsley, P., Hartshorn, M. J., Mimmack, M. M., Gileadi, U., Pearce, S. R., Gallagher, M. P., Gill, D. R., Hubbard, R. E., and Higgins, C. F. (1990). Structural model of ATP-binding proteins associated with cystic fibrosis, multidrug resistance and bacterial transport. *Nature* **346:**362–365.
48. Ames, G. F., Mimura, C. S., and Shyamala, V. (1990). Bacterial periplasmic permeases belong to a family of transport proteins operating from *Escherichia coli* to human: Traffic ATPases. *FEMS Microbiol. Rev.* **6:**429–446.
49. Ames, G. F.-L., Mimura, C. S., Holbrook, S. R., and Shyamala, V. (1992). Traffic ATPases: A superfamily of transport proteins operating from Escherichia coli to humans. *Adv. Enzymol.* **65:**1–47.
50. Devaux, P. F. (1991). Static and dynamic lipid asymmetry in cell membranes. *Biochemistry* **30:**1163–1173.
51. Ambudkar, S. V., Lelong, I. H., Zhang, J., Cardarelli, C. O., Gottesman, M. M., and Pastan, I. (1992). Partial purification and reconstitution of the human multidrug-resistance pump: Characterization of the drug-stimulatable ATP hydrolysis. *Proc. Natl. Acad. Sci. USA* **89:**8472–8476.
52. Groen, B. H., Berden, J. A., and van Dam, K. (1990). Differentiation between leaks and slips in oxidative phosphorylation. *Biochim. Biophys. Acta* **1019:**121–127.
53. Gill, D. R., Hyde, S. C., Higgins, C. F., Valverde, M. A., Mintenig, G. M., and Sapulveda, F. V. (1992). Separation of drug transport and chloride channel functions of the human multidrug resistance P-glycoprotein. *Cell* **71:**23–32.
54. Cole, S. P. C., Bhardwaj, G., Gerlach, J. H., Mackie, J. E., Grant, C. E., Almquist, K. C., Stewart, A. J., Kurz, E. U., Duncan, A. M. V., and Deeley, R. G. (1992). Overexpression of a transporter gene in a multidrug-resistant human lung cancer cell line. *Science* **258:**1650–1654.
55. Chin, K.-V., Pastan, I., and Gottesman, M. M. (1993). Function and regulation of the human multidrug resistance gene. *Adv. Cancer Res.* **60:**157–180.
56. Abraham, E. H., Prat, A. G., Gerweck, L., Seneveratne, T., Arceci, R. J., Kramer, R., Guidotti, G., and Cantiello, H. F. (1993). The multidrug resistance (mdr1) gene product functions as an ATP channel. *Proc. Natl. Acad. Sci. USA* **90:**312–316.
57. Reisin, I. L., Prat, A. G., Abraham, E. H., Amara, J., Gregory, R., Ausiello, D. A., and Cantiello, H. F. (1992). Expression of CFTR results in the appearance of an ATP channel. *46th Meeting, Society of General Physiologists* (abstr.).
58. Arias, I. M. (1993). Is the multidrug resistance an ATP channel? *Hepatology* **18:**216–222.
59. Kadmon, M., Kluenemann, C., Boehme, M., Ishikawa, T., Gorgas, K., Otto, G., Herfarth, C., and Keppler, D. (1993). Inhibition by cyclosporin A of adenosine triphosphate-dependent transport from the hepatocyte into bile. *Gastroenterology* **104:**1507–1514.
60. Nishida, T., Che, M., Gatmaitan, Z., and Arias, I. M. (1992). Evidence for a single ATP and membrane potential-dependent taurocholate transport system in rat liver canalicular membrane vesicles. *Hepatology* **16:**149A (abstr.).
61. Kast, C., Stieger, B., and Meier, P. J. (1992). Electrogenic and ATP-dependent taurocholate transport exhibit distinct subcellular distributions in rat hepatocytes. *Hepatology* **16:**148A (abstr.).
62. Fernandez-Checa, J. C., Ookhtens, M., and Kaplowitz, N. (1993). Selective induction by phenobarbital of the electrogenic transport of glutathione and organic anions in rat liver canalicular membrane vesicles. *J. Biol. Chem.* **268:**10836–10841.
63. Lin, S.-H. (1990). Liver plasma membrane ecto-ATPase. *Ann. N.Y. Acad. Sci* **603:**394–400.
64. Lin, S.-H., and Guidotti, G. (1989). Cloning and expression of a cDNA coding for a rat liver plasma membrane ecto-ATPase. The primary structure of the ecto-ATPase is similar to that of the human biliary glycoprotein I. *J. Biol. Chem.* **364:**14408–14414.
65. Lin, S.-H., Culic, O., Flanagan, D., and Hixson, D. C. (1991). Immunochemical characterization of two isoforms of rat liver ecto-ATPase that show an immunological and structural identity with a glycoprotein cell-adhesion molecule with Mr 105,000. *Biochem. J.* **278:**155–161.
66. Najjar, S. M., Accili, D., Philippe, N., Jernberg, J., Margolis, R., and Taylor, S. I. (1993). pp120/ecto-ATPase, an endogenous substrate of the insulin receptor tyrosine kinase, is expressed as two variably spliced isoforms. *J. Biol. Chem.* **268:**1201–1206.
67. Culic, O., Huang, Q. H., Flanagan, D., Hixson, D., and Lin, S. H. (1992). Molecular cloning and expression of a new rat liver cell-CAM105 isoform. Differential phosphorylation of isoforms. *Biochem. J.* **285:**47–53.
68. Che, M., Nishida, T., Gatmaitan, Z., and Arias, I. M. (1992). A nucleoside transporter is functionally linked to ectonucleotidases in rat liver canalicular membrane. *J. Biol. Chem.* **267:**9684–9688.
69. Cheung, P. H., Luo, W., Qiu, Y., Zhang, X., Earley, K., Millirons, P., and Lin, S.-H. (1993). Structure and function of C-CAM1. The first immunoglobulin domain is required for intracellular adhesion. *J. Biol. Chem.* **268:**24303–24310.
70. Obrink, B. (1991). C-CAM (cell-CAM 105)—a member of the growing immunoglobulin superfamily of cell adhesion proteins. *Bioessays* **13:**227–234.
71. Hohmann, J., Kowalewski, H., Vogel, M., and Zimmermann, H. (1993). Isolation of a Ca^{2+} or Mg^{2+}-activated ATPase (ecto-ATPase) from bovine brain synaptic membranes. *Biochim. Biophys. Acta* **1152:**146–154.
72. Sippel, C. J., Suchy, F. J., Ananthanarayanan, M., and Perlmutter, D. H. (1993). The rat liver ecto-ATPase is also a canalicular bile acid transport protein. *J. Biol. Chem.* **268:**2083–2091.
73. Sippel, C. J., McCollum, M. J., and Perlmutter, D. H. (1994). Bile acid transport by the rat liver canalicular bile acid transport/ecto-ATPase protein is dependent on ATP but not on its own ecto-ATPase activity. *J. Biol. Chem.* **269:**2820–2826.
74. Kast, C., Stieger, B., Winterhalter, K. H., and Meier, P. J. (1994). Hepatocellular transport of bile acids. Evidence for distinct subcellular localizations of electrogenic and ATP-dependent taurocholate transport in rat hepatocytes. *J. Biol. Chem.* **269:**5179–5186.
75. von Dippe, P., Amoui, M., Alves, C., and Levy, D. (1993). Na^+-dependent bile acid transport by hepatocytes is mediated by a protein similar to microsomal epoxide hydrolase. *Am. J. Physiol.* **264:**G528–G534.
76. Alves, C., von Dippe, P., Amoui, M., and Levy, D. (1993). Bile acid transport into hepatocyte smooth endoplasmic reticulum vesicles is mediated by microsomal epoxide hydrolase, a membrane protein exhibiting two distinct topological orientations. *J. Biol. Chem.* **268:**20148–20155.
77. Hortensteiner, S., Vogt, E., Hagenbuch, B., Meier, P. J., Amrhein, N., and Martinoia, E. (1993). Direct energization of bile acid transport into plant vacuoles. *J. Biol. Chem.* **268:**18446–18449.
78. Ishikawa, T. (1992). The ATP-dependent glutathione S-conjugate pump. *Trends Biochem. Sci.* **17:**463–468.

79. Kuipers, F., Radominska, A., Zimniak, P., Little, J. M., Havinga, R., Vonk, R. J., and Lester, R. (1989). Defective biliary secretion of bile acid 3-O-glucuronides in rats with hereditary conjugated hyperbilirubinemia. *J. Lipid Res.* **30:**1835–1845.
80. De Vries, M. H., Redegeld, F. A. M., Koster, A. S., Noordhoek, J., De Haan, J. G., Oude Elferink, R. P. J., and Jansen, P. L. M. (1989). Hepatic, intestinal and renal transport of 1-naphthol-β-D-glucuronide in mutant rats with hereditary-conjugated hyperbilirubinemia. *Naunyn-Schmiedebergs Arch. Pharmacol.* **340:**588–592.
81. Kobayashi, K., Komatsu, S., Nishi, T., Hara, H., and Hayashi, K. (1991). ATP-dependent transport for glucuronides in canalicular plasma membrane vesicles. *Biochem. Biophys. Res. Commun.* **176:** 622–626.
82. Ishikawa, T., Mueller, M., Kluenemann, C., Schaub, T., and Keppler, D. (1990). ATP-dependent primary active transport of cysteinyl leukotrienes across liver canalicular membrane. Role of the ATP-dependent transport system for glutathione S-conjugates. *J. Biol. Chem.* **265:**19279–19286.
83. Zimniak, P., Awasthi, S., and Awasthi, Y. C. (1993). Phase III detoxification system. *Trends Biochem. Sci.* **18:**164–165.
84. Inoue, M., Akerboom, T. P., Sies, H., Kinne, R., Thao, T., and Arias, I. M. (1984). Biliary transport of glutathione S-conjugate by rat liver canalicular membrane vesicles. *J. Biol. Chem.* **259:**4998–5002.
85. Nicotera, P., Moore, M., Bellomo, G., Mirabelli, F., and Orrenius, S. (1985). Demonstration and partial characterization of glutathione disulfide-stimulated ATPase activity in the plasma membrane fraction from rat hepatocytes. *J. Biol. Chem.* **260:**1999–2002.
86. Nicotera, P., Baldi, C., Svensson, S.-A., Larsson, R., Bellomo, G., and Orrenius, S. (1985). Glutathione S-conjugates stimulate ATP hydrolysis in the plasma membrane fraction of rat hepatocytes. *FEBS Lett.* **187:**121–125.
87. Kobayashi, K., Sogame, Y., Hayashi, K., Nicotera, P., and Orrenius, S. (1988). ATP stimulates the uptake of S-dinitrophenylglutathione by rat liver plasma membrane vesicles. *FEBS Lett.* **240:**55–58.
88. Kunst, M., Sies, H., and Akerboom, T. P. M. (1989). ATP-stimulated uptake of S-(2,4-dinitrophenyl)glutathione by plasma membrane vesicles from rat liver. *Biochim. Biophys. Acta* **983:**123–125.
89. Kitamura, T., Jansen, P., Hardenbrook, C., Kamimoto, Y., Gatmaitan, Z., and Arias, I. M. (1990). Defective ATP-dependent bile canalicular transport of organic anions in mutant (TR^-) rats with conjugated hyperbilirubinemia. *Proc. Natl. Acad. Sci. USA* **87:** 3557–3561.
90. Kobayashi, K., Sogame, Y., Hara, H., and Hayashi, K. (1990). Mechanism of glutathione S-conjugate transport in canalicular and basolateral rat liver plasma membranes. *J. Biol. Chem.* **265:**7737–7741.
91. Akerboom, T. P. M., Narayanaswami, V., Kunst, M., and Sies, H. (1991). ATP-dependent S-(2,4-dinitrophenyl)glutathione transport in canalicular plasma membrane vesicles from rat liver. *J. Biol. Chem.* **266:**13147–13152.
92. Nishida, T., Hardenbrook, C., Gatmaitan, Z., and Arias, I. M. (1992). ATP-dependent organic anion transport system in normal and TR^- rat liver canalicular membranes. *Am. J. Physiol.* **262:**G629–G635.
93. Nishida, T., Gatmaitan, Z., Roy-Chowdhury, J., and Arias, I. M. (1992). Two distinct mechanisms for bilirubin glucuronide transport by rat bile canalicular membrane vesicles. Demonstration of defective ATP-dependent transport in rats (TR^-) with inherited conjugated hyperbilirubinemia. *J. Clin. Invest.* **90:**2130–2135.
94. Goeser, T., Nakata, R., Braly, L. F., Sosiak, A., Campbell, C. G., Dermietzel, R., Novikoff, P. M., Stockert, R. J., Burk, R. D., and Wolkoff, A. W. (1990). The rat hepatocyte plasma membrane organic anion binding protein is immunologically related to the mitochondrial F1 adenosine triphosphatase beta subunit. *J. Clin. Invest.* **86:**220–227.
95. Board, P. G. (1981). Transport of glutathione S-conjugate from human erythrocytes. *FEBS Lett.* **124:**163–165.
96. Awasthi, Y. C., Misra, G., Rassin, D. K., and Srivastava, S. K. (1983). Detoxification of xenobiotics by glutathione S-transferases in erythrocytes: The transport of the conjugate of glutathione and 1-chloro-2,4-dinitrobenzene. *Br. J. Haematol.* **55:**419–425.
97. LaBelle, E. F., Singh, S. V., Srivastava, S. K., and Awasthi, Y. C. (1986). Dinitrophenyl glutathione efflux from human erythrocytes is primary active ATP-dependent transport. *Biochem. J.* **238:**443–449.
98. LaBelle, E. F., Singh, S. V., Ahmad, H., Wronski, L., Srivastava, S. K., and Awasthi, Y. C. (1988). A novel dinitrophenylglutathione-stimulated ATPase is present in human erythrocyte membranes. *FEBS Lett.* **228:**53–56.
99. Sharma, R., Gupta, S., Singh, S. V., Medh, R. D., Ahmad, H., LaBelle, E. F., and Awasthi, Y. C. (1990). Purification and characterization of dinitrophenylglutathione ATPase of human erythrocytes and its expression in other tissues. *Biochem. Biophys. Res. Commun.* **171:**155–161.
100. Singhal, S. S., Sharma, R., Gupta, S., Ahmad, H., Zimniak, P., Radominska, A., Lester, R., and Awasthi, Y. C. (1991). The anionic conjugates of bilirubin and bile acids stimulate ATP hydrolysis by S-(dinitrophenyl) glutathione ATPase of human erythrocytes. *FEBS Lett.* **281:**255–257.
101. Sharma, R., Gupta, S., Ahmad, H., Ansari, G. A. S., and Awasthi, Y. C. (1990). Stimulation of a human erythrocyte membrane ATPase by glutathione conjugates. *Toxicol. Appl. Pharmacol.* **104:**421–428.
102. Saxena, M., Singhal, S. S., Awasthi, S., Singh, S. V., LaBelle, E. F., Zimniak, P., and Awasthi, Y. C. (1992). Dinitrophenyl S-glutathione ATPase purified from human muscle catalyzes ATP hydrolysis in the presence of leukotrienes. *Arch. Biochem. Biophys.* **298:**231–237.
103. Awasthi, Y. C., Singhal, S. S., Gupta, S., Ahmad, H., Zimniak, P., Radominska, A., Lester, R., and Sharma, R. (1991). Purification and characterization of an ATPase from human liver which catalyzes ATP hydrolysis in presence of the conjugates of bilirubin, bile acids and glutathione. *Biochem. Biophys. Res. Commun.* **175:**1090–1096.
104. Zimniak, P., Ziller, S. A., III, Panfil, I., Radominska, A., Wolters, H., Kuipers, F., Sharma, R., Saxena, M., Moslen, M. T., Vore, M., Vonk, R., Awasthi, Y. C., and Lester, R. (1992). Identification of an anion-transport ATPase that catalyzes glutathione conjugate-dependent ATP hydrolysis in canalicular plasma membranes from normal rats and rats with conjugated hyperbilirubinemia (GY mutant). *Arch. Biochem. Biophys.* **292:**534–538.
105. Board, P., Nishida, T., Gatmaitan, Z., Che, M., and Arias, I. M. (1992). Erythrocyte membrane transport of glutathione conjugates and oxidized glutathione in the Dubin–Johnson syndrome and in rats with hereditary hyperbilirubinemia. *Hepatology* **15:**722–725.
106. LaBelle, E. F., Singh, S. V., Srivastava, S. K., and Awasthi, Y. C. (1986). Evidence for different transport systems for oxidized glutathione and S-dinitrophenyl glutathione in human erythrocytes. *Biochem. Biophys. Res. Commun.* **139:**538–544.
107. Awasthi, S., Singhal, S. S., Srivastava, S. K., Zimniak, P., Bajpai, K. K., Saxena, M., Sharma, R., Ziller, S. A., Frenkel, E., Singh, S. V., He, N. G., and Awasthi, Y. C. (1994). Adenosine triphosphate-dependent transport of doxorubicin, daunomycin and vinblastine in human tissues by a mechanism distinct from the P-glycoprotein. *J. Clin. Invest.* **93:**958–965.
108. Sexana, M., Sharma, R., Singhal, S. S., Ahmad, H., and Awasthi, Y. C. (1991). The effect of doxorubicin on the transport of glutathione conjugates from human erythrocytes. *Biochem.* Arch. **7:**285–292.
109. Crawford, J. M., and Gollan, J. L. (1993). Bilirubin metabolism and the pathophysiology of jaundice. In *Diseases of the Liver*, 7th ed. (L. Schiff and E. R. Schiff, eds.), Lippincott, Philadelphia, pp. 42–84.
110. Chowdhury, J. R., Chowdhury, N. R., Wolkoff, A. W., and Arias, I. M. (1994). Heme and bile pigment metabolism. In *The Liver: Biology and Pathology*, 3rd ed. (I. M. Arias, N. Fausto, W. B. Jakoby, D. Schachter, and D. Shafritz, eds.), Raven Press, New York, pp. 471–504.

111. Berk, P. D. (1985). Bilirubin metabolism and hereditary hyperbilirubinemias. In *Bockus Gastroenterology,* 4th ed. (J. E. Berk, ed.), Saunders, Philadelphia, pp. 2732–2797.
112. Dubin, I. N. (1958). Chronic idiopathic jaundice. A review of fifty cases. *Am. J. Med.* **24:**268–292.
113. Rosenthal, P., Kabra, P., Blanckaert, N., Kondo, T., and Schmid, R. (1981). Homozygous Dubin–Johnson syndrome exhibits a characteristic serum bilirubin pattern. *Hepatology* **1:**540 (abstr.).
114. Weiss, J. S., Gautam, A., Lauff, J. J., Sunderberg, M. W., Jatlow, P., Boyer, J. L., and Seligson, D. (1983). The clinical importance of protein-bound fraction of serum bilirubin in patients with hyperbilirubinemia. *N. Engl. J. Med.* **309:**147–150.
115. Cohen, L., Lewis, C., and Arias, I. M. (1972). Pregnancy, oral contraceptives, and chronic familial jaundice with predominantly conjugated hyperbilirubinemia (Dubin–Johnson syndrome). *Gastroenterology* **62:**1182–1190.
116. Shani, M., Seligsohn, U., Gilon, E., Sheba, C., and Adam, A. (1970). Dubin–Johnson syndrome in Israel. I. Clinical, laboratory, and genetic aspects of 101 cases. *Q. J. Med.* **39:**549–567.
117. Seligsohn, U., Shani, M., Ramot, B., Adam, A., and Sheba, C. (1970). Dubin–Johnson syndrome in Israel. II. Association with factor-VII deficiency. *Q. J. Med.* **39:**569–584.
118. Bar-Meir, S., Baron, J., Seligson, U., Gottesfeld, F., Levy, R., and Gilat, T. (1982). 99mTc-HIDA cholescintigraphy in Dubin–Johnson and Rotor syndromes. *Radiology* **142:**743–746.
119. Klatskin, G., and Conn, H. D. (1993). *Histophatology of the Liver,* Oxford University Press, London.
120. Kitamura, T., Alroy, J., Gatmaitan, Z., Inoue, M., Mikami, T., Jansen, P., and Arias, I. M. (1992). Defective biliary excretion of epinephrine metabolites in mutant (TR^-) rats: Relation to the pathogenesis of black liver in the Dubin–Johnson syndrome and Corriedale sheep with an analogous excretory defect. *Hepatology* **15:** 1154–1159.
121. Hunter, F. M., Sparks, R. D., and Flinner, R. L. (1964). Hepatitis with resulting mobilization of hepatic pigment in a patient with Dubin–Johnson syndrome. *Gastroenterology* **47:**631.
122. Swartz, H. M., Sarna, T., and Varma, R. R. (1979). On the nature and excretion of the hepatic pigment in the Dubin–Johnson syndrome. *Gastroenterology* **76:**958–964.
123. Shani, M., Gilon, E., Ben-Ezzer, J., and Sheba, C. (1970). Sulfobromophthalein tolerance test in patients with Dubin–Johnson syndrome and their relatives. *Gastroenterology* **59:**842–847.
124. Schoenfield, L. J., McGill, D. B., Hunton, D. B., Foulk, W. T., and Butt, H. R. (1963). Studies of chronic idiopathic jaundice (Dubin–Johnson syndrome). I. Demonstration of hepatic excretory defect. *Gastroenterology* **44:**101–111.
125. Berk, P. D., Osola, L. M., and Jones, E. A. (1986). Specific defects in hepatic storage and clearance of bilirubin. In *Bile Pigments and Jaundice* (J. D. Ostrow, ed.), Dekker, New York, pp. 279–316.
126. Kappas, A., Sassa, S., Galbraith, R. A., and Nordmann, Y. (1989). The porphyrias. In *The Metabolic Basis of Inherited Disease* (C. R. Scriver, A. L. Beaudet, W. S. Sly, and D. Valle, eds.), McGraw–Hill, New York, pp. 1305–1365.
127. Kaplowitz, N., Javitt, N., and Kappas, A. (1972). Coproporphyrin I and III excretion in bile and urine. *J. Clin. Invest.* **51:**2895–2899.
128. Wolkoff, A. W., Cohen, L. E., and Arias, I. M. (1973). Inheritance of the Dubin–Johnson syndrome. *N. Engl. J. Med.* **288:**113–117.
129. Cornelius, C. E. (1986). Comparative bile pigment metabolism in vertebrates. In *Bile Pigments and Jaundice* (J. D. Ostrow, ed.), Dekker, New York, pp. 601–647.
130. Jansen, P. L. M., Peters, W. H., and Lamers, W. H. (1985). Hereditary chronic conjugated hyperbilirubinemia in mutant rats caused by defective hepatic anion transport. *Hepatology* **5:**573–579.
131. Kuipers, F., Enserink, M., Havinga, R., van der Steen, A. B. M., Hardonk, M. J., Fevery, J., and Vonk, R. J. (1988). Separate transport systems for biliary secretion of sulfated and unsulfated bile acids in the rat. *J. Clin. Invest.* **81:**1593–1599.
132. Oude Elferink, R. P. J., Ottenhoff, R., Liefting, W., de Haan, J., and Jansen, P. L. M. (1989). Hepatobiliary transport of glutathione and glutathione conjugate in rats with hereditary hyperbilirubinemia. *J. Clin. Invest.* **84:**476–483.
133. Oude Elferink, R. P. J., Ottenhoff, R., Radominska, A., Hofmann, A. F., Kuipers, F., and Jansen, P. L. M. (1991). Inhibition of glutathione-conjugate secretion from isolated hepatocytes by dipolar bile acids and other organic anions. *Biochem.* J. **274:**281–286.
134. Huber, M., Guhlmann, A., Jansen, P. L. M., and Keppler, D. (1987). Hereditary defect of hepatobiliary cysteinyl leukotriene elimination in mutant rats with defective hepatic anion excretion. *Hepatology* **7:**224–228.
135. Jansen, P. L. M., Van Klinken, J.-W., Van Gelder, M., Ottenhoff, R., and Oude Elferink, R. P. J. (1993). Preserved organic anion transport in mutant TR^- rats with a hepatobiliary secretion defect. *Am. J. Physiol.* **265:**G445–G452.
136. Leier, I., Jedlitschky, G., Buchholz, U., and Keppler, D. (1994). Characterization of the ATP-dependent leukotriene C4 export carrier in mastocytoma cells. *Eur. J. Biochem.* **220:**599–606.
137. Pikula, S., Hayden, J. B., Awasthi, S., Awasthi, Y. C., and Zimniak, P. (1994). Organic anion-transporting ATPase of rat liver. I. Purification, regulation by phosphorylation, and photoaffinity labeling. *J. Biol. Chem.* **269:**27566–27573.
138. Pikula, S., Hayden, J. B., Awasthi, S., Awasthi, Y. C., and Zimniak, P. (1994). Organic anion-transporting ATPase of rat liver. II. Functional reconstitution of active transport and regulation of phosphorylation. *J. Biol. Chem.* **269:**27574–27579.
139. Schaub, T., Ishikawa, T., and Keppler, D. (1991). ATP-dependent leukotriene export from mastocytoma cells. *FEBS Lett.* **279:**83–86.
140. Kondo, T., Miyamoto, K., Gasa, S., Taniguchi, N., and Kawakami, Y. (1989). Purification and characterization of glutathione disulfide-stimulated Mg^{2+}-ATPase from human erythrocytes. *Biochem. Biophys. Res. Commun.* **162:**1–8.
141. Kondo, T., Kawakami, Y., Taniguchi, N., and Beutler, E. (1987). Glutathione disulfide-stimulated Mg^{2+}-ATPase of human erythrocyte membranes. *Proc. Natl. Acad. Sci. USA* **84:**7373–7377.
142. Zimniak, P., Ziller, S., III, Radominska, A., Blaauw, M., Wolters, H., Kuipers, F., Saxena, M., Singhal, S., Vonk, R., Awasthi, Y. C., and Lester, R. (1992). Photoaffinity labeling and partial purification of a transporting glutathione conjugate-dependent ATPase from canalicular rat liver plasma membranes. *FASEB J.* **6:**A117 (abstr.).
143. Roelofsen, H., Ottenhoff, R., Oude Elferink, R. P., and Jansen, P. L. (1991). Hepatocanalicular organic-anion transport is regulated by protein kinase C. *Biochem. J.* **278:**637–641.

27

The Pathophysiology of Diarrhea

Joseph H. Sellin

27.1. INTRODUCTION

The gut faces the daily challenge of transforming an intake highly variable in volume and composition into a manageable solution, from which it extracts the necessary nutrients, electrolytes, minerals, and water, simultaneously excluding bacteria, antigens, and toxins. An input of 8–9 liters is predictably reduced to an input of 100–200 cm^3/day. Specific segmental transport pathways represent the traffic routes that ions, nutrients, and other solutes necessarily follow. The traffic is regulated at several different levels, including neural, hormonal, immune, and luminal. Normally this traffic runs smoothly, but when the regulation does go awry, diarrhea may result.

The multiple levels of control generally serve the gut well. But when the intestine is challenged by poorly absorbable luminal contents, diminished absorptive capacity, or stimulation of secretion by toxins or inflammatory mediators, excessive fluid accumulates in the intestinal lumen. The intestine can adapt to increased intestinal fluid losses. The colon is capable of increasing its absorptive capacity two- to threefold in the face of increased small bowel losses. Carbohydrate malabsorption in the small bowel may be compensated for by colonic salvage of short-chain fatty acids. Adaptation after intestinal resection clearly occurs.

Rapid strides have been made in understanding the molecular basis of intestinal transport. However, correlating these fundamental molecular events with the clinical pathophysiology of diarrhea remains a challenge. This chapter will provide an overview of the mechanisms of ion transport and regulation of fluid movement and then examine how these models can be applied to specific diarrheal diseases, including cholera, inflammatory bowel disease, and AIDS-related diarrhea.

27.2. CLASSIFICATIONS OF DIARRHEA

The categorization of diarrheas into secretory and osmotic has served as a useful heuristic device in our understanding of the pathophysiologic factors that determine the development of clinical symptoms. By calculating whether electrolytes can account for the osmolality of fecal water, one can determine what underlying mechanisms are involved. Recent clinical studies have more carefully substantiated the ability of this approach to distinguish secretory and osmotic types of diarrhea.[1] However, it is relatively uncommon to find a case of pure osmotic or secretory diarrhea. Instead, most diarrheas tend to be multifactorial; changes in epithelial structure, absorptive function, and motility may commonly outweigh either osmotic factors or electrogenic anion secretion as the determinants of a diarrheal illness. With some diarrheas it still remains difficult to implicate any specific factors and only broad generalities can be applied. In this chapter, we will emphasize the secretory mechanisms involved in diarrhea, with the caveat that consideration necessarily has to be given to other contributing factors.

27.3. BASIC CONSIDERATIONS IN ION TRANSPORT

27.3.1. Segmental Heterogeneity

Two types of segmental heterogeneity are important considerations in modeling intestinal transport function. Significant differences exist along the crypt–villus axis in the small intestine or the crypt–surface axis in the colon (Figure 27.1). Crypt cells exhibit predominantly secretory characteristics; villus and surface cells are absorptive.[2] As epithelial cells migrate up from the base of the crypt, they acquire additional transporters. In the small bowel, Na–nutrient-coupled transporters are primarily present in the mature villus cells, but absent from crypt cells.[3] A similar pattern exists for the apical Na–H exchanger.[4] This may have important clinical implications. If a pathological process destroys primarily surface cells, the absorptive capacity of the epithelium will be impacted to a much greater extent than the secretory capacity. This selective damage of the epithelium could shift the balance of ion transport within the intestine. A similar effect may be obtained if there is a more rapid turnover of epithelial cells, with a preponderance of relatively immature cells appearing farther up the villus than normal.[2,5] However, this dichotomy between absorptive surface cells and secretory crypt cells should not be considered absolute. Surface cells probably retain some secretory capacity.[6]

Joseph H. Sellin • Departments of Medicine and Integrative Biology, The University of Texas–Houston Medical School, Houston, Texas 77030.
Molecular Biology of Membrane Transport Disorders, edited by S.G. Schultz *et al.* Plenum Press, New York, 1996

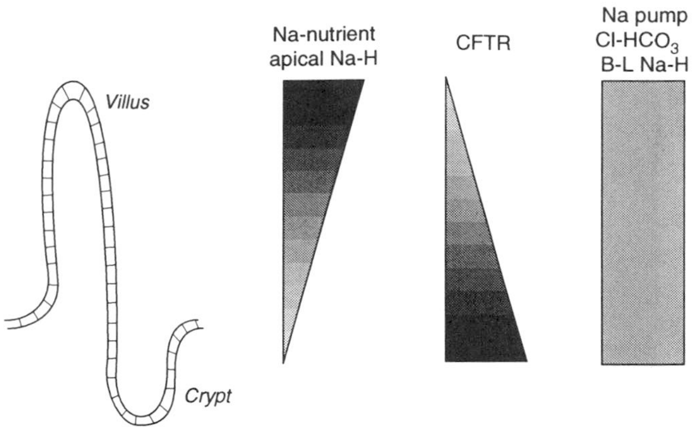

Figure 27.1. Segmental heterogeneity along the crypt–villus axis. Specific transporters exhibit a differential expression along the crypt–villus axis. This may account for the geographical localization of absorptive function toward the villus and secretory function in the crypt.

A second type of heterogeneity exists along the length of the gut. Different regions of the intestine exhibit a distinct array of transporters and significant variability in permeability. Some of these differences have long been recognized, such as between absorptive mechanisms in the distal colon and the jejunum, whereas others have been appreciated more recently. It is now apparent that the cecum, proximal colon, and distal colon exhibit distinctly different transporters.[7,8] Small intestinal transport is dominated by Na–nutrient-coupled absorption and electroneutral Na absorption mediated by Na–H exchange. The distal colon of rabbit and human demonstrates the classic amiloride-inhibitable electrogenic Na transport pathway. Rabbit proximal colon has a primarily electroneutral NaCl absorptive pathway, with limited electrogenic transport of either Na or Cl. The cecum absorbs Na avidly by an electrogenic process that is not readily blocked by amiloride; this cecal Na absorption may be mediated by an apical nonselective cation channel.[9] Human colon exhibits a similar segmental heterogeneity of transport.[10] The variation in transport characteristics is indicative of diverse physiologic function and perhaps pathologic response. Clearly the epithelial response to a particular agonist will be constrained by the presence or absence of specific sets of transporters. Although the segmental differences in Na absorptive mechanisms have been more carefully delineated, differences in secretory responses are also present (see below).

Permeability. Another level of segmental heterogeneity within the gut involves permeability, the relative "tightness" or "leakiness" of the intestine (Figure 27.2). The images of a

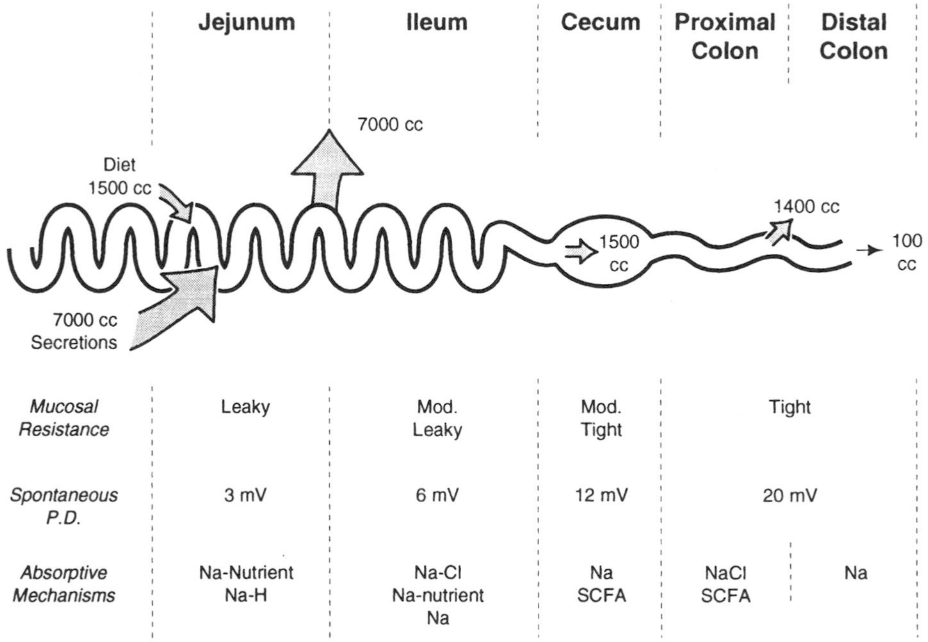

Figure 27.2. Overview of intestinal fluid balance. Eight to nine liters of fluid flows into intestine. Salivary, gastric, biliary, pancreatic, and intestinal secretions make up the bulk of this amount. Most intestinal fluid is absorbed in the small bowel, with approximately 1500 cm^3 of fluid crossing the ileocecal valve. The colon extracts most of this fluid, leaving 100 to 200 cm^3 of still water daily. In progression down the intestine, it becomes progressively "tighter." Spontaneous potential difference measured *in vitro* demonstrates a corresponding rise. Absorptive mechanisms in each segment of the gut differ, whereas chloride secretion is found throughout the intestine. [From Sellin, J. H. Intestinal electrolyte absorption and secretion. In *Gastrointestinal Disease,* 5th ed. (Sleisenger and Fordtran, eds.), Saunders, Philadelphia, Volume 1, p. 955.]

relatively impermeable or porous membrane are not physiologically precise, but are quite useful. Leaky epithelia like the jejunum generally transfer large amounts of isotonic fluid, while tight epithelia usually effect net solute transfer against a gradient. The rectum and distal colon are the tightest regions of the intestines, but only moderately tight when compared to frog skin or bladder.

The "tightness" of an epithelium is determined by the paracellular shunt pathway; *in vitro* electrical parameters predict the tightness of an epithelium, as exhibited by high transepithelial resistance and voltage. The paracellular shunt conductance provides the mechanism for maintaining or dissipating the transcellular potential difference; thus, in a leaky epithelium, the relatively high shunt conductance tends to short-circuit the potential difference generated by transcellular transport.

The anatomic basis of "tightness" is most likely the intercellular tight junction, or zona occludens (ZO), a zone of 0.1 to 0.6 μm where the lateral membranes of adjacent cells are in close apposition. Freeze-fracture electron microscopy reveals "kiss sites," an interlaced net of strands and grooves. The number of ZO strands tends to correlate with resistance to passive ion flow and epithelial resistance. There may be a series of fixed negative charges associated with the ZO, which account for the cation selectivity of the epithelium. Differences between ZOs of crypt and villus cells may lead to a greater permeability to passive ion movement in the crypts.[11,12]

ZOs no longer can be thought of as static structures. There is increasing evidence that ZOs respond to a variety of changes in the transport status of the epithelium. Stimulation of Na–glucose cotransport reduces the resistance of ZOs two- to threefold, with corresponding dilations of the paracellular space within the ZO. This is associated with increased permeability of the epithelium to multiple small molecules, suggesting an increase in the "leakiness" of the epithelium.[11–13] Bacterial toxins also modify the paracellular pathway; these permeability changes may be important in the pathophysiology of infectious diarrhea. Whether predetermined (genetic) differences in permeability play a role in the development of chronic diarrheas or inflammatory bowel disease is an important, but as yet unanswered question. Thus, there are several levels of heterogeneity that must be considered within the intestine. The interplay among these regional differences, either along the crypt–villus axis or the length of the gut, raises important issues for the molecular biologist, the physiologist, and the clinician in understanding the etiology of diverse diarrheas.

27.3.2. Basics of Electrogenic Chloride Secretion

A common mechanism of electrogenic chloride secretion exists in several epithelia, including the small intestine, colon, cornea, and trachea. The general applicability of this model to different organs and different diseases has provided an important focus for research in epithelial biology. A detailed description of this model can be found in Chapter 22. What distinguishes (and complicates) the understanding of, and the correlation between, secretion on the basic level and diarrhea on the clinical level is the array of regulatory factors, including luminal contents, enteric neurons, immunomodulators, hormonal and paracrine inputs, and intestinal motility. More than with any other transporting epithelia, in the intestines these factors beyond the epithelium complicate our understanding and challenge our ability to successfully model the sequence of transport events leading to diarrhea. However, even with these considerations, all approaches to diarrhea and intestinal transport must begin with the consensus model that has been developed over the last several years. This model of secretion entails both discrete basolateral entry steps into the cell and an exit step across the apical membrane. The basolateral entry step couples the uphill movement of Cl to Na entry and is electroneutral. In contrast to electroneutral NaCl absorption, which is generally mediated by dual antiporters exchanging Na for H and Cl for HCO_3, this pathway is mediated by a single carrier that binds three ionic species in the ratio 1 Na : 1 K : 2 Cl and is inhibited by the loop diuretics bumetanide and furosemide.[14–16]

Cl entry into the cell is propelled by the electrochemical gradient favoring Na entry across the basolateral membrane (Figure 27.3). Na then exits via the Na pump, resulting in no net transcellular Na movement. Cl accumulates in the cell above its electrochemical equilibrium, then exits the cell through an apical chloride channel into the intestinal lumen.

Potassium entering through the electroneutral carrier is recycled across the basolateral membrane through K channels.[16,17] Although this may seem like an inefficient and futile recycling, linkage of K to NaCl influx serves several purposes. Basolateral K exit electrically balances the large Cl flux across the apical membrane. Continued K exit across the

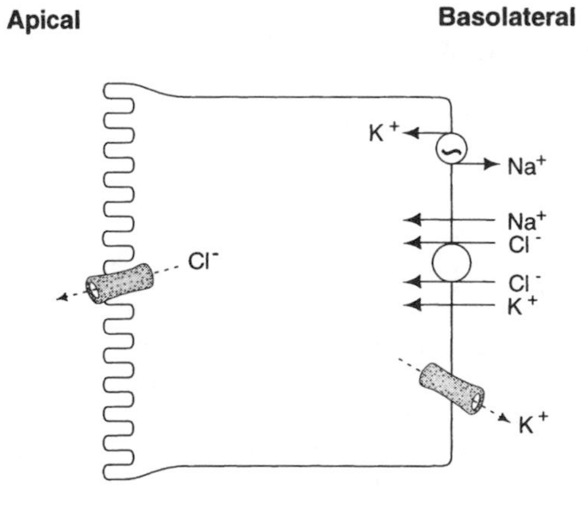

Figure 27.3. Chloride secretion. Discrete basolateral entry steps and apical exit steps are integral to chloride secretion. A carrier couples the movement of sodium, potassium, and chloride in a 1:1:2 stoichiometry and permits chloride to accumulate in the cell above its electrochemical equilibrium. Chloride exits the cell across the apical membrane via a chloride channel. The sodium and potassium that entered with the chloride are recycled by the sodium pump and a basolateral potassium channel, respectively. [Sellin, J. H. Intestinal electrolyte absorption and secretion. In *Gastrointestinal Disease,* 5th ed. (Sleisenger and Fordtran, eds.) Saunders, Philadelphia, Volume 1, p. 960.]

basolateral membrane prevents the dissipation of the electrochemical driving force for apical Cl secretion.[16]

Molecular Basis for Specific Transport Processes

As is the case in much of biological research, identification at the molecular level of specific transport proteins and characterization of the electrophysiological behavior of individual transport proteins have become central goals in membrane biology. Elegant descriptions of this progress will be found elsewhere in this volume. We will focus briefly on the impact that molecular biology has had on our understanding of two specific transporters that are central to intestinal transport processes, CFTR and NHEs. In both cases, molecular tools have provided insights into the complexities and subtleties of intestinal transport.

CFTR. The discovery of CFTR as the underlying molecular defect in cystic fibrosis (CF) has, in one sense, clarified our understanding of intestinal secretion, but, in another, has highlighted our gaps of knowledge. Electrophysiological fingerprints exist for several different types of chloride channels in secretory epithelia, including the intestine.[18,19] These may respond uniquely or in combination to different agonists including cAMP, Ca^{2+}, and volume. CFTR has been shown to be the structural equivalent of a specific chloride conductance and thus integral to the process of Cl secretion.

Although it is clear that CF is associated with, among other abnormalities, a defect in AMP-stimulated Cl secretion, it has not been a straightforward exercise to relate the genetic defect to a specific electrophysiological abnormality at the level of a single chloride channel. There was an initial focus on a relatively large (40–50 pS), outwardly rectifying channel that was stimulated by cAMP.[20,21] However, expression experiments in a variety of systems produced a different channel, with a lower unitary conductance (<10 pS), no rectification, and differing pharmacological inhibitory profile.[22,23]

The relation between the small, linear channel and whole-cell recordings supports its role as CFTR-dependent Cl channel. Demonstration of its role in intact epithelia is a necessary step in fully characterizing its physiologic role, especially in relation to different Cl channels and other transport processes. It has been suggested that the outwardly rectified Cl channel may be related to the volume-sensitive Cl channel, but further characterization and subcellular localization will be necessary before the physiologic function of the 40- to 50-pS channel can be determined.[24]

Given the hypothesis that CF represents a defect in cAMP-stimulated secretion and that CF airway epithelia respond normally to Ca-mediated agonists, the expectation existed that a similar pattern of response would be seen in intestinal tissues. This has not been the case. Although freshly isolated normal intestinal epithelia respond to Ca-related agonists with an expected increase in Cl secretion, small intestine or colon from CF patients is insensitive to either Ca^{2+} ionophores or phorbol esters as well as cAMP-mediated agonists, suggesting a different defect in stimulus–secretion coupling in the CF gut.[25–27] It is clear that cAMP- and Ca-mediated stimulatory pathways may be intertwined, but the explanation for these results in CF intestinal tissues remains unclear. To resolve this will require careful characterization of apical Ca^{2+}-mediated Cl channels and delineation of the mechanisms of intestinal secretion induced by Ca^{2+}-related agonists. Unlike airway epithelia, the gut may have a dearth of Ca^{2+}-sensitive apical Cl channels. If, as has been suggested, in the colon, Ca^{2+}-mediated secretagogues operate primarily by activating basolateral K channels rather than a specific apical Cl channel, then the observations in CF intestine may be explained. In normal colon, activation of the basolateral K channels indirectly increases the driving forces for Cl secretion through cAMP-dependent Cl channels. However, in CF intestine, the lack of apical Cl channels may negate any effect of the activation of basolateral K channels.[28]

With identification of the gene, the gene product, and its biophysical fingerprint, investigators have reached the critical stage where they can begin to define the cellular anatomy and physiology of CFTR. Immunochemical studies have localized CFTR to the apical membrane of T84 cells, as would be expected for a Cl conductance.[29] Determination of CFTR mRNA by *in situ* hybridization showed a remarkable segmental heterogeneity with decreasing gradients of expression along the crypt–villus axis and along a proximal–distal axis.[30,31] This vertical distribution of message along the crypt–villus axis is consistent with the generalized transport model localizing the site of chloride secretion to the crypts. The implications of finding decreasing CFTR message as one moves down the gut are less clear. CFTR exhibits specific glycosylation patterns as it moves from the endoplasmic reticulum to the Golgi apparatus; it may also require specific intracellular proteins (e.g., heat-shock proteins) to function as chaperones in this intracellular trafficking.[32]

Theories about the defect in CF have been continuously modified as our knowledge about CFTR increases. Several lines of evidence suggest that there is a complex processing of normal CFTR in the epithelial cell and that the critical defect in CF may be in intracellular trafficking of the protein. The membrane expression of CFTR is temperature-sensitive.[33] There are suggestions that there may be differences in the posttranslational processing of CFTR, such as glycosylation, depending on the epithelial cell type.[34] Such differences may be important in determining different functional or regulatory properties of CFTR. In a particularly intriguing study, Morris *et al.* demonstrated that the polarization of an epithelial cell may be a necessary factor in expressing the Cl conductance associated with CFTR. The HT29 intestinal cell line, depending on culture conditions, can exhibit characteristics of either a polarized or nonpolarized epithelium. Only the polarized cells exhibited the small linear cAMP-sensitive Cl channel, although both the differentiated and undifferentiated cells had equivalent amounts of CFTR mRNA and protein.[35] This observation suggests that expression of the characteristic Cl channel behavior depends on a series of factors beyond message and protein. Differences in expression of Ca-sensitive Cl conductances in polarized and nonpolarized epithelial cells raise similar issues.[28] A caveat is in order: results of several of the recent approaches to characterizing channels may be subject to subtle but critical variables such as cell culture techniques and variations on patch-clamp methodology; it is imperative to relate these findings to perhaps less sophisticated but more physiological approaches to fully assess their function. The development of knockout animal models with predominant intestinal symptoms may provide the opportunity to more thoroughly evaluate the physiologic function of CFTR.[36] Determining the role of CFTR in acidification of intracellular organelles and colonic mucus secretion may now be possible with these animal models.

27.3.3. NHEs: Sodium–Hydrogen Exchangers

Na–H exchange activity can be demonstrated in essentially every cell type. Exchange of extracellular Na for intracellular H has been an integral component of most models of electroneutral NaCl absorption in the intestine. Studies employing either intestinal perfusion techniques in humans or basic membrane vesicle methodology have been consistent with an apical Na–H antiporter.[37–39] Further studies comparing Na–H exchange in apical and basolateral membrane vesicle preparations suggested that there may be a family of antiporters with different kinetic and regulatory features. Apical Na–H exchanger activity is more resistant to inhibition by amiloride than the basolateral exchanger.[40,41] Also, NHE activity in intestinal brush border membranes does not obviously exhibit the pH-sensitive allosteric modification site associated with the basolateral Na–H exchanger.[40,42] Intestinal Na–H antiporters mediating NaCl transport clearly exhibited different responses to a variety of agonists when compared to Na–H exchangers in other epithelia. For example, intestinal Na–H exchange activity is inhibited by cyclic nucleotide-dependent or Ca-dependent protein kinases; in other tissues, these kinases increase Na–H exchange activity.[43]

Four isoforms of Na–H exchangers have been cloned.[44–47] NHE-1 is found in both epithelial and nonepithelial cells. It most likely represents the ubiquitous "housekeeper" involved in regulation of intracellular pH, cell volume, and growth.[48] In polarized epithelia, NHE-1 is localized to the basolateral membrane, consistent with its suggested housekeeping role rather than a mediator of Na absorption. Comparatively little is known about NHE-2 and NHE-4, which may have significant homologies to each other. NHE-2 appears to be relatively specific for the intestine.

NHE-3 is one, if not *the,* apical Na–H exchanger that mediates electroneutral Na absorption in the intestine. NHE-3 is expressed predominantly in intestinal epithelia that exhibit apical Na–H exchange activity. It is localized to the villus cells in the small intestine and to the surface cells in the colon, consistent with the vertical heterogeneity of transport function.[4] Glucocorticoids stimulate Na absorption; NHE-3, but not NHE-1, is upregulated by glucocorticoids, consistent with its role in vectorial Na transport.[49] The ability to correlate changes in Na transport with changes in NHE isoforms will permit a penetrating insight into the mechanisms involved in regulating Na transport and additional cellular homeostatic functions.

27.3.4. Bicarbonate

In contrast to the elegant descriptions of the molecular basis of Cl secretion or Na–H exchange, the mechanisms of bicarbonate transport remain problematic. Bicarbonate secretion occurs in the ileum and colon. It is one of the major transport mechanisms of intestinal epithelia. Nevertheless, achieving a reasonable understanding of bicarbonate secretion remains one of the most vexing problems for gastrointestinal physiologists.

The problem is manyfold. To a certain extent, bicarbonate secretion is a chameleon. Clinical observation suggests that in significant secretory diarrheas such as cholera, the principal anion lost in the stool is bicarbonate.[50] Yet, as is obvious from even a cursory scan of the literature, cholera toxin or its surrogates stimulate Cl secretion *in vitro.* The reason for this remains elusive, but Cl and bicarbonate secretion may be intertwined in a yet-to-be-determined mechanism and the experimental results may be model-dependent.

Bicarbonate is a transported species, a metabolic product, and a probable regulator of intracellular processes. Changes in $pH/pCO_2/HCO_3$ may have profound effects on the transport of other ionic species, but there is no uniform pattern. Mucosal bicarbonate exposure to gallbladder epithelium stimulates NaCl absorption; in the intestine, increasing $[HCO_3]$ and pH inhibit salt absorption.[51,52] Recent studies employing measurements of intracellular pH have suggested that intracellular bicarbonate concentrations may modulate basal rates of Cl secretion in the rat colon. Similar studies have raised the possibility that pCO_2 in addition to pH may be a regulator of Na–H exchange.[53,54] Thus, at one level, bicarbonate can be viewed as a regulator of ion transport.

Changes in acid–base balance may affect bicarbonate secretion specifically and ion transport in general, but attempts to delineate the specific component(s) that regulate transport frequently yield conflicting data.[51,55]

Secreted bicarbonate may originate either in the subepithelial (serosal) space or intracellularly. Bicarbonate secretion may be either electrogenic or electroneutral. Proton absorption may mimic bicarbonate secretion. Separation of absorptive and secretory fluxes of bicarbonate has demonstrated discrete transport pathways and regulatory mechanisms for these unidirectional fluxes in the intestine. In fact, what had initially been thought to be an inhibition of bicarbonate secretion by epinephrine is more likely to be a stimulation of bicarbonate absorption.[56–59] Given the confounding variability of the techniques employed, it is not surprising that there is no clear consensus on a model for bicarbonate secretion.

In synthesizing observations from intestine and other bicarbonate-secreting epithelia (e.g., bladder, pancreas), several specific hypotheses emerge regarding both bicarbonate entry and exit steps across the epithelium (Figure 27.4). Bicarbonate may enter across the basolateral membrane either by diffu-

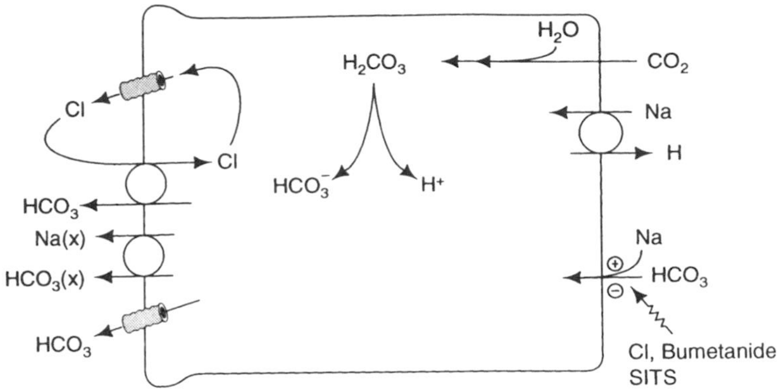

Figure 27.4. Bicarbonate secretion. Multiple mechanisms have been proposed for both basolateral entry and apical exit of bicarbonate in the intestinal epithelial cell. Basolateral entry may be mediated by diffusion of CO_2, Na–H exchange, or a Na-coupled bicarbonate transporter. Proposed apical exit steps include (1) a cyclic nucleotide-sensitive anion conductance that permits electrodiffusion of HCO_3 out of the cell, (2) a Na_y/HCO_{3x} symporter ($x>y$), or (3) coordinated function of a Cl/HCO_3 exchanger with a Cl-conductance, such that Cl is recycled while there is net movement of HCO_3.

sion of CO_2 or by an ion-specific transport mechanism. Bicarbonate secretion is a function of increasing serosal HCO_3,[57,59,60] Na-dependent,[56,60,61] inhibited by serosal [Cl],[57,61] and partially blocked by bumetanide.[57] These findings are consistent with the movement of an ionic species across the basolateral membrane via specific transporters. Differential effects of the pharmacological inhibitors SITS and amiloride are consistent with two different basolateral entry mechanisms, Na–H exchange and Na-coupled bicarbonate entry.[60] In contrast, an integral step in several models of bicarbonate secretion is the diffusion of CO_2 into the cell with subsequent conversion to HCO_3 by carbonic anhydrase. The inhibition of bicarbonate secretion by acetazolamide as found in pancreas and other epithelia[54,62] is consistent with this hypothesis. However, inhibition of secretion by carbonic anhydrase inhibitors is not a universal finding.[56,63] The preponderance of studies focusing on intestinal bicarbonate secretion have concluded that the secreted anion originates from serosal bicarbonate or CO_2 rather than intracellular CO_2.[56,57,60]

The apical exit step(s) for bicarbonate have not been specifically delineated. A consensus model of bicarbonate secretion for pancreas and bladder involves the synchronized function of an apical chloride channel (CFTR) with a Cl–HCO_3 exchanger. In this model, the "classic" electrogenic Cl secretory apparatus provides luminal Cl which is recycled across the apical membrane in exchange for intracellular HCO_3. The individual components of this model are certainly present in both small intestine and colon. The finding that luminal Cl stimulates HCO_3 secretion[57,61,64,65] is consistent with this theory; however, luminal chloride is clearly not an absolute requirement for HCO_3 secretion.[59]

A bicarbonate conductance on the apical membrane has been considered by several investigators. Patch-clamp studies in T84 cells indicate that HCO_3 will traverse Cl channels, albeit significantly less efficiently (P-HCO_3/P-Cl = 0.44).[66] Significant bicarbonate permeation through chloride channels in a pancreatic ductular carcinoma cell line has been observed.[67] In airway epithelial cells, cAMP stimulates an electrogenic HCO_3 secretion which is inhibited by the channel blocker DPC.[68] In rabbit ileum, lowering external [Cl] reveals a cyclic nucleotide-mediated electrogenic secretion of bicarbonate.[60] Similar findings have been reported in *Amphiuma* small intestine and turtle bladder; electrogenic bicarbonate secretion occurs in Cl-free solutions.[69,70] Thus, bicarbonate may exit the cell through an apical membrane anion conductance; whether it is specifically CFTR or a distinct channel remains to be determined. Clearly such a model is mechanistically simpler and therefore appealing; however, its feasibility may depend on more exacting determinations of relative permeabilities of different anions, intracellular activities, and electrochemical gradients.

An additional model of HCO_3 secretion depends on the geographical separation of antiporter systems along the crypt–villus axis in the ileum (Figure 27.5). Although villus cells

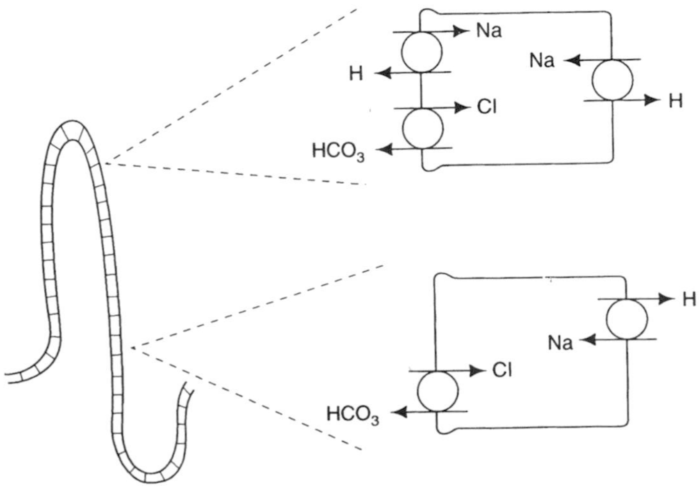

Figure 27.5. The segmental distribution of Na–H and Cl–HCO_3 exchangers along the crypt–villus axis in the small intestine may permit the net movement of HCO_3.

possess apical and basolateral Na–H exchangers, crypt cells apparently exhibit only basolateral Na–H exchangers. In contrast, Cl–HCO_3 exchangers are found on apical, but not basolateral, membranes of both crypt and villus cells.[71,72]

This separation provides a mechanism for villus cells to absorb NaCl, but for crypt cells to secrete HCO_3. Thus, this segmental heterogeneity along the crypt–villus axis may provide a mechanism for different transport functions; this pathway may account for electroneutral HCO_3 secretion.

Although intestinal HCO_3 secretory mechanisms are central to secretory diarrheas, methodological and conceptual barriers have precluded a full understanding of this process. Unlike other areas of transport, bicarbonate secretion will require an integrative, as well as a molecular, approach.

27.4. REGULATION OF ION TRANSPORT

The elaborate array of intestinal membrane transporters is finely regulated to carry out routine physiologic function. Given the multiplicity of threats to this equilibrium, the gut generally responds successfully and maintains its homeostasis. Occasionally the response may be unsuccessful or inappropriate, leading to clinically significant symptoms.

Regulation of ion transport occurs on three different levels: extracellular, intracellular, and homocellular. There are a complex arcana of luminal, neural, hormonal, paracrine, and inflammatory mediators that supply the extracellular signals for intestinal epithelial cells. The borders between these discrete systems have become increasingly blurred. Not only is there a constant interaction among these systems, but it may be difficult to categorize a specific agonist. For example, pancreatic cholera is mediated by islet cell tumors producing large amounts of vasoactive intestinal peptide (VIP). These VIPomas are considered a classically "hormonally" mediated diarrhea. However, VIP is not normally found in the adult pancreas; in contrast, its physiologic role in the gut is primarily as a peptidergic neurotransmitter. Thus, one must recognize a certain arbitrariness in classifying a particular agonist as hormone or neurotransmitter, when, in reality, it may function differently in different clinical settings. Given the multitude of interconnections, it may be more appropriate to view these individual mediators collectively as part of a superregulatory agency. Inside the cell, extracellular information is processed and translated into specific alterations of transporters through a series of second messengers. Finally, entry and exit of transported species across the two borders of the cell must be coordinated to maintain cellular integrity.

27.4.1. Extracellular Regulation

The extracellular signals may be divided into those that stimulate Na and water absorption, i.e., absorbagogues (Table 27.1), and those that either inhibit Na absorption or stimulate Cl secretion, i.e., secretagogues (Table 27.2). The net balance between the absorptive and secretory stimuli from all sources will determine rates of ion and fluid transport. Quite obviously, if secretory stimuli predominate, diarrhea will ensue.

Table 27.1. Intestinal Absorbagogues

Endogenous absorbagogues	Pharmacologic agents
Aldosterone	Steroids
Glucocorticoids	Octreotide
α-Adrenergic agonists	Cyclooxygenase inhibitors
Enkephalins	Lithium
Somatostatin	Clonidine (α-2 agonist)
Angiotensin	Propranolol
Peptide YY	Opiates
Neuropeptide Y	Berberine
Prolactin	

Whether there is a clinical correlate for a hyperabsorptive pattern is unclear.

27.4.1.1. Hormonal Regulation

Numerous hormones may alter ion transport *in vitro,* but only a few have demonstrable physiologic effects: mineralocorticoids, glucocorticoids, insulin, and, perhaps, secretin. Others may have an effect only when associated with a hormone-producing tumor, such as VIP or serotonin. Often, it is unclear whether a specific hormone has a significant physiologic role.

27.4.1.2. Neural Regulation

Neural regulation of epithelial function is critical for fluid and electrolyte transport. The extrinsic innervation of the gut has two basic components: (1) cholinergic stimulation of secretion, predominantly via vagal input, and (2) adrenergic stimulation of absorption via prevertebral and sympathetic ganglia. Tonic sympathetic and parasympathetic input modulates ion transport. The presence of a cholinergic secretory tone is supported by ablation and inhibitor experiments.[73,74] Loss of sympathetic regulatory mechanisms occurs in the "diabetic diarrhea" associated with diabetic autonomic neuropathy. "Replacement" therapy with α-2 adrenergic agonists may be therapeutic.[75]

The enteric nervous system (ENS) itself adds several labyrinthine layers of complex innervation to the neural regulation of the gut.[76–80] The ENS independently integrates the neural activity of the muscles and blood vessels of the gut, in addition to the epithelium, primarily through its two hubs, the ganglionated plexuses in the submucosal (Meissner's) and myenteric (Auerbach's) regions. As in any neural arc, there are sensory, inter-, and motor neurons. Relatively little is known about the sensory input into the ENS, but one would surmise that likely factors would be the luminal environment or volume. The sensory neurons may be capsaicin-sensitive. Endocrine and paracrine cells interspersed among epithelial cells may function as auxiliary sensors for the ENS. Interneurons appear to be primarily cholinergic, while motor neurons are cholinergic and VIPergic. These classifications represent the principal neurotransmitter for these neurons, but each type also releases additional neuroactive substances.[79,80] Peptidergic neurons may release VIP, CCK, bombesin, and other neu-

Table 27.2. Endogenous Secretagogues (Intestinal Secretagogues)

Agonist	Intracellular mediator	Source[a]
Prostaglandins	cAMP	Immune, mesenchymal
Bradykinin	cAMP	Immune
Arachidonic acid	cAMP	Immune, cell membranes
VIP	cAMP	ENS/VIPoma
Secretin	cAMP	Paracrine
Peptide histidine isoleucine	cAMP	ENS
Platelet-activating factor	cAMP	Immune
Oxygen-derived free radicals	cAMP	Immune
Adenosine	cAMP	Immune
Leukotrienes	??	Immune
Acetylcholine	Ca	ENS
Serotonin	Ca	ENS
Histamine	Ca	ENS, carcinoid
Substance P	Ca	Immune
Neurotensin	Ca	ENS
Atrial natriuretic peptide	cGMP	
Calcitonin/CGRP[a]	??	ENS, MCT
Gastrin	??	Paracrine
GIP	??	??
Motilin	??	??
Bombesin	??	??

[a]ENS, enteric nervous system; MCT, medullary carcinoma of the thyroid; CGRP, calcitonin-gene related peptide.

roactive agents. Determining the physiologic effects of a mix of agonists can be exponentially more difficult than tracking the effects of a single agonist. The ultimate target cells for these neurons include, quite obviously, epithelial, vascular, and muscle cells, but, in addition, other neurons, paracrine cells, and inflammatory and subepithelial cells.

Agents may function either as neurotransmitters or as neuromodulators. In contrast to the synaptic target of neurotransmitters, neuromodulators act at a presynaptic site on either the source neuron or another neuron. Neuromodulators fine-tune the neuronal circuits through feedback effects on their own release or by altering the release of other neurotransmitters of neighboring neurons.[76,77] The ability to map the neuronal circuitry of the gut has, to some extent, outpaced our understanding of the biologic significance of specific neuroactive agents.

27.4.1.3. Immunologic Regulation

Immunocytes in the lamina propria and their ever-increasing array of cytokines and other soluble products play an important coordinated role with the ENS and the paracrine–endocrine network in regulating fluid and electrolyte transport (Figure 27.6).

The number and type of inflammatory cells residing in the lamina propria, and thus their regulatory input, vary depending on the underlying state of the gut. Under normal, noninflamed conditions, the majority of immunocompetent cells in the lamina propria are T lymphocytes (60%), with smaller numbers of plasma cells and B lymphocytes (25–30%), macrophages (8–10%), with scattered mast cells and polymorphonuclear cells, usually eosinophils. Inflammatory changes obviously increase the number of intestinal immunocytes. Acute bacterial infections result in an increase in PMNs; in contrast, celiac sprue is associated with an increase in lymphocytes. In parasitic infections, the intestinal mast cell population expands dramatically.[81–83] In inflammatory bowel disease, all components of the immune system are activated with a particular increase in IgG-secreting cells. Thus, not all inflammatory changes will be the same and the effect on the epithelial cell's transport function may vary depending on which specific population of immunocytes is increased during "inflammation."

Inflammatory mediators released by immunocytes provide the linkage to changes in epithelial function. The eicosanoids (prostaglandins and leukotrienes) are a family of 20-carbon (from the Greek *eikosa,* meaning 20) oxygenated metabolites of arachidonic acid and are central to the secretory process associated with inflammation. Membrane phospholipids are converted to arachidonic acid and subsequently metabolized to cyclooxygenase (prostaglandins) and lipoxygenase (leukotrienes) products. There are complex differences in the biological actions within the families of leukotrienes and prostaglandins.[84] Within the normal intestine, more than 95% of gut prostaglandins arise from inflammatory cells in the submucosa. However, prostaglandins are preferentially metabolized by epithelial cells.[85,86] The secretory response of most other inflammatory mediators is linked, at some stage, to prostaglandin production. Bradykinin liberates arachidonic acid and stimulates prostaglandin production. Prostaglandins stimulate other subepithelial cells, particularly enteric neurons. Prostaglandins also directly affect the epithelial cell.[25,81,87–89]

cAMP is the principal intracellular second messenger for prostaglandins,[90–92] although there is evidence to support

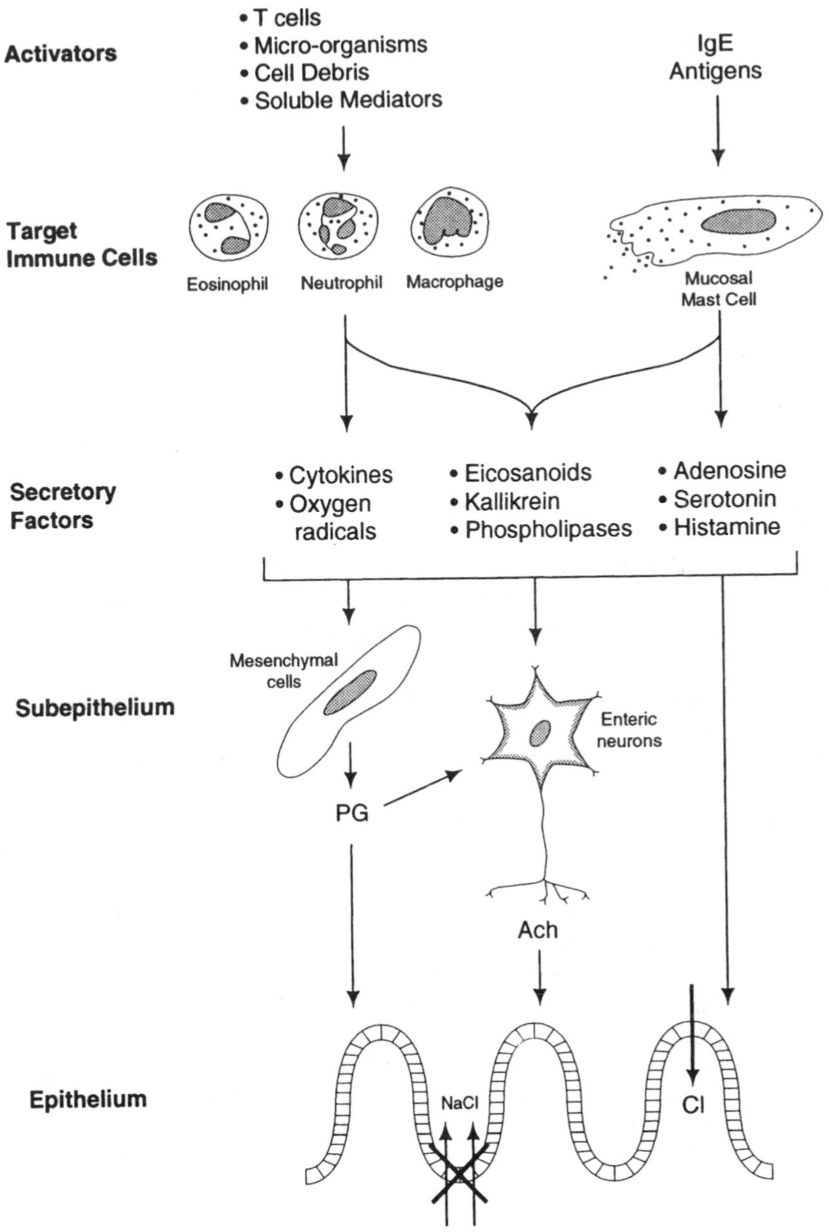

Figure 27.6. Immune responses. Multiple activators can stimulate target immune cells. These intestinal immune cells can release an array of secretory factors, which may act either directly on the epithelium or indirectly by stimulating the mesenchymal cells or enteric neurons to release prostaglandins or acetylcholine. [Adapted from Hinterleitner, T. A., and Powell, D. W. (1991). Immune system control of intestinal ion transport. *Proc. Soc. Exp. Biol. Med.* **197:**249.]

a role for intracellular calcium.[93,94] In addition to inhibiting electroneutral NaCl absorption and stimulating electrogenic Cl secretion, eicosanoids have significant effects on intestinal motility and blood flow. In some *in vitro* systems, prostaglandins contribute to a basal secretory tone; cyclooxygenase inhibitors such as indomethacin or aspirin increase basal rates of Na and Cl absorption.[95,96]

Mucosal mast cells are integral to several inflammatory processes, particularly parasitic infections and allergic reactions.[87] Because of their strategic location in close proximity to enteric neurons, blood vessels, and epithelial cells, mast cell release of histamine, serotonin, and adenosine may have important localized effects.[97] These three agents are all potent secretagogues, either by direct effects on the epithelial cells or by indirect neural stimulation and prostaglandin release.[87,98,99] Interestingly, like prostaglandins, these mediators also tend to affect intestinal motility. Although the purine nucleoside adenosine is most commonly recognized as a component of the energy store ATP, it is a potent agonist in its own right, an inflammatory mediator, acting on multiple organ systems. It stimulates electrogenic Cl secretion by activating both Cl and K channels. Adenosine may also act through cAMP as a second messenger, but there is some evidence suggesting that it exerts its intracellular effects independent of cyclic nucleotides or intracellular calcium.[100]

Several other novel secretagogues associated with inflammation have been identified. Oxidants such as superoxides, hydrogen peroxide, and hydroxyl radicals released from neutrophils stimulate electrogenic chloride secretion.[88,101] Chemotactic peptides that enhance neutrophil migration also stimulate secretion.[102] Cytokines such as interleukins 1 and 3 have a secretory effect.[103] Arachidonic acid, platelet-activating factor (PAF), substance P, kallikreins, and bradykinin may be important factors in the secretory response to inflammation.[81]

27.4.1.4. Systemic Factors in Regulation of Ion Transport

Acid–base balance modulates intestinal electrolyte transport, particularly electroneutral NaCl absorption, both *in vivo* and *in vitro*. Changes consistent with metabolic acidosis stimulate absorption, while changes of metabolic alkalosis tend to inhibit absorption.[51,104,105] Although it is probable that intracellular pH is the controlling factor, limited studies examining the governing intracellular factors have been performed. A recent report suggests that intracellular pCO_2, rather than pH, modulates Na–H exchange in the apical membrane of rat distal colon.[54] Intracellular HCO_3 affects basal rates of chloride secretion in the same tissue.[53] Correlations among different methodologies (e.g., vesicle and short-circuit techniques) will be necessary to clarify the mechanisms involved.

Intestinal blood flow and volume status may alter ion transport. Both active absorption and secretion are associated with increased intestinal blood flow.[106,107] This may simply reflect the tissue response to increased metabolic activity resulting from active transport. Decreased intravascular volume, as may be expected in hemorrhage or shock, elicits a series of adaptive responses that increase fluid absorption. Cardiopulmonary mechanoreceptors and carotid baroreceptors increase sympathetic input into the ENS which results in a decrease in secretion. Angiotensin II, antidiuretic hormone, and atrial natriuretic peptide may also have a role in regulating fluid transport under these conditions.

The metabolic status of the gut has an impact on its transport capacity. It is not surprising that a well-fed gut transports more effectively; what has become more apparent is the segmental heterogeneity of fuel preferences. Although the entire intestinal tract utilizes glucose, the small bowel epithelium prefers glutamine and the colon short-chain fatty acids, particularly butyrate.[108–110]

27.4.1.5. Luminal Factors Regulating Epithelial Transport

27.4.1.5a. Bacteria. The potential for bacterial, viral, or parasitic modification of gut function is evident. One of the distinguishing features of the intestine, compared to other epithelia, is its continual intimate contact with a multitude of complex bacteria. The interaction between bacteria and the apical membrane of colonic and small intestinal epithelial cells is an area of intense interest. Invasive bacteria, quite obviously, wreak havoc with the barrier function of the gut. But there are more subtle interactions that alter transport, permeability, and barrier functions.

Bacterial toxins, such as those produced by *Vibrio cholerae* or *E. coli*, target specific apical membrane receptors with subsequent sabotage of the cell's transport regulatory mechanisms through uncontrolled production of cAMP or other second messengers (Table 27.3). This results in electrogenic Cl secretion, the basis for the archetypical secretory diarrheas. However, bacteria possess additional virulence factors, beyond these classic enterotoxins, that may have specific interactions with the apical membrane.[111] Certain bacteria adhere closely to the mucosa, but neither produce an enterotoxin nor invade the epithelium. However, these effacing, adherent bacteria, such as the enteropathogenic strains of *E. coli* (EPEC), cause alterations of the apical membrane and decrease the transepithelial resistance of either native tissue or an epithelial monolayer.[112] The decrease in resistance may be related to cytoskeletal rearrangement secondary to contraction of the perijunctional ring. The use of discrete bacterial mutations permits a correlation between specific bacterial adherence factors, such as intimin, and subsequent cytoskeletal rearrangement.[113,114] The range of modulatory effects of bacteria on epithelial function may be considerable.[115]

Table 27.3. Luminal Secretagogues

Bacterial enterotoxins	
V. cholerae (CT)	cAMP
V. cholerae ZOT	?
V. cholerae ACE	?
E. coli (heat labile)	cAMP
Salmonella	cAMP
Campylobacter jejuni	cAMP
Aeromonas	cAMP
E. coli (heat stable)	cGMP
Y. enterocolitica	cGMP
C. perfringens	?
C. difficile (A)	?
Miscellaneous	
Bile salts	cAMP/Ca
Long-chain fatty acids	cAMP/Ca
Laxatives	?

27.4.1.5b. Bile Acids. In addition to bacteria, other luminal factors may have a profound effect on epithelial function. Small bowel malabsorption of bile acids or oral intake of pharmacologic amounts of certain bile acids cause diarrhea. Only dihydroxy bile acids with the hydroxyl groups in the α, but not β, position cause diarrhea. This seemingly obscure detail of bile acid biochemistry indicates why chenodeoxycholate (3α, 7α), but not ursodeoxycholate (3α, 7β), causes diarrhea. The interactions of bile acids with the colonic epithelium have been extensively investigated, but uncertainties remain about the specific mechanisms of bile acid-induced diarrhea. Bile acids have a detergent effect on epithelial membranes at high concentrations but this is of limited physiological or clinical relevance. At lower concentrations bile acids stimulate chloride secretion by increasing either cAMP or intracellular calcium.[116,117]

Long-chain fatty acids within the colonic lumen originate from triglycerides digested by pancreatic lipase but malabsorbed within the small bowel. These fatty acids have effects similar to those of bile acids. Hydroxylated fatty acids are more potent secretagogues than their corresponding nonhydroxy fatty acid; they may originate either from colonic bacterial metabolism or given therapeutically as the active component of castor oil.[116,117]

27.4.1.5c. SCFAs. Short-chain fatty acids (SCFAs) are important luminal factors that serve a very different role. They originate from the bacterial metabolism of carbohydrates, proteins, and fiber not absorbed in the small intestine. The luminal concentration of SCFAs is approximately 100 mM; thus, they are the predominant luminal anion in the large intestine. Although initially presumed to be an etiologic factor in the diarrhea associated with carbohydrate malabsorption, it is now clear that SCFAs are rapidly absorbed from the colon and enhance Na and fluid absorption. The mechanisms of SCFA transport remain uncertain. Because SCFAs are weak electrolytes, there are two potentially transportable species, the protonated acid or the ion.

Investigators employing flux measurements have obtained results consistent with diffusion of the protonated SCFA, generally correlated with the availability of luminal protons (Figure 27.7). In proximal colon, changes in apical Na–H activity correspond to changes in SCFA transport.[118] In distal colon, inhibition of an apical K/H-ATPase inhibits SCFA absorption.[119] Changes in luminal pH, with creation of a transepithelial pH gradient, alter SCFA transport in a manner consistent with diffusion of a weak electrolyte.[120] In contrast, studies employing apical membrane vesicles from human ileum and rat colon have demonstrated an SCFA–HCO_3 anion exchanger.[121,122] The reason for the discrepancies between

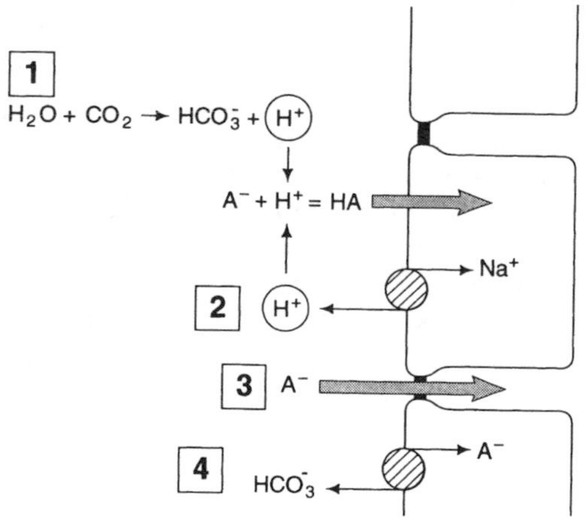

Figure 27.7. Models of short-chain fatty acid absorption. Several potential absorptive pathways exist for SCFA in the colon. (1) Generation of luminal CO_2, of either bacterial or cellular origin, may provide H^+ to combine with the ionized SCFA (A^-) within an acid microclimate, forming the protonated SCFA (HA), which would readily diffuse across the apical membrane. (2) Luminal H^+, supplied by Na^+–H^+ exchange, would also allow for formation of HA. (3) A^- may diffuse across the tight junction. (4) Anion exchange mechanisms may permit movement of the ionized species (A^-) across the apical membrane. There is considerable evidence to support mechanisms (2) and (4). As with bicarbonate, there is little evidence for diffusion of SCFA in the ionized form across the epithelium. [From Sellin, J. H. and Duffey, M. E. Mechanisms of intestinal chloride absorption. In *Textbook of Secretory Diarrhea* (E. Lebenthal and M. E. Duffey, eds.), Raven Press, New York.]

these two methodological approaches is unclear, but obviously has important implications for further research.[123] SCFAs have been employed as a probe to investigate the regulatory mechanisms involved in controlling cell volume and intracellular pH in several different cell systems[124–126]; their effect on the colonic epithelium is unclear, but may have important implications for the interactions between the luminal environment and the epithelium.

27.4.1.5d. Osmotic Effects. Within the gastrointestinal tract, the stomach is relatively impermeable and capable of maintaining an osmotic gradient between lumen and subepithelial space. However, as one moves distally, there is no evidence to suggest that either the small intestinal or colonic epithelium can maintain an osmotic gradient. Given the vagaries of dietary intake and gastric emptying, the duodenum and upper jejunum are normally subject to major fluid shifts as they adjust osmotically to hypertonic or hypotonic foods and liquids. Normal absorptive processes serve to reduce the osmolar load in the intestine. However, the continued presence of nonabsorbable solute within the intestinal lumen can negate functioning absorptive pathways in the distal gut. This provides the basis for osmotic diarrheas.

Carbohydrates, particularly disaccharides, are a common source of nonabsorbable solutes. Because disaccharides require specific brush border membrane enzymes for conversion to simple sugars prior to absorption, the lack of a specific disaccharidase results in malabsorption. The most frequent clinical example is lactose intolerance, in which the glucose-galactose disaccharide is not metabolized because of a lactase deficiency. The continued osmotic load, in addition to the loss of nutrient-coupled Na absorption, results in increased fluid within the small bowel lumen.

However, the physiology of carbohydrate-induced osmotic diarrhea is complicated by the fact that what may be a nonabsorbable solute in the ileum may be converted into an absorbable solute by colonic bacteria. Disaccharides and most other carbohydrates malabsorbed by the small bowel are converted to SCFAs once they cross the ileocecal valve and encounter the colonic flora. SCFAs are rapidly absorbed by the colon (see above). Thus, depending on the rate of conversion to SCFAs and colonic capacity for SCFA absorption, small bowel fluid loss may be compensated for by colonic fluid absorption, independent of any hormonal regulation by steroids. However, the degree of carbohydrate malabsorption may be sufficient to either overwhelm the SCFA absorptive capacity of the colon or exceed the ability of the colonic bacteria to metabolize carbohydrates, permitting the continued osmotic effect of an unabsorbed solute.[127] In contrast to carbohydrates, the other commonly encountered poorly absorbable solutes, calcium, magnesium, and sulfate, are minimally affected by the colon.[128]

27.4.2. Intracellular Mediators

The epithelial cell is exposed to a barrage of potential stimuli. Translating these external messages into an orderly, comprehensive "internal language" permits the cell to regulate its transport function. Adenylate cyclase, guanylate cyclase, intracellular calcium, and the inositol phosphate–diacyl glycerol cascade function as the second messengers inside the cell that perform this regulatory function. Epithelia and other tissues share a common regulatory mechanism. Gastric acid secretion, pancreatic enzyme secretion, and hormonal effects all depend on basically similar systems. More detailed descriptions of the second messenger systems can be found elsewhere.[128–131]

The second messengers have a common effect on transport function that involves both an inhibition of absorption and a stimulation of secretion. Na–H exchange and electroneutral NaCl absorption are inhibited by these second messengers in intestinal epithelia, in contrast to the regulatory response in other tissues. There is no demonstrable effect on Na conductances in the colon, again contrasting to other epithelia. None of the second messengers alter Na-coupled nutrient transport. Thus, only one of multiple absorptive pathways is affected by the second messengers.

Secretion is enhanced by increases in both apical chloride and basolateral K conductances associated with electrogenic chloride secretion. There is evidence to suggest that the cyclic nucleotides primarily affect apical chloride channels, whereas the Ca-related pathways target basolateral K channels. The pattern of response to second messengers in a specific intestinal segment is determined in large part by which of the potentially targeted transporters are present. For example, in rabbit distal colon, only electrogenic chloride secretion is stimulated; in proximal colon, which has limited chloride secretory capacity, the primary effect is antiabsorptive. In ileum, there is both inhibition of NaCl absorption and stimulation of chloride secretion.

27.4.3. Homocellular Regulation

Simply stated, what enters a polarized epithelial cell on one side must exit the cell on the other side at a similar rate or the cell will suffer dire consequences. The cell will either shrink or explode owing to a rapid change in ionic content and osmolality. The concept of homocellular regulation has been applied most commonly to absorption, but necessarily also applies to secretion. For example, when the Na–glucose transporter is maximally stimulated, the amount of Na flowing through the cell each minute is several orders of magnitude greater than the basal cell Na content. The rate limiting step for absorptive processes is generally at the apical membrane. Changes in rates of Na entry initiate a series of coordinated changes in the Na pump, basolateral K conductances, and the apical entry mechanisms themselves so that the intracellular environment is minimally disrupted. Potential regulators of this dialogue between membranes include cell volume, stretch-sensitive ion channels, and specialized intracellular pools of either Na or Ca.[132] Given the vicissitudes in luminal osmolarity and content, the changes between fasting and feeding, the intestine must be prepared for large and rapid changes in the rates of ion and nutrient transport. The ability of the cell to fine-tune discrete events at its opposite borders allows it to function effectively in this environment.

27.5. CASE STUDIES OF DIARRHEA

Sophisticated models of epithelial ion secretion have clearly facilitated our understanding of the underlying pathophysiology of diarrheal diseases. However, in moving from a generalized model to an individual disease, the specific details of that disease highlight both the strengths and the shortcomings of the more general model. We will consider three diarrheal diseases that have broad clinical applicability: cholera, inflammatory bowel disease, and cryptosporidiosis. In each case, the attempt to bridge the gap between general and the specific demonstrates the intricacies and subtleties of intestinal regulation of fluid transport.

27.5.1. Cholera, the Paradigm of Secretory Diarrhea

Over the last two decades, cholera has been considered the paradigm of secretory diarrheas. A specific agonist (cholera toxin) targets the epithelial cell specifically, and without inducing any structural damage, captures the cellular machinery for ion transport and by subverting a critical regulatory step, causes uncontrolled secretion (Figure 27.8). This model has served as an invaluable heuristic tool for understanding both intracellular second messenger systems and the electrophysiology of ion transport and has led to the development of important therapeutic advances. Early *in vitro* studies with cholera toxin, focusing specifically on the response of the epithelial cell, established the broad outlines of the consensus model of electrogenic chloride secretion and related increases in intracellular cAMP to specific changes in ion transport. The realization that cholera did not inhibit nutrient-coupled Na absorption in the small intestine formed the basis for development of oral rehydration therapy, which remains a major triumph of modern medicine in the Third World. Cholera continues to pose challenges for understanding the cellular mechanics of secretion such as intracellular trafficking; however, it also is becoming clearer that multiple effects beyond the ep-

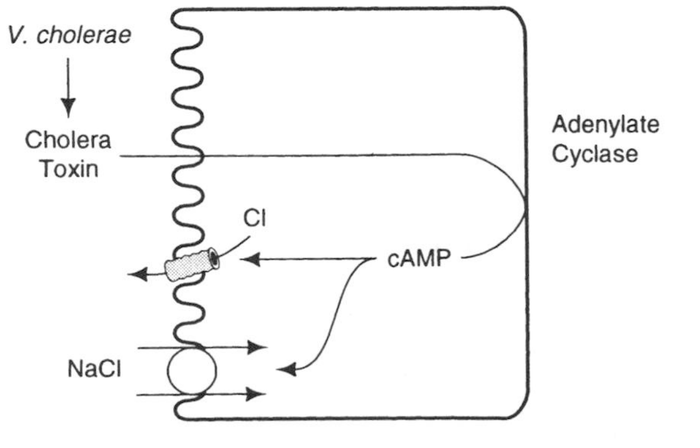

Figure 27.8. Cholera, the classic view. Original descriptions of the mechanism(s) of cholera-induced diarrhea focused on an enterotoxin-induced increase in cAMP in intestinal epithelial cells.

ithelial cell itself may be important in producing the clinical picture of cholera.

Because of the increasing appreciation of the complexities of cholera, our understanding of its effects on the epithelium is "less complete" now than a decade ago. An inkling of that complexity could be seen in the differences between clinical cholera, a bicarbonate-wasting diarrhea, and the *in vitro* model of cholera, electrogenic Cl secretion. This major discrepancy still awaits an adequate explanation.

27.5.1.1. Intracellular Processing of Toxin

Although cholera toxin binds to the apical membrane, its epithelial site of action, the adenylate cyclase, is on the basolateral membrane. The broad outline of how this occurs has been elegantly detailed. The B subunits of the toxin bind to specific monosialoganglioside molecules in the plasma membrane; the A subunit of the toxin is internalized and reduced, releasing the A1 subunit. At the basolateral membrane, the ADP ribosylase activity of A1 targets the α subunit of Gs, the G protein that activates adenylate cyclase; once modified, the α subunit of Gs binds to the catalytic subunit of adenylate cyclase and leaves it "switched on." But how cholera toxin, or its activated component, crosses the cell to reach its target has remained very much a "black box" phenomenon until recently; as the methodological and conceptual tools for studying intracellular trafficking have improved, new insights into the cellular biology of cholera toxin have been obtained. Some investigators have found evidence that cholera toxin activates apical G proteins; one of these may be the α subunit of Gs. However, this still leaves the problems of how the protein traverses the cell.

Studies employing a polarized epithelial monolayer to investigate the transcellular movement of cholera toxin have identified several specific steps between binding to the apical membrane and activation of adenylate cyclase. The A1 subunit of the toxin is taken into the cell by a non-clathrin-coated endocytotic vesicle pathway (potocytosis) and subsequently reduced. There is a critical step in the early endocytotic pathway prior to the reduction of A1 that may be inhibited by the fungal metabolite Brefeldin A,[133] which has served as a useful probe of vesicular transport; Brefeldin A inhibits vesicular passage through the *trans*-Golgi network by interfering with specific vesicle-associated proteins including coating-associated proteins (COPs) and ADP-ribosylating factors (ARFs), but there may be additional sites of action.[134] It is unclear whether the observed inhibition of cholera toxin by Brefeldin A represents an effect on the *trans*-Golgi network directly, or alternatively, an effect of Brefeldin A on the endocytotic pathway by an ARF/COP-independent mechanism. Nambiar *et al.*[135] have suggested that an intact Golgi apparatus is necessary for the toxin to affect intracellular trafficking and/or processing. Cholera toxin did not exhibit cytotoxicity in cell lines that exhibited a Brefeldin A-resistant Golgi apparatus. A temperature-sensitive step later in the vesicular pathway is necessary for the action of apical, but not basolateral, cholera toxin.[136] Thus, information is now emerging about how luminal cholera toxin finds its basolateral target. Whether the vesicular pathway being defined for cholera may serve as a model for other luminal agonists remains to be determined.

27.5.1.2. Multiple Toxins

The identification of multiple cholera toxins has broadened the potential range of epithelial effects. In the course of vaccine development studies, molecularly engineered strains of cholera that did not possess the classic cholera toxin were shown to cause a mild diarrhea. *In vitro* electrical studies demonstrated that these engineered strains elicited a rapid increase in tissue conductance. Histological studies showed a decrease in the complexity of the ZO strand complexity and decreased epithelial exclusion of electron-dense horseradish peroxidase.[137] These observations implied that a second cholera toxin, termed ZOT (zona occludens toxin), has a direct effect on epithelial permeability. In conjunction with studies on enteroadherent *E. coli* and *C. difficile* toxin A,[138] this suggests that there may be a class of toxins that exerts their effect primarily by altering barrier function and epithelial permeability rather than stimulating electrogenic Cl secretion. However, it is clear, based on size, molecular sequence, and cytotoxic effects, that ZOT and *C. difficile* toxin A are very different. A recent report has suggested the presence of a third cholera toxin that induces fluid accumulation in rabbit ileal loops. Molecular biological studies revealed a predicted protein sequence similar to a family of ion-transporting ATPases.[139] Finally, *V. cholerae* also produces a neuraminidase that serves as a "booster," potentiating the toxin effect by increasing the number of monosialoganglioside receptors for cholera toxin.[140] Aspects of the clinical spectrum of cholera may need to be reassessed in the light of this recent proliferation of enterotoxins.

27.5.1.3. Multiple Targets

Initial studies focused on the action of cholera on the epithelial cell, but more recent investigations have suggested that there may be multiple targets involving other cellular components of the gut. Cholera toxin may target epithelial neuroendocrine cells, eliciting a release of serotonin and perhaps other neurohormonal factors. Serotonin then stimulates the release of prostaglandins, most probably from epithelial cells. Although pharmacological doses of prostaglandins are associated with an increase of intracellular cAMP, physiologic levels of eicosanoids may utilize Ca-mediated pathways to stimulate secretion. Indomethacin and kentanserin, the serotonin antagonist, can decrease cholera-induced secretion whereas calcium channel blockers can abolish secretion in *in vivo* models.[141–143] An additional pathway for cholera toxin-mediated secretion may involve the ENS. Neurotransmitter inhibition by tetrodotoxin and similar agents has been shown to block *in vivo* secretion induced by cholera toxin.[142,143] In contrast, tetrodotoxin does not exhibit a similar inhibition *in vitro*.[144] Selective destruction of the myenteric plexus with benzalkonium chloride inhibits the secretory response to cholera toxin of isolated jejunal loops, suggesting that a reflex loop of the ENS involving the myenteric plexus is integral to the secretory effect of the toxin.[145] In addition to its effect on electrolyte transport, cholera toxin also stimulates mucus secretion[146,147]; this action of cholera toxin may be mediated by a capsaicin-sensitive neural pathway rather than a direct effect on mucin-producing cells.[148,149]

Varying results arising from differing methodologies raise the concomitant issue of the most appropriate approach to study the effects of cholera toxin. *In vivo* models of cholera toxin-induced secretion apparently target predominantly the villus epithelia, because of marginal mixing of luminal contents and minimal distention, resulting in a selective increase in adenylate cyclase activity in this portion of the epithelium. In contrast, in the Ussing chamber, cholera toxin has a universal effect on adenylate cyclase activity in both the crypts and villi. The *in vitro* model may correspond more closely to the clinical situation in which *V. cholerae* attack the luminal surface of both villus and crypt cells.[150] However, the conventional "stripped" epithelium used in flux studies may be lacking critical components of the ENS that determine the overall clinical response. Quite obviously, the multiplicity of stimulatory pathways is somewhat model-dependent; this probably reflects the complexity of the clinical reality of cholera infection which may involve more than direct stimulation of the epithelial cell. As in other diarrheas, even this most "classical" of secretory diarrheas may stimulate multiple arms of the superregulatory neuro-immuno-hormonal system that modulates intestinal ion transport.

27.5.1.4. Motility

The consideration of extraepithelial effects necessarily leads to the issue of motility in the genesis of the diarrhea of cholera. Cholera toxin elicits an increase in migrating action-potential complexes (MAPC), a single moving ring contraction of the small intestine. MAPCs may be an electrophysiologic equivalent of a "peristaltic rush," and therefore an indicator of transit. MAPCs are under both neural and hormonal control; they may originate from cholera toxin stimulation of enterochromaffin cells and enteric neurons. Thus, specific neural and hormonal antagonists (e.g., tetrodotoxin, indomethacin) may alter the motility response. Other luminal agonists, including *E. coli* toxins, cathartics, and bile salts, also stimulate MAPCs.[151] Alterations in motility clearly occur secondary to cholera and its toxins, but the relative contribution to diarrhea may be modest compared to the direct secretory effects.[152]

27.5.1.5. Histologic Damage

A basic catechism of the pathophysiology of cholera is that there is no epithelial damage. However, a recent report suggested that careful inspection by light and scanning electron microscopy in a rat ligated loop model demonstrates mucosal edema, villous shortening, and destruction of enterocytes.[153] Whether this is a primary action of cholera or secondary epithelial response to secretion or distention remains to be determined.[153]

27.5.1.6. Role of Colon

Although the effects of cholera toxin on small bowel transport may be somewhat model-dependent, there is no doubt that there is a significant effect. Somewhat surprisingly, the effect of cholera on the colon has not been clearly characterized. In canine models of cholera, the effects of toxin on the colon have been rather minimal.[154,155] However, in rat cecum, cholera toxin reverses net absorption of electrolytes to secre-

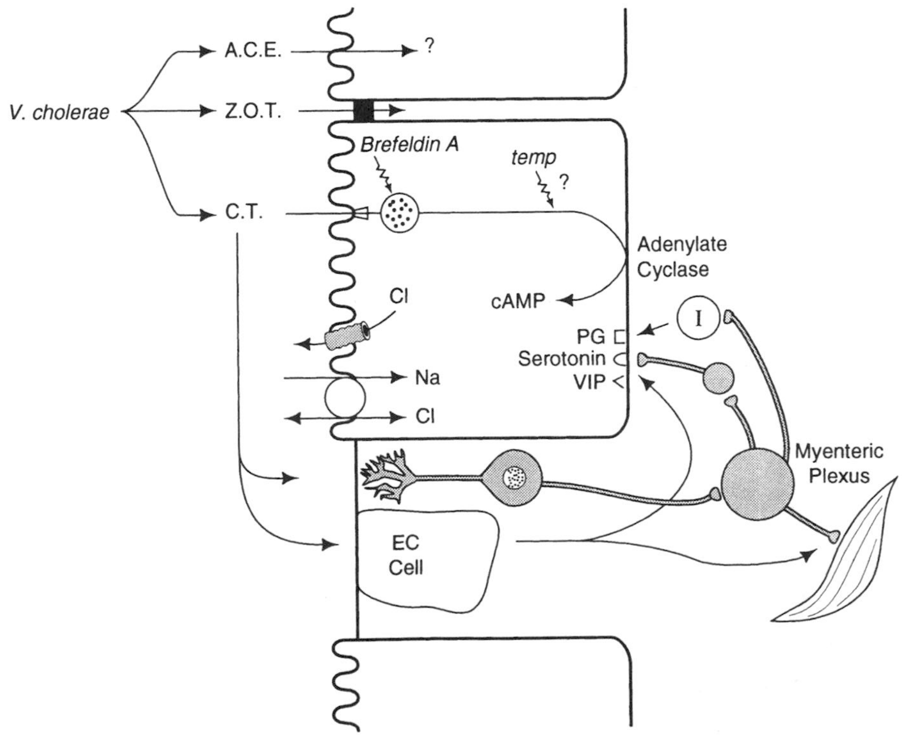

Figure 27.9. Cholera, the updated view. *V. cholerae* produces three toxins (CT, ZOT, ACE) with at least three different targets. CT's voyage across the epithelial cell is complex, with several discrete steps, before it reaches its ultimate target, the adenylate cyclase system in the basolateral membrane. CT has additional targets including enterochromaffin cells (EC) and sensory neurons in the epithelium. This initiates a cascade of events involving the enteric nervous system, primarily via the myenteric plexus, and the subepithelial immunologically reactive cells (I). Release of additional mediators including prostaglandins, serotonin, and VIP affect both the intestinal muscle and the epithelial cell.

tion.[156] In clinical cholera, measurement of fluid movement can be limited by technical considerations. Speelman *et al.* employed total colonic perfusion to demonstrate minimal water absorption during acute cholera, concluding that there was significant impairment of transport function during acute cholera. However, there was little improvement in absorptive function during convalescent studies.[157] Ramakrishna and Mathan, using rectal dialysis, found minimal fluid absorption during cholera and other acute diarrheas, but with a significant shift toward absorption during convalescence. Interestingly, there was a decrease in SCFA excretion during cholera and other acute diarrheas and rectal instillation of SCFAs had a proabsorptive effect during the acute stage of the diarrheas. This raises several interesting issues. Much like bicarbonate, SCFAs have been shown to stimulate electroneutral NaCl absorption in the colon.[159] However, Na absorption in the distal colon is typically mediated by electrogenic Na absorption rather than coupled NaCl absorption. Therefore, the mechanism of stimulation of Na absorption is not immediately evident. There are multiple possibilities for the decreased SCFA excretion, including dietary restriction or change in bacterial flora capable of SCFA formation. Nevertheless, this points out the complexities of the mechanisms of colonic salvage of small intestinal fluid loss and raises the possibility that delivery of SCFAs to the colon in acute diarrheas may augment the oral rehydration therapies that have proved so clinically successful in treating cholera and other secretory diarrheas. Although transport models would predict specific segmental responses to cholera toxin in different regions of the colon and surrogate secretagogues have been studied, surprisingly little data are available about cholera's effect on the colon.

27.5.2. Inflammatory Bowel Disease

Diarrhea is a prominent symptom of inflammatory bowel disease (IBD). It may be a dominant factor in clinical management, especially with colonic disease. However, until relatively recently, there had been little research into the etiology of the diarrhea associated with IBD. In general, the severity of diarrhea has correlated with the degree of inflammation. Clinically, the classic "paint can" approach of measuring stool volume and electrolytes to separate out secretory and osmotic diarrheas yields little meaningful information and is rarely employed.[160–162] The etiology is usually deemed "multifactorial" and presumed to result from a combination of the loss of absorptive function from injured (or resected) epithelium; inflammation and ulceration, with concomitant exudation of protein and fluid; changes in intestinal motility with an emphasis on loss of the reservoir capacity of the distal colon; and, finally, the additional factor of malabsorption of nutrients must be added into the assessment of small bowel Crohn's disease.

27.5.2.1. Inflammatory Mediators

As research into the etiology and pathophysiology of the inflammatory process(es) involved in IBD progressed, increased efforts were focused on the relationship between the epithelium and the arcana of subepithelial cells and secretory stimuli. By identifying factors associated with increased intestinal inflammation and demonstrating that these products are capable of inducing secretion *in vitro,* investigators were able to establish a mechanistic link between inflammation and secretory diarrhea. In this way, IBD has become an important, clinically relevant model for examining immune regulation of epithelial transport and the interaction among the immune, neural, paracrine, and endocrine component of the enteric superregulatory system.

Because prostaglandins have been demonstrated to cause active secretion in small intestine and colon[163] and increased levels of prostaglandins occur in inflamed colon,[164] prostaglandins were one of the factors initially implicated in the secretory processes of IBD. However, it rapidly became apparent that the underlying pathophysiology is infinitely more complex. Therapeutic interventions based on inhibition of prostaglandin production were either ineffective or exacerbated the disease, therefore necessitating a thorough reassessment.[165]

As the arcana of cells and cytokines involved in intestinal inflammation became more fully appreciated, the potential suspects involved in the secretion and their modus operandi were continually expanded. Conceptually, inflammatory mediators may have several effects that result in active secretion. In the simplest model, an inflammatory mediator directly stimulates an epithelial cell. However, it is now apparent that a pure effect on an epithelial cell may be a relatively rare event. It is more probable that mediators have multiple effects at different levels within the neuro-immuno-endo-paracrine system, setting off a cascade of events. Other levels of interaction also occur including activation of enteric neurons, stimulation of subepithelial cells, enhanced release of arachidonic acid metabolites from cell membranes, alteration of blood flow, and changes in intestinal permeability. Chronic inflammation may lead to a "primed" epithelial or mesenchymal response. With tissue damage, there may be a relative increase in the proportion of crypt cells, which are presumed to have a relatively greater secretory function. Thus, an agonist may have multiple targets in eliciting a characteristic response; although the epithelial cell is the most obvious, the observed effect may be caused by an initial stimulation of a fibroblast or enteric neuron.

Because of the paradoxical clinical effect of cyclooxygenase inhibition, the potential role of leukotrienes was investigated. A series of arachidonic acid metabolites and lipoxygenase products elicit chloride secretion in colonic epithelial cells.[166] Multiple agonists (bradykinin, complement, interleukin 1, endotoxin, fMLP) liberate arachidonic acid from cell membranes. Subepithelial cells are the primary source of eicosanoids in normal intestine, but PMNs may be a major source of eicosanoids in an inflamed intestine. As our understanding of the complexities of the inflammatory process in the intestine increases, more potential mediators such as oxygen radicals may be implicated.

27.5.2.2. Crypt Abscess

Crypt abscesses, collection of polymorphonuclear leukocytes (PMNs) within a colonic crypt, resulting from PMN migration across the intestinal epithelium, is a classic finding in ulcerative colitis, Crohn's disease, and other inflammatory conditions in the intestine. By employing a reductionist model of PMN transmigration across a monolayer of T84 cells, Madara's group has defined several salient features of the interaction between PMNs and the epithelium and also characterized a novel mechanism of inflammatory induced secretion.

PMNs can be stimulated to cross the epithelium by bacterial chemoattractant peptides such as fMLP. PMNs cross the epithelium by impaling intercellular tight junctions. Analogous to PMN–endothelial interactions, specific adhesive interactions between the epithelial cell and the leukocyte are required for transmigration to occur. The use of selective monoclonal antibodies can define the specific interactions required: the leukocyte integrins CD18 and CD11b are integral to transmigration, but not CD11a or CD11c.[167] This transmigration causes transepithelial resistance to fall and the paracellular permeability to inert solutes to increase.[168] Eicosanoids may modify the PMN's ability to permeate the epithelium.[169] Interestingly, the fall in resistance could be resolved into two components, one unrelated to junctional permeability. This nonjunctional resistance change was associated with a neutrophil-derived secretory product that increased transepithelial short-circuit current and Cl secretion.[170] Further studies have now characterized the neutrophil-derived secretory factor: as neutrophils traverse the tight junction into the crypt lumen, they release 5′-AMP, which is a weak secretagogue. However, an apical membrane ecto-5′-nucleotidase converts 5′-AMP to adenosine, which functions as a potent direct secretagogue on the apical membrane. Thus, there is an exceedingly complex interplay between the neutrophil and epithelial cell membrane in the process of transmigration and stimulation of secretion.[171]

Unraveling the basis of inflammation, at the molecular level, has major implications for therapeutic strategies. Novel approaches have been developed on the basis of insights developed about the mechanisms of immune mediation and a broad approach to colonic physiology. For example, trials of fish oils and/or their derivative polyunsaturated fatty acids (e.g., eicosapentaenoic acid) have been based on the observation that these diets elicit a significant reduction of 5-lipoxygenase products generated from monocytes and neutrophils and may therefore reduce the inflammatory response in the colon. Pharmaceutical development of specific leukotriene inhibitors follows the same goal.[172,173] Because of a potential neural input, lidocaine enemas have been employed in ulcerative colitis.[174] SCFAs have been shown to be therapeutic in ulcerative colitis, presumably on a metabolic basis. The finding that interleukin 1 is elevated in IBD formed the basis for a trial of the receptor antagonist for this interleukin. As our insights into the specific interactions between cytokines and epithelial cells expand, we can expect that similar therapeutic trials based on "surgically" interfering with specific steps in the inflammatory pathway will be developed.

27.5.2.3. Intestinal Permeability

If frank ulcerations of the epithelium occur during the course of ulcerative colitis or Crohn's disease, it is quite obvious that the "permeability" of the gut will be increased, with relatively free transudation of proteins and fluid into the lumen. In the other direction, the loss of barrier function may promote bacterial translocation and absorption of endotoxin and bacterial chemotactic factors, which will inevitably continue the inflammatory cascade.

There is, however, perhaps a more subtle alteration of intestinal permeability that may play a role in IBD. Several investigators have found that inert markers that presumably would cross the epithelium only by a paracellular route, e.g., lactulose, mannitol, PEG, exhibit increased absorption in patients with Crohn's disease. Perhaps even more intriguing is the observation that healthy relatives of patients with Crohn's disease also demonstrate a similar increased permeability, suggesting that this may be a preclinical marker of the disease.[175,176]

Attempts to further define these alterations in permeability have not necessarily clarified the picture. A series of inert markers of different diameters and shapes did not permeate the intestine in a predictable manner in Crohn's disease patients.[176,177] Although there may be individual relatives with demonstrable increases in intestinal permeability, it has been more difficult to show statistically significant increases in permeability in groups of relatives.[178] This may reflect the fact that only a subset of relatives of IBD patients (<20%) are destined to develop the disease. At this point, prospective studies are required to adequately assess the importance of increased permeabilities in asymptomatic relatives.[176,179]

27.5.2.4. Motility

The motility changes associated with chronic inflammation have not been as intensively studied as the effects on fluid transport. Because cytokines and other inflammatory mediators may target muscle cells as well as epithelial cells, it is not surprising that significant changes exist.[180] In mild to moderate ulcerative colitis, there are demonstrable differences in colonic motility. Patients with ulcerative colitis had more frequent, low-amplitude contractions compared to controls. These propagating contractions invariably elicited an antegrade movement of luminal contents and caused rapid movement from the transverse to the sigmoid colon.

There is a decrease in the force of muscle contraction *in vitro* in circular colonic muscle strips from patients with ulcerative colitis compared to those with either cancer or diverticular disease. This suggested that the decreased contraction may be secondary to alteration of the cell contractile machinery.[181]

In colonic muscle obtained from animals with experimental colitis, there is a decrease in phosphorylation of a 20-kDa myosin light chain, consistent with a decrease in the cross-bridge cycling rates associated with inflammation.[182] Cytokines are recognized as growth factors; indeed, the release of inflammatory mediators such as PDGF may stimulate growth of intestinal muscles.[183] Although there are significant alterations in both motility function and intestinal muscle structurally, it is unclear whether these changes have any impact on fluid transport, whether they are a secondary reactive process or simply an epiphenomenon.

27.5.2.5. Sick Cell Syndrome

The interest in identifying inflammatory mediators associated with either ulcerative colitis or Crohn's disease often has an accompanying aim: characterization of the effect of these mediators on ion transport in normal epithelium. Because prostaglandins and leukotrienes stimulate electrogenic chloride secretion, these observations have often been generalized to imply that the diarrhea associated with IBD is the result of electrogenic chloride secretion. Further observations that a series of inflammatory mediators have also been shown to stimulate chloride secretion in normal epithelia have strengthened this assumption.

However, studies that have examined ion transport in models of colonic inflammation do not support this general model. Rather than finding an epithelium actively secreting chloride, the general finding has been one of a "sick" epithelium that is unable to sustain its normal absorptive function. These studies have demonstrated major changes in active Na absorption; interestingly, the inflammatory effect has been seen with electroneutral NaCl absorption, cecal Na absorption, and amiloride-inhibitable electrogenic Na absorption in the distal colon.[184–187] Inhibition of electroneutral NaCl absorption may be related to an increase in intracellular second messengers secondary to inflammatory mediators. However, it is more difficult to relate changes in electrogenic Na absorption to second messengers.

Although there must be a concern about the model-dependency of any findings, Sandle *et al.* have shown basically similar results in distal colonic samples obtained from patients with ulcerative colitis or Crohn's colitis. Compared to normal controls, the IBD colons exhibited decreased electrogenic Na absorption and significant increases in both apical membrane and total tissue conductance. Mucosal inflammation also was associated with decreased activity of the basolateral Na pump.[188] These findings suggest that it may be appropriate to focus on how intestinal inflammation alters either electrogenic or electroneutral Na absorption rather than on the stimulation of Cl secretion.

27.5.3. AIDS

The AIDS pandemic has provided an unfortunate opportunity to reassess the complex interrelationships among potential infectious agents, the intestinal epithelium, the mucosal immune system, and fluid and electrolyte transport by the gut. Diarrhea is a prominent AIDS symptom that has striking geographical variability. Although incapacitating diarrhea is a frequent, late finding in patients with AIDS in the United States or Europe, it occurs more frequently and earlier in the course of the disease in less developed countries. For example, in Africa, diarrhea is a more prominent early finding in the course of HIV infection; the descriptive clinical term *Slim disease* refers to the associated nutritional consequences, which have a devastating impact on the clinical course of the patient.

A Pandora's box of bacteria, viruses, fungi, and protozoans has been newly recognized as potential diarrheal agents in AIDS.[189] The identification of organisms in either tissue or stool specimens has raised difficult and intriguing questions. Is the identification of a unique organism in the gut of a patient with AIDS and diarrhea sufficient to establish it as a causative agent? Are these agents pathogenic only in immunosuppressed patients or do they also cause diarrhea in immunocompetent patients? Is there a difference in the pathophysiology between immunocompetent and immunosuppressed patients? And, finally, what are the mechanisms of fluid secretion elicited by these agents? Achieving a fuller understanding of the factors involved in AIDS diarrhea will be an important challenge to both clinicians and basic scientists interested in epithelial function and transport.

The number of potential infectious agents continues to expand (Table 27.4). The length of this list serves to emphasize that HIV-associated diarrhea is not one disease, but may be several. Although the identification of an unusual organism either in a biopsy or in a stool sample may suggest a causal relation, there are other possibilities to consider. A specific organism may be an innocent bystander or may simply reflect the severity of the intestinal disease or the depression of immune surveillance but not be an etiologic factor. HIV itself may be a pathogenic factor in the development of the intestinal manifestations of AIDS. Presence of the HIV virus in intestinal epithelia, but with no other demonstrable agents, in patients with AIDS-related diarrhea raises the possibility of a specific AIDS enteropathy, but critical studies establishing HIV as a diarrheal agent remain to be done.

27.5.3.1. Cryptosporidiosis

The evolution of our understanding of cryptosporidiosis over the last decade can serve as a paradigm for the complexities in both clinical medicine and epithelial biology that AIDS diarrhea presents. In the pre-AIDS era, cryptosporidiosis was recognized as an uncommon cause of infectious diarrhea, acquired through contact with farm animals or in day-care centers. Cryptosporidiosis was quickly recognized as a frequent infectious agent found in the stool of HIV-positive patients with diarrhea. In 1993, the municipal water supply of Milwaukee became contaminated with cryptosporidiosis causing a massive epidemic of diarrhea that crippled the city, afflicting immunocompetent as well as immunosuppressed patients.[190] In immunocompetent patients cryptosporidiosis causes a relatively short-lived (7–10 days), noninflammatory diarrhea with an uneventful recovery.

The clinical course in HIV-positive patients is quite different. There appears to be a correlation between the degree of immune suppression and the course of the cryptosporidial infection. The disease will generally resolve in patients with CD4 lymphocyte count (an indicator of immune status) greater than 200, whereas it becomes persistent in those with a CD4 count less than 200.[189,191]

Cryptosporidiosis in immunocompromised patients may have a spectrum of presentations: (1) tropical sprue-like picture, with malabsorption and a low D-xylose; (2) ileocolitis; and (3) a cholera-like picture with profuse watery diarrhea (D. Kotler, personal communication). The disease is usually unremitting and essentially unresponsive to antibiotic therapy.

Table 27.4. Pathogens in HIV-Associated Diarrhea[a]

Pathogen	Small bowel, duodenum, jejunum	Colon, with or without terminal ileum
Bacteria	*Campylobacter* *Salmonella* ? *Escherichia coli* ? Overgrowth	*Campylobacter* *Salmonella* *Shigella* *Yersinia* *Clostridium difficile* *Chlamydia* Spirochetes
Mycobacteria	*Mycobacterium avium* complex *M. tuberculosis*	*M. avium* complex
Viruses	Rotavirus Norwalk virus and similar agents Cytomegalovirus	Adenovirus ? HIV Herpes simplex virus
Parasites	*Giardia* *Isospora* ? *Blastocystis* *Cryptosporidium* *Microsporidium* *Cyclospora*	*Entamoeba histolytica* ? *Cryptosporidium*

[a]From Wanke, C. A. (1993). *N. Engl. J. Med.* **329:**1946–1954.

It is unclear what accounts for the different symptom complexes. Both clinical and experimental data support a role for malabsorption, as estimated by impaired D-xylose absorption. In common with other malabsorptive diarrheal diseases, such as viral gastroenteritis or tropical sprue, histologic examination has shown reduced villus height and crypt hyperplasia.[192,193] The morphological changes themselves may not be an adequate explanation of the diarrhea. Most AIDS patients with cryptosporidiosis have normal duodenal villus architecture. Several morphological abnormalities such as flattening of the villi occur in about 25% of AIDS-related cryptosporidiosis and are associated with the intensity of the infection, as estimated by the percentage of mucosa covered by organisms.[194] Changes in villus architecture may develop pari passu with advancing HIV infection independent of clinical development of diarrhea; similarly, rodents infected with *C. parvum* develop a similar set of morphological changes, but no diarrhea.[195,196]

Development of an animal model for infectious cryptosporidiosis (the normal neonatal piglet) may allow for a better understanding of the underlying pathophysiology. Argenzio *et al.* demonstrated significant villous atrophy with an increase in lamina propria inflammatory cells in the ileum. Although basal rates of fluid and electrolyte absorption were not altered during infection, there was a significant decrease in glucose-coupled Na absorption from both jejunum and ileum.[197] These findings are more consistent with a malabsorptive diarrhea rather than a toxigenic diarrhea. However, these immunocompetent animals had a rapid and uncomplicated recovery; therefore, these results are more relevant to understanding the disease in healthy rather than immunocompromised patients.[197] In a modification of this model, colostrum-deprived neonatal pigs were studied both *in vivo* and *in vitro* with somewhat different results. In the colostrum-deprived animals, in addition to the changes in villus architecture and glucose-coupled Na absorption, there was a decrease in basal rates of electroneutral NaCl absorption coupled with increased PGE_2 levels in infected piglet ileum compared to controls. The depressed NaCl absorption was reversed by indomethacin treatment, suggesting that the change in basal transport rates was not related to a structural defect, but, rather, a change in regulatory input to the epithelia.[198] The colostrum-deprived animals were clearly sicker, despite an identical inoculum, raising the possibility of a protective role for passive immunity or for another component of colostrum.

The massive choleralike diarrhea seen in some patients with cryptosporidiosis cannot be easily ascribed to a malabsorptive state; however, the search for an enterotoxin has proved elusive. Available data are somewhat conflicting. A recent report found that stool filtrates from cryptosporidiosis-infected calves elicited electrogenic Cl secretion in human jejunum *in vitro.*[197] Clearly, isolation of a specific toxin is necessary for adequate assessment of this phenomenon.

Cryptosporidiosis serves as a paradigm for AIDS-related diarrhea. Although there may be an enterotoxin involved in the clinical manifestations of cryptosporidiosis, it has not been a simple task to isolate such a factor. Cryptosporidiosis may induce architectural changes in small bowel structure, but it is not clear that this is either a necessary or sufficient insult to cause diarrhea. Therefore, like so many other infectious agents, the specific mechanisms that lead to cryptosporidiosis-induced diarrhea remain enigmatic. The role of immunosuppression associated with AIDS in modulating the course of crytposporidiosis serves to emphasize that the interaction between an infectious agent and the epithelial cell alone cannot adequately explain the clinical manifestations of this diarrhea. Factors beyond the epithelial cell involving the superregulatory immuno-neuro-endo-paracrine system must be the determinants of the clinical spectrum of cryptosporidiosis. At present there is minimal information about how the barrier function of the intestine may be altered by immunocytes, but this will become an increasingly critical puzzle to decipher.

REFERENCES

1. Eherer, A. J., and Fordtran, J. S. (1992). Fecal osmotic gap and pH in experimental diarrhea of various causes. *Gastroenterology* **103:** 545–551.
2. Welsh, M. J., Smith, P. L., Fromm, M., *et al.* (1982). Crypts are the site of intestinal fluid and electrolyte secretion. *Science* **218:**1219–1221.
3. Hopfer, J. (1987). Membrane transport mechanisms for hexoses and amino acids in the small intestine. In *Physiology of the Gastrointestinal Tract* (L. R. Johnson, ed.), Raven Press, New York, pp. 1499–1526.
4. Bookstein, C., DePaoli, A. M., Xie, Y., *et al.* (1987). Na/H exchangers, NHE-1 and NHE-3, of rat intestine: Expression and localization. *J. Clin. Invest.* **93:**106–113.
5. Fedorak, R. N., Chang, E. B., Madara, J. L., *et al.* (1987). Intestinal adaptation to diabetes: Altered Na-dependent nutrient absorption in streptozocin-treated chronically diabetic rats. *J. Clin. Invest.* **79:** 1571–1578.
6. Diener, M., Rummel, W., Mestres, P., *et al.* (1989). Single chloride channels in colon mucosa and isolated colonic enterocytes of the rat. *J. Membr. Biol.* **108:**21–30.
7. Sellin, J. H., Oyarzabal, H., and Cragoe, E. J. (1988). Electrogenic sodium absorption in rabbit cecum in vitro. *J. Clin. Invest.* **81:** 1275–1283.
8. Hatch, M., and Freel, R. W. (1988). Electrolyte transport across the rabbit caecum in vitro. *Pfluegers Arch.* **411:**333–338.
9. Sellin, J. H., and DeSoignie, R. (1984). Rabbit proximal colon: A distinct transport epithelium. *Am. J. Physiol.* **246:**G603–G610.
10. Sellin, J. H., and DeSoignie, R. (1987). Ion transport in human colon in vitro. *Gastroenterology* **93:**441–448.
11. Madara, J. L. (1990). Contributions of the paracellular pathway to secretion, absorption, and barrier function of the small intestine. In *Textbook of Secretory Diarrhea* (E. Lebenthal and M. Duffey, eds.), Raven Press, New York, pp. 125–238.
12. Madara, J. L. (1988). Tight junction dynamics: Is paracellular transport regulated? *Cell* **53:**497–498.
13. Pappenheimer, J. R., and Reiss, K. Z. (1987). Contribution of solvent drag through intercellular junctions to absorption of nutrients by the small intestine of the rat. *J. Membr. Biol.* **100:**123–135.
14. Dharmsathaphorn, K., Mandel, K. G., Masui, H., *et al.* (1985). Vasoactive intestinal polypeptide-induced chloride secretion by a colonic epithelial cell line: Direct participation of a basolaterally localized Na–K–Cl cotransport system. *J. Clin. Invest.* **75:**462–471.
15. O'Grady, S. M., Palfry, H. C., and Field, M. (1987). Characteristics and functions of Na–H–Cl cotransport in epithelial tissues. *Am. J. Physiol.* **253:**C177–C192.
16. Halm, D. R., and Frizzell, R. A. (1990). Intestinal chloride secretion. In *Textbook of Secretory Diarrhea* (M. Duffey and E. Lebenthal, eds.), Raven Press, New York, pp. 47–58.
17. Halm, D. R., and Frizzell, R. A. (1986). Active K transport across rabbit distal colon: Relation to Na absorption and Cl secretion. *Am. J. Physiol.* **251:**C252–C267.
18. Cliff, W. H., and Frizzell, R. A. (1990). Separate Cl conductances activated by cAMP and Ca in Cl secretory epithelial cells. *Proc. Natl. Acad. Sci. USA* **87:**4956–4960.
19. Frizzell, R. A., and Cliff, W. H. (1992). Chloride channels: No common motif. *Curr. Biol.* **2:**285–287.
20. Frizzell, R. A., Rechkemmer, G., and Shoemaker, R. L. (1986). Altered regulation of airway epithelial cell chloride channel in cystic fibrosis. *Science* **233:**558–560.
21. Welsh, M. J. (1986). An apical-membrane chloride channel in human tracheal epithelium. *Science* **232:**1648–1650.
22. Kartner, N. J., Hanrahan, W., Jensen, T. J., *et al.* (1991). Expression of the cystic fibrosis gene in non-epithelial invertebrate cells produces a regulated anion conductance. *Cell* **64:**681–691.
23. Bear, C. E., Duguay, F., Naismith, A. L., *et al.* (1991). Cl channel activity in Xenopus oocytes expressing the cystic fibrosis gene. *J. Biol. Chem.* **266:**19142–19145.
24. Fuller, C. M., and Benos, D. J. (1992). CFTR! *Am. J. Physiol.* **263:** C267–C268.
25. Berschneider, H. M., Knowles, M. R., Azizkhan, R. G., *et al.* (1988). Altered intestinal chloride transport in cystic fibrosis. *FASEB J.* **2:**2625–2629.
26. Loughlin, E. V., Hunt, D. M., Gaskin, K. J., *et al.* (1991). Abnormal epithelial transport in cystic fibrosis jejunum. *Am. J. Physiol.* **260:**G758–G763.
27. Goldstein, J. L., Nash, N. T., Al-Bazzazt, F., *et al.* (1988). Rectum has abnormal ion transport but normal cAMP-binding proteins in cystic fibrosis. *Am. J. Physiol.* **254:**C719–C724.
28. Anderson, M. P., Sheppard, D. N., Berger, H. A., *et al.* (1992). Chloride channels in the apical membrane of normal and cystic fibrosis airways and intestinal epithelial cells. *Am. J. Physiol.* **263:**L1–L14.
29. Cohn, J. A., Nairn, A. C., Marino, C. R., *et al.* (1992). Characterization of the cystic fibrosis transmembrane conductance regulator in a colonocyte cell line. *Proc. Natl. Acad. Sci. USA* **89:**2340–2344.
30. Trezise, A. E. O., and Buchwald, M. (1991). In vivo cell-specific expression of the cystic fibrosis transmembrane regulator. *Nature* **353:**434–437.
31. Strong, T. V., Boehm, K., and Collins, F. S. (1994). Localization of cystic fibrosis transmembrane conductance regulator mRNA in the human gastrointestinal tract by in situ hybridization. *J. Clin. Invest.* **93:**347–354.
32. Yang, Y., Janisch, S., Cohn, J. A., *et al.* (1993). The common variant of cystic fibrosis transmembrane conductance regulator is recognized by hsp70 and degraded in a pre-Golgi nonlysosomal compartment. *Proc. Natl. Acad. Sci. USA* **90:**9480–9484.
33. Demming, G. M., Anderson, M. P., Amara, J. F., *et al.* (1992). Processing of mutant cystic fibrosis transmembrane conductance regulator is temperature-sensitive. *Nature* **358:**761–764.
34. Sarkadi, B., Bauzon, D., Huckle, W. R., *et al.* (1992). Biochemical characterization of the cystic fibrosis transmembrane conductance regulator in normal and cystic fibrosis epithelial cells. *J. Biol. Chem.* **267:**2087–2095.
35. Morris, A. P., Cunningham, S. A., Benos, D. J., *et al.* (1992). Cellular differentiation is required for cAMP but not Ca^{2+}-dependent Cl^- secretion in colonic epithelial cells expressing high levels of cystic fibrosis transmembrane conductance regulator. *J. Biol. Chem.* **267:** 5575–5583.
36. Snouwaert, J. N., Brigman, K. K., Latour, A. M., *et al.* (1992). An animal model for cystic fibrosis made by gene targeting. *Science* **257:**1083–1088.
37. Turnberg, L., Fordtran, J. S., Carter, N., *et al.* (1970). Mechanism of bicarbonate absorption and its relationship to sodium transport in human jejunum. *J. Clin. Invest.* **49:**557–567.
38. Turnberg, L., Bieberdorf, F., Morawski, S., *et al.* (1970). Interrelationships of chloride, bicarbonate, sodium, and hydrogen transport to the human ileum. *J. Clin. Invest.* **49:**557–567.
39. Murer, H., Hopfer, U., and Kinne, R. (1976). Sodium/proton antiport in brush border membrane vesicles isolated from rat small intestine and kidney. *Biochem. J.* **154:**597–604.
40. Knickelbein, R. G., Aronson, P. S., and Dobbins, J. W. (1990). Characterization of Na/H exchangers on villus cells in rabbit ileum. *Am. J. Physiol.* **259:**G802–G806.

41. Sundaram, U., Knickelbein, R. G., and Dobbins, J. W. (1991). Mechanisms of intestinal secretions: Effect of cyclic AMP on rabbit ileal villus and crypt cells. *Proc. Natl. Acad. Sci. USA* **88:** 6249–6253.
42. Musch, M. W., Drabik-Arvans, D., Rao, M. C., *et al.* (1992). Bethanechol inhibition of chicken intestinal brush border Na/H exchange; role of protein kinase C and other calcium-dependent processes. *J. Cell. Physiol.* **152:**362–371.
43. Semrad, C. E., and Chang, E. G. (1987). Calcium mediated cyclic AMP inhibition of Na–H exchange in small intestine. *Am. J. Physiol.* **252:**C315–C322.
44. Sardet, C., Franchi, A., and Pouyssegur, J. (1989). Molecular cloning, primary structure and expression of the human growth factor activatable Na/H antitransporter. *Cell* **56:**271–280.
45. Orlowski, J., Kandasamy, R. A., and Shull, G. E. (1992). Molecular cloning of putative members of the NHE exchanger gene family. cDNA cloning, deduced amino acid sequence, and mRNA tissue expression of the rat NHE exchanger NHE-1 and two structurally related proteins. *J. Biol. Chem.* **267:**9331–9339.
46. Tse, C. M., Ma, A. I., Yang, V. W., *et al.* (1991). Molecular cloning and expression of a cDNA encoding the rabbit ileal villus cell basolateral membrane Na–H exchanger. *EMBO J.* **10:**1957–1967.
47. Tse, C.-M., Brant, S. R., Walker, M. S., *et al.* (1992). Cloning and sequencing of a rabbit cDNA encoding an intestinal and kidney-specific Na^+/H^+ exchanger isoform (NHE-3). *J. Biol. Chem.* **267:** 9340–9346.
48. Rothstein, A. (1989). The Na/H exchange system in cell pH and volume control. *Rev. Physiol. Biochem. Pharmacol.* **112:**235–257.
49. Yun, C. H. C., Gurubhagavatula, S., Levine, S. A., *et al.* (1993). Glucocorticoid stimulation of ileal Na^+ absorptive cell brush border Na^+/H^+ exchange and association with an increase in message for NHE-3, an epithelial exchanger isoform. *J. Biol. Chem.* **268:** 206–211.
50. Hubel, K. A. (1974). The mechanism of bicarbonate secretion in rabbit ileum exposed to choleragen. *J. Clin. Invest.* **53:**964–970.
51. DeSoignie, R., and Sellin, J. H. (1990). Acid–base regulation of ion transport in rabbit ileum in vitro. *Gastroenterology* **99:**1332–1341.
52. Heintze, K., Petersen, K.-U., and Wood, J. R. (1981). Effects of bicarbonate on fluid and electrolyte transport by guinea pig and rabbit gallbladder: Stimulation of absorption. *J. Membr. Biol.* **62:**175–181.
53. Dagher, P. C., Balsam, L., Weber, J. T., *et al.* (1992). Modulation of chloride secretion in the rat colon by intracellular bicarbonate. *Gastroenterology* **103:**120–127.
54. Dagher, P. C., Egnor, R. W., and Charney, A. N. (1993). Effect of intracellular acidification on colonic NaCl absorption. *Am. J. Physiol.* **264:**G569–G575.
55. Vaccarezza, S. G., and Charney, A. N. (1988). Acid–base effects of ileal sodium chloride absorption. *Am. J. Physiol.* **254:**G329–G333.
56. Smith, P. L., Cascairo, M. A., and Sullivan, S. K. (1985). Sodium dependence of luminal alkalinization by rabbit ileal mucosa. *Am. J. Physiol.* **249:**G358–G368.
57. Sellin, J. H., and DeSoignie, R. (1989). Regulation of bicarbonate transport by rabbit ileum. pH stat studies. *Am. J. Physiol.* **257:** G607–G615.
58. Field, M., and McColl, I. (1973). Ion transport in rabbit ileal mucosa. III. Effects of catecholamines. *Am. J. Physiol.* **225:**852–857.
59. Sheerin, H. E., and Field, M. (1975). Ileal HCO_3 secretion: Relationship to Na and Cl transport and effect of theophylline. *Am. J. Physiol.* **228:**1065–1074.
60. Minhas, B. S., Sullivan, S. K., and Field, M. (1993). Bicarbonate secretion in rabbit ileum: Electrogenicity, ion dependence, and effects of cycline nucleotides. *Gastroenterology* **105:**1617–1629.
61. White, J. F., and Imon, M. A. (1982). Intestinal HCO_3 secretion in Amphiuma: Stimulation by mucosal Cl and serosal Na. *J. Membr. Biol.* **68:**207–214.
62. Binder, H. J. (1992). The gastroenterologist's osmotic gap: Fact or fiction. *Gastroenterology* **103:**702–704.
63. Sullivan, S. K., and Smith, P. L. (1986). Bicarbonate secretion by rabbit proximal colon. *Am. J. Physiol.* **250:**G475–G483.
64. Hubel, K. A. (1967). Bicarbonate secretion in rat ileum and its dependence on intraluminal chloride. *Am. J. Physiol.* **213:**1409–1413.
65. Hubel, K. A. (1969). Effect of luminal chloride concentration on bicarbonate secretion in rat ileum. *Am. J. Physiol.* **217:**40–45.
66. Halm, D. R., and Frizzell, R. A. (1992). Anion permeation in an apical membrane chloride channel of a secretory epithelial cell. *J. Gen. Physiol.* **99:**339–366.
67. Hanrahan, J. W., and Tabcharani, J. A. (1989). Possible role of outwardly rectifying anion channels in epithelial transport. *Ann. N.Y. Acad. Sci.* **574:**30–43.
68. Smith, J. J., Welsh, M. J., and Baccam, D. N. (1992). cAMP stimulates bicarbonate secretion across normal, but not cystic fibrosis airway epithelia. *J. Clin. Invest.* **89:**1148–1153.
69. Rich, A., Dixon, T. E., and Clausen, C. (1991). Electrogenic bicarbonate secretion in the turtle bladder: Apical membrane conductance characteristics. *J. Membr. Biol.* **119:**241–252.
70. Imon, M. A., and White, J. R. (1981). The effect of theophylline on intestinal bicarbonate transport measured by pH stat in Amphiuma. *J. Physiol. (London)* **321:**343–354.
71. Knickelbein, R. A., Aronson, P. S., and Dobbins, J. W. (1988). Membrane distinction of sodium–hydrogen and chloride–bicarbonate exchangers in crypt and villus cell membranes with rabbit ileum. *J. Clin. Invest.* **82:**2158–2163.
72. Sundaram, U., Knickelbein, R. G., and Dobbins, J. W. (1991). pH regulation in ileum: Na–H and Cl–HCO_3 exchange in isolated crypt and villus cells. *Am. J. Physiol.* **260:**G440–G449.
73. Hubel, K. A. (1976). Intestinal ion transport: Effect of norepinephrine, pilocarpine and atropine. *Am. J. Physiol.* **231:**2552–2557.
74. Morris, A. I., and Turnberg, L. A. (1980). The influence of a parasympathetic agonist and antagonist on human intestinal transport in vivo. *Gastroenterology* **79:**861–866.
75. Chang, E. B., Bergenstal, E. M., and Field, M. (1985). Diarrhea of streptozotocin-treated rats. Loss of adrenergic regulation of intestinal fluid and electrolyte transport. *J. Clin. Invest* **75:**1666–1670.
76. Tapper, E. J. (1983). Local modulation of intestinal ion transport by enteric neurons. *Am. J. Physiol.* **244:**G457–G468.
77. Tapper, E. J., Powell, D. W., and Morris, S. M. (1978). Cholinergic–adrenergic interactions on intestinal ion transport. *Am. J. Physiol.* **235:**E402–E409.
78. Makhlouf, G. M. (1990). Neural and hormonal regulation of function in the gut. *Hosp. Pract.* **25:**59–78.
79. Cooke, H. J. (1987). Neural and humoral regulation of small intestinal electrolyte transport. In *Physiology of the Gastrointestinal Tract* (L. R. Johnson, ed.), Raven Press, New York, pp. 1307–1351.
80. Cooke, H. J. (1991). Hormones and neurotransmitters regulating intestinal ion transport. In *Diarrheal Diseases* (M. Field, ed.), Elsevier, Amsterdam, pp. 23–48.
81. Sartor, R. B., and Powell, D. W. (1991). Mechanisms of diarrhea in intestinal inflammation and hypersensitivity. In *Diarrheal Diseases* (M. Field, ed.), Elsevier, Amsterdam, pp. 75–114.
82. Castro, G. A. (1989). Gut immunophysiology: Regulatory pathways within a common mucosal immune system. *NIPS* **4:**59–64.
83. Perdue, M. H., Ramage, J. K., Burget, D., *et al.* (1989). Intestinal mucosal injury is associated with mast cell activation and leukotriene generation during Nippostrongylus-induced inflammation in the rat. *Dig. Dis. Sci.* **34:**724–731.
84. Madara, J. L., Patapoff, T. W., Gillece-Castro, B., *et al.* (1993). 5′ adenosine monophosphate is the neutrophil derived paracrine factor that elicits Cl secretion from T84 intestine epithelial monolayers. *J. Clin. Invest.* **91:**2320–2325.

85. Lawson, L. D., and Powell, D. W. (1987). Bradykinin-stimulated eicosanoid synthesis and secretion by rabbit ileal components. *Am. J. Physiol.* **252:**G783–G790.
86. Craven, P. A., and DeRubertis, F. R. (1983). Patterns of prostaglandin synthesis and degradation in isolated superficial and proliferative colonic epithelial cells compared to residual colon. *Prostaglandins* **225:**583–604.
87. Perdue, N. H., Masson, S., Wershil, B. K., *et al.* (1991). Role of mast cells in ion transport abnormalities associated with intestinal anaphylaxis. Correction of the diminished secretory response in genetically mast cell-deficient W/W^v mice by bone marrow transplantation. *J. Clin. Invest.* **87:**687–693.
88. Bern, J. M., Sturbaum, C. W., Karayalcin, S. S., Berschneider, H. M., Wachsman, J. T., and Powell, D. W. (1989). Immune system control of rat and rabbit colonic electrolyte transport. *J. Clin. Invest.* **83:**1810–1820.
89. Hinterleitner, T. A., and Powell, D. W. (1991). Immune system control of intestinal ion transport. *Proc. Soc. Exp. Biol. Med.* **197:** 249–260.
90. Gaginella, T. S. (1990). Eicosanoid-mediated intestinal secretion. In *Textbook of Secretory Diarrhea* (M. Duffey and E. Lebenthal, eds.), Raven Press, New York, pp. 15–30.
91. Kimberg, D. V., Field, M., Gershon, E., *et al.* (1974). Effects of prostaglandins and cholera enterotoxin on intestinal mucosal cyclic AMP accumulation. Evidence against an essential role for prostaglandins in the action of toxin. *J. Clin. Invest.* **53:**941–949.
92. Kimberg, D. V., Field, M., Johnson, J., *et al.* (1971). Stimulation of intestinal mucosal adenyl cyclase by cholera enterotoxin and prostaglandins. *J. Clin. Invest.* **50:**1218–1230.
93. Beubler, E., Bukhave, E., and Rask-Madsen, J. (1986). Significance of calcium for the prostaglandin E_2-mediated secretory response to 5-hydroxytryptamine in the small intestine of the rat in vivo. *Gastroenterology* **90:**1972–1977.
94. Bukhave, K., and Rask-Madsen, J. (1980). Saturation kinetics applied to in vitro effects of low prostaglandin E2 and F2a concentrations on ion transport across human jejunal mucosa. *Gastroenterology* **78:**32–42.
95. Smith, P. L., Blumberg, J. B., Stoff, J. S., *et al.* (1981). Antisecretory effects of indomethacin on rabbit ileal mucosa in vitro. *Gastroenterology* **80:**356–365.
96. Clarke, L. L., and Argenzio, R. A. (1990). NaCl transport across equine proximal colon and the effect of endogenous prostanoids. *Am. J. Physiol.* **259:**G62–G69.
97. Stead, R. H., Tomicka, M., Quinonez, G., *et al.* (1987). Intestinal mucosal mast cells in normal and nematode-infected rat intestines are in intimate contact with peptidergic nerves. *Proc. Natl. Acad. Sci. USA* **84:**2975–2979.
98. Wang, Y.-Z., Cooke, H. J., Su, H.-C., *et al.* (1990). Histamine augments colonic secretion in guinea pig distal colon. *Am. J. Physiol.* **258:**G432–G439.
99. Castro, G. A., Harari, Y., and Russell, D. (1987). Mediators of anaphylaxis-induced ion transport changes in small intestine. *Am. J. Physiol.* **253:**G540–G548.
100. Barrett, K. E., Cohn, J. A., Huott, P. A., *et al.* (1990). Immune-related intestinal Cl-secretion. Effect of adenosine on the T-84 cell line. *Am. J. Physiol.* **258:**C902–C912.
101. Karayalcin, S. S., Sturbaum, C. W., Wachsman, J. T., *et al.* (1990). Hydrogen peroxide stimulates rat colonic prostaglandin production and alters electrolyte transport. *J. Clin. Invest.* **86:**60–68.
102. Barrett, T. A., Musch, M. W., and Chang, E. B. (1990). Chemotactic peptide effects on intestinal electrolyte transport. *Am. J. Physiol.* **259:**G947–G954.
103. Chang, E. B., Musch, M. W., and Mayer, L. (1990). Interleukins 1 and 3 stimulate anion secretion in chicken intestine. *Gastroenterology* **98:**1518–1524.
104. Charney, A. N., and Feldman, G. M. (1984). Systemic acid–base disorders and intestinal electrolyte transport. *Am. J. Physiol.* **247:**G1–G12.
105. Charney, A. N., and Egnor, R. W. (1991). NaCl absorption in the rabbit ileum. Effect of acid–base variables. *Gastroenterology* **100:** 403–409.
106. Mailman, D. (1982). Blood flow and intestinal absorption. *Fed. Proc.* **41:**2096–2100.
107. Grange, D. N., Richardson, P. D. I., Kvietys, P. R., *et al.* (1980). Intestinal blood flow. *Gastroenterology* **78:**837–863.
108. Penn, D., and Lebenthal, E. (1990). Intestinal mucosal energy metabolism—A new approach to therapy for gastrointestinal disease. *J. Pediatr. Gastroenterol. Nutr.* **10:**1–4.
109. Bugaut, M. (1987). Absorption and metabolism of short chain fatty acids in the digestive tract of mammals. *Comp. Biochem. Physiol.* **86B:**439–472.
110. Rhoads, J. M., Keku, E. O., Quinn, J., *et al.* (1991). L-Glutamine stimulates jejunal sodium and chloride absorption in pig rotavirus enteritis. *Gastroenterology* **100:**683–691.
111. Ashkenzai, S., Cleary, T. G., and Pickering, L. K. (1990). Bacterial toxins associated with diarrheal diseases. In *Textbook of Secretory Diarrhea* (M. Duffey and E. Lebenthal, eds.), Raven Press, New York, pp. 255–272.
112. Dytoc, M., Gold, B., Louie, M., *et al.* (1993). Comparison of Helicobacter pylori and attaching-effacing Escherichia coli adhesion to eukaryotic cells. *Infect. Immun.* **61:**448–456.
113. Canil, C., Rosenshine, I., Ruschkowski, S., *et al.* (1993). Enteropathogenic Escherichia coli decreases the transepithelial electrical resistance of polarized epithelial monolayers. *Infect. Immun.* **61:**2755–2762.
114. Tai, Y. H., Gage, T. P., McQueen, C., *et al.* (1989). Electrolyte transport in rabbit cecum I. Effect of RDEC-1 infection. *Am. J. Physiol.* **256:**G721–G726.
115. Boedeker, E. C. (1994). Adherent bacteria: Breaching the mucosal barrier? *Gastroenterology* **106:**255–256.
116. Binder, H. J. (1980). Pathophysiology of bile acid and fatty acid-induced bacteria. In *Disturbances in Intestinal Salt and Water Transport* (M. Field, ed.), American Physiological Society, Bethesda, pp. 159–178.
117. Dharmsathaphorn, K., Huott, P. A., Vongkovit, P., *et al.* (1989). Chloride secretion induced by bile salts. *J. Clin. Invest.* **84:**945–953.
118. Sellin, J. H., and DeSoignie, R. (1990). Short-chain fatty acid absorption in rabbit colon in vitro. *Gastroenterology* **99:**676–683.
119. von Engelhardt, W., Burmester, M., Hansen, K., *et al.* (1993). Effects of amiloride and ouabain on short-chain fatty acid transport in guinea pig large intestine. *J. Physiol. (London)* **460:**455–466.
120. Sellin, J. H., DeSoignie, R., and Burlingame, S. (1993). Segmental differences in short-chain fatty acid transport in rabbit colon: Effect of pH and Na. *J. Membr. Biol.* **136:**147–158.
121. Harig, J. M., Soergel, K. H., Barry, J. A., *et al.* (1991). Transport of propionate by human ileal brush border membrane vesicles. *Am. J. Physiol.* **260:**G776–G782.
122. Mascolo, N., Rajendran, V. M., and Binder, H. J. (1991). Mechanisms of short-chain fatty acid uptake by apical membrane vesicles of rat distal colon. *Gastroenterology* **101:**331–338.
123. Sellin, J. H. (1994). Methodological approaches to study colonic short chain fatty acid transport. In *Short-Chain Fatty Acids* (H. J. Binder, ed.), Kluwer, Dordrecht, pp. 71–82.
124. Grinstein, S., Goetz, J. D., Furuya, W., *et al.* (1984). Amiloride-sensitive Na–H exchange in platelets and leukocytes: Detection by electronic sizing. *Am. J. Physiol.* **247:**C293–C298.
125. Grinstein, S., and Furuya, W. (1986). Characterization of the amiloride-sensitive Na–H antiport of human neutrophils. *Am. J. Physiol.* **247:**C283–C291.

126. Rowe, W. A., Blackmon, D. L., and Montrose, M. (1993). Propionate activates multiple ion transport mechanisms in the HT29-18-Cl human colon cell line. *Am. J. Physiol.*, **265:**G564–G571.
127. Hammer, H. F., Santa Ana, C. A., Schiller, L. R., *et al.* (1989). Studies of osmotic diarrhea induced in normal subjects by ingestion of polyethylene glycol and lactulose. *J. Clin. Invest.* **84:**1056–1062.
128. Donowitz, M., and Welsh, M. J. (1990). Regulation of mammalian small intestinal electrolyte secretion. In *Physiology of the Gastrointestinal Tract* (L. R. Johnson, ed.), Raven Press, New York, pp. 191–208.
129. DeJonge, H. R., and Rao, M. C. (1990). Cyclic nucleotide-dependent kinases. In *Textbook of Secretory Diarrhea* (M. Duffey and E. Lebenthal, eds.), Raven Press, New York, pp. 191–208.
130. Rao, M. C., and DeJonge, H. R. (1990). Ca and phospholipid-dependent protein kinases. In *Textbook of Secretory Diarrhea* (M. Duffey and E. Lebenthal, eds.), Raven Press, New York, pp. 209–223.
131. Chang, E. B., and Rao, M. C. (1991). Intracellular mediators of intestinal electrolyte transport. In *Diarrheal Diseases* (M. Field, ed.), Elsevier, Amsterdam, pp. 49–72.
132. Schultz, S. G., and Hudson, R. L. (1991). Biology of sodium-absorbing epithelial cells: Dawning of a new era. In *Handbook of Physiology—The Gastrointestinal System IV* (S. G. Schultz, ed.), American Physiological Society, pp. 45–81.
133. Lencer, W. I., Delp, C., Neutra, M. R., *et al.* (1992). Mechanism of cholera toxin action on a polarized human epithelial cell line: Role of vesicular traffic. *J. Cell Biol.* **117:**1197–1209.
134. Morris, A. P., and Frizzell, R. A. (1994). Vesicle targeting and ion secretion in epithelial cells: Implication for cystic fibrosis. *Annu. Rev. Physiol.* **56:**371–397.
135. Nambiar, M., Oda, T., Chen, C., *et al.* (1993). Involvement of the Golgi region in the intracellular trafficking of cholera toxin. *J. Cell. Physiol.* **154:**222–228.
136. Lencer, W. I., deAlmeida, J. B., Moe, S., *et al.* (1993). Entry of cholera toxin into polarized human intestinal epithelial cells. *J. Clin. Invest.* **92:**2941–2951.
137. Fasano, A., Baudry, B., Pumplin, D. W., *et al.* (1991). Vibrio cholerae produces a second enterotoxin, which affects intestinal tight junctions. *Proc. Natl. Acad. Sci. USA* **88:**5242–5246.
138. Hecht, G., Pothoulakis, J. T., LaMont, J. T., *et al.* (1988). Clostridium difficile toxin A perturbs cytoskeletal structure and tight junction permeability of the cultured human intestine. *J. Clin. Invest.* **82:**1516–1524.
139. Trucksis, M., Galen, J. E., Michalski, J., *et al.* (1993). Accessory cholera enterotoxin (ACE), the third toxin of a Vibrio cholerae virulence cassette. *Proc. Natl. Acad. Sci. USA* **90:**5267–5271.
140. Fasano, A., Baudry, B., Galen, J., *et al.* (1992). Role of Vibrio cholerae neuraminidase in the function of cholera toxin. *Infect. Immun.* **60:**406–415.
141. Beubler, E., Kollar, G., Saria, A., *et al.* (1989). Involvement of 5-hydroxytryptamine, prostaglandin E_2, and cyclic adenosine monophosphate in cholera toxin-induced fluid secretion in the small intestine of the rat in vivo. *Gastroenterology* **96:**368–376.
142. Cassuto, J., Jodal, M., and Lundgren, O. (1982). The effect of nicotinic and muscarinic receptor blockade on cholera toxin induced intestinal secretion in cats and rats. *Acta Physiol. Scand.* **114:**573–577.
143. Cassuto, J., Siewert, A., Jodal, M., *et al.* (1983). The involvement of intramural nerves in cholera toxin induced intestinal secretion. *Acta Physiol. Scand.* **117:**195–202.
144. Moriarty, K. J., Higgs, N. B., Woodford, M., *et al.* (1989). An investigation of the role of possible neural mechanisms in cholera toxin-induced secretion in rabbit ileal mucosa in vitro. *Clin. Sci.* **77:**161–166.
145. Jodal, M., Holmgren, S., Lundgren, O., *et al.* (1993). Involvement of the myenteric plexus in the cholera toxin-induced net fluid secretion in the rat small intestine. *Gastroenterology* **105:**1286–1293.
146. Forstner, J. G., Boomi, N. W., Fahim, R. E. F., *et al.* (1981). Cholera toxin stimulates secretion of immunoreactive mucin. *Am. J. Physiol.* **240:**G10–G16.
147. Njoku, O. O., and Leitch, G. J. (1983). Separation of cholera enterotoxin-induced mucus secretion from electrolyte secretion in rabbit ileum by acetazolamide, colchicine, cycloheximide, cytochalasin B and indomethacin. *Digestion* **27:**174–184.
148. Moor, B. A., Sharkey, K. A., and Mantle, M. (1993). Neural mediation of cholera toxin-induced mucin secretion in the rat small intestine. *Am. J. Physiol.* **265:**G1050–G1056.
149. Lencer, W. I., Reinhart, F. D., and Neutra, M. R. (1990). Interaction of cholera toxin with cloned human goblet cells in monolayer culture. *Am. J. Physiol.* **258:**G96–G102.
150. Field, M., and Semrad, C. E. (1993). Toxigenic diarrheas, congenital diarrheas, and cystic fibrosis: Disorders of intestinal ion transport. *Annu. Rev. Physiol.* **55:**631–655.
151. Mathias, J. R., and Clench, M. H. (1989). Alterations of small intestine motility by bacteria and their enterotoxins. In *Handbook of Physiology—The Gastrointestinal System I,* American Physiological Society, Bethesda, pp. 1159–1177.
152. Read, N. W. (1991). The role of motility in diarrheal disease. In *Diarrheal Diseases* (M. Field, ed.), Elsevier, Amsterdam, pp. 173–185.
153. Schirgi-Degen, A., Pabst, M. A., Klimpfinger, M., *et al.* (1993). Histopathological effects of cholera toxin in the rat jejunum in vivo. *Gastroenterology* **104:**A279.
154. Sack, R. B., Carpenter, C. C. J., Steenburg, R. W., *et al.* (1966). Experimental cholera. A canine model. *Lancet* **2:**206–297.
155. Carpenter, C. C. J., Sack, R. B., Feeley, J. C., *et al.* (1968). Site and characteristics of electrolyte loss and effect of intraluminal glucose in experimental canine cholera. *J. Clin. Invest.* **47:**1210–1220.
156. Donowitz, M., and Binder, H. J. (1976). Effect of enterotoxins of Vibrio cholerae, Escherichia coli, and Shigella dysenteriae Type 1 on fluid and electrolyte transport in the colon. *J. Infect. Dis.* **134:**135–143.
157. Speelman, P., Butler, T., Kabir, I., *et al.* (1986). Colonic dysfunction during cholera infection. *Gastroenterology* **91:**1164–1170.
158. Ramakrishna, B. S., and Mathan, V. I. (1993). Colonic dysfunction in acute diarrhea: The role of luminal short chain fatty acids. *Gut* **34:**1215–1218.
159. Binder, H. J., and Mehta, P. (1990). Characterization of butyrate-dependent electroneutral Na-Cl absorption in the rat distal colon. *Pfluegers Arch.* **417:**365–369.
160. Smiddy, F. G., Gregory, S. D., Smith, I. B., *et al.* (1960). Faecal loss of fluid, electrolytes and nitrogens in colitis before and after ileostomy. *Lancet* **1:**14–19.
161. Schilli, R., Breuer, R. I., Kelin, F., *et al.* (1982). Comparison of the composition of faecal fluid in Crohn's disease and ulcerative colitis. *Gut* **23:**326–332.
162. Farmer, R. G., Hawk, W. A., and Turnbull, R. B. (1975). Clinical patterns in Crohn's disease: A statistical study of 615 cases. *Gastroenterology* **68:**627–635.
163. Al-Awqati, W., and Greenough, W. B. (1972). Prostaglandins inhibit intestinal sodium transport (abstract). *Nature* **238:**26–27.
164. Ligumsky, M., Karmeli, F., Sharon, P., *et al.* (1981). Enhanced thomboxane A2 and prostacyclin production by cultured rectal mucosa in ulcerative colitis and its inhibition by steroids and by sulfasalazine. *Gastroenterology* **88:**444–449.
165. Bjarnason, I., Giueseppe, Z., Smith, T., *et al.* (1987). Nonsteroidal anti-inflammatory drug-induced inflammation in humans. *Gastroenterology* **93:**480–489.
166. Kaufman, H. J., and Taubin, H. L. (1987). Nonsteroidal anti-inflammatory drugs activate quiescent inflammatory bowel disease. *Ann. Intern. Med.* **107:**513–516.
167. Parkos, C. A., Delp, C., Arnaout, M. A., *et al.* (1991). Neutrophil migration across a cultured intestinal epithelium. *J. Clin. Invest.* **88:**1605–1612.

168. Nash, S., Stafford, J., and Madara, J. L. (1987). Effects of polymorphonuclear leukocyte transmigration on the barrier function of cultured intestinal epithelial monolayers. *J. Clin. Invest.* **80:**1104–1113.
169. Colgan, S. P., Parkos, C. A., Delp, C., *et al.* (1993). Neutrophil migration across cultured intestinal epithelial monolayers is modulated by epithelial exposure to IFN-gamma in a highly polarized fashion. *J. Cell Biol.* **120:**785–798.
170. Parkos, C. A., Colgan, S. P., Delp, C., *et al.* (1992). Neutrophil migration across a cultured epithelial monolayer elicits a biphasic resistance response representing sequential effects on transcellular and paracellular pathways. *J. Cell Biol.* **117:**757–764.
171. Madara, J. L., Patapoff, T. W., Gillece-Castro, B., *et al.* (1993). 5′ adenosine monophosphate is the neutrophil derived paracrine factor that elicits Cl secretion from T84 intestinal epithelial monolayers. *J. Clin. Invest.* **91:**2320–2325.
172. Stenson, W. F., Cort, D., Rodgers, J., *et al.* (1992). Dietary supplementation with fish oil in ulcerative colitis. *Ann. Intern. Med.* **116:**609–614.
173. Rask-Madsen, J., Bukhave, K., Laursen, L. S., *et al.* (1991). 5-Lipoxygenase inhibitors for the treatment of inflammatory bowel disease. *Agents Action Spec. Conf.* C37–C46.
174. Bjorch, S., Dahlstrom, L., Johansson, L., *et al.* (1992). Treatment of the mucosa with local anesthetics in ulcerative colitis. *Agents Action Spec. Conf.* C60–C72.
175. Hollander, D., Vadheim, C. M., Brettholz, E., *et al.* (1986). Increased intestinal permeability in patients with Crohn's disease and their relations. *Ann. Intern. Med.* **105:**883–885.
176. Hollander, D. (1993). Permeability in Crohn's disease: Altered barrier functions in healthy relatives. *Gastroenterology* **104:**1848–1851.
177. Katz, K. D., Hollander, D., Vadheim, C. M., *et al.* (1989). Intestinal permeability in patients with Crohn's disease and their healthy relatives. *Gastroenterology* **97:**927–931.
178. May, G. R., Sutherland, L. R., and Meddings, J. B. (1993). Is small intestinal permeability really increased in relatives of patients with Crohn's disease? *Gastroenterology* **104:**1627–1632.
179. Ruttenberg, D., Young, G. O., Wright, J. P., *et al.* (1992). PEG-400 excretion in patients with Crohn's disease, their first-degree relatives and healthy volunteers. *Dig. Dis. Sci.* **37:**705–708.
180. Reddy, S. N., Bazzocchi, G., Chan, S., *et al.* (1991). Colonic motility and transit in health and ulcerative colitis. *Gastroenterology* **101:**1298–1306.
181. Snape, W. J., Jr., Williams, R., and Hyman, P.E. (1991). Defect in colonic smooth muscle contraction in patients with ulcerative colitis. *Am. J. Physiol.* **261:**G987–G991.
182. Xie, Y. N., Gerthoffer, W. T., Reddy, S. N., *et al.* (1992). An abnormal rate of actin myosin cross-bridge cycling in colonic smooth muscle associated with experimental colitis. *Am. J. Physiol.* **262:**G921–G926.
183. Blennerhassett, M. G., Vignjevic, P., Vermillion, D. L., *et al.* (1992). Inflammation causes hyperplasia and hypertrophy in smooth muscle cells of rat small intestine. *Am. J. Physiol.* **262:**G1041–G1046.
184. Ciancio, M. J., Vitiritti, L., Dhar, A., *et al.* (1992). Endotoxin-induced alterations in rat colonic water and electrolyte transport. *Gastroenterology* **103:**1437–1443.
185. Bell, C. J., Gall, D. G., and Wallace, J. L. (1993). Disruption of colonic electrolyte transport in experimental colitis. *Gastroenterology* **104:**A235.
186. Dalal, V., and Butzner, J. D. (1993). Acute colitis inhibits short-chain fatty acid absorption in the proximal colon of the rabbit. *Gastroenterology* **104:**A241.
187. Sellin, J. H., and Oyarzabal, H. (1988). Carrageenan-induced colitis alters ion transport in rabbit colon in vitro. In *Inflammatory Bowel Disease, Current Status and Future Approach* (R. P. MacDermott, ed.), Excerpta Medica, Amsterdam, pp. 391–396.
188. Alberts, B., Bray, D., Lewis, J., *et al.* (eds.) (1991). *Molecular Biology of the Cell,* Garland, New York, p. 300.
189. Wanke, C. A. (1993). Case 51-1993: A man with AIDS and diarrhea. *N. Engl. J. Med.* **329:**1946–1954.
190. Terry, S. (1993). Drinking water comes to a boil. *New York Times Magazine,* September 26.
191. Flanigan, T., Whalen, C., Turner, J., *et al.* (1992). Cryptosporidium infection and CD4 counts. *Ann. Intern. Med.* **116:**840–842.
192. Soave, R., Danner, R. I., Honig, C. I., *et al.* (1984). Cryptosporidiosis in homosexual men. *Ann. Intern. Med.* **100:**504–511.
193. Fayer, R., and Ungar, B. L. P. (1986). Cryptosporidium spp. and cryptosporidiosis. *Microbiol. Rev.* **50:**458–483.
194. Genta, R. M., Chappel, C. L., White, A. C., *et al.* (1993). Duodenal morphology and intensity of infection in AIDS-related intestinal cryptosporidiosis. *Gastroenterology* **105:**1769–1775.
195. Greenson, J. K., Belitsos, P. C., Yardley, J. H. AIDS enteropathy: Occult enteric infections and duodenal mucosal alterations in chronic diarrhea. *Ann. Intern. Med.* **114:**336–372.
196. Sears, C. L., and Guerrant, R. L. (1994). Cryptosporidiosis: The complexity of intestinal pathophysiology. *Gastoenterology* **106:** 252–254.
197. Argenzio, R. A., Liacos, J. A., Levy, M. L., *et al.* (1990). Villous atrophy, crypt hyperplasia, cellular infiltration and impaired glucose-Na absorption in enteric cryptosporidiosis of pigs. *Gastroenterology* **98:**1129–1140.
198. Argenzio, R. A., Lecce, J., and Powell, D. W. (1993). Prostanoids inhibit intestinal NaCl absorption in experimental porcine cryptosporidiosis. *Gastroenterology* **104:**440–447.
199. Guarino, A., Canani, R. B., Pozio, E., *et al.* (1994). Enterotoxic effect of stool supernatant of cryptosporidium-infected calves on human jejunum. *Gastroenterology* **106:**28–34.

28

The Myasthenic Syndromes

Henry J. Kaminski and Robert L. Ruff

28.1. INTRODUCTION

This chapter deals with disorders of the neuromuscular junction including myasthenia gravis (MG), Lambert–Eaton myasthenic syndrome (LEMS), and congenital disorders of neuromuscular transmission (CDNT). The pathophysiologies of these disorders lie at the presynaptic nerve terminal and the specialized structures located on the postsynaptic skeletal muscle membrane. The myasthenic syndromes are of general interest because an understanding of the pathophysiology at the neuromuscular junction may provide insight into disorders at other neural synapses as, for example, may occur in certain epileptic disorders. In addition, understanding the consequence of specific mutations in the nicotinic acetylcholine receptor (AChR), as occurs in some of the CDNT, may provide insights into the structure–function relationships of the AChR.

28.2. OVERVIEW OF NEUROMUSCULAR PHYSIOLOGY

Motor nerve fibers are myelinated and the main ionic channel responsible for propagation of action potentials in a saltatory fashion down the axon is the neuronal sodium channel.[1] When a nerve fiber is demyelinated, additional channel types appear on the demyelinated membrane.[1] The distal portion of a motor nerve fiber branches to provide a single nerve terminal to each of the many muscle fibers that are innervated by that nerve fiber. The terminal branches of the nerve fiber, which are each up to 100 μm long, are unmyelinated.[2] The unmyelinated terminal branches of the motor nerve fibers contain delayed rectifier and inward rectifier potassium channels as well as sodium channels.[1] Therefore, the amplitude and duration of the action potential in the terminal nerve fibers are controlled by potassium channels as well as by sodium channels.

28.2.1. The Neuromuscular Junction Is a Chemical Synapse with Acetylcholine as the Transmitter

Acetylcholine (ACh) is stored in vesicles within the nerve terminal.[2–6] The ACh-containing vesicles are aligned near release sites in the nerve terminal, where the vesicles will fuse with the presynaptic nerve terminal membrane.[2] The vesicles may be anchored to the nerve terminal cytoskeleton by actin.[7,8] The fusion of synaptic vesicles with the nerve terminal allows vesicles to empty their contents into the synaptic cleft.[9–11] The release sites lie in direct opposition to the tops of the secondary synaptic folds of the postsynaptic muscle membrane.[2,12,13]

28.2.2. Transmitter Release Requires Calcium Influx

There is a first- to second-order relationship between calcium current into the nerve terminal and the size of the endplate current elicited in the muscle membrane.[12–14] The calcium channels responsible for the calcium entry that triggers ACh release are located in the active zones or release sites.[15,16] These calcium channels which trigger ACh release are distributed as two parallel double rows with approximately five channels per row, a spacing of about 20 nm between rows and a spacing of about 60 nm between the double rows of channels at the active zone. It appears that the calcium channels are the particles seen on freeze-fracture electron micrographs that define the active zones visualized by freeze-fracture electron microscopy. The concentration of calcium channels at the active zones enables the calcium concentration to reach 100 to 1000 μM quickly in the regions where vesicle fusion occurs.[12,13] The delay between calcium entry and vesicle release is only about 100 μsec.[14]

There are many types of calcium channels that can be distinguished by their sensitivity to toxins or blocking agents (see Table 28.1).[17–20] The calcium currents in motor nerve terminals of reptiles and birds are sensitive to both dihydropyridines and ω-conotoxin.[18,21,22] The nerve terminal calcium currents in the mammalian neurohypophysis are also sensitive to both dihydropyridines and ω-conotoxin.[20] Therefore, both N- and L-type calcium channels appear to be present at avian, reptilian, and mammalian nerve terminals. However, ACh release at mammalian motor nerve terminals is exquisitely sensitive to funnel-web spider toxin[23] and relatively insensitive to dihydropyridines and ω-conotoxin, which suggest that P-type

Henry J. Kaminski • Department of Neurology, Case Western Reserve University School of Medicine, Cleveland Veterans Affairs Medical Center, University Hospitals of Cleveland, Cleveland, Ohio 44106. **Robert L. Ruff** • Departments of Neurology and Neuroscience, Case Western Reserve University School of Medicine, Cleveland Veterans Affairs Medical Center, University Hospitals of Cleveland, Cleveland, Ohio 44106.

Molecular Biology of Membrane Transport Disorders, edited by S.G. Schultz *et al.* Plenum Press, New York, 1996

Table 28.1. Calcium Channel Sensitivity to Blocking Agents[a]

	Sensitivity to blocking agent		
Channel type	DHP	ω-CTX	FTX
L	+[b]	−	−
N	−	+	−
P	−	−	+
T	−	−	−

[a]DHP, dihydropyridines; ω-CTX, ω-conotoxin; FTX, funnel-web spider venom toxin.
[b]+, sensitive; −, insensitive.

calcium channels may be the channel type that is responsible for the calcium entry which triggers transmitter release at mammalian motor nerve terminals. However, N-type calcium channels are probably also present on mammalian motor nerve terminals.[22,24,25]

In the vertebrate neuromuscular junction, a normal nerve terminal action potential does not fully activate the nerve terminal calcium channels because the duration of the action potential is $\leq$ 1 msec and the nerve terminal calcium channels are activated with a time constant of $\geq$ 1.3 msec.[21,22] Increasing the duration of the nerve terminal action potential by blocking delayed rectifier potassium channels with tetraethylammonium or 3,4-diaminopyridine (3,4-DAP) will increase calcium entry and increase ACh release.[18,21,22]

28.2.3. Calcium Entry Triggers Synaptic Vesicle Fusion

The initial step in vesicle fusion appears to be that vesicles located close to release sites in the nerve terminal must move closer to the nerve terminal membrane so that membrane fusion can start. The approximation of synaptic vesicles and the presynaptic nerve terminal membrane may be opposed by electrostatic forces due to the similar polarity of surface charges on the nerve terminal and vesicle membranes. Calcium may bind to the membrane surfaces and neutralize the negative surface charges, thereby removing a restraint to membrane fusion.[26] Calcium may also open calcium-activated cationic channels and the entry of cations may also reduce the negative surface charges on the synaptic vesicle and nerve terminal membranes.[27]

In addition to neutralizing repulsive charges, calcium may also trigger conformational changes in large molecules that allow synaptic vesicles to detach from the cytoskeleton and that actively trigger membrane fusion. Synapsin is a protein located in presynaptic nerve terminals and on the cytoplasmic face of synaptic vesicles. Synapsin is phosphorylated by calcium–calmodulin protein kinase type II (protein kinase C). Therefore, calcium will stimulate phosphorylation of synapsin. The dephosphorylated form of synapsin cross-links synaptic vesicles to cytoskeletal elements such as actin and blocks synaptic transmission.[14] Calcium entry may trigger phosphorylation of synapsin and allow synaptic vesicles to release from cytoskeletal attachments. Synaptotagmin is another synaptic vesicle-associated protein that is phosphorylated by protein kinase C which may be associated with synaptic vesicle fusion to the nerve terminal membrane.[28,29] Synexin is another nerve terminal protein that may stimulate vesicle fusion with the nerve terminal membrane. According to the "hydrophobic bridge hypothesis,"[30] calcium enters the nerve terminal and binds to monomeric synexin, which activates the synexin and results in the formation of polymers of activated synexin. The activated synexin binds to acidic phospholipids on the nerve terminal membrane and on synaptic vesicles, which initiates membrane fusion. At the onset of membrane fusion in vesicle exocytosis, a fusion pore is formed.[31] The fusion pore is an aqueous channel, about 10–40 nm in diameter, that has the conductance of a large ionic channel.[32,33] The fusion pore rapidly expands as the entire synaptic vesicle fuses with the presynaptic nerve terminal membrane.[9,10,32]

28.2.4. ACh Diffuses across the Synaptic Cleft to Activate AChRs

Each synaptic vesicle fusion releases about 10,000 ACh molecules into the synaptic cleft.[34] ATP is also released by synaptic vesicle fusion and the released ATP may modulate transmitter release and postsynaptic transmitter sensitivity.[35] An action potential propagating into the nerve terminal stimulates the fusion of between 50 and 300 synaptic vesicles (i.e., the normal quantal content is between 50 and 300).[6] The diffusion of ACh across the synaptic cleft is very rapid because of the small distance to be traversed and the relatively high diffusion constant for ACh.[36] Acetylcholine esterase (ChE) in the basal lamina of the postsynaptic membrane accelerates the decline in concentration of ACh in the synaptic cleft as does diffusion of ACh out of the cleft.[37] Inactivation of ChE prolongs the duration of action of ACh on the postsynaptic membrane and slows the decay of the ACh-induced endplate current.[5] The concentration of ChE is approximately 3000 molecules/μm^2 postsynaptic membrane,[37] which is about five- to eightfold lower than the concentration of AChRs.[38]

28.2.5. Postsynaptic Membrane Specialization

Figure 28.1 shows an electron micrograph of a rat neuromuscular junction. The nerve terminal branches lie in depressions of the postsynaptic membrane referred to as primary synaptic clefts. The space between the nerve terminal and the postsynaptic membrane is about 50 nm.[2] The postsynaptic membrane area is increased by folding into secondary synaptic clefts or folds.

The AChRs are concentrated at the tops of the secondary synaptic folds.[39] The concentration of AChRs at the endplate is about 15,000 to 20,000 receptors/μm^2.[38] Away from the endplate the concentration of AChRs is about 1000-fold lower with a slight increase in AChR density at the tendon ends of the muscle fibers.[40] The relatively high concentration of AChRs at the endplate in part results because the muscle nuclei near the endplate preferentially produce the mRNA that encodes for AChR subunits.[41,42]

AChRs continually turn over, with old receptors internalized and degraded. The removed receptors are replaced with

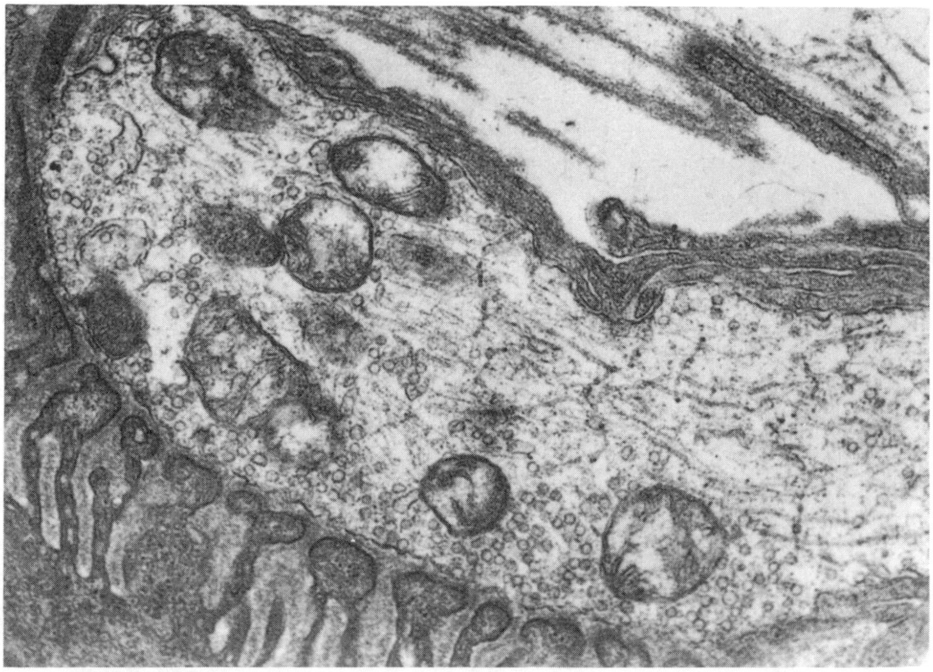

Figure 28.1. Electron micrograph of a rat neuromuscular junction showing junctional folds and synaptic vesicles (see text for description).

new receptors. The AChRs are not recycled. Early in development the half-life of AChRs is short, about 13–24 hr.[43] At a mature endplate the half-life of AChRs is about 8 to 11 days. Cross-linking of receptors by antibodies dramatically shortens AChR half-life by accelerating internalization of the receptors.[44,45]

The AChRs are firmly anchored to the cytoskeleton. A 43-kDa protein is closely associated with the AChRs and during early development the 43-kDa protein appears to be responsible for clustering of AChRs.[39] Other cytoskeletal proteins that may anchor the AChRs include β-spectrin, dystrophin, dystrophin-related protein (also called utropin), and actin.[39,41,42,46] The signaling between nerve and muscle that triggers AChR clustering at the contact point between nerve and muscle is still being resolved. Neural cell adhesion molecule (NCAM) is diffusely present on the muscle membrane surface before innervation, becomes restricted to the endplate region after innervation, and may serve to direct nerve terminal growth cones during initial innervation and during reinnervation after denervation.[47]

The ChE, located in the basal lamina, covers the secondary synaptic folds. The concentration of ChE in the secondary synaptic folds is sufficiently high so that most of the ACh that enters a synaptic cleft is hydrolyzed. Consequently, the secondary synaptic folds act like sinks to terminate the action of ACh and to limit the extent that AChRs can be activated more than once in response to ACh released by a nerve terminal action potential.[48]

Sodium channels are also concentrated in the endplate region. In rat and human fast twitch and garter snake twitch muscle fibers, the sodium current densities are about five- to tenfold higher immediately adjacent to the endplate than at regions away from the endplate (Figure 28.2).[39,49–54] The increased sodium current density falls off rapidly with distance reaching the background level at 100–200 μm from the endplate. When antibodies or labeled toxins were used to study the distribution of sodium channels, the density of channels was greater at the endplate than in the surrounding extrajunctional region and the sodium channels were concentrated in the depths of the secondary synaptic folds.[39,49,55] The sodium channels are immobile in the membrane.[56] Ankyrin may link the sodium channels to the cytoskeleton.[39] The increased density of sodium channels may serve to raise the safety factor for neuromuscular transmission.[51–54,57,58] In addition, a higher concentration of sodium channels may be needed because at the endplate two action potentials must be generated, one traveling toward each tendon end of the muscle fiber.

The density of AChRs in the endplate membrane is about 10,000 channels/μm^2.[59–61] The density of sodium channels varies with fiber type.[52–54] Fast twitch fibers have about 500–550 sodium channels/μm^2 on the endplate membrane and slow twitch fibers have about 100–150 sodium channels/μm^2.

28.2.6. The Safety Factor for Neuromuscular Transmission

The safety factor (SF) for neuromuscular transmission can be defined as

$$SF = \frac{EPP}{E_{AP} - E_M}$$

where EPP is the endplate potential amplitude, E_M is the membrane potential, and E_{AP} is the threshold potential for initiating

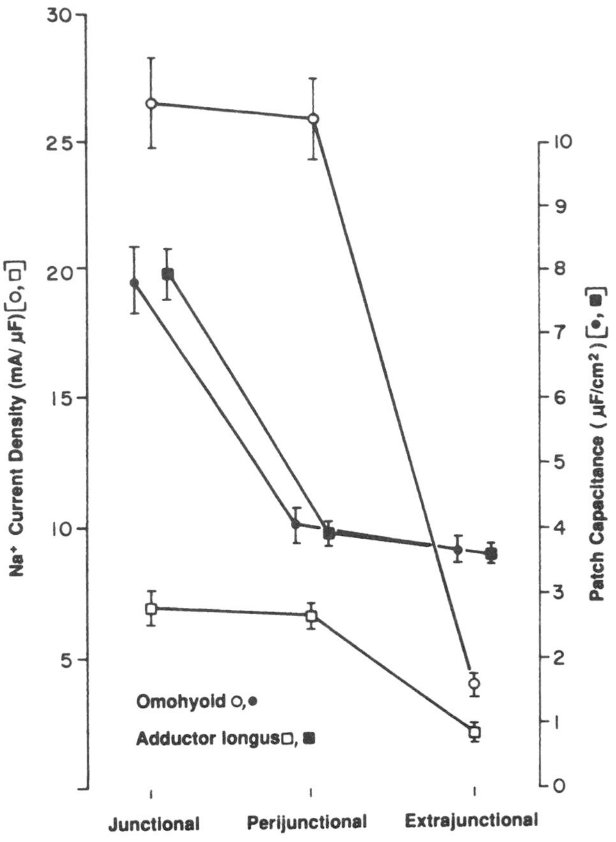

Figure 28.2. Sodium current density (normalized to patch capacitance; unfilled symbols) and patch capacitance (filled symbols) on the endplate (junctional), the endplate boundary (perijunctional), and > 200 μm away from the endplate boundary (extrajunctional) regions of fast twitch omohyoid (circles) and slow twitch adductor longus (squares) fibers. The symbols indicate mean ± S.E. for 12–23 measurements from seven omohyoid and seven adductor longus fibers. The current densities of omohyoid fibers were significantly greater than the adductor longus current densities for each of the three regions studied ($p < 0.001$). The junctional or perijunctional current densities of either adductor longus or omohyoid fibers were significantly greater than the extrajunctional current densities ($p < 0.001$). The junctional and perijunctional current densities were not significantly different for either adductor longus or omohyoid fibers. The patch capacitances on the endplates were significantly larger than on perijunctional or extrajunctional membrane for omohyoid and adductor longus fibers ($p < 0.001$). At each of the three regions the sodium current density was larger on omohyoid fibers ($p < 0.001$), but the patch capacitance of adductor longus and omohyoid fibers for each region were similar.

an action potential.[59] Several factors appear to contribute to increase the SF for neuromuscular transmission of fast compared to slow twitch mammalian skeletal muscle fibers.[62] The nerve terminal morphology and transmitter release properties of axons innervating fast and slow twitch fibers are different. The nerve terminals for fast twitch fibers appear to be larger than those innervating slow twitch fibers.[63] The quantal contents of synapses on rodent fast twitch fibers are larger than for slow twitch fibers.[62,64]

The postsynaptic sensitivities of fast twitch fibers are also greater than slow twitch fibers. Endplates on fast twitch fibers depolarize more in response to quantitative iontophoresis of ACh.[65] The increased endplate sensitivity to ACh is probably not related to fast twitch fiber endplates having more AChRs[59] or different types of AChRs.[65] However, the AChRs on fast and slow twitch fibers differ appreciably in their sensitivities to several nicotinic receptor blockers.[66] Consequently, the increased sensitivity of the endplates of fast twitch fibers to ACh may result from differences in the agonist sensitivities of AChRs when they are incorporated in endplate membranes of fast or slow twitch fibers.

Fast twitch endplates have a higher concentration of sodium channels compared to slow twitch endplates (Figure 28.2).[51–54,58] The increased concentration of sodium channels on and near the endplates of fast twitch fibers may provide another way to increase the SF for neuromuscular transmission of fast twitch fibers.[51–54,57,58,67] In addition, the increased sodium current on fast twitch fibers may be needed because fast twitch fibers have a higher mechanical threshold than slow twitch fibers. Strength-duration studies found that mammalian fast twitch fibers required larger-amplitude depolarization to initiate contraction than slow twitch fibers.[68]

The differences in synaptic transmission for fast and slow twitch fibers may be in response to the different functional properties of fast and slow twitch motor units. Slow twitch motor units in mammals *in vivo* are tonically active at rates of about 10 Hz.[69] Under these conditions, transmitter depletion and other factors may not appreciably compromise neuromuscular transmission.[62,70–72] The effective safety factor for slow twitch motor units from rat soleus muscle at a steady-state firing rate of 10 Hz is about 1.8.[62] In contrast, fast twitch motor units are phasically active at rates of 40 Hz or higher.[69] The higher firing rates of fast twitch motor units make them more susceptible to compromise of neuromuscular transmission caused by depression of transmitter release from depletion of readily releasable vesicles of transmitter.[62,70–72] In addition, at higher rates of stimulation, potassium may accumulate in the extracellular space around muscle fibers resulting in membrane depolarization and depolarization-induced sodium channel inactivation.[73] Muscle fibers in predominantly fast twitch muscles may be more susceptible to reduction in membrane excitability because the capillary networks in these muscles are not well developed so that potassium may not be rapidly cleared from the extracellular space.[51,52,74,75] The SF for neuromuscular transmission for fibers from the fast twitch rat extensor digitorum longus muscle stimulated at 40 Hz drops from 3.7 for the first stimulation to 2.0 after 200 stimuli.[62]

28.3. STRUCTURE AND FUNCTION OF THE AChR

28.3.1. The AChR Is an Allosteric Pentameric Glycoprotein

The AChR is composed of four subunits.[76–79] In mammals the channel is known to exist in at least two isoforms. The mature or "innervated" form of the AChR is composed of two α subunits and one copy of each of the β, δ, and ε subunits.[78,80–82] The fetal or "denervated" form of the AChR has a γsubunit in place of the εsubunit. There are at least two isoforms of the human αsubunit.[83] Studies in which cDNA associated with specific subunits was injected into *Xenopus* oocytes indicated that few if any functional channels were produced if the α, β, or δ subunit was omitted.[81,84] Although the δ subunit can substitute for the γ or ε subunits, AChRs are more efficiently made from a combination of α, β, δ, and γ or ε subunits.[85,86] Muscle fibers express AChRs before there is any contact between motor axons and the muscle fibers.[87,88] AChRs collect into clusters prior to neuronal contact. Contact between a neurite growth cone and a muscle fiber enhances AChR clustering in the vicinity of the contact. Before and after the initial contacts of nerve fibers on a muscle fiber, both the fetal- and adult-type AChRs are expressed with the fetal-type receptor predominating. Early in development muscle fibers receive synaptic contacts from many axons at several points along a fiber. Later in development, at about the stage of synaptic rejection when a twitch fiber supports only one synapse per fiber, the fiber suppresses expression of the γ subunit and upregulates expression of the ε subunit.[78,80–82,89] Mature twitch fibers express the ε subunit in much greater abundance than the γ subunit so that > 97% of the AChRs are the mature isoform.[88] In mature innervated twitch muscle fibers the nuclei near the endplate transcribe AChR subunits more actively than other nuclei.[41,42] After denervation the γ subunit is again expressed and non-endplate nuclei express AChR subunits. Consequently, fetal-type receptors reappear on denervated muscle fibers at and away from the endplate. After denervation, AChRs appear diffusely over the muscle surface, but remain in relative abundance at the endplate.[88] The only exception to this pattern is found in extraocular muscle, in which the γ subunit is expressed in the adult muscle.[89a]

Each α subunit contains one binding site for ACh. The binding sites exist at the interface between the α subunit and the δ subunit or either the γ subunit or ε subunit. AChR channel opening usually requires that two ACh molecules bind to the AChR.[90] There is positive cooperativity for ACh binding.[91] The dissociation constant is about 50 μM with one ACh molecule bound, and about 10 μM for a receptor in the closed state with two ACh molecules bound.[91] In humans, there are two α-subunit isoforms.[83] The significance of these isoforms is not presently understood. The difference in the binding affinity for the first and second ACh molecule could result from: a conformational change induced by binding of the first ACh molecule, a difference in the subunits surrounding each α subunit, or perhaps the slight difference between the two α-subunit isoforms.[91–93]

28.3.2. Differences between AChR Isoforms

Adult- and fetal-type AChRs can be distinguished by their single-channel conductances and channel open times. Adult-type channels have shorter mean open times and a single-channel conductance that is about 50% larger than is found with fetal-type channels.[87,88] Mishina *et al.*[81] demonstrated that differences between adult- and fetal-type AChR channels in mammals could largely be accounted for by the replacement of the γ with the ε subunit. The AChR has kinase A, kinase C, and tyrosine kinase phosphorylation sites.[94,95] Phosphorylation and other forms of posttranslational modification can alter properties of the AChR.[87,96] In particular, subunit phosphorylation appears to regulate agonist-induced desensitization.[94,95] Both subunit substitution and posttranslational modification may contribute to developmental and denervation-induced changes in AChR properties in amphibians and reptiles. Of note, avian and probably fish species do not show a developmental or denervation-induced change in AChR properties and both avian and fish adult AChRs contain the γ subunit, not the ε subunit.[88] It is not known if both γ and ε subunits are expressed by amphibians or reptiles.[85,87,88,97,98]

Electrophysiological studies suggest that mature tonic fiber synapses in reptiles and amphibians have both adult- and fetal-type AChR.[85,97,98] If mature tonic fibers in reptiles and amphibians express both γ and ε subunits, then perhaps multi-innervated mammalian extraocular muscle fibers, which resemble tonic fibers, also express γ and ε subunits.[99,100] The mRNA for the γ subunit as well as the ε subunit is present in extraocular muscle, which supports the hypothesis that fetal-type as well as adult-type AChRs are found in extraocular muscle fibers.[101,102]

28.3.3. Ion Passage through the AChR

The AChR forms a cation-selective channel that is relatively nonselective among cations compared to other voltage-gated cation-selective channels, such as sodium, calcium, and potassium channels.[76,79,103–105] The selectivity of ionic channels is based partly on the size of the pore. The AChR ion pore at its narrowest point is approximately 6.5 Å, which is considerably larger than that of the voltage-gated ion channels which are very ion selective such as the sodium channel.[103] Ions do not pass through the channel freely in a bulk flow fashion; instead, they bind to specific sites within the pore and traverse the channel by moving from one site to the next. To traverse the narrowest portion of the channel, called the selectivity filter, an ion must be stripped of accompanying water molecules, which requires energy.[104] In order for an ion to traverse the AChR channel, it must pass through the selectivity filter and the free energy of the ion bound to a binding site within the channel must be less than or similar to the free energy of the hydrated ion.

The structure of the binding sites within the pore is critical to ion passage.[105] If the interaction between a binding site and ion is too weak, the ion will not pass through the channel. If the interaction is too strong, then the ion will remain fixed at the binding site and effectively block the channel. Several noncompetitive AChR inhibitors appear to block current flow by binding to and remaining at sites within the pore.[90] The charge of the amino acid side chains which line the pore and which form the inner and outer vestibule contribute to the ion selectivity of the channel.[77,79,104,105] The binding of ACh to the channel likely leads to a change in pore structure which allows ions to traverse the channel.[106,107] In addition, ACh binding may also trigger a change in the properties of ion binding sites within the pore.[108]

28.3.4. Molecular Structure of the AChR Subunits

There is homology in the amino acid sequences among the subunits of the AChR.[80,81,84,88] Each of the subunits appears to have its N- and C-terminal regions in the extracellular space. In addition, each subunit contains four αhelices, called M1 to M4, that probably span the membrane.[76,109] The extracellular portions of the subunits consisting of the N- and C-terminal regions and the region between M2 and M3 form a large extracellular vestibule that surrounds the channel extracellular orifice. The regions between M1 and M2 and between M3 and M4 form a smaller vestibule around the intracellular orifice of the ion channel.[76,106,107]

28.3.5. ACh Binding Site

The α-subunit portion of the ACh binding site is located close to a reducible disulfide bond that is formed between cysteines at positions 192 and 193. Reduction of the disulfide bond greatly alters ACh binding affinity.[90] The binding site for α-bungarotoxin is also located on the N-terminal region of the α subunit, close to, but distinct from, the ACh binding site. Amino acid substitutions in the N-terminal region could therefore alter the binding affinity of ACh, other nicotinic agonists, or competitive inhibitors such as α-bungarotoxin or d-tubocurarine.[110]

28.3.6. Lining of the Ion Pore

Several experiments suggest that the M2 segments of the subunits line the ion channel. Noncompetitive AChR inhibitors, which enter and block the AChR ionic channel, bind to specific amino acids in the M2 segments of specific subunits such as the serine in the M2 segment of the δ subunit at position 262 (δ-262).[111,112] Permanently charged local anesthetic analogues such as QX222 block AChR channels by entering the extracellular orifice of the channel and binding to a site within the channel that is close to the intracellular orifice.[113,114] Based on the work with noncompetitive inhibitors, the serine at δ-262 and amino acids on comparable positions of the other subunits appeared to be good candidates for the local anesthetic binding site. Leonard *et al.*[115] used site-directed mutagenesis of mouse AChR subunits to study the effects of interchanging polar and nonpolar amino acids on the binding affinity of QX222 for the AChR channel. The amino acid substitutions they studied were exchanging the serine at δ-262 and the analogous serine on the α subunit with a nonpolar alanine, and replacing a nonpolar phenylalanine on the β subunit with a polar serine. Since QX222 is positively charged, the affinity of QX222 for its channel binding site should be related to the number of polar and nonpolar amino acids near the binding site. As the number of serines decreased, there was a decrease in the residence time for QX222 in the channel and consequently a decrease in the equilibrium binding affinity of QX222. Receptors with three serines replaced with alanines displayed a decrease in outward single-channel currents, which would be consistent with the hypothesis that reducing the complement of polar amino acids near the intracellular orifice of the pore would inhibit intracellular cations from entering and traversing through the channel. The effects of the amino acid substitutions in M2 on the binding affinity of QX222 for the channel and on the direction of the current rectification support the hypothesis that the M2 segments contribute to the lining of the ion channel.[108]

28.3.7. Effect of Site-Directed Mutagenesis on Ion Passage

Experiments using site-directed mutagenesis demonstrated that the net amount of negative charges in the intra- and extracellular vestibules surrounding the AChR ion pore strongly affect the size of the single-channel currents. Negative charges in the vestibules attract cations, thereby increasing the concentration of current-carrying ions at the pore orifices. Imoto *et al.*[116] observed that in an extracellular solution with low divalent cation concentrations, the bovine AChR channel (formed with the γ subunit) had a smaller conductance for inward currents than the *Torpedo* AChR channel. Replacement of the δ subunit of *Torpedo* with the bovine δ subunit produced a chimeric channel with conductance similar to the channel composed entirely of bovine subunits. The experimenters then-constructed δ-subunit chimeras consisting of sections of the *Torpedo* and bovine δ subunits. They found that a region comprising the M2 segment and the extracellular segment between M2 and M3 was the key factor in determining the conductance of channels made with a chimeric δ subunit. The segment between M2 and M3 is part of the extracellular, or outer, vestibule surrounding the ion pore. Figure 28.3 shows that the outer vestibule of the *Torpedo* δ subunit contains one net positive charge, whereas the outer vestibule of the bovine δ subunit contains three net positive charges.

Imoto *et al.*[117] went on to make a series of site-directed mutations of *Torpedo* subunits in the inner and outer vestibules surrounding M2. Figure 28.3 indicates the locations (1–6) of the mutations. They found that changing the charge in the inner vestibule altered the conductance for outward-flowing currents and changing the charge of the outer vestibule altered the channel conductance for inward-flowing currents. Increasing the net negative charge in a vestibule increased the single-channel current.

As mentioned above, the adult form of the bovine AChR with the ε subunit has a larger single-channel conductance than the bovine channel with the γ subunit. Figure 28.3 shows that the inner vestibule of the γ subunit has two net positive charges compared to one net positive charge for the ε subunit. The outer vestibule of the bovine γ subunit has two net positive charges compared to no net charge for the ε subunit.

28.3.8. Effect of Site-Directed Mutagenesis on Open Time

Two lines of evidence suggest that the amino acid sequence in M2 may play a role in regulating the stability of the

```
                       INNER            OUTER VESTIBULE
                     VESTIBULE

Torpedo   α    M1⇒  DSG,EK  ⇐M2⇒  ELIPSTSSAVPLIGK  ⇐M3
          β    M1⇒  DAG,EK  ⇐M2⇒  DKVPETSLSVPIIIR  ⇐M3
          γ    M1⇒  QAGGQK  ⇐M2⇒  QKVPETSLNVPLIGK  ⇐M3
          δ    M1⇒  ESG,EK  ⇐M2⇒  QRLPETALAVPLIGK  ⇐M3

Bovine    γ    M1⇒  KAGGQK  ⇐M2⇒  KKVPETSQAVPLISK  ⇐M3
          ε    M1⇒  QAGGQK  ⇐M2⇒  QKTPETSLSVPLLGG  ⇐M3
          δ    M1⇒  DCG,EK  ⇐M2⇒  KRLPATSMAIPLIGK  ⇐M3
                    1   23        45  6
```

Figure 28.3. Amino acid sequences of the inner and outer vestibules that surround the M2 segment of several different subunits from *Torpedo* electric organ and bovine skeletal muscle. The single-letter abbreviations for important amino acids are: alanine (A; nonpolar), arginine (R; positive charge), aspartate (D; negative charge), glutamate (E; negative charge), glutamine (Q; polar), isoleucine (I; nonpolar), lysine (K; positive charge), phenylalanine (F; nonpolar), serine (S; polar), and valine (V; nonpolar). The positively charged amino acids are double underlined and the negatively charged amino acids are singly underlined. Numbers 1–6 indicate locations of amino acid substitutions to change the net charge in the inner or outer vestibule of a subunit.

open state and hence the open time of the channel. One of the prime differences in mammalian AChR channels built with the ε subunit compared to channels built with the γ subunit is that the channel with the ε subunit has a shorter open time.[81] While the difference in open times could be attributed to several differences between the γ and ε subunits, let us consider the differences in the M2 segments between the γ and ε subunits (Figure 28.4).[80] There are two amino acid differences. The first is that an alanine in the γ subunit is replaced by a serine in the ε subunit. The second difference is that a valine in the γ subunit is replaced by an isoleucine. Alanine, valine, and isoleucine are nonpolar amino acids; serine is a polar amino acid. Hence, the switch of alanine for serine changes the polarity of the M2 segment, whereas the exchange of valine for isoleucine does not change the polarity of the M2 segment. We suggest that replacing a nonpolar amino acid with a polar amino acid in the M2 segment in a critical region near the cytoplasmic orifice of the channel may destabilize the open state resulting in shorter open times. Support for this suggestion comes from the work of Leonard *et al.*,[115] who found that when a nonpolar phenylalanine in the β subunit was replaced by a polar serine, the open time of the channel was reduced (Figure 28.5). When a polar serine in the α or δ subunit was replaced by a nonpolar alanine, the channel open time increased (Figure 28.5).[115] Though circumstantial, the evidence supports the hypothesis that the presence of polar versus nonpolar amino acids near the cytoplasmic end of M2 may regulate the channel open time.

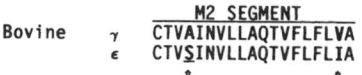

```
                    M2 SEGMENT
Bovine   γ   CTVAINVLLAQTVFLFLVA
         ε   CTVSINVLLAQTVFLFLIA
                ⇑              ⇑
```

Figure 28.4. Amino acid sequences of the M2 segments of the bovine γ and ε subunits. The arrows refer to amino acid differences. S indicates the polar serine. The other amino acids are nonpolar.

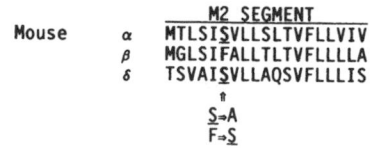

```
                   M2 SEGMENT
Mouse    α   MTLSISVLLSLTVFLLVIV
         β   MGLSIFALLTLTVFLLLLA
         δ   TSVAISVLLAQSVFLLLIS
                  ⇑
                 S⇒A
                 F⇒S
```

Figure 28.5. Amino acid sequences of the M2 segments of the mouse α, β, and δ subunits. The arrow refers to site of amino acid substitutions. S indicates the polar serine. The other amino acids are nonpolar.

28.4. CONGENITAL DISORDERS OF NEUROMUSCULAR TRANSMISSION

28.4.1. Introduction

CDNTs are commonly referred to as congenital myasthenias because of their clinical similarity to immune-mediated MG. Patients often present with fatigable weakness and a propensity for extraocular (EOM) involvement, and electromyography (EMG) usually reveals a decremental response to repetitive stimulation. However, antibodies toward the AChR are not present, manifestations of the illness may be traced to birth, and symptoms respond poorly to immunosuppressive treatments. The CDNTs should be differentiated from neonatal and juvenile MG, which are immune-related (see Table 28.2). Neonatal MG occurs in 10 to 20% of infants of mothers with MG because of transfer of anti-AChR antibodies across the placenta.[118] Juvenile MG refers to the autoimmune form of MG with onset usually before age 17 (arbitrarily defined by most studies) and accounts for at least 10% of autoimmune MG cases. Its pathogenesis is the same as the adult form of the disease. Less than 100 cases of CDNT have been described in detail, but this may be a substantial underestimate of the actual frequency of these disorders because of the extensive evaluation required for diagnosis and their possible confusion with MG. The CDNTs are caused by a variety of defects at the neuromuscular junction and provide unique insights into mechanisms of neurotransmitter release, AChR function, and ACh metabolism. The following sections attempt to categorize the CDNTs on the basis of the presumed physiological defect of neuromuscular transmission and how the defect leads to clinical manifestations.

28.4.2. Presynaptic Disorders

28.4.2.1. Abnormal Packaging of Transmitter

The patients described below share an abnormality of quantal release and bear clinical similarity to cases of "familial infantile myasthenia," which were described prior to the advent of sophisticated electrophysiological techniques and electron microscopy.

Engel *et al.*[119] reported a case that strongly emphasizes the role of synaptic vesicles in producing the EPP. A 23-year-old woman had episodic generalized and bulbar fatigable weakness from birth. Her symptoms responded partially to anti-ChE medications. Physical examination demonstrated small stature, generalized and bulbar weakness, decreased muscle bulk, and

Table 28.2. Contrasting Features of CDNTs and Juvenile MG

	CDNT	Juvenile MG
Fluctuating strength	Yes	Yes
Length of complaints	Commonly date to birth	1–2 years
Response to immunosuppressives	Poor	Good
Associated autoimmune disorders	None	Common
Family history	May be +	Usually −
Tensilon test	+ or −	Usually +
CMAP	May show repeated potentials to single stimulation	Normal
AChR antibodies	−	Usually +

fluctuating ophthalmoparesis. A decremental response of the compound muscle action potential (CMAP) was found with 2 Hz repetitive stimulation which improved with neostigmine treatment. Miniature endplate potential (MEPP) amplitudes were normal, but the EPP at 1 Hz stimulation indicated a marked reduction of quantal content. Elevation of extracellular calcium led to a normal increase in quantal content indicating that quantal release mechanisms were normal. Electron micrographs of the postsynaptic membrane demonstrated normal morphology and AChR density as determined by labeling with peroxidase-labeled α-bungarotoxin. The nerve terminal, however, had a reduced synaptic vesicle density. The anatomic abnormality suggests that the small EPPs resulted from a reduction of readily releasable vesicles containing ACh.

Synaptic vesicle density at the nerve terminal is dependent on two processes: (1) the recycling of vesicles present at the nerve terminal and (2) the synthesis and transport of new vesicles from the cell body. After fusion of synaptic vesicles with the presynaptic membrane, endocytosis of surface membrane occurs with the aid of a protein, clathrin, which forms a coat around the vesicle. After internalization the vesicles shed their coats and fuse to form endosomes. New synaptic vesicles are presumably formed from the endosomes.[8] Since a decrement of CMAP is evident with repetitive stimulation which is presumably dependent to a greater extent on recycling, a defect of recycling is possible. Engel *et al.*[119] hypothesized that since synaptic vesicle density was depleted in the rested state, axonal transport is most likely to be defective. However, a defect of axonal transport may be expected to produce greater clinical and pathological manifestations.

Several patients with deficits of ACh processing have been characterized.[120,121] In 1979 Engel and colleagues described an 18-year-old male who as an infant had ptosis and poor feeding. As he grew older, he fatigued easily and suffered episodic apneas after crying or febrile illnesses. Anti-ChE medications improved his symptoms. The patient's sister had similar difficulties, and three siblings had died of apneic spells. Light microscopy of muscle from this patient and two clinically similar but unrelated patients demonstrated normal muscle. There was no evidence of complement or IgG deposition at the endplates. AChR and ChE content at the endplate were normal. The only anatomical abnormality was a 60% increase in the number of synaptic vesicles. Repetitive stimulation at 2 Hz produced a decrement of the CMAP. Microelectrode studies of an intercostal muscle biopsy revealed normal resting MEPP amplitudes, but after 10 Hz stimulation MEPP amplitude declined. Since MEPP amplitude decreased only after repetitive stimulation, an abnormality of ACh resynthesis is suggested. Such a defect could lie in the (1) choline transporter, (2) ACh resynthesis of choline acetyltransferase, or (3) packaging of ACh into vesicles, the net effect being that vesicles contain less releasable ACh. Desensitization of the AChR is an unlikely explanation because ChE inhibitors would be expected to worsen such a defect.

In a study of three clinically similar patients, Mora *et al.*[121] found a reduction in synaptic vesicle size in intercostal muscle at rest. MEPP amplitude was normal but decreased after repetitive stimulation in each of the patients. With stimulation the vesicle size in patients increased or was unchanged, while controls showed a decrease or did not change. The normal MEPP at rest indicates that quantal content is normal. Further, these results indicate that vesicle size is not related to the ACh concentration of the vesicle. In controls and patients, the vesicle density near the nerve terminal synaptic membrane decreased after stimulation. Mora *et al.*[121] investigated the effect of hemicholinium, which competitively inhibits choline uptake by the nerve terminal, on MEPP amplitudes. Although MEPP amplitude was reduced in controls after drug treatment, no further reduction in amplitude occurred in the patients, and synaptic vesicle size did change in patients or controls. Since synaptic vesicles contain H^+, Mg^{2+}, Ca^{2+}, ATP, GTP, and proteoglycans,[8] it is not surprising that ACh content is not the sole determinant of vesicle size. The decline in MEPP amplitude with stimulation suggests an abnormality of resynthesis of functional synaptic vesicles. A defect of resynthesis of ACh could be related to any of the three reasons mentioned in the previous paragraph. The lack of effect of hemicholinium treatment would indicate that choline uptake was already maximally affected by the underlying abnormality. This deficit would occur because of a primary abnormality of choline uptake which is not worsened by hemicholinium or a defect downstream to choline uptake which limits synaptic vesicle filing with ACh to a greater extent than does the addition of hemicholinium.

Despite the clinical and electrophysiological similarities among these patients, they apparently have different underlying abnormalities. The two groups appear to have defects of ACh processing, while the other has a deficit of production of synaptic vesicles. These cases demonstrate the necessity of detailed electrophysiological and anatomical studies.

28.4.2.2. Impaired Vesicle Release

An abnormality of quantal release has been described in two patients. Bady *et al.*[122] described a child who had generalized weakness from birth, but without bulbar weakness, and moderate mental retardation. Administration of guanidine hydrochloride improved muscle tone and head control. The patient's only sibling was normal, but six of the father's ten siblings died of unknown causes prior to age 7. An EMG showed a decrement at 1 and 3 Hz stimulation of CMAP but an incremental response with 20 Hz stimulation. Guanidine hydrochloride led to improvement in strength. Unfortunately, no microelectrode studies could be performed. Vincent *et al.*[123] described a 15-year-old boy with reduced quantal content despite normal MEPPs, normal ChE staining, and α-bungarotoxin-labeling. The defect in these patients was not defined but may have resulted from an abnormality of presynaptic calcium channels as is thought to occur in LEMS.

28.4.3. Acetylcholinesterase Deficiency

Several patients described with ChE deficiency demonstrate the delicate balance of factors that regulate neuromuscular transmission.[124,125] ChE concentration and kinetics, ACh diffusion, and the electrophysiological characteristics of the AChR determine the time course and amplitude of the MEPP. Several characteristics of the neuromuscular junction lead to the synchronous opening of many AChRs that is required to produce the rapid rising phase of the MEPP. The release of ACh into the small area of the synaptic cleft produces local concentrations well in excess of the K_d of the AChR. ChE is found at five to eight times lower concentrations than AChR on the postsynaptic membrane, and the turnover time of ChE (the time for the enzyme to hydrolyze an ACh molecule) approximates the rise time of the MEPP. Therefore, the ACh initially binds to receptor before it is hydrolyzed or diffuses from the synaptic cleft. The concentration of ACh must drop quickly so that repeated binding and AChR channel opening does not occur. During the falling phase of the MEPP, ACh hydrolysis and diffusion occurs more rapidly than AChR channel closure, and therefore, the time course of the falling phase is determined by the single-channel open time of the AChR.

Using cyto- and immunochemical methods, Engel *et al.*[125] demonstrated a severe deficiency of ChE from the neuromuscular junction of a 14-year-old boy. From birth he had ptosis, fatigable weakness, and had experienced a bout of respiratory insufficiency associated with a viral illness. Physical examination demonstrated scoliosis, decreased muscle bulk, and generalized weakness. Anti-ChE did not improve his symptoms. Significant inhibition of ChE in normal individuals may lead to receptor desensitization and the clinical correlate, "cholinergic crisis," which was not observed in this patient. Stains for ChE demonstrated an absence of ChE (see Figure 28.6), but the muscle was otherwise normal by routine histochemical staining. α-Bungarotoxin-labeling demonstrated normal AChR density at intercostal endplates, but slightly reduced in the biceps. In some muscles junctional folds were degenerated. Other patients with ChE deficiency have demonstrated greater degeneration of junctional folds and a decrease in AChR density.[124]

EMG provided important insights into the physiological consequences of ChE deficiency. A single stimulus evoked two or more CMAPs, and intracellular microelectrode studies demonstrated a prolonged MEPP duration and a low-normal MEPP amplitude. Maintenance of the EPP above the level for action potential generation during the refractory period of the action potential resulted in triggering of multiple action potentials. The ChE deficiency led to an increase in the ACh in the synaptic cleft, and illustrates that diffusion is not sufficient to eliminate ACh from the synaptic cleft.

Further investigations revealed several secondary abnormalities,[124,125] which explain why these patients have a decremental response to repetitive stimulation and fatigable weakness typical of MG. Many nerve terminals were smaller than normal covering only a portion of the postsynaptic membrane. The quantal content of the EPP was reduced in keeping with the decreased size of the nerve terminal. The changes in structure of the synapses may have functioned to increase diffusion of ACh away from the postsynaptic membrane which would decrease endplate depolarization and prevent AChR desensitization. Since the AChR density was normal, the reduced MEPP amplitude would indicate a receptor abnormality or reduced ACh concentration in the vesicles. The latter suggestion appears more likely. A reduction of vesicle ACh may serve to offset the lack of ChE and allow diffusion to rapidly terminate the local action of ACh. The decreased ACh content may also be related to a decrease in choline reuptake because of decreased hydrolysis of ACh in the synaptic cleft. Although these changes may have been adaptive, they lead to a decrease in safety factor for action potential generation.

28.4.4. Abnormal Development of the Postsynaptic Membrane

Several patients have been described with a reduction in the number and depth of secondary synaptic clefts.[126,127] Wokke *et al.*[127] described two siblings in their sixth and seventh decades who had ptosis from childhood and had gradually developed generalized weakness, loss of muscle bulk, and ophthalmoparesis. In contrast, the two brothers reported by Smit *et al.*[126] presented at birth with weakness, ptosis, and contractures. By age 3 their strength had improved. All four patients responded to anti-ChE, and repetitive stimulation produced a decremental response. MEPP amplitudes were reduced, but quantal content was normal. AChR densities assessed by α-bungarotoxin-labeling were reduced explaining the reduction in MEPP amplitude.

The primary defect in these patients appears to be an abnormality of the formation of the synaptic clefts. The synaptic clefts of these patients resembled those of fetal muscle. Therefore, the underlying disturbance may be of developmental regulation. In immature muscle, junctional folds develop after AChRs have begun to be inserted into the membrane. The invagination of the membrane increases surface area and may serve to increase the capacity of the membrane to accommodate AChR insertion. The cause of the AChR reduction may be impaired insertion of receptors into the membrane. The prolonged stability of adult-type AChR appears to be related to the presence of the ε subunit and its association with other synaptic proteins.[128] Disruption of the structure of the end-

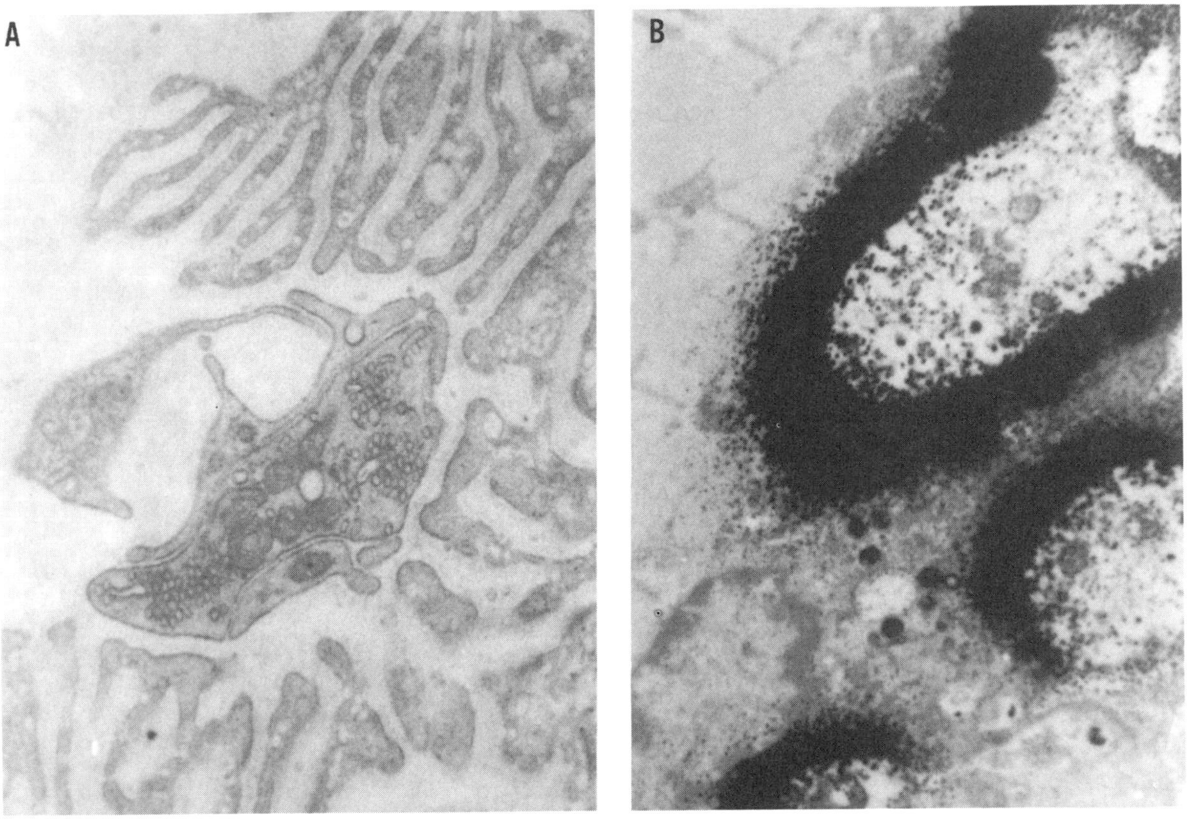

Figure 28.6. AChE localization in patient with AChE deficiency (A) and control (B). (From Engel *et al.*[125] with permission of the *Annals of Neurology*.)

plate in this disorder could lead to an increased degradation of AChR. Decreased synthesis may also account for the reduction in AChR density.

28.4.5. Defects of AChR Function

Several patients with CDNTs have been described who appear to have defects of AChR function. In no patient has the actual defect been elucidated on a molecular level, but electrophysiological data coupled with mutational analysis of the AChR (described above) may be used to postulate where an abnormality may exist in the amino acid sequence.

28.4.5.1. High-Conductance Fast Channel Syndrome

A 9-year-old girl had delayed motor development and fatigable weakness since infancy.[129] Her symptoms responded only partially to anti-ChE medication. EMG revealed a decremental response but only after exercise. A younger sister had similar findings. Routine histochemical studies of the muscle were normal, and electron microscopy revealed a significant reduction in area of the junctional folds and clefts. The density of AChR was normal, and ChE was present at the endplate. MEPP amplitude was large, and the rate of decay was short. Channel conductance was increased as determined by noise analysis and conductance plot while the channel open time reduction was found by noise analysis.

The most likely explanation of the findings in this case is an alteration in the AChR. The increase in conductance may be explained by a mutation that results in an increase in the net negative charge of the external surface of the AChR channel. The shorter channel open time could result if the mutation destabilized the open state conformation of the AChR. A mutation similar to that produced by Leonard *et al.*[115] substituting a polar amino acid for a nonpolar in the M2 region of the β subunit will lead to a reduction in channel open time.

The mechanism by which the abnormality of AChR results in a reduction in the safety factor for neuromuscular transmission is not clear. Anatomical studies of the postsynaptic region demonstrated some degeneration and immaturity of the junctional folds which could lead to a reduction in endplate potential. These degenerative changes could have resulted in part from the increased influx of calcium through the larger conductance channel. The variation in clinical muscle involvement suggests that muscle groups differ in their ability to adapt to an alteration in AChR conductance and open time.

28.4.5.2. Short Channel Open Time and AChR Deficiency

A single patient with an AChR deficiency and decreased open time has been described.[130] At birth the girl required mechanical ventilation, was hypotonic, and diffusely weak. Her fatigable weakness improved with anti-ChE. She was delayed in achieving normal motor developmental milestones. EMG showed a decremental response with repetitive stimulation. Routine histochemical studies revealed a type I fiber preponderance. Morphological studies demonstrated an enlargement of the endplate region, and some junctional folds were simplified. Bungarotoxin binding revealed a decrease in AChR density. MEPP amplitude was low but increased with neostigmine treatment. The AChR open time was decreased while the conductance was normal. The reduced MEPP amplitude was probably secondary to fewer AChR. The shorter channel open time and MEPP duration suggest a mutation that destabilized the open state, and a mutation similar to that described above may be present. The abnormality of the AChR in this patient also led to a decrease in AChR density, which could have resulted from an increase in degradation or decreased synthesis.

28.4.5.3. Low Conductance, Increased Open Time, and AChR Deficiency

Two patients have been described with electrophysiological abnormalities consistent with a mutation of the ε subunit.[131] A 48-year-old woman had generalized and EOM weakness since age 18 months. After a respiratory arrest on exposure to a muscle relaxant at age 35, she was diagnosed with MG after a positive response to edrophonium and a decremental response was identified by EMG. After several years of treatment with anti-ChE medications with good response, the treatment became ineffective and was discontinued. A 16-year-old patient had a significantly worse clinical presentation despite similar *in vitro* electrophysiological studies. She had myasthenic manifestations since birth and was wheelchair-dependent by age 13. Physical examination revealed diffuse weakness with relative sparing of bulbar and ocular muscles. Quanidine improved her symptoms transiently. Both patients had a type I fiber preponderance. The endplate morphology in the first patient was normal, but the younger patient had many degenerating junctional folds and mitochondria, and necrotic nuclei. Both patients had reduced numbers of AChR. In both patients the MEPP amplitudes were decreased. In the younger patient the MEPP decay time and noise analysis data indicated the presence of two channel types, one with a normal channel open time and the other with a markedly prolonged channel open time. Analysis of channel characteristics of the first patient indicated a single channel with a markedly prolonged open time. In both patients the channel conductances were decreased.

Structural analysis using antibodies directed toward epitopes on the AChR demonstrated a failure of binding of antibodies directed toward amino acid residues 360 to 369 of the ε subunit in either patient. In the first patient monoclonal antibodies directed toward ε residues 364 to 373 did not bind. Therefore, the presumed mutations in these two patients appear not to be identical. The region against which the antibodies were directed lie in the cytoplasmic loop of the ε subunit between the M3 and M4 membrane-spanning domains. From the present analysis one cannot determine if the primary mutation lies in this region or has caused a change in tertiary structure of this region, and from mutagenesis studies it is not clear how an abnormality of this region would lead to the observed channel characteristics. These patients also had a significant reduction in AChR density. Gu *et al.*[132] identified amino acid residues 106 and 115 of the rat and mouse ε subunit to be critical in the assembly of the AChR. Alterations in these sites lead to a decreased expression of AChR on the cell surface. It is possible that the mutation in the ε subunit of these patients also affects AChR assembly.

28.4.5.4. Slow Channel Syndrome

The patients described with this syndrome have presented from infancy to adulthood with intermittent weakness on the backdrop of a slowly progressive myopathy.[133,134] Involvement appears to be greatest in the upper extremities and trunk compared to the lower limbs. Weakness does not improve with anti-ChE treatment, and among some, anti-ChE medications actually worsen strength. EMG shows decremental responses to repetitive stimulation, and single electrical stimulation produces repetitive CMAPs, as seen in ChE-deficient patients. Microelectrode studies have demonstrated prolonged MEPP duration and a decreased amplitude. The electrophysiological studies resemble those of AChE deficiency; however, ChE stains reveal the presence of functional ChE. Since ChE is normal, the prolongation of the MEPP most likely results from a prolongation of the open time of the AChR. The presumed increased open time would prolong the depolarization of the membrane beyond the refractory period for action potential generation and produce the observed repetitive CMAPs.

The only anatomical abnormality identified is degeneration of junctional folds (Figure 28.7), but AChR density is normal. The degeneration of the junctional folds presumably accounts for the decreased safety factor for neuromuscular transmission. The cause of the membrane damage has been hypothesized to result from the prolonged depolarization of the endplate. Prolonged depolarization could lead to increased influx of calcium and activation of calcium-sensitive proteases. The hypothesis remains speculative.

28.4.5.5. Abnormality of ACh and AChR Interaction

Since birth a 21-year-old had suffered severe fatigable generalized weakness which responded partially to anti-ChE.[135] EMG revealed a decremental response to 2 Hz stimulation. Histochemical studies were normal, endplate morphology by electron microscopy was normal, and AChR density was normal or slightly increased compared to controls. MEPP amplitudes were decreased. No reduction in AChR density or synaptic vesicle abnormality was identified to explain the MEPP reduction. ACh-induced endplate current fluctuations were used to study AChR kinetic properties. These studies found a normal single-channel conductance, but noise power spectrum indicated complex channel closure. Three explanations may be considered to explain the current fluctuation data: (1) two populations of AChRs with different open times may be pres-

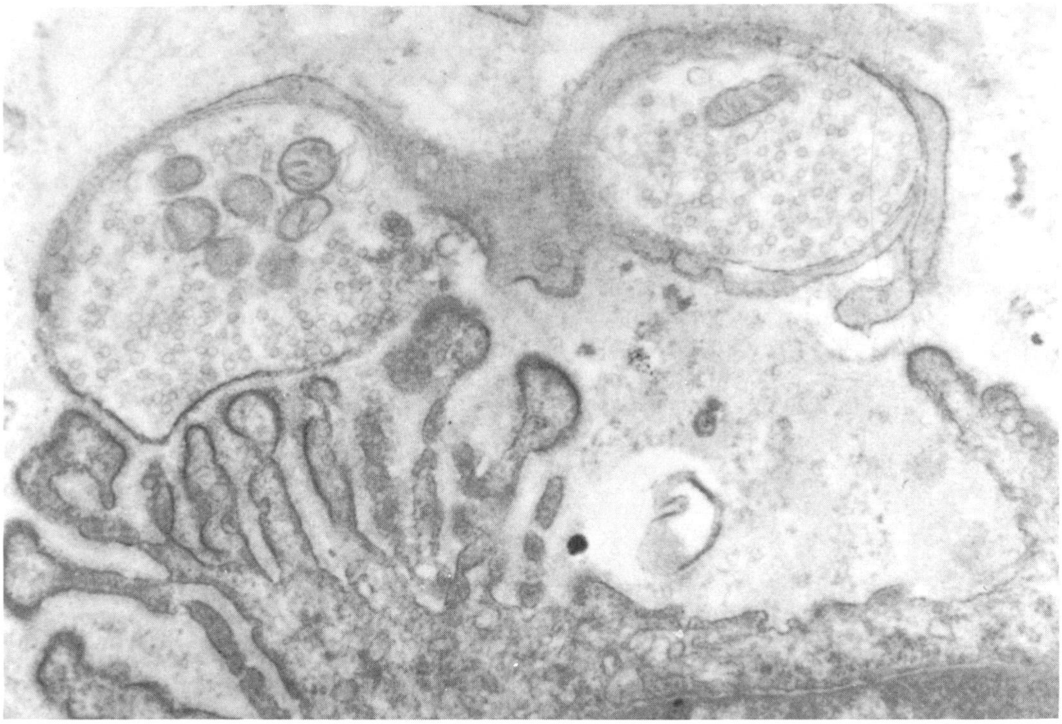

Figure 28.7. Neuromuscular junction from patient with slow channel syndrome showing degenerated junctional folds. (From Engel *et al.*[251] with permission of the *Annals of the New York Academy of Sciences.*)

ent, (2) the open ion channel may be transiently blocked by ACh or transiently blocked in some other way, and (3) a decrease in the rate that ACh dissociates from the receptor or a receptor abnormality so that closure of the channel is no longer the rate-limiting step. Since the channel recordings revealed normal single-channel conductance, the first possibility would not explain the low MEPP amplitude. A block of the channel by agonist would not explain the clinical improvement with anti-ChE treatment and increase in MEPP amplitude with neostigmine would not be expected. Reduced affinity of the receptor for ACh could explain the decreased MEPP in the setting of normal AChR conductance and a normal number of AChRs, and the improvement with anti-ChE. Mutational analysis of the *Torpedo* and mouse AChR demonstrates that the replacement of cystine residues 192 or 193 of the α subunit by a serine or tyrosine 190 by phenylalanine would lead to a reduction of ACh affinity by 10- to 50-fold.[135,136] The ACh binding site appears to be close to peptide loops which include tryptophan 86 and 149 and tyrosine 93 and 151. This syndrome may be explained by mutations in the loops that form the ACh binding site but do not inhibit α-bungarotoxin binding.

28.4.5.6. AChR with Altered *d*-Tubocurarine Binding

Morgan-Hughes *et al.*[137] described a 32-year-old man with a 14-month history of fluctuating generalized and bulbar weakness. The patient's parents were first cousins and two siblings had died during infancy. EMG showed a slight decrement in CMAP after 5 Hz stimulation. Light microscopic analysis of a triceps biopsy revealed tubular aggregates. Electron microscopy of some endplates revealed reduced synaptic vesicles and short, irregular junctional folds. Microelectrode recordings were not performed. The AChR density was reduced as measured by α-bungarotoxin binding. The AChR affinity for binding of *d*-tubocurarine was increased compared to normal and myasthenic controls. Presumably, a change in AChR structure led to altered sensitivity to *d*-tubocurarine and a reduction of AChR density and secondary alterations of the endplate. Mutagenesis studies have demonstrated that binding of antagonists may be modified without altering ACh binding.[136]

28.4.6. Identification of Patients with CDNT

Distinction between patients with autoimmune MG and forms of CDNT is of utmost importance for the appropriate care of the patient. The treatment of MG often involves long-term treatment with immunosuppressive agents, plasmapheresis, or thymectomy, which would not be expected to benefit patients with CDNT and may be harmful.

Several clinical and diagnostic features (see Table 28.2) should alert the physician that a CDNT may be present.[138,139] Since MG and LEMS are rare in childhood, the physician should consider a CDNT in an infant or child with weakness and an EMG that shows a decremental or incremental response to repetitive stimulation. A long history of muscle weakness, often to early childhood, also supports a CDNT. Patients with MG often have been symptomatic for a year or more prior to diagnosis and may have been misdiagnosed several times, but their complaints tend to be of more recent onset than seen in CDNT. A diagnosis of MG in a family member may also serve as a clue to the presence of a CDNT. A poor, or oversensitive, response to anti-ChE treatment is seen in some CDNTs, but improvement is also consistent with the diagnosis. A positive edrophonium test is consistent with acquired or congenital diseases of neuromuscular transmission.

The EMG will confirm a disorder of neuromuscular transmission and repetitive stimulations, which are not routinely performed if not requested, are imperative. An incremental response usually indicates the presence of LEMS so that a search for a neoplasm or autoimmune disorder should be initiated, although two CDNT patients with an EMG response similar to LEMS have been described. The single most helpful finding may be the repetitive CMAP observed after a single nerve stimulation. This finding suggests maintenance of the endplate depolarization beyond the refractory period for action potential generation. The only acquired condition that could produce a prolonged EPP is organophosphate poisoning or overdose with anti-ChE agents.

All disorders of CDNT lack serum antibodies directed toward the AChR; however, roughly 10% of generalized myasthenics and 50% of ocular myasthenics lack AChR antibodies (described below). The majority of seronegative myasthenics probably have an autoimmune disorder and respond to standard therapy of MG, but if the features described above exist in a seronegative patient, then a CDNT should be strongly considered.

After history, physical examination, EMG, and serological tests, including AChR antibody titers, have been performed, and the diagnosis of a CDNT is considered likely, then the patient should be referred to a center able to perform specialized testing.[139] A muscle biopsy is required in such situations, and electron microscopic analysis of the neuromuscular junction which allows for characterization of postsynaptic anatomy, quantitative estimates of AChR density, and synaptic vesicle morphology and concentration is necessary. Microelectrode studies of endplate currents are helpful to further delineate the defect. Also, sufficient quantities of muscle should be removed and frozen at −70°C to allow for future studies, including genetic analysis.

28.5. MYASTHENIA GRAVIS

28.5.1. Autoimmune Pathogenesis

MG is one of the few diseases that fulfill strict criteria for an immune-mediated disorder. First, the purported antigen can be administered to an animal and induce a disease similar to the human form. This was first demonstrated by Patrick and Lindstrom.[140] They injected purified electric eel AChR with Freund's adjuvant in rabbits and observed the development of generalized weakness which improved after edrophonium administration, antibodies toward the AChR, and a decremental response to repetitive stimulation. Second, antibodies may be passively transferred to induce the disease. Lindstrom *et al.*[141] and Toyka *et al.*[142] have produced MG in laboratory animals by administration of AChR antibodies. Finally, immunoglobulin can be identified at the presumed site of the disease. IgG is found bound to the neuromuscular junction in the myasthenic (see Figure 28.8), and AChR isolated from myasthenic tissue has IgG bound to it.[143]

AChR antibodies appear to produce the neuromuscular transmission defect in MG by (1) binding to the AChR and affecting its function (2) accelerating the degradation rate of AChR and thereby lowering the concentration of AChR, and (3) causing complement-mediated lysis of the muscle endplate.[144,145] The first mechanism appears to be relatively unimportant. Antibodies have been identified that block binding of α-bungarotoxin and therefore may affect ACh-induced opening of the ion channel.[146,147] However, these antibodies appear to be only a small percentage of total AChR antibodies and their effect on MEPP is small. In addition, binding of myasthenic antibody to AChR does not appear to affect channel conductance or open time. However, monoclonal antibodies have been produced that modify AChR electrophysiological characteristics.[146] Antibody can cross-link AChR and increase its degradation rate, and the ability of AChR antibody to increase degradation appears to correlate with clinical manifestations to some degree.[146] By far the most important effect of AChR antibody is complement-mediated destruction of the neuromuscular junction. Engel and Fumagalli[145] have convincingly demonstrated deposition of complement components and activation of the lytic phase at the neuromuscular junction. The C9 component of complement is associated with degraded membranous material and the abundance of C9 correlates with destruction of junctional folds. Loss and simplification of junctional folds leads to loss of AChR-rich membrane and probably a decrease in Na channel density. Alteration in the architecture of the junctional folds decreases the amount of membrane surface available for AChR insertion and may affect AChR turnover. All of these factors would lead to a decrease in the safety factor for neuromuscular transmission.

The endplate does appear to respond to the loss of AChR. Choline acetyltransferase activity and ACh release are increased at the myasthenic endplate.[148] These alterations could lead to an increase in the endplate potential. The transcription of AChR subunit genes is increased in response to experimental MG in animals,[149] but the increased synthesis does not appear to lead to sufficient synthesis of AChR to balance the loss produced by the disease. It is interesting to note that experimental MG does not induce transcription of the γ-subunit gene, unlike denervation.[149] Therefore, MG does not appear to produce a functional denervation of muscle.

AChR antibodies are polyclonal IgG subclasses that bind to many different sites on the AChR.[144,146] Investigators using monoclonal antibodies toward the AChR have defined immunogenic regions on the α subunit of the AChR. The majority of myasthenic antibodies and monoclonal antibodies raised against native AChR bind to the so-called main immunogenic region (MIR) which is located on the extracellular loop of the

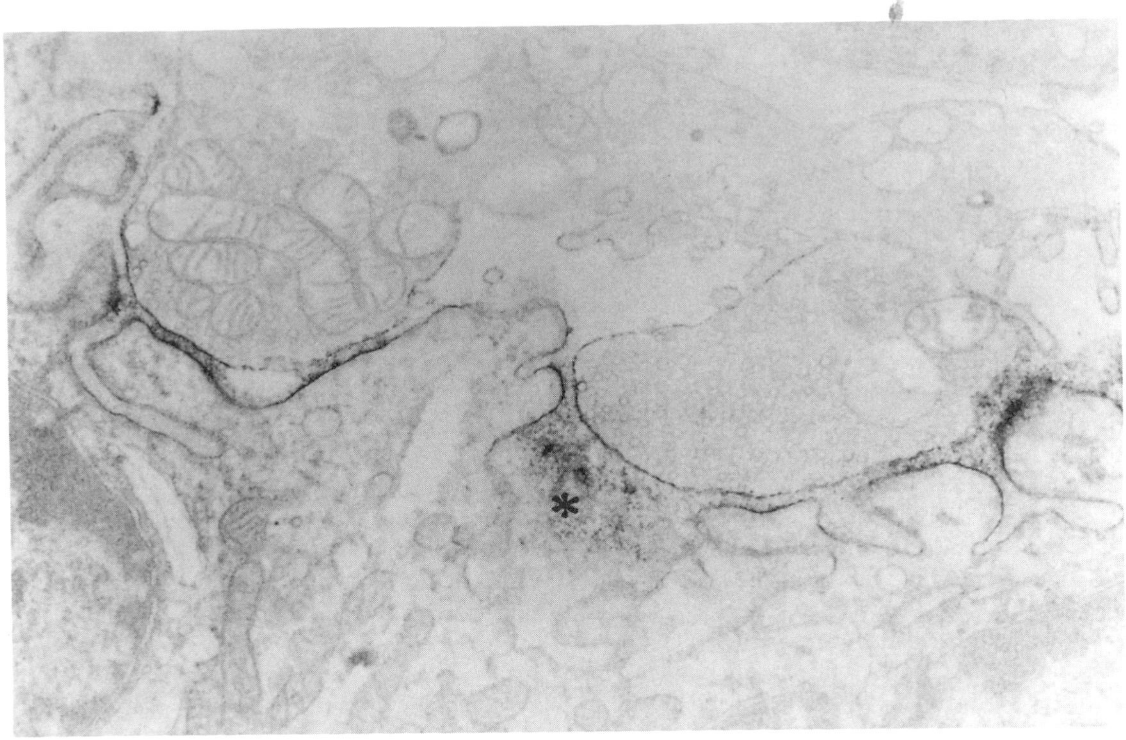

Figure 28.8. IgG localization with peroxidase-labeled protein A at endplate of patient with MG. The junctional folds are degenerated. (From Engel[144] with permission of Alan R. Liss, Inc.)

AChR α subunit in the region of amino acids 61 to 76. Vincent and colleagues[147] have identified a greater frequency of antibody binding to a particular region of the MIR in patients with thymoma compared to other myasthenics. Variations in binding characteristics of antibodies may reflect differences in the ontogeny of the disorder in subgroups of patients and partially explain differences in clinical characteristics and response to treatment.

Autoantibody production in MG is a T-cell-dependent process and a breakdown in tolerance toward self-antigens appears to be the primary abnormality in MG.[150,151] The thymus plays a key role in tolerance induction to self antigens and in responsiveness of lymphocytes to foreign antigens. During development, bone marrow stem cells appear in the thymic subcapsular epithelium where a random process of gene rearrangements occurs in the regions that will ultimately code for the T-cell antigen receptor. The immature T cells pass through the thymic cortex and those that recognize self major histocompatibility complex (MHC) antigens pass through to the medulla. The majority of immature T cells that do not recognize self MHC antigens die and are engulfed by macrophages. During this stage in thymic cortex, T cells that would react toward self antigens are also removed, although the mechanisms by which this occurs are unknown. Sequestration of antigens from the immune system may also be important in preventing autoimmune attack. Once in the medulla the T cells differentiate into helper and suppressor cells and eventually are released to the periphery.[151]

The autoantibody synthesis in MG is under the influence of autoreactive T cells.[150,151] Helper T cells are found in patients with MG which can increase the production of anti-AChR antibodies, and suppressor T cells are present which can decrease antibody production. These T cells are AChR specific. T cells bind AChR only after it is processed by proteolytic cleavage and associated with an MHC molecule by antigen-presenting cells (B cell, macrophage, or dendritic cell). The requirement for MHC association restricts the number of epitopes that a T cell can recognize. The AChR α subunit contains the majority of T-cell recognition sites although other subunits are recognized by T cells from myasthenics. The epitopes on the α subunit recognized by the T cell are distinct from the MIR recognized by antibodies. Similar to antibody production, the T-cell response is polyclonal and varies among myasthenics.[146] The T-cell-dependent nature of MG explains its association with HLA antigens.[152–154] An increased association of HLA-A1, B8, and DRw3 (and B12 antigen in Japan) occurs among female myasthenics under 40 without thymoma. HLA-A3, B7, and DRw2 (and A10 in Japan) occur with greater frequency in males after 40 without thymoma. Restriction fragment length polymorphism analysis has demonstrated a link between MG and HLA-DQ which suggests that HLA site is close to the coding region for the epitope of the AChR binding site. Development of experimental allergic MG in animals is also linked to the MHC sites.[150]

The development of a T-cell-dependent autoimmune disease is caused by a loss of self tolerance.[150,151,155] How a loss of tolerance develops in MG is not understood, but several pieces of data indicate that thymic abnormalities are important.[156] First, thymectomy improves the clinical course of MG. Second, pathological changes occur in 90% of myasthenics, lymphoid follicular hyperplasia of the thymus occurs in 80%, and neoplasia in the remainder. Some patients have a normal or atrophic gland. Compston *et al.*[154] found that differences in thymic histology correlated with certain clinical features. Patients with thymoma had no sex or HLA associations, high titers of anti-AChR antibody, striated muscle antibodies, and a low frequency of associated autoimmune diseases. Third, AChR antibody-producing cells may be isolated from thymus, bone marrow, and peripheral blood, but a greater proportion of antibody is spontaneously produced by thymic cells.[151] These observations may explain the therapeutic benefits of thymectomy and that thymectomy does not lead to complete resolution of the disorder. Finally, the thymus contains proteins that are antigenically similar, or identical, to the AChR. Myoid cells, which express muscle proteins and form muscle fibers in tissue culture, are present in normal and hyperplastic thymuses of myasthenics, but not in thymomas.[157,158] Cultured myoid cells and thymuses of myasthenics express transcripts of AChR subunits,[158] and some AChR antibodies and α-bungarotoxin will bind to myoid cells. Epithelial cells from thymomas bind AChR antibodies and express transcripts of the α subunit, but functional AChRs are not expressed. The expression of antigenically similar proteins to the AChR in the thymus would provide a source of antigen for autosensitization of T cells toward the skeletal muscle AChR.[156]

28.5.2. Possible Explanations for Differential Involvement of Muscles by MG

Most muscle disorders demonstrate a differential involvement of muscle groups, regardless of whether they are genetic, metabolic, or autoimmune in etiology. Polymyositis affects proximal musculature and frequently affects the myocardium. Duchenne/Becker muscular dystrophy disproportionately affects fast contracting musculature, except for EOM.[159] Clinical observations demonstrate a variation in involvement of muscle groups among myasthenics.[160] Ocular, bulbar, neck extensor, and proximal limb musculature are affected to a greater extent than thigh and distal limb muscles. The basis for this differential involvement is no doubt the inherent differences in physiological properties of the muscles, their ability to adapt to disease-induced damage, and the nature of the disease itself.

The differential involvement of muscle by MG is exemplified best by the preponderance of ocular manifestations. Weakness of EOM is the first sign of MG in the majority of patients and ultimately occurs in 90% of myasthenics. The greater frequency of ocular manifestations is dependent on factors unique to EOM (see Table 28.3) and other features that may aid in the explanation of variations in extremity muscle involvement (for complete review see Ruff *et al.*,[99] Kaminski *et al.*[100]) First, slight EOM weakness will sufficiently misalign the visual axes to produce symptoms. Second, the high firing frequencies of EOM motor units may increase EOM susceptibility to fatigue. Third, physiological properties of EOM fibers may make them more susceptible to the block of muscular excitation produced by MG. Finally, neuromuscular epitopes unique to EOM may exist, and thus antibodies in some myasthenics could be targeted specifically to EOM postsynaptic membrane.

Only slight weakness of an EOM is necessary to cause symptoms.[161] The visual axes must be aligned precisely, otherwise a slight discrepancy in retinal images occurs, and symptoms, such as diplopia, develop. Extraocular motoneuron activity is primarily efferently coded with little continuous proprioceptive feedback. The ocular motor system may be particularly susceptible to MG because its central control mechanism, utilizing primarily visual feedback, cannot adapt quickly to asymmetric or variable weakness of the EOMs. In contrast, during limb movements continuous correction of motoneuron activity occurs by feedback of position, velocity, and muscle tension. A slight reduction in force generation of extremity muscles, as could occur in the early phases of MG, may be compensated for prior to symptom development.

To understand the physiologic characteristics that make EOM susceptible to neuromuscular transmission failure, a brief introduction to EOM anatomy and physiology is necessary. EOM contains two basic muscle fiber types (this is an oversimplification; for a complete review see Spencer and Porter,[162] Ruff *et al.*[99]). Approximately 80% of EOM fibers have a single neuromuscular junction and are termed singly innervated fibers (SIFs). Morphological, histochemical, and physiological studies suggest that the SIFs most closely resemble fast-twitch extremity muscle fibers. In response to a single nerve stimulation, twitch fibers produce a synchronized contraction and a propagated action potential. The SIFs possess a single en plaque neuromuscular synapse (Figure 28.9) similar to extremity muscle endplates. The remaining EOM fibers, the multiply innervated fibers (MIFs), receive multiple synapses per fiber (see Figures 28.9 and 28.10). MIFs have been studied best in nonmammalian species, and other than EOM, are present only in pharyngeal, tensor tympani, and stapedius muscles of humans. In amphibians and reptiles, the MIFs can be divided into tonic and intermediate fibers. Tonic fibers are distinguished by three physiological characteristics. First, they cannot propagate action potentials. Second, they maintain a graded contraction to depolarization, and the amplitude of the contraction is proportional to the degree of membrane depolarization. Third, tonic fibers contract and relax very slowly compared with twitch fibers. The MIFs possess en plaque and en grappe endplates. Intermediate fibers

Table 28.3. Differences between Extraocular and Extremity Skeletal Muscle

	EOM	Extremity
Proprioceptive feedback for motor control	Little	Prominent
Intermediate and tonic fibers	Present	Absent
Typical motor unit firing frequency	>150 Hz	<50 Hz
Maximal motor unit firing frequency	~400 Hz	~100 Hz
Fetal-type AChR mRNA expression	Yes	Little to none

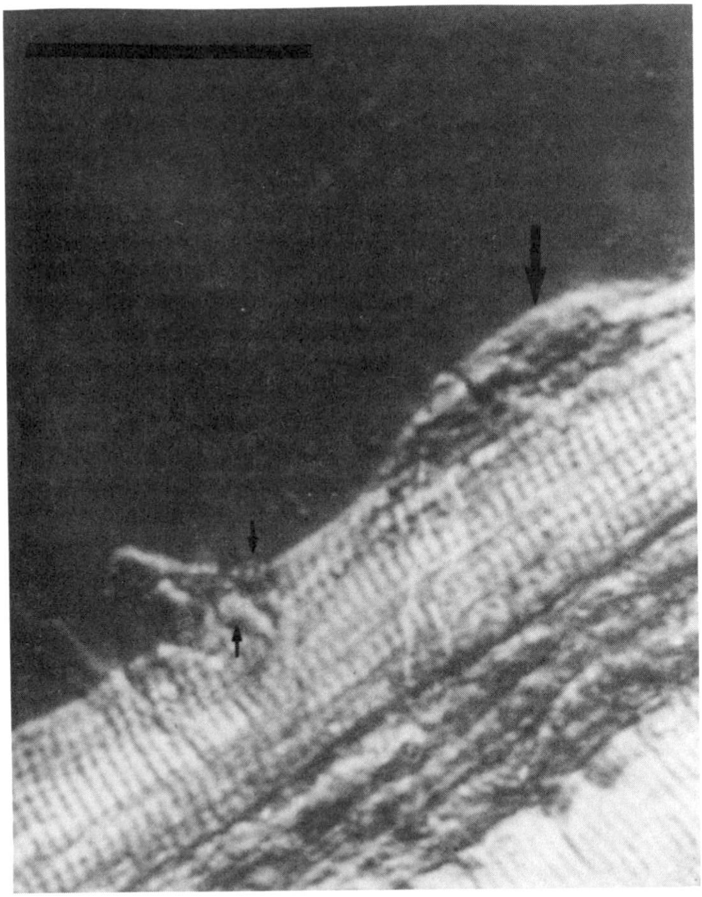

Figure 28.9. Rat EOM fiber with two en grappe synapses (small arrow) and an en plaque synapse (large arrow). Scale bar = 50 μm.

share properties of both twitch and tonic fibers. Human EOM appears to possess both tonic and intermediate fibers.

EOM twitch fibers may be more susceptible than extremity twitch fibers to MG for several reasons. EOM twitch motor units operate at higher motor unit firing frequencies, which may make EOM particularly susceptible to fatigue. The firing frequencies of fast EOM motor units during saccades exceed 400 Hz, while the maximum firing frequencies of extremity motor units do not exceed 100 to 200 Hz. In normal EOM, the safety factor is sufficient to prevent fatigue during repeated saccades or maintained eccentric gaze. EOM twitch fibers appear to possess certain physiological properties that may make them more susceptible to neuromuscular blockade. EOM twitch fibers have less prominent secondary synaptic folds and therefore may have fewer AChR and Na^+ channels on their postsynaptic membrane. In addition, the mean quantal content is decreased in twitch EOM fibers compared with extremity muscle.[163] A reduction in AChR, Na^+ channels, and quantal content would reduce the safety factor for neuromuscular transmission of EOM twitch fibers relative to extremity twitch fibers. It is not known if significant variations in AChR and Na^+ channel density and quantal release occur among extremity muscles and account for variations in extremity muscle involvement by MG. However, proximal and distal muscles differ in their response to repetitive stimulation,[164] which reflects differences in the safety factor for neuromuscular transmission.

Assuming MIFs are similar to nonmammalian tonic fibers, they would be predisposed to MG for structural and physiological reasons. Junctional folds are sparse to nonexistent in the tonic fibers of reptiles and the AChR receptor density is lower by a factor of 1.3 to 1.5. With a lower density of AChR, MG would compromise synaptic transmission at the tonic EOM synapse more than at a twitch synapse. Force generation in tonic fibers is directly proportional to the membrane depolarization caused by the synaptic potentials. Hence, unlike the twitch fibers, no safety factor exists for neuromuscular transmission in tonic fibers, and any reduction of synaptic depolarization could lead to symptomatic weakness.

Evidence exists that MIFs are preferentially involved by MG. Available data are compatible with the hypothesis that MIFs play a role in maintaining gaze at an intended position and small eye movements from midposition while MIFs are responsible for rapid eye movements, termed *saccades.* Saccades are spared in myasthenics,[165] and this finding suggests that MIFs are preferentially affected by the disease. This view is further supported by a computer model of eye movements of myasthenics.[166] In this model, the ocular motility disorders of myasthenics could be simulated by a defect in the component of EOM force generation that is responsible for holding the eyes fixed with sparing of the fibers that generate a saccade. In addition, tests of stapedius muscle function, which contains MIFs, demonstrate abnormalities in 84% of myasthenics.[167]

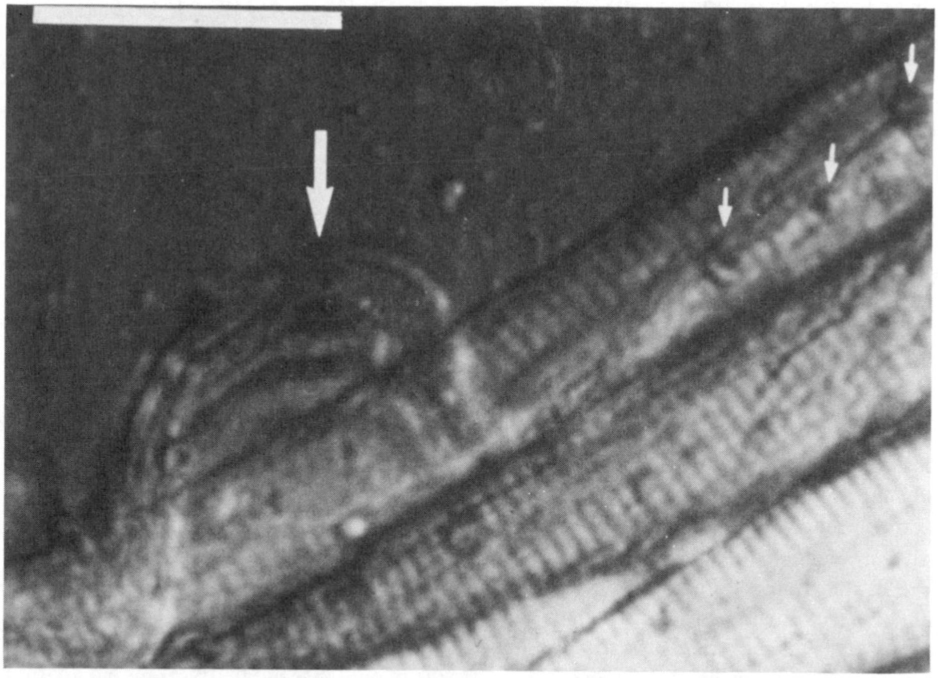

Figure 28.10. En grappe synapses distributed along single rat EOM fiber (small arrows). Large arrow points to nerve fiber. Scale bar = 50 μm. (From Kaminski *et al.*[100] with permission of *Neurology*.)

Some studies suggest that the sera of ocular myasthenics reacts more strongly with ocular muscle antigen. Oda[168] found a subgroup of seronegative ocular myasthenics who were positive when tested against EOM-derived antigen. By immunohistochemistry, two sera reacted only with en grappe synapses of EOM, four with both en plaque and en grappe synapses, and two with en plaque synapses of EOM and extremity muscle. However, others[169] have not identified differential binding of ocular myasthenic sera to EOM antigens.

Neuromuscular epitopes unique to EOM exist, and thus antibodies in some myasthenics could be targeted specifically to EOM postsynaptic membrane. Studies of snake tonic muscle fibers have shown both fetal- and adult-type AChR at the neuromuscular junction.[97,98] Twitch fibers contain only the adult-type channels. Bovine and murine EOM express mRNA of the γ subunit of the fetal-type AChR.*[101,102] The presence of fetal-type channels at the tonic fiber synapses may facilitate tonic force generation because the channels have a longer mean open time which would tend to produce a more uniform spread of depolarization along the fiber.

Serum from myasthenic patients contains antibodies that react selectively with fetal-type or extrajunctional AChRs, which are not expected to be found on normal adult extremity muscle. T cells from myasthenics also react against fetal-type AChR epitopes.[170] Schuetze and Role[88] studied the effects of myasthenic serum on AChR in developing rat muscle. Myasthenic serum blocked the fetal-type AChR. Others[147] have confirmed that myasthenic serum reacts against extrajunctional AChR. Detailed study of antibody reactivity demonstrated two determinants on AChR, one common to extrajunctional and junctional AChR, the other unique to extrajunctional AChR. Using monoclonal antibodies, Vincent and colleagues[147] have found that some myasthenic antibodies bind only to the α subunit of the extrajunctional AChR. Therefore, the presence of the γ subunit appears to induce a conformational change which exposes an antigenic site on the α subunit. Tzartos *et al.*[171] found that myasthenic serum contained antibodies directed at regions of the γ subunit, although this did not correlate with ocular symptoms. The presence of the fetal-type AChR in EOM provides an antigenic site for differential autoimmune attack.

MG provides a particularly good example of how the clinical manifestations of a muscle disorder cannot be explained simply by the primary disease but requires understanding of the function of the individual muscle and its physiological properties.

28.5.3. Diagnosis

The prevalence rates of MG range from 5 to 142 per million with incidence estimates of 22 to 100.[172,173] Women in their third and fourth decades are affected more frequently than men, while the incidence of MG rises in both sexes with age.[172,173] In contrast to earlier studies, Phillips *et al.*[172] found the incidence and prevalence of MG to be higher among American blacks than whites. Oriental populations differ in the clinical characteristics of the disease. The peak incidence

*The MIFs and a subset of SIFs, of rat EOM express the adult and fetal AChR at their mature endplates.[89] Interestingly, the levator palpebrae, which is frequently affected by MG and shares anatomic similarity with EOM, does not express the fetal AChR.[109a]

of MG in Japanese and Chinese populations occurs before age 20, and ocular myasthenia is more common than in Caucasians.[174,175]

As described, the majority of patients present with ocular manifestations, roughly equally distributed between ptosis and diplopia.[176] Within 6 months of presentation, slightly more than one-half of patients develop generalized disease, and three-quarters of patients will have bulbar or extremity weakness within the first year. After 3 years of onset, only 6% of ocular myasthenics develop generalized disease. Rare ocular myasthenics develop generalized disease decades after initial presentation. In 15% of myasthenics clinical involvement is restricted to the EOM.[177] Remissions occur in 10% of patients usually in the first year.

The differential diagnosis of ocular MG is relatively limited.[161,177] Ocular MG may mimic cranial neuropathies or intrinsic brain-stem dysfunction, but several features aid in distinguishing these disorders. Since MG never involves the pupil, isolated ptosis is easily differentiated from Horner's syndrome and third cranial nerve dysfunction. Unilateral ptosis may be a sign of a third cranial nerve abnormality, but if it develops painlessly over several days, MG is the most likely diagnosis. Alternating or recurrent ptosis is only caused by MG. Isolated abduction weakness caused by MG may mimic sixth nerve palsy but can be differentiated by fatigability or involvement of other muscles. MG may produce isolated medial rectus weakness causing a "pseudo-internuclear ophthalmoplegia." Patients with a true internuclear ophthalmoplegia usually have spared convergence and other signs of brain-stem disease. Graves' ophthalmopathy[178] produces a limitation of ocular movement but can be differentiated from MG by the absence of ptosis and presence of proptosis. Radiologic imaging demonstrates enlargement of EOM. However, one must bear in mind that MG and Graves' may coexist. Kearns' syndrome, a mitochondrial disorder, leads to progressive ophthalmoplegia, which may be difficult to distinguish from patients with long-standing ocular MG.[161,177] Ancillary diagnostic testing, as described below, may be necessary to distinguish these disorders with confidence.

Despite the frequent occurrence of ocular manifestations, the presentation of MG may be extremely varied.[176] A fifth of patients present with prominent manifestations of dysarthria and dysphagia. Because of palatal muscle weakness, speech becomes nasal and regurgitation of liquids through the nose occurs. Facial muscle weakness produces a smooth face and the smile becomes straight. Respiratory muscle weakness may occur early in MG. Patients commonly take rapid shallow breaths, and their condition is confused with hysteric hyperventilation. This presentation is particularly dangerous because of the risk of respiratory arrest. Isolated limb weakness occurs in fewer than 10% of patients, and rarely isolated weakness of certain muscles, such as the neck extensors, may occur. The diagnosis of MG must be seriously questioned, if ocular manifestations do not develop at some time during the disease.[177]

Confusion of generalized MG with other neuromuscular disorders must be guarded against. Polymyositis, inclusion body myositis, and late-onset muscular dystrophies can be distinguished from MG by the presence of elevated serum creatine phosphokinase, abnormal EMG, and characteristic muscle pathology. Endocrine disorders, such as hyper- and hypothyroidism, glucocorticoid excess, and adrenal insufficiency produce generalized extremity weakness but usually are differentiated by associated clinical findings and characteristic laboratory abnormalities.[178] Autoimmune thyroid diseases may coexist with MG and complicate the diagnosis of both conditions. Amyotrophic lateral sclerosis and MG may mimic each other; however, EMG should distinguish the disorders. LEMS and MG rarely coexist,[179,180] and the two may be difficult to differentiate clinically. Just as the fatigable weakness of MG should not be confused with psychiatric disease, vague symptoms of "tiredness," which represent underlying depression, should not be considered evidence of MG. However, depression may occur in undiagnosed myasthenics, in particular, in older age groups among whom depression is common. MG occurs in association with hyperthyroidism, rheumatoid arthritis, systemic lupus erythematosus, and, possibly, pernicious anemia, and the diagnosis of MG should be considered in individuals who have these disorders and present with weakness.

Improvement in strength after intravenous injection of edrophonium is the hallmark of MG. Opinions as to the exact manner in which edrophonium should be administered vary,[181] but in general, a total of 10 mg of edrophonium is given over approximately 5 min and a response is expected within 5 min. If unequivocal improvement occurs in a single weakened muscle, the test is considered positive. The test is most useful if improvement in ptosis or the strength of an EOM is demonstrated, because of the objective nature of this response. Difficulties arise in evaluation of the test results when attempting to evaluate improvement in limb strength or bulbar function.[182] Injection of saline and edrophonium in a double-blind fashion is not useful given the characteristic muscarinic side effects that occur. False-positive edrophonium tests have been described with motoneuron disease, LEMS, and intracranial mass lesions.[177,182–184] False-negative tests are relatively common, and repeated tests are of value. A single positive edrophonium test is strong support for the diagnosis of MG.

A radioimmunoassay, which uses human extremity muscle as an antigenic source, is used for clinical detection of anti-AChR antibodies. Antibodies are present in up to 80% of patients, but only 50% of ocular myasthenics possess anti-AChR antibodies by such tests.[185,186] Levels of antibodies correlate poorly with clinical status. False-positive detection of AChR antibodies may occur in patients with LEMS and patients treated with penicillamine.

Antistriational muscle antibodies may be useful as a marker for thymoma. However, false-positive and -negative tests are common, and their use should be restricted to patients between the ages of 20 and 50 to be clinically useful.[187] Antistriational antibodies may be found in LEMS and other autoimmune and paraneoplastic diseases.

Patients with seronegative MG may represent a special subset of acquired autoimmune MG. Birmanns *et al.*[188] divided 12 seronegative myasthenics into two groups. Seven had generalized weakness and a clear response to immunosuppressive treatments, while five had oculobulbar manifestations that were relatively benign and in the three who received immunotherapy, treatment was not affective. From the nine patients tested, peripheral blood lymphocytes did not synthesize anti-AChR antibodies when stimulated with pokeweed mitogen or Epstein–Barr virus. Mossman *et al.*[189] found that immunoglobulin preparations from eight seronegative myasthenics (six generalized and two ocular myasthenics) infused into mice led to impairments of neuromuscular transmission and a small reduction of AChR in the diaphragm. This study suggests that antibodies not directed at the AChR may lead to typical MG. In a study of 221 patients, Soliven *et al.*[190] identified 41 seronegative patients. No significant differences in clinical characteristics were identified between seronegative and positive patients.

The EMG is critical in the diagnosis of MG. A decremental response of the CMAP with 3 Hz stimulation can be identified in at least 74% of myasthenics.[173,186] Ocular myasthenics despite lack of generalized manifestations may also demonstrate a decremental response, but in a lower percentage than generalized myasthenics.[177] Single fiber EMG is the most sensitive test for detecting abnormalities consistent with MG. However, its usefulness is compromised by a high false-positive rate and requirement for specialized training. Therefore it should be used only by experienced electromyographers in the diagnosis of patients without other definitive evidence of MG.[186]

28.5.4. Treatment

Proper care of every newly diagnosed myasthenic begins with a thorough discussion of the natural history, treatment options, and pathogenesis of the disease. Patients should receive a list of medications that may exacerbate myasthenic manifestations and emphasize that every health practitioner they encounter must be informed of their condition. Because of their special needs, patients should be encouraged to educate physicians caring for their other medical conditions about MG. Follow-up visits should include reinforcement of these issues, and education of family members and friends is vital. Psychological adjustment is difficult to a chronic illness in which definite predictions of outcome cannot be made and patients fear the development of incapacitating weakness. These fears may limit an individual's functional improvement despite significant resolution of myasthenic weakness. The physician should attempt to strike a balance between the difficulties in treating MG and the hope of complete remission. Support group services provided by the Myasthenia Gravis Foundation are of particular benefit to the newly diagnosed myasthenic.

ChE inhibitors, which retard the hydrolysis of ACh at the neuromuscular junction, are the primary treatment of MG. ChE therapy must be tailored to the needs of each patient. Pyridostigmine bromide and neostigmine bromide are the most commonly used ChE inhibitors. Initial therapy begins at doses of 30 to 60 mg of pyridostigmine with individual doses of greater than 120 mg rarely being effective. Dosing intervals of pyridostigmine are set at 3 to 6 hr depending on symptoms. Timed-release forms are available but tend to have erratic absorption, although they may be useful in some patients. ChE inhibitors affect muscarinic synapses and therefore may lead to cholinergic side effects. Gastrointestinal complaints of nausea, vomiting, diarrhea, and cramps are most common but may be controlled with administration of atropine or glycopyrrolate. Pyridostigmine tends to have a lower frequency of gastrointestinal side effects. Respiratory secretions may be increased, which further complicates the management of patients with dysphagia and breathing difficulties. Rarely, patients may develop mental confusion from ChE therapy. The development of ChE inhibitor-induced weakness ("cholinergic crisis") is frequently discussed but is rarely encountered. The use of intravenous edrophonium to determine if muscle weakness is secondary to MG or ChE treatment is unreliable.[177,181,191] If cholinergic weakness is seriously considered, then ChE therapy should be temporarily discontinued and improvement observed rapidly.

Corticosteroids are begun in patients with generalized weakness that is not significantly improved by ChE inhibitors. Disagreement exists as to the optimal manner in which to initiate corticosteroid therapy.[176,191–194] Some advocate high-dose prednisone (60 to 80 mg) treatment until clinical improvement occurs followed by alternate-day steroid treatment with gradual tapering as tolerated.[192] Fifty percent of patients develop a worsening of strength in the first month, usually within the first few days, of treatment. In the experience of the University of Virginia group,[192] 80% of patients may expect significant improvement or total remission on such a regimen during the first year of therapy; however, only slightly less than 20% will be able to stop steroid treatment. Gradual initiation of corticosteroids probably decreases the frequency of exacerbation of weakness but delays the onset of improvement.[194,195]

Thymectomy is the generally accepted treatment for patients with thymoma and with generalized myasthenia.[196] Rates of stable remission after thymectomy vary widely ranging from 15%[193] to 64%,[197] and some reports describe significant improvement occurring in over 90% of thymectomized patients.[197] Comparison among studies is difficult because of variations in surgical approach (transcervical versus transsternal), time of follow-up, and population characteristics[191]; however, the general consensus among experts is that thymectomy is an effective therapy.[196] With improvements in surgical technique, anesthesia, and respiratory care, operative morbidity and mortality approaches zero in recent reviews. When thymectomy should be performed is unclear. One report suggests that thymectomy may be used as initial therapy and obviate the need for corticosteroid treatment.[197] Most reviews[195] indicate that thymectomy is best performed within the first few years of diagnosis, but thymectomy appears to produce improvement regardless of when it is performed.[198] No clear relationship among age, sex, thymic pathology, or severity of disease and clinical response to thymectomy has been observed.[176,191–193]

Cytotoxic medications, such as azathioprine and cyclophosphamide, are used for treatment of patients with severe myasthenia that has not responded to thymectomy and corticosteroids or with complications from corticosteroids.[191] Although studies are difficult to compare because of their uncontrolled nature, azathioprine appears to be effective when used in isolation and with corticosteroids.[199,200] Improvement is gradual and may continue for 2 years after initiation of treatment. Rates of side effects appear to be similar to those of corticosteroids. Increased rates of lymphoma in patients treated with azathioprine have not been clearly confirmed among myasthenics.[199] Perez *et al.*[201] used cyclophosphamide to treat 42 myasthenics and found more than 50% to be asymptomatic after 1 year. Side effects of alopecia, leukopenia, anorexia, and skin discoloration occurred frequently. Cyclosporine suppresses T-helper-cell-dependent function, has been proven to maintain tolerance in transplant patients, and experimental autoimmune MG can be suppressed by treatment with cyclosporine. Cyclosporine has been used to treat myasthenics previously treated with thymectomy and corticosteroids with benefit.[202]

Plasma exchange was first used for the treatment of MG in 1976[203] and is useful for production of rapid improvement. Plasma exchange is used in myasthenic crisis, to optimize muscle function prior to thymectomy, and in rare patients as chronic therapy. Although plasma exchange reduces immunoglobulin levels rapidly, rebound occurs in days and may lead to clinical worsening if concurrent immunosuppressive treatment is not used.[204]

Less well established treatments designed to modulate or suppress immune function have been attempted for treatment of some myasthenics. Intravenous IgG (IVIg) has been used in myasthenics with severe disease and poor responses to other treatments.[205–207] IVIg therapy appears to improve strength rapidly, within 5 days of initiation, and lower anti-AChR antibodies, but the response is short-lived and not uniformly observed.[208] IVIg therapy may serve as an alternative to plasma exchange in individuals with poor vascular access or contraindications to plasma exchange.[207] The presumed mechanism of immunosuppression of IVIg is modulation of antiidiotypic antibodies.[205] Durelli *et al.*[209] treated 12 thymectomized patients with total-body irradiation and 5 had significant clinical improvement for 2 years after treatment.

Care of the purely ocular myasthenic has its own special requirements.[177] The patient and physician should appreciate that purely ocular manifestations are primarily a cosmetic problem and do not warrant treatment with significant complications. Ptosis crutches or ptosis tape may be helpful in some patients. Alternating eye patching will alleviate diplopia. Attempts may be made to treat ocular manifestations with ChE inhibitors, but these rarely produce fully satisfactory results. Corticosteroid therapy, although effective, should not be initiated in patients with purely ocular myasthenia unless the patient is fully aware of the numerous side effects of such treatment.

28.6. LAMBERT–EATON MYASTHENIC SYNDROME

28.6.1. Introduction

LEMS is characterized by weakness and fatigability of the trunk and proximal limb muscles. Mild or transient EOM symptoms may be present in about three-fourths of patients. About 80% of patients complain of symptoms related to the autonomic nervous system including decreased salivation, impaired lacrimation, reduced sweating, orthostatic hypotension, and impotence. Impaired pupillary constriction to light may be present on physical examination. The characteristic feature of the weakness in LEMS is that strength is most severely reduced in rested muscles and strength increases over a period of seconds if the patient is able to initiate voluntary contraction. Tendon reflexes may be markedly diminished or absent in rested muscles and then increased after a brief maximal voluntary contraction.[210–213] Cerebellar ataxia associated with subacute cerebellar degeneration is occasionally associated with LEMS.[213,214]

LEMS occurs more frequently in men than in women with a ratio of about 5 to 1.[212,215,216] Approximately three-fourths of men and one-fourth of women have an associated malignancy, with malignancy being more common in LEMS patients over the age of 40.[212,216] Approximately 80% of the associated carcinomas are small-cell lung cancers.[212,216,217] Overall, about two-thirds of patients with LEMS have an associated carcinoma. The disease onset in the cancer-associated LEMS is usually greater than 40 years of age and in the non-cancer-associated LEMS can be at any age. Nonneoplastic LEMS can be associated with other autoimmune diseases.[212,218]

28.6.2. Electrophysiologic Findings

EMG shows low-amplitude motor unit potentials that increase in amplitude with continued activity.[211] The amplitude of the CMAP evoked by a single nerve stimulus to rested muscle is extremely small. After a brief period of maximal voluntary activity, the CMAP amplitude increases. Stimulation at 10 Hz or higher frequencies usually produces a progressive increase of the CMAP amplitude. Similar stimulation in normal subjects has little effect on the CMAP amplitude.[211] In contrast, stimulation at about 5 to 10 Hz in patients with MG produces a decremental response.[211] Single fiber EMG shows increased jitter and frequent blocking. The abnormalities improve after stimulation or voluntary contraction. In contrast, in MG, blocking increases with activity or stimulation.[219]

The neuromuscular transmission defect in LEMS is that too few synaptic vesicles are released in response to nerve stimulation so that the quantal content of the endplate potential is abnormally low.[215,220] The content of ACh and the activity of choline acetyltransferase are normal in LEMS nerve terminals.[221] The relationship between the number of vesicles released in response to nerve stimulation and extracellular calcium is disturbed in LEMs indicating that calcium entry into the nerve terminals is compromised.[222] MEP amplitudes are normal indicating that the amount of ACh in the synaptic vesicles and the postsynaptic sensitivity are normal in LEMS.[215,220]

The difference between the decremental response seen in MG and the incremental responses seen in LEMS to repetitive stimulation can be understood in the following context. With repetitive stimulation, there are two competing forces acting on the nerve terminal. Stimulation will tend to deplete the pool of readily releasable synaptic vesicles. This depletion effect will tend to reduce transmitter release by reducing the number of vesicles that are released in response to a nerve terminal action potential. The opposite is that with repeated stimulation, calcium can accumulate within the nerve terminal, thereby increasing the probability that a given vesicle will fuse with the nerve terminal membrane. In a normal nerve terminal, the effect of depletion of readily releasable synaptic vesicles predominates, so that with repeated stimulation, the number of vesicles released or quantal content will decrease. In MG, because of reduced postsynaptic sensitivity, the safety factor for neuromuscular transmission is greatly compromised. Consequently, when the quantal content is diminished during repetitive stimulation, the size of the postsynaptic endplate potential can drop below that necessary to elicit an action potential and hence repetitive stimulation can result in transmission failure in a fraction of the synapses which is seen as the decremental response in MG. In LEMS, very few vesicles are released so that depletion of vesicles is not a prominent effect. With repeated stimulation, the calcium concentration in the nerve terminal can rise high enough to stimulate synaptic vesicle fusion for a sufficient number of synaptic vesicles to result in a suprathreshold endplate potential. In this context, it is possible to understand the decremental response in MG and the incremental response in LEMS.[77,105]

28.6.3. Autoimmune Pathogenesis

The autoimmune etiology of LEMS is suggested by: (1) the response of the nonneoplastic form to glucocorticoids,[223] (2) nonneoplastic LEMS is associated with other autoimmune disorders,[212,224,225] (3) plasmapheresis and immunosuppressive therapy improve both neoplastic and nonneoplastic LEMS,[226] and (4) IgG from LEMS patients injected into rodents reproduces the electrophysiological features of LEMS.[224]

A striking structural feature of LEMS is that the density of active zone particles is markedly reduced.[227] Injection of IgG from LEMS patients into mice also results in depletion of active zone particles (Figure 28.11).[228] The active zone particles are thought to be associated with the calcium channels responsible for transmitter release or possibly the calcium channels themselves.[15] The LEMS IgG binds to active zones at the nerve terminal.[144,229,230] The antigenic modulation that results in depletion of the active zone particles requires that the LEMS IgG be able to cross-link active zone particles.[231] As was true for AChRs and MG, cross-linking of channels by antibodies results in clustering of the active zone particles[228] followed by a depletion of active zone particles which probably results from accelerated internalization of the calcium channels that is insufficiently compensated for by increased channel production in the nerve cell.[144,229,230]

Passive transfer of LEMS to mice produced by injection of LEMS IgG results in impaired transmitter release at the recipient's neuromuscular junction. In addition, the LEMS IgG impairs the function of calcium channels isolated from small-cell lung cancer cells[232] and calcium channels in other tissues.[233,234] The LEMS IgG binds directly to several classes of calcium channels including N-type,[235–238] L-type,[235,239,240] and P-type calcium channels.[235] Based on our current understanding of transmitter release at mammalian motor nerve terminals (see above), it appears that the antibodies directed against P-type calcium channels may be the most important in compromising transmitter release at the nerve terminal.[23,235] Another possible target for LEMS antibodies is the protein synaptotagmin which is associated with both synaptic vesicles and nerve terminal calcium channels. Antibodies induced to synaptotagmin can reduce the quantal content of the endplate potential and produce a transmission defect resembling that seen in LEMS.[241] An interesting feature of the LEMS antibodies directed against calcium channels is that at least a fraction of the antibodies appear to recognize the β subunit of the calcium channel, which exists only in the cytoplasm and does not appear to have transmembrane or extracellular domains.[242] It is not clear how antibodies directed against a cytoplasmic protein can gain entry into the cell. It is interesting to note, however, that many other antibodies in paraneoplastic syndromes are directed against cytoplasmic rather than surface proteins.[242] From a diagnostic standpoint, antibodies directed against N-type calcium channels are the most sensitive serologic test for LEMS with significantly elevated titers found in approximately 80% of patients with neoplastic LEMS but only about 30% of nonneoplastic LEMS.[236]

28.6.4. Diagnosis

The diagnosis of LEMS is made based on clinical and EMG data with serologic testing adding support.[243,244] A high suspicion of an underlying malignancy, particularly small-cell lung cancer, should be maintained in patients with LEMS above the age of 40 and in patients who complain of a severe deep aching pain in association with the proximal muscle weakness and fatigability and autonomic impairment (Dr. Ed Lambert, personal communication).

The differential diagnosis includes acquired MG, Guillain–Barre syndrome, polyradiculopathy, peripheral neuropathy, polymyositis, botulism, magnesium poisoning, and in younger patients a CDNT. Rare patients may have both LEMS and MG.[218] The reason for the antibodies against calcium channels and other elements of the nerve terminal in LEMS is not clear. However, small-cell lung cancer cells can express either L-, N-, or P-type calcium channels.[217]

28.6.5. Treatment

In patient with LEMS associated with cancer, treatment of the underlying tumor may improve symptoms.[245] Anti-ChE drugs have limited usefulness.[211] Guanidine increases the quantal content of the endplate potential[246] and can improve symptoms in LEMS.[211,212] However, guanidine can also cause ataxia, paresthesias, cardiac arrhythmia, and renal failure.[247]

3,4-DAP prolongs the duration of the presynaptic action potential by blocking delayed rectifier potassium channels.[248] The prolonged action potential increases calcium entry into

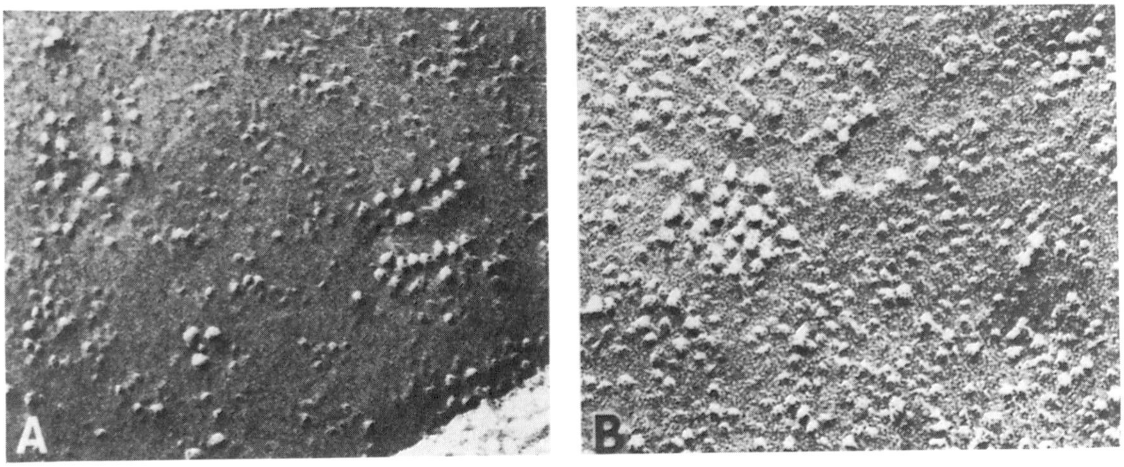

Figure 28.11. Comparison of active zones of normal mice (A) and mice treated with LEMS IgG (B). In LEMS antibody-treated mice, stereometric analysis demonstrated a decrease in normal active zones, an aggregation of particles in abnormal active zones, and clusters of large membrane particles. (From Engel *et al.*[243] with permission of the *Annals of the New York Academy of Sciences.*)

the nerve terminal which increases quantal release.[34] 3,4-DAP in doses of up to 100 mg per day can relieve the motor and autonomic symptoms of LEMS.[249] Plasmapheresis can be effective in LEMS, but it is costly.[226] High-dose intravenous IgG therapy can produce temporary improvement in symptoms.[250] Azathioprine, glucocorticoids, or both can improve the symptoms of LEMS. The two drugs may act synergistically to reduce the motor and autonomic manifestations of LEMS.[226] A combination of alternate-day corticosteroid therapy and azathioprine 1.5 to 2 mg/kg per day can be used to treat nonneoplastic LEMS. 3,4-DAP can be used in conjunction with corticosteriods and azathioprine.

28.7. SUMMARY

This chapter has discussed disorders of the neuromuscular junction. With the application of modern electrophysiological methods, electron microscopy, and molecular biological techniques, great strides in the understanding of CDNT, MG, and LEMS have occurred. Detailed analysis of the CDNT is providing insights into normal neuromuscular transmission, the complex interaction of the presynaptic nerve terminal and postsynaptic muscle membrane, and AChR function. Because of knowledge of the AChR and advances in immunology, MG has become the best-characterized autoimmune disorder. Immunomodulating therapies have led to such significant improvement in the survival of myasthenics that the term *gravis,* is no longer appropriate. Investigation of paraneoplastic syndromes, such as LEMS, has shed light on the immune system's response to cancer. Further investigation of these three disorders will aid in the elucidation of the events that lead to normal neuromuscular transmission.

ACKNOWLEDGMENTS. This work is supported by the Office of Research and Development, Medical Research Service of the Department of Veterans Affairs. H.J.K. is supported by National Institutes of Health grant (EY-00332-01) and an Osserman fellowship from the Myasthenia Gravis Foundation. R.L.R. receives support from the Muscular Dystrophy Association.

REFERENCES

1. Black, J. A., Kocsis, J. D., and Waxman, S. G. (1990). Ion channel organization of the myelinated fiber. *Trends Neurosci.* **13:**48–54.
2. Salpeter, M. M. (1987). *The Vertebrate Neuromuscular Junction* Liss, New York.
3. Katz, B. (1966). *Nerve Muscle and Synapse,* McGraw–Hill, New York.
4. Katz, B., and Miledi, R. (1972). The statistical nature of the acetylcholine potential and its molecular components. *J. Physiol. (London)* **244:**665–699.
5. Katz, B., and Miledi, R. (1973). The binding of acetylcholine to receptors and its removal from the synaptic cleft. *J. Physiol. (London)* **231:**549–574.
6. Katz, B., and Miledi, R. (1979). Estimates of quantal content during chemical potentiation of transmitter release. *Proc. R. Soc. London* **205:**369–378.
7. Hall, Z. W., Lubit, B. W., and Schwartz, J. H. (1981). Cytoplasmic actin in postsynaptic structures at the neuromuscular junction. *J. Cell Biol.* **90:**789–792.
8. Hall, Z. W. (1992). The nerve terminal. In *An Introduction to Molecular Neurobiology* (Z. W. Hall, ed.), Sinauer Associates, Sunderland, MA, pp. 148–180.
9. Heuser, J. E., and Reese, T. S. (1973). Evidence for recycling of synaptic vesicle membrane during transmitter release at the frog neuromuscular junction. *J. Cell Biol.* **57:**315–344.
10. Heuser, J. E., Reese, T. S., Dennis, M. J., Jan, Y., Yan, L., and Evans, L. (1979). Synaptic vesicle exocytosis captured by quick freezing and correlated with quantal transmitter release. *J. Cell Biol.* **81:**275–300.
11. Peper, K., Dreyer, F., Sandri, C., and Akert, K. (1974). Structure and ultrastructure of the frog motor endplate. A freeze-etching study. *Cell Tissue Res* **149:**437–455.

12. Augustine, G. J., Adler, E. M., and Charlton, M. P. (1991). The calcium signal for transmitter secretion from presynaptic nerve terminals. *Ann. N.Y. Acad. Sci.* **635:**365–381.
13. Smith, S. J., and Augustine, G. J. (1988). Calcium ions, active zones and synaptic transmitter release. *Trends Neurosci.* **10:**458–464.
14. Llinas, R. R. (1991). Depolarization release coupling: An overview. *Ann. N.Y. Acad. Sci.* **635:**3–17.
15. Engel, A. G. (1991). Review of evidence for loss of motor nerve terminal calcium channels in Lambert–Eaton myasthenic syndrome. *Ann. N.Y. Acad. Sci.* **635:**246–258.
16. Pumplin, D. W., Reese, T. S., and Llinas, R. (1981). Are the presynaptic membrane particles calcium channels? *Proc. Natl. Acad. Sci. USA* **78:**7210–7213.
17. Cherksey, B. D., Sugimori, M., and Llinas, R. R. (1991). Properties of calcium channels isolated with spider toxin, FTX. *Ann. N.Y. Acad. Sci.* **635:**80–89.
18. Lindgren, C. A., and Moore, J. W. (1991). Calcium current in motor nerve endings of the lizard. *Ann. N.Y. Acad. Sci.* **635:**58–69.
19. Nowycky, M. C., Fox, A. P., and Tsien, R. W. (1985). Three types of neuronal calcium channel with calcium agonist sensitivity. *Nature* **316:**440–443.
20. Nowycky, M. C. (1991). Two high-threshold Ca^{2+} channels contribute Ca^{2+} for depolarization–secretion coupling in the mammalian neurohypophysis. *Ann. N.Y. Acad. Sci.* **635:**45–57.
21. Stanley, E. F., and Cox, C. (1991). Calcium channels in the presynaptic nerve terminal of the chick ciliary ganglion giant synapse. *Ann. N.Y. Acad. Sci.* **635:**70–79.
22. Stanley, E. F. (1993). Presynaptic calcium channels and the transmitter release mechanism. *Ann. N.Y. Acad. Sci.* **681:**368–372.
23. Protti, D. A., Sanchez, V. A., Cherksey, B. D., Sugimori, M., Llinas, R. R., and Uchitel, O. D. (1993). Mammalian neuromuscular transmission blocked by funnel web toxin. *Ann. N.Y. Acad. Sci.* **681:**405–407.
24. Catterall, W. A., De Jongh, K., Rotman, E., Hell, J., Westenbroek, R., Dubel, S. J., and Snutch, T. P. (1993). Molecular properties of calcium channels in skeletal muscle and neurons. *Ann. N.Y. Acad. Sci.* **681:**342–355.
25. Wray, D., and Porter, V. (1993). Calcium channel types at the neuromuscular junction. *Ann. N.Y. Acad. Sci.* **681:**356–367.
26. Niles, W. D., and Cohen, F. S. (1991). Video-microscopy studies of vesicle–planar membrane adhesion and fusion. *Ann. N.Y. Acad. Sci.* **635:**273–306.
27. Ehrenstein, G., Stanley, E. F., Pocotte, S. L., Jia, M., Iwasa, K. H., and Krebs, K. E. (1991). Evidence of a model of exocytosis that involves calcium-activated channels. *Ann. N.Y. Acad. Sci.* **635:**297–306.
28. Perin, M. S., Fried, V. A., Mignery, G., Jahn, R., and Südhof, T. C. (1990). Phospholipid binding by a synaptic vesicle protein homologous to the regulatory region of protein kinase C. *Nature* **345:**260–263.
29. Perin, M. S., Brose, N., Jahn, R., and Südhof, T. C. (1991). Domain structure of synaptotagmin (p65). *J. Biol. Chem.* **266:**623–629.
30. Pollard, H. B., Rojas, E., Pastor, R. W., Rojas, E., Guy, H. R., and Burns, A. L. (1991). Synexin: Molecular mechanism of calcium-dependent membrane fusion and voltage-dependent calcium-channel activity. Evidence in support of the "hydrophobic bridge hypothesis" for exocytotic membrane fusion. *Ann. N.Y. Acad. Sci.* **635:**328–351.
31. Zimmerberg, J., Curran, M., and Cohen, F. S. (1991). A lipid/protein complex hypothesis for exocytotic fusion pore formation. *Ann. N.Y. Acad. Sci.* **681:**307–317.
32. Almers, W., Breckenridge, L. J., Iwata, A., Lee, A. K., Spruce, A. E., and Tse, F. W. (1991). Millisecond studies of single membrane fusion events. *Ann. N.Y. Acad. Sci.* **635:**318–327.
33. Chandler, D. E. (1991). Membrane fusion as seen in rapidly frozen secretory cells. *Ann. N.Y. Acad. Sci.* **635:**234–245.
34. Miledi, R., Molenaar, P. C., and Polak, R. L. (1983). Electrophysiological and chemical determination of acetylcholine release at the frog neuromuscular junction. *J. Physiol. (London)* **334:**245–254.
35. Etcheberrigaray, R., Fielder, J. L., Pollard, H. B., and Rojas, E. (1991). Endoplasmic reticulum as a source of Ca^{2+} in neurotransmitter secretion. *Ann. N.Y. Acad. Sci.* **635:**90–99.
36. Land, B. R., Harris, W. V., Salpeter, E. E., and Salpeter, M. M. (1984). Diffusion and binding constants for acetylcholine derived from the falling phase of miniature endplate currents. *Proc. Natl. Acad. Sci. USA* **81:**1594–1598.
37. McMahan, U. J., Sanes, J. R., and Marshall, L. M. (1978). Cholinesterase is associated with the basal lamina at the neuromuscular junction. *Nature* **271:**172–174.
38. Land, B. R., Salpeter, E. E., and Salpeter, M. M. (1981). Kinetic parameters for acetylcholine interaction in intact neuromuscular junction. *Proc. Natl. Acad. Sci. USA* **78:**7200–7204.
39. Flucher, B. E., and Daniels, M. P. (1989). Distribution of Na^{+} channels and ankyrin in neuromuscular junctions is complementary to that of acetylcholine receptors and the 43 kD protein. *Neuron* **3:**163–175.
40. Kuffler, S. W., and Yoshikami, D. (1975). The distribution of acetylcholine sensitivity at the post-synaptic membrane of vertebrate skeletal twitch muscles: Iontophoretic mapping in the micron range. *J. Physiol. (London)* **244:**703–730.
41. Merlie, J. P., and Sanes, J. R. (1985). Concentration of acetylcholine receptor mRNA in synaptic regions of adult muscle fibers. *Nature* **317:**66–68.
42. Sanes, J. R., Johnson, Y. R., Kotzbauer, P. T., Mudd, J., Hanley, T., Martinou, J.-C., and Merlie, J. P. (1991). Selective expression of an acetylcholine receptor-lacZ transgene in synaptic nuclei of adult muscle fibers. *Development* **113:**1181–1191.
43. Salpeter, M. M., and Loring, R. H. (1985). Nicotinic acetylcholine receptors in vertebrate muscle: Properties, distribution and neural control. *Prog. Neurobiol.* **25:**297–325.
44. Kao, I., and Drachman, D. (1977). Myasthenic immunoglobulin accelerates acetylcholine receptor degradation. *Science* **196:**526.
45. Merlie, J. P., Heinemann, S., and Lindstrom, J. M. (1979). Acetylcholine receptor degradation in adult rat diaphragms in organ culture and the effect of anti-acetylcholine receptor antibodies. *J. Biol. Chem.* **254:**6320–6327.
46. Martinou, J. C., Falls, D. I., Fischback, G. D., and Merlie, J. P. (1991). Acetylcholine receptor-inducing activity stimulates expression of the epsilon-subunit gene of the muscle acetylcholine receptor. *Proc. Natl. Acad. Sci. USA* **88:**7669–7673.
47. Rutishauser, U., Grumet, M., and Edelman, G. M. (1983). Neural cell adhesion molecule mediates initial interactions between spinal cord neurons and muscle cells in culture. *J. Cell Biol.* **97:**145–152.
48. Colquhoun, D., and Sakmann, B. (1985). Fast events in single-channel currents activated by acetylcholine and its analogues at the frog muscle end-plate. *J. Physiol. (London)* **369:**501–557.
49. Haimovich, B., Schotland, D. L., Fieles, W. E., and Barchi, R. L. (1987). Localization of sodium channel subtypes in rat skeletal muscle using channel-specific monoclonal antibodies. *J. Neurosci.* **7:**2957–2966.
50. Roberts, W. M. (1987). Sodium channels near end-plates and nuclei of snake skeletal muscle. *J. Physiol. (London)* **388:**213–232.
51. Ruff, R. L. (1992). Na current density at and away from end plates on rat fast- and slow-twitch skeletal muscle fibers. *Am. J. Physiol.* **262:**C229–C234.
52. Ruff, R. L., and Whittlesey, D. (1992). Na^{+} current densities and voltage dependence in human intercostal muscle fibres. *J. Physiol. (London)* **458:**85–97.

53. Ruff, R. L., and Whittlesey, D. (1993). Na^+ currents near and away from endplates on human fast and slow twitch muscle fibers. *Muscle Nerve* **16:**922–929.
54. Ruff, R. L., and Whittlesey, D. (1993). Comparison of Na^+ currents from type IIa and IIb human intercostal muscle fibers. *Am. J. Physiol.* **265:**C171–C177.
55. Angelides, K. J. (1986). Fluorescently labelled Na^+ channels are localized and immobilized to synapses of innervated muscle fibres. *Nature* **321:**63–66.
56. Almers, W., Stanfield, P. R., and Stühmer, W. (1983). Lateral distribution of sodium and potassium channels in frog skeletal muscle: Measurements with a patch-clamp technique. *J. Physiol. (London)* **321:**63–66.
57. Caldwell, J. H., Campbell, D. T., and Beam, K. G. (1986). Sodium channel distribution in vertebrate skeletal muscle. *J. Gen. Physiol.* **87:**907–932.
58. Milton, R. L., Lupa, M. T., and Caldwell, J. H. (1992). Fast and slow twitch skeletal muscle fibres differ in their distributions of Na channels near the endplate. *Neurosci. Lett.* **135:**41–44.
59. Banker, B. Q., Kelly, S. S., and Robbins, N. (1983). Neuromuscular transmission and correlative morphology in young and old mice. *J. Physiol. (London)* **339:**355–375.
60. Fertuck, H. C., and Salpeter, M. M. (1976). Quantitation of junctional and extrajunctional receptors by electron microscope autoradiography after 125-I-labeled alpha-bungarotoxin binding at mouse motor endplates. *J. Cell Biol.* **69:**144–158.
61. Porter, C. W., and Barnard, E. A. (1975). The density of cholinergic receptors at the endplate postsynaptic membrane: Ultrastructural studies in two mammalian species. *J. Membr. Biol.* **20:**31–49.
62. Gertler, R. A., and Robbins, N. (1978). Differences in neuromuscular transmission in red and white muscles. *Brain Res.* **142:**255–284.
63. Padykula, H. A., and Gauthier, G. F. (1970). The ultrastructure of the neuromuscular junctions of mammalian red, white and intermediate skeletal muscle fibers. *J. Cell Biol.* **46:**27–41.
64. Tonge, D. A. (1974). Chronic effects of botulinum toxin on neuromuscular transmission and sensitivity to acetylcholine in slow and fast skeletal muscle of the mouse. *J. Physiol. (London)* **241:**127– 139.
65. Sterz, R., Pagala, M., and Peper, K. (1983). Postjunctional characteristics of the endplates in mammalian fast and slow muscles. *Pfluegers Arch.* **398:**48–54.
66. Storella, R. J., Riker, W. F., and Baker, T. (1985). d-Tubocurarine sensitivities of fast and slow neuromuscular system of the rat. *Eur. J. Pharmacol.* **118:**181–184.
67. Caldwell, J. H., and Milton, R. L. (1988). Sodium channel distribution in normal and denervated rodent and snake skeletal muscle. *J. Physiol. (London)* **401:**145–161.
68. Laszewski, B., and Ruff, R. L. (1985). The effects of glucocorticoid treatment on excitation–contraction coupling. *Am. J. Physiol.* **248:**E363–E369.
69. Hennig, R., and Lømo, T. (1985). Firing patterns of motor units in normal rats. *Nature* **314:**164–166.
70. Kelly, S. S., and Robbins, N. (1986). Sustained transmitter output by increased transmitter turnover in limb muscles of old mice. *J. Neurosci.* **6:**2900–2907.
71. Lev-Tov, A. (1987). Junctional transmission in fast- and slow-twitch mammalian muscle units. *J. Neurophysiol.* **57:**660–671.
72. Lev-Tov, A., and Fishman, R. (1986). The modulation of transmitter release in motor nerve endings varies with the type of muscle fiber innervated. *Brain Res.* **363:**379–382.
73. Ruff, R. L., Simoncini, L., and Stühmer, W. (1988). Slow sodium channel inactivation in mammalian muscle: A possible role in regulating excitability. *Muscle Nerve* **11:**502–510.
74. Ruff, R. L., Simoncini, L., and Stühmer, W. (1987). Comparison between slow sodium channel inactivation in rat slow and fast twitch muscle. *J. Physiol. (London)* **383:**339–348.
75. Ruff, R. L., Simoncini, L., and Stühmer, W. (1988). The possible role of slow sodium channel inactivation in regulating membrane excitability in mammalian skeletal muscle. In *Contributions to Contemporary Neurology: A Tribute to Joseph M. Foley* (J. Conomy and R. B. Daroff, eds.), Butterworth, Boston, pp. 153–170.
76. Guy, H. R., and Hucho, F. (1987). The ion channel of nicotinic acetylcholine receptor. *Trends Neurosci.* **10:**318–322.
77. Ruff, R. L. (1986). Ionic channels II. Voltage- and agonist-gated and agonist-modified channel properties and structure. *Muscle Nerve* **9:**767–786.
78. Witzemann, V., Barg, B., Criado, M., Stein, E., and Sakmann, B. (1989). Developmental regulation of five subunits specific mRNAs encoding acetylcholine receptor subtypes in rat muscle. *FEBS Lett.* **242:**419–424.
79. Dani, J. A. (1989). Site-directed mutagenesis and single-channel currents define the ionic channel of the nicotinic acetylcholine receptor. *Trends Neurosci.* **12:**125–130.
80. Takai, T., Noda, M., Mishina, M., Shimizu, S., Furutani, Y., Kayano, T., Ikeda, T., Tai, K., Takahashi, H., Takahashi, T., Kuno, M., and Numa, S. (1985). Cloning, sequencing and expression of cDNA for a novel subunit of acetylcholine receptor from calf muscle. *Nature* **315:**761–764.
81. Mishina, M., Takai, T., Imoto, K., Noda, M., Takahashi, T., Numa, S., Methfessl, C., and Sakmann, B. (1986). Molecular distinction between fetal and adult forms of muscle acetylcholine receptor. *Nature* **321:**406–411.
82. Witzemann, V., Barg, B., Nishikawa, Y., Sakmann, B., and Numa, S. (1987). Differential regulation of muscle acetylcholine receptor γ- and ε-subunit mRNAs. *FEBS Lett.* **223:**104–112.
83. Morris, A., Beeson, D., Jacobson, L., Baggi, F., Vincent, A., and Newsom-Davis, J. (1991). Two isoforms of the muscle acetylcholine rector of alpha-subunit are translated in the human cell line TE671. *FEBS Lett.* **295:**116–118.
84. Mishina, M., Kurosaki, T., Tobimatsu, T., Morimoto, Y., Noda, M., Yamamoto, T., Terao, M., Lindstrom, J., Takahashi, T., Kuno, M., and Numa, S. (1984). Expression of functional acetylcholine receptor from cloned cDNAs. *Nature* **307:**604–608.
85. Henderson, L. P., and Brehm, P. (1989). The single-channel basis for the slow kinetics of synaptic currents in vertebrate slow muscle fibers. *Neuron* **2:**1399–1405.
86. Brehm, P., and Henderson, L. (1988). Regulation of acetylcholine receptor channel function during development of skeletal muscle. *Dev. Biol.* **129:**1–11.
87. Brehm, P. (1989). Resolving the structural basis for developmental changes in muscle ACh receptor function: It takes nerve. *Trends Neurosci.* **12:**174–177.
88. Schuetze, S. M., and Role, L. W. (1987). Developmental regulation of nicotinic acetylcholine receptors. *Annu. Rev. Neurosci.* **10:**403–457.
89. Noda, M., Furutani, Y., Takahashi, H., Toyosato, M., Tanabe, T., Shimizu, S., Kikyotani, S., Kayano, T., Hirose, T., Inayama, S., and Numa, S. (1983). Cloning and sequence analysis of calf cDNA and human genomic DNA encoding alpha-subunit precursor of muscle acetylcholine receptor. *Nature* **305:**818–823.

89a. Kaminski, H. J., Kusner, L. L., and Block, C. H. (1996). Expression of acetylcholine receptor isoforms at extraocular muscle endplates. *Invest. Opthamol. Vis. Sci.*, **37:**345–351.

90. Karlin, A., Kao, P. N., and DiPaola, M. (1986). Molecular pharmacology of the nicotinic acetylcholine receptor. *Trends Pharmacol. Sci.* **4:**304–308.
91. Sigworth, F. J., and Sine, S. M. (1987). Data transformations for improved display and fitting of single-channel dwell time histograms. *Biophys. J.* **52:**1047–1054.
92. Sine, S. M., and Steinbach, J. H. (1984). Activation of a nicotinic acetylcholine receptor. *Biophys. J.* **45:**175–185.

93. Auerback, A. B., and Sachs, F. (1984). Single channel currents from acetylcholine receptors in embryonic chick muscle. Kinetic and conductance properties of gaps within bursts. *Biophys. J.* **45:** 187–198.
94. Qu, Z., Moritz, E., and Huganir, R. L. (1990). Regulation of tyrosine phosphorylation of the nicotinic acetylcholine receptor at the rat neuromuscular junction. *Neuron* **2:**367–378.
95. Safran, A., Provenzano, C., Sagi-Eisenberg, R., and Fuchs, S. (1990). Phosphorylation of membrane-bound acetylcholine receptor by protein kinase C: Characterization and subunit specificity. *Biochemistry* **29:**6730–6734.
96. Rozental, R. (1991). In vitro denervation of frog skeletal muscle: Expression of several conductance classes of nicotinic receptors. *Neurosci. Lett.* **133:**65–67.
97. Dionne, V. E. (1989). Two types of nicotinic acetylcholine receptor channels at slow fibre end-plates of the garter snake. *J. Physiol. (London)* **409:**313–331.
98. Ruff, R. L., and Spiegel, P. (1990). Ca sensitivity and AChR currents of twitch and tonic snake muscle fibers. *Am. J. Physiol.* **259:**C911–C919.
99. Ruff, R. L., Kaminski, H. J., Maas, E., and Spiegel, P. (1989). Ocular muscles: Physiology and structure–function correlations. *Bull. Soc. Belg. Ophthalmol.* **237:**321–352.
100. Kaminski, H. J., Maas, E., Spiegel, P., and Ruff, R. L. (1990). Why are eye muscles frequently involved by myasthenia gravis? *Neurology* **40:**1663–1669.
101. Horton, R. M., Manfredi, A. A., and Conti-Tronconi, B. M. (1993). The "embryonic" gamma subunit of the nicotinic acetylcholine receptor is expressed in adult extraocular muscle. *Neurology* **43:** 983–986.
102. Kaminski, H. J., Fenstermaker, R., and Ruff, R. L. (1991). Adult extraocular and intercostal muscle express the gamma-subunit of fetal AChR. *Biophys. J.* **59:**444a.
103. Dwyer, T. M., Adams, D. J., and Hille, B. (1980). The permeability of the endplate channel to organic cations in frog muscle. *J. Gen. Physiol.* **75:**469–492.
104. Dani, J. A., and Eisenman, G. (1987). Monovalent and divalent cation permeation in acetylcholine receptors channel. *J. Gen. Physiol.* **89:**959–983.
105. Ruff, R. L. (1986). Ionic channels: I. The biophysical basis for ion passage and channel gating. *Muscle Nerve* **9:**675–699.
106. Toyoshima, C., and Unwin, N. (1988). Ion channel of acetylcholine receptor reconstructed from images of postsynaptic membranes. *Nature* **336:**247–250.
107. Unwin, N., Toyoshima, C., and Kubalek, E. (1988). Arrangement of the acetylcholine receptor subunits in the resting and desensitized states, determined by cryoelectron microscopy of crystallized torpedo postsynaptic membranes. *J. Cell Biol.* **107:**1123–1138.
108. Kaminski, H. J., and Ruff, R. L. (1993). Insights into possible skeletal muscle nicotinic acetylcholine receptor (AChR) changes in some congenital myasthenias from physiological studies, point mutations, subunit substitutions of the AChR. *Ann. N.Y. Acad. Sci.* **681:**435–450.
109. Chavez, R. A., and Hall, Z. W. (1992). Expression of fusion proteins of the nicotinic acetylcholine receptor from mammalian muscle identifies the membrane-spanning regions in the α and δ subunits. *J. Cell Biol.* **116:**385–393.
109a. Kaminski, H. J., Kusner, L. L., Nash, K. V., and Ruff, R. L. (1995). The γ-subunit of the acetylcholine receptor is not expressed in the levator palpebrae superioris. *Neurology* **45:**516–518.
110. Neumann, D., Barchan, D., Horowitz, M., Kochva, E., and Fuchs, S. (1989). Snake acetylcholine receptor: Cloning of the domain containing the four extracellular cysteines of the α-subunit. *Proc. Natl. Acad. Sci. USA* **86:**7255–7259.
111. Giraudat, J., Dennis, M., Heidmann, T., Chang, J.-Y., and Changeux, J.-P. (1986). Structure of the high-affinity binding site for noncompetitive blockers of the acetylcholine receptor: Serine-262 of the δ subunit is labeled [^{3}H]chlorpromazine. *Proc. Natl. Acad. Sci. USA* **83:**2719–2723.
112. Hucho, F., Oberthür, W., and Lottspeich, F. (1986). The ion channel of the nicotinic acetylcholine receptor is formed by the homologous helices M II of the receptor subunits. *FEBS Lett.* **205:**137–142.
113. Ruff, R. A. (1977). A quantitative analysis of local anesthetic alteration of miniature endplate current fluctuations. *J. Physiol. (London)* **264:**89–124.
114. Ruff, R. L. (1982). The kinetics of local anesthetic blockade of endplate channels. *Biophys. J.* **37:**625–631.
115. Leonard, R. J., Labarce, C. G., Charnet, P., Davidson, N., and Lester, H. A. (1988). Evidence that the M2 membrane-spanning region lines the ion channel pore of the nicotinic receptor. *Science* **242:**1578–1581.
116. Imoto, K., Methfessel, C., Sakmann, B., Mishina, M., Mori, Y., Konno, T., Fukuda, K., Kurasaki, M., Bujo, H., Fujita, Y., and Numa, S. (1986). Location of a δ subunit region determining ion transport through the acetylcholine receptor channel. *Nature* **324:** 670–674.
117. Imoto, K., Busch, C., Sakmann, B., Mishina, M., Konno, T., Nakai, J., Bujo, H., Mori, Y., Fukuda, K., and Numa, S. (1988). Rings of negatively charged amino acids determine the acetylcholine receptor channel conductance. *Nature* **305:**645–648.
118. Morel, E., Eymard, B., Vernet-der Garabedian, B., Pannier, C., Dulac, O., and Bach, J. F. (1988). Neonatal myasthenia gravis: A new clinical and immunological appraisal on 30 cases. *Neurology* **38:** 138–142.
119. Engel, A. G., Walls, T. J., Nagel, A., and Uchitel, O. (1990). Newly recognized congenital myasthenic syndromes: I. Congenital paucity of synaptic vesicles and reduced quantal release. II. High conductance fast-channel syndrome. III. Abnormal acetylcholine receptor (AChR) interaction with acetylcholine. IV. AChR deficiency and short channel open time. *Prog. Brain Res.* **84:**125–137.
120. Engel, A. G., Lambert, E. H., Mulder, D. M., Torres, C. F., Sahashi, K., Bertorini, T. E., and Whitaker, J. N. (1979). Investigations of 3 cases of a newly recognized familial, congenital myasthenic syndrome. *Trans. Am. Neurol. Assoc.* **104:**8–11.
121. Mora, M., Lambert, E. H., and Engel, A. G. (1987). Synaptic vesicle abnormality in familial infantile myasthenia. *Neurology* **37:** 206–214.
122. Bady, B., Chauplannaz, G., and Carrier, H. (1987). Congenital Lambert–Eaton myasthenic syndrome. *J. Neurol. Neurosurg. Psychiatry* **50:**476–478.
123. Vincent, A., Newsom-Davis, J., Wray, D., Shillito, P., Harrison, J., Betty, M., Beeson, D., Mills, K., Palace, J., Molenaar, P., and Murray, N. (1993). Clinical and experimental observations in patients with congenital myasthenic syndromes. *Ann. N.Y. Acad. Sci.* **681:** 451–460.
124. Hutchinson, D. O., Engel, A. G., Walls, T. J., Nakano, S., Camp, S., Taylor, P., Harper, C. M., and Brengman, J. M. (1993). The spectrum of congenital end-plate acetylcholinesterase deficiency. *Ann. N.Y. Acad. Sci.* **681:**469–486.
125. Engel, A. G., Lambert, E. H., and Gomez, A. R. (1977). A new myasthenic syndrome with end-plate acetylcholinesterase deficiency, small nerve terminals, and reduced acetyline release. *Ann. Neurol.* **1:**315–330.
126. Smit, L. M. E., Hageman, G., Veldman, H., Molenaar, P. C., Oen, B. S., and Jennekens, F. G. I. (1988). A myasthenic syndrome with congenital paucity of secondary synaptic clefts: CPSC syndrome. *Muscle Nerve* **11:**337–348.
127. Wokke, J. H. J., Jennekens, F. G. I., Molenaar, P. C., Van Den Oord, C. J. M., Oen, B. S., and Busch, H. F. M. Congenital paucity of secondary synaptic clefts (CPSC) syndrome in adult sibs. *Neurology* **39:**648–654.

128. Gu, Y., Franco, A., Gardner, P. D., Lansman, J. B., Forsayeth, J. R., and Hall, Z. W. (1990). Properties of embryonic and adult muscle acetylcholine receptors transiently expressed in COS cells. *Neuron* **5:**147–157.
129. Engel, A. G., Uchitel, O. D., Walls, T. J., Nagel, A., Harper, C. M., and Bodensteiner, J. (1993). Newly recognized congenital myasthenic syndrome associated with high conductance and fast closure of the acetylcholine receptor channel. *Ann. Neurol.* **34:**38–47.
130. Engel, A. G., Nagel, A., Walls, T. J., Harper, C. M., and Waisburg, H. A. (1993). Congenital myasthenic syndromes: I. Deficiency and short open-time of the acetylcholine receptor. *Muscle Nerve* **16:** 1284–1292.
131. Engel, A. G., Hutchinson, D. O., Nakano, S., Murphy, L., Griggs, R. C., Gu, Y., Hall, Z. W., and Lindstrom, J. (1993). Myasthenic syndromes attributed to mutations affecting the epsilon subunit of the acetylcholine receptor. *Ann. N.Y. Acad. Sci.* **681:**496–508.
132. Gu, Y., Camacho, P., Gardner, P., and Hall, Z. W. (1991). Identification of two amino acid residues in the ε-subunit that promote mammalian muscle acetylcholine receptor assembly in COS cells. *Neuron* **6:**879–887.
133. Oosterhuis, H. J. G. H., Newsom-Davis, J., Wokke, J. H. J., Molenaar, P. C., Weerden, T. V., Oen, B. S., Jennekens, F. G. I., Veldman, H., Vincent, A., Wray, D. W., Prior, C., and Murray, N. M. F. (1987). The slow channel syndrome. Two new cases. *Brain* **110:** 1061–1079.
134. Engel, A. G., Lambert, E. H., Mulder, D. M., Torres, C. F., Sahashi, K., Bertorini, T. E., Whitaker, J. N. (1982). A newly recognized congenital myasthenic syndrome attributed to a prolonged open time of the acetylcholine-induced ion channel. *Ann. Neurol.* **11:** 553–569.
135. Uchitel, O., Engel, A. G., Wals, T. J., Nagel, A., Atassi, M. Z., and Bril, V. (1993). Congenital myasthenic syndromes: II. Syndrome attributed to abnormal interaction of acetylcholine with its receptor. *Muscle Nerve* **16:**1293–1301.
136. Changeux, J.-P., Devillers-Thiery, A., Galzi, J.-L., and Bertrand, D. (1992). New mutants to explore nicotinic receptor functions. *Trends Pharmacol. Sci.* **13:**299–301.
137. Morgan-Hughes, J. A., Lecky, B. R. F., Landon, D. N., and Murray, N. M. F. (1981). Alterations in the number and affinity of junctional acetylcholine receptors in a myopathy with tubular aggregates. A newly recognized receptor defect. *Brain* **104:** 279–295.
138. Kaminski, H. J., and Ruff, R. L. (1992). Congenital disorders of neuromuscular transmission. *Hosp. Pract.* **39:**73–86.
139. Engel, A. G. (1993). The investigation of congenital myasthenic syndromes. *Ann. N.Y. Acad. Sci.* **681:**425–434.
140. Patrick, J., and Lindstrom, J. (1973). Autoimmune response to acetylcholine receptor. *Science* **180:**871–872.
141. Lindstrom, J. M., Engel, A. G., Seybold, M. E., Lennon, V. A., and Lambert, E. H. (1976). Pathological mechanisms in experimental autoimmune myasthenia gravis. II. Passive transfer of experimental autoimmune myasthenia gravis in rats with anti-acetylcholine receptor antibodies. *J. Exp. Med.* **144:**739–753.
142. Toyka, K. V., Drachman, D. B., Griffin, D. E., and Pestronk, D. (1977). Myasthenia gravis study of humoral immune mechanisms by transfer to mice. *N. Engl. J. Med.* **296:**125–131.
143. Engel, A. G., Lambert, E. H., and Howard, F. M. (1977). Immune complexes (IgG and C3) at the motor end-plate in myasthenia gravis. Ultrastructure and light microscopic localization and electrophysiological correlations. *Mayo Clin. Proc.* **52:**267–280.
144. Engel, A. G. (1987). The molecular biology of end-plate diseases. In *The Vertebrate Neuromuscular Junction* (M. M. Salpeter, ed.), Liss, New York, pp. 361–424.
145. Engel, A. G., and Fumagalli, G. (1982). Mechanisms of acetylcholine receptor loss from the neuromuscular junction. *Ciba Found. Symp.* **90:**197–224.
146. Schönbeck, S., Chrestel, S., and Hollfeld, R. (1990). Myasthenia gravis: Prototype of the antireceptor autoimmune diseases. *Int. Rev. Neurobiol.* **32:**175–200.
147. Vincent, A. (1987). Disorders affecting the acetylcholine receptor: Myasthenia gravis and congenital myasthenia. *J. Recept. Res.* **7:** 599–616.
148. Molenaar, P. C. (1990). Synaptic adaptation in diseases of the neuromuscular junction. *Prog. Brain Res.* **84:**145–149.
149. Asher, O., Neumann, D., Witzemann, V., and Fuchs, S. (1990). Acetylcholine receptor gene expression in experimental autoimmune myasthenia gravis. *FEBS Lett.* **261:**231–235.
150. Steinman, L. (1990). Immunogenetic mechanisms in myasthenia gravis. *Prog. Brain Res.* **84:**117–124.
151. Smiley, J. D., and Moore, S. E. Jr. (1988). Molecular mechanisms of autoimmunity. *Am. J. Med. Sci.* **295:**478–496.
152. Vincent, A., and Newsom-Davis, J. (1982). Acetylcholine receptor antibody characteristics in myasthenia gravis. I. Patients with generalized myasthenia or disease restricted to ocular muscles. *Clin. Exp. Immunol.* **49:**257–265.
153. Limburg, P. C., The, T. C., Hummel-Teppel, E., and Oosterhuis, H. (1983). Anti-acetylcholine receptor antibodies in myasthenia gravis. I. Relation to clinical parameters in 250 patients. *J. Neurol. Sci.* **58:**357–370.
154. Compston, D. A. S., Vincent, A., Newsom-Davis, J., and Batchelor, J. R. (1980). Clinical, pathological, HLA antigen, and immunological evidence for disease heterogeneity in myasthenia gravis. *Brain* **103:**579–601.
155. Sprent, J. (1993). The thymus and T cell tolerance. *Ann. N.Y. Acad. Sci.* **681:**5–15.
156. Wekerle, H. (1993). The thymus in myasthenia gravis. *Ann. N.Y. Acad. Sci.* **681:**47–55.
157. Muller-Hermelink, H. K., Marx, A., Geuder, K., and Kirchner, T. (1993). The pathological basis of thymoma-associated myasthenia gravis. *Ann. N.Y. Acad. Sci.* **681:**56–65.
158. Kaminski, H. J., Fenstermaker, R. A., Abdul-Karim, F. W., Clayman, J., and Ruff, R. L. (1993). Acetylcholine receptor subunit gene expression in thymic tissue. *Muscle Nerve* **16:**1332–1337.
159. Kaminski, H. J., Al-Hakim, M., Leigh, R. J., Katirji, M. B., and Ruff, R. L. (1992). Extraocular muscle are spared in advanced Duchenne dystrophy. *Ann. Neurol.* **32:**586–588.
160. Simpson, J. (1960). Myasthenia gravis: A new hypothesis. *Scott. Med. J.* **5:**419–436.
161. Leigh, R., and Zee, D. (1991). *The Neurology of Eye Movements.* Davis, Philadelphia.
162. Spencer, R. F., and Porter, J. D. (1988). Structural organization of the extraocular muscles. In *Neuroanatomy of the Oculomotor System.* (J. Buttner-Ennever, ed.), Elsevier, Amsterdam, pp. 33–79.
163. Kim, Y., Zahm, D., Liu, H., and Johns, T. (1982). Safety margin of neuromuscular transmission in rat extraocular muscle. *Soc. Neurosci.* **8:**616.
164. Oh, S. J., and Eslami, N. (1987). Eight-to-ten percent decremental response is not the normal limit for all muscles. *Ann. N.Y. Acad Sci.* **505:**851–853.
165. Yee, R. D., Cogna, D. G., Zee, D. S., Baloh, R. W., and Honrubia, V. (1976). Rapid eye movements in myasthenia gravis. II. Electrooculographic analysis. *Arch. Ophthalmol.* **94:**1465–1472.
166. Abel, L., Dell'Osso, L. F., Schmidt, D., and Daroff, R. B. (1980). Myasthenia gravis: Analog computer model. *Exp. Neurol.* **68:** 378–389.
167. Kramer, L. D., Ruth, R. A., Johns, M. E., and Saunders, D. B. (1981). A comparison of stapedial reflex fatigue with repetitive stimulation and single-fiber EMG in myasthenia gravis. *Ann. Neurol.* **9:**531–536.
168. Oda, K. (1993). Differences in acetylcholine receptor–antibody interactions between extraocular and extremity muscle fibers. *Ann. N.Y. Acad. Sci.* **681:**238–255.

169. Hayashi, M., Kida, K., Yamada, I., Matsuda, H., Tsuneishi, M., and Tamura, O. (1989). Differences between ocular and generalized myasthenia gravis: Binding characteristics of anti-acetylcholine receptor antibody against bovine muscles. *J. Neuroimmunol.* **21:** 227–233.
170. Protti, M. P., Manfredi, A. A., Howard, J. F., and Conti-Tronconi, B. M. (1991). T cells in myasthenia gravis specific for embryonic acetylcholine receptor. *Neurology* **41:**1809–1814.
171. Tzartos, S., Seybold, M., and Lindstrom, J. (1982). Specificities of antibodies to acetylcholine receptors in sera from myasthenia gravis patients measured by monoclonal antibodies. *Proc. Natl. Acad. Sci. USA* **79:**188–192.
172. Phillips, L., Torner, J., Anderson, M., and Cox, G. (1992). The epidemiology of myasthenia gravis in central and western Virginia. *Neurology* **42:**1888–1893.
173. Christensen, P., Jensen, T., Tsiropoulos, I., Sorensen, T., Kjaer, M., Hojer-Pedersen, E., Rasmussen, M., Lehfeldt, E., and de Fine Olivarius, B. (1993). Incidence and prevalence of myasthenia gravis in western Denmark 1975 to 1989. *Neurology* **43:**1779–1783.
174. Uona, M. (1980). Clinical statistics of myasthenia gravis in Japan. *Int. J. Neurol.* **14:**87–99.
175. Chiu, H.-C., Vincent, A., Nemsom-Davis, J., Hsieh, K.-H., and Hung, T.-P. (1987). Myasthenia gravis: Population differences in disease expression and acetylcholine receptor antibody titers between Chinese and Caucasians. *Neurology* **37:**1854–1857.
176. Grob, D., Arsura, E. L., Brunner, N. G., and Namba, T. (1987). The course of myasthenia gravis and therapies affecting outcome. *Ann. N.Y. Acad. Sci.* **505:**472–499.
177. Daroff, R. B. (1980). Ocular myasthenia: Diagnosis and therapy. In *Neuro-ophthalmology* (J. Glaser, ed.), Mosby, St. Louis, pp. 62–71.
178. Kaminski, H. J., and Ruff, R. L. (1989). Neurologic complications of endocrine diseases. *Neurol. Clin.* **7:**489–508.
179. Taphoorn, M. J. B., Van Duijn, H., and Wolters, E. C. H. (1988). A neuromuscular transmission disorder: Combined myasthenia gravis and Lambert Eaton syndrome in one patient. *J. Neurol. Neurosurg. Psychiatry* **51:**880–882.
180. Oh, S. J., Dweyer, D. S., and Bradley, R. J. (1987). Overlap myasthenic syndrome: Combined myasthenia gravis and Eaton–Lambert syndrome. *Neurology* **37:**1411–1414.
181. Daroff, R. B. (1986). The office tensilon test for ocular myasthenia gravis. *Arch. Neurol.* **43:**843–844.
182. Oh, S. J., and Cho, H. K. (1990). Edrophonium responsiveness not necessarily diagnostic of myasthenia gravis. *Muscle Nerve* **13:** 187–191.
183. Dirr, L. Y., Donofrio, P. D., Patton, J. F., and Troost, B. T. (1989). A false-positive edrophonium test in a patient with a brainstem glioma. *Neurology* **39:**865–867.
184. Moorthy, G., Behrens, M. M., Drachman, D. B., Kirkham, T. H., Knox, D. L., Miller, N. R., Slamovitz, T. L., and Zinreich, S. J. (1989). Ocular pseudomyasthenia or ocular myasthenia "plus": A warning to clinicians. *Neurology* **39:**1150–1154.
185. Kelly, J. J., Daube, J. R., Lennon, V. A., Howard, F. M., and Younge, B. R. (1982). The laboratory diagnosis of mild myasthenia gravis. *Ann. Neurol.* **12:**238–242.
186. Oh, S. J., Kim, D. E., Kuruoglu, R., Bradley, R. J., and Dwyer, D. (1992). Diagnostic sensitivity of the laboratory tests in myasthenia gravis. *Muscle Nerve* **15:**720–724.
187. Lanska, D. J. (1991). Diagnosis of thymoma in myasthenics using anti-striated muscle antibodies: Predictive value and gain in diagnostic certainty. *Neurology* **41:**520–524.
188. Birmanns, B., Brenner, T., Abramsky, O., and Steiner, I. (1991). Seronegative myasthenia gravis: Clinical features, response to therapy and synthesis of acetylcholine receptor antibodies *in vitro. J. Neurol. Sci.* **102:**184–189.
189. Mossman, S., Vincent, A., and Newsom-Davis, J. (1986). Myasthenia gravis without acetylcholine receptor antibody: A distinct disease entity. *Lancet* **1:**116–119.
190. Soliven, B. C., Lange, D. J., Penn, A. S., Younger, D., Jaretzki, A., Lovelace, R. E., and Rowland, L. P. (1988). Seronegative myasthenia gravis. *Neurology* **38:**514–517.
191. Rowland, L. P. (1980). Controversies about the treatment of myasthenia gravis. *J. Neurol. Neurosurg. Psychiatry* **43:**644–659.
192. Pascuzzi, R. M., Coslett, H. B., and Johns, T. R. (1984). Long-term corticosteroid treatment of myasthenia gravis: Report of 116 patients. *Ann. Neurol.* **15:**291–298.
193. Beghi, E., Antozzi, C., Batocchi, A. P., Cornelio, F., Cosi, V., Evoli, A., Lombardi, M., Mantegazzi, R., Monticelli, M. L., Piccolo, G., Tonali, P., Trevisan, D., and Zarrelli, M. (1991). Prognosis of myasthenia gravis: A multicenter follow-up study. *J. Neurol. Sci.* **106:**213–220.
194. Seybold, M., and Drachman, D. (1974). Gradually increasing doses of prednisone in myasthenia gravis: Reducing the hazards of treatment. *N. Engl. J. Med.* **290:**81–84.
195. Durelli, L., Maggi, G., Casadio, C., Ferri, R., Rendine, S., and Bergamini, L. (1991). Actuarial analysis of the occurrence of remissions following thymectomy for myasthenia gravis in 400 patients. *J. Neurol. Neurosurg. Psychiatry* **54:**406–411.
196. Lanska, D. J. (1990). Indications for thymectomy in myasthenia gravis. *Neurology* **40:**1828–1829.
197. Olanow, C. W., Wechsler, A. S., Sirotkin-Roses, M., Stajich, J., and Roses, A. D. (1987). Thymectomy as primary therapy in myasthenia gravis. *Ann. N.Y. Acad. Sci.* **505:**595–606.
198. Olanow, C. W., Lane, R. J. M., and Roses, A. D. (1982). Thymectomy in late-onset myasthenia gravis. *Arch. Neurol.* **39:**82–83.
199. Hohlfeld, R., Michels, M., Heininger, K., Besinger, U., and Toyka, K. V. (1988). Azathioprine toxicity during long-term immunosuppression of generalized myasthenia gravis. *Neurology* **38:** 258–261.
200. Witte, A. S., Cornblath, D. R., Parry, G. J., Lisak, R. P., and Schatz, N. J. (1984). Azathioprine in the treatment of myasthenia gravis. *Ann. Neurol.* **15:**602–605.
201. Perez, M., Buot, W. L., Mercado-Danguilan, C., Bagabaldo, Z. G., and Renales, L. D. (1981). Stable remissions in myasthenia gravis. *Neurology* **31:**32–37.
202. Tindall, R. S. A., Phillips, J. T., Rollins, J. A., Wells, L., and Hall, K. (1993). A clinical therapeutic trial of cyclosporine in myasthenia gravis. *Ann. N.Y. Acad. Sci.* **681:**539–551.
203. Pinching, A. J., Peters, D. K., and Newsom-Davis, J. (1976). Remission of myasthenia gravis following plasma exchange. *Lancet* **2:**1373–1376.
204. Seybold, M. E. (1987). Plasmapheresis in myasthenia gravis. *Ann. N.Y. Acad. Sci.* **505:**584–587.
205. Ferrero, B., Durelli, L., Cavallo, R., Dutto, A., Aimo, G., Pecchio, F., and Bergamasco, B. (1993). Therapies for exacerbation of myasthenia gravis. The mechanism of action of intravenous high-dose immunoglobulin. *Ann. N.Y. Acad. Sci.* **681:**563–566.
206. Gajdos, P., Outin, H., Elkharray, D., Brunel, D., Rohan-Chabot, P. D., Raphael, J. C., Goulon, M., Goulon-Goeau, C., and Geursen, R. G. (1984). High-dose intravenous gammaglobulin for myasthenia gravis. *Lancet* **1:**406–407.
207. Gajdos, P., Outin, H. D., Morel, E., Raphael, J. C., and Goulon, M. (1987). High-dose intravenous gamma globulin for myasthenia gravis: An alternative to plasma exchange. *Ann. N.Y. Acad. Sci.* **505:**842–844.
208. Uchiyama, M., Ichikawa, Y., Takaya, M., Moriuchi, J., Shimizu, H., and Arimori, S. (1987). High-dose gammaglobulin therapy of generalized myasthenia gravis. *Ann. N.Y. Acad. Sci.* **505:**868–871.
209. Durelli, L., Ferrio, M. F., Urgesi, A., Poccardi, G., Ferrero, B., and Bergamini, L. (1993). Total body irradiation for myasthenia gravis: A long-term follow-up. *Neurology* **43:**2215–2221.

210. Lambert, E. H., Eaton, L. M., and Rooke, E. D. (1956). Defect of neuromuscular transmission associated with malignant neoplasm. *Am. J. Physiol.* **187:**612–613.
211. Lambert, E. H., Rooke, E. D., Eaton, L. M., and Hodgson, C. H. (1961). Myasthenic syndrome occasionally associated with bronchial neoplasm: Neurophysiologic studies. In *Myasthenia Gravis* (H. R. Viets, ed.), Thomas, Springfield, pp. 362–410.
212. O'Neill, J. H., Murray, N. M. F., and Newsom-Davis, J. (1988). The Lambert–Eaton myasthenic syndrome. A review of 50 cases. *Brain* **111:**577–596.
213. Rooke, E. D., Eaton, L. M., Lambert, E. H., and Hodgson, C. H. (1960). Myasthenia and malignant intrathoracic tumor. *Med. Clin. North Am.* **44:**977–988.
214. Satoyoshi, E., Kowa, H., and Fukunaga, N. (1973). Subacute cerebellar degeneration in Eaton–Lambert syndrome with bronchogenic carcinoma. *Neurology* **23:**764–768.
215. Elmqvist, D., and Lambert, E. H. (1968). Detailed analysis of neuromuscular transmission in a patient with the myasthenic syndrome sometimes associated with bronchogenic carcinoma. *Mayo Proc. Clin.* **43:**689–713.
216. Lambert, E. H., and Lennon, V. A. (1982). Neuromuscular transmission in nude mice bearing oat-cell tumors from Lambert–Eaton myasthenic syndrome. *Muscle Nerve* **5:**S39–S45.
217. Oguro-Okamoto, M., Griesman, G. E., Wieben, E. D., Slaymaker, S. J., Snutch, T. P., and Lennon, V. A. (1992). Molecular diversity of neuronal-type calcium channels identified in small cell lung carcinoma. *Mayo Clin. Proc.* **67:**1150–1159.
218. Newsom-Davis, J., Leys, K., Vincent, A., Ferguson, I., Modi, G., and Mills, K. (1991). Immunological evidence for the co-existence of the Lambert–Eaton myasthenic syndrome and myasthenia gravis in two patients. *J. Neurol. Neurosurg. Psychiatry.* **54:**452–453.
219. Trontelj, J. V., and Stalberg, E. (1990). Single motor endplates in myasthenia gravis and LEMS at different firing rates. *Muscle Nerve* 14:226–232.
220. Lambert, E. H., and Elmqvist D. (1971). Quantal components of end-plate potentials in the myasthenic syndrome. *Ann. N.Y. Acad. Sci.* **183:**183–199.
221. Molenaar, P. C., Newsom-Davis, J., Polak, R. L., Vincent, A., and Murray, N. (1982). Eaton–Lambert syndrome: Acetylcholine and choline acetyltransferase in skeletal muscle. *Neurology* **32:**1062–1065.
222. Cull-Candy, S. G., Meledi, R., Trautmann, A., and Uchitel, O. D. (1980). On the release of transmitter at normal, myasthenia gravis and myasthenic syndrome affected human end-plates. *J. Physiol. (London)* **299:**621–638.
223. Streib, E. W., and Rothner, A. D. (1981). Eaton–Lambert myasthenic syndrome: Long-term treatment of three patients with prenisone. *Ann. Neurol.* **10:**448–453.
224. Lang, B., Newsom-Davis, J., Wray, D. W., and Vincent, A. (1981). Autoimmune aetiology for myasthenic (Lambert–Eaton) syndrome. *Lancet* **2:**224–226.
225. Lennon, V. A., Lambert, E. H., Whittingham, S., and Fairbainks, V. (1982). Autoimmunity in the Lambert–Eaton myasthenic syndrome. *Muscle Nerve* **5:**S21–S25.
226. Newsom-Davis, J., and Murray, N. M. F. (1984). Plasma exchange and immunosuppressive drug treatment in the Lambert–Eaton myasthenic syndrome. *Neurology* **34:**480–485.
227. Fukunaga, H., Engel, A. G., Osame, M., and Lambert, E. H. (1982). Paucity and disorganization of presynaptic membrane active zones in the Lambert–Eaton myasthenic syndrome. *Muscle Nerve* **5:**686–697.
228. Fukunaga, H., Engel, A. G., Lang, B., Newsom-Davis, B., and Vincent, A. (1983). Passive transfer of Lambert–Eaton myasthenic syndrome IgG from man to mouse depletes the presynaptic membrane active zones. *Proc. Natl. Acad. Sci. USA* **80:**7636–7640.
229. Fukuoka, T., Engel, A. G., Lang, B., Newsom-Davis, J., Prior, C., and Wray, D. W. (1987). Lambert–Eaton myasthenic syndrome: I. Early morphologic effects of IgG on the presynaptic membrane active zones. *Ann. Neurol.* **22:**193–199.
230. Fukuoka, T., Engel, A. G., Lang, B., Newsom-Davis, J., and Vincent, A. (1987). Lambert–Eaton myasthenic syndrome. II. Immunoelectron microscopy localization of IgG at the mouse motor end-plate. *Ann. Neurol.* **22:**200–211.
231. Nagel, A., Engel, A. G., Lang, B., Newsom-Davis, J., and Fukuoka, T. (1988). Lambert–Eaton syndrome IgG depletes presynaptic membrane active zone particles by antigenic modulation. *Ann. Neurol.* **24:**552–558.
232. Lang, B., Vincent, A., Murray, N. M. F., Newsom-Davis, J. (1989). Lambert–Eaton myasthenic syndrome: Immunoglobulin G inhibition of Ca^{2+} flux in tumor cells correlates with disease severity. *Ann. Neurol.* **25:**265–271.
233. Kim, Y. I., and Neher, E. (1988). IgG from patients with Lambert–Eaton syndrome blocks voltage-dependent calcium channels. *Science.* **239:**405–408.
234. Peers, C., Lang, B., Newsom-Davis, J., and Wray, D. W. (1987). Selective action of Lambert–Eaton myasthenic syndrome antibodies on calcium channels in a rodent neuroblastoma × glioma hybrid cell line. *J. Physiol. (London)* **421:**293–308.
235. Lang, B., Johnston, I., Leys, K., Elrington, G., Marqueze, B., Leveque, C., Martin-Moutot, N., Seagar, M., Hoshino, T., Takahashi, M., Sugimori, M., Cherksey, B. D., Llinas, R., and Newsom-Davis, J. (1993). Autoantibody specificities in Lambert–Eaton myasthenic syndrome. *Ann. N.Y. Acad. Sci.* **681:**382–391.
236. Lennon, V. A., and Lambert, E. H. (1989). Antibodies bind solubilized calcium channel–omega-conotoxin complexes from small cell lung carcinoma: A diagnostic aid for Lambert–Eaton myasthenic syndrome. *Mayo Clin. Proc.* **64:**1498–1504.
237. Leys, K., Lang, B., Johnston, I., and Newsom-Davis, J. (1991). Calcium channel autoantibodies in the Lambert–Eaton myasthenic syndrome. *Ann. Neurol.* **29:**307–314.
238. Sher, E., Gotti, C., Canal, N., Scopetta, C., Piccolo, G., Evoli, A., and Clementi, F. (1989). Specificity of calcium channel autoantibodies in Lambert–Eaton myasthenic syndrome. *Lancet* **2:**640–643.
239. Blandino, J. K. W., and Kim, Y. I. (1993). Lambert–Eaton syndrome IgG inhibits dihydropyridine-sensitive, slowly inactivating calcium channels in bovine adrenal chromaffin cells. *Ann. N.Y. Acad. Sci.* **681:**394–397.
240. Viglione, M. P., Blandino, J. K. W., and Kim, Y. I. (1993). Effects of Lambert–Eaton syndrome serum and IgG on calcium and sodium currents in small-cell lung cancer cells. *Ann. N.Y. Acad. Sci.* **681:**418–421.
241. Takamori, M., Hamada, T., Komai, K., Takahashi, M., and Yoshida, A. (1994). Synaptotagmin can cause an immune-mediated model of Lambert–Eaton myasthenic syndrome in rats. *Ann. Neurol.* **35:**74–80.
242. Rosenfeld, M. R., Wong, E., Dalmau, J., Manley, G., Egan, D., Posner, J. P., Sher, E., and Furneaux, H. M. (1993). Sera from patients with Lambert–Eaton myasthenic syndrome recognize the β-subunit of Ca^{2+} channel complexes. *Ann. N.Y. Acad. Sci.* **681:** 408–411.
243. Engel, A. G., Fukuoka, T., Lang, B., Newsom-Davis, J., Vincent, A., and Wray, D. W. (1982). Lambert–Eaton myasthenic syndrome IgG: Early morphologic effects and immunolocalization at the motor end-plate. *Ann. N.Y. Acad. Sci.* **505:**333–345.
244. Engel, A. G. (1986). Myasthenic syndromes. In *Myology* (A. G. Engel and B. Q. Banker, eds.), McGraw–Hill, New York, pp. 1955–1990.
245. Chalk, C. H., Murray, N. M. F., Newsom-Davis, J., O'Neill, J. H., and Spiro, S. G. (1990). Response of the Lambert–Eaton myas-

thenic syndrome to treatment of associated small-cell lung carcinoma. *Neurology* **40:**1552–1556.
246. Otsuka, M., and Endo, M. (1960). The effect of guanidine on neuromuscular transmission. *J. Pharmacol. Exp. Ther.* 128 :273–282.
247. Cherington, M. (1976). Guanidine and germaine in Lambert–Eaton syndrome. *Neurology* **26:**944–946.
248. Saint, D. A. (1989). The effects of 4-aminopyridine and tetraethylammonium on the kinetics of transmitter release at the mammalian neuromuscular synapse. *Can. J. Physiol. Pharmacol.* **67:**1045–1050.
249. McEvoy, K. M., Windebank, A. J., Daube, J. R., and Low, P. A. (1989). 3,4-Diaminopyridine in the treatment of Lambert–Eaton myasthenic syndrome. *N. Engl. J. Med.* **321:**1567–1571.
250. Bird, S. J. (1991). Clinical and electrophysiologic improvement in the Lambert–Eaton syndrome with intravenous immunoglobulin therapy. *Muscle Nerve* **14:**913–914.
251. Engel, A. G., Lambert, E. H., Mulder, D. M., Gomez, M. R., Whitaker, J. H., Hart, Z., and Sahashi, K. (1981). Recently recognized congenital myasthenic syndromes: A) Endplate acetylcholine (ACh) esterase deficiency, B) putative abnormality of the ACh induced ion channel, C) putative defect of ACh resynthesis or mobilization—Clinical features, ultrastructure and cytochemistry. *Ann. N.Y. Acad. Sci.* **377:**614–639.

29

Genesis of Cardiac Arrhythmias
Roles of Calcium and Delayed Potassium Channels in the Heart

Robert S. Kass

29.1. INTRODUCTION

Electrical activity underlies the control of the frequency, strength, and duration of contraction of the heart. During the cardiac cycle, a regular rhythmic pattern must be established in time-dependent changes in ionic conductances in order to ensure events that underlie normal cardiac function. Electrical impulses, originating at the sinoatrial (SA) node, are conducted throughout the atria until they converge at the atrioventricular (AV) node, pass through the bundle of His and the Purkinje fiber conducting system, and eventually excite the working myocardial cells of both ventricles. Coordination of these cellular events is ensured by the unique electrical properties of the different cell types of the heart as well as the junctional connections that join them. Rapid conduction in the His–Purkinje network ensures uniform and synchronous contraction of the ventricle; slow conduction through the AV node coordinates atrial and ventricular contraction. Pacemaker activity of the SA node determines heart rate and latent pacemaker properties of both the AV node and the Purkinje fiber network serve as important pacing reserves.

Current flow through a large number of ion channels, exchange mechanisms, and pumps underlies and coordinates these electrical signals and alteration of the critical balance of these multiple current pathways can lead to disruptive, often fatal, rhythm disturbances: the cardiac arrhythmias. Although the cardiac arrhythmias form a complicated and diverse group reflecting the complexities of the ionic mechanisms underlying the electrical activity of the human heart, a surprisingly large number of rhythm disturbances are caused either directly or indirectly by mechanisms that prolong the duration of the action potential of the working myocardium. This chapter focuses on this class of arrhythmia and on the cellular and molecular mechanisms that underlie them.

29.2. REPOLARIZATION OF THE CARDIAC ACTION POTENTIAL: CRUCIAL ROLES OF CALCIUM AND POTASSIUM CHANNELS

In general, inward movement of positive ions depolarizes cells (makes the intracellular membrane more positive than the extracellular membrane) and outward flow of positive charge repolarizes the cell (makes the intracellular membrane more negative than the extracellular membrane). Action potentials of the heart are long lasting, and particularly in the ventricle, they are characterized by a period of slowly changing and maintained depolarization, the action potential plateau, a period of activity that separates phases 2 and 3 of the electrical response. The plateau is crucial in determining the strength and duration of contraction of the myocardium, in setting the proper relationship between systolic contraction and diastolic filling times, and in providing a cardioprotective window in which reexcitation cannot take place via voltage-dependent sodium or calcium channels. During the plateau, the net flow of current is small because of the inherent high input resistance of the cell membrane during this crucial period. Weidmann[1] first demonstrated this property of the cardiac action potential plateau by showing that the conductance of the plateau phase was lower than during diastole. This characteristic nonlinear relationship between conductance and membrane potential is the result of the unique properties of background inwardly-rectifying potassium channels. The important functional consequence of this property of the background channels is twofold: (1) small net currents are needed to maintain the plateau which is highly energy efficient for the cell and (2) small changes in net plateau current can markedly influence the time course of the plateau. Repolarization begins when the net current during the plateau becomes outward: either by an increase in an outward current component or a decrease in an inward current component. The

Robert S. Kass • Department of Pharmacology, College of Physicians and Surgeons, Columbia University, New York, New York 10032.
Molecular Biology of Membrane Transport Disorders, edited by S.G. Schultz *et al.* Plenum Press, New York, 1996

repolarization process and its underlying controlling mechanisms have been reviewed and summarized by Noble.[2,3]

29.3. THE IMPORTANCE OF POTASSIUM CHANNEL CURRENTS IN REGULATION OF THE ACTION POTENTIAL DURATION

Initial attempts to explain the ionic mechanisms responsible for cardiac action potentials relied heavily on the pioneering work of Hodgkin and Huxley in nerve.[4] It seemed reasonable that modification of the gating kinetics of Na and K conductances might be responsible for the slow time course that characterizes responses in the heart. However, it was not until Noble[3] incorporated the unique nonlinear "rectifying" properties of the background potassium conductance into a model, that the proper high input impedance characteristic of the plateau phase of the action potential could be accounted for. Because of the sparse experimental data available, this model of the action potential was oversimplified; it failed to include important time-dependent ionic currents now known to be crucial to the action potential: calcium channel currents, transient outward currents, potassium currents, and currents carried via exchangers and pumps. Nevertheless, this work pointed to the importance of potassium channel currents in controlling the duration of the cardiac action potential.

29.4. DELAYED POTASSIUM CHANNEL CURRENTS IN THE HEART: MULTIPLE CHANNEL TYPES

Noble and Tsien[5] provided the first quantitative investigation of time-dependent outward currents that activate over the voltage range of and with a time course similar to the Purkinje fiber action potential plateau. Because the underlying conductance(s) of these currents activate slowly and with distinctively slow kinetics, the conductances were referred to as "delayed rectifier" conductances.

Two components of potassium-sensitive current were identified and found to be activated at voltages positive to −50 mV. Although the currents were sensitive to the external potassium ion concentration, they were not perfectly potassium-selective as judged by the equilibrium potentials of the two current components. Consequently, the names I_{x1} and I_{x2} were chosen to identify the components. The first component, I_{x1}, the most potassium-selective component, activated with time constants that were voltage-dependent and on the order of 0.05 to 0.5 sec, depending on membrane potential. Additionally, the fully activated current–voltage relationship for this component rectified in the inward direction. That is, it limited the amount of current that could flow in outward direction at more positive voltages. The second component, I_{x2}, was characterized by extremely slow activation kinetics, less potassium selectivity, a relatively linear instantaneous current–voltage relationship, and less selectivity for potassium ions. Both current components were used to successfully reconstruct the measured currents[5] as well as the Purkinje fiber action potential.[6] These findings were confirmed in a number of cardiac preparations by other groups.[7,8] In most cases, the kinetics of activation and deactivation of delayed rectifier components were shown to be complex, and the interpretation of the data as representative of two individual current components was preserved.

Bennett *et al.*[9] reinvestigated the delayed potassium current of the calf Purkinje fiber and specifically designed experiments to discriminate between single and multiple channel types. As in previous work, the delayed potassium currents were best described by multiple exponential functions of time. Evidence for a single population of channels was based on measurement of the potassium-sensitivity of the amplitudes of the individual current components which were identified by kinetic analysis. In a second kinetic approach, measurement of the envelope of tails recorded at the holding potential after prepulses to a more positive voltage but of variable duration, showed a time course consistent with one population of channels. The results of this study suggested that the second component of delayed rectification reported by Noble and Tsien[5] could have been a reflection of changes in potassium ion concentration that occurred in restricted spaces during the extremely long depolarizing pulses needed to activate it.

Balser and Roden,[10] investigating the lanthanum-sensitivity of delayed potassium current in isolated guinea pig ventricular cells, found that the envelope of tails accompanying depolarizing pulses of variable duration did not share the same time course as the development of current during a maintained depolarizing voltage pulse, and that exposure to lanthanum corrected this discrepancy. Balser *et al.*[11] then analyzed the kinetic properties of delayed currents measured in the presence of La^{3+} and confirmed multiple state-kinetics for this type of channel, but these experiments raised the possibility that, at least in the guinea pig ventricle, multiple types of delayed rectifier channels existed.

This possibility was confirmed by Sanguinetti and Jurkiewicz[12] who, in an attempt to resolve the mechanism of action of the benzenesulfonamide antiarrhythmic drug E-4031, a putative class III agent, found that currents measured during depolarizing test pulses appeared to be less sensitive to this compound than tail currents measured after termination of the test pulses (Figure 29.1). Importantly, as was the case with lanthanum, the application of E-4031 abolished the difference between the time course of pulse current and the envelope of tails following pulses of variable duration (envelope of tails test).

Sanguinetti and Jurkiewicz[12] concluded that delayed rectification in guinea pig ventricular cells consists of two components: one that activates rapidly but then spontaneously inactivates with depolarization (I_{Kr}) and a second, approximately tenfold larger component, which is lanthanum- and E-4031-insensitive and does not inactivate, which is labeled I_{Ks} because of its slow kinetics. I_{Ks} is the dominant component of I_K recorded positive to voltages of +20 mV in the ventricle. It is less prominent in the atria.[13] Thus, this pharmacological evidence, to a certain extent, supports the view first proposed by Noble and Tsien[5] of multiple delayed rectifier channel types in

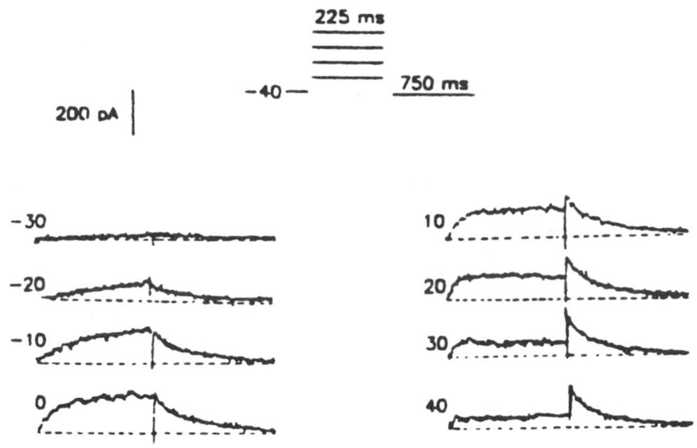

Figure 29.1. E-4031-sensitive current recorded in guinea pig ventricular cells. Currents were recorded in the presence and absence of drug and the difference traces were obtained by digital subtraction. Note the rapid kinetics and rectification that occurs at pulse voltages positive to +20 mV. Note also that tail amplitudes do not rectify. (From Ref. 12.)

the heart, although the detailed properties of I_{Ks} and I_{Kr} differ considerably from the properties of I_{x1} and I_{x2} as originally described by Noble and Tsien.[5]

Delayed potassium currents have also been reported in mammalian SA and AV nodes.[8,14–18] Initial studies of nodal I_K using multicellular and single-cell nodal preparations reported current similar to I_{Ks} of ventricular preparations: slow, voltage-dependent activation, no inactivation, largely potassium-selective, and multiple exponential activation.[8,14,15,19] More extensive analysis of the nodal cell delayed rectifier current by Shibasaki[20] revealed a major I_K component in the rabbit node that was characterized by marked inward rectification in both whole-cell and single-channel records resembling I_{Kr} in its kinetic and rectifier properties. More recent work suggests that even nodal tissue can consist of multiple delayed rectifier components and that the relative importance of each current type varies not only with tissue but also with species.[21]

Novel delayed rectifier activity described in human atrium may in fact be carried by current through yet a distinctly different channel population[22] and shows the necessity of identification of the functionally relevant channel activity in the normal and pathologic human heart. Inactivating potassium channel proteins have been cloned from human hearts[22–26] but the pharmacology, gating, and voltage-dependence of the cloned channels differ from native I_{Kr} channels of the guinea pig.

29.5. NEUROMODULATION OF CARDIAC I_K

Cardiac electrical and mechanical activity is modulated extensively by the neuroendocrine system: heart rate and contractile activity are particularly sensitive to catecholamines. Increase in norepinephrine release with high sympathetic tone speeds heart rate and strengthens the force of contraction (Figure 29.3). Key to proper electrical and mechanical response to sympathetic stimulation is the control of action potential duration (APD) via modulated cardiac K channels. Changes in heart rate must be accompanied by concomitant shortening of the action potential to ensure a proper temporal relationship between diastolic filling and systolic ejection. Because delayed rectifier channels are such major determinants of APD, their neurohormonal regulation which has been shown to be closely linked to control of APD[27] is key to cardiac function.

29.6. β-ADRENERGIC STIMULATION

Changes in heart rate and contractile activity are linked to the well-known stimulatory effects of β-adrenergic stimulation on L-type calcium channel activity which has been investigated extensively by many groups[28] and is summarized below. Additionally, I_{Ks} is also under marked sympathetic control. Tsien *et al.*[29] first showed that cAMP can increase I_K in calf Purkinje fiber, and since that work many groups have shown that β-adrenergic agonists, cAMP and its analogues, and phosphodiesterase inhibitors increase I_{Ks} in Purkinje fibers and other multicellular preparations[27,30–34] as well as in isolated cardiac myocytes (Figure 29.2).[35–39] Because of the dramatic enhancement of I_{Ks} by sympathetic stimulation (up to sixfold increases in I_{Ks} magnitude have been reported,[27] it is very likely that this channel plays a major role in the normal physiologically important control of action potential duration by catecholamines and, perhaps, underlies rhythm disturbances that occur when sympathetic control of heart electrical activity is dysfunctional (see below).

29.7. MULTIPLE ROLES OF CALCIUM MOVEMENT DURING THE ACTION POTENTIAL PLATEAU

Although at least four types of calcium channels have been identified based on their kinetics, inactivation properties, and specific pharmacology, it is the dihydropyridine-sensitive L-type channel that is crucial to regulation of the cardiac action potential plateau and regulation of contractile activation.[40] It has been known since the pioneering work of Reuter[41] that calcium influx during the plateau phase of the action potential is closely linked with activation of the contractile proteins as well as with control of the action potential duration. L-type calcium channels are modulated by catecholamines and by calcium channel antagonists, specific drugs that regulate L-chan-

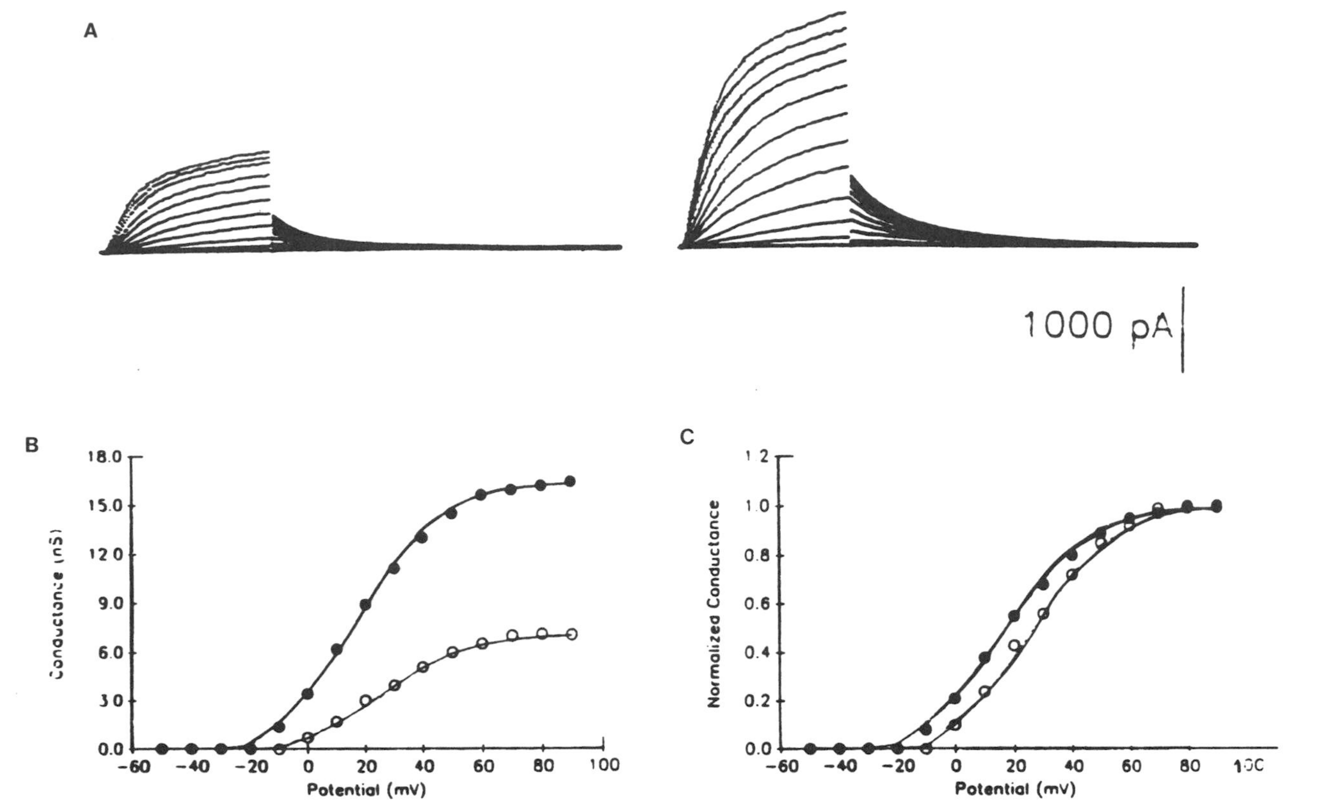

Figure 29.2. Stimulation of protein kinase A shifts delayed rectifier gating kinetics. Total delayed rectifier currents measured before (left, A) and after (right, A) exposure of cell to isoproterenol (1 μM). Panels B and C show raw (B) and normalized (C) activation curves before and after isoproterenol treatment. Normalized activation is shifted by about −10 mV. (From Ref. 72.)

nel gating in such a manner that both the action potential plateau and accompanying twitch can either be potentiated or diminished.[40] Thus, the inward movement of calcium ions serves not only as an electrical signal, but provides a key second messenger, the calcium ion itself, to modulate several aspects of cellular function.

L-type calcium channels inactivate in a manner that depends both on calcium and on membrane potential.[42] Inactivation of channel activity is accelerated with calcium entry through open pores and with membrane depolarization. Since the duration of the action potential plateau depends on the balance between inward and outward membrane currents, the du-

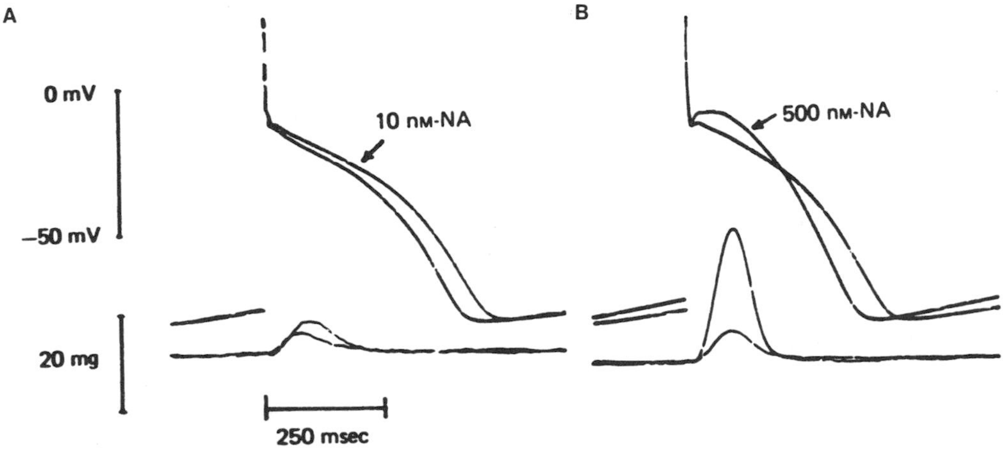

Figure 29.3. Concentration-dependent control of calf Purkinje fiber action potential and twitch tension by norepinephrine (NA): dual interactions with potassium and calcium channels. The action potential prolongation in the presence of 10 nM NA and action potential shortening by 500 nM NA were shown to be the result of concentration-dependent enhancement of L-type calcium and delayed potassium channel currents. (From Ref. 27.)

ration of the plateau period should be affected both by calcium entry and by changes in the amplitude of this phase of the action potential. Kass and Tsien[43] demonstrated that this was in fact the case by studying the effects of calcium concentration and of membrane depolarization on membrane currents and action potential duration in the calf Purkinje fiber, and Kass and Wiegers[27] showed that action potential duration was altered by catecholamines in a manner that was consistent with the interactions between the two catecholamine-sensitive currents: (L-type) calcium and delayed potassium channels.

Thus, the duration of the action potential plateau will be affected by agents and interventions that target at least these two channel types. Agents that block potassium channel currents tend to prolong the action potential. Agents that inhibit calcium entry tend to reduce the duration of the action potential. However, besides these electrical effects, the additional consequences of perturbations in calcium entry must be taken into account when the action potential duration changes. Thus, action potential prolongation will be associated with increased calcium influx, whereas shortening of the action potential reduces calcium influx and thus cellular calcium loading.

29.8. DEFINING THE PROBLEM: TWO CLASSES OF CALCIUM-DEPENDENT ARRHYTHMIAS

Related in part to the dual roles of calcium ions in controlling the time course of the electrical response as well as in providing a transmembrane pathway for calcium influx, two distinct classes of calcium-dependent rhythm disturbances exist: (1) electrical activity driven directly by calcium entry and (2) activity controlled by intracellular calcium concentration and thus indirectly driven by calcium entry. Early afterdepolarizations (EADs) are a type of arrhythmic electrical activity that is generated during the action potential plateau, before repolarization begins.[44] This class of triggered activity fits definition 1 as shown by January and colleagues.[45] Using drugs that specifically interact with L-type calcium channel gating, these investigators have shown that induction of EADs requires (1) lengthening and/or flattening of the action potential plateau phase and (2) the subsequent voltage-dependent recovery of inactivated (L-type) calcium channels. These observations are interpreted in terms of a so-called "window" of voltages over which L-type channels can provide conducting pathways for calcium entry during prolonged depolarization. Thus, action potential prolongation in this critical voltage range will allow L channel window current to provide a continuous pathway for calcium entry. This window current, if sufficiently large, can thus directly generate excitatory electrical activity via the inward current they conduct. It is this directly driven activity that can be classified as one type of EAD (Figure 29.4).

This cycling of channels from unavailable (inactivated) to available (closed, but not inactivated) states normally occurs at more negative (diastolic) voltages. However, under conditions in which the action potential plateau is sufficiently prolonged, this cycling occurs and, in addition, transitions from closed to open (reactivated channels) states can occur and drive electrical responses. The voltage over which these responses occurs is on the order of −30 to −40 mV. Interventions that change L-channel gating and/or numbers of available channels such as sympathetic nerve stimulation or administration of calcium antagonists, will affect EADs driven by this mechanism.

A second type of arrhythmic disturbance can occur over more negative voltages. These events occur at diastolic voltages and are triggered by preceding electrical activity. This type of activity is generally referred to as "delayed afterdepolarization" (DAD) because of the close coupling between preceding action potential activity and the excitatory event that follows with a delay.[46] The underlying current that drives this type of arrhythmogenic event is not thought to be carried di-

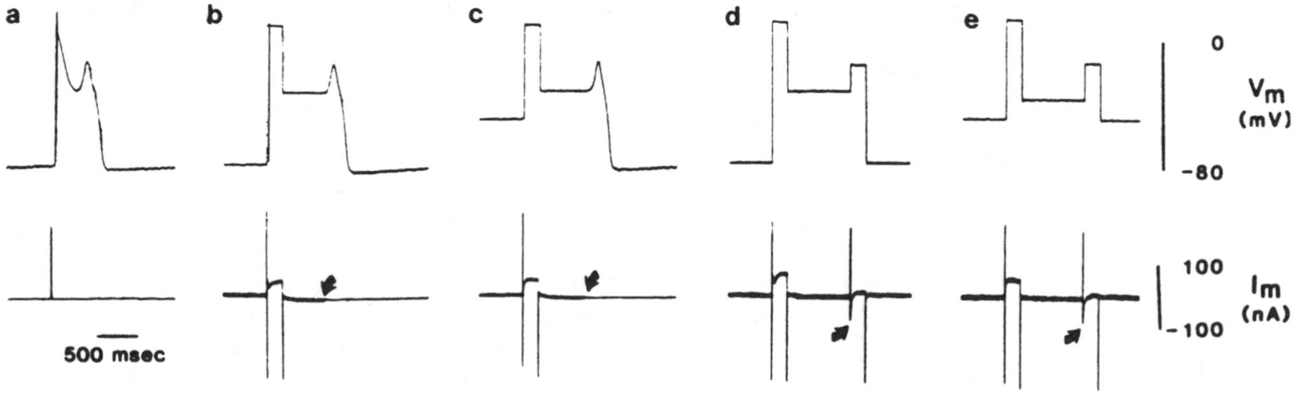

Figure 29.4. An inward current is shown to underlie electrical depolarization caused by prolonging the action potential of a sheep cardiac Purkinje fiber (A). Records of membrane potential (upper traces) and current (lower traces) were obtained in a Purkinje fiber continuously exposed to Bay K8644 in order to enhance L-type calcium channel current. The secondary depolarization in (A) is an EAD, which, as shown in (B) and (C), can be generated after returning to current clamp mode in cells that had been held at −75 mV (B) or −50 mV (C) suggesting that sodium channel currents are not responsible for this response. Inward currents revealed under voltage clamp in (D) and (E) occur at times that are consistent with the EADs of (B) and (C) recorded under current clamp conditions. These results provide strong evidence that the current underlying EADs is a dihydropyridine-sensitive calcium channel current. (From Ref. 45.)

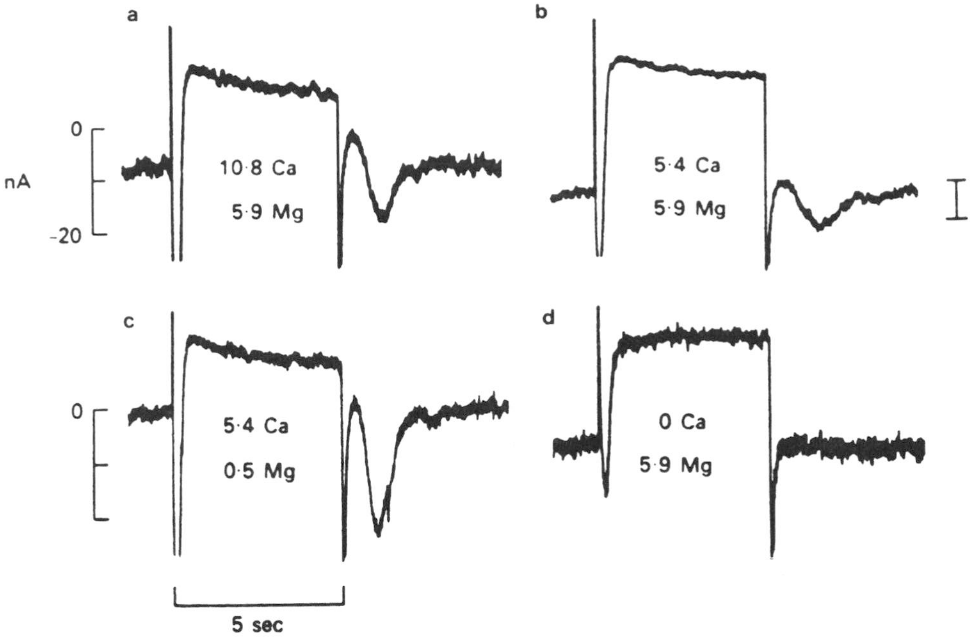

Figure 29.5. Transient inward current induced by prolonged depolarizing voltage steps applied to a calf Purkinje fiber is sensitive to extracellular calcium ions. Ionic currents recorded from a calf Purkinje fiber exposed to strophanthidin exhibited characteristic oscillatory membrane currents following prolonged (5 sec) steps to −32 mV. The panels show the effects of varying calcium and magnesium in the extracellular solution on the amplitude of this oscillatory current which was named "TI" for transient inward current. Calcium potentiates and magnesium antagonizes the response. Subsequent work showed that the trigger for this oscilatory event was release of calcium from intracellular stores (From Ref. 48.)

rectly by calcium ions, but, instead, is the result of a calcium-regulated conductance and/or exchange mechanism that is modulated by release of calcium from intracellular stores (Figure 29.5).[47,48] Although this activity is not driven directly by current that flows during the plateau phase of the action potential, it is potentiated by action potential prolongation most likely because of additional filling of intracellular calcium stores that is promoted by the prolongation of the plateau. Thus, although generated by fundamentally different mechanisms, both EADs and DADs are exacerbated by action potential prolongation and can contribute to arrhythmias associated with this change in action potential configuration. Because of the well-known importance of K^+ channel currents in controlling action potential duration, regulation of these currents is closely associated with the genesis of both of these types of arrhythmias. The remainder of this chapter will focus on the roles of K^+ channel currents in the genesis of two members of this class of arrhythmic disturbance: (1) the acquired and (2) congenital prolonged QT syndrome.

29.9. ACQUIRED LQTS: IS THERE DRUG INDUCTION OR GENETIC SUSCEPTIBILITY?

The term "acquired LQTS" is used broadly to describe conditions in which patients, treated with drugs designed to prevent rhythm disturbances caused by reentry via action potential prolongation, develop additional dysfunction such as EADs or DADs that are induced directly or indirectly by the lengthened action potential of the ventricle. In the simplest cases, these drug-induced arrhythmias are the result of a concentration-dependent blockade of target K channels that causes depolarization-induced electrical disturbances as described above. Such drug action can occur in a "reverse use-dependent" manner, the basis for which has been reviewed by Kass and Freeman.[49] An additional possible underlying cause of acquired LQTS is a latent genetic tendency toward exaggerated drug-induced K channel blockade. In this view, acquired LQTS could be related to the congenital form of this disorder which has clearly been shown to have a genetic basis. In either case, however, acquired LQTS most likely results from the blockade or inhibition of normally functioning K channel activity and the arrhythmic activity that results is most likely related to one of the two calcium-dependent pathways described above. In this view, I_{Ks} channels are likely to be responsive to β-stimulation and contribute to action potential shortening in the ventricle. Thus, in the case of acquired LQTS, it is not surprising that administration of isoproterenol is therapeutic.

Drugs designed to prolong cardiac action potentials through K channel inhibition, the so-called Class III antiarrhythmics, all have a tendency to exacerbate acquired LQTS. With the confirmation of the existence of multiple types of delayed potassium channels that contribute to the control of cardiac action potential duration, there has been considerable renewed interest in the development of specific pharmacologic regulators of these component channels to control the duration of action potentials in the heart under different physiological and pathological conditions. Because of the distinct kinetic and rectifying properties of I_{Kr} and I_{Ks}, these currents will have differential effects on the action potential at different membrane potentials and heart rates, but in all cases, risk of the development of calcium-dependent arrhythmias via the prolonged action potential exists. Understanding the pharmacology of cardiac delayed K^+ channel currents is essential to an understanding of the causes and prevention of these disorders as well as defining the dosages that limit drug action to therapeutic and not toxic effects. The pharmacology of these channels has been reviewed recently by Sanguinetti,[50] Kass and Freeman,[49] and Anumonwo *et al.*[21]

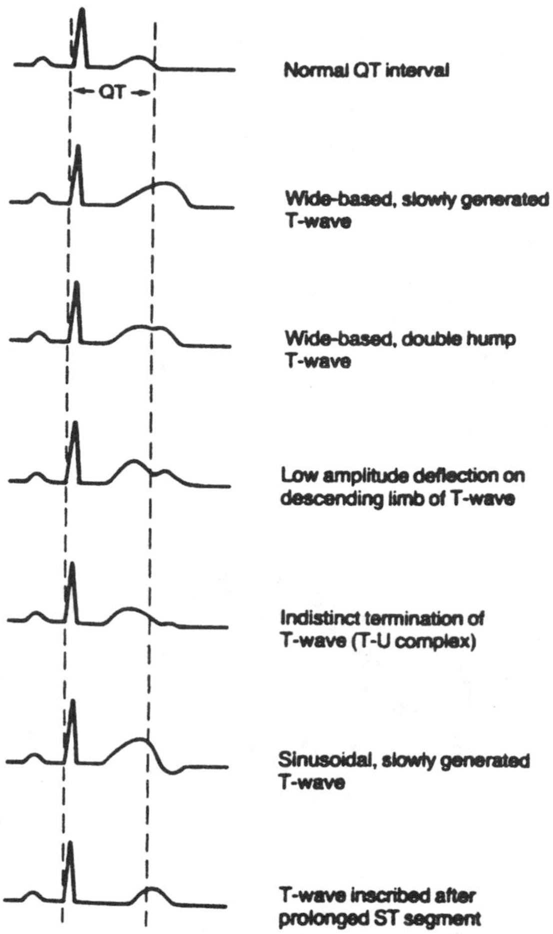

Figure 29.6. Examples of EKGs that show prolonged QT intervals typical of the prolonged QT syndrome. (From Ref. 52.)

29.10. CONGENITAL LQTS: A MOLECULAR HYPOTHESIS

There is considerable interest linking genetic information about the long QT syndrome (LQTS) to a gene defect related to expression of a cardiac ion channel.[51,52] The idiopathic LQTS is an infrequently occurring disease in which the electrocardiogram is characterized by an unusually long period of ventricular repolarization (QT interval) (Figure 29.6). The genetic basis of this disorder has been confirmed,[53,54] linking its basis to the short arm of chromosome 11 and thus to the Harvey RAS gene. However, it is not yet clear whether the disorder is linked to a genetic defect in the structure and/or regulation of an ion channel or whether more than one gene may be involved in generating the disease phenotype.[54,55] Interestingly, an important distinction between acquired and congenital LQTS concerns the distinct actions of sympathetic stimulation in each condition. As described above, β-stimulation is therapeutic in the case of acquired LQTS, but it is well known that elevation of sympathetic tone is lethal in patients with the congenital form of the disorder.

What could be the basis of this difference? Because of the importance of I_{Ks} in controlling the ventricular action potential duration, its robust modulation by β-adrenergic stimulation, and with the failure of the QT interval to adjust (during exercise) in patients with congenital LQTS, the I_{Ks} channel is a strong candidate for the molecular basis of this disease. To date, most cloned channels that have delayed rectifier properties are channels that inactivate and thus are not likely candidates for I_{Ks}. Inactivating potassium-selective channels fall in a class of channels first isolated from experiments with the fruit fly *Drosophila* by both the Jan[56] and Pongs[57] laboratories. These channels, the so-called *Shaker* A-type channel, were so-named because of the phenotype associated with the fruit fly mutation. Based on homology with *Shaker* K^+ channels, several types of voltage-gated K^+ channels have now been cloned from rat and human heart (Figure 29.7).[22–24,26,58] All of these channels have *Shaker*-like properties in that they activate quickly and inactivate over different time scales. Thus, it is not likely that these channel proteins underlie the very slowly activating delayed rectifier channel discussed above, but they could be candidates for several important heart K^+ channels including the transient outward current channel and the I_{Kr} channel.

The functional properties of I_{Ks} are not well-correlated with properties reported for channels cloned from the *Shaker* family described above, but instead with properties of a unique channel, structurally very different from *Shaker*-type channels. This channel, originally cloned from rat kidney,[59] uterus,[60] human genomic DNA,[59] and also neonatal mouse hearts,[61] is a protein of 129–130 amino acids with only one transmembrane-spanning domain, and no homology with other cloned channels.[62] The putative channel protein is called minK because of its small size.[62] Like ventricular I_K,[63–65] the current expressed by *Xenopus* ooctyes injected with mRNA encoding minK is blocked by the antiarrhythmic drug clofilium and by high external concentrations of Ba^{2+} and TEA.[61,66–68]

Recently, I_{minK} has been expressed in mammalian cells transiently transfected with the cDNA encoding the channel protein (Figure 29.8).[69,70] Additionally, immunofluorescence was used to identify minK channel protein in guinea pig SA node and ventricular cells.[69,70] Thus, of all channels cloned to date, minK is the most likely candidate to underlie the heart's slow delayed K^+ current I_{Ks}. Now that the gene that encodes minK has been linked to chromosome 21,[71] it remains to be seen whether this

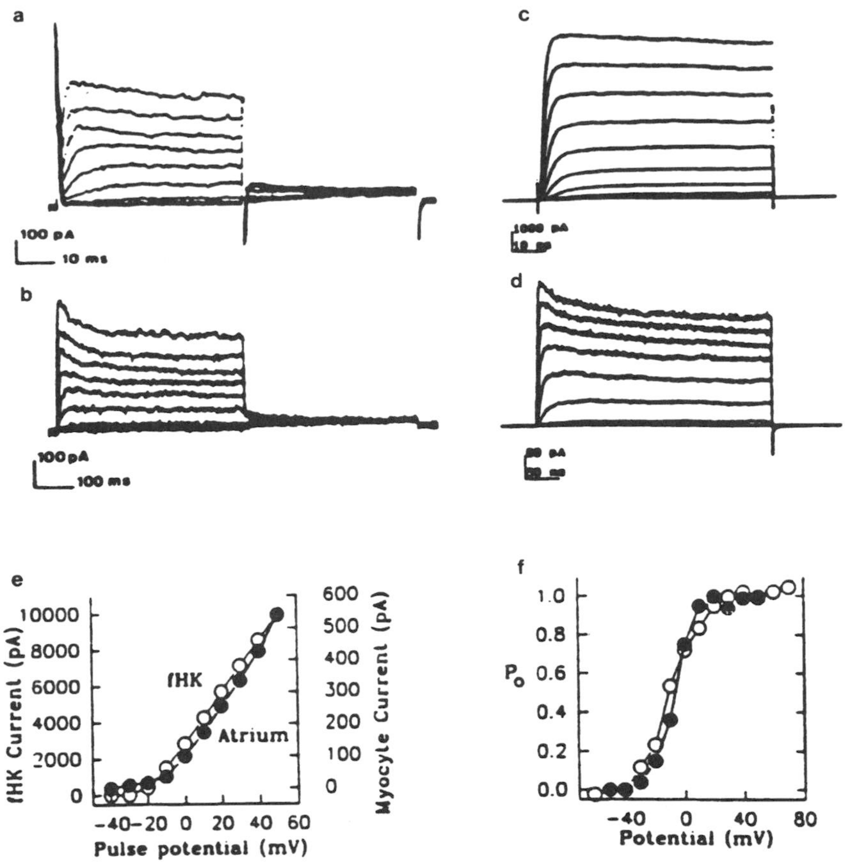

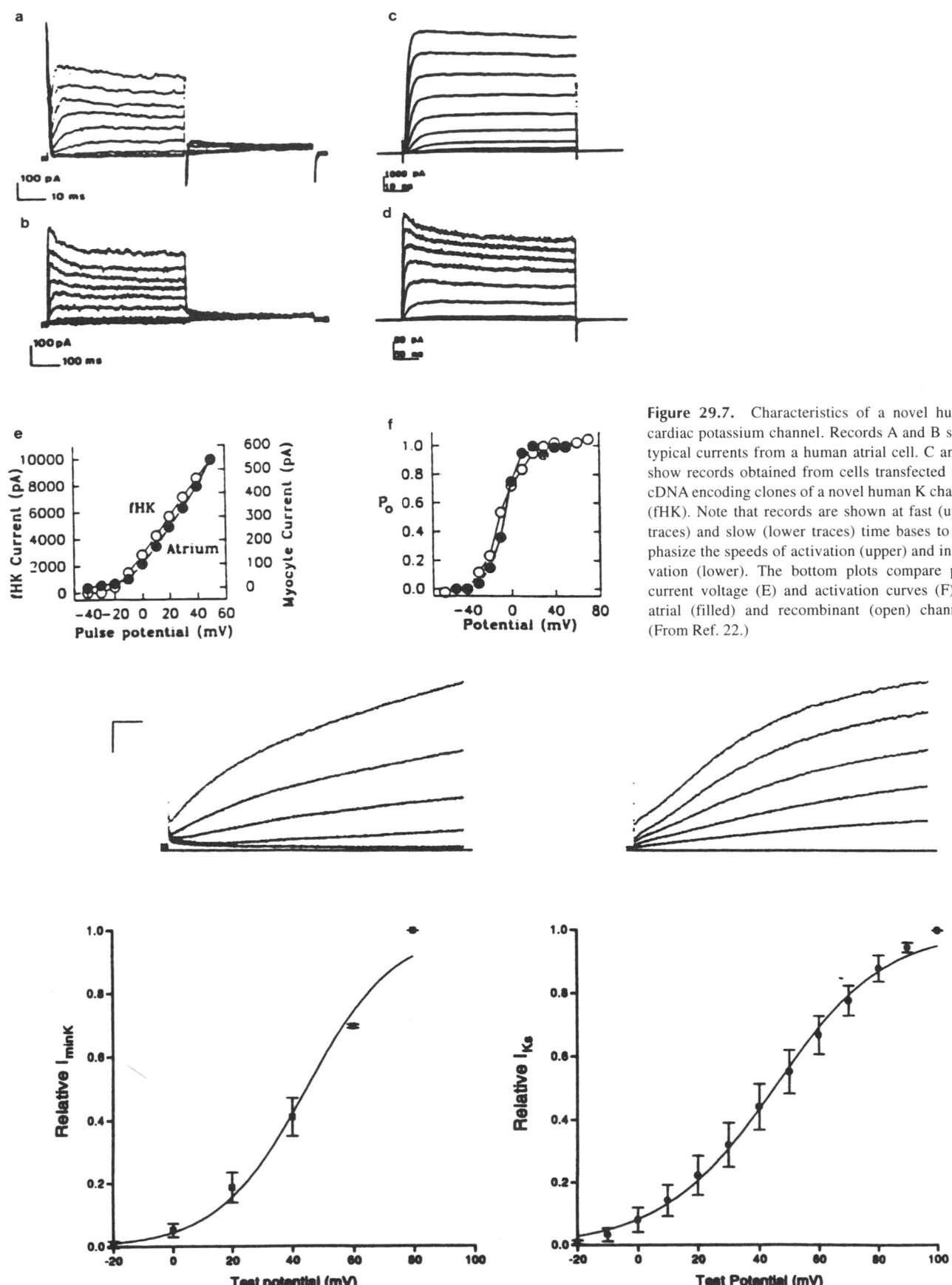

Figure 29.7. Characteristics of a novel human cardiac potassium channel. Records A and B show typical currents from a human atrial cell. C and D show records obtained from cells transfected with cDNA encoding clones of a novel human K channel (fHK). Note that records are shown at fast (upper traces) and slow (lower traces) time bases to emphasize the speeds of activation (upper) and inactivation (lower). The bottom plots compare peak current voltage (E) and activation curves (F) for atrial (filled) and recombinant (open) channels. (From Ref. 22.)

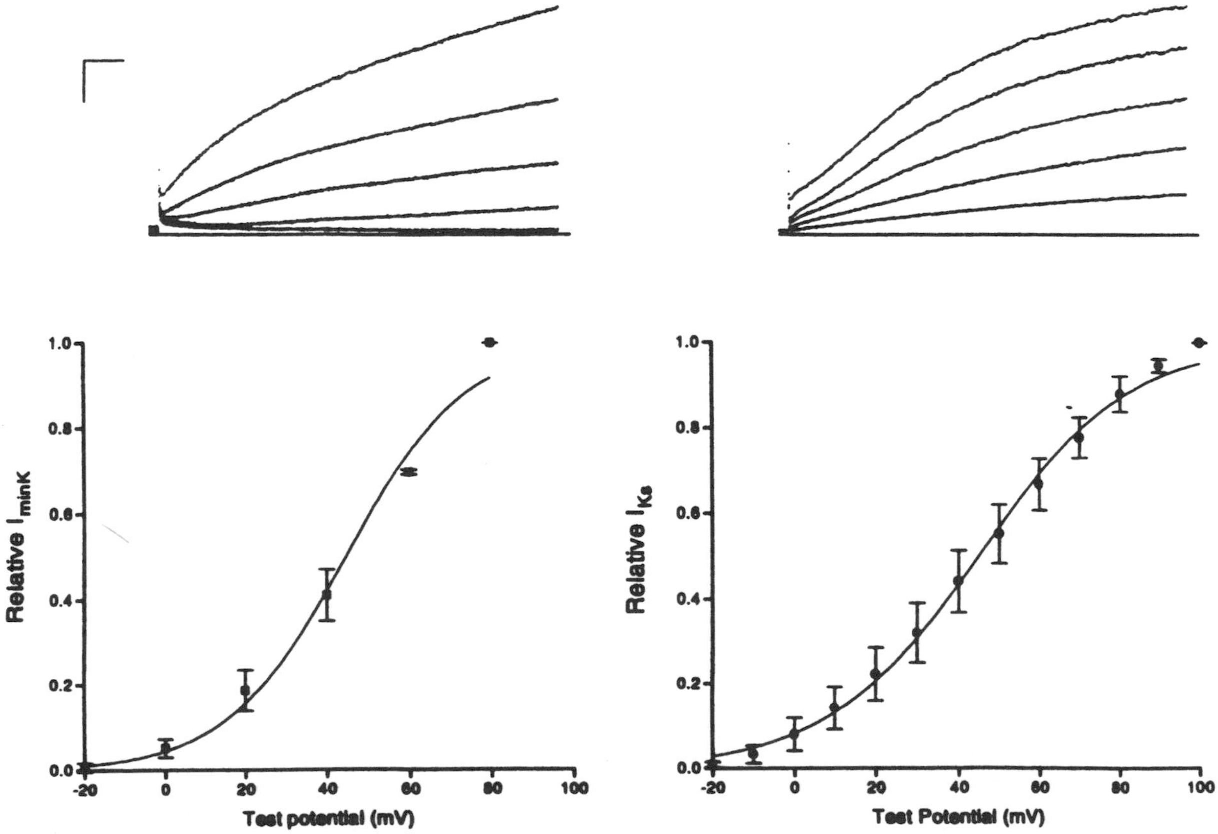

Figure 29.8. Comparison of slow potassium current recorded from guinea pig ventricular cells (right) with channel activity recorded from HEK 293 cells that have been transfected with a synthetic gene encoding rat minK.[70]

gene and, in turn, channel are in fact causally related to congenital LQTS. Dysfunction of the channel protein, its expression, or its regulation by PKA all could contribute to this disorder.

Clearly, research into the molecular pharmacology of this and other cardiac potassium channels is essential to unravel the mechanisms that underlie this and other genetic disorders of the heart.

REFERENCES

1. Weidmann, S. (1951). Effect of current flow on the membrane potential of cardiac muscle. *J. Physiol. (London)* **115:**227–236.
2. Noble, D. (1990). Ionic mechanisms in normal cardiac activity. In *Cardiac Electrophysiology* (J. Jalife and D. P. Zipes, eds.), Saunders, Philadelphia, pp. 163–172.
3. Noble, D. (1978). *The Initiation of the Heartbeat,* Oxford University Press (Clarendon), London.
4. Hodgkin, A. L., and Huxley, A. F. (1952). A quantitative description of membrane current and its application to conduction and excitation in nerve. *J. Physiol. (London)* **117:**500–544.
5. Noble, D., and Tsien, R. W. (1969). Outward membrane currents activated in the plateau range of potentials in cardiac Purkinje fibres. *J. Physiol. (London)* **200:**205–231.
6. McAllister, R. E., Noble, D., and Tsien, R. W. (1975). Reconstruction of the electrical activity of cardiac Purkinje fibres. *J. Physiol. (London)* **251:**1–59.
7. McDonald, T. F., and Trautwein, W. (1978). The potassium current underlying delayed rectification in cat ventricular muscle. *J. Physiol. (London)* **274:**217–245.
8. Noma, A., and Irisawa, H. (1976). A time and voltage-dependent potassium current in the rabbit sinoatrial node cell. *Pfluegers Arch.* **366:**251–258.
9. Bennett, P. B., McKinney, L. C., Kass, R. S., and Begenisich, T. (1985). Delayed rectification in the calf cardiac Purkinje fiber. *Biophys. J.* **48**:553–567.
10. Balser, J. R., and Roden, D. M. (1988). Lanthanum-sensitive current contaminates Ik in guinea pig ventricular myocytes. *Biophys. J.* **53:** 642a.
11. Balser, J. R., Bennett, P. B., and Roden, D. M. (1990). Time-dependent outward current in guinea pig ventricular myocytes. Gating kinetics of the delayed rectifier. *J. Gen. Physiol.* **96:**835–863.
12. Sanguinetti, M. C., and Jurkiewicz, N. K. (1990). Two components of cardiac delayed rectifier K^+ current. Differential sensitivity to block by class III antiarrhythmic agents. *J. Gen. Physiol.* **96:**195–215.
13. Sanguinetti, M. C., and Jurkiewicz, N. K. (1991). Delayed rectifier outward K current is composed of two currents in guinea pig atrial cells. *Am. J. Physiol.* **260:**H393–H399.
14. DiFrancesco, D., Noma, A., and Trautwein, W. (1979). Kinetics and magnitude of the time-dependent potassium current in the rabbit sinoatrial node. *Pfluegers Arch.* **381:**271–279.
15. Kokubun, S., Nishimura, M., Noma, A., and Irisawa, H. (1982). Membrane currents in the rabbit atrioventricular node cell. *Pfluegers Arch.* **393:**15–22.
16. Hauswirth, O., Noble, D., and Tsien, R. W. (1969). The mechanism of oscillatory activity at low membrane potentials in cardiac Purkinje fibres. *J. Physiol. (London)* **200:**255:265.
17. Irisawa, H. (1978). Comparative physiology of the cardiac pacemaker mechanism. *Physiol. Rev.* **58:**461–498.
18. Brown, H. F. (1982). Electrophysiology of the sinoatrial node. *Physiol. Rev.* **62:**505–530.
19. Nakayama, T., Kurachi, Y., Noma, A., and Irisawa, H. (1984). Action potential and membrane currents of single pacemaker cells of the rabbit heart. *Pfluegers Arch.* **402:**248–257.
20. Shibasaki, T. (1987). Conductance and kinetics of delayed rectifier potassium channels in nodal cells of the rabbit heart. *J. Physiol. (London)* **387:**227–250.
21. Anumonwo, J. M. B., Freeman, L. C., Kwok, W. M., and Kass, R. S. (1992). Delayed rectification in single cells isolated from guinea pig sino-atrial node. *Am. J. Physiol.* **262:**H921–H925.
22. Fedida, D., Wible, B., Wang, Z., Fermini, B., Faust, F., Nattel, S., and Brown, A. M. (1993). Identity of a novel delayed rectifier current from human heart with a cloned K channel current. *Circ. Res.* **73:** 210–216.
23. Po, S., Snyders, D. J., Baker, R., Tamkun, M. M., and Bennett, P. B. (1992). Functional expression of an inactivating potassium channel cloned from human heart. *Circ. Res.* **71:**732–736.
24. Snyders, D. J., Roberds, S. L., Knoth, K. M., Bennett, P. B., and Tamkun, M. M. (1992). Block of a cloned human cardiac delayed rectifier by class III antiarrhythmic agents. *Biophys. J.* **61:**151a.
25. Roberds, S. L., and Tamkun, M. M. (1991). Cloning and tissue-specific expression of five voltage-gated potassium channel cDNAs expressed in rat heart. *Proc. Natl. Acad. Sci. USA* **88:**1798–1802.
26. Po, S., Roberds, S., Snyders, D. J., Tamkun, M. M., and Bennett, P. B. (1993). Heteromultimeric assembly of human potassium channels: Molecular basis of a transient outward current? *Circ. Res.* **72:**1326–1336.
27. Kass, R. S., and Wiegers, S. E. (1982). The ionic basis of concentration-related effects of noradrenaline on the action potential of calf cardiac Purkinje fibres. *J. Physiol. (London)* **322:**541–558.
28. Tsien, R. W. (1987). Calcium currents in heart cells and neurons. In *Neuromodulation: The Biochemical Control of Neuronal Excitability* (L. K. Kaczmarek and I. B. Levitan, ed.), Oxford University Press, London, pp. 206–242.
29. Tsien R. W., Giles, W., and Greengard, P. (1972). Cyclic AMP mediates the effects of adrenaline on cardiac Purkinje fibers. *Nature* **240:**181–183.
30. Bennett, P. B., McKinney, L. C., Begenisch, T., and Kass, R. S. (1986). Adrenergic modulation of the delayed rectifier potassium channel in calf cardiac Purkinje fibers. *Biophys. J.* **49:**839–848.
31. Brown, H. F., and Noble, S. J. (1974). Effects of adrenaline on membrane currents underlying pacemaker activity in frog atrial muscle. *J. Physiol. (London)* **238:**51–52.
32. Carmeliet, E., and Mubagwa, K. (1986). Changes by acetylcholine of membrane currents in rabbit cardiac Purkinje fibres. *J. Physiol. (London)* **371:**201–217.
33. Pappano, A. J., and Carmeliet, E. (1979). Epinephrine and the pacemaking mechanism at plateau potentials in sheep cardiac Purkinje fibers. *Pfluegers Arch.* **382:**17–26.
34. Umeno, T. (1984). β-actions of catecholamines on the K-related currents of the bullfrog atrial muscle. *Jpn. J. Physiol.* **34:**513–528.
35. Walsh, K. B., Begenisich, T., and Kass, R. S. (1988). Beta-adrenergic modulation in the heart: Evidence for independent regulation of K and Ca channels. *Biophys. J.* **53:**460a.
36. Walsh, K. B., and Kass, R. S. (1988). Regulation of a heart potassium channel by protein kinase A and C. *Science* **242:**67–69.
37. Duchatelle-Gourdon, I., Hartzell, H. C., and Lagrutta, A. A. (1989). Modulation of the delayed rectifier potassium current in frog cardiomyocytes by β-adrenergic agonists and magnesium. *J. Physiol. (London)* **415:**251–274.
38. Harvey, R. D., and Hume, J. R. (1989). Autonomic regulation of delayed rectifier K^+ current in mammalian heart involves G proteins. *Am. J. Physiol.* **257:**H818–H823.
39. Yazawa, K., and Kameyama, M. (1990). Mechanism of receptor-mediated modulation of the delayed outward potassium current in guinea-pig ventricular myocytes. *J. Physiol. (London)* **421:**135–150.
40. Kokubun, S., Prod'hom, B., Becker, C., Porzig, H., and Reuter, H. (1986). Studies on Ca channels in intact cardiac cells: Voltage-dependent effects and cooperative interactions of dihydropyridine enantiomers. *Mol. Pharmacol.* **30**(6):571–584.

41. Reuter, H. (1984). Electrophysiology of calcium channels in the heart. In *Perspectives in Cardiovascular Research* (L. H. Opie, ed.), Raven Press, New York, pp. 43–51.
42. Kass, R. S., and Sanguinetti, M. C. (1984). Calcium channel inactivation in the cardiac Purkinje fiber. Evidence for voltage- and calcium-mediated mechanisms. *J. Gen. Physiol.* **84:**705–726.
43. Kass, R. S., and Tsien, R. W. (1976). Control of action potential duration by calcium ions in cardiac Purkinje fibers. *J. Gen. Physiol.* **67:** 599–617.
44. Busch, A. E., Varnum, M. D., North, R. A., and Adelman, J. P. (1992). An amino acid mutation on a potassium channel that prevents inhibition by protein kinase C. *Science* **255:**1705–1707.
45. January, C. T., and Riddle, J. M. (1989). Early afterdepolarizations: Mechanism of induction and block. A role for L-type calcium current. *Circ. Res.* **64:**977–990.
46. Ferrier, G. R., and Moe, G. K. (1973). The effect of calcium on acetylstrophanthidin-induced transient depolarizations in canine Purkinje tissue. *Circ. Res.* **33:**508–515.
47. Kass, R. S., Tsien, R. W., and Weingart, R. (1978). Ionic basis of transient inward current induced by strophanthidin in cardiac Purkinje fibres. *J. Physiol. (London)* **281:**209–226.
48. Kass, R. S., Lederer, J. W., Tsien, R. W., and Weingart, R. (1978). Role of calcium ions in transient inward currents and aftercontractions induced by strophanthidin in cardiac Purkinje fibres. *J. Physiol. (London)* **281:**187–208.
49. Kass, R. S., and Freeman, L. C. (1993). Potassium channels in the heart: Cellular, molecular, and clinical implications. *Trends Cardiovasc. Med.* **3:**149–159.
50. Sanguinetti, M. C. (1992). Modulation of potassium channels by antiarrhythmic and antihypertensive drugs. *Hypertension* **19:**228–236.
51. Moss, A. J., Schwartz, P. J., Crampton, R. S., Tzivoni, D., Locati, E. H., MacCluer, J., Hall, W. J., Weitkamp, K., Vincent, M., Garso, A., Robinson, J. L., Benhorin, J., and Choi, S. (1991). The long QT syndrome: Prospective longitudinal study of 328 families. *Circulation* **84:**1136–1144.
52. Moss, A. J., and Robinson, J. (1992). Clinical features of the idiopathic long QT syndrome. *Circulation* **85**(Suppl I):I140–I144.
53. Keating, M., Atkinson, D., Dunn, C., Timothy, K., Vincent, G. M., and Leppert, M. (1991). Linkage of a cardiac arrhythmia, the long QT syndrome, and the Harvey ras-1 gene. *Science* **252:**704–706.
54. Keating, M., Atkinson, D., Dunn, C., Timothy, K., Vincent, G. M., and Leppert, M. (1993). Evidence of genetic heterogeneity in the long QT syndrome. *Science* **260:**1960–1961.
55. Benhorin, J., Kalman, Y. M., Medina, A., Towbin, J., Rave-Harel, N., Dyer, T. D., Blangero, J., MacCluer, J. W., and Krem, B. (1993). Evidence of genetic heterogeneity in the long QT syndrome. *Science* **260:**1960–1961.
56. Papazian, D. M., Schwarz, T. L., Tempel, B. L., Jan, Y. N., and Jan, L. Y. (1987). Cloning of genomic and complementary DNA from Shaker, a putative potassium channels gene from Drosophila. *Science* **237:**749–753.
57. Pongs, O., Kecskemethy, N., Muller, R., Krah-Jentgens, I., Baumann, A., Kiltz, H. H., Canal, I., Llamazares, S., and Ferrus, A. (1988). Shaker encodes a family of putative potassium channel proteins in the nervous system of Drosophila. *EMBO* **7:**1087–1096.
58. Roberds, S. L., Knoth, K. M., Po, S., Blair, T. A., Bennett, P. B., Hartshorne, R. P., Snyders, D. J., and Tamkun, M. M. (1993). Molecular biology of the voltage-gated potassium channels of the cardiovascular system. *J. Cardiovasc. Electrophysiol.* **4:**68–80.
59. Murai, T., Kazikuza, A., Takumi, T., Ohkubo, H., and Nakanishi, S. (1989). Molecular cloning and sequence analysis of human genomic DNA encoding a novel membrane protein which exhibits a slowly activating potassium channel activity. *Biochem. Biophys. Res. Commun.* **161:**176–181.
60. Pragnell, M., Snay, K. J., Trimmer, J. S., MacLusky, N. J., Naftolin, F., Kaczmarek, L. K., and Boyle, M. (1990). Estrogen induction of a small, putative K channel mRNA in rat uterus. *Neuron* **4:**807–812.
61. Honore, E., Attali, B., Romey, G., Heurteaux, C., Ricard, P., Lesage, F., Lazdunski, M., and Barhanin, J. (1991). Cloning, expression, pharmacology, and regulation of a delayed rectifier K channel in mouse heart. *EMBO* **10:**2805–2811.
62. Kaczmarek, L. K. (1991). Voltage-dependent potassium channels: minK and Shaker families. *New Biol.* **3:**315–323.
63. Kass, R. S., Arena, J. P., and Walsh, K. B. (1990). Measurement and block of potassium channel currents in the heart: Importance of channel type. *Drug Dev. Res.* **19:**115–127.
64. Arena, J. P., and Kass, R. S. (1988). Block of heart potassium channels by clofilium and its tertiary analogs: Relationship between drug structure and type of channel blocked. *Mol. Pharmacol.* **34:**60–66.
65. Hirano, Y., and Hiraoka, M. (1986). Changes in K^+ currents induced by Ba^{2+} in guinea pig ventricular muscles. *Am. J. Physiol.* **251:** H24–H33.
66. Ertel, E. A., Smith, M. M., Leibowitz, M. D., and Cohen, C. J. (1994). Isolation of myocardial L-type calcium channel gating currents with the spider toxin omega-Aga-IIIA. *J. Gen. Physiol.* **103:**731–753.
67. Hausdorff, S. F., Goldstein, S. A. N., Rushin, E. E., and Miller, C. (1991). Functional characterization of a minimal K channel expressed from a synthetic gene. *Biochemistry* **30:**3341–3346.
68. Goldstein, S. A. N., and Miller, C. (1991). Site-specific mutations in a minimal voltage-dependent K^+ channel alter ion selectivity and open-channel block. *Neuron* **7:**403–408.
69. Freeman, L. C., and Kass, R. S. (1993). minK:expression in mammalian (HEK 293) cells and immunolocalization in guinea pig heart. *Biophys. J.* **64:**A341.
70. Freeman, L. C., and Kass, R. S. (1993). Expression of a minimal K channel protein in mammalian cells and immunolocalization in guinea pig heart. *Circ. Res.* **73:**968–973.
71. Chevillard, C., Attali, B., Florian, L., Fontes, M., Barhanin, J., Lazdunski, M., and Mattei, M. (1993). Localization of a potassium channel gene (KCNE1) to 21q22.1–q22.2 by in situ hybridization and somatic cell hybridization. *Genomics* **15:**243–245.
72. Walsh, K. B., and Kass, R. S. (1991). Distinct voltage-dependent regulation of a heart delayed IK by protein kinases A and C. *Am. J. Physiol.* **261:**C1081–C1090.

30

Cystic Fibrosis

Michael J. Welsh

30.1. INTRODUCTION

Cystic fibrosis (CF) is a genetic disease caused by mutations in a single gene encoding the cystic fibrosis transmembrane conductance regulator (CFTR). CF is the most common life-threatening autosomal recessive disease of the Caucasian population. It primarily affects children and young adults, but with improved treatments many people with CF are living into adulthood. Recent advances in research on CF and CFTR have dramatically increased our understanding of the abnormal epithelial electrolyte transport that is the hallmark of the disease, the structure and function of the CFTR Cl^- channel, and the way that mutations cause dysfunction of CFTR. The knowledge from the basic research has led to the proposal and testing of several new strategies to treat the disease.

30.2. CLINICAL MANIFESTATIONS OF CYSTIC FIBROSIS

Approximately one in every 2500 infants in the United States is born with CF and it is the predominant cause of severe chronic lung disease in children and young adults.[1] Patients have a life expectancy of only 20–30 years.

Although CF is caused by mutations in a single gene, it has a complex clinical phenotype that involves a number of organs (Table 30.1). Today, disease of the pulmonary airways is the major cause of morbidity and is responsible for 90–95% of the mortality. Chronic bronchitis and bronchiectasis result from chronic colonization of the airways and recurrent infections, particularly with *Pseudomonas.* The course of the lung disease is punctuated by increasingly frequent exacerbations ending in respiratory failure. The mainstay of treatment for the lung disease has been intensive antibiotic use and a program of postural drainage with chest percussion. As the disease progresses, frequent hospitalizations are required and in end-stage disease, lung transplantation is sometimes attempted. Although such treatments have alleviated some symptoms of the disease and delayed its progress, the underlying lung disease remains untreated.

Exocrine pancreatic insufficiency occurs in approximately 85% of patients. Disease of the pancreas appears to begin with obstruction of ducts by inspissated secretions followed by tissue destruction. As the disease progresses, fibrosis replaces the acinar tissue. With extensive disease the pancreatic destruction can also lead to endocrine pancreatic failure and diabetes. In the past, pancreatic failure was a major cause of malnutrition. However, with the administration of pancreatic enzymes and the improvement of nutrition the clinical sequelae of pancreatic failure have been lessened.

The most prominent manifestation of gastrointestinal disease is meconium ileus, which occurs in 5–15% of patients. The intestinal obstruction may result from altered fluid and electrolyte transport by the intestinal epithelium. Small-bowel obstruction may also occur beyond the newborn period, and has been called "meconium ileus equivalent." In such cases obstruction typically occurs in the terminal ileum.

Evidence of liver disease is variable, but as patients survive into adulthood it is observed more frequently and has occasionally led to liver failure requiring transplantation. Focal biliary cirrhosis is observed with inspissation of secretions within the small bile ducts. In more severe disease, the biliary system is more extensively involved. Cholelithiasis is observed in 12% of older patients. The pathogenesis is not certain.

In the genitourinary tract the vas deferens, body of the epididymis, and seminal vesicles are frequently obstructed, fibrotic, or completely absent. As a result, approximately 97% of male patients with CF are sterile. Physical examination shows that most patients with CF have enlarged submandibular, sublingual, and submucosal salivary glands, although the enlargement is usually asymptomatic.

The abnormal electrolyte content of sweat is a characteristic of CF. The elevated sweat Cl^- has been, and continues to be, a mainstay of the diagnosis. Although the increased sweat Cl^- is not usually associated with disease, it can cause fluid

Michael J. Welsh • Howard Hughes Medical Institute, Departments of Internal Medicine and Physiology and Biophysics, University of Iowa College of Medicine, Iowa City, Iowa 52242.

Molecular Biology of Membrane Transport Disorders, edited by S.G. Schultz *et al.* Plenum Press, New York, 1996

Table 30.1. Common Clinical Abnormalities in CF

Respiratory tract	Mucus obstruction of the airways, infection, bronchiectasis
Sweat gland	Increased Cl^- and Na^+ content of sweat
Pancreas	Obstruction of the ducts—fibrosis
Gastrointestinal tract	Meconium ileus, small-bowel obstruction
Genitourinary tract	Male sterility, atrophy of vas deferens
Hepatobiliary system	Biliary cirrhosis, cholelithiasis
Salivary glands	Enlargement, dilated ducts

and electrolyte abnormalities during heat stress or during salt loss associated with vomiting or diarrhea.

30.3. ABNORMALITIES OF EPITHELIAL ELECTROLYTE TRANSPORT

Because the clinical manifestations of the disease seemed so varied, an appreciation of the biologic defect was slow to develop. However, during the early 1980s, physicians and scientists began to appreciate that CF was a disease that primarily affected epithelia; in each of the organs affected, the primary site of involvement was an epithelium. This appreciation led to studies that focused on the physiologic defect in epithelia and generated a unifying hypothesis about the basic defect. Work performed in the mid-1980s showed that electrolyte transport was abnormal in several CF epithelia. Some of the very first appreciation that CF epithelia were Cl^- impermeable came through studies of the sweat gland duct epithelia by Dr. Paul Quinton. At about the same time, Drs. Michael Knowles, Richard Boucher, and their colleagues recognized electrolyte transport abnormalities in the airways. When it has been possible to perform similar studies on other organs, related abnormalities have been discovered. Here I will first describe electrolyte transport abnormalities in airway epithelia, because they have been investigated most extensively. Then other organs are discussed.

30.3.1. Electrolyte Transport by Airway Epithelia

Electrolyte transport by the airway epithelium controls, in part, the quantity and composition of the respiratory tract fluid. Thus, it helps effect normal mucociliary clearance, the pulmonary defense mechanism that removes inhaled particulate material from the lungs. In CF, the mucociliary clearance process is defective, probably as a result of the abnormal electrolyte transport. The CF defect in electrolyte transport may alter the quantity or properties of respiratory tract fluid and possibly the properties of the mucos, so that the mucociliary defense system is impaired.

Human airway epithelium generates a transepithelial electrical potential difference (V_t) of approximately −10 to −30 mV with the lumen electrically negative relative to the submucosal surface.[2–4] This value is in good agreement with values measured *in vivo*.[5–7] The absolute value of voltage decreases progressively as one moves from proximal to distal airways. Transepithelial electrical resistance ranges from 150 to 1000 ohms-cm² in proximal airways and again decreases as one moves to more distal airways. When the epithelium is studied *in vitro* in Ussing chambers and transepithelial voltage is clamped to 0 mV, the resulting current (the short-circuit current) is accounted for by the absorption of Na^+ from the mucosal to the submucosal surface and by secretion of Cl^- from the submucosal to the mucosal surface. In proximal human airway epithelium the majority of the current results from active Na^+ transport whereas in some animal models the major component is Cl^- secretion. The absolute values of Na^+ and Cl^- transport vary from species to species and depend on the neurohumoral environment.[8]

Proximal airway epithelium is a pseudostratified columnar epithelium composed of ciliated cells, goblet cells, and basal cells.[9] The distal airway epithelia are predominantly composed of ciliated and nonciliated bronchiolar (CLARA) cells.[10] In the proximal epithelium, studies with intracellular microelectrodes suggest that the ciliated surface cells are most likely to be the ones responsible for both Na^+ absorption and Cl^- secretion.[11]

The cellular mechanism of Cl^- transport is shown in Figure 30.1 (for reviews see Refs. 8, 12–15). Net transepithelial Cl^- secretion results from the coordinated activity of several transporters located at the apical and basolateral membranes. Cl^- enters the cell across the basolateral membrane on a Na^+–K^+–$2Cl^-$ cotransporter. This transporter is electrically neutral and is driven by the movement of Na^+ down its electrochemical gradient. As a result, Cl^- is accumulated in the cell above electrochemical equilibrium. This transporter is inhibited by a variety of loop diuretics, including furosemide and bumetanide. Na^+, which enters the cell coupled to Cl^- and K^+, is recycled via the activity of the Na^+/K^+-ATPase. This enzyme provides the energy for transepithelial Cl^- secretion by

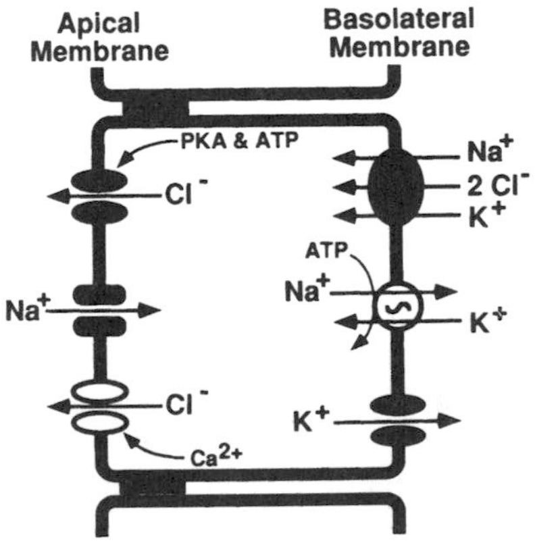

Figure 30.1. Diagram of the cellular mechanism of electrolyte transport by airway epithelia. Figure shows apical and basolateral membrane of airway epithelia. Shaded Cl^- channel in apical membrane represents CFTR which is activated by PKA and ATP. Note that CFTR, the Ca^{2+}-dependent Cl^- channel, and the amiloride sensitive Na^+ channel appear to occur in the apical membrane of the same cell type. The basolateral membrane contains a Na^+–K^+–$2Cl^-$ cotransporter, the Na^+–K^+ ATPase, and at least one type of K^+ channel.

maintaining a low intracellular Na^+ concentration. The Na^+/K^+-ATPase also accumulates K^+ inside the cell. K^+ exits passively moving across the basolateral membrane through one or more types of K^+ channel. The basolateral K^+ conductance and the K^+ gradient hyperpolarize the cell, providing an electrical driving force for Cl^- and Na^+ movement at the apical membrane. The Cl^- which is accumulated in the cell above electrochemical equilibrium exits passively through an apical membrane Cl^- channel, moving down a favorable electrochemical gradient. As discussed below, there are at least two types of apical membrane Cl^- channel: those activated when intracellular levels of cAMP increase thereby stimulating cAMP-dependent protein kinase (PKA) and those activated by an increase in the intracellular Ca^{2+} concentration. The channels activated by PKA-dependent phosphorylation are CFTR Cl^- channels (see below). Regulation of apical membrane Cl^- channels controls, in part, the rate of transepithelial Cl^- secretion. Transepithelial Na^+ transport occurs when Na^+ enters the cell passively through an apical membrane Na^+ channel, moving down a favorable electrochemical gradient. Na^+ then exits across the basolateral membrane as a result of the activity of the Na^+/K^+-ATPase.

Chloride secretion can be controlled at several steps, but one of the most important ones for determining the rate of transepithelial transport is the activity of apical membrane Cl^- channels. A variety of agonists have been observed to increase the two main intracellular second messengers, cAMP and intracellular Ca^{2+}. In addition, some agents stimulate protein kinase C, which also regulates the apical Cl^- channels by phosphorylation. Much less is known about how apical membrane Na^+ channels are regulated.

In addition to regulation of transepithelial electrolyte transport by Cl^- and Na^+ channels in the apical membrane, basolateral membrane transporters are regulated. Indirect evidence suggests regulation of the Na^+–K^+–$2Cl^-$ entry step and it is likely that there are more than one basolateral K^+ channel, including channels regulated by phosphorylation, channels regulated by Ca^{2+}, and possibly other K^+ channels. The coordinated activity of the apical and basolateral ion channels is key to effective transepithelial electrolyte transport without producing large changes in cell volume.

Studies of fluid transport in cultured human airway epithelia suggest that the rate of fluid transport is sufficiently large that it could easily modify the respiratory tract fluid. Under basal conditions active, Na^+ absorption appears to be responsible for fluid absorption.[16–18] When the epithelium is treated with amiloride and agonists that increase cellular levels of cAMP, fluid secretion ensues.

In vitro and *in vivo* studies of transepithelial electrolyte transport indicate that there are two main abnormalities in airway epithelia from patients with CF.[3–5,19–27] First, CF airway epithelia lack cAMP-stimulated Cl^- secretion. Microelectrode studies indicate that the apical membrane is relatively Cl^- impermeable and that the apical Cl^- conductance fails to increase when cAMP levels increase. In contrast, Ca^{2+}-activated Cl^- secretion remains intact in CF airway epithelia.

The second major abnormality in CF airway epithelia is an increase in Na^+ absorption caused by an increased apical membrane Na^+ permeability. As a result, the rate of transepithelial Na^+ absorption under short-circuit conditions is two- to threefold greater in CF than in non-CF epithelia. Nearly all of the increased Na^+ permeability can be attributed to Na^+ transport through amiloride-sensitive apical membrane Na^+ channels. It is thought that the increased Na^+ transport and decreased Cl^- permeability combine to produce respiratory tract fluid that is abnormal in either quantity or composition.

30.3.2. Sweat Gland

The sweat gland is composed of two regions, the secretory coil and the reabsorptive duct. The secretory coil produces nearly isotonic fluid. As the sweat passes up through the reabsorptive duct, which is relatively water impermeable, Na^+ and Cl^- are reabsorbed leaving a hypotonic sweat to emerge at the surface of the skin.

In the secretory coil, active Cl^- transport drives fluid and electrolyte secretion (Figure 30.2).[28–33] It appears that there are two different cells involved in secretion by the secretory

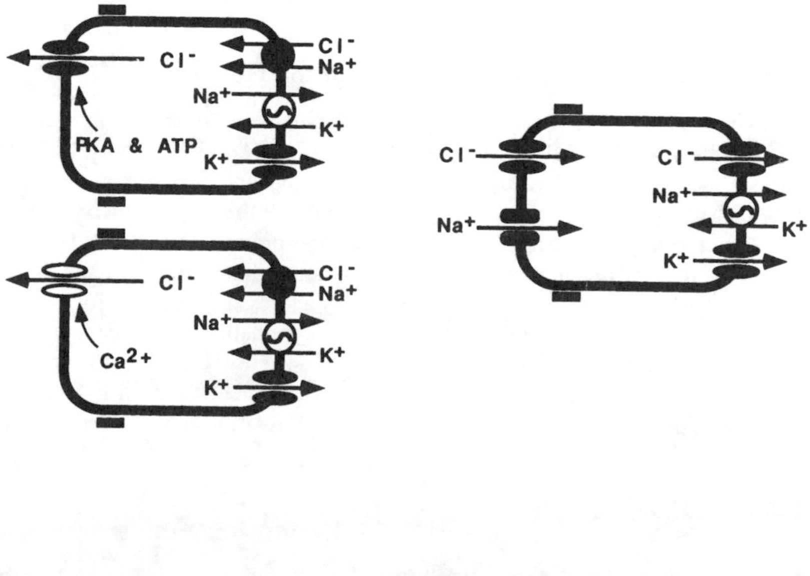

Figure 30.2. Diagram of cellular mechanisms of electrolyte transport by the sweat gland reabsorptive duct (left) and secretory coil (right). Note that in the secretory coil there are two different cell types involved in electrolyte transport. In one cell type, the CFTR Cl^- channel, indicated by the shaded channel in the apical membrane, is regulated by phosphorylation in response to increases in cellular levels of cAMP. In a second cell type, an increase in cell Ca^{2+} concentration opens a different apical membrane Cl^- channel. In the reabsorptive duct, Cl^- transport occurs through CFTR Cl^- channels located in both the apical and basolateral

coil: clear cells and dark cells. One cell type appears to respond to cholinergic agonists which increase Ca^{2+} and stimulate secretion, whereas the other cell type appears to respond to β-adrenergic agonists which increase intracellular levels of cAMP. At present the relationship between the morphologic appearance and the function is not clear. What is clear is that in the CF secretory coil, only cAMP-stimulated secretion is abnormal. The cellular mechanisms of cAMP-stimulated Cl^- secretion and Ca^{2+}-stimulated Cl^- secretion appear to conform to a model similar to that described for the airway epithelium except that the two appear to occur in different cell types.

Absorption of electrolytes by the sweat gland duct appears to be driven by Na^+ absorption with Na^+ entering the cell across the apical membrane, through an amiloride-sensitive Na^+ channel. Na^+ that exits across the basolateral membrane in exchange for K^+ via the Na^+/K^+-ATPase. K^+ that enters the cell is predominantly recycled via basolateral membrane K^+ channels. These two processes establish the ion concentration and voltage gradients that drive Cl^- absorption. Cl^- channels at both cell membranes appear to be controlled by cAMP but under basal conditions appear to be open, possibly because the sweat gland ducts have less specific phosphatase activity specific for CFTR or because they have increased cAMP-stimulated activity under baseline conditions. Thus, Cl^- absorption appears to occur through Cl^- channels located in both the apical membrane and the basolateral membrane of the sweat gland duct epithelia. In CF, conductive Cl^- transport at both apical and basolateral membranes is absent or markedly reduced.[34–39]

There is less information about the abnormal electrolyte transport in CF pancreatic duct epithelium, primarily because of the difficulty in obtaining appropriate tissue. In the pancreatic ducts, apical membrane Cl^- channels are thought to work in parallel with Cl^-–HCO_3^- exchangers to secrete the HCO_3^- rich fluid. Cl^- is thought to enter the lumen through Cl^- channels. It is then recycled back across the apical membrane into the cell in exchange for intracellular HCO_3^-. The Cl^- channel responsible for apical Cl^- transport has properties similar to those of phosphorylation-regulated Cl^- channels in airway epithelia and to the CFTR Cl^- channel.[40–43] The loss of cAMP-stimulated fluid secretion in CF pancreatic ducts is thought to produce fluid that is abnormally dehydrated.[44] As a result the ducts become plugged by thick secretions.

30.3.3. Abnormal Intestinal Electrolyte Transport

The most consistent abnormality observed in CF intestine has been the loss of cAMP-stimulated Cl^- secretion.[27,45–53] The cellular mechanism of intestinal Cl^- secretion appears to be similar to that of airway epithelium, with Cl^- entry at the basolateral membrane via a Na^+–K^+–$2Cl^-$ cotransport process and Cl^- exit across the apical membrane via phosphorylation-regulated Cl^- channels. The most extensive studies addressing this model have come from studies of Cl^--secreting epithelial cell lines derived from intestinal tumors, including T84 cells, HT29 cells, and CaCo2 cells. In these cells, studies of apical Cl^- channels show a small channel with a linear *I–V* relationship and a small single-channel conductance. These properties are similar to those observed in airway epithelial Cl^- channels and similar to the CFTR Cl^- channel.

30.3.4. Apparent Anomalies in Cl⁻ Transport

At first glance, the failure of *absorption* of Cl^- in the sweat gland duct and the failure of *secretion* of Cl^- in intestine and airway epithelium seem contradictory. However, both processes are readily explained by the lack of a Cl^- channel. In airway and intestinal epithelia the Cl^- channel provides a pathway for Cl^- secretion, whereas in the sweat gland duct it provides a pathway for absorption. Because Cl^- movement through the Cl^- channel is driven solely by the electrochemical gradient, the difference in electrochemical gradient provides the driving force for absorption in one epithelium and secretion in the other.

Another tissue-specific difference that was at first confusing was the difference in Ca^{2+}-stimulated Cl^- secretion in intestine and in airway epithelium. Ca^{2+}-stimulated Cl^- secretion was intact in airway epithelium but appeared to be defective in intestinal epithelium. Although this observation initially suggested that there might be tissue-specific regulation of apical Cl^- channels in CF, it now appears that the observation is explained by the lack of apical membrane Ca^{2+}-activated Cl^- channels in at least some intestinal epithelia, rather than their defective regulation. That an increase in cell Ca^{2+} could stimulate Cl^- secretion in intestine appears to be related to the fact that an increase in cell Ca^{2+} can activate basolateral K^+ channels thereby increasing the driving force for Cl^- exit through phosphorylation-activated apical membrane Cl^- channels, some of which are open under baseline conditions. The lack of Ca^{2+}-stimulated Cl^- secretion in CF intestine is related not to the difference in Ca^{2+}-dependent regulation, but rather to the lack of phosphorylation-regulated apical Cl^- channels.[27]

30.4. THE CFTR GENE

In 1989, the gene that encodes CFTR was identified and cloned.[54–56] Four lines of evidence have unequivocally established that the gene encoding CFTR is the one involved in CF. First, the gene was mapped to the correct chromosomal location by positional cloning.[54] Second, an appropriate pattern of tissue expression was observed. That is, the gene was expressed in epithelia involved by the disease.[55] Third, mutations were identified that were present on CF chromosomes, but not normal chromosomes.[56,57] Fourth, expression of normal, or wild-type CFTR, but not mutated CFTR in CF epithelia corrected the cAMP-stimulated Cl^- transport defect.[58–60]

The ability to detect mutations has allowed potentially affected individuals to be screened and has provided the capability of prenatal diagnosis. Most mutations are rare or at most account for a few percent of CF chromosomes. However, the most common mutation, the deletion of phenylalanine at position 508 (ΔF508), accounts for approximately 70% of CF chromosomes.[56,61] The CF gene includes approximately 230 kb of

DNA and contains at least 27 exons. The encoded mRNA is about 6.5 kb long and can be detected by Northern blotting in a variety of affected tissues. The CFTR protein is comprised of 1480 amino acids.[55] There is some evidence suggesting the possibility of multiple initiation sites and possibly some alternative splicing, but the physiologic significance is uncertain.

More than 400 sequence variations have been detected in the CFTR gene in the 4 years since it was identified.[57,62] Of these, about 350 are presumed to be pathologic mutations whereas the others are likely benign sequence variations. All forms of mutations have been identified including amino acid substitutions, missense mutations, nonsense mutations, frame shifts due to insertions and deletions, and mRNA splicing mutations. There are a few single amino acid deletions and one large in-frame deletion. As of this writing, no promoter mutations have been described. The distribution of mutations in different regions of the gene appears to be nonrandom. There are clearly mutation hotspots and different substitutions for a single base pair can be found in many places in the gene. This is particularly true for NBD1 which has a high density of mutations.

30.5. CFTR

30.5.1. Predicted Structure of CFTR

Identification of the primary sequence of CFTR allowed the prediction of potential structure. In large part, initial predictions were based on a comparison of CFTR with a family of proteins named the traffic ATPases[63,64] or ABC transporters.[65] As shown in Figure 30.3, the 1480 amino acids were proposed to be folded into five domains[55]: two membrane-spanning domains (MSDs), each composed of six transmembrane segments; an R domain, which contains several consensus phosphorylation sequences; and two nucleotide-binding domains (NBDs), which were predicted to interact with ATP. Sequence similarity in the NBDs, the prediction of two MSDs, and the overall topology were features that placed CFTR in the traffic ATPase/ABC transporter family. This family includes periplasmic permeases in prokaryotes, such as the histidine and maltose transport systems; STE6, involved in the secretion of a-mating factor in yeast; and P-glycoprotein, which is responsible for pumping a variety of chemotherapeutic agents of cells. The R domain, with its many potential phosphorylation sites and multiple charged amino acids, is a unique feature of CFTR not shared by other members of the traffic ATPase/ABC transporter family.

30.5.2. Localization of CFTR

Not surprisingly, CFTR has primarily been localized to the epithelia involved by the disease. Initial Northern blot analysis of CFTR mRNA suggested that it was expressed in airway, pancreas, and intestinal epithelia.[55] Subsequent *in situ* hybridization to detect CFTR transcripts and the use of antibodies to immunolocalize CFTR provided a more detailed identification of its localization.[66–81] In the lung, CFTR was observed in serous cells in the submucosal glands and in the surface epithelium of the airways. In the pancreas, CFTR is primarily observed in the small pancreatic ducts. In the intestine, it is primarily found in the crypt epithelium, with an especially predominant localization in the Brunner's glands. CFTR has also been observed in the sweat gland duct and the salivary glands, consistent with the observation that CF affects these epithelia. With one exception, wherever CFTR is expressed at sufficiently high levels to be detected by immunocytochemistry, it has been localized to the apical cell membrane of the epithelia. The one exception is the sweat gland duct epithelium where it may be localized at both apical and basolateral membranes.

30.5.3. Topology of CFTR

The topology proposed at the time that the primary sequence of CFTR was discovered has proven to accurately predict the topology of the protein. In Figure 30.3, the sites that have been identified to lie on either the external or the internal surface of the membrane are indicated. Studies with antibodies in permeabilized and unpermeabilized cells have placed the loop between M1 and M2 on the extracellular surface and

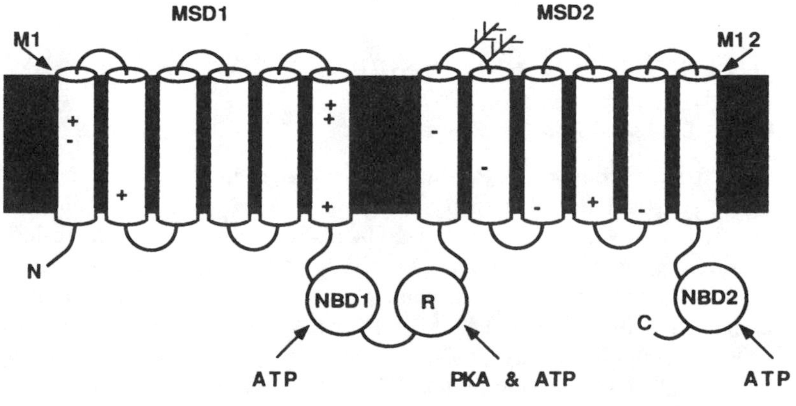

Figure 30.3. Model showing the proposed domain structure of CFTR. MSD refers to the membrane spanning domains, NBD refers to nucleotide-binding domains, and R refers to R domain. The membrane is represented as the shaded area. The location of charged residues within the putative transmembrane sequences are indicated by plus and minus symbols. Glycosylation sites between M6 and M7 are indicated by the branched structures.

the carboxy-terminus on the intracellular surface.[73] As will be described below, functional studies also placed the R domain and the NBDs on the internal surface of the membrane. Because there is no signal sequence, the amino-terminus is probably intracellular. CFTR is a membrane glycoprotein[59] and Asp894 and Asp900 between M7 and M8 are glycosylated; thus, they lie on the extracellular surface.[82] Studies in which the original glycosylation sites at Asp894 and Asp900 were removed and glycosylation sites were inserted at other regions of the proteins confirmed that the extracellular domains shown in Figure 30.3 do, in fact, lie on the extracellular surface.[83] Thus, the data indicate that there are two MSDs, each composed of six membrane-spanning sequences.

Studies of CFTR expressed in heterologous cells indicate that CFTR likely functions as a monomer.[84] Using a variety of different CFTR variants created by site-directed mutagenesis, different solubilization and membrane preparation techniques, and different antibodies, no evidence could be obtained that CFTR was a dimer. Moreover, there is no evidence that any member of the traffic ATPase/ABC transporter family functions as a unit composed of more than two MSDs and two NBDs. Thus, although with negative data it is impossible to rule out function as a multimer, all current data suggest CFTR function as a monomer.

Evidence that CFTR is located within the apical membrane has come from two kinds of studies. In one study an antibody directed against an extracellular epitope of CFTR was able to immunolabel the protein in unpermeabilized cells.[73,75] In contrast, antibodies directed against intracellular epitopes could only label the protein after the cells had been permeabilized. In other studies, CFTR was biochemically labeled from the extracellular surface using a hydrazide reagent to biotinylate CFTR.[85]

CFTR is also located beneath the apical membrane within intracellular vesicles. Labeling of cell-surface CFTR suggested only half of the protein was available for labeling and that the other half was intracellular.[85] Preliminary studies suggest that much of the intracellular CFTR may be located within clathrin-coated vesicles.[86–89] Studies of cells expressing recombinant CFTR suggest that CFTR in endosomes is functionally active and that it recycles back and forth between the subapical region of the cell and the cell surface. The significance of this process is uncertain.

30.6. CHLORIDE CHANNEL FUNCTION OF CFTR

30.6.1. Evidence That CFTR Is a Cl- Channel

After the cloning of CFTR and deduction of its primary sequence, speculation abounded concerning its potential function. The observation that cAMP-regulated Cl^- permeability is lost from the apical membrane of CF epithelia suggested that CFTR might encode a Cl^- channel. However, the primary sequence did not resemble that of any known ion channels. As a result, a number of hypotheses emerged suggesting that CFTR might in some way regulate Cl^- channels. However,

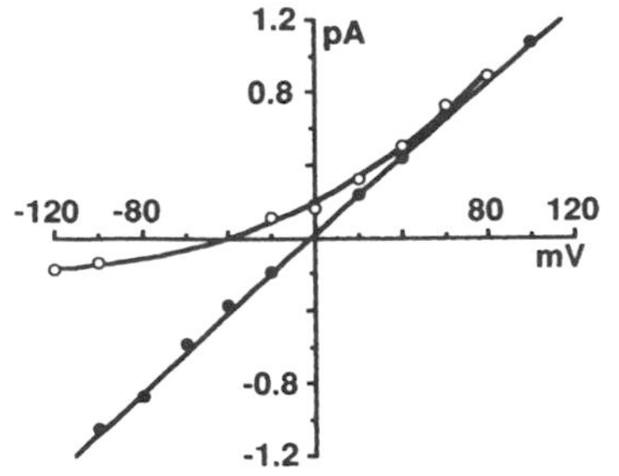

Figure 30.4. Current-voltage relationship for a single CFTR Cl^- channel. Pippette contained 140 mM Cl^-; bath Cl^- concentration was 139 mM (●) or 14 mM (○) (From Berger *et al.*,[93] with permission).

we now know that CFTR is itself a Cl^- channel. Evidence from several types of study support this conclusion.

The first evidence that CFTR was a Cl^- channel came from studies in which CFTR was expressed in cells that did not normally contain cAMP-regulated Cl^- channels and expressed little or no endogenous CFTR. When CFTR has been expressed in a number of mammalian and nonmammalian cells, Cl^- channels have been detected.[90–102] These channels share several common biophysical properties including relatively linear *I–V* relationship with a small unitary conductance of 5–10 pS (Figure 30.4). The channels select Cl^- over cations and are more permeable to Cl^- than to I^- as described below. They also share a unique regulation, requiring both phosphorylation by PKA or PKC, plus the presence of ATP to open. As indicated above, CFTR and the phosphorylation-regulated Cl^- channels that are defective in CF are both located in the apical membrane. Colocalization was a required criterion to support the hypothesis that CFTR is a Cl^- channel.

The regulatory and biophysical properties of Cl^- currents in cells expressing recombinant CFTR and in epithelial cells expressing endogenous CFTR and in the apical membrane of Cl^- secretory epithelia are the same. There is little, if any, time-dependent voltage effect on open state probability.

The first definitive evidence that CFTR is a Cl^- channel came from studies of CFTR variants containing site-directed mutations in the MSDs. CFTR contains a number of charged residues in the MSDs. It was hypothesized that if CFTR is a Cl^- channel, then altering the identity of the amino acids in the part of the channel that forms the pore might change the properties of the channel.[91] Mutation of two of the positively charged amino acids, Lys95 and Lys335, to negatively charged amino acids altered the anion selectivity. With both mutations the permeability sequence was converted from $Br^- > Cl^- > I^-$ to $I > Br^- > Cl^-$. The two mutations were not, however, equivalent. Unlike the changes in permeability, only K335E changed the conductivity sequence. Mutation of two other basic residues, R347E and R1030E, did not change the selectivity sequence. It is interesting that in the topological model of CFTR shown in Figure 30.3, both Lys95 and Lys335 are predicted to lie toward the outer half of the channel whereas Arg347 and Arg1030 lie toward the inner half of the channel. Since the initial study there have been several studies in which residues in the MSDs have been observed to change the biophysical properties of the channels.

Additional studies showed that incorporation of purified CFTR into planar lipid bilayers produced Cl^- channel.[103] When CFTR was purified from SF9 insect cells, reconstituted into proteoliposomes, and fused with bilayers, Cl^- channels were generated that had properties similar to those of Cl^- channels studied in cells expressing recombinant CFTR and cells expressing endogenous CFTR. These data support the conclusion that CFTR is a Cl^- channel and argue against the requirement for loosely associated factors for regulation or function of the channel. Similar results have been obtained when membrane vesicles from cells expressing high-level CFTR have been fused with bilayers.[103,104]

30.6.2. The MSDs Contribute to the Formation of the Cl^- Channel Pore

Knowledge of the topology of CFTR shown in Figure 30.3 implicates the MSDs in formation of the Cl^- channel pore. More direct evidence comes from the study of Cl^- channel variants created by site-directed mutagenesis. As indicated above, mutation in M1(K95D), and mutation in M6(K335E), altered the anion selectivity sequence. Additional evidence that M1 forms part of the channel pore came from a study using sulfhydryl-reactive reagents to bind to and block CFTR.[105] A series of residues in M1 were individually mutated to cysteine which will interact with the sulfhydryl reagents. Each of the residues from 91 to 99 was individually mutated to cysteine. When these mutants were tested with the sulfhydryl reagents, they blocked G91C, K95C, and Q98C. This result is consistent with the observation that K95D alters anion selectivity and suggests that these mutants may line the pore as part of an αhelix where they are accessible to a watery environment. Studies of voltage-dependent block of CFTR by diphenylamine-2-carboxylate (DPC) have also identified residues in M6 and M12 that contribute to the pore.[106,107] The data were also consistent with an α-helical structure for the transmembrane sequences.

R117H, a CF mutation, also changed the conductive properties of the channel.[108] Although this region did not change anion selectivity, it did induce sensitivity to external pH and reduced single-channel conductance from 7.86 to 6.76 pS. Another mutation located toward the putative external surface of the channel pore, R334W, also drastically reduced single-channel conductance to a value at which it would not be accurately measured.

Mutation Arg347, which is predicted to lie toward the intracellular surface of the channel in M6, has also been observed to alter the properties of the channel.[108,109] R347P, which is a CF mutation, has a reduced single-channel conductance of 2.34 pS. In addition, mutation of R347 to Asp or His altered the multiion pore behavior of the channel. Recent studies clearly indicate that M1 and M6 contribute to the formation of the Cl^- channel pore. The data are also consistent with the notion that these residues form αhelices which generate a pore. If CFTR functions as a monomer, as the data suggest, then clearly other portions of the sequence must contribute to the formation of the pore. In voltage-gated K^+, Na^+, and Ca^{2+} channels, a hydrophobic sequence between the fifth and sixth membrane-spanning sequence of a subunit appears to contribute to the pore, perhaps as a β sheet. However, in contrast to these channels, in which hydrophobic residues led to the prediction of the sequence that might contribute to the pore, no similar sequence has been identified in CFTR.

The construction of CFTR Cl^- channels from MSD–NBD motifs is reminiscent of other ion channels that are built from repeated structural motifs such as voltage-gated Na^+, Ca^{2+}, and K^+ channels. Therefore, the hypothesis was tested that one MSD–NBD from CFTR might be sufficient to form a functional Cl^- channel.[110] When an MSD1–NBD1-R domain "half channel" (D836X) was expressed in heterologous cells, Cl^- channels were formed that had conductive properties surprisingly similar to those of wild-type CFTR. However, channel regulation differed. Although phosphorylation increased activity, channels opened without phosphorylation. MgATP was also required for channel activity, although it more potently stimulated activity in D836X than in wild-type CFTR. Data from the regulation of D836X, plus the migration of D836X on sucrose density gradients suggest that D836X may function as a multimer, perhaps forming a dimer. The observation that D836X channels had the same anion-to-cation permeability, anion-selectivity sequence, single-channel conductance, linear *I–V* relationship, and lack of voltage-dependent activation and inactivation suggest that the amino-terminal portion of CFTR contains all of the sequences required to build a Cl^- channel pore.

If sequences in MSD1 are sufficient to produce a Cl^- channel pore, what then is the function of MSD2? There are several possibilities. First, sequences within MSD2 might also contribute to the Cl^- channel pore. Although there is no sequence similarity between MSD1 and MSD2, there might be structural similarity so that MSD1 sequences could substitute for MSD2 sequences to form the pore. In addition, studies with DPC suggest that MSD2 contribute to the pore.[106] Second, MSD2 sequences might stabilize the three-dimensional structure of the channel, perhaps assisting in the arrangement of MSD sequences into a functional pore or in shielding some MSD1 sequences from the lipid bilayer. Although large amounts of D836X protein were produced in the cells, the number of functional Cl^- channels that were generated was very small. Thus, D836X channels were not as efficient as wild-type protein at generating functional channels.

30.7. REGULATION OF CFTR BY THE R DOMAIN

Figure 30.5 shows the amino acid sequence of the R domain of CFTR from a number of species. The R domain was originally defined as the sequence encoded by exon 13. The R domain contains a number of consensus PKA phosphorylation sites. PKA favors the consensus phosphorylation sequence R-R/K-X-S*/T* > R-X-X-S*/T* = R-X-S*/T*, with X being any amino acid and the phosphoacceptor site indicated by an

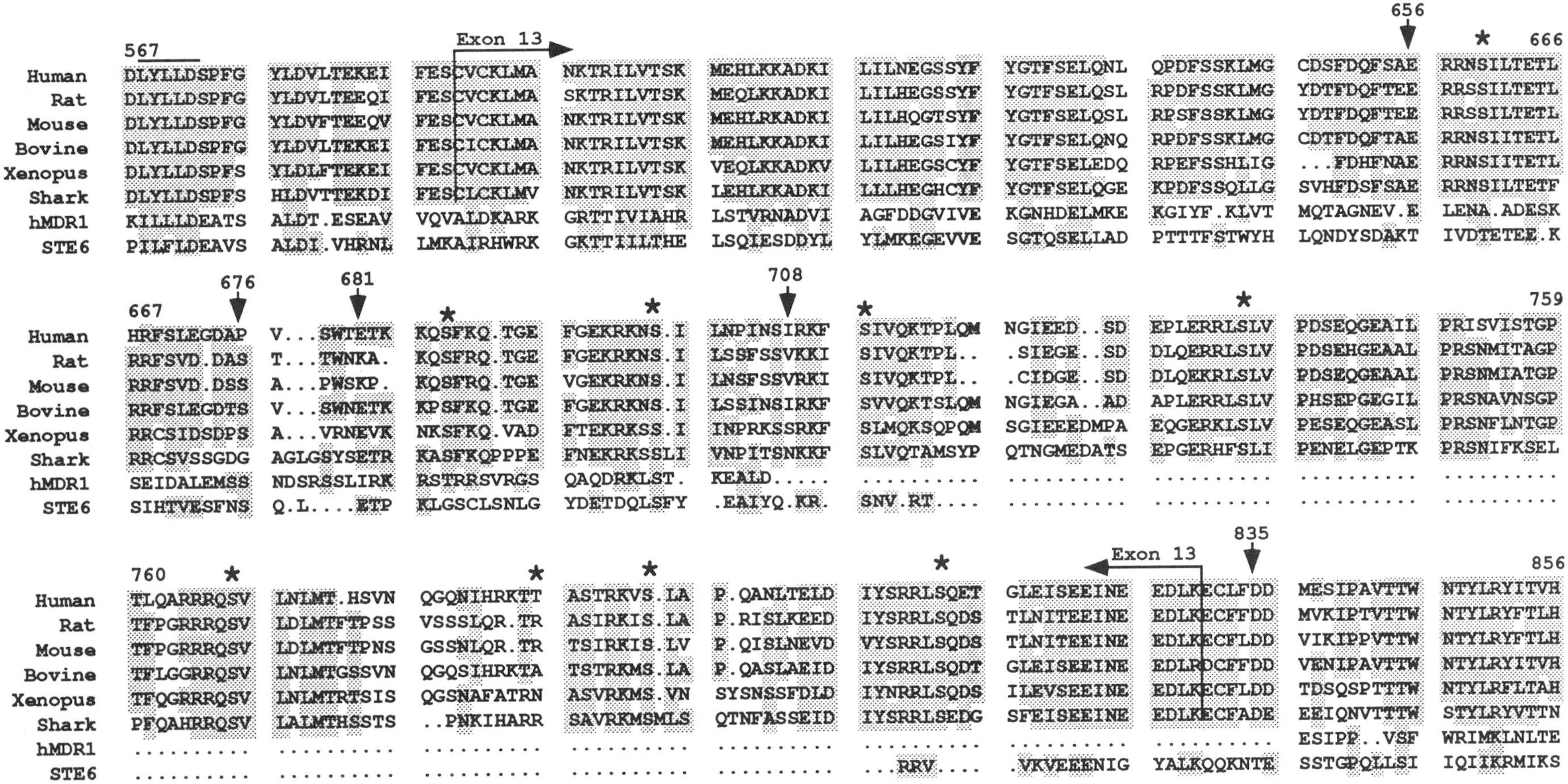

Figure 30.5. Amino acid alignment of Exon 13 and surrounding sequences of human CFTR with five different species of CFTR, human MDR1, and yeast STE6. Amino acid sequences are shown in single-letter code. Identical and conservatively substituted residues are shaded. Dots represent shifts in the amino acid sequence necessary for the "best" alignment. The Walker B motif in NBD1 is indicated by an overline and the nine PKA phosphorylation sites located within the R domain are labeled with asterisks. (Adapted from ref. 118, with permission.)

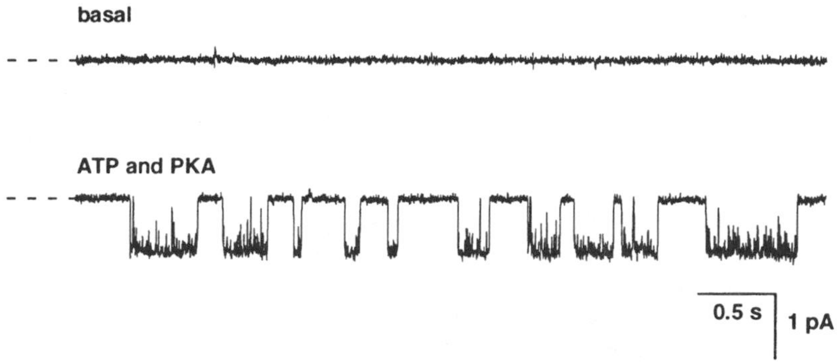

Figure 30.6. Effect of PKA and ATP on CFTR Cl^- channel activity. Data are currents at −90 mV from an excised, inside-out patch. ATP (1 mM) and PKA (75 nM) were present during second trace. Dashed line indicates zero current level. (Courtesy of Mark R. Carson and Michael J. Welsh.)

asterisk. There are ten "classic" R-R/K-X-S*/T* PKA consensus sequences within human CFTR: eight serines plus one threonine within the R domain (these are indicated by asterisks in Figure 30.5) and one serine just prior to NBD1. Although the R domain is the site of the greatest amino acid sequence divergence in CFTR cloned from different species, the consensus phosphorylation sites are for the most part conserved.

These observations about the sequence of the R domain, as well as the observation that cAMP-stimulated Cl^- secretion is lost in CF epithelia, suggested that the CFTR Cl^- channel may be regulated by PKA-dependent phosphorylation. That suggestion was supported by patch-clamp studies of recombinant CFTR. In cell-attached patches of membrane, an increase in cellular levels of cAMP reversibly activated CFTR Cl^- channels. In cell-free inside-out patches of membrane, addition of the catalytic subunit of PKA (in the presence of ATP) activated single channels.[93,98] An example is shown in Figure 30.6. CFTR could also be activated by phosphorylation by protein kinase C, although the absolute level of activity was less than that observed with PKA. There may also be interactions between phosphorylation by PKA and PKC, because it has been reported that phosphorylation with PKC increases the effect of subsequent phosphorylation by PKA. It appears that at least two different isoforms of PKC are able to phosphorylate and activate the channel; both a Ca^{2+}-dependent and a Ca^{2+}-independent form have been implicated.[111]

cAMP-dependent stimulation of Cl^- secretion in intact cells is reversible; once the agonist is removed, the rate of secretion returns to basal values and subsequent readdition of agonists can once again stimulate Cl^- secretion. This observation implies that phosphorylation is reversible and suggests that specific phosphatases dephosphorylate CFTR. When CFTR Cl^- channels are excised from cells expressing endogenous or recombinant CFTR, the rate at which the effect of phosphorylation is reversed (run down) is variable, depending on the cell type.[112] This result suggests that phosphatases which regulate CFTR may in some cases attach to the membrane patch or in some cases may diffuse away. Studies of CFTR in excised patches of membrane show that PP2A, a ubiquitous enzyme found in the cytoplasm of many types of cells, is capable of dephosphorylating CFTR.[111] Addition of PP2A to CFTR that has been phosphorylated *in vitro* by PKA caused a dephosphorylation that was blocked by the phosphatase inhibitors okadaic acid and calyculin-A. When PP2A was added to the cytosolic surface of excised membrane patches containing CFTR Cl^- channels, it also inhibited current to near basal values. After PP2A was removed and PKA readded, the channels reopened.

In contrast to the effect of PP2A, neither PP1 nor PP2B dephosphorylated CFTR nor did they close CFTR Cl^- channels that had been activated by PKA and ATP. These results suggested that the effect of PP2A was specific. Alkaline phosphatase may also dephosphorylate and inactivate CFTR, although reports have been conflicting and additional studies are required to investigate the effect and specificity of alkaline phosphatases.[98,111] Although PP2A and alkaline phosphatase can dephosphorylate and inactivate CFTR, it is likely that other phosphatases will also regulate phosphorylation. This suggestion is supported by the observation that okadaic acid does not completely prevent dephosphorylation and inactivation of CFTR.

30.7.1. Identification of Phosphorylated Sites in CFTR

The study of CFTR that has been phosphorylated by an increase in cellular levels of cAMP *in vivo* and the phosphorylation of CFTR by PKA *in vitro* has begun to identify specific sequences that are phosphorylated.[113–116] Two-dimensional tryptic phosphopeptide mapping of wild-type CFTR and variants in which each of the potential phosphoserines was mutated individually to alanine has identified a number of serines that are phosphorylated. *In vitro* studies show that several sites (S660, S700, S712, S737, S768, S795, and S813) are phosphorylated. Of the three putative phosphorylation sites that were not phosphorylated, S422 lies outside the R domain and S686 and T788 are not conserved in other species (Figure 30.5). In addition, phosphoamino acid analysis indicates that only serines are phosphorylated; there is no evidence of phosphotyrosine or phosphothreonine.

When similar studies were done under *in vivo* conditions in which labeling was stimulated by addition of cAMP agonists, four major phosphopeptides were identified, corresponding to residues S660, S737, S795, and S813. Additional studies using CFTR from the T84 intestinal epithelial cell line or using a peptide containing part of the R domain, found that

S700 was also phosphorylated *in vivo* after stimulation by cAMP agonists.

30.7.2. Functional Consequences of Mutating Phosphorylation Sites in CFTR

Study of CFTR variants that contain phosphoserines mutated to alanine has begun to provide some insight into how phosphorylation regulates CFTR.[113–115] When the four *in vivo* phosphorylation sites (S660, S737, S795, and S813) were all simultaneously mutated to alanine (S-Quad-A), there was a large decrease in open state probability. This result suggested that phosphorylation of these sites controls the activity of CFTR. However, the fact that the S-Quad-A mutant remained responsive to cAMP seemed surprising because the mutation removed four major *in vivo* phosphorylation sites. Subsequent studies showed that an S-Oct-A mutant and S-Dec-A mutant in which eight or all ten of the potential phosphorylation sites were mutated to alanine continued to exhibit PKA-stimulated channel activity. This result suggested that there may be residual phosphorylation sites in CFTR.

Thus, it appears that there may be residual phosphorylation that can regulate the channel, but that the level of phosphorylation at any one site is too low to allow detection and identification of the corresponding phosphopeptide. The possibility that cAMP is having some effect other than phosphorylation seems unlikely because the channels were activated only by addition of PKA and ATP to excised membrane patches and because the mutants containing multiple serine-to-alanine mutations were still phosphorylated *in vivo*. An interesting question is whether nonclassic sites are involved in the physiologic regulation of CFTR. At present it is not possible to answer this question with certainty, but the preponderance of phosphorylation on S660, S737, S795, and S813 and the large decrease in open state probability observed when these serines are mutated to alanine suggest that they may well be the major *in vivo* phosphorylation sites.

One mechanism by which phosphorylation may open the CFTR Cl^- channel is by addition of negative charge to the R domain. An assessment of this hypothesis came from studies in which phosphoserines were replaced by negatively charged residues.[114,115] When four or five of the serines were mutated to aspartate, channel activity was only observed after addition of cAMP agonists. However, when six, seven, or eight serines were mutated to aspartate, channels were found to be open, even without addition of cAMP. Similar studies were performed in which the serines were mutated to glutamic acid.

These results suggest that negative charge in the R domain can open the CFTR Cl^- channel. The data presented above suggest that phosphorylation of the R domain opens the CFTR Cl^- channel. To test the hypothesis that the R domain has a negative effect that keeps the channel closed, the effect of deleting the R domain was tested. As shown in Figure 30.5 the R domain was originally defined as the residues encoded by exon 13. However, comparison of the amino acid sequence of human CFTR of several different species and with that of two other members of the traffic ATPase/ABC transporter family, MDR and STE6 (Figure 30.5), suggested that CFTR contains a segment of over 120 amino acids that is not found in any other member of the family. Therefore, the effect of deletion of that part of the R domain was tested.[117,118] When residues 708–813 were deleted, CFTR formed channels that were open even in the absence of phosphorylation by PKA. However, subsequent addition of cAMP *in vivo* or application of PKA to CFTR in excised membrane patches produced a further increase in activity. Subsequent studies showed that more extensive deletion of the R domain failed to produce Cl^- channels (Figure 30.5). The increase in activity observed on addition of PKA to CFTR lacking residues 708–835 resulted from phosphorylation of at least one additional site within the protein: S660. When the S660A mutation was made in CFTR lacking much of the R domain, the PKA-dependent increase in P_o was abolished. That data suggest that S660 is involved in regulation of the R domain function, whereas S700 may not be involved in all cell types.

One interpretation of how the R domain regulates CFTR is shown in Figure 30.7. This interpretation suggests that the R domain has a negative influence, keeping the channel closed. In CFTR, inhibition of channel function by the R domain can be relieved in three ways: by phosphorylation, by addition of negatively charged amino acids, or by deleting a

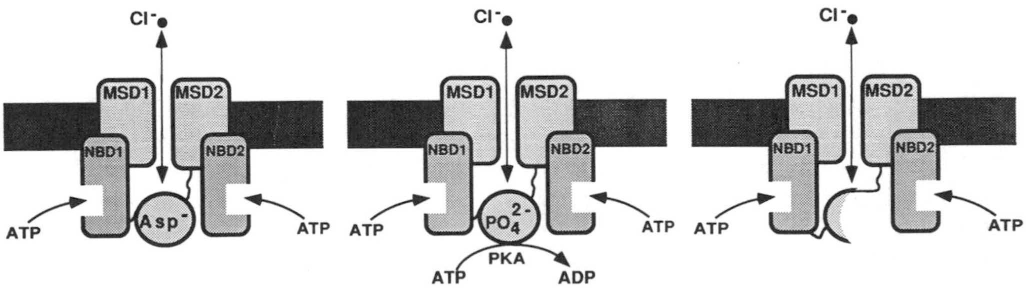

Figure 30.7. Model of regulation of CFTR Cl^- channel by the R domain. The channel can be opened by three different interventions: by phosphorylation of the R domain, by substitution of phosphoserines by aspartates, or by deletion of part of the R domain. In all three cases ATP is required for channel activity. (Adapted from ref. 1, with permission.)

large part of the R domain. The ability of phosphorylation and negatively charged residues to open the channel suggests that electrostatic interactions within the protein may be important for channel activation. Perhaps charge–charge interactions between the R domain and another part of the protein alter the position of the R domain, thereby opening the channel. Alternatively, charge-induced conformational changes in the R domain may open the channel. In either case, it is unclear whether the important factor for channel activation is the number of charges or charge density.

However, the interpretation given above is probably overly simplistic based on the observation that deletion of the R domain produced Cl^- channels that had a lower open-state probability than wild-type channels studied after phosphorylation. If the R domain only had a negative effect, one would expect that its removal might produce a Cl^- channel that had a P_o similar to or even greater than that of phosphorylated wild-type CFTR. Thus, in addition to negative effects, the presence of the R domain most likely has positive effects on channel function.

Studies of deletions within the R domain begin to suggest "functional" boundaries for the R domain. The R domain was originally defined as those amino acids encoded by exon 13. Such exon–intron boundaries provide a convenient way to define the R domain but that does not necessarily reflect the functional domain. In fact, as shown in Figure 30.5 most of the variation in sequence between species of CFTR occurs in the second half of exon 13. Moreover, there is sequence similarity between CFTR, MDR, and STE6 within the initial portion of the R domain; such sequence similarities may reflect an extension of the structural and functional region of the NBDs.

One interpretation of the effect of deletion mutations and the sequence analysis shown in Figure 30.5 is that the unique segment of 127 amino acids from the second half of exon 13 defines, at least in part, a "functional" domain within CFTR. However, other parts of exon 13 may also be involved as evidenced by the fact that phosphorylation of S660 increases activity in a channel lacking much of the R domain. Thus, the actual "functional" R domain may consist of amino acids 708–835 plus additional parts of the protein defined by its overall three-dimensional structure.

It is also interesting that relatively few missense mutations associated with CF mapped to the R domain, whereas other parts of the molecule with important functional roles appear to have many more CF-associated mutations. Perhaps the observation that mutation of any one or indeed all of the major *in vivo* phosphorylation sites does not completely abolish channel activity means that the R domain of CFTR is somewhat plastic within the structure of CFTR and can accommodate structural alteration without measurable effect on function. This would support a model arguing that regulation of the R domain is brought about by charge–charge interactions and that these interactions are degenerate. In other words, charge may be important in regulation, but its precise location, at least in the primary sequence, is less critical.

Finally, one might ask why regulation of CFTR by the R domain is so complex. There are several possibilities. It might be that multiple phosphorylation sites provide a mechanism by which graded phosphorylation could provide graded activity and, hence, graded regulation of transepithelial Cl^- secretion. It is also possible that different sites within CFTR provide substrates for different protein kinases and phosphatases thereby providing an additional mechanism for regulation. Finally, there may be interactions between the phosphorylation sites and a pattern of hierarchical phosphorylation that is important in control of the channel.

30.8. REGULATION OF CFTR BY THE NBDs

When the sequence of CFTR was first identified, the presence of the NBDs seemed a puzzling feature. If CFTR were a Cl^- channel, why should the channel contain NBDs which in related family members hydrolyze ATP? The presence of the NBDs initially suggested that CFTR might be a protein involved in active transport, rather than forming a channel in which ions move passively down their electrochemical gradients. However, several studies have now shown that the interaction of ATP with the NBDs regulates the activity of CFTR channels.

Because the NBDs were predicted to interact with ATP, Anderson *et al.*[92] tested the hypothesis that ATP would regulate channel activity. They found that in channels that had been activated by PKA, ATP was required on the cytosolic surface for activity. When ATP was removed, the channel closed. Readdition of ATP reactivated the channel (Figure 30.6). These results indicated that ATP regulates the channel, but only when it has first been phosphorylated by PKA. A variety of additional studies showed that ATP-dependent regulation was not the result of reversible phosphorylation and dephosphorylation. It appears that hydrolysis is required for CFTR Cl^- channel activity. This conclusion is supported by the fact that nonhydrolyzable nucleotides and Mg-free ATP are unable to substitute for ATP in activating the channel. The nucleoside triphosphate potency is, however, rather broad (at 1 mM relative to ATP): ATP (1.00) > AMP-CPP (0.75) > GTP (0.65) > ITP (0.49) $\geq$ UTP (0.42) > CTP (0.25). This broad specificity contrasts with the high specificity for ATP observed for a number of kinases and the Na^+/K^+-ATPase. ATP increased Cl^- channel activity in a dose-dependent manner with an EC_{50} of approximately 250 μM. The broad nucleoside triphosphate specificity and the high EC_{50} are similar to results that have been obtained with other members of the traffic ATPase/ABC transporter family.

Several studies have shown that as the concentration of intracellular ATP increases, the probability of the channel being open (P_o) increases (Figure 30.8). Studies of single, phosphorylated CFTR Cl^- channels in excised, inside-out membrane patches have examined the kinetics of gating at different ATP concentrations.[119] As the ATP concentration increased, the mean closed time decreased but mean open time did not

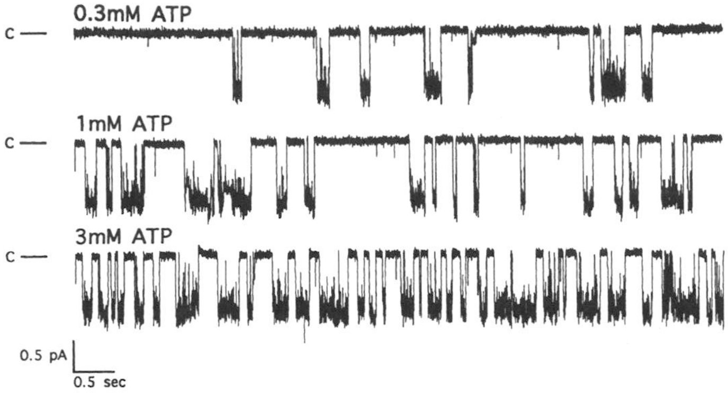

Figure 30.8. Single-channel currents recorded from an excised, inside-out patch. The bath (cytosolic) solution contained the indicated concentrations of ATP. Before obtaining the first trace, the channel had been phosphorylated with PKA and ATP. Downward reflections represent channel openings. The holding potential was −120 mV. (From ref. 119, with permission.)

change. Analysis of the data using the maximum likelihood method and the log-likelihood ratio test suggested that channel behavior could be described by a model containing one open and two closed states:

$$C_1 \underset{\alpha_1}{\overset{\beta_1}{\rightleftharpoons}} C_2 \underset{\alpha_2}{\overset{\beta_2}{\rightleftharpoons}} O \qquad (30.1)$$

This equation represents a minimal model of the channel gating that is sufficient to explain the data. However, it is likely that with future studies, more complex models will be required. In addition, in this model the transition from one state to a neighboring state may represent several steps that cannot be discriminated at the present time. Using this model the effect of ATP could be attributed to an increase in the β_1 transition. As the concentration of ATP increased, β_1 increased. In contrast, transitions from C2 to C1 and between C2 and the open state were not significantly altered by ATP. This model gives the appearance of bursts of activity as is shown in Figures 30.6 and 30.8. As the ATP concentration increased, the interval between bursts of opening is shortened with no apparent change in the duration of bursts or in the closed times within bursts.

ADP inhibited ATP-supported channel activity through what appeared to be a competitive interaction. When studied at the single-channel level, ADP decreased β_1 without altering α_1, α_2, or β_2.[119]

These observations taken in the context of a molecule that contains NBDs suggested that ATP interacts with CFTR through the NBDs. Studies of ATP binding showed that a synthetic peptide containing 67 amino acids from NBD1 and a larger peptide containing a substantial amount of NBD1 sequence bound nucleotide analogues.[120–122] Intact CFTR studied in the membranes of F9 insect cells was also specifically labeled by a photoactivable ATP analogue (γ-^{32}P-β-azidoadenosine 5′-triphosphate, 8-N_3 ATP).[123] Both ATP and GTP prevented photolabeling with half-maximal inhibition at approximately 1 mM. ADP and ANP-PNP prevented photolabeling, but at much higher concentrations, whereas ANP did not inhibit, even at concentrations up to 100 mM. These results support the conclusion that the NBDs of CFTR interact directly with nucleotides at concentrations similar to those that regulate channel activity.

An interesting feature of the CFTR Cl^- channel is that it contains two NBDs. Studies of CFTR variants show that mutations designed to disrupt the function of one or the other of the NBDs decrease the potency of ATP and reduce net activity measured as P_o.[124] Studies of the effect of increasing concentrations of ATP generated data that suggest cooperative kinetics. Thus, it was suggested that two NBDs may interact. The first evidence that the two NBDs are not functionally equivalent came from the study of the effect of ADP. The inhibitory effect of ATP was abolished or reduced by mutation of several residues in NBD2. In contrast, some mutations in NBD1 did not alter the inhibitory effect of ADP. Those data suggested that ADP inhibits CFTR by competing with ATP and that competition occurs at NBD2. Thus, the divergence of amino acid sequence between the two NBDs is paralleled by divergence in their function.

Recent studies have suggested that the two NBDs may have additional distinct functions in controlling channel activity. Insight into distinct functions came initially from the study of the hydrolysis-resistant ATP analogue 5′-adenosine (β,γ-imino) triphosphate (ANP-PNP).[92,125–128] ANP-PNP was not able to open Cl^- channels on its own. However, in the presence of ATP, ANP-PNP increased P_o. The increase in P_o occurred as a result of an increase in the duration of bursts of activity. That is, once CFTR opened into a burst of activity [the transitions between C_2 and O in Eq. (30.1)], it exited from the burst more slowly. The effect of ANP-PNP was only observed in the presence of PKA, suggesting that only channels in a high state of phosphorylation were susceptible to the effect.

Additional studies showed that VO_4, BeF_3, and PP_i produced similar effects, preventing delayed exit from the open bursting state.[129,130] The observations were interpreted to suggest that ATP hydrolysis may be required not only to open the channel but also to actively close the channel.

Assignment of distinct functions to specific NBDs came from studies in which conserved motifs in NBD1 and NBD2 were mutated. In CFTR the two NBDs share sequence similarity and certain conserved regions such as the Walker A

(GXXGXGKT/S) motif (where X is any amino acid). The single-channel activity was measured in CFTR variants which contained mutations in the absolutely conserved Walker A motif lysine in either NBD1 (K464A), NBD2 (K1250A and K1250N), or both NBDs simultaneously (K464A/K1250A).[128] Studies in related proteins suggest that such mutations slow the rate of ATP hydrolysis. These mutations did not alter the conductive properties of the channel or the requirement for phosphorylation and the ATP to open the channel. However, all of the mutations decreased P_o. Mutations in NBD1 decreased the frequency of bursts of activity, whereas mutations in NBD2 and mutations in both NBDs simultaneously produced prolonged bursts of activity, as well as decreased the frequency of bursts. These results were not attributed to altered binding of nucleotide because none of the mutants studied had reduced 8-N_3ATP binding. These and additional data were interpreted to suggest that the two NBDs had distinct functions in channel gating: ATP hydrolysis at NBD1 initiates a burst of activity, and hydrolysis at NBD2 terminates a burst.

30.9. MECHANISM BY WHICH CF-ASSOCIATED MUTATIONS CAUSE CHANNEL DYSFUNCTION

Knowledge that CFTR is a Cl^- channel helps to explain cAMP-regulated Cl^- secretion in the epithelia that express CFTR and the loss of Cl^- permeability in CF. At the present, there appear to be at least four mechanisms by which mutations can disrupt CFTR function. These are diagrammed in Figure 30.9.

30.9.1. Class I Mutations: Defective Protein Production

Mutations throughout the CFTR gene are predicted to produce premature termination because of splice site abnormalities, frameshifts via insertions or deletions, and nonsense mutations. In some cases such as R553X, the mutation results in unstable mRNA and no detectable protein.[131] In other cases, a truncated protein or an abnormal protein containing deleted or novel amino acid sequences may be produced. However, such proteins are often unstable and usually would be expected to be degraded relatively rapidly or have little function. Thus, all mutations in this class are expected to produce little or no full-length protein and thereby a loss of CFTR Cl^- channel function in affected epithelia.

30.9.2. Class II Mutations: Defective Protein Processing

Several CF-associated mutations fail to traffic to the correct cellular location. This includes the most common mutation, deletion of phenylalanine at residue 508 (ΔF508).[82,132] In the case of CFTR, the failure of protein to progress through the biosynthetic pathway can be followed by assessing its state of glycosylation. In recombinant cells, CFTR containing the ΔF508 mutation fails to mature to the fully glycosylated form. This is shown in Figure 30.10. Band A represents the nascent protein that is not glycosylated. Band B represents protein with a pattern of "core" glycosylation consistent with processing in the endoplasmic reticulum. Band C represents the mature protein that has undergone a pattern of glycosylation consistent with processing in the Golgi complex. Figure 30.10 shows that the ΔF508 protein does not have a pattern consistent with traffic through the Golgi complex. ΔF508 protein undergoes core glycosylation in the endoplasmic reticulum and then is degraded. As a result, the ΔF508 protein cannot be detected at the cell surface.[69,70,75] Studies in CF airway epithelial cells, submucosal glands, and sweat gland duct cells have shown that the ΔF508 protein is either missing or present in much reduced amounts at the apical membrane.

The reason why mutant forms of CFTR fail to be transported correctly has not been established. Experience with other proteins suggests that the mutation prevents CFTR from adopting its correct conformation, that is, the nascent protein folds incorrectly and is recognized as abnormal by cellular "quality control" mechanisms which mark the protein for degradation rather than movement to the plasma membrane.[82,121] Studies of wild-type and ΔF508 protein in recombinant cells suggest that both forms of the protein interact with HSP70 and with calnexin.[133–135] The interaction with these chaperones may be prolonged for the ΔF508 protein, but it is not clear

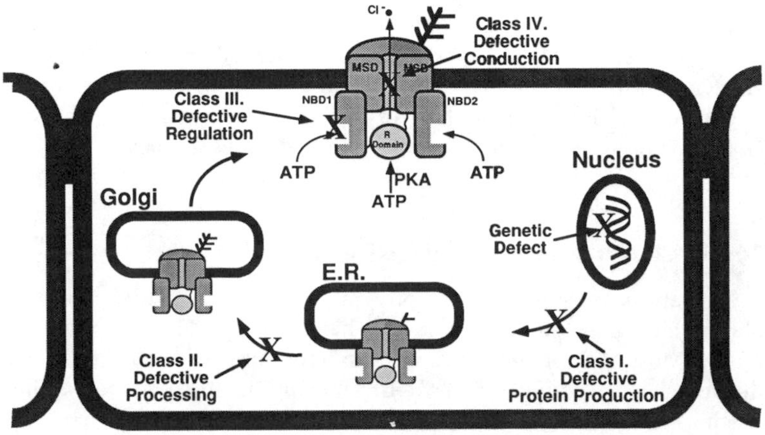

Figure 30.9. Diagram of the biosynthesis and function of CFTR in an epithelial cell and of mechanisms of dysfunction associated with CF mutations. (From ref. 146, with permission.)

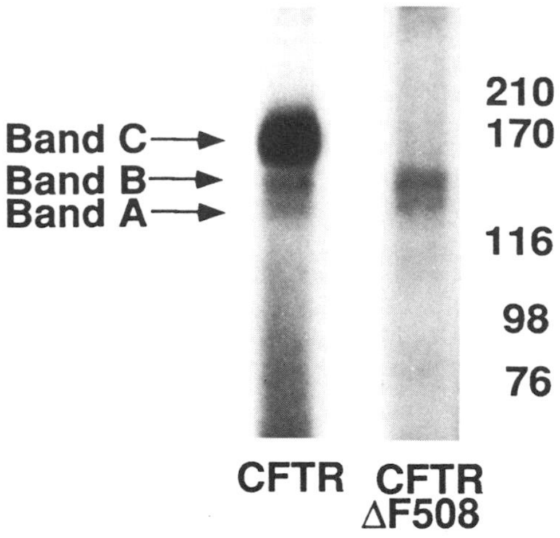

Figure 30.10. Immunoprecipitation of CFTR and CFTRΔF508. Immunoprecipitates were phosphorylated with PKA and [^{32}P]ATP and separated by gel electrophoresis. The unglycosylated protein was labeled band A. The core glycosylated form is band B, and the fully glycosylated form is band C. Recombinant CFTR or CFTRΔF508 was expressed in 3T3 fibroblasts. (Courtesy of Dr. Lynda S. Ostedgaard and Michael J. Welsh.)

whether these or other chaperones are critical to recognition and retention of the ΔF508 mutant in the endoplasmic reticulum and then its targeting to the degradation pathway.

Although CFTRΔF508 is mislocalized in cells grown in 37°C, when the incubation temperature is reduced to 23 to 30°C some of the mutant protein escapes from the endoplasmic reticulum, is fully glycosylated in the Golgi complex, and is delivered to the cell membrane.[136] Presumably the folding process is able to occur, at least partially, at the reduced temperature. Once delivered to the cell membrane, CFTRΔF508 retains some function: the Cl^- conductive properties and regulation by phosphorylation appear to be intact. Most reports suggest that the absolute activity of the ΔF508 mutant is reduced to approximately one-third that of wild-type CFTR.[94,95,136] However, some studies suggest that activity may be normal.[137]

Recent reports suggest that the processing defect may not be an all-or-none defect. Some mutants, including the P574H and A455E mutants, also show misprocessing, but a small amount of the protein does progress through the Golgi complex and does produce channels that have normal or even greater than normal activity.[138] Additional class II mutations might conceivably be identified or processed through the Golgi complex yet fail to be transported to the apical membrane.

30.9.3. Class III Mutations: Defective Regulation

Analysis of some mutant proteins that do reach the plasma membrane shows that several have mutations in the NBDs. Most of these mutants appear to have normal regulation of channel activity. Some NBD mutants in this class (such as G551D) have very little function. In some (such as S1255P) ATP is less potent at stimulating activity. In others (such as G551S, G1244E, G1349D) the absolute activity is reduced.[94,124,139] Defective regulation of these mutants results in decreased Cl^- channel activity and is likely to be responsible for the defective epithelial Cl^- permeability in patients bearing these mutations.

30.9.4. Class IV Mutations: Defective Conduction

The MSDs are the site of a number of CF-associated missense mutations. Mutations in MSD1, including R117H, R334W, R347P, and R347H, affect arginine residues located in putative membrane-spanning sequences. When these mutant CFTRs are expressed in heterologous epithelial cells, all are correctly processed, are present in the apical membrane, and generate cAMP-regulated apical membrane Cl^- currents. Moreover, regulation by cAMP-dependent phosphorylation and by intracellular ATP appear to be similar to that of wild-type CFTR. Nevertheless, the amount of current is reduced in each. The explanation for why these proteins generate less Cl^- current has come from an analysis of single channels in excised patches of membrane.[108,109] In each mutant, the rate of ion flow through a single open channel is reduced. In addition, at least for R117H, the amount of time that the channel is opened is reduced to about one-third that of wild-type CFTR. Mutations in R347 also appear to alter the multiion pore behavior of the channel so that a decreased number of ions in the pore results in a lower rate of ion flow through the channel.

30.9.5. Relationship between Genotype and Clinical Phenotype

The classification of CFTR mutations into four classes based on biochemical dysfunction is beginning to provide some insight into clinical phenotype. However, a precise understanding of the relationship is difficult for several reasons. First, the clinical phenotype can vary substantially and at least some of the variability is likely related to environmental factors. Second, variability in clinical phenotype may result from differences in the genetic background and the influence of other genes. Third, many patients are compound heterozygotes and it is difficult to predict clinical abnormalities resulting from two different CF-associated mutations. Fourth, at least in some cases, CF mutations may occur on different backgrounds. This has been most clearly shown for the R117H mutation which appears to have occurred twice in history, occurring on the background of CFTR which has different splicing efficiencies.[140] In one case the R117H occurs in a gene in which splicing is relatively efficient (90% full-length transcripts) and in another case the mutation occurs in a gene in which splicing is relatively inefficient (roughly 10%) leading to deletion of an exon 9 and thereby loss of function. Fifth, it is possible that some patients, particularly those with apparently anomalous clinical phenotypes, may bear more than one mutation within the CFTR gene, only one of which has been identified.

Interestingly, such an example may be a CF patient reported to have both the ΔF508 and the R553Q mutation on the same chromosome.[141] Although this patient had both pancreatic insufficiency and lung disease, the sweat Cl^- concentration was close to the normal range. In an independent study identifying second-site revertants of the ΔF508 mutation using STE6–CFTR chimeras in yeast, R553M and, to a lesser extent, R553Q partially reverted the localization and functional effects of the ΔF508 mutation.[142] Thus, one can speculate that the R553Q mutation partially suppressed the ΔF508 Cl^- transport defect in sweat glands in that patient. Identification of other second-site suppressors may provide further insight into CFTR structure and function.

At present the clinical variable that most readily discriminates different phenotypes is the state of pancreatic function.[1,143] Most patients have pancreatic failure, requiring ingestion and supplemental pancreatic enzymes; such patients are referred to as pancreatic insufficient. However, some patients retain significant pancreatic function and require little or no pancreatic enzyme supplementation; they are pancreatic sufficient. Class I and II mutant proteins are missing from the correct cellular location which would be expected to have a severe effect on mutation. Indeed, to date, all Class I and II mutations have been associated with a severe pancreatic insufficient phenotype with the exception of P574H and A455E, in which there is residual processing of the mutant protein to the mature form. In contrast, all Class III and IV mutants are correctly processed. Some retain significant residual function, while others have little measurable activity. In Class IV mutants in which channel regulation appears to be intact but ion flow through an open Cl^- channel is reduced, residual activity may be sufficient to confer a pancreatic sufficient phenotype. Some individuals with Class III mutations (e.g., 551S) are also pancreatic sufficient[144] whereas most others (e.g., G551D) are pancreatic insufficient.[145] Presumably this difference relates to the degree of abnormal regulation, but its molecular basis is not yet understood. Since CF is a recessive disease, patients with a pancreatic insufficient mutation on one chromosome and a pancreatic sufficient mutation on the other have a pancreatic sufficient phenotype. That is, the dominant mutation provides sufficient CFTR Cl^- channel function to confer pancreatic sufficiency.

30.10. SUMMARY

Understanding of the biology of CF and CFTR has increased dramatically during the last 5 years. Yet in many respects our knowledge remains superficial. That is certainly the case for our understanding of the structure and function of the protein. A particularly important problem will be to determine the structure of the different domains and how they associate to govern the opening and closing of the channel. It will also be important to learn how CFTR influences the function of other ion channels. Much also remains to be discovered about how CF mutations cause protein dysfunction. One problem that is in particular need of investigation is to tie together knowledge of the clinical manifestations of the disease with knowledge of CFTR function and dysfunction and with knowledge of the physiology of epithelia. That is, we need to know how the cellular and molecular defects cause the disease. This area of research is difficult because it demands that investigators apply a cellular, molecular, and physiologic approach to understanding the function of an entire organ and organism.

Finally, an important goal of research in CF is to develop more rational approaches to therapy. Although the survival of patients has increased dramatically in the last 30 years, current treatments are not directed at the basic underlying defect. As a result, CF remains a life-threatening and usually lethal disease. Recent attempts to develop therapies based on the physiologic abnormalities and on the molecular defects give hope that in the future CF may not have to be a lethal disease.

REFERENCES

1. Welsh, M. J., Tsui, L.-C., Boat, T. F., and Beaudet, A. L. (1995). Cystic fibrosis. In *The Metabolic and Molecular Basis of Inherited Disease,* 7th ed. (C. R. Scriver, A. L. Beaudet, W. S. Sly, and D. Valle, eds.), McGraw-Hill, New York pp. 3799–3876.
2. Knowles, M., Murray, G., Shallal, J., Askin, F., Ranga, V., Gatzy, J., and Boucher, R. (1984). Bioelectric properties and ion flow across excised human bronchi. *J. Appl. Physiol.* **56:**868–877.
3. Knowles, M. R., Stutts, M. J., Spock, A., Fischer, N., Gatzy, J. T., and Boucher, R. C. (1983). Abnormal ion permeation through cystic fibrosis respiratory epithelium. *Science* **221:**1067–1070.
4. Widdicombe, J. H., Welsh, M. J., and Finkbeiner, W. E. (1985). Cystic fibrosis decreases the apical membrane chloride permeability of monolayers cultured from cells of tracheal epithelium. *Proc. Natl. Acad. Sci. USA* **82:**6167–6171.
5. Knowles, M. R., Buntin, W. H., Bromberg, P. A., Gatzy, J. T., and Boucher, R. C. (1982). Measurements of transepithelial electric potential differences in the trachea and bronchi of human subjects *in vivo. Am. Rev. Respir. Dis.* **126:**108–112.
6. Knowles, M. R., Carson, J. L., Collier, A. M., Gatzy, J. T., and Boucher, R. C. (1981). Measurements of nasal transepithelial electric potential differences in normal human subjects *in vivo. Am. Rev. Respir. Dis.* **124:**484–490.
7. Widdicombe, J. H., Coleman, D. L., Finkbeiner, W. E., and Tuet, I. K. (1985). Electrical properties of monolayers cultured from cells of human tracheal mucosa. *J. Appl. Physiol.* **58:**1729–1735.
8. Welsh, M. J. (1987). Electrolyte transport by airway epithelia. *Physiol. Rev.* **67:**1143–1184.
9. Breeze, R. G., and Wheeldon, R. B. (1977). The cells of the pulmonary airways. *Am. Rev. Respir. Dis.* **116:**705–777.
10. Plopper, C. G., Hill, L. L., and Mariassy, A. T. (1980). Ultrastructure of the nonciliated bronchiolar epithelial (Clara) cell of mammalian lung III. A study of man with comparison of 15 mammalian species. *Exp. Lung Res.* **1:**171–180.
11. Welsh, M. J., Smith, P. L., and Frizzell, R. A. (1983). Chloride secretion by canine tracheal epithelium. III. Membrane resistances and electromotive forces. *J. Membr. Biol.* **71:**209–218.
12. Basbaum, C., and Welsh, M. J. (1994). Defense mechanisms and immunology: Mucous secretion and ion transport in airways. In *Textbook of Respiratory Medicine,* 2nd ed. (J. Nadel, ed.), Saunders, Philadelphia, pp. 323–344.
13. Boucher, R. C., Knowles, M. R., Stutts, M. J., and Gatzy, J. T. (1983). Epithelial dysfunction in cystic fibrosis lung disease. *Lung* **161:** 1–17.
14. Frizzell, R. A., Halm, D. R., Rechkemmer, G., and Shoemaker, R. L. (1986). Chloride channel regulation in secretory epithelia. *Fed. Proc.* **45:**2727–2731.

15. Liedtke, C. M. (1992). Electrolyte transport in the epithelium of pulmonary segments of normal and cystic fibrosis lung. *FASEB J.* **6:** 3076–3084.
16. Smith, J. J., and Welsh, M. J. (1993). Fluid and electrolyte transport by cultured human airway epithelia. *J. Clin. Invest.* **91:**1590–1597.
17. Welsh, M. J., Widdicombe, J. H., and Nadel, J. A. (1980). Fluid transport across canine tracheal epithelium. *J. Appl. Physiol. Respir. Environ. Exercise Physiol.* **49:**905–909.
18. Jiang, C., Finkbeiner, W. E., Widdicombe, J. H., McCray, P. B. Jr., and Miller, S. S. (1993). Altered fluid transport across airway epithelium in cystic fibrosis. *Science* **262:**424–427.
19. Knowles, M., Gatzy, J., and Boucher, R. (1981). Increased bioelectric potential difference across respiratory epithelia in cystic fibrosis. *N. Engl. J. Med.* **305:**1489–1495.
20. Gowen, C. W., Lawson, E. E., Gingras-Leatherman, J., Gatzy, J. T., Boucher, R. C., and Knowles, M. R. (1986). Increased nasal potential difference and amiloride sensitivity in neonates with cystic fibrosis. *J. Pediatr.* **108:**517–521.
21. Knowles, M., Gatzy, J., and Boucher, R. (1983). Relative ion permeability of normal and cystic fibrosis nasal epithelium. *J. Clin. Invest.* **71:**1410–1417.
22. Alton, E. W. F. W., Currie, D., Logan-Sinclair, R., Warner, J. O., Hodson, M. E., and Geddes, D. M. (1990). Nasal potential difference: A clinical diagnostic test for cystic fibrosis. *Eur. Respir. J.* **3:** 922–926.
23. Boucher, R. C., Cotton, C. U., Gatzy, J. T., Knowles, M. R., and Yankaskas, J. R. (1988). Evidence for reduced Cl^- and increased Na^+ permeability in cystic fibrosis human primary cell cultures. *J. Physiol. (London)* **405:**77–103.
24. Widdicombe, J. H. (1986). Cystic fibrosis and beta-adrenergic response of airway epithelial cell cultures. *Am. J. Physiol.* **251:** R818–R822.
25. Cotton, C. U., Stutts, M. J., Knowles, M. R., Gatzy, J. T., and Boucher, R. C. (1987). Abnormal apical cell membrane in cystic fibrosis respiratory epithelium. An in vitro electrophysiologic analysis. *J. Clin. Invest.* **79:**80–85.
26. Boucher, R. C., Cheng, E. H., Paradiso, A. M., Stutts, M. J., Knowles, M. R., and Earp, H. S. (1989). Chloride secretory response of cystic fibrosis human airway epithelia. Preservation of calcium but not protein kinase C- and A-dependent mechanisms. *J. Clin. Invest.* **84:**1424–1431.
27. Anderson, M. P., and Welsh, M. J. (1991). Calcium and cAMP activate different chloride channels in the apical membrane of normal and cystic fibrosis epithelia. *Proc. Natl. Acad. Sci. USA* **88:**6003–6007.
28. Sato, K., and Sato, F. (1984). Defective beta adrenergic response of cystic fibrosis sweat glands in vivo and in vitro. *J. Clin. Invest.* **73:** 1763–1771.
29. Sato, K., and Sato, F. (1988). Relationship between quin2-determined cytosolic $[Ca^{2+}]$and sweat secretion. *Am. J. Physiol.* **254:** C310–C317.
30. Sato, K., Saga, K., and Sato, F. (1988). Membrane transport and intracellular events in control and cystic fibrosis eccrine sweat glands. In *Cellular and Molecular Basis of Cystic Fibrosis* (G. Mastella and P. M. Quinton, eds.), San Francisco Press, San Francisco, pp. 171–185.
31. Takemura, T., Sato, F., Saga, K., Suzuki, Y., and Sato, K. (1991). Intracellular ion concentrations and cell volume during cholinergic stimulation of eccrine secretory coil cells. *J. Membr. Biol.* **119:**211–219.
32. Reddy, M. M., Bell, C. L., and Quinton, P. M. (1992). Evidence of two distinct epithelial cell types in primary cultures from human sweat gland secretory coil. *Am. J. Physiol.* **262:**C891–C898.
33. Reddy, M. M., and Quinton, P. M. (1992). Electrophysiologically distinct cell types in human sweat gland secretory coil. *Am. J. Physiol.* **262:**C287–C292.
34. Quinton, P. M., and Bijman, J. (1983). Higher bioelectric potentials due to decreased chloride absorption in the sweat glands of patients with cystic fibrosis. *N. Engl. J. Med.* **308:**1185–1189.
35. Quinton, P. M. (1983). Chloride impermeability in cystic fibrosis. *Nature* **301:**421–422.
36. Quinton, P. M. (1990). Cystic fibrosis: A disease in electrolyte transport. *FASEB J.* **4:**2709–2717.
37. Bijman, J., and Quinton, P. (1987). Permeability properties of cell membranes and tight junctions of normal and cystic fibrosis sweat ducts. *Pfluegers Arch.* **408:**505–510.
38. Reddy, M. M., and Quinton, P. M. (1989). Localization of Cl^- conductance in normal and Cl^- impermeability in cystic fibrosis sweat duct epithelium. *Am. J. Physiol.* **257:**C727–C735.
39. Reddy, M. M., and Quinton, P. M. (1992). cAMP activation of CF-affected Cl^- conductance in both cell membranes of an absorptive epithelium. *J. Membr. Biol.* **130:**49–62.
40. Gray, M. A., Harris, A., Coleman, L., Greenwell, J. R., and Argent, B. E. (1989). Two types of chloride channel on duct cells cultured from human fetal pancreas. *Am. J. Physiol.* **257:**C240–C251.
41. Gray, M. A., Greenwell, J. R., and Argent, B. E. (1988). Secretin-regulated chloride channel on the apical plasma membrane of pancreatic duct cells. *J. Membr. Biol.* **105:**131–142.
42. Gray, M. A., Pollard, C. E., Harris, A., Coleman, L., Greenwell, J. R., and Argent, B. E. (1990). Anion selectivity and block of the small-conductance chloride channel on pancreatic duct cells. *Am. J. Physiol.* **259:**C752–C761.
43. Gray, M. A., Plant, S., and Argent, B. E. (1993). cAMP-regulated whole cell chloride currents in pancreatic duct cells. *Am. J. Physiol.* **264:**C591–C602.
44. Kopelman, H., Durie, P., Gaskin, K., Weizman, Z., and Forstner, G. (1985). Pancreatic fluid secretion and protein hyperconcentration in cystic fibrosis. *N. Engl. J. Med.* **312:**329.
45. Taylor, C. J., Baxter, P. S., Hardcastle, J., and Hardcastle, P. T. (1988). Failure to induce secretion in jejunal biopsies from children with cystic fibrosis. *Gut* **29:**957–962.
46. Berschneider, H. M., Knowles, M. R., Azizkhan, R. G., Boucher, R. C., Tobey, N. A., Orlando, R. C., and Powell, D. W. (1988). Altered intestinal chloride transport in cystic fibrosis. *FASEB J.* **2:**2625–2629.
47. Veeze, H. J., Sinaasappel, M., Bijman, J., Bouquet, J., and De-Jonge, H. R. (1991). Ion transport abnormalities in rectal suction biopsies from children with cystic fibrosis. *Gastroenterology* **101:** 398–403.
48. Goldstein, J. L., Shapiro, A. B., Rao, M. C., and Layden, T. J. (1991). In vivo evidence of altered chloride but not potassium secretion in cystic fibrosis rectal mucosa. *Gastroenterology* **101:**1012–1019.
49. De-Jonge, H. R., Van den Berghe, N., Tilly, B. C., Kansen, M., and Bijman, J. (1989). (Dys)regulation of epithelial chloride channels. *Biochem. Soc. Trans.* **17:**816–818.
50. De-Jonge, H. R., Bijman, J., and Sinaasappel, M. (1987). Relation of regulatory enzyme levels to chloride transport in intestinal epithelial cells. *Pediatr. Pulmonol. Suppl.* **1:**54–57.
51. Tabcharani, J. A., Low, W., Elie, D., and Hanrahan, J. W. (1990). Low-conductance chloride channel activated by cAMP in the epithelial cell line T84. *FEBS Lett.* **270:**157–164.
52. Bear, C. E., and Reyes, E. F. (1992). cAMP-activated chloride conductance in the colonic cell line, Caco-2. *Am. J. Physiol.* **262:** C251–C256.
53. Frizzell, R. A., and Halm, D. R. (1990). Chloride channels in epithelial cells. In *Current Topics in Membranes and Transport,* 37th ed., Academic Press, New York, pp. 247–282.
54. Rommens, J. M., Iannuzzi, M. C., Kerem, B.-S., Drumm, M. L., Melmer, G., Dean, M., Rozmahel, R., Cole, J. L., Kennedy, D., Hidaka, N., Zsiga, M., Buchwald, M., Riordan, J. R., Tsui, L.-C., and Collins, F. S. (1989). Identification of the cystic fibrosis gene: Chromosome walking and jumping. *Science* **245:**1059–1065.

55. Riordan, J. R., Rommens, J. M., Kerem, B.-S., Alon, N., Rozmahel, R., Grzelczak, Z., Zielenski, J., Lok, S., Plavsic, N., Chou, J.-L., Drumm, M. L., Iannuzzi, M. C., Collins, F. S., and Tsui, L.-C. (1989). Identification of the cystic fibrosis gene: Cloning and characterization of complementary DNA. *Science* **245:**1066–1073.
56. Kerem, B.-S., Rommens, J. M., Buchanan, J. A., Markiewicz, D., Cox, T. K., Chakravarti, A., Buchwald, M., and Tsui, L.-C. (1989). Identification of the cystic fibrosis gene: Genetic analysis. *Science* **245:**1073–1080.
57. Tsui, L.-C. (1992). Mutations and sequence variations detected in the cystic fibrosis transmembrane conductance regulator (CFTR) gene: A report from the Cystic Fibrosis Genetic Analysis Consortium. *Hum. Mutat.* **1:**197–203.
58. Rich, D. P., Anderson, M. P., Gregory, R. J., Cheng, S. H., Paul, S., Jefferson, D. M., McCann, J. D., Klinger, K. W., Smith, A. E., and Welsh, M. J. (1990). Expression of cystic fibrosis transmembrane conductance regulator corrects defective chloride channel regulation in cystic fibrosis airway epithelial cells. *Nature* **347:**358–363.
59. Gregory, R. J., Cheng, S. H., Rich, D. P., Marshall, J., Paul, S., Hehir, K., Ostedgaard, L., Klinger, K. W., Welsh, M. J., and Smith, A. E. (1990). Expression and characterization of the cystic fibrosis transmembrane conductance regulator. *Nature* **347:**382–386.
60. Drumm, M. L., Pope, H. A., Cliff, W. H., Rommens, J. M., Marvin, S. A., Tsui, L.-C., Collins, F. S., Frizzell, R. A., and Wilson, J. M. (1990). Correction of the cystic fibrosis defect in vitro by retrovirus-mediated gene transfer. *Cell* **62:**1227–1233.
61. Kerem, E., Corey, M., Kerem, B. S., Rommens, J., Markiewicz, D., Levison, H., Tsui, L.-C., and Durie, P. (1990). The relation between genotype and phenotype in cystic fibrosis—analysis of the most common mutation (delta F508). *N. Engl. J. Med.* **323:**1517–1522.
62. Tsui, L.-C. (1992). The spectrum of cystic fibrosis mutations. *Trends Genet.* **8:**392–398.
63. Ames, G. F.-L., Mimura, C. S., and Shyamala, V. (1990). Bacterial periplasmic permeases belong to a family of transport proteins operating from *Escherichia coli* to human: Traffic ATPases. *FEMS Microbiol. Rev.* **75:**429–446.
64. Mimura, C. S., Holbrook, S. R., and Ames, G. F. (1991). Structural model of the nucleotide-binding conserved component of periplasmic permeases. *Proc. Natl. Acad. Sci. USA* **88:**84–88.
65. Hyde, S. C., Emsley, P., Hartshorn, M. J., Mimmack, M. M., Gileadi, U., Pearce, S. R., Gallagher, M. P., Gill, D. R., Hubbard, R. E., and Higgins, C. F. (1990). Structural model of ATP-binding proteins associated with cystic fibrosis, multidrug resistance and bacterial transport. *Nature* **346:**362–365.
66. Trezise, A. E., and Buchwald, M. (1991). *In vivo* cell-specific expression of the cystic fibrosis transmembrane conductance regulator. *Nature* **353:**434–437.
67. Trezise, A. E. O., Romano, P. R., Gill, D. R., Hyde, S. C., Sepúlveda, F. V., Buchwald, M., and Higgins, C. F. (1992). The multidrug resistance and cystic fibrosis genes have complementary patterns of epithelial expression. *EMBO J* **11:**4291–4303.
68. Trezise, A. E., Linder, C. C., Grieger, D., Thompson, E. W., Meunier, H., Griswold, M. D., and Buchwald, M. (1993). CFTR expression is regulated during both the cycle of the seminiferous epithelium and the oestrous cycle of rodents. *Nature Genet.* **3:**157–164.
69. Engelhardt, J. F., Yankaskas, J. R., Ernst, S. A., Yang, Y., Marino, C. R., Boucher, R. C., Cohn, J. A., and Wilson, J. M. (1992). Submucosal glands are the predominant site of CFTR expression in the human bronchus. *Nature Genet.* **2:**240–248.
70. Kartner, N., Augustinas, O., Jensen, T. J., Naismith, A. L., and Riordan, J. R. (1992). Mislocalization of delta F508 CFTR in cystic fibrosis sweat gland. *Nature Genet.* **1:**321–327.
71. Cohn, J. A., Melhus, O., Page, L. J., Dittrich, K. L., and Vigna, S. R. (1991). CFTR: Development of high-affinity antibodies and localization in sweat gland. *Biochem. Biophys. Res. Commun.* **181:**36–43.
72. Marino, C. R., Matovcik, L. M., Gorelick, F. S., and Cohn, J. A. (1991). Localization of the cystic fibrosis transmembrane conductance regulator in pancreas. *J. Clin. Invest.* **88:**712–716.
73. Denning, G. M., Ostedgaard, L. S., Cheng, S. H., Smith, A. E., and Welsh, M. J. (1992). Localization of cystic fibrosis transmembrane conductance regulator in chloride secretory epithelia. *J. Clin. Invest.* **89:**339–349.
74. Crawford, I., Maloney, P. C., Zeitlin, P. L., Guggino, W. B., Hyde, S. C., Turley, H., Gatter, K. C., Harris, A., and Higgins, C. F. (1991). Immunocytochemical localization of the cystic fibrosis gene product CFTR. *Proc. Natl. Acad. Sci. USA* **88:**9262–9266.
75. Denning, G. M., Ostedgaard, L. S., and Welsh, M. J. (1992). Abnormal localization of cystic fibrosis transmembrane conductance regulator in primary cultures of cystic fibrosis airway epithelia. *J. Cell Biol.* **118:**551–559.
76. Dalemans, W., Hinnrasky, J., Slos, P., Dreyer, D., Fuchey, C., Pavirani, A., and Puchelle, E. (1992). Immunocytochemical analysis reveals differences between the subcellular localization of normal and DPhe508 recombinant cystic fibrosis transmembrane conductance regulator. *Exp. Cell Res.* **201:**235–240.
77. Puchelle, E., Gaillard, D., Ploton, D., Hinnrasky, J., Fuchey, C., Boutterin, M.-C., Jacquot, J., Dreyer, D., Pavirani, A., and Dalemans, W. (1992). Differential localization of the cystic fibrosis transmembrane conductance regulator in normal and cystic fibrosis airway epithelium. *Am. J. Respir. Cell Mol. Biol.* **7:**485–491.
78. Morris, A. P., Cunningham, S. A., and Frizzell, R. A. (1993). CFTR targeting in epithelial cells. *J. Bioenerg. Biomembr.* **25:**21–26.
79. Cohn, J. A., Nairn, A. C., Marino, C. R., Melhus, O., and Kole, J. (1992). Characterization of the cystic fibrosis transmembrane conductance regulator in a colonocyte cell line. *Proc. Natl. Acad. Sci. USA* **89:**2340–2344.
80. Hoogeveen, A. T., Keulemans, J., Willemsen, R., Scholte, B. J., Bijman, J., Edixhoven, M. J., De-Jonge, H. R., and Galjaard, H. (1991). Immunological localization of cystic fibrosis candidate gene products. *Exp. Cell Res.* **193:**435–437.
81. McCray, P. B., Jr., Bettencourt, J. D., Bastacky, J., Denning, G. M., and Welsh, M. J. (1993). Expression of CFTR and a cAMP-stimulated chloride secretory current in cultured human fetal alveolar epithelial cells. *Am. J. Respir. Cell Mol. Biol.* **9:**578–585.
82. Cheng, S. H., Gregory, R. J., Marshall, J., Paul, S., Souza, D. W., White, G. A., O'Riordan, C. R., and Smith, A. E. (1990). Defective intracellular transport and processing of CFTR is the molecular basis of most cystic fibrosis. *Cell* **63:**827–834.
83. Chang, X.-B., Hou, Y.-X., Jensen, T. J., and Riordan, J. R. (1994). Mapping of cystic fibrosis transmembrane conductance regulator membrane topology by glycosylation site insertion. *J. Biol. Chem.* **269:**18572–18575.
84. Marshall, J., Fang, S., Ostedgaard, L. S., O'Riordan, C. R., Ferrara, D., Amara, J. F., Hoppe, H., Scheule, R. K., Welsh, M. J., Smith, A. E., and Cheng, S. H. (1994). Stoichiometry of recombinant cystic fibrosis transmembrane conductance regulator in epithelial cells and its functional reconstitution into cells in vitro. *J. Biol. Chem.* **269:**2987–2995.
85. Prince, L. S., Tousson, A., and Marchase, R. B. (1993). Cell surface labeling of CFTR in T84 cells. *Am. J. Physiol.* **264:**C491–C498.
86. Bradbury, N. A., Jilling, T., Berta, G., Sorscher, E. J., Bridges, R. J., and Kirk, K. L. (1992). Regulation of plasma membrane recycling by CFTR. *Science* **256:**530–532.
87. Bradbury, N. A., Jilling, T., Kirk, K. L., and Bridges, R. J. (1992). Regulated endocytosis in a chloride secretory epithelial cell line. *Am. J. Physiol.* **262:**C752–C759.
88. Bradbury, N. A., Cohn, J. A., Venglarik, C. J., and Bridges, R. J. (1994). Biochemical and biophysical identification of cystic fibrosis transmembrane conductance regulator chloride channels as components of endocytic clathrin-coated vesicles. *J. Biol. Chem.* **269:**8296–8302.

89. Lukacs, G. L., Chang, X.-B., Kartner, N., Rotstein, O. D., Riordan, J. R., and Grinstein, S. (1992). The cystic fibrosis transmembrane regulator is present and functional in endosomes. Role as a determinant of endosomal pH. *J. Biol. Chem.* **267:**14568–14572.
90. Anderson, M. P., Rich, D. P., Gregory, R. J., Smith, A. E., and Welsh, M. J. (1991). Generation of cAMP-activated chloride currents by expression of CFTR. *Science* **251:**679–682.
91. Anderson, M. P., Gregory, R. J., Thompson, S., Souza, D. W., Paul, S., Mulligan, R. C., Smith, A. E., and Welsh, M. J. (1991). Demonstration that CFTR is a chloride channel by alteration of its anion selectivity. *Science* **253:**202–205.
92. Anderson, M. P., Berger, H. A., Rich, D. P., Gregory, R. J., Smith, A. E., and Welsh, M. J. (1991). Nucleoside triphosphates are required to open the CFTR chloride channel. *Cell* **67:**775–784.
93. Berger, H. A., Anderson, M. P., Gregory, R. J., Thompson, S., Howard, P. W., Maurer, R. A., Mulligan, R., Smith, A. E., and Welsh, M. J. (1991). Identification and regulation of the cystic fibrosis transmembrane conductance regulator-generated chloride channel. *J. Clin. Invest.* **88:**1422–1431.
94. Drumm, M. L., Wilkinson, D. J., Smit, L. S., Worrell, R. T., Strong, T. V., Frizzell, R. A., Dawson, D. C., and Collins, F. S. (1991). Chloride conductance expressed by ΔF508 and other mutant CFTRs in Xenopus oocytes. *Science* **254:**1797–1799.
95. Dalemans, W., Barbry, P., Champigny, G., Jallat, S., Dott, K., Dreyer, D., Crystal, R. G., Pavirani, A., Lecocq, J. P., and Lazdunski, M. (1991). Altered chloride ion channel kinetics associated with the ΔF508 cystic fibrosis mutation. *Nature* **354:**526–528.
96. Bijman, J., Dalemans, W., Kansen, M., Keulemans, J., Verbeek, E., Hoogeveen, A., De-Jonge, H., Wilke, M., Dreyer, D., Lecocq, J. P., Pavirani, A., and Scholte, B. (1993). Low-conductance chloride channels in IEC-6 and CF nasal cells expressing CFTR. *Am. J. Physiol.* **264:**L229–L235.
97. Bear, C. E., Duguay, F., Naismith, A. L., Kartner, N., Hanrahan, J. W., and Riordan, J. R. (1991). Cl^- channel activity in *Xenopus* oocytes expressing the cystic fibrosis gene. *J. Biol. Chem.* **266:**19142–19145.
98. Tabcharani, J. A., Chang, X.-B., Riordan, J. R., and Hanrahan, J. W. (1991). Phosphorylation-regulated Cl^- channel in CHO cells stably expressing the cystic fibrosis gene. *Nature* **352:**628–631.
99. Kartner, N., Hanrahan, J. W., Jensen, T. J., Naismith, A. L., Sun, S., Ackerley, C. A., Reyes, E. F., Tsui, L.-C., Rommens, J. M., Bear, C. E., and Riordan, J. R. (1991). Expression of the cystic fibrosis gene in non-epithelial invertebrate cells produces a regulated anion conductance. *Cell* **64:**681–691.
100. Rommens, J. M., Dho, S., Bear, C. E., Kartner, N., Kennedy, D., Riordan, J. R., Tsui, L.-C., and Foskett, J. K. (1991). cAMP-inducible chloride conductance in mouse fibroblast lines stably expressing the human cystic fibrosis transmembrane conductance regulator. *Proc. Natl. Acad. Sci. USA* **88:**7500–7504.
101. Cliff, W. H., Schoumacher, R. A., and Frizzell, R. A. (1992). cAMP-activated Cl channels in CFTR-transfected cystic fibrosis pancreatic epithelial cells. *Am. J. Physiol.* **262:**C1154–C1160.
102. Cunningham, S. A., Worrell, R. T., Benos, D. J., and Frizzell, R. A. (1992). cAMP-stimulated ion currents in Xenopus oocytes expressing CFTR cRNA. *Am. J. Physiol.* **262:**C783–C788.
103. Bear, C. E., Li, C., Kartner, N., Bridges, R. J., Jensen, T. J., Ramjeesingh, M., and Riordan, J. R. (1992). Purification and functional reconstitution of the cystic fibrosis transmembrane conductance regulator (CFTR). *Cell* **68:**809–818.
104. Tilly, B. C., Winter, M. C., Ostedgaard, L. S., O'Riordan, C., Smith, A. E., and Welsh, M. J. (1992). Cyclic AMP-dependent protein kinase activation of cystic fibrosis transmembrane conductance regulator chloride channels in planar lipid bilayers. *J. Biol. Chem.* **267:**9470–9473.
105. Akabas, M. H., Kaufmann, C., Cook, T. A., and Archdeacon, P. (1994). Amino acid residues lining the chloride channel of the cystic fibrosis transmembrane conductance regulator. *J. Biol. Chem.* **269:**14865–14868.
106. McDonough, S., Davidson, N., Lester, H. A., and McCarty, N. A. (1994). Novel pore-lining residues in CFTR that govern permeation and open-channel block. *Neuron* **13:**623–634.
107. McCarty, N. A., McDonough, S., Cohen, B. N., Riordan, J. R., Davidson, N., and Lester, H. A. (1993). Voltage-dependent block of the cystic fibrosis transmembrane conductance regulator Cl^- channel by two closely related arylaminobenzoates. *J. Gen. Physiol.* **102:**1–23.
108. Sheppard, D. N., Rich, D. P., Ostegaard, L. S., Gregory, R. J., Smith, A. E., and Welsh, M. J. (1993). Mutations in CFTR associated with mild disease from Cl^- channels with altered pore properties. *Nature* **362:**160–164.
109. Tabcharani, J. A., Rommens, J. M., Hou, Y. X., Chang, X.-B., Tsui, L.-C., Riordan, J. R., and Hanrahan, J. W. (1993). Multi-ion pore behaviour in the CFTR chloride channel. *Nature* **366:**79–82.
110. Sheppard, D. N., Ostedgaard, L. S., Rich, D. P., and Welsh, M. J. (1994). The amino-terminal portion of CFTR forms a regulated Cl^- channel. *Cell* **76:**1091–1098.
111. Berger, H. A., Travis, S. M., and Welsh, M. J. (1993). Regulation of the cystic fibrosis transmembrane conductance regulator Cl^- channel by specific protein kinases and protein phosphatases. *J. Biol. Chem.* **268:**2037–2047.
112. Haws, C., Finkbeiner, W. E., Widdicombe, J. H., and Wine, J. J. (1994). CFTR in Calu-3 human airway cells: Channel properties and role in cAMP-activated Cl^- conductance. *Am. J. Physiol.* **266:**L502–L512.
113. Cheng, S. H., Rich, D. P., Marshall, J., Gregory, R. J., Welsh, M. J., and Smith, A. E. (1991). Phosphorylation of the R domain by cAMP-dependent protein kinase regulates the CFTR chloride channel. *Cell* **66:**1027–1036.
114. Rich, D. P., Berger, H. A., Cheng, S. H., Travis, S. M., Saxena, M., Smith, A. E., and Welsh, M. J. (1993). Regulation of the cystic fibrosis transmembrane conductance regulator Cl^- channel by negative charge in the R domain. *J. Biol. Chem.* **268:**20259–20267.
115. Chang, X. B., Tabcharani, J. A., Hou, Y. X., Jensen, T. J., Kartner, N., Alon, N., Hanrahan, J. W., and Riordan, J. R. (1993). Protein kinase A (PKA) still activates CFTR chloride channel after mutagenesis of all 10 PKA consensus phosphorylation sites. *J. Biol. Chem.* **268:**11304–11311.
116. Picciotto, M. R., Cohn, J. A., Bertuzzi, G., Greengard, P., and Nairn, A. C. (1992). Phosphorylation of the cystic fibrosis transmembrane conductance regulator. *J. Biol. Chem.* **267:**12742–12752.
117. Rich, D. P., Gregory, R. J., Anderson, M. P., Manavalan, P., Smith, A. E., and Welsh, M. J. (1991). Effect of deleting the R domain on CFTR-generated chloride channels. *Science* **253:**205–207.
118. Rich, D. P., Gregory, R. J., Cheng, S. H., Smith, A. E., and Welsh, M. J. (1993). Effect of deletion mutations on the function of CFTR chloride channels. *Recept. Chan.* **1:**221–232.
119. Winter, M. C., Sheppard, D. N., Carson, M. R., and Welsh, M. J. (1994). Effect of ATP concentration on CFTR Cl^- channels: A kinetic analysis of channel regulation. *Biophys. J.* **66:**1398–1403.
120. Thomas, P. J., Shenbagamurthi, P., Ysern, X., and Pedersen, P. L. (1991). Cystic fibrosis transmembrane conductance regulator: Nucleotide binding to a synthetic peptide. *Science* **251:**555–557.
121. Thomas, P. J., Shenbagamurthi, P., Sondek, J., Hullihen, J. M., and Pedersen, P. L. (1992). The cystic fibrosis transmembrane conductance regulator. Effects of the most common cystic fibrosis-causing mutation on the secondary structure and stability of a synthetic peptide. *J. Biol. Chem.* **267:**5727–5730.
122. Hartman, J., Huang, Z., Rado, T. A., Peng, S., Jilling, T., Muccio, D. D., and Sorscher, E. J. (1992). Recombinant synthesis, purification, and nucleotide binding characteristics of the first nucleotide binding domain of the cystic fibrosis gene product. *J. Biol. Chem.* **267:**6455–6458.

123. Travis, S. M., Carson, M. R., Ries, D. R., and Welsh, M. J. (1993). Interaction of nucleotides with membrane-associated cystic fibrosis transmembrane conductance regulator. *J. Biol. Chem.* **268:**15336–15339.
124. Anderson, M. P., and Welsh, M. J. (1992). Regulation by ATP and ADP of CFTR chloride channels that contain mutant nucleotide-binding domains [published erratum appears in *Science* 1992 **258:** 1719]. *Science* **257:**1701–1704.
125. Quinton, P. M., and Reddy, M. M. (1992). Control of CFTR chloride conductance by ATP levels through non-hydrolytic binding. *Nature* **360:**79–81.
126. Carson, M. R., and Welsh, M. J. (1993). 5′-Adenylylimidodiphosphate does not activate CFTR chloride channels in cell-free patches of membrane. *Am. J. Physiol.* **265:**L27–L32.
127. Hwang, T.-C., Nagel, G., Nairn, A. C., and Gadsby, D. C. (1994). Regulation of the gating of CFTR Cl channels by phosphorylation and ATP hydrolysis. *Proc. Natl. Acad. Sci. USA* **91:**4698–4702.
128. Carson, M. R., Travis, S. M., and Welsh, M. J. (1995). The two nucleotide-binding domains of CFTR have distinct functions in controlling channel activity. *J. Biol. Chem.* **270:**1711–1717.
129. Baukrowitz, T., Hwang, T.-C., Nairn, A. C., and Gadsby, D. C. (1994). Coupling of CFTR Cl^- channel gating to an ATP hydrolysis cycle. *Neuron* **12:**473–482.
130. Gunderson, K. L., and Kopito, R. R. (1994). Effects of pyrophosphate and nucleotide analogs suggest a role for ATP hydrolysis in cystic fibrosis transmembrane regulator channel gating. *J. Biol. Chem.* **269:**19349–19353.
131. Hamosh, A., Trapnell, B. C., Zeitlin, P. L., Montrose-Rafizadeh, C., Rosenstein, B. J., Crystal, R. G., and Cutting, G. R. (1991). Severe deficiency of cystic fibrosis transmembrane conductance regulator messenger RNA carrying nonsense mutations R553X and W1316X in respiratory epithelial cells of patients with cystic fibrosis. *J. Clin. Invest.* **88:**1880–1885.
132. Gregory, R. J., Rich, D. P., Cheng, S. H., Souza, D. W., Paul, S., Manavalan, P., Anderson, M. P., Welsh, M. J., and Smith, A. E. (1991). Maturation and function of cystic fibrosis transmembrane conductance regulator variants bearing mutations in putative nucleotide-binding domains 1 and 2. *Mol. Cell Biol.* **11:**3886–3893.
133. Yang, Y., Janich, S., Cohn, J. A., and Wilson, J. M. (1993). The common variant of cystic fibrosis transmembrane conductance regulator is recognized by hsp70 and degraded in a pre-Golgi nonlysosomal compartment. *Proc. Natl. Acad. Sci. USA* **90:**9480–9484.
134. Pind, S., Riordan, J. R., and Williams, D. B. (1994). Participation of the endoplasmic reticulum chaperone calnexin (p88, IP90) in the biogenesis of the cystic fibrosis transmembrane conductance regulator. *J. Biol. Chem.* **269:**12784–12788.
135. Ward, C. L., and Kopito, R. R. (1994). Intracellular turnover of cystic fibrosis transmembrane conductance regulator: Inefficient processing and rapid degradation of wild-type and mutant proteins. *J. Biol. Chem.* **269:**25710–25718.
136. Denning, G. M., Anderson, M. P., Amara, J., Marshall, J., Smith, A. E., and Welsh, M. J. (1992). Processing of mutant cystic fibrosis transmembrane conductance regulator is temperature-sensitive. *Nature* **358:**761–764.
137. Li, C., Ramjeesingh, M., Reyes, E., Jensen, T., Chang, X., Rommens, J. M., and Bear, C. E. (1993). The cystic fibrosis mutation (ΔF508) does not influence the chloride channel activity of CFTR. *Nature Genet.* **3:**311–316.
138. Sheppard, D. N., Ostedgaard, L. S., Winter, M. C., and Welsh, M. J. (1994). Mechanism of dysfunction of two nucleotide-binding domain mutations in CFTR that are associated with pancreatic sufficiency. *EMBO J.* **14:**101–109.
139. Smit, L. S., Wilkinson, D. J., Mansoura, M. K., Collins, F. S., and Dawson, D. C. (1993). Functional roles of the nucleotide-binding folds in the activation of the cystic fibrosis transmembrane conductance regulator. *Proc. Natl. Acad. Sci. USA* **90:**9963–9967.
140. Chu, C.-S., Trapnell, B. C., Curristin, S., Cutting, G. R., and Crystal, R. G. (1993). Genetic basis of variable exon 9 skipping in cystic fibrosis transmembrane conductance regulator mRNA. *Nature Genet.* **3:**151–156.
141. Dörk, T., Wulbrand, U., Richter, T., Neumann, T., Wolfes, H., Wulf, B., Maab, G., and Tümmler, B. (1991). Cystic fibrosis with three mutations in the cystic fibrosis transmembrane conductance regulator gene. *Hum. Genet.* **87:**441–446.
142. Teem, J. L., Berger, H. A., Ostedgaard, L. S., Rich, D. P., Tsui, L.-C., and Welsh, M. J. (1993). Identification of revertants for the cystic fibrosis ΔF508 mutation using STE6-CFTR chimeras in yeast. *Cell* **73:**335–346.
143. Kristidis, P., Bozon, D., Corey, M., Markiewicz, D., Rommens, J., Tsui, L.-C., and Durie, P. (1992). Genetic determination of exocrine pancreatic function in cystic fibrosis. *Am. J. Hum. Genet.* **50:**1178–1184.
144. Strong, T. V., Smit, L. S., Turpin, S. V., Cole, J. L., Hon, C. T., Markiewicz, D., Petty, T. L., Craig, M. W., Rosenow, E. C., Tsui, L.-C., Iannuzzi, M. C., Knowles, M. R., and Collins, F. S. (1991). Cystic fibrosis gene mutation in two sisters with mild disease and normal sweat electrolyte levels. *N. Engl. J. Med.* **325:**1630–1634.
145. Cutting, G. R., Kasch, L. M., Rosenstein, B. J., Zielenski, J., Tsui, L.-C., Antonarakis, S. E., and Kazazian, H. H. J. (1990). A cluster of cystic fibrosis mutations in the first nucleotide-binding fold of the cystic fibrosis conductance regulator protein. *Nature* **346:**366–369.
146. Welsh, M. J., and Smith, A. E. (1993). Molecular mechanisms of CFTR chloride channel dysfunction in cystic fibrosis. *Cell* **73:** 1251–1254.

31

Familial Periodic Paralysis

Louis Ptáček and Robert C. Griggs

31.1. INTRODUCTION

The periodic paralyses have traditionally been divided into hypokalemic, hyperkalemic, normokalemic, and paramyotonic forms.[1] Over the past decade, a combination of electrophysiologic and molecular biologic studies have clarified the classification of the disease (Table 31.1). It has become apparent that there are two broad categories of disease: hypokalemic periodic paralysis and hyperkalemic periodic paralysis. All forms of periodic paralysis are either autosomal dominantly inherited or occur as sporadic cases that are probably the result of new mutations. Hyperkalemic periodic paralysis usually results from a disorder of the skeletal muscle, voltage-gated sodium channel. The molecular alterations have been defined for most cases.[2,3] It is becoming clear that a number of disorders once considered separate entities are in fact allelic to hyperkalemic periodic paralysis including: paramyotonia congenita[4,5] and most recently, a form of myotonia without periodic paralysis that is potassium-sensitive.[6,7] There are, however, a small proportion of patients with hyperkalemic periodic analysis that is not allelic.[8] The gene causing hypokalemic periodic paralysis has also been identified.[9,10] Three mutations have been identified to date and lead to phenotypically similar disorders. Remarkably, transient weakness also characterizes a disorder of the chloride channel.[11] This chapter describes the clinical and historical features of the periodic paralyses, delineates the laboratory abnormalities that have been identified in patients, reviews genetic and molecular studies in the disorders, relates molecular abnormalities to phenotypic variation, and summarizes current treatment for symptoms of the disorders.

Table 31.1. Genetic Classification of the Periodic Paralyses and Myotonias

Sodium channel diseases—chromosome 17q locus (*SCN4A*)
Hyperkalemic periodic paralysis (with and without myotonia)
Normokalemic periodic paralysis
Paramyotonia congenita
Myotonia without periodic paralysis
Calcium channel disease—chromosome 1q locus (*CACNL1A3*)
Hypokalemic periodic paralysis
Chloride channel diseases—chromosome 7q locus (*HUMCLC1*)
Thomsen's (autosomal dominant) myotonia congenita
Becker's (autosomal recessive) myotonia congenita
Unknown
Chondrodystrophic myotonia[97] (autosomal dominant)
Periodic paralysis with cardiac dysrhythmias (Andersen syndrome)

31.2. DISEASE CLASSIFICATION

31.2.1. Sodium Channel Disorders of Muscle

31.2.1.1. Clinical Phenotypes

31.2.1.1a. Hyperkalemic Periodic Paralyses. Patients who had hyperkalemia rather than hypokalemia during attacks of weakness were first recognized in 1951.[12] Patients with hyperkalemic periodic paralysis have the onset of symptoms in infancy or early childhood: frequently myotonia is clinically evident in infancy and episodic weakness develops subsequently. Attacks are frequent, last much shorter periods of time than those of hypokalemic periodic paralysis (often 30 min to 2–3 hr) but are similarly precipitated by rest after exercise and by emotional stress. Five years later, Gamstorp described a large kindred of patients and introduced the term *adynamia episodica hereditaria.*[13] In her patients and many subsequent reports, hyperkalemia is not present during many attacks of weakness. Potassium levels tend to rise at the onset of weakness but are often normal and occasionally fall to hypokalemic levels. The disease is defined by the patient's response to potassium rather than by the absolute potassium level during attacks. Patients are initially of normal strength during attack-free intervals, but develop persistent interattack weakness after years of attacks.[14] Myotonia is often but not invariably evident during attack-free intervals.

Laboratory evaluation frequently discloses slightly elevated serum potassium at times when patients are not reporting attacks and the creatine kinase level is often elevated. Other serum chemistries are normal. Electromyography often discloses myotonia (Figure 31.1); muscle histology is usually abnormal and may show degenerating fibers, vacuoles, or atrophic fibers (Figure 31.2).

Louis Ptáček • Department of Neurology, Human Molecular Biology and Genetics, The University of Utah, Salt Lake City, Utah 84132.
Robert C. Griggs • Department of Neurology, University of Rochester, Rochester, New York 14642.

Molecular Biology of Membrane Transport Disorders, edited by S.G. Schultz *et al.* Plenum Press, New York, 1996

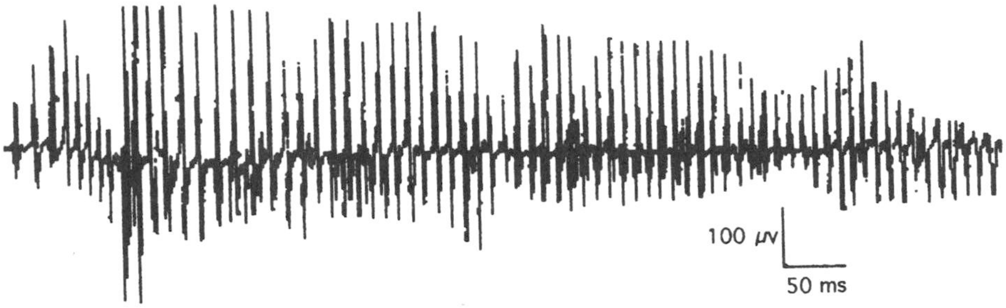

Figure 31.1. Myotonic discharges recorded from the right first dorsal interosseous muscle in a 42-year-old woman with myotonic dystrophy. Repetitive discharges at 60–100 Hz are shown. Both amplitude and frequency wax and wane. (Figure provided by J. S. Lou, MD, PhD, EMG Laboratory, Department of Neurology, University of Rochester.)

31.2.1.1b. Paramyotonia Congenita. Eulenburg originally described paramyotonia congenita noting paradoxical myotonia (myotonia that worsens with activity), cold exacerbation of myotonia, and attacks of weakness that are provoked by cold or can occur without cold exposure.[15] Paradoxical myotonia contrasts with the expected improvement with exercise seen in most forms of classical myotonia. Interattack weakness commonly develops after years of attacks much as in other forms of periodic paralyses.

Laboratory studies are normal in many patients with paramyotonia congenita. Electromyography usually demonstrates myotonia. Muscle histology may show degenerating fibers, vacuoles, or atrophic fibers. The cold-sensitivity that is seen in paramyotonia congenita can be quantified using electromyography with standard protocols.[16]

Patients with paramyotonia congenita are often potassium-sensitive. As with hyperkalemic periodic paralysis, weakness can sometimes also be provoked by *hypo*kalemic challenge.[1,17]

31.2.1.1c. Normokalemic Periodic Paralysis. Most cases described as "normokalemic periodic paralysis" have been potassium-sensitive and therefore are indistinguishable from hyperkalemic periodic paralysis.[14] Rare kindreds have been reported where neither elevating nor lowering the serum potassium provoked attacks. Controversy exists concerning the relationship of this disorder to hyperKPP. It is not clear whether these patients are distinct from the population of hyperKPP patients.

31.2.1.1d. Potassium-Activated Myotonia. This sodium channel disorder is of particular interest because of the potassium-sensitive myotonia that these individuals demonstrate.[18] Their symptoms are therefore episodic although they

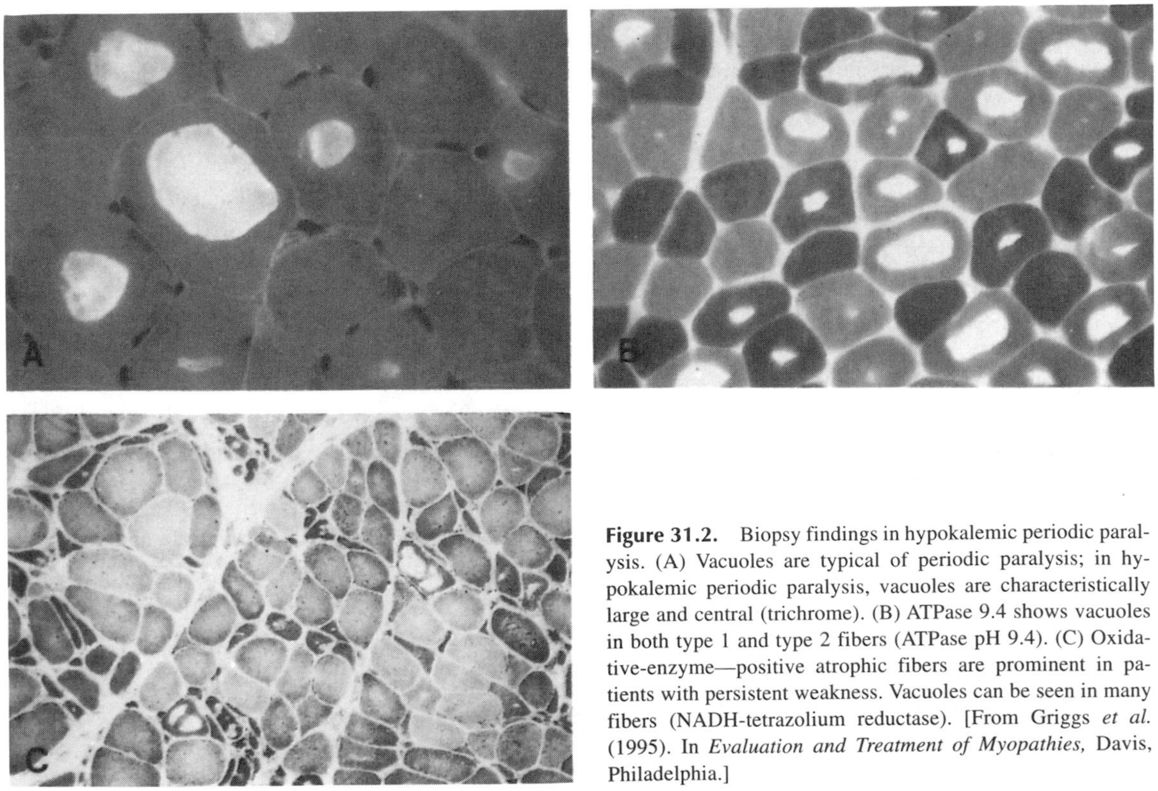

Figure 31.2. Biopsy findings in hypokalemic periodic paralysis. (A) Vacuoles are typical of periodic paralysis; in hypokalemic periodic paralysis, vacuoles are characteristically large and central (trichrome). (B) ATPase 9.4 shows vacuoles in both type 1 and type 2 fibers (ATPase pH 9.4). (C) Oxidative-enzyme—positive atrophic fibers are prominent in patients with persistent weakness. Vacuoles can be seen in many fibers (NADH-tetrazolium reductase). [From Griggs *et al.* (1995). In *Evaluation and Treatment of Myopathies,* Davis, Philadelphia.]

do not have periodic paralysis. The myotonia of these patients is markedly improved by lowering potassium. The myotonia is not temperature-sensitive.

31.2.1.2. Genetics and Molecular: Sodium Channel Disorders *(SCN4A)*

Physiologic abnormalities characterized *in vitro* in hyperKPP patient muscle first suggested a skeletal muscle sodium channel gene as the site of defect in these patients.[19,20] Linkage analysis with genetic probes for sodium channels supported this hypothesis even before the genomic localization of this gene was known.[21,22] This finding proved that hyperKPP must be caused by the mutations in the sodium channel gene or in some other gene that resides very close to this sodium channel gene. The disease allele in one hyperKPP family is not linked to the sodium channel locus[8] suggesting that a second hyperKPP gene is present. However, a large majority of hyperKPP disease alleles are linked to the sodium channel gene.

Genetic linkage analysis subsequently showed that the nonmyotonic form of hyperKPP is also linked to the sodium channel on chromosome 17.[23] This finding suggested that the myotonic and nonmyotonic forms of hyperKPP are allelic disorders.

In vitro study of muscle from paramyotonia congenita patients showed sodium conductance abnormalities that were similar to those seen in hyperKPP.[24] These electrophysiologic similarities, along with clinical similarities, led to the hypothesis that paramyotonia congenita and hyperKPP are allelic disorders. The first genetic support of this hypothesis was demonstration of genetic linkage of this disease locus to the sodium channel gene on chromosome 17.[25]

Despite the lack of episodic weakness in these patients, the potassium-sensitive and acetazolamide-responsive myotonia suggested that this disorder was also allelic to hyperKPP and paramyotonia congenita.[18] This hypothesis was also supported by genetic linkage to the chromosome 17 sodium channel locus.[26] Muscle from these patients has not been studied *in vitro*.

31.2.1.2a. Sodium Channel Structure. A family of genes encoding voltage-gated sodium channel α subunits has been cloned from Drosophila,[27,28] electric eel,[29] rat,[30–33] and human.[34–36] These cDNAs represent genes that share considerable DNA homology with one another. The functional sub-

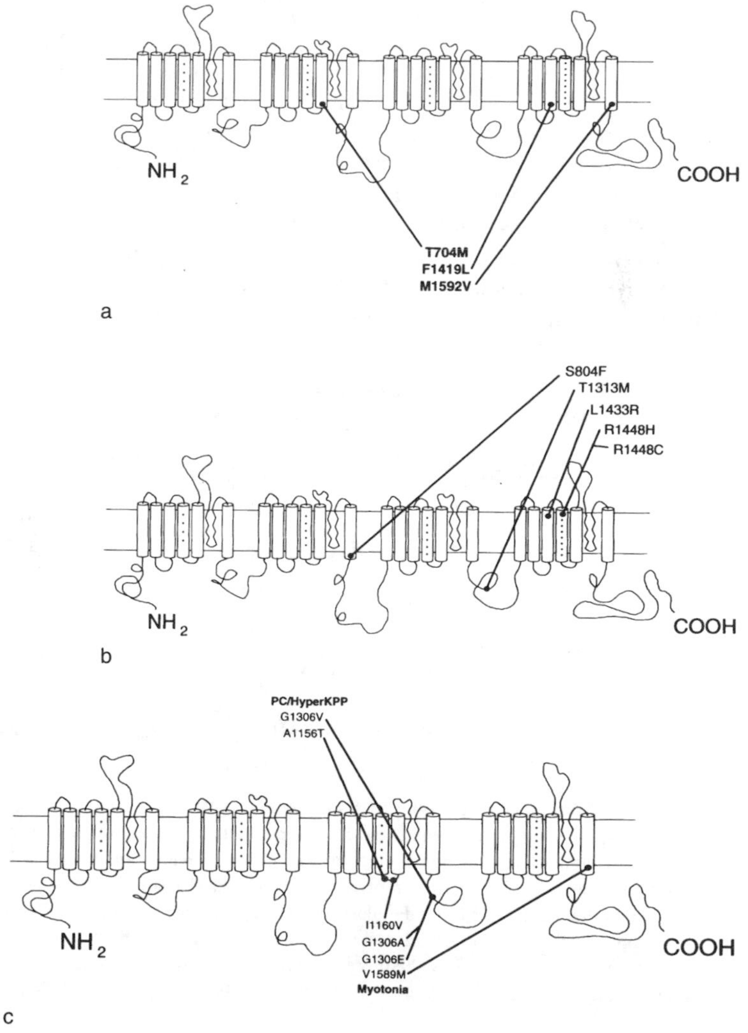

Figure 31.3. Sodium channel structure with mutations. The sodium channel protein is represented in schematic form with 24 putative membrane-spanning segments (cylinders). They are grouped into four domains of six segments each. The fourth segment in each domain contains positively charged arginine and lysine residues ("+"). The cytoplasmic loop connecting the third and fourth domains is the putative inactivation gate. (a) The three mutations causing pure hyperKPP are shown. T704M and M1592V occur in humans. The F1419L mutation occurs in American quarter horses with hyperKPP. (b) Five pure temperature-sensitive sodium channel mutations have been demonstrated in paramyotonia congenita. (c) Two sodium channel mutations have been demonstrated in patients who have a hybrid (PC and hyperKPP) phenotype. Additionally, four mutations have been associated with myotonia in the absence of any periodic paralysis.

units of these sodium channels are homologous to potassium and calcium channels; together they form a large family of ion channel genes. Voltage-dependent sodium channels are membrane proteins that consist of a single large polypeptide of ~260 kDa containing 1800 to 2000 amino acids and 25 to 30% carbohydrate by weight. The proteins encoded by these genes show a high degree of evolutionary conservation in regions presumed to be critical for channel function.[37–39] The conserved regions include areas thought to function in voltage-sensitive gating, inactivation, and ion selectivity. The striking sequence homology among species ranging from Drosophila to humans suggests extraordinary genetic pressure for conservation of sodium channel function. There are four regions of internal homology in sodium channels (domains 1 to 4) each encompassing 225 to 325 amino acids (Figure 31.3). Analysis of these repeat domains has revealed at least six hydrophobic segments within each domain. These segments (S1 to S6) are putative transmembrane helices and are located at conserved regions in each domain.

The S4 helix contains a repeating motif with a positively charged amino acid at every third position. The high charge density of this helix suggests that it may function as a "voltage sensor," thereby playing a central role in voltage-dependent channel activation and inactivation, perhaps through mechanical shifts of this positively charged region in response to membrane depolarization (Figure 31.4a). Such movement may confer a conformational change in the protein resulting in channel pore opening (Figure 31.4b). Site-directed mutagenesis experiments and the development of antibodies to certain portions of this sodium channel protein have begun to define the specific functions of some of the sodium channel regions. For example, altering the intercytoplasmic loop between domains 3 and 4 alters the kinetics of inactivation, implying that this part of the channel serves as the inactivation gate of the sodium channel.[40,41] In addition to the α subunit, mammalian sodium channels may contain one or more smaller β subunits. They serve, in part, to modulate channel kinetics *in vivo*.[42,43]

31.2.1.2b. Sodium Channel Mutations

HyperKPP with myotonia. Two distinct mutations have been identified in patients with hyperKPP and established the sodium channel gene as the disease gene in these families[2,3] (Figure 31.3a). These two mutations account for a large majority of the hyperKPP families.[6,44,45] The two mutations (T704M and M1592V) occur in putative membrane-spanning segments of the sodium channel and are predicted to reside very close to the cytoplasmic side of the membrane. HyperKPP is recognized as a pathological condition in American quarter horses and a mutation in the horse homologue of the skeletal muscle voltage-gated sodium channel has been characterized.[46] It corresponds to F1419L, the homologous mutation in the human gene (Figure 31.3a). This mutation has not yet been seen in human patients with this disease.

HyperKPP without myotonia. Interestingly, though this phenotype breeds true in families and is distinct from the myotonic form of hyperKPP, mutation analysis demonstrated that these patients have one of the same mutations (Figure 31.3a) described above.[2] The reason for the phenotypic difference between families with and without myotonia is not understood.

The fact that both phenotypes breed true in disease families suggests that the modulating factor must be cosegregating with the disease allele. Two possibilities exist: (1) There may be a distinct gene closely linked to the sodium channel gene that somehow modulates sodium channel function, or (2) there may be a normal alteration within the sodium channel itself (polymorphism) that does not cause abnormal sodium channel function but does interact with the T704M mutation in such a way as to modulate its abnormal function. Several polymorphisms are known to exist in the sodium channel gene.[3,47] In either case, the modulating factor would be tightly

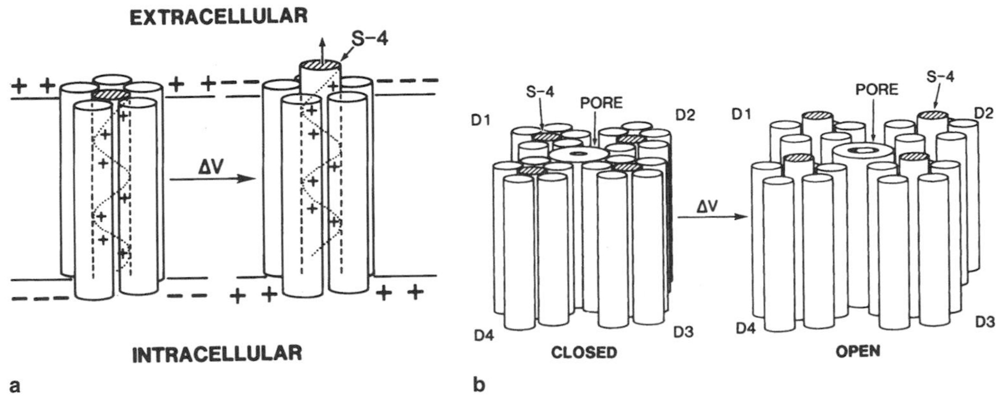

Figure 31.4. Sodium channel voltage sensor model. (a) The S4 segments of sodium channels contain a repeating motif of positively charged amino acids (arginine or lysine) at every third position separated by two neutral amino acids. The S4 segment is shaded in the single domain shown here and moves slightly, under the influence of electrostatic forces, when the membrane is depolarized. The S4 segment of domain 4 is the site of two of the recognized mutations in patients with paramyotonia congenita. (b) This model shows the four domains (D1 to D4) of the sodium channel arranged around the ion pore in the membrane. After depolarization, channel opening likely results from a conformational shift of the protein. Channel closing is thought to result from a "ball-valve" mechanism where the cytoplasmic loop between domains D3 and D4 falls into the ion pore, thus blocking it. In this model, repolarization would then result in the protein assuming its closed conformation and release of the inactivation gate. [From Ptáček *et al.* (1993). Genetics and physiology of the myotonic muscle disorders. *N. Engl. J. Med.* **328:**482–489, with permission.]

linked (genetically) to the T704M mutation. Then the mutation and the modulating factor would cosegregate and the resulting phenotype would be consistent with any one family. Clinical and electrophysiological characterization of additional families with various phenotypes and molecular alterations may reveal other examples where one mutation yields different family-specific phenotypes. Exploration of the electrophysiological phenotype and its association with the clinical phenotypes may shed light on sodium channel structure–function relationships that are not currently understood.

Paramyotonia congenita. Five recognized sodium channel mutations are associated with a pure temperature-sensitive phenotype of myotonia and periodic paralysis (S804F, T1313M, L1433R, R1448H, R1448C; see Figure 31.3b). Two of these mutations occur at the same amino acid (1448) in the S4 segment. These two mutations replace this arginine residue with either cysteine or histidine and therefore neutralize this highly conserved S4 positive charge [the charge on histidine (pKa = 6) will be dependent on its local environment, but is not likely to have a strong positive charge]. Although the exact functional consequences of neutralizing Arg1448 in *hSkM1* can only be revealed by a detailed electrophysiological study of the mutant channel itself, some insight can be gained from studies of the effects of S4 segment mutations in other voltage-dependent sodium and potassium channels. Site-directed mutagenesis experiments involving the S4 segment of the rat brain II sodium channel and the *Shaker* B potassium channel support the hypothesis that this segment does act as the voltage sensor in the biophysical mechanism of activation.[40,48] Substitution of a neutral amino acid for individual positively charged residues in S4 of domains 1 and 2 of the rat brain II channel decreases the steepness of the voltage-dependence of activation and shifts the midpoint of the activation curve along the voltage axis.[40] Mutations neutralizing charges in the *Shaker* potassium channel S4 have a less predictable effect, but also shift both the slope of the activation curve and the location of the curve on the voltage axis. In addition, S4 mutations in the *Shaker* potassium channel indicate coupling between channel activation and inactivation.

A characteristic of the paramyotonia congenita phenotype is the presence of both cold-induced myotonia and intermittent weakness, suggesting that these mutations can, under different physiological circumstances, lead to either repetitive sodium channel activation or failure of activation and paralysis. Although repetitive activity could result from either a shift in the voltage-dependence of activation to more negative potentials or of inactivation to more positive potentials, the persistent sodium currents reported in electrophysiological recordings from paramyotonia congenita muscle[24] suggest that channel inactivation may be the process that is primarily affected. It has been shown that failure of sodium channel inactivation, even in a small percentage of the channel population, can lead to membrane depolarization, inactivation of the other normal sodium channels, and muscle paralysis.[49] Given the effects of S4 mutations in *Shaker,* it is also possible that the mutations described here could affect both inactivation and activation. However, since these mutations insert different amino acids than the substitution studied at the comparable location in *Shaker,* this point can only be resolved through the study of the mutated paramyotonia congenita channel expressed *in vitro.*

A third mutation occurs in the adjacent S3 segment and introduces a positively charged arginine for the normal leucine. The T1313M mutation occurs in the sodium channel "inactivation gate," the cytoplasmic loop connecting domains 3 and 4. This mutation probably interferes, through some steric alteration, with the ability of this gate to interact normally with its docking site. The S804F mutation occurs neither in the "voltage sensor" nor in the "inactivation gate" of the sodium channel. Rather, this mutation resides in the S6 segment of domain 2 in a region predicted to be near the cytoplasmic surface of the membrane.[47] Unfortunately, results of potassium provocation in these patients are not reported and it is therefore possible that patients with this mutation will eventually be reclassified as a mixed hyperKPP/paramyotonia congenita phenotype. McClatchey and her colleagues suggested that the substitution of this polar serine with the larger, nonpolar phenylalanine at the cytoplasmic face of segment S6 of domain 2 opposite the 3–4 interdomain might indicate a role for this part of the channel in receiving the inactivation gate as it pivots to block the channel pore. This is an analogous situation to that suggested for the two human hyperKPP mutations that are also predicted to be located on the cytoplasmic face of the membrane.[14] The location of these five paramyotonia congenita sodium channel mutations is distinct from the two reported hyperKPP mutations and this molecular heterogeneity is likely to be responsible for differences in the clinical presentation of these two syndromes.

The temperature sensitivity of symptoms is an interesting aspect of the paramyotonia congenita phenotype and we can speculate as to how this relates to sodium channel defects. Structural changes related to the mutations reported here might alter the relative energy levels of various conformations associated with different sodium channel gating modes. If the factors stabilizing such states differ in their entropic contributions between mutant and wild-type channels, changes in temperature may differentially affect the preferred channel conformation in each case, and lower temperatures might serve to stabilize the mutant channel in a state associated with an abnormal gating mode. This might be anticipated if, for example, hydrophobic interactions play a prominent role in stabilizing a particular conformation. Alternatively, the mutations may render the channel abnormally sensitive to other cellular processes that are themselves temperature sensitive, such as phosphorylation or G-protein interaction. Ultimately, the correlation between the mutations described here and the pathophysiology of excitation in paramyotonia congenita will require the expression and analysis of human skeletal muscle sodium channels in which this mutation has been introduced.

Mutants with both temperature- and potassium-sensitivity. Two sodium channel mutations (A1156T, G1306V) have been identified in patients with both potassium- and cold-sensitive myotonia and episodic weakness[4,47,50,51] (Figure 31.3c). The relationship of both temperature-and potassium-sensitive phenotype in the same patients remains a mystery. Electrophysiological study of these mutants may lead to understanding of this relationship. Because patients harboring these two mutations are rare among periodic paralysis patients, only a few patients have been subjected to rigorous electrophysiological study. It will be important to study additional families with these mutations to determine whether they consistently produce this phenotype. As noted in the preceding subsection, patients with the S804F mutation have not been subjected to potassium challenge and may ultimately be reclassified as having the hyperKPP/paramyotonia congenita phenotype once that has been clarified.

Sodium channel myotonia without associated weakness. Patients with one of these four mutations (Figure 31.3c) never have episodic weakness. The myotonia caused by the I1160V and V1589M mutations is worsened by potassium administration.[6,7] Potassium-sensitivity has not yet been assessed in patients with the G1306A and G1306E mutations.[52] The reason for muscle membrane hyperexcitability in all of these patients in the absence of any weakness is not known. One possibility is that these mutations result in physiologic abnormalities that are more minor than the myotonia in other families. An attenuated physiologic abnormality might be sufficiently severe to cause myotonia but not dramatic enough to result in inactivation of most or all of the skeletal muscle sodium channels. Alternatively, if these mutations result in altered activation of sodium channels without affecting channel inactivation and without causing persistent sodium conductance, then muscle cells might be hyperexcitable while retaining their ability to inactivate and reactivate completely. This might yield regenerative action potentials (myotonia) without inactivating all of the voltage-gated sodium channels in muscle (periodic paralysis). The explanation for the behavior of these mutations may be much clearer once they are expressed *in vitro* and physiological data are available.

31.2.1.3. Physiology

31.2.1.3a. Experiments Using Patient Muscle Tissue

Hyperkalemic periodic paralysis. In vitro electrophysiological studies have suggested that sodium channel genes are excellent candidates for the site of the defect in hyperKPP. Such studies have demonstrated altered inactivation and alternate gating modes of sodium channels in hyperKPP.[19,20] These investigators found persistent current across hyperKPP muscle membrane after depolarization. This current could be completely and reversibly blocked by tetrodotoxin, a potent and selective sodium channel blocker. This work led to the genetic analysis described above using sodium channel markers for initial genetic localization. Patch-clamp recordings from cultured hyperKPP myotubes supported this idea by demonstrating a defect in the voltage-dependent inactivation of sodium channels in which elevation of extracellular K^+ favored an aberrant gating mode in a small fraction of the channels.[49] Abnormal inactivation in a subpopulation of channels would lead to persistent membrane depolarization, leading to inactivation of normal sodium channels and, ultimately, to membrane inexcitability manifested clinically as transient paresis.

Paramyotonia congenita. In paramyotonia congenita, cooling of skeletal muscle leads first to hyperexcitability and then, in some cases, to paralysis. *In vitro,* skeletal muscle from paramyotonia congenita patients is electrophysiologically normal at 37°C but depolarizes and becomes electrically unexcitable when the temperature is lowered to 27°C.[24] Like muscle from hyperKPP patients, the depolarization seen in muscle from paramyotonia congenita patients is caused by a persistent current that can be completely and reversibly blocked by tetrodotoxin. Along with the clinical phenotype of temperature-sensitive myotonia and periodic paralysis, and the genetic linkage data, the electrophysiological data further supported the hypothesis that the molecular alteration may reside in a skeletal muscle sodium channel gene.

Tahmoush and his colleagues performed patch-clamp, cell-attached, single sodium channel studies on myotubes aneurally cultured from paramyotonia congenita muscle biopsies. The biopsies were obtained from two patients with the T1313M mutation and from two normal controls. These studies were conducted at both 22 and 34°C. Single sodium channel studies of paramyotonia congenita and normal myotubes showed similar mean open time, amplitude, and time to first latency at both temperatures. However, for paramyotonia congenita myotubes at 22°C, openings occurred throughout the 100-msec test pulse while paramyotonia congenita and normal myotubes at 34°C showed only openings clustered within the first 20 msec of the test pulse.[53] They concluded that the T1313M mutation produces a temperature-sensitive inactivation defect where single sodium channels would go from an open to closed state instead of an open to inactivated state. This would permit reopenings and would result in repetitive membrane depolarization and reexcitation (myotonia).[53]

Sodium channel myotonia. Electrophysiological studies were performed on biopsied muscle samples obtained from patients with the G1306E, G1306V, G1306A mutations. Patch-clamp recordings revealed an increase in the time constant of fast sodium channel inactivation in addition to increased late channel openings compared to wild-type muscle.[52] No difference was noted in the single-channel conductance of mutants versus wild-type. In patients with one of these mutations, the severity of the myotonia phenotype correlated well with the size and charge of substituted side chains in the mutants at amino acid 1306, supporting the hypothesis in which the normal glycines at positions 1306 and 1307 were proposed as the hinge in the hinge–lid model proposed by West *et al.*[54]

31.2.1.3b. Expression Studies of Described Mutations

In vitro *expression of the hyperKPP mutations.* The T704M mutation occurs in a highly conserved portion of the sodium channel. This led investigators to attempt *in vitro* expression of this mutation using a full-length cDNA for the rat skeletal muscle channel before a full-length human sodium channel cDNA was available. The region harboring this mutation is completely conserved between the rat and human homologues. Two groups successfully pursued this strategy and

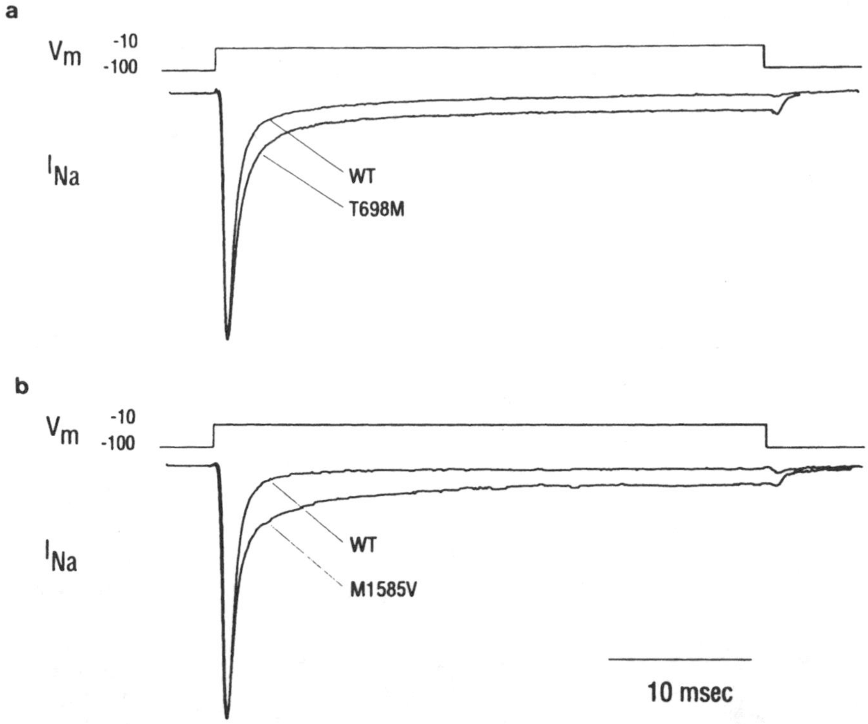

Figure 31.5. Wild-type and mutant currents recorded in HEK 293 cells. Whole-cell currents that have been normalized to the peak amplitude are shown. The HEK 293 cells have been transfected with a wild-type rat sodium channel clone. Others have been transfected with the rat clone after site-directed mutagenesis to insert one of the two human hyperKPP mutations into the analogous site in the rat channel gene. The currents for the two mutants (T698M and M1585V, corresponding to the human T704M and M1592V mutants, respectively) are superimposed with wild-type responses. Both mutants demonstrate persistent sodium conductance when compared to wild-type. [From Cannon and Strittmatter. (1993). Functional consequences of sodium channel mutations identified in families with periodic paralysis. *Neuron* **10:**317–326, with permission.]

reported slightly different results. Cannon and Strittmatter[55] made the rat sodium channel mutation (T698M) that corresponds to the human T704M hyperKPP mutation and used it in transfection experiments in HEK 293 cells in culture. Their work showed that this mutation disrupted inactivation without affecting the time course of the onset of the sodium current or the single-channel conductance (Figure 31.5). Cummins *et al.*[56] expressed the same rat muscle sodium channel with the T698M mutation in HEK 293 cells. Patch-clamp recordings demonstrated that this mutation shifts the voltage dependence of activation by 10–15 mV in the negative direction (Figure 31.6). While this shift in the activation curve was not reported by Cannon and Strittmatter,[55] data were not presented for potentials more negative than −40 mV. Also, Cannon and Strittmatter commented that a smaller sample size and the concern of possible artifacts introduced by series resistance errors while recording large whole-cell currents contribute to possibly overlooking a shift of this magnitude.[55] The other main discrepancy in the data from the two labs relates to the amplitude of the persistent current. Cannon and Strittmatter observed the relative amount of persistent current to be greater (8%) than that observed by Cummins *et al.* (3%). The reason for this discrepancy is not entirely clear. One possible explanation is that Ca^{2+} inhibits the steady-state current since Cummins *et al.* used no Ca^{2+} buffers in their recording solution while Cannon and Strittmatter used a recording solution that contained 5 mM EGTA. The possibility that calcium ions might modulate the persistent current in intriguing and will require further study. Both groups noted that there was no change in single-channel conductance for the mutant and neither noted a consistent potassium dependence of the abnormalities.[55,56] More recently, electrophysiological study of the T704M mutation in *hSkM1* showed a small increase in noninactivating current, especially seen at positive voltages, and a hyperpolarizing shift in the peak current–voltage relationship. These alterations appear to be more severe than the homologous mutation expressed in the rat channel *rSkM1*.[57]

Cannon and Strittmatter also transfected HEK 293 cells with the rat skeletal muscle sodium channel cDNA into which the rat homologue (M1585V) of the human M1592V mutation had been introduced.[55] They noted that this mutation also disrupted inactivation without affecting single-channel conductance. No shift in the activation curve was seen although the same limitations noted above may have obscured such a shift if it were present. Interestingly, cultured myotubes from a patient with this same mutation were studied previously. One important difference was noted between the noninactivating sodium channels in hyperKPP myotubes and the mutant constructs of the rat sodium channel expressed in HEK 293 cells. In the human myotubes, the proportion of sodium channels gating in the noninactivating mode increased when the extracellular potassium concentration was raised from 0 to 10 mM.[49] This behavior correlates well with the clinical observation that attacks of paralysis occur in patients with hyperKPP when the serum potassium concentration is elevated.[14] No potassium-dependence of the persistent current was noted when this mutation was expressed in HEK 293 cells.[55] Possibilities for this discrepancy include differences in posttranslational modification between HEK 293 cells and myotubes, or the absence of required cofactors of the sodium channel α subunit. The high degree of conservation between the human and rat α subunits makes it unlikely that a species difference accounts for the difference in potassium sensitivity, but this awaits confirmation by expression of human constructs.

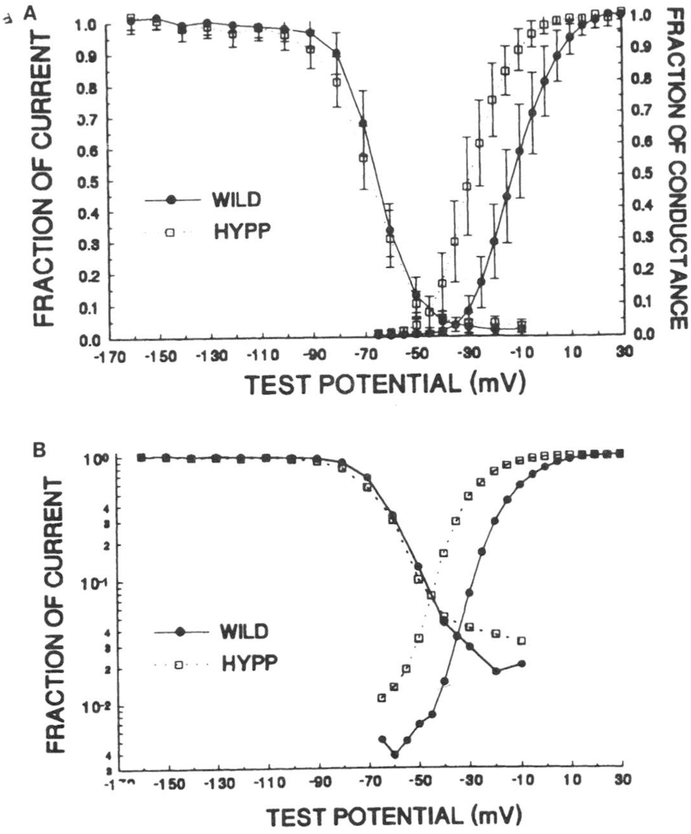

Figure 31.6. Steady-state inactivation properties of hyperKPP mutant. (A) The steady-state inactivation curves for T698M (open squares) and wild-type (closed circles) channels are shown to the left. Current is plotted as a fraction of peak current. The activation curves for T698M and wild-type currents are also shown. (B) Data from panel A plotted on a semilog scale to show the overlap region of the activation and steady-state inactivation curves in greater detail. [From Cummins *et al.* (1993). Functional consequences of a sodium channel mutation causing hyperkalemic periodic paralysis. *Neuron* **10**:667–678, with permission.]

In vitro *expression of the PC mutations. In vitro* study of tsA201 cells transfected with the R1448H and R1448C *hSkM1* mutants showed only small effects on the activation of sodium channels and slowed activation with less voltage dependence and an enhanced rate of recovery from inactivation.[58] The authors suggested a critical role for the S4 helix of domain 4 in coupling of activation and inactivation in this sodium channel.

Yang and colleagues transiently expressed five cold-sensitive mutations (R1448H, R1448C, L1433R, T1313M, A1156T) in tsA201 cells and examined sodium currents using whole-cell patch clamp (Figure 31.7). All five of the mutant constructs exhibited normal activation kinetics but showed (1) reduced rate and voltage-dependence of inactivation, (2) altered voltage-dependence of steady-state inactivation, and (3) increased rate and decreased voltage-dependence of recovery from inactivation. Interestingly, steady-state properties of inactivation differed among these mutants. Hyperpolarizing shifts of steady-state inactivation were seen for the R1448 mutants while depolarizing shifts were seen with the other mutants.[57]

In vitro *expression of sodium channel myotonia mutants.* The V1589M *hSkM1* mutant was expressed in HEK 293 cells by Mitrovic and colleagues and studied using a standard whole-cell electrophysiological recording. Single-channel recordings were also studied using cell-attached patches. They showed an increased sodium steady-state to peak current ratio and increased rate of recovery from inactivation. Single-channel recording demonstrated mutant channels to have a higher probability of short isolated late openings with bursts of late openings.[59] Extracellular potassium had no effects on wild-type or mutant currents; the authors postulated that the potassium-sensitivity of the myotonia in patients with this mutation was related to the associated membrane depolarization favoring the occurrence of late openings in the mutant channel. The increased sodium steady-state to peak current ratio for this mutant was 3–4%,[59] less than the increase seen by Cannon and

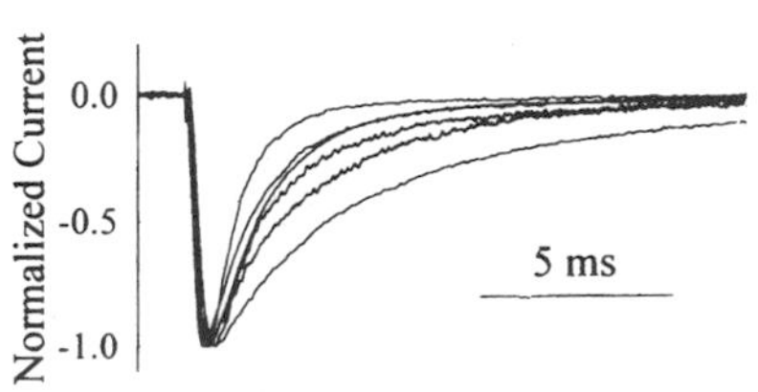

Figure 31.7. Abnormal sodium current in PC muscle. Superimposed currents at −10 mV, normalized to the same peak value. The clones are listed in order of their effects on the rate of inactivation with wild type being fastest and R1448C being the slowest at this voltage (from top to bottom: Wildtype, R1448H, T1313M, A1156T, L1433R, R1448C). [From Yang *et al.* (1994). Sodium channel mutations in paramyotonia congenita exhibit similar biophysical phenotypes *in vitro*. *Proc. Natl. Acad. Sci. USA* **91**:12785–12789, with permission.]

Strittmatter[55] in the hyperKPP M1592V mutant (7.5%). This may be the explanation for the similar physiologic phenotypes of these two mutants yielding different clinical phenotypes. The smaller V1589M inactivation defect may cause hyperexcitability alone. A larger defect, like that seen with M1592V, may yield hyperexcitability initially (myotonia) but be sufficiently dramatic to lead ultimately to sodium channel inactivation, inexcitability, and paralysis.

31.2.2. Calcium Channel Disorder of Muscle

31.2.2.1. Clinical Features: Hypokalemic Periodic Paralysis

The clinical syndrome of hypokalemic periodic paralysis was first described in 1885.[60] It was not until 1934, however, that the association of hypokalemia with the syndrome of periodic paralysis was first identified.[61] In the typical patient, an apparently normal person suddenly awakens with severe weakness of the limbs. There is no alteration of consciousness and the limbs are painless. Patients are often unable to walk and at times may be totally quadriplegic. Attacks of weakness typically resolve over 3 to 4 hr but may occasionally persist for as long as 24 hr.[11] Once the attack has resolved, patients are again completely normal. The weakness that occurs during attacks of paralysis is usually confined to the limbs; facial and respiratory muscle weakness can occasionally occur. Extraocular muscle weakness is rarely, if ever, observed. During attacks of severe weakness, patients lose their muscle stretch reflexes. Patients may have subjective sensory symptoms but tests of sensation are invariably normal. Myotonia cannot be detected by electromyographic examination but eyelid myotonia (Figure 31.8) is frequently present even during attack-free intervals.

Attacks of weakness in hypoKPP have their onset typically in the second decade but can begin as early as age 3 or 4. Almost invariably, attacks of familial hypoKPP begin before age 30. The appearance of typical attacks of hypokalemic weakness in patients after age 30 is usually indicative of either thyrotoxic hypoKPP or a secondary cause of hypokalemia leading to periodic paralysis.

Within families, the severity of periodic attacks is greater in males than females. Attacks usually occur either while the patient is sleeping at night or at other times when the patient has rested following exercise. Attacks are precipitated by maneuvers that lower serum potassium such as high-carbohydrate and high-sodium-containing meals and by emotional stress. Gentle exercise of an affected limb will restore strength to that limb more rapidly than an unexercised limb.

After years of attacks of paralysis, interattack weakness will appear and progress in severity, particularly in patients who are untreated. Attack frequency is often reported to diminish with age but is frequently replaced by disabling weakness that ultimately may progress to the point of wheelchair confinement. Patients are often unaware of this weakness early in the course of their illness. They are, moreover, equally unaware of the fact that they are having major fluctuations in strength during times when they consider their strength to be normal. During attack-free intervals, and early in the course of the illness, physical examination is usually normal. The one exception is the presence of eyelid myotonia (Figure 31.8).

Episodes of weakness occurring in patients with hypoKPP are characteristically associated with a low serum potassium level. In a small proportion of patients, the serum potassium declines but remains within normal limits.[62] Blood studies in patients with periodic paralysis are unrevealing during attack-free intervals. Muscle biopsy is invariably abnormal during attacks when vacuoles are numerous (Figure 31.2). It is usual for the biopsy to remain abnormal during attack-free intervals showing either vacuoles or atrophic changes (Figure 31.2). Vacuoles have been shown by electron microscopy to be dilated sarcoplasmic reticulum, possibly as a result of water movement across the membrane in response to ion fluxes across the sarcolemma.

Approximately two-thirds of patients with hypoKPP have a family history of a disorder in which case it is invariably autosomal dominant. Approximately one-third of patients occur as sporadic cases; sporadic cases can then transmit the disease to subsequent generations and result from new mutations.

31.2.2.2. Genetics and Molecular: Calcium Channel Disorders *(CACNL1A3)*

The shared clinical features of periodic paralysis seen in both hypo- and hyperKPP suggested the possibility that these disorders might be allelic. This hypothesis was tested in several large hypoKPP pedigrees using linkage analysis. Recombinants were noted between the disease allele in several hypoKPP families and the sodium channel gene locus on chromosome 17.[63,64] This demonstrated that a distinct gene causes hypoKPP in these families. The homogeneity seen in the clinical presentation of hypoKPP families suggests that a single gene may be responsible for this disorder. However, it does not rule out the possibility that more than one gene can lead to the hypoKPP phenotype. Recently, a hypoKPP was reported to be genetically linked to a locus on chromosome 1q[9,65] near a gene encoding a dihydropyridine-sensitive, voltage-gated calcium channel. No recombinants were noted between the disease allele in the reported families and an intragenic marker for the calcium channel gene. While it is possible that another hypoKPP gene exists, there is, to date, no genetic evidence supporting that hypothesis.

31.2.2.2a. Calcium Channel Structure. The skeletal-muscle DHP receptor is a complex of five subunits (α_1, α_2, β, γ, δ). The α_1 subunit forms the structural channel (Figure 31.9). DHP receptors serve two functions, excitation–contraction (EC) coupling and voltage-gated calcium conductance.[66] When expressed *in vitro,* this protein induces calcium currents that are voltage-sensitive and can be blocked by dihydropyridines.[67,68] In addition, this protein is critical for EC coupling in skeletal muscle. Muscular dysgenesis occurs as an autoso-

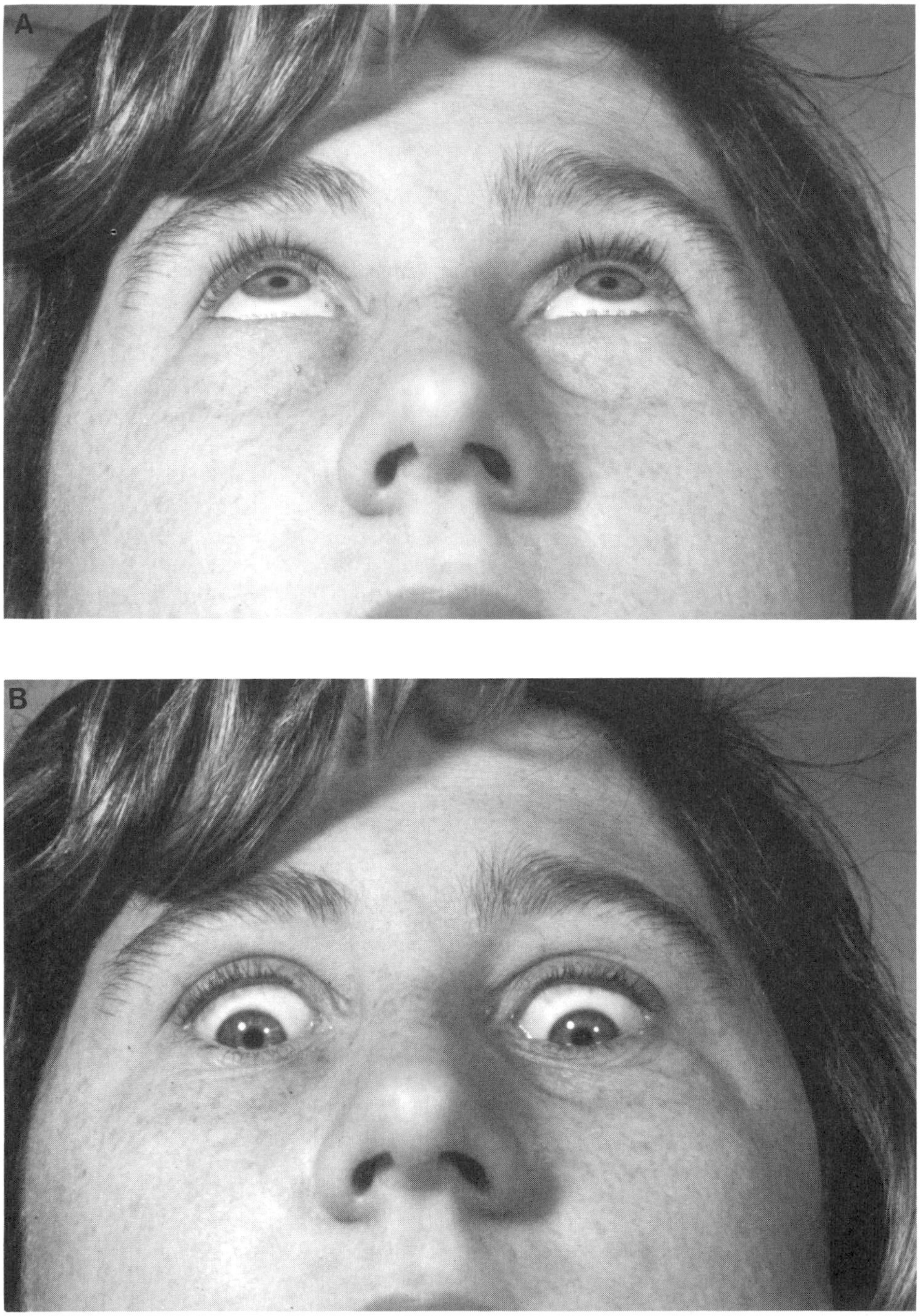

Figure 31.8. Eyelid myotonia in a patient with periodic paralysis. (A) The patient is asked to look up. (B) After looking up, the patient is asked to look down. Normally, the eyelid would come down to the top of the iris. Eyelid myotonia prevents this so that a rim of sclera is seen between the eyelid and the iris.

mal recessive mutation in mice resulting from a single base pair deletion in the DHP receptor that prevents its expression.[69–73] Muscular dysgenesis is a lethal mutation resulting in complete absence of skeletal muscle contraction related to the failure of depolarization of the transverse tubular membrane to trigger calcium release from the sarcoplasmic reticulum. EC coupling can be restored to muscular dysgenesis myotubes growing in culture when they are transfected with a wild-type DHP receptor cDNA-containing vector.[67]

The S4 segments of this voltage-gated calcium channel are thought to function as voltage sensors for both EC coupling and calcium conductance. Movements of these charged parts of the proteins could, for example, occur in response to depolarization and lead to alterations of the protein conformation. This, in turn, leads to interaction of the cytoplasmic loop between domains 2 and 3 with the ryanodine receptor in such a way as to lead to opening of a calcium channel in the sarcoplasmic reticulum (Figure 31.10). Opening of this distinct

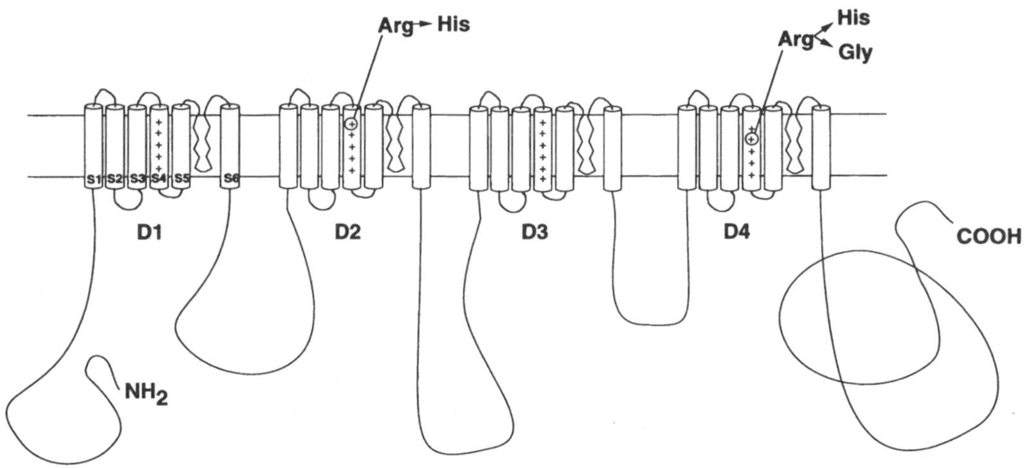

Figure 31.9. Calcium channel α_1 subunit with hypoKPP mutations. The calcium channel protein depicted here shares considerable homology with the sodium channel α subunit. There are four homologous domains each containing six putative membrane-spanning segments (S1–S6). The fourth segment in each domain has a number of positively charged arginine or lysine residues and, like the sodium channel, is thought to function as a voltage sensor for the protein. The locations of the three recognized hypoKPP mutations are shown and all involve an S4 arginine residue.

channel allows calcium from the sarcoplasmic reticulum to move into the cytoplasm where it interacts with the contractile proteins to effect muscle contraction.

31.2.2.2b. DHP Receptor/Calcium Channel Mutations in HypoKPP. Genetic linkage of the hypoKPP disease allele to the *CACNL1A3* locus led to a search for mutations to prove that this DHP receptor/calcium channel was indeed the hypoKPP gene. Two mutations were identified and occurred in adjacent nucleotide positions in the region of the gene encoding the S4 segment of domain 4 (Figure 31.9). One of these mutations accounted for a large proportion of the disease

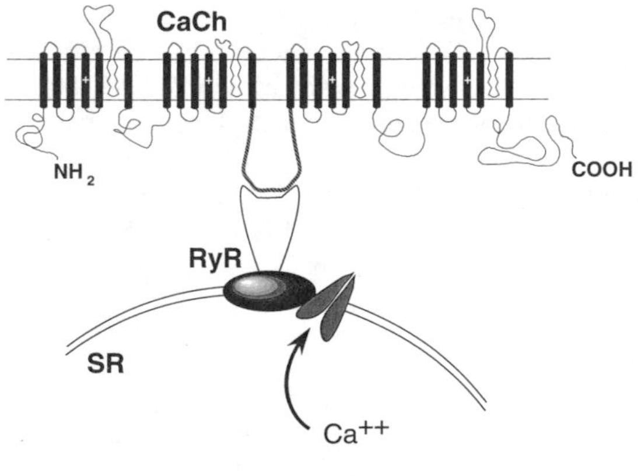

Figure 31.10. Calcium channel and ryanodine receptor model. This skeletal muscle calcium channel (CaCh) is known to interact with the ryanodine receptor (RyR) which resides in the sarcoplasmic reticulum (SR) membrane. The contribution of the CaCh to calcium flux across the sarcolemma is small. A depolarization of the muscle cell membrane is sensed by the CaCh which, through a physical interaction with the RyR, leads to opening of a calcium slow release channel in the SR. It is the movement of calcium from the SR into the cytosol that leads to contraction of muscle.

(about one-third of more than 30 hypoKPP families) and arose as a *de novo* mutation in one patient with sporadic disease. The second mutation was found in only one family.[9] Identification of these mutations therefore established the DHP-sensitive calcium channel as the hypoKPP gene. Subsequently, a third mutation, occurring in the S4 segment of domain 2, has been reported[10] (Figure 31.9).

31.2.2.3. Physiology—Hypokalemic Periodic Paralysis

In contrast to hyperKPP and PC, muscle fibers in hypoKPP are permanently depolarized (by 10–15 mV) but membrane conductance is normal between attacks. *In vitro* study of patient muscle on exposure to a 1 mM potassium solution led to depolarization to −50 mV and to muscle inexcitability.[74] How these mutations in hypoKPP lead to the disease phenotype is not known. Understanding of the pathophysiology awaits *in vitro* expression and physiologic characterization of the mutant proteins.

31.2.3. Chloride Channel Periodic Paralysis

31.2.3.1. Clinical Phenotypes

31.2.3.1a. Becker's Myotonia. This is an autosomal recessive disorder that is associated not only with myotonia but also with transient weakness that can eventually be "warmed up" to normal strength. This weakness is thus analogous to typical myotonia which lessens with repeated muscle contraction. The transient weakness often results in major disability and suggests a diagnosis of either a periodic paralysis or muscular dystrophy. The diagnosis is evident clinically by careful evaluation of patients because strength is characteristically normal on the initial voluntary contraction and then diminishes rapidly. Repeated muscle contraction then results in a return to normal strength. Muscles are generally large or normal in size, in contrast to the atrophy seen in muscular dystrophy.

31.2.3.1b. Thomsen's Disease. This disorder is characterized by autosomal dominant inheritance and painless myotonia. Patients do not experience weakness nor can weakness be detected on examination. The myotonia does, however, "warm up" with repeated activity such that it may be totally abolished. This warm-up contrasts with the worsening of myotonia (paradoxical myotonia) seen with sodium channel myotonia. Electromyography invariably shows myotonia in all forms of chloride channel myotonia.

31.2.3.2. Genetics and Molecular: Chloride Channel Disorders *(HUMCLC1)*

In vitro physiologic studies of patient muscle implicated abnormal chloride conductance in Thomsen's disease[75] and led to genetic linkage experiments using markers near the human skeletal muscle voltage-gated chloride channel locus. Tight linkage supported the hypothesis that this chloride channel is the site of defect in this disease,[76] further implicating this chloride channel gene as the site of defect in these families.

Because of the implication of chloride channels in the mouse model of this disorder[77] and the known chloride channel mutation in the mouse model of autosomal recessive myotonia,[78] mutational analysis of the chloride channel gene *HUMCLC* was performed directly. This led to identification of chloride channel mutations in some of these patients.[79]

31.2.3.2a. Chloride Channel Structure. Voltage-gated chloride channel genes have been cloned more recently[78,80,81] than sodium channels and much less is known about their structure and molecular properties. The chloride channel protein is about 1000 amino acids in length and 90–100 kDa. Hydropathy plots predict 12 (or 13) membrane-spanning segments (Figure 31.11). Biophysical data led to a kinetic model in which chloride channels are thought to form homodimers, the so-called double-barreled channel.[82,83]

31.2.3.2b. Chloride Channel Mutations

Becker's disease. An autosomal recessive myotonic mouse was known to have altered chloride conductance.[77] This led Steinmeyer and colleagues to first show genetic linkage of the *ClC-1* locus to the ADR mouse disease locus and then to identify a transposon insertion in the mouse *ClC-1* gene.[78] Subsequently, a mutation was noted in the chloride channel of humans with the same autosomal recessive myotonia phenotype. This mutation resulted from a phenylalanine-to-cysteine change in the eighth putative transmembrane domain of the chloride channel.[79] Heine and colleagues[84] identified a patient homozygous for a 4-bp deletion in exon 3 of the human *ClC-1* gene.

Thomsen's disease. To date, two mutations have been identified to cause the autosomal dominant myotonia congenita phenotype. The G230E mutation was noted in three of four families studied by George and colleagues[85] and occurs in a part of the chloride channel protein thought to be an extracellular loop between putative membrane-spanning segments D3 and D4. They postulated that the negatively charged glutamic acid residue introduces an electrostatic force that affects conductance of chloride through the channel and/or ion selectivity. The second, a proline-to-leucine substitution at position 480, resides in the loop connecting the ninth and tenth putative membrane-spanning segments.[86] This mutation was noted in descendants of the Julius Thomsen, who was afflicted with the disease and the first to characterize the typical autosomal dominant form of myotonia congenita that has come to bear his name.

31.2.3.3. Physiology

Two mutations found in autosomal dominant myotonia congenita patients (P480L, G230E) were introduced into the human ClC-1 cDNA and used to transfect COS-7 cells.[86] Wild-type human ClC-1 was used to transfect control cells. Both mutations abolished chloride currents in the physiological voltage range. Both mutant channels and wild-type channels were coexpressed by injection of a constant amount of wild-type cRNA and increasing amounts of mutant cRNA into *Xenopus* oocytes. With both mutants, chloride currents were suppressed in a dose-dependent manner. The P480L mutant had a more pronounced effect than the G230E mutant. Channel properties were altered when wild-type was coexpressed with G230E but not with P480L, suggesting that the G230E mutant forms heterooligomers with wild-type. The changes included slower activation at positive potentials. Iodide block was reduced in the positive voltage range.

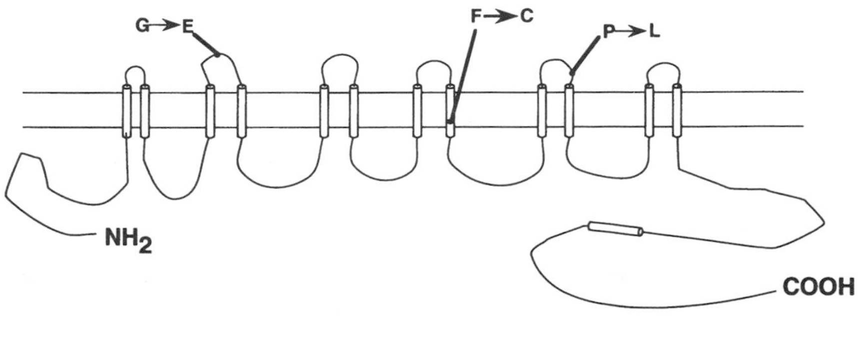

Figure 31.11. Chloride channel. The voltage-gated chloride channel is shown and contains 12 putative membrane-spanning segments. These proteins are thought to form homomultimers. The location of three recognized missense mutations causing myotonia congenita are shown.

31.2.4. Disorders of Unknown Pathogenesis

Clinical Phenotypes

Thyrotoxic Periodic Paralysis. Thyrotoxic periodic paralysis usually appears as an acquired, sporadic disorder that resolves with treatment of the underlying thyrotoxicosis; it will recur if thyrotoxicosis recurs. Attacks of thyrotoxic periodic paralysis are invariably associated with hypokalemia and resemble those of familial hypokalemic periodic paralysis. Thyrotoxic periodic paralysis occurs most frequently in Oriental adults, although it can occur in all races. Despite the higher incidence of thyrotoxicosis in women, less than 5% of cases of thyrotoxic periodic paralysis occur in women. As many as 10% of thyrotoxic Oriental males may develop thyrotoxic periodic paralysis; the proclivity to develop the syndrome when thyrotoxic has been transmitted as an autosomal dominant disorder.[1]

The diagnosis of thyrotoxicosis is often clinically inapparent but is suspected when a patient over the age of 30 first develops typical periodic paralysis. Examination during attack-free intervals often demonstrates proximal weakness from the thyrotoxicosis. Other signs of thyrotoxicosis may be completely lacking. The usual laboratory indicator of thyrotoxicosis, an elevation of thyroxine, is often lacking. Elevations in triiodothyronine are sometimes present but normal thyroid hormone levels may be found. In such instances a depression of TSH and increased radioactive iodine uptake by the thyroid gland are the sole evidence for thyrotoxicosis.[1,92]

Periodic Paralysis with Cardiac Arrhythmias. Patients with this distinctive disorder have a potassium-sensitive periodic paralysis and frequent ectopic ventricular premature beats, most commonly bigeminy or bidirectional tachycardia. The recognition of a form of periodic paralysis associated with dysmorphic features including short stature, hypertelorism, low-set ears, mandibular hypoplasia, and clinodactyly (Figure 31.12) is credited to Andersen.[87] Only subsequently, however, was the nature of the periodic paralysis recognized.[88] Cardiac arrhythmia detected by physical examination is often the pre-

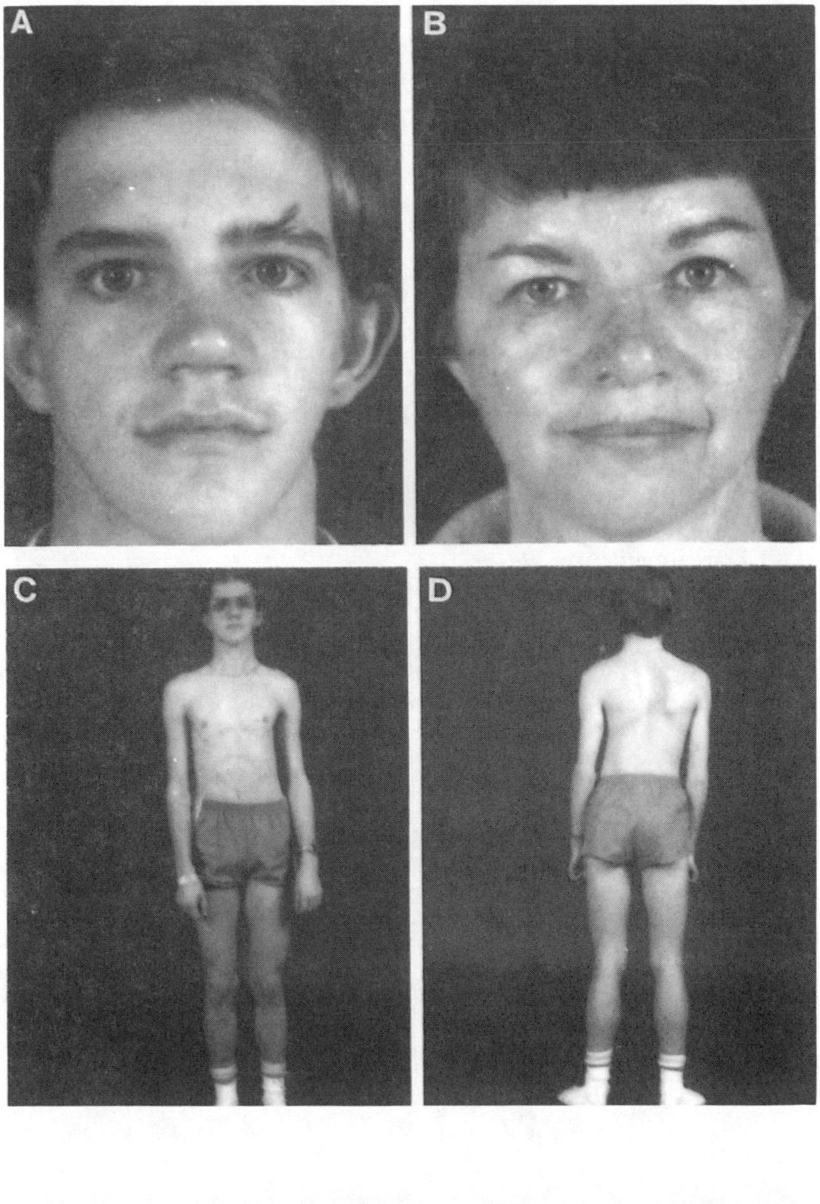

Figure 31.12. Patients with potassium-sensitive periodic paralysis and cardiac arrhythmia. (A) Sixteen-year-old with characteristic facial features: low-set ears, hypoplastic mandible. This patient had frequent attacks of weakness, persistent interattack weakness, and ventricular ectopy. (Reprinted with permission from *Annals of Neurology* **35**:326–330, 1994.) (B) Mother of patient in A. She had only two episodes of weakness, slight proximal weakness, and ventricular ectopy. (C) The patient has proximal weakness and atrophy. (D) Scoliosis was prominent. [From Griggs *et al.* (1995). In *Evaluation and Treatment of Myopathies,* Davis, Philadelphia.]

senting manifestation of the disorder. The cardiac arrhythmia and periodic paralysis are mirror images of each other in terms of provocative and therapeutic tests: raising serum potassium precipitates weakness but normalizes the electrocardiogram whereas lowering serum potassium improves strength but worsens the electrocardiographic abnormalities. While the majority of cases are potassium-sensitive, occasional patients can develop weakness with hypokalemia.

31.2.5. Episodic Ataxias

This group of disorders does not have true periodic paralysis but patients frequently state that they are "weak" or "paralyzed" when they misinterpret their inability to perform motor activity.[89] Typically a sudden loss of balance renders the patient unable to stand or to walk. Remarkably, the pathogenesis of one of the episodic ataxias has been shown to result from mutations in a potassium channel[90]; other episodic movement disorders may well be the result of an ion channel dysfunction.[91]

31.3. IMPLICATIONS FOR FUTURE WORK

31.3.1. Diagnosis

Identification of mutations in hyperKPP and paramyotonia congenita now makes direct testing for these disorders feasible. This is more efficient than predictive testing using linkage to a genetic marker and allows precise diagnosis in a majority of patients. There are, however, some patients with these clinical disorders in whom mutations have not yet been identified. Further work may elucidate additional sodium channel mutations causing these phenotypes. In families where no sodium channel mutations are found, linkage analysis can be used to test the hypothesis that the disease in such families is caused by a different gene. One such example already exists.[8]

31.3.2. Treatment

The attacks of all forms of periodic paralysis respond to treatment (Table 31.2). Myotonia seen with the sodium and chloride channel disorders requires different therapies than those used for episodic weakness (Table 31.3). Acute paralyses can be treated and once the diagnosis is established, prophylactic treatment can usually be devised to prevent attacks. The best regimen(s) has not been established nor is it clear that prevention of attacks will forestall the development of the disabling interattack weakness that patients develop. A multicenter, randomized controlled trial currently in progress is seeking to define the best approach to preventive treatment for attacks of both hypo- and hyperkalemic periodic paralysis.

31.3.2.1. Hypokalemic Periodic Paralysis

Acute attacks of hypoKPP can be aborted by the administration of oral potassium preparations.[1] Intravenous potassium treatment should be avoided if possible since the solutions used to dilute the potassium often worsen hypokalemia. Attack prevention strategies include dietary modification with a low-carbohydrate, low-sodium diet. Attack prevention usually requires the administration of a carbonic anhydrase inhibitor such as acetazolamide or dichlorphenamide. A small proportion of patients are dramatically worsened by carbonic anhydrase inhibitors[62] and may have a genetically distinct disorder. Triamterene has proved effective for attack prevention in such patients.[62] Both acetazolamide and dichlorphenamide can produce some improvement in persistent interattack weakness; it is not known whether such weakness can be totally prevented. No agent has returned strength to normal once severe weakness has developed.

31.3.2.2. Thyrotoxic Periodic Paralysis

Thyrotoxic periodic paralysis responds to correction of thyrotoxicosis. Carbonic anhydrase inhibitors worsen the symptoms of these patients. Beta-blockade treatment of thyrotoxicosis usually eliminates periodic paralysis providing time for specific antithyroid treatment to become effective. In patients with no other signs of thyrotoxicosis, β-blockade may be the only treatment necessary.[92]

The mechanism of action of carbonic anhydrase inhibitors for hypoKPP has not been established. Since there is no sulfonamide-inhibited carbonic anhydrase in muscle,[1] it is probable that a systemic metabolic effect is responsible for attack prevention. The metabolic acidosis produced may under-

Table 31.2. Summary of Treatment Strategies for Periodic Paralysis

	Response to acetazolamide	Other medication	Diet
Sodium channel disorder			
Hyperkalemic periodic paralysis	Yes	Thiazides	High potassium
Myotonia	Yes	Mexiletine	
Calcium channel			
Hypokalemic periodic paralysis	Yes	Spironolactone Triamterene	Low carbohydrate Low sodium
Chloride channel periodic paralysis	Yes	Mexiletine	
Disorders of uncertain etiology			
Thyrotoxic periodic paralysis	No	β-blockade	

Table 31.3. Summary of Treatment Strategies for Myotonia

	Response to acetazolamide	Other medications
Sodium channel	Yes	Mexiletine
Chloride channel		
Thomsen's disease	No	Quinine, procaineamide, phenytoin
Becker's disease	Yes	Mexiletine, tocainide

lie the salutary effects since the ingress of potassium to muscle is lessened in subjects treated with acetazolamide.[93]

31.3.2.3. Hyperkalemic Periodic Paralysis

Acute, individual attacks are mild and seldom require treatment. Patients themselves often discover that carbohydrate-containing food promptly improves weakness. An alternative strategy includes the administration of β-adrenergic agonists such as salbutamol. When used as an inhalant, it is convenient and relatively safe. Those cases with cardiac arrhythmia must be carefully excluded before such treatment is considered. Prophylactic treatment of hyperKPP includes kaliopenic diuretics such as thiazides or carbonic anhydrase inhibitors. Whether the regular use of these agents will prevent the development of interattack weakness is not known. Since patients often decline long-term treatment that will prevent attacks, it is important to determine whether such prophylactic treatment is of value in improving or preventing the persistent weakness that often develops after many years.

31.3.2.4. Paramyotonia Congenita

Most patients with paramyotonia congenita have sufficiently symptomatic myotonia that it overshadows their infrequent attacks of episodic weakness. Patients seldom require medication for prevention of attacks of weakness; acetazolamide is occasionally useful for prevention of weakness or treatment of myotonia but has also precipitated acute weakness in certain patients.[94]

Paramyotonia congenita and acetazolamide-responsive myotonia congenita are both responsive to carbonic anhydrase inhibitors in terms of their effect on myotonia. Since potassium administration dramatically worsens myotonia in many of these patients, the kaliopenic effect of carbonic anhydrase inhibitors probably underlies their therapeutic effect in myotonia as well as in periodic paralysis.[95] Paramyotonia congenita and acetazolamide-responsive myotonia congenita both respond well to mexilitine and tocainide.[96]

31.3.2.5. Becker's Myotonia

The weakness of autosomal recessive myotonia congenita is markedly improved by the orally absorbed, methylated lidocaine derivative mexilitine.[14] Some patients derive benefit from acetazolamide but the response is variable.[14]

31.4. SUMMARY

The molecular defects of the familial periodic paralyses have recently been discovered: (1) hypokalemic periodic paralysis is caused by mutations in the dihydropyridine receptor (calcium channel); (2) hyperkalemic periodic paralysis is caused by mutations in the α subunit of the voltage-gated sodium channel. Most of the once confusing number of different periodic paralyses such as normokalemic periodic paralysis, paramyotonia congenita, and myotonic periodic paralysis have proved to be allelic variations of the sodium channel. Remarkably, defects of still other channels can have clinical overlap with the periodic paralyses. Thus, patients with myotonia congenita caused by chloride channel mutations can have features of intermittent weakness. Furthermore, the potassium channel mutations associated with episodic ataxia can produce clinical symptoms that resemble periodic paralysis.

The basis for the marked differences in phenotype resulting from mutations in different portions of ion channels is still incompletely explained in periodic paralysis. *In vitro* expression systems are beginning to unravel the nature of the defects in terms of cellular physiology. Such studies also may pave the way for design of better treatments that can modify ion channel dysfunction. It also remains unclear why patients with periodic paralysis have only intermittent symptoms, typically less than 1% of the time, despite the persistent gene defect and alteration in channel function. Nonetheless, an understanding of the gene lesions in the various disorders has already permitted improved classification and more reliable diagnosis. Successful treatment is already available for many patients with these disorders.

REFERENCES

1. Riggs, J. E., and Griggs, R. C. (1979). Diagnosis and treatment of the periodic paralyses. *Clin. Neuropharmacol.* **4:**123–138.
2. Ptáček, L. J., George, A. L., Griggs, R. C., Tawil, R., Kallen, R. G., Barchi, R. L., Robertson, M., and Leppert, M. F. (1991). Identification of a mutation in the gene causing hyperkalemic periodic paralysis. *Cell* **67:**1021–1027.
3. Rojas, C. V., Wang, J., Schwartz, L. S., Hoffman, E. P., Powell, B. R., and Brown, R. H. (1991). A met-to-val mutation in the skeletal muscle Na^+ channel α-subunit in hyperkalaemic periodic paralysis. *Nature* **354:**387–389.
4. McClatchey, A., Van den Bergh, P., Pericak-Vance, M., Raskind, W., Verellen, C., McKenna-Yasek, D., Rao, K., Haines, J. L., Bird, T., Brown, R. H., Jr., and Gusella, J. F. (1992). Temperature-sensitive mutations in the III–IV cytoplasmic loop region of the skeletal muscle sodium channel gene in paramyotonia congenita. *Cell* **68:** 769–774.
5. Ptáček, L. J., George, A. L., Barchi, R. L., Griggs, R. C., Riggs, J. E., Robertson, M., and Leppert, M. F. (1992). Mutations in an S4 segment of the adult skeletal muscle sodium channel gene cause paramyotonia congenita. *Neuron* **8:**891–897.
6. Ptáček, L. J., Griggs, R. C., Tawil, R., Meola, G., McManis, P., Mendell, J., Harris, C., Barohn, R., Spitzer, R., Santiago, F., and Leppert, M. F. (1994). Sodium channel mutations in acetazolamide-respon-

sive myotonia congenita, paramyotonia congenita and hyperkalemic periodic paralysis. *Neurology* **44:**1500–1503.
7. Heine, R., Pika, U., and Lehmann-Horn, F. (1993). A novel SCN4A mutation causing myotonia aggravated by cold and potassium. *Hum. Mol. Genet.* **2:**1349–1353.
8. Wang, J., Shou, J., Todorovic, S. M., Feero, W. G., Barany, F., Conwit, R., Hausmanowa-Petrusewicz, I., Fidzianska, A., Arahata, K., Wessel, H. B., Sillen, A., Marks, H. G., Hartlage, P., Galloway, G., Ricker, K., Lehmann-Horn, F., Hayakawa, H., and Hoffman, E. P. (1993). Molecular genetic and genetic correlations in sodium channelopathies: Lack of founder effect and evidence for a second gene. *Am. J. Hum. Genet.* **52:**1074–1084.
9. Ptáček, L. J., Tawil, R., Griggs, R. C., Engel, A. G., Layzer, R. B., Kwieciński, H., McManis, P., Santiago, F., Moore, M., Fouad, G., Bradley, P., and Leppert, M. F. (1994). Dihydropyridine receptor mutations cause hypokalemic periodic paralysis. *Cell* **77:**863–868.
10. Jurkat-Rott, K., Lehmann-Horn, F., Elbaz, A., Heine, R., Gregg, R. G., Hogan, K., Powers, P. A., Lapie, P., Vale-Santos, J., Weissenbach, J., and Fontaine, B. (1994). A calcium channel mutation causing hypokalemic periodic paralysis. *Hum. Mol. Genet.* **3:**1415–1419.
11. Griggs, R. C., and Ptáček, L. J. (1992). The periodic paralyses. *Hosp. Pract.* **27:**123–137.
12. Tyler, F. H., Stephens, F. E., Gunn, F. D., and Perkoff, G. T. (1951). Studies in disorders of muscle. VII. Clinical manifestations and inheritance of a type of periodic paralysis without hypopotassemia. *J. Clin. Invest.* **30:**492–502.
13. Gamstorp, I., (1956). Adynamia episodica hereditaria. *Acta Paediatr.* **45**(Suppl. 108):1–126.
14. Ptáček, L. J., Johnson, K. J., and Griggs, R. C. (1993). Mechanisms of disease: Genetics and physiology of the myotonic muscle disorders. *N. Engl. J. Med.* **328:**482–489.
15. Eulenburg, A. (1886). Über eine familiäre, durch 6 Generationen verfolgbare Form Congenitaler Paramyotonie. *Zentralbl. Neurol.* **5:** 265–272.
16. Jackson, C. E., Barohn, R. J., and Ptáček, L. J. (1994). Paramyotonia congenita: Abnormal short-exercise test following cooling in the absence of weakness and improvement after mexiletine therapy. *Muscle Nerve* **17:**763–768.
17. Layzer, R. B., Lovelace, R. E., and Rowland, L. P. (1967). Hyperkalemic periodic paralysis. *Arch. Neurol.* **16:**455.
18. Trudell, R. G., Kaiser, K. K., and Griggs, R. C. (1987). Acetazolamide responsive myotonia congenita. *Neurology* **37:**488–491.
19. Lehmann-Horn, F., Kuther, G., Ricker, K., Grafe, P., Ballanyi, K., and Rüdel, R. (1987). Adynamia episodica hereditaria with myotonia: A non-inactivating sodium current and the effect of extracellular pH. *Muscle Nerve* **10:**363–374.
20. Rüdel, R., Ruppersberg, J. P., and Spittelmeister, W. (1989). Abnormalities of the fast sodium current in myotonic dystrophy, recessive generalized myotonia, and adynamia episodica. *Muscle Nerve* **12:**281– 287.
21. Fontaine, B., Khurana, T. S., Hoffman, E. P., Bruns, G. A. P., Haines, J. L., Trofatter, J. A., Janson, M. P., Rich, J., McFarlane, H., McKenna Yasek, D., Romano, D., Gusella, J. F., and Brown, R. H., Jr. (1990). Hyperkalemic periodic paralysis and the adult muscle sodium channel gene. *Science* **250:**1000–1002.
22. Ptáček, L. J., Tyler, F., Trimmer, J. S., Agnew, W. S., and Leppert, M. (1991). Analysis in a large hyperkalemic periodic paralysis pedigree supports tight linkage to a sodium channel locus. *Am. J. Hum. Genet.* **49:**378–382.
23. Ebers, G. C., George, A. L., Barchi, R. L., Ting-Passador, S. S., Kallen, R. G., Lathrop, G. M., Beckman, J. S., Hahn, A. F., Brown, W. F., Campbell, R. D., and Hudson, A. J. (1991). Paramyotonia congenita and hyperkalemic periodic paralysis are linked to the adult muscle sodium channel gene. *Ann. Neurol.* **30:**810–816.
24. Lehmann-Horn, F., Rüdel, R., and Ricker, K. (1987). Membrane defects in paramyotonia congenita (Eulenburg). *Muscle Nerve* **10:**633–641.
25. Ptáček, L. J., Trimmer, J. S., Agnew, W. S., Roberts, J. W., Petajan, J. H., and Leppert, M. (1991). Paramyotonia congenita and hyperkalemic periodic paralysis map to the same sodium channel gene locus. *Am. J. Hum. Genet.* **49:**851–854.
26. Ptáček, L. J., Tawil, R., Griggs, R. C., Storvick, D., and Leppert, M. F. (1992). Linkage of atypical myotonia congenita to a sodium channel locus. *Neurology* **42:**431–433.
27. Salkoff, L., Butler, A., Wu, A., Scavarda, N., Giffen, K., Ifune, C., Goodman, R., and Mandel, G. (1987). Genomic organization and deduced amino acid sequence of a putative sodium channel gene in *Drosophila. Science* **237:**744–749.
28. Loughney, K., Kreber, R., and Ganetzky, B. (1989). Molecular analysis of the para locus, a sodium channel gene in drosophila. *Cell* **58:** 1143–1154.
29. Noda, M., Shimizu, S., Tanabe, T., Takai, T., Kayanao, T., Ikeda, T., Takahashi, H., Nakayama, H., Kanaoka, Y., Miniamino, N., Kangawa, K., Natsuo, H., Raftery, M. A., Hirose, T., Inayama, S., Hayashida, H., Miyata, T., and Numa, S. (1984). Primary structure of *Electrophorus electricus* sodium channel deduced from cDNA sequences. *Nature* **312:**121–127.
30. Noda, M., Ikeda, T., Kayanao, T., Suzuki, H., Takeshima, H., Kurasaki, M., Takahashi, H., and Numa, S. (1984). Existence of distinct sodium channel messenger RNA's in rat brain. *Nature* **320:** 188–192.
31. Trimmer, J. S., Cooperman, S. S., Tomiko, S. A., Zhou, J., Crean, S. M., Boyle, M. B., Kallen, R. G., Sheng, Z., Barchi, R. L., Sigworth, F. J., Goodman, R. H., Agnew, W. S., and Mandel, G. (1989). Primary structure and functional expression of a mammalian skeletal muscle sodium channel. *Neuron* **3:**33–49.
32. Rogart, R. B., Cribbs, L. L., Muglia, L. K., Kephart, D. D., and Kaiser, M. W. (1989). Molecular cloning of a putative tetrodotoxin-resistant rat heart sodium channel isoform. *Proc. Natl. Acad. Sci. USA* **86:**8170–8174.
33. Kallen, R. G., Sheng, Z. H., Yang, J., Chen, L., Rogart, R. B., and Barchi, R. L. (1990). Primary structure and expression of a sodium channel characteristic of denervated and immature rat skeletal muscle. *Neuron* **4:**233–242.
34. George, A. L., Komisarof, J., Kallen, R. G., and Barchi, R. L. (1990). Primary structure of the adult human skeletal muscle voltage-dependent Na^+ channel. *Ann. Neurol.* **31:**131–137.
35. Wang, J. Z., Rojas, C. V., Zhou, J., Schwartz, L. S., Nicholas, H., and Hoffman, E. P. (1992). Sequence and genomic structure of the human adult skeletal muscle sodium channel α-subunit gene on 17q. *Biochem. Biophys. Res. Commun.* **182:**794–801.
36. McClatchey, A. I., Liu, C. S., Wang, J., Hoffman, E. P., Rojas, C., and Gusella, J. (1992). Genomic structure of the human skeletal muscle sodium channel gene. *Hum. Mol. Genet.* **1:**521–527.
37. Numa, S., and Noda, M. (1986). Molecular structure of sodium channels. *Ann. N. Y. Acad. Sci.* **479:**338–355.
38. Barchi, R. L. (1988). Probing the molecular structure of the voltage-dependent sodium channel. *Annu. Rev. Neurosci.* **11:**455–495.
39. Trimmer, J. S., and Agnew W. S. (1989). Molecular diversity of voltage-sensitive Na channels. *Annu. Rev. Physiol.* **51:**401–418.
40. Stühmer, W., Conti, F., Suzuki, H., Wang, X., Noda, M., Yahagi, N., Kubo, H., and Numa, S. (1989). Structural parts involved in activation and inactivation of the sodium channel. *Nature* **339:**597–603.
41. Moorman, J. R., Kirsch, G. E., Brown, A. M., Joho R. H. (1990). Changes in sodium channel gating produced by point mutations in a cytoplasmic linker. *Science* **250:**688–691.
42. Isom, L. L., De Jongh, K. S., and Catterall, W. A. (1994). Auxiliary subunits of voltage-gated ion channels. *Neuron* **12:**1183–1194.
43. Isom, L. L., De Jongh, K. S., Patton, D. E., Reber, B. F. X., Offord, J., Charbonneau, H., Walsh, K., Goldin, A. L., and Catterall, W. A. (1992). Primary structure and functional expression of the β_1 subunit of the rat brain sodium channel. *Science* **256:**839–842.

44. Feero, W. G., Wang, J., Barany, F., Zhou, J., Todorovic, S. M., Conwit, R., Galloway, G., Hausmanowa-Petrusewocz, I., Fidzianska, A., Arahata, K., Wessel, H. B., Wadelius, C., Marks, H. G., Hartlage, P., Hayakawa, H., and Hoffman, E. P. (1993). Hyperkalemic periodic paralysis: Rapid molecular diagnosis and relationship of genotype to phenotype in 12 families. *Neurology* **43:**668–673.
45. Ptáček, L. J., Gouw, L., Kwiecinski, H., McManis, P. G., Mendell, J., George, A. L., Barchi, R. L., Robertson, M., and Leppert, M. F. (1993). Sodium channel mutations in hyperkalemic periodic paralysis and paramyotonia congenita. *Ann. Neurol.* **33:**300–307.
46. Rudolph, J. A., Spier, S. J., Byrns, G., Rojas, C. V., Bernoco, D., and Hoffman, E. P. (1992). Periodic paralysis in quarter horses: A sodium channel mutation disseminated by selective breeding. *Nature Genet.* **2:**144–147.
47. McClatchey, A. I., McKenna-Yasek, D., Cros, D., Worthen, H. G., Kuncl, R. W., DeSilva, S. M., Cornblath, D. R., Gusella, J. F., and Brown, R. H., Jr. (1992). Novel mutations in families with unusual and variable disorders of the skeletal muscle sodium channel. *Nature Genet.* **2:**148–152.
48. Papazian, D. M., Timpe, L. C., Jan, Y. N., and Jan, L. Y. (1991). Alteration of voltage-dependence of *Shaker* potassium channel by mutations in the S4 sequence. *Nature* **349:**305–310.
49. Cannon, S. C., Brown, R. H., and Corey, D. P. (1991). A sodium channel defect in hyperkalemic periodic paralysis: Potassium-induced failure of inactivation. *Neuron* **6:**619–626.
50. Van den Bergh, P., Van de Wyngaert, F., and Brucher, J.-M. (1991). Potassium sensitivity in 'pure' paramyotonia congenita. *Neurology* **41**(Suppl. 1):419–420.
51. DeSilva, S. M., Kuncl, R. W., Griffin, J. W., Cornblath, D. R., and Chavoustie, S. (1990). Paramyotonia congenita or hyperkalemic periodic paralysis? Clinical and electrophysiological features of each entity in one family. *Muscle Nerve* **13:**21–26.
52. Lerche, H., Heine, R., Pika, U., George, A. L., Mitrovic, N., Browatzki, M., Weiss, T., River-Bastide, M., Franke, C., Lomonaco, M., Ricker, K., and Lehmann-Horn, F. (1993). Human sodium channel myotonia: Slowed channel inactivation due to substitutions for a glycine within the III–IV linker. *J. Physiol. (London)* **470:**13–22.
53. Tahmoush, A. J., Zhang, P., Hyslop, T. M., Heiman-Patterson, T. D., Schaller, K. L., and Caldwell, J. H. (1993). Thr;raMet substitution and inactivation defect in muscle Na channel in a family with paramyotonia congenita (PC). *Neurology* **43:**1441.
54. West, J. W., Patton, D. E., Scheuer, T., Wang, Y., Goldin, A. L., and Catterall W. A. (1992). A cluster of hydrophobic amino acid residues required for fast Na^+-channel inactivation. *Proc. Natl. Acad. Sci. USA* **89:**10910–10914.
55. Cannon, S. C., and Strittmatter, S. M. (1993). Functional consequences of sodium channel mutations identified in families with periodic paralysis. *Neuron* **10:**317–326.
56. Cummins, T. R., Zhou, J., Sigworth, F. J., Ukomadu, C., Stephan, M., Ptáček, L. J., and Agnew, W. S. (1993). Functional consequences of a sodium channel mutation causing hyperkalemic periodic paralysis. *Neuron* **10:**667–678.
57. Yang, N., Ji, S., Zhou, M., Ptáček, L. J., Barchi, R. L., Horn, R., and George, A. L. (1994). Sodium channel mutations in paramyotonia congenita exhibit similar biophysical phenotypes *in vitro. Proc. Natl. Acad. Sci. USA* **91:**12785–12789.
58. Chahine, M., George, A. L., Zhou, M., Ji, S., Sun, W., Barchi, R. L., and Horn, R. (1994). Sodium channel mutations in paramyotonia congenita uncouple inactivation from activation. *Neuron* **12:**281–294.
59. Mitrovic, N., George, A. L., Heine, R., Wagner, S., Pika, U., Hartlaub, U., Zhou, M., Lerche, H., Fahlke, C., and Lehmann-Horn, F. (1994). Potassium-aggravated myotonia: The V1589M mutation destabilizes the inactivated state of the human muscle sodium channel. *J. Physiol. (London)* **478:**395–402.
60. Westphal, C. (1885). Über einen merkwurdigen fall von periodischer lahmung aller vier extemitaten mit gleichzeitigem erloschen der elektrischen erregbarkeit wahrend der lahmung. *Klin. Wochenschr.* **22:** 489–491.
61. Biemond, A., and Daniels, A. P. (1934). Familial periodic paralysis and its transition into spinal muscular atrophy. *Brain* **57:**90–108.
62. Torres, C. F., Griggs, R. C., Moxley, R. T., and Bender, A. N. (1981). Hypokalemic periodic paralysis exacerbated by acetazolamide. *Neurology* **31:**1423–1428.
63. Fontaine, B., Trofatter, J., Rouleau, G. A., Khurana, T. S., Haines, J., Brown, R., and Gusella, J. (1991). Different gene loci for hyperkalemic and hypokalemic periodic paralysis. *Neuromuscular Disorders* **1:**235–238.
64. Casley, W., Allon, M., Cousin, H. K., Ting, S. S., Crackower, M. A., Hashimoto, L., Cornelis, F., Beckmann, J. S., Hudson, A. J., and Ebers, G. C. (1992). Exclusion of linkage between hypokalemic periodic paralysis (HOKPP) and three candidate loci. *Genomics* **14:** 493–494.
65. Fontaine, B., Vale-Santos, J., Jurkat-Rott, K., Reboul, J., Plassart, E., Rime, C.-S., Elbaz, A., Heine, R., Guimaraes, J., Weissenbach, J., Baumann, N., Fardeau, M., and Lehmann-Horn, F. (1994). Mapping of the hypokalaemic periodic paralysis (HypoPP) locus to chromosome 1q31-32 in three European families. *Nature Genet.* **6:**267–272.
66. Tanabe, T., Beam, K. G., Adams, B. A., Niidome, T., and Numa, S. (1990). Regions of the skeletal muscle dihydropyridine receptor critical for excitation–contraction coupling. *Nature* **346:**567–569.
67. Tanabe, T., Beam, K. G., Powell, J. A., and Numa, S. (1988). Restoration of excitation–contraction coupling and slow calcium current in dysgenic muscle by dihydropyridine receptor complementary DNA. *Nature* **336:**134–139.
68. Perez-Reyes, E., Kim, H. S., Lacerda, A. E., Horne, W., Wei, X., Rampe, D., Campbell, K., Brown, A. M., and Birnbaumer, L. (1989). Induction of calcium currents by the expression of the α_1 subunit from the dihydropyridine receptor of skeletal muscle. *Nature* **340:** 233–236.
69. Gluecksohn-Waelsch, S. (1963). Lethal genes and analysis of differentiation. *Science* **142:**1269–1276.
70. Powell, J. A., and Fambrough, D. M. (1973). Electrical properties of normal and dysgenic mouse skeletal muscle in culture. *J. Cell. Physiol.* **82:**21–38.
71. Knudson, C. M., Chaudhari, N., Sharp, A. H., Powell, J. A., Beam, K. G., and Campbell, K. P. (1989). Specific absence of α_1 subunit of the dihydropyridine receptor in mice with muscular dysgenesis. *J. Biol. Chem.* **264:**1345–1348.
72. Beam, K. G., Knudson, C. M., and Powell, J. A. (1986). A lethal mutation in mice eliminates the slow calcium current in skeletal muscle cells. *Nature* **320:**168–170.
73. Chaudhari, N. (1992). A single nucleotide deletion in the skeletal muscle-specific calcium channel transcript of muscular dysgenesis (*mdg*) mice. *J. Biol. Chem.* **267:**25636–25639.
74. Rüdel, R., Lehmann-Horn, F., Ricker, K., and Küther, G. (1984). Hypokalemic periodic paralysis: In vitro investigation of muscle fiber membrane parameters. *Muscle Nerve* **7:**110–120.
75. Lipicky, R. J., and Bryant, S. H. (1973). A biophysical study of the human myotonias. In *New Developments in Electromyography and Clinical Neurophysiology* (J. E. Desmedt, ed.), Karger, Basel, pp. 451–463.
76. Abdalla, J. A., Casley, W. L., Cousin, L., Hudson, A. J., Murphy, E. G., Cornélis, F. C., Hashimoto, L., and Ebers, G. C. (1992). Linkage of Thomsen disease to the T-cell-receptor Beta (TCRB) locus on chromosome 7q35. *Am. J. Hum. Genet.* **51:**579–584.
77. Mehrke, G., Brinkmeier, H., and Jockusch, H. (1988). The myotonic mouse mutant ADR: Electrophysiology of the muscle fiber. *Muscle Nerve* **11:**440–446.
78. Steinmeyer, K., Klocke, R., Ortland, C., Gronemeirer, M., Jockusch,

H., Gründer, S., and Jentsch, T. J. (1991). Inactivation of muscle chloride channel by transposon insertion in myotonic mice. *Nature* **354:**304–308.

79. Koch, M. C., Steinmeyer, K., Lorenz, C., Ricker, K., Wolf, F., Otto, M., Zoll, B., Lehmann-Horn, F., Grzeschik, K.-H., and Jentsch, T. J. (1992). The skeletal muscle chloride channel in dominant and recessive human myotonia. *Science* **257:**797–800.
80. Jentsch, T. J., Steinmeyer, K., and Schwartz, G. (1990). Primary structure of *Torpedo marmorata* chloride channel isolated by expression cloning of *Xenopus* oocytes. *Nature* **348:**510–514.
81. Thiemann, A., Grunder, S., Pusch, M., and Jentsch, T. J. (1992). A chloride channel widely expressed in epithelial and non-epithelial cells. *Nature* **356:**357–360.
82. Hanke, W., and Miller, C. (1983). Single chloride channels from *Torpedo* electroplax. *J. Gen. Physiol.* **82:**25–45.
83. Bauer, C. K., Steinmeyer, K., Schwarz, J. R., and Jentsch, T. J. (1991). Completely functional double-barreled chloride channel expressed from a single *Torpedo* cDNA. *Proc. Natl. Acad. Sci. USA* **88:**11052–11056.
84. Heine, R., George, A. L., Pika, U., Deymeer, R., Rüdel, R., and Lehmann-Horn, F. (1994). Proof of nonfunctional muscle chloride channel in recessive myotonia congenita (Becker) by detection of a 4 base pair deletion. *Hum. Mol. Genet.* **3:**1123–1128.
85. George, A. L., Crackower, M. A., Abdalla, J. A., Hudson, A. J., and Ebers, G. C. (1993). Molecular basis of Thomsen's disease (autosomal dominant myotonia congenita). *Nature Genet.* **3:**305–310.
86. Steinmeyer, K., Lorenz, C., Pusch, M., Koch, M., and Jentsch, T. J. (1994). Multimeric structure of ClC-1 chloride channel revealed by mutations in dominant myotonia congenita (Thomsen). *EMBO J.* **13:**737–743.
87. Andersen, E. D., Krasilnikoff, P. A., and Overvad, H. (1971). Intermittent muscular weakness, extrasystoles, and multiple developmental anomalies. *Acta Paediatr. Scand.* **60:**559–564.
88. Tawil, R., Ptáček, L. J., Pavlakis, S. G., DeVivo, D. C., Penn, A. S., and Griggs, R. C. (1994). Andersen's syndrome: Potassium-sensitive periodic paralysis, ventricular ectopy and dysmorphic features. *Ann. Neurol.* **35:**326–330.
89. Gancher, S. T., and Nutt, J. G. (1986). Autosomal dominant episodic ataxia: A heterogeneous syndrome. *Movement Disorders* **1:**239–253.
90. Browne, D. L., Gancher, S. T., Nutt, J. G., Brunt, E. R. P., Smith, E. A., Kramer, P., and Litt, M. (1994). Episodic ataxia/myokymia syndrome is associated with point mutations in the human potassium channel gene, KCNA1. *Nature Genet.* **8:**136–140.
91. Griggs, R. C., and Nutt, J. G. (1995). Episodic ataxias as channelopathies. *Ann. Neurol.* **37:**285–286.
92. Griggs, R. C., Bender, A. N., and Tawil, R. (1996). A puzzling case of periodic paralysis. *Muscle Nerve,* **19:**362–364.
93. Riggs, J. E., Griggs, R. C., and Moxley, R. T., III. (1984). Dissociation of glucose and potassium arterial–venous differences across the forearm by acetazolamide. *Arch. Neurol.* **41:**35–38.
94. Riggs, J. E., Griggs, R. C., and Moxley, R. T. (1977). Acetazolamide induced weakness in paramyotonia congenita. *Ann. Intern. Med.* **86:**169–173.
95. Riggs, J. E., Griggs, R. C., Moxley, and Lewis, E. D. (1981). Acute effects of acetazolamide in hyperkalemic periodic paralysis. *Neurology* **31:**725–729.
96. Streib, E. W. (1986). Successful treatment with tocainide of recessive generalized congenital myotonia. *Ann. Neurol.* **19:**501.
97. Schwartz, O., and Jampel, R. S. (1962). Congenital blepharophimosis associated with a unique generalized myopathy. *Arch. Ophthalmol.* **68:**52–57.

32

Disorders of Renal Tubular Transport Processes

W. Brian Reeves and Thomas E. Andreoli

32.1. INTRODUCTION

The glomeruli of individuals with normal renal function deliver, on an average, about 180 liters of plasma ultrafiltrate to the nephrons each day. Included in this vast volume of filtrate are 25,000 mEq of sodium, 4300 mEq of bicarbonate, 700 mEq of potassium, 180 g of glucose, and 10 g of calcium. Indeed, for many of the constituents of the glomerular ultrafiltrate, such as water and sodium, the daily filtered load greatly exceeds the total body content of these substances. One of the primary functions of the nephron, then, is to reabsorb the bulk of the water and solutes presented to it. This process, involving the regulated actions of a wide variety of both active and passive transport pathways proceeding in spatially discrete segments of the nephron, results in a final urine that contains only a small fraction of the original volume and of certain solutes, while other solutes may be present in higher content than in the original glomerular filtrate. That is, certain solutes are avidly absorbed by the nephron while others may undergo net secretion. It also must be emphasized that the resorptive and secretory processes in the kidney are under strict physiologic regulation. Thus, while an individual on a normal diet may excrete only about 1% of the filtered water, and 5–15% of the filtered potassium in the urine, these values can increase tenfold in the setting of increased water or potassium intake. These changes in renal tubular transport may occur in response to hormonal stimuli, such as antidiuretic hormone or aldosterone, or local factors, such as changes in the filtered load or serum concentration of a solute.

Given the importance of renal tubular transport to the maintenance of fluid and solute homeostasis, it is not surprising that clinical disorders may result when these transport processes either are defective or are not regulated appropriately in response to physiologic stimuli. Such clinical disorders may arise from either the excessive losses of substances in the urine, such as water in diabetes insipidus, the deposition of unreabsorbed solutes in the kidney, such as cystine in cystinuria, or the excessive accumulation of substances that are normally secreted into the urine, such as protons in distal renal tubular acidosis.[1] Table 32.1 lists some of the disorders referable to abnormalities in renal tubular transport processes. The following sections will review briefly the clinical, pathophysiologic, and, where applicable, molecular aspects of some of the more common renal tubular disorders.

32.2. AMINOACIDURIA

It has long been known that the proximal tubule reabsorbs amino acids from the tubular fluid. The accumulation of amino acids in the cell at higher concentration than in the tubular fluid places an active transport step at the luminal membrane.[2,3] Samarzija and Fromter,[4] using double perfusion micropuncture techniques and electrophysiologic measurements, examined the kinetics of amino acid reabsorption in the rat proximal tubule. They observed a depolarization of the luminal membrane during amino acid transport and, from competition studies, were able to identify five classes of amino acid transporters in the luminal membrane.

Both sodium-dependent[5–8] and sodium-independent[9] amino acid uptake pathways have been characterized in the kidney. Neutral amino acid transport appears to involve at least three separate transport systems,[10] one that transports all neutral amino acids, one specific for imino acids, and one for the β-amino acids. The acidic and basic amino acid groups each have their own transport system.[11,12] At least one amino acid transporter, a sodium-independent transporter for neutral and dibasic amino acids, has been cloned from the kidney.[9,13,14]

Aminoaciduria, the presence of an excess amount of one or more amino acids in the urine, results either from an increase in the filtered load of an amino acid which overwhelms the tubular transport capacity, or from an abnormality in tubular amino acid transport. The latter presumably occurs due to either a defect in one of the specific amino acid transporters, increased back diffusion of amino acids across the luminal membrane of the proximal tubule, or a generalized failure of

W. Brian Reeves and Thomas E. Andreoli • Division of Nephrology and Department of Internal Medicine, University of Arkansas College of Medicine, Little Rock, Arkansas 72205.

Molecular Biology of Membrane Transport Disorders, edited by S.G. Schultz *et al.* Plenum Press, New York, 1996

Table 32.1. Disorders of Renal Tubular Transport

Disorder	Transport defect	Molecular defect	Clinical features
Cystinuria	Renal and GI cystine and dibasic amino acid transport	Mutant dibasic amino acid transporter (D2H)	Renal calculi
Hartnup disorder	Renal and GI neutral amino acid transport	Unknown	Usually benign; pellagra-like syndrome in some
Iminoglycinuria	Renal imino acid transport	Unknown	Benign
Renal glycosuria	Renal glucose transport	Unknown (mutant SGLT2 ?)	Glucosuria, benign
Nephrogenic diabetes insipidus	Water reabsorption	Mutant V_2 receptor Mutant AQP2 protein	Vasopressin-resistant polyuria, hypernatremia
Renal tubular acidosis—distal	Proton secretion	Absent H^+-ATPase Absent H^+/K^+-ATPase (?)	Metabolic acidosis, nephrocalcinosis
Renal tubular acidosis—proximal	Bicarbonate reabsorption	Unknown	Metabolic acidosis, Fanconi syndrome
Hereditary hypophosphatemia	Renal phosphate reabsorption	Unknown	Hypophosphatemia, osteomalacia
Renal hypouricemia	Renal urate transport	Unknown	Benign

proximal tubular transport, perhaps related to abnormal energy metabolism.[1] The latter may account for the multiple defects in the transport of amino acids, phosphate, glucose, and protons seen in the Fanconi syndrome.

32.2.1. Cystinuria

Cystinuria, the most common heritable tubular defect in amino acid reabsorption, is defined by increased excretion of cystine in the urine and is characterized, clinically, by the formation of calculi in the urinary tract.[15] Defects in the transport of cystine and the dibasic amino acids can be demonstrated in both the kidney and intestinal epithelium of patients with cystinuria leading to the suggestion that a single transport protein expressed in both tissues is affected in this disorder. As noted above, a protein has been cloned from rat[13] adn rabbit[16] kidney that mediates high-affinity transport of dibasic amino acids as well as cystine. The chromosomal location of the human ortholog of this gene (*D2H*) has also been determined.[14] Linkage studies[17] in multiple families afflicted with cystinuria localized the disease to chromosome 2p, the same region as *D2H*. Moreover, mutations in the *D2H* gene have been identified in a number of families with cystinuria.[18] Expression of one of these mutants, M467T, yielded a protein with markedly decreased transport activity. These studies demonstrate that a defect in the D2H dibasic amino acid transporter plays a proximate role in the pathogenesis of cystinuria.

The disorder is transmitted as an autosomal recessive trait, although in some families, heterozygotes may have elevated urinary amino acid levels but no clinical disease. The incidence of the disease is about 1 in 7000 births.[19] The disease typically presents in the second and third decades of life as renal colic. Cystine, the least soluble amino acid, forms radiopaque stones which may cause flank pain, hematuria, urinary tract obstruction, subsequent infection, and possibly renal insufficiency. The diagnosis should be considered in any individual with renal calculi and is easily confirmed by measuring cystine excretion in a 24-hr urine sample. The mainstay of treatment is to increase the urine volume in order to reduce the cystine concentration below about 300 mg/liter. This may require the ingestion of 3 to 4 liters of water daily. The intake of two glasses of water in the middle of the night is recommended to avoid saturation of the urine during the night when urine flow is low.[20] Alkalinization of the urine with oral bicarbonate or citrate supplementation may also help to increase the solubility of cystine in the urine. Penicillamine, mercaptopropionylglycine, and captopril may be helpful in managing those patients who continue to form stones in spite of increased fluid intake and urinary alkalinization.[15] These agents form mixed disulfides with cystine thereby reducing the concentration of free cystine in the urine. The effectiveness of these drugs is limited by their side effects.

32.2.2. Hartnup Disorder

Hartnup disorder is another disorder of both renal and intestinal amino acid transport.[21] The disorder is transmitted as an autosomal recessive trait and is defined by the excretion of excessive amounts of neutral amino acids, i.e., alanine, serine, threonine, valine, leucine, isoleucine, phenylalanine, tyrosine, tryptophan, histidine, glutamine, and asparagine, in the urine. The normal rate of excretion of other amino acids, such as taurine, proline, hydroxyproline, lysine, ornithine, and arginine, separates Hartnup disorder from generalized hyperaminoaciduria.[21] While most individuals expressing this disorder experience no clinical manifestations, some patients develop a pellagralike syndrome.[22] Clinical manifestations can include a photosensitive rash[23] and neurologic abnormalities[24] including ataxia, mental retardation, and hyperreflexia. In patients with neurologic symptoms, the EEG may show nonspecific slowing and neuroradiologic studies may show a variety of nonspecific abnormalities such as cortical atrophy and cerebral calcifications.[24,25] The pellegralike syndrome is believed to be the result of a deficiency of tryptophan, from both enhanced urinary losses and decreased intestinal absorption, which is a precursor for niacin.[21] The diagnosis is based on the pattern of excretion of amino acids in the urine. The treatment of symptomatic patients consists of dietary supplementation with either niacin or nicotinamide, 50 to 300 mg/day. The protein responsible for neutral amino acid has not been identified and the molecular basis for Hartnup disorder remains to be determined.

32.2.3. Iminoglycinuria

Iminoglycinuria is a rare disorder in which the renal clearance of proline, hydroxyproline, and glycine is increased.[26] Iminoglycinuria is inherited as an autosomal recessive trait. The condition is completely benign and requires no treatment.

32.3. RENAL GLYCOSURIA

In the presence of normal blood glucose levels, the renal tubule is capable of reabsorbing virtually all of the filtered glucose load such that glucose is not present in the voided urine. Renal glycosuria refers to a condition in which glucose appears in the urine in the face of normal blood glucose concentrations. This disorder results from a selective defect in sodium-coupled glucose absorption in the proximal tubule.[27] Several features define this disorder. First, patients display normal glucose tolerance and normal fasting blood glucose concentrations, clearly distinguishing them from patients with diabetes who have glycosuria related to an increased filtered glucose load. The defect is specific for glucose as other sugars, such as fructose and galactose, do not appear in the urine. Finally, the transport defect, unlike the aminoacidurias discussed above, is limited to the kidney—intestinal glucose transport is normal and extrarenal carbohydrate metabolism is normal. The disorder is generally believed to be transmitted as an autosomal dominant trait.[28]

The relationship between glucose transport and the filtered glucose load is depicted in Figure 32.1. At physiologic filtered glucose loads, the rate of reabsorption matches the rate of filtration. As the filtered load is increased, however, a point is reached when filtration begins to exceed reabsorption and small amounts of glucose appear in the urine. This point is referred to as the minimal renal glucose threshold, F_{minG}. As the filtered glucose load is increased further, reabsorption reaches a maximum rate (T_m) and all additional filtered glucose appears quantitatively in the urine. Glucose titration studies in patients with renal glycosuria indicate that renal glycosuria is a physiologically heterogeneous disorder. Some patients, with so-called Type A glycosuria, demonstrate a reduction in both T_m and K_{minG}. This pattern may arise from a decrease in the functional number of glucose transporters. In Type B glycosuria, the maximal rate of glucose transport is normal, but K_{minG} is reduced, leading to an increased amount of splay in the glucose titration curve. This pattern may arise from alterations in the affinity of the glucose transporter for glucose.[27] Finally, a single patient has been reported with a complete absence of glucose reabsorption (Type O glycosuria). This patient may represent the homozygous state for Type A glycosuria since both parents displayed modest glycosuria.[29]

The proximal tubule is the primary site for glucose reabsorption by the nephron.[2] The coupling of proximal tubular glucose and amino acid absorption to Na^+ absorption has already been mentioned. Electrophysiologic studies in rat, rabbit, and frog kidney all show that, on addition of glucose to

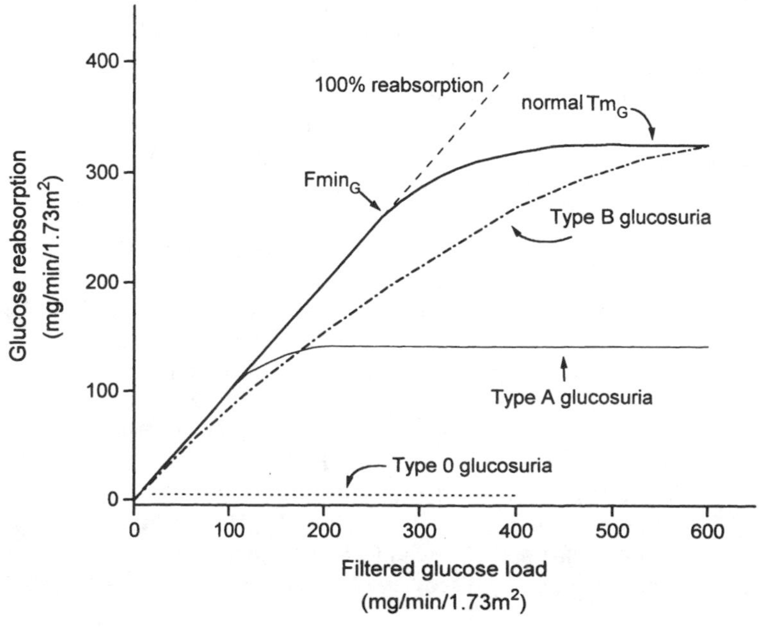

Figure 32.1 Glucose titration curve depicting the relation between filtered glucose load (GFR × plasma glucose) and glucose reabsorption. The dashed line represents theoeretical 100% reabsorption. The heavy solid line represents normal individuals. Titration curves for patients with Type A, Type B, and Type O glucosuria are indicated. F_{minG} is the minimal renal glucose threshold for normal individuals; T_{mG} is the maximal glucose transport rate for normal individuals. (Redrawn with modifications from Ref. 29.)

luminal fluids, apical membranes depolarize.[30] The depolarization occurs because the Na^+–glucose cotransporter transfers a positive charge into the cell, and thus is electrogenic. The Na^+–glucose cotransporter is specific for the D-stereoisomers of glucose, galactose, and α-methyl-D-glucoside.[31,32] The Na^+–glucose cotransporter has little affinity for cations other than sodium.

The rate of glucose transport by the early proximal tubule is greater than in late proximal segments. Barfuss and Schafer,[33] studying perfused rabbit proximal tubules, and Turner and Moran,[34] studying rabbit brush border membrane vesicles, found axial differences in the affinities of the Na^+–glucose carrier for glucose. In the proximal straight tubule, the K_m for D-glucose is 5–20 times lower than in the proximal convoluted tubule. Moreover, the transporter in the early proximal tubule (cortical origin) has a 1:1 sodium-to-glucose stoichiometry,[35] whereas the transporter in the straight segment (medullary origin) has a 2:1 stoichiometry.[36] This 2:1 stoichiometry couples the energy from two sodium ions moving down their electrochemical gradient, thus enabling the transporter to establish a cellular-to-extracellular glucose concentration ratio of 40,000, in comparison to a maximum ratio of 200 with a 1:1 Na^+:glucose stoichiometry. The 2 Na^+:1 glucose transporter is therefore well suited to the straight segment, where tubular fluid glucose concentrations have already been reduced by glucose absorption in the convoluted tubule.

Hediger *et al.* have cloned a Na^+–glucose cotransporter (SGLT1) from rabbit jejunum[37] and human intestine.[38] The cDNA codes for proteins of 662 and 664 amino acids, respectively, and a molecular mass of 73 kDa. Analysis of kidney mRNA using the intestinal Na^+–glucose cotransporter cDNA as a probe indicates that the kidney contains a transporter identical to that in the intestine.[39] Western and Northern blot analysis indicates that SGLT1 is located predominantly in the outer medulla.[40] A second Na^+–glucose cotransporter, SGLT2, has been cloned from human kidney.[41] This protein shares 59% homology to SGLT1. When expressed in *Xenopus* oocytes, SGLT2 mediates low-affinity Na–glucose cotransport with a sodium-to-glucose coupling ratio of 1:1. Moreover, *in situ* hybridization revealed high levels of SGLT2 message in the S1 segment of the proximal tubule.[42] Thus, it appears that SGLT2 may represent the low-affinity, high-capacity sodium–glucose cotransporter in the early proximal tubule, while SGLT1 may represent the high-affinity, low-capacity transporter of the proximal straight tubule.

Since SGLT1 is present in both small intestine and kidney and intestinal glucose transport is normal in renal glycosuria, it is unlikely that mutations in SGLT1 are responsible for renal glycosuria. In contrast, mutations of SGLT1 can lead to congenital glucose–galactose malabsorption, a condition characterized by severe diarrhea.[27,43] It seems reasonable to speculate that mutations in the gene for SGLT2 may underlie at least some cases of renal glycosuria. To date, however, the molecular basis for this disorder remains unknown.

Clinically, renal glycosuria is a rather benign condition that does not require specific treatment. The diagnosis is most often made only after the discovery of glucose on routine urinalysis. Occasionally, patients may present with symptoms of polyuria. The most important point in making a correct diagnosis is to avoid the erroneous diagnosis, and treatment, of diabetes mellitus.

32.4. NEPHROGENIC DIABETES INSIPIDUS

Nephrogenic diabetes insipidus (NDI) can be either a familial or acquired disorder in which renal tubular cells are unresponsive to vasopressin.[44] The clinical hallmarks of the syndrome, polyuria and polydipsia, are the direct consequence of this tubular unresponsiveness to vasopressin, since patients with NDI are unable to concentrate urine effectively, even with extreme degrees of volume contraction. The clinical picture in dehydrated patients with NDI is one of volume contraction, hypernatremia, hyperthermia, polyuria, vomiting, constipation, and failure to thrive. Because of the relatively nonspecific nature of symptoms in early stages of the disease, the disorder may be difficult to identify in the first few months of life. Polyuria, in particular, may frequently be absent in infancy due presumably to dehydration, hypovolemia, and a reduced glomerular filtration rate. Mental and physical retardation may accompany hereditary NDI diabetes insipidus, but children with the disorder may have normal intelligence and physical maturation. Observations suggest that the sequelae of repeated bouts of hypernatremia lead to mental impairment in familial NDI.

In normal individuals, the plasma osmolality is maintained remarkably constant, in the range 285–295 mOsm/kg H_2O, despite wide variations in water or solute intake. The key elements involved in regulating water homeostasis are illustrated in Figure 32.2. The solid lines indicate mechanisms activated by changes in effective extracellular fluid (ECF) osmolality, and consequently changes in cell volume; the dashed lines indicate mechanisms activated by changes in effective circulating volume; and the dotted lines indicate negative feedback limbs.

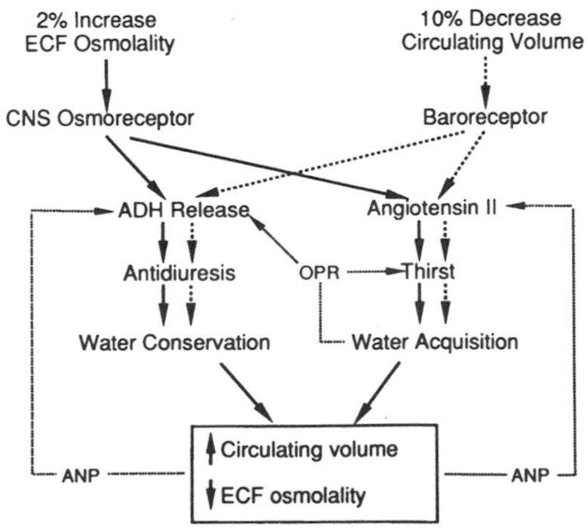

Figure 32.2. A schematic view of the elements involved in water homeostasis. (From Ref. 45, with permission.)

Figure 32.2 illustrates the sensor elements that adjust water balance. The osmoreceptors, both for vasopressin release and for thirst, respond to small changes in effective ECF osmolality, while baroreceptors respond to changes in effective circulating volume. As little as a 2% increase in effective ECF osmolality causes shrinkage of osmoreceptor cells and stimulation of both vasopressin release from the posterior pituitary and thirst. A second way of stimulating both vasopressin release and thirst involves volume-mediated stimuli which can operate independently of changes in plasma osmolality. When the effective circulating volume is reduced by approximately 10%, these volume-dependent mechanisms stimulate vasopressin release. Volume contraction also stimulates thirst by way of local angiotensin II release. Thus, water acquisition involves two distinct mechanisms, osmotic and nonosmotic, each of which operates at a different sensitivity. Moreover, both osmotic as well as nonosmotic stimuli evoke two distinct responses, water conservation by antidiuresis and water acquisition via thirst.

The major site at which vasopressin modulates renal concentrating power is the collecting tubule, where the hormone increases strikingly the permeability to water. In the presence of vasopressin, the increase in water permeability of the collecting tubule augments osmotic water flow from tubular lumen into a hypertonic medullary interstitium, thus increasing the final urine osmolality.[45] Morphological studies of water flow in the collecting tubule have confirmed that, in the absence of vasopressin, the apical plasma membrane is the rate-limiting site for osmotic water flow[46,47] and that vasopressin increases the water permeability of this membrane.[47,48]

The effects of vasopressin on transport processes in renal epithelia are mediated primarily by the intracellular second messenger cAMP.[45] Vasopressin binds to specific receptors, the V_2 receptor, on basolateral membrane surfaces of hormone-responsive epithelial cells and activates membrane-associated adenylate cyclase to catalyze cAMP generation from ATP (Figure 32.3). Adenylate cyclase is a multicomponent enzyme system in which the catalytic subunit is under regulation by two GTP binding proteins, G_s and G_i.[45] In tissues in which hormone action is mediated by cAMP, the hormone–receptor complex activates the G_s subunit of the adenylate cyclase enzyme. The activated, GTP-bound G_s then stimulates the catalytic subunit of adenylate cyclase to produce more cAMP. The vasopressin V_2 receptor has been cloned from rat and human tissue.[49,50] The V_2 receptor gene encodes a 370-amino-acid protein with a transmembrane topography which is characteristic of members of the G-protein-coupled superfamily of receptors. The mRNA for the V_2 receptor is found mainly in the kidney.[49,50] *In situ* hybridization histochemistry revealed localization of the V_2 receptor mRNA in a pattern corresponding to the collecting ducts and medullary thick ascending limb of Henle (mTAL).[50]

A series of structural studies (reviewed in Ref. 44) have led to the formulation of the "shuttle" hypothesis for the effects of vasopressin on apical membrane water permeability. This hypothesis holds that water channel proteins are inserted into the membrane from a cytosolic vacuolar reservoir in response to vasopressin stimulation. These channels are subsequently retrieved from the apical membrane by endocytosis when vasopressin is withdrawn (Figure 32.3). The recent identification of the vasopressin-responsive water channel (AQP2) and the development of immunologic and molecular probes for the channel have allowed direct testing of the shuttle hypothesis.[51] In isolated perfused collecting duct segments, both water permeability and the distribution of AQP2 protein were measured during vasopressin stimulation and withdrawal.[52] In the basal state, water permeability was low and the majority of AQP2 was in intracellular vesicles. During vasopressin stimulation, water permeability increased and the labeling of AQP2 in the apical membrane increased fourfold. Following removal of vasopressin, both the water permeability and the distribution of AQP2 returned to the basal state, thus providing direct evidence supporting the shuttle hypothesis.

32.4.1. Pathogenesis of NDI

Vasopressin-resistant hyposthenuria or isosthenuria may be caused by three types of disturbances[44]:

1. The renal tubules may be vasopressin-responsive and negative free water formation may be normal, but the urine osmolality approaches isotonicity because the volume of isotonic fluid entering the loop of Henle is increased. This mechanism may account for the isosthenuria of chronic renal disease.
2. The renal tubule may be vasopressin-responsive but the renal countercurrent multiplication and exchange systems fail to generate an appropriately hypertonic medullary interstitium; thus, the osmotic driving force for negative free water formation is either reduced or abolished. Such a disturbance may contribute to the vasopressin-unresponsive concentrating defect in a

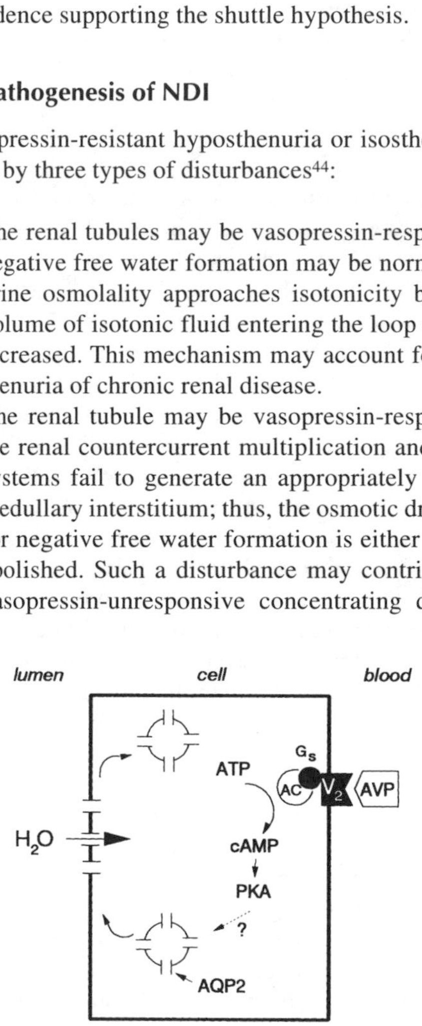

Figure 32.3. A schematic model of the mechanism of the hydroosmotic effect of vasopressin in principal cells of the collecting duct. Binding of arginine vasopressin (AVP) to the V_2 receptor stimulates cAMP formation via G_s activation of adenyl cyclase (AC) and results in activation of cAMP-dependent protein kinase A (PKA). Aquaporin2 (AQP2)-containing subapical vacuoles subsequently fuse with the apical membrane thereby increasing the water permeability of the membrane. The steps controlling the delivery to and retrieval from the apical membrane of AQP2 remain poorly understood.

number of disorders, including: sickle-cell disease; hypokalemia and hypercalcemia, where renal papillary Na^+ and total solute concentrations are reduced; and obstructive uropathy.

3. The volume of isotonic fluid entering the loop of Henle and the countercurrent multiplication and exchange systems are normal, but the renal tubule is unresponsive to vasopressin.

The diagnosis of NDI should be applied to disorders where renal tubular unresponsiveness to vasopressin, without disturbances either in solute delivery to the loop of Henle or in countercurrent multiplication or exchange processes, is responsible for polyuria and hyposthenuria. Thus, as discussed below, NDI may be related to an inability of vasopressin to raise cellular cAMP concentrations, to an inability of cAMP to increase the water permeability of luminal membranes, or to a combination of these two disorders.

32.4.1.1. Familial NDI

Familial NDI is usually an X-linked disorder with variable degrees of expression in heterozygous females. In normal individuals or patients with central diabetes insipidus, exogenous vasopressin increases the rate of urinary cAMP excretion. In the majority of patients with familial NDI, however, vasopressin or dDAVP have little or no effect on the cAMP excretion rate.[44,53] The specificity of defect is demonstrated by the ability of epinephrine to stimulate cAMP excretion normally in these individuals.[53] Thus, the defect is specific to vasopressin responses and appears to reside at a "precyclic AMP" locus.

The lack of response to vasopressin in these patients appears to be restricted to those responses mediated by the V_2 receptor. V_1 receptor-mediated effects, such as vasoconstriction, ACTH release, and renal prostaglandin production, are preserved in patients with familial NDI. In addition, the V_2 receptor defect appears to be generalized and not restricted to the renal action of vasopressin. Extrarenal effects mediated by V_2 receptors include an increase in von Willebrand factor and Factor VIII, and a fall in diastolic blood pressure and stimulation of renin release. In most patients with familial NDI, these extrarenal V_2-mediated responses are absent, indicating a generalized defect in the V_2 receptor signal transduction pathway.[53–56] Obligate carriers of the disorder, e.g., mothers of affected patients, demonstrate intermediate extrarenal responses to the administration of the selective V_2 agonist dDAVP. Occasional patients have been reported in whom extrarenal V_2-mediated responses are preserved.[57,58]

Finally, the recent cloning of the V_2 receptor has provided new insights into the cause of NDI. The NDI gene locus has been mapped to the Xq28 region on the distal long arm of the X chromosome.[59–61] Jans *et al.*[62] created a series of somatic cell hybrid carrying portions of the human X chromosome and demonstrated that cells carrying the Xq28 locus expressed active V_2 receptors. Likewise, using probes to the cloned human V_2 receptor gene, Lolait *et al.*[50] localized the V_2 receptor gene to the Xq27–28 regions. This colocalization of the V_2 receptor gene and the NDI gene supports the view that NDI may result from a defect in the V_2 receptor gene. In fact, a variety of mutations in the V_2 receptor gene have now been identified in a number of families with NDI.[63–66] In one family, a frameshift mutation at codon 247 resulting in a premature termination codon at codon 270 was found in affected males and heterozygous females. The resulting receptor protein is truncated in the third cytoplasmic loop and is presumably unable to transduce hormone-receptor binding to G-protein activation.[63] Another mutation causing a change in amino acid 113 from arginine to tryptophan resulted in a receptor that had both a 20-fold reduction in its affinity for vasopressin and a reduced level of protein expression compared to the normal receptor.[67] Finally, while the majority of familial NDI is transmitted as an X-linked disorder, autosomal recessive varieties may occur rarely.[68] In at least two families with the autosomal recessive form of NDI, mutations in the AQP2 water channel have been identified.[69,70] In one of these cases, expression of the mutant AQP2 channel in *Xenopus* oocytes demonstrated that the protein was nonfunctional.[70]

Thus, defects in either the vasopressin receptor or the apical membrane water channel, the final effect of of vasopressin's hydroosmotic effect, may cause familial NDI.

32.4.1.2. Acquired NDI

Vasopressin-resistent hyposthenuria associated with otherwise normal or nearly normal renal function may occur as a complication of drug therapy or in association with systemic diseases. Appropriately termed acquired NDI, this condition is to be distinguished from the rare familial disorder described earlier. Table 32.2 summarizes the more common causes of acquired vasopressin-resistant polyuria.[44]

Table 32.2. Causes of Acquired Vasopressin-Resistant Polyuria[a]

Osmotic diuresis
Primary polydipsia
Impaired countercurrent exchange
Chronic interstitial renal disease
Obstructive uropathy
Multiple myeloma
Amyloidosis
Sjogren syndrome
Hypercalcemia
Hypokalemia
Diuretics
Sickle-cell anemia
Impaired collecting duct response to vasopressin
Lithium
Methoxyflurane anesthesia
Demeclocycline
Colchicine
Vinblastine
Cyclophosphamide
Amphotericin B

[a]From Ref. 44.

32.4.2. Treatment of NDI

In cases of acquired NDI, efforts should be directed toward correcting the underlying cause. There is no specific therapy for the disorder. Adequate hydration, easily achieved by oral intake in children and adults, but sometimes requiring parenteral supplementation in infants, is essential to prevent the damaging effects of hypernatremia and circulatory collapse, particularly in children. Polyuria may be minimized by reducing solute intake. This is accomplished in infants by limiting the intake of low-sodium formula to only that which is necessary to supply adequate caloric intake.[71] Additional fluid is provided as water or fruit juice on a frequent basis (every 1–2 hr during the day and three times during the night) until children are old enough to satisfy their own thirst.[71]

Neither arginine vasopressin nor any of its analogues has any effect on the disease. Obviously, drugs that stimulate endogenous vasopressin release, such as clofibrate, or that enhance the tubular activity of vasopressin, such as chlorpropamide, are also ineffective.[45]

Nonsteroidal antiinflammatory drugs (NSAID) reduce the urine volume and the free water clearance in children with NDI.[44] Indomethacin is the agent that has been used most frequently. NSAIDs appear to work by reducing the delivery of solute to the distal tubule rather than by alleviating the prostaglandin antagonism of the tubular action of vasopressin.

Diuretics such as chlorothiazide are useful therapeutic agents. Chlorothiazide-induced reductions in urine volume are referable to mild volume depletion and an attendant increase in the proximal fractional absorption of glomerular filtrate. This increase in isosmotic proximal tubular fluid absorption results in a decrease in the volume of fluid delivered to the distal nephron, the site of defective water absorption, and therefore, to a reduction in urine volume. Furthermore, once a sodium deficiency is achieved by the diuretic agent, antidiuresis persists without further drug administration as long as the sodium deficiency is maintained by salt restriction. When salt losses are restored, polyuria rapidly returns.

Treatment of NDI with thiazide diuretics commonly causes hypokalemia, which may aggravate further the renal concentrating defect. Amiloride, a potassium-sparing diuretic, when used in combination with thiazide diuretics may limit potassium losses and have and additive effect to the thiazide in terms of reducing urine volume.[72,73] Amiloride also appears to be particularly effective for the treatment of polyuria in patients receiving lithium therapy (Table 32.2). By blocking sodium channels in the collecting duct cell, amiloride may limit the entry of lithium into these cells.[74]

32.5. RENAL TUBULAR ACIDOSIS

Renal tubular acidosis (RTA) refers to a group of conditions (Table 32.3) that are characterized by a hyperchloremic metabolic acidosis resulting from impaired renal acid excretion. The metabolism of proteins and nucleic acids produces about 50–100 mEq of nonvolatile acids, such as phosphoric and sulfuric acid, daily. These acids are buffered almost immediately by bicarbonate thereby consuming body bicarbonate stores. This process, if left unattended, would eventually exhaust completely the body supplies of both bicarbonate and nonbicarbonate buffers. However, by excreting an equivalent amount of protons, the kidney generates new bicarbonate and prevents the depletion of body bicarbonate and nonbicarbonate buffer stores. In addition to the 50–100 mEq of bicarbonate that is utilized to buffer nonvolatile acids, approximately 4300 mEq of bicarbonate is removed each day from the ECF by glomerular filtration. If this filtered bicarbonate load were not reabsorbed by the nephron, bicarbonate depletion and metabolic acidosis would quickly ensue. Thus, the nephron has two main chores with respect to bicarbonate reabsorption: first, it must completely *reclaim* the large amount of bicarbonate which is filtered by the glomerulus, and, second, it must *regenerate* new bicarbonate to replace that which was titrated by metabolic nonvolatile acid production. The former process occurs mainly in the proximal tubule while the latter occurs in the distal tubule, specifically, in the cortical and medullary collecting ducts. The failure of these processes results in proximal or distal renal tubular acidosis, respectively.

Table 32.3. Classification of Renal Tubular Acidosis

Hypokalemic RTA
Proximal (Type II) RTA
Distal (Classic, Type I) RTA
Hyperkalemic RTA
Voltage-dependent RTA
Generalized distal nephron dysfunction (Type IV RTA)

32.5.1. Proximal Tubule Bicarbonate Transport

The proximal tubule reabsorbs about 80–90% of the filtered bicarbonate. The reabsorption of bicarbonate by the proximal tubule is driven by the active secretion of protons across the apical cell membrane.[75] The mechanism of proximal tubule bicarbonate reabsorption is depicted in Figure 32.4.

Directly coupled Na^+/H^+ exchange in the proximal tubular brush border[76] is responsible for most proton secretion in the proximal tubule. The mammalian Na^+/H^+ exchanger is electroneutral. The kinetic characteristics of the Na^+/H^+ exchanger conform to a simple, saturable single-site model, consistent with a stoichiometry of one proton for one sodium.[76,77] The exchanger is inhibited reversibly by amiloride,[78] and is stimulated by intracellular acidosis. The stimulatory effect of internal protons, mediated by an internal activator site rather than increased occupancy of the internal transport site,[79] helps to account for the increase in Na^+/H^+ exchange in response to intracellular acidosis.[80,81]

Studies indicate that the apical and basolateral membranes of cultured kidney cells contain different forms of Na^+/H^+ exchangers with different affinities for amiloride.[82,83]

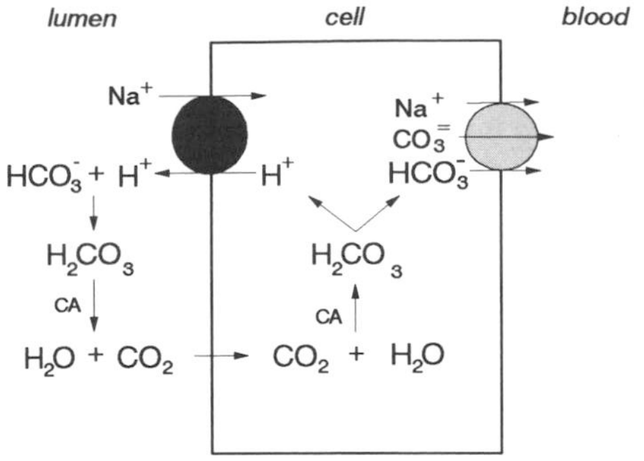

Figure 32.4. Mechanism of bicarbonate reabsorption in the proximal tubule. See text for explanation. Ca, carbonic anhydrase.

The apical Na^+/H^+ exchanger is involved in urinary acidification and has a low affinity for amiloride while the basolateral exchanger is presumed to serve a "housekeeping" role and has a high affinity for amiloride. An Na^+/H^+ exchanger was first cloned from human genomic DNA.[84] The clone NHE-1 encodes a 99-kDa protein which contains ten membrane-spanning regions. The sensitivity of NHE-1 to amiloride[84] and the basolateral localization[85] suggest that it represents the "housekeeping" Na^+/H^+ exchanger.

Other isoforms of the Na^+/H^+ exchanger have also been cloned. One of these, NHE-3, has been shown to be located in the brush border membrane of proximal tubule cells[86] and may be the relatively amiloride-resistant Na^+/H^+ exchanger responsible for proximal tubular acidification.

In addition to Na^+/H^+ exchange, a small portion of proximal tubular H^+ secretion is mediated by an H^+-ATPase.[75]

The protons secreted across the apical membrane are supplied by the dissociation of carbonic acid (H_2CO_3) to H^+ and HCO_3^-. The HCO_3^- that is formed leaves the cell across the basolateral membrane via an electrogenic $Na^+/HCO_3^-/CO_3^{2-}$ cotransporter with a 1:1:1 stoichiometry.[87,88] This transporter is believed to mediate most of the basolateral bicarbonate transport in the proximal tubule.

As indicated in Figure 32.4, protons that are secreted into the tubule lumen combine with filtered HCO_3^- to form H_2CO_3. In the presence of carbonic anhydrase, an abundant brush border enzyme, H_2CO_3 rapidly dehydrates to H_2O and CO_2. The CO_2 is freely permeable to the tubular cell and, on entering, is rehydrated back to H_2CO_3 in a reaction also catalyzed by carbonic anhydrase. This H_2CO_3 then serves as the source for further H+ secretion and HCO_3^- reabsorption. The absence of carbonic anhydrase activity, either as a genetic disorder[89] or from pharmacologic inhibition, can result in proximal RTA.

Even though most of the H^+ secretion in the proximal tubule is mediated via "passive" Na^+/H^+ exchange, H^+ secretion nonetheless requires the expenditure of metabolic energy. Na^+/H^+ exchange is an example of a "secondary" active transport process. Specifically, the movement of H^+ against a pH gradient is driven by the chemical gradient for Na^+ to enter the cell. The Na^+ gradient, however, is established and maintained by the action of the basolateral membrane Na^+,K^+-ATPase. The transport of many other solutes in the proximal tubule is coupled to sodium and dependent, ultimately, on Na^+,K^+-ATPase activity. The Fanconi syndrome refers to a generalized defect in proximal tubule function which leads to impaired reabsorption of bicarbonate, amino acids, glucose, and phosphate.[90] At least some forms of this disorder result from impaired ATP production and consequent decreased Na^+,K^+-ATPase activity.

The reabsorption of bicarbonate in the proximal tubular exhibits a "T_m" pattern similar to that discussed earlier for proximal glucose transport.[75] Specifically, as the serum bicarbonate concentration, and filtered load of bicarbonate, is increased, the reabsorption of bicarbonate initially increases almost quantitatively (Figure 32.5). As the bicarbonate concentration is increased, however, a threshold is reached when filtration begins to exceed reabsorption and bicarbonate begins to appear in the urine. As the filtered bicarbonate is increased further, reabsorption reaches a maximum rate (T_m) and all additional filtered bicarbonate appears quantitatively in the urine. This relationship between filtered bicarbonate and proximal bicarbonate transport is pertinent to understanding the clinical features and response to treatment of patients with proximal RTA.[75] Patients with proximal RTA have a reduced T_m for bicarbonate transport (Figure 32.5). Therefore, at normal serum bicarbonate levels the filtered bicarbonate load exceeds the transport capacity of the proximal tubule. Some of the unabsorbed bicarbonate is subsequently reabsorbed in the distal tubule but its limited capacity is easily exceeded such that bicarbonate appears in the urine. This bicarbonate loss leads to a reduction in the serum bicarbonate concentration and, hence, filtered bicarbonate load. Without treatment, the serum bicarbonate falls until the filtered load equals the maximum transport capacity for the proximal tubule and little bicarbonate escapes proximal reabsorption. At this point, since distal tubule acidification mechanisms are intact in proximal RTA, the urine pH will fall. Patients will then maintain

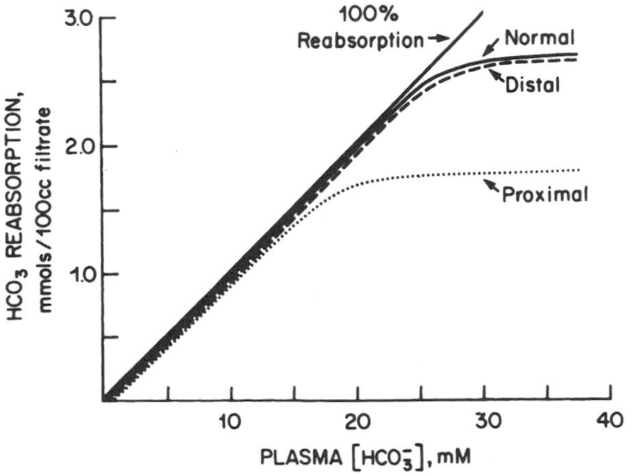

Figure 32.5. The reabsorption of filtered bicarbonate as a function of the plasma bicarbonate concentration. In patients with proximal RTA, the threshold at which bicarbonate appears in the urine is reduced. This reduction in the bicarbonate threshold is reflection of a decrease in the maximal rate of bicarbonate transport (T_m) by the proximal tubule. (From Ref. 75, with permission.)

acid–base balance—able to reclaim the filtered bicarbonate and regenerate sufficient bicarbonate to cope with ongoing nonvolatile acid production—albeit at a lower steady-state serum bicarbonate concentration. However, if one attempts to raise the serum bicarbonate concentration toward normal by supplying bicarbonate, as soon as the filtered load exceeds the T_m the additional bicarbonate will appear in the urine and the urine pH will rise (>5.5). For this reason, it is difficult, without using large amounts of bicarbonate, to normalize the serum bicarbonate of a patient with proximal RTA.

32.5.2. Distal Tubule Proton Transport

The distal tubule is responsible for the reabsorption of the 5–10% of filtered bicarbonate that normally escapes proximal reabsorption. In addition, the distal tubule secretes approximately 50–100 mEq/day of acid to match the amount produced by systemic metabolism.

The mechanism of distal tubule bicarbonate reabsorption, like in the proximal tubule, is mediated by the secretion of H^+ across the apical cell membrane (Figure 32.6). Unlike the proximal tubule, however, the majority of H^+ secretion in the distal tubule is mediated by an active, electrogenic H^+-ATPase, rather than an electroneutral Na^+/H^+ exchanger.[75,91] The active transporter of H^+ by this pump allows proton secretion to occur even in the face of large pH gradients and is responsible for the ability of the kidney to lower the urinary pH to <5.0. The H^+-ATPase in mammalian kidney has been purified and extensively characterized.[92] The kidney H^+-ATPase is a vacuolar ATPase, similar to that found in endosomes and other intracellular organelles. The H^+-ATPase pump directly couples the extrusion of H^+, against both chemical and electrical gradients, to the hydrolysis of ATP. The activity of the kidney H^+-ATPase is inhibited by *N*-ethylmaleimide but not by oligomycin, an inhibitor of the mitochondrial H^+-ATPase.[93] The kidney H^+-ATPase is a protein complex containing subunits of 70, 58, 31, and 17 kDa.[94,95] Immunolocalization studies of the H^+-ATPase indicate it is present in the apical membrane of α-intercalated cells.[96]

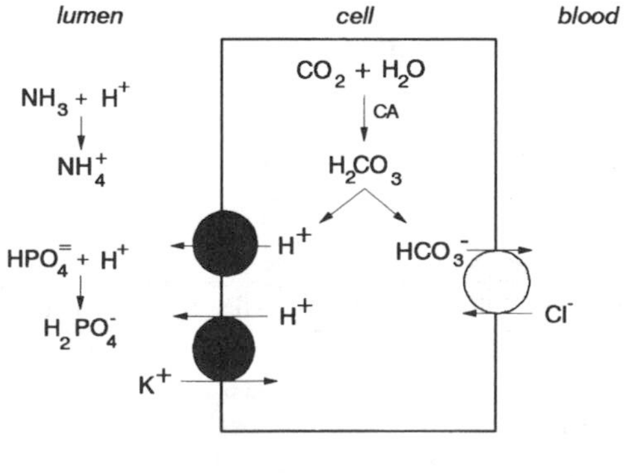

Figure 32.6. A schematic view of proton secretion by the α-intercalated cell of the collecting duct. Protons are secreted across the apical membrane via the H^+-ATPase and H^+/K^+-ATPase where they titrate luminal buffers. Bicarbonate exit across the basolateral membrane is mediated by a Cl^-/HCO_3^- exchanger.

A second proton pump, the H^+/K^+-ATPase, has also been implicated in renal acid excretion. This pump, which is similar to the gastric H^+/K^+-ATPase, catalyzes the electroneutral exchange of luminal potassium for cellular protons. The H^+/K^+-ATPase is inhibited by vanadate and by omeprazole. The contribution of H^+/K^+-ATPase to overall distal acid excretion is unclear. In isolated perfused tubules, a contribution can be demonstrated only in tubules from hypokalemic animals, a setting known to increase the expression of the transporter.[97] However, chronic treatment with vanadate can produce a hypokalemic acidification defect in rats suggesting that the H^+/K^+-ATPase does play a significant role in distal acidification.[98] Inhibition of the H^+/K^+-ATPase by dietary vanadate has been suggested as a possible cause for the high rate of distal renal tubular acidosis in humans and water buffalo in northern Thailand, an area with high soil vanadium content.[99] As in the proximal tubule, protons for apical secretion are derived from the dissociation of carbonic acid. The bicarbonate that is produced by this reaction leaves the intercalated cell across the basolateral membrane. Thus, the secretion of protons results in the generation of an equivalent number of bicarbonate ions. The exit of bicarbonate across the basolateral membrane is mediated by a Cl^-/HCO_3^- exchanger.[75]

32.5.3. Role of Urinary Buffers in Renal Acidification

The availability of buffers in the luminal fluid greatly influences the capacity of the distal tubule to secrete acid. In the proximal tubule, filtered bicarbonate ions effectively buffer the secreted protons such that the luminal pH in the proximal tubule remains above 6.9. However, in the distal tubule, where little bicarbonate normally reaches, other buffer systems must be used to allow proton secretion to proceed. In the complete absence of urinary buffers, for example, the H^+ content of urine with a pH of 5.0 is only 0.01 mEq/liter. In order to excrete the 50–100 mEq daily acid load in a reasonable urine volume (2 liters), the urine pH, in the absence of buffers, would have to be reduced to between 2 and 3. The presence of nonbicarbonate buffers in the urine allows the distal tubule to secrete the required daily acid load without reducing the pH to less than 5.0. The major urinary buffers can be divided into two groups; titratable buffers, in which the buffer pair remain within the luminal fluid, and nontitratable buffers, in which one of the buffer pair is free to diffuse into or out of the luminal fluid. The major titratable buffer in the urine is HPO_4^{2-} with the formation of titratable acidity resulting from the titration of HPO_4^{2-} to $H_2PO_4^-$ ($pK_a = 6.8$). The major nontitratable buffer is ammonia with the formation of nontitratable acidity resulting from the protonation of NH_3 to NH_4^+. NH_3 is freely diffusible across cell membranes and enters the luminal fluid by nonionic diffusion. The secretion of protons by the distal tubule causes the NH_3 to be converted to NH_4^+. Because NH_4^+ is relatively impermeable to the collecting duct cells, it becomes "trapped" in the urine and is excreted along with its proton. Increased renal ammoniagenesis and the increased excretion of NH_4^+ (nontitratable acidity) accounts almost entirely for the enhanced net acid excretion during chronic metabolic acidosis. In contrast, inadequate production of ammonium contributes to the pathogenesis of renal tubular acidosis as seen in generalized distal nephron dysfunction (Type IV RTA).

Finally, mineralocorticoids influence distal nephron acidification. In the cortical collecting duct, mineralocorticoids stimulate proton secretion indirectly. Since the H^+-ATPase is electrogenic, the rate of pumping is affected by the apical membrane potential. By stimulating sodium absorption in principle cells of the CCT, aldosterone increases the lumen-negative transepithelial voltage and increases electrogenic H^+ secretion by the α-intercalated cells.[100] In addition, aldosterone increases proton secretion through sodium-independent mechanisms in the medullary collecting duct.[101]

32.5.4. Proximal RTA

32.5.4.1. Clinical Features

Proximal RTA can present as an isolated defect in proximal tubule bicarbonate reabsorption or, more commonly, as a generalized defect in proximal tubular function. The latter disorder, the Fanconi syndrome, also features glycosuria, aminoaciduria, phosphaturia, and uricosuria.[90] Table 32.4 lists some of the causes of proximal RTA. The diagnosis is initially suspected in a patient with a hyperchloremic metabolic acidosis. The primary causes of a hyperchloremic metabolic acidosis are related to either gastrointestinal bicarbonate losses or renal bicarbonate losses (RTA). A helpful clue in distinguishing between these two possibilities comes from the measurement of the urinary anion gap (UAG). This gap, defined as $[Na^+ + K^+]_{urine} - [Cl^-]_{urine}$, has been proposed as an indirect measure of the ammonium content of the urine since ammonium, NH_4^+, represents the major unmeasured cation in the urine.[102] A UAG of 0 corresponds to the daily excretion of approximately 80 mmole of NH_4^+. However, as mentioned earlier, the excretion of nontitratable acidity, i.e., ammonium, increases dramatically in response to metabolic acidosis. This increase in an unmeasured cation will result in a UAG that is negative, i.e., the urine Cl^- exceeds the sum of urine Na^+ and K^+. In metabolic acidosis caused by defects in renal acidification (RTA), the urine ammonium content is not increased. Thus, the finding of a negative UAG in the setting of hyperchloremic metabolic acidosis suggests an external bicarbonate loss (or acid gain) while a positive UAG suggests a renal bicarbonate loss.[103]

As discussed earlier, patients with proximal RTA generally present in a steady-state acidosis in which the serum bicarbonate concentration is at or below their threshold value for proximal bicarbonate reabsorption (Figure 32.5). Accordingly, the urine pH may be low (<5.5) and the urine will contain little bicarbonate. Subsequent supplementation with bicarbonate, however, leads to bicarbonaturia (fractional excretion of $HCO_3^- > 10–15\%$) and an alkaline urine.[104] Patients with Fanconi syndrome will also have other evidence of abnormal proximal tubular dysfunction such as glycosuria with a normal blood glucose concentration, hypophosphatemia, hypouricemia, hyperphosphaturia, and aminoaciduria. In contrast to distal RTA (see below), nephrocalcinosis and nephrolithiasis are uncommon in proximal RTA. Hypokalemia may be present, particularly in patients treated with bicarbonate. The hypokalemia is related primarily to renal potassium losses induced by the presence of nonabsorbable anions (HCO_3^-) in the collecting duct. Osteomalacia resulting from hypophosphatemia may develop in patients with the Fanconi syndrome.[75,90] Table 32.5 lists some of the diagnostic studies in different types of RTA.

Table 32.4. Disorders Associated with Proximal RTA[a]

Isolated RTA
- Familial
- Carbonic anhydrase deficiency
  - Pharmacologic—acetazolamide
  - Congenital

Generalized (Fanconi syndrome)
- Familial
- Secondary
  - Heritable and systemic diseases
    - Cystinosis
    - Tyrosinemia
    - Hereditary fructose intolerance
    - Galactosemia
    - Glycogen storage disease
    - Wilson disease
    - Multiple myeloma
    - Amyloidosis
    - Paroxysmal nocturnal hemoglobinuria
    - Balkan nephropathy
    - Medullary cystic disease
  - Vitamin D resistance and deficiency
  - Drugs and toxins
    - Cadmium
    - Uranium
    - Lead
    - Mercury
    - Outdated tetracycline
    - Aminoglycosides
    - Toluene
    - Maleic acid

[a]Compiled from Refs. 75, 90, 104.

32.5.4.2. Treatment of Proximal RTA

The goal of therapy is to normalize the serum bicarbonate concentration and systemic pH. The treatment of any un-

Table 32.5. Diagnostic Studies in RTA

	Proximal RTA	Classic dRTA	Generalized dRTA
Serum K^+	Low	Low	High
Urine pH	<5.5	>5.5	<5.5
Urine anion gap	Low	Low	Low
Fanconi syndrome	Common	No	No
Fe_{HCO3}[a]	>10%[b]	<5%	<5%
Nephrocalcinosis	No	Common	No

[a]FE_{HCO_3}, fractional excretion of bicarbonate.
[b]Elevated FE_{HCO_3} in proximal RTA is during bicarbonate loading.

derlying disorder should also be attempted. The mainstay of therapy for the metabolic acidosis is the administration of exogenous alkali such as bicarbonate (Table 32.6). Since such treatment aggravates the urinary bicarbonate losses, massive supplementation may be required to achieve and maintain a normal serum bicarbonate concentration. Indeed, the degree of difficulty in correcting the acidosis may be a clue in distinguishing proximal RTA from distal RTA. The bicarbonaturia may be diminished by treatment with thiazide diuretics. This maneuver is based on the observation that extracellular volume depletion increases the threshold for proximal tubular bicarbonate reabsorption.

32.5.5. Distal RTA

Defects in the secretion of acid by the distal tubule could result through several potential mechanisms: a decrease in the functional number of either H^+-ATPase or H^+,K^+-ATPase; decreased sodium transport by principal cells of the collecting duct; aldosterone deficiency or resistance; decreased renal ammoniagenesis; or increased back leak of H^+ across the tubule epithelium.[91,105] Table 32.7 presents a classification of distal RTA according to these mechanisms along with clinical examples of each.

32.5.5.1. Hypokalemic dRTA

32.5.5.1a. Clinical Features. A first step in the differential diagnosis of distal RTA is the measurement of the serum potassium concentration (Table 32.5). Patients with classic distal RTA (also referred to as Type I RTA) have hypokalemia, hyperchloremic metabolic acidosis, and an inability to lower the urine pH below 5.5. Nephrocalcinosis and nephrolithiasis are common.[104] This disorder should be suspected in any patient with hyperchloremic metabolic acidosis and an elevated urine pH. The distinction between dRTA and gastrointestinal bicarbonate losses, which can also cause a high urine pH via increased ammonium excretion, can be aided by measurement of the UAG (see above). Classic distal RTA is rarely seen as an isolated disorder. Table 32.8 lists many of the disorders associated with distal RTA.

32.5.5.1b. Pathophysiology. Classic distal RTA has been attributed to either a proton secretory defect, such as a decrease in either H^+-ATPase or H^+,K^+-ATPase, or to a "gradient" defect resulting in increased back leak of protons.

Table 32.6. Forms of Alkali Replacement[a]

Agent	HCO_3^- equivalents
Shohl's solution (citric acid and sodium citrate)	1 mEq/ml
Polycitra (citric acid and Na^+/K^+ citrate)	2 mEq/ml
Baking soda	50–60 mEq/tsp
$NaHCO_3$	3.9 mEq/325-mg tablet
K-Lyte	25- and 50-mEq tablets

[a]Modified from Ref. 75.

Evidence for a defect in H^+-ATPase comes from an immunohistochemical study of one patient with distal RTA associated with Sjogren syndrome which revealed the absence of H^+-ATPase in the collecting duct.[106] There is no direct evidence, as yet, for a decrease in H^+,K^+-ATPase in distal RTA. However, the findings that vanadate, an inhibitor of H^+,K^+-ATPase, can cause hypokalemic distal RTA in rats,[98] and that distal RTA is endemic in areas with high soil vanadium content[99] are consistent with the possibility that decreases in H^+,K^+-ATPase may result in distal RTA. The best characterized, but probably least common, mechanism for distal RTA is increased back leak of protons causing a "gradient" defect. In this disorder, the secretion of protons is normal but steep pH gradients cannot be maintained because of the diffusion of protons back through the leaky epithelium into the blood. The distal RTA caused by amphotericin B is the best, and perhaps only, clinical example of this type of defect.[107]

32.5.5.1c. Treatment. Even though the daily bicarbonate losses in distal RTA are rather small (50–100 mEq/day), these patients never achieve a steady-state acid–base balance. This is in contrast to patients with proximal RTA who, even though bicarbonate losses can be massive, eventually come into an acid–base balance. Thus, untreated patients tend to have an inexorable erosion of the bicarbonate and nonbicarbonate buffer supplies resulting in acidosis and osteopenia. The goal of treatment is restoration and maintenance of body bicarbonate stores. Since the ongoing bicarbonate losses are modest, most patients will respond to 2–3 mEq kg per day of alkali replacement (Table 32.6).

32.5.5.2. Hyperkalemic dRTA

32.5.5.2a. Clinical Features. Two forms of distal RTA characterized by hyperkalemia are recognized: voltage-dependent RTA and generalized distal nephron dysfunction. The latter is sometimes referred to as Type IV RTA—we fa vor the more descriptive name. Voltage-dependent RTA occurs following urinary tract obstruction, in patients with sickle-cell disease, and as a side effect of potassium-sparing diuretics such as amiloride and triamterene. The features of voltage-dependent RTA include: inability to acidify the urine to pH<5.5, normal or increased plasma aldosterone concentrations, impaired H^+ secretion as reflected by a low urine P_{CO_2}, and an abnormal response to sulfate infusion.[91] The response to sulfate infusion has been used to gauge the effects of sodium and voltage on H^+ secretion in the cortical collecting duct; however, its value in providing any mechanistic information has been questioned.[75]

Patients with generalized distal nephron dysfunction usually present with hyperchloremic metabolic acidosis, mild renal insufficiency, and hyperkalemia which is out of proportion to the degree of renal insufficiency. The urinary excretion of ammonium is decreased. In contrast to classic distal RTA or voltage-dependent RTA, patients with generalized distal nephron dysfunction can lower the urine pH below 5.5. The

Table 32.7. Classification of Distal Tubule Acidification Defects[a]

Defect	Descriptive terminology	Clinical examples
Decreased or defective H^+-ATPase	Secretory defect	Sjogren syndrome
Decreased or defective H^+,K^+-ATPase	Secretory defect	Vanadate toxicity (?)
Decreased sodium transport in CCT	Voltage defect	Amiloride, lithium
Hypoaldosteronism	Generalized defect	Diabetes
Impaired ammoniagenesis	Generalized defect	Interstitial renal disease
Increased leak of H^+	Gradient defect	Amphotericin B

[a]Adapted from Ref. 105.

secretion of H^+, then, is *qualitatively* normal, but is limited *quantitatively* by the availability of ammonium as a urinary buffer. Potassium secretion by the cortical collecting duct is impaired as measured by the transtubular potassium gradient (TTKG).[75] This has been attributed ti low aldosterone levels, either from primary hypoaldosteronism or hyporeninemia, or a decreased response to aldosterone.[108] This disorder is particularly common in patients with diabetic nephropathy. Patients receiving NSAIDs or angiotensin-converting enzyme inhibitors are also at risk for developing generalized distal nephron dysfunction. Some of the disorders associated with generalized distal nephron dysfunction are listed in Table 32.9.

32.5.5.2b. Pathophysiology. Several factors contribute to the development of metabolic acidosis in hypoaldosteronism. As noted in an earlier section, aldosterone stimulates proton secretion in the collecting duct through both sodium-dependent and -independent mechanisms. Second, the hyperkalemia that results from hypoaldosteronism suppresses ammonia synthesis in the proximal tubule thereby reducing the buffering capacity of the urine. Third, both H^+ secretion and ammoniagenesis may be limited by the reduction in nephron mass associated with the mild to moderate renal insufficiency that usually accompanies this disorder. Finally, hyperkalemia and renal interstitial disease may impair the accumulation of ammonium within the renal medulla.

32.5.5.2c. Treatment. The decision to treat patients with generalized distal nephron dysfunction depends on the severity of the acidosis and hyperkalemia. Many patients have mild disease and do not require specific treatment. Because hyperkalemia is potentially life-threatening, the presence of hyperkalemia, more than acidosis, often dictates the need for therapy. In addition, correction of the hyperkalemia often improves the metabolic acidosis by increasing ammonium delivery to the collecting duct. Mineralocorticoid replacement is useful, particularly in patients with primary hypoaldosteronism. Patients with hyporeninemic hypoaldosteronism may

Table 32.8. Conditions Associated with Classic Distal RTA

Primary
Secondary
Hypercalciuria and nephrocalcinosis
Hyperparathyroidism
Vitamin D intoxication
Idiopathic hypercalcivria
Autoimmune diseases
Cryoglobulinemia
Sjogren syndrome
Chronic active hepatitis
Systemic lupus erythematosus
Primary biliary cirrhosis
Polyarteritis nodosa
Thyroiditis
Drugs and toxins
Amphotericin B
Lithium
Toluene
Vanadate
Tubulointerstitial diseases
Amyloidosis
Obstructive uropathy
Renal transplantation
Balkan nephropathy
Heritable diseases
Sickle-cell anemia
Medullary cystic disease
Medullary sponge kidney
Carbonic anhydrase deficiency
Ehlers–Danlos syndrome
Fabry disease
Marfan syndrome
Hereditary elliptocytosis

Table 32.9. Disorders Associated with Generalized Distal Nephron Dysfunction

Primary hypoaldosteronism
Addison disease
Congenital enzyme defects
21-Hydroxylase deficiency
3-β-Hydroxydehydrogenase deficiency
Familial methyl oxidase deficiency
Chronic heparin therapy
Hyporeninemic hypoaldosteronism
Diabetes
Urinary tract obstruction
Analgesic nephropathy
Gout
Sickle-cell anemia
Multiple myeloma
Renal transplantation
Systemic lupus erythematosus
Drugs
Angiotensin-converting enzyme inhibitors
Nonsteroidal antiinflammatory drugs
Amiloride
Cyclosporin
Spironolactone
Triamterene

also respond to mineralocorticoid supplementation. Generally, though, initial therapy for hyperkalemia in these patients consists of cation-exchange resins (Kayexalate) or treatment with a loop diuretic and high-salt diet to induce a kaliuresis.

REFERENCES

1. Friedman, A. L., and Chesney, R. W. (1993). Isolated renal tubular disorders. In *Diseases of the Kidney,* 5th ed. (R. W. Schrier and C. W. Gottschalk, eds.), Little, Brown, Boston, pp. 611–634.
2. Reeves, W. B., and Andreoli, T. E. (1993). Tubular sodium transport. In *Diseases of the Kidney,* 5th ed. (R. W. Schrier and C. W. Gottschalk, eds.), Little, Brown, Boston, pp. 139–179.
3. Barfuss, D. W., and Schafer, J. A. (1979). Active amino acid absorption by proximal convoluted and proximal straight tubules. *Am. J. Physiol.* **236:**F149–F155.
4. Samarzija, I., and Fromter, E. (1982). Electrophysiologic analysis of rat renal sugar and amino acid transport. I. Basic phenomema. *Pfluegers Arch.* **393:**179–188.
5. Fass, S. J., Hammerman, M. R., and Sactor, B. (1977). Transport of amino acids in renal brush border membrane vesicles. Uptake of the neutral amino acid L-alanine. *J. Biol. Chem.* **252:**583–587.
6. Murer, H., Leopolder, A., Kinne, R., and Burckhardt, G. (1980). Recent observations on the proximal tubular transport of acid and basic amino acids by rat renal proximal tubular brush border vesicles. *Int. J. Biochem.* **12:**222–230.
7. Doyle, F. A., and McGivan, J. D. (1992). The bovine renal epithelial cell line NBL-1 expresses a broad specificity Na^+-dependent neutral amino acid transport system similar to that in bovine renal brush border membrane vesicles. *Biochim. Biophys. Acta* **1104:**55–59.
8. Doyle, F. A., and McGivan, J. D. (1992). Reconstitution and identification of the major Na^+-dependent neutral amino acid transport protein from bovine renal brush-border membrane vesicles. *Biochem. J.* **281:**95–101.
9. Tate, S. S., Yan, N., and Undenfriend, S. (1992). Expression cloning of a Na^+-independent neutral amino acid transporter from rat kidney. *Proc. Natl. Acad. Sci. USA* **89:**1–5.
10. Samarzija, I., and Fromter, E. (1982). Electrophysiologic analysis of rat renal sugar and amino acid transport. III. Neutral amino acids. *Pfluegers Arch.* **393:**199–209.
11. Samarzija, I., and Fromter, E. (1982). Electrophysiologic analysis of rat renal sugar and amino acid transport. IV. Basic amino acids. *Pfluegers Arch.* **393:**210–214.
12. Samarzija, I., and Fromter, E. (1982). Electrophysiologic analysis of rat renal sugar and amino acid transport. V. Acidic amino acids. *Pfluegers Arch.* **393:**215–219.
13. Wells, R. G., and Hediger, M. A. (1992). Cloning of a rat kidney cDNA that stimulates dibasic and neutral amino acid transport and has sequence similarity to glucosidases. *Proc. Natl. Acad. Sci. USA* **89:**5596–5600.
14. Lee, W. S., Wells, R. G., Sabbag, R. V., Mohandas, T. K., and Hediger, M. A. (1993). Cloning and chromosomal localization of a human kidney cDNA involved in cystine, dibasic, and neutral amino acid transport. *J. Clin. Invest.* **91:**1959–1965.
15. Segal, S., and Thier, S. O. (1995). Cystinuria. In *The Metabolic and Molecular Bases of Inherited Disease,* 7th ed. (C. R. Scriver, A. L. Beaudet, W. S. Sly, and D. Valle, eds.), McGraw–Hill, New York, pp. 3581–3606.
16. Bertrand, J., Werner, A., Moore, M. L., Strange, G., Markovich, D., Biber, J., Testar, X., Zorzano, A., Palacin, M., and Murer, H. (1992). Expression cloning of a cDNA from rabbit kidney cortex that induces a single transport system for cystine and dibasic and neutral amino acids. *Proc. Natl. Acad. Sci. USA* **89:**5601–5605.
17. Pras, E., Arber, N., Aksentijevich, I., Katz, G., Schapiro, J. M., Prosen, L., Gruberg, L., Harel, D., Liberman, U., Weissenbach, J., Pras, M., and Kastner, D. L. (1994). Localization of a gene causing cystinuria to chromosome 2p. *Nature Genet.* **6:**415–419.
18. Calonge, M. J., Gasarine, P., Chillaron, J., Chillon, M., Gallucci, M., Roussard, F., Zelante, L., Testar, X., Dallapiccola, B., DiSilverio, F., Barcelo, P., Estivill, X., Zorzano, A., Nunes, V., and Palacin, M. (1994). Cystinuria caused by mutations in rBAT, a gene involved in the transport of cystine. *Nature Genet.* **6:**420–425.
19. Levy, H. L. (1973). Genetic screening. In *Advances in Human Genetics* (H. Harris and K. Hirschhorn, eds.), Plenum Press, New York, pp. 1–82.
20. Dent, C. E., Friedmann, M., Green, H., and Watson, L. C. A. (1965). Treatment of cystinuria. *Br. Med. J.* **1:**403–405.
21. Levy, H. L. (1995). Hartnup disorder. In *The Metabolic and Molecular Bases of Inherited Disease,* 7th ed. (C. R. Scriver, A. L. Beaudet, W. S. Sly, and D. Valle, eds.), McGraw–Hill, New York, pp. 3629–3642.
22. Scriver, C. R., Mahon, B., Levy, H. L., Clow, C. L., Reade, T. M., Kronick, J., Lemieux, B., and Laberge, C. (1987). The Hartnup phenotype: Mendelian transport disorder, multifactorial disease. *Am. J. Hum. Genet.* **40:**401–411.
23. Haim, S., Gilhar, A., and Cohen, A. (1978). Cutaneous manifestations associated with aminoaciduria. Report of two cases. *Dermatologica* **156:**244–249.
24. Mori, E., Yamadori, A., Tsutsumi, A., and Kyotani, Y. (1989). Adult onset Hartnup disease presenting with neuropsychiatric symptoms but without skin lesions. *Clin. Neurol.* **29:**687–690.
25. Erly, W., Castillo, M., Foosaner, D., and Bonmati, C. (1991). Hartnup disease: MR findings. *AJNR* **12:**1026–1031.
26. Chesney, R. W. (1995). Iminoglycinuria. In *The Metabolic and Molecular Bases of Inherited Disease,* 7th ed. (C. R. Scriver, A. L. Beaudet, W. S. Sly, and D. Valle, eds.), McGraw–Hill, New York, pp. 3643–3653.
27. Desjeux, J.-F., Turk, E., and Wright, E. (1995). Congenital selective Na^+ D-glucose cotransport defects leading to renal glycosuria and congenital selective intestinal malabsorption of glucose and galactose. In *The Metabolic and Molecular Bases of Inherited Disease,* 7th ed. (C. R. Scriver, A. L. Beaudet, W. S. Sly, and D. Valle, eds.), McGraw–Hill, New York, pp. 3563–3580.
28. DeMarchi, S., Cecchin, E., Basile, A., Proto, G., Donadon, W., Jengo, A., Schinella, D., Jus, A., Villalta, D., De Paoli, P., Santini, G., and Tesio, F. (1984). Close genetic linkage between HLA and renal glycosuria. *Am. J. Nephrol.* **4:**280–286.
29. Oemar, B. S., Byrd, D. J., and Brodehl, J. (1987). Complete absence of tubular glucose absorption: A new type of renal glucosuria. *Clin. Nephrol.* **27:**156–160.
30. Biagi, B. A., Kubota, T., Sohtell, M., and Giebisch, G. (1981). Intracellular potentials in rabbit proximal tubules perfused *in vitro. Am. J. Physiol.* **240:**F200–F208.
31. Aronson, P. S., and Sacktor, B. (1975). The Na^+ gradient-dependent transport of D-glucose in renal brush border membranes. *J. Biol. Chem.* **250:**6032–6036.
32. Kinne, R., Murer, H., Kinne-Saffran, E., Thees, M., and Sachs, G. (1975). Sugar transport by renal plasma membrane vesicles. *J. Membr. Biol.* **21:**375–382.
33. Barfuss, D. W., and Schafer, J. A. (1981). Differences in active and passive glucose transport along the proximal nephron. *Am. J. Physiol.* **236:**F149–F156.
34. Turner, R. J., and Moran, A. (1982). Heterogeneity of sodium-dependent D-glucose transport sites along the proximal tubule: Evidence from vesicle studies. *Am. J. Physiol.* **242:**F406–F412.

35. Turner, R. J., and Moran, A. (1982). Stoichiometric studies of renal outer critical brush border membrane D-glucose transporter. *J. Membr. Biol.* **67:**73–80.
36. Turner, R. J., and Moran, A. (1982). Further studies of proximal tubular brush border membrane D-glucose transport heterogeneity. *J. Membr. Biol.* **70:**37–43.
37. Hediger, M. A., Coady, M. J., Ikeder, T. S, and Wright, E. M. (1987). Expression cloning and cDNA sequencing of the Na^+/glucose co-transporter. *Nature* **330:**379–381.
38. Hediger, M. A., Turk, E., and Wright, E. M. (1989). Homology of the human intestinal Na^+/glucose and *Escherichia coli* Na^+/proline cotransporters. *Proc. Natl. Acad. Sci. USA* **86:**5748–5753.
39. Coady, M. J., Pajor, A. M., and Wright, E. M. (1990). Sequency homologies among intestinal and renal Na^+/glucose cotransporters. *Am. J. Physiol.* **259:**C605–C612.
40. Pajor, A. M., Hirayama, B. A., and Wright, E. M. (1992). Molecular evidence for two renal Na^+/glucose cotransporters. *Biochim. Biophys. Acta* **1106:**216–219.
41. Wells, R. G., Pajor, A. M., Kanai, T., Turk, E., Wright, E. M., and Hediger, M. A. (1992). Cloning of a human kidney cDNA with similarity to the sodium–glucose cotransporter. *Am. J. Physiol.* **263:** F459–F467.
42. Kanai, Y., Lee, W. S, You, G., Brown, D., and Hediger, M. A. (1994). The human kidney low affinity Na^+/glucose cotransporter SGLT2. Delineation of the major renal resorptive mechanism for D-glucose. *J. Clin. Invest.* **93:**397–405.
43. Turk, E., Zabel, B., Mundlos, S., Dyer, J., and Wright, E. (1991). Glucose/galactose malabsorption: A defect in the Na^+/glucose co-transporter. *Nature* **350:**354–356.
44. Reeves, W. B., and Andreoli, T. E. (1995). Nephrogenic diabetes insipidus. In *The Metabolic and Molecular Bases of Inherited Disease,* 7th ed. (C. R. Scriver, A. L. Beaudet, W. S. Sly, and D. Valle, eds.), McGraw–Hill, New York.
45. Reeves, W. B., and Andreoli, T. E. (1992). The posterior pituitary and water metabolism. In *Williams Textbook of Endocrinology,* 8th ed. (J. D. Wilson and D. W. Foster, eds.), Saunders, Philadelphia, pp. 311–356.
46. Ganote, C. E., Grantham, J. J., Moses, H. L., Burg, M. B., and Orloff, J. (1968). Ultrastructural studies of vasopressin effect in isolated perfused renal collecting tubules of the rabbit. *J. Cell Biol.* **36:**355–362.
47. Grantham, J. J., Ganote, C. E., Burg, M. B., and Orloff, J. (1968). Paths of transtubular water flow in isolated renal collecting tubules. *J. Cell Biol.* **41:**562–570.
48. Schafer, J. A., and Andreoli, T. E. (1972). Cellular constraints to diffusion: The effect of antidiuretic hormone on water flows in isolated mammalian collecting tubules. *J. Clin. Invest.* **51:**1264–1270.
49. Birnbaumer, M., Seibold, A., Gilbert, S., Ishido, M., Barberis, C., Antaramian, A., Brabet, P., and Rosenthal, W. (1992). Molecular cloning of the receptor for human antidiuretic hormone. *Nature* **357:** 333–335.
50. Lolait, S. J., O'Carroll, A.-M., McBride, O. W., Konig, M., Morel, A., and Brownstein, M. J. (1992). Cloning and characterization of a vasopressin V2 receptor and possible link to nephrogenic diabetes insipidus. *Nature* **357:**336–340.
51. Nielsen, S., and Agre, P. (1995). The aquaporin family of water channels in kidney. *Kidney Int.* **48:**1057–1068.
52. Nielsen, S., Chou, C. L., Marples, D., Christensen, E. I., Kishore, B. K., and Knepper, M. A. (1995). Vasopressin increases water permeability of kidney collecting duct by producing translocation of aquaporin-CD water channels to plasma membrane. *Proc. Natl. Acad. Sci. USA* **92:**1013–1017.
53. Bichet, D. G., Razi, M., Arthus, M.-F., Lonergan, M., Tittley, P., Smiley, R. K., Rock, G., and Hirsch, D. J. (1989). Epinephrine and dDAVP administration in patients with congenital nephrogenic diabetes isipidus. Evidence for a pre-cyclic AMP V2 receptor defective mechanism. *Kidney Int.* **36:**859–867.
54. Knoers, N., Brommer, E. J. P., Willems, H., van Oost, B. A., and Monens, L. A. H. (1990). Fibrinolytic responses to 1-desamino-8-D-arginine-vasopressin in patients with congenital nephrogenic diabetes insipidus. *Nephron* **54:**326–331.
55. D'Avanzo, M., Toraldo, R., Fazzone, A., Papa, M. L., Santinelli, R., Tolone, C., and Lafusco, F. (1991). Factor VIII response to vasopressin in nephrogenic diabetes insipidus. *J. Pediatr.* **119:**504–510.
56. Kobrinsky, N. L., Doyle, J. J., Israels, E. D., Winter, J. J., Cheang, M. S., Walker, R. D., and Bishop, A. J. (1985). Absent factor VIII response to synthetic vasopressin analogue (DDAVP) in nephrogenic diabetes insipidus. *Lancet* **8441:**1293–1294.
57. Brenner, B., Seligsohn, U., and Hochberg, Z. (1988). Normal response of factor VIII and von Willebrand factor to 1-deamino-8-D-arginine vasopressin in nephrogenic diabetes insipidus. *J. Clin. Endocrinol. Metab.* **67:**191–197.
58. Moses, A. M., Miller, J. L., and Levine, M. A. (1988). Two distinct pathophysiological mechanisms in congenital nephrogenic diabetes insipidus. *J. Clin. Endocrinol. Metab.* **66:**1259–1265.
59. Knoers, N., van der Heyden, H., van Oost, B. A., Monnens, L., Willems, J., and Ropers, H. H. (1988). Linkage of X-linked nephrogenic diabetes insipidus with DAS52, a polymorphic DNA marker. *Nephron* **50:**187–193.
60. Knoers, N., van der Heyden, H., van Oost, B. A., Monners, L., Ropers, H. H., and Willems, J. (1988). Nephrogenic diabetes insipidus: Close linkage with markers from the distal long arm of the human X chromosome. *Hum. Genet.* **80:**31–36.
61. Kambouris, M., Dlouhy, S. R., Trofatter, J. A., Conneally, P. M., and Hodes, M. E. (1988). Localization of the gene for X-linked nephrogenic diabetes insipidus to Xq28. *Am. J. Med. Genet.* **29:**239–245.
62. Jans, D. A., van Oost, B. A., Ropers, H. H., and Fahrenholz, F. (1990). Derivatives of somatic cell hybrids which carry the human gene locus for nephrogenic diabetes insipidus (NDI) express functional vasopressin renal V2-type receptors. *J. Biol. Chem.* **265:** 15379–15383.
63. Rosenthal, W., Seibold, A., Antaramian, A., Lonergan, M., Arthus, M.-F., Hendy, G. N., Birnbaumer, M., and Bichet, D. G. (1992). Molecular identification of the gene responsible for congenital nephrogenic diabetes insipidus. *Nature* **359:**233–236.
64. Oksche, A., Dickson, J., Schulein, R., Seyberth, H. W., Muller, M., Rascher, W., Birnbaumer, M., and Rosenthal, W. (1994). Two novel mutations in the vasopressin V2 receptor gene in patients with congenital nephrogenic diabetes insipidus. *Biochem. Biophys. Res. Commun.* **205:**552–557.
65. Holtzman, E. J., Kolakowski, L. F., Jr., Geifman-Holtzman, O., O'Brien, D. G., Rasoulpour, M., Guillot, A. P., and Ausiello, D. A. (1994). Mutations in the vasopressin V2 receptor gene in two families with nephrogenic diabetes insipidus. *J. Am. Soc. Neph.* **5:**169–176.
66. Knoers, N. V., van den Ouweland, A. M., Verdijk, M., Monnens, L. A., and van Oost, B. A. (1994). Inheritance of mutations in the V2 receptor gene in thirteen families with nephrogenic diabetes insipidus. *Kidney Int.* **46:**170–176.
67. Birnbaumer, M., Gilber, S., and Rosenthal, W. (1994). An extracellular congenital nephrogenic diabetes insipidus mutation of the vasopressin receptor reduces cell surface expression, affinity for ligand, and coupling to the Gs/adenylyl cyclase system. *Mol. Endocrinol.* **8:**886–894.
68. Langley, J. M., Balfe, J. W., Selander, T., Ray, P. N., and Clarke, J. T. R. (1991). Autosomal recessive inheritance of vasopressin-resistant diabetes insipidus. *Am. J. Med.* **38:**90–93.
69. Deen, P. M. T., Verdijk, M., Knoers, N., Wieringa, B., Monnens, L. A., van Os, C. H., and van Oost, B. A. (1994). Requirement of human renal water channel aquaporin-2 for vasopressin-dependent concentration of urine. *Science* **264:**92–95.

70. van Lieburg, A. F., Verdijk, M. A., Knoers, N. V., van Essen, A. J., Proesmans, W., Mallmann, R., Monnens, L. A., van Oost, B. A., van Os, C. H., and Deen, P. M. (1994). Patients with autosomal nephrogenic diabetes insipidus homozygous for mutations in the aquaporin 2 water-channel gene. *Am. J. Hum. Genet.* **55:**648–652.
71. Bergstein, J. M. Nephrogenic diabetes insipidus. In *Textbook of Pediatrics* (R. E. Behrman, R. M. Kliegman, W. E. Nelson, and V. C. Vaughan III, eds.), Saunders, Philadelphia.
72. Alon, U., and Chan, J. C. (1985). Hydrochlorothiazide–amiloride in the treatment of congenital nephrogenic diabetes insipidus. *Am. J. Nephrol.* **5:**9–13.
73. Knoers, N., and Monnens, L. A. H. (1990). Amiloride–hydrochlorothiazide versus indomethacin–hydrochlorothiazide in the treatment of nephrogenic diabetes insipidus. *J. Pediatr.* **117:**499–503.
74. Battle, D. C., von Riotte, A. B., Gaviria, M., and Grupp, M. (1985). Amelioration of polyuria by amiloride in patients receiving long-term lithium therapy. *N. Engl. J. Med.* **312:**408–411.
75. DuBose, T. D., Jr., and Alpern, R. J. (1995). Renal tubular acidosis. In *The Metabolic and Molecular Bases of Inherited Disease,* 7th ed. (C. R. Scriver, A. L. Beaudet, W. S. Sly, and D. Valle, eds.), McGraw–Hill, New York, pp. 3655–3689.
76. Murer, H., Hopfer, V., and Kinne, R. (1976). Sodium/protein antiport in brush-border membrane vesicles isolated from rat small intestine and kidney. *Biochem. J.* **154:**597–601.
77. Kinsella, J. L., and Aronson, P. S. (1980). Properties of the Na^+–H^+ exchange in renal microvillus membrane vesicles. *Am. J. Physiol.* **238:**F401–F410.
78. Aronson, P. S. (1983). Mechanisms of active H^+ secretion in the proximal tubule. *Am. J. Physiol.* **245:**F647–F654.
79. Aronson, P. S., Nee, J., and Suhm, M. A. (1982). Modifier role of internal H^+ in activating the Na^+–H^+ in renal microvillus vesicles. *Nature* **299:**161–163.
80. Boron, W. F., and Boulpaep, E. L. (1983). Intracellular pH regulation in the renal proximal tubule of the salamander. Na^+–H^+ exchange. *J. Gen. Physiol.* **81:**29–41.
81. Tsai, C. J., Ives, H. E., Alpern, R. J., Yee, V. J., Warnock, D. G., and Rector, F. C., Jr. (1984). Increased Vmax for Na^+/H^+ antiporter activity in proximal tubule brush border vesicles from rabbits with metabolic acidosis. *Am. J. Physiol.* **247:**F339–F346.
82. Casavola, V., Helmle-Kolb, C., and Murer, H. (1989). Separate regulatory control of apical and basolateral Na^+/H^+ exchange in renal epithelial cells. *Biochem. Biophys. Res. Commun.* **165:**833–837.
83. Haggerty, J. G., Agarwal, N., Reilly, R. F., Adelberg, E. A., and Slayman, C. W. (1988). Pharmacologically different Na/H antiporters on the apical and basolateral surfaces of cultured porcine kidney cells (LLC-PK1). *Proc. Natl. Acad. Sci. USA* **85:**6797–6801.
84. Sardet, C., Franchi, A., and Pouyssegur, J. (1989). Molecular cloning, primary structure, and expression of the human growth factor-activatable Na^+/H^+ antiporter. *Cell* **56:**271–280.
85. Helmle-Kolb, C., Counillon, L., Roux, D., Pouyssegur, J., Mrkic, B., and Murer, H. (1993). Na/H exchange activities in NHE1-transfected OK-cells: Cell polarity and regulation. *Pfluegers Arch.* **425:**34–41.
86. Biemesderfer, D., Pizzonia, J., Abu-Alfa, A., Exner, M., Reilly, R., Igarashi, P., and Aronson, P. S. (1993). NHE3: A Na^+/H^+ exchanger isoform of renal brush border. *Am. J. Physiol.* **265:**F736–F742.
87. Alpern, R. J. (1985). Mechanism of basolateral membrane $H^+/OH^-/HCO_3^-$ transport in the rat proximal convoluted tubule. A sodium-coupled electrogenic process. *J. Gen. Physiol.* **86:**613–623.
88. Sasaki, S., Shiigai, T., Yoshiyama, N., and Takeuchi, J. (1987). Mechanism of bicarbonate exit across basolateral membrane of rabbit proximal straight tubule. *Am. J. Physiol.* **21:**F11–F20.
89. Sly, W. S., Whyte, M. P., Sundaram, V., Tashian, R. E., Hewett-Emmett, D., Guibaud, P., Vainsel, M., Baluarte, H. J., Gruskin, A., Al-Mosawi, M., Sakati, N., and Ohlsson, A. (1985). Carbonic anhydrase II deficiency in 12 families with autosomal recessing syndrome of osteopetrosis with renal tubular acidosis and cerebral calcification. *N. Engl. J. Med.* **313:**139–143.
90. Bergeron, M., Gougoux, A., and Vinay, P. (1995). The renal Fanconi syndrome. In *The Metabolic and Molecular Bases of Inherited Disease,* 7th ed. (C. R. Scriver, A. L. Beaudet, W. S. Sly, and D. Valle, eds.), McGraw–Hill, New York, pp. 3691–3704.
91. Sabatini, S., and Kurtzman, N. A. (1991). Pathophysiology of the renal tubular acidoses. *Semin. Neph.* **11:**202–211.
92. Gluck, S. L., Nelson, R. D., and Lee, B. S. (1993). Properties and regulation of the renal vacuolar H^+-ATPase and H^+-K^+-ATPase. *Curr. Opin. Neph. Hypertension* **2:**715–724.
93. Diaz-Diaz, F. D., LaBelle, E. F., Eaton, D. C., and DuBose, T. D., Jr. (1986). ATP-dependent proton transport in human renal medulla. *Am. J. Physiol.* **20:**F297–F305.
94. Gluck, S., and Caldwell, J. (1987). Immunoaffinity purification and characterization of vacuolar H^+ ATPase from bovine kidney. *J. Biol. Chem.* **262:**15780–15784.
95. Gluck, S., and Bastani, B. (1991). The biochemistry of distal urinary acidification in health and disease. In *Acid–Base Balance* (N. G. DeSanto and G. Capasso, eds.), Editoriale Bios, Consenza, Italy, pp. 21–34.
96. Brown, D., Hirsch, S., and Gluck, S. (1988). Localization of a proton-pumping ATPase in rat kidney. *J. Clin. Invest.* **82:**2114–2126.
97. Wingo, C. S. (1989). Active proton secretion and potassium absorption in the rabbit outer medullary collecting duct. *J. Clin. Invest.* **84:**361–369.
98. Dafnis, E., Spohn, M., Lonis, B., Kurtzman, N. A., and Sabatini, S. (1992). Vanadate causes hypokalemic distal renal tubular acidosis. *Am. J. Physiol.* **262:**F449–F453.
99. Sitprija, V., Tungsanga, K., Eiam-Ong, S., Leelhaphunt, N., and Scriboonlue, P. (1990). Renal tubular acidosis, vanadium and buffaloes. *Nephron* **54:**97–98.
100. O'Neil, R. G., and Helman, S. I. (1977). Transport characteristics of renal collecting tubules: Influences of DOCA and diet. *Am. J. Physiol.* **233:**F544–F551.
101. Stone, D. S., Seldin, D. W., Kokko, J. P., and Jacobson, H. R. (1983). Mineralocorticoid modulation of rabbit medullary collecting duct acidification. A sodium-independent acidification. *J. Clin. Invest.* **72:**77–85.
102. Goldstein, M. B., Bear, R., Richardson, R. M. A., Marsden, P. A., and Halperin, M. L. (1986). The urine anion gap: A clinically useful index of ammonium excretion. *Am. J. Med. Sci.* **292:**198–202.
103. Batlle, D. C., Hizon, M., Cohen, E., Gutterman, C., and Gupta, R. (1988). The use of the urinary anion gap in the diagnosis of hyperchloremic metabolic acidosis. *N. Engl. J. Med.* **318:**594–599.
104. Rothstein, M., Obialo, C., and Hruska, K. A. (1990). Renal tubular acidosis. *Endocrinol. Metab. Clin. North Am.* **19:**869–886.
105. Arruda, J. A. L., and Cowell, G. (1994). Distal renal tubular acidosis: Molecular and clinical aspects. *Hosp. Pract.* **29:**75–88.
106. Cohen, E. P., Bastani, B., Cohen, M. R., Kolner, S., Hemkin, P., and Gluck, S. L. (1992). Absence of H^+-ATPase in cortical collection tubules of a patient with Sjogren's syndrome and distal renal tubular acidosis. *J. Am. Soc. Neph.* **3:**264–269.
107. Douglas, J. B., and Healy, J. K. (1969). Nephrotoxic effects of amphotericin B, including renal tubular acidosis. *Am. J. Med.* **46:**154–162.
108. DeFronzo, R. A. (1980). Hyperkalemia and hyporeninemic hypoaldosteronism. *Kidney Int.* **17:**118–125.

Index